Catastrophic Events and Mass Extinctions: Impacts and Beyond

Edited by

Christian Koeberl
Institute of Geochemistry, University of Vienna
Althanstrasse 14
A-1090 Vienna
Austria

and

Kenneth G. MacLeod
Department of Geological Sciences
University of Missouri
Columbia, Missouri 65221
USA

356

Geological Society of America
3300 Penrose Place
P.O. Box 9140
Boulder, Colorado 80301-9140
2002

Published by The Geological Society of America, Inc.
3300 Penrose Place, P.O. Box 9140, Boulder, Colorado 80301
www.geosociety.org

Printed in U.S.A.

GSA Books Science Editor Abhijit Basu
Cover graphics production by Margo Good

Library of Congress Cataloging-in-Publication Data

Catastrophic events and mass extinctions : impacts and beyond / edited by Christian Koeberl and Kenneth G. MacLeod.
p. cm. — (Special Paper ; 356)
Includes bibliographical references and index.
ISBN 1-8137-2356-6
1. Catastrophic events and mass extinctions. I. Koeberl, Christian II. MacLeod, Kenneth G. III. Special papers (Geological Society of America) ; 356.

QE527.7 .M36 2001
551.1′16—dc21 00-052071

Cover: This illustration shows a polished and etched section of an iron meteorite juxtaposed with the silhouette of a dinosaur skeleton. It is meant to emphasize two distinctly different, yet, for the topic of the meeting and this book, intimately interrelated branches of the earth sciences. Planetary science (meteoritics) and paleontology share a common denominator—in this case, impacts. The basis for the design is similar to the concept behind Rubik's Cube®: the idea that lining up all the elements of the puzzle will show that the pieces can, in fact, fit together. Concept and design: Dona Jalufka, Vienna.

10 9 8 7 6 5 4 3 2 1

Contents

Dedication vii
Graham Ryder, 1949–2002

Preface ix

The Big Picture

1. ***Meteors and comets in ancient Mexico*** 1
Ulrich Köhler

2. ***Impact lethality and risks in today's world: Lessons for interpreting Earth history*** 7
Clark R. Chapman

3. ***Tracers of the extraterrestrial component in sediments and inferences for Earth's accretion history*** 21
Frank T. Kyte

Chicxulub and the Cretaceous-Tertiary Boundary Deposits

4. ***Geophysical constraints on the size and structure of the Chicxulub impact crater*** 39
Joanna Morgan, Mike Warner, and Richard Grieve

5. ***Micro-Raman and optical identification of coesite in suevite from Chicxulub*** 47
Elena Lounejeva, Mikjail Ostroumov, and Gerardo Sánchez-Rubio

6. ***Distribution of Chicxulub ejecta at the Cretaceous-Tertiary boundary*** 55
Philippe Claeys, Wolfgang Kiessling, and Walter Alvarez

7. ***Generation and propagation of a tsunami from the Cretaceous-Tertiary impact event*** 69
T. Matsui, F. Imamura, E. Tajika, Y. Nakano, and Y. Fujisawa

8. ***Mass wasting of Atlantic continental margins following the Chicxulub impact event*** 79
R.D. Norris and J.V. Firth

9. ***Sequence stratigraphy and sea-level change across the Cretaceous-Tertiary boundary on the New Jersey passive margin*** 97
Richard K. Olsson, Kenneth G. Miller, James V. Browning, James D. Wright, and Benjamin S. Cramer

10. ***Complex tsunami waves suggested by the Cretaceous-Tertiary boundary deposit at the Moncada section, western Cuba*** 109
R. Tada, Y. Nakano, M.A. Iturralde-Vinent, S. Yamamoto, T. Kamata, E. Tajika, K. Toyoda, S. Kiyokawa, D. Garcia Delgado, T. Oji, K. Goto, H. Takayama, R. Rojas-Consuegra, and T. Matsui

11. ***Cretaceous-Tertiary boundary sequence in the Cacarajicara Formation, western Cuba: An impact-related, high-energy, gravity-flow deposit*** 125
Shoichi Kiyokawa, Ryuji Tada, Manuel Iturralde-Vinent, Eiichi Tajika, Shinji Yamamoto, Tetsuo Oji, Youichiro Nakano, Kazuhisa Goto, Hideo Takayama, Dora Garcia Delgado, Consuelo Diaz Otero, Reinaldo Rojas-Consuegra, and Takafumi Matsui

12. ***Multiple spherule layers in the late Maastrichtian of northeastern Mexico*** 145
Gerta Keller, Thierry Adatte, Wolfgang Stinnesbeck, Mark Affolter, Lionel Schilli, and José Guadalupe Lopez-Oliva

13. *Two anomalies of platinum group elements above the Cretaceous-Tertiary boundary at Beloc, Haiti: Geochemical context and consequences for the impact scenario* 163
Doris Stüben, Utz Kramar, Zsolt Berner, Jörg-Detlef Eckhardt, Wolfgang Stinnesbeck, Gerta Keller, Thierry Adatte, and Klaus Heide

14. *Cretaceous-Tertiary boundary at Blake Nose (Ocean drilling Program Leg 171B): A record of the Chicxulub impact ejecta* 189
Francisca Martínez-Ruíz, Miguel Ortega-Huertas, Inmaculada Palomo, and Jan Smit

15. *Global occurrence of magnetic and superparamagnetic iron phases in Cretaceous-Tertiary boundary clays* 201
N. Bhandari, H.C. Verma, C. Upadhyay, Amita Tripathi, and R.P. Tripathi

16. *Cretaceous-Tertiary profile, rhythmic deposition, and geomagnetic polarity reversals of marine sediments near Bjala, Bulgaria* 213
Anton Preisinger, Selma Aslanian, Franz Brandstätter, Friedrich Grass, Herbert Stradner, and Herbert Summesberger

17. *Paleoenvironment across the Cretaceous-Tertiary transition in eastern Bulgaria* 231
Thierry Adatte, Gerta Keller, S. Burns, K.H. Stoykova, M.I. Ivanov, D. Vangelov, Utz Kramar, and Doris Stüben

Biological and Chemical Changes Across the Cretaceous-Tertiary Boundary

18. *Cretaceous-Tertiary boundary planktic foraminiferal mass extinction and biochronology at La Ceiba and Bochil, Mexico, and El Kef, Tunisia* 253
Ignacio Arenillas, Laia Alegret, José A. Arz, Carlos Liesa, Alfonso Meléndez, Eustoquio Molina, Ana R. Soria, Esteban Cedillo-Pardo, José M. Grajales-Nishimura, and Carmen Rosales-Domínguez

19. *Distribution pattern of calcareous dinoflagellate cysts across the Cretaceous-Tertiary boundary (Fish Clay, Stevns Klint, Denmark): Implications for our understanding of species-selective extinction* 265
Jens Wendler and Helmut Willems

20. *Abrupt extinction and subsequent reworking of Cretaceous planktonic foraminifera across the Cretaceous-Tertiary boundary: Evidence from the subtropical North Atlantic* 277
Brian T. Huber, Kenneth G. MacLeod, and Richard D. Norris

21. *Faunal changes across the Cretaceous-Tertiary (K-T) boundary in the Atlantic coastal plain of New Jersey: Restructuring the marine community after the K-T mass-extinction event* 291
William B. Gallagher

22. *Giant ground birds at the Cretaceous-Tertiary boundary: Extinction or survival?* 303
Eric Buffetaut

23. *Dinosaurs that did not die: Evidence for Paleocene dinosaurs in the Ojo Alamo Sandstone, San Juan Basin, New Mexico* 307
James E. Fassett, Robert A. Zielinski, and James R. Budahn

24. *Sulfur isotopic compositions across terrestrial Cretaceous-Tertiary boundary successions* 337
Teruyuki Maruoka, Christian Koeberl, Jason Newton, Iain Gilmour, and Bruce F. Bohor

25. *Natural fullerenes from the Cretaceous-Tertiary boundary layer at Anjar, Kutch, India* 345
G. Parthasarathy, N. Bhandari, M. Vairamani, A.C. Kunwar, and B. Narasaiah

26. *Organic geochemical investigation of terrestrial Cretaceous-Tertiary boundary successions from Brownie Butte, Montana, and the Raton Basin, New Mexico* 351
A.F. Gardner and I. Gilmour

Permian-Triassic Boundary

27. *End-Permian mass extinctions: A review* 363
Douglas H. Erwin, Samuel A. Bowring, and Jin Yugan

28. *Permian-Triassic boundary in the southwestern United States: Hiatus or continuity?* 385
Walter Alvarez and Diane O'Connor

29. *Extent, duration, and nature of the Permian-Triassic superanoxic event* 395
Paul B. Wignall and Richard J. Twitchett

30. *Abruptness of the end-Permian mass extinction as determined from biostratigraphic and cyclostratigraphic analyses of European western Tethyan sections* 415
Michael R. Rampino, Andreas Prokoph, Andre C. Adler, and Dylan M. Schwindt

31. *Permian-Triassic boundary in the northwest Karoo Basin: Current stratigraphic placement, implications for basin development models, and the search for evidence of impact* 429
P. John Hancox, D. Brandt, W.U. Reimold, C. Koeberl, and J. Neveling

32. *Chemical signatures of the Permian-Triassic transitional environment in Spiti Valley, India* 445
A.D. Shukla, N. Bhandari, and P.N. Shukla

33. *Synchronous record of $\delta^{13}C$ shifts in the oceans and atmosphere at the end of the Permian* 455
Mark A. Sephton, Cindy V. Looy, Ruben J. Veefkind, Henk Brinkhius, Jan W. De Leeuw, and Henk Visscher

Other Extinctions and Boundaries

34. *Late Ordovician extinction: A Laurentian view* 463
William B.N. Berry, Robert L. Ripperdan, and Stanley C. Finney

35. *Late Devonian sea-level changes, catastrophic events, and mass extinctions* 473
Charles A. Sandberg, Jared R. Morrow, and Willi Ziegler

36. *Impact-generated carbonate accretionary lapilli in the Late Devonian Alamo Breccia* 489
John E. Warme, Matthew Morgan, and Hans-Christian Kuehner

37. *Continental Triassic-Jurassic boundary in central Pangea: Recent progress and discussion of an Ir anomaly* 505
Paul E. Olsen, Christian Koeberl, Heinz Huber, Alessandro Montanari, Sarah J. Fowell, Mohammed Et-Touhami, and Dennis V. Kent

38. *Dating the end-Triassic and Early Jurassic mass extinctions, correlative large igneous provinces, and isotopic events* 523
József Pálfy, Paul L. Smith, and James K. Mortensen

39. *Sea-level changes and black shales associated with the late Paleocene thermal maximum: Organic-geochemical and micropaleontologic evidence from the southern Tethyan margin (Egypt-Israel)* 533
Robert P. Speijer and Thomas Wagner

Case Studies

40. *Sedimentary record of impact events in Spain* 551
Enrique Díaz-Martínez, Enrique Sanz-Rubio, and Jésus Martínez-Frías

41. *Postglacial impact events in Estonia and their influence on people and the environment* 563
Anto Raukas

Experiments, Methods, and Model Calculations

42. *Mineralogical investigations of experimentally shocked dolomite: Implications for the outgassing of carbonates* 571
Roman Skála, Jana Ederová, Pavel Matějka, and Friedrich Hörz

43. *How strong was impact-induced CO_2 degassing in the Cretaceous-Tertiary event? Numerical modeling of shock recovery experiments* 587
Boris A. Ivanov, Falko Langenhorst, Alexander Deutsch, and Ulrich Hornemann

44. *Laboratory impact experiments versus natural impact events* 595
Paul S. DeCarli, Emma Bowden, Adrian P. Jones, and G. David Price

45. *Comparison of the osmium and chromium isotopic methods for the detection of meteoritic components in impactites: Examples from the Morokweng and Vredefort impact structures, South Africa* 607
Christian Koeberl, Bernhard Peucker-Ehrenbrink, Wolf Uwe Reimold, Alex Shukolyukov, and Günter W. Lugmair

46. *Numerical modeling of the formation of large impact craters* 619
Boris A. Ivanov and Natalia A. Artemieva

47. *Multiple stages of condensation in impact-produced vapor clouds* 631
Detlef de Niem

Astronomical Studies

48. *On the completeness of the discovery rate of the potentially hazardous asteroids* 645
Eric W. Elst

49. *Possible sources of Earth crossers* 651
Nina A. Solovaya and Eduard M. Pittich

50. *Measurement of the lunar impact record for the past 3.5 b.y. and implications for the Nemesis theory* 659
Richard A. Muller

51. *Role of the galaxy in periodic impacts and mass extinctions on the Earth* 667
Michael R. Rampino

52. *Solar system linked to a gigantic interstellar cloud during the past 500 m.y.: Implications for a galactic theory of terrestrial catastrophism* 679
Carlos A. Olano

Impacts and Extinction Mechanisms

53. *Grazing meteoroids could ignite continental-scale fires* 685
Vladimir V. Svetsov

54. *Atmospheric erosion and radiation impulse induced by impacts* 695
V.V. Shuvalov and N.A. Artemieva

55. *Toxins produced by meteorite impacts and their possible role in a biotic mass extinction* 705
M.V. Gerasimov

56. *Modeling long-term climatic effects of impacts: First results* 717
Thomas Luder, Willy Benz, and Thomas F. Stocker

Index 731

Dedication

Photo by D. Jalufka.

Graham Ryder (1949–2002)

This volume is dedicated to the memory of Graham Ryder, who passed away much too soon. He died on January 5, 2002, as a result of complications from cancer of the esophagus diagnosed in November 2001. Since 1983, Graham was a staff scientist at the Lunar and Planetary Institute in Houston, Texas, United States, where he specialized in petrology and geochemistry of lunar rocks, working extensively with the collection of Apollo rocks at the NASA Johnson Space Center. For more than a decade, he became very interested in impact craters on Earth and their importance for the evolution of the planet.

Graham was a native of Wales, United Kingdom, and he received his B.Sc. from the University of Wales (Swansea; 1970) and his Ph.D. from Michigan State University (1974) on the petrology of igneous rocks. Subsequently, he did postdoctoral studies at the Harvard-Smithsonian Astrophysical Observatory in Cambridge, Massachusetts, and then worked with Northrop Services Inc. in the Lunar Curatorial Facility at the NASA Johnson Space Center. His main research dealt with the petrography and petrology of igneous rocks and impact melt rocks on the moon, and he was instrumental in providing new evidence for the so-called "late heavy bombardment" on the moon. Besides explaining the history of the lunar crust, he produced detailed catalogs and guides to the Apollo lunar sample collections.

Besides being an eminent scientist, Graham was a quick-witted and helpful colleague and friend with an encyclopedic knowledge, which was not confined to scientific topics. During dull moments on a weeklong field trip to impact deposits in Nevada and Utah in April 2001, he enlivened the proceedings by reciting—verbatim—entire episodes of the TV series "Monty Python." His passion for soccer led him to play in various teams and forge long-lasting friendships. His sharp and incisive, but always witty comments invigorated many a scientific conference, and he was often instrumental in bringing debaters "back to Earth."

Graham was planning to do detailed research on the new drill cores from the Chicxulub impact structure that are now, in 2002, becoming available. He was also the senior editor on the previous book (GSA Special Paper 307, *The Cretaceous-Tertiary Event and Other Catastrophes in Earth History*) of the informal series, of which the present volume is a part. He will be missed by his family and many friends, and our science will be poorer without him.

Christian Koeberl

Geological Society of America
Special Paper 356
2002

Preface

Christian Koeberl
Institute of Geochemistry, University of Vienna, Althanstrasse 14, A-1090 Vienna, Austria
Kenneth G. MacLeod
Department of Geological Sciences, University of Missouri, Columbia, Missouri 65211 USA

INTRODUCTION

During the 1980s and early 1990s there was lively debate in the geological community regarding the cause of the mass extinction that marks the end of the Cretaceous Period. The now-famous paper by Alvarez et al. (1980) is the reference cited in many subsequent papers as having implicitly signaled the start of the current debate. Alvarez et al. reported geochemical evidence for a large asteroid or comet impact event at the stratigraphic level of the Cretaceous-Tertiary (K-T) boundary in Italy; both their discovery and their conclusion were reached independently at about the same time by J. Smit, on the basis on the study of rocks in southern Spain. The crux of the original arguments is that the concentrations of some siderophile elements (most prominently iridium) are as much as four orders of magnitude higher in the thin clay layer at the K-T boundary than in strata above and below the boundary clay or in rocks of the Earth's crust in general. Although rare in crustal rocks, these siderophile elements have high abundances in many extraterrestrial objects (and in subcrustal rocks on Earth). The stratigraphic coincidence of a high concentration of potentially extraterrestrial material with the virtually complete turnover in the microfossil assemblage (i.e., the K-T mass extinction) was interpreted as indicating that a large asteroid or comet nucleus hit the Earth ca. 65 Ma and that that impact caused extreme, global environmental stress leading to the mass extinction.

This proposal caused a significant stir in the geological and paleontological community, and an interdisciplinary meeting was organized, "Large Body Impacts and Terrestrial Evolution: Geological, Climatological, and Biological Implications." The meeting was held in Snowbird, Utah, October 19–22, 1981, and is now commonly referred to as "Snowbird I." The proceedings were published in 1982 as a Geological Society of America Special Paper (Silver and Schultz, 1982). The debate was lively at that meeting. The initial geochemical, stratigraphic, and paleontological data in favor of the impact scenario were augmented by arguments from a small group of planetary scientists familiar with impact as one of the main surface-forming and surface-modifying process on the Moon, Mars, Mercury, and Venus. A significant number of geologists and paleontologists ("anti-impactors") argued that in many places the rock record did not seem to match the predictions of the impact hypothesis and that earthbound forcing mechanisms could more simply and completely explain observed patterns than the impact hypothesis. The debate can be portrayed as an echo of the nineteenth century arguments between uniformitarianism and catastrophism, or, alternatively and perhaps more accurately, as an illustration of biases against invoking catastrophic or external events to explain earth history until less dramatic, internal options have been discarded. Regardless, it was clear that more rigor and resolution were necessary to distinguish among competing hypotheses. Outstanding problems included the reality of large-scale impacts on Earth and how to recognize them, the accurate characterization of mass extinctions, and the meaning of observed first and last appearances relative to the inferred timing of extinctions and originations. In short, this first meeting got the scientists and media alike interested in pursuing these studies more seriously.

The impact hypothesis received a major boost in 1984 when Bohor and coworkers of the U.S. Geological Survey announced their discovery of shocked minerals (especially shocked quartz) in the K-T boundary layer in western North America. This finding provided strong evidence that a large impact event in continental rocks coincided with the K-T boundary. Such shocked minerals have been found associated only with hypervelocity impacts and nuclear bomb explosions, and shocked minerals would not be subject to the same types of diagenetic modification that could compromise geochemical profiles. Unfortunately, features superficially similar to planar deformation features can be formed in a variety of geological settings. Although detailed petrographical work can easily lead to a clear identification of planar deformation features, the ambiguity, coupled with positive evidence of large-scale volcanism near the K-T boundary and lack of an impact crater of the

Koeberl, C., and MacLeod, K.G., 2002, Preface, *in* Koeberl, C., and MacLeod, K.G., eds., Catastrophic Events and Mass Extinctions: Impacts and Beyond: Boulder, Colorado, Geological Society of America Special Paper 356, p. ix–xii.

right size and age, led some to support a volcanic explanation for observations across the K-T boundary. The K-T debate echoed an earlier similar debate that began in the 1920s concerning an impact versus volcanic origin for a variety of unusual structures (cryptoexplosive features) around the world.

Repeating on long time scales, large impacts have generally not been incorporated into a uniformitarian view of earth history, i.e., cryptoexplosive features or mass-extinction boundaries. One important early contribution of the K-T debate was a reexamination of rates and time scales that are involved in the discussion of the role of impacts in terrestrial events. Earthquakes, volcanic eruptions, and landslides are locally devastating during periods of years to centuries and are integrated into interpretations of earth history. Large meteorite impacts have not been observed during the last few millennia, and they have tended to be neglected when thinking of events that shape the Earth. Small meteorites have been observed frequently, but over the centuries scientists have failed to make the connection to impact events by applying the same principle used for extrapolating the frequency of volcanic eruptions and earthquakes: i.e., large and devastating ones occur less often than small events. In other words, there is no conflict between uniformitarianism and meteorite impact. The geological evolution of the Earth is the result of more or less continuous processes combined with the effects of a large number of individual events and catastrophes of various magnitudes summed over a long time span. Over long periods of time, impact is a common phenomenon on Earth.

Impacts are also dramatic, and interest in the K-T boundary continued to grow. A second interdisciplinary conference, "Global Catastrophes in Earth History: An Interdisciplinary Conference on Impacts, Volcanism, and Mass Mortality" was held October 20–23, 1988, again in Snowbird, Utah ("Snowbird II"). Snowbird II was not only a truly interdisciplinary meeting, but also one at which intense (and often personal) debates dominated the stage. Opinions were polarized. Proponents of the impact hypothesis advanced physical and chemical evidence for impact, as well as stratigraphic and geochemical data suggesting catastrophic and globally synchronous ecological perturbations at the K-T boundary. Opponents of the impact hypothesis forwarded counter examples and offered alternative interpretations of siderophile element anomalies, shock metamorphism, and other data derived from impact-cratering studies. It was also noted that existence of an impact structure was a prominent prediction of the impact hypothesis, but that a suitable structure had yet to be discovered. As with the previous meeting, Snowbird II crystallized areas of disagreement and led to more rigorous examination of impact signatures, stratigraphic data (particularly reworking), and a search for the missing K-T boundary impact crater. The proceedings were published as Geological Society of America Special Paper 247 (Sharpton and Ward, 1990).

Six years later, the next Snowbird meeting was held and both the venue and majority opinion had changed. Snowbird III took place in Houston, Texas, February 9–12, 1994, and most participants accepted evidence for a large impact event 65 Ma. In the early 1990s, after years of intense search, a subsurface structure (which had already been suggested as a possible impact structure in 1982) centered near the town of Chicxulub on the Yucatan Peninsula in Mexico, had been generally confirmed as an impact crater and became the leading candidate for the K-T impact site. One of the main points of contention at the Houston meeting was the size of the Chicxulub structure; estimates of its diameter ranged from ~200 to 300 km. At either end of that range, Chicxulub is one of the largest impact structures currently known on Earth. A second theme of the meeting concerned how to link evidence for an impact with changes in the fossil record. Notable discussions included the results of a blind test of foraminiferal distributions across the K-T boundary, sampling biases and statistical tests of uncertainties in estimated stratigraphic ranges, and studies of possible microfossil reworking and vertebrate taphonomy. The proceedings of that meeting were published as Geological Society of America Special Paper 307 (Ryder et al., 1996). Consensus seemed to have been reached that the Chicxulub impact was real and was a significant ultimate cause of at least some of the biological changes across the boundary, but details of Chicxulub structure, more proximate links between impact and extinction, and the relevance of insights and techniques drawn from K-T studies to other boundaries remained unresolved and actively disputed.

In 2000, six years after the third Snowbird meeting, the time seemed ripe to attempt to build on this tradition of the previous meetings. The first meeting dealt mainly with the then-controversial hypothesis that a large-scale impact 65 Ma was responsible for the end-Cretaceous mass extinction. The second meeting was mainly concerned with the evidence (e.g., shock metamorphism) that such a large impact event happened. At the third meeting (in Houston), the discussion centered on the Chicxulub impact structure as the long-sought K-T boundary impact crater. This fourth meeting was designed to try to quantify the question if (and how) short-term, high-energy events influence biological evolution on the Earth. To this end, an international, interdisciplinary conference, "Catastrophic Events and Mass Extinctions: Impacts and Beyond," was held at the University of Vienna, Vienna, Austria, July 9–12, 2000 (Fig. 1).

The Vienna meeting was cosponsored by the University of Vienna, the Lunar and Planetary Institute (Houston), the Barringer Crater Company (USA), the European Science Foundation Impact Programme, the Federal Ministry of Education, Science, and Culture (Austria), the Geological Survey of Austria, the Vienna Convention Bureau, the International Association of Geochemistry and Cosmochemistry, Finnegan, Micromass, and Cameca Instruments. The International Organizing and Program Committee included Walter Alvarez (University of California, Berkeley, United States), Eric Buffetaut (Centre National de la Recherche Scientifique [CNRS], Paris, France), Douglas H. Erwin (Smithsonian Institution, Washington, D.C., United States), Iain Gilmour (Open University, Milton Keynes,

Figure 1. The "veterans" who have participated in at least three of the four "Snowbird" meetings (1981, 1988, 1994, and/or 2000). Front row, left to right: D. Nichols, K. Johnson, F. Kyte, A. Montanari, J. Smit, W. Alvarez, and C. Chapman. Middle row, left to right: A. Preisinger, S. D'Hondt, P. Ward, C. Koeberl, R. Muller, G. Ryder, W.U. Reimold, and M. Rampino. Back row, left to right: C. Pillmore, P. Claeys, B. Sharpton, and I. Gilmour. Photo by D. Jalufka in Vienna, July 12, 2000.

United Kingdom), Christian Koeberl (University of Vienna, Austria), W. Uwe Reimold (University of the Witwatersrand, Johannesburg, South Africa), Graham Ryder (Lunar and Planetary Institute, Houston, Texas, United States), Charles A. Sandberg (U.S. Geological Survey, Denver, Colorado, United States), Jan Smit (Free University, Amsterdam, The Netherlands), and Peter D. Ward (University of Washington, Seattle, Washington, United States). The local organizing committee consisted of Christian Koeberl, Franz Brandstätter (Natural History Museum, Vienna, Austria), Heinz Huber (University of Vienna, Austria), Wolfgang Kiesl (University of Vienna, Austria), Gero Kurat (Natural History Museum, Vienna, Austria), Anton Preisinger (Technical University, Vienna, Austria), and Hans Peter Schönlaub (Geological Survey, Vienna, Austria).

The Vienna meeting was attended by 221 people from 33 different countries from all over the world, ranging from Argentina to the United States, from Belarus to South Africa. Thus, it was the largest one in this informal series of meetings; the 1981 meeting attracted 101 attendees, the 1988 meeting attracted 184 people, and the 1994 meeting had 190 participants. In 2000, the largest contingency came from the United States (60 participants), followed by Germany, Austria, England, France, Italy, and Japan. Of the total number of participants, 34 were students. Many participants came from developing countries in Europe and elsewhere around the world. Contributions received from a variety of sources funded 38 travel grants to support participants from these countries (researchers and students), as well as students and a few other participants from other countries. Three of four suggested geological field trips were well subscribed and took part as planned. There was a four-day premeeting field trip to the Ries and Steinheim impact structures in Germany and K-T boundary sites in Austria, a four-day postmeeting field trip to the Carnic Alps in Austria (e.g., Permian-Triassic boundary), and a five-day postmeeting field trip to the classical K-T boundary and late Eocene impact clastic layers in central Italy.

The main focus of the Vienna meeting was the question if and how short-term, high-energy events influence biological evolution on the Earth. Thus, a wide variety of topics targeted included the following: (1) K-T boundary: the marine record; (2) K-T boundary: the terrestrial record; (3) K-T boundary melange; (4) the Chicxulub impact structure; (5) snowball Earth–late Eocene record; (6) Permian-Triassic boundary; (7) carbon cycle–early Paleocene record; (8) catastrophic event markers; (9) causes of extinctions; (10) impact volcanism–environmental effects; and (11) impacts and beyond. The poster session included 118 presentations on topics across this variety of themes. Oral presentations were of two types and limited to a single session. Main topics were reviewed by speakers who gave 20-minute oral reviews. In addition, authors of selected volunteered abstracts presented 5-minute summaries of their findings. In total, 65 oral presentations were given. This format allowed for presentation of a variety of topics and results and plenty of time for discussion during the three-day meeting.

The meeting and the proceedings volume clearly show that the K-T boundary event still concerns many researchers; about half of all papers dealt with various aspects of that important boundary. However, the Permian-Triassic extinction is increasingly being scrutinized by a growing number of researchers with a variety of perspectives and analytical techniques. No smoking gun of any sort has yet been found at the Permian-Triassic extinction horizon, even though the extinctions were apparently more severe than those at the K-T boundary. The situation across many other boundaries and events is at least as unresolved. We look forward to progress, if not resolution, by the time of the next meeting.

The success of the Vienna meeting is reflected in the papers published in this proceedings volume. Of 64 submitted manuscripts, 56 were accepted for publication in what is now the fourth volume on the interdisciplinary study of the relationship of impacts and other high-energy events to extinctions. The schedule for this volume was demanding: manuscripts were submitted in October and November of 2000, and all reviews were returned to the authors by January of 2001. The last of the final revised manuscripts were received by the editors in mid-February of 2001. The publication of this book, and the rapid pace of reviewing, revising, editing, and revising again would have been impossible without the full cooperation of the reviewers and authors. Many went beyond their duty to ensure timely evaluations and revisions of these manuscripts. We are especially grateful to those who reviewed more than one manuscript, and who may have reread a paper after revision. We also want to acknowledge those authors who stayed on schedule even when faced with one of Earth's high-energy events, a major earthquake. We hope that this book is not only a timely product that reflects the state of the knowledge in a still under-explored field, but that the chapters herein will stand the test of time and become benchmarks on par with many of the contributions to the previous three volumes.

REFERENCES CITED

Alvarez, L., Alvarez, W., Asaro, F., and Michel, H.V., 1980, Extraterrestrial cause for the Cretaceous-Tertiary extinction: Science, v. 208, p. 1095–1108.

Ryder, G., Fastovsky, D., and Gartner, S., eds., 1996, The Cretaceous-Tertiary event and other catastrophes in earth history: Geological Society of America Special Paper 307, 569 p.

Sharpton, V., and Ward, P., eds., 1990, Global catastrophes in earth history: An interdisciplinary conference on impacts, volcanism, and mass mortality: Geological Society of America Special Paper 247, 631 p.

Silver, L., and Schultz, P., 1982, Geological implications of impacts of large asteroids and comets on the Earth: Geological Society of America Special Paper 190, 528 p.

Printed in the U.S.A.

Geological Society of America
Special Paper 356
2002

Meteors and comets in ancient Mexico

Ulrich Köhler*
Institut für Völkerkunde, Universität Freiburg, Werderring 10, 79085 Freiburg, Germany

ABSTRACT

Meteors and comets were of great importance to the native inhabitants of Mexico. Meteors were alternatively viewed as arrows of stellar gods, as their cigar butts, and even as their excrement. The arrows could hit animals or people and were feared when walking at night. Comets were conceived as smoking stars and as bad omens, e.g., announcing the death of a ruler. In the north of Yucatan, Maya oral traditions report of a lake that originated when a huge meteorite hit the ground.

INTRODUCTION

Since the identification of the Chicxulub impact structure, the area of northern Yucatan, Mexico, is famous among geologists for a spectacular extraterrestrial impact. This impact of a bolide occurred at the Cretaceous-Tertiary boundary ca. 65 Ma. The impact structure has a diameter of ~200 km (see Morgan, Warner, and Grieve, this volume), and such an enormous size is suggestive of great repercussions on Earth. The bolide is thought to have caused or contributed to the extinction of dinosaurs and other species, thus indirectly leading to the rise of mammals and finally humans. Several papers of this volume deal with that impact at Chicxulub, emphasizing its importance as object of research.

It is still debated whether that extraterrestrial object was a comet or a meteorite. Because the latter are sometimes an effect of the former, they are treated together in this chapter. However, in contrast to the Chicxulub structure, I discuss not real events, but ideas of people about comets, meteors, and meteorites. The focus here is on the ideas of the natives of ancient Mexico, but some concepts of present-day native groups of the same area are included.

The principal sources are the following: (1) codices, i.e., native books of pictoral writing, (2) Spanish chronicles of the sixteenth and seventeenth centuries, and (3) information from present-day indigenous groups in ethnographic reports. As an anthropologist and historian, I present a picture according to the point of view of these disciplines.

Unfortunately, the early sources are not precise in defining meteors and comets. Although *meteoro, aerolito*, and *estrella fugaz* in Spanish clearly describe meteors, most early and many recent sources employ the word *cometa* for comets and meteors. These two phenomena have a different appearance in the sky. A comet looks like a small lengthy cloud and can be seen as a slow-moving bright spot in the sky for weeks. In contrast, a meteor (also called a fireball) is characterized by rapid movement and can normally be seen only for a few seconds. Meteors have an average speed of 18 km/s (Carlisle, 1995, p. 125). If a meteor survives passage through the atmosphere and arrives on Earth, that object is called a meteorite. The different appearances of comets and meteors give us clues to identify some of the meteors that are mentioned in the sources by the name of *cometa*. Whenever there is a mention of a running *cometa* or of fast movement, we can be sure that the reference pertains to a meteor or a shower of meteors.

COMETS: SMOKING STARS

What were the ideas about comets in ancient Mexico? What images were associated with them? In Europe we have the image of a star with a tail; in China it is a star in the form of a broom; in Mexico, however, it is a smoking star. In Aztec it is called *citlalin popoca* and in Yucatec Maya it is called *budzil ek'*. The Aztecs feared comets as bad omens announcing the death of a prince or king or foreshadowing incipient war or famine (Sahagún, 1950–69, vol. VII, p. 13). Unequivocal data

*E-mail: ulrich.koehler@ethno.uni-freiburg.de

Köhler, U., 2002, Meteors and comets in ancient Mexico, *in* Koeberl, C., and MacLeod, K.G., eds., Catastrophic Events and Mass Extinctions: Impacts and Beyond: Boulder, Colorado, Geological Society of America Special Paper 356, p. 1–6.

for specific comets have survived only for the time following the conquest by the Spanish. The appearance of Halley's comet in 1531 is mentioned in several sources. Codex Telleriano-Remensis (1955, folio 44, obverse) gives an illustration of the comet (Fig. 1). It can also be seen in Codex Vaticanus A (1979, folio 88, obverse), wherein the picture shows the comet and an eclipse of the sun that occurred in that same year. The normal rendering of a comet in these two codices is shown in Figure 2.

There is an interesting chronicle of a central Mexican village, the *Anales de Tecamachalco* (1981), in which six comets between 1556 and 1581 are listed; four, those of 1556, 1577, 1580, and 1581, were also registered in the Old World (Baldet, 1950). Comparing the data, it becomes evident that the first worldwide observation of the comets of 1569 and 1580 was in Tecamachalco. In the Old World they were spotted a few days later. The credibility of the source from that native village in central Mexico is supported by the mention of two eclipses of the sun, which allowed us to check whether the information of the chronicle is correct. The dates given, January 25, 1571, and November 2, 1575, when referenced to the famous *Canon der Finsternisse* of von Oppolzer (1969), are correct to the very day. We may thus conclude that the dates given for the first appearance of comets are just as accurate.

Figure 1. Halley's comet in 1531, shown together with eclipse of sun, from Codex Telleriano-Remensis (1995).

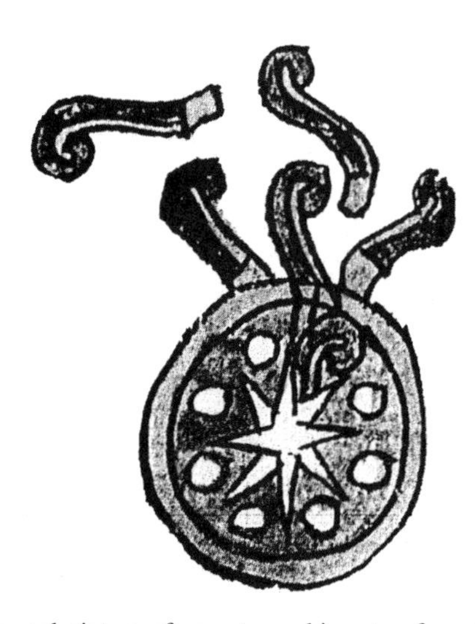

Figure 2. Normal picture of comet, smoking star, from Codex Telleriano-Remensis (1995).

The reappearance of Halley's comet in 1607 is reported by the Indian chronicler Chimalpahin (1963–65, vol. II, p. 63) as well as by his Spanish counterpart Torquemada (1969, vol. I, p. 85). Both observed it personally from September 21 onward, i.e., 10 days after its first being spotted in the Old World.

Many present-day native groups of Mexico view comets as smoking stars, just as among the Aztecs. Slight variations are the images of a burning star, or a light-beam star. Omens related to the appearance of a comet are just as pessimistic as in the sixteenth century, and foreshadow war, pestilence, and famine. The appearance of Halley's comet in 1910, reported in several ethnographies (see Köhler, 1989, p. 292), was followed by the Mexican Revolution, a civil war that lasted seven years. This calamity helped to revive ancient fears related to comets.

METEORS: STELLAR ARROWS

The Aztec concept of meteors clearly conveys the idea of a fast movement. The Aztec term is *citlalin tlamina* (star shoots), a concept almost identical with the English expression of a shooting star. Among the Maya of Yucatan, sixteenth century dictionaries list three images: running star, running witch, and cigarette of witch, the latter conveying the image of a falling cigarette butt. However, who are the stars or celestial bodies that shoot with arrows or spears? Among the Aztecs, these were the sun god, *Nahui Olin*, and the Venus god, *Tlahuizcalpantecutli*. In Figure 3 the sun god can be seen sitting on his throne: in his left hand he has spears, and in the other a spear thrower; the beams around his body clearly identify him as sun god. This identification is also given by his calendar name *Nahui Olin*, which can be seen below his throne. The figure resembling a St. Andrew's cross stands for the day sign *Olin*, and the four dots for *nahui*, four. In this illustration the sun god is swallowing a stream of blood that comes from the beheaded body of a sacrificed quail. The head of another quail can be seen in the open mouth of the Earth goddess, who is typically depicted as a monster in the form of a crocodile. The Earth, where humans live, where people work in their fields and build their houses and cities, is located on and within the extension of this open mouth, which may suddenly close and devour them at any time.

Figure 3. Sun god on his throne (Codex Borgia, 1976); in upper right is moon with rabbit in its midst, in lower right is Earth goddess, depicted as head of crocodile. Sun god is fed with blood of quail, Earth goddess is fed with quail head (Seler, 1902–24, vol. IV, p. 628).

High above the surface of the Earth, the moon can be seen in the night sky, showing in its midst a rabbit. This is how people in central Mexico generally interpreted the darker parts on the surface of the moon.

Nahui Olin was not the first sun. According to the Aztecs and their neighbors, there have been four previous suns. Each of them resided over a world that was destroyed in a cosmic catastrophe. These catastrophes did not always result in mass extinction; the results were sometimes transformations, i.e., of humans into animals. The means of destruction of the four previous worlds were jaguars, wind, rains of fire, and water. More than a dozen sources on these cyclical catastrophes have survived (Moreno de los Arcos, 1967). They coincide mostly with the types of catastrophes, but some give a different sequence. Because the sequence is of little importance for the general message of the myth I present only one prominent version (Lehmann, 1938, p. 322–327). At the end of the first cosmic period the sun perished, and jaguars devoured people. At the end of the second previous sun, a strong destructive wind appeared; the sun and people were blown away, the latter partly transformed into monkeys (Fig. 4). At the end of the following period, a rain of fire poured down on Earth, and the sun and people were burned, except some who were transformed into turkeys. During the last catastrophe the sky collapsed, water destroyed the world, and people were drowned or changed into fishes. It is interesting that, in contrast to European Darwinian ideas on evolution, in Aztec myth apes and monkeys do not precede humans, but descend from them.

This sequence of catastrophes was not considered to have come to its end. The Aztecs were convinced that our sun and world would perish, just like the previous ones, and they knew by what kind of catastrophic event this would happen in the future. The destruction would be brought about by earthquakes (Lehmann, 1938, p. 62); i.e., the ferocious terrestrial monster seen in Figure 3 would cause the destruction and annihilation of human beings.

The other prominent star that shoots arrows was Venus. To the Aztecs, Venus was a fierce and threatening warrior, invariably depicted with spears and spear thrower (*atlatl*), as well as a shield. In the Codices Borgia (1976, p. 53–54), Cospi (1968, p. 9–11), Dresdensis (1975, p. 46–50), and Vaticanus B (1972, p. 80–84), he can be seen in a sequence of five pictures that represent the five phases of the Venus cycle. During each phase he is hitting or killing a different being or phenomenon, such as the maize god, the water goddess, a jaguar, or a throne. Figure 5 (from Codex Borgia, 1976, p. 53) shows how he is wounding the water goddess *Chalchiuhtlicue*, who can be identified easily because she is standing in water, in which a shellfish and a turtle can be seen. Blood is streaming from her left leg and from the two animals.

EFFECTS OF METEORITES

The Aztecs considered falling stars to be dangerous; if these stellar arrows hit an animal or human, an *ocuili*, a maggot or caterpillar, would be left in the wound. Therefore, people protected themselves when walking at night and they refrained from eating animals wounded by a shooting star (Sahagún,

Figure 4. Destruction of world by wind and transformation of surviving people into monkeys, according to Codex Vaticanus A (1979, folio 6, obverse; see also Seler, 1902–24, vol. IV, p. 51).

1950–69, vol. VII, p. 13). Two impressive meteors are mentioned in early historic sources. The first refers to the year 1489, and Codex Telleriano-Remensis (1995, folio 39, reverse) shows a snake-like being (Fig. 6). It can also be seen in Codex Vaticanus A (1979, folio 88, obverse). Observing it closely, it becomes evident that it has no split tongue. Hence, according to Aztec iconographic rules, it cannot be a snake. Its mouth is similar to the mouth of butterflies, as depicted in Aztec codices. We may thus conclude that a caterpillar is depicted. The spines on the body also point toward such identification, because there are many species of caterpillars in Mexico that have such spines on their body. This identification of the object depicted is in congruence with the information just mentioned, i.e., that a stellar shot results in caterpillars. Codex Telleriano-Remensis (1995, folio 39, reverse) contains a written commentary in

Figure 5. Venus god shooting at water goddess, wounding her as well as shellfish and turtle, from Codex Borgia (1976; see also Seler, 1902–24, vol. II, p. 375).

Figure 6. Big meteor in 1489, depicted as caterpillar with spines, from Codex Telleriano-Remensis (1995).

which the object is called a *cometa*, yet the additional description as "running" identifies it clearly as a meteor.

The second instance is reported from the eve of the conquest by the Spanish, between 1515 and 1519; this meteor was considered as a bad omen, and caused great fear. Sahagún (1950–69, vol. VIII, p. 18), the Franciscan friar, described it as a *cometa* that appeared before dusk and ran from west to east; it split into three parts and gave the impression of showering sparks (Fig. 7). This description convincingly identifies it as a shower of meteors.

There is no uniform concept of meteors and meteorites among present-day indigenous groups of Mexico, although to many they are both considered to be parts of the same phenomenon. The image most frequently encountered now is that of excrement of stars. In certain places the idea is held that this excrement arrives in the form of obsidian blades, and occasionally meteors and meteorites are interpreted as arrow points of gods. This seems reminiscent of the Aztec concept.

The Nahua of the Sierra de Zongolica of eastern Mexico bring together the ideas of caterpillars and excrement. According to them, where a meteorite arrives on Earth, black caterpillars will appear, each about 3 cm long, and hundreds of them are intertwined and form a heap in the size of a hand. The Nahua have two words for them: *citlalocuile* (star caterpillars), and *citlalcuitlatl* (star excrement).

No uniform concept regarding the effects of falling stars could be detected among present-day Indians. There are pessimistic, optimistic, and ambivalent expectations. Negative ideas were recorded among the Popoloca of Veracruz, the Tzotzil of Zinacantan in Chiapas, and the Jacalteca, just across the border in Guatemala. Some Popoloca say that if one points at a falling star, incurable wounds will erupt in that arm. In Zinacantan, a falling star may cause a swollen leg. According to the Jacalteca, a falling star that bursts near the house causes sickness, and if it bursts over the house, it results in death. Positive expectations are rare; among the Totonac in the Sierra Madre Oriental, however, there are stones said to have been sent by stars and these have therapeutic qualities. According to some members of this native group, four arrow-shooting stars are located at the corners of the sky, and from there they protect humanity.

That meteors and meteorites are conceived as possible *naguals* of humans may be considered an ambivalent attitude. A *nagual* is a companion spirit of a person in the form of which certain people can act at their will. Frequent *naguals* are jaguar, fox, eagle, owl, snake, but also whirlwind or meteor. The extrahuman power derived from them may be employed for any purpose. Therefore, negative effects are expected if the *naguals* are employed by a witch, and positive ones if employed by a curer. For example, a curer of the Mixe in Oaxaca explained to me that when the soul of his patient is imprisoned behind a rock, he smashes that rock with his meteor and thus frees the soul of the sick person, who will recover once the soul is back in his body. (For references to individual ethnographic accounts, see Köhler, 1989.)

Figure 7. Shower of meteors between 1515 and 1519 (Sahagún 1950–69, vol. VIII, Fig. 57).

FINAL QUESTION

According to the sixteenth century *Relaciones histórico-geográficas de la gobernación de Yucatán* (1983), near Chahuac-Ha there is a lake called *Yocah Ek'*, which means "star has pierced." The local Maya Indians explained that it received its name because a star fell into it with massive rains (*Relaciones histórico-geográficas de la gobernación de Yucatán*, 1983, v. II, p. 33). Could that be a vague reminiscence of the impact at Chicxulub? The answer is up to the reader.

REFERENCES CITED

Note: Some of the references cited here are primary sources, such as facsimile editions of native documents, that are traditionally referenced by the name of the document, rather than by name of editor, translator, or compiler. All codices listed are facsimile editions.

Anales de Tecamachalco, 1981, Crónica local y colonial en idioma Nahuatl, 1398 y 1590: México, D.F., Editorial Innovación, 101 p.

Baldet, F., 1950, Liste générale des comètes, de l'origine à 1948: Paris, Annuaire du Bureau des Longitudes pour l'an 1950, 118 p.

Carlisle, D.B., 1995, Dinosaurs, diamonds, and things from outer space: The great extinction: Stanford, California, Stanford University Press, 241 p.

Chimalpahin, D. de San Anton Muñon, 1963–65, Die Relationen Chimalpahin's zur Geschichte Méxicos, v. I, II [edited by G. Zimmermann]: Hamburg, Cram, de Gruyter & Company, 401 p.

Codex Borgia, 1976, Codices Selecti, Volume 58: Graz, Akademische Druck- und Verlagsanstalt, 76 p.

Codex Cospi, 1968, Codices Selecti, Volume 18: Graz, Akademische Druck- und Verlagsanstalt, 27 p.

Codex Dresdensis, 1975, Codices Selecti, Volume 54: Graz, Akademische Druck- und Verlagsanstalt, 74 p.

Codex Telleriano-Remensis, 1995, edited by Eloise Quiñones Keber: Austin, University of Texas Press, 102 p.

Codex Vaticanus A (3738), 1979, Codices Selecti, Volume 65: Graz, Akademische Druck- und Verlagsanstalt. 192 p.

Codex Vaticanus B (3773), 1972, Codices Selecti, Volume 36: Graz, Akademische Druck- und Verlagsanstalt. 96 p.

Köhler, U., 1989, Comets and falling stars in the perception of Mesoamerican Indians, *in* Aveni, A.F., ed., World archaeoastronomy: Cambridge, Cambridge University Press, p. 289–299.

Lehmann, W., translator and editor, 1938, Die Geschichte der Königreiche von Colhuacan und Mexico: Stuttgart, W. Kohlhammer, 391 p.

Moreno de los Arcos, R., 1967, Los cinco soles cosmogónicos: México, D.F., Estudios de Cultura Náhuatl, v. 7, p. 183–210.

Oppolzer, T.R. von, 1969, Canon of Eclipses [Canon der Finsternisse, Wien 1887], Gingrich, O., translator: New York, Dover Publications, 376 p.

Relaciones Histórico-geográficas de la gobernación de Yucatán, 1983, Volumes 1 and 2: México, D.F., Universidad Nacional Autónoma de México, 939 p.
Sahagún, B. de, 1950–69, Florentine Codex, Volumes 1 through 12 [Anderson, A.J.O., and Dibble, Ch.E., eds. and translators]: Santa Fe, School of American Research, 1736 p.
Seler, E., 1902–24, Gesammelte Abhandlungen zur Amerikanischen Sprach- und Altertumskunde, Volumes 1 through 5: Berlin, Ascher, 4041 p.
Torquemada, J. de, 1969, Monarquía Indiana: México, Porrúa, D.F., 2025 p.

Manuscript Submitted September 15, 2000; Accepted by the Society March 22, 2001

Printed in the U.S.A.

Geological Society of America
Special Paper 356
2002

Impact lethality and risks in today's world: Lessons for interpreting Earth history

Clark R. Chapman*
*Southwest Research Institute, Suite 426, 1050 Walnut Street
Boulder, Colorado 80302, USA*

ABSTRACT

There is a modern-day hazard, threatening the existence of civilization, from impacts of comets and asteroids larger than ~1.5 km diameter. The average annual world fatality rate is similar to that due to significant accidents (e.g., airplane crashes) and natural disasters (e.g., floods), although impact events are *much* more rare and the deaths per impact event are *much* greater. (Smaller, more frequent impacts can cause regional catastrophes from tsunamis of unprecedented scale at intervals similar to the duration of recorded human history.) As the telescopic Spaceguard Survey census of Near Earth Asteroids advances, numerical simulations of the dynamic and collisional evolution of asteroids and comets have become robust, defining unambiguously past rates of impacts of larger, more dangerous cosmic bodies on Earth. What are very tiny risks for impacts during a human lifetime become certainties on geologic time scales. Widely reported errors in predictions of possible impacts during the next century have no bearing on the certainty that enormous impacts have happened in the past. The magnitudes and qualitative features of environmental consequences of impacts of objects of various sizes are increasingly well understood. Prime attributes of impacts, not duplicated by any other natural processes, are (1) extreme suddenness, providing little opportunity for escape and no chance for adaptation, (2) globally pervasive, and (3) unlimited potential (for Cretaceous-Tertiary [K-T] boundary-scale impacts and larger) for overwhelming destruction of the life-sustaining characteristics of the fragile ecosphere, notwithstanding the rather puny evidence for impacts in the geologic record. A civilization-ending impact would be an environmental and human catastrophe of unprecedented proportions. The K-T-scale impacts, of which there must have been at least several during the Phanerozoic (past 0.5 b.y.), are 1000 times more destructive. No other plausible, known natural (or human) processes can approach such catastrophic potential. The largest impacts must have caused mass extinctions in the fossil record; other natural processes could not have done so. Perspectives concerning both the potential modern-day destructive potential of impacts and conceivable, almost miraculous refugia in our own world provide a new gestalt for thinking about past cataclysms.

*E-mail: cchapman@boulder.swri.edu

Chapman, C. R., 2002, Impact lethality and risks in today's world: Lessons for interpreting Earth history, *in* Koeberl, C., and MacLeod, K.G., eds., Catastrophic Events and Mass Extinctions: Impacts and Beyond: Boulder, Colorado, Geological Society of America Special Paper 356, p. 7–19.

INTRODUCTION

The idea that cosmic impacts on the Earth have played a significant, or even dominant, role in mass extinctions (and subsequent explosive radiation of new species) has evolved from widespread skepticism to substantial acceptance during the two decades since publication of the Alvarez et al. (1980a) hypothesis concerning the Cretaceous-Tertiary (K-T) boundary. However, there remain pockets of nonacceptance as well as a wide spectrum of opinions about the degree to which impacts have influenced evolution. Even among those who fully accept a role for impact in the K-T boundary extinctions, views range all the way from belief that the only substantiated case of an impact playing a role in mass extinction was the Chicxulub impact 65 Ma, which is viewed as just hastening the demise of already stressed populations, to the hypothesis of Raup (1991), that all mass extinctions, large and small, could have been caused by sudden environmental changes due to impacts.

As the Alvarez hypothesis was researched, awareness grew among scientists and the public of the modern-day risk to civilization from cosmic impacts. Once the purview of science fiction, it has become widely accepted (e.g., the report of an independent Task Force set up by the British Government; Atkinson, 2000) that the threat of a calamitous impact ranks among other hazards meriting national and international attention and consideration of preventative measures. While the chances of impact with an asteroid larger than 1.5 km diameter (deemed sufficient to threaten modern civilization; Chapman and Morrison, 1994) are very small, about one chance in several hundred thousand per year, the potential consequences are so enormous (perhaps the deaths of a quarter of the world's population) that the annualized fatality rate is similar to fatality rates associated with other natural hazards, such as floods and earthquakes (Fig. 1).

Research on the impact hazard, especially during the 1990s, has yielded a voluminous literature on the numbers and physical traits of the impactors, on the physical and environmental effects of impacts, and even on the potential response of human society to an imminent impact or to the aftermath of one. With the perspective from modern research on the impact hazard, the issues faced by historical geologists trying to understand the role of impacts during Earth history can be viewed from a new gestalt.

However, the modern-day impact hazard is not well understood by the public, by policy makers, and even by most scientists. Both the extremely low probabilities and the extremely great consequences of impact tax our intuition and common sense, because they are so far beyond the realm of our personal, or even historical, experience. Therefore, before applying insights from studies of the modern impact hazard to the historical record (the subject of a following section), I first introduce the impactor population and what is known about the consequences of impacts. I then discuss the issues of risk perception and uncertainties in impact prediction in order to demonstrate that they have no bearing on the certainty that impacts with unparalleled ecological consequences happened in the past. I then discuss the implications for the role of impacts in Earth history from lessons learned in the study of the impact hazard.

The following assertions about past mass extinctions are justified later herein.

It is virtually certain that several other impacts have occurred during the Phanerozoic (past 0.5 b.y.) that had at least the energy and potential ecological consequences of the Chicxulub K-T boundary impact, and that many other impacts have occurred with potential global consequences nearly as great. This is not a hypothesis: it is an inescapable fact derived from robust knowledge of asteroids and comets.

There is no other plausible, known kind of natural calamity that can possibly approach asteroid and/or comet impact in terms of the suddenness of the onset of devastating global consequences. (I exclude human devastations such as nuclear war, as well as other conceivable but unlikely disasters; e.g., a nearby supernova.) I assert that this suddenness, ranging from minutes to months, greatly magnifies the devastation compared with any other equally profound geologic, oceanic, or meteorological catastrophe.

The largest impacts during the Phanerozoic must have caused mass extinctions and, conversely, no other known plausible mechanism can approach the magnitude of consequences of such impacts. Therefore, the largest mass extinctions must have been caused by impact. (Only if required evidence of such impacts is missing from the geologic record must one then turn to the unlikely alternative explanations, e.g., a nearby supernova or explosion of an unexpectedly stupendous supervolcano.)

What is commonly accepted among impact hazard researchers as the threshold size of asteroid that could terminate civilization as we know it (1.5 km diameter) is more energetic than the explosive force of the world's combined arsenals of nuclear weapons by a factor of ~20. Yet the magnitude of the K-T boundary impactor (10–15 km diameter), and each of the several other equivalent or larger impacts that must have occurred since the Phanerozoic, is equivalent to a thousand civilization-ending impacts, all occurring simultaneously. The miracle is that anything survived. Perhaps the best way to visualize mass extinctions is to try to imagine the refugia that might exist for us, and for various species of animals and plants, in our modern world after it has been utterly devastated by an unimaginably colossal holocaust.

THE POPULATION OF IMPACTORS

Geologists have traditionally invoked the uniformitarian concept that continuous geologic processes observable today can account for what is observed in the geologic record of the past. In recent decades, a reasonable balance has been achieved between this two-century-old tradition and the important role

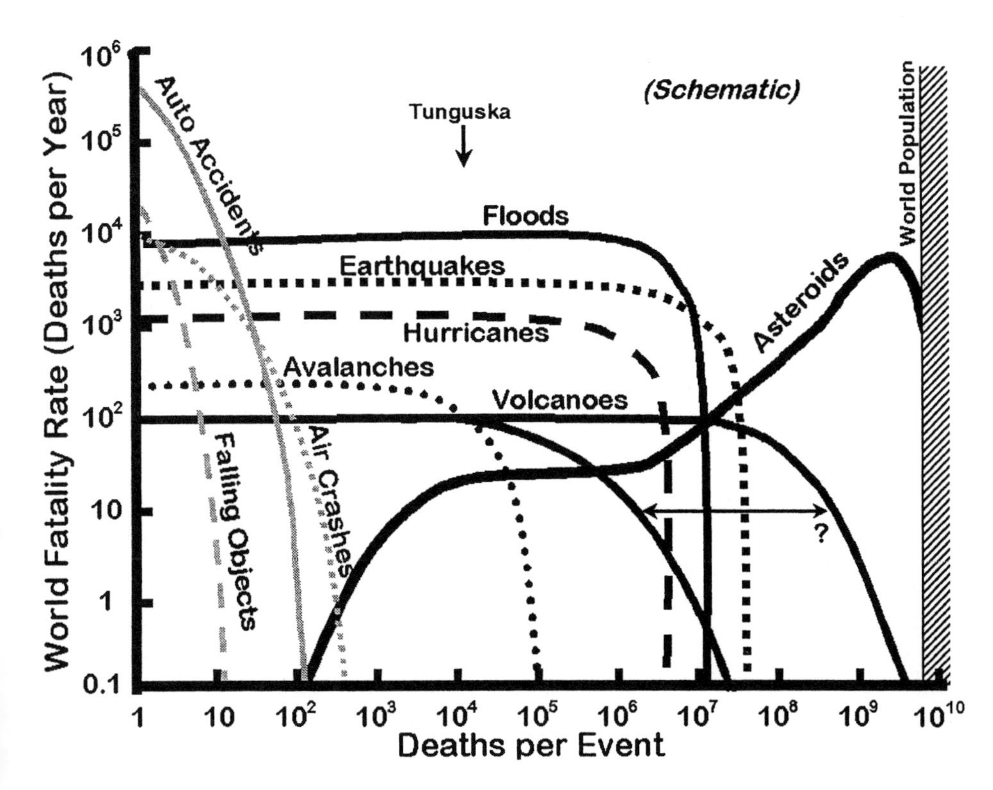

Figure 1. Schematic illustration of approximate average annual worldwide fatality rate for various kinds of accidents and natural disasters of various magnitudes. Accidents generally kill only a few people at one time, although accidents involving large transportation vehicles (e.g., aircraft) can kill hundreds. Natural disasters, as a class, are far less deadly than automobile accidents, but the largest among them can be far more deadly per event. Natural disasters comparable in lethality to Tunguska-class impacts (which occur every century or two; see downward pointing arrow) are about two orders of magnitude more frequent than such impacts. Most natural disasters have natural upper limits to lethality, because they are confined to particular geographic localities and/or are limited by physics or by strength of Earth's crustal materials in maximum magnitude. The upper limit in lethality of explosive volcanism is less well known; two alternative limiting curves are shown (see double-headed arrow), one extending roughly to consequences of impact of a 1–2-km-diameter asteroid. Only asteroids and comets have no upper bound in size and could, conceivably, eradicate the human species. In terms of fatality rate, asteroids dominate over all other natural disasters combined for individual events that kill more than 10^8 people at once, and are maximized for civilization-threatening impacts that can kill 10^9 people or more.

of episodic, even catastrophic, geologic processes. However, when geoscientists turn their attentions away from their specialties, old habits can reemerge. Frequently during this conference ("Catastrophic Events and Mass Extinctions: Impacts and Beyond"), speakers from geological and paleontological backgrounds made statements such as, "for this particular mass extinction, there is no reason to invoke an ad hoc impact from the heavens." They apparently miss the point, developed robustly over the past 70 yr, that cosmic impacts—despite their rare, catastrophic manifestations on Earth that concern us here—are part of an ongoing, continuous process that is observable today and has necessarily operated during the past history of the Earth.

Impact hazard researchers currently direct most attention to telescopic searches for Near Earth Asteroids (defined as those so-called Earth-crossing asteroids [including dead comets] whose orbits cross the Earth's orbit, in the sense that their closest and farthest distances from the Sun include 1 Astronomical Unit, the mean distance of the Earth from the Sun, plus the so-called Amors, which get as close to the Sun as 1.3 times the Earth's distance from the Sun). Depending on details of counting, currently more than 1550 are known, even though the first asteroid in an orbit that actually crosses the Earth's orbit was not found until 1932 and fewer than 20 Near Earth Asteroids were known as recently as the early 1970s. Currently, numerous telescopes equipped with modern detectors systematically scan the skies. Astronomers assemble data on detections and calculate orbits for these bodies. About one-half of all Earth approachers larger than 1 km diameter have been cataloged, as well as large numbers of smaller bodies ranging to the size of a small house.

These are just the largest, and potentially most dangerous, of a vast complex of interplanetary objects and particles in Earth-approaching orbits, ranging in size from enormous asteroids like Eros (34 km long; Veverka et al., 2000) and still larger comets, to rocks and dust particles. The basic physics of how this debris is created (by hypervelocity collisions among the debris) and how it is lost (by collision with the Sun or planets, and other loss mechanisms) has been understood for a long time. For example, the collisional cascade that creates and maintains the population of smaller bodies was explained by Piotrowski (1953) and Dohnanyi (1969); modern research has made changes that only specialists could care about. Fundamentally, the asteroids and smaller debris orbit around the solar system (inside of Jupiter's orbit) in a way that, despite some regularities, generates essentially random encounters and collisions among themselves at speeds of many kilometers per second. From the well-known mechanical properties of the common materials of which the debris is composed (rock, ice, carbonaceous mud, metal), the objects inevitably are broken by such collisions into smaller fragments and dust.

The resulting size distribution (numbers as a function of size) from multihundred kilometer asteroids and comets down to dust grains is well known and essentially invariant (cf. Durda et al., 1998). Dust grains are abundant and the Earth's large cross section continuously sweeps them up, as anyone can observe on a clear dark night: a meteor flashes through the upper atmosphere, as viewed from one location on the ground, every few minutes. Dust detectors on spacecraft confirm the widespread distribution of such grains throughout interplanetary space. Impacts of meter-sized bodies are much more rare, but are routinely observed by downward-looking satellites searching (primarily) for signs of military activity (Nemtchinov et al., 1997), and occasionally by ordinary human beings, as stunningly brilliant fireballs. For example, a 5-m-diameter impactor shone 10 times brighter than the Sun, as observed from the Yukon, when it struck in January 2000 (Brown et al. 2000), yielding some precious meteorite fragments.

Impactors several tens of meters and larger are too uncommon to strike regularly during a human lifetime, although the 15-Mt-equivalent Tunguska event in 1908 is well documented (probably caused by an asteroid ~50 m in diameter). However, objects of these sizes passing within a few million kilometers of the Earth and the Moon are regularly discovered by the telescopic scanning programs, especially by the Spacewatch Program on Kitt Peak, Arizona, which is optimized for discovering smaller bodies. The sampling becomes a complete census for Earth-approaching bodies larger than ~7 km diameter, not counting rare comets that can approach from the darkness of the outer solar system.

It is purely a matter of random chance, equivalent to rolling dice, when an impact will happen—whether a faint meteor streaking across the sky or a dinosaur-killing impact—but the average frequencies of impacts of objects of different sizes is well known and has not significantly changed since Shoemaker's (1983) review. Subtle regularities cause only slight departures from purely random chance. Specialists debate the exact numbers of bodies of specific sizes. However, differences are rarely greater than a factor of a few, and are often less than a factor of two. For example, it was long estimated that the number of Earth-approaching asteroids larger than 1 km diameter might be ~1500. During the last few years, there has been a well-publicized debate (e.g., Rabinowitz, 2000; Bottke et al., 2000; Stuart, 2001) about whether that number is really as low as 700 or as high as 1200. The answer has potentially important political consequences, such as whether NASA can reach its committed goal to find 90% of such objects by 2008 (Pilcher, 1998) without building more, larger search telescopes. Such arguments are inconsequential, however, in the context of impact catastrophes past or future.

Not only are the numbers and impact frequencies of interplanetary objects of all sizes well known today, but today's samplings and census are known to have been generally unchanged during the past 3.5 b.y. The physics of these bodies and their collisional evolution is well understood, and must have been as applicable in the past as today. Furthermore, understanding the sources and sinks of these bodies and their dynamics (e.g., how they move through the solar system on time scales ranging to billions of years) has developed remarkably in the past decade due to the advent of inexpensive, very fast computers. Although Kepler's Laws were never in doubt, the dynamic systematics of the entire complex of asteroids and comets has become well understood only during the past five years. Furthermore, examination of the cratering record on the Earth and terrestrial planets—and especially on the Moon—has demonstrated the continuity through the past 3.5 b.y. of impact processes. Specialists are interested in minor variations in impact rates and in the shape of the size distribution. However, since the late heavy bombardment ended ca. 3.8 Ga, the impact rates at all sizes have never varied by more than factors of a few, except (probably) for brief, transient showers of modest magnitude that have made a negligible contribution to the cumulative record of craters.

Table 1 translates the known, largely invariant size distribution of interplanetary projectiles, and their rates of colliding with Earth, into some relevant chances of impact by bodies of three interesting sizes: a small asteroid 200 m across capable of creating a devastating tsunami unprecedented in historical times; a civilization-ending impactor 2 km in diameter; and a K-T boundary extinctor (10–15 km in diameter). Regarding the consequences of such impacts, how do we know what will happen if the Earth is struck by a 200 m, 2 km, or >10 km body?

CONSEQUENCES OF IMPACTS

Environmental effects

Studies of the modern-day impact hazard have greatly augmented our understanding of the consequences of impacts, probably more so than have analogous studies of the physical, chemical, environmental, and biological effects of giant impacts in a K-T context. Studies of the modern-day hazard (cf. Adushkin and Nemchinov, 1994) have usually focused on the dangerous objects that are most likely to strike, those ranging from producers of giant tsunamis (~200 m diameter, ~1000 Mt explosive yield) to the civilization enders (~2 km diameter, 10^5 Mt), which involve modest extrapolations from weapons tests and the Tunguska event. A reality check, the impact of Comet Shoemaker-Levy 9 into Jupiter in 1994 (roughly equivalent to the ~2 km terrestrial case because of the much higher impact velocity at Jupiter), was extensively researched and applied to the Near Earth Asteroids hazard (cf. Boslough and Crawford,

TABLE 1. CHANCES OF EVENT HAPPENING IN SPECIFIED DURATION

	Human-Scale	Historical		Geological		Planetary
Object (diameter)	1 yr	100 yr	10,000 yr	1 m.y.	100 m.y.	4 b.y.
Cretaceous–Tertiary extinctor (10–15 km)	10^{-8}	10^{-6}	10^{-4}	1%	50%	100%
Civilization ender (2 km)	10^{-6}	10^{-4}	1%	50%	100%	100%
Huge tsunami (200 m)	10^{-4}	1%	50%	100%	100%	100%

1997). Such research has also provided a guide for extrapolating to the far more energetic case of the K-T boundary impact. The latest, most comprehensive review of the environmental consequences of impacts, ranging from 20 m to 20 km diameter ($1–10^9$ Mt), is that of Toon et al. (1997).

The first salient fact is that the impact of a cosmic body with Earth, whether at 15 or 25 km/s (or sometimes greater speeds for comets), essentially causes an explosion: an instantaneous conversion of the kinetic energy of the impactor into fragmentation, comminution, and cratering of the substrate; heating, melting, and vaporization of the projectile and target materials; kinetic energy of cratering ejecta; seismic shock waves penetrating the planet; and other types of destructive energy. Precisely how the kinetic energy is partitioned into the various forms of energy is the subject of continuing research, but we only have to look at the now many-decades-old sites of nuclear weapons tests to understand the general idea. Modern computer codes reliably reproduce the weapons tests and, based on sound physics, can be extrapolated robustly to the energy scales of civilization-ending impacts and perhaps even to mass extinctors.

The second salient fact is that a very significant fraction of the energy from an impact is dissipated in the ecosphere, the thin shell of air, water, and surface rocks and soils whose constancy sustains and nurtures life. The Earth as a planetary body has been unfazed by any impacts subsequent to the colossal interplanetary collision that is believed to have formed our Moon. The geologic record is only marginally perturbed by even the largest post-late heavy bombardment impacts: witness the general obscurity of the famous clay layer at the K-T boundary. The boundary is readily recognized by the permanent change in the diversity of species, but it is not prominent as a geologic feature (the centimeter-scale layer is dwarfed by ordinary sedimentation and erosion and by faulting and other pervasive effects of tectonism and volcanism). However, our thin ecosphere is exceptionally subject to damage and instantaneous modification by events of these magnitudes, even if only a tiny fraction of the kinetic energy of the impactor is partitioned into the atmosphere during the bolide phase (passage of the impactor through the atmosphere), during the explosion, and during the subsequent ejecta plume phase.

A final fact about consequences of impacts is that those that exceed the relevant threshold sizes (dependent on the particular consequence) necessarily distribute their consequences globally: while the greatest damage is obviously at ground zero, the stratosphere is badly polluted with dust on a global scale from impacts exceeding 10^5 Mt (1 km diameter), glowing ejecta are distributed globally from impacts exceeding 10^8 Mt (15 km diameter), and even seismic shock waves may reach moderately damaging proportions on a global scale for impacts of 10^8 Mt scale (K-T level). Even much smaller impacts (e.g., by a 200 m impactor), if into the ocean, can cause devastation thousands of kilometers away due to the efficient transmission of energy to great distances by tsunamis (Ward and Asphaug, 2000). In normal times, the distributive character of air and water is what lubricates our world, maintains chemical balance, and sustains life. In times of catastrophe, however, which overwhelm the modest mass of the atmosphere and ocean and their thermal and chemical balances, these media distribute poisons, sun-darkening dust and aerosols, and meteorological and climatological consequences around the globe. Rebound from past catastrophes that have afflicted civilization (e.g., World War II) have often depended on some portions of the planet remaining unaffected by the localized or regional devastation, thus serving as nuclei of recovery. In the case of a sufficiently large impact, there are essentially no unaffected refugia where life continues normally.

Consider the Comet Shoemaker-Levy 9 impact into Jupiter in 1994: with the kinetic energies roughly that of a civilization-ending impact on Earth, the largest comet fragments created immense, black patches in Jupiter's stratosphere (certainly appreciably dimming the sunlight beneath); several of them exceeded the size of the entire planet Earth and persisted for months (Chapman, 1995).

Precisely what dominant environmental consequences arise from impacts is less certain than the generalizations just listed. Certainly the vagaries of weather forecasting and of other contemporary forecasts of environmental scenarios (e.g., global warming) engender an understandable skepticism among the public about the predictive sciences (cf. Sarewitz et al., 2000). However, the magnitude of a major impact is so enormous compared with the environmental perturbations resulting from twentieth and twenty-first century civilization, that the kinds of uncertainties that plague the other predictions are overwhelmed. Furthermore, because there are so many separate phenomena, the synergies among them, which are difficult to model, probably lead to conditions appreciably worse than the simple addition of their separate effects. If one or two of them are less effective than initially calculated, there remain numerous other damaging consequences. For example, estimates of the production of nitric acid, once thought to be a primary environmental effect of a K-T-scale impact, have more recently waned even as sulfuric acid has received greater attention due to the probable anhydrite-rich substrate near Chicxulub (Pope et al., 1994).

The complete suite of consequences for a 2 km impactor and for a 10–15 km impactor, primarily as gleaned from the comprehensive review of Toon et al. (1997), is summarized in Table 2, supplemented in some cases by insights from other, more recent work. I have left out less significant, more localized damage (e.g., blast effects near ground zero), less well understood effects (general toxicity of the environment and effects on ocean chemistry), and secondary and long-lasting effects.

Although there are significant uncertainties in some of these results, the inevitability of most of the effects within the range of impactor scales we are considering is assured. Several of the effects may independently range from global deterioration of the biosphere (for 2 km impactors) to massive destruction of the biosphere (for K-T-scale impacts). Some of the effects are complementary, e.g., the dramatic cooling effects of impact winter would be moderated near ocean shores due to the ocean's heat capacity (Covey et al., 1994); however, these are the regions that would be inundated and scoured by tsunamis. The tabulated consequences acting in concert (along with other effects not yet fully evaluated), and extended by the less certain, longer term consequences for the chemistry and temperature of the atmosphere and the ocean, would make life on Earth following a big impact horrific.

Civilization-ending impact

The consequences of a civilization-ending impact can dwarf the environmental effects of historical environmental catastrophes, such as the so-called year without summer due to the massive Tambora volcanic eruption in 1815 as well as nuclear winter scenarios envisioned to result from all-out nuclear war (discounting the immediate and long-lasting radiation effects of the latter). An impact is far more efficient than nuclear war (or volcanic explosions) at polluting the stratosphere, despite the fact that other kinds of damage are far more concentrated in one locality in an impact. The most dramatic consequence for modern civilization seems to be the prospect that all agriculture would be lost for a year. Given ongoing episodes of Third World starvation that occur even under the optimized international food-distribution systems in stable times, it seems likely that a sudden impact by a kilometer-scale comet or 2 km asteroid would lead to mass starvation of a sizeable fraction of the world's population.

The end-game of such a scenario naturally involves highly uncertain speculation about the longer term response of the ecosphere, of corporate, national, and international infrastructures, and of the global economic system. Some commentators view civilization as inherently fragile. Human beings have moved away from nature and lack knowledge about survival in the absence of manufactured goods and retail stores. Technology has become highly specialized and is generally inaccessible and incomprehensible to nonspecialists. American society proved to be astonishingly vulnerable to terrorist acts in late 2001, which had objective consequences comparable to one month of automobile traffic fatalities. The network of interdependencies among nations is fragile, even absent a global calamity. A breakdown of social order (like that postulated in the aftermath of a comet strike in *Lucifer's Hammer*; Niven and Pournelle, 1977) is viewed by some as inevitable, probably leading to conflicts and wars on local to global scales (and modern warfare

TABLE 2. MAGNITUDES OF SEVERAL KINDS OF ENVIRONMENTAL CONSEQUENCES FOR TWO SIZES OF IMPACTORS

Chief environmental consequences of impacts	Civilization ender (2 km)	Cretaceous–Tertiary extinctor (10–15 km)
Fires ignited by fireball and/or reentering ejecta	Fires ignited only within hundreds of kilometers of ground zero.	Fires ignited globally; global firestorm assured (Wolbach et al., 1988).
Stratospheric dust obscures sunlight	Sunlight drops to "very cloudy day" (nearly globally); global agriculture threatened by summertime freezes.	Global night; vision is impossible. Severe, multi-year "impact winter."
Other atmospheric effects: sulfate aerosols, water injected into stratosphere, ozone layer destruction, nitric acid, smoke.	Sulfates and smoke augment effects of dust; ozone layer may be destroyed.	Synergy of all factors yields decade-long winter. Approaches level that would acidify oceans (more likely by sulfuric acid than nitric acid).
Earthquakes	Significant damage within hundreds of kilometers of ground zero.	Modest to moderate damage globally.
Tsunamis	Shorelines of proximate ocean flooded inland tens of kilometers.	Primary and secondary tsunami flood most shorelines ~100 km inland, inundating low-lying areas worldwide.

has become *very* dangerous). Such fragility could easily lead from an impact catastrophe to the death of most of the world's population and a long-lasting Dark Ages.

However, other commentators believe that civilization is robust. Frequently, the human spirit rises to meet challenges that seem overwhelming. Cooperation rather than social disintegration seems more likely to some. There are technological refugia (e.g., bomb shelters) and other forms of mitigating the disaster, especially if there is some warning (e.g., food supplies to outlast the darkness could be grown and stored, given a decade's warning, and thanks to the Spaceguard Survey, warning of an impact years to decades in advance is increasingly likely). Human history has demonstrated society's ability to recover from such holocausts as plague and World War II (although less affected peoples and nations contributed to recovery from World War II, which might not be the case in a truly global catastrophe).

Extrapolation to K-T scale impact

With increasing size of impactor, the magnitude of the catastrophe grows toward the scale of a K-T boundary event (with a thousand times the destructive energy of the civilization-threatening event just discussed), and the certainty that, not only would civilization collapse, but the human species would be rendered extinct. Who could survive? Even a well-trained survivalist, capable of living off desert lands in perpetuity, would be overwhelmed by months (not to mention years) of trying to survive in a burned, denuded, bitterly cold, perpetually dark, and poisoned environment. To even try to live off the land and the dregs of a destroyed civilization, an individual would have to have survived (in some deep cave or other shelter somewhere) the initial calamity of a global firestorm, global earthquakes, and other immediate traumas of the impact. How would land-based animals, or complex plants, be any more successful at surviving? Oceanic life would be buffered from the fire, but would still be subject to changes in chemistry and, eventually, temperature, which would be pervasively distributed throughout the waters, with adequate refugia being even more difficult to imagine.

From the perspective of a typical individual (human, animal, plant), survivability from a K-T-scale event is impossible to imagine. However, lessons from the aftermath and recovery of local populations following the Mount St. Helen's eruption and evidence that some species have evolved accidental protections from otherwise highly lethal environments (including extreme temperatures and even high doses of radiation), suggest why the K-T event did not doom all life larger than microbial. To understand such survivability, one must concentrate on *exceptional* environments, most readily imagined by thinking of the world we know, including its special environmental niches and microclimates. Presumably analogous circumstances existed in the past.

An example of such an exceptional environmental niche might be a small herd happening to be next to a thermal spring, and thus luckily in a much better position to survive a multiyear winter than most individuals of the same species; especially so if the spring happens to be deep within a cave where the lucky herd avoided being scorched during the initial postimpact firestorm and was thermally buffered from the multiyear winter. If the cave with springs happened, also, to be on a far-offshore island perhaps shielded from the glowing ejecta by a thick overcast at the time of impact, a small ecosystem of animals and plants might have temporarily survived. All would be lost if that island were subsequently submerged by the impact-generated tsunami or by storms generated by catastrophic meteorological changes. But perhaps not; perhaps the island is perched high above sea level and/or is very far from ground zero. This hypothetical concatenation of lucky circumstances puts the lucky herd a few steps up the ladder of potential survival, although many more environmental challenges must still be overcome to assure long-lasting survival and repopulation of the species. Through such fortuitous circumstances in an exceptional refugium, one can imagine that small reproductive groups might permit the survival of certain lucky species, even if 99.99999% of individual species members have died. That is presumably what happened 65 Ma.

PERCEPTIONS OF RISK

The impact hazard has received some bad press in recent years, giving the subject a certain "Chicken Little" unreality. To underscore the robustness of my central message, I address the issues that affect individual (and society's) perception of the risks associated with the impact hazard. Among the most important are the following.

1. The failure to grasp the meaning of low-level probabilities or of randomness.
2. The fact that ordinarily negligible errors can overwhelm the "signal" of a low-probability event, requiring exceptional procedures for handling calculations and reporting of low-probability events.
3. The failure to understand that scientific research (in this arena, especially) is an ever-improving process and that retracted predictions of impacts or near misses are the usual outcomes of this research, and generally do not imply that mistakes have been made.

In the literature of the psychology of risk perception (e.g., see Cole, 1998), it is commonplace that the human brain finds it inherently difficult to grasp the meaning of probabilities outside of the range of our practical experience. The 1 in 649 739 chance of being dealt a royal flush in poker (not to mention winning a national lottery) is lower than the chance that the Earth will be struck by a civilization-ending asteroid next year. Few gamblers could imagine worrying about the end of everything and everyone they know and love while they still harbor a real hope of beating the odds. People also fear that they may die by several other frightening causes less likely than that of being killed by an impact catastrophe, including death by a wild animal, lightning, or tornado. Companies, governments, and citizens apply great pressure for increased airline safety, despite the fact that an individual American is more likely to die as a result of an asteroid impact than by jetliner crash. Extremely dangerous activities (far exceeding dangers from airplanes or asteroids), however, such as smoking or driving automobiles, or leading a sedentary life, are readily tolerated and rationalized.

Another common confusion involves misunderstanding that the typical waiting time until the next impact (a few hundred thousand years for the end-civilization impact) justifies current inaction. (A related, common confusion familiar to geologists is the layperson's expectation that one can ignore the possibility of a flood because "the hundred-year flood just happened two years ago.") The impact could happen just as readily next year as in some particular year tens of thousands of years from now.

The history of widely publicized impact scares during the past decade may be leading to a "boy who cries wolf" skepticism about the robustness of astronomers' observations and calculations about impact probabilities. Despite attempts to improve, regularize, and simplify the reporting of inherently difficult to understand results to the public (e.g., through de facto adoption of the Torino scale [Binzel, 2000], analogous to the Richter scale for earthquakes, to categorize predictions of possible future impacts), there continue to be headlines about dangerous impacts in the next decades, generally immediately followed by what are perceived as retractions. Several factors, beyond the commonplace hyperbole and misreporting by news media, contribute to these unfortunate perceptions.

Consider what is happening in interplanetary space and in astronomical observatories. The Earth is in a cosmic shooting gallery, although space is very big, so nothing consequential hits Earth very often. During the past three decades, and especially during the past five years, astronomers have begun to scan the skies for asteroids, especially the ones more likely to hit (e.g., not asteroids in the main asteroid belt, most of which are safely there "forever" and all of which are safely there for millennia). Near Earth Asteroids are found as an unknown, uncharted star on a photographic plate or, more recently, on a charge-coupled device (CCD) image. They are confirmed when, after several exposures, they are found to be moving at an appropriate rate (not as fast as an airplane or satellite, but not so slowly as a main-belt asteroid or distant comet) during the course of the night. After observations over the course of a few weeks (provided skies are clear and the patch of sky is in the coverage area of one of the photographic search telescopes), positions of the object are established well enough to calculate an approximate orbit.

While most such preliminary orbits do not permit the asteroid to come anywhere near the Earth in the foreseeable future (in which case the future impact probability is exactly zero), a small fraction of such orbits, especially when propagated forward in time a few decades, include the Earth in the large volume of space that is within the very broad error bars associated with the preliminary orbit. The chances of impact may even be smaller than the chance of a random, thus far undiscovered object hitting the Earth, but at least there is now a known date or dates in the future when such a specific object could conceivably hit; it thus bears monitoring in the future.

After more weeks of additional observations of this still-threatening object, or possibly after discovery of a preexisting observation of it in an archive (that had not previously been successfully linked with other observations to compute a preliminary orbit), the preliminary orbit can be refined and the error bars reduced. In most such cases, the refined orbits no longer include the Earth within uncertainties, and the probability of impact goes to zero. Very occasionally the refined orbit narrows down to a zone that still includes the Earth, and the probability of impact goes up, perhaps to better than 1 chance in a 1 000 000 (for a 1-km-diameter asteroid) or 1 chance in 10 000 (for a 100 m body), which merits moving it from 0 on the Torino scale (meaning roughly equivalent to the background chance of unknown asteroids striking the Earth) to 1 (events meriting careful monitoring). Such cases have been happening a couple of times a year lately, and they may happen more frequently as search techniques advance.

A Torino scale rating of 1 (or higher) generates considerable interest in the media and within the astronomical community. An automatic review of the calculations by a Working Group of the International Astronomical Union (IAU) commences, and observers around the world focus on the potential impactor with urgency, generating new observations or discoveries of archived observations. Commonly, within a few days, the refined data shrink the error bars and an accurate orbit can now be computed. Almost always, the chance of impact reverts to exactly zero and an "all clear" is announced, which the media, having just published news of an impact possibility a few days earlier, tend to call a retraction. The possibility exists, however, although it has never happened yet and is not likely to, that the accurate orbit predicts—now with much higher likelihood, perhaps certainty—a future impact. That, after all, is the purpose of the search. We already know that there is only a one in a few thousand chance of impact of a kilometer-sized body sometime this century, so we expect that refined orbits of new discoveries will continue to move toward zero probability impact. However, there are bound to be a few cases a year in the intermediate stage of orbit improvement that temporarily swing as high as 1 on the Torino scale, meriting attention for a while.

The normal routine described here illustrates why media discussion (e.g., "it is not going to hit after all") misrepresents the Spaceguard search process, although there have been surprises and even mistakes. A surprise occurred in October–November 2000, when an asteroid was calculated to have an astonishingly high 1 in 500 chance of impacting the Earth 30 years hence. The body was faint, hence small, but plausibly of Tunguska size, hence meriting a 1 on the Torino scale. The IAU, following its mandated 72 hour review process, reported confirmation of the calculation; unfortunately, just hours later an earlier observation was found, proving that the impact would not happen. The news media had a field day with the "correction." Further investigation revealed that the object was, in all probability, a highly reflective old booster rocket from the early 1970s. Not only is it hollow, but it is much smaller than had been estimated, and constitutes no danger at all if it is to hit the Earth, which, indeed, seems likely to happen within some thousands of years. Its surprisingly Earth-like orbit would be a strange one for a real asteroid, but typical of space junk. In the future, astronomers are likely to be more aware of the possibility of being confused by space debris.

Much of the skepticism about astronomers' predictions is the legacy of an actual mistake made in 1998 (cf. Chapman, 2000), when an internationally respected astronomer announced that a civilization-ending asteroid, 1997 XF11, would come spectacularly close to the Earth in 2028, "virtually certain" to pass within the orbit of the Moon but nominally only 40000 km away, implying an impact probability as high as 0.1%. The calculations were faulty. Data archived by the astronomer during several previous months were sufficient to calculate an impact probability of essentially zero (about 1 chance in 10^{42}), but he was excited and failed to check his results with colleagues before issuing a press information statement that generated headlines around the world. Once again, astronomers rushed to their archived images and found positions for 1997 XF11 that showed it to be in an orbit such that it could not possibly hit the Earth, but would actually pass 2.5 times farther away than the Moon in 2028. Unlike the nominal process described here, this time the original prediction was just wrong.

An unappreciated reality affecting predictions of very low probability occurrences is that the probability of making an error in calculating such a probability is much larger than the probability itself. Ordinary human care, resulting perhaps in 99% reliability, doesn't suffice when trying to reduce the already extremely tiny chances of an airliner accident, or in assuredly calculating a low-probability asteroid impact. In the operations arena, the engineering discipline of surety systems analysis has been devised to build in safeguards against even the extremely low probability concatenation of improbable events that after the fact analysis often shows to be the cause of rare accidents, e.g., airliner crashes or the Three Mile Island nuclear accident. Surety involves "out of the box" thinking about exceptionally unusual circumstances, human factors analysis, and multiple closed-loop redundancies.

In asteroid astronomy, similar procedures must be implemented to avoid cries of "wolf!" At the time of the 1998 mistaken announcement, given the known impact probabilities, it was much more likely that the astronomer had made a mistake than that the newly implemented Spaceguard Survey had already found an asteroid, large enough to destroy human civilization, with a significant chance of striking within our lifetimes. Indeed, the astronomer was mistaken. The calculation-checking procedures of the IAU were subsequently developed, in part, to minimize the chances of future mistakes. Henceforth, we may hope that reported possibilities of future impacts are at least objective, even if they will almost certainly quickly evolve to zero.

In conclusion, the widespread dissension within the astronomical community concerning issues of impact probabilities and the outright skepticism sometimes expressed in the media are an inevitable result of misunderstandings over how to understand and communicate about unfamiliarly tiny probabilities. They in no way should be taken to undercut the robust understanding of how often the Earth is likely to be struck by cosmic projectiles of various sizes.

There is a related analogy relevant to how geologists and paleontologists, facing rare crises in Earth history, should evaluate evidence in the geologic record. Given the unimaginably grotesque consequences of large asteroid impacts, which have certainly happened, as well as the range of lesser but nonetheless dramatic catastrophes occasionally posed by volcanism, tectonics, and potential climatological instabilities, we must step "out of the box" of our normal world and think realistically about how biological populations and ecosystems might have been affected by such rare disasters. The rules are different at such times from anything we have personally witnessed or can even easily imagine.

UNDERSTANDING CRISES IN EARTH HISTORY

Comparisons of natural hazards

The first lesson for historical geology from studies of how the impact hazard affects our modern world is to understand the almost unfathomable differences in scale of impacts of various sized asteroids. Even the "small" ones have enormous consequences beyond our experiences. The 1908 Tunguska impact unleashed an explosive energy equal to more than a thousand Hiroshima bombs and only a few times less than the largest ever bomb test. Tunguska devastated $\sim$1000 km^2 of Siberian forest or $\sim$0.001% of the land area of the Earth. In contrast, the energy of the K-T boundary impact was 10^7 times greater than Tunguska; one could think of every 1000 km^2 land unit on our planet being allocated 500 times the energy that leveled the Tunguska region. Actually the destructive processes change with scale of impact and the consequences vary with distance from ground zero, but clearly, even if the comparative destructive efficiencies are extremely low, our fragile ecosphere has to absorb an enormous amount of destructive energy within an hour or two of a K-T-scale impact.

Figure 1 is a highly schematic representation of the comparative consequences of various kinds of accidents and natural disasters, represented by human lethality. The vertical axis represents the annualized world fatality rate from various types of accidents and disasters; the more serious sources of death plot higher on the graph. The horizontal axis (deaths per event) depicts an important qualitative difference between the various accidents and disasters. Automobile accidents kill many people; they happen frequently, but generally kill only a few at a time. Accidents involving buses, trains, ocean vessels, and airplanes have the potential for killing many more people at a time, and occasionally they do, which is why their curves extend somewhat to the right. While natural disasters, like a small avalanche or a minor earthquake, can kill just a few people, many deaths from natural disasters result from rather rare, big events. For example, between 10^5 and 2×10^6 people died in each of the 11 worst natural disasters (chiefly earthquakes, floods, and cyclones) during the period 1900–1987 (Munich Reinsurance Company, 1988), even though many years passed with no natural disasters even approaching these rates of lethality.

The impact hazard represents another jump toward extremely high lethality per event, but extreme rarity. Averaged over time, the lethality (height on the diagram) is comparable with many other individual kinds of natural disasters, although less than for some kinds of accidents. (War, famine, and especially disease greatly exceed both natural disasters and accidents as the chief killers.) Qualitatively, the impact hazard is very different from anything else plotted: it is the only hazard capable of killing hundreds of millions of people, or even the entire world population, in one event.

Of course, nuclear war has been hypothesized as having the potential to reach this level of death and destruction. However, it presumably has no relevance for understanding past mass extinctions. Conceivably, some virulent disease could break out and decimate, or even eradicate, the human species; this also is probably not relevant to understanding mass extinctions because diseases are normally species specific and are not easily spread among numerous species, although breakdowns of ecological systems could conceivably magnify the consequences of such an outbreak. A nearby astrophysical disaster (supernova) cannot be completely ruled out, although it would be very unlikely.

Most geophysical natural hazards necessarily have natural upper bounds to their catastrophic potential. For example, Chinnery and North (1975, p. 1198) stated "There are good reasons for believing that there must be an upper bound to earthquake M_o values, due to the geometry of seismic zones and the strength of crustal material." The only possible competitor for asteroid impacts is volcanism. It has been argued that monstrous volcanic explosions (cf. Rampino et al., 1988), dwarfing those recorded during human history but occasionally recognizable in the geologic record, could approach the magnitude of a kilometer-scale asteroid impact. This topic deserves further research (see double-headed arrow in Fig. 1 indicating conservative and liberal possibilities for the magnitude of large volcanic events), but it also seems unlikely to apply to mass extinctions. There are inherent limitations, imposed by the strength of the Earth's crustal rocks, in the possible magnitude to which pent-up volcanic energy can rise before breaking through. Therefore, there must be an upper limit to the magnitude of a volcanic explosion; the Toba event of ca. 75 ka, recorded in the geologic record, may be as big as they get, and no mass extinction was associated with that.

The asteroid and comet size distributions, however, continue to larger sizes without end. While only a few Earth-approaching asteroids currently exceed the size of the K-T boundary impactor (none of them can strike the Earth in the near future, although Earth approachers are replenished on time scales of millions to tens of millions of years), an unknown comet could arrive with a warning of only months or a year, and it could have an immense size. Comet Hale-Bopp, prominent in the sky in 1997, was estimated to have a diameter of at least 25 km and perhaps as large as 70 km. It came within the Earth's orbit, although on the other side of the solar system. Had it struck, with its energy of tens to hundreds of K-T boundary impactors at once, it might have sterilized our planet of all but microbial life. Thus no hazard other than cosmic impacts has the possibility of conceivably eradicating humanity in a single event. Fortunately, the odds are very small that such an event will happen soon.

Some perspectives on the past from today

Looking to Earth history, however, extremely small odds during a human lifetime become virtual certainties on a time scale of geologic epochs. The odds of any of the three examples of impactors (Table 1; 200 m, 2 km, 10–15 km) striking during a year—the usual temporal measure for human hazards—are very small, ranging from 10^{-4} to 10^{-8}. However, all of them are *certain* to happen on geologic time scales. There have been repeated impacts resulting in huge tsunamis during Earth's history, and one may even have struck during human history (conceivably contributing to one or more of the great flood myths). A "civilization ender" is likely to strike a couple of times every 10^6 yr (which means ~100 of them since the time of the K-T boundary). They have necessarily caused "bad years" for most species dependent, directly or indirectly, on a summer season. K-T scale impactors have surely struck several to a dozen times during the Phanerozoic, and it is natural to try to associate the worst crises in Earth history with those randomly timed but irrefutable cataclysmic events of the past. What is rare with almost negligible chances on a human time scale (thus permitting international society to largely ignore this threat to its very existence) becomes a certain fact in the context of interpreting the paleontologic record. Impacts cannot be ignored: they have happened, and the larger among them were unimaginably devastating.

There are some ways of thinking about mass-extinction events that can be seen as unrealistic if viewed from the perspective of a modern-day catastrophe. We must especially heed the variety of things that can happen within lengthy durations that are unresolvable in the geologic record. We must not attribute to the global ecosystem, but rather to exceptional refugia, the characteristics that permitted some species to survive a mass extinction. The following anecdotes from discussions at the 2000 Snowbird conference exemplify how we must change our thinking.

The difference in time scales relevant to the survival of a species in the face of a sudden, global, environmental catastrophe compared with that resolvable in the geologic record is profound. The survivability of animals may depend on migrations over enormous distances taking just weeks or months, time scales orders of magnitude shorter than the precision of dating the stratigraphic age of fossils.

A speaker at this conference suggested that subfreezing temperatures lasting months would be incompatible with the survival of certain reptiles. But that would not be true if a few reptiles survived next to a thermal hot springs in a favorably located cave. One must guard against attributing to the environments of a few exceptional refugia the average conditions of the Earth during a global environmental crisis.

There is a tendency to confuse killing or survival of a species with general death or survival of individuals during a crisis. Thus one speaker discussed the theoretical possibility that small carnivorous dinosaurs might have been able to survive on mammals, lizards, and other species that made it through the extinction. In all probability, however, this is not even theoretically possible: in a devastated world, where virtually every individual mammal presumably was killed, the survivors that enabled continuance of some mammal species were probably small groups in totally exceptional refugia, hardly a findable food source for some carnivores stumbling blindly through the darkness.

Importance of sudden changes for mass extinctions

Traditionally, mass extinctions have been ascribed to various changes in the environment that evolve extremely slowly compared with the sudden events (impacts, volcanic explosions) that I have discussed. Sea-level changes, chemical and thermal changes to the oceans, global warming, glaciations, and hotspot volcanic outpourings have all have traditionally been interpreted to evolve over durations ranging from tens of thousands of years to millions of years. Even recently hypothesized runaway geophysical processes commence on time scales that are long compared with the characteristic time scales of impact devastation, i.e., minutes to years. To me, it seems obvious that a sudden event (happening on a time scale, like months, that is short compared with the lifetime of an individual animal or plant) would be a far more potent cause of mass death and a possible mass extinction than changes, almost no matter how great, that evolve over centuries, millennia, or even millions of years. Here, a modern-day perspective is helpful.

A disaster, in human terms, is necessarily something that happens during a day, or perhaps over months or a year, but never over decades or centuries. After all, 100% of human beings now alive will die during the next 120 years or so, but that is considered normal, not a catastrophe. A powerful hurricane that strikes Florida can be a major natural disaster, but if waters rise and flood Florida during the next half-century (perhaps resulting from global warming), then people and enterprises can calmly move out of Florida at the rate that they moved in during the past half-century; it would be one of the usual ebbs and flows of economic and societal change, not a catastrophe.

As we look at Earth history, we must realize that species will be much more seriously affected by a catastrophe that is short compared with the reproductive cycle of individuals, and globally pervasive, two unique attributes of impacts. Some of the most powerful effects of impacts are over within the first few hours; most of the others are over within a few years. While much longer lasting effects will certainly ensue, and are recognized in the post-K-T geologic record, they are, like other slowly acting environmental changes, of little consequence to mass extinctions, no matter how much they may inhibit recovery and radiation of new species. For impacts over certain thresholds (that vary depending on the specific consequence; see Table 2), the effects are global in extent, notwithstanding possibilities that small refugia may be less affected. If all individuals starve, freeze, and die within a year of an impact holocaust, how will the species reproduce and survive? Adaptation to such radical environmental shocks is practically impossible.

The onset of an ice age is something that can be adapted to. (Walls of ice never arrive suddenly in suburban New Jersey, as depicted in Thornton Wilder's play *The Skin of Our Teeth.*) Seas don't suddenly regress, dramatically decreasing certain ecological niches worldwide, within the lifetimes of aquatic species. Species can migrate and/or develop new behaviors. Even if competition results in stresses and lowered population numbers, the survival of small breeding populations within such evolving ecosystems seems much more likely than in the instantly scorched, but frozen Earth aftermath of an impact. One of the most popular causes for mass extinctions discussed at the 2000 Vienna conference are episodes millions of years long of enhanced hotspot volcanism in certain localities on Earth. I cannot understand why anyone would regard such localized formation of a volcanic province like the Deccan Traps as conceivably resulting in a mass extinction. What are the killing mechanisms from such a slowly evolving process on the opposite side of the planet? Localized volcanism enhanced by factors of many compared with the modern rate may show up prominently in the geologic record, but the modest global ecological ramifications would be readily adapted to by migrations, evolutionary change, and other long-term responses.

Understanding that slow-acting climatological changes are impotent as causes of mass extinctions, some researchers have hypothesized that there are possibilities for natural, rapid instabilities on Earth, including sudden melting and destruction of polar ice caps, great landslides on continental shelfs, and dramatic changes in the carbon dioxide budget. Such events could possibly stress populations in ways not readily responded to, but even they are much slower acting and less dramatic in their consequences than are impacts. They rely on such factors as rising sea levels (which fail to severely affect habitats far from shorelines) and changing climates. Yet none of them transmit their devastating effects at the speed of many kilometers per second, spreading around the globe in a couple of hours, and none of them can be as globally and suddenly effective in changing the climate as the instantaneous and efficient injection of dust and aerosols into the stratosphere, greatly dimming or blocking out the Sun around the entire globe within a matter of weeks and lasting for many months to many years.

Even despite recent advances (S.A. Bowring, this conference), resolvable time scales concerning ancient events in the geologic record are long compared with human time scales. It is understandable, therefore, that geologists try to measure and think about environmental changes over such resolvable times. However, by imagining a multikilometer asteroid impact occurring today, in our modern built-up and natural world, we become much more aware of the amazingly sudden and profound changes that would present dramatic obstacles to survivability.

Huge impacts, which were nearly instantaneous in their globally devastating effects, have certainly occurred several times since the Precambrian. Their potency in causing the nearly instantaneous collapse of ecosystems (within minutes to months) dramatically exceeds any other suggested mechanism for mass extinction. The smoking guns (like extant, nonsubducted craters) become less likely to remain in the geologic record as we search back in time, but should not be required as evidence for impact, given the inevitability that the monster impacts have occurred. Other evidences of the K-T boundary impact, including the famous iridium excess, are not necessary outcomes of all major impacts (e.g., iridium content varies among impactors, and survival of projectile material is problematic, depending on the velocity and angle of impact). The impacts have occurred and have the unique attribute of sudden, global simultaneity. I think it is no coincidence that, as the techniques for making temporal measurements improve, the time scales associated with the largest mass extinctions (like the Permian-Triassic extinction; D.H. Erwin et al., this volume) continually shrink.

The huge impacts were so instantly awful, they must have left a paleontological record, and must have caused mass extinctions of some scale. It is a testimony to the resilience of life that, through localized, exceptional circumstances, breeding populations survived so that enough species managed to make it through the year-long frozen night of terror and death. It then becomes problematic that any other gradualistic geologic or environmental process could have played such a significant role, if any at all, in mass extinctions. If a total lack of evidence (e.g., of a layer of shocked dust) requires searching for another cause in the case of a particular mass extinction, only then are we compelled to turn to other improbable but still instantaneous causes, such as an immense volcanic explosion or supernova.

Raup's idea that the record of extinction reflects the cosmic impactor size distribution, and that impacts may be the cause of essentially all mass extinctions, was actually first enunciated in 1980 (Alvarez et al., 1980b, p. 2):

> It is reasonable to assume that the Permian-Triassic (P-T) and K-T extinctions were caused by large Earth-crossers, while lesser extinctions may have been caused by more numerous smaller asteroids. If so, the severity vs. frequency should relate to the size vs. number of Earth-crossing objects.

From the perspective of modern research on the impact hazard, it seems even more likely now that impacts have been the dominant cause of mass extinctions during the Phanerozoic.

ACKNOWLEDGMENTS

This work was supported by the NASA Near Earth Objects Programs Office at the Jet Propulsion Laboratory and by a Presidential Discretionary Internal Research and Development grant from Southwest Research Institute.

REFERENCES CITED

Adushkin, V.V., and Nemchinov, I.V., 1994, Consequences of impacts of cosmic bodies on the surface of the earth, *in* Gehrels, T., ed., Hazards due to comets and asteroids: Tucson, University of Arizona Press, p. 721–778.

Alvarez, L., Alvarez, W., Asaro, F., and Michel, H.V., 1980a, Extraterrestrial cause for the Cretaceous-Tertiary extinction: Science, v. 208, p. 1095–1108.

Alvarez, L.W., Dyson, F., Frosch, R.A., Hunter, P., Meinel, A., Naugle, J., Niehoff, J., Oliver, B.M., Sadin, S.A., and Yardley, J.F., 1980b, Project Spacewatch, *in* Report of New Directions Symposium: Woods Hole, Massachusetts, NASA Advisory Council.

Atkinson, H.H. (Chairman), 2000, UK Government Report of Task Force on Potentially Hazardous Near Earth Objects: British National Space Centre, London (http://www.nearearthobjects.co.uk/index.cfm).

Binzel, R.P., 2000, The Torino Impact Hazard Scale: Planetary and Space Science, v. 48, p. 297–303.

Boslough, M.B.E., and Crawford, D.A., 1997, Shoemaker-Levy 9 and plume-forming collisions on Earth, *in* Remo, J.L., ed., Near-Earth Objects: The United Nations Conference: Annals of the New York Academy of Sciences, v. 822, p. 236–282.

Bottke, W.F., Jedicke, R., Morbidelli, A., Petit, J.-M., and Gladman, B., 2000, Understanding the distribution of Near-Earth Asteroids: Science, v. 288, p. 2190–2194.

Brown, P.G., and 21 others, 2000, The fall, recovery, orbit, and composition of the Tagish Lake meteorite: A new type of carbonaceous chondrite: Science, v. 290, p. 320–325.

Chapman, C.R., 1995, What if? . . . , *in* Spencer, J.R., and Mitton, J., eds., The great comet crash: Cambridge, Cambridge University Press, p. 103–108.

Chapman, C.R., 2000, The asteroid/comet impact hazard: Homo sapiens as dinosaur?, *in* Sarewitz, D., Pielke, R.A., Jr., and Byerly, R., Jr., eds., Prediction: Science, decision making, and the future of nature: Washington, D.C., Island Press, p. 107–134.

Chapman, C.R., and Morrison, D., 1994, Impacts on the earth by asteroids and comets: Assessing the hazard: Nature, v. 367, p. 33–40.

Chinnery, M.A., and North, R.G., 1975, The frequency of very large earthquakes: Science, v. 190, p. 1197–1198.

Cole, K.C., 1998, Calculated risks: Skeptical Inquirer, v. 22, n. 5, p. 32–36.
Covey, C., Thompson, S.L., Weissman, P.R., and MacCracken, M.C., 1994, Global climatic effects of atmospheric dust from an asteroid or comet impact on Earth: Global and Planetary Change, v. 9, 263–273.
Dohnanyi, J.W., 1969, Collisional model of asteroids and their debris: Journal of Geophysical Research, v. 74, p. 2531–2554.
Durda, D.D., Greenberg, R., and Jedicke, R., 1998, Collisional models and scaling laws: A new interpretation of the shape of the main-belt asteroid size distribution: Icarus, v. 135, p. 431–440.
Munich Reinsurance Company (Münchener Rückversicherungs-Gesellschaft), 1988, World Map of Natural Hazards: Munich, 36 p.
Nemtchinov, I.V., Jacobs, C., and Tagliaferri, E., 1997, Analysis of satellite observations of large meteoroid impacts, *in* Remo, J.L., ed., Near-Earth Objects: The United Nations Conference: Annals of the New York Academy of Sciences, v. 822, p. 303–317.
Niven, L., and Pournelle, J., 1977, Lucifer's hammer: Chicago, Playboy Press, 494 p.
Pilcher, C., 1998, Testimony before House Subcommittee on Space and Aeronautics, http://www.house.gov/science/pilcher_05-21.htm.
Piotrowski, S.I., 1953, The collisions of asteroids: Acta Astronautica, ser. A, v. 6, p. 115–138.
Pope, K.O., Baines, K.H., Ocampo, A.C., and Ivanov, B.A., 1994, Impact winter and the Cretaceous/Tertiary extinctions: Results of a Chicxulub asteroid impact model: Earth and Planetary Science Letters, v. 128, p. 716–725.
Rabinowitz, D., Helin, E., Lawrence, K., and Pravdo, S., 2000, A reduced estimate of the number of kilometre-sized near-Earth Asteroids: Nature, v. 403, p. 165–166.
Rampino, M.R., Stothers, R.B., and Self, S., 1988, Volcanic winters: Annual Review of Earth and Planetary Science, v. 16, p. 73–99.
Raup, D.M., 1991, Bad genes or bad luck?: New York, Norton, 210 p.
Sarewitz, D., Pielke, R.A., Jr., and Byerly, R., Jr., editors., 2000, Prediction: Science, decision making, and the future of nature: Washington, D.C., Island Press, 405 p.
Shoemaker, E.M., 1983, Asteroid and comet bombardment of the earth: Annual Review of Earth and Planetary Science, v. 11, p. 461–494.
Stuart, J.S., 2001, A Near-Earth Asteroid population estimate from the LINEAR survey: Science, v. 294, p. 1691–1693.
Toon, O.B., Zahnle, K., Morrison, D., Turco, R.P., and Covey, C., 1997, Environmental perturbations caused by the impacts of asteroids and comets: Reviews of Geophysics, v. 35, p. 41–78.
Veverka, J., and 32 others, 2000, NEAR at Eros: Imaging and spectral results: Science, v. 289, p. 2088–2097.
Ward, S.N., and Asphaug, E., 2000, Asteroid impact tsunami: A probabilistic hazard assessment: Icarus, v. 145, p. 64–78.
Wolbach, W.S., Gilmour, I., Anders, E., Orth, C.J., and Brooks, R.R., 1988, Global fire at the Cretaceous-Tertiary boundary, Nature, v. 334, p. 665–669.

Manuscript Submitted December 1, 2000; Accepted by the Society March 22, 2001

Printed in the U.S.A.

Geological Society of America
Special Paper 356
2002

Tracers of the extraterrestrial component in sediments and inferences for Earth's accretion history

Frank T. Kyte*
Center for Astrobiology, Institute of Geophysics and Planetary Physics, University of California, Los Angeles, California 90095-1567, USA

ABSTRACT

The study of extraterrestrial matter in sediments began with the discovery of cosmic spherules during the HMS *Challenger* Expedition (1873–1876), but has evolved into a multidisciplinary study of the chemical, physical, and isotopic study of sediments. Extraterrestrial matter in sediments comes mainly from dust and large impactors from the asteroid belt and comets. What we know of the nature of these source materials comes from the study of stratospheric dust particles, cosmic spherules, micrometeorites, meteorites, and astronomical observations.

The most common chemical tracers of extraterrestrial matter in sediments are the siderophile elements, most commonly iridium and other platinum group elements. Physical tracers include cosmic and impact spherules, Ni-rich spinels, meteorites, fossil meteorites, and ocean-impact melt debris. Three types of isotopic systems have been used to trace extraterrestrial matter. Osmium isotopes cannot distinguish chondritic from mantle sources, but provide a useful tool in modeling long-term accretion rates. Helium isotopes can be used to trace the long-term flux of the fine fraction of the interplanetary dust complex. Chromium isotopes can provide unequivocal evidence of an extraterrestrial source for sediments with high concentrations of meteoritic Cr.

The terrestrial history of impacts, as recorded in sediments, is still poorly understood. Helium isotopes, multiple Ir anomalies, spherule beds, and craters all indicate a comet shower in the late Eocene. The Cretaceous-Tertiary boundary impact event appears to have been caused by a single carbonaceous chondrite projectile, most likely of asteroid origin. Little is known of the impact record in sediments from the rest of the Phanerozoic. Several impact deposits are known in the Precambrian, including several possible megaimpacts in the Early Archean.

INTRODUCTION

The discovery of anomalous iridium concentrations in Cretaceous-Tertiary (K-T) boundary sediments (Alvarez et al., 1980) sparked an explosion of research into the study of impacts, biotic extinction events, and relationships between extraterrestrial phenomena and terrestrial ecosystems. One field that has seen enormous growth has been the detection, characterization, and understanding of the extraterrestrial component in sediments. More than a decade ago I wrote a paper on the extraterrestrial component in sediments (Kyte, 1988) that is now hopelessly outdated. This latest paper brings much of this topic up to date. Like the earlier paper, this is meant to be a general (although not comprehensive) review of what has been learned

*E-mail: kyte@igpp.ucla.edu

Kyte, F.T., 2002, Tracers of the extraterrestrial component in sediments and inferences for Earth's accretion history, *in* Koeberl, C., and MacLeod, K.G., eds., Catastrophic Events and Mass Extinctions: Impacts and Beyond: Boulder, Colorado, Geological Society of America Special Paper 356, p. 21–38.

about extraterrestrial matter in sediments. It focuses on the different methods that have been developed to detect an extraterrestrial component and the types of interpretations that can be made using these methods. As much as possible, these topics are presented in a historical perspective. There is a brief discussion of the major implications of this work, based on my views. For a discussion of the identification of meteoritic components in impact melt rocks, breccias, and other impactites, see Koeberl (1998).

HISTORICAL BEGINNINGS

The study of this extraterrestrial component in sediments began with the discovery of cosmic spherules in deep-sea sediments collected during the expedition of the HMS *Challenger* (1873–1876). Murray and Renard (1883, 1891) found that the magnetic fraction of some sediments contained what they called "cosmic dust." These were black magnetic spherules, rarely as large as 200 μm, and brown spherules, typically 500 μm. Most of the magnetic fraction could be attributed to volcanic or other terrestrial sources. However, the spherules were unusual; the black magnetic variety had metallic cores and the brown spherules bore a striking resemblance to chondrules, a common component of chondritic meteorites. Murray and Renard (1891, p. 333–336) noted that "while they are universally distributed, they are more abundant in regions where the accumulation of the deposit is relatively slow, and most abundant where the rate of deposition is reduced to a minimum, *viz.* in the deepest water far removed from continental land." They concluded that these were "extraterrestrial bodies allied to meteorites, and in all probability thrown off by them in their passage through the Earth's atmosphere." In this work, the scientists of the HMS *Challenger* recognized two important facts: (1) cosmic spherules formed by atmospheric ablation of meteoritic material and (2) the concentration of extraterrestrial material was highest in the most slowly accumulating sediments.

Murray and Renard (1891) cited other work of their time describing possible "cosmic dust." None appeared to be so conclusive in their results as the *Challenger* work, but they noted that Nordenskjold (1881, cited in Murray and Renard, 1891) collected dust in Greenland in deposits of "Krykonit." These deposits, which occur in lakes on the Greenland ice sheet, are now known to contain some of the best-preserved concentrates of cosmic spherules yet discovered (Maurette et al., 1986). In the succeeding century nearly all work on extraterrestrial matter in sediments concentrated on cosmic spherule studies. This work was mainly descriptive, or focused on new types of deposits in which cosmic spherules could be found, such as Pleistocene beach sands (Marvin and Einaudi, 1967) and Paleozoic salt deposits (Mutch, 1966). One of the first applications of the newly designed electron microprobe was to directly measure the Ni content of magnetic cosmic spherules (Castaing and Fredriksson, 1958), providing the first strong chemical evidence supporting their cosmic origin.

Cosmic spherule science took a major step forward in the 1970s and early 1980s with new collections and refined analyses that directly linked them to specific types of meteorites. This work was largely spearheaded by Don Brownlee and coworkers who were analyzing the source materials of cosmic spherules, i.e., interplanetary dust. In the mid-1970s NASA U-2 aircraft were first used routinely to collect dust in the stratosphere (e.g., Brownlee, 1985). This dust was found to have similarities to some types of chondritic meteorites. Large volume collections of cosmic spherules included use of a magnetic sled dragged on the ocean floor to obtain millions of specimens (Brownlee et al., 1979). Neutron activation analyses of individual spherules (Ganapathy et al., 1978) added Ir to the list of elements indicating a cosmic origin. Scanning electron microscope (SEM) and electron microprobe analyses of stony spherules (the brown spherules of Murray and Renard, 1891) proved that the major element chemistry was similar to that of chondrites and that unmelted relict grains of meteorite minerals could survive atmospheric ablation (Blanchard et al., 1981). The final proof that cosmic spherules had to be of extraterrestrial origin was from the detection of cosmogenic ^{53}Mn, a short-lived isotope that can only have formed by irradiation in space (Nishiizumi, 1983).

In addition to the work on cosmic spherules, a number of workers attempted to use chemical methods to determine the flux of cosmic matter to marine sediments. Analyses of the Ni content of deep-sea sediments and Antarctic ice (Pettersson and Rotschi, 1950; Pettersson and Fredriksson, 1958; Bonner and Lourenco, 1965) yielded what are now known to be erroneously high values. Barker and Anders (1968) used two platinum group elements (PGEs), Ir and Os, to obtain the first reasonable estimate of the long-term flux of extraterrestrial matter. They showed that the concentration of Ir was inversely proportional to the sedimentation rate of slowly accumulating deep-sea clays (Fig. 1). From this relationship they derived an estimate of 60 000 t/yr of cosmic matter accreted to the Earth, a value that has not changed by a factor of two over the past three decades.

The first attempt to use the Barker and Anders (1968) data to measure the accumulation rate of a sediment deposit, based on its Ir concentration, was by Alvarez et al. (1980). The results of this experiment are now almost legend. They found that the Ir concentration of sediments from the K-T boundary were far too high to be from the background influx of cosmic debris, and thus this experiment was a failure; however, it was one of the more spectacular failures in the history of Earth sciences. Alvarez et al. (1980) explained the anomalous Ir concentrations as being due a sudden influx of extraterrestrial matter at the end of the Cretaceous. Their hypothesis was that this influx was due to the impact of an asteroid or comet, 10 km in diameter, and that the K-T boundary clay was ejecta from this impact. The most exciting aspect of their hypothesis was the speculation that this impact was likely responsible for the mass extinctions recorded at the K-T boundary. The cause of these extinctions had been the subject of controversy for the preceding century, and

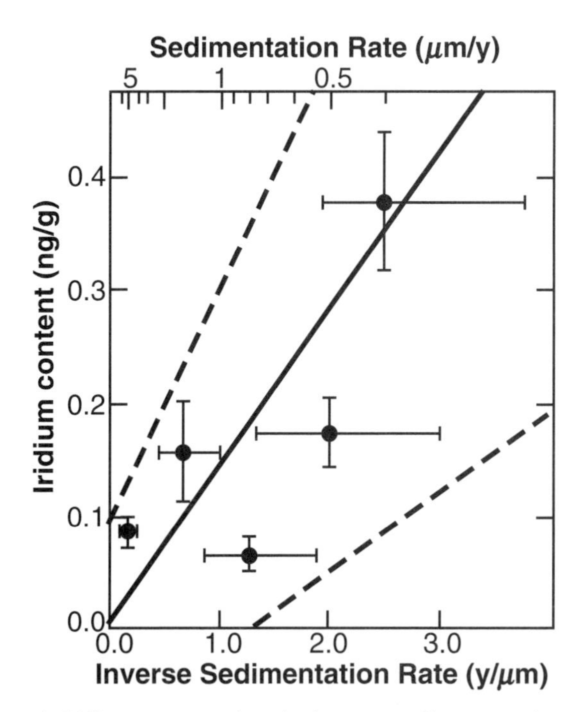

Figure 1. Iridium concentrations in deep-sea sediments are inversely correlated with sediment accumulation rates. Solid line is correlation based on five sediment samples. Dashed lines are 2σ uncertainties. After Barker and Anders (1968).

the Alvarez et al. (1980) discovery provided the first extinction hypothesis that was backed up by reproducible physical data and that had testable predictions. Although not precisely 100 yr after the initial discovery of cosmic dust, the discovery of the K-T boundary Ir anomaly, along with modern analyses of cosmic spherules and interplanetary dust, could be considered the start of the second century of research on the extraterrestrial component in sediments.

SOURCES OF EXTRATERRESTRIAL MATTER

The influx of extraterrestrial materials to the Earth is dominated by two size fractions: submillimeter interplanetary dust and impacting asteroids and comets (Fig. 2; Kyte and Wasson, 1986). The dust is derived from collisions in the asteroid belt and from materials expelled from the surface of comets. To some degree, both the large projectiles and the finest dust have the same two source regions. The modern flux of interplanetary dust, as directly measured from impacts on a satellite collector (Love and Brownlee, 1993), is 30000 ± 10 000 t/yr (revised from 40000; Engrand and Maurette, 1998). The flux of large impactors is less well known, and must be estimated from analyses of crater statistics and astronomical observations of populations of comets and Earth-crossing asteroids (e.g., Wetherill and Shoemaker, 1982). However, it is generally accepted that most of the mass of extraterrestrial matter accreted to the Earth over geological time scales is from the rare, largest impactors.

The actual composition of these source materials, particularly over long time scales, is uncertain. The modern flux of interplanetary dust could be dominated by recent impacts in the asteroid belt or by a few large, dusty comets. Because dust particles have dynamic lifetimes of only about 10^5 yr before they spiral into the Sun (Love and Brownlee, 1993), the dust complex can be dominated by just a few source objects at any given time. Although it is generally accepted that meteorites are derived from asteroid belt objects, the well-classified meteorites are probably from no more than 20 asteroids (Dodd, 1989). Astronomical observations of the spectral reflectance of asteroids show that they appear to vary in composition with radial distance from the Sun (Fig. 3; Bell et al., 1989; Shearer et al., 1998). It appears that the outer portion of the asteroid belt is populated by highly carbonaceous asteroids (types P and D) that may be unlike typical meteorites. Direct sampling of comets is limited to flybys of comet Halley, but these have shown that, in addition to the high concentrations of ice in this comet, the dust contains about 30% carbonaceous matter (e.g., Jessberger and Kissel, 1989).

Interplanetary dust is distinctly different, on average, from typical meteorites. Interplanetary dust particles (IDPs) collected from the stratosphere sample the 5–50 µm fraction of the dust complex. These are mostly fine-grained chondritic particles (Brownlee, 1985). The remainder of the dust is typically either melted particles or individual mineral grains. The carbon content of the smaller chondritic IDPs (~10 µm), those least heated by atmospheric entry, is typically several times that of the most carbon-rich meteorites (e.g., Brownlee et al., 1997a), so these objects are distinct from most meteorites. Possibly these particles are from the outer asteroid belt or from comets. Larger unmelted dust particles (50–500 µm) have been recovered as unmelted micrometeorites from Antarctic melt ice (Engrand and Maurette, 1998). These particles are also much more carbon rich than carbonaceous chondrite meteorites and petrographically they most resemble CM and CR chondrites, two relatively rare meteorite groups.

Several attempts have been made to link IDPs to either asteroid or cometary sources (e.g., Klöck et al., 1989; Bradley and Brownlee, 1991). The fine-grained IDPs can generally be classified as belonging to one of two types, smooth or porous (Brownlee, 1985). The smooth IDPs are typically dominated by hydrated clays (e.g., saponite, crondstedite) and are from an object that underwent aqueous alteration (Fig. 4A). The porous IDPs (Fig. 4, B and C) are composed mainly of very fine grained, anhydrous phases, and have been found to contain unusual materials called GEMS (Fig. 4D), an acronym for "glass with embedded metal and sulfides." Because of their apparently amorphous silicates, and possibly heavy radiation damage, these GEMS are considered a likely candidate for unprocessed interstellar materials (Bradley et al., 1999). Detailed studies of the He content of IDPs has been used to model relative velocities of a large number of IDPs in an attempt to distinguish asteroidal from cometary sources (Joswiak et al., 2000). This

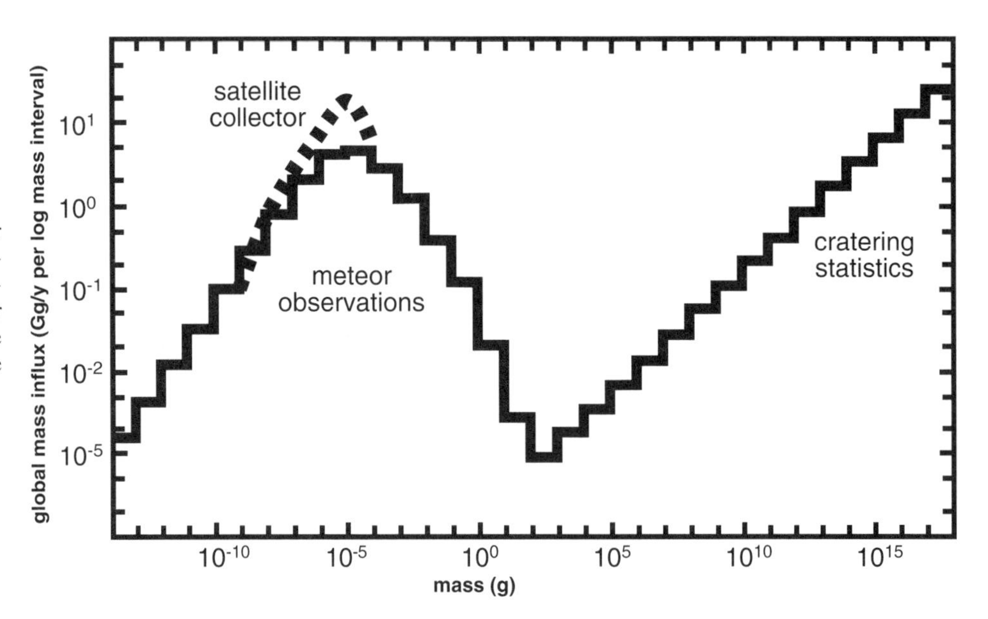

Figure 2. Influx of extraterrestrial matter to Earth is dominated by two size fractions, interplanetary dust and large asteroids and comets. Histogram is after Kyte and Wasson (1986). Dashed curve is adapted from Love and Brownlee (1993).

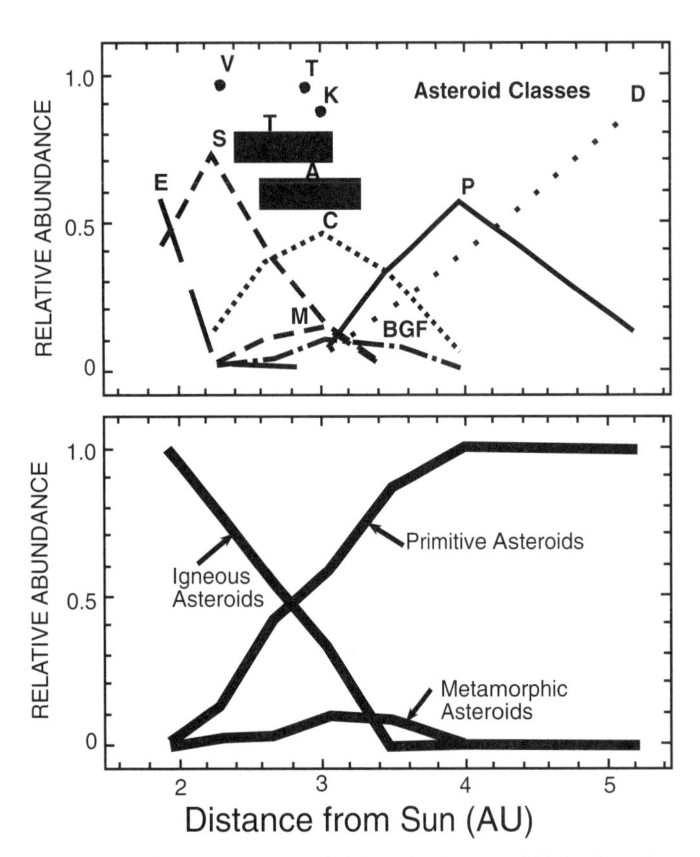

Figure 3. Reflectance spectra of objects in the asteroid belt show that there are several types of asteroids. The outer asteroid belt appears to be dominated by icy carbonaceous chondrites that may not be common in meteorite collections (after Bell et al., 1989).

work found that the asteroidal IDPs, having velocities <14 km/s, were dominated by compact particles with hydrated silicates. Cometary IDPs, having velocities >18 km/s, were mostly anhydrous, porous types containing GEMS. These data probably present the best criteria for distinguishing cometary from asteroidal materials in interplanetary dust.

TRACERS OF EXTRATERRESTRIAL MATTER IN SEDIMENTS

The extraterrestrial component in sediments can be traced by three distinct methods: chemical, physical, and isotopic. The most common chemical tracers are the PGEs, of which Ir is the most frequently studied. More recently, fullerenes have been described as potential tracers that may survive even large-body impacts. Physical tracers include actual meteorites, cosmic spherules, impact spherules, spinels derived from either of these rocks, and ocean-impact melt debris. Three isotopic systems, using isotopes of Os, He, and Cr, have been successfully applied to tracing extraterrestrial matter. Each of these has been successful in examining different aspects of the accretion of meteoritic materials to the Earth. These three types of tracers are discussed in the following.

Chemical tracers

Siderophile elements. Siderophile elements are those that would be concentrated in a metallic phase, if one were present in a rock. Siderophile elements can be useful tracers of extraterrestrial matter because they are highly depleted in the Earth's crust as a result of both the core-mantle separation early in Earth history (Chou, 1978), and the subsequent mantle-crust fractionation. Siderophiles include the PGEs (Ru, Rh, Pd, Os, Ir, and

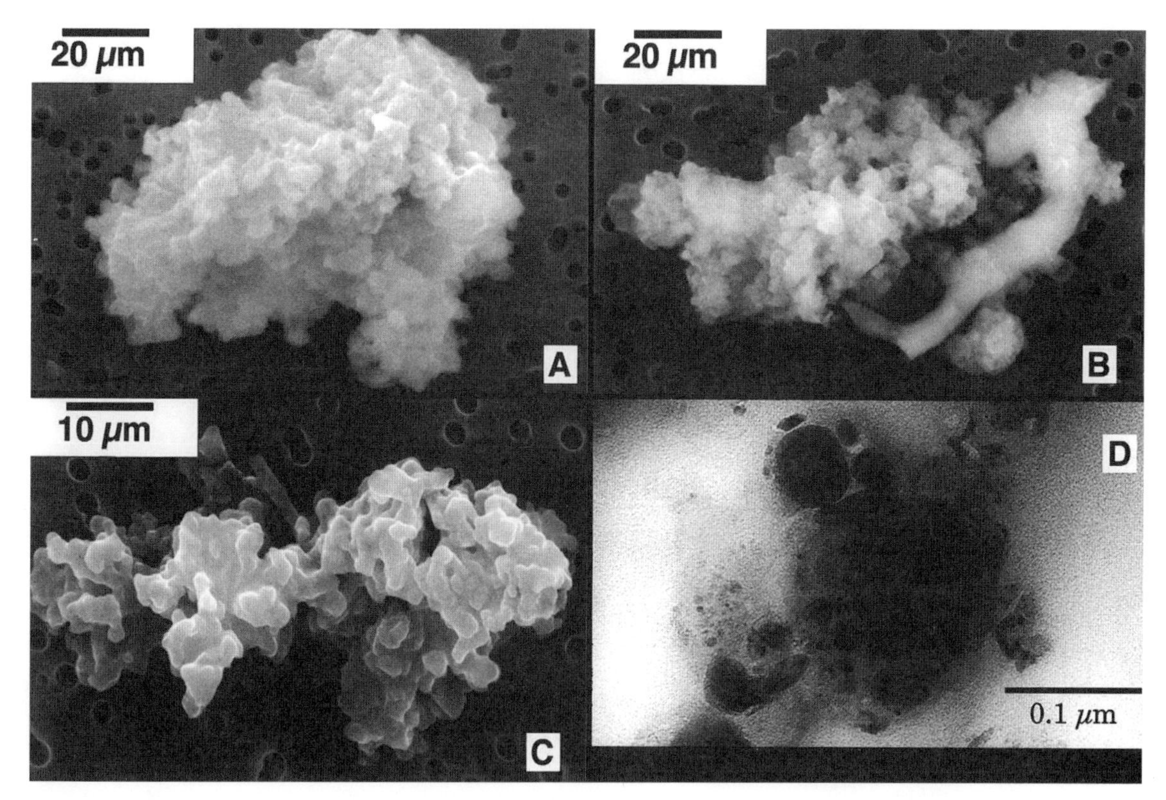

Figure 4. Interplanetary dust particles are routinely collected in the stratosphere. Low-velocity dust particles are probably derived from asteroids. These are commonly compact particles with hydrated silicates (A). High-velocity particles are probably cometary. These are commonly anhydrous porous aggregates (B and C) with GEMS (D; glass with embedded sulfides). Long object in B is probably glass. A–C are secondary electron images. D is transmission electron microscope image. Photos courtesy of D.E. Brownlee.

Pt) and Au, Re, Ni, Co, and Fe. Although it is not typically considered a siderophile in meteorites, Cr is also highly depleted in the crust relative to its concentration in chondritic meteorites, and it can also be a useful measure of extraterrestrial matter. Chromium isotopic studies are discussed in the following. Of the siderophiles, Ir has proven the most commonly used and effective tracer of extraterrestrial matter due to its strong depletion in crustal rocks relative to chondrites (by a factor of $\sim 10^4$; Palme et al., 1978) and its relatively easy detection by neutron activation analysis at extremely low concentrations. Recent work in the application of inductively coupled plasma mass spectroscopy to PGE analyses has seen the development of methods for rapid determination of PGEs with detection limits similar to those previously obtainable only by radiochemical neutron activation analyses (Ravizza and Pyle, 1997).

The first successful attempt at using siderophiles as a measure of extraterrestrial matter was that of Barker and Anders (1968), who measured Ir and Os concentrations in slowly accumulating deep-sea clays. They used this to calculate an extraterrestrial matter accretion rate that is little changed to this date (e.g., Kyte and Wasson, 1986; Peucker-Ehrenbrink, 1996). The Barker and Anders (1968) study is attributed as the basis for the Ir measurements performed by Alvarez et al. (1980) that led to the discovery of the K-T boundary Ir anomaly. Shortly after the discovery of the K-T Ir anomaly, other workers found anomalies for Os (Smit and Hertogen, 1980), as well as other PGEs, Re, and Au (Ganapathy, 1980; Kyte et al., 1980). The other PGEs were all found to have interelemental abundances roughly similar to those in chondrites. This was considered to be another strong argument for a meteoritic signature in K-T boundary sediments.

In succeeding years, a number of workers used abundances of a number of PGEs and siderophiles to argue for an extraterrestrial provenance for K-T boundary sediments (e.g., Smit and Kyte, 1984; Gilmore et al., 1984; Kyte et al., 1985). Although this work showed that interelement abundances for siderophiles were similar to those in chondritic meteorites, there were fractionations noted by other workers. For example, Tredoux et al. (1989) argued that PGE abundances were different between Northern and Southern Hemisphere K-T boundary sites, and that this might best be explained by a mantle-derived source for the PGEs. Evans et al. (1993) argued that differences in the ratio of Ru/Ir between K-T sites in North America and Europe was a primary feature, related to the relative volatilities of these two elements and their condensation from an impact-generated vapor plume. Evans et al. (1993) argued that the Ru/Ir ratio should increase with increasing distance from the impact crater. In retrospect, while it is now relatively clear that the PGEs in K-T boundary sediments were probably largely derived from the Chicxulub projectile, the specific physical and chemical processes involved in their transfer to sediments, as well as their final preservation in K-T boundary clays, are still poorly understood.

Immediately following the Alvarez et al. (1980) discovery, additional Ir anomalies were linked to other likely impact deposits. Anomalous Ir was inadvertently discovered by Crocket and Kuo (1979), actually predating the K-T boundary discovery, in a sample of late Pliocene sediment from USNS *Eltanin* sediment core E13-3. However, the significance of this anomaly was not recognized until Kyte et al. (1981) studied the core in greater detail. Subsequent work has now thoroughly documented this as a signal from the Eltanin asteroid impact (e.g., Gersonde et al., 1997). Ganapathy (1982) found an Ir anomaly in late Eocene sediments closely associated with microtektite deposits. The confirmation of this anomaly as being stratigraphically below the microtektite horizon (Sanfilippo et al., 1985) led to the understanding that late Eocene clinopyroxene spherules were from a separate impact event, actually predating the impact that produced the microtektites (Glass et al., 1985; Glass and Burns, 1987).

The excitement surrounding the possible extraterrestrial cause of the K-T boundary extinctions led to speculation that many other mass extinctions might have been caused by impacts. When Raup and Sepkoski (1984) found an apparent periodicity in the frequency of mass extinctions, several astronomical mechanisms were proposed that could cause periodic perturbations of the Oort Cloud, a symmetric cloud of comets extending to interstellar distances (Weissman, 1985). These mechanisms for causing periodic comet showers included perturbations by a dark companion star to the Sun (Whitmire and Jackson, 1984; Davis et al., 1984), giant molecular clouds encountered as the solar system oscillates through the galactic plane (Rampino and Stothers, 1984), and an unobserved tenth planet (Whitmire and Matese, 1985).

This excitement was largely followed by disappointment as extensive searches of numerous extinction horizons (summarized by Orth et al., 1985; Kyte, 1988) found no strong evidence to link them to extraterrestrial factors. Reports of an Ir anomaly at the Permian-Triassic (P-T) boundary (Xu et al., 1985) lent some hope after earlier work proved negative (Asaro et al., 1982; Alekseev et al., 1983), but subsequent work failed to confirm this (Clark et al., 1986; Zhou and Kyte, 1988). Retallack et al. (1998) described deformed and possibly shocked quartz in P-T sediments, but a possible extraterrestrial event at this greatest mass extinction remains unproven. Several reports of spherules in Late Devonian sediments (e.g., Wang, 1992; Claeys and Cassier, 1994) been were intriguing, but detection of any clearly defined extraterrestrial component, such as a strong Ir anomaly, has remained elusive (e.g., Mc Ghee et al., 1985; Claeys et al., 1996). In my opinion, no mass extinction other than that at the K-T boundary has been conclusively linked to a strong extraterrestrial signature.

A significant extraterrestrial component has been detected in several Precambrian deposits. Late Precambrian ejecta found in Australian shales that contain shocked minerals and rocks are probably derived from the Acraman impact structure (Gostin et al., 1986; Williams, 1986). These deposits have anomalous Ir and other PGEs (Gostin et al., 1989; Wallace et al., 1990). Small Ir anomalies have been found in at least two of four known spherule deposits from the Hamersley basin of western Australia (Simonson et al., 1998), and anomalous PGEs in addition to Ir have been found in a Late Archean spherule layer in South Africa (Simonson et al., 2000). These reports all lend significant credence to interpretations of an impact origin for these multiple spherule deposits.

Perhaps the most spectacular deposits with a large extraterrestrial component are found in the Early Archean (3.2–3.5 Ga) Barberton Greenstone Belt, South Africa. Lowe et al. (1989) identified at least four spherule beds with thicknesses ranging from ~10 to 100 cm. They cited several criteria that distinguish these beds from typical volcanic and clastic sediments, including (1) wide geographic distribution in a variety of depositional environments, (2) relict quench textures, (3) absence of juvenile volcaniclastic debris, and (4) extreme enrichment of Ir and other PGEs. However, some workers argue for a terrestrial origin, possibly related to volcanism and gold mineralization (e.g., Koeberl et al., 1993; Koeberl and Reimold, 1995). Perhaps the strongest arguments, by Koeberl and Reimold (1995), are the extreme enrichments of Ir in these rocks and the proximity of some spherule bed localities to rocks with strong gold mineralization. A few individual samples from within these beds contain Ir concentrations equal to or higher than those in chondrites. The average content of extraterrestrial matter in these beds (based on PGE abundances) is probably 10%–20% by weight (e.g., Koeberl and Reimold, 1995; Shukolyukov et al., 2000a). These are very high concentrations, but comparable extremes have been observed in K-T boundary sediments (on a carbonate-free basis; e.g., Kyte et al., 1980; Schmitz, 1988), and in materials concentrated from K-T boundary sediments (Robin et al., 1993; Kyte, 1998).

In a detailed study of PGE abundances in one of these spherule beds (bed S4), Kyte et al. (1992) found strong correlations between Ir, Pt, Os, Pd, and Au in a set of 15 samples that had absolute PGE concentrations that varied by an order of magnitude. Average elemental ratios based on correlations for Os/Ir and Pt/Ir were ~0.8 times those ratios in CI chondrites (Fig. 5). This is a remarkable agreement, considering analytical uncertainties of at least 10% and the fact that these rocks had undergone early hydrothermal metamorphism that completely replaced primary impact minerals with chlorite and quartz. They found that Pd was depleted and Au was strongly depleted relative to chondrites, Pd/Ir and Au/Ir being 0.41 and 0.02 times CI, respectively. Kyte et al. (1992) pointed out that of these five siderophiles, Au and to a lesser extent Pd, are known to be the

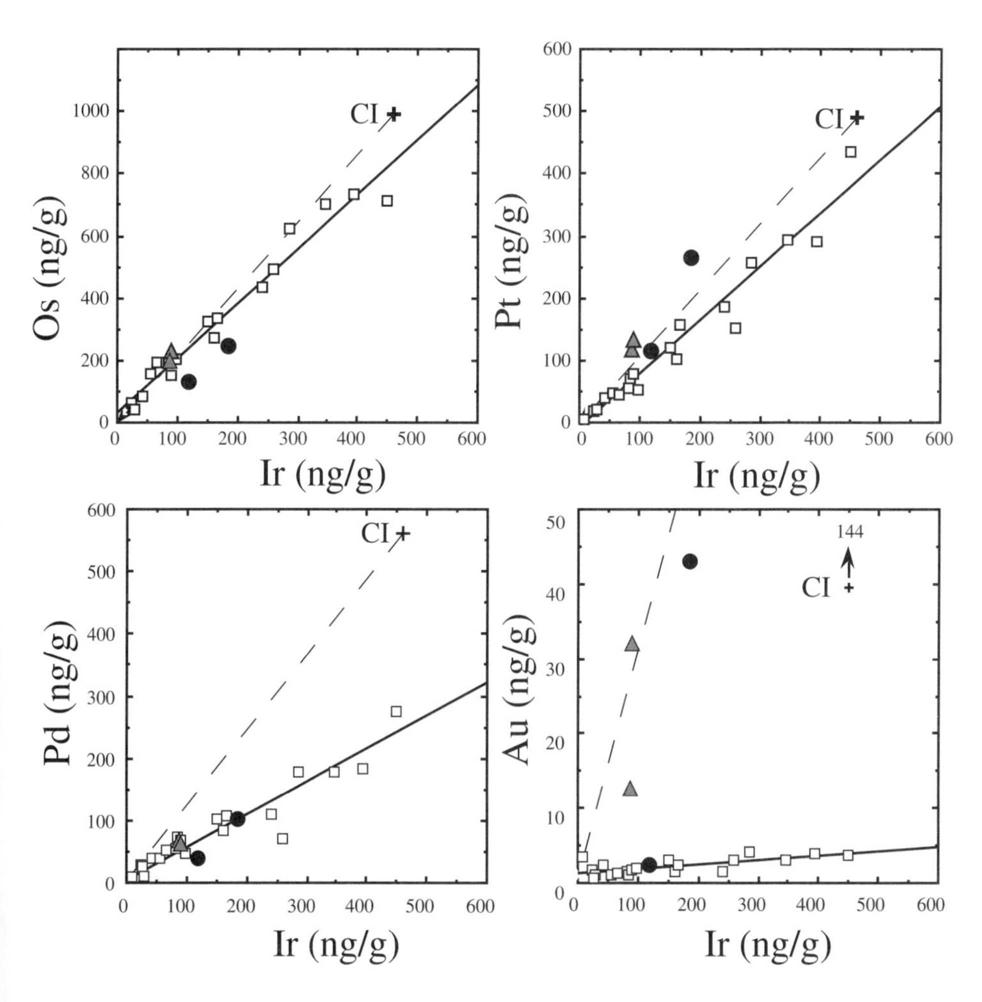

Figure 5. Plots of Os, Pt, Pd, and Au vs. Ir in samples from three Early Archean spherule beds: 15 samples from bed S4 (open squares) are strongly correlated (solid line). Ratios for Os/Ir and Pt/Ir are close to those ratios in CI chondrites (dashed line and cross). Samples from beds S3 (black circles) and S2 (gray triangles) also have abundances similar to those in chondrites, except for Au in one sample. After Kyte et al. (1992).

most mobile during alteration. They argued that this pattern of high concentrations of Ir, Os, and Pt at chondritic abundances with depleted Pd and Au is consistent with an original chondritic source that had been altered by secondary processes. This is the opposite of the pattern that one would expect due to gold mineralization, which would result in deposits highly enriched in Au and Pd relative to other siderophiles.

This pattern of Au and Pd enrichment that Kyte et al. (1992) predicted should be caused by mineralization was observed in a study of another Early Archean spherule bed. Reimold et al. (2000) reported detailed analyses of PGEs and Au in a number of samples from the Princeton Mine, South Africa. These included a number of samples from spherule bed S2 as well as samples from rocks that were heavily mineralized. Most of their spherule bed samples had chondritic interelement abundances for Ir, Ru, Rh, and Pt, but variable abundances for Pd and Au. The latter elements were both either somewhat depleted or enriched relative to chondritic interelement abundances. In contrast, the rocks that had undergone high levels of Au mineralization had more fractionated PGE abundances (relative to chondrites) at much lower concentrations than in the spherule beds. These mineralized samples also had moderate enrichment of Pd (~5 times) and extreme enrichment of Au (to 10^4 times) relative to the other PGEs. Reimold et al. (2000) remained highly skeptical of an extraterrestrial origin for this PGE anomaly, noting the extreme enrichment of Ir in a few samples, as well the fact that a few samples that did not contain spherules also had high Ir concentrations. However, arguments of terrestrial versus extraterrestrial origin for these beds may soon be moot. Recent analyses of Cr isotopes have confirmed the extraterrestrial nature of the siderophile anomaly, at least in bed S4 (Shukolyukov et al., 2000a). This method should soon resolve this question in the other spherule beds, and allow future studies to work on unraveling the nature of extraterrestrial and terrestrial processes that cause such large PGE enrichments.

Fullerenes. Fullerenes (C_{60} and C_{70}) have been reported in K-T boundary sediments (Heymann et al., 1994; Becker, 2000) and in breccias from the 1.85 Ga Sudbury impact structure (Becker et al., 1994). The origin of these fullerenes was originally attributed to K-T boundary wildfires (Heymann et al., 1994) or formation within the impact plume (Becker et al., 1994). However, ^{3}He has now been reported in fullerenes from both of these deposits (Becker et al., 1996; Becker, 2000), and it has been suggested that the fullerenes might also be of extraterrestrial origin. If so, fullerenes would represent a new type of tracer of extraterrestrial matter. However, another study of the Sudbury breccia (Mukhopadhyay et al., 1998) contradicted these results, and concluded that all ^3He in these rocks could be attributed to decay of ^6Li. Further work is needed to evaluate the potential of this method before it can be accepted as a routine tracer of extraterrestrial matter in sediments.

Physical tracers

Cosmic and impact spherules. These two types of spherules are discussed together because in some cases it is not possible to distinguish them. Spheroidal material has been described in numerous settings, but much of this work includes spherules of unproven meteoritic or even impact origin. For this reason, in this section I concentrate on well-documented occurrences of spherules that contain a significant meteoritic component.

Cosmic spherules are derived primarily from atmospheric ablation of interplanetary dust grains. Murray and Renard (1983) originally distinguished two types, black magnetic spherules and brown chondrule-like spherules. The chondrule-like spherules are now referred to as stony, and have a roughly chondritic major element composition (e.g., Brownlee et al., 1997b). These are by far the most abundant type of cosmic spherule. The black spherules are composed of oxides of Fe (magnetite, wüstite, hematite), often with a core of metallic NiFe. Although they clearly had a metallic precursor, a large fraction of the black magnetic type is almost certainly derived from chondritic metal, rather than from iron meteorites (Herzog et al., 1999). Thus, the bulk composition of most of the materials that make up the cosmic spherules is chondritic in origin. Detailed analyses of the major element chemistry of the stony spherules shows that they have compositions that are on average similar to type CM chondrites (Brownlee et al., 1997b), although they are depleted in volatiles (Na, S) due to atmospheric heating and depleted in siderophiles by a less understood mechanism, possibly due to separation of immiscible phases during melting in the atmosphere.

Cosmic spherules are now known from a wide range of environments and ages. These include recent deep-sea sediments, Pleistocene beach sands (Marvin and Einaudi, 1967), mid-Cenozoic clays (Taylor and Brownlee, 1991; Kyte and Bostwick, 1995), Jurassic hardgrounds (Czajkowski, 1987; Jehanno et al., 1988), Silurian and Permian salt deposits (Mutch, 1966), and Proterozoic red sandstones (Kettrup et al., 2000). It is now well established that silicate spherules are poorly preserved in deep-sea sediments and new types of spherules have been discovered in less corrosive environments, such as blue lakes in Greenland (Maurette et al., 1986; Robin, 1988; Taylor and Brownlee, 1991). The most spectacular and well-preserved cosmic spherule collections have come from a water well at the South Pole (Taylor et al., 2000). This collection includes a full range of particles, from totally melted spherules (including types never described before) through unmelted micrometeorites, and represents the best collection to date of samples from the coarse fraction of the interplanetary dust complex.

Most impact spherules probably do not contain a significant extraterrestrial component, so, although they may be tracers of impact events, they are not necessarily tracers of meteoritic materials. Microtektite deposits are well known for having extremely low Ir concentrations (e.g., Schmidt et al., 1993). The common interpretation of this is that there is an extremely low concentration of meteoritic matter in the ejecta. Thus, the microtektites are composed almost entirely of terrestrial impact target materials. Glass and Koeberl (1999) reported Ir concentrations as high as 0.3 ng/g in late Eocene microtektites, but this is <0.1% of the concentration in chondritic meteorites (~580 ng/g on a volatile-free basis).

Late Eocene clinopyroxene-bearing spherules are associated with a significant Ir anomaly (e.g., Sanfilippo et al., 1985; Glass and Koeberl, 1999; Vonhof and Smit, 1999). Unfortunately, no analyses of Ir have been reported for the clinopyroxene spherules. However, Glass and Koeberl (1999) cited a Ni/Ir ratio of 10^6 from R. Ganapathy (1983, personal commun.). Glass and Koeberl (1999) measured Ni concentrations up to 4 mg/g in clinopyroxene spherules from Ocean Drilling Program Site 689, so one can infer that clinopyroxene spherules have an Ir content of ~4 ng/g. This is a significant concentration of Ir, but still <1% of chondritic abundances and an insufficient amount to account for the Ir deposited at this site. The peak concentration of clinopyroxene spherules found at Site 689 by Glass and Koeberl (1999) was <30 spherules per gram of sediment. These small particles (~100 μm) cannot possibly be the carrier of the Ir in sediments in which Montanari et al. (1993) found bulk Ir concentration of 0.16 ng/g.

Many types of spherules are known in K-T boundary sediments, but most are probably not the principal carrier of Ir and other meteoritic components. Large spherules composed of a tektite-like glass occur in regional deposits near the Chicxulub impact structure, in Haiti, and in the base of tsunami deposits along the Gulf of Mexico (e.g., Smit 1999). These tektites may also be the source of the lower clay layer of the impact couplet in the North American interior (e.g., Bohor and Glass, 1995). They are not a significant carrier of the global Ir anomaly (e.g., Rocchia et al., 1996). The main carrier of the extraterrestrial component in K-T boundary sediments is in the global ejecta, which compose the uppermost layer of sediments in regional deposits near Chicxulub (Smit, 1999). This layer is also mostly composed of spherules, and Smit (1999) estimated the spherule concentration to be ~20 000/cm^2. At least three types of spherules have been described from the global K-T layer (Montanari et al., 1983; Smit et al., 1992). Spherules composed largely of glauconite or potassium feldspar have similar dendritic textures, which Smit et al. (1992) suggested may be derived from clinopyroxenes. Spherules containing unaltered clinopyroxene have only been observed in the K-T boundary from the North Pacific Deep Sea Drilling Project (DSDP) Site 577. Smectitic spherules containing an Ni-rich, magnesioferrite spinel are known from a number of sites around the world (e.g., Kyte and Smit, 1986; Kyte and Bostwick, 1995).

Only the spinel-bearing spherules are clearly established as an important carrier of the siderophile-element anomaly in K-T boundary sediments. Montanari et al. (1983) measured 68 ng/g Ir in magnetic concentrates of spherules from Petriccio, Italy. This was nearly an order of magnitude greater than the 8 ng/g Ir of the bulk sediment. By contrast, K-feldspar spherules contained 1 ng/g Ir and glauconitic spherules were <2 ng/g Ir. Smit and Kyte (1984) also showed that the Italian spinel-bearing spherules had element/Ir ratios for Pt, Au, and Pd that were within a factor of two of those ratios in chondritic meteorites. The conclusion was that these spherules contained $\sim 10\%$ chondritic materials. In a similar study of spinel-bearing spherules from North Pacific DSDP Site 577, Robin et al. (1993) analyzed individual spherules. Their average concentration of 84 ng/g Ir is comparable to, but somewhat higher than, the values measured in Italy. Robin et al. (1983) found that a few individual spherules had concentrations as high as 610 ng/g Ir, comparable to levels in chondritic meteorites.

There is some disagreement as to the actual source of the spinel-bearing spherules. Robin et al. (1993) argued that they are derived directly from meteoritic materials by atmospheric ablation related to localized impact events. Rocchia et al. (1996) proposed that this might have resulted from reentry of projectile material that was ricocheted from the Chicxulub crater during a low-angle impact event. Whereas a direct meteoritic source may be the only reasonable interpretation of the very Ir-rich particles, Kyte and Bostwick (1996) argued that the regional variability of spinel compositions and textures are best explained by formation of spherules within the impact plume, possibly by condensation from the impact vapor. Competing hypotheses of formation of spinel-bearing spherules as individual ablation droplets (Gayraud et al., 1996) or vapor condensates (e.g., Kyte and Bostwick, 1985; Ebel and Grossman, 1999) have yet to be resolved. However, an undisputed fact is that these particles are an important carrier of meteoritic materials in K-T boundary sediments.

Several other occurrences of spherules that are likely carriers of a significant extraterrestrial component have been documented. Spinel-bearing spherules were produced by the late Pliocene oceanic impact of the Eltanin asteroid (Margolis et al., 1991; discussed further in the following). Spinel-bearing spherules from a late Eocene Ir anomaly in sediments from Massignano, Italy, were described as ablation particles formed from a comet impact (Pierrard et al., 1998). However, there are no specific Ir data on these spherules, and it is not clear that it is possible to distinguish them from clinopyroxene spherules found in other late Eocene deposits (e.g., Glass and Koeberl, 1999; Vonhof and Smit, 1999), which cannot be the main carrier of Ir in these deposits. Unusual spinel-bearing spherules from a Jurassic hardground (Jehanno et al., 1988) have Ir contents averaging 300 ng/g and include individuals with unusual, compound spherule shapes. Presumably they formed by accretion of multiple droplets in flight. The only known analogue to this is compound spherules formed by the Eltanin impact (Margolis et al., 1991). Jehanno et al. (1988) inferred that the Jurassic spherules all formed in a single large accretionary event. Spherules are also known in several layers in Proterozoic and Late Archean sediments, several of which contain small Ir anomalies (e.g., Simonson et al., 2000; Simonson and Harnik, 2000). Perhaps the most spectacular spherule deposits are those found in Early Archean sediments (e.g., Lowe et al., 1989), which are discussed herein.

Spinels. Ni-rich magnesioferrite spinels are known to occur in many types of spherules, including cosmic spherules (e.g., Robin et al., 1992) and impact spherules such as at the K-T boundary (Smit and Kyte, 1984), in Eltanin ejecta (Margolis et al., 1991), and even in an Archean spherule bed (Byerly and Lowe, 1994). These spinels are often considered to be high-temperature phases that have survived diagenetic alteration and record processes involved in formation and evolution of at least one K-T boundary component (e.g., Kyte and Smit, 1986; Robin et al., 1992; Kyte and Bostwick, 1995). Very little is known about the trace element chemistry of these spinels. Bohor et al. (1986) measured 29 ± 11 ng/g Ir in spinels from the K-T boundary at Carvaca, Spain, less than the bulk Ir content of the Caravaca boundary clay (52 ng/g; Bohor et al., 1986). Therefore, although the spinel has an extraterrestrial component, it cannot be the principal carrier of Ir at this site. Ni, Cr, and a significant fraction of other major elements (e.g., Mg, Fe, Al) in these spinels may be derived directly from meteoritic precursors.

Unfortunately, the silicate portion of the spherules from which these spinels crystallized does not survive alteration. At best, the silicates are preserved as clays with relict textures (e.g., Montanari et al., 1983; Robin et al., 1993), but they often fall apart during sample processing and only individual spinels can be extracted from sediment. In the latter case, spinels can be used as a proxy tracer of extraterrestrial matter that was originally contained in spherules.

At many K-T boundary sites, individual spinel grains have served as useful tracers of the physical debris deposited by the impact event (Bohor et al., 1986; Zhou et al., 1991; Robin et al., 1991; Rocchia et al., 1996). An important point raised by Robin et al. (1991) is that in K-T boundary sediments, the distribution of physical tracers such as spinels can be quite different from chemical tracers such as Ir. Whereas the stratigraphic distribution of spinels can be influenced only by physical processes such as reworking or bioturbation, Ir can be mobilized by diagenesis and can diffuse through sediment pore waters (e.g., Wallace et al., 1990; Colodner et al., 1992; Evans et al., 1993). At different K-T boundary sites, Robin et al. (1991) found that spinels are always restricted to a narrow stratigraphic interval, while the Ir signal is always dispersed, probably as a result of postdepositional chemical processes (Fig. 6). These results provide a strong argument for a single, geologically brief (<100 yr), accretionary event at the K-T boundary that deposited physical ejecta and was followed by chemical diffusion during diagenesis.

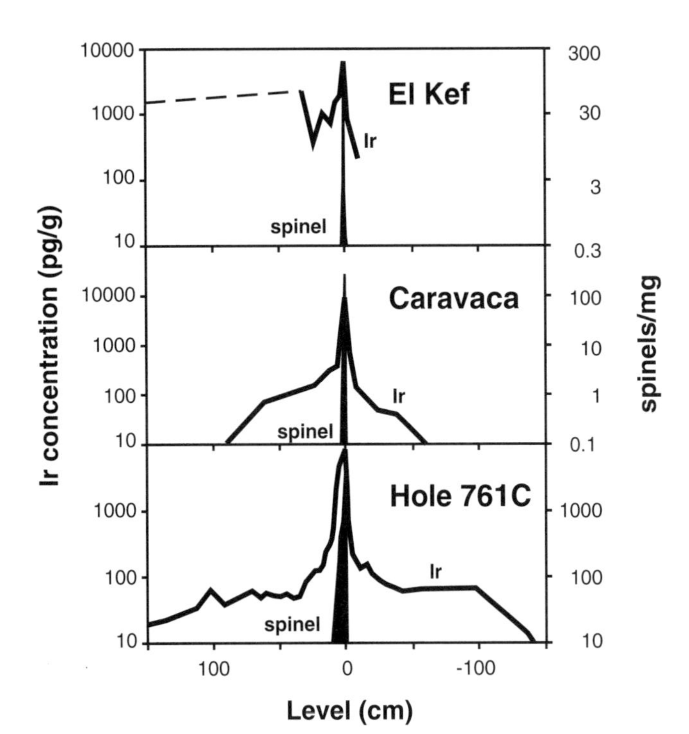

Figure 6. Ir anomaly in Cretaceous-Tertiary boundary sediments can be quite broad, extending into sediments below horizon. However, physical ejecta such as spinels are restricted to very thin stratigraphic interval. After Robin et al. (1993).

The stratigraphic distribution of spinels is being examined in LL44-GPC3, a large-diameter piston core of pelagic clay from the central North Pacific. This core contains a relatively complete record of sediment accumulation since the Late Cretaceous (Kyte et al., 1993). Preliminary results from Robin et al. (1999) indicate that at least five horizons contain anomalous concentrations of spinels. One of these is at the K-T boundary, two are from the middle Cenozoic, and two are from the Pleistocene. Robin et al. (1999) attributed these to impact events, but provided no specific criteria to distinguish these as impact spherules, as opposed to cosmic spherules. At least one of the middle Cenozoic horizons, at 11.2 m depth, also contains abundant spherules that were studied by Kyte and Bostwick (1995) and interpreted as cosmic spherules. This is also nearly coincident with a peak in extraterrestrial ^{3}He, centered at 11.4 m depth in LL44-GPC3 (Farley, 1995). One of these middle Cenozoic, spinel-rich horizons probably correlates with the spinel-rich horizon in Massignano, Italy (Pierrard et al., 1998). Whether these spinel horizons are impact deposits, or the result of other types of accretionary events, they are valuable tracers of the flux of extraterrestrial matter to deep-sea sediments.

Meteorites. Recovery of actual meteorites provides the best information about the composition of extraterrestrial matter accreted to Earth, because they can be directly compared to potential source materials. Meteorites and fossil meteorites have been described in a number of deposits. Meteorites have been recovered from two hypervelocity impacts, the Eltanin oceanic and the K-T boundary events. Micrometeorites are now routinely recovered in cosmic spherule collections (Engrand and Maurette, 1998; Taylor et al., 2000).

The first fossil meteorite found in ancient sediments was Brunflo, a heavily altered 10 cm chondrite found in Ordovician limestone (Thorslund et al., 1981, 1984). Relict textures are well preserved, but nearly all the precursor minerals in this meteorite were replaced by calcite, clays, and other trace phases. Only chromites appear to retain their original chemistry, and these are used to infer that this meteorite was an H-type, ordinary chondrite. Schmitz et al. (1996) recovered 13 additional fossil meteorites from these strata, and proposed that the accumulation rate of these meteorites and Ir in associated sediments is higher than in modern sediments and that this is evidence of a large collision in the asteroid belt during the Ordovician.

Numerous meteorite fragments are found in the ejecta layer of the Eltanin impact event (Kyte and Brownlee, 1985; Gersonde et al., 1997; Kyte, 2002a, 2002b). These are typically a few millimeters in size, but one individual as large as 1.5 cm was recovered. Although most of the coarse ejecta from the Eltanin impact is melted asteroidal materials (discussed in the following), ~7% survived as unmelted meteorites. The meteorites recovered in deposits of the Eltanin impact are polymict breccias composed mainly of plagioclase and pyroxene with compositions similar to those found in basaltic meteorites; specifically the howardites and the silicates of mesosiderites. Because of relatively high Ir concentrations in the Eltanin impact melt (Kyte and Brownlee, 1985; Kyte, 2000c), the Eltanin asteroid is inferred to be related to mesosiderites, meteorites containing both basaltic and metallic components (e.g., Mittlefehldt et al., 1998). These meteorite fragments provide conclusive evidence that the Eltanin projectile was from an asteroidal, rather than a cometary source.

Meteoritic materials have also been described in K-T boundary sediments. Robin et al. (1993) inferred that irregular particles with spinel-rich rims and high Ir concentrations from the K-T boundary at DSDP Site 577 may be partially melted meteoritic debris. These particles are ~250 µm in size, and no relict meteorite textures or minerals have been described, but Robin et al. (1993) pointed out that similar objects occur as partially melted meteorites in cosmic spherule collections. At a nearby locality in the western North Pacific, DSDP Site 576, Kyte (1998) recovered a 2.5 mm fossil meteorite in the K-T boundary. This specimen has chondritic concentrations of Ir, Fe, and Cr, as well as relict mineral textures that were interpreted to be from olivine, metal, sulfides, and a fine-grained clay matrix. The clay mineral saponite, which is common in some carbonaceous chondrites (Zolensky et al., 1993) and IDPs (Joswiak et al., 2000), was also found. On the basis of these data, Kyte (1998) inferred that the K-T bolide had characteristics similar to those of many carbonaceous chondrites, with olivine, metal and sulfide in a fine-grained, phyllosilicate-rich matrix. On the basis of the abundances of these phases in this small specimen, it was inferred to be most like the CO, CV, and CR groups of chondrites. Notably, because the K-T meteorite appeared to be from a compact, hydrated object, rather than a porous, anhydrous one, Kyte (1998) favored an asteroidal source over a cometary one for the K-T projectile. This is consistent with interpretations of provenance for stratospheric IDPs discussed herein (Joswiak et al., 2000).

Ocean impact melt products. Deep-ocean impacts of small asteroids or comets can produce a unique class of meteoritic debris. This is because smaller projectiles cannot penetrate several kilometers of water. In effect, the projectile will not mix with silicates from the ocean floor, and melt products will be composed almost entirely of meteoritic materials. There may also be a similar type of meteoritic materials that are produced by objects that cannot penetrate the atmosphere (i.e., atmospheric impacts), but no extraterrestrial materials from such an event have been documented. A classic example of an atmospheric impact is the Tunguska event of 1908. No unequivocal evidence of an extraterrestrial deposit from this event is known (Zahnle, 1996). Chyba et al. (1993) proposed that the Tunguska explosion was caused by catastrophic disruption of a small stony asteroid in the upper troposphere. The ability of a projectile to penetrate the atmosphere and impact the ocean is a function of its strength, size, velocity, and impact angle (Chyba et al., 1993). For stony asteroids, minimum diameters of a few hundred meters may be required for projectiles to penetrate the atmosphere and impact the ocean surface with cosmic velocities.

Numerical simulations of deep-ocean impacts (Artemieva and Shuvalov, 2002) provide some limits on the size of a projectile that will not mix with the ocean floor during a deep-ocean impact. For a vertical impact at asteroidal velocities (~20 km/s), mixing is only likely when the projectile diameter is greater than 1/2 of the water depth. For oblique impacts, even larger projectiles will not mix with the ocean floor. Given the typical water depths of 4–5 km in deep-ocean basins, asteroidal projectiles with diameters as large as 2 or 3 km may commonly produce silicate ejecta composed only of meteoritic materials and seawater salts. However, the compressed water column beneath the projectile can still disrupt and shock metamorphose the ocean floor. Given that 60% of the Earth's surface is covered by oceanic lithosphere and 500 m projectiles hit the Earth on 10^5 yr time scales (Wetherill and Shoemaker, 1982), there must be hundreds of oceanic impact deposits in the sediment record to be discovered.

The only known case of a deep-ocean impact event is the late Pliocene impact of the Eltanin asteroid. This was originally discovered as an Ir anomaly in a sediment core from the sub-Antarctic Pacific (Kyte et al., 1981). Ejecta from this impact is now known from at least eight sediment cores across at least 500 km of the ocean floor (Kyte et al., 1988; Gersonde et al., 1997; Kyte, 2002a). Most of the ejecta is a vesicular melt rock composed mostly of asteroidal materials (Fig. 7, A–C; Kyte and Brownlee, 1985). The only terrestrial component in the ejecta is a few percent Na, K, and Cl from seawater salts, the least volatile component of the impact target. The salt component is clear evidence that the asteroid actually mixed with a seawater target. The high vesicularity of the melt is presumably produced by water vapor escaping from the melt. On the basis of the composition of meteorite fragments that compose several percent of the ejecta, as well as the content of Ir, Ni, and Fe in the melt, Kyte (2002c) inferred that the original asteroid was a meteoritic basalt with ~4 wt% metal. Another melt produced by this impact formed spherules that are only a trace component of the ejecta (Fig. 7, D and E; Margolis et al., 1991). The spherules are not vesicular and contain only ~5 ng/g Ir, as opposed to 187 ng/g in the vesicular melt (Kyte, 2002c). Therefore these spherules must have formed under different conditions than the bulk of the ejecta, but the causes of these differences are unknown. It is conceivable that if ejecta from this impact is ever discovered at sites far removed from ground zero, it may be dominated by spheroidal rather than vesicular debris.

The highly vesicular characteristics of oceanic impact melts have probably hindered their discovery in sediments, because they may be confused with volcanic ash. One possible diagnostic criterion could be a very high density of very small vesicles that produce a rough surface texture (Fig. 7A). In polished section, vesicular melt from a chondritic source should contain abundant olivine, an interstitial glass, and an oxide phase, as in cosmic spherules (e.g., Brownlee, 1985). Qualitative energy dispersive analyses in an SEM would easily detect Ni, an element unlikely to be from a terrestrial source.

Isotopic tracers

Three isotopic systems have been used successfully to trace meteoritic matter in sediments. Each system has proven useful in measuring different aspects of the accretion of extraterrestrial materials on Earth.

Re-Os system. This system is based on the decay of ^{187}Re to ^{187}Os (half life, $t_{1/2}$ = 45.6 b.y.), and the differing geochemical behaviors of Re and Os (Faure, 1986). The potential of this isotopic system as a tracer of extraterrestrial matter was first recognized by Turekian (1982) at the first Snowbird Conference. While mantle rocks are expected to have an osmium isotopic composition, measured as ^{187}Os/^{186}Os, similar to that in chondritic meteorites, crustal rocks should have very different values. This is because during melting in the upper mantle, Re behaves as an incompatible element and has been enriched in derivative melts relative to Os, which is concentrated in residual mantle solids. Thus, crustal rocks have Re/Os ratios much greater than those in the mantle or in chondrites, and the decay of the excess ^{187}Re has led to high levels of ^{187}Os in the crust. Turekian (1982) predicted that if K-T boundary sediments were enriched in extraterrestrial matter, they should have ^{187}Os/^{186}Os ratios near unity, the value in chondrites, whereas crustal rocks should have ^{187}Os/^{186}Os at least an order of magnitude higher.

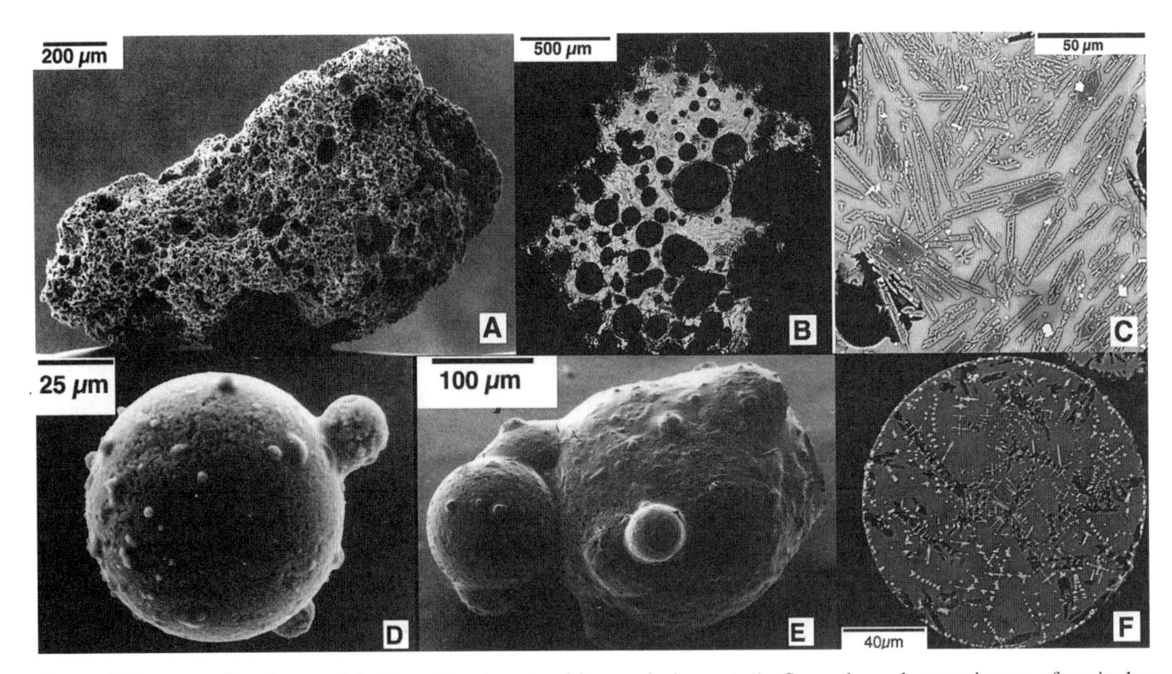

Figure 7. Images of melt materials from Eltanin asteroid oceanic impact. A: Secondary electron image of vesicular impact melt-rock particle. B: Backscatter electron image of polished section of vesicular impact melt-rock particle. C: Portion of B showing glass with olivine crystals. Small white grains are chromite. D: Secondary electron image of spherule from Eltanin impact. These spherules commonly have multiple small droplets attached to their surfaces. E: Secondary electron image of highly compound spherule from Eltanin impact. F: Backscatter electron impact of polished section of Eltanin impact spherule. Skeletal spinels (white) are imbedded in glass groundmass (gray). Black holes are probably from olivine that has dissolved.

Luck and Turekian (1983) confirmed this prediction when they found $^{187}Os/^{186}Os$ ratios of 1.3 and 1.6 in K-T boundary clays from the Raton basin, Colorado, and Stevns Klint, Denmark, respectively. In the same study they found that manganese nodules, which should reflect the seawater composition, had $^{187}Os/^{186}Os$ ratios ranging from ~6 to 8. Although this was the first isotopic evidence to support the K-T boundary extinction hypothesis, it remained ambiguous because it left open the possibility that K-T boundary PGEs were from mantle sources, which should have Os isotopic abundances similar to those in chondritic meteorites. The discovery of enhanced Ir concentrations in aerosols from Kilauea, Hawaii (Zoller et al., 1983), provided an alternate potential source for the Ir and Os, and lent considerable credence to a volcanic alternative to the K-T boundary impact model.

Osmium isotopes have also proven useful in estimating the long-term flux of extraterrestrial matter to the Earth (Esser and Turekian, 1988; Peucker-Ehrenbrink, 1996). Simple chemical methods, such as accumulation rates of Ir or Os in marine sediments (Barker and Anders, 1968; Kyte and Wasson, 1986), essentially provide an upper limit on the meteoritic flux, because some fraction of the PGEs in sediments is terrestrial. Os isotopes make it possible to model terrestrial contributions from crustal and mantle sources. The flux of 37000 ± 13000 t/yr of extraterrestrial matter in Pacific sediments, based on Os isotopes (Peucker-Ehrenbrink, 1996), is probably the best estimate available to date. In addition to its application to the flux of meteoritic matter, the Os isotopic compositions of sediments and seawater have proven to be a useful measure of a number of terrestrial processes (e.g., Peucker-Ehrenbrink et al., 1995; Ravizza and Pyle, 1997).

Helium isotopes. Because of their high surface areas relative to their mass, interplanetary dust particles have high concentrations of solar-wind implanted He (Rajan et al., 1977), and thus $^3He/^4He$ ratios much higher than in any terrestrial materials (Nier and Schlutter, 1990). Much of this He survives atmospheric entry and is deposited in deep-sea sediments. Ozima et al. (1984) found an approximately inverse relationship between ^3He concentrations and the accumulation rate of Pacific surface sediments, a relationship that was attributed to the influx of interplanetary dust. Ozima et al. (1984) calculated an influx rate of only 2000 t/yr, at least an order of magnitude lower than other influx estimates (e.g., Barker and Anders, 1968; Kyte and Wasson, 1986; Love and Brownlee, 1993; Peucker-Ehrenbrink, 1996). This is because the ^3He is most concentrated in the finest fractions of the interplanetary dust complex and is degassed from larger particles during atmospheric entry (e.g., Joswiak et al., 2000). Although the ^3He content of sediments is not a measure of the total influx of extraterrestrial matter, it is very useful because it is a tracer of the fine fraction of the interplanetary dust complex.

Farley (1995) showed that much of the ^3He flux is retained in pelagic clay sediments from the central north Pacific that are as old as 70 Ma. He also found an apparent peak both in the concentration and flux of ^{3}He in middle Cenozoic sediments. This peak in ^{3}He also corresponds to high concentrations of cosmic spherules and Ni-rich spinels (Kyte and Bostwick, 1985; Robin et al., 1999) in the same core (LL44-GPC3). Because of the low stratigraphic resolution of North Pacific clays, Farley et al. (1998) measured the ^{3}He flux in a high-resolution carbonate sequence from the Massignano quarry, Italy. They found a five-fold increase in the ^{3}He flux over an interval of ~2 m.y. in the late Eocene (Fig. 8). These sediments in Massignano also contain at least one Ir anomaly (and possibly two), shocked minerals, and Ni-rich spinel that are related to impacts in the late Eocene (Montanari et al., 1993; Clymer et al., 1996; Pierrard et al., 1998). The two largest impact structures known in the Cenozoic, Popigai and Chesapeake Bay, are also roughly coincident with this time interval (Bottomley et al., 1997; Koeberl et al., 1996). Farley et al. (1998) concluded that the only possible source for this increase in ^{3}He flux was an increase in the interplanetary dust in the inner solar system. Because dust particles have a dynamic lifetime of only ~10^5 yr before they spiral into the Sun, they would need to be continuously replenished to produce an anomaly for 2 m.y. Farley et al. (1998) concluded that the only reasonable source was a comet shower, which could have a duration of about 3 m.y. (Hut et al., 1987). With this discovery, Farley et al. (1998) presented the first physical evidence that required a perturbation of the Oort Cloud of comets. This was not detected from major large-body impacts, but through tracers from fine-grained dust. Mukhopadhyay et al. (2001) measured ^{3}He in a high-resolution section of pelagic limestone from Gubbio, Italy, providing the first detailed profile of the accretion rate of interplanetary dust from the early Maastrichtian to the middle Eocene. Although there appear to be variations in the flux of dust by factors of two or more, they found no evidence of new comet showers, as had been identified in the late Eocene (Farley et al., 1998).

Chromium isotopes. Extraterrestrial material can contain two types of Cr isotopic anomalies. One of these is based on the decay of ^{53}Mn to ^{53}Cr ($t_{1/2}$ = 3.7 m.y.). Although ^{53}Mn was present in the early solar system, it rapidly decayed to extinction. The decay products of ^{53}Mn were first discovered in refractory inclusions from the Allende meteorite, a CV-type carbonaceous chondrite (Birck and Allegre, 1985), and have now been identified in a variety of solar system objects (Lugmair and Shukolyukov, 1998; Birck and Allegre, 1988). These studies have shown that ^{53}Mn was heterogeneously distributed in the early solar system and that different types of meteorites can have distinct ^{53}Cr/^{52}Cr ratios (Lugmair and Shukolyukov, 1998; Shukolyukov and Lugmair, 1999). Because the Earth homogenized long after the ^{53}Mn decayed, all terrestrial rocks should have identical ^{53}Cr/^{52}Cr ratios. The isotopic variations in meteorites are measured as the deviations of the ^{53}Cr/^{52}Cr ratios from the standard terrestrial ^{53}Cr/^{52}Cr ratio and are usually expressed in ε units (1 ε is 1 part in 10^4, or 0.01%). Thus, by definition, the standard terrestrial ^{53}Cr/^{52}Cr is ≡ 0 ε. The other type of Cr isotopic anomaly is an excess of ^{54}Cr known to occur in carbonaceous chondrites (Papanastassiou, 1986; Podosek et al., 1997).

High-precision measurements of ^{53}Cr/^{52}Cr ratios in ordinary chondrites, differentiated meteorites, and Martian meteorites (SNCs) show that all these extraterrestrial materials have an excess of ^{53}Cr relative to terrestrial rocks (Lugmair and Shukolyukov, 1998), and thus a positive value for ε ^{53}Cr for all meteorites, except the carbonaceous chondrites (Fig. 9). The high-precision measurement of this ratio relies on a second-order fractionation correction that is based on a normalization

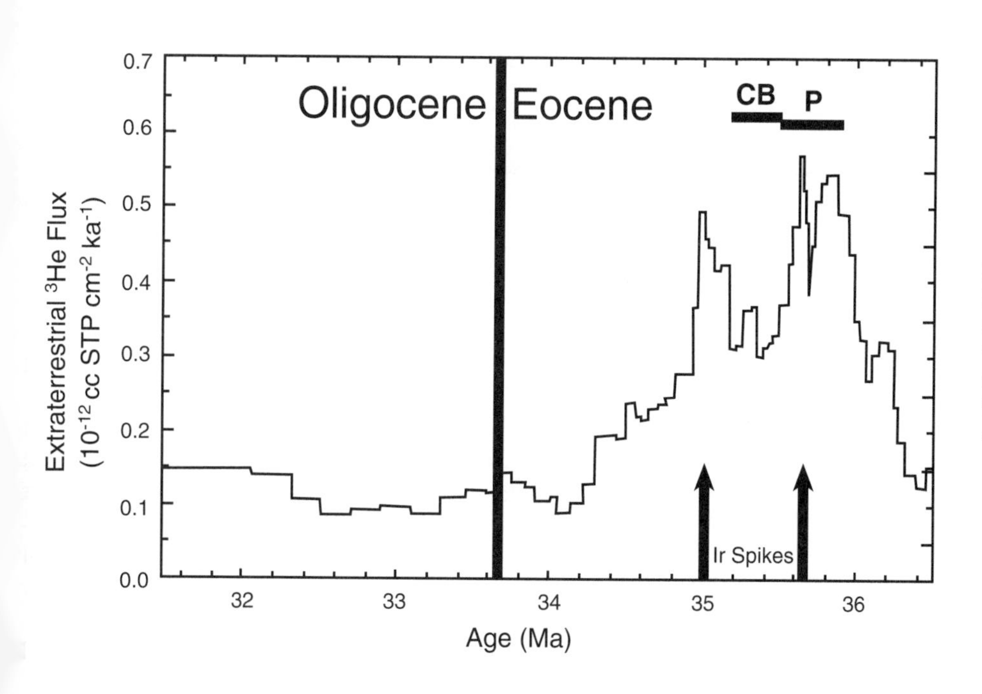

Figure 8. Peak in flux of ^{3}He is found in late Eocene sediments from Massignano, Italy. This is roughly coincident with at least one Ir anomaly, spherule deposits, and two large impact craters, Chesapeake Bay (CB) and Popigai (P). After Farley et al. (1998).

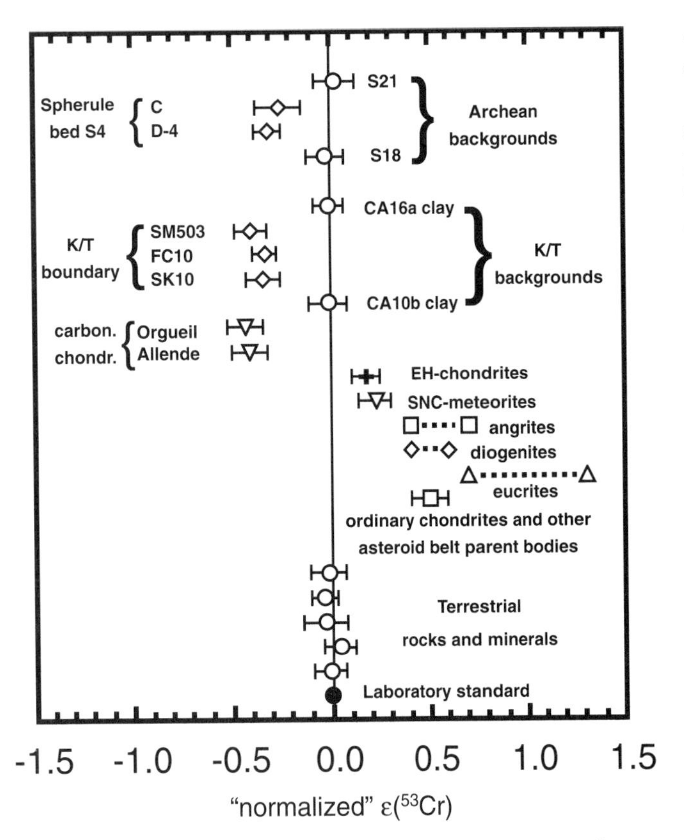

Figure 9. All terrestrial materials, including background sediments near impact deposits, have Cr isotopic compositions with $\varepsilon\,^{53}Cr = 0$. Most meteorites have positive $\varepsilon\,^{53}Cr$ values, reflecting excess of ^{53}Cr. Carbonaceous chondrites have normalized $\varepsilon\,^{53}Cr$ that is negative, reflecting excess of both ^{53}Cr and ^{54}Cr relative to terrestrial samples. Samples from Cretaceous-Tertiary (K-T) boundary and Early Archean spherule bed have negative normalized $\varepsilon\,^{53}Cr$, indicating confirmed extraterrestrial component similar to that in carbonaceous chondrites. After Shukolyukov et al. (2000a).

to ^{54}Cr. Therefore, when Cr isotopes are measured in carbonaceous chondrites, which have an excess of ^{54}Cr, this fractionation correction results in a normalized $\varepsilon\,^{53}Cr$ that has negative values relative to terrestrial rocks (Fig. 9). Thus, high precision Cr isotopic analyses can distinguish meteoritic materials from terrestrial rocks and carbonaceous chondrites from other meteorite groups (Shukolyukov et al., 2000a).

Shukolyukov and Lugmair (1998) applied this method to K-T boundary sediments from Caravaca, Spain, and Stevns Klint, Denmark. As expected, background sediments from Caravaca have $\varepsilon\,^{53}Cr$ values indistinguishable from the terrestrial value of zero. The K-T boundary clays, however, have normalized $\varepsilon\,^{53}Cr$ of ~ -0.4 and are indistinguishable from the carbonaceous chondrites Orgueil and Allende (Fig. 9). It took 18 years from the initial report of anomalous Ir in K-T boundary sediments, but this method provided the first unequivocal isotopic proof of an extraterrestrial component in the K-T boundary. In addition, this work showed that the K-T projectile had both anomalous ^{53}Cr and ^{54}Cr and was probably related to the carbonaceous chondrites, a result consistent with interpretations of relict textures in a fossil K-T meteorite (Kyte, 1998).

Chromium isotopic data have been obtained on Early Archean spherule bed S4 (Shukolyukov et al., 2000a), which also has a negative $\varepsilon\,^{53}Cr$ (Fig. 9). This deposit is now clearly identified as containing Cr from an extraterrestrial source such as a carbonaceous chondrite. Preliminary reports (Shukolyukov et al., 2000b) from another spherule bed (S3) indicate that it may have a similar source.

The Cr isotopic method has enormous potential as a tracer of extraterrestrial matter. The main limitation of this technique is that a significant fraction of the Cr in a sample (perhaps at least 30%) must be meteoritic. Thus, Cr isotopes are not useful as a tracer of the background flux of extraterrestrial matter, but they are useful for ejecta deposits with high Ir and Cr concentrations. One intriguing possibility is that Cr isotopes might be able to identify ejecta from a projectile with low concentrations of Ir. For example, differentiated meteorites such as howardites, eucrites, and diogenites have Ir concentrations at levels typically <1% of the concentration in chondrites (Mittlefehldt et al., 1998), but Cr concentrations 10–100 times higher than in typical sediments. Ejecta from the impact of a eucritic asteroid would be virtually undetectable using siderophile elements, but well within the range of possibility using Cr isotopic analyses.

CONCLUSIONS

Much has been learned in the past two decades about the accretion history of the Earth. The interplanetary dust complex is becoming very well characterized. The finest fraction of this, the stratospheric IDPs, are mostly derived from chondritic materials. IDPs that are most likely from asteroidal sources generally are compact and have hydrated minerals, whereas IDPs from cometary sources appear to mostly be porous anhydrous aggregates with GEMS. Cosmic spherule collections and associated micrometeorites from the coarse fraction of the dust complex have characteristics that are compatible with an origin mostly from asteroidal material similar to carbonaceous chondrite meteorites.

The long-term flux of interplanetary dust has been relatively constant for much of the past 70 m.y., except for a pulse in the late Eocene recorded by high concentrations of ^{3}He in marine sediments. This is coincident with one or more Ir anomalies, at least two spherule deposits, and two large impact structures. Because this pulse of dust persists for at least 2 m.y., this is believed to be the signature of a comet shower, triggered by a perturbation of the Oort Cloud of comets. Although it is well established that the Eocene-Oligocene transition was marked by high extinction rates (Raup and Sepkoski, 1984), and rapid climate change (e.g., Zachos et al., 1996), these phenomena have not been linked to this apparent comet shower by any strong physical data.

Deposits of the major impact event at the K-T boundary are now extremely well characterized. The fallout layer from the global ejecta probably contained 5%–10% chondritic material, and one carrier of this component is spinel-bearing spherules. Analyses of a small fossil meteorite, as well as the isotopic composition of Cr in K-T boundary sediments, point to a projectile similar to carbonaceous chondrites. Physical debris (i.e., Ni-rich spinels) in the global fallout is restricted to a single layer, and there is no strong evidence to support any hypothesis other than a single, geologically instantaneous accretionary event. This observation, in addition to the apparent lack of an increased flux of ^{3}He at the K-T boundary, are strong arguments against a comet shower at 65 Ma. That the K-T meteorite is more similar to compact, hydrated IDPs than to anhydrous porous IDPs is also reason to suspect an asteroidal, rather than a cometary source for the K-T projectile.

Very little is known about the meteoritic flux through the rest of the Phanerozoic. Exhaustive searches of a number of extinction horizons in sediments have failed to find conclusive evidence of an extraterrestrial component. At this time, the only great mass extinction that can be confidently linked to a major impact event is that at the K-T boundary. Considerably more work is needed in this area.

A number of potential impact deposits ranging in age from 0.6 to 3.4 Ga have been identified in the Precambrian. Most spectacular among these are a number of thick spherule deposits in the Barberton Greenstone Belt, South Africa. Chromium isotopes have conclusively shown that at least one of these contains ~20% meteoritic material with an isotopic composition similar to that of CV chondrites. If other beds can be proven to contain high concentrations of extraterrestrial matter, they may indicate an extensive record of megaimpacts, possibly by projectiles considerably larger than the K-T impactor.

ACKNOWLEDGMENTS

I thank I. McDonald and B. Schmitz for helpful reviews. This work was supported by NASA grant NAG5-9411.

REFERENCES CITED

Alekseev, A.S., Barsukova, L.D., Kolesov, G.M., Nazarov, M.A., and Grigoryan, A.G., 1983, The Permian-Triassic boundary event: Geochemical investigation of the Transcaucasia section [abs.]: Lunar and Planetary Science, v. 14, p. 3–4.

Alvarez, L.W., Alvarez, W., Asaro, F., and Michel, H.V., 1980, Extraterrestrial cause for the Cretaceous–Tertiary extinction: Science, v. 208, p. 1095–1108.

Artemieva, N.A., and Shuvalov, V.V., 2002, Shock metamorphism on the ocean floor (numerical simulations): Deep Sea Research II, v. 49, p. 959–968.

Asaro, F., Alvarez, L.W., Alvarez, W., and Michel, H.V., 1982, Geochemical anomalies near the Eocene/Oligocene and Permian/Triassic boundaries: Geological Society of America Special Paper 190, p. 517–528.

Barker, J.L., and Anders, E., 1968, Accretion rate of cosmic matter from iridium and osmium contents of deep-sea sediments: Geochimica et Cosmochimica Acta, v. 32, p. 627–645.

Becker, L., 2000, Fullerenes, noble gases and the flux of extraterrestrial debris to the surface of the earth over geologic time: First Astrobiology Conference, Mountain View, California, Abstracts, p. 30.

Becker, L., Bada, J.L., Winans, R.E., Hunt, J.E., Bunch, T.E., and French, B.M., 1994, Fullerenes in the 1.85-billion-year-old Sudbury impact structure: Science, v. 265, p. 642–644.

Becker, L., Poreda, R.J., and Bada, J.L., 1996, Extraterrestrial helium trapped in fullerenes in the Sudbury impact structure: Science, v. 272, p. 249–252.

Bell, J.F., Davis, D.R., Hartman, W.K., and Gaffey, M.J., 1989, Asteroids: The big picture, *in* Binzel, R.P., Gehrels, R., and Matthews, M.S., eds., Asteroids II: Tucson, University of Arizona Press, p. 921–945.

Birck, J.-L., and Allègre, C.J., 1985, Evidence for the presence of ^{53}Mn in the early solar system: Geophysical Research Letters, v. 12, p. 745–748.

Birck, J.-L., and Allègre, C.J., 1988, Manganese-chromium isotope systematics and development of the early solar system: Nature, v. 331, p. 579–584.

Bohor, B.F., Ford, E.E., and Ganapathy, R., 1986, Magnesioferrite from the Cretaceous-Tertiary boundary, Caravaca, Spain: Earth and Planetary Science Letters, v. 81, p. 57–66.

Bohor, B.F., and Glass, B.P., 1995, Origin and diagenesis of K/T impact spherules—from Haiti to Wyoming and beyond: Meteoritics, v. 30, p. 182–198.

Bonner, F.T., and Laurenco, A.S., 1965, Nickel content in Pacific Ocean cores: Nature, v. 207, p. 933–935.

Bottomley, R., Grieve, R.A.F., York, D., and Masaitis, V., 1997, The age of the Popigai impact event and its relationship to events at the Eocene/Oligocene boundary: Nature, v. 388, p. 365–368.

Bradley, J.P., Brownlee, D.E., 1991, An interplanetary dust particle linked directly to type-CM meteorites and an asteroidal origin: Science, v. 251, p. 549–552.

Bradley, J.P., Keller, L.P., Snow, T.P., Hanner, M.S., Flynn, G.J., Gezo, J.C., Clemett, S.J., Brownlee, D.E., and Bowey, J.E., 1999, An infrared spectral match between GEMS and interstellar grains: Science, v. 285, p. 1716–1718.

Brownlee, D.E., 1985, Cosmic dust: Collection and research: Annual Review of Earth and Planetary Sciences, v. 13, p. 147–173.

Brownlee, D.E., Pilachowski, L.B., and Hodge, P.W., 1979, Meteorite mining on the ocean floor [abs.]: Lunar and Planetary Science, v. 10, p. 157–158.

Brownlee, D.E., Joswiak, D., and Bradley, J.P., 1997a, Vesicular carbon in strongly heated IDPs [abs.]: Lunar and Planetary Science, v. 27, Abstract #1585, CD-ROM.

Brownlee, D.E., Bates, B., and Schramm, L., 1997b, The elemental composition of stony cosmic spherules: Meteoritics and Planetary Science, v. 32, p. 157–175.

Byerly, G.R., and Lowe, D.R., 1994, Spinel from Archean impact spherules: Geochimica et Cosmochimica Acta, v. 58, p. 3469–3486.

Castaing, R., and Fredriksson, K., 1958, Analyses of cosmic spherules with and X-ray microanalyzer: Geochimica et Cosmochimica Acta, v. 14, p. 114–117.

Chou, C.-L., 1978, Fractionation of siderophiles elements in the earth's upper mantle: Proceedings of the 9th Lunar and Planetary Science Conference, p. 219–230.

Chyba, C.F., Thomas, P.J., and Zahnle, K.J., 1993, The 1908 Tunguska explosion: Atmospheric disruption of a stony asteroid: Nature, v. 361, p. 40–44.

Claeys, P., and Casier, J.-G., 1994, Microtektite-like impact glass associated with the Frasnian-Famennian boundary mass extinction: Earth and Planetary Science Letters, v. 122, p. 303–315.

Claeys, P., Kyte, F.T., Herbosh, A., and Casier, J.-G., 1996, Geochemistry of the Frasnian Famennian boundary: Mass extinctions, anoxic oceans, and microtektite layer, but not much iridium?, *in* Ryder, G., Fatovsky, D., and Gartner, S., eds., The Cretaceous-Tertiary boundary and other catastrophes in Earth history: Geological Society of America Special Paper 307, p. 491–504.

Clark, D.L., Wang, C., Orth, C.J., and Gilmore, J.S., 1986, Conodont survival and low iridium abundances across the Permian-Triassic boundary in south China: Science, v. 233, p. 984–986.

Clymer, A.K., Bice, D.M., and Montanari, A., 1996, Shocked quartz from the late Eocene: Impact evidence from Massignano, Italy: Geology, v. 24, p. 483–486.

Colodner, D.C., Boyle, E.A., Edmond, J.M., and Thomson, J., 1992, Post-depositional mobility of platinum, iridium, and rhenium in marine sediments: Nature, v. 358, p. 402–404.

Crocket, J.H., and Kuo, H.Y., 1979, Sources for gold, palladium, and iridium in deep-sea sediments: Geochimica et Cosmochimica Acta, v. 43, p. 831–842.

Czajkowski, J., 1987, Cosmo and geochemistry of the Jurassic hardgrounds [Ph.D. thesis]: San Diego, University of California, 418 p.

Davis, M., Hut, P., and Muller, R.A., 1984, Extinction of species by periodic comet showers: Nature, v. 308, p. 715–717.

Dodd, R.T., 1989, Unique find from Antarctica: Nature, v. 338, p. 296–297.

Ebel, D.S., and Grossman, L., 1999, Condensation in a model Chicxulub fireball [abs.]: Lunar and Planetary Science, v. 30, Abstract #1906, CD-ROM.

Engrand, C., and Maurette, M., 1998, Carbonaceous micrometeorites from Antarctica: Meteoritics and Planetary Science, v. 33, p. 565–580.

Esser, B.K., and Turekian, K.K., 1988, Accretion rate of extraterrestrial particles determined from osmium isotope systematics of Pacific pelagic clay and manganese nodules: Geochimica et Cosmochimica Acta, v. 52, p. 1383–1388.

Evans, N.J., Gregoire, D.C., Goodfellow, W.D., McInnes, B.I., Miles, N., and Veizer, J., 1993, Ru/Ir ratios at the Cretacous-Tertiary boundary: Implications for PGE source and fractionation within the ejecta cloud: Geochimica et Cosmochimica Acta, v. 57, p. 3149–3158.

Farley, K.A., 1995, Cenozoic variations in the flux of interplanetary dust recorded by He-3 in a deep-sea sediment: Nature, v. 376, p. 153–156.

Farley, K.A., Montanari, A., Shoemaker, E.M., and Shoemaker, C.S., 1998, Geochemical evidence for a comet shower in the Late Eocene: Science, v. 280, p. 1250–1253.

Faure, G., 1986, Principles of isotope geology (second edition): New York, John Wiley & Sons, 589 p.

Ganapathy, R., 1980, A major meteorite impact on the earth 65 million years ago: Evidence from the Cretacous-Tertiary boundary clay: Science, v. 209, p. 921–923.

Ganapathy, R., 1982, Evidence for a major meteorite impact on the earth 34 million years ago: Implications for Eocene extinctions: Science, v. 216, p. 885–886.

Ganapathy, R., Brownlee, D.E., and Hodge, P.W., 1978, silicate spherules from deep-sea sediments: Confirmation of extraterrestrial origin: Science, v. 201, p. 1119–1121.

Gayraud, J., Robin, E., Rocchia, R., and Froget, L., 1996, Formation conditions of oxidized Ni-rich spinel and their relevance to the K/T boundary event: Geological Society of America Special Paper 307, p. 425–443.

Gersonde, R., Kyte, F.T., Bleil, U., Diekmann, B., Flores, J.A., Gohl, K., Grahl, G., Hagen, R., Kuhn, G., Sierro, F.J., Völker, D., Abelmann, A., and Bostwick, J.A., 1997, Geological record and reconstruction of the late Pliocene impact of the Eltanin asteroid in the Southern Ocean: Nature, v. 390, p. 357–363.

Gilmore, J.S., Knight, J.D., Orth, C.J., Pillmore, C.L., and Tschudy, R.H., 1984, Trace element patterns at a non-marine Cretaceous-Tertiary boundary: Nature, v. 307, p. 224–228.

Glass, B.P., and Burns, C.A., 1987, Late Eocene crystal-bearing spherules: Two layers or one?: Meteoritics, v. 22, p. 265–279.

Glass, B.P., and Koeberl, C., 1999, Ocean Drilling Project Hole 689B spherules and upper Eocene microtektite and clinopyroxene-bearing spherule strewn fields: Meteoritics and Planetary Science, v. 34, p. 197–208.

Glass, B.P., Burns, C.A., Crosbie, J.R., and DuBois, D.L., 1985, Late Eocene North American microtektites and clinopyroxene-bearing spherules: Journal of Geophysical Research, v. 90, p. D175–D196.

Gostin, V.A., Haines, P.W., Jenkins, R.J.F., Compston, W., and Williams, I.S., 1986, Impact ejecta horizon within late Precambrian shales, Adelaide geosyncline, South Australia: Science, v. 233, p. 198–200.

Gostin, V.A., Keays, R.R., and Wallace, M.W., 1989, Iridium anomaly from the Acraman impact ejecta horizon: Impacts can produce sedimentary iridum peaks: Nature, v. 340, p. 542–544.

Herzog, G.F., Xue, S., Hall, G.S., Nyquist, L.E., Shih, C.Y., Wiesmann, H., Brownlee, D.E., 1999, Isotopic and elemental composition of iron, nickel, and chromium in type I deep-sea spherules: Implications for origin and composition of the parent micrometeoroids: Geochimica et Cosmochimica Acta, v. 63, p. 1443–1457.

Heymann, D., Chibante, L.P.F., Brooks, R.R., Wolbach, W.S., and Smalley, R.E., 1994, Fullerenes in the Cretaceous-Tertiary boundary layer: Science, v. 265, p. 645–647.

Hut, P., Alvarez, W., Elder, W.P., Hansen, T., Kauffman, E.G., Keller, G., Shoemaker, E.M., and Weissman, P.R., 1987, Comet showers as a cause of mass extinctions: Nature, v. 329, p. 118–126.

Jehanno, C., Boclet, D., Bonte, Ph., Castellarin, A., and Rocchia, R., 1988, Identification of two populations of extraterrestrial particles in a Jurassic hardground of the southern Alps: Proceedings of the 18th Lunar and Planetary Science Conference, p. 625–630.

Jessberger, E.K., and Kissel, J., 1989, The compositions of comets, *in* Atreya, S.K., Pollack, J.B., and Matthews, M.S., eds., Origin and evolution of planetary and satellite atmospheres: Tuscon, Univiversity of Arizona Press, p. 167–191.

Joswiak, D.J., Brownlee, D.E., Pepin, R.O., and Schlutter, D.J., 2000, Characteristics of asteroidal and cometary IDPs obtained from stratospheric collectors: Summary of measured He release temperatures, velocities and descriptive [abs.]: Lunar and Planetary Science, v. 31, Abstract #1500, CD-ROM.

Kettrup, D., Deutsch, A., Pihlaja, P., and Pesonen, L.J., 2000, Fossil micrometeorites from Finland: Basic features, scientific potential, and characteristics of the Mesoproterozoic host rocks, *in* Gilmour, I., and Koeberl, C., eds., Impacts and the early earth: Heidelberg, Germany, Springer-Verlag, Lecture Notes in Earth Sciences, v. 92, p. 215–227.

Klöck, W., Thomas, K.L., McKay, D.S., and Palme, H., 1989, Unusual olivine and pyroxene composition in interplanetary dust and unequilibrated ordinary chondrites: Nature, v. 339, p. 126–128.

Koeberl, C., 1998, Identification of meteoritic components in impactites, *in* Grady, M.M., Hutchison, R., McCall, G.H., and Rothery, D.A., eds., Meteorites: Flux with time and impact effects: Geological Society [London] Special Publication 140, p. 133–153.

Koeberl, C., and Reimold, W.U., 1995, Early Archean spherule beds in the Barberton Mountain Land, South Aftica: No evidence for impact origin: Precambrian Research, v. 74, 1–33.

Koeberl, C., Reimold, W.U., and Boer, R.H., 1993, Geochemistry and mineralogy of early Archean spherule beds, Barberton Mountain Land, South Aftica: Evidence for origin by impact doubtful: Earth and Planetary Science Letters, v. 119, p. 441–452.

Koeberl, C., Poag, C.W., Reimold, W.U., and Brandt, D., 1996, Impact origin of the Chesapeake Bay structure and the source of the North American tektites: Science, v. 271, p. 1263–1266.

Kyte, F.T., 1988, The extraterrestrial component in marine sediments: Paleoceanography, v. 3, p. 235–247.

Kyte, F.T., 1998, A meteorite from the Cretaceous-Tertiary boundary: Nature, v. 396, p. 237–239.

Kyte, F.T., 2002a, Iridium concentrations and abundances of meteoritic ejecta from the Eltanin impact in sediment cores from *Polarstern* expedition ANT XII/4: Deep Sea Research II, v. 49, p. 1049–1061.

Kyte, F.T., 2002b, Unmelted meteoritic debris collected from Eltanin Ejecta in *Polarstern* Cores from Expedition ANT XII/4: Deep Sea Research II, v. 49, p. 1063–1071.

Kyte, F.T., 2002c, Composition of Impact Melt Debris from the Eltanin Impact Strewn Field, Bellingshausen Sea: Deep Sea Research II, v. 49, p. 1029–1047.

Kyte, F.T., and Bostwick, J.A., 1995, Magnesioferrite spinel in Cretaceous-Tertiary boundary sediments of the Pacific basin: Hot, early condensates of the Chicxulub impact?: Earth and Planetary Science Letters, v. 132, p. 113–127.

Kyte, F.T., and Brownlee, D.E., 1985, Unmelted meteoritic debris in the Late Pliocene Ir anomaly: Evidence for the impact of a nonchondritic asteroid: Geochimica et Cosmochimica Acta, v. 49, p. 1095–1108.

Kyte, F.T., and Smit, J., 1986, Regional variations in spinel compositions: An important key to the Cretaceous-Tertiary event: Geology, v. 14, p. 485–487.

Kyte, F.T., and Wasson, J.T., 1986, Accretion rate of extraterrestrial matter: Iridium deposited 33 to 67 million years ago: Science, v. 232, p. 1225–1229.

Kyte, F.T., Heath, G.R., Leinen, M., and Zhou, L., 1993, Cenozoic sedimentation history of the central North Pacific: Inferences from the elemental geochemistry of core LL44-GPC3: Geochimica et Cosmochimica Acta, v. 57, p. 1719–1740.

Kyte, F.T., Zhou, Z., and Wasson, J.T., 1980, Siderophile-enriched sediments from the Cretaceous-Tertiary boundary: Nature, v. 288, p. 651–656.

Kyte, F.T., Zhou, Z., and Wasson, J.T., 1981, High noble metal concentrations in a late Pliocene sediment: Nature, v. 292, p. 417–420.

Kyte, F.T., Smit, J., and Wasson, J.T., 1985. Siderophile interelement variations in the Cretaceous-Tertiary boundary sediments from Caravaca, Spain: Earth and Planetary Science Letters, v. 73, p. 183–195.

Kyte, F.T., Zhou, L., and Wasson J.T., 1988, New evidence on the size and possible effects of a late Pliocene oceanic impact: Science, v. 241, p. 63–65.

Kyte, F.T., Zhou, L., and Lowe D.R., 1992, Noble metal abundances in an early Archean impact deposit: Geochimica et Cosmochimica Acta, v. 56, p. 1365–1372.

Love, S.G., and Brownlee, D.E., 1993, A direct measurement of the terrestrial mass accretion rate of cosmic dust: Science, v. 262, p. 550–553.

Lowe, D.R., Byerly, G.R., Asaro, F., and Kyte, F.T., 1989, Geological and geochemical evidence for a record of 3,400 Ma-old terrestrial meteorite impacts: Science, v. 245, p. 959–962.

Luck, J.M., and Turekian, K.K., 1983, Osmium-187/Osmium-186 in manganese nodules and the Cretaceous-Tertiary boundary: Science, v. 222, p. 613–615.

Lugmair, G.W., and Shukolyukov, A., 1998, Early solar system timescales according to ^{53}Mn-^{53}Cr systematics: Geochimica et Cosmochimica Acta, v. 62, p. 2863–2886.

Margolis, S.V., Claeys, P., and Kyte, F.T., 1991, Microtektites, microkrystites and spinels from a Late Pliocene asteroid impact in the Southern Ocean: Science, v. 251, p. 1594–1597.

Marvin, U.B., and Einaudi, M.T., 1967, Black, magnetic spherules from Pleistocene and recent beach sands: Geochimica et Cosmochimica Acta, v. 31, p. 1871–1884.

Maurette, M., Hammer, C., Brownlee, D.E., Reeh, N., and Thomsen, H.H., 1986, Placers of cosmic dust in the blue ice lakes of Greenland: Science, v. 233, p. 869–872.

Mc Ghee, G.R., Orth, C.J., Gilmore, J.S., and Olsen, E.J., 1984, No geochemical evidence for an asteroidal impact at Late Devonian mass extinction horizon: Nature, v. 308, p. 629–631.

Mittlefehldt, D.W., McCoy, T.J., Goodrich, C.A., and Kracher, A., 1998, Nonchondritic meteorites from asteroidal bodies, *in* Papike, J.J., ed., Planetary materials: Reviews in Mineralogy, v. 36, p. 4-1–4-195.

Montanari, A., Hay, R.L., Alvarez, W., Asaro, F., Michel, H.V., Alvarez, L.W., and Smit, J., 1983, Spheroids at the Cretaceous-Tertiary boundary are altered impact droplets of basaltic composition: Geology, v. 11, p. 668–671.

Montanari, A., Asaro, F., Michel, H.V., and Kennett, J.P., 1993, Iridium anomalies of late Eocene age at Massignano (Italy) and ODP Site 689B (Maude Rise, Antarctic): Palaios, v. 8, p. 420–437.

Mukhopadhyay, S., Farley, K.A., and Montanari, A., 2001, A 35 Myr record of helium in pelagic limestones from Italy: Implications for interplanetary dust accretion from the early Maastrichtian to the middle Eocene: Geochimica et Cosmochimica Acta, v. 65, p. 653–669.

Mukhopadhyay, S., Farley, K.A., Montanari, A., and Ahrens, T.J., 1998, Extraterrestrial ^{3}He in the sedimentary record [abs.]: Lunar and Planetary Science, v. 29, Abstract #1535, CD-ROM.

Murray, J., and Renard, A.F., 1883, On the microscopic characters of volcanic ashes and cosmic dust, and their distribution in deep-sea deposits: Proceedings of the Royal Society of Edinburgh, v. 12, p. 474–495.

Murray, J., and Renard, A.F., 1891, Deep sea deposits, *in* Report on the scientific results of the H.M.S. Challenger during the years 1873–1876 [reprinted 1965]: New York, Johnson Reprint Corporation, 327 p.

Mutch, T.A., 1966, Abundances of magnetic spherules in Silurian and Permian salt samples: Earth and Planetary Science Letters, v. 1, p. 325–329.

Nier, A.O., and Schlutter, D.J., 1990, Helium and neon isotopes in stratospheric particles: Meteoritics, v. 25, p. 263–267.

Nishiizumi, K., 1983, Measurement of ^{53}Mn in deep sea iron and stony spherules: Earth and Planetary Science Letters, v. 63, p. 223–228.

Orth, C.J., Gilmore, J.S., and Knight, J.D., 1985, A search for evidence of large body Earth impacts associated with biological crisis zones in the fossil record [abs.]: Lunar and Planetary Science, v. 16, p. 631–632.

Orth, C.J., Attrep, M., Jr., Mao, X.Y., Kauffman, E.G., Diner, R., and Elder, W.P., 1988, Iridium abundance maxima in the upper Cenomanian extinction interval: Geophysical Research Letters, v. 15, p. 346–349.

Ozima, M., Takayanagi, M., Zashu, S., and Amari, S., 1984, High ^{3}He/^{4}He ratio in ocean sediments: Nature, v. 311, p. 448–450.

Palme, H., Janssens, M., Takahashi, H., Anders, E., and Hertogen, J., 1978, Meteoritic material at five large impact craters: Geochimica et Cosmochimica Acta, v. 42, p. 313–323.

Papanastassiou, D.A., 1986, Chromium isotopic anomalies in the Allende meteorite: Astrophysical Journal, v. 308, p. L27–L30.

Pettersson, H., and Fredriksson, K., 1958, Magnetic spherules in deep-sea deposits: Pacific Science, v. 12, p. 71–81.

Pettersson, H., and Rotschi, H., 1950, Nickel content of deep-sea deposits: Nature, v. 166, p. 308–310.

Peucker-Ehrenbrink, B., 1996, Accretion of extraterrestrial matter during the last 80 million years and its effect on the marine osmium isotope record: Geochimica et Cosmochimica Acta, v. 60, p. 3187–3196.

Peucker-Ehrenbrink, B., Ravizza, G., and Hofmann, A.W., 1995, The marine ^{187}Os/^{186}Os record of the past 80 million years: Earth and Planetary Science Letters, v. 130, p. 155–167.

Pierrard, O., Robin, E., Rocchia, R., and Montanari, A., 1998, Extraterrestrial Ni-rich spinel in upper Eocene sediments from Massignano, Italy: Geology, v. 26, p. 307–310.

Podosek, F.A., Ott, U., Brannon, J.C., Neal, C.R., Bernatowicz, T.J., Swan, P., and Mahan, S.E., 1997, Thoroughly anomalous chromium in Orgueil: Meteoritics and Planetary Science, v. 32, p. 617–627.

Rajan, R.S., Brownlee, D.E., Tomandl, D., Hodge, P.W., Farrar, H., and Britten, R.A., 1977, Detection of ^{4}He in stratospheric particles gives evidence of extraterrestrial origin: Nature, v. 267, p. 133–134.

Rampino, M.R., and Stothers, R.B., 1984, Terrestrial mass extinctions, cometary impacts and the sun's motion perpendicular to the galactic plane: Nature, v. 308, p. 709–712.

Raup, D.M., and Sepkoski, J.J., 1984, Periodicity of extinctions in the geologic past: Proceedings of the National Academy of Science, USA, v. 81, p. 801–805.

Ravizza, G., and Pyle, D., 1997, PGE and Os isotopic analyses of single sample aliquots with NiS fire assay preconcentration: Chemical Geology, v. 141, p. 251–268.

Retallack, G.J., Seyedolali, A., Krull, E.S., Holser, W.T., Ambers, C.P., and Kyte, F.T., 1998, Search for evidence of impact at the Permian-Triassic boundary in Antarctica and Australia: Geology, v. 26, p. 979–982.

Robin, E., 1988, Des poussieres cosmiques dans les cryconites du Greenland: Nature, origine et applications [Ph.D. thesis]: Paris, University d'Orsay, 131 p.

Robin, E., Boclet, D., Bonte, Ph., Froget, L., Jehanno, C., and Rocchia, R., 1991, The stratigraphic distribution of Ni-rich spinels in Cretaceous-Tertiary boundary rocks at El Kef, (Tunisia), Caravaca (Spain) and Hole 761C (Leg 122): Earth and Planetary Science Letters, v. 107, p. 715–721.

Robin, E., Bonte, Ph., Froget, L., Jehanno, C., and Rocchia, R., 1992, Formation of spinels in cosmic objects during atmospheric entry: A clue to the Cretaceous-Tertiary boundary event: Earth and Planetary Science Letters, v. 108, p. 181–190.

Robin, E., Froget, L., Jéhanno, C., and Rocchia, R., 1993, Evidence for a K/T impact event in the Pacific Ocean: Nature, v. 363, p. 615–617.

Robin, E., Pierrard, O., Lefevre, I., and Rocchia, R., 1999, A search for extraterrestrial spinel in pelagic sediments from the central North Pacific [abs.]: Geological Society of America Abstracts with Programs, v. 31, no. 7, p. A63.

Rocchia, R., Robin, E., Froget, L., and Gayraud, J., 1996, Stratigraphic distribution of extraterrestrial markers at the Cretaceous-Tertiary boundary in the Gulf of Mexico area: Implications for the temporal complexity of the event, *in* Ryder, G., Fatovsky, D., and Gartner, S., eds., The Cretaceous-Tertiary boundary and other catastrophes in Earth history: Geological Society of America Special Paper 307, p. 279–286.

Sanfilippo, A., Riedel, W.R., Glass, B.P., and Kyte, F.T., 1985, Late Eocene microtektites and radiolarian extinctions on Barbados: Nature, v. 314, p. 613–615.

Schmidt, G., Zhou, L., and Wasson, J.T., 1993, Iridium anomaly associated with the Australian tektite-producing impact: Masses of the impactor and of the Australasian tektites: Geochimica et Cosmochimica Acta, v. 57, p. 4851–4859.

Schmitz, B., 1988, Origin of microlayering in worldwide distributed Ir-rich marine Cretaceous/Tertiary boundary clays: Geology, v. 16, p. 1068–1072.

Schmitz, B., Lindstrom, M., Asaro, F., and Tassinari, M., 1996, Geochemistry of meteorite-rich marine limestone strata and fossil meteorites from the lower Ordovician at Kinnekulle, Sweden: Earth and Planetary Science Letters, v. 145, p. 31–48.

Shearer, C.K., Papike, J.J., and Rietmeijer, F.J.M., 1998, The planetary sample suite and environments of origin, *in* Papike, J.J., ed., Planetary materials: Reviews in Mineralogy, v. 36, p. 1-1–1-28.

Shukolyukov, A., and Lugmair, G.W., 1998, Isotopic evidence for the Cretaceous-Tertiary impactor and its type: Science, v. 282, p. 927–929.

Shukolyukov, A., and Lugmair, G.W., 1999, The ^{53}Mn-^{53}Cr isotope systematics of the enstatite chondrites [abs.]: Lunar and Planetary Science, v. 30, Abstract #1093, CD-ROM.

Shukolyukov, A., Kyte, F.T., Lugmair, G.W., Lowe, D.R., and Byerly, G.R., 2000a, The oldest impact deposits on Earth: First confirmation of an extraterrestrial component, *in* Gilmour, I., and Koeberl, C., eds., Impacts and the early earth: Heidelberg, Germany, Springer-Verlag, Lecture Notes in Earth Sciences, v. 92, p. 99–116.

Shukolyukov, A., Kyte, F.T., Lugmair, G.W., Lowe, D.R., and Byerly, G.R., 2000b, Early Archean spherule beds: Confirmation of impact origin: Meteoritics and Planetary Science, Abstracts, v. 35, p. A146–A147.

Simonson, B.M., and Harnik, P., 2000, Have distal impact ejecta changed through geologic time?: Geology, v. 28, p. 975–978.

Simonson, B.M., Davies, D. Wallace, M., Reeves, S., and Hassler, S.W., 1998, Iridium anomaly but no shocked quartz from late Archean microkrystite layer: Oceanic impact ejecta?: Geology, v. 26, p. 195–198.

Simonson, B.M., Koeberl, C., McDonald, I., and Reimold, W.U., 2000, Geochemical evidence for an impact origin for a Late Archean spherule layer, Transvaal Supergroup, South Africa: Geology, v. 28, p. 1103–1106.

Smit, J., 1999, The global stratigraphy of the Cretaceous-Tertiary boundary impact ejecta: Annual Review of Earth and Planetary Sciences, v. 27, p. 75–113.

Smit, J., Alvarez, W., Montanari, A., Swinburne, N., van Kempen, T.M., Klaver, G.T., and Lustenhouwer, W.J., 1992, "Tektites" and microkrystites at the Cretaceous Tertiary boundary: Two strewn fields, one crater?: Proceedings of Lunar and Planetary Science, v. 22, p. 87–100.

Smit, J., and Hertogen, J., 1980, An extra-terrestrial event at the Cretaceous-Tertiary boundary: Nature, v. 285, p. 198–200.

Smit, J., and Kyte, F.T., 1984, Siderophile-rich magnetic spheroids from the Cretaceous-Tertiary boundary in Umbria, Italy: Nature, v. 310, p. 403–405.

Taylor, S., Brownlee, D.E., 1991, Cosmic spherules in the cosmic record: Meteoritics, v. 26, p. 203–211.

Taylor, S., Lever, J.H., and Harvey, R.P., 2000, Numbers, types and compositions of an unbiased collection of cosmic spherules: Meteoritics and Planetary Science, v. 35, p. 651–666.

Thorslund, P., and Wickman, F.E., 1981, Middle Ordovician chondrite in fossiliferous limestone from Brunflo, central Sweden: Nature, v. 289, p. 285–286.

Thorslund, P., Wickman, F.E., and Nystrom, J.A., 1984, The Ordovician chondrite from Brunflo, central Sweden. 1. General description and primary minerals: Lithos, v. 17, p. 87–100.

Tredoux, M., DeWitt, M.J., Hart, R.J., Lindsay, N.M., Verhagen, B., and Sellschop, J.P.F., 1989, Chemostratigraphy across the Cretaceous-Tertiary boundary and a critical assessment of the iridium anomaly: Journal of Geology, v. 97, p. 585–605.

Turekian, K.K., 1982, Potential of ^{187}Os/^{186}Os as a cosmic versus terrestrial indicator in high iridium layers of sedimentary strata, *in* Silver, L.T., and Schultz, P.H., eds., Geological implications of impacts and large asteroids and comets on the Earth: Geological Society of America Special Paper 190, p. 243–249.

Vonhof, H.B., and Smit, J., 1999, Late Eocene microkrystites and microtektites at Maude Rise (ODP Hole 689B; Southern Ocean) suggest a global extension of the ~35.5 Ma Pacific impact ejecta strewn field: Meteoritics and Planetary Science, v. 34, p. 747–756.

Wallace, M.W., Gostin, V.A., and Keays, R.R., 1990, Acraman impact ejecta and host shales: Evidence for low-temperatuure mobilization of iridium and other platinoids: Geology, v. 18, p. 132–135.

Wang, K., 1992, Glassy microspherules (microtektites) from an Upper Devonian Limestone: Science, v. 256, p. 1546–1549.

Weissman, P.R., 1985, Cometary dymanics: Space Science Reviews, v. 41, p. 299–349.

Wetherill, G.W., and Shoemaker, E.M., 1982, Collision of astronomically observable objects with the earth: Geological Society of America Special Paper 190, p. 1–13.

Whitmire, D.P., and Jackson, A.A., 1984, Are periodic mass extinctions driven by a distant solar companion?: Nature, v. 308, p. 713–715.

Whitmire, D.P., and Matese, J.J., 1985, Periodic comet showers and planet X: Nature, v. 313, p. 36–38.

Williams, G.E., 1986, The Acraman impact structure: Source of ejecta in the late Precambrian shales, South Australia: Science, v. 233, p. 200–203.

Xu, D., Ma, S., Chai, Z., Mao, X., Sun, Y., Zhang, Q., and Yang, Z., 1985, Abundance variation of iridium and trace elements at the Permian/Triassic boundary at Shangsi in China: Nature, v. 314, p. 154–156.

Zachos, J.C., Quinn, T.M., and Salamy, K.A., 1996, High-resolution (10^4 years) deep-sea foraminiferal stable isotope records of the Eocene-Oligocene climate transition: Paleoceanography, v. 11, p. 251–256.

Zahnle, K., 1996, Tunguska: Leaving no stone unburned: Nature, v. 383, p. 674–675.

Zhou, L., and Kyte, F.T., 1988, The Permian-Triassic boundary event: A geochemical study of three Chinese sections: Earth and Planetary Science Letters, v. 90, p. 411–421.

Zhou L., Kyte F.T., and Bohor B.F., 1991, Cretaceous/Tertiary boundary of DSDP Site 596, South Pacific: Geology, v. 19, p. 694–697.

Zolensky, M., Barrett, R., and Browning, L., 1993, Mineralogy and composition of matrix and chondrule rims in carbonaceous chondrites: Geochimica et Cosmochimica Acta, v. 57, p. 3123–3148.

Zoller, W.H., Parrington, J.R., and Phelan-Kotra, J.M., 1983, Iridium enrichment in airborne particles from Kilauea volcano: January 1983: Science, v. 222, p. 1118–1120.

Manuscript Submitted December 7, 2000; Accepted by the Society March 22, 2001

Printed in the U.S.A

Geological Society of America
Special Paper 356
2002

Geophysical constraints on the size and structure of the Chicxulub impact crater

Joanna Morgan*
Mike Warner*
Department of Earth Science and Engineering, Imperial College, London SW7 2BP, UK
Richard Grieve*
Earth Science Sector, NRCan, 601 Booth Street, Ottawa, Ontario K1A OE8, Canada

ABSTRACT

Clear images of impact craters on other planetary bodies reveal a progressive change in crater morphology with increasing crater size. Attempts to make direct comparisons between extraterrestrial and terrestrial craters have been hindered by the lack of pristine craters on Earth, particularly in the larger size range. This deficiency in ground truth data has also slowed our progress in understanding the cratering process for large impacts. However, the buried Chicxulub crater in Mexico now provides a pristine example of a large impact crater on Earth. Early structural models across Chicxulub were extremely divergent. They illustrate that we do not know how peak rings are formed, how stratigraphic uplifts are related to topographic peak rings, or how these morphological elements are related spatially to allogenic impact breccias and melt rocks. New reflection and refraction seismic data helped improve constraints on structural models of Chicxulub, and led to a better understanding of the cratering process for large impacts. There is now general agreement that the transient cavity at Chicxulub was 80–110 km in diameter. Impact-related structures within the target rocks are clearly observed in the reflection data, but there remains disagreement on the interpretation of some of these structural elements. If the outermost significant inward-facing asymmetric scarp locates the crater rim, Chicxulub has a diameter of 180–195 km. However, if the outermost topographic high locates the crater rim, Chicxulub probably has a crater diameter of between 250 and 270 km. Chicxulub has been interpreted as having the morphology of a peak ring crater and a multiring basin. There is no consensual model for the formation of rings in multiring basins. Once such a model is agreed upon, we will be in a better position to categorize the morphology of Chicxulub.

INTRODUCTION

Structure of large impact craters

Remotely sensed images of other planets and moons reveal that the morphology of impact craters change as crater size increases (Fig. 1). The smallest craters (so-called simple craters) are bowl shaped (Fig. 1A). As crater size increases we observe a dramatic change in surface morphology: simple craters evolve into broad and shallow complex craters (Fig. 1, B–D) (Alexopoulos and McKinnon, 1994). The smallest complex craters are central-peak craters (Fig. 1B), in which a central mound

*E-mails: Morgan, j.v.morgan@ic.ac.uk; Warner, m.warner@ic.ac.uk; Grieve, rgrieve@nrcan.gc.ca

Morgan, J., Warner, M., and Grieve, R., 2002, Geophysical constraints on the size and structure of the Chicxulub impact crater, *in* Koeberl, C., and MacLeod, K.G., eds., Catastrophic Events and Mass Extinctions: Impacts and Beyond: Boulder, Colorado, Geological Society of America Special Paper 356, p. 39–46.

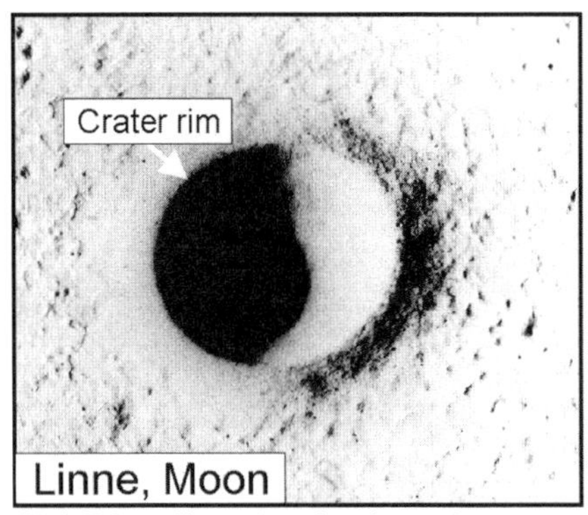

A) 2.5-km-diameter simple crater

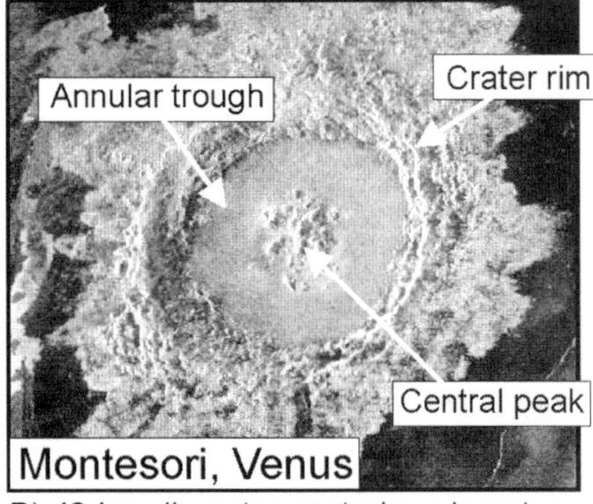

B) 43-km-diameter central-peak crater

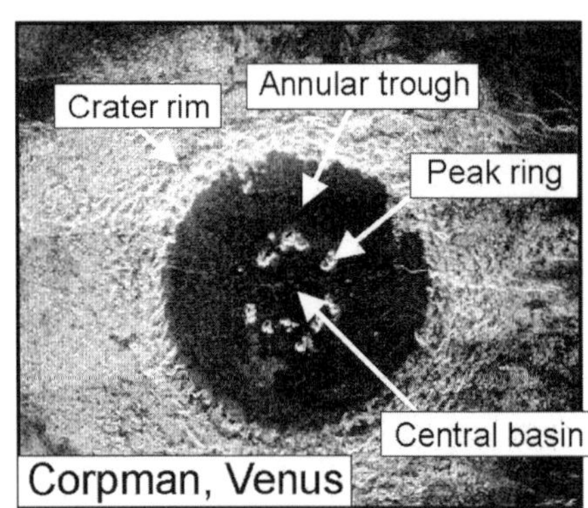

C) 45-km-diameter peak-ring basin

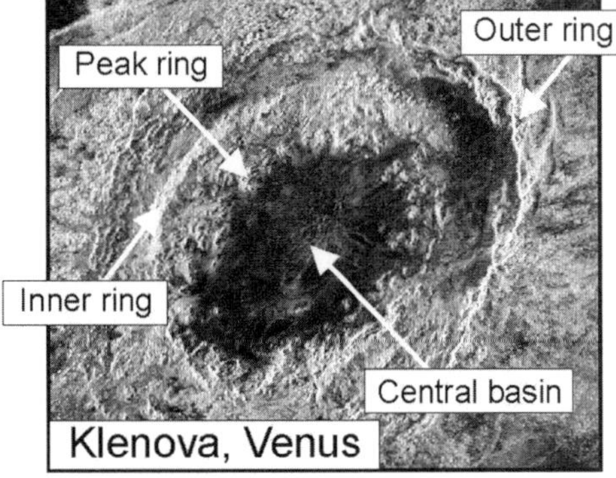

D) 145-km-diameter multi-ring basin

Figure 1. Morphological progression of impact craters with increasing crater size. A: Simple lunar bowl-shaped crater. B: Venusian central peak crater. C: Venusian peak ring crater. D: Venusian multiring basin. Peak rings are roughly circular rings of rugged hills and massifs that stand above flat crater floor. Crater rim in central peak and peak ring craters is outer edge of terrace zone. In multiring basins, two or more widely spaced, inward-facing, asymmetric scarps are outside of central basin.

protrudes through the relatively flat visible crater floor. With increasing size the central peak evolves to a roughly circular ring of hills and massifs (Fig. 1C), forming so-called peak ring craters. Some craters have both a peak ring and central peak. On some planets and moons, the largest craters are multiringed, defined as having at least two rings formed by inward-facing asymmetric scarps (Fig. 1D). Not all multiring basins contain a peak ring.

Images of extraterrestrial craters do not typically provide information on the subsurface structure and lithological characteristics associated with a particular crater morphology. These data can currently only be supplied by the study of terrestrial impact craters. The Earth is the most endogenically active of all the terrestrial planets, and thus has preserved only a small sample of the population of impact craters (~160 to date) acquired over geologic time. Many of these known craters are eroded, buried, or tectonically deformed. The steep size distribution of planetary crater populations results in many fewer large craters than small craters (Grieve, 1998). This lack of ground truth data has hindered our progress in understanding the cratering process in large impacts.

There are several examples of well-preserved simple craters on Earth, the best known being Meteor Crater in Arizona. There are also many examples of complex craters, where there is significant stratigraphic uplift in the crater center (as depicted in Fig. 2D). Some well-known examples are Vredefort in South Africa (Reimold and Gibson, 1996), Haughton in Canada (Grieve, 1988), and Mjølnir in the Barents Sea (Tsikalas et al., 1998). Where observed, the autochthonous rocks that form the crater floor in complex craters are pervaded by pseudotachylytes, possibly suggesting that they behaved hydrodynamically during impact (Grieve et al., 1981; Spray, 1997). Several large terrestrial craters have been interpreted as being peak ring craters. Within the Popigai crater in Siberia (~100 km in diameter), there is a 50-km-diameter ring of uplifted basement rocks (Masaitis, 1994) that may be the eroded remnant of a topographic peak ring. Structural models of the Ries crater in Germany (25 km in diameter) (Pohl et al., 1977) show uplifted basement rocks around a deeply excavated central zone. These uplifted rocks do not appear to have the same character as the topographic peak rings observed on other planetary bodies, and may not be a direct analogue. On the basis of gravitational scaling laws, we would predict that craters on Earth larger than ~25 km should be peak ringed (Pike, 1985); however, we lack ground truth data to confirm it. Two large craters, Manicouagan (100 km in diameter) and Vredefort (250–300 km in diameter), should be peak ring craters (or possibly multiringed basins). Both craters have been eroded to reveal narrow zones of stratigraphic uplift, and at Manicouagan, the central uplift is surrounded by impact melt. It is not obvious how these stratigraphic uplifts might link to a broad topographic peak ring.

On the basis of the extraterrestrial and terrestrial observa-

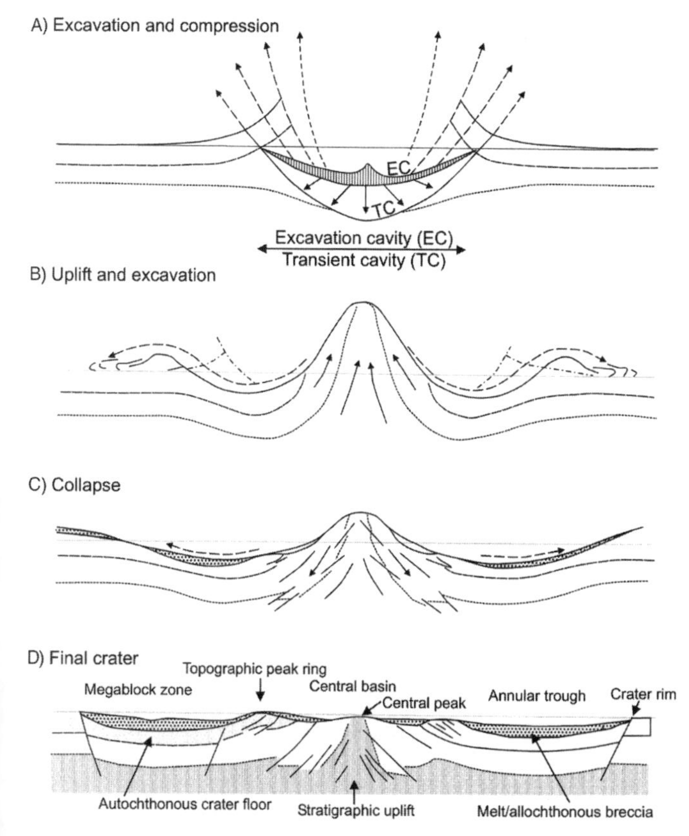

Figure 2. Generic conceptual model for collapse of large impact crater. A: Target rocks are compressed and pushed down to form transient cavity (TC); shading indicates melted rocks, above which target is excavated as defined by excavation cavity (EC). B: Base of central transient cavity moves upward as excavation continues. C: Uplifted rim of transient crater collapses inward and downward as central uplift collapses downward and outward. D: Final crater has narrow stratigraphic uplift in crater center, central uplift collapses to form topographic peak ring, and transient crater rim collapses to form megablock and/or terrace zone. Melt and allochthonous breccias are within central basin and annular trough, and rocks within autochthonous crater floor are pervaded with pseudotachylytes.

tions noted here, models for the formation of complex craters all show upward movement within the crater center during collapse (see Fig. 2). During the initial stages of impact a transient cavity is formed (Fig. 2A). Gravitational forces act to collapse this cavity, but collapse can only occur if the yield strength and internal friction of the target rocks are significantly reduced (Melosh, 1977; McKinnon, 1978). In this model, the center of the transient cavity starts to collapse gravitationally before excavation has ceased (Fig. 2B). When excavation has ceased, the uplifted rim collapses inward and downward to form a terrace or megablock zone, while the uplifted central zone collapses downward and outward to form a peak ring (Fig. 2, C–D). Although most workers agree with this generic model for large crater collapse, several issues remain poorly defined: (1) the detailed kinematics of collapse of the central uplift and uplifted transient cavity rim (specifically how much lateral and vertical movement is involved), and how that collapse changes as central peak craters evolve into peak ring craters; (2) the mechanism for weakening the rocks to allow hydrodynamic flow; (3) the mechanism for forming rings in multiring basins; and (4) the relationship of specific morphologic elements (central peaks and peak rings) with specific impact lithologies (melt sheets and autochthonous and allogenic impact breccias).

Because craters on other planetary bodies show a morphological progression with increasing size, we would expect the same progression on Earth. Multiring basins, however, are not present on all planetary bodies. Their existence on Venus, which has gravity similar to that of Earth, favors their occurrence on Earth. However, without knowing how rings are formed, or which properties of the target are critical to ring formation, we cannot be sure that they should occur on Earth. At Vredefort, a series of anticlinal and synclinal structures exist to 150 km from the crater center (McCarthy et al., 1990). At the Sudbury crater in Canada several rings of pseudotachylytes are present outside the outcrop of the impact melt rocks (Spray and Thompson, 1995). These features may indicate that these craters are multiringed basins, but it is not certain that these structural rings correspond directly to the morphological rings observed around, for example, the large lunar multiring basins (Grieve and Therriault, 2000).

Character of Chicxulub

The identification of the Chicxulub crater (Hildebrand et al., 1991; Pope et al., 1991) was an exciting discovery. Chicxulub is the most pristine large impact crater in its size range known on Earth. The crater has been slowly buried on a tectonically quiet carbonate platform, leaving the impact basin relatively intact. Chicxulub offers us an opportunity to improve our understanding of the cratering process. Structural models of the Chicxulub crater have been developed using onshore wells, and potential field, seismic reflection, and topographic data (Fig. 3). These models have been guided by the somewhat conflicting observations at other complex terrestrial craters.

In the Pilkington et al. (1994) model (Fig. 3A), the central uplift is narrow, and is surrounded by a thick melt sheet, matching observations at the Manicouagan crater (Grieve and Head, 1983). The peak ring overlies the melt, and is formed as impact breccia is sloughed off from a rebounding central uplift. The melt sheet does not extend significantly beyond the peak ring. Pilkington et al. (1994) proposed that Chicxulub was a 180-km-diameter peak ring crater on the basis that they observed no signature in the gravity data outside of this diameter.

The Espindola et al. (1995) model (Fig. 3B) is derived principally from potential field data. The peak ring has no clear gravitational signal (Brittan et al., 1999), so they do not include one in their model. The crater diameter is limited to ~200 km, and a central zone of high density is interpreted as either a melt sheet or a central uplift. This model includes significant east-west dip in the basement rocks across the crater.

In the Sharpton et al. (1996) model (Fig. 3C) the central uplift is broad. The peak ring is produced by subvertically uplifted basement rocks and is overlain by impact melt rocks. A significant fraction of the melt is outside the peak ring, in the annular trough. The annular ring of uplifted basement matches observations at the Popigai crater (Masaitis, 1994). Sharpton et al. (1996) proposed that Chicxulub was a 260–300-km-diameter multiring basin, on the basis of the observation of a subtle gravity ring of ~280 km diameter.

From seismic data and hydrocode modelling, Morgan et al. (2000) proposed that the peak ring at Chicxulub was formed as the outward-collapsing central uplift overrode the inward-collapsing transient cavity rim. They proposed, on the basis of velocity data, that there is a narrow zone of stratigraphic uplift with a concave-upward top, and that the melt rocks are mainly within the central basin interior to the peak ring. On the basis of inferred faulting in the target rocks, the crater is argued to be 195 km in diameter and to have the morphology of a multiring basin (Morgan et al., 1997; Morgan and Warner, 1999).

Gravity, magnetic, topographic, and seismic data across Chicxulub

The Chicxulub crater is detectable with gravity data: average gravity values are low across a roughly circular region that is 180–200 km in diameter, and there is a local gravity high in the central region of the crater (Hildebrand et al., 1991; Camargo-Zanoguera and Suárez-Reynosa, 1994). Gravity profiles taken in different directions from the nominal crater center (89.54°W, 21.3°N) are highly variable (Fig. 4A), and there are local and regional gravity anomalies superimposed on the crater-related gravity. There have been claims and counterclaims as to the existence of subtle gravity rings at distances >100 km from the crater center. In Figure 4A we show three gravity profiles, at bearings of 120°, 180°, and 220° from the center. On the first two of these profiles there are clear local gravity highs at radii of ~128 and 137 km, respectively, whereas on the third there is no gravity signature at a radial distance >100 km. Sharpton et al. (1993) averaged a number of radial profiles, and obtained a gravity high at ~139 km radius (their fourth gravity ring). However, it remains unproven which of these features are directly related to the Chicxulub crater, and what those features mean in terms of crater structure.

All potential field modeling is inherently nonunique, and an infinite number of density profiles can reproduce any given observed anomaly. Models can be better defined through direct sampling, enabling specific rock types to be assigned specific densities. At Chicxulub, the density of the target rocks and impactites remains largely unknown: the limited number of density measurements made within the Universidad Nacional Autónoma de México (UNAM) and Petroleos Mexicanos (PeMex) wells show a wide range of densities (V. Sharpton, 1998, personal commun.). Within 90 km of the crater center, the samples are limited to <1.6 km depth. This leaves us with no means to estimate density values across the crater, particularly within the central regions of the crater. It is unlikely, therefore, that we can use gravity data alone to argue for a particular structural model. It is generally agreed that any topographic rims are likely to have been eroded prior to burial (Pilkington et al., 1994; Sharpton et al., 1996). It is thus unlikely that we can use

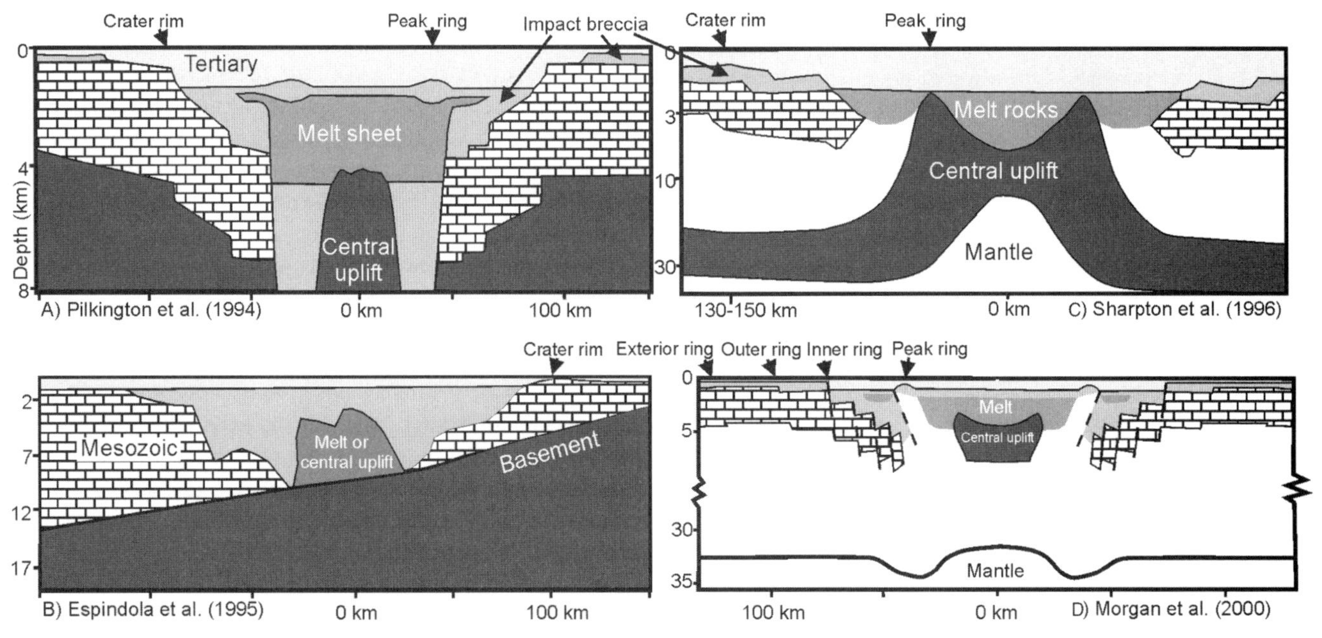

Figure 3. Structural models for Chicxulub impact crater, redrawn from (A) Pilkington et al. (1994), (B) Espindola et al. (1995), (C) Sharpton et al. (1996), and (D) Morgan et al. (2000). Note varying vertical exaggeration.

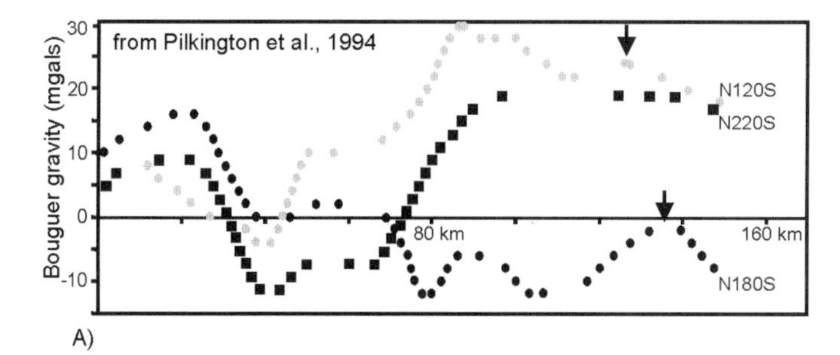

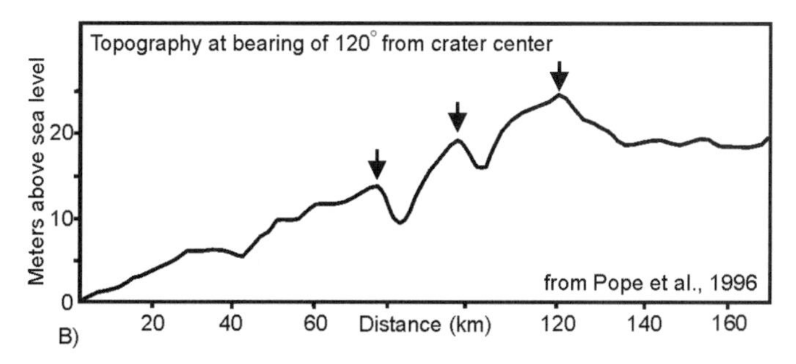

Figure 4. A: Bouguer gravity anomalies from Pilkington et al. (1994). Zero represents approximate crater center; profiles are at bearings of 120°, 180°, and 220°. B: Topographic data redrawn from Pope et al. (1996). Arrows indicate local highs at 75, 95, and 120 km from crater center.

gravity data to assert whether Chicxulub has a peak ring or multiring basin morphology.

A strong magnetic anomaly, ~90 km in diameter, of several hundred nanoteslas is observed across Chicxulub (Cornejo-Toledo and Hernandez-Osuna, 1950; Hildebrand et al., 1991; Camargo-Zanoguera and Suárez-Reynoso, 1994). Magnetic signatures at impact craters can originate from suevitic impact breccias (as at Ries), melt rocks (as at Sudbury), and within the stratigraphic uplift (as at Vredefort). Pilkington and Hildebrand (2000) modeled the Chicxulub data with two source bodies: one with a diameter of 90 km and average source depth of 2 km, and a deeper layer with a diameter of ~40 km and average source depth of 5 km. These two bodies are interpreted as originating within a melt sheet and central uplift, respectively. Magnetic properties of rocks tend to vary more than those of density, and there are several parameters that can be varied when forward modeling these data. This usually makes such models less reliable than those obtained using gravity data. Direct sampling will be required to validate the Pilkington and Hildebrand (2000) model.

Pope et al. (1996) published onshore topographic data (Fig. 4B), and proposed that the current topography reflects the original (and now buried) crater morphology. At a bearing of 120° from the crater center, these data show subtle local highs at crater radii of ~75, 95, and 120 km. The averaged values for these highs are ~75, 103, and 129 km. Pope (1997) and Sharpton (1997) both agreed that the topographically highest point represents the crater rim, suggesting a crater diameter of ~260 km. However, the existence of these topographic highs is disputed: in a similar study Connors et al. (1996) did not detect concentric topographic features at the radii proposed by Pope et al. (1996).

By far the most powerful geophysical technique for mapping subsurface structure is seismic reflection profiling. Reflection data across Chicxulub have been converted to depth using velocity data obtained from traveltime tomographic inversions (Christeson et al., 1999). These data reveal a postimpact basin in which high-frequency reflections deepen to 1–1.5 km within the central region of the crater (Fig. 5). Within this basin there is an ~80-km-diameter topographic ring, which appears to be analogous to peak rings observed on other planetary bodies. Outside this central basin, reflections from preimpact target rocks are observed between 0.6 and 3 km. These target rocks show concentric highs, produced by faulting within the target, at average radii of 75, 97.5, and 125 km (see faults in Fig. 5) (Morgan and Warner, 1999). These highs occur at remarkably similar radii to those within the topographic data (Fig. 4B).

In Figure 5, the reflectors in the central basin between 0 and 1.5 km depth are interpreted as Tertiary, and those outside the basin between 0.6 and 3 km are interpreted as Mesozoic. These interpretations are based on comparisons with wells onshore, and are in agreement with regional interpretations of PeMex (A. Camargo-Zanoguera, 1997, personal commun.). The Tertiary section is 1–1.2 km thick in PeMex wells S1, C1, and Y6. The thickening and deepening of the Mesozoic section from east to west along Chicx-A1 is also observed onshore between PeMex wells Y4 and Y5a. The westerly dip of the Mesozoic section along Chicx-A is consistent with PeMex wells in Campeche (a few hundred kilometers west of the Yucatan), where ~3.5-km-deep wells bottom in Upper Cretaceous rocks (Grajales-Nishimura et al., 2000). The velocity data parallel the reflection data: as the postimpact basin deepens, so do the velocity contours. The postimpact deep-water Tertiary marls have a significantly lower velocity than the adjacent shallow-water platform carbonates.

The Mesozoic rocks are not aligned across the postimpact basin. If the eastern Mesozoic reflectors were extended westward they would pass beneath the Mesozoic reflectors on the western margin. This suggests there may have been some postimpact readjustment, the section west of the basin moving upward relative to the east. This is reflected in the particularly deep annular trough in the eastern part of the basin, where the inferred presence of high-energy deposits supports the hypothesis of a period of rapid subsidence (see Fig. 5).

CURRENT STATUS OF GEOPHYSICAL INTERPRETATIONS

Transient crater diameter

One of the most important parameters at Chicxulub, in terms of determining the volume of ejecta and volatiles placed

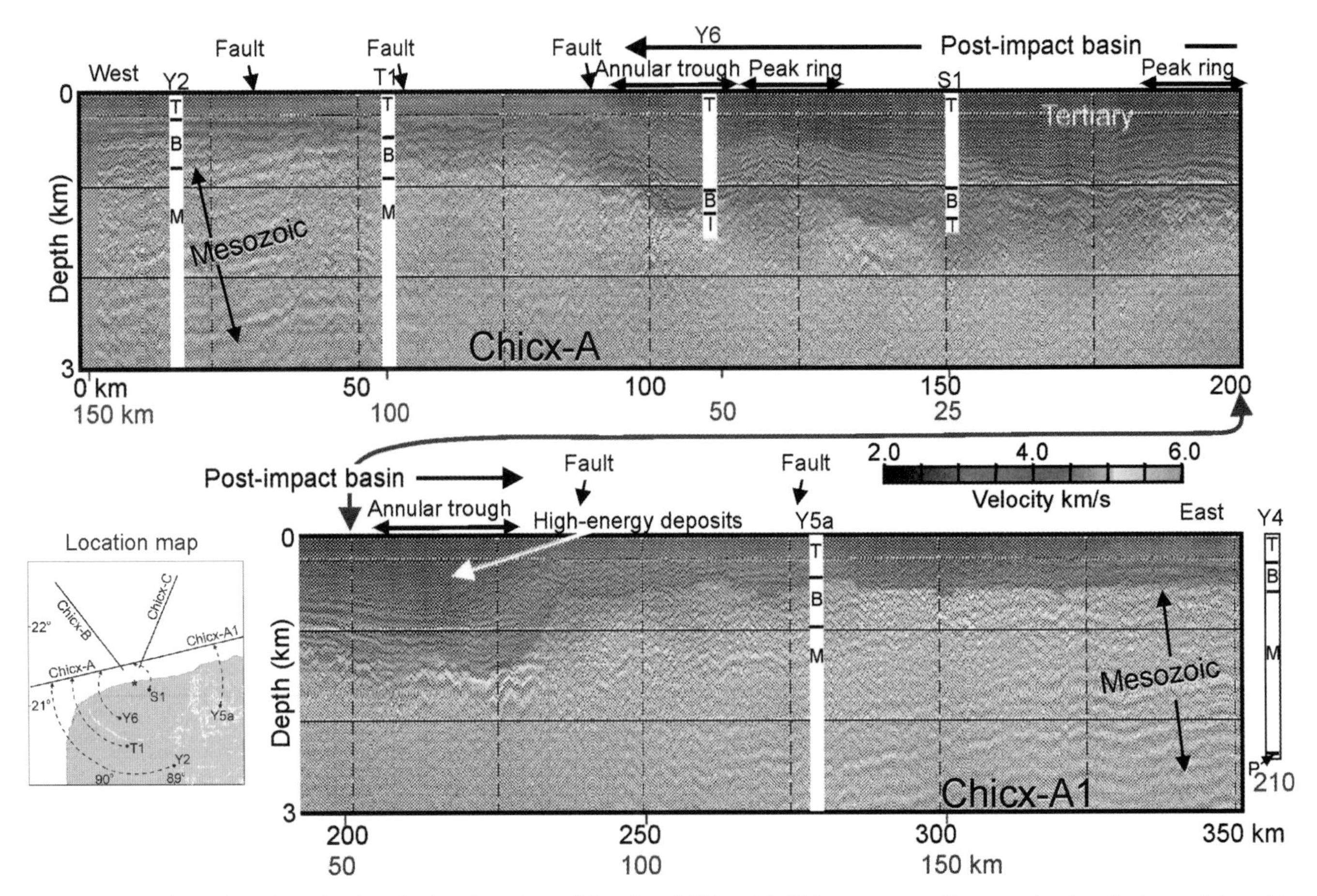

Figure 5. Unmigrated marine seismic reflection data along Chicx-A and Chicx-A1. Color represents P-wave seismic velocity model used to convert reflection time to depth. Horizontal scale in blue represents distance in kilometers along profile; horizontal scale in red represents radial distance in kilometers. For display purposes, two profiles overlap, as indicated by red arrows. Lithologies within onshore wells (from Ward et al., 1995; Sharpton et al., 1996) have been projected onto marine seismic data at their equivalent radial distances from crater center (map shows projection). T is Tertiary, M is Mesozoic, B is impact breccia, I is impact melt breccia, P is Paleozoic. Faults indicate locations at which we infer significant faulting in Mesozoic rocks along Chicx-A. Peak ring appears as local high within Tertiary basin.

in the atmosphere by the impact, is the diameter of the transient cavity (see Fig. 2A). The size of this cavity relates directly to the dimensions of the excavation cavity and to the energy of impact. Early estimates for the size of the transient cavity varied by an order of magnitude, but estimates have converged. Kring (1995), Morgan et al. (1997), and Hildebrand et al. (1998) estimated the transient cavity to be between 80 and 110 km in diameter. The Mesozoic reflectors disappear at ~85 km diameter and ~8 km depth. This delineates the maximum size of the collapsed excavation cavity (Morgan et al., 1997). The seismic data reveal the location of the target rocks after collapse (Fig. 2D), and we can use these data to attempt to reconstruct the original transient cavity (Fig. 2A). The variation between individual estimates occurs because authors make different assumptions about the kinematics of collapse. Hildebrand et al. (1998) assumed that the target rocks collapsed at an angle of ~60° to obtain a 80-km-diameter transient cavity; Morgan et al. (1997) assumed 30°–45° to obtain a diameter of 90–105 km; V. Sharpton (1999, personal commun.) assumes an angle of ≤30° to obtain a transient cavity ~110 km in diameter. To better define the size of the transient cavity, we must improve our understanding of the mechanics of large-crater collapse. Some preliminary attempts to do so, with numerical modeling using hydrocodes, have recently been published (O'Keefe and Ahrens, 1999; Melosh and Ivanov, 1999; Morgan et al., 2000).

Final crater diameter

The seismic reflection data image the target rocks clearly. Concentric structural features within the target rocks are likely to be impact induced. In terms of structural models of the crater, however, the interpretation of these features remains controversial. The outermost consistent feature is a small thrust (~50 m throw) at ~125 km radius (Morgan and Warner, 1999). Such a thrust might be responsible for the topographic high and gravity ring seen at similar radii (Fig. 4). If we define the diameter of a crater as the topographically highest point, then Chicxulub would probably have a diameter of 250–270 km, as argued by

Pope (1997) and Sharpton (1997). We say "probably" because it is not clear from the marine seismic reflection data that the target rocks here are topographically higher than the outer ring at ~195 km diameter. Morgan et al. (1997) argued that the crater rim is the most significant ring with an inward-facing scarp, following definitions used for large basins on the Moon. They used a crater diameter of ~195 km diameter, where there is a normal fault in the target rocks with a throw of ~400 m. Hildebrand et al. (1998) agreed that this normal fault locates the crater rim, but preferred a slightly smaller diameter of 180 km. The difference occurs principally because: (1) Hildebrand et al. (1998) used the location of the fault scarp, whereas Morgan et al. (1997) used the topographically highest point (there is a bulge in the target rocks that is outboard of the scarp), and (2) Hildebrand et al. (1998) did not include seismic data along the Chicx-C profile, on which the fault scarp is significantly farther out.

Crater morphology

Most agree that Chicxulub has a peak ring, and is, therefore, either a peak ring crater (as suggested by Pilkington et al., 1994; Pope et al., 1996; Hildebrand et al., 1998) and analogous to Corpman crater (Fig. 1C), or it is a multiring basin, analogous to Klenova (Fig. 1D). By definition, Chicxulub is a multiring basin if more than one of the fault scarps at ~250 km, ~195 km, and ~145 km diameter (Fig. 5) were visible as separate topographic rings after impact (ignoring any removal by backwash), as advocated by Sharpton et al. (1996) and Morgan et al. (1997). Hildebrand et al. (1998) did not believe that either of the fault scarps at 145 km or 250 produced a distinct ring. Regardless of whether Chicxulub has more than one topographic ring (apart from the peak ring), we still cannot be sure that its structural rings are direct analogues of topographic rings in extraterrestrial large multiring basins (Grieve and Therriault, 2000).

Future

A three-dimensional seismic tomographic investigation of Chicxulub is planned in 2003; onshore International Continental Drilling Project and offshore Ocean Drilling Program drilling are planned in the near future. Measurements of velocity, density, and magnetic properties within core samples will be used to calibrate our potential-field and velocity modeling. Continued geophysical investigation will thus enable better definition of the structure of the Chicxulub crater, and refine our understanding of the cratering process.

ACKNOWLEDGMENTS

The Chicxulub seismic experiment was funded by the Natural Environment Research Council through the British Institutions Reflection Profiling Syndicate (BIRPS), the Leverhulme Trust, the National Science Foundation, the BIRPS Industrial Associates Programme, the Royal Commission for the Exhibition of 1851, and the Royal Society (London). The seismic reflection data were acquired by Geco-Prakla and processed by Bedford Interactive Processing Services. Velocity data were supplied by Gail Christeeson. The assistance of Petroleos Méxicanos and the Universidad Nacional Autónoma de México in planning and permitting the survey is gratefully acknowledged. We thank B. Ivanov and W. Poag for their reviews. This is Geological Survey of Canada contribution 2000208.

REFERENCES CITED

Alexopoulos, J.S., and McKinnon, W.B., 1994, Large impact craters and basins on Venus, with implications for ring mechanics on the terrestrial planets, *in* Dressler, B.O., Grieve, R.A.F., and Sharpton, V.L., eds., Large meteorite impacts and planetary evolution: Geological Society of America Special Paper 293, p. 29–50.

Brittan, J., Morgan, J.V., Warner, M.R., and Marin L., 1999, Near-surface seismic expression of the Chicxulub impact crater, *in* Dressler, B.O., and Sharpton, V.L., eds., Large meteorite impacts and planetary evolution II: Geological Society of America Special Paper 339, p. 281–290.

Camargo-Zanoguera, A., and Suárez-Reynoso, G., 1994, Evidencia sísmica del cráter impacto de Chicxulub: Boletín de la Asociación Mexicana de Geofísicos de Exploración, v. 34, p. 1–28.

Christeson, G.L., Buffler, R.T., and Nakamura, Y., 1999, Upper crustal structure of the Chicxulub impact crater from wide-angle ocean bottom seismograph data, *in* Dressler, B.O., and Sharpton, V.L., eds., Large meteorite impacts and planetary evolution II: Geological Society of America Special Paper 339, p. 291–305.

Connors, M., Hildebrand, A.R., Pilkington, M., Oritz-Aleman, C., Chavez, R.E., Urrutia-Fucugauchi, J., Graniel-Castro, E., Camara-Zi, A., and Vasquez, J., 1996, Yucatán karst features and the size of Chicxulub crater: Geophysical Journal International, v. 127, p. F11–F17.

Cornejo-Toledo, A., and Hernandez-Osuna, A., 1950, Las anomalias gravimetricas en la cuenca salina del istmo, planicie costera de Tabasco, Campeche y Penisula de Yucatan: Boletín de la Asociación Mexicana de Geólogos Petroleros, v. 2, p. 453–460.

Espindola, J.M., Mena, M., de La Fuente, M., and Campos-Enriquez., J.O., 1995, A model of the Chicxulub impact structure (Yucatan, Mexico) based on its gravity and magnetic signatures: Physics of the Earth and Planetary Interiors, v. 92, p. 271–278.

Grajales-Nishimura, J.M., Cedillo-Pardo, E., Rosales-Domínguez, C., Moran-Zenteno, D.J., Alvarez, W., Claeys, P., Ruíz-Morales, J., García-Hernández, J., Padilla-Avila, P., and Sánchez-Ríos, A., 2000, Chicxulub impact: The origin of reservoir seal facies in the southeastern Mexico oil fields: Geology, v. 28, p. 307–310.

Grieve, R.A.F., Robertson, P.B., and Dence, M.R., 1981, Constraints on the formation of ring impact structures, based on terrestrial data, *in* Schultz, P.H., and Merrill, R.B., eds., Multi-ring basins: Proceedings of the Lunar and Planetary Science Conference, v. 12A, p. 37–57.

Grieve, R.A.F., and Head, J.W., 1983, The Manicouagan impact structure: An analysis of its original dimensions and form: Journal of Geophysical Research, v. 88, p. 807–818.

Grieve, R.A.F., 1988, The Haughton impact structure: Summary and synthesis of the results of the HISS Project: Meteoritics, v. 23, p. 249–254.

Grieve, R.A.F., 1998, Extraterrestrial impacts on Earth: The evidence and consequences, *in* Grady, M.M., Hutchinson, R., McCall, G.J.H., and Rotherby, D.A., eds., Meteorites: Flux with time and impact effects: Geological Society [London] Special Publication 140, p. 105–132.

Grieve, R., and Therriault, A., 2000, Vredefort, Sudbury, Chicxulub: Three of a kind?: Annual Review of Earth and Planetary Sciences, v. 28, p. 305–338.

Hildebrand, A.R., Penfield, G.T., Kring, D., Pilkington, M., Carmargo, A., Jacobsen, S.B., and Boynton, W., 1991, Chicxulub Crater: A possible Cretaceous-Tertiary boundary impact crater on the Yucatán peninsula, Mexico: Geology, v. 19, p. 867–871.

Hildebrand, A.R., Pilkington, M., Ortiz-Aleman, C., Chavez, R., Urrutia-Fucugauchi, J., Connors, M., Graniel-Castro, E., Camara-Zi, A., Halpenny, J., and Niehaus, D., 1998, Mapping Chicxulub crater structure with gravity and seismic reflection data, *in* Grady, M.M., Hutchinson, R., McCall, G.J.H., and Rotherby, D.A., eds., Meteorites: Flux with time and impact effects: Geological Society [London] Special Publication 140, p. 155–176.

Kring, D.A., 1995, The dimensions of the Chicxulub crater and impact melt sheet: Journal of Geophysical Research, v. 100, p. 16979–16986.

Masaitis, V., 1994, Impactites from Popigai crater, *in* Dressler, B.O., Grieve, R.A.F., and Sharpton, V.L., eds., Large meteorite impacts and planetary evolution: Geological Society of America Special Paper 293, p. 153–162.

McCarthy, T.S., Stanistreet, I.G., and Robb, L.J., 1990, Geological studies related to the origin of the Witwatersrand Basin and its mineralization: An introduction and a strategy for research and exploration: South African Journal of Geology, v. 93, p. 1–4.

McKinnon, W.B., 1978, An investigation into plastic failure in crater modification: Proceedings of the 9th Lunar and Planetary Science Conference, p. 3965–3973.

Melosh, H.J., 1977, Crater modification by gravity: A mechanical analysis of slumping, *in* Roddy, D.J., Pepen, R.O., and Merill, R.B., eds: Impact and explosion cratering: New York, Pergamon Press, p. 1245–1260.

Melosh, H.J., and Ivanov, B.A., 1999, Impact crater collapse: Annual Review of Earth and Planetary Sciences, v. 27, p. 385–415.

Morgan, J.V., Warner, M.R., and The Chicxulub Working Group, 1997, Size and morphology of the Chicxulub impact crater: Nature, v. 390, p. 472–476.

Morgan, J.V., and Warner, M.R., 1999, The third dimension of a multi-ring impact basin: Geology, v. 27, p. 407–410.

Morgan, J.V., Warner, M.R., Collins, G.R., Melosh, H.J., and Christeson, G.L., 2000, Peak ring formation in large impact craters: Geophysical constraints from Chicxulub: Earth and Planetary Science Letters, v. 183, p. 347–354.

O'Keefe, J.D., and Ahrens, T., 1999, Complex craters: Relationship of stratigraphy and rings to impact conditions: Journal of Geophysical Research, v. 104, p. 27091–27104.

Pike, R.J., 1985, Some morphologic systematics of complex impact structures: Meteoritics, v. 20, p. 49–68.

Pilkington, M., Hildebrand, A., and Ortiz-Aleman, C., 1994, Gravity and magnetic field modelling and structure of the Chicxulub crater, Mexico: Journal of Geophysical Research, v. 99, p. 13147–13162.

Pilkington, M., and Hildebrand, A.R., 2000, Three-dimensional magnetic imaging of the Chicxulub crater [abs.]: Lunar Planetary Science, v. 31, 1190–1191.

Pohl, J., Stöffler, D., Gall, H., and Ernston, K., 1977, The Ries impact crater, *in* Roddy, D.J., Pepin, R.O., and Merrill, R.B., eds., Impact and explosion cratering: New York, Pergamon Press, p. 343–404.

Pope, K.O., Ocampo, A., and Duller, D., 1991, Mexican site for the K/T crater?; Nature, v. 351, p. 105.

Pope, K.O., Ocampo, A.C., Kinsland, G.L., and Smith, R., 1996, Surface expression of the Chicxulub crater: Geology, v. 24, p. 527–530.

Pope, K.O. 1997, Surface expression of the Chicxulub crater: Reply: Geology, v. 25, p. 568–569.

Reimold, W.U., Gibson, R.L., 1996, Geology and evolution of the Vredefort impact structure, South Africa: Journal African Earth Science, v. 23, p. 125–162.

Sharpton, V.L., Burke, K., Camargo-Z., A., Hall, S.A., Lee, S., Marin, L.E., Suarez-R., G., Quezada-M., J.M., Spudis, P.D., and Urrutia-F., J., 1993, Chicxulub multi-ring impact basin: Size and other characteristics derived from gravity analysis: Science, v. 261, p. 1564–1567.

Sharpton, V.L., Marin, L.E., Carney, J.L., Lee, S., Ryder, G., Schuraytz, B.C., Sikora, P., and Spudis, P.D., 1996, Model of the Chicxulub impact basin, *in* Ryder, G., Fastovsky, D., and Gartner, S., eds., The Cretaceous–Tertiary event and other catastrophes in Earth history: Geological Society of America Special Paper 307, p. 55–74.

Sharpton, V.L., 1997, Surface expression of the Chicxulub crater: Comment: Geology, v. 25, p. 567–568.

Spray, J.G., and Thompson, L., 1995, Friction melt distribution in a multi-ring impact basin: Nature, v. 373, p. 130–132.

Spray, J.G., 1997, Superfaults: Geology, v. 25, p. 305–308.

Spudis, P.D., 1993, The geology of multi-ring impact basins: The moon and other planets: Cambridge, Cambridge University Press, 263 p.

Tsikalas, F., Gudlaugsson, S.T., Eldholm, O., and Faleide, J.I., 1998, Integrated geophysical analysis supporting the impact origin of the Mjølner structure, Barents Sea: Tectonophysics, v. 289, p. 257–280.

Ward, W., Keller, G., Stinnesbeck, W., and Adatte, T., 1995, Yucatan subsurface stratigraphy: Implications and constraints for the Chicxulub impact: Geology, v. 23, p. 873–876.

Manuscript Submitted October 17, 2000; Accepted by the Society March 22, 2001

Geological Society of America
Special Paper 356
2002

Micro-Raman and optical identification of coesite in suevite from Chicxulub

Elena Lounejeva*
Instituto de Geología, Universidad Nacional Autónoma de México, Ciudad Universitaria, Código Postal 04510, México, D.F., México
Mikjail Ostroumov*
Universidad Michoacana de San Nicolas de Hidalgo, Instituto de Investigaciones Metalúrgicas, Departamento de Geología y Mineralogía, Edifício "U", Ciudad Universitaria, Apartado Postal 52J, Código Postal 58000, Morelia, Michoacán, México
Gerardo Sánchez-Rubio*
Instituto de Geología, Universidad Nacional Autónoma de México, Ciudad Universitaria, Código Postal 04510, México, D.F., México

ABSTRACT

In this work we report the presence of coesite in suevite from the Chicxulub impact structure. We studied polished thin sections from sample Y6N14, which represents the fallback suevite breccia from the depth interval 1208–1211 m of the Yucatan-6 well. Micro-Raman spectroscopy was selected as the most adequate technique to search for high-pressure silica polymorphs. More than 60 shocked quartz grains were subjected to Raman study. Spectral data were compared with synthetic coesite as well as with published data for quartz and coesite. In four cases characteristic spectra of coesite displaying Raman shifts of ~521, 271, 178, and 119 cm^{-1} were obtained in shocked quartz grains.

In one case the coesite-bearing quartz grains form an aplite-like aggregate. The three other grains were found in one fragment of partially melted crystalline basement rock. Optical microscopy reveals that coesite-bearing quartz grains contain comparable petrographic shock-metamorphic features, such as mosaicism, brownish domains, low birefringence, planar fractures, and/or planar deformation features.

The coesite occurs as shapeless polycrystalline aggregates up to 50 µm or as fine, well-rounded, brownish crystals <10 µm in size, with positive relief relative to quartz, situated either along some planar deformation features, or arranged in short curved chains. We infer that the quartz grains where coesite was identified were subjected to shock pressures between 30 and 35 GPa.

INTRODUCTION

The Chicxulub crater, in the northern part of the Yucatan Peninsula, Mexico, is one of the largest terrestrial multiring impact structures on Earth (Fig. 1). It is ~195 km in diameter (Morgan et al., 1997) and its age is 65 Ma (Swisher et al., 1992). The structure is formed by massive impact melt rocks of dacitic composition overlain by impact breccias containing clasts of

*E-mails: Lounejeva, elenal@servidor.unam.mx; Ostroumov, ostroum@zeus.umich.mx; Sánchez-Rubio, gesaru@servidor.unam.mx

Lounejeva, E., Ostroumov, M., and Sánchez-Rubio, G., 2002, Micro-Raman and optical identification of coesite in suevite from Chicxulub, *in* Koeberl, C., and MacLeod, K.G., eds., Catastrophic Events and Mass Extinctions: Impacts and Beyond: Boulder, Colorado, Geological Society of America Special Paper 356, p. 47–54.

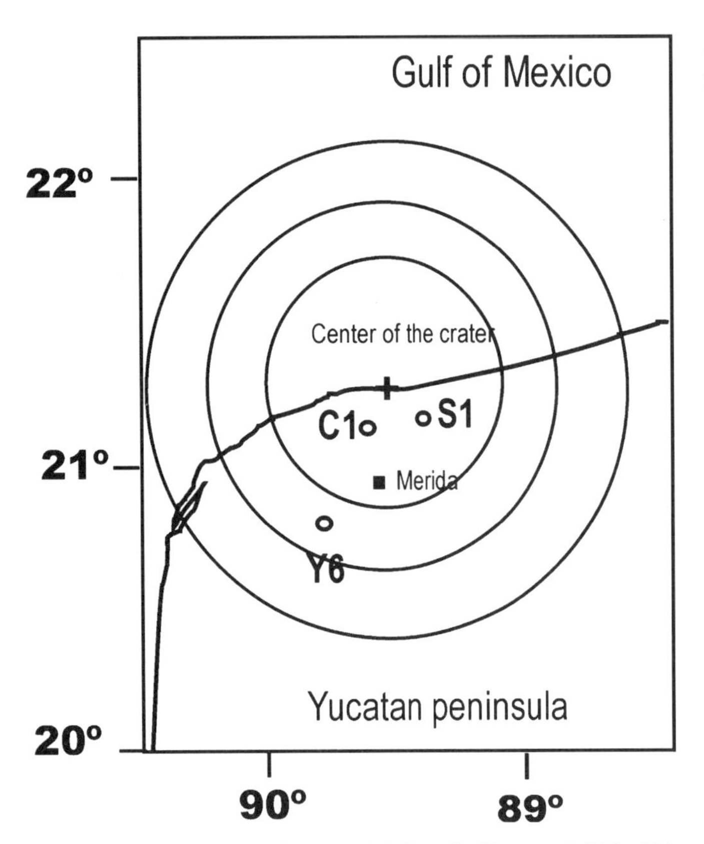

Figure 1. Location map of PEMEX drill wells Yucatan-6 (Y6), Chicxulub 1 (C1) and Sacapuc (S1). Chicxulub crater rings as defined by gravity model are shown (simplified from Hildebrand et al., 1991).

pre-Paleocene marine sedimentary units and crystalline rocks of basement affinity; this sequence is buried by several hundred meters of Cenozoic marine platform sediments.

A number of publications deal with specific petrological aspects of impactites from Chicxulub (Hildebrand et al., 1991; Quezada-Muñeton et al., 1992; Sharpton et al., 1992; Cedillo-Pardo et al., 1994; Koeberl et al., 1994; Shuraytz et al., 1994; Kettrup and Deustch, 2000; Vera-Sánchez, 2000; Rebolledo-Vieyra et al., 2000; and references therein). Nevertheless, there is no record, to our knowledge, of high-pressure phases in the Chicxulub impact breccias. Coesite (Boslough et al., 1995) and stishovite (McHone et al., 1989) were discovered in samples from the Cretaceous-Tertiary (K-T) boundary at Raton basin, New Mexico, but so far only shocked quartz, and none of the high-pressure quartz polymorphs, has been reported from the Chicxulub crater rocks.

Different silica polymorphs, such as partially disordered quartz, diaplectic silica glass, coesite, and stishovite, may form from quartz depending on the pressure and temperature to which the rocks were subjected. Monoclinic coesite, with silicon in tetrahedric coordination, and tetragonal stishovite, with silicon in octahedric coordination, are the densest phases. The pressures required for their formation, at least 3.5 GPa and 8.5 GPa, respectively (Stishov and Popova, 1961), are difficult to achieve under lithostatic crustal conditions, but are easily reached during shock events. Stishovite is formed during the shock compression stage, whereas coesite forms during pressure release (Stöffler and Langenhorst, 1994). Both phases are commonly used as indicators of thermal and pressure history and represent mineralogical evidence for an impact event. Coesite is also known from nonimpact environments of deep-seated metamorphic rocks (Boyer et al., 1985; Liou et al., 1998) or volcanic diatremes (Novgorodov et al., 1990), usually forming inclusions of large single crystals within high-pressure minerals such as pyrope or diamond. The absence of ultrahigh-pressure metamorphic rocks and the mineral assemblage in the Chicxulub breccias rules out a steady-state formation scenario.

The main goal of our research project was to search for high-pressure silica polymorphs. Preliminary reports were given by Gómez et al. (1997) and Lounejeva et al. (2000).

SAMPLE SETTING

Sample Y6N14 represents the depth interval 1208–1211 m from the Yucatan-6 well, located about 50 km southwest of the center of the crater (Fig. 1), and inside the crater rim that was defined as between 60 and 80 km by Morgan and Warner (1999). This core comes from the middle part of a layer of fallback suevite (i.e., a polymict impact breccia containing more than 15 vol% glass that fell inside the crater) that is ~250 m thick in this well. The suevitic breccia sampled from the upper 100 m (Y6N13) is carbonate rich, but poor in shocked basement and melt fragments; the suevite sampled 40 m below (Y6N15) contains abundant shocked basement clasts and silicate melt material (S. Heuschkel, 1998, personal commun.), and overlies breccias with melt matrix rich in basement clasts.

For our study we used material from the Y6N14 core, which is housed by the Institute of Geology (Universidad Nacional Autónoma de México). We studied 10 covered thin sections and 2 slabs from an area of ~70 cm^2 in 3 subsamples of the core to determine the general petrographic features. For the Raman study we prepared three polished thin sections from subsample Y6N14-4 as representative of the entire depth interval, and one polished thin section from subsample Y6N14-6, macrosopically distinguished from the others by its generally bluish color. It was not possible to establish the exact depth interval of the subsamples.

PETROGRAPHY

Sample Y6N14 was classified as a suevitic breccia because it contains abundant impact melt glasses (Hildebrand et al., 1991; Sharpton et al., 1992).

The subrounded elongate fragments composed of green and beige clayish minerals are the most abundant in the breccia (35–40 vol%), and we agree with others' opinions that they must represent altered impact glass particles, because in most cases the fluidal texture is preserved. Furthermore, we have observed undigested felsic shocked clasts, anhydrite aggregates, a diaplectic mineral, and brownish isotropic material with fluidal texture that could be fresh (unaltered) glass in some of these fragments.

The suevite breccia is also rich in shocked basement clasts <1 cm (20–40 vol%). These clasts were the main target of our study. The precise nature of the basement clasts was difficult to identify by petrographic observations because most of them were partially fused and altered; thus, they can only be described as being derived either from granitic or metamorphic rocks.

The granitic group shows remnants of granular quartz and feldspar embedded in a matrix of what is probably devitrified silicic melt. The shape of these fragments varies from subangular to amoeboid, in this case outlined by matrix pockets. The matrix shows either cryptocrystalline or opaque domains in different proportions, or it is formed by fibrous or ghost textures of acicular feldspar crystals (<20 μm) floating in a cryptocrystalline groundmass. Well-rounded aggregates of anhydrite may also be present in these fragments.

The metamorphic group of fragments displays subparallel light and dark bands. In some cases the light bands (about 60 vol%) are composed of felsic minerals that show well-developed spherulitic texture with acicular crystals up to 300 μm. The dark colored bands are formed of euhedral prismatic opaque crystals that might comprise some shock metamorphosed Fe-Mg—rich silicate. We assume that these fragments were derived from gneisses. In other cases the subparallel light colored domains (20–40 vol%) are formed by polycrystalline quartz aggregates with shear strain that may have formed previous to impact. The dark domains are composed of cryptocrystalline brown or opaque groundmass. We believe that these fragments derived from quartzite or quartzitic schist. The remaining quartz and feldspar grains in the basement clasts often display plastic deformation features, such as characteristic planar deformation features, mosaicism, low birefringence, wavy extinction, brownish spots, and shear displacements. Single shocked quartz grains and aplite-like aggregates form a minor component of the breccia.

The sedimentary target rocks are also represented in the breccia by fragments (to 20 vol%) of limestone, anhydrite, calcite, and dolomite. The limestone appears as dark, well-rounded, fossil-rich fragments, normally with reaction rims. Anhydrite forms aggregates (<10 vol%) of elongated prismatic crystals, and some calcite fragments (~5 vol%) are distinguished by a feathery texture, whereas dolomite forms rare aggregates (<1 vol%).

Subsample Y6N14-6 is unusual due to the exceptional abundance of sedimentary fragments, biogenic material, and altered melt fragments with strongly elongate opaque fluidal textures. The feathery textured calcite and anhydrite aggregates form >20 vol%, similar to core Y6N13.

The breccia matrix (30–65 vol%) seems to be composed mostly of micritic calcite, but a mixture of rounded K-feldspar, plagioclase, and quartz grains, all <10 μm in size, was also detected in the cryptocrystalline material around the calcite grains (S. Heuschkel, 1998, personal commun.).

Chlorite, calcite, anhydrite, gypsum, and hematite were identified among the secondary minerals; the presence of other phyllosilicates is inferred.

METHODOLOGY

The optical identification of both high-pressure silica polymorphs, either coesite or stishovite, is difficult because of their typical very fine grained occurrence. During the past decade the traditional material-consuming X-ray diffraction method for identification of high-pressure silica phases has given way to nondestructive spectroscopic methods, such as magic angle sample spinning nuclear magnetic resonance (MASS NMR) (Yang et al., 1986) and Raman spectroscopy (Boyer et al., 1985). It is generally accepted that Raman scattering is well suited for the study of disordered materials, particularly for disordered silica, as well as its high-pressure polymorphs and synthetic glasses (McMillan, 1992). With the development of micro-Raman techniques, the Raman microprobe (RMP) emerged as an important tool in mineralogical research and, in particular, in the case of in situ studies of high-pressure polymorphs (Wopenka et al., 1996). For example, coesite was detected in situ by Raman spectroscopy in impactites from the Vredefort impact structure; the impactites were found to have a fairly high abundance of coesite and stishovite (Halvorson and McHone, 1992).

For the study we used Jobin-Yvon T64000 (λ = 514.5 nm) and Bruker (λ = 1064 nm) instruments to obtain simple and Fourier transform Raman spectra, respectively. In both cases a microscope with 2 μm spatial resolution was used; the beam diameter was 10–20 μm, and the beam power was 200–250 mW. The spectral slit width was 2.5 cm^{-1} and the wavelength accuracy was 1 cm^{-1}. Integration time constants in the range 30–60 s were used to eliminate the experimental fluctuations. The spectral data were corrected for background noise and the exact shift positions were defined using the GRAMS 386c program. Coesite powder, synthesized at pressure, $P \approx 50$ kbar and temperature, $T \approx 1000$ °C using a Konak-type high-pressure cell at the Institute of High Pressure Physics of Russian Academy of Sciences, was used as a reference sample. The phase purity of this sample was confirmed by comparing its X-ray diffraction pattern (X-ray powder diffraction instrument SIEMENS 5000) with ASTM 14-654 (Table 1). The micro-Raman spectra of the Chicxulub samples and reference materials were measured under identical conditions.

More than 60 promising areas for detection of silica polymorphs, i.e., quartz grains with shock-metamorphic features, were selected on thin sections for our micro-Raman spectroscopic study. As many as five scans were done on each grain. The spectral data for the Chicxulub samples were compared both with those obtained for the synthetic coesite and with

TABLE 1. X-RAY DIFFRACTION DATA FOR THE SYNTHETIC COESITE USED AS REFERENCE MATERIAL

d_{hkl} ASTM 14-654	Intensity (%) ASTM14-654	d_{hkl} MEASURED	Intensity (%) MEASURED
3.43	30	3.43	40
3.09	100	3.10	100
2.76	8	2.77	8
2.69	10	2.70	10
2.33	4	2.34	4
2.29	8	2.30	7
2.18	4	2.19	5
2.03	6	2.03	6
1.84	4	1.84	4
1.79	8	1.80	8
1.79	8	1.79	8
1.71	12	1.72	10
1.70	10	1.70	10
1.58	6	1.58	6
1.545	10	1.55	8
1.501	4	1.501	4
1.418	<1	1.418	<1
1.407	4	1.407	4
1.345	12	1.346	12
1.321	2	1.321	3
1.285	6	1.285	6

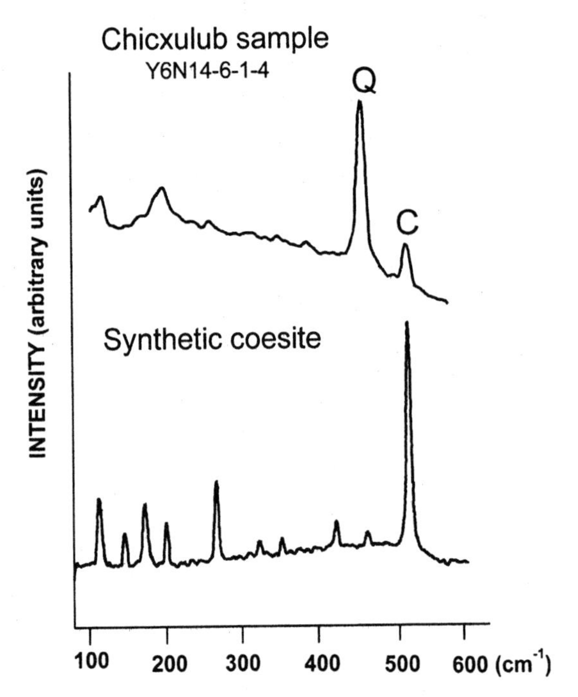

Figure 2. Micro-Raman spectrum of synthetic coesite and one of several spectra obtained from coesite-bearing quartz aggregate (number 4) from subsample Y6N14-6, thin section 1, from Chicxulub. Main characteristic quartz (Q) and coesite (C) peaks are labeled. See Table 2 for Raman shift data and Figures 3, 5, and 6 for petrographic details.

previously published data for coesite and quartz (Sharma et al., 1981; Scott and Porto, 1967). We returned to the optical examination of thin sections after the spectroscopic study.

RESULTS

Raman study

The majority of the studied quartz grains are composed of common alpha quartz, which is readily identified by its intense Raman line at 464 cm^{-1}. In four cases coesite and disordered quartz were also detected. The presence of disordered quartz was inferred by measuring the half-width of the band at 464 cm^{-1}. Further studies of disordered quartz are ongoing.

The coesite was identified from its characteristic Raman line at ~521 cm^{-1}. Both of the strongest bands, at 521 cm^{-1} for coesite and at 464 cm^{-1} for alpha quartz, have been assigned to a symmetric Si-O-Si stretching vibration (A1 vibration mode; see Sharma et al., 1981). These Raman lines are sufficient to distinguish unambiguously coesite from quartz. Representative Raman spectra for one of the coesite-bearing quartz grains as well as for synthetic coesite are shown in Figure 2. Details of spectral data for Raman shifts are given in Table 2. The intensity of Raman peaks is in arbitrary units assuming random orientation of the grains.

Coesite-bearing quartz

In the thin section from subsample Y6N14-6, coesite-bearing quartz grains were detected by Raman microprobe in only one fragment; this was a melt-free aplite-like fragment with triple junctions among quartz grains (Fig. 3). The three other cases in which coesite was detected were found in the same rock fragment from subsample Y6N14-4 (thin section 2). This subangular fragment, ~1 cm in size, contains many small quartz grains and an opaque phase embedded in a cryptocrystalline brownish groundmass (Fig. 4). We observed other similar rock fragments in this subsample, which might be mostly fused and altered granitic remnants, including that where coesite was detected by previous Raman study (fragment 31 from slab Y6N14-4; Table 2).

All coesite-bearing quartz grains contain similar shock-metamorphic features, such as mosaicism or wavy extinction, brownish domains, low birefringence, planar fractures and/or planar deformation features (Fig. 5).

The phase inferred to be coesite forms <10 vol% of the host quartz grains. It appears as shapeless polycrystalline aggregates up to 50 µm or forms small (<10 µm in diameter), well-rounded, brownish crystals with slightly positive relief with respect to quartz, and it is situated either along some planar deformation features, or arranged in short curved chains (Fig. 6).

DISCUSSION AND CONCLUSIONS

The coesite identified in sample Y6N14 of the suevite breccia from the Chicxulub crater is similar to that observed in

TABLE 2. RAMAN SHIFTS (cm^{-1}) FOR CHICXULUB AND REFERENCED MINERALS

Natural quartz	Natural coesite	Synthetic coesite	Coesite-bearing quartz from Chicxulub		
			Present study		***
*	**	Powder	Subsample Y6N14-6 Thin section 1 Quartz grain 4	Subsample Y6N 14-4 Thin section 2 Quartz grain 17	Subsample Y6N 14-4 Slab 1 Fragment 31
	116s	115	116		119
128s			124	127	128
	151m	150	151		151
	176	175	175		178
207s,bd	204s	202	203	205	205
264m			262	264	
	269s	268	272		272
	326m	326	324		328
356m	355m	355	352	356	356
395m,sh			390	395	
401w					405
	427s	425	425		427
464vs	466m	466	461	464	464
	521vs	519	520		520
696w				695	
	785	785			
795w,sh			794	795	
807w			804	806	
	815w	815			
	837w	837	834		
	1036w	1036			
1066w,sh	1065w	1065	1064	1065	
1083w			1081	1082	
	1144	1144			
1161w	1164	1164	1160	1162	
1231w			1232	1230	

*Scott and Porto, 1967
**Sharma et al., 1981
***Lounejeva et al., 2000
Note: w—weak; s—strong; vs—very strong; bd—broad; sh—shoulder; m—moderate.

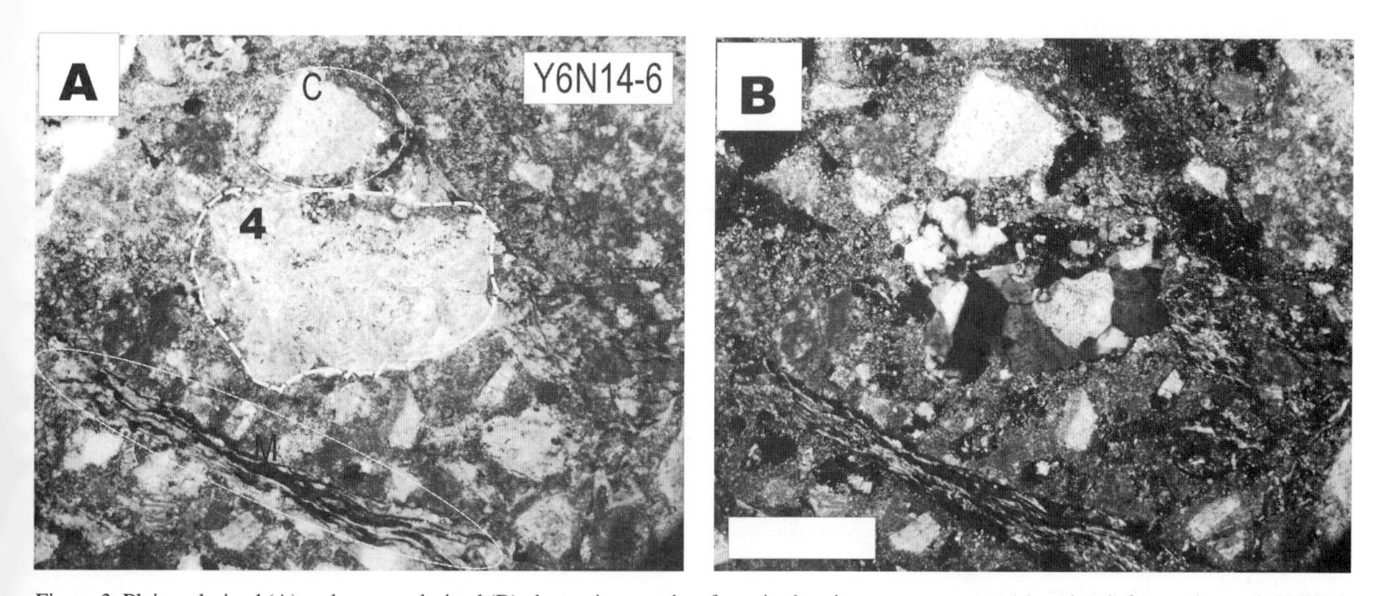

Figure 3. Plain polarized (A) and cross-polarized (B) photomicrographs of coesite-bearing quartz aggregate (number 4) from subsample Y6N14-6, thin section 1. Clast environment includes fragments of impact melt altered glass (M) and feathery textured calcite (C). Note absence of reaction rim around quartz aggregate. Scale bar is ~0.5 cm. See also Figures 5 and 6.

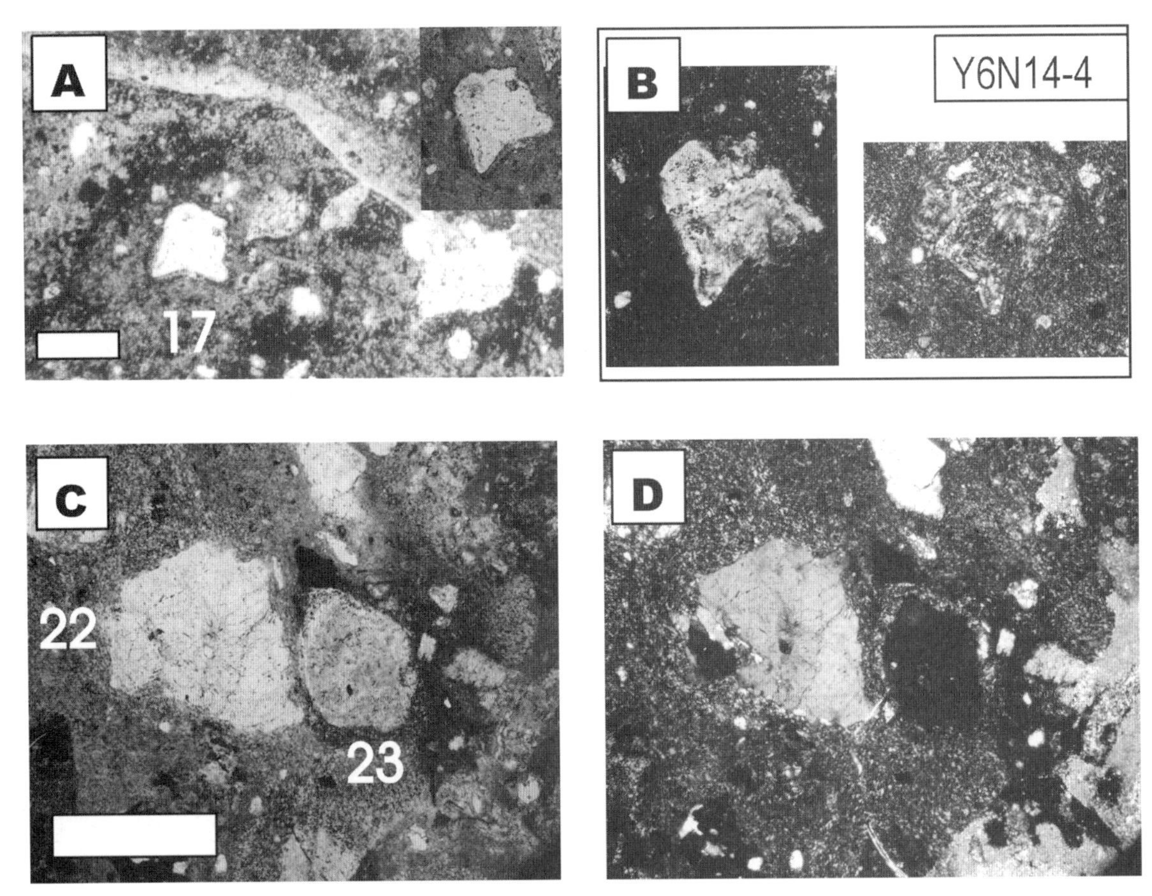

Figure 4. Photomicrographs of mostly fused and altered fragment from subsample Y6N14-4, thin section 2. Quartz grains where coesite was detected by Raman study are embedded in cryptocrystalline brownish groundmass of fragment. A, B: One horn-shaped grain with mosaic extinction (17); C, D: One grain with very low birefringence and reaction rim (23) and another subangular grain with curved features around central hole (22). A, C: Plane polarized light. B, D: Cross-polarized light. Scale bars are ~100 μm. See also Figures 5 and 6.

impact craters with crystalline nonporous targets (Stöffler and Langenhorst, 1994). Thus we believe that the detected coesite is a result of the shock transformation of quartz. The pressure-temperature conditions of coesite formation are estimated as follows. The preshock temperatures of crystalline nonporous target rocks may be lower than 250°C, according to the maximum excavation depth of the Chicxulub crater (12 km; see Morgan et al., 1997). At these conditions the planar deformation features could form at pressures between 5 and 35 GPa. The coesite, as considered, has an even greater pressure stability range, between 3.5 and 50 GPa. Nevertheless, if both coesite and abundant planar deformation features are present in the same quartz grain, the pressure can be determined to be between 30 and 35 GPa (Grieve et al., 1996). The upper limit is confined by the transformation of crystalline quartz into diaplectic glass, usually associated with coesite formed at higher pressures. The assumed pressure close to 35 GPa is in good agreement with the presence of shock-disordered quartz detected by the Raman study and with the absence of diaplectic glass. The postshock temperature range assigned to the estimated pressure range is 170–300 °C. Shock-metamorphosed crystalline basement debris, including the coesite-bearing quartz, may be mixed with impact melts and incorporated into the suevite breccia as air-transported ejecta. Previous studies, based on measurements of the orientation of planar deformation features in shocked quartz, also indicate that the Y6N14 suevite breccia contains target materials derived from several shock pressure zones, corresponding to varying distances from the center of the structure (Sharpton et al., 1992). Further studies of these samples, e.g., measurement of orientation of planar deformation features in coesite-bearing quartz, are required to better document the pressure-temperature history of these rocks.

We used micro-Raman spectroscopy, followed by optical mineralogy, to unambiguously demonstrate, for the first time, the presence of coesite in an impact breccia from the Chicxulub structure. Our observations of coesite in suevitic breccia from Chicxulub indicate that the pressure in the investigated shocked quartz grains reached at least 30 GPa, in agreement with the presence of other shock-metamorphic effects in these samples.

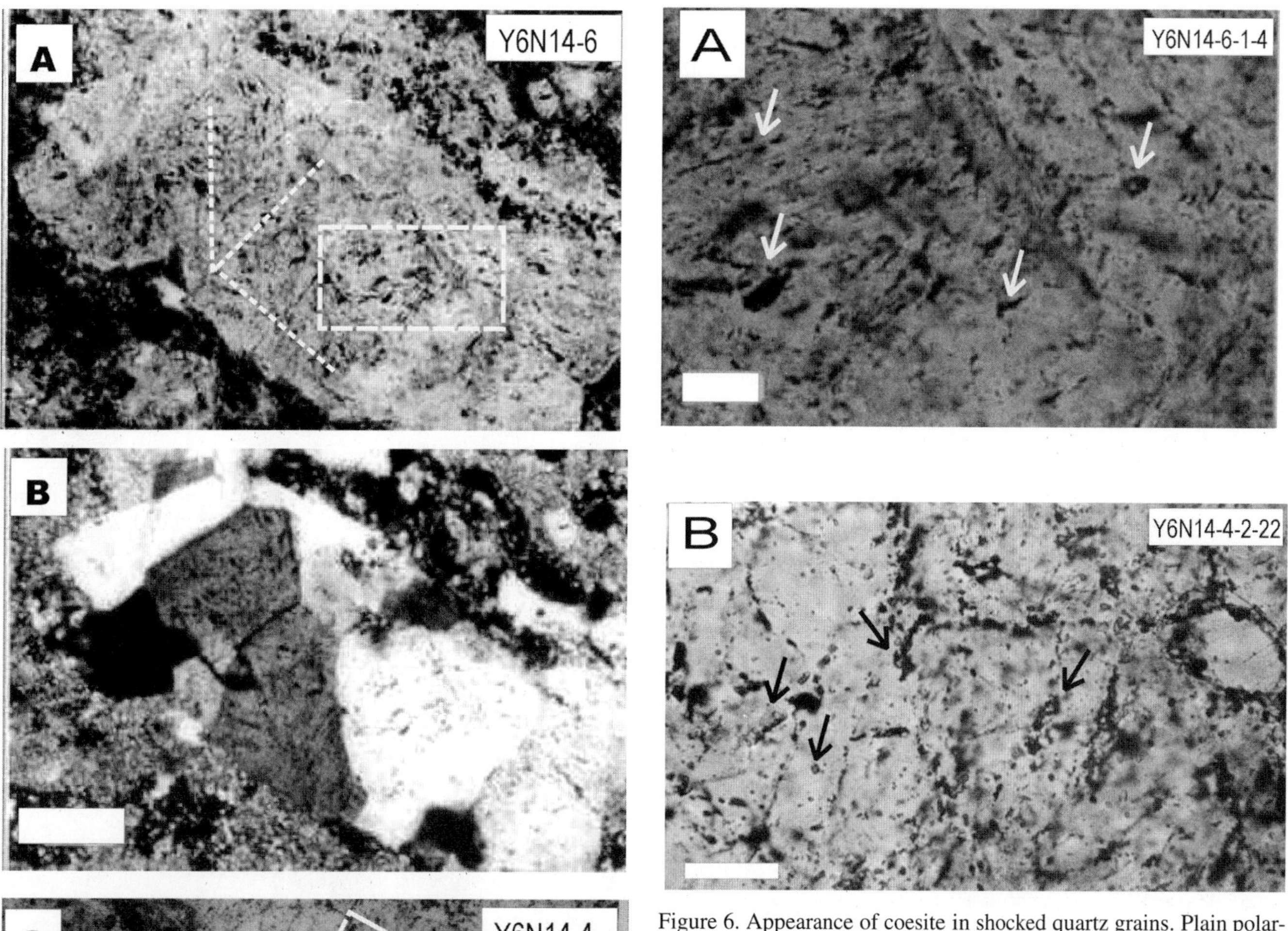

Figure 5. Enlarged view of coesite-bearing quartz grains. Plane polarized (A) and cross-polarized (B) photomicrographs of quartz aggregate from subsample Y6N14-6, thin section 1. C: Plane light photomicrograph of quartz grains 22 and 23 from subsample Y6N14-4, thin section 2. Dashed lines indicate some general directions of shock planar features development. Frames mark areas enlarged in Figure 6. Scale bars are ~40 µm.

Figure 6. Appearance of coesite in shocked quartz grains. Plain polarized photomicrographs of (A) one of grains from quartz aggregate (number 4) from subsample Y6N14-6, thin section 1, and (B) grain 22 from subsample Y6N14-4, thin section 2. Arrows indicate coesite occurring as single crystals and as aggregates. Scale bars are ~25 µm.

ACKNOWLEDGMENTS

We thank E. Faulques, University of Nantes, for technical assistance in the Raman study; J.T. Vásquez-Ramirez and M.A. Reyes-Salas, Institute of Geology (Universidad Nacional Autónoma de México), for the preparation of thin sections and some microprobe analyses, respectively; and the Institute of High Pressure Physics of the Russian Academy of Sciences for providing a synthetic coesite sample. We thank B. Martiny for the English revision and F. Ortega-Gutiérrez, R. Gibson, E. Libowitzky, and C. Koeberl for their constructive critical reviews of the manuscript.

REFERENCES CITED

Boslough, M.B., Cygan, R.T., and Izett, G.A., 1995, NMR spectroscopy of quartz from the K/T boundary: Shock-induced peak broadening, dense glass, and coesite [abs.]: Lunar and Planetary Science, v. 26, p. 149–150

Boyer, H., Smith, D., Chopin, C., and Lasnier, B., 1985, Raman microprobe determinations of natural and synthetic coesite: Physics and Chemistry of Minerals, v. 12, p. 45–48.

Cedillo-Pardo, E., Claeys, P., Grajales-Nishimura, J.M., and Alvarez, W., 1994, New mineralogical constraints on the nature of target rocks at the Chicxulub crater [abs.], *in* New developments regarding the K/T Event and other catastrophes in Earth history: Houston, Texas, Lunar and Planetary Institute, LPI Contribution No. 825, p. 20–21.

Gómez, R., Quintana, M., Ridaura, R., Marquina, V., Marquina, M.L., Jiménez, M., Aburto, S., Sánchez-Rubio, G., and Lounejeva, E., 1997, Evidence of high pressure polymorphs in the materials extracted from the Chicxulub basin: Microscope IR study: Boston, Material Research Society Program and Abstracts, Fall Meeting, p. 583.

Grieve, R.A.F., Langenhorst, F., and Stöffler, D., 1996, Shock metamorphism of quartz in nature and experiment. 2. Significance in geoscience: Meteoritics and Planetary Science, v. 31, p. 6–35.

Halvorson, K., and McHone, J.F., 1992, Vredefort coesite confirmed with Raman spectroscopy [abs.]: Lunar and Planetary Science, v. 23, p. 477–478.

Hildebrand, A.R, Penfield, G.T., Kring, D.A., Pilkington, M., Camárgo-Zaragoza, A., Jacobsen, S.B., and Boynton, W.V., 1991, Chicxulub crater: A possible Cretaceous/Tertiary boundary impact crater on the Yucatan Peninsula, Mexico: Geology, v. 19, p. 867–871.

Kettrup, B., and Deutsch, A., 2000, Composition and variability of the crystalline basement at the Chicxulub target site deduced from geochemical-petrographical data of clasts in impactites [abs.], *in* Catastrophic events and mass extinctions: Impacts and beyond: Houston, Texas, Lunar and Planetary Institute, LPI Contribution No. 1053, p. 92–93.

Koeberl, C., Sharpton, V.L., Schuraytz, B.C., Shirey, S.B., Blum, J.D., and Marín, L.E., 1994, Evidence for a meteoritic component in impact melt rock from Chicxulub structure: Geochimica et Cosmochimica Acta, v. 58, p. 1679–1684.

Liou, J.G., Zhang, R.Y., Ernst, W.G., Rumble, D., III, and Maruyama, S., 1998, High-pressure minerals from deeply subducted metamorphic rocks, *in* Hemley, R.J., ed., Ultrahigh-pressure mineralogy: Physics and chemistry of the earth's deep interior: Washington, D.C., Mineralogical Society of America, Reviews in Mineralogy, v. 37, p. 33–96.

Lounejeva, E., Ostroumov, M., and Sánchez-Rubio, G., 2000, Polimorfos de alta presión de sílice en las impactitas de Chicxulub, México: Resultados de espectrometría Raman: Revista Mexicana de Ciencias Geológicas, v. 17, no. 2, p. 138–142.

McHone, J.F., Nieman, R.A., Lewis, C.F., and Yates, A.M., 1989, Stishovite at the Cretaceous-Tertiary Boundary, Raton, New Mexico: Science, v. 243, p. 1182–1184.

McMillan, P.F., 1992, A Raman spectroscopic study of shocked single crystalline quartz: Physics and Chemistry of Minerals, v. 19, p. 71–79.

Morgan, J., Warner, M., and the Chicxulub working group [Brittan, J., Buffler, R., Camargo, A., Christeson, G., Denton, P., Hildebrand, A., Hobbs, R., Macintyre, H., Mackenzie, G., Maguire, P., Marin, L., Nakamura, Y., Pilkington, M., Sharpton, V., Snyder, D., Suarez, G., and Trejo, A., 1997, Size and morphology of the Chicxulub impact crater: Nature, v. 390, p. 472–476.

Morgan, J., and Warner, M., 1999, Chicxulub: The third dimension of a multiring impact basin: Geology, v. 27, p. 407–410.

Novgorodov, P.G., Bulanov, G.P., Pavlova, L.A., Mikhaylov, V.N., Ugarov, V.V., Shebanin, A.P., and Argunov, K.P., 1990, Inclusions of potassic phases, coesite and omphacite in coated diamond crystals from the Mir pipe (in Russian): Moscow, Doklady Academi Nauk SSSR, v. 10, p. 439–443.

Quezada-Muñeton, J.M., Marín, L.E., Sharpton, V.L., Schuraytz, B.C., and Ryder, G., 1992, The Chicxulub impact structure: Shock deformation and target composition [abs.]: Lunar and Planetary Science, v. 23, p. 1121–1122.

Rebolledo-Vieyra, M., Vera-Sánchez, P., Urrutia-Fucugauchi, J., and Marín, L.E., 2000, Physical characteristics of deposition of impact breccias and Pan-African basement affinities of Chicxulub crater [abs.], *in* Catastrophic events and mass extinctions: Impacts and beyond: Houston, Texas, Lunar and Planetary Institute, LPI Contribution No. 1053, p. 180–181.

Schuraytz, B.C., Sharpton, V.L, and Marín, L., 1994, Petrology of impact melt rocks at the Chicxulub multi-ring basin, Yucatán, México: Geology, v. 22, p. 868–872.

Scott J.F., and Porto S.P., 1967, Longitudinal and transverse optical lattice vibrations in quartz: Physical Reviews, v. 161, p. 903–910.

Sharma, S.K., Mammine, J.F., and Nicol, M.F., 1981, Raman investigation of ring configuration in vitreous silica: Nature, v. 292, p. 140–141.

Sharpton, V.L., Dalrymple, B.G., Marín, L.E., Ryder, G., Schuraytz, B.C., and Urrutia-Fucugauchi, J., 1992, New links between the Chicxulub impact structure and the Cretaceous/Tertiary boundary: Nature, v. 359, p. 819–821.

Stishov, S.M., and Popova, S.V., 1961, A new modification of silica: Geochemistry, v. 10, p. 923–926.

Stöffler, D., and Langenhorst, F., 1994, Shock metamorphism of quartz in nature and experiment. 1. Basic observation and theory: Meteoritics, v. 29, p.155–181.

Swisher, C.C., Grajales-Nishimura, J.M., Montanari, A., Margolis, S.V., Claeys, P., Alvarez, W., Renne, P., Cedillo-Pardo, E., Maurrasse, F., Curtis, G.H., Smit, J., and McWilliams, M.O., 1992, Coeval $^{40}Ar/^{39}Ar$ ages of 65.0 million years ago from Chicxulub crater melt rock and Cretaceous-Tertiary boundary tektites: Science, v. 257, p. 954–958.

Vera-Sánchez, P., 2000, Caracterización geoquímica de las unidades basales del bloque de Yucatán y su afinidad con unidades similares en el Golfo de México [M.S. thesis]: México, D.F., Universidad Nacional Autónoma de México, 105 p.

Wopenka, B., El Goresy, A., Chen, M., and Sharp, T., 1996, In situ identification of naturally occurring HP polymorphs: Wadsleyite, ringwoodite, majorite and apatite [abs.]: Terra Nova, v. 8, p. 25 (Abstract supplement no. 2).

Yang, Wang-Hong, Kirkpatrick, R.J., Vergo, N., and McHone, J., 1986, Detection of high-pressure silica polymorphs in whole-rock samples from Meteor crater, Arizona, impact sample using solid-state silicon-29 nuclear magnetic resonance spectroscopy: Meteoritics, v. 21, p. 117–124.

Manuscript Submitted October 10, 2000; Accepted by the Society March 22, 2001

Printed in the U.S.A.

Geological Society of America
Special Paper 356
2002

Distribution of Chicxulub ejecta at the Cretaceous-Tertiary boundary

Philippe Claeys*
Department of Geology, Vrije Universiteit Brussel, Pleinlann 2, B-1050, Brussels, Belgium
Wolfgang Kiessling*
Natural History Museum, Invalidenstr. 43 D-10099, Berlin, Germany
Walter Alvarez*
Department of Earth and Planetary Science, University of California, Berkeley, California 94720-4767, USA

ABSTRACT

The mineralogical, sedimentological, and geochemical information in a large database on the Cretaceous-Tertiary (K-T) boundary is used to document the distribution of impact debris derived from the Chicxulub crater. The database is coupled with a geographic information system (GIS) allowing the plotting of the information on a latest Cretaceous paleogeographic map. The database will be available in part on the internet in the near future, and contains data from 345 K-T boundary sites worldwide. However, relatively few sites are known in South America, Australia, Africa, and in the high latitudes. Major disturbances of sedimentation, such as massive debris flows, failure of platform margins, or significant erosion of Upper Cretaceous layers, occur throughout the Gulf of Mexico. Mass wasting of material also took place in the western and eastern parts of the Atlantic Ocean. Almost 100 K-T boundary sites analyzed for Ir recorded the positive anomaly; Ir is spread homogeneously throughout the world, but is diluted at proximal sites because of the high volume of sediment that was in suspension in the Gulf of Mexico after the impact. Shocked quartz is more common, and maybe larger in size west of the crater. The main advantage of the database is to provide a convenient method to manage the huge amount of data available in the literature, and to reveal patterns or characteristics of the data. The database can help refine the variables used in mathematical models and documents the origin, transport, and deposition of ejecta during a cratering event.

INTRODUCTION

The Cretaceous-Tertiary (K-T) boundary impact event produced a broad range of ejecta material. Impact debris can be easily located in many Late Cretaceous and early Tertiary sections, thanks to their close association with the K-T boundary mass extinction. It is difficult to argue against the fact that the ejecta debris and geochemical signal found in the K-T boundary layer resulted from a unique event, i.e., the formation of the Chicxulub crater in Yucatan. In the Gulf of Mexico region, millimeter-sized spherules, some with a preserved glass core, occur at the K-T boundary (Izett et al., 1990; Sigurdsson et al., 1991; Smit, 1999). The glass is linked to Chicxulub impactites by its major and trace element compositions, isotopic signa-

*E-mail: Claeys, phclaeys@vub.ac.be; Kiessling, kiess@pal.uni-erlangen.de; Alvarez, platetec@socrates.berkeley.edu

Claeys, P., Kiessling, W., and Alvarez, W., 2002, Distribution of Chicxulub ejecta at the Cretaceous-Tertiary boundary, *in* Koeberl, C., and MacLeod, K.G., eds., Catastrophic Events and Mass Extinctions: Impacts and Beyond: Boulder, Colorado, Geological Society of America Special Paper 356, p. 55–68.

tures, and radiometric age (Blum et al., 1993; Swisher et al., 1992). The crater impactites and several K-T layers contain shocked zircons dated as ca. 540 Ma by U-Pb methods (Kamo and Krogh, 1995; Krogh et al., 1993). This is the age of the Pan-African basement, which underlies the Yucatan carbonate platform (López Ramos, 1975). The study of Alvarez et al. (1990) in Italy and the results of the geochemical part of a K-T blind test carried out in 1994 clearly demonstrated that the K-T Ir anomaly is unique in the last 10 m.y. of the Cretaceous.

The Chicxulub crater is one of the few craters on Earth with a well-identified and complete ejecta sequence. Ejecta blanket material covers most of today's Yucatan, and proximal ejecta that is deposited within 3000 km from the impact site is found all over the Gulf of Mexico region, and in parts of North America. Distal ejecta occurs all over the world at the K-T boundary. It is possible to document variations in type, composition, and concentration of impact debris with distance from the source crater. Chicxulub is thus an ideal case study to understand the formation, transport, and distribution of various ejecta products during a large impact. Study of the ejecta material at different sites also provides information on the chemical and physical interactions between ejecta and the atmosphere during transport. Knowledge of the ejecta distribution pattern can be used to refine existing mathematical models for cratering and ejecta production on Earth and other rocky planets. Eventually, understanding the distribution of ejecta at the K-T boundary will help to identify ejecta debris from other craters and to link impact particles found in the sedimentary record with a specific source crater. Before these studies can be carried out, careful and extensive documentation of the Chicxulub ejecta is required. This can best be accomplished with the support of a database, coupled with a geographic information system tied to the Late Cretaceous geography. This chapter is largely a survey of K-T boundary ejecta distribution. The new database approach and the focus on geographic patterns rather than vertical successions make this chapter complementary to the extensive review of Smit (1999).

DATABASE

Computer databases offer the opportunity to organize, review, and evaluate large amounts of data easily and more objectively than other data collection means. Our database (named KTbase) is a relational database designed to combine geochemical, mineralogical, sedimentological, and paleontological information from all currently known K-T boundary sites. The development of KTbase started in May 1999, and is based on the information extracted from the literature. It is designed for Visual dBase 5.5, but is also available in Microsoft Access format. The main table in KTbase is a locality-based summary table mostly containing information on the kind of data available on each site. Basic lithological, chemical, and mineralogical data are also included in the main table. For each important entry a link to the reference table is available. The paleontological data require several additional tables that are also linked to the main table. KTbase currently summarizes data from 348 K-T boundary sites. A K-T boundary site is defined as an area where at least late Maastrichtian and/or Danian sediments are preserved. The paleontological part of the database has been described in Kiessling and Claeys (2001), and is not discussed here. Interested colleagues are invited to use additional parts of the database. However, the paleontological part of the database will not be available until it has been published.

The part of the database containing K-T boundary geochemical, sedimentological, and mineralogical information is based on the earlier unpublished database developed at the University of California, Berkeley, by Walter Alvarez and colleagues. It listed 113 localities with iridium anomalies, 28 localities with occurrences of shocked quartz, 54 localities with spherules, and 18 localities with Ni-rich spinel. Whereas the Berkeley database was largely text based, the KTbase contains numerous quantitative fields.

A major difference from the former database is the 20 km distance criterion for K-T sites to be included separately. In this chapter, a K-T site may thus be composed of several outcrops, separated by <20 km. The criterion was applied to obtain a more homogenous spatial distribution of sites in the database. This implies that data from an area of $\sim$315 km^2 are summarized for each site in KTbase. Consequently, the number of sites with a recorded Ir anomaly is reduced in Ktbase, because many well known but closely spaced sections in North America and the Italian Apennines are now combined. There are 85 sites known on a global scale to have a significant iridium anomaly; 49 sites yield shocked quartz, 54 sites contain spherules, and 21 sites with Ni-rich spinel are known. A total of 101 of 345 K-T boundary sites contain ejecta material.

A locality-based entry format was defined for the database main table in connection with a geographic information system (GIS). This main table presents an overview of the K-T sites studied and lists basic data and information, the details of which are contained in a set of related tables. The relevant fields for this study are (1) maximum iridium concentration reported from the boundary clay (in ppt); (2) integrated iridium concentration in the boundary clay when available (in ng/cm^2); (3) maximum diameter of shocked quartz grains in the boundary clay (in µm); (4) abundance of shocked quartz grains in the boundary clay (in grains/cm^2); (5) presence of suspect tsunami and mass-wasting deposits; (6) presence of Ni-rich magnesioferrite spinel; (7) maximum concentration of Ni-rich spinel; (8) presence of impact spherules; (9) maximum diameter of spherules; and (10) presence and abundance of soot.

KTbase is essentially a compilation of the data available in the literature, combined with the authors' experiences of the problem. To be complete, the database must be maintained and regularly updated. The aim of our database is first to gather and then to render the existing information more approachable, manageable, and objective. The database and the presentation of the data on a Late Cretaceous paleogeographic map (Scotese and Golonka, 1992) can also point out possible characteristics or ordering of the data, which then need to be further investigated in the field or in the laboratory.

In KTbase, quantitative information is provided wherever possible and as precisely as possible. Unfortunately, not all authors present their data in the same manner and/or with the same rigor. The database format reflects the average quality of published data extracted from the literature. This is clearly illustrated by the concentration of Ir. In the K-T layer Ir is ideally reported in flux per square centimeter of sediment (ng/cm). However, most commonly in the literature the Ir concentration is given in parts per billion or parts per trillion (ppb or ppt = ng/g or pg/g). Only 30 Ir flux values are stored in KTbase, whereas 67 precise Ir maximum concentration measurements are available. Thus our analysis focuses on the maximum Ir concentration in the boundary clay, although, when available, integrated values are also considered. This information (e.g., flux or concentration of other platinum group elements [PGE]) is stored in a memo field together with additional data and remarks concerning the site.

Despite these drawbacks, some patterns emerge. The main advantage of the database is to point out these possible patterns and allow scientists to test them in a more constrained manner. In a multidisciplinary field such as the K-T boundary, it also rapidly provides the specialist in one discipline with both a global perspective of the problem and with detailed information in an unfamiliar field. Another significant advantage of the database is to indicate clearly what needs to be done and where, in order to further develop the understanding of the ejecta distribution problem. As clearly shown in Figure 1, the worldwide distribution of studied K-T sequences is strongly biased toward Europe and North America. Africa, Australia, South America, and the high latitudes (>60°) appear as open fields of research on the K-T event, especially in term of impact tracers.

In this chapter we first focus our discussion on the sedimentology and the quality of K-T sites for ejecta studies, then examine the broad distribution of chaotic sedimentary units at the K-T boundary. We then present and discuss the distribution patterns of Ir and shocked quartz.

EVALUATION OF K-T BOUNDARY SECTIONS

Completeness criteria

In the database, a K-T boundary site is defined as a locality with at least the late Maastrichtian or the Danian preserved. So far, 345 K-T boundary sites are recorded in KTbase. The completeness of K-T sections is often debated (see discussion in Kiessling and Claeys, 2001). Biostratigraphic correlation and precision between marine and continental location can be problematic. In this case, we decided to rely on the presence of impact products to define precisely the K-T boundary, and to evaluate stratigraphic completeness. The timing of impact debris deposition ranges from about an hour for the proximal ejecta curtain material (Alvarez et al., 1995) to several years for the finest Ir-rich dust to settle from the upper atmosphere (Toon et al., 1982) and through the water column. In a section, the record of an event as geologically short as the deposition of ejecta debris is probably one of the best indications (if not the best) of stratigraphic completeness available. In addition, deposition of ejecta is more likely to be synchronous than the global first occurrence of a newly evolved species. In addition, the ejecta material criterion eliminates the possible problems of biostratigraphic correlation between marine and continental sections. Although strictly speaking the ejecta criterion only applies to the completeness of the K-T layer, nearly all of the reported sites containing impact debris are also biostratigraphically complete.

On the basis of this convention, nearly 30% of all localities in KTbase are stratigraphically complete; i.e., 101 K-T sites contain ejecta debris, and can thus be considered as true K-T boundaries for the purpose of this study (Fig. 1). Another 15% do not seem to contain the Chicxulub ejecta, but would be considered complete in terms of biostratigraphy, because they contain the latest Maastrichtian and earliest Danian zones. They are classified as fairly complete sections for the purpose of this study and are also shown in Figure 1. Most of them have never been checked for impact tracers. This gives a first hint of the research possibility opened by KTbase.

At almost 60% of the localities in KTbase the completeness is either unknown or a significant sedimentation break is evident. These so-called incomplete sites cannot be used to study ejecta distribution patterns, but they might be used to understand some aspects of the K-T extinction (Kiessling and Claeys, 2001).

Depositional environment

Sites from all paleoenvironments are considered in KTbase. The great majority of K-T boundary sites are located in the marine realm (84%); 33% are in deep water and 48% are in shallow-marine environments. The other 3% are labeled only as marine sites without any bathymetric information. The terrestrial environment represents 16% of the sites. Comparing Figures 1 and 2, it is obvious that the sites considered incomplete in term of ejecta distribution are mostly from shallow-water depositional environments.

K-T regression

Several authors advocate a significant regression at or just prior to the K-T boundary (Archibald, 1996a, 1996b; Keller and Stinnesbeck, 1996; Schmitz et al., 1992). Figure 3 indicates that this regression does not seem to be globally recorded. Looking at the sea-level curve, the K-T regression appears very minor compared to the one clearly marked ~7 m.y. later at the Selandian-Thanetian boundary (Haq et al., 1988; Hardenbol et al., 1998). In addition, a good part of the data for a pre-K-T

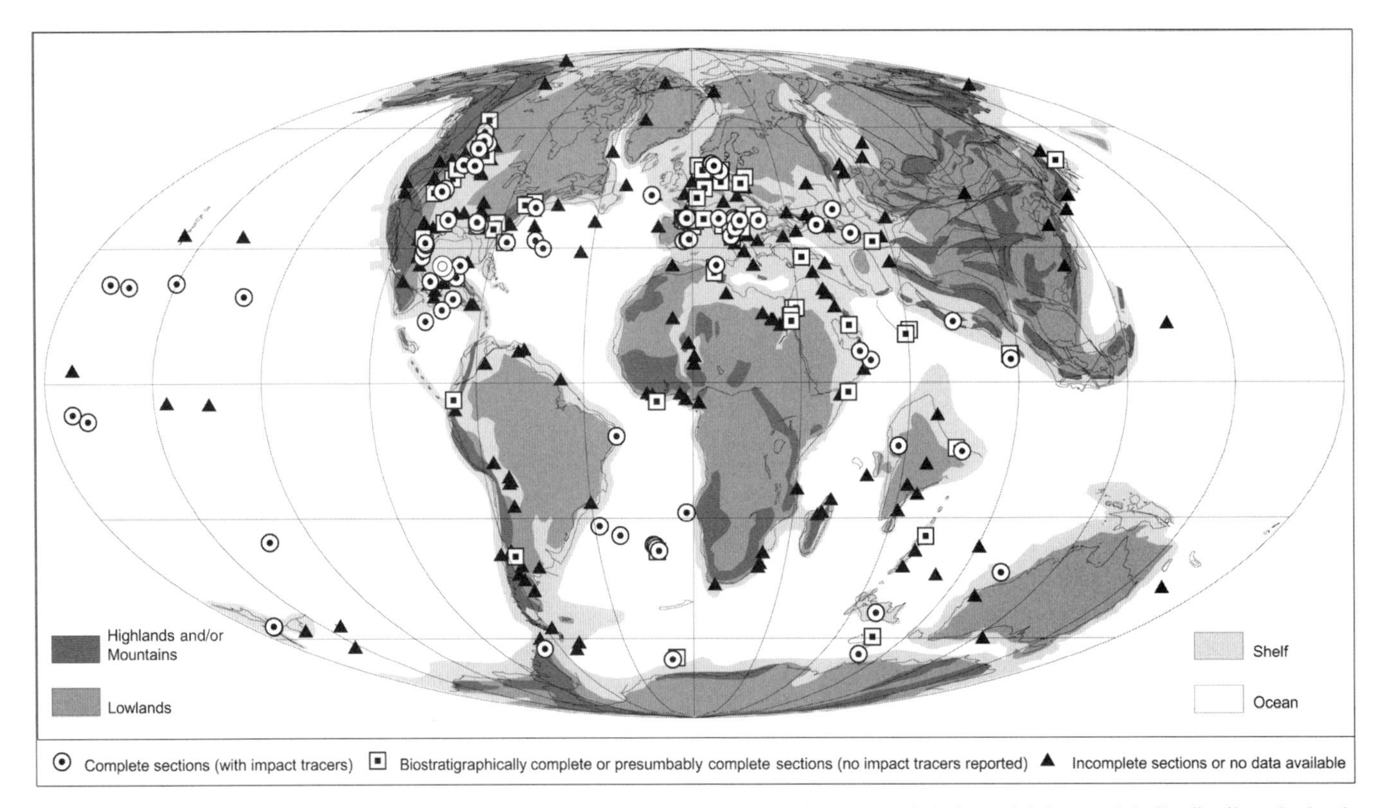

Figure 1. Completeness of Cretaceous-Tertiary (K-T) sections: 101 K-T sites that have recorded ejecta debris are globally distributed, clearly supporting worldwide deposition of Chicxulub ejecta. Paleogeographic reconstruction (J. Golonka, 1997, personal commun.) presented here is based on modified version of Late Cretaceous map of Scotese and Golonka (1992.)

boundary regression came from the Braggs section in Alabama (Baum and Vail, 1988; Donovan et al., 1988; Habib et al., 1992). This Gulf of Mexico locality is <1500 km from the Chicxulub crater, in a region where the K-T boundary is marked by a clastic sequence several meters thick that has proximal ejecta at its base (Smit, 1999). The displacement of material and large earthquakes involved in the formation of the 200-km-diameter Chicxulub crater must have severely affected the sedimentology of the Gulf of Mexico and triggered major sediment disturbances. Gravity flows and tsunami-related deposition are present throughout the Gulf of Mexico (Bourgeois et al., 1988; Bralower et al., 1998; Smit, 1999; Smit et al., 1992, 1996). Therefore, the sediments interpreted at Bragg as a transgressive sequence tract overlying a sequence boundary (Baum and Vail, 1988; Donovan et al., 1988) may instead reflect the deposition of debris flows triggered by tsunami waves related to the nearby Chicxulub impact, as suggested by Pitakpaivan et al. (1994) (Fig. 4). These coarse sediments thus have no relation to eustatic sea level. KTbase also supports the view of Speijer and van der Zwaan (1996) that the complex pattern of sea-level fluctuations for the late Maastrichtian evoked by some is based on unconstrained and contradictory data. It is thus more likely that the pattern observed in Figure 3 outside the Gulf of Mexico region reflects local regression (Keller et al., 1998) opposed by local transgressions (Pardo et al., 1999), even in tectonically stable areas.

TSUNAMI AND MASS-WASTING DEPOSITION

Effect on regional sedimentation

Paleogeographic reconstruction shows that, contrary to the tectonically active Caribbean region, the morphology of the Gulf of Mexico has not changed much since the Late Cretaceous (Acton et al., 2000; Pindell, 1994). The impact took place on the platform, displacing and/or pushing away huge volumes of sediments, leading to major wave disturbance in deeper water. Very large earthquakes were also generated for an extended period of time after the event (Covey et al., 1994; O'Keefe and Ahrens, 1991). Seismic shaking induced the fracturing and collapse of the unstable parts of the Yucatan platform margins even at a distance of several hundred kilometers from the crater (Grajales-Nishimura et al., 2000). As a result, huge tsunami waves and massive debris flows formed in the Gulf of Mexico, locally eroding the Upper Cretaceous strata, and reworking, transporting, and redepositing sediments. Bourgeois et al. (1988) estimated that at Brazos River (Texas) the waves were at least 50–100 m high. Crashing onshore, the huge tsunami

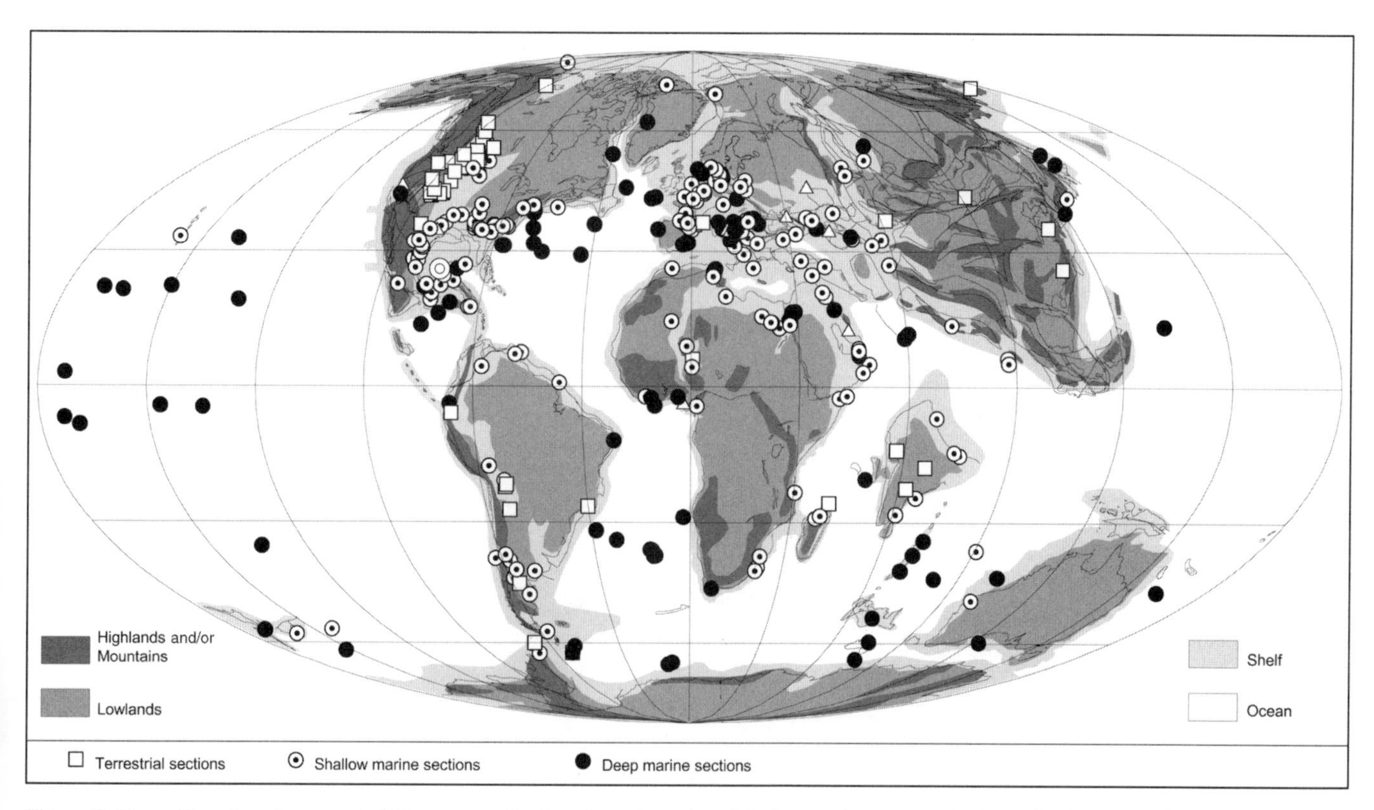

Figure 2. Depositional environment of Cretaceous-Tertiary boundary sites. Marine environments dominate by far; terrestrial sections are only conspicuous in North America.

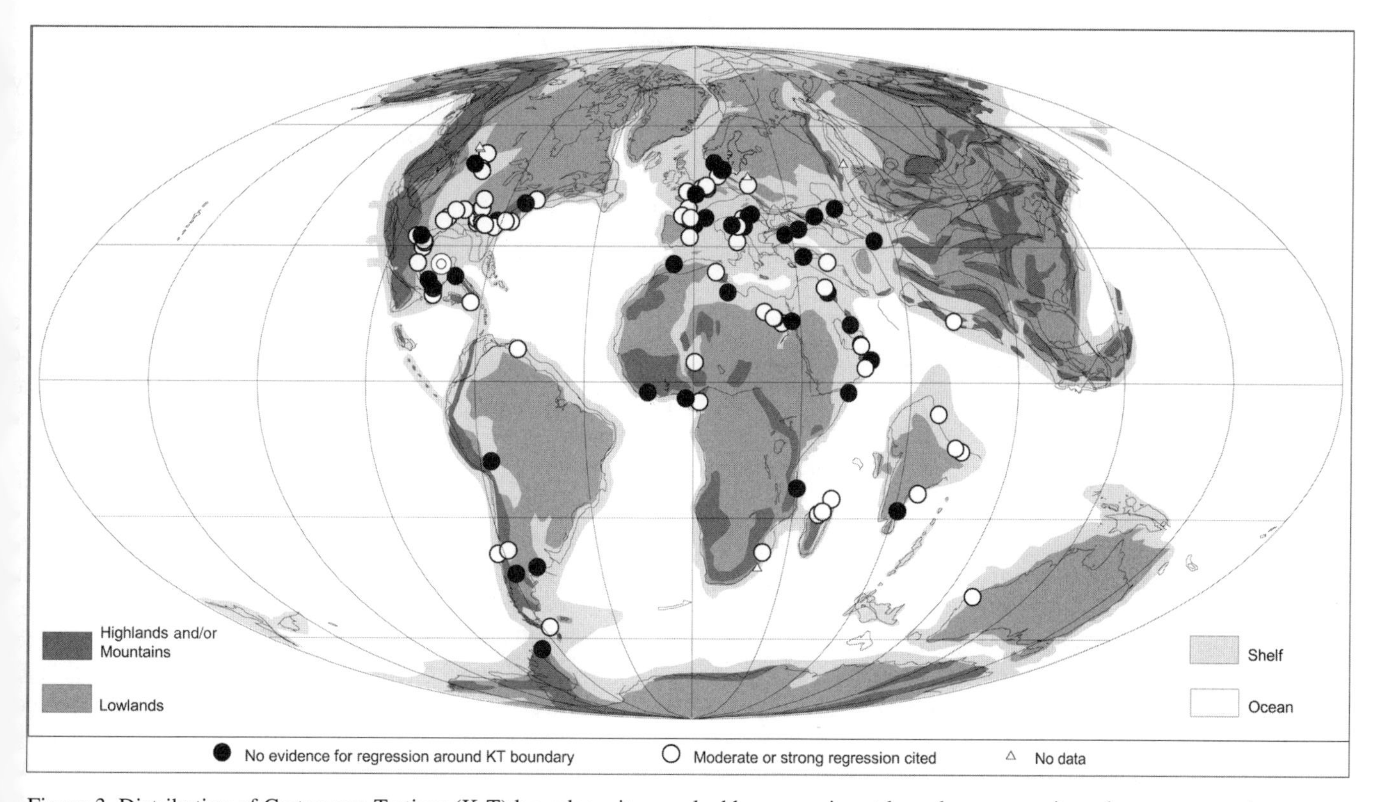

Figure 3. Distribution of Cretaceous-Tertiary (K-T) boundary sites marked by regression at boundary versus sites where no regression or even transgression is indicated. Only shallow-marine sites with sedimentological data are shown.

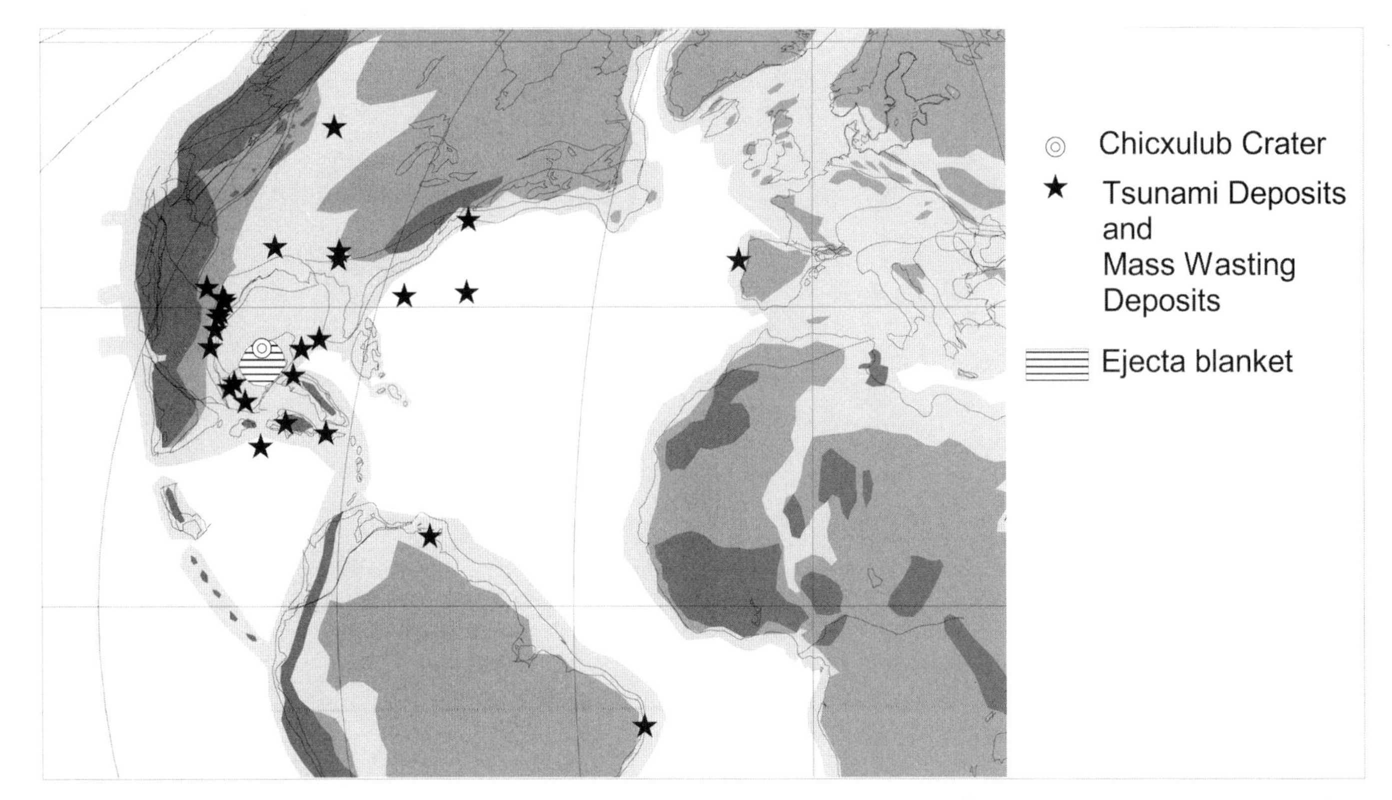

Figure 4. Major disturbance of sedimentation at Cretaceous-Tertiary boundary is recorded in Gulf of Mexico region and in Atlantic. Ejecta blanket surrounding crater, and first discovered in Belize up to 350 km from crater (Ocampo et al., 1996), is shown by hachured pattern.

waves had a strong erosive power, and as they retreated they dragged back large amounts sediment to the deep water (Kruge et al., 1994). Figure 4 indicates the location of K-T boundary chaotic sediments, such as breccias, coarse clastic units, debris flows, and tsunamites related to the Chicxulub impact event. It seems that this catastrophic sedimentation also extended to parts of the Atlantic Ocean.

Proximal ejecta blanket

On the Yucatan platform, from the crater rim all the way to Belize, the K-T boundary is composed of breccias. At the crater rim, it is formed of suevite overlying a Bunte Breccia-like unit (Urrutia-Fucugauchi et al., 1996). This polymict breccia is composed essentially of blocks of carbonate and evaporites from the upper part of the Chicxulub target rock. Near the Mexican-Belize border the breccia is composed of a >20-m-thick diamictite formed essentially of dolomite blocks. This unit formed on the shallow-water carbonate platform and is recognized all over Yucatan, from Campeche to Valladolid, around Chetumal, and in northern Belize (hachured pattern in Fig. 4). The mode of formation of this unit is still unclear. It could have been formed as a ground-hugging flow, according to the Oberbeck (1975) model of ballistic ejection and secondary cratering. This model was successfully advocated by Hörz (1982; Hörz et al., 1983) for the Bunte Breccia around the Ries crater.

Failure of the Yucatan platform margins

The stratigraphy and distribution of the large sediment gravity flows resulting from the collapse of the continental margins around the Chicxulub crater were examined in detail by Bralower et al. (1998) and nicknamed the "K-T boundary cocK-Tail." Evidence for erosion caused by the passage of the flows, or deposition of K-T cocK-Tail material, are widespread from the margin of the Yucatan platform to the north, between Yucatan and Florida to the Nicaragua Rise and the Venezuela basin to the south. Farther west, offshore of the Cretaceous platform in the Mexican states of Tabasco and Chiapas, an ~40-m-thick breccia is below the ejecta-rich K-T boundary sequence (Grajales-Nishimura et al., 2000; Montanari et al., 1994). This breccia is 170 m thick in the oil-producing region of the Campeche bank (Grajales-Nishimura et al., 2000). Similar breccias are also reported from Guatemala and Cuba (Fourcade et al., 1997, 1998; Kiyokawa et al., 1999; Tada et al., 1999). In the entire region, the breccia is formed of decametric to millimetric blocks of limestone, and the fossil content indicates that they originated from an Upper Cretaceous shallow-water carbonate platform environment. Where found in deep-water settings, this chaotic deposit fines upward and resembles a major single-unit debris flow. The breccia is interpreted to be the result of fracturing and collapse of the Yucatan platform margin, caused by the Chicxulub-induced seismic shaking (Bralower et al., 1998; Grajales-Nishimura et al., 2000).

Tsunamites in the Gulf of Mexico

Material displaced by the crater formation, the failure of the margins, and huge earthquakes induced the formation of major tsunamis in the enclosed basin of the Gulf of Mexico. It is difficult to separate the effects of major mass flow and that of tsunamis on sedimentation across the K-T boundary. The breaking of the huge tsunamis (at least 100 m high, according to Bourgeois et al., 1988) on shore or in the shallow water eroded and dislocated large volumes of sediment, which then traveled as large masses or flows to the deeper water environment.

From Alabama to the northern part of the Mexican state of Vera Cruz, the K-T boundary is marked by a coarse clastic sequence interpreted as rapidly deposited by the combination of high-energy tsunamis and triggered gravity flows (Smit et al., 1996). However, this explanation is contested (for a complete discussion see Smit, 1999). A turbidite origin has also been proposed (Bohor, 1996). Others estimate that this unit resulted from channelized deposition taking place over several hundred thousand years due to rapid sea-level changes (Stinnesbeck et al., 1993). One of the characteristics of this sequence is its remarkable homogeneity over more than 2500 km, which clearly reflects unusual sedimentary processes. The sequence can be subdivided into four units (Smit, 1999).

The lower unit is a sandstone composed of ejecta material such as impact spherules, some with a preserved glass core, shocked grains, and limestone fragments. This unit also contains rip-up clasts of the underlying Upper Cretaceous Mendez marls. Unit two is a coarse carbonate-cemented sandstone with lithic clasts, quartz grains, foraminifera debris, and plant fragments. This unit is ejecta free, except for some reworked material at its base. Smit et al. (1996) reported repetitive changes of paleocurrent direction. Unit II fines upward and the transition to unit III is gradual and marked by the appearance of silt layers. The amount of sand diminishes upward. Unit IV is composed essentially of size-graded silt and appears often more lithified than the underlying units (Smit, 1999). It contains reworked fine ejecta, including shocked grains and Cretaceous fossils, and is enriched in Ir.

Debris flows and tsunamis outside the Gulf of Mexico

Sediment disturbances are also reported outside the Gulf of Mexico region. South and east of Yucatan the situation appears confused. In Cuba, Tada et al. (2000) interpreted the 180-m-thick Peñalver Formation as a K-T boundary deposit formed by debris flow and tsunamis. In Cuba, the 500-m-thick Cacarajicara Formation is viewed as related to giant mass-flow deposits triggered by the Chicxulub impact (Kiyokawa et al., 2000). The two units contain impact ejecta material such as shocked quartz and altered glass, and appear to be formed of an accumulation of chaotic clastic carbonate material from various sources. It is difficult to clearly differentiate the precise sedimentological processes taking place in this region. At K-T time, these Cuba sites must have been located very at close to the crater to explain such an input of material, capable of forming sections several hundred meters thick. Another possibility is that all the material originated from the collapse of the southern Yucatan platform margins, as observed to a lesser extent in the Campeche area (Grajales-Nishimura et al., 2000). In comparison, the K-T boundary debris-flow deposits found at nearby Beloc (Haiti) and at Ocean Drilling Program (ODP) site 1001 are <10 m thick (Maurasse and Sen, 1991; Bralower et al., 1998; Smit, 1999).

The Chicxulub impact also triggered large submarine slope failures in both the western and eastern North Atlantic. At ODP Site 171B on Blake Nose in the western Atlantic north of Florida, the K-T boundary is marked by an ~20-cm-thick clastic unit underlying the classic ejecta Ir succession (Smit et al., 1997). Seismic data in the area show sediment disturbance near the K-T boundary (Klaus et al., 1997).

At Bass River (New Jersey), more than 2500 km from Chicxulub, a biostratigraphically complete upper Maastrichtian—Paleocene sequence is interrupted by a 12-cm-thick spherule bed and reworked clay clasts of various sizes (Olsson et al., 1997). This sequence is equivalent but much thinner than that found throughout the Gulf of Mexico region. The basal contact of the spherule layer is nonerosional, indicating that the spherules settled quietly on a soft surface undergoing sedimentation (Olsson et al., 1997). This contrasts with the Gulf of Mexico sequence, where the presence of rip-up clasts of the underlying upper Maastrichtian marls reworked in the spherule bed indicate more energetic deposition and cutting of the underlying sediments. At Bass River, the upper contact of the spherule bed is sharp. The overlying 6-cm-thick clay-clast layer contains Cretaceous foraminifera and nannofossils indicating that they were eroded, transported, and redeposited on the spherule bed (Olsson et al., 1997). This unit was interpreted by Olsson et al. (1997) to have originated from erosive action of impact-triggered tsunami or megastorms affecting the North Atlantic. It appears to be a single event with waves strong enough to affect the Bass River middle shelf at a depth of ~100 m. Norris et al. (2000) identified K-T boundary age mass-flow deposits associated with ejecta in Deep Sea Drilling Project (DSDP) Sites 387 and 386 on the Bermuda Rise, ~2500-2800 km from Chicxulub. These deposits can be correlated with a distinctive acoustic reflector known across the North Atlantic, from Puerto Rico to the Grand Bank of Canada (Norris et al., 2000). In the eastern Atlantic, at DSDP Hole 398D on the abyssal plain offshore Portugal, an interval of 70 cm below the K-T boundary is slumped (Norris et al., 2000). These authors interpreted it as possibly reflecting a single mass-failure event predating the arrival of the overlying ejecta material, caused by slope failure along the Portuguese coast by direct effect of ground motion or tsunamites across the Atlantic. In the western Atlantic the mass-wasting material originated in shallow-water environments and was transported over a significant distance; in the eastern Atlantic the slump is locally derived and reworked (Norris et al., 2000), which might perhaps be viewed as reflecting the variation in the intensity of the ground motion disturbance with distance from the crater.

It appears that sediment disturbance across the boundary is also present in the Poty quarry of northern Brazil (Albertão et al., 1994), and probably in Venezuela. Coarse- to medium-grained sandstones overlying silty shales mark the K-T boundary interval in eastern Venezuela (Helenes and Somoza, 1999). Although detailed sedimentological data are lacking, the overall stratigraphic context is similar to that reported at Gulf of Mexico tsunamite sites. The Pernambuco region of northeastern Brazil seems to mark the southern extension of Chicxulub-induced disturbance of K-T sedimentation. The tsunamis, slope failure, or major earthquakes do not seem to have extended south of the equator. At ODP and DSDP sites in the southern Atlantic, the K-T boundary appears to be marked only by the classic thin Ir-rich clay layer.

EJECTA MATERIAL

Iridium distribution

The first reported and probably most characteristic feature of the K-T boundary is an enrichment in Ir. Several studies have demonstrated that the K-T boundary clay also contains high concentrations in the other PGE (Pt, Pd, Os, Ru, Re), Ni, Cr, Co, and gold (Evans et al., 1993a, 1993b; Kyte et al., 1980). However, KTbase contains essentially Ir data; most authors have reported the K-T boundary geochemical anomaly mainly in terms of Ir concentration (Fig. 5).

The database shows that the Ir anomaly was detected in 85 K-T boundary sites worldwide, in a broad range of depositional environments from deep marine to continental settings. KTbase allows the comparison of Ir data determined at different laboratories, using different methods. We discussed, in the presentation of the database, the problem of reporting Ir data in flux versus maximum concentration. Despite this difficulty and the various analytical techniques utilized, the reported Ir data are generally in good agreement. The maximum concentration ranges from 0.1 to >87 ppb.

Outside the Gulf of Mexico, there is no correlation between Ir concentration and distance from the impact site. Local conditions, such as sedimentation rate, lateral sediment redistribution, bioturbation, and diagenesis, can probably account for the difference in Ir concentration reported, even between geographically close sites. In some K-T sites, for example in the U.S. Western Interior, the Ir anomaly is concentrated in a thin (<1 cm) interval. At other locations, bioturbation, reworking, diagenesis, and chemical diffusion can cause the remobilization of Ir and its spread over as much as several meters of section. Small Ir peaks recorded in the early Paleocene can also be attributed to redeposition of eroded Ir-rich K-T boundary material. In that case, the Ir is likely to be associated with reworked Cretaceous microfossils (Pospichal, 1996). Robin et al. (1991) discussed the diffusion of Ir compared to other denser and thus less mobile ejecta components such as the Ni-rich magnesioferrite spinels. Meteorite-rich dust and vapor from the impacting bolide and target rock were transported to the upper atmosphere by the fireball rising from the crater. It thus appears that, after the impact, the Earth was engulfed in a homogeneous cloud of vapor and dust particles. The database does not support the idea that some areas of the Earth, e.g., the high latitudes, were less affected by this dust cloud than tropical regions. Several southern high-latitude sites (i.e., Seymour Island or Kerguelen Plateau) display fairly high Ir concentrations. KTbase thus confirms that the Ir anomaly is global and homogeneous (Fig. 5). It is detected in all known parts of the Late Cretaceous world, including in high latitudes.

The only region to show a particular Ir pattern is the proximal Gulf of Mexico. At nine sites around the Gulf of Mexico, the concentration of Ir determined in the upper part (units III and IV of Smit, 1999) of the K-T sequence ranges between 0.25 and 1 ng/g. These concentrations are clearly lower than those determined at more distal sites (Fig. 5). This is a direct reflection of the highly unusual sedimentation taking place close to the Chicxulub crater. At the base, in the ejecta-rich unit I, the Ir concentration varies extensively between samples, reaching 0.2 ng/g. This is probably due to the presence of some minor meteoritic components included in the ejecta. No anomalous Ir concentration is detected in unit II, which is mainly formed of local coarse sand material eroded and transported by the combined effect of tsunami waves and debris flows. The silt layers at the base of unit III contain more Ir than the underlying sands (Fig. 6). In the fining-upward successions of units III and IV the finer layers are systematically enriched in Ir. The highest Ir concentration occurs in the size-graded fine silts of unit four (Fig. 6).

At the time of deposition of units III and IV, weeks to months after the K-T impact (Smit, 1999), Ir-rich material was raining down from the atmosphere and slowly settling through the water column. Although the carrier is still unknown (Schmitz, 1988; Schmitz et al., 1990), it appears that Ir is associated with the very fine fraction of the sediment and thus settled slowly. The sand interlayered with silt in unit III represents the last pulses of coarse material being deposited. Their rapid deposition had a diluting effect on the Ir sedimentation. The fining-upward sequence reflects the progressive decrease in frequency and magnitude of the tsunamis and debris-flow deposition (Smit, 1999). However, at that time, the waters of the Gulf of Mexico probably still contained much fine material in suspension, stirred up by the sedimentological disturbances induced directly or indirectly by the impact. The finer silts in unit III and especially in unit IV represent periods of less active sedimentation. During this time, finer sediments settled slowly,

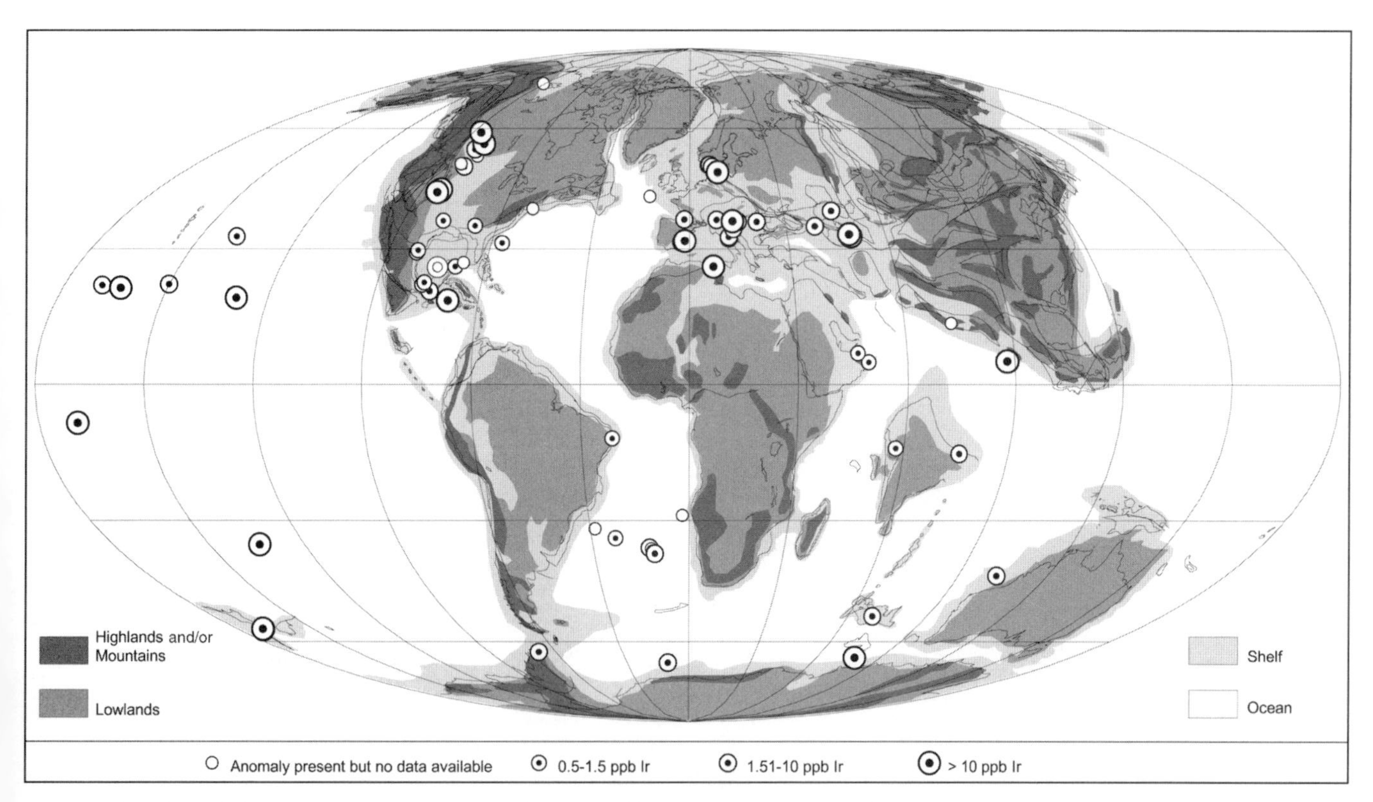

Figure 5. Location of Cretaceous-Tertiary boundary sites marked by positive Ir anomaly. Size of dot is proportional to maximum concentration of Ir (in ppb; or ng/g). Analyzed sites located around Gulf of Mexico contain <1.5 ppb Ir because of Ir dilution effect by high volume of disturbed sediment in suspension in Gulf of Mexico. Outside this region there is no correlation between concentration and distance from impact site.

incorporating the Ir. In the upper part of the K-T sequence, the Ir enrichment is thus diluted over almost 30 cm, rather than concentrated in a thin 1 cm layer as at more distal sites. This dilution effect is responsible for the lower Ir concentrations determined when single silt layers are analyzed. However, if the Ir data are integrated over the entire thickness of units III and IV, the Ir flux is comparable to the highest value detected outside the Gulf of Mexico region.

Shocked mineral distribution

The high-pressure shock wave generated by a meteorite impact produces deformations in minerals. In quartz, at dynamic pressures above 5 Gpa, several sets of very fine glassy silicate lamellae known as planar deformation features are created. Such shocked quartz grains were discovered in the K-T boundary sediments by Bohor et al. (1984). Shocked quartz distribution appears to be widespread (Bohor et al., 1987) (Fig. 7). Detailed studies at K-T sites in the U.S Western Interior yielded abundant quartz grains, commonly >300 μm (Bohor, 1990; Izett, 1990; Pillmore and Flores, 1987). These U.S. Western Interior sites were 2200–4200 km from the Chicxulub crater. ODP or DSDP cores recovered at distances as far as 10 000 km from Chicxulub crater (e.g., ODP Site 596) in the southwest Pacific, west of the crater, also yielded numerous shocked quartz grains (Bostwick and Kyte, 1996; Kyte et al., 1996). Abundance of shocked quartz in the K-T unit reaches >1000 grains per cm^2 (Bostwick and Kyte, 1996).

The largest shocked quartz grains are found in the U.S. Western Interior, where they reach sizes >500 μm. The only location outside the Western Interior where large shocked quartz grains are reported is in the Poty quarry in northern Brazil (Albertão et al., 1994). However, no precise statistics are given for this site. In the Pacific, a few grains are as large as 150 μm, but the majority are below 100 μm, the average being ~30 μm (Bostwick and Kyte, 1996). It is difficult to compare the searches for shocked material made at different sites, by different authors, using different techniques. The size difference could be an effect of the amount of sample available for study, and/or the amount of time invested in the search. Much larger volumes of sediments can be processed from the Western Interior outcrops than from deep-sea cores. Shocked quartz is also reported in the K-T clay layer in Italy, Spain, France, and Denmark (Bohor et al., 1987). However, at these outcrops, shocked quartz concentrations appear lower and their size smaller than reported at equivalent distances (~6000–8000 km) to the west in the Pacific Ocean (Bohor and Izett, 1986; Montanari, 1991). Outcrops in Europe also provide ample material for shocked

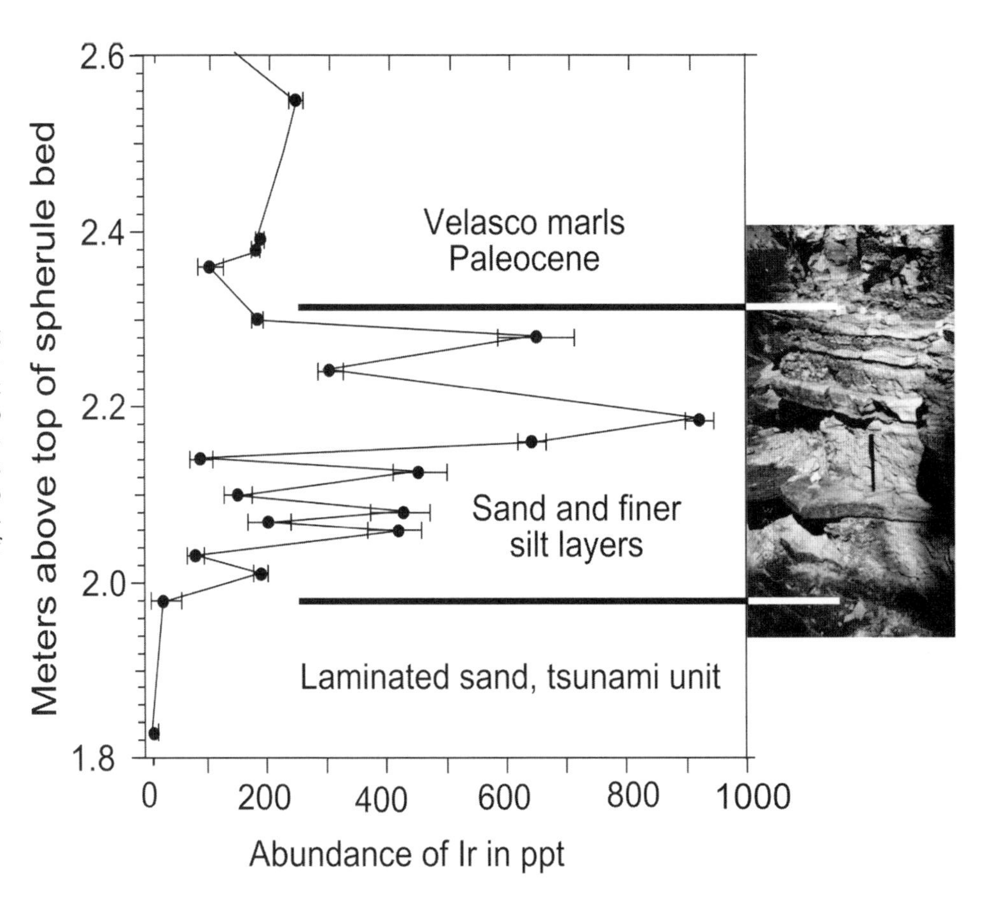

Figure 6. Ir profile across alternating sand and silt layers forming fining-upward sequence of units III and IV at Cretaceous-Tertiary boundary in arroyo El Mimbral (northeastern Mexico) illustrating progressive enrichment in Ir as sediment fines upward from sand to finer silt. Lower Ir content systematically characterizes coarser material of sequence.

mineral searches: no grains >300 μm have been reported there.

Unfortunately, many reports of shocked quartz failed to mention clearly the range and average sizes as well as precise concentration. This information is only available for K-T boundary sites on the Pacific plate, and for some in the U.S. Western Interior. More rigor in the way shocked minerals are characterized is required to extract quantitative meaningful information about the global distribution of this type of ejecta. On the basis of the available data, we suggest that there is a pattern of higher abundance, and possible larger sizes west of the Chicxulub crater, as first pointed out by Alvarez et al. (1995). Alvarez et al. attributed this asymmetric distribution of the shocked quartz to the rotation of the Earth, which affected differently the ballistic trajectory and orbit of the eastbound and westbound particles (see Fig. 1 in Alvarez, 1996; Alvarez et al., 1995). The shocked grain pattern agrees with the oblique impact hypothesis of Schultz and D'Hondt (1996). The impact structure asymmetries suggest a trajectory from the southeast to the northwest at an angle <30° from the horizontal (Schultz and D'Hondt, 1996). This predicts that most of the shocked material was ejected preferentially toward the northwest. The large shocked quartz found in the Brazilian Poty quarry is problematic for both hypotheses. Unfortunately, there is too little information on how common or rare such grains are in this section, which is the only shocked mineral-bearing K-T outcrop reported in South America.

In the Gulf of Mexico region, distribution of shocked quartz is complex. In northeastern Mexico sites, shocked grains are associated with the basal ejecta-rich unit of the K-T sequence (Smit et al., 1992). Shocked quartz grains are rare in this unit, compared to impact glass spherules and limestone fragments. The uppermost fine-grained unit IV, the K-T cocKTail of Bralower et al. (1998), also contains shocked grains. There are no reports of quartz with planar deformation features in the coarser clastic units II and III of the sequence. In Beloc, Haiti, shocked quartz occurs in the spherule layer and associated with Ir in the upper part of the sequence (Leroux et al., 1995). Leroux et al. reported shocked quartz abundance of 104 grains/cm^2 in the spherule layer, much higher than that found in the same level in northeastern Mexico. Shocked quartz grains are also found in southern Mexico, in the ejecta-rich layers above the breccia in Bochil and El Guayal, and in the bentonite layer sealing the oil-producing breccia on the Campeche bank (Claeys et al., 1996; Grajales-Nishimura et al., 2000; Montanari et al., 1994).

Rare shocked quartz is present in the diamictite-like dolomite ejecta breccia that forms the distal part of the ejecta blanket in Yucatan (Pope et al., 1999). The ejecta blanket also contains small, highly altered basement fragments, similar to those found in the Chicxulub suevite inside the crater. In this unit, crater products appear to be greatly diluted by the local dolomitic material. However, shocked grains are more common

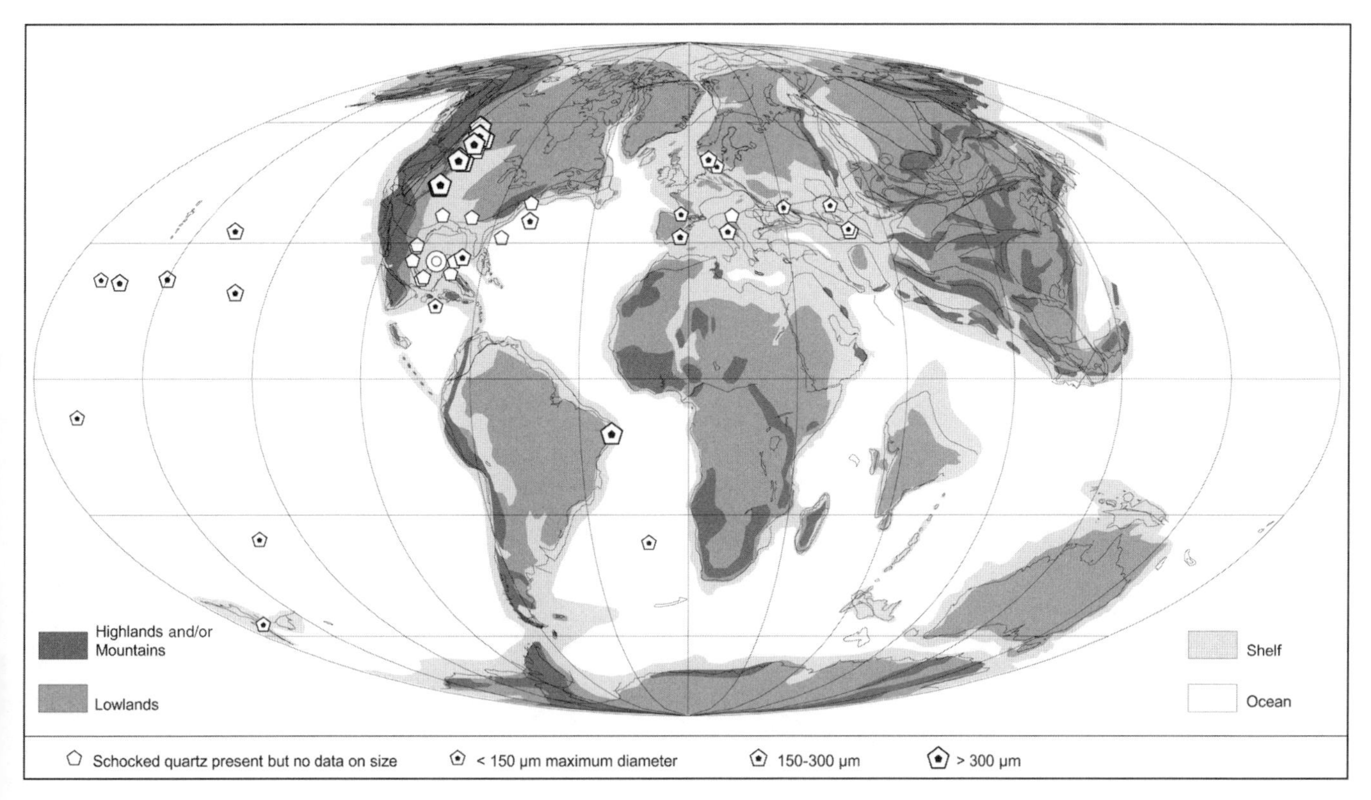

Figure 7. Grain-size distribution of shocked quartz grains at Cretaceous-Tertiary (K-T) boundary. K-T layer in U.S. Western Interior contains largest quartz with planar deformation features (PDF) reported so far. Only maximum size is represented; figure does not show abundance of shocked quartz grains. Data suggest that at equivalent distance, quartz with PDFs are more common to west of crater.

in the suevite and polymict breccia of wells UNAM 5 and UNAM 6, drilled at the crater rim (Claeys et al., 2000).

CONCLUSIONS

1. There are 101 K-T boundary sites (some representing multiple outcrops), all over the world, that contain ejecta debris. This represents nearly 30% of all the sites entered in KTbase. Of the sites that contain latest Maastrichtian and earliest Danian biozones, 15% (i.e., more than 50 sites) have not been investigated for ejecta material.

2. K-T sites formed in shallow-water depositional environments are more commonly considered incomplete in terms of ejecta debris than deeper marine ones.

3. KTbase does not support a global regression at the K-T boundary. The hypothesis of a K-T regression stems in parts from the misinterpretation of impact-related coarse clastic or debris-flow units deposited in the Gulf of Mexico region. These coarse sandy units do not represent a transgressive sequence tract overlying a sequence boundary and are not related to a K-T boundary sea-level change.

4. The Chicxulub impact affected sedimentation within the Gulf of Mexico region and the Atlantic, probably all the way to offshore Portugal. This effect is reflected by the presence of coarse clastic units and/or breccia in deep-water settings and/or by the erosion of Upper Cretaceous sediments from depositional settings usually not prone to unconformities. The precise sedimentological mechanisms are not fully understood: for example, it is not clear if debris flows are created by the seismic wave or ground shaking or by the tsunami waves generated by the impact. It is also possible that in some places, massive debris flows generated tsunami waves. Most likely all these processes acted together, leading to the chaotic sedimentation and erosion occurring at or near the K-T boundary in the Gulf of Mexico and the North Atlantic.

5. The positive Ir anomaly has been recorded at 85 sites and appears to have been spread homogeneously all around the globe. Concentration does not vary systematically with distance from the crater. At proximal sites, the Ir concentration is diluted by the high amount of sediment put in suspension in the Gulf of Mexico's water after the impact.

6. Shocked quartz grains appear more abundant and larger west of the Chicxulub crater, although the absolute size factor may strongly depend on the amount of material available for study. Nevertheless, as proposed by Alvarez et al. (1995) and Bostwick and Kyte (1996), the maximum grain size of quartz

grains with planar deformation features seems larger in the Pacific than at sites located at equivalent distances from the crater in Europe.

7. KTbase demonstrates that a significant effort is needed to improve our knowledge of K-T boundary sites in South America, Africa, Australia, and the high latitudes (>60°).

Preliminary examination of the K-T database in terms of the ejecta debris distribution and K-T boundary sedimentation shows that it is probably the most convenient and user-friendly method to compile and sort the huge amount of literature on the subject. Plotting the current data on paleogeographic maps indicates zones where further studies are required. The occurrence of impact debris must also be reported in a clear and quantitative manner (e.g., abundance per cm^2 of sediments) before the distribution pattern of impact products can be understood.

The main advantage of the database is to point toward potential trends or characteristics in the data, which can then be investigated further. The information extracted from the database coupled with mathematical models will permit the documentation of the origin, transport, and deposition of ejecta material during cratering events. It is our goal to have the ejecta debris part of the database available through the internet in the near future.

ACKNOWLEDGMENTS

Kiessling was supported by a postdoctoral fellowship of the Graduate Program "Evolutionary Transformations and Mass Extinctions" held at the Museum of Natural History in Berlin. We thank Henning Dypvik and an anonymous reviewer for corrections and suggestions that significantly improved this manuscript.

REFERENCES CITED

Acton, G.D., Galbrun, B., and King, J.W., 2000, Paleolatitude of the Caribbean Plate since the Late Cretaceous, *in* Leckie, R.M., Sigurdsson, H., Acton, G.D., and Draper, G., eds., Proceedings of the Ocean Drilling Program, Scientific Results, Volume 165: Washington, D.C., U.S. Government Printing Office, p. 149–173.

Albertao, G.A., Koutsoukos, A.A.M., Regali, M.P.S., Attrep, M., Jr., and Martins, P.P., Jr., 1994, The Cretaceous-Tertiary boundary in southern low-latitude regions: Preliminary study in Pernambuco, northeastern Brazil: Terra Nova, v. 6, p. 366–375.

Alvarez, W., 1996, Trajectories of ballistic ejecta from the Chicxulub crater, *in* Ryder, G., Fastovski, D., and Gartner, S., eds., The Cretaceous-Tertiary event and other catastrophes in Earth history: Geological Society of America Special Paper 307, p. 141–150.

Alvarez, W., Asaro, F., and Montanari, A., 1990, Iridium profile for 10 million years across the Cretaceous-Tertiary boundary at Gubbio (Italy): Science, v. 250, p. 1700–1702.

Alvarez, W., Claeys, P., and Kieffer, S.W., 1995, Emplacement of KT boundary shocked quartz from Chicxulub crater: Science, v. 269, p. 930–935.

Archibald, J.D., 1996a, Dinosaur extinction and the end of an era: What the fossils say, *in* Bottjer, D.J., and Bambach, R.K., eds., Critical moments in paleobiology and earth history series: New York, Columbia University Press, 237 p.

Archibald, J.D., 1996b, Testing extinction theories at the Cretaceous-Tertiary boundary using the vertebrate fossil record, *in* MacLeod, N., and Keller, G., eds., Cretaceous-Tertiary mass extinctions biotic and environmental changes: New York, W.W. Norton & Company, p. 373–397.

Baum, G.R., and Vail, P.P., 1988, Sequence stratigraphic concepts applied to Paleogene outcrops, Gulf and Atlantic basins, *in* Wilgus, C.K., Kendall, C.G.S.C., Posamentier, H.W., and Ross, C.A., eds., Sea-level changes: An integrated approach: Society of Economic Paleontologists and Mineralogists Special Publication, v. 42 p. 309–327.

Blum, J.D., Chamberlain, C.P., Hingston, M.P., Koeberl, C., Marin, L.E., Schuraytz, B.C., and Sharpton, V.L., 1993, Isotopic comparison of K/T boundary impact glasses with melt rock from the Chicxulub and Manson impact structures: Nature, v. 364, p. 325–327.

Bohor, B.F., 1990, Shocked quartz and more: Impact signatures in Cretaceous/Tertiary boundary clays, *in* Sharpton, V.L., and Ward, P.D., eds., Global catastrophes in Earth history: Geological Society of America Special Paper 247, p. 335–342.

Bohor, B.F., 1996, A sediment gravity flow hypothesis for siliciclastic units at the K/T boundary, Northeastern Mexico, *in* Ryder, G., Fastovski, D., and Gartner, S., eds., The Cretaceous-Tertiary event and other catastrophes in Earth history: Geological Society of America Special Paper 307 p. 183–196.

Bohor, B.F., Foord, E.E., Modreski, P.J., and Triplehorn, D.M., 1984, Mineralogic evidence for an impact event at the Cretaceous-Tertiary boundary: Science, v. 224, p. 867–869.

Bohor, B.F., and Izett, G.A., 1986, Worldwide size distribution of shocked quartz at the K/T boundary: Evidence for a North American impact site [abs.]: Lunar and Planetary Science, v. 17, p. 68–69.

Bohor, B.F., Modreski, P.J., and Foord, E.E., 1987, Shocked quartz in the Cretaceous-Tertiary boundary clays: Evidence for a global distribution: Science, v. 236, p. 705–709.

Bostwick, J.A., and Kyte, F.T., 1996, The size and abundance of shocked quartz in Cretaceous-Tertiary boundary sediments from the Pacific basin, *in* Ryder, G., Fastovski, D., and Gartner, S., eds., The Cretaceous-Tertiary event and other catastrophes in Earth history: Geological Society of America Special Paper 307, p. 403–416.

Bourgeois, J., Hansen, T.A., Wiberg, P.L., and Kauffman, E.G., 1988, A tsunami deposit at the Cretaceous-Tertiary boundary in Texas: Science, v. 241, p. 567–570.

Bralower, T.J., Paull, C.K., and Leckie, R.M., 1998, The Cretaceous-Tertiary boundary cocktail: Chicxulub impact triggers margin collapse and extensive sediment gravity flows: Geology, v. 26, p. 331–334.

Claeys, P., Montanari, A., Smit, J., Alvarez, W., and Vega, F., 1996, KT boundary megabreccias in southern Mexico: Geological Society of America Abstracts with Programs, v. 28, no. 7, p. A181.

Claeys, P., Grajales-Nishimura, J.M., and Salge, T., 2000, Variability and distribution of the Chicxulub suevite [abs.]: Eos (Transactions, American Geophysical Union), v. 81, p. 801.

Covey, C., Thompson, S.L., Weissman, P.R., and Maccracken, M.C., 1994, Global climatic effects of atmospheric dust from an asteroid or comet impact on Earth: Global and Planetary Change, v. 9, p. 263–273.

Donovan, A.D., Baum, G.R., Blechschmidt, G.L., Loutit, T.S., Pflum, C.E., and Vail, P.R., 1988, Sequence stratigraphic setting of the Cretaceous-Tertiary boundary in central Alabama, *in* Wilgus, C.K., Kendall, C.G.S.C., Posamentier, H.W., and Ross, C.A., eds., Sea-level changes: An integrated approach: Society of Economic Paleontologists and Mineralogists Special Publication, v. 42, p. 299–307.

Evans, N.J., Gregoire, D.C., Goodfellow, W.D., McInnes, B.I., Miles, N., and Veizer, J., 1993a, Ru/Ir ratios at the Cretaceous-Tertiary boundary: Implications for PGE source and fractionation within the ejecta cloud: Geochimica et Cosmochimica Acta, v. 57, p. 3149–3158.

Evans, N.J., Gregoire, D.C., Grieve, R.A.F., Goodfellow, W.D., and Veizer, J., 1993b, Use of platinum-group elements for impactor identification: Terrestrial impact craters and Cretaceous-Tertiary boundary: Geochimica et Cosmochimica Acta, v. 57, p. 3737–3748.

Fourcade, E., Alonzo, M., Barrillas, M., Bellier, J.-P., Bonneau, M., Cosillo, A., Cros, P., Debrabant, P., Gardin, S., Masure, E., Philip, J., Renard, M., Rocchia, R., and Romero, J., 1997, La limite Crétacé/Tertiaire dans le Sud-Ouest du Petén (Guatemala): Earth and Planetary Sciences, v. 325, p. 57–64.

Fourcade, E., Rocchia, R., Gardin, S., Bellier, J.-P., Debrabant, P., Masure, E., Robin, E., and Pop, W.T., 1998, Age of the Guatemala breccias around the Cretaceous-Tertiary boundary: Relationships with the asteroid impact on the Yucatan: Earth and Planetary Science Letters, v. 327, p. 47–53.

Grajales-Nishimura, J.M., Cedillo-Pardo, E., Rosales-Dominguez, C., Moràn-Zenteno, D.J., Alvarez, W., Claeys, P., Ruiz-Morales, J., Garcia-Hermandez, J., Padilla-Avila, P., and Sànchez-Rios, A., 2000, Chicxulub impact: The origin of reservoir and seal facies in the southeastern Mexico oil fields: Geology, v. 28, p. 307–310.

Habib, D., Moshkovitz, S., and Kramer, C., 1992, Dinoflagellate and calcareous nanofossil response to sea-level change in Cretaceous-Tertiary boundary sections: Geology, v. 20, p. 164–168.

Haq, B.U., Hardenbol, J., and Vail, P.R., 1988, Mesozoic and Cenozoic chronostratigraphy and cycles of sea-level change, *in* Wilgus, C.K., Hastings, B.S., Kendall, C.G.S.C., Posamentier, H.W., and Ross, C.A., eds., Sea-level changes: An integrated approach: Society of Economic Paleontologists and Mineralogists Special Publication, v. 42, p. 71–108.

Hardenbol, J., Thierry, J., Farley, M.B., Jacquin, T., de Graciansky, P.-C., and Vail, P.R., 1998, Mesozoic and Cenozoic sequence chronostratigraphic framework of European basins, *in* de Graciansky, P.-C., Hardenbol, J., Jacquin, T., and Vail, P.R., eds., Mesozoic and Cenozoic sequence stratigraphy of European basins: SEPM (Society for Sedimentary Geology) Special Publications, v. 60, p. 3–13.

Helenes, J., and Somoza, D., 1999, Palynology and sequence stratigraphy of the Cretaceous of eastern Venezuela: Cretaceous Research, v. 20, p. 447–463.

Hörz, F., 1982, Ejecta of the Ries Crater, Germany, *in* Silver, L.T., and Schultz, P.H., eds., Geological implications of impacts of large asteroids and comets on Earth: Geological Society of America Special Paper 190, p. 39–55.

Hörz, F., Ostertag, R., and Rainy, D.A., 1983, Bunte Breccia of the Ries: Continuous deposits of large impact craters: Reviews of Geophysics and Space Physics, v. 21, p. 1667–1725.

Izett, G.A., 1990, The Cretaceous/Tertiary boundary interval, Raton Basin, Colorado and New Mexico, and its content of shock-metamorphosed minerals: Evidence relevant to the K/T boundary impact-extinction theory: Geological Society of America Special Paper 249, p. 1–100.

Izett, G.A., Maurrasse, F.J.-M.R., Lichte, F.E., Meeker, G.P., and Bates, R., 1990, Tektites in Cretaceous-Tertiary boundary rocks on Haiti: U.S. Geological Survey Open-File Report, v. 90-635, p. 1–31.

Kamo, S.L., and Krogh, T.E., 1995, Chicxulub crater source for shocked zircon crystals from the Cretaceous-Tertiary boundary layer, Sasketchewan: Evidence from new U-Pb data: Geology, v. 23, p. 281–284.

Keller, G., and Stinnesbeck, W., 1996, Near-K/T age of elastic deposits from Texas to Brazil: Impact, volcanism and/or sea-level lowstand: Terra Nova, v. 8, p. 277–285.

Keller, G., Adatte, T., Stinnesbeck, W., Stüben, D., Kramar, U., Berner, Z., Li, L., and Perch-Nielsen, K.V.S., 1998, The Cretaceous-Tertiary transition on the shallow Saharan Platform of Southern Tunisia: Geobios, v. 30, p. 951–975.

Kiessling W., and Claeys, P., 2001, A geographic database to the KT boundary, *in* Buffetaut, E., and Koeberl, C., eds., Geological and biological effects of impact events: Berlin, Springer-Verlag, Lecture Notes in Earth Sciences, p. 33–140.

Kiyokawa, S., Tada, R., Matsui, T., Tajika, E., Takayama, H., and Iturralde-Vinent, M.A., 1999, Extraordinary thick K/T boundary sequence, Cacarajicara Formation, Western Cuba [abs.]: Lunar and Planetary Science, v. 30, Abstract #1577, CD-ROM.

Kiyokawa, S., Tada, R., Oji, T., Tajika, E., Nakano, Y., Goto, K., Yamamoto, S., Takayama, H., Toyoda, K., Rojas, R., Garcia, D., Iturralde-Vinent, M.A., and Matsui, T., 2000, More than 500 m thick K/T boundary sequence: Cacarajicara formation Western Cuba. Impact related giant flow deposit [abs.] *in* Catastrophic events and mass extinctions: Impacts and beyond: Houston, Texas, Lunar and Planetary Institute, LPI Contribution No. 1053, p. 100–101.

Klaus, A., Norris, R.D., Smit, J., Kroon, D., Martinez-Ruiz, F., and LEG 171B Scientific Party, 1997, Impact-induced K-T boundary mass wasting across the Blake Nose, W. North Atlantic: Evidence from seismic reflection and core data [abs.]: Eos (Transactions, American Geophysical Union), v. 78, p. 371.

Krogh, T.E., Kamo, S.L., Sharpton, V.L., Marin, L.E., and Hildebrand, A.R., 1993, U-Pb ages of single shocked zircons linking distal K/T ejecta to the Chicxulub crater: Nature, v. 366, p. 731–734.

Kruge, M.A., Stankiewicz, B.A., Crelling, J.C., Montanari, A., and Bensley, D.F., 1994, Fossil charcoal in Cretaceous-Tertiary boundary strata: Evidence for catastrophic firestorm and megawave: Geochimica and Cosmochimica Acta, v. 58, p. 1393–1397.

Kyte, F.T., Bostwick, J.A., and Zhou, L., 1996, The Cretaceous-Tertiary boundary on the Pacific plate: Composition and distribution of impact debris, *in* Ryder, G., Fastovski, D., and Gartner, S., eds., The Cretaceous-Tertiary event and other catastrophes in Earth history: Geological Society of America Special Paper 307, p. 389–402.

Kyte, F.T., Zhou, Z., and Wasson, J.T., 1980, Siderophile-enriched sediments from the Cretaceous-Tertiary boundary: Nature, v. 288, p. 651–656.

Leroux, H., Rocchia, R., Froget, L., Orue-Etxebarria, X., Doukhan, J.-C., and Robin, E., 1995, The K/T boundary at Beloc (Haiti): Compared stratigraphic distributions of the boundary markers: Earth and Planetary Science Letters, v. 131, p. 255–268.

López Ramos, E., 1975, Geological summary of the Yucatán Peninsula, *in* Nairn, A.E.M., and Stehli, F.G., eds., The Gulf of Mexico and the Caribbean. 3. The Ocean Basins and Margins: New York, Plenum, p. 257–282.

Maurrasse, F.J.-M.R., and Sen, G., 1991, Impacts, tsunamis, and the Haitian Cretaceous-Tertiary boundary layer: Science, v. 252, p. 1690–1693.

Montanari, A., 1991, Authigenesis of impact spheroids in the K/T boundary clay from Italy: New constraints for high-resolution stratigraphy of terminal Cretaceous events: Journal of Sedimentary Petrology, v. 61, p. 315–339.

Montanari, A., Claeys, P., Asaro, F., Bermudez, J., and Smit, J., 1994, Preliminary stratigraphy and iridium and other geochemical anomalies across the KT boundary in the Bochil section (Chiapas, Southeastern Mexico) [abs.], *in* New developments regarding the KT event and other catastrophes in Earth history: Houston, Texas, Lunar and Planetary Institute, Contribution No. 825, p. 84–85.

Norris, R.D., Firth, J., Blusztajn, J.S., and Ravizza, G., 2000, Mass failure of the North Atlantic margin triggered by the Cretaceous/Paleocene bolide impact [abs.], *in* Catastrophic events and mass extinctions: Impact and beyond, Vienna, Austria: Houston, Texas, Lunar and Planetary Institute, Contribution No. 1053, p. 152–153.

Oberbeck, V.R., 1975, The role of ballistic erosion and sedimentation in lunar startigraphy: Reviews of Geophysics and Space Physics, v. 13, p. 337–362.

Ocampo, A.C., Pope, K.O., and Fisher, A.G., 1996, Ejecta blanket deposits of the Chicxulub crater from Albion island, Belize, *in* Ryder, G., Fastovski. D., and Gartner, S., eds., The Cretaceous-Tertiary event and other catastrophes in Earth history: Geological Society of America Special Paper 307, p. 75–88.

O'Keefe, J.D., and Ahrens, T.J., 1991, Tsunamis from giant impacts on solid planets [abs.]: Lunar and Planetary Science, v. 21, p. 997–998.

Olsson, R.K., Miller, K.G., Browning, J.V., Habib, D., and Sugarman, P.J.,

1997, Ejecta layer at the Cretaceous-Tertiary boundary, Bass River, New Jersey (Ocean Drilling Program Leg 174AX): Geology, v. 25, p. 759–762.

Pardo, A., Adatte, T., Keller, G., and Oberhänsli, H., 1999, Paleoenvironmental changes across the Cretaceous-Tertiary boundary at Koshak, Kazakhstan, based on planktic foraminifera and clay mineralogy: Palaeogeography, Palaeoclimatology, Palaeoecology, v. 154, p. 247–273.

Pillmore, C.L., and Flores, R.M., 1987, Stratigraphy and depositional environments of the Cretaceous-Tertiary boundary clay and associated rocks, Raton basin, New Mexico and Colorado, *in* Fassett, J.E., and Rigby, J.K., eds., The Cretaceous-Tertiary boundary in the San Juan and Raton Basins, New Mexico and Colorado: Geological Society of America Special Paper 209, p. 111–130.

Pindell, J.L., 1994, Evolution of the Gulf of Mexico and the Caribbean, *in* Donovan, S.K., and Jackson, T.A, eds., Caribbean geology: An introduction: Kingston, Jamaica, University of the West Indies Publishers Association, p. 13–39.

Pitakpaivan, K., Byerly, G.R., and Hazel, J.E., 1994, Pseudomorphs of impact spherules from the Cretaceous-Tertiary boundary section at Shell Creek, Alabama: Earth and Planetary Science Letters, v. 124, p. 49–56.

Pope, K.O., Ocampo, A.C., Fisher, A.G., Alvarez, W., Fouke, B.W., Webster, C.L., Vega, F.J., Smit, J., Fritsche, A.E., and Claeys, P., 1999, Chicxulub impact ejecta from Albion Island, Belize: Earth and Planetary Science Letters, v. 170, p. 351–364.

Pospichal, J.J., 1996, Calcareous nannoplankton mass-extinctions at the Cretaceous-Tertiary boundary, *in* Ryder, G., Fastovski, D., and Gartner, S., eds., The Cretaceous-Tertiary event and other catastrophes in Earth history: Geological Society of America Special Paper 307, p. 335–360.

Robin, E., Boclet, D., Bonte, P., Froget, L., Jehanno, C., and Rocchia, R., 1991, The stratigraphic distribution of Ni-rich spinels in Cretaceous-Tertiary boundary rocks at El-Kef (Tunisia), Caravaca (Spain) and Hole-761C (Leg-122): Earth and Planetary Science Letters, v. 107, p. 715–721.

Schmitz, B., 1988, Origin of microlayering in worldwide distributed Ir-rich marine Cretaceous Tertiary boundary clays: Geology, v. 16, p. 1068–1072.

Schmitz, B., Kyte, F.T., Bohor, B.F., and Triplehorn, D.M., 1990, Comments and replies on "Origin of microlayering in worldwide distributed Ir-rich marine Cretaceous/Tertiary boundary clays": Geology, v. 18, p. 87–94.

Schmitz, B., Keller, G., and Stenvall, O., 1992, Stable isotope and foraminiferal changes across the Cretaceous-Tertiary boundary at Stevns Klint, Denmark: Arguments for long-term oceanic instability before and after bolide-impact event: Palaeogeography, Palaeoclimatology, Palaeoecology, v. 96, p. 233–260.

Schultz, P.H., and D'hondt, S., 1996, Cretaceous-Tertiary (Chicxulub) impact angle and its consequences: Geology, v. 24, p. 963–967.

Scotese, C.R., and Golonka, J., 1992, PALEOMAP Paleogeographic Atlas: Arlington, Texas, University of Texas at Arlington Publication, PALEOMAP Progress Report, v. 20, 34 p.

Sigurdsson, H., D'Hondt, S., Arthur, M.A., Bralower, T.J., Zachos, J.C., Van Fossen, M., and Channell, J.E.T., 1991, Glass from the Cretaceous/Tertiary boundary in Haiti: Nature, v. 349, p. 482–487.

Smit, J., 1999, The global stratigraphy of the Cretaceous-Tertiary boundary impact ejecta: Annual Review of Earth and Planetary Sciences, v. 27, p. 75–113.

Smit, J., Montanari, A., Swinburne, N.H.M., Alvarez, W., Hildebrand, A.R., Margolis, S.V., Claeys, P., Lowrie, W., and Asaro, F., 1992, Tektite-bearing, deep-water clastic unit at the Cretaceous-Tertiary boundary in northeastern Mexico: Geology, v. 20, p. 99–103.

Smit, J., Rocchia, R., Robin, E., and the ODP 171B shipboard party, 1997, Preliminary iridium analysis from a graded spherule layer at the K/T boundary and Late Eocene ejecta from ODP sites 1049, 1052, 1053 Blake Nose, Florida: Geological Society of America Abstracts with Programs, v. 29, no. 6, p. A141.

Smit, J., Roep, T.B., Alvarez, W., Montanari, A., Claeys, P., Grajales Nishimura, J.M., and Bermudez, J., 1996, Coarse-grained, clastic sandstone complex at the KT boundary around the Gulf of Mexico: Deposition by tsunami waves induced by the Chicxulub impact, *in* Ryder, G., Fastovsky, D., and Gartner, S., eds., The Cretaceous-Tertiary boundary event and other catastrophes in Earth history: Geological Society of America Special Paper 307, p. 151–182.

Speijer, R.P., and van der Zwaan, G.T., 1996, Extinction and survivorship of southern Tethyan benthic foraminifera across the Cretaceous/Paleocene boundary, *in* Hart, M.B., ed., Biotic recovery from mass extinction events: Geological Society [London] Special Publication 102, p. 343–371.

Stinnesbeck, W., Barbarin, J.M., Keller, G., Lopez-Oliva, J.G., Pivnik, D.A., Lyons, J.B., Officer, C.B., Adatte, T., Graup, G., Rocchia, R., and Robin, E., 1993, Deposition of channel deposits near the Cretaceous-Tertiary boundary in northeastern Mexico: Catastrophic or "normal" sedimentary deposits?: Geology, v. 21, p. 797–800.

Swisher, C.C., Grajales-Nishimura, J.M., Montanari, A., Margolis, S.V., Claeys, P., Alvarez, W., Renne, P., Cedillo-Pardo, E., Maurrasse, F.J.-M.R., Curtis, G.H., Smit, J., and McWilliams, M.O., 1992, Coeval ^{40}Ar/^{39}Ar ages of 65.0 million years ago from Chicxulub Crater melt rock and Cretaceous-Tertiary boundary tektites: Science, v. 257, p. 954–958.

Tada, R., Matsui, T., Iturralde-Vinent, M.A., Oji, T., Tajika, E., Kiyokawa, S., Garcia, D., Okada, H., Hasegawa, T., and Toyoda, K., 1999, Origin of a giant event deposit in northwestern Cuba and its relation to K/T boundary impact [abs.]: Lunar and Planetary Science, v. 30, Abstract #1534, CD-ROM.

Tada, R., Takayama, H., Iturralde Vinent, M., Tajika, E., Oji, T., Kiyokawa, S., Garcia, D., Okada, H., and Toyoda K., 2000, A giant tsunami deposit at Cretaceous-Tertiary boundary in Cuba [abs.], *in* Catastrophic events and mass extinctions: Impact and beyond: Lunar and Planetary Institute Contribution, no. 1023, p. 226–227.

Toon, O.B., Pollack, J.B., Ackerman, T.P., Turco, R.P., McKay, C.P., and Liu, M.S., 1982, Evolution of an impact-generated dust cloud and its effects on the atmosphere, *in* Silver, L.T., and Schultz, P.H., eds., Geological implications of impacts of large asteroids and comets on Earth: Geological Society of America Special Paper 190, p. 187–200.

Urrutia-Fucugauchi, J., Marin, L., and Trejo-Garcia, A., 1996, UNAM Scientific drilling program of Chicxulub impact structure: Evidence for a 300 kilometer crater diameter: Geophysical Research Letters, v. 23, p. 1565–1568.

MANUSCRIPT SUBMITTED NOVEMBER 28, 2000; ACCEPTED BY THE SOCIETY MARCH 22, 2001

Printed in the U.S.A.

Geological Society of America
Special Paper 356
2002

Generation and propagation of a tsunami from the Cretaceous-Tertiary impact event

T. Matsui*
Department of Complexity Science and Engineering, Graduate School of Frontier Science, University of Tokyo, 7-3-1 Hongo, Bunkyo-ku, Tokyo 113-0033, Japan
F. Imamura*
Disaster Control Research Center, Graduate School of Engineering, Tohoku University, Aoba 06, Sendai 980-8579, Japan
E. Tajika*
Y. Nakano*
Department of Earth and Planetary Science, Graduate School of Science, University of Tokyo, 7-3-1 Hongo, Bunkyo-ku, Tokyo 113-0033, Japan
Y. Fujisawa*
Technical Research Institute, Obayashi Corporation, Shimo-kiyose 4-640, Kiyose 204-0011, Tokyo, Japan

ABSTRACT

We have studied the mechanism of tsunami generation by meteorite impact on a shallow ocean at 65 Ma and modeled the propagation of that tsunami in the Gulf of Mexico. We found that the water flow into and out of the crater cavity causes most of the tsunami. The height of the wave coming out of the crater is controlled by the depth of the shallow-water region surrounding the crater. We show that the lower the flow velocity in the shallow-water region, the lower the wave height, and the longer the oscillation period. If the depth of the sea above the Yucatan platform was 200 m at the end of the Maastrichtian, the maximum tsunami wave height and period at the rim of the crater are estimated to be ~50 m and 10 h, respectively.

Using these results, we simulated the propagation of the K-T impact-generated tsunami in the Gulf of Mexico. There are two types of tsunami; the receding wave and the rushing wave. The receding wave traveled across the entire gulf within 10 h of the impact. Tsunamis attacked the coast as a leading negative wave. The rushing wave flowed with a height of more than 200 m and reached the coastal area of North America. It ran up over the land and crossed the Mississippi embayment, a distance of more than 300 km. The averaged runup was more than 150 m, but it reached a height of 300 m near the Rio Grande embayment.

*E-mails: Matsui, matsui@k.u-tokyo.ac.jp; Imamura, imamura@tsunami2.civil.tohoku.ac.jp; Tajika, tajika@eps.s.u-tokyo.ac.jp; Nakano, nakano@sys.eps.s.u-tokyo.ac.jp; Fujisawa, Fujisawa@tri.obayashi.co.jp

Matsui, T., Imamura, F., Tajika, E., Nakano, Y., and Fujisawa, Y., 2002, Generation and propagation of a tsunami from the Cretaceous-Tertiary impact event, *in* Koeberl, C., and MacLeod, K.G., eds., Catastrophic Events and Mass Extinctions: Impacts and Beyond: Boulder, Colorado, Geological Society of America Special Paper 356, p. 69–77.

INTRODUCTION

A gigantic meteorite impact occurred at the Yucatan Peninsula ca. 65 Ma and resulted in the formation of the Chicxulub crater (e.g., Hildebrand et al., 1991); hereafter, we call this the K-T (Cretaceous-Tertiary) impact event. After the discovery of the Chicxulub crater on the Yucatan Peninsula (Hildebrand et al., 1991), studies of the K-T impact event focused on the environmental consequences of this gigantic meteorite impact. One example of such studies concerns the K-T boundary tsunami deposits found in and around the Gulf of Mexico (e.g., Bourgeois et al., 1988; Smit et al., 1996), although their origin is still controversial (e.g., Bohor, 1996). A Japanese-Cuban joint research group reported the discovery of an ~180-m-thick layer in Cuba, which consists of a lower grain flow unit overlain by an upper homogenite unit, the age of which is definitively defined as that of the K-T boundary (Takayama et al., 2000). The homogenite is a thick, normally graded, but otherwise structureless deposit formed by a tsunami in a deep-sea environment (Cita et al., 1996). This discovery supports tsunami generation by the K-T impact.

There are some studies on numerical simulation of a tsunami caused by a small asteroid impact into deep water (e.g., Hills et al., 1994; Crawford and Mader, 1998; Ward and Asphaug, 2000). There has been, however, no study of a tsunami generated by a large asteroid impact into shallow water, as in the case of the K-T impact. Here, we study the generation mechanism of the K-T impact tsunami and its propagation within the Gulf of Mexico.

MODEL

Tsunami generation

The diameter of the impactor that formed the Chicxulub crater has been estimated as ~10 km, based on the total amount of Ir in the K-T boundary layer (Alvarez et al., 1980). The energy released by a 10 km meteorite impact with a velocity of several tens of kilometers/second on the Earth would correspond to a TNT explosion energy of 10^8 to 10^9 Mt. The diameter and depth of the crater formed by this impact are ~180 km and several kilometers, respectively (Morgan et al., 1997; Hildebrand et al., 1998).

In association with this impact, shock waves would have propagated through the atmosphere, ocean, crust, and mantle, and both the seawater and seafloor at the impact point would have been vaporized instantaneously. Shock waves caused by the entry of the meteorite into the atmosphere and ejecta from the crater could have both induced tsunamis. Hereafter we refer to the first of these as the shock-wave-induced tsunami and to the second, i.e., the wave formed at the front of the ejecta curtain, as the rim wave. According to Gault and Sonett (1982), the rim wave is generated and propagates away from the cavity as shown in Figure 1. As a more gradual process, the surrounding seawater may have flowed back into the crater. Then the crater would have been overfilled with water, and a central water column would form above the crater. The height of the water column would be dependent on the depth of the ocean around the crater, because the flow rate of water into the crater from the surrounding ocean is controlled by the water depth. The elevated water column would then collapse and could eventually propagate outward from the crater site as a tsunami.

We also expect the formation of landslides at the margin of the Yucatan platform. The basal unit of the type locality for the Peñalver Formation in Cuba is several tens of meters thick and shows features characteristic of a sediment gravity flow (Takyama et al., 2000). Such gravity flow units are common for the Peñalver Formation at other localities and for the Cacarajicara Formation in Cuba (Kiyokawa et al., 2000). These observations suggest that large-scale landslides occurred at the margin of the Yucatan platform just after the K-T impact.

We consider four stages of tsunami generation by the K-T impact: (1) the shock-wave-induced tsunami associated with high air pressure and wind generated by the passage of the meteorite through the atmosphere, (2) the rim wave formed at the front of the ejecta curtain, (3) the crater-generated tsunami, caused by movement of water to fill and flow out of the crater cavity after crater formation (hereafter called receding and rushing waves, respectively), and (4) the landslide-generated tsu-

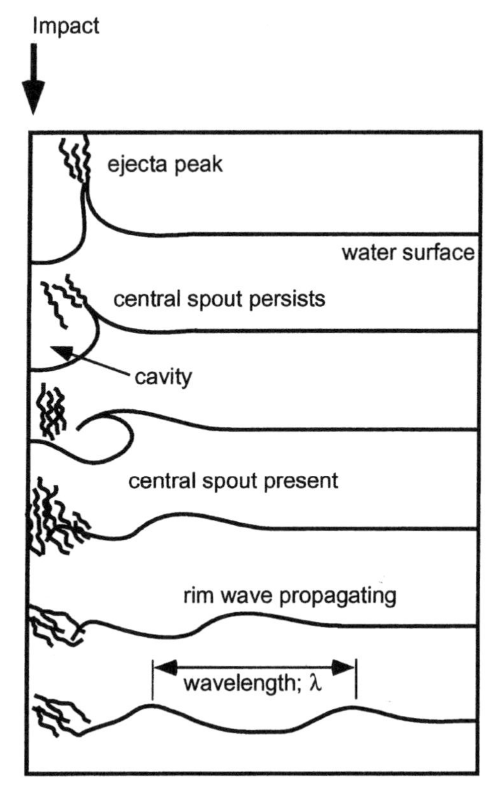

Figure 1. Wave formation sequence in case of impact into shallow water (modified from Gault and Sonett, 1982). Rim wave is generated and propagated away from cavity.

nami, caused by landslides at the margin of the Yucatan platform. The wave generation by the first mechanism is negligible, because waves generated by strong winds are dispersive and rapidly damped during propagation over a long distance. Waves generated by the second mechanism were modeled as the initial condition of the rim wave, wave height and length being estimated using the hydraulic experimental results of Gault and Sonett (1982). Waves generated by the third and fourth mechanisms might have been the most devastating, and were simulated based on shallow-water (nonlinear long wave) theory.

Paleobathymetry

In order to simulate tsunami generation and propagation due to the K-T impact, we assumed a probable paleobathymetry of the region at the time of the K-T impact. As shown herein, the shape of the crater does not have a significant effect on the generation and propagation of tsunami. Hence, we simply assume that the crater shape just after the impact was almost like the present shape of the Chicxulub crater (e.g., Morgan et al., 1997; Hildebrand, 1997). We also assume that the Yucatan Peninsula was covered with shallow water (<200 m deep) at the end of the Maastrichtian (Sohl et al., 1991). For the reconstruction of paleobathymetry of the Gulf of Mexico and the Caribbean region, we adopted the tectonic reconstruction model by Ross and Scotese (1988), modified slightly based on the field work in Cuba by Takayama et al. (2000). In Figure 2 we show the reconstructed ocean-floor geometry in the Gulf of Mexico (after Ross and Scotese, 1988) and the location of the Chicxulub crater. A cross section along the line A-A′ in Figure 2 is shown in Figure 3.

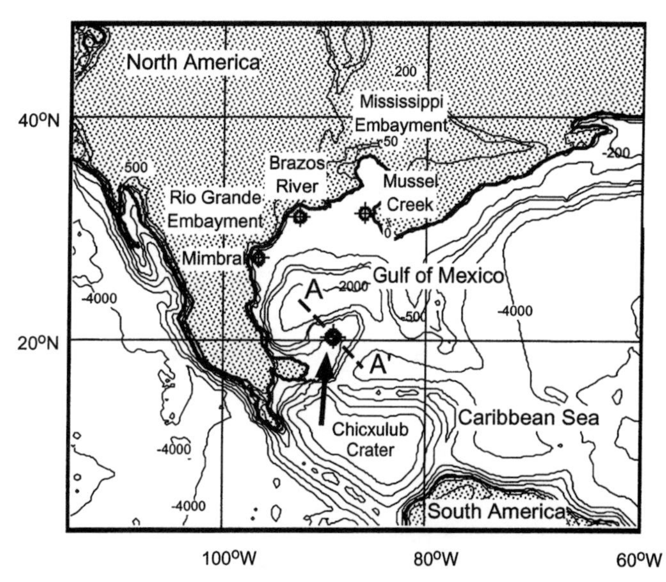

Figure 2. Reconstructed geometry of Gulf of Mexico at time of Cretaceous-Tertiary (K-T) impact, based on model by Ross and Scotese (1988). Locations of Chicxulub crater and some possible K-T boundary tsunami deposits around Gulf of Mexico are also shown.

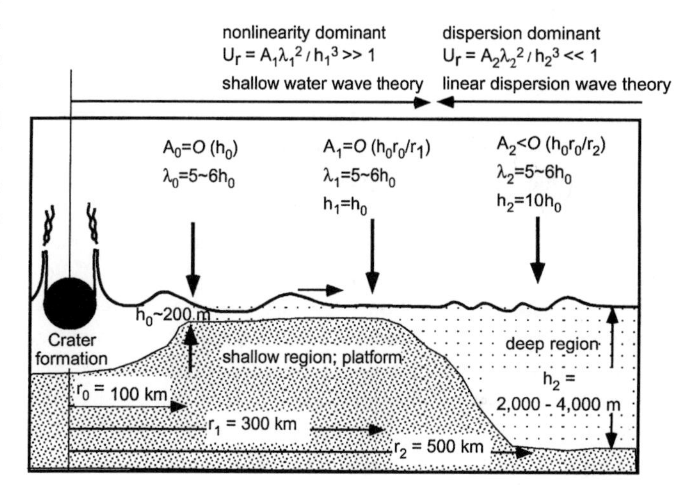

Figure 3. Cross section of topography of Chicxulub crater and Yucatan platform along line A-A′ in Figure 2. Schematic view of propagation of rim wave in near and far field and order estimate of wave amplitude are also shown. *A*, λ, *h*, *r*, and Ur represent amplitude, wavelength, depth, distance, and Ursell parameter, respectively.

NUMERICAL PROCEDURE

We used the nonlinear long-wave theory (e.g., Imamura, 1996) to simulate the movement of water on the Yucatan platform and in the Gulf of Mexico. In this section we describe briefly the basic equations and numerical procedures used in this study.

Equations for tsunami generation and propagation

We applied the nonlinear long-wave theory integrated over a layer with nonhorizontal bottom and interface as the governing equation for the numerical model. We assume hydrostatic pressure distribution, and uniform density and velocity distributions in each layer. For tsunami generation by landslide, the seawater is modeled as an upper layer for wave generation and propagation, and the landslide is modeled as a lower layer in Figure 4. The Navier-Stokes equations of mass and momentum continuity are integrated in each layer, with the kinetic and dynamic conditions at a free surface and an interface (Imamura and Imteaz, 1995).

The governing equations of the upper layer for a one-dimensional problem are expressed as follows. The equation of the mass conservation is given by:

$$\frac{\partial(\eta_1 - \eta_2)}{\partial t} + \frac{\partial M_1}{\partial x} = 0, \tag{1}$$

where x is the horizontal coordinate, t is time, η_1 is the difference between the water surface and the still water depth, η_2 is the vertical displacement of the bottom, and subscripts 1 and 2 indicate the upper and lower layers, respectively (for details, see Fig. 4). M is the discharge in the x direction given by the integration of velocity over a layer, which is defined by:

$$M = \int_{-h}^{\eta} u dz. \quad (2)$$

The momentum equation is given by:

$$\frac{\partial M_1}{\partial t} + \frac{\partial}{\partial x}\left(\frac{M_1^2}{D_1}\right) + gD_1 \frac{\partial \eta_1}{\partial x} - L_x = 0, \quad (3)$$

where L_x is the drag force induced by a lower layer at the flow front, and is defined by:

$$L_x = \frac{1}{2} C_D \frac{1}{h} \frac{A_z}{A_c} \left(\frac{M}{D} - V\right)\left|\frac{M}{D} - V\right|, \quad (4)$$

where D is the total depth ($= h + \eta$), h is the still water depth, A_z is the projecting area of a lower layer flow at the flow front, A_c is the grid cell area of the water at the flow front, C_D is the drag-force coefficient (as proposed by Raney and Bulter, 1976), and V is the velocity of sea-bottom displacement.

Landslide in a lower layer

We can assume two stages for a landslide; slumping and debris flows. The first stage, slumping, is modeled by a circular arc failure along a shear plane, similar to the model by Okusa and Yoshimura (1981) for displacement of the sea bottom under an oceanic wave. There are two types of slumping, toe slip and base slip. It is usually assumed that the surface of failure has a circular arc profile for both types (Imamura and Gica, 1996). In this study we use base slip for the surface of failure. Initiation of motion occurs when a sudden ground quake or external disturbance alters the balance of the internal resisting force and driving force. For the case of base slip, we can neglect the resisting force. If these conditions are satisfied, the tongue at the front of a landslide would start to move and flow down to the slope.

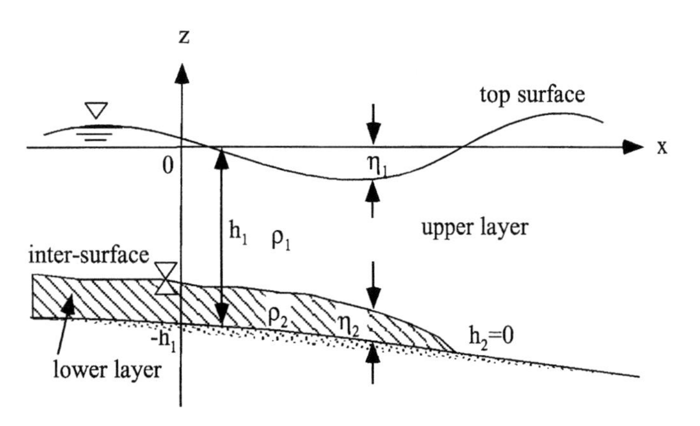

Figure 4. Two-layer model for movement of water and landslide (see text for discussion).

When landslide acceleration becomes negligibly small, debris flows become important. We use a different model of a lower layer for this stage. Here, the mass conservation equation for the lower layer is given by:

$$\frac{\partial \eta_2}{\partial t} + \frac{\partial M_2}{\partial x} = 0, \quad (5)$$

and the momentum equation becomes

$$\frac{\partial M_2}{\partial t} + \frac{\partial}{\partial x}\left(\frac{M_2^2}{D_2}\right) + gD_2 \left\{\alpha\left(\frac{\partial \eta_1}{\partial x} + \frac{\partial h_1}{\partial x} - \frac{\partial \eta_2}{\partial x}\right) + \frac{\partial \eta_2}{\partial x} - \frac{\partial h_1}{\partial x}\right\} + g\eta_2(\alpha - 1)\frac{\partial h_1}{\partial x} + R_x + L_x = 0, \quad (6)$$

where

$$R_x = \frac{g\left(\frac{M}{D} - V\right)\left|\frac{M}{D} - V\right|}{C^2(h + \eta)}, \quad (7)$$

wherein ρ is the fluid density, $\alpha = \rho_1/\rho_2$, R_x is the bottom friction, and C is Chezy's roughness coefficient; other variables are as described for equations 3 and 4.

In addition, we considered two nonlinear interactions between the upper and lower layers. One is a change of η_1 due to η_2 in equation 1 for the upper layer. The other is a gravitational force caused by the upper layer, which affects the lower layer (see equation 6).

Numerical scheme and computational condition

We used the staggered leap-frog scheme for the linear terms and the up-wind scheme for the nonlinear terms in this study. The leap-frog scheme is a central difference scheme with a truncation error of the second order. In order to set the boundary conditions easily, the leap-frog scheme assumes that the computation point for the water surface does not coincide with the point for discharge and is therefore shifted by a half-time step and spatial grid. This is called the staggered scheme. Although the accuracy of the up-wind difference scheme is lower than that of the leap-frog scheme, we used it to make the computation stable for nonlinear terms. By using these schemes, we can resolve the water surface and discharge in upper and lower layers explicitly for each time step, reducing the CPU time in the computation. The computational region covered the Gulf of Mexico and the Caribbean Sea, as shown in Figure 2; the data were digitized with a spatial resolution of 2.5 km. The time step was changed to satisfy the stability condition, as explained in the following section.

The condition of the tsunami front is important for estimating the runup height and inundation area. In this study we used the moving boundary condition by employing the staggered leap-frog scheme, in which a water level and discharge are alternatively calculated. Assuming that a water level is already computed at each computation cell, we can compare the levels of the water surface and the bottom of the next landward cell. If the water level is higher than the latter, the water may flow into the lowland cell, meaning that a runup of tsunami proceeds.

Stability condition

Nonlinear terms in the momentum equation and the interactions between the two layers make it difficult to derive a stability condition using the Von Neumann method (e.g., Imamura, 1996). Moreover, the Courant-Friedrichs-Lewys (CFL) condition, which is normally used in the numerical scheme for wave propagation, is not directly applicable to the present case because a representative wave celerity (the phase velocity) cannot be uniquely determined. There are two celerities for progressive and reflected waves derived from the governing equation for two-layer flow. The analytical solution for linearized governing equations provides $C_1 = \sqrt{gh_1(1 + \alpha\beta)}$ and $C_2 = \sqrt{gh_2(1 - \alpha)/(1 + \alpha\beta)}$, where $\alpha = \rho_1/\rho_2$ and $\beta = h_2/h_1$. A stability condition is determined by selecting some arbitrary spatial grid interval Δx and a time step Δt. According to the above solution, the model is shown to be stable up to a certain limit of $\Delta x/\Delta t$, which varies with α and β. For example, in the case of $\alpha = 0.5$ and $\beta = 4.0$, the celerity of the upper layer C_2 controls the stability criteria. However, for the case of $\alpha = 0.4$ and $\beta = 1.0$, C_1 controls the stability criteria. We cannot determine a stability condition a priori. At each numerical step, we need to compare the maximum value of C_1 and C_2 in order to derive the stability condition.

NUMERICAL RESULTS

We discuss, in succession, the behavior of the rim wave, the tsunami caused by the flow into and out of the crater (the crater-generated tsunami), and the propagation of the crater-generated tsunamis and landslide-generated tsunamis in the Gulf of Mexico.

Tsunami generation

Behavior of the rim wave. According to breaking wave criteria, the amplitude of the rim wave is estimated to be on the order of the mean shelf water depth (Dean and Dalrymple, 1991). In the near field for a centered wave system, the higher frequency components will be dispersed backward and the wave amplitude will decrease with distance traveled due to turbulence and bottom friction. In addition, the wave amplitude will also decrease as a function of $1/r$ (r is the distance from the center of the crater). We can use the Ursell parameter ($= A\lambda^2/h^3$, where A is amplitude is λ, wave length, and h is water depth) to estimate the importance of nonlinearity (the second term in equation 3) relative to dispersion effects (the fourth term in equation 3). As shown in Figure 3, the values for the Ursell parameters are O(10) at the edge of the platform and O(1/10) in the deep water. This indicates that the effect of nonlinearity is dominant on the platform, whereas the dispersion effect becomes significant in deep water. It further suggests that breaking in shallow water does not amplify the wave height. The amplitude of the rim wave would be ~10 m at the edge of the platform and of several meters in deep water 500 km from the impact center.

Receding and rushing waves. We used nonlinear long-wave theory with dispersion effects to simulate the movement of water on the Yucatan platform. As shown in Figure 3, the water depth above the Yucatan platform around the crater was much shallower than the crater depth. Therefore, the crater shape does not strongly affect tsunami generation; i.e., the flow rate into crater cavity does not depend on the depth of the crater, but depends on water depth of the Yucatan platform. In Figure 5, the results of a numerical simulation of water movement in and around the crater are shown. This numerical simulation demonstrates that the water movement generates the receding and rushing waves. The water that flowed into the crater cavity accumulates to the point where the crater cavity is overfilled, which then generates the rushing wave outward as this accumulation collapses. In Figure 6, we show the maximum height of the water levels at the center and rim of the crater, and the period of oscillation of the water movement. The maximum height achieved by the water column within the crater is dependent on the water depth above the Yucatan platform. We found that the amplitude and oscillation period of the wave going out of the crater is controlled by the depth of the shallow-water region surrounding the crater. The lower the inward flow velocity in the shallow-water region, the lower the wave height of the rushing wave and the longer the oscillation period. If the water depth of the Yucatan platform was 200 m at the end of the Maastrichtian, the wave height and period at the edge of the crater are estimated as 50 m and 10 h, respectively. We modeled the numerical simulation for the cases with different crater depths. As shown in Figure 6, the depth of the crater affects the period of the crater-generated tsunami, but affects minimally the wave height.

Tsunami propagation

Crater-generated tsunami. We obtained the wave amplitude and period of receding and rushing waves as a function of the crater size and the water depth around the crater (Fig. 6). Using these results as initial conditions, we simulated propagation of the tsunami across the Gulf of Mexico followed by coastal runup.

Numerical calculations indicate that the coastal region of North America was attacked by two types of tsunamis; a receding wave and a rushing wave (Fig. 7). No measurable wave was found ahead of the receding wave, which suggests that the rim wave must have dispersed quickly during propagation to the shore. The receding wave traveled across the entire gulf within 10 h after the impact. The Mimbral and Brazos River

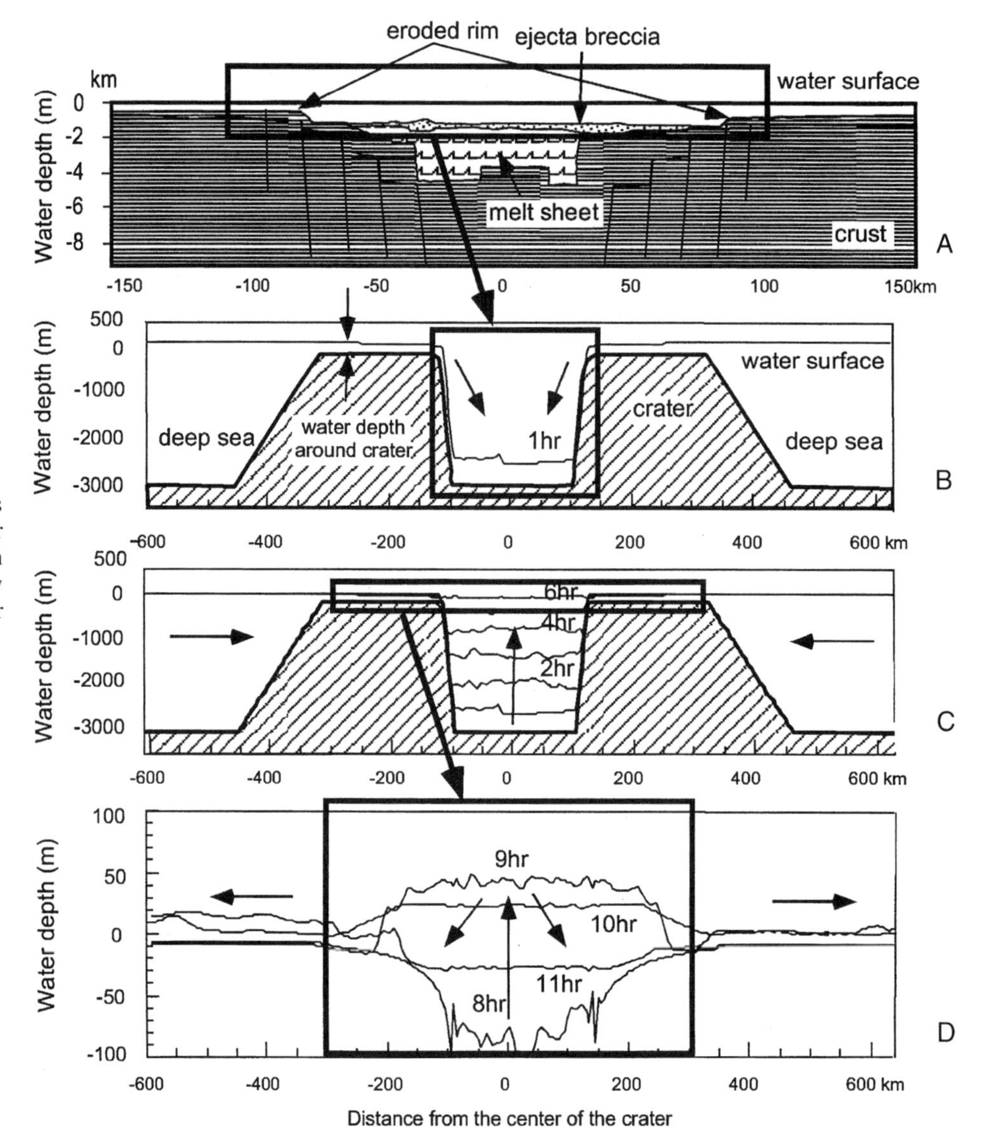

Figure 5. Time series of water mass movement flowing into and out of crater cavity. Because shallow-water region around crater strongly reduces inflow rate of water, it takes ~8 h to fill crater cavity for this case.

localities were exposed after ~10 h. The rushing wave, having a height of more than 200 m, then attacked the coast of North America and was reflected back, followed by significant wave oscillations having a periodicity of 1–2 h. Bourgeois et al. (1988) estimated the tsunami heights to have been 50–100 m high from the conditions requested for deposition of the K-T sandstone complex at the Brazos River, Texas, site. The wave height at the Brazos River locality in Figure 7 is consistent with this observation.

Calculations indicate that the tsunami runup inundated North America to 300 km beyond the Mississippi embayment (Fig. 8). The tsunami reached 300 m above sea level near the Rio Grande embayment, whereas the average runup height is in excess of 150 m.

Tsunami generated by landslide. We assume that a landslide occurred on the northern margin of the Yucatan platform just north of the impact point (Fig. 2). The location, size, and direction of the landslide are chosen arbitrarily. The area of the landslide is assumed to be 140 km (in east-west direction) by 75 km (in north-south direction). The thickness of the landslide layer along the north-south direction is assumed to be constant (100 m) for the southern region (0–20 km) and to decrease linearly toward the north (75–20 km). The landslide is assumed to be moving due north. We could simulate other cases, but in this chapter we just show the results for the parameters chosen to compare with the case of the crater-generated tsunami.

In the coastal regions of North America, the amplitude of the tsunami generated by a landslide is shown to be much smaller than that of the crater-generated tsunami (Fig. 9; cf. Fig. 7). However, the amplitude is highly dependent on the thickness of the sliding layer. If the layer were 10 times thicker than that assumed in this study, as suggested by Hildebrand (1997), the amplitude of the tsunami generated by the landslide is shown to be much larger and closer to that for the crater-

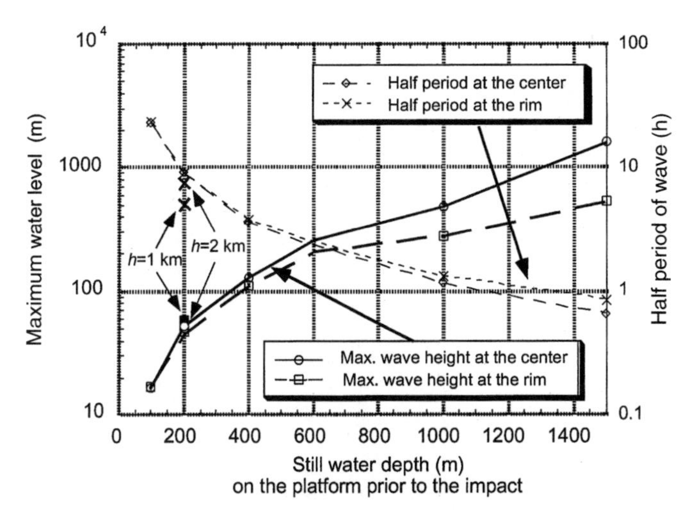

Figure 6. Relation of maximum water level, period of crater-generated tsunami at center and rim of crater, and still water depth of Yucatan platform. Numerical results for different crater depth models are also shown.

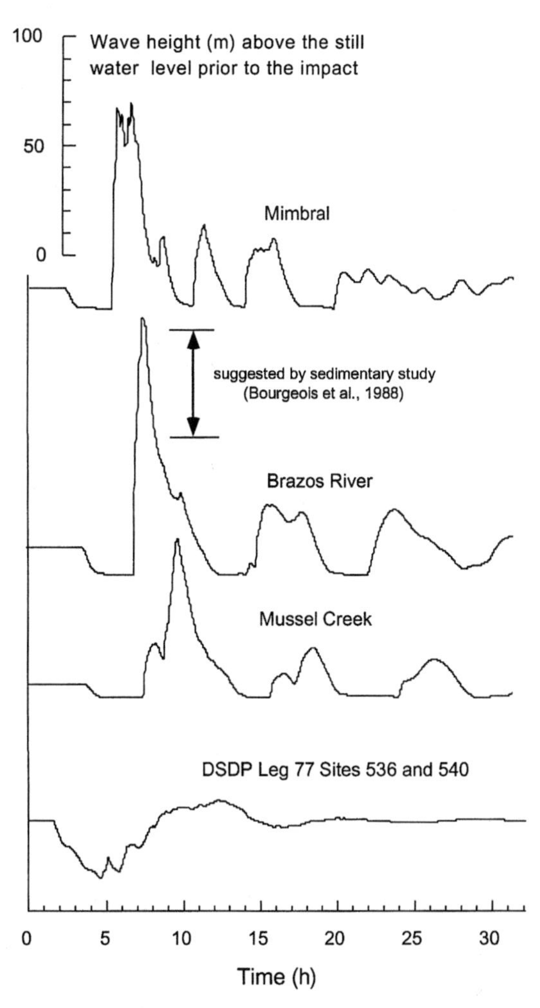

Figure 7. Temporal variations of water surface level at several locations in and around Gulf of Mexico for crater-generated tsunami. DSDP is Deep Sea Drilling Project.

generated tsunami; however, the first arrival in this case is a positive wave. This is the most significant difference between the crater-generated tsunami and the landslide-generated tsunami.

CONCLUSIONS

The wave height of the crater-generated tsunami is dependent on the depth of water covering the Yucatan platform at that time (Fig. 6), although the depth is difficult to estimate exactly. In this study we assumed that the depth of water was 200 m on the Yucatan platform, although the depth might have been much shallower. If the water was only 100 m deep, the maximum water level attained in the crater is reduced to ~40% of the maximum obtained for the 200 m case (Fig. 6). The energy of a wave is proportional to the square of the amplitude of that wave. Therefore, the magnitude of the tsunami for the 100 m case would be roughly ~17% that of the 200 m case, and the height of tsunami waves in costal regions would be 40% of the 200 m case.

Our calculations indicate there are significant differences between the crater-generated tsunamis and the landslide-generated tsunamis (cf. Figs. 7 and 9). This applies to their amplitudes in coastal regions as well as their directions and attenuations. The landslide-generated tsunami produces a rushing wave (first arrival), whereas the crater-generated tsunami produces a receding wave. This might be reflected in paleocurrent directions in the K-T sandstone complexes around the Gulf of Mexico (e.g., Smit et al., 1996). It is interesting to note that the study of the Moncada Formation in Cuba (Tada et al., this volume) suggests a receding wave as the first arrival.

Another contrasting feature of the two types of tsunami is that the amplitude of the crater-generated tsunami attenuates, whereas that of the landslide-generated tsunami does not attenuate. This is because the crater-generated tsunami runs up over the plain around the Gulf of Mexico, thus undergoing significant attenuation. In contrast, the landslide-generated tsunami is too small to run up, and thus undergoes little attenuation.

The magnitude of a landslide-generated tsunami is highly dependent on the area and thickness of the landslide. A seismic survey of the slope of the Yucatan platform is necessary to determine the landslide parameters; with this information, the generation and propagation of landslide-generated tsunami can be modeled more precisely. It is important to study the sedimentation mechanisms of the deposits formed on the floor of the crater in order to help clarify the amount of water that entered the crater. The continuously cored drill hole into the Chicxulub impact crater planned by the International Continental Scientific Drilling Program will provide a unique opportunity to do this.

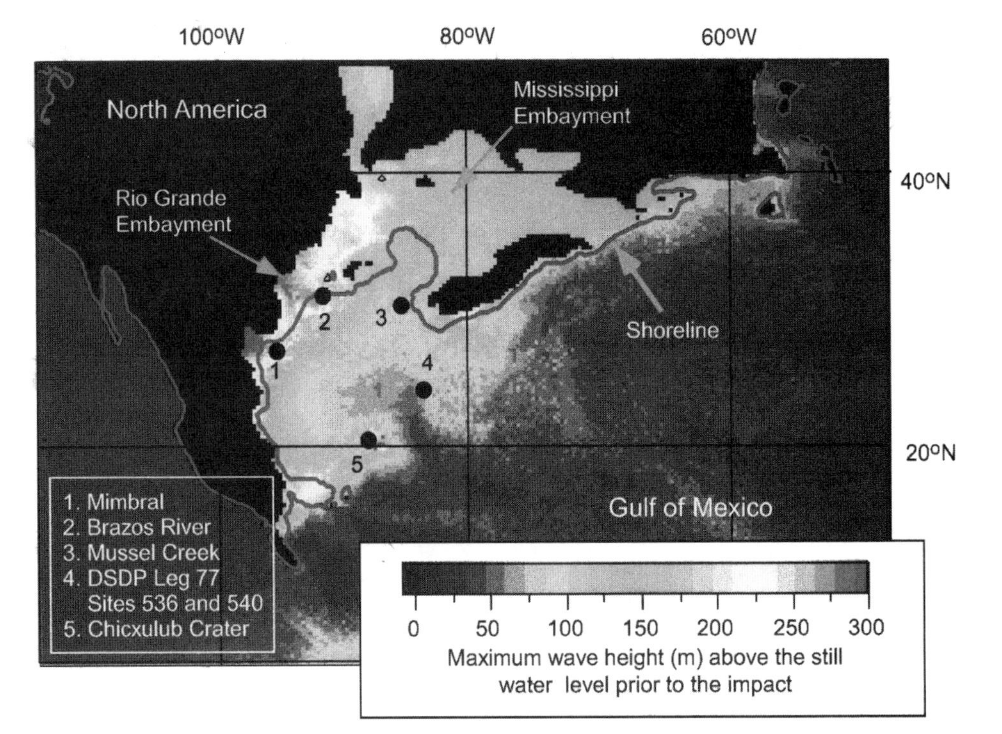

Figure 8. Inundation map of tsunami runup over plain around Gulf of Mexico. DSDP is Deep Sea Drilling Project.

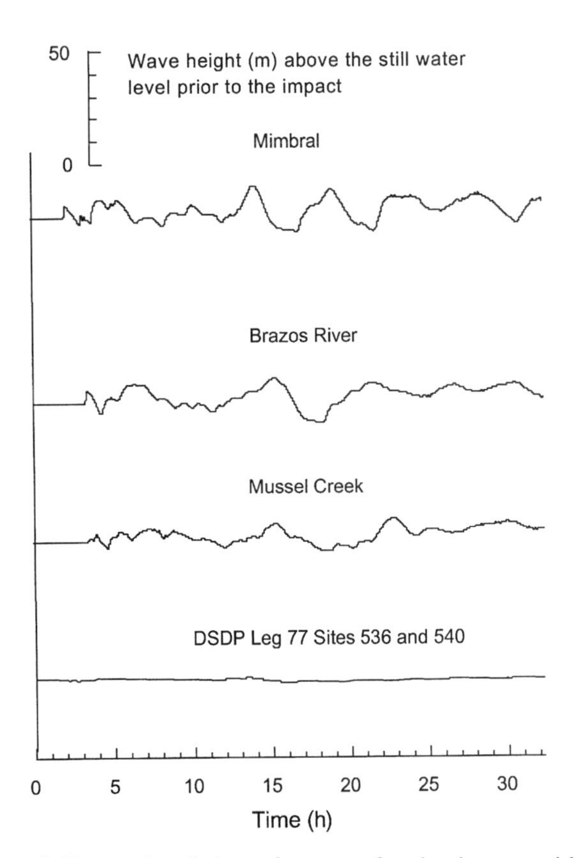

Figure 9. Temporal variations of water surface level at several locations in and around Gulf of Mexico for tsunami caused by landslide. DSDP is Deep Sea Drilling Project.

ACKNOWLEDGMENTS

We appreciate the critical comments and suggestions of A.R. Hildebrand, B. Simonson, and W.U. Reimold. We thank A.L. Moore for his efforts in improving our English, and the editor C. Koeberl for his helpful suggestions and efforts in handling the manuscript. This study was partially supported by grant-in-aids for scientific research of the Japan Society for the Promotion of Science (no. 11691116) and by research funds donated to University of Tokyo by I. Okawa, M. Iizuka, and K. Ihara.

REFERENCES CITED

Alvarez, L.W., Alvarez, W., Asaro, F., and Michel, H.V., 1980, Extraterrestrial cause for the Cretaceous-Tertiary extinction: Science, v. 208, p. 1095–1108.

Bohor, B.F., 1996, A sediment gravity flow hypothesis for siliciclastic units at the K/T boundary, northeastern Mexico, *in* Ryder, G., Fastovsky, D., and Gartner, S., eds., The Cretaceous-Tertiary event and other catastrophes in Earth history: Geological Society of America Special Paper 307, p. 183–195.

Bourgeois, J., Hansen, T.A., Wiberg, P.L., and Kauffman, E.G., 1988, A tsunami deposit at the Cretaceous-Tertiary boundary in Texas: Science, v. 241, p. 567–570.

Cita, M.B., Camerlenghi, A., and Rimoldi, B., 1996, Deep-sea tsunami deposits in the eastern Mediterranean: New evidence and depositional models: Sedimentary Geology, v. 104, p. 155–173.

Crawford, D.A., and Mader, C., 1998, Modeling asteroid impact and tsunami: Science of Tsunami Hazards, v. 16, p. 21–31.

Dean, R.G., and Dalrymple, R.A., 1991, Water wave mechanics for engineers and scientists: Singapore: World Scientific, 353 p.

Gault, D.E., and Sonett C.P., 1982, Laboratory simulations of pelagic asteroidal impact: Atmospheric injection, benthic topography, and the surface wave radiation field, *in* Silver, L.T., and Schultz, P.H., eds., Geological implications of impacts of large asteroids and comets on the earth: Geological Society of America Special Paper 190, p. 69–92.

Hildebrand, A.R., 1997, Contrasting Chicxulub Crater structural models: What can seismic velocity studies differentiate?: Journal of Conference Proceedings, v. 1, p. 37–46.

Hildebrand, A.R., Penfield, G.T., Kring, D.A., Pilkington, M., Camargo, Z.A., Jacobsen, S.B., and Boynton, W.V., 1991, Chicxulub crater: A possible Cretaceous/Tertiary boundary impact crater on the Yucatán Peninsula, Mexico: Geology, v. 19, p. 867–871.

Hildebrand, A.R., Pilkington, M., Ortiz-Aleman, C., Chavez, R.E., Urrutia-Fucugauchi, J., Connors, M., Graniel-Castro, E., Camara-Z., A., Halpenny, I.A., and Nichaus, D., 1998, Mapping Chicxulub crater structure with gravity and seismic reflection data, *in* Grady, M.M., Hutchison, R., McCall, G.J.H., and Rothery, D.A., eds., Meteorites: Flux with time and impact effects: Geological Society [London] Special Publication 140, p. 153–173.

Hills, J.G., Memchinov, I.V., Popov, S.P., and Teterev A.V., 1994, Tsunami generated by small asteroid impacts, *in* Gehrels, T., ed., Hazards due to comets and asteroids: Tucson, University of Arizona Press, p. 779–789.

Imamura, F., 1996, Review of tsunami simulation with a finite difference method, *in* Yeh, H., Liu, P., and Synolakis, C., eds., Long-wave runup models: Singapore, World Scientific, p. 25–42.

Imamura, F., and Gica, E.C., 1996, Numerical model for wave generation due to subaqueous landslide along a coast: A case of the 1992 Flores tsunami, Indonesia: Science of Tsunami Hazards, v. 14, p. 13–28.

Imamura, F., and Imteaz, M.M.A., 1995, Long waves in two-layers: Governing equations and numerical model: Science of Tsunami Hazards, v. 13, p. 3–24.

Kiyokawa, S., Tada, R., Oji, T., Tajika, E., Nakano, Y., Goto, K., Yamamoto, S., Takayama, H., Toyoda, K., Rojas, R., Gracia, D., Iturralde-Vinent, M.A., and Matsui, T., 2000, More than 500m thick K/T boundary sequence: Cacarajicara formation, Western Cuba. Impact related giant flow deposits [abs.], *in* Catastrophic events and mass extinction: Impacts and beyond: Houston, Texas, Lunar and Planetary Institute, LPI Contribution No. 1053, p. 100–101.

Morgan, J.V., Warner, M., and the Chicxulub Working Group, 1997, Size and morphology of the Chicxulub impact crater: Nature, v. 390, p. 472–476.

Okusa, S., and Yoshimura, M., 1981, Possibility of submarine slope failure due to waves (in Japanese): Tokai University Bulletin, v. 14, p. 227–234.

Raney, D.C., and Butler, H.L., 1976, Landslide generated water wave model: Journal of Hydraulic Research Division, American Society of Civil Engineering, v. 102, no. HY9, p. 1269–1281.

Ross, M.J., and Scotese, C.R., 1988, A hierarchical tectonic model of the Gulf of Mexico and Caribbean region: Tectonophysics, v. 155, p. 139–168.

Smit, J., Roep, Th.B., Alvarez, W. Claeys, P., Grajales-Nishimura, J.M., and Bermudez, J., 1996, Coarse-grained, clastic sandstone complex at the K/T boundary around the Gulf of Mexico: Deposition by tsunami waves induced by the Chicxulub impact?, *in* Ryder, G., Fastovsky, D., and Gartner, S., eds., The Cretaceous-Tertiary event and other catastrophes in Earth history: Geological Society of America Special Paper 307, p. 151–182.

Sohl, N.F., Martinez, R.E., Salmeron-Urena, P., and Soto-Jaramillo, F., 1991, Upper Cretaceous, *in* Salvador, A., ed., The Gulf of Mexico Basin: Boulder, Colorado, Geological Society of America, Geology of North America, v. J, p. 205–244.

Takayama, H., Tada, R., Matsui, T., Iturralde-Vinent, M.A., Oji, T., Tajika, E. Kiyokawa, S., Garcia, D., Okada, H., Hasegawa, T., and Toyoda, K., 2000, Origin of the Peñalver Formation in northwestern Cuba and its relation to K/T boundary impact event: Sedimentary Geology, v. 135, p. 295–320.

Ward, S.N., and Asphaug, E., 2000, Asteroid impact tsunami: A probabilistic hazard assessment: Icarus, v. 145, p. 64–78.

Manuscript Submitted October 10, 2000; Accepted by the Society March 22, 2001

Geological Society of America
Special Paper 356
2002

Mass wasting of Atlantic continental margins following the Chicxulub impact event

R.D. Norris*
Woods Hole Oceanographic Institution, Woods Hole, Massachusetts 02543-1541, USA
J.V. Firth*
Ocean Drilling Program, 1000 Discovery Drive, College Station, Texas 77845-9547, USA

ABSTRACT

The Chicxulub impact 65 Ma triggered massive submarine failure of continental margins around the North Atlantic. Slumped sediments associated with impact ejecta and geochemical tracers of the bolide are present on the Blake Plateau, the mid-Atlantic continental slope and rise, Bermuda Rise, and the Iberian Abyssal Plain more than 6000 km from the impact crater. Evidence from deep-sea drilling and seismic stratigraphy suggests that much of the eastern seaboard of North America and at least parts of the eastern margin of the North Atlantic must have failed catastrophically because of the ~10–13 magnitude earthquake associated with the impact event, and created one of the largest, composite mass-wasting deposits on Earth. We infer that mass failure of the eastern margin of North America can account for elevated extinction rates and delayed recovery of North American invertebrates compared to other places during the Cretaceous-Paleogene mass extinction. Slumping on the margins also redeposited large volumes of carbonates into the deep sea below the Cretaceous carbonate compensation depth (CCD), giving the false impression of a drop in the CCD during the Maastrichtian.

INTRODUCTION

The Cretaceous-Paleogene (K-P) boundary is associated with a 40%–70% species-level extinction that has widely been attributed to the climatological effects of a large body impact (Alvarez et al., 1980). The Chicxulub crater on the Yucatan Peninsula in northeastern Mexico is generally considered to be the primary impact site (Hildebrand and Boynton, 1990) and has been dated at the K-P boundary by radiometric methods (Swisher et al., 1992). The Chicxulub extraterrestrial impact released energy equivalent to more than 100 $\times$ 10^6 Mt of TNT (Morrison et al., 1994; Toon et al., 1997) and triggered one of the five largest mass extinctions known in Earth's history (Sepkoski, 1986; but see MacLeod, 1996; MacLeod et al., 1997 for an opposing view).

Conventionally, much of the immediate damage in the first hours following impact has been attributed to the direct effects of the blast around the Gulf of Mexico, the passage of tsunamis precipitated by the impact in shallow water over the Yucatan platform, and the effects of ejecta fallout within a few thousand kilometers of the crater. These events are followed in the succeeding days and weeks by the longer term effects of global wildfires, dust loading in the atmosphere, and collapse of oceanic and terrestrial food chains. For example, the occurrence of redeposited sediments (Bourgeois et al., 1988; Yancey, 1996) in the Gulf coastal plain, groove markings on the K-P bedding surface (Olsson et al., 1996), and sand deposits within deep-water facies in Mexico and Central America (Smit et al., 1992, 1996; Bohor, 1996; O'Campo et al., 1996) suggest erosion produced by bolide-generated tsunamis, although others have in-

*E-mails: Norris, RNorris@whoi.edu; Firth, firth@ODPmail.tamu.edu

Norris, R.D., and Firth, J.V., 2002, Mass wasting of Atlantic continental margins following the Chicxulub impact event, *in* Koeberl, C., and MacLeod, K.G., eds., Catastophic Events and Mass Extinctions: Impacts and Beyond: Boulder, Colorado, Geological Society of America Special Paper 356, p. 79–95.

terpreted some of these features as a result of sea-level lowstand (e.g., Stinnesbeck and Keller, 1996). Bralower et al. (1998) documented slope failure associated with the K-P boundary at numerous sites in the Gulf of Mexico and the Caribbean. They proposed that mass wasting involved both large-scale margin collapse of the Yucatan platform and gravity flows that distributed sediment throughout the western Caribbean. The far-field effects of tsunamis and impact fallout are poorly documented, but could include erosion by tsunamis in the Atlantic (Albertão et al., 1994) and Pacific basins. Loading of unstable sediments by impact fallout might cause local sediment failure as much as several thousand kilometers from the crater.

The direct seismic effects of the impact are dependent on the efficiency of transfer in energy from the impact into ground motion. For example, a 12.8 magnitude quake could be generated if all of the Chicxulub bolide's energy was converted into elastic waves, whereas a magnitude 10 earthquake would result from a more realistic conversion efficiency (Toon et al., 1997). Computer simulations with a fairly simple planetary model suggest vertical ground motion of >15 m within 100 km of the crater (Boslough et al., 1996), which is more than enough to account for the large-scale collapse of the Yucatan platform and gravity slides along the Gulf coast and throughout the Caribbean and Gulf of Mexico (Bralower et al., 1998). Olsson et al (1996) documented normal faulting in Cretaceous sediments just below the K-P boundary that are overlain by unfaulted Danian sediments, and proposed that the normal faulting was induced by impact-related earthquakes.

As with the long-distance consequences of impact fallout and tsunamis, the far-field effects of seismicity associated with the impact are poorly known. Computer simulations suggest that seismic effects could be nearly as large at the antipode (in the southern Indian Ocean) as at the site of the impact on the Yucatan (Boslough et al., 1996). In addition, these same models suggest that surface ground motion could be in excess of 1 m within 7000 km of the impact crater. Large radial displacements are also associated with large peak strain amplitudes and suggest that the impact earthquakes could have easily produced enough strain in bottom sediments to cause mass flows of unstable sediments around North America, northern South America, and much of the eastern North Atlantic. It is not surprising that large-scale mass wasting has been reported from the Upper Cretaceous in Baja California (Busby et al., 1999), the Blake Plateau (Klaus et al., 2000), and the central Bermuda Rise more than 2500 km from the crater (Norris et al., 2000a). Possible tsunami deposits have also been reported from coastal Brazil (Albertão et al., 1994).

Here we show that mass flows in Upper Cretaceous sediments are even more widespread in the western North Atlantic than has been previously documented, as well as being present on the Iberian continental margin (Fig. 1) more than 6000 km from Chicxulub. We believe that slope failure was triggered by ground motion produced by the impact, because the resulting mass flows originated just before the arrival of impact debris on the seafloor. We have previously shown seismic and coring evidence for large-scale slope failure on the Blake Plateau (Klaus et al., 2000). We have also shown that there is considerable evidence from the central Bermuda Rise that mass flows originating from the continental margin of North America reached as much as 1400 km offshore (Norris et al., 2000a, 2000b). Furthermore, our previous work on Bermuda Rise suggests that a prominent acoustic reflector in the western North Atlantic, Horizon A*, is likely to be at least partly related to mass wasting at the K-P boundary. Most previous workers have suggested that Horizon A* (which has been mapped throughout the northwestern Atlantic) is related to an increase in the carbonate compensation depth (CCD) in the Maastrichtian that allowed calcareous ooze to be deposited on the Bermuda Rise for the first time in the Late Cretaceous (Tucholke and Vogt, 1979b; Barrera and Savin, 1999). This interpretation obviously hinges on whether the carbonate sediments are actually pelagic deposits. Here we argue that the reconstructions of the Atlantic CCD are biased by calcareous mass-flow deposits that introduced carbonate well below the CCD in sufficient amounts to leave a preserved record. We also address the possible effects of large-scale mass wasting on the marine biota and global climate.

MASS WASTING IN THE WESTERN NORTH ATLANTIC

There are eight deep-sea sites in the western North Atlantic that preserve biostratigraphically complete K-P boundary sequences (Fig. 1). Many other sites have drilled through the boundary, but have either failed to recover boundary sediments or have had incomplete boundary sections. Of those sites with relatively complete boundary sections, five Deep Sea Drilling Program (DSDP) and Ocean Drilling Program (ODP) sites (DSDP Sites 603 and 605 and ODP Sites 1049, 1050, and 1052) are on the North American continental margin; the remainder are located on the Bermuda Rise (DSDP Sites 385–387) (Fig. 2). Cores are supplemented with seismic reflection profiles that have been used to map acoustic facies (such as Horizon A*) over much of the western North Atlantic. Acoustic horizons have been mapped mostly by correlation to a small number of drill sites and by tracing reflectors to substantial distances from their point of known age (Tucholke, 1979, 1981; Tucholke and Mountain, 1979). In other cases, reflectors may be picked on the basis of their acoustic character; in essence the relationship of a given reflector to other parts of the seismic stratigraphy (Mountain and Miller, 1992).

Evidence for margin collapse on the Blake Plateau

Ocean drilling on the Blake Nose east of Florida (Norris et al., 1998) has provided some of the first strong evidence for large-scale mass wasting of K-P boundary age in the North Atlantic (Klaus et al., 2000). Upper Maastrichtian calcareous ooze and chalk is highly deformed immediately below the K-P

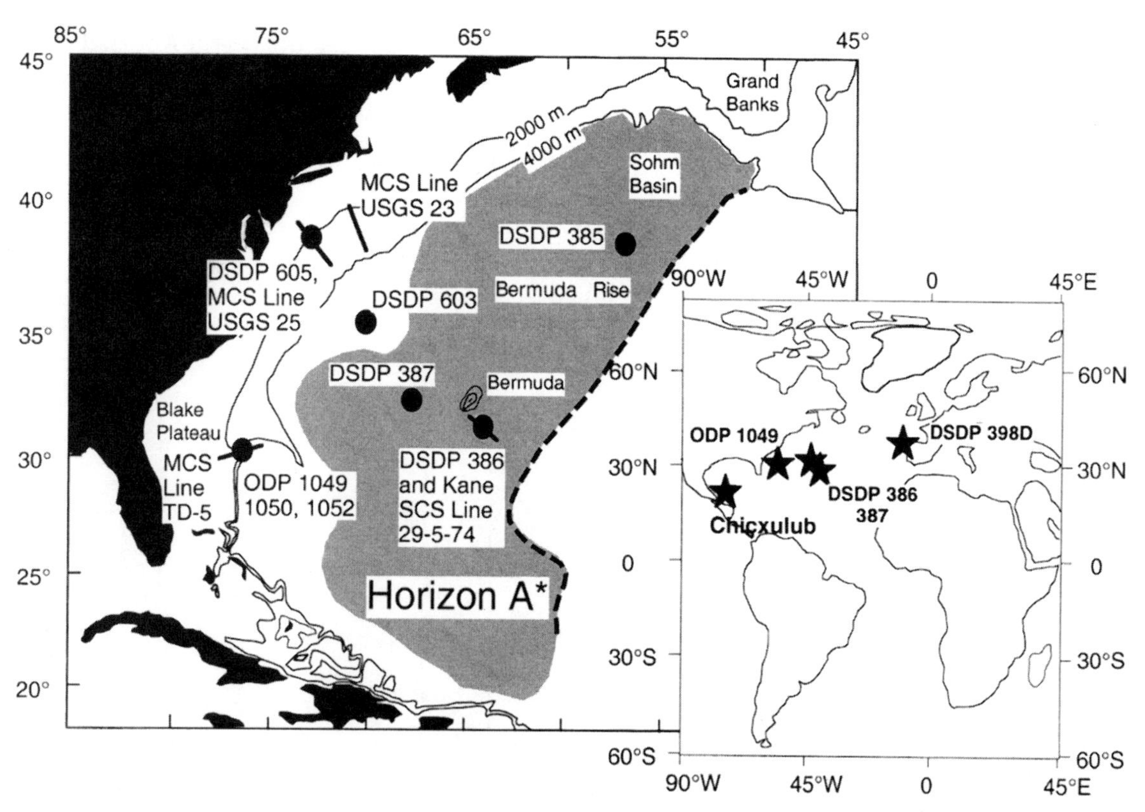

Figure 1. Geographic distribution of drill sites and seismic evidence for Atlantic Cretaceous-Paleogene (K-P) boundary slope failure deposits. Gray stippled pattern shows distribution of acoustic Horizon A* correlated to K-P boundary mass-wasting deposits. Horizon A* is largely absent along southeastern margin of North America, probably because of erosion during Oligocene (ca. 33 Ma), but has been recognized in Deep Sea Drilling Project (DSDP) Site 603. Reflector equivalent with Horizon A* has been identified in slope settings at DSDP Site 605 and Ocean Drilling Program Site 1049. Mass-wasting deposits have been identified as far east as DSDP Hole 398D, offshore Portugal. SCS = Single Channel Seismic Line, MCS = Multichannel Seismic Line.

boundary at ODP Sites 1049, 1050, and 1052 located along a bathymetric transect between 1300 m (ODP Site 1052) and 2670 m (ODP Site 1049). The K-P boundary is marked by an ejecta bed at Site 1049 (Norris et al., 1998, 1999; Klaus et al., 2000), where flat-lying Paleocene ooze overlies 16.8 m of strongly deformed Maastrichtian ooze (Fig. 2). Bedding is inclined at 15°–80° and exhibits microfaulting and pull-apart structures throughout the Maastrichtian (Norris et al., 1998). At Site 1050, deformation is most severe in the lower 30 m of the Maastrichtian, where the chalk displays spectacular recumbent folds. The overlying chalk at Site 1050 shows well-developed physical property cycles that have been shown to correspond to the orbital precession cycle (21 k.y.) (MacLeod and Huber, 2001) and are very slightly deformed in comparison to the highly folded and faulted chalk at the base of the sequence. In contrast, deformation is much more intense in the upper 84 m of the Maastrichtian at ODP Site 1052 than at greater depths. The relatively undeformed sequence at ODP Site 1052 has bedding with dips to 20° from horizontal and is cut periodically by high-angle faults with slickensided surfaces (Norris et al., 1998; Klaus et al., 2000). It appears that the drill core may have cut through large, rotated fault blocks at ODP Site 1052 and penetrated a basal decollement at ODP Site 1050, where relatively intact Maastrichtian sediment sifted downslope over highly deformed sediment.

Seismic data show that the K-P boundary has an irregular, hummocky surface over chaotic reflectors and steeply dipping reflectors that are characteristic of rotated fault blocks (Fig. 3). Interpretations of the seismic line suggest that Maastrichtian sediments are extensively deformed all across Blake Nose and have broken up into fault-bounded blocks as much as several kilometers long in the vicinity of ODP Site 1052 (Klaus et al., 2000).

Deformation on Blake Nose involved relatively consolidated chalk buried to depths of ~200 m below the Cretaceous seafloor (Fig. 3). The brittle deformation and preservation of slickensided surfaces, microfaulting, and pull-apart structures all suggest that the chalk was consolidated before it was deformed. The magnitude and depth of penetration of mass movement strongly suggest that deformation was unrelated to loading of surficial unconsolidated sediments by a few centimeters of impact ejecta. Likewise, the 1300–2600 m paleowater depth

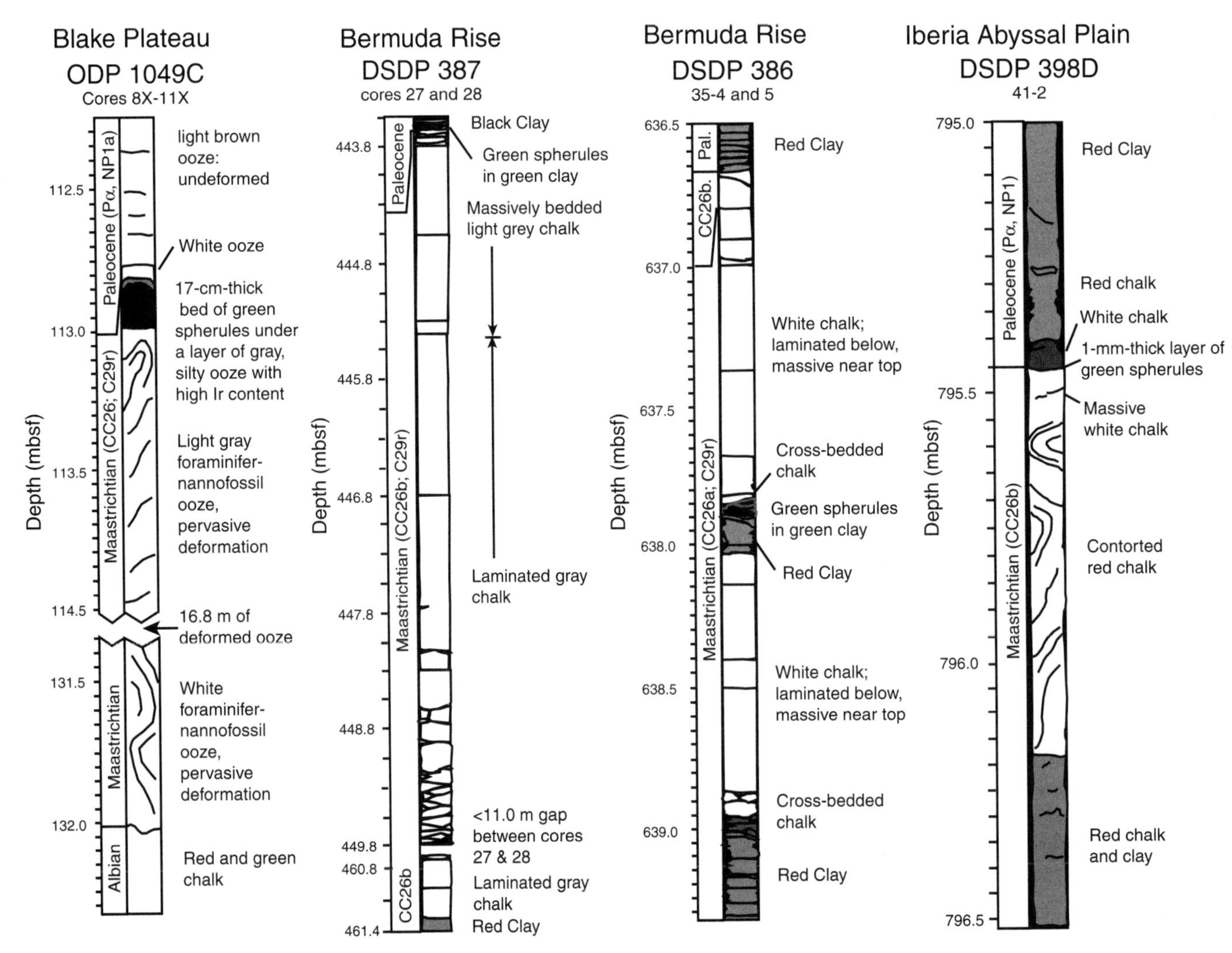

Figure 2. Summary of Cretaceous-Paleogene stratigraphy in four sites from west to east transect across North Atlantic. Deformation in Maastrichtian on Blake Nose (west) and Iberian Abyssal Plain (east) is due to slumping that preceded deposition of flat-lying impact spherule deposits and Paleocene chalk or ooze. Deep-water sequences on Bermuda Rise consist of thick calcareous turbidite deposits that display graded bedding, cross-lamination, and allochthonous fossils. Clastic turbidites (Deep Sea Drilling Project Site 603) suggest sources of sediment different from calcareous turbidites shown here.

of the sediments during the Cretaceous suggests that they should not have been greatly affected by the passage of tsunamis on their way to the continental shoreline. Therefore, Klaus et al. (2000) concluded that mass wasting was likely to have been triggered by the seismic effects of the impact.

K-P boundary on the mid-Atlantic continental slope

DSDP Site 605 (38°44N, 72°36W, water depth 2194 m) was drilled on the continental slope offshore of New Jersey (Fig. 1). Drilling recovered a thick sequence of green, highly bioturbated, Paleocene and Maastrichtian chalk (van Hinte and Wise, 1987). The K-P boundary (in core 605-66-1, 759.8 mbsf [meters below seafloor]) is brecciated but mostly biostratigraphically complete, although it might be missing part of the earliest Paleocene as well as the boundary clay (Smit and van Kempen, 1987). A total of 56.9 m of middle and upper Maastrichtian chalk was recovered between the K-P boundary and the bottom of the hole. The chalk is generally only weakly deformed, showing small normal faults and bedding tilted 10°–20° from horizontal. The shipboard party describing the core suggested that the core might have penetrated rotated fault blocks in the Maastrichtian section (van Hinte and Wise, 1987). However the sequence appears to be in correct stratigraphic order, suggesting that there has been no repetition of section.

A multichannel seismic reflection profile across DSDP Site 605 (U.S. Geological Survey [USGS] line 25) shows a pair of well-developed reflectors at 3.72 and 3.77 s (two-way traveltime). The upper of these reflectors has been correlated to an impedance contrast associated with an unconformity at the Danian-Selandian boundary (ca. 61 Ma). Olsson and Wise

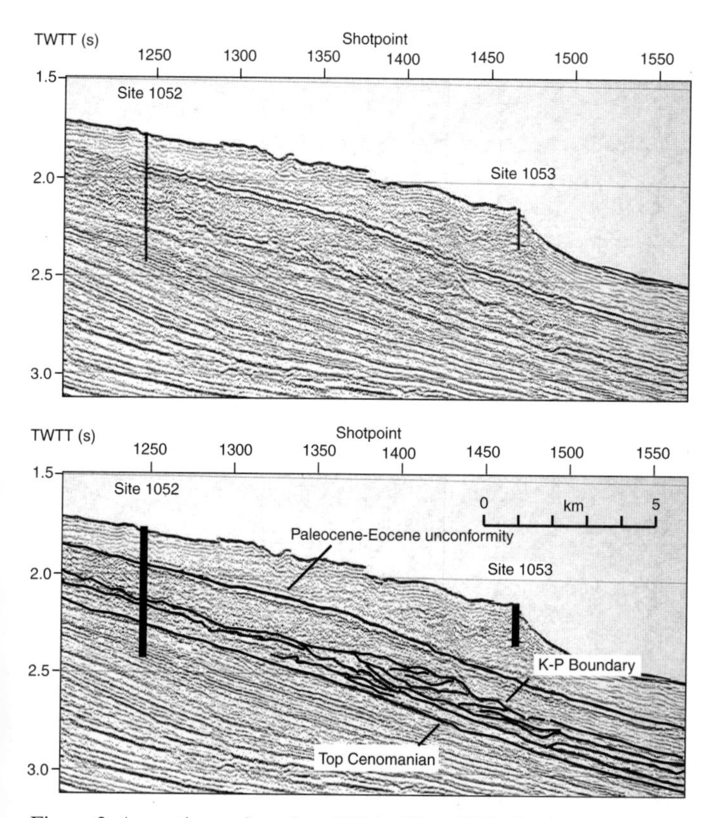

Figure 3. Acoustic stratigraphy of Blake Nose (U.S. Geological Survey line TD-5) showing inferred slumping in Maastrichtian chalk between Ocean Drilling Program Sites 1052 and 1053 (after Norris et al., 1998; Klaus et al., 2000). Uninterpreted line is above, interpreted line is below. Chaotic reflectors within Maastrichtian sequence suggest large (~1 km) slide blocks below Cretaceous-Paleogene (K-P) boundary that we interpret as result of mass wasting related to Chicxulub impact. TWTT is two-way traveltime.

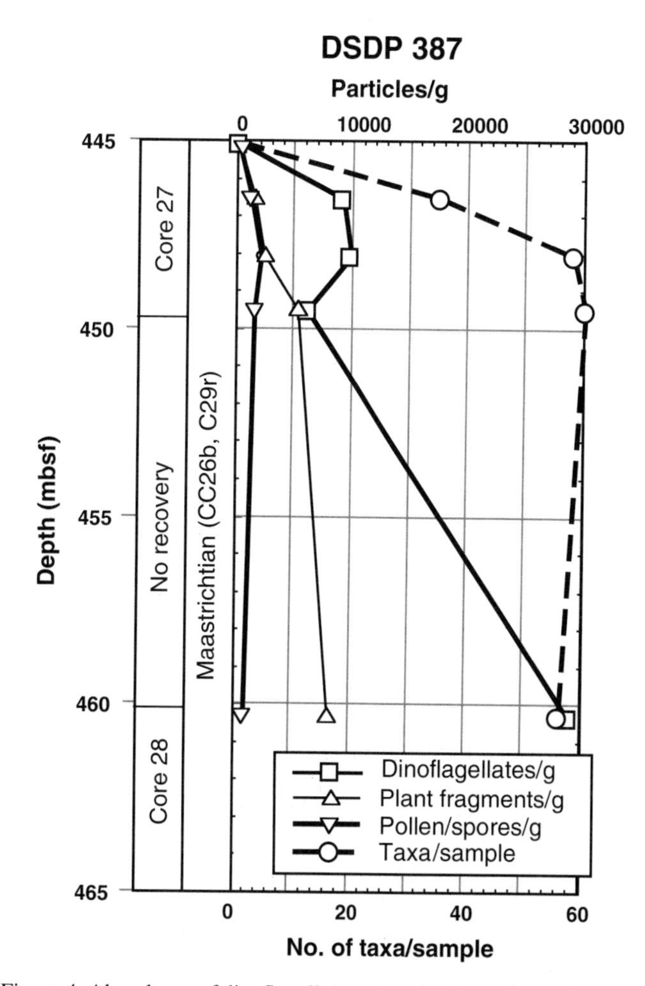

Figure 4. Abundance of dinoflagellates, plant debris, pollen and spores, and total dinoflagellate taxa in chalk bed at Deep Sea Drilling Project (DSDP) Site 387. Decrease in abundance of each component toward top of core suggests that sediments are size graded. Shallow-water origin for parts of dinoflagellate assemblage suggest that chalk was derived by mass flow from continental margin.

(1987) identified the Danian-Selandian hiatus as the likely source of reflector A*. However, the reflector at 3.77 s corresponds to the K-P boundary and is both stronger and more continuous than that at 3.72 s. We suggest that the deeper of these two reflectors is actually correlative with Horizon A* as described from the Bermuda Rise (Tucholke, 1979; Tucholke and Vogt, 1979a).

Seismic stratigraphy of USGS line 25 shows that reflectors below Horizon A* have a wavy, discontinuous character, in contrast to the more parallel character of reflectors in the overlying Paleocene section. The shipboard party interpreted the seismic facies as indicating slumped section in the early Maastrichtian and Campanian (below the deepest penetration of DSDP Site 605) (van Hinte and Wise, 1987, p. 282; their Fig. 4). The wavy, hummocky section intersects Horizon A* between shotpoints 3650 and 3750 on USGS line 25 ~ 14 km southeast of DSDP Site 605 (van Hinte and Wise, 1987, p. 281; their Fig. 3). Disrupted reflectors have also been reported from just below the K-P boundary in nearby multichannel seismic (MCS) line USGS 23, where the deformation spans an interval ~400–500 m thick in the continental rise (Max et al., 1999).

K-P mass wasting in the deep western North Atlantic

Sediment slumped off the continental margin should have accumulated on the abyssal plain in the western North Atlantic. Post-Oligocene erosion by the deep western boundary current has removed most traces of the K-P boundary along the base of the continental slope (Fig. 1; Tucholke, 1981). Many drill sites also have gaps in core recovery across the K-P boundary that limit their usefulness to this study. However, the K-P boundary is mostly preserved at four DSDP sites on the Bermuda Rise and the continental rise (Fig. 1). These sites include DSDP Site 385 (37°22N, 60°09W, water depth 4936 m), DSDP Site 386 (31°11N, 64°14W, water depth 4783 m), DSDP Site 387 (32°19N, 67°40W, water depth 5118 m), and DSDP Site 603 (35°29N, 70°01W, water depth 4633 m). DSDP Site 603 is located on the continental rise offshore of the mid-Atlantic continental margin, where it forms a deep-water end member of the New Jersey transect (van Hinte and Wise, 1987). DSDP Site 385 is located at the base of the Vogel Seamount ~800 km southeast of the continental rise (Tucholke and Vogt, 1979a). DSDP Sites 387 and 386 are on the central Bermuda Rise and are 700 and 1200 km offshore of the continental escarpment, and 2560 and 2830 km from the Chicxulub crater, respectively (Tucholke and Vogt, 1979a). All the sites were below the CCD (below which chalk does not usually accumulate on the sea floor) for most of the Late Cretaceous and Paleocene (Tucholke, 1981). However, at three sites (DSDP Sites 385–387) the K-P boundary is marked by chalk that is overlain and underlain by pelagic clay (Fig. 2). The K-P chalk has been generally interpreted as pelagic sediment that accumulated during a brief increase in the depth of the CCD during the late Maastrichtian (Tucholke and Vogt, 1979b). However, the sedimentary structures, fossil content, and geochemistry show that the chalk represents a mass-wasting deposit derived from the North American margin.

Stratigraphy of DSDP Site 603

The K-P boundary at DSDP 603 is marked by rusty brown to olive-green siltstone that lacks calcareous and organic microfossils. Accordingly, there are no direct biostratigraphic determinations of the age of the sediments across the K-P boundary. However, Klaver et al. (1987) reported green spherules in a 5-cm-thick section of light green, cross-bedded sandstone. The spherules are 0.5–1 mm in diameter and composed of smectite. Relict bubbles filled with secondary minerals are present within the spherules, as has been observed in altered tektites from other K-P boundary localities (e.g., review by Smit, 1999). The high concentrations of Ni, Co, and As in the spherules and overlying turbidite siltstone are consistent with the enrichment of these elements observed in the K-P boundary deposits from other areas (Klaver et al., 1987).

Restudy of the core in the ODP repository shows that the stratigraphic sequence below the spherule bed consists of rusty brown to dark olive-green silty claystone, micaceous and glauconite sandstone, and silty sandstone. Well-defined graded beds typical of turbidity current deposition are present, some with cross-bedded or plane-laminated bases suggestive of distal turbidites. There is little evidence of large-scale deformation of the section, but there is enough fracturing in the core (presumably drilling disturbance) that faulting may have been overlooked. The silty sandstone overlying the spherule bed is divided into two fining-upward sequences that are each ~70 cm thick. These sandstones are in turn overlain by greenish-gray radiolarian siltstone of late Paleocene age (van Hinte and Wise, 1987).

Lithostratigraphy of sites on the Bermuda Rise

DSDP Site 385, Vogel Seamount. The sedimentary sequence at DSDP Site 385 consists mostly of light yellowish-brown zeolitic clay of Paleogene age overlying gray-brown to light greenish-gray clay of Maastrichtian age (Tucholke and Vogt, 1979a). Less than ~33% of the K-P boundary interval was recovered; the rest was lost in 3–10 m core gaps. Calcareous microfossils are present throughout the boundary section, but organic microfossils are absent. The youngest sediments are dark olive-gray-green calcareous ooze in the lowest 25 cm of core 385-11-2. Our observations show that the well-preserved foraminifer assemblage is typical of lower Danian foraminifer biozone P1a. The top of this chalk forms a sharp contact with overlying claystone and suggests the presence of an unconformity or the top of a turbidite. A similar, but more dissolved, foraminifer assemblage occurs throughout core 385-12, but Danian planktic foraminifera are mixed with fragmentary Cretaceous planktic foraminifera and abundant calcareous and agglutinated benthic foraminifera. The shipboard science party identified core 385-12 as Maastrichtian age (Tucholke and Vogt, 1979a), but this is probably because of the abundance of reworked Cretaceous sediment. We also observed rare, small *Cruciplacolithus* sp., among abundant Cretaceous nannofossils, down through core 385-12. This nannofossil datum corroborates the foraminiferal age of early (but not earliest) Danian for this core.

We have not found foraminifera typical of the earliest Danian Pα Zone, or any well-defined ejecta bed. Small flakes of green clay and carbonate are present in samples 385-12-2 (24–26 cm) and 385-13-1 (140–142 cm), but it is not clear whether these are pieces of diagenetically altered tektites because none of the pieces are very large. *Thoracosphaera* cysts are abundant in light gray chalk intervals in samples 385-12-1 (114–116 cm) and 385-12-2 (24–26 cm) and suggest the presence of reworked lower Danian sediment.

The K-P boundary probably occurs in the ~10 m coring gap between cores 12 and 13 (222–231 mbsf). The foraminifer assemblage in core 385-13 is composed largely of dissolved Maastrichtian planktic foraminifera and a mixture of calcareous and agglutinated benthic species. Inoceramid prisms occur in a sample from core 385-13-4 (40–42 cm), suggesting an early-middle Maastrichtian age. In agreement with our data, the upper-middle Maastrichtian boundary was identified by the shipboard science party (Tucholke and Vogt, 1979a) as occurring between sections two and three in core 385-13 (~234.6 mbsf) based upon calcareous nannofossil biostratigraphy. There is some downhole contamination of the core; we discovered Paleocene planktic foraminifera (the genus *Acarinina*, typical of biozone P4, ca. 59 Ma or younger) in a sample from 385-13-2 (76–78 cm).

Danian calcareous ooze in cores 385-11 and 385-12 is likely to represent turbidites reworked from the flanks of the Vogel Seamount. The light gray beds containing *Thoracosphaera* cysts in core 385-12 have sharp contacts with overlying and underlying brown, calcareous claystone. A series of these light gray calcareous oozes is interbedded with the less calcareous brown claystones, suggesting mass transport from well above the CCD. This rhythmic bedding is also found in the upper Maastrichtian sediments present in core 385-13 and suggests that most of the calcareous sediments have been reworked in a series of downslope transport events. Although these sediments suggest that much of the section is affected by redeposition, the absence of lowermost Danian sediments and ejecta suggests that DSDP 385 cannot be used to define the distribution of K-P boundary mass-wasting deposits. At the same time, the evidence for downslope transport in the Cretaceous and Danian suggests that the presence of carbonate sediments at this site may not have much significance for the depth of the CCD during K-P boundary time.

DSDP Site 387, Central Bermuda Rise. The chalk at DSDP Site 387 is size graded from a foraminifer chalk at the base to a nannofossil chalk at the top (Fig. 2). A ~1-mm-thick layer of green spherules similar to the impact ejecta found on Blake Nose (Norris et al., 1999; Smit, 1999; Martínez-Ruiz et al., 2001) occurs in a layer of green clay at the very top of the chalk sequence. The entire chalk unit was not recovered during drilling; 6.9 m of sediment was recovered, but much of the chalk sequence was lost in coring gaps. When all the gaps in core recovery are considered, the chalk could be as thick as 17.6 m between the top of the chalk unit at 443.8 mbsf and the top of the youngest Cretaceous red clay at 461.4 mbsf.

The upper 1.5 m of chalk is a pale yellowish-gray nannofossil chalk that displays no sedimentary structures, and the remainder of the chalk is light to medium gray and shows horizontal partings. Samples of chalk washed through a 38 µm screen yielded only rare, extremely small planktic foraminifera (McNulty, 1979), but our data show that the largest of these were ~50–75 µm larger in the lowest part of the chalk than in the uppermost part of the chalk; therefore, the chalk is weakly size graded.

The entire chalk unit contains calcareous nannofossils typical of the latest Maastrichtian *Micula prinsii* zone (CC26b). We did not find species diagnostic of older parts of the Cretaceous, in agreement with earlier work by Okada and Thierstein (1979). The planktic foraminifera are also typical species of the latest Maastrichtian. They are represented primarily by extremely small, but well-preserved, individuals of the genera *Heterohelix*, *Globigerinelloides*, *Globotruncanella*, and *Hedbergella* (Fig. 5), whereas species that usually grow quite large, such as globotruncanids and rugoglobigerinids, are notably absent or very rare.

Palynologic analysis of the chalk at DSDP Site 387 revealed a surprising kerogen assemblage, consisting of abundant dinoflagellate cysts and common terrestrial plant debris and pollen, which is most typical of neritic environments (Habib and Miller, 1989). The kerogen also shows a weak grading in abundance through the recovered chalk sequence (Fig. 4). The dinoflagellate assemblage, furthermore, is similar in abundance, diversity, and species content to those found in eastern North American shelf environments (Benson, 1976; Firth, 1987; Habib and Miller, 1989). The latest Maastrichtian species *Palynodinium grallator*, *Disphaerogena carposphaeropsis*, and *Manumiella seelandica* are all present in core 27, whereas the late Maastrichtian species *Deflandrea galeata* and *Thalassiphora pelagica* are present in cores 27 and 28. High abundances of *Manumiella seelandica* have been reported worldwide in shelf environments just below the Cretaceous-Tertiary boundary, and have been interpreted as indicating marginal marine to shallow shelf environments (Nøhr-Hansen and Dam, 1997). This species comprises between 4% and 7% of the dinoflagellate cyst assemblage at DSDP Site 387. In addition, *Areoligera/Glaphyrocysta* spp. and *Spiniferites* spp. compose up to 18.7% and 35% of the assemblage, respectively. These two species groups also typically have their highest abundances in neritic environments (Brinkhuis and Zachariasse, 1988). The base of the chalk in core 28 has more common large-sized cysts, such as *Areoligera*, *Glaphyrocysta*, and *Palaeocystodinium*, whereas the top of the chalk in core 27 has more common smaller cysts, such as *Senegalinium*, *Piercites*, *Kallosphaeridium*, and *Impagidinium*. This indicates a weak size grading of the dinoflagellate cysts, similar to that observed with the foraminifera.

DSDP Site 386, Southeast Bermuda Rise. The chalk at DSDP Site 386 was described by Norris et al. (2000a). There are two discrete chalk layers, 0.90 m and 1.15 m thick, separated by 17 cm of red and green pelagic clay (Fig. 6). Both chalk beds are cross-laminated at the base, horizontally laminated in the middle, and have very fine grained, structureless tops. The base of each chalk layer is composed of granular carbonate and rare minute planktic foraminifera that grade upward into very fine grained nannofossil chalk. Exotic nannofossils (*Nannoconus* sp.) are present in the chalk sequence in addition to a typical uppermost Maastrichtian nannoflora. Organic microfossils are absent in this sequence.

The upper chalk bed can be firmly tied to the K-P boundary by its lithological and geochemical composition. A 5-cm-thick layer of green spherules, identical to the impact ejecta on Blake Nose, is present at the base of the upper chalk bed at DSDP Site 386 (Fig. 7). The spherule layer contains small amounts of angular limestone and chert fragments similar to those in the ejecta bed at ODP Site 1049 on Blake Nose (Martínez Ruiz et al., 2001; Norris et al., 1999) and in DSDP Site 603 (Klaver et al., 1987). The spherules resemble a fine green sand and have formed load casts into the underlying green claystone. Individual grains range to ~1 mm diameter. A chondritic ratio of Pt to Ir is also present in the cross-bedded chalk at the base of the upper chalk layer (Fig. 6 and Norris et al., 2000a). The meteoritic Pt/Ir signature is clearly associated with the spherule horizon at the base of the upper chalk bed and strongly supports an impact origin and K-P age for the chalk sequence.

Failure of the eastern North Atlantic margin

Few complete K-P boundary records have been recovered in the deep eastern North Atlantic. However, DSDP Hole 398D on the Iberian Abyssal Plain has a well-preserved K-P boundary

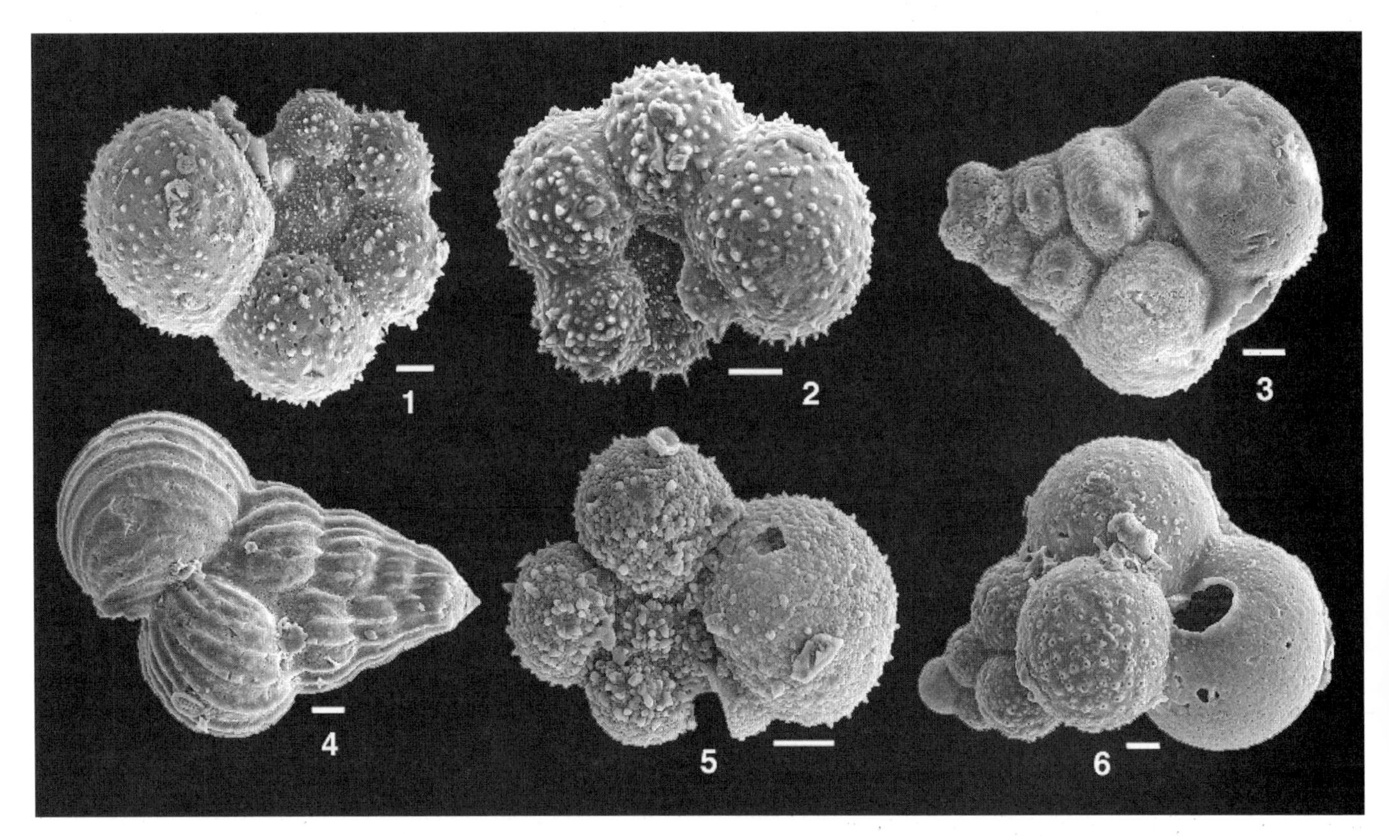

Figure 5. Minute planktic foraminifera recovered from turbidite chalk at Deep Sea Drilling Project (DSDP) Site 387 (core 28-1, 135 cm). Foraminifera represent largest specimens available and are juveniles of species common in fine fraction of Maastrichtian planktic foraminifera assemblages. Typical large species, such as globotruncanids and rugoglobigerinids, are absent or represented only by rare juveniles. Exclusive presence of small foraminifera suggests that chalk at DSDP Site 387 is not typical Maastrichtian pelagic chalk. Scale bars are 10 μm. 1, *Hedbergella* sp.; 2, *Hedbergella* sp.; 3, *Heterohelix glabrans*; 4, *Pseudoguembelina costulata*; 5, *Globigerinelloides subcarinatus*; 6, *Guembelitria cretacea*.

sequence ~80 km from the continental rise off Portugal and only 20 km south of the Vigo Seamount (Sibeut et al., 1979). The Upper Cretaceous and basal Paleogene sediments consist of red calcareous claystone and red to yellow nannofossil chalk. In general, calcareous microfossils are poorly preserved in the Upper Cretaceous, but large, partly dissolved foraminifera are present and represent species typical of Maastrichtian pelagic assemblages. Section 398D-41-2 contains an overall poor to moderately preserved nannofossil assemblage with *Micula murus* and *Micula prinsii*, indicative of the latest Maastrichtian nannofossil zone CC26b. Organic microfossils are absent in this section.

The lowermost Danian sediments contain planktic foraminifera indicative of the earliest Paleocene Pα. Biozone (Iaccarino and Premoli Silva, 1979). Nannofossils are indicative of zone NP1 (Blechschmidt, 1979), and common, well-preserved *Thoracosphaera* cysts are found at 39 cm within section 398D-41-2, immediately above the K-P boundary. Stable isotope analysis of bulk carbonate delineates a −1.3‰ shift in $\delta^{13}C$ precisely at the biostratigraphic K-P boundary that is consistent in magnitude and direction with stable isotope results from other, better known K-P boundary sections (Létolle, 1979). Furthermore, our observations show that a 1-mm-thick layer of green spherules is associated with the biostratigraphic K-P boundary (Fig. 8). As elsewhere in the North Atlantic, the spherules are thought to represent impact ejecta that has been diagenetically altered to clay minerals. Therefore, the biostratigraphy, isotope stratigraphy, and presence of spherules suggest that the K-P boundary is reasonably complete and can be identified precisely at 42 cm depth in section 398D-41-2 (795.44 mbsf).

Maastrichtian and Danian sediments at DSDP Site 398D are generally flat lying save for an interval ~70 cm thick immediately below the K-P boundary (Fig. 8). There, the chalk and calcareous claystone are slumped into a series of recumbent folds. The base of the slumped interval is laminated and has a sharp contact with underlying bioturbated chalk. The laminations are probably not sedimentary structures, but appear to represent stretched bedding along the sole of the slump. The slumped sediments show somewhat chaotic structure near the base. The rest of the slump contains a set of two or three recumbent folds before the red chalk grades into the overlying structureless white chalk. The structureless chalk is ~12 cm

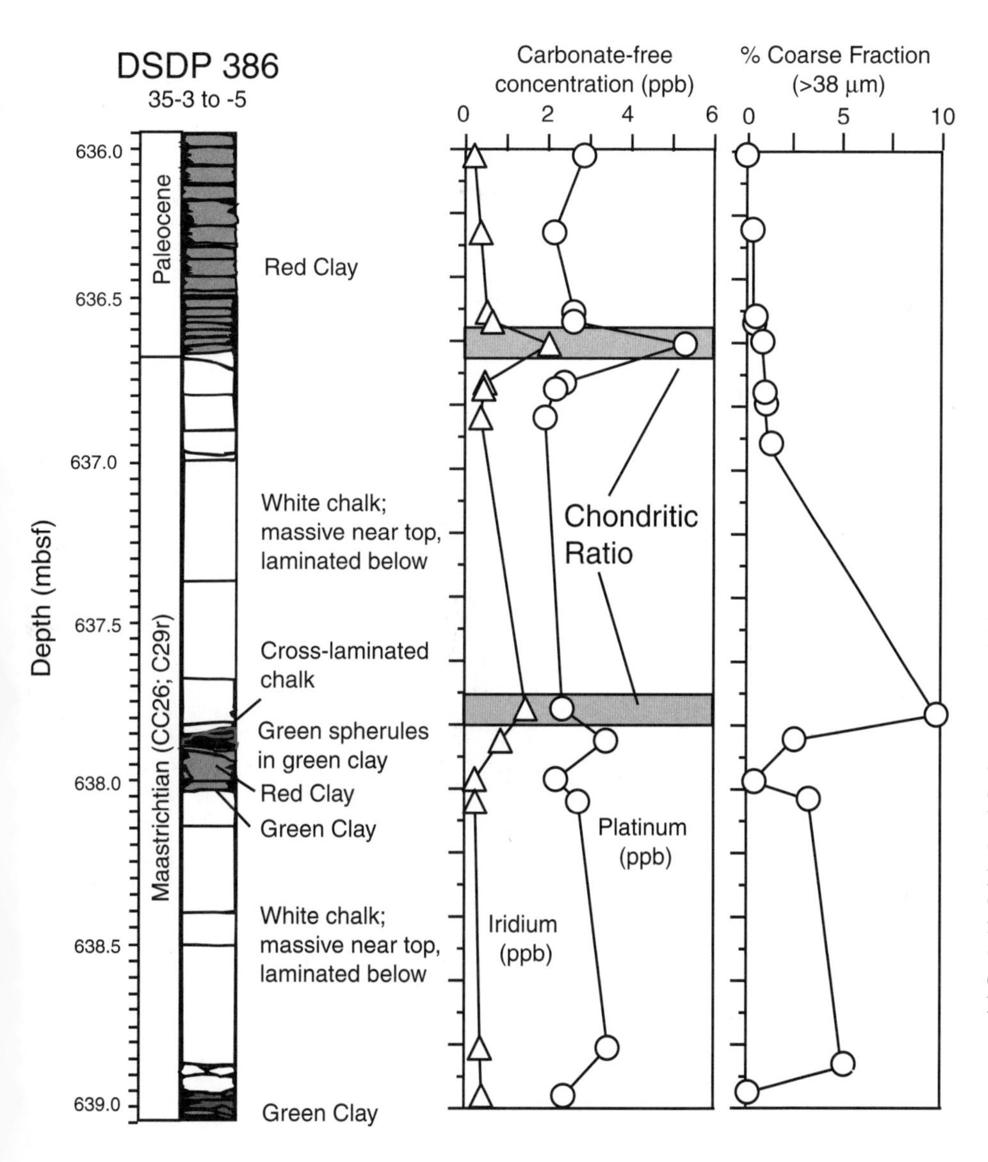

Figure 6. Sediment geochemistry, and % coarse fraction of Cretaceous-Paleogene mass-wasting deposits at Deep Sea Drilling Project (DSDP) Site 386 on Bermuda Rise, western North Atlantic (after Norris et al., 2000a). Both chalk beds display classic evidence for mass flows, including size grading from coarse-grained bases to fine-grained tops, lamination produced by high flow velocities (see Fig. 7) and sharp erosion surfaces at their bases. Platinum and iridium have peaks in red clay at top of upper chalk (bed 5) showing that chalk accumulated before all fine-grained meteoritic component of impact dust had settled to ocean floor. Pt/Ir ratio approaches that seen in chondrites in two places (gray bands), demonstrating meteoritic source. Percent coarse fraction (derived by weighing dry sample before and after washing through 38 μm screen) reflects graded bedding of each calcareous turbidite. Record of percent coarse fraction is biased in two ways: (1) base of upper turbidite (sample at 637.85 mbsf [meters below seafloor]) is partially cemented, which may result in overestimate of original percent coarse fraction, and (2) sample in spherule bed (at 637.9 mbsf) underestimates percent coarse fraction because most of tektites have been altered to clay minerals.

thick and is penetrated by dark brown or greenish, back-filled burrows filled with Paleocene sediment (Sibeut et al., 1979). Our studies show that planktic foraminifera in the folded chalk are not size sorted, and lack uniquely shallow-water species, whereas the nannofossils in the folded chalk are poor to moderately preserved, and slightly more diverse than the poorly preserved assemblages in the underlying sediments. This slump is the only such deposit within the upper Maastrichtian at this site. Overlying Paleocene sediment is not deformed.

The basal part of the Paleocene immediately above the spherule layer consists of 2–3 cm of burrow-mottled white or light gray chalk similar to that at the top of the underlying slumped Maastrichtian chalk. Burrows filled with greenish-brown Paleocene sediment penetrate into the top of the white chalk above the spherule bed. The spherule bed and the K-P boundary both occur within the white chalk.

DISTRIBUTION AND INTERPRETATION OF K-P MASS-WASTING DEPOSITS

Evidence from sites on the North American continental slope and rise, Bermuda Rise, and the eastern North Atlantic suggests that mass-wasting deposits of K-P boundary age are widespread throughout the North Atlantic. We believe that slope failure occurred at a number of points along the continental margin of eastern North America and probably on the Bahamian platform. Diverse sources of remobilized sediment help to explain the variations in carbonate content and microfossil composition between the mass-flow deposits at the various sites. There may also have been multiple generations of mass flows in the weeks or months following the Chicxulub impact, either because of aftershocks on faults released by the

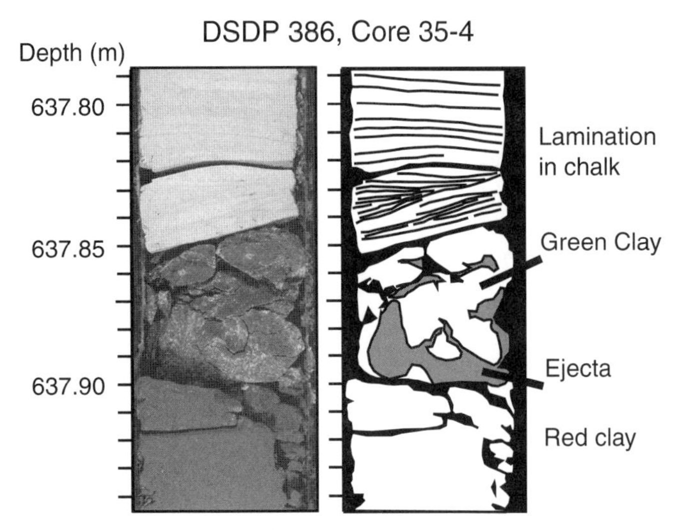

Figure 7. Spherule bed in Deep Sea Drilling Project (DSDP) Site 386 (after Norris et al., 2000a). Sand-sized spherules have formed load casts into underlying green clay and are overlain by cross-bedded chalk. Spherule bed and overlying chalk show classic evidence for decreasing flow energies that include (1) load casts in spherule bed, (2) decreasing grain size from spherule bed into chalk, and (3) upward transition from cross-bedded chalk to plane-laminated chalk.

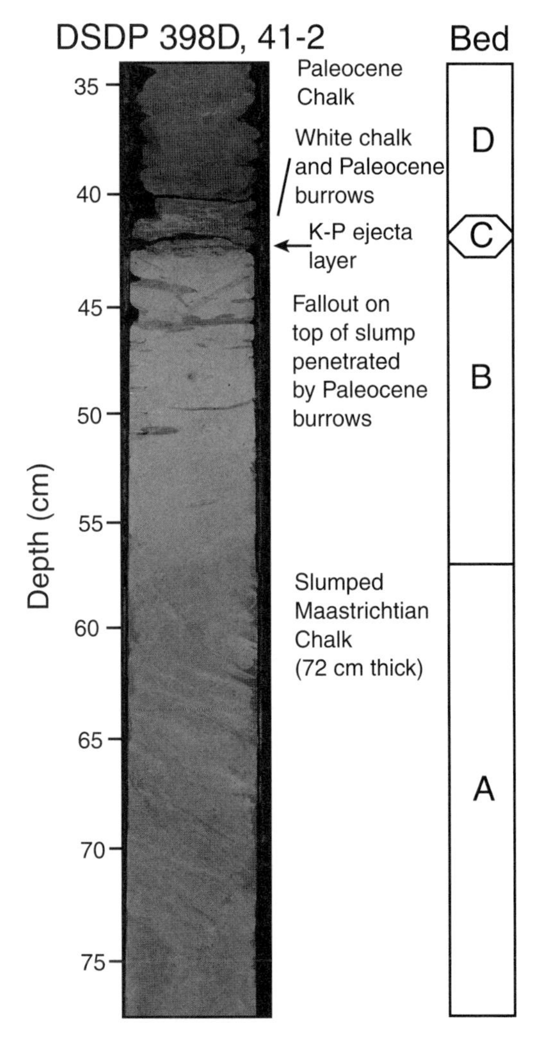

Figure 8. Evidence for slope failure in Iberian Abyssal Plain off Portugal. Core photograph of Cretaceous-Paleogene (K-P) boundary in Deep Sea Drilling Project (DSDP) Hole 398D showing folded, slumped chalk (bed A) of latest Cretaceous age, very fine grained white chalk fluidized by slump (bed B) immediately below K-P ejecta layer (bed C), and undisturbed, brown chalk containing earliest Paleocene microfossils (bed D). Occurrence of mass-flow deposits just below ejecta bed strongly suggests mass failure of Portuguese margin synchronous with Chicxulub impact.

initial impact-generated earthquake, or by failure of poorly consolidated sediments deposited by the initial gravity flows.

We have previously shown seismic and coring data that suggest widespread slope failure of the Blake Plateau that is synchronous with the K-P boundary. Failure evidently occurred along a decollement as much as 200 m subbottom that broke the overlying consolidated sediments into kilometer-scale blocks (Klaus et al., 2000). These blocks fragmented downslope, resulting in pervasive deformation in the Maastrichtian at the deepest site on Blake Nose. Much of the deformation was complete by the time impact ejecta began to rain through the water column, although some secondary slides may have redeposited impact ejecta in some cases.

There is also evidence for slope failure north of the Blake Plateau on the New Jersey margin. The sequence penetrated at DSDP Site 605 is comparable in many respects to that drilled at Sites 1050 and 1052 on the Blake Plateau. At all three sites, the K-P boundary is biostratigraphically complete, although the short duration of magnetochron C29r or planktic foraminifer biozones suggests there may be some missing sediment. In each case, the Paleocene sedimentary sequences are relatively undisturbed, and the Maastrichtian is largely intact, but cut by high-angle faults and small fractures. Bedding, particularly in Sites 1052 and 605, is frequently tilted from horizontal. Strata in DSDP Site 605 alternate between nearly horizontal bedding and tilted bedding, suggesting that the core may have cut through a series of rotated fault blocks rather than being related to downslope creep of the entire section during the Cenozoic. Unfortunately, DSDP site 605 was not drilled deep enough to penetrate the base of the Maastrichtian, so we cannot tell if there is a basal decollement like that found in ODP Site 1050.

The seismic stratigraphy downdip of DSDP Site 605 is also broadly consistent with the style of deformation seen on the Blake Plateau. The reflector stratigraphy of the New Jersey transect suggests that most deformation is concentrated in the early Maastrichtian and Campanian, as it is on Blake Nose. Wavy and hummocky reflectors below Horizon A* suggest that the most extensive deformation occurred throughout the Maastrichtian sequence ~14 km to the southeast of DSDP Site 605 (van Hinte and Wise, 1987). Disrupted reflectors have also been reported below the K-P boundary on MCS line USGS 23, near USGS line 25. Max et al. (1999) interpreted the disrupted reflectors in line USGS 23 as evidence of massive submarine slope failure on the U.S. continental rise at the K-P boundary. Similar, massive deformation is visible in MCS line TD-5 downdip of ODP Site 1052 on Blake Nose, and probably represents the breakup of large slide blocks where the slope begins to level out (Klaus et al., 2000). Reprocessing the seismic data from USGS line 25 may improve the resolution of these slide blocks and help evaluate the distribution of slumped sediments on the New Jersey margin.

The distal equivalents of submarine slides derived from places such as Blake Nose or the New Jersey slope are not preserved along much of the continental rise. However, the sequence seen at DSDP Site 603 may represent the deep-water turbidites that should have been derived from transport off the shelf and slope. The absence of calcareous microfossils in K-P boundary sediments at DSDP Site 603 (Klaver et al., 1987; van Hinte and Wise, 1987) suggests that sediment was remobilized either from nearshore deposits that bypassed the shelf and slope during transport into deep water or from preexisting turbidites deposited higher up on the continental rise. Decomposition of organic matter in the turbidites may have contributed to the dissolution of calcareous fossils or the turbidites may have carried little carbonate in the first place. Turbidites derived from more calcareous facies on the shelf and slope apparently bypassed DSDP Site 603 on their way to the Bermuda Rise.

The seismic stratigraphy at DSDP Site 603 suggests that the K-P mass-failure deposits were just one of a long sequence of turbidites deposited before and after the K-P boundary. Horizon A* and the K-P Boundary occur only 60 m below a widespread erosion surface of Oligocene age (Horizon A^u), but both reflectors are nearly horizontal and are underlain by generally flat-lying, parallel reflectors typical of turbidite facies (van Hinte and Wise, 1987). Hence, it seems that turbidites are not unique to the K-P boundary on the continental rise.

Offshore, on the Bermuda Rise, we currently have only a record of chalk turbidites without the clastic mass-flow deposits seen at DSDP Site 603. The sedimentary structures and size grading of the microfossils in both DSDP Sites 386 and 387 demonstrate that chalk layers at both sites were rapidly deposited from gravity flows, such as turbidity currents, followed by slower accumulation of very fine grained sediment suspended in the water column. The progression within the chalk beds from cross-bedding to plane lamination (e.g., Fig. 7), and then to fine-grained, structureless tops is classic evidence for the decreasing flow energies associated with turbidity current deposition. Shallow-water dinoflagellates as well as pollen and plant debris in the chalk at DSDP 387 are consistent with transport of sediment from the continental shelf (Norris et al., 2000a). The absence of appreciable quantities of clastics in the chalks at DSDP Sites 386 and 387 suggests that the turbidites currently known from the Bermuda Rise may have been derived from a different source region along the continental slope, such as Blake Plateau or the Bahama platform, than the silty sandstones at DSDP Site 603.

The sizes of planktic foraminifera in sites on Bermuda Rise and on Blake Nose provide additional evidence for offshore transport of K-P boundary mass flows (Fig. 9). Assemblages of Cretaceous planktic foraminifera are normally skewed toward small size classes (63–150 μm), but also contain appreciable numbers of specimens in the size classes between 250 and >400 μm. However, the Blake Nose samples within the ejecta bed and overlying basal Danian sediments are skewed toward the largest size classes and contain very few specimens smaller than 150–125 μm (Norris et al., 1999; Huber et al., this volume). In contrast, the chalk turbidites at DSDP Site 387 contain few, if any, specimens of Cretaceous planktic foraminifera larger than 125 μm (Figs. 5 and 9).

The simplest explanation for the near absence of small foraminifera on Blake Nose and their exclusive presence on Bermuda Rise is offshore transport of fine-grained sediment during the K-P event. We suggest that slope failure on Blake Nose and elsewhere along the southeastern margin of North America suspended large quantities of fine-grained carbonate (Fig. 10). The coarse fractions containing large foraminifera were redeposited close to their source. Some of these large foraminifera were mixed into impact ejecta, either as the ejecta settled through the water column, or when the ejecta was redeposited from higher up the slope. Bioturbation or continued settling of the resuspended sediment plume mixed large Cretaceous foraminifera

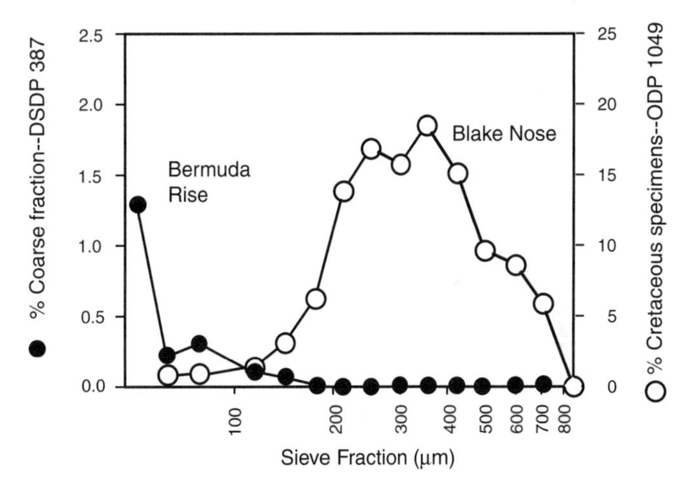

Figure 9. Size distribution of Cretaceous planktic foraminifera from continental margin (Blake Nose, open circles; Ocean Drilling Program [ODP] Hole 1049C, 8X-5, 85 cm, 112.95 m below seafloor) and basinal site (Bermuda Rise, filled circles; Deep Sea Drilling Project [DSDP] Site 387; 29R-2, 63–65 cm). Continental margin sample includes all Cretaceous planktic foraminifera as discussed by Norris et al. (1999). Sample from Bermuda Rise includes fine-grained clastic particles as well as foraminifera, but visual observation shows that foraminifera track size distribution of entire sample. Typical Maastrichtian faunas are dominated by small individuals, but also contain appreciable numbers of large (>300–400 μm) individuals. In contrast, skewed size distributions of foraminifera on Blake Nose and Bermuda Rise suggest foraminifer assemblages were winnowed of their fine-grained component on Blake Nose, and these small specimens were transported offshore to DSDP Site 387 in turbidity currents.

above the ejecta bed, where it was subsequently mixed with Paleogene sediments on the continental margin. The fine-grained nannofossils and minute foraminifera were suspended during the initial slope failure and transported offshore in density currents (Fig. 10).

The volumes of resuspended sediments transported onto Bermuda Rise must have been very large. The chalk bed at DSDP Site 387 is as thick as 17.6 m when gaps in core recovery are accounted for. DSDP Site 387 is more than 700 km from the present continental slope, and DSDP Site 386, where the turbidite bed are each ~1.2 m thick, is nearly 1200 km from the continental margin. Both sites may have penetrated the chalk beds where the turbidites filled in submarine valleys and exaggerated the regional thickness of the mass-wasting deposits. Coring gaps also allow for the possibility that the chalk deposit at DSDP 387 may consist of several stacked carbonate turbidites, rather than a single bed. The record at DSDP Site 386 shows that there were at least two mass flows at or near the K-P boundary (Norris et al., 2000a). Apparently, slope failure not only generated a mixture of clastic and carbonate turbidites at different points along the continental margin, but also produced a series of mass flows that became stacked on top of one another in some places.

The relative ages of multiple mass flows are suggested by the stratigraphy and geochemistry of the chalk turbidites at DSDP Site 386 (Fig. 11). We have previously interpreted the K-P deposits there as indicating two separate mass-flow deposits, the first released before the ballistic arrival of impact ejecta, and the second originating with hours to days of the impact, after most ejecta had reached the seafloor (Fig. 10). The two turbidites are separated by a slump of pelagic claystone derived from a local topographic high on the seabed (Fig. 11; Norris et al., 2000a). The area around DSDP Site 386 had considerable bottom relief in the Cretaceous (Fig. 12), and the submarine ridges may have served to funnel turbidites derived from the abyssal plain to the west. The nearby ridges and abyssal hills could be the source of the red clay that was apparently remobilized by the passage and deposition of the first of the chalk turbidites. We speculate that the first chalk turbidite (Fig. 11, unit B) represents a mass failure associated with the impact-generated earthquake, whereas the second chalk turbidite (Fig. 11, unit D) may represent a flow triggered by tsunamis in the Atlantic or by aftershocks on faults released by the impact event. Paleocene sediments representing the final fallout of very

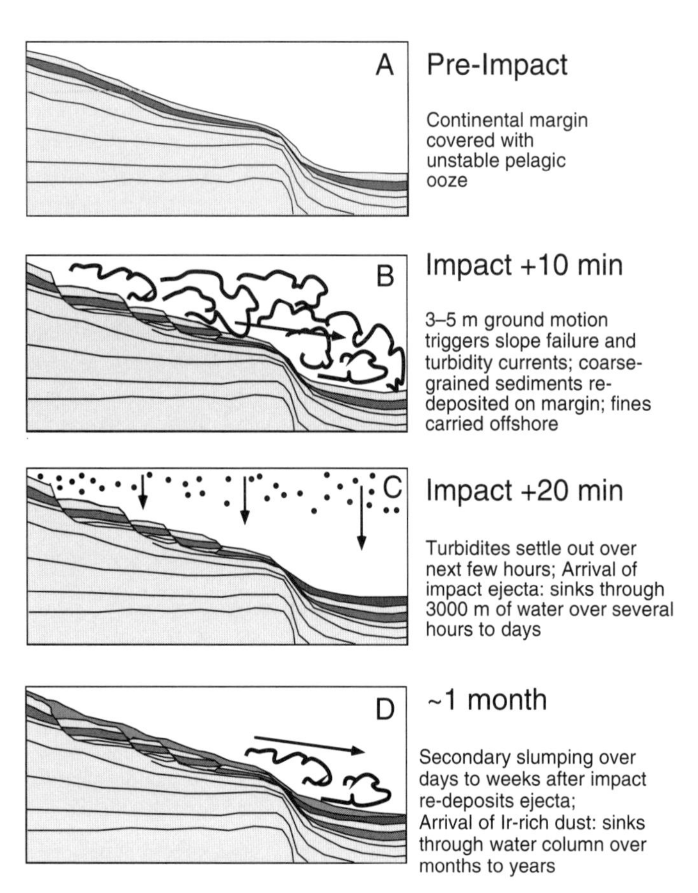

Figure 10. Model for sequence of events surrounding mass wasting of Atlantic continental margin. Slumping precedes arrival of impact ejecta as consequence of disruption caused by impact-generated earthquake, and time required for impact ejecta to sink through 2–4 km of seawater. Secondary slides created by aftershocks on continental margin faults, or by remobilization of previously deposited mass-flow deposits, result in complex stratigraphy of multiple mass-flow deposits.

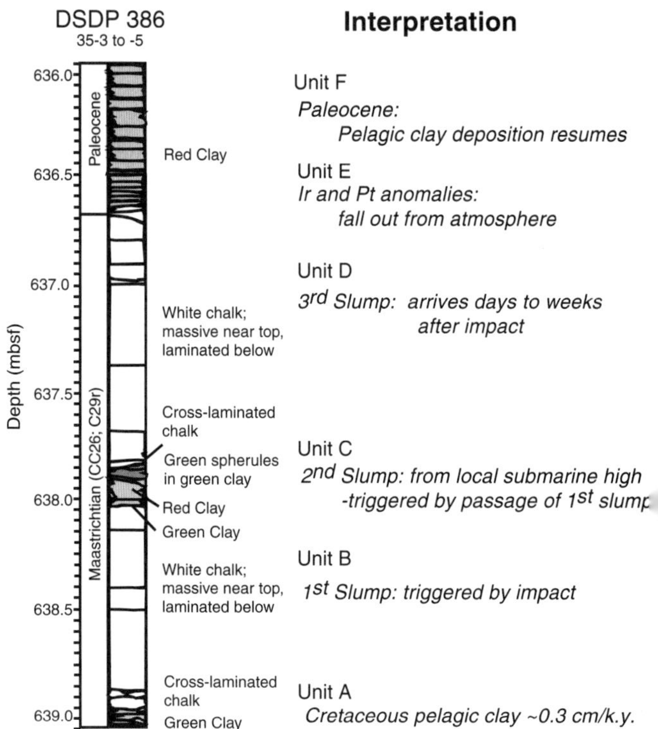

Figure 11. Interpretation of depositional units at Deep Sea Drilling Project (DSDP) Site 386 (mbsf is meters below seafloor).

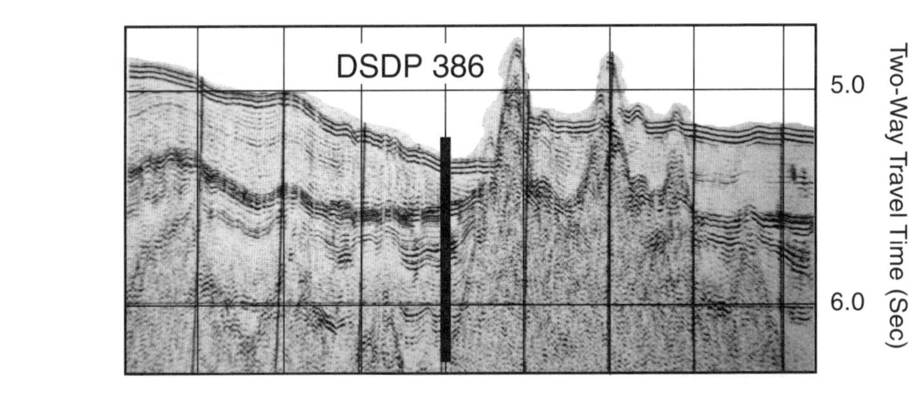

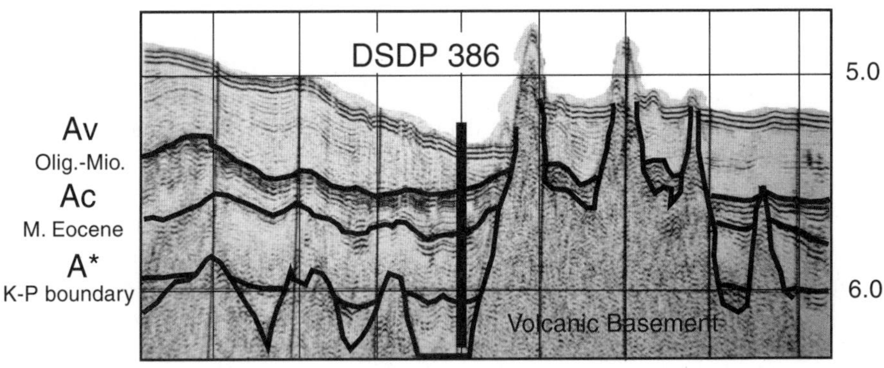

Figure 12. Seismic stratigraphy in vicinity of Deep Sea Drilling Project (DSDP) Site 386. Steep bottom topography at Cretaceous-Paleogene (K-P) boundary (indicated by basin-filling nature of Horizon A*) may account for presence of multiple mass-wasting deposits at DSDP Site 386 by contributing slumps of pelagic clay following erosion by passing turbidite (to form slump 2 in Fig. 11).

fine grained meteoric debris account for the Ir and Pt anomalies (Fig. 11, unit E) and are overlain by claystone recording the return to normal, extremely slow, pelagic clay sedimentation (Fig. 11, unit F).

EXTENT OF MASS WASTING IN THE NORTH ATLANTIC—EVIDENCE FROM IBERIA

Across the Atlantic, the record in DSDP Hole 398D provides additional evidence that mass wasting preceded the arrival of impact ejecta on the seabed. We interpret the record to reflect a single mass-failure event immediately predating the arrival of impact ejecta on the seafloor. The slump preserved below the K-P boundary at DSDP Hole 398D (Fig. 8, bed A) must have been locally derived because it contains microfossils and sediment similar to that preserved below the slump. The microfossil assemblages lack uniquely shallow-water species, unlike the turbidite sequence at DSDP Site 387. The folded sediment in bed A did not travel long distances, unlike the deposits at DSDP Sites 386 and 387, because it did not become size graded or develop sedimentary structures indicative of a fluidized gravity flow. The structureless white chalk overlying the slump (Fig. 8, bed B) probably represents material that was suspended during slope failure and settled out before arrival of impact ejecta (Fig. 8, bed C) on the seafloor. Sediment failure must have stopped within a day or two after the impact because the spherule layer, interpreted as impact ejecta, is undeformed on top of the slump deposit. Sedimentation resumed in the earliest Paleocene (Fig. 8, bed D) and burrowing organisms introduced Paleocene sediment into the ejecta layer (Fig. 8, bed C) and the upper part of the slumped Maastrichtian chalk (Fig. 8, bed B).

The spherule bed is buried by ~2 cm of white chalk at the base of bed D. Greenish sediments filling burrows in the chalk have $\delta^{13}C$ ratios typical of overlying reddish-brown Paleocene sediment, whereas the white chalk has a $\delta^{13}C$ chemistry similar to underlying Cretaceous sediments (Létolle, 1979). The chalk may represent the last fine-grained sediment suspended in the water column during the initial slump. The slump must have occurred just before the fallout of impact ejecta from the atmosphere. Coarse, sand-sized ejecta particles must have settled through the water column to arrive on the seabed before the fine-grained nannofossil plume completely sank to the bottom. Alternatively, the arrival of the impact ejecta on the seafloor may have remobilized previously deposited white chalk and become buried by a small flow of this material to form the first sediments of the Paleogene. Bioturbation has largely obscured any sedimentary structures that may have once existed in the chalk overlying the spherule bed. However, the record in DSDP Hole 398D leaves little doubt that slumping must have occurred at the same time as the Chicxulub impact event. Our data suggest that the impact caused slope failures along the Portuguese coast whether by the direct effects of ground motion produced by the impact earthquake or the secondary effects of tsunamis crossing the Atlantic.

ACOUSTIC EVIDENCE FOR MASS-SLOPE FAILURE

Seismic data from Blake Nose and the New Jersey continental margin (near DSDP Site 605) suggest that there was large-scale gravity sliding in both regions during the Late Cretaceous. Wavy and hummocky reflectors suggestive of gravity slides are present just below the K-P boundary southeast of DSDP Site 605 between shotpoints 3600 and 3800 on USGS seismic reflection line 25 (van Hinte and Wise, 1987; p. 281, their Fig. 3). We believe that the strong reflector immediately overlying the hummocky sequence of reflectors in the Cretaceous is correlative with Horizon A* on both the Bermuda Rise (where the A series of reflectors was first named) and on Blake Nose.

Tucholke (1979) and Tucholke and Mountain (1979) mapped the distribution of Horizon A* over much of the Bermuda Rise and later workers extended the distribution of the reflector to the New England Seamount chain (Swift et al., 1986) and in the Sohlm Abyssal Plain south of the Grand Banks (Swift et al., 1986; Ebinger and Tucholke, 1988). The eastern extent of the reflector is ~1300 km from the continental slope of eastern North America and 1400–1500 km from the present shoreline (Tucholke and Mountain, 1979). The eastern limits of Horizon A* may reflect either the actual disappearance of the chalk turbidites or the loss of the acoustic impedance contrast between the chalk and increasingly carbonate-rich sediments deposited close to the mid-Atlantic Ridge crest. It is frequently difficult to correlate Horizon A* from the Bermuda Rise up onto the continental margin because of widespread erosion along the base of the continental slope and rise during the Oligocene. Accordingly, there have been substantial differences of opinion in the location of Horizon A* in seismic reflection profiles that cross the mid-Atlantic margin (e.g., cf. Kitgord and Grow, 1980, and Schlee and Grow, 1980).

We believe Horizon A* reflects the upper surface of K-P boundary mass-wasting deposits over large portions of the western North Atlantic. Our evidence is mainly from seismic correlations to DSDP Sites 386 and 387 where there is a direct tie between the acoustic horizons and the biostratigraphy (e.g., Fig. 12). However, the turbidites may not be resolvable by acoustic means everywhere on the Bermuda Rise, either because they are too thin or because they are not present on former topographic highs on the seafloor. In our interpretation, the K-P mass flows should occur only in former submarine basins because gravity flows generally will not ride over topographic highs. Neal Driscoll (Woods Hole Oceanographic Institution, 2000, personal commun.) noted that sequences equivalent with Horizon A* on the southern Bermuda Rise often show mounded acoustic character suggestive of hemipelagic sedimentation. Although Horizon A* is often associated with turbidite facies in the Sohm Abyssal Plain, Bermuda Rise, and the continental rise near DSDP Site 603, it is clear that the entire deep western North Atlantic was not covered with mass-flow deposits. Nonetheless, the region affected by mass flows is enormous, representing an area of as much as $\sim 3.9 \times 10^6$ km^2. Our calculations do not take into account mass flows that may have occurred in the eastern Atlantic or in the Caribbean and Pacific. Therefore we believe that mass-wasting deposits generated immediately after the K-P impact represent by far the most widespread series of flows that were caused by a single event on the Earth.

K-P MASS WASTING AND THE CCD

Tucholke and Vogt (1979b) used the distribution of carbonate sediments in backtracked Atlantic DSDP sites to estimate changes in the depth of the CCD from the Late Jurassic to the present (Fig. 13). Their reconstruction (and later compilations of Arthur and Dean, 1986; Barrera and Savin, 1999; Frank and Arthur, 1999) suggests that the North Atlantic CCD was at ~3 km depth during the Campanian and early Maastrichtian, plunged to more than 5.2 km depth in the late Maastrichtian, and then rebounded to ~4 km depth in the Paleocene. In contrast, the CCD in the Pacific, Indian Ocean, and South Atlantic rose through the Maastrichtian to 3–4 km by the K-P boundary (Barrera and Savin, 1999). The overall ~1 km increase in depth of the North Atlantic CCD between the Late Cretaceous and the Paleogene is well supported, but the sharp ~2.2 km increase in the Maastrichtian is entirely dependent upon data from three localities, DSDP Sites 385, 386, and 387.

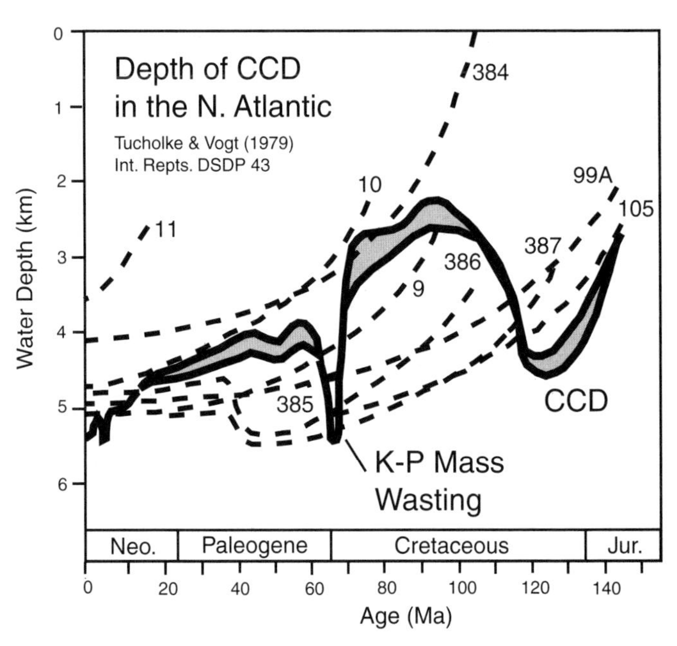

Figure 13. Reconstruction of Jurassic to recent carbonate compensation depth (CCD) (modified from Tucholke and Vogt, 1979a). We suggest that dramatic increase in depth of CCD in latest Maastrichtian is artifact of introduction of large quantities of pelagic carbonates below CCD as result of Cretaceous-Paleogene (K-P) mass wasting. Deep Sea Drilling Project (DSDP) Sites 385–387 represent only stratigraphic control on CCD drop in Maastrichtian. Other sites (DSDP Sites 9 and 99A) do not have calcareous sediments across K-P boundary.

We believe the abundant evidence for mass wasting of Maastrichtian carbonates in all three of these sites sweeps away any support for an abrupt fall in the North Atlantic CCD at the close of the Cretaceous. The preservation of carbonates in each site owes much to the massive size of the K-P slope failure, which introduced sufficient carbonate below the CCD that most could be preserved despite corrosive bottom waters. The widespread introduction of massive quantities of fine-grained carbonates into the deep sea may have caused an increase in the CCD and created a marked change in carbonate chemistry of the North Atlantic Ocean. However, we have no direct evidence for the extent of this change in the North Atlantic CCD because all our present records come from sites contaminated with mass flows.

BIOTIC EFFECTS OF K-P SLOPE FAILURE

Slope failures associated with the impact may have had substantial effects on the marine biota. Devastation wrought by slope failure may account for the delayed recovery of North American molluscan assemblages compared to faunas elsewhere in the world (Jablonski, 1998). The combination of slumping on the slope and continental rise coupled with extensive erosion of the shelf by tsunamis should have been particularly severe around the North American margin. Endemic North American species and those without pelagic larvae had particularly high extinction rates compared to plankton-feeding species during the Cretaceous-Tertiary mass extinction (Hansen, 1980; Jablonski, 1987, 1998).

CONCLUSIONS

K-P boundary sites on the Atlantic margin of North America, the Bermuda Rise, and offshore Portugal display sedimentological, paleontological, and geochemical evidence for massive slope failure at the Cretaceous-Paleogene boundary. Sites on Blake Plateau, offshore Florida, and on the New Jersey slope show evidence for slumping of the upper several hundred meters of Maastrichtian sediments. Deformation appears to stop at the K-P boundary, because overlying early Danian sediments are frequently biostratigraphically complete and unbroken by the extensive deformation seen below the boundary. Seismic reflection profiles on Blake Plateau display evidence for kilometer-scale slide blocks, and profiles on the New Jersey margin show chaotic reflectors suggestive of mass flow of Maastrichtian sediments below the K-P boundary. Distal mass-flow deposits are in places clastic deposits (DSDP Site 603) and in other places carbonates (DSDP Sites 386 and 387), suggesting different source areas and points of failure along the continental margin. The presence of two chalk turbidites at DSDP Site 386 further suggests that there may have been multiple phases of mass wasting during and shortly after the Chicxulub impact. The distribution of acoustic reflector A*, tied to the K-P boundary at DSDP Sites 386 and 387, suggests that there was slope failure along much of the eastern North American margin. Across the Atlantic, DSDP Hole 398D shows that slumping immediately preceded the arrival of impact ejecta on the seafloor.

Our results show that large impact events such as the Chicxulub impact leave a distinctive sedimentary record in the deep sea that includes the widespread, massive deposition of shallow-marine, calcareous sediments below the CCD, and geochemical signatures of the impacting bolide. How the impact event triggered slope failure is unclear, but the absence of impact ejecta in the lowest of two chalk turbidites in DSDP Site 386, and the presence of a slump just below impact ejecta in DSDP Hole 398D, suggests that slope failure occurred before ejecta had reached the ocean bottom. The extensive slumping of consolidated Maastrichtian sediments on Blake Plateau suggests that deformation was produced by earthquakes associated with the initial impact or slip on previously quiescent faults released by the impact. Tsunamis generated by the impact or as a consequence of the initial slope failure may have caused further devastation and later phases of slumping.

Passive margins around the Atlantic, eastern Africa, Australia, and New Zealand should preserve particularly good records of seismic events associated with large marine or terrestrial impacts, because impact-generated turbidites will not have been subducted or greatly deformed by tectonic processes. In the North Atlantic, large impacts in the Chesapeake Bay ca. 35 Ma (Poag et al., 1994; Poag, 1996) and on the Canadian margin at 49 Ma (Jansa and Pe-Piper, 1987; Aubry et al., 1990) are likely to have precipitated large-scale slope collapse. Unfortunately present drill holes from the deep western North Atlantic contain large gaps in core recovery from these time periods that prevent a definitive test of this idea. Still, it may be that the deep-sea record has been sculpted by impact events more than is generally appreciated.

ACKNOWLEDGMENTS

Our work has been supported by grants from the National Science Foundation and the U.S. Science Support Program of the Ocean Drilling Program. We thank the Ocean Drilling Program for core sampling and photography, Tim Bralower, Adam Klaus, and Neal Driscoll for discussions that improved the manuscript, Lu-Ping Zou and Ellen Rosen for sample preparation, and Tracy Frank, Dick Buffler, and Ken MacLeod for excellent, detailed reviews.

REFERENCES CITED

Albertão, G.A., Koutsoukos, E.A.M., Regali, M.P.S., Attrep, M., Jr., and Martins, P.P., Jr., 1994, The Cretaceous–Tertiary boundary in southern low-latitude regions: Preliminary study in Pernambuco, northeastern Brazil: Terra Nova, v. 6, p. 366–375.

Alvarez, L.W., Alvarez, W., Asaro, F., and Michel, H.V., 1980, Extraterrestrial causes of the Cretaceous–Tertiary extinction: Science, v. 208, p. 1095–1108.

Arthur, M.A., and Dean, W.E., 1986, Cretaceous paleoceanography of the western North Atlantic Ocean, *in* Vogt, P.R., ed., The Western North Atlantic region: Boulder, Colorado, Geological Society of America, Geology of North America, v. M, p. 617–630.

Aubry, M.-P., Gradstein, F.M., and Jansa, L.F., 1990, The late early Eocene Montagnais bolide: No impact on biotic diversity: Micropaleontology, v. 36, p. 164–172.

Barrera, E., and Savin, S.M., 1999, Evolution of late Campanian-Maastrichtian marine climates and oceans: Geological Society of America Special Paper 332, p. 245–281.

Benson, D.G., 1976, Dinoflagellate taxonomy and biostratigraphy at the Cretaceous-Tertiary boundary, Round Bay, Maryland: Tulane Studies in Geology and Paleontology, v. 12, p. 169–233.

Blechschmidt, G., 1979, Biostratigraphy of calcareous nannofossils: Leg 47B, Deep Sea Drilling Project, *in* Sibeut, J.-C., Ryan, W.B.F., et al., eds., Initial Reports of the Deep Sea Drilling Project, Volume 47B: Washington, D.C., U.S. Government Printing Office, p. 327–360.

Bohor, B.F., 1996, A sediment gravity flow hypothesis for siliciclastic units at the K/T boundary, northeastern Mexico, *in* Ryder, G., Fastovsky, D., and Gartner, S., eds., The Cretaceous-Tertiary event and other catastrophes in Earth history: Boulder, Colorado, Geological Society of America Special Paper 307, p. 183–195.

Boslough, M.E., Chael, E.P., Trucano, T.G., Crawford, D.A., and Campbell, D.L., 1996, Axial focusing of impact energy in the Earth's interior: A possible link to flood basalts and hotspots, *in* Ryder, G., Fastovsky, D., and Gartner, S., eds., The Cretaceous-Tertiary event and other catastrophes in Earth history: Boulder, Colorado, Geological Society of America Special Paper 307, p. 541–550.

Bourgeois, J., Hansen, T.A., Wiberg, P.L., and Kaufman, E.G., 1988, A tsunami deposit at the Cretaceous-Tertiary boundary in Texas: Science, v. 241, p. 567–570.

Bralower, T.J., Paull, C.K., and Leckie, R.M., 1998, The Cretaceous–Tertiary boundary cocktail: Chicxulub impact triggers margin collapse and extensive sediment gravity flows: Geology, v. 26, p. 331–334.

Brinkhuis, H., and Zachariasse, W.J., 1988, Dinoflagellate cysts, sea level changes and planktonic foraminifers across the Cretaceous–Tertiary boundary at Al Haria, northwest Tunisia: Marine Micropaleontology, v. 13, p. 153–191.

Busby, C.J., Blikra, L.H., and Renne, P.R., 1999, Coastal landsliding and catastrophic sedimentation triggered by bolide impact: Geological Society of America Abstracts with Programs, v. 31, no. 6, p. A42.

Colley, S., Thomson, J., Wilson, T.R.S., and Higgs, N.C., 1984, Post-depositional migration of elements during diagenesis in brown clay and turbidite sequences in the North East Atlantic: Geochimica et Cosmochimica Acta, v. 48, p. 1223–1235.

Ebinger, C.J., and Tucholke, B.E., 1988, Marine geology of Sohm Basin, Canadian Atlantic Margin: American Association of Petroleum Geologists Bulletin, v. 72, p. 1450–1468.

Firth, J.V., 1987, Dinoflagellate biostratigraphy of the Maastrichtian to Danian interval in the U.S. Geological Survey Albany core, Georgia: Palynology, v. 11, p. 199–216.

Frank, T.D., and Arthur, M.A., 1999, Tectonic forcings of Maastrichtian ocean-climate evolution: Paleoceanography, v. 14, p. 103–117.

Habib, D., and Miller, J.A., 1989, Dinoflagellate species and organic facies evidence of marine transgression and regression in the Atlantic Coastal Plain: Palaeogeogrpahy, Palaeoclimatology, Palaeoecology, v. 74, p. 23–47.

Hansen, T.A., 1980, Influence of larval dispersal and geographic distribution on species longevity in neogastropods: Paleobiology, v. 6, p. 193–207.

Hildebrand, A.R., and Boynton, W.V., 1990, Proximal Cretaceous/Tertiary boundary impact deposits in the Caribbean: Science, v. 248, p. 843–847.

Iaccarino, S., and Premoli Silva, I., 1979, Paleogene planktonic foraminiferal biostratigraphy of DSDP Hole 398D, Leg 47B, Vigo Seamount, Spain, *in* Sibeut, J.-C., Ryan, W.B.F., et al., eds., Initial Reports of the Deep Sea Drilling Project, Volume 47B: Washington, D.C., U.S. Government Printing Office, p. 237–253.

Jablonski, D.J., 1998, Geographic variation in the molluscan recovery from the end-Cretaceous extinction: Science, v. 279, p. 1327–1330.

Jablonski, D.J., 1987, Heritability at the species level: Analysis of geographic ranges of Cretaceous mollusks: Science, v. 238, p. 360–363.

Jansa, L.F., and Pe-Piper, G., 1987, Identification of an underwater extraterrestrial impact crater: Nature, v. 327, p. 612–614.

Keating, B.H., and Helsley, C.E., 1979, Magnetostratigraphy of Cretaceous sediments from DSDP 386, *in* Tucholke, B.E., Vogt, P.R., et al., eds., Initial Reports of the Deep Sea Drilling Project, Volume 43: Washington, D.C., U.S. Government Printing Office, p. 781–784.

Kitgord, K.D., and Grow, J.A., 1980, Jurassic seismic stratigraphy and basement structure of western Atlantic magnetic quiet zone: American Association of Petroleum Geologists Bulletin, v. 64, p. 1658–1680.

Klaus, A., Norris, R.D., Kroon, D., and Smit, J., 2000, Impact-induced K-T Boundary mass wasting across the Blake Nose, western North Atlantic: Geology, v. 28, p. 319–322.

Klaver, G.T., van Kempen, T.M.G., Bianchi, F.R., and van der Gaast, S.J., 1987, Green spherules as indicators of the Cretaceous/Tertiary boundary in Deep Sea Drilling Project Hole 603B, *in* van Hinte, J.E., and Wise, S.W., eds., Initial Reports of the Deep Sea Drilling Project, Volume 93: Washington, D.C., U.S. Government Printing Office, p. 1039–1060.

Létolle, R., 1979, Oxygen 18 and carbon 13 isotopes from bulk carbonate samples, Leg 47B, Initial Reports of the Deep Sea Drilling Project, Volume 47: Washington, D.C., U.S. Government Printing Office, p. 493–496.

MacLeod, K.G., and Huber, B.T., 2001, The Maastrichtian record at Blake Nose (western Atlantic) and implications for global palaeoceanographic and biotic changes, *in* Kroon, D., Norris, R.D., and Klaus, A., eds., Western North Atlantic Palaeogene and Cretaceous palaeoceanography: Geological Society [London] Special Publication 183, p. 49–72.

MacLeod, N., 1996, Nature of the Cretaceous–Tertiary planktonic foraminiferal record: Stratigraphic confidence intervals, signor-Lipps effect, and patterns of survivorship, *in* MacLeod, N., and Keller, G., eds., Cretaceous-Tertiary mass extinctions: Biotic and environmental changes: New York, W.W. Norton, p. 85–138.

MacLeod, N., Rawson, P.F., Forey, P.L., Banner, F.T., Boudagher-Fadel, M.K., Bown, P.R., Burnett, J.A., Chambers, P., Culver, S., Evans, S.E., Jeffrey, C., Kaminski, M.A., Lord, A.R., Milner, A.C., Milner, A.R., Morris, N., Owen, E., Rosen, B.R., Smith, A.B., Taylor, P.D., Urquhart, E., and Young, J.R., 1997, The Cretaceous-Tertiary biotic transition: Journal of the Geological Society, London, v. 154, p. 265–292.

Martínez-Ruiz, F., Ortega-Huertas, M., Palomo-Delgado, I., and Smit, J., 2001, K-T boundary spherules from Blake Nose (ODP Leg 171B) as a record of the Chicxulub ejecta deposits, *in* Kroon, D., Norris, R., and Klaus, A., eds., Western North Atlantic Palaeogene and Cretaceous palaeoceanography: Geological Society [London] Special Publication 183, p. 149–163.

Max, M.D., Dillon, W.P., Nishimura, C., and Hurdle, B.G., 1999, Sea-floor methane blow-out and global firestorm at the K-T boundary: Geo-Marine Letters, v. 18, p. 285–291.

Max, M.D., Pellanbarg, R.E., and Hurdle, B.G., 1997, Methane hydrate, a special clathrate: Its attributes and potential: Naval Research Laboratory, v. NRL/MR/r101-97-7926, 74 p.

McNulty, C.L., 1979, Smaller Cretaceous foraminifers of Leg 43, Deep Sea Drilling Project, *in* Tucholke, B.E., and Vogt, P.R., eds., Initial reports of the Deep Sea Drilling Project, Volume 43: Washington, D.C., U.S. Government Printing Office, p. 487–505.

Morrison, D., Chapman, C.R., and Slovic, P., 1994, The impact hazard, *in* Gehrels, T., ed., Hazards due to comets and asteroids: Tucson, University of Arizona Press, p. 59–92.

Mountain, G.S., and Miller, K.G., 1992, Seismic and geologic evidence for early Paleogene deep-water circulation in the western North Atlantic: Paleoceanography, v. 7, p. 423–439.

Nøhr-Hansen, H., and Dam, G., 1997, Palynology and sedimentology across a new marine Cretaceous–Tertiary boundary section on Nuussuap, West Greenland: Geology, v. 25, p. 851–854.

Norris, R.D., Firth, J., Blusztajn, J., and Ravizza, G., 2000a, Mass failure of the North Atlantic margin triggered by the Cretaceous/Paleogene bolide impact: Geology, v. 28, p. 1119–1122.

Norris, R.D., Klaus, A., and Kroon, D., 2000b, Middle Eocene deep water, the late Paleocene thermal maximum and continental slope mass wasting during the Cretaceous/Paleogene impact, *in* Kroon, D., Norris, R., and Klaus, A., eds., Paleogene and Cretaceous palaeoceanography of the western North Atlantic, Volume 2000: Geological Society [London] Special Publication 183, p. 23–48.

Norris, R.D., Huber, B.T., and Self-Trail, J., 1999, Synchroneity of the K-T oceanic mass extinction and meteorite impact: Blake Nose, western North Atlantic: Geology, v. 27, p. 419–422.

Norris, R.D., Kroon, D., Klaus, A., et. al., 1998, Blake Nose Paleoceanographic Transect, Western North Atlantic, Procedings of the Ocean Drilling Program, Part A: Initial Reports, v. 171B: College Station, Texas, Ocean Drilling Program, 749 p.

O'Campo, A.C., Pope, K.O., and Fischer, A.G., 1996, Ejecta blanket deposits of the Chicxulub Crater from Albion Island, Belize, *in* Ryder, G., Fastovsky, D., and Gartner, S., eds., The Cretaceous-Tertiary Event and other catastrophes in Earth history: Boulder, Colorado, Geological Society of America Special Paper 307, p. 75–88.

Okada, H., and Thierstein, H.R., 1979, Calcareous nannoplankton—Leg 43, Deep Sea Drilling Project, *in* Tucholke, B.E., Vogt, P.R., et al., eds., Initial reports of the Deep Sea Drilling Project, Volume 43: Washington, D.C., U.S. Government Printing Office, p. 507–573.

Olsson, R.K., Lui, C., and van Fossen, M., 1996, The Cretaceous–Tertiary catastrophic event at Millers Ferry, Alabama, *in* Ryder, G., Fastovsky, D., and Gartner, S., eds., The Cretaceous–Tertiary event and other catastrophes in Earth history: Boulder, Colorado, Geological Society of America Special Paper. 307, p. 263–277.

Olsson, R.K., and Wise, S.W., 1987, Upper Maestrichtian to middle Eocene stratigraphy of the New Jersey slope and coastal plain, *in* van Hinte, J.E., and Wise, S.W., eds., Initial Reports of the Deep Sea Drilling Project, Volume 93: Washington, D.C., U.S. Government Printing Office, p. 1343–1365.

Poag, C.W., 1996, The Chesapeake Bay bolide impact: A convulsive event in Atlantic Coastal Plain evolution: Sedimentary Geology, v. 108, p. 45–90.

Poag, C.W., Powars, D.S., Poppe, L.J., and Mixon, R.B., 1994, Meteoroid mayhem in Ole Virginny: Source of the North American tektite strewn field: Geology, v. 22, p. 691–694.

Schlee, J.S., and Grow, J.A., 1980, Seismic stratigraphy in the vicinity of the COST No. B-3 well, *in* Scholle, P.A., ed., Geological studies of the COST No. B-3 well: Washington, D.C., U.S. Government Printing Office, U.S. Geological Survey Circular, v. 833, p. 111–116.

Sepkoski, J.J., Jr., 1986, Phanerozoic overview of mass extinction, *in* Raup, D.M., and Jablonski, D., eds., Patterns and processes in the history of life: Berlin, Springer-Verlag, p. 277–295.

Sibeut, J.-C., Ryan, W.B.F., et al., 1979, Initial Reports of the Deep Sea Drilling Program: Washington, D.C., U.S Government Printing Office, v. 47B, 1115 p.

Smit, J., 1999, The global stratigraphy of the Cretaceous–Tertiary boundary impact ejecta: Annual Reviews of Earth and Planetary Sciences, v. 27, p. 75–113.

Smit, J., Montanari, A., Swinburne, N., Alvarez, W., Hildebrand, A.R., Margolis, S.V., Claeys, P., Lowrie, W., and Asaro, F., 1992, Tektite-bearing, deep-water clastic unit at the Cretaceous–Tertiary boundary in northeastern Mexico: Geology, v. 20, p. 99–103.

Smit, J., Roep, T.B., Alvarez, W., Montanari, A., Claeys, P., Grajales-Nishimura, J.M., and Bermudez, J., 1996, Coarse-grained, clastic sandstone complex at the K/T boundary around the Gulf of Mexico: Deposition by tsunami waves induced by the Chicxulub impact?, *in* Ryder, G., Fastovsky, D., and Gartner, S., eds., The Cretaceous–Tertiary event and other catastrophes in Earth history: Boulder, Colorado, Geological Society of America Special Paper 307, p. 151–182.

Smit, J., and van Kempen, T.M.G., 1987, Planktonic foraminifers from the Cretaceous/Tertiary boundary at Deep Sea Drilling Project Site 605, North Atlantic, *in* van Hinte, J.E., and Wise, S.W., eds., Initial Reports of the Deep Sea Drilling Project, Volume 93: Washington, D.C., U.S. Government Printing Office, p. 549–553.

Stinnesbeck, W., and Keller, G., 1996, K/T boundary coarse-grained siliciclastic deposits in northeastern Mexico and northeastern Brazil: Evidence for mega-tsunami or sea-level changes?, *in* Ryder, G., Fastovsky, D., and Gartner, S., eds., The Cretaceous–Tertiary event and other catastrophes in Earth history: Boulder, Colorado, Geological Society of America Special Paper 307, p. 211–226.

Swift, S.A., Ebinger, C.J., and Tucholke, B.E., 1986, Seismic stratigraphic correlation across the New England Seamounts, western North Atlantic Ocean: Geology, v. 14, p. 346–349.

Swisher, C.C., III, and 11 others, 1992, Coeval $^{40}Ar/^{39}Ar$ ages of 65.0 million years ago from Chicxulub crater melt rock and Cretaceous–Tertiary boundary tektites: Science, v. 257, p. 954–958.

Thomson, J., Rothwell, R.G., and Higgs, N.C., 1994, Development of reduction haloes under calcareous and volcaniclastic turbidites in the Lau Basin (Southwest Pacific), Proceedings of the Ocean Drilling Program, Scientific Results, Volume 135: College Station, Texas, Ocean Drilling Program, p. 151–162.

Toon, O.B., Turco, R.P., and Covey, C., 1997, Environmental perturbations caused by the impacts of asteroids and comets: Reviews of Geophysics, v. 35, p. 41–78.

Tucholke, B.E., 1979, Relationships between acoustic stratigraphy and lithostratigraphy in the western North Atlantic Basin, *in* Tucholke, B.E., Vogt, P.R., et al., eds., Initial Reports of the Deep Sea Drilling Project, Volume 43: Washington, D.C., U.S. Government Printing Office, p. 827–846.

Tucholke, B.E., 1981, Geologic significance of seismic reflectors in the deep western North Atlantic Basin: Society of Economic Paleontologists and Mineralogists Special Publication, v. 32, p. 23–37.

Tucholke, B.E., and Mountain, G.S., 1979, Seismic stratigraphy, lithostratigraphy and paleosedimentation patterns in the North Atlantic Basin, *in* Talwani, M., Hay, W., and Ryan, W.B.F., eds., Deep Drilling Results in the Atlantic Ocean: Continental margins and paleoenvironment, Maurice Ewing Symposium, Volume 3: Washington, D.C., American Geophysical Union, p. 58–86.

Tucholke, B.E., and Vogt, P.R., editors, 1979a, Initial Reports of the Deep Sea Drilling Project: Washington, D.C., U.S. Government Printing Office, v. 43, 1115 p.

Tucholke, B.E., and Vogt, P.R., 1979b, Western North Atlantic: Sedimentary evolution and aspects of tectonic history, *in* Tucholke, B.E., Vogt, P.R., et al., eds., Initial Reports of the Deep Sea Drilling Project, Volume 43: Washington, D.C., U.S. Government Printing Office, p. 791–825.

van Hinte, J.E., and Wise, S.W., editors, 1987, Initial Reports of the Deep Sea Drilling Project, Volume 93: Washington, D.C., U.S. Government Printing Office, 469 p.

Yancey, T.E., 1996, Stratigraphy and depositional environments of the Cretaceous–Tertiary boundary complex and basal Paleocene section, Brazos River, Texas: Transactions of the Gulf Coast Association of Geological Societies, v. 66, p. 433–442.

Zachos, J.C., and Arthur, M.A., 1986, Paleoceanography of the Cretaceous–Tertiary boundary event: Inferences from stable isotopic and other data: Paleoceanography, v. 1, p. 5–26.

Manuscript Submitted November 6, 2000; Accepted by the Society March 22, 2001

Printed in the U.S.A.

Geological Society of America
Special Paper 356
2002

Sequence stratigraphy and sea-level change across the Cretaceous-Tertiary boundary on the New Jersey passive margin

Richard K. Olsson
Kenneth G. Miller
James V. Browning
James D. Wright
Benjamin S. Cramer
Department of Geological Sciences, Rutgers University, Piscataway, New Jersey 08854, USA

ABSTRACT

In the New Jersey coastal plain the Cretaceous-Tertiary (K-T) boundary is within an unconformity-bounded Navesink depositional sequence (ca. 69.1-64.5 Ma). At the Bass River, New Jersey, borehole, a 2.2 m.y. hiatus separates the Navesink sequence from underlying Campanian sequences, and an ~1.5 m.y. hiatus separates Danian zone P1a from zone P1c and younger sequences. A 6-cm-thick spherule layer that contains shocked minerals and an iridium anomaly marks the K-T boundary at this site. Benthic foraminiferal biofacies and biostratigraphy indicate that sedimentation was continuous across the K-T boundary. During deposition of the Navesink sequence, relative sea level fell from 100–150 m above present sea level in the lower part of the sequence (transgressive systems tract) to ~50 m (highstand systems tract) at the K-T boundary.

Three significant events are inferred from the Navesink depositional record: (1) an ~5°C warming of sea-surface temperatures perhaps related to the main outpouring of the Deccan Traps in India that began ~500 k.y. and ended ~22 k.y. before the K-T boundary; (2) the K-T event caused by an asteroid impact at Chicxulub, Mexico; and (3) a tsunami event immediately following the ballistic fallout of tektites from the Chicxulub ejecta vapor cloud, possibly triggered by massive slumping on the Atlantic slope. There is no relationship between these events and sea-level change during deposition of the Navesink sequence.

INTRODUCTION

Sea-level change has played a prominent role in interpreting the cause of mass extinctions either by regression reducing marine habitat or by transgression leading to widespread anoxia (Hallam and Wignall, 1997, 1999; Hallam, 2000). According to these authors, no general pattern to extinction at the major extinction boundaries has been unequivocally shown to be related either to regression or transgression. The Cretaceous-Tertiary boundary (K-T) has the additional consideration that it is associated with the Chicxulub asteroid impact. Nevertheless, Hallam and Wignall (1997) pointed out that there was evidence for major sea-level change at the K-T boundary that needed further evaluation. A general viewpoint summarized by Hallam

Olsson, R.K., Miller, K.G., Browning, J.V., Wright, J.D., and Cramer, B.S., 2002, Sequence stratigraphy and sea-level change across the Cretaceous-Tertiary boundary on the New Jersey passive margin, *in* Koeberl, C., and MacLeod, K.G., eds., Catastrophic Events and Mass Extinctions: Impacts and Beyond: Boulder, Colorado, Geological Society of America Special Paper 356, p. 97–108.

(1990) is that the K-T boundary marks a global regression followed by a transgression. This viewpoint is based on the fact that most shallow-water stratigraphic sections have a hiatus at the K-T boundary. Studies of Alabama sections have interpreted a hiatus at the K-T boundary as being caused by an impact-generated tsunami event and not due to regression (Olsson et al., 1996; Smit et al., 1996). Another general viewpoint is that continuous sedimentation across the K-T boundary is found only in deeper water pelagic facies (see Olsson and Liu, 1993, for discussion). It is difficult to determine relative sea-level changes in deep-water paleoenvironments, because sea-level changes generally do not cause significant changes in sediment and biota, as they do in shallower water paleoenvironments.

Inferred sea-level change across the K-T boundary is based primarily on studies of shallow-water deposits, and can be controversial. For example, based on a section in the coastal plain at Braggs, Alabama, the EXXON group (Donovan et al., 1988) placed the K-T boundary in the lowermost Clayton Formation (Pine Barren Member) within sequence TA 1.1 on the EXXON coastal onlap chart (Haq et al., 1987, 1988). They placed the sequence boundary at an unconformity separating the Clayton Formation from the Prairie Bluff Chalk (upper Maastrichtian); thus, they inferred a sea-level fall that predated slightly the K-T boundary. However, Habib et al. (1992) and Olsson and Liu (1993) showed that the K-T boundary was misplaced at Braggs. At the Braggs section, Olsson et al. (1996) identified Danian zone Pα (*Parvularugoglobigerina eugubina*) in the basal beds of the Clayton Formation just above the Prairie Bluff Chalk and 0.9 m below where Donovan et al. (1988) placed the K-T boundary. According to Olsson et al. (1996), the surface separating the Maastrichtian Prairie Bluff Chalk from the Danian Clayton Formation is not a sequence boundary. Rather, this surface was eroded by a giant tsunami wave that was generated by the K-T asteroid impact at Chicxulub, Mexico. Using benthic foraminiferal biofacies, Olsson et al. estimated paleodepth as ~30 m across the K-T boundary and concluded that sea level was at a lowstand. However, due to limited exposure, the K-T boundary sections in Alabama do not show the entire stratigraphic extent of the K-T sequence and its relationship with sequences above and below, which is important to understanding sequence stratigraphy and long-term sea-level change.

The New Jersey coastal plain is regarded as a classic passive margin and has received much study since the 1970s, including extensive seismic reflection surveys, petroleum exploration wells, Deep Sea Drilling Project (DSDP) and Ocean Drilling Project (ODP) drill sites, and coastal plain drilling. The New Jersey coastal plain contains a marine record of the Cenomanian to the Miocene and is ideally situated to study sequence stratigraphy and sea-level change. The New Jersey Coastal Plain Drilling Project (ODP Legs 150X and 174AX, Miller et al., 1994, 1996, 1997, 1998, 1999b) represents the onshore component of the New Jersey Sea-Level Transect, which included the Ocean Drilling Program (ODP) shelf and slope drilling on Legs 150X (Mountain et al., 1996) and 174A (Austin et al., 1998). The objective of this drilling program was to study sea-level history and depositional sequences on a classic passive margin, where these events should be unambiguously expressed in the geologic record.

Leg 150X drilling focused on the Miocene, Oligocene, and Eocene sea-level cycles and their corresponding depositional sequences. Studies showed that depositional sequences can be recognized by their bounding unconformities, by lithologic criteria, and by benthic foraminiferal biofacies that reflect changes in relative sea level (Browning et al., 1996; Miller et al., 1998). Hiatuses separating depositional sequences range in age from 100 to 500 k.y. in the Oligocene and 100 k.y. to 1 m.y. in the Miocene, and are correlated by the New Jersey $\delta^{18}O$ isotope record to periods of glaciation (Miller et al., 1998). Hiatuses in the Eocene range from 500 k.y. to ~ 2 m.y., but correlation to glacial events is uncertain (Browning et al., 1996).

The Ancora and Bass River boreholes (Fig. 1) were drilled during Leg 174AX. They were cored continuously, penetrated the K-T boundary, and bottomed in the upper Cenomanian. Seven unconformity-bounded sequences were identified in the Cenomanian to Maastrichtian section (Miller et al., 1998). In the lowermost Cenomanian-Turonian sequence, deposited in in-

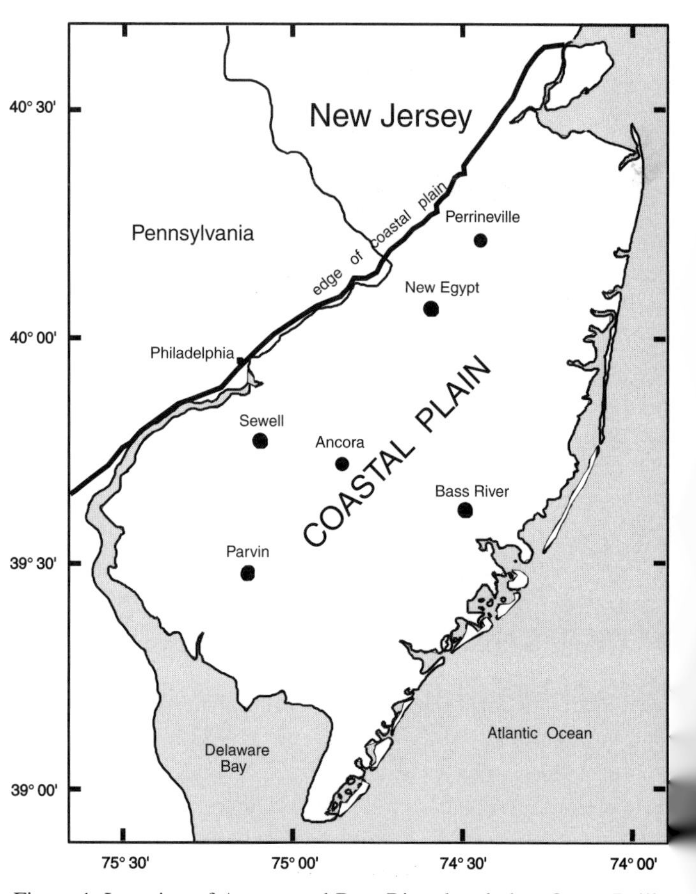

Figure 1. Location of Ancora and Bass River boreholes (Ocean Drilling Program Leg 174AX) and other Cretaceous-Tertiary boundary sections mentioned in text.

ner shelf paleodepths, five parasequences were identified on the basis of well-preserved benthic foraminifera and lithologic criteria (Sugarman et al., 1999). In summary, the New Jersey Coastal Plain Drilling Project demonstrated that coastal plain sequences preserve detailed records for studies of eustasy and that contrasting facies and other environmental indicators, including well-preserved benthic foraminifera, allow for clear identification of water-depth variations within sequences. New Jersey coastal plain drilling also recovered spectacular records of deposition at the K-T boundary.

In the downdip Bass River borehole (Fig. 2), a 6-cm-thick spherule layer separates lowermost Danian deposits (zone P0) from uppermost Maastrichtian deposits (*Micula prinsii* zone) (Olsson et al., 1997). Reworked spherules occur above the K-T boundary at the updip Ancora borehole, but no distinct spherule layer is present. Nevertheless, an interval of clay clasts that has been interpreted as tsunami derived occurs immediately above the spherule layer at Bass River and at the K-T boundary at Ancora (Olsson et al., 2000). The Ancora section is considered biostratigraphically complete. Thus, the New Jersey coastal plain appears ideal for assessing sea-level change across the K-T boundary. Here we report on the depositional sequences in the Maastrichtian and the Danian to obtain a longer term view of sea-level change during ~10 m.y. prior to and following the K-T event.

METHODS

Weighed samples for foraminiferal analysis were washed on a 63 µm sieve, dried, and separated into three size fractions for counting (63 µm, 125 µm, 250 µm). Foraminifera were well preserved and their chambers generally free of sediment, which in most cases allowed concentration of foraminifera by flotation procedures. Samples were split with a microsplitter to a man-

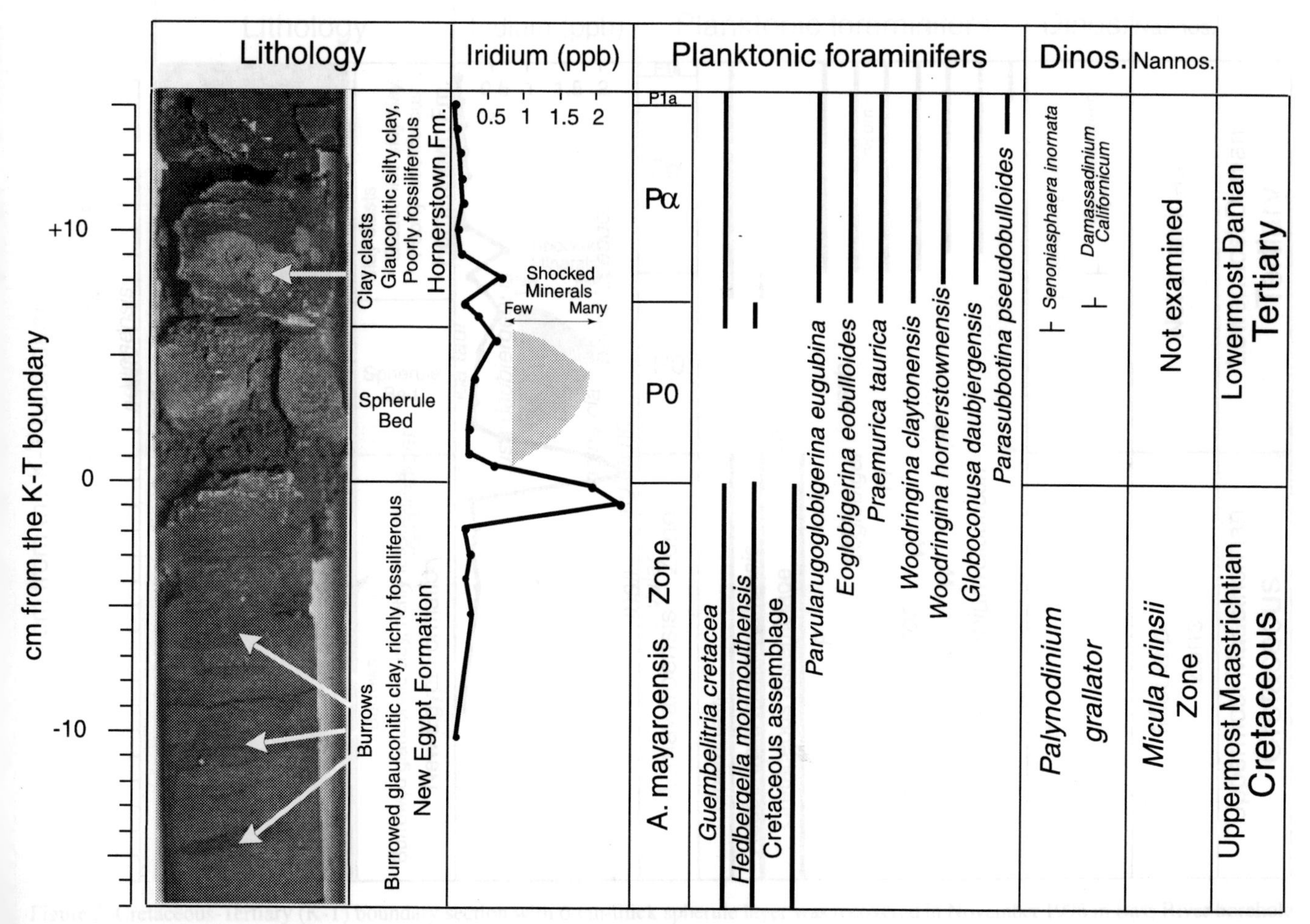

Figure 2. Cretaceous-Tertiary (K-T) boundary section with 6-cm-thick spherule layer was recovered in November 1996 in Bass River borehole (New Jersey Coastal Plain Drilling Project, Ocean Drilling Program ODP Leg 174AX). Shocked minerals (identified by Izett, 1997, personal commun.) and elevated iridium values (measurements by Asaro, 1998, personal commun.) are associated with spherule layer. Spherule bed is overlain by 10 cm zone that includes clay clasts that are surrounded by Danian sediments. Clay clasts are interpreted as being emplaced by tsunami activity associated with boundary event. Planktonic foraminifera, dinoflagellates, and calcareous nannofossils provide biostratigraphic resolution in K-T boundary section (Olsson et al., 1997). Fm.—Formation.

ageable size and at least 300 specimens were counted in each size fraction. The total number of species was summed for each size fraction and then totaled for each sample; in most instances more than 1000 specimens were counted. Benthic foraminiferal abundance data were analyzed to help estimate paleowater depth. The great number of the species *Buliminella carseyae* skewed the results, so their abundance data were eliminated from the data matrix. The dataset was normalized to percentages and Q-mode factor analysis was used to compare variations among the samples. Three factors, explaining 80.0% of the faunal variation, were extracted using a Varimax Factor rotation using Systat 5.2.1 on a Macintosh microcomputer. These factors were interpreted as biofacies.

The paleoslope model for Campanian and Maastrichtian benthic foraminifera of the New Jersey coastal plain (Olsson and Nyong, 1984) was used for estimating paleodepth for the foraminiferal biofacies.

Clean specimens of the planktonic foraminifer *Rugoglobigerina* sp. and the benthic foraminifer *Anomalinoides midwayensis* were picked from the >125 μm size fraction. Stable isotope values of foraminifera were measured in stable isotope laboratories at the University of Maine and Rutgers University. The analyses from the University of Maine laboratory were made on a VG Prism II mass spectrometer using an IsoCarb automated carbonate preparation system. At Rutgers University, the measurements were made on a Micromass Optima mass spectrometer using a Multiprep automated carbonate preparation system. Samples of specimens were reacted in 100% phosphoric acid at 90°C. Oxygen and carbon isotopic values in Tables 1 and 2 are reported relative to Vienna Peedee belemnite by normalizing the NBS-19 or NBS-20 standards to the values reported in Coplen et al. (1983). The standard deviations (1σ) of the standards (minimum of 6 standards measured with each run of 30 samples) are 0.06‰ and 0.05‰ for $\delta^{18}O$ and $\delta^{13}C$, respectively.

We obtained 23 Sr isotopic measurements from foraminifer tests (~4–6 mg) at the Bass River borehole. Sediments adhering to the tests were removed by ultrasonically cleaning for 1–2 s. Samples were dissolved in 1.5 N HCL, centrifuged, and introduced into ion-exchange columns. Standard ion exchange techniques were used to separate the strontium (Hart and Brooks, 1974) and samples were analyzed on a VG Sector mass spectrometer at Rutgers University. Internal precision on the sector for the data set (Table 3) averaged 0.000011, and the external precision is ~0.000020–0.000030 (Oslick et al., 1994). NBS 987 is measured for these analysis at 0.710255 (2σ standard deviation 0.000008, n = 22) normalized to $^{87}Sr/^{86}Sr$ of 0.71194.

TABLE 1. BASS RIVER

Sample depth ft (m)	*Anomalinoides midwayensis* $\delta^{13}C$	*Anomalinoides midwayensis* $\delta^{18}O$
1260.28 (384.13)	0.79	−1.518
1260.35 (384.15)	0.95	−1.52
1260.41 (384.17)	1.067	−1.174
1260.41 (384.17)	0.917	−1.42
1260.51 (384.20)	0.894	−1.075
1260.57 (384.22)	1.13	−1.22
1260.64 (384.24)	1.25	−1.16
1260.71 (384.26)	1.35	−1.18
1260.77 (384.28)	1.229	−1.181
1261.07 (384.37)	1.16	−1.26
1261.50 (384.50)	1.224	−1.22
1262.00 (384.65)	1.133	−1.212
1263.00 (384.96)	1.083	−1.347
1263.50 (385.11)	0.922	−1.322
1264.00 (385.26)	1.056	−1.416
1265.10 (385.60)	0.592	−1.388
1266.00 (385.87)	0.554	−1.212
1267.00 (386.18)	0.428	−1.19
1268.00 (386.48)	0.512	−1.309
1269.00 (386.79)	0.175	−1.075
1271.00 (387.40)	0.45	−1.08
Sample depth ft (m)	***Rugoglobigerina* $\delta^{13}C$**	***Rugoglobigerina* $\delta^{18}O$**
1260.55 (384.31)	1.88	−1.83
1260.62 (384.33)	2.09	−2.11
1260.68 (384.35)	2.2	−2.3
1260.78 (384.38)	2.19	−2.23
1260.85 (384.40)	2.09	−1.99
1260.90 (384.42)	2.24	−2.16
1260.91 (384.42)	2.22	−2.06
1261.00 (384.45)	2.25	−2.16
1261.50 (384.50)	2.34	−2.76
1261.82 (384.70)	2.37	−2.65
1262.00 (384.65)	2.28	−3.22
1262.50 (384.90)	2.37	−3.02
1263.00 (384.96)	2.27	−2.88
1263.50 (385.11)	2.34	−2.9
1264.00 (385.26)	2.21	−2.87
1265.10 (385.60)	1.84	−2.89
1266.00 (385.87)	1.43	−1.98
1267.00 (386.18)	1.61	−2.22
1268.00 (386.48)	1.31	−1.99
1269.00 (386.79)	1.49	−1.79
1271.00 (387.40)	1.75	−2.18

UNCONFORMITIES IN THE MAASTRICHTIAN AND DANIAN SECTIONS

Sugarman et al. (1995) established that the only sequence represented in the Maastrichtian of the New Jersey coastal plain was the Navesink sequence. The Navesink sequence is characterized by shallowing-upward lithofacies, from a basal clastic sequence-bounding unconformity, to transgressive clayey, glauconite-rich deposits (the Navesink Formation), to silt-rich, glauconitic highstand deposits (the Red Bank—New Egypt Formation). Above the K-T boundary the highstand deposits (Hornerstown Formation) are more glauconite rich, but are silty without significant clay content. A prominent basal unconformity separates the Navesink sequence from an upper Campanian sequence (the Marshalltown sequence). Using strontium isotope dating on outcrops, Sugarman et al. (1995) estimated that the duration of the hiatus separating the two sequences was ~3 m.y. Miller et al. (1999a), using additional strontium isotope data from the Bass River borehole, estimated the duration of

TABLE 2. ANCORA

Sample depth ft (m)	*Anomalinoides midwayensis* δ13C	*Anomalinoides midwayensis* δ18O
618.40 (188.53)	0.68	−1.28
618.50 (188.56)	0.94	−1.21
619.00 (188.71)	0.83	−1.48
619.50 (188.87)	0.89	−1.45
620.00 (189.02)	0.94	−1.47
621.00 (189.32)	0.82	−1.38
622.00 (189.63)	0.67	−1.32
623.00 (189.93)	0.18	−1.15
624.00 (190.24)	0.06	−1.05
625.00 (190.54)	0.19	−1.12
626.00 (190.85)	0.26	−0.91
627.00 (191.15)	0.27	−0.94
628.00 (191.46)	0.34	−0.89
630.50 (192.22)	0.27	−0.97
631.00 (192.37)	0.36	−0.80
636.00 (193.90)	0.49	−0.93
641.00 (195.42)	0.781	−1.051
646.00 (196.95)	0.539	−1.146
651.00 (198.47)	0.596	−0.925
Sample depth ft (m)	***Rugoglobigerina* δ13C**	***Rugoglobigerina* δ18O**
618.40 (188.53)	2.40	−2.92
618.50 (188.56)	2.15	−2.82
619.00 (188.71)	1.763	−3.109
619.50 (188.87)	1.58	−2.67
620.00 (189.02)	1.72	−3.46
621.00 (189.32)	1.91	−3.37
622.00 (189.63)	2.02	−2.80
623.00 (189.93)	1.341	−2.519
624.00 (190.24)	1.33	−2.10
625.00 (190.54)	1.38	−2.48
626.00 (190.85)	1.56	−2.14
627.00 (191.15)	1.81	−2.54
628.00 (191.46)	1.93	−2.42
630.50 (192.22)	1.61	−2.33
631.00 (192.37)	1.77	−2.25
636.00 (193.90)	2.15	−2.23
641.00 (195.42)	1.86	−2.11
646.00 (196.95)	1.26	−2.47
651.00 (198.47)	1.61	−1.84

TABLE 3. BASS RIVER STRONTIUM

Sample depth ft (m)	87Sr/86Sr	Error
1260.60 (384.32)	0.707915	13
1260.73 (384.36)	0.707893	20
1260.90 (384.42)	0.707889	18
1261.40 (384.57)	0.707876	13
1261.75 (384.67)	0.707885	4
1262.25 (384.83)	0.707878	10
1262.50 (384.90)	0.707896	7
1263.00 (385.06)	0.707900	6
1263.25 (385.13)	0.707894	20
1263.50 (385.21)	0.707895	5
1263.75 (385.28)	0.707884	6
1264.00 (385.36)	0.707908	5
1264.50 (385.51)	0.707900	4
1265.00 (385.67)	0.707874	16
1266.00 (385.97)	0.707893	4
1266.50 (386.12)	0.707884	10
1267.00 (386.28)	0.707872	10
1267.50 (386.43)	0.707872	13
1269.00 (417.37)	0.707897	6
1269.50 (387.04)	0.707875	5
1270.00 (387.19)	0.707892	7
1274.50 (388.56)	0.707909	6
1280.00 (390.24)	0.707882	32

the hiatus as 2.2 m.y. (71.3-69.1 Ma) (Fig. 3). They correlated the sequence boundary with synchronous $\delta^{18}O$ increases in deep-water benthic and low-latitude surface-dwelling planktonic foraminifera, and speculated that changes in paleobathymetry of 30–40 m based on benthic foraminifera across the Navesink-Marshalltown sequence boundary were caused by development of a small to moderate-sized Antarctic ice sheet during the early Maastrichtian (Miller et al., 1999a).

The top of the Navesink sequence is identified by a prominent unconformity within the lower Danian (Fig. 3). In the Ancora and Bass River boreholes the sequence boundary is between planktonic foraminiferal zones P1a and P1c; zone P1b is absent at Bass River and is represented by only a thin interval at Ancora (hiatus ca. 64.5–63 Ma at Bass River and ca. 64–63 Ma at Ancora based on the time scale of Berggren et al., 1995). The sediments deposited in zone P1c are clearly transgressive, because they contain middle to outer shelf benthic foraminifera and abundant planktonic foraminifera and overlie a zone P1a inner to middle shelf benthic assemblage containing rarer planktonic foraminifera. The duration of the hiatus separating the two sequences is estimated as ~1.5 m.y. at Bass River and 1.0 m.y. at Ancora using the time scale of Berggren et al. (1995).

BENTHIC FORAMINIFERAL BIOFACIES AND PALEODEPTH ACROSS THE K-T BOUNDARY

Factor analysis of the pooled dataset identifies three benthic biofacies in the Ancora and Bass River boreholes (Figs. 4–7). Biofacies 1, which occurs in the lower part of the sections analyzed in both boreholes, is characterized by relatively high abundances of the species *Gyroidinoides imitata, Anomalinoides midwayensis, Quadrimorphina allomorphinoides*, and *Pseudouvigerina seligi*. Biofacies 1 is replaced upsection in both boreholes by biofacies 3 and is characterized by the species *Anomalinoides* cf. *welleri, Corphyostoma plaitum, Pseudouvigerina seligi*, and *Pulsiphonina prima*. Biofacies 2 is found in a thin interval spanning the K-T boundary in both boreholes. The species *Alabamina midwayensis, Anomalinoides acuta, P. prima*, and *Tappanina selmensis* characterize this biofacies. It is notable that taxa such as *A. midwayensis, A. acuta*, and *Anomalinoides welleri* that characterize Danian Midway benthic assemblages (Berggren and Aubert, 1975) in the Gulf and Atlantic coastal plains first appear in biofacies 2 in the uppermost Maastrichtian (Figs. 4 and 6). Thus, they appear to have evolved in inner shelf environments in the latest Maastrichtian.

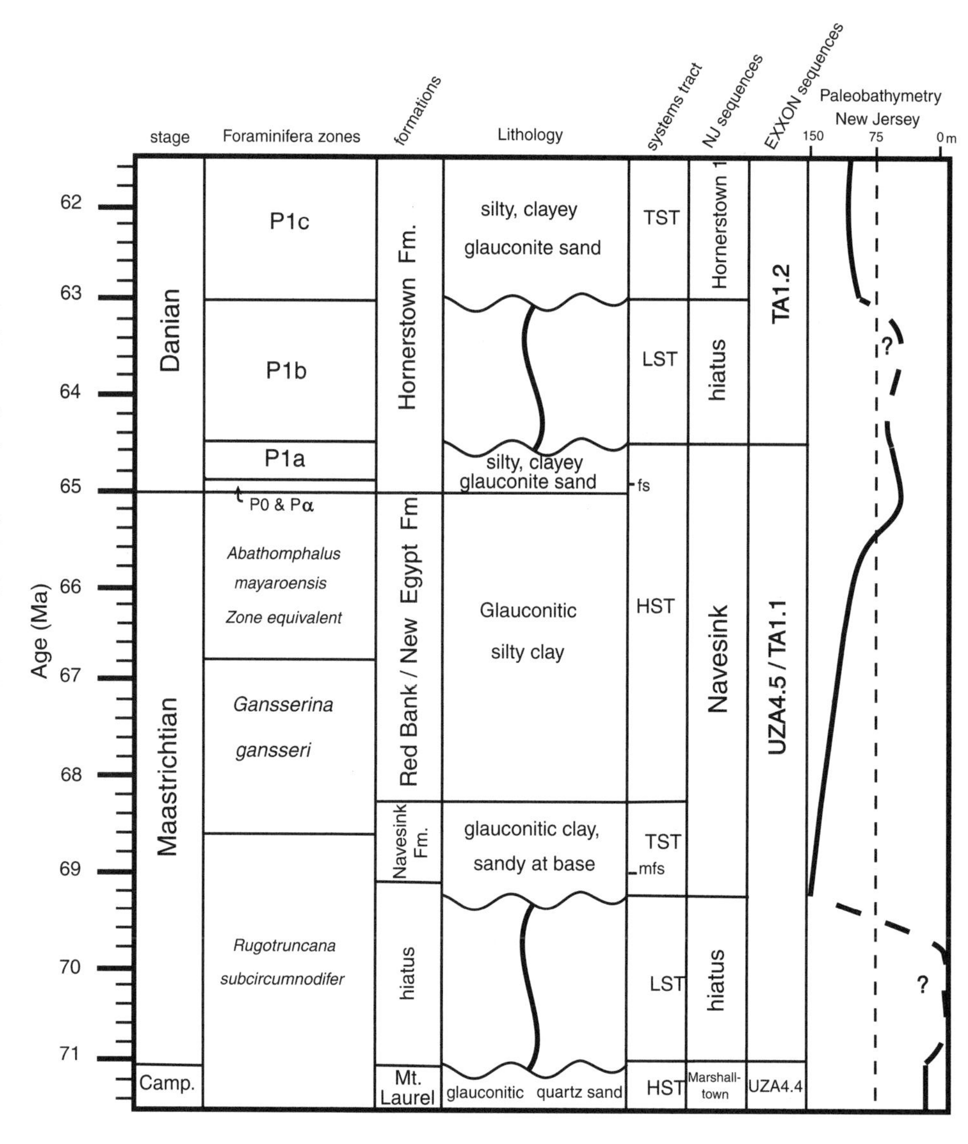

Figure 3. Navesink depositional sequence at Bass River borehole showing paleobathymetry based on benthic foraminifera. Note that the Navesink sequence spans the Cretaceous-Tertiary (K-T) boundary and is clearly separated from sequences above and below by prominent unconformities and their respective hiatuses. Navesink sequence contains Cretaceous Navesink and New Egypt Formations and lower portion of Hornerstown Formation. Portion of Hornerstown Formation (glauconite-rich unit) above Navesink sequence can be further separated into three sequences based on biostratigraphy and benthic foraminiferal biofacies. LST is lowstand, TST is transgressive, HST is highstand, msf is maximum flooding surface, fs is flooding surface.

Bass River borehole

In the downdip Bass River borehole, biofacies 1 (386.28–388.10 m) planktonic foraminifera compose as much as 85% of the assemblage, averaging 78% (Fig. 5). Paleodepth is estimated as middle to outer neritic, using the paleoslope model of Olsson and Nyong (1984). Planktonic foraminifera average ~73% of the assemblage within biofacies 3 (386.03–384.35 m). Paleodepth is estimated as middle neritic using the paleoslope model of Olsson and Nyong (1984). Water depth gradually shoaled during deposition of biofacies 3. Biofacies 3 correlates with a significant warming trend in sea-surface temperatures, as indicated by decreased $\delta^{18}O$ values of the shallow-dwelling planktonic foraminifer *Rugoglobigerina* (Fig. 5).

A sharp decrease in the planktonic foraminiferal component to ~40% of the assemblage occurs in the Maastrichtian part of biofacies 2 (384.32–384.0 m). This decrease in percent planktonics and the transition from biofacies 3 to 2 occurs immediately below the K-T boundary (384.22 m); little change is associated with the boundary. Due to the planktonic foraminiferal extinctions at the K-T boundary, the diversity and percentage of planktonic foraminifera is very low in the lowermost Danian, but it gradually increases to ~30% of the assemblage at the top of the sequence. Paleodepth is estimated to have lowered to inner shelf depths below the K-T boundary and continued at these depths above the K-T boundary to the sequence boundary.

Ancora borehole

In the updip Ancora borehole (Fig. 1), planktonic foraminifera, particularly adult specimens, are not as plentiful (averaging 63%) due to the shallower environment of deposition. Due to the shallower paleodepth, benthic foraminiferal abundances differ somewhat from abundances at Bass River. Nevertheless, factor analysis identifies the same succession of biofacies that is identified at Bass River (Fig. 7). The paleodepth of biofacies 1 (192.37–189.93 m) is estimated as middle neritic using the paleoslope model of Olsson and Nyong (1984). The paleodepth of biofacies 3 (189.63–188.87 m) is estimated as inner to middle neritic. As at Bass River, biofacies 3 correlates with a significant warming trend in sea-surface temperatures indicated by decreased $\delta^{18}O$ values of the shallow-dwelling planktonic foraminifer *Rugoglobigerina* (Fig. 7). The percentage of planktonic foraminifera in biofacies 2 decreases across the K-T boundary, from 55% in the Maastrichtian to 1.7% in the lowermost Danian, increasing to 6% at the top of the analyzed section (Fig. 7). The paleodepth of biofacies 2 (188.71–188.18 m) is estimated as inner neritic using the paleoslope model of Olsson and Nyong (1984).

DISCUSSION

Biostratigraphic and paleomagnetic data show that a stratigraphically complete K-T section is present in the New Jersey coastal plain (Fig. 2). Sequence stratigraphy indicates that the K-T event occurred within a highstand systems tract of the Navesink sequence. The Navesink sequence is equivalent to sequence cycles UZA 4.5 and TA1.1 (the base of TA1.1, as defined, is not a sequence boundary) on the EXXON coastal onlap chart (Haq et al., 1988). On the basis of our data relative sea level is estimated to have fallen from ~100–150 m above present at the beginning of the Navesink sequence to <50 m prior to the K-T boundary and remained low during final deposition of the sequence in the Danian. The paleodepth across the K-T boundary in New Jersey is similar to the paleodepth estimated at Millers Ferry, Alabama (Olsson et al., 1996), which was also situated in a shallow shelf setting at K-T boundary time. As in New Jersey, relatively little change occurred in the composition of benthic foraminiferal assemblages across the K-T boundary.

Three significant events occurred during deposition of the Navesink sequence in New Jersey. The earliest event occurred ~500 k.y. before the K-T boundary and involved a significant warming of sea-surface temperatures. The $\delta^{18}O$ record of the shallow-dwelling planktonic foraminifer *Rugoglobigerina* shows a sharp negative shift in oxygen isotope values of nearly 1%, indicating a warming of surface waters by ~4 °C; peak warming during the trend was ~5 °C (Figs. 5 and 7). On the basis of estimated sedimentation rates at Bass River, this interval of warming ended ~22 k.y. before the K-T boundary. The warming trend is also evident in the benthic foraminiferal $\delta^{18}O$ record and correlates with biofacies 3 (Figs. 5 and 7). We suggest that benthic foraminiferal changes from biofacies 3 to biofacies 1 at Ancora and Bass River are in part due to warming of bottom waters. At Bass River the warming trend started near the base of C29R (Fig. 5). We support the hypothesis (Barrera and Savin, 1999) that the warming trend was due to an increase

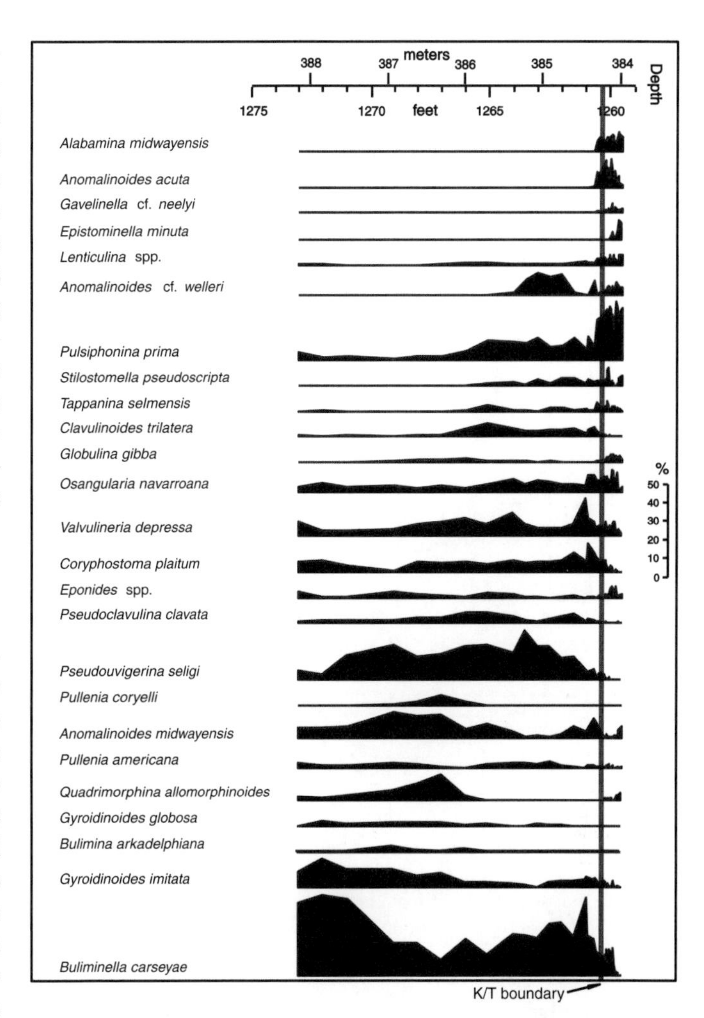

Figure 4. Relative abundances of benthic foraminifera in Maastrichtian and lower Danian Navesink sequence, Bass River borehole. These results are data used in factor analysis shown in Figure 5. K-T, Cretaceous-Tertiary.

in greenhouse gases related to the main outpouring of the Deccan Traps in India, which started near the base of subchron C29R (Courtillot et al., 1986; Hansen, et al., 1996).

A positive shift in $^{87}Sr/^{86}Sr$ ratios occurs at Bass River near the base of Subchron C29R and correlates with the latest Maastrichtian warming trend (Fig. 5). High-resolution (50 k.y. scale) Maastrichtian variations in $^{87}Sr/^{86}Sr$ observed at Bass River show a remarkably similar pattern to coeval sections at Bidart, France, and El Kef, Tunisia (Vonhof and Smit, 1997). However, $^{87}Sr/^{86}Sr$ values measured at Bass River (~0.707 890) are offset from the values of Vonhof and Smit (1997) (0.707 830) by ~0.000 040 ppm. Similar Sr isotopic offsets (few tens of ppm) have been noted in high-resolution Sr isotopic studies of Pliocene-Pleistocene sediments and ascribed to adhering of clay minerals to the carbonates; we speculate that a similar effect influenced the Bass River Maastrichtian section. The difference between our values and those of Vonhof and Smit cannot be ascribed to an interlaboratory calibration problem because we

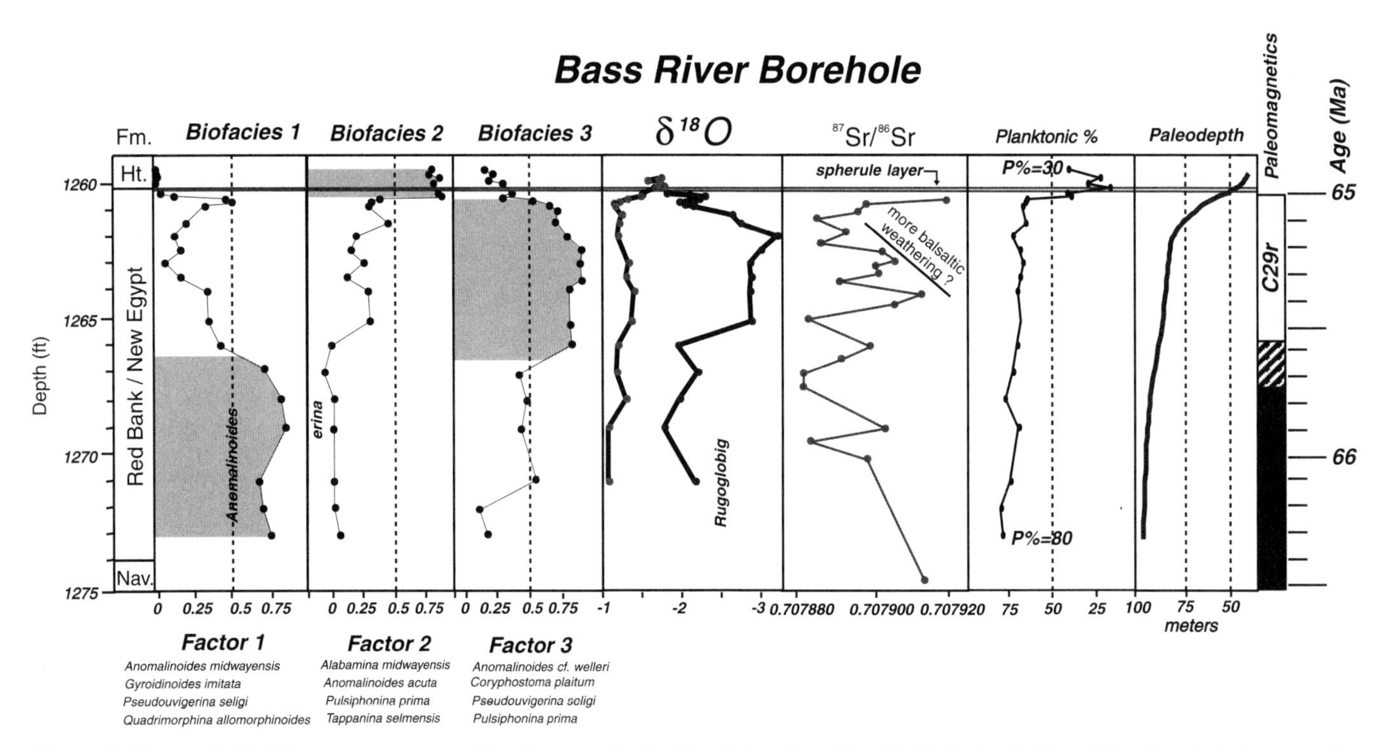

Figure 5. Three principal factors are separated by factor analysis, identifying three benthic biofacies in Bass River borehole (see text for explanation). Note that biofacies 3 is associated with negative shift in $\delta^{18}O$ isotope values indicating warming trend. Decrease of as much as 1% in ^{18}O values on shells of Cretaceous planktonic foraminifer *Rugoglobigerina* indicates significant warming trend of as much as 5°C in near sea-surface temperatures that began ~500 k.y. and ended ~22 k.y. before K-T boundary. Warming trend that starts near lower boundary of chron C29R is believed related to main outpouring of Deccan Traps in India. Benthic biofacies and planktonic foraminiferal percentages indicate that sea level fell from highstand of ~150 m in Maastrichtian to ~50 m prior to Cretaceous-Tertiary (K-T) boundary event. Sea level remained low across K-T boundary. Fm.—Formation, Ht.—AQP, Nav.–AQP.

note a similar offset between Maastrichtian $^{87}Sr/^{86}Sr$ values at Bass River versus coeval New Jersey coastal plain sections (Sugarman et al., 1995). The variations noted by Vonhof and Smit (1997) include a sharp decrease in chron C29r that they ascribe to increased weathering of basaltic rocks associated with the Deccan Traps. This strontium isotope change also correlates with this global warming trend. We conclude that Maastrichtian Sr isotopic values at Bass River must be affected by minor diagenesis, although the general global pattern of Sr isotopic variations is preserved. Diagenesis is minor because the stable isotopic values from the Maastrichtian at Bass River (Fig. 5) are similar in amplitude and pattern to coeval global values, the specimens are generally well preserved, and the Sr isotopic offset is relatively small.

Li and Keller (1998a, 1998b) identified a latest Maastrichtian (chron C29R) warming trend in $\delta^{18}O$ values of benthic foraminifera at ODP Site 525, but due to the lack of a clear warming trend in $\delta^{18}O$ values of planktonic foraminifera in this interval, they ruled out a link to the Deccan Traps volcanism. However, the study by Kucera and Malmgren (1998) of Site 525 and other South Atlantic sites showed that this warming trend affected surface waters, based on the poleward migration of the low-latitude planktonic foraminifer *Contusotruncana contusa*. Barrera and Savin (1999) concluded in their study of $\delta^{18}O$ records that intermediate and deep waters in the South and North Atlantic, Indian, and Pacific Oceans warmed globally by 3–4 °C between 65.5 and 65.3 Ma and then cooled slightly ca. 65.2 Ma. They suggested that this increase in marine temperatures correlated with the main episode of eruptions of the Deccan Traps that may have led to greenhouse global warming. Olsson et al. (2000, 2001) correlated a poleward migration in the North and South Atlantic Oceans of the subtropical planktonic foraminifera *Pseudotextularia elegans* with this warming trend. We conclude that a late Maastrichtian global warming event occurred in earliest subchron C29R (ca. 65.5 Ma).

The second event, the most significant, is the K-T event, which very briefly masked deposition of the Navesink sequence by condensation and fallout of tektites from the ejecta vapor cloud generated by the impact of the K-T asteroid at Chicxulub, Yucatan, Mexico (Hildebrand et al., 1991; Alvarez, 1996; Olsson et al., 1997). The 6-cm-thick layer of tektites that settled on the New Jersey seafloor left impressions in the soft mud that are interpreted as original, but possibly enhanced by post-depositional compaction. These impressions are preserved in the Bass River K-T boundary core (Fig. 8). The 6-cm-thick K-T boundary spherule layer that occurs as a thin layer in an otherwise continuous section at Bass River is testimony to this instantaneous event. Shocked minerals and the K-T iridium

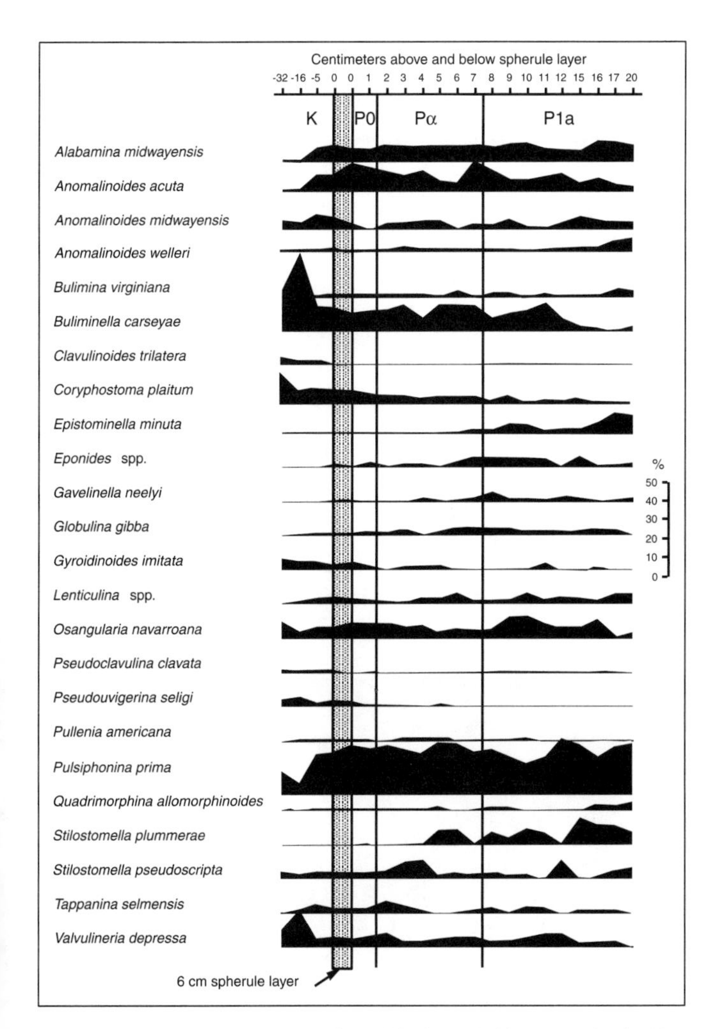

Figure 6. Relative abundance of benthic foraminiferal species in biofacies 2 across Cretaceous-Tertiary (K-T) boundary in Bass River borehole. Typical Danian species such as *Alabamina midwayensis, Anomalinoides acuta*, and *Anomalinoides welleri* first appear in uppermost Maastrichtian with lowering of sea level, suggesting that they migrated from inner shelf environments. Note that benthic foraminifera were not affected by K-T impact event.

anomaly are associated with the spherule layer that appears to be an in situ deposit at Bass River (Fig. 2). The higher concentration of iridium at the base of the spherule layer is believed to be due to the postdepositional downward diffusion of iridium through pore water in the highly permeable and porous spherule layer; the underlying Maastrichtian clayey silts act as an aquitard. Benthic foraminifera apparently were little affected by this event because biofacies 2 continues immediately above the spherule layer (Fig. 5). Planktonic foraminifera underwent a mass extinction; the only survivors appear to be the well-known taxa *Guembelitria cretacea, Hedbergella monmouthensis*, and *Hedbergella holmdelensis*. The Bass River record unequivocally links the extinctions of marine plankton to the impact (Olsson et al., 1997; Norris et al., 1999).

The third event, which immediately followed the deposition of tektites, was a tsunami. The tsunami generated in the Gulf of Mexico by the impact at Chicxulub probably would have been blocked by the shallow Florida platform and prevented from crossing into the North Atlantic. However, a tsunami may have been triggered by slope failure on the New Jersey margin due to earthquakes generated by the Chicxulub impact (Klaus et al., 2000; Norris et al., 2000; Olsson et al., 2000). Small clay clasts, ranging from a few centimeters to millimeters in size and containing Cretaceous foraminifera and calcite-replaced tektites (possibly original calcite tektites), occur directly above the spherule layer at Bass River and form a marker for the K-T boundary throughout the New Jersey coastal plain where the spherule bed is absent (Fig. 9). We have noted this clast layer at Ancora, Bass River, and a well at Parvin, and outcrops at New Egypt, Perrineville, and Sewell (Fig. 1). Cretaceous planktonic foraminifera that are found in the Danian occur within clay clasts and are confined to a 10-cm-thick zone above the spherule layer. The clay clasts are surrounded by Danian sediment containing typical Danian planktonic foraminiferal species and dinoflagellates (Olsson et al., 1997) that indicate that the clasts are confined to zone P0 to the lowermost part of zone P1a (Fig. 10).

At Bass River a concentrated layer of echinoid fecal pellets (Fig. 11) that is interpreted as a condensed interval occurs in zone P1a. This interval probably represents a flooding surface and may be a parasequence boundary in the upper part of the Navesink sequence. This condensed interval may correspond to a flooding surface that Donovan et al. (1988) associated with the K-T boundary at Braggs, Alabama, and to a flooding surface identified by Olsson et al. (1996) at Millers Ferry, Alabama.

CONCLUSIONS

The K-T boundary occurs in a highstand systems tract within an unconformity-bounded depositional sequence, the Navesink sequence in New Jersey. Sea level fell to ~50 m prior to the K-T boundary and remained low for the remainder of the sequence. Three significant events occurred during deposition of the Navesink sequence. (1) A global warming event possibly brought on by the main outpouring of the Deccan Traps in India began ~500 k.y. and ended ~22 k.y. prior to the K-T boundary. (2) The K-T boundary event at Chicxulub, Mexico led to fallout of tektites on the New Jersey margin (the spherule layer at Bass River) and the mass extinction of marine calcareous plankton. (3) A possible tsunami event, triggered by mass slumping of the New Jersey outer continental margin or elsewhere, may have immediately followed the K-T boundary event.

Sea-level changes do not appear to be a significant cause or influence on any of these three events.

ACKNOWLEDGMENTS

Drilling of the Bass River and Ancora boreholes (Ocean Drilling Program Leg 174AX) was supported by National Science

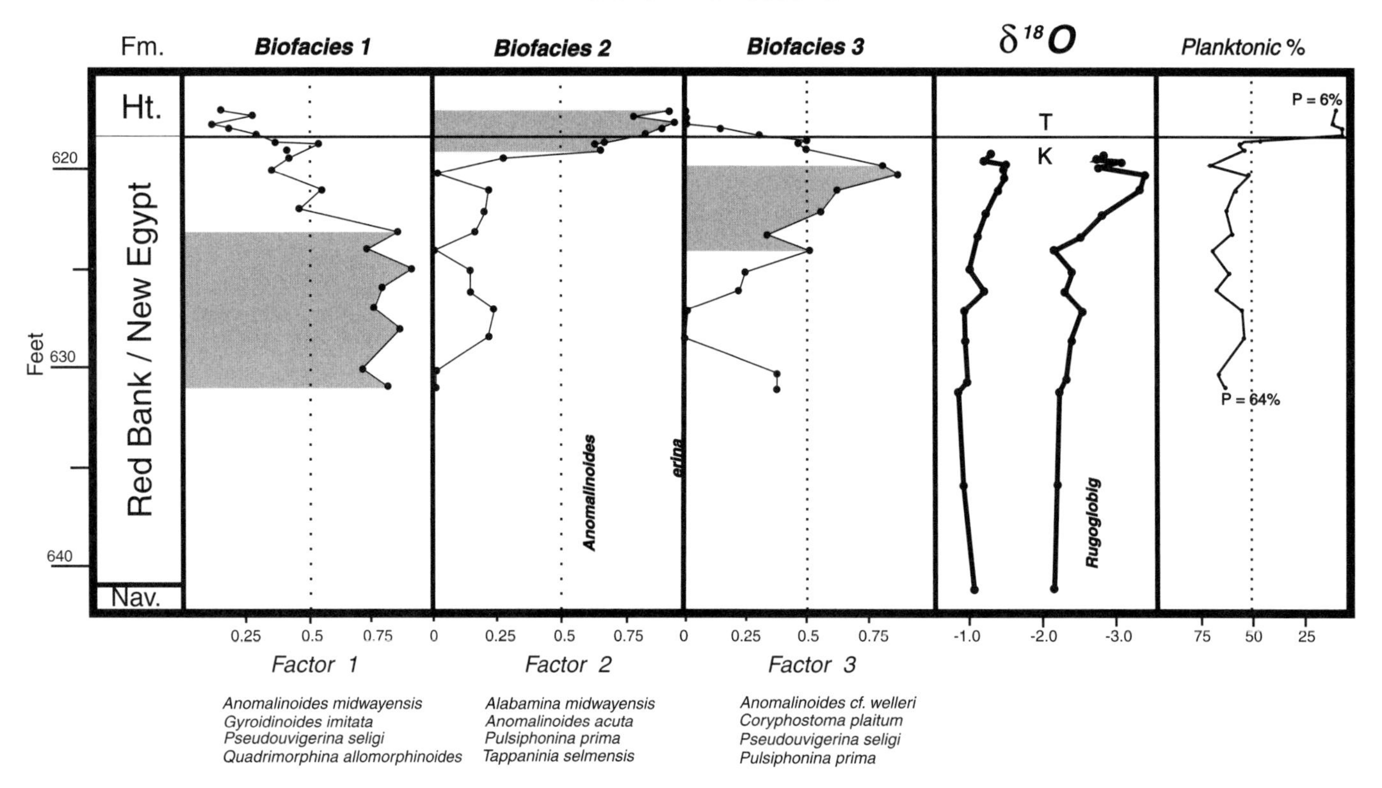

Figure 7. Factor analysis identifies same succession of three biofacies that are identified at Bass River (see Fig. 5 and text for explanation). Note that factor 3, as at Bass River, is associated with negative shift in $\delta^{18}O$ isotope values of Cretaceous planktonic foraminifer *Rugoglobigerina*, indicating warming trend. Decrease of as much as 1% in ^{18}O values on shells of Cretaceous planktonic foraminifer *Rugoglobigerina* indicates significant warming trend of as much as 5°C in near sea-surface temperatures that began ~500 k.y. and ended ~22 k.y. before Cretaceous-Tertiary boundary.

Foundation grants (EAR-94-17108 and EAR-97-08664 to Miller), by the New Jersey Geological Survey, and by the U.S. Geological Survey. We thank the science staff and many who assisted during drilling and recovery of cores at Bass River and Ancora. B. Huber, M. Leckie, and K. MacLeod provided helpful reviews.

REFERENCES CITED

Alvarez, W., 1996, Trajectories of ballistic ejecta from the Chicxulub Crater, *in* Ryder, G., Fastovsky, D., and Gartner, S., eds., The Cretaceous-Tertiary event and other catastrophes in Earth history: Geological Society of America Special Paper 307, p. 141–150.

Austin, J.A., Christie-Blick, N., Malone, M.J., and 25 others, 1998, Continuing the New Jersey Mid-Atlantic Sea-Level Transect: Proceedings of the Ocean Drilling Program, Initial Reports, v. 174A: College Station, Texas, Ocean Drilling Program, 324 p.

Berggren, W.A., and Aubert, J., 1975, Paleocene benthonic foraminiferal biostratigraphy, paleobiogeography and paleoecology of Atlantic-Tethyan regions: Midway-type fauna: Palaeogeography, Palaeoclimatology, Palaeoecology, v. 18, p. 73–192.

Berggren, W.A., Kent, D.V., Swisher, C.C., III, and Aubry, M.-P., 1995, A revised Cenozoic geochronology and chronostratigraphy, *in* Berggren, W.A., et al., eds., Geochronology, time scales, and global stratigraphic correlation: SEPM (Society for Sedimentary Geology) Special Publication, v. 54, p. 129–212.

Barrera, E., and Savin, S.M., 1999, Evolution of late Campanian-Maastrichtian marine climates and oceans, *in* Barrera, E., and Johnson, E., eds., Evolution of the Cretaceous ocean-climate system: Geological Society of America Special Paper 332, p. 245–282.

Browning, J.V., Miller, K.G., and Pak, D.K., 1996, Global implications of lower to middle Eocene sequence boundaries on the New Jersey coastal plain: The icehouse cometh: Geology, v. 24, p. 639–642.

Coplen, T.B., Kendall, C., and Hopple, J., 1983, Comparison of stable isotope reference sections: Nature, v. 302, p. 236–238.

Courtillot, V., Besse, J., Vandamme, D., Montigny, R., Jaeger, J-J., and Cappetta, H., 1986, Deccan flood basalts at the Cretaceous/Tertiary boundary?: Earth and Planetary Science Letters, v. 80, p. 361–374.

Donovan, A.D., Baum, G.R., Blechschmidt, G.L., Loutit, L.S., Pflum, C.E., and Vail, P.R., 1988, Sequence stratigraphic setting of the Cretaceous-Tertiary boundary in central Alabama, *in* Wilgus, C.K., and five others, Sea-level changes: An integrated approach: Society of Economic Paleontologists and Mineralogists Special Publication, v. 42, p. 299–307.

Habib, D., Moskovitz, S., and Kramer, C., 1992, Dinoflagellate and calcareous nannofossil response to sea-level change in Cretaceous-Tertiary boundary sections: Geology, v. 20, p. 165–168.

Hallam, A., 1990, Eustatic sea-level change and the K-T boundary: Geotimes, August, p. 18, 23.

Hallam, A., 2000, Mass extinctions and sea-level changes, *in* Catastrophic events and mass extinctions: Impacts and beyond: Houston, Texas, Lunar and Planetary Institute, LPI Contribution No. 1053, p. 65.

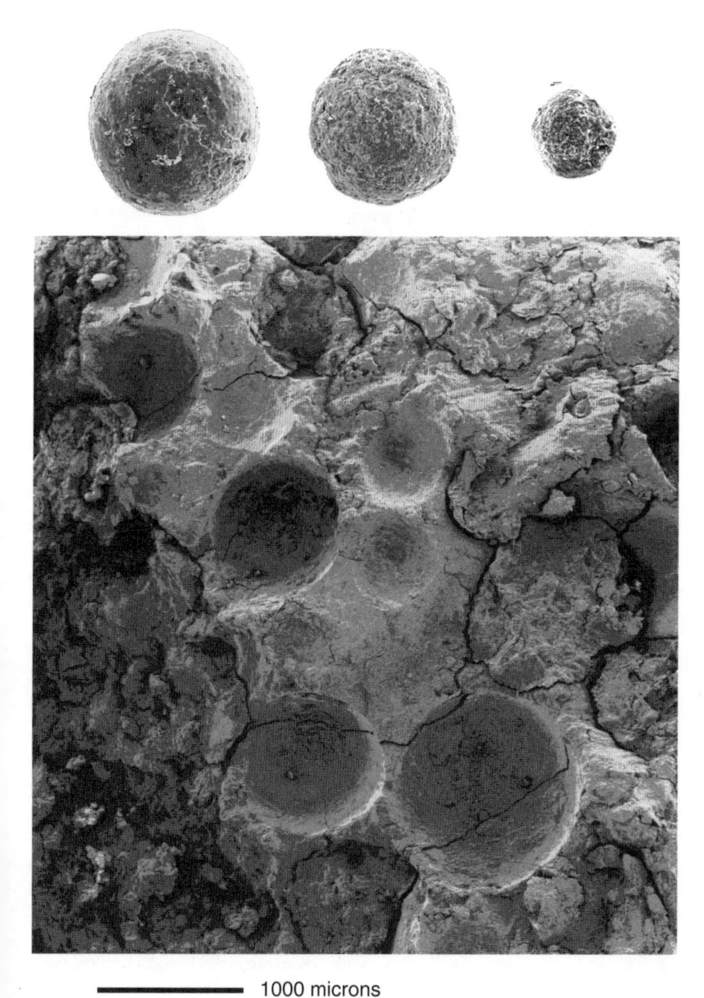

Figure 8. Impressions made by tektites when they settled on surface of Maastrichtian glauconitic clays at Cretaceous-Tertiary boundary time, Bass River borehole. Altered tektites (spherules) are shown at top.

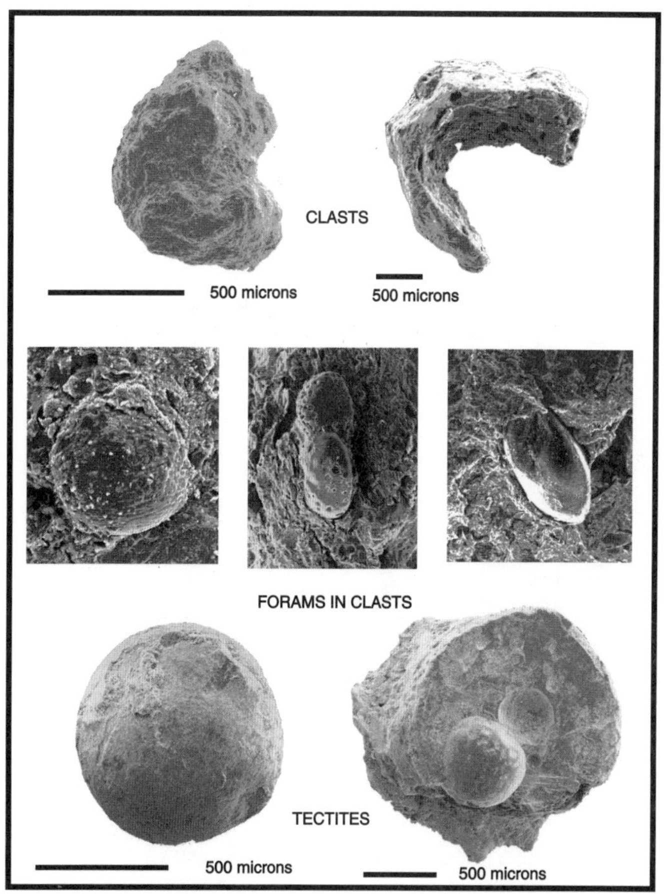

Figure 9. Above spherule layer, small clay clasts, containing calcite-replaced tektites and Cretaceous foraminifera and dinoflagellates, occur in lower 10 cm of Danian. Clasts are interpreted as ripups that were transported by tsunami activity shoreward from deeper part of Maastrichtian seafloor. Note internal globules, typical of Cretaceous-Tertiary boundary tektites, in tectite on right. From left to right ultimate chamber of *Heterohelix globulosa*, gavelinid, and *Alabamina midwayensis* are identified in clay clasts.

Hallam, A., and Wignall, P.B., 1997, Mass extinctions and their aftermath: Oxford, Oxford University Press, 328 p.

Hallam, A., and Wignall, P.B., 1999, Mass extinctions and sea-level changes: Earth-Science Reviews, v. 48, p. 217–250.

Hansen, H.J., Toft, P., Mohabey, D.M., and Surkar, A., 1996, Lameta age: Dating the main pulse of the Deccan Traps volcanism, in National Symposium Deccan Flood Basalts, India: Gondwana Geology Magazine, v. 2, p. 365–374.

Haq, B.U., Hardenbol, J., and Vail, P.R., 1987, Chronology of fluctuating sea levels since the Triassic: Science, v. 235, p. 1156–1167.

Haq, B.U., Hardenbol, J., and Vail, P.R., 1988, Mesozoic and Cenozoic chronostratigraphy and cycles of sea-level change, *in* Wilgus, C.K., and five others, Sea-level changes: An integrated approach: Society of Economic Paleontologists and Mineralogists Special Publication, v. 42, p. 72–108.

Hart, S.R., and Brooks, C., 1974, Clinopyroxene-matrix partitioning of K, Rb, Cs, and Ba: Geochimica et Cosmochimica Acta, v. 38, p. 1799–1806.

Hildebrand, A.R., Penfield, G.T., Kring, D.A., Pilkington, M., Camargo, Z.A., Jacobsen, S.B., and Boynton, W.V., 1991, Chicxulub Crater: A possible Cretaceous Tertiary boundary impact crater on the Yucatán Peninsula, Mexico: Geology, v. 19, p. 867–871.

Klaus, A., Norris, R.D., Kroon, D, and Smit, J., 2000, Impact-induced mass wasting at the K-T boundary: Blake Nose, western North Atlantic: Geology, v. 28, p. 319–322.

Kucera, M., and Malmgren, B.J., 1998, Terminal Cretaceous warming event in the mid-latitude South Atlantic Ocean: Evidence from poleward migration of *Contusotruncana contusa* (planktonic foraminifera) morphotypes: Palaeogeography, Palaeoclimatology, Palaeoecology, v. 138, p. 1–15.

Li, L., and Keller, G., 1999a, Maastrichtian climate, productivity and faunal turnovers in planktic foraminifera in South Atlantic DSDP Sites 525 and 21: Marine Micropaleontology, v. 33, p. 55–86.

Li, L., and Keller, G., 1999b, Abrupt deep-sea warming at the end of the Cretaceous: Geology, v. 26, p. 995–998.

Miller, K.G., et al., 1994, Initial Reports, Ocean Drilling Program, Leg 150X: College Station, Texas, Ocean Drilling Program, 59 p.

Miller, K.G., Liu, C., Browning, J.V., Pekar, S.F., and 9 others, 1996, Cape May site report: Proceedings of the Ocean Drilling Program, Initial Reports, v. 150X (supplement): College Station, Texas, Ocean Drilling Program, 28 p.

Miller, K.G., Rufolo, S., Sugarman, P.J., Pekar, S.F., Browning, J.V., and Gwynn, D. W., 1997, Early to middle Miocene sequences, systems tracts, and benthic foraminiferal biofacies, New Jersey coastal plain: Scientific

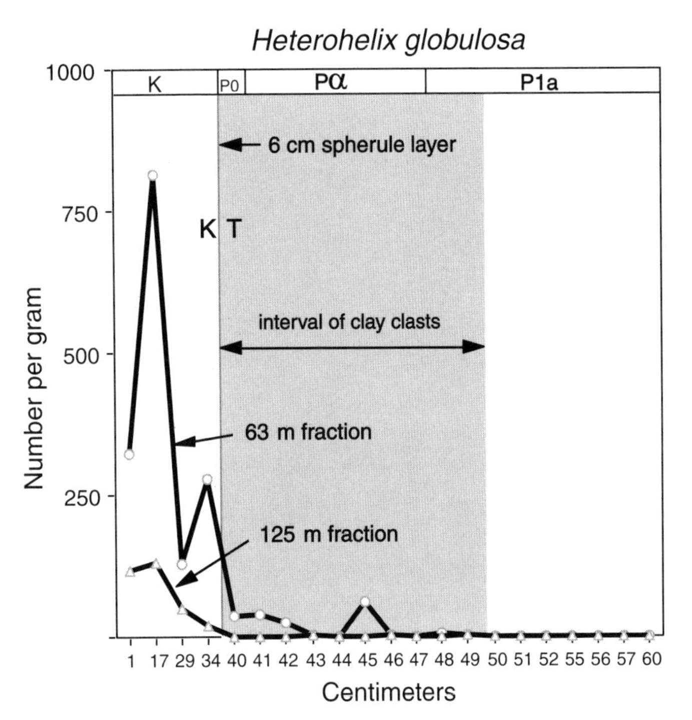

Figure 10. Cretaceous planktonic foraminifer, *Heterohelix globulosa*, is most abundant foraminifer in clay clasts. This species is absent above zone of clay clasts, indicating that its presence in Danian comes from clay clasts.

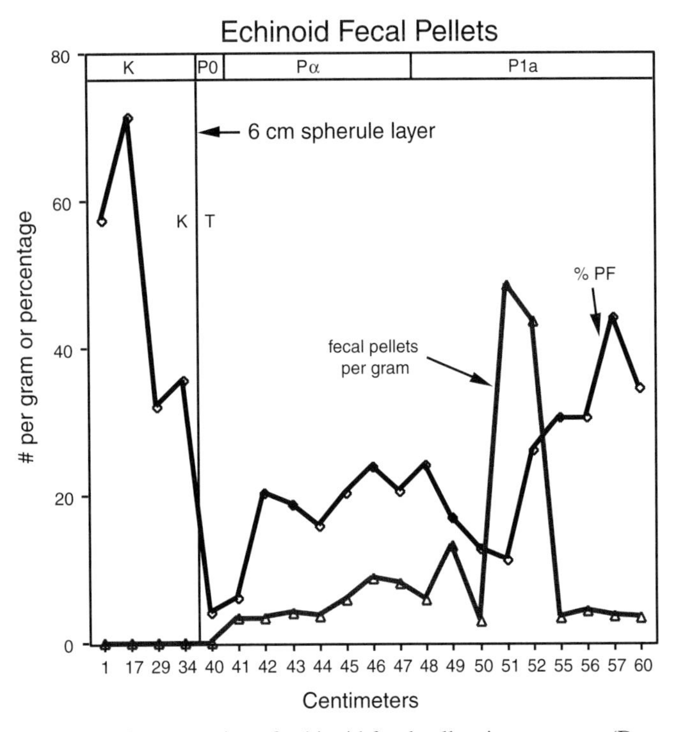

Figure 11. Concentration of echinoid fecal pellets in upper part (Danian) of Navesink sequence in Bass River borehole. Peak concentration of fecal pellets is interpreted as condensed interval, suggesting flooding surface of parasequence. See text for explanation.

Results, Ocean Drilling Program, Leg 150X: College Station, Texas, Ocean Drilling Program, p. 169–186.

Miller, K.G., Sugarman, P.J., Browning, J.V., and 16 others, 1998, Bass River Site report, *in* Scientific results, Ocean Drilling Program, Leg 174AX: College Station, Texas, Ocean Drilling Program, 39 p.

Miller, K.G., Barrera, E., Olsson, R.K., Sugarman, P.J., and Savin, S.M., 1999a, Does ice drive early Maastrichtian eustasy?: Geology, v. 27, p. 783–786.

Miller, K.G., Sugarman, P.J., Browning, J.V., and 16 others, 1999b, Ancora Site: Proceedings of the Ocean Drilling Program, Initial Reports, Volume 174AX (Suppl.), p. 1–65 [Online]. Available from World Wide Web: ⟨http://www-odp.tamu.edu/publications/174AXSIR/VOLUME/CHAPTERS/.

Mountain, G.S., Miller, K.G., Blum, P., et al., 1994, Initial Reports, Ocean Drilling Program, Volume 150: College Station, Texas, Ocean Drilling Program, 885 p.

Norris, R.D., Huber, B.T., and Self-Trail, J., 1999, Synchroneity of the K-T oceanic mass extinction and meteorite impact: Blake Nose, western North Atlantic: Geology, v. 27, p. 419–422.

Norris, R.D., Firth, J., Blusztajn, J.S., and Ravizza, G., 2000, Mass failure of the North Atlantic margin triggered by the Cretaceous-Paleogene bolide impact: Geology, v. 28, p. 1119–1122.

Olsson, R.K., and Nyong, E.E., 1984, A paleoslope model for Campanian-lower Maestrichtian foraminifera of New Jersey and Delaware: Journal of Foraminiferal Research, v. 14, p. 50–68.

Olsson, R.K., and Liu, C., 1993, Controversies on the placement of Cretaceous-Paleogene boundary and the K/P mass extinction of planktonic foraminifera: Palaios, v. 7, p. 127–139.

Olsson, R.K., Liu, C., and Van Fossen, M., 1996, The Cretaceous-Tertiary catastrophic event at Millers Ferry, Alabama, *in* Ryder, G., Fastovsky, D., and Gartner, S., eds., The Cretaceous-Tertiary event and other catastrophes in Earth history: Geological Society of America Special Paper 307, p. 263–277.

Olsson, R.K., Miller, K.G., Browning, J.V., Habib, D., and Sugarman, P.J., 1997, Ejecta layer at the Cretaceous-Tertiary boundary, Bass River, New Jersey (Ocean Drilling Program Leg 174AX): Geology, v. 25, p. 759–762.

Olsson, R.K., Wright, J.D., Miller, K.G., Browning, J.V., and Cramer, B.S., 2000, Cretaceous-Tertiary boundary events on the New Jersey continental margin, *in* Catastrophic events and mass extinctions: Impacts and beyond: Houston, Texas, Lunar and Planetary Institute, LPI Contribution No. 1053, p. 160.

Olsson, R.K., Wright, J.D., and Miller, K.G., 2001, Paleobiogeography of *Pseudotextularia elegans* during the latest Maastrichtian global warming event: Journal of Foraminiferal Research, v. 31, p. 275–282.

Oslick, J.F., Miller, K.G., Feigenson, M.D., and Wright, J.D., 1994, Testing Oligocene-Miocene strontium isotopic correlations: Relationships with an inferred glacioeustatic record: Paleoceanography, v. 9, p. 427–443.

Smit, J., Roep, Th.B., Alvarez, W., Montanari, A., Claeys, P., Grajales-Nishimura, J.M., and Bermudez, J., 1996, Coarse-grained, clastic sandstone complex at the K-T boundary around the Gulf of Mexico: Deposition by tsunami waves induced by the Chicxulub impact?, *in* Ryder, G., Fastovsky, D., and Gartner, S., eds., The Cretaceous-Tertiary event and other catastrophes in Earth history: Geological Society of America Special Paper 307, p. 151–182.

Sugarman, P.J., Miller, K.G., Burky, D., and Feigenson, M.D., 1995, Uppermost Campanian-Maastrichtian strontium isotopic, biostratigraphic, and sequence stratigraphic framework of the New Jersey Coastal Plain: Geological Society of America Bulletin, v. 107, p. 19–37.

Sugarman, P.J., Miller, K.G., Olsson, R.K., and six others, 1999, The Cenomanian/Turonian carbon burial event, Bass River, New Jersey, USA: Journal of Foraminiferal Research, v. 29, p. 438–452.

Vonhof, H.B., and Smit, J., 1997, High-resolution late Maastrichtian-early Danian oceanic $^{87}Sr/^{86}Sr$ record: Implications for Cretaceous-Tertiary boundary events: Geology, v. 25, p. 347–350.

MANUSCRIPT SUBMITTED OCTOBER 5, 2000; ACCEPTED BY THE SOCIETY MARCH 22, 2001

Printed in the U.S.A.

Geological Society of America
Special Paper 356
2002

Complex tsunami waves suggested by the Cretaceous-Tertiary boundary deposit at the Moncada section, western Cuba

R. Tada
Y. Nakano
Department of Earth and Planetary Science, Graduate School of Science, University of Tokyo, 7-3-1 Hongo, Tokyo 113-0033, Japan
M.A. Iturralde-Vinent
Museo Nacional de Historia Natural, Obispo no. 61, Plaza de Armas, La Habana 10 100, Cuba
S. Yamamoto
T. Kamata
E. Tajika
Department of Earth and Planetary Science, Graduate School of Science, University of Tokyo, 7-3-1 Hongo, Tokyo 113-0033, Japan
K. Toyoda
Graduate School of Environmental Earth Science, Hokkaido University, N17 W8, Sapporo 060-0810, Japan
S. Kiyokawa
Department of Earth and Planetary Sciences, Faculty of Sciences, Kyushu University 33 Hakozaki, Fukuoka 812-8581, Japan
D. Garcia Delgado
Instituto de Geologia y Paleontologia, Via Blanca y Linea del Ferrocarril San Miguel del Padron, La Habana 11 000, Cuba
T. Oji
K. Goto
H. Takayama*
Department of Earth and Planetary Science, Graduate School of Science, University of Tokyo, 7-3-1 Hongo, Tokyo 113-0033, Japan
R. Rojas-Consuegra
Museo Nacional de Historia Natural, Obispo no. 61, Plaza de Armas, La Habana 10 100, Cuba
T. Matsui
Department of Complexity Science and Engineering, Graduate School of Frontier Science, University of Tokyo, 7-3-1 Hongo, Bunkyo-ku, Tokyo 113-0031, Japan

ABSTRACT

The Moncada Formation in western Cuba is an ~2-m-thick weakly metamorphosed complex characterized by repetition of calcareous sandstone units that show overall upward fining and thinning. The Moncada Formation contains abundant

*Present address: NHK Japan Broadcasting Corporation, Nagoya Office 1-13-3 Higashisakura, Higashiku, Nagoya 461-8725, Japan.

Tada, R., Nakano, Y., Iturralde-Vinent, M.A., Yamamoto, S., Kamata, T., Tajika, E., Toyoda, K., Kiyokawa, S., Garcia Delgado, D., Oji, T., Goto, K., Takayama, H., Rojas-Consuegra, R., and Matsui, T., 2002, Complex tsunami waves suggested by the Cretaceous-Tertiary boundary deposit at the Moncada section, western Cuba, *in* Koeberl, C., and MacLeod, K.G., eds., Catastrophic Events and Mass Extinctions: Impacts and Beyond: Boulder, Colorado, Geological Society of America Special Paper 356, p. 109–123.

shocked quartz, altered vesicular impact-melt fragments, and altered and deformed greenish grains of possible impact glass origin. In addition, a high iridium (~0.8 ppb) peak is identified at the top of the formation. Together with the biostratigraphically estimated age, between late Maastrichtian and early Paleocene, this evidence supports a Cretaceous-Tertiary (K-T) boundary origin for the deposit. The Moncada Formation has ripple cross-laminations at several horizons that indicate north-south–trending paleocurrent directions with reversals. Changes in detrital provenance corresponding to paleocurrent reversals are also recognized. These characteristics are similar to K-T boundary sandstone complexes reported from the Gulf of Mexico region, and strongly support a K-T boundary tsunami origin for the Moncada Formation. The pattern of paleocurrent reversals in the Moncada Formation suggests that tsunami waves were not simple alternations of a single beat, but rather alternations of double beats following the first wave that came from the south. In addition, the maximum grain-size variation within each unit suggests the presence of higher frequency waves superimposed on the lower frequency waves. Thus, our results suggest that K-T impact tsunami waves had a complex rhythm that was caused either by reflections and diffractions of waves or by multiple tsunami waves created by multiple gravity-flows triggered by seismic shocks of the impact.

INTRODUCTION

Since the discovery of the Cretaceous-Tertiary (K-T) boundary impact crater at Chicxulub, Yucatan (Hildebrand et al., 1991), the focus of K-T boundary research has shifted toward exploring the environmental consequences of the K-T impact (e.g., Ryder et al., 1996; D'Hondt et al., 1994). A giant tsunami is one of the probable consequences, and possible K-T boundary tsunami deposits have been reported at many localities in the marginal part of the Gulf of Mexico (e.g., Bourgeois et al., 1988; Smit et al., 1992, 1996; Smit, 1999). However, their origin is still controversial and an impact-induced megaturbidite origin has also been advocated (Bohor, 1996).

The Late Cretaceous to early Tertiary sedimentary sequence is widely distributed throughout the Cuban fold belt, and extraordinarily thick (to 700 m), coarse-grained event deposits, usually dated as upper Maastrichtian, have been known for some time in the sections of central and western Cuba (Pszczolkowski, 1986; Iturralde-Vinent, 1992). Although their megaturbidite origin, and a possible relation with the K-T bolide impact, was advocated by Pszczolkowski (1986), not enough evidence was presented to demonstrate a megaturbidite origin, and the relation with the K-T boundary impact remained controversial, because no conclusive evidence was presented to support their K-T boundary age (e.g., Iturralde-Vinent, 1992). Takayama et al. (2000) reported latest Maastrichtian nannofossils, *Micula prinsii*, from a mudstone intraclast in the deposit near Havana, Cuba (the Peñalver Formation). Together with their discovery of shocked quartz from the same deposit, these authors concluded that the deposit was formed in association with the K-T bolide impact. They also argued that the deposit consists of a lower gravity-flow unit and an upper homogenite (a deep-sea tsunami deposit) unit, and proposed the occurrence of deep-sea tsunamis at K-T boundary sites in western Cuba.

Although most of the K-T boundary deposits in western and central Cuba are characterized by thick, coarse-grained deposits (Takayama et al., 2000, Iturralde-Vinent et al., 2000, Kiyokawa et al., 2000), a relatively thin (~2 m) calcareous sandstone of probable K-T boundary age is present in the Los Organos belt of the Guaniguanico terrane in western Cuba (Iturralde-Vinent, 1995). Such a thin (compared to other K-T boundary sites in western Cuba) deposit needs to be explained, and the answer may give a clue to solving the origin and extent of tsunami waves at the K-T boundary.

In this chapter we describe the occurrence of this 2-m-thick deposit, present evidence to support its association with the K-T boundary impact, and discuss the origin of the deposit and its implication for the nature of tsunamis. The observation suggests a complex behavior of tsunami waves immediately following the impact in the eastern margin of the Yucatan platform.

GEOLOGICAL SETTING OF WESTERN CUBA

The present Cuban fold belt contains five geotectonic units; the autochthonous Bahamian platform, the allochthonous southwestern Cuba terranes (the Guaniguanico, Escambray, and Pinos terranes), the northern ophiolite and Placetas belts, the Cretaceous volcanic arc, and neoautochthonous postorogenic sedimentary rocks of latest Eocene to recent age (Iturralde-Vinent, 1994, 1995, 1998). The Guaniguanico terrane is located in Pinar del Rio Province of western Cuba, where it occurs as a stack of north- to northwest-thrusted sheets that were emplaced onto the Gulf of Mexico autochthon (Iturralde-Vinent, 1994), probably as a result of collision of the extinct Cretaceous volcanic arc caused by the opening of the rift basins in the western part of the Yucatan basin during the late Paleocene to middle Eocene (Rosencrantz, 1990). The Guaniguanico terrane is exposed on the north side of the Pinar fault; the tectonically upper horizons are exposed to the north (Figure. 1). North and east portions of the terrane are tectonically covered by the Bahia Honda allochthon, which is composed of ophiolites, Cretaceous volcaniclastics, and latest Cretaceous to Eocene sedimentary rocks (Iturralde-Vinent, 1994, 1998). The deformation and emplacement of all thrust sheets took place between the late Paleocene and middle Eocene (Bralower and Iturralde-Vinent, 1997).

The allochthonous nature of the Guaniguanico terrane has been discussed in the Cuban literature, and its original location is estimated as the western margin of the Yucatan platform (Iturralde-Vinent, 1994, 1998). It is also considered that the sedimentary sequences of the Guaniguanico terrane originated as a part of the continental margin of the Yucatan platform (Maya block) and the western Caribbean Sea basin (Iturralde-Vinent, 1994, 1998) (Fig. 2). The Moncada section (the site studied in this chapter) occurs in the Los Organos belt, which consists of the thrust sheet at the tectonically lowest position, and includes sedimentary strata of early Middle Jurassic to middle Eocene age (Fig. 1).

LOCALITY AND STRATIGRAPHIC SETTING OF THE MONCADA SECTION

The Moncada section is located 18 km to the west of the town of Viñales, along the road to Pons, just at the intersection with the village of Moncada. The section is exposed on both sides of a roadcut immediately to the west of the intersection, where a >40-m-thick sequence of Cretaceous to lower Tertiary strata are exposed. The strata dip gently east; the general strike and dip are N15°-40°E and 8°-12°E, respectively.

Iturralde-Vinent (1995) described a 2-m-thick calcareous sandstone from this locality, which was referred to as the La Güira Member of the Ancón Formation. Bralower and Iturralde-Vinent (1997) paleontologically dated the La Güira Member as early Paleocene (older than NP4). Additional stratigraphic observation in the area suggests that this 2-m-thick

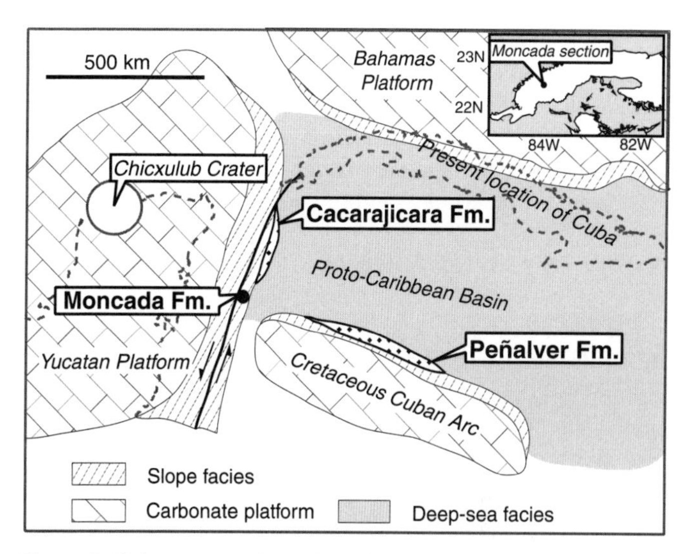

Figure 2. Paleogeotectonic setting of Moncada Formation and other Cretaceous-Tertiary (K-T) boundary deposits in western Cuba at time of K-T impact. Location of Moncada section is also shown in upper right corner of figure. Fm.—Formation.

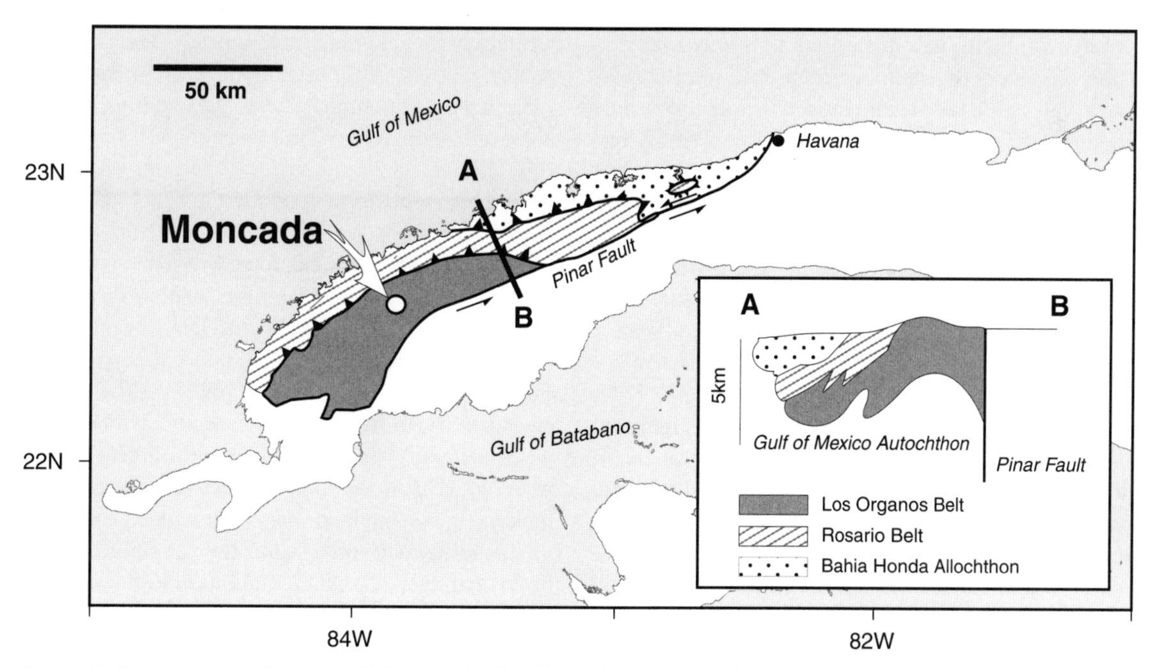

Figure 1. Geologic map of western Cuba showing locality of Moncada section (study site).

calcareous sandstone should be defined as an independent stratigraphic unit, which crops out only in the Moncada section, so it is named the Moncada Formation. The ~2-m-thick Moncada Formation is exposed near the western end of the roadcut. It disconformably overlies thick-bedded, grayish-black, micritic limestone of the Pons Formation, which yields microfossils of Albian-Cenomanian age. It is conformably overlain by well-bedded, light violet, gray, and greenish-gray micritic slaty limestone of the Ancón Formation, which dates from the late early Paleocene (older than NP4) to the earliest Eocene (P6a) (Bralower and Iturralde-Vinent, 1997). Diaz Otero et al. (2000) reported a mixed microfossil assemblage from the Moncada Formation, the ages ranging from Aptian to late Maastrichtian. Together with the early Paleocene age of the Ancón Formation, this result suggests an age between late Maastrichtian and early Paleocene for the Moncada Formation.

SAMPLES AND METHODS

The entire sequence of the Moncada Formation, as well as the strata immediately above and below it, were sampled almost continuously, using a motor-powered cutter. Sedimentary structures were observed and described on the cut surfaces. Rough estimation of the current directions was conducted for cross-laminated intervals in field based on observation of the differently orientated cut surfaces. In addition, orientated samples were taken for such intervals to estimate current directions based on three-dimensional observations of the samples in the laboratory. Each sample, generally representing 5–10 cm of stratigraphic interval, was polished and sedimentary structures were described. Thin sections were prepared for most of the samples to examine grain composition and the maximum grain size. The maximum grain size of detrital silicate and/or limestone and chert fragments was measured in 1 cm intervals throughout the sequence. Subsamples, each representing 0.5–2-cm-thick intervals, were taken from most of the samples, pulverized, and subjected to X-ray powder diffraction and X-ray fluorescence (XRF) analyses to determine mineral and major element compositions (Tables 1 and 2). Sample preparation and analytical methods for XRD and XRF analyses are the same as those described in Irino and Tada (2000). The reproducibility of XRD measurements is better than ±30% for smectite, illite, and chlorite, ±15% for calcite and plagioclase, and ±10% for quartz. The 2σ relative standard deviation of the XRF measurements is ±0.6%–0.8% for SiO_2, TiO_2, Al_2O_3, Fe_2O_3, K_2O, and CaO, and 1.0%–1.2% for MgO, P_2O_5, and loss on ignition, and ±1.4%–1.6% for MnO and Na_2O. Identification of smectite and chlorite was carried out for all samples based on glycerol treatment that shifts the 14 Å peak of smectite to ~17.9 Å, and heat treatment at 550°C for 1 h that intensifies the 14.2 Å peak of chlorite and slightly shifts it to 13.8 Å (Moore and Reynolds, 1997). The results suggest that the 14 Å peak is attributed to either smectite or chlorite, whereas the 7 Å peak is attributed to chlorite. Coexistence of chlorite and smectite is identified in some samples. The smectite peak height is corrected for chlorite contribution based on the observation that the 14 Å peak of chlorite is approximately the same in height as its 7 Å peak.

The iridium (Ir) content in selected subsamples was determined by radiochemical neutron activation analysis (RNAA) by irradiation in the JRR-3M reactor of the Japan Atomic Energy Research Institute (JAERI) at a neutron flux of 1.2×10^{14} $cm^{-2}s^{-1}$ for 6 h. Chemical separation was done with Srafion NMRR resin prior to counting on high-resolution Ge detectors at the Central Institute of Isotope Science, Hokkaido University. Recovery of the Ir carrier was 50%–70%, as determined by reactivation. The RNAA procedures were essentially the same as those of Kyte et al. (1992). For the standard, appropriate amounts of diluted solution from Ir standard solution (SIGMA Chemical Co.) were weighed into the quartz tubes immediately after preparation, dried, and then sealed. Our result for SARM7 certified reference material (MINTEK, South African Bureau of Standard) is 73 ± 6 ppb; the reference value of SARM7 from Potts et al. (1992) is 74 ± 12 ppb with 95% confidence intervals. No Ir peaks were observed in blank sample measurement.

MONCADA FORMATION

The Moncada Formation at the Moncada section is 184–191 cm thick, and consists of a calcareous sandstone complex; there are alternations of calcareous claystone and very fine sandstone in the uppermost part of the formation (Fig. 3). The Moncada Formation disconformably overlies grayish-black, bedded micritic limestone of the Pons Formation with a slightly undulating erosional surface (Fig. 4A). The Moncada Formation is conformably overlain by marly limestone of the Ancón Formation with a gradational contact. The sequence is weakly metamorphosed (to pumpellyite facies) and contains some weak shear planes subparallel to the bedding. However, primary fabric and sedimentary structures are still preserved.

Calcareous sandstone complex

The calcareous sandstone complex is characterized by repetition of five sandstone units, with an upward decrease in unit thickness from 93 to 9 cm. The lower two units show a distinct upward fining, whereas the upper three units do not show a clear upward fining (Figure 3). These sandstone units are numbered from 1 to 5 in ascending order. Boundaries between the units are gradational and no erosional contacts are observed. The lower part of each unit is characterized by a thicker, coarser-grained, very thin-bedded, olive-green to light olive, slightly calcareous sandstone compared to the upper part that is characterized by alternations of thinner and finer-grained, parallel to ripple cross-laminated, light gray, calcareous sandstone and grayish-black muddy drapes and/or flasers.

The lower part of unit 1 is composed of faintly stratified

TABLE 1. XRD PEAK INTENSITIES OF CONSTITUENT MINERALS IN BULK SAMPLES

Sample No.	Position (m)	Smectite (cps)	Illite (cps)	Chlorite (cps)	Quartz (cps)	Plagioclase (cps)	Calcite (cps)	Pyroxene (cps)	Rhodochrosite (cps)	Hematite (cps)
Mn32.5	229	0	0	0	175	0	2806	0	58	0
Mn32	227.5	0	6	6	146	0	2582	0	65	10
Mn32-2	227.1	0	0	0	136	0	2710	0	55	0
Mn32-1	225	0	0	0	198	0	2983	0	68	0
Mn31-5	221.6	0	0	0	160	0	2700	0	72	0
Mn31-4	219.8	0	0	0	176	15	2719	0	56	0
Mn31-3	218.4	0	31	0	158	0	2774	0	59	0
Mn31-2	217	0	0	0	150	0	2718	0	74	0
Mn31-1	215.5	0	0	0	138	0	2916	0	68	0
Mn30E-5	214.2	0	0	0	138	0	3056	0	76	0
Mn30E-2	210.9	0	0	0	75	0	2938	0	76	0
Mn30E-1	209.6	0	0	0	58	0	3143	0	68	0
Mn30E-0	208.5	0	0	0	113	0	2894	0	66	0
Mn30D-3	207.5	0	0	0	44	0	3077	0	68	0
Mn30D-2	206.4	0	0	0	45	0	3294	0	63	0
Mn30D-1	205	0	0	0	73	0	3256	0	74	0
Mn30C-3	202.7	0	0	0	51	0	3756	0	73	0
Mn30C	201	0	0	0	58	15	2826	0	72	0
Mn30C-2	199.9	0	0	0	44	0	3359	0	70	0
Mn30C-1	198.2	0	0	0	66	23	3159	0	82	0
Mn30B-6	197	0	0	0	77	28	3149	0	73	0
Mn30B-5	195.9	0	0	0	76	47	3130	0	75	0
Mn30B-4	195.1	0	0	0	82	41	3133	0	72	0
Mn29-8	191.5	0	0	39	291	151	2732	0	64	0
Mn30B	190.5	14	0	0	301	151	2611	0	52	0
Mn29-7	190.5	0	0	69	434	211	2610	0	74	0
Mn29U	190.3	0	0	38	315	179	2186	0	62	0
Mn29-6	189.7	7	0	31	851	486	1997	0	73	0
Mn29-5	189.2	57	0	59	559	378	1735	0	39	0
Mn29-4	188.3	43	44	76	342	306	2719	0	59	0
Mn29-2	187.1	48	0	0	1266	364	1964	0	47	0
Mn29L	186.7	6	0	42	155	265	1570	0	46	0
Mn29-1	186.5	22	0	33	1060	353	1703	0	50	0
Mn28-7	184	145	58	0	540	206	1660	0	38	0
Mn28-6	182.8	153	66	0	548	206	2100	0	41	0
Mn28-5	181.3	97	58	0	536	170	1905	0	51	0
Mn28	181	34	42	21	614	180	1524	0	44	0
Mn28-4	180.1	77	64	38	607	180	1816	0	49	0
Mn28-3	179.2	128	97	53	710	220	1102	0	34	0
Mn28-2	178.3	78	64	49	653	201	1815	0	38	30
Mn28-1	177.1	36	36	34	740	224	2163	0	51	25
Mn27	173	9	28	43	661	228	1396	0	59	46
Mn26	169	22	44	67	676	230	909	0	42	51
Mn25	161.5	21	14	22	737	185	2543	0	45	0
Mn24	158.5	31	18	19	628	269	1792	0	43	50
Mn23	152	20	0	0	418	336	2127	0	48	42
Mn22.5	148.5	32	0	0	512	399	2234	65	48	30
Mn22	144	21	0	0	538	350	2253	64	43	0
Mn21	136	73	64	0	538	263	441	105	19	66
Mn20	129	133	72	0	424	185	194	85	19	58
Mn19	124.5	31	0	0	688	268	2534	52	52	31
Mn18	120.5	23	0	0	778	326	1459	107	40	55
Mn17	113	70	28	0	732	180	1588	43	36	0
Mn16	111	107	28	0	973	175	936	56	41	29
Mn15	100	118	38	0	1062	198	755	62	27	48
Mn14	94	119	25	0	618	195	1067	62	35	43
Mn13	89	22	0	0	425	194	2528	0	55	0
Mn12	85	14	0	0	283	177	2581	0	52	0
Mn11	79	62	0	0	521	252	1866	0	41	69
Mn10.5	75	22	0	0	387	167	3417	0	54	0
Mn9	63	22	22	35	1004	98	1598	38	49	46
Mn8	58.5	0	14	35	1212	96	1408	55	40	31
Mn7	50	23	38	57	872	132	1230	46	33	59
Mn5	35.5	0	33	56	916	131	572	44	25	0
Mn3	11	0	0	72	1384	205	767	52	47	0
Mn2	5	0	0	81	1212	226	996	0	62	0
Mn1	−3	0	0	0	71	16	2834	0	67	0

*Peak intensity of each mineral is normalized by the peak height of a standard quartz sample × 10000.
#Smectite peak height is corrected for chlorite contribution.

TABLE 2. MAJOR ELEMENT COMPOSITION OF BULK SAMPLES DETERMINED BY X-RAY FLUORESCENCE ANALYSIS

Sample	Position (cm)	SiO_2 (wt%)	TiO_2 (wt%)	Al_2O_3 (wt%)	Fe_2O_3 (wt%)	MnO (wt%)	MgO (wt%)	CaO (wt%)	Na_2O (wt%)	K_2O (wt%)	P_2O_5 (wt%)	L.O.I. (wt%)	Total (wt%)
Mn32-2	227.1	5.37	0.05	1.37	0.60	0.06	0.45	49.70	n.d.	0.09	0.07	40.30	98.06
Mn32-1	225	6.55	0.07	1.81	0.71	0.06	0.51	48.71	n.d.	0.20	0.10	39.58	98.30
Mn32-0	223.3	5.47	0.07	1.53	0.64	0.06	0.47	49.70	n.d.	0.11	0.09	40.41	98.55
Mn31-5	221.6	6.01	0.08	1.75	0.76	0.07	0.53	48.32	n.d.	0.18	0.10	40.07	97.87
Mn31-4	219.8	6.84	0.08	1.86	0.77	0.07	0.52	49.32	n.d.	0.25	0.10	39.71	99.52
Mn31-3	218.4	7.18	0.08	2.02	0.85	0.07	0.56	49.20	n.d.	0.22	0.11	39.43	99.72
Mn31-1	215.5	5.41	0.06	1.51	0.65	0.06	0.49	48.79	n.d.	0.18	0.09	40.55	97.79
Mn30E-5	214.2	4.79	0.05	1.25	0.54	0.06	0.47	49.83	n.d.	0.10	0.08	40.81	97.98
Mn30E-4	213.1	5.24	0.06	1.38	0.59	0.06	0.52	50.56	n.d.	0.13	0.09	40.03	98.66
Mn30E-3	212	3.85	0.04	1.03	0.44	0.06	0.46	52.14	n.d.	0.04	0.07	41.43	99.56
Mn30E-2	210.9	3.39	0.04	0.97	0.43	0.05	0.51	52.32	n.d.	0.03	0.08	41.76	99.58
Mn30E-0	208.5	4.72	0.05	1.25	0.52	0.05	0.51	49.82	n.d.	0.08	0.08	41.00	98.08
Mn30D-2	206.4	2.60	0.03	0.83	0.36	0.05	0.46	51.95	n.d.	0.01	0.07	42.22	98.58
Mn30D-1	205	2.47	0.03	0.68	0.29	0.04	0.41	51.53	n.d.	0.03	0.06	42.29	97.83
Mn30C-3	202.7	2.15	0.02	0.62	0.24	0.05	0.38	51.93	n.d.	n.d.	0.05	42.49	97.93
Mn30C-2	199.9	2.28	0.03	0.65	0.26	0.05	0.43	53.14	n.d.	n.d.	0.06	42.47	99.37
Mn30B-4	195.1	4.95	0.06	1.32	0.58	0.05	0.47	48.83	n.d.	0.02	0.09	40.55	96.92
Mn30B-3	194	5.47	0.06	1.33	0.60	0.06	0.46	48.58	0.02	0.01	0.09	40.30	96.98
Mn30B-2	192.7	9.25	0.10	2.27	1.11	0.05	0.71	45.34	0.15	0.02	0.11	37.67	96.78
Mn29-9	191.9	9.46	0.11	2.46	1.21	0.05	0.84	45.97	0.28	0.01	0.12	37.34	97.85
Mn29-5	189.2	34.97	0.47	9.69	4.20	0.05	4.19	21.12	2.20	0.61	0.04	18.88	96.42
Mn29-4	188.3	29.28	0.66	8.87	4.21	0.05	4.34	24.90	1.33	1.22	0.04	22.23	97.13
Mn28-7	184	37.49	0.61	10.26	4.82	0.08	5.73	18.42	1.04	2.11	0.06	17.56	98.18
Mn28-6	182.8	39.84	0.58	11.07	4.97	0.05	6.30	17.20	1.01	2.34	0.06	16.51	99.93
Mn28-5	181.3	36.66	0.49	9.56	4.16	0.05	5.11	20.80	0.97	2.17	0.06	18.91	98.94
Mn28-4	180.1	39.77	0.55	10.56	4.62	0.07	5.93	17.54	1.06	2.42	0.06	16.41	98.99
Mn28-3	179.2	47.20	0.68	12.90	5.69	0.06	7.51	10.35	1.10	2.97	0.07	11.25	99.78
Mn28-2	178.3	42.39	0.57	10.94	4.81	0.06	6.29	15.94	1.14	2.39	0.07	15.11	99.71
Mn28-1	177.1	36.88	0.42	8.18	3.53	0.05	4.17	22.97	1.28	1.57	0.05	19.85	98.95
Mn27	173	43.52	0.57	10.70	4.71	0.05	6.68	16.07	1.88	2.03	0.07	14.25	100.53
Mn26	169	48.08	0.64	12.32	5.47	0.07	8.09	11.06	1.75	2.49	0.06	10.92	100.95
Mn25	161.5	35.41	0.33	6.50	2.74	0.06	4.28	27.04	1.27	1.09	0.06	22.64	101.42
Mn24	158.5	39.91	0.46	9.08	3.94	0.05	5.47	20.69	2.04	1.29	0.05	17.99	100.97
Mn23	152	31.13	0.33	7.37	2.72	0.07	3.06	30.05	2.35	0.47	0.07	24.20	101.82
Mn22.5	148.5	33.61	0.39	7.68	3.24	0.09	3.91	28.25	2.04	0.23	0.11	21.98	101.53
Mn22	144	32.38	0.37	7.37	3.04	0.08	3.26	29.87	1.96	0.31	0.10	23.15	101.89
Mn21	136	50.67	0.77	14.48	6.07	0.06	8.44	7.24	1.71	3.27	0.06	8.21	100.98
Mn20	129	51.26	0.90	15.57	6.86	0.06	8.38	5.01	1.48	3.96	0.05	6.58	100.11
Mn19	124.5	34.20	0.35	7.22	2.96	0.10	2.67	29.43	1.74	0.38	0.13	22.67	101.85
Mn18	120.5	41.21	0.47	9.40	3.91	0.08	3.68	23.27	2.23	0.37	0.13	17.17	101.92
Mn17	113	43.25	0.53	10.26	4.36	0.07	5.82	17.87	1.39	1.82	0.06	15.93	101.36
Mn16	111	49.23	0.56	11.36	4.95	0.05	6.38	12.52	1.42	2.02	0.05	12.22	100.76
Mn15	100	51.15	0.61	12.13	5.22	0.11	7.47	10.30	1.39	2.29	0.06	10.82	101.55
Mn14	94	46.80	0.59	11.59	4.90	0.12	6.28	14.26	1.67	1.83	0.05	13.50	101.59
Mn13	89	24.88	0.21	5.06	1.86	0.12	1.72	36.82	1.48	0.29	0.11	29.05	101.60
Mn12	85	20.22	0.20	4.41	1.76	0.12	2.00	39.94	1.10	0.11	0.13	31.86	101.85
Mn11	79	33.94	0.39	8.67	3.83	0.06	4.45	25.30	1.69	0.70	0.06	21.31	100.40
Mn10.5	75	21.80	0.18	4.53	1.78	0.13	2.28	38.96	0.98	0.24	0.12	30.63	101.63
Mn9	63	43.42	0.49	9.21	4.27	0.08	7.79	17.87	0.85	1.82	0.05	15.85	101.70
Mn8	58.5	45.73	0.46	8.64	3.86	0.08	6.59	18.55	0.98	1.45	0.04	15.55	101.93
Mn7	50	48.18	0.65	12.02	5.88	0.07	10.03	10.14	1.05	2.57	0.07	10.58	101.24
Mn5	35.5	48.58	0.67	12.36	5.63	0.06	9.79	9.82	0.98	2.40	0.05	10.72	101.06
Mn3	11	50.39	0.59	11.03	4.63	0.05	8.14	12.00	1.38	1.05	0.05	12.08	101.39
Mn2	5	48.18	0.60	11.14	4.75	0.06	9.04	11.78	1.46	1.01	0.04	12.61	100.67
Mn1	−3	2.61	0.01	0.48	0.15	0.07	0.60	53.50	0.07	n.d.	0.03	42.49	100.01

and normally graded, very coarse to medium-grained sandstone with occasional intercalations of 1–2-cm-thick granule-rich layers in the basal part. The sandstone is composed of flattened, olive-green grains and angular, whitish, vesicular fragments. The amount of whitish, vesicular fragments decreases upward, and small amounts of angular, dark red, hematitic mudstone fragments are in the middle part. Granules are predominantly composed of flat and rounded fragments of light gray micritic limestone, gray cherty limestone, and grayish-black chert, probably derived from the underlying Pons Formation. Granules of angular, whitish, vesicular fragments are also present, especially in the basal part. The flat limestone and chert fragments show imbrication structure in places (Fig. 4A). Under the petrographic microscope, olive-green, flattened grains are replaced

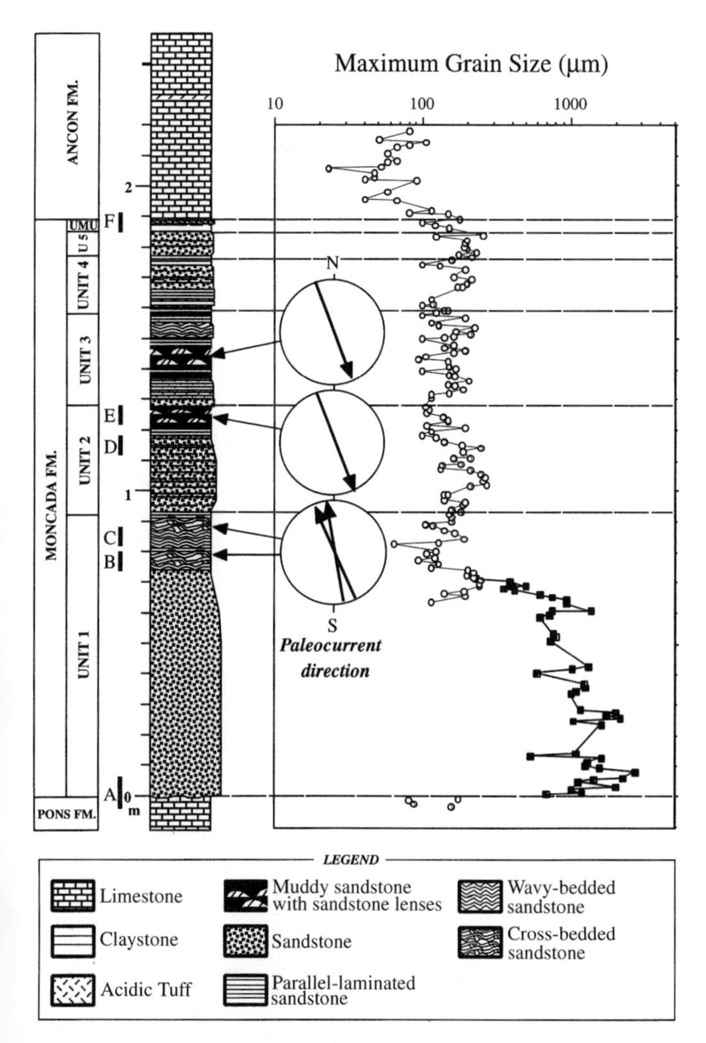

Figure 3. Columnar section of Moncada Formation. UMU means uppermost unit. Also shown on right side of column is maximum size of detrital silicate grains (open circle) and limestone and chert lithics (solid square), and paleocurrent directions measured at positions indicated by arrows. Positions of photographs A–F in Figure 4 are indicated on left side of column. Fm.—Formation.

by chlorite and subordinate amount of granular pumpellyite, whereas the whitish vesicular fragments are grayish, dusty-looking, and composed of smectite with small amount of clinopyroxene microcrystals.

The upper part of unit 1 is characterized by flaser bedding (Fig. 4B) and wavy bedding (Fig. 4C). The ripple cross-laminated part consists of fine-grained, light greenish-gray, calcareous sandstone, which alternates with thin, grayish-black flaser and/or drapes of very fine grained, muddy sandstone. Under the petrographic microscope, the ripple cross-bedded sandstone is composed of recrystallized calcite grains, and small amounts of detrital plagioclase and quartz. The grayish-black muddy part is composed of clayey micritic matrix and opaque wisps of probable hematite, and small amounts of very fine grained detrital quartz and plagioclase, micritic limestone fragments, and foraminiferal skeletons. Although partly recrystallized, cross-laminations are still preserved. The estimated paleocurrent direction is consistently from S17° ± 10°E (Fig. 3).

Units 2, 3, and 4 are composed of thin-bedded to parallel-laminated (0.5–2 cm thick), pale olive, medium- to fine-grained, calcareous sandstone with intercalations of millimeter-thick, pale greenish-yellow lamina (Fig. 4D) in the lower part, and wavy- to lenticular-bedded and ripple cross-laminated, light gray, finer grained, calcareous sandstone with grayish-black, very fine, sandy mudstone drapes in the upper part (Fig. 4E), although only parallel laminations are observed in unit 4. Under the petrographic microscope, the pale olive sandstone is predominantly composed of grayish, dusty, and occasionally vesicular fragments (corresponding to whitish vesicular fragments), pale yellowish-green, transparent stringers, highly deformed and flattened grains of probable chlorite, micritic limestone fragments, and subordinate amounts of detrital quartz, chert, and plagioclase grains. The maximum detrital grain size varies within each thin bed; finer grain sizes are in its margins (Fig. 3). The composition of the light gray sandstone and grayish-black sandy mudstone in the upper part is basically the same as in unit 1, although foraminiferal skeletons become more common in these units. Light gray, wavy beds and lenses tend to be recrystallized; however, some of the ripple cross-laminations are still preserved. The estimated paleocurrent direction within two horizons in the upper part of units 2 and 3 is from N20° ± 15° W and N10° ± 15° W, respectively, which is approximately reversed from the directions observed in unit 1 (Fig. 3).

Unit 5 is composed of thin-bedded, light olive-brown, fine- to very fine grained, calcareous sandstone, separated by a millimeter-thick, pale greenish-yellow lamina. Under the petrographic microscope, the sandstone is composed of brownish-gray, dusty, vesicular fragments, micritic limestone and chert fragments, and other sedimentary lithics, including hematitic, silty, mudstone, and sandstone. No ripple cross-laminations or thin, parallel laminations are recognized in the upper part of this unit.

Calcareous claystone and very fine sandstone alternation (uppermost unit)

A 3–5-cm-thick unit of light colored, calcareous claystone and dark colored, very fine calcareous sandstone alternations (Fig. 4F) overlies the sandstone complex. A 1–4.5-cm-thick bed of light gray to dusty yellow, calcareous, sandy claystone containing thin, discontinuous lamina of light gray, calcareous, fine sandstone covers a slightly undulating surface of the underlying sandstone complex. This claystone is overlain by 1-cm-thick, olive-gray, fine sandstone, which is characterized by a yellowish rim at its upper and lower boundaries. The lower boundary of this sandstone bed is sharp, whereas the upper boundary is slightly bioturbated. Under the petrographic microscope, the light colored claystone consists of micritic calcite matrix with

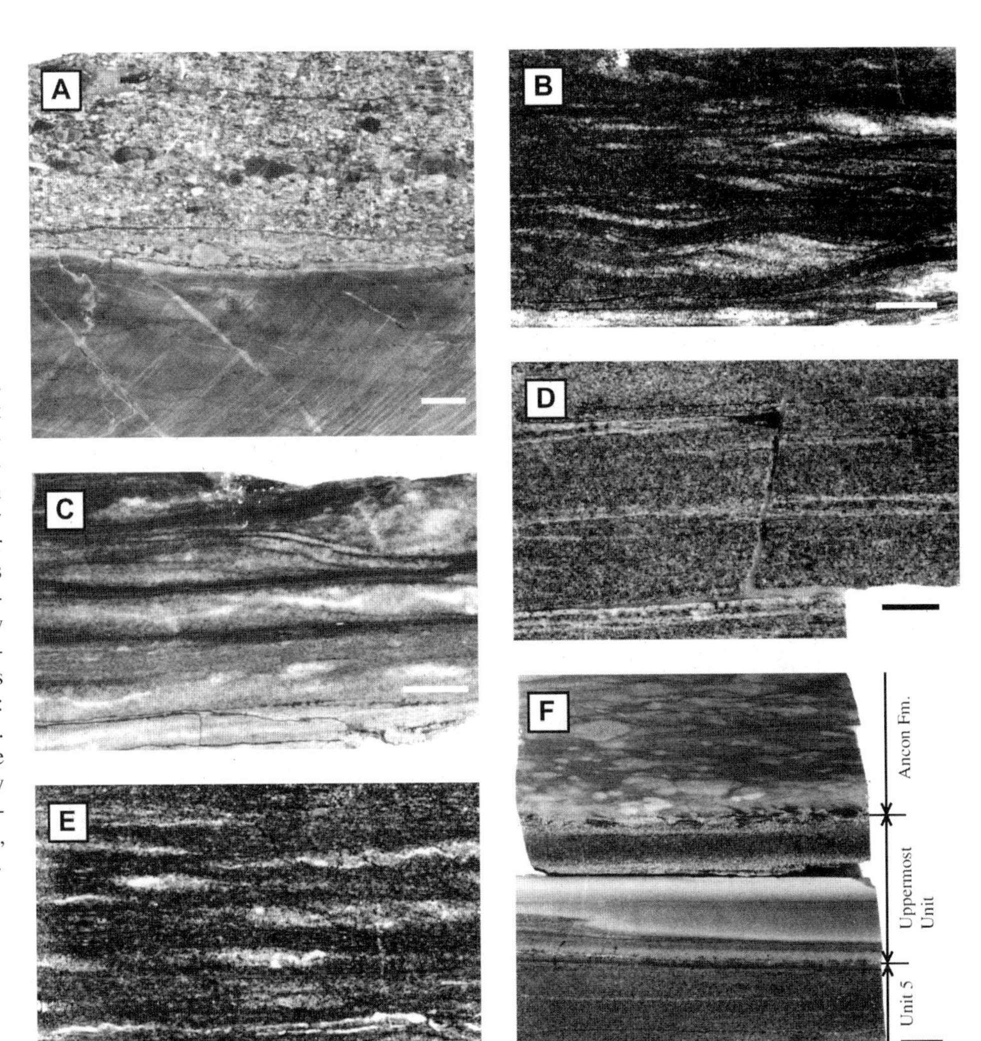

Figure 4. Photographs of typical lithology of Moncada Formation. Scale bar in lower right corner in each photograph is 1 cm. A: Slightly undulating erosional contact between grayish-black limestone of Pons Formation and calcareous sandstone of Moncada Formation. Flat granules in basal part of Moncada Formation show imbrication. B: Flaser bedding with ripple cross-laminations in upper part of unit 1 (Mn22-11). C: Wavy bedding with thin mud drape in upper part of unit 1 (Mn22-12). D: Thin beds of pale olive-gray medium to fine sandstone with thin, pale greenish-yellow lamina in unit 2 (Mn22-17). E: Lenticular beds with muddy sandstone drapes in upper part of unit 2 (Mn22-19). F: Uppermost unit of Moncada Formation. Light gray claystone with thin sandstone lamina in its lower part covers slightly undulating surface of sandstone complex, and it is covered by 1-cm-thick, olive-gray sandstone. Fm.—Formation.

chert fragments and detrital quartz floating in its lower part, whereas the olive-gray sandstone consists of micritic limestone fragments, recrystallized calcite grains, brownish, dusty, vesicular fragments, and detrital quartz and plagioclase. The upper and lower margin of the bed, where the color is yellowish, is extensively cemented by pale yellowish-green, transparent chlorite.

Mineral and major element composition

The petrography of the Moncada Formation suggests the presence of systematic compositional variations within each unit, associated with variation in sedimentary structures and grain size. In order to characterize this compositional variation semiquantitatively, mineral and major element compositions were examined.

Calcite, quartz, plagioclase, and clay minerals, such as chlorite, illite, and smectite, are the major constituents of the Moncada Formation; pumpellyite and hematite are present in trace amounts. Within each unit, the calcite and plagioclase contents are lower in the lower part and gradually increase in the upper part, whereas those of quartz and clay minerals are higher in the lower part and gradually decrease in the upper part (Fig. 5). An exception is unit 4, where this trend is less clear. In the uppermost unit, dark colored, very fine calcareous sandstone is enriched in calcite, chlorite, and illite, whereas light colored calcareous claystone is enriched in quartz and plagioclase.

Major element compositions also show significant variation. SiO_2, TiO_2, Al_2O_3, Fe_2O_3, MgO, and K_2O are enriched in the lower part and decrease upward within each unit, whereas CaO, MnO, Na_2O, and P_2O_5 show the opposite trend (Fig. 5). These patterns reflect dilution effects by $CaCO_3$. However, even after normalization to Al_2O_3, a systematic compositional variation within each unit is still observable, especially for Fe_2O_3/Al_2O_3, MgO/Al_2O_3, K_2O/Al_2O_3, and TiO_2/Al_2O_3 ratios (Fig. 5).

The intervals with higher contents of quartz and clay minerals are less calcareous, and generally coincide with intervals with higher Fe_2O_3/Al_2O_3, MgO/Al_2O_3, K_2O/Al_2O_3, and TiO_2/Al_2O_3 ratios. These intervals also correspond to coarse-grained

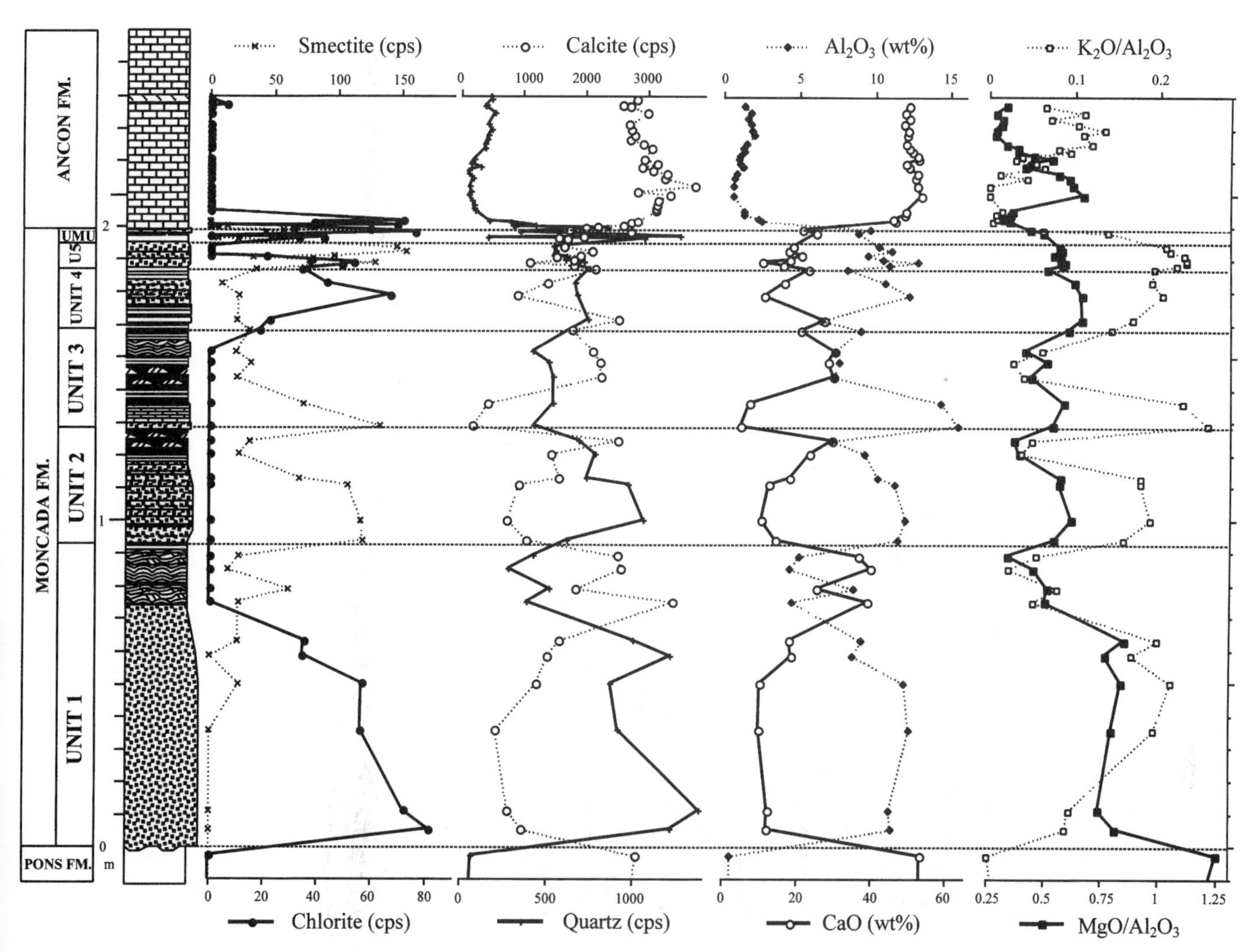

Figure 5. Diagrams showing vertical variations of mineral and chemical compositions in Moncada Formation. Legend for columnar section is same as in Figure 3. Peak intensity of each mineral is expressed as cps (count per second), which is normalized to condition that peak height of standard quartz sample is 10000 cps. Fm.—Formation.

olive-green to pale olive sandstone that is enriched in whitish, vesicular fragments and/or flattened, olive-green grains. According to petrographic observations in conjunction with energy-dispersive spectrometry (EDS) analysis, flattened, olive-green grains are replaced by chlorite, illite, and pumpellyite, whereas vesicular fragments are replaced by smectite. Considering the chemical composition of these grains, higher Fe_2O_3/Al_2O_3, MgO/Al_2O_3, and K_2O/Al_2O_3 ratios are attributable to the flattened, olive-green grains.

The intervals with higher contents of calcite and plagioclase generally coincide with intervals with higher CaO, MnO, Na_2O, and P_2O_5 contents, and correspond to medium- to fine-grained, light gray, calcareous sandstone, alternating with grayish-black, very fine grained, sandy mudstone. Higher concentrations of Na_2O are attributable to plagioclase, whereas higher concentrations of CaO and MnO are attributable to calcite.

There are also variations in the mineral and major element compositions between the units. Most conspicuous is the distribution of chlorite versus smectite: i.e., chlorite content is high in units 1 and 4 and nearly absent in units 2 and 3, whereas smectite and illite show the opposite trend. In unit 5, the chlorite content is high in the lower half, whereas smectite content is high in the upper half. Although less conspicuous, plagioclase shows a trend similar to that of smectite, with slightly higher concentrations in units 2 and 3. This trend is also observed in a slightly lower MgO/Al_2O_3 ratio in units 2 and 3 (Fig. 5).

BASAL PART OF THE ANCÓN FORMATION

The basal 0.5 m of the Ancón Formation consists of dark gray, flaggy limestone; the rest of the formation is characterized by light violet, slaty limestone. Dark gray, flaggy limestone of the Ancón Formation conformably overlies the Moncada Formation with a gradational contact (Fig. 3). The basal 0.5 cm of the limestone is a strong dusty yellow that gradually fades out within next 8 cm interval. Flattened burrows of ~1 cm in diameter are abundant, especially in the basal 3 cm interval. Stylolites develop every 0.2–1.5 cm, which causes the flaggy appearance. Under the petrographic microscope, the flaggy limestone consists of a micritic matrix with a small amount of poorly preserved, small (30–60 μm in diameter), foraminiferal skeletons scattered in the matrix. Although we could not identify these foraminifera, their small size suggests earliest Danian age (P0 zone). Detrital grains, such as quartz and feldspar, are very rare and <50 μm.

PROBABLE RELATION OF THE MONCADA FORMATION WITH K-T BOUNDARY IMPACT

In order to explore the possible relation of the Moncada Formation with the K-T boundary impact, we determined the Ir concentrations in 26 samples (Table 3; Fig. 6). As is obvious from Figure 6, the Ir concentration in the grayish-black limestone of the underlying Pons Formation is ~10 ppt, which is within the range of ordinary crustal rocks (<100 ppt; after Koeberl, 1998). The Ir concentration in the lower to middle part of the sandstone complex is also low, below 50 ppt, whereas it increases to 138 ppt in the upper part of unit 5 where the sandstone color becomes slightly brownish. The Ir concentration further increases to 450 ppt in the 1-cm-thick light gray claystone, and decreases to 220 ppt in the 1-cm-thick olive-gray sandstone in the uppermost unit. A second Ir peak of 820 ppt occurs in the basal 1 cm of the Ancón Formation, where the color of the limestone is dusty yellow. The Ir concentration remains relatively high, between 335 and 396 ppt, to 13 cm above the base of the Ancón Formation, then decreases to 117 ppt within the next 3 cm interval. However, this level is still high compared to ordinary crustal rocks, especially taking into account the dilution effects of $CaCO_3$ in these rocks (Fig. 6). The Ir peak position at the top of the sandstone complex and a peak Ir concentration of 820 ppt in the Moncada section is comparable to the Ir peak concentration reported from shallow-water K-T boundary sites in the Gulf of Mexico region (e.g., Smit et al., 1996).

TABLE 3. IR ABUNDANCE

Sample	Position (cm)	Concentration (ppt)
Mn32-2	227.05	13 ± 14
Mn31	213.00	32 ± 13
Mn30B-5	195.90	117 ± 14
Mn30B-4	195.05	191 ± 21
Mn30B-3	194.00	201 ± 11
Mn30B-2	192.70	335 ± 13
Mn29-9	191.90	376 ± 33
Mn29-8	191.45	396 ± 23
Mn29-7	190.50	663 ± 17
Mn29U	190.25	815 ± 31
Mn29-6	189.65	616 ± 17
Mn29-5	189.15	279 ± 14
Mn29M	188.30	217 ± 24
Mn29-4	188.30	160 ± 17
Mn29-3	187.50	194 ± 27
Mn29-2	187.10	380 ± 22
Mn29-L	186.70	448 ± 22
Mn29-1	186.45	390 ± 15
Mn28-7	183.95	138 ± 12
Mn28-6	182.75	88 ± 9
Mn28-5	181.30	122 ± 11
Mn28	181.00	46 ± 36
Mn28-2	178.30	75 ± 8
Mn8	58.50	50 ± 12
Mn2	5.00	21 ± 3
Mn1	−3.00	10 ± 4

We also searched for impact ejecta within the Moncada Formation. Quartz grains with planar deformation features (PDF) are commonly found throughout the formation (Fig. 7A). Our preliminary semiquantitative observation suggests that ~18% of detrital quartz grains have PDFs. The crystallographic orientations of PDFs in quartz from the Moncada Formation were measured using a universal stage with five axes (Montanari and Koeberl, 2000). As shown in Figure 8, the orientations strongly support their impact shock origin (e.g., Stöffler and Langenhorst, 1994). As for olive-green, flattened, and deformed grains, it is difficult to identify the original texture, which could be indicative of their origin. However, considering that they are replaced by chlorite and pumpellyite, the original material should have been an unstable material, such as impact glass enriched in Mg, Fe, and Ca.

Thin-section observation in conjunction with EDS analysis of angular, whitish, vesicular fragments suggests that they are composed of a mixture of smectite and small clinopyroxene crystals that show quench textures (Fig. 7B). Such quench textures of clinopyroxene are only found in microkrystites, impact melt rocks, and deep-sea basalt (Glass and Burns, 1988; Schuraytz et al., 1994; Smit, 1999; Bryan, 1972). Together with the highly vesicular nature and relatively large size of the grains, these whitish vesicular fragments are most likely of impact melt origin.

The presence of a significant peak in the Ir abundance in the uppermost unit of the Moncada Formation, as well as the presence of shocked quartz and vesicular glass fragments with pyroxene quench textures throughout the formation, strongly argues for a genetic relation of the Moncada Formation with meteorite impact. Together with its biostratigraphically estimated age of between late Maastrichtian and early Paleocene and its proximity to the Chicxulub crater, this evidence strongly suggests a K-T boundary impact origin for the Moncada Formation.

DEPOSITIONAL MECHANISM OF THE MONCADA FORMATION

Although it seems evident that the deposition of the Moncada Formation was associated with the K-T boundary impact, its genetic relation with the impact is not well understood. Detailed examination of sedimentary structures, as well as grain size and composition, should give some clues for the depositional mechanism.

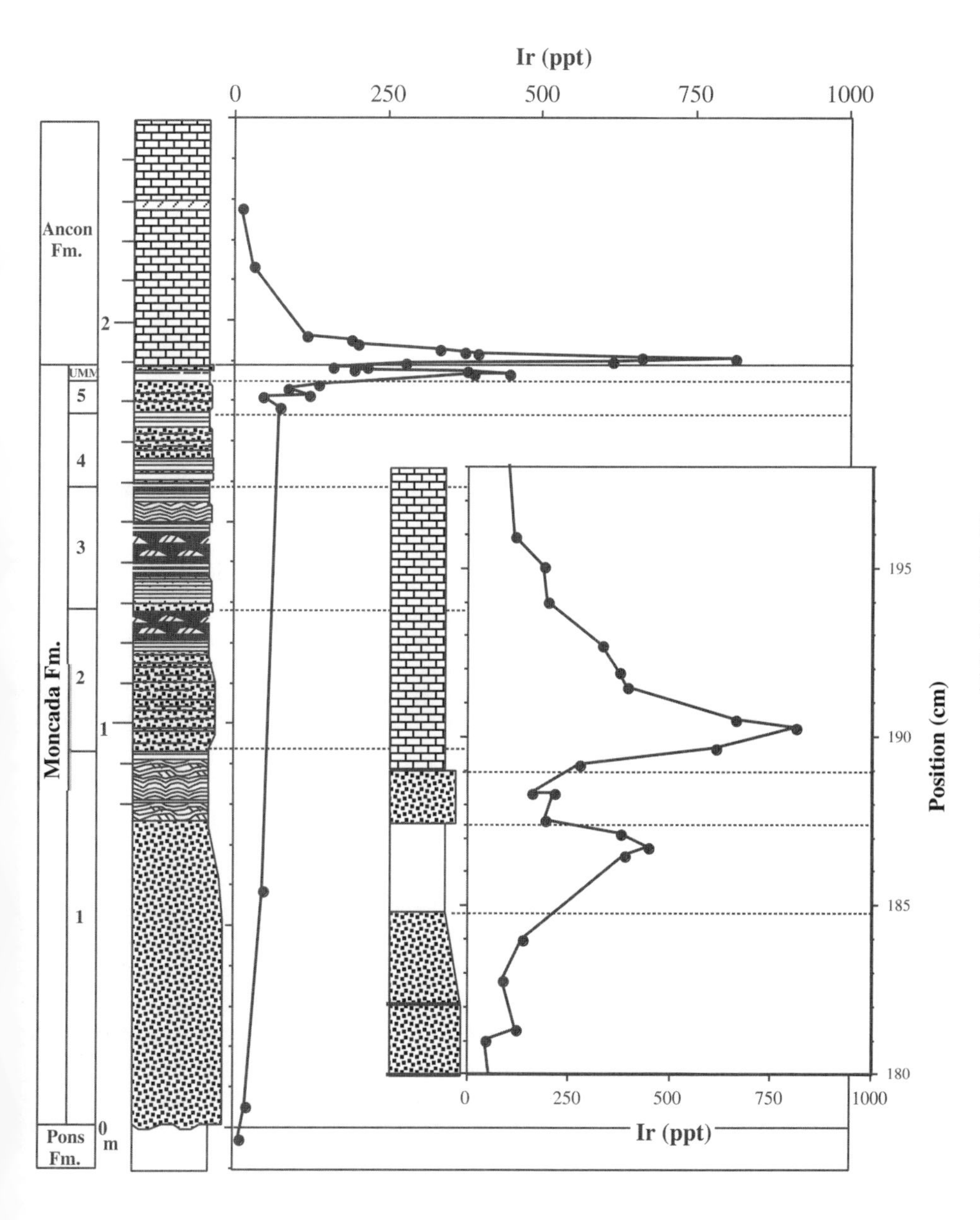

Figure 6. Variation of Ir concentration across Moncada Formation. Magnification of 180–198 cm interval, covering uppermost unit (UMU), is inserted in right lower corner. Legend for columnar section is same as in Figure 3. Fm.—Formation.

The sandstone complex of the Moncada Formation is characterized by repetition of sandstone units with an overall upward decrease in grain size and unit thickness. Each unit shows systematic upward changes in sedimentary structure, from thin parallel beds, to parallel laminations, to flaser and/or lenticular bedding with ripple cross-laminations. These sedimentary structures suggest deposition from a flowing current. The maximum grain sizes for thin, parallel-bedded sandstone, parallel-laminated sandstone, and flaser-bedded sandstone are between ~0.2 and 0.3 mm, ~0.22 and 0.12 mm, and ~0.15 and 0.1 mm, respectively, suggesting that changes in these sedimentary structures reflect a gradual decrease in current speed. It is likely that thin, parallel beds to parallel laminations represent flat beds in an upper flow regime, whereas ripple cross-laminations represent ripples in a lower flow regime (e.g., Friedman and Sanders, 1978). Thus, the systematic changes in sedimentary structures and an upward decrease in grain size, especially in units 1 and 2, most likely reflect an upward decrease in flow speed within the units. The pattern of the maximum grain-size variation within a unit becomes more symmetrical in units 3, 4, and 5. Several centimeter-scale oscillations in the maximum grain size are observable within each unit, probably reflecting shorter scale oscillations in current speed.

The paleocurrent direction estimated from ripple cross-laminations is unidirectional within individual units, whereas reversing paleocurrent direction is observed between units 1 and 2. The direction is unchanged between units 2 and 3. The paleocurrent direction is from S17° ± 10°E in unit 1 and from N15° ± 20°W in units 2 and 3. These paleocurrent directions should be corrected for crustal rotation, and our preliminary paleomagnetic measurement of the Moncada Formation (three samples) indicates counterclockwise rotation of ~22°. After

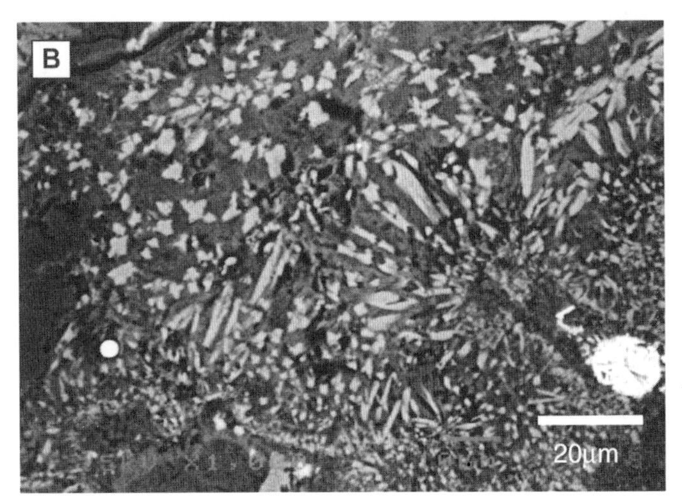

Figure 7. Thin-section photographs of ejecta materials in Moncada Formation. A: Detrital quartz grain (240 μm) with two sets of planar deformation features (PDF) from unit 5 (Mn22-28). Each set of PDFs is parallel to ω(23°). Cross-polarized light. B: Whitish, vesicular fragments with clinopyroxene quench textures. Light gray, blade-like, fine crystals are clinopyroxene. Gray matrix is smectite, which is probably altered from impact glass. Spherical aggregate of bright needle-like grains on right side is composed of goethite, dark gray grain on left side is quartz, and gray grain on lower left is chlorite. Back-scattered electron image.

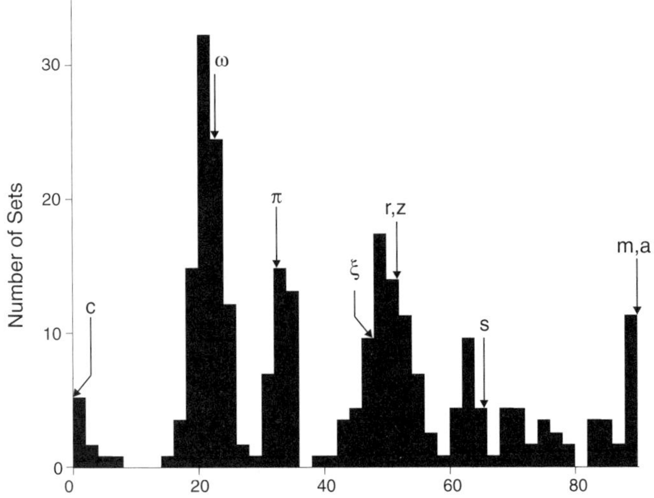

Figure 8. Histogram showing variation in angle between c-axis and pole to planar deformation features (PDF) in shocked quartz grains obtained from Moncada Formation. Crystallographic orientations of c(0°), ω(22.95°), π(32.42°), ξ(47.73°), r, z(51.79°), s(65.56°) and m, a(90°) are prominent, indicating their impact-derived origin (Stöffler and Langenhorst, 1994). We measured 303 sets in 187 grains. Samples were selected from 58–64 cm (unit 1), 77–85 cm (unit 1), 113–120 cm (unit 2), 152–158 cm (unit 3), 168–171 cm (unit 4), and 180–185 cm (unit 5).

correcting for the crustal rotation, the current direction for the Moncada Formation is S5°W and N7°E, nearly parallel to the eastern margin of the Yucatan platform.

The mineral and major element compositions within each unit also vary systematically (Fig. 5), probably reflecting an upward decrease in current speed. Variations in mineral and chemical compositions, in conjunction with petrographic observation, suggest that coarse grains are dominated by ejecta materials, such as altered, vesicular, impact-melt fragments and shocked quartz grains, and flattened olive-green grains of possible altered impact glass, whereas material reworked from underlying substrates, such as calcareous microfossils, detrital plagioclase, and sedimentary lithic fragments, is predominantly fine grained. Further examination revealed that mineral and chemical compositions in the lower (and coarser) part of individual units, especially the relative abundance of smectite and illite versus chlorite, are different between units 1, 4, and 5, and units 2 and 3. This may reflect changes in provenance in response to the changes in paleocurrent direction; unit 1, which is characterized by high chlorite content, has an opposite current direction compared to units 2 and 3, which are characterized by high smectite and illite contents. Assuming that chlorite and smectite represent flattened, olive-green grains of possible altered impact glass and altered, vesicular, impact-melt fragments, respectively, possible altered impact glass and altered, vesicular, impact-melt fragments might have been derived from southern and northern directions, respectively. If this interpretation is correct, units 4 and 5 were also derived from the north.

It is difficult to estimate the bathymetry of the Moncada Formation at the time of impact, because of the lack of late Maastrichtian strata underneath the Moncada Formation, and poor preservation of depth-diagnostic microfossils in the overlying Ancón Formation. However, the lower part of the Ancón Formation is very fine grained, calcareous, and free of turbidites, suggesting a hemipelagic to pelagic environment of deposition. Considering that the calcium carbonate compensation depth in the Atlantic during the early Paleocene has been <3.3 km (van Andel, 1975), and that the Ancón Formation is highly calcareous, the depositional depth of the Moncada Formation should have been shallower than 3.3 km. The paleotectonic reconstruction suggests that the Los Organos belt was located immediately to the east of the eastern margin of the present Yucatan block and was scraped off this margin in the course of northward movement of the Cretaceous volcanic arc during the late Paleocene to early Eocene (Iturralde-Vinent, 1994, 1998). According to Rosencrantz (1990), the eastern margin of Yucatan (the Yucatan borderland) is an offshore continuation of the Yucatan platform, and Paleogene-Late Cretaceous deep-water clastics were found in offshore wells to the east of Belize. Although the bathymetry of these deep-water clastics is not clear, they most likely represent a slope facies. Because the Los Organos belt is considered to have been located immediately to the east of the Yucatan borderland, the Moncada Formation should have been deposited in a slope setting with a probable paleodepth between 0.2 and 3.3 km.

The depositional mechanism for each unit should explain the unidirectional flow with a gradual upward decrease in flow speed. The mechanism should also explain the paleocurrent direction subparallel to the Yucatan platform margin, as well as paleocurrent reversal between the units. Turbidity currents are one possibility. Longitudinal paleocurrent direction with paleocurrent reversal between different units could be explained by turbidites filling a trench-type basin plain (e.g., Hesse, 1974). However, a gradational contact at the base of each unit (except unit 1), the symmetric maximum grain-size-variation patterns, especially in units 4 and 5, the presence of centimeter-scale oscillations of maximum grain size within each unit, and the lack of hemipelagic claystone at the top of each unit, taken together, are inconsistent with a turbidite origin for individual units. Moreover, a trench-type basin-plain setting required to explain longitudinal paleocurrent direction with reversals is inconsistent with the continental slope setting as estimated here, and a systematic upward decrease in unit thickness without any interruptions by hemipelagic deposition are difficult to explain by a turbidite origin. Explanation that the entire sandstone complex represents a single turbidite bed is also inconsistent with paleocurrent reversals within the sandstone complex.

K-T boundary sandstone complexes as thick as 9 m deposited on shelf to upper slope environments are reported from multiple sites in the Gulf of Mexico region, and they are interpreted as being formed as a result of tsunamis generated by the K-T boundary impact (Bourgeois et al., 1988; Smit et al., 1996; Smit, 1999). The sandstone complexes in Mexico (Smit et al., 1996; Smit, 1999) are especially similar to the sandstone complex of the Moncada Formation with respect to such characteristics as (1) repetition of upward-fining units, (2) reversing current directions between the units, (3) systematic upward decreases in unit thickness, and (4) the high concentration of Ir at the top of the sandstone complex. In addition, the number of units in the Moncada sandstone complex is comparable to those in the Gulf of Mexico region (as many as seen, according to Smit et al., 1996). The first three characteristics strongly support a tsunami origin for the sandstone complex of the Moncada Formation, as was proposed for the K-T sandstone complexes in Mexico (Smit et al., 1996).

Major differences between the sandstone complex of the Moncada Formation and those of Mexico are the patterns of paleocurrent reversals and the paleocurrent directions with respect to the paleoshoreline. Smit et al (1996) described paleocurrent reversals between every subsequent unit of the K-T sandstone complex in La Lajilla, Mexico, whereas paleocurrent direction is reversed every two units in the Moncada Formation. Although paleocurrent direction is not estimated directly for units 4 and 5, paleocurrent reversals are estimated between units 3 and 4, according to changes in the clay mineral assemblage and the MgO/Al_2O_3 ratio. Smaller scale oscillations in the maximum grain size observed within each unit suggest the presence of two to six higher frequency waves superimposed on larger amplitude, lower frequency waves. Thus, the pattern of current direction changes in the Moncada Formation is more complex than in Mexico. Such a complex tsunami wave rhythm could have been created by local reflection of tsunami waves due to the complex paleogeography of the region. Alternatively, multiple gravity flows triggered by seismic shocks of the impact (e.g., Takayama et al., 2000) might have caused multiple tsunami waves with a complex rhythm. The north-south-trending paleocurrent directions subparallel to the eastern margin of the Yucatan platform can be explained by the relatively deep depth of the studied site that prevented the wave direction from becoming parallel to the shoreline direction.

COMPARISON WITH OTHER K-T BOUNDARY DEPOSITS IN WESTERN CUBA

K-T boundary deposits at other sites in western Cuba are characterized by extremely thick (to 700 m) deposits, including a lower coarse-grained debris-flow unit and an upper calcarenite to calcilutite unit that shows normal grading (e.g., Takayama et al., 2000). The Moncada Formation, however, is only 2 m thick. The continental slope depositional setting of the Moncada Formation and its proximity to the southern gateway between the Yucatan platform and the Cretaceous volcanic arc may explain such a large difference in thickness. The absence of a basal debris-flow unit in the Moncada Formation, and the lack of a Turonian to Maastrichtian sedimentary sequence underneath it, suggests that the Turonian to Maastrichtian sequence was eroded by the debris flow before deposition of the Moncada Formation. Kiyokawa et al. (2000) demonstrated that the lower part of the Cacarajicara Formation is composed of an ~300-m-thick gravity-flow unit that is characterized by breccias of limestone and chert similar in appearance and age to those of the Upper Cretaceous sedimentary sequence in the Guaniguanico terrane. Iturralde-Vinent et al. (2000) estimated that the Cacarajicara Formation in the Rosario belt of the Guaniguanico terrane was deposited in the lower slope to basinal environment on the eastern frank of the Yucatan platform. Thus, it is likely that Moncada was located on the slope between the Yucatan platform to the west and the Rosario belt to the east, and was actually part of the source area for the debris-flow unit of the

Cacarajicara Formation. On the basis of this interpretation, we consider that the debris flow triggered by the K-T boundary impact removed the Upper Cretaceous sedimentary sequence from the middle to upper slope of the eastern Yucatan margin, including Moncada, transported the sediment downslope, and deposited the thick K-T boundary deposits such as the Cacarajicara Formation in the deeper part of the proto-Caribbean basin. Consequently, no deposit corresponding to the debris-flow unit of the Cacarajicara Formation accumulated in Moncada.

The upper unit of K-T boundary deposits at other sites in western Cuba is composed of as much as 400 m of normally graded calcarenite to calcilutite, which Takayama et al. (2000) interpreted as a homogenite (a deep-sea tsunami deposit). The calcarenite part is 100–300 m thick at these localities, whereas the calcareous sandstone at the Moncada section is only 2 m thick. Our on-going high-resolution study of homogenite in the Peñalver Formation near Havana revealed slight but consistent oscillations in grain size and composition within the calcarenite part that repeated five to six times. This oscillation may suggest repeated agitation of the upper water column by tsunami waves, or lateral injection of the coarser material to the water column by backwash. For this reason, we consider that the calcarenite part of the homogenite was deposited at about the same time as the sandstone complex in Moncada. The difference in their thicknesses can be explained by the difference in their depositional depths. Assuming that the suspended sediment cloud created by tsunami agitation was widespread in the deeper part of the sea, its thickness should have increased with increasing water depth. Thus, thick calcarenite was deposited by vertical settling in the basinal sites, whereas the thin calcareous sandstone complex was deposited under the influence of tsunami waves in shallow sites.

The calcilutite part of these homogenite beds is 30–100 m thick (Takayama et al., 2000; Kiyokawa et al., 2000), and is interpreted as being deposited from a low-concentration sediment suspension cloud within several days to weeks after the impact. However, calcareous claystone at the top of the Moncada Formation is only a few centimeters thick. This difference needs to be explained, especially if the sediment cloud prevailed throughout the water column within the proto-Caribbean basin. The proximity of the studied site to the southern gateway, in addition to its shallow bathymetry compared to other K-T boundary sites in western Cuba, could be one explanation. The surface current (western boundary current) flowing through the gateway between the Yucatan platform and the Cuban volcanic arc and continuing along the eastern margin of the Yucatan platform could have cleaned the upper ~500 m of the water column, especially in the southern part of the proto-Caribbean basin, and thus prevented deposition of thick calcilutite at the study site. Alternatively, it is possible that the sediment suspension cloud was restricted to the deeper part of the proto-Caribbean basin.

CONCLUSIONS

The Moncada Formation in western Cuba is a K-T boundary sandstone complex that is characterized by (1) abundant ejecta materials such as shocked quartz and altered impact-melt fragments throughout the formation, (2) the presence of an Ir-rich claystone at the top of the complex, (3) repetition of fining-upward sandstone units with a gradual upward decrease in thickness, (4) the occurrence of current ripples showing reversing paleocurrent direction between the units, and (5) the lack of hemipelagic clays between the units. These characteristics are similar to other K-T sandstone complexes in Mexico that are interpreted to have been formed by tsunamis caused by the impact. On the basis of these characteristics in conjunction with its biostratigraphically estimated age, between late Maastrichtian and early Paleocene, it is concluded that the Moncada sandstone complex was formed by repeated tsunami waves caused by the K-T impact.

Although similarities between the K-T sandstone complex of Moncada and those in Mexico are obvious, there are several differences. First, the inferred tsunami waves are not simple oscillatory waves, but had more complex rhythms, the first wave moving from south to north, and the subsequent two waves from north to south. Second, shorter scale grain-size oscillations within each unit are recognized, suggesting higher frequency waves superimposed on the lower frequency waves. Third, the depositional site was not on the shelf but on the slope; thus, the bathymetry was probably deeper than most of the other K-T boundary sandstone complexes in the Gulf of Mexico region. The lack of thick gravity-flow and homogenite-type deposits at Moncada probably reflects its slope setting, where the Turonian to Maastrichtian limestone and chert sequence was eroded, most likely by gravity flows triggered by the seismic shocks of the impact, and its proximity to the southern gateway that allowed influx of clear surface water. Better estimates of the bathymetry are needed to allow a more precise estimation of the magnitude and mode of the tsunami waves caused by the K-T boundary impact.

ACKNOWLEDGMENTS

We thank the Agencia Medio Ambiente in Cuba for their support of the field survey in Cuba. This research was made possible thanks to an agreement between the Department of Earth and Planetary Science, University of Tokyo, the Museo Nacional de Historia Natural (Agencia del Medio Ambiente) de Cuba, and the Instituto de Geologia del Ministerio del Industria Basica. We acknowledge especially the support provided to the field research in Cuba by Mitsui & Co., Ltd., as well as their manager in Havana, A. Nakata. We also thank C. Koeberl, B. Bohor, and P. Claeys, who critically read the manuscript and gave us many valuable suggestions, and J. Compton, who helped improve our English. The neutron activation analysis was carried out at the Inter-University Laboratory for the joint use of JAERI

(Japan Atomic Energy Research Institute) facilities. The survey was supported by grants-in-aid for scientific research of the Japan Society for the Promotion of Science (no. 11691116) and also by research funds donated to the University of Tokyo by NEC Corporation, I. Ohkawa, M. Iizuka., and K. Ihara.

REFERENCES CITED

Bohor, B.F., 1996, A sediment gravity-flow hypothesis for siliciclastic units at the K/T boundary, northeastern Mexico, *in* Ryder, G., Fastovsky, D., and Gartner, S., eds., The Cretaceous–Tertiary event and other catastrophes in Earth history: Geological Society of America Special Paper 307, p. 183–195.

Bourgeois, J., Hansen, T.A., Wiberg, P.L., and Kaufmann, E.G., 1988, A tsunami deposit at the Cretaceous–Tertiary boundary in Texas: Science, v. 241, p. 567–570.

Bralower, T.J., and Iturralde-Vinent, M.A., 1997, Micropaleontological dating of the collision between the North American plate and the Greater Antilles Arc in western Cuba: Palaios, v. 12, p. 133–150.

Bryan, W.B., 1972, Morphology of quench crystals in submarine basalts: Journal of Geophysical Research, v. 29, p. 5812–5819.

D'Hondt, S., Pilson, M.E., Sigurdsson, H., Hanson, A.F., Jr., and Carey, S., 1994, Surface-water acidification and extinction at the Cretaceous–Tertiary boundary: Geology, v. 22, p. 983–986.

Diaz Otero, C., Iturralde-Vinent, M.A., and Garcia Delgado, D., 2000, The Cretaceous–Tertiary Boundary "Cocktail" in Western Cuba, Greater Antilles [abs.], *in* Catastrophic events and mass extinctions: Impact and beyond: Houston, Texas, Lunar and Planetary Institute, LPI Contribution No. 1053, p. 37–38.

Friedman, G.M., and Sanders, J.E., 1978, Principles of sedimentology: New York, John Wiley and Sons, 792 p.

Glass, B.P., and Burns, C.A., 1988, Microkrystites: A new term for impact-produced glassy spherules containing primary crystallites [abs.]: Lunar and Planetary Science Conference, 18th, Houston, Texas, Lunar and Planetary Institute, p. 455–458.

Hesse, R., 1974, Long-distance continuity of turbidites: Possible evidence for an Early Cretaceous trench-abyssal plain in the East Alps: Geological Society of America Bulletin, v. 85, p. 859–870.

Hildebrand, A.R., Penfield, G.T., Kring, D.A., Pilkington, M., Camargo, Z.A., Jacobsen, S.B., and Boynton, W.V., 1991, Chicxulub crater: A possible Cretaceous/Tertiary boundary impact crater on the Yucatán Peninsula, Mexico: Geology, v. 19, p. 867–871.

Irino T., and Tada, R., 2000, Quantification of aeolian dust (Kosa) contribution to the Japan Sea sediments and its variation during the last 200 ky: Geochemical Journal, v. 34, p. 59–93.

Iturralde-Vinent, M.A., 1992, A short note on the Cuban late Maastrichtian megaturbidite (an impact-derived deposit?): Earth and Planetary Science Letters, v. 109, p. 225–228.

Iturralde-Vinent, M.A., 1994, Cuban geology: A new plate-tectonic synthesis: Journal of Petroleum Geology, v. 17, p. 39–70.

Iturralde-Vinent, M.A., 1995, Sedimentary geology of Western Cuba, in SEPM Congress on Sedimentary Geology, 1st Field Trip Guidebook: St. Petersburg, Florida, (SEPM) Society for Sedimentary Geology, 21 p.

Iturralde-Vinent, M., 1998, Sinopsis de la constitución geológica de Cuba, *in* Melgarejo, J.C., and Proenza, J.A., eds., Geología y metalogenia de Cuba: Una introducción: Acta Geológica Hispanica, v. 33, nos. 1–4, p. 9–56.

Iturralde-Vinent, M.A., Garcia, D., Diaz, C., Rojas, R., Tada, R., Takayama, H., and Kiyokawa, S., 2000, The K/T boundary impact layer in Cuba: Update of an international project [abs.], *in* Catastrophic events and mass extinctions: Impact and beyond: Houston, Texas, Lunar and Planetary Institute, LPI Contribution No. 1053, p. 76–77.

Kiyokawa, S., Tada, R., Oji, T., Tajika, E., Nakano, Y., Goto, K., Yamamoto, S., Takayama, H., Toyoda, K., Rojas, R., Garcia, D., Iturralde-Vinent, M.A., and Matsui, T., 2000, More than 500 m thick K/T boundary sequence; Cacarajicara Formation, western Cuba [abs.], *in* Catastrophic events and mass extinctions: Impact and beyond: Houston, Texas, Lunar and Planetary Institute, LPI Contribution No. 1053, p. 100–101.

Koeberl, C., 1998, Identification of meteoritical components in impactites, *in* Grady, M.M., Hutchison, R., McCall, G.J.H., and Rothery, D.A., eds., Meteorites: Flux with time and impact effects: Geological Society [London] Special Publication 140, p. 133–152.

Kyte, F.T., Zhou, L., and Lowe, D.R., 1992, Noble metal abundances in an Early Archean impact deposit: Geochimica et Cosmochimica Acta, v. 56, p. 1365–1372.

Montanari, A., and Koeberl, C., 2000, Impact stratigraphy: The Italian record, *in* Bhattacharji, S., Friedman, G.M., Neugebauer H.J., and Seilacher, A., eds., Lecture Notes in Earth Sciences, Volume 93: Berlin, Springer-Verlag, p. 295–300.

Moore, D.M., and Reynolds, R.C., Jr., 1997, X-ray diffraction and the identification and analysis of clay minerals: Oxford, Oxford University Press, 378 p.

Potts, P.J., Tindle, A.G., and Webb, P.C., 1992, Geochemical reference material compositions: Rocks, minerals, sediments, soils, carbonates, refractories and ores used in research and industry: Caithness/Boca Raton, Lousiana, Whittles Publishing/CRC Press Inc., 313 p.

Pszczolkowski, A., 1986, Megacapas del maestrichtiano en Cuba occidental y central: Bulletin of the Polish Academy of Earth Science, v. 34, p. 81–94.

Rosencrantz, E., 1990, Structure and tectonics of the Yucatán basin, Caribbean Sea, as determined from seismic reflection studies: Tectonics, v. 9, p. 1037–1059.

Ryder, G., Fastovsky, D., and Gartner, S., editors, 1996, The Cretaceous–Tertiary event and other catastrophes in Earth history: Geological Society of America Special Paper 307, 569 p.

Schuraytz, B.C., Sharpton, V.L., and Marin, L.E., 1994, Petrology of impact-melt rocks at the Chicxulub multiring basin, Yucatán, Mexico: Geology, v. 22, p. 868–872.

Sigurdsson, H., D'Hondt, S., Arthur, M.A., Bralower, T.J., Zachos, J.C., van Fossen, M., and Channell, J.E.T., 1991, Glass from the Cretaceous/Tertiary boundary in Haiti: Nature, v. 349, p. 482–487.

Smit, J., 1999, The global stratigraphy of the Cretaceous–Tertiary boundary impact ejecta: Annual Reviews of Earth and Planetary Science, v. 27, p. 75–113.

Smit, J., Montanari, A., Swinburne, N.H.M., Alvarez, W., Hildebrand, A.R., Margolis, S.V., Claeys, P., Lowrie, W., and Asaro, F., 1992, Tektite-bearing, deep-water clastic unit at the Cretaceous–Tertiary boundary in northeastern Mexico: Geology, v. 20, p. 99–103.

Smit, J., Roep, Th.B., Alvarez, W., Claeys, P., Grajales-Nishimura, J.M., and Bermudez, J., 1996, Coarse-grained, clastic sandstone complex at the K/T boundary around the Gulf of Mexico: Deposition by tsunami waves induced by the Chicxulub impact?, *in* Ryder, G., Fastovsky, D., and Gartner, S., eds., Cretaceous–Tertiary event and other catastrophes in Earth history: Geological Society of America Special Paper 307, p. 151–182.

Stöffler, D., and Langenhorst, F., 1994, Shock metamorphism of quartz in nature and experiment. 1. Basic observation and theory: Meteoritics, v. 29, p. 155–181.

Takayama, H., Tada, R., Matsui, T., Iturralde-Vinent, M.A., Oji, T., Tajika, E., Kiyokawa, S., Garcia, D., Okada, H., Hasegawa, T., and Toyoda, K., 2000, Origin of the Peñalver Formation in northwestern Cuba and its relation to K/T boundary impact event: Sedimentary Geology, v. 135, p. 295–320.

Van Andel, T.J., 1975, Mesozoic/Cenozoic calcite compensation depth and the global distribution of calcareous sediments: Earth and Planetary Science Letters, v. 26, p. 187–194.

Manuscript Submitted October 16, 2000; Accepted by the Society March 22, 2001

Printed in the U.S.A.

Geological Society of America
Special Paper 356
2002

Cretaceous-Tertiary boundary sequence in the Cacarajicara Formation, western Cuba: An impact-related, high-energy, gravity-flow deposit

Shoichi Kiyokawa
Kyushu University, Department of Earth and Planetary Sciences, 6-10-1 Hakozaki, Higashiku, Fukuoka 812-8581, Japan
Ryuji Tada
Department of Earth and Planetary Science, Graduate School of Science, University of Tokyo, 7-3-1 Hongo, Tokyo 113-0033, Japan
Manuel Iturralde-Vinent
Museo Nacional de Historia Natural, Obispo No. 61, Plaza de Armas, La Habana Vieja 10100, Cuba
Eiichi Tajika, Shinji Yamamoto, Tetsuo Oji, Youichiro Nakano, and Kazuhisa Goto
Department of Earth and Planetary Science, Graduate School of Science, University of Tokyo, 7-3-1 Hongo, Tokyo 113-0033, Japan
Hideo Takayama
NHK Japan Broadcasting Corporation, Nagoya Office, 1-13-3 Higashisakura, Higashiku, Nagoya 461-8725, Japan
Dora Garcia Delgado
Consuelo Diaz Otero
Instituto de Geologia y Paleontologia Via Blanca y Linea del Ferrocarril San Miguel del Padron, La Habana 11000, Cuba
Reinaldo Rojas-Consuegra
Museo Nacional de Historia Natural, Obispo No. 61, Plaza de Armas, La Habana 10100, Cuba
Takafumi Matsui
Department of Complexity Science and Engineering, Graduate School of Frontier Science, University of Tokyo, 7-3-1 Hongo, Bunkyo-ku, Tokyo 113-0031, Japan

ABSTRACT

The Cacarajicara Formation of western Cuba is a more than 700-m-thick calcareous clastic sequence that contains shocked quartz throughout, and spherules. Three members are recognized. The lower member consists of limestone and chert boulders, and disconformably overlies Cretaceous deep-water turbidite. It is characterized by: (1) a grain-supported fabric with only a small amount of matrix, (2) 5–15 cm, well-sorted clasts and occasional boulders, (3) reversely graded, discoidal or rectangular boulders showing a preferred orientation, (4) abundant shallow- and deep-water carbonate clasts in a well-mixed fabric, (5) direct contact between adjacent clasts, and (6) hydrostatic deformation within a black clay matrix. This evidence suggests that the lower member was deposited under conditions of high-density and high-speed laminar flow. The middle member consists of upward graded, massive to well-bedded,

Kiyokawa, S., Tada, R., Iturralde-Vinent, M., Tajika, E., Yamamoto, S., Oji, T., Nakano, Y., Goto, K., Takayama, H., Garcia Delgado, D., Diaz Otero, C., Rojas-Consuegra, R., and Matsui, T., 2002, Cretaceous-Tertiary boundary sequence in the Cacarajicara Formation, western Cuba: An impact-related, high-energy, gravity-flow deposit, *in* Koeberl, C., and MacLeod, K.G., eds., Catastrophic Events and Mass Extinctions: Impacts and Beyond: Boulder, Colorado, Geological Society of America Special Paper 356, p. 125–144.

homogeneous calcarenite. Unusual fluid-escape structures in the thick calcarenite suggest that this member formed by high-density turbidity suspension. The upper member consists of fine calcarenite mudstone; there is no evidence of bioturbation. We infer that it was deposited from a dilute, low-density suspension.

On the basis of these criteria, the Cacarajicara Formation is interpreted to be a single hyperconcentrated flow that was formed by high-energy and high-speed concentrated flow. The south-southeast paleocurrent direction suggests that this high-energy flow originated on the Yucatan platform and was triggered by the Chicxulub impact. We propose that a gigantic flow deposit was induced by earthquake-generated collapse of the Yucatan platform margin owing to ballistic flow from the Chicxulub impact.

INTRODUCTION

Impact-related sedimentary deposits have been described at Cretaceous-Tertiary (K-T) boundary sites in the Yucatan Peninsula region (e.g., Pope et al., 1999). In addition, coarse clastic sequences of impact origin have been reported from marginal areas of the adjoining Gulf of Mexico (e.g., Smit, 1999), and are inferred tsunami or gravity flow deposits (Bourgeois et al., 1988; Bohor, 1996; Smit et al., 1996). In the central Gulf of Mexico and Caribbean Sea, debris-flow deposits at some Ocean Drilling Program sites have been interpreted as originating from a collapsed platform margin (Alvarez et al., 1992; Bralower et al., 1998). Proximal deposits of impact origin have also been discovered east and south of the Yucatan Peninsula. The nearest such outcrop is on Albion Island, Belize, ~350 km south of the Chicxulub crater (Ocampo et al., 1996; Pope et al., 1999). The East Yucatán oil field, which is located 350–600 km from the crater, contains a thick layer of impact ejecta and clasts (Grajales-Nishimura et al., 2000). Deep core data from the crater have revealed detailed characteristics (Hildebrand et al., 1991; Sharpton et al., 1996). However, the record from the continental shelf to the oceanic basin on the east and south side of the Yucatan Peninsula is still poorly known.

Western Cuba consists of a Paleogene orogenic belt formed by interaction between the North American and Caribbean plates (Pindell and Barrett, 1990; Ross and Scotese, 1988; Gordon et al., 1997). The orogenic belt in western Cuba contains Late Cretaceous to early Tertiary sequences that include a thick, well-preserved late Maastrichtian section consisting of the Cacarajicara and Peñalver Formations (Pszczolkowski, 1986; Bohor and Seitz, 1990; Piotrowska, 1993; Iturralde-Vinent, 1992, 1994a, 1994b, 1996). The depositional environments and trigger events for these unusually thick formations have been discussed, and several origins have been proposed, including orogenic-volcanic debris deposits, impact-related ejecta, tsunami deposits, and earthquake-related megaturbidites (e.g., Palmer, 1945; Pszczolkowski, 1986; Bohor and Seitz, 1990; Iturralde-Vinent, 1992; Takayama et al., 2000). End-Cretaceous plate reconstructions (Iturralde-Vinent, 1994a) indicate that the study site was situated near the eastern margin of the Yucatan Peninsula, and may preserve proximal evidence related to the K-T impact.

In this chapter we describe the lithologic characteristics of the unusually thick Cacarajicara Formation, focusing especially on the lower member, which contains a thick clastic sequence. We also discuss the possible depositional mechanisms responsible for this thick sequence.

REGIONAL GEOLOGIC SETTING

From Cretaceous to Eocene time, the Cuban island arc moved northeastward along the southeast margin of the Yucatan Peninsula (Fig. 1) (Pindell et al., 1988; Ross and Scotese, 1988). Fragments of the Yucatan block were detached and moved at least 350 km by left-lateral transform motion in the Paleocene to early Eocene and formed the Guaniguanico terrane (Rosencrantz, 1990; Iturralde-Vinent, 1994b). The Cretaceous Cuban island arc finally collided with the North American continent and formed a collisional fold and thrust belt (Iturralde-Vinent, 1994a, 1994b; Gordon et al., 1997). The orogeny in western Cuba was characterized by collision and strike-slip deformation during the Paleocene and early Eocene and resulted in very complex geology (Gordon et al., 1997).

Western Cuba consists of the Pinos and Guaniguanico terranes (Iturralde-Vinent, 1994a; Kerr et al., 1999). The Pinos terrane is a medium pressure–medium temperature metamorphic sialic complex. The Guaniguanico terrane contains ophiolite-bearing thrust-nappe sequences of Paleocene to early Eocene age (e.g., Cajalbana ophiolite: Pszczolkowski, 1994), and was strongly affected by Paleocene–middle Eocene strike-slip deformation during the opening of the Yucatan basin (Rosencrantz, 1990). The Guaniguanico terrane consists of several tectonic belts, one of which is the Rosario belt (Iturralde-Vinent, 1996). The Rosario stratigraphic sections include continental slope and rise hemipelagic deposits of Late Jurassic to Late Cretaceous age, as well as Paleocene and Eocene foreland deposits (Pszczolkowski, 1978, 1994; Iturralde-Vinent, 1994b; Bralower and Iturralde-Vinent, 1997). The Guaniguanico terrane is overlain by ophiolites and volcano-sedimentary rocks of the Bahía Honda–Matanzas allochthon. Well-preserved K-T

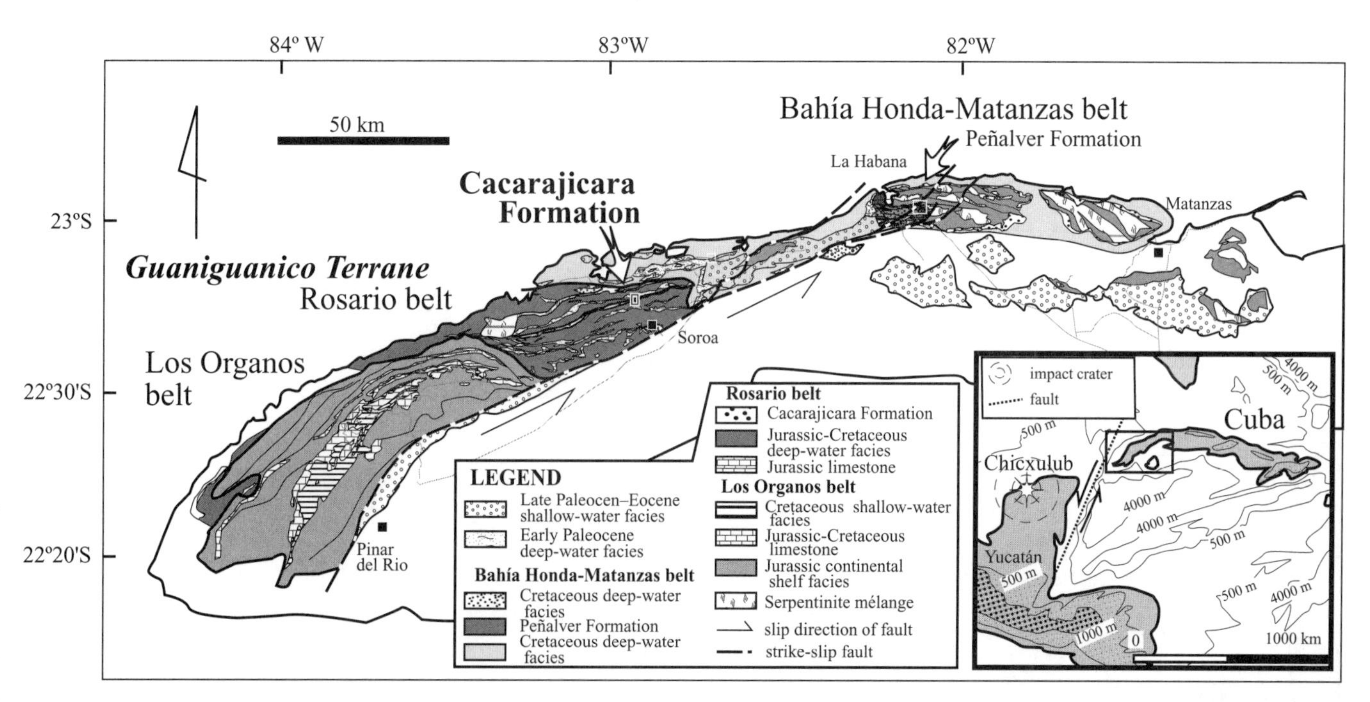

Figure 1. Geologic map of western Cuba, showing Guaniguanico terrane and location of Cacarajicara Formation outcrops (after Pushcharovsky, 1989). Map on right shows location of western Cuba and Yucatan Peninsula, with respect to buried Chicxulub impact crater and topographic centerlines. Dashed circles represent proposed crater diameters: 180 km (Hildebrand et al., 1991) and 300 km (Sharpton et al., 1992, 1996).

boundary sequences in western Cuba include the Moncada Formation in the Los Organos belt (Tada et al., this volume), the Cacarajicara Formation in the Rosario belt, and the Peñalver Formation in the Bahía Honda–Matanzas allochthon (Takayama et al., 2000).

The study area is located near Loma Cornelia, along the road from Soroa to Bahía Honda, ~10 km north of the town of Soroa. There are excellent outcrops along an adjacent river, where we conducted mapping and sampling (Fig. 2). We sampled along roadcuts and river bluffs continuously every 5 m vertically, collecting the clasts and matrix in the breccia sequences. Thin sections were prepared, and grain separation was accomplished by dissolution in hydrochloric acid and magnetic separation. More than 100 thin sections of rock chips, and 100 acid-dissolution samples were used to determine the grain type and size, and grain-boundary conditions, by using a petrographic microscope, energy-dispersive spectrometer (EDS; JEOL 5400), and electron probe microanalyzer (EPMA; JXA-8800M) (for details see Yokoyama et al., 1993).

CACARAJICARA FORMATION

The Cacarajicara Formation is present only in the Rosario belt as nearly continuous strips 0.5–1.0 km wide and to 100 km long in an east-west direction (Fig. 1). Latest Maastrichtian rudists, foraminifera, and nannofossils occur in the matrix and in fossil-bearing limestone fragments, but no Paleocene fossils have ever been reported (Pszczolkowski, 1986; Iturralde-Vinent, 1992).

CROSS SECTION

The Cacarajicara Formation dips 70°–80° to the north, as indicated by the thin pebbly layer in the calcarenite (Fig. 3); the pressure-solution cleavage preserved in the calcarenite dips 30° to the north. The lower boundary with the Lower Cretaceous Polier Formation is partly disturbed by a reverse fault that dips 60° south and has steeply plunging lineations. This disconformable (erosional) contact is preserved in the hanging wall of the fault. The upper boundary was disturbed by strike-slip and shear deformation and is overlain by the Paleocene Ancon Formation. The Cacarajicara Formation is underlain by the Santonian?-Campanian-Maastrichtian Moreno Formation or older units at some other localities (Pszczolkowski, 1978, 1994).

LITHOSTRATIGRAPHY

The Cacarajicara Formation is an ~700-m-thick, upward-fining, homogeneous, calciclastic sequence. It can be subdivided into the Lower Breccia, Middle Calcarenite, and Upper Lime Mudstone Members (Fig. 4). The boundaries between the members are gradual; there are no sharp lithologic discontinuities except for the contacts with Eocene sediments.

In order to understand the depositional mechanism for the

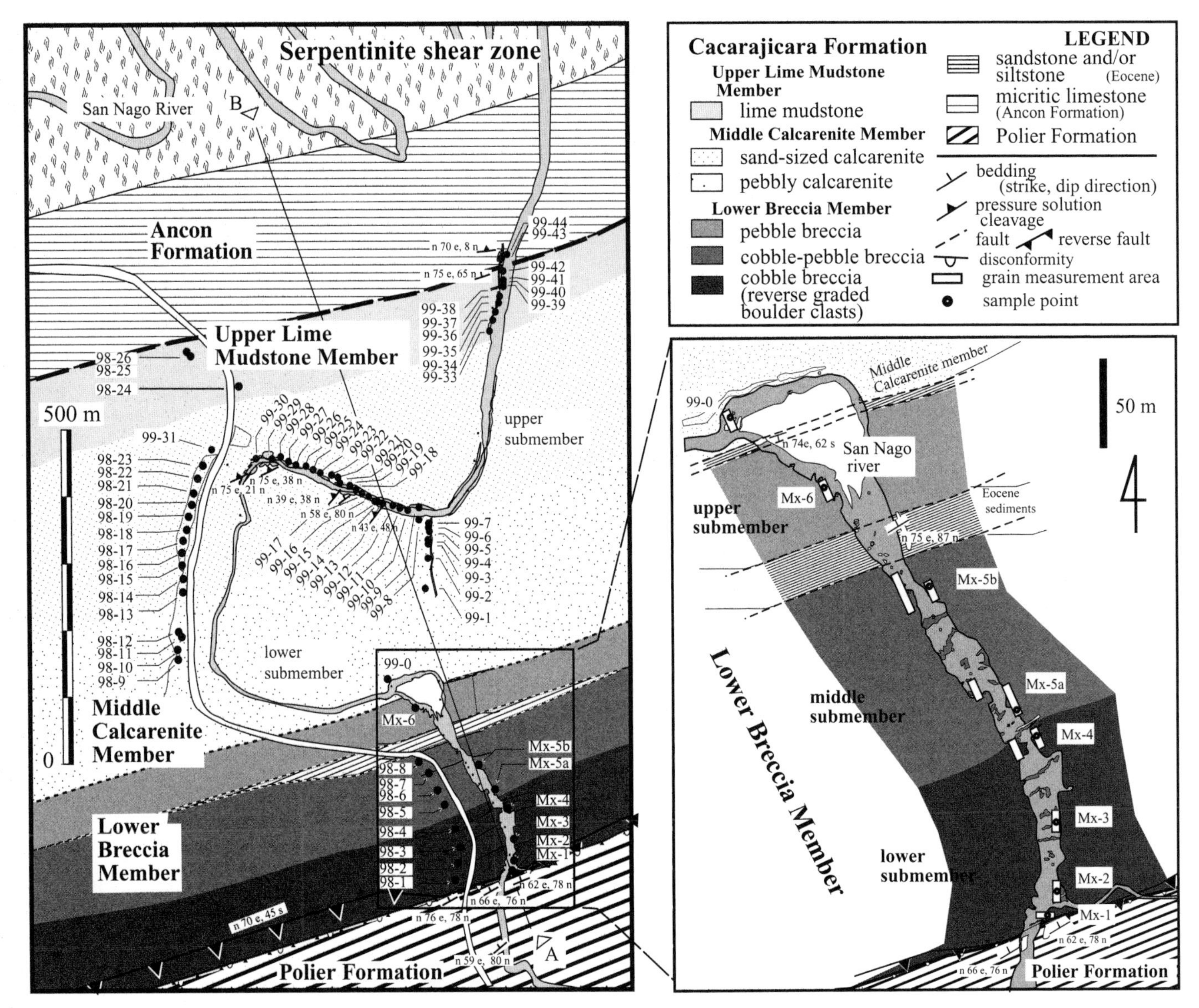

Figure 2. Geologic map of Cacarajicara Formation along San Nago River (left). Geologic map of Lower Breccia Member (right). Black dots and numbers are sampling points and sample number.

Cacarajicara Formation, we focused our study on the detailed sedimentary structures and lithologic characteristics, including grain composition, size, shape, and boundaries, and matrix components. We used microscopic point counts for the Middle Calcarenite and the Upper Lime Mudstone Members and outcrop point counts for the Lower Breccia Member. Grain-size variations were determined from 200 counts at one locality in each member.

Lower Breccia Member

The Lower Breccia Member is more than 250 m thick and is composed of cobble- to pebble-size clasts consisting of shallow- and deep-water limestones, black chert, bedded chert, and greenish shale (Fig. 5, A, B, and C). The breccia is well sorted and grain supported, with only a small amount of matrix. Discoidal or rectangular boulders floating among the cobble-pebble clasts are conspicuous (Fig. 5, B and C).

The Lower Breccia Member is subdivided into the lower, middle, and upper submembers (Fig. 6). The lower submember contains the cobble- to boulder-size breccia and larger boulder clasts (Fig. 5B). The black clay matrix contains some shocked quartz; the matrix in the basal part of this submember contains spherules. The middle submember contains the pebble- to cobble-size breccia; there are no larger boulder clasts (Fig. 5C). The upper submember contains pebble-size breccia; black chert pebbles are more homogeneous in size than in the middle and lower submembers.

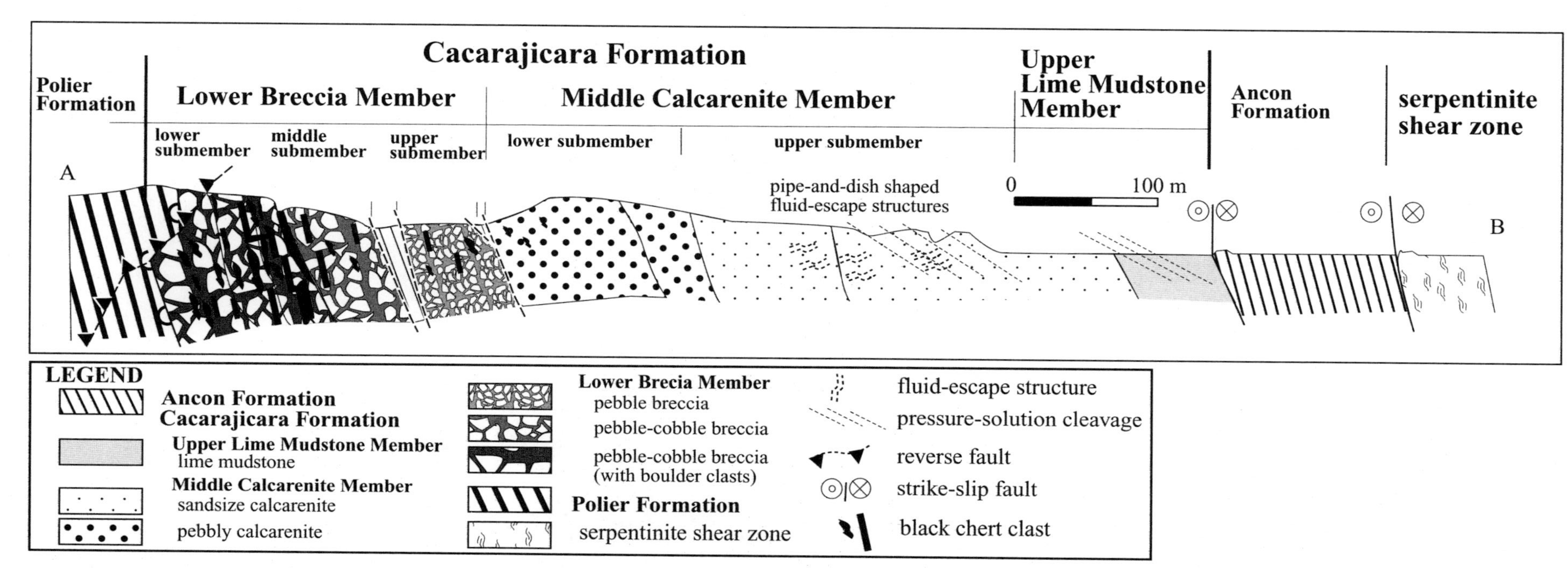

Figure 3. Cross section of Cacarajicara Formation.

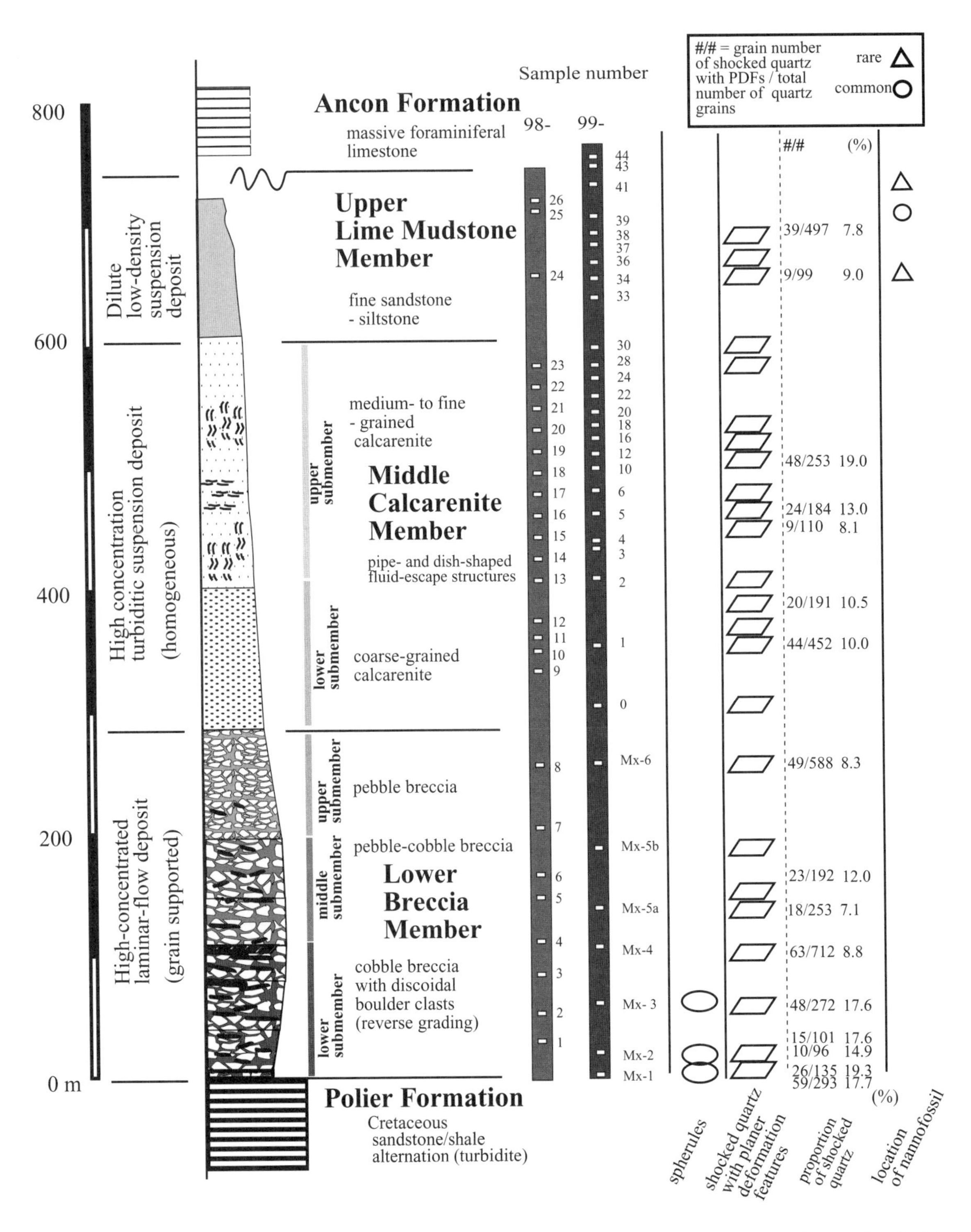

Figure 4. Lithologic column for Cacarajicara Formation. Squares and line of sample numbers indicate points in thin sections where proportion of shocked versus unshocked quartz was counted. Occurrences of spherules, shocked quartz, and nannofossils are shown in right columns. PDF is planar deformation feature.

Figure 5. A: Overview of Lower Breccia Member. Continuous outcrop is preserved along Nago River. B: Basal part of Lower Breccia Member. Breccia contains discoidal and/or rectangular clasts of black chert, greenish shale, and gray limestone. Pebble size fraction mainly contains rounded limestone clasts. Matrix in these clasts is very poorly preserved. C: Lower submember of Lower Breccia Member. Note black chert and greenish shale particles preserved in pebbly limestone clast. D: Pebbly calcarenite of lower Middle Calcarenite Member. White margin of black chert clast is replaced by carbonate. E: Elongated fluid-escape pipe structures in homogeneous calcarenite in upper Middle Calcarenite Member. Weak subhorizontal foliation is pressure-solution cleavage. F: Fluid-escape vein structures in homogeneous calcarenite of upper Middle Calcarenite Member. Diffuse webby laminations (slanting to right) of fluid-escape structures are well preserved in this sequence, and resemble structures seen in Peñalver Formation. Lens cap, 55 mm for scale. G: Rectangular bedded red chert clast in uppermost lower submember of Lower Breccia Member. Hammer, 35 cm for scale. H: Black chert clast in upper part of lower submember of Lower Breccia Member. Scale is film cap in center. I: Black siliceous sandstone clast in upper part of lower submember of Lower Breccia Member. Sandstone is 1 m thick and more than 10 m long. Note that there is no fault in this calcarenite. (Figure 5 continues on p. 132–133.)

Grain composition. More than 50% of the clasts in the Lower Breccia Member are calcareous angular or rounded limestone. The angular limestone is composed of rudist limestone, massive coarse calcarenite, foraminiferal limestone, and micritic limestone with black chert layers. Small amounts of oolitic limestone and calcareous algal mat limestone are also present. The rounded limestone has a more eroded surface than the angular limestone, and is composed of micritic limestone, foraminiferal limestone, and dolostone, 3–10 cm in diameter (Fig. 7).

Chert represents 40% of the pebble- to cobble-size clasts, which consist mainly of black chert (Fig. 5, B, C, and H), black chert associated with foraminiferal limestone, and red radiolarian chert (Fig. 5G). The black chert clasts are similar in lithology to the black chert in the Cretaceous deep-water Pons Formation of the Guaniguanico terrane.

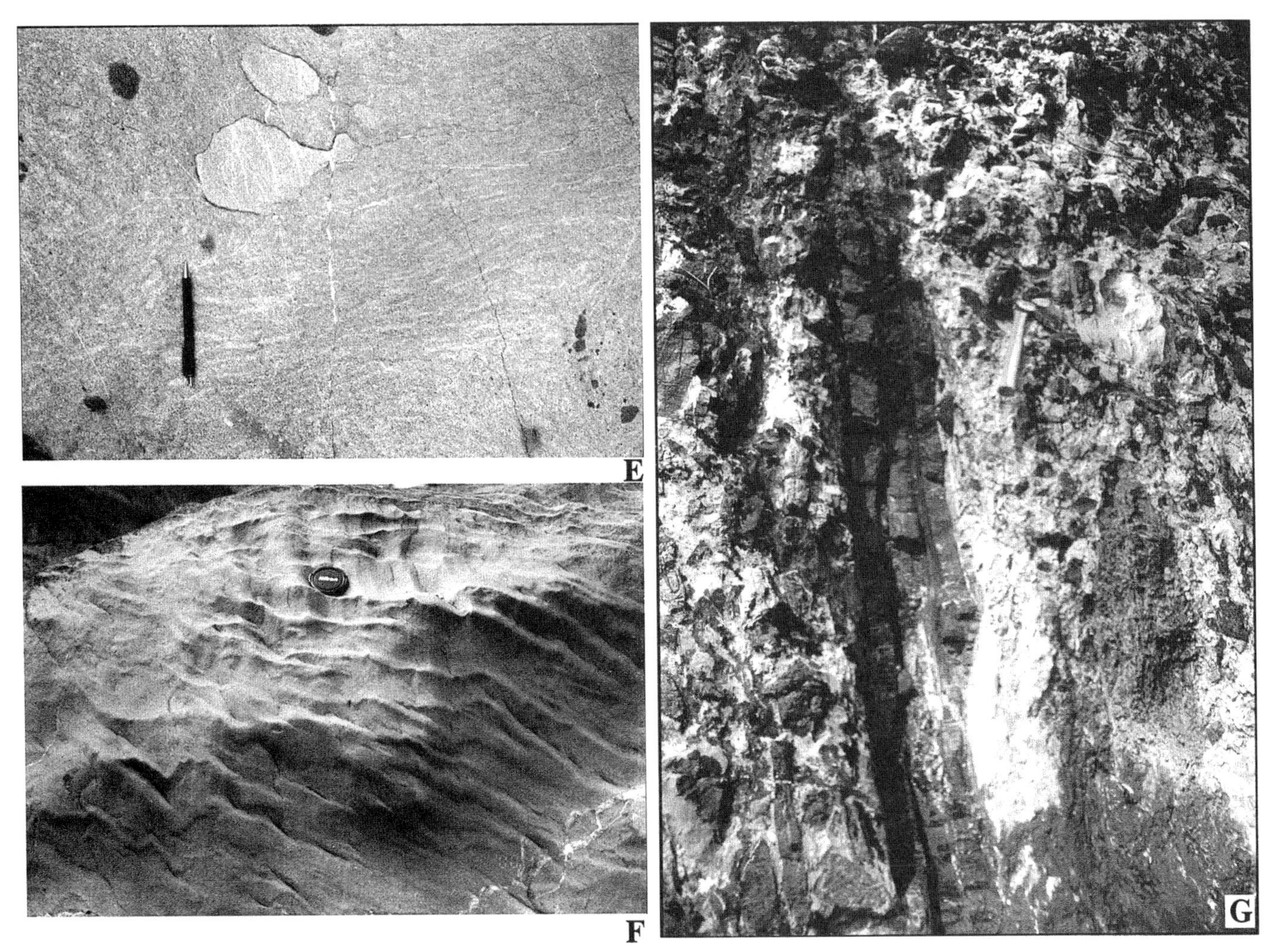

Figure 5. (*continued*)

Minor clast components include green shale (Fig. 5G), green altered volcanic rock, trachytic basalt (Fig. 8A), vesicular basic rocks (Fig. 8B), quartz-rich sandstones (Fig. 5I) and pelitic and psammitic schists. The minor grain compositions imply a variable source area within a volcanic and orogenic region.

Grain size and shape. There are two modes of grain sizes in the lower submember of the Lower Breccia Member (Fig 6). The smaller clasts are ~3–15 cm and no size grading is obvious (Fig. 5C); this range composes 60%–70% of this member by volume. The larger clasts are 30–300 cm (average long axis 40 cm; Fig. 5G), and compose to 5% by volume; this clast size range exhibits reverse grading (Fig. 6). The uppermost part of the lower submember contains the largest clasts, including some that are >2-m-long, discoidal to rectangular black chert, and others that are >10-m-long, 80-cm-thick, black, less calcareous sandstone blocks (Fig. 5, G, H, and I). The upper and middle submembers lack large boulder clasts in the gradually formed fining-upward sequences.

The largest angular clasts in the Lower Breccia Member were not affected by abrasion. The cobble- to pebble-size carbonate clasts are subrounded and subangular, and have homogeneous compositions. The minor clast components, such as green shale, schist, and volcanic rocks, are mostly <5 cm in diameter and well rounded. They were probably reworked long before deposition in the Cacarajicara Formation.

Grain-boundary condition. The clasts are mostly in direct contact with each other; harder angular grains are partly abraded into softer, rounded limestone clasts (Figs. 5, B and C, and 8D). There are many hydrofractured limestone clasts in the matrix alongside the larger clasts. Cracked chert clasts, matrix clay intrusion structures, and highly brecciated material form a jigsaw puzzle-like mosaic or texture preserved along the limestone clast margins (Figs. 8F and 9, A and B). Some black chert is fractured, and sharp cracks are filled by the general matrix clay (Fig. 8, E and F). Along narrow cracks in the large discoidal boulders, dark brown clay matrix contains brecciated pebbly micritic limestone (Fig. 9A).

Figure 5. (*continued*)

Matrix. The matrix composes <1–3 vol% of the Lower Breccia Member, and fills the very narrow spaces between the closely packed clasts. This matrix is composed of dark brown to black clay and some sand-size carbonate fragments. According to EPMA element mapping, there are many dolomite fragments in this clay matrix. The matrix is composed mainly of brown to dark brown clay, and small amounts of sand-size carbonate, siliceous, and opaque grains (Fig. 9C). The basal part in this member exhibits an especially low content of limestone grains. Sand-size carbonate grains in the clay matrix increase upsection throughout the Lower Breccia Member.

Middle Calcarenite Member

The ~300-m-thick Middle Calcarenite Member is composed of well-sorted, massive, homogeneous calcarenite, which exhibits upward fining and distribution grading. A partly preserved pebbly bed and parallel laminations are rarely present as bedding in this homogeneous calcarenite. The boundary between the Lower Breccia and the Middle Calcarenite Members is a fault contact within the Eocene turbidite sequence. However, the rock gradually changed to fining upward. Except for the chert grains, the grain composition resembles the upper submember of the Lower Breccia and the lowest Middle Calcarenite Members.

The Middle Calcarenite Member can be subdivided into upper and lower submembers. The lower submember is composed of coarse granular carbonate calcarenite with pebble-size clasts (Fig. 5D). The calcareous sand grains are mainly composed of limestone fragments of rudists, algal mats, and foraminifers. The pebble-size clasts are fragments of black chert, greenish shale, volcanic rocks, and schist. The percentage of pebbly clasts gradually decreases upward.

The upper submember is composed of massive, coarse- to medium-grained, calcareous calcarenite, with a well-sorted and very homogeneous fabric (Fig. 5, E and F). The sand grains are mainly composed of micritic limestone fragments and foraminiferal skeletons, and a smaller amount of larger bioclasts. The massive, coarse calcarenite contains many unusual fluid-escape structures similar to those reported in the Peñalver Formation near Havana (Bronnimann and Rigassi, 1963; Takayama et al., 2000). Fluid-escape structures in this submember are mainly pillar, pipe, and spiral shaped; there are some dish-shaped structures (Fig. 5, E and F).

Grain composition. The calcarenite of the Middle Calcarenite Member consists of more than 85%–90% calcareous grains, including two types of fragments: shallow-water bioclastic limestone, comprising rudists, stromatolites, and oolite, and deep-water fragments of micritic limestone and foraminiferal limestone (Fig. 7). The lower submember mostly contains rudist, bioclastic, and foraminifer limestone fragments, and the upper submember mainly contains foraminiferal grains and micritic limestone (Fig. 7). The remaining 15%–10% of the grains are quartz and feldspar (6%–10%), chlorite-smectite-replaced volcanics or serpentinite (3%–4%), and other minerals (Fig. 7). Single-crystal piemontite, indicative of low- to medium-grade

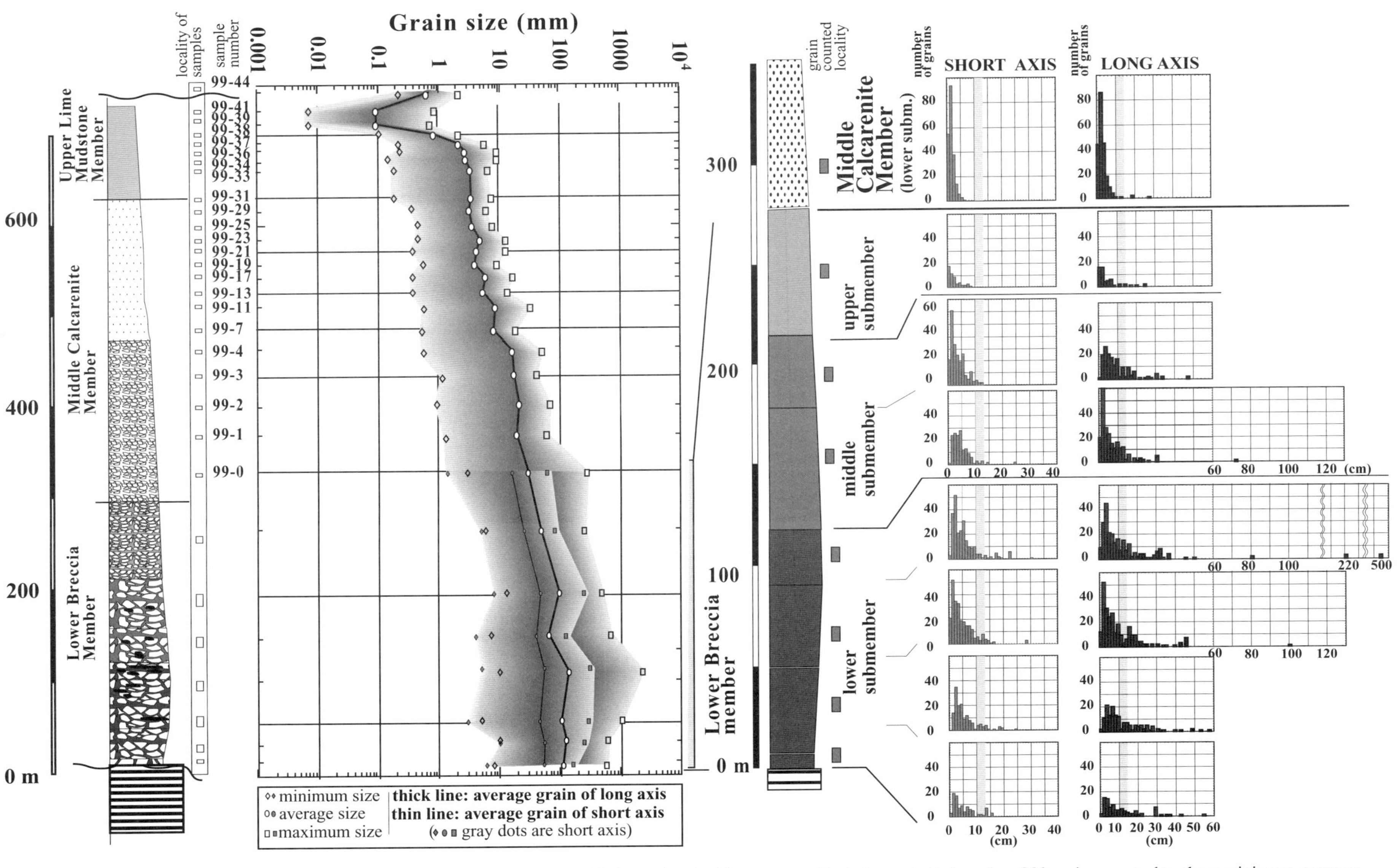

Figure 6. Left: Grain-size distribution of Cacarajicara Formation. There are 33 data points in this sequence. Each data point is based on 200 grains counted to show minimum, average, and maximum size data. Thin-section method used above Middle Calcarenite Member and field-observation method used for Lower Breccia Member. Width of shadow shows grain-size variation. Square and rhombus points show largest and smallest grain sizes, respectively. Darkest part with circle point shows average of grain-size variation. Note that thin and thick lines of Lower Breccia Member show average grain size on short axis and fluctuations on long axis. Right: Detailed grain-size variation diagram of Lower Breccia Member. Note that clasts are of two types: major 3–7 cm size group and minor boulder size group. Sizes of boulder clasts increase from lower to upper in lower submember of Lower Breccia Member.

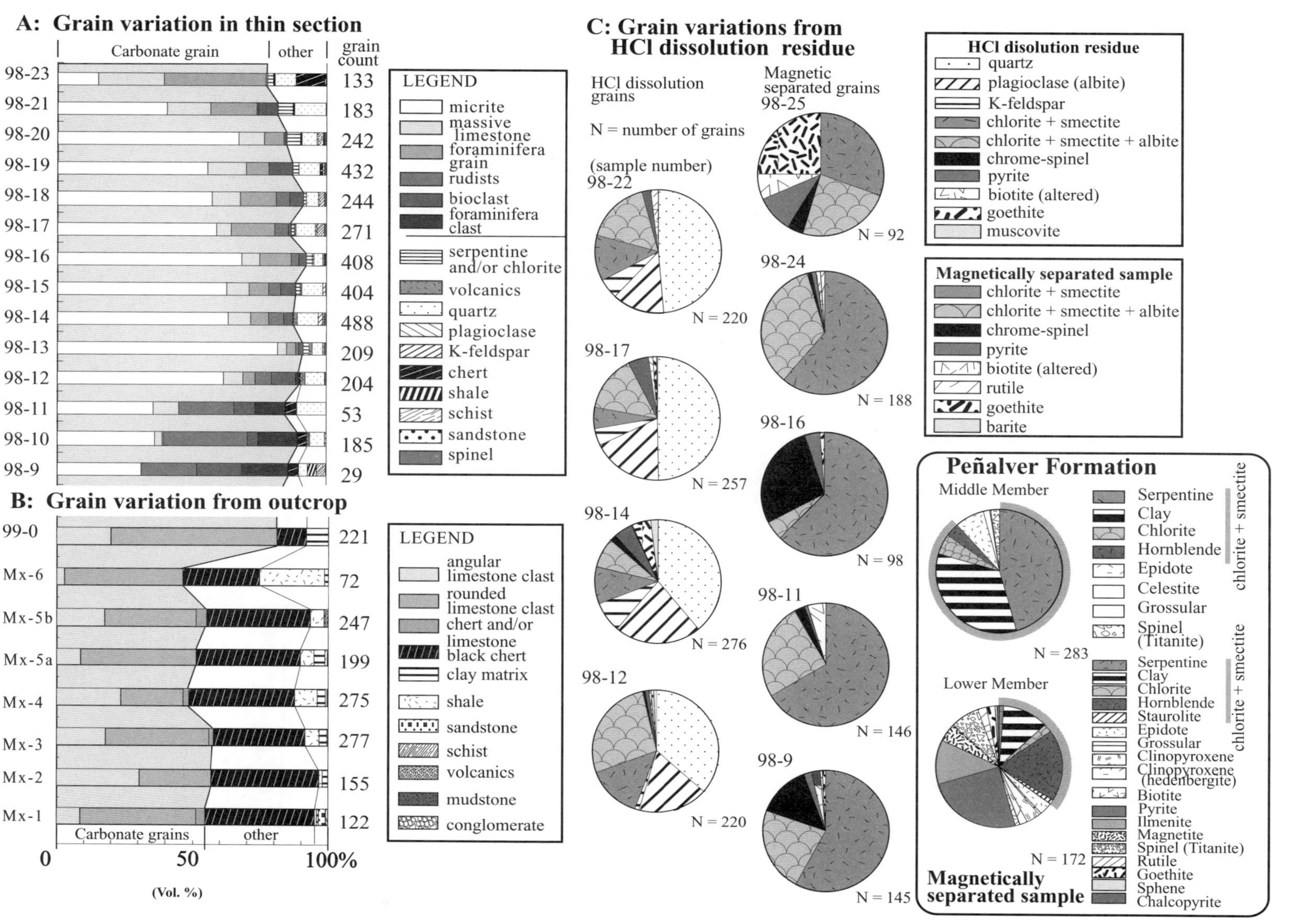

Figure 7. Thin-section and outcrop-scale grain-type variation of nondissolved samples from throughout Cacarajicara Formation (left). Grain composition of HCl dissolution and magnetically separated residue (right). Box on right shows data comparable with Peñalver Formation, which has more variable grains. Considering metamorphic alteration, composition of grains between Cacarajicara and Peñalver Formations is very similar.

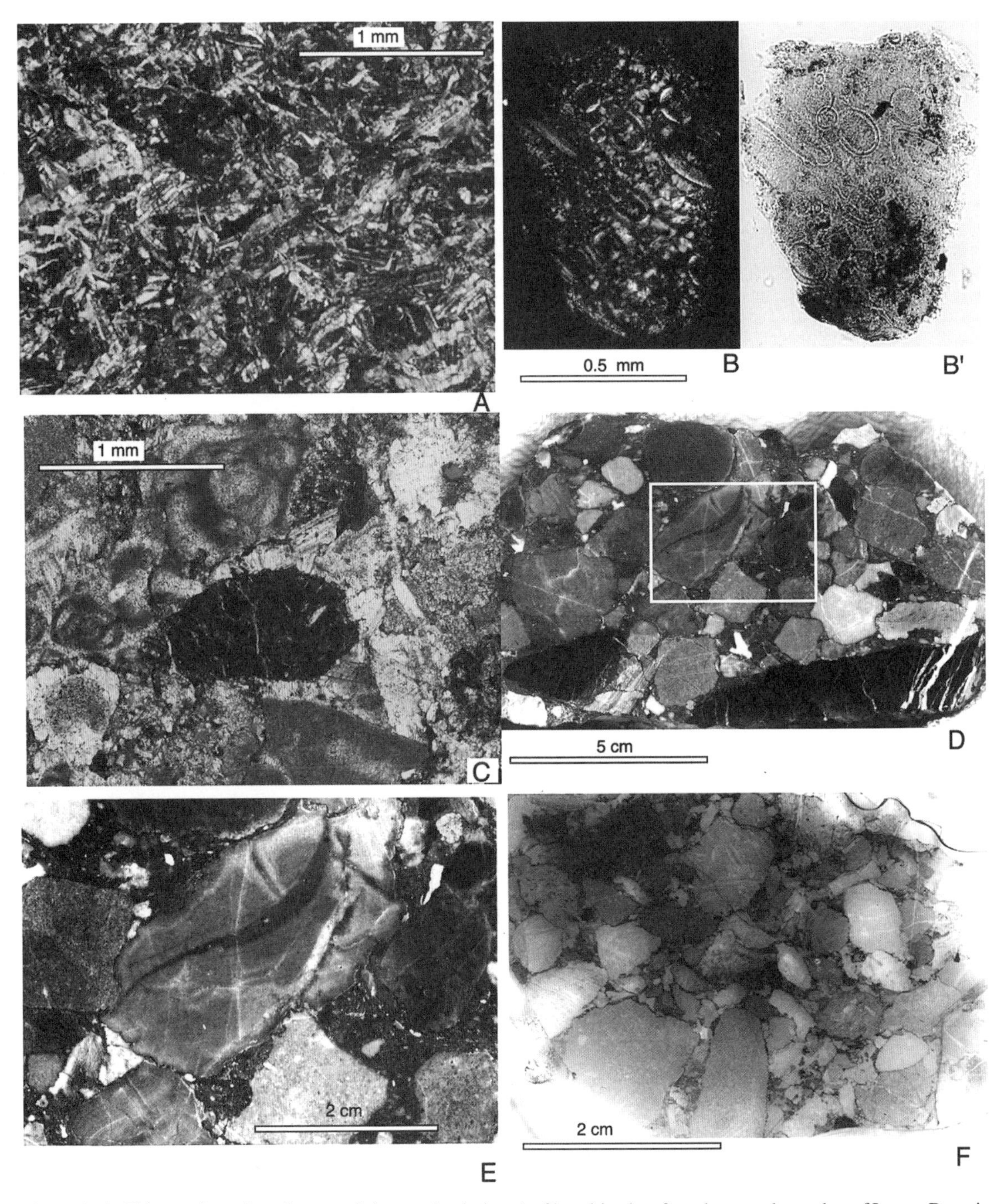

Figure 8. A: Thin-section microphotograph (crossed polarizers) of basaltic clast from lower submember of Lower Breccia Member. B: Thin section of vesicular shard in Middle Calcarenite Member: B, crossed nicols; B', parallel polarizers. Note that grains are filled by chlorite, although some portions of shard are isotropic and may be mostly glass. C: Thin section of well-rounded volcanic grains in coarse calcarenite from Middle Calcarenite Member (plain light). D: Polished section of pebbly clast from Lower Breccia Member. Varied clast types preserved in poor matrix. Rectangle shows E. E: Muddy matrix fills fracture in green chert. F: Polished section of coarse calcarenite of lower Middle Calcarenite Member.

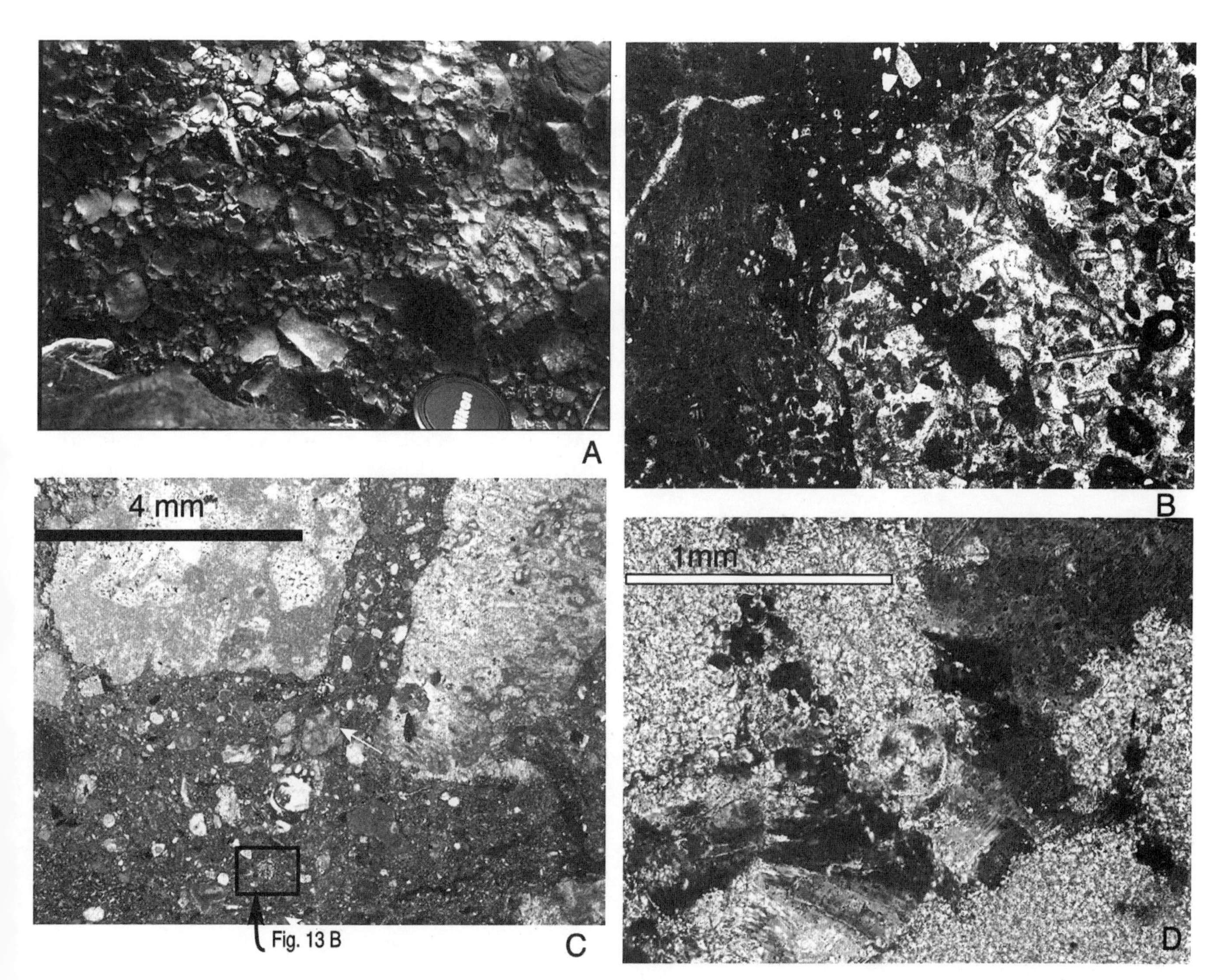

Figure 9. A: Pebbly limestone with black matrix along 3-m-long, disk-shaped clast. Note that limestone float in black matrix has been affected by hydrofracturing. Lens cap, 5.5 cm for scale. B: Thin-section view of margin of limestone grain. Intrusion of matrix preserved in limestone grain in upper submember of Lower Breccia Member (parallel polarizers). C: Thin-section view of matrix from basal part of Lower Breccia Member. Isolated round quartz spherules are preserved. (Square shows more detailed Fig. 13B) D: Microcarbonate pisolith in thin section of medium calcarenite. Pisolith forms radial pattern crystal of carbonate with very thin, spherical shell.

metamorphism, and zircon and chrome spinel are present in the matrix in small amounts.

Calcareous grains compose >80 vol% of both the Cacarajicara and Peñalver Formations, but the relative presence of associate minerals differs significantly. On the basis of the magnetically separated samples from each formation, the Cacarajicara Formation preserved only smectite and chlorite grains, but the Peñalver Formation, especially the middle member, contains serpentine, clay, epidote, chlorite, hornblende, and grossular (Fig. 7). This difference probably reflects the low-grade metamorphism of the Rosario belt. In addition, minerals such as serpentinite, clay, and hornblende may have decomposed or altered to chlorite-smectite minerals (Fig. 7).

Grain size and shape. The calcarenite in this member is well sorted, and the carbonate grains are mainly angular, whereas the volcanic, schist, and shale grains are well rounded. The rounded noncarbonate grains may have a reworked origin. The lower submember contains large fragments of rudist, bioclastic, and foraminiferal limestone. The upper submember contains small grains of foraminifer skeletons and micritic limestone. This size fractionation may have resulted from depositional sorting of the original grains.

Matrix. Matrix is rarely present in the well-sorted calcarenite, and when present it consists of very fine calcareous clay in the narrow spaces between sand-size grains. Calcite pisoliths, 200–300 μm and having very thin skin, are well preserved in the matrix of medium-fine calcarenite. They appears as 10 or more grains seen in 3 × 4 cm thin sections, and compose <10% of the matrix.

Upper Lime Mudstone Member

The Upper Lime Mudstone Member is more than 100 m thick, and consists of homogeneous, massive, calcareous, fine calcarenite and lime mudstone. The lithology gradually changes from that of the underlying Middle Calcarenite Member. The upper boundary with the Paleocene Ancon Formation is a fault contact, and the exact boundary is not observable. This member consists of calcareous, fine calcarenite and lime mudstone with some foraminifer skeletons, shallow-origin bioclasts, and black carbon fragments that might be wood. Foraminifer skeletons are very rare in the uppermost part of this member. Poorly preserved sedimentary structures are present, such as cross-lamination, bioturbation, or fluid-escape structures, in addition to faint parallel bedding. The overlying Ancon Formation contains many foraminifers and exhibits a bioturbation texture.

There are some nannofossils in the limestone (Fig. 4), including *Micula dexussata*, which ranges from Coniacian to Maastrichtian; no Paleocene fossil was found in this member.

PALEOCURRENTS

Imbricated clast structures indicative of paleocurrents are present in the Lower Breccia Member; the discoidal and/or rectangular clasts show flow orientation. We measured the dips and strikes of clasts >40 cm in greatest diameter. The estimated current orientation is corrected for folding; the fold plane dips 80° north with zero plunge of the axis (Fig. 10). The area of Cacarajicara Formation distribution is 100 km long and the lithology is very uniform, and we estimate that the folding plunge in this area is very shallow and not involved in crustal rotation. The disk-shaped clasts preserve imbricated structures, and the paleocurrent evidently flowed from north-northwest to south-southeast in current terms (Fig. 10).

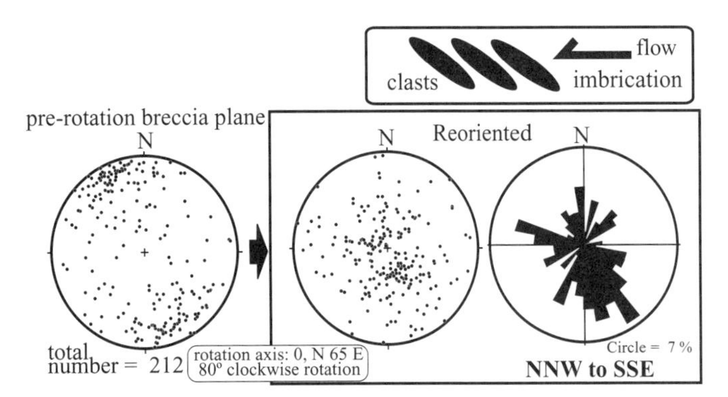

Figure 10. Paleocurrent direction for Lower Breccia Member. Plot shows dip and strike of disk-shaped clast, which preserved similar direction. Note data reoriented 80° clockwise by N65E horizontal axis.

IMPACT EVIDENCE

Shocked quartz and spherules of probable impact origin (e.g., Bohor and Glass, 1995; French, 1998) are present in the Cacarajicara Formation. Carbonate pisoliths are also preserved in the calcarenite, which might be a microcarbonate pisolith (Pope et al., 1999).

Analytical methods

Isolation of grains. We fragmented ~100 samples of calcarenite, weighing ~50 g each, to a grain size of 1 cm to few millimeters. The samples were then treated with dilute hydrochloric acid for ~30 min and after complete dissolution of the carbonate material, the residue was magnetically separated into magnetic, weak magnetic, and nonmagnetic grains. Thin sections of the three kinds of residues (~1–3 g) were prepared, and these thin sections were used for counting and measuring grains under optical and electron microscopes. The composition of the grains was determined by EDS or EPMA.

The nonmagnetic residue of shocked quartz was treated with silica-saturated hydrofluorosilicic acid for three days, so that other minerals were completely dissolved. After making a thin section of the new residue, we produced a grain map of the thin section in a slide scanner. These scanner images make it easy to identify the normal and deformed quartz. Shocked quartz was identified by the presence of more than two sets of planar deformation features (PDFs: Fig. 11, A, B, and C), which have distinct angles from the c-axis in the quartz (e.g., Stöffler and Langenhorst, 1994; French, 1998). When looking for deformed quartz on a map, we try to measure the PDFs, which are characterized by at least two sets of straight planar features. We exclude the spaced and planar fracture (PF) structures formed at this stage. Measuring the orientation of the PDFs of quartz in the thin section was performed on a universal stage (e.g., Montanari and Koeberl, 2000), and then plotted in frequency distribution histograms.

We processed the magnetically separated residue to obtain spherules. Spherules were searched for under a stereomicroscope, then the spherule surfaces were observed by electron microscopy. At this stage, we picked up only smooth spherules. After we examined the spherule surfaces, each grain was glued to a slide glass and we made half-cut polished sections for observation by EDS and EPMA. Microfossils such as radiolaria, ostracoda, and foraminifera were distinguished at this stage. The radiolaria, which are totally replaced with goethite, have round tests with well-regulated small openings and are very common in this formation. The ostracoda and foraminifera were preserved only as lattice shapes replaced by oblong clay minerals.

Shocked quartz

Shocked quartz is found throughout the Cacarajicara Formation (Fig. 4). PDF orientation has been used as a shock ba-

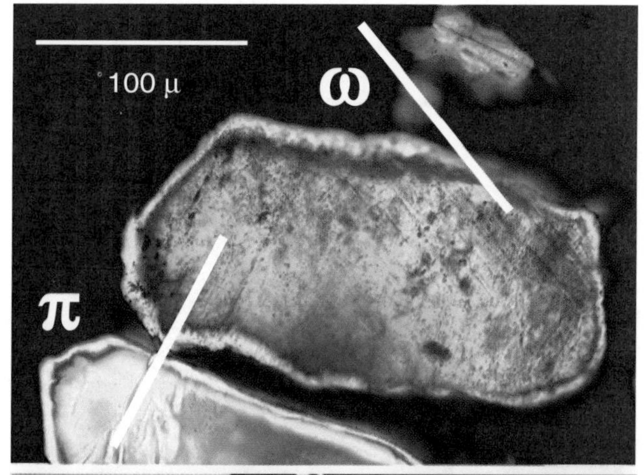

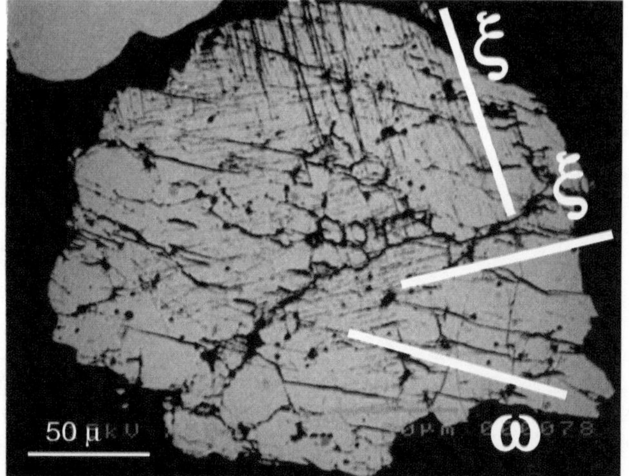

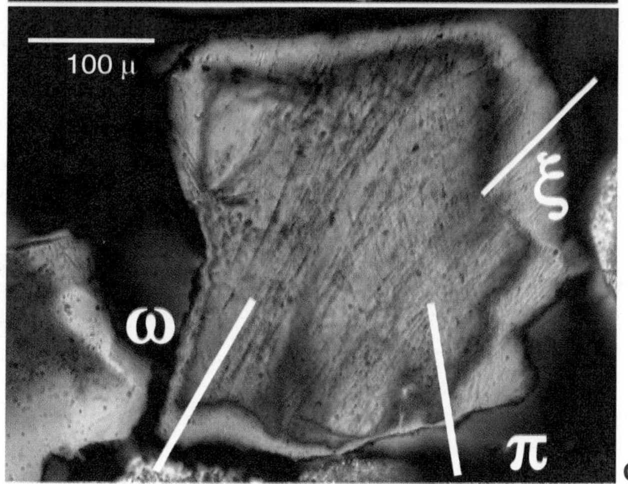

Figure 11. A: Planar deformation features (PDFs) in shocked quartz (crossed nicols) from Middle Calcarenite Member (sample 98-18). PDF shows ω and π planes. B: Backscattered electron (BSE) image of shocked quartz in Middle Calcarenite Member (sample 99-4). PDF shows ω and ζ planes, as measured on universal stage by microscope. C: PDFs in shocked quartz (crossed nicols) in Lower Breccia Member (sample Mx-1). PDF shows π, ζ, and ω planes.

rometer to measure the intensity and distribution of shock pressures (Grieve and Robertson, 1976; Dressler and Sharpton, 1997) Most of the PDFs in the shocked quartz from this formation are concentrated at the ω (23°), π (32°), ζ (48°), and γ (52°) positions from the c-axis of quartz crystals (Fig. 12). The presence of the π plane with a number of different orientation planes suggests that the shocked quartz in the Cacarajicara Formation was affected by more than 16 GPa of pressure.

The abundance of shocked quartz with PDFs in the Cacarajicara Formation varies from <20% to 7% of total quartz grains in the acid-dissolved residue (Fig. 4). The matrix in the basal part of the Lower Breccia Member and the fine sand-size calcarenite in the upper Middle Calcarenite Member preserve shocked quartz in abundances as high as 15%–20% of the total number of quartz grains.

Spherules

The identity of impact-related spherules in the Cacarajicara Formation is still not known with certainty. However, the matrix in the basal part of the Lower Breccia Member contains three types of spherules. The first type of spherule is gray, 200–300 μm in diameter, and has a very smooth surface and partly elongated shape (Fig. 13A). Based on backscattered electron images and EDS analysis, this type of spherule is composed of homogeneous clay minerals, mainly smectite. It may be an altered glass spherule. The second type of spherule consists of isolated silica spheroids preserved in the clay matrix within the basal Lower Breccia Member (Fig. 8B). These spherules are well rounded and are formed of a quartz crystal with radical extinc-

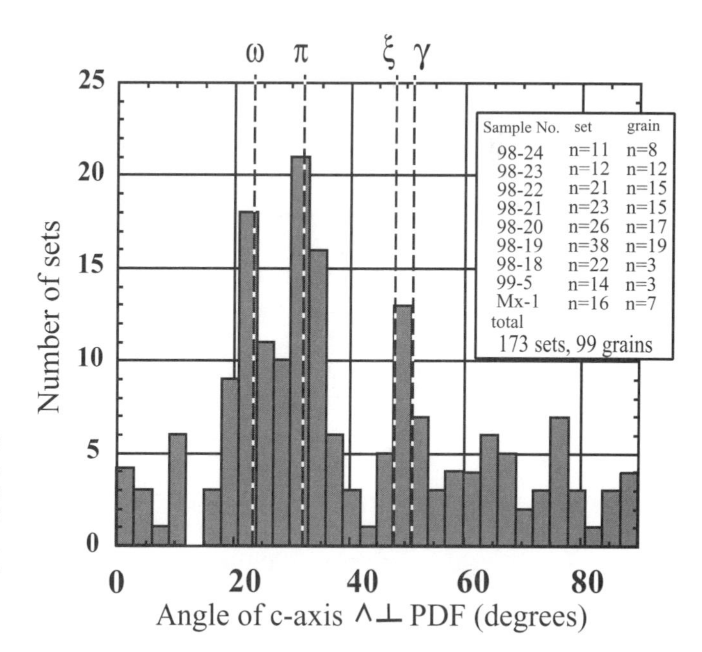

Figure 12. Shocked-quartz planar deformation feature (PDF) orientation diagram.

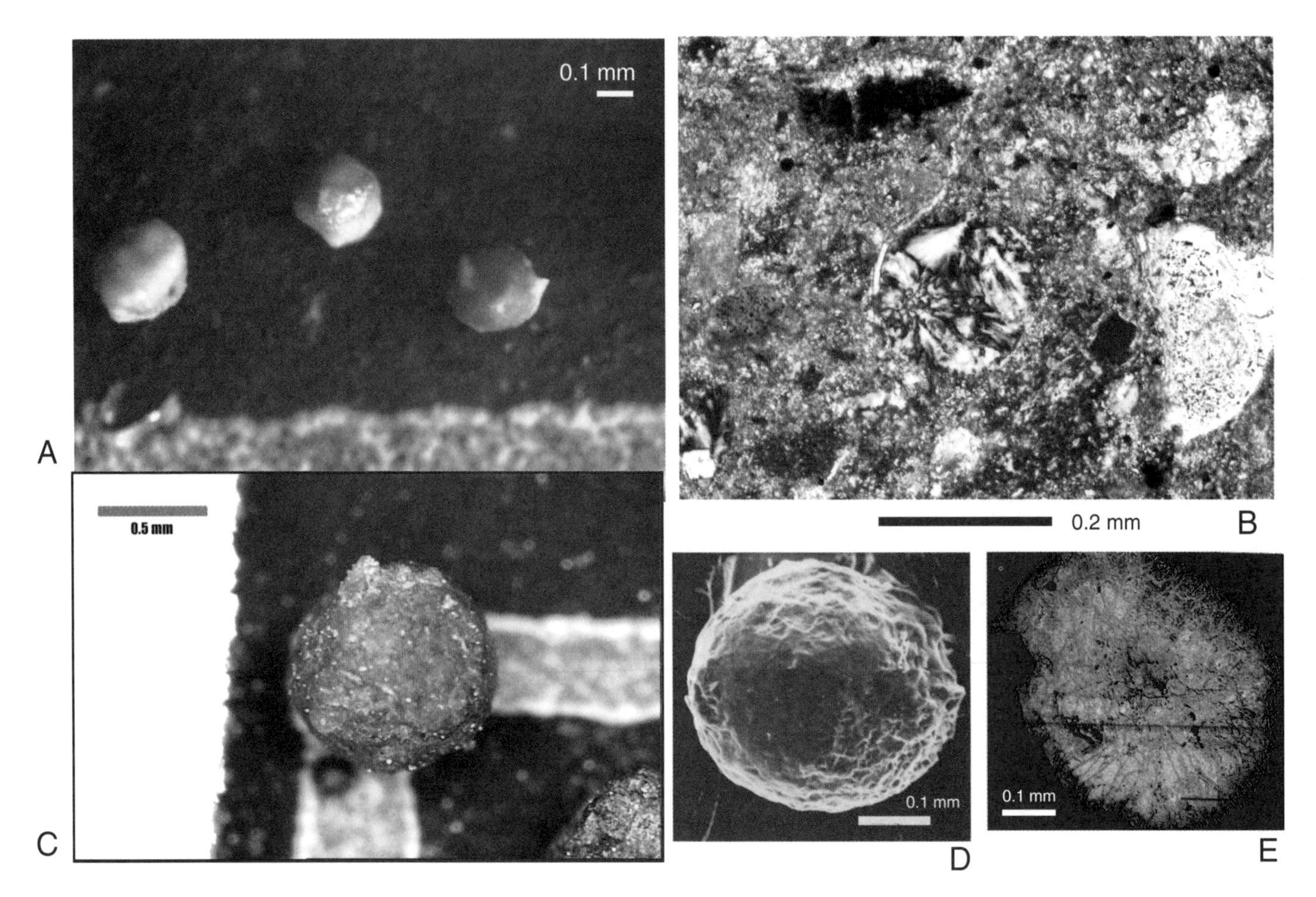

Figure 13. A: Gray spherules in matrix of Lower Breccia Member (0.3 mm in diameter, sample Mx-1). Note smectite in each grain. B: Silica-rich spherules in matrix of Lower Breccia Member (0.3 mm diameter). C: Large goethite spherules (sample Mx-1). D: Scanning electron microscope (SEM) image of gray spherule surface (sample Mx-1). E: SEM image of section of goethite spherules (sample Mx-1). Note dendritic texture preserved and overprinted by goethite.

tion (Fig. 13B). The third spherule type consists of goethite, which is distinguished by some larger (500–1000 µm in diameter) and other smaller (200–300 µm in diameter) spherules (Fig. 8C). The larger spherules are similar in size to those in the Beloc samples (Izett, 1991; Bohor and Glass, 1995). The surface of the goethite spherules is a smooth and irregular plane. The insides of large goethite spherules are very homogeneous and partly preserve a dendritic texture (Fig. 13E) that resembles the texture seen in the second type of spherule that has been entirely replaced by goethite (Bohor and Glass, 1995).

Carbonate pisoliths

The carbonate pisoliths are well preserved in matrix of the middle-fine calcarenite of the Middle Calcarenite Member (Fig. 9D). The matrix is formed of simple 200–300-µm-diameter calcite crystals within a fibrous radial structure. These pisoliths have 1-µm-thick calcite wall that forms perfect spheres. This is a simpler crystal than the laminated ooids or microfossils, and it is not preserved in the other carbonate clasts or carbonate grains. Similar pisoliths are also contained in the matrix within the middle member of the Peñalver Formation, which was not metamorphosed. Calcite pisoliths in the spherule layer on Albion Island are 7–10 mm in diameter, larger than in the Cacarajicara Formation. However, a radial structure with 1–2-mm-thick shells is also present (Pope et al., 1999). The carbonate pisoliths in the Middle Calcarenite Member might have been formed as accretional carbonate tuff from impact-induced calcite vapor.

DISCUSSION

Here we review the origin, transportation, and depositional mechanisms of the Cacarajicara Formation. The upward-fining sequence in this formation was likely formed by one gravity flow event. The sedimentary characteristics, however, are very different from those seen in normal turbidite and debris-flow deposits. We have focused on the flow characteristics of each member of this thick formation in order to infer flow transportation mechanisms, and relate these to the Chicxulub impact.

Transportation mechanisms

Lower Breccia Member. The clasts in the Lower Breccia Member have shallow-water, deep-water, and orogenic belt origins. Many limestone and chert breccias resemble those in underlying Cretaceous formations. The thick mixed-clast member may represent a collapse of the continental margin of the Yucatan platform. We proposed that these thick clasts are transported by laminar flow for the following reasons.

The cobble- to pebble-size clasts in the Lower Breccia Member are mostly in direct contact with each other and are similar to those in other grain-flow deposits (Bagnold, 1954). Such clasts were affected by the high dispersive pressure of grain collisions. The grain-boundary conditions around the clay matrix, however, preserved hydrofractured clasts, cracked chert clasts, and matrix intrusion structures within limestone clasts. The pebble-size carbonate present in the matrix alongside large boulders, which can be regarded as flow shadows, is preserved as a highly fractured fabric. There is no evidence that the other cobble- to boulder-size clasts were similarly influenced by this flow shadow. These features imply a high pore-pressure condition in this matrix, and the large clasts were fractured by the hydrostatic conditions.

The flow behavior of grain-supported sediments is changed by flow speed (Pierson and Costa, 1987). Grain-supporting mechanisms, such as the turbulence, dispersive stress, and fluidization, increase with flow velocity. In this way, with increasing pore pressure of the flow matrix, the grains floated and were transported long distances (e.g., Shanmugam, 1996). The Lower Breccia Member contains the following characteristics: (1) the reverse-graded and imbricated fabrics of the boulder clasts show the shear distribution of flow; (2) direct contact between each clast implies a high dispersive pressure condition; and (3) many matrix intrusions and a mosaic or jigsaw puzzle texture alongside limestone clasts also indicate high pore pressures. These features suggest that the Lower Breccia Member formed under laminar flow conditions in a very high speed dilatant situation (Lowe, 1976; Todd, 1989; Shanmugam, 1996).

Middle Calcarenite Member. The homogeneous and well-sorted calcarenite with deep- and shallow-origin rocks implies that the Middle Calcarenite Member formed under extremely mixed, strongly turbulent conditions. Sustained steady or quasi-steady current and upward migration of a depositional flow were required to form such a thick, massive calcarenite with a homogeneous fabric and large-scale grading. This process is referred to as high-density turbidity suspension (Kneller and Branney, 1995). The diffused flow laminations in the massive and well-sorted facies also suggest that the grains belonged to a strong suspension (deposit) environment, rather than a flow (shear) environment. The many fluid-escape structures also suggest that the deposit was formed by fast loading of high-density sediments under overpressured conditions. This evidence implies that the member formed from a high-density turbidity suspension.

Upper Lime Mudstone Member. The fine-grained Upper Lime Mudstone Member was probably deposited from a dilute sedimentary suspension. The gradually changing grain size and composition from those seen in the Middle Calcarenite Member suggest that the Upper Lime Mudstone Member was formed by a continuous, more dilute suspension.

Takayama et al. (2000), however, proposed "homogenite," which is formed by a tsunami, for the middle member of the Peñalver Formation. The Middle Calcarenite Member preserves sedimentary features that are similar to those in the Peñalver Formation. A tsunami produces mixing and turbidity in ocean sediments and may occur with or slightly after gravity sliding and ejecta flow. The very thick high-density turbidity suspension might have been produced tsunami-induced remixing. However, it is unknown whether the Cacarajicara Formation was influenced by a tsunami wave.

Relationship to the Chicxulub impact. The presence of shocked quartz and microspherules indicates that the Cacarajicara Formation resulted from an impact. The paleocurrent distribution inferred from the boulder imbrication in the Lower Breccia Member shows a north-northwest to south-southeast direction in modern terms. The Rosario belt belongs to an orogenic terrane, and the basic structure is top-to-the-north vergence with east-west-striking folds and thrusts (Gordon et al., 1997). It is difficult to rotate bedding 180° inside such an orogenic belt, so we propose that this paleocurrent flow came from the depositional area. When the Guaniguanico terrane of western Cuba is restored to its original orientation at the K-T interval, the Cacarajicara Formation was situated on the southeastern margin of the Yucatan Peninsula (Pindell et al., 1988; Ross and Scotese, 1988). In this position, the south-southeast direction of paleocurrents is consistent with the former location of the Yucatan continental margin.

Sedimentation model

Clastic rocks of the Cacarajicara Formation mainly originated from the collapse of the Yucatan platform (Fig. 14). The deep-water limestone and black chert, which are main components of this formation, represent continental slope deposition, whereas shallow-water limestone such as rudist and oolite limestones and dolomite represent platform margin deposition. The volcanic and metamorphic rocks are also present in the San Cayetano Formation and the Roble Member of the Polier Formation, which were also located on the continental margin or slope along the Yucatan platform.

The inferred sedimentation of the Cacarajicara Formation, which records laminar flow, high-turbidite suspension, and diluted suspension, resembles the hyperconcentrated flow that has been associated with high-energy and high-speed flows (Pierson and Costa, 1987). These sedimentary characteristics of the Cacarajicara Formation are similar to those of deposits produced by a subaqueous eruption (White, 2000). Subaqueous-eruption flow deposits are also composed of three layers; in

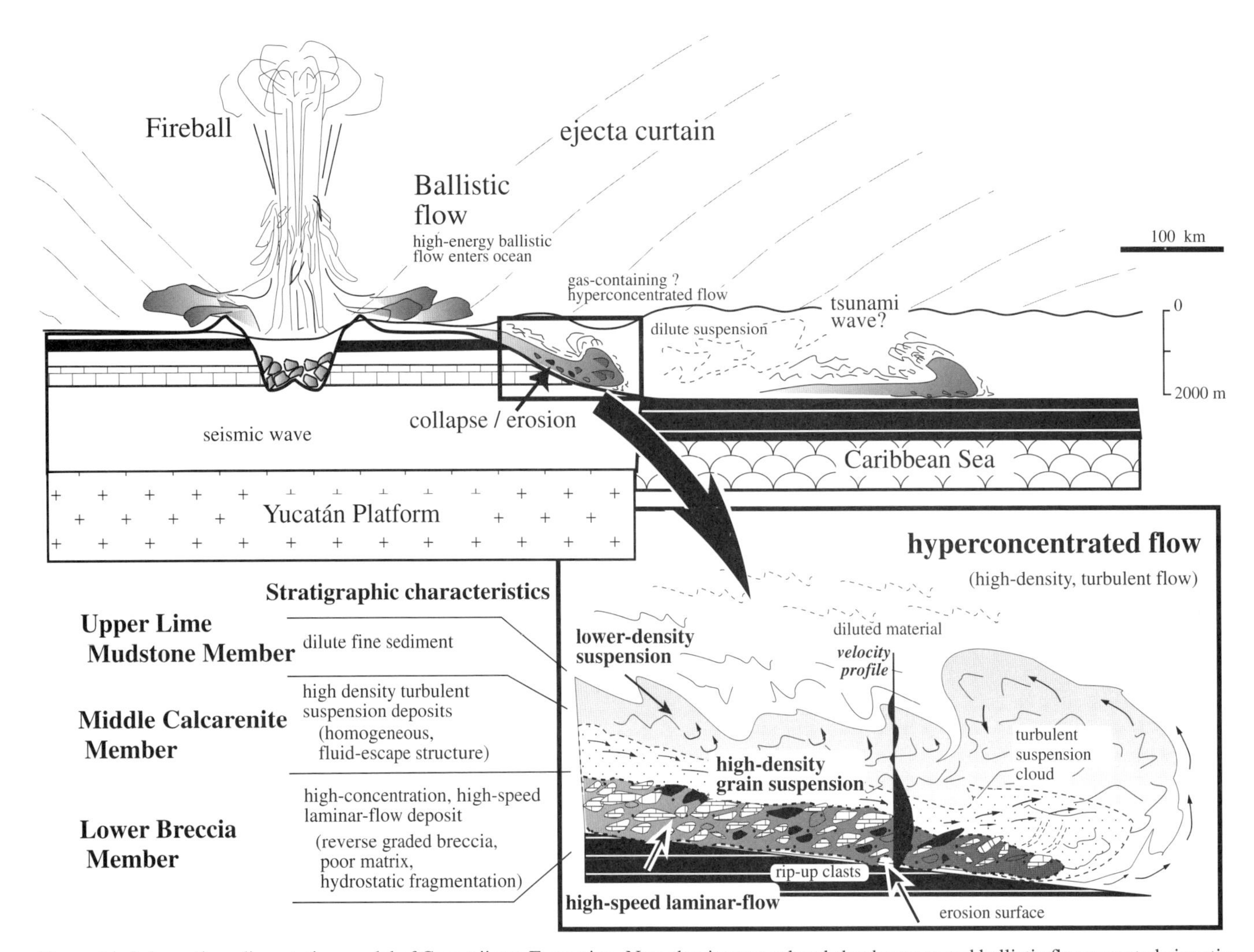

Figure 14. Schematic sedimentation model of Cacarajicara Formation. Note that impact-related shock waves and ballistic flows created gigantic hyperconcentrated flow. Flow consists of three different flow units that formed three sedimentary facies sequences: high-speed laminar-flow clasts, high-density turbulent-suspension deposits, and low-density, diluted, fine material. Middle and upper sequences might have been affected by tsunami wave mixing.

ascending order, these are a grain-supported high concentration of rocks that is identified with high-concentration laminar flow, a massive to stratified, graded, volcaniclastic sandstone layer that formed from a high-concentration turbidite flow, and a fine-grained tuffaceous layer that formed a dilute low-density suspension (Mueller and White, 1993). The subaqueous eruption flows are products of high-energy flow and suspension deposition in the deep sea. The stratified middle sequence of the subaqueous eruption flow, however, is different from that of the Cacarajicara Formation, which preserves a homogeneous calcarenite sequence. The massive, homogeneous calcarenite sequence of the Cacarajicara Formation has a more mixed and suspended character.

One possibility is that an extraordinary gravity flow was triggered by an impact-related earthquake, keeping in mind that the estimated collapse locality of the Yucatan platform margin is more than 500 km from the impact site. The arrival of the seismic wave (at 5–7 km/s) would occur ~2 min after the arrival of impact ballistic flows (the initial blast is 30 km/s and main body of the vapor cloud is 10 km/s or less) (Melosh, 1989). This suggests that the seismic shock wave and ballistic flows arrived almost simultaneously.

As described here, the Cacarajicara Formation is a high-energy flow deposit that was formed by platform collapse resulting from an impact-related seismic wave and ballistic flow. This flow transported the shallow- (continental shelf) and deep-water (continental slop) sediments of the Yucatan platform margin to the deep Yucatan ocean basin. As evidenced by the thick homogeneous calcarenite, such a gigantic flow would have been very turbulent. An immense tsunami resulting from the impact might have mingled with this turbulence to form the very thick homogeneous calcarenite sequence.

CONCLUSIONS

The Cacarajicara Formation is a >700-m-thick calcareous sedimentary sequence that fines upward. This formation contains Maastrichtian fossils (foraminifera), and impact-related shocked quartz and spherules. Three members are identified in this formation. The Lower Breccia Member consists of a matrix-poor breccia sequence formed by high-energy lamina-flow. The Middle Calcarenite Member is characterized by well-sorted homogeneous calcarenite that was deposited from a high-concentration sediment suspension. The Upper Lime Mudstone Member is composed of massive, fine calcarenite-lime mudstone, which was deposited from a dilute suspension. The gradual transition between the three members of the Cacarajicara Formation can be explained by high-energy hyperconcentrated flow. Alternatively, the upper half of the formation can be interpreted as a deep-sea tsunami deposit (homogenite) similar to that in the upper half of the Peñalver Formation. The presence of shallow- and deep-water facies clasts, shocked quartz, spherules, and a south-southeast-trending paleocurrent suggest that this flow was formed by Chicxulub impact-related, seismically induced gravity flow combined with impact ballistic flow. Therefore, the Cacarajicara Formation is one of thickest sequences resulting from the gigantic Chicxulub impact-induced flow.

ACKNOWLEDGMENTS

We thank Agencia Medio Ambiente, Cuba, for their support of field survey in Cuba. This research was made possible by an agreement between the Department of Earth and Planetary Sciences, the University of Tokyo, and the Museo Nacional de Historia Natural (Agencia del Medio Ambiente) de Cuba and el Instituto de Geologia del Ministerio del Industria Basica. Y. Saito, K. Yokoyama, and colleagues of the National Science Museum of Japan are acknowledged for their suggestions and for using the energy-dispersive spectrometer and electron probe microanalyzer analysis. A. Taira provided useful discussions and provided helpful suggestions. We acknowledge the important support provided to field research in Cuba by Mitsui & Co., Ltd., as well as their manager in Havana, A. Nakata. We thank Christian Koeberl, Michael Rampino, and an unknown reader for constructive and insightful reviews. The survey was supported by research funds donated to University of Tokyo by NEC Corp., I. Ohkawa, M. Iizuka, and K. Ihara. This study was partly supported by a grant-in-aid (11691116) for scientific research from the Japan Society for the Promotion of Science.

REFERENCES CITED

Alvarez, W., Smit, J., Lowrie, W., Asaro, F., Margolis, S.V., Claeys, P., Kastner, M., and Hildebrand, A.R., 1992, Proximal impact deposits at the Cretaceous-Tertiary boundary in the Gulf of Mexico: A restudy of DSDP Leg 77 Site 536 and 540: Geology, v. 20, p. 697–700.

Bagnold, R.A., 1954, Experiments on a gravity-free dispersion of large solid spheres in a Newtonian Fluid under shear: Proceedings of the Royal Society, London, ser. A, v. 225, p. 49–63.

Bohor, B.F., and Seitz, R., 1990, Cuban K/T catastrophe: Nature, v. 344, p. 593.

Bohor, B.F., and Glass, B.P., 1995, Origin and diagenesis of K/T impact spherules: From Haiti to Wyoming and beyond: Meteoritics, v. 30, p. 182–198.

Bohor, B.F., 1996, Sediment gravity flow hypothesis for siliciclastic units at the K/T boundary, northeastern Mexico, *in* Ryder G., Fastovsky D., and Gartner, S., eds., The Cretaceous-Tertiary event and other catastrophes in Earth history: Geological Society of America Special Paper 307, p. 183–195.

Bourgeois, J., Hansen, T.A., Wiberg, L., and Kauffman, E.G., 1988, A tsunami deposit at the Cretaceous-Tertiary boundary in Texas: Science, v. 241, p. 567–570.

Bralower T.J., and Iturralde-Vinent M., 1997, Micropaleontological dating of the collision between the North American Plate and the Greater Antilles Arc in Western Cuba: Palaios, v. 12, p. 133–150.

Bralower, T.J., Paull, C.K., and Leckie, R.M., 1998, The Cretaceous-Tertiary boundary cocktail: Chicxulub impact triggers margin collapse and extensive sediment gravity flows: Geology, v. 26, p. 331–334.

Bronnimann P., and Rigassi, D., 1963, Contribution to the geology and paleontology of the city of La Habana, Cuba, and its surroundings: Eclogae Geologicae Helvetiae, v. 56, p. 193–480.

Dressler, B.O., and Sharpton, V.L., 1997, Breccia formation at a complex impact crater: Slate Island, Lake Superior, Ontario Canada: Tectonophysics, v. 275, p. 285–311.

French, B.M., 1998. Traces of catastrophe: A handbook of shock-metamorphic effects in terrestrial meteorite impact structures: Houston, Texas, Lunar and Planetary Institute, LPI Contribution No. 954, 120 p.

Gordon, M.B., Mann, P., Caceres, D., and Flores, R., 1997, Cenozoic tectonic history of the North America-Caribbean plate boundary zone in western Cuba: Journal of Geophysical Research, v. 102, p. 10055–10082.

Grajales-Nishimura, J.M., Cedillo-Pardo, E., Rosales-Dominguez, C., Moran-Zenteno, D.J., Alvarez W., Claeys P., Ruiz-Morales J., Garcia-Hernandez J., Padilla-Avila P., and Sanchez-Rios, A., 2000, Chicxulub impact: The origin of reservoir and seal facies in the southeastern Mexico oil fields: Geology, v. 28, p. 307–310.

Grieve, R.A., and Robertson, P.B., 1976, Variations in shock deformation at the Slate Islands impact structure, Lake Superior, Canada: Contribution to Mineralogy and Petrology, v. 58, p. 37–49.

Hildebrand, A.R., Penfield, G.T., Kring. D.A., Pilkington, M., Camargo, Z.A., Jacobsen, S.B., and Boynton, W.V., 1991, Chicxulub crater: A possible Cretaceous/Tertiary boundary impact crater on the Yucatan Peninsula, Mexico: Geology, v. 19, p. 867–871.

Iturralde-Vinent, M.A., 1992, A short note on the Cuban late Maastrichtian megaturbidite (and impact-derived deposit?): Earth and Planetary Science Letters, v. 109, p. 225–228.

Iturralde-Vinent, M.A., 1994a, Cuban geology: A new plate-tectonic synthesis: Journal of Petroleum Geology, v. 17, p. 39–70.

Iturralde-Vinent, M.A., 1994b, Interrelationship of the terranes in western and central Cuba: Comment: Tectonophysics, v. 234, p. 345–348.

Iturralde-Vinent, M.A., 1996, Introduction to Cuban geology and tectonics, *in* Iturralde-Vinent, M.A., ed., Ofiolitas y arcos volcanicos de Cuba (Cuban ophiolites and volcanic arcs): Miami, Florida, International Geological Correlation Programme 364, p. 3–23.

Izett, G.A., 1991, Tectites in Cretaceous-Tertiary Boundary rocks on Haiti and their bearing on the Alvarez impact extinction hypothesis: Journal of Geophysical Research, v. 96, p. 20879–20905.

Kerr, A.C., Iturralde-Vinent, M.A., Saunders, A.D., Babbs, T.L., and Tarney, J., 1999, A new plate tectonic model of the Caribbean: Implications from a geochemical reconnaissance of Cuban mesozoic volcanic rocks: Geological Society of America Bulletin, v. 111, p. 1581–1599.

Kneller, B.C., and Branney, M.J., 1995, Sustained high-density turbidity currents and the deposition of thick massive sands: Sedimentology, v. 42, p. 607–616.

Lowe, D.R., 1976, Grain flow and grain flow deposits: Journal of Sedimentary Petrology, v. 46, p. 188–199.

Melosh, H.J., 1989, Impact cratering: A geologic process: Oxford, UK, Oxford University Press, 245 p.

Montanari, A., and Koeberl, C., 2000. Impact stratigraphy: The Italian record, *in* Bhattacharji, S., Friedman, G.M., Neugebauer H.J., and Seilacher, A., eds., Lecture notes in earth sciences, Volume 93: Berlin, Springer-Verlag, 364 p.

Mueller, W., and White, J.D.L., 1993, Felsic fire-fountaining beneath Archean seas: Pyroclastic deposits of the 2730 Ma Hunter Mine Group, Quebec, Canada: Journal of Volcanology Geothermal Research, v. 54, p. 117–134.

Oberbeck, V.R., 1975, The role of ballistic erosion and sedimentation in lunar stratigraphy: Reviews of Geophysics Space Physics, v. 13, p. 337–362.

Ocampo, A.C., Pope, K.O., and Fischer, A.G., 1996, Ejecta blanket deposits of the Chicxulub crater from Albion Island, Belize, *in* Ryder G., Fastovsky D., and Gartner, S., eds., The Cretaceous-Tertiary event and other catastrophes in Earth history: Geological Society of America Special Paper 307, p. 75–88.

Palmer, R.H., 1945, Outline of the geology of Cuba: Journal of Geology, v. 53, p. 1–34.

Pierson, T.C., and Costa, J.E., 1987 A rheologic classification of subaerial sediment-water flows, *in* Costa, J.E., and Wieczorek, G.F., eds., Debris flows/avalanches: Process, recognition, and mitigation: Reviews in Engineering Geology, v. 7, p. 1–12.

Pindell, J.L., and Barrett, S.F., 1990, Geologic evolution of the Caribbean region: A plate-tectonic perspective, *in* Dengo, G., and Case, J.E., eds., The Caribbean Region: Boulder, Colorado, Geological Society of America, Geology of North America, v. H, p. 405–432.

Pindell, J.L., Cande, S.C., Pitman, W.C., III, Rowley, D.B., Dewey, J.F., Labercque, J., and Haxby, W., 1988, A plate-kinematic framework for models of Caribbean evolution: Tectonophysics, v. 155, p. 121–138.

Piotrowska, K., 1993. Interrelationship of the terranes in western and central Cuba: Tectonophysics, v. 220, p. 273–282.

Pope, K.O., Ocampo, A.C., Fischer, A.G., Alvarez, W., Fouke, B.W., Webster, C.L., Vega, F.J., Smit, J., Fritsche, A.E., and Claeys, P., 1999, Chicxulub impact ejecta from Albion Island, Belize: Earth and Planetary Science Letters, v. 170, p. 351–364.

Pszczolkowski, A., 1978, Geosynclinal sequences of the Cordillera de Guaniguanico in western Cuba: Their lithostratigraphy, facies development and paleogeography: Acta Geological Polish, v. 28, p. 1–96.

Pszczolkowski, A., 1986, Megacapsa del maestrichitiano de Cuba occidental y central: Bulletin of Polish Academic of Earth Science, v. 34, p. 81–87.

Pszczolkowski, A., 1994, Interrelationship of the terranes in western and central Cuba: Comment: Tectonophysics, v. 234, p. 339–344.

Pushcharovsky, Y., 1989, Geology of Cuba, Explanatory note to the 1:250 000 geological map of Cuba: Moscow, Geological Institute of the USSR Academy of Sciences, Nauka, 55 p.

Rosencrants, E., 1990, Structure and tectonics of the Yucatan basin, Caribbean Sea, as determined from seismic reflection studies: Tectonics, v. 9, p. 1037–1059.

Ross, M.I., and Scotese, C.R., 1988, A hierarchical tectonic model of the Gulf of Mexico and Caribbean region: Tectonophysics, v. 155, p. 139–168.

Shanmugam, G., 1996, High-density turbidity currents: Are they sandy debris flows?: Journal of Sedimentary Research, v. 66, p. 2–10.

Sharpton, V.L., Dalrymple, G.B., Marin, L.E., Ryder, G., Schuraytz, B.C., and Urrutia-Fucugauchi, J., 1992, New links between the Chicxulub impact structure and the Cretaceous/Tertiary boundary: Nature, v. 359, p. 819–821.

Sharpton, V.L., Martin, L.E., Carney, J.L., Ryder, G.S., Scjiraytz, C.S, Sikora, P., and Spudis, P.D., 1996, A model of the Chicxulub impact basin based on evaluation of geophysical data, well logs and drill core samples, *in* Ryder, G., Fastovsky, D., and Gartner, S., eds., The Cretaceous-Tertiary event and other catastrophes in Earth history: Geological Society of America Special Paper 307, p. 55–74.

Smit, J., 1999, The global stratigraphy of the Cretaceous-Tertiary boundary impact ejecta: Annual Reviews of Earth and Planetary Science, v. 27, p. 75–113.

Smit, J., Reop, R.B., Alvarez, W., Montanari, A., Claeys, P., Grajales-Nishimura, J.M., and Bermudez, J., 1996, Coarse-grained, clastic sandstone complex at the K/T boundary around the Gulf of Mexico: Deposition by tsunami waves induced by the Chicxulub impact?, *in* Ryder, G., Fastovsky, D., and Gartner, S., eds., The Cretaceous-Tertiary event and other catastrophes in Earth history: Geological Society of America Special Paper 307, p. 151–182.

Stöffler, D., and Langenhorst, F., 1994, Shock metamorphism of quartz in nature and experiment. 1. Basic observation and theory: Meteoritics, v. 29, p. 155–181.

Takayama, H., Tada, R., Matsui, T., Iturralde-Vinent, M.A., Oji, T., Tajika, E., Kiyokawa, S., Garcia, D., Okada, H., Hasegawa, T., and Toyoda, K., 2000, Origin of the Peñalver Formation in northwestern Cuba and its relation to K/T boundary impact event: Sedimentary Geology, v. 135, p. 295–320.

Todd, S.P., 1989, Stream-driven high-density gravelly traction carpets: Possible deposits in the Trabeg Conglomerate Formation, SW Ireland and theoretical considerations of their origin: Sedimentology, v. 36, p. 513–530.

White, J.D.L., 2000, Subaqueous eruption-fed density currents and their deposits: Precambrian Research, v. 101, p. 87–109.

Yokoyama, K., Matsubara, S., Saito, Y., Tiba, T., and Kato, A., 1993. Analyses of natural minerals by energy-dispersive spectrometer: Bulletin of National Science Museum of Tokyo, ser. C. v. 10, p. 115–126.

Manuscript Submitted October 27, 2000; Accepted by the Society March 22, 2001

Geological Society of America
Special Paper 356
2002

Multiple spherule layers in the late Maastrichtian of northeastern Mexico

Gerta Keller*
Department of Geosciences, Princeton University, Princeton, New Jersey, 08544 USA
Thierry Adatte
Geological Institute, University of Neuchâtel, Neuchâtel, Switzerland
Wolfgang Stinnesbeck
Geological Institute, University of Karlsruhe, Karlsruhe, Germany
Mark Affolter
Lionel Schilli
Institut de Géologie, University of Neuchâtel, Neuchâtel, Switzerland
Jose Guadalupe Lopez-Oliva
Facultad de Ciencias de la Tierra, Universidad Autónoma de Nuevo Leon, Linares, Nuevo Leon, Mexico

ABSTRACT

The discovery of as many as 4 spherule layers within 10 m of pelagic marls below the sandstone-siltstone complex and Cretaceous-Tertiary (K-T) boundary in the La Sierrita area of northeastern Mexico reveals a more complex K-T scenario than previously imagined. These spherule layers were deposited within pelagic marls of the Mendez Formation; the oldest layer is as much as 10 m below the K-T boundary. The marls are of latest Maastrichtian calcareous nannofossil *Micula prinsii* zone and planktic foraminiferal zone CF1 (*Plummerita hantkeninoides*) age; the latter spans the last 300 k.y. of the Maastrichtian. The oldest spherule layer was deposited near the base of zone CF1 and marks the original spherule-producing event. This is indicated by the presence of a few marl clasts and benthic foraminifera that are frequently surrounded by welded glass, and many welded spherules with schlieren features, indicating that deposition occurred while the glass was still hot and ductile. It is possible that some, or all, of the three stratigraphically younger spherule layers have been reworked from the original spherule deposit, as suggested by the common marl clasts, terrigenous input, reworked benthic and planktic foraminifera, and clusters of agglutinated spherules. These data indicate that at least one spherule-producing event occurred during the late Maastrichtian and provide strong evidence for multiple catastrophic events across the K-T transition.

*E-mail: gkeller@princeton.edu

Keller, G., Adatte, T., Stinnesbeck, W., Affolter, M., Schilli, L., and Lopez-Oliva, J.G., 2002, Multiple spherule layers in the late Maastrichtian of northeastern Mexico, *in* Koeberl, C., and MacLeod, K.G., eds., Catastrophic Events and Mass Extinctions: Impacts and Beyond: Boulder, Colorado, Geological Society of America Special Paper 356, p. 145–161.

INTRODUCTION

In 1990, glass spherule deposits were discovered in sediments at the Cretaceous-Tertiary (K-T) boundary at Beloc, Haiti, and subsequently similar glass spherule deposits were discovered in K-T deposits at Mimbral and many other localities in northeastern Mexico (Izett et al., 1990; Smit et al., 1992, 1996; Stinnesbeck et al., 1993; Keller et al., 1997). The chemical similarity of the glass spherules with melt rock in subsurface samples at Chicxulub led many workers to interpret an impact origin for the glass spherules and Chicxulub as the impact crater (Izett et al., 1990; Sigurdsson et al., 1991; Koeberl, 1993; Koeber et al., 1994).

However, on the basis of the Haiti sections, some workers suggested that volcanism (Lyons and Officer, 1992), two impacts (Leroux et al., 1995), or one impact and one volcanic event (Jéhanno et al., 1992) produced the stratigraphically separated glass spherules, shocked quartz, and iridium anomaly. Recent analyses of new and more complete K-T boundary sections at Beloc, Haiti, revealed that the glass spherules were deposited in the early Danian *Parvularugoglobigerina eugubina* zone ~100 k.y. after the K-T boundary (Stinnesbeck et al., 2000; Keller et al., 2001). A spherule deposit in the *P. eugubina* zone was also observed above a thick breccia in the Caribe section of Guatemala (Fourcade et al., 1998, 1999; Keller and Stinnesbeck, 2000). In addition, a new K-T section at Coxquihui, Mexico, also revealed deposition of a spherule layer within the early Danian *P. eugubina* zone (Stinnesbeck et al., 2002). In all three localities, and in addition to the well-known K-T Ir anomaly, a small but significant iridium anomaly was also observed immediately above the early Danian spherule layer, and rare shocked minerals are also present. Although it is possible that the early Danian spherule layer and Ir anomalies in Haiti, Guatemala, and Mexico are reworked from older original deposits, an early Danian impact event cannot be ruled out.

An impact event at the K-T boundary is generally accepted and well documented based on a global iridium anomaly, shocked quartz, and Ni-rich spinels (Alvarez et al., 1992; Robin et al., 1992; Rocchia et al., 1996), the extinction of all tropical and subtropical planktic foraminifera and many calcareous nannofossil species (see summaries in D'Hondt et al., 1996; Keller et al., 1995; Pospichal, 1996). It is also widely accepted that the Chicxulub crater represents this K-T boundary impact event, based on the stratigraphic position of the glass spherules near the K-T boundary in sections in Haiti and Mexico (Alvarez et al., 1992; Smit et al., 1992, 1996), their chemical similarity with melt rock in subsurface cores at Chicxulub (Izett et al., 1990; Sigurdsson et al., 1991; Blum and Chamberlain, 1992; Blum et al., 1993; Koeberl, 1993; Koeberl et al., 1994), and an $^{40}Ar/^{39}Ar$ age of ca. 65 Ma of the spherules and melt rock (Izett et al., 1990; Swisher et al., 1992; Dalrymple et al., 1993). However, the age and origin of the glass spherules are still disputed.

In a continuing effort to determine the age and origin of the glass spherules in northeastern Mexico, a major study was undertaken by us and a group of students to map and examine the sandstone-siltstone complex and underlying late Maastrichtian sediments over 50 km in the La Sierrita area of northeastern Mexico (Fig. 1, Lindenmaier, 1999; Schulte, 1999; Schilli, 2000, Affolter, 2000). The results are surprising: of more than 24 sections examined below the sandstone-siltstone complex, nearly all contain as many as 4 distinct spherule layers, often separated by several meters of pelagic marls of the Mendez Formation (Fig. 2). Here we report the stratigraphic results of one of these sections at Loma Cerca, and address specifically: (1) the age of the spherule layers; (2) the likelihood that the oldest layer is the original spherule-producing event and that the other layers are reworked; (3) environmental changes associated with the spherule event(s); and (4) the implications of this discovery for impact scenarios.

Lithology of spherule-bearing deposits

Cretaceous-Tertiary (K-T) boundary sequences in the La Sierrita area form a northwest-southeast-trending series of low-lying hills that consist of Maastrichtian marls of the Mendez Formation. The top of these hills is formed by a resistant sandstone-siltstone complex that varies in thickness from 0.20 cm to 4 m in the La Sierrita sections (Figs. 2 and 3) and 8 m in the Peñon section (Stinnesbeck et al., 1996). A small iridium anomaly is generally present beginning at the top of this depositional complex and reaches peak abundance in the overlying early Danian marls (Smit et al., 1992; Keller et al., 1997). Some workers suggested that this sandstone-siltstone complex (including the spherule layer at its base) and Ir anomaly represent a megatsunami deposit generated by the Chicxulub impact, and was deposited over a few hours or days (Smit et al., 1992, 1996; Bralower et al., 1998). However, this interpretation is no longer tenable based on the presence of several horizons of trace fossils within this sedimentary complex.

Evidence of trace fossils in the sandstone-siltstone complex is present in numerous outcrops. In the past workers typically subdivided this depositional complex into three distinct sedimentary units that represent distinct depositional events. At the base is unit 1, the spherule layer that is generally not bioturbated. Above the spherule layer is unit 2, a sandstone that is mostly not bioturbated (Fig. 3), but generally contains a few J-shaped, 5–10-cm-long, spherule-filled burrows near the base (Fig. 4, A and B) that are often truncated by overlying sand layers. Ekdale and Stinnesbeck (1998, p. 593) interpreted these burrows as having been "excavated following deposition of the first sand layers and then filled with spherules, scoured, and overlain by more of unit II sand." The uppermost unit 3 consists of alternating sandstone, siltstone, and shale that contain abundant trace fossils (Fig. 3). These trace fossils were abundantly illustrated and discussed in Ekdale and Stinnesbeck (1998) and include *Chondrites*, *Ophiomorpha*, *Planolites*, and *Zoophycos*. Additional illustrations are shown in Figure 4 (C–E). The burrows of unit 3 indicate that sediment deposition occurred epi-

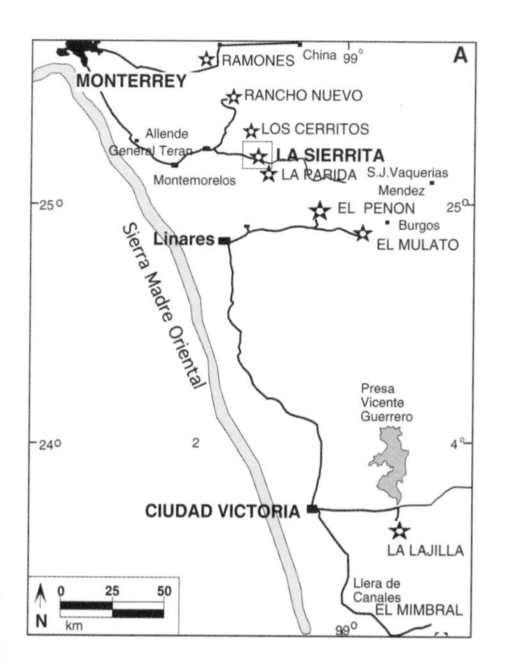

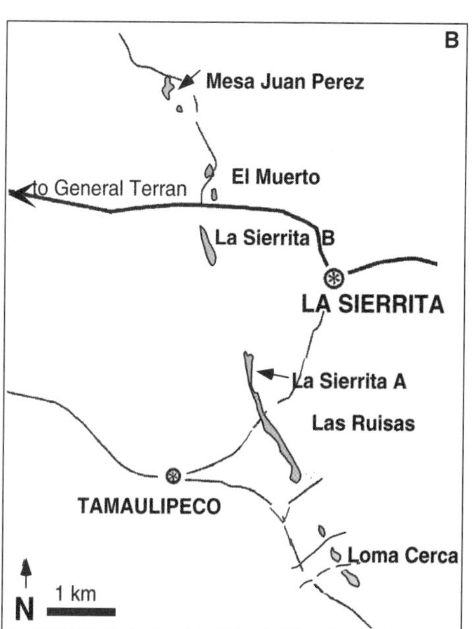

Figure 1. A: Location map of Cretaceous-Tertiary boundary sections with spherule-rich deposits below sandstone-siltstone complex in northeastern Mexico. Stars mark localities of sections. B: Location map of La Sierrita area with low-lying northeast-southwest–trending hills of Mendez Formation topped by sandstone-siltstone complex. There are 2–4 spherule layers within upper 10 m of pelagic marls of Mendez Formation. Series of lithologs of these sections are illustrated in Figure 2.

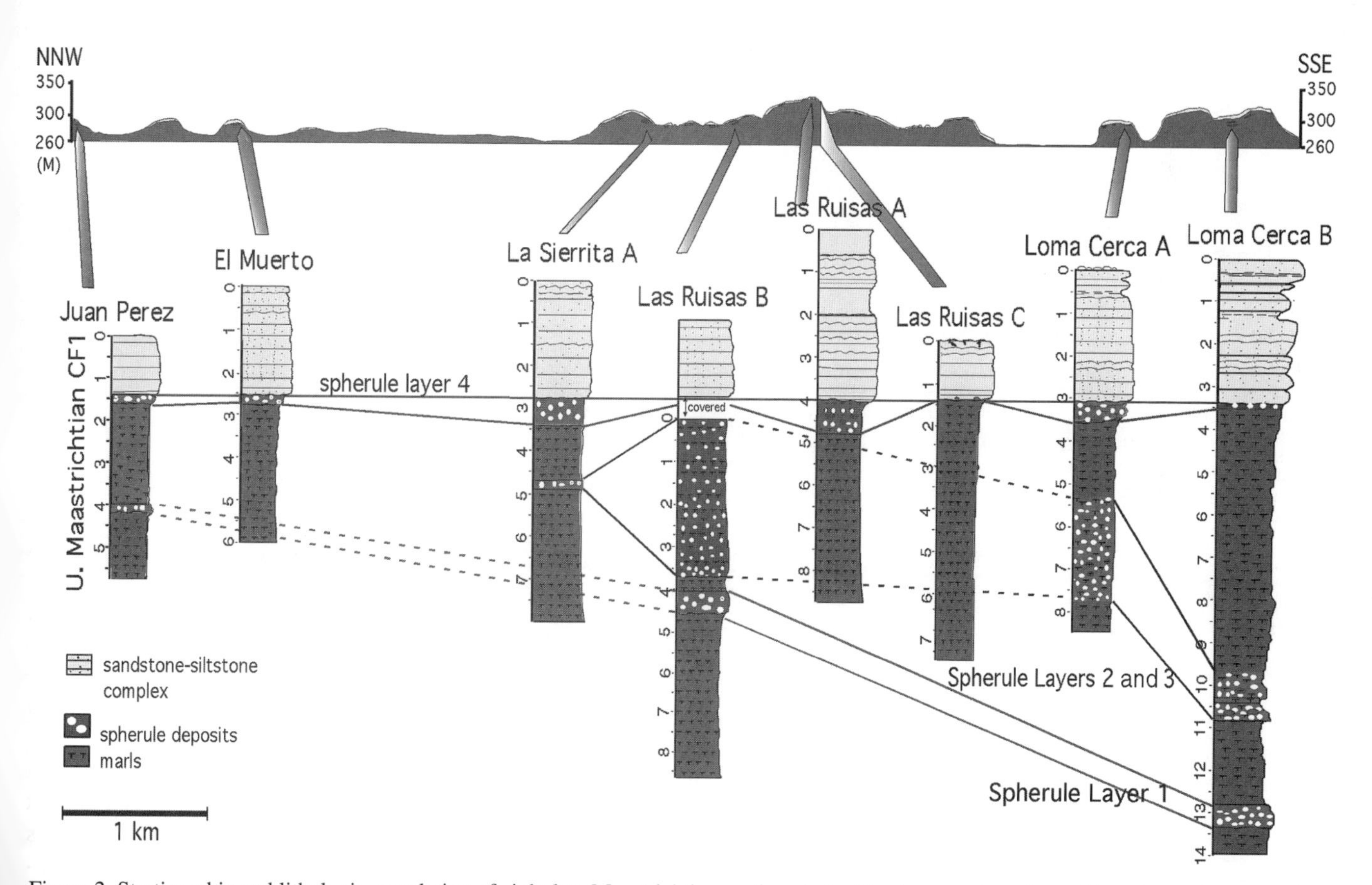

Figure 2. Stratigraphic and lithologic correlation of eight late Maastrichtian sections in La Sierrita area, from Loma Cerca in southwest to Mesa Juan Perez in northeast (see Fig. 1B). Topographic relief illustrates positions of these sections from sandstone-siltstone complex at top to Mendez Formation marl that makes up slopes of hills. Each section contains spherule layer directly below sandstone-siltstone complex and as many as three additional spherule layers in 10 m of pelagic marls below it. Age of oldest spherule layer at Loma Cerca B predates Cretaceous-Tertiary boundary by ~270 k.y. Numbers on lithological columns indicate meters. Solid lines mark correlation of spherule layers. Dashed correlation lines indicate where particular spherule layer is missing in some sections.

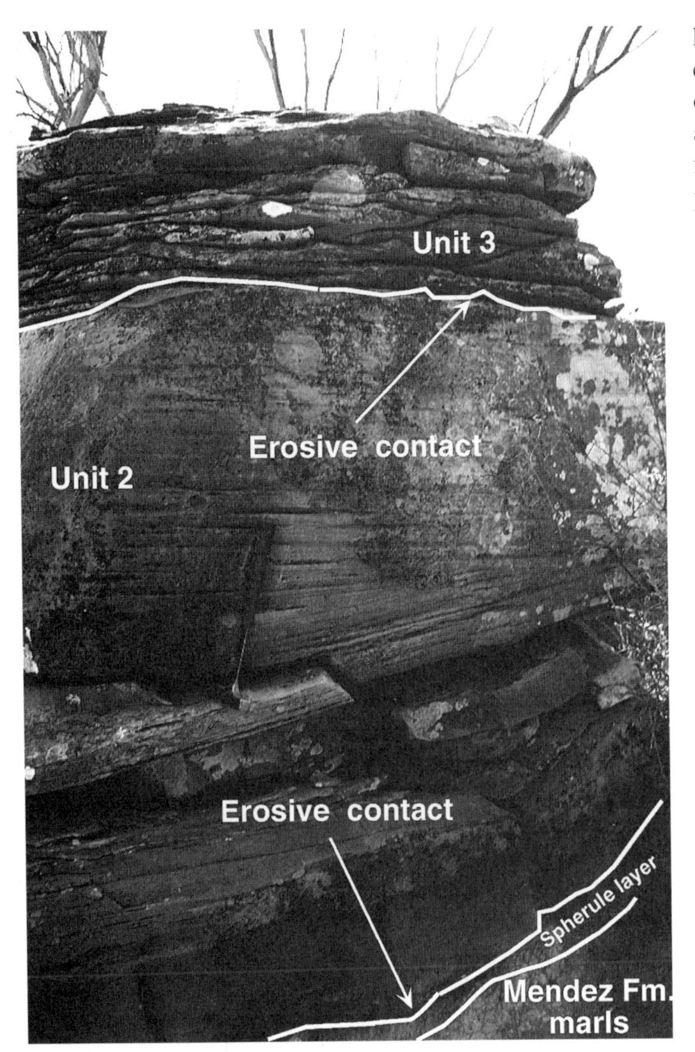

Figure 3. Sandstone-siltstone complex at La Sierrita A showing marl of Mendez Formation and spherule layer (unit 1) at base, sandstone (unit 2) in middle, and overlying alternating sand, silt, shale layers (unit 3). Trace fossils are rare in unit 2 and very common in unit 3. Hammer, 30 cm, for scale. Fm.—Formation.

sodically and that burrowing occurred during periods of deposition, not after deposition had ceased.

Trace-fossil evidence indicates that the entire sandstone-siltstone complex of units 2 and 3 was deposited over a long period of time during which periods of normal pelagic sedimentation and colonization of the ocean floor alternated with periods of erosion followed by rapid deposition. The entire sandstone-siltstone complex (units 1–3) could therefore not have been deposited by a tsunami over a period of hours to days, as suggested by Smit et al. (1992, 1996). However, this evidence does not question the reality of a K-T boundary impact, or even the possibility that the rippled upper beds (20–25 cm) of the sandstone-siltstone complex beneath the iridium anomaly could have been deposited by a tsunami event. These data, however, contradict the interpretation that the spherule layer (unit 1) is directly related to the planktic foraminiferal extinctions at the K-T boundary and the Ir anomaly at the top of the burrowed sandstone-siltstone complex. The K-T boundary and the spherule event are temporally separated by the trace-fossil horizons of units 2 and 3 in the sandstone-siltstone complex in northeastern Mexico.

Throughout the La Sierrita area, a spherule layer (unit 1) that varies from a few centimeters to 1 m thick typically underlies the sandstone-siltstone complex (Fig. 2). As many as 3 additional spherule layers are present in the top 10 m of the underlying marls in more than 24 sections examined (Figs. 2 and 5A). These spherule layers may be separated by 2–6 m of pelagic marls, indicating that deposition occurred in a normal marine environment. The contact between the stratigraphically lowermost spherule layer and overlying marl is often sharp, suggesting little erosion (Fig. 5B). The Loma Cerca B section is one of these sequences located on the western flank of the Loma Cerca hill, ~40 km east of Montemorelos. The top of the hill is capped by a 3.5-m-thick sandstone-siltstone complex that has a 5–10-cm-thick spherule layer at its base. More than 15 m of marls of the Mendez Formation are exposed below it; there are two closely spaced 50-cm-thick spherule layers between 6.8 and 7.5 m. Both of these layers show decreasing spherule abundance upward and mark two depositional events. Another 50-cm-thick spherule layer, without upward decreasing abundance, is present between 9.5 and 10 m below the base of the sandstone-siltstone complex. Thus four spherule layers are present in these late Maastrichtian Mendez marls. No iridium anomaly is associated with any of the four late Maastrichtian spherule layers, and shocked quartz grains are rare.

Methods

In the field, samples were collected at 10–20 cm intervals for biostratigraphic, mineralogical, and stable isotope analyses. For biostratigraphic analysis, samples were processed for foraminifera using standard laboratory techniques (Keller et al., 1995). The washed residue (>63 μm size fraction) was examined for planktic foraminifera and a species census was taken to evaluate the age of the sediments. Changes in species populations indicative of climatic variations, reworking and transport of species, and hiatuses were evaluated based on quantitative analysis of ~250–300 individuals per sample. Microfossils are abundant, although generally recrystallized, in the Mendez marls and exhibit no evidence of preservational bias due to carbonate dissolution or breakage. For hard marls, thin sections were also made and examined to counter possible bias in washed residues. No major bias was observed. Thin sections were also made of the spherule layers and examined for glass and to identify the nature of the spherules, reworked clasts, terrigenous influx, and transported foraminifera. The biostratigraphic zonal scheme of Pardo et al. (1996) and Li and Keller (1998b) was used where the last 300 k.y. of the Maastrichtian or the upper interval of chron 29r below the K-T boundary, are

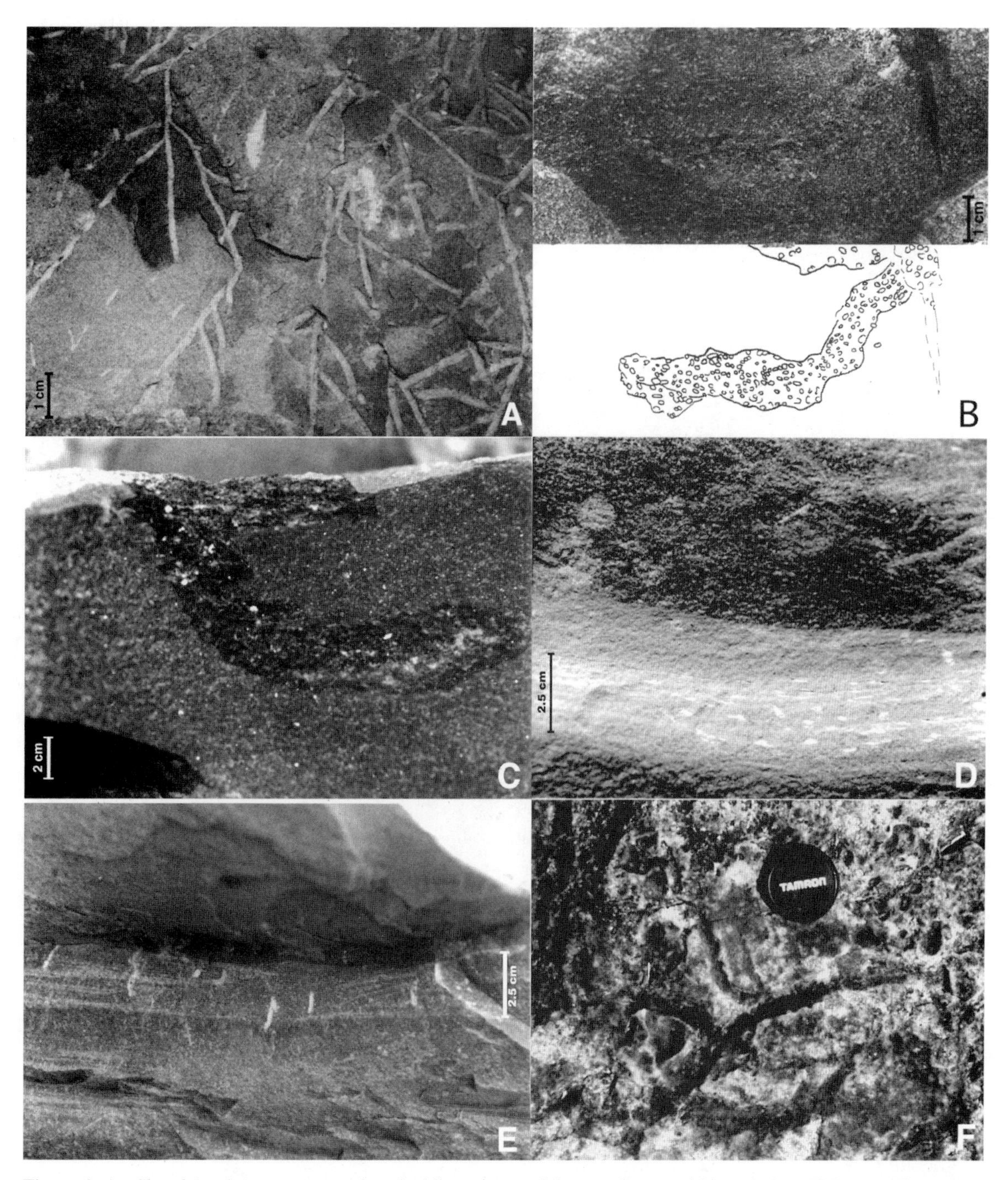

Figure 4. A: *Chondrites* burrows exposed on bedding planes of fine sandstone within unit 3 at El Peñon. These deep deposit-feeding burrows consist of branching and dominantly horizontal tunnels. Scale bar is 1 cm (from Ekdale and Stinnesbeck, 1998). B, C: Spherule-filled burrows in sandstone near base of unit 2 at El Peñon and Rancho Canales, and outline of burrow in photograph of B. Note that burrows are filled with dark colored, unbroken, calcite spherules and truncated at top by scour and overlying sand. Burrows of this nature are 0.5–1.5 cm in diameter and 6–10 cm long. D: *Chondrites* burrows exposed on bedding planes of fine silt in unit 3. Such burrows are common in fine-grained silt or shale layers of unit 3 in many outcrops. E: Vertical shafts of Chondrites burrows; some are truncated by scour and overlying sand. Such burrows are common in fine-grained silt and shale layers of unit 3 in many outcrops. F: *Ophiomorpha* burrows exposed on bedding planes within sandstone near top of unit 3 at La Sierrita. Such burrows are abundant within unit 3 and also at top of unit 3 in many outcrops Lens cap, 50 mm, for scale.

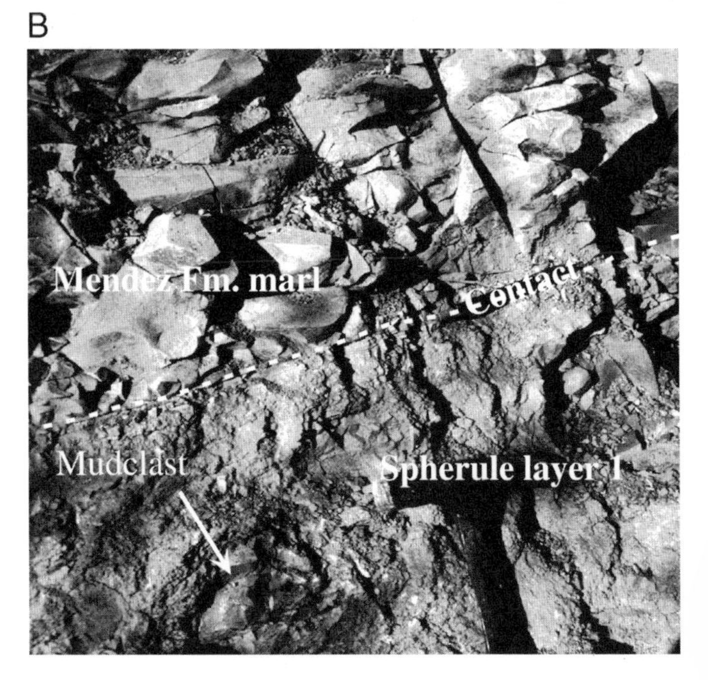

Figure 5. A: Spherule layer 1 at Loma Cerca B is within marls of Mendez Formation and 10 m below sandstone-siltstone complex. This spherule layer is nearly 1 m thick and was deposited within *Micula prinsii* zone and near base of *Plummerita hantkeninoides* zone (CF1); latter spans last 300 k.y. of Maastrichtian. Sediment accumulation rates suggest that this spherule layer was deposited ~270 k.y. before Cretaceous-Tertiary boundary. B: Spherule layer 1 showing sharp upper contact with overlying marl. This suggests gradational transition. Mud clasts are rare in spherule layer 1 and there is generally no upward fining of spherules. Fm.—Formation.

identified by the total range of *Plummerita hantkeninoides*, and the calcareous nannofossil zone *Micula prinsii* spans the last 450 k.y. of the Maastrichtian.

Stable isotope analyses were done on bulk-rock samples at the stable isotope laboratory of the University of Bern, Switzerland, using a VG Prism II ratio mass spectrometer equipped with a common acid bath. The results are reported relative to the Vienna Peedee belemnite standard reference material with a standard error of 0.1‰ for oxygen and 0.05‰ for carbon. Clay and whole-rock mineralogical analyses were done at the Geological Institute of the University of Neuchatel, Switzerland, based on X-ray diffraction (XRD) (SCINTAG XRD 2000 diffractometer) using the method of Kuebler (1987) and Adatte et al. (1996).

RESULTS

Age of spherule layers

Biostratigraphic investigations based on planktic foraminifera indicate that the top 10 m of marls in the Mendez Formation at Loma Cerca B were deposited within the latest Maastrichtian *Plummerita hantkeninoides* zone CF1, which spans the last 300 k.y. of the Maastrichtian (Pardo et al., 1996). This age is supported by the presence of the *Micula prinsii* zone calcareous nannofossil assemblage that marks the last 450 k.y. of the Maastrichtian. Faunal assemblages within the CF1 interval are diverse and typical of the latest Maastrichtian low-latitude Tethys, as characterized by the presence of *P. hantkeninoides* (fragile apical spines frequently broken), the dominance of biserial species, and generally low abundance of globotruncanids (Fig. 6). Species populations within the marls show no evidence of transport and reworking, repetition due to slumps, or missing intervals due to major erosion at the base of the spherule layers within the marls. However, there are occasional outcrops in the La Sierrita area where there are minor local slumps, a few meters in lateral extent. At Loma Cerca B, a short interval may be missing at the unconformity at the base of the sandstone-siltstone complex, as indicated by the undulating erosional surface the spherules filling the troughs, and the presence of commor

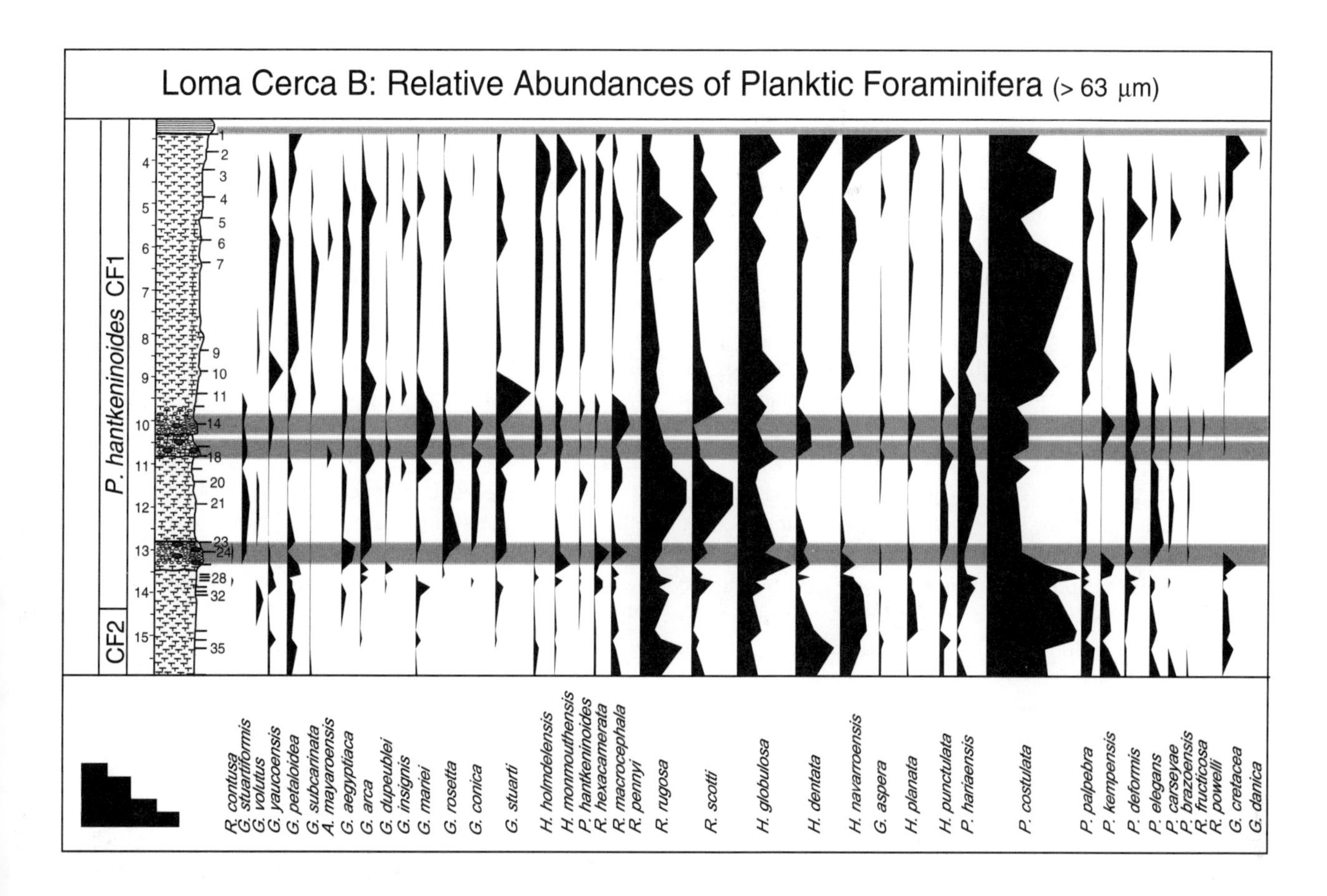

Figure 6. Relative abundances of late Maastrichtian planktic foraminifera in Mendez marls below sandstone-siltstone complex. Note that species diversity and relative abundances are characteristic of Tethyan assemblages and that there is no indication of chaotic deposition or major reworking in Mendez marls. However, anomalous abundance peaks in globotruncanids at bases of spherule layers 2 and 3 indicate influx of reworked specimens. Deposition of Mendez marls from 1 m below spherule layer 1 to top occurred within latest Maastrichtian *Plummerita hantkeninoides* (CF1) zone that was deposited during last 300 k.y. of Maastrichtian.

clasts. Erosion of the uppermost Maastrichtian is also suggested by the absence of the characteristic end-Maastrichtian increase in biserial species. However, the missing interval appears to be minor, as indicated by the lithological variations of more than 12 sections in the region (Fig. 2). In addition, the presence of common *Guembelitria cretacea* in the upper 4 m of the Mendez Formation marl (Fig. 6) is characteristic of the latest Maastrichtian in the Tethys (Abramovich et al., 1998).

The CF1-CF2 boundary is 11 m below the sandstone-siltstone complex and 1 m below the first spherule layer. On the basis of paleomagnetic and biostratigraphic correlations at Agost, Spain, the base of zone CF1 is about 300 k.y. below the K-T boundary (Pardo et al., 1996). Assuming that most of the CF1 zone interval is present at Loma Cerca B, the average sediment accumulation rate of the marl was 3 cm/k.y. This compares favorably with 2 cm/k.y. at El Kef and 4 cm/k.y. for marls at Elles, Tunisia, for the same interval (Li et al., 1999; Abramovich and Keller, 2002) and suggests relatively continuous sedimentation at Loma Cerca B. On the basis of these sediment accumulation rates, the oldest spherule layer was deposited about 270 k.y. before the K-T boundary. The second and third layers were deposited ~215 k.y. and ~210 k.y. before the K-T boundary, respectively, assuming that there was little erosion prior to deposition. Although these are approximate ages with uncertainties due to possibly discontinuous sedimentation, they provide the best age estimates consistent with biostratigraphy at this time.

Estimating the depositional age of spherule layer 4 has additional uncertainty due to the unknown length of time it took to deposit the sandstone-siltstone complex. This deposit must predate the K-T boundary by an unknown, although short, interval, as indicated by the multiple horizons of trace fossils in units 2 and 3. Each renewed colonization of the ocean floor could have taken place within a few years. If we assume that no significant erosion and time lapses occurred between repeated influxes of siliciclastic sediments alternating with pelagic sedimentation, then spherule layer 4 could have been deposited only a few thousand years before the K-T boundary.

Evidence for original deposition of late Maastrichtian marls

Until the discovery of multiple spherule layers within the late Maastrichtian Mendez marls in sections throughout northeastern Mexico, there was no question that these marls represented original pelagic sedimentation (Smit et al., 1992, 1996; Stinnesbeck et al., 1996; Keller et al., 1994, 1997). However, the age and origin of these multiple spherule deposits within the marls and the far-reaching implications of a late Maastrichtian age for the current impact–mass-extinction scenario require thorough evaluation of all possible depositional scenarios. In particular, it has been suggested that the multiple spherule layers may be due to slumping and repeated sections as a result of seismic activity induced by the Chicxulub impact at the K-T boundary (Soria et al., 2001). Although we find no evidence to support this scenario in our survey of numerous outcrops over more than 50 km, we detail our observations in the following.

Faunal evidence: planktic foraminifera. Biostratigraphic and quantitative planktic foraminiferal analyses reveal normal pelagic deposition within a subtropical to tropical Tethyan ocean with late Maastrichtian faunal assemblages similar to those documented throughout the Tethys (Li and Keller, 1998b; Luciani, 1997; Abramovich et al., 1998). There is no faunal evidence of significant reworking within the marls of Loma Cerca (Fig. 3), or any of the other numerous sections examined. There is no faunal evidence of large-scale slumps, repeated sections due to slumping, or chaotic deposition as a result of storm deposits (e.g., impact tsunami, Stinnesbeck et al., 2001). Whether the faunal and biostratigraphic data of Loma Cerca B is viewed alone, or within the context of more than 24 northeastern Mexico sections, or even within the context of the larger Tethyan ocean from the Negev of Israel to Mexico, these data can not be interpreted as having been compromised by slumps, repetition of sections, or major reworking within the marls. This is indicated by the overall similarity of the fossil assemblages, the similarity in the relative abundances of most species, and the relative abundance changes within species populations (e.g., globotruncanids, guembelirids, roguglobigerinids, and heterohelicids) at Loma Cerca B. All of these proxies reflect similar environmental changes in other Tethys sections with known undisturbed and continuous sedimentation records of the last 500 k.y. of the late Maastrichtian (Pardo et al., 1996; Abramovich et al., 1998; Abramovich and Keller, 2002; Luciani, 1997).

Faunal evidence: benthic foraminifera. The relative abundance of benthic species is an index of paleodepth. In neritic environments, benthic species are abundant and diverse, and dominate foraminiferal assemblages. At bathyal depths benthic species are less diverse and decrease to <5% with planktic foraminifera dominating (95%). Peak abundances at bathyal depths are therefore generally due to downslope transport, reworking, or dissolution. In addition, many benthic species are specific in their depth habitats, neritic species generally being restricted to shelf areas and bathyal species being restricted to slope areas at depths deeper than 300 m (Morkhoven et al., 1986; Keller, 1992). At Loma Cerca B, benthic species average between 2% and 5% in the Mendez Formation marls, which indicates normal pelagic deposition in a relatively deep water environment (Fig. 7). They thus provide no evidence for significant reworking during marl deposition. However, peak increases in benthic abundances to 10% and 23% at the bases of spherule layers 2 and 3, respectively, indicate that significant reworking occurred in these intervals. Apparently no significant reworking was associated with deposition of spherule layer 1. Earlier studies have shown that spherule layer 4 at the base of the sandstone-siltstone complex also contains common transported shallower shelf species, indicating reworking (Keller et al., 1994, 1997).

Most of the benthic species present at Loma Cerca B and other Sierrita sections examined are known to live in upper bathyal and outer neritic environments (Table 1), and there are no shallower neritic species in any of the samples examined. This indicates that deposition occurred in an upper bathyal environment at depths of ~300–500 m. All of the species present in the Mendez Formation are also commonly present in outer neritic to upper bathyal depths in late Maastrichtian to early Danian sections at Caravaca, Spain, the Negev, Israel, and El Kef in Tunisia (Keller, 1992). However, in the Mexican sections benthic species are very rare (2%–5% relative to planktic species), which indicates that the Mexican sections were deposited in a deeper upper bathyal environment than those in Spain, the Negev, or Tunisia.

Lithological evidence. More than 48 K-T boundary transitions have been examined in Mexico spanning a distance of more than 300 km. More than 24 of these sections are in the La Sierrita area (Fig. 1), where the late Maastrichtian was examined in detail and where individual spherule layers are exposed for several tens of meters along the hillslopes without evidence of major slumps and faults. However, in some isolated sections minor local slumps spanning a few meters were observed. In general, the various spherule layers can be observed in many outcrops spanning an area of more than 50 km (Fig. 2). That not all four spherule layers are present in all sections indicates a variable degree of erosion and reworking, and it may reflect the topographic relief of the seafloor. This is also indicated by the variable thickness or occasional absence of the sandstone-siltstone complex (Lindenmaier, 1999; Schulte, 1999; Affolter, 2000; Schilli, 2000). The regional continuity of pelagic marls and presence of spherule layers within the last 300 k.y. of the Maastrichtian zone CF1 rules out the possibility that deposition occurred as a result of slump deposits, or any other type of significantly disturbed deposition. However, it does not rule out the possibility that spherule layers 2, 3, and 4 derived from a single spherule-producing event in the late Maastrichtian (spherule layer 1) via periodic reworking and redeposition.

Mineralogic evidence. X-ray diffraction (XRD) and granulometric analyses of the insoluble size fraction provide further

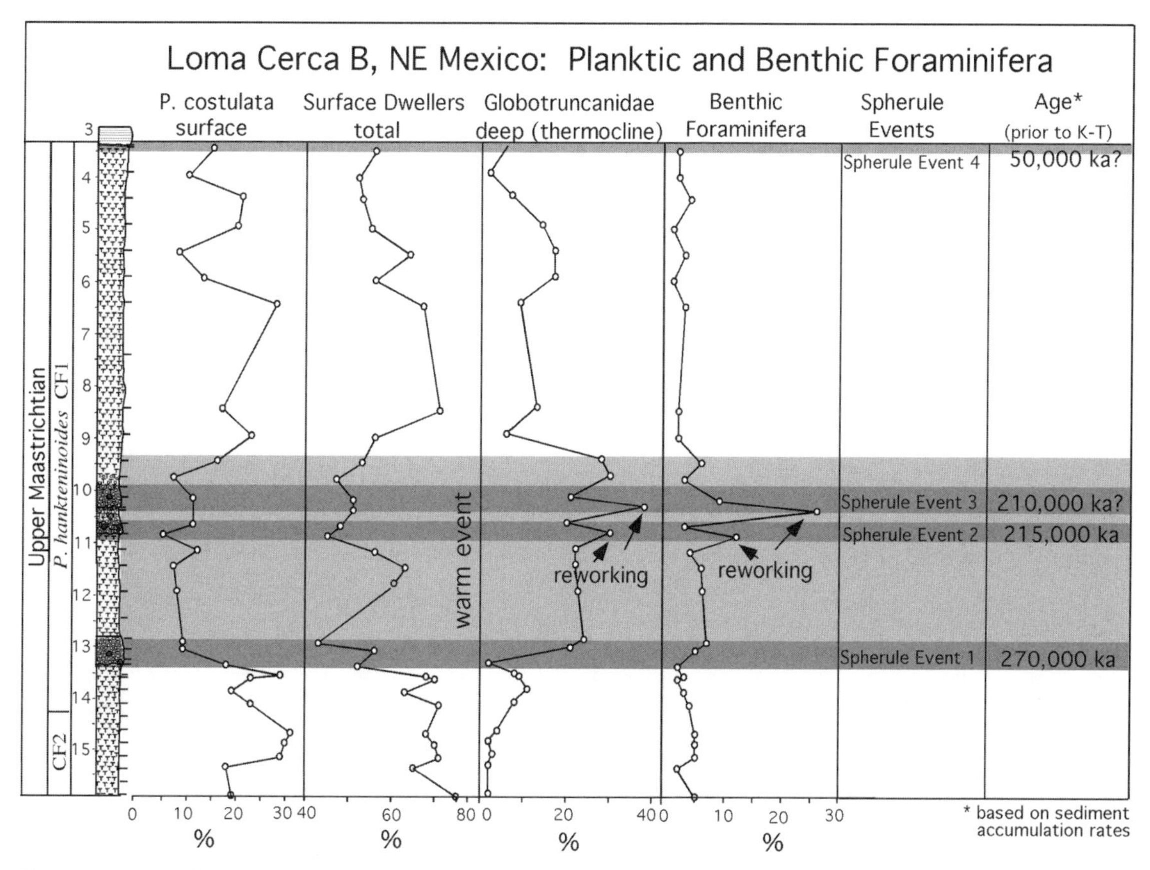

Figure 7. Foraminiferal proxies as indicators of climate change and reworking. Note that reworked sediments are restricted to short anomalous peaks in globotruncanids and benthic foraminifera at bases of spherule layers 2 and 3. Planktic foraminifera are depth ranked based on carbon and oxygen isotopes (see Li and Keller, 1998b). At Loma Cerca B, as well as Tethys ocean in general, late Maastrichtian surface mixed layer is characterized by common to abundant surface dwellers (e.g., rugoglobigerinids and heterohelicids), whereas subsurface (thermocline) dwellers are dominated by globotruncanids (Abramovich et al., 1998; Li and Keller, 1998a, 1998b; Luciani, 1997). At Loma Cerca B, increased abundance of globotruncanids indicates climate warming beginning with deposition of spherule layer 1 and ending ~1.5 m above spherule layer 3. Age of this warming coincides with major warm pulse in stable isotope record of middle latitude Deep Sea Drilling Project Site 525 (Li and Keller, 1998a), and also with major Deccan volcanism between 65.4 and 65.2 Ma (Hoffman et al., 2000).

evidence for normal pelagic sedimentation. Pelagic marls tend to be uniformly fine grained (<16 μm); calcite and phyllosilicates are the most abundant components. This reflects normal pelagic sedimentation under weak hydrodynamic conditions. At Loma Cerca B, insoluble grain-size fractions average ~50% in the 0–4 μm range and 50% in the 4–16 μm range (Fig. 8). Just below the sandstone-siltstone complex and spherule layer 4 a fine-grained interval marks a bentonite layer that has not yet been dated. XRD analyses of whole-rock marls indicate the same average compositions (45%–50% calcite, 12%–18% quartz, 5%–10% plagioclase, 25%–30% phyllosilicates) for the entire marl sequence at Loma Cerca B, as well as Mesa Juan Perez, 10 km to the northeast (Stinnesbeck et al., 2001). Similar values were previously reported for the Mendez marls at other northeastern Mexico sections where no additional spherule layers were observed (Adatte et al., 1996). Only the insoluble residue of the spherule-rich layers has a larger (16–150 μm) grain size present due to the spherules. In addition, there is a small (~10%) peak in the larger (16–150 μm) grain size apparent at the CF1-CF2 boundary that marks the presence of a bioturbated interval with trace fossils at Loma Cerca B and may represent a short hiatus. These uniform mineralogical values, either within the marls of the Loma Cerca B section or within the marls across the region (Stinnesbeck et al., 2001), do not support a scenario of slumps, reworking, and sliding movements induced by seismic waves as a result of a K-T boundary impact, as suggested by Soria et al. (2001).

Stable isotopes. Marls of the Mendez Formation are diagenetically altered and therefore measured stable isotopes do not represent original $\delta^{18}O$ values. In contrast, $\delta^{13}C$ values are little affected by recrystallization processes because pore waters have low concentrations of carbon (Magaritz, 1975; Brand and

TABLE 1. BENTHIC FORAMINIFERA FROM MARLS AND SPHERULE LAYERS AT LOMA CERCA B, MEXICO

Species	Depositional environment
Cibicicoides dayi (White)	Primarily bathyal
Cibicidoides succedens	Outer neritic and upper bathyal
Coryphostoma midwayensis (Cushman)	Outer neritic and upper bathyal
Bolivinoides draco (Marsson)	Outer neritic and upper bathyal
Coryphostoma incrassata (Reuss)	Outer neritic and upper bathyal
Bolivina incrassata gigantea (Wicker)	Outer neritic and upper bathyal
Bulimina faragraensis Le Roy	Outer neritic and upper bathyal
Prebulimina cushmani (Sandidgei)	Outer neritic and upper bathyal
Praebulimina reussi	Outer neritic and upper bathyal
Dorothia oxycona (Reuss)	Outer neritic and upper bathyal
Gaudryina pyramidata Cushman	Outer neritic and upper bathyal
Gyroidinoides planulata	Outer neritic and upper bathyal
Gyroidinioides subangulataus (Plummer)	Outer neritic and upper bathyal
Gyroidinoides globulosus	Outer neritic and upper bathyal
Cassidulina globosa	Outer neritic and upper bathyal
Uvigerina maqfiensis Le Roy	Outer neritic and upper bathyal
Trifarina esnaensis Le Roy	Outer neritic and upper bathyal
Spiroplectamina dentata	Outer neritic and upper bathyal
Stensioina beccariiformis	Outer neritic and upper bathyal
Lenticulina muensteri	Neritic to bathyal
Anomalinoides welleri	Outer neritic and upper bathyal
Anomalinoides acuta	Outer neritic and upper bathyal

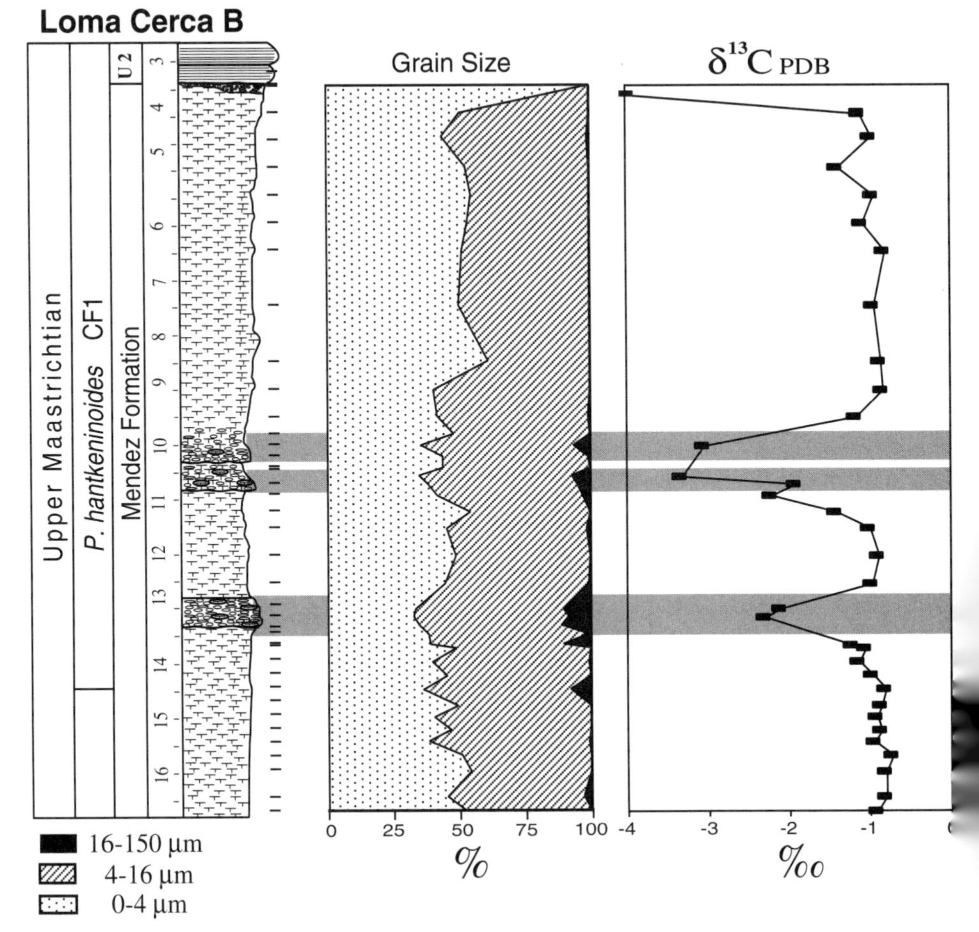

Figure 8. Grain-size distribution and δ^{13}C record at Loma Cerca B. Note that marly sediments are uniform; no major changes are apparent in lithology or stable isotopes. Only variations in these patterns are seen in spherule layers that differ by their larger grain sizes and lighter δ^{13}C values as result of non-biogenic calcite infilling of spherule voids. PDB is Peedee belemnite.

Veizer, 1980; Schrag et al., 1995). In this study, $\delta^{13}C$ values of bulk carbonate were analyzed in order to test whether sediments show major isotopic trends that would indicate the presence of K-T sediments, or chaotic values due to mixing of sediments. No such trends are observed. Bulk-rock $\delta^{13}C$ data show relatively uniform values of $-1‰$ throughout the Maastrichtian marls (Fig. 8). These values and stable trends are comparable to other low-latitude sections with late Maastrichtian marls (e.g., Zachos et al., 1989, 1994; Keller and Lindinger, 1989), and suggest pelagic deposition. Lower $\delta^{13}C$ values are only observed in the spherule layers and the bentonite at the top of the section. These highly negative values are likely due to the presence of significant quantities of nonbiogenic calcite that now infills most of the original spherule cavities (Fig. 9A). Although these $\delta^{13}C$ data provide no firm evidence for normal pelagic deposition of the Mendez marls, they provide no evidence against such an interpretation in the form of chaotic or random fluctuations.

Evidence for original spherule-producing event

Although the arguments presented here show that we can rule out the multiple spherule layers within the marls being due to slumps or repeated stacking of sections with each containing the original spherule layer, questions remain of when, where, and how the original spherule deposition occurred. One of the most challenging tasks is to determine whether one or more of the spherule layers represent the original spherule-producing event(s). Alternatively, all or some of the spherule layers may be redeposited from an earlier, unknown, spherule-producing event. Although none of the four spherule layers at Loma Cerca B or any of the other localities (Schilli, 2000; Affolter, 2000) is associated with an Ir anomaly, there is strong evidence that the oldest late Maastrichtian spherule layer 1 at Loma Cerca B, as well as at several other northeastern Mexico sections (e.g., Mesa Juan Perez, Las Ruisas B, Fig. 2), represents the original spherule-producing event, whereas others may have been re-

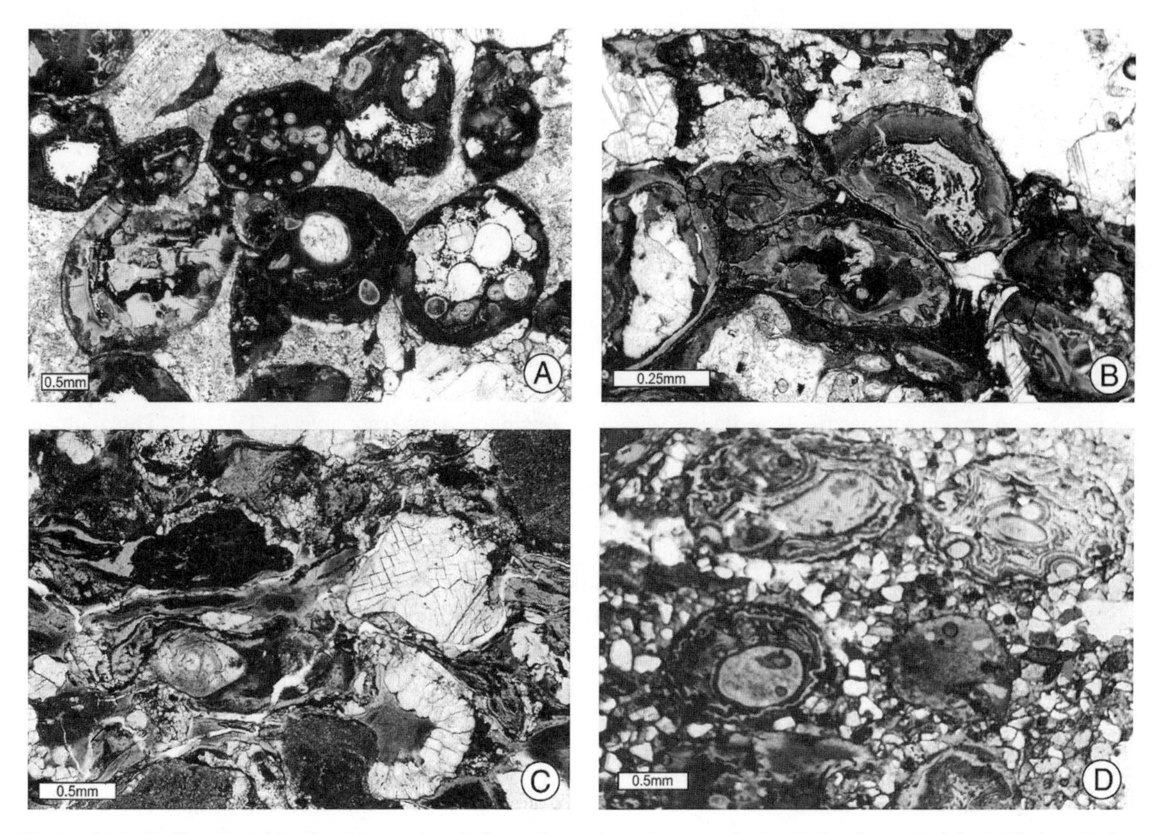

Figure 9. A–D: Stratigraphically oldest spherule layer 1 consists almost entirely of closely packed devitrified vesicular spherules and glass fragments to 5 mm and 7 mm, respectively, in diameter within matrix of blocky calcite (A). Spherules are often compressed and welded together with concave-convex contacts and dark schlieren features (B). Another feature of spherule layer 1 resembles polymict microbreccia where dominant component consists of angular to flaser-like shards and welding or plastic deformation of glass around more rigid constituents, such as carbonate clasts and occasional shells of benthic foraminifera (C). These features suggest that deposition of glass spherules occurred while glass was still hot and ductile during transport and primary deposition. In contrast, matrix of spherule layers 2–4 consists largely of marl clasts, and in spherule layer 4 of terrigenous influx (sand; D).

worked. There are several lines of evidence that lead us to this conclusion.

Nature of spherule layer 1. Thin sections of the spherule layers indicate that the oldest spherule layer 1 (shown in Fig. 5, A and B) consists almost entirely of closely packed devitrified vesicular spherules and altered glass fragments up to 5 mm and 7 mm in diameter, respectively, with a blocky calcite matrix (Fig. 9A). There are very few broken spherules. Spherules are often compressed and welded together with concave-convex contacts and dark schlieren features (Fig. 9B). In some sections, spherule layer 1 resembles a polymict microbreccia, the dominant component of which consists of angular to flaser-like shards with welding or plastic deformation of glass around more rigid constituents, such as carbonate clasts and occasional shells of benthic foraminifera (Fig. 9C). These features suggest that deposition of the glass spherules occurred while the glass was still hot and ductile during transport and primary deposition.

Faunal evidence against major reworking in spherule layer 1. Most benthic foraminifera, as well as the planktic foraminifers (e.g., globotruncanids) have robust, thick-shelled tests that preferentially survive carbonate dissolution, transport, and reworking. Therefore, sharp anomalous abundance peaks in these species groups are proxies for poor preservation (dissolution) or transport and sediment redeposition. At Loma Cerca B there is no evidence of carbonate dissolution, and no significant reworking of species was observed in either benthic or planktic foraminifera in spherule layer 1. Quantitative faunal analysis indicates that benthic foraminifera in the Mendez marls average between 2% and 5%, and there is no increase at the base or within spherule layer 1 (Fig. 7). In addition, there are no transported shallower neritic species present that would indicate downslope transport. Globotruncanids are only 2% of the total foraminiferal population near the base of the section and rise to 11% prior to the first spherule layer. However, they decreased to 1% at the base of spherule layer 1 and abruptly increased to 20% at the top of spherule layer 1 (Fig. 7).

This abundance increase is sustained and long term, continuing through the 2 m of marls between spherule layers 1 and 2, through spherule layers 2 and 3, and well into the marls above. There is no evidence of test abrasion, breakage of specimens, elimination of fragile planktic species, and increase of benthic species that would suggest that the increased abundance of globotruncanids in spherule layer 1, or throughout the interval of globotruncanid dominance, is due to reworking and transport. These faunal data are in agreement with observations based on grain-size analyses, mineralogy, and lithology. The reason for the globotruncanid abundance may be found in the other constituents of the planktic foraminiferal assemblage. Relative abundance changes indicate that the increase in globotruncanids is coupled with a significant increase in warm-water surface dwellers (e.g., *Rugoglobigerina rugosa* and *R. scotti*, Fig. 6) that suggests climate warming. Similar peaks in warm-water species in the latest Maastrichtian are observed in sections across latitudes (e.g., Keller, 1993; Schmitz et al., 1992; Luciani, 1997; Nederbragt, 1998; Li and Keller, 1998a, 1998b; Pardo et al., 1999). We conclude that (1) the welded nature of spherules in spherule layer 1, (2) the near absence of benthic and planktic foraminifera, clasts, and terrigenous influx in the spherule matrix, and (3) the absence of peak abundance in transported benthic species and globotruncanids all point toward spherule layer 1 as the time of deposition of the original spherule-producing event. The presence of this spherule layer near the base of zone CF1, which has been dated as 65.3 Ma, and extrapolation based on sediment accumulation rates suggest that deposition of spherule layer 1 occurred ~270 k.y. prior to the K-T boundary and that deposition of this spherule layer coincided with the onset of the global warm event documented between 65.4 and 65.2 Ma (Li and Keller, 1998a).

Evidence for reworked spherule layers 2–4

It is easier to argue that spherule layers are reworked because we generally assume that deposition of impact or volcanic spherules is likely to be accompanied by seismic disturbance, slumping, or tsunami waves. A significant transported component would be expected if, as a result of a sea-level lowstand, a shallow-water spherule deposit is reworked and transported into deeper waters. Spherule layers 2, 3, and 4 appear to fall within the latter category for the following reasons.

Nature of spherule layers 2–4. These spherule layers contain abundant devitrified glass spherules and fragments, and occasional welded or agglutinated spherules with fluidal textures similar to spherule layer 1. However, they differ in that the matrix contains a chaotic mixture of irregularly shaped marl clasts to several centimeters in size, lithic fragments, and benthic and planktic foraminifera. Terrigenous input (sand) is most abundant in spherule layer 4 (Fig. 9D). This indicates transport, albeit over a relatively short distance as suggested by the size and angularity of marl clasts and sand grains. The strongest evidence for transport and redeposition of spherule layers 2–4 comes from planktic and benthic foraminiferal assemblages.

Faunal evidence for redeposition. Spherule layers 2 and 3 show anomalous peaks in globotruncanids of 10% and 18% above background values at the base of these spherule layers (Fig. 5). Benthic foraminifera also show anomalous peaks 10% and 23% above background values at the base of spherule layers 2 and 3, respectively. Because these abundance peaks are due to concentration of large, robust species that readily survived transport, they most likely mark intervals of transport, reworking, and redeposition associated with spherule layers 2 and 3. The absence of shallower neritic species within these spherule layers suggests that reworking and transport occurred from outer shelf to upper bathyal depths. Insufficient material was available from spherule layer 4 for a quantitative estimate. However, previous investigations of this spherule layer revealed the presence of abundant transported benthic foraminifera and marl clasts (Keller et al., 1994, 1997).

Paleoclimate at Loma Cerca during the last 300 k.y. of the Maastrichtian

During the latest Maastrichtian, sediment deposition in northeastern Mexico occurred in a tectonically active region (Laramide event), as indicated by the clay mineralogy and the high detrital content. The climate was probably arid, although active tectonism that caused increased erosion from uplifted areas overprinted climate signals. However, planktic foraminiferal assemblages indicate that a typical subtropical to tropical Tethyan environment prevailed.

Maastrichtian climate changes are well documented based on high-resolution stable isotope records from low and middle latitudes. In the southern middle to high latitudes, temperature records based on stable isotopes indicate that climate cooled by ~7–8°C during the Maastrichtian and had reached a minimum 500 k.y. before the K-T boundary (Barrera, 1994; Li and Keller, 1998a). Beginning at 65.45 Ma the climate rapidly warmed, resulting in a 3–4°C increase in bottom and surface waters in middle to high latitudes, and reached a maximum by 65.30 Ma. During the last 100 k.y. of the Maastrichtian, climate cooled gradually by 3°C (Fig. 10). These strong climate changes are evident in faunal assemblages across latitudes. In middle and high latitudes, there is a brief incursion of subtropical and tropical species (e.g., rugoglobigerinids) during the maximum warming (Schmitz et al., 1992; Pardo et al., 1999). At the same time, globotruncanids that lived at thermocline depth doubled their populations in low latitudes. However, by K-T boundary time a more temperate cool climate prevailed again. Globotruncanids responded to this cooling by decreased abundance

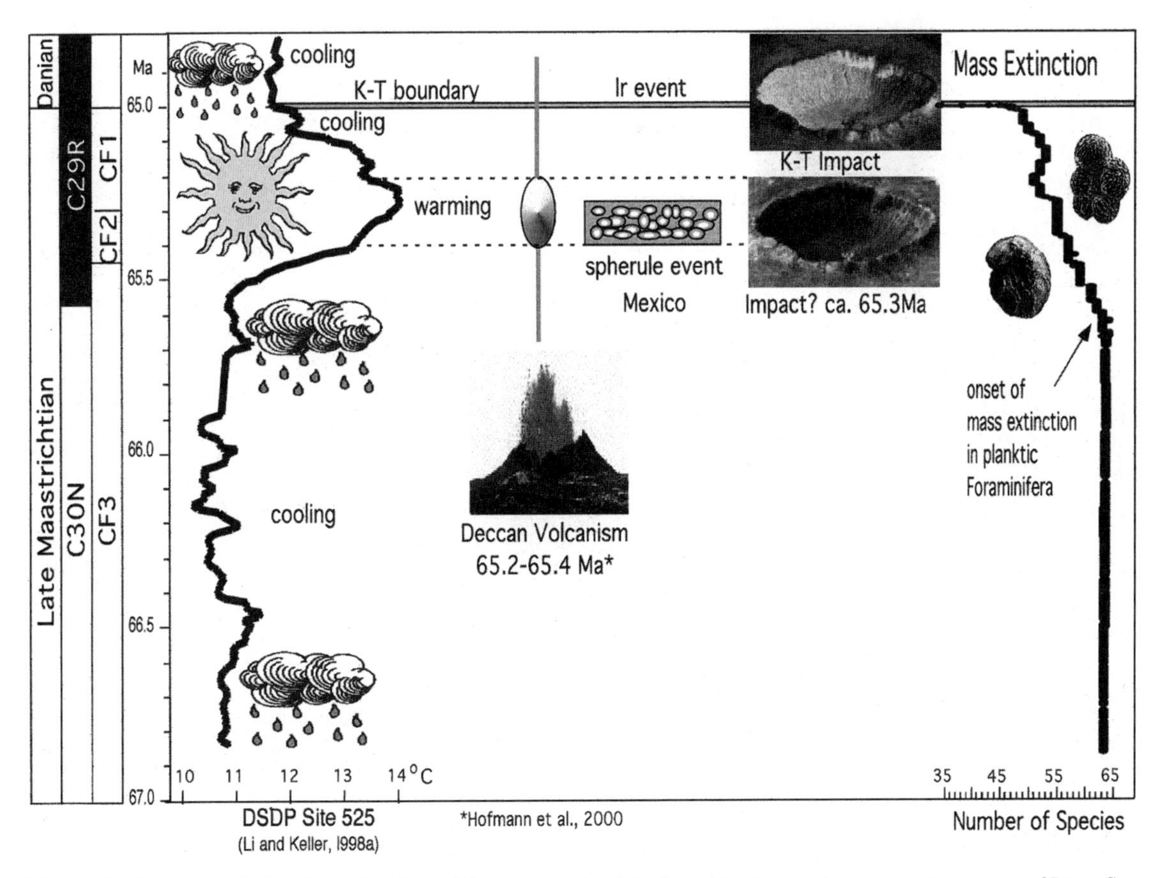

Figure 10. Sequence of climatic, volcanic, and impact events plotted next to deep-water temperature curve of Deep Sea Drilling Project (DSDP) Site 525 (data from Li and Keller, 1998a). Note that global warming of 3–4°C occurred between 65.4 and 65.2 Ma, coincident with latest estimates of major Deccan volcanism (Hoffman et al., 2000). New discovery of spherule layers in late Maastrichtian Mendez marls indicates that original spherule deposition event occurred ~270 k.y. before Cretaceous-Tertiary (K-T) boundary and is also associated with major warming. These data indicate that latest Maastrichtian global climate warming was triggered by Deccan volcanism, and at least locally (in Central and North America) exacerbated by an impact (spherule layer 1). Spherules from Mexico and Haiti were originally linked to Chicxulub impact based on chemical evidence of glass spherules. If this conclusion still holds in view of present multiple spherule layers, then the Chicxulub impact predates the K-T boundary by ~270 k.y. Regardless of age of Chicxulub, there is now strong evidence for multiple impacts, one at K-T boundary and one in late Maastrichtian ca. 65.3 Ma. Extinctions in planktic foraminifera began during maximum cooling at 65.5 Ma and gradually continued through last 0.5 m.y. of Maastrichtian, ending with rapid extinction of all tropical and subtropical species at or near K-T boundary.

to <5% of the foraminiferal assemblage in the >63 µm size fraction, and to <10% in the >150 µm size fraction (Nederbragt, 1991, 1998; Schmitz et al., 1992; Keller et al., 1995, 2001; Abramovich et al., 1998; Abramovich and Keller, 2002). Relative abundance fluctuations in planktic foraminiferal species are thus excellent environmental proxies for changes in water-mass stratification and climate.

At Loma Cerca faunal assemblages show clear indications of major climate changes during the last 300–400 k.y. of the Maastrichtian, as seen in the relative abundance changes of surface and thermocline dwellers (Fig. 7). For example, the relative abundance changes of the dominant surface dweller, *Pseudoguembelina costulata*, and those of globotruncanids show opposite trends. The same trends are amplified by the strong relative abundance changes in the overall population of surface dwellers that reflect major climate fluctuations.

Prior to deposition of the first spherule layer, *P. costulate* dominated, averaging 30%, and the total group of surface dwellers averaged 70% of the planktic foraminiferal population. At the same time, tropical and subtropical intermediate dwellers (globotruncanids) averaged 2% and rose to 10% near the base of CF1. The remainder consisted of intermediate-dwelling heterohelicids (e.g., *Heterohelix globulosa, H. dentata, H. planata*) that were ecological generalists able to thrive in variable conditions across latitudes. The extremely low abundance of species restricted to the tropical-subtropical environment (2%) in zone CF2 reflects high-stress conditions as a result of cool temperatures and correlates with the global cooling trend that reached a maximum ca. 65.5 Ma (Li and Keller, 1998a).

Spherule layer 1 coincides with a major faunal change that continued through spherule layers 2 and 3. During this interval, globotruncanids averaged a constant 20%–25%, except for two brief peaks of nearly 30% and 40% at the bases of spherule layers 2 and 3, respectively. Although it is tempting to dismiss this long-term abundance maximum in tropical and subtropical globotruncanid species as being due to reworking, there is little support for this interpretation. As noted herein, only the anomalous peaks in globotruncanids and benthic foraminifera at the bases of spherule layers 2 and 3 indicate reworking. The remainder of the spherule samples, including the samples from the marls and marly limestone between the spherule layers, contain typical warm-climate, low-latitude Tethyan assemblages dominated by globorotalids, rugoglobigerinids, and low-oxygen-tolerant heterohelicids. This strongly suggests that the spherule-producing event near the base of zone CF1 coincides with, or is causally related to, the onset of the short-term rapid global warming that has been dated as occurring between 65.4 to 65.3 Ma (Li and Keller, 1998a). This intriguing possibility must be explored further in additional sections in Mexico. Oxygen isotope analyses of Loma Cerca and another nearby section have proved useless because carbonate is strongly diagenetically altered.

In the 6 m of marls between spherule layers 3 and 4 at the base of the sandstone-siltstone complex, planktic foraminiferal assemblages suggest climate cooling with increasing abundance of ecological generalists (heterohelicids) near the top of the section (Fig. 3). Increasing dominance of heterohelicids (e.g., *H. globulosa, H. dentata, H. planata, H. navarroensis*) generally characterizes Tethyan assemblages during the last 100–200 k.y. of the Maastrichtian (Keller et al., 1995, 2001; Abramovich et al., 1998; Luciani, 2002). That this increase begins only near the top of the section just below the sandstone-siltstone complex at Loma Cerca indicates that deposition of the sandstone-siltstone complex occurred near the end of the Maastrichtian.

K-T extinction scenario revised

The new discovery of an older late Maastrichtian spherule layer deposited in pelagic marls about 270 k.y. prior to the K-T boundary in northeastern Mexico, and the associated faunal evidence of significant climate warming correlates well with the global warming of 3–4°C in surface and deep waters between 65.4 and 65.2 Ma (Li and Keller, 1998a), a major pulse in Deccan volcanism between 65.4 and 65.2 Ma (Hofmann et al., 2000), and the onset of mass extinctions in planktic foraminifera ca. 65.5 Ma (Fig. 10; Abramovich et al., 1998; Abramovich and Keller, 2001). These data indicate a complex interplay of climate change, volcanism, and impacts that caused the end-Cretaceous mass extinction.

This multievent mass-extinction scenario began ~400–500 k.y. before the K-T boundary at the transition that marks the end of the long-term Maastrichtian global cooling and onset of rapid global warming (Fig. 10). Maximum global warming between 65.4 and 65.2 Ma was probably linked to increased atmospheric CO_2 due to volcanic eruptions (Hoffman et al., 2000). The spherule event documented in Central America coincided with this warm event and exacerbated the already extreme climatic and environmental changes. During the last 100 k.y. of the Maastrichtian, climate cooled rapidly by 2–3°C across latitudes. The K-T boundary event coincided with a relatively cool climate that continued through the early Danian (Fig. 10). These extreme climatic fluctuations resulted in severe biotic stresses for tropical and subtropical planktic foraminifera. A gradual decrease in diversity began ca. 65.5 Ma, their combined relative abundance dropped to ~5%–10% of the total population, and their demise was met at or near the K-T boundary.

Chixculub impact or multiple impact events?

Glass spherules in Cretaceous-Tertiary boundary sections in Central America and the Caribbean are generally interpreted as melt droplets of target rocks dispersed by the Chicxulub impact in Yucatan, Mexico. The critical supporting evidence is based on the stratigraphic position of the glass spherules near the K-T boundary, their chemical similarity with melt rock in subsurface cores at Chicxulub, and an $^{40}Ar/^{39}Ar$ age of ca. 65 Ma of the spherules and melt rock (Izett et al., 1990; Swisher et al., 1992; Dalrymple et al., 1993). However, at the time these conclusions were reached only one spherule layer was known at the base of the sandstone-siltstone complex in northeastern Mexico. Four new discoveries have since been made that challenge this interpretation. (1) As many as 4 spherule layers within 10 m of pelagic marls below the sandstone-siltstone complex and K-T boundary in northeastern Mexico have been discovered (this study; Stinnesbeck et al., 2001). (2) At Beloc, Haiti, the spherule layer and Ir anomaly above it are within early Danian shales of the *P. eugubina* zone (Stinnesbeck et al., 2000; Keller et al., 2001; Stüben et al., this volume). (3) At Coxquihui, east-central Mexico, a thick spherule deposit and Ir anomaly above it also occur in the early Danian *P. eugubina* zone (Stinnesbeck et al., 2002). (4) Above a thick breccia deposit at El Caribe in Guatemala, spherules and an Ir anomaly occur in shales of the early Danian *P. eugubina* zone (Fourcade et al., 1998, 1999; Keller and Stinnesbeck, 1999).

In view of these discoveries, it remains to be determined whether the Chicxulub crater is the source of all of these glass spherules, as previously concluded on the basis of chemical similarity of the glass spherules and melt rock from Chicxulub cores. If this conclusion stands, then the Chicxulub event must predate the K-T boundary by ~270–300 k.y. Regardless of the age of Chicxulub, current spherule data from late Maastrichtian marls and the K-T iridium anomaly worldwide strongly support a multiple impact scenario with a spherule-producing event (but no iridium anomaly) in the latest Maastrichtian (65.3 Ma), and an impact event at the K-T boundary (65.0 Ma). Accumulating data of spherules and Ir anomaly in the *P. eugubina* zone of Mexico, Haiti, and Guatemala indicates that there may have been a third impact event in the early Danian.

ACKNOWLEDGMENTS

This study would not have been possible without the dedicated work of four Masters degree students, Lionel Schilli and Mark Affolter from the University of Neuchâtel, Switzerland, whose study area is discussed here, and Peter Schulte and Falk Lindenmaier from the University of Karlsruhe, Germany, who worked on an adjacent area finding yet more sections with multiple spherule layers. We also thank Doris Stueben, University of Karlsruhe, and Steve Burns, University of Bern, for their efforts in analyzing trace elements and stable isotopes, and Abdel Aziz Tantawy for analyzing the calcareous nannofossils. These studies were supported in part by Deutsche Forschungsgemeinschaft grants Sti 128/2-3 and Sti 128/2-4 and Mexico's Conacyt grant E 120.561.

REFERENCES CITED

Abramovich, S., and Keller, G., 2002, Planktic foraminiferal population changes during the late Maastrichtian at Elles, Tunisia: Palaeogeography, Palaeoecology, Palaeoclimatology (in press).

Abramovich, S., Almogi-Labin, A., and Benjamini, C., 1998, Decline of the Maastrichtian pelagic ecosystem based on planktic foraminifera assemblage change: Implications for the terminal Cretaceous faunal crisis: Geology, v. 26, no. 1, p. 63–66.

Adatte, T., Stinnesbeck, W., and Keller, G., 1996, Lithostratigraphic and mineralogic correlations of near K/T boundary clastic sediments in northeastern Mexico: Implications for origin and nature of deposition, *in* Ryder, G., Fastovsky D., and Gartner, S., eds., The Cretaceous-Tertiary event and other catastrophes in Earth history: Geological Society of America Special Paper 307, p. 211–226.

Affolter, M., 2000, Etude des depots clastiques de la limite Cretace-Tertiaire dans la region de la Sierrita, Nuevo Leon, Mexique [M.S. thesis]: Neuchatel, Switzerland, University of Neuchatel, Geological Institute, 133 p.

Alvarez, W., Smit, J., Lowrie, W., Asaro, F., Margolis, S.V., Claeys, P., Kastner, M., and Hildebrand, A., 1992, Proximal impact deposits at the K/T boundary in the Gulf of Mexico: A restudy of DSDP Leg 77 Sites 536 and 540: Geology, v. 20, p. 697–700.

Barrera, E., 1994, Global environmental changes preceeding the Cretaceous-Tertiary boundary: Early-late Maastrichtian transition: Geology, v. 22, p. 877–880.

Blum, J.D., and Chamberlain, C.P., 1992, Oxygen isotope constraints on the origin of impact glasses from the Cretaceous-Tertiary boundary: Science, v. 257, p. 1104–1107.

Blum, J.D., Chamberlain, C.P., Hingston, M.P., Koeberl, C., Marin, L.E., Schuraytz, B.C., and Sharpton, V.L., 1993, Isotopic comparison of K-T boundary impact glass with melt rock from the Chicxulub and Manson impact structures: Nature, v. 364, p. 325–327.

Bralower, T., Paull, C.K., and Leckie, R.M., 1998, The Cretaceous-Tertiary boundary cocktail: Chicxulub impact triggers margin collapse and extensive sediment gravity flows: Geology, v. 26, p. 331–334.

Brand, U., and Veizer, J., 1980, Chemical diagensis of a multicomponent carbonate system. 1. Trace elements: Journal of Sedimentary Petrology, v. 50, p. 1219–1236.

Dalrymple, G.B., Izett, G.A., Snee, L.W., and Obradovich, J.D., 1993, ^{40}Ar/^{39}Ar age spectra and total fusion ages of tektites from Cretaceous-Tertiary boundary sedimentary rocks in the Beloc formation, Haiti: U.S. Geological Survey Bulletin 2065, 20 p.

D'Hondt, S., Herbert, T.D., King, J., and Gibson, C., 1996, Planktic foraminifera, asteroids and marine production: Death and recovery at the Cretaceous-Tertiary boundary, *in* Ryder, G., Fastovsky, D., and Gartner, S., eds., The Cretaceous-Tertiary event and other catastrophes in Earth history, Geological Society of America Special Paper 307, p. 303–318.

Ekdale, A.A., and Stinnesbeck, W., 1998, Ichnology of Cretaceous-Tertiary (K/T) boundary beds in northeastern Mexico: Palaios, v. 13, p. 593–602.

Fourcade, E., Rocchia, R., Gardin, S., Bellier, J-P., Debrabant, P., Masure, E., Robin, E., and Pop, W.T., 1998, Age of the Guatemala breccias around the Cretaceous-Tertiary boundary: Relationships with the asteroid impact on the Yucatan: Comptes Rendus de l'Académie des Sciences, Serie 2, Sciences de la Terre et des Planetes, v. 327, p. 47–53.

Fourcade, E., Piccioni, L., Escribá, J., and Rosselo, E., 1999, Cretaceous stratigraphy and palaeoenvironments of the Southern Petén Basin, Guatemala: Cretaceous Research, v. 20, p. 793–811.

Hoffman, C., Feraud, G., and Courtillot, V., 2000, ^{40}Ar/^{39}Ar dating of mineral separates and whole rocks from the Western Ghats lava pile: Further constraints on duration and age of Deccan traps: Earth and Planetary Science Letters, v. 180, p. 13–27.

Izett, G., Maurasse, F.J.-M.R., Lichte, F.E., Meeker, G.P., and Bates, R., 1990, Tektites in Cretaceous/Tertiary boundary rocks on Haiti: U.S. Geological Survey Open-File Report 90–635, 122 p.

Jéhanno, C., Boclet, D., Froget, L., Lambert, B., Robin, E., Rocchia, R., and Turpin, L., 1992, The Cretaceous-Tertiary boundary at Beloc, Haiti: No evidence for an impact in the Caribbean area: Earth and Planetary Science Letters, v. 109, p. 229–241.

Keller, G., 1992, Paleoecologic response of Tethyan benthic foraminifera to the Cretaceous-Tertiary boundary transition: Studies in benthic foraminifera: BENTHOS '90, Sendai, 1990, Tokai University Press, p. 77–91.

Keller, G., and Lindinger, M., 1989, Stable isotope, TOC and $CaCO_3$ records across the Cretaceous-Tertiary boundary at El Kef, Tunisia: Palaeogeography, Palaeoclimatology, Palaeoecology, v. 73, p. 243–265.

Keller, G., and Stinnesbeck, W., 1999, Ir and the K/T boundary at El Caribe, Guatemala: International Journal of Earth Sciences, v. 88, p. 844–852.

Keller, G., Stinnesbeck, W., and Lopez-Oliva, J.G., 1994, Age, deposition and biotic effects of the Cretaceous/Tertiary boundary event at Mimbral, NE Mexico: Palaios, v. 9, p. 144–157.

Keller, G., Li, L., and MacLeod, N., 1995, The Cretaceous/Tertiary boundary stratotype section at El Kef, Tunisia: How catastrophic was the mass extinction?: Palaeogeography, Palaeoclimatology, Palaeoecology, v. 119, p. 221–254.

Keller, G., Lopez-Oliva, J.G., Stinnesbeck, W., and Adatte, T., 1997, Age, stratigraphy and deposition of near K/T siliciclastic deposits in Mexico: Relation to bolide impact?: Geological Society of America Bulletin, v. 109, p. 410–428.

Keller, G., Adatte, T., Stinnesbeck, W., Stueben, D., and Berner, Z., 2001, Age, chemo- and biostratigraphy of Haiti spherule-rich deposits: A multi-event K-T scenario: Canadian Journal of Earth Sciences v. 38, p. 197–227.

Koeberl, C., 1993, Chicxulub crater, Yucatan: Tektites, impact glasses, and the geochemistry of target rocks and breccias: Geology, v. 21, p. 211–214.

Koeberl, C., Sharpton, V.L., Schuraytz, B.C., Shirley, S.B., Blum, J.D., and Marin, L.E., 1994, Evidence for a meteoric component in impact melt rock from the Chicxulub structure: Geochimica et Cosmochimica Acta, v. 56, p. 2113–2129.

Kuebler, B., 1987, Cristallinite de l'illite, methods normaisees de preparations, methods normalisees de measures: Neuchâtel, Suisse, Cahiers Institut de Geologie, sér. ADX, v. 1, p. 1–12.

Leroux, H., Rocchia, R., Froget, L., Orue-Etxebarria, X., Doukhan, J., and Robin, E., 1995, The K/T boundary of Beloc (Haiti): Compared stratigraphic distributions of boundary markers: Earth and Planetary Science Letters, v. 131, p. 255–268.

Li, L., and Keller, G., 1998a, Abrupt deep-sea warming at the end of the Cretaceous: Geology, v. 26, p. 995–998.

Li, L., and Keller, G., 1998b, Diversification and extinction in Campanian-Maastrichtian planktic foraminifera of northwestern Tunisia: Eclogae Geologicae Helvetiae, v. 91, p. 75–102.

Li, L., Keller, G., and Stinnesbeck, W., 1999, The late Campanian and Maastrichtian in northwestern Tunisia: Paleoenvironmental inferences from lithology, macrofauna and benthic foraminifera: Cretaceous Research, v. 20, p. 231–252.

Lindenmaier, F., 1999, Geologie und geochemie an der Kreide/Tertiär-Grenze im Nordosten von Mexiko: Karlsruhe, Germany, Diplomarbeit, Universität Karlsruhe, Institute fur Regionale Geologie, 90 p.

Lopez-Oliva, J.G., and Keller, G., 1996, Age and stratigraphy of near-K/T boundary clastic deposits in northeastern Mexico, *in* Ryder, G., Fastovsky, D., and Gartner, S., eds., The Cretaceous-Tertiary event and other catastrophes in Earth history: Geological Society of America Special Paper 307, p. 227–242.

Luciani, V., 1997, Planktonic foraminiferal turnover across the Cretaceous-Tertiary boundary in the Vajont valley (Southern Alps, northern Italy): Cretaceous Research, v. 18, p. 799–821.

Luciani, V., 2002, High resolution planktonic foraminiferal analysis from the Cretaceous/Tertiary boundary at Ain Settara (Tunisia): Evidence of an extended mass extinction: Palaeogeography, Palaeoclimatology, Palaeoecology (in press).

Lyons, J.B., and Officer, C.B., 1992, Mineralogy and petrology of the Haiti Cretaceous-Tertiary section: Earth and Planetary Science Letters, v. 109, p. 205–242.

Magaritz, M., 1975, Sparitization of pelleted limestone: A case study of carbon and oxygen isotopic composition: Journal of Sedimentary Petrology, v. 45, p. 599–603.

MacLeod, N., and Keller, G., 1991, How complete are Cretaceous/Tertiary boundary sections?: Geological Society of America Bulletin, v. 103, p. 1439–1457.

Morkhoven, F.P.C.M., Berggren, W.A., and Edwards, A., 1986, Cenozoic cosmopolitan deep-water benthic foraminifera: Bulletin Centres Recherche Exploration-Production Elf-Aquitaine, Pau, France, Memoir 11, 421 p.

Nederbragt, A., 1991, Late Cretaceous biostratigraphy and development of Heterohelicidae (planktic foraminifera): Micropaleontology, v. 37, p. 329–372.

Nederbragt, A., 1998, Quantitative biogeography of late Maastrichtian planktic foraminifera: Micropaleontology, v. 44, p. 385–412.

Pardo, A., Ortiz, N., and Keller, G., 1996, Latest Maastrichtian and K/T boundary foraminiferal turnover and environmental changes at Agost, Spain, *in* MacLeod, N., and Keller, G., eds., The Cretaceous-Tertiary boundary mass extinction: Biotic and environmental events: New York, W.W. Norton and Co., p. 155–176.

Pardo, A., Adatte, T., Keller, G., and Oberhaensli, H., 1999, Paleoenvironmental changes across the Cretaceous-Tertiary boundary at Koshak, Kazakhstan, based on planktic foraminifera and clay mineralogy: Palaeogeography, Palaeoclimatology, Palaeoecology, v. 154, p. 247–273.

Pospichal, J.J., 1996, Calcareous nannoplankton mass extinction at the Cretaceous-Tertiary boundary: An update, *in* Ryder, G., Fastovsky, D., and Gartner, S., eds., The Cretaceous-Tertiary event and other catastrophes in Earth history: Geological Society of America Special Paper 307, p. 335–360.

Robin, E., Bonté, Ph., Froget, L., Jéhanno, C., and Rocchia, R., 1992, Formation of spinels in cosmic objects during atmospheric entry: A clue to the Cretaceous-Tertiary boundary event: Earth and Planetary Science Letters, v. 108, p. 181–190.

Rocchia, R., Robin, E., Froget, L., and Gayraud, J., 1996, Stratigraphic distribution of extraterrestrial markers at the Cretaceous-Tertiary boundary in the Gulf of Mexico area: Implications for the temporal complexity of the event, *in* Ryder, G., Fastovsky, D., and Gartner, S., eds., The Cretaceous-Tertiary event and other catastrophes in Earth history: Geological Society of America Special Paper 307, p. 279–286.

Schilli, L., 2000, Etude de la limite K-T dans la région de la Sierrita, Nuevo Leon, Mexique [M.S. thesis]: Neuchatel, Switzerland, University of Neuchatel, Geological Institute, 138 p.

Schmitz, B., Keller, G., and Stenvall, O., 1992, Stable isotope and foraminiferal changes across the Cretaceous/Tertiary boundary at Stevns Klint, Denmark: Arguments for longterm oceanic instability before and after bolide impact: Palaeogeography, Palaeoclimatology, Palaeoecology, v. 96, p. 233–260.

Schrag, D.P., DePaolo, D.J., and Richter, F.M., 1995, Reconstructing past sea surface temperatures: Correcting for diagenesis of bulk marine carbon: Geochimica et Cosmochimica Acta, v. 59, p. 2265–2278.

Schulte, P., 1999, Geologisch-sedimentologische Untersuchungen des Kreide/Tertiär (K/T)-Übergangs im Gebiet zwischen La Sierrita und El Toro, Nuevo Leon, Mexiko: Karlsruhe, Germany, Diplomarbeit, Universität Karlsruhe, Institute fur Regionale Geologie, 134 p.

Sigurdsson, H., Bonté, P., Turpin, L., Chaussidon, M., Metrich, N., Steinberg, M., Pradel, P., and D'Hondt, S., 1991, Geochemical constraints on source region of Cretaceous/Tertiary impact glasses: Nature, v. 353, p. 839–842.

Smit, J., Montanari, A., Swinburne, N.H.M., Alvarez, W., Hildebrand, A., Margolis, S.V., Claeys, P., Lowrie, W., and Asaro, F., 1992, Tektite bearing deep-water clastic unit at the Cretaceous-Tertiary boundary in northeastern Mexico: Geology, v. 20, p. 99–103.

Smit, J., Roep, T.B., Alvarez, W., Montanari, A., Claeys, P., Grajales-Nishimura, J.M., and Bermúdez, J., 1996, Coarse-grained, clastic sandstone complex at the K/T boundary around the Gulf of Mexico: Deposition by tsunami waves induced by the Chicxulub impact, *in* Ryder, G., Fastovsky, D., and Gartner, S., eds., The Cretaceous-Tertiary event and other catastrophes in Earth history: Geological Society of America Special Paper 307, p. 151–182.

Soria, A.R., Llesa, C.L., Mata, M.P., Arz, J.A., Alegret, L., Arenillas, I., and Meléndez, A., 2001, Slumping and a sandbar deposit at the Cretaceous-Tertiary boundary in the El Tecolote section (northeastern Mexico): An impact-induced sediment gravity flow: Geology, v. 29, p. 231–234.

Stinnesbeck, W., Barbarin, J.M., Keller G., Lopez-Oliva, J.G., Pivnik, D.A., Lyons, J.B., Officer, C.B., Adatte, T., Graup, G., Rocchia, R., and Robin, E., 1993, Deposition of channel deposits near the Cretaceous-Tertiary boundary in northeastern Mexico: Catastrophic or "normal" sedimentary deposits: Geology, v. 21, p. 797–800.

Stinnesbeck, W., Keller, G., Adatte, T., Stüben, D., Kramar, U., Berner, Z., Desremaux, C., and Moliere, E., 2000, Beloc, Haiti, revisited: Multiple events across the Cretaceous-Tertiary transition in the Caribbean?: Terra Nova, v. 11, p. 303–310.

Stinnesbeck, W., Schulte, P., Lindenmaier, F., Adatte, T., Affolter, M., Schilli, L., Keller, G., Stueben, D., Berner, Z., Kramar, U., and Lopez-Oliva, J.G., 2001, Late Maastrichtian age of spherule deposits in northeastern Mexico: Implication for Chicxulub scenario: Canadian Journal of Earth Sciences, v. 38, p. 229–238.

Stinnesbeck, W., Keller, G., Schulte, P., Stueben, D., Berner, Z., Kramar, U., and Lopez-Oliva, J.G., 2002, The Cretaceous-Tertiary (K/T) boundary transition at Coxquihui, state of Veracruz, Mexico: Evidence for an early Danian impact event?: International Journal of Earth Sciences (in press).

Swisher, C.C., and 11 others, 1992, Coeval $^{40}Ar/^{39}Ar$ ages of 65 million years ago from Chicxulub crater melt rock and Cretaceous-Tertiary boundary tektites: Science, v. 257, p. 954–958.

Zachos, J.C., Arthur, M.A., Dean, W.E., 1989, Geochemical evidence for suppression of pelagic marine productivity at the Cretaceous/Tertiary boundary: Nature, v. 337, p. 61–64.

Zachos, J.C., Stott, L.D., and Lohmann, K.C., 1994, Evolution of early Cenozoic marine temperatures: Paleoceanography, v. 9, p. 353–387.

Manuscript Submitted November 23, 2000; Accepted by the Society March 22, 2001

Printed in the U.S.A.

Geological Society of America
Special Paper 356
2002

Two anomalies of platinum group elements above the Cretaceous-Tertiary boundary at Beloc, Haiti: Geochemical context and consequences for the impact scenario

Doris Stüben*
Utz Kramar
Zsolt Berner
Jörg-Detlef Eckhardt
Institut für Mineralogie und Geochemie, Universität Karlsruhe, 76128 Karlsruhe, Germany
Wolfgang Stinnesbeck
Geologisches Institut, Universität Karlsruhe, 76128 Karlsruhe, Germany
Gerta Keller
Department of Geosciences, Princeton University, Princeton, New Jersey, 08544 USA
Thierry Adatte
Institut de Géologie, 11 Rue Emile Argand, 2007, Neuchâtel, Switzerland
Klaus Heide
Institut für Geowissenschaften, Universität Jena, Burgweg 11, 07749 Jena, Germany

ABSTRACT

A detailed geochemical investigation of an expanded Cretaceous-Tertiary (K-T) boundary section near Beloc (B3), Haiti, reveals a complex pattern of sedimentation of multiple origins as a result of erosional, biogenic, volcanic, and impact events. Carbonate-rich uppermost Maastrichtian sediments with high excess rates for Cu, Zn, and Sr (biogenic origin) indicate high productivity ($\delta^{13}C$) and warm temperatures ($\delta^{18}O$). These sediments are overlain by Paleocene (early Danian zone P1a) spherule-rich clayey layers that indicate lower productivity, lower temperatures, and high input of glass and biogenic carbonate. Reworked Maastrichtian sediments are mixed with spherule-rich layers. This spherule-rich deposit is topped by a thin layer rich in Fe that also contains an Ir-dominated anomaly of platinum group elements (PGE) with an almost chondritic abundance pattern, which appears to be the result of a cosmic influx. Monotonous limestones above this interval reflect recovery to normal pelagic sedimentation, which is interrupted by a second PGE anomaly in an Fe-rich clayey layer in the middle part of zone P1a. All PGEs are enriched in this interval and the PGE pattern is basalt like, suggesting a volcanic source. Both PGE anomaly horizons coincide with productivity and temperature changes.

*E-mail: doris.stueben@bio-geo.uni-karlsruhe.de

Stüben, D., Kramar, U., Berner, Z., Eckhardt, J.-D., Stinnesbeck, W., Keller, G., Adatte, T., and Heide, K., 2002, Two anomalies of platinum group elements above the Cretaceous-Tertiary boundary at Beloc, Haiti: Geochemical context and consequences for the impact scenario, *in* Koeberl, C., and MacLeod, K.G., eds., Catastrophic Events and Mass Extinctions: Impacts and Beyond: Boulder, Colorado, Geological Society of America Special Paper 356, p. 163–188.

INTRODUCTION

The origin and genesis of Cretaceous-Tertiary (K-T) boundary sediment layers are still discussed, although most researchers interpret them to be impact derived (Sutherland, 1994; Glasby and Kunzendorf, 1996; Montanari and Koeberl, 2000, and references herein). Central America and the Caribbean play a key role in this debate. In this region, K-T boundary sediments contain glass spherules, shocked quartz, anomalous Ir concentrations (e.g., Alvarez et al., 1980; Officer and Drake, 1985; Bohor and Seitz, 1990; Hildebrand and Boyton, 1990; Hildebrand et al., 1991; Koeberl et al., 1994), and evidence of biotic mass extinction (Maurasse and Sen, 1991; Lamolda et al., 1997; Keller et al., 2001) that are thought to have been generated by a large impact event at Chicxulub in Yucatan, Mexico (Izett et al., 1990; Izett, 1991; Sigurdsson et al., 1991a; Koeberl and Sigurdsson, 1992; Koeberl, 1992; Smit et al., 1992; Kring et al., 1994). According to this scenario, a siliciclastic sequence, which can be intermittently followed over several hundreds of kilometers in northeastern Mexico, was the result of an impact-generated tsunami or gravity flow. Partially altered glass spherules in Mexican and Haitian K-T sediments are interpreted as microtektites or impact glasses, and igneous rocks with andesitic composition in subsurface cores at Chicxulub are considered to be impact melt rocks. A sequence of events is recognized in the K-T boundary transition in the Caribbean area, including an impact, volcanism, and climate and sea-level changes that explain the K-T sediments deposited in the Caribbean area (Keller and Stinnesbeck, 1996; Keller et al., 1997, 2001, this volume; Stinnesbeck et al., 2001).

Stinnesbeck et al. (1999) suggested a multiple-event scenario across the K-T boundary from sections at Beloc, Haiti. Of major interest in these sections is a spherule-rich deposit, which was originally interpreted as a series of volcanogenic turbidites (Maurrasse, 1982). However, geochemical studies revealed that the different layers contain evidence of an impact event (platinum group element [PGE] anomalies, Ni-rich spinels, glass spherules, shocked quartz grains) (Izett et al., 1990, 1991; Izett, 1991; Sigurdsson et al., 1991a, 1991b; Leroux et al., 1995).

Earlier studies of Beloc sections documented a complex sedimentation history, including two Ir enhancements above the spherule-rich deposit, one of them located in an Ir-rich Fe-oxide horizon (Lyons and Officer, 1992; Jéhanno et al., 1992; Leroux et al., 1995). Maurrasse and Sen (1991) and Lamolda et al. (1997) reported the first early Danian species near the top of the spherule layer at 32 cm above the base and 30 cm below the Ir enrichment. Sigurdsson et al. (1991b) reported the first Danian species at 110 cm above the base of the spherule layer, whereas Keller et al. (2001) reported the first Danian species at the base of the spherule layer. The differences in the biostratigraphic results of these reports are primarily due to the localities studied: K-T outcrops along the road are tectonically faulted and sheared and, in consequence, often reduced in thickness. Maurrasse and Sen (1991), Stinnesbeck et al. (2001), and Keller et al. (2001) described outcrops on the steep hillside below the road to Beloc. These outcrops are less affected by faulting and more expanded than the road sections, and provide better age data. This study provides a detailed geochemical and mineralogical investigation of the most complete of these new sections at Beloc, Haiti (section B3), in which two PGE anomalies were reported (Stinnesbeck et al., 1999). Specifically, this study concentrates on the following. (1) A detailed chemostratigraphy study of the Beloc 3 (also known as B3) section was done in order to identify the source and the depositional history of the accumulated material. Special attention is paid to the geochemical discrimination of extraterrestrial, volcanogenic, and terrigenous-detrital signals. (2) A geochemical investigation was made of spherules that accumulated in several layers and their provenance and postdepositional alteration.

ANALYTICAL METHODS

Material and sample preparation

We selected 42 samples for geochemical investigations from an expanded K-T boundary transition at Beloc (Figs. 1 and 2; see Keller et al., 2001, for details of location). Sample locations with the corresponding labels are marked on the lithological profiles by horizontal dashes. The sediment samples were dried, crushed, finely ground, and homogenized in an agate mill and dried at 105°C prior to analysis.

Major and trace elements

Major and trace elements were determined by energy-dispersive (EDXRF) and wavelength-dispersive X-ray fluorescence (WDXRF) analysis. We analyzed 14 trace elements (Ni, Cu, Zn, Ga, As, Rb, Sr, Y, Zr, Nb, Cd, Sb, Ba, and Pb) by EDXRF. For these determinations an aliquot of ~5 g from each of the 42 samples was used as bulk powder in Polystyrol containers with a 6 µm Mylar window for measurement. The samples were measured three times with a SPECTRACE 5000 X-ray analyzer using Al, Pd-, and Cu primary filters to optimize the excitation of elements emitting low, intermediate, and high X-ray energies, respectively. For details of analytical procedures and detection limits for bulk-rock powder samples, see Kramar (1997).

Major elements and some of the minor elements (V, Cr, and Ni) were determined by wavelength-dispersive WDXRF. For these samples fused glass disks were prepared from a mixture of 0.5 g ignited bulk powder, 0.5 g SiO_2, and 4 g SPECTROMELT. The fused disks were analyzed with an SRS 303 AS wavelength-dispersive X-ray spectrometer with Rh-tube excitation. Major elements were evaluated by a fundamental parameter calibration procedure, whereas trace elements were determined using a combined Compton and intensity matrix correction procedure.

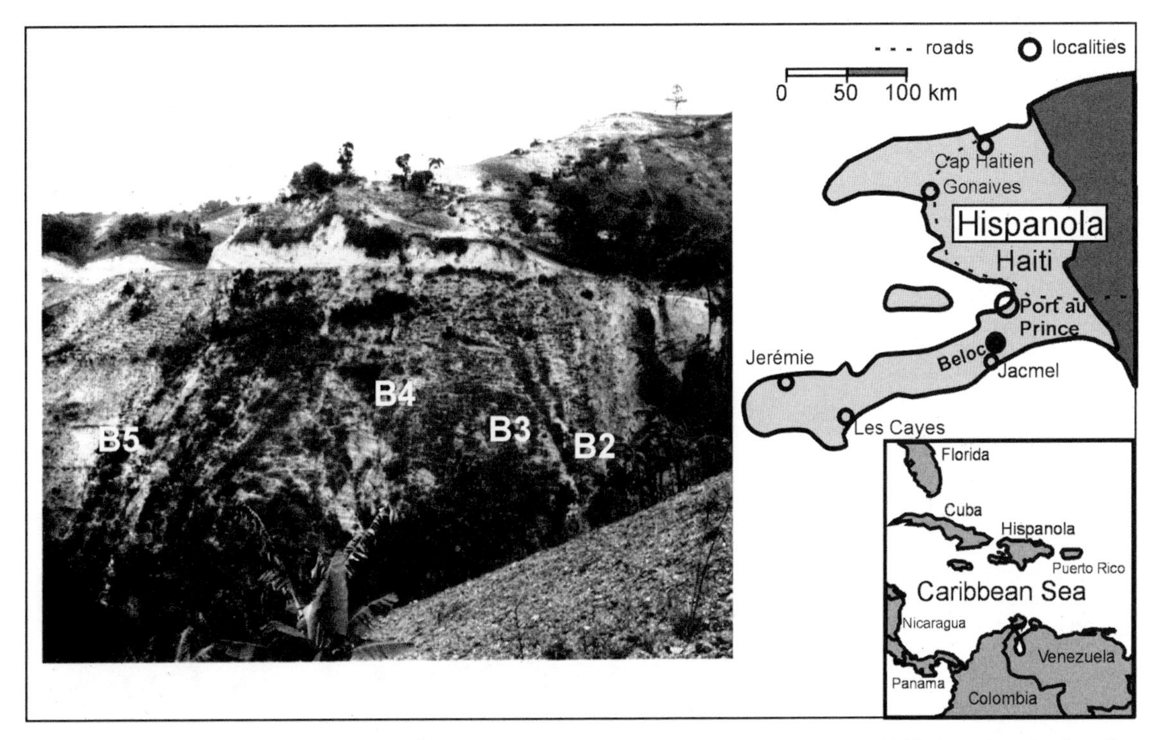

Figure 1. Sketch map of topographic setting of Beloc on island of Haiti and photo of exposed hill slope showing location of Beloc sections B2, B3, and B4 (map is redrawn from Planet Erde multimedia CD, Bertelsmann, 1997, ISBN 3-575-09509-4).

Wavelength-dispersive electron microprobe analyses were conducted by a Cameca SX50 microprobe to determine the main composition of spherules. Accelerating voltage was set to 15–20 kV and counting times of 120–150 s were used. Automatic calibration was performed by means of standard reference minerals. Detection limits are between 500 and 1000 mg/kg. Sulfur contents were evaluated in energy-dispersive mode, detection limits being above 1000 ppm for this element.

Rare earth elements

We selected 10 samples to evaluate the rare earth element (REE) patterns of the bulk material of typical lithological units of the profile. In addition, three samples were prepared of hand-picked spherules consisting of yellowish glass, dark colored glass, and spherules with smectite-like composition, in order to evaluate possible differences in the REE patterns of the different types of spherules in the spherule-rich deposit. After complete digestion with concentrated HF-$HClO_4$(5:1), samples were taken up in HNO_3 (0.15 M). The ratio between sample weight and solution was ~1:1000. Element concentrations were measured by standard inductively coupled plasma-mass spectrometry (ICP-MS) technique (Fisons Plasmaquad 2 turbo+), using an REE multielement standard solution (High-Purity Standard, Charleston, South Carolina) for external calibration and Rh and Bi as internal standards. Analytical accuracy was tested by means of the U.S. Geological Survey Standards SCo-1 (Cody Shale) and SGR-1 (shale). For most elements the measured concentrations are within ±6 relative% of certified values, and precision was better than 3%.

PGEs

We selected 28 samples across the profile for PGE analysis. The very low contents impose a combined procedure of preconcentration and matrix elimination prior to the detection with ICP-MS. This was achieved by fire assay with nickel sulfide collection on parts of thoroughly homogenized samples weighing ~50g. After drying at 105°C, sample material mixed with 30 g of sodium carbonate, 100 g of sodium tetraborate, 10 g of nickel powder, 7.5 g of sulfur and 5–10 g of diatomite was fused for 1 h at 1140°C. The resulting melt segregates by liquation, leading to a Ni-sulfide phase (Ni button or regulus), which collects and concentrates the PGEs. The cooled regulus was crushed and its bulk (NiS) dissolved in concentrated HCl, without attacking the precious metals and their alloys. After filtration through a polytetrafluorethylene (PTFE) membrane filter, the part of the residue containing the PGEs was dissolved in a mixture of hydrogen peroxide and concentrated HCl. Undissolved particles were retained by filtration through a white band quality paper filter. The solution containing the PGEs was slowly evaporated to dryness and taken up with 1% HNO_3 in 10 mL flasks. All chemicals used in the fire-assay step were of reagent grade, those used in the digestion were of suprapure

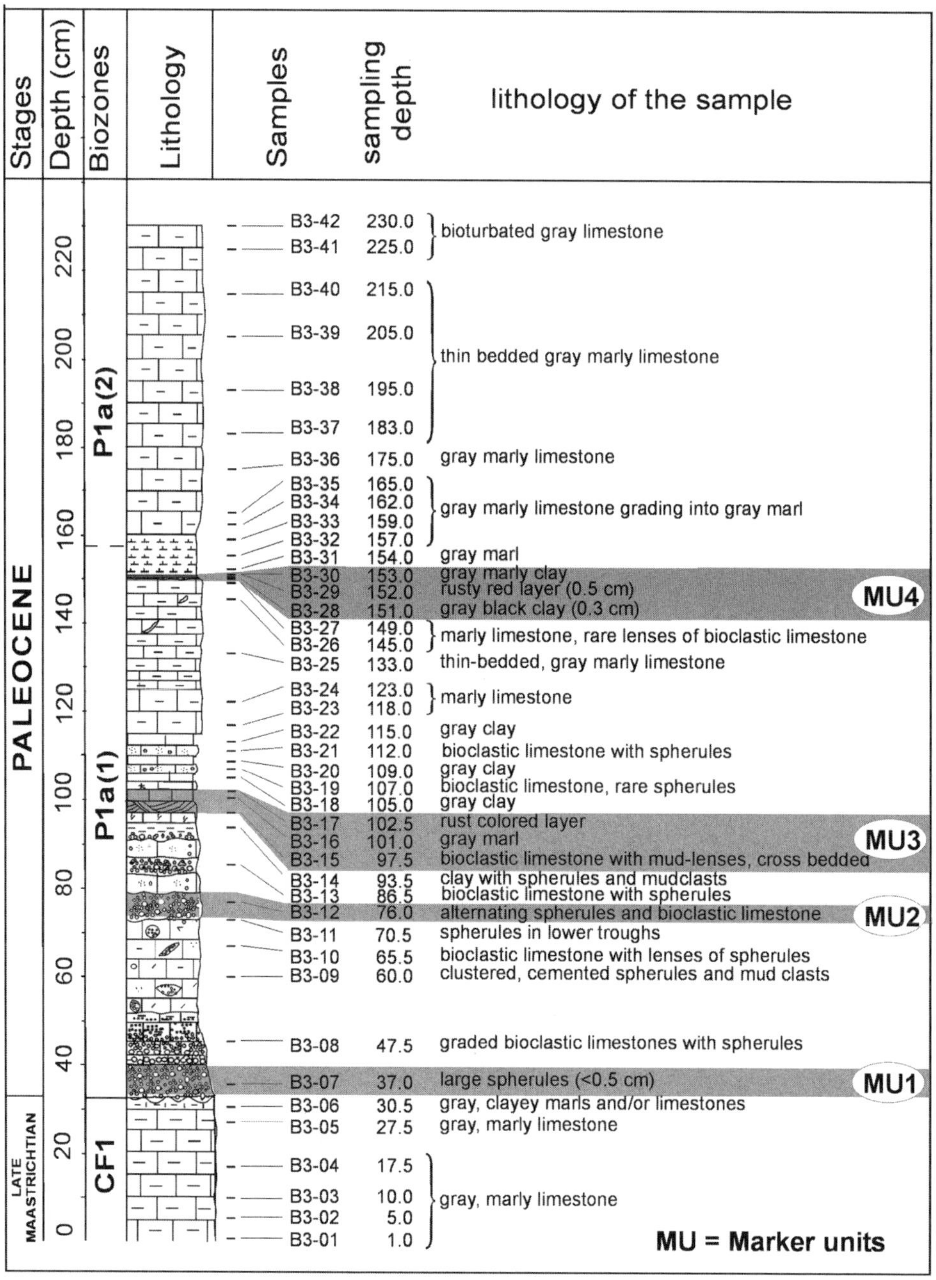

Figure 2. Description of Beloc 3 section giving stratigraphic unit, lithology, sample location, sampling depth, and lithological description of each subsample. (Lithology, stratigraphy, and biozones are redrawn from Stinnesbeck et al., 1999.)

grade. ICP-MS analysis was performed in a similar way as described here, using PGE multielement standard Claritas (SPEX Industries, Grasbrunn, Germany) and Tm and Bi as internal standards.

Accuracy was checked by means of WPR-1 (Canada Centre for Mineral and Energy Technology, CANMET) and SARM-7 (South African Committee for Certified Reference Materials) certified reference materials. The measured contents were found to be within 11% of the certified values, while precision, estimated by the standard deviation of replicate analyses (3σ) was in the range 7%–10%. Based on reference samples analyzed during several years in the lab, recovery was estimated to be better than 85%, which is in the range of the efficiency usually obtained by NiS fire assay (Date et al., 1987; Reddi et al., 1994; Zereini et al., 1994). Detection limits as determined from the average of the blanks were 0.09 ng/g Ir, 0.1 ng/g Rh, 0.4 ng/g Pd, and 0.4 ng/g Pt. These values are mainly dependent on the NiS fire assay; instrumental detection limits of the ICP-MS being approximately two orders of magnitude lower.

Stable isotopes (carbon and oxygen)

Carbon and oxygen isotope ratios in the carbonate fraction of finely ground bulk samples were determined by means of an automated carbonate preparation system connected on-line to an isotope ratio mass spectrometer (MultiPrep and Optima, both from Micromass UK, Ltd.). The preparation line is based on the standard method by measuring the isotope ratios in CO_2 released by reaction with 100% phosphoric acid. Samples are dissolved in individual vials, eliminating the risk of cross-contamination from one sample to the next, a requirement crucial for accurate and precise isotope ratio measurements on samples as small as <10 µg of calcite. Gas from a CO_2 pressure bottle, calibrated and certified by Messer Griesheim (Germany), was used as reference gas. Instrumental precision was better than 0.008‰ for $\delta^{13}C$ and <0.015‰ for $\delta^{18}O$. Accuracy was checked in each of the analytical batches by running the carbonate standard NBS19. Results were in the range of the certified values within ±‰ for $\delta^{13}C$ and ±‰ for $\delta^{18}O$. Isotope values are reported relative to the Vienna Peedee belemnite (VPDB) standard for both carbon and oxygen.

Bulk-rock mineralogy

For mineralogical characterization of the samples, X-ray diffraction (XRD) analyses on whole-rock samples were carried out. We ground ~5 g of dried rock (60°C) to a homogeneous powder of particle sizes <40 µm (Kübler, 1983). The powdered material (800 mg) was pressed at 20 bar into a powder holder and analyzed by a SCINTAG XRD 2000 diffractometer, using standard semiquantitative techniques based on external standardization (Klug and Alexander, 1974; Kübler, 1983, 1987).

Degassing experiments

For the degassing experiments glass spherules were separated by hand-picking under binocular (samples B3–7 and B3–12). The sample weight (10–30 mg, representing 10–20 spherules) is limited with respect to the degassing rate and the content of volatiles. The evolved gas analyses (EGA) were carried out using high-temperature mass spectrometry as described in detail in Heide (1974) and Stelzner and Heide (1996). The degassing process occurs far from equilibrium conditions. Reverse reactions between the volatiles and the melt thus cannot occur. The samples are heated under vacuum of 10^{-4} Pa to 10^{-3} Pa at a rate of 10 K/min to 1500°C; hence interactions between the evolved gases were minimized and a determination of the primary volatile species is possible. Turbomolecular pumps are used to eliminate hydrocarbons in the background of the vacuum devices. Analyses of the volatiles were carried out using a quadrupol mass spectrometer (QMA 125, Balzers) operated in rapid scan mode or in a multiple ion detection mode during the entire heating period. The measurements yield ion currents at distinct mass numbers (m/z 1–100; at a mass resolution of ~1 amu), which are proportional to the partial pressure of the various volatiles escaping from the melt during the heat treatment. Fragmentation of molecules occurs during ionization in the ion source of the mass spectrometer. Isotopic abundances and background contributions were taken into consideration for the interpretation of the gas release curves. To test the system background before and after a cycle, measurements were carried out in a blank experiment (i.e., degassing profiles of the equipment without samples). Detection limits depend on sample size and gas species; herein 0.1–1 ppm for CO_2 and SO_2 and ~100 ppm for H_2O.

GEOLOGICAL SETTING AND LITHOLOGY

The Beloc 3 (B3) section is located ~1 km north of Beloc (Fig. 1) on a steep slope, 20–30 m below the road. More than 100 m of Late Cretaceous and 30 m of Danian deposits are horizontally layered and exposed along this hillslope. The B3 section analyzed for this study includes 30 cm of latest Maastrichtian sediments, the K-T boundary, and 200 cm of Paleocene sediments (Stinnesbeck et al., 2001) (Fig. 3). Latest Maastrichtian pelagic marly limestones contain radiolarians, calcispheres, sponge spicules, ostracods, benthic foraminifera, and abundant planktic foraminifera and disconformably underlie spherule-rich deposits. An undulating erosional surface characterizes the lithological contact that separates the spherule-rich deposit from the latest Maastrichtian pelagic marly limestone. The spherule-rich deposit contains abundant evidence of reworked sediments, such as subrounded clasts of limestone, mudstone, and wackestone, some of which contain reworked early late Maastrichtian planktic foraminifera. The K-T boundary is placed at the disconformity at the base of the spherule-rich deposit based on the abundant presence of Danian planktic foraminifera in the spherule-rich deposit (Keller et al., 2001). Maurrasse and Sen (1991) and all other authors also place the K-T boundary at the lithological change between the limestones and the base of the spherule-rich deposit.

The spherule-rich deposit (30–100 cm) consists of 11 lithologically distinct layers (Stinnesbeck et al., 2001). These layers are characterized by changes in abundance of spherules and bioclastic debris and are separated by erosional contacts and reworking, as indicated by grain-size gradation and erosional disconformities. Four distinct lithological units (marker units, MU) can be recognized within the section, and these are interlayered with marls or marly limestones (Fig. 3). MU1 forms the basal 10–20 cm of the spherule-rich deposit just above the previously defined K-T boundary and is characterized by abundant spherules altered to blocky calcite and smectite. MU2 is located 70–80 cm upsection and is characterized by the presence of abundant black glass spherules with altered rims (to ~60%–70% relative abundance), visible with the naked eye in the outcrop. MU3 is near the top of the spherule-rich deposit at 100–102 cm from the base, and consists of a 2-cm-thick gray-green shale containing a thin rust-colored layer. MU4 is 150 cm above the base of the spherule-rich deposit and is marked by a rust-colored layer associated with volcanic ash between layers of micritic limestone rich in pelagic microfossils (e.g.,

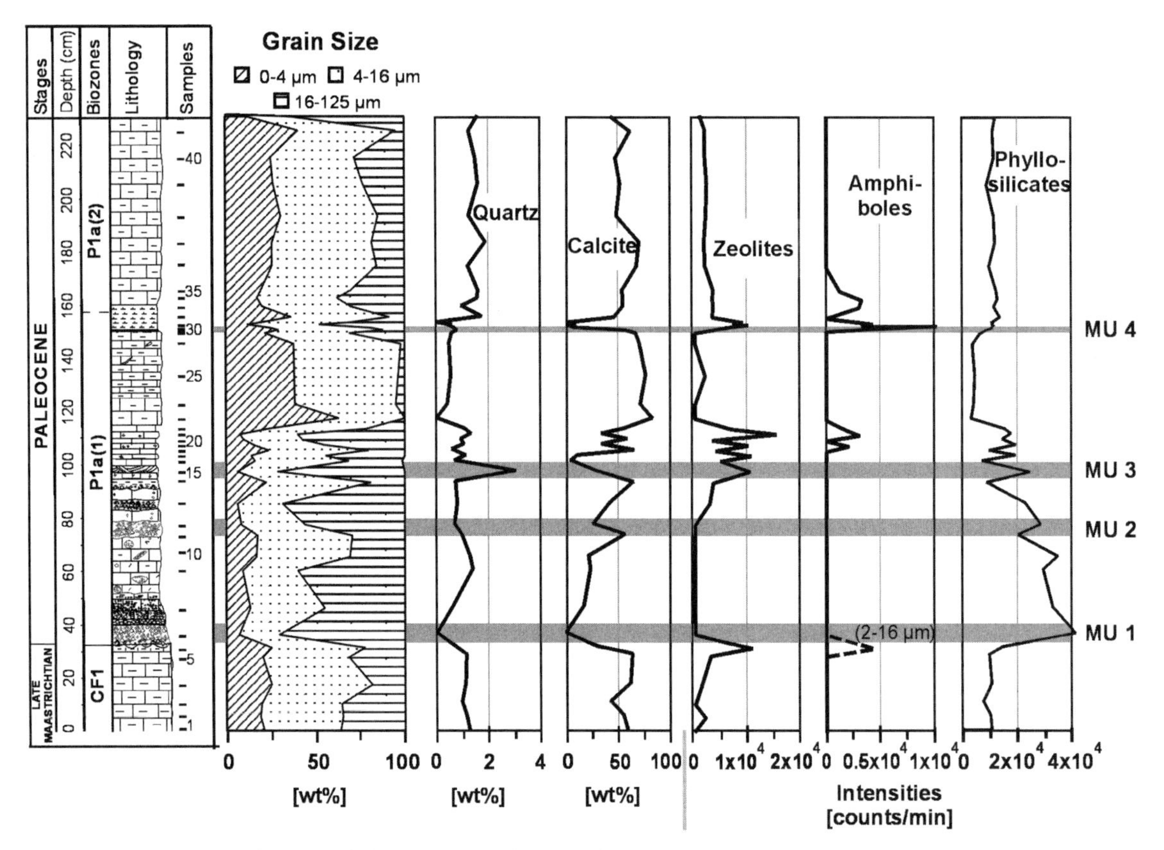

Figure 3. Stratigraphy of Beloc B3 Cretaceous-Tertiary (K-T) section showing grain size, quartz, calcite, zeolite, amphibole, and phyllosilicate distribution. (Lithology, stratigraphy, and biozones are redrawn from Stinnesbeck et al., 1999.)

planktic foraminifera and the calcareous nannofossil *Thoracosphaera*).

RESULTS AND DISCUSSION

Mineralogical results

Bulk-rock data of the Beloc section indicate that normal pelagic conditions dominated by biogenic calcite and low detrital quartz characterize the Maastrichtian (Fig. 3). A possible volcanic pulse is indicated in the uppermost Maastrichtian sediment layer (sample B3–6), as suggested by the presence of amphiboles (hornblende, basaltic type) and increased amount of zeolite (heulandite-clinoptilolite and phillipsite). Basaltic hornblende occurs in a large variety of volcanic rocks varying in composition from basalts to trachytes (Deer et al., 1992). Kring et al. (1994) interpreted hornblende in K-T boundary sediments at Beloc as reflecting a direct volcanic input or the erosion of volcanic rocks. Phillipsite predominates in volcanogenic and clayey sediments, as a diagenetic alteration product of basaltic glass (Kastner and Stonecipher, 1978). At Beloc, clinoptilolite is present throughout the section and derives either from chert and/or opal, or from alteration of basaltic glass (Hataway and Sachs, 1965; Weaver, 1968). The occurrence of zeolites (clinoptilolite and harmotome) at Beloc was reported by several authors in connection with altered glass spherules (Lyons and Officer, 1992; Jéhanno et al., 1992; Bohor and Glass, 1995), but note that maxima in hornblende coincide with high zeolite and smectite contents, suggesting that zeolites (especially phillipsite) mainly derive from alteration of basaltic volcanic material. There is no doubt that the smectite rinds around the glass cores of spherules are a direct alteration product of the spherule glass. However, the mineral association zeolite, smectite, and amphibole, characteristic for several distinct layers (e.g., at 30 cm in the topmost part of the Maastrichtian, at 10–20 cm above MU3, and in MU4) is not always associated with spherules in this section. Consequently, smectites and zeolites are not necessarily an alteration product of the spherules (Bohor and Glass, 1995), but could also be indicative of episodic volcanic inputs throughout the section.

The spherule-rich layers within the spherule-rich deposit (MU1 and MU2) are generally characterized by a high amount of phyllosilicates (mostly smectite, which is probably derived from glass alteration), variable calcite, and low quartz content. The absence of significant detrital input into these layers is consistent with an interpretation of reworked particles (e.g., spherules, glass, planktic foraminifera) derived from a pelagic environment (e.g., upper slope), rather than intensive erosion of an adjacent emergent area.

MU3 and MU4, which are characterized by significant PGE anomalies (see following) are rich in zeolites and MU4 may coincide with a volcanic influx, as indicated by a high amphibole content. The marly limestone layers located between the two PGE anomalies show mineralogical assemblages dominated by high calcite, and reflect normal pelagic conditions.

A volcaniclastic contribution was also observed by Jéhanno et al. (1992) and Leroux et al. (1995) in their Beloc section ~70–80 cm above the spherule-rich deposit and in association with shocked minerals and an Ir anomaly of 28 ng/g. Revisiting the Beloc section of Maurrasse and Sen (1991) and Lamolda et al. (1997), we found hornblende and increased zeolite abundances in the upper part of the section in association with their reported Ir anomaly.

Major and trace elements

Concentrations of major and trace elements in the Beloc K-T section show considerable variations that indicate fluctuation in the source and nature of the deposited materials, and may also reflect postdepositional changes (Tables 1 and 2). The geochemistry of marine sediments depends on a series of factors, the most important being the biogenic particle flux (productivity) and terrigenous flux. The latter represents a mixture of wet-deposited and eolian-transported material, downslope clastic input near continental margins, volcaniclastic and hydrothermal flux in areas with volcanic activities, and accumulation of authigenic phases. High productivity results in accumulation of biogenic carbonate, opal, and C_{org} dominated sediments (Leinen et al., 1986). Calcium, Mg, Sr, and Cd are incorporated into biogenic carbonate. Manganese, traces of other metals (e.g., V, Mo), and REEs tend to be adsorbed onto organic matter. The major parts of Ti, Al, Cr, K, and Fe are bound in the lattices of minerals of terrigenous debris (Murray and Leinen, 1993). Within this group Ti is by far the most immobile element during weathering (Schroeder et al., 1997). The small amount of Ti involved in the biogeochemical cycle (Orians et al., 1990) is negligible. Authigenic phases (e.g., zeolites, authigenic carbonates, manganese coatings, and hydrogenous phases) play an important role in the accumulation of trace metals in sediments. The major part of Ba is removed from the water column by precipitation as sulfate and is enriched at continental margins (Torres et al., 1996). Alteration can lead to a considerable redistribution of the elements, but it is generally restricted to the volcanogenic material, and is less important in the more resistant detrital components and in authigenic phases that are generally in equilibrium with the depositional environment. Postburial diagenesis is unlikely to change the ratio between the accumulated metals, although it may redistribute some of the trace elements among the different mineral phases on a local scale (Leinen and Stakes, 1979).

CaO is the main component for biogenic carbonate sedimentation and varies from 4 wt% (in layers dominated by clays and terrigenous and/or volcanic material) to 50 wt% (nearly pure calcite) with a clear negative correlation to Zr and TiO_2 (Fig. 4). Because Zr and Ti are nearly immobile during weathering or diagenetic alteration (e.g., Barkatt et al., 1984; Zielinski, 1982, 1985) and are introduced into the sediments exclusively by detrital accumulation, they can be used as proxies for the amount of terrigenous and/or volcanogenic components within the sediment.

From the base of the section to the paleontologically defined Maastrichtian-Paleocene boundary at 30 cm (Keller et al., 2001), CaO concentrations in the marly limestones decrease slightly (from ~35 to 25 wt%). Zirconium and TiO_2 contents (Fig. 4) remain nearly constant at low levels (average 27 µg/g Zr; 0.2 wt% TiO_2), indicating a very low and approximately constant terrigenous component in the sediment during the latest Maastrichtian. Within the spherule-rich layer of MU1, Zr and TiO_2 concentrations increase (~70 µg/g Zr; ~0.4 wt% TiO_2), whereas the Rb content decreases from 10 to 2 µg/g. These elements remain relatively constant in the interval from 30 to 65 cm. CaO also shows constant values of ~20 wt% in this interval, except for one sample at 35 cm, where the value drops to 5 wt% (Fig. 4). In general, this part of the section is less carbonate rich than the late Maastrichtian, but shows the highest phyllosilicate content (Fig. 4).

In the upper part of the spherule-rich deposit (65–100 cm), these elements display rapidly changing compositions of a factor of 2 for CaO and Zr, indicating abrupt changes in the sediment mineralogy (increasing calcite to the detriment of phyllosilicates). These rapid changes can be explained by reworking and redeposition of older Maastrichtian sediments within the early Danian. At 102 cm, MU3 contains a thin, rust-colored Fe-rich layer (to 19 wt% Fe_2O_3; Fig. 4) enriched in most of the trace elements examined (e.g., TiO_2, Zr, Ni, Cu, Zn, Cd, Sn, and Sb), and very low in CaO (~5 wt%) (Figs. 4 and 5). This rust-colored layer is overlain by a 50-cm-thick marly limestone, which is low in TiO_2, Zr, Rb, Cu, Zn, Sn, and Sb, and indicates very low terrigenous input and high carbonate accumulation.

Dark gray clay and a second Fe-rich horizon (to 20 wt% Fe_2O_3) are observed between 150 and 152 cm (MU4). These layers are clearly enriched in Cu, Zn, Ni, As, Pb, Cd, and Sb (Fig. 5). Compared with the marly limestone between the two rust-colored layers, the marls overlying the dark gray clays of MU4 (152–239 cm) show higher concentrations of terrigenous or volcanic materials (e.g., Rb, Zr, and Ti). Rubidium content increases continuously and Ti and Zr increase slightly upsection.

Strontium content varies from 96 to 950 µg/g within the section, whereas Ba ranges from 200 to 6500 µg/g. Within the Beloc section, Ba shows very low concentrations (~200 µg/g) in the marly limestone of the latest Maastrichtian, whereas Sr displays intermediate concentrations (500–600 µg/g). Both elements are evidently enriched (950 µg/g Sr and 6455 µg/g Ba) at the Maastrichtian-Paleocene boundary. The overlying spherule-rich layers and bioclastic limestones (30–100 cm above the base) are lower in Sr (200–300 µg/g) and higher in

TABLE 1. MAJOR ELEMENT COMPOSITION OF BULK SAMPLES

Sample #	Sampling depth (cm)	Na_2O (wt%)	MgO (wt%)	Al_2O_3 (wt%)	SiO_2 (wt%)	P_2O_5 (wt%)	CaO (wt%)	K_2O (wt%)	TiO_2 (wt%)	MnO (wt%)	Fe_2O_3* (wt%)	LOI (wt%)	Total (wt%)
BE3-42	230	0.33	2.15	4.69	34.32	0.17	30.07	0.62	0.30	0.04	2.99	25.63	101.33
BE3-41	225	0.30	1.79	4.07	42.03	0.14	27.09	0.55	0.26	0.03	2.81	23.04	102.12
BE3-40	215	0.38	2.10	4.87	33.42	0.15	28.89	0.67	0.34	0.07	3.57	24.91	99.36
BE3-39	205	0.34	1.85	4.72	31.10	0.13	31.80	0.57	0.30	0.04	3.12	27.01	100.99
BE3-38	195	0.24	1.39	3.74	16.27	0.09	40.47	0.41	0.22	0.08	2.21	33.21	98.34
BE3-37	183	0.23	1.69	3.92	20.69	0.12	38.15	0.44	0.23	0.08	2.53	32.32	100.39
BE3-36	175	0.26	1.43	3.82	17.29	0.10	40.66	0.48	0.24	0.08	2.04	31.96	98.36
BE3-35	165	0.32	2.04	5.12	23.20	0.12	35.50	0.63	0.30	0.06	3.16	29.79	100.24
BE3-34	162	0.33	2.12	5.54	23.30	0.11	34.51	0.70	0.32	0.05	3.08	29.05	99.10
BE3-33	159	0.32	2.27	5.60	25.08	0.17	33.96	0.65	0.31	0.05	3.50	29.26	101.18
BE3-32	157	0.35	3.15	7.01	29.61	0.13	30.01	0.68	0.35	0.03	3.78	25.50	100.60
BE3-31	154	0.36	3.46	7.71	32.94	0.12	27.63	0.65	0.35	0.03	4.15	23.79	101.20
BE3-30	153	0.79	5.87	13.20	51.13	0.08	8.45	0.43	0.51	0.01	11.09	8.18	99.75
BE3-29	152	1.05	3.33	11.09	41.83	0.12	7.92	0.33	0.50	0.07	18.91	9.03	94.17
BE3-28	151	0.32	2.16	5.97	27.26	0.11	31.40	0.51	0.31	0.05	3.75	26.81	98.63
BE3-27	149	0.08	0.98	1.69	18.20	0.10	42.88	0.15	0.11	0.07	1.20	35.58	101.05
BE3-26	145	0.07	0.66	1.08	8.42	0.06	49.52	0.14	0.07	0.10	0.67	39.64	100.43
BE3-25	133	0.11	0.97	1.74	9.85	0.08	47.03	0.22	0.13	0.08	1.09	37.72	99.01
BE3-24	123	0.05	0.70	0.90	5.90	0.05	49.14	0.11	0.06	0.09	0.71	39.74	97.47
BE3-23	118	N.D.	N.D.	N.D.	N.D.	N.D.	N.D.	N.D.	N.D.	N.D.	N.D.		
BE3-22	115	0.08	1.17	2.57	15.37	0.05	44.21	0.18	0.12	0.08	1.37	35.78	100.98
BE3-21	112	0.17	1.85	5.17	25.28	0.06	35.07	0.23	0.17	0.04	2.25	30.48	100.78
BE3-20	109	0.13	2.18	4.51	20.85	0.06	37.96	0.20	0.17	0.05	2.14	32.55	100.81
BE3-19	107	0.09	3.44	6.77	33.03	0.05	28.87	0.25	0.31	0.01	4.17	24.94	101.94
BE3-18	105	0.07	1.54	3.29	15.56	0.08	41.99	0.21	0.11	0.06	1.12	35.59	99.62
BE3-17	102.5	0.05	2.88	6.90	39.40	0.05	10.38	0.21	0.45	0.03	15.88	23.37	99.62
BE3-16	101	0.10	2.84	5.80	28.27	0.08	30.90	0.37	0.25	0.06	2.52	27.75	98.93
BE3-15	97.5	0.06	4.37	7.22	33.31	0.02	16.95	0.21	0.35	0.03	4.35	32.90	99.77
BE3-14	93.54	0.11	3.57	6.32	28.82	0.07	28.08	0.34	0.21	0.04	2.55	31.12	101.24
BE3-13	86.5	N.D.	N.D.	N.D.	N.D.	N.D.	N.D.	N.D.	N.D.	N.D.	N.D.	N.D.	
BE3-12	76	0.03	6.00	6.58	30.78	0.04	25.78	0.10	0.26	0.09	5.01	25.08	99.76
BE3-11	70.5	0.04	4.78	5.84	27.05	0.04	31.25	0.15	0.25	0.04	3.96	27.38	100.79
BE3-10	65.5	0.04	7.35	9.36	42.10	0.02	16.65	0.16	0.39	0.01	6.31	18.38	100.78
BE3-09	60	0.04	6.37	8.99	43.29	0.02	15.69	0.18	0.45	0.03	6.70	18.28	100.02
BE3-08	47.5	<0.02	8.09	11.72	50.26	0.02	10.74	0.13	0.52	0.01	7.77	14.25	103.53
BE3-07	37	N.D.	8.26	11.72	51.00	0.01	5.30	0.10	0.53	0.04	7.96	10.82	95.74
BE3-06	30.5	0.54	3.20	10.16	45.02	0.11	19.82	0.58	0.49	0.03	3.76	18.20	101.93
BE3-05	27.5	0.28	1.56	3.58	23.96	0.13	36.59	0.45	0.25	0.07	2.42	30.71	100.00
BE3-04	17.5	0.21	1.00	2.49	46.81	0.07	26.96	0.32	0.19	0.03	1.70	21.76	101.54
BE3-03	10	N.D.	N.D.	N.D.	N.D.	N.D.	N.D.	N.D.	N.D.	N.D.	N.D.	N.D.	
BE3-02	5	0.25	1.28	3.09	26.31	0.10	36.28	0.37	0.21	0.06	2.10	30.00	100.08
BE3-01	1	0.25	1.50	3.26	31.69	0.09	33.77	0.42	0.23	0.06	2.22	27.23	100.72

Note: Major element composition and LOI of the bulk samples determined by Wavelength Dispersive X-ray Fluorescence spectrometry (WDXRF). Concentrations of major elements in the Beloc Cretaceous-Tertiary section B3 show considerable variations, which primarily indicate fluctuation in the source and nature of the deposited materials, but may also reflect postdepositional changes.
*All Fe as Fe_2O_3
N.D. = not determined; LOI = loss on ignition.

Ba concentrations (450–800 mg/g) than the Maastrichtian marly limestone. The two rust-colored layers of MU3 and MU4 and adjacent clays are enriched in both elements (to 920 mg/g Sr and 6000 μg/g Ba) (Fig. 4). Above the gray clay of MU4, Sr and Ba contents decrease gradually to the top of the section. Barium shows a strong and Sr a moderate positive correlation to Zr and TiO_2 in the interval above 90 cm (Fig. 4).

The abundances of Sr and Ba in sediments are controlled by various factors, including terrigenous influx, biogenic flux, hydrothermal input, and sedimentation rates (Schroeder et al., 1997). To evaluate the part of the elements bound to the diffuse terrigenous component, the so-called excess concentrations of various elements (El*) were calculated (Murray and Leinen, 1993, 1996; Schroeder et al., 1997). Assuming the same element to Ti ratio in the sample as in a representative reference material, a theoretically expected content for each element can be calculated. The excess concentration (El*) is the deviation between the actually found content and the expected concentration of the respective element:

$$El^* = El_{total} - [Ti_{sample}{}^* (El_{NASC}/Ti_{NASC})], \quad (1)$$

where El_{total} is the bulk concentration of the element; Ti_{sample}, the concentration of Ti in the sample; El_{NASC} and El_{PAAS} are the concentrations of the element in the NASC (North Ameri-

TABLE 2. MINOR ELEMENT CONTENTS AND STABLE ISOTOPE DATA OF BULK SAMPLES

Sample #	Sampling depth (cm)	Cr (µg/g)	Ni (µg/g)	Cu (µg/g)	Zn (µg/g)	As (µg/g)	Rb (µg/g)	Sr (µg/g)	Y (µg/g)	Zr (µg/g)	Nb (µg/g)	Ba (µg/g)	Pb (µg/g)	$\delta^{13}C$ ‰ VPDB	$\delta^{18}O$ ‰ VPDB
BE3-42	230	48	32	83	101	3	20	619	35	91	3	1993	<5	−0.27	−4.00
BE3-41	225	47	32	59	60	<3	16	561	22	79	<2	1199	<5	0.11	−3.41
BE3-40	215	51	35	73	90	3	20	612	24	90	3	2547	<5	−0.45	−3.73
BE3-39	205	47	28	81	84	<3	15	655	27	94	3	2011	<5	−0.39	−3.24
BE3-38	195	62	34	61	66	<3	12	734	33	85	10	1579	<5	−0.17	−2.75
BE3-37	183	37	26	46	68	<3	14	639	33	92	<2	1508	<5	0.41	−1.48
BE3-36	175	72	29	57	66	<3	14	760	34	89	13	1715	<5	0.27	−1.41
BE3-35	165	68	35	89	110	3	17	800	29	97	12	3000	7	0.04	−1.98
BE3-34	162	56	34	77	114	3	24	752	41	110	7	2792	<5	−0.42	−2.61
BE3-33	159	50	35	70	97	3	18	750	37	105	4	3000	<5	−0.21	−1.78
BE3-32	157	73	55	72	87	<3	22	763	13	108	13	3093	<5	−1.46	−1.83
BE3-31	154	53	58	77	84	<3	15	784	17	109	10	3899	<5	−0.86	−1.75
BE3-30	153	29	103	105	149	8	10	666	4	116	10	5273	7	−1.68	−1.19
BE3-29	152	54	209	145	152	21	7	792	5	62	4	5479	12	−0.37	−2.83
BE3-28	151	73	42	60	80	<3	14	865	17	97	12	5371	<5	0.39	−2.08
BE3-27	149	25	15	32	50	<3	6	618	11	81	<2	796	<5	0.71	−0.82
BE3-26	145	30	20	21	36	N.D.	6	630	9	66	<2	690	<5	0.74	−0.93
BE3-25	133	58	25	33	42	<3	10	663	13	69	3	941	<5	0.86	−1.06
BE3-24	123	37	16	22	29	<3	7	683	11	67	<2	452	<5	0.77	−0.62
BE3-23	118	N.D.	N.D.	15	18	<3	N.D.	N.D.	0	N.D.	N.D.	410	<5	0.64	−0.69
BE3-22	115	38	20	20	37	<3	7	735	18	81	4	1305	<5	0.42	−0.77
BE3-21	112	57	21	N.D.	N.D.	N.D.	7	889	12	109	3	N.D.	N.D.	−0.15	−0.78
BE3-20	109	46	31	23	48	<3	4	617	12	94	<2	1400	<5	−0.50	−0.93
BE3-19	107	59	47	35	73	<3	12	955	2	118	3	2181	<5	−0.56	−2.16
BE3-18	105	47	19	20	32	<3	5	632	17	84	N.D.	1285	<5	−1.04	−1.16
BE3-17	102.5	53	103	41	97	49	11	754	N.D.	98	N.D.	2017	16	−0.27	−3.75
BE3-16	101	40	34	29	45	<3	7	701	N.D.	92	<2	2642	<5	−0.29	−1.97
BE3-15	97.5	40	38	32	89	<3	3	557	3	114	<2	5939	6	−1.05	−3.77
BE3-14	93.5	45	23	20	33	<3	9	706	20	112	3	1463	<5	−0.49	−1.12
BE3-13	86.5	N.D.	N.D.	73	90	3	N.D.	N.D.	0	N.D.	N.D.	2547	<5	−0.21	−2.06
BE3-12	76	32	23	22	30	<3	<2	324	9	82	<2	584	<5	−5.05	−4.06
BE3-11	70.5	56	25	19	22	<3	<2	392	6	86	<2	671	<5	−1.72	−1.41
BE3-10	65.5	42	25	22	31	<3	5	213	2	89	4	466	<5	−5.65	−4.03
BE3-09	60	39	25	20	24	<3	11	261	N.D.	94	<2	876	<5	−0.09	−1.64
BE3-08	47.5	40	42	24	30	<3	4	191	N.D.	97	2	854	6	−4.95	−3.77
BE3-07	37	38	82	21	43	<3	6	140	N.D.	93	3	1457	<5	−3.34	−1.16
BE3-06	30.5	61	33	50	60	<3	19	1154	12	130	11	6455	<5	−0.37	−2.51
BE3-05	27.5	51	20	N.D.	N.D.	N.D.	13	711	34	93	<2	N.D.	N.D.	0.56	−2.09
BE3-04	17.5	45	24	39	51	<3	11	610	11	54	4	194	<5	1.53	−2.71
BE3-03	10	N.D.	N.D.	51	64	N.D.	N.D.	N.D.	0	N.D.	N.D.	326	<5	1.25	−3.02
BE3-02	5	N.D.	N.D.	55	69	<3	N.D.	N.D.	0	N.D.	N.D.	435	<5	0.64	−2.04
BE3-01	1	57	28	49	117	<3	14	726	19	75	9	384	<5	1.46	−2.09

Note: Minor element contents and stable isotope data of the bulk samples (minor elements determined by Energy Dispersive X-ray Flourescence [EDXRF] and Wave Dispersive X-ray Flourescence [WDXRF] and C and O isotopes by Isotope Ratio Mass Spectrometry). The minor element contents show considerable variations, which primarily indicate fluctuation in the source, reworking, and primary productivity. C and O isotopes mainly reflect climatic changes.
N.D. = not determined; VPDB = Vienna Peedee belemnite.

can Shale Composite) and the PAAS (Post-Archaean Average Australian Shale), respectively, and Ti_{NASC} is the concentration of Ti in NASC. Because the eolian input in the Caribbean originates mainly from North America, data for the NASC (Gromet et al., 1984) were used instead of PAAS (Taylor and McLennan, 1985), except where no NASC data were available (e.g., for Cu and Zn).

To allow an easier comparison of the excess concentrations of different elements, the ratio of excess to bulk concentrations were calculated:

$$R_{El} = El^*/El_{total}. \quad (2)$$

This ratio expresses the relative amount of an element that is not bound to the detrital components of the sediment. Accuracy of the estimates depends on the composition of the local terrestrial rocks, which were exposed to weathering, but variations in the average composition of the source rocks are probably much lower than the variations observed within the Beloc section.

The El^*/El_{total} ratios for Cu, Zn, Sr, and Ba (Fig. 6) and for Ca confirm that the major part of these elements was not introduced into the sediments by detrital minerals. However, characteristic units with specific slightly different features can be distinguished.

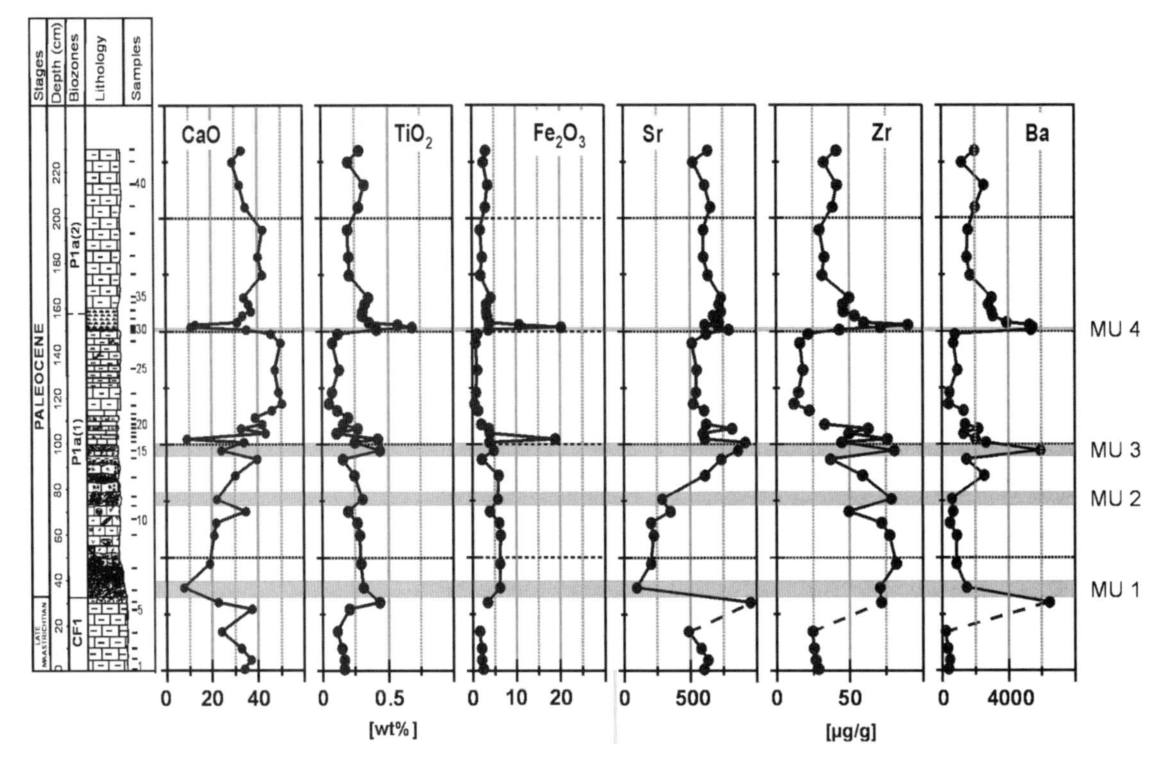

Figure 4. Distribution of CaO, TiO_2, Fe_2O_3, Sr, Zr, and Ba content along Cretaceous-Tertiary (K-T) section Beloc B3. (Lithology, stratigraphy, and biozones are redrawn from Stinnesbeck et al., 1999.)

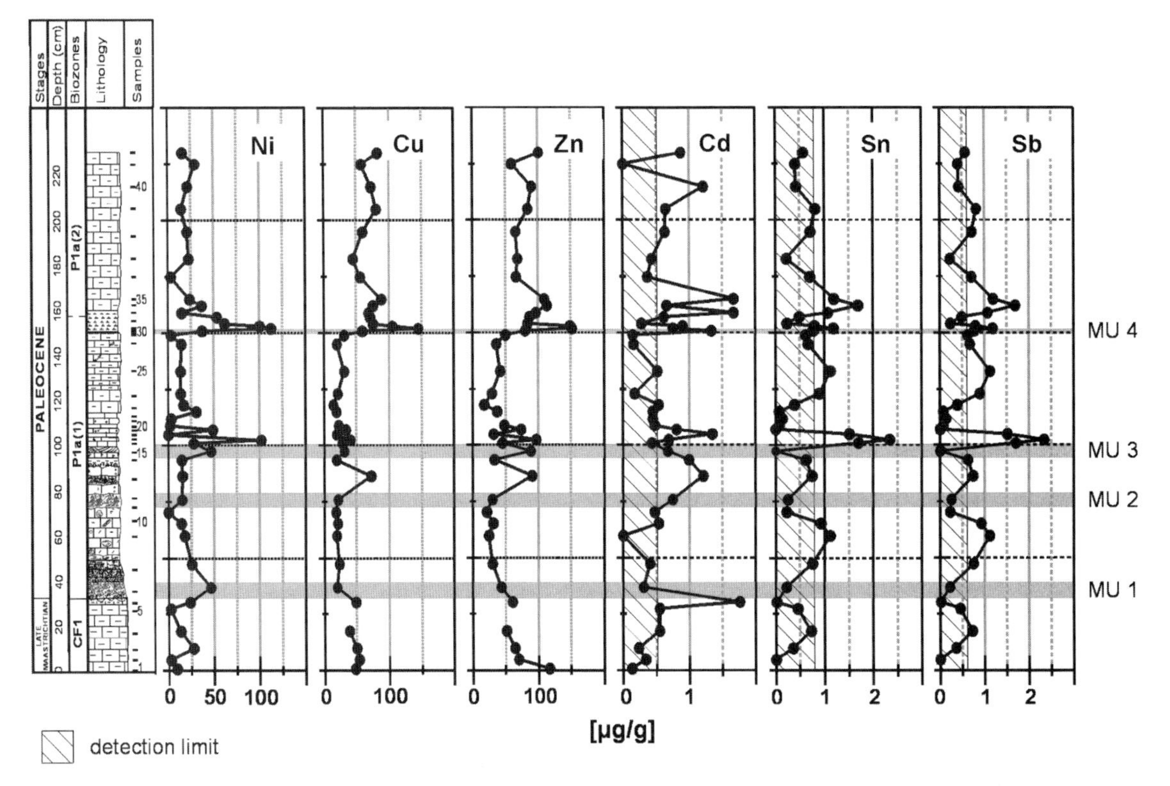

Figure 5. Distribution of Ni, Cu, Zn, Cd, Sn, and Sb content along K-T section Beloc B3. (Lithology, stratigraphy, and biozones are redrawn from Stinnesbeck et al., 1999.)

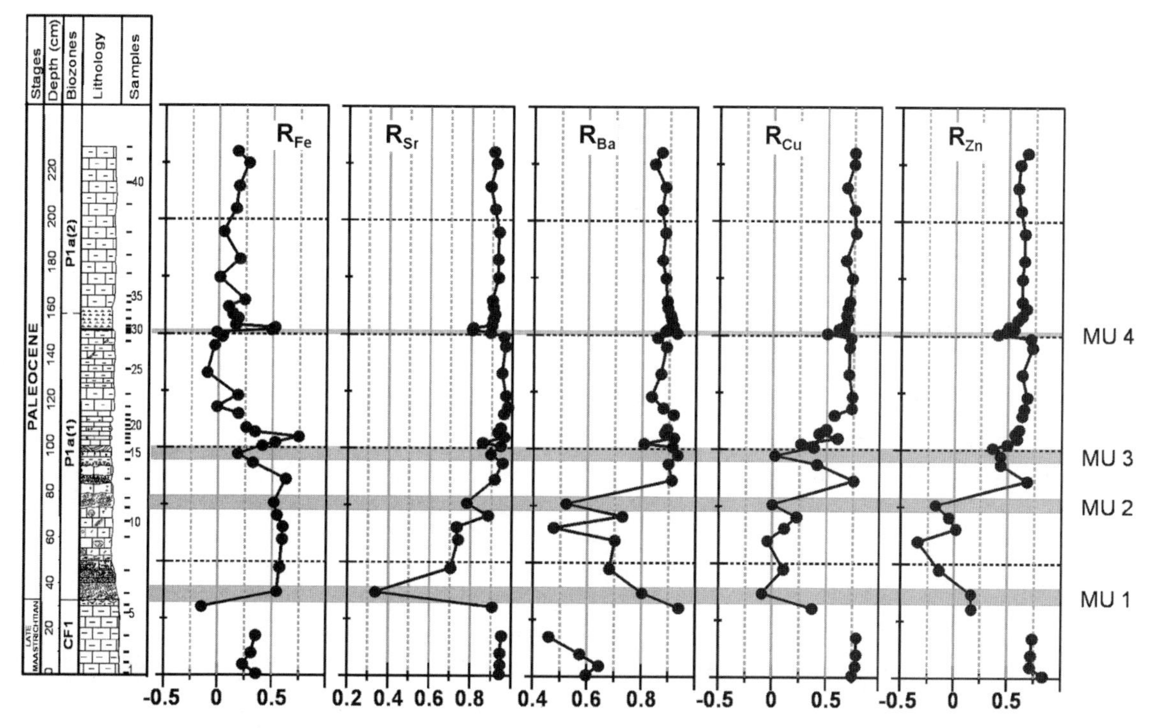

Figure 6. Excess ratios (R) of Fe, Sr, Ba, Cu, and Zn along Cretaceous-Tertiary (K-T) section Beloc B3. Excess ratios $R_{El} = El^*/El_{total}$ are estimates for ratio of part of element $El^* = El_{total} - [Ti_{sample}^* (El_{NASC}/Ti_{NASC})]$ not bound to detrital phase and bulk concentration El_{total}. R values can be interpreted as follows. $R \gtrsim 0.8$: only minor part of respective element is bound to detrital fraction of sediment; $R \sim 0.2$–0.8: significant part of element content is of detrital origin; $R \sim 0$: practically, entire amount of element is bound to detrital part of sediment; $R < 0$: detrital part of sediment derived from rocks with lower element/Ti ratios than North American Shale Composite (NASC) or Post-Archaean Average Australian Shale (PAAS) and/or part of respective element was lost during sedimentation or weathering. (Lithology, stratigraphy, and biozones are redrawn from Stinnesbeck et al., 1999.)

The Maastrichtian has a monotonous excess/bulk ratio, ~70% of the Fe and half of the Ba content being of terrigenous origin. Strontium, Cu, and Zn are controlled by incorporation in biogenic carbonates.

At the top of the Maastrichtian marly limestones (below the base of the spherule-rich deposit), increasing amounts of terrigenous and/or volcanic Fe, Cu, and Zn are indicated by the respective R values ($R_{Fe} = -0.2$; $R_{Cu} = 0.4$; $R_{Zn} = 0.2$; Fig. 6). The nondetrital fraction of Sr remains at high levels ($R_{Sr} \sim 0.8$) and that of Ba increases (R_{Ba} 0.4–0.95). A few centimeters above MU1 and to the top of the black spherule layer of MU2, Fe reaches maximum excess values ($R_{Fe} \sim 0.6$), except for the two Fe-rich horizons of MU3 and MU4. R_{Cu} and R_{Zn} fluctuate around 0. The nondetrital fraction of Sr shows a considerable drop in MU1 and increases upward to the top of MU2 ($R_{Sr} \sim 0.3$–0.8). Nonterrigenous Ba (Fig. 6) decreases in MU1 and shows considerable fluctuations to the top of MU2 ($R_{Ba} \sim 0.4$–0.7). Between MU1 and the top of MU2 reworking of different sediments is indicated by the high and variable contribution of the terrigenous and/or volcanic flux to the Cu, Zn, Sr, and Ba fractions together with a high and nearly constant excess rate of Fe.

At the base of both layers enriched in PGEs (MU3 and MU4) a slight increase in the fraction enriched in detrital components is observed for Cu, Zn, Sr, and Ba (Fig. 6).

In the limestone layers between MU3 and MU4, and in the marls above MU4, nearly all Fe can be assigned to the terrigenous and/or volcanic fraction with a small additional component increasing to the top of the profile. Copper, Zn, Sr, and Ba remain at nearly constant high excess rates, indicating accumulation of these elements by high biogenic carbonate sedimentation, by scavenging from the water column, and/or enriched authigenic minerals. The increase in detrital input as indicated by the R values of Sr, Ba, Cu, and Zn coincides with the influx of volcanic material.

To examine the possibility of a volcaniclastic input into the sediments the discrimination diagram proposed by Andreozzi et al. (1997) was used. Based on primary genetic differences in respect of the content and postdepositional behavior of some of the first-row transition elements (V, Ni, Cr) and immobile elements (Zr, TiO_2, Al_2O_3), Andreozzi et al. proposed a diagram in which volcaniclastic and normal terrigenous sediments define two separate areas and thus can be easily distinguished. Except for the two rust-colored, iron-rich layers (samples B3-17 and B3-29), all samples from the spherule-rich deposit (MU1, MU2, MU3, layers between) and from MU4 are much closer to the field of the volcaniclastic deposits than are those from CF1, from Pla(1) above the spherule-rich deposit, and from Pla(2) (Fig. 7), suggesting that a volcaniclastic component was admixed to the entire spherule-rich deposit and MU4. However, it is not clear if remelting of crustal material by impact would confer a geochemical behavior (specifically higher mobility) to the transition metals, similar to that empirically observed in volcaniclastic deposits.

Platinum group elements

The concentrations of PGEs display variations of nearly two orders of magnitude (Table 3; Fig. 7). The lowest PGE concentrations, corresponding to background values for marine carbonates, are observed in the Maastrichtian part (basal 30 cm) of the section with 0.050 ng/g Ir, 0.2 ng/g Rh, <0.5 ng/g Pt, and 1 ng/g Pd. Above the K-T disconformity Ir increases slightly to 0.2 ng/g, whereas Pt, Rh, and Pd remain at low levels. In the rust-colored layer of MU3 (at 102 cm), Ir is enriched up to a factor of 20 compared to local Maastrichtian background values, whereas the other PGEs are only slightly enriched. Iridium concentrations tail below and above this marker unit (Fig. 7). In the limestones above MU3, Ir decreases to 0.2–0.3 ng/g, but is still slightly higher as compared to the Maastrichtian sediments. A second, Pt- and Pd-dominated, PGE anomaly is observed in the rust-colored layer and gray clay of MU4 (Fig. 7). Iridium is only enriched by a factor of 2 as compared with the marls below and above, whereas Pt is enriched to 6 ng/g and Pd to 9 ng/g. The chondrite-normalized PGE pattern of these samples is similar to that of ocean-floor basalts (e.g., Greenough and Fryer, 1990), Hawaiian basalts (Crocket and Kabir, 1988), or rift-related acid volcanics (Borg et al., 1987). The chondrite-normalized PGE pattern of the marly layers of normal pelagic sedimentation (17.5 and 195 cm, but also some of the samples belonging to MU2 and MU3) shows a distinct negative Pt anomaly. Because the residence time of Pt in the oceanic reservoir is much higher than for Pd (i.e., seawater is relatively enriched in Pt), a negative Pt anomaly is considered to reflect sedimentation in equilibrium with seawater (Tredoux et al., 1989).

Normalized to chondrite (McDonough and Sun, 1995) and plotted in order of decreasing melting points, the PGE pattern of MU3 is roughly chondritic (Fig. 9) and contrasts with the PGE pattern of the rest of the section. Chondritic PGE ratios are usually interpreted as indicating an extraterrestrial origin (Kyte et al., 1980; Ganapathy, 1980). The peak of the Ir anomaly is centered on a clayey layer strongly enriched in Fe-oxihydroxides. Because PGEs are readily scavenged by ferromanganese phases (at pH 7–9), their distribution in oxidized marine sediments is dominated by Fe-Mn oxihydroxide minerals (Anbar et al., 1996; Stüben et al., 1999), although other possibilities—such as precipitation with bacterial iron oxides, sulfides, and concentration from seawater into sediments—are considered (Wallace et al., 1990; Colodner et al., 1992). How-

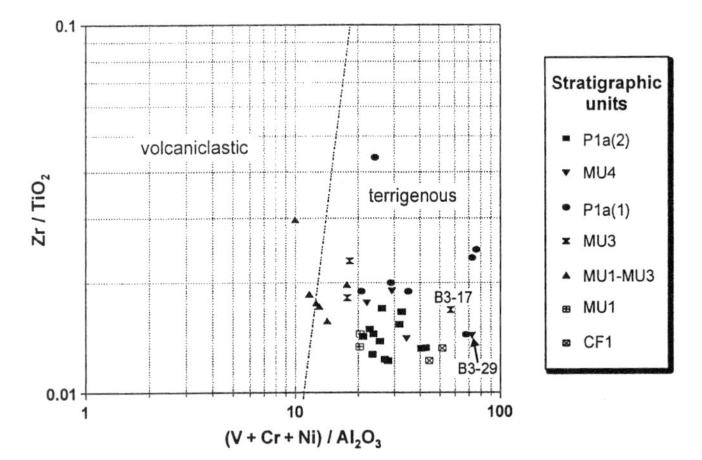

Figure 7. Discrimination diagram for sediments of volcaniclastic vs. terrigenous origin for different stratigraphic layers (according to Andreozzi et al., 1997). CF1 = *Plummerita hantkenoides* (CF1) zone of latest Maastrichtian; P1a(1) = lower *P. eugubina* (P1a[1]) zone of early Danian; P1a(2) = upper *P. eugubina* (P1a[2]) zone of early Danian.

TABLE 3. PLATINUM GROUP ELEMENT CONTENTS OF THE BULK SAMPLES

Sample #	Sampling depth (cm)	Ir (ng/g)	Pt (ng/g)	Rh (ng/g)	Pd (ng/g)
BE3-38	195	0.22	0.73	0.11	2.82
BE3-36	175	0.23	0.52	0.12	2.85
BE3-35	165	0.20	0.20	0.03	2.44
BE3-34	162	0.30	1.12	0.12	2.95
BE3-32	157	0.50	0.49	0.05	8.88
BE3-31	154	0.46	N.D.	N.D.	N.D.
BE3-30	153	0.29	0.78	0.02	7.54
BE3-29	152	0.56	6.19	0.11	8.10
BE3-28	151	0.14	0.11	0.00	2.92
BE3-26	145	0.19	N.D.	0.02	0.80
BE3-25	133	0.23	0.08	0.03	3.26
BE3-24	123	0.15	N.D.	0.02	1.15
BE3-22	115	0.27	4.93	0.05	2.42
BE3-19	107	0.59	N.D.	0.03	0.10
BE3-18	105	0.55	0.11	0.30	1.43
BE3-17	102.5	1.00	2.13	0.14	2.51
BE3-16	101	0.92	N.D.	0.05	N.D.
BE3-15	97.5	1.00	N.D.	0.09	0.06
BE3-14	93.5	0.66	0.39	0.04	2.04
BE3-12	76	0.28	0.19	0.09	0.63
BE3-10	65.5	0.26	N.D.	0.04	0.55
BE3-09	60	0.21	0.20	0.05	N.D.
BE3-08	47.5	0.20	N.D.	0.01	1.13
BE3-07	37	0.18	2.45	0.08	1.08
BE3-06	30.5	0.10	0.66	0.04	2.18
BE3-04	17.5	0.05	0.04	0.22	1.10
BE3-03	10	0.05	N.D.	0.40	1.40
BE3-01	1	0.09	0.61	0.18	1.82

Note: Platinum group element (PGE) contents of the bulk samples as determined by NiS fire assay and Inductively Coupled Plasma-Mass Spectrometry (ICP-MS). Neighboring samples were determined in different batches. The PGE contents display background values in the Maastrichtian, slightly elevated concentration in the Danian, and 2 PGE anomalies: an Ir-dominated one (samples Be3-14 to Be3-19) and a Pd-dominated one (samples Be3-29 to Be3-32).

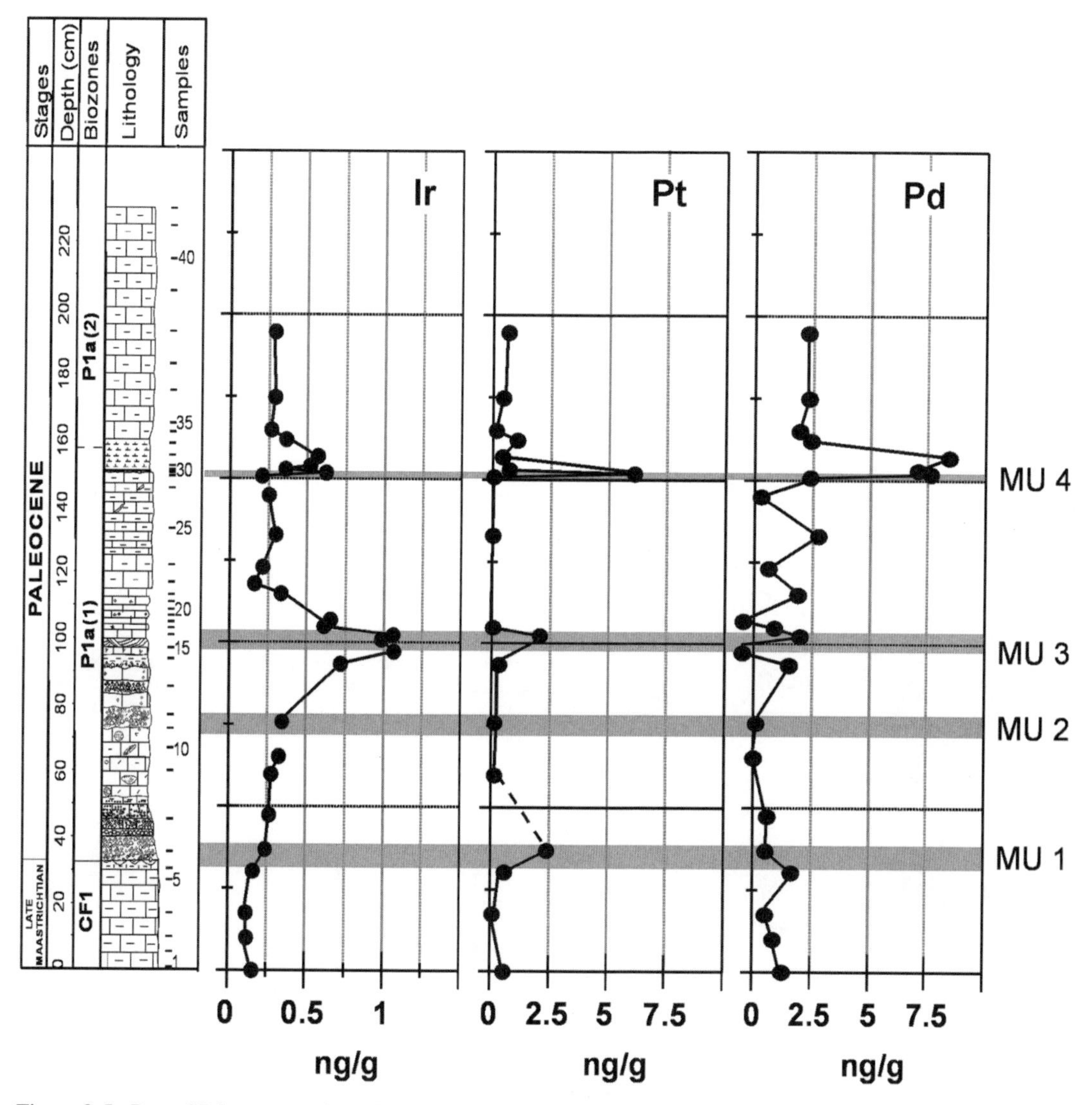

Figure 8. Ir, Pt, and Pd contents along Cretaceous-Tertiary (K-T) section Beloc B3 showing two different platinum group element anomalies for units MU3 and MU4 (Lithology, stratigraphy, and biozones are redrawn from Stinnesbeck et al., 1999.)

ever, unlike the PGE pattern of MU3, Fe-oxihydroxides and hydrogenous Mn crusts are enriched in Pt relative to the chondrite-normalized Pd contents (Stüben et al., 1999).

Below and above MU3, the anomalously high Ir contents show a gradual transition into the surrounding sediments over more than 50 cm (Fig. 8). Similar situations in several K-T sections have been mentioned (e.g., Crocket et al., 1988; Kyte et al., 1996; Rocchia et al., 1996; Rocchia and Robin, 1998; Smit, 1999), but the origin of such tailings is not clear. Considering possible primary depositional mechanisms resulting in such a profile, they would require prolonged periods of Ir input, if scenarios such as volcanic activity (Alvarez et al., 1980; Crocket et al., 1988) or a ring of projectile debris around the Earth following the impact of a huge extraterrestrial body (Rocchia and Robin, 1998) are envisaged. However, a number of other possibilities have been suggested to explain the gradual and sometimes symmetrical shape of the Ir anomalies, including erosion of ultramafic source rocks (Orth et al., 1988), gravitational or ballistic sorting of the ejecta during deposition (Evans et al., 1995), postdepositional redistribution during diagenesis or weathering (Wallace et al., 1990; Colodner et al., 1992), microbial activity (Dyer et al., 1989), carbonate dissolution (Rocchia et al., 1990), and physical mixing by bioturbation (Evans et al., 1995; Smit, 1999).

The diffuse extension of the Ir concentrations above the MU3 is probably mainly a consequence of the long residence time of the Ir in the oceanic reservoir (to 20 k.y.; Anbar et al., 1996) as a result of a meteorite impact, as suggested by Rocchia and Robin (1998). However, a different explanation is necessary to explain the tailing of the Ir values into the underlying

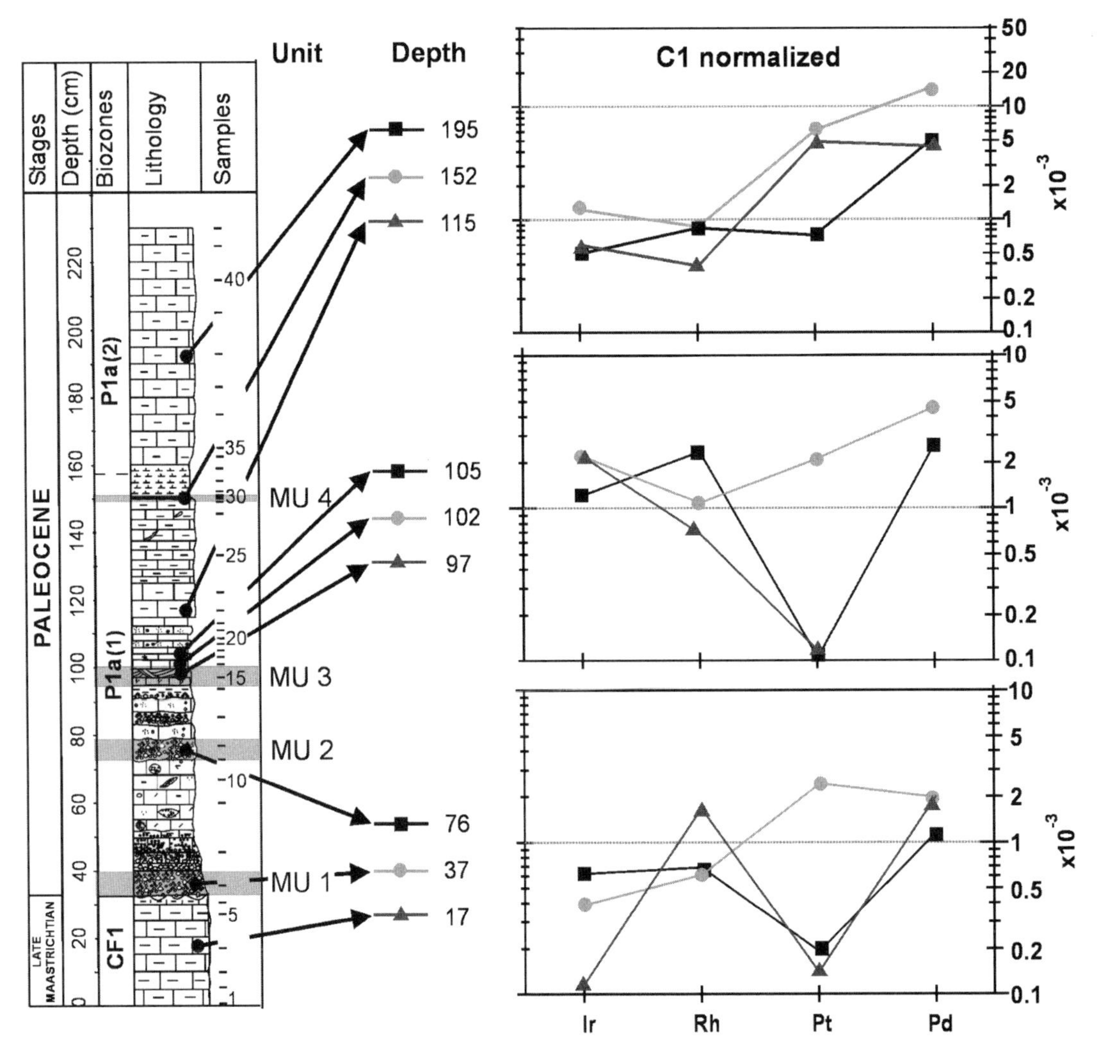

Figure 9. Chondrite-normalized platinum group element (PGE) patterns of defined sediment horizons: Maastrichtian marls (17 cm) of lower section, marker units MU1 (37 cm), MU2 (76 cm), MU3 (97 cm, 102.5 cm; Ir anomaly), MU4 (152 cm; upper PGE anomaly), and Paleocene marls of upper section (105 cm, 115 cm, and 195 cm). Normalization C1 values are from McDonough and Sun (1995).

sediments. Despite their known low mobility, several observations seem to support the possibility of a postdepositional chemical redistribution of the PGEs, as a consequence of redox changes in the sediments (Wallace et al., 1990; Colodner et al., 1992; Anbar et al., 1996). The behavior of the individual elements is slightly different in this respect, but the conclusions are equivocal and only little experimental work was carried out to substantiate these observations (Dai et al., 2000). For example, while some (Colodner et al., 1992; Evans et al., 1994; Anbar et al., 1996) relate the remobilization of Ir to reduction of oxic sediments driven by degradation of organic matter, others (e.g., Bowles, 1986; Wallace et al., 1990; Sawlowicz, 1993) consider that PGEs are carried as chloride or organic complexes by highly oxidized acidic fluids and are redeposited when they interact with low-Eh brines. A postdepositional remobilization of the PGEs cannot be undoubtedly demonstrated in the section examined.

Palladium shows the highest chondrite normalized values compared to other PGEs ($>1 \times 10^{-3}$), except for the reworked horizon at 37 cm. For example, in sample B3-17 (MU3, red layer at 102.5 cm), the chondrite-normalized values for Ir and Pt are much lower ($\sim 2 \times 10^{-3}$ for Ir and Pt) than for Pd ($\sim 5 \times 10^{-3}$) (Fig. 9), indicating that the material was depleted more in Ir and Pt than in Pd (accepting an extraterrestrial source). Such a fractionation pattern (presuming that it is due to postdepositional mobilization) would not be in agreement with the general mobility of the PGEs during diagenesis and weathering, which generally appears to decrease approximately in order Pd $\gg$ Pt $>$ Rh $>$ Ru $>$ Os $>$ Ir (Westland, 1981). Regardless of the mechanisms considered, the postdepositional mobilization of PGEs should be accompanied by the redistribution of many transition metals (Wallace et al., 1990; Colodner et al., 1992). Such redistribution cannot be unambiguously proved in the Beloc 3 section. In both MU3 and MU4, the highest Ir contents are related to thin Fe-rich layers. Elements such as As and Sb, which are typically bound to Fe-oxihydroxides and are readily removed if a redox front moves through the sediment, show very sharp maximums in the Fe-rich layers, without any signs of mobilization. Nevertheless, covariation of the Ir content with some of the heavy metals (e.g., Cu, Zn, Pb) across the MU3 horizon can be demonstrated. However, this seems more likely to be a primary feature controlled by the mineral composition, because the higher metal contents coincide with the presence of spherules and correlate with the amount of their alteration products (smectites, zeolites). We think that, rather than postdepositional remobilization, the tailing of the Ir content below MU3 is related to sediment reworking, as a result of a high-energy depositional environment, as indicated by biostratigraphic, sedimentologic, and granulometric data (Stinnesbeck et al., 1999; Keller et al., 2001). Nevertheless, a postdepositional mobilization of the PGEs cannot be completely excluded.

The slight deviation of the Ir anomaly from a typical chondritic pattern suggests that other influxes, such as volcanic emission or some other magmatic input, may also have been contributing factors. Dunites associated with a dismembered ophiolite suite of Upper Cretaceous age on Jamaica, and a suite of ultramafic rocks forming the lower part of a plutonic complex of island arc affinity of mid-Cretaceous age on Tobago, can be considered possible sources for PGEs in the Caribbean area. From both localities, Ni-Cu mineral assemblages enriched in PGEs associated with a primary stage of metallization, and individual PGE mineral phases were reported (PtS, grains composed of Rh, Ir, Pt, Cu, As, and S, native copper enriched in Pt and Pd, to 23 wt% and 15 wt%, respectively) (Scott et al., 1999).

Rare earth elements

The most important factor controlling the REE content of terrigenous sediments is the mineralogical composition and the provenance (McLennan, 1989; Condie, 1991). The REEs are chiefly transported with particulate matter or adsorbed phases. Because they are not easily fractionated during transport, sedimentation, diagenesis, and weathering, they generally reflect the chemistry of their source. Argillaceous rocks therefore frequently display a pattern that is very close to the average of the upper continental crust, being enriched in the light REEs and presenting an almost ubiquitous negative Eu anomaly (McLennan, 1989). Deviations from the crustal average are primarily due to specific mineralogical composition. Among these, heavy minerals (e.g., apatite, monazite, allanite, zircon) can easily dominate the REE pattern of sedimentary rocks because of their high concentrations in REEs (e.g., McLennan, 1989; Ohr et al., 1994). Plagioclase preferentially concentrates divalent Eu during igneous processes, and sedimentary rocks enriched in plagioclase (e.g., volcanogenic graywackes) will inherit this pattern. Low to moderate contents of amphiboles generally do not alter substantially the average pattern of the sedimentary rocks. Quartz and carbonate, due to their very low REE contents, have mostly a dilution effect, generally lowering the REE contents well below that of average shales. In foraminiferal tests, the slight enrichment of the middle REEs leads to a roof-shaped pattern (Palmer, 1985; Winter et al., 1997). Marls with a higher proportion of biogenic carbonate will have relatively low chondrite-normalized La/Sm and Gd/Yb ratios. Seawater is characterized by a distinct negative Ce anomaly, and authigenic phases formed in equilibrium with seawater will inherit this feature (Fleet, 1984). Distribution patterns similar to seawater have been reported in limestones and calcareous oozes (Taylor and McLennan, 1985; Pattan et al., 1995; Sethi et al., 1998). Because of the low mobility of the REEs, diagenetic alteration and weathering of the detrital, biogenic, and authigenic components will not affect substantially the original distribution of the lanthanides in the sediment. However, stronger remobilization may occur during the alteration of glass fragments, spherules, and volcanogenic minerals (e.g., Elliot, 1993).

In the bulk samples investigated, REE contents are 5–15 times the chondritic values, with moderate enrichments of the light over the middle REEs ($La_{Ch\text{-}N}/Sm_{Ch\text{-}N} = 1.9–3.8$) and flat heavy REE patterns ($Gd_{Ch\text{-}N}/Yb_{Ch\text{-}N} = 0.9–1.3$) (Table 4; Fig. 10). Compared with the NASC, the total REE contents are ~4 times lower (excepting the more depleted sample from MU1) and the samples are slightly depleted in the light lanthanides relative to the heavy lanthanides ($La_{NASC}/Yb_{NASC} = 0.4–0.7$). The generally low REE contents relative to NASC are due to variable amount of carbonate admixed, while the depletion of the relatively more mobile light REEs could be connected to a slight remobilization during late diagenesis (Hannigan and Basu, 1998).

The chondrite-normalized REE patterns observed throughout the section can be conveniently subdivided into three groups, types 1–3.

Type 1 displays negative Ce and Eu anomalies ($Ce/Ce^* = 0.58$ and $Eu/Eu^* = 0.81$, respectively) and is typical for the Maastrichtian marly limestones. The pattern can be explained by the mixing of detrital clay with carbonate material. The clay fraction confers to the samples a typical crustal pattern with a negative chondritic Eu anomaly, while the carbonate lowers the REE contents of the bulk samples and imprints a negative Ce anomaly (Pattan et al., 1995).

Type 2 is characterized by the absence of a Ce anomaly and is encountered in the samples of the spherule-rich deposit (including the base of MU3, at 97.5 cm). This feature is not restricted to the spherule-rich layers of the spherule-rich deposit (MU1 and MU2), but is also found in the interlayered marls (at 65 and 86 cm). The chondrite-normalized Eu anomalies show

TABLE 4. RARE EARTH ELEMENT CONTENTS IN BULK SAMPLES AND SPHERULES

Sample number	Sampling depth (cm)	Sc (µg/g)	Y (µg/g)	La (µg/g)	Ce (µg/g)	Pr (µg/g)	Nd (µg/g)	Sm (µg/g)	Eu (µg/g)	Gd (µg/g)	Tb (µg/g)	Dy (µg/g)	Ho (µg/g)	Er (µg/g)	Tm (µg/g)	Yb (µg/g)	Lu (µg/g)
Bulk samples																	
B3-32	157.0	12.6	14.4	10.1	10.4	1.93	8.52	1.87	0.81	2.35	0.37	2.43	0.53	1.62	0.24	1.58	0.23
B3-29	152.0	15.4	8.5	5.1	8.5	1.32	6.18	1.54	1.03	1.87	0.27	1.79	0.37	1.13	0.17	1.25	0.18
B3-28	151.0	11.5	15.2	10.0	10.8	1.96	8.65	1.90	1.26	2.38	0.37	2.38	0.52	1.59	0.25	1.57	0.22
B3-16	101.0	9.0	10.4	8.7	11.9	1.78	7.51	1.65	0.74	1.94	0.29	1.89	0.39	1.23	0.19	1.27	0.18
B3-15	97.5	16.4	4.8	4.4	7.9	1.07	4.35	0.99	0.10	1.06	0.17	1.11	0.23	0.74	0.12	0.93	0.14
B3-13	86.5	10.1	6.4	5.0	10.9	1.48	6.25	1.49	0.13	1.56	0.25	1.57	0.31	0.95	0.16	1.08	0.16
B3-12	76.0	14.1	7.5	8.2	19.4	2.19	8.91	1.97	0.51	1.98	0.30	1.91	0.37	1.14	0.18	1.27	0.19
B3-10	65.5	12.5	5.0	5.5	12.7	1.62	6.65	1.56	0.37	1.50	0.24	1.52	0.29	0.88	0.16	1.24	0.19
B3-07	37.0	15.9	2.5	2.2	5.1	0.69	2.99	0.71	0.30	0.70	0.11	0.63	0.12	0.37	0.07	0.47	0.08
B3-03	10.0	8.0	17.1	12.0	12.4	2.24	9.73	2.01	0.61	2.61	0.41	2.74	0.58	1.81	0.28	1.64	0.25
Spherules																	
B3-12	Dark	21.4	26.8	21.2	46.0	5.37	23.3	4.80	1.35	4.70	0.68	4.42	0.92	2.77	0.43	2.86	0.42
B3-12	Yellow	21.4	26.4	21.2	45.6	5.17	22.6	4.73	1.23	4.29	0.71	4.34	0.83	2.59	0.36	2.77	0.38
B3-12	Smectite	12.4	1.2	0.9	2.0	0.27	1.20	0.27	0.17	0.24	0.04	0.23	0.05	0.13	0.02	0.14	0.02

Note: Rare earth element contents in bulk samples and spherules, determined by inductively coupled plasma-mass spectrometry (ICP-MS). The rare earth element (REE) contents of the bulk samples indicate depositional changes and fluctuation in the source and nature of the deposited materials. The REE contents of glass spherules are 20 times higher than those of smectite spherules.

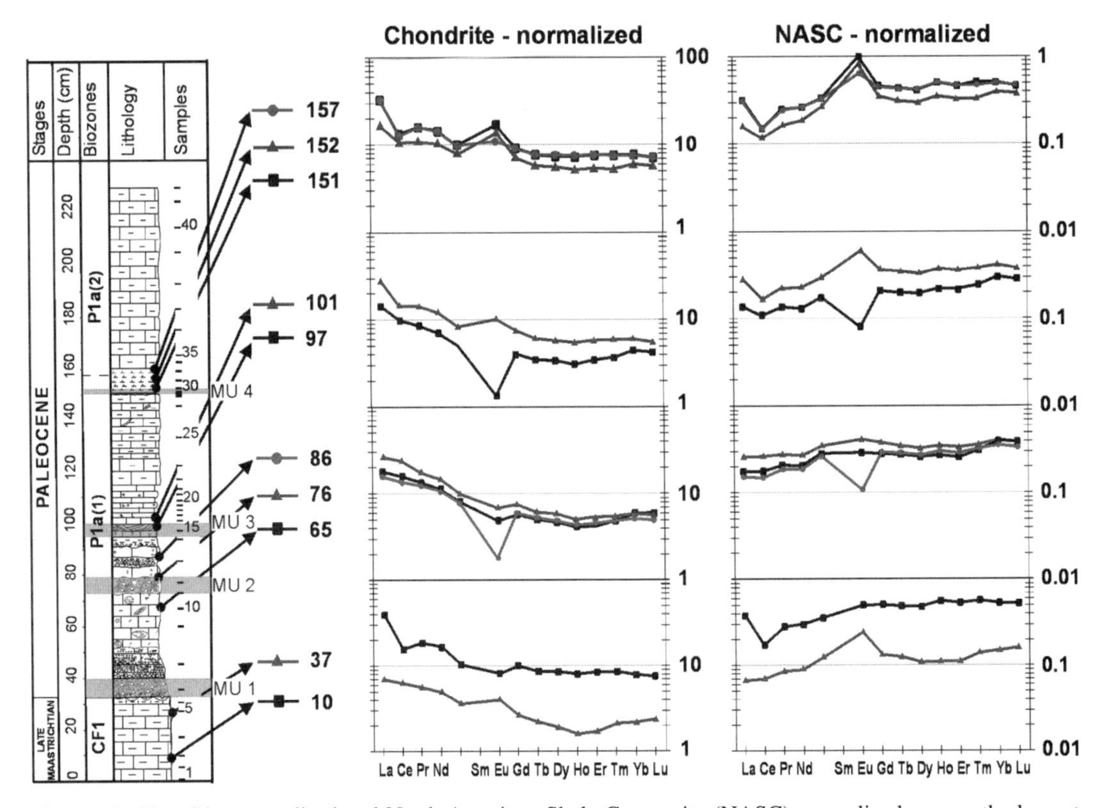

Figure 10. Chondrite-normalized and North American Shale Composite (NASC) normalized rare earth element (REE) patterns of distinct sediment layers of different horizons of Cretaceous-Tertiary (K-T) section, Beloc. Normalization values are from McDonough and Sun (1995, chondrites) and Gromet et al. (1984, NASC).

an apparently systematic change upsection, with slightly positive values at MU1 (Eu/Eu* = 1.31), and progressively lower values upsection (from Eu/Eu* = 0.73–0.80 to 0.30 in MU3). The absence of a Ce anomaly in the spherule-rich deposit, in contrast to the sequences below and above, could indicate that these sediments were not formed in equilibrium with seawater, but were deposited faster. However, a change in the redox state of the seawater may be an alternative explanation for the lack of Ce anomalies (see following). The formation of secondary mineral phases (smectites and/or palagonite, zeolites, opal) from glass has certainly affected the original distribution of the REEs in the spherule-rich deposit to some extent. However, a direct relationship between the composition of the smectitic spherules and that of the known glass varieties could not be demonstrated (see following). However, it is noteworthy that the REE pattern of the bulk samples B3-7 (MU1, with a slight chondritic negative Eu anomaly) and B3-12 (MU2) can be closely modeled by the mixing of glass and smectite with a composition similar to that of the spherules analyzed from sample B3-12 in a proportion of 10% to 90% and 35% to 65%, respectively (Fig. 11). (Because of the relatively low CaO contents of these samples and the low concentrations of REEs in calcite, the influence of carbonate was disregarded.) The modeling is based on mass-balance acceptances, and was solved by successive iterations until the best fit was obtained. Guy et al. (1999) reported that REE concentrations in clays resulting from seawater-basalt interaction are close to that of unaltered basalt, while concentrations in zeolites are 10 times lower. However, the fate of REEs during the alteration of volcanic ash and glass material to smectite and zeolite (regardless of the origin of glass) is not well understood. Available data are not sufficient or conclusive (Hopf, 1993; Martin-Barajas and Lallier-Verges, 1993; Terakado and Nakajima, 1995; Christidis, 1998) and preclude further interpretation of our REE data with respect to glass alteration.

Type 3 is restricted to the gray-green shale of MU3 (101 cm) and to the samples around MU4. Compared to type 1, type 3 is characterized by a positive Eu anomaly (Eu/Eu* = 1.18–1.85) in addition to the negative Ce anomaly (Ce/Ce* = 0.53–0.79). The Eu anomaly is considerably higher in MU4 (Eu/Eu* = 1.81–1.85), where it may be related to higher feldspar contents or to its alteration products. However, in the absence of mineralogical proof the development of a positive Eu anomaly may be interpreted as a result of diagenetic reprecipitation, promoted by compaction and reduced pore waters enriched in Eu^{2+} entering an oxidizing environment (MacRae et al., 1992). This mechanism is supported by the presence of the oxidized, Fe-rich layers at or in close proximity to the samples with the highest Eu anomalies. The REE pattern of the gray-green clay of MU3 (at 101 cm) is almost identical to that of the samples from MU4. Except for the positive Eu anomaly, the total REE content and the presence of a characteristic negative Ce anomaly are similar to that found in the clay fraction of the Fish Clay horizon of the K-T transition at Stevns Klint, Denmark (Tredoux et al., 1989; Elliot, 1993). The negative Ce anomaly was interpreted by Elliott (1993) to be the result of possible equilibration with seawater; the Mg-smectites he investigated were formed authigenically from a volcanic glass precursor.

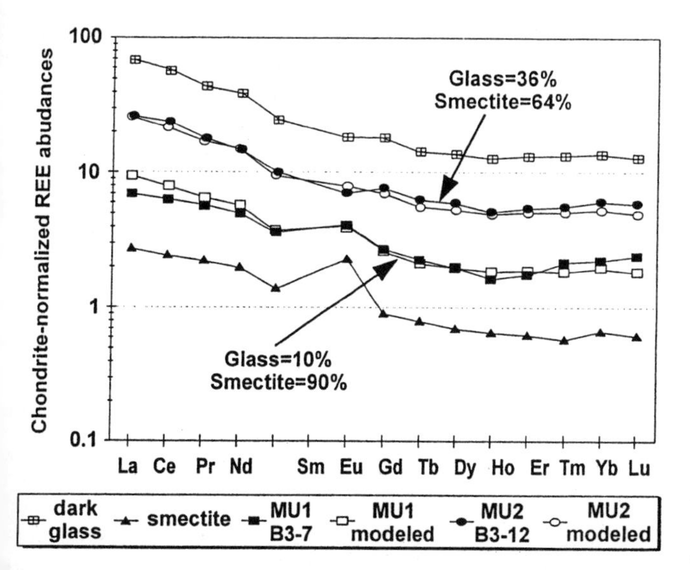

Figure 11. Simulated rare earth element (REE) pattern of bulk samples from MU1 (B3-07) and MU2 (B3-12) by mixing glass and smectite spherules with compositions similar to those analyzed from sample B3-12.

Several authors reported systematic stratigraphic trends in the distribution of Ce anomalies. Such trends were interpreted to reflect changes in the oceanic redox conditions (e.g., Hu et al., 1988; Shields and Stille, 1998) or in the tectonic depositional environment (Murray et al., 1990). According to Shields and Stille (1998), the correlation of the Ce anomaly with other proxy parameters (Sr and Nd isotope ratios, $\delta^{13}C$ of carbonate and organic fraction, primary sedimentological features) proves that the Ce anomaly reflects primary changes in the redox state of coeval seawater. On the basis of the good correlation between the Ce anomaly and $\delta^{13}C$ of carbonate ($r = -0.86$) and assuming that no suboxic or anoxic diagenesis has taken place, we conclude that the seawater was less oxygenated during the sedimentation of the spherule-rich deposit as compared to the sequences below and above.

Stable isotopes

During diagenesis, primary biogenic carbonate can be dissolved and replaced by secondary calcite with a different isotopic composition. Oxygen isotopic values of bulk carbonate are more sensitive to postdepositional alteration than carbon isotopic ratios. An increase in temperature caused by sediment burial and possibly different isotopic composition of pore fluids are the most important factors altering the primary isotopic signal of biogenic calcite (Killingley, 1983; Jenkyns et al., 1994; Schrag et al., 1995; Mitchell et al., 1997). Generally, both of these factors tend to lower the initial oxygen isotopic values. Consequently, if a part of the carbonate was affected by recrystallization, the excursions toward more negative $\delta^{18}O$ values may represent diagenetically altered isotopic values, whereas higher values should be regarded as minimum values.

Bulk $\delta^{13}C$ values range between 0.37‰ and 0.65‰ in the latest Maastrichtian marly limestones and decline to −3.34‰ at the base of MU1 with a further decline to −4.85‰ above MU1 (Table 2; Fig. 12). This negative $\delta^{13}C$ shift in the early Danian P1a(1) sediments reflects the drop in primary produc-

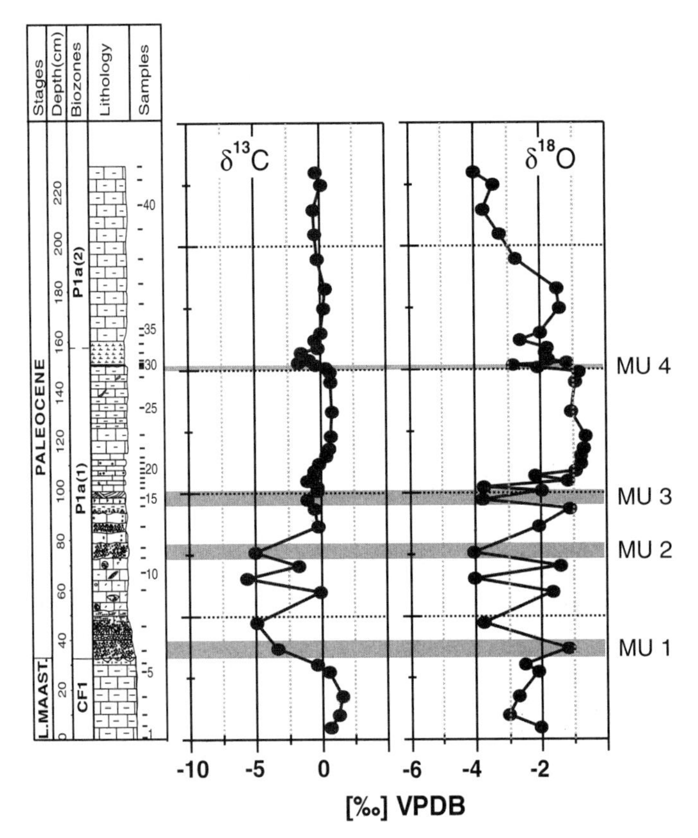

Figure 12. Carbon and oxygen isotope data of carbonate fraction in section Beloc B3 with oscillating pattern for both isotopes within spherule-rich deposit, negative δ^{13}C shift in early Danian, and with values stabilized at -1.0‰ δ^{18}O between MU3 and MU4 and gradual decrease to -4.0‰ near top of section. VPDB is Vienna Peedee belemnite. (Lithology, stratigraphy, and biozones are redrawn from Stinnesbeck et al., 1999.)

tivity across the K-T boundary, though the shift at Beloc is larger (3.3–4.9) than in other tropical marine sequences (~2‰–3‰) (Keller and Lindinger, 1989; Zachos et al., 1985, 1989). Between MU1 and 2, δ^{13}C values oscillate (-0.09‰ to -5.65‰) and may reflect mechanical reworking of Cretaceous sediments into Danian deposits. The original signal also may have been overprinted by diagenetic effects from the sparry calcite matrix that cements the spherule debris. In the limestone between MU2 and MU3, δ^{13}C values vary between 0‰ and -1.0‰ with a 0.5‰ excursion associated with the Ir-dominated PGE anomaly. In the limestone above MU3, δ^{13}C values increase from -1.0‰ to $+0.5$‰ and remain stable to MU4, when they drop by more than 2‰ (Fig. 12). Consequently, δ^{13}C values narrowly fluctuate around 0‰. The two negative excursions associated with the PGE anomalies may reflect short-term changes in primary productivity, although because they coincide with very low to nearly absent calcite, they may be artifacts of the sedimentary record.

Oxygen isotope values range between -4.05‰ and -0.77‰; Maastrichtian values vary between -2.0‰ and -3.0‰ (Table 2). Between MU1 and MU3 (spherule-rich deposit and bioclastic limestone,Fig. 12), δ^{18}O values oscillate between -4.0‰ and -2.0‰, probably due to the presence of Cretaceous reworked sediments and diagenetic alteration. In the pelagic limestone, between MU3 and MU4, values stabilize between -0.5‰ and -1.0‰ and gradually decrease to -4.0‰ near the top of the section. Diagenetic alteration of the carbonate as well as reworked sediments in the lower part of the section (zone P1a[1]) impede the use of the δ^{18}O bulk record as a temperature record, but some trends in the *Parvularugoglobigerina eugubina* zone are apparent. For example, temperature appears to have been relatively cool between MU3 and MU4 (upper P1a, lower *P. eugubina* zone) and gradually warmed above MU4 (P1a[2], upper *P. eugubina* zone). Temperature trends between MU1 and MU3 are obscured by abundant reworked sediments, but suggest overall cool climates (Fig. 12).

Chemical, mineralogical, and isotopic correlations

The hierarchical clusters deduced from the Spearman rank correlation matrix (Fig. 13) of minerals, stable isotopes, and major and trace elements display a clear grouping that reflects the mineralogy and chemistry of the parent material and chemical mobility of the elements. One group, consisting of $CaCO_3$, CaO, MnO, and loss on ignition (LOI), corresponds to marine biogenic or chemical carbonate precipitation. The second group combines minerals and elements of mainly terrigenous, detrital, and/or volcanic origin (SiO_2, TiO_2, Al_2O_3, Fe_2O_3, MgO, Na_2O, Zr, Nb, phyllosilicates, and Ga). Barium is closely correlated with zeolites and to a lesser extent with Sr and weakly with Ir. The close correlation of zeolites and Ba may be caused by the release of Ba and Sr from chemical weathering of feldspars followed by adsorption on clay and Fe oxides and coprecipitation into the sediment. Thus variations in biogenic carbonate deposition are overprinted by this effect, and should be used with caution as paleo-sea-level proxies (Stoll and Schrag, 1996, 1998). The association of Pd with Br, which is probably scavenged from seawater, may be an indication of higher solubility of Pd in seawater compared to the other PGEs in the profile (Sawlowicz, 1993; Stüben et al., 1999). The weak association of Pt and Ni is a reflection of the enrichment of Pt and Ni in the gray clays and the upper rust-colored layer of MU4.

Spherules

Three types of spherules (dark glass, yellow glass, and smectite spherules) were separated from the spherule-rich deposit in MU2 (sample B3-12) and analyzed by electron microprobe for major elements and by ICP-MS for REEs. The major element chemistry indicates two different types of glassy spherules and a significant chemical difference between the glass and smectite spherules (Tables 4 and 5).

The black glass is relatively homogeneous, with an andesitic to dacitic composition, similar to that found by Izett (1991).

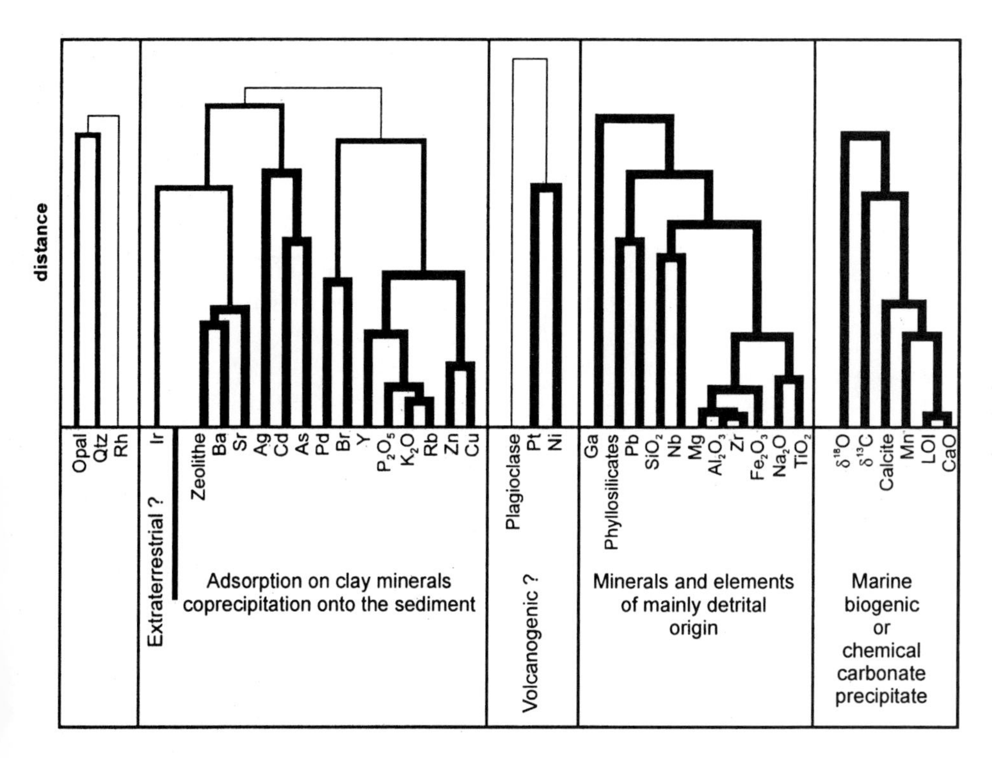

Figure 13. Hierarchical cluster analyses of Spearman rank correlation matrix (Rock, 1988) of minerals, stable isotopes, and major and trace elements from Beloc B3 showing five different groups. LOI is loss on ignition.

TABLE 5. MAJOR ELEMENT COMPOSITION OF SPHERULES FROM SAMPLE B3-12 (MICROPROBE DATA)

		FeO (wt%)	MnO (wt%)	SiO_2 (wt%)	MgO (wt%)	K_2O (wt%)	CaO (wt%)	TiO_2 (wt%)	Na_2O (wt%)	Al_2O_3 (wt%)	Total (wt%)
4 black glass spherules (18 points)	avg.	5.31	0.15	66.85	2.75	1.53	5.38	0.65	1.96	14.94	99.52
	stdv.	0.36	0.05	1.28	0.20	0.11	0.54	0.05	0.37	0.29	
	min.	4.79	0.06	64.61	2.39	1.33	4.79	0.54	1.51	14.48	
	max.	5.92	0.22	68.94	3.11	1.75	6.58	0.74	2.63	15.47	
	Q1	4.99	0.12	66.12	2.65	1.45	5.06	0.63	1.68	14.77	
	Q3	5.52	0.19	67.51	2.91	1.61	5.44	0.68	2.32	15.06	
2 yellow glass spherules (8 points)	avg.	5.37	0.15	60.47	3.16	1.40	11.45	0.65	2.78	14.21	99.63
	stdv.	0.60	0.07	4.49	0.55	0.42	5.29	0.06	0.30	0.82	
	min.	4.59	0.06	51.07	2.60	0.63	8.24	0.55	2.13	12.65	
	max.	6.22	0.27	64.33	4.35	1.98	23.13	0.73	3.11	15.19	
	Q1	4.95	0.11	59.60	2.80	1.30	8.30	0.61	2.72	13.81	
	Q3	5.78	0.18	63.92	3.31	1.66	11.37	0.69	2.95	14.79	
Coating of 3 smectitic spherules (10 points)	avg.	5.07	0.01	61.98	3.92	1.13	1.06	0.81	0.05	9.82	83.84
	stdv.	1.81	0.01	4.41	1.42	0.49	0.29	0.13	0.02	3.25	
	min.	0.63	0.00	53.79	0.02	0.24	0.64	0.64	0.01	0.76	
	max.	7.73	0.02	68.13	4.92	1.89	1.49	1.11	0.08	11.75	
	Q1	4.75	0.00	60.65	4.01	0.93	0.80	0.73	0.03	10.08	
	Q3	5.84	0.01	65.50	4.68	1.41	1.25	0.84	0.06	11.39	
3 smectitic spherules (9 points)	avg.	4.81	0.01	65.26	4.64	1.10	0.88	0.34	0.06	11.12	88.23
	stdv.	0.41	0.02	3.21	0.40	0.43	0.16	0.13	0.02	0.83	
	min.	4.31	0.00	59.61	3.95	0.15	0.67	0.15	0.03	9.83	
	max.	5.32	0.06	69.58	5.19	1.56	1.16	0.49	0.08	11.98	
	Q1	4.41	0.00	63.25	4.45	0.99	0.78	0.30	0.04	10.66	
	Q3	5.15	0.01	67.82	4.93	1.29	0.93	0.45	0.07	11.82	

Note: Glass, Ca-rich and Ca-poor, and two types of smectites, rims as weathering products of glass spherules, which are slightly enriched in TiO_2 compared to the glass, and spherules depleted in TiO_2 compared to the glass.
avg. = average, stdv. = standard deviation, min. = minimum, max. = maximum, Q1 = first quartile, Q3 = third quartile.

Koeberl and Sigurdsson (1992), Koeberl (1993). Sulfur was not detected in the black glass fragments. The yellow glass is largely zoned, showing flow structures with varying CaO contents (the CaO content varies between 8 and 23 wt% along these structures) within one single grain with a significant spatial correlation of Ca and S (Fig. 14A). Sulfur content, $\delta^{34}S$ values, and B isotopes of yellow glass fragments are thought to result from inclusion of evaporitic material (Chaussidon et al., 1996), although Koeberl (1993) showed that B isotopes are incompatible with an evaporitic source and trace elements are not in agreement with any combination of materials from Chicxulub analyzed by him at that time. Based on oxygen isotope data, Blum and Chamberlain (1992) also rejected the involvement of a sulfate-rich evaporitic material, and argued for the mixture of carbonate and silicate rocks. The oxygen, Sr, and Nd isotopic composition of the Haitian glass spherules proved to be practically identical with those of Chicxulub melt rocks (Blum et al., 1993). Recalculating the uncontaminated composition of the glass by removing the excess CaO, a source rock of dacitic composition similar to the black glass results.

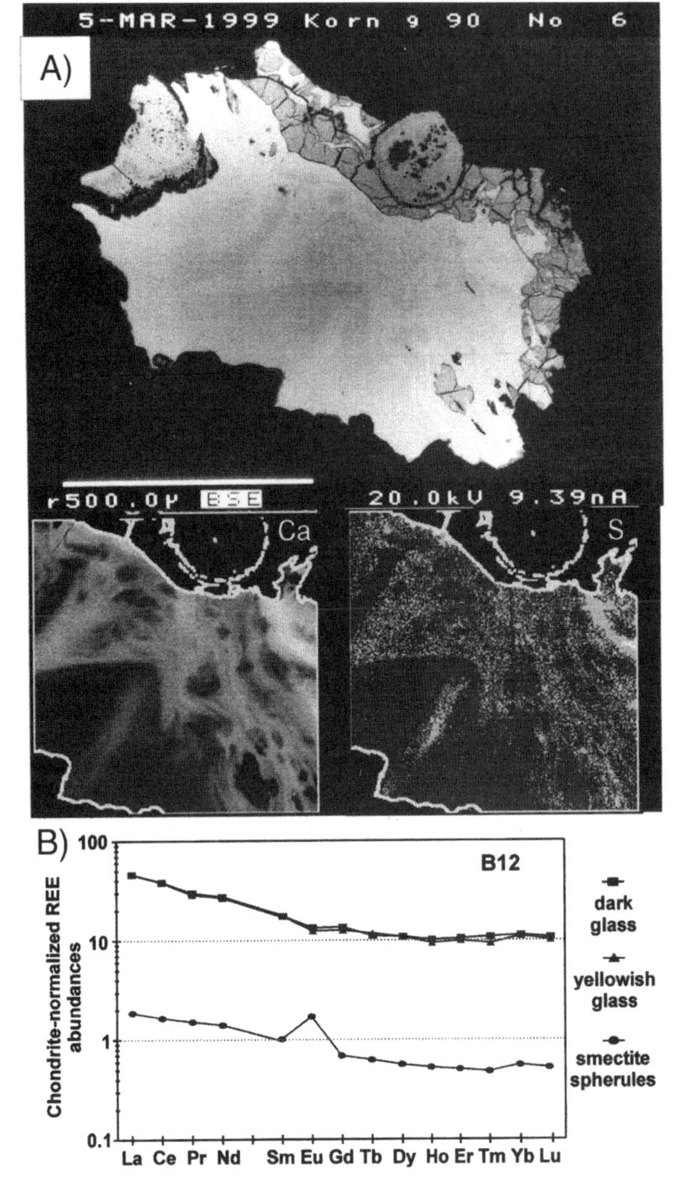

Figure 14. A: Microphotographs of Ca-rich glass spherule (sample B3-12) from spherule-rich deposit at Beloc B3, showing backscatter electron image and distribution of Ca and S within spherule. Yellow glass is largely zoned, showing flow structures with varying CaO contents. B: Chondrite-normalized rare earth element (REE) patterns of two glass spherules and one smectite spherule from spherule-rich deposit at Beloc B3 (sample B3-12).

For the degassing experiments two spherule samples were taken from B3-7 (MU1) and B3-12 (MU2). The interpretation of the evolved gas analyses (EGA curves) is based on the partial pressure curves of the evolved gas species. Only the gas-release curve above 1000°C is due to degassing of the melt and is considered to carry a primary signal. The spherule (B3-12) in MU2 is slightly altered, as demonstrated by water release already well below 1000°C, which can be assigned to the degradation of some alteration products. Distinct spikes in the partial pressure curve well above 1000°C correspond to SO_2, which supports the uptake of SO_2 from evaporites into the silicate melt, as argued by Chaussidon et al. (1996) on the basis of the isotopic composition of the sulfur. Dark colored, fresh glass spherules were analyzed for comparison from a spherulitic bed in the Beloc 1 section that correlates with MU2. The degassing profiles of this sample are almost completely free of any of the gas species considered, and hence characteristic of tektites.

The EGA curve of the spherule B3-7 shows no SO_2 degassing, but a spontaneous water release in a relatively narrow range (between 1000–1100°C), which is more characteristic for degassing of volcanic glasses; on the contrary, impact-derived glasses (e.g., glassy part of suevites, zhamanshinite) typically lose their low water content over a considerably larger temperature range (Heide, 1974; Stelzner and Heide, 1996). Because EGA curves do not permit an exact quantitative evaluation of the amount of water released, these data do not necessarily conflict with the very low water concentrations measured by Koeberl (1992).

The chondrite normalized REE patterns of the glass spherules analyzed resemble the light REE-enriched patterns (La/Yb = 4.2) described by previous workers (Koeberl and Sigurdsson, 1992), but without the negative Eu anomaly found by Koeberl (1993). The REE concentrations in the yellow glass are slightly lower (Fig. 14B; Table 4). Thus contamination of an andesitic to dacitic glass melt by evaporites or carbonates as described by Sigurdsson et al. (1991a) is probable. The REE pattern of the glass spherules does not provide any firm evidence for a volcanic or impact origin, but it allows us to address the problem of the origin of the smectitic and/or palagonitic spherules that were considered by several authors to represent alteration products of these glassy precursors (e.g., Kring and Boynton, 1991; Sigurdsson et al., 1991a; Koeberl and Sigurdsson, 1992; Bohor and Glass, 1995).

In the Beloc 3 section two types of smectites (or palagonites, as defined in Lyons and Officer, 1992; Bohor and Glass, 1995) can be distinguished morphologically: (1) smectite as coatings of glass spherules and (2) smectite spherules as pseudomorphs of impactites, like those described by Izett (1991). These two types also differ significantly in their chemistry. Compared to black and yellow Haiti glass, elements that can be mobilized by alteration are depleted in both types. Na_2O and CaO are nearly completely removed, while SiO_2, K_2O, FeO, and Al_2O_3 are only partly remobilized. Similar element losses were reported in the palagonitized rims of the Haitian glass spherules (Bohor and Glass, 1995). Palagonite is considered to be an incipient phase in the alteration of glass, with a considerably higher Si/Al ratio compared to smectites, into which they transform over time. TiO_2 is immobile during alteration, as are REEs, to a lesser degree. Nevertheless, the smectite spherules are strongly depleted in TiO_2 and REEs relative to both types of glasses, whereas the coatings are TiO_2 enriched by ~20 wt%, compared to the glass nucleus. Although there is increasing evidence that REEs can be efficiently removed during hydrothermal alteration (e.g., Hopf, 1993; Bach and Irber, 1998; Christidis, 1998), the shape of the chondrite-normalized pattern should remain unchanged (Sigurdsson et al., 1991a; Guy et al., 1999). Therefore, even if the low REE contents of the smectite spherules could be attributed to alteration and leaching under a continental weathering regime, implying very high water/rock ratios, the explanation of the positive Eu anomaly in the pattern of the altered spherule analyzed is still not straightforward, and it would be even more problematic to interpret if we consider the negative Eu anomalies of the glass analyzed by Koeberl (1993). Leaching experiments (Bach and Irber, 1998) with hydrothermally altered diabase (to 50% lower REE concentrations than the surrounding, weakly altered rocks) showed significant losses of REE and yielded leaching solutions with negative Eu anomalies compared to the host rock. Nevertheless, the derivation of the smectite spherules from the types of glass material already known is not convincing because the loss of Ti still cannot be readily explained. Thus the smectites probably represent two different genetic types: the smectite coatings are alteration products of the adjacent glass, whereas the smectite spherules, showing very low REE and Ti concentrations, are probably authigenic fillings of voids. None of the glass materials described until now, including the high Si-K glass of Koeberl and Sigurdsson (1992), could have produced by alteration the smectitic spherules analyzed here. In our opinion, the alteration of the glass spherules probably started in the marine sedimentation environment, but may have been completed only after the uplift of the section above sea level, as suggested by Bohor and Glass (1995).

DEPOSITIONAL HISTORY AT BELOC

The age and biostratigraphy of the Beloc section presented here are discussed elsewhere (Keller et al., 2001). The section displays a complex interaction of different sedimentary, volcanic, and cosmogenic sources resulting from a multievent scenario (Fig. 15). The latest Maastrichtian marly limestones indicate a time of relatively high productivity marked by high $\delta^{13}C$ values and high excess rates for Cu, Zn, and Sr, biogenic deposition, and moderately warm temperatures compared with the early Danian. Compared to NASC, REE contents are lower, which is typical for marine carbonates. In particular, NASC-normalized REE patterns show a Ce anomaly, reflecting seawater patterns (Rollinson, 1993). The distribution of the PGEs within the Maastrichtian limestone is similar to that of other marine precipitates (Stüben et al., 2001). Mineralogical bulk-rock data indicate normal pelagic conditions dominated by biogenic calcite, phyllosilicates, opal, and low detrital quartz contents. The presence of hornblende just below the unconformity at the top of the Maastrichtian suggests a volcanic influx due to volcanic activity or to erosion of volcanic rocks.

A spherule-rich deposit marks the earliest Danian *P. eugubina* zone Pla(l) (32–97 cm). The base of the spherule layer of MU1 overlies an undulating erosive surface of the marly limestone enriched in zeolites and amphiboles and marks the paleontologically defined K-T disconformity (e.g., based on the first appearance of Danian species). The spherule-rich deposit is lithologically heterogeneous and is characterized by alternating layers rich in spherules (MU1 and MU2), clayey sediments, and bioclastic limestone. Clasts of reworked Cretaceous sediments are common and suggest intermittent high-energy environments, as indicated by the cross-bedded layer at the top of this interval.

Surface productivity dropped abruptly across this lithological contact and temperatures cooled, as suggested by isotopic data as well as by excess ratios of some trace elements. Due to intensive reworking the entire interval between MU1 and MU3 displays considerable variations in trace element contents, excess ratios, and isotopic data, alternating between values typical for pre-K-T and post-K-T sediments. Detrital minerals are nearly absent here, and this indicates that reworked particles were derived from a pelagic environment (e.g., upper slope), rather than from erosion of an adjacent emerged area.

The cross-bedded layer that marks the top of the high-energy depositional regime of the spherule-rich deposit is overlain by gray-green shale with a thin rust-colored oxidized layer (MU3), which contains the highest values of the Ir-dominated PGE anomaly. Of the PGEs only Ir shows a steady increase from background values in the latest Maastrichtian and the early Danian spherule-rich deposit to 0.6 ng/g in the cross-bedded layer, reaching a maximum of 1 ng/g in the clay and rust-colored layer. The gradual increase in Ir (to 0.6 ng/g) cannot be readily explained by postdepositional remobilization, by decreasing sediment accumulation rates, or by progressive exposure and erosion of an Ir-rich source rock. The Ir anomaly coincides with a possible volcanic influx, as indicated by high zeolite and amphibole contents (Figs. 3 and 8), and decreased calcite content and with drastically lowered sedimentation rates.

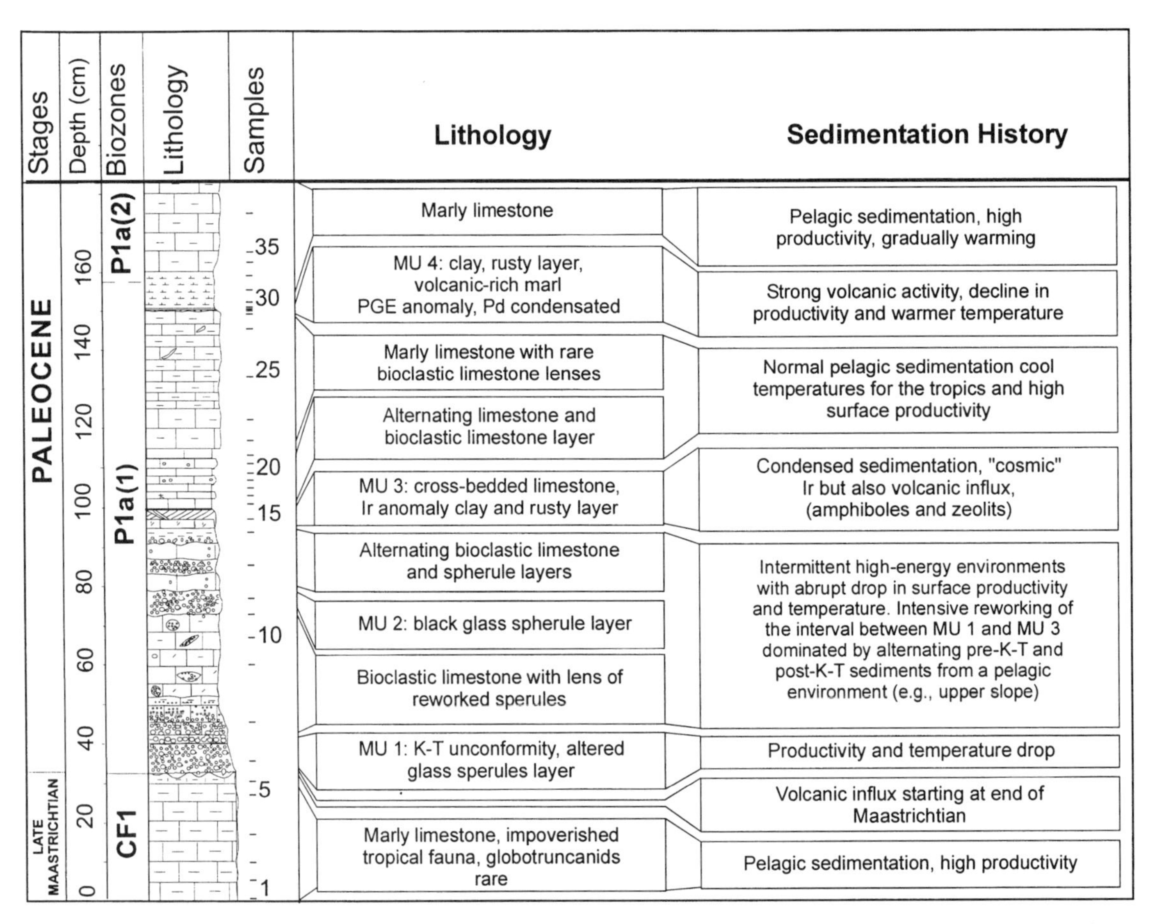

Figure 15. Reconstruction of depositional history of Beloc B3 section based on lithology, mineralogy, sedimentology, and chemostratigraphy. (Lithology, stratigraphy, and biozones are redrawn from Stinnesbeck et al., 1999.) K-T is Cretaceous-Tertiary.

The maximum Ir values within the rust-colored layer (MU3) display a roughly chondritic PGE pattern and may reflect a cosmic overprint on a long-lasting period of volcanic activity. The peak of the Ir anomaly shows a tailing into the surrounding sediments, possibly caused by reworking, diffusion, or bioturbation. The occurrence of glass and smectite spherules below the Ir anomaly may also be considered as indicative of impact material. However, the association of the PGE anomalies with volcanic material (including glass shards) indicates a magmatic contribution and contradicts a pure impact source of the material.

Alternating bioclastic limestones and limestones above the Ir anomaly (105–150 cm, Fig. 8) grade into marly limestones, reflecting normal pelagic sedimentation, as also expressed by carbon and oxygen isotopes and excess values of trace elements. Stable isotopes in this interval indicate very cool temperatures for the tropics and high surface productivity (Fig. 15). This rapid return to nearly pre-K-T values in surface productivity within the lower part of the early Danian zone P1a(1) (lower part of *P. eugubina* zone) is not observed in Tunisian sections (e.g., El Kef) (Keller and Lindinger, 1989), where pre-K-T values are reached in the later Danian zone P1c, but is consistent with low-latitude open-marine environments (Zachos et al., 1989). This may reflect the more stable environment of the Caribbean platform at this time.

Above this interval, MU4 consists of a thin oxidized rust-colored layer and a clay- and clayey marl-layer rich in volcanic glass, amphiboles, and zeolites and is enriched in PGE, particularly in Pd (Fig. 15). The PGE pattern of this anomaly is more compatible with oceanic flood basalts than with a cosmic origin (Greenough and Fryer, 1990). Because Haiti is part of the oceanic flood basalt province within the Caribbean plate, which developed 90 Ma, this PGE anomaly thus may be related with the younger volcanic activity in the Dominican Republic and Costa Rica, which is estimated to have occurred between 69 and 63 Ma (Kerr et al., 1996, 1997; Sinton et al., 1998). Denudation of ultramafic rocks with occasionally very high contents of PGE of middle to Upper Cretaceous age on Jamaica and Tobago (Scott et al., 1999) could also be considered to be responsible for these higher concentrations. Thus, for this anomaly a magmatic origin appears probable. Normalized REE patterns with a strong positive Eu anomaly support this interpretation.

Stable isotope data in MU4 suggest a temporary decline in productivity and warmer temperatures. Above MU4, marly limestones reveal relatively stable high productivity, as also indicated by high calcite contents, C isotopes, and trace element excess rates. Oxygen isotope data show a decreasing trend up to the top of profile, suggesting a gradual warming at that time.

CONCLUSIONS

Detailed geochemical and mineralogical studies of the Beloc section in Haiti revealed the following.

An Ir anomaly is present within a rust-colored clayey layer (MU3) above the spherule-rich deposit. Cosmic and volcanic inputs are reflected by mineral phases, impactites, and chondrite-normalized PGE patterns.

A Pd-dominated PGE anomaly is present 45 cm above the first Ir anomaly and is associated with a second rust-colored layer (MU4) composed of clay, amphiboles, zeolites, and volcanic glass shards. This layer and PGE anomaly are due to volcanic activity, as indicated by mineral composition and chondrite-normalized PGE and REE patterns.

The chemical variations within the three types of spherules found in the lower marker units can be explained by post-depositional alteration and/or by the lithological heterogeneity of the target material.

ACKNOWLEDGMENTS

This study was supported by the Deutsche Forschungsgemeinschaft (grants Sti128/2-1 and Sti128/2-2, and Stu 169/10-1), the National Science Foundation (grant OCE-9021338), the Petroleum Research Fund (grant 2670-AC8), and the Swiss National Fund (grant 8220-028367). We thank Jan Smit and R. Rocchia for reviewing the paper.

REFERENCES CITED

Adatte, T., Stinnesbeck, W., and Keller, G., 1996, Lithostratigraphic and mineralogical correlations of near K/T boundary clastic sediments in northeastern Mexico: Implications for origin and nature of deposition, *in* Ryder, G., Fastovsky, D., and Gartner, S., eds., The Cretaceous-Tertiary event and other catastrophes in Earth history: Geological Society of America Special Paper 307, p. 211–226.

Alvarez, L.W., Alvarez, W., Asaro, F., and Michel, H.V., 1980, Extraterrestrial cause for the Cretaceous-Tertiary extinction experimental results and theoretical interpretation: Science, v. 208, p. 1095–1108.

Alvarez, L.W., Alvarez, W., Asaro, F., and Michel, H.V., 1982, Current status of the impact theory for the terminal Cretaceous extinction, *in* Silver, L.T., and Schultz, P.H., eds., Geological implications of impacts of large asteroids and comets on the Earth: Geological Society of America Special Paper 190, p. 305–315.

Anbar, A.D., Wasserburg, G.J., Papanastassiou, D.A., and Andersson, P.S., 1996, Iridium in natural waters: Science, v. 273, p. 1524–1528.

Andreozzi, M., Dinelli, E., and Tateo, F., 1997, Geochemical and mineralogical criteria for the identification of ash layers in the stratigraphic framework of a foredeep: The early Miocene Mt. Cervarola Sandstones, northern Italy: Chemical Geology, v. 137, p. 23–39.

Bach, W., and Irber, W., 1998, Rare earth element mobility in the oceanic lower sheeted dyke complex: Evidence from geochemical data and leaching experiments: Chemical Geology, v.151, p. 309–326.

Barkatt, A., Boulos, M., Barkatt, A., Samsapour, W., Boroomand, M., and Macedo, P., 1984, The chemical durability of tektites laboratory study and correlation with long-term corrosion behaviour: Geochimica et Cosmochimica Acta, v. 48, p. 361–371.

Blum, J.D., and Chamberlain, C.P., 1992, Oxygen isotope constraints on the origin of impact glasses from the Cretaceous-Tertiary boundary: Science, v. 257, p. 1104–1107.

Blum, J.D., Chamberlain, C.P., Hingston, M.P., Koeberl, C., Marin, L.E., Schuraytz, B.C., and Sharpton, V.L., 1993, Isotopic comparison of K/T boundary impact glass with melt rock from the Chicxulub and Manson structures: Nature, v. 364, p. 325–327.

Bohor, B.F., and Glass, B.P., 1995, Origin and diagenesis of K/T impact spherules: From Haiti to Wyoming and beyond: Meteoritics, v. 30, p. 182–198.

Bohor, B.F., and Seitz, R., 1990, Cuban K/T catastrophe: Nature, v. 433, p. 593.

Borg, G., Tredoux, M., Maiden, K., Sellschop, J., and Wayward, O., 1987, PGE- and Au-distribution in rift-related volcanics, sediments and stratabound Cu/Ag ores of middle Proterozoic age in Central SWA/ Namibia, *in* Prichard, H., Potts, P., Bowles, J., and Cribb, S., eds., Geo-Platinum 87: London and New York, Elsevier, p. 303–317.

Bowles, J.F.W., 1986, The development of platinum-group minerals in laterites: Economic Geology, v. 81, p. 1278–1285.

Chaussidon, M., Sigurdsson, H., and Métrich, N., 1996, Sulfur and boron isotope study of high-Ca impact glasses from the K/T boundary: Constraints on source rocks, *in* Ryder, G., Fastovsky, D., and Gartner, S., eds., The Cretaceous-Tertiary event and other catastrophes in Earth history: Geological Society of America Special Paper 307, p. 253–262.

Christidis, G.E., 1998, Comparative study of the mobility of major and trace elements during alteration of an andesite and rhyolite to bentonite in the islands of Milos and Kimolos, Aegean, Greece: Clays and Clay Minerals, v. 46, p. 379–399.

Colodner, D.C., Boyle, E.A., Edmond, J.M., and Thomson, J., 1992, Post-depositional mobility of platinum, iridium, and rhenium in marine sediments: Nature, v. 358, p. 402–404.

Condie, K.C., 1991, Another look at rare earth elements in shales: Geochimica et Cosmochimica Acta, v. 55, p. 2527–2531.

Crocket, J., and Kabir, A., 1988, PGE in Hawaiian basalt: Implication of hydrothermal alteration on PGE mobility in volcanic fluids, *in* Prichard, H., Potts, P., Bowles, J., and Cribb, S., eds., Geoplatinum: London and New York, Elsevier, p. 259.

Crocket, J.H., Officer, C.B., Wezel, F.C., and Johnson, G.D., 1988, Distribution of noble metals across the Cretaceous/Tertiary boundary at Gubbio, Italy: Iridium variation as a constraint on the duration and nature of Cretaceous/ Tertiary boundary events: Geology, v. 16, p. 77–88.

Dai, X., Chai, Z., Mao, X., and Ouyang, H., 2000, Sorption and desorption of iridium by coastal sediment: Effects of iridium speciation and sediment components: Chemical Geology, v. 166, p. 15–22.

Date, A.R., Davis, A.E., and Cheung, Y.Y., 1987, The potential of fire assay and inductively coupled plasma source mass spectrometry for the determination of platinum group elements in geological materials: Analyst, v. 112, p. 1217–1222.

Deer, W.A, Howie, R.A., and Zussman, J., 1992, An introduction to the rock-forming minerals (second edition): London, Longman Scientific and Technical, 696 p.

Dyer, B.D., Lyalikova, N.N., Murray, D., Doyle, M., Kolesov, G.M., and Krumbein, W., 1989, Role for microorganisms in the formation of iridium anomalies: Geology, v. 17, p. 1036–1039.

Elderfield, H., and Greaves, M.J., 1982, The rare earth elements in seawater: Nature, v. 296, p. 214–219.

Elliot, W.C., 1993, Origin of the Mg-smectite at the Cretaceous/Tertiary (K/T) boundary at Stevns Klint, Denmark: Clays and Clay Minerals, v. 41, p. 442–452.

Evans, N.J., and Chai, C.F., 1997, The distribution and geochemistry of platinum-group elements as event markers in the Phanerozoic: Palaeogeography, Palaeoclimatology, Palaeoecology, v. 132, p. 373–390.

Evans, D.M., Buchanan, D.L., and Hall, G.E.M., 1994, Dispersion of platinum, palladium and gold from the main sulphide zone, Great Dyke, Zimbabwe: Institution of Mining and Metallurgy, Transactions, Section B: Applied Earth Science, v. 103, p. B57–B67.

Evans, N.J., Ahrens, T.J., and Gregoire, D.C., 1995, Fractionation of ruthenium from iridium at the Cretaceous-Tertiary boundary: Earth and Planetary Science Letters, v. 134, p. 141–153.

Fleet, A.J., 1984, Aqueous and sedimentary geochemistry of the rare earth elements, *in* Henderson, P., ed., Rare earth element geochemistry: Amsterdam, Elsevier, Developments in Geochemistry 2, p. 343–373.

Ganapathy, R., 1980, A major meteorite impact on the Earth 65 million years ago: Evidence from the Cretaceous-Tertiary boundary clay: Science, v. 209, p. 921–923.

Glasby, G.P., and Kunzendorf, H., 1996, Multiple factors in the origin of the Cretaceous/Tertiary boundary: The role of environmental stress and Deccan Trap volcanism: Geologische Rundschau, v. 85, p.191–210.

Greenough, J.D., and Fryer, B.J., 1990, Indian Ocean basalts, *in* Duncan, R.A., and Backmann, J., et al., eds., Proceedings of the Ocean Drilling Program, Scientific Results, Leg 115: College Station, Texas, Ocean Drilling Program, p. 71–84.

Gromet, L.P., Dymek, R.F., Haskin, L.A., and Korotev, R.L., 1984, The "North American shale composite": Its compilation, major and trace element characteristics: Geochimica et Cosmochimica Acta, v. 48, p. 2469–2482.

Guy, C., Daux, V., and Schott, J., 1999, Behaviour of rare earth elements during seawater/basalt interactions in the Mururoa Massif: Chemical Geology, v. 158, p. 21–35.

Hannigan, R., and Basu, A.R., 1998, Late diagenetic trace element remobilization in organic-rich black shales of the Taconic Foreland Basin of Québec, Ontario and New York, *in* Schieber, J., Zimmerle, W., and Sethi, P.S., eds., Shales and mudstones 2: Stuttgart, Germany, Schweizerbart'sche Verlagsbuchhandlung (Nägele u. Obermüller), Petrography, Petrophysics, Geochemistry, and Economic Geology, p. 209–234.

Hataway, J.C., and Sachs, P.L., 1965, Sepiolite and clinoptilolite from the mid-Atlantic Ridge: American Mineralogist, v. 50, p. 852–867.

Heide, K., 1974, Untersuchung der Hochvakuumentgasung bei dynamischer Temperaturänderung bis 1200°C von natürlichen Gläsern unterschiedlicher Genese: Chemie der Erde, v. 33, p. 195–214.

Hildebrand, A.R., Penfield, G.T., Kring, D.A., Pilkington, M., Camargo, A., Jacobson, Z.S.B., and Channell, J.E.T., 1991, Glass from the Cretaceous/Tertiary boundary in Haiti: Nature, v. 349, p. 482.

Hildebrand, A.R., and Boyton, W.V., 1990, Proximal Cretaceous-Tertiary boundary impact deposits in the Caribbean: Science, v. 248, p. 843.

Hopf, S., 1993, Behaviour of rare earth elements in geothermal systems of New Zealand: Journal of Geochemical Exploration, v. 47, p. 333–357.

Hu, X., Wang, Y.L., and Schmitt, R.A., 1988, Geochemistry of sediments on the Rio Grande Rise and the redox evolution of the South Atlantic Ocean: Geochimica et Cosmochimica Acta, v. 52, p. 201–207.

Izett, G.A., 1991, Tektites in Cretaceous-Tertiary boundary rocks on Haiti and their bearing on the Alvarez impact extinction hypothesis: Journal of Geophysical Research, v. 96, p. 20879–20905.

Izett, G.A., Maurrasse, F.J.-M.R., Lichte, F.E., Meeker, G.P., and Bates, R., 1990, Tektites in Cretaceous-Tertiary boundary rocks on Haiti: U.S. Geological Survey Open-File Report 90–635, 31 p.

Izett, G.A., Dalrymple, G.B., and Snee, L.W., 1991, $^{40}Ar/^{39}Ar$ age of Cretaceous/Tertiary boundary tektites from Haiti: Science, v. 252, p. 1539–1542.

Jéhanno, C., Boclet, D., Froget, L., Lambert, B., Robin, E., Rocchia, R., and Turpin, L., 1992, The Cretaceous-Tertiary boundary at Beloc, Haiti: No evidence for an impact in the Caribbean area: Earth and Planetary Science Letters, v. 109, p. 229–241.

Jenkyns, H.C., Gale, A.S., and Corfield, R.M., 1994, Carbon- and oxygen-isotope stratigraphy of the English Chalk and Italian Scaglia and its paleoclimatic significance: Geological Magazine, v. 131, p. 1–34.

Kastner, M., and Stonecipher, S.A., 1978, Zeolites in pelagic sediments of the Atlantic, Pacific and Indian Oceans, *in* Sand, L.B., and Mumpton, F.A., eds., Natural zeolithes occurrence, properties, use: Oxford, Pergamon Press, p. 199–220.

Keller, G., and Lindinger, M., 1989, Stable isotope, TOC and $CaCO_3$ record across the Cretaceous/Tertiary boundary at El Kef, Tunisia: Palaeogeography, Palaeoclimatology, Palaeoecology, v. 73, p. 243–265.

Keller, G., and Stinnesbeck, W., 1996, Sea level changes, clastic deposits and megatsunamis across the Cretaceous/Tertiary boundary, *in* MacLeod, N., and Keller, G., eds., The Cretaceous-Tertiary boundary mass extinction: Biotic and environmental events: New York, Norton Press, p. 415–449.

Keller, G., Lopez-Oliva, J.G., Stinnesbeck, W., and Adatte, T., 1997, Age, stratigraphy and deposition of near K/T siliciclastic deposits in Mexico: Relation to bolide impact?: Geological Society of America Bulletin, v. 109, p. 410–428.

Keller, G., Adatte, T., Stinnesbeck, W., Stüben, D., and Berner, Z., 2001, Age, chemo- and biostratigraphy of Haiti spherule-rich deposits: A multi-event K-T scenario: Canadian Journal of Earth Sciences, v. 38, p. 197–227.

Kerr, A.C., Tarney, J., Marriner, G.F., Nivia, A., Klaver, G.T., and Saunders, A.D., 1996, The geochemistry and tectonic setting of late Cretaceous Caribbean and Colombian volcanism: Journal of South American Earth Sciences, v. 9, p. 111–120.

Kerr, A.C., Marriner, G.F., Tarney, J., Nivia, A., Saunders, A.D., Thirlwall, M.F., and Sinton, C.W., 1997, Cretaceous basaltic terranes in western Colombia: Elemental, chronological and Sr-Nd isotopic constraints on petrogenesis: Journal of Petrology, v. 38, p. 677–702.

Killingley, J.S., 1983, Effects of diagenetic recrystallization on $^{18}O/^{16}O$ values of deep-sea sediments: Nature, v. 301, p. 594–597.

Klug, H.P., and Alexander, L., 1974, X-ray diffraction procedures for polycrystalline and amorphous materials (first and second editions): New York, John Wiley and Sons, 960 p.

Koeberl, C., 1992, Water content of glasses from the K/T boundary, Haiti: An indication of impact origin: Geochimica et Cosmochimica Acta, v. 56, p. 4329–4332.

Koeberl, C., 1993, Chicxulub Crater, Yucatan: Tektites, impact glasses, and the geochemistry of target rocks and breccias: Geology, v. 21, p. 211–214.

Koeberl, C., and Sigurdsson, H., 1992, Geochemistry of impact glasses from the K/T boundary in Haiti: Relation to smectites and new types of glass: Geochimica et Cosmochimica Acta, v. 56, p. 2113–2129.

Koeberl, C., Sharpton, V.L., Schuraytz, B.C., Shirey, S.B., Blum, J.D., and Marin, L.E., 1994, Evidence for a meteoritic component in impact melt rock from the Chicxulub structure: Geochimica et Cosmochimica Acta, v. 58, p. 1679–1884.

Kramar, U., 1997, Advances in energy-dispersive x-ray fluorescence: Journal of Geochemical Exploration, v. 58, p. 73–80.

Kring, D.A., and Boynton, W.V., 1991, Altered spherules of impact melt and associated relic glass from the K/T boundary sediments in Haiti: Geochimica et Cosmochimica Acta, v. 55, p. 1737–1742.

Kring, D.A., Hildebrand, A.R., and Boynton, W.V., 1994, Provenance of mineral phases in the Cretaceous-Tertiary boundary sediments exposed on the southern peninsula of Haiti: Earth and Planetary Science Letters, v 128, p. 629–641.

Kübler, B., 1983, Dosage quantitatif des minéraux majeurs des roches sédimentaires par diffraction X: Cahier de l'Institut de Géologie de Neuchatel Suisse, sér. ADX, 1.1 and 1.2, 15 p.

Kübler, B., 1987, Cristallinité de l'illite, méthodes normalisées de préparations méthodes normalisées de mesures: Cahiers de l'Institut de Géologie de Neuchatel, Suisse, sér. ADX, 3, 21 p.

Kyte, F.T., Zhiming, Z., and Wasson, J.T., 1980, Siderophile-enriched sedi ments from the Cretaceous-Tertiary boundary: Nature, v. 288, p. 651–656.

Kyte, F.T., Bostwick, J.A., and Zhou, L., 1996, The Cretaceous-Tertiary boundary on the Pacific plate: Composition and distribution of impact debris, *in* Ryder, G., Fastovsky, D., and Gartner, S., eds., The Cretaceous-Tertiary event and other catastrophes in Earth history: Geological Society of America Special Paper 307, p. 389–401.

Lamolda, M., Aguado, R., Maurrasse, F.T.-M.R., and Peryt, D., 1997, El transito Cretacico–Terciario en Beloc, Haiti: Registro micropaleontológico e implicaciones bioestratigraficas: Geogaceta, v. 22, p. 97–100.

Leinen, M., and Stakes, D., 1979, Metal accumulation rate in the central Pacific during Cenozoic time: Geological Society of America Bulletin, v. 90, p. 357–375.

Leinen, M., Cwienk, D., Heath, G.R., Biscaye, P.E., Kolla, V., Thiede J., and Dauphin, J.P., 1986, Distribution of biogenic silica and quartz in recent deep-sea sediments: Geology, v. 14, p. 199–203.

Lerbekmo, J.F., Sweet, A.R., and Davidson, R.A., 1999, Geochemistry of the Cretaceous–Tertiary (K-T) boundary interval: South-central Saskatchewan and Montana: Canadian Journal of Earth Sciences, v. 36, p. 717–724.

Leroux, H., Rocchia, R., Froget, L., Orue-Etxebarria, X., Doukhan, J., and Robin, E., 1995, The K/T boundary of Beloc (Haiti): Compared stratigraphic distributions of boundary markers: Earth and Planetary Science Letters, v. 131, p. 255–268.

Li, L., Keller, G., Adatte, T., and Stinnesbeck, W., 2001, Late Cretaceous sea level changes in Tunisia: A multi-disciplinary approach: Journal of the Geological Society [London] v. 157, p. 447–458.

Lyons, J.B., and Officer, C.B., 1992, Mineralogy and petrology of the Haiti Cretaceous/Tertiary section: Earth and Planetary Science Letters, v. 109, p. 205–224.

MacRae, N.D., Nesbitt, H.W., and Kronberg, B.I., 1992, Development of a positive Eu anomaly during diagenesis: Earth and Planetary Science Letters, v. 109, p. 585–591.

Martin-Barajas, A., and Lallier-Verges, E., 1993, Ash layers and pumice in the central Indian basin: Relationship to the formation of manganese nodules: Marine Geology, v. 115, p. 307–329.

Maurrasse, F.J.-M.R., 1982, Survey of the geology of Haiti, *in* Guide to the field excursions in Haiti: Miami, Florida, Miami Geological Society, v. 130, 103 p.

Maurrasse, F.J.-M.R., and Sen, G., 1991, Impacts, tsunamis, and the Haitian Cretaceous-Tertiary boundary layer: Science, v. 252, p. 1690–1693.

McDonough, W.F., and Sun, S.-S., 1995, The composition of the earth: Chemical Geology, v. 120, p. 223–253.

McLennan, S.M., 1989, Rare earth elements in sedimentary rocks: Influence of provenance and sedimentary processes, *in* Lipin, B.R., and McKay, G.A., eds., Geochemistry and mineralogy of rare earth elements: Washington, D.C., Mineralogical Society of America, Reviews in Mineralogy, v. 21, p. 169–200.

Mitchell, S.F., Ball, J.D., Crowley, S.F., Marshall, J.D., Paul, C.R.C., Veltkamp, C.J., and Samir, A., 1997, Isotope data from cretaceous chalks and foraminifera: Environmental or diagenetic signals?: Geology, v. 25, p. 691–694.

Montanari, A., and Koeberl, C., 2000, Impact stratigraphy: Heidelberg, Germany, Springer-Verlag, 364 p.

Murray, R.W., and Leinen, M., 1993, Chemical transport to the seafloor of the equatorial Pacific Ocean across a latitudinal transect at 135° W: Tracking sedimentary major, trace and rare earth element fluxes at the equator and the Intertropical Convergence Zone: Geochimica et Cosmochimica Acta, v. 57, p. 4141–4163.

Murray, R.W., and Leinen, M., 1996, Scavenged excess aluminum and its relationship to bulk titanium in biogenic sediment from central equatorial Pacific Ocean: Geochimica et Cosmochimica Acta, v. 60, p. 3869–3878.

Murray, R.W., Buchholtz ten Brink, M.R., Jones, D.L., Gerlach, D.C., and Russ G.P., III, 1990, Rare earth elements as indicators of different marine depositional environment in chert and shale: Geology, v. 18, p. 268–293.

Officer, C.B., and Drake, C.L., 1985, Terminal Cretaceous environmental events: Science, v. 277, p. 1161–1167.

Ohr, M., Halliday, A.N., and Peacor, D.R., 1994, Mobility and fractionation of rare earth elements in argillaceous sediments: Implications for dating diagenesis and low-grade metamorphism: Geochimica et Cosmochimica Acta, v. 58, p. 289–312.

Orians, K.J., Boyle, E.A., and Bruland, K.W., 1990, Dissolved titanium in the open ocean: Nature, v. 348, p. 322–325.

Orth, C.J., Quintana, L.R., Gilmore, J.S., Barrick, J.E., Haywa, J.N., and Spesshardt, S.A., 1988, Pt-metal anomalies in the lower Mississippian of southern Oklahoma: Geology, v. 16, p. 627–603.

Palmer, M.R., 1985, Rare earth elements in foraminifera tests: Earth and Planetary Science Letters, v. 73, p. 285–298.

Pattan, J.N., Rao, Ch.M., Higgs, N.C., Colley, S., and Parthiban, G., 1995, Distribution of major, trace and rare-earth elements in surface sediments of the Wharton Basin, Indian Ocean: Chemical Geology, v. 121, p. 201–215.

Reddi, G.S., Rao, C.R.M., Rao, T.A.S, Lakshmi, S.V., Prabhu, R.K., and Mahalingam, T.R., 1994, Nickel sulphide fire assay: ICPMS method for the determination of platinum group elements—A detailed study on the recovery and losses at different stages: Fresenius Journal of Analytical Chemistry, v. 348, p. 350–352.

Rocchia, R., and Robin, E., 1998, The stratigraphic distribution of iridium at the Cretaceous–Tertiary boundary of El Kef, Tunisia: Bulletin de la Société Géologique de France, v. 169, p. 515–526.

Rocchia, R., Boclet, D., Bonté, Ph., Jéhanno, C., Chen, Y., Courtillot, V., Mary, C., and Wezel, F., 1990, The Cretaceous-Tertiary boundary at Gubbio revisited: Vertical extent of the Ir anomaly: Earth and Planetary Science Letters, v. 99, p. 206–219.

Rocchia, E., Robin, E., Froget, L., and Gayraud, J., 1996, Stratigraphic distribution of extraterrestrial markers at the Cretaceous-Tertiary boundary in the Gulf of Mexico area: Implications for the temporal complexity of the event, *in* Ryder, G., Fastovsky, D., and Gartner, S., eds., The Cretaceous-Tertiary event and other catastrophes in Earth history: Geological Society of America Special Paper 307, p. 279–302.

Rock, N.M.S., 1988, Numerical geology: Berlin, Springer-Verlag, Lecture Notes in Earth Sciences, v. 18, 427 p.

Rollinson, H., 1993, Using geochemical data: London, Longman, 352 p.

Sawlowicz, Z., 1993, Iridium and other platinum-group elements as geochemical markers in sedimentary environments: Palaeogeography, Palaeoclimatology, Palaeoecology, v. 104, p. 253–270.

Schrag, D.P., DePaolo, D.J., and Richter, F.M., 1995, Reconstructing past sea surface temperatures: Correcting for diagenesis of bulk marine carbonate: Geochimica et Cosmochimica Acta, v. 59, p. 2265–2278.

Schroeder, J.O., Murray, R.W., Leinen, M., Pflaum, R.C., and Janacek, T.R., 1997, Barium in equatorial Pacific carbonate sediment: Terrigenous, oxide, and biogenic associations: Paleoceanography, v. 12, no. 1, p. 125–146.

Scott, P.W., Jackson, T.A., and Dunham, A.C., 1999, Economic potential of the ultramafic rocks of Jamaica and Tobago: Two contrasting geological settings in the Caribbean: Mineralium Deposita, v. 34, p. 718–723.

Sethi, P.S., Hanningan, R.E., and Leithold, E.L., 1998, Rare-earth element chemistry of Cenomanian-Turonian shales of the North American Greenhorn Sea, Utah, *in* Schieber, J., Zimmerle, W., and Sethi, P.S., eds., Shales and mudstones: Stuttgart, Germany, Schweizerbart'sche Verlagsbuchhandlung (Nägele u. Obermüller), p. 195–208.

Shields, G., and Stille, P., 1998, Stratigraphic trends in cerium anomaly in authigenic marine carbonates and phosphates: Diagenetic alteration or seawater signals? [abs.]: Goldschmidt Conference, Toulouse, Mineralogical Magazine, v. 62A, p. 1387–1388.

Sigurdsson, H., D'Hondt, S., Arthur, M.A., Bralower, T.J., Zachos, J.C., Van Fossen, M., and Channell, J.E.T., 1991a, Glass from the Cretaceous/Tertiary boundary in Haiti: Nature, v. 349, p. 482–487.

Sigurdsson, H., Bonté, Ph., Turpin, L., Chaussidon, M., Metrich, N., Steinberg,

M., Pradel, Ph., and D'Hondt, S.D., 1991b, Geochemical constraints on source region of Cretaceous/Tertiary impact glasses: Nature, v. 353, p. 839–842.

Sinton, C.W., Duncan, R.A., Storey, M., Lewis, J., and Estrada, J.J., 1998, An oceanic flood basalt province within the Caribbean plate: Earth and Planetary Science Letters, v. 155, p. 221–235.

Smit, J., 1999, The global stratigraphy of the Cretaceous-Tertiary boundary impact ejecta: Annual Reviews in Earth and Planetary Science, v. 27, p. 75–113.

Smit, J., Montanari, A., Swinburne, N.H.M., Alvarez, W., Hildebrand, A., Margolis, S.V., Claeys, P., Lowrie, W., and Asaro, F., 1992, Tektite-bearing deep-water clastic unit at the Cretaceous-Tertiary boundary in northeastern Mexico: Geology, v. 20, p. 99–103.

Stelzner, Th., and Heide, K., 1996, The study of weathering products of meteorites by means of evolved gas analysis: Meteoritics and Planetary Science, v. 31, p. 249–254.

Stinnesbeck, W., Barbarin, J.M., Keller, G., Lopez-Oliva, J.G., Pivnik, D., Lyons, J., Officer, C., Adatte, T., Graup, G., Rocchia, R., and Robin, E., 1993, Deposition of channel deposits near the Cretaceous/Tertiary boundary in northeastern Mexico: Catastrophic or "normal" sedimentary deposits?: Geology, v. 21, p. 797–800.

Stinnesbeck, W., Keller, G., Adatte, T., Stüben, D., Kramar, U., Berner, Z., Desremeaux, C., and Molière, E., 1999, Beloc, Haiti, revisited: Multiple events across the Cretaceous-Tertiary transition in the Caribbean: Terra Nova, v. 11, p. 303–310.

Stinnesbeck, W., Schulte, P., Lindenmaier, F., Adatte, T., Affolter, M., Schilli, L., Keller, G., Stüben, D., Berner, Z., Kramar, U., Burns, S., and Lopez-Oliva, J.G., 2001, Late Maastrichtian age of spherule deposits in northeastern Mexico: Implication for Chicxulub scenario: Canadian Journal of Earth Sciences, v. 38, p. 1–10.

Stoll, H.M., and Schrag, D.P., 1996, Evidence for glacial control of rapid sea level changes in the Early Cretaceous: Science, v. 272, p. 1171–1174.

Stoll, H.M., and Schrag, D.P., 1998, Effects of Quaternary sea level cycles on strontium in seawater: Geochimica et Cosmochimica Acta, v. 62, p. 1107–1118.

Stüben, D., Glasby, G.P., Eckhardt, J.-D., Berner, Z., Mountain, B.W., and Usui, A., 1999, Enrichments of platinum-group elements in hydrogeneous, diagenetic and hydrothermal marine manganese and iron deposits: Exploration and Mining Geology, v. 8, p. 1–15.

Stüben, D., Kramar, U., Berner, Z., Stinnesbeck, W., Keller, G., and Adatte, T., 2001, Trace elements, stable isotopes, and clay mineralogy of the Elles II K/T profile: Indications for sealevel fluctuations and primary productivity: Palaeogeography, Palaeoclimatology, Palaeoecology, special volumne (in press).

Sutherland, F.L., 1994, Volcanism around K/T boundary time: Its role in an impact scenario for the K/T extinction event: Earth Science Reviews, v. 36, p. 1–26.

Taylor, S.R., and McLennan, S.M., 1985, The continental crust: Its composition and evolution: Oxford, Blackwell, 312 p.

Terakado, Y., and Nakajima, W., 1995, Characteristics of rare-earth elements, Ba, Sr and Rb abundances in natural zeolithes: Geochemical Journal, v. 29, p. 337–345.

Torres, M.E., Brumsack, H.J., Bohrmann, G., and Emeis, K.C., 1996, Barite fronts in continental margin sediments: A new look at barium remobilization in the zone of sulfate reduction and formation of heavy barites in diagenetic fronts: Chemical Geology, v. 127, p. 125–139.

Tredoux, M., DeWitt, M.J., Hart, R.J., Lindsay, N.M., Verhagen, B., and Sellschop, J.P.F., 1989, Chemostratigraphy across the Cretaceous-Tertiary boundary and a critical assessment of the iridium anomaly: Journal of Geology, v. 97, p. 585–605.

Wallace, M.W., Gostin, V.A., and Keays, R.R., 1990, Acraman impact ejecta and host shales: Evidence for low-temperature mobilization of iridium and other platinoids: Geology, v. 18, p. 132–135.

Weaver, C.E., 1968, Mineral facies in the Tertiary of the continental shelf and Blake Plateau: Southeastern Geology, v. 9, p. 57–63.

Westland, A.D., 1981, Inorganic chemistry of the platinum-group elements, *in* Cabri, L.J., ed., Platinum-group elements: Mineralogy, geology and recovery: Canadian Institute of Mining and Metallurgy, Special Issue, v. 23, p. 5–18.

Winter, B.L., Johnson, C.M., Clark, D.L., 1997, Geochemical constraints on the formation of Late Cenozoic ferromanganese micronodules from the central Arctic Ocean: Marine Geology, v. 138, p. 149–169.

Zachos, J.C., Arthur, M.A., Thunell, R.C., Williams, D.F., and Tappa, E.J., 1985, Stable isotope and trace element geochemistry of carbonate sediments across the Cretaceous/Tertiary boundary at Deep Sea Drilling Project Hole 577, Leg 86, *in* Heath, G.R., Burckle, L.H., et al., eds, Initial reports of the Deep Sea Drilling Project, Volume 86: Washington, D.C., U.S. Government Printing Office, p. 513–532.

Zachos, J.C., Arthur, M.A., and Dean, W.E., 1989, Geochemical evidence for suppression of pelagic marine productivity at the Cretaceous/Tertiary boundary: Nature, v. 337, p. 61–64.

Zereini, F., Skerstupp, B., and Urban, H., 1994, Comparison between the use of sodium and lithium tetraborate in platinum-group element determination by nickel sulphide fire-assay: Geostandards Newsletter, v. 18, p. 105–109.

Zielinski, R.A., 1982, The mobility of uranium and other elements during alteration of rhyolite ash to montmorilonite: A case study in the Troublesome Formation, Colorado, U.S.A.: Chemical Geology, v. 35, p. 185–204.

Zielinski, R.A., 1985, Element mobility during alteration of silicic ash to kaolinite: A study of tonstein: Sedimentology, v. 32, p. 567–579.

Manuscript Submitted October 10, 2000; Accepted by the Society March 22, 2001

Printed in the U.S.A.

Geological Society of America
Special Paper 356
2002

Cretaceous-Tertiary boundary at Blake Nose (Ocean Drilling Program Leg 171B): A record of the Chicxulub impact ejecta

Francisca Martínez-Ruiz
Instituto Andaluz de Ciencias de la Tierra (CSIC-UGR), Facultad de Ciencias, Campus Fuentenueva, 18002 Granada, Spain
Miguel Ortega-Huertas
Inmaculada Palomo
Departamento de Mineralogía y Petrología, Facultad de Ciencias, Campus Fuentenueva, 18002 Granada, Spain
Jan Smit
Department of Sedimentary Geology, Vrije Universiteit, de Boelelaan 1085, 1081HV Amsterdam, The Netherlands

ABSTRACT

The Ocean Drilling Program (ODP) included as one of its Leg 171B objectives the recovery of a detailed record of the Cretaceous-Tertiary (K-T) events at Blake Nose (northwest Atlantic). This aim was successfully achieved with sections across the K-T boundary recovered at Sites 1049, 1050, and 1052, and a thick spherule bed recovered at ODP Site 1049. This spherule bed varies from 7 to 17 cm in thickness at the three different holes drilled at Site 1049, and occurs at the biostratigraphic boundary between the Cretaceous and the Paleocene. Mineralogical and geochemical analyses of the Blake Nose spherule bed reveal that it is mainly composed of smectite derived from the alteration of a precursor material, mostly glass. Also present in minor proportions are dolomite, quartz, zeolites, and trace amounts of rutile and some lithic fragments. Different types of spherules, dark green, pale yellow, and light green, that can be related to different precursors were observed in the Blake Nose spherule bed. Transmission electron microscope observations showed that smectite directly replaced the original material and that dark green spherules originated from a Si-rich precursor, whereas pale yellow spherules originated from a more Ca-rich precursor. The chemical composition of the spherule-bed material at Blake Nose shows little evidence for a significant extraterrestrial contribution, suggesting that the spherule-bed material was mainly derived from the alteration of target-rock-derived material from Chicxulub crater. In addition, rare earth element C1-normalized patterns also suggest that this material was derived from upper crustal rocks.

Martínez-Ruiz, F., Ortega-Huertas, M., Palomo, I., and Smit, J., 2002, Cretaceous-Tertiary boundary at Blake Nose (Ocean Drilling Program Leg 171B): A record of the Chicxulub impact ejecta, *in* Koeberl, C., and MacLeod, K.G., eds., Catastrophic Events and Mass Extinctions: Impacts and Beyond: Boulder, Colorado, Geological Society of America Special Paper 356, p. 189–199.

INTRODUCTION

The Chicxulub structure was first reported by Penfield and Camargo (1981) as a large buried impact crater, and it was suggested at that time that it could be the site of the Cretaceous-Tertiary (K-T) impact event (Byars, 1981). However, little research was devoted to this structure until evidence for thick ejecta layers and larger shocked quartz grains in the United States interior (e.g., Bohor, 1990; Sharpton et al., 1990), and evidence for tsunami deposits in Brazos River, Texas (Bourgeois et al., 1988; Smit and Romein, 1985), turned attention to a possible buried crater near the Gulf of Mexico. In the early 1990s, evidence for the temporal link of the Chicxulub structure to the K-T mass-extinction event (e.g., Hildebrand et al., 1991; Kring and Boynton, 1992; Izett, 1991; Izett et al., 1991; Sharpton et al., 1992; Swisher et al., 1992) concentrated intense research on this structure and revitalized the issue of a buried K-T impact crater to further confirm the Alvarez hypothesis (Alvarez et al., 1980). Gravity measurements and drill-core data from Chicxulub, as well as the discovery of new K-T boundary outcrops in the Gulf of Mexico area, reinforced the hypothesis of the K-T boundary impact at this site (e.g., Sigurdsson et al., 1991, 1992; Sharpton et al., 1992, 1993, 1994; Blum et al., 1993; Koeberl et al., 1994; Smit et al., 1992a, 1992b). The location of the impact would also explain the different nature of proximal and distal K-T boundary deposits and some features such as spherule size and Ir concentration (e.g., Smit, 1999). The Ocean Drilling Program (ODP) also addressed this line of research by including the K-T boundary sediments in the Gulf of Mexico and the North American Atlantic margin in its drilling objectives. At the Blake Nose Plateau, ODP Leg 171B included as one of its objectives the recovery of a detailed record of the K-T events. This aim was successfully achieved and K-T boundary materials were recovered at ODP Site 1049 in three adjacent holes: 1049A (30°08.5436′N, 76°06.7312′W), 1049B (30 + 08.5423′N, 76 + 06.7264′W), and 1049C (30°08.5370′N, 76°06.7271′W). The K-T boundary is marked in Holes 1049A, 1049B, and 1049C by a single bed of spherules, 17, 7, and 9 cm thick, respectively, capped by a limonitic layer. The excellent Cretaceous-Tertiary (K-T) boundary interval recovered provided evidence of the deposition of K-T impact-generated material at this location.

SAMPLES AND METHODS

During ODP Leg 171B five sites were cored at Blake Nose (Fig. 1); sections across the K-T boundary recovered at Sites 1049, 1050, and 1052 (Norris et al., 1998). At ODP Site 1049, a spherule bed occurs at the biostratigraphic boundary between the Cretaceous and the Paleocene. Analytical work has therefore focused on sediments from the K-T boundary interval recovered at this site because there is a complete record of the K-T event here. The uppermost Maastrichtian sediments comprise light gray nannofossil-foraminifer ooze (*Abathomphalus mayaroensis* zone and *Micula prinsii* zone) that is slumped (Norris et al., 1998, 1999; Klaus et al., 2000). A sharp contact separates this ooze from an overlying bed of spherical and oval-shaped spherules. This spherule bed varies from 7 to 17 cm in thickness at the three holes drilled at Site 1049 (Fig. 2), which suggests reworking of the ejecta material (Klaus et al., 2000). Despite this, the spherule bed confirms that the impact-generated material from the Chicxulub crater is well preserved at the Blake Nose Plateau. The spherule bed is capped by a 1–3-mm-thick orange limonitic layer, overlain by lowermost Paleocene ooze with a foraminiferal assemblage indicative of the P-alpha zone (Norris et al., 1998, 1999). The limonitic layer was initially (during on-board analysis) considered to be a candidate for the so-called fireball layer, but the usual extraterrestrial markers, such as Ni-rich spinels and a strong iridium anomaly, are conspicuously absent. The spherule bed and Cretaceous and Tertiary sediments were sampled in sections 1049A-17X-2 and 1049B-8H-2, by continuous sampling every 2 cm between 20 cm above and 20 cm below the boundary bed and samples spaced 2 cm above and below this interval. Because similar results were obtained from K-T boundary sediments from Holes 1049A and 1049B, only data from Hole 1049A are reported in tables and figures (see Table 1 for location of samples). Spherules were hand-picked under a stereomicroscope with a dry brush. Mineralogical and geochemical analyses of bulk samples and representative hand-picked spherules were done using the following methods.

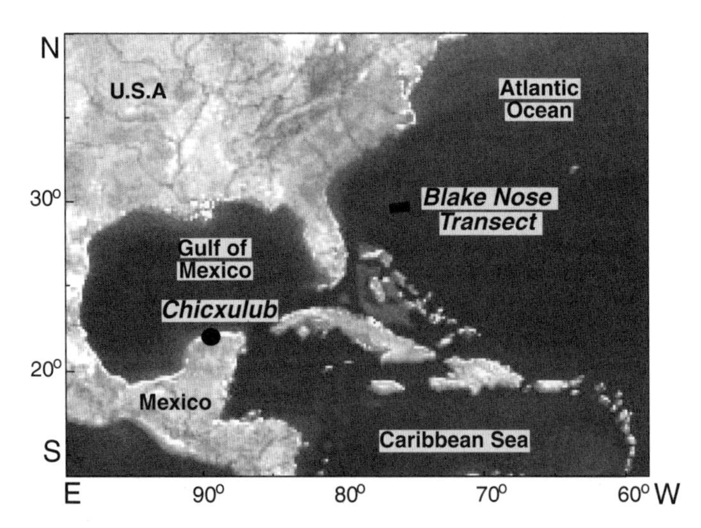

Figure 1. Location of Ocean Drilling Program (ODP) Leg 171B Blake Nose drilling transect.

X-ray diffraction

For bulk mineralogy analyses, samples were packed in Al holders for X-ray diffraction (XRD). For clay mineral analyses, the carbonate fraction was removed using acetic acid, starting the reaction at a very low concentration (0.1 N) and increasing

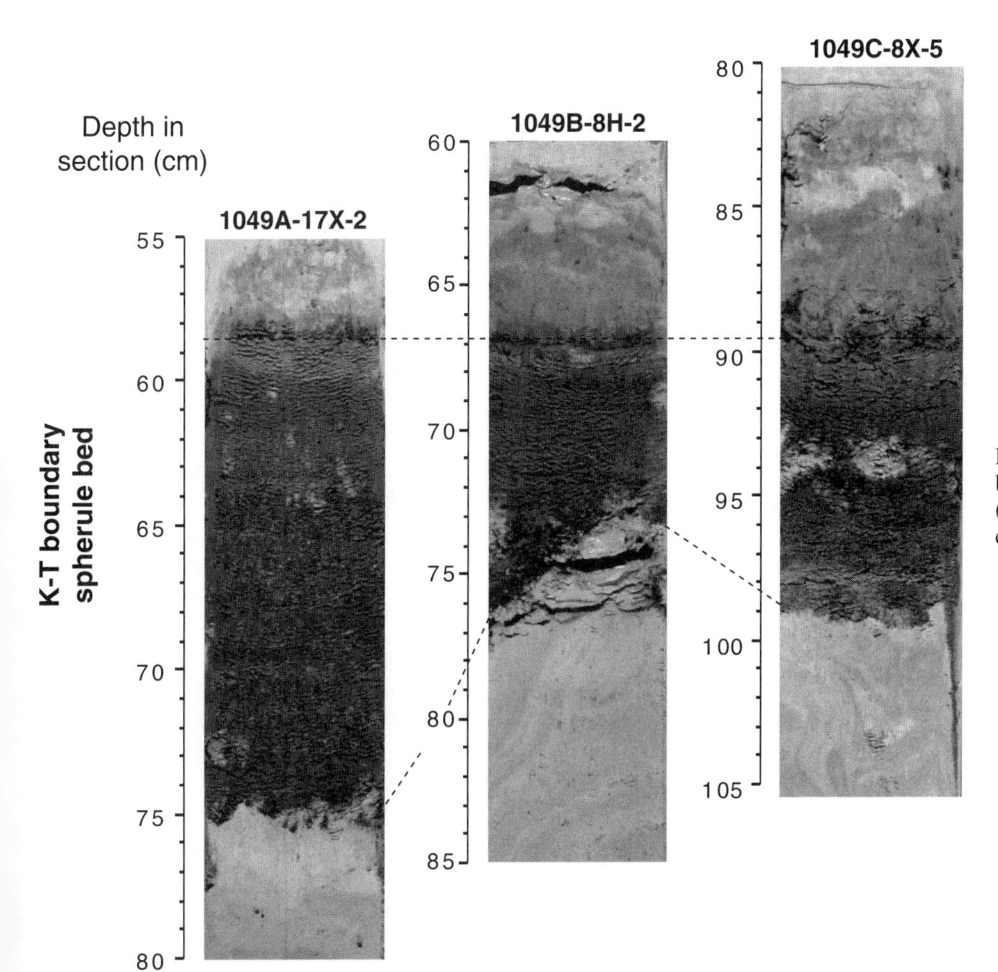

Figure 2. Core photographs of spherule bed that marks Cretaceous-Tertiary (K-T) boundary interval at three holes drilled at Site 1049.

to 1 N. Clays were deflocculated by successive washing and the <2 μm fraction was separated by centrifuging. The clay fraction was smeared onto glass slides for XRD. Diffractograms were obtained using a Philips PW 1710 diffractometer with Cu-Kα radiation. Scans were run from 2° to 64° 2θ for bulk samples and untreated clay preparations, and from 2° to 30° 2θ for glycolated, heated, and dimethyl-sulfoxide-treated samples. Semiquantitative analyses were performed on integrated peak areas using a specific computer program for the diffractometer used (Nieto et al., 1989). The estimated semiquantitative analysis error is 5%.

Electron microscopy

Morphological studies on bulk samples and hand-picked spherules were performed using binocular microscope and scanning electron microscopy (SEM; Zeiss DSM 950). Quantitative microanalyses of clay minerals were obtained by transmission electron microscopy (TEM, Philips CM-20 equipped with an EDAX microanalysis system). Quantitative analyses were obtained in scanning TEM mode only from particle edges using a 70 D diameter beam with a 200 × 1000 D scanning area and a short counting time to avoid alkali loss (Nieto et al., 1996). Smectite formulas were normalized to 11 oxygens.

Inductively coupled plasma-mass spectrometry and atomic absorption spectrometry

Samples from the spherule bed were cleaned of Cretaceous clasts under a stereomicroscope and dried, homogenized, and ground in an agate mortar for chemical analyses by inductively coupled plasma-mass spectrometry (ICP-MS) and atomic absorption spectrometry (AAS). Rb, Sr, Ba, V, Cr, Co, Ni, Cu, Zr, Hf, Mo, Pb, U, Th, and rare earth elements (REE) were analyzed by ICP-MS, and Al, K, Fe, Mn, Ca, and Mg were analyzed by AAS. Analyses were performed on bulk samples following sample digestion with HNO_3 + HF of 0.100 g of sample powder in a Teflon-lined vessel at high temperature and pressure, evaporation to dryness, and subsequent dissolution in 100 ml of 4 vol% HNO_3. ICP-MS instrument measurements were performed in triplicate using a Perkin Elmer Sciex Elan-5000 spectrometer with Rh as internal standard, and AAS

TABLE 1. X-RAY DIFFRACTION DATA OF THE CRETACEOUS-TERTIARY BOUNDARY INTERVAL AT HOLE 1049A

Core, section, interval (cm)	Depth (mbsf)	Clay minerals	Quartz (wt%)	Calcite (wt%)	Smectite (wt%)	Illite (wt%)	Kaolinite (wt%)
17X-2, 001–003	125.32	28	<5	69	83	8	9
17X-2, 004–006	125.35	28	<5	69	82	12	6
17X-2, 008–010	125.39	24	5	71	63	30	7
17X-2, 012–014	125.43	25	<5	74	56	20	24
17X-2, 016–018	125.47	27	<5	70	61	17	22
17X-2, 020–022	125.51	30	5	65	49	25	26
17X-2, 024–026	125.55	30	5	65	56	25	19
17X-2, 028–030	125.59	28	7	65	63	22	15
17X-2, 032–034	125.63	28	7	65	65	20	15
17X-2, 035–037	125.66	30	<5	66	60	29	11
17X-2, 038–040	125.69	25	<5	72	83	14	<5
17X-2, 040–042	125.71	28	<5	69	90	<5	8
17X-2, 042–044	125.73	26	<5	72	85	11	<5
17X-2, 044–046	125.75	19	<5	77	85	9	6
17X-2, 046–048	125.77	9	<5	87	84	9	7
17X-2, 048–050	125.79	11	<5	85	86	6	8
17X-2, 050–052	125.81	12	<5	84	80	13	7
17X-2, 052–054	125.83	11	<5	85	84	9	7
17X-2, 054–056	125.85	14	<5	83	88	5	7
17X-2, 056–058	125.87	12	<5	85	89	5	6
17X-2, 060–062	125.91	75	<5	12	99	<5	<5
17X-2, 062–064	125.93	92	<5	6	99	<5	<5
17X-2, 064–066	125.95	90	<5	8	98	<5	<5
17X-2, 066–068	125.97	91	<5	7	98	<5	<5
17X-2, 068–070	125.99	90	<5	8	96	<5	3
17X-2, 070–072	126.01	95	<5	5	98	<5	<5
17X-2, 072–074	126.03	91	<5	7	99	<5	<5
17X-2, 074–076	126.05	91	<5	7	99	<5	<5
17X-2, 076–078	126.07	13	<5	84	98	<5	<5
17X-2, 078–080	126.09	23	<5	75	77	15	8
17X-2, 080–082	126.11	22	<5	75	80	14	6
17X-2, 082–084	126.13	22	<5	74	82	10	8
17X-2, 084–086	126.15	21	<5	76	80	12	8
17X-2, 086–088	126.17	21	<5	76	84	13	<5
17X-2, 088–090	126.19	19	<5	79	83	10	7
17X-2, 090–092	126.21	22	<5	76	77	15	8
17X-2, 092–094	126.23	28	<5	68	45	40	15
17X-2, 096–098	126.27	30	<5	67	72	<5	24
17X-2, 100–102	126.31	30	<5	68	62	23	15
17X-2, 104–106	126.35	31	<5	68	30	30	40
17X-2, 108–110	126.39	29	<5	68	55	26	19
17X-2, 112–114	126.43	30	<5	67	50	22	28
17X-2, 116–118	126.47	28	<5	68	43	17	40
17X-2, 120–122	126.51	28	<5	68	48	12	40
17X-2, 124–126	126.55	31	<5	67	46	30	24
17X-2, 128–130	126.59	27	<5	69	68	19	13
17X-2, 132–134	126.63	28	<5	68	83	11	6
17X-2, 136–138	126.67	24	<5	75	66	13	21
17X-2, 140–142	126.71	24	<5	75	48	35	17
17X-2, 144–146	126.75	25	<5	75	81	11	8
17X-2, 148–150	126.79	20	<5	79	82	10	8

Note: Main mineral components are in the bulk sediments (clay minerals, quartz, and calcite) and clay minerals proportions are in the <2 μm fraction (smectite, illite, and kaolinite). Shaded area corresponds to the Cretaceous-Tertiary boundary layer; mbsf: meters below sea floor.

analyses were carried out with a Perkin Elmer 5100 ZL spectrometer. The quality of the analyses was monitored with laboratory and international standards from the U.S. Geological Survey (USGS). ICP-MS precision and accuracy was better than ±2% and ±5% for analyte concentrations of 50 and 5 ppm in the rock, respectively. AAS analytical error was <2%.

CRETACEOUS AND TERTIARY SEDIMENTS

The ooze immediately below the spherule bed contains planktonic foraminifera and calcareous nannofossils of late Maastrichtian age; fossils are abundant and well preserved (Norris et al., 1998, 1999). The burrow-mottled ooze overlying the spherule bed (Fig. 2) contains some reworked Cretaceous planktonic foraminifera, but typical early Danian species are also present (Norris et al., 1998, 1999). These uppermost Cretaceous and lowermost Danian materials are composed of carbonates, clay minerals, and quartz (Table 1), and minor quantities of feldspars and traces of heavy minerals and pyrite. Feldspar recognized by XRD is always <5 wt%, and trace minerals were only identified by SEM. Calcite is the dominant carbonate phase, but small quantities of dolomite are occasionally present in the burrow-mottled ooze. Clay mineral assemblages consist of smectite, which is dominant, kaolinite, and illite (Table 1). Cretaceous sediments are slump folded, although the overlying K-T boundary stratigraphy is undisturbed. The deformation of Cretaceous sediments is a general feature at proximal ejecta sites, related to the seismic energy input from the Chicxulub impact, some of it induced before the emplacement of the ejecta from this impact (Alvarez et al., 1992; Smit, 1999; Norris et al., 2000).

Although precise chemical stratigraphy of the uppermost Cretaceous sediments cannot be established due to slumping, their chemical composition is very homogeneous (Fig. 3), except for some variations related to detrital mineral abundances and redox conditions (for the chemical data, see Martínez-Ruiz et al., 2001a). Thus, some potassium-concentration fluctuations can be related to illite abundance. The slight decrease in Mn and Fe contents, and the slight increase in redox-sensitive element concentration, could be related to a change in redox conditions at the end of the Maastrichtian. The lowermost Danian stratigraphy is undisturbed. Some changes observed in the burrow-mottled ooze are mainly related to diagenetic alteration. The Mn content increases above the boundary bed, indicating diagenetic remobilization of Mn, and the Mn peak marks the penetration of the oxidation front. The REE concentrations are depleted in the spherule bed, but they are slightly enriched in the burrow-mottled ooze, which suggests reprecipitation of REEs mobilized from the spherule bed. Smit et al. (1997) reported the maximum Ir concentration in Blake Nose sediments just above the spherule bed, which may suggest that diagenetic remobilization may also have affected extraterrestrial elements, although Ni-rich spinels are abundant above the spherule bed as well (R. Rocchia and E. Robin, 1997, personal commun.).

K-T BOUNDARY BED

The spherule bed at Blake Nose (Fig. 2) consists of a coarse, graded, and poorly cemented unit. It is mostly composed of spherules, but contains Cretaceous foraminifera and clasts. Mineralogical analyses reveal that the spherule bed mostly consists of clays and minor proportions of calcite (Table 1), partially derived from the Cretaceous material. Dolomite, quartz, and zeolites are also present in minor proportions, and trace amounts of rutile, biotite, and some lithic fragments. Clays are mostly smectites; occasional traces of illite and kaolinite are probably derived from contamination by Cretaceous material. The contact of the spherule bed with sediments above and below is very sharp, suggesting very rapid deposition. The presence of Cretaceous materials within the spherule bed strongly supports downslope transport of the spherule bed material.

Spherules

Stereomicroscope and SEM observations reveal that the morphologies of the Blake Nose spherules are mainly perfect spheres with lesser proportions of oval spherules that contain bubble cavities. Sizes usually range from 100 μm to 1000 μm. Different types of spherules have been distinguished on the basis of color, morphology, and surface texture. They are light green, dark green, or pale yellow, with nodular, smooth or rough (Fig. 4) surfaces (Martínez-Ruiz et al., 2001b).

Diagenetic alteration and precursor material

X-ray diffraction scans of oriented samples reveal that the spherules are mainly composed of smectite (Table 2); there is some evidence for preserved unaltered glass relics. Moreover, TEM microanalyses show some compositional variations between dark green spherules and pale yellow spherules (Table 2). Smectites from dark green spherules are richer in Fe (Table 2), and Si/Al usually ranges from 3.1 to 3.3; however, in some Si-rich areas, the Si/Al range of 3.5 to 5 (Fig. 5A) does not correspond to a true smectite composition but to the altering glass. This suggests that a Si-rich glass has been their precursor. Smectites from pale yellow spherules have lower Si/Al, usually ranging from 2.1 to 2.5, and originated from a Ca-rich material, the Ca/Si ratio being ~6. Some calcite crystals are observed in the Ca-rich matrix that could be an original, unaltered phase. Light green spherules smectites have lower Si/Al (2–2.5) than those from dark green spherules.

The compositional differences are probably derived from different precursor glass types. Two end-member types of glass, black andesitic and CaO-poor, and honey-colored CaO-rich, are present in the proximal K-T ejecta from Haitian sections, and Mimbral, Mexico (Izett, 1991; Izett et al., 1991; Sigurdsson et al., 1991; Smit et al., 1992b; Koeberl and Sigurdsson, 1992). The differences between Blake Nose spherules therefore seem to indicate that they were derived from the alteration of compositionally different impact glasses. Variations in octahedral cations (Table 2) support a compositionally variable precursor because impact-generated glass was only briefly melted, so there was not enough time for the elements to mix and homogenize (e.g., Alvarez et al., 1992).

The smectite morphologies observed by TEM in this study are similar to those of smectites originated from the alteration of volcanic glass (e.g., De la Fuente et al., 2000) (Fig. 5A), implying that the smectite directly replaced the original glass phase. Direct formation of smectites from glass material is also consistent with the presence of altered rims where there seems

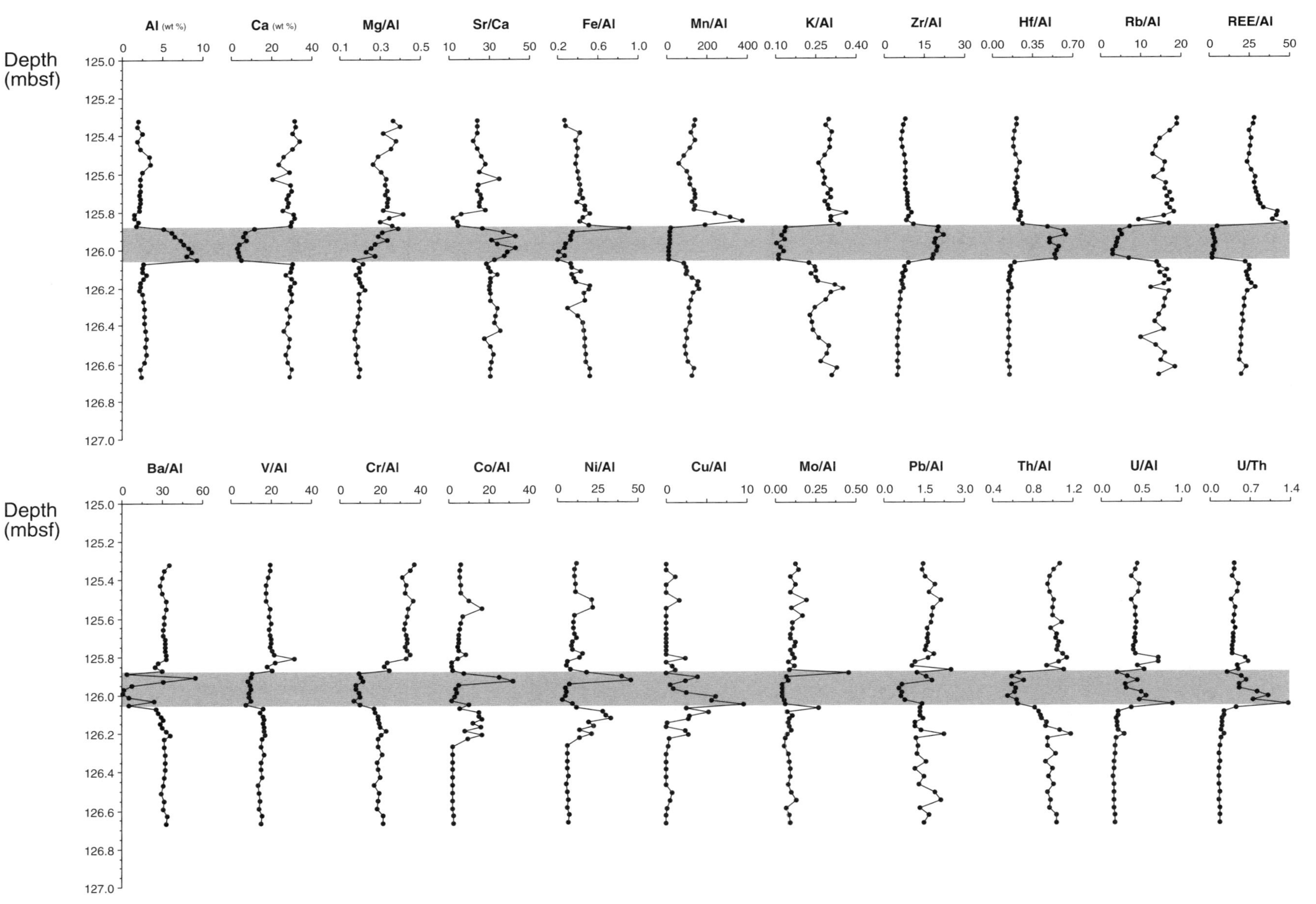

Figure 3. Geochemical data from Cretaceous-Tertiary (K-T) boundary interval at Hole 1049A. Plots show Ca and Al concentrations (wt%), Sr/Ca and Th/U ratios, Fe, K, and Mg concentrations normalized to Al and trace element/Al weight ratio ($\times 10^4$) vs. depth (for chemical data, see Martínez-Ruiz et al., 2001a). REE, rare earth elements; MBSF, meters below seafloor.

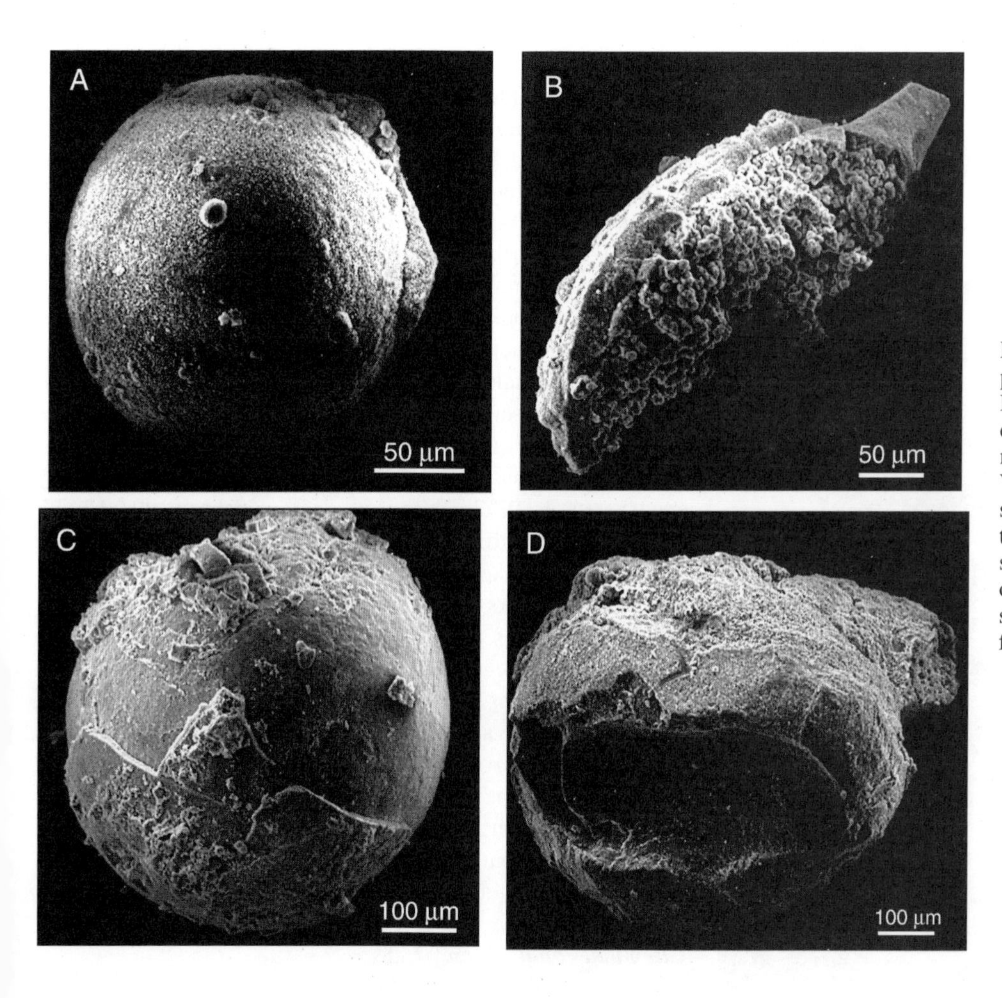

Figure 4. Scanning electron microscopy photographs of smectite spherules from Blake Nose at Site 1049. A: View of dark green spherule showing spherical morphology and nodular surface. B: View of dark green spherule with drop-shape morphology and filled with smectite aggregates. C: View of pale yellow spherule showing its spherical morphology. D: View of light green spherule showing its morphology and rough surface.

TABLE 2. REPRESENTATIVE ANALYTICAL ELECTRON MICROSCOPY DATA FROM SMECTITE SPHERULES OF THE CRETACEOUS-TERTIARY BOUNDARY BED AT HOLE 1049A

Samples	Si (a. f. u.)	Al^{IV} (a. f. u.)	Al^{VI} (a. f. u.)	Mg (a. f. u.)	Fe (a. f. u.)	Ti (a. f. u.)	Σ^{VI} (a. f. u.)	K (a. f. u.)	Ca (a. f. u.)	Na (a. f. u.)	$\Sigma^{Int.}$ (a. f. u.)
Light green spherules	3.80	0.20	1.47	0.42	0.23	0.03	2.15	0.04	0.02	0.05	0.11
	3.80	0.20	1.55	0.43	0.10	0.04	2.12	0.05	0.05	0.10	0.20
	3.71	0.29	1.53	0.55	0.09	0.06	2.23	0.06	0.02	0.21	0.29
	3.89	0.11	1.44	0.44	0.15	0.05	2.08	0.05	0.03	0.12	0.20
	3.65	0.35	1.36	0.61	0.16	0.12	2.25	0.12	0.02	0.37	0.51
	3.76	0.24	1.28	0.51	0.35	0.00	2.14	0.16	0.01	0.15	0.32
	3.70	0.30	1.50	0.47	0.19	0.00	2.16	0.07	0.04	0.15	0.26
	3.71	0.29	1.21	0.57	0.33	0.00	2.11	0.09	0.04	0.25	0.38
	3.75	0.25	1.38	0.47	0.27	0.02	2.14	0.06	0.19	0.19	0.44
	3.53	0.47	1.28	0.58	0.17	0.09	2.12	0.11	0.15	0.19	0.45
Dark green spherules	3.57	0.43	0.71	0.49	0.81	0.04	2.05	0.49	0.02	0.18	0.69
	3.67	0.35	0.77	0.41	0.81	0.04	2.03	0.41	0.00	0.12	0.53
	3.66	0.34	0.84	0.48	0.71	0.03	2.06	0.22	0.02	0.25	0.49
	3.58	0.42	0.83	0.53	0.68	0.04	2.08	0.29	0.03	0.27	0.59
	3.77	0.23	0.91	0.44	0.64	0.03	2.02	0.29	0.02	0.08	0.39
Pale yellow spherules	3.64	0.36	1.35	0.50	0.37	0.00	2.22	0.13	0.00	0.07	0.20
	3.64	0.36	1.18	0.39	0.48	0.02	2.07	0.20	0.02	0.16	0.38
	3.69	0.31	1.20	0.39	0.48	0.03	2.10	0.21	0.00	0.16	0.37
	3.56	0.44	1.04	0.48	0.48	0.00	2.00	0.06	0.11	0.48	0.65
	3.76	0.24	1.24	0.40	0.40	0.01	2.05	0.05	0.05	0.13	0.23

Note: Smectite formulae normalized to $O_{10}(OH)_2$. a.f.u. = atoms per formulae unit.

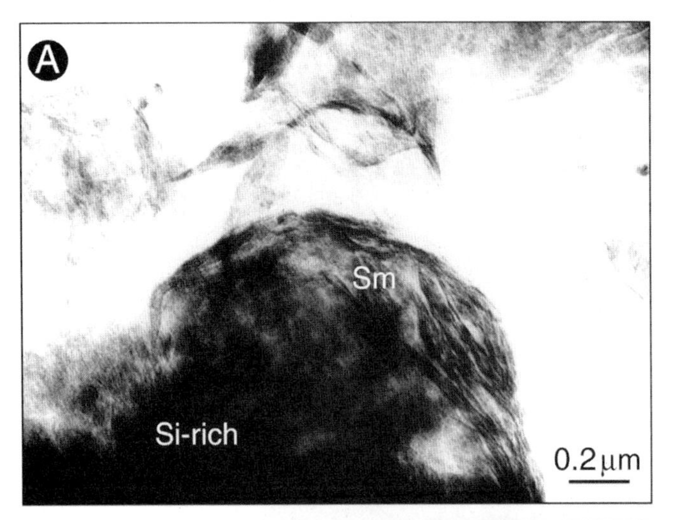

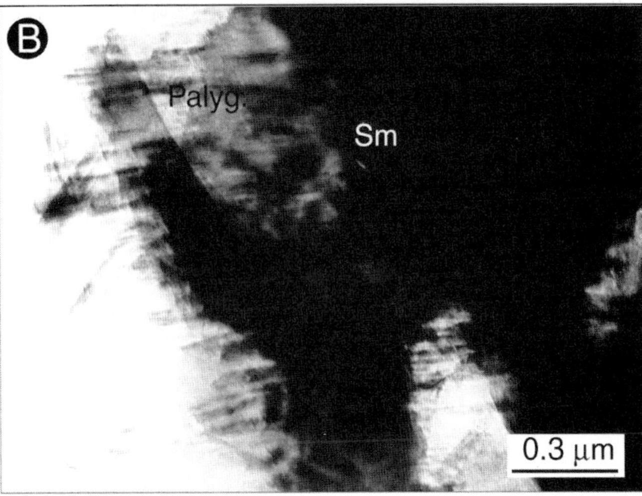

Figure 5. Transmission electron microscopy micrographs of smectites from dark green spherules (Site 1049). A: Smectites formed from high-silica material, probably altering glass. B: Higher resolution micrograph showing authigenic palygorskite fibers.

to be a morphological and chemical gradation from glass material to smectites (Fig. 5A).

During the smectite-forming diagenetic alteration of the Blake Nose spherules, zeolites and palygorskite were also formed, requiring a Si-rich source and alkaline conditions. In general, Si-rich environments favor the precipitation of chain-structure silicates (Beck and Weaver, 1978). Smectites can be a precursor of fibrous clays (Singer, 1979). Just like the low-temperature alteration of basalts (Velde, 1985), the diagenetic alteration of the precursor glass involved the expulsion of fibrous clay and zeolite-forming elements. In the present case, when the original Si-rich glass was altered to smectite, the latter did not incorporate all the available silica, thereby producing the Si-rich environments required for the formation of zeolites and palygorskite. In addition, some calcite may also have been derived from diagenetic reactions leading to clay authigenesis, such as reactions including the formation of palygorskite from smectite and dolomite (Jones and Galán, 1988). This suggestion is supported by TEM observations that reveal that the palygorskite formed from a smectite precursor (Martínez-Ruiz et al., 2001b) (Fig. 5B) and that dolomite might have been abundant in the ejecta material, as it is in other K-T ejecta deposits such as those from Albion Island, Belize (Ocampo et al., 1996; Pope et al., 1999). Furthermore, XRD and SEM data also show the presence of dolomite in the spherule bed. Thus, except for the calcite, which could be derived from either Cretaceous material or authigenic reactions, dolomite and calcite could be partially derived as original phases from carbonate target rocks. Carbonate material present in other K-T ejecta deposits supports this proposal (Smit et al., 1992a). The existence of Ca-rich and Si-rich phases is consistent with the preimpact target stratigraphy (Swisher et al., 1992; Blum et al., 1993; Koeberl, 1993; Hough et al., 1998). All this evidence suggests that the K-T boundary material from Blake Nose spherule layer was derived from Chicxulub target rocks.

The spherules from Blake Nose are comparable to spherules from other locations on the North America Atlantic margin, such as Bass River (Olsson et al., 1997) and Deep Sea Drilling Project Sites 390b and 603B (Klaver et al., 1987), and therefore they probably all represent the same diagenetically altered impact ejecta from the Chicxulub crater. Spherules from numerous sections in eastern Mexico, such as El Mibral and La Lajilla, are also similar, but moreover contain a preserved impact glass core (e.g., Smit et al., 1992b).

Chemical composition

Element concentration within the spherule bed (Fig. 3) is mainly affected by two factors: (1) a difference in composition, in particular an increase in Al and a decrease in Ca and K contents, compared with overlying and underlying sediments, and (2) diagenetic alteration under reducing conditions (Martínez-Ruiz et al., 2001a) favoring the remobilization of Fe and Mn, which diffused upward and reprecipitated upon encountering oxygenated pore waters. Differences in the Eh stability field of these elements made them uncouple during diagenetic alteration. Fe was reprecipitated, forming the limonitic layer capping the spherule bed, and Mn was diffused further upward and reprecipitated in the burrow-mottled ooze.

The REE concentrations significantly decrease in the spherule bed relative to overlying and underlying sediments. Originally, some assumed that REEs are relatively immobile during diagenetic alteration, and explained the low REE concentration in K-T boundary sediments as resulting from an impact in oceanic crust (e.g., Smit and ten Kate, 1982; Hildebrand and Boynton, 1987). However, REE mobilization during diagenesis has been demonstrated (e.g., Nesbitt, 1979; Taylor and McLennan, 1988), and therefore diagenetic remobilization could explain the low REE abundances in the K-T boundary sediments, as demonstrated by Izett (1990), who compared the REE composition of impact glass cores with the smectite rims.

Although it has also been suggested that REEs may not reflect the nature of the progenitor material (Izett, 1990), we suggest that at Blake Nose the K-T boundary sediments show C1-normalized REE patterns that can be considered informative. In certain diagenetic environment REE are leached during alteration (e.g., Zelinski, 1982) but C1-normalized patterns remain similar to the parent glass. The latter seems to be the case for the Blake Nose spherule bed material because its REE patterns are similar to those of upper crustal rocks (McLennan, 1989) and to Cretaceous and Tertiary sediments (Fig. 6), and Haitian glass cores (Izett, 1990). Koeberl and Sigurdsson (1992) reported the REE C1-normalized patterns of smectites, derived from Haitian impact glasses, being almost flat. At Blake Nose the C1-normalized patterns are not flat, and cannot be attributed solely to alteration, but instead suggest inheritance from upper crustal rocks.

Regarding extraterrestrial elements, Cr, Co, Ni, and Ir appear in low concentrations in the spherule bed at Blake Nose. A slight Ir enrichment was only reported above the spherule bed (Smit et al., 1997). Although Co and Ni concentrations (Martínez-Ruiz et al., 2001a) are not as high as in some more distal sections (e.g., Caravaca and Agost sections, Smit and ten Kate, 1982; Martínez-Ruiz et al., 1999), both elements are enriched in the upper part of the spherule bed, suggesting possible extraterrestrial contamination. However, little evidence for significant extraterrestrial contribution is observed at Blake Nose (absence of Ir enrichment and Ni-rich spinels), suggesting that the spherule-bed material mainly originated from the alteration of target-rock-derived material, as also suggested by the REE composition.

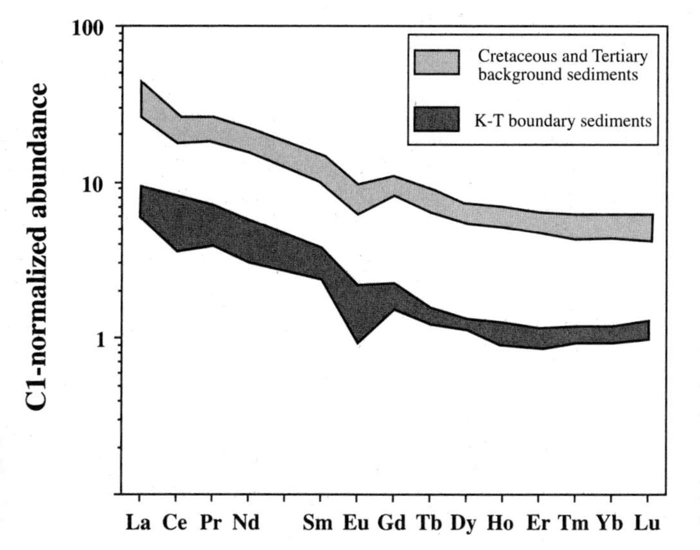

Figure 6. Rare earth element abundances normalized to C1 (Anders and Ebihara, 1982) from Blake Nose sediments (Hole 1049A). Analyzed samples are listed in Table 1.

CONCLUSIONS

The biostratigraphic K-T boundary at Blake Nose is marked by a coarse, poorly cemented bed composed mostly of spherules. These spherules are morphologically and compositionally similar to spherules from different locations on the North American Atlantic margin, and the Gulf of Mexico and the Caribbean, and all of them represent the same diagenetically altered impact ejecta. Mineralogical and geochemical analyses of the Blake Nose spherule bed reveal that this bed is mainly composed of smectite (derived from the alteration of the original precursors), minor proportions of calcite and dolomite, and other trace components such as quartz, rutile, zeolites, and lithic fragments. Different types of spherules were observed in the Blake Nose spherule bed: dark green, pale yellow, and light green spherules, which can be related to different precursors. TEM observations showed that smectite directly replaced the original material and that dark green spherules originated from a Si-rich precursor, whereas pale yellow spherules originated from a Ca-rich precursor. This result is in agreement with the variation of impact glass compositions reported in other K-T boundary sections around the Gulf of Mexico. Prior to smectite formation the precursor of the Blake Nose spherules was probably compositionally similar to the impact glasses reported at Haitian sections. Other mineral phases that originated during alteration, such as palygorskite and zeolites, also indicate a very Si-rich environment. In addition, the occurrence of palygorskite suggests the presence of dolomite in the original precursor material. Some dolomite and calcite within the spherule bed may represent original phases. The chemical composition of the spherule-bed material at Blake Nose does not show a significant extraterrestrial contribution, suggesting that the spherule-bed material mainly originated from the alteration of target-rock-derived material. Major chemical changes accompanied the diagenetic alteration of glass to smectite, the REE concentrations being significantly depleted during this alteration. Low Eh conditions also led to trace element remobilization. Fe and Mn were the most significantly mobilized elements, diffusing upward and reprecipitating upon encountering oxygenated pore waters. Diagenetic alteration is therefore the main control of the geochemical profiles across the K-T boundary at Blake Nose. However, REE C1-normalized patterns suggest that this material was derived from upper crustal rocks.

ACKNOWLEDGMENTS

We thank the Ocean Drilling Program (ODP) Leg 171B Shipboard Scientific Party and the crew of the *Joides Resolution* for assistance with the samples and data, the Bremen ODP Core Repository for assistance during the sampling party, and the Centro de Instrumentación Científica CIC (University of Granada, Spain) for the use of analytical facilities. We also thank J. Scarrow for reading the manuscript and two anonymous reviewers for constructive suggestions. This work was supported by Projects PB96-1429, BET2000-1493, and REN2000-0798 of the Ministerio de Ciencia y Tecnologia, Spain, and Research Group RNM0179 of the Junta de Andalucia, Spain.

REFERENCES CITED

Alvarez, L.W., Alvarez, W., Asaro, F., and Michel, H.V., 1980, Extraterrestrial cause for the Cretaceous/Tertiary extinction: Science, v. 208, p. 1095–1108.

Alvarez, W., Smit, J., Lowrie, W., Asaro, F., Margolis, S.V., Claeys, P., Kastner, M., and Hildebrand, A.R., 1992, Proximal impact deposits at the Cretaceous-Tertiary boundary in the Gulf of Mexico: A restudy of DSDP Leg 77 Sites 536 and 540: Geology, v. 20, p. 697–700.

Anders, E., and Ebihara, H., 1982, Solar-system abundances of the elements: Geochimica et Cosmochimica Acta, v. 46, p. 2362–2380.

Beck, K.C., and Weaver, C.E., 1978, Miocene of the S.E. United States: A model for chemical sedimentation in a peri-marine environment: Reply: Sedimentary Geology, v. 21, p. 154–157.

Blum, J.D., Chamberlain, C.P., Hingston, M.P., and Koeberl, C., 1993, Isotopic composition of K/T boundary impact glass with melt rock from Chicxulub and Manson impact structures: Nature, v. 364, p. 325–327.

Bohor, B.F., 1990, Shocked quartz and more: Impact signatures in Cretaceous/Tertiary boundary clays, *in* Sharpton, V.L., and Ward, P.D. eds., Global catastrophes in Earth history: An interdisciplinary conference on impacts, volcanism, and mass mortality: Geological Society of America Special Paper 247, p. 335–342.

Bourgeois, J., Hansen, T.A., Wiberg, P.L., and Kauffman, E.G., 1988, A tsunami deposit at the Cretaceous-Tertiary boundary in Texas: Science, v. 241, p. 567–570.

Byars, C., 1981, Mexican site may be link to dinosaurs disappearance: Houston Chronicle, December 13, p. 1, 18.

De la Fuente, S., Cuadros, J., Fiore, S., and Linares, J., 2000, Electron-microscopy study of volcanic tuff alteration to illite-smectite under hydrothermal conditions: Clays and Clay Minerals, v. 48, p. 339–350.

Hildebrand, A.R., and Boynton, W.V., 1987, The K/T impact excavated oceanic mantle: Evidence from REE abundances [abs.]: Lunar and Planetary Science Conference, 17th, Houston, Texas, Lunar and Planetary Institute, p. 427–428.

Hildebrand, A.R., Pewnfield, G.T., Kring, D.A., Pilkington, M., Camargo, Z.A., Jacobsen, S.B., and Boynton, W.V., 1991, Chicxulub Crater: A possible Cretaceous-Tertiary boundary impact crater on the Yucatán Peninsula, Mexico: Geology, v. 19, p. 867–871.

Hough, R.M., Wright, I.P., Sigurdsson, H., Pillinger, C.T., and Gilmour, I., 1998, Carbon content and isotopic composition of K/T impact glasses from Haiti: Geochimica et Cosmochimica Acta, v. 62, p. 1285–1291.

Izett, G.A., 1990, The Cretaceous/Tertiary boundary interval, Raton Basin, Colorado and New Mexico, and its content of shock-metamorphosed minerals: Evidence relevant to the K/T boundary impact-extinction theory: Geological Society of America Special Paper 249, 100 p.

Izett, G.A., 1991, Tektites in Cretaceous/Tertiary boundary rocks on Haiti and their bearing on the Alvarez impact extinction hypothesis: Journal of Geophysical Research, v. 96, p. 20879–20905.

Izett, G.A., Dalrymple, G.B., and Snee, L.W., 1991, $^{40}Ar/^{39}Ar$ Age of Cretaceous-Tertiary boundary tektites from Haiti: Science, v. 252, p. 1539–1541.

Jones, B.F., and Galan, E., 1988, Sepiolite and palygorskite, *in* Bailey, S.W., ed., Hydrous phyllosilicates (exclusive of micas): Washington, D.C., Mineralogical Society of America, Reviews in Mineralogy, v. 19, p. 631–674.

Klaus A., Norris R.D., Kroon D., and Smit J., 2000, Impact-induced mass wasting at the K-T boundary: Blake Nose, western North Atlantic: Geology, v. 28, p. 319–322.

Klaver, G.T., van Kempen, T.M.G., Bianchi, F.R., and van der Gaast, S.J., 1987, Green spherules as indicators of the Cretaceous/Tertiary boundary in Deep Sea Drilling Project Hole 603B, *in* van Hinte, J.E., and Wise, S.W., Jr., eds., Initial reports of the Deep Sea Drilling Project, Volume 93: Washington, D.C., U.S. Government Printing Office, p. 1039–1056.

Koeberl, C., 1993, Chicxulub Crater, Yucatan: Tektites, impact glasses, and the geochemistry of target rocks and breccias: Geology, v. 21, p. 211–214.

Koeberl, C., and Sigurdsson, H., 1992, Geochemistry of impact glasses from the K/T boundary in Haiti: Relation to smectites and a new type of glass: Geochimica et Cosmochimica Acta, v. 56, p. 2113–2129.

Koeberl, C., Sharpton, V.L., Schuraytz, B.C., Shirey, S.B., Blum, J.D., and Marin, L. E., 1994, Evidence for a meteoritic component in impact melt rock from the Chicxulub structure: Geochimica et Cosmochimica Acta, v. 58, p. 1679–1684.

Kring, D.A., and Boynton, W.V., 1992, Petrogenesis of an augite-bearing melt rock in the Chicxulub structure and its relationship to K/T impact spherules in Haiti: Nature, v. 358, p. 141–144.

Martínez-Ruiz, F., Ortega-Huertas, M., and Palomo, I., 1999, Positive Eu anomaly development during diagenesis of the K/T boundary ejecta layer in the Agost section (SE Spain): Implications for trace-element remobilization: Terra Nova, v. 11, p. 290–296.

Martínez-Ruiz, F., Ortega-Huertas, M., Kroon, D., Smit, J., Palomo, I., and Rocchia, R., 2001a, Geochemistry of the Cretaceous-Tertiary boundary at Blake Nose (ODP Leg 171B): Geological Society [London] Special Publication 183, p. 131–148.

Martínez-Ruiz, F., Ortega-Huertas, Palomo, I., and Smit, J., 2001b, K/T boundary spherules from Blake Nose (ODP Leg 171B) as a record of the Chicxulub ejecta deposits: Geological Society [London] Special Publication 183, p. 149–161.

McLennan, S.M., 1989, Rare earth elements in sedimentary rocks: Influence of provenance and sedimentary processes, *in* Lipin, B.R., and McKay, G.A., eds., Geochemistry and mineralogy of rare earth elements: Washington, D.C., Mineralogical Society of America, Reviews in Mineralogy, v. 20, p. 169–200.

Nesbitt, H.W., 1979, Mobility and fractionation of rare earth elements during weathering of a granodiorite: Nature, v. 279, p. 206–210.

Nieto, F., Lopez-Galindo, A., and Peinado-Fenoll, E., 1989, Programa de recogida de datos del difractometro de rayos X: Granada, Spain, Universidad de Granada, 55 p.

Nieto, F., Ortega-Huertas, M., Peacor, D.R., and Arostegui, J., 1996, Evolution of illite/smectite from early diagenesis through incipient metamorphism in sediments of the Basque-Cantabrian basin: Clays and Clay Minerals, v. 44, p. 304–323.

Norris, R.D., Kroon, D., Klaus, A., and others, 1998, Initial reports, Ocean Drilling Program, Leg 171B: College Station, Texas, Ocean Drilling Program, 749 p.

Norris R.D., Huber B.T., and Self-Trail, J., 1999, Synchroneity of the K-T oceanic mass extinction and meteorite impact: Blake Nose, western North Atlantic: Geology, v. 27, p. 419–422.

Norris, R.D., Firth, J., Blusztajn, J.S., and Ravizza, G., 2000, Mass failure of the North Atlantic margin triggered by the Cretaceous-Paleogene bolide impact: Geology, v. 28, p. 1119–1122.

Ocampo, A.C., Pope, K.O., and Fischer, A.G., 1996, Ejecta blanket deposits of the Chicxulub crater from Albion Island, Belize, *in* Ryder, G., Fastovsky, D., and Gartner, S., eds., The Cretaceous-Tertiary event and other catastrophes in Earth history: Geological Society of America Special Paper 307, p. 75–88.

Olsson, R.K., Miller, K.G., Browning, J.V., Habib, D., and Sugarman, P.J., 1997, Ejecta layer at the Cretaceous-Tertiary boundary, Bass River, New Jersey (Ocean Drilling Program Leg 174AX): Geology, v. 25, p. 759–762.

Penfield, G.T., and Camargo, Z.A., 1981, Definition of a major igneous zone in the Central Yucatan platform with aeromagnetics and gravity: Society of Exploration Geophysicists, 51st Annual Meeting, Tulsa, Oklahoma, Abstracts, p. 37.

Pope, K.O., Ocampo, A.C., Fischer, A.G., Alvarez, W., Fouke, B.W., Webster, C.L., Vega, F.J., Smit, J., Fritsche, A.E., and Claeys, P., 1999, Chicxulub impact ejecta from Albion Island, Belize: Earth and Planetary Science Letters, v. 170, p. 351–364.

Sharpton, V.L., Schuraytz, B.C., and Jones, J., 1990, Arguments favoring a single continental impact at the KT boundary: Meteoritics, v. 25, p. 408–409.

Sharpton, V.L., Dalrymple, G.B., Marin, L.E., Ryder, G., Schuraytz, B.C., and Urrutia-Fucugauchi, J., 1992, New links between the Chicxulub impact structure and the Cretaceous/Tertiary boundary: Nature, v. 359, p. 819–821.

Sharpton, V.L., Burke, K., Camargo-Zanoguera, A., Hall, S.A., Lee D.S., Marin, L.E., Suarez-Reynoso, G., Quezada J.M., Spudis, P.D., and Urrutia-Fucugauchi, J., 1993, Chicxulub multi-ring impact basin: Size and other characteristics derived from gravity analyses: Science, v. 261, p. 1564–1567.

Sharpton, V.L., Marin, L.E., and Schuraytz, B.C., 1994, The Chicxulub multi-ring impact basin: Evaluation of geophysical data, well logs, and drill core samples [abs.], *in* New developments regarding the KT event and other catastrophes in Earth history: Houston, Texas, Lunar and Planetary Institute, LPI Contribution No. 825, p. 108–110.

Sigurdsson, H., D'Hont, S., Arthur, M.A., Bralower, T.J., Zachos, J.C., van Fossen, M., and Channell, E.T., 1991, Glass from the Cretaceous/Tertiary boundary in Haiti: Nature, v. 349, p. 482–487.

Sigurdsson, H., D'Hont, S., and Carey, S., 1992, The impact of the Cretaceous/Tertiary bolide on evaporite terrane and generation of major sulfuric acid aerosol: Earth and Planetary Science Letters, v. 109, p. 543–559.

Singer, A., 1979, Palygorskite in sediments: Detrital, diagenetic or neoformed—A critical review: Geological Rundschau, v. 68, p. 996–1008.

Smit, J., 1999, The global stratigraphy of the Cretaceous-Tertiary boundary impact ejecta: Annual Review of Earth and Planetary Sciences, v. 27, p. 75–113.

Smit, J., and ten Kate, W.G.H.Z., 1982, Trace element patterns at the Cretaceous-Tertiary boundary: Consequences of a large impact: Cretaceous Research, v. 3, p. 307–332.

Smit, J., and Romein, A.J.T., 1985, A sequence of events across the Cretaceous-Tertiary boundary: Earth and Planetary Science Letters, v. 74, p. 155–170.

Smit, J., Alvarez, W., Montanari, A., Swinburne, N., Kempen, T.M.V., Klaver, G.T., and Lustenhouwer, W.J., 1992a, Tektites and microkrystites at the KT boundary: Two strewn fields, one crater: Proceedings of Lunar and Planetary Science v. 22, p. 87–100.

Smit, J., Montanari, A., Swinburne, N.H.S., Alvarez, W., Hildebrand, A.R., Margolis, S.V., Claeys, P., Lowrie, W., and Asaro, F., 1992b, Tektite-bearing, deep-water clastic unit at the Cretaceous-Tertiary boundary in northeastern Mexico: Geology, v. 20, p. 99–103.

Smit, J., Rocchia, R., Robin, E., and ODP Leg 171B Shipboard Party, 1997, Preliminary iridium analyses from a graded spherule layer at the K/T boundary and late Eocene ejecta from ODP Sites 1049, 1052, 1053, Blake Nose, Florida: Geological Society of America Abstracts with Programs, v. 29, no. 6, p. A141.

Swisher, C.C., Grajales, N.J.M., Montanari, A., Margolis, S.V., Claeys, P., Curtis, G. H., Smit, J., and McWilliams, M.O., 1992, Coeval $^{40}Ar/^{39}Ar$ ages of 65.0 million years from Chicxulub melt rock and Cretaceous-Tertiary boundary tektites: Science, v. 257, p. 954–958.

Taylor, S.R., and McLennan, S.M., 1988, The significance of the rare earths in geochemistry and cosmochemistry, *in* Gschneidner, K.A., Jr., and Eying, L., eds., Handbook on the physics and chemistry of rare earths, Volume 11: Amsterdam, Elsevier, p. 485–578.

Velde, B., 1985, Clay minerals: A physical-chemical explanation of their occurrence: Amsterdam, Elsevier, 198 p.

Zelinski, R.A., 1982, The mobility of uranium and other elements during alteration of rhyolite ash to montmorillonite: A case study in the Troublesome Formation, Colorado, U.S.A.: Chemical Geology, v. 35, p. 185–204.

MANUSCRIPT SUBMITTED OCTOBER 5, 2000; ACCEPTED BY THE SOCIETY MARCH 22, 2001

Printed in the U.S.A.

Geological Society of America
Special Paper 356
2002

Global occurrence of magnetic and superparamagnetic iron phases in Cretaceous-Tertiary boundary clays

N. Bhandari*
Physical Research Laboratory, Navrangpura, Ahmedabad 380009, India
H.C. Verma
C. Upadhyay
Department of Physics, Indian Institute of Technology, Kanpur 208016, India
Amita Tripathi
R.P. Tripathi
Department of Physics, Jai Narain Vyas University, Jodhpur 342001, India

ABSTRACT

The iron mineralogy of the Cretaceous-Tertiary (K-T) boundary clays from four different sites (Gubbio, Turkmenistan, Anjar, and Meghalaya) has been determined using Mössbauer spectroscopy. At all four sites the K-T boundary samples show the presence of oxide and/or oxyhydroxide phases of iron, often in the form of particles of a few nanometers in size, which exhibit superparamagnetic behavior. The abundance of these iron phases across the boundary correlates fairly well with the iridium content, which is considered to be the geochemical signature of the impact of an extraterrestrial body. The association of the nanoparticles of iron minerals with iridium, their global occurrence within the K-T boundary layer regardless of the local depositional environment, and their absence above and below the K-T boundary layer indicate that their formation is related to the K-T impact. At certain sites, iron phases are present adjacent to the K-T boundary layer, but they do not contain nanoparticles, whereas the K-T boundary layer in the same section has nanosized oxidized iron minerals. The observations indicate that these iron oxide and/or oxyhydroxide phases were formed in the environmental conditions created by the impact.

INTRODUCTION

The Cretaceous-Tertiary (K-T) boundary in the marine sediments is generally characterized by a thin (~2 cm) layer of clay. Geochemical anomalies in the Cretaceous-Tertiary (K-T) boundary clays have been attributed to the impact of a large bolide on Earth, 65 Ma (Alvarez et al., 1980). Several studies have been made to determine the nature of the impactor, climatic conditions arising due to the impact, and their biological effects. On the basis of the isotopic composition of chromium ($^{53}Cr/^{52}Cr$), the impactor appears to belong to the carbonaceous chondrite group of meteorites (Shukolyukov and Lugmair, 1998). Meteorites, in general, have high concentrations of iron ($\geq$20%) in the form of silicates, metal, magnetite, and other iron-bearing minerals. In addition, the terrestrial rocks at the impact site would have been subjected to extremely high temperature and pressure conditions and chemically altered during and after the impact. Accordingly, iron phases in K-T boundary clays must contain some information about the impact or conditions following the impact.

*E-mail: bhandari@prl.ernet.in

Bhandari, N., Verma, H.C., Upadhyay, C., Tripathi, A., and Tripathi, R.P., 2002, Global occurrence of magnetic and superparamagnetic iron phases in Cretaceous-Tertiary boundary clays, *in* Koeberl, C., and MacLeod, K.G., eds., Catastrophic Events and Mass Extinctions: Impacts and Beyond: Boulder, Colorado, Geological Society of America Special Paper 356, p. 201–211.

Although extensive work has been carried out on different aspects of K-T boundary clays, only scattered information concerning its iron mineralogy is available (Thorpe et al., 1994). Realizing that a study of the chemical state of iron is useful in characterizing the geochemical conditions prevalent at the time of deposition, we have carried out ^{57}Fe Mössbauer spectroscopy of K-T boundary clays. This technique is well suited for such a study because of the characteristic spectral shapes and distinct Mössbauer parameters for each kind of iron-containing species. For example, a paramagnetic iron complex gives rise to a quadrupole doublet in the Mössbauer spectrum, whereas a hyperfine magnetic field (HMF) existing in magnetically ordered materials causes a six-line pattern with characteristic splitting (Gütlich, 1975). The Mössbauer parameters measured from the spectrum are isomer shift (IS), quadrupole splitting (QS), and HMF (B). Whereas IS is related to the total electron density overlapping with the ^{57}Fe nucleus, QS gives the deviation from the cubic symmetry in the charge distribution around the nucleus. The value of B gives the extent to which the sample is ordered magnetically. New features appear in the spectrum when the particle size in a magnetic system is so small (few nanometers) that each particle is a single magnetic domain. The direction of magnetization in nanosized single domain particles fluctuates rapidly among the easy axes of magnetization, giving zero average HMF during the measurement time (superparamagnetic relaxation). This causes a temperature-dependent collapse of the usual six-line Mössbauer spectrum in magnetic systems to a doublet or a singlet (Haneda, 1987; Mørup et al., 1980; Gangopadhyay et al., 1992). Because of the distribution in particle size, the Mössbauer spectrum of a natural fine particle system often shows a superposition of large number of components with reduced HMF and a superparamagnetic doublet with a parabolic-shaped baseline. At low temperatures (i.e., low thermal energy), the superparamagnetic effects are reduced and the characteristic six-line patterns reappear. Mössbauer spectroscopy can be effectively used to study the nanosized magnetic iron phases because the response time of these measurements is very small (10^{-8} s).

With a view to understanding the mineralogy of the iron-bearing phases, we have carried out a detailed Mössbauer study of four K-T boundary sites extending from Italy to eastern India. The results of this work, providing evidence of the occurrence of nanoparticles of oxide/oxyhydroxide phases of iron at the K-T boundary, were reported in Bhandari et al. (2000) and are described in detail here. Wdowiak et al. (2001) reported similar Mössbauer study of 10 European and American K-T boundary sites and have shown the presence of nanosize hematite-goethite, consistent with our studies (Bhandari et al., 2000). All the Mössbauer data (Bhandari et al., 2000; Wdowiak et al., 2001), taken together, provide clear evidence that the nanosize iron oxide and/or oxyhydroxide phases are globally present at the K-T boundary.

SELECTION OF SAMPLES

Samples collected from four well-documented K-T boundary sections, Gubbio (Italy), Turkmenistan, Anjar (Kutch, western India), and Meghalaya (eastern India), have been studied. Their geographical locations and the stratigraphic positions are shown in Figure 1. All the samples analyzed here were deposited ca. 65 Ma. The Gubbio, Turkmenistan, and Meghalaya sections comprise marine sediments, whereas the Anjar sediments were deposited in a lacustrine environment. Some of these samples (Gubbio, Meghalaya, and Anjar) were collected by one of us (Bhandari) from 1985 to 1995. The samples from the Turkmenistan section were made available to us by M.A. Nazarov, U.S.S.R Academy of Sciences (Alekseev et al., 1986). The surface material was removed from the exposed sections, and fresh samples collected from the interior were used for the Mössbauer investigations. Their chemical analyses and lithostratigraphy were reported earlier (Bhandari et al., 1994a, 1994b, 1995, 1996; Alekseev et al., 1986). The Gubbio (Bottaccione) section in Italy is the well-known deep-marine site, where the iridium anomaly was first discovered (Alvarez et al., 1980). The Um Sohryngkew River section in Meghalaya is a near-coastal marine section in eastern India where an iridium anomaly and Ni rich spinels have been found (Bhandari et al., 1994a, 1994b; Bhandari, 1998; Robin et al., 1997) in the *Micula prinsii* zone. Anjar is a continental lake section in the Deccan volcano-sedimentary sequence in Kutch, western India, where three closely spaced iridium-rich layers have been identified (Bhandari et al., 1995, 1996). The Turkmenistan section shows the presence of an iridium anomaly (Alekseev et al., 1986) and shocked quartz grains (Badjukov et al., 1986). Sediments above and below the iridium-rich layers were also examined to understand the nature and variation of iron phases across the boundary. The stratigraphic locations as well as iron and iridium contents of the samples (Bhandari et al., 1994a, 1994b, 1995, 1996; Alekseev et al., 1986) studied in the present work are given in Table 1.

EXPERIMENTAL TECHNIQUES

We have used ^{57}Fe Mössbauer spectroscopy in transmission mode in the present work. The samples were collected after removal of the surface material from the exposed sections. In all except the Turkmenistan boundary samples, the chocolate colored boundary layer was mechanically separated from the adhering sediments. These samples have been studied earlier for their chemical composition (Bhandari et al., 1994a, 1994b, 1995, 1996; Alekseev et al., 1986). Iridium in Anjar, Meghalaya, and Gubbio samples were measured in our laboratory by neutron activation analyses, followed by radiochemical purification and gamma ray spectrometry. Neutron irradiation was carried out at the Cirus reactor of the Bhabha Atomic Research Center, Bombay. The detection limit of iridium concentration is 10 pg/g and the precision of measurement, including the

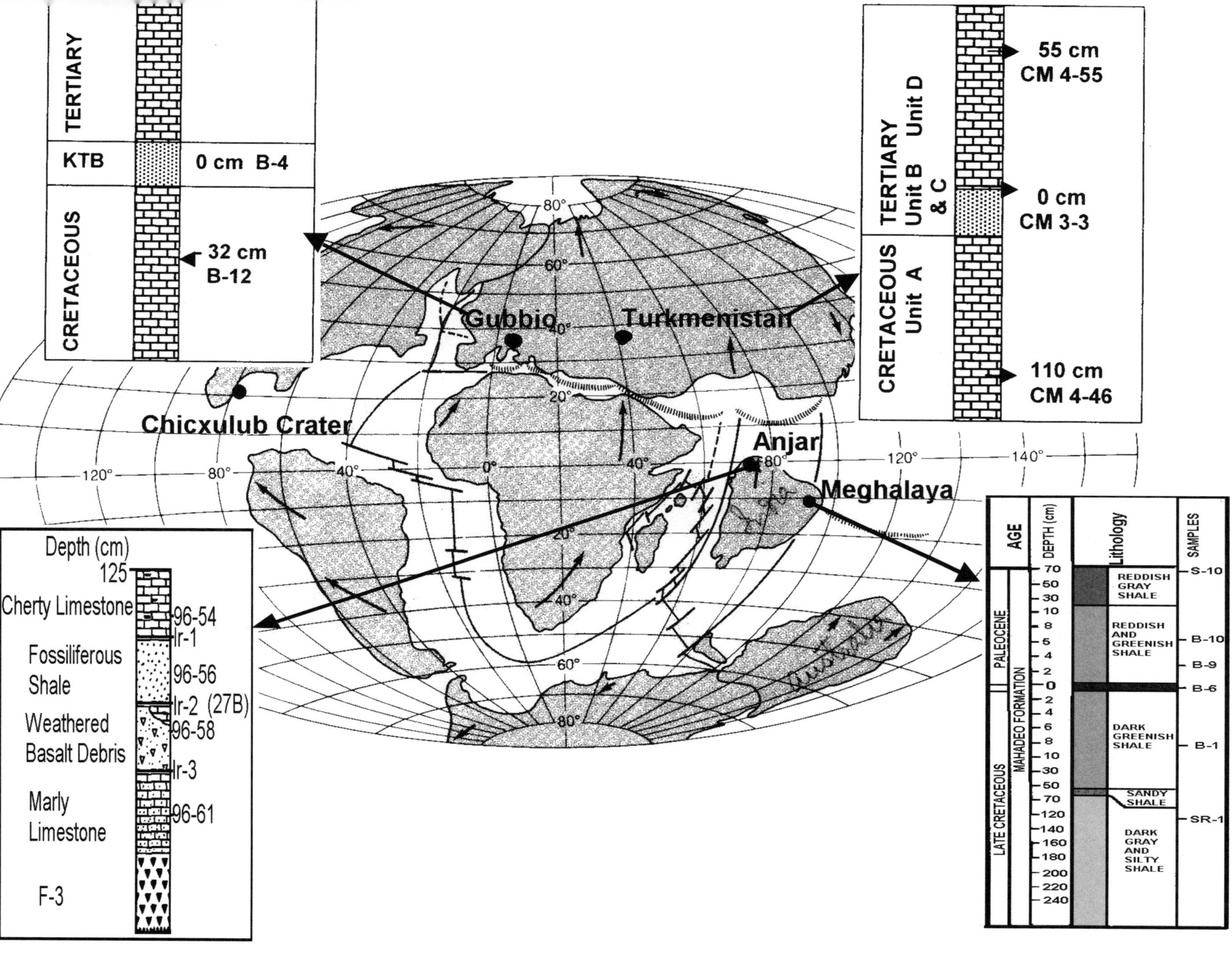

Figure 1. Geographical locations (at Cretaceous-Tertiary boundary time, KTB) of various sites studied. Lithostratigraphy around K-T boundary and stratigraphic positions of these samples are shown in insets. Further details of samples are given in Table 1.

TABLE 1. SAMPLE DETAILS OF VARIOUS CRETACEOUS-TERTIARY BOUNDARY SECTIONS

Location	Sample code	Nature of sediments	Stratigraphic level (cm)*	Ir concentration† (pg/g)	Fe concentration† (wt%)
Meghalaya	S10	Shale	+82	N.M.	2.5
	B10	Shale	+7.5 ± 2.5	125 ± 1.6	3.86
	B9	Shale	+4 ± 1	299 ± 9	3.95
	B6	Limonitic layer	0	7800 ± 450	13.31
	B1	Shale	−10.5 ± 2.5	132 ± 2.7	3.39
	SR1	Shale	−170	7.5 ± 1	6.41
Anjar	96-54	Cherty LS	+35	113	1.27
	96-56	Limestone	+15	95	N.M.
	27b	Limonitic layer	0	1271	37.4
	96-58	Grey Shale	−5	100	1.28
	96-61	Shale	−35	78	2.38
Gubbio	B4	Limonitic layer	0	5200	5.36
	B12	Limestone	−32	<50	0.32
Turkmenistan	C4-55	Marly Clay	+40	~100	N.M.
	C3-3	Dark layer	0	~45000	N.M.
	C4-46	Marl	−110	~100	N.M.

*Stratigraphic level is measured from the base of the iridium-rich layer.
†Bhandari et al. (1994a, 1994b, 1995, 1996; Alekseev et al., 1986). In some cases, the Ir and Fe concentrations are given for other samples collected from the same level. N.M.: Not measured.

statistical errors, is 5%–10%, depending on the concentration. The chemical procedure and counting details are given in Bhandari et al. (1994b). The chemical composition for Turkmenistan samples was taken from Alekseev et al. (1986). For Mössbauer spectroscopy, the samples were not given chemical or any other treatment because it can potentially change their chemical form. Each sample was crushed into fine powder and absorbers of nearly uniform thickness were made by pressing 50–100 mg of this powder between two transparent tapes fixed at the open ends of a copper ring (inner diameter 12 mm). A radioactive source of ^{57}Co embedded in Rh matrix (obtained from Inter University Center, Calcutta) was fixed to an electromechanical transducer that imparted a periodic motion, resulting in a linear change in velocity between its maximum and minimum values. The Doppler shifted gamma photons passing through the fixed absorber were counted as a function of the source velocity. The Mössbauer spectra were taken at room temperature and at 100 K (LT, or low temperature) using a cryogenic system. In some cases we recorded the spectra at 25 K using a liquid helium cooled system. The Mössbauer parameters, i.e., isomer shift (IS), quadrupole splitting (QS), and the effective hyperfine magnetic field (HMF), were calculated from the spectra. The data were analyzed in two different ways. In the first approach, discrete doublets and sextets with Lorentzian peaks were fitted to the spectra using a least squares code. In the second approach, a near continuous variation of HMF was assumed and probability of each HMF was calculated using Window's (1971) approach. The isomer shift and quadrupole splitting reported here have an uncertainty of about 0.04 mm/s. The relative areas given are uncertain within ~5%.

X-ray diffraction (XRD) patterns were recorded using Cu K_α radiation using a Seifert Iso-Debyflex 2002 diffractometer. The d-values were calculated from the peak positions and were compared with the standard values for different possible minerals to get the information about the minerals present.

RESULTS AND DISCUSSION

Iridium-rich layers

The Mössbauer spectra of the samples from the most iridium-rich limonitic layers of the four K-T boundary sections (Meghalaya, Anjar, Gubbio, and Turkmenistan) are shown in Figure 2. The p-B distributions (indicating probability p in the interval dB of the hyperfine magnetic field B) obtained from the low-temperature spectra are also included in Figure 2. The Mössbauer parameters (IS, QS, and B) deduced from the discrete fitting of the spectra taken at various temperatures are given in Table 2. The K-T boundary spectra from different sites have several features in common. All of them show presence of iron in magnetically ordered phases. The first spectrum from the top is for Gubbio, which is a well-studied K-T boundary site having a thin (~2 cm) boundary layer. The Mössbauer spectrum at room temperature as well as at 100 K is a superposition of a well-split sextet and a doublet. As seen from Table 2, the HMF of the sextet is 522 kOe at 100 K, which corresponds to hematite. The HMF at room temperature is 498 kOe, which is somewhat smaller than the standard value of 522 kOe. Wdowiak et al. (2001) have obtained a similar Mössbauer spectrum of the Gubbio K-T boundary layer, and attributed the sextet at room temperature with similar HMF to hematite. The lower HMF compared to the standard value for hematite may be due to the collective magnetic excitation induced in the fine particle hematite (Haneda, 1987). Apart from the sextet, there

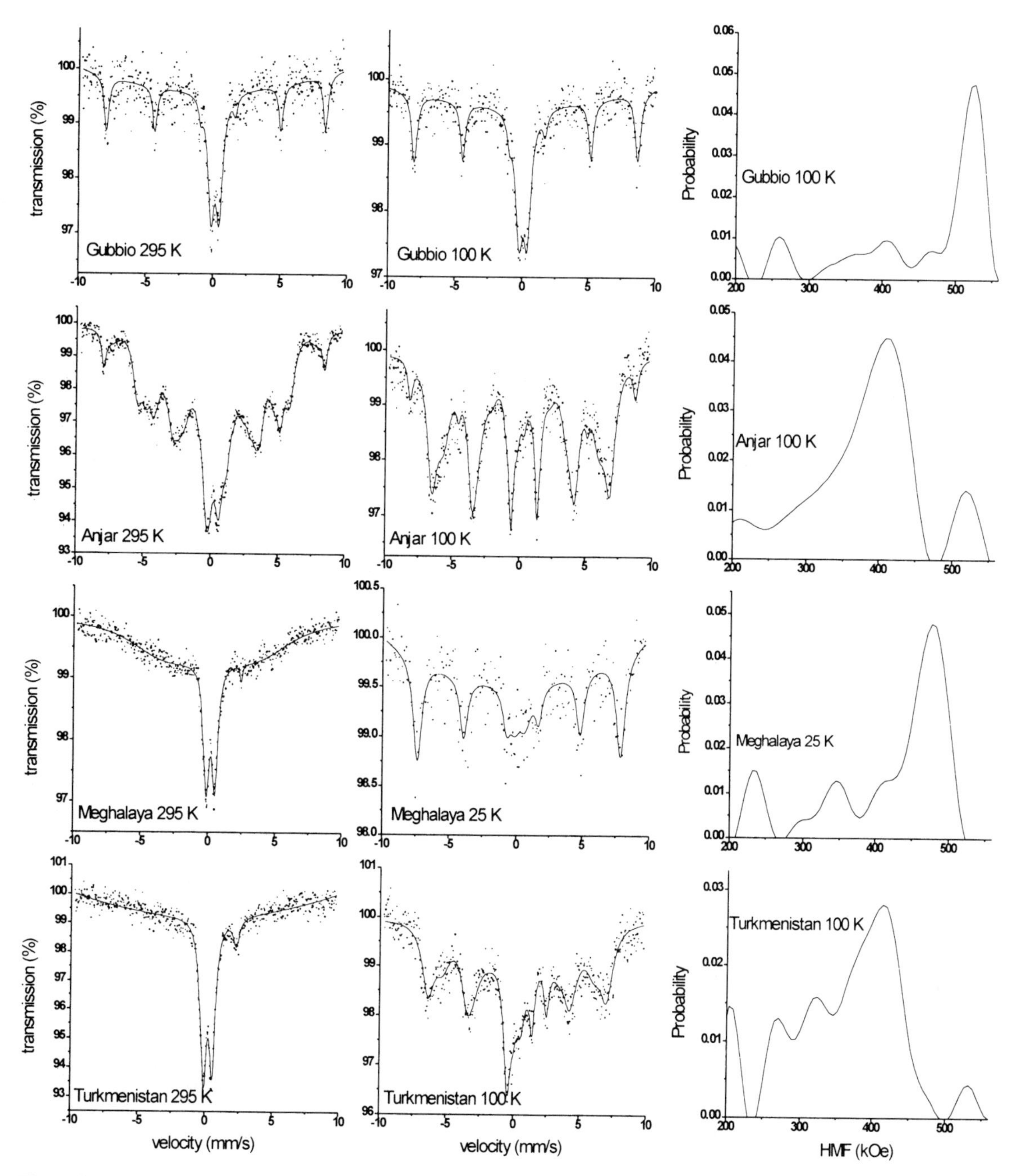

Figure 2. Mössbauer spectra of various Cretaceous-Tertiary (K-T) boundary clays at 295 K (left) and at low temperature (middle). Right column shows p-B distributions (indicating probability p in interval dB of magnetic hyperfine field B) obtained for various K-T layers at low temperature.

is a doublet with isomer shift and quadrupole splitting, which indicates the presence of an Fe^{3+} phase.

The K-T boundary samples from Anjar show the presence of several iron phases. The room-temperature spectrum appears as a typical partially relaxed spectrum when the particles have a size distribution in nanometer range. The presence of a small amount of α-Fe_2O_3 is clearly indicated by the sharp first and sixth line of the characteristic sextet of small amplitude (Fig. 2). The p-B curve at 100 K shows a peak at 522 kOe, but most of the magnetic phases have an HMF of 415 kOe and below (Fig. 2). The broad asymmetric peak centered at ~415 kOe suggests the existence of nanoparticles of varying size and indicates that superparamagnetism, although considerably subdued, is not completely quenched at 100 K. This HMF indicates the possible presence of goethite (FeOOH) and maghemite (γ-Fe_2O_3). The discrete peak fitting reveals the presence of a minor

TABLE 2. MÖSSBAUER PARAMETERS OF THE CRETACEOUS-TERTIARY BOUNDARY SAMPLES

Sample code	Description	Temperature (K)	IS (mm/s)	QS (mm/s)	B* (kOe)	a	Iron phase†
B4	Gubbio	295	0.28	0.61		61	P, M
			0.39	−0.06	498	39	
		100	0.23	0.61		55	P, M
			0.48	−0.03	522	45	
27b	Anjar	295	0.34	0.77		16	P, M, S
			0.42	−0.07	510	5	
			0.25	0.11	473	7	
			0.39	−0.09	350	17	
			0.41	−0.11	310	11	
			0.44	−0.03	272	8	
			0.30	−0.02	194	34	
			1.47	−1.08	133	3	
		100	0.50	1.93		6	P, M, S
			0.20	0.46		3	
			0.40	−0.01	523	6	
			0.38	−0.09	415	32	
			0.33	−0.03	363	38	
			0.36	0.15	232	16	
B6	Meghalaya	295	0.33	0.63		21	P, S, Fe^{2+}
			1.14	2.80		1	
			0.25	0.00	248	78	
		25	0.36	0.78		20	P, M
			0.44	−0.12	470	80	
B6	Annealed#	100	0.44	1.03		22	P, M
			0.40	−0.11	509	45	
			0.38	−0.12	475	33	
C3-3	Turkmenistan	295	0.35	0.62		83	P, S, Fe^{2+}
			1.04	2.68		17	
		100	0.38	0.57		10	P, S, Fe^{2+}
			0.58	1.92		5	
			1.20	2.87		5	
			0.49	−0.04	415	32	
			0.41	0.11	352	24	
			0.73	0.19	227	24	

*Missing B value means absence of hyperfine magnetic field. Iron is either in paramagnetic phase or superparamagnetic phase (magnetic but fine particles).
†P = paramagnetic, S = superparamagnetic, M = magnetic.
#The sample was annealed at 300 °C.

iron phase with a quadrupole splitting ~0.46 mm/s and isomer shift ~0.20 mm/s, with respect to the natural iron. These parameters correspond to iron in Fe^{3+} state.

In Meghalaya, the K-T boundary spectrum at room temperature is not split magnetically (Fig. 2). It shows a doublet with very broad absorption at baseline. This is a typical spectrum of superparamagnetic particles. The blocking temperature for all the particles is below room temperature so that magnetic splitting does not occur, but the indication of their fine particle nature is given by the shape of the baseline. Cooling the system gives the magnetically split spectrum. We recorded the Mössbauer spectra at 100 K and 25 K. The spectrum and the p-B distribution shown in Figure 2 correspond to 25 K. It shows a single dominant phase at ~475 kOe. This HMF with the negative quadrupole splitting suggests goethite as the dominant iron-bearing phase. The presence of other minor magnetic phases is suggested by the humps at lower magnetic fields. Like Gubbio, this sample also gives indication of Fe^{3+} phases. A very small amount of Fe^{2+} is also indicated in the discrete fitting. A part of this sample was heated at 300°C for a few hours in nitrogen atmosphere. The Mössbauer spectrum of this preheated sample shows the formation of hematite (Table 2). This confirms our assignment of the original phase to goethite because only goethite can convert into hematite at such a relatively low temperature.

A much broader p-B distribution is observed in the Turkmenistan spectrum at 100 K, showing persistent superparamagnetic effects indicating the presence of ultrafine particles. The dominant peak is at 415 kOe as in Anjar, indicating goethite, but other magnetic phases may also be present in smaller amounts. A small amount of hematite is also possible. The Mössbauer parameters obtained from the discrete fitting shows the presence of both Fe^{3+} and Fe^{2+} components in paramagnetic phases.

The XRD studies of all the samples show several sharp and intense peaks corresponding to calcite, quartz, and illite, and several smaller and broader peaks that support the broad assignment of iron phases made here. These XRD spectra indicate the presence of goethite in Meghalaya and Anjar samples and minor hematite in Gubbio and Anjar. The Turkmenistan sample had some faint peaks corresponding to iron phases, but poor signal to noise ratio made it difficult to identify the minerals present. The unusually large width of these peaks is consistent with the presence of fine particles in the sample.

An estimate of the particle size of the iron components can be made from the Mössbauer spectra because the particle size is related to the blocking temperature at which the sextets collapse to doublets. The larger the particle size, the higher the blocking temperature; it also depends on the magnetic anisotropy of the material. For α-Fe_2O_3, a blocking temperature of 300 K corresponds to a particle size of ~10 nm (Van der Geissen, 1968). The observation of a well-developed sextet in the Gubbio spectrum at room temperature suggests that the particle size may be $\geq$10 nm. For Anjar the room temperature spectrum is a superposition of the well-grown α-Fe_2O_3 and other magnetic phases that are split, at least partially. Consequently, the average particle size of iron-bearing minerals in Anjar samples is estimated to be $\leq$10 nm. The blocking temperatures in Meghalaya and Turkmenistan samples are between room temperature and 100 K, indicating that the particles may be significantly smaller than 10 nm.

It is important to note that the sedimentary facies occurring at different sites are different, although they represent approximately the same time bracket including the 65 Ma K-T event. Lithologically, the Gubbio section is limestone, the Meghalaya and Anjar sections are shales, and the Turkmenistan section is a marly clay (Table 1). In terrestrial sediments, the phases of iron generally depend on the lithology of the local sediments. The concentration of iron is, in general, very low in limestones and it is present in the form of the carbonate minerals siderite, ankerite, or dolomite. Shales have siderite and pyrite in relatively larger quantities. In the K-T boundary layers, however, we find that these common phases are absent and the oxide phases of iron appear prominently at all the sites regardless of the lithology of the local sediments.

Samples above and below the K-T boundary layer

A number of samples collected from stratigraphic levels above and below the iridium-rich layers at all the four sections were studied. Table 3 gives their Mössbauer parameters. Figure 3 shows the 100 K Mössbauer spectra of Anjar and Meghalaya samples taken across a vertical sequence including the K-T boundary layer. The solid lines represent the fitted curve according to the parameters given in Table 3. We have also included K-T boundary spectra in this figure for comparison.

The Mössbauer spectra of Anjar exhibit magnetic particles only at the K-T boundary. The samples above and below the K-T boundary show only a quadrupole doublet, characteristic of paramagnetic minerals. The samples above and below the K-T boundary have similar Mössbauer parameters, indicating that a lithologically similar sedimentary sequence continues across the boundary. The abrupt change in mineralogy at K-T boundary thus indicates that the boundary layer consists of extraneous material that was suddenly emplaced in a short time.

In Meghalaya we studied samples at close intervals from the iridium-rich K-T boundary layer. These samples show magnetically split sextets in addition to certain doublets. However, the relative intensity of the sextet part is strongest in the layer that contains the largest amount of iridium, and weakens (as compared to the doublet part) away from this layer. We noted a broad band (~70 cm) of sediments in Meghalaya section having a small amount of excess iridium on which a relatively sharp and strong iridium peak is superposed (Bhandari et al., 1994b). The intensity of the sextet shows strong correlation with the abundance of iridium in the samples. Furthermore, the samples, which are devoid of iridium, do not show the presence of any oxide phase, either magnetic or superparamagnetic. The correlation of abundance of oxide phases with that of iridium (Fig. 4) is discussed in detail in the next section.

The samples from above and below the K-T boundary in the Turkmenistan section follow the trend seen at Anjar, although the Mössbauer absorption is weak because the Turkmenistan section has mainly marly clay. These samples contain only the paramagnetic Fe^{3+} and Fe^{2+}, while the K-T boundary layer contains magnetic phases (in fine particle form) together with a small amount of paramagnetic phases (Table 3). The Gubbio samples, above and below the boundary, also had small Mössbauer absorption, as expected from its limestone character. These observations are consistent with the conclusion that the nanosized particles of magnetic iron phases present in the K-T boundary layers in different sections have a common origin and are different from phases present in the samples just above and below the K-T boundary layer.

Correlation of iron phases with iridium content

The K-T layer in various sections is identified by enhanced iridium concentrations and certain other geochemical and mineral markers. The integrated iridium concentration at all the K-T boundary sites has been found to vary within a narrow range, but vertical profiles of iridium differ from place to place. In general, the thickness of the iridium-rich layer is more than that expected during the short period of impact ejecta fallout. We have also seen that magnetic or superparamagnetic phases of iron oxides and oxyhydroxides have larger abundance in this layer in all the four sections studied. At Anjar, Gubbio, and Turkmenistan, the samples above and below the K-T boundary have no magnetic particles and their iridium content is also low. The magnetic fraction as well as the iridium concentration change more gradually in the Meghalaya section. The concentrations of iridium in the samples studied in this work are given in Table 1. To look for a possible correlation between the iridium content and the magnetic phases, we have calculated the relative area under the sextets relative to the total absorption area for the Meghalaya samples at 100 K, which qualitatively

TABLE 3. MÖSSBAUER PARAMETERS OF OFF-BOUNDARY SAMPLES AT DIFFERENT STRATIGRAPHIC LEVELS RELATIVE TO THE KTB LAYER AT 100 K

Section	Sample	Stratigraphic level* (cm)	IS (mm/s)	QS (mm/s)	B (kOe)	Relative area %	Iron phases†
Anjar	96-54	+35	0.35	0.48		100	P
	96-56	+15	0.33	0.43		100	P
	96-58	−5	0.34	0.41		100	P
	96-61	−35	0.28	0.51		100	P
Meghalaya	S10	+82	0.29	0.87		67	P, Fe^{2+}
			1.28	2.36		33	
	B10	+7.5 ± 2.5	0.25	0.74		77	P, M
			0.54	0.01	381	23	
	B9	+4 ± 1	0.30	0.81		45	P, M
			0.52	−0.03	425	55	
	B1	−10.5 ± 2.5	0.34	0.66		81	P, Fe^{2+}
			1.11	2.80		19	
Turkmenistan	C4-55	+40	0.31	0.63		70	P, Fe^{2+}
			1.09	2.57		30	
	C4-46	−110	0.24	0.72		100	P

*The stratigraphic level is the height from the base of the iridium rich layer.
IS = isomer shift; QS = quadrupole splitting; B = hyperfine magnetic field.
†P = paramagnetic; S = superparamagnetic; M = magnetic. Missing B value means absence of hyperfine magnetic field. Iron is either in paramagnetic phase or superparamagnetic phase (magnetic but fine particles).

gives the magnetic fraction present in the sample. The iron content, magnetic fraction, and the iridium concentration in the Meghalaya section, after normalization, are plotted against the stratigraphic level in Figure 4. The three profiles are roughly parallel, suggesting a fair degree of correlation between them. Therefore, we conclude that a correlation exists between the abundance of magnetic iron phases and iridium content at the K-T boundary.

Implications of oxide and oxyhydroxide iron phases

The results of the Mössbauer investigations of the K-T boundary layer from the four different basins, located far away from each other (Fig. 1), reveal the common presence of hematite and goethite regardless of the local lithology, and as the major iron-bearing minerals. Although these oxides are known to occur in terrestrial sediments and are generally formed due to weathering of iron-bearing minerals, their association with iridium, global occurrence within the K-T boundary layer regardless of the local depositional environment (marine or continental), and their absence above and below the K-T boundary layer indicate that their formation is related to the K-T event. The absence of pyrite, siderite, and other iron minerals in the K-T boundary layer that can weather into the oxide and/or oxyhydroxide phases indicates that they are not the products of weathering of local sediments but have been emplaced independently in this form. Their origin may, therefore, be related to the impact-induced processes.

Primary or secondary origin of superparamagnetic phases. An important question is whether the observed oxide phases are primary in the sense that they were formed in the impact-induced processes (in the vapor plume or just after their fallout in the extreme environmental conditions; e.g., high temperatures, acid rains) that might have prevailed for a few years or these are secondary phases, i.e., formed by postdepositional alteration of sediments. Several observations, summarized in the following, favor a primary origin.

The Anjar intertrappean section has intercalations of shale with marly limestone in which nanosized oxide phases have not been found. If these phases are a product of terrestrial weathering of iron-bearing minerals such as pyrite and siderite, they should be present everywhere. Only the iridium-bearing K-T boundary layer contains the oxidized phases, and in the rest of the section iron is present in paramagnetic phases.

In Meghalaya, the oxide and/or oxyhydroxide phases in the vertical section correlate with the iridium concentration (Fig. 4). It is impossible to visualize selective weathering dependent on the iridium concentration, and therefore this observation supports a common source for iridium and the oxide and/or oxyhydroxide phases. In the Turkmenistan K-T boundary layer we still see a ferrous component. It is unlikely that it did not convert into a ferric phase while the major part became hematite or goethite. This observation suggests that postdepositional alterations have not transformed the iron-bearing minerals into oxide phases. It is difficult to visualize common terrestrial sedimentary processes, operating at all places independent of the geographical or geological environment, that lead to the formation of oxide and oxyhydroxide nanophases at the same time globally.

There are now 13 sites distributed over the Earth where these oxide and oxyhydroxide phases have been found in the boundary layer if we include the work of Wdowiak et al., 2001.

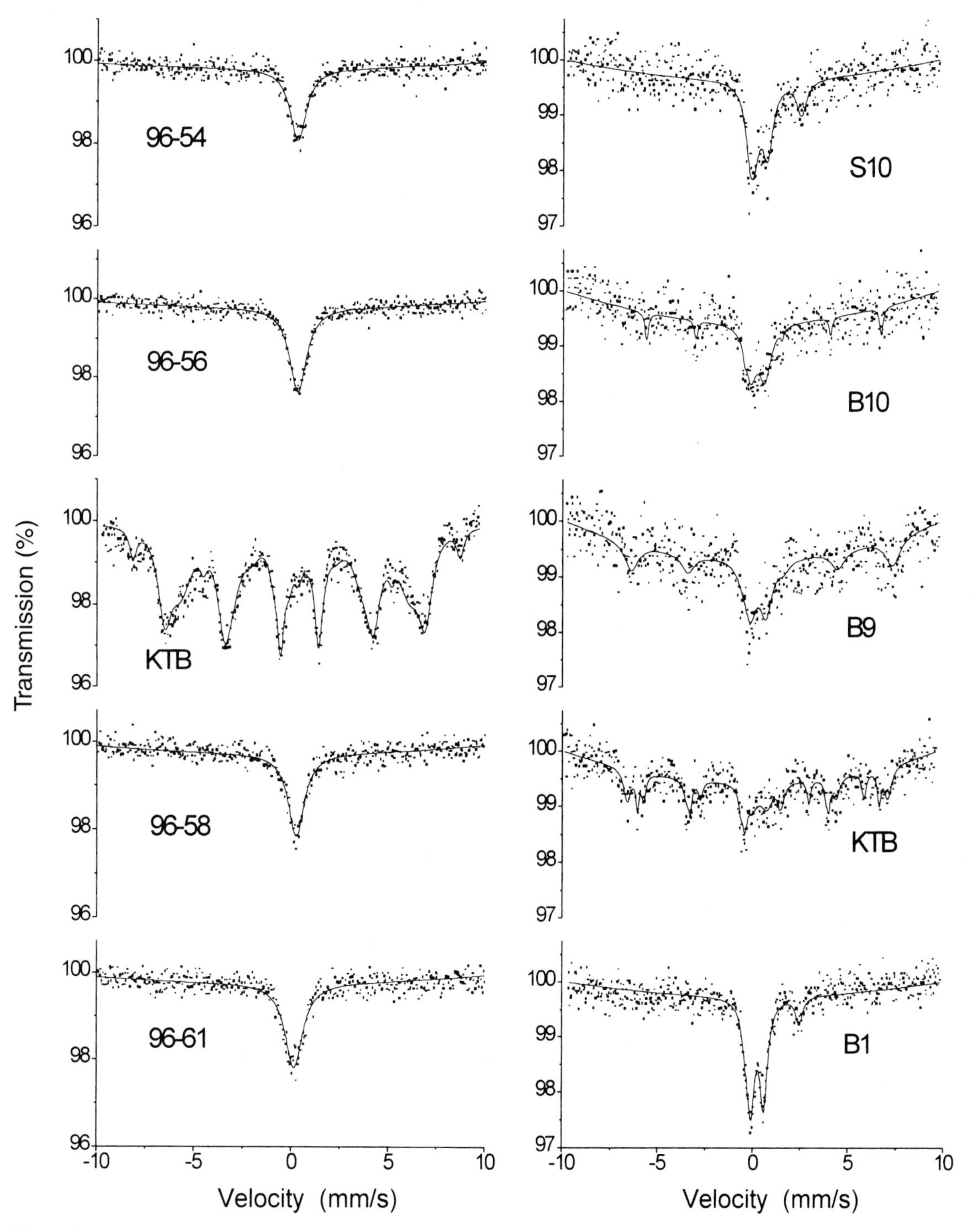

Figure 3. Mössbauer spectra of samples at different stratigraphic distances from Cretaceous-Tertiary (K-T) boundary in Anjar (left) and Meghalaya sections (right), recorded at 100 K.

In view of these observations, it appears that the oxide and oxyhydroxide phases were formed before being deposited in the sediments.

Nanoparticles are known to form in vapor condensates during rapid cooling, as revealed by laboratory experiments (Stephens, 1978). This observation favors the formation of nanoparticle iron phases observed at the K-T boundary during condensation of the high-temperature impact vapor plume.

Possible mechanism of deposition of the impact-ejecta material in sediments. The K-T boundary ejecta layer at most sites seems to have hematite, goethite, and/or other iron oxides as the major iron-bearing minerals. There are two possibilities

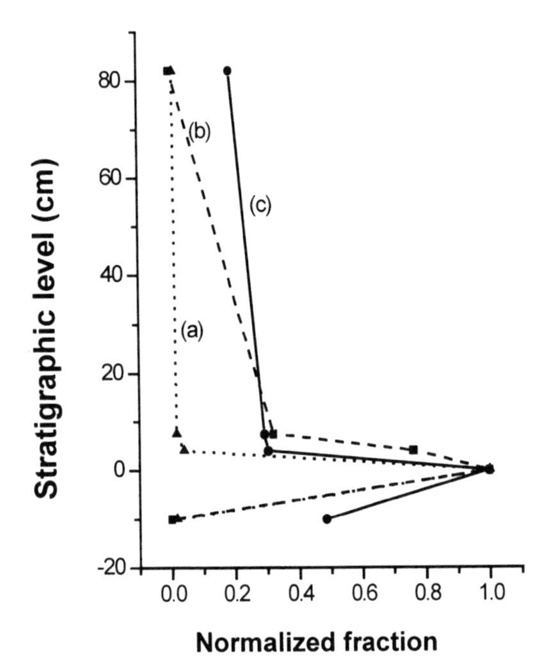

Figure 4. Vertical profiles of (a) iridium content, (b) magnetic fraction, and (c) iron content in different samples of Meghalaya section as function of stratigraphic distance from iridium-rich (K-T boundary) layer. Magnetic fraction has been obtained from area under sextets compared to total absorption area at 100 K. All values are normalized relative to corresponding value at iridium-rich layer.

for their formation. (1) These phases were formed in the impact vapor plume (Melosh, 1989, 1991; Zahnle and Sleep, 1997), or (2) these oxide and/or oxyhydroxides may be the weathered products of the iron phases formed in the plume before being incorporated in the sediments. The Mössbauer studies have a bearing on these two alternatives.

In the Gubbio K-T boundary spectrum (Fig. 2; Table 2), we see a prominent paramagnetic Fe^{3+} phase together with hematite. This phase is not present in the layers deposited a few centimeters above or below the iridium-rich layer. The Fe^{3+} phase could be a precursor to hematite in the weathering chain and, because the deposition of the K-T boundary layer was quite fast, we can assume that it was not completely weathered into hematite. The amount of hematite observed suggests that a significant amount of iron might have been in the form of phases that were prone to weathering, and that the harsh environmental conditions (high temperatures, acid rains) created by the impact were conducive to extremely fast weathering of these phases.

In the Meghalaya site, Fe^{3+} minerals are present throughout the ~20 cm vertical section (B1–B10,Table 3). This implies that the marine sediment was already receiving these minerals before most of the impact ejecta was deposited. This is corroborated by the fact that these magnetic phases correlate with the iridium concentration (Fig. 4). In Turkmenistan, a minor ferrous component has survived, whereas in Anjar all the iron minerals were converted into oxides and oxyhydroxides. In summary, the results discussed herein show that iron in nanoparticle, superparamagnetic phases probably formed in the impact vapor plume, and were altered to the oxide and/or oxyhydroxide phases to different stages in the weathering chain, depending on the local environmental conditions prevalent following the impact. The iron-bearing minerals present in the K-T layer (hematite, goethite, or limonite) suggest that most of the primary iron minerals that formed in the vapor plume were susceptible to weathering, and the geochemical conditions created by the impact were quite effective for weathering.

CONCLUSIONS

Mössbauer studies of the K-T boundary sections from four different sites, extending from Italy to eastern India, show sharp variations in iron mineralogy across the boundary layer. The boundary layer is characterized by the presence of nanoparticles of iron in oxide and/or oxyhydroxide phases. These superparamagnetic and magnetic iron phases occur in the K-T boundary layers in all the sites studied. The concentrations of iron and some other siderophile elements are also enhanced in the K-T boundary layer. The magnetically ordered iron phases show a correlation with the iridium content, indicating that they have a common origin. These phases or their precursors could have been formed in impact-related processes, e.g., during condensation of the impact vapor plume. The observations can be explained by assuming that after the large bolide impact, the ejecta (including condensates of the vapor plume) dispersed in the atmosphere and was deposited globally. The prevalent atmospheric and local geochemical conditions quickly altered the iron phases to different degrees before they were incorporated in the sediments.

ACKNOWLEDGMENTS

We are grateful to M.A. Nazarov for providing us with the samples of the Turkmenistan section. We thank C. Koeberl and B. Hofmann for useful suggestions for improvement of this paper and A.D. Shukla for help in preparation of this manuscript.

REFERENCES CITED

Alekseev, A.S., Barsukova, L.D., Kolesov, G.M., Nazarov, M.A., and Amanniyazov, N., 1986, Cretaceous/Tertiary event: Iridium distribution in Turkmenia sections [abs.]: Lunar and Planetary Science, v. 27, p. 9–10.

Alvarez, L.W., Alvarez, W., Asaro, F., and Michel, H.V., 1980, Extraterrestrial cause of the Cretaceous/Tertiary extinction: Science, v. 208, p. 1095–1108.

Badjukov, D.D., Nazarov, M.A., and Suponeva, I.V., 1986, Shocked quartz grains from K/T Boundary sediments [abs.]: Lunar and Planetary Science, v. 27, p. 18–19.

Bhandari, N., 1998, Astronomical and terrestrial causes of physical, chemical and biological changes at geological boundaries: Proceedings of Indian Academy of Sciences, v. 107, p. 251–263.

Bhandari, N., Shukla, P.N., and Castagnoli, G.C., 1994a, Geochemistry of some K/T sections in India: Palaeogeography, Palaeoclimatology, Palaeoecology, v. 104, p. 199–211.

Bhandari, N., Gupta, M., Pandey, J., and Shukla, P.N., 1994b, Chemical profiles in K/T boundary section of Meghalaya, India: Cometary, asteroidal or volcanic: Chemical Geology, v. 113, p. 45–60.

Bhandari, N., Shukla, P.N., Ghevariya, Z.G., and Sundaram, S.M., 1995, Impact did not trigger Deccan volcanism: Evidence from Anjar K/T boundary intertrappean sediments: Geophysical Research Letters, v. 22, p. 433–436.

Bhandari, N., Shukla, P.N., Ghevariya, Z.G., and Sundaram, S.M., 1996, K/T boundary layer in Deccan intertrappeans at Anjar, Kutch, *in* Ryder, G., Fastovsky, D., and Gartner, S., eds., The Cretaceous-Tertiary event and other catastrophes in Earth history: Geological Society of America Special Paper 307, p. 417–424.

Bhandari, N., Verma, H.C., Tripathi, A., Upadhyay, C., and Tripathi, R.P., 2000, Mössbauer spectroscopy of K/T boundary clays: Characteristics of iron bearing minerals, *in* Catastrophic events and mass extinctions: Impacts and beyond: Houston, Texas, Lunar and Planetary Institute, LPI Contribution No. 1053, p. 12–13.

Gangopadhyay, S., Hadjipanayis, G.C., Dale, B., Sorensen, C.M., Klabunde, K.J., Papaefthymiou, V., and Kostikas, A., 1992, Magnetic properties of ultrafine iron particles: Physical Review B, v. 45, p. 9778–9787.

Gütlich, P., 1975, Mössbauer spectroscopy in chemistry, *in* Gonser, U., ed., Mössbauer spectroscopy: Berlin, Springer-Verlag, p. 53–96.

Haneda, K., 1987, Recent advances in the magnetism of fine particles: Canadian Journal of Physics, v. 65, p. 1233–1243.

Melosh, H.J., 1989, Impact cratering: A geologic process: New York, Oxford University Press, 245 p.

Melosh, H.J., 1991, Atmospheric impact process: Advances in Space Research, v. 11, p. 87–93.

Mørup, S., Dumesic, J.A., and Topsøe, H., 1980, Magnetic microcrystals, *in* Cohen, R.L., ed., Application of Mössbauer spectroscopy: New York, Academic Press, p. 1–53.

Robin, E., Rocchia, R., Bhandari, N., and Shukla, P.N., 1997, Cosmic imprints in the Meghalaya K/T section; International Conference on Isotopes in the Solar System, Physical Research Laboratory, Ahmedabad, India, Abstracts, p. 95–96.

Stephens, J.R., 1978, Laboratory condensation of chondritic minerals, *in* Thermodynamics and kinematics of dust formation in space medium: Houston, Texas, Lunar and Planetary Institute, LPI Contribution No. 330, p. 30–33.

Shukolyukov, A., and Lugmair, G.W., 1998, Isotopic evidence for the Cretaceous-Tertiary impactor and its type: Science, v. 282, p. 927–929.

Thorpe, A.N., Senftle, F.E., May, L., Barkatt, A., Adel-Hadadi, M.A., Marbury, G.A. Izett, G.S., and Maurrasse, F.R., 1994, Comparison of the magnetic properties and Mössbauer analysis of glass from the Cretaceous-Tertiary boundary, Beloc, Haiti, with tektites: Journal of Geophysical Research, v. 99, p. 10881–10886.

Van der Giessen, A.A., 1968, Chemical and physical properties of iron (II) oxide and hydrate: Phillips Research Reports, Supplement, no. 12, p. 31–46.

Wdowiak, T.J., Armendarez, L.P., Agresti, D.G., Wade, M.L., Wdowiak, S.Y., Claeys, P. and Izett, G., 2001, Presence of an iron-rich nanophase material in the upper layer of the Cretaceous-Tertiary boundary clay: Meteoritics and Planetary Science, v. 36, p. 123–133.

Window, B., 1971, Hyperfine field distributions from Mössbauer spectra: Journal of Physics, E: Scientific Instruments, v. 4, p. 401–402.

Zahnle, K.J. and Sleep, N.H., 1997, Impacts and the early evolution of life, *in* Thomas, P.G., Chyba, C.F., and McKay, C.P., eds., Comets and the origin and evolution of life: New York, Springer-Verlag, p.175–208.

Manuscript Submitted October 5, 2000; Accepted by the Society March 22, 2001

Printed in the U.S.A.

Geological Society of America
Special Paper 356
2002

Cretaceous-Tertiary profile, rhythmic deposition, and geomagnetic polarity reversals of marine sediments near Bjala, Bulgaria

Anton Preisinger*
Selma Aslanian
Institute of Mineralogy, Crystallography and Structural Chemistry, Technical University of Vienna, Getreidemarkt 9, A-1060 Vienna, Austria
Franz Brandstätter
Museum of Natural History, Burgring 7, A-1010 Vienna, Austria
Friedrich Grass
Atomic Institute of the Austrian Universities, Schüttelstrasse 115, A-1020 Vienna, Austria
Herbert Stradner
Geological Survey, Rasumofskygasse 23, A-1030 Vienna, Austria
Herbert Summesberger
Museum of Natural History, Burgring 7, A-1010 Vienna, Austria

ABSTRACT

In 1991, a Cretaceous-Tertiary (K-T) boundary was discovered in Bulgaria in marine sediments on the coast of the Black Sea near the city of Bjala, 35 km south of Varna. At the Bjala 2b site, rhythmic sedimentation under hemipelagic conditions from the Late Cretaceous (30 m below the K-T boundary) to the early Paleogene (70 m above the K-T boundary) resulted in the deposition of 100 m of marly limestone and intercalated marl. In these strata, 200 marly limestone beds correspond to precessional Milankovitch cycles of an average duration of 22.5 k.y. An absolute geological time scale over 4 m.y. is obtained from the number of limestone beds of the measured magnetic polarity zones (chronozones 29R to 27R). The reference age (65.0 Ma) is assumed to correspond to the level with the maximum Ir and Ni-rich spinel content in the K-T profile.

The process of geomagnetic polarity reversals is relatively short (5–10 k.y.) compared to the geomagnetic polarity zones between the reversals (0.2–1.2 m.y.). The sediments show $CaCO_3$ minima of the marly limestone beds and maxima of magnetic susceptibility in the geomagnetic polarity reversals. In addition to the magnetic minerals such as greigite, extraterrestrial ferrimagnetic spinels composed of Ni-rich magnesioferrites with high Cr contents were found at the polarity reversals. These extraterrestrial spinels are similar in their composition to the K-T spinels of the K-T impact at Chicxulub, but without etched pits on the octahedral faces of the spinel crystals.

The K-T boundary profiles at Bjala are characterized by fallout sedimented within a short time, overlapped by boundary clay sedimented over a long time, and some partially reworked sediments. Names of extinct species, of survivors, and first occurrences of new species at Bjala 2b (chronozone 29R) are given.

*Mailing address: Lerchengasse 23/2/9, A-1080 Vienna, Austria.
E-mail: apreisin@mail.zserv.tuwien.ac.at

Preisinger, A., Aslanian, S., Brandstätter, F., Grass, F., Stradner, H., and Summesberger, H., 2002, Cretaceous-Tertiary profile, rhythmic deposition, and geomagnetic polarity reversals of marine sediments near Bjala, Bulgaria, *in* Koeberl, C., and MacLeod, K.G., eds., Catastrophic Events and Mass Extinctions: Impacts and Beyond: Boulder, Colorado, Geological Society of America Special Paper 356, p. 213–229.

INTRODUCTION

At the end of the Cretaceous, 65 Ma (Swisher et al., 1992), a carbonaceous chondritic body (Shukolyukov and Lugmair, 1998) collided with the Earth in the vicinity of Chicxulub, on the Yucatan Peninsula, Mexico (Hildebrand et al., 1991; Sharpton et al., 1993). The extraterrestrial body was suggested to have hit Earth at an oblique angle from the southeast (Schultz and D'Hondt, 1996). At the time of the impact, Yucatan was a 3-km-thick platform of calcite, dolomite, and anhydrite and was covered by a shallow sea (Ward et al., 1995). The impact caused hot and warm fireballs, an ejecta curtain, a megatsunami, earthquakes, shock waves, and melting of rock (Pope et al., 1997; Preisinger, 2000). Material ejected from the Chicxulub impact was globally distributed, and has been found in ~100 Cretaceous-Tertiary (K-T) boundary sites, in ocean sediments as well as on land (see Claeys et al., 2000).

Continuous K-T boundary sections were investigated at different sites in the Mediterranean area of the Tethys, from Spain to Bulgaria; e.g., in Barranco del Gredera near Caravaca, southeastern Spain (Smit and Hertogen, 1980; Lindinger, 1988); in the Scaglia Rossa of the Apennines at Gubbio (Alvarez et al., 1980) and 25 km north of Gubbio at Cerbara, Italy (Preisinger et al., 1996); at Elendgraben, Gosau (Preisinger et al., 1986) and at Knappengraben, Gams, Austria (Stradner et al., 1987); at El Kef, Tunisia (Smit, 1982; Lindinger, 1988); and on the Black Sea coast near Bjala, Bulgaria (Preisinger et al., 1993a, 1993b; Rögl et al., 1995, 1996). The K-T profile and the rhythmic depositions of marine sediments at Bjala, Bulgaria (herein Bjala 2b) were investigated in detail and the results are presented here.

METHODS

In continuation of previous studies, an additional 516 samples were analyzed by biostratigraphic, mineralogical, geochemical, and magnetic methods using the following: (1) manual, microscopically checked separation of the samples in intervals from a few mm to 1.0 cm; (2) mechanical disintegration in water and separation of different fractions (>20 μm, 20-8 μm, < 8 μm); (3) determination of the carbonate content in bulk and in fractionated samples by means of the carbonate bomb (Müller and Gastner, 1971); (4) quantitative phase determination by X-ray analysis of bulk samples and fractions; and (5) extraction of spinels with a strong permanent Co-Sm magnet from water suspensions of clay collected from the K-T boundary profile. After decalcifying the samples with 7% HCl, the water suspension was passed through a polyethylene tube squeezed by two strong Co-Sm magnets. To prevent plugging of the tube during the separation, a string of 1-mm-diameter was inserted between the magnets. The operation was repeated 10 times. The particles collected between the magnets were washed with ~500 ml of distilled water. The magnets were removed, and the particles washed into a beaker with distilled water, transferred to a stoppered pipette tip, and dried at 50°C. (6) In addition, experimental etching of K-T spinels was done with concentrated H_2SO_4; (7) mineralogical analysis of the fractionated samples was done by scanning electron microscopy (JEOL JSM-6400 equipped with a KEVEX energy dispersive system, operating conditions 15 kV, 1.5 nA, and CAMSCAN 24 plus EDS Link system); (8) grain-size analysis of Cretaceous and Tertiary decalcified marly limestone beds was done with a particle-size analyzer using an X-ray Sedigraph 5000 ET; and (9) $\delta^{13}C_{carb}$ and $\delta^{18}O$ measurements of calcareous nannofossils (<20 μm) and of planktonic foraminifera were done (H. Oberhänsli, Alfred-Wegener Institute, Potsdam, Germany, 1995, personal commun.)

After sample preparation by means of an automatic instrument of the Finnigan system, mass-spectrometric measurements of $\delta^{18}O$ and $\delta^{13}C$ were done by Oberhänsli (1996) using a Finnigan MAT 251 (University of Bremen), and a VG Prism (Max-Planck Institute of Chemistry, Mainz).

Eighty-three measurements of magnetic susceptibility at the magnetic polarity reversals were performed by the following procedure: the original samples were powdered, mixed with camphor (0.4 g sample with 0.07 g camphor), and pressed to pills. The susceptibility was measured at 293K with a superconductivity quantum interference device (SQUID) magnetometer (Schwartz and Foner, 1977) by A. Vostner, Atomic Institute of the Austrian Universities, Vienna.

Paleomagnetic measurements were made of 170 oriented cores of the marly limestone. These were collected by means of a gasoline-driven drill at an average stratigraphic spacing of 40 cm. After thermal treatment of the oriented cores, polarity measurements gave a lucid, reliable picture of the normal and reversed zones in paleodeclination and paleoinclination (Mauritsch and Scholger, *in* Preisinger et al., 1993a).

Special attention was paid to the boundary clay, which is a few millimeters to 2 cm thick, as well as to the sediments above and below it. A combination of X-ray diffractometry, chemical analysis, microscopy, and scanning electron microscopy (SEM), isotope analysis, and instrumental neutron-activation analysis (INAA) was applied. The mineralogy of bulk and decalcified samples was determined by an automatic powder diffractometer (Philips PW1710, using Cu-K radiation and graphite monochromatization, 40 kV, 40 mA, and step scanning; 1 step 0.02°/1.00 s).

The quantity of spinels in the magnetically isolated fractions was determined by X-ray diffraction analysis with an automatic powder diffractometer (Philips X'Pert-MPD). The analysis was performed by theta-/theta-geometry, silicon sample holder spinning, and step scanning (1 step 0.04°/40 s). The mechanically separated fractions were analyzed by SEM (CAMSCAN 24 plus EDS Link system). The magnetically isolated fractions were investigated by SEM (JEOL JSM-6400 equipped with a KEVEX energy-dispersive system [EDS], operating conditions: 15 kV, 1 nA beam current). The iridium and trace elements were determined by INAA. The samples were welded into 5 mm high-purity quartz (Suprasil) tubes and irradiated for 120 h at a flux of 8.10^{13}n/cm^{2}s^1 at the ASTRA reactor (Seibersdorf, Austria), together with a Cerbara K-T standard and a Stevns Klint Fish Clay standard with a recommended value of 32 ± 2 ppb (6.25 wt%) Ir (Gwozdz et al., 1992). (Note that the red boundary clay of the Cerbara K-T boundary was collected by G. Gottardi, F. Grass, and A. Preisinger in 1984. About 300 g of the red boundary clay was dried at 120°C for 12 h, ground in an agate mortar until all the powder passed a 280 μm Perlon sieve, and tested for homogeneity by X-ray. The sample was calibrated in a round-robin by S. Meloni by radio-chemical neutron activation analysis (RNAA), and by P. Bonte and F. Grass by INAA, giving a mean value of 13 Ir measurements of 6.54 ± 0.21 ppb (3.2 wt%), using Ir and PCC1 as primary standards.) After 6 weeks decay, the samples and standards were measured with a HPGe detector (30% relative efficiency) having a resolution of 1.81 keV for the 1332 keV-peak of ^{60}Co. The Stevns Klint Fish Clay standard evaluated by the Cerbara K-T standard gave 30.8 ± 1.8 ppb (5.9 wt%) Ir. Thus the accuracy of our Ir measurements is ~6% and the precision is ~8%.

GEOLOGY OF THE K-T SECTIONS AT BJALA, BULGARIA

In 1991, the K-T boundary was discovered in marine sediments on the coast of the Black Sea near the city of Bjala, Bulgaria, 35 km south of Varna (Preisinger et al., 1991). The continuous K-T boundary sections at Bjala are located in the Luda Kamchiya tectonic unit, a synclinorium filled with Upper Cretaceous and Paleogene sediments (Bončev, 1974). The Luda Kamchiya tectonic unit is situated between the Balkan chain and the Moesian platform (Rögl et al., 1995, 1996). The K-T profiles at Bjala (Fig. 1) show a continuous deposition of marine sediments from the upper Maastrichtian into the Danian. The K-T boundaries at Bjala 2b, 2c, and 3 are characterized by fallout and boundary clay, followed by some partially reworked sediments (Preisinger et al., 1993b). The K-T boundary section Bjala 2b is 250 m long, and located at 42°52′40″N, and 27°53′58″E (Fig. 2) within a paleodistance of 9500 km from the impact crater at Chicxulub. The section was formed in the Tethys at a paleolatitude of ~29°N at the shelf edge (water depth ~300 m). Marly limestone with intercalated marl was

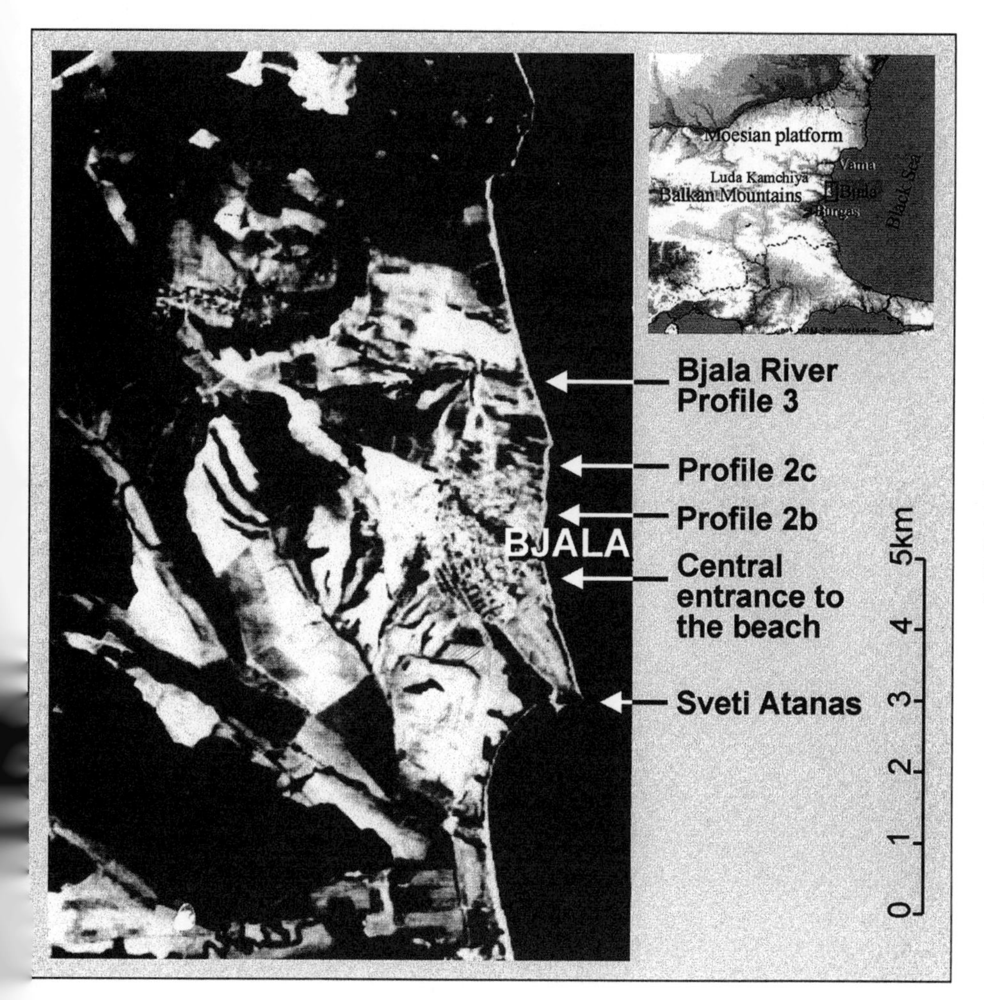

Figure 1. Satellite photograph (Landsat 5 from September 24, 1990, scale 1:50 000) of Cretaceous-Tertiary (K-T) boundary sites near Bjala at the coast of Black Sea. Profiles Bjala 2b, 2c, and 3 are indicated. Inset: Simplified geological map of Bulgarian part of Black Sea coast.

sedimented over a vertical range of 100 m under hemipelagic conditions from the Late Cretaceous (30 m below the K-T boundary) to the early Paleogene (70 m above the K-T boundary).

The paleomagnetic profile Bjala 2b shows the chronozones from 30N to 26R (Fig. 2). The marly limestone beds show occasional minor disturbances; however the sequence of the beds could be analyzed. At 27 m south of the K-T boundary of Bjala 2b, which is marked by a red-painted iron rod (Fig. 2), the connection between the blocks was shifted by a major fault. This was verified by the analysis of the carbonate contents of the beds (Preisinger et al., 1993a). The K-T boundary is located in chronozone 29R.

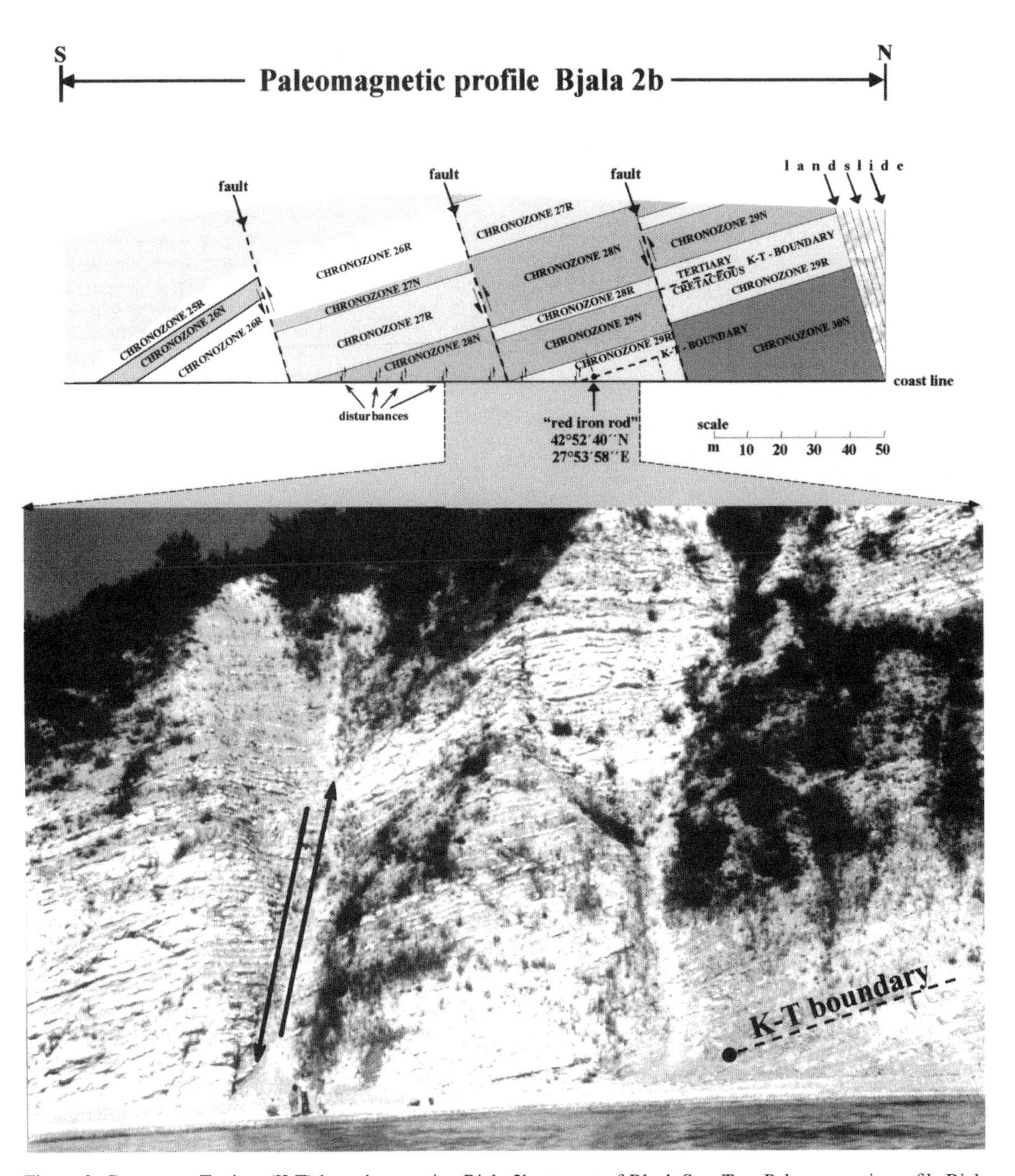

Figure 2. Cretaceous-Tertiary (K-T) boundary section Bjala 2b at coast of Black Sea. Top: Paleomagnetic profile Bjala 2b. Bottom: Photograph of part of Bjala 2b section from seaside. K-T boundary is marked by a red-painted iron rod (black dot).

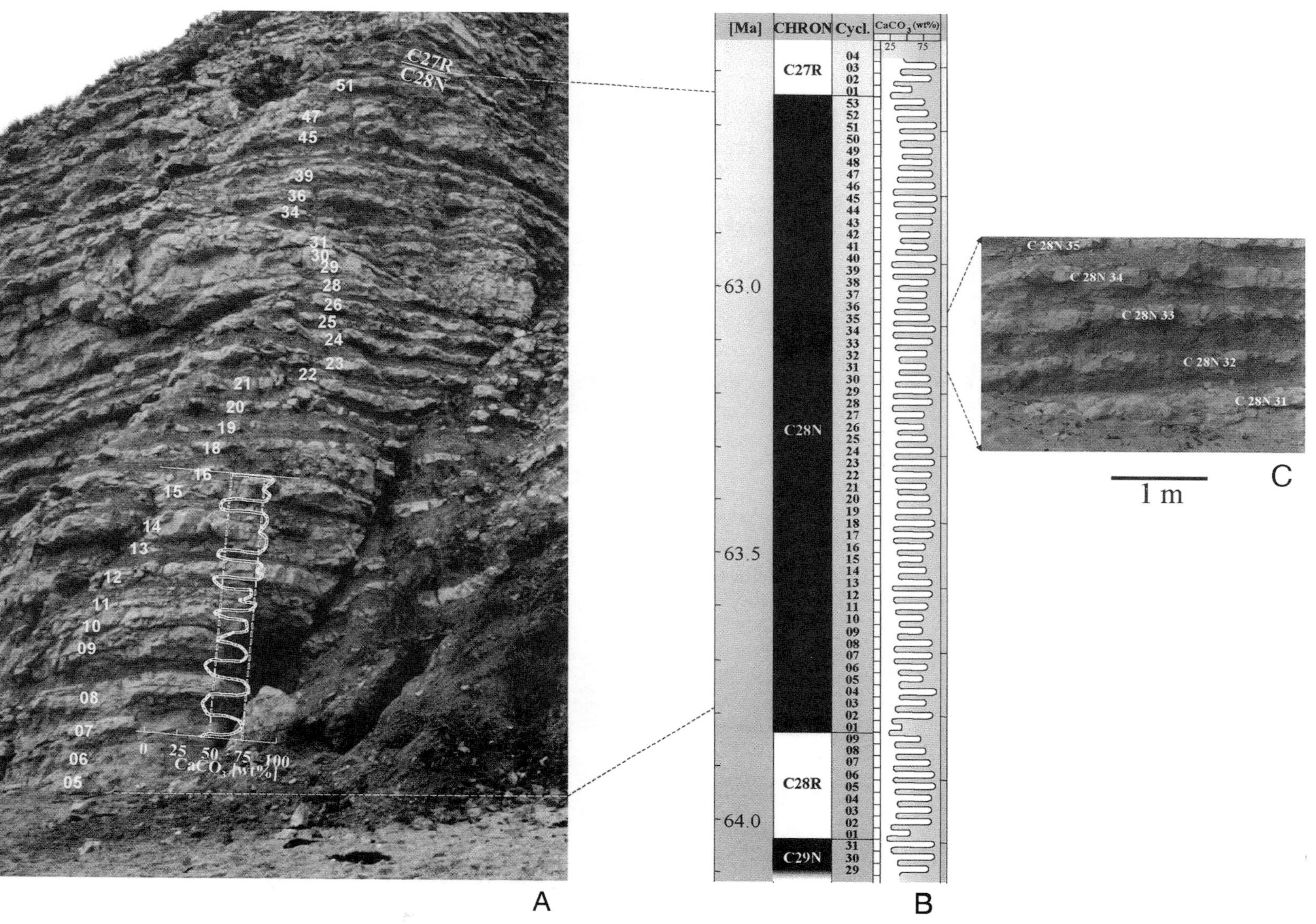

Figure 3. A: Photograph of marly limestone beds of chronozones 28N and the polarity reversal 28N-27R at Bjala 2b. B: Precessional Milankovitch cycles of chron 28R (9 cycles) and chron 28N (53 cycles) are correlated with marly limestone beds and numbered from 01 to 09, and from 01 to 53, respectively. C: Photograph of five marly limestone beds corresponding to precessional Milankovitch cycles numbered from 31 to 35 in chronozone 28N.

GEOLOGICAL TIME SCALE OF THE K-T PROFILE BJALA 2B

Magnetostratigraphy and Milankovitch cycles

In relation to the K-T boundary as a reference, the absolute geological age of the marly limestone beds has been determined by using the correlation of the magnetic polarity chronozones with the number of rhythmic marly limestone beds. The thickness of these beds corresponds to the time span of a single Milankovitch cycle of the climatic precession parameter (Berger, 1978) with a measured periodicity of 22.5 k.y. (Fig. 3). The present 23 k.y. period of solar insolation on Earth results from the combined effects of precession of the axis of rotation and the elliptical orbit of Earth. Due to the shorter Earth-Moon distance 65 Ma, the 23 k.y. precession period was calculated to 22.5 k.y. by Berger et al. (1989). At Bjala 2b, the combination of the paleomagnetic polarities with the number of Milankovitch cycles gives an absolute geological time scale spanning 4 m.y. A comparison of our results with the stratigraphic time scales (Harland et al., 1990; Cande and Kent, 1995) is shown in Figure 4. All three time scales are related to the reference date of the K-T boundary, 65.0 Ma (Swisher et al., 1992).

The duration of chron 29R within the Paleocene corresponds to 270 ka, and the duration of the total chron 29 has

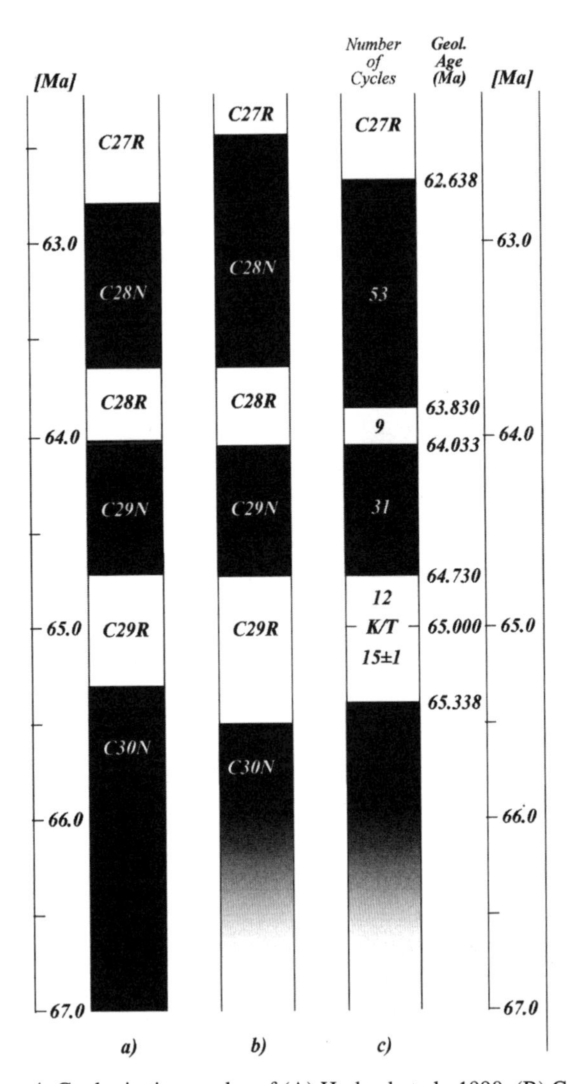

Figure 4. Geologic time scales of (A) Harland et al., 1990, (B) Cande and Kent, 1995, and (C) this study: Number of Milankovitch cycles of Bjala 2b permits calculation of geological time spent for sedimentation in each chronozone. K-T is Cretaceous-Tertiary boundary.

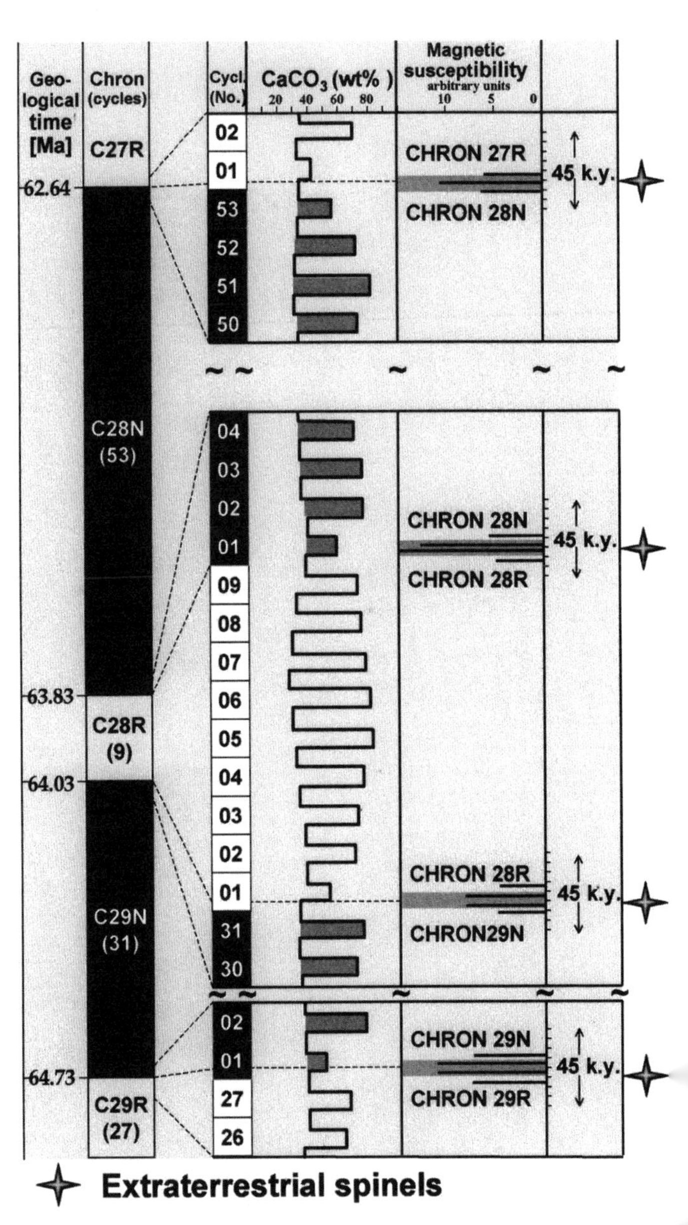

Figure 5. Four magnetic polarity reversals at Bjala 2b. $CaCO_3$ content of Milankovitch cycles shows minima at reversals. Magnetic susceptibility shows maxima at reversals connected with occurrence of extraterrestrial spinels.

TABLE 1. COMPARISON OF EXTRATERRESTRIAL POLARITY REVERSAL SPINELS AND CRETACEOUS-TERTIARY SPINELS

	Cation distribution in $A^{[4]}B_2^{[6]}O_4$ spinels										
	$A^{[4]}$			$B^{[6]}$							O
Crystal size	[Fe^{3+}	Mn^{2+}	Fe^{2+}]	[Mg^{2+}	Ni^{2+}	Al^{3+}	Cr^{3+}	Ti^{4+}	Fe^{2+}	Fe^{3+}]	O^{2-}
3.6 µm	0.43	0.01	0.56	0.29	0.14	0.15	0.18	0.02	0.02	1.20	4
7.0 µm	0.54	0.00	0.46	0.44	0.10	0.19	0.33	0.01	0.01	0.92	4
10.0 µm	0.84	0.02	0.14	0.75	0.09	0.24	0.54	0.01	0.01	0.36	4
Spot No.*											
1	0.62	0.00	0.38	0.48	0.14	0.23	0.34	0.01	0.01	0.79	4
2	0.65	0.02	0.33	0.54	0.12	0.27	0.28	0.01	0.01	0.77	4
3	0.53	0.02	0.45	0.37	0.16	0.20	0.27	0.00	0.00	1.00	4
4	0.60	0.00	0.40	0.42	0.19	0.19	0.21	0.01	0.01	0.97	4
	MEAN VALUES										
	MgO	Al_2O_3	TiO_2	Cr_2O_3	MnO	FeO	NiO	Total			
Extraterrestrial	10.1	5.0	0.6	13.5	0.4	66.3	4.1	100			
Cretaceous-Tertiary	11.1	4.9	0.6	9.3	0.8	68.0	5.3	100			

*K-T spinels at Caravaca, Spain. No. 1–4 indicate spots in Figure 6B

been found to be 608 ± 23 ka, which is in good agreement with 270 ± 17 ka and 603 ± 26 ka, respectively, at Deep Sea Drilling Project Site 529 in the South Atlantic (D'Hondt et al., 1996).

GEOMAGNETIC POLARITY REVERSALS

The geomagnetic polarity reversals are relatively short (5–10 k.y.) compared to the constant polarity intervals between the reversals (0.2–1.2 m.y.). The carbonate content of the marly limestone beds within the constant polarity intervals is in the range of 65–80 wt% $CaCO_3$, whereas it is significantly lower, 45–55 wt% $CaCO_3$, during the time of the polarity reversals. Magnetic susceptibility is significantly higher during the geomagnetic polarity reversals due to the occurrence of extraterrestrial spinels (Fig. 5).

EXTRATERRESTRIAL SPINELS

Ni-rich magnesioferrite spinels with variable Cr contents were found by magnetic separation at all magnetic polarity reversals at the K-T profile of Bjala 2b. These spinels have compositions similar to those of spinels reported from the crust of meteorites (Gayraud, 1995). The compositions of the geomagnetic polarity reversal spinels are also similar to those of K-T spinels (Table 1). However, K-T spinels have etched pits, whereas polarity reversal spinels have none (Fig. 6), because an H_2SO_4 aerosol existed in the stratosphere only after the K-T event (Pope et al., 1997). Similar extraterrestrial Ni-rich magnesioferrites were also found at the magnetic polarity reversal between chronozone 29R and 30N in the pelagic K-T profile at Cerbara, Italy.

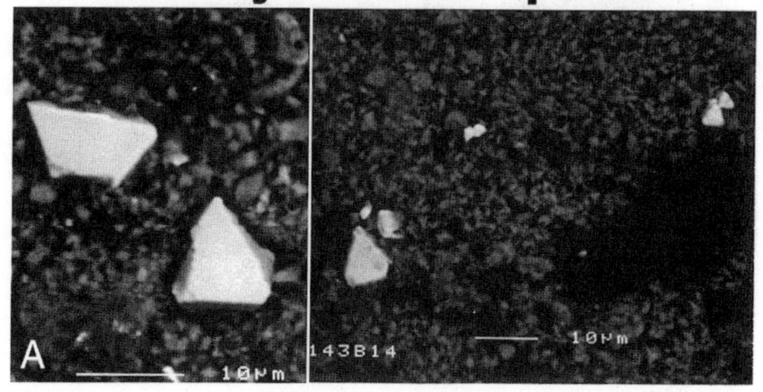

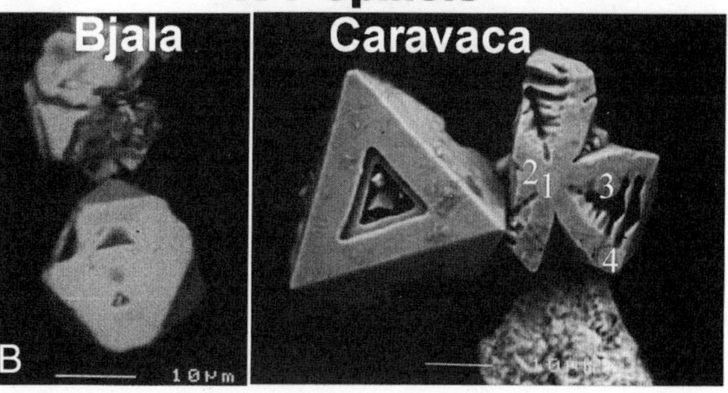

Figure 6. Comparison of geomagnetic polarity reversal spinels with etched Cretaceous-Tertiary (K-T) spinels at Bjala, Bulgaria (A), and Caravaca, Spain (B).

TABLE 2. QUANTITATIVE PHASE ANALYSES OF BULK SAMPLES AND FRACTIONS OF CRETACEOUS-TERTIARY PROFILE BJALA 2b

Distance from Cretaceous-Tertiary (cm)	Bulk samples			Fractions								
	Carb.	Silicates		>20 μm			8–20 μm			<8 μm		
	$CaCO_3$ (wt%)	$\overset{3}{\infty}$ (wt%)	$\overset{2}{\infty}$ (wt%)	$CaCO_3$ (wt%)	$\overset{3}{\infty}$ (wt%)	$\overset{2}{\infty}$ (wt%)	$CaCO_3$ (wt%)	$\overset{3}{\infty}$ (wt%)	$\overset{2}{\infty}$ (wt%)	$CaCO_3$ (wt%)	$\overset{3}{\infty}$ (wt%)	$\overset{2}{\infty}$ (wt%)
+29.0–+33.0	72.9	7.1	20.0	29.9	2.9	6.0	19.1	2.0	4.7	23.9	2.2	9.3
+22.0–+29.0	72.5	8.6	18.9	37.4	4.7	8.2	14.0	2.0	3.4	21.1	1.9	7.3
+16.0–+21.0	69.7	9.1	21.2	29.2	4.0	6.8	14.6	2.5	5.7	25.9	2.6	8.7
+16.0–+18.0	67.9	10.4	21.7	26.7	4.5	5.8	14.0	3.0	6.8	27.2	2.9	9.1
+9.2– +9.5	74.4	8.7	16.9	13.9	2.4	1.8	9.6	1.7	1.8	50.9	4.6	13.3
+6.0– +8.0	72.6	7.9	19.5	18.0	3.0	2.2	14.1	2.0	3.0	40.5	2.9	14.3
+4.0– +6.0	69.7	9.7	20.6	31.0	4.3	7.1	11.6	1.9	6.1	27.1	3.5	7.4
+3.5– +5.0	60.2	10.8	29.0	23.1	4.8	7.4	11.4	2.2	4.7	25.7	3.8	16.9
+3.5– +4.0	52.2	15.9	31.9	16.1	2.6	4.8	7.8	3.1	2.0	28.3	10.2	25.1
+2.7– +3.2	11.7	28.7	59.6	1.1	1.5	1.4	0.2	0.8	0.6	10.4	26.4	57.6
+1.5– +2.0	1.6	30.9	67.5	*0.1	2.5	1.8	*0.1	2.1	3.8	1.4	26.3	61.9
+1.0– +1.5	*<0.1	29.8	70.2	*<0.1	6.1	4.3	*<0.1	6.0	11.3	*<0.1	17.7	54.6
+0.5– +1.0	*<0.1	33.2	66.8	*<0.1	7.6	11.0	*<0.1	11.2	12.6	*0.1	14.4	43.2
0.0– +0.5	1.7	38.1	60.2	1.0	13.5	12.8	<0.1	13.4	16.7	0.7	11.2	30.7
0.0– −1.0	71.2	8.6	20.2	17.3	2.8	3.3	8.3	2.0	3.8	45.6	3.8	13.1
−1.0– −2.0	74.2	7.8	18.0	17.3	1.9	4.2	4.7	1.1	1.3	52.2	4.8	12.5
−2.5– −3.8	68.6	9.1	22.3	18.8	3.1	3.8	8.7	1.7	3.8	41.1	4.3	14.7
−3.8– −5.5	69.9	8.4	21.7	17.9	2.3	3.6	9.4	1.7	2.8	42.6	4.4	15.3
−5.5– −9.0	70.2	9.8	20.0	22.3	3.6	3.9	14.8	2.4	4.5	33.1	3.8	11.6
−10.0–−12.5	74.1	7.8	18.1	31.8	3.5	6.6	10.0	1.3	2.1	32.3	3.0	9.4
−12.0–−14.0	72.2	9.8	18.0	28.1	4.2	5.3	11.0	1.6	2.5	33.1	4.0	10.2
−13.5–−16.5	69.0	9.2	21.8	29.8	4.0	6.9	9.3	1.5	2.5	29.9	3.7	12.4
−26.0–−28.0	70.6	7.6	21.8	27.1	3.2	8.1	10.7	1.3	2.3	32.8	3.1	11.4
−68.0–−70.0	63.7	8.7	27.6	27.6	4.1	8.4	10.6	1.8	4.2	25.5	2.8	15.0
−71.0–−76.0	69.1	9.5	21.4	24.9	3.6	5.1	16.9	2.8	7.7	27.3	3.1	8.6

Note: The values of sheet silicates ($\overset{2}{\infty}$) are calculated as difference to 100 wt%.
*Mg calcite. $\overset{3}{\infty}$ = quartz + feldspar; $\overset{2}{\infty}$ = sheet silicates. Carb. = Carbonates

The first experimental investigation of extraterrestrial Ni-rich spinels at magnetic polarity reversals was documented by Preisinger et al. (2000). Ni-rich spinels in marine sediments are not only markers for big impacts, but also markers for a weak terrestrial magnetic field.

The change of magnetic polarity is related to decreases in the intensity of Earth's magnetic field (Merrill et al., 1998). In this weakened magnetic field, screening of Earth from cosmic radiation and particles is drastically reduced. That cosmic radiation may lead to a higher mutation rate of organisms is well known. The magnetic polarity reversals are thus thought to be important for the evolution of new species.

K-T BOUNDARY PROFILE BJALA 2B

Fallout and boundary clay

We split 25 samples from the K-T boundary profile Bjala 2b into 3 fractions (>20 μm, 20 − 8 μm, <8 μm). The bulk samples as well as the different fractions were investigated by X-ray analysis and SEM. The mineral composition, $CaCO_3$, quartz + feldspar ($\overset{3}{\infty}$), and sheet silicates ($\overset{2}{\infty}$) of the bulk samples and their fractions are given in Table 2. Figure 7 shows the quantitative phase analysis of carbonates, calcareous nannoplankton (<20 μm) and foraminifera (>20 μm), as well as silicates, quartz + feldspar ($\overset{3}{\infty}$), and sheet silicates ($\overset{2}{\infty}$) in the range of 20 cm below the K-T boundary (C −20) to 30 cm above it (T +30). C designates Cretaceous, and T designates Tertiary.

At the K-T boundary there is a radical change in the composition of the long-term state of the Cretaceous sediment: The calcite content decreases to a minimum (from 70 wt% to <0.1 wt% within <0.2 mm) in <10 yr, and increases again in sediment layers 2.5 cm above this minimum with a bloom of surviving nannoplankton (e.g., *Thoracosphaera operculata*; Fig. 8A). In the fraction <20 μm, the calcareous cysts of these dinoflagellates have an average diameter of 18 μm; smaller specimens are only 14 μm. The first occurrences (FO) of new species after the K-T impact are as follows: *Woodringina sp.* (T +2 cm), *Biantholithus sparsus* (T +3 cm), and *Parvularugoglobigerina eugubina* (T +10 cm). In the calcite gap, sheet silicates ($\overset{2}{\infty}$) and framework silicates ($\overset{3}{\infty}$), quartz + feldspar, are dominant. Up to T +4 cm, small amounts of authigenic Mg-containing calcites (12 mol % $MgCO_3$) in solid solution are observed (Fig. 8B). Framboidal iron sulfides and greigite (Fig. 8C), produced by the activity of sulfate reducing bacteria, occur between C −0.4 cm and T +4 cm.

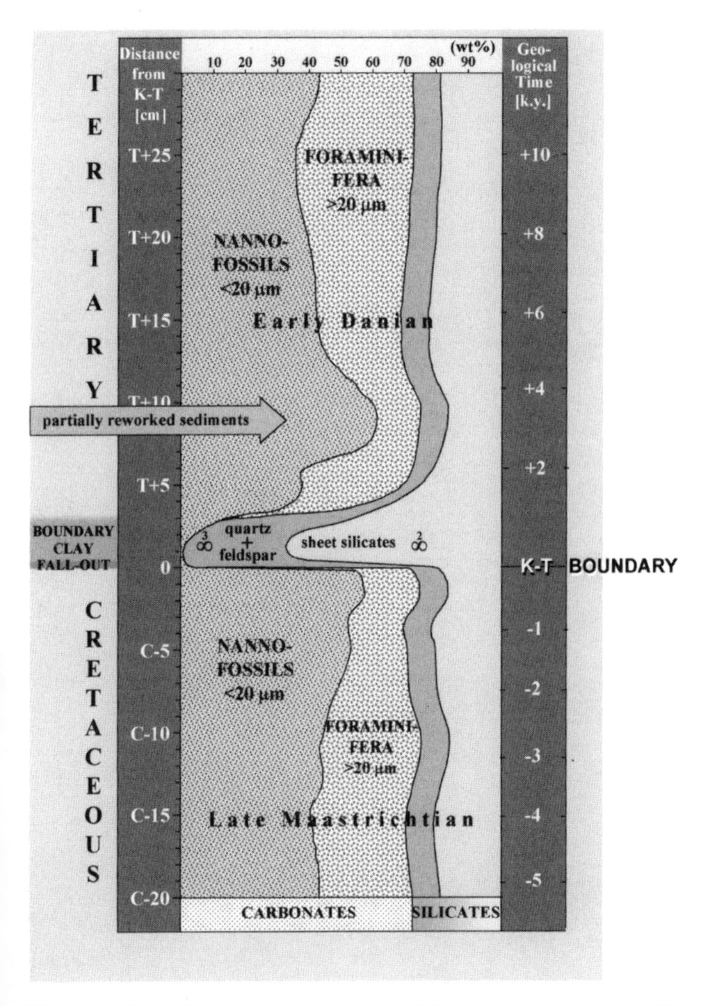

Figure 7. Quantitative phase analysis of Cretaceous-Tertiary (K-T) profile Bjala 2b. Carbonate is calcite; silicates are framework silicates, quartz + feldspar ($\overset{3}{\infty}$), and sheet silicates, smectite, illite, mixed-layer minerals, and kaolinite ($\overset{2}{\infty}$).

Figure 8. A: Dinoflagellate *Thoracospharaera operculata*, survivor after Cretaceous-Tertiary (K-T) event. B: Mg-calcite crystals, authigenically formed after K-T event. C: Framboidal iron sulfides, authigenically formed by activity of sulfate-reducing bacteria.

The $\delta^{13}C$ and $\delta^{18}O$ data for carbonates >20 µm and <20 µm (Fig. 9) in the same range as in Figure 7 show lower values of $\delta^{13}C$ and higher values of $\delta^{18}O$ in the Tertiary, which indicates a lower bioproductivity and a lower water temperature in the Tertiary. The strong negative signal of $\delta^{13}C$ is due to authigenically formed Mg calcites and mixing of surface water with deep water caused by the megatsunami.

Figure 10 shows the grain-size distribution of the carbonate and silicate fractions. These fractions (>20 µm, 8–20 µm, and <8 µm) illustrate the successive deposition of transported grains according to size. In the K-T boundary profile at Bjala 2b the Ir, Cr, and other siderophile elements (Preisinger et al., 1993a) show asymmetric curves (Fig. 11). These can be interpreted as the result of two different processes (Eder and Preisinger, 1987): a short-term component (fallout) and a long-term component (boundary clay). The maximum of these distribution curves and the maximum of K-T spinels were used as reference peaks for pinpointing the K-T boundary.

The fallout is characterized by shocked quartz (curtain ejecta), magnesioferrite spinels (condensed from the hot fireball in the mesosphere and etched by sulfuric acid in the stratosphere), glasses (condensed from the warm fireball and decomposed in the sediment to sheet silicate), reaction products of sulfuric acid (celestine, gypsum), and amorphous materials (soot and organic matter). The amount of fallout material (~3 mm thickness) depends mainly on the distance and direction from the location of the impact event.

The 3-mm-thick fallout layer is overlapped by 2.8 cm of boundary clay composed mainly of material transported from the coast to the place of sedimentation. The boundary clay was produced mainly by the megatsunami caused by the impact.

The paleocoastline north of the Bjala 2b section was reached by the tsunami within 12 h after the impact and the resulting wave flooded the coastal area to an estimated height of 50 m (Matsui et al., 1999).

The boundary clay consists mostly of sheet silicates, quartz, and small amounts of feldspar transported from the flooded coastal area into the sea. A small amount of the boundary clay consists of surviving nannoplankton, such as *Thoracosphaera operculata*, Mg-calcite, and framboidal iron sulfides (Fig. 8, A–C).

K-T spinels

The K-T spinels are discussed separately because of the importance attached to their origin. These Ni-rich K-T spinels contain Ir and Cr, the extraterrestrial origin of which has been proved (Shukolyukov and Lugmair, 1998).

We investigated 197 spinel crystals from the K-T profile Bjala 2b by SEM. The majority of the spinels are euhedral. Most single crystals have octahedral shapes; some have a narrow (110) or (100) face. Only a few crystals are twinned. The size of the octahedra is between 1 and 20 μm; the average is 6 μm. More than half of the investigated crystals of sizes greater than 5 μm have surfaces with etched pits on the (111) face, rarely on the (100) face. Some crystals have a growing peak on the (100) face (Fig. 12).

X-ray analysis of the magnetically separated samples shows them to consist of spinel and framboidal greigite in addition to quartz and accessory amounts of feldspar and mixed-layer clay. The spinel content varies in the K-T boundary section. In T +0.2 cm to T +0.6 cm, T +0.2 cm to C −0.2 cm, and K-T to C −0.4 cm, the samples contain ~0.0004 wt% spinel. In the T +0.1 cm to T +0.3 cm interval they contain ~0.003 wt% spinel. The composition of the spinels from sections T +1.6 cm to C −0.4 cm is given in Table 3. The spinels show high mean values of NiO (~6 wt%) and Cr_2O_3 (~7.8 wt%); the highest Cr_2O_3 content (11 wt%) is from T +0.1 cm to T +0.3 cm.

All the spinels described in this work are magnesioferrites with the space group Fd3m, and the structural formula $8(A^{[4]}B_2{}^{[6]}O_4)$; the lattice parameter of the Bjala K-T spinels is a = 8.356 Å (Aslanian et al., 1996). The cation distribution of the spinels is partially inverse: The substitution of Fe^{3+} in the 6 coordination is indicated by a well-defined negative correlation between FeO and Al_2O_3 + Cr_2O_3 (Fig. 13).

The trace element distribution pattern of the magnetically isolated fractions exhibits some peculiar features. In the interval T +0.2 cm to T +0.6 cm and T +0.1 cm to T +0.3 cm, Ir is enriched in comparison to the raw samples by a factor of 12 (±2) (Table 4). The spinel concentration peak corresponds to the maximum of the Ir content found by INAA. It may therefore be assumed that about half of the Ir is contained in the K-T spinels. The investigated K-T spinels from Bjala 2b are similar to the Ir-bearing K-T spinels from the profile in Caravaca, Spain, identified by Bohor et al. (1986).

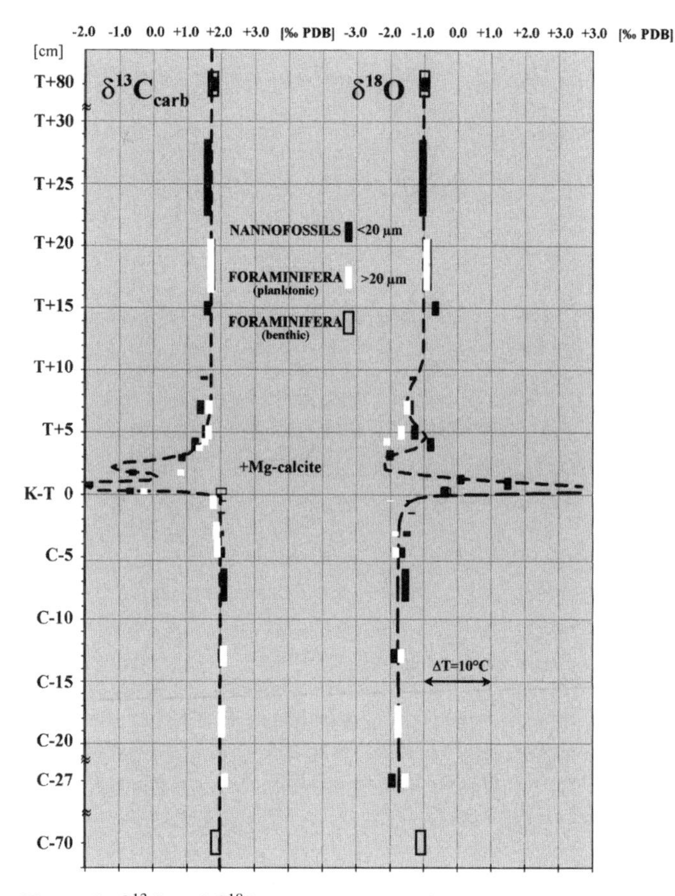

Figure 9. $\delta^{13}C$ and $\delta^{18}O$ measurements of carbonates >20 μm and <20 μm, showing rapid decrease of $\delta^{13}C$ after impact. $\delta^{13}C$ value is generally lower in Tertiary (T) than in Cretaceous (K). $\delta^{18}O$ indicates lower water temperature in Tertiary time. PDB is Peedee belemnite.

Etching experiments on K-T spinels with concentrated H_2SO_4 in two time steps of 20 min each also show a faster decrease of Cr, Ni, Fe, Co, and Zn, while the Ir content decreases slowly. This behavior of Ir in the spinels renders it possible for Ir and the platinum metals to have acted in the hot fireball as condensation nuclei on which Ni-Cr spinels were later condensed. The assumption that condensation occurs in the mesosphere was borne out by experimental investigations by Gayraud et al. (1996). According to these investigations, the formation of such Ni-rich spinels does not occur under conditions typical for Earth's crust. The etched crystal surfaces of most of the spinels could be a consequence of the high H_2SO_4 content in the stratosphere after the impact.

A comparison of spinels of the K-T profile Bjala 2b with those from other K-T sites in the Mediterranean area, such as Caravaca, Spain, and Cerbara, Italy, which are situated at different distances from the impact crater, but approximately in a line of the same direction, is given in Table 5. The mean

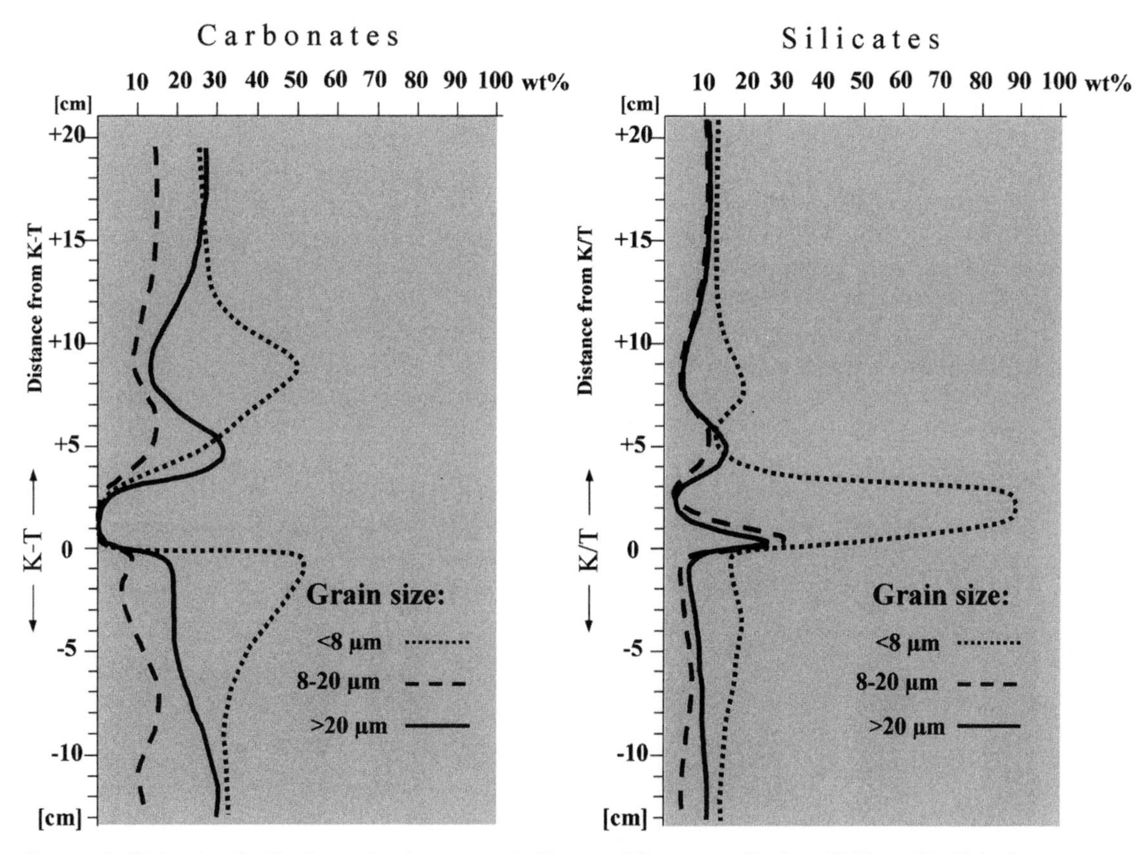

Figure 10. Grain-size distributions of carbonates and silicates of Cretaceous-Tertiary (K-T) profile Bjala 2b.

values of the grain sizes and the weight percentages of spinels are reciprocal to the distance from the impact site.

Partially reworked sediments

The partially reworked sediments are a consequence of sea-level changes. The sea level fell after the K-T impact due to the formation of ice sheets in the Northern and Southern Hemispheres. After melting of the ice, the sea level rose again, reaching the former coastal areas. From these coasts and the backshore areas, the fallout and Cretaceous fossils were washed into the sea. The reworked sediments consist of transported marine sediments, some K-T spinels, a certain amount of Ir and Cr, as well as reworked Cretaceous microfossils and nannofossils (Figs. 7, 10, and 11). The second maximum in these figures is not indicative of a second impact, but clearly shows a partial reworking event.

STRATIGRAPHIC PROFILE OF BJALA 2B

At the K-T boundary, all the species of planktonic foraminifera larger than 150 µm became extinct. Only small ones like *Hedbergella, Guembelitria*, and *Zeanvigerina* are survivors. With calcareous nannoplankton species, the situation is similar: most of them became extinct and only a small group of species survived, e.g., *Thoracosphaera operculata* and *Braarudosphaera bigelowii*, which bloomed at the beginning of the Danian. Only 6 calcareous nannoplankton genera and 3 of 15 genera of planktonic foraminifera survived the K-T event at Bjala 2b. Some extinct species, the survivors, and the first occurrences of new species in the Danian are given in Figure 14.

Very large (1–3 mm) planktonic foraminifera such as *Globotruncana* and *Guembelina* occur in the *Abathomphalus mayaroensis* zone of chronozone 29R. Evidently the final Cretaceous planktonic fauna became extinct while in full development. Only species smaller than 150 µm, such as *Guembelitria cretacea*, survived the K-T crisis (Smit, 1982). The bioproduction of these smaller planktonic foraminifera of the early Paleocene shows minor values compared to the bioproduction of larger species in the Cretaceous under similar ecological conditions.

The calculation of the absolute geological age by means of precessional Milankovitch cycles (22.5 k.y.) based on the date of the K-T boundary (65.0 Ma) gives an absolute time scale for biostratigraphic evolution. The mean accumulation rate connects the sediment thickness with the absolute geological age (Fig. 15). The mean $CaCO_3$ accumulation rate decreases rapidly from 9.5 g/cm^2/k.y. before the K-T event to 4.8 g/cm^2/k.y.

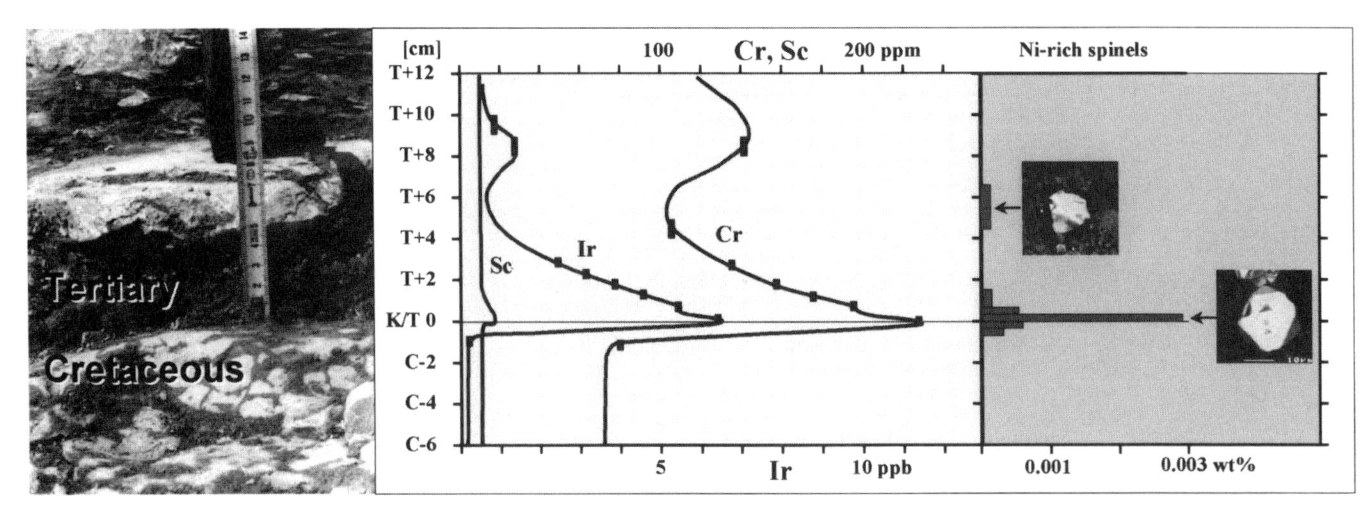

Figure 11. Trace elements Ir, Cr, and Sc showing anomaly of Ir and Cr at Cretaceous-Tertiary (K-T) boundary, and K-T spinels of profile Bjala 2b, as well as second peak of reworked sediment above it.

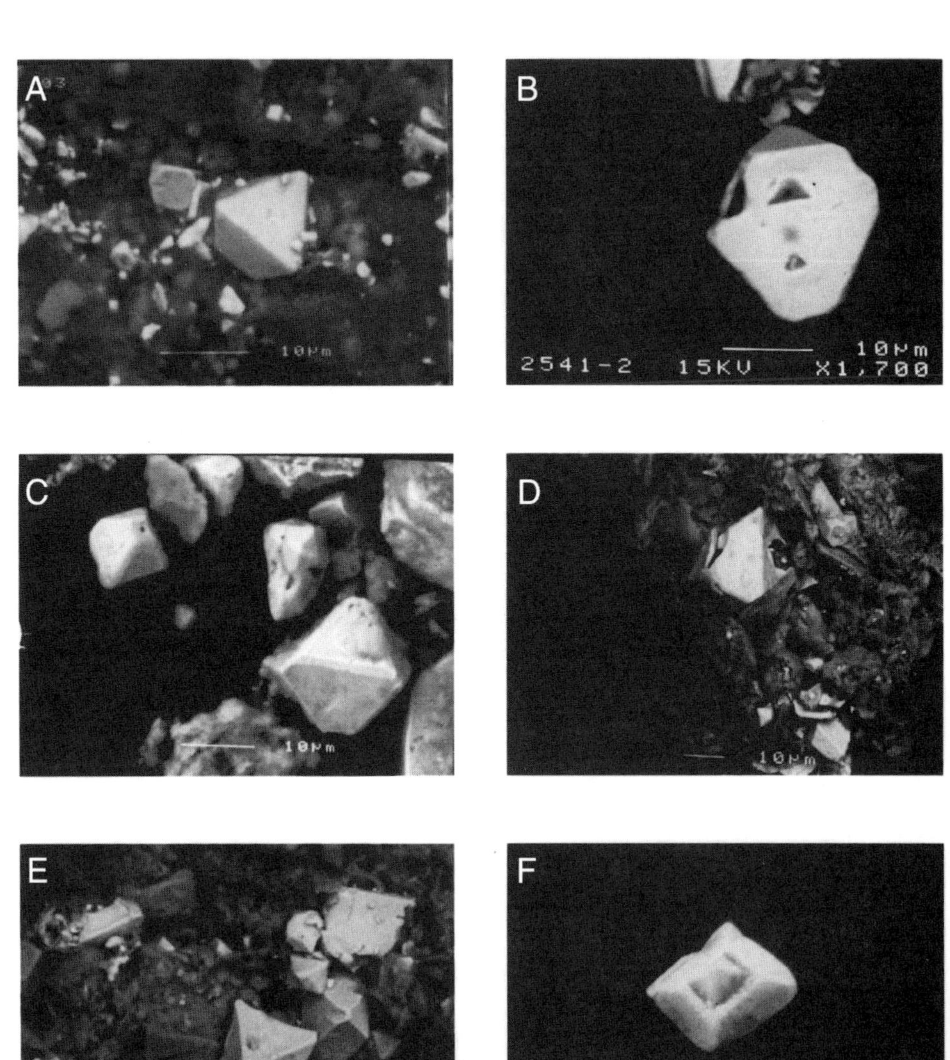

Figure 12. Morphology and growth features of selected magnesioferrite spinels of Cretaceous-Tertiary (K-T) profile Bjala 2b, Bulgaria. A: Spinel twin on {111} with triangular pits on (111). B: Pit on (100) in upper left corner. C–E: Crystals exhibiting characteristic growth defects at edges and corners. F: Octahedral crystal overgrown by second individual crystal in parallel orientation.

TABLE 3. ENERGY-DISPERSIVE SPECTROMETRY ANALYSIS OF SPINELS FROM THE CRETACEOUS-TERTIARY BOUNDARY, BJALA 2B

Components	T +0.6–T +1.6 (cm) wt%	T +0.2–T +0.6 (cm) wt%	T +0.1–T +0.3 (cm) wt%	T +0.2–C-0.2 (cm) wt%	K-T–C-0.4 (cm) wt%
MgO	7.5	8.7	9.6	11.1	9.2
Al_2O_3	4.0	5.3	5.4	6.9	5.5
TiO_2	0.5	0.6	0.5	0.8	0.7
Cr_2O_3	6.8	7.2	11.2	9.7	7.6
MnO	1.1	0.9	0.9	0.8	1.0
FeO	74.1	69.4	65.9	65.3	70.1
NiO	6.0	7.3	5.7	5.4	5.2
ZnO	<0.1	0.6	0.8	<0.1	0.7
Total	100.0	100.0	100.0	100.0	100.0
n	13.0	18.0	29.0	29.0	6.0
⟨Ø⟩ (μm)	4.8	6.8	6.0	9.1	7.4

Note: n, number of analyzed crystals.
FeO, calculated from Fe_{total}.
⟨Ø⟩, average diameter.

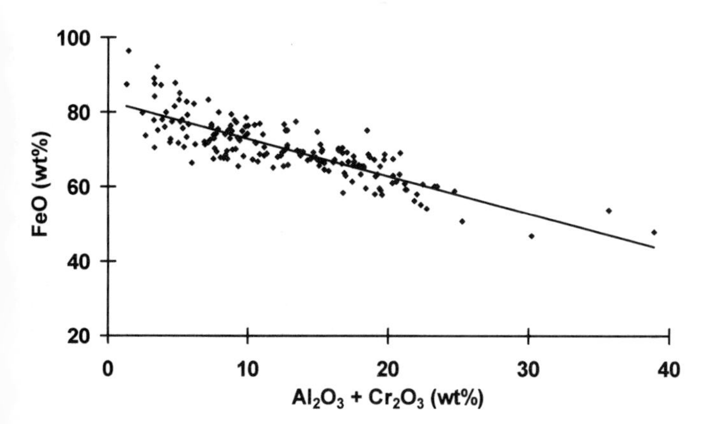

Figure 13. Correlation diagram of FeO (wt%) vs. Al_2O_3 + Cr_2O_3 (wt%) in Cretaceous-Tertiary spinels of Bjala 2b.

within 300 k.y., then slowly to 4.2 g/cm^2/k.y. over a time span of 2.5 m.y. (Fig. 15). Similar results were found in K-T profiles in the South Atlantic (Herbert and D'Hondt, 1990; D'Hondt et al., 1996), in pelagic sediments of the Scaglia Rossa, Italy (Montanari and Koeberl, 2000), and in the Caribbean Sea (Louvel and Galbrun, 2000).

CLIMATIC CHANGES ACROSS THE K-T-BOUNDARY PROFILE

During the Maastrichtian, just after the end of the paleomagnetic normal superchron (80–115 Ma), there was an extreme situation in the Tethys (comparable to the present situation of the western Pacific) produced by hotspot activity and sea-floor spreading. Thus, an ice-free state in the northern polar sea was generated with very high sea level (+340 m, Haq et al., 1987), an atmospheric CO_2 content four times higher than today, and an extreme absolute humidity of 16 g/m^3 (Hay, 1997). Furthermore, the temperature gradient from the equator to the poles was ~22° (D'Hondt and Arthur, 1996). An open east-west gateway (e.g., Indonesian-Australian passage, Himalayan Southeast Asian passage, Tethyan passage, Central American passage [Hay et al., 1996]) allowed free east-west circulation. A deep warm-water current circulated from the equator to the poles. Long-term climate changes were produced by solar insolation (precessional Milankovitch cycles), atmospheric greenhouse gases (especially H_2O and CO_2), paleogeographic changes (land-sea distribution), vegetation (low albedo of forests at high latitudes) (Broecker, 1995). Sudden short-term changes in climate occurred after the K-T event, caused by the effects of a tsunami, shielding of sunlight by solid particles (dust), wildfire, and a short-term sea-level change by glaciation and deglaciation. A climate change had taken place during the late Maastrichtian because of a general lowering of the sea level and an increase of the total shelf area. In addition to short-term changes after the K-T event, a higher erosion rate can be deduced from the higher silicate input at Bjala (Fig. 16). The sedimentary regimes induced by the precessional cycles became more pronounced.

CONCLUSIONS

The K-T profile at Bjala 2b exposes a sequence of rhythmic hemipelagic sediments spanning 4 m.y. An absolute geological time scale could be established by counting the number of sedimentary cycles from chronozones 30N to 27R. Each limestone bed corresponds to one precessional Milankovitch cycle of an average duration of 22.5 k.y. The K-T boundary profile is in chronozone 29R, which comprises 27 ± 1 cycles, 15 ± 1 below and 12 above the K-T boundary. The polarity reversal between chronozones 29R and 29N is located 6.70 m above the K-T boundary.

The time span between the K-T impact and the end of

TABLE 4. INSTRUMENTAL NEUTRON-ACTIVATION ANALYSIS, DATA FOR BULK SAMPLES AND MAGNETIC FRACTIONS OF CRETACEOUS-TERTIARY BOUNDARY, BJALA 2B

	T +0.2–T +0.6 (cm)		T +0.1–T +0.3 (cm)		T +0.2–C-0.2 (cm)	
Elements	Bulk sample	Magnetic fraction	Bulk sample	Magnetic fraction	Bulk sample	Magnetic fraction
Sc (ppm)	10.0	61.0	15.1	29.7	28.9	12.7
Cr (ppm)	98.2	1120	159.4	4286	278.0	1614
Fe (%)	2.6	17.9	4.0	9.1	6.3	3.6
Co (ppm)	16.6	207.5	32.6	87.3	30.0	25.0
Ni (ppm)	145.0	8858	242.0	2874	233.0	800
Zn (ppm)	724.0	9856	637.0	4719	750.0	1542
Ir (ppb)	3.0	40.8	5.5	57.1	1.4	3.2
Th (ppm)	6.0	12.3	8.7	7.3	15.9	7.9

Note: The bulk sample contents are calculated for decalcified samples.

TABLE 5. AVERAGE COMPOSITION OF CRETACEOUS-TERTIARY SPINELS BY ENERGY-DISPERSIVE SPECTROMETRY ANALYSIS

Components	Bjala, Bulgaria		Cerbara, Italy			Caravaca, Spain	
	n = 197 wt%	σ	n = 24 wt%$^{\times}$	n = 22 wt%$^{+}$	$\sigma^{\times,+}$	n = 115 wt%	σ
MgO	8.8	3.7	9.3	9.8	4.8	9.7	3.4
Al_2O_3	4.7	2.9	3.0	4.6	2.5	6.0	2.4
TiO_2	0.7	1.6	1.4	0.3	0.3	0.6	0.5
Cr_2O_3	7.8	4.8	3.5	8.1	5.1	9.2	4.1
MnO	0.9	0.8	1.5	0.7	1.0	0.6	0.4
FeO	70.2	7.9	75.8	69.5	7.3	67.9	7.8
NiO	6.0	2.6	5.6	6.0	2.8	5.3	2.2
Total	99.1		100.1	99.0		99.3	
⟨Ø⟩ (μm)	6.0		7.0	9.4		22.5	
Spinel (wt%)	0.004		0.08	0.02		0.11	
D (km)	9500		8100	8100		6700	

Note: n, number of analyzed crystals.
wt%, average value of the component.
×, spinels in the red layer (7,5 mm).
+, spinels in the green layer (2,5 mm).
σ, standard deviation.
$\sigma^{\times,+}$, standard deviation of spinels of the total layer (10 mm).
FeO, calculated from Fe_{total}.
⟨Ø⟩, the average diameter.
Spinel, the content of spinels, calcualted for decalcified samples.
D, the palaeodistance from the impact crater at Chicxulub, Mexico.

chronozone 29R can be calculated to 270 k.y. Most of the calcareous nannoplankton genera and planktonic foraminifera genera became extinct through the K-T event. Only 6 calcareous nannoplankton genera and 3 of 15 genera of planktonic foraminifera survived the K-T event at Bjala 2b. After the K-T event, the first occurrence of new species of nannoplankton and foraminifera is in chronozone 29R.

The short-term fallout at the K-T boundary of Bjala 2b is characterized by shocked quartz grains and anomalies of extraterrestrial Ir, Cr, and other siderophile elements, as well as K-T spinels (Ni-rich magnesioferrites with variable Cr contents and etched pits). The longer term sedimentation, evidenced by the boundary clay at the K-T profile of Bjala 2b, consists mainly of sheet silicates, quartz and feldspar, small amounts of Mg-calcites, and framboidal iron sulfides as well as surviving nannoplankton and foraminifera. The accumulation in the Tertiary between T +5 cm and T + 10 cm is evidenced by considerable reworking of fallout with Ir, Cr, and K-T spinels, boundary clay, and Cretaceous microfossils.

The changes in magnetic polarities in the range of chronozones 30N to 27R were found to correspond to minima of the $CaCO_3$ content and maxima of magnetic susceptibility and extraterrestrial spinels. The intensity of Earth's magnetic field decreases significantly during the change of the dipole axis of Earth, so that screening against cosmic radiation and particles is reduced.

The presence of extraterrestrial Ni-rich magnesioferrites during the time span of the polarity change indicates a higher influx of cosmic particles and radiation on Earth. This may lead to a higher mutation rate of organisms. Therefore the polarity reversals are thought to be important for the evolution of new species.

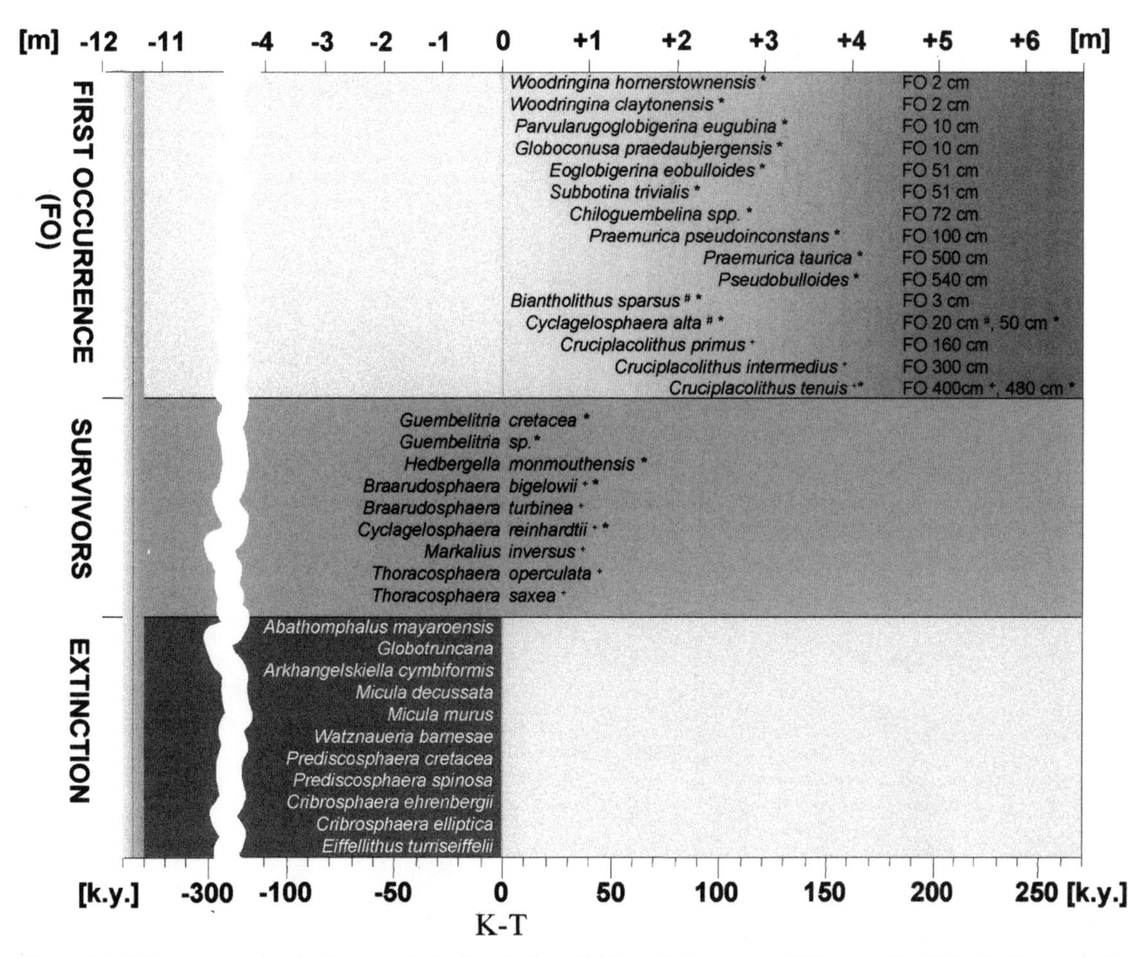

Figure 14. Calcareous nannoplanktons and planktonic foraminifera of chronozone 29R at profile Bjala 2b. Nannoplankton: asterisk indicates as determined in Rögl et al. 1996. Plus sign indicates as determined in Preisinger et al., 1993a, 1993b. Pound sign indicates as determined in Ivanov and Stoykova, 1994. Foraminifera: asterisk indicates as determined in Rögl et al., 1996.

The presence of Ni-rich magnesioferrite is not only a marker for impact events, but also for a minimum in the intensity of the Earth's magnetic field.

ACKNOWLEDGMENTS

This report was supported by the Austrian Ministry of Science and Research BMWF contract GZ.45134, under the auspices of the Austrian Academy of Sciences, and the Austrian Science Foundation FWF Project 12643-GEO. We thank A.G. Fischer and D.T. King Jr. for their careful reviews.

REFERENCES CITED

Alvarez, L.W., Alvarez, W., Asaro, F., and Michel, H.V., 1980, Extraterrestrial cause for the Cretaceous-Tertiary extinction: Science, v. 208, p. 1095–1108.

Aslanian, S., Preisinger, A., and Petras, L., 1996, Spinel formation in the mesosphere of the earth after Cretaceous/Tertiary impact: International Union of Crystallography 17th Congress, Seattle, Washington, Collected Abstracts, p. C-335.

Berger, A., 1978, Long term variations of daily insolution and quaternary climatic changes: Journal of Atmospheric Science, v. 35, n. 2, p. 2362–2367.

Berger, A., Loutre, M.F., and Dehant, V., 1989, Astronomical frequencies for pre-Quaternary palaeoclimate studies: Terra Nova, v. 1, p. 474–479.

Bohor, B.F., Foord, E.E., and Ganapathy, R., 1986, Magnesioferrite from the Cretaceous-Tertiary boundary, Caravaca, Spain: Earth and Planetary Science Letters, v. 81, p. 57–66.

Bončev, E., 1974, Regions of alpine folding: The Bulgarian Carpathian-Balkan area, the Balkanides, *in* Mahel, M., ed., Tectonics of the Carpathian Balkan Regions: Explanations to the tectonic map of the Carpathian-Balkan regions and their foreland: Bratislava, Slovakia, Geological Institute "Dionyz Stur," p. 307–308.

Broecker, W., 1995, The glacial world according to Wally (revised edition): New York, Eldigio Press, 318 p.

Cande, S.C., and Kent, D.V., 1995, Revised calibration of the geomagnetic polarity timescale for the Late Cretaceous and Cenozoic: Journal of Geophysical Research, v. 100, p. 6093–6095.

Claeys, P., Kiessling, W., and Alvarez, W., 2000, Global distribution of Chicxulub ejecta: Catastrophic events and mass extinctions: Impacts and

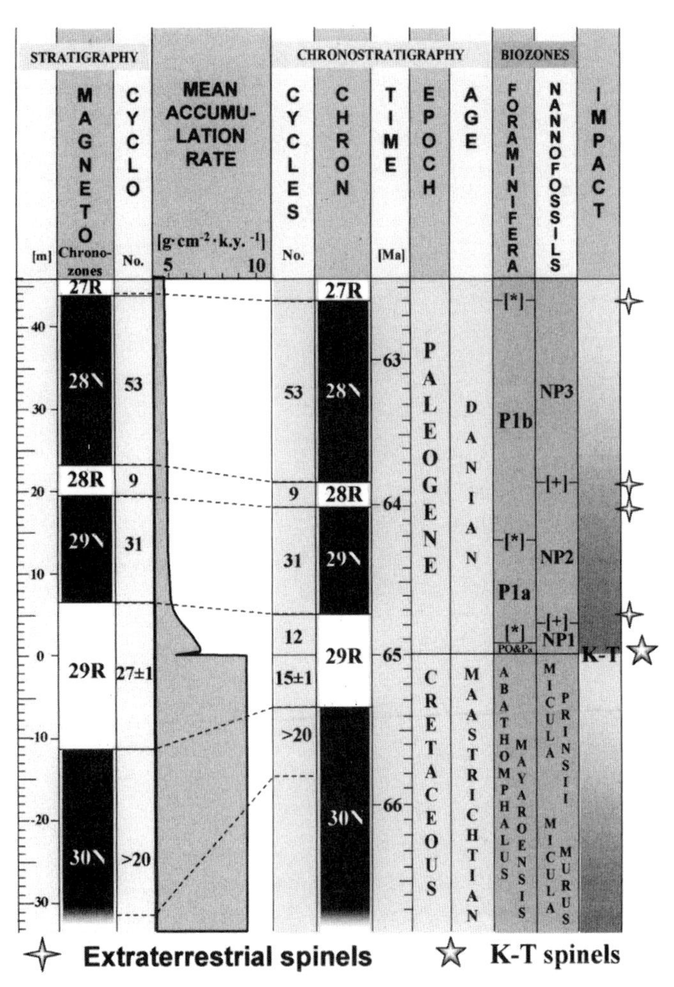

Figure 15. Stratigraphic profile Bjala 2b. Left to right: distance in meters measured above and below Cretaceous-Tertiary (K-T) boundary. Magnetostratigraphy shows chronozones determined by H.J. Mauritsch and R. Scholger. Cyclostratigraphy based on number of Milankovitch cycles. Mean $CaCO_3$ accumulation rate of hemipelagic sediment. Chronostratigraphy correlating magnetozones with absolute geological time. Biozones after F. Rögl (*) and K. Stoykova (+). Occurrences of spinels are marked on right.

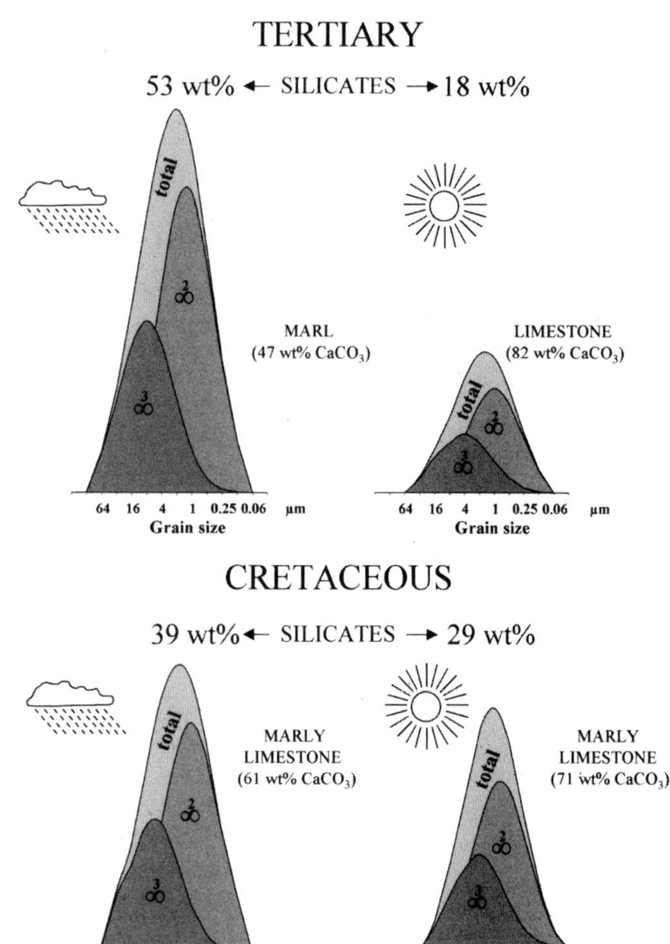

Figure 16. Grain-size distributions of transported silicates in Tertiary (limestone and marl) and in Cretaceous (marly limestones). Total area corresponds to indicated weight percent of silicates: $\overset{3}{\infty}$ indicates quartz + feldspar, $\overset{2}{\infty}$ indicates sheet silicates.

beyond: Houston, Texas, Lunar and Planetary Institute, LPI Contribution No. 1053, p. 26–27.

D'Hondt, S., and Arthur, M., 1996, Late Cretaceous oceans and the cool tropic paradox: Science, v. 271, p. 1838–1841.

D'Hondt, S., Herbert, T.D., King, J., and Gibson, C., 1996, Planktic foraminifera, asteroids, and marine production: Death and recovery at the Cretaceous-Tertiary boundary, *in* Ryder, G., Fastovsky, D., and Gartner, S., eds., The Cretaceous-Tertiary event and other catastrophes in Earth history: Boulder, Colorado, Geological Society of America Special Paper 307, p. 303–317.

Eder, G., and Preisinger, A., 1987, Zeitstruktur globaler Ereignisse veranschaulicht an der Kreide/Tertiär Grenze: Naturwissenschaften, v. 74, p. 35–37.

Gayraud, J., 1995, Synthèse du spinelle nickélifère: études cinétique, thermodynamique et application à l'étude de la limite Crétacé-Tertiaire [Ph.D.thesis]: Paris, Université de Paris-sud, 303 p.

Gayraud, J., Robin, E., Rocchia, R., and Froget, L., 1996, Formation conditions of oxidized Ni-rich spinel and their relevance to the K/T boundary event, *in* Ryder, G., Fastovsky, D., and Gartner, S., eds., The Cretaceous-Tertiary event and other catastrophes in Earth history: Boulder, Colorado, Geological Society of America Special Paper 307, p. 425–443.

Gwozdz, R., Hansen, H.J., and Rasmussen, K.L., 1992, 40 kg sample of Fish Clay from Stevns Clint, Denmark: Meteoritics, v. 27, p. 228–234.

Haq, B.U., Hardenbol, J., Vail, P.R., 1987, Chronology of fluctuating sea levels since the Triassic: Science, v. 235, p. 1156–1167.

Harland, W.B., Armstrong, R.L., Cox, A.V., Craig, L.E., Smith, A.G., and Smith, D.G., 1990, A geologic time-scale: Cambridge, Cambridge University Press, 246 p.

Hay, W.W., De Conto, R.M., and Wold, Ch.N., 1997, Climate: Is the past the key to the future?: Geologische Rundschau, v. 86, p. 471–491.

Herbert, T.D., and D'Hondt, S.L., 1990, Precessional climate cyclicity in Late Cretaceous-Early Tertiary marine sediments: A high resolution chronometer of Cretaceous-Tertiary boundary events: Earth and Planetary Science Letters, v. 99, p. 263–275.

Hildebrand, A.R., Penfield, G.T., Kring, D.A, Pilkington, M., Camargo Z.A., Jacobsen, S.B., and Boynton, W.V., 1991, Chicxulub crater: A possible Cretaceous/Tertiary boundary impact crater on the Yucatan Peninsula, Mexico: Geology, v. 19, p. 867–871.

Ivanov, M., and Stoykova, K., 1994, Cretaceous/Tertiary boundary in the area of Bjala, eastern Bulgaria: Biostratigraphic results: Geologica Balcanica, v. 24, no. 6, p. 3–22.

Lindinger, M., 1988, The Cretaceous/Tertiary boundaries of El Kef and Caravaca: Sedimentological, geochemical and clay mineralogical aspects [Ph.D.thesis]: Zürich, Swiss Federal Institute of Technology, 253 p.

Louvel, V., and Galbrun, B., 2000, Magnetic polarity sequences from downhole measurements in ODP Holes 998B and 1001A, Leg 165, Caribbean Sea: Marine Geophysical Researches, v. 21, no. 6, p. 561–577.

Matsui, T., Imamura, F., Tajika, E., Nakamo, Y., and Fujisava, Y., 1999, K/T impact tsunami: Lunar and Planetary Science, v. 30, Abstract #1527, CD-ROM.

Merrill, R.T., McElhinny, M.W., and McFadden, P.L., 1998, The magnetic field of the Earth: London, Academic Press, International Geophysics Series, v. 63, 531 p.

Montanari, A., and Koeberl, C., 2000, Impact stratigraphy: The Italian record, *in* Bhattacharji, S., Friedman, G.M., Neugebauer H.J., and Seilacher, A., eds., Lecture notes in Earth Sciences, v. 93: Berlin, Springer-Verlag, 364 p.

Müller, G., and Gastner, M., 1971, The "Karbonat-Bombe", a simple device for determination of the carbonate content in sediments, soils, and other materials: Neues Jahrbuch für Mineralogie, v. 10, p. 466–469.

Oberhänsli, H., 1996, Klimatische und ozeanographische Veränderungen im Eozän: Zeitschrift deutsche geologische Gesellschaft, v. 147, p. 303–413.

Pope, K.O., Baine, K.H., Ocampo, A.C., and Ivanov, B.A., 1997, Energy, volatile production, and climatic effects of the Chicxulub Cretaceous-Tertiary impact: Journal of Geophysical Research, v. 102, p. 21 645–21 664.

Preisinger, A., 2000, Die Kreide/Tertiär-Grenze, *in* Events und Evolution, Katastrophen und Entwicklung in der Erdgeschichte: Geoschule Paybach, Barbara-Gespräche 1997, v. 4, p. 57–72.

Preisinger, A., Zobetz, E., Gratz, A.J., Lahodynsky, R., Becke, M., Mauritsch, H.J., Eder, G., Grass, F., Rögl, F., Stradner, H., and Surenian, R., 1986, The Cretaceous/Tertiary boundary in the Gosau Basin, Austria: Nature, v. 322, p. 794–799.

Preisinger, A., Aslanian, S., Grass, F., Mauritsch, H.J., and Stoykova, K., 1991, Cretaceous-Tertiary boundary on the coast of the Black Sea (Bulgaria) [abs.]: Event markers in the earth history, Joint Meeting of IGCP Projekt 216: Global Biological Events in Earth History, IGCP Projekt 293: Geochemical Event Markers in the Phanerozoic, IGCP Projekt 303: Precambrian/Cambrian Event Stratigraphy, Calgary, Alberta, Canada, p. 303.

Preisinger, A., Aslanian, S., Stoykova, K., Grass, F., Mauritsch, H.J., and Scholger, R., 1993a, Cretaceous/Tertiary boundary sections on the coast of the Black Sea near Bjala (Bulgaria): Palaeogeography, Palaeoclimatology, Palaeoecology, v. 104, p. 219–228.

Preisinger, A., Aslanian, S., Stoykova, K., Grass, F., Mauritsch, H.J., and Scholger, R., 1993b, Cretaceous/Tertiary boundary sections in the East Balkan area, Bulgaria: Geologica Balcanica, v. 23, no. 5, p. 3–13.

Preisinger, A., Aslanian, S., Brandstätter, F., and Grass, F., 1996, Distribution of K/T-spinels in the Mediterranean area: International Workshop Postojna, Slovenia, The Role of Impact Processes in the Geological and Biological Evolution of Planet Earth, Geology of West Slovenia Field Guide, Abstracts, p. 65.

Preisinger, A., Aslanian, S., Brandstätter, F., and Grass, F., 2000, Geomagnetic polarity reversals from Chron 29R to Chron 27N in marine sediments near Bjala, Bulgaria: Lunar and Planetary Science, v. 31, Abstract #1985, CD-ROM.

Rögl, F., Von Salis, K., Preisinger, A., Aslanian, S., and Summesberger, H., 1995, A continuous Cretaceous/Paleogene boundary section near Bjala, Bulgaria: International Subcommission on Paleogene Stratigraphy, Newsletter no. 5, p. 11–17.

Rögl, F., Von Salis, K., Preisinger, A., Aslanian, S., and Summesberger, H., 1996, Stratigraphy across the Cretaceous/Paleogene boundary near Bjala, Bulgaria, *in* Jardine, S., Klasz, J., and de Delenay, J.-P., eds., Geologie de l'Afrique et de l'Atlantique sud: Actes Colloques Angers, 1994, Elf Aquitaine Edition, Angers, France, p. 673–683.

Schultz, P.H., and D'Hondt, S., 1996, Cretaceous-Tertiary (Chicxulub) impact angle and its consequences: Geology, v. 24, p. 963–976.

Schwartz, B.B., and Foner, S., 1977, Superconductor applications: SQUIDS and Machines: New York, Plenum Press, 321 p.

Sharpton, V.L., Burke, K., Camargo-Zanoguera, A., Hall, S.A., Lee, D.S., Marin, L.E., Suares-Reynoso, G., Quezada-Muneton, J.M., Spudis, P.H., and Urrutia-Fucugauchi, J., 1993, Chicxulub multiring impact basin: Size and other characteristics derived from gravity analysis: Science, v. 261, p. 1564–1567.

Shukolyukov, A., and Lugmair, G. W., 1998, Isotope evidence for the Cretaceous Tertiary impactor and its type: Science, v. 282, p. 927–929.

Smit, J., 1982, Extinction and evolution of planktonic foraminifera at the Cretaceous/Tertiary boundary after a major impact, *in* Silver, L.T., and Schultz, P.H., eds., Geological implications of impacts of large asteroids and comets on the earth: Geological Society of America Special Paper 190, p. 329–352.

Smit, J., and Hertogen, T, 1980, An extraterrestrial event at the Cretaceous/Tertiary boundary: Nature, v. 200, p. 198–200.

Stradner, H., Eder, G., Grass, F., Lahodynsky, R., Mauritsch, H.J., Preisinger, A., Rögl, F., Surenian, R., Zeiss, W., and Zobetz, E., 1987, New K/T boundary sites in the Gosau Formation of Austria [abs.]: Terra Cognita, v. 7, p. 212.

Swisher, C.C., III, Grajales-Nishimura, J.M., Montanari, A., Margolis, S.V., Claeys, P., Alvarez, W., Renne, P., Cedillo-Pardo, E., Maurrasse, F.J.-M.R., Curtis, G.H., Smit, J., and McWilliams, M.O., 1992, Coeval ^{40}Ar/^{39}Ar Ages of 65.0 million years ago from Chicxulub crater melt rocks and K/T boundary tektites: Science, v. 257, p. 954–958.

Ward, W.C., Keller, G., Stinnesbeck, W., and Adatte, T., 1995, Yucatan subsurface stratigraphy: Implications and constrains for the Chicxulub impact: Geology, v. 23, p. 873–876.

Manuscript Submitted October 25, 2000; Accepted by the Society March 22, 2001

Geological Society of America
Special Paper 356
2002

Paleoenvironment across the Cretaceous-Tertiary transition in eastern Bulgaria

Thierry Adatte*
Institut de Géologie, Université Neuchâtel, CH-2007 Neuchâtel, Switzerland
Gerta Keller
Department of Geosciences, Princeton University, Princeton, New Jersey 08544, USA
S. Burns*
Geologisches Institut, Universität Bern, Baltzerstrasse 1, 3012 Bern, Switzerland
K.H. Stoykova
Geological Institute of Bulgarian Academy of Sciences, 113 Sofia, Bulgaria
M.I. Ivanov
D. Vangelov
Sofia University St. Kliment Okhridski, 1000 Sofia, Bulgaria
Utz Kramar
Doris Stüben
Institut für Petrographie und Geochemie, Universität Karlsruhe, D-76128 Karlsruhe, Germany

ABSTRACT

The Cretaceous-Tertiary (K-T) transition in eastern Bulgaria (Bjala) was analyzed in terms of lithology, mineralogy, stable isotopes, trace elements, and planktic foraminifera. The sequence represents a boreal-Tethyan transitional setting, spans from the last 300 k.y. of the Maastrichtian (zone CF1) through the early Danian (zones P0-Plc), and contains several short hiatuses. It differs from low-latitude Tethyan sequences primarily by lower diversity assemblages, pre-K-T faunal changes, a reduced K-T $\delta^{13}C$ shift, and the presence of two clay layers with platinum group element anomalies. The first clay layer marks the K-T boundary impact event, as indicated by an iridium anomaly (6.1 ppb), the mass extinction of tropical and subtropical planktic foraminifera, and cooling. The second clay layer is stratigraphically within the upper *Parvularugoglobigerina eugubina* (Pla) zone and contains a small Ir enrichment (0.22 ppb), a major Pd enrichment (1.34 ppb), and anomalies in Ru (0.30 ppb) and Rh (0.13 ppb) that suggest a volcanic source.

INTRODUCTION

Cretaceous-Tertiary (K-T) boundary studies have often concentrated on documenting the geochemical anomalies and the pattern of planktic foraminiferal species extinctions immediately below and above this boundary event in low latitudes (Fig. 1). Most of these studies reveal a major mass extinction of all tropical and subtropical species at or near the K-T boundary and iridium anomalies that have been variously attributed to the sole effects of an impact (e.g., Smit, 1990; D'Hondt et

*E-mail: thierry.adatte@geo.unine. Present address, Burns: Department of Geosciences, 233 Morril Science Center, University of Massachusetts, Amherst, Massachusetts 01003.

Adatte, T., Keller, G., Burns, S., Stoykova, K.H., Ivanov, M.I., Vangelov, D., Kramar, U., and Stüben, D., 2002, Paleoenvironment across the Cretaceous-Tertiary transition in eastern Bulgaria, *in* Koeberl, C., and MacLeod, K.G., eds., Catastrophic Events and Mass Extinctions: Impacts and Beyond: Boulder, Colorado, Geological Society of America Special Paper 356, p. 231–251.

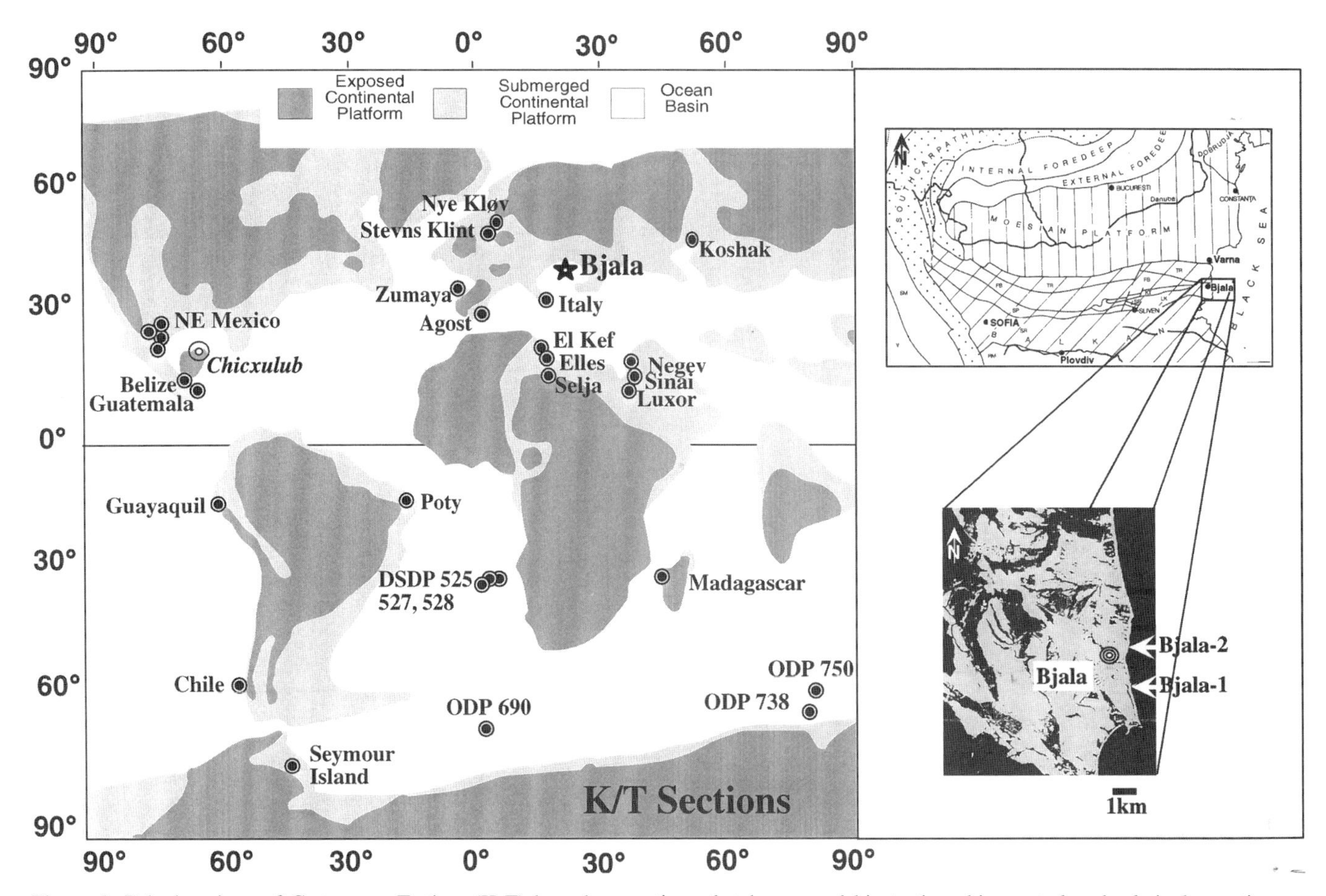

Figure 1. Paleolocations of Cretaceous-Tertiary (K-T) boundary sections that have good biostratigraphic control and relatively continuous sediment records. Note that current knowledge of K-T boundary event is primarily from sections in low-latitude northern Tethys, including Mexico, Spain, and Tunisia. Few K-T sections are currently known from middle and high latitudes. Bjala sections of Bulgaria provide critical information of this event in transitional environment between northern Tethys and boreal sea. Inset shows satellite photograph of Black Sea coast with Bjala locality, and simplified tectonic map of South Carpathian arc (modified from Preisinger et al., 1993a; Rögl et al., 1996). Ocean basins are white, continental platform is gray, continents are black (modified from MacLeod and Keller, 1991a). ODP is Ocean Drilling Program, DSDP is Deep Sea Drilling Project.

al., 1996; Apellaniz et al., 1997; Olsson, 1997), or long-term environmental changes followed by an impact (e.g., Keller, 1988, 1996; Luciani, 1997; Abramovich et al., 1998). Few studies have attempted to evaluate some aspects of the K-T event on a regional or global scale, including studies of hiatus distributions (MacLeod and Keller, 1991a, 1991b), species survivorship, and records of pre-K-T extinctions (MacLeod and Keller, 1994; Abramovich et al., 1998; Pardo et al., 1999). The absence of more comprehensive integrated summary studies is largely because most K-T sections are separated by large distances (Fig. 1), and direct comparisons across latitudes are difficult due to little known regional effects, particularly in middle to high latitudes.

The mass-extinction pattern of planktic foraminifera in high latitudes is quantitatively documented from a few localities in the southern ocean (e.g., Deep Sea Drilling Project [DSDP] Sites 690 and 738, Keller, 1993) and Northern Hemisphere (e.g., Denmark and Kazakhstan, Schmitz et al., 1992; Keller et al., 1993; Pardo et al., 1999; Fig. 1). These data suggest a diminished mass-extinction effect compared with low latitudes primarily because tropical and subtropical species are absent and assemblages are dominated by small species that tolerate environmental fluctuations, including *Guembelitria cretacea, G. danica, G. trifolia, G. dammula, Heterohelix globulosa, H. navarroensis, H. planate, Hedbergella holmdelensis, H. monmouthensis*, and *Globigerinelloides aspera* (MacLeod and Keller, 1994). Most of these species are also observed in Danian sediments of low latitudes, and between 4 and 12 species have been considered as survivors (e.g., Canudo, 1997; Keller, 1997; Masters, 1997; Olsson, 1997; Orue-etxebarria, 1997; Smit and Nederbragt, 1997; Luciani, 1997). Stable isotopes measured on some of these species in Danian sediments have been found to record Danian values, also suggesting that they are survivors (Barrera and Keller, 1994; Keller et al., 1993; MacLeod and Keller, 1994). Huber (1996) attributed the presence of these species in Danian sediments to reworking, citing relatively unchanged stable isotope values of *Globigerinelloides multispinus* across the K-T boundary at Site 738 as evidence (see also Huber et al., 1994; Keller and MacLeod, 1994). We do not argue that Danian sediments may contain reworked Maastrichtian species, as indeed any sedimentary interval may, but we disagree that the consistent presence of these small species in sections across latitudes can be attributed to reworking, and that the Danian isotopic signal of species can be explained as artifact of reworking.

We suggest that these species are K-T survivors for two major reasons. (1) They are ecological generalists able to survive the K-T environmental changes, whereas the larger complex tropical and subtropical species are not. (2) Environmental effects of the K-T impact diminished into higher latitudes, as suggested by $\delta^{13}C$ values that indicate a relatively minor decrease in primary productivity in high latitudes as compared with low latitudes (e.g., Keller and Lindinger, 1989; Zachos et al., 1989; Keller et al., 1993; Barrera and Keller, 1994).

Environmental changes and the nature of the mass extinction between the extremes of the northern boreal sea (Denmark and Kazakhstan) and the low-latitude Tethys are still relatively unknown. This shortcoming is because K-T sequences with nearly continuous sedimentation and good microfossil preservation are very rare and currently only reported from Bjala in eastern Bulgaria (Preisinger et al., 1993a, 1993b; Rögl et al., 1996). We chose to study the Bjala sections in order to obtain a quantitative record that would allow evaluation of the mass extinction and environmental changes in this transitional environment and permit comparison with both low- and high-latitude records. To this end we analyzed the biostratigraphy, stable isotopes, bulk rock compositions, clay minerals, trace element abundances, and quantitative changes among planktic foraminiferal assemblages.

PREVIOUS STUDIES IN BULGARIA

The K-T boundary in eastern Bulgaria was first recognized on the basis of planktic foraminifera from boreholes (Juranov and Dzhuranov, 1983), and a subsequent search for outcrops in the vicinity of Bjala revealed a relatively complete K-T transition based on calcareous nannoplankton (Stoykova and Ivanov, 1992; Ivanov and Stoykova, 1994; Sinnyovsky and Stoykova, 1995). Geochemical and planktic foraminiferal data were given in Preisinger et al. (1993a, 1993b) and Rögl et al. (1996).

In those publications the K-T biostratigraphy is based primarily on calcareous nannoplankton, the most complete sequences containing the *Micula prinsii* zone in the topmost 14–17 m of the Maastrichtian, and zone NP1, spanning the basal 4 m of the Danian. The K-T boundary was recognized on the basis of a 2–3-cm-thick dark clay layer containing an Ir anomaly of 6.1 ppb (Preisinger et al., 1993a). A second, smaller Ir and Co enrichment was recognized 7–8 cm above this interval in a marly limestone layer, and was considered reworked. The first Danian nannofossil species *Biantholithus sparsus* and *Cyclagellosphaera alta* are reported from this marly limestone layer that also contains blooms of *Thoracosphaera operculata* and *Braarudosphaera bigelowii* (Preisinger et al., 1993a, 1993b; Ivanov and Stoykova, 1994). Rögl et al. (1996) reported the range of four Maastrichtian planktic foraminifer species (*Hedbergella monmouthensis, Guembelitria cretacea, Racemiguembelina intermedia, Abathomphalus mayaroensis*) and six undifferentiated genera in the 90 cm below the K-T boundary. They observed the first Danian species (*Woodringina hornerstownensis, W. claytonensis, Globoconusa predaubjergensis*) 2 cm above the base of the K-T clay layer.

GEOLOGICAL SETTING AND LOCATION

The studied outcrops are exposed along the Black Sea coast, close to the town of Bjala (Fig. 1). Tectonically, this area belongs to the Luda Kamchia unit of the Stara Planica zone, which is part of the High Balkan mountain range (Ivanov, 1983, 1988). The Bjala area was a well-differentiated basin characterized by rhythmic sedimentation of hemipelagic marls and marly limestones during the Late Cretaceous. Deposition of more detrital and turbiditic sediment began in the early Paleocene and appears to reflect the first pulse of alpine tectonic activity. Subsequent tectonic activity resulted in the numerous nappes, thrust folds, and faults that can now be observed in the area (Ivanov, 1988).

Bjala-1 is located 800 m north of a trench leading down from the town of Bjala to the beach (Figs. 1 and 2). This section corresponds to the Bjala 2b section of Ivanov and Stoykova (1994) and Preisinger (1994), and can be observed laterally over 15–20 m. This locality represents the best exposure and structurally least disturbed outcrop of the K-T transition known to date in Bulgaria. The area surrounding Bjala-1 is strongly tectonized with steep east-west faults and vertical displacements of 10–20 m. The beds dip 20°–30° southwest (Figs. 2A and 3A). About 2 km to the north is the Bjala-2 section (Figs. 1 and 3B), located along the beach close to the Bjala River outlet (near the Bjala-2 locality of Ivanov and Stoykova, 1994; Preisinger, 1994). This section is more tectonically disturbed than Bjala-1, and small-scale faults cut the K-T boundary clay layer, making it difficult to trace laterally over any distance.

METHODS

We collected 107 samples at Bjala-1 (Fig. 3A) at 15–20 cm intervals for the first 5 m of the section, 5–10 cm intervals for the K-T transition, and 20–30 cm intervals for the 8 m of lower Paleocene sediments. Due to the tectonic disturbance at the Bjala-2 outcrop, only 15 samples were collected at that locality (Fig. 3B). In the field, the sections were measured and the lithology was examined and described with particular emphasis on structural disturbance (faults and folds), bioturbation, trace fossils, macrofossils, and erosion surfaces (e.g., undulating surfaces, clasts, truncated trace-fossil burrows).

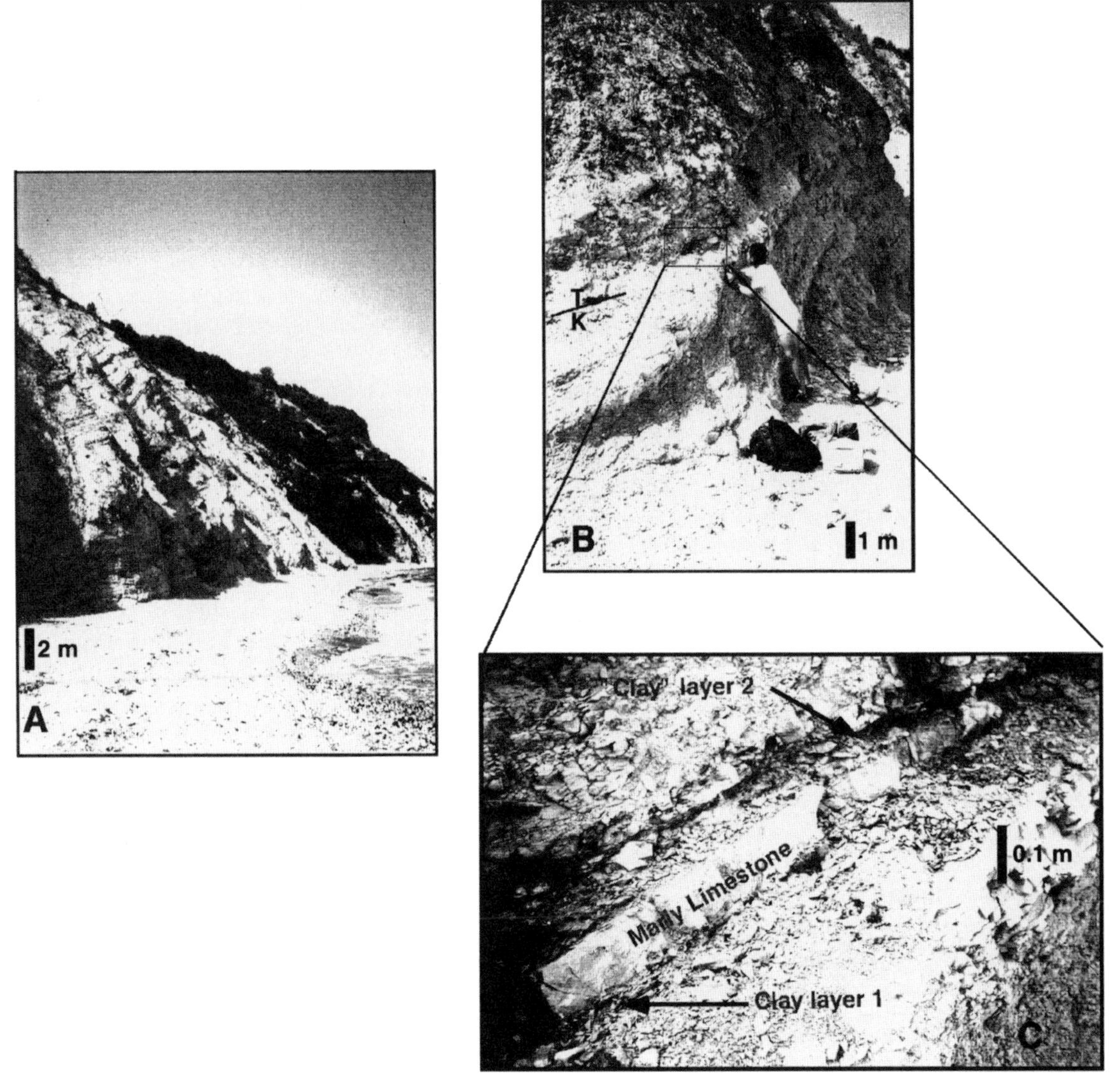

Figure 2. Photographs of Bjala-1 outcrop. A: Alternating marl and marly limestone layers of zone CF1, which spans last 300 k.y. of Maastrichtian. B: View of Cretaceous-Tertiary (K-T) boundary transition at Bjala-1. C: K-T boundary at Bjala-1 showing clay layer 1 (K-T) and clay layer 2 separated by 7–10-cm-thick marly limestone layer.

For foraminiferal studies, samples were processed following the standard method of Keller et al. (1995). Sediments were washed over a 63 μm sieve for the Maastrichtian samples and <38 μm sieve for the early Danian samples. The 38 μm sieve size was used for the early Danian because the first evolving Tertiary species in the boundary clay are usually smaller than 63 μm (see Keller et al., 1995). If necessary, foraminifera were further cleaned by ultrasonic agitation for 10–15 s, until a clean residue was obtained, and then washed again. The samples were oven dried at 50 °C. Planktic foraminifera are abundant and relatively well preserved, although recrystallized. From each sample, ~250–300 specimens were picked from random sample splits (using a microsplitter) from two size fractions, >63 μm and >150 μm. The two size fractions were analyzed in order to obtain statistically significant representations of the smaller and larger species populations. The entire sample was searched for rare species and these were included in the species range distributions and species richness data. The 38–63 μm size fraction was examined for small Danian species, but none were observed in the Bjala sections.

Major faunal assemblage changes noted in the quantitative analysis of washed residues were also examined in thin sections in order to evaluate potential preservation effects. Because identification of species in thin sections is more difficult than in washed residues, no quantitative abundance data were obtained from these samples. However, the overall species assemblages were evaluated on the basis of relative abundances of easily identifiable groups, such as heterohelicids, globotruncanids, and rugoglobigerinids, and compared with the quantitative counts.

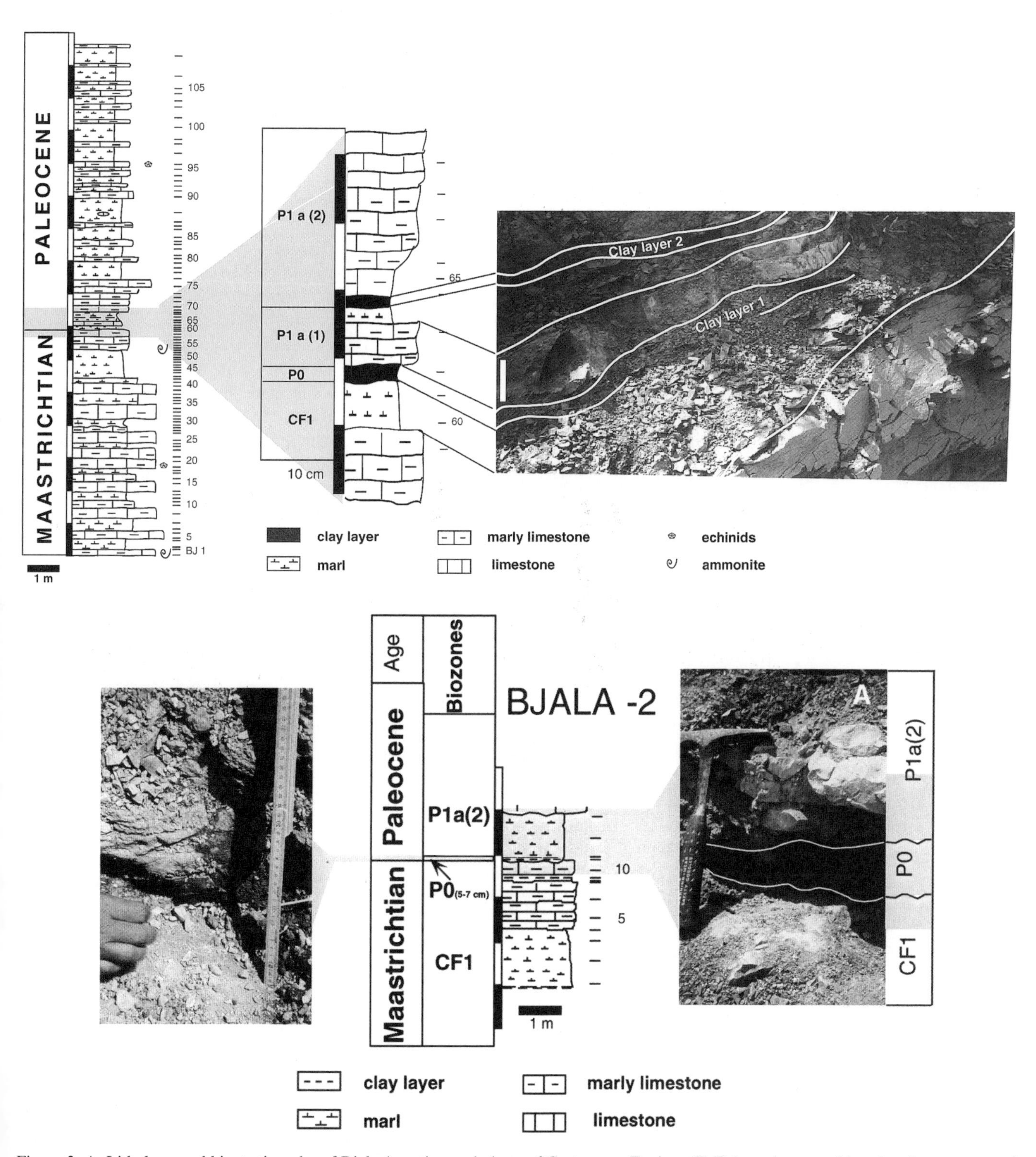

Figure 3. A. Lithology and biostratigraphy of Bjala-1 section and photo of Cretaceous-Tertiary (K-T) boundary transition showing presence of boundary clay layer 1 and second clay layer within zone Pla, ~10 cm above it. Note that two clay layers are separated by 7–10-cm-thick marly limestone layer with undulose erosional contact that marks hiatus. B: Lithology and biostratigraphy of Bjala-2 section. K-T boundary clay layer (P0) is ~2–3 cm thick (right side), but varies due to infilling of depressions and forms lenses as thick as 5–7 cm (left side). Note that this clay layer is between marly limestones with undulating, irregular erosional contacts that mark hiatuses at K-T boundary (CF1-P0) and in early Danian *P. eugubina* zone (P0-Pla[2]).

Whole-rock and clay mineral analyses were conducted at the Geological Institute of the University of Neuchatel, Switzerland, based on X-ray diffraction (XRD) analyses (SCINTAG XRD 2000 diffractometer). Sample processing followed the procedure outlined by Kübler (1987) and Adatte et al. (1996). Whole-rock compositions were determined by XRD (SCINTAG XRD 2000 diffractometer) based on methods described by Klug and Alexander (1974) and Kübler (1983). This method for semiquantitative analysis of the bulk-rock mineralogy (obtained by XRD patterns of random powder samples) uses external standards. The intensities of selected XRD peaks that characterize each clay mineral in the <2 μm size fraction (e.g., chlorite, mica, kaolinite, smectite) were measured for semiquantitative estimates. Therefore, clay minerals are given in relative percent abundances without correction factors. The percent smectite is estimated by using the method of Moore and Reynolds (1989).

We selected 10 samples for trace element analysis across the K-T transition. The very low trace element contents required preconcentration, and matrix elimination prior to detection with inductively coupled plasma-mass spectrometry (ICP-MS) was achieved by fire assay with NiS collection as described in detail in Cubelic et al. (1997). ICP-MS analysis was performed using platinum group element (PGE) multielement standards Claritas (SPEX Industries Grasbrunn, Germany) and Tm and Bi as internal standards. Accuracy was checked by means of WPR-1 and SARM-7 standard reference materials. Based on reference samples analyzed in the laboratory at the University of Karlsruhe over several years, recovery was estimated to be better than 85%, which is in the range of the efficiency usually obtained by NiS fire assay (Reddi et al., 1994; Zereini et al., 1994). Detection limits are 0.05 ng/g Ir, 0.1 ng/g Rh, 0.4 ng/g Pd, and 0.4 ng/g Pt. Detection limits are mainly dependent on blanks of the NiS fire assay.

Stable isotope analyses were conducted on the bulk-rock carbonate samples at the stable isotope laboratory of the University of Bern, Switzerland, using a VG Prism II ratio mass spectrometer equipped with a common acid bath (H_3PO_4). The results are reported relative to the Vienna Peedee belemnite standard reference material with a standard error of 0.1‰ for $\delta^{18}O$ and 0.05‰ for $\delta^{13}C$.

LITHOLOGY

At the Bjala-1 locality, ~16 m of relatively undisturbed alternating gray marl and light yellow to beige marly limestone layers are exposed (Fig. 3A). The lower 5.5 m of the section are characterized by alternating marl and marly limestone layers of approximately equal thickness (30–40 cm). The upper surfaces of the marly limestone layers are undulating, frequently bioturbated (*Chondrites, Thalassinoides*), and occasionally slightly glauconitic. These surfaces are indications of nondeposition, or possibly erosion. Between 67 cm and 150 cm below the K-T boundary, the sediments are more marly and interbedded with discrete 2–3-cm-thick calcareous layers. Overlying this interval is a 60-cm-thick white to gray marly limestone with an undulating, bioturbated upper surface (Fig. 3A). This surface is overlain by a 7-cm-thick gray bioturbated marl with an undulating upper surface that marks the top of the Maastrichtian and suggests erosion prior to deposition of the K-T boundary clay. The boundary clay consists of a 2–3-cm-thick, dark gray to black clay layer that contains an Ir anomaly of 6.1 ppb (Preisinger et al., 1993a) and is labeled "clay layer 1" (Fig. 3A).

Above the K-T boundary clay layer is a 7–10-cm-thick, white to beige marly limestone layer with undulating surfaces at the bottom and top suggesting erosion, a conclusion supported by a strongly bioturbated (*Chondrites*) upper surface. This limestone layer is laterally continuous, although variable in thickness. Above this limestone layer is a 1–2-cm-thick marl layer, followed by a 2-cm-thick clayey marl layer which we labeled "clay layer 2" (Fig. 3A). The overlying sediments consist of alternating marl and marly limestone layers with occasional thin clayey intercalations. The upper part of the section consists of 30–50-cm-thick marl layers that are interbedded with thin, light gray, marly limestone layers.

The Bjala-2 section is significantly tectonically disturbed, and therefore only the K-T transition interval was examined. The first 1.20 m of the section consist of light gray marl (Fig. 3B). Above the marl are 1.1 m of marly limestone layers having strongly undulating and bioturbated surfaces. A 1–3-cm-thick dark brown to dark gray clay layer represents the K-T boundary clay and contains a 2.1 ppb Ir anomaly. The thickness of this clay layer is variable and may reach 6 cm, infilling depressions and cracks in the underlying marly limestone. A 1.05-m-thick gray marl layer with echinoid fragments overlies the K-T boundary clay. Above the marl is a light beige limestone with an undulating upper surface (Fig. 3B).

BIOSTRATIGRAPHY

Macrofossils are rare at Bjala-1. An unidentifiable ammonite mold was collected 60 cm below the K-T boundary. Ivanov (1993) noted common ammonites at 10 m below the K-T boundary, only very rare specimens in the last 10 m of the Maastrichtian, and the last ammonite at 40 cm below the K-T boundary. Echinoid fragments were observed in the lower 3.5 m of the section and in the Danian at 12 m (Fig. 3A).

In this study the biostratigraphy of the Cretaceous-Tertiary transition in the Bjala sections is based on quantitative analysis of planktic foraminifera. We use the biozonation of Keller (1993) and Keller et al. (1995), shown in Figure 4, in comparison with the zonal scheme and paleomagnetic correlation of Berggren et al. (1995), and the calcareous nannofossil zonation of Martini (1971). Zones P0 and Pα in the Berggren et al. (1995) zonal scheme are equivalent to zones P0 and Pla of Keller et al. (1995). However, it is difficult to correlate these two zonal schemes above this interval for reasons of methodology. Zones Pla, Plb, and Plc of Berggren et al. (1995) are

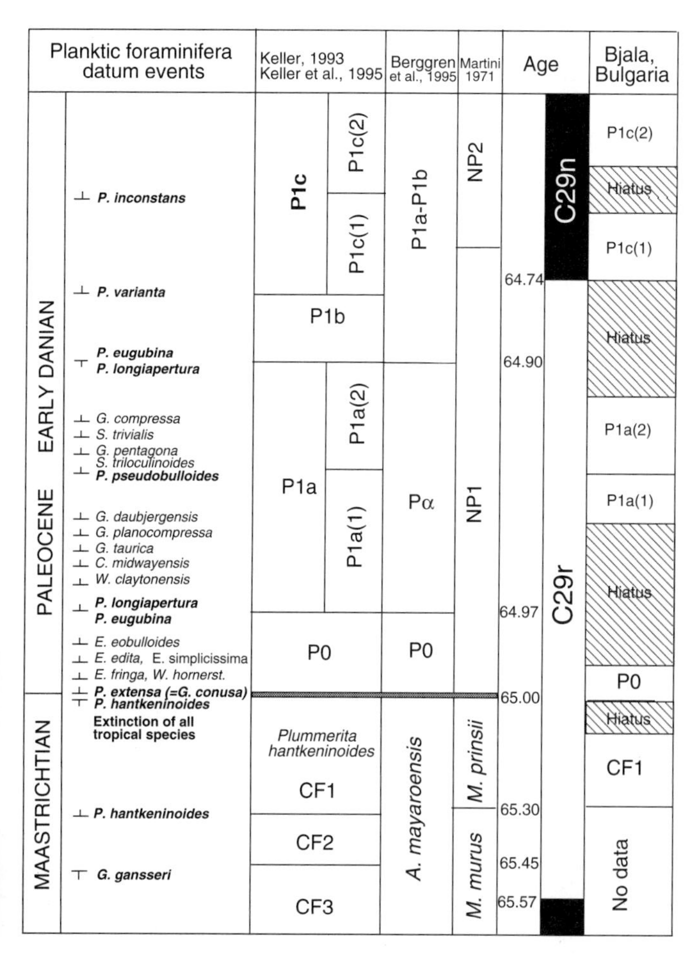

Figure 4. Late Maastrichtian–early Paleocene plankton zones and time scale. Correlation of commonly used planktic foraminiferal and calcareous nannofossil zonations as applied at Bjala, Bulgaria. Note presence of short intrazonal hiatuses recognized in upper part of CF1, at P0-Pla(1), Pla(2)-Plc, and Plc intervals. Lithologic observations (e.g., undulating erosional surfaces) indicate that additional short hiatuses may be present. See Figure 3A for rock unit symbols.

based on the successive first appearances of *Subbotina triloculinoides, Globanomalina compressa*, and *Praemurica inconstans*, all of which they reported to originate well above the extinction of *Parvularugoglobigerina eugubina*. However, *S. triloculinoides* and *G. compressa* have been observed to first appear in the upper part of the range of *P. eugubina* (see MacLeod and Keller, 1991a, 1991b; Keller et al., 1995). The difference in the reported first appearances is apparently due to the size fraction analyzed. Keller (1988, 1993; Keller et al., 1995, 2001) documented that early Danian species are very small (38–63 μm) and generally do not reach sizes larger than 100 μm until zone Plc (Fig. 4), ~500 k.y. after the K-T boundary. When only the >100 μm size fraction is analyzed, the first appearances of these species are delayed as a function of size, although these species are present earlier in the smaller size fraction (both 38–63 μm and >63 μm size fractions are analyzed in this study).

There is generally good agreement between nannofossil and planktic foraminiferal zones (e.g., Henriksson, 1993; Berggren et al., 1995; Pardo et al., 1996; Li and Keller, 1998a). The latest Maastrichtian *Micula prinsii* zone is an excellent marker species that first appears near the base of paleomagnetic chron 29R, ~500 k.y. before the K-T boundary. The *M. prinsii* zone thus encompasses planktic foraminiferal zones CF1 and CF2, which span the last 450 k.y. of the Maastrichtian (Li and Keller, 1998a). Zone CF1 spans the last 300 k.y. of the Maastrichtian (Pardo et al., 1996). The early Danian zone NP1 spans foraminiferal zones P0, Pla, and Plb of Keller (1993) and Keller et al. (1995), and zone NP2 begins near the base of zone Plc (Fig. 4). The stratigraphic resolution of calcareous nannoplankton is therefore not comparable to that of planktic foraminifera in the early Danian.

MAASTRICHTIAN

The nearly 7 m of Maastrichtian marl and marly limestone layers exposed at Bjala-1 contain diverse, abundant planktic foraminiferal assemblages indicative of zone CF1, and calcareous nannofossils (*Micula prinsii* zone; Ivanov and Stoykova, 1994). *Plummerita hantkeninoides*, the index species for zone CF1, is present but rare (Fig. 4). The undulating surface at the top of the Maastrichtian suggests the presence of a short hiatus in both Bjala-1 and Bjala-2 sections. However, very little of the latest Maastrichtian zone CF1 may be missing, as suggested by the thickness of this zone (7 m), compared to Tunisian sections (6 m, Abramovich and Keller, 2002). Other short hiatuses may be present in the lower part of the section, as indicated by the lithological changes and undulating and glauconitic surfaces. Current biostratigraphic resolution is insufficient to detect these short hiatuses.

A total of 53 species were identified, although most faunal assemblages contain between 30 and 40 species (Fig. 5). This species richness is significantly lower than the species diversity generally observed in the Tethys (e.g., 55–65 species in Tunisia, Israel, Italy, Spain; Keller et al., 1995; Apellaniz et al., 1997; Luciani, 1997; Abramovich et al., 1998), and reflects the location of Bjala in the northern Tethys. The assemblages observed are generally more characteristic of middle than low latitudes, as indicated by the high abundances of *Globotruncana arca* (to 40% in >150 μm) and *Heterohelix globulosa* (Fig. 6; see also Malmgren, 1991; Nederbragt, 1998; Li and Keller, 1998a), rarity of *P. hantkeninoides* and low-latitude globotruncanids, and the relatively low abundance of rugoglobigerinids (5%–10%). For most of zone CF1, species populations fluctuated, but in the 67 cm below the K-T boundary the relative abundance of all large-sized species decreased significantly and permanently (Fig. 6). All tropical and subtropical species disappeared at or below the K-T boundary. In contrast, the small (63–150 μm) ecological generalist species (Keller, 1993, 1996) increased during this time, suggesting a major environmental change and increased stress preceding the K-T boundary.

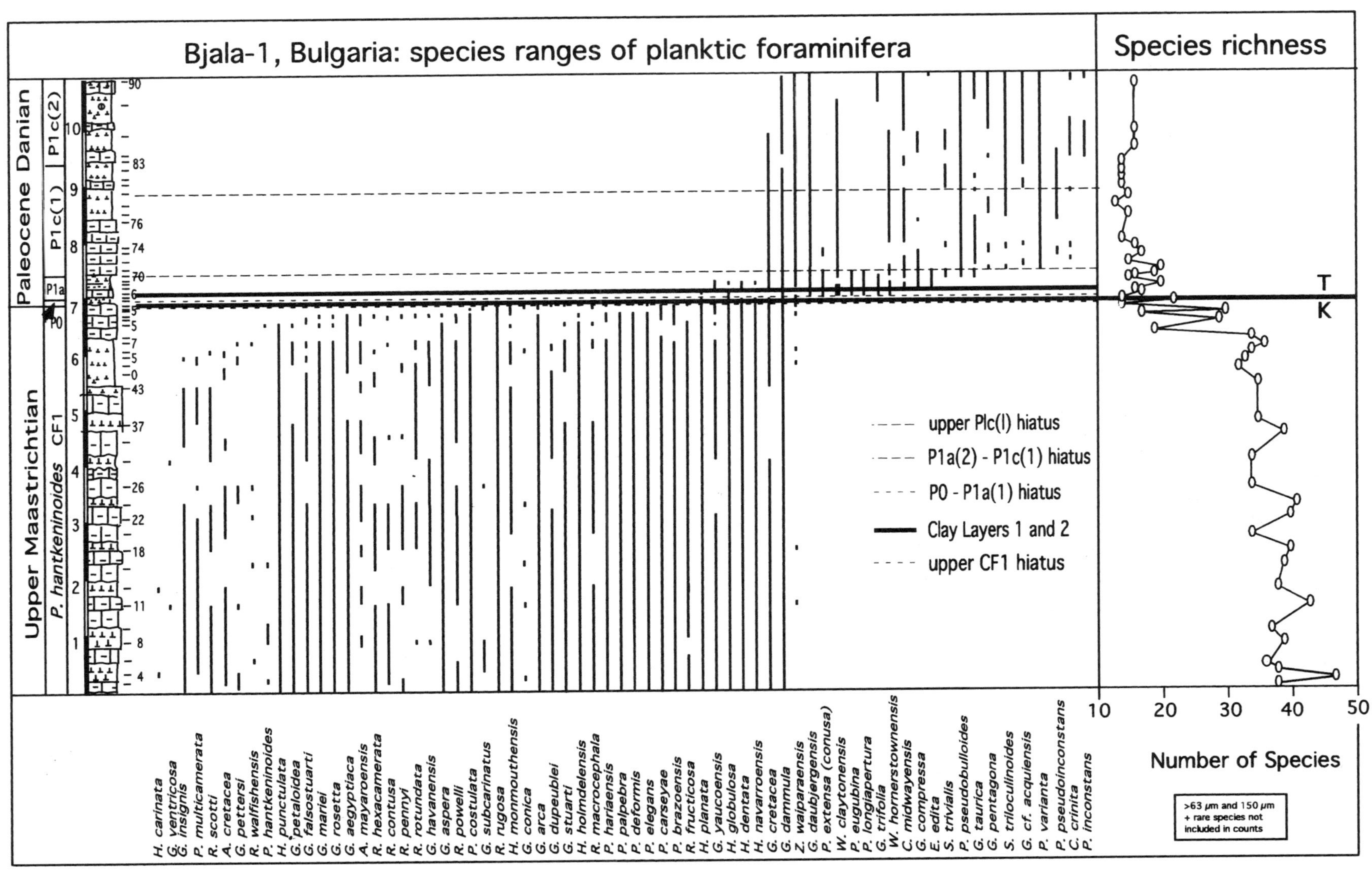

Figure 5. Species richness and biostratigraphic ranges of species across Cretaceous-Tertiary (K-T) boundary at Bjala-1, Bulgaria. Note decreased species richness in upper half of zone CF1, followed by rapid, but variable decrease in the 67 cm below K-T boundary. Decreased diversity reflects environmental changes and increasing biotic stress that resulted in local species disappearances and migrations. See Figure 3A for rock unit symbols.

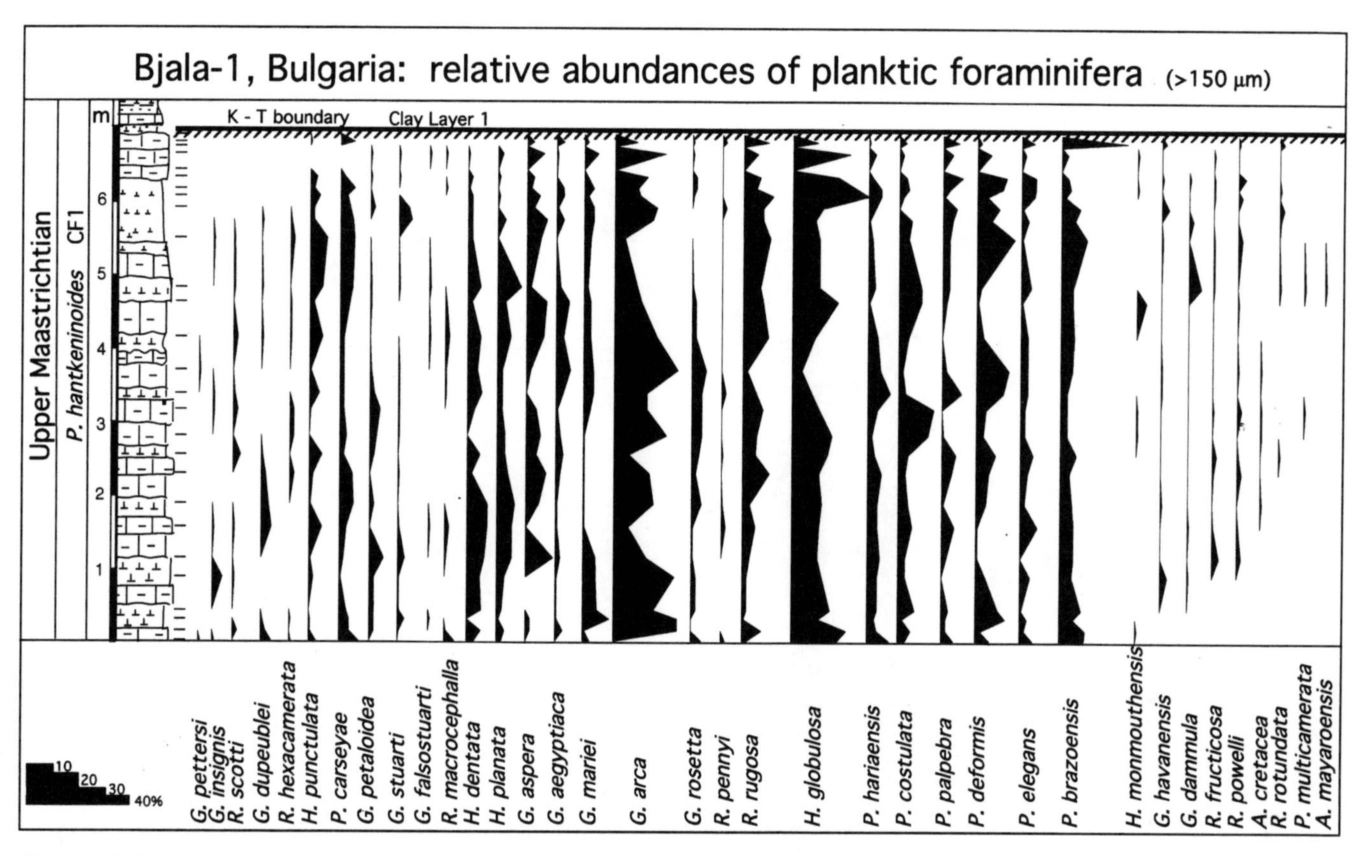

Figure 6. Relative abundance changes in species populations of planktic foraminifera (>150 µm) during last 300 k.y. of Maastrichtian at Bjala-1, Bulgaria. Note high abundance of *Globotruncana arca* and *Heterohelix globulosa*, species that thrived in relatively cool middle latitudes. Major environmental change is indicated by general decrease in all larger sized species in the top 67 cm of Maastrichtian interval. See Figure 3A for rock unit symbols.

The most unique aspect of the latest Maastrichtian Bjala-1 fauna is the dominance (60%–80%) of *Guembelitria dammula*, which has not been reported elsewhere (Fig. 7). *Guembelitria dammula* is a very high spired and large triserial species that has a morphology similar to that of the smaller *G. danica*, which is common in northern boreal seas (e.g., Denmark, Kazakstan; Keller et al., 1993; Pardo et al., 1999). *G. dammula* has been observed in late Maastrichtian sediments of Madagascar, which also suggests that this species is endemic to higher latitudes (Abramovich, 2000, written commun.). Quantitative data from other localities are needed before environmental affinities of *G. dammula* can be interpreted.

Species richness

Species census data provide an estimate of the number of species present in a faunal assemblage. In this study, the species richness of each sample at Bjala-1 is based on three types of records: (1) the number of species present in a count of 250–300 individuals in the >150 µm size fraction of a random sample split (using a microsplitter), (2) the number of species present in a count of 250–300 individuals in the >63 µm size fraction of a random sample split (using a microsplitter), and (3) a careful search of the remaining sample residues of the >63 µm and >150 µm size fractions for rare species that were not observed in either of the two counts. The species richness curve in Figure 5 thus reflects the union of these three species tallies for each sample and therefore is not necessarily the same as the plotted species ranges, which are based on the two counts.

Species richness averaged 40 species in the lower 3.5 m of the Maastrichtian zone CF1, and decreased to an average of 35 species in the upper 3 m (Fig. 5). A major drop in species richness occurred in the top 67 cm of the Maastrichtian, coincident with decreased abundance of large species (Fig. 6). However, species variability is high (14–30 species) in this interval and suggests that at least part of the decrease is due to preservation bias. Thin sections from this interval show the presence of impoverished species assemblages dominated by small (63–150 µm) ecological generalists (e.g., small biserial species and guembelitria), similar to those observed in the >63 µm faunal counts, whereas species with larger morphologies (>150 µm) are generally few to rare (e.g., globotruncanids, rugoglobigerinids, and large heterohelicids). This suggests that a significant faunal change occurred prior to the K-T boundary that differentially affected more specialized large species, leading to decreased populations, whereas the inferred ecological generalist

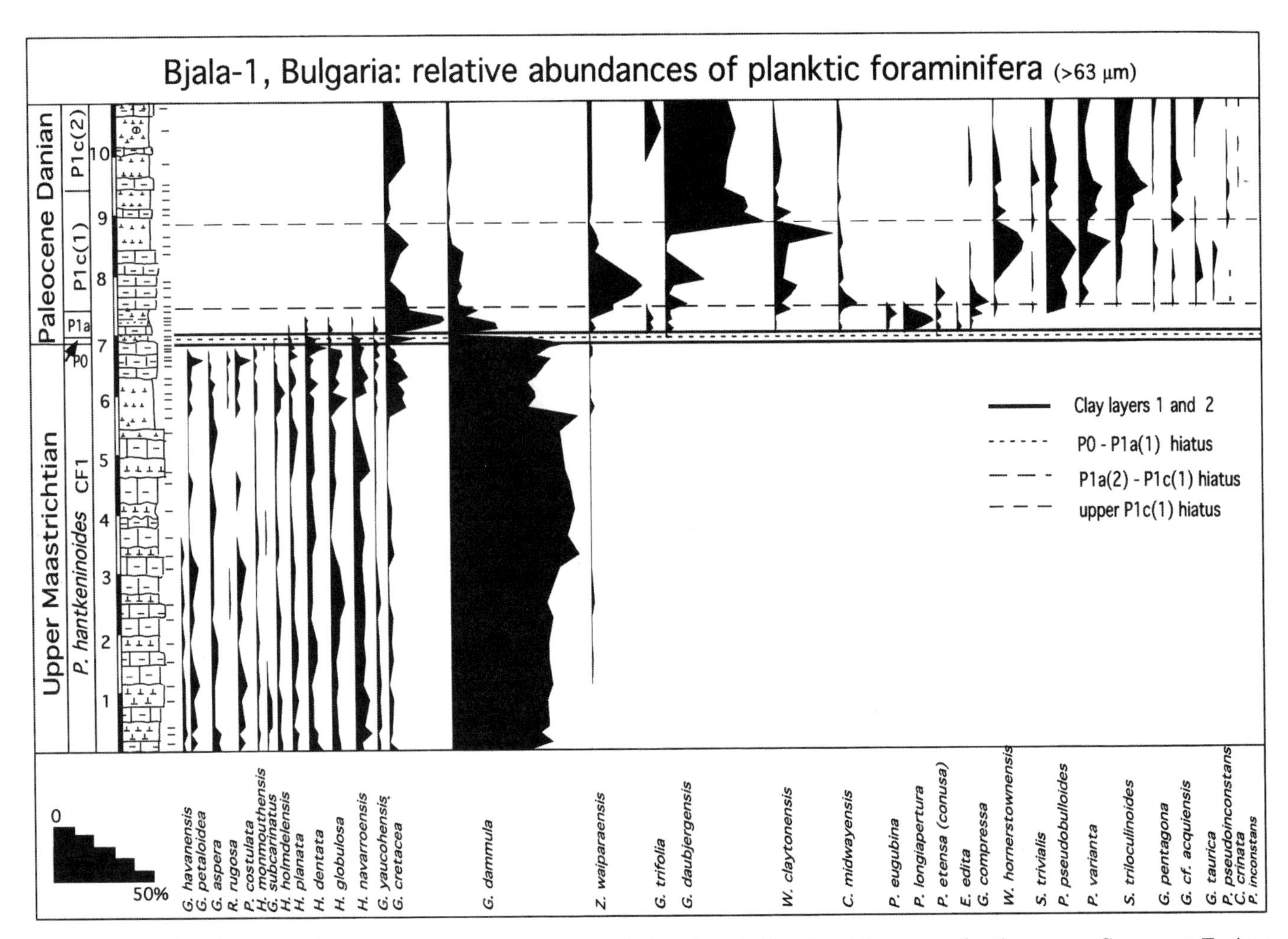

Figure 7. Relative abundance changes in species populations of planktic foraminifera in >63 µm size fraction across Cretaceous-Tertiary (K-T) boundary at Bjala-1, Bulgaria. Note that this size fraction contains only ecological generalists, many of which are known to be K-T survivors. What is unique to Bulgaria is dominance of triserial species *Guembelitria dammula* in late Maastrichtian. Ecological generalists thrived during biotic crisis that led to terminal decline of larger sized planktic foraminifera at end of Maastrichtian. Abrupt faunal changes in early Danian mark hiatuses. See Figure 3A for rock unit symbols.

species remained unaffected or thrived. The high species richness variability in this interval may reflect local species disappearances and reappearances due to migration from the Tethys during climate fluctuations.

K-T BOUNDARY

At Bjala-1, the K-T boundary clay consists of a 2–3-cm-thick clay layer (labeled clay layer 1) that contains a 6.1 ppb Ir enrichment (Preisinger et al., 1993a). This clay layer 1 contains a planktic foraminiferal assemblage dominated by *Guembelitria* (75% *G. dammula*) and common small biserial Maastrichtian species (*Heterohelix globulosa, H. dentata, H. navarroensis*, and *H. planata*, Fig. 8). A few large and possibly reworked Maastrichtian species are also present. Rögl et al. (1996) reported the first occurrence of Danian species *Woodringina hornerstownensis, W. claytonensis*, and *Globoconusa predaubjergensis* 2 cm above the base of the clay layer. In the overlying marly limestone layer, we observed abundant very small Danian planktic foraminifera, including *Parvularugoglobigerina extensa* (formerly *Globoconusa conusa), Globoconusa daubjergensis, G. predaubjergensis, Woodringina claytonensis, W. hornerstownensis*, abundant *Guembelitria dammula*, and *Parvularugoglobigerina eugubina* and *P. longiapertura*, the index species for the early Danian zone Pla (Figs. 4 and 8). However, the simultaneous first appearance of these species immediately above clay layer 1 indicates an interval of nondeposition or a short hiatus between the clay and overlying marly limestone (zone P0-Pla[1]); most of the lower part of zone Pla (*P. eugubina* zone) missing. A hiatus is also suggested by the abrupt lithologic change from clay to marly limestone with undulating surface.

Only one clay layer was observed at the nearby Bjala-2 section. This clay layer represents the boundary clay, as indicated by the presence of an Ir anomaly of 2.03 ppb. As at Bjala-

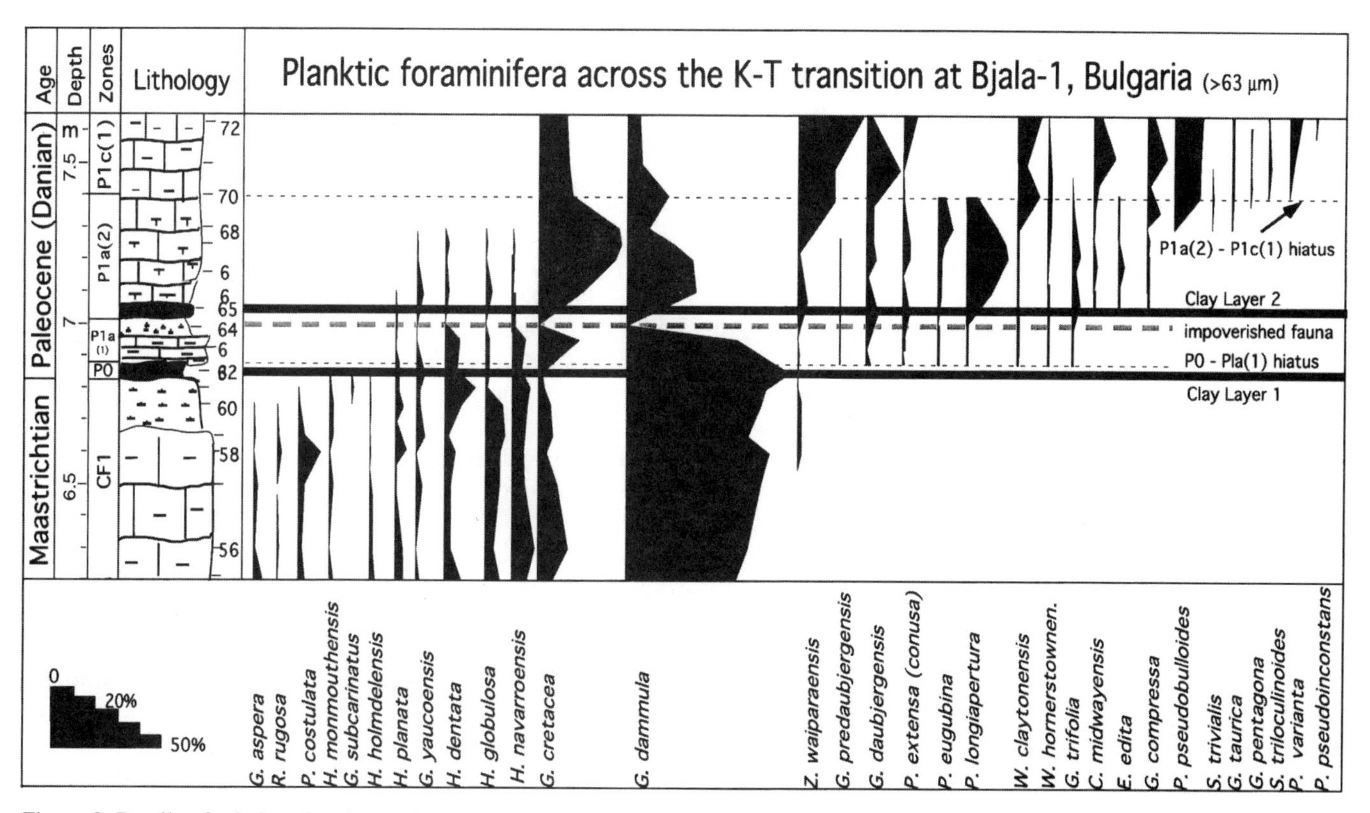

Figure 8. Details of relative abundance changes in planktic foraminifera across Cretaceous-Tertiary (K-T) boundary at Bjala-1, Bulgaria, in >63 μm size fraction. Note dominance of *Guembelitria dammula* and its decreased abundance in early Danian, and unusually high abundance of *Heterohelix navarroensis* in early Danian. First appearance of eight new Danian species above boundary clay layer marks hiatus; most of lower part of *P. eugubina* zone, or Pla(1) is missing. Note also that second clay layer in upper part of *P. eugubina* zone, or Pla(2), also marks hiatus, as indicated by abrupt disappearance of *P. eugubina* and *P. longiapertura*, and lithological change and undulating erosional surface. See Figure 3A for rock unit symbols.

1, this clay layer contains small Maastrichtian heterohelicids, guembelitrids, and hedbergellids, as well as rare reworked globotruncanids and rugoglobigerinids; however, it contains no Danian species. The first Danian species are observed in the overlying marl, which contains a well-developed early Danian assemblage with specimens predominantly in the >63 μm size fraction. The assemblage includes *P. eugubina, P. longiapertura, G. conusa, G. daubjergensis, G. dammula, G. cretacea, P. pseudobulloides*, and *C. midwayensis*. This species assemblage and the relatively large size of the specimens indicate zone Pla(2) (upper part of *P. eugubina* zone). Zone Pla(1) is characterized by generally smaller size (>63 μm) and absence of *P. pseudobulloides*. Thus at Bjala-2, biostratigraphy indicates that an early Danian hiatus spans the interval from the boundary clay to Pla(2).

DANIAN

A second clay, clay layer 2, is present in the early Danian zone Pla of Bjala-1. This clay layer, separated from the boundary clay by thin limestone and marl layers (Fig. 3A), contains small Ir (0.22 ppb), Ru (0.30 ppb), and Rh (0.13 ppb) anomalies and a larger Pd anomaly (1.34 ppb). Although this clay layer was not mentioned by Preisinger et al. (1993a, 1993b), it is clearly visible in their photo of the section. The small Ir and Co anomalies noted by Preisinger et al. (1993a) are present in the marly limestone layer between the two clay layers (A. Preisinger, 2000, personal commun.). The lithology, biostratigraphy, and geochemistry of these two clay layers indicate that they are separate depositional events.

In clay layer 2, *G. compressa, Chiloguembelina midwayensis*, and *Eoglobigerina edita* first appear, followed by *Parasubbotina pseudobulloides, Subbotina trivialis*, and *G. taurica* (Fig. 8). This assemblage marks the upper part of the *P. eugubina* zone or Pla(2). Early Danian planktic foraminifera are more abundant and larger than in the interval below, particularly *P. longiapertura* and *P. eugubina*, whereas *Guembelitria dammula* decreased. These abundance and test size changes are characteristics of the upper part of the *P. eugubina* zone or Pla(2). A short hiatus is suggested by this abrupt faunal and lithological change from clay to limestone.

The Pla(2) assemblage continued for 27 cm (sample 70), where the Pla index species *P. eugubina* and *P. longiapertura* disappeared coincident with the first appearance of *G. penta-*

gona, Subbotina triloculinoides, and the Plc index species *Parasubbotina varianta* (Fig. 8). The juxtaposition of the Pla and Plc index species suggests another condensed interval or short hiatus. The abrupt faunal change at this interval is also evident in the decreased relative abundance in triserial species (*G. cretacea, G. dammula*) and subsequent increased abundance in biserial species (*Zeauvigerina waiparaensis, Woodringina claytonensis, Chiloguembelina midwayensis*), as well as *G. daubjergensis G. compressa*, and *P. pseudobulloides* (Figs. 7 and 8). Triserial species continued to decrease in Plc(1), and all biserial and trochospiral species nearly disappeared coincident with the onset of dominant *G. daubjergensis* just below the Plc(1)-Plc(2) boundary, which is marked by the first appearance of *P. inconstans*. An abrupt lithologic change from marl to limestone with an undulating surface marks this faunal change and suggests another short hiatus.

Interpretation

Planktic foraminiferal biostratigraphy and assemblage changes indicate that the early Danian at Bjala is very condensed and probably underwent significant erosion or intermittent periods of nondeposition at the P0-Pla, Pla-Plc, and Plc(1)-Plc(2) intervals. An incomplete sedimentary record for the Pla zone at Bjala-1 is also suggested by the fact that this zone interval spans only 40 cm, as compared with more than 4 m at El Kef and Elles in Tunisia (Keller, 1988; Karoui-Yakoub et al., 2002). Hiatuses are frequently observed in early Danian sections at these stratigraphic intervals in marine sections (MacLeod and Keller, 1991a, 1991b; Keller and Stinnesbeck, 1996).

Previous workers have argued that eastern Bulgaria has one of the most continuous sediment records across the K-T transition (Preisinger et al., 1993a, 1993b; Ivanov and Stoykova, 1994; Rögl et al., 1996). Our study indicates that these sections are far from complete. However, because each hiatus spans less than a biozone, they cannot be detected by biostratigraphy alone, and that is probably why they were not been detected earlier. In this study, the hiatuses were recognized by field observations (e.g., lithological changes, erosional surfaces, bioturbation), high-resolution sampling, and quantitative faunal analysis. In earlier studies, Ivanov and Stoykova (1994) considered the Bjala-1 (their Bjala 2b) section complete on the basis of the presence of early Danian calcareous nannofossil zones. However, because nannofossil zone NP1 corresponds to planktic foraminiferal zones P0, Pla(1), Pla(2), and Plb, and zone NP2 encompasses more than Plc(1) and Plc(2) (Fig. 4), the short early Danian hiatuses within these intervals could not be detected by nannofossil biostratigraphy. Preisinger et al. (1993a, 1993b) and Ivanov and Stoykova (1994) cited the presence of *Braarudosphaera* and *Thoracosphaera* blooms in the limestone immediately above the K-T boundary clay as evidence for continuous sedimentation. However, these blooms are not restricted to the basal Danian, but are characteristic of the early Danian zones Pla(1)-Plc(1) (Keller and v. Salis Perch-Nielsen, 1995; Keller et al., 2001).

MINERALOGY

Bulk rock

In the Bjala-1 section, sediments are generally dominated by calcite (50%–75%), phyllosilicates (10%–40%), and quartz (8%–20%), and there are sporadic occurrences of K-feldspar (0%–3%) and plagioclase (0%–2%, Fig. 9). The first 4.5 m of the analyzed uppermost Maastrichtian do not show significant fluctuations in the bulk-rock composition. With the exception of sample BJ3, plagioclase (low albite) averaged 2%. In the top 2 m of the Maastrichtian, calcite shows an overall increase from a mean value of 70% to 80%. These data suggest a brief period of proportionally higher calcite at the top of zone CF1, ~100 k.y. prior to the K-T boundary.

At the K-T boundary, calcite content dropped abruptly from 78% to 3% and detrital influx increased (Fig. 9). Immediately above the clay layer, calcite content increased to pre-K-T values (80%). In complete K-T transitions, calcite content remained low from zone P0 through at least the lower part of zone Plb (e.g., Zachos et al., 1989; Keller and Lindinger, 1989). The rapid return to higher values at Bjala-1 may reflect the hiatus at the P0-Pla(1) boundary that is recognized by abrupt changes in lithology and planktic foraminiferal assemblages. Calcite also decreased (from 84% to 66%) in the second clay layer and at the base of zone Plc (from 91% to 72%). These intervals coincide with increased detrital influx, abrupt changes in lithology, and faunal assemblages that mark short hiatuses. In the overlying marly limestone, calcite increased to 80%–87%, then decreased gradually to 55%–60%: at the same time, detrital minerals, such as quartz (10%–15%), phyllosilicates (20%–30%), and plagioclase (2%–3%) increased. This enhanced detrital influx prevailed into zone Pld (>20%) and reflects increased erosion due to tectonic activity in the area.

CLAY MINERALS

At Bjala-1, the main clay phases are smectite, illite, chlorite, and kaolinite. Random illite-smectite mixed layers (I/S) with a high percentage of smectite layers and smectite are the predominant clay minerals. We refer to I/S with a high percentage of smectite layers (85%) as smectite (Fig. 10). The constant presence of smectite throughout the section indicates the absence of a strong diagenetic overprint due to burial. The smectite presence implies that the clay minerals are not transformed, and therefore are mostly of detrital origin, and reflect local uplift and/or variations in weathering processes and soil formation in the bordering continental areas (Chamley, 1989; Weaver, 1989). Two types of smectite are present in the Bjala sediments: (1) an almost pure Mg smectite, crystallized and characterized by a high percentage of expandable layers

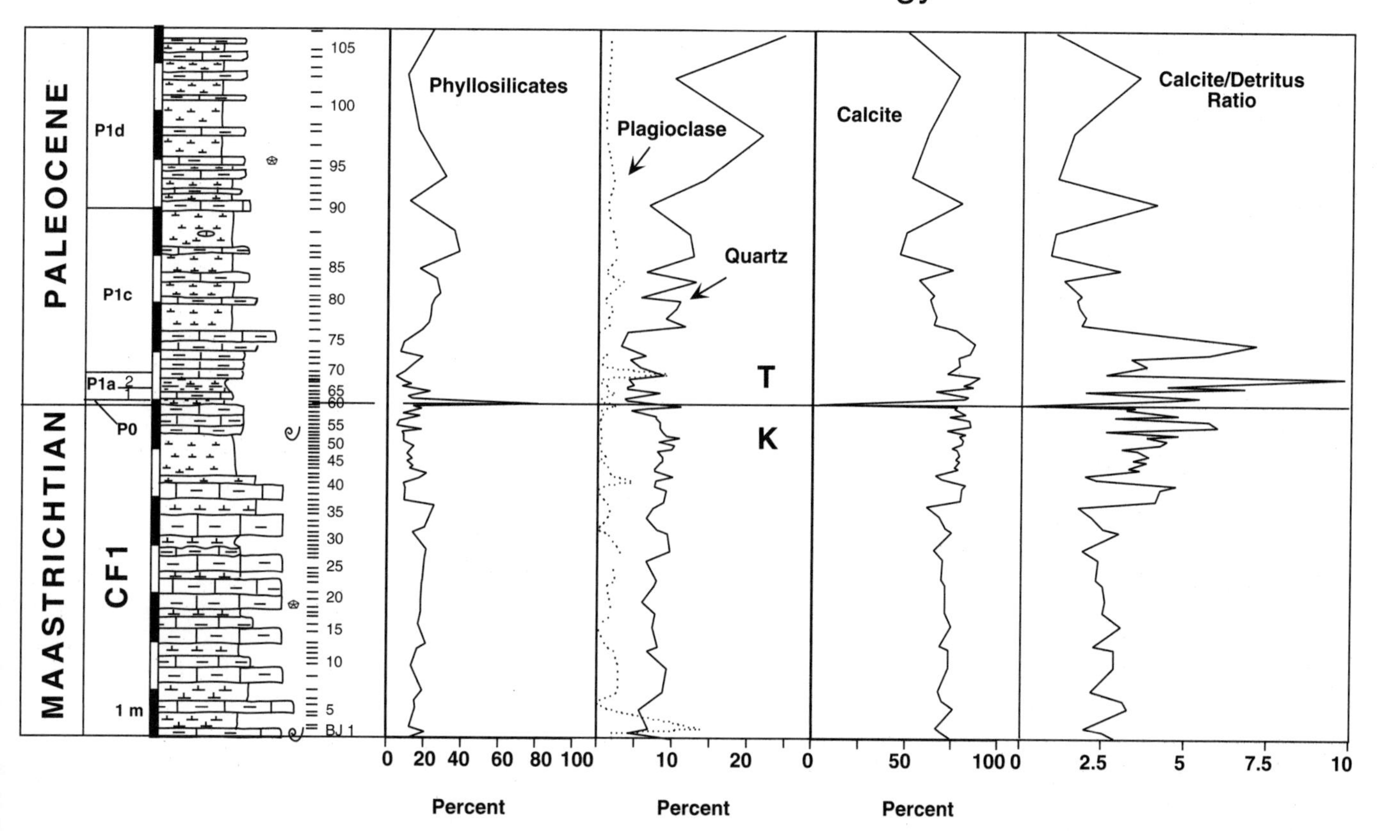

Figure 9. Bulk-rock composition at Bjala-1, Bulgaria. Note that detritus includes quartz, plagioclase, and phyllosilicates. Dominant sedimentary components are calcite and phyllosilicates with minor influx of quartz and plagioclase. Gradually increasing calcite/detritus ratio in upper part of CF1 is followed by decrease just below Cretaceous-Tertiary (K-T) boundary that suggests enhanced erosion and possible hiatus. Note that calcite is abundant directly above zone P0, contrary to more complete K-T sections (e.g., El Kef, Tunisia), and indicates presence of hiatus. See Figure 3A for rock unit symbols.

(>95%), and (2) a chemically more variable smectite characterized by broader XRD peaks and a lower percent of expandable layers (<80%). Environmental scanning electron microscope (ESEM) and energy dispersive X-ray (EDX) analyses of the type 1 (sample 65 from clay layer 2) reveal a weblike morphology and show that the major element is a typical Mg smectite (Si, Al, Mg, with minor Fe and K). Type 2 smectite is characterized by higher Al and K contents (sample 62 from K-T clay layer 1). The Mg smectite may be derived from a volcanic precursor (Elliot et al., 1989; Elliot, 1993), whereas smectite type 2 may be derived from alteration of soils under temperate to semiarid climate conditions (Gaucher, 1981; Hillier, 1995). Cool annual temperatures, low precipitation, and a short time for soil evolution restrict weathering and enhance the direct flux of illite and chlorite eroded from parent rocks. Note that detrital smectite in marine sediments increases during sea-level highstand periods, whereas chlorite, mica, and kaolinite increase during lowstand periods.

Smectite is abundant at the base of the section (80%), but decreased gradually to 60% by 3.5 m and increased to 80% by 6 m (Fig. 10). Rhythmic alternations between marl and marly limestone layers coincide with discrete, but repeated, changes in clay mineral assemblages. Marly limestones contain slightly increased smectite and decreased illite and chlorite (Fig. 10). The overall evolution of the clay mineral assemblages can be shown by the smectite/chlorite + illite (S/IC) ratio, which confirms the trends indicated by the bulk calcite/detritus ratio.

Interpretation

The S/IC ratio and the presence of kaolinite in the lower half of the Maastrichtian interval (Fig. 9) indicate enhanced erosion (due to low sea level and/or tectonic activity) under relatively warm and humid conditions. More temperate and arid conditions and possibly higher sea level prevailed shortly before (1–2 m below) the K-T boundary. The sharp decrease in the S/IC ratio near the top of the Maastrichtian marks a lower sea level and increased detrital input under cool and arid conditions (abundant illite and chlorite) that culminated in the boundary clay layer 1. Above the K-T boundary, the increased

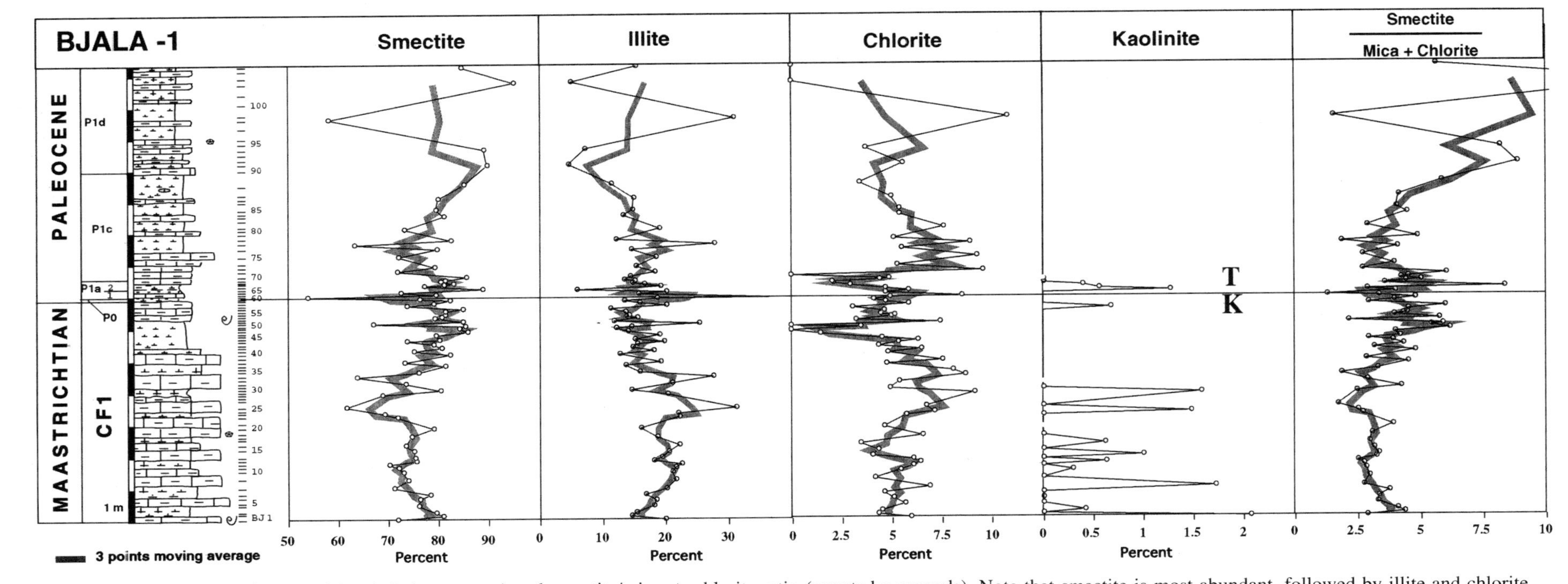

Figure 10. Clay mineral composition (relative percent) and smectite/mica + chlorite ratio (counts by seconds). Note that smectite is most abundant, followed by illite and chlorite, and reflect overall temperate-cool to arid conditions. Minor amounts of kaolinite in lower part of zone CF1 may indicate episodes of slightly warmer and humid conditions. Sharp decrease in smectite/chlorite + illite (S/IC) ratio near top of Maastrichtian marks lower sea level and increased detrital input under cool-arid conditions (abundant illite and chlorite) that culminated in boundary clay layer 1. Similar decrease in S/IC ratio is observed in lower Plc zone. See Figure 3A for rock unit symbols.

S/IC values in clay layer 2 could be due to deposition during a time of high sea level with increasing seasonality, or enhanced alteration of volcanic rocks. At the base of zone Plc, the significantly decreased S/IC and calcite/detritus ratios (Figs. 9 and 10) suggest a sharp sea-level drop. However, the change from a smectite-rich facies to a predominantly illite-chlorite assemblage may be due to increased erosion from a closer source (due to uplift or lower sea level). Beginning in the middle part of zone Plc (Figs. 9 and 10), the detrital input and smectite increased gradually and sediments changed from marl and marly limestone layers to more siliciclastic sediments. This suggests increased tectonic activity in the region, although alteration of volcanic rocks may have contributed to the increased smectite.

STABLE ISOTOPES

During the last 300 k.y. of the Maastrichtian at Bjala-1 (zone CF1), bulk carbonate $\delta^{13}C$ values narrowly fluctuated (1.9‰–2.3‰) for the first 5 m, and slightly increased in the upper meter of the Maastrichtian. At the K-T boundary, there is a relatively small (0.6‰) $\delta^{13}C$ negative shift in clay layer 1, and a small 0.2‰ decrease in clay layer 2. The $\delta^{13}C$ values continued to decrease through the early Danian into zone Plc (Figs. 11 and 12). Assuming that these values are original and that the shape of the curve has not been compromised by hiatuses, the $\delta^{13}C$ values suggest relatively high productivity during the Maastrichtian zone CF1 with a slight increase prior to the K-T boundary. The unusually small (0.6‰) $\delta^{13}C$ shift at the K-T boundary, as compared with 2‰–3‰ in low latitudes, indicates that the characteristic productivity crash did not extend to this middle latitude locality. However, productivity appears to have continued to decrease through the early Danian. A similar $\delta^{13}C$ pattern has been observed in other middle- and high-latitude localities (Keller et al., 1993; Barrera and Keller, 1994).

During the last 300 k.y. of the Maastrichtian at Bjala-1, $\delta^{18}O$ values fluctuated between -2.1‰ and -2.6‰, then decreased to a steady -2‰ in the 1.5 m below the K-T boundary (Fig. 11). In the 10 cm below the K-T boundary, the $\delta^{18}O$ values increased from -2‰ to -1.5‰ in the boundary clay and increased to -1.3‰ in clay layer 2 (Fig. 11). Thereafter, $\delta^{18}O$ values narrowly fluctuated between -1.1‰ and -1.4‰ in zone Plc(1). Assuming that these $\delta^{18}O$ trends reflect paleoenvironmental conditions, they suggest a relatively warm but fluctuating climate with increased cooling near the end of the Maastrichtian. Support for rapid cooling coincident with the K-T boundary and continuing into the early Danian is provided by the marked increase in illite and chlorite (e.g., increased detrital influx). These climate trends are consistent with those observed in low and high latitudes (Zachos et al., 1989; Barrera and Keller, 1994).

DISCUSSION

Paleoenvironment

Planktic foraminiferal assemblages of Bjala-1, eastern Bulgaria, reflect their depositional environment between the Tethys and northern boreal sea and are most similar to middle-latitude assemblages, as indicated by their relatively low species diversity (32–40 species), high abundance of *Globotruncana arca* and *Heterohelix globulosa*, and low abundance of species endemic to low latitudes (Li and Keller, 1998a; Nederbragt, 1998). A strong boreal influence is reflected by the dominance of *Guembelitria dammula*, which has been observed in faunal assemblages from Madagascar (S. Abramovich, 2000, written commun.).

The K-T mass extinction in planktic foraminifera at Bjala is similar to that in low latitudes in that all morphologically complex, large (>150 μm) species disappeared at or before the K-T boundary and only small, morphologically simple species survived (e.g., guembelitrids, heterohelicids, and hedbergellids). However, there are major differences in the patterns of pre-K-T species disappearances. About one-third of the large morphotypes that are known to range up to the K-T boundary in low latitudes disappeared within 1 m below the boundary clay at Bjala-1. Because most of these species are rare, their early disappearance may be partly due to the Signor-Lipps effect, and partly to increasingly unfavorable environmental conditions.

Stable isotope and clay mineral data indicate a cooling climate correlative with the disappearance of tropical and subtropical species. Evidence for late Maastrichtian climate changes were recorded from middle to high latitudes in the Northern and Southern Hemispheres (e.g., Barrera, 1994; Kucera and Malmgren, 1998; Li and Keller, 1998a, 1998b). At DSDP Site 525 in the middle-latitude South Atlantic, a 3–4°C climate warming occurred in deep and surface waters between 65.2 and 65.4 Ma, followed by a 3°C cooling during the last 100 k.y. of the Maastrichtian (Li and Keller, 1998b). At Bjala-1, this warming may be reflected by high calcite content and the presence of kaolinite in the lower 5 m of zone CF1, the maximum abundance in globotruncanids (43%), and maximum average species richness (40 species). The end-Maastrichtian cooling is reflected by the decreased relative abundances of all larger species (>150 μm), decreased abundance of globotruncanids (10%–28%), and increased abundance of small (63–150 μm) morphologically simple taxa (Figs. 5 and 6), increased detrital minerals, chlorite, and mica at the expense of calcite and smectite, respectively.

Humid Tethys-cool middle latitudes

Climate and sea-level changes at Bjala can best be interpreted in comparison with clay mineral data from K-T sections of the Tethys and northern middle to high latitudes. Overall,

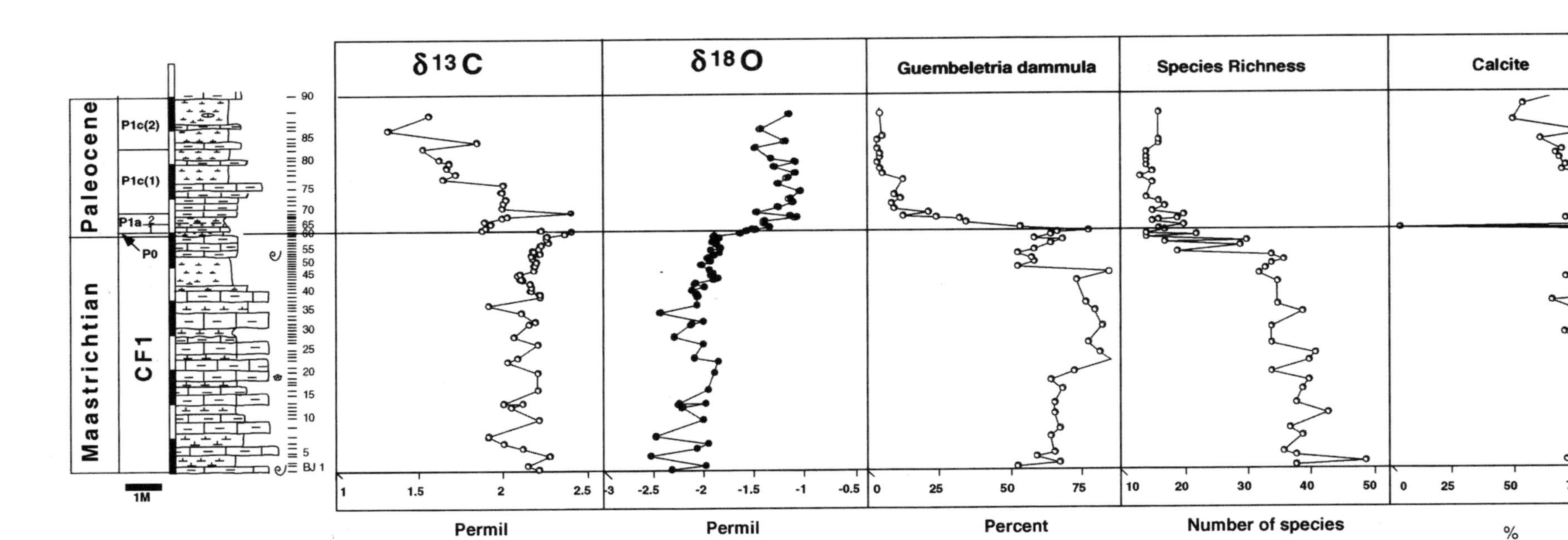

Figure 11. Stable isotope analyses ($\delta^{13}C$ and $\delta^{18}O$) of bulk rock at Bjala-1, correlated with *Guembelitria dammula* populations, planktic foraminiferal species richness, and calcite content. Note gradual increase in $\delta^{13}C$ values in late Maastrichtian, relatively small (~0.6‰) negative shift at Cretaceous-Tertiary (K-T) boundary, and major decrease in early Danian all suggest long-term changes in productivity and absence of sudden productivity crash at K-T boundary. $\delta^{18}O$ data indicate that cooling began in latest Maastrichtian (upper part of CF1) and continued across K-T boundary into early Danian.

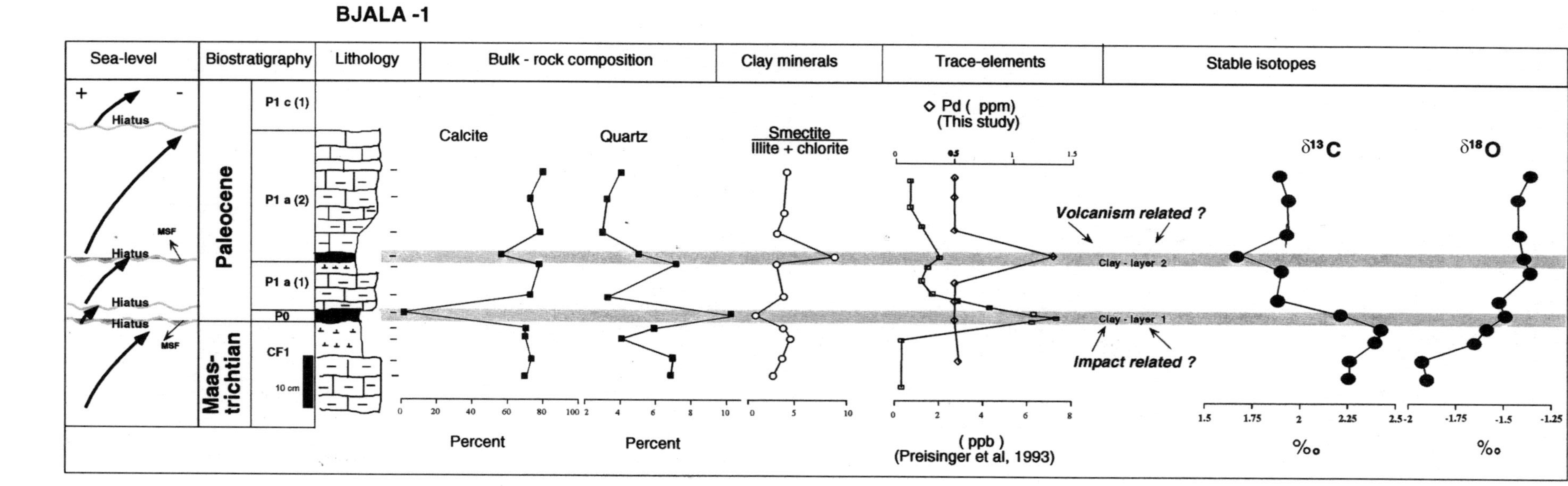

Figure 12. Summary of changing parameters across Cretaceous-Tertiary (K-T) transition at Bjala-1. Relatively small negative $\delta^{13}C$ shift and increasing $\delta^{18}O$ values mark environmental changes at Bjala that are more similar to higher latitudes than low-latitude Tethys. Two clay layers are present; K-T boundary clay with relatively high Ir anomaly marks cosmic event, whereas Pd-dominated early Danian clay layer in *P. eugubin* a zone suggests volcanic origin. See Figure 3A for rock unit symbols. MSF: Maximum flooding surface.

these data indicate that the low-latitude Tethys region underwent a time of increased humidity (Chamley, 1989), whereas the middle and higher latitudes underwent cooling (Adatte et al., 2001) coeval with the late Maastrichtian to early Danian climate cooling observed in the oxygen isotope record. For example, in Tunisia kaolinite increased from the late Maastrichtian into the early Danian, reflecting overall increased humidity. Smectite peaked just below the K-T boundary and again near the base of zone Pla and marks sea-level rises linked with drier climatic conditions (Adatte et al., 2002). Throughout the early Danian a series of warm humid periods alternated with seasonal temperate periods in the tropical and subtropical Tethys, from Tunisia to Haiti (Keller et al., 2001). However in higher latitudes, such as the east boreal paratethys of Kazakstan, the predominance of mica and smectite and minor amounts of kaolinite indicate a relatively cool and seasonally variable climate during the late Maaastrichtian and a cooler climate during the earliest Danian (Pardo et al., 1999). Similar cool climatic conditions are also inferred from clay minerals and stable isotopes for the Denmark K-T sections (Elliott, 1989, 1993; Schmitz et al., 1992). The source of the Mg smectite in the boundary clay of Stevns Klint was interpreted as volcanic, deposited under a temperate seasonal climate.

Productivity crash limited to low latitudes

The $\delta^{13}C$ record across the K-T boundary transition at Bjala-1 is very different from that of coeval records from low latitudes. Global stable isotope records across the K-T boundary in marine sediments show a major negative $\delta^{13}C$ excursion in surface waters of 2‰–3‰ in low latitudes (Zachos et al., 1989, 1992; Keller and Lindinger, 1989; Hsü and McKenzie, 1990), but only 0.5‰–1.0‰ in middle to high latitudes, such as Denmark, Kazakhstan (Schmitz et al., 1992; Keller et al., 1993; Oberhänsli et al., 1998), and DSDP Sites 738, 690, and 527 (Shackleton et al., 1994; Barrera and Keller, 1994). In low latitudes, low $\delta^{13}C$ values persisted through zones Pla and Plb; recovery is recorded in Plc, ~500 k.y. after the K-T boundary (Keller, 1988; D'Hondt et al., 1996). However, in middle to high latitudes $\delta^{13}C$ values continued to decrease through zones Pla and Plb and to the base of zone Plc (Fig. 11). This decreasing $\delta^{13}C$ effect into higher latitudes indicates that the productivity crash was highest in low latitudes and diminished into higher latitudes. This latitudinal difference in surface productivity is likely due to increased upwelling and proximity to a deep-water source in higher latitudes, which may also explain the high rate of species survivorship. The Bjala-1 section defines the productivity crash in the Tethys ocean as south of ~30°N (Fig. 1), in agreement with the conclusion reached by Pardo et al. (1999) based on a summary of faunal assemblages.

Multiple events

Bjala-1 also differs from most other K-T boundary sequences in that it has two clay layers. A single clay layer generally enriched in iridium of cosmic origin characterizes the K-T boundary globally. At Bjala-1 and Bjala-2, the K-T boundary has Ir anomalies of 6.1 ppb and 2.3 ppb, respectively, and coincides with the mass extinction in planktic foraminifera. However, there is a second clay layer stratigraphically within the upper part of zone Pla (*P. eugubina*). This clay layer has a high of Mg smectite content, a small iridium enrichment (0.22 ppb), a major enrichment in palladium (Pd, 1.34 ppb), and anomalies in Ru (0.30 ppb) and Rh (0.13 ppb, Fig. 11). These PGE anomalies suggest a probable volcanic source for this early Danian event. Volcanic rocks in Bulgaria are of Santonian and early Campanian age. However, volcanism related to the Balkan tectonic activity occurred in the early Paleocene in northern Turkey and in Georgia (Senel, 1991; Tzankov and Burchfiel, 1993).

A second event in the early Danian zone Pla with similar PGE anomalies has been documented in a number of sections, including Haiti (Stinnesbeck et al., 2000; Keller et al., 2001; Stueben et al., this volume), Guatemala (Fourcade et al., 1998; Keller and Stinnesbeck, 1999), and Mexico (W. Stinnesbeck, 2000, written commun.). The origin of this second event is still speculative. In Haiti, the PGE anomaly is present in a clayey layer rich in Fe, and has a basalt-like pattern, suggesting a volcanic source (Stinnesbeck et al., 2000; Stueben et al., this volume). In Mexico, a small Ir anomaly of 0.5 ppb correlates with the highest Pt content (1.7 ng/g^{-1}) and minor enrichments of Rh and Pd. The Pt/Ir ratio is nearly chondritic and suggests a cosmic origin (Stinnesbeck, 2000, written commun.). Thus there is evidence of a cosmic event and a volcanic event in the early Danian zone Pla. These data suggest that, in addition to the K-T boundary event, at least one and possibly two major events resulting in PGE anomalies occurred in the early Danian Pla (*P. eugubina*) zone. Further studies are needed to determine the origin and nature of these events.

CONCLUSIONS

The K-T sections at Bjala, Bulgaria, provide critical information that help define the extent and distribution of the environmental effects and the K-T boundary mass extinction across latitudes.

1. The K-T mass extinction in planktic foraminifera at Bjala is similar to that in low latitudes in that all morphologically complex, large (>150 μm) species disappeared at or before the K-T boundary. However, Bjala differs in its pre-K-T decreased species richness, decreased abundance of all low-latitude species, and increased abundance of ecological generalists, all of which suggest unfavorable environmental conditions.

2. Clay mineral data indicate that the low-latitude Tethys region underwent a time of increased humidity during the late Maastrichtian, whereas the middle and higher latitudes under-

went cooling. Temperate to cool conditions prevailed across latitudes during the early Danian.

3. Oxygen isotope data indicate that climate cooling occurred during the last 100 k.y. of the Maastrichtian and accelerated across the K-T boundary, with generally cool climate across latitudes in the early Danian.

4. There is a decreasing $\delta^{13}C$ effect into higher latitudes across the K-T boundary, which indicates that the productivity crash was highest in low latitudes and diminished into higher latitudes. This latitudinal difference in surface productivity is likely due to increased upwelling and proximity to a deep-water source in higher latitudes, which may also explain the possibly high rates of species survivorship in these regions.

5. There are two clay layers with PGE anomalies at Bjala-1 that suggest multiple events. Clay layer 1 marks the K-T boundary and contains an Ir anomaly of 6.1 ppb. Clay layer 2 is in the early Danian *P. eugubina* (Pla) zone and contains a small Ir anomaly, and a major Pd anomaly that suggests a volcanic origin.

ACKNOWLEDGMENTS

We thank A. Preisinger for long discussions and M. Kucera and K. MacLeod for many helpful comments and suggestions. This study was supported by Binational Science Foundation grant 1954059 (Keller), Deutsche Forschungsgemeinschaft grants Sti 128/2-3 and Sti 128/2-4 (Stinnesbeck) and STÜ169/21-2 (Stüben), and the Swiss National Fund grant 8220-028367 (Adatte).

REFERENCES CITED

Abramovich, S., Almogi-Labin, A., and Benjamini, C., 1998, Decline of the Maastrichtian pelagic ecosystem based on planktic foraminiferal assemblage change: Implication for the terminal Cretaceous faunal crisis: Geology, v. 26, p. 63–66.

Abramovich, S., and Keller, G., 2002, High stress upper Maastrichtian paleoenvironment: Inference from planktic foraminifera in Tunisia: Palaeogeography, Palaeoclimatology, Palaeoeology (in press).

Adatte, T., Stinnesbeck, W., and Keller, G., 1996, Lithostratigraphic and mineralogic correlations of near K/T boundary clastic sediments in northeastern Mexico: Implications for origin and nature of deposition: Geological Society of America Special Paper 307, p. 211–226.

Adatte, T., Keller, G., and Stinnesbeck, W., 2002, Late Cretaceous to Early Paleocene climate and sea-level fluctuations: The Tunisian record: Palaeogeography, Palaeoclimatology, Palaeoeology (in press).

Apellaniz, E., Baceta, J. I., Bernaola-Bilbao, G., Nunez-Betelu, K., Orue-Etxebarria, X., Payros, A., Pujalte, V., Robin, E., and Rocchia, R., 1997, Analysis of uppermost Cretaceous-lowermost Tertiary hemipelagic successions in the Basque Country (western Pyrenees): Evidence for a sudden extinction of more than half planktic foraminifer species at the K/T boundary: Bulletin de la Société Géologique de France, v. 168, p. 783–793.

Barrera, E., 1994, Global environmental changes preceding the Cretaceous-Tertiary boundary–early-late Maastrichtian transition: Geology, v. 22, p. 877–880.

Barrera, E., and Keller, G., 1999, Foraminiferal stable isotope evidence for gradual decrease of marine productivity and Cretaceous species survivorship in the earliest Danian: Paleoceanography, v.5, p. 867–890.

Barrera, E., and Keller, G., 1994, Productivity across the Cretaceous-Tertiary boundary in high latitudes: Geological Society of America Bulletin, v. 106, p. 1254–1266.

Berggren, W.A., Kent, D.V., Swisher, C.C., and Aubry, M.P., 1995, A revised Cenozoic geochronology and chronostratigraphy: SEPM (Society for Sedimentary Geology), v. 54, p. 129–213.

Canudo, J.I., 1997, El Kef blind test I results: Marine Micropaleontology, v. 29, p. 73–76.

Chamley, H., 1989, Clay sedimentology: Berlin, Springer-Verlag, Berlin, 623 p.

Cubelic, M., Peccoroni, R., Schäfer, J., Eckhardt, J.-D., Berner, Z., and Stüben, D., 1997, Verteilung verkehrsbedingter Edelmetallimmissionen in Böden, UWSF–Z: Umweltchemicalische Ökotoxycology, v. 9, p. 249–258.

D'Hondt, S., Herbert, T.D., King, J., and Gibson, C., 1996, Planktic foraminifera, asteroids and marine production: Death and recovery at the Cretaceous-Tertiary boundary, *in* Ryder, G., Fastovsky, D. and Gartner, S., eds., The Cretaceous-Tertiary event and other catastrophes in Earth history: Geological Society of America Special Paper 307, p. 303–318.

Elliot, C.W., Aronson J.L., Millard, H.T., and Gierlowski-Kordesch E., 1989, The origin of clay minerals at the Cretaceous/Tertiary boundary in Denmark: Geological Society America Bulletin, v. 101, p. 702–710.

Elliot, C.W., 1993, Origin of the Mg smectite at the Cretaceous/Tertiary (K/T) boundary at Stevns Klint, Denmark: Clays and Clay Minerals, v. 41, p. 442–452.

Fourcade, E., Rocchia, R., Gardin, S., Bellier, J.P., Debrabant, P., Masure, E., and Robin, E., 1998, Age of the Guatemala breccias around the Cretaceous-Tertiary boundary: Relationships with the asteroid impact on the Yucatan: Comptes Rendus de l'Académie des Sciences, Série 2. Sciences de la Terre et des Planètes, v. 327, p. 47–53.

Gaucher, G., 1981, Les facteurs de la pédogenèse, Lelotte, G., édition/publication, Dison, Belgium, 730 p.

Henriksson, A.S., 1993, Biochronology of the terminal Cretaceous calcareous nannofossil zone of Micula prinsii: Cretaceous Research, v. 14, p. 59–68.

Hillier, S., 1995, Erosion, sedimentation and sedimentary origin of clays, *in* Velde, B., ed., Origin and mineralogy of clays, clays and the environment: Berlin, Heidelberg, New York, Springer-Verlag, p. 162–219.

Hsü, K.J., and McKenzie, J.A., 1990, Carbon-isotope anomalies at era boundaries: Global catastrophes and their ultimate cause, *in* Sharpton, V.L., and Ward, P.D., eds., Global catastrophes in Earth history: An interdisciplinary conference on impacts, volcanism, and mass mortality: Geological Society of America Special Paper 247, p. 61–70.

Huber, B.T., 1996, Evidence for planktonic foraminifer reworking versus survivorship across the Cretaceous-Tertiary boundary at high latitudes: Implications for origin and nature of deposition, *in* Ryder, G., Fastovsky, D. and Gartner, S., eds., The Cretaceous-Tertiary event and other catastrophes in Earth history: Geological Society of America Special Paper 307, p. 319–334.

Huber, B.T., Liu, C., Olsson, R.K., and Berggren, W.A., 1994, Comment on "The Cretaceous-Tertiary boundary transition in the Antarctic ocean and its global implications", by G. Keller: Marine Micropaleontology, v. 24, p. 91–99.

Ivanov, M., 1993, Uppermost Maastrichtian ammonites from the uninterrupted Upper Cretaceous-Paleogene section at Bjala (east Bulgaria): Geologica Balcanica, v. 23, 54 p.

Ivanov, M.I., and Stoykova, K., 1994, Cretaceous/tertiary boundary in the area of Bjala, eastern Bulgaria: Biostratigraphical results: Sofia, Bulgaria, Geologica Balcanica, v. 24, p. 3–22.

Ivanov, Z., 1983, Apercu général sur l'évolution géologique et structurale des Balkanides, *in* Ivanov, Z., and Nikolov, T., eds., Guide de l'excursion, Réunion extraordinaire de la Société Géologique de France en Bulgarie: Sofia, Bulgaria, Presse Universitaire, p. 3–28.

Ivanov, Z., 1988, Apercu général sur l'évolution géologique et structurale du massif des Rhodopes dans le cadre des Balkanides: Bulletin de la Société Géologique de France, v. 8, p. 227–240.

Juranov, S.G., and Dzhuranov, S.G., 1983, Planktonic foraminiferal zonation of the Paleocene and the lower Eocene in part of East Balkan Mountains: Sofia, Bulgaria, Geologica Balcanica, v. 13, p. 59–73.

Karoui-Yakoub, N., Zaghbib-Turki, N., and Keller G., 2002, The Cretaceous-Tertiary (K-T) mass extinction in planktic foraminifera at Elles I and El Melah, Tunisia: Palaeogeography, Palaeoclimatology, Palaeoecology (in press).

Keller, G., 1988, Extinction, survivorship and evolution of planktic foraminifera across the Cretaceous/Tertiary boundary at El Kef, Tunisia: Marine Micropaleontology, v. 13, p. 239–263.

Keller, G., 1993, The Cretaceous/Tertiary boundary transition in the Antarctic Ocean and its global implications: Marine Micropaleontology, v. 21, p. 1–45.

Keller, G., 1996, The Cretaceous-Tertiary mass extinction in planktonic foraminifera: Biotic constraints for catastrophe theories, *in* MacLeod, N., and Keller, G., eds., The Cretaceous/Tertiary boundary mass extinction: Biotic and environmental changes: New York, W.W. Norton and Co., p. 49–84.

Keller, G., 1997, Analysis of El Kef blind test I: Marine Micropaleontology, v. 29, p. 89–94.

Keller, G., and Lindinger, M., 1989, Stable isotope, TOC and $CaCO_3$ records across the Cretaceous-Tertiary boundary at El Kef, Tunisia: Palaeogeography, Palaeoclimatology, Palaeoecology, v. 73, p. 243–265.

Keller, G., and MacLeod, N., 1994, Reply to comment on "The Cretaceous-Tertiary boundary transition in the Antarctic Ocean and its global implications" by Huber et al.: Marine Micropaleontology, v. 24, p. 101–118.

Keller, G., and Von Salis Perch-Nielsen, K., 1995, Cretaceous-Tertiary (K-T) mass extinction: Effect of global change on calcareous microplankton, *in* Steven, M., ed., Effects of past global change on life: Studies in geophysics: Washington, D.C., National Academy Press, p. 72–92.

Keller, G., and Stinnesbeck, W., 1996, Sea-level changes, clastic deposits, and megatsunamis across the Cretaceous-Tertiary boundary, *in* MacLeod, N., and Keller, G., eds., The Cretaceous/Tertiary boundary mass extinction: Biotic and environmental changes: New York, W.W. Norton and Co., p. 415–450.

Keller, G., and Stinnesbeck, W., 1999, Ir and the K/T boundary at El Caribe, Guatemala: International Journal of Earth Sciences, v. 88, p. 844–852.

Keller, G., Barrera, E., Schmitz, B., and Mattson, E., 1993, Gradual mass extinction, species survivorship and long-term environmental changes across the Cretaceous/Tertiary boundary in high latitudes: Geological Society of America Bulletin, v. 105, p. 979–997.

Keller, G., Li, L., and MacLeod, N., 1995, The Cretaceous/Tertiary boundary stratotype section at El Kef, Tunisia: How catastrophic was the mass extinction? Palaeogeography, Palaeoclimatology, Palaeoecology, v. 119, p. 221–254.

Keller, G., Adatte, T., Stinnesbeck, W., Stüben, T., Kramar, U. and Berner, Z., 2001, Age and biostratigraphy of Haiti spherule-rich deposits: A multi-event K-T scenario: Canadian Journal of Earth Sciences, v. 38(2), p. 197–227.

Klug, H.P., and Alexander, L., 1974, X-ray diffraction procedures for polycrystalline and amorphous materials (first and second editions): New York, John Wiley and Sons, Inc., 210 p.

Kübler, B., 1983, Dosage quantitatif des minéraux majeurs des roches sédimentaires par diffraction X: Cahiers de l'Institut de Géologie de Neuchâtel, Série ADX, v. 1, 12 p.

Kübler, B., 1987, Cristallinite de l'illite, méthodes normalisées de préparations, méthodes normalisées de mesures: Neuchâtel, Suisse, Cahiers de l'Institut de Géologie, Série ADX, v. 1, 13 p.

Kucera, M., and Malmgren, B.A., 1998, Terminal Cretaceous warming event in the mid-latitude south Atlantic Ocean: Evidence from poleward migration of Contusotruncana contusa (planktonic foraminifera) morphotypes: Palaeogeography, Palaeoclimatology, Palaeoecology, v. 138, p. 1–15.

Li, L., and Keller, G., 1998a, Maastrichtian climate, productivity and faunal turnovers in planktic foraminifera in South Atlantic DSDP Sites 525A and 21: Marine Micropaleontology, v. 33, p. 55–86.

Li, L., and Keller, G., 1998b, Abrupt deep-sea warming at the end of the Cretaceous: Geology, v. 26, p. 995–998.

Luciani, V., 1997, Planktonic foraminiferal turnover across the Cretaceous-Tertiary boundary in the Vajont Valley (southern Alps, northern Italy): Cretaceous Research, v. 18, p. 799–821.

MacLeod, N., and Keller, G., 1991a, Hiatus distribution and mass extinction at the Cretaceous/Tertiary boundary: Geology, v. 19, p. 497–501.

MacLeod, N., and Keller, G., 1991b, How complete are Cretaceous/Tertiary boundary sections? A chronostratigraphic estimate based on graphic correlation: Geological Society of America Bulletin, v. 103, p. 1439–1457.

MacLeod, N., and Keller, G., 1994, Mass extinction and planktic foraminiferal survivorship across the Cretaceous-Tertiary boundary: A biogeographic test: Paleobiology, v. 20, p. 143–177.

Malmgren, B.A., 1991, Biogeographic patterns in terminal Cretaceous planktonic foraminifera from Tethyan and warm transitional waters: Marine Micropaleontology, v. 18, p. 73–99.

Martini, E., 1971, Standard Tertiary and Quaternary calcareous nannoplankton zonations, *in* Farinacci, A., ed., Proceedings 2nd Planktonic Conference: Roma, Tecnoscienza, p. 739–785.

Masters, Bruce A., 1997, El Kef blind test II results: Marine Micropaleontology, v. 29, p. 77–79.

Millot, G., 1970, Geology of clays: Berlin, Springer-Verlag, 499 p.

Moore, D., and Reynolds, R., 1989, X-ray-diffraction and the identification and analysis of clay-minerals: Oxford, New York, Oxford University Press, 332 p.

Nederbragt, A.J., 1998, Quantitative biogeography of late Maastrichtian planktic foraminifera: Micropaleontology, v. 44, p. 385–412.

Oberhaensli, H., Keller, G., Adatte, T., and Pardo, A., 1998, Diagenetically and environmentally controlled changes across the K/T transition at Koshak, Mangyshlak (Kazakstan): Bulletin de la Société Géologique de France, v. 169, p. 493–501.

Olsson, R.K., 1997, El Kef blind test III results: Marine Micropaleontology, v. 29, p. 80–84.

Orue-etxebarria, X., 1997, El Kef blind test IV results: Marine Micropaleontology, v. 29, p. 85–88.

Pardo, A., Ortiz, N., and Keller, G., 1996, Latest Maastrichtian foraminiferal turnover and its environmental implications at Agost, Spain, *in* MacLeod, N., and Keller, G., eds., Cretaceous-Tertiary mass extinction: New York, W.W. Norton and Co., p. 139–172.

Pardo, A., Adatte, T., Keller, G. and Oberhänsli, H., 1999, Paleoenvironmental changes across the Cretaceous-Tertiary boundary at Koshak, Kazakhstan, based on planktic foraminifera and clay mineralogy: Palaeogeography, Palaeoclimatology, Palaeoeology, v. 154, p. 247–273.

Preisinger, A., Aslanian, S., Stoykova, K., Grass, F., Mauritsch, H.J., and Scholger, R., 1993a, Cretaceous/Tertiary boundary sections in the East Balkan area, Bulgaria: Geologica Balcanica, v. 23, p. 3–13.

Preisinger, A., Aslanian, S., Stoykova, K., Grass, F., Mauritsch, H.J., and Scholger, R., 1993b, Cretaceous/Tertiary boundary sections on the coast of the Black Sea near Bjala (Bulgaria): Palaeogeography, Palaeoclimatology, Palaeoecology, v. 104, p. 219–228.

Preisinger, A., 1994, Cyclostratigraphy of the KT boundary section on the coast of the Black Sea near Bjala (Bulgaria), *in* New developments regarding the KT event and other catastrophes in Earth history: Houston, Texas, Lunar and Planetary Institute, LPI Contribution No. 825, p. 90.

Reddi, G.S., Rao, C.R.M., Rao, T.A.S, Lakshmi, S.V., Prabhu, R.K., and Mahalingam, T.R., 1994, Nickel sulphide fire assay—ICPMS method for the determination of platinum group elements: A detailed study on the recovery and losses at different stages: Fresenius Journal of Analytical Chemistry, v. 348, p. 350–352.

Rögl, F., Von Salis, K., Preisinger, A., Aslanian. S., and Summesberger, H. 1996, Stratigraphy across the Cretaceous/Paleogene boundary near Bjala

Bulgaria: Bulletin des Centres de Recherches Exploration-Production Elf Aquitaine, Mémoire 16, p. 673–683.

Schmitz, B., Keller, G., Stenvall, O., 1992, Stable isotope and foraminiferal changes across the Cretaceous-Tertiary boundary at Stevns Klint, Denmark: Arguments for long-term oceanic instability before and after bolide-impact event: Palaeogeography, Palaeoclimatology, Palaeoecology, v. 96, p. 233–260.

Senel, M., 1991, Paleocene-Eocene sediments interbedded with volcanics within the Lycian nappes, Faralya Formation: Bulletin of the Mineral Research and Exploration Institute of Turkey, v. 113, p. 1–14.

Shackleton, N.J., Hall, M.A., and Boersma, A., 1984, Oxygen and carbon isotope data from Leg 74 foraminifers, Initial Reports of the Deep Sea Drilling Project, Volume 74: Washington, D.C., U.S. Government Printing Office, p. 599–612.

Sinnyovsky, D.S., and Sroykova, K.H., 1995, Cretaceous/Tertiary boundary in the Emine Flysch Formation, east Balkan, Bulgaria: Nannofossil evidences: Comptes Rendus Académie Bulgare des Sciences, v. 48, p. 45–48.

Smit, J., 1990, Meteorite impact, extinctions and the Cretaceous/Tertiary boundary: Geologie en Mijnbouw, v. 69, p. 187–204.

Smit, J., and Nederbragt, A.J., 1997, Analysis of the El Kef blind test II: Marine Micropaleontology, v. 29, p. 94–100.

Stinnesbeck, W., Keller, G., Adatte, T., Lopez-Oliva, J.G., and MacLeod, N., 1996, Cretaceous-Tertiary boundary clastic deposits in northeastern Mexico: Impact tsunami or sea-level lowstand?, *in* MacLeod, N., and Keller, G., eds., Cretaceous-Tertiary mass extinctions: Biotic and environmental changes: New York, W.W. Norton and Co., p. 471–518.

Stinnesbeck, W., Keller, G., Adatte, T., Stüben T., Kramar, U., Berner Z., Desremeaux, C., and Moliere, E., 2000, Beloc, Haiti, revisited: Multiple events across the KT boundary in the Caribbean: Terra Nova, v. 11, p. 303–310.

Stoykova, K., and Ivanov, M.I., 1992, An interrupted section across the Cretaceous/Tertiary boundary at the town of Bjala, Black Sea coast (Bulgaria): Comptes Rendus, Académie Bulgare des Sciences, v. 45, p. 61–64.

Tzankov, T., Burchfiel, B.C., 1993, Alpine tectonic evolution of Bulgaria: American Association of Petroleum Geologists Bulletin, v. 77, p. 1671.

Weaver, C.E., 1989, Clays, muds and shales: Amsterdam, Elsevier, Developments in Sedimentology, v. 44, 819 p.

Zachos, J.C., Arthur, M.A., and Dean, W.E., 1989, Geochemical evidence for suppression of pelagic marine productivity at the Cretaceous/Tertiary boundary: Nature, v. 337, p. 61–67.

Zachos, J.C., Aubry, M.P., Berggren, W.A., Ehrendorfer, T., Heider, F., and Lohmann, K.C., 1992, Chemobiostratigraphy of the Cretaceous/Paleocene boundary at Site 750, southern Kerguelen Plateau, *in* Wise, S.W., Jr., Schlich, R., et al., eds., Scientific Results, Proceedings of the Ocean Drilling Program: College Station, Texas, Ocean Drilling Program, v. 120, p. 961–977.

Zereini, F., Skerstupp, B., and Urban, H., 1994, Comparison between the use of sodium and lithium tetraborate in platinum-group element determination by nickel sulphide fire-assay: Geostandards Newsletter, v. 18, p. 105–109.

Manuscript Submitted November 23, 2000; Accepted by the Society March 22, 2001

Geological Society of America
Special Paper 356
2002

Cretaceous-Tertiary boundary planktic foraminiferal mass extinction and biochronology at La Ceiba and Bochil, Mexico, and El Kef, Tunisia

Ignacio Arenillas*
Laia Alegret
José A. Arz
Carlos Liesa
Alfonso Meléndez
Eustoquio Molina
Ana R. Soria
Departamento de Ciencias de la Tierra, Universidad de Zaragoza, Campus Plaza San Francisco, Zaragoza, E-50009, Spain
Esteban Cedillo-Pardo
José M. Grajales-Nishimura
Carmen Rosales-Domínguez
Exploración y Producción, YNF, Instituto Mexicano del Petróleo, Eje Lázaro Cárdenas #152, Mexico D.F., MEX-07730, Mexico

ABSTRACT

Micropaleontology studies across the Cretaceous-Tertiary (K-T) boundary from sections at La Ceiba, Bochil, Mexico, and El Kef, Tunisia, suggest a close cause and effect relationship between the Chicxulub impact and the K-T planktic foraminiferal mass extinction. The K-T planktic foraminiferal biostratigraphy and assemblage turnover in Mexico was examined and the approximate deposition timing of K-T-related material (clastic unit) was estimated. On the basis of established biomagnetochronologic calibrations, the first appearance datum (FAD) of *Parvularugoglobigerina longiapertura* occurred ~3.5–5 k.y. after the K-T boundary, and the FADs of *Parvularugoglobigerina eugubina*, *Eoglobigerina simplicissima*, and *Parasubbotina pseudobulloides* occurred ~15–17.5 k.y., ~28–31 k.y., and ~45–55 k.y., respectively, after the K-T boundary. According to estimated average sedimentation rates and estimated age, the K-T red layer at El Kef was probably formed in <20 yr and the deposition of the K-T clastic unit in the Gulf of Mexico was geologically instantaneous. The last appearance of most Maastrichtian species is just below the K-T impact-generated bed, clearly implying a catastrophic planktic foraminiferal mass extinction.

*E-mail: ias@posta.unizar.es

Arenillas, I., Alegret, L., Arz, J.A., Liesa, C., Meléndez, A., Molina, E., Soria, A.R., Cedillo-Pardo, E., Grajales-Nishimura, J.M., and Rosales-Domínguez, C., 2002, Cretaceous-Tertiary boundary planktic foraminiferal mass extinction and biochronology at La Ceiba and Bochil, Mexico, and El Kef, Tunisia, *in* Koeberl, C., and MacLeod, K.G., eds., Catastrophic Events and Mass Extinctions: Impacts and Beyond: Boulder, Colorado, Geological Society of America Special Paper 356, p. 253-264.

INTRODUCTION

The Cretaceous-Tertiary (K-T) boundary Global Stratotype Section and Point (GSSP) was officially defined at the El Kef section (Tunisia) at the base of a clay layer anomalously rich in iridium (Cowie et al., 1989). The placement of the K-T boundary in Europe and in North Africa is clear because the base of the boundary clay and the impact evidence coincide with the mass extinction of Maastrichtian planktic foraminifera. Most specialists agree that planktic foraminifera at the K-T boundary underwent a mass extinction. However, there is still a controversy about whether the extinction occurred instantaneously or more gradually. Nevertheless, the main planktic foraminiferal extinction coincides biostratigraphically with the K-T boundary in addition to the impact evidence at Tethyan sections (Alvarez et al., 1980; Smit and Hertogen, 1980; Smit, 1982), where the placement of the K-T boundary is unambiguous.

The placement of the K-T boundary in the Gulf of Mexico is more difficult because the deposits in this area are more complex and the boundary is marked by a clastic deposit, the nature and timing of which are controversial. The base of the clastic deposit is usually marked by a layer rich in millimeter-size microspherules, normally interpreted as altered microtektites from the Chicxulub impact crater (Smit et al., 1992a, 1992b, 1996). For this reason, most specialists place the K-T boundary at the base of the clastic bed (Hansen et al., 1987; Bourgeois et al., 1988; Smit et al., 1992b, 1994, 1996). However, others place it above the clastic deposits because the boundary clay and Ir anomaly—one of the main criteria used for K-T boundary placement—are also above them (Keller et al., 1993, 1994a; Stinnesbeck et al., 1993, 1994; Stinnesbeck and Keller, 1996). These last authors interpret that the main planktic foraminiferal extinction also occurs in this horizon, suggesting that the Chicxulub impact and K-T mass extinction do not coincide. Using sedimentological, micropaleontological, and paleoichnological criteria, they proposed that the clastic deposits are the result of successive sea-level lowstands over several thousands of years.

The timing of clastic deposition must be clarified in order to establish whether the clastic deposits were deposited in one event (Smit et al., 1996) or several different events (Stinnesbeck et al., 1993; Keller et al., 1994a; Stinnesbeck and Keller, 1996). Previous studies of planktic foraminiferal biostratigraphy at the most expanded and continuous K-T boundary sections in Spain (Caravaca, Agost, Zumaya) and Tunisia (El Kef, Aïn Settara, Elles) helped to evaluate the continuity and thickness of the K-T sections and correlate biozones and sections (Molina et al., 1998; Arenillas et al., 2000a). In this chapter we consider the K-T boundary at La Ceiba and Bochil in order to examine K-T planktic foraminiferal biostratigraphy and assemblage turnover in Mexico and approximately calibrate the timing of clastic deposition. We compared the findings with the El Kef stratotype (Tunisia) in order to analyze the planktic foraminiferal extinction pattern across the K-T boundary and its possible cause and effect relationship to the Chicxulub impact event.

LOCATION AND STRATIGRAPHY

La Ceiba, Mexico

The La Ceiba section crops out ~7 km south of La Ceiba (Avila Camacho), along the road from La Ceiba to Tlaxcalantongo, in the State of Veracruz (central-eastern Mexico). Here, the K-T boundary is marked by a >1-m-thick clastic unit, intercalated between two pelagic marly units (Méndez and Velasco Formations). Arz et al. (2001) divided this clastic unit into four different subunits according to their texture and architectural characteristics. (1) A basal subunit of calcareous marls is rich in shocked quartz and millimeter-size spherules, altered to clay minerals, most likely representing microtektites (Smit et al., 1992a, 1992b, 1996) as well as bioclasts of shallow-water origin. (2) A second subunit is a 25-cm-thick body of medium-grained sandstone of tabular geometry and a slightly channeled base that displays abundant parallel lamination. (3) The third subunit is composed of a single body of medium- to fine-grained sandstone that has tabular geometry; some internal erosive surfaces separate ~15–20-cm-thick tabular strata, and the subunit exhibits parallel- and cross-lamination, trough cross-stratification, current ripples, and climbing ripples. (4) An upper subunit is a tabular body of fine-grained sandstones, exhibiting parallel and low-angle cross-lamination, asymmetric ripples, and burrow traces on the top. The second and third subunits could correspond to the Unit II in Smit et al. (1996) and fourth subunit to the Unit III. Just above the last clastic bed is a 20-cm-thick subunit of marl, clay, or silt alternating with millimeter-size fine sands. This subunit has Paleocene planktic foraminifera and must be included in the Velasco Formation.

The clastic unit displays a general fining upward, similar to a turbidite sequence (i.e., tends to fine upward) (Fig 1). The sedimentological features support an impact-generated sediment gravity flow at lower bathyal depths, deeper than 1000 m according to benthic foraminiferal assemblages. The clastic unit was deposited under a high sedimentation rate in upper flow regimes and placed in a single-pulse event as turbidites (Arz et al., 2001).

Bochil, Mexico

The Bochil section is ~9 km northeast of the town of Bochil along the road to the PEMEX Soyalo-1 well, in the State of Chiapas (southern Mexico). Here, the K-T boundary has an upward-fining clastic sequence that can be subdivided into three main subunits (Fig 1): (1) a >60-m-thick basal clast-supported coarse carbonate breccia containing blocks of as much as 2 m in diameter, followed by (2) a 4-m-thick medium-grained breccia, including a 2-m-thick bed containing round calcareous objects in a whitish matrix (altered ejecta?), and (3) an ~2-m-thick sandstone to claystone subunit. This section is similar to the one in Guayal, Tabasco, described by Grajales-Nishimura et al. (2000).

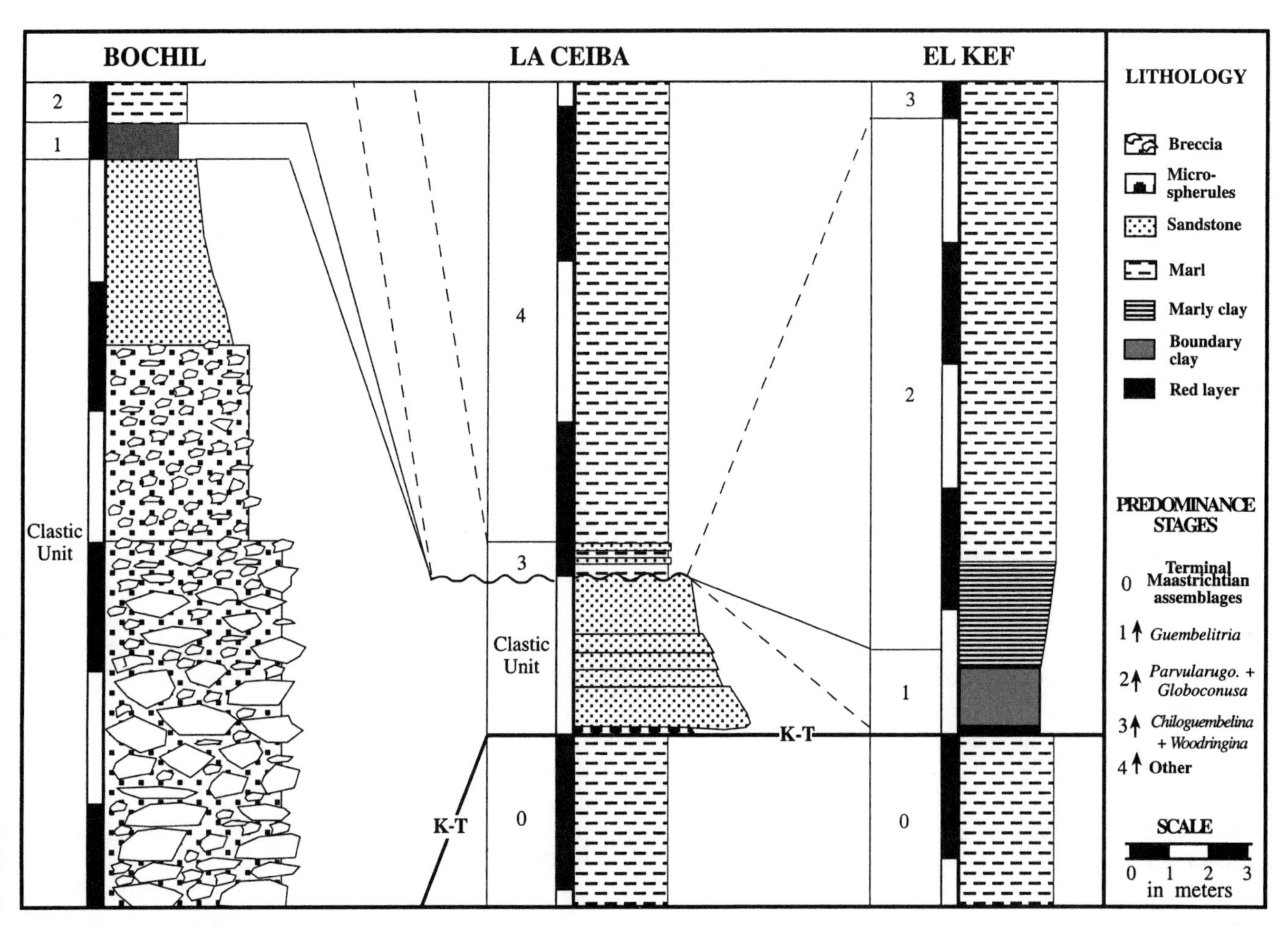

Figure 1. Stratigraphic columns of Bochil, La Ceiba, and El Kef sections across Cretaceous-Tertiary (K-T) boundary, showing planktic foraminiferal predominance stages.

The source of the particles forming the breccia and sandstone units is variable. The lower part of the breccia of subunit 1 is made of clasts containing rudist fragments and larger foraminifera such as *Vaughanina cubensis*, *Orbitoides media*, *Asterorbis aguayoi*, *Sulcoperculina globosa*, *Smoutina* spp., and *Aktinorbitoides* spp., typical of platform margins (Rosales-Domínguez et al., 1997). Some fragments appear to be rip-ups of Maastrichtian sediments eroded during the deposition of clasts from the platform. The clasts of the upper part of this subunit contain alveolinids such as *Chubbina jamaicensis* and miliolids typical of lagoonal environments. Even though the matrix is very scarce, we identified *Globotruncana mariei*, *Rugoglobigerina* spp., *Heterohelix* spp., and the larger foraminifera *Chubbina jamaicensis* and *Orbitoides media*. The breccia of subunit 2 comprises fragments with mostly alveolinids and miliolids, similar to the upper part of subunit 1. The sandstone of subunit 3 has many fragments of loose benthic foraminifera, altered glass fragments, and shocked quartz (Grajales-Nishimura et al., 2000). An Ir anomaly was found in the clay on the top part of this subunit (Montanari et al., 1994).

We interpret that the basal coarse breccia (subunit 1) represents a gravity-flow deposit formed by seismic shacking, triggered by the impact at Chicxulub. Fragments of the platform margin were deposited first, followed by fragments from the inner platform, lagoon environments. Subunits 2 and 3 were originated by a combination of mechanisms, including ballistic sedimentation (ejecta deposits), later reworked and mixed with local material by the backwash of the tsunamis. The Ir-bearing uppermost part of subunit 3 represents the waning stage of turbulent flow when sediments were deposited by the settling of fine-grained suspensions. Stratigraphic and micropaleontological studies support this interpretation.

El Kef, Tunisia

The El Kef section is ~7 km west of El Kef, northwestern Tunisia. This section was officially designated the Cretaceous-Paleogene (K-P) boundary GSSP (Cowie et al., 1989). The K-T boundary was defined at the base of a 50–60-cm-thick marly clay layer (Fig 1), which is intercalated between Maastrichtian and Danian pelagic marly sediments (El Haria Formation). The clay layer has a drop in $CaCO_3$, a maximum of organic carbon, and a negative excursion in $\delta^{13}C$ (Keller and Lidinger, 1989). A 2–3-mm-thick rust-red layer at the base of this clay unit marks the boundary event. This red layer also has an Ir anomaly, crystalline microspherules (altered microkrystites), Ni-rich spinels, and shocked minerals (Smit and Klaver, 1981; Smit, 1982; Robin et al., 1991; Robin and Rocchia, 1998).

The red lamina represents the direct fallout layer of the impact event (Smit, 1982). Small microkrystites identified at El Kef and in other Tethyan sections are finely crystallized, altered to K-feldspar or goethite spherules (Smit and Klaver, 1981; Smit et al., 1992a; Martínez-Ruiz et al., 1997), and were probably derived from the Chicxulub impact. Biostratigraphic data indicate that the red layer with microkrystites from Tethyan sections and the microtektite layer from Gulf Coast sections are geologically isochronous (Arenillas et al., 2000d). Moreover, K-T boundary impact glass in Beloc (Haiti) and El Mimbral (Mexico) sections and melt rock from the Chicxulub impact structures are geochemically and isotopically similar (Sigurdsson et al., 1991; Smit et al., 1992b; Blum et al., 1993), indicating a similar origin. The $^{40}Ar/^{39}Ar$ dating shows that the Chicxulub crater melt rock and K-T boundary microtektites have the same age and that the age of the K-T boundary is 65 Ma (Swisher et al., 1992).

BIOSTRATIGRAPHY AND ASSEMBLAGE TURNOVER

We considered five biozones: *Abathomphalus mayaroensis*, *Plummerita hantkeninoides* (Cretaceous), *Guembelitria cretacea*, *Parvularugoglobigerina eugubina*, and *Parasubbotina pseudobulloides* (Tertiary). The base of each of these biozones is placed at the FAD of the eponymous species, except for the base of *G. cretacea*, which was placed at the last appearance datum (LAD) of *A. mayaroensis*, precisely at the K-T boundary (Molina et al., 1996). Because *P. eugubina* and *P. longiapertura* are considered synonyms by some authors (Smit, 1982; Berggren et al., 1995), the *P. longiapertura* FAD has frequently been used to situate the top of the P0, which is not completely equivalent to the *P. eugubina* biozone used by Arenillas and Arz (2000). These five biozones were recognized at El Kef. At La Ceiba, a hiatus affects the lower part of the Danian (Arz et al., 2001), including the *Guembelitria cretacea* and *Parvularugoglobigerina eugubina* biozones and the lower part of the *Parasubbotina pseudobulloides* biozone. At Bochil, we only studied the lower part of the *G. cretacea* biozone (approximately the P0 zone and the lower part of P1 zone).

Terminal Maastrichtian planktic foraminiferal assemblages were very abundant and diverse at La Ceiba (Gulf of Mexico) and El Kef (Tunisia, Tethys). There are 67 species at El Kef and 63 species at La Ceiba (Arenillas et al., 2000b; Arz et al., 2001). One species (1.6%) at El Kef disappears in the last meter of the Maastrichtian, 46 (74.1%) species extinct at the K-P boundary, and 15 (24.2%) range into the earliest Danian (Arenillas et al., 2000b). At La Ceiba, nearly all Maastrichtian planktic foraminiferal species are found in the last Maastrichtian sample, with no support for a gradual mass extinction pattern in the terminal Cretaceous (Arz et al., 2001). At Bochil, the contact between upper Maastrichtian hemipelagic marl and limestone and basal coarse carbonate breccia (subunit 1) could not be observed. As a result, we could not analyze the planktic foraminiferal biostratigraphy and extinction pattern across the uppermost part of the Maastrichtian.

This catastrophic mass extinction is the largest and most sudden extinction event in the history of planktic foraminifera. It also consistently coincides with the possible catastrophic effects caused by the impact of a large asteroid (Smit, 1982; Molina et al., 1998). Furthermore, there is an evolutionary radiation of planktic foraminifera in the earliest Danian. This radiation always begins above the K-T boundary and never below, which is very consistent with the impact theory. Figure 2 shows planktic foraminiferal ranges from subtropical-temperate sections (Molina et al., 1998; Arz, 2000; Arenillas, 2000) and indicates the probable worldwide model of planktic foraminiferal mass extinction across the K-T boundary. The only species that went extinct just below the K-T boundary were probably *Archaeoglobigerina cretacea* and *Gublerina acuta*. Other Cretaceous species, such as *Archaeoglobigerina blowi* or *Contusotruncana walfischensis*, seem to disappear in some sections before the K-T boundary. However, these pre-K-T disappearances may be local or the remaining Signor-Lipps effect (Molina et al., 1998; Arz, 2000; Arenillas et al., 2000c).

Danian planktic foraminifera evolve sequentially in continuous sections of the K-T boundary. Because these gradual Tertiary species appear over an extensive stratigraphic interval in the lower Danian from El Kef, this section is considered one of the most continuous and expanded marine K-T boundary sections known at the Tethys (Cowie et al., 1989; Keller et al., 1995). A comparison of Tethyan and Gulf coast sections helps to identify four stages in the planktic foraminiferal population in the lowermost Danian (see Arenillas et al., 2000a, 2000c). The lowermost assemblages were dominated successively by *Guembelitria* (stage 1), *Parvularugoglobigerina* and *Globoconusa* (stage 2), *Chiloguembelina* and *Woodringina* (stage 3), and *Eoglobigerina*, *Parasubbotina*, *Praemurica*, and *Globanomalina* (stage 4). Figure 3 shows the correlation between planktic foraminiferal stages and biozones. The infaunal benthic foraminiferal species underwent a general Lazarus effect across stages 1 and 2 (Alegret et al., 2002). The identification of these stages may help to quantify the size of the hiatus across the K-T boundary because they do not involve problematic taxonomic species assignments.

All stages were identified in the El Kef stratotype (Fig. 4; Arenillas et al., 2000b). However, stages 1 and 2 were not identified at La Ceiba (Fig. 5; Arz et al., 2001) and stages 3 and 4 were not studied at Bochil. The simultaneous first appearances of Tertiary planktic foraminifera at La Ceiba mark a hiatus that affects the lower part of the Danian, impeding investigation of the FADs and LADs of possible Maastrichtian survivors and

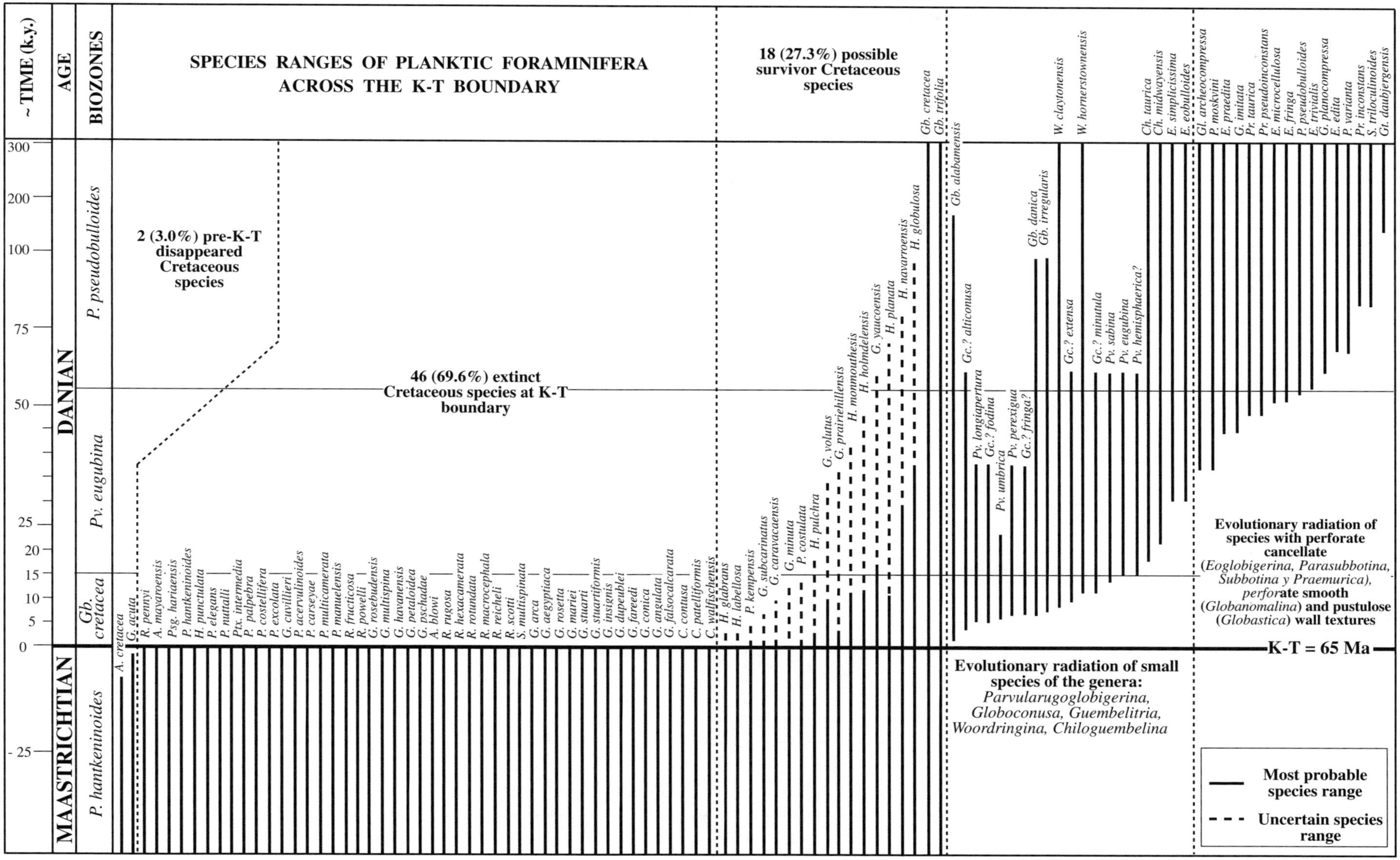

Figure 2. Planktic foraminiferal ranges from subtropical-temperate sections by Arenillas (2000), Arz (2000), and Molina et al. (1998), and probable Cretaceous-Tertiary (K-T) planktic foraminiferal extinction and evolutionary models.

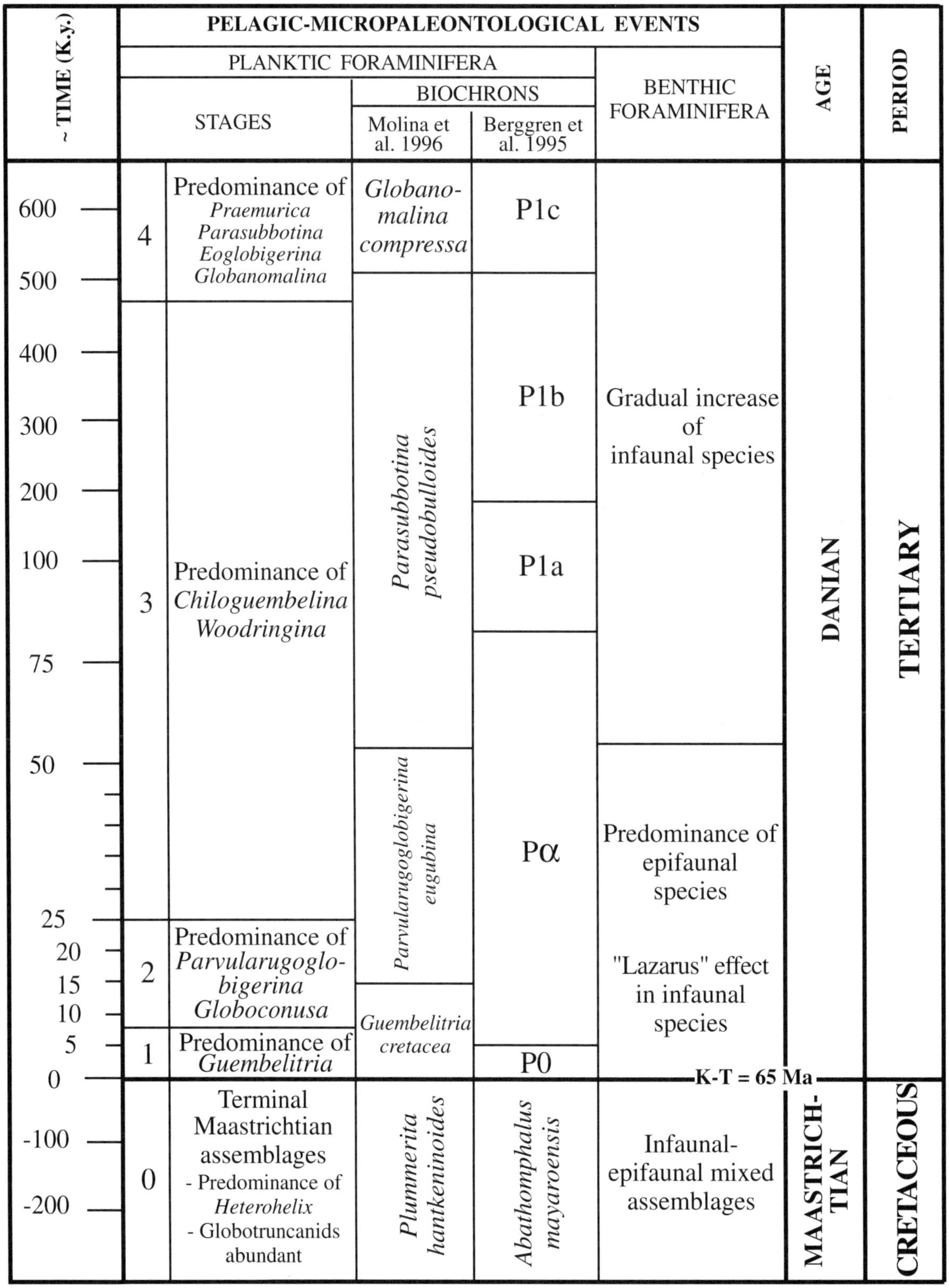

~ TIME (K.y.)	PELAGIC-MICROPALEONTOLOGICAL EVENTS: PLANKTIC FORAMINIFERA – STAGES		BIOCHRONS – Molina et al. 1996	BIOCHRONS – Berggren et al. 1995	BENTHIC FORAMINIFERA	AGE	PERIOD
600, 500	4	Predominance of *Praemurica Parasubbotina Eoglobigerina Globanomalina*	*Globanomalina compressa*	P1c	Gradual increase of infaunal species	DANIAN	TERTIARY
400, 300	3	Predominance of *Chiloguembelina Woodringina*	*Parasubbotina pseudobulloides*	P1b			
200, 100				P1a			
75				Pα			
50			*Parvularugoglobigerina eugubina*		Predominance of epifaunal species		
25, 20, 15, 10	2	Predominance of *Parvularugoglobigerina Globoconusa*			"Lazarus" effect in infaunal species		
5	1	Predominance of *Guembelitria*	*Guembelitria cretacea*	P0			
0	K-T = 65 Ma						
-100, -200	0	Terminal Maastrichtian assemblages - Predominance of *Heterohelix* - Globotruncanids abundant	*Plummerita hantkeninoides*	*Abathomphalus mayaroensis*	Infaunal-epifaunal mixed assemblages	MAASTRICHTIAN	CRETACEOUS

Figure 3. Correlation among planktic foraminiferal predominance stages, biozones, and benthic foraminiferal events.

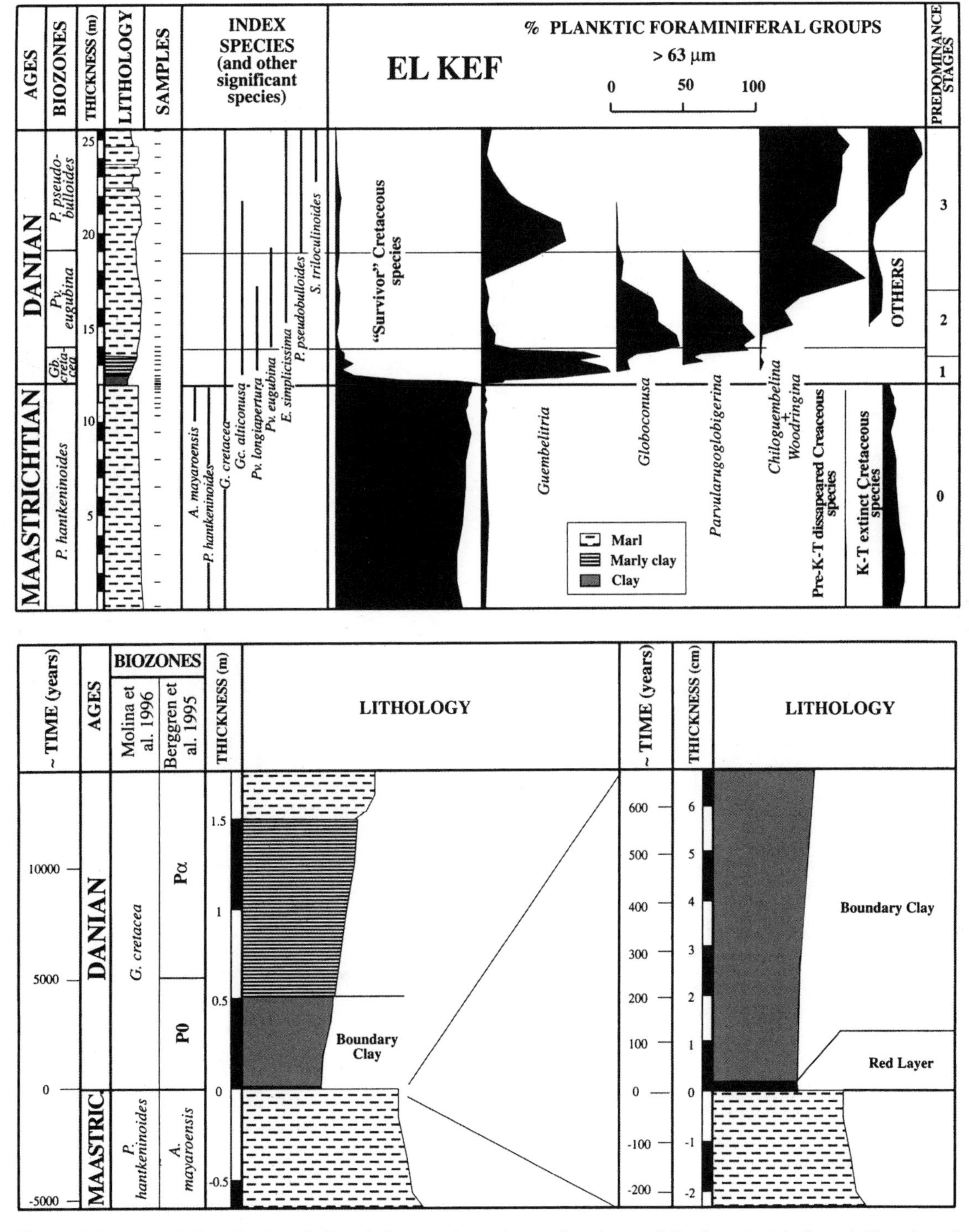

Figure 4. Ranges of planktic foraminifera index species, relative abundance of Danian planktic foraminifera faunal groups, and stages at El Kef section in size fractions larger than 63 μm. Probable duration (in years) of El Kef boundary clay and red layer.

new Danian species. A shorter hiatus has been also identified at other Mexican sections such as El Mimbral and El Mulato (López-Oliva and Keller, 1996; López-Oliva et al., 1998). However, the two first Danian stages were identified at Bochil just above the clastic sequence (Fig. 6). Here, stage 1 and the P0 biozone (lower part of *G. cretacea* biozone) approximately coincide with the clay—anomalously rich in Ir—at the top of impact-derived breccia and sandstone.

BIOMAGNETOCHRONOLOGY

The biomagnetostratigraphic correlation of planktic foraminifera with geomagnetic polarity scales was used to develop a geologic time scale and, ultimately, a magnetobiochronology framework. Using this we approximated the sedimentation rate of the sections and absolute age of the FADs of index taxa (Kent, 1977; Smit et al., 1996). The K-T boundary is near the

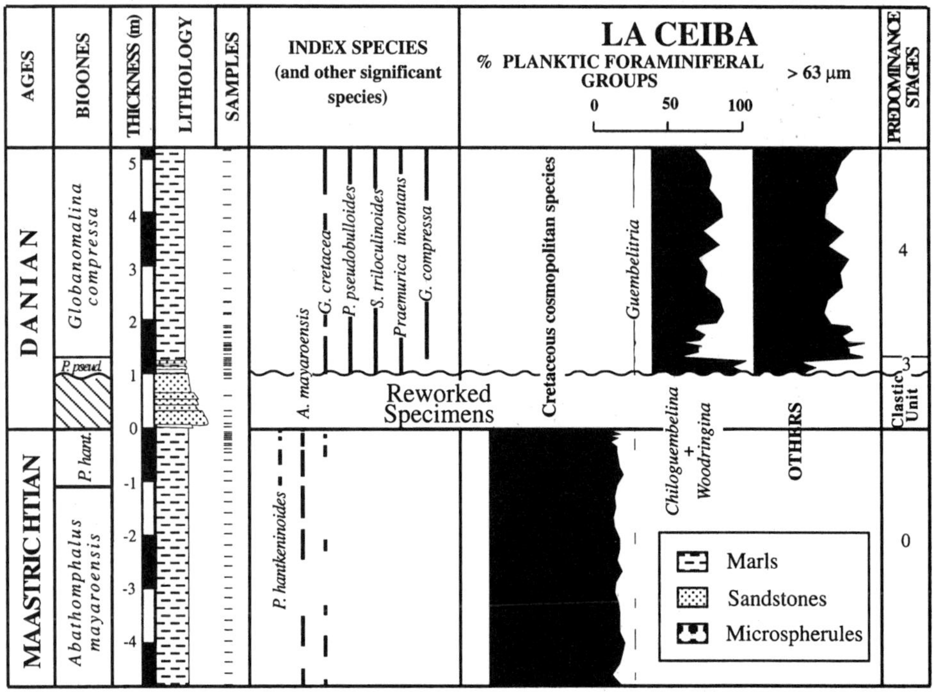

Figure 5. Species ranges of planktic foraminifera index species, relative abundance of Danian planktic foraminifera faunal groups, and stages at La Ceiba section in size fractions larger than 63 µm.

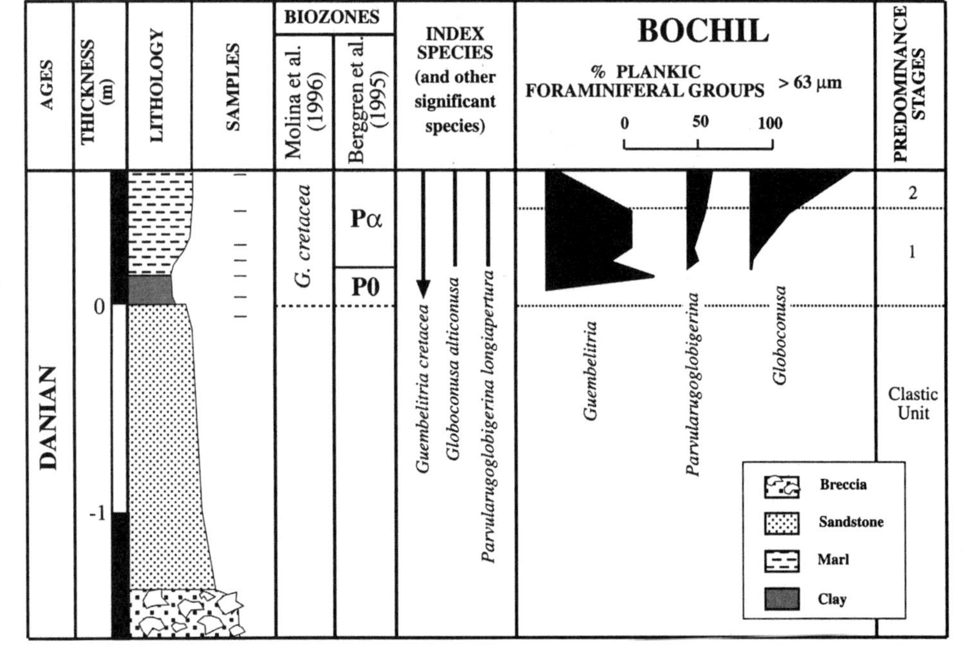

Figure 6. Ranges of planktic foraminifera index species, relative abundance of Danian planktic foraminifera faunal groups, and stages at Bochil section in size fractions larger than 63 µm.

upper middle part of C29r (Lowrie and Alvarez, 1981) and is ca. 65 Ma (Swisher et al., 1992). According to biomagnetochronology studies by Robaszynski and Caron (1995), Gradstein et al. (1995), and Berggren et al. (1995), the base of C29r is 578 k.y. before the K-T boundary, ca. 65.578 Ma, and the top of C29r is 255 k.y. after the K-T boundary, ca. 64.745 Ma. The base of the three first Tertiary biozones is between the K-T boundary and the top of C29r. A biomagnetostratigraphic calibration of the K-T events is possible by correlating the magnetostratigraphic studies of the most expanded and continuous sections in Spain, such as Caravaca, Agost, and Zumaya (Roggenthen, 1976; Smit, 1982; Groot et al., 1989): the sedimentation rates in the basal part of the Danian were ~2.00 cm/k.y., 0.85 cm/k.y., and 1.31 cm/k.y., respectively. The sedimentation rates in the upper part of the Maastrichtian at Caravaca and Agost were ~2.02 cm/k.y. and 1.00 cm/k.y. respectively. Nevertheless, the sedimentation rate in the boundary clay of the three sections must be very similar, probably <0.8 cm/k.y., because the boundary clay has a similar thickness in the three sections (Kaiho et al., 1999; Arenillas, 2000).

It is obvious that the sedimentation rate fluctuated with time. The boundary clay sedimentation was slower than normal Maastrichtian and Danian pelagic marly sediments in Spain, Tunisia, and Mexico. This is due to the decrease in oceanic productivity and flux of biogenic components (Keller and Lidinger, 1989; Arz et al., 1999). However, the rust-red layer sedimentation rate could be greater than that of the boundary clay, because the former includes worldwide dispersed ejecta (iridium-rich dust, shocked minerals, and microspherules). The sedimentation rates in the clastic unit at both Bochil and La Ceiba was evidently much larger. We can only extrapolate the estimated time using the available magnetochronologic data, but the sedimentation rates and absolute ages estimated here must be of the same order of magnitude (see Smit, 1982).

Biomagnetostratigraphic calibration of Agost and Caravaca suggests that the FAD of *Plummerita hantkeninoides* was ~300 k.y. before the K-T boundary (Pardo et al., 1996; Arz, 2000). According to our data, the last pre-K-T extinctions of *Archaeoglobigerina cretacea* and *Gublerina acuta* probably occurred in the last 25 k.y. of the Maastrichtian. Biomagnetostratigraphic calibration of Agost, Zumaya, Caravaca, and Gubbio (Arenillas, 1998, 2000) suggests that the FAD of *P. longiapertura* (top of P0) was ~3.5–5 k.y. after the K-T boundary, and the FADs of *P. eugubina, Eoglobigerina simplicissima, and P. pseudobulloides* were ~15–17.5 k.y., ~28–31 k.y., and ~45–55 k.y., respectively, after the K-T boundary (Figs. 2 and 3). According to these data, the sedimentation rate in the basal part of Danian at El Kef is ~14.9 cm/k.y. At Bochil, the sedimentation rate could be ~3.4 cm/k.y. from the base of clay topping the clastic and breccia sequence.

DISCUSSION

The K-T stratotype definition of El Kef (Cowie et al., 1989) implies that the K-T boundary is at the base of the clay that includes the red-rust layer containing the Ir anomaly and microspherules. For this reason, the K-T boundary in other sections should also be placed at the base of the layer with the impact evidence, because this K-T horizon is the most isochronous worldwide (Smit et al., 1996). The closer the studied section is to the Chicxulub crater, the clearer is the impact evidence. The energy transmitted by the Chicxulub impact was able to collapse continental margins and generate large tsunami waves that affected Gulf Coast shelf sedimentation (Bourgeois et al., 1988; Bralower et al., 1998). Unstable deposits with microtektites were mobilized from the shelf, forming sediment gravity currents toward the slope and deep basin (Bohor, 1996).

The coarse ejecta (e.g., breccia, microtektites) and the Ir anomaly are separated by a controversial sandstone unit in the Gulf of Mexico. Arz et al. (2001) have suggested that sandstone deposits at La Ceiba were deposited in a single-pulse event as turbidites at lower bathyal depths. This sedimentation model is similar to the one proposed by Bohor (1996) for other Gulf Coast sections. If all these deposits were directly or indirectly caused by the bolide impact mentioned in Alvarez et al. (1980), then the clastic breccia, microspherules, and tsunami-derived sandstone unit are equivalent to the K-T red-rust layer from the El Kef stratotype. According to the criteria used at El Kef, the K-T boundary at La Ceiba must be placed at the base of the clastic (microspherules) layer because it is equivalent to the base of the boundary clay at El Kef (Arz et al., 2001). Similarly, the K-T boundary at Bochil should be placed at the base of the K-T clastic complex just below the impact-generated polymict debris flow.

The identification of the stages 1 and 2 at Bochil, just above the polymict breccia and sandstone complex, is significant because it supports the following two views.

1. The K-T boundary and mass extinction are just below the clastic unit. According to the sedimentological characteristics of the breccia and sandstone bed, the Maastrichtian specimens in the clastic unit are obviously reworked. Therefore, the last appearance of the mostly indigenous Maastrichtian species coincides with the base of the clastic unit at La Ceiba (and probably at Bochil) and the K-T boundary must be placed at this horizon (Arz et al., 2001).

2. The planktic foraminiferal evolutionary radiation also occurs just above clastic unit in the Gulf of Mexico. At El Kef, stage 1 is located just above the K-T boundary and red-rust layer, which confirms that the Gulf Coast clastic unit is equivalent to the Tethyan and the worldwide red layer. Moreover, the planktic foraminiferal evolution model and the different stages across the basal part of Danian are similar worldwide.

How long did the red layer deposition last? The El Kef red layer is 2–3 mm thick. Assuming a constant sedimentation rate (14.9 cm/k.y.), the red layer deposition would have taken ~13–20 yr. However, $CaCO_3$ in the clay layer and red layer is lower than in the upper Maastrichtian and lower Danian marls, indicating that the sedimentation rate fluctuated across the K-T boundary. At El Kef, the average $CaCO_3$ in the red layer and the clay layer is respectively 3 and 10 wt%, and 35 wt% in the lowermost Danian marl. If the depositional flux of the residual mineral matrix was constant (Kaiho et al., 1999), the sedimentation rate in the El Kef red layer and clay layer would be ~10 and 10.7 cm/k.y., respectively. In this case, the red layer deposition would have taken 20–30 yr. However, this is uncertain because the depositional flux of extraterrestrial and impact-derived material was larger in the red layer than the uppermost Maastrichtian and lowermost Danian. The duration of the red layer deposition at El Kef was probably <20 yr (Fig. 4), i.e., geologically instantaneous.

How long did the clastic unit deposition last? Fine-grained Ir-rich particles settle more slowly through the atmosphere and water column than coarse ejecta (Smit et al., 1996). Ir concentration must have increased after the high-energy episode represented by the coarse ejecta (breccia, microspherules, and shocked minerals) and megatsunami clastic unit. This suggests that the clastic unit, including polymict breccia, microspherules (microtektites), and megatsunami-derived sandstone, was deposited between the K-T boundary and red layer or boundary clay. If the deposition of the red layer occurred during the first years of the Danian, we can assume that the K-T clastic unit formed almost instantaneously.

Stinnesbeck et al. (1993) and Keller et al. (1994b) suggested that the clastic units were deposited over a long period of time. However, this cannot have been the case because they are characterized by a high sedimentation rate in upper flow regimes (Bohor, 1996; Arz et al., 2001), suggesting that K-T sandstones were deposited rapidly over a very short period. According to Bralower et al. (1998), much of the impact-derived material would have accumulated within hours or days after the impact. The gravity-flow deposit throughout the Gulf of Mexico basin was probably the result of destabilization of unconsolidated shelf sediments after the huge accumulation of impact- and tsunami-generated sediments above the Gulf Coast shelf and associated seismic events. Consequently, the main clastic deposition may have lasted only days or weeks. If the last indigenous specimens of most Maastrichtian planktic foraminifera are only found just below the K-T impact-generated bed, we conclude that their mass-extinction model was clearly catastrophic. We suggest that there is a close relation between the K-T planktic foraminiferal mass extinction and the Chicxulub impact.

CONCLUSIONS

A micropaleontological study across the K-T boundary from the El Kef, Tunisia, and La Ceiba and Bochil, Mexico, sections was used to examine K-T planktic foraminiferal biostratigraphy and assemblage turnover in Mexico and calibrate the timing of clastic deposition. The El Kef section is one of the most continuous and expanded marine K-T boundary sections known at the Tethys, because the gradual Tertiary species appear over an extensive stratigraphic interval at the lower part of the Danian. All stages were identified in the El Kef stratotype, including the peak of *Guembelitria* (stage 1). However, stages 1 and 2 were not identified at La Ceiba, indicating a hiatus in the lower Danian. However, the identification of stages 1 and 2 at Bochil support two views: (1) the K-T boundary and mass extinction are placed below the impact-generated polymict debris-flow and sandstone complex, and (2) the planktic foraminiferal evolutionary radiation also occurs just above impact-generated clastic complex in the Gulf of Mexico. Consequently, our micropaleontological and sedimentological evidence is consistent with the K-T impact theory and the impact on the Yucatan Peninsula.

On the basis of calculated average sedimentation rates and estimated age, the duration of the K-T red layer deposition at El Kef is probably <20 yr. The impact-generated clastic unit in the Gulf of Mexico was deposited between the K-T boundary and the red layer or boundary clay. If the red layer deposition occurred during the first years of the Danian, we can assume that the K-T clastic unit was geologically instantaneous. If the last certain indigenous specimens of most of Maastrichtian species are only found just below the K-T impact-generated bed, it appears that the planktic foraminiferal mass extinction was catastrophic.

ACKNOWLEDGMENTS

We thank B. Schmitz and J. Smit for reviews of the manuscript, and Morris Villarroel (McGill University, Canada) for correcting the English. Laia Alegret is grateful to the Spanish Ministerio de Educación y Cultura for the predoctoral grant FP98. This research was funded by Ministerio de Ciencia y Tecnología, Dirección General de Investigación (Spain) project BTE 2001-1809 and Universidad de Zaragoza (Spain) project UZ2001-CIEN-01 (no. 221-153). IMP researchers recognize the contribution from the FIES-9575-1 Project and the D.0002 Project from YNF Program (Mexico).

REFERENCES CITED

Alegret, L., Arenillas, I., Arz, J.A., and Molina, E., 2002, Eventoestratigrafía del límite Cretácico/Terciario en Aïn Settara (Tunicia): disminución de la productividad y oxigenación oceánicas: Revista Mexicana de Ciencias Geológicas, v. 19, no. 2 (in press).

Alvarez, L.W., Alvarez, W., Asaro, F., and Michel, H.V., 1980, Extraterrestrial cause for the Cretaceous-Tertiary extinction: Science, v. 208, p. 1095–1108.

Arenillas, I., 1998, Biostratigrafía con foraminíferos planctónicos del Paleoceno y Eoceno inferior de Gubbio (Italia): Calibración biomagnetoestratigráfica: Neues Jahrbuch für Geologie and Paläontologie, Monatshefte, v. 5, p. 299–320.

Arenillas, I., 2000, Los foraminíferos planctónicos del Paleoceno-Eoceno inferior: Sistemática, bioestratigrafía, cronoestratigrafía y paleoceanografía [Ph.D. thesis]: Zaragoza, Spain, Prensas Universitarias de Zaragoza, 513 p.

Arenillas, I., and Arz, J.A., 2000, *Parvularugoglobigerina eugubina* type-sample at Ceselli (Italy): Planktic foraminiferal assemblage and lowermost Danian biostratigraphic implications: Rivista Italiana de Paleontologia e Stratigrafia, v. 106, p. 379–390.

Arenillas, I., Arz, J.A. and Molina, E., 2000a, Spanish and Tunisian Cretaceous-Tertiary boundary sections: A planktic foraminiferal biostratigraphic comparison and evolutive events: Geologiska Föreningens i Stockholm Förhandlingar, v. 122, p. 11–12.

Arenillas, I., Arz, J.A., Molina, E., and Dupuis, C., 2000b, An independent test of planktic foraminiferal turnover across the Cretaceous/Paleogene (K/P) boundary at El Kef, Tunisia: Catastrophic mass extinction and possible survivorship: Micropaleontology, v. 46, p. 31–49.

Arenillas, I., Arz, J.A., Molina, E., and Dupuis, C., 2000c, The Cretaceous/Paleogene (K/P) boundary at Aïn Settara, Tunisia: Sudden catastrophic mass extinction in planktic foraminifera: Journal of Foraminiferal Research, v. 30, p. 46–62.

Arenillas, I., Alegret, L., Arz, J.A., Meléndez, A., Molina, E., and Soria, A.R., 2000d, Secuencia estratigráfica y eventos evolutivos de foraminíferos en el tránsito Cretácico-Terciario: Geotemas, v. 2, p. 25–28.

Arz, J.A., 2000, Los foraminíferos planctónicos del Campaniense y Maastrichtiense: Bioestratigrafía, cronoestratigrafía y eventos paleoecológicos [Ph.D. thesis]: Zaragoza, Spain, Prensas Universitarias de Zaragoza, 419 p.

Arz, J.A., Arenillas, I., Molina, E., and Dupuis, C., 1999, La extinción en masa de foraminíferos planctónicos en el límite Cretácico/Terciario (K/T) de Elles (Túnez): Los efectos tafonómico y "Signor-Lipps": Revista de la Sociedad Geológica de España, v. 12, no. 2, p. 251–268.

Arz, J.A., Arenillas, I., Soria, A.R., Alegret, L., Grajales-Nishimura, J.M., Liesa, C., Mélendez, A., Molina, E., and Rosales, M.C., 2001, Micropaleontology and sedimentology across the Cretaceous/Paleogene boundary at La Ceiba (Mexico): Impact-generated sediment gravity flows: Journal of South American Earth Sciences, v. 14, no. 5, p. 505–519.

Berggren, W.A., Kent, D.V. Swisher, C.C., and Aubry, M.P., 1995, A revised Cenozoic geochronology and chronostratigraphy, *in* Berggren, W.A., Kent, D.V., Aubry, M.P., and Hardenbol, J., eds., Geochronology, time scales and stratigraphic correlation: Society of Economic Paleontologists and Mineralogists Special Publication, v. 54, p. 129–212.

Blum, J.D., Chamberlain, C.P., Hingston, M.P., Koeberl, C., Marín, L.E., Schuraytz, B.C., and Sharpton, V.L., 1993, Isotopic comparison on K/T boundary impact glass with melt rock from the Chicxulub and Manson impact structures: Nature, v. 364, p. 325–327.

Bohor, B.F., 1996, A sediment gravity flow hypothesis for siliciclastic units at the K/T boundary, northeastern Mexico, *in* Ryder, G., Fastovsky, D., and Gartner, S., eds., The Cretaceous-Tertiary event and other catastrophes in Earth history: Geological Society of America Special Paper 307, p. 183–195.

Bourgeois, J., Hansen, T.A., Wiberg, P.L., and Kauffman, E.G., 1988, A tsunami deposit at the Cretaceous-Tertiary boundary in Texas: Science, v. 241, p. 567–570.

Bralower, T.J., Paull, C.K., and Leckie, R.M., 1998, The Cretaceous-Tertiary boundary cocktail: Chicxulub impact triggers margin collapse and extensive sediment gravity flows: Geology, v. 26, p. 331–334.

Cowie, J.W., Zieger, W., and Remane, J., 1989, Stratigraphic commission accelerates progress, 1984–1989: Episodes, v. 112, p. 79–83.

Gradstein, F.M., Agterberg, F.P., Ogg, J.G., Hardenbol, J., Van Veen, P., Thierry, J., and Huang, Z., 1995, A Triassic, Jurassic and Cretaceous time scale, *in* Berggren, W.A., Kent, D.V., Aubry, M.P., and Handerbol, J., eds., Geochronology, time scales and stratigraphic correlation: Society of Economic Paleontologists and Mineralogists Special Publication, v. 54, p. 95–126.

Grajales-Nishimura, J.M., Cedillo-Pardo, E., Rosales-Domínguez, C., Morán-Zenteno, D.J., Alvarez, W., Claeys, P., Ruíz-Morales, J., García-Hernández, J., Padilla-Avila, P., and Sanchez-Ríos, A., 2000, Chicxulub impact: The origin of reservoir and seal facies in the southeastern Mexico oil fields: Geology, v. 28, p. 307–310.

Groot, J.J., De Jonge, R.B.G., Langereis, C.G., Ten Kate, W.G.H.Z., and Smit, J., 1989, Magnetostratigraphy of the Cretaceous-Tertiary boundary at Agost (Spain): Earth and Planetary Science Letters, v. 94, p. 385–397.

Hansen, T., Farrand, R.B., Montgomery, H.A., Billman, H.G., and Blechschmidt, G., 1987, Sedimentology and extinction patterns across the Cretaceous-Tertiary boundary interval in East Texas: Cretaceous Research, v. 8, p. 229–252.

Kaiho, H., Kajiwara, Y., Tazaki, K., Ueshima, M., Takeda, N., Kawahata, H., Arinobu, T., Ishiwatari, R., Hirai, A., and Lamolda, M.A., 1999, Oceanic primary productivity and dissolved oxygen levels at the Cretaceous/Tertiary boundary: Their decrease, subsequent warming and recovery: Paleoceanography, v. 14, p. 511–524.

Keller, G., and Lidinger, M., 1989, Stable isotope, TOC and $CaCO_3$ records across the Cretaceous/Tertiary boundary at El Kef, Tunisia: Palaeogeography, Palaeoclimatology, Palaeoecology, v. 73, p. 243–265.

Keller, G., MacLeod, N., Lyons, J.B., and Officer, C.B., 1993, Is there evidence for Cretaceous-Tertiary boundary-age deep-water deposits in the Caribbean and Gulf of Mexico?: Geology, v. 21, p. 776–780.

Keller, G., Stinnesbeck, W., and López-Oliva, J.G., 1994a, Age, deposition and biotic effects of the Cretaceous/Tertiary boundary event at Mimbral, NE Mexico: Palaios, v. 9, p. 144–157.

Keller, G., Stinnesbeck, W., Adatte, T., López-Oliva, J.G., and MacLeod, N., 1994b, The KT boundary clastic deposits in northeastern Mexico as product of noncatastrophic geologic processes?, *in* Keller, G., Stinnesbeck, W., Adatte, T., MacLeod, N., and Lowe, D.R., eds., Field guide to Cretaceous-Tertiary boundary sections in northeastern Mexico: Houston, Texas, Lunar and Planetary Institute, LPI Contribution No. 827, p. 65–94.

Keller, G., Li, L., and MacLeod, N., 1995, The Cretaceous/Tertiary boundary stratotype sections at El Kef, Tunisia: How catastrophic was the mass extinction?: Palaeogeography, Palaeoclimatology, Palaeoecology, v. 119, p. 221–254.

Kent, D.V., 1977, An estimate of the duration of the faunal change at the Cretaceous-Tertiary boundary: Geology, v. 5, p. 769–771.

López-Oliva, J.G., and Keller, G., 1996, Age and stratigraphy of near-K/T boundary siliciclastic deposits in Northeastern Mexico, *in* Ryder, G., Fastovsky, D., and Gartner, S., eds., The Cretaceous-Tertiary event and other catastrophes in Earth history: Geological Society of America Bulletin Special Paper 307, p. 227–242.

López-Oliva, J.G., Keller, G., and Stinnesbeck, W., 1998, El límite Cretácico/Terciario (K/T) en el Noreste de México: Extinción de los foraminíferos planctónicos: Revista Mexicana de Ciencias Geológicas, v. 15, no. 1, p. 109–113.

Lowrie, W., and Alvarez, W., 1981, One hundred years of geomagnetic polarity history: Geology, v. 9, p. 392–397.

Martínez-Ruiz, F., Ortega-Huertas, M., Palomo, I., and Acquafredda, P., 1997, Quench textures in altered spherules from the Cretaceous-Tertiary boundary layer at Agost and Caravaca, SE Spain: Sedimentary Geology, v. 113, p. 137–147.

Molina, E., Arenillas, I., and Arz, J.A., 1996, The Cretaceous/Tertiary boundary mass extinction in planktic foraminifera at Agost, Spain: Revue de Micropaléontologie, v. 39, p. 225–243.

Molina, E., Arenillas, I., and Arz, J.A., 1998, Mass extinction in planktic foraminifera at the Cretaceous/Tertiary boundary in subtropical and temperate latitudes: Bulletin de la Société Géologique de France, v. 169, p. 351–363.

Montanari, S., Claeys, P., Asaro, F., Bermudez, J., and Smit, J., 1994, Preliminary stratigraphy and iridium and other geochemical anomalies across the KT boundary in the Bochil section (Chiapas, southeastern Mexico), *in* KT event and other catastrophes in Earth history: Houston, Texas, Lunar and Planetary Institute, LPI Contribution No. 825, p. 84–85.

Pardo, A., Ortiz, N., and Keller, G., 1996, Latest Maastrichtian and Cretaceous-Tertiary boundary foraminiferal turnover and environmental changes at Agost, Spain, *in* MacLeod N., and Keller, G., eds., Cretaceous-Tertiary mass extinctions: Biotic and environmental changes: New York, Norton and Co., p. 139–171.

Robaszynski, F., and Caron, M., 1995, Foraminiferes planctoniques du Crétacé:

Commentaire de la zonation Europe-Mediterranée: Bulletin de la Société Géologique de France, v. 166, no. 6, p. 681–692.
Robin, E., and Rocchia, R., 1998, Ni-rich spinel at the Cretaceous-Tertiary boundary of El Kef, Tunisia: Bulletin de la Société Géologique de France, v. 169, no. 3, p. 365–372.
Robin, E., Boclet, D., Bonte, P., Froget, L., Jehanno, C., and Rocchia, R., 1991, The stratigraphic distribution of Ni-rich spinels in Cretaceous-Tertiary boundary rocks at El Kef (Tunisia), Caravaca (Spain) and Hole 761C (Leg 122): Earth and Planetary Science Letters, v. 107, p. 715–721.
Roggenthen, W.M., 1976, Magnetic stratigraphy in the Paleocene: A comparison between Spain and Italy: Memorie de la Societa Geologica Italiana, v. 15, p. 73–82.
Rosales-Dominguez, M.C., Bermúdez-Santana, J., and Aguilar-Piña, M., 1997, Mid and Upper Cretaceous foraminiferal assemblages from the Sierra de Chiapas, southeastern Mexico: Cretaceous Research, v. 18, p. 697–712.
Sigurdsson, H., D'Hondt, S., Arthur, M.A., Bralower, T.J., Zachos, J.C., Van Fossen, M., and Channell, E.T., 1991, Glass from the Cretaceous/Tertiary boundary in Haiti: Nature, v. 349, p. 482–487.
Smit, J., 1982, Extinction and evolution of planktonic foraminifera after a major impact at the Cretaceous/Tertiary boundary, *in* Silver, L.T., and Schultz, P.H., eds., Geological implications of impacts of large asteroid and comets on the earth: Geological Society of America Special Paper 190, p. 329–352.
Smit, J., and Hertogen, J., 1980, An extraterrestrial event at the Cretaceous-Tertiary boundary: Nature, v. 285, p. 198–200.
Smit, J., and Klaver, G., 1981, Sanidine spherules at the Cretaceous-Tertiary boundary indicate a large impact event: Nature, v. 292, p. 47–49.
Smit, J., Alvarez, W., Montanari, A., Swinburne, N.H.M., Van Kempen, T.M., Klaver, G.T., and Lustenhouwer, W.J., 1992a, "Tektite" and microkrystites at the Cretaceous-Tertiary boundary: Two strewnfields, one crater?: Proceedings, Lunar and Planetary Science Conference, v. 22, p. 87–100.
Smit, J., Montanari, A., Swinburne, N.H.M., Alvarez, W., Hildebrand, A.R., Margolis, S.V., Claeys, Ph., Lowrie, W., and Asaro, F., 1992b, Tektite-bearing, deep-water clastic unit at the Cretaceous-Tertiary boundary in northeastern Mexico: Geology, v. 20, p. 99–103.
Smit, J., Roep, T.B., Alvarez, W., Claeys, P., and Montanari, A., 1994, Deposition of channel deposits near the Cretaceous-Tertiary boundary in northeastern Mexico: Catastrophic or "normal" sedimentary deposits?: Comment: Is there evidence for Cretaceous-Tertiary boundary age deep-water deposits in the Caribbean and Gulf of Mexico?: Reply: Geology, v. 22, p. 953–959.
Smit, J., Roep, T.B., Alvarez, W., Montanari, A., Claeys, P., Grajales-Nishimura, J.M., and Bermudez, J., 1996, Coarse-grained, clastic sandstone complex at the K/T boundary around the Gulf of Mexico: Deposition by tsunami waves induced by the Chicxulub impact?, *in* Ryder, G., Fastovsky, D., and Gartner, S., eds., The Cretaceous-Tertiary event and other catastrophes in Earth history: Geological Society of America Special Paper 307, p. 151–182.
Stinnesbeck, W., and Keller, G., 1996, K/T boundary coarse-grained siliciclastic deposits in northeastern Mexico and northeastern Brazil: Evidence for mega-tsunami or sea-level changes?, *in* Ryder, G., Fastovsky, D., and Gartner, S., eds., The Cretaceous-Tertiary event and other catastrophes in Earth history: Geological Society of America Special Paper 307, p. 197–209.
Stinnesbeck, W., Barbarin, J.M., Keller, G., López-Oliva, J.G., Pivnik, D.A., Lyons, J.B., Officer, C.B., Adatte, T., Graup, G., Rocchia, R., and Robin, E., 1993, Deposition of channel deposits near the Cretaceous-Tertiary boundary in northeastern Mexico: Catastrophic or "normal" sedimentary deposits?: Geology, v. 21, p. 797–800.
Stinnesbeck, W., Keller, G., Adatte, T., and MacLeod, N., 1994, Deposition of channel deposits near the Cretaceous-Tertiary boundary in northeastern Mexico: Catastrophic or "normal" sedimentary deposits?: Reply: Is there evidence for Cretaceous-Tertiary boundary age deep-water deposits in the Caribbean and Gulf of Mexico?: Geology, v. 22, p. 955–956.
Swisher, C.C., Grajales-Nishimura, J.M., Montanari, A., Cedillo-Pardo, E., Margolis, S.V., Claeys, P., Alvarez, W., Smit, J., Renne, P., Maurrasse, F.J.-M.R., and Curtis, G.H., 1992, Chicxulub crater melt-rock and K-T boundary tektites from Mexico and Haiti yield coeval $^{40}Ar/^{39}Ar$ ages of 65 Ma: Science, v. 257, p. 954–958.

Manuscript Submitted October 10, 2000; Accepted by the Society March 22, 2001

Geological Society of America
Special Paper 356
2002

Distribution pattern of calcareous dinoflagellate cysts across the Cretaceous-Tertiary boundary (Fish Clay, Stevns Klint, Denmark): Implications for our understanding of species-selective extinction

Jens Wendler*
Helmut Willems
Geological Department, University of Bremen, P.O. Box 330440, 28334 Bremen, Germany

ABSTRACT

The distribution patterns of calcareous dinoflagellate cysts were studied in the classic Cretaceous-Tertiary (K-T) boundary section of Stevns Klint, Denmark, focusing mainly on the response of the cyst association to an abrupt environmental catastrophe. A major part of the Fish Clay, which covers the K-T boundary at its base and is exposed in the investigated section, contains fallout produced by an asteroid impact. Calcareous dinoflagellate cysts are the best-preserved remains of carbonate-producing phytoplankton in this layer. The potential of this group of microfossils for the analysis of survival strategies and extinction patterns has been underestimated. The cyst species of the investigated section can be grouped into four assemblages that represent victims, survivors, opportunists, and specially adapted forms. The victims (Pithonelloideae) were an extremely successful group throughout the Upper Cretaceous, but were restricted to the narrow outer shelf. This restriction minimized their spatial distribution, which generally should be large to facilitate escape from unfavorable conditions. Spatial restriction optimized the population decrease by mass mortality, disabling a successful recovery. In contrast, the survivors that became the dominating group in the Danian had a wide spatial range from the shelf environment to the oceanic realm. A unique calcareous dinocyst assemblage in the Fish Clay shows that even under the stressed conditions immediately following the impact event, some species flourished due to special adaptation or high ecological tolerance. The ability of these dinoflagellate species to form calcareous resting cysts in combination with their generally wide spatial distribution in a variety of environments appears to be the main reason for a low extinction rate at the K-T boundary as opposed to the high extinction rate of other phytoplankton groups, such as the coccolithophorids.

INTRODUCTION

The Stevns Klint, Denmark, section represents one of the classic Cretaceous-Tertiary (K-T) boundary sections and has been extensively studied during the twentieth century (e.g., Christensen et al., 1973; Birkelund and Håkansson, 1982; Schmitz et al., 1992). The boundary clay in this section, the so-called Fish Clay, consists for a major part of products of impact fallout at its base (Alvarez et al., 1980; Kastner et al., 1984; Smit, 1999) and can be considered to represent the abrupt

*E-mail: wendler@uni-bremen.de

Wendler, J., and Willems, H., 2002, Distribution pattern of calcareous dinoflagellate cysts across the Cretaceous-Tertiary boundary (Fish Clay, Stevns Klint, Denmark): Implications for our understanding of species-selective extinction, *in* Koeberl, C., and MacLeod, K.G., eds., Catastrophic Events and Mass Extinctions: Impacts and Beyond: Boulder, Colorado, Geological Society of America Special Paper 356, p. 265–275.

change following an impact event that had global consequences, probably the Chicxulub impact (Hildebrand et al., 1991).

The extinction event at the end of the Cretaceous is still hotly debated because of the controversy of paleontologic data that obviously reflect two different scenarios, i.e., a long-term faunal and floral change causing extinction of taxa throughout the upper Maastrichtian, and the sudden impact catastrophe. A comprehensive discussion on this aspect was given by Ward (1995). The gradual decline in species richness of ammonites (Marshall and Ward, 1996) and planktic foraminifera during the Maastrichtian was probably related to long-term global climate change, characterized by strong cooling and sea-level regression (Keller et al., 1995; Keller, 2000). Evidence from molluscs (Marshall and Ward, 1996), foraminifera (e.g., Smit, 1982; Olsson and Liu, 1993; Molina et al., 1998) and coccolithophorids (Pospichal et al., 1992; Pospichal, 1994), however, shows that the major extinction at the K-T boundary was related to a sudden catastrophe rather than culmination of long-term environmental stress. Fossil groups with low extinction rates are important for understanding the exact extinction mechanism of the impact, because their association changes may provide information on migration events and survival strategies and can improve our knowledge of the ecological effects of the catastrophic events. Most calcareous plankton groups were strongly affected by extinction at the K-T boundary. Organic-walled dinoflagellate cysts, however, reflect no extinction, but show characteristic lateral distribution changes that can be used to reconstruct variations in sea-surface temperatures (Brinkhuis et al., 1998).

In this study we focus on the distribution patterns of calcareous cyst-producing dinoflagellates (Calciodinelloideae) over the K-T boundary in the Stevns Klint section; they show no accelerated extinction rate (Hildebrand-Habel et al., 1999). The study aims at ecologically interpreting the pattern of selective extinction of these organisms over the boundary, which is clearly related to the impact event.

The subfamily Calciodinelloideae (Fensome et al., 1993) comprises all dinoflagellates that form calcareous resting cysts as part of their life cycle. These fossilizable cysts are mostly spherical in shape and do not show any signs of paratabulation, although some species reflect clearly the peridinioid plate pattern (paratabulation) of the corresponding motile dinoflagellates. Calcareous dinoflagellate cysts (dinocysts) show a high diversity of forms throughout the Cretaceous. Their recent descendants are autotrophic organisms, so we assume that Cretaceous dinocysts are also remains of phytoplanktonic organisms, reflecting conditions within the photic zone. Therefore, they can be a tool for the reconstruction of conditions in the upper water column, an area strongly influenced by atmospheric change, and thus play a key role in determining the reaction of the ocean to both short-term catastrophes and long-term climate change.

This study represents the first high-resolution analysis of calcareous dinocysts across the Fish Clay. Calcareous dinocysts of the stratigraphic interval containing the K-T boundary were studied by Kienel (1994) in various paleogeographic positions of the Boreal realm, including Stevns Klint; however, Kienel did not sample the Fish Clay. Studies of the calcareous dinocysts throughout the Maastrichtian to Paleocene worldwide are scarce and generally only include investigations of the southern Atlantic Ocean (Fütterer, 1984, 1990; Hildebrand-Habel et al., 1999) and the Boreal realm (Willems, 1994, 1995, 1996). In many cases, a hiatus prevents detailed studies exactly at the boundary.

MATERIAL AND METHODS

The sampled section is located ~500 m northeast of Højerup "gamle kirke" at Stevns Klint, near Copenhagen, Denmark (Fig. 1.) Details of the lithostratigraphy of the Fish Clay were given in Christensen et al. (1973); foraminiferal biostratigraphy and isotope stratigraphy were provided by Schmitz et al. (1992). Four main lithofacies types can be distinguished: (1) white Maastrichtian limestone, rich in bryozoan fragments; (2) reddish and black to gray, pyritic, laminated clay (Fish Clay); (3) yellowish, highly burrowed, and intensively cemented limestone (Cerithium Limestone); and (4) white Danian limestone, very rich in bryozoans. Preservation of the calcareous dinoflagellate cysts is good in the Maastrichtian limestone and very good in the Bryozoan Limestone. In contrast, the carbonate grains of the Cerithium Limestone (sample Fi9) and the uppermost part of the Fish Clay (sample Fi8) are strongly altered. Preservation of carbonate grains in the lower Fish Clay appears to have been selective; planktic foraminifera show signs of dissolution (see Schmitz et al., 1992) whereas dinocysts are particularly well preserved.

Samples were taken with a microcarve at 1 cm sample spacing, and ~2 g of sediment were disintegrated in water (clay samples). The limestone samples were disintegrated by repeated freezing and thawing in a saturated solution of sodium sulfate. The disintegrated clay samples were treated with clay dispersant (Rewoquat) for 24 h to remove clay particles during washing. The material was wet sieved using mesh sizes of 15 µm, 20 µm, 75 µm, and 125 µm, and then dried.

Counts were done by optical microscopy (120× magnification) using weighted splits of 1–2.8 mg of the 20–75 µm grain-size fraction, which encompasses the size range of most specimens of calcareous dinoflagellate cysts. Exceptions are large Maastrichtian specimens, which were analyzed in the size fraction >75 µm, and the lower size limit of Danian species, which were analyzed qualitatively in smear slides of the 15–20 µm size fraction, using a CamScan CS 44 scanning electron microscope (SEM). Due to the dominance of one species in the Maastrichtian, the presence of rare species in samples Fi1 to Fi4 could only be assessed qualitatively by SEM analyses of

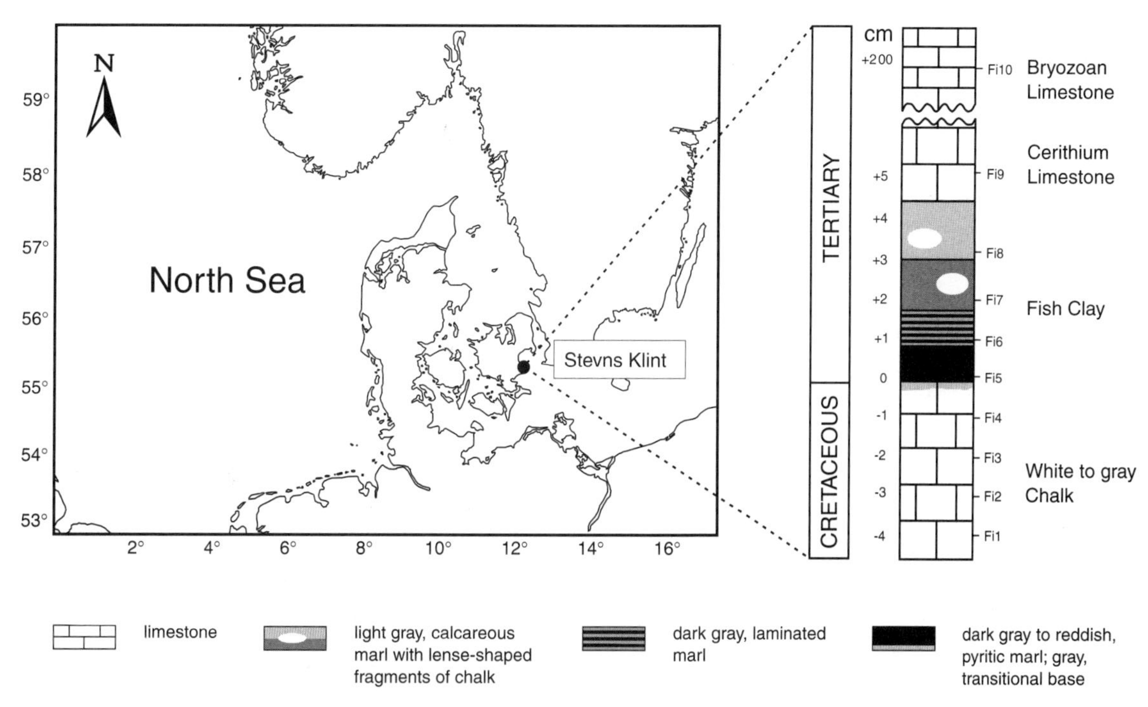

Figure 1. Geographic location and lithological profile of investigated section, including sample positions.

the 20–28 μm size fraction, where these are enriched. This fraction was dry sieved from 4 mg of the 20–75 μm fraction (qualitative evaluation of results: 1–10 specimens per SEM stub = present [<0.5% of the dinocyst assemblage]; more than 10 specimens per SEM stub = common). Each specimen counted in the splits under the light microscope was picked, excluding the optically determinable Pithonelloideae, and mounted on a stub for SEM studies for taxonomic determination. The taxonomy follows the classification of Fensome et al. (1993) and Young et al. (1997). The carbonate content of each sample was measured in order to evaluate the comparability of dinocyst counts of the clay to the counts in the limestones. Measurement was carried out using a carbonate bomb. (Carbometer mod. 23: 1 g of crushed sample is placed in a small pressure vessel along with a plastic boat containing 6N hydrochloric acid. The vessel, which has a manometer attached, is sealed and the carbonate is reacted with the acid. The amount of pressure registered by the gauge is proportional to the carbonate content.). Due to strong variations in carbonate content, the results of the cyst counts only represent a semiquantitative estimate of the concentration of specimens per sample.

The use of cathodoluminescence to distinguish reworked Maastrichtian from Danian specimens was tested using a cold cathode luminescence system 8200 MK II coupled with an Olympus BH-2 optical microscope. However, there is no difference in luminescence color and/or intensity between Maastrichtian and Danian carbonates. All material is stored at the Division of Historical Geology and Paleontology of the University of Bremen.

RESULTS

The carbonate content and grain-size distribution (Table 2) of the investigated samples are given in Figure 2 to illustrate the significant lithologic variations throughout the profile. The basal Fish Clay shows an abrupt breakdown in carbonate content and contains an increased amount of coarse grains, which are predominantly noncarbonate. Toward the top of the clay, the carbonate content increases gradually and the grain-size distribution indicates upward fining. The deposition of coarse grains (mainly fish remains, spherules, and corroded foraminifera) at the base of the Fish Clay is followed by a clay that almost lacks carbonate grains >125 μm, but shows a first carbonate recovery in the fraction <125 μm.

The Cretaceous chalk and the Danian Bryozoan Limestone have similar carbonate contents but show considerable difference in the percentage of the coarse fraction, the Danian having higher values. The percentages of the dinocyst-bearing fraction 20–125 μm, however are approximately equal.

Although the cyst counts of these limestones can be compared to each other due to the lithological similarity, the cyst amount counted in the clay samples is biased incalculably by selective carbonate dissolution (foraminifera) and dilution by noncarbonate.

Stratigraphic distribution of cyst assemblages in the studied section

The 15 common species of calcareous dinocysts of the studied material (Fig. 3) were grouped into 4 assemblages ac-

TABLE 1. ASSEMBLAGES OF CALCAREOUS DINOFLAGELLATE CYSTS

Assemblage	Species	Samples	Comments
P assemblage (Pithonella)	*Pithonella sphaerica* Kaufmann *Pithonella discoidea* Willems *Pithonella ovalis* Kaufmann	Fi1, Fi2, Fi3, Fi4 Fi8*, Fi9*, Fi10*	Most abundant species, formerly known as calcispheres
D assemblage (common in the Danian)	*Operculodinella operculata* Kienel *Operculodinella costata* Kienel *Operculodinella reticulata* Kienel *Lentodinella danica* Kienel *Orthotabulata obscura* Kienel *Calcicarpinum tetramurus* Kienel	Fi1, Fi2, Fi3, Fi4 Fi8*, Fi9*, Fi10	*Operculodinella operculata* was formerly described as *Thoracosphaera operculata* Fütterer Assemblage is named for its dominance in the Danian but is also present in the Maastrichtian
B assemblage (Background)	*Orthopithonella* sp. *Rhabdothorax* spp. (2 species)	Fi1–Fi10	Present throughout the section; increased abundance in the Fish Clay
C assemblage (Clay-specific)	*Orthopithonella collaris* (n. sp.) *Pirumella lepidota* Keupp *Pentadinellum vimineum* Keupp	Fi5, Fi6, Fi7	Pulse-like occurrence exclusively in the Fish Clay

*probably reworked specimens.

TABLE 2. GRAIN-SIZE FRACTIONS AND CARBONATE CONTENT

Sample no.	Fraction <20 μm (wt%)	Fraction 20–125 μm (wt%)	Fraction >125 μm (wt%)	Carbonate (wt%)
Fi1	59.1	27.8	13.1	95
Fi2	60.4	29.1	10.5	95
Fi3	73.0	17.0	10.0	95
Fi4	62.3	26.4	10.3	95
Fi5a	49.4	27.3	23.3	77
Fi5b	29.9	26.2	42.9	32
Fi6	44.9	45.5	9.6	43
Fi7	61.3	36.7	2.0	58
Fi8	63.8	30.0	6.2	77
Fi9	35.0	33.6	31.4	91
Fi10	32.1	25.2	42.7	96

cording to their stratigraphic distribution (Table 1), which were given the abbreviations P (*Pithonella* spp.), D (dominant in the Danian), C (clay specific), and B (background). The distribution patterns of these four assemblages are illustrated in Figure 4. Cyst counts are given in Table 3.

The uppermost Maastrichtian samples (Fi1–Fi4) are characterized by a typical Upper Cretaceous dinocyst association, ~99% of which consists of species of the P assemblage. This group represents the dominant element of the Boreal microflora since the Albian. Cyst abundances are ~1000 cysts/mg (fraction 20–75 μm). The remaining 1% of the cyst association is formed by the D and B assemblages. Approaching the uppermost Maastrichtian, abundances of the P assemblage decrease.

Cyst abundances are extremely low in the Fish Clay (Fi5–Fi7) compared to the Maastrichtian chalk due to the abrupt, complete disappearance of the P and D assemblages. The lowermost 4 mm of the reddish layer contain almost no cysts. The C assemblage is present just above the base of the clay layer. Specimens are exceptionally well preserved, in contrast to planktic foraminifera and coccoliths. The occurrence of a new species, *Orthopithonella collaris* n.sp. (Wendler et al., 2001), is particularly prominent because it is restricted to the K-T boundary clay. The B assemblage shows increased abundance in the Fish Clay. No calcareous dinocysts were found in the <20 μm size fraction.

Abundances of calcareous dinocysts remain low in the entire clay succession. Toward the top (Fi8), the C assemblage disappears and the species of the B assemblage decrease in number while the P and D assemblages reappear. Preservation of specimens, especially those of the P assemblage, is mostly poor in this upper part of the clay. No data could be obtained from the Cerithium Limestone due to extremely poor preservation.

In the Bryozoan Limestone (Fi10), the total abundances of cysts are ~80 cysts/mg (fraction 20–75 μm). The D assemblage represents ~75% of the association. The size fraction 15–20 μm contains *Operculodinella operculata* (1.4% of that fraction). There is a decrease in abundance of the B assemblage. The C assemblage is absent. Very few specimens (3% of the dinocyst association) of the P assemblage can be found. The good preservation provides no clear hints that these specimens were reworked.

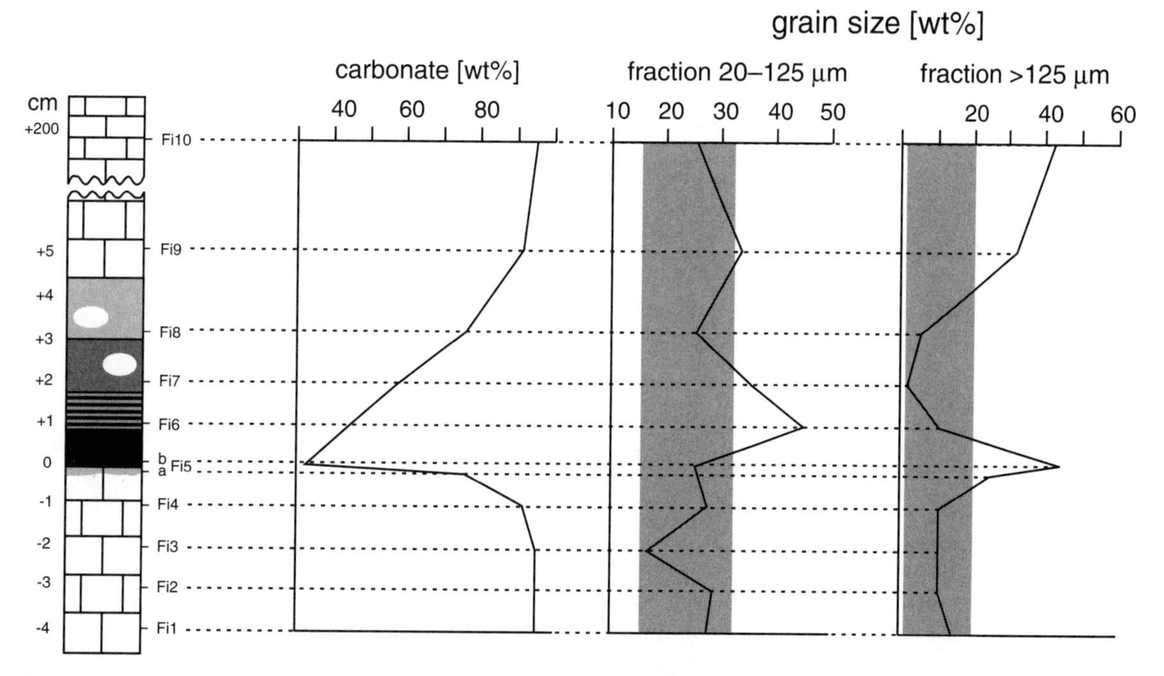

Figure 2. Carbonate content and grain-size distribution of investigated samples; gray areas of diagrams mark common ranges in percentage throughout Cretaceous. Note significantly higher amount of coarse fraction in Danian sample. See Figure 1 for rock symbols.

DISCUSSION

Duration of deposition of the boundary clay

The Fish Clay is generally considered to be a condensed layer (Schmitz, 1990; MacLeod and Keller, 1991). The amount of time represented by this layer is, however, still debated. An evaluation of the duration of deposition, which is certainly variable due to local circumstances, is important for understanding the mechanisms of ecological change that are associated with the impact.

We consider the reddish and dark gray layers as representing the first decades or centuries following the impact because of their mineralogical particularities:, i.e., featuring only traces of quartz as terrigenous component, an almost pure smectitic clay mineral assemblage, and spherules of probable impact origin (Kastner et al., 1984). Alvarez et al. (1980) pointed out that, in contrast to the terrigenous clays found in the chalks below and above the Fish Clay, the completely different clay mineral assemblage of this layer is likely the product of alteration of fallout material. Had the time of deposition been in the order of millennia, the continual terrestrial input and reworking would have diluted the comparably small amount of fallout. Furthermore, upward fining in the sediments covering the reddish and dark gray layers suggests a fast, continuous sedimentation under decreasing energy. Accordingly, the presence of non-Maastrichtian carbonate producers, causing a gradual increase in carbonate content during deposition of the fallout, indicates a fast recovery of an initially low carbonate production. This supposition is corroborated by the results of Pope et al. (1994), who modeled the impact-winter cooling and a subsequent warming pulse, which took place during the discussed interval, to have had a duration of only decades. The warming pulse was recognized to occur at the base of the Fish Clay by Brinkhuis et al. (1998), using dinoflagellate-based sea-surface temperature reconstruction. However, on the basis of biostratigraphical results, Brinkhuis et al. considered the period of warming to have had a longer duration, as much as 10 k.y., which is inconsistent with the conclusions drawn from the mineralogical and sedimentological characteristics of the basal Fish Clay discussed here.

Distribution patterns of calcareous dinocysts

Victims. The four characteristic assemblages of calcareous dinocysts show specific reactions to the impact catastrophe. Most strikingly, the dominance of the Pithonelloideae ended abruptly with the onset of clay deposition. These species, which were characteristic for the Boreal Realm throughout the Upper Cretaceous, were clearly victims of the K-T event. It is ambiguous whether the P assemblage underwent a weak recovery in the Danian, which, according to Kienel (1994), ends at the top of the *Cruciplacolithus intermedius* nannofossil zone. In the Geulhemmerberg K-T boundary section (Limburg, southeastern Netherlands) (Willems, 1996) no break at all in the dominance of the species of the P assemblage can be found. This is most

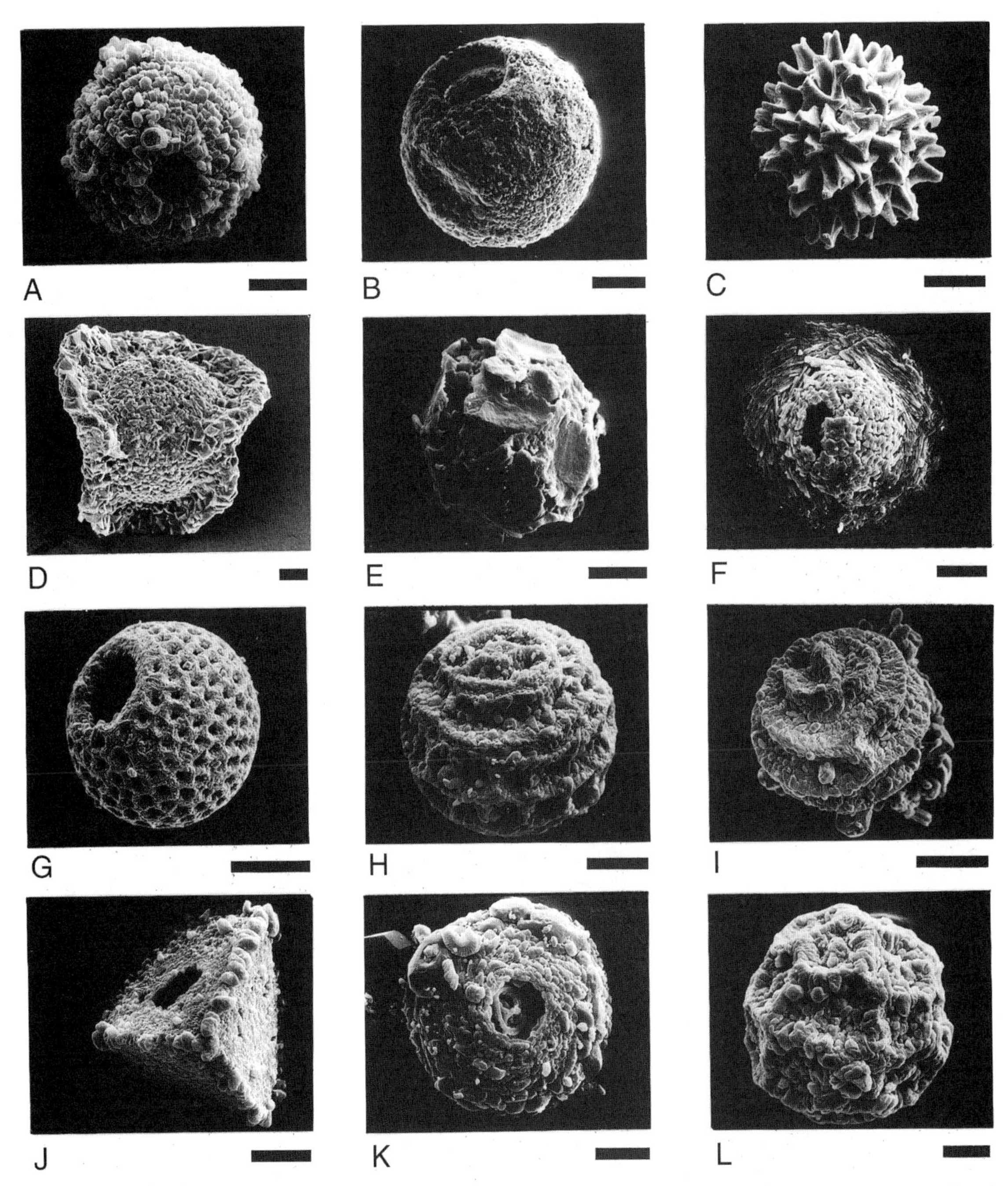

Figure 3. Common calcareous dinoflagellate cysts of Stevns Klint Cretaceous-Tertiary (K-T) boundary section. All scale bars are 10 µm. A: *Pithonella sphaerica*, main species of P assemblage, sample Fi1. B: *Orthopithonella* sp., B assemblage, sample Fi6. C: *Rhabdothorax* sp., B assemblage, sample Fi5. D: *Orthopithonella collaris* (spec. nov., Wendler et al., 2001), C assemblage, sample Fi5. E: *Pirumella lepidota*, C assemblage, sample Fi5. F: *Pentadinellum vimineum*, C assemblage, sample Fi5. G: *Operculodinella operculata*, D assemblage, sample Fi3. H: *Operculodinella reticulata*, D assemblage, sample Fi10. I: *Operculodinella costata*, D assemblage, sample Fi10. J: *Calcicarpinum tetramurus*, D assemblage, sample Fi10. K: *Lentodinella danica*, D assemblage, sample Fi10. L: *Orthotabulata obscura*, D assemblage, sample Fi10.

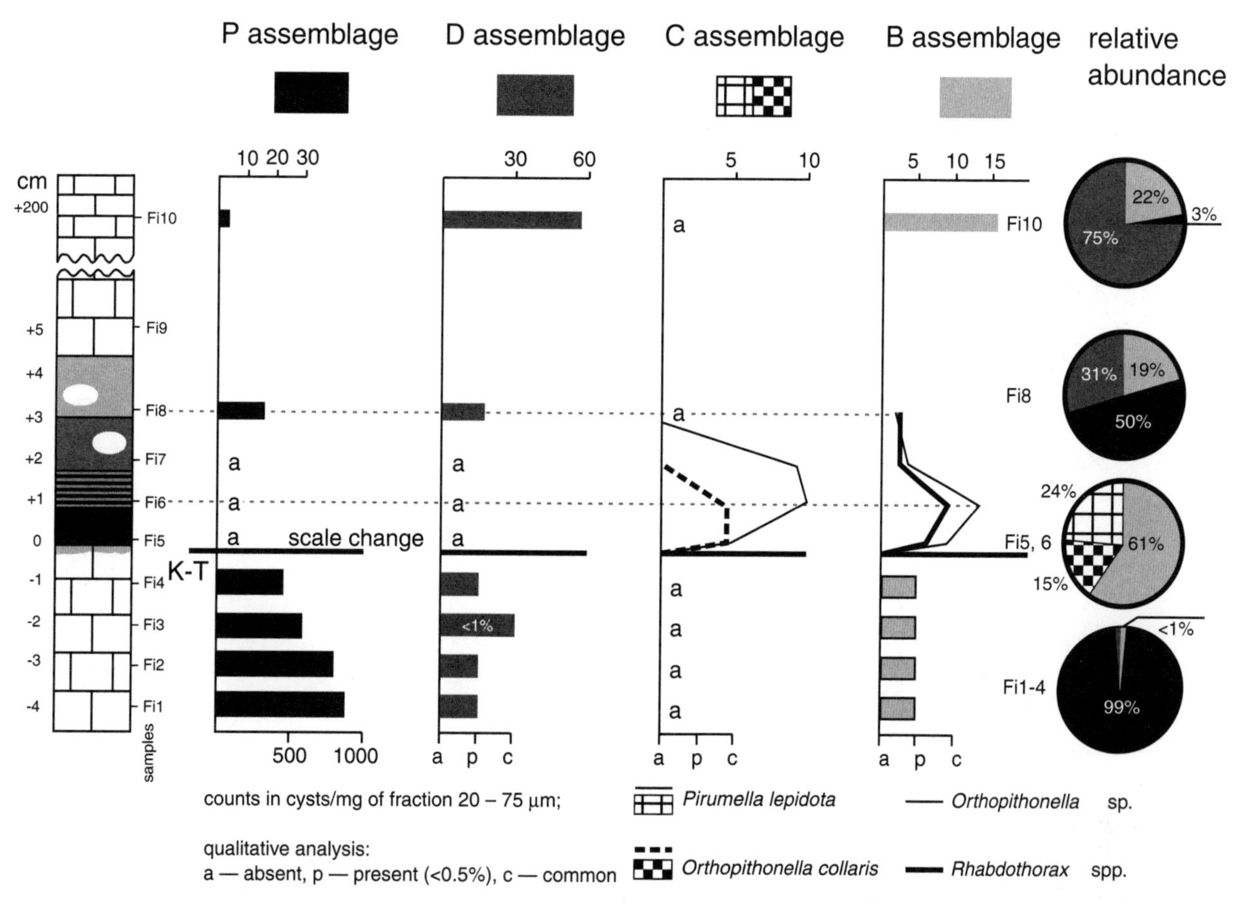

Figure 4. Qualitative and semiquantitative distribution patterns of four assemblages (described in Table 1) of calcareous dinocyst species throughout Fish Clay section. See Figure 1 for rock symbols.

likely due to reworking. In the Stevns Klint section, a few specimens of this assemblage can also be found above the K-T boundary. Those are likely reworked, although their mostly poor preservation could also be due to increased cementation, i.e., diagenetic overprinting. Despite the possibility of a weak and short-lived recovery, this assemblage is an example of the sudden end of an era of extremely successful Cretaceous organisms initiated by the impact catastrophe.

Because changing distribution patterns of Pithonelloideae are known to reflect sea-level change (Dali-Ressot, 1987; Keupp, 1991; Zügel, 1994), the decrease in number toward the end of the Maastrichtian can be interpreted as a reaction to the long-term global eustatic sea-level fall (Haq et al., 1987; Hardenbol et al., 1998). This major regression resulted in a pronounced sea-level lowstand during the late Maastrichtian to early Danian, and reduced the area of epicontinental shallow seas (Håkansson and Thomsen, 1999). In addition, short-term regression, estimated to have been ~70–100 m in the Tethys (Keller, 1988), is reflected by the uppermost Maastrichtian associations of planktic foraminifera and dinoflagellates (Brinkhuis and Zachariasse, 1988; Schmitz et al., 1992).

One reason for the extinction of the P assemblage organisms may be their restriction to the outer shelf environment (e.g., Villain, 1981; Zügel, 1994), which was progressively reduced during sea-level fall. Because of their restriction to a shrinking environmental niche, the P assemblage organisms were strongly affected by the environmental catastrophe; they could not escape into other environments, where they might have found niches and survived.

Furthermore, the ability to form a resting cyst as part of the life cycle appears to be an important general factor in dinoflagellate survival, as suggested by Brinkhuis et al. (1998) for the dinoflagellates that form organic-walled cysts. The production of long-term resting cysts can be considered a survival strategy (Lewis et al., 1999). Experimental data (Dale, 1983; Lewis et al., 1999) on resting cysts confirm a wide range of survival time, between 12 and 66 months for different dinoflagellate species, coastal and fresh-water species having the longest dormancy periods. These differences could account for selective extinction, although little is known about the ability of dinoflagellates to adjust their dormancy period because certain natural environmental conditions could never be simulated in experiments. Experiments performed by Griffis and Chapman (1989, 1990) showed that light blackout exceeding three

TABLE 3. COUNTS OF CALCAREOUS DINOFLAGELLATE CYSTS

Sample no.	Counted split [mg] of 20–75 μm fraction	Counted cysts (total)	*Pithonella sphaerica*	*Pithonella ovalis*	*Pithonella discoidea*	*Rhabdothorax* spp.	*Orthopithonella* sp.
Fi1	1.4	1280	1230	10	30	1	2
Fi2	0.9	772	720	10	40	0	1
Fi3	0.7	436	400	4	20	1	1
Fi4	0.8	428	350	2	70	0	0
Fi5b	2.8	64	0	0	0	18	24
Fi6	2.2	74	0	0	0	18	28
Fi7	1.0	28	0	0	0	3	4
Fi8	1.2	48	24	0	0	5	4
Fi9	no count						
Fi10	1.2	92	3	0	0	10	11

Sample no.	Split [mg] of 20–75 μm fraction	*Orthotabulata obscura*	*Operculodinella* spp.	*Calcicarp. tetramurus*	*Lentodinella danica*	*Pirumella lepidota*	*Orthopithonella collaris*
Fi1	1.4	1	1	2	3	0	0
Fi2	0.9	0	0	0	1	0	0
Fi3	0.7	1	2	3	4	0	0
Fi4	0.8	1	1	2	2	0	0
Fi5b	2.8	0	0	0	0	10	12
Fi6	2.2	0	0	0	0	22	8
Fi7	1.0	0	0	0	0	9	0
Fi8	1.2	0	8	0	7	0	0
Fi9	no count						
Fi10	1.2	17	16	11	24	0	0

Qualitative analysis of the 20–28 μm size fraction					*Pithonella sphaerica* >75 μm	
Sample no.	*Orthotabulata obscura*	*Operculodinella* spp.	*Calcicarpinum tetramurus*	*Lentodinella danica*	*Pithonella sphaerica*	split [mg] of >75 μm fraction
Fi1	p	p	p	p	770	2.2
Fi2	p	p	p	p	660	3.3
Fi3	p	c	c	c	312	3.3
Fi4	p	p	p	p	265	2.8

p—present (~0.5%); c—common

months and the effect of acid rain on the growth of cells are critical factors for survival of the motile dinoflagellate stage. We hypothesize that the species of the P assemblage, in addition to their disadvantage of diminished spatial distribution, had a rapid life cycle with a short resting stage. Thus they were subject to extinction because their dormancy period was not long enough to survive the impact winter and prolonged period of unfavorable conditions.

Survivors. The D and B assemblages were not significantly affected by the impact and survived into the Tertiary. The first appearance of species of the D assemblage in the late Maastrichtian at Stevns Klint and in a section from the South Atlantic Ocean (Hildebrand-Habel et al., 1999) most likely reflects a change in the cyst association as a consequence of long-term environmental change and evolution.

The disappearance of the D assemblage in the K-T boundary layer could be the result of environmental stress or of selective dissolution. The latter seems unlikely, because the assemblage contains a wide range of cyst sizes, various wall thicknesses, and all wall types. Therefore, the absence of the D assemblage in the Fish Clay is interpreted to mainly reflect an ecological signal. It may indicate that these species underwent instant regional migration due to a major facies shift at the investigated location. However, the species of the D assemblage seem to have lived in a wide range of environments; they are known from shelf environments (Kienel, 1994; Willems, 1996; and this study) and the oceanic realm (Hildebrand-Habel et al., 1999). Such an extended habitat can provide a large variety of ecological niches, increasing the chance of survival for at least part of the species' population. Even if 99.9% of the population dies, the species can still survive (Smit, 1999, 2000).

Adapters and opportunists. In contrast to the species of the D and P assemblages, the species of the C and B assemblages present in the Fish Clay seem to have adapted to the special environmental conditions following the impact, or they were opportunists. The spiny cysts of the B assemblage might be compared to the recent calcareous dinoflagellate species *Scrippsiella trochoidea* (von Stein), which forms similar spiny cysts and is exclusively found in coastal environments (Janofske, 2000, and references therein), where unstable conditions are typical. *Scrippsiella trochoidea* (von Stein) is considered a euryhaline species and is occasionally described from brackish environments. Lewis et al. (1999, p. 352) found in a study including *Scrippsiella* sp. that "the species to have the greatest longevity are amongst the most common members of coastal phytoplankton assemblages." This means that the coastal environment could have been a reservoir of cysts that are capable of surviving a long period of unfavorable conditions at the K-T boundary (Lewis et al., 1999). This idea is corroborated by the increased presence of spiny cysts within the Fish Clay; these probably represent species that successfully pioneered the postimpact marine environment.

Special adaptation to the environment established during Fish Clay deposition must be assigned particularly to the C assemblage. The species *Orthopithonella collaris* and *Pentadinellum vimineum* with their distinctive reduced paratabulation and *Pirumella lepidota* are exotic elements in the dinocyst association of the Boreal Realm. *Pentadinellum vimineum* represents a group of reduced paratabulated species that shows a specific distribution pattern throughout the Cretaceous, a pattern characterized by restricted temporal ranges. They were observed to occur sporadically in short-term events related to ingressions of warm Tethyan water masses into the Boreal Realm during the Lower to Middle Cretaceous (Keupp, 1991; Neumann, 1999). Zügel (1994) stated that the development of paratabulated cysts, which he found to be dominating in coastal areas, must be related to short-term, regional, ecological events.

Possible reasons for the C assemblage pulse occurrence in the Fish Clay are (1) the advantage in survival strategy, probably connected with the development of stress-controlled, ecological morphotypes, or (2) a migration of the species into the studied area. A comparable pulse of organic-walled dinoflagellate cysts exactly at the K-T boundary was discussed by Brinkhuis et al. (1998). They discovered a short-term dominance of an equatorial species in the lower parts of the Fish Clay, and interpreted it as representing a migration of a tropical species into higher latitudes due to the aerosol-forced, post-impact warming event. Too little is known about the ecological affinities of the C assemblage species to check if such an interpretation also applies for this group of dinoflagellates. Although it is possible that the pulse-like occurrence in the Fish Clay was caused by a migration event, the most straightforward explanation is that the C as well as the B assemblages represent a pioneer flora of robust species that flourished during the time of diminished carbonate production, which was later replaced by a newly developed ecosystem.

The dinocyst association of the Fish Clay is similar to the calcareous dinocyst assemblages found in the C-clay of the Geulhemmerberg K-T section (Willems, 1996). According to paleogeographic distribution patterns, including those of the Pithonelloideae, this association was interpreted as indicative of maximum transgression. Considering (1) that reworking played a major role in the P assemblage distribution in the Stevns Klint and Geulhemmerberg sections, (2) the pulse-like appearance of the C assemblage in the Fish clay, and (3) the distinct breakdown of the P assemblage in the clay, we interpret the C and B assemblage equivalents of the Geulhemmerberg section to reflect a unique ecological situation rather than a sea-level-controlled facies shift.

Although the Fish Clay is an exceptional example of breakdown of carbonate production that is clearly related to an impact and subsequent substantial extinctions, there is a striking analogy between its dinocyst distribution and that of cyclic sediments throughout the mid-Cenomanian, interpreted as orbitally forced sedimentary cycles (Gale et al., 1999). Reduced paratabulated species occur in high relative abundance in certain marls of the chalk-marl couplets of the Anglo-Paris basin, while simultaneously, an almost complete disappearance of Pithonelloideae can be observed (Wendler, 2001). These short-term disruptions of the chalk deposition can be related to climatically forced stratification events on the basis of paleoceanographic models (e.g., Mitchell and Carr, 1998). If the similar dinocyst distribution patterns of this long-term climate change on the one hand and the impact catastrophe on the other hand were caused by the same oceanographic changes, stratification was perhaps another important local factor of the extinction mechanism related to the impact.

CONCLUSIONS

Two main ecological factors were apparently important for the survival over the K-T boundary of the dinoflagellates investigated in this study. First the ability to form a resting cyst as part of the life cycle appears to be an important advantage for dinoflagellate survival. Second, the distribution of species over large areas and a wide range of ecosystems were crucial for survival. A combination of both factors appears to be the main reason for a low extinction rate at the K-T as opposed to the high extinction rate of other phytoplankton groups, such as the coccolithophorids.

The P assemblage distribution shows that even though some specimens occur above the K-T boundary, and may represent a short recovery, the complete disappearance of taxa at the base of the boundary clay is an abrupt biotic response clearly related to the impact. Our definition of an impact-related extinction should not necessarily require the instant disappearance of a taxon, but rather accept the impact catastrophe as an event that initiated a substantial ecological change, which allowed for limited survival of some victims.

ACKNOWLEDGMENTS

We thank I. Wendler, A. Vink, and K.A.F. Zonneveld for stimulating discussions and review of the manuscript, and H. Brinkhuis and I. Arenillas for helpful comments. The investigation was financially supported by the German Science Foundation (Deutsche Forschungsgemeinschaft, project Wi 725/12).

REFERENCES CITED

Alvarez, L.W., Alvarez, W., Asaro, F., and Michel, H.V., 1980, Extraterrestrial cause for the Cretaceous-Tertiary extinction: Science, v. 208, 1095–1108.

Birkelund, T., and Håkansson, E., 1982, The terminal Cretaceous extinction in Boreal shelf seas: A multicausal event: Geological Society of America Special Paper 190, p. 373–384.

Brinkhuis, H., and Zachariasse, W.J., 1988, Dinoflagellate cysts, sea level changes, and planktonic foraminifers across the Cretaceous/Tertiary boundary at El Haria, northwest Tunisia: Marine Micropaleontology, v. 13, p. 153–191.

Brinkhuis, H., Bujak, J.P., Smit, J., Versteegh, G.J.M., and Visscher, H., 1998, Dinoflagellate-based sea surface temperature reconstructions across the Cretaceous-Tertiary boundary: Palaeogeography, Palaeoclimatology, Palaeoecology, v. 141, p. 67–83.

Christensen, L., Fregerslev, S., Simonsen, A., and Thiede, J., 1973, Sedimentology and depositional environment of Lower Danian fish clay from Stevns Klint, Denmark: Bulletin of the Geological Society of Denmark, v. 22, p. 193–212.

Dale, B., 1983, Dinoflagellate resting cysts: "benthic plankton", *in* Fryxell, G.A., ed., Survival strategies of the algae: Cambridge, Cambridge University Press, p. 69–134.

Dali-Ressot, M.-D., 1987, Les Calcisphaerulidae des terrains Albien à Maastrichtien de Tunesie centrale (J. Bireno et J. Bou el Ahneche): Intérets systématique, stratigraphique et paléogéographique [Ph.D. thesis]: Tunis, University of Tunis, 191 p.

Fensome, R.A., Norris, G., Sarjeant, W.A.S., Taylor, F.J.R., Wharton, D.I., and Williams, G.L., 1993, A classification of living and fossil dinoflagellates: Micropalaeontology, Special Publication, v. 7, 351 p.

Fütterer, D.K., 1984, Pithonelloid calcareous dinoflagellates from the Upper Cretaceous and Cenozoic of the Southeastern Atlantic Ocean, Deep Sea Drilling Project, Leg 74, *in* von Huene, R., Aubouin, J., et al., Initial Reports of the Deep Sea Drilling Project, Volume 74: Washington, D.C., U.S. Government Printing Office, p. 533–541.

Fütterer, D.K., 1990, Distribution of calcareous dinoflagellates at the Cretaceous-Tertiary boundary of Queen Maud Rise, Eastern Weddell Sea, Antarctica (ODP Leg 113), *in* Barker, P.F., Kennett, J.P., et al., Scientific Results, Ocean Drilling Program, Leg 113: College Station, Texas, Ocean Drilling Program, p. 533–548.

Gale, A.S., Young, J.R., Shackleton, N.J., Crowhurst, S.J., and Wray, D.S., 1999, Orbital tuning of Cenomanian marly chalk successions: Towards a Milankovitch time-scale for the Late Cretaceous: Royal Society of London Philosophical Transactions, ser. A, v. 357, p. 1815–1829.

Griffis, K., and Chapman, D.J., 1989, Survival of phytoplankton under prolonged darkness: Implications for the Cretaceous/Tertiary boundary darkness hypothesis: Palaeogeography, Palaeoclimatology, Palaeoecology, v. 67, p. 305–314.

Griffis, K., and Chapman, D.J., 1990, Modeling Cretaceous/Tertiary boundary events with extant photosynthetic plankton: Effects of impact-related acid rain: Lethaia, v. 23, p. 379–383.

Håkansson, E., and Thomsen, E., 1999, Benthic extinction and recovery patterns at the K/T boundary in shallow water carbonates, Denmark: Palaeogeography, Palaeoclimatology, Palaeoecology, v. 154, p. 67–85.

Haq, B.U., Hardenbol, J., and Vail, P.R., 1987, Chronology of fluctuating sea level since the Triassic: Science, v. 235, p. 1156–1166.

Hardenbol, J., Thierry, J., Farley, M.B., Jacquin, T., de Graciansky, P.-C., and Vail, P.R., 1998, Mesozoic and Cenozoic sequence chronostratigraphic framework of European basins, *in* de Graciansky, P.-C., Hardenbol, J., Jacquin, T., and Vail, P.R., eds., Mesozoic and Cenozoic sequence stratigraphy of European basins: SEPM (Society for Sedimentary Geology) Special Publication, v. 60, chart 1.

Hildebrand, A.R., Penfield, G.T., Kring, D.A., Pilkington, M., Camargo, A., Jacobsen, S.B., and Boynton, W.V., 1991, Chicxulub crater: A possible Cretaceous/Tertiary boundary impact crater on the Yucatan peninsula, Mexico: Geology, v. 19, p. 867–871.

Hildebrand-Habel, T., Willems, H., and Versteegh, G.J.M., 1999, Variations in calcareous dinoflagellate associations from the Maastrichtian to Middle Eocene of the western South Atlantic Ocean (São Paulo Plateau, DSDP Leg 39, Site 356): Review of Palaeobotany and Palynology, v. 106, p. 57–87.

Janofske, D., 2000, *Scrippsiella trochoidea* and *Scrippsiella regalis*, nov. comb. (Peridiniales, Dinophyceae): A comparison: Journal of Phycology, v. 36, p. 178–189.

Kastner, M., Asaro, F., Michel, H.V., Alvarez, W., and Alvarez, L.W., 1984, The precursor of the Cretaceous-Tertiary boundary clays at Stevns Klint, Denmark, and DSDP hole 465A: Science, v. 226, p. 137–143.

Keller, G., 1988, Biotic turnover in benthic foraminifera across the Cretaceous/Tertiary boundary at El Kef, Tunisia: Palaeogeography, Palaeoclimatology, Palaeoecology, v. 66, p. 153–171.

Keller, G., 2000, The K/T boundary mass extinction: Year 2000 assessment: Abstracts of the Spring Meeting (AMICO 2000) Astronomische Gesellschaft und Deutsche Geologische Gesellschaft, Schriftenreihe der Deutschen Geologischen Gesellschaft, v. 11, p. 28–30.

Keller, G., Li, L., and MacLeod, N., 1995, The Cretaceous/Tertiary boundary stratotype section at El Kef, Tunisia: How catastrophic was the mass extinction?: Palaeogeography, Palaeoclimatology, Palaeoecology, v. 119, p. 221–254.

Keupp, H., 1991, Kalkige Dinoflagellatenzysten aus dem Eibrunner Mergel (Cenoman-Turon-Grenzbereich) bei Abbach/Süddeutschland: Berliner Geowissenschaftliche Abhandlungen, ser. A, v. 134, p. 127–145.

Kienel, U., 1994, Die Entwicklung der kalkigen Nannofossilien und der kalkigen Dinoflagellaten-Zysten an der Kreide/Tertiär-Grenze in Westbrandenburg im Vergleich mit Profilen in Nordjütland und Seeland (Dänemark) [Ph.D. thesis]: Berliner Geowissenschaftliche Abhandlungen, ser. E, v. 12, 87 p.

Lewis, J., Harris, A.S.D., Jones, K.J., and Edmonds, R.L., 1999, Long-term survival of marine planktonic diatoms and dinoflagellates in stored sediment samples: Journal of Plankton Research, v. 21, p. 343–354.

MacLeod, N., and Keller, G., 1991, Hiatus distribution and mass extinctions at the Cretaceous/Tertiary boundary: Geology, v. 19, p. 497–501.

Marshall, C.R., and Ward, P.D., 1996, Sudden and gradual molluscan extinction in the Latest Cretaceous of Western European Tethys: Science, v. 274, p. 1360–1363.

Mitchell, S.F., and Carr, I.T., 1998, Foraminiferal response to mid-Cenomanian (Upper Cretaceous) palaeoceanographic events in the Anglo-Paris Basin (Northwest Europe): Palaeogeography, Palaeoclimatology, Palaeoecology, v. 137, p. 103–125.

Molina E., Arenillas I., and Arz, J.A., 1998, Mass extinction in planktic foraminifera at the Cretaceous/Tertiary boundary in subtropical and temperate latitudes: Bulletin de la Société Géologique de France, v. 169, no. 3, p. 351–363.

Neumann, Chr., 1999, Diversitäts- und Häufigkeitsmuster kalkiger Dinoflagellatenzysten aus dem Alb der Bohrung Kirchrode II (zentrales Niedersächsisches Becken, NW-Deutschland) und ihre möglichen Steuerungsmechanismen [Ph.D. thesis]: Berliner Geowissenschaftliche Abhandlungen, ser. E, v. 31, 79 p.

Olsson, R.K., and Liu, C., 1993, Controversies on the placement of Cretaceous-Paleogene boundary and the K/P mass extinction of planktonic foraminifera: Palaios, v. 8, p. 127–139.

Pope, K.O., Baines, K.H., Ocampo, A.C., and Ivanov, B.A., 1994, Impact winter and the Cretaceous/Tertiary extinctions: Results of a Chicxulub asteroid impact model: Earth and Planetary Science Letters, v. 128, p. 719–725.

Pospichal, J.J., and Bralower, T.J., 1992, Calcareous nannofossils across the Cretaceous/Tertiary boundary, Site 761, Northwest Australian margin, *in* von Rad, U., Haq, B.U., et al., Proceedings of the Ocean Drilling Program, Scientific Results, v. 122, p. 735–747.

Pospichal, J.J., 1994, Calcareous nannofossils at the K-T boundary, El Kef: No evidence for stepwise, gradual, or sequential extinctions: Geology, v. 22, p. 99–102.

Schmitz, B., 1990, Origin of microlayering in worldwide distributed Ir-rich marine Cretaceous/Tertiary boundary clays: Reply on comment: Geology v. 18, p. 93–94.

Schmitz, B., Keller, G., and Stenwall, O., 1992, Stable isotope and foraminiferal changes across the Cretaceous-Tertiary boundary at Stevns Klint, Denmark: Arguments for long-term oceanic instability before and after bolide impact event: Palaeogeography, Palaeoclimatology, Palaeoecology, v. 96, p. 233–260.

Smit, J., 1982, Extinction and evolution of planktonic foraminifera after a major impact at the Cretaceous/Tertiary boundary: Geological Society of America Special Paper 190, p. 329–352.

Smit, J., 1999, The global stratigraphy of the Cretaceous-Tertiary boundary impact ejecta: Annual Review of Earth and Planetary Sciences, v. 27, p. 75–113.

Smit, J., 2000, Global mass-extinction at the KT boundary: Abstracts of the Spring Meeting (AMICO 2000) Astronomische Gesellschaft und Deutsche Geologische Gesellschaft, Schriftenreihe der Deutschen Geologischen Gesellschaft, v. 11, p. 46–47.

Ward, P.D., 1995, After the fall: Lessons and directions from the K/T debate: Palaios, v. 10, p. 530–538.

Wendler, J., 2001, Reconstruction of astronomically-forced cyclic and abrupt paleoecological changes in the Upper Cretaceous Boreal Realm based on calcareous dinoflagellate cysts [Ph.D. thesis]: Berichte, Fachbereich Geowissenschaften, Universität Bremen, no. 183, 149 p.

Wendler, J., Wendler, I., and Willems, H., 2001, *Orthopithonella collaris* sp. nov., a new calcareous dinoflagellate cyst from the K/T boundary (Fish Clay, Stevns Klint/Denmark): Review of Palaeobotany and Palynology, v. 115, p. 69–77.

Willems, H., 1988, Kalkige Dinoflagellaten-Zysten aus der oberkretazischen Schreibkreide-Fazies N-Deutschlands (Coniac bis Maastricht): Senckenbergiana Lethaea, v. 68, nos. 5/6, p. 433–477.

Willems, H., 1994, New calcareous dinoflagellates from the Upper Cretaceous white chalk of northern Germany: Review of Palaeobotany and Palynology, v. 84, p. 57–72.

Willems, H., 1995, *Praecalcigonellum duopylum* n.sp., a new calcareous dinoflagellate cyst from the lowermost Danian (*Biantholithus sparsus* Zone) of the Geulhemmerberg Section (South Limburg, The Netherlands): Neues Jahrbuch Geologische und Paläontologische Abhandlungen, v. 198, p. 141–152.

Willems, H., 1996, Calcareous dinocysts from the Geulhemmerberg K/T boundary section (Limburg, SE Netherlands): Geologie en Mijnbouw, v. 75, p. 215–231.

Villain, J.-M., 1981, Les Calcisphaerulidae: Intérêt stratigraphique et paléoécologique: Cretaceous Research, v. 2, p. 435–438.

Young, J.R., Bergen, J.A., Bown, P.R., Burnett, J.A., Fiorentino, A., Jordan, R.W., Kleijne, A., van Niel, B.E., Romein, A.J.T., and Von Salis, K., 1997, Guidelines for coccolith and calcareous nannofossil terminology: Palaeontology, v. 40, p. 875–912.

Zügel, P., 1994, Verbreitung kalkiger Dinoflagellaten-Zysten im Cenoman/Turon von Westfrankreich und Norddeutschland: Courier Forschungs-Institut Senckenberg, v. 176, 159 p.

MANUSCRIPT SUBMITTED OCTOBER 5, 2000; ACCEPTED BY THE SOCIETY MARCH 22, 2001

Printed in the U.S.A.

Geological Society of America
Special Paper 356
2002

Abrupt extinction and subsequent reworking of Cretaceous planktonic foraminifera across the Cretaceous-Tertiary boundary: Evidence from the subtropical North Atlantic

Brian T. Huber*
Department of Paleobiology, MRC: NHB-121, Smithsonian Institution, Washington, D.C. 20560, USA
Kenneth G. MacLeod*
Department of Geological Sciences, University of Missouri, Columbia, Missouri 65211, USA
Richard D. Norris*
Woods Hole Oceanographic Institute, Woods Hole, Massachusetts 02543, USA

ABSTRACT

An impact ejecta bed containing shocked quartz and diagenetically altered tektite spherules coincides exactly with biostratigraphic placement of the Cretaceous-Tertiary (K-T) boundary in three drill cores recovered from Ocean Drilling Program Site 1049 (located in the subtropical North Atlantic Ocean). Both the bracketing pelagic ooze and the ejecta bed are undisturbed at Site 1049, allowing detailed examination of the expression of the boundary event in an open ocean setting. The youngest Cretaceous sediments contain a diverse assemblage of well-preserved upper Maastrichtian Tethyan microfossils. The overlying ejecta bed varies laterally in thickness, has sharp lower and upper contacts, and contains features (e.g., presence of a foraminiferal grainstone layer at its base and large chalk clasts in its middle, and dominance of poorly sorted coarse grains throughout) that suggest it was deposited by one or several mass-flow events. The oldest Danian ooze contains abundant, tiny planktonic foraminifera characteristic of the early Danian Pα Zone as well as common, large Cretaceous individuals. The lowermost Danian P0 Zone (assemblage dominated by *Guembelitria cretacea*) is apparently absent. This absence could reflect restriction of the P0 assemblage to shallower settings, slow sedimentation rates coupled with bioturbation mixing Tertiary forms into (and thus obscuring) the P0 Zone, or an interval of erosion or nondeposition. Cretaceous species decline and last occur in the first several meters of section above the ejecta bed. This pattern could be interpreted as evidence for gradual extinction above the impact bed, but thin-section observations, relative abundance counts, size-distribution analyses, and comparison with species extinctions at other K-T sections demonstrate sudden extinction, nearly all post-K-T occurrences of Cretaceous planktonic foraminiferal species being explained as the result of sediment reworking.

*E-mails: Huber, huber.brian@nmnh.si.edu; MacLeod, macleodk@missouri.edu; Norris, rnorris@whoi.edu

Huber, B.T., MacLeod, K.G., and Norris, R.D., 2002, Abrupt extinction and subsequent reworking of Cretaceous planktonic foraminifera across the Cretaceous-Tertiary boundary: Evidence from the subtropical North Atlantic, *in* Koeberl, C., and MacLeod, K.G., eds., Catastrophic Events and Mass Extinctions: Impacts and Beyond: Boulder, Colorado, Geological Society of America Special Paper 356, p. 277–289.

INTRODUCTION

One of the long-standing controversies in the Cretaceous-Tertiary (K-T) boundary debate is whether the terminal Cretaceous extinctions were abrupt and coincident with physical evidence for bolide impact, or gradual and linked to Earth-bound environmental changes. Although there is abundant evidence that a large bolide impact occurred at the K-T boundary (e.g., Ryder et al., 1996) and that the K-T turnover is exceptionally sharp relative to other Phanerozoic extinctions (Ryder, 1996), some workers maintain that the Chicxulub impact event was not the major cause of the end-Cretaceous extinctions (e.g., Keller, 1996; Archibald, 1996; Sarjent, 1996; Pardo et al., 1999).

Differing interpretations of the planktonic foraminiferal biostratigraphic record have been the focus of much of this controversy. Some authors argue that changes in sea level, temperature, and global volcanism led to prolonged extinctions before and after the K-T boundary (e.g., MacLeod and Keller, 1994; Keller and Stinnesbeck, 1996; Keller et al., 1998; Li and Keller, 1998; Pardo et al., 1999). Others suggest that the pattern of gradual preboundary extinction among these organisms can be explained as a combination of sampling or taxonomic artifacts and low-level, background extinctions, while the post-boundary occurrences mostly result from sediment reworking (e.g., Olsson and Liu, 1993; Huber, 1996). The reworking hypothesis is strongly supported by recognition of offsets between the strontium and carbon isotopic composition of trans-K-T boundary Cretaceous species and co-occurring in situ Danian species (Zachos et al., 1992; Huber, 1996; MacLeod and Huber, 1996; Kaiho et al., 1999; MacLeod et al., 2001).

A remarkably complete and well-preserved K-T boundary sequence was recovered in the three holes drilled at Ocean Drilling Program (ODP) Site 1049, which is located in the subtropical North Atlantic and ~1600 km from the Chicxulub impact site (Fig. 1). Between uppermost Maastrichtian and lowermost Danian nannofossil ooze in these three holes, there is a 9–17-cm-thick layer of asteroid impact ejecta that was presumably derived from the Chicxulub impact site (Norris et al., 1998, 1999; Klaus et al., 2000). The ejecta bed is capped by a 1–3-mm-thick red limonitic layer that is coincident with a zone of low iridium enrichment and is bracketed by intervals of higher Ir concentration above and below (Smit et al., 1997). The uppermost Maastrichtian sediments reveal evidence of soft-sediment deformation, perhaps resulting from seismicity caused by the Chicxulub impact (Klaus et al., 2000; Norris and Firth, this volume). The lowermost Danian sequence reveals no physical or biostratigraphic evidence of deformation or discontinuous sedimentation, yet it contains 36 species of Cretaceous planktonic foraminifera. In this study we focused on analyses of changes in planktonic foraminiferal population structure, stratigraphic distributions, and shell size/mass ratios across the boundary succession to evaluate evidence for post-K-T reworking.

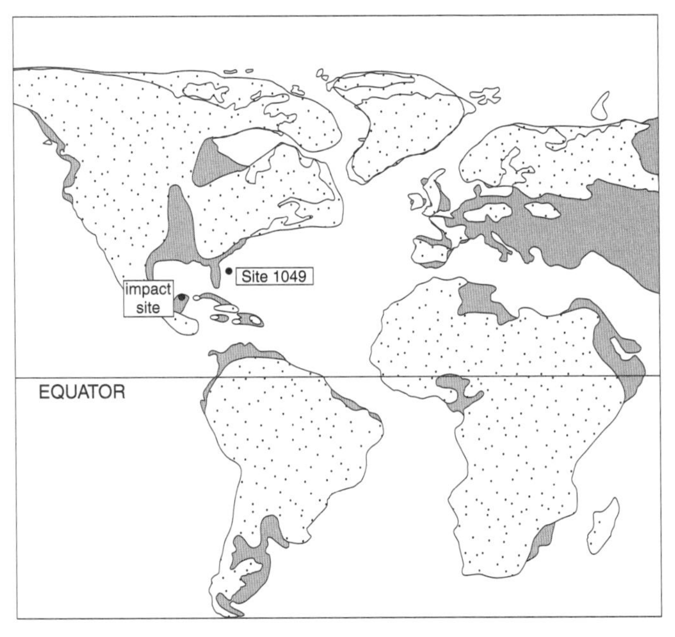

Figure 1. Paleogeography for 65 Ma showing approximate locations of Site 1049 and Chicxulub impact crater. Three holes drilled ~30 m apart at Site 1049 yielded complete Cretaceous-Tertiary sequences containing ejecta bed varying from 9 to 17 cm in thickness. Continental positions were generated using PGIS/Mac™ software package and land and sea distributions are modified from Barron (1987).

MATERIAL AND METHODS

The analyzed samples are from section 1049C-8X-5, between 112.10 and 113.15 m below seafloor (mbsf). The spacing of samples used in abundance and size/mass determinations is between 6 and 10 cm. Bulk samples ranging from 0.5 to 1.0 cm^3 were disaggregated in tap water, washed over a 38 µm sieve, and dried in an oven at ~50°C. The size/mass ratio of Cretaceous species in Danian sediments was determined by pouring the samples through a nest of sieves ranging from 125 to 500 µm in size and separated at 1/4 phi size intervals. All Cretaceous planktonic foraminifera were picked from each size interval and weighed on a microbalance. Because nonforaminiferal grains compose <2% of the total >125 µm size fraction, the residue without Cretaceous specimens closely approximates the proportion of Danian planktonic foraminifera in each size fraction. This residue was also weighed to determine the total mass of foraminifera in each sample.

Numerical abundance counts were based on identification of all Cretaceous specimens picked from the >125 µm size intervals. Relative abundance estimates of the Danian species were made from visual observation of the >38 µm size fraction. Taxonomic concepts for Cretaceous species are the same as those used in MacLeod et al. (2000) for Deep Sea Drilling Project Site 390A, which was drilled at the same location as Site 1049. Taxonomic concepts for Danian species and Danian biozone definitions are based on those used in Olsson et al. (1999).

Three thin sections were prepared from a continuous 15.2 × 3.6 × 1.3 cm slab of sediment that was removed from the K-T interval of Hole 1049C. Thin sectioning was accomplished by building a Plexiglas epoxy case around the slab of sediment, impregnating the slab with Epo-Tech 301 epoxy in a vacuum chamber, cutting the block lengthwise, dividing one half of one length into thirds, and grinding these into thin sections to the petrographic thickness of calcite (20 μm or less). The remaining sediment block was reimpregnated and attached to a 3 mil clear polyester plastic sheet on one side and a 3-mm-thick Plexiglas sheet on the other. Some minor disturbance of the ejecta bed occurred during this procedure as a result of dehydration and development of shrinkage cracks in the slab of sediment and swelling of the porous, smectite-rich spherule layer when the epoxy was added (Fig. 2).

GEOLOGIC SETTING

The K-T boundary was cored at three sites along a depth transect extending from 1344 to 2670 m water depth on Blake Nose during ODP Leg 171B, but only the deepest site (Site 1049) yielded an impact-derived ejecta bed. The oldest sediments on Blake Nose are Jurassic and Lower Cretaceous platform carbonates that were deposited on the rifted margin of North America (Benson et al., 1978). These shallow-water limestones are overlain by a relatively thin succession of hemipelagic and pelagic carbonates that range from late Aptian through late Eocene in age. Capping the Cretaceous-Eocene sequence is a thin layer of manganiferous nodules and sand containing Pleistocene to Holocene foraminifera (Norris et al., 1998). The shallow burial depths account for the remarkably good preservation of biogenic calcite throughout most of the Upper Cretaceous-Eocene sequence at all of the Leg 171B drill sites.

Soft-sediment deformation features such as contorted bedding, microfaults, and variably angled dips are pervasive in the upper Maastrichtian core sections. Such deformation features are absent from the overlying Danian sediments. Analysis of Maastrichtian sediment cores and seismic reflection data along the Leg 171B depth transect led Klaus et al. (2000) as well as Norris and Firth (this volume) to conclude that mass wasting occurred across Blake Nose as a result of the Chicxulub impact event. Despite this evidence for sediment deformation, the middle through upper Maastrichtian sequence at Site 1049 is biostratigraphically complete and there is no evidence for stratigraphic repetition (Norris et al., 1999; Self-Trail, 2001).

Benthic foraminifera from the Maastrichtian-Danian sequence at Site 1049 compose <2% of the total foraminiferal assemblage. Their rarity and dominance by species that are not found at upper slope or shelfal depths suggest that this site occupied middle to upper bathyal paleodepths during K-T boundary time (Norris et al., 1998).

RESULTS

Maastrichtian sediments at Site 1049 are composed of foraminifer-nannofossil chalk and ooze and contain very little detrital material (Fig. 2A). Ooze immediately below the ejecta bed contains well-preserved planktonic foraminiferal (Fig. 3A) and calcareous nannofossil assemblages from the *Abathomphalus mayaroensis* and *Micula prinsii* zones, respectively, and have been assigned to the latest Maastrichtian portion of magnetic polarity chron 29R (Norris et al., 1998). Although Pardo et al. (1999) identified *Plummerita hantkenoides* as a planktonic foraminiferal marker species for the uppermost Maastrichtian, its absence at Site 1049 is consistent with its absence from deep-sea sediments worldwide, suggesting that this species is restricted to shallower water biofacies. Nonetheless, the presence of *Pseudoguembelina hariaensis* within 1.5 m below the ejecta bed (Table 1) confirms assignment of this interval to the uppermost Maastrichtian, because this species is restricted to the upper *A. mayaroensis* zone elsewhere (Nederbragt, 1991; Li and Keller, 1998).

Petrology

The ejecta bed is in sharp contact with the underlying ooze. It is predominantly composed of unconsolidated, circular to ovoid spherules (Fig. 3B) that range to 3 mm in size and vary in color from dark green to pale yellow. X-ray diffraction analyses indicate that the originally glassy spherules have been diagenetically altered to smectite (Martínez-Ruiz et al., 2001). Rare spherules composed of calcite have been observed in thin section. Large Cretaceous planktonic foraminifera are concentrated in discrete layers within the basal 1 cm of the ejecta bed (Fig. 2B), and are abundant and randomly distributed through the rest of the bed. Thin-section and macroscopic observations of the core reveal that the ejecta bed contains little fine-grained matrix and, apart from the foraminiferal grainstone, is poorly sorted throughout and lacks any sedimentary structures. Chalk clasts to 1 cm in diameter, grains of euhedral dolomite that reach 0.1 mm in length, and rare grains of shocked quartz (Fig. 2D) and euhedral zeolite occur in the middle and upper portions of the ejecta bed. An echinoid spine found in the middle of the ejecta bed (Fig. 2C) is considered exotic because none were found in the pelagic carbonate below and above. The thickness of the ejecta bed is 17 cm at Hole 1049A, 13 cm at Hole 1049B, and 9 cm at Hole 1049C (see Klaus et al., 2000, for color images). At the top of the ejecta bed is a 1–3-mm-thick orange, limonitic layer that contains flat goethite concretions (Fig. 2E). Lateral variation in the thickness of the limonitic layer is consistent with evidence for upward diffusion and precipitation of trace elements in the upper part of the ejecta bed (Martínez-Ruiz et al., 2001).

Immediately above the ejecta bed is a 3–7-cm-thick dark, burrow-mottled, clay-rich ooze that contains well-preserved,

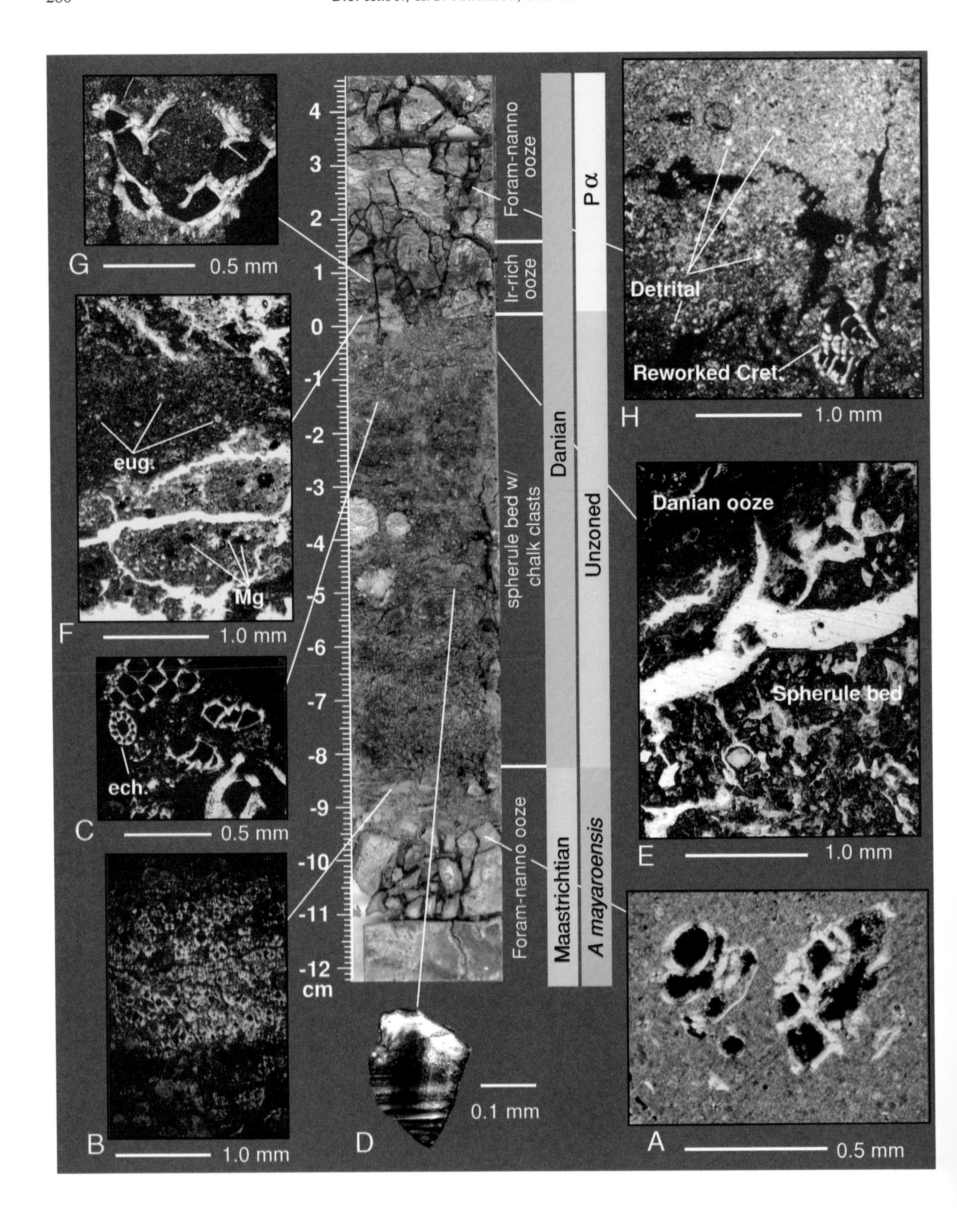

G
0.5 mm
F
eug.
Mg
1.0 mm
C
ech.
0.5 mm
B
1.0 mm
4
3
2
1
0
-1
-2
-3
-4
-5
-6
-7
-8
-9
-10
-11
-12
cm
D
0.1 mm
Foram-nanno ooze
Ir-rich ooze
spherule bed w/ chalk clasts
Foram-nanno ooze
Danian
Maastrichtian
P α
Unzoned
A mayaroensis
H
Detrital
Reworked Cret.
1.0 mm
E
Danian ooze
Spherule bed
1.0 mm
A
0.5 mm

Figure 2. Epoxy block and thin sections across Cretaceous-Tertiary boundary in Hole 1049C revealing changes in lithology, sediment fabric, and species distributions. A: Uppermost Maastrichtian chalk shows large Cretaceous planktonic foraminifera floating in a matrix of nannofossil ooze. B: Basal ejecta bed shows size sorting of Cretaceous planktonic foraminifera into grainstone, suggesting winnowing and lateral transport. C: Presence of occasional echinoid spines may indicate derivation of some sediment from shallower depth. D: Shocked quartz occurs within middle and upper levels of ejecta bed. E: Contact between top of spherule bed and basal Danian ooze shows no evidence of sediment winnowing or presence of hardground, which might be expected if this contact were disconformable. F: Presence of tiny specimens of *Parvulorugoglobigerina eugubina* (eug.) in basal Danian sediments indicates absence of lowermost Danian zone P0. Opaque grains within 2 mm above top of ejecta bed are Mg spinels (J. Smit, 2000, personal commun.). G: Different colored matrix within shell of *Contusotruncana contusa* compared to surrounding matrix reveals that this specimen has been reworked. H: Larger size and greater abundance of detrital clastic sediment in lower Danian vs. Cretaceous chalk indicates increased downslope sediment transport.

although strongly recrystallized, minute planktonic foraminifer species characteristic of the lower Danian Pα Zone (Fig. 3C). The contact between this layer and the underlying ejecta bed is sharp (Fig. 2E). Smit et al. (1997) reported Ir concentrations of 1.3 ppb within the burrow-mottled ooze, in contrast to values of <0.06 ppb at the base of the ejecta bed and 3 cm above its top. Detrital grains including quartz, pyroxene, and mica are more abundant and larger in the mottled interval than in the ooze below the ejecta bed (Fig. 2, F and H). Opaque minerals identified as Mg-rich spinels are concentrated in the lower part of this bed (Fig. 2F) and are randomly dispersed in the sediment matrix for several millimeters above the contact. A specimen of the Maastrichtian species *Contusotruncana contusa* that contains an infilling matrix of a different color than the surrounding matrix was observed in thin section within 1 cm above the limonitic layer (Fig. 2G). Scanning electron microscope (SEM) observation of the internal matrix of another trans-K-T species found in Danian sediments, *Heterohelix globulosa*, reveals an assemblage of Cretaceous calcareous nannofossil species not considered to be survivors of the K-T extinction event (Fig. 4). No Danian calcareous nannofossil species occur within the shell of this specimen, despite their abundant presence in the surrounding Danian sediments.

The sample from within 0.5 cm above the ejecta bed is dominated by *Guembelitria cretacea*, but it also contains rare *Parvularugoglobigerina extensa* and *Parvularugoglobigerina eugubina*. The presence of the latter species and absence of *Chiloguembelina morsei* or *Chiloguembelina midwayensis* indicate that this sample correlates with the lowermost Pα Zone.

The absence of the basal Danian *Guembelitria cretacea* Zone (P0 Zone) at all deep-sea sites, including Site 1049, has previously been attributed to the temporary absence or rare

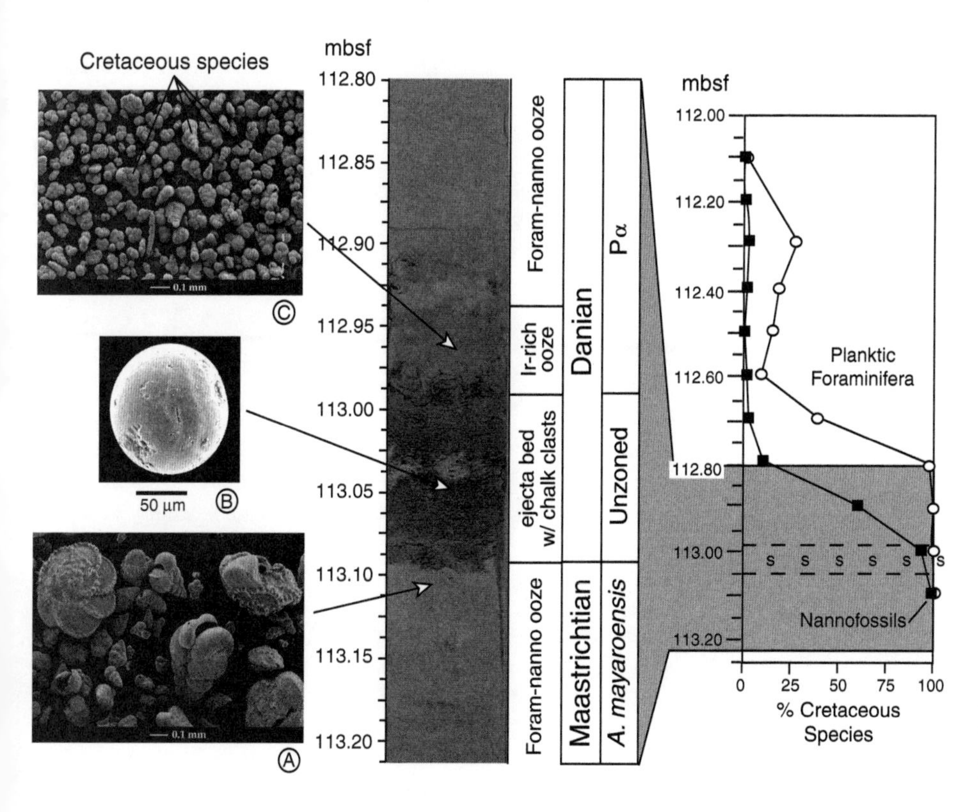

Figure 3. Comparison of pre-Cretaceous-Tertiary (K-T) (A) and post-K-T (C) planktonic foraminiferal assemblages and example of spherule (B) from K-T boundary interval of Ocean Drilling Program Hole 1049C. Relative abundances of planktonic foraminifera (wt% Cretaceous species in >125 µm fraction) and calcareous nannofossils (counts of 450 specimens/sample at 1600× along random smear slide traverse) are from Norris et al. (1999). Abrupt and dramatic turnover in planktonic foraminifera coincident with K-T impact event is evident from comparison of Cretaceous assemblage from 1 cm below top of Maastrichtian chalk (A) with basal Danian assemblage from 2 cm above ejecta bed (C). Scanning electron microscope images shown were prepared using randomly poured >38 µm sieved residues from two sample levels. Note that Danian assemblage is strikingly smaller in size and lower in species diversity than Cretaceous assemblage, and that shells of Danian species are much simpler in morphology. Also note that Cretaceous species that occur in Danian (denoted by K) are relatively rare (mbsf is meters below sea floor).

TABLE 1. PERCENTAGE ABUNDANCE DETERMINATIONS FOR CRETACEOUS PLANKTONIC FORAMINIFERA ACROSS THE CRETACEOUS-TERTIARY INTERVAL OF OCEAN DRILLING PROGRAM HOLE 1049C

HOLE 1049C	Top Zone	Zone	*Abathomphalus mayaroensis*	*Archaeoglobigerina cf. blowi*	*Contusotruncana contusa*	*Globigerinelloides prairiehillensis*	*Globigerinelloides subcarinatus*	*Globotruncana aegyptiaca*	*Globotruncana arca*	*Globotruncana dupueblei*	*Globotruncana esnehensis*	*Globotruncana falsostuarti*	*Globotruncana insignis*	*Globotruncana orientalis*	*Globotruncana rosetta*	*Globotruncana sp.*	*Globotruncanella minuta*	*Globotruncanella petaloidea*	*Globotruncanita stuarti*	*Globotruncanita stuartiformis*	*Gublerina acuta*	*Guembelina cretacea*	*Hedbergella monmouthensis*
8X-5, 0–2	112.10	P1a				2.3	2.3		6.8				4.5	6.8		9.1		4.5		2.3	2.3	P	
8X-5, 10–12	112.20	Pα			1.9		5.7						3.8			11.3	1.9	7.5				P	
8X-5, 20–22	112.30	Pα			2.2	2.2	4.3		10.9				2.2			13.0		6.5		2.2		P	
8X-5, 30–32	112.40	Pα		2.6		2.6	5.3		5.3				2.6	5.3		15.8		5.3		2.6		P	
8X-5, 40–42	112.50	Pα		7.9		2.6			15.8				2.6	10.5		2.6		10.5		5.3		P	
8X-5, 50–52	112.60	Pα		1.9	1.9	3.8	1.9		13.5				1.9	5.8		5.8		1.9		1.9	1.9	P	
8X-5, 60–61	112.70	Pα			2.3	2.3	2.3		4.7				1.2	2.3		11.6	2.3	5.8		2.3		P	
8X-5, 70–71	112.80	Pα			0.7	8.5	1.4		3.5		2.8		1.4			18.4	1.4	0.7		0.7	0.7	P	0.7
8X-5, 80–80.5	112.90	Pα			0.8	3.1	1.6		3.1			1.6		1.6		14.7	0.8	3.1		3.1	0.8	P	5.4
8X-5, 88–88.5	112.96	Pα	1.4		1.4	1.4			4.1					1.4		18.9	0.0	1.4	5.4	6.8		P	4.1
8X-5, 105–106	113.16	*A. may.*		0.2	0.5	3.9	1.4	0.2	2.3		0.2		0.7	1.6		17.0	1.4	1.1		2.8	0.7	P	0.7
8x-6, 85–87	114.45	*A. may.*	0.4	5.4	0.8	3.5	2.7	0.4	5.3	0.6	4.8	1.2	2.8		3.9	11.4	1.6	1.0		3.2	2.2	P	0.8

HOLE 1049C	*Heterohelix globulosa*	*Heterohelix navarroensis*	*Heterohelix planata*	*Heterohelix punctulata*	*Heterohelix semicostata*	*Heterohelix cf. sphenoides*	*Laeviheterohelix glabrans*	*Planoglobulina acervulinoides*	*Planoglobulina carseyae*	*Planoglobulina multicamerata*	*Pseudoguembelina costulata*	*Pseudoguembelina hariaensis*	*Pseudoguembelina palpebra*	*Pseudotextularia elegans*	*Pseudotextularia intermedia*	*Psuedotextularia nuttalli*	*Racemiguembelina fructicosa*	*Rugoglobigerina hexacamerata*	*Rugoglobigerina milamensis*	*Rugoglobigerina rugosa*	*Schackoina multispina*	*Trinitella scotti*	Total specimen no.
8X-5, 0–2	34.1	9.1				2.3	2.3			2.3			2.3		2.3	2.3						2.3	44
8X-5, 10–12	28.3	11.3		7.5							5.7		1.9		1.9			1.9		5.7		3.8	53
8X-5, 20–22	19.6	10.9		2.2							2.2		4.3		2.2	6.5	4.3					4.3	46
8X-5, 30–32	23.7	7.9									2.6		2.6		2.6	5.3	5.3					2.6	38
8X-5, 40–42	21.1	5.3					2.6				5.3		2.6							5.3			38
8X-5, 50–52	25.0	5.8	1.9		3.8	1.9				1.9	1.9		3.8		1.9	1.9	1.9			3.8	1.9		52
8X-5, 60–61	37.2	7.0				4.7		1.2			1.2		1.2		2.3			1.2		5.8		1.2	86
8X-5, 70–71	27.7	3.5	1.4			3.5	1.4	0.7		0.7	2.1		2.8	0.7	1.4	4.3	3.5	1.4		1.4	0.7	1.4	141
8X-5, 80–80.5	24.8	7.0	1.6			3.1					1.6		1.6		0.8	3.9	7.0	0.8		7.8		0.8	129
8X-5, 88–88.5	20.3	2.7	1.4					1.4		1.4			4.1	1.4	1.4	1.4	12.2			5.4		1.4	74
8X-5, 105–106	29.8	6.2	0.2	0.7		4.6	0.5	0.5	0.2	0.5	0.5	0.2	2.8	0.9	2.1	0.9	3.2	0.9	0.5	5.7	P	4.6	436
8x-6, 85–87	26.2	4.7	1.1			1.2	1.2	0.5		0.3	1.1	0.6	2.2	1.6		2.1	3.8	1.4					382

P = present in <125 μm fraction.

occurrence in shallow-marine sections of *P. eugubina*, the nominate marker for the overlying Pα Zone (Norris et al., 1999). However, the P0 Zone could be missing because of a slow deep-sea sedimentation rate, which would have increased the likelihood that the seafloor surface was exposed to the effects of bioturbation and current winnowing. Although thin-section observation of the contact between the ejecta bed and basal Danian sediments at Site 1049 reveals no evidence of sedimentary winnowing or development of a hardground surface (Fig. 2E), we cannot unequivocally determine how much, if any, of the lowermost Danian may be missing from Site 1049. However, the presence of Mg spinels and an Ir anomaly in basal Paleocene sediments suggest that the boundary section is complete.

A burrowed contact marks the base of an overlying 5–15-cm-thick, white foraminiferal-nannofossil ooze that also contains a Pα Zone foraminiferal assemblage and fossils typical of calcareous nannofossil biozone NP1 (Norris et al., 1999; Self-Trail, 2001). In sharp contact with the top of the white interval is greenish-gray nannofossil foraminiferal ooze, which is the final bed of the studied sequence. The base of the P1a Zone, identified at the level of the extinction of *P. eugubina*, is placed within this greenish-gray interval, 88.5 cm above the top of the ejecta bed (Table 1).

Relative abundance changes

Of the 37 species identified in the uppermost Maastrichtian sample studied, 33 species (89%) occur in the overlying Pα Zone sediments and 19 species (51%) range into Zone P1a (Table 2). The lowermost 20 cm of the Pα Zone is dominated by Cretaceous taxa, but by the top of this zone, the Cretaceous component represents <1% of the total assemblage (Fig. 3). A similar, but more rapid, pattern of gradual replacement of Cretaceous by Danian taxa at this site has been observed for calcareous nannofossils (Norris et al., 1999; Self-Trail, 2001).

The most abundant and consistently occurring Cretaceous species in the Danian sediments, such as *Heterohelix globulosa*, tend to be the most abundant species below the K-T ejecta bed. Similarly, the rare species with sporadic distributions in Maastrichtian sediments, such as *Globotruncana aegyptiaca* (see also MacLeod et al., 2000, Table 8.4), also tend to be rare in Danian sediments. Some Cretaceous species, such as *Racemiguembelina fructicosa*, have erratic changes in abundance in the Danian sediments. Although this species usually composes <5% of upper Maastrichtian planktonic foraminiferal assemblages and is consistently present in all samples throughout its upper Maastrichtian range (e.g., MacLeod et al., 2000), its abundance is >12% in one Danian sample but it is absent from others.

Mass/size ratios

Comparison between the masses of Cretaceous planktonic foraminiferal shells per size fraction from below versus above

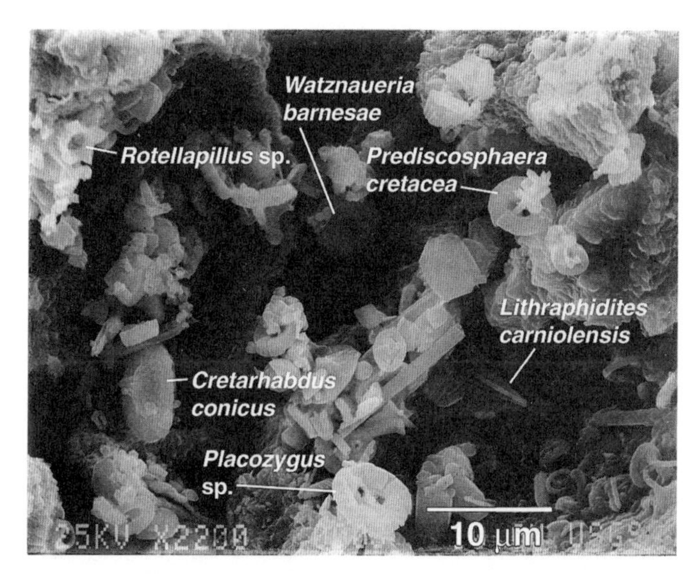

Figure 4. Scanning electron micrograph of calcareous nannofossils from within shell of *Heterohelix globulosa*, trans-Cretaceous-Tertiary (K-T) planktonic foraminifer found in sample 1049C-8X-5, 75.0–75.5 cm, 13.5 cm above top of spherule bed. Only Cretaceous nannofossil species that lived during Maastrichtian occur in matrix, despite abundant presence of Danian species in surrounding sediments, and none are considered as K-T survivor species in the sense of Pospichal (1996). Calcareous nannofossil species identifications were done by J. Self-Trail (2001, personal commun.).

the K-T ejecta bed reveals a significant change in size distributions. The largest percentage of specimens in the upper Maastrichtian sample from 113.16 mbsf occurs in the smallest size fractions (Fig. 5). This is similar to the size distribution of living populations, which tend to be dominated by small individuals (Berger, 1971). However, the Danian samples analyzed yield a size distribution with the greatest percentage of specimens in the >212 µm size fractions. The most extreme departure from the expected size distribution occurs in the sample from 112.70 mbsf, ~28 cm above the top of the ejecta bed, with nearly 60% of the assemblage comprising the >500 µm size fraction.

Plots of the percentage mass of Cretaceous specimens relative to the combined mass of Cretaceous and Danian specimens document the prolonged presence of Cretaceous specimens in the larger size fractions and gradual increase in Danian species beginning in the smallest size fractions (Fig. 6). In the sample from within 1 cm above the ejecta bed (112.98 mbsf), Danian species are only present in the <125 µm size fraction. Danian species are included in 4% of the 125–150 µm size interval 8 cm above this sample (112.90 mbsf), but 100% of the >150 µm size fraction is composed of Cretaceous specimens only. Within 28 cm above the ejecta bed (112.70 mbsf) Danian species dominate the <212 µm size fraction sample, but Cretaceous specimens compose 100% of the >300 µm fraction. Cretaceous specimens continue to represent 100% of the >300 µm fraction to 78 cm above the ejecta bed (112.20 mbsf) and dominate that fraction throughout the Pα Zone and into the lower P1a Zone.

DISCUSSION

Redeposition of the ejecta bed

The composition of the spherule bed at Site 1049 suggests an impact origin for most of the nonbiogenic grains (Norris et al., 1998, 1999; Martinez-Ruiz et al., 2001), but the bed does not seem to solely represent material deposited directly after settling through the water column. The sedimentary features of the spherule bed are more consistent with deposition from one or several mass-flow events. Evidence for this interpretation includes the presence of: (1) a scoured basal contact with the underlying Maastrichtian chalk; (2) a size-sorted foraminiferal grainstone in the lowermost 1 cm, immediately above a layer of spherules; (3) large foraminifera scattered throughout the bed; (4) an exotic echinoid spine and large chalk clasts in the middle and upper portion of the bed; and (5) a slight decrease in grain size in the uppermost portion of the bed (Fig. 2).

Because it occurs precisely at the K-T boundary, it is tempting to interpret the mass flow(s) as being related to impact seismicity (e.g., Klaus et al., 2000). However, the sedimentology of the K-T succession at Site 1049 suggests a more complex history. Immediately after the impact, but before there was time for the ejecta to settle to the seafloor, impact seismicity probably caused slumping of the Maastrichtian ooze (Klaus et al., 2000). The redeposited chalk and ooze probably predated gravity settling of the tektites through ~2 km of water by a several hours or less. The source for the planktonic foraminifera in the ejecta bed seems to have been upslope equivalents of the underlying upper Maastrichtian ooze, because all species identified (including those in the chalk clasts) occur in underlying upper Maastrichtian samples. There is no evidence at Site 1049 for reworking of older Cretaceous material into the boundary bed as seen in the Caribbean area (cf. Bralower et al., 1998), which likely reflects details of the local geology and the location of Site 1049 on the protected side of the Florida carbonate platform. Absence from the ejecta bed of shelf-dwelling benthic foraminifera suggests that the source of reworked specimens was from below the shelf-slope break.

The tektites could have been remobilized following their rapid accumulation on already disturbed sediments, or they could have originated by gravity flows from the continental shelf and slope following impact-generated aftershocks or tsunamis generated by mass gravity flows elsewhere on the continental slope following margin collapse (e.g., Olsson et al., this volume). The different grain sizes in various parts of the ejecta bed could reflect deposition of ejecta by multiple plumes of sediment coming off Blake Plateau, some of which would have flowed directly down Blake Nose (delivering relatively coarse sediments) and others bypassing Blake Nose but delivering finer grained ejecta in laterally spreading plumes (Norris and

TABLE 2. VISUAL RELATIVE ABUNDANCE ESTIMATES FOR DANIAN PLANKTONIC FORAMINIFERA IN HOLE 1049C (>38 μm FRACTION)

HOLE 1049C	Top Depth (mbsf)	Zone	*Guembelitria cretacea*	*Woodringina? sp.*	*Globanomalina archeocompressa*	*Parasubbotina aff. pseudobulloides*	*Parvarugoglobigerina eugubina*	*Parvularugoglobigerina alabamensis*	*Woodringina hornerstownensis*	*Chiloguembelina midwayensis*	*Parvularugoglobigerina extensa*	*Woodringina claytonensis*	*Eoglobigerina eobulloides*	*Parasubbotina pseudobulloides*	*Globoconusa daubjergensis*	*Chiloguembelina morsei*	*Eoglobigerina edita*	*Globanomalina planocompressa*	*Praemurica taurica*	*Subbotina trivialis*
8X-5, 0–2	112.10	P1a							R			R	C	C	P	P	F	F		R
8X-5, 10–12	112.20	Pα					P		C	C		C	F	A	F	R	C	C	F	
8X-5, 20–22	112.30	Pα	R				R		C	C		C	F	F	R	R	F	F	R	
8X-5, 30–32	112.40	Pα	C				C		A	C	R	C	F	C	R	R	C	F	R	
8X-5, 40–42	112.50	Pα	C				C		A	C	R	C	F	C	R	R	R	R	P	
8X-5, 50–52	112.60	Pα	C				F		C	C	R	C	F	F	R	R	R	R	P	
8X-5, 60–61	112.70	Pα	C		?		C		A	A	F	F	C	F	F	F	R	R	F	
8X-5, 70–71	112.80	Pα	C		R	R	C	R	R	C	F	F	F	F	?					
8X-5, 80–80.5	112.90	Pα	C		R	R	C	P	R	P	F									
8X-5, 88–88.5	112.96	Pα	C	P	R	P	P	P												
8X-5, 105–106	113.16	*A. may.*	P																	

A = abundant (>15%); C = common (5%–15%); F = few (1%–5%); R = rare < 1%); P = present; ? = uncertain; mbsf = meters below sea floor.

Firth, this volume). The foraminiferal grainstone at the base of the ejecta bed may reflect either sorting produced by the initial slump of the upper Maastrichtian section or a different plume of ejecta and carbonate sediment than the coarser grained middle and upper parts of the ejecta bed. Lateral variation in thickness of the ejecta bed could be explained by mass-flow accumulation on the irregular, slumped surface of Maastrichtian ooze or lateral variation in mass-flow sediment volume.

The uppermost 1–3 mm of the spherule bed may have been deposited by direct settling of impact dust several weeks to many months after the impact event. This possibility is suggested by the fine-grained nature of this interval and concentration of iron oxide and iridium, which probably originated from vaporization of the K-T asteroid, although diagenetic mobilization of some elements has modified original depositional patterns (Martinez-Ruiz et al., 2001).

Evidence for Danian reworking

The thin section and SEM observations show that some Cretaceous taxa in the Danian are reworked (e.g., exotic internal matrix; Figs. 2G and 4), and reworking of Cretaceous deposits during the early Danian would explain peculiarities in the stratigraphic and size distribution of Cretaceous assemblages present above the K-T boundary (Norris et al., 1999; Norris and Firth, this volume). Unlike living and in situ fossil foraminiferal assemblages, which have progressively greater abundance in smaller size fractions (Berger, 1971), the distribution of Cretaceous taxa in the Danian shows (1) a peak abundance above the smallest size fraction, with some samples exhibiting extreme departures from the expected pattern (e.g., 112.98, 112.70, 112.10 mbsf in Fig. 5); (2) the modes of the size-distribution plots vary from sample to sample in the Danian interval; and (3) bimodal size distributions in some samples, suggesting that the foraminiferal shells were derived from multiple source beds. In addition, some species that are consistently present in low abundance in Maastrichtian samples have sporadic and highly variable abundance in the Danian samples, and several species that have been accepted as likely victims of the K-T event (e.g., *Globotruncanita stuartiformis, Globotruncana arca, Globotruncana insignis, Racemiguembelina fructicosa, Pseudotextularia elegans*) by those espousing gradual post-K-T foraminiferal extinctions (e.g., Keller, 1996; MacLeod and Keller, 1994; Pardo et al., 1999) are abundant above the Site 1049 ejecta bed. Complementary studies of the $^{87}Sr/^{86}Sr$ ratios of Cretaceous and Tertiary foraminifera tests from Hole 1049C showed that Cretaceous foraminifera occurring above the boundary have the isotopic signature of the upper Maastrichtian and did not live with the co-occurring Tertiary forms; i.e., they are reworked (MacLeod et al., 2001). The greater abundance and larger size of detrital grains in the ooze above the ejecta bed compared to below (Fig. 2) suggest that winnowing and downslope transport was more active during the earliest Danian than in the latest Maastrichtian.

Norris and Firth (this volume) showed that turbidites of K-T boundary age occur in the Bermuda Rise, where they consist largely of calcareous nannofossils and extremely small (<63 μm) planktonic foraminifera. Apparently slumping of the continental margin suspended large quantities of upper Maastrichtian sediments, some of which was transported off the margin into the deep sea. Size sorting within these resuspended sediments separated the fine fraction, dominated by calcareous nannofossils and small foraminifera, from the coarse fraction, dominated by large foraminifera and rock debris. The coarse

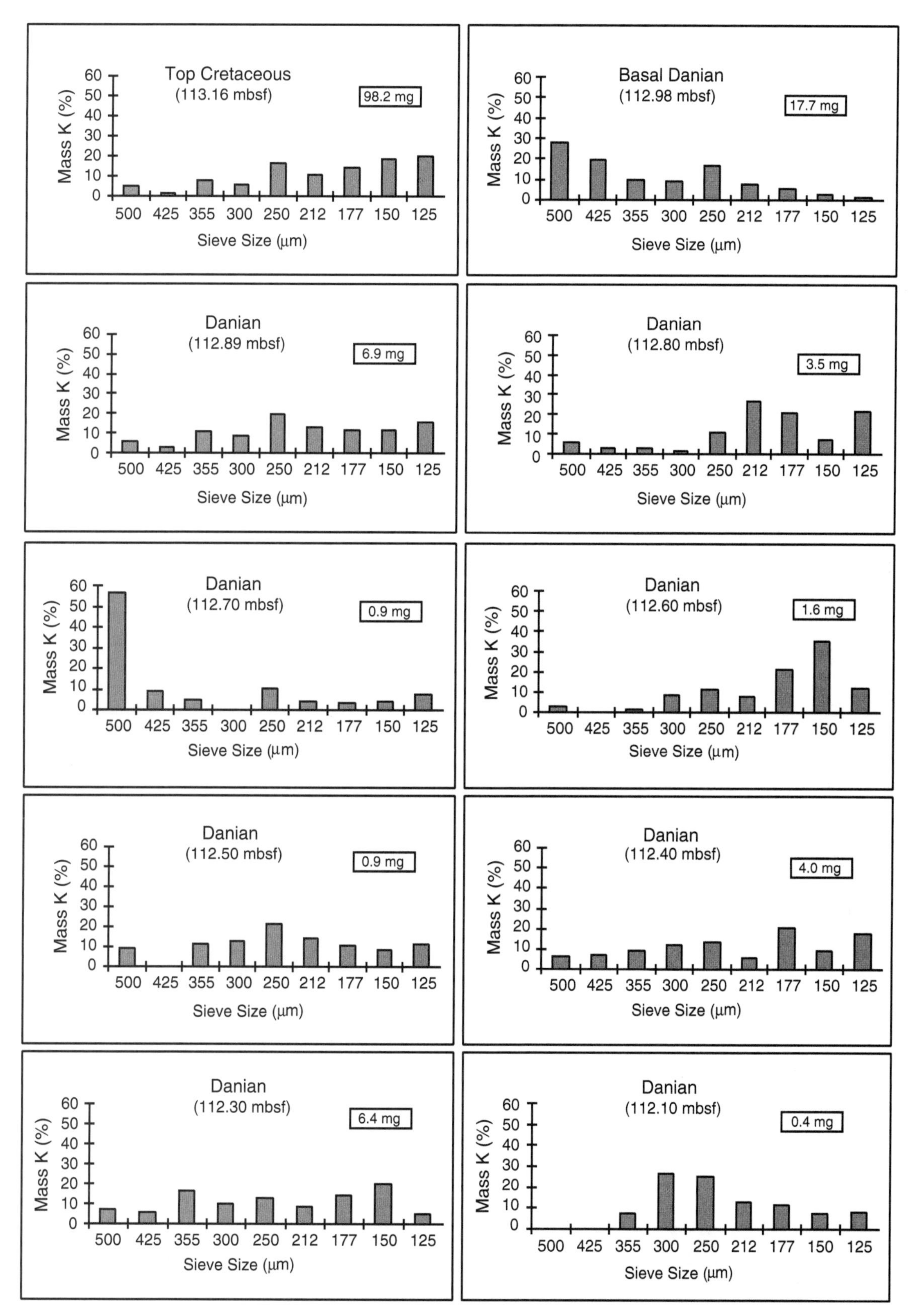

Figure 5. Weight percentage of Cretaceous (K) specimens from 1/4 phi sieve size intervals relative to total >125 μm fraction mass of Cretaceous specimens (shown in boxes) for uppermost Cretaceous and Danian samples. Note that bulk of mass of foraminifera occurs in <300 μm size fraction in Cretaceous sample (113.16 m below seafloor, mbsf), which is typical for Cenozoic and modern assemblages. Samples from above Cretaceous-Tertiary boundary have greatest mass of Cretaceous specimens concentrated in largest and intermediate size fractions.

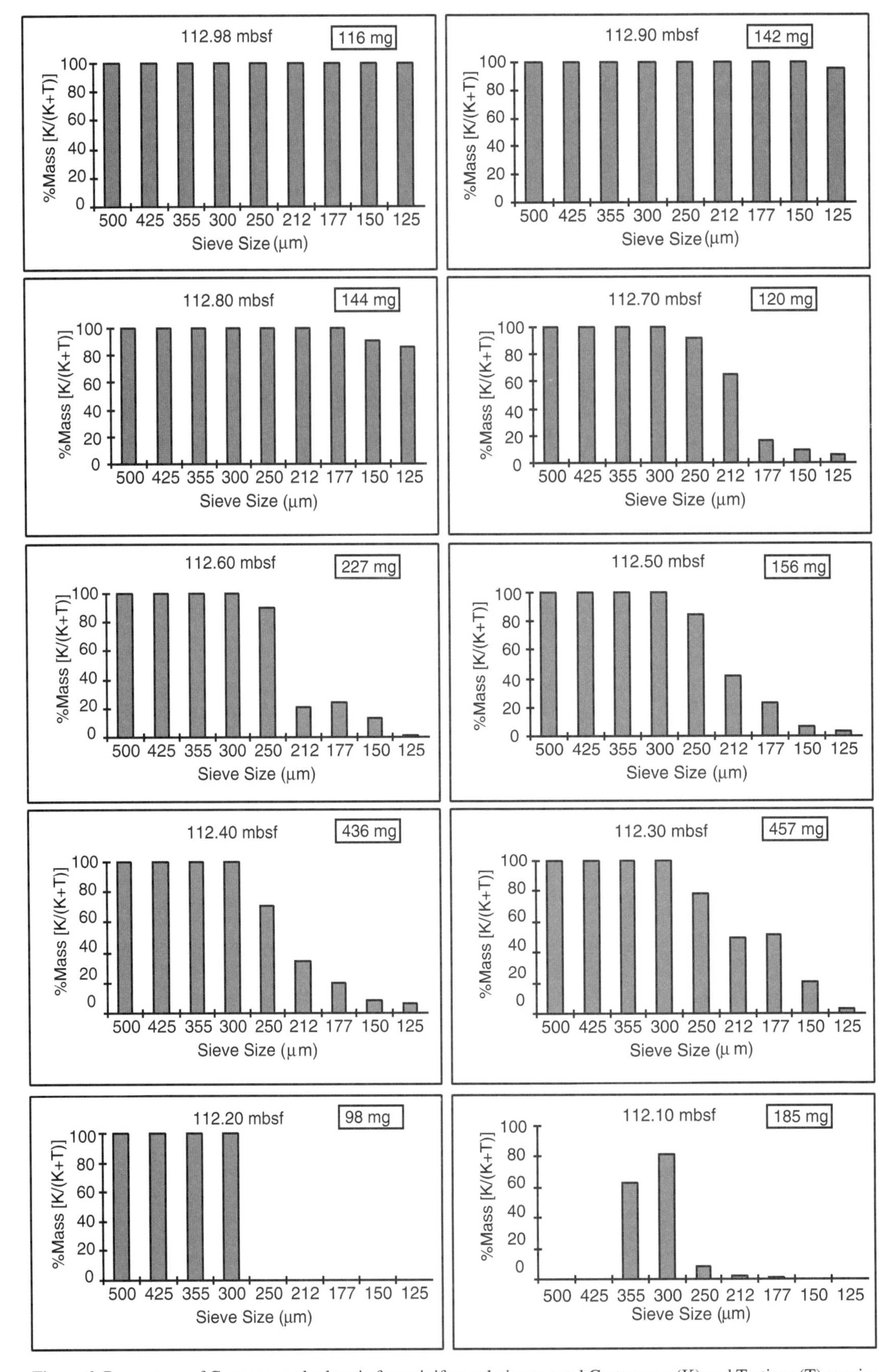

Figure 6. Percentage of Cretaceous planktonic foraminifera relative to total Cretaceous (K) and Tertiary (T) species in each of 1/4 phi size fractions above 125 μm. Note that Danian planktonic foraminifera are initially absent from all >125 μm size intervals then gradually increase in abundance beginning in smallest size fractions (mbsf is meters below sea floor).

fraction was mostly redeposited on or adjacent to the continental margin, and the fine fraction was carried offshore, which contributed to the near absence of small Cretaceous foraminifera in lowermost Paleocene sediments at ODP Site 1049.

Trans-K-T survivors

Only three Cretaceous species are considered to be definite survivors of the impact event, because their shell morphology has been linked to earliest Danian descendent species (Olsson et al., 1999). Two of these three species, *Guembelitria cretacea* and *Hedbergella monmouthensis*, were identified at Site 1049 (Fig. 7). If, on the basis of evidence presented here, all other Cretaceous species found in Danian sediments are considered reworked from underlying upper Maastrichtian sediments, then 35 of 37 species present (95%) in the topmost Maastrichtian sample studied at Site 1049 became extinct at the K-T boundary.

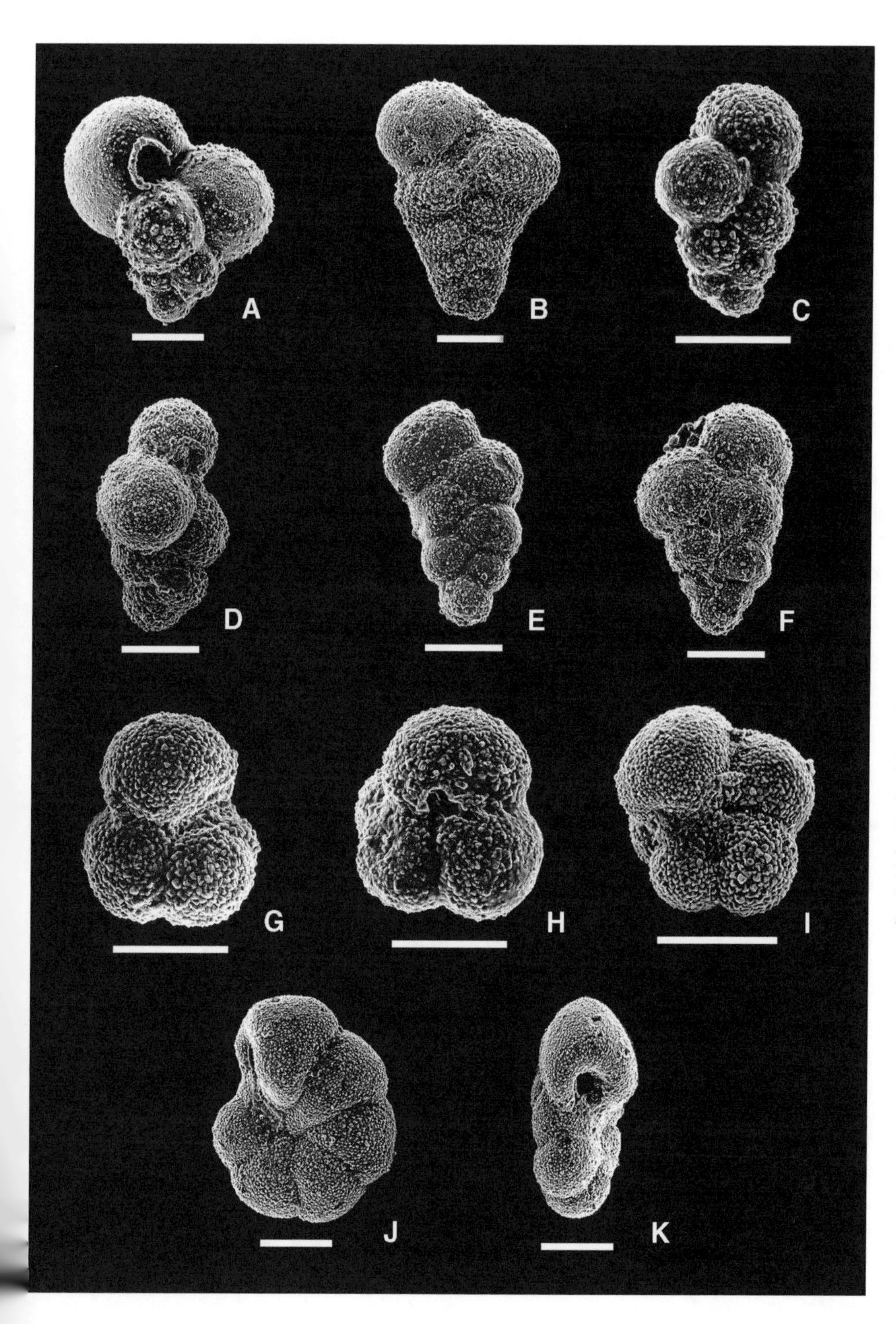

Figure 7. Planktonic foraminifera from lower Danian of Site 1049. Scale bars represent 30 μm. A: *Guembelitria cretacea* (1049C-8X-5, 87–89 cm). B: Side view of *Heterohelix navarroensis* (1049C-8X-5, 87–89 cm). C: Side view of *Woodringina claytonensis* (1049C-8X-5, 78–80 cm). D: Edge view of *Woodringina claytonensis* (1049C-8X-5, 78–80 cm). E: Side view of *Chiloguembelina* sp. (1049A-17X-2, 64–66 cm). F: Side view of *Chiloguembelina midwayensis* (1049C-8X-5, 78–80 cm). G: Umbilical view of *Parvularugoglobigerina alabamensis* (1049C-8X-5, 78–80 cm). H: Umbilical view of *Parvularugoglobigerina extensa* (1049C-8X-5, 78–80 cm). I: Umbilical view of *Eoglobigerina eobulloides* (1049C-8X-5, 87–89 cm). J: Umbilical and (K) edge view of *Parvularugoglobigerina eugubina* (1049C-8X-5, 87–89 cm).

Although not considered as ancestral to any Danian planktonic foraminiferal lineages, the Cretaceous species *Heterohelix globulosa* is consistently more abundant than any other trans-K-T species (Table 1) and is also abundant in the smallest (>38 μm) size fractions relative to co-occurring Danian species, suggesting that it also may have survived the bolide impact event. However, SEM analysis of calcareous nannofossils from within the shell of a *H. globulosa* found in Danian sediments reveals the presence of Cretaceous species and the absence of Danian species (Fig. 4), which indicates that the observed specimen was reworked. Carbon isotope ratios of Danian specimens of *H. globulosa* should be compared with co-occurring Danian species to resolve this uncertainty. This approach has been effectively used to identify trans-K-T reworking at several deep-sea sites (e.g., Zachos et al., 1992; Huber, 1996) and it was used to suggest that *H. globulosa* was a K-T survivor in the Brazos River, Texas, K-T sequence (Barrera and Keller, 1990).

CONCLUSIONS

A single 9–17-cm-thick meteorite impact ejecta bed containing shocked quartz and diagenetically altered tektite spherules marks the K-T boundary at three holes drilled at ODP Site 1049. Several features of the ejecta bed, including (1) irregularity of its basal contact with the underlying chalk, (2) presence of a foraminiferal grainstone in the lowermost 1 cm, (3) occurrence of clasts of chalk to 1 cm in diameter, (4) poor size sorting of the impact debris, and (5) variable thickness, suggest that the ejecta bed was probably emplaced during a slump or turbidity flow triggered by sediment overloading during the rapid accumulation of ejecta higher up on the continental slope.

Thin-section study reveals a sharp contact between the top of the ejecta bed and the overlying calcareous ooze containing abundant, tiny planktonic foraminifera characteristic of the early Danian Pα Zone. It is not clear why the lowermost Danian P0 Zone is absent at Site 1049. A hiatus is one possibility, but there is no physical evidence for an unconformity at this contact (e.g., scour or hardground surface). It is also possible that the P0 Zone is facies controlled (Norris et al., 1999). The absence of this zone might have indirectly resulted from the extremely slow sedimentation rates that would have been characteristic of deep-sea pelagic carbonate environments immediately after the mass extinction of a large proportion of the calcareous plankton biomass. Slow accumulation rates in combination with bioturbation could have resulted in mixing thorough enough to obliterate the distinction between the P0 and Pα zones.

Foraminifer distribution analysis and thin-section study indicate that the post-K-T occurrences of nearly all Cretaceous planktonic foraminiferal species are the result of reworking and that the extinction rate of foraminifera at the boundary was very high (95%) and abrupt. The rarity of small specimens of most Cretaceous species found in Danian sediments and the presence of trans-K-T specimens in Danian sediments that contain an exotic or unequivocally Cretaceous internal matrix provide the strongest evidence that they were redeposited. The bias toward intermediate and larger sizes among these trans-K-T species is unlikely to be an artifact of in situ winnowing of an assemblage once dominated by smaller specimens, because small Danian taxa constitute nearly the entire <125 μm fraction of the total assemblage. Instead, the widespread blanket of Cretaceous sediment underlying the Danian pelagic ooze would have been easily eroded and resuspended by lateral and downslope current activity. Such reworking of older sediments is commonly found in continental slope settings today.

ACKNOWLEDGMENTS

We gratefully acknowledge reviews by H.B. Vonhof and R.K. Olsson, calcareous nannofossil identifications by J. Self-Trail, thin-section preparations by D.A. Dean, and the Ocean Drilling Program for providing samples. This research was supported by grants from the Smithsonian Scholarly Studies Program to B. Huber, and the Joint Oceanographic Institutions–U.S. Science Support Program to K. MacLeod and R. Norris.

REFERENCES CITED

Archibald, J.D., 1996, Dinosaur extinction and the end of an era: What the fossils say: New York, Columbia University Press, 237 p.

Barrera, E., and Keller, G., 1990, Foraminiferal stable isotope evidence for a gradual decrease of marine productivity and Cretaceous species survivorship in the earliest Danian: Paleoceanography, v. 5, p. 867–890.

Barron, E.J., 1987, Global Cretaceous paleogeography-International Geologic Correlation Program Project 191: Palaeogeography, Palaeoclimatology, Palaeoecology, v. 59, p. 207–216.

Benson, W.E., Sheridan, R.E., and others, 1978, Initial Reports of the Deep Sea Drilling Project, Volume 44: Washington, D.C., U.S. Government Printing Office, p. 1–1005.

Berger, W.H., 1971, Planktonic foraminifera: Sediment production in an oceanic front: Journal of Foraminiferal Research, v. 1, p. 95–118.

Bralower, T.J., Paull, C.K., and Leckie, R.M., 1998, The Cretaceous-Tertiary boundary cocktail: Chicxulub impact triggers margin collapse and extensive sediment gravity flows: Geology, v. 26, p. 331–334.

Huber, B.T., 1996, Evidence for planktonic foraminifer reworking vs. survivorship across the Cretaceous/Tertiary boundary at high latitudes, *in* Ryder, G., Fastovsky, D., and Gartner, S., eds., The Cretaceous-Tertiary event and other catastrophes in Earth history: Geological Society of America Special Paper 307, p. 319–334.

Kaiho, K., Kajiwara, Y., Tazaki, K., Ueshima, M., Takeda, N., Kawahata, H., Arinobu, T., Ishiwatari, R., Hirai, A., and Lamolda, M.A., 1999, Oceanic primary productivity and dissolved oxygen levels at the Cretaceous/Tertiary boundary: Their decrease, subsequent warming, and recovery: Paleoceanography, v. 14, p. 511–524.

Keller, G., 1996, The Cretaceous-Tertiary mass extinction in planktonic foraminifera: Biotic constraints for catastrophe theories, *in* MacLeod, N., and Keller, G., eds., Cretaceous-Tertiary mass extinctions: New York, W.W. Norton and Co., p. 49–84.

Keller, G., Li, L., Stinnesbeck, W., and Vicenzi, E., 1998, The K/T mass extinction, Chicxulub and the impact-kill effect: Bulletin de la Société Géologique de France, v. 169, p. 485–491.

Keller, G., and Stinnesbeck, W., 1996, Sea-level changes, clastic deposits, and megatsunamis across the Cretaceous-Tertiary boundary, *in* MacLeod, N., and Keller, G., eds., Cretaceous-Tertiary mass extinctions: New York, W.W. Norton and Co., p. 415–449.

Klaus, A., Norris, R.D., Kroon, D., and Smit, J., 2000, Impact-induced mass wasting at the K-T boundary: Blake Nose, western North Atlantic: Geology, v. 28, p. 319–322.

Li, L., and Keller, G., 1998, Maastrichtian climate, productivity and faunal turnovers in planktic foraminifera in South Atlantic DSDP sites 525A and 21: Marine Micropaleontology, v. 33, p. 55–86.

MacLeod, K.G., and Huber, B.T., 1996, Strontium isotopic evidence for extensive reworking in sediments spanning the Cretaceous/Tertiary boundary at ODP Site 738: Geology, v. 24, p. 463–466.

MacLeod, K.G., Huber, B.T., and Ducharme, M.L., 2000, Paleontological and geochemical constraints on changes in the deep ocean during the Cretaceous greenhouse interval, *in* Huber, B.T., MacLeod, K.G., and Wing, S.L., eds., Warm climates in Earth history: Cambridge, Cambridge University Press, p. 241–274.

MacLeod, K.G., Huber, B.T., and Fullagar, P.D., 2001, Evidence for a small (~0.000030) but resolvable increase in seawater $^{87}Sr/^{86}Sr$ ratios across the Cretaceous-Tertiary boundary: Geology, v. 29, p. 303–306.

MacLeod, N., and Keller, G., 1994, Comparative biogeographic analysis of planktic foraminiferal survivorship across the Cretaceous/Tertiary (K/T) boundary: Paleobiology, v. 20, p. 143–177.

Martínez-Ruiz, F., Ortega-Huertas, M., Palomo-Delgado, I., and Smit, J., 2001, Diagenetic alteration of K/T boundary spherules at Blake Nose (ODP Leg 171B): Implications for the preservation of the impact record, *in* Kroon, D., Norris, R.D., and Klaus, A., eds., Western North Atlantic Palaeogene and Cretaceous palaeoceanography: Geological Society [London] Special Publication 183, p. 149–161.

Nederbragt, A.J., 1991, Late Cretaceous biostratigraphy and development of Heterohelicidae (planktic foraminifera): Micropaleontology, v. 37, p. 329–372.

Norris, R.D., Huber, B.T., and Self-Trail, J., 1999, Synchroneity of the K-T oceanic mass extinction and meteorite impact: Blake Nose, western North Atlantic: Geology, v. 27, p. 419–422.

Norris, R.D., Firth, J., Blusztajn, J.S., and Ravizza, G., 2000, Mass failure of the North Atlantic margin triggered by the Cretaceous-Paleogene bolide impact: Geology, v. 28, p. 1119–1122.

Norris, R.D., Kroon, D., and Klaus, A., and Shipboard Scientific Party, 1998, Proceedings of the Ocean Drilling Program, Initial Reports, Leg 171B: College Station, Texas, Ocean Drilling Program, p. 1–749.

Olsson, R.K., Berggren, W.A., Hemleben, C., and Huber, B.T., eds., 1999, Atlas of Paleocene planktonic foraminifera: Smithsonian Contributions to Paleobiology, v. 85, 252 p.

Olsson, R.K., and Liu, C., 1993, Controversies on the placement of Cretaceous-Paleogene boundary and the K/P mass extinction of planktonic foraminifera: Palaios, v. 8, p. 127–139.

Pardo, A., Adatte, T., Keller, G., and Oberhänsli, H., 1999, Paleoenvironmental changes across the Cretaceous-Tertiary boundary at Koshak, Kazakhstan, based on planktic foraminifera and clay mineralogy: Palaeogeography, Palaeoclimatology, Palaeoecology, v. 154, p. 247–273.

Pospichal, J.J., 1996, Calcareous nannoplankton mass extinction at the Cretaceous/Tertiary boundary: An update, *in* Ryder, G., Fastovsky, D., and Gartner, S., eds., The Cretaceous-Tertiary event and other catastrophes in Earth history: Geological Society of America Special Paper 307, p. 335–360.

Ryder, G., Fastovsky, D., and Gartner, S., eds., 1996, The Cretaceous-Tertiary event and other catastrophes in Earth history: Geological Society of America Special Paper 307, 569 p.

Sarjent, W.A.S., 1996, Dinosaur extinction: Sudden or slow, cataclysmic or climatic?: Geoscience Canada, v. 23, p. 161–164.

Self-Trail, J.M., 2001, Biostratigraphic subdivision and correlation of upper Maastrichtian sediments from the Atlantic Coastal Plain and Blake Nose, Western Atlantic, *in* Kroon, D., Norris, R.D., and Klaus, A., eds., Western North Atlantic Palaeogene and Cretaceous palaeoceanography: Geological Society [London] Special Publication 183, p. 93–110.

Smit, J., Rocchia, R., and Robin, E., 1997, Preliminary iridium analysis from a graded spherule layer at the K/T boundary and late Eocene ejecta from ODP sites 1049, 1052, 1053, Blake Nose, Florida. Geological Society of America Abstracts with Programs, v. 29, no. 6, p. 141.

Zachos, J.C., Berggren, W.A., Aubry, M.-P., and Mackensen, A., 1992, Chemostratigraphy of the Cretaceous/Paleocene boundary at Site 750, southern Kerguelen Plateau, *in* Wise, S.W., Jr., Schlich, R., eds., et al., Proceedings of the Ocean Drilling Program, Scientific Results, Volume 120: College Station, Texas, Ocean Drilling Program, p. 961–977.

MANUSCRIPT SUBMITTED DECEMBER 1, 2000; ACCEPTED BY THE SOCIETY MARCH 22, 2001

Printed in the U.S.A.

Geological Society of America
Special Paper 356
2002

Faunal changes across the Cretaceous-Tertiary (K-T) boundary in the Atlantic coastal plain of New Jersey: Restructuring the marine community after the K-T mass-extinction event

William B. Gallagher
New Jersey State Museum, Trenton, New Jersey 08625-0530, and Department of Geological Sciences, Rutgers University, Piscataway, New Jersey 08855, USA

ABSTRACT

The inner Atlantic coastal plain of the eastern United States reveals exposures of Maastrichtian and Danian deposits that contain fossil assemblages from the Late Cretaceous and early Paleocene. Recent fossil discoveries in this interval are reported, and placed in the context of Cretaceous-Tertiary (K-T) faunal changes. The Inversand Pit at Sewell, New Jersey, is the last active marl mine in the region, and the exposure there is an important reference section for the many significant discoveries of invertebrate and vertebrate fossils produced by the marl mining industry at its zenith. Changes in planktonic populations across the K-T boundary are related to Maastrichtian-Danian marine ecosystem community reorganization and demonstrate changes in abundance of dominant marine invertebrates in successive fossil assemblages. Marine invertebrates with nonplanktotrophic larval stages were the most common fossils preserved in the Danian sediments in this region.

INTRODUCTION

The northern Atlantic coastal plain of eastern North America figured prominently in the early development of the science of paleontology in North America, and although the fossiliferous deposits of the Atlantic coastal plain have been investigated for two centuries, new data and discoveries are still emerging from these classic sections. The inner coastal plain of New Jersey (Fig. 1), Delaware, and Maryland is underlain by deposits of Late Cretaceous and early Tertiary age; these deposits (Table 1) contain an abundant record of life from these intervals.

The biostratigraphic framework of these deposits was defined by R. Olsson and his students in a series of studies focusing on the micropaleontology of the Cretaceous-Tertiary (K-T) boundary interval (Koch and Olsson, 1977; Olsson, 1987, 1989). They demonstrated a major reduction in the diversity of planktonic foraminifera in the northern Atlantic coastal plain, as elsewhere at the K-T boundary (Olsson and Liu, 1993). Moreover, the K-T boundary ejecta layer has been located in a downdip well section in southern New Jersey (Olsson et al., 1997), and stable carbon isotope data from the Danian section of this borehole indicate that primary productivity in the ocean was significantly reduced for 3 m.y. after the K-T boundary (Olsson et al., 2000). My goal here is to investigate how changes in the macrofossil faunas below and above the K-T boundary may be related to reorganization of marine food webs in earliest Tertiary time.

Gallagher, W.B., 2002, Faunal changes across the Cretaceous-Tertiary (K-T) boundary in the Atlantic coastal plain of New Jersey: Restructuring the marine community after the K-T mass-extinction event, *in* Koeberl, C., and MacLeod, K.G., eds., Catastrophic Events and Mass Extinctions: Impacts and Beyond: Boulder, Colorado, Geological Society of America Special Paper 356, p. 291–301.

Figure 1. Outcrop belt of Late Cretaceous–Paleocene sediments in Atlantic coastal plain of New Jersey (X is location of Inversand Pit). Inset shows position of New Jersey in United States.

TABLE 1. STRATIGRAPHY OF UPPER CRETACEOUS-LOWER TERTIARY DEPOSITS, NEW JERSEY ATLANTIC COASTAL PLAIN

Period	Epoch	Stage	Formation
Tertiary	Paleocene	Thanetian	Vincentown*
		Danian	Hornerstown*
Cretaceous	Late		Tinton
			Red Bank
		Maastrichtian	Navesink*
			Mount Laurel*
			Wenonah
			Marshalltown*
		Campanian	Englishtown
			Woodbury
			Merchantville

*Indicates units sampled for bulk matrix (dm^3) bioclast counts; see Table 2.

RATIONALE

It is necessary to establish the large-scale paleoecological dynamics of mass extinctions; specifically, the nature of community reorganization during and after mass-extinction events, how trophic structure changes, and how the environmental effects of large-body impacts might have ecological consequences. Diversity counts have given us the first approximation of the effects of bolide impacts on ecosystems of that time. However, this approach has some fundamental problems; diversity counts essentially equate the disappearance of small, rare, endemic taxa with the extinction of large, abundant, widespread taxa. Because the obvious interest in mass-extinction studies is the disappearance of previously dominant taxa, which usually translates into the loss of biomass among the common widespread species, we need to look at relative abundances rather than mere absolute occurrence, as in diversity-count studies. To do this, we can apply the basic methods of modern field ecology, studying densities within quadrants or volumes.

We recognize the limitations of the fossil record, but the fossil record can be surprisingly forthcoming when the right questions are asked. Thus we can expect to determine, e.g., whether changes in planktonic diversity at the K-T boundary had any effect on the dominant shelled benthos of latest Cretaceous and earliest Tertiary time. Another interesting line of inquiry involves how mass extinctions affect the fundamental nature of marine food webs. Some insight into these issues can be gained by controlled sampling of specific sections around the K-T boundary, and comparing the composition of the marine fauna in the same environments before and after the K-T event.

METHODOLOGY

In this study, population-density data of marine organisms were obtained by counting representatives of the most abundant taxa in cubic decimeter bulk samples and by mapping of areal surfaces (in square meter grids) of specific fossiliferous beds in the Maastrichtian and Danian sections of the Upper Cretaceous–lower Tertiary outcrop belt in the northern Atlantic coastal plain of eastern North America.

Fossils in the K-T section of the New Jersey coastal plain tend to occur in well-defined layers or concentrations that provide most of the specimens known from this interval. Seven fossiliferous concentrations were for sampled for bulk matrix; these ranged in age from late Campanian to Thanetian, crossing the K-T boundary in this region (Table 1). Samples were taken at the best available exposures of the fossil assemblages over a distance of 112 km along the strike of the outcrop belt (Fig. 1), within a stratigraphic thickness of <100 m. Numbers of the most common bioclasts found in the fossil assemblages were converted to percentages of total bioclasts found in sampled matrix (Table 2). Comparison of these fossiliferous horizons is then used to interpret paleoecological change and trophic dynamics across the K-T boundary.

DESCRIPTION OF SECTIONS

Some of the most fossiliferous deposits in the Atlantic coastal plain are the beds of glauconite that were widely mined as fertilizer in the nineteenth century. The marl pits produced many of the earliest discoveries of dinosaurs and other vertebrate fossils in North America. At Haddonfield in Camden County, New Jersey, *Hadrosaurus foulkii* was discovered in an old marl pit that yielded the most complete skeleton of a dinosaur known at that time (Leidy, 1858). The marl deposits have yielded numerous specimens for private and professional collectors.

The glauconitic marl beds of the Atlantic coastal plain represent the basal transgressive system tract of a series of sedimentary cycles that reflect changing sea levels coupled with subsidence of the North American eastern continental margin since the Jurassic (Olsson, 1989). The typical or idealized cycle consists of a basal glauconitic unit that contains the maximum flooding surface, a superjacent clay or silt that represents the highstand tract (deposited in shallower water than the basal transgressive unit), and a sandy unit that may contain lowstand tract deposits at the top. The sequence is then repeated by the next cycle of glauconite, clay, and sand. Four such cycles can be recognized in the northern Atlantic coastal plain in the Upper Cretaceous through Paleocene deposits, although the idealized pattern of glauconite, clay, and sand is not always present everywhere in the coastal plain. The maximum marine faunal diversity is usually located in the basal glauconitic beds; these represent the most widespread full marine deeper water inner to mid-shelf environments (Olsson, 1989).

Fossils of Maastrichtian age are found in the Monmouth Group, which comprises, in ascending order, the Mount Laurel, Navesink, Red Bank, and Tinton Formations (Table 1). The Campanian-Maastrichtian boundary is now placed at the Mount Laurel–Navesink formational contact (Miller et al., 1999). The Mount Laurel is a sandy regressive unit containing abundantly fossiliferous lag deposits (Gallagher et al., 2000). The superjacent Navesink Formation is a transgressive deposit of greensand marl that contains several oyster bed assemblages. Typical Maastrichtian specimens found in the Navesink beds are the abundant bivalves *Exogyra costata, Pycnodonte convexa*, and *Agerostrea mesenterica*, belemnite pens (*Belemnitella americana*), ammonites (*Baculites ovatus*), the small brachiopod *Choristothyris plicata*, shark teeth of several species (*Scapanorhyncus texanus, Squalicorax pristodontus, Cretolamna appendiculata*), and scrappy remains of larger vertebrates, including turtles, crocodiles, plesiosaurs, mosasaurs, and dinosaurs (Gallagher, 1993).

Miller et al. (1999) interpreted the Mount Laurel–Navesink contact as a sequence boundary caused by glacioeustatic sea-level fall, and concluded that continental glaciation was driving sea-level change in the Late Cretaceous. Just below the contact, a concentration of fossils contains abundant vertebrate remains, primarily chondrichthyan and mosasaur material, as well as an invertebrate assemblage dominated by inoceramids and ammonites (Gallagher et al., 2000). This bed has yielded some noteworthy finds, including a hollow coelurosaurian long bone shaft, a giant coelacanth coronoid (Schwimmer et al., 1994), the first lungfish jaw plate known from the Upper Cretaceous of North America, and a possible Cretaceous multituberculate incisor (Grandstaff et al., 2000).

In the northern end of the outcrop belt, the Red Bank and Tinton Formations have limited areal extent in Monmouth County before pinching out to the southeast. The Red Bank Formation is a silty to sandy unit, representing an influx of

TABLE 2. MOST ABUNDANT MARINE INVERTEBRATE SPECIES BY FOSSIL ASSEMBLAGE, NEW JERSEY CRETACEOUS-TERTIARY STRATIGRAPHIC INTERVAL

Assemblage	Age (Stage)	Dominant species	Bioclasts (%/dm³)	Reproductive type
Vincentown Limesand	Thanetian	*Coscinopleura digitata* (bryozoa)	81	Nonplanktotroph
Oleneothyris biozone	Danian-Thanetian	*Oleneothyris harlani* (brachiopod)	99	Nonplanktotroph
Middle Hornerstown	Danian	*Peridonella dichotoma*; (sponge)	40	Nonplanktotroph
		Oleneothyris harlani	30	Nonplanktotroph
Basal Hornerstown	Danian-Maastrchtian	*Pycnodonte dissimilaris* (oyster)	78	Planktotroph
Middle Navesink shellbed	Maastrichtian	*Pycnodonte convexa* (oyster)	98	Planktotroph
Upper Mount Laurel Fm.	Late Campanian	*Inoceramus vanuxemi* (bivalve)	33	Planktotroph
Marshalltown shellbed	Early Late Campanian	*Exogyra ponderosa* (oyster)	99	Planktotroph

sediment from the north, and it has yielded Maastrichtian fossils, primarily the bivalve *Cucullaea* and shark teeth. The Tinton Formation is a coarse glauconitic quartz sand containing an oyster-dominated fauna typical of the Cretaceous greensands (Minard et al., 1969).

In the southern part of the outcrop belt, the Navesink Formation is directly overlain by the Hornerstown Formation, the bulk of which is Danian in age. The Hornerstown Formation is a persistent glauconitic sand that crops out from Monmouth County in the northeast across New Jersey diagonally to the southwest into the Delmarva Peninsula. The Hornerstown Formation is a dusky green, heavily bioturbated, nearly pure glauconite sand. Paleocene deposits are combined into the Rancocas Group, including the Hornerstown and the Vincentown Formations. The Vincentown, in contrast to the Hornerstown, shows a wide variability in its composition both vertically and along strike; at its base it is a glauconitic quartz sand, grading upward into a silty quartz sand. At some places along strike the upper part of the formation consists of a patchy carbonate deposit known as the Vincentown limesand.

The mining of greensand marl was a widespread industry in this region in the nineteenth century, and numerous fossils were obtained from the extensive excavations into the marl deposits surrounding the K-T boundary (Gallagher, 1997). During the nineteenth century, glauconite was used as a long-term soil conditioner to enhance the yield of crops. The marl mining industry declined and almost disappeared entirely in the earlier part of the twentieth century. Glauconite is still mined at the Inversand Pit, the last commercial greensand mine in existence, for use as a water conditioner. Glauconite is a hydrous potassium aluminum iron silicate with a high cation exchange capacity; it is formed at very low sedimentation rates in reducing marine environments, partly as the result of bioturbation by marine burrowing organisms. The exposures at the Inversand Pit (Fig. 2) are the best available outcrops of the Maastrichtian-Danian section in New Jersey, and the section at the pit is an important reference section for the latest Cretaceous and Paleocene in this part of the world. Other, less complete exposures are available in stream bank outcrops and as a result of temporary excavations.

AGE LIMITATIONS ON THE K-T BOUNDARY IN NEW JERSEY

The precise relationship of the Navesink and the Hornerstown Formations in this area has been a minor controversy; U.S. Geological Survey workers have favored an angular unconformity between the two formations at their contact (Minard et al., 1969; Owens et al., 1970), and others have proposed that the contact between the Navesink and the Hornerstown is more accurately characterized as a disconformity or paraconformity (Gallagher et al., 1986; Gallagher, 1993). The precise boundary is smeared out by bioturbation and the contact is usually more obvious from a distance. The contact is essentially between two glauconitic units, the overlying Hornerstown Formation representing a transgressive, slightly deeper water depositional environment. There may be missing section here, certainly in comparison to the downdip Bass River Core section, which has a complete K-T section in the subsurface, including an impact ejecta layer (Olsson et al., 1997).

Estimates of the age involved in this section have been extremely variable; the glauconites are difficult to date in absolute years. Owens and Sohl (1973) used potassium-argon dating on the glauconites, and obtained an age of 61.1 Ma for the K-T boundary in New Jersey; dates for the Navesink are 59–60 Ma. Because the currently accepted figure for the K-T boundary is 65.4 ± 0.1 Ma (Obradovich, 1993) and the Navesink is unquestionably Maastrichtian in age, these K-Ar dates are not reliable. Gallagher and Parris (1996) reported several lines of evidence for the age of these deposits, including an attempt to date the Navesink-Hornerstown contact at the Inversand Pit using rubidium-strontium radiometric dating. This effort yielded an age of 53 Ma for the Navesink and 54 Ma for the overlying Hornerstown, obviously an unsatisfactory result. Gallagher and Parris (1996) concluded that the glauconites of the New Jersey coastal plain are an open system unsuitable for radiometric dating because of their geochemical activity, and that the best means of dating these beds is through traditional biostratigraphic methods. For example, a specimen of the ammonite *Discoscaphites gulosus* from the Navesink Formation at Inversand confirms that this formation must be correlative at least in part to the uppermost ammonite zone of the Fox Hills Formation in the American Western Interior, the *Jeletzkytes nebrascensis* zone (Kennedy et al., 2000). This zone is upper Maastrichtian in stratigraphic position (Obradovich, 1993).

Stable strontium isotopes from fossil oysters (*Pycnodonte dissimilaris*) were employed to try to provide an age estimate for the basal Hornerstown main fossiliferous layer (known as the MFL bed), but these also supplied an anomalously young date (Gallagher and Parris, 1996). That may be due to the strontium spike at the K-T boundary hypothesized to be caused by extensive continental weathering (O'Keefe and Ahrens, 1989); however, mineralization is also a possible source of the anomalously high values, because many of the MFL oysters have bornite coatings. Elsewhere and lower in the same section, Sugarman et al. (1995) obtained stable strontium age estimates of 65–66 Ma for fossil oysters from the Navesink Formation, supporting a late Maastrichtian age for this bed.

Planktonic biostratigraphy is probably the best way to date these beds, but unfortunately the exposed glauconites at the Inversand Pit are leached of carbonate, so updip correlation from downdip nonleached borehole sections and noncarbonate microfossils are the only way to determine relative ages. The nannoplankton data of Self-Trail and Bybell (1995) from a borehole in nearby Clayton, New Jersey, is unfortunately compromised by missing section around the critical stratigraphic interval of greatest interest, the Navesink-Hornerstown contact. Planktonic foraminifera from the Bass River core show the

Figure 2. Exposures of Maastrichtian through Miocene section at Inversand Pit, Sewell, Gloucester County, New Jersey. Four-wheel drive vehicle for scale.

presence of the P0 zone above the K-T ejecta layer (Olsson et al., 1997); this zone is not present in updip sections. Dinoflagellate biostratigraphy indicates the presence of late Maastrichtian dinoflagellate fossils in the basal Hornerstown and the uppermost Navesink at the Inversand Pit (Koch and Olsson, 1977). This evidence suggests that at most any erosional interval was probably confined to the very latest Maastrichtian and the P0-Pα zone of the earliest Paleocene. Zone P1$_a$ is present in the basal Hornerstown (Olsson, 1987), so the missing interval here is at least 100 k.y., the length of the absent earliest Paleocene planktonic foraminiferal zones (Berggren and Norris, 1999).

To summarize, radiometric dating has failed to provide any reliable dates for the K-T section in the New Jersey coastal plain. Stable strontium isotope age estimation and ammonite biostratigraphy suggest a late Maastrichtian age for the Navesink Formation. Basal Paleocene planktonic foraminiferal zones are missing, indicating a gap of at least 0.1 m.y. in the record. However, even if the section is not perfectly complete, enough of the latest Cretaceous and earliest Paleocene faunas are present to make some determination of the larger scale pattern in K-T faunal turnover.

FAUNAL CHANGES IN THE K-T SECTION

Three well-defined fossiliferous layers are present at the Inversand Pit, ranging in age from the Maastrichtian oyster bank assemblage in the Navesink Formation at the bottom of the pit through the basal Hornerstown layer to the Danian layer in the middle part of the Hornerstown Formation (Fig. 3). The fossil faunas bridge the K-T boundary, and in New Jersey the pit provides our best view in outcrop of this critical stratigraphic interval with its record of paleoecological change. The Navesink shell bed at the base of the pit here is characterized by a diversity of molluscs (26 species; see faunal list in Appendix) dominated by oysters, including *Exogyra costata, Pycnodonte convexa, Agerostrea falcata*, and *Gryphaeostrea vomer*. The large arc shell *Cucullaea antrosa* occurs sporadically in the upper part of the Navesink Formation. The small brachiopod *Choristothyris plicata* is encountered rarely. Vertebrates found in the Navesink Formation include rare shark teeth, *Enchodus ferox* teeth and jaw pieces, turtles, mosasaurs, and dinosaurs.

However, although there is a lithostratigraphic break at the contact in the pit exposure, the basal fossiliferous layer (MFL) in the lower Hornerstown Formation contains the last remains of Cretaceous organisms. The MFL in the base of the Hornerstown Formation yields a diverse molluscan fauna (26 species; see Appendix). The most common is the thin-shelled fragile oyster *Pycnodonte dissimilaris*, which can be seen in abundance in outcrop (10–12 valves/dm^3 in the basal MFL), but which is very difficult to collect intact because of its delicacy. The next most abundant invertebrate is the clam *Cucullaea vulgaris*. Among the gastropods, *Turritella vertebroides* and *Pyropsis trochiformis* are the most common. There are also several ammonite species, mostly in the form of broken and worn steinkerns (Kennedy et al., 1995).

Vertebrates from the MFL are more diverse than in the other two layers, and include a variety of chondrichthyians, bony fish, turtles, crocodiles, mosasaurs, and birds, for a total of more than 30 species. Certain elements of this assemblage

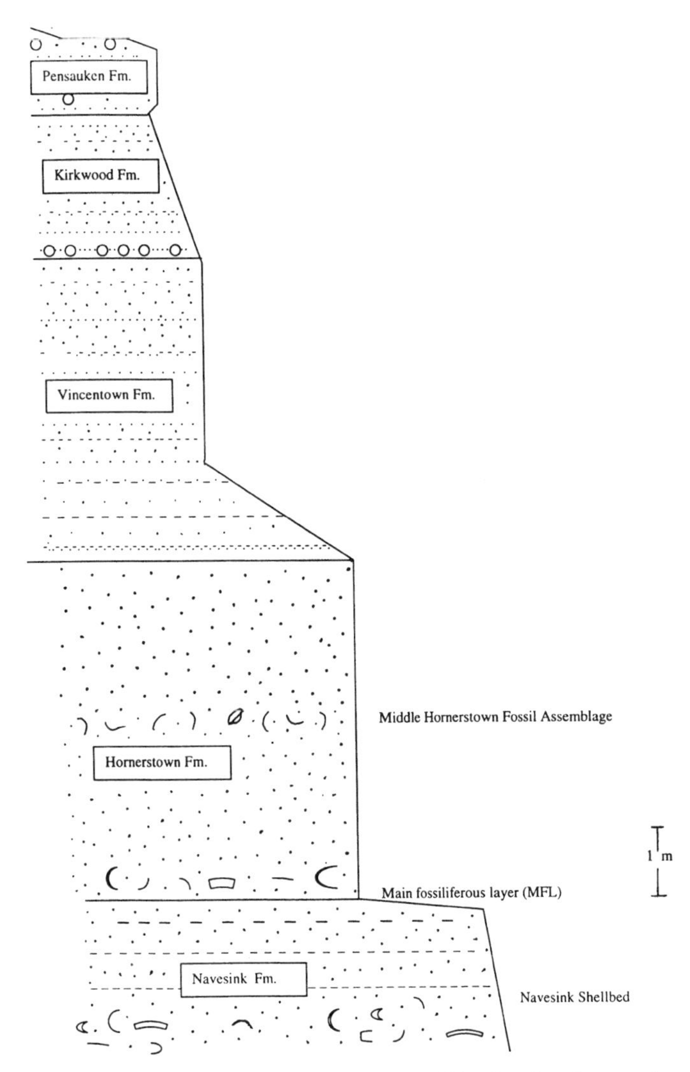

Figure 3. Stratigraphic section at Inversand Pit, showing three major fossil concentrations exposed at excavation. Dash pattern is clay, dot pattern is sand, open circles indicate gravel, and curved lines show fossil concentrations. Profile of section reflects typical slope seen in mining operations at Inversand Pit. Fm.—Formation.

(the shark *Squalicorax pristodontus, Enchodus*, the mosasaurs) are characteristic Cretaceous forms, and other taxa (the turtles and crocodiles) are less distinctively Mesozoic (see Appendix).

One possible way to interpret the basal Hornerstown MFL is as a condensed section lag, an accumulation of remanié fossils that represent protracted time averaging over the K-T boundary. Estimates of the time necessary to produce this concentration are 300 k.y. (Gallagher, 1993). An interpretation of this deposit as an entirely reworked assemblage is not consistent with the many complete and partial skeletons of crocodilians and turtles that have been found at this horizon, especially in comparison to other vertebrate concentrations in the New Jersey coastal plain (see Gallagher, 1993, for data). Some of these specimens are extremely delicate in nature, for example an articulated bird wing (Olson and Parris, 1987). However, many of the Cretaceous forms are found as single isolated elements or as worn and abraded internal molds. A possible alternative hypothesis is that the MFL fossil bed is the product of tsunami mixing and erosion, creating the formational contact and producing a lag of Cretaceous fossils, over which the remains of Paleocene survivors accumulated during protracted mortality in the wake of the K-T environmental disturbance. Preliminary results of mapping and microstratigraphic studies of the MFL reveal clay rip-up clasts, small quartz pebbles, and stacked, imbricated clusters of single oyster valves at the base of the MFL; these are indicative of sediment disturbance. There is evidence elsewhere in eastern North America for impact-induced disturbance and mixing of sediment, including slumping and mass wasting along the continental margin (Bralower et al., 1998; Norris et al., 2000).

Above this horizon rare crocodile remains or an occasional shark tooth may be found, but the section is essentially barren for 2 m until another concentration of fossils occurs. This fossil layer is significant because it differs markedly from the underlying beds. Molluscs are reduced in size, abundance, and diversity (6 species). The most abundant bioclasts are sponges, brachiopods, and solitary corals (Table 2), but each group is represented by just one species. Fossil distribution in this layer is more irregular, with a varying thickness of 2–3 m in the middle of the Hornerstown Formation. Vertebrates are rare; small lamnid shark teeth are the most frequent specimens.

In other sites there is a concentration of the large terebratulid brachiopod *Oleneothyris harlani* at and above the level of the Hornerstown-Vincentown formational contact (Feldman, 1977; Gallagher, 1991, 1993). Densities may reach hundreds of single and articulated brachiopod valves per cubic decimeter at localities such as Shingle Run in Monmouth County, New Jersey (Gallagher, 1993). This remarkable density of brachiopods occurs sporadically where the same level is exposed along strike as far to the south as southeastern North Carolina. The brachiopods first appear in the middle of the glauconitic Hornerstown Formation and are also found at the Hornerstown-Vincentown formational contact and into the basal part of the Vincentown. The distribution of *O. harlani* is thus facies independent, because it is present in the deeper water glauconite and in the shallower water sand of the Vincentown. In the upper part of the Thanetian Vincentown Formation, glauconite disappears altogether and is replaced by quartz sands and patchy occurrences of bryozoan limestone. Diversity rebounds to pre-Danian levels, especially in the limestone beds at the top of the formation (Gallagher, 1993); these indurated limesands are interpreted here as a patch reef ecosystem. The glauconitic Aquia Formation of equivalent Thanetian age in Maryland also displays a more diverse marine invertebrate fauna dominated by molluscs. A gigantic species of *Cucullaea* (*C. gigantea*) occurs in some Aquia sites (Table 3). The Aquia Formation is a glauconitic unit correlated with the Vincentown (Olsson et al. 1988), more comparable in depositional environment to the subjacent glauconitic beds of Maastrichtian and Danian age.

TABLE 3. DWARFING AND GIGANTICISM IN THE *CUCULLAEA* CLADE IN THE LATE CRETACEOUS-PALEOCENE SECTION OF THE NORTHERN ATLANTIC COASTAL PLAIN

Species	Formation	Age	Average Length (cm)	Average Height (cm)	Average Width (cm)
*C. antrosa**	Navesink	Maastrichtian	5.8	5.2	4.7
C. vulgaris†	Basal Hornerstown	Maastrichtian-Danian	3.7	3.6	2.9
C. saffordi#	Middle Hornerstown	Danian	3.0	2.6	2.6
*C. gigantea***	Aquia	Thanetian	13.1	10.9	11.6

Note: Measurements made on *Cucullaea* internal casts as follows; length was measured along the commissure; height was measured from the tip of the beak to the commissure; and width was measured acoss the widest point of both valves. NJSM = New Jersey State Museum collection numbers.
*NJSM numbers 7507, 7594, 9696, 9882, 9885, 9887, 9888; n = 14.
†NJSM number 11322; n = 21.
#NJSM number 10863; n = 22.
**NJSM numbers 10485, 14981; n = 4.

To summarize, there is a typical Mesozoic marine community at the bottom of the section in the Navesink Formation, a possibly reworked last Cretaceous assemblage in the MFL, and an unusual Danian fauna in the middle of the Hornerstown Formation, giving us some insight into the nature of faunal changes across the K-T boundary in this part of the world (see Appendix).

DISCUSSION

Mass extinctions in the marine realm are often marked by reductions in the diversity and abundance of plankton. Diversity reduction among the plankton is documented for the Frasnian-Fammenian event (McGhee, 1996), the Permian-Triassic mass extinction (Erwin, 1994), and the K-T boundary (Olsson and Liu, 1993). The effects of these population crashes in the base of the food chain should reverberate upward to higher levels of the trophic pyramid; for confirmation of this hypothesis, we seek evidence in the fossil record of mass extinctions.

Across the K-T boundary in eastern North America, there is evidence for a resetting of the evolutionary clock in Danian marine ecosystems, which are dominated by faunas of Paleozoic aspect. Widespread diverse Cretaceous molluscan assemblages are replaced temporarily in the Danian by a sponge-brachiopod-coral-dominated assemblage (Gallagher, 1991, 1993). In particular, fossil concentrations in the K-T section of the Atlantic coastal plain show a diverse oyster bank ecosystem replaced by a brachiopod-dominated shell gravel, wherein large terebratulids of one species (*Oleneothyris harlani*) can form as much as 99%+ of the bioclasts (Table 2; Gallagher, 1991, 1993). The *Oleneothyris* zone can be traced in Paleocene outcrops from New Jersey in the north into southeastern North Carolina. A sponge (*Peridonella dichotoma*) and a solitary coral (*Flabellum mortoni*) often occur with the brachiopods (Table 2). Dwarfed bivalves form an insignificant proportion of this fauna.

Sponges are minimalist organisms, able to filter to 10–20 $\times 10^3$ times their own body volume of water per day (Rigby, 1983). They specialize in extricating smaller bits of nutrients from the water column. In mass extinctions such as the K-T event, the kinds of plankton (e.g., foraminifera) left after the event are generally smaller than the larger and more ornate forms found before the extinction. Therefore sponges are liable to be favored as survivors because they are more effective at filtering out small particles and processing large amounts of water for these particles. In an ocean where the primary planktonic food resources for suspension feeders were reduced to fewer, smaller forms, this would be a distinct advantage. Sponges are capable of regenerating themselves from a strained liquid of their cells, and can reproduce by budding as well (Rigby, 1983).

Brachiopods are also minimalist organisms. They are able to survive at lower oxygen levels, and in general have lower food requirements and lower metabolism than other exoskeletonized benthos (Thayer, 1981). The abundance of a single species such as *Oleneothyris harlani* over a wide area after the K-T event suggests that it was an opportunistic species that could thrive despite the reduction in planktonic diversity and productivity characteristic of the Danian ocean in this region (Olsson and Liu, 1993; Olsson et al., 2000).

CONCLUSIONS

Hypothesized effects of large-body impacts, such as reduction in solar radiation reaching Earth's surface, acid rain, and near destruction of the ozone layer, would all have negative effects on the plankton population. As Olsson and Liu (1993) showed, only three species of planktonic foraminifera survived the K-T boundary. The resulting Danian marine community would have to be adapted to the profound change in plankton populations, especially in waters closer to the Chicxulub impact site in eastern North America. Such a pattern is demonstrated in the minimalist assemblage of the Danian of the Atlantic coastal plain.

The plankton crash may have affected those organisms that reproduced by means of planktotrophic larvae. Normally, wide dispersal patterns and large numbers of planktotrophic larvae were favored during times of high planktonic productivity; but when the planktonic populations declined severely at the K-T boundary, those organisms that depended upon a developmental stage in the plankton were severely affected (Gallagher, 1991). This explains why ammonites, which were probably planktotrophic reproducers (Sharigeta, 1993), went extinct, while nautiloids, which lay lecithotrophic eggs on the sea bottom, survived (Landman et al., 1983). Brachiopods are larval brooders (Thayer, 1981), while sponges can reproduce asexually (Rigby, 1983).

Dwarfing among Danian benthos may be due to a decrease in trophic resources in the post K-T ocean. Danian representatives of Maastrichtian molluscan genera were generally smaller, if they survived (e.g., *Pycnodonte, Cucullaea*; Table 3). However, some reversion to larger size occurred later in the Thanetian, e.g., with the appearance of the biggest member of the *Cucullaea* clade, *C. gigantea* (Table 3). Russian echinoids were shown to dwarf across the K-T boundary (Markov and Soloviev, 1997), and reduction in molluscan body size was noted across the K-T boundary in the Brazos River sections of Texas (Hansen et al., 1984).

The niche of duraphagous predator was an important trophic level in the Cretaceous marine food web: shell-crushing carnivores took advantage of the rich food resources of Cretaceous mollusks and crustaceans. Chimaerids, rays, certain sharks, various bony fish species, some turtles, and a few mosasaurs (*Globidens, Carinidens*) have jaws or dentition adapted for shell crushing.

In contrast, marine predators of the Paleocene tend to be more generalized carnivores such as crocodiles or lamnid sharks. Duraphagy survived as a niche, but during the Danian opportunistic predators with broader tastes predominated (Gallagher, 1993). The scarce resources of the early Paleocene ocean affected specialized predators.

This pattern of extinction and survival across the K-T boundary provides a model for the collapse of marine ecosystems under stresses generated by large-scale environmental disturbance, such as an asteroid impact. Such a pattern should be zonally distributed, reflecting decreasing intensity of impact effects farther away from the crater. Preliminary proof for this comes from new Late Cretaceous fossil discoveries in eastern North America; coelacanths (Schwimmer et al., 1994) and lungfish of Late Cretaceous age were destroyed in this part of the world, but survived at the antipodes of the impact in the India-Australia region (Carroll, 1988). This idea is also supported by the distribution of other well-known living fossils such as *Neotrigonia*, various plant families, and marsupials, and is the topic of further research now in progress.

ACKNOWLEDGMENTS

I thank David C. Parris of the New Jersey State Museum and the Inversand Company of Clayton, New Jersey, for their encouragement of this work. I also thank the numerous people who have helped in this effort over the years, through assistance in the field, useful discussions, and donation of specimens, including James Barnett, Joseph and Sandra Camburn, Gudni Fabian, Ned Gilmore, Barbara Grandstaff, Paul Hanczaryk, Margaret Martinson, Richard Olsson, Earle Spamer, John Rebar, and Chris Storck. Reviews by Claudia Johnson, Richard Norris, and Kenneth MacLeod greatly improved this paper, and I thank them for their thoughtful comments. This project was supported through the Horace G. Richards Fund of the Friends of the New Jersey State Museum.

Appendix: Faunal list for Inversand Pit, Mantua Township, Gloucester County, New Jersey

The following are adapted and updated from Gallagher et al. (1986).

Navesink Formation (Maastrichtian): Invertebrates

Porifera
Cliona cretacea Fenton and Fenton

Brachiopoda
Choristothyris plicata (Say)

Bryozoa
Encrusting bryozoa indet.

Bivalvia
Cucullaea antrosa Morton
C. neglecta Gabb
C. vulgaris Morton
Glycimeris mortoni (Conrad)
Agerostrea nasuta (Morton)
Pycnodonte convexa (Say)
Gryphaeostrea vomer Morton
Exogyra costata Say
Trigonia mortoni Whitfield
Crassatellites vadosus (Morton)
Liopistha protexta (Conrad)
Spondylus (Dianchora) echinata (Morton)
Pachycardium spillmani (Conrad)
Solyma cf. *lineolatus* Conrad
Lithophaga ripleyana Gabb

Gastropoda
Lunatia halli Gabb
Gyrodes petrosus (Morton)
Turritella cf. *vertebroides*
Anchura cf. *abrupta* Conrad
Anchura pennata (Morton)
Pyrifusus macfarlandi Whitfield
Turbinopsis curta Whitfield
Volutomorpha ponderosa Whitfield

Nautiloidea
Eutrophoceras dekayi Morton

Ammonoidea
Baculites ovatus Say
Discoscaphites conradi (Morton)

Echinodermata
Hemiaster sp.

Navesink Formation (Maastrichtian): Vertebrates

Chondrichthyes
Squalicorax pristodontus (Agassiz)
Odontaspis sp.

Osteichthyes
Enchodus ferox Leidy
Anomaeodus phaseolus (Hay)
large teleost indet. (*Xiphactinus*?)

Chelonia
Peretresius ornatus Leidy
Cheloniidae indet.

Squamata
Halisaurus platyspondylus Marsh
Mosasaurus sp.
Mosasaurus maximus Cope
Prognathodon rapax (Hay)

Dinosauria
Hadrosaurus minor Marsh
Lambeosaurine? indet

Basal Hornerstown Formation (main fossiliferous layer): Invertebrates

Brachiopoda
Terebratulina atlantica Morton

Bivalvia
Nuculana stephensoni Richards
Cucullaea vulgaris Morton
Glycimeris mortoni (Conrad)
Gervilliopsis ensiformis (Conrad)
Pycnodont dissimilaris
Gryphaeostrea vomer Morton
Veniella conradi Morton
Cardium tenuistriatum Whitfield
Etea delawarensis (Gabb)
Crassatellites vadosus (Morton)
Panopea decisa Conrad
Lithophaga ripleyana Gabb
Xylophagella irregularis (Gabb)

Gastropoda
Lunatia halli Gabb
Gyrodes abyssinus Morton
Turritella vertebroides Morton
Anchura abrupta Conrad
Volutoderma ovata Whitfield
Turbinella subconica Gabb
T. parva Gabb
Pyropsis trochiformis (Tuomey)
Acteon cretacea Gabb

Nautiloidea
Eutrephoceras dekayi (Morton)

Ammonoidea
Baculites sp.
Pachydiscus (*Neodesmoceras*) sp.
Sphenodiscus lobatus (Tuomey)

Crustacea
cf. *Hoploparia* sp.

Basal Hornerstown Formation (main fossiliferous layer): Vertebrates

Chondrichthyes
Edaphodon stenobyrus (Cope)
E. mirificus Leidy
Ischyodus thurmanni Pictet and Campiche
Squatina sp.
Squalicorax pristodontus (Morton)
Cretolamna appendiculata (Agassiz)
Odontaspis cuspidata (Agassiz)
Elasmobranchia indet.—vertebrae and teeth
Rhombodus levis Cappetta and Case
Rhinoptera sp.
Myliobatis cf. *leidyi* Hay
Batomorpha indet.—vertebrae

Osteichthyes
Acipenser cf. *albertensis* Lambe
Enchodus ferox Leidy
cf. *Bananogmus* sp.
Teleost indet.—vertebrae

Chelonia
Taphrosphys sulcatus Leidy
cf. *Bothremys* sp.
Adocus beatus Leidy
Agomphus turgidus Cope
Dollochelys atlantica (Zangerl)
Osteopygis emarginatus Cope
Peretresius cf. *emarginatus* Leidy

Crocodylia
Hyposaurus rogersii Owen
Diplocynodon sp.
cf. *Procaimanoidea* sp.
Bottosaurus harlani Meyer
Thoracosaurus neocesariensis

Squamata
Mosasauridae indet.
Mosasaurus sp.

Aves
Telmatornis priscus Marsh
Paleotringa littoralis Marsh
Graculavis velox Marsh
Tithostonyx glauconiticus Olson and Parris
Aves indet.

Middle Hornerstown Formation Fossil Assemblage (Danian): Invertebrates

Porifera
Peridonella dichotoma Gabb

Cnidaria
Flabellum mortoni Vaughan

Brachiopoda
Oleneothyris harlani, var. *manasquani* Stenzel

Bivalvia
Cucullaea macrodonta Whitfield
Ostrea glandiformis Whitfield
Crassatellites cf. littoralis Conrad
Caryatis veta Whitfield

Gastropoda
cf. *Volutacorbis* sp.

Nautloidea
cf. *Aturia* sp.

Middle Hornerstown Formation Fossil Assemblage (Danian): Vertebrates

Chondrichthyes
Edaphodon agassizi (Buckland)
Odontaspis sp.
Otodus obliquus (Agassiz)
Paleocarcharodon sp.

Chelonia
Dollochelys sp.

Crocodylia
Hyposaurus rogersii Owen

Vincentown Formation (Thanetian): Invertebrates

Bryozoa
Encrusting bryozoa indet.

Bivalvia
Polorthus tibialis (Morton)

REFERENCES CITED

Berggren, W.A., and Norris, R.D., 1999, Biostratigraphy, *in* Olsson, R.K., Hemleben, C., Berggren, W.A., and Huber, B., eds., Atlas of Paleocene planktonic foraminifera: Smithsonian Contributions to Paleobiology, no. 85, p. 8–10.

Bralower, T.J., Paul, C.K., and Leckie, R.M., 1998, The Cretaceous-Tertiary boundary cocktail: Chicxulub impact triggers margin collapse and extensive sediment gravity flows: Geology, v. 26, p. 331–334.

Carroll, R.L., 1988, Vertebrate paleontology and evolution: New York, W.H. Freeman and Company, 698 p.

Erwin, D.H., 1994, The Permo-Triassic extinction: Nature, v. 367, p. 231–236.

Feldman, H.R., 1977, Paleoecology and morphologic variations of a Paleocene terebratulid brachiopod (*Oleneothyris harlani*) from the Hornerstown Formation of New Jersey: Journal of Paleontology, v. 51, p. 86–107.

Gallagher, W.B., 1991, Selective extinction and survival across the Cretaceous/Tertiary boundary in the northern Atlantic Coastal Plain: Geology, v. 19, p. 967–970.

Gallagher, W.B., 1993, The Cretaceous/Tertiary mass extinction event in the northern Atlantic Coastal Plain: The Mosasaur, v. 5, p. 75–154.

Gallagher, W.B., 1997, When dinosaurs roamed New Jersey: New Brunswick, New Jersey, Rutgers University Press, 176 p.

Gallagher, W.B., Parris, D.C., and Spamer, E.E., 1986, Paleontology, biostratigraphy, and depositional environments of the Cretaceous-Tertiary transition in the New Jersey Coastal Plain: The Mosasaur, v. 3, p. 1–35.

Gallagher, W.B., and Parris, D.C., 1996, Age determinations for Late Cretaceous dinosaur sites in the New Jersey coastal plain: Sixth North American Paleontological Convention, Washington, D.C., Paleontological Society Special Publication Number 8, Abstracts of Papers, p. 133.

Gallagher, W.B., Camburn, J., Camburn, S., Albright, S.S., and Hanzcaryk, P.A., 2000, Taphonomy of a Maastrichtian fossil assemblage in the Mount Laurel Formation of New Jersey: Geological Society of America Abstracts with Programs, v. 32, p. A19.

Grandstaff, B.S., Gallagher, W.B., Shannon, K., and Parris, D.C., 2000, New discoveries of Late Cretaceous mammals in eastern North America: Society of Vertebrate Paleontology Abstracts with Program, Annual Meeting, Mexico City, Mexico, v. 60, p. 46A.

Hansen, T.A., Farrand, R., Montgomery, H., and Billman, H., 1984, Sedimentology and extinction patterns across the Cretaceous-Tertiary boundary interval in East Texas, *in* Yancey, T.E., The Cretaceous-Tertiary boundary and Lower Tertiary of the Brazos River Valley: San Antonio, Texas, Field Trip for American Association of Petroleum Geologists and Society of Economic Paleontologists and Mineralogists Annual Meeting, Guidebook, p. 21–36.

Kennedy, W.J., Johnson, R.O., and Cobban, W.A., 1995, Upper Cretaceous ammonite faunas of New Jersey, *in* Baker, J.E.B., ed., Contributions to the paleontology of New Jersey: Proceedings of the Geological Association of New Jersey 12th Annual Meeting, William Patterson College of New Jersey, Wayne, New Jersey, p. 24–55.

Kennedy, W.J., Landman, N.H., Cobban, W.A., and Johnson, R.O., 2000, Additions to the ammonite fauna of the Upper Cretaceous Navesink Formation of New Jersey: American Museum of Natural History Novitates 3306, 30 p.

Koch, R.C., and Olsson, R.K., 1977, Dinoflagellate and planktonic foraminiferal biostratigraphy of the uppermost Cretaceous of New Jersey: Journal of Paleontology, v. 51, p. 480–491.

Landman, N.H., Rye, D.M., Shelton, K.L., 1983, Early ontogeny of *Eutrephoceras* compared to Recent *Nautilus* and Mesozoic ammonites: Evidence from shell morphology and light stable isotopes: Paleobiology, v. 9, p. 269–279.

Leidy, J., 1858, Remarks concerning *Hadrosaurus foulkii*: Proceedings of the Academy of Natural Sciences of Philadelphia, v. 10, p. 215–218.

Markov, A.V., and Soloviev, A.N., 1997, Echinoids at the Cretaceous-Paleogene Boundary, *in* Rozanov, A.Y., Vickers-Rich, P., and Tassell, C., eds., Evolution of the biosphere: Launceston, Australia, Records of the Queen Victoria Museum, no. 104, p. 35–37.

McGhee, G.R., 1996, The Late Devonian mass extinction: The Frasnian/Famenian crisis: New York, Columbia University Press, 303 p.

Miller, K.G., Barrera, E., Olsson, R.K., Sugarman, P.J., and Savin, S.M., 1999, Does ice drive early Maastrichtian eustacy?: Geology, v. 27, p. 783–786.

Minard, J.P., Owens, J.P., Sohl, N.F., Gill, H.E., and Mello, J.F., 1969, Cretaceous-Tertiary boundary in New Jersey, Delaware, and eastern Maryland: U.S. Geological Survey Bulletin 1274-H, 33 p.

Norris, R.D., Firth, J., Blusztajn, J.S., and Ravizza, G., 2000, Mass failure of the North Atlantic margin triggered by the Cretaceous-Paleogene bolide impact: Geology, v. 28, p. 1119–1122.

Obradovich, J.D., 1993, A Cretaceous time scale, *in* Caldwell, W.G.E., and Kauffman, E.G., eds., Evolution of the Western Interior Basin: Geological Association of Canada, Special Paper 39, p. 379–396.

O'Keefe, J.D., and Ahrens, T.J., 1989, Impact production of CO_2 by the Cretaceous/Tertiary extinction bolide and the resultant heating of the earth: Nature, v. 338, p. 247–249.

Olson, S.L., and Parris, D.C., 1987, The Cretaceous birds of New Jersey: Smithsonian Contributions to Paleobiology 63, 22 p.

Olsson, R.K., 1987, Cretaceous stratigraphy of the Atlantic Coastal Plain, Atlantic Highlands of New Jersey: Geological Society of America Centennial Field Guide, Northeastern Section, p. 87–90.

Olsson, R.K., 1989, Depositional sequences in the Cretaceous post-rift sediments on the New Jersey Atlantic margin: Marine Geology, v. 90, p. 113–118.

Olsson, R.K., Gibson, T.G., Hansen, H.J., and Owen, J.P., 1988, Geology of the northern Atlantic coastal plain: Long Island to Virginia, *in* Sheridan, R.E., and Grow, J.A., eds., The Atlantic Continental Margin, U.S.: Geological Society of America, Geology of North America, v. I-2, p. 87–105.

Olsson, R.K., and Liu, C., 1993, Controversies on the placement of Cretaceous-Paleogene boundary and the K/P mass extinction of planktonic foraminifera: Palaios, v. 8, p. 127–139.

Olsson, R.K., Miller, K.G., Browning, J.V., Habib, D., and Sugarman, P.J., 1997, Ejecta layer at the Cretaceous-Tertiary boundary, Bass River, New Jersey (Ocean Drilling Program, Leg 174AX): Geology, v. 25, p. 759–762.

Olsson, R.K., Wright, J.D., Miller K.G., Browning, J.V., and Cramer, B.S., 2000, Cretaceous-Tertiary boundary events on the New Jersey continental margin, *in* Catastrophic events and mass extinctions: Impacts and beyond: Houston, Texas, Lunar and Planetary Institute, LPI Contribution No. 1053, p. 160–161.

Owens, J.P., Minard, J.P., Sohl, N.F., and Mello J.F., 1970, Stratigraphy of the outcropping post-Magothy Upper Cretaceous formations in southern New Jersey and northern Delmarva Peninsula, Delaware and Maryland: U.S. Geological Survey Professional Paper 674, 60 p.

Owens, J.P., and Sohl, N.F., 1973, Glauconites from New Jersey-Maryland coastal plain: Their K-Ar Ages and application in stratigraphic studies: Geological Society of America Bulletin, v. 84, p. 2811–2838.

Rigby, J.K., 1983, Introduction to the Porifera, *in* Rigby, J.K., and Stearn, C.W., eds., Sponges and spongiomorphs: Knoxville, University of Tennessee, Department of Geological Sciences, Studies in Geology 7, p. 1–11.

Schwimmer, D.R., Stewart, J.D., and Williams, G.D., 1994, Giant fossil coelacanths of the Late Cretaceous in the eastern United States: Geology, v. 22, p. 503–506.

Self-Trail, J.M., and Bybell, L.M., 1995, Cretaceous and Paleogene calcareous nannofossil biostratigraphy of New Jersey *in* Baker, J.E.B., ed., Contributions to the paleontology of New Jersey: Proceedings of the Geological Association of New Jersey Twelfth Annual Meeting, The William Patterson College of New Jersey, Wayne, New Jersey, p. 102–139.

Sharigeta, Y., 1993, Post-hatching early life history of Cretaceous ammonoidea: Lethaia, v. 26, p. 133–145.

Sugarman, P.J., Miller, K.G., Bukry, D., and Feigenson, M.D., 1995, Uppermost Campanian-Maastrichtian strontium isotopic, biostratigraphic, and sequence stratigraphic framework of the New Jersey Coastal Plain: Geological Society of America Bulletin, v. 107, p. 19–37.

Thayer, C.W., 1981, Ecology of living brachiopods, *in* Dutro, J.T., and Boardman, R.S., eds., Lophophorates: Knoxville, University of Tennessee, Department of Geological Sciences, Studies in Geology 5, p. 110–126.

MANUSCRIPT SUBMITTED OCTOBER 5, 2000; ACCEPTED BY THE SOCIETY MARCH 22, 2001

Printed in the U.S.A.

Geological Society of America
Special Paper 356
2002

Giant ground birds at the Cretaceous-Tertiary boundary: Extinction or survival?

Eric Buffetaut*
Centre National de la Recherche Scientifique, 16 cour du Liégat, 75013 Paris, France

ABSTRACT

A family of giant, flightless ground birds, the Gastornithidae, has been known for a long time from the early Tertiary (late Paleocene to middle Eocene) of both Europe and North America. The giant ground bird *Gargantuavis* was recently described from the Upper Cretaceous (probably early Maastrichtian) of France. The question may therefore be asked whether there is any close phylogenetic relationship between *Gargantuavis* and the Gastornithidae, which would suggest survival of giant birds across the Cretaceous-Tertiary boundary. A close anatomical comparison, however, reveals that *Gargantuavis* is a much more primitive bird than the Gastornithidae, and that they do not belong to the same lineage, resemblances probably being due to convergent adaptation to a similar mode of life. Although it cannot be demonstrated at the moment that *Gargantuavis* became extinct at the Cretaceous-Tertiary (K-T) boundary, this is a distinct possibility. If this is the case, the mass extinction of the K-T boundary will appear to have stopped early giantism and flightlessness in birds; these were followed by renewed and similar giantism and flightlessness in a different group of birds in the Paleocene.

INTRODUCTION

What happened to birds at the Cretaceous-Tertiary (K-T) boundary is one of the least clear effects of the mass extinction that took place at that time, although it is obvious that birds survived the K-T catastrophe. According to Feduccia (1995, 1996, 1999), there was a mass extinction of archaic birds, followed by a Tertiary radiation of modern birds during the Tertiary. Although this view of a bottleneck in bird evolution at the K-T boundary has been accepted by some paleontologists (Chatterjee, 1997), it has been challenged by others (Chiappe, 1995), who think that the fossil record of birds does not warrant such an assumption, and by molecular biologists (Cooper and Penny, 1997), who argue that modern bird groups originated well before the end of the Cretaceous, and that there was a mass survival of birds at the K-T boundary. Much of the controversy appears to arise from the inadequate fossil record of birds both just before and just after the K-T boundary, which makes it difficult to reconstruct what may have happened to them at that time with any accuracy. In this chapter I discuss one aspect of the problem, i.e., what happened to giant flightless birds at the K-T boundary, on the basis of newly discovered Late Cretaceous remains of such birds, which can be compared with better known giant birds from the Paleocene and Eocene.

EARLY TERTIARY AND LATE CRETACEOUS GIANT GROUND BIRDS

The occurrence of giant, flightless birds in the early Tertiary was first reported in 1855, when a tibiotarsus of such a bird was discovered by Gaston Planté in the basal Eocene of Meudon, ncar Paris, and subsequently described as *Gastornis parisiensis* by Hébert (1855). Later discoveries (see Buffetaut, 1997a, for a review) showed that *Gastornis* also occurred in

*E-mail: Eric.Buffetaut@wanadoo.fr.

Buffetaut, E., 2002, Giant ground birds at the Cretaceous-Tertiary boundary: Extinction or survival?, *in* Koeberl, C., and MacLeod, K.G., eds., Catastrophic Events and Mass Extinctions: Impacts and Beyond: Boulder, Colorado, Geological Society of America Special Paper 356, p. 303–306.

late Paleocene beds in the eastern Paris basin (Lemoine, 1878, 1881), and that a similar giant bird was present in the early Eocene of North America (Cope, 1876; Matthew and Granger, 1917), and in the middle Eocene of Europe (Fischer, 1962; Berg, 1965). A similar bird is known from the Eocene of China (Hou, 1980). The North American form was named *Diatryma* (see Andors, 1988, 1992, for a revision), but recent work has shown that *Diatryma* and *Gastornis* are very closely allied and referable to a single family, the Gastornithidae (Martin, 1992), and are in all likelihood congeneric (Buffetaut, 1997b, 2000). Those early Tertiary giant flightless birds from Europe and North America are referred to as *Gastornis* herein.

The known stratigraphic range of *Gastornis* in Europe is from the Thanetian to the middle Eocene, whereas it appears to be restricted to the early Eocene in North America. The earliest occurrence of *Gastornis* is from the Thanetian of Walbeck in Germany (Weigelt, 1939), but no description of the Walbeck material has been published. Good *Gastornis* material is known from the slightly later deposits of Cernay and Berru near Reims, France (Martin, 1992; Buffetaut, 1997b). The osteology of the Gastornithidae is now well known, because a large amount of material is available from both Europe and North America, and a fairly complete skeleton has been found in the lower Eocene of Wyoming (Matthew and Granger, 1917). *Gastornis* was a large, to 2 m tall, large-headed, and massively built flightless bird. Biogeographic and stratigraphic evidence suggests that gastornithids may have originated in Europe and then spread to North America during a phase of faunal interchange in the earliest Eocene (Buffetaut, 1997b).

For a long time, the only known Cretaceous flightless birds were *Hesperornis* and *Baptornis* from the Late Cretaceous Niobrara Chalk of Kansas, first described by Marsh (1880). These relatively large forms were highly specialized marine birds with reduced forelimbs and hindlimbs adapted to swimming, quite unlike terrestrial flightless birds. The first undoubted Cretaceous flightless ground bird to be described was *Patagopteryx deferrariisi* from the Upper Cretaceous of Patagonia (Alvarenga and Bonaparte, 1992; Chiappe, 1996), which was only the size of a chicken. It has been suggested that some of the so-called "feathered dinosaurs" from the Lower Cretaceous of China, especially *Caudipteryx*, may in fact be secondarily flightless birds (Feduccia, 1999; Jones et al., 2000), but this hypothesis is still controversial (Zhou and Wang, 2000); in any case, *Caudipteryx* was a relatively small animal (~50 cm high at the hip). The avian status of the definitely flightless but controversial Alvarezsauridae is disputed and they are not considered here.

Remains of a giant Cretaceous terrestrial bird were first reported from the Upper Cretaceous of France (Buffetaut et al., 1995). This form was subsequently described as *Gargantuavis philoinos* by Buffetaut and Le Loeuff (1998) on the basis of less fragmentary material. *Gargantuavis* remains are known from several localities in southern France that are referred to the early Maastrichtian, or possibly late Campanian. Although the available material is still scanty, consisting of remains of the synsacrum, pelvis, and femur, *Gargantuavis* appears to have been a robustly built bird weighing as much as 140 kg (i.e., the weight of an adult ostrich; see Buffetaut and Le Loeuff, 1998). *Gargantuavis* may have been endemic to Europe; no remains of such a large ground bird have been reported from anywhere else.

One of the questions that arise following the discovery of *Gargantuavis philoinos* is whether there could be a phylogenetic link between this giant Late Cretaceous bird and the Paleocene gastornithids. The stratigraphic gap separating their known records, from the early Maastrichtian to the Thanetian, is at least 10 m.y., but, in view of the paucity of the vertebrate fossil record for the early Paleocene, the absence of intervening fossils could easily be explained by the imperfection of the fossil record. If *Gargantuavis* and *Gastornis* could be shown to be closely related, these giant birds would provide an unusual example of a group of large terrestrial vertebrates surviving across the K-T boundary.

GARGANTUAVIS AND *GASTORNIS*: A COMPARISON

At first sight, there are resemblances between what is known of *Gargantuavis* and the corresponding parts of the much better known *Gastornis*. Both have a massively built and broad pelvis, which sets them apart from fast-running ground birds such as the ostrich. (However, it is reminiscent of some of the more sturdily built moas, such as *Emeus crassus*.) In both, the acetabulum is in anterior position relative to the total length of both the synsacrum and the pelvis. (This may be linked to the position of the center of gravity of the bird. In *Gastornis* at least, the head was very large, and an anteriorly located acetabulum may have been advantageous for the balance of the body. The anterior part of the body is unknown in *Gargantuavis*.)

One of the main differences between *Gargantuavis* and the Gastornithidae is the smaller number of fused vertebrae forming the synsacrum in the Cretaceous form. Although it may be slightly incomplete posteriorly, there were apparently 10 fused vertebrae in the synsacrum of *Gargantuavis*, which is comparable to the number in other Cretaceous birds such as *Baptornis* and *Ichthyornis*, but definitely less than in the Gastornithidae. In the skeleton (AMNH 6165) described as *Diatryma steini* by Matthew and Granger (1917), the number of fused vertebrae in the synsacrum is at least 16, according to Andors (1988). The first synsacral vertebra of *Gargantuavis* has a nearly circular concave anterior articular surface, unlike that of *Gastornis*, which is more saddle shaped, as is usual in modern birds. A further difference is in the relative position of the acetabulum. The anterior part of the pelvis is relatively short in both *Gargantuavis* and *Gastornis*, but this is much more striking in *Gargantuavis*, in which the acetabulum is located only a short distance posterior to the anterior end of the synsacrum, at the level of the third and fourth transverse processes of the synsacral vertebrae. In *Gastornis*, the preacetabular part of the synsacrum is markedly longer.

In *Gastornis*, the ilia extend medially to cover much of the synsacrum region, although apparently they do not really meet dorsally; according to Fischer (1962), in the anterior part they abut against a median ridge formed by the neural spines of the synsacral vertebrae, and in the posterior part they are separated by a relatively broad median section formed by the transverse processes of the synsacral vertebrae. In *Gargantuavis*, the neural spines of the synsacral vertebrae are fused to form a median ridge, but the ilia do not reach it and are separated from it by a broad space (Buffetaut and Le Loeuff, 1998).

The femur of *Gargantuavis* differs from that of *Gastornis* mainly in the shape of the trochanteric crest located lateral to the articular head. In *Gastornis*, as in most other large flightless birds, this crest forms a rather prominent, proximally directed, and sharply pointed process. In *Gargantuavis*, the trochanteric crest is smoothly rounded, and there is no clear indication of the oblique line issuing from the trochanter on the anterior face of the shaft, which is present in *Gastornis*.

Most of the characters separating *Gargantuavis* from *Gastornis*, especially in the synsacrum and pelvis, indicate that *Gargantuavis* is more primitive than the Gastornithidae. Although a precise phylogenetic and systematic placement of *Gargantuavis* is difficult to establish because of the incompleteness of the available material, the various primitive features of its pelvic region suggest that it is not an ornithurine (Buffetaut and Le Loeuff, 1998), although according to Feduccia (1999), it might be an archaic ornithurine. Resemblances with another flightless, but much smaller, Cretaceous bird, *Patagopteryx*, from the Upper Cretaceous of Patagonia (Alvarenga and Bonaparte, 1992; Chiappe, 1996), have been noted (Buffetaut and Le Loeuff, 1998). *Patagopteryx* was considered a nonornithurine by Chiappe (1996).

The position of gastornithids in bird classification has been the subject of much discussion since the discovery of the first remains of *Gastornis* in the mid-nineteenth century (Buffetaut, 1997a). Both Martin (1992) and Andors (1992) placed the Gastornithidae in an order of their own, the Gastornithiformes. According to Andors (1992), the Gastornithiformes are related to the Anseriformes rather than to the Gruiformes, as formerly maintained by many. Whatever their exact phylogenetic and systematic position, they belong to the Neornithes and are clearly more advanced than the archaic *Gargantuavis*.

CONCLUSIONS

Despite some resemblances that can be ascribed to convergent evolution in similarly adapted, robustly built large ground birds, *Gargantuavis* and *Gastornis* appear to differ from each other in important respects. *Gargantuavis* apparently belongs to an archaic group of birds, whereas the Gastornithidae are clearly more advanced. There is no indication of close phylogenetic links between the Late Cretaceous *Gargantuavis* and the early Tertiary Gastornithidae.

Whether *Gargantuavis* disappeared at the K-T boundary is uncertain. All the available material is from deposits that apparently antedate the K-T boundary by several million years, and at the moment it cannot be demonstrated that *Gargantuavis* survived until the end of the Maastrichtian. Nothing really resembling it has ever been reported from the Tertiary.

What appears clear, however, is that *Gargantuavis* is not a Cretaceous representative of the Gastornithidae. No evidence of gastornithids is known from the Cretaceous, and they probably represent an early Tertiary lineage of giant ground birds that developed after the extinction of the dinosaurs at the K-T boundary. Although they were long considered as carnivorous birds feeding on small mammals (see Witmer and Rose, 1991), Andors (1988, 1992) reinterpreted them as large folivorous birds. Whatever their diet, it has generally been accepted that such giant ground birds could only evolve after the disappearance of the dinosaurs had left the continents devoid of any large tetrapods. As expressed by Martin (1992, p. 106), "the extinction of the dinosaurs and the small size of the early mammals reduced the threat to ground-nesting birds and probably made possible an increased avian expansion into terrestrial habitats." The discovery of *Gargantuavis*, a giant ground bird that lived among dinosaurs, has shown that the disappearance of the dinosaurs was not a prerequisite for the evolution of such large flightless birds (although it may have played a part in the evolution of the early Tertiary gastornithids).

What can be said at the moment is that there is no evidence for the survival of a lineage of giant flightless birds across the Cretaceous-Tertiary boundary. *Gargantuavis* in the Late Cretaceous and the Gastornithidae in the early Tertiary should be considered as results of convergent evolution in two different groups of birds. If it can ultimately be shown that *Gargantuavis* was among the victims of the mass extinction at the K-T boundary, this would suggest that, in the case of giant ground birds, extinction "set the clock back" and resulted in renewed giantism and flightlessness that gave rise to the Gastornithidae.

ACKNOWLEDGMENTS

I thank Christian Koeberl (University of Vienna) for inviting me to give a paper at the symposium on Catastrophic Events and Mass Extinctions in Vienna. Comparisons between *Gargantuavis* and *Diatryma* were made possible by E.S. Gaffney and A.V. Andors (American Museum of Natural History). Useful suggestions were provided by reviewers M.J. Benton, A. Feduccia, and C. Koeberl.

REFERENCES CITED

Alvarenga, H., and Bonaparte, J.F., 1992, A new flightless landbird from the Cretaceous of Patagonia, *in* Campbell, K., ed., Papers in avian paleontology honoring Pierce Brodkorb: Natural History Museum of Los Angeles County Science Series, v. 36, p. 51–64.

Andors, A.V., 1988, Giant groundbirds of North America (Aves, Diatrymidae) [Ph.D. thesis]: New York, Columbia University, 577 p.

Andors, A.V., 1992, Reappraisal of the Eocene groundbird Diatryma (Aves, Anserimorphae), *in* Campbell, K., ed., Papers in avian paleontology honoring Pierce Brodkorb: Natural History Museum of Los Angeles County Science Series, v. 36, p. 109–125.

Berg, D.E., 1965, Nachweis des Riesenlaufvogels *Diatryma* im Eozän von Messel bei Darmstadt/Hessen: Notizblatt des hessischen Landesamtes für Bodenforschung, v. 93, p. 68–72.

Buffetaut, E., 1997a, L'oiseau géant *Gastornis*: Interprétation, reconstitution et vulgarisation de fossiles inhabituels dans la France du XIXe siècle: Bulletin de la Société Géologique de France, v. 168, p. 805–811.

Buffetaut, E., 1997b, New remains of the giant bird *Gastornis* from the Upper Palaeocene of the eastern Paris Basin and the relationships between *Gastornis* and *Diatryma*: Neues Jahrbuch für Geologie und Paläontologie, Monatshefte, v. 3, p. 179–190.

Buffetaut, E., 2000, Are *Gastornis* and *Diatryma* congeneric? [abs.]: Vertebrata Palasiatica, v. 38, supplement, p. 3.

Buffetaut, E., and Le Loeuff, J., 1998, A new giant ground bird from the Upper Cretaceous of southern France: Journal of the Geological Society, London, v. 155, p. 1–4.

Buffetaut, E., Le Loeuff, J., Mechin, P., and Mechin-Salessy, A., 1995, A large French Cretaceous bird: Nature, v. 377, p. 110.

Chatterjee, S., 1997, The rise of birds: Baltimore, The Johns Hopkins University Press, 312 p.

Chiappe, L.M., 1995, The first 85 million years of bird evolution: Nature, v. 378, p. 349–355.

Chiappe, L.M., 1996, Late Cretaceous birds of southern South America: Anatomy and systematics of Enantiornithes and *Patagopteryx deferrariisi*, *in* Arratia, G., ed., Contributions of southern South America to vertebrate paleontology: Münchner Geowissenschaftliche Abhandlungen, v. 30, p. 203–244.

Cooper, A., and Penny, D., 1997, Mass survival of birds across the Cretaceous-Tertiary boundary: Molecular evidence: Science, v. 275, p. 1109–1113.

Cope, E.D., 1876, On a gigantic bird from the Eocene of New Mexico: Proceedings of the Academy of Natural Sciences of Philadelphia, v. 28, p. 10–11.

Feduccia, A., 1995, Explosive evolution in Tertiary birds and mammals: Science, v. 267, p. 637–638.

Feduccia, A., 1996, The origin and evolution of birds: New Haven, Yale University Press, 420 p.

Feduccia, A. 1999, The origin and evolution of birds (second edition): New Haven, Yale University Press, 466 p.

Fischer, K., 1962, Der Riesenlaufvogel *Diatryma* aus der eozänen Braunkohle des Geiseltales: Hallesches Jahrbuch für mitteldeutsche Erdgeschichte, v. 4, p. 26–33.

Hébert, E., 1855, Note sur le tibia du *Gastornis parisiensis*: Comptes Rendus de l'Académie des Sciences de Paris, v. 40, p. 579–582.

Hou, L., 1980, New form of the Gastornithidae from the Lower Eocene of the Xichuan, Honan: Vertebrata Palasiatica, v. 16, p. 111–115.

Jones, T.D., Farlow, J.O., Ruben, J.A., Henderson, D.M., and Hillenius, W.J., 2000, Cursoriality in bipedal archosaurs: Nature, v. 406, p. 716–718.

Lemoine, V., 1878, Recherches sur les oiseaux fossiles des terrains tertiaires inférieurs des environs de Reims: Reims, France, Keller, 69 p.

Lemoine, V., 1881, Recherches sur les oiseaux fossiles des terrains tertiaires inférieurs des environs de Reims. Deuxième partie: Reims, France, Matot-Braine, p. 75–170.

Marsh, O.C., 1880, Odontornithes: A monograph on the extinct toothed birds of North America: United States Geological Exploration of the Fortieth Parallel: Washington D.C., U.S. Government Printing Office, 201 p.

Martin, L.D., 1992, The status of the Late Paleocene birds *Gastornis* and *Remiornis, in* Campbell, K., ed., Papers in avian paleontology honoring Pierce Brodkorb: Natural History Museum of Los Angeles County Science Series, v. 36, p. 97–108.

Matthew, W.D., and Granger, W., 1917, The skeleton of *Diatryma*, a gigantic bird from the Lower Eocene of Wyoming: Bulletin of the American Museum of Natural History, v. 37, p. 307–326.

Weigelt, J., 1939, Die Aufdeckung der bisher ältesten tertiären Säugetierfauna Deutschlands: Nova Acta Leopoldina, v. 7, p. 515–528.

Witmer, L.M., and Rose, K.D., 1991, Biomechanics of the jaw apparatus of the gigantic Eocene bird *Diatryma*: Implications for diet and mode of life: Paleobiology, v. 17, p. 95–120.

Zhou, Z., and Wang, X., 2000, A new species of *Caudipteryx* from the Yixian Formation of Liaoning, northeast China: Vertebrate Palasiatica, v. 38, p. 111–127.

Manuscript Submitted October 5, 2000; Accepted by the Society March 22, 2001

Geological Society of America
Special Paper 356
2002

Dinosaurs that did not die: Evidence for Paleocene dinosaurs in the Ojo Alamo Sandstone, San Juan Basin, New Mexico

James E. Fassett*
U.S. Geological Survey, 552 Los Nidos Drive, Santa Fe, New Mexico 87501, USA
Robert A. Zielinski
U.S. Geological Survey, P.O. Box 25046, MS 973, Denver Federal Center, Denver, Colorado 80225, USA
James R. Budahn
U.S. Geological Survey, P.O. Box 25046, MS 974, Denver Federal Center, Denver, Colorado 80225, USA

ABSTRACT

Palynologic and paleomagnetic data confirm a Paleocene age for the Ojo Alamo Sandstone (and its contained dinosaurs) throughout the San Juan Basin of New Mexico. The recently reported discovery of 34 skeletal elements from a single hadrosaur in the Ojo Alamo provides unequivocal evidence that these bones were not reworked from underlying Cretaceous strata. Geochemical studies of samples from several single-dinosaur-bone specimens from the Paleocene Ojo Alamo Sandstone and the underlying Late Cretaceous (Campanian) Kirtland Formation show that mineralized bones from these two rock units contain distinctly different abundances of uranium and rare-earth elements and demonstrate that Cretaceous and Paleocene bones were mineralized at different times when mineralizing fluids had distinctly different chemical compositions. These findings indicate that the dinosaur bone from the Paleocene Ojo Alamo is indigenous and not reworked.

These data show that a relatively diverse assemblage of dinosaurs survived the end-Cretaceous asteroid-impact extinction event of 65.5 Ma. The San Juan Basin's Paleocene dinosaur fauna is herein named the Alamoan fauna. Magnetic-polarity chronology shows that these survivors lived for about one million years into the Paleocene and then became extinct around 64.5 Ma. We suggest that a plausible survival mechanism for this Lazarus fauna may have been the large numbers of buried dinosaur eggs, laid just before the asteroid impact occurred. These buried eggs would have provided a safe haven for developing dinosaur embryos for the first one to two years after the impact, thereby making it possible for them to survive the worst of the impact's early devastation.

OJO ALAMO SANDSTONE

Previous studies

The Ojo Alamo Sandstone has been a controversial rock unit from the time the name Ojo Alamo Beds was coined by Barnum Brown (1910) to label dinosaur-bearing rocks near Ojo Alamo Arroyo (Fig. 1A) in the southern San Juan Basin of New Mexico. Almost every characteristic of the Ojo Alamo (i.e., age, rock-stratigraphic definition, and the nature of its upper and lower contacts) continues to be vigorously debated. Discussions

*E-mail: jimgeology@qwest.net

Fassett, J.E., Zielinski, R.A., and Budahn, J.R., 2002, Dinosaurs that did not die: Evidence for Paleocene dinosaurs in the Ojo Alamo Sandstone, San Juan Basin, New Mexico, *in* Koeberl, C., and MacLeod, K.G., eds., Catastrophic Events and Mass Extinctions: Impacts and Beyond: Boulder, Colorado, Geological Society of America Special Paper 356, p. 307–336.

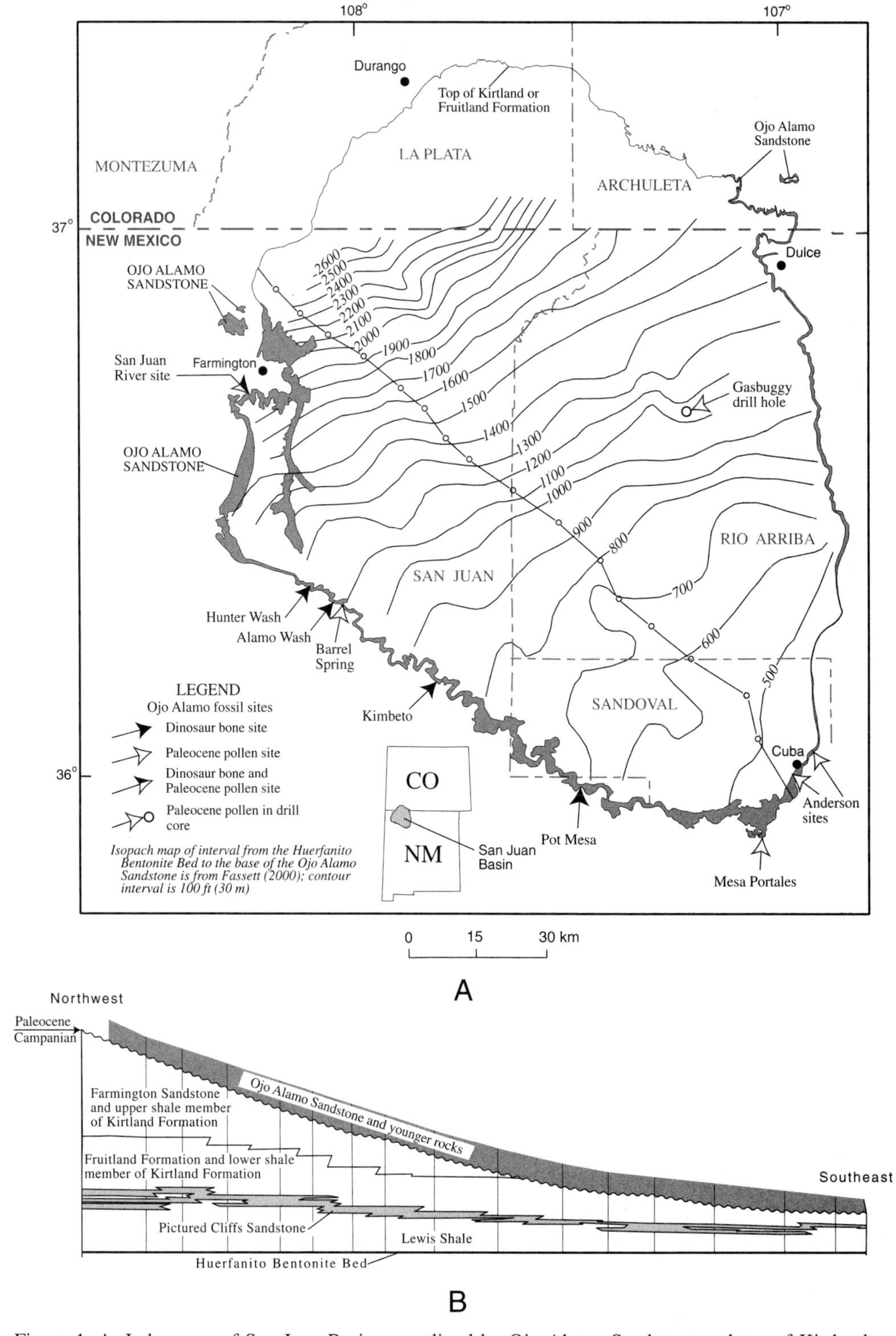

Figure 1. A: Index map of San Juan Basin as outlined by Ojo Alamo Sandstone and top of Kirtland or Fruitland Formation. Significant fossil localities for pollen or dinosaur bone and isopach map showing thickness of interval from Huerfanito Bentonite Bed to base of Ojo Alamo Sandstone are shown. Ojo Alamo type area is between Hunter Wash and Barrel Spring. B: Northwest-trending stratigraphic cross section across San Juan Basin showing interval from Huerfanito Bentonite Bed through Ojo Alamo Sandstone (line of cross section is shown in A). Modified from Figure A3-3 of Fassett (2000).

of the Ojo Alamo controversy by Fassett (1973) and Fassett et al. (1987) are summarized in the following.

Soon after Brown (1910) first used the name Ojo Alamo to refer to dinosaur-bearing beds near Alamo Wash (Fig. 1A), Bauer (1916) conducted a detailed geologic study of these strata throughout the type area between Hunter Wash and Barrel Spring (Fig. 1A), and formally defined the Ojo Alamo Sandstone there as a conglomeratic sandstone containing lenses of shale, siltstone, and nonconglomeratic sandstones. Bauer found that the Ojo Alamo throughout much of this area generally consisted of a relatively thin (average 2–3 m) lower conglomerate, a middle shalier unit as much as 25 m thick, and an upper conglomerate to 15 m thick (Fig. 2). Bauer (1916) noted, however, that the lithologic composition of the Ojo Alamo was variable and that at some places, notably to the west near Hunter Wash and in the east near Barrel Spring, the middle shalier unit was absent and the formation consisted of a massive conglomeratic sandstone bed containing only a few, thin, scattered shale layers. It is the middle, generally shalier part of the Ojo Alamo that contains most of the dinosaur fossils in the type area.

U.S. Geological Survey (USGS) geologists, principally Reeside (1924) and Dane (1936), mapped the Ojo Alamo in the west and south parts of the San Juan Basin, as defined by Bauer (1916). Baltz (1967) further extended the name along the east side of the basin northward to the New Mexico state line by applying it to rocks in that area previously mapped as the basal sandstone of the Animas Formation by Dane (1946, 1948). Fassett and Hinds (1971) completed the mapping of the Ojo Alamo in the basin by extending the name into the Colorado part of the basin (Fig. 1A). In the course of this work, the Ojo Alamo was found to vary in its stratigraphic composition in exposures around the periphery of the basin, ranging from four massive sandstone beds separated by shaly layers totaling more than 120 m thick in the northwest near the town of Farmington (Fig. 1A), to a single sandstone bed a few meters thick at other places around the south and east sides of the basin. The Ojo Alamo is

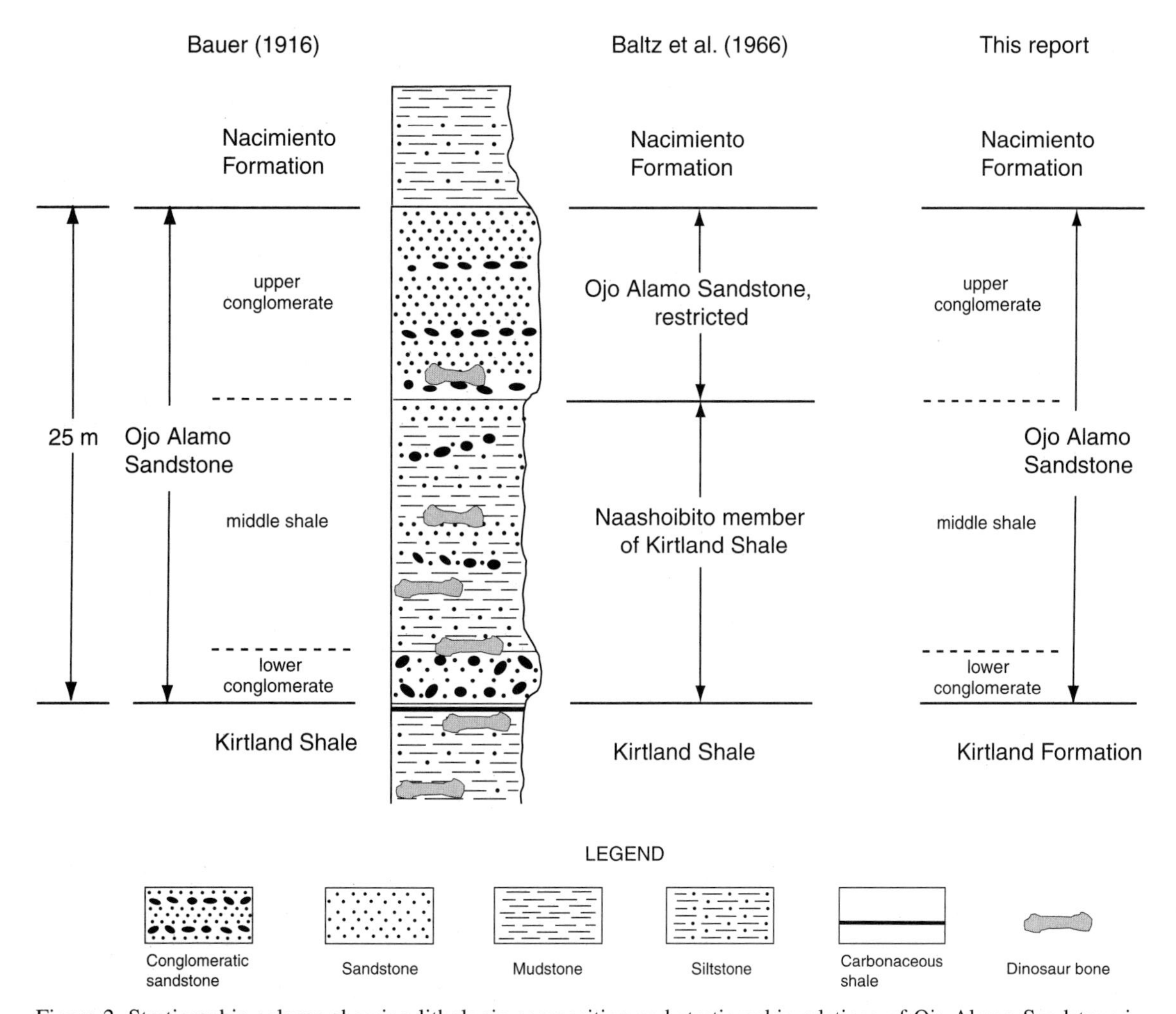

Figure 2. Stratigraphic column showing lithologic composition and stratigraphic relations of Ojo Alamo Sandstone in type area in southern San Juan Basin (Fig. 1A). Figure shows original definition of Ojo Alamo by Bauer (1916), revised definition of Baltz et al. (1966), and definition used in this chapter.

today recognized as a multistoried, channel sandstone complex consisting of interbedded sandstones and mudstones representing stream-channel and overbank sediments deposited by high-energy braided streams flowing from the north or northwest (Fassett and Hinds, 1971; Fassett, 1985; Powell, 1972, 1973; Sikkink, 1987). Channel sandstone beds are generally tabular, discontinuous, and frequently overlapping. In places, upper story channel sandstones have deeply eroded bases that in places have cut through intervening shaly overbank deposits and into lower story channel sandstone deposits. In those places where the Ojo Alamo forms massive sandstone cliffs tens of meters thick, the cliff face, on close examination, is found to consist of a complex of stacked stream-channel deposits separated by conglomerate or coarse-grained sandstone layers or thin, clay-drape layers. Channel sandstones are typically separated by intervening mudstones and siltstones that accumulated as overbank deposits adjacent to braided stream channels. Conglomerates are abundant in the Ojo Alamo in the northwest part of the basin, closer to the source areas, and conglomerate clasts become smaller and less abundant southward and eastward across the basin. (See Fassett et al. [1987] for detailed descriptions of the Ojo Alamo at the various fossil localities shown in Fig. 1A.)

The Ojo Alamo Sandstone was assigned a Cretaceous age when it was first defined by Bauer (1916) because of its dinosaur fauna, but in spite of these Cretaceous index fossils, Reeside (1924, p. 32) assigned a Tertiary (?) age to the Ojo Alamo as follows: "In view of the wide differences in opinion expressed by various students as to the correct assignment of this whole group of related formations, the Ojo Alamo Sandstone and Animas formation are herein classified as Tertiary (?)." Reeside was thus the first geologist to (tacitly) suggest that the Ojo Alamo and its dinosaur fauna were actually Tertiary in age, basing this age designation partly on fragmentary plant fossils from the Ojo Alamo identified by Knowlton (*in* Reeside, 1924, p. 31, 32) that "appear to be Tertiary." More significant for Reeside, however, was his correlation of the dinosaur-bearing Animas Formation in the northern San Juan Basin with the Ojo Alamo Sandstone to the south, because Knowlton (1924) had assigned an unequivocal Tertiary age to the Animas on the basis of its flora. Most geologists working in the San Juan Basin after Reeside's 1924 publication rejected his Tertiary (?) age designation for the Ojo Alamo, suggesting that at least the lower, dinosaur-bearing part of the Ojo Alamo was Cretaceous.

Anderson (1960) was the first geologist to use palynology to try and resolve the conflicting data regarding the age of the Ojo Alamo. In his innovative study, he described and identified palynomorphs from a suite of rock samples from within, above, and below the Ojo Alamo Sandstone in the vicinity of Cuba, New Mexico (Fig. 1A). In his conclusions, Anderson (1960, p. 13) wrote the following (italics ours).

> Paleobotanical evidence suggests that most of the Ojo Alamo Sandstone is Tertiary, but the basal part may be either Cretaceous or Paleocene. The dinosaurs taken from the middle shale unit of the Ojo Alamo Sandstone [in the type area] have been considered Montanan in age; the discrepancy may be explained by assuming that the dinosaur bones and fragments have been reworked or misidentified. *Alternatively, pre-Lance-type dinosaurs persisted into a "Tertiary environment."*

Anderson subsequently collected rock samples from the Ojo Alamo and adjacent strata in the type area, near Barrel Spring and Alamo Wash (Fig. 1A), and concluded (*in* Baltz et al., 1966) that palynomorphs from the Ojo Alamo Sandstone and overlying Nacimiento Formation were similar to the Paleocene palynomorphs from these same rock units near Cuba. The Ojo Alamo collection from the Barrel Spring area was from the middle of the upper conglomeratic member of the Ojo Alamo Sandstone (above the dinosaur-rich, middle shale member) and thus for the first time directly confirmed the age of at least the upper part of the Ojo Alamo in its type area as Paleocene. Baltz et al. (1966) also demonstrated that the Ojo Alamo and the overlying Nacimiento Formation intertongued near Barrel Spring, providing further confirmation of the Paleocene age of at least the upper part of the Ojo Alamo because of the Puercan mammalian fossils known to occur nearby in the lowermost part of the Nacimiento.

Baltz et al. (1966) recommended that the problem of the mixed age of the Ojo Alamo Sandstone be resolved by restricting the name Ojo Alamo to the uppermost conglomeratic sandstone bed of Bauer's (1916) Ojo Alamo and renaming the middle shale and lower conglomerate of Bauer's Ojo Alamo the Naashoibito Member of the Kirtland Shale. Thus, the "restricted Ojo Alamo," containing no unreworked dinosaurs, was unequivocally Paleocene in age and the newly named "Naashoibito Member"—now a dinosaur-bearing member of the Kirtland Shale—was Cretaceous in age. This redefinition of the Ojo Alamo removed the last of Brown's (1910) "Ojo Alamo Beds" from the Ojo Alamo Sandstone. Figure 2 shows the Ojo Alamo Sandstone in the type area as defined by Bauer (1916) and as later redefined by Baltz et al. (1966).

Several papers in a publication on the Cretaceous and Tertiary rocks of the southern Colorado Plateau (Fassett, 1973; Clemens, 1973; Powell, 1973) rejected the redefinition of the Ojo Alamo Sandstone by Baltz et al. (1966) and recommended a return to Bauer's original definition. Powell (1973) suggested retaining the Naashoibito name for the lower conglomerate and dinosaur-bearing middle shale unit of Bauer's Ojo Alamo and further suggested that the upper conglomerate of Bauer (restricted Ojo Alamo of Baltz et al., 1966) be given a new name; the Kimbeto Member of the Ojo Alamo Sandstone. Subsequent papers used either the Baltz et al. (1966) redefinition of the Ojo Alamo or the original definition of Bauer (1916) to describe this unit, resulting in a confusing and untidy stratigraphic situation that continues to this day. We offer a solution to this problem in the following.

Fassett and Hinds (1971) published palynological analyses by R.H. Tschudy (USGS, Denver, Colorado) on samples collected from Mesa Portales in the southeast part of the basin (Fig. 1A). Tschudy's data indicated that all of the strata assigned to the Ojo Alamo Sandstone at Mesa Portales were Paleocene, confirming Anderson's (1960) findings. Tschudy (1973) published palynologic data from samples from the core of the Gasbuggy (GB-1) drill hole (Fig. 1A) that penetrated the Ojo Alamo Sandstone at depths of 1060–1122 m (3480–3680 ft on geophysical logs from that drill hole). These data indicated that the Ojo Alamo Sandstone was Tertiary in age in its entirety in this drill core and that the underlying Fruitland Formation was Campanian. Tschudy also concluded that a hiatus representing most of Maastrichtian time was present at the base of the Ojo Alamo in this core. Numerous rock samples have since been collected from the dinosaur-bearing part of the Ojo Alamo (of Bauer, 1916) in the Ojo Alamo type area for palynologic analysis, but thus far no samples have yielded diagnostic palynomorphs.

Fassett (1982) and Fassett et al. (1987) examined all available data for the dinosaur and palynomorph localities known in the San Juan Basin and conditionally concluded that the dinosaur fauna in the Ojo Alamo type area and elsewhere in the basin is Paleocene in age. Those reports also rejected the redefinition of the Ojo Alamo Sandstone by Baltz et al. (1966) in favor of retaining the original definition of Bauer (1916). Fassett and Lucas (2000) presented pollen data provided by Douglas J. Nichols (1994, written commun.) from a coaly, carbonaceous shale layer, 3 m beneath a large hadrosaur femur and 12 m above the base of the Ojo Alamo Sandstone, at a locality in the northern San Juan Basin (discussed in detail in the following). These data indicate that the Ojo Alamo and its contained dinosaur fossil there is Paleocene in age; Fassett and Lucas (2000, p. 229) concluded, "some dinosaurs in the San Juan Basin survived the 'terminal' end-Cretaceous asteroid impact event."

Lithologic components

The Ojo Alamo Sandstone is a multistoried, compound, stratigraphic unit containing abundant interbeds of conglomeratic sandstone, sandstone, siltstone, mudstone, and, rarely, carbonaceous mudstone. Figure 1A shows that the Ojo Alamo is present only in the southern part of the basin and is missing throughout most of the Colorado part of the basin. Sandstone beds range from very coarse grained to fine grained and are arkosic and poorly sorted; sand grains are generally subangular to poorly rounded. Conglomerate clasts range from grit to cobble size and clasts diminish in size from north to south and from west to east across the basin; clasts are generally composed of well-rounded quartzite, chert (mostly jasper), andesite, and fine-grained volcanic rock fragments (Powell, 1972). Powell (1972, p. 66) noted: "Toward the south, the volcanic rock pebbles rapidly decrease in number until the formation is practically a chert-quartzite pebble conglomerate." Conglomerates are essentially absent in the east and southeast parts of the basin. The Ojo Alamo contains large silicified logs at nearly all localities. None of the lithologic layers of the Ojo Alamo Sandstone have extensive lateral continuity, thus the numbers and thicknesses of the various beds that together make up the Ojo Alamo are different at nearly every location. The variable lithology of this formation is apparent even in the relatively small type area of the Ojo Alamo, between Hunter Wash and Barrel Spring (Fig. 1A), as shown on measured sections in that area by Bauer (1916) and Baltz et al. (1966).

Figure 3 shows eight stratigraphic columns from seven

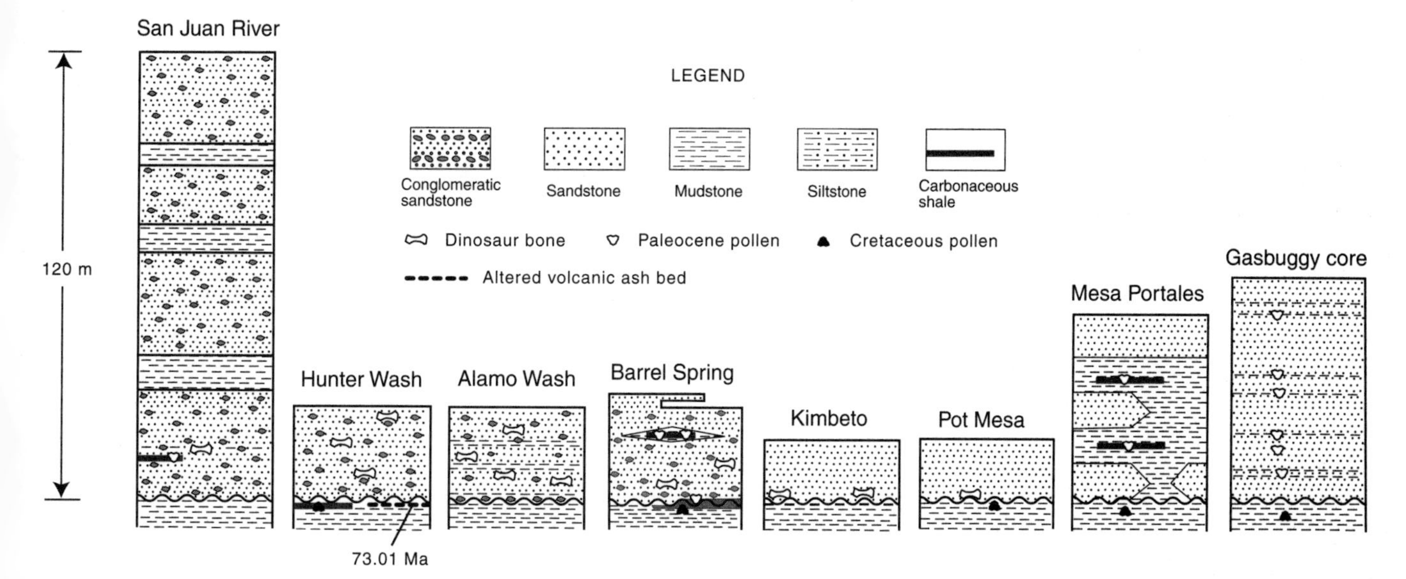

Figure 3. Seven stratigraphic columns showing nature of Ojo Alamo Sandstone at localities, shown in Figure 1A, where pollen or dinosaur bone have been found. Ojo Alamo is underlain by Kirtland or Fruitland Formation and overlain by Nacimiento Formation and its full thickness is shown at all localities except for GB-1, where upper part of Ojo Alamo is not shown. Modified from Fassett (1987).

measured outcrop sections and the Gasbuggy (GB-1) drill core in the New Mexico part of the San Juan Basin (the localities for each column are shown in Fig. 1A). These sections show the variable nature of the Ojo Alamo at these localities and clearly show that the lithology and thickness of the Ojo Alamo in its type area are not typical of the Ojo Alamo throughout the basin. These columns also show that the conglomeratic component of the Ojo Alamo diminishes between the Barrel Spring site and the Kimbeto site.

Figure 4 shows a comparison of the Ojo Alamo Sandstone in exposures in the type area and in the subsurface at the Fannin Government 1 drill hole, 4.5 km to the northeast. The lower conglomerate of the type area is visible on the geophysical log of the Fannin hole and is essentially the same thickness at both locations. This figure shows that the upper part of the Ojo Alamo also maintains about the same thickness from outcrop to drill hole, but the lithologic constituents vary. In the drill hole, the Ojo Alamo consists of an upper massive sandstone bed (containing two thin mudstone layers) separated from the lower conglomerate by ~3 m of mudstone, whereas the outcrops of the Ojo Alamo consist of an upper conglomerate, a middle, predominantly shaly unit (containing lensing sandstone and siltstone beds), and a lower conglomerate. Figure 4 shows that the upper conglomerate in the type area has no clear counterpart in the Fannin drill hole.

Figure 5 is a northeast-trending, subsurface cross section from near the Ojo Alamo Sandstone type area to the northeast part of the San Juan Basin showing the extreme variability of the lithologic constituents of the Ojo Alamo in the 16 drill holes shown. (The correlation of drill hole 1 on this cross section, the Fannin Government 1 hole, to the outcrop in the type area is shown in Fig. 4.) The length of the cross section is 106 km and the average distance between drill holes is ~7.6 km. The Ojo Alamo is thinnest (25 m) in drill hole 1 at the southwest end of the section near the type area and is thickest (~115 m) in drill hole 9. The Ojo Alamo consists of 5 sandstone beds separated by 4 relatively thin mudstone layers in hole 9, 4 sandstone beds separated by 3 mudstone beds in hole 1, a single massive sandstone bed ~30 m thick in hole 14, and 5 sandstone beds separated by 4 relatively thick mudstone beds in hole 7.

This variability in the lithologic composition of the Ojo Alamo reflects its origin as a high-energy, braided-stream deposit laid down on a relatively flat erosion surface as stream channels rapidly migrated laterally across the basin area in early Paleocene time. In some areas, major channel systems maintained the same general geographic location, resulting in thicker and more massive sandstone deposits, whereas in other areas, channels migrated relatively rapidly over large areas, resulting in thinner and larger numbers of channel-sandstone deposits interlayered with overbank mudrock deposits. The environment of deposition of the Ojo Alamo was described and illustrated in detail in Fassett (1985.)

Validity of Naashoibito member definition

The redefinition of the Ojo Alamo Sandstone in its type area by Baltz et al. (1966) was primarily aimed at removing the lower dinosaur-bearing part of the Ojo Alamo Sandstone from

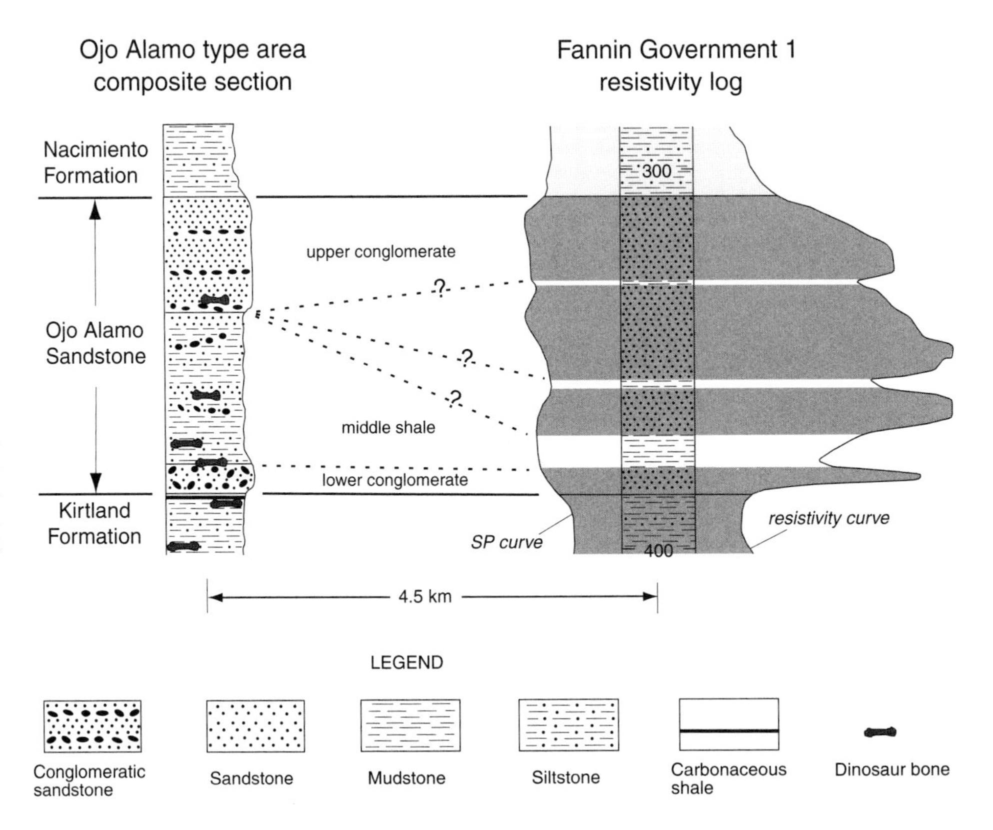

Figure 4. Correlation diagram showing relations between Ojo Alamo Sandstone outcrops in Ojo Alamo type area (Fig. 1A) and in subsurface on geophysical log of Fannin Government 1 drill hole, 4.5 km downdip to northeast. SP is spontaneous potential. Shaded area inside log traces is sandstone; white areas represent mudstone interbeds; lithologies are based on geophysical-log interpretation. Depths below surface shown on log of Fannin hole are in feet (300 ft = 91 m, 400 ft = 122 m).

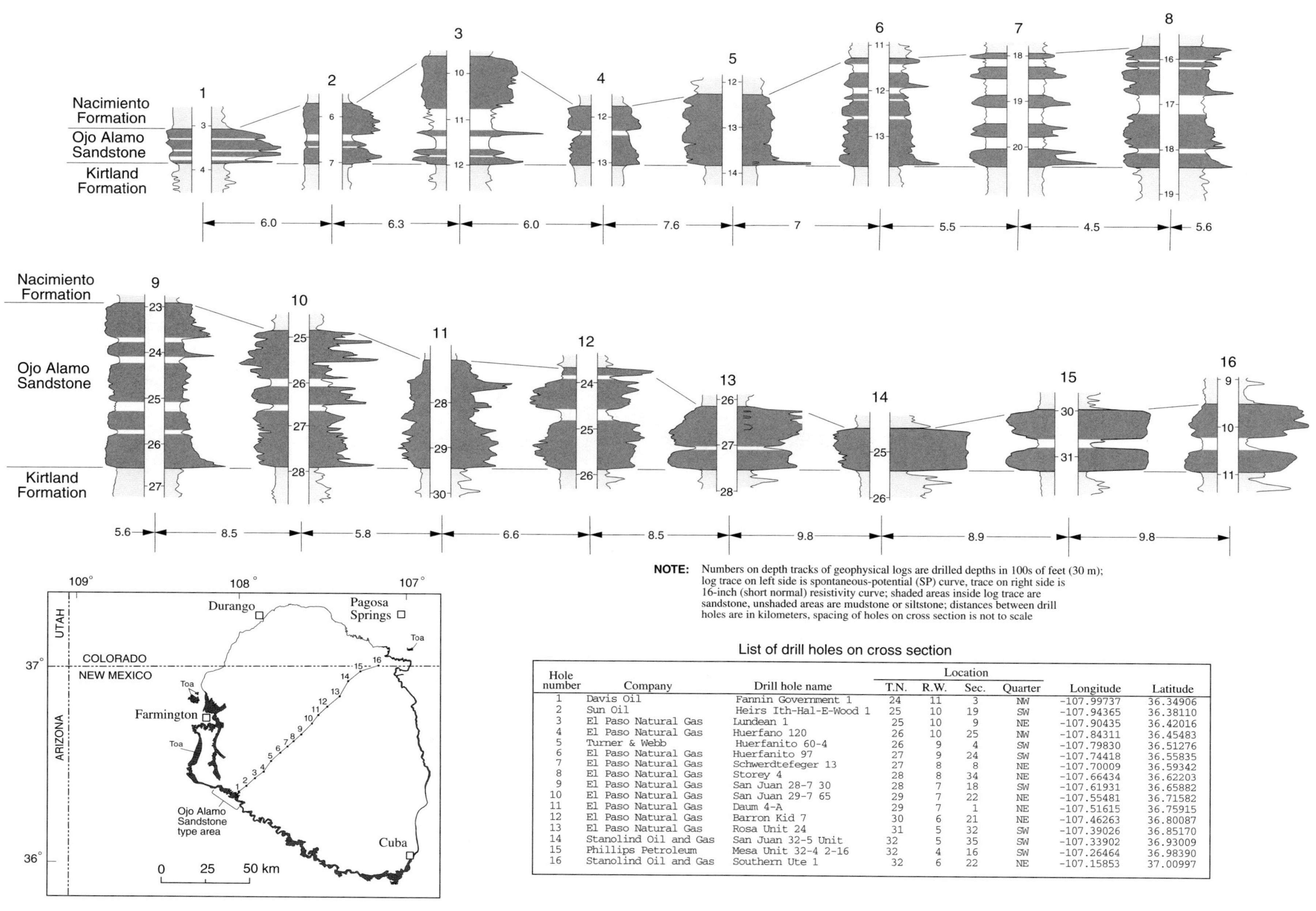

List of drill holes on cross section

Hole number	Company	Drill hole name	Location T.N.	R.W.	Sec.	Quarter	Longitude	Latitude
1	Davis Oil	Fannin Government 1	24	11	3	NW	-107.99737	36.34906
2	Sun Oil	Heirs Ith-Hal-E-Wood 1	25	10	19	SW	-107.94365	36.38110
3	El Paso Natural Gas	Lundean 1	25	10	9	NE	-107.90435	36.42016
4	El Paso Natural Gas	Huerfano 120	26	10	25	NW	-107.84311	36.45483
5	Turner & Webb	Huerfanito 60-4	26	9	4	SW	-107.79830	36.51276
6	El Paso Natural Gas	Huerfanito 97	27	9	24	SW	-107.74418	36.55835
7	El Paso Natural Gas	Schwerdtfeger 13	27	8	8	NE	-107.70009	36.59342
8	El Paso Natural Gas	Storey 4	28	8	34	NE	-107.66434	36.62203
9	El Paso Natural Gas	San Juan 28-7 30	28	7	18	SW	-107.61931	36.65882
10	El Paso Natural Gas	San Juan 29-7 65	29	7	22	NE	-107.55481	36.71582
11	El Paso Natural Gas	Daum 4-A	29	7	1	NE	-107.51615	36.75915
12	El Paso Natural Gas	Barron Kid 7	30	6	21	NE	-107.46263	36.80087
13	El Paso Natural Gas	Rosa Unit 24	31	5	32	SW	-107.39026	36.85170
14	Stanolind Oil and Gas	San Juan 32-5 Unit	32	5	35	SW	-107.33902	36.93009
15	Phillips Petroleum	Mesa Unit 32-4 2-16	32	4	16	SW	-107.26464	36.98390
16	Stanolind Oil and Gas	Southern Ute 1	32	6	22	NE	-107.15853	37.00997

Figure 5. Northeast-trending geophysical-log cross section across San Juan Basin showing variability of Ojo Alamo Sandstone in subsurface of San Juan Basin. Geophysical log traces of Ojo Alamo are taken from stratigraphic cross-section A-A′ of Fassett (2000, Plate 1). Shaded areas inside log traces are sandstone; white areas represent mudstone interbeds; lithologies are based on geophysical log interpretation.

the formation (as originally defined), thereby restricting the name to the upper part that was thought to be unequivocally Paleocene in age. This redefinition, however, conflicts with the American Commission on Stratigraphic Nomenclature (1961, 1970) and the North American Commission on Stratigraphic Nomenclature (1983), which state that rock-stratigraphic units must be defined on the basis of lithologic characteristics and not biochronologic criteria. The 1983 code states (p. 856) that "Inferred geologic history, depositional environment, and *biological sequence* have no place in the definition of a lithostratigraphic unit, which must be based on composition and other lithic characteristics" (our italics). Bauer's (1916) definition of the Ojo Alamo Sandstone was based on the lithologic similarities between the lower and upper conglomerates, which have no lithologic similarity to the underlying Kirtland Formation. Thus, the placement of the conglomeratic Naashoibito member in the nonconglomeratic Kirtland Shale cannot be defended in terms of defining rock stratigraphic units on the basis of composition and other lithic characteristics.

Furthermore, the Naashoibito Member is not a lithologically consistent rock unit, even within the type area between Hunter Wash and Barrel Spring. The Naashoibito member was essentially defined to include the rocks between the base of the upper conglomerate and the base of the lower conglomerate. However, both of these have limited lateral continuity, which leads to difficulty and uncertainty in defining them at various localities. We therefore recommend that the names "Naashoibito Member of the Kirtland Shale" and "Ojo Alamo restricted" of Baltz et al. (1966) be abandoned, and that the name Ojo Alamo Sandstone be restored to the usage of Bauer (1916) and as further extended by Reeside (1924), Dane (1936), Baltz (1967), and Fassett and Hinds (1971). Figure 2 shows our definition of the Ojo Alamo Sandstone in the Ojo Alamo type area.

Consequently, we do not support the recent extension of the Naashoibito name to the Betonnie Tsosie Wash area (Kimbeto area of Fig. 1A), as recommended by Lucas and Sullivan (2000a) in their definition of the Naashoibito Member of the Kirtland Formation as being the dinosaur-bearing part of a rock unit previously mapped as Ojo Alamo Sandstone. Lucas and Sullivan (2000a) showed a cross section in the Betonnie Tsosie area (their Fig. 3) <~3.2 km long and consisting of seven measured sections. The lithologic components of their Naashoibito Member of the Kirtland Formation, as depicted on this cross section, are extremely variable, consisting of 100% sandstone at locality A and five sandstone beds separated by multiple mudstone layers at locality E; each of the other five sections shows the Naashoibito Member as containing a distinctly different assortment of sandstone and mudstone beds. The Naashoibito Member has no rock-stratigraphic consistency in the Betonnie Tsosie study area. Although Lucas and Sullivan (2000a) referred to this unit as the "Naashoibito Member of the Kirtland Formation" throughout their report, they concluded with the following paragraph (p. 100).

> This also indicates that the original definition of the Ojo Alamo Sandstone by Bauer (1916) may be the most useful (mappable) definition of a formation-rank unit for basinwide (sic) recognition. The presence of a mud-dominated Naashoibito Member is very localized in the west-central San Juan Basin; elsewhere, the lower part of the Ojo Alamo Sandstone is sandstone dominated. Thus, a basin-wide stratigraphy of the Ojo Alamo Sandstone should recognize it as a formation-rank unit composed locally of a lower, Naashoibito Member of Baltz et. al. (1966) (lower conglomerate and middle shale of Bauer, 1916) and an upper, Kimbeto Member of Powell (1973) (upper conglomerate of Bauer, 1916).

Thus, Lucas and Sullivan (2000a) recommended (as we do) a return to the original definition of the Ojo Alamo Sandstone by Bauer (1916), even though throughout their paper they referred to the Naashoibito as a member of the Kirtland Formation: because of this ambiguity, their intent is unclear. Although we mostly agree with Lucas and Sullivan (2000a), as quoted here, we do not agree that the names "Naashoibito" and "Kimbeto" be used as local member names for the Ojo Alamo in the west-central San Juan Basin. We suggest that, where appropriate, parts of the Ojo Alamo may be referred to informally using terms such as "lower conglomerate" or "middle shale," following Bauer (1916).

Basal contact

The nature of the basal contact of the Ojo Alamo Sandstone is not everywhere clear on the outcrops, primarily because of the lensing nature of the formation's sandstone beds. This is especially troublesome in those places where a lowermost sandstone bed pinches out and the overlying mudstone layer directly overlies mudstones of the underlying Kirtland Formation. Some workers, observing such relations in the outcrops, concluded that in those areas the Ojo Alamo intertongued with the underlying Kirtland. Reeside (1924), however, found that a substantial unconformity was present at the base of the Ojo Alamo. He reached this conclusion because of the thinning of the underlying Kirtland Formation beneath the Ojo Alamo as seen on a series of measured sections around the west and south sides of the basin. However, later workers (e.g., Dane, 1936) concluded that the (apparent) intertonguing of the basal Ojo Alamo and underlying Kirtland demonstrated that there was no unconformity at this contact. Fassett and Hinds (1971, p. 26–31) presented the first detailed subsurface study of the Kirtland–Ojo Alamo contact throughout the San Juan Basin and concluded that Reeside was correct in stating that a substantial unconformity was present at that contact.

Challenges to the unconformable nature of the basal Ojo Alamo contact (as described by Fassett and Hinds, 1971) have been offered by a number of workers, principally Butler et al. (1977) and Lindsay et al. (1978, 1981, 1982), who argued that biostratigraphic and paleomagnetic evidence across the Kirtland–Ojo Alamo contact in the southern San Juan Basin demonstrated that there was no hiatus at the contact and that depo-

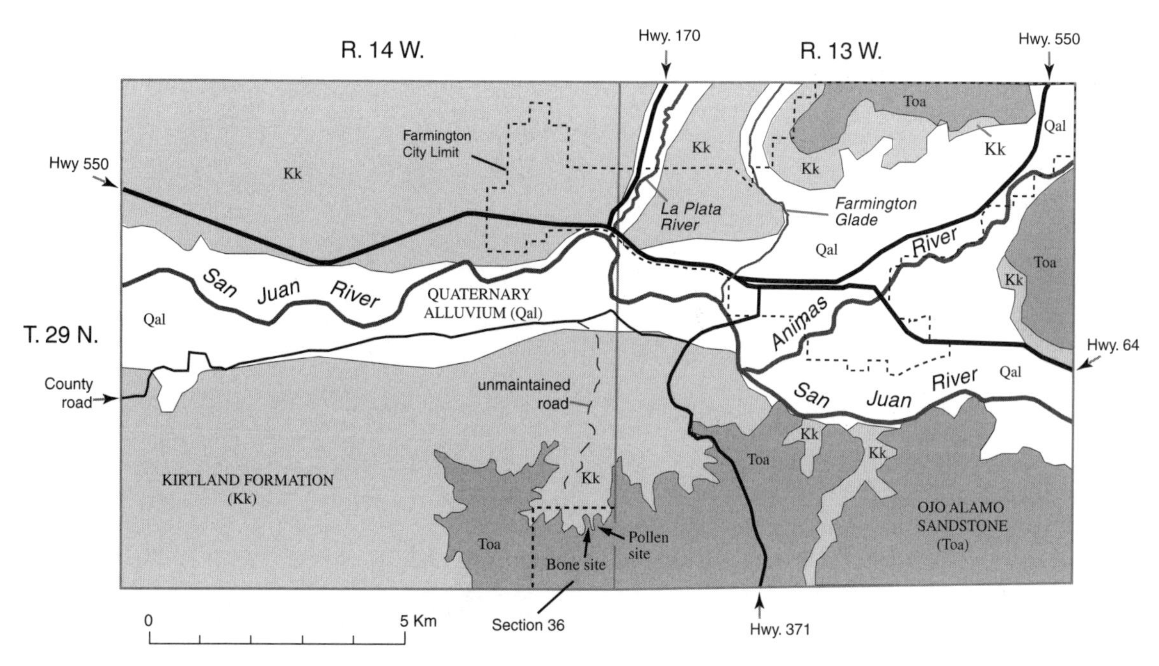

Figure 6. Geologic map of Farmington, New Mexico, area showing location of San Juan River dinosaur bone and pollen sites. Geology is modified from O'Sullivan and Beikman (1963).

sition across this contact had been continuous. Several papers in Fassett and Rigby (1987) directly addressed the nature of the basal Ojo Alamo Sandstone contact and reached the conclusion that there was a substantial unconformity at the base of the Ojo Alamo Sandstone. For example, Fassett (1987, p. 14) estimated that the hiatus separating the Ojo Alamo from the underlying Kirtland represented at least 6 m.y. The duration of the hiatus is now known precisely; radiometric age dates from samples of altered volcanic ash beds in the Kirtland Formation ($^{40}Ar/^{39}Ar$ single-crystal sanidine dates) were reported by Fassett and Steiner (1997), Fassett and Lucas (2000), and Fassett (2000). The hiatus ranges from 8 m.y. in the Hunter Wash area to ~5 m.y. at the San Juan River site. The duration of the hiatus of ~5 m.y. at the San Juan River site is based on the Fruitland-Kirtland section being ~230 m thicker there (Fig. 1) than at Hunter Wash, and the average rate of deposition (undecompacted rock rate) for these rocks is ~75 m/m.y. (Fassett, 2000). Figure 1A shows the locations of these sites and isopach lines showing the thinning of the rocks underlying the Ojo Alamo southeastward across the basin; Figure 1B shows this thinning in cross section.

OCCURRENCES OF PALEOCENE POLLEN UNDERLYING DINOSAUR BONE

San Juan River site

The San Juan River site was discussed by Fassett and Lucas (2000) and a slightly modified version of that discussion is as follows. In 1983 a large dinosaur bone was discovered in the Ojo Alamo Sandstone in the bluffs south of the San Juan River, ~5 km southwest of Farmington, New Mexico (Figs. 1 and 6). This was the first discovery of a dinosaur bone in the Ojo Alamo in the northern part of the San Juan Basin. The site is located in the NE1/4 Sec. 36, T. 29 N., R. 14 W., New Mexico Principal Meridian, San Juan County, New Mexico (Fig. 6). The bone was excavated in the summer of 1983 by a field party headed by Mike O'Neill of the Farmington District Office, Bureau of Land Management, and volunteers Sid Ashe, Dave Thomas, and Brad Peterson. The excavation was conducted under the auspices of the New Mexico Bureau of Mines and Mineral Resources in Socorro, New Mexico. Figure 7A is a photograph of the bone in place as it was first discovered. The bone was ~15 m above the base of the Ojo Alamo in a conglomeratic sandstone that was less well cemented than the overlying, massive, vertical-cliff-forming part of the Ojo Alamo. The softer rock layer containing the bone formed an undercut bench beneath the overlying massive cliff of Ojo Alamo Sandstone (Fig. 7B). This specimen was transported to Albuquerque, where it is now on display in the Geology Museum of the University of New Mexico. In 1999, it was discovered that a slab of Ojo Alamo Sandstone directly overlying the dinosaur bone collection site had fallen away from the cliff face, probably as a result of frost wedging during the previous winter; the site was destroyed (Fig. 8). Figure 9 (A and B) shows the bone in place and the prepared and mounted specimen in about the same orientation: the position of the bone in B is rotated ~45° toward the viewer relative to A. The area of the bone outlined by a white dotted line and containing a white X helps to orient the bone in the figure. The outlined area shows damage by recent erosion, but

A

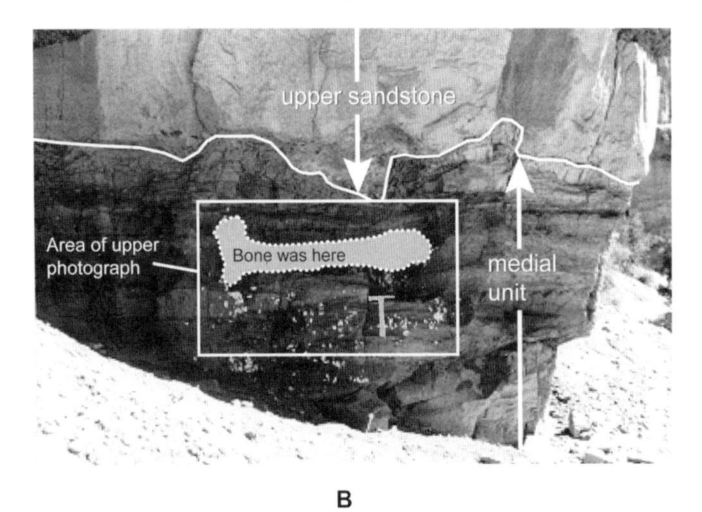

B

Figure 7. A: Photograph of large dinosaur bone in place in Ojo Alamo Sandstone at San Juan River site (photograph from F. Michael O'Neill, Bureau of Land Management, Albuquerque, New Mexico); site location is shown in Figures 1A and 6. B: Photograph of bone site of A showing overlying massive cliff of Ojo Alamo Sandstone and undercut nature of locality from which bone was excavated. White spots are plaster from jacketing bone as it was excavated. Rock hammer is 28.4 cm long.

A

B

Figure 8. Two views of San Juan River dinosaur bone site. A: Site in 1984, 1 yr after bone was excavated; arrow shows place from which bone was excavated. B: Same site in 1999 after slab of Ojo Alamo fell from cliff face above bone collection site. Note rubble from recent rock fall in lower right area of photograph. The small shrub, visible at the base of the columnar-rock column in both photographs, is about 1 m high.

otherwise the surface of the specimen is in remarkably pristine condition, as seen in Figures 9B and 10.

This right femur (Fig. 10) has a maximum length of 1310 mm, a maximum proximal width of 370 mm, and a maximum distal width of 330 mm. The shaft is nearly straight and only widens slightly near the proximal and distal ends. The femoral head is well developed, hemispherical, and offset medially. The large and prominent greater trochanter makes up the proximo-lateral end of the bone. A small, lesser trochanter projects anteriorly from the antero-lateral margin of the greater trochanter. The large fourth trochanter is a rounded flange of bone on the postero-medial edge at about the middle of the shaft. The distal condyles are prominent and rounded postero-distally. The medial condyle is the larger of the two, and they are separated by sulci anteriorly and posteriorly. The cortical surface of the bone is very well preserved, and shows numerous fine striae and rugosities for muscle attachment. All visible damage to the bone appears to have been the result of recent weathering on the outcrop, principally the area outlined in white containing a white X in Figure 10D.

The bone is the right femur of a hadrosaur (cf. Lull and Wright, 1942, p. 90, Fig. 24, Plate 6; Brett-Surman, 1975, Plate 7; Weishampel and Horner, 1990, p. 551–552, Fig. 26.10). However, it does not provide a more precise identification than Hadrosauridae, because the genus- and species-level taxonomy of hadrosaurs is based on cranial features. The San Juan River hadrosaur femur is ~8%–12% longer than the longest hadrosaur femora listed by Lull and Wright (1942, their Table 5), those of *Kritosaurus* and *Saurolophus*, 1140 and 1150 mm long, respectively. However, it is shorter than the femur of the largest

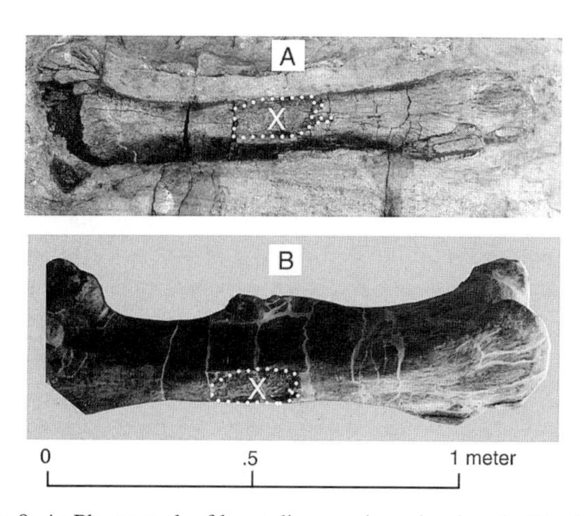

Figure 9. A: Photograph of large dinosaur bone in place in Ojo Alamo Sandstone at San Juan River site; site location is shown in Figures 1A and 6. B: Bone after preparation and mounting. Photograph is shown in about same position as bone in place; note area of damage to bone by recent erosion (outlined by dotted white line and containing white X) in A and B. View in B shows bone rotated ~45° toward viewer from view in A.

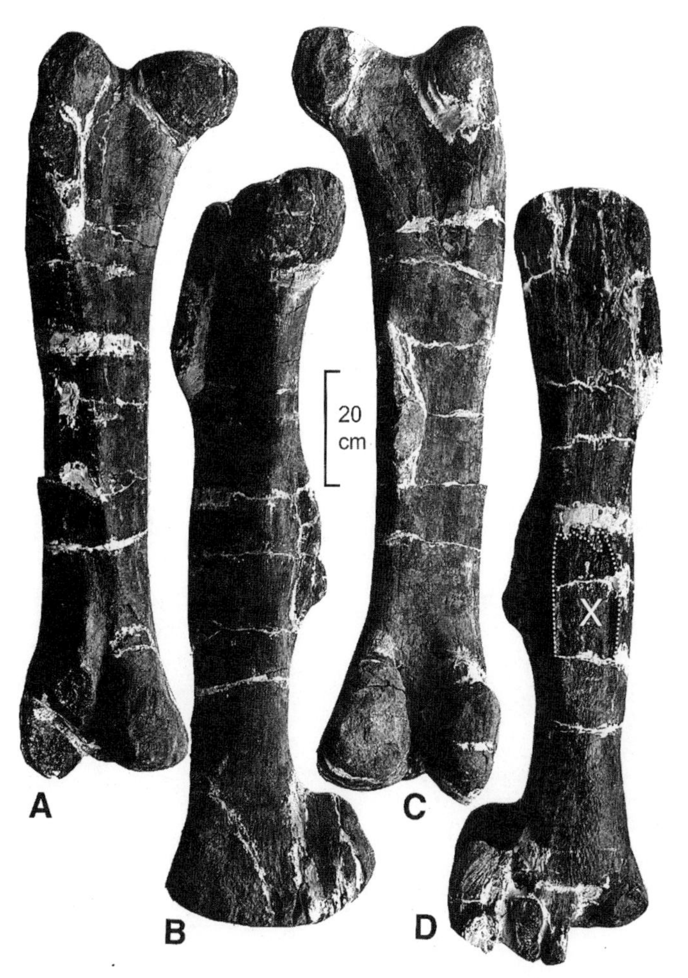

Figure 10. Right femur of hadrosaurian dinosaur from Ojo Alamo Sandstone at San Juan River site showing anterior (A), medial (B), posterior (C), and lateral (D) views. Note area of damage to bone by recent erosion (outlined by dotted white line containing white X in D); damaged area is also seen in Figure 9. Photographs by H. Foster, courtesy of F. Michael O'Neill, Bureau of Land Management, Albuquerque, New Mexico.

hadrosaur, *Shantungosaurus giganteus* from the Upper Cretaceous of China, which is ~1650 mm long (Hu, 1973).

The photograph of the bone in place (Figs. 7A and 9A) was published in Fassett et al. (1987, their Fig. 3B) but at that time the age of the Ojo Alamo Sandstone had not been directly determined at this site, or at any outcrop in the northern part of the San Juan Basin. In 1985, a coaly, carbonaceous shale bed was discovered in the Ojo Alamo Sandstone ~160 m east of the dinosaur bone locality (Figs. 6 and 11), 3 m stratigraphically below the level from which the bone had been excavated. Three samples of this coaly bed (Fig. 11B) were collected and analyzed for their pollen and spore content by Douglas J. Nichols (1994, written commun.) and found to contain a diverse palynomorph assemblage (Table 1), including *Brevicolporites colpella* and *Momipites tenuipolis* (photographs 3, 4, and 12–14, Fig. 12), indicating a Paleocene age for these rocks. Figure 12 shows photographs from Anderson (1960) of some of the palynomorphs identified from the San Juan River site. *M. tenuipolis* was also reported by Anderson (1960) in his Nacimiento 1 and 2 florules collected from the Nacimiento Formation near Cuba (Fig. 1A), 30 cm and 35 m, respectively, above the top of the Ojo Alamo Sandstone. Anderson (*in* Baltz et al., 1966) also reported the presence of this Paleocene index fossil from the upper part of the Ojo Alamo Sandstone near Barrel Spring. Tschudy (1973) reported the presence of *M. tenuipolis* in the Ojo Alamo Sandstone in the Gasbuggy core (Fig. 1A), at Mesa Portales (*in* Fassett and Hinds, 1971), and near Barrel Spring (*in* Fassett et al., 1987).

Figure 13A is a composite stratigraphic column showing the stratigraphy of the lower part of the Ojo Alamo Sandstone at the San Juan River site. Figure 13B is a stratigraphic column of the entire thickness of the Ojo Alamo Sandstone as measured by Reeside (1924) near the San Juan River and shows the relative position of the bone-bearing lowermost sandstone bed of the Ojo Alamo in this section. The data from the San Juan River site demonstrate that the large hadrosaur femur found there was preserved in rocks of Paleocene age. Because this is a single bone, however, the question of possible reworking from the underlying Kirtland Formation of Cretaceous age must be addressed. We find that possibility highly unlikely because of the following. (1) The base of the Ojo Alamo Sandstone is a planar surface in this area; there are no local topographic highs in the underlying Cretaceous strata extending 15 m upward into the Ojo Alamo to the level where the dinosaur bone was found, and the isopach lines of Figure 1A also indicate a relatively flat surface at the base of the Ojo Alamo in the vicinity of the San Juan River site. (2) Because the gradient of the pre-Ojo Alamo erosion surface sloped to the south, uppermost Kirtland strata

A

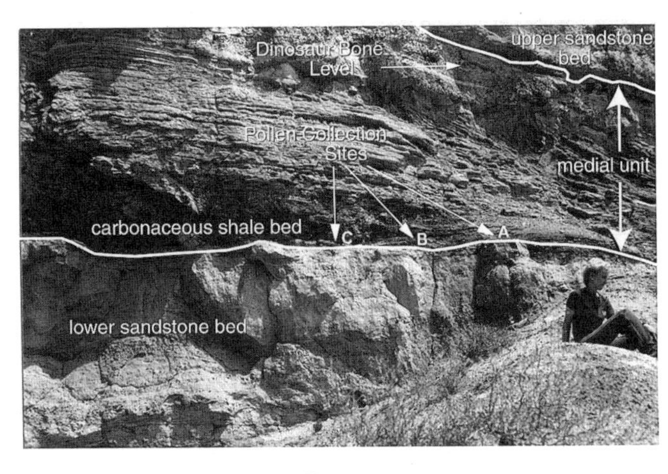

B

Figure 11. A: San Juan River pollen site in cliff of Ojo Alamo Sandstone; view is looking east from dinosaur bone site of Figure 7; distance between bone and pollen sites is ~160 m. Medial unit can be traced continuously along outcrop from bone site to pollen site. Location is shown in Figures 1A and 6. Medial unit is about 3.5 m thick in the area of lower photo. B: San Juan River pollen site showing coaly, carbonaceous shale bed from which samples A, B, and C were collected for palynologic evaluation; bed is at base of medial unit of Ojo Alamo Sandstone. Stratigraphic level of bone collected from top of this medial unit (Fig. 7), 160 m to west is shown.

must have been at the same level as the bone at the San Juan River site several kilometers to the northwest; however, given the size of this bone and its silicified weight of more than 130 kg, it would have been practically impossible for this bone to have been transported intact even a few meters, not to mention tens of kilometers southward to the San Juan River site. (3) The bone (Figs. 9B and 10) has a pristine outer surface with no abrasions or scratches, and all of its delicate features are intact. There is thus no evidence of significant movement of this bone.

We suggest that this animal lived in early, but not earliest, Paleocene time and died near the place where this silicified femur was found. As the corpse decayed, predators and river currents disarticulated the skeleton, dispersing the lighter elements, and leaving this large massive bone behind to be quickly buried and mineralized in place.

Barrel Spring site

No palynomorphs have been found in the middle shale unit that contains numerous dinosaur fossils in the Ojo Alamo Sandstone in the type area (Fig. 1A); however, a palynomorph assemblage (collection 1) from the upper part of the Ojo Alamo was reported by Anderson (in Baltz et al., 1966, p. D17) from "a bed of lignitic shale enclosed between the upper and lower parts of the restricted Ojo Alamo Sandstone . . . about one-eighth mile north of Barrel Spring." (See Fig. 14 for the collection-site localities and Fig. 20 for the stratigraphic positions of pollen collections in the type area of the Ojo Alamo Sandstone.) Baltz et al. (1966, p. D17) stated that this collection "contains forms that are known to occur also in the Fort Union Formation and other formations generally considered to be of Paleocene age" but then concluded that "The palynology does not directly fix the age of the restricted Ojo Alamo Sandstone, which, however, appears to be Paleocene on the basis of its gradational and intertonguing relation with the Nacimiento."

Tschudy (in Fassett et al., 1987, p. 27) listed and discussed palynomorphs recovered from a carbonaceous mudstone lens in the upper part of the Ojo Alamo ~500 m east of Barrel Spring, and concluded "This assemblage is clearly of Paleocene age." (It is not clear if this sample site is the same one reported by Baltz et al. [1966], but it may not be, based on the different directions and distances given from Barrel Spring.) Tschudy stated (Fassett et al., 1987, p. 27) that the assemblage of Palynomorphs from a carbonaceous mudstone bed a few meters below the base of the Ojo Alamo Sandstone near Barrel Spring was Cretaceous in age and "is latest Campanian or early Maastrichtian."

Additional rock samples for palynologic analysis were collected in 1985 in the Barrel Spring area. One of these samples was from the same mudstone bed in the upper Ojo Alamo Sandstone that was analyzed by Tschudy. D.J. Nichols (1994, written commun.) agreed with Tschudy's findings that palynomorphs from this rock unit indicated a Paleocene age (sample 24-3C, Table 2). A second sample analyzed by Nichols from a coaly, carbonaceous mudstone bed, <1 m beneath the base of the Ojo Alamo (Fig. 15), yielded abundant palynomorphs, including Paleocene and Cretaceous forms (sample 24-5, Table 2). The palynomorph assemblage from this sample is remarkably similar to the assemblage found at the San Juan River site and in the upper part of the Ojo Alamo Sandstone northeast of Barrel Spring (Table 2), except that it contains two species of the Cretaceous index fossil *Proteacidites*. This pollen assemblage is undoubtedly Paleocene. The *Proteacidites* specimens in this assemblage must be reworked from underlying Cretaceous strata: the reworking of some Cretaceous palynomorphs into this Paleocene assemblage is not unexpected. The early Paleocene swamp in which indigenous Paleocene pollen was accumulat-

TABLE 1. PALYNOMORPHS IDENTIFIED FROM THE OJO ALAMO SANDSTONE AT THREE LOCALITIES, SAN JUAN RIVER SITE, SAN JUAN BASIN, NEW MEXICO

Sample 25Ga	Sample 25Gb	Sample 25Gc
Arecipites reticulatus	*Arecipites reticulatus*	*Arecipites reticulatus*
	Arecipites sp.	
		Azolla cretacea
	Brevicolporites colpella	
		Chenopodipollis sp.
		Cicatricosisporites spp.
Corollina torosa (incl. monads and tetrads)	*Corollina torosa* (incl. monads and tetrads)	*Corollina torosa* (incl. monads and tetrads)
Cupanieidites sp.		
Cupuliferoidaepollenites minutus	*Cupuliferoidaepollenites minutus*	
	Cyathidites minor	
Fraxinoipollenites variabilis	*Fraxinoipollenites variabilis*	*Fraxinoipollenites variabilis*
	Laevigatosporites sp.	*Laevigatosporites* sp.
Momipites inaequalis	*Momipites inaequalis*	*Momipites inaequalis*
Momipites tenuipolus	*Momipites tenuipolus*	*Momipites tenuipolus*
Nyssapollenites sp.		*Nyssapollenites* sp.
		"Palaeoisoetes" sp.
		"Paliurus" triplicatus
	Pandaniidites typicus	*Pandaniidites typicus*
Pityosporites sp.	*Pityosporites* sp.	*Pityosporites* sp.
Taxodiaceaepollenites hiatus	*Taxodiaceaepollenites hiatus*	
	Tricolpites sp.	
	Triporoletes novomexicanum	
Ulmipollenites krempii	*Ulmipollenites krempii*	*Ulmipollenites krempii*
	U. tricostatus	*U. tricostatus*

Note: Palynomorphs identified by D.G. Nichols, U.S. Geological Survey, Denver, Colorado (1994 written commun.). Sample sites are 12 m above the base of the Ojo Alamo Sandstone and 3 m below the level of a large hadrosaur femur collected from the San Juan River site. Photographs of the hadrosaur femur are shown in Figures 7, 9, 10; sample-site localities are shown in Figure 6.

ing was located on an erosion surface (peneplain) on Kirtland Formation strata of Campanian age, and Cretaceous pollen could easily have been transported laterally a few meters to a few tens of meters across this surface in wind-blown dust and deposited in the Paleocene swamp.

Another rock sample was collected from the lower part of the same carbonaceous, coaly shale bed at the Barrel Spring site ~3 m below the base of the Ojo Alamo Sandstone (Fig. 15). (This may be the same site from which the Cretaceous assemblage came that was reported on by Tschudy [*in* Fassett et al., 1987].) The palynomorphs identified from this sample are listed in Table 2 (sample 0430200), and D.J Nichols (2000, written commun.) concluded that the assemblage is Late Cretaceous; probably upper Campanian to lower Maastrichtian. Additional samples from the same level in this bed (a few meters below the base of the Ojo Alamo) a few hundred meters northwest of the 0430200 sample site also yielded Campanian to lower Maastrichtian palynomorphs (D.J. Nichols, 2000, written commun.).

Figure 15 is a photograph of the Barrel Spring site showing the location of the 1985 sample site near the base of the Ojo Alamo Sandstone on the left, and the location of the April 2000 collection site, nearly 3 m beneath the base of the Ojo Alamo, on the right. Barrel Spring is located ~200 m east (left of photo) of the 1985 sample site at the head of a short, narrow, side canyon where the spring issues from the lower part of the Ojo Alamo Sandstone (Figs. 14 and 16). The spring is not labeled on the Alamo Mesa East 1/24 000 USGS topographic quadrangle, but it is in a short side canyon that is clearly visible on this map in the west-central part of sec. 16, T. 24 N., R. 11 W. The two pollen sites shown in Figure 15 are in the east central part of sec. 17, T. 24 N., R. 11 W.

Figure 16 is a photograph of a trench that was excavated through the rock interval from the April 2000 sample site to the projected level of the 1985 sample site, just beneath the base of the Ojo Alamo. The trench was cut to attempt to locate a lithologic break marking the Cretaceous-Tertiary (K-T) boundary between the Cretaceous and Paleocene pollen-sample sites. (The K-T boundary must be present within the 2–3 m interval between the two sample localities shown in Figs. 15 and 16.) The April 2000 sample site was from the coaly layer that forms the prominent black band on the exposed Kirtland rock face ~3 m below the base of the Ojo Alamo Sandstone on the right side of the photograph in Figure 15. The only distinct lithologic break that is visible in this exposure is at the top of the coaly, carbonaceous shale layer from which the April 2000 sample was collected; thus, the K-T interface is tentatively placed at the top of this coaly layer.

Eight additional samples of carbonaceous shale were subsequently collected from sample sites directly underlying the Ojo Alamo Sandstone. Three of these samples were from near the top of the trench shown in Figure 16 and the other samples were from sites between the trench site and Barrel Spring. None of these samples contained diagnostic palynomorphs (D.J. Nichols, 2000, written commun.).

It is interesting to note that the K-T boundary at Barrel

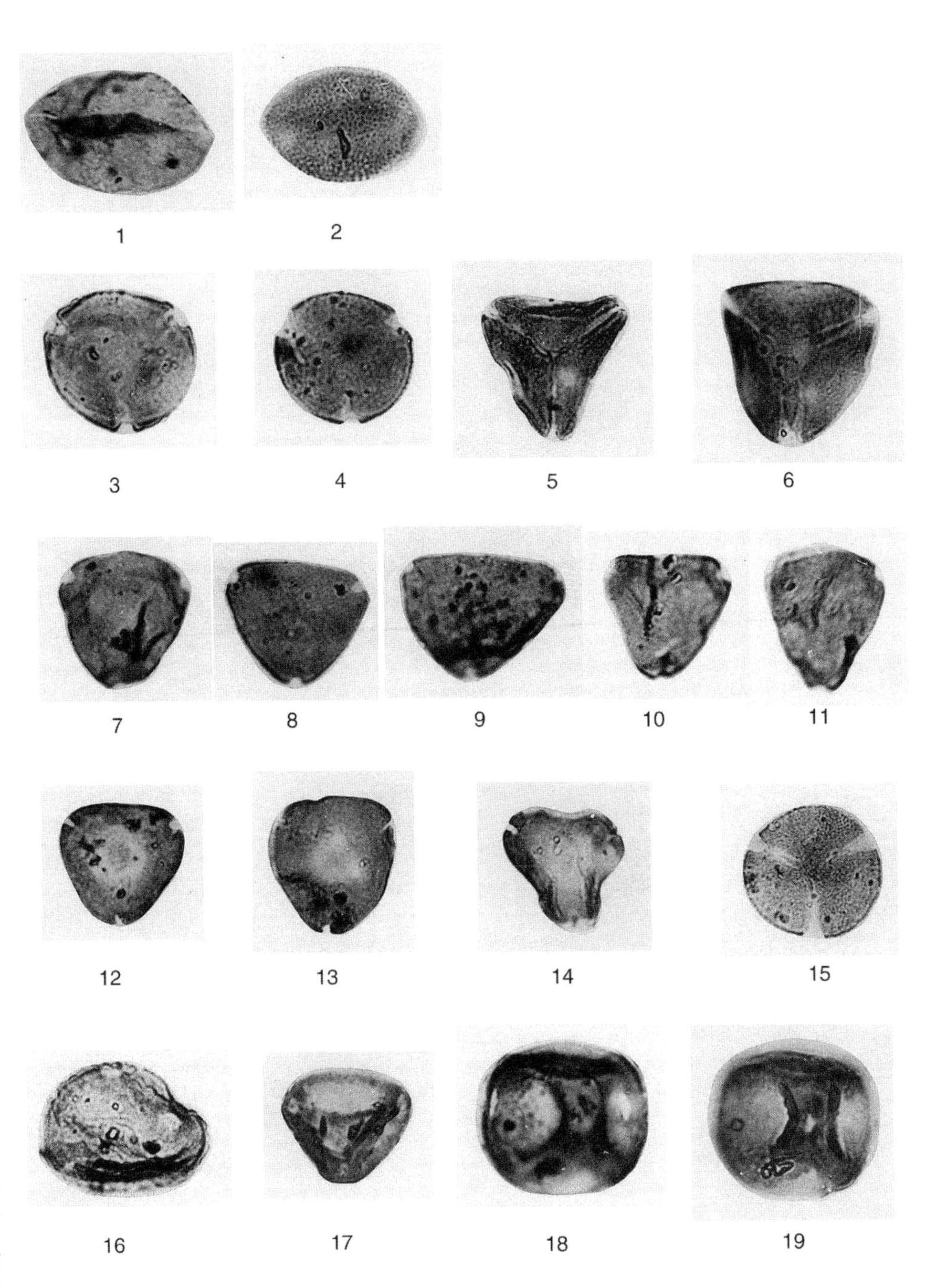

Figure 12. Photographs of palynomorphs from Paleocene rock samples collected at sites near Cuba, New Mexico, in southeast San Juan Basin (from Anderson, 1960). These same forms were identified by D.J. Nichols (USGS, Denver, Colorado) in rock samples from Ojo Alamo Sandstone at San Juan River pollen site (Figs. 1 and 6) and Barrel Spring site (Figs. 1 and 14). Anderson's Ojo Alamo florule 1 is from base of Ojo Alamo, Ojo Alamo florule 2 is in middle of Ojo Alamo, Nacimiento florule 1 is 30 cm above top of Ojo Alamo, and Nacimiento florule 2 is 35 m above top of Ojo Alamo. Following plates and figures refer to Anderson (1960). 1, *Arecipites reticulates* (Plate 7, Fig. 2; Nacimiento 1 florule, 14 × 23 µm). 2, *Arecipites reticulatus* (Plate 8, Fig. 3; Nacimiento 2 florule, 14 × 21 µm). 3, *Brevicolporites colpella* (Plate 6, Fig. 11; Ojo Alamo 2 florule, 21 µm). 4, *Brevicolporites colpella* (Plate 6, Fig. 12; Ojo Alamo 2 florule, 20 µm). 5, *Cupanieidites* sp. (Plate 8, Fig. 10; Nacimiento 2 florule, diameter 20 µm). 6, *Cupanieidites* sp. (Plate 8, Fig. 11; Nacimiento 2 florule, diameter 22 µm). 7, *Momipites inaequalis* (Plate 7, Fig. 13; Nacimiento 1 florule, diameter 19 µm). 8, *Momipites inaequalis* (Plate 6, Fig. 7;: Ojo Alamo 2 florule, diameter 24 µm. 9, *Momipites inaequalis* (Plate 6, Fig. 8: Ojo Alamo 2 florule, diameter 22 µm). 10, *Momipites inaequalis* (Plate 6, Fig. 9; Ojo Alamo 2 florule, diameter 19 µm). 11, *Momipites inaequalis* (Plate 6, Fig. 10; Ojo Alamo 2 florule, diameter 16 µm). 12, *Momipites tenuipolus* (Plate 7, Fig. 14; Nacimiento 1 florule, diameter 17 µm). 13, *Momipites tenuipolus* (Plate 8, Fig. 14; Nacimiento 2 florule, diameter 18 µm). 14, *Momipites tenuipolus* (Plate 8, Fig. 15; Nacimiento 1 florule, diameter 16 µm). 15, *Tricolpites* sp. (Plate 6, Fig. 18; Ojo Alamo 2 florule, diameter 25 µm). 16, *Ulmipollenites krempii* (*Ulmoideipites krempii* of Anderson) (Plate 4, Fig. 12; Ojo Alamo 1 florule, diameter 23 µm). 17, *Ulmipollenites tricostatus* (*Ulmoideipites tricostatus* of Anderson) (Plate 7, Fig. 8; Nacimiento 1 florule, diameter 17 µm). 18, *Ulmipollenites tricostatus* (*Ulmoideipites tricostatus of Anderson*) (Plate 8, Fig. 8; Nacimiento 2 florule, diameter 22 µm). 19, *Ulmipollenites tricostatus* (*Ulmoideipites tricostatus* of Anderson) (Plate 8, Fig. 9; Nacimiento 2 florule, diameter 22 µm).

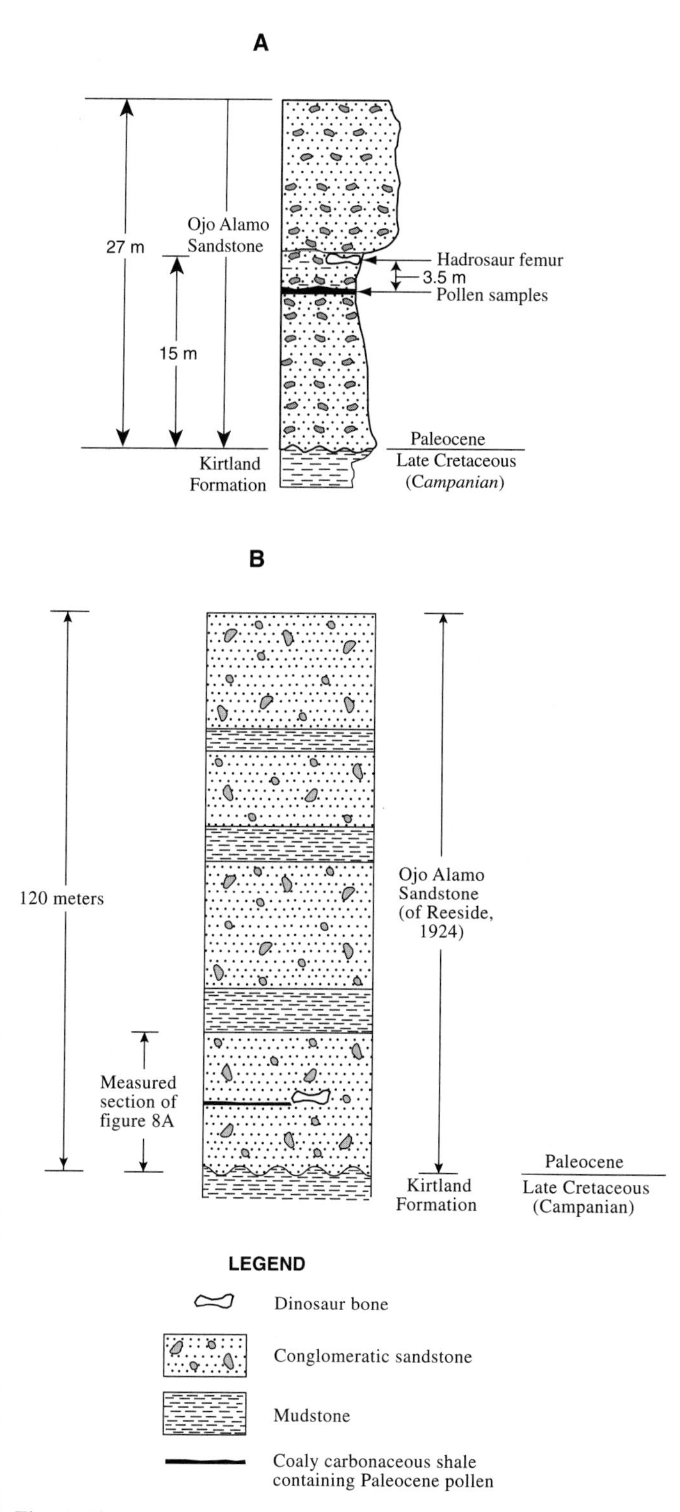

Figure 13. A: Composite stratigraphic column of Ojo Alamo Sandstone at San Juan River site showing relative stratigraphic positions of dinosaur bone and pollen sample collection sites. (Locations of sites shown in Fig. 6; photographs of bone site in Figs. 7 and 8; photograph of pollen sites is in Fig. 11.) B: Stratigraphic column showing full thickness of Ojo Alamo Sandstone near San Juan River as measured and described by Reeside (1924) and showing relative positions of dinosaur bone and pollen samples.

Spring is located within the uppermost bed of the rock-stratigraphic Kirtland Formation and not precisely at the Kirtland-Ojo Alamo contact; thus the Kirtland there is both Cretaceous and Tertiary in age. There is little doubt that the 8 m.y. unconformity is at the K-T boundary at this exposure; thus the carbonaceous shale bed containing Paleocene palynomorphs, <1 m below the base of the Ojo Alamo, must have been deposited on the pre-Ojo Alamo erosion surface in early (but not earliest) Paleocene time. Fassett and Hinds (1971, p. 33) found a similar relationship at Mesa Portales (Figs. 1A and 3) in the southeast part of the basin, where Paleocene-age uppermost Kirtland mudstone overlies similar looking Cretaceous mudstone. Fassett and Hinds pointed out that the K-T boundary is probably present at the base of the Ojo Alamo Sandstone in most parts of the basin, but in other places, such as at Mesa Portales and the Barrel Spring site, the first sediments deposited in early Paleocene time (as the San Juan Basin began to subside again, following an episode of uplift) were mudstones deposited shortly before Ojo Alamo fluvial channels reached those areas. It seems clear that the K-T boundary, where it is not at the base of the Ojo Alamo, is not far below the base of this rock unit, but without detailed, closely spaced sampling for palynological analysis, this contact may be virtually unidentifiable in places solely on the basis of lithology. Fassett and Hinds (1971, p. 33) noted, however, that in some places in the basin "a bed of rusty-brown friable coarse-grained, in some places conglomeratic, sandstone as much as several inches thick is present below and near the base of the lowermost sandstone bed of the Ojo Alamo. This conglomeratic sandstone may be a lag deposit and could mark the Tertiary-Cretaceous boundary in some areas. Unfortunately, this sandstone bed is usually not well exposed, and it is difficult to trace." No such bed has been found at the Barrel Spring locality.

The identification of Paleocene palynomorphs in the uppermost part of the Kirtland Formation at the Barrel Spring locality (Fig. 14) provides compelling evidence that all of the many dinosaur bones that have been discovered in the Ojo Alamo Sandstone in the type area are Paleocene in age, as posited by Reeside (1924), Anderson (1960), Fassett (1982, 2000), Fassett et al. (1987, 2000), and Fassett and Lucas (2000). One of the difficulties in assessing the significance of dinosaur fossils in the Ojo Alamo Sandstone has been the incorrect and/or uncertain assignment of stratigraphic provenance for some of the specimens described and collected from the Kirtland and Ojo Alamo, especially some of the older collections. Hunt and Lucas (1992) reviewed the known literature regarding dinosaur bone discoveries from the Fruitland and Kirtland strata in the San Juan Basin and attempted to place all of the described specimens in their proper stratigraphic setting. They used the stratigraphic nomenclature of Baltz et al. (1966) (Fig. 2 herein). Their assignment of dinosaur fossils to the Naashoibito Member of the Kirtland Formation is therefore equivalent to assigning them to the Ojo Alamo Sandstone as we have defined that formation herein.

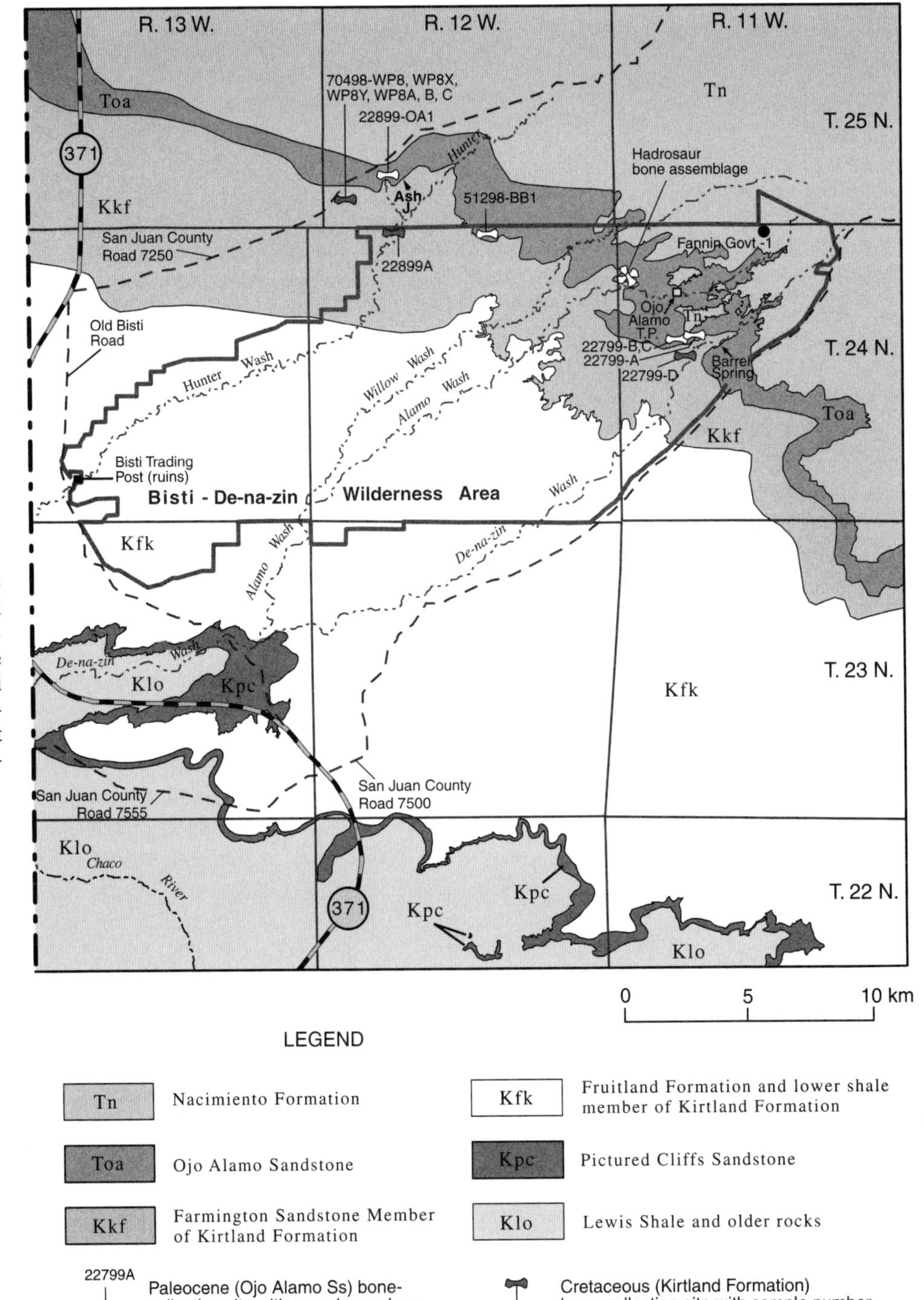

Figure 14. Geologic map of Ojo Alamo Sandstone type area (Fig. 1A). Altered volcanic ash bed and dinosaur bone collection sites are shown. Sites where samples were collected for pollen and spore analyses are all in vicinity of Barrel Spring (see Figs. 15 and 16 for exact locations of critical sites). Map is modified from Fassett (2000, Fig. 9).

The work of Hunt and Lucas (1992) has been updated, principally by Lucas et al. (2000) and Williamson (2000). The following list of dinosaurs from the Ojo Alamo Sandstone (Naashoibito Member of the Kirtland Formation) from Lucas et al. (2000, p. 88), is considered to be the most accurate now available. This list includes dinosaurs from the Kimbeto site (Fig. 1A), and all but one of these identifications is based on a single bone specimen. Some of the older collections may contain additional Ojo Alamo dinosaurs, but only the specimens that could unquestionably be placed within the Ojo Alamo Sandstone are listed: ornithomimid, indeterminate; dromaeosaurid, indeterminate; saurornithoidids, indeterminate; ?*Alber-*

TABLE 2. PALYNOMORPHS IDENTIFIED FROM SAN JUAN RIVER AND BARREL SPRING SITES, SAN JUAN BASIN, NEW MEXICO

Samples 25Ga, b, c, composite list from San Juan River site, 1985, 12 m above base O.A., Paleocene	Sample 24-5 from Barrel Spring site, 1985, <1 m below base of O.A., Paleocene	Sample 24-3C, from northeast of Barrel Spring, 1985, ~15 m above base O.A., Paleocene	Sample 04302000, from Barrel Spring site, April 2000, 3 m below base O.A., Campanian–early Maastrichtian
	Araucariacites australis		
Arecipites reticulatus	*Arecipites reticulatus*		*Arecipites reticulatus*
Arecipites sp.		*Arecipites* sp.	
Azolla cretacea	*Azolla cretacea*		
Brevicolporites colpella		*Brevicolporites colpella*	
		Cercidiphyllites sp.	
Chenopodipollis sp.			
Cicatricosisporites spp.			
Corollina torosa (incl. Monads and tetrads)	*Corollina torosa*	*Corollina torosa*	
Cupanieidites sp.	*Cupanieidites* sp.	*Cupanieidites* sp.	
Cupuliferoidaepollenites minutus	*Cupuliferoidaepollenites minutus*	*Cupuliferoidaepollenites minutus*	
Cyathidites minor			*Cyathidites* sp.
	Dyadonapites reticulatus		*Dyadonapites reticulatus*
	Ghoshispora sp.		
Fraxinoipollenites variabilis		*Fraxinoipollenites variabilis* /*Ghoshispora* sp.	
Laevigatosporites sp.	*Laevigatosporites* sp.	*Laevigatosporites* sp.	
	Liliacidites leei		*Liliacidites leei*
	Liliacidites sp. of Anderson		
Momipites inaequalis	*Momipites inaequalis*	*Momipites inaequalis*	
Momipites tenuipolus	*Momipites tenuipolus*	*Momipites tenuipolus*	
	Momipites sp.	*Momipites* sp.	
	Osmundacidites wellmannii		
Nyssapollenites sp.			
"Palaeoisoetes" sp.	*"Palaeoisoetes"*	*"Palaeoisoetes"* sp.	
"Paliurus" triplicatus			
Pandaniidites typicus	*Pandaniidites typicus*	*Pandaniidites typicus*	*Pandaniidites typicus*
Pityosporites sp.	*Pityosporites* sp.	*Pityosporites* sp.	*Pityosporites* spp.
		Psilastephanocolpites sp.	
		"Quercus" explanata	
		Rectosulcites latus	
		Syncolporites minimus	
	Proteacidites retusus		*Proteacidites retusus*
	Proteacidites thalmannii		*Proteacidites thalmannii*
	Rhoipites sp.		
Taxodiaceaepollenites hiatus	*Taxodiaceaepollenites hiatus*		
	Tetraporina sp.		
		Tricolpites anguloluminosus	
	Tricolpites? sp. cf. *Gunnera*		
Tricolpites sp.	*Tricolpites* spp.		
			Tricolpites interangulus
Triporoletes novomexicanum			
Ulmipollenites krempii	*Ulmipollenites krempii*	*Ulmipollenites krempii*	
U. tricostatus		*U. tricostatus*	*Ulmipollenites ("Ulmoideipites") tricostatus*

Note: Palynomorphs identified by D.G. Nichols, U.S. Geological Survey, Denver, Colorado (1994, 1998, 2000, written communs.). O.A. = Ojo Alamo Sandstone, sample collection sites in Figures 1, 6, 14, and 15.

tosaurus sp., cf.; *Tyrannosaurus* sp.; *Alamosaurus sanjuanensis*; ankylosaurid, indeterminate; nodosaurids, indeterminate; *Torosaurus* cf. *T. latus*; *Pentaceratops*; hadrosaurids, indeterminate.

We discuss a method to uniquely identify the formation of origin of some of the older, uncertain, specimens in a following section. One of the hadrosaur collections consists of 34 skeletal elements and is discussed in more detail.

Lucas et al. (2000, p. 88) characterized this assemblage as "a limited, but distinctive dinosaur fauna" and stated "three opinions have been advanced as to its precise age. . . . The Naashoibito [Ojo Alamo] fauna is of Maastrichtian age, or Lancian in terms of vertebrate biochronology. . . . The Naashoibito [Ojo Alamo] fauna is late Campanian, discounting the identifications of Lancian-age indicators such as Torosaurus and Tyrannosaurus. . . . The Naashoibito [Ojo Alamo] dinosaurs [are] Paleocene in age."

Lucas et al. characterized the three opinions as conventional, reasonable, and unreasonable, respectively, but refrained from suggesting specifically which of the three opinions they considered to be correct. Photographs of some of the dinosaur bone specimens collected from the Ojo Alamo Sandstone and listed here are in the following: *Tyranosaurus*, Lucas and Sullivan (2000a, Fig. 5K); *Alamosaurus sanjunanensis*, Lucas and

Figure 15. Photograph showing location of two Barrel Spring sample collection sites for pollen and spore analysis; view is looking southwest in De-na-zin Wash. Trench labeled in right center of photograph is shown in Figure 16. Mouth of small side canyon leading to Barrel Spring is on left side of photograph. Distance between base of Ojo Alamo Sandstone and prominent black coaly bed in area where arrow points to upper end of trench is 3 m thick, for scale.

Sullivan (2000b, Figs. 2 and 6); hadrosaurid, Williamson (2000, Fig. 13). Drawings of Nodosaurid elements from the Ojo Alamo are in Ford (2000, Figs. 7 and 13). Photographs of other Ojo Alamo specimens are in Lucas et al. (1987), as follows: *Albertosaurus* and *Tyranosaurus* (Fig. 3), *Alamosaurus sanjuanensis* (Fig. 4, A and B), nodosaurid (Fig. 4, D and E), *ankylosaurid* (Fig. 4, F and G), *Taurosaurus* (Fig. 5, A–D), and *Pentaceratops* (Fig. 5, E and F).

OTHER SITES

Pot Mesa site

The Pot Mesa site (Figs. 1 and 3) was discussed by Fassett et al. (1987, p. 28, 29). They reported that the definition of the Ojo Alamo at Pot Mesa was controversial because Dane (1936) had mapped the Ojo Alamo as a single sandstone bed, whereas Scott et al. (1980) had mapped the Ojo Alamo as two sandstone beds; an upper, more continuous sandstone bed (Dane's [1936] Ojo Alamo) and a lower, lensing sandstone bed that Dane had mapped as part of the Kirtland Formation. The unique identification of all of the lithologic components of the Ojo Alamo Sandstone in the eastern part of the basin, where it does not contain its characteristic conglomerates that are so distinctive to the west, can be difficult in places, such as Pot Mesa where the Ojo Alamo is relatively thin. Without the aid of biochronologic evidence in an area such as Pot Mesa, there can be no positive identification of lensing sandstone beds that underlie the more continuous upper sandstones of the Ojo Alamo Sandstone. Nevertheless, Fassett et al. (1987) were convinced, on the basis of physical stratigraphy, that the Scott et al. (1980) mapping was correct and that the Ojo Alamo consisted of two sandstone beds at Pot Mesa (Fig. 17).

A rock sample collected from a carbonaceous siltstone ~1.5 m above the top of the lowermost sandstone bed (Fig. 17) included in the Ojo Alamo by Scott et al. (1980) and Fassett et al. (1987) was analyzed by D.J. Nichols (1994, written commun.), who concluded that "Based on the diverse and well-preserved assemblage, the sample is clearly of latest Cretaceous age." On the basis of this evidence, we conclude that the lower, less continuous, sandstone bed is part of the Kirtland Formation and that Dane's (1936) mapping of the Ojo Alamo Sandstone at Pot Mesa was correct.

Dinosaur bones in the Ojo Alamo at the Pot Mesa site were discussed by Fassett et al. (1987), who incorrectly assigned some of those fossils to the Ojo Alamo; the correct placement is shown in Figure 17. A photograph of the large bone in the basal part of the Ojo Alamo (as redefined at Pot Mesa herein) is shown as it was discovered by Fassett et al. (1987, Fig. 11). This specimen has been collected and given the number NMTOA-2. Even though this specimen was fragmented by weathering, the outer cortical surfaces appeared to be unabraded by transport.

Kimbeto site

The discovery of dinosaur bone in the Ojo Alamo Sandstone in the Kimbeto area (Lucas and Sullivan, 2000d) has extended the area of such occurrences 30 km southeast of the Ojo Alamo type area and fills in a large gap between the Barrel Spring and Pot Mesa sites (Figs. 1A and 3). Photographs of the Kimbeto fossils are in Lucas and Heckert (2000). An organic-rich mudstone bed was collected from within the Ojo Alamo

Figure 16. Photograph of trench cut into outcrop of upper part of Kirtland Formation in Barrel Spring area, showing pollen collection levels. Area of trench is on right side of photograph in Figure 15. Cretaceous-Tertiary interface is tentatively placed at top of coaly carbonaceous shale layer in lower part of trench. Alcove where Barrel Spring is located is visible in upper left corner of this photograph. Hammer is 29.8 cm long.

below the level of a dinosaur bone collected from the Kimbeto site (illustrated in Lucas and Sullivan, 2000d, Fig. 8K) and D.J. Nichols (1998, written commun.) stated that the samples contained a sparse assemblage of palynomorphs indicating deposition in a freshwater pond or boggy area, but that none of the species identified were age diagnostic.

REWORKING QUESTION

Physical evidence

Some earlier workers, confounded by the presence of dinosaurs, Cretaceous index fossils, in a rock unit that was apparently Paleocene in age, suggested that these fossils were reworked from older, Cretaceous strata (see discussion in Clemens, 1973). Clemens (1973, p. 160) addressed the reworking question in the type area of the Ojo Alamo and concluded the following.

> Most of the fossils [in the Ojo Alamo] are isolated bones, albeit some of them are large or not particularly durable elements to be moved by a high-energy stream. Also, we do not know if, "a few fragmentary vertebrae of a very large carnivorous dinosaur . . . collected by J.B. Reeside, Jr. . . . (Gilmore 1919, p. 67)" were articulated. However, even their association in the same area argues against reworking. Subsequent collecting by the author has yielded evidence indicating that at least some of the vertebrate fossils from the Ojo Alamo sandstone, unit 3, [middle shale unit of Fig. 2 of this report] have not been reworked. Though the possibility of reworking of a few vertebrate fossils cannot be fully excluded, the chances of reworking of large numbers of bones are slight.

The question as to whether Ojo Alamo dinosaur bones were all reworked was largely resolved by Hunt and Lucas (1991), who described an assemblage of dinosaur bones found in the lower Ojo Alamo Sandstone in Alamo Wash (Fig. 14). A photograph of those bones in place is shown in Figure 18A. Hunt and Lucas (1991, p. 28–30) reported the following.

> The partial hadrosaur skeleton (NMMNH P-19147) consists of about 34 elements, including 22 ribs or partial ribs, 2 dorsal vertebrae, 5 neural spines, 1 ossified tendon, 1 sacrum (poorly preserved), 1 scapula and 2 pubes (Figs. 1, 2A–B). As there is no duplication of body parts and no other evidence (e.g. size differences) to suggest the mixing of more than one skeleton, we assume that all the bones represent one individual.
>
> The surface texture of the bones is well preserved and is covered with elongate fractures indicative of surface weathering (Behrensmeyer, 1978). Many of the ribs preserve pre-fossilization breaks, but most of the damage to the larger bones (pubes, scapula) is due to recent weathering. The bones were scattered over an area of approximately 13 m^3 (Fig. 1). This partial skeleton lacks most of the vertebral column, all limb elements and the skull (Fig. 3).

Figure 18B (Fig. 3 of Hunt and Lucas, 1991) shows the placement of the dinosaur bones described in the preceding, in a drawing of the skeleton of a hadrosaur. This concentration of the skeletal parts of a single hadrosaur from the lower part of the Ojo Alamo Sandstone is interpreted as unequivocal evidence that these animals lived in Paleocene time. It thus seems likely to us that most, if not all, of the other single dinosaur bones found in the Ojo Alamo Sandstone in the San Juan Basin also represent animals that lived in Paleocene time and are not reworked Cretaceous dinosaur bones. Evidence based on geochemical analyses of dinosaur bone supports this conclusion.

Geochemistry of dinosaur bones

Geochemical analyses of dinosaur bone samples were conducted as part of this study to provide additional evidence that dinosaur bones found in the Ojo Alamo Sandstone had not been reworked from underlying Cretaceous strata. Because of the 8 m.y. hiatus at the K-T interface in the Hunter Wash area (decreasing to 5 m.y. at the San Juan River site) and because the Cretaceous Kirtland Formation had a distant, southwest source

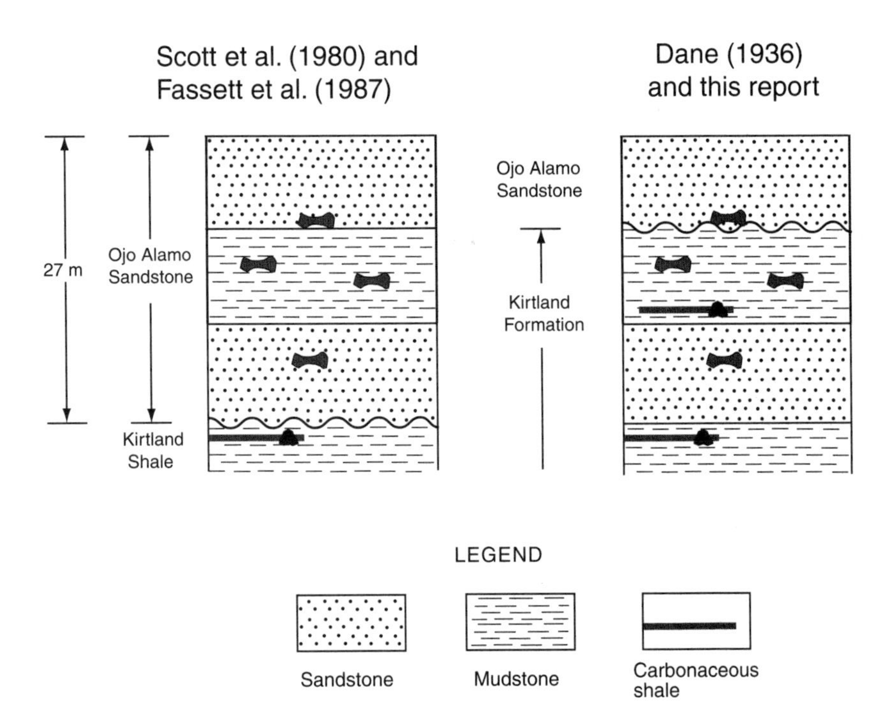

Figure 17. Stratigraphic columns at Pot Mesa locality (Fig. 1A) showing different interpretations of Ojo Alamo Sandstone and Kirtland Formation.

area and the Paleocene Ojo Alamo Sandstone had a nearby north to northwest source area (Fassett, 1985), we thought it reasonable to assume that the groundwater circulating around the bones of recently deceased and buried dinosaurs would have contained distinctly different percentages of dissolved minerals in Cretaceous (Kirtland) time versus Paleocene (Ojo Alamo) time. If this hypothesis proved correct, then the trace element assemblages in mineralized dinosaur bone from the Kirtland Formation would be different from the trace element assemblages present in dinosaur bone from the Ojo Alamo Sandstone. A key element of this hypothesis was the presumption that mineralized dinosaur bone, reworked from Cretaceous into Paleocene strata, would retain its Cretaceous trace element signature and thus be readily recognizable as reworked material. Even though we were not sure at the beginning of this study that reworked Cretaceous bone couldn't be remineralized after reworking into Paleocene sediments to give it a Paleocene geochemical signature, we hoped that our study would shed light on the location of trace elements in the bone matrix and on the manner in which the elements were geochemically bound, and that there would be evidence of remineralization of the bone samples.

In early 1998, dinosaur bones were collected in the field from two localities near Hunter Wash (Figs. 1 and 14). One sample (051298-BB1) was a fragment of a large, massive, sauropod femur ~6 m above the base of the Ojo Alamo Sandstone east of Hunter Wash (photographs of this specimen are in Lucas and Sullivan, 2000b, Fig. 2, E and F); a second sample (070498-WP8) was a hand-sized fragment of a frill bone from a ceratopsian dinosaur ~9 m below the top of the Kirtland Formation west of Hunter Wash (see Fig. 14 for sample localities). The ceratops frill bone (070498-WP8) was from a collection of bone fragments of several different animals preserved as channel-lag material at the base of a sandstone bed in the upper part of the Kirtland. Samples of silicified wood from the Ojo Alamo and the Kirtland were also collected for analysis to determine if trace element compositions from fossil wood would be distinctly different for Cretaceous and Paleocene material. The instrumental neutron activation analysis (INAA) data for the two mineralized dinosaur bone samples showed distinct differences in the abundances of several elements; thus we were encouraged to broaden our geochemical study of mineralized dinosaur bone above and below the K-T boundary in the San Juan Basin. Chemical differences between Cretaceous and Paleocene silicified wood samples did not appear to be significant; therefore, no additional silicified-wood analyses were done.

In late 1998–early 1999, additional bone samples were collected in the field from the Kimbeto, Barrel Spring, and Hunter Wash areas (Figs. 1A and 14); six from the Ojo Alamo Sandstone and three from the Kirtland Formation. Four additional specimens were selected for analysis from the assortment of bone samples collected earlier at the 070498-WP8 sample site (Fig. 14); these specimens were a small turtle bone, part of a dinosaur vertebra, and two small dinosaur limb bones. Tables 3 and 4 identify all of the bone samples analyzed in this report and describe sampling methods. A core from the hadrosaur femur collected from the Ojo Alamo Sandstone at the San Juan River site was provided to us for geochemical study by Gary

A

B

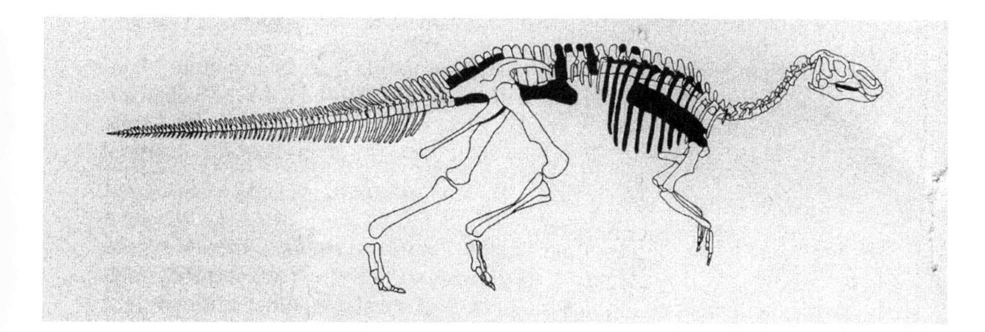

Figure 18. A: Photograph of dinosaur bone assemblage before excavation in lower part of Ojo Alamo Sandstone at Alamo Wash locality of Hunt and Lucas (1991). Hammer is 25 cm long. Photograph provided by Spencer G. Lucas, New Mexico Museum of Natural History and Science, Albuquerque, New Mexico. B: Drawing of skeleton of hadrosaur (from Hunt and Lucas, 1991, Fig. 3) showing positions of bones (shaded) collected from bone assemblage shown in photograph 18A. Length of skeleton is 13.5 m.

Smith, Associate Professor, Department of Earth and Planetary Science, University of New Mexico, Albuquerque. The core, ~1.2 cm in diameter and ~7 cm long, was labeled sample FBHF (Farmington Bluffs hadrosaur femur).

With the exception of sample FBHF and the large sauropod femur (051298BB1), both from the Ojo Alamo Sandstone, most bone specimens were small, ranging from a few centimeters in maximum dimension to hand size. Specimens were gently washed under a stream of water prior to sampling but were otherwise unprocessed. A few grams of material were sampled from each bone specimen. In most cases (14 of 18 samples) material was chipped from an outside surface, sheared from a protruding edge, or broken from the end of the long dimension using a single sharp blow with a steel chisel. Assorted small chips of each sample totaling ~0.2–0.6 g were loaded into a sealable polyethylene vial and submitted for INAA. This procedure minimized sample handling and possible associated contamination, and provided material that was assumed to be reasonably representative of the outer surface of each bone sample.

Sample FBHF from the San Juan River site was sampled by coring the specimen with a stainless steel, 1.25-cm-diameter core barrel to a depth of 7 cm. The outermost, 1.5-cm-long segment of this core was sawed length-wise with a thin stainless steel diamond saw, and half of it was crushed in a ceramic jaw crusher, pulverized, and homogenized in a ceramic ball mill, and an aliquot (0.5 g) submitted for INAA. The three sawed samples, 9498-WP8X and 9498-WP8Y from the Kirtland Formation (Table 3) and 22799B from the Ojo Alamo Sandstone (Table 4), were sawed in half with a stainless steel diamond saw. Half of each sawed bone was then crushed in a ceramic jaw crusher, pulverized, and homogenized in a ceramic ball mill, and aliquots (0.5 g) were submitted for INAA.

Fossilized bone samples were irradiated for 8 h in the Lazy Susan facility of the USGS-TRIGA reactor having a flux of

TABLE 3. CHEMICAL COMPOSITION OF DINOSAUR BONE SAMPLES FROM THE KIRTLAND FORMATION, SAN JUAN BASIN, NEW MEXICO

Sample number	22799 D	22899 A	72598 6C	70498 WP8	70498 WP8A	70498 WP8B	70498 WP8C	9498 WP8X	9498 WP8Y
Meters below base of O.A.	6.1	54.9	18.3	9.1	9.1	9.1	9.1	9.1	9.1
Bone type and how sampled	limb chipped	limb chipped	turtle chipped	frill chipped	vertebra chipped	limb chipped	turtle chipped	limb sawed	limb sawed
				Major and minor elements (wt%)					
Ca	30.2	28.5	26.8	26.7	23.6	10.9	19.5	30.9	22.3
Na	0.26	0.55	0.40	0.39	0.77	0.23	0.33	0.32	0.32
Sr	0.27	0.23	0.13	0.26	0.17	0.72	1.4	0.37	0.46
Ba	0.09	0.16	0.06	3.0	0.24	30.8	22.7	4.7	9.2
Fe	0.29	1.78	0.16	0.44	1.66	0.40	0.73	0.60	0.71
				Trace elements (ppm)					
Sc	0.3	2.6	6.6	4.6	4.9	1.6	3.4	4.3	4.9
Co	5.4	6.3	0.6	5.1	6.5	1.5	2.1	10.9	9.3
Zn	62.2	107	324	266	255	116	261	158	233
As	5.8	131	1.1	5.5	16.9	3.7	12.2	15.2	12.3
Sb	0.2	0.6	0.4	0.8	1.0	0.5	0.8	2.5	1.6
U	26.6	2.1	38.3	22.1	13.9	2.3	44.5	45.1	24.5
Th	0.03	1.3	0.2	0.9	2.7	0.03	0.5	0.2	0.5
Hf	0.06	0.20	0.6	0.3	2.2	0.3	0.4	0.2	0.3
Ta	0.02	0.06	0.02	0.1	0.2	0.01	0.05	0.04	0.1
La	93.1	531	1030	1160	1010	472	1000	670	1310
Ce	130.0	814	2090	2150	1630	349	1250	833	2320
Nd	70.6	487	1290	1190	780	69.1	405	350	1260
Sm	18.1	95.9	290	259	163	9.9	67.8	63.9	224
Eu	5.6	22.8	72.6	48.2	38.5	3.8	16.0	12.6	51.4
Gd	36.0	91.8	444	284	187	18.2	67.9	78	320
Tb	5.4	12.9	67.3	40.4	27.4	4.3	12.9	10.3	40.9
Ho	5.7	14.8	92.2	38.9	35.2	6.3	13.2	11.3	36.2
Tm	1.9	3.2	30.4	8.5	8.5	2.7	4.0	5.2	9.3
Yb	10.2	16.0	178	45.8	48.0	17.1	24.8	31.0	48.5
Lu	1.4	2.0	25.4	5.7	6.6	2.4	3.2	4.3	5.7
Sum REE	378	2091	5610	5231	3934	955	2865	2070	5626
La_{cn}/Yb_{cn}	6.2	22.4	3.9	17.1	14.2	20.9	27.2	14.6	18.2

Note: Additional elements (K, Rb, Cs, Cr, Zr, W, Se, Ni, Au, Br, Mo) are not tabulated because reported analytical precision resulting from interelement interferences, fission yield corrections, and/or counting statistics typically exceeds 15% (relative standard deviation). O.A. — Ojo Alamo Sandstone; cn — chondrite-normalized abundance.

$2.5 \cdot 10^{12}$ n/cm^{2}s. Two different synthetic composite quartz standards (CQS) served as primary monitors and two USGS standard reference materials (SRM), BHVO-2 and BCR-1, were used to measure the accuracy of the analyses. Following decay periods of 6–7 days, 14–16 days, and 65–70 days, the samples, monitors, and SRM were counted twice for 1–2 h, 2–3 h, and 4–5 h, respectively, on both coaxial- and planar-type detectors. This ensures that at least two determinations are made on the elemental abundances reported. Analytical precision of the analysis is determined by propagating the counting statistic errors for a nuclide in the monitor and the sample. Typically, $<1\%$ counting statistic errors are associated with the monitor because of the enhanced levels of the various elements doped. The precision for most reported elements is better than $\pm 15\%$ (relative standard deviation, RSD) based on counting statistics. (Na, Ba, Sc, Co, U, La, Sm, Eu, and Lu = 1%–2%; Ca, Sr, Fe, Ce, Tb, and Yb = 2%–5%; Zn, Sb, Th, Nd, Gd, and Ho = 5%–10%; As, Hf, Ta, and Tm = 10%–15%.) Accuracy is similar to precision, based on replicate simultaneous analyses of standard reference materials. For a few unreported elements (K, Rb, Cs, Cr, Zr, W, Se, Ni, Au, Br) analytical precision is poorer because of poor counting statistics, or uncertainties in additional corrections for interelement interferences or U fission products. Additional details on the INAA method can be found in Baedecker and McKown (1987).

The chemical compositions of fossil bone samples from the Kirtland Formation (Table 3) and the Ojo Alamo Sandstone (Table 4) record the cumulative effects of diagenesis on original bone-mineral substrate composed of fine-grained hydroxyapatite, $Ca_5(PO_4)_3OH$. Present concentrations of trace and minor elements are primarily the result of addition by (1) isomorphic substitution for calcium (Sr, Ba, Zn, U, rare earth elements [REE]), or phosphorous (As) in the apatite structure; (2) ion exchange uptake on crystal surfaces (Sr, Na, Ba); and (3) precipitation of secondary minerals that replace apatite or fill open canals (osteons) in the bone matrix (Williams, 1989; Pate et al. 1989). Secondary minerals identified by X-ray diffraction in these samples include barite, calcite, and goethite.

TABLE 4. CHEMICAL COMPOSITION OF DINOSAUR BONE SAMPLES FROM THE OJO ALAMO SANDSTONE, SAN JUAN BASIN, NEW MEXICO

Sample number	22799 A	22799 B-1	22799 B-2	22799 C	22899 OA1	72598 1C	72598 4C	51298 BB1	FBHF
Meters above base of O.A.	6.1	8.2	8.2	6.1	7.3	2.4	1.8	6.1	15
Bone type and how sampled	limb chipped	limb chipped	limb sawed	limb chipped	limb chipped	limb chipped	femur chipped	femur chipped	femur cored
				Major and minor elements (wt%)					
Ca	28.8	32.7	26.5	30.4	33.9	29.0	31.3	33.4	14.1
Na	0.26	0.40	0.28	0.36	0.18	0.42	0.30	0.33	0.22
Sr	0.13	0.18	0.14	0.11	0.15	0.14	0.14	0.20	0.07
Ba	0.06	0.05	0.10	0.03	0.05	0.30	0.41	0.10	0.03
Fe	0.53	0.24	11.9	0.03	0.23	1.9	1.4	0.62	19.9
				Trace elements (ppm)					
Sc	10.5	10.2	1.7	2.2	0.99	2.9	2.1	7.5	8.5
Co	4.3	4.8	27.3	0.6	50.5	16.0	0.5	5.0	106
Zn	27	147	59	363	62	116	144	161	100
As	8.1	8.4	10.0	2.1	7.3	3.5	2.1	6.0	2.7
Sb	1.6	1.0	16.6	0.3	0.7	0.5	0.5	0.8	0.3
U	232	720	822	436	530	247	166	834	33.2
Th	0.1	0.1	N.D.	0.3	0.05	0.8	0.06	0.2	1.6
Hf	0.3	0.2	0.5	0.5	0.2	0.8	0.04	0.5	1.9
Ta	0.03	0.01	0.2	0.3	0.03	0.1	0.01	0.07	0.4
La	28.1	319	568	1120	265	189	19.1	456	119
Ce	18.3	525	949	2370	293	167	9.0	750	335
Nd	4.5	302	409	1420	101	71.7	7.6	419	299
Sm	1.2	68.7	90	415	23.7	17.1	3.1	96.3	65.7
Eu	0.44	16.7	24.7	90.4	11.0	5.5	0.82	24.0	19.8
Gd	2.2	96.3	137	467	57.4	38.1	4.8	143	105
Tb	0.56	14.4	18.0	75.2	7.7	6.6	1.0	21.4	12.1
Ho	1.8	17.8	N.D.	89.2	8.7	10.1	2.1	23.1	14.8
Tm	1.9	7.1	7.0	16.2	2.2	4.4	1.2	7.4	4.7
Yb	11.9	43.4	36.3	97.8	13.3	27.2	8.1	42.5	25.5
Lu	2.0	6.4	5.3	13.9	1.9	3.8	1.3	6.3	3.6
Sum REE	72.9	1417	2244	6174	785	541	58.1	1989	1004
La_{cn}/Yb_{cn}	1.6	5.0	10.6	7.7	14.0	4.7	1.6	7.2	3.2

Note: Additional elements (K, Rb, Cs, Cr, Zr, W, Se, Ni, Au, Br, Mo) are not tabulated because reported analytical precision resulting from interelement interferences, fission yield corrections, and/or counting statistics typically exceeds 15% (relative standard deviation). O.A. — Ojo Alamo Sandstone; cn — chondrite-normalized abundance; N.D. — not detected.

Chemical variability within each sample suite reflects variable abundance of secondary minerals, and variable uptake and/or exchange of elements from mineral-rich groundwater during bone diagenesis. During earliest diagenesis the rate of recrystallization and void filling in originally permeable bone material determines the effective duration of interaction with throughgoing groundwater. As interior portions of bone become more isolated from flux of groundwater, further uptake of elements is restricted to outer surfaces or linings of unfilled cavities, resulting in inhomogeneous distribution of added elements. Outer surfaces of young fossil bone tend to be enriched in trace elements, but are also potentially more susceptible to later leaching (Rae and Ivanovich, 1986). On the basis of these previous observations, bone exteriors collected and compared in this study should contain different concentrations of elements than interior portions. Preliminary chemical analyses of 7-cm-long cores in the large femur samples FBHF and BB1 indicate generally lower concentrations of trace elements at depth, but considerable variation in the effect. U is more pervasive at depth than REEs.

Despite the chemical variability of the bone samples, there remain distinct differences in the concentration of U and REEs between the two sample suites (Table 5). Concentrations of U in the Ojo Alamo samples are typically an order of magnitude higher than in the Kirtland samples. Such differences are large enough to emerge from the within-suite variability and the analytical precision for U ($\pm 2\%$, 1σ). In contrast to U, the total abundance of REEs tends to be higher in Kirtland samples, primarily because of a greater abundance of light REEs (La, Ce, Nd). This translates into a more steeply sloped REE pattern when data for each sample are normalized to chondritic abundances (Fig. 19, A and B). Steeper sloped REE patterns are represented in the tables as a higher ratio of La_{cn}/Yb_{cn} (cn is chondrite-normalized value).

The contrasting U content and REE patterns in Ojo Alamo and Kirtland bones reflect differences in the U and REE composition of throughgoing groundwater during bone diagenesis. Contrasting groundwater composition results from water-rock interaction along different paleohydrologic flow paths and under different geochemical conditions. For example, higher U

TABLE 5. SUMMARY STATISTICS FOR CHEMICAL PARAMETERS DISTINGUISHING KIRTLAND FORMATION BONES FROM OJO ALAMO SANDSTONE BONES

	Ojo Alamo (nine samples)					Kirtland (nine samples)				
Chemical parameter	Minimum	Maximum	Median	Mean	Standard deviation	Minimum	Maximum	Median	Mean	Standard deviation
U	33.2	822	436	447	298	2.1	45.1	24.5	24.4	16.3
La_{cn}/Yb_{cn}	1.6	14	5.0	6.2	4.2	3.9	27.2	17.1	16.1	7.5
Sum REE	58.1	6174	1004	1587	1883	378	5626	2865	3196	2000

Note: cn — chondrite-normalized abundance; REE is rare earth element.

content in bones from the Ojo Alamo sandstone suggests that pore waters were enriched in dissolved U. High dissolved U could indicate a nearby upland source with abundant labile U and/or a solution chemistry (oxidizing, carbonate rich) particularly favorable for dissolution and transport of uranium (VI). The different REE patterns in the two suites of bone likewise suggest that Kirtland Formation groundwater was more enriched in light REEs.

The geologic history of deposition of the rocks adjacent to the K-T interface in the San Juan Basin (as described in Fassett, 1985) supports these conclusions. The Cretaceous Kirtland Formation consists of fine- to medium-grained sandstone, siltstone, and mudstone that was delivered to the basin area by low-energy, northeasterly flowing streams from a source area ~1000 km to the southwest. The gradient of the paleoslope over which those streams flowed must have been extremely low. In contrast, the Ojo Alamo Sandstone (deposited 8 m.y. after the uppermost Kirtland strata in the southern San Juan Basin) consists of cobble to pebble conglomerate (andesitic to the north), coarse-grained, immature arkosic sandstone, and lesser amounts of siltstone and mudstone. The Ojo Alamo was deposited by very high energy, south-flowing, braided streams carrying sediment into the basin area from a rapidly rising source terrain <100 km from the northern part of the basin. Groundwater chemistry would have been distinctly different at 73 Ma, in late Kirtland time, from that ca. 65 Ma, in Ojo Alamo time. Sikkink (1987, p. 82) referred to "a low-grade disseminated deposit of uranium [in the Ojo Alamo] near Cuba, New Mexico" (Fig. 1A) attesting to an enrichment of dissolved U in groundwater moving through the Ojo Alamo in Paleocene time.

The concentrations of U and REEs in most of the bone samples (Tables 3–5) are very high compared to the crustal average for U of 2.8 ppm (Taylor and McLennan, 1985) and the REE content of a North American shales composite (167 ppm; Henderson, 1984). Strong partitioning of dissolved U and REE into apatite is the result of isomorphous substitution for Ca. An additional mechanism for U enrichment from groundwater is precipitation of insoluble U (IV) in response to locally reducing conditions (Finch, 1967). Transient, chemically reducing environments are established in bone tissue during early degradation of the organic collagen matrix.

Rapid uptake of REEs by apatite is supported by experiments that indicate as much as 70% of Ca site substitution by

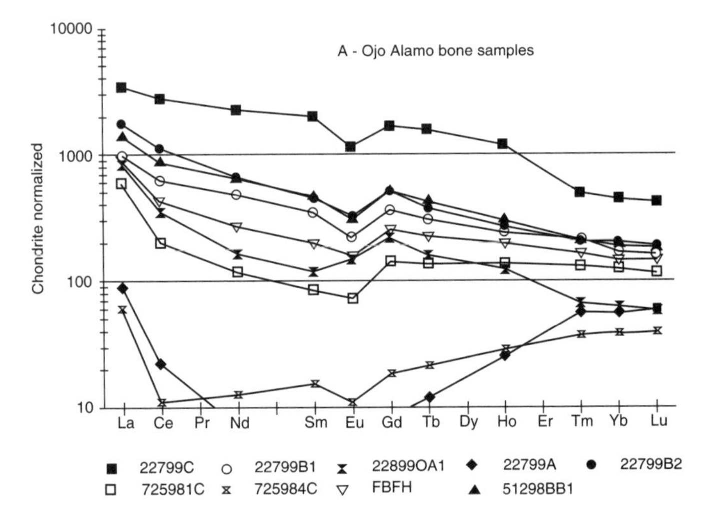

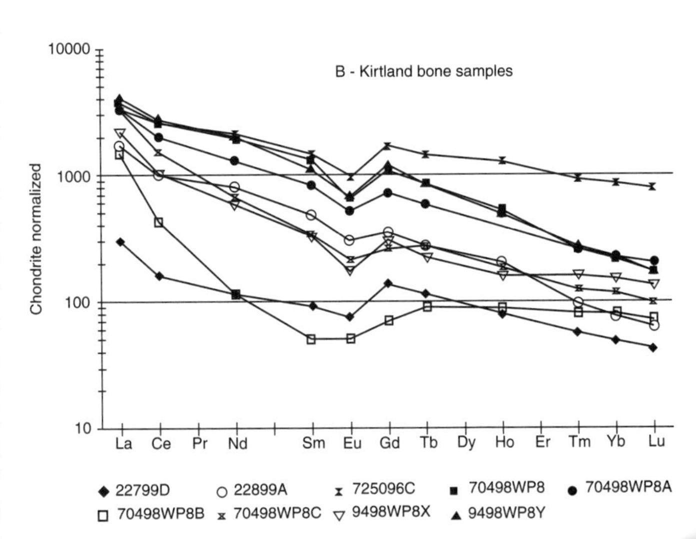

Figure 19. Chondrite-normalized rare earth element patterns of fossil bone samples from (A) Ojo Alamo Sandstone and (B) Kirtland Formation, San Juan Basin, New Mexico. Values used in normalization are La = 0.311, Ce = 0.813, Nd = 0.603, Sm = 0.196, Eu = .074 Gd = 0.26, Tb = 0.047, Ho = 0.0718, Tm = .0326, Yb = 0.21 and Lu = 0.0323, and were derived from multiplying 1.32 times CI values of Anders and Grevesse (1989) to correct to volatile-free basis. Sample localities are shown in Figures 1 and 14. Specimen numbers are keyed to Tables 3 and 4.

REEs within 24 h (Valsami-Jones et al., 1996). The REE content of Quaternary bones indicates that uptake is rapid, and differences in REE composition are attributed to different pore-water environments (Williams, 1988). Rapid uptake during early diagenesis is indicated by U-series ages of Quaternary bone that are compatible with independent age estimates by ^{14}C or stratigraphy (Szabo, 1980; Rae and Ivanovich, 1986).

Additional support for early diagenetic uptake of REEs and U in the bones of this study is that chemical differences between the two suites are preserved, despite their close geographic and stratigraphic proximity, and despite their largely shared post-Kirtland alteration history. If chemical signatures of early diagenesis are preserved in mineralized bone from the Kirtland Formation, they should also be preserved in any bone reworked from the Kirtland into the overlying Ojo Alamo. The REE composition of fossil bone has been used to identify reworked bone in the Triassic of southwest England (Trueman and Benton, 1997). That the REE composition and U content of Ojo Alamo bone are clearly different from those of underlying Kirtland bone (only a few meters below Ojo Alamo bone) strongly indicates that Ojo Alamo bone has not been reworked from the Kirtland.

MAGNETIC POLARITY OF THE OJO ALAMO SANDSTONE

The first studies of the magnetic polarity of the Ojo Alamo Sandstone in the San Juan Basin were published in a series of papers by Butler et al. (1977) and Lindsay et al. (1978, 1981, 1982), and are collectively referred to here as the Butler and Lindsay reports. These studies reported the presence in the southern San Juan Basin of a thin interval of reversed magnetic polarity in the lowermost part of the Ojo Alamo Sandstone overlain by a much thicker interval of normal polarity, including the dinosaur bone-bearing part of the Ojo Alamo. The Butler and Lindsay reports identified the reversed polarity interval in the lower part of the Ojo Alamo as polarity chron C29r and the overlying normal polarity interval as chron C29n. The reversed polarity interval directly underlying the Ojo Alamo in the Kirtland Shale was labeled in the reports as also being C29r based on the assumption that there was no unconformity at the K-T boundary; however, this upper Kirtland reversed interval is really C32r, as demonstrated by Fassett and Steiner (1997). Lindsay et al. (1981, p. 431) stated the following.

> Therefore, the extinction of dinosaurs in the San Jun Basin occurred slightly later than the faunal change recognized as the Cretaceous/Tertiary boundary in the marine deposits at Gubbio, Italy. Conversely, if the marine deposits at Gubbio correctly identify the Cretaceous/Tertiary boundary, there are earliest Tertiary dinosaurs in the San Juan Basin.
>
> The weight of the present evidence, therefore, indicates that the K-T boundary in marine and terrestrial sediments is not synchronous.

Butler (1985) reassessed the normal polarity interval (C29n) identified in the Ojo Alamo Sandstone in the southern San Juan Basin in the reports cited herein, and concluded, on the basis of new studies of the mineralogy of the magnetic minerals in the Ojo Alamo, that the previously identified normal polarity interval in the Ojo Alamo (C29n) did not reliably represent Paleocene remnant magnetism, but rather was a present-day, normal-field overprint and should be removed from the magnetic polarity column in the southern San Juan Basin. Butler further concluded that the dinosaur-bearing part of the Ojo Alamo was actually in the lower part of chron C29r and thus there was no longer a lack of synchroneity between the extinction of the dinosaurs in the San Juan Basin and the marine, end-Cretaceous extinctions at Gubbio, Italy.

Fassett and Steiner (1997, Fig. 2), however, in an independent study of the magnetic polarity of the Kirtland Formation and Ojo Alamo Sandstone in the Hunter Wash area (Figs. 1 and 14) found a thin, reversed magnetic polarity interval at the base of the Ojo Alamo Sandstone overlain by a thicker, normal polarity interval that included the dinosaur-bearing part of the Ojo Alamo. Fassett and Steiner labeled these polarity chrons as C29r and C29n and concluded that they represented certain Cretaceous (or early Paleocene) remnant magnetization of the polarities. Figure 20 shows the magnetic polarity of the Ojo Alamo Sandstone at Hunter Wash as determined by Fassett and Steiner (1997). Because the upper part of chron C29r and all of chron C29n are early Paleocene (Cande and Kent, 1992, 1995), the Ojo Alamo at Hunter Wash must be Paleocene in its entirety; thus, these data provide independent corroborating evidence for the Paleocene age of the Ojo Alamo and its dinosaur fauna in the San Juan Basin. Lindsay et al. (1981, p. 431) were correct in their initial alternative conclusion "Conversely, if the marine deposits at Gubbio correctly identify the Cretaceous/Tertiary boundary, there are earliest Tertiary dinosaurs in the San Juan Basin," but they were incorrect in concluding "The weight of the present evidence, therefore, indicates that the Cretaceous-Tertiary boundary in marine and terrestrial sediments is not synchronous."

AGE AND DURATION OF PALEOCENE DINOSAURS OF THE SAN JUAN BASIN

Biochronology

Prior to our study, many workers (e.g., Clemens, 1973; Hunt and Lucas, 1992; Lucas et al., 2000) have debated whether the dinosaur fauna of the Ojo Alamo was distinctly different from that of the underlying Kirtland Formation and whether these rock units were separated by a hiatus of significant duration. Clemens (1973, p. 161) coined the name "Alamo Wash local fauna" for "the fossils collected from the Ojo Alamo Sandstone and upper part of the Kirtland Shale in the vicinity of the old Ojo Alamo Trading Post," thus implying there was little or no biochronologic break between the Ojo Alamo Sandstone and

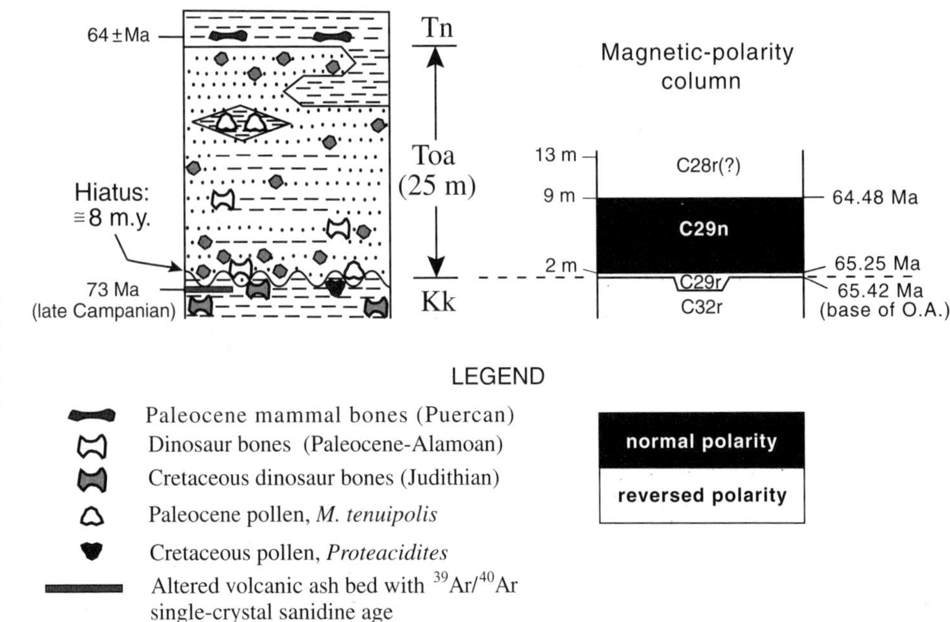

Figure 20. Composite stratigraphic column of Ojo Alamo Sandstone and adjacent strata in Ojo Alamo type area and magnetic polarity of Ojo Alamo and Kirtland Formation as determined in Hunter Wash area by Fassett and Steiner (1997). Ages shown for C29r-C29n and C29n-C28r magnetic polarity reversals are modified from Cande and Kent (1995). Kk = Kirtland Formation, Toa = Ojo Alamo Sandstone, Tn = Nacimiento Formation.

at least the upper part of the Kirtland Shale. Later workers, e.g., Hunt and Lucas (1992, p. 235) have used the name "Ojo Alamo local fauna" to refer just to the dinosaurs found in the Ojo Alamo Sandstone and have stated that this fauna is of Lancian age, whereas the upper Kirtland fauna is of Judithian age, indicating a significant biochronologic break. Lucas et al. (2000) concluded that the "conventional" view now is that the dinosaurs of the Ojo Alamo Sandstone (their Naashoibito) are of Maastrichtian age, or Lancian in terms of vertebrate biochronology: however, they stated that a reasonable alternative is that the Ojo Alamo fauna is late Campanian, indicating continuing uncertainty regarding the age of this fauna.

Our study has demonstrated that the dinosaur fauna of the Ojo Alamo Sandstone is early, but not earliest, Paleocene, and is separated from the underlying Kirtland Formation by an 8 m.y. hiatus in the Hunter Wash area. Therefore, this fauna cannot be correctly referred to as Lancian, which is a latest Cretaceous age designation. We therefore suggest that the Paleocene dinosaur fauna of the Ojo Alamo Sandstone be named the Alamoan fauna.

Figure 20 contains a composite stratigraphic column for the Ojo Alamo Sandstone in its type area summarizing the data relating to the age of the Ojo Alamo dinosaurs in that part of the San Juan Basin. The age of this fauna is fairly well defined on the basis of the evidence summarized in Figure 20. If the age of the K-T boundary is 65.5 ± 0.1 Ma (Obradovich and Hicks, 1999), the base of the Ojo Alamo in the San Juan Basin must be a few hundred thousand years younger. Because Paleocene (Puercan) fossils occur in strata that closely overlie and intertongue with the top of the Ojo Alamo, the age for the top of the Ojo Alamo cannot be more than ~1–2 m.y. younger. Thus, on the basis of biochronologic evidence, the Ojo Alamo must represent a span of time of ~1 or 2 m.y.

Magnetochronology

Cande and Kent (1992, 1995) provided revised ages for the geomagnetic time scale and stated in their 1995 report that the age of the Cretaceous-Tertiary boundary is 65.0 Ma. We place the age of the K-T boundary at 65.5 ± 0.1 Ma (the date published by Obradovich and Hicks, 1999) in order to be consistent with radiometric ages obtained by Obradovich for ash beds in the Hunter Wash area of the San Juan Basin, one of which is shown in Figure 20. We thus add 0.5 m.y. to the Cande and Kent (1995) age of 64.745 Ma for the C29r-C29n magnetic polarity reversal, making the age of this reversal 65.245 Ma. Cande and Kent give the duration of C29r as 0.833 m.y. and C29n as 0.769 m.y.

Figure 20 contains the magnetic polarity column from Fassett and Steiner (1997) for the Ojo Alamo Sandstone in the Hunter Wash area. This figure shows that C29r is only ~2 m thick and C29n is ~9 m thick. The rate of deposition (undecompacted rock rate) for polarity chron C29n is thus equal to 9 m divided by 0.769 m.y. or 11.8 m/m.y., and each 1 m of rock represents 0.085 m.y. Thus, the base of the Ojo Alamo Sandstone at Hunter Wash has an age of 65.25 Ma plus 0.170 m.y., or 65.42 Ma. The amount of missing Paleocene time at the K-T interface in the San Juan Basin is thus 65.50–65.42 Ma, or 0.08 m.y. The duration of the survival time for Paleocene dinosaurs would thus have been ~1 m.y. (to about the top of chron C29n) and the time of final extinction was ca. 64.5 Ma.

IMPLICATIONS OF PALEOCENE DINOSAURS

Two thought-provoking questions come to mind regarding our conclusion that not all dinosaurs became extinct at the end of the Cretaceous. (1) How did dinosaurs in the San Juan Basin, living in an area just 2500 km away from the end-Cretaceous impact site at the Chicxulub crater, survive such a globally devastating event? (2) Why did the surviving dinosaurs in the San Juan Basin die out 1 m.y. after the end-Cretaceous impact event?

A response to these questions at this time can be only speculative at best, but we suggest some possible answers. Kring (2000, p. 2–3) presented a concise summary of the regional and global effects of the impact and the estimated duration of those effects and stated "In the immediate vicinity of the crater, the shock wave, air blast, and heat produced by the impact explosion killed many plants and animals. The air blast, for example, flattened any forests within a 1,000–2000 km region . . . Within a few hundred kilometers of the Chicxulub crater, the thick blanket of ejecta was sufficient to exterminate life." Kring asserted that globally, dust in the atmosphere may have made it too dark to see for from one to six months and too dark for photosynthesis for two months to one year, seriously disrupting continental food chains and decreasing continental surface temperatures. He further speculated that aerosols in the atmosphere, acid rain, and greenhouse warming could have adversely affected conditions on Earth from a few years to decades.

It would seem that dinosaurs—not just in the San Juan Basin area but to the northernmost extent of the Western Interior area in what is now Canada—even if some survived the initial blast effects, would have had a difficult time surviving the first year's events of an impact that would have effectively wiped out the food supply for as much as a year after the impact occurred. We suggest that a possible survival mechanism for the San Juan Basin dinosaurs may have been the thousands of eggs that had been laid in the days just before the impact occurred. It is not known how long it took for dinosaur eggs to hatch, but eggs of the Komodo Dragon, one of the largest living reptiles on Earth, take about eight and a half months to hatch. If the hatching time for dinosaurs can be scaled up to one or possibly two years, based on their much larger body size, then it is not unreasonable to think that a few, newly hatched dinosaurs coming out of their shells a year or two after the impact event might have survived the grueling conditions existing on Earth at that time. At least these embryonic dinosaurs would have weathered the worst of the postimpact devastation buried in the sand and kept snug in their shells up and down the Western Interior. Moreover, the survivors need not have had to survive in the San Juan Basin area, so close to the impact site. The survivors could well have hatched farther north, where the impact effects were less severe, and plant refugia existed, and then their progeny could have ultimately migrated south to the San Juan Basin area as conditions there became more hospitable.

The question as to why the surviving dinosaurs in the San Juan Basin area became extinct in another million years or so has no easy answer. We can only speculate that their genetic diversity may have been greatly diminished by their being offspring of a very few original hardy survivors of the K-T impact event, and thus they would have been susceptible to natural stresses such as disease, global climate change, or other unknown events.

SUMMARY AND CONCLUSIONS

In this study we discussed the presence of a large, pristine, unabraded hadrosaur femur in the Ojo Alamo Sandstone underlain by Paleocene palynomorphs at the San Juan River site (see also Fassett et al., 2000; Fassett and Lucas, 2000). In addition, we show that a dinosaur bone assemblage, representing the remains of one animal, has been preserved in the Ojo Alamo Sandstone that is underlain by Paleocene palynomorphs nearby at the Barrel Spring site. Geochemical data show that all of the analyzed samples of single dinosaur bones from the Ojo Alamo Sandstone have a chemical signature that is distinctly different from that of Cretaceous bone specimens from the underlying Kirtland Formation. This geochemical evidence provides corroboration that Ojo Alamo dinosaur bone is not reworked from older rocks. On the basis of magnetic polarity chronology, we can quantify the duration of the survival of the San Juan Basin dinosaurs into the Paleocene as ~1 m.y. and the final extinction having occurred ca. 64.5 Ma.

Our geochemical data also provide a unique geochemical identifier for dinosaur bone mineralized in Ojo Alamo time, thus it is now possible to place earlier bone collections of uncertain stratigraphic provenance as having come from the Kirtland Formation or Ojo Alamo Sandstone by doing INAA analysis to see if they match Paleocene or Cretaceous geochemical patterns. This technique will also permit the identification of dinosaur bone as definitely reworked from the Cretaceous, should any be found in the Ojo Alamo in the future. We are in the process of conducting additional chemical analyses of dinosaur bone samples from Cretaceous and Tertiary strata in the San Juan Basin to make this database more robust.

The only cogent argument remaining that might cloud our conclusions that dinosaur fossils from the Ojo Alamo Sandstone represent survivors of the K-T extinction event is that somehow the Paleocene palynomorphs found in the Ojo Alamo Sandstone at numerous sites in the San Juan Basin (Figs. 1A and 3) are really Cretaceous forms that have been incorrectly attributed to the Paleocene.

Prior to the discovery of the K-T impact-event fallout layer in the Raton basin by Orth et al. (1981, 1982) and subsequently by other workers at numerous other sites in the Western Interior of North America, the possibility of some overlap of Paleocene and Cretaceous index palynomorphs from boundary rocks in the Western Interior could not have been denied. Now that the K-T impact fallout layer has been found as a thin, but discrete physical rock layer that represents the global K-T geochron in several areas of the Western Interior, palynologists are able to relate their Cretaceous and Tertiary palynomorphs precisely to positions above or below the rock-stratigraphic K-T boundary layer. Recent palynological studies of the strata above and below this rock layer at numerous sites throughout the Western Interior of North American have confirmed that there is no significant overlap of key Cretaceous and Tertiary palynomorphs across the K-T boundary in the Western Interior (J.D. Nichols, 2000, personal commun.).

More to the point, detailed studies in the Raton basin by Orth et al. (1981, 1982), Pillmore et al. (1984), and Fleming (1990) confirmed that the Cretaceous palynomorph *Proteacidites* never occurs above the fallout layer (except for rare, reworked specimens), and *Momipites tenuipolis* has never been found nearer than 15 m above the fallout layer and never below the fallout layer. The Raton basin is essentially at the same latitude as the San Juan Basin and only ~150 km east of it; thus it is extremely unlikely that plants producing Paleocene index pollen forms were present earlier (in Late Cretaceous time) in the San Juan Basin than in the Raton basin due to differing regional climatic or other conditions.

On the basis of all currently available data from throughout the Western Interior of North America, we conclude that the Paleocene index fossils identified from the Ojo Alamo Sandstone in the San Juan Basin are Paleocene in age and thus are bona fide Paleocene index fossils. With no existing contravening data regarding the age of the Ojo Alamo Sandstone and its contained, in place dinosaurs, we conclude that dinosaurs did not all become extinct at the end of the Cretaceous.

ACKNOWLEDGMENTS

Some of the information contained in this paper was presented at the First Annual Symposium on the Dinosaurs of New Mexico held at the New Mexico Museum of Natural History and Science in Albuquerque on April 29, 2000, and at the IMPACT2000 meeting held in Vienna, Austria, in July 2000. Discussions with many colleagues at those meetings contributed greatly to improving the focus and quality of this report. We thank palynologist Douglas J. Nichols, U.S. Geological Survey (USGS), Denver, Colorado, for analysis of many of the rock samples collected from the San Juan River, Barrel Spring, and other sites in the San Juan Basin and for the identification and age determination of the palynomorphs extracted from those samples; this work was important in confirming the Paleocene age of the Ojo Alamo Sandstone. F. Michael O'Neill of the Bureau of Land Management, Albuquerque, New Mexico, was responsible for the excavation and preservation of the large hadrosaur femur from the San Juan River site, and without his efforts to collect this unique specimen, this report would not have been as broadly based. We also thank Isabel Brownfield and Rhonda Driscoll, USGS, Denver, for help with sample preparation for the geochemical analysis of San Juan Basin dinosaur bones. Fred Peterson and W.R. Keefer, USGS, Denver, and Jeff Grossman, USGS, Reston, Virginia, reviewed the manuscript for the USGS and provided insightful comments that greatly improved the final product. Eric Buffetaut, Centre National de la Recherche Scientifique, Paris, and an anonymous reader reviewed the manuscript for the Geological Society of America and provided valuable comments that improved the paper. The radiometric dates and the paleomagnetic data and most of the palynologic analyses of rocks adjacent to the Cretaceous-Tertiary boundary in the San Juan Basin, that provide the basic geochronologic backbone of this report, were obtained through a USGS Gilbert Fellowship awarded to Fassett in 1987.

REFERENCES CITED

American Commission on Stratigraphic Nomenclature, 1961, Code of stratigraphic nomenclature: American Association of Petroleum Geologists Bulletin, v. 45, p. 645–665.

American Commission on Stratigraphic Nomenclature, 1970, Code of stratigraphic nomenclature [second edition]: Tulsa, Oklahoma, American Association of Petroleum Geologists, 45 p.

Anders, E., and Grevesse, N., 1989, Abundances of the elements: Meteoritic and solar: Geochimica et Cosmochimica Acta, v. 53, p. 197–214.

Anderson, R.Y., 1960, Cretaceous–Tertiary palynology, eastern side of the San Juan Basin, New Mexico: New Mexico Bureau of Mines and Mineral Resources Memoir 6, 59 p.

Baedecker, P.A., and McKown, D.M, 1987, Instrumental neutron activation analysis of geochemical samples, *in* Baedecker, P.A., ed., Methods for geochemical analysis: U.S. Geological Survey Bulletin 1770, p. H1–H14.

Baltz, E.H., 1967, Stratigraphy and regional tectonic implications of part of Upper Cretaceous and Tertiary rocks east-central San Juan Basin New Mexico: U.S. Geological Survey Professional Paper 552, 101 p.

Baltz, E.H., Ash, S.R., and Anderson, R.Y., 1966, History of nomenclature and stratigraphy of rocks adjacent to the Cretaceous–Tertiary boundary, Western San Juan Basin, New Mexico: U.S. Geological Survey Professional Paper 524-D, 23 p.

Bauer, C.M., 1916 (1917), Stratigraphy of a part of the Chaco River valley: U.S. Geological Survey Professional Paper 98-P, p. 271–278.

Behrensmeyer, A.K., 1978, Taphonomic and ecologic information from bone weathering: Paleobiology, v. 4, p. 150–162.

Brett-Surman, M.K., 1975, The appendicular anatomy of hadrosaurian dinosaurs [M.A. thesis]: Berkeley, University of California, 70 p.

Brown, B., 1910, The Cretaceous Ojo Alamo beds of New Mexico with description of the new dinosaur genus *Kritosaurus*: American Museum of Natural History Bulletin, v. 28, p. 267–274.

Butler, R.F., 1985, Mineralogy of magnetic minerals and revised magnetic polarity stratigraphy of continental sediments, San Juan Basin, New Mexico: Journal of Geology, p. 535–554.

Butler, R.F., Lindsay, E.H., Jacobs, L.L., and Johnson, N.M., 1977, Magnetostratigraphy of the Cretaceous–Tertiary boundary in the San Juan Basin, New Mexico: Nature, v. 267, p. 318–323.

Cande, S.C., and Kent, D.V., 1992, A new geomagnetic polarity time scale for the Late Cretaceous and Cenozoic: Journal of Geophysical Research, v. 97, no. B10, p. 13917–13951.

Cande, S.C., and Kent, D.V., 1995, Revised calibration of the geomagnetic polarity time scale for the Late Cretaceous and Cenozoic: Journal of Geophysical Research, v. 100, p. 6093–6095.

Clemens, W.A., 1973, The roles of fossil vertebrates in interpretation of Late Cretaceous stratigraphy of the San Juan Basin, New Mexico, *in* Fassett, J.E., ed., Cretaceous and Tertiary rocks of the Colorado Plateau: Four Corners Geological Society Memoir, 1973, p. 154–167.

Dane, C.H., 1936 (1937), Geology and fuel resources of the southern part of the San Juan Basin, New Mexico. 3. The La Ventana-Chacra Mesa coal field: U.S. Geological Survey Bulletin 860-C, p. 81–161.

Dane, C.H., 1946, Stratigraphic relations of Eocene, Paleocene, and latest Cretaceous formations of eastern side of San Juan Basin New Mexico: U.S. Geological Survey Oil and Gas Investigations Preliminary Chart 24.

Dane, C.H., 1948, Geology and oil possibilities of the eastern side of San Juan Basin, Rio Arriba County, New Mexico: U.S. Geological Survey Oil and Gas Investigations Preliminary Map 78, scale 1:63 360, 1 sheet.

Fassett, J.E., 1973, The saga of the Ojo Alamo Sandstone; or the rock-stratigrapher and the paleontologist should be friends, *in* Fassett, J.E., ed., Cretaceous and Tertiary rocks of the Colorado Plateau: Four Corners Geological Society Memoir, 1973, p. 123–130.

Fassett, J.E., 1982, Dinosaurs in the San Juan Basin, New Mexico, may have survived the event that resulted in creation of an iridium-enriched zone near the Cretaceous–Tertiary boundary, *in* Silver, L.T., and Schultz, P.H., eds., Geological implications of impacts of large asteroids and comets on the earth: Geological Society of America Special Paper 190, p. 435–447.

Fassett, J.E., 1985, Early Tertiary paleogeography and paleotectonics of the San Juan Basin area, New Mexico and Colorado, *in* Flores, R.M., and Kaplan, S.S., eds., Cenozoic paleogeography of the west-central United States *in* Proceedings, Rocky Mountain Paleogeography Symposium 3: Rocky Mountain Section, Society of Economic Paleontologists and Mineralogists, p. 317–334.

Fassett, J.E., 1987, The age of the continental, Upper Cretaceous, Fruitland Formation and Kirtland Shale based on a projection of ammonite zones from the Lewis Shale, San Juan Basin, New Mexico and Colorado, *in* Fassett, J.E., and Rigby, J.K., Jr., eds., The Cretaceous–Tertiary boundary in the San Juan and Raton Basins, New Mexico and Colorado: Geological Society of America Special Paper 209, p. 5–16.

Fassett, J.E., 2000, Geology and coal resources of the Upper Cretaceous Fruitland Formation, San Juan Basin, New Mexico and Colorado, *in* Kirschbaum, M.A., Roberts, L.N.R., and Biewick, L.R.H., Geologic assessment of coal in the Colorado Plateau: Arizona, Colorado, New Mexico, and Utah, Chapter Q: U.S. Geological Survey Professional Paper 1625-B, CD-ROM.

Fassett, J.E., and Hinds, J.S., 1971, Geology and fuel resources of the Fruitland Formation and Kirtland Shale of the San Juan Basin, New Mexico and Colorado: U.S. Geological Survey Professional Paper 676, 76 p.

Fassett, J.E., and Lucas, S.G., 2000, Evidence for Paleocene dinosaurs in the Ojo Alamo Sandstone, San Juan Basin, New Mexico, *in* Lucas, S.G., and Heckert, A.B., eds., Dinosaurs of New Mexico: Albuquerque, New Mexico Museum of Natural History and Science Bulletin 17, p. 221–230.

Fassett, J.E., Lucas, S.G., and O'Neill, F.M., 1987, Dinosaurs, pollen and spores, and the age of the Ojo Alamo Sandstone, San Juan Basin, New Mexico: Geological Society of America Special Paper 209, p. 17–34.

Fassett, J.E., Lucas, S.G., Zielinski, R.A., and Budahn, J.R., 2000, Compelling new evidence for Paleocene dinosaurs in the Ojo Alamo Sandstone, San Juan Basin, New Mexico and Colorado, USA, *in* Catastrophic events and mass extinctions: Impacts and beyond: Houston, Texas, Lunar and Planetary Institute, LPI Contribution No. 1053, p. 45–46.

Fassett, J.E., and Rigby, J.K., Jr., eds., 1987, The Cretaceous-Tertiary boundary in the San Juan and Raton Basins, New Mexico and Colorado: Geological Society of America Special Paper 209, 200 p.

Fassett, J.E., and Steiner, M.B., 1997, Precise age of C33n-C32r magnetic polarity reversal, San Juan Basin, New Mexico and Colorado, *in* Anderson, O.J., et al., eds., Mesozoic geology and paleontology of the Four Corners Region: New Mexico Geological Society Guidebook 48, p. 239–247.

Finch, W.I., 1967, Geology of epigenetic uranium deposits in sandstone in the United States: U.S. Geological Survey Professional Paper 538, 121 p.

Fleming, R.F., 1990, Palynology of the Cretaceous-Tertiary boundary interval and Paleocene part of the Raton Formation, Colorado and New Mexico [Ph.D. thesis]: Boulder, University of Colorado, 283 p.

Ford, T.L., 2000, A review of ankylosaur osteoderms from New Mexico and a preliminary review of ankylosaur armor, *in* Lucas, S.G., and Heckert, A.B., eds., Dinosaurs of New Mexico: Albuquerque, New Mexico Museum of Natural History and Science Bulletin 17, p. 157–176.

Gilmore, C.W., 1919, Reptilian faunas of the Torreon, Puerco, and underlying Cretaceous formations of San Juan County, New Mexico: U.S. Geological Survey Professional Paper 119, 71 p.

Henderson, P., 1984, Rare earth element chemistry: New York, Elsevier, 510 p.

Hu, C., 1973, A new hadrosaur from the Cretaceous of Chucheng, Shantung: Acta Geologica Sinica, v. 2, p. 179–202.

Hunt, A.P., and Lucas, S.G., 1991, An associated Maastrichtian hadrosaur and a Turonian ammonite from the Naashoibito Member, Kirtland Formation (Late Cretaceous: Maastrichtian), northwestern New Mexico: New Mexico Journal of Science, v. 31, no. 1, p. 27–35.

Hunt, A.P., and Lucas, S.G., 1992, Stratigraphy, paleontology and age of the Fruitland and Kirtland Formations (Upper Cretaceous), San Juan Basin, New Mexico, *in* Lucas, S.G., Kues, B.S., Williamson, T.E., and Hunt, A.P., eds., San Juan Basin IV: New Mexico Geological Society Guidebook 43, p. 217–239.

Knowlton, F.H., 1924, Flora of the Animas Formation: U.S. Geological Survey Professional Paper 134, p. 71–114.

Kring, D.A., 2000, Impact events and their effect on the origin, evolution, and distribution of life: GSA Today, v. 10, no. 8, p. 1–7.

Lindsay, E.H., Jacobs, L.L., and Butler, R.F., 1978, Biostratigraphy and magnetostratigraphy of Paleocene terrestrial deposits, San Juan Basin, New Mexico: Geology, v. 6, p. 425–429.

Lindsay, E.H., Butler, R.F., and Johnson, N.M., 1981, Magnetic polarity zonation and biostratigraphy of Late Cretaceous and Paleocene continental deposits, San Juan Basin, New Mexico: American Journal of Science, v. 281, p. 390–435.

Lindsay, E.H., Butler, R.F., and Johnson, N.M., 1982, Testing of magnetostratigraphy in Late Cretaceous and early Tertiary deposits, San Juan Basin, New Mexico, *in* Silver, L.T., and Schultz, P.H., eds., Geological implications of impacts of large asteroids and comets on the earth: Geological Society of America Special Paper 190, p. 435–447.

Lucas, S.G., and Heckert, A.B., eds., 2000, Dinosaurs of New Mexico: Albuquerque, New Mexico Museum of Natural History and Science Bulletin 17, 230 p.

Lucas, S.G., and Sullivan, R.M., 2000a, Stratigraphy and vertebrate biostratigraphy across the Cretaceous-Tertiary boundary, Betonnie Tsosie Wash, San Juan Basin, New Mexico, *in* Lucas, S.G., and Heckert, A.B., eds., Dinosaurs of New Mexico: Albuquerque, New Mexico Museum of Natural History and Science Bulletin 17, p. 95–103.

Lucas, S.G., and Sullivan, R.M., 2000b, The Sauropod dinosaur Alamosaurus from the Upper Cretaceous of the San Juan Basin, New Mexico, *in* Lucas, S.G., and Heckert, A.B., eds., Dinosaurs of New Mexico: Albuquerque, New Mexico Museum of Natural History and Science Bulletin 17, p. 147–156.

Lucas, S.G., Heckert, A.B., and Sullivan, R.M., 2000, Cretaceous dinosaurs in New Mexico, *in* Lucas, S.G., and Heckert, A.B., eds., Dinosaurs of New Mexico: Albuquerque, New Mexico Museum of Natural History and Science Bulletin 17, p. 83–90.

Lucas, S.G., Mateer, N.J., Hunt, A.P., and O'Neill, M.O., 1987, Dinosaurs, the age of the Fruitland and Kirtland Formations, and the Cretaceous-Tertiary boundary in the San Juan Basin, New Mexico, *in* Fassett, J.E., and Rigby,

J.K., Jr., eds., The Cretaceous-Tertiary boundary in the San Juan and Raton Basins, New Mexico and Colorado: Geological Society of America Special Paper 209, p. 35–50.

Lull, R.S., and Wright, N.E., 1942, Hadrosaurian dinosaurs of North America: Geological Society of America Special Paper 40, 241 p.

North American Commission on Stratigraphic Nomenclature, 1983, North American stratigraphic code: American Association of Petroleum Geologists Bulletin, v. 67, p. 841–875.

Obradovich, J.D., 1993, A Cretaceous time scale, *in* Caldwell, W.G.E., and Kauffman, E.G., eds., Evolution of the Western Interior Basin: Geological Association of Canada Special Paper 39, p. 379–396.

Obradovich, J.D., and Hicks, J.F., 1999, A review of the isotopic calibration points for the geomagnetic polarity time scale, in the interval 83 to 33 ma (c34n to c13n): Geological Society of America Abstracts with Programs, v. 31, n. 7, p. A71.

Orth, C.J., Gilmore, J.S., Knight, J.D., Pillmore, C.L., Tschudy, R.H., and Fassett, J.E., 1981, An iridium abundance anomaly at the palynological Cretaceous-Tertiary boundary in northern New Mexico: Science, v. 214, p. 1341–1343.

Orth, C.J., Gilmore, J.S., Knight, J.D., Pillmore, C.L., Tschudy, R.H., and Fassett, J.E., 1982, Iridium abundance measurements across the Cretaceous/Tertiary boundary in the San Juan and Raton Basins of northern New Mexico, *in* Silver, L.T., and Schultz, P.H., eds., Geological implications of impacts of large asteroids and comets on the earth: Geological Society of America Special Paper 190, p. 423–433.

O'Sullivan, R.B., and Beikman, H.M., 1963, Geology, structure, and uranium deposits of the Shiprock quadrangle, New Mexico and Arizona: U.S. Geological Survey Miscellaneous Geological Investigations Map I-345, scale 1:250000, 2 sheets.

Pate, D.F., Hutton, J.T., and Norrish, K., 1989, Ionic exchange between soil solution and bone: Toward a predictive model: Applied Geochemistry, v. 4, p. 303–316.

Pillmore, C.L., Tschudy, R.H., Orth, C.J., Gilmore, J.S., and Knight, J.D., 1984, Geologic framework of nonmarine Cretaceous–Tertiary boundary sites, Raton Basin, New Mexico and Colorado: Science, v. 223, p. 1180–1183.

Powell, J.S., 1972, The Gallegos Sandstone (formerly Ojo Alamo Sandstone) of the San Juan Basin, New Mexico [M.S. thesis]: Tucson, University of Arizona, 130 p.

Powell, J.S., 1973, Paleontology and sedimentation models of the Kimbeto Member of the Ojo Alamo Sandstone, *in* Fassett, J.E., ed., Cretaceous and Tertiary rocks of the Colorado Plateau: Four Corners Geological Society Memoir, p. 111–122.

Rae, A.M., and Ivanovich, M., 1986, Successful application of uranium series dating of fossil bone: Applied Geochemistry, v. 1, p. 419–426.

Reeside, J.B., Jr., 1924, Upper Cretaceous and Tertiary formations of the western part of the San Juan Basin of Colorado and New Mexico: U.S. Geological Survey Professional Paper 134, p. 1–70.

Scott, G.R., Mytton, J.W., and Schneider, G.B., 1980, Geologic map of the Star Lake Quadrangle, McKinley County, New Mexico: U.S. Geological Survey Miscellaneous Field Studies Map MF-1248, scale 1:24000, 1 sheet.

Sikkink, P.G.L., 1987, Lithofacies relationships and depositional environment of the Tertiary Ojo Alamo Sandstone and related strata, San Juan Basin, New Mexico and Colorado, *in* Fassett, J.E., and Rigby, J.K., Jr., eds., The Cretaceous–Tertiary boundary in the San Juan and Raton Basins, New Mexico and Colorado: Geological Society of America Special Paper 209, p. 81–104.

Szabo, B.J., 1980, Results and assessment of uranium series dating of vertebrate fossils from Quaternary alluviums in Colorado: Arctic Alpine Research, v. 12, p. 95–100.

Taylor, S.R., and McLennan, S.M., 1985, The continental crust: Its composition and evolution: Oxford, Blackwell Scientific Publications, 312 p.

Trueman, C.N., and Benton, M.J., 1997, A geochemical method to trace the taphonomic history of reworked bones in sedimentary settings: Geology, v. 25, no. 3, p. 263–266.

Tschudy, R.H., 1973, The Gasbuggy core; a palynological appraisal, *in* Fassett, J.E., ed., Cretaceous and Tertiary rocks of the Colorado Plateau: Four Corners Geological Society Memoir, p. 123–130.

Valsami-Jones, E., Ragnarsdottir, K.V., Crewe-Read, N.O., Mann, T., Kemp, A.J., and Allen, G.C., 1996, An experimental investigation of the potential of apatite as radioactive and industrial waste scavenger, *in* Bottrells, S.H., ed., Fourth International Symposium on the Geochemistry of the Earth's surface: Yorkshire, United Kingdom, University of Leeds, p. 686–689.

Weishampel, D.B., and Horner, J.R., 1990, Hadrosauridae, *in* Weishampel, D.B., Dodson, P., and Osmólska, H., eds., The Dinosauria: Berkeley, University of California Press, p. 534–561.

Williams, C.T., 1988, Alteration of chemical composition of fossil bones by soil processes and ground water, *in* Grupe, G., and Herrmann, B., eds., Trace elements in environmental history: Berlin, Springer-Verlag, p. 27–40.

Williams, C.T., 1989, Trace elements in fossil bone: Applied Geochemistry, v. 4, p. 247–248.

Williamson, T.E., 2000, Review of *Hadrosauridae* (Dinosauria, Ornithischia) from the San Juan Basin, New Mexico, *in* Lucas, S.G., and Heckert, A.B., eds., Dinosaurs of New Mexico: Albuquerque, New Mexico Museum of Natural History and Science Bulletin 17, p. 191–213.

Manuscript Submitted October 23, 2000; Accepted by the Society March 22, 2001

Printed in the U.S.A.

Geological Society of America
Special Paper 356
2002

Sulfur isotopic compositions across terrestrial Cretaceous-Tertiary boundary successions

Teruyuki Maruoka
Christian Koeberl
Institute of Geochemistry, University of Vienna, Althanstrasse 14, A-1090 Vienna, Austria
Jason Newton
Institute of Geochemistry, University of Vienna, Althanstrasse 14, A-1090 Vienna, Austria, and Department of Earth Sciences, University of California, Santa Cruz, California 95064, USA
Iain Gilmour
Planetary Science Research Institute, Open University, Milton Keynes MK7 6AA, UK
Bruce F. Bohor
U.S. Geological Survey, MS 926A, Denver Federal Center, Denver, Colorado 80225-0046, USA

ABSTRACT

Isotopic compositions of sulfur and concentrations of sulfur and carbon have been measured for sedimentary rocks across terrestrial Cretaceous-Tertiary (K-T) boundary successions that originated from flood-plain and backswamp environments. Organic carbon contents are relatively constant below the boundary (26 cm for the Dogie Creek section and 9 cm for the Brownie Butte section), but change abruptly at the K-T boundary. At the K-T event, a high input of sulfate to the freshwater wetlands might have resulted from the melt ejecta and/or acid rain. We interpret the low ratio of organic C to nonorganic S at the melt ejecta layer and sulfur isotopic data as consistent with this hypothesis; however, additional analyses across a thicker interval are necessary to rule out alternative hypotheses.

INTRODUCTION

Most sulfides present in sediments originate from H_2S produced by sulfate-reducing bacteria (e.g., Berner, 1984). This process is accompanied by a fractionation of the sulfur isotopes that results in sulfide being ^{34}S depleted with respect to the parent sulfate. The extent of isotopic fractionation between sulfate and sulfide is dependent on the kinetics of the sulfate reduction reaction (e.g., Kaplan and Rittenberg, 1964; Chambers and Trudinger, 1979), which, in turn, is affected by the environment of deposition. Consequently, the isotopic composition of sulfide can be used as a proxy for environmental conditions such as the temperature of water, sulfate abundance, and types of electron source.

Holser et al. (1988) observed that $\delta^{34}S$ values of sulfides and sulfate in marine sediments have not changed much since the end of the Cretaceous. However, there have been relatively few studies of sulfur across the Cretaceous-Tertiary (K-T) boundary. Schmitz et al. (1988) reported $\delta^{34}S$ values of $-31.5‰$ in metal-rich pyrite spherules that occur in the basal Fish Clay from the marine K-T succession at Stevns Klint, Denmark. Heymann et al. (1998) observed native sulfur in a spherule-bearing unit from marine K-T successions around the Gulf of Mexico with $\delta^{34}S$ values of $-24.9‰$. Both these $\delta^{34}S$ values are similar to those observed in Late Cretaceous sedimentary rocks, and suggest that the geological record provides no indication of major perturbations in the microbiological utilization of sulfur across the K-T boundary. However, Kajiwara and Kaiho (1991, 1992) observed a temporary and marked increase in pyrite $\delta^{34}S$ values across the K-T boundary in the

Maruoka, T., Koeberl, C., Newton, J., Gilmour, I., and Bohor, B.F., 2002, Sulfur isotopic compositions across terrestrial Cretaceous-Tertiary boundary successions, *in* Koeberl, C., and MacLeod, K.G., eds., Catastrophic Events and Mass Extinctions: Impacts and Beyond: Boulder, Colorado, Geological Society of America Special Paper 356, p. 337–344.

marine succession at Kawaruppu, Hokkaido, Japan, and suggested an anoxic depositional environment immediately after the K-T boundary as an explanation for the variation in $\delta^{34}S$ values.

Holmes and Bohor (1994) observed an increase in $\delta^{34}S$ values in the terrestrial K-T boundary claystone unit at the Sugarite site in the Raton Basin, New Mexico (Pillmore and Flores, 1987). The $\delta^{34}S$ value increases by 3.5‰ and the $\delta^{13}C$ value of the organic matter is also ~2‰–3‰ heavier in the boundary claystone unit (fireball and melt ejecta layer) than in the surrounding coal. They suggested impact-derived carbon and sulfur from vaporized marine target rocks as the origin of the isotopically heavier S and C in the boundary claystone. The terrestrial environment is suitable for the detection of an impact-derived sulfate input because the concentrations of sulfate in terrestrial environments are generally low relative to marine environments, and the perturbation effects are minimal in terrestrial environments. However, it is important to distinguish isotopically heavy S originating from vaporization of target rocks from that associated with environmental change caused by the K-T event. In this study we determined S contents and $\delta^{34}S$ values in sedimentary rocks across two K-T boundary sequences and used these data to examine paleoenvironmental conditions of terrestrial successions across the K-T boundary from the Western Interior of the United States.

SAMPLES

Dogie Creek, Wyoming

We analyzed 17 samples from 26 cm below the boundary to 16 cm above. They include Upper Cretaceous carbonaceous shale, melt ejecta layer (boundary claystone), fireball layer (magic layer), and lower Tertiary clay-shale and lignite (Bohor et al., 1987).

Brownie Butte, Montana

A series of 11 samples in 2 cm intervals from 9 cm below the boundary to 12 cm above was analyzed (Bohor et al., 1984). At the time of the terminal Cretaceous event, this locality was part of a large area consisting of flood plains and backswamps, allowing the deposition of sandstones interlayered with coal, mudstone, carbonaceous shale, and siltstone (Tschudy, 1970).

EXPERIMENTAL METHODS

Mass spectrometry

Concentrations and isotopic compositions of sulfur were measured in bulk sediments and 5 M HCl-treated residues using a helium-gas continuous flow isotope ratio mass spectrometer (CF-IR-MS; Micromass Optima; Giesemann et al., 1994). The concentrations of carbon were measured using a thermal conductivity detector (TCD; Carlo Erba) with a combustion furnace, followed by measurement in the mass spectrometer. The samples were weighed into 12 × 5 mm tin capsule with a mixture of V_2O_5 and SiO_2 to promote full combustion (Yanagisawa and Sakai, 1983).

The $\delta^{34}S$ values were calculated relative to Canyon Diablo Troilite (CDT) based on comparison with the analyses of two silver sulfide standards of International Atomic Energy Agency (IAEA) (IAEA-S-1, −0.3‰ CDT; IAEA-S-2, +22‰ CDT; Coplen and Krouse, 1998) and determined with a measured precision of ±0.4‰ (1σ). The reference sulfides were also used in the calibration of the sulfur contents. The contents of S were determined with a precision of ~±5 rel%. An elemental standard (Acetanilide Standard, ThermoQuest Italia S.p.A.) was used in the calibration of the C contents. The contents of C were determined with a precision of ~1 rel% for 2 mg C and 10 rel% for 0.2 mg C.

Acid treatment

For each sample, we treated ~100–200 mg of powdered rock for 30 min (including 20 min for centrifugation) in ~5 mL of 5 M HCl to eliminate sulfate and carbonate in the samples. Treatment of pure pyrite and gypsum test samples resulted in 86% and 16% recovery yields, respectively. For some samples, the sulfate was collected in the form of $BaSO_4$ by precipitation from the solutions through the addition of ~5 mL of $BaCl_2$ (2 M) to the solutions.

RESULTS AND DISCUSSION

Concentrations of carbon and sulfur together with isotopic compositions of sulfur in bulk samples and 5 M HCl residues are given in Table 1. Figure 1 shows $\delta^{34}S$ values against C/S ratio in the 5 M HCl residues. The $\delta^{34}S$ values are positively correlated with the C/S ratios, indicating that the organic-rich rocks contain abundant organic matter with respect to S. The organic sulfur has a high C/S ratio and high $\delta^{34}S$ value and disulfide sulfur (acid nonvolatile sulfur, which mainly consists of pyrite) with a low $\delta^{34}S$ value. The $\delta^{34}S$ value of the 5 M HCl residue of LC-86-1-A (−1.95 ± 0.46) is the same as that of the bulk sample (−1.64 ± 0.50), although the C/S values vary from 29.0 to 6.1 (Fig. 2). The $\delta^{34}S$ value of the 5 M HCl residue of BB-3 (−1.16 ± 0.13) is the same as that of the bulk sample (−0.97 ± 0.15), although the C/S values vary from 20.3 to 12.0 (Fig. 2). These observations indicate that the presence of organic sulfur does not severely affect the $\delta^{34}S$ in the samples having these ranges of the C/S ratio (i.e., <30 for Dogie Creek samples and <20 for Brownie Butte samples). However, the $\delta^{34}S$ values of the 5 M HCl residue with C/S ratios exceeding these values become higher than that of the bulk sample due to the effect of organic sulfur (Fig. 3; e.g., 5 M HCl residues of LC-86-K and L, residue and bulk sample of BB-6, BB-7, and BB-8). Therefore, the $\delta^{34}S$ values with C/S ratios of less than the values mentioned here can be regarded as repre-

TABLE 1. CONCENTRATION OF CARBON AND SULFUR AND ISOTOPIC COMPOSITION OF SULFUR IN CRETACEOUS-TERTIARY BOUNDARY SEDIMENTS

Sample	Depth from KTB		Description	Bulk analysis			Analysis after 5 M HCl			Sulfate	
	Bottom (cm)	Top (cm)		C (wt%)	S (wt%)	$\delta^{34}S$ (‰)	C (wt%)	S (wt%)	$\delta^{34}S$ (‰)	S Yield (wt%)	$\delta^{34}S$ (‰)
Dogie Creek											
LC-86-1O	14	16	Carbonaceous shale	2.54 ± 0.28	0.219 ± 0.031	−3.87 ± 0.39	2.58 ± 0.32	0.134 ± 0.008	−5.79 ± 0.37	n.d	
LC-86-1N	8	11	Lignite	23.17 ± 1.37	0.631 ± 0.037	0.49 ± 0.38	27.19 ± 1.47	0.576 ± 0.099	0.68 ± 0.55	n.d	
LC-86-1M	4	6	Clay-shale	5.43 ± 0.43	2.618 ± 0.302	−3.80 ± 0.34	4.92 ± 1.23	1.805 ± 0.209	−5.52 ± 0.74	0.216	−2.82 ± 0.13
LC-86-1L	2	4	Clay-shale	1.15 ± 0.16	0.051 ± 0.012	0.57 ± 0.56	1.04 ± 0.08	0.024 ± 0.003	3.17 ± 0.59	n.d	
LC-86-1K	0	2	Clay-shale	0.99 ± 0.17	0.051 ± 0.002	−0.30 ± 0.51	1.02 ± 0.14	0.020 ± 0.003	2.87 ± 0.50	n.d	
Fireball Layer			Smectitic claystone	4.57	1.462 ± 0.033	−4.60 ± 0.79	4.09 ± 0.22	1.256 ± 0.030	−6.58 ± 0.26	0.078	−5.89
Melt Ejecta Layer			Kaolinitic claystone	2.92	1.318 ± 0.033	−3.33 ± 0.65	3.36 ± 0.16	0.864 ± 0.088	−4.99 ± 0.20	0.049	−4.67 ± 0.22
LC-86-1H	−2	0	Carbonaceous shale	0.50 ± 0.10	0.054 ± 0.015	−2.85 ± 0.57	0.54 ± 0.01	0.039 ± 0.001	−4.54 ± 0.34	n.d	
LC-86-1G	−4	−2	Carbonaceous shale	0.60 ± 0.10	0.206 ± 0.059	−2.53 ± 0.86	0.71 ± 0.11	0.115 ± 0.004	−3.24 ± 0.76	0.009	−2.17
LC-86-1F	−6	−4	Carbonaceous shale	0.56 ± 0.02	0.062 ± 0.004	−2.61 ± 0.28	0.71 ± 0.07	0.044 ± 0.002	−2.19 ± 0.26	n.d	
LC-86-1E	−10	−8	Carbonaceous shale	0.64 ± 0.11	0.085 ± 0.009	−2.44 ± 0.59	0.72 ± 0.05	0.043 ± 0.002	−2.51 ± 0.08	n.d	
LC-86-1D	−14	−12	Carbonaceous shale	1.71	0.416 ± 0.258	−2.32 ± 0.38	0.81 ± 0.17	0.136 ± 0.011	−4.06 ± 0.64	0.009	−1.67 ± 0.03
LC-86-1C	−18	−16	Carbonaceous shale	0.72 ± 0.11	0.105 ± 0.000	−3.39 ± 0.25	0.83 ± 0.04	0.070 ± 0.003	−3.54 ± 0.42	n.d	
LC-86-1B	−22	−20	Carbonaceous shale	0.65 ± 0.15	0.175 ± 0.018	−0.54 ± 0.29	0.67 ± 0.08	0.029 ± 0.001	0.07 ± 0.46	0.012	−1.25 ± 0.10
LC-86-1A	−26	−24	Carbonaceous shale	0.50 ± 0.03	0.083 ± 0.007	−1.64 ± 0.50	0.62 ± 0.03	0.021 ± 0.001	−1.95 ± 0.46	n.d	−1.62
Brownie Butte											
BB-11	10	12	Gray shale	0.70 ± 0.10	0.246 ± 0.008	−3.15 ± 0.29	0.40 ± 0.07	0.069 ± 0.004	−3.69 ± 0.19	n.d	
BB-10	8	10	Gray shale	0.68 ± 0.02	0.204 ± 0.003	−2.99 ± 0.39	0.56 ± 0.10	0.047 ± 0.002	−1.33 ± 0.20	n.d	
BB-9	6	9	Carbonaceous shale	9.92 ± 1.22	0.899 ± 0.017	1.20 ± 0.11	8.41 ± 0.10	0.275 ± 0.002	5.52 ± 0.16	0.291	−1.58 ± 0.11
BB-8	4	6	Lignite	38.45 ± 3.36	1.463 ± 0.021	5.40 ± 0.23	38.98 ± 1.00	0.828 ± 0.028	12.91 ± 0.67	n.d	
BB-7	2	4	Lignite	44.50 ± 3.29	2.177 ± 0.014	6.00 ± 0.35	48.76 ± 5.44	1.204 ± 0.275	11.10 ± 0.68	n.d	
BB-6	0	2	Lignite	20.72 ± 8.35	1.587 ± 0.245	1.72 ± 0.23	15.74 ± 1.04	0.737 ± 0.024	3.82 ± 0.26	0.292	−3.36 ± 0.49
BB-14			Fireball	4.19 ± 0.41	0.377 ± 0.016	−1.85 ± 0.39	n.d.				
BB-13			Melt ejecta	0.75	0.344 ± 0.010	−0.83 ± 0.30	0.82 ± 0.43	0.113 ± 0.019	−0.19 ± 0.34	0.041	−1.30 ± 0.04
BB-5	−2	0	Gray shale	0.81	0.236 ± 0.004	−1.62 ± 0.29	0.60 ± 0.06	0.082 ± 0.002	−2.25 ± 0.42	0.042	−2.49 ± 0.17
BB-4	−4	−2	Carbonaceous shale	4.50 ± 0.89	0.491 ± 0.002	−1.39 ± 0.26	3.16 ± 0.11	0.221 ± 0.013	−1.77 ± 0.16	n.d	
BB-3	−6	−4	Carbonaceous shale	4.01 ± 0.28	0.334 ± 0.010	−0.97 ± 0.15	3.33 ± 0.32	0.164 ± 0.006	−1.16 ± 0.13	n.d	
BB-2	−8	−6	Carbonaceous shale	3.94 ± 0.49	0.281 ± 0.003	−0.18 ± 0.43	3.50 ± 0.29	0.098 ± 0.009	0.93 ± 0.23	n.d	
BB-1	−9	−8	Carbonaceous shale	4.08 ± 0.63	0.325 ± 0.010	1.43 ± 0.26	3.29 ± 0.28	0.107 ± 0.005	1.86 ± 0.10	0.034	0.65

Note: n.d. — not determined; KTB — Cretaceous-Tertiary boundary

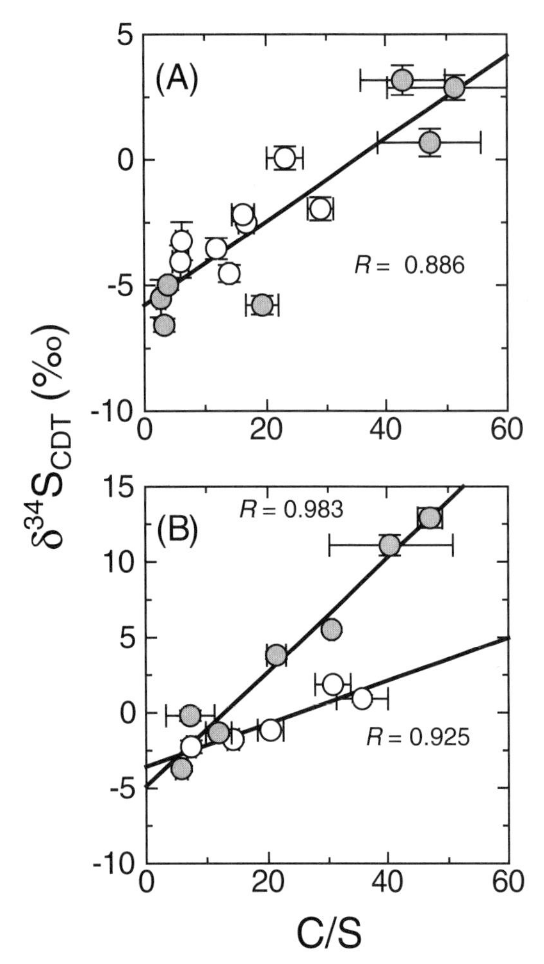

Figure 1. Isotopic compositions of sulfur against C/S ratio in 5 M HCl residue of sediments from (A) Dogie Creek and (B) Brownie Butte. Open and gray circles represent data below and above Cretaceous-Tertiary boundary, respectively. Solid lines represent regression lines. For Brownie Butte samples, two lines can be obtained from data of post- and pre-boundary sediments. Isotopic compositions are given as deviation from that of Canyon Diablo Troilite (CDT).

senting the δ^{34}S values of nonorganic sulfur. For all samples, except for the coal-bearing samples (i.e., Dogie Creek sample LC-86-N and Brownie Butte samples BB-6, BB-7, and BB-8), the C/S ratios of the bulk samples are less than these critical values. Therefore, the δ^{34}S values of the bulk samples of noncoal sediments can be used for our discussion.

Weathering causes the oxidation of sulfide to sulfate, such as barite and celestite, as suggested by Schmitz (1989). For most of the samples with a C/S ratio lower than the limit stated in the preceding, no difference of the δ^{34}S values between the 5 M HCl residue and the bulk is observed, suggesting that the oxidation did not severely affect the δ^{34}S values. However, the δ^{34}S values of the 5 M HCl residue of such samples as LC-86-D, LC-86-H, LC-86-M, and LC-86-O, fireball, and melt eject layer, for the Dogie Creek samples, are different from those of the bulk samples. These differences could indicate that the oxidation of sulfide affected the δ^{34}S values of sulfide for these samples. Although some fraction of the sulfur might be lost during weathering, the δ^{34}S value of the initial sulfide may be restricted between that of the residual component (i.e., bulk sample) and that of the lost component. The latter component can be represented by the δ^{34}S value of the acid-volatile sulfate (Table 1), because some of the acid-volatile sulfate might be dissolved during weathering. However, because the δ^{34}S values of the acid-volatile sulfate are generally similar to those of the bulk samples (Fig. 3), the loss of sulfur is probably not significant and should not affect the δ^{34}S values of the bulk samples.

C and S concentrations and S isotope compositions across the boundary

Below the K-T boundary. Concentrations of total organic carbon (calculated using the carbon content in the 5 M HCl residue and mass losses during 5 M HCl treatment; Table 1) and total sulfur (equivalent to the sulfur content in the bulk sediment) are shown in Figure 4.

The concentrations of organic carbon are relatively constant below the K-T boundary (0.72% ± 0.07% from 26 cm to 2 cm below the boundary for the Dogie Creek sediments; 3.32% ± 0.14% from 9 cm to 2 cm for the Brownie Butte sediments). Organic carbon burial efficiency (the ratio of the carbon burial rate and the carbon flux to the sediment surface) was shown to be a function of sedimentation rate (Henrichs and Reeburgh, 1987). At higher sedimentation rates, labile and metabolizable organic material undergoes a shorter period of oxic and suboxic degradation in the surface of sediments. Therefore, the constancy of the organic carbon concentration (i.e., the ratio of the carbon burial rate to the sedimentation rate) implies that the carbon flux should be expressed as a function of sedimentation rate. We have no reason to assume that the carbon flux is affected by the sedimentation rate, because of the difference of origin between the carbon flux and the sediments. The former originates from the biogenic activity and the latter is dominated by the influx of lithic material. Therefore, it is likely that both parameters, carbon flux and sedimentation rate, were constant over the short interval analyzed.

At the K-T boundary (melt ejecta layer and fireball layer). The ratios of organic C to bulk S in the melt-ejecta layer (2.88 ± 0.15 for Dogie Creek site and 2.57 ± 1.35 for Brownie Butte site) are similar to the mean C/S value for the marine sediments (1.8 ± 0.5 for Devonian to Tertiary marine shale; Raiswell and Berner, 1986). In normal (noneuxinic) marine sediments, where dissolved sulfate and iron minerals are abundant, organic matter is the major control on pyrite formation (Berner, 1984). In contrast, pyrite formation in nonmarine, freshwater sediments is not significantly limited by the abundance of organic matter, because abundant organic matter is available in freshwater environments (Berner and Raiswell, 1984). Therefore, the low C/S ratios, similar to the marine sediments, indicate that the sulfate concentrations in the water of the wetlands

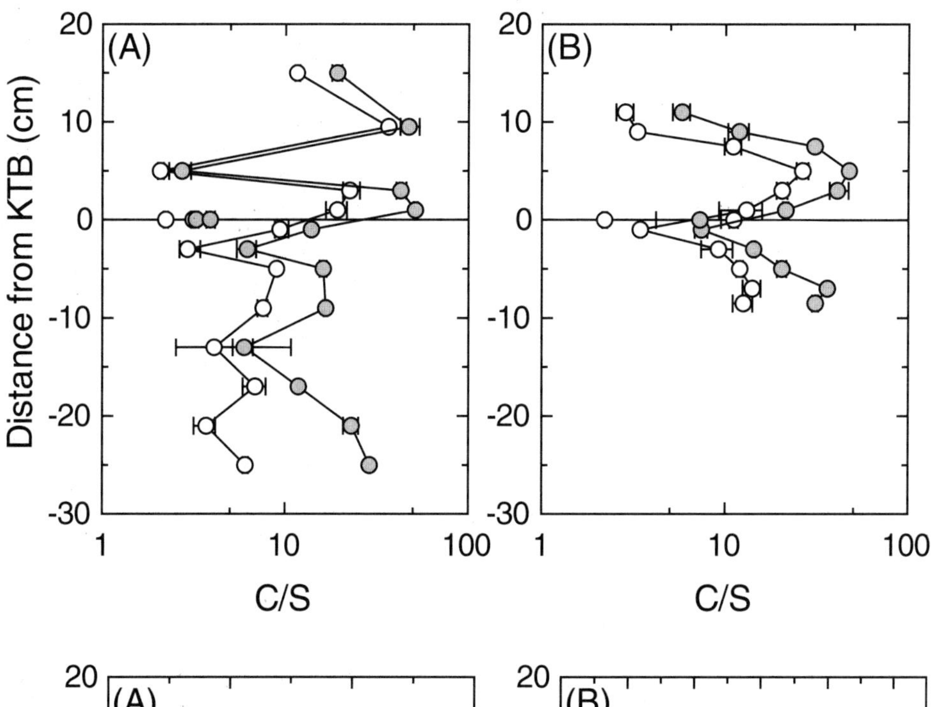

Figure 2. Ratio of organic carbon to sulfur across Cretaceous-Tertiary (K-T) boundary (KTB) at (A) Dogie Creek and (B) Brownie Butte. Open and gray circles represent data for bulk sample and 5 M HCl residue, respectively. Distances from K-T boundary are given from center of boundary layer.

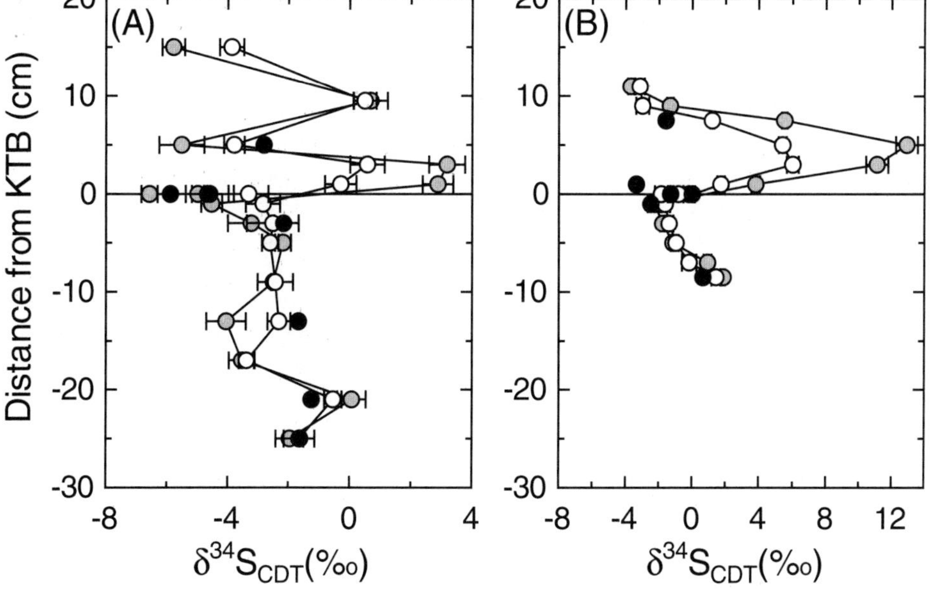

Figure 3. Isotopic composition of sulfur across Cretaceous-Tertiary (K-T) boundary (KTB) at (A) Dogie Creek and (B) Brownie Butte sites. Open, gray, and closed circles represent data for bulk sample, 5 M HCl residue, and acid-volatile sulfate, respectively. Distances from K-T boundary are given from center of boundary layer. CDT is Canyon Diablo Troilite.

at the period of melt ejecta deposition might be much higher relative to those at other periods. The accumulation of sulfides in lake sediments increases as the sulfate concentration increases (e.g., Nriagu and Coker, 1983; Fry, 1986). The melt ejecta and/or acid rain induced by the K-T impact event (Sigurdsson et al., 1992) may have led to such a higher concentration in the wetlands.

An increase of Fe^{2+} ions also causes the high concentration of pyritic sulfur. However, a higher amount of Fe^{2+} ions may be accompanied with a higher $\delta^{34}S$ value, as stated in detail in the following. Therefore, it cannot explain the low C/S ratio at the boundary.

Sheehan and Fastovsky (1992) noted only a minor extinction of freshwater taxa (10%) at the K-T boundary. This observation appears to be inconsistent with the acid rain scenario. However, acid rain does not have to cause the extinction of freshwater animals, if the acid addition did not exceed the ability of acid-neutralization of the wetlands. Moreover, olivine, which might have formed during the K-T impact, and subsequently altered to clay minerals (Evans et al., 1994), can act as an acid buffer (cf. Schuiling et al., 1986). During dissolution of olivine in acid, Mg^{2+} ions are replaced by H^+, yielding $Si(OH)_4$ monomers and Mg^{2+} ions in solution. Because the rate of neutralization depends on the geometrical surface area of the olivine grains (e.g., Jonckbloedt, 1998), small or porous grains of the condensed mafic droplets induced by the impact may

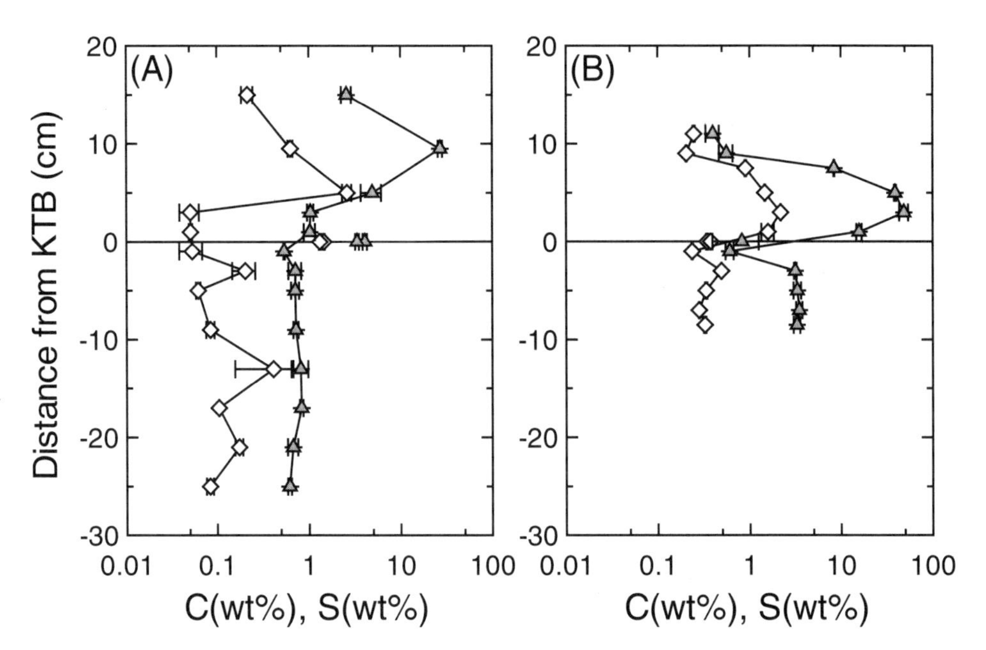

Figure 4. Concentration profiles of organic carbon and total sulfur across Cretaceous-Tertiary (K-T) boundary (KTB) at (A) Dogie Creek and (B) Brownie Butte. Triangles and diamonds represent data for carbon and sulfur, respectively. Distances from K-T boundary are given from center of boundary layer.

improve neutralization. Therefore, neutralization by olivine grains might help to prevent the extinction of freshwater animals. During neutralization, anions, such as sulfate, should remain in solution. The scenario of olivine-assistant neutralization is consistent with the high input of sulfate from melt ejecta and/or acid rains.

Above the K-T boundary. Concentrations of organic carbon and total sulfur reached a maximum after the K-T boundary (Fig. 4). The variation of carbon concentrations should reflect changes in the environmental factors, such as deposition rate, oxygen concentration in wetland waters, organic input, and water temperature. However, because the distance above the K-T boundary of the Dogie Creek section to the maximum in total S concentration (4–6 cm above the boundary; Fig. 4) is different from that of the organic C concentrations (8–11 cm above the boundary; Fig. 4), the environmental conditions that caused the variation of sulfur and organic concentrations might have been different.

The maximum concentration of nonorganic sulfur corresponds with minimum $\delta^{34}S$ values above the K-T boundary (including the boundary layer) for the Dogie Creek sediments. The maxima at the melt-ejecta and fireball layer and at 4–6 cm above the boundary (LC-86-1-M) were accompanied by a minimum of the $\delta^{34}S$ values; i.e., values are negatively correlated with the concentration of sulfur.

Under conditions in which the concentration of Fe^{2+} ions is not limited by the formation of pyrite, the higher concentration of sulfate ions results in a higher residual fraction of sulfate, reflecting the higher isotopic fractionation between sulfate and microbiological H_2S (the lower $\delta^{34}S$ values). Sulfate reduction rate, R, can be expressed using the concentration of sulfate, [S], as $1/R = a/[S] + b$, where a and b are constants ($a, b > 0$) (Boudreau and Westrich, 1984). The fraction of reduced sulfate, F, can be obtained as $F = (\Sigma Rdt)/[S]_o$, where $[S]_o$ represents the initial sulfate concentration in the water incorporated into the sediment in which bacterial reduction occurs. The final amount of reduced sulfide will be controlled by the initial reduction rate so that the fraction of reduced sulfide is roughly correlated to $R \times \Delta t/[S]_o = \Delta t/(a + b[S]_o)$, where Δt (time) is the interval of the deposition. Consequently, an increase in the initial sulfate concentration would cause a decrease in the fraction of reduced sulfate, reflected by an increase in isotopic fractionation between reduced products, H_2S, and initial sulfate (the decrease of $\delta^{34}S$ values in pyrite). This process causes a negative correlation between sulfide concentrations and $\delta^{34}S$ values. As dissolved Fe^{2+} ions are derived from the reductive dissolution of oxide phases (Canfield et al., 1996), water with saturated Fe^{2+} ions would have been anoxic during the period with S-$\delta^{34}S$ negative correlation. Therefore, reduction under anoxic conditions is likely to result in the negative correlation between S and $\delta^{34}S$ observed above the K-T boundary. The bottom of the wetland should be anoxic, inferred from the presence of sulfur reduced by sulfate-reducing bacteria; however, the saturation of Fe^{2+} may require anoxia of some fraction of the non-bottom water to supply enough Fe^{2+} ions.

The variation of the abundance of Fe^{2+} cannot cause S-$\delta^{34}S$ negative correlation. An increase of Fe^{2+} ions promotes pyrite formation from H_2S before it reoxidizes to sulfate. Therefore, the increase in Fe^{2+} ions leads to an increase in the reduction speed, and therefore to a decrease in the amount of residual sulfate. The lower amount of residual sulfate causes the smaller isotopic fractionation between pyrite and initial sulfate, and therefore the higher $\delta^{34}S$ in the pyritic sulfur. The high amount of reactive Fe^{2+} should cause the high concentration and high $\delta^{34}S$ value of pyritic sulfur. Although parameters such as abundance of organic matter and oxygen also affect the reduction speed, their variations also may cause S-$\delta^{34}S$ positive correlation as the variation of the abundance of Fe^{2+} should cause.

After the concentration maximum of S and organic C of Dogie Creek sediments above the K-T boundary, no negative correlation between $\delta^{34}S$ values and concentration of sulfur is observed. Therefore, the anoxic conditions may be short lived and of extraordinary nature.

CONCLUSIONS

We have presented data for sulfur contents and isotopic compositions across two K-T boundary successions. From these results, we derived the following observations and conclusions.

The concentrations of organic carbon were constant below the K-T boundary (26 cm at the Dogie Creek and 9 cm at Brownie Butte sites), suggesting constancy of such parameters as depositional rate, oxygen concentration, and organic input to the wetland.

At the K-T boundary, sulfates in wetlands water might be much more abundant than those before and after the K-T boundary, which causes low C/S ratios, similar to those observed in marine sediments. The high input of sulfate to the wetlands may have been caused by the melt ejecta and/or acid rain, induced by the Chicxulub impact event.

Just above the K-T boundary at the Dogie Creek site, the concentrations of nonorganic sulfur are negatively correlated with the $\delta^{34}S$ values, suggesting that the $\delta^{34}S$ values might not be controlled by the concentration of Fe^{2+} ions. This means that the wetlands water at the Dogie Creek site might have been anoxic for the period of deposition.

After the maximum concentration of organic C and nonorganic S, the anoxic conditions of the wetland water might have ceased and returned to normal conditions.

ACKNOWLEDGMENTS

This work was supported by the Austrian Science Foundation project Y58-GEO (to Koeberl). We thank T. Lyons, K. MacLeod, B. Schmitz, C. Pilmore, and J. Hatch for their helpful comments and critical reviews.

REFERENCES CITED

Berner, R.A., 1984, Sedimentary pyrite formation: An update: Geochimica et Cosmochimica Acta, v. 48, p. 605–615.

Berner, R.A., and Raiswell, R., 1984, C/S method for distinguishing freshwater from marine sedimentary rocks: Geology, v. 12, p. 365–368.

Bohor, B.F., Foord, E.E., Modreski, P.J., and Triplehorn, D.M., 1984, Mineralogic evidence for an impact event at the Cretaceous-Tertiary boundary: Science, v. 224, 867–869.

Bohor, B.F., Triplehorn, D.M., Nichols, D.J., and Millard, H.T., Jr., 1987, Dinosaurs, spherules, and the "magic" layer: A new K-T boundary clay site in Wyoming: Geology, v. 15, p. 896–899.

Boudreau, B.P., and Westrich, J.T., 1984, The dependence of bacterial sulfate reduction on sulfate concentration in marine sediments: Geochimica et Cosmochimica Acta, v. 48, p. 2503–2516.

Canfield, D.E., Lyons, T.W., and Raiswell, R., 1996, A model for iron deposition to euxinic black sea sediments: American Journal of Science, v. 296, p. 818–834.

Chambers, L.A., and Trudinger, P.A., 1979, Microbiological fractionation of stable sulfur isotopes: A review and critique: Geomicrobiology Journal, v. 1, p. 249–293.

Coplen, T.B., and Krouse, H.R., 1998, Sulphur isotope data consistency improved: Nature, v. 392, p. 32.

Evans, N.J., Gregoire, D.C., Goodfellow, W.D., Miles, N., and Veizer J., 1994, The Cretaceous-Tertiary fireball layer, ejecta layer and coal seam: Platinum-group element content and mineralogy of size fractions: Meteoritics, v. 29, p. 223–235.

Fry, B., 1986, Stable sulfur isotopic distributions and sulfate reduction in lake sediments of the Adirondack Mountains, New York: Biogeochemistry, v. 2, p. 329–343.

Giesemann, A., Jäger, H.-J., Norman, A.L., Krouse, H.R., and Brand, W.A., 1994, On-line sulfur-isotope determination using an elemental analyzer coupled to a mass spectrometer: Analytical Chemistry, v. 66, p. 2816–2819.

Henrichs, S.M., and Reeburgh, W.S., 1987, Anaerobic mineralization of marine sediment organic matter: Rate and the role of anaerobic processes in the oceanic carbon economy: Geomicrobiology Journal, v. 5, p. 191–237.

Heymann, D., Yancey, T.E., Wolbach, W.S., Thiemens, M.H., Johnson, E.A., Roach, D., and Moecker, S., 1998, Geochemical markers of the Cretaceous-Tertiary boundary event at Barazos River, Texas, USA: Geochimica et Cosmochimica Acta, v. 62, p. 173–181.

Holmes, C.W., and Bohor, B.F., 1994, Stable isotope (C, S, N) distributions in coals spanning the Cretaceous/Tertiary boundary in the Raton Basin, Colorado and New Mexico [abs.]: U.S. Geological Survey Circular 1107, p. 141.

Holser, W.T., Schidlowski, M., Mackenzie, F.T., and Maynard, J-B., 1988, Geochemical cycles of carbon and sulfur, *in* Gregor, C.B, Garrels, R.M., Mackenzie, F.T., and Maynard, J.B., eds., Chemical cycles in the evolution of the earth: New York, John Wiley and Sons, p. 105–173.

Jonckbloedt, R.C.L., 1998, Olivine dissolution in sulphuric acid at elevated temperatures: Implications for the olivine process, an alternative waste acid neutralizing process: Journal of Geochemical Exploration, v. 62, p. 337–346.

Kajiwara, Y., and Kaiho, K., 1991, Sulfur isotopic data from the Cretaceous/Tertiary boundary sediments in the eastern Hokkaido, Japan: Annual Report of the Institute of Geoscience, University of Tsukuba, v. 17, p. 68–73.

Kajiwara, Y., and Kaiho, K., 1992, Oceanic anoxia at the Cretaceous/tertiary boundary supported by the sulfur isotopic record: Palaeogeography, Palaeoclimatology, Palaeoecology, v. 99, p. 151–162.

Kaplan, I.R., and Rittenberg, S.C., 1964, Microbiological fractionation of sulphur isotopes: Journal of General Microbiology, v. 34, p. 195–212.

Nriagu, J.O., and Coker, R.D., 1983, Sulphur in sediments chronicles past changes in lake acidification: Nature, v. 303, p. 692–694.

Pillmore, C.L., and Flores, R.M., 1987, Stratigraphy and depositional environments of the Cretaceous-Tertiary boundary clay and associated rocks, Raton Basin, New Mexico and Colorado, *in* Fassett, J.E., and Rigby, J.K., Jr., eds., The Cretaceous-Tertiary boundary in the San Juan and Raton Basins, New Mexico and Colorado: Geological Society of America Special Paper 209, p. 111–129.

Raiswell, R., and Berner, R.A., 1986, Pyrite and organic matter in Phanerozoic normal marine shales: Geochimica et Cosmochimica Acta, v. 50, p. 1967–1976.

Schmitz, B., 1989, Recent formation of baraite and celestite in weathering Cretaceous-Tertiary boundary clays: International Geological Congress, 28th, Washington, D.C., Abstracts, v. 3, p. 51.

Schmitz, B., Andersson, P., and Dahl, J., 1988, Iridium, sulfur isotopes and rare earth elements in the Cretaceous-Tertiary boundary clay at Stevns Klint, Denmark: Geochimica et Cosmochimica Acta, v. 52, p. 229–236.

Schuiling, R.D., and van Herk, J., and Pietersen, H.S., 1986, A potential process for the neutralization of waste acids by reaction with olivine: Geologie en Mijnbouw, v. 65, p. 243–246.

Sheehan, P.M., and Fastovsky, D.E., 1992, Major extinctions of land-dwelling vertebrates at the Cretaceous-Tertiary boundary, eastern Montana: Geology, v. 20, p. 556–560.

Sigurdsson, H., D'Hondt, S., and Carey, S., 1992, The impact of the Cretaceous/Tertiary bolide on evaporate terrane and generation of major sulfuric acid aerosol: Earth and Planetary Science Letters, v. 109, p. 543–559.

Tschudy, R.H., 1970, Palynology of the Cretaceous-Tertiary boundary in the northern Rocky Mountain and Mississippi Embayment regions: Geological Society of America Special Paper 127, p. 65–111.

Yanagisawa, F., and Sakai, H., 1983, Thermal decomposition of barium sulfate-vanadium pentaoxide-silica glass mixtures for preparation of sulfur dioxide in sulfur isotope ratio measurements: Analytical Chemistry, v. 55, p. 985–987.

Manuscript Submitted November 3, 2000; Accepted by the Society March 22, 2001

Printed in the U.S.A.

Geological Society of America
Special Paper 356
2002

Natural fullerenes from the Cretaceous-Tertiary boundary layer at Anjar, Kutch, India

G. Parthasarathy
National Geophysical Research Institute, Hyderabad-500 007, India
N. Bhandari*
Physical Research Laboratory, Ahmedabad-380 009, India
M. Vairamani
A.C. Kunwar
B. Narasaiah
Indian Institute of Chemical Technology, Hyderabad-500 007, India

ABSTRACT

We report here the presence of fullerenes in the iridium-rich Cretaceous-Tertiary (K-T) boundary layer in the intertrappean beds of Anjar within the Deccan Volcanic Province. Fullerenes (C_{60}) have been identified in or near the three iridium-rich horizons from this section by high-resolution electron-impact ionization mass spectrometry, Fourier transform infrared (FT-IR), and ^{13}C-nuclear magnetic resonance spectroscopic techniques. Fullerenes are absent in six other iridium-poor horizons of the same section. The association of high-pressure–high-temperature forms of fullerenes with high iridium concentrations suggests that fullerenes can be used as geochemical indicators of an extraterrestrial impact.

INTRODUCTION

Fullerenes have been synthesized in the laboratory by gas-phase chemistry at high temperatures (>1300 K; Kroto et al., 1985). However, there are also a few locations on Earth where natural fullerenes have been found. These include some low-pressure metamorphic rocks such as shungite from Russia (Buseck et al., 1992; Parthasarathy et al., 1998), clays at the Cretaceous-Tertiary (K-T) boundary (Heymann et al., 1994a, 1994b, 1996, 1998), claystone from the Permian-Triassic (P-T) boundary (Chijiwa et al., 1999), the Sudbury impact structure (Becker et al., 1994), and in a micrometeorite crater formed on a space platform (Di Brozolo et al., 1994). Here we report the first occurrence of fullerenes from the intertrappean sediments of Kutch, on the Indian subcontinent, that are believed to be associated with the K-T boundary.

The first natural occurrence of fullerenes was reported by Buseck et al. (1992) in ca. 2 Ga Precambrian carbon-rich rocks of Shunga (Russia) called shungite. Subsequently, other workers found fullerenes in rocks that underwent singular geological events such as lightning strikes (Daly et al., 1993), and shock-produced impact-generated breccias from the Sudbury meteorite crater (Becker et al., 1994). Fullerenes presumably produced in wildfires at the K-T boundary (Heymann et al., 1996) and from samples of P-T boundary sections (Chijiwa et al., 1999) have been reported. The occurrence of fullerenes and an increase in carbon soot content are considered to be important geochemical markers of the K-T boundary (Heymann et al., 1998). Mita and Shimoyama (1999) reported ~26 polycyclic aromatic hydrocarbons from naphthalene to coronene from the K-T boundary sediments at Kawaruppu, Hokkaido, Japan, indicating a possible combustion origin of these hydrocarbons

*E-mail: bhandari@prl.ernet.in.

Parthasarathy, G., Bhandari, N., Vairamani, M., Kunwar, A.C., and Narasaiah, B., 2002, Natural fullerenes from the Cretaceous-Tertiary boundary layer at Anjar, Kutch, India, *in* Koeberl, C., and MacLeod, K.G., eds., Catastrophic Events and Mass Extinctions: Impacts and Beyond: Boulder, Colorado, Geological Society of America Special Paper 356, p. 345–350.

within the K-T boundary clays. However, no fullerenes were found at these sites. Becker et al. (1996, 2000a, 2000b) reported trapped noble gases in K-T boundary fullerenes, indicating their extraterrestrial origin. Although fullerenes have been reported to form by high-temperature benzene flames, they have not been detected in modern-day wildfires (Becker et al., 2000b). Despite the multiple and poorly understood processes that may produce fullerenes in nature, there appears to be an association of these compounds with the K-T and P-T boundaries. This fact, combined with the otherwise rare occurrence of these carbonaceous materials in nature, makes fullerenes potential chemical markers for certain geological boundaries. In this chapter we investigate whether fullerenes are present in the intertrappean sediment beds at Anjar, India, and whether their presence can help us confirm that these beds represent a continental K-T boundary within the Deccan flood basalt area (Bhandari et al., 1995; Shukla et al., 1997).

GEOLOGICAL SETTING

The Anjar volcano-sedimentary sequence is located at the western periphery of the Deccan flood basalt province and consists of nine lava flows and at least four intertrappean beds (Ghevariya, 1988; Bhandari et al., 1995). The third intertrappean bed is well developed (~6 m thick) and consists of cherty limestone, shale, and mudstone (Fig. 1). Three thin limonitic layers are present in the lower 1.5 m of the intertrappean bed that was exposed in pit BG-1 (Shukla et al., 1997). The limonitic layers have high concentrations of iridium (650–1333 pg/g) and osmium (650–2230 pg/g) (Bhandari et al., 1996) compared to <100 pg/g measured in other horizons of this pit. On the basis of the geochronological, geochemical, paleomagnetic, and paleontological data (Venkatesan et al., 1996; Bhandari, 1998), it is believed that the limonitic layers were deposited during or close to the K-T boundary event. The deposition rate of sediments in the intertrappean beds is much higher than in the contemporaneous marine sections; therefore, they offer a better time resolution. A study of the intertrappean sediments should allow the sequence of events that occurred around the K-T boundary and their relation to Deccan volcanism to be better defined.

EXPERIMENTAL METHODS AND RESULTS

Nine samples from the third intertrappean bed, collected from pit BG1 during 1994, were analyzed in this study. Their stratigraphic locations with respect to the three iridium-rich layers (Br-1, Br-2, Br-3) are shown in Figure 1 (Shukla et al., 1997). Sample 964 was located just below the iridium-rich brown layer (BR-1), sample L is the limonitic layer (BR-2) containing the highest concentration of iridium in this section, and samples Anjar-J and M represent the third iridium-rich layer (Br-3). Anjar-M was collected from a location laterally displaced with respect to that of Anjar-J by about 3 m.

The carbonaceous matter was extracted from the powdered samples by using a standard acid-digestion method (Wedeking et al., 1983). All the samples were treated at room temperature with 12 N HCl for 24 h followed by 60% HF for 18 h. Thereafter the samples were heated in 60% HF at 70 °C for 1 h. The resulting solid residue was washed with ~1 L of double-distilled water and dried overnight at 80 °C. The carbon-rich fraction so obtained was analyzed by Fourier transform infrared (FT-IR) spectroscopy. A fraction of the carbon-rich residue was also treated with toluene or d6-benzene for the extraction of fullerenes, which were then characterized by mass spectroscopy and ^{13}C-nuclear magnetic resonance (NMR) spectroscopy. Only four samples, Anjar-J and M (Ir layer, Br-3), L (Ir layer, Br-2), and 964 (just below Ir layer Br-1) yielded some extractable fullerenes (Fig. 1). The amounts of the carbonaceous matter extracted from these three samples were ~1.3%–3.8%. The other samples yielded only neo-formed fluoride residues as fluorite crystals, and sample 965 had a noticeable amount of pyrite.

The electron impact ionization mass spectra were obtained with a VG AUTOSPEC mass spectrometer and an OPUS. V3.IX data system. The conditions of measurements were as follows: source temperature, 250 °C; electron energy, 70 eV; trap current, 200 mA; resolution, 1000; mass range scanned, 300–900 amu; scan time, 5 s per decade of mass; interscan delay, 0.5 s. Samples were loaded in a quartz cup, introduced through direct inlet probe, and instantly heated to 250 °C. The mass spectrometer was calibrated to 900 amu with the reference sample of high-purity perfluorokerosene. The samples were scanned between 300 and 900 amu to improve the sensitivity of detection of C_{60} and C_{70}. The spectra, given in Figure 2, clearly show peaks corresponding to C_{60}^{++} and C_{60}^{+} at 360 amu (for doubly charged ions) and 720 amu, respectively, and ions corresponding to C_{70}^{+} are not present. Although a single scan gave enough signal to identify C_{60}, an average of 20 scans was used for better precision during analysis. C_{60} was resolved into four peaks at 720, 721, 722, and 723 amu, corresponding to $^{12}C_{60}^{+}$, $^{12}C_{59}\,^{13}C^{+}$, $^{12}C_{58}\,^{13}C_2^{+}$, and $^{12}C_{57}\,^{13}C_3^{+}$, respectively, as schematically shown in Figure 3. It is interesting to note that peaks corresponding to the addition of as many as 3 oxygen atoms (mass to charge ratio, m/z = 736, 752 and 768) are observed in some of the samples. These peaks might be due to the oxidized fullerenes, because they are very susceptible to oxidation. T. Bunch (2001, personal commun.) has also observed the presence of oxidized fullerenes in samples that were accidentally exposed to oxidizing conditions during sample preparation. However, we have not observed these peaks in other samples, which were prepared under similar experimental conditions. Their origin is therefore not yet fully understood. In order to avoid any contamination, we did not use any synthetic fullerenes for estimating the concentration of fullerenes in various samples. Therefore, our estimate of fullerene content in the samples (Fig. 1) is approximate and based only on the intensity of the peaks at the mass to charge ratio m/z = 720.

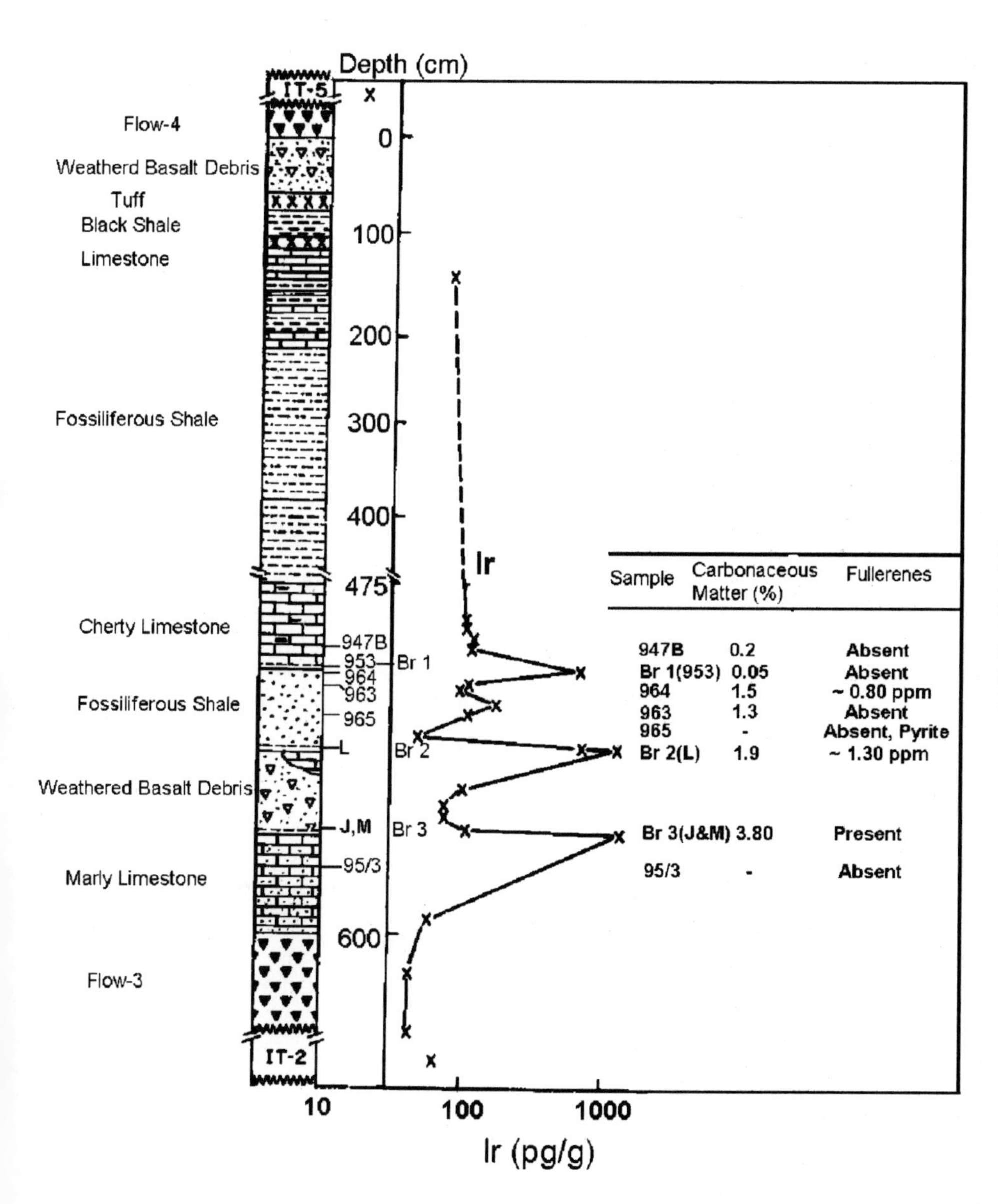

Figure 1. Stratigraphic locations of samples analyzed for fullerenes with respect to iridium-rich layers (Br-1, Br-2, Br-3; Bhandari et al., 1996). Content of carbonaceous matter and approximate concentration of fullerenes at various horizons are given. Details of samples L, J, M and others are given in text.

Because we have not observed any peaks beyond C_{60}, we have not attempted to search for higher mass fullerenes. The highest concentration of fullerenes was found in sample L, which contains the highest average concentration of iridium (1287 pg/g) (Bhandari et al., 1996).

Because the high-pressure–high-temperature forms of C_{60} fullerenes are insoluble in toluene or benzene (Kozlov and Yakashi, 1995), we have carried out FT-IR spectroscopic studies directly on the carbon-rich residues of the samples of 964, L, and Anjar-J to confirm their presence. The standard KBr pellet method was employed and the IR-transmission spectra were collected in the frequency range 400–4000 cm^{-1} at room temperature, by using a Biorad-FT-IR spectrometer. All the samples exhibit identical spectral bands with minor change in intensities. IR bands at 1430, 1180, 575, and 525 cm^{-1}, which are characteristic of the C_{60} fullerenes F_{1U} vibrational mode (Kozlov and Yakashi, 1995), have been observed in all three samples. In addition, a few vibrational bands at 2920, 2850, and 1380 cm^{-1}, which are characteristic of aliphatic CH, CH_2, and CH_3 groups, have also been observed. Well-resolved peaks at ~740 and 509 cm^{-1} in these samples indicate the presence of a high-pressure–high-temperature form of fullerenes. The presence of this high-pressure–high-temperature form of fullerenes (probably subjected to 3 GPa and 800 K; Kozlov and Yakashi, 1995) suggests an impact origin, probably during the K-T boundary impact.

We also carried out ^{13}C-NMR spectroscopic studies on the powdered carbonaceous material to confirm the presence of fullerenes. The ^{13}C-NMR spectrum was recorded in benzene-d6 at 125 MHz on a Varian INOVA 500 NMR spectrometer. We collected 3200 free induction decays using 30° pulse width and 15 s relaxation delay. The ^{13}C-NMR spectrum (with an acceptable signal to noise ratio) exhibits clearly the presence of a single NMR line at 143.28 ppm for sample L (Fig. 4) and

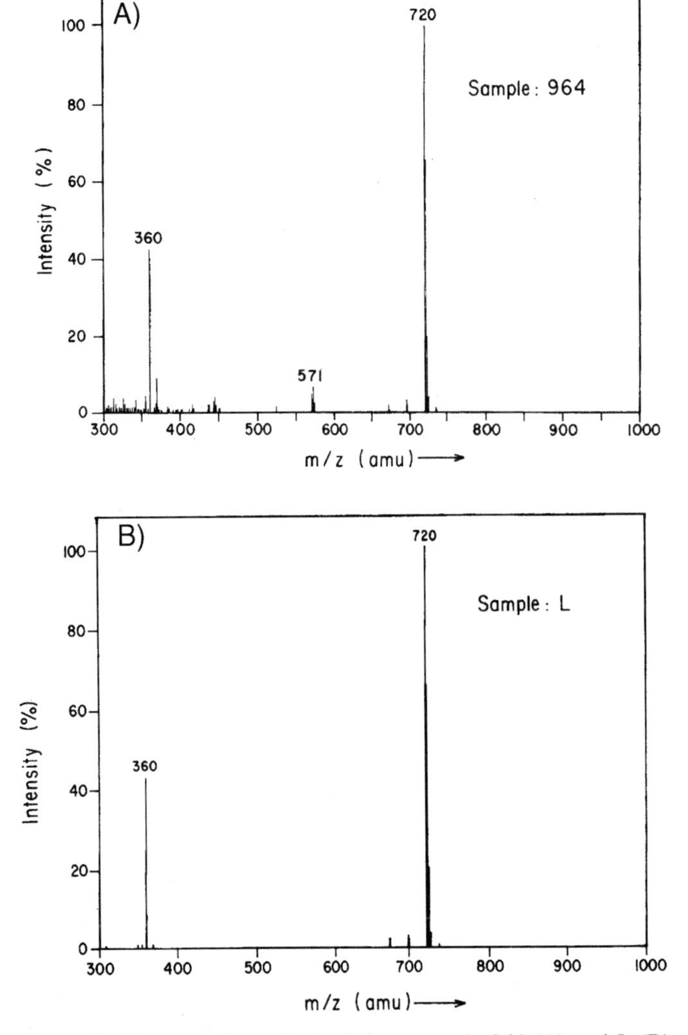

Figure 2. Mass spectrum obtained from sample 964 (A) and L (B) showing peaks at m/z = 360 and 720, characteristic of C_{60} ions.

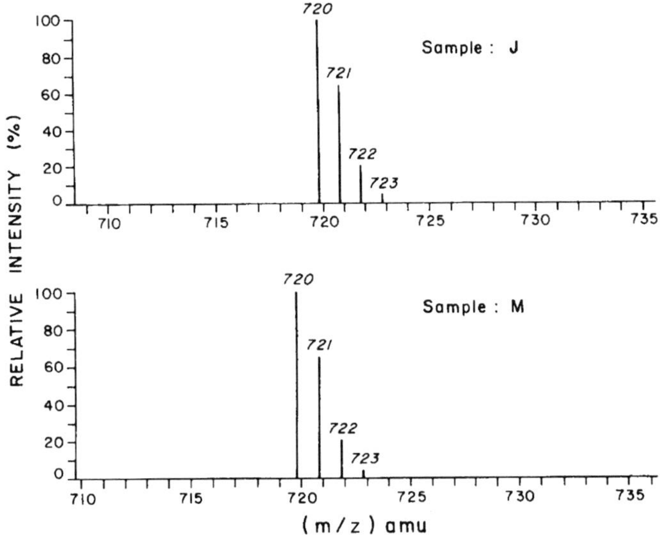

Figure 3. High-resolution mass spectra around peak at m/z = 720 for samples Anjar-J and Anjar-M, which are resolved into four peaks at 720, 721, 722, and 723 amu. Data are systematically shifted slightly toward lower values of m/z.

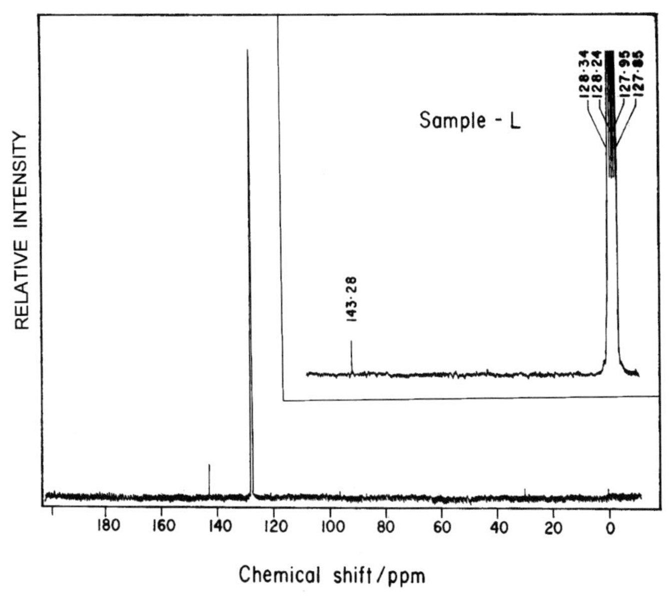

Figure 4. ^{13}C-NMR (nuclear magnetic resonance) spectrum showing single NMR line at 143.278 ppm for sample L, corresponding to sp^2 carbons. Peaks at ~128, which can be resolved into six peaks (shown in inset) are due to d6-benzene solvent.

143.24 ppm for sample 964, corresponding to the sp^2 carbons of the C_{60} fullerenes (Ajie et al., 1990). The peaks expected for C_{70} at 130.9, 145.4, 147.4, 148.1, and 150.7 ppm are not observed in any of the samples, corroborating the results obtained by mass spectroscopic studies. The concentration of C_{60} fullerene is approximate, but appears to be higher than the values found in K-T boundary samples at other locations (Heymann et al., 1996), and lower than some natural samples such as Karelian shungite, which was estimated to be ~0.01–0.1 wt%. The C_{70} abundance is found to be between one-third and one-fifth relative to C_{60} in a few K-T boundary clays, whereas it is absent at many other locations, e.g., at Woodside Creek (Heymann et al., 1996, 1998). Shungites of Karelia (Parthasarathy et al., 1998) and the P-T boundary (Chijiwa et al., 1999) also show an absence of C_{70}. It has been suggested that the absence of C_{70} may be due to weathering and differential oxidation, as indicated by the results of Chibante et al. (1993), although the experiments of Heymann et al. (1996) disagree with this explanation. Becker et al. (1994) suggested that the survival of fullerenes in the Sudbury crater is due to the presence of diagenetic sulfides, indicating low oxygen and high sulfur concentrations. In such an environment, fullerenes were protected from oxidation by their surrounding sulfide-silicate matrix. However, in the Anjar samples, despite the noticeable sulfur content in samples 965, just above Br-2 (Fig. 1), C_{70} is not found. We therefore believe that C_{70} is absent at Anjar because it was not deposited, rather than having been later lost by oxidation. The occasional presence of C_{70} at the K-T boundary, therefore, still remains to be fully understood in terms of its production and preservation.

DISCUSSION

Several mechanisms for origin of fullerenes at the K-T boundary have been proposed. These include (1) survival of fullerenes that were already present in the K-T impactor. Trapped ^{3}He in fullerenes, reported by Becker et al. (2000a), support this mechanism. Pentagonal-shaped fullerenes-like nanoparticles 2–10 nm in diameter were found in the Allende meteorite by Harris et al. (2000), suggesting that these nanoparticles could be carriers of planetary gases. (2) Another mechanism is the formation of fullerenes during combustion of carbonaceous material in wildfires created by the K-T impact (Heymann et al., 1994a, 1994b, 1996, 1998). However, the absence of fullerenes in modern wildfires reported by Becker et al. (2000a) makes this mechanism improbable (3) Heymann et al. (1998) suggested that fullerenes may have formed in the hot atmospheric plume of the Chicxulub crater (Smit, 1999). Heymann et al. (1996) estimated that the mean global C_{60} concentration at the K-T boundary is 14 ng/cm^2, which corresponds to a mean C_{60} content of 0.41 ppm in the soot. They suggested that these values are consistent with the formation of fullerenes by wildfires than by carbon chemistry in the hot atmospheric plume of the Chicxulub impact (Heymann et al., 1996). Whether these wildfires were impact related or were caused by volcanism remains to be established. It is difficult to get further insight into the mechanism of formation of fullerenes, but we make use of the association of fullerenes with K-T boundary to understand the peculiar stratigraphic distribution of iridium at Anjar, where there are three iridium-rich horizons (Bhandari et al, 1996). The question has arisen whether these represent three independent events of deposition or whether the three layers have been enriched in iridium due to secondary processes such as downward fluid mobilization during the emplacement of the upper basaltic flow F4 (Courtillot et al., 2000). Because fullerenes and iridium are both insoluble in water, it is unlikely that their coexistence in three different layers (Fig. 1) is due to fluid mobilization. The presence of any other impact markers, e.g., planar deformation features in quartz or the existence of Ni-rich spinels, could have been useful in understanding the depositional history of the Anjar K-T boundary section, but their search has not yielded any positive results (Courtillot et al., 2000). The present experimental studies reveal the presence of trace amounts of the high-pressure form of fullerenes C_{60}, indicating that they are formed by impact, and their association with iridium-rich horizons supports their association with the K-T boundary event.

ACKNOWLEDGMENTS

We are grateful to Govardhan Mehta, K.V. Raghavan, and H.K. Gupta for providing facilities and support for this work. We thank A.D. Shukla for his help in preparation of the samples. This chapter was improved substantially by the constructive comments of Ted Bunch, Mark Sephton, and C. Koeberl.

REFERENCES CITED

Ajie, H., Alvaris, M.M., Anz, S.J., Beck, R.D., Diederich, F., Fost, K., Poulous, I.R.O., Huffman, D.R., Krafschmer, W., Rubin, Y., Shriber, K.E., Sensharma, U., and Whepten, R.L., 1990, Characterization of the soluble all carbon molecules C_{60} and C_{70}: Journal of Physical Chemistry, v. 94, p. 8630–8633.

Becker, L., Bada, J.L., Winans, R.E., Hunt, J.E., Bunch, T.E., and French, B.M., 1994, Fullerenes in the 1.85 billion year old Sudbury impact structure: Science v. 265, p. 642–645.

Becker, L., Poreda, R.J., and Bunch, T.E., 2000a, Fullerenes: An extraterrestrial carbon carrier phase for noble gases: Proceedings of the National Academy of Sciences, v. 97, p. 2979–2983.

Becker, L., Poreda, R.J., and Bunch, T.E., 2000b, The origin of fullerenes in the 65 Myr old Cretaceous/Tertiary 'K/T' boundary [abs.]: Lunar and Planetary Sciences, v. 31, p. 1832.

Becker, L., Poreda, R.J., and Bada, J.L., 1996, Extraterrestrial helium trapped in fullerenes in the Sudbury Impact Structure: Science, v. 272, p. 249–252.

Bhandari, N., 1998, Astronomical and terrestrial causes of physical, chemical and biological changes at geological boundaries: Proceedings of the Indian Academy of Sciences: Earth and Planetary Sciences, v. 107, p. 251–263.

Bhandari, N., Shukla, P.N., Ghevariya, Z.G., and Sundaram, S.M., 1995, Impact did not trigger Deccan volcanism: Evidence from Anjar K/T boundary intertrappean sediments: Geophysical Research Letters, v. 22, p. 433–436.

Bhandari, N., Shukla, P.N., Ghevariya, Z.G., and Sundaram, S.M., 1996, K/T boundary layer in Deccan Intertrappeans at Anjar, Kutch, *in* Ryder, G., Fastovsky, D., and Gartner, S., eds., The Cretaceous-Tertiary event and other catastrophes in Earth history: Geological Society of America Special Paper 307, p. 417–424.

Buseck, P.R., Tsipursky, S.J., and Hettich, R., 1992, Fullerenes from the geological environment: Science, v. 257, p. 215–217.

Chibante, L.P.F., Pan, C., Pierson, M.L., Haufler, R.E., and Heymann R., 1993, Rate of decomposition of C_{60} and C_{70} heated in air and the attempted characterization of the products: Carbon, v. 31, p.185–193.

Chijiwa, T., Arai, T., Sugai, T., Shinohara, H., Kumazawa, M., Takano, M., and Kawakami, S., 1999, Fullerenes found in the Permo-Triassic Mass extinction period: Geophysical Research Letters, v. 26, p. 767–770.

Courtillot, V., Gallet, Y., Rocchia, R., Feraud, G., Robin, E., Hofmann, C., Bhandari, N., and Ghevariya, Z.G., 2000, Cosmic markers, ^{40}Ar/^{39}Ar dating and paleomagnetism of the KT sections in the Anjar area of the Deccan large igneous province: Earth and Planetary Science Letters, v. 182, p. 137–156.

Daly, T.K., Buseck, P.R., Williams, P., and Lewis, C.F., 1993, Fullerenes from the fulgurite: Science, v. 259, p. 1599–1601.

Di Brozolo, F.R., Bunch, Th.E., Fleming, R.H., and Macmillan, J., 1994, Fullerenes in an impact crater on the LDEF spacecraft: Nature, v. 369, p. 37–40.

Ghevariya, Z.G., 1988, Intertrappean dinosaurian fossils from Anjar area, Kachchh District, Gujarat: Current Science, v. 57, p. 248–251.

Harris, P.J.F., Vis, R.D., and Heymann, D., 2000, Fullerene-like nanostructures in the Allende meteorite: Earth and Planetary Science Letters, v. 183, p. 355–359.

Heymann, D., Chibante, L.P.F., Brooks, P.R., Wolbach, W.S., and Smalley, R.E., 1994a, Fullerenes in the K/T boundary layer: Science, v. 265, p. 645–647.

Heymann, D., Wolbach, W.S., Chibante, L.P.F., Brooks, R.R., and Smalley, R.E., 1994b. Search for extractable fullerenes in clays from the Cretaceous/Tertiary boundary of the Woodside Creek and Flaxbourne river sites, New Zealand: Geochimica et Cosmochimica Acta, v. 58, p. 3531–3534.

Heymann, D., Chibante, L.P.F., Brooks, R.R., Wolbach, W.S., Smit, J., Korochantsev, A., Nazarov, M.A., and Smalley, R.E., 1996, Fullerenes of possible wildfire origin in Cretaceous-Tertiary boundary sediments, *in* Ryder, G., Fastovsky, D., and Gartner, S., eds., The Cretaceous-Tertiary event and other catastrophes in Earth history: Geological Society of America Special Paper 307, p. 453–464.

Heymann, D., Yancey, T.E., Wolbach, W.S., Thiemens, M.H., Johnson, E.A., Roach, D., and Moecker, S., 1998, Geochemical markers of the Cretaceous-Tertiary Boundary event at Brazos River, Texas, U.S.A.: Geochimica et Cosmochimica Acta, v. 62, p. 173–181.

Kozlov, M.E., and Yakushi, K., 1995, Optical properties of high-pressure phase of C_{60} fullerene: Journal of Physics (Condensed Matter), v. 7, p. L209–216.

Kroto, H.W., Heath, J.R., O'Brien, S.C., Curl, R.F., and Smalley, R.E., 1985, C_{60}: Buckminsterfullerene: Nature, v. 318, p. 162–163.

Mita, H., and Shimoyama, A., 1999, Characterization of n-alkanes, pristane and phytane in the K/T boundary sediments at Kawaruppu, Hokkaido, Japan: Geochemical Journal, v. 33, p. 285–294.

Parthasarathy, G., Srinivasan, R., Vairamani, M., Ravikumar, K., and Kunwar, A.C., 1998, Occurrence of natural fullerenes in low grade metamorphosed Proterozoic shungite from Karelia: Geochimica et Cosmochimica Acta, v. 62, p. 3541–3544.

Shukla, P.N., Shukla, A.D., and Bhandari, N., 1997, Geochemical characterisation of the Cretaceous-Tertiary Boundary sediments at Anjar, India: Palaeobotanist, v. 46, p. 127–132.

Smit, J., 1999, The global stratigraphy of the Cretaceous-Tertiary boundary impact ejecta: Annual Reviews of Earth and Planetary Sciences, v. 27, p. 75–113.

Venkatesan, T.R., Pande, K., and Ghevariya, Z.G., 1996, $^{40}Ar/^{39}Ar$ ages of Anjar Traps, western Deccan Province (India) and its relation to the Cretaceous-Tertiary boundary events: Current Science, v. 70, p. 990–996.

Wedeking, K.W., Hayes, J.M., and Matzigkeit, U., 1983, Procedures of organic geochemical analysis, *in* Schopf, J.W., ed., Earth's earliest biosphere: Its origin and evolution: New Jersey, Princeton University Press, p. 428–441.

Manuscript Submitted October 5, 2000; Accepted by the Society March 22, 2001

Printed in the U.S.A.

Geological Society of America
Special Paper 356
2002

Organic geochemical investigation of terrestrial Cretaceous-Tertiary boundary successions from Brownie Butte, Montana, and the Raton Basin, New Mexico

A.F. Gardner
I. Gilmour*
Planetary and Space Sciences Research Institute, The Open University, Milton Keynes MK7 6AA, UK

ABSTRACT

We have used coupled carbon and nitrogen isotope measurements together with molecular level isotope analysis of biomarker molecules to assess environmental changes across two nonmarine Cretaceous-Tertiary (K-T) boundary successions in the Western Interior of North America. Measurement of the carbon isotope compositions of individual molecules greatly increases the usefulness of biomarkers as indicators of specific sources.

The terrestrial organic matter carbon isotope record apparently reflects the 2‰–3‰ isotope excursion observed in the marine foraminiferal record, and the nitrogen isotope record contains a small positive excursion immediately above the boundary. Biomarker analysis indicates that the two major organic matter inputs were higher plants and algae. However, the relative contributions of these two sources vary above the K-T boundary toward a greater input from algae and bacteria. Biomarker and isotopic evidence for the presence of a methane-oxidizing bacterial community is also recorded immediately above the boundary, suggesting the onset of partially anoxic conditions in freshwater ecosystems.

INTRODUCTION

Since the initial discovery of an iridium enrichment at the Cretaceous-Tertiary (K-T) boundary at Gubbio, Italy (Alvarez et al., 1980), a substantial body of evidence has been amassed for a K-T impact, including enrichments in other elements, shocked minerals, impact-derived glasses, and proximal impact deposits from an apparent crater centered at Chicxulub on the Yucatan Peninsula in Mexico (Ganapathy, 1980; Smit and Klaver, 1981; Bohor et al., 1987b; Gilmour and Anders, 1989; Hildebrand et al., 1991; Kring and Boynton, 1991, 1992; Sigurdsson et al., 1991; Sharpton et al., 1992). However, the effects of the impact and, in particular, whether these were severe and widespread enough to explain the extinction patterns observed at the K-T boundary remain controversial. Forcing mechanisms proposed have included rock dust thrown into the atmosphere resulting in worldwide darkness and cooling (e.g., Alvarez et al., 1980; Toon et al., 1982; Pollack et al., 1983), and it is now apparent that an impact into a shallow-marine carbonate platform, such as that present at Chicxulub, would release CO_2 by shock-pressure–induced devolatization of $CaCO_3$ (O'Keefe and Ahrens, 1989) and increased atmospheric pCO_2 resulting in a major perturbation of Earth's climate. In addition, the discovery of soot in the basal boundary clay layer at Woodside Creek, New Zealand, and at other K-T sequences together with the presence of pyrogenic hydrocarbons (Wol-

*E-mail: I.Gilmour@open.ac.uk

Gardner, A.F., and Gilmour, I., 2002, Organic geochemical investigation of terrestrial Cretaceous-Tertiary boundary successions from Brownie Butte, Montana, and the Raton Basin, New Mexico, *in* Koeberl, C., and MacLeod, K.G., eds., Catastrophic Events and Mass Extinctions: Impacts and Beyond: Boulder, Colorado, Geological Society of America Special Paper 356, p. 351–362.

bach et al., 1985, 1988, 1990; Simoneit and Beller, 1987; Gilmour et al., 1989; Venkatesan and Dahl, 1989) is compelling evidence for widespread fires caused either by the energy from the impact or reentering ejecta (Anders et al., 1986; Melosh et al., 1990). The presence of soot in the atmosphere would add to the reduction of light at the surface as well as contributing CO_2 and other gases to the atmosphere, adding to an already environmentally stressed situation (Crutzen, 1987) and a global bioproductivity collapse.

Previous stable isotopic studies of carbon associated with K-T boundary sites have identified a 1‰–2‰ negative shift in the planktonic foraminiferal $\delta^{13}C$ record (e.g., Buchardt and Jørgensen, 1979; Boersma and Shackleton, 1981; Perch-Nielsen et al., 1982; Zachos and Arthur, 1986; Zachos et al., 1989). This shift has been interpreted as evidence for depressed marine bioproductivity resulting in ^{12}C being returned to the inorganic reservoir faster than it was utilized by photosynthesis (Hsü and McKenzie, 1985). The reduced carbon record is marked by the sharp increases in elemental carbon and soot concentrations over Cretaceous background values (Wolbach et al., 1988, 1990; Gilmour et al., 1989; Wolbach and Anders, 1989). The organic carbon record at marine sites is more variable with consistent enrichments in organic carbon abundance in the boundary clay and with changes in $\delta^{13}C_{org}$ values that can be resolved on a millimeter scale (Gilmour et al., 1990; Wolbach et al., 1990) although no significant isotopic or concentration changes are observed between the upper Maastrichtian and lower Paleocene records (Meyers, 1992). These small-scale variations in $\delta^{13}C_{org}$ have been interpreted as further evidence of reduced surface-water productivity in that they have been attributed to preferential fixation of ^{12}C by plankton blooms colonizing nutrient-rich surface waters as environmental conditions improved after the impact (Hollander et al., 1993a).

Few studies have been made of organic carbon in terrestrial sequences or of nitrogen isotope variations. Schimmelman and Deniro (1984) observed oscillations in $\delta^{13}C$ and $\delta^{15}N$ values at York Canyon, on the western side of the Raton Basin in New Mexico, United States. Gilmour et al. (1990) observed a marked positive shift in $\delta^{15}N$ values of total sediment nitrogen at Woodside Creek boundary clay, although the background nitrogen values were measured in limestone that was extremely low in organic carbon.

Studies of palynomorphs and plant megafossils across the K-T boundary of the Western Interior in North America suggest sudden and traumatic vegetation disturbance and extinction patterns that are heterogeneous (e.g., Hickey, 1981; Tschudy et al., 1984; Tschudy and Tschudy, 1986; Wolfe and Upchurch, 1986, 1987; Johnson, 1992), apparently representing an ecological catastrophe. Geochemical evidence for such a catastrophe would provide important additional information on the possible effects of the Chicxulub impact. We have therefore examined the organic geochemistry of two nonmarine K-T boundary sites Raton (New Mexico) and Brownie Butte (Montana). These two locations represent two different paleogeographical and paleoecological settings at the time of the impact.

The measurement of the carbon and nitrogen isotope compositions of sedimentary organic matter can provide clues for the reconstruction of paleoenvironments, but are also equivocal because of the wide variety of possible organic matter inputs to a particular depositional environment. However, the analysis of organic compounds and their stable isotope signatures has become an increasingly powerful tool in the reconstruction of biogeochemical processes and past depositional environments, and in determining the sources of extractable sedimentary lipids (Freeman et al., 1990; Collister et al., 1992). The carbon and nitrogen isotopic compositions of organic matter in samples across the K-T boundary from the two localities studied were determined and the distribution patterns of organic compounds and the carbon isotope compositions of specific biomarker molecules analyzed. In particular we have examined the distribution and carbon isotopic composition of hopanes, a bacterial biomarker. In the bacteria, bacteriohopanepolyols are amphiphilic membrane biochemicals that serve a regulating and rigidifying function (Jahnke et al., 1999). Their hydrocarbon skeletons are extremely refractory and resist biodegradation to become incorporated into sedimentary organic matter (Sinninghe Damste et al., 1995), providing an opportunity to examine bacterial inputs to sedimentary organic matter across the K-T boundary.

SAMPLES AND EXPERIMENTS

Samples

The continental K-T sites of the U.S. Western Interior are composed of two layers (Bohor et al., 1987a, 1987b; Hildebrand and Boynton, 1990). Above the uppermost Cretaceous material is a kaolinitic claystone (10–30 mm thick) that contains spherules believed to represent a distal impact ejecta blanket. This is overlain by a smectitic layer (2–3 mm thick), which contains shock-metamorphosed minerals, enrichments in trace metals (including an Ir anomaly of 14 ppb at Brownie Butte, 1.3 ppb at Raton), and soot (Hildebrand and Wolbach, 1989). The boundary is marked by an abrupt change in the abundance of fern spores relative to pollen (Tschudy et al., 1984; Nichols et al., 1986; Tschudy and Tschudy, 1986). At Brownie Butte a series of 12 samples in ≤2 cm intervals from 9 cm below the boundary to 14 cm above were analyzed. At Raton, the 13-mm-thick kaolinitic ejecta layer overlies Cretaceous shale. On top of this shale is the 3-mm-thick smectite (fireball) layer, which is overlain by coal. A series of 16 samples of varying intervals to 3 cm thick was taken across the boundary, which at the time of the terminal Cretaceous event was part of a large area consisting of flood plains and backswamps, allowing the deposition of sandstones interlayered with coal, mudstone, carbonaceous shale, and siltstone.

Acid demineralization and determination of carbon and nitrogen isotope composition

The samples were subjected to acid treatment to remove silicates using a cycle of alternating 9 M HCl followed by 10 M/1 M HF/HCl at room temperature. Any neoformed fluorides were removed using 20% $AlCl_3$. The resulting black or dark brown residue contains elemental carbon, kerogen (in variable amounts reflected in the color), and acid-insoluble minerals such as rutile. The total carbon content of the residues was determined by placing a weighed sample (~300 μg) with excess CuO in a sealed quartz tube. The sample was then combusted at ~850°C for a minimum of 2 h. Removal of gases other than CO_2 was carried out under vacuum and the amounts of CO_2 measured using a capacitance manometer. Carbon isotope measurements were made on a VG SIRA 24 mass spectrometer to a precision of ±0.1‰. The total nitrogen content of each residue was determined by sealing a weighed sample (~30 μg) in a quartz glass tube and combusting at ~950°C. Conversion of any nitrogen oxide gases to N_2, and CO, CH_4, and higher hydrocarbons to CO_2 were carried out using CuO and a Pt catalyst in a vacuum system prior to isotopic and concentration determination on a static vacuum mass spectrometer, to a precision of ±0.5‰ (Wright et al., 1988; Boyd and Pillinger, 1990).

The carbon and nitrogen isotopic results are reported as per mil deviations (‰) relative to the international standards, Peedee belemnite (PDB) for carbon, and atmospheric N_2 for nitrogen.

Gas chromatography–mass spectrometry

Soluble organic compounds were isolated from the samples using solvent extraction with a 50:50 mixture of dichloromethane and methanol. A saturated hydrocarbon fraction was prepared from the total solvent extract by silica gel column chromatography. Compound detection and identification was performed by gas chromatography–mass spectrometry (GC-MS) using a Hewlett Packard 5890 gas chromatograph interfaced with a 5971 mass selective detector. Analyses were by on-column injection onto an HP5 capillary column (50 m × 0.32 mm × 0.17 μm). Following a 10 min period at 25°C the GC oven was programmed from 25 to 220°C at 5°C min^{-1} and then from 220 to 300°C at 10°C min^{-1}. The final temperature was held for 12 min.

Gas chromatography–isotope ratio–mass spectrometry

Samples were introduced as solutions in isooctane into the split-splitless injector of a Varian 3400 gas chromatograph linked via a combustion interface to a Finnigan MAT Delta S/GC isotope ratio mass spectrometer (IRMS). Data acquisition and reduction were performed using Finnigan MAT Isodat software (version 5.2). For a detailed description of the principles behind the instrumental design of the GC-IRMS system, see Hayes et al. (1990). Gas chromatography was performed on a 25 m × 0.32 mm × 0.5 μm BPX5 capillary column (SGE Ltd.). The GC oven was held at 40°C for 2 min and subsequently programmed from 40 to 300°C at 5°C min^{-1}. The final temperature was held for 6 min.

Carbon isotopic compositions were determined in triplicate, allowing mean values and standard deviations to be calculated. All carbon isotopic ratios are expressed relative to the international PDB standard.

RESULTS AND DISCUSSION

Carbon and nitrogen abundance and isotopic composition of organic matter

Uppermost Cretaceous organic carbon concentrations are typically ~2.6% at Brownie Butte, and decrease markedly in the lower boundary clay layer to 0.1% (Fig. 1A). The carbon content increases in the upper boundary layer to 1.8% before rising sharply to ~32% in the lowermost Tertiary layer. The carbon content gradually decreases above this layer until a marked drop to 0.1% at 8 cm above the boundary coincident with a change in lithology from coal to gray shale. The same general distribution in carbon content is recorded from the Raton (Berwind Canyon) section (Fig. 2A), although background Cretaceous levels are lower (typically 0.4%) and the carbon content of the ejecta layer was only 0.04%.

Figure 1B shows the $\delta^{13}C_{org}$ values of total carbon for the Brownie Butte section. There is a shift of −2.1‰ to −28.3‰ in the layer immediately above the uppermost boundary layer. Above this level $\delta^{13}C_{org}$ values become progressively heavier, returning to Cretaceous values at +8 cm and more ^{13}C enriched above this. The Raton section (Fig. 2B) also shows a negative isotopic shift, $\delta^{13}C_{org}$ values becoming 1.7‰ lighter immediately above the boundary.

At Brownie Butte, organic nitrogen concentrations determined from the nitrogen content of the HF/HCl residues (Fig. 1C) decrease from typical Cretaceous levels of 700 ppm to ~300 ppm in the uppermost boundary layer, followed by a marked increase to 10000 ppm in the overlying layer at the base of the coal. The nitrogen content declines gradually throughout the coal bed, falling rapidly to very small amounts in the gray shale. The Raton section (Fig. 2C) shows a decrease in nitrogen concentration in the lower boundary layer to <10 ppm, followed by a three-fold increase over Cretaceous values to 300 ppm in the layer above.

The nitrogen isotopic data are shown in Figures 1D and 2D. At Brownie Butte there is a small negative shift in $\delta^{15}N$ within the boundary layers coincident with the drop in nitrogen concentration followed by a positive shift of 2‰–3‰ in the overlying lowest Tertiary sample. Overall, there is a ^{15}N enrichment in the lowermost Tertiary of 1‰–2‰ compared to the background Cretaceous values. A similar profile is observed at

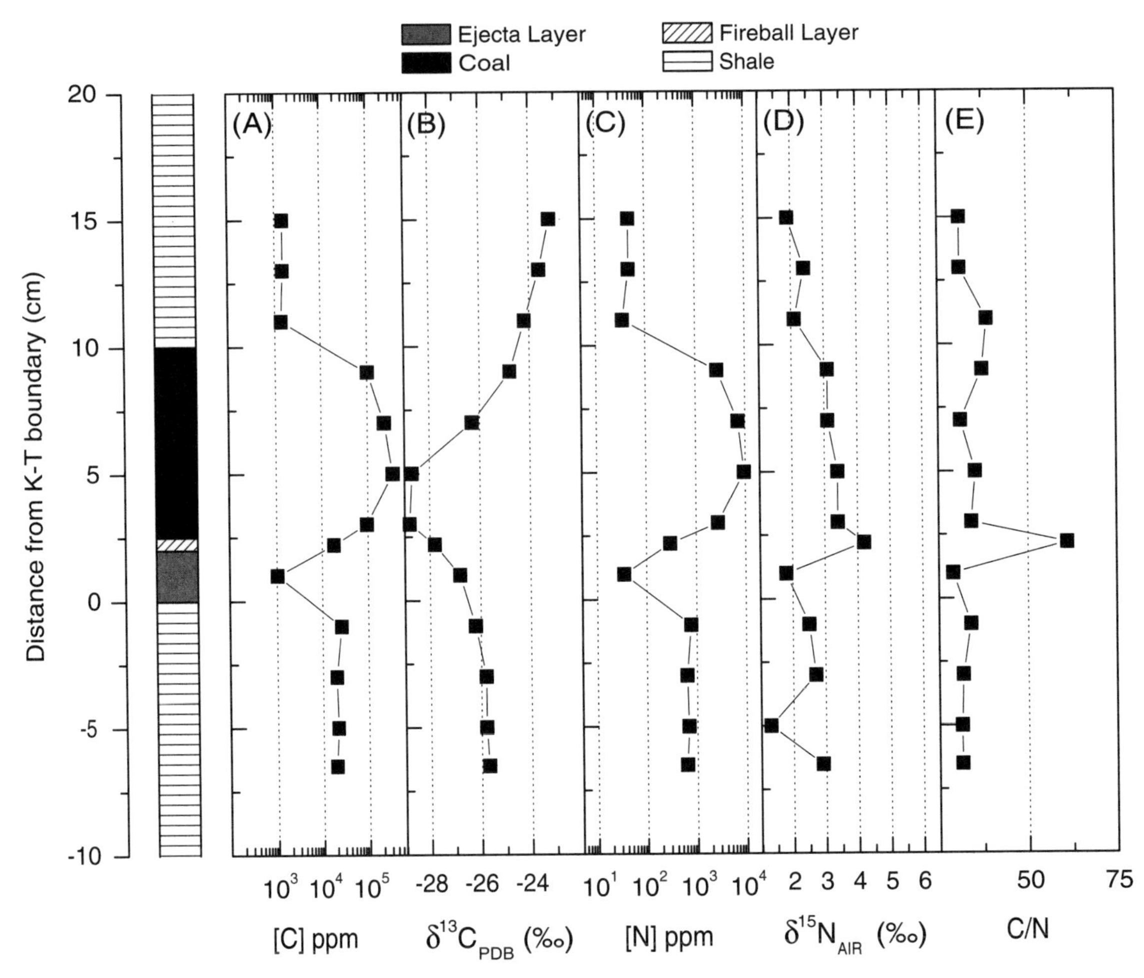

Figure 1. Carbon and nitrogen isotope compositions and abundances across Cretaceous-Tertiary (K-T) boundary at Brownie Butte, Montana. PDB is Peedee belemnite.

Raton; Cretaceous and Tertiary levels remain constant at $\sim+1‰$, the lowermost boundary layer has a $\delta^{15}N$ value of $\sim 0‰$, and the two lowermost Tertiary layers are $\sim 4‰$ heavier than background values. Apparently, at both sites there is an ^{15}N enrichment overlying the uppermost boundary layer.

Isotope excursions and biotic crises

The palynologic K-T boundary of the Western Interior of North America consistently occurs at the uppermost boundary clay horizon (Tschudy and Tschudy, 1986) and is represented by the abrupt regional disappearance of diagnostic Late Cretaceous angiosperm pollen taxa. The changes are similar in both the northern (encompassing Brownie Butte) and southern (encompassing Raton) parts of the Western Interior, although different taxa disappear at the boundary, reflecting the different paleoecological settings of the two areas (Tschudy and Tschudy, 1986). The boundary sequences have several discrete palynological phases. Immediately above the uppermost boundary layer a barren organic-rich zone usually occurs that is devoid of palynomorphs and is overlain by an interval rich in fern spores that has been interpreted as representing the dominance of fern plant species following the impact (Tschudy et al., 1984; Spicer et al., 1985; Wolfe and Upchurch, 1987). Overlying the fern spike are phases indicating angiosperm recolonization and recovery (Wolfe and Upchurch, 1987).

Under normal depositional conditions the total organic content of these fresh-water swamp sequences will contain inputs of both terrigenous-derived plant material and aquatic algal contributions, and the two sites studied have different paleoecological settings (e.g., Pillmore et al., 1984; Fastovsky and Dott, 1986; Tschudy and Tschudy, 1986; Fastovsky and McSweeney, 1987). At both sites organic carbon concentration increases immediately above the uppermost boundary layer and carbon isotopic compositions become relatively ^{13}C depleted. In both cases this increase in organic carbon content has led to the formation of a coal. Schimmelman and Deniro (1984) observed a similar trend at York Canyon in the Raton Basin. At Brownie Butte the organic carbon content increases immediately above the boundary and the ^{13}C-depleted zone persists

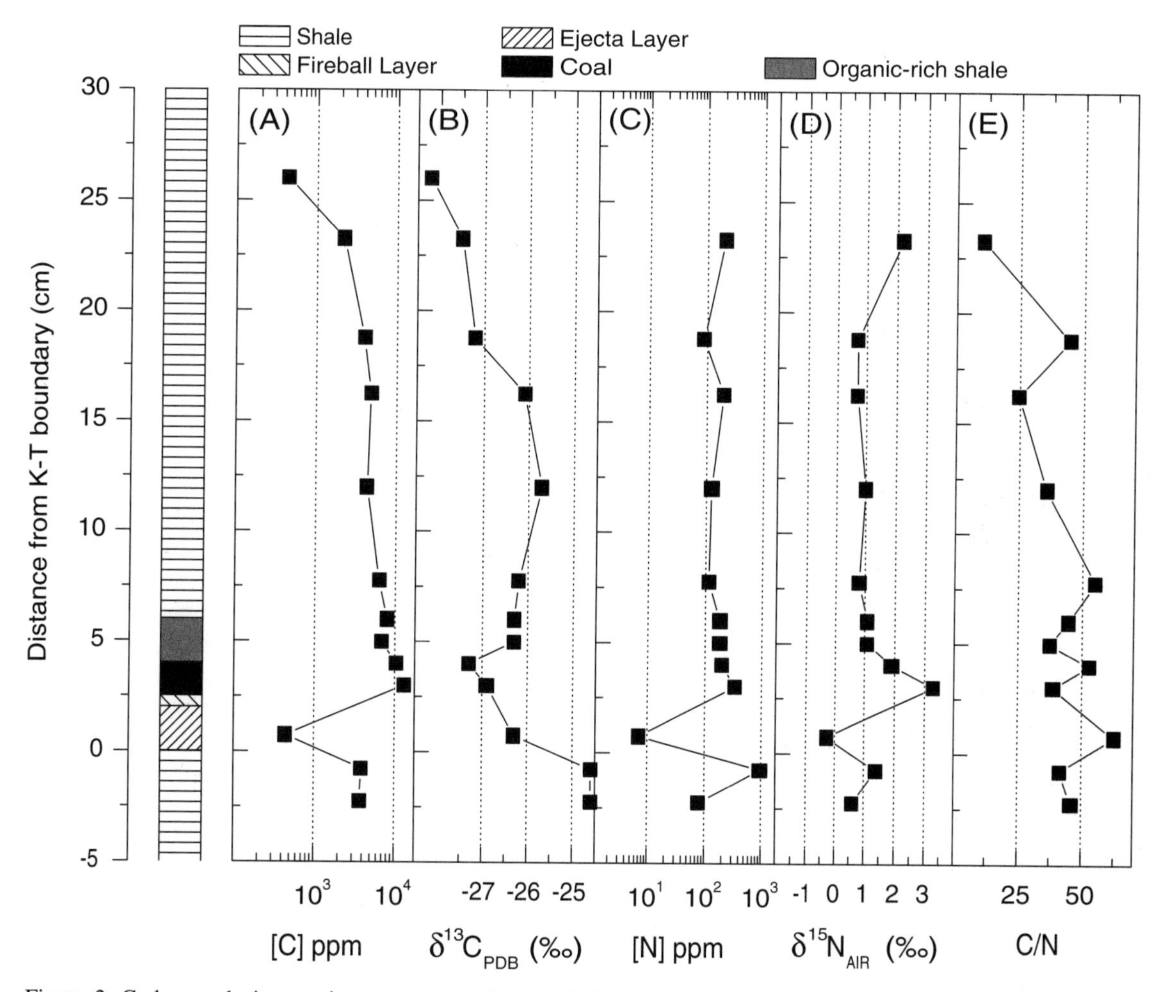

Figure 2. Carbon and nitrogen isotope compositions and abundances across Cretaceous-Tertiary (K-T) boundary at Berwind Canyon, Raton Basin, New Mexico. PDB is Peedee belemnite.

into the Tertiary for ~8 cm. One possible explanation for the pattern is that the isotopic shifts are the result of lithological or palynofloral changes.

Coalification, the process that transforms plant remains to a black, lustrous solid organic fossil fuel, was originally thought to involve the full degradation of plant remains and their subsequent reconstitution (Hatcher and Clifford, 1997). However, the general consensus today is that the process involves the selective preservation of resistant plant components followed by minor reorganization of the biopolymers that remain (Hatcher and Clifford, 1997). It is therefore possible that the negative carbon and positive nitrogen isotope shifts reflect the loss of more ^{13}C- and ^{14}N-rich organic components as a result of the coalification process. Such enriched components might be expected to be ~5‰ different from the bulk isotopic composition at most (Park and Epstein, 1961), and mass-balance calculations indicate that this would require the loss of ~40% of such a component to produce the isotopic shift observed. Furthermore, at the Raton site, the more negative $\delta^{13}C$ values persist into the lower Tertiary while at Brownie Butte they appear to be restricted to the lower few centimeters of the coal.

Alternatively, Schimmelman and Deniro (1984), who also observed a negative shift in $\delta^{13}C$ that persisted well into the lower Tertiary at York Canyon, proposed the same mechanism that has been cited to explain the negative isotopic shifts in marine carbonates; i.e., a decrease in photosynthetic activity in ocean surface waters accompanied by excess oxidation of organic matter resulting in a lowering of the $\delta^{13}C$ values of oceanic bicarbonate (Hsü and McKenzie, 1985). They suggested that the negative $\delta^{13}C$ values at York Canyon reflected the gradual reequilibration of marine and atmospheric CO_2 occurring over several millennia following the impact. Such a model is an appealing one because it provides a unified explanation for the disruption of both the marine and terrestrial carbon cycles. However, the model is difficult to support from the determination of bulk organic carbon and nitrogen isotope measurements alone, because such measurements can be influenced by changes in the organic matter contributions to the depositional environment from which the samples are derived.

Other contributors to atmospheric CO_2 as a result of the

Chicxulub impact also need to be considered. Wildfires, either as a direct result of the impact or caused by reentering ejecta, would inject substantial amounts of isotopically light CO_2 into the atmosphere (Wolbach et al., 1985, 1988; Gilmour et al., 1989); soot and elemental carbon, the primary evidence for wildfires, have a mean global carbon isotopic composition of $-25.8‰ \pm 0.6‰$, and CO_2 produced would also be isotopically light (Gleason and Kyser, 1984). The breakdown of calcium carbonate in the target rocks of the Late Cretaceous carbonate platform in the vicinity of the Chicxulub crater would have resulted in the production of isotopically heavy CO_2 (O'Keefe and Ahrens, 1989); the carbon dioxide liberated from fluid inclusions in K-T impact melt glasses has an isotopic composition of $\sim -2‰$ (Hough et al., 1998). Either of these sources of atmospheric CO_2 may have influenced the isotopic composition of carbon in terrestrial ecosystems. Ivany and Salawitch (1993) calculated that it would require ~25% of the world's above-ground biomass to be combusted to produce the observed shift in $\delta^{13}C$ of marine carbonate, and that this was in keeping with the amount of soot observed in K-T boundary clays (Wolbach et al., 1988; Gilmour et al., 1989). Vegetation growing within the area of combustion-produced CO_2 can have $\delta^{13}C$ values several per mil lighter than normal (Gleason and Kyser, 1984), so that one might expect terrestrial ecosystems at the K-T boundary to reflect any changes in the isotopic composition of atmospheric CO_2. However, Ivany and Salawitch (1993) did not include a contribution to atmospheric pCO_2 levels from carbonate in their calculations.

No widespread changes in the $\delta^{13}C_{org}$ record between the upper Maastrichtian and lower Paleocene have been observed at marine sites (Meyers and Simoneit, 1990; Meyers, 1992), and the few high-resolution studies of marine organic carbon across the K-T boundary reveal an inconsistent pattern (Gilmour et al., 1987, 1990; Wolbach et al., 1990). The concomitant changes in $\delta^{13}C_{org}$ org and $\delta^{13}C_{carb}$ at Woodside Creek, New Zealand, have been interpreted as indicating unusual plankton blooms and bacterially dominated degradation of organic matter following decreased photosynthesis (Hollander et al., 1993a).

The terrestrial palynofloral record provides evidence of significant regional biotic perturbations following the Chicxulub impact. Wolfe and Upchurch (1986) interpreted changes in leaf flora across the K-T boundary as evidence of significant climatic perturbations, including a brief low-temperature excursion and increased precipitation. It seems, therefore, that the preserved isotopic signatures could reflect the responses of local environments to climatic and other environmental changes, or disruption of the global carbon cycle as a result of decreased photosynthesis.

Nitrogen isotopic evidence for anoxia

The response of the local paleoenvironments to major biotic perturbations is apparently reflected in the record of nitrogen abundance and isotopic composition at the sites studied. Both the Brownie Butte and Raton records show increased nitrogen concentrations over Upper Cretaceous values immediately above the boundary. However, there is no apparent change in the C/N ratio (Figs. 1E and 2E) between Upper Cretaceous and lower Tertiary values, although the boundary layers show higher C/N ratios, presumably reflecting the lesser amount of organic relative to elemental carbon in these layers (Hildebrand and Wolbach, 1989). There are no significant net increases or decreases in nitrogen input across the K-T boundary at these localities. Nitrogen isotopic composition of organic matter, however, becomes more ^{15}N enriched by several per mil in the lowermost Tertiary layer. Schimmelman and Deniro (1984) found a similar shift in $\delta^{15}N$ values at York Canyon.

The marine organic nitrogen record across the K-T boundary is sparse. Gilmour et al. (1990) found an increase in total sediment nitrogen and a positive shift in $\delta^{15}N$ at Woodside Creek, New Zealand, but the background values in the Upper Cretaceous limestones and Tertiary marls are extremely low and do not have $\delta^{15}N$ values typical of marine organic matter.

The enrichment in ^{15}N in the two study sites may reflect the onset of anaerobic conditions in these paleoenvironments as a result of decaying and rotting vegetation killed by the impact. Velinksy et al. (1991) found that dissolved ammonium in modern marine environments where anoxic bottom waters develop is strongly enriched in ^{15}N. In sedimentary environments organic matter from the strongly stratified waters of the Eocene Green River Formation are also enriched in ^{15}N (Collister et al., 1992), as are kerogens from the Tertiary Mulhouse basin in France, where molecular biomarker evidence also indicates the development of anaerobic conditions (Hollander et al., 1993b). These enrichments result from isotopic effects associated with the uptake, oxidation, and reduction of nitrogen species within the water column and are characteristic of systems in which high concentrations of NH_4^+ develop in anaerobic bottom waters (Hollander et al., 1993b). The enrichments in $\delta^{15}N$ values at Brownie Butte and Raton indicate that anaerobic conditions may have prevailed after the devastation of the terrestrial ecosystem by the impact at Chicxulub.

Saturated hydrocarbon biomarkers

The remnants of some carbon-based compounds from the soft parts of animals and plants can still be recognized in sedimentary material, providing us with chemical fossils or biomarkers (Chaffee et al., 1986). Various types of biomarkers exist, some derived from specific organisms, others providing a more general picture of organic matter inputs to depositional environments. Normal alkanes (straight-chain saturated hydrocarbons) belong to the latter category of biomarkers and provide a useful tool for the delineation of different organic matter sources: different inputs can result in a change in the distribution of the carbon chains.

Previous studies of chemical fossils at the K-T boundary

have been restricted to the examination of marine sections (Meyers and Simoneit, 1990; Yamamoto et al., 1996) or have focused on particular problems, such as evidence for fire (Venkatesan and Dahl, 1989). Yamamoto et al. (1996) examined the organic compound distributions from the lowermost Danian section at the extremely well preserved Geulhemmerberg, Netherlands, K-T section. Their observations of highly functionalized organic compounds demonstrated the potential that biomarker studies may have for unraveling some of the environmental consequences of the Chicxulub impact.

A suite of four of the samples from Brownie Butte and six from the Raton section that were examined for bulk carbon and nitrogen isotope measurements were selected for a biomarker study. Figure 3 shows the distribution of n-alkanes across the K-T boundary at Brownie Butte. The uppermost Cretaceous sample from 2–4 cm below the boundary (Fig. 3A) contains a predominance of odd-numbered n-alkanes in the range n-C_{25} to n-C_{31}. Long-chain odd-numbered n-alkanes in the range n-C_{27}^{+} typically originate from higher plant waxes; other organic matter sources such as algae or bacteria were apparently minimal, as evidenced by the relatively lower abundance of n-alkanes in the n-C_{15} to n-C_{25} region. The lowermost Tertiary sample that immediately overlies the boundary clay layer displays a bimodal distribution with both n-C_{25} to n-C_{31} higher plant-derived n-alkanes and n-C_{14} to n-C_{18} n-alkanes together with the isoprenoid hydrocarbons pristane and phytane. The latter group of compounds most likely represents the input of organic matter from algal and/or bacterial sources; n-C_{17} is a common algal biomarker and n-C_{17}/n-C_{27} = 0.6, indicating a substantial algal/bacterial contribution relative to higher plants. Early studies indicated that pristane and phytane could be derived from the side chain of chlorophyll under variable conditions of oxidation in the paleoenvironment (Brooks et al., 1969; Powell and McKirdy, 1973). However, the side chain of tocopherols and lipids of various groups of bacteria, including methanogens, have also been proposed as possible sources (Goossens et al., 1984; ten Haven et al., 1985; Rowland, 1990). By the base of the coal (2–4 cm above the boundary, Fig. 3C), higher plant n-alkanes begin to dominate the distribution again; i.e., n-C_{17}/n-C_{27} = 0.12, although pristane and phytane are still clearly present. The shale overlying the coal continues this higher plant-dominated n-alkane distribution (n-C_{17}/n-C_{27} = 0.27).

The distributions of n-alkanes in the Raton section are shown in Figure 4. As at Brownie Butte, the uppermost Cre-

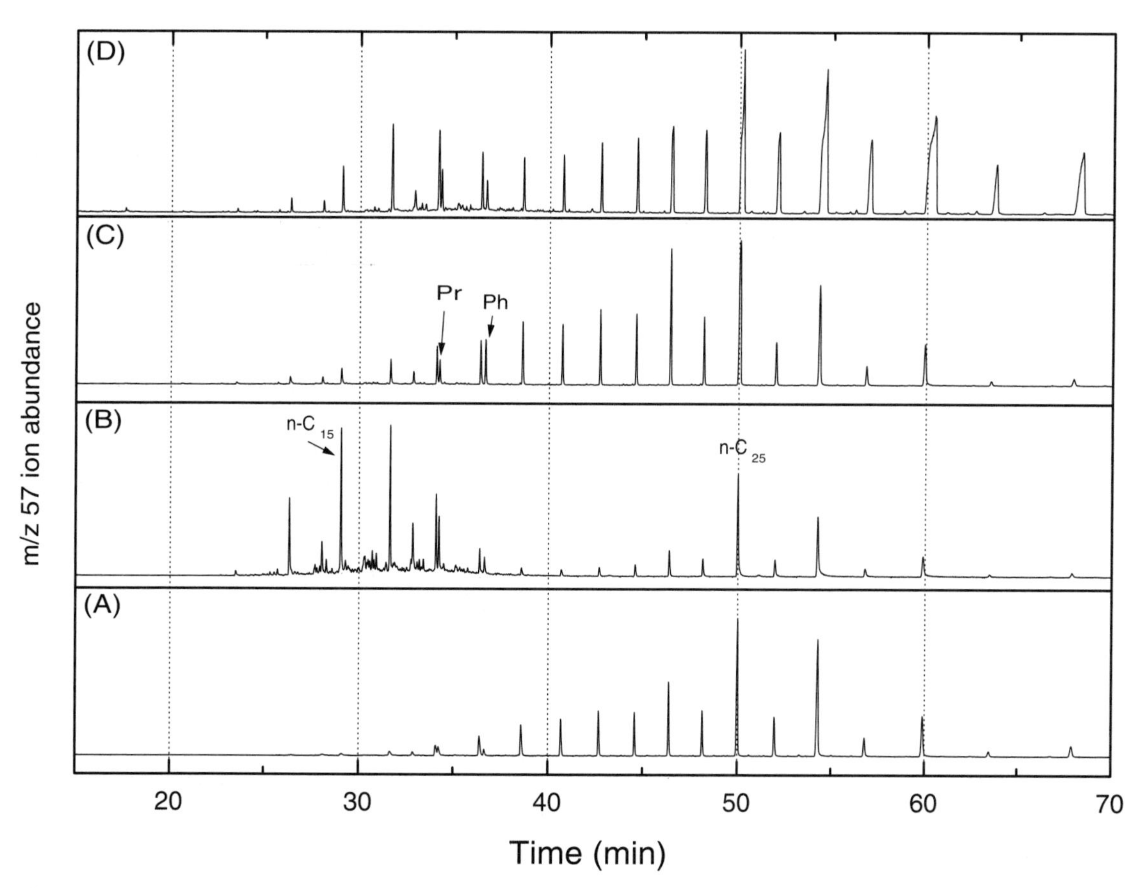

Figure 3. Mass chromatograms (m/z 57) of saturated hydrocarbons showing distribution of n-alkanes across Cretaceous-Tertiary boundary at Brownie Butte, Montana. A: 4–2 cm below boundary. B: 0–2 cm above boundary, including upper and lower boundary clays. C: 2–4 cm above boundary. D: 6–8 cm above boundary.

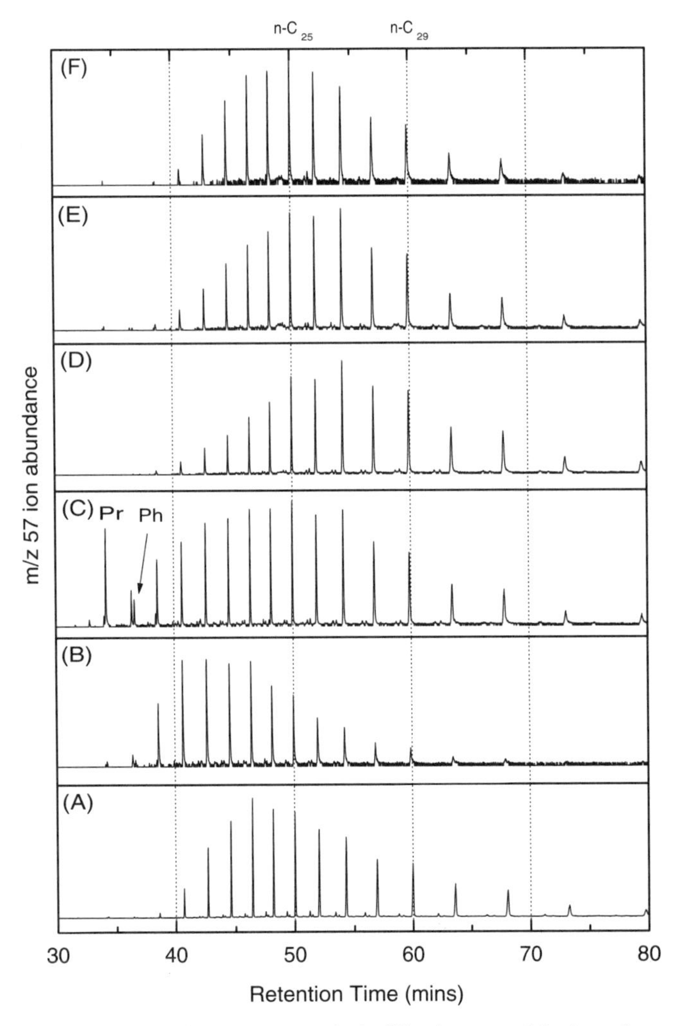

Figure 4. Mass chromatograms (m/z 57) of saturated hydrocarbons showing distribution of n-alkanes across Cretaceous-Tertiary boundary at Berwind Canyon, Raton Basin, New Mexico. A: 4–2 cm below boundary. B: 2–0 cm below boundary. C: 0–2 cm above boundary. D: 2–4 cm above boundary. E: 4–6 cm above boundary. F: 6–8 cm above boundary.

taceous samples (Fig. 4, A and B) contain some evidence for higher plant organic inputs in the form of n-C_{27} to n-C_{31} n-alkanes. However, the n-alkane profiles are dominated by maxima at ~n-C_{23}, and the relative abundance of n-alkanes in the C_{19}–C_{25} region is much higher than at Brownie Butte, suggesting that the dominant organic matter input may have been algal. However, as at Brownie Butte, the lowermost Tertiary sample (Fig. 4C) contains a marked increase in the relative abundances of n-C_{17}, pristine, and phytane. The n-alkane profiles above the boundary (Fig. 4, D–F) are fairly similar, with n-alkane maxima at n-C_{27} (Fig. 4, D and E) or n-C_{25} (Fig. 4F), suggesting that higher plant sources became increasingly significant contributors.

At both K-T sites, the lowermost Tertiary samples contain evidence for variations in the organic matter inputs to their respective depositional environments in that there is a marked increase in the relative abundance of n-C_{17}, pristine, and phytane. These coincide with the shifts observed in bulk organic matter $\delta^{13}C$ and $\delta^{15}N$ values; however, the isotopic changes appear to persist into the Tertiary, most notably for carbon. Changes in the organic matter input above the boundary are in agreement with the palynofloral evidence for a biotic crisis. Given the marked change in the ratio of angiosperm pollen to fern spores recorded in Western Interior K-T sections, it might be expected that the organic geochemical record would preserve such a change. The saturated hydrocarbon profiles are unlikely to distinguish between higher plant groups; instead they indicate a relative increase in organic matter derived from algal or bacterial sources immediately above the boundary.

Cyclic hydrocarbon biomarkers

Hopanes are a class of cyclic hydrocarbon biomarkers (triterpanes) with a widespread occurrence in recent and ancient sediments and in petroleum (Ensminger et al., 1974). Geological samples typically contain hopanes in the range C_{27} and C_{29}–C_{35}.

The distribution of hopanes in the suite of samples from Brownie Butte is shown in Figure 5. Several hopanes were identified in the uppermost coal sample (Fig. 5D); the rest of the coal contained fewer hopanes (Fig. 5, C and D), being dominated by $C_{27\beta\beta}$. This hopane is common to all of the Brownie Butte samples, including the uppermost Cretaceous sample. The hopane distributions for the Raton samples are shown in Figure 6. No hopanes are apparent in the uppermost Cretaceous samples, but a range of C_{27}–C_{30} hopanes is recorded immediately above the boundary clay; the hopanes are also present further up in the succession as the lithology changes from coal to an organic-rich shale.

With several hopanes detected in samples from both Brownie Butte and Raton, it would be reasonable to suggest the presence of increased microbial activity in the lowermost Tertiary from this evidence alone because they are common bacterial biomarkers. Three hopanes, $C_{27\alpha\beta}$, $C_{29\alpha\beta}$, and $C_{30\alpha\beta}$, are common to both localities. The sources of C_{27}, C_{29}, and C_{30} hopanes include autotrophic cyanobacteria, heterotrophic bacteria, chemotrophs, and methanotrophic bacteria. However, it is possible to investigate the possible origin of these compounds further if their carbon isotope compositions are determined by GC-IRMS. Table 1 lists the carbon isotope compositions of individual hopanes from the Raton samples. All of the hopanes analyzed gave extremely negative $\delta^{13}C$ values ranging from −45.8‰ to −60.3‰. The significant ^{13}C depletion of hopanes compared to pristane and phytane suggests a different habitat for the prokaryotic biota that biosynthesized these compounds. Negative $\delta^{13}C$ values for hopanes have been reported previously for hopanes from similar lithologies. Schoell et al. (1994) examined the hopane content of Tertiary brown coals from southern China and reported $\delta^{13}C$ values of between −41.7‰

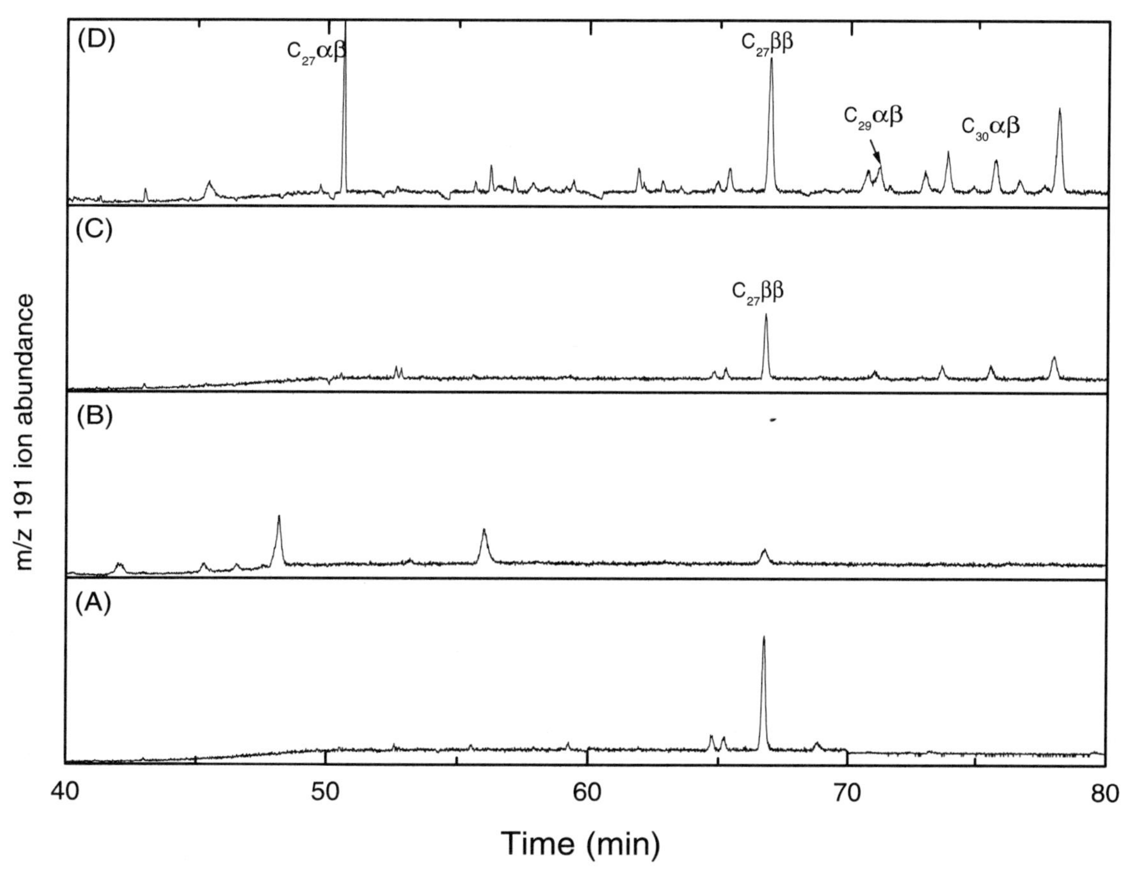

Figure 5. Mass chromatograms (m/z 191) showing distribution of hopanes at Brownie Butte, Montana. A: 4–2 cm below boundary. B: 0–2 cm above boundary, including upper and lower boundary clays. C: 2–4 cm above boundary. D: 6–8 cm above boundary.

and −58.8‰. They concluded that the negative $\delta^{13}C$ values indicated that a methane-utilizing process was involved, suggesting the recycling of methanogenically produced methane by methanotrophic bacteria; the methane produced by methanogenic bacteria is typically extremely ^{13}C depleted. However, methanogen-derived hopanes are generally less ^{13}C depleted than the methane these organisms produce; consequently it is the recycling of methane by methane-oxidizing bacteria (methanotrophs) that is thought to be the principal process by which methane is recycled in lakes (Cicerone and Oremland, 1988). Nonmarine-derived ^{13}C-depleted C_{27} and C_{29} hopanes were reported by Freeman et al. (1990) in the Messel shale and by Collister et al. (1992) in the Green River Formation. In both cases either methanotropic or chemotrophic bacteria were suggested as the probable origin. The highly negative $\delta^{13}C$ values for the C_{27}–C_{30} hopanes observed in this study therefore suggest a source from a methane-oxidizing bacterial community.

CONCLUSIONS

We have used coupled carbon and nitrogen isotope measurements together with molecular level isotope analysis of biomarker molecules to contribute to the assessment of environmental conditions at two nonmarine K-T boundary sites in the Western Interior of North America. Several conclusions can be drawn from this study.

1. The organic matter carbon isotope record for terrestrial sites apparently reflects the 2‰–3‰ isotope excursion observed in the marine foraminiferal record.

2. The nitrogen isotope record contains a small positive excursion immediately above the boundary; however, no significant changes in nitrogen input to the sedimentary environment could be discerned.

3. Both the carbon and nitrogen isotope excursions coincide with a marked change in lithology from shale below the boundary to an overlying coal. While the coal is an indicator of a relative increase in organic matter accumulation, the coalification process also results in the selective preservation of organic matter, which may, in part, account for the isotope variations observed in bulk organic matter.

4. At the two localities studied saturated hydrocarbons (n-alkanes) indicated a significant organic matter input from higher plants and algae. However, above the K-T boundary algal and/or bacterial contributions become more significant. Upward in

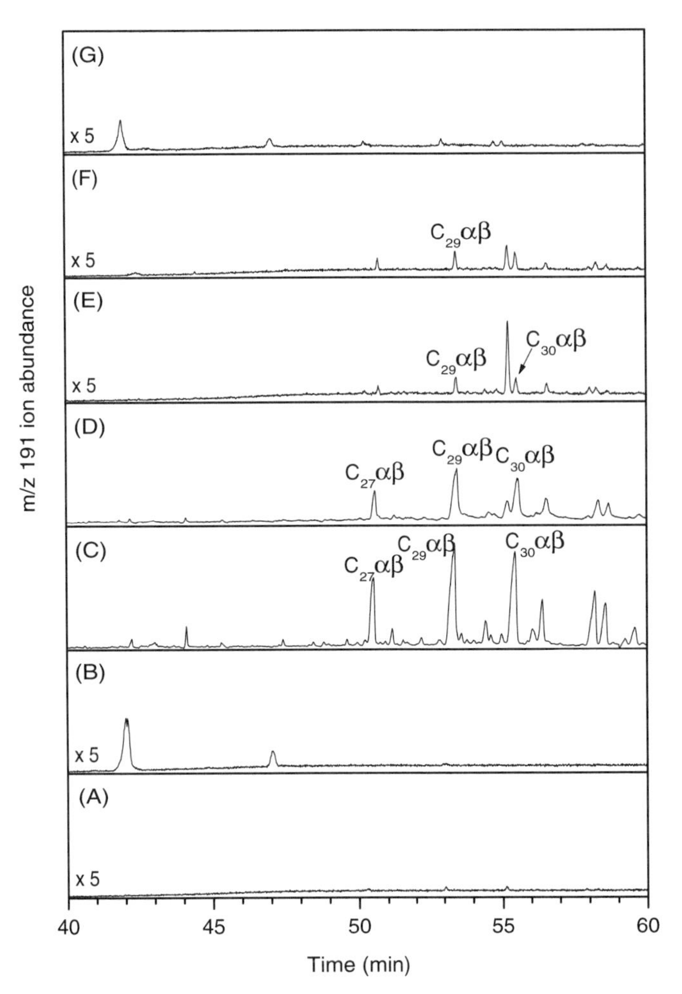

Figure 6. Mass chromatograms (m/z 191) showing distribution of hopanes at Berwind Canyon, Raton Basin, New Mexico. A: 4–2 cm below boundary. B: 2–0 cm below boundary. C: 0–2 cm above boundary. D: 2–4 cm above boundary. E: 4–6 cm above boundary. F: 6–8 cm above boundary. G: 8–10 cm above boundary.

TABLE 1. CARBON ISOTOPE COMPOSITIONS OF BIOMARKERS FROM BERWIND CANYON, RATON BASIN

Biomarker	Distance from Cretaceous Tertiary boundary	
	2–4 cm $\delta^{13}C_{PDB}$ (‰)	0–2 cm $\delta^{13}C_{PDB}$ (‰)
Hopane $C_{27\alpha\beta}$	−57.1	−51.0
Hopane $C_{29\alpha\beta}$	−60.3	−54.9
Hopane $C_{30\alpha\beta}$	−51.1	−45.8
Pristane	−28.4	−27.3
Phytane	−32.4	−27.7

Note: Distance is above boundary. PDB is Peedee belemnite.

the successions, higher plants again appear to be the dominant source of n-alkanes.

5. The coals overlying the boundary at both localities contain biomarker evidence for the presence of prokaryotic organisms in the form of bacterially derived hopanes. These hopanes were extremely ^{13}C depleted, suggesting a source from methane-oxidizing bacteria.

ACKNOWLEDGMENTS

Bruce Bohor provided the Brownie Butte samples used in this study. This work was supported by a Royal Society university research fellowship to Gilmour, a Natural Environment Research Council (NERC) studentship to Gardner and NERC grant GR9/199. We are grateful to Jim Carter, NERC Scientific Services, for confirmation of hopane identities. We thank Wendy Wolbach, Ken MacLeod, and an anonymous reader for reviews.

REFERENCES CITED

Alvarez, L.W., Alvarez, W., Asaro, F., and Michel, H.V., 1980, Extraterrestrial cause for the Cretaceous-Tertiary extinction: Science, v. 208, p. 1095–1108.

Anders, E., Wolbach, W.S., and Lewis, R.S., 1986, Cretaceous extinctions and wildfires: Science, v. 234, p. 261–264.

Boersma, A., and Shackleton, N.J., 1981, Oxygen- and carbon-isotope variations and planktonic-foraminifer depth habitats, Late Cretaceous to Paleocene, Central Pacific, Deep Sea Drilling Project Sites 463 and 465, *in* Thiede, J., and Vallier, T. L., eds., Initial reports of the Deep Sea Drilling Project: Washington, D.C., U.S. Government Printing Office, p. 513–526.

Bohor, B.F., Modreski, P.J., and Foord, E.E., 1987a, Shocked quartz in the Cretaceous-Tertiary boundary clays: Evidence for a global distribution: Science, v. 236, p. 705–709.

Bohor, B.F., Triplehorn, D.M., Nichols, D.J., and Millard, H.T., 1987b, Dinosaurs, spherules, and the "magic" layer: A new K-T boundary clay site in Wyoming: Geology, v. 15, p. 896–899.

Boyd, S.R., and Pillinger, C.T., 1990, Determination of the abundance and isotope composition of nitrogen within organic compounds: A sealed tube technique for use with static vacuum mass spectrometers: Measurement Science and Technology, v. 1, p. 1176–1183.

Brooks, J.D., Gould, K., and Smith, J.W., 1969, Isoprenoid hydrocarbons in coal and petroleum: Nature, v. 222, p. 257–259.

Buchardt, B., and Jørgensen, N.O., 1979, Stable isotope variations at the Cretaceous/Tertiary boundary in Denmark, *in* Christensen, W.K., and Birkelund, T., eds., Cretaceous Tertiary boundary events symposium: Copenhagen, University of Copenhagen, p. 54–61.

Chaffee, D.S., Hoover, R.B., Johns, R.B., and Schweighardt, F.K., 1986, Biological markers extractable from coal, *in* Johns, R.B., ed., Biological markers in the sedimentary record: Methods in geochemistry and geophysics: Amsterdam, Elsevier, p. 311–346.

Cicerone, R.J., and Oremland, R.S., 1988, Biogeochemical aspects of atmospheric methane: Global Biogeochemical Cycles, v. 2, p. 299–327.

Collister, J.W., Summons, R.E., Lichtfouse, E., and Hayes, J.M., 1992, An isotopic biogeochemical study of the Green River oil shale: Organic Geochemistry, v. 19, p. 265–276.

Crutzen, P.J., 1987, Acid rain at the K/T boundary: Nature, v. 330, p. 108–109.

Ensminger, A., van Dorsselaer, A., Spyckerelle, C., Albrecht, P., and Ourisson, G., 1974, Pentacyclic triterpenes of the hopane type as ubiquitous geo chemical markers: Origin and significance, *in* Tissot, B.P., and Bienner, F., eds., Advances in Organic Geochemistry 1973, p. 245–260.

Fastovsky, D.E., and Dott, R.H., 1986, Sedimentology, stratigraphy, and extinctions during the Cretaceous-Paleogene transition at Bug-Creek, Montana: Geology, v. 14, p. 279–282.

Fastovsky, D.E., and McSweeney, K., 1987, Paleosols spanning the Cretaceous-Paleogene transition, Eastern Montana and Western North Dakota: Geological Society of America Bulletin, v. 99, p. 66–77.

Freeman, K.H., Hayes, J.M., Trendel, J.-M., and Albrecht, P., 1990, Evidence from carbon isotope measurements for diverse origins of sedimentary hydrocarbons: Nature, v. 343, p. 254–256.

Ganapathy, R., 1980, A major meteorite impact on the earth 65 million years ago: Evidence from the Cretaceous-Tertiary boundary: Science, v. 209, p. 921–923.

Gilmour, I., and Anders, E., 1989, Cretaceous-Tertiary boundary event: Evidence for a short time scale: Geochimica et Cosmochimica Acta, v. 53, no. 2, p. 503–511.

Gilmour, I., Orth, C.J., and Brooks, R.R., 1987, Carbon at a new K-T boundary site in New Zealand: Meteoritics, v. 22, p. 385–388.

Gilmour, I., Wolbach, W.S., and Anders, E., 1989, Major wildfires at the Cretaceous-Tertiary boundary, *in* Clube, S.V.M., ed., Catastrophes and evolution: Astronomical foundations: Cambridge, Cambridge University Press, p. 195–213.

Gilmour, I., Wolbach, W.S., and Anders, E., 1990, Early environmental effects of the terminal Cretaceous impact, *in* Sharpton, V. L., and Ward, P., eds., Global catastrophes in Earth history; An interdisciplinary conference on impacts, volcanism, and mass mortality: Geological Society of America Special Paper 247, p. 383–390.

Gleason, J.D., and Kyser, T.K., 1984, Stable isotope compositions of gases and vegetation near naturally burning coal: Nature, v. 307, p. 254–257.

Goossens, H., De Leeuw, J.W., and Shenck, P.A., 1984, Tocopherols as likely precursors of pristane in ancient sediments: Nature, v. 312, p. 440–442.

Hatcher, P.G., and Clifford, D.J., 1997, The organic geochemistry of coal: From plant materials to coal: Organic Geochemistry, v. 27, p. 251–274.

Hayes, J.M., Freeman, K.H., Popp, B.N., and Hoham, C.H., 1990, Compound-specific isotopic analyses: A novel tool for reconstruction of ancient biogeochemical processes: Organic Geochemistry, v. 16, no. 4–6, p. 1115–1128.

Hickey, L.J., 1981, Land plant evidence compatible with gradual, not catastrophic, change at the end of the Cretaceous: Nature, v. 292, no. 5823, p. 529–531.

Hildebrand, A.R., and Boynton, W.V., 1990, Proximal Cretaceous-Tertiary boundary impact deposits in the Caribbean: Science, v. 248, p. 843–847.

Hildebrand, A.R., Penfield, G.T., Kring, D.A., Pilkington, M., Camargo, A., Jacobsen, S.B., and Boynton, W.V., 1991, Chicxulub Crater: A possible Cretaceous-Tertiary boundary impact crater on the Yucatan peninsula, Mexico: Geology, v. 19, p. 867–871.

Hildebrand, A.R., and Wolbach, W.S., 1989, Carbon and chalcophiles at a nonmarine K/T boundary: Joint investigation of the Raton section, New Mexico: Lunar and Planetary Science, v. 20, p. 414–415.

Hollander, D.J., McKenzie, J.A., and Hsü, K.J., 1993a, Carbon isotope evidence for unusual plankton blooms and fluctuations of surface water CO_2 in "Strangelove" ocean after terminal Cretaceous event: Palaeogeography, Palaeoclimatology, Palaeoecology, v. 104, p. 229–237.

Hollander, D.J., Sinninghe Damsté, J.S., Hayes, J.M., De Leeuw, J.W., and Huc, A.Y., 1993b, Molecular and bulk isotopic analysis of organic matter in marls of the Mulhouse Basin (Tertiary, Alsace, France): Organic Geochemistry, v. 20, p. 1253–1263.

Hough, R.M., Wright, I.P., Sigurdsson, H., Pillinger, C.T., and Gilmour, I., 1998, Carbon content and isotopic composition of K/T impact glasses from Haiti: Geochimica et Cosmochimica Acta, v. 62, no. 7, p. 1285–1291.

Hsü, K.J., and McKenzie, J.A., 1985, A strangelove ocean in the earliest Tertiary, *in* Sundquist, E.T., and Broecker, W.S., eds., The carbon cycle and atmospheric CO_2: Washington, D.C., American Geophysical Union, p. 487–492.

Ivany, L.C., and Salawitch, R.J., 1993, Carbon isotopic evidence for biomass burning at the K-T boundary: Geology, v. 21, p. 487–490.

Jahnke, L.L., Summons, R.E., Hope, J.M., and Des Marais, D.J., 1999, Carbon isotopic fractionation in lipids from methanotrophic bacteria II: The effects of physiology and environmental parameters on the biosynthesis and isotopic signatures of biomarkers—Correlation of the hopanoids from extant methylotrophic eria with their fossil analogues: Geochimica et Cosmochimica Acta, v. 63, no. 1, p. 79–93.

Johnson, K.R., 1992, Leaf-fossil evidence for extensive floral extinction at the Cretaceous Tertiary boundary, North-Dakota, USA: Cretaceous Research, v. 13, no. 1, p. 91–117.

Kring, D.A., and Boynton, W.V., 1991, Altered spherules of impact melt and associated relic glass from the K/T boundary sediments in Haiti: Geochimica et Cosmochimica Acta, v. 55, no. 6, p. 1737–1742.

Kring, D.A., and Boynton, W.V., 1992, Petrogenesis of an augite-bearing melt rock in the Chicxulub structure and its relationship to K/T impact spherules in Haiti: Nature, v. 358, p. 141–144.

Melosh, H.J., Schneider, N.M., Zahnle, K.J., and Latham, D., 1990, Ignition of global wildfires at the Cretaceous/Tertiary boundary: Nature, v. 343, p. 251–254.

Meyers, P.A., 1992, Changes in organic carbon stable isotope ratios across the K/T boundary: Global or local control?: Chemical Geology, v. 101, p. 283–291.

Meyers, P.A., and Simoneit, B.R.T., 1990, Global comparisons of organic matter in sediments across the Cretaceous Tertiary boundary: Organic Geochemistry, v. 16, no. 4–6, p. 641–648.

Nichols, D.J., Jarzen, D.M., Orth, C.J., and Oliver, P.Q., 1986, Palynological and iridium anomalies at Cretaceous-Tertiary boundary, South-central Saskatchewan: Science, v. 231, p. 714–717.

O'Keefe, J.D., and Ahrens, T.J., 1989, Impact production of CO_2 by the Cretaceous/Tertiary extinction bolide and the resultant heating of the Earth: Nature, v. 338, p. 247–249.

Park, R., and Epstein, S., 1961, Metabolic fractionation of C^{13} and C^{12} in plants: Plant Physiology, v. 36, p. 133–138.

Perch-Nielsen, K., McKenzie, J., and He, Q., 1982, Biostratigraphy and isotope stratigraphy and the "catastrophic" extinction of calcareous nanoplankton at the Cretaceous/Tertiary boundary, *in* Silver, L.T., and Schultz, P.H., eds., Geological implication of impacts of large asteroids and comets on the Earth: Geological Society of America Special Paper 190, p. 353–371.

Pillmore, C.L., Tschudy, R.H., Orth, C.J., Gilmore, J.S., and Knight, J.D., 1984, Geologic framework of nonmarine Cretaceous-Tertiary boundary sites, Raton Basin, New Mexico and Colorado: Science, v. 223, p. 1180–1183.

Pollack, J.B., Toon, O.B., Ackerman, T.P., McKay, C.P., and Turco, R.P., 1983, Environmental effects of an impact-generated dust cloud: Implications for the Cretaceous-Tertiary extinctions: Science, v. 219, p. 287–289.

Powell, T.G., and McKirdy, D.M., 1973, Relationship between ratio of pristane to phytane, crude oil composition and geological environment in Australia: Nature, v. 243, p. 37–39.

Rowland, S.J., 1990, Production of acyclic isoprenoid hydrocarbons by laboratory maturation of methanogenic bacteria: Organic Geochemistry, v. 15, p. 9–16.

Schimmelmann, A., and DeNiro, M.J., 1984, Elemental and stable isotope variations of organic matter from a terrestrial sequence containing the Cretaceous/Tertiary boundary at York Canyon, New Mexico: Earth and Planetary Science Letters, v. 68, p. 392–398.

Schoell, M., Simoneit, B.R.T., and Wang, T.G., 1994, Organic geochemistry and coal petrology of Tertiary brown coal in the Zhoujing mine, Baise Basin, South China. 4. Biomarker sources inferred from stable carbon isotope compositions of individual compounds: Organic Geochemistry, v. 21, p. 713.

Sharpton, V.L., Dalrymple, G.B., Marin, L.E., Ryder, G., Schuraytz, B.C., and Urrutia-Fucugauchi, J., 1992, New links between the Chicxulub impact structure and the Cretaceous/Tertiary boundary: Nature, v. 359, p. 819–821.

Sigurdsson, H., D'Hondt, S., Arthur, M.A., Bralower, T.J., Zachos, J.C., van Fossen, M., and Channell, E.T., 1991, Glass from the Cretaceous/Tertiary boundary in Haiti: Nature, v. 349, p. 482–487.

Simoneit, B.R.T., and Beller, H.R., 1987, Lipid geochemistry of Cretaceous/Tertiary boundary sediments, hole 605, Deep Sea Drilling Project Leg 93, and Stevns Klint, Denmark, *in* van Hinte, J.E., and Wise, S.W., eds., Initial Reports of the Deep Sea Drilling Project: Washington, D.C., U.S. Government Printing Office, p. 1211–1215.

Sinninghe Damste, J.S., Van Duin, A.C.T., Hollander, D., Kohnen, M.E.L., and De Leeuw, J.W., 1995, Early diagenesis of bacteriohopanepolyol derivatives: Formation of fossil homohopanoids: Geochimica et Cosmochimica Acta, v. 59, no. 24, p. 5141–5158.

Smit, J., and Klaver, G., 1981, Sanidine spherules at the Cretaceous-Tertiary boundary indicate a large impact event: Nature, v. 292, p. 47–49.

Spicer, R.A., Burnham, R.J., Grant, P., and Glicken, H., 1985, *Pityrogramma calomelanos*, the primary post-eruption colonizer of Volcán Chichonal, Chiapas, Mexico: American Fern Journal, v. 75, p. 1–5.

ten Haven, H.L., de Leeuw, J.W., and Schenck, P.A., 1985, Organic geochemical studies of a Messinian evaporitic basin, northern Apennines (Italy). 1. Hydrocarbon biological markers for a hypersaline environment: Geochimica et Cosmochimica Acta, v. 49, p. 2181–2191.

Toon, O.B., Pollack, J.B., Ackerman, T.P., Turco, R.P., McKay, C.P., and Liu, M.S., 1982, Evolution of impact generated dust cloud and its effects on the atmosphere, *in* Silver, L.T., and Schultz, P.H., eds., Geological implications of impacts of large asteroids and comets on the earth: Geological Society of America Special Paper 190, p. 187–200.

Tschudy, R.H., Pillmore, C.L., Orth, C.J., Gilmore, J.S., and Knight, J.D., 1984, Disruption of the terrestrial plant ecosystem at the Cretaceous-Tertiary boundary, Western Interior: Science, v. 225, p. 1030–1032.

Tschudy, R.H., and Tschudy, B.D., 1986, Extinction and survival of plant life following the Cretaceous/Tertiary boundary event, Western Interior, North America: Geology, v. 14, p. 667–670.

Velinksy, D.J., Fogel, M.L., Todd, J.F., and Tebo, B.M., 1991, Isotopic fractionation of dissolved ammonium at the oxygen-hydrogen sulfide interface in anoxic waters: Geophysical Research Letters, v. 18, p. 649–652.

Venkatesan, M.I., and Dahl, J., 1989, Organic geochemical evidence for global fires at the Cretaceous-Tertiary boundary: Nature, v. 338, p. 57–60.

Wolbach, W.S., and Anders, E., 1989, Elemental carbon in sediments: Determination and isotopic analysis in the presence of kerogen: Geochimica et Cosmochimica Acta, v. 53, no. 7, p. 1637–1647.

Wolbach, W.S., Gilmour, I., and Anders, E., 1990, Major wildfires at the Cretaceous-Tertiary boundary, *in* Sharpton, V.L., and Ward, P., eds., Global catastrophes in Earth history: An interdisciplinary conference on impacts, volcanism, and mass mortality: Geological Society of America Special Paper 247, p. 391–400.

Wolbach, W.S., Gilmour, I., Anders, E., Orth, C.J., and Brooks, R.R., 1988, Global fire at the Cretaceous-Tertiary boundary: Nature, v. 334, p. 665–669.

Wolbach, W.S., Lewis, R.S., and Anders, E., 1985, Cretaceous extinctions: Evidence for wildfires and search for meteoritic material: Science, v. 230, p. 167–170.

Wolfe, J.A., and Upchurch, G.R., 1986, Vegetation, climatic and floral changes at the Cretaceous-Tertiary boundary: Nature, v. 324, p. 148–152.

Wolfe, J.A., and Upchurch, G.R., 1987, Leaf assemblages across the Cretaceous Tertiary boundary in the Raton Basin, New Mexico and Colorado: Proceedings of the National Academy of Sciences of the United States of America, v. 84, p. 5096–5100.

Wright, I.P., Boyd, S.R., Franchi, I.A., and Pillinger, C.T., 1988, High-precision determination of nitrogen stable isotope ratios at the sub-nonomole level: Journal of Physics E: Scientific Instrumentation, v. 21, p. 865–875.

Yamamoto, M., Ficken, K., Baas, M., Bosch, H.J., and deLeeuw, J.W., 1996, Molecular palaeontology of the earliest Danian at Geulhemmerberg (the Netherlands): Geologie en Mijnbouw, v. 75, no. 2–3, p. 255–267.

Zachos, J.C., and Arthur, M.A., 1986, Paleooceanography of the Cretaceous/Tertiary boundary event: Inferences from stable isotopic and other data: Paleooceanography, v. 1, p. 26–35.

Zachos, J.C., Arthur, M.A., and Dean, W.E., 1989, Geochemical evidence for suppression of pelagic marine productivity at the Cretaceous/Tertiary boundary: Nature, v. 337, p. 61–64.

Manuscript Submitted December 4, 2000; Accepted by the Society March 22, 2001

Printed in the U.S.A.

Geological Society of America
Special Paper 356
2002

End-Permian mass extinctions: A review

Douglas H. Erwin*
Department of Paleobiology, MRC-121, National Museum of Natural History, Washington, D.C. 20560, USA
Samuel A. Bowring
Department of Earth, Atmospheric and Planetary Sciences, Massachusetts Institute of Technology, 77 Massachusetts Avenue, Cambridge, Massachusetts 02130, USA
Jin Yugan
Nanjing Institute of Geology and Palaeontology, Chinese Academy of Sciences, Nanjing 210008, China

ABSTRACT

Two mass extinctions brought the Paleozoic to a close: one at the end of the Guadalupian, or middle Permian (ca. 260 Ma), and a more severe, second event at the close of the Changhsingian Stage (ca. 251.6 Ma). Here we review work over the past decade that defines the probable causes of the mass extinction, and evaluate several extinction hypotheses. The marine extinctions were selective; epifaunal suspension feeders were more affected than other clades, although significant variations occurred even among the filter feeders. In southern China, the Changhsingian marine extinction was nearly catastrophic, occurring in <0.5 m.y. On land, vertebrates, plants, and insects all underwent major extinctions. The event coincides with (1) a drop of $\delta^{13}C$ in carbonates, from $\sim+2‰$ to $-2‰$ in both marine and terrestrial sections; (2) the eruption of the massive Siberian continental flood basalts; and (3) evidence of shallow-water marine anoxia, and perhaps deep-water anoxia. Although the cause of the extinction remains unclear, a series of constraints on speculation have been established in the past few years. Leading contenders for the cause are the climatic effects, including acid rain and global warming, possibly induced by the eruption of the Siberian flood basalts; and marine anoxia. An extraterrestrial impact is consistent with the geochronological and paleontological data from southern China and elsewhere, and some possible evidence for impact has recently been advanced.

INTRODUCTION

The role of mass extinctions and subsequent biotic recoveries in determining the course of the history of life has become widely appreciated in the past two decades. At least five major mass extinctions occurred during the past 540 m.y., and while many paleontologists accept an extraterrestrial impact as the primary cause of the end-Cretaceous mass extinction, other causes are much less well understood. A full understanding of mass extinctions requires an integration of the details of the fossil record, changes in the chemistry of oceans and atmospheres, the tempo of extinction, and the distinction between triggers and mechanisms. The end-Permian mass extinctions are the most profound in the past 540 m.y., and although our understanding of these extinctions has advanced considerably in the past decade, we remain far from understanding either the trigger or mechanisms of extinction.

Many once widely accepted views of the end-Permian extinction have now been repudiated. For example, new data have led to rejection of older views of a single prolonged extinction (Teichert, 1990), perhaps peaking in the latest Permian (Erwin, 1993). Careful analysis has demonstrated the presence of two

*E-mail: Erwin.Doug@nmnh.si.edu.

Erwin, D.H., Bowring, S.A., and Yugan, J., 2002, End-Permian mass extinctions: A review, *in* Koeberl, C., and MacLeod, K.G., eds., Catastrophic Events and Mass Extinctions: Impacts and Beyond: Boulder, Colorado, Geological Society of America Special Paper 356, p. 363–383.

discrete pulses of extinction, the first near the Capitanian-Wuchiapingian boundary, and the second at the close of the Changhsingian Stage (Fig. 1) (Stanley and Yang, 1994; Jin et al., 1994). The development of a reliable conodont biostratigraphy, coupled with chemostratigraphy and a reliable temporal framework from geochronology, has produced a detailed picture of events during the Permian-Triassic (P-Tr) transition (Fig. 2). Less is known of the earlier extinction phase, in part because development of a reliable global chronostratigraphy is pending.

The information developed over the past decade constrains the possible explanations of this event, and using this information we evaluate a number of proposed causes of the latest Permian or Changhsingian mass extinction. The classic argument that the extinction is a result of the formation of the supercontinent of Pangea has long been untenable (Erwin, 1993). Other hypotheses have been rejected by the rapidity of the extinction, now shown to be <0.5 m.y. at marine sections in southern China (Bowring et al., 1998; Jin et al., 2000a). Similarly, other explanations invoked a global lowstand in sea level, based in part on the apparent paucity of marine boundary sections across the P-Tr boundary (Erwin, 1993). Although regression may have been involved in the relatively poorly known end-Guadalupian event, sequence stratigraphic analysis has shown that a transgression began in late Changhsingian time and extended across the boundary into the Early Triassic (Wignall and Hallam, 1997).

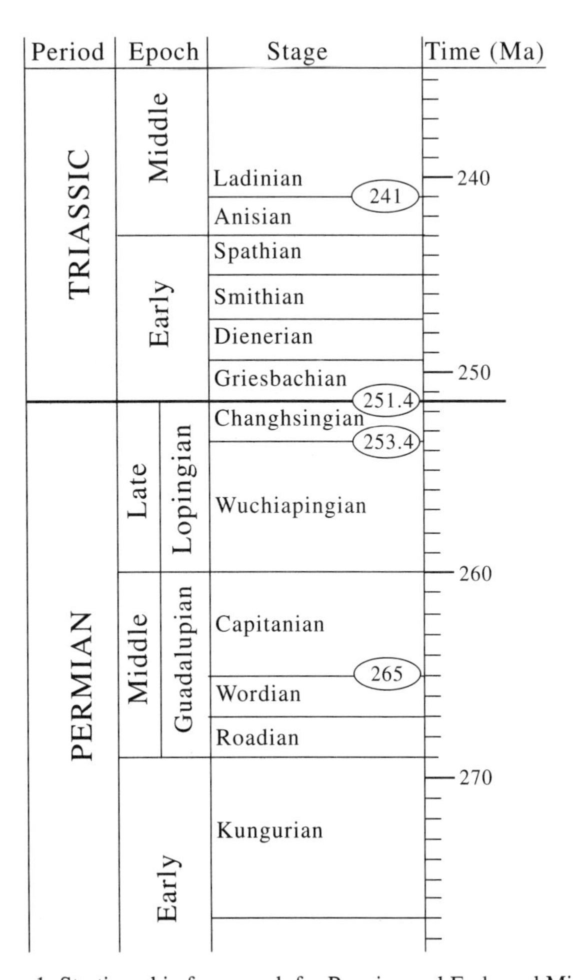

Figure 1. Stratigraphic framework for Permian and Early and Middle Triassic, including geochronologic date control. Established geochronology is in circles; interpolated time scale is to right. Permian dates are from Bowring et al. (1998) and Ladinian date is from Brack et al. (1996).

In this chapter we review the stratigraphic framework and then turn to the pattern of extinction through the Late Permian and the geological context. This review focuses largely on developments since earlier reviews (Erwin, 1993, 1994; Wignall and Hallam, 1997). We close with a brief discussion of the aftermath of the extinction, the biotic recovery in the Early Triassic.

BIOSTRATIGRAPHIC FRAMEWORK

The Late Permian has been the subject of more detailed biostratigraphic analysis than the Early Triassic, yet widespread marine regression and biotic provincialism have made the Late Permian the most problematic part of the P-Tr boundary sequence. Earlier reliance upon ammonoid, brachiopod, and fusulinid foraminiferal biozones has now largely been replaced by higher resolution conodont biostratigraphy through the P-Tr boundary (e.g., Sweet et al., 1992; Mei et al., 1998). Although some difficulties still persist over precise conodont identifications, conodont biostratigraphy has produced important advances in global correlations of marine sections, including fewer latitudinal, facies, and biogeographic problems, although direct Tethyan-Boreal-Gondwanan correlations remain difficult. Attempts to correlate Permian successions of conodont zones in warm water and cold water are particularly encouraging. For example, the Salt Range sequence of the Permian was reestablished as a link between the Tethyan and Gondwanan realms. Work by the Permian Subcommission of the International Union of Geological Sciences has produced a widely accepted biostratigraphic framework for the Permian (Fig. 1) in which conodont zones define stage boundaries (Jin et al., 1997). Detailed discussions of global and regional biostratigraphic correlation (Sweet et al., 1992; Dickins et al., 1997; Jin et al., 1998; Lucas and Yin, 1998; Yin et al., 2000) have been supplemented by detailed chemostratigraphy and magnetostratigraphy for many critical sections. The sharp shift in carbon isotopes at the P-Tr boundary provides additional support for biostratigraphic correlations.

Correlations between the marine and terrestrial realm, however, have proven far more difficult. The first appearance of the *Lystrosaurus* vertebrate assemblage has long been taken as the P-Tr boundary, particularly in the relatively abundant faunas of southern Africa (Rubidge, 1995; Ward et al., 2000). The overlap between *Lystrosaurus* and the underlying *Dicynodon* zone in South Africa has increased the problems in precisely identifying the position of the P-Tr boundary (Smith,

Marine sites

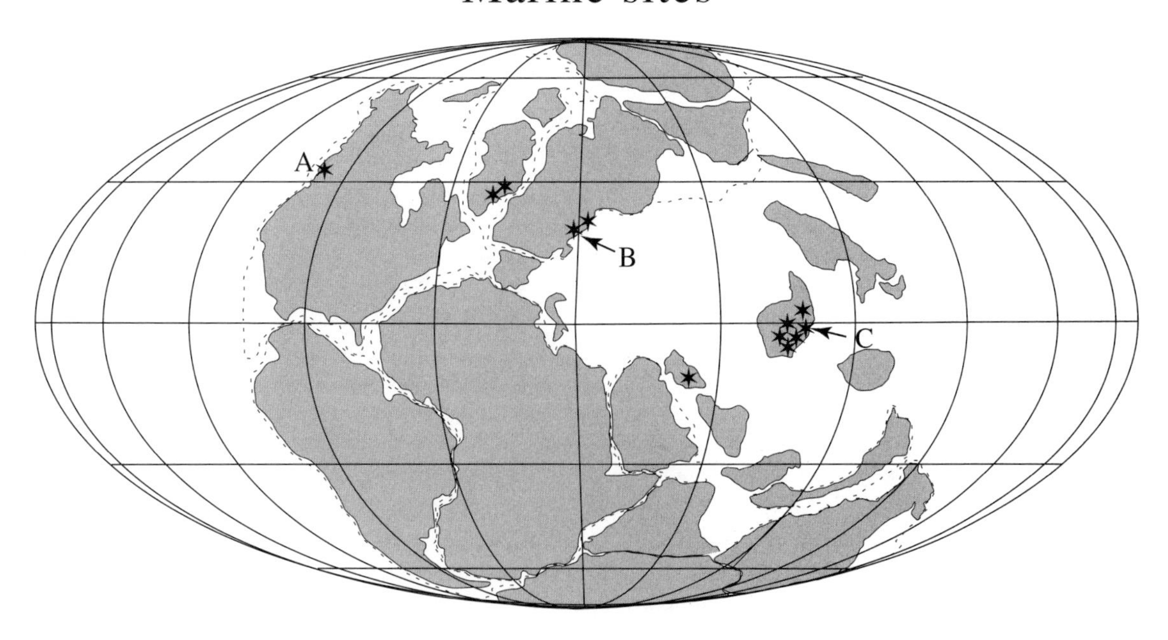

Terrestrial sites

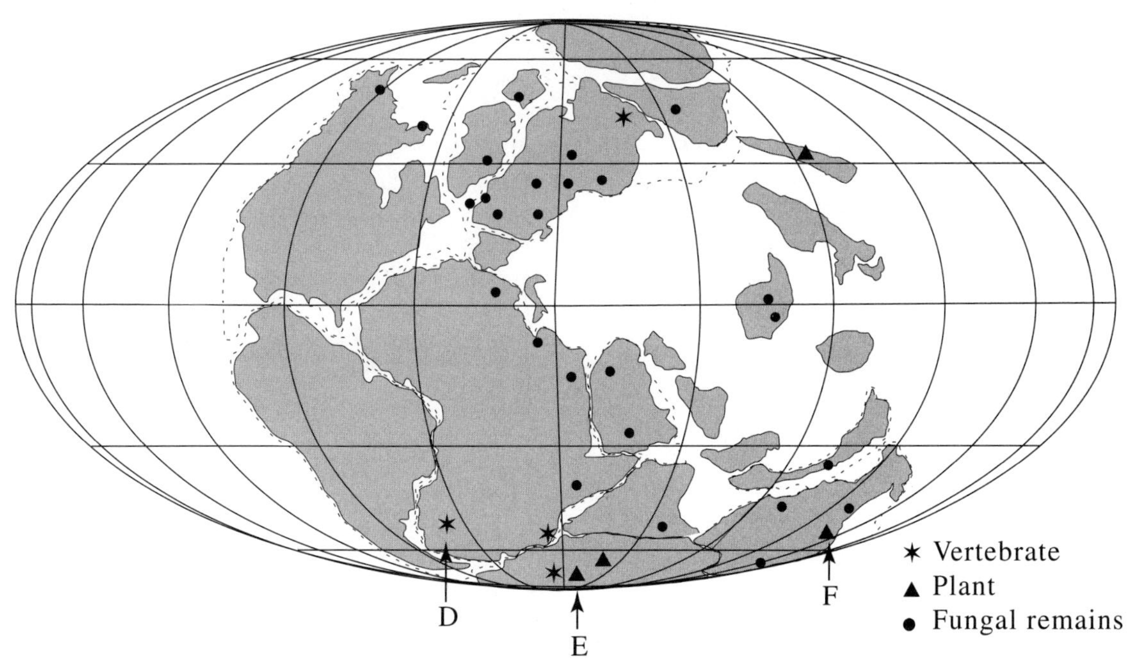

Figure 2. Marine sites: Distribution of important marine Permian-Triassic (P-Tr) boundary sites on Late Permian continental reconstruction. Note paucity of sites outside southern China, where more than 50 putative P-Tr boundary sites have been described. Marine sites: A—terranes in British Columbia, Canada; B—Garternkofel core in western Austria; C—southern China. Terrestrial sites: Distribution of terrestrial P-Tr boundary information on Late Permian continental reconstruction, showing distribution of vertebrate, plant, and fungal remains. D—Karoo deposits of South Africa; E—plant and vertebrate material from Antarctica; F—plant deposits from eastern Australia. Data were compiled from variety of sources cited in text.

1995; MacLeod et al., 2000; Ward et al., 2000). However, the wide distribution of many tetrapod genera across Pangea has aided interregional correlations, and the co-occurrence of amphibians with ammonoids in some sections has strengthened correlation between marine and terrestrial sections (Lozovsky, 1998). Chemostratigraphy (e.g., MacLeod et al., 2000), magnetostratigraphy (Jin et al., 2000b), pollen (Visscher et al., 1996), and a fungal spike at the boundary have all served as the basis of intraregional correlation, as has the disappearance of coal (e.g., Retallack et al., 1996; Faure et al., 1995). The strongest evidence for a close correlation between marine and terrestrial events comes from sections in East Greenland where terrestrial palynomorphs are found with marine fossils across the Permian-Triassic boundary (Twitchett et al., 2001). An abrupt shift in $\delta^{13}C_{carbonate}$ and changes in fossil abundance and diversity occur about 24 m below the first appearance of the diagnostic earliest Triassic conodont *Hindeodus parvus*. This establishes that the marine and terrestrial disturbances are roughly synchronous. Despite these advances, placement of the boundary in some areas remains difficult (e.g., Australia; Foster et al., 1998).

EXTINCTION PATTERNS AND SELECTIVITY

Patterns of extinction and survival have been analyzed on global, regional, and local scales for the marine realm and on global and regional scales for plants and terrestrial vertebrates.

Marine

The end-Permian mass extinction marks the demise of distinctive Paleozoic communities dominated by epifaunal, suspension-feeding clades, including articulate brachiopods, crinoids, blastoids, tabulate and rugose corals, and stenolaemate bryozoans (summarized in Erwin, 1993; Wignall and Hallam, 1997). Clades that dominate modern shallow-marine ecosystems, including bivalve and gastropod molluscs, arthropods, and nautiloid cephalopods, underwent far less extinction. Some relatively minor Permian groups, including blastoids and the last few trilobites, disappeared completely, while others, notably the echinoids, survived with only a few species. Even within these broad clades, however, considerable variation in extinction selectivity has been identified, and clues to the causes of this severe biotic crisis may be within such selective patterns.

Sepkoski's (1992) compendium of marine families and genera remains the best global view of patterns of extinction and survival, despite continuing questions about the reliability of some of the stratigraphic data. The compendium provides a benchmark for more detailed investigations of individual clades. Overall, Sepkoski's data show a 54% extinction of marine families, with generic disappearances of 58% at the Capitanian, and 67% for the Changhsingian (Stanley and Yang, 1994). From the 1979 estimates of 52% loss of families and 17% of orders, Raup (1979) applied the rarefaction curves from modern echinoids to estimate the species extinction as 96% for the entire Late Permian interval; rarefaction based on 65% loss of genera yielded an estimated species loss of 88%. These results were criticized by Stanley and Yang (1994), in part because the extinctions were clearly taxonomically selective, rather than random, as assumed by a rarefaction approach. Their revised estimate of global marine extinction, based on observed generic extinctions, is 76%–84%, although Stanley and Yang conceded that this is an underestimate. (Stanley and Yang subtracted the effects of background extinction from their estimates, but also underestimated the duration of the Lopingian.) Calculated species extinctions in southern China (Yang et al., 1993), the only region where data are available at this level, are much closer to Raup's original result than to Stanley and Yang's revised estimate.

Different patterns of extinction are also well documented among reef communities. Although many well-developed middle Permian reefs disappeared at the close of the Capitanian as their marine basins were eliminated (e.g., west Texas, Zechstein), other diverse reefs persisted through the Late Permian in southern China (Fan et al., 1990; Reinhardt, 1988), including probably the youngest Paleozoic coral reef (Shen et al., 1998), in Greece (Flügel and Reinhardt, 1989; see also Flügel, 1994, for discussion), and as calcisponge-dominated reefs in Tunisia (Toomey, 1991). Thus, reefs were not completely eliminated by the first phase of the Permian extinctions, as previously thought.

The highly selective patterns of extinction received a new explanation from Knoll et al.'s (1996) analysis of differences in metabolic capability. Using Sepkoski's (1992) generic data set, Knoll et al. recorded 65% extinction during the Capitanian, and 81% extinction during the Changhsingian among groups characterized by a heavy, calcified skeleton, gas-permeable surfaces rather than gills, weak internal circulation, and low metabolic rates. This group includes tabulate and rugose corals, stenolaemate bryozoans, brachiopods, and blastoid and crinoid echinoderms. In contrast, clades with active gills, active circulatory systems, and relatively high metabolic rates exhibit 49% and 38% extinction percentages during these two intervals. This group includes most molluscs, arthropods, conodonts, and other vertebrates. Knoll et al. (1996) proposed that this pattern of extinction selectivity matches that expected for extinctions induced by hypercapnia, or high concentrations of carbon dioxide (see following discussion).

The first pulse of extinction occurred at the end of the Capitanian, and what some had viewed as a statistical artifact of a poor fossil record now appears to be a discrete episode (Stanley and Yang, 1994; Jin et al., 1994; Shen and Shi, 2000, for brachiopods; Wang and Sugiyama, 2000, for corals), although the duration of this episode and whether it is simultaneous across the globe remain unexamined. Stanley and Yang (1994) used three statistical tests to determine whether the apparent Capitanian extinction phase could have been produced by preservational bias; the results do not support a preservational artifact. Blastoid echinoderms became extinct at this point, and most tabulate and rugose corals became extinct; only a few survived into the Wuchiapingian before disappearing during the Changhsingian (Wang and Sugiyama, 2000). Crinoids and fusulinid foraminifera nearly disappeared

This first crisis was coincident with a major global sea-level fall in the late Guadalupian that drained the shallow epicontinental seas of the Gondwana, North America, and Boreal realms. The endemic faunas in these epicontinental seas were extinguished as their habitats were eliminated, but the magnitude of the extinction may be exaggerated by limited records of Lopingian faunas. In the Tethys, benthic groups that flourished on carbonate platforms, such as the corals, fusulinids, and bryozoans, underwent a significant decline, but the survivors did not produce any new families during the Lopingian. Turnover occurred among ammonoids and conodonts, although diversity remained essentially the same during the Lopingian. The number of brachiopod families did not decrease significantly, although all the dominant elements of the Guadalupian brachiopod faunas disappeared. Nonfusulinid foraminifera, bivalves, and gastropods did not show distinctive changes at the family and generic levels (Jin et al., 1994).

Ammonoids have been particularly well studied because of their biostratigraphic utility. A global generic analysis of Permian ammonoids shows only a 44% (14 of 32 genera) extinction at the close of the Capitanian, within the range of other stage level boundaries, and far less than the 87% at the close of the Wuchiapingian. Of 33 genera, 32 (97%) disappeared at the end-Changhsingian (Zhou et al., 1996). This difference seems to suggest that the end-Capitanian was not as large as the end-Permian event, at least among ammonoids. However, the reliability of these data is uncertain. For example, Guadalupian ammonoid genera seemingly disappeared slightly later than most benthic faunas. In southern Hunan Province, China, the Guadalupian genera *Paraceltites*, *Roadoceras*, *Strigogoniatites*, *Cibolites*, *Altudocera*, *Doulingoceras*, and *Neogeoceras* extend to the very basal part of the Wuchiapingian, the *Clarkina postbitteri* zone, and Zhou et al. (1996) assigned them to the Wuchiapingian. However, they arguably disappeared as part of the end-Guadalupian event, significantly increasing the apparent rate of ammonoid extinction. In contrast, Yang and Wang (2000) referred the ammonoids from the very basal part of the Wuchiapingian to the Guadalupian fauna. Consequently they found the extinction rates of family, genera, and species to be 69%, 91%, and 100%, respectively, in southern China. Most genera that originated in the Wordian and Capitanian did not persist into the Lopingian; almost all genera that disappeared during the Wuchiapingian and Changhsingian also originated during these two stages (Zhou et al., 1996).

The P-Tr boundary section at Meishan, Zhejiang Province, southern China, has been intensively studied, and is the global boundary stratotype; detailed investigations have revealed the faunal content of each bed. In addition, the carbon isotopes have been well studied, and many of the volcanic ash beds in the sequence have been precisely dated (Bowring et al., 1998; see following). These data allowed Jin et al. (2000a) to evaluate the pattern of extinction within a well-defined chronostratigraphic and geochemical framework. By employing two statistical analyses of the extinction data, Jin et al. were able to test proposals for as many as three pulses of extinction at this locality (e.g., Yang et al., 1993; Yin and Tong, 1998).

The claim of multiple extinction horizons includes a first horizon corresponding to the disappearance of shallow-marine forms. The second proposed extinction pulse occurs in the middle of the boundary beds and is marked by the disappearance of Permian conodonts and the appearance of the first Triassic elements, as well as widespread marine anoxia. The third proposed extinction horizon corresponds to the disappearance of Permian relicts, including the final Permian brachiopods, and the initial diversification of Triassic faunal elements (Yang et al., 1993; Yin and Tong, 1998). Yin and Tong correlated the shift in $\delta^{13}C$ to the second extinction level, and suggested that a minute iridium anomaly, denoting the onset of the extinction, precedes the lower extinction horizon. These extinction levels correspond to inferred sequence boundaries, which, in combination with the carbon isotopic shifts, allow both regional and global correlation to other sections.

The development of statistical techniques based upon actual occurrence data has greatly aided interpretation of apparent extinction patterns (e.g., Meldahl, 1990; Marshall, 1994; Marshall and Ward, 1996). Due to preservational effects (the Signor-Lipps [1982] effect), the final occurrence of a taxon will precede the actual extinction. By analyzing fossil occurrences leading up to the apparent extinction horizon, the difference between the last known occurrence and the probable level of extinction may be estimated. This estimate places statistical constraints on discussions of extinction pattern.

Jin et al. (2000a) assembled data on the Meishan fauna, using the volcanic ash beds to construct a composite section of 64 horizons between the five local sections. In 162 genera, 333 species were recorded among foraminifera, fusulinids, radiolarians, rugosan corals, bryozoans, brachiopods, bivalves, cephalopods, gastropods, ostracods, trilobites, conodonts, fish, calcareous algae, and others. Of the 333 species, 161 disappeared below the P-Tr boundary beds (the extinction rate is <33% for each of these horizons). Confidence intervals were calculated for the 93 genera (265 species) with multiple occurrences. When the analysis was performed based on rock thickness, the 95% confidence intervals extend well above the boundary, reflecting the well-established depositional hiatus at the boundary beds. Using the dated ash beds, depositional rates were calculated and rock thickness converted to time. The confidence interval analysis was rerun; the 50% confidence intervals for all 93 genera are consistent with a sudden extinction at 251.4 Ma. The most reasonable interpretation of the data is a sudden extinction at 251.4 Ma followed by the gradual disappearance of a small number of surviving genera over the following million years. To test for higher or lower extinction horizons, the analysis was rerun with the 38 genera that cross the boundary, and for the genera having confidence intervals that did not reach the P-Tr boundary. No support was found for a stepwise extinction. Stepwise, gradual, and catastrophic extinctions are expected to produce different patterns of decline (Meldahl, 1990). Jin et al. (2000a) plotted the age of last occurrence versus stratigraphic abundance for all 162 genera, and the results support a sudden extinction at 251.4 Ma followed by a gradual decline in surviving species.

Correlation to the dates from Meishan has allowed other groups to infer accumulation rates and assess the duration of the extinction. For the East Greenland section, Twitchett et al. (2001) estimated an extinction duration of between 10 and 60 k.y. Analysis of apparent Milankovitch cycles in the Gartnerkofel-1 core from western Austria suggest the extinction occurred in <60 k.y., with the carbon shift in <30 k.y. (Rampino et al., 2000).

Terrestrial

The Karoo basin of South Africa remains the key area for understanding terrestrial vertebrate extinction patterns through the P-Tr interval. Excellent fossiliferous sections crop out throughout the basin, and sediment accumulation rates were evidently high. Although precise correlation to marine rocks remains uncertain, a generally accepted correlation places the Oudeberg Member of the Balfour Formation and the *Cistecephalus* biozone as uppermost Guadalupian, with the *Dicynodon* biozone encompassing the entire Lopingian (Tatarian) (Rubidge, 1995). The P-Tr boundary occurs within a zone of overlap between *Dicynodon* and *Lystrosaurus* (Hotton, 1967; Smith, 1995; King and Jenkins, 1997), as supported by recent $\delta^{13}C$ chemostratigraphy (MacLeod et al., 2000). King (1991) recorded 85 reptilian genera (including therapsids) from the *Cistecephalus* zone, 16 from the *Dicynodon* zone and 23 from the *Lystrosaurus* zone of the earliest Triassic (including new forms). Unfortunately, identification of many of these genera is problematic, even with well-preserved and well-prepared material. Rubidge's (1995) detailed biostratigraphic analysis documents the extinction of 9 of 26 reptilian genera (including 8 genera of therapsids) between the *Cistecephalus* and *Dicynodon* zones. Of the 44 genera in the *Dicynodon* zone, 7 have been recovered from the overlying *Lystrosaurus* zone (at least one additional genus, a cynodont, must also have survived because the lineage persists into the Triassic; C. Sidor, 2001, personal commun.). The redescription of the amphibian *'Lydekkerina' puterilli* Broom from the *Lystrosaurus* zone of South Africa, previously thought to be a lydekkerinid or juvenile rhinesuchid, as a paedomorphic rhinesuchid (Shiskin and Rubidge, 2000) means that the most common amphibians of the Early Triassic are all small or medium sized, which Shiskin and Rubidge argued is consistent with increased aridity relative to the Permian.

Correlative with the overlap zone between *Dicynodon* and *Lystrosaurus*, Ward et al. (2000) reported a rapid change in sedimentary style from meandering to braided river systems (see Smith, 1995), similar to the pattern reported from the P-Tr boundary in the Ural Mountains of Russia (Newell et al., 1999). This change in sedimentary style suggests a rapid die-off of terrestrial vegetation, a more arid climate, and rapid increases in sedimentation rate (Newell et al., 1999; Ward et al., 2000). Earlier work in the Karoo (e.g., Smith, 1995) suggested that the change in sedimentary style was diachronous across the basin, induced by uplift of the Cape Fold Belt, producing a more arid climate and a shift in vegetation. Thus, the conclusions of Ward et al. (2000) are highly dependant upon both their new correlations and the age of initiation of uplift relative to the extinction.

Similar evidence for a warmer and drier climate has been described in China, Australia, and Antarctica. Wang (1993, 2000) documented multiple charcoal horizons, desertification, and a decline in plant remains through the extensive Upper Permian red beds of the North China block, with a decimation of the flora at the P-Tr boundary. In both the Sydney basin, Australia (Retallack, 1999), and sections in Antarctica (Retallack and Krull, 1999), latest Permian swamps of the *Glossopteris* flora record a seasonal, humid, cold temperate climate. Although there is no difference in inferred paleolatitude, Early Triassic paleosols indicate a warmer climate with higher sedimentation rates (Retallack, 1999). The change in sedimentary style and inferred shift in climate correlate to the carbon isotopic shift at the P-Tr boundary, plant extinctions, and an increase in fungal remains. The record from Australia suggests that the change from the *Glossopteris*-dominated Permian floras to the *Dicroidium*-dominated earliest Triassic floras was abrupt (Retallack, 1995), and Retallack (1999) interpreted this as evidence of a post-apocalyptic greenhouse.

The global nature of the environmental disturbance is indicated by the widespread preservation of apparent fungal remains at the P-Tr boundary. Visscher et al. (1996) reported abundant fungal remains from the terrestrial component of palynomorph assemblages in marine rocks from the western Alps and from southern Israel, and summarized evidence of a similar horizon from the Canadian Arctic through paleo-Tethys to southern China, India, Malaysia, and Australia. This fungal spike is reminiscent of the increased abundance of fern spores immediately after the Cretaceous-Tertiary (K-T) boundary, but actually begins before the marine P-Tr extinction. The geochronologic results from the Meishan section (Bowring et al., 1998), in conjunction with the record of fungal spores from this section (Ouyang and Utting, 1990), establish that the increase in fungal remains began gradually below the P-Tr boundary and the associated isotopic shift, and decreased abruptly above the boundary. This pattern confirms the suggestion of Visscher et al. (1996) that the duration of the fungal anomaly was more than 1 m.y. Visscher et al. (1996) argued that this pattern represents an extensive die-back of terrestrial vegetation and ecosystem collapse. Megafloral evidence is consistent with an extinction of Permian conifers at the family level in Europe, and a temporary increase in bryophyte abundance in the earliest

Triassic. Further evidence of the magnitude of the terrestrial crisis comes from the complete lack of coal deposition during the Early Triassic (Faure et al., 1995; McLoughlin et al., 1997; Retallack et al., 1996; Veevers et al., 1994).

GEOLOGICAL CONTEXT

Here we consider the general geological setting of the end-Permian mass extinction, including stable isotopes, magnetostratigraphy, geochronology, and the climate record. To avoid repetition, elements common to several extinction hypotheses are discussed here, including changes in sea level and the general issue of marine anoxia. Aspects specific to single extinction hypotheses are discussed in the following section.

Stable isotopes

The Late Permian to Early Triassic interval encompassed some of the most dramatic shifts in carbon isotopes since the late Neoproterozoic and Cambrian. An early synthesis for six Tethyan sections (Baud et al., 1989) described a gradual decline from ~ +4‰ during the Guadalupian and a sharp fall to −1‰ near the Dzulfian-Dorashamian boundary (now approximately equal to the Wuchiapingian-Changhsingian boundary), a recovery to +4‰ in the basal Dorashamian, and a gradual decline to ~ −2‰ at the P-Tr boundary. The Early Triassic data are fairly stable at ~ +0.5‰ to +1‰. In addition to the short-term shifts associated with the extinction, the overall pattern involves an abrupt drop of ~3‰ from the Late Permian into the Early Triassic, indicating a major, long-term shift in the carbon cycle. In other words, there was no Early Triassic recovery to Permian values following the extinction.

Subsequent work has considerably elaborated this picture. At the Wuchiaping-Changhsing transition, two sections in southern China reveal a shift from ~ +4.5‰–5.0‰ during the Wuchiapingian to +2.0‰–3.0‰ during deposition of the Changhsing Formation (Shao et al., 2000). Shao et al. reported a single very negative point at 0‰ just above the Wuchiapingian-Changhsingian boundary at Matan in Guangxi Province, but additional data are required to confirm this result.

A negative shift in carbon isotopes is a consistent marker of the P-Tr interval in marine carbon (Holser and Magaritz, 1987; Holser et al., 1989; Baud et al., 1989), marine organic carbon (Magaritz et al., 1992; Wang et al., 1994; Isozaki, 1997; Krull et al., 2000), and terrestrial organic carbon (Morante, 1996; Retallack and Krull, 1999), tooth apatite (Thackery et al., 1990), and pedogenic carbon (MacLeod et al., 2000; Krull and Retallack, 2000), although the correlation between marine and terrestrial sections is often based on the assumption of a single global shift that must be at the P-Tr boundary. (See de Wit et al. [2002] for a criticism of this claim.) The apparent abruptness of the shift may vary between sections, in part because of differences in sediment accumulation rates. The global extent of the isotopic shift is confirmed by analyses of shallow-water carbonates from seamounts in Panthallasa (Musashi et al., 2001).

The abrupt shift in carbonate carbon isotopes coincident with the peak extinction at the Meishan section (Bowring et al., 1998; Jin et al., 2000a) and seen globally in marine sections has been interpreted not as a whole-ocean shift (e.g., involving both surface and deep waters; Erwin, 1993), but as a composite of a transient, shallow-water shift associated with a decline in productivity superimposed on a long-term shift of ~2‰. The abrupt negative spike in <500 k.y. (Bowring et al., 1998) and associated changes in organic carbon (e.g., Wang et al., 1994) are too rapid to be explained by a whole-ocean shift. The most reasonable explanation is a reduction in productivity with a drop in surface-water $\delta^{13}C$ coupled with a shift in the carbon cycle. The rapid loss of terrestrial vegetation may have been at least partly responsible for the transient negative shift in carbon isotopes at the P-Tr boundary (e.g., Visscher et al., 1996; Newell et al., 1999; Ward et al., 2000).

Several complexities affect the analysis of the carbon isotopic record from marine carbonates. Scholle (1995) described processes that may alter carbon isotopic signals, and cautioned that the common assumption that the signal reflects a single, global event may be unwarranted. In many cases independent biostratigraphic data have been used to identify the boundary, although there are many other cases, particularly in sections in Australia and Antarctica, where the chemostratigraphy has been used in isolation. Krull et al. (2000) reported $\delta^{13}C_{org}$ data from the Little Ben Sandstone in New Zealand, and documented a shift of ~7‰, from −25‰ to −32‰, with a briefer transient to −38‰ where they place the P-Tr boundary; however, there is no biostratigraphic support for this assignment. Krull et al. (2000) proposed the release of methane hydrates as the only reasonable source of such depleted carbon, perhaps from a submarine slide or warming of permafrost. In support of this contention, they charted the latitudinal variations in the isotopic shift, from −7‰ to −10‰ in higher latitudes, and from −4‰ to −2‰ in equatorial latitudes (their point for southern China was based on incorrect data; results in Jin et al. [2000a] place the southern China shift with comparable latitudinal data). However, Krull et al. (2000) included analyses from Gruszyzynski et al. (1989), which Mii et al. (1997) showed to be diagenetic artifacts, a problem that may affect some other high-latitude points reported from this group. Removal of these data eliminates all of the Northern Hemisphere high-latitude points, except those reported from west Spitsbergen by Wignall et al. (1998). Thus, confirmation of the latitudinal gradient suggested by Krull et al. (2000) requires additional paleomagnetically defined high-latitude data and additional checks for diagenetic complications, as described by Mii et al. (1997) and Scholle (1995).

Krull and Retallack (2000) presented further support for release of methane during this interval in $\delta^{13}C_{org}$ analyses of paleosols from sections through the P-Tr boundary in Antarctica. Unusual variation in $\delta^{13}C_{org}$ in depth profiles of Early

Triassic paleosols compared with Permian and modern soils, and values as low as −42‰ imply significant isotopic fractionation. They argued that the most plausible explanation is derivation of most of the organic material from methanogenesis. Specifically, this requires that atmospheric methane must have been sufficiently abundant when the soils were formed to support large populations of methanotrophs, and there must have been preferential preservation of the methanotrophic carbon relative to soil carbon to preserve the isotopic signal. In light of the short atmospheric residence time of methane (>10 yr), their data seem to require a continuing methane input into the atmosphere in the earliest Triassic.

The shift in organic carbon described by Krull et al. (2000) is plausibly related to release of methane, but more subtle shifts in isotopic values may reflect other shifts in sources, and thus complicate interpretation of the results. Foster et al. (1998) showed that the organic carbon record in Australia could be affected by the input of organic matter with a different isotopic values, specifically terrestrial material in the case they studied. Krull and Retallack (2000) argued, however, that the mechanism proposed by Foster et al. (whom they did not cite) will not influence bulk $\delta^{13}C_{org}$ unless much of the organic matter is derived from a depleted source.

A rapid shift in sulfur isotopes of sulfides has also been documented through the uppermost Permian. At the Sasayama section in Japan, the $\delta^{34}S$ values from pyrite range from −35‰ to −25‰ through the middle Permian (Kajiwara et al., 1994), but shift rapidly to ∼ −15‰ ± 5‰ beginning near the base of what Kajiwara called the middle-Upper Permian boundary, although whether this corresponds to the Capitanian-Wuchiapingian or younger is unclear. From the P-Tr into the earliest Triassic, values remain at −15‰ except for a brief negative excursion to as much as −40‰ at the assumed P-Tr boundary. Sulfates, in contrast, evidently remained at ∼ −10‰. Kajiwara et al. (1994) argued that this is persuasive evidence for an anoxic interval at the P-Tr boundary. Thus, as with the carbon record, Kajiwara et al. found a long-term shift in sulfur isotopes from the Permian to the Triassic, and a transient spike at the P-Tr boundary, although unlike the carbon isotopes, the sulfur spike is in the opposite sense of the general shift from the Permian to Triassic. Broecker and Peacock (1999) suggested that the carbon and sulfur isotopes indicate a major change in the burial of organic matter between the Permian and Triassic, associated with a disruption in terrestrial ecosystems and the introduction of less efficient processing of organic matter in marine ecosystems. Kaiho et al. (2001) report highly resolved $\delta^{34}S$ sulfate date from the upper portion of Bed 24 at the Meishan section, with a sharp drop from about 20‰ to about 4‰ at the boundary with Bed 25. They argue that this reflects a global negative shift in sulfur isotopes, from injection of a large volume of isotopically light sulfur. They attribute this to an impact event, with oxidation of reduced sulfur. This model is further discussed under impact hypotheses.

The strontium isotopic composition of seawater also changed dramatically through the Late Permian. Martin and Macdougall (1995) showed a minimum in $^{87}Sr/^{86}Sr$ in the Capitanian (0.7070). The Late Permian rise in the strontium ratio was fairly rapid, and Martin and Macdougall (1995) invoked increased continental weathering due to a change in the riverine flux and a global shift in the $^{87}Sr/^{86}Sr$ ratio, perhaps associated with global warming. The recent analysis by Kaiho et al. (2001) of the Meishan section includes a sharp drop in strontium isotopes from 0.715 to 0.708 in the uppermost part of Bed 24, followed by an increase to as high as 0.733 in Bed 27. This suggests a shift from continental to mantle sources, and then a return to a continental source. While the long-term shift documented by Martin and Macdougall (1995) may plausibly reflect a change in weathering regimes (and corresponds to the McLoughlin et al. 1997 qualitative analysis of increasing red beds), it may reflect mantle-derived ejecta. Kaiho et al. (2001) favor an impact origin for this strontium.

Magneotostratigraphy

Magnetic reversals are a key tool for correlation, but have been difficult to apply to the P-Tr boundary because of the frequent reversals, and resetting of the magnetic signal in many key sections. The long Permian-Carboniferous reversed polarity superchron ended during the Capitanian Stage of the middle Permian, and is succeeded by the Illawarra mixed polarity superchron (Menning, 1995). These two events provide important benchmarks in the correlation of Late Permian rocks across biogeographic boundaries (Jin et al., 2000b). Menning and Jin (1998) suggested that as many as 13 reversals may have occurred during the Late Permian and thus the paleomagnetic record is best used in conjunction with biostratigraphic and chemostratigraphic data. Where such data exist, however, the magnetostratigraphy may help to define the continuity of the sections.

The composite record for the P-Tr boundary (Ogg and Steiner, 1991) indicates a normal interval during at least the late Changhsingian, with short reversal during the very latest Changhsingian, followed by normal polarity through most of the overlying Griesbachian stage of the Early Triassic (Steiner et al., 1989; Menning, 1995). Results matching this pattern, and thus confirming continuity of sections to the level of magnetostratigraphic resolution, have come from southern China (summarized in Jin et al., 2000b), the southern Alps (Scholger et al., 2000), and elsewhere, although some correlation problems continue to plague the details (e.g., Orchard and Krystyn, 1998).

Geochronology

Precise geochronology is essential for understanding the causes of extinction, especially where triggering events such as the eruption of the Siberian Traps or a bolide impact have been proposed. In particular, precise geochronology allows for the testing of global synchroneity of an extinction event as well as determining the tempo of extinction and recovery and/or rapid shifts in isotopic signals across a boundary. The type section at Meishan is rich in volcanic ash beds and thus has been the site of numerous attempts to define the age and duration of the end-Permian extinction (Claoue-Long et al., 1991; Renne et al., 1995; Bowring et al., 1998; Mundil et al., 2001) using both Ar-Ar and U-Pb geochronology. In addition, both U-Pb (Kamo et al., 1996, 2000; Fedorenko et al., 2000; Campbell et al., 1992) and Ar-Ar geochronology (Renne et al., 1995; Basu et al., 1995) have been applied in order to determine the age of the Siberian Traps and evaluate their temporal coincidence with the P-Tr boundary. Despite the large amount of geochronological data, there is still some disagreement regarding the age and duration of the P-Tr boundary. In part, the debate over the age of the boundary reflects difficulties in comparing U-Pb and Ar-Ar geochronology.

U-Pb zircon geochronology takes advantage of two independent decay schemes (^{235}U-^{207}Pb and ^{238}U-^{206}Pb), allowing for evaluation of closed-system behavior; inheritance of older zircon components and diffusive Pb loss are the two most common causes of open-system behavior. It has become apparent that there is significant bias between Ar-Ar and U-Pb dates, with U-Pb dates being $\sim$1% older; this discrepancy has been attributed in part to uncertainty in the ^{40}K decay constant (Renne et al., 1995, 1998a; Min et al., 2000, 2001). Unfortunately, simple corrections are complicated by the fact that the values of the fluence monitor in many Ar laboratories (usually the Fish Canyon Tuff) have changed over the past decade. A recent U-Pb geochronological study of the Fish Canyon Tuff (Schmitz and Bowring, 2001) highlights this problem. Zircon from the Fish Canyon Tuff yield a crystallization age of 28.5 $\pm$ 0.035 Ma, a minimum of 400–500 k.y. older than the generally accepted Ar-Ar date of sanidine from the same sample. This $>$1% discrepancy is thought to be the result of a combination of uncertainty in the decay constants for K and the possibility of magmatic residence time for the zircons. Thus considerable care must be taken when comparing dates from the two decay systems. This systematic bias is much larger than calculated uncertainties associated with dates of individual volcanic layers. Geochronological studies of the extinction must now be taken to new levels, with multiple ages for volcanic rocks in stratigraphic succession a high priority. Thus, considerable care must be taken when comparing dates from the two decay systems.

The first published U-Pb geochronology from Meishan was the SHRIMP ion-probe study of Claoue-Long et al. (1991), who dated an ash bed (bed 25) just below the paleontologically defined P-Tr boundary and concluded that its age was 251.1 $\pm$ 3.4 Ma. Campbell et al. (1992) published a SHRIMP date of 248 $\pm$ 4 Ma for zircons from the Noril'sk-1 intrusion, a sill that intrudes the lower third of the Siberian Traps, and concluded that within error, the flood basalts and the Permo-Triassic boundary were synchronous. Renne et al. (1995) compared high-precision Ar-Ar geochronology of sanidine and plagioclase feldspars from bed 25 at Meishan with Ar-Ar results from the Noril'sk-1 intrusion and concluded that they were synchronous, ca. 250 Ma. Bowring et al. (1998) presented U-Pb geochronology from a series of ash beds at Meishan and two other localities in southern China that they interpreted to indicate that the extinction occurred in $<$500 k.y., ca. 251.7-251.4 Ma. Mundil et al. (2001) questioned the results of Bowring et al. (1998), and suggested an age for the boundary that is closer to 253 or 254 Ma. While much effort has been expended to define the age and duration of the boundary, it is clear that more work will be needed. Geochronological data from marine sections other than those in southern China and terrestrial sections such as the Karoo basin in South Africa are essentially nonexistent. One notable exception where both Ar-Ar and U-Pb dating methods have been applied are rocks from the Delaware basin of northwest Texas and southeastern New Mexico, where Renne et al. (1998b) reported a ^{40}Ar/^{39}Ar date of 251.0 $\pm$ 0.2 Ma on langbeinite (a potassium-rich evaporite mineral) from the Salado Formation (Wuchiapingian). If we apply an approximate bias of 1%, the date would be ca. 254 Ma, consistent with the results of Bowring et al. (1998). In addition, Renne et al. (1996) reported both conventional and SHRIMP U-Pb dates of 248–250 Ma from volcanic ashes in the stratigraphically younger Quartermaster Formation in Texas. The Quartermaster Formation has been viewed as latest Permian on the basis of biostratigraphy. Clearly more data are needed to evaluate global synchroneity and duration of the end-Capitanian and end-Changhsingian extinction.

Sea level

For decades the apparent widespread absence of marine P-Tr boundary sections from much of the world led to the acceptance of the occurrence of a widespread marine regression beginning in the Capitanian and extending until the earliest Triassic (Newell, 1967; Dickins, 1983; Holser and Magaritz, 1987; Ross and Ross, 1987; Erwin, 1993). This perspective was largely based on North American and Russian sections. Over the past decade, detailed studies of many P-Tr boundary sections, improved conodont biostratigraphy for reliable placement of the boundary, and application of sequence stratigraphic approaches has overturned this view, although distinguishing between global and regional sea-level signals remains difficult in some regions. For example, new sequence stratigraphic (Mertmann, 2000) and conodont data indicate that the Chhidru Formation in the Salt Range sections of Pakistan must be early Changhsingian in age, rather than Wuchiapingian (cf. Wignall and Hallam, 1993; Hallam and Wignall, 1999). This revision places the onset of the transgression in the mid-Changhsingian. Sections in Italy also show transgression beginning in the Changhsingian and continuing into the Triassic (Wignall and Hallam, 1992), and similar results were reported from Spitsbergen (Hallam and Wignall, 1999).

Sections from throughout China indicate a significant lowstand near the Guadalupian-Wuchiapingian boundary, followed by generally rising sea level through the Lopingian (Chen et al., 1998). Yang et al. (1993; see also Wignall and Hallam, 1993; Chen et al., 1998) demonstrated that the major transgression, previously described as beginning in the basal Griesbachian, actually began during the latest Changhsingian, but prior to the extinction. Slightly different views were provided by Zhang et al. (1996), who identified three third-order sequence boundaries, at the lower and upper Changhsingian boundaries and the basal Griesbachian, and by Tong et al. (1999) who identified two distinct transgressive systems tracts during the Changhsingian, but a slight regression during the latest Changhsingian. Whether these differences reflect local variations or differences in interpretation is unclear, but the general global picture that results is the onset of transgression in lower to mid-Changhsingian time, and extending through the P-Tr boundary into the Early Triassic.

Climate record

The movement of Pangea northward through the Permian helped end the Permian-Carboniferous glaciation by the close of the Early Permian, and marked a transition to much warmer climates that persisted through the Mesozoic (Parrish, 1995). The continued northward movement of Pangea led to a gradual warming and drying of many continental regions. McLoughlin et al. (1997) compiled sedimentologic indicators of this climate change, including the spread of red beds, but emphasized that more rapid climatic changes near the P-Tr boundary were superimposed on this long-term trend. These changes had substantial effects on both marine and terrestrial biota (e.g., Archibold and Shi, 1996, western Pacific brachiopods; Wang, 1993, 2000, plants from northern China; McLoughlin et al., 1997, Antarctic plants). The sedimentological record in South Africa, Australia, and Antarctica, and the plant record in northern China raise the possibility of a very rapid global warming during the earliest Triassic, possibly associated with the extinction.

Marine anoxia

Many arguments for both shallow- and deep-water anoxia have been advanced (Wignall and Hallam, 1992, 1993, 1997; Kajiwara et al., 1994; Isozaki, 1995, 1997; Kakuwa, 1996), and the existence of at least some degree of marine anoxia is now widely accepted, although interpretations of the causes and the relationship to the P-Tr mass extinction have varied widely.

In the deep sea, radiolarian siliceous mudstones and claystones through the P-Tr boundary interval in Japan have provided some of the most detailed records of deep-sea anoxia, including a prolonged sharp shift in sulfur isotopes. Although the boundary is difficult to unambiguously locate in these sections, Kajiwara et al. (1994) interpreted the sedimentological evidence as anoxia associated with a stagnant, stratified ocean. Isozaki (1995, 1997) proposed a global, deep-sea anoxic episode based on the occurrence of black chert and carbonaceous claystone with framboidal pyrite; biostratigraphically useful radiolarians and conodonts bracket the anoxic interval (see Kakuwa, 1996). He interpreted a gray siliceous claystone from the ?Early Triassic of the Cache Creek terrane in British Columbia, Canada, as evidence of the global extent of this anoxic episode.

Arguments in favor of shallow-water anoxia, possibly associated with the marine transgression, were developed by Paul Wignall and Tony Hallam and colleagues. Their initial studies of the earliest Triassic of the western United States, northern Italy (Wignall and Hallam, 1992), and Pakistan and China (Wignall and Hallam, 1993) invoked laminated sediments, apparently dysaerobic groups such as lingulid brachiopods and pectins such as *Claraia*, pyrite, and geochemical signatures. They argued that the Changhsingian extinction coincides with the appearance of anaerobic and dysaerobic facies. This theme was further developed in later papers (Wignall et al., 1995), including detailed analyses of high-latitude sections in Spitsbergen (Wignall and Twitchett, 1996; Wignall et al., 1998) and a comparison between high-latitude and low-latitude sections (Wignall and Twitchett, 1996). Trace fossils and additional geochemical proxies, including pyrite and sulfur/carbon ratios, were applied to bolster the evidence for anoxia. One of the difficulties in confirming this hypothesis is that many of the indicators of anoxia, including laminated sediments and the disappearance of active bioturbation, could simply reflect the extinction rather than anoxia, and some of the pyrite is clearly diagenetic and may reveal little about oxygen levels in the water column. The evidence appears to favor shallow-water anoxia coincident with the extinction in many sections, and extending into earliest Triassic sediments.

EVALUATION OF EXTINCTION HYPOTHESES

Extraterrestrial impact

The growing evidence of a rapid, even catastrophic, extinction at the P-Tr boundary continues to stimulate those favoring an extraterrestrial impact as the cause of the extinction. To date no generally accepted evidence of such an impact has been published, although Bowring et al. (1998) and Jin et al. (2000a) noted that the geochronology, stable isotopes, and marine paleontological data are largely consistent with an impact scenario. However, early analyses of boundary clays at Meishan (Zhejiang Province) and the Wachapo Mountains (Guizhou Province), China, revealed <0.5 ppb of iridium. Trace element geochemistry supported a volcanic origin for the clays (Asaro et al., 1982), which has been born out by many subsequent studies, although studies by some Chinese geologists (summarized in Xu et al., 1985, 1989) continued to support a possible extraterrestrial source of iridium. Xu Daoyi's group proposed a stony achondritic meteorite as the impactor to account for the low iridium abundances (Xu et al., 1989). Repeated attempts to replicate these reports of iridium have not met with success (Clark et al., 1986; Orth, 1989; Orth et al., 1990; Zhou and Kyte, 1988). The Gartnerkofel-1 core in western Austria contains two minor peaks in iridium abundance, the lower of which corresponds with the $\delta^{13}C$ minima at the top of the Tesero Oolite, but above the P-Tr boundary; the second anomaly occurs ~40 cm above the Tesero Oolite and is associated with pyrite, and a trace element chemistry is inconsistent with a chondritic meteorite (Orth et al., 1990).

Shocked quartz provides another indication of impact. In their study of two Antarctic and one Australian P-Tr boundary sections, Retallack et al. (1998) published evidence of what they interpreted as shocked quartz. The grains are rare, much smaller than those at K-T boundary sites, and do not clearly contain characteristic planar deformation features. Iridium anomalies are reported in the picogram range (rather than as ppt or ppb), and occur below the apparent boundary. Retallack et al. accept that this is not convincing evidence for an impact. (The stratigraphic placement of the P-Tr boundary in the Antarctic localities was also questioned by Isbell and Askin, 1999).

Perhaps the most suggestive evidence of an extraterrestrial component associated with the P-Tr boundary is the presence of helium and argon trapped in fullerenes (C_{60} and C_{70}) from sections in Japan and southern China (Becker et al., 2001; a section from Hungary was also sampled, but with no significant recovery). Fullerenes were also reported from a P-Tr boundary section in central Japan, although they were linked to production in terrestrial wildfires and preservation in anoxic marine sediments (Chijiwa et al., 1999). Becker et al. (2001) have documented a variety of fullerenes containing trapped helium and argon, similar to reports from the K-T boundary. The ^3He abundances are similar to the Murchison and Sudbury meteorites, and the $^3He/^{36}Ar$ ratios support an origin in the planetary nebula. Moreover, Becker et al. (2001) suggest that only a star or a collapsing gas cloud could produce the observed partial pressures of helium in the fullerenes: all of these lines of evidence, they suggest, support an extraterrestrial source for the fullerenes, and thus an impact at the P-Tr boundary. They propose that an object of 9 $\pm$ 3 km would have been sufficient to produced the observed ^3He. The Becker et al. analysis has been criticized by Farley and Mukhopadhyay (2001) and Isozaki (2001). Farley and Mukhopadhyay (2001) searched for ^3He in samples we provided from Bed 25 at the Meishan boundary section and from the Shangshi boundary section. In neither sample could they identify a ^3He signal consistent with the results of Becker et al. (2001). Becker and Poreda (2001) responded to Farley and Mukhopadhyay (2001) with more details of their approach and a criticism of Farley and Mukhopadhyay's techniques. They also claim evidence of spatial variation in $\delta^{34}C$ through the section, and suggest a similar variability in fullerenes may explain the failure of Farley and Mukhopadhyay to replicate the results of Becker et al. (2001). Despite this, reproduction of experimental results was absolutely critical to validation of the impact at the Cretaceous-Tertiary boundary. The absence of confirmation of the Becker et al. results is troubling. Isozaki (2001) noted that the placement of the boundary at the Sasayama section is difficult and notes that the samples analyzed by Becker et al. (2001) came from a point about 0.8 m below the actual boundary. Although stimulating, this hypothesis clearly requires further verification.

Additional suggestive evidence of impact at the Meishan locality comes from Ni-rich-Fe-Si-Ni particles at the top of Bed 24, evidently produced by condensation from a post-impact vapor cloud (Kaiho et al., 2001). Volcanic eruptions tend to be Ni-poor, with temperatures too low to form Fe-Si-Ni grains (Kaiho et al., 2001). Whether the Siberian flood basalts, where eruption rates are believed to have been an order of magnitude higher than most continental flood basalts, were also Ni-poor is unclear. The shifts in sulfur and strontium isotopes have been interpreted as impact related (Kaiho et al., 2001). Kaiho et al. estimate the maximum diameter of the crater at 600 to 1200 km (and thus the asteroid diameter at 30 to 60 km) based on the volume of vaporized reduced sulfur. They acknowledge that the object may have been much smaller if sulfur was derived from impact-induced volcanism. Finding an impact structure convincingly dated to the P-Tr boundary would also support an impact scenario. Mory et al. (2000a) suggested that a possible impact structure near Woodleigh in the Carnavon basin of Western Australia might be associated with the P-Tr extinction. Their evidence includes an apparent multiring structure with an outermost diameter of 120 km, shock-induced planar deformed quartz, and other impact features; however, there are no precise age data. Mory et al. (2000a) claimed that the structure was filled with Lower Jurassic lacustrine deposits, and that shale clasts contain palynomorphs from the Sakmarian (Early Permian). Rb-Sr geochronology of biotite reported by Mory et al. (2000a) yield 835 Ma dates corresponding to widespread basement ages. Apatite fission-track ages suggest a regional thermal peak in the range 280–250 Ma, but it is unclear whether this relates to the Woodleigh impact structure (Mory et al., 2000a). Although emphasizing the need for further study, Reimold and Koeberl (2000) questioned the reliability of the data, particularly the shock deformation structures, and suggested that the structure may be only 40 km wide. In response, Mory et al. (2000b) provided additional evidence for shocked quartz and the structure of the impact feature, but also noted unpublished data suggesting a possible Late Devonian to Early Carboniferous age for the impact. Although additional study of the structure seems warranted, at present there is no evidence to link it to the end-Permian mass extinction.

The association between the Siberian flood basalts and the P-Tr extinction has raised the issue of whether impacts can trigger continental flood basalt eruptions, and even superplume initiation (e.g., Rampino and Strothers, 1988; Alt et al., 1988). As discussed in the following, there is considerable doubt that the Siberian volcanism is plume related. Gilkson (1999) estimated that a 300-km-diameter impactor with 10%–50% mantle melting could produce 0.22 $\times$ 10^6 km^3 of basaltic magma, about an order of magnitude less than the Deccan flood basalts at the K-T boundary, and far less than the Siberian flood basalts. Melosh (2000) reached similar conclusions, and noted that there is no evidence of impact-induced volcanism from the Earth or any other planetary body. These studies suggest that impact was an unlikely trigger for the Siberian volcanism.

Abbas et al. (2000) suggested that the double extinction during the Late Permian was caused by the collision of Earth with weakly interacting massive particles (WIMPS), a class of postulated dark matter. They argued that accumulation of WIMPS in the core would produce massive heating, initiating a superplume that would rise toward the crust, causing eruption of flood basalt. Abbas et al. (2000) postulated that accompanying changes in tectonics and oceanic circulation would trigger anoxia, and possibly release of gas hydrates; they suggested that eruption of the superplume would take 5 m.y., but the source of this estimate is obscure. In the Abbas et al. (2000) model the initial extinction is due to increased carcinogenesis and anoxia consequent to the extinctions. Among other difficulties, there is no evidence that carcinogenicity is an effective cause of extinction.

Oceanic overturn and CO_2 poisoning

Widespread Late Permian seafloor carbonate marine cements and microbial precipitates led Grotzinger and Knoll (1995) to note a strong similarity to carbonates from the Proterozoic. They suggested that an increased upwelling of calcium-rich, anoxic deep waters produced waters with increased alkalinities and promoted the formation of these unusual deposits. Knoll et al. (1996) combined this insight with additional geochemical data in a model where overturn of anoxic deep waters liberated large volumes of carbon dioxide, leading to the preferential extinction of susceptible clades via carbon dioxide poisoning, or hypercapnia. In their original model, a preextinction, stagnant ocean sequestered carbon by burial of organic material from photosynthesis in surface waters, reducing atmospheric carbon dioxide; glaciation induced overturn of a stagnant ocean, introducing large volumes of carbon dioxide into the atmosphere.

This hypothesis, however, lumps large clades that exhibit distinct patterns of extinction and survival. For example, the life habits, environmental preferences, and resulting extinction patterns of prosobranch gastropods vary substantially (Erwin, 1990). The euomphalids are essentially sessile as adults and are ecologically closer to brachiopods than other gastropods: it is not surprising that they became extinct. While this example strengthens the claims of Knoll et al. (1996), whether they have carried out the analysis at the appropriate taxonomic level remain unclear. In addition, Knoll et al. (1996) appeared to conflate different physiological patterns, although whether this invalidates the results is unclear. The absence of any evidence for Late Permian glaciation is another difficulty with the Knoll et al. (1996) hypothesis, as is the considerable uncertainty over whether stagnant deep oceans can persist as long as proposed by Knoll et al. (1996). A global climate simulation by Hotinski et al. (2001) with a Permian paleography was designed to test this model. Hotinski et al. (2001) concluded that although low oxygen levels could build up in the deep sea, they would be insufficient to produced the high levels of preoverturn carbon dioxide required by the Knoll et al. (1996) model without very high levels of phosphate concentration. Hallam and Wignall (1997) also noted that the Early Triassic evidence for global warming and increased anoxia of the oceans, rather than oxygenation, are both exactly opposite to the trends suggested by Knoll et al. (1996), and Isozaki (1997) emphasized that the persistence of apparent deep-sea anoxia at two deep-sea sections in Japan and Canada is inconsistent with an oceanic overturn at the P-Tr boundary. The Japanese section, and less certainly the Canadian section, suggests persistence of the anoxia well into the Early Triassic. Whereas such evidence tends to reject the mechanism proposed by Knoll et al. (1996), the pattern of extinction selectivity is independent of the mechanism they proposed to explain it.

Suggestive evidence for unusual oceanographic conditions at the P-Tr boundary comes from a peculiar carbonate deposit in eastern Sichuan Province, southern China. At several localities an ~1-m-thick anomalous carbonate caps the shallow-water crinoidal limestones that form part of a reef complex at the top of the Changhsing Formation. The crust is overlain by Lower Triassic shales and micrites (Kershaw et al., 1999). The genesis of this carbonate has been disputed, and karst, calcrete, and a microbial origin have all been proposed. Kershaw et al. (1999) discounted the calcrete and karst interpretations, and suggested that the carbonate crust is either microbial, with recrystallization having largely destroyed diagnostic microbial features, or an inorganic carbonate precipitate. Two of their four likely solutions involve CO_2-rich waters.

Marine anoxia and transgression

An alternative anoxia model was advanced by Wignall, Hallam, and colleagues (Wignall and Hallam, 1992, 1993; Wignall et al., 1995, 1998; Wignall and Hallam, 1997; Hallam, 1994; Wignall and Twitchett, 1996).: they proposed that a latest Permian-earliest Triassic transgression brought a dysaerobic to anoxic layer onshore, the extinction coinciding with the appearance of the dysaerobic waters at that locality. Wignall and Hallam (1993) noted that this implies that the extinction should be diachronous, with the extent of the anoxia, rather than its duration as the effective extinction agent through the reduction of oxygenated shallow-marine habitats. Possible diachroneity of the extinction has been suggested by conodont biostratigraphy of a variety of sections (Wignall et al., 1995). Wignall et al. noted that the earliest appearance of dysaerobic facies is in the *Hindeotus latidentatus* conodont zone of the Changhsingian in basinal sections in Kashmir and possibly southern China. In contrast, in the shallow-water section in the Italian Alps anoxia and extinction occur at the base of the overlying *Hindeotus parvus* conodont zone. In the Salt Range of Pakistan anoxia did not arrive in until the *Clarkina carinata* zone of the Early Triassic. These conclusions depend on sedimentologic criteria for identifying anoxia, and the reliability of the biostratigraphic correlation. Wignall and Hallam (1993) suggested that the Siberian Traps may have been the cause of the extinction, via production of extensive warm saline bottom waters. Wignall and Twitchett (1996), however, suggested that the link between the extinction, anoxia, and the Siberian Traps was through a decline in oceanic circulation due to reduction or elimination of the latitudinal temperature gradient. Contributory factors may have been the decrease in oxygen solubility in water as temperature rose.

Isozaki's (1997) superanoxia model links the onset of the stagnant ocean to the Capitanian extinction episode, and links the Changhsingian event to the climax of the superanoxia in the deep sea through unspecified changes in ocean dynamics. This prolonged stratified ocean thus lasts into the lower Anisian, or ~20 m.y., with a superanoxic ocean lasting from within the Changhsingian to the Dienerian, and then waning through the Smithian and Spathian. An alternative interpretation of the Japanese record by Kakuwa (1996) interprets the Griesbachian carbonaceous mudstone as representing dysaerobic but not anoxic conditions, the accompanying pyrite being diagenetic. He also indicated that the decline in radiolarian abundance (an index of marine productivity) occurred during the Changhsingian, but before the production of the black chert. In contrast to Isozaki (1997), Kakuwa (1996) interpreted the Griesbachian carbonaceous mudstone as evidence of phytoplankton blooms and consequent spread of anoxia in epicontinental seas that influenced the deep sea. This hypothesis should be testable via organic geochemistry and biomarker studies. For example, organic biomarker evidence from the relatively shallow water Phosphoria Formation (mid-Late Permian) of Montana is consistent with anoxia along a chemocline separating normal marine surface waters from saline, evidently anoxic bottom waters; anoxia and salinity appear to have increased during the Late Permian (Dahl et al., 1993). Similar studies have yet to be carried out in deep-water P-Tr sections.

While some degree of deep- and shallow-water anoxia may have occurred during this interval, Isozaki's data suggest that the deep-water anoxia began before the marine mass extinction. Neither of these models seem capable of explaining the shift in carbon isotopes, nor can they plausibly produce the terrestrial extinctions. Thus, if anoxia played any role in the extinction it appears to have been a subsidiary one, and we must look elsewhere for the primary cause of the extinction.

Methane release

A possible role for methane release, possibly triggered by marine regression, was first suggested by Erwin (1993) on the basis of simple models of the carbon isotopic shifts. Growing evidence for marine transgression rather than regression raised substantial questions about this model, as did the suggestion that the shift in carbon isotopes involved both a long-term shift and a transient shift associated with a productivity decline at the extinction horizon, rather than a whole-ocean shift. Variants of the methane hypothesis have been invoked (e.g., Morante, 1996).

Krull and Retallack (2000; and see Krull et al., 2000) suggested involvement of methane release based on the very light organic carbon record from New Zealand and other regions, and $\delta^{13}C_{org}$ profiles in soils from the Early Triassic of Antarctica. Krull et al. (2000) also described a latitudinal gradient in $\delta^{13}C_{org}$, with high values in higher latitudes. Such isotopically depleted values reported are difficult to explain without a methanotrophic source. Krull et al. (2000) claimed that there is not a single highly depleted signal, as suggested by most previous authors, but several, which would require persistent input of methane over a long interval (>10 k.y.). Krull and Retallack (2000) suggested that a drop in sea level initially destabilized clathrates while the Siberian Traps further exacerbated greenhouse warming. In their model, this in turn destabilized polar clathrates, continuing the positive feedback and climatic destabilization. This model requires an isotopic shift coincident with the lowstand in sea level, but numerous studies suggest that the lowstand occurred well before the onset of extinction. The existence of a large volume of polar clathrates seems unlikely in view of the global warming since the mid-Permian. At present, while methane release remains a plausible explanation of the carbon shift, the only apparent mechanism of releasing sufficient marine clathrates is a substantial warming of the oceans, presumably by the Siberian Traps (Bowring et al., 1998).

Siberian flood basalt volcanism

The Siberian flood basalts are one of the two largest continental flood basalt (CFB) provinces. With an estimated volume of 2–3 $\times$ 10^6 km^3, the complex includes both intrusive and extrusive rocks and ranges from 100 to 3000 m thick, covering an area almost two-thirds the size of the continental United States. In the Noril'sk area 45 flows in 11 sequences with a total thickness of 3700 m have been identified (Renne and Basu, 1991).

Campbell et al. (1992) first drew attention to the coincidence in the ages of Siberian flood basalt eruption and the age of the P-Tr extinction (with relatively large errors). They used zircons to date the Noril'sk intrusion to 248 $\pm$ 4 Ma; this date overlapped the date for bed 25 at Meishan obtained by Claoue-Long et al. (1991) of 251.1 $\pm$ 3.4 Ma. Campbell et al. (1992) proposed an extinction model involving injection of large amounts of sulfur dioxide aerosols into the upper atmosphere with attendant global cooling and growth of polar ice caps with attendant drops in sea level. On the basis of available paleomagnetic data, Campbell et al. (1992) also speculated that the eruption of the basalts occurred in ~600 k.y. Renne et al. (1995)

published high-precision Ar-Ar data from both the Noril'sk gabbro and ash beds from Meishan and showed that they were the same age, within uncertainties, ca. 250 ± 0.2 Ma.

Kamo et al. (1996) obtained high-precision, single-grain, U-Pb geochronological data on both zircon and badellyite from the ore-bearing Noril'sk-1 intrusion that yielded an age of 251.2 ± 0.3 Ma. In addition, Kamo et al. (2000) and Fedorenko et al. (2000) reported and discussed, respectively, U-Pb geochronological results from the Maymecha-Kotuy area that confirm the suggestion that the entire sequence erupted in <1 m.y. From near the base of the sequence in the Maymecha-Kotuy area, perovskite from a melanephelinite gave a date of 252.1 ± 0.4 Ma and two lavas near the top yielded U-Pb zircon dates of 251.1 ± 0.5 Ma. Basu et al. (1995) reported a $^{40}Ar/^{39}Ar$ date of 253.3 ± 2.6 Ma from a lava stratigraphically below the one dated as 252.1 ± 0.4 Ma. The age and duration of the Siberian Traps overlap the extinction based on the U-Pb dates obtained at Meishan (Bowring et al., 1998), with the oldest dates ca. 252.1 Ma and the youngest ca. 251.1 Ma. In contrast, if the age of the extinction is as old as 253–254 Ma, as proposed by Mundil and Ludwig (1998), the majority of the volcanism is distinctly younger. This discrepancy is a matter that must be resolved with better data on the age of the boundary both in China and in other localities. It is clear that we need to know the age and duration of both the Siberian Traps and the P-Tr boundary to better than 200 k.y. to test for contemporaneity. However, the evident coincidence between the formation of the Siberian flood basalts and the mass extinction should be considered in any model for the extinction. Although paleomagnetic results from the Noril'sk area suggest that the eruption occurred during a normal magnetic polarity interval, inferred to be the earliest Triassic (Lind et al., 1994), the reversal history during the P-Tr boundary is sufficiently complex (M. Steiner, 2000, personal commun.), that paleomagnetic inferences about age of eruption are uncertain.

The close temporal association between the dates for the P-Tr boundary in southern China and the flood basalt in Siberia has spurred many hypotheses about possible links between these events (Rampino and Strothers, 1988; Campbell et al., 1992; Renne et al., 1995; Visscher et al., 1996; Kozur, 1998). The coincidence between the age of the Deccan flood basalts in India and the K-T boundary and the more recent demonstration of apparent coincidence in the formation of the central Atlantic magmatic province and the end-Triassic mass extinction (Marzoli et al., 1999; Palfy et al., 2000) have strengthened suggestions of a general link between mass extinctions and CFB eruptions—a faint echo of the enthusiasm for impacts. Wignall (2001) reexamined the postulated correlation between flood basalts and mass extinctions. Improved dating suggests the correlation only exists for three large and one smaller biotic crisis, and that even in these cases the eruptions appear to postdate the main phase of the extinctions. Further, Wignall questioned many of the postulated causal links between the eruptions and mass extinctions, suggesting the most plausible connection was a runaway greenhouse effect, perhaps involving a release of gas hydrates.

Several possibilities have been advanced to link a massive eruption of flood basalt to the mass extinctions in both marine and terrestrial ecosystems. Campbell et al. (1992) proposed the release of sulfate aerosols, perhaps augmented by release of sulfates from evaporates encountered during ascent of the magma as a source of acidic aerosols. Renne et al. (1995) suggested the release of sulfate aerosols, perhaps augmented by release of sulfates from evaporates encountered during ascent of the magma, as a source of acidic aerosols. In the Renne et al. (1995) model, global cooling was triggered by stratospheric sulfate aerosols as well as ice-sheet formation, particularly in the region lifted by the Siberian plume; the glaciation would in turn have produced marine regression. The marine extinctions result from both global cooling and regression. In this model the brief cooling episode would be followed by global warming resulting from build up of volcanic carbon dioxide.

A more involved model (Veevers et al., 1994) begins with formation of a mantle plume coincident with the end of the Kiaman paleomagnetic superchron and production of a large volume of carbon dioxide during eruption of the flood basalts. The resulting global greenhouse triggers most extinctions, lowers organic productivity, and leads to oxidation of surface organic carbon to shift the carbon isotopes. The anoxia is explained as a result of reduced ocean circulation due to global warming. A greenhouse climate would directly lower marine oxygen levels simply because of the drop in oxygen solubility as water temperature increases. The remainder of the hypothesis is independent of the cause of the Siberian Traps, and this model is not inconsistent with available data, although the postulated changes in oceanic circulation due to global warming are dependant on the (unknown) causes of Permian oceanic circulation. Another issue that requires further attention is the potential role of volcanic sulfates and carbon dioxide. Visscher et al. (1996) suggested that the fungal spike reflects destruction of terrestrial ecosystems by acid rain produced by the release of sulfur and chlorine from the Siberian volcanism. This model appears to be consistent with the terrestrial data, although whether it is consistent with the marine extinction patterns remains unknown.

Several of these models make specific assumptions about the source of the Siberian Traps. Golonka and Bocharova (2000) support the view, originally suggested by Morgan (1981), linking the Siberian flood basalts to the Jan Mayen hotspot near Iceland via a Middle Triassic track through the Yenisei-Khtanga trough south of the Taimyr Peninsula. Nevertheless, a number of questions have been raised about a plume origin for the Siberian volcanism (earlier work summarized in Erwin, 1993). Although Renne et al. (1995) proposed 1–3 km uplift in a core region of 500 km, Czamanske et al. (1998) argued that subsidence, rather than uplift, occurred during the formation of the flood basalt and was conclusive evidence against a plume origin. Czamanske et al. argued instead for partial melting associated with lithospheric shear and extension, although alternative models of the interaction between the plume and the upper mantle (e.g., Arndt, 2000) remain possibilities. Courtillot et al. (1999) argued for a close correlation between continental flood basalts and continental breakup and noted that the Siberian flood basalts and the probably Guadalupian-age Emeishan flood basalt in China are the only major episodes not associated with continental breakup. They suggested that the Siberian flood basalts represent a failed rifting event and offered as evidence the enormous deep sedimentary basin in western Siberia, which they suggest opened in the latest Permian to Early Triassic. Further support of the Courtillot et al. (1999) model comes from comparative analysis of elemental compositions of CFBs, which has demonstrated distinct differences between most CFBs and Siberia, CAMP, and a South African CFB (Puffer, 2001). The latter are very similar to volcanic arcs, suggesting they reflect reactivation of arc or backarc sources, consistent with the Courtillot et al. views of the Siberian flood basalt. Tanton and Hager (2000) developed a somewhat different model for Siberian magmatism that is consistent with the geological observations that rule out uplift associated with the magmatism (Fedorenko et al., 1996; Czamanske et al., 1998). They propose a model in which precursor melt intrudes and heats the lithospheric mantle, lowers its viscosity, and increases its density as melt turns to eclogite. This results in foundering and removal of lithospheric mantle. Consequently, a large volume of mantle melt would be produced in a very short time interval without a large amount of surface uplift. The widely assumed link between the flood basalt and a mantle plume does not appear to be supported.

The evidence for a rapid interval of global warming during the earliest Triassic has been linked to carbon dioxide released from the eruptions; however, with little hope of determining the volume of carbon dioxide released, this may remain merely an appealing, but untestable, suggestion. The effects of increased atmospheric carbon dioxide would be pervasive, whatever the source. In addition to global warming, the increase would change surface-ocean pH and carbonate chemistry (Wolf-Gladrow et al., 1999), and experimental evidence has shown that it reduces calcification rates in corals, coralline algae, and calcifying phytoplankton (Riebesell et al., 2000). The apparent temporal overlap between the eruption of the Siberian basalt and the mass extinction remains intriguing, and some part of these explanations may be correct. However, establishing a close causal link between the eruption and the patterns of differential marine and terrestrial extinction remains a task for future research. In addition, a detailed chronology of the Siberian flood basalts to compare with that of the extinction is critical for a full evaluation of this hypothesis.

EXTINCTION MECHANISMS AND SYNTHESIS

The available geological, geochemical, and paleontological data establish a set of constraints on mechanisms for the Changhsingian extinction (Bowring et al., 1998, 1999). These observations must be accommodated within any acceptable model for the extinction. (1) There is widespread evidence for shallow anoxia and some evidence for deep-water anoxia; the deep-water anoxia corresponds in part to a positive shift in carbon isotopes during the Lopingian that could reflect the burial of a massive volume of organic carbon. (2) The $\delta^{13}C$ excursion occurs in both marine and terrestrial sections, and in the marine sections in China corresponds within precision with the primary extinction horizon, although finer scale analyses are warranted. (3) The age of the extinction in southern China and the eruption of the major phase of the Siberian flood basalts are coincident within experimental error. (4) The marine extinction occurred in <500 k.y. and during a rise in sea level, not a regression. (5) There is no evidence for latest Permian glaciation. (6) Evidence suggestive of rapid global warming has been accumulating from Russia, Australia, and possibly South Africa. (7) The increase in fungal spores, indicating a disturbance in terrestrial ecosystems, begins before the marine extinction in many sections, including southern China, and has a duration exceeding 1 m.y. (8) Possible, although not yet conclusive, evidence of impact has recently been advanced in the form of fullerenes containing trapped helium and argon gases with ratios indicating an extraterrestrial origin. (9) The early onset of the fungal spike and the deep-sea anoxia suggest the possibility that disruption began on land and in the deep sea before shallow-marine ecosystems were affected.

Although recent research has dramatically reduced the range of viable possibilities, the cause, or causes, of the end-Permian mass extinction remain unclear. The principal disagreements between the various hypotheses described herein involve whether the growing evidence for rapid earliest Triassic greenhouse is correct, and if so, its origin and relationship to the extinction; whether the shift in carbon isotopes preceded or was isochronous with the mass extinction; and whether the magnitude of the transient spike in light carbon at the P-Tr boundary can be explained solely by a drop in productivity, or requires the input of additional light carbon, from methane hydrates, terrestrial organic, or an extraterrestrial (cometary) source.

Among the many remaining unanswered issues are whether there is a causal link between the Capitanian and Changhsingian extinction pulses; and if the terrestrial and marine extinction pulses during the Changhsingian were isochronous. The presence of a sharp carbon isotopic shift in South African and other terrestrial sections is compelling, but not yet conclusive evidence for isochroneity. Carbon isotopic evidence is similarly taken to demonstrate the isochroneity between the well-studied marine extinctions in southern China and those elsewhere, particularly around Pangea. Other issues are whether, at a fine scale, the carbon isotopic shift precedes or lags the peak of marine extinction, and what are the latitudinal and onshore-offshore (i.e. shallow to deep) gradients in carbon isotopes across the P-Tr boundary. Can we test different hypotheses for the causes of surface and deep-water anoxia? The pattern of transgression and the link to the appearance of anoxic sediments remains unclear. Some Chinese workers see the major transgression beginning in latest Changhsingian time, yet this would require a lag between the onset of transgression and the appearance of shallow-water anoxic facies beyond that suggested by Wignall and Hallam (1993; see also Wignall et al., 1995, Wignall and Twitchett, 1996).

Most explanations for mass extinctions involve a single triggering event, although often with a variety of subsequent proximal causes of extinction. Although this approach is intuitively satisfying, and such hypotheses are easier to test, there is no a priori reason why mass extinction could not result from a more complex web of causality (Erwin, 1993; see also MacLeod et al., 2000). The evident rapidity of this extinction adds credence to the search for a single causal force. The report by Becker et al. (2001) may be the first valid extraterrestrial signal associated with the P-Tr boundary, although it will require considerable additional confirmation. The evidence discussed here is largely consistent with an extraterrestrial impact, but does not require one, and may be consistent with other extinction scenarios. The strong correlation between the timing of eruption of the Siberian flood basalts and the extinction continues to strengthen suggestions of a link between these events, although the causal connection remains unclear and is possibly difficult to test. Against this correlation, however, is the evidence from China and elsewhere that the increase in fungal spores and apparent disruption of terrestrial ecosystems began before the marine extinction, and possibly before the onset of the Siberian eruptions.

POSTEXTINCTION BIOTIC RECOVERY

From the perspective of the history of life, one of the most intriguing aspects of the end-Permian mass extinction remains the delay in the onset of recovery in most groups during the earliest Triassic (Erwin, 1998). With the exception of ammonoids, which diversified quickly during the earliest Triassic, other marine groups exhibit a survival interval of low-diversity, cosmopolitan assemblage of generalists until near the close of the Early Triassic, perhaps 5 m.y. after the P-Tr boundary. These assemblages often contain the brachiopod *Lingula* and the bivalve *Claraia* in large numbers. The common association of these two genera with anoxic conditions led Hallam (1991, 1994) to invoke persistent anoxia as a cause of the delayed recovery. In the western United States, for example, low-diversity, low-complexity assemblages are common, often with particular species in very high abundance (Schubert and Bottjer, 1995; Woods and Bottjer, 2000). Many of these taxa exhibit nearly cosmopolitan distribution and low provinciality during the Early Triassic. By the Spathian, the final substage of the Early Triassic, new taxa began to appear, and marine communities achieved a more normal aspect by the Anisian (Middle Triassic), although stromatolites are still found in the Spathian (Schubert and Bottjer, 1995; whether these are truly biogenic stromatolites or inorganic precipitates requires further study.) Wignall et al. (1998) noted that reasonably diverse faunas, by Early Triassic standards, are found in Spitsbergen and suggest that recovery may have begun earlier in high latitudes. This area is clearly one where considerable additional research is required; there is far too little information on the biogeographic architecture of recovery. Although reefs constructed with a metazoan framework appear to be largely missing from the Early Triassic (Flügel, 1994), calcimicrobial mounds and biostromes have been recorded from the Griesbachian and Smithian-Spathian in the Nanpanjiang basin of Guizhou, China (Lehrmann, 1999).

Many lineages are known from Middle to Upper Permian rocks, and again from latest Lower Triassic to Middle Triassic deposits, but are missing from the latest Permian and Lower Triassic. Such Lazarus taxa (Jablonski, 1986), first described for gastropods (Batten, 1973), reveal the ecological and preservational complexities of this recovery and that it affected a variety of taxa (Erwin, 1993). In contrast to earlier reports, claims of extensive Lazarus lineages among reef organisms appear to be have been due to inadequate systematic treatment (Flügel, 1994). Among gastropods, ~30% of Permian genera became Lazarus taxa (Erwin, 1996). The significance of this effect should not be overestimated, because preservational bias clearly accounts for some missing taxa; e.g., the many lineages only known from silicified specimens are unlikely to be recovered from Early Triassic rocks, because no appropriate silicified assemblages are known from the Early Triassic (Erwin, 1996). The missing lineages reappear from the Smithian into the Middle Triassic, essentially coincident with the diversification of new taxa in a variety of marine clades. The lineages must have persisted during the missing interregnum, but whether population sizes were too small to ensure recovery or species migrated to as-yet undiscovered refugia remains a controversy (although the utility of oceanic islands as refugia has been overestimated; Erwin, 1993). Wignall and Benton (1999) criticized Erwin's analysis and attempted to discriminate between these alternatives, albeit with inadequate methods. They counted the total number of shallow-marine formations, and concluded that because Changhsingian and Griesbachian units were approximately equal in number, deficiencies in the fossil record were unlikely to be significant. Their analysis is essentially meaningless because they failed to quantify the presence of formations with silicified faunas.

The recovery of terrestrial communities was evidently similarly delayed. Following the demise of the conifer-dominated floras at the P-Tr boundary, lycopsids dominated the Early Triassic. The weedy lycopsid *Isoetes* was widespread, and a single earliest Triassic species gave rise to 4 additional genera and 11 species in the Early Triassic (Retallack, 1997). Lycopsids also dominated European floras during the Early Triassic (Looy et al., 1999). Coals are unknown from the Early Triassic (McLoughlin et al., 1997; Retallack and Krull, 1999; Veevers et al.,

1994). There is evidence for an earliest Triassic greenhouse climate from Australia (Retallack, 1999), Antarctica (Retallack and Krull, 1999; although McLoughlin et al., 1997, suggested that in the Lambert graben, red beds do not appear until ~100 m above the P-Tr boundary, in ?Anisian deposits), and Russia (Newell et al., 1999). A study of the Karoo in South Africa (Ward et al., 2000) documented a change in sedimentary style to braided streams at the P-Tr boundary, plausibly interpreted as the result of rapid loss of vegetation. Although further study is required, macrofloral, pollen, soil, sedimentologic, and carbon isotopic records all suggest an interval of low-diversity plant assemblages in a warm and arid environment through the Early Triassic. Pollen evidence suggests that conifers returned to dominance near the end of the Early Triassic in Europe, where there was a rapid transition to new floral assemblages, perhaps in 0.5 m.y. (Looy et al., 1999). A similar pattern has been described from Australia, where floras did not return to Permian levels of biodiversity until the Middle Triassic (Retallack, 1995; Retallack and Krull, 1999).

The two leading explanations for the delayed onset of recovery are: (1) continuing environmental perturbation, or dampening, from the P-Tr boundary through the Smithian (Hallam, 1991; Woods and Bottjer, 2000); or (2) a delay simply reflecting the ecological disturbance associated with the magnitude of the extinction (Erwin, 1996, 1998). If the former is correct, we should, in principle, be able to identify environmental factors inhibiting biotic recovery, and the recovery of marine and terrestrial ecosystems might occur at the same point in time. Hallam (1991) suggested that the prolonged delay in recovery during the Early Triassic reflected a persistent anoxic marine layer associated with the Early Triassic transgression, although some of the evidence in that paper, including the cerium enrichment patterns, is of questionable utility (Erwin, 1993). There is evidence for anoxia through the Early Triassic, although largely in the earliest and latest parts of the stage. Wignall and Hallam (1993) documented apparent dysaerobic shallow-water marine facies through much of the Griesbachian in the Salt Range of Pakistan and southern China, and Twitchell and Wignall (1996) found a similar pattern based on trace fossils in Early Triassic sections in northern Italy. Further evidence of the unusual nature of the Early Triassic is in reports of inorganic carbonate cements in outer slope deposits of the Smithian Union Wash Formation in California (Woods et al., 1999). However, similar deposits are not found throughout the Early Triassic, suggesting that the occurrence of such conditions was at best episodic. Tosk and Andersson (1988) described a diverse foraminiferal assemblage from dysaerobic to anaerobic environments in the Thaynes Formation of Idaho (Spathian).

ACKNOWLEDGMENTS

We thank Ken MacLeod and Tony Hallam for helpful reviews. Support from the National Aeronautics and Space Administration (NASA) Exobiology Program and the NASA Astrobiology Institute, the Ministry of Science and Technology of China Basic Research project G2000077705, and the Santa Fe Institute is gratefully acknowledged.

REFERENCES CITED

Abbas, S., Abbas, A., and Mohanty, S., 2000, Anoxia during the Late Permian binary mass extinction: Current Science, v. 78, p. 1290–1292.

Alt, D., Sears, J.W., and Hyndman, D.W., 1988, Terrestrial maria: The origins of large basalt plateaus, hotspot tracks and spreading ridges: Journal of Geology, v. 96, p. 647–662.

Archibold, N.W., and Shi Guanrong, 1996, Western Pacific Permian marine invertebrate palaeobiogeography: Australian Journal of Earth Sciences, v. 43, p. 635–641.

Arndt, N., 2000, Hot heads and cold tails: Nature v. 407, p. 458–460.

Asaro, F., Alvarez, L.W., Alvarez, W., and Michel, H.V., 1982, Geochemical anomalies near the Eocene/Oligocene and Permian/Triassic boundaries, *in* Silver, L.T., and Schultz, P.H., eds., Geological implications of impact hypothesis of large asteroids and comets on the earth: Geological Society of America Special Paper 190, p. 517–528.

Basu, A.R., Poreda, R.J., Renne, P.R., Teichmann, F., Vasiliev, Y.R., Sobolev, N.V., and Turrin, B.D., 1995, High-^{3}He plume origin and temporal-spatial evolution of the Siberian flood basalts: Nature, v. 269, p. 822–825.

Batten, R.L., 1973, The vicissitudes of the Gastropoda during the interval of Guadalupian-Ladinian time, *in* Logan, A., and Hills, L.V., eds., The Permian and Triassic systems and their mutual boundary: Calgary, Canadian Society of Petroleum Geologists, Memoir 2, p. 596–607.

Baud, A., Magaritz, M., and Holser, W.T., 1989, Permian-Triassic of the Tethys: Carbon isotope studies: Sonderdruck aus Geologische Rundschau, v. 78, p. 649–677.

Becker, L., Poreda, R.J., Hunt, A.G., Bunch, T.E., and Rampino, M., 2001, Impact event at the Permian-Triassic boundary: Evidence from extraterrestrial noble gases in fullerenes: Science, v. 291, p. 1530–1533.

Becker, L., and Poreda, R.J., 2001, An extraterrestrial impact at the Permian-Triassic Boundary?: Reply: Science, v. 293, p. 2343a.

Bowring, S.A., Erwin, D.H., Jin Yugan, Martin, M.W., Davidek, K.L., and Wang Wei, 1998, U/Pb zircon geochronology and tempo of the end-Permian mass extinction: Science, v. 280, p. 1039–1045.

Bowring, S.A., Erwin, D.H., and Isozaki, Y., 1999, The tempo of mass extinction and recovery: The end-Permian example: Proceedings of the National Academy of Sciences of the United States of America, v. 96, p. 8827–8828.

Brack, P., Mundil, R., Oberli, F., Meiser, M., and Rieber, H., 1996, Biostratigraphic and radiometric age data question the Milankovitch characteristics of the Latemar cycles (southern Alps, Italy): Geology, v. 24, p. 371–375.

Broecker, W.S., and Peacock, S., 1999, An ecologic explanation for the Permo-Triassic carbon and sulfur isotope shifts: Global Biogeochemical Cycles, v. 13, p. 1167–1172.

Campbell, I.H., Czamanske, G.K., Fedorenko, V.A., Hill, R.I., and Stepanov, V., 1992, Synchronism of the Siberian traps and the Permian-Triassic boundary: Science, v. 258, p. 1760–1763.

Chen Zhongqiang, Jin Yugan and Shi Guanrong, 1998, Permian transgression-regression sequences and sea-level changes of south China: Proceedings of the Royal Society of Victoria, v. 110, p. 345–367.

Chijiwa, T., Arai, T., Sugai, T., Shinohara, H., Kumazawa, M., Takano, M., and Kawakami, S., 1999, Fullerenes found in the Permo-Triassic mass extinction period: Geophysical Research Letters, v. 26, p. 767–770.

Claoue-Long, J.C., Zhang Zichao, Ma Guogan, and Du Shaohua, 1991, The age of the Permian-Triassic boundary: Earth and Planetary Science Letters, v. 105, p. 182–190.

Clark, D.J., Wang, C.Y., Orth, C.J., and Gilmore, J.S., 1986, Conodont survival and low iridium abundances across the Permian-Triassic boundary in South China: Science, v. 233, p. 984–986.

Courtillot, V., Jaupart, C. Manighetti, I., Tapponnier, P., and Besse, J., 1999, On causal links between flood basalts and continental breakup: Earth and Planetary Science Letters, v. 166, p. 177–195.

Czamanske, G.K., Gurevitch, A.B., Fedorenko, V., and Simonov, O., 1998, Demise of the Siberian plume: Paleogeographic and paleotectonic reconstruction from the prevolcanic and volcanic record, north-central Siberia: International Geological Review, v. 40, p. 95–115.

Dahl, J., Maldowan, J.M., and Sundararaman, P., 1993, Relationship of biomarker distribution to depositional environment: Organic Geochemistry, v. 20, p. 1001–1017.

de Wit, M.J., Ghosh, J.G., de Villiers, S., Rakotosolofo, N., Alexander, J., Tripathi, A., and Looy, C., 2002. Multiple organic carbon isotope reversals across the Permo-Triassic boundary of Terrestrial Gondwanan sequences: Clues to extinction patterns and delayed ecosystem recovery. Journal of Geology, v. 110, p. 227–240.

Dickins, J.M., 1983, Permian to Triassic changes in life: Memoirs of the Association of Australasian Paleontologists, v. 1, p. 297–303.

Dickins, J.M., Yang Zunyi, Yin Hongfu, Lucas, S.G., and Acharyya, S.K., 1997, Late Palaeozoic and Early Mesozoic circum-Pacific events and their global correlation: Cambridge, Cambridge University Press, 245 p.

Erwin, D.H., 1990, Carboniferous-Triassic gastropod diversity patterns and the Permo-Triassic mass extinction: Paleobiology, v. 16, p. 187–203.

Erwin, D.H., 1993, The great Paleozoic crisis: New York, Columbia University Press, 327 p.

Erwin, D.H., 1994, The Permo-Triassic extinction: Nature, v. 367, p. 231–236.

Erwin, D.H., 1996, Understanding biotic recoveries: Extinction, survival and preservation during the End-Permian mass extinction, *in* Jablonski, D., Erwin, D.H., and Lipps, J.H., eds., Evolutionary paleobiology: Chicago, University of Chicago Press, p. 398–418.

Erwin, D.H., 1998, The end and the beginning: Recoveries from mass extinctions: Trends in Ecology and Evolution, v. 13, p. 344–349.

Fan Jiasong, Rigby, J.K, Qi Jingwen, 1990, The Permian reefs of South China and comparisons with the Permian Reef Complex of the Guadalupe Mountains, West Texas and New Mexico: Brigham Young University Geological Studies, v. 36, p. 15–55.

Farley, K.A., and Mukhopadhyay, S., 2001, An extraterrestrial impact at the Permian-Triassic Boundary?: Comment: Science, v. 293, p. 2343a.

Faure, K., de Wit, M.J., and Wills, J.P., 1995, Late Permian global coal hiatus linked to ^{13}C-depleted CO_2 flux into the atmosphere during the final consolidation of Pangea: Geology, v. 23, p. 507–510.

Fedorenko, V., Czamnske, G., Zen'ko, T., Budahn, J., and Siems, D., 2000, Field and geochemical studies of the melilite-bearing Arydzhangsky Suite, and an overall perspective on the Siberian alkaline-ultramafic flood basalt volcanic rocks: International Geology Review, v. 42, p. 769–804.

Flügel, E., 1994, Pangean shelf carbonates: Controls and paleoclimatic significance of Permian and Triassic reefs, *in* Klein, G.D., ed., Pangea: Paleoclimate, tectonics and sedimentation during accretion, zenith, and breakup of a supercontinent: Geological Society of America Special Paper 288, p. 247–266.

Flügel, E., and Reinhardt, J., 1989, Uppermost Permian reefs in Skyros (Greece) and Sichuan (China): Implications for the Late Permian mass extinction event: Palaios, v. 4, p. 502–518.

Foster, C.B., Logan, G.A., and Summons, R.E., 1998, The Permian-Triassic boundary in Australia: Where is it and how is it expressed?: Proceedings of the Royal Society of Victoria, v. 110, p. 247–266.

Gilkson, A.Y., 1999, Oceanic mega-impacts and crustal evolution: Geology, v. 27, p. 387–390.

Golonka, J., and Bocharova, N.Y., 2000, Hot spot activity and the break-up of Pangea: Palaeogeography, Palaeoclimatology, Palaeoecology, v. 161, p. 49–69.

Grotzinger, J.P., and Knoll, A.H., 1995, Anomalous carbonate precipitates: Is the Precambrian the key to the past?: Palaios, v. 10, p. 578–596.

Gruszczynski, M., Halas, S. Hoffman, A., and Malkowski, K., 1989, A brachiopod calcite record of the oceanic carbon and oxygen level isotope shifts at the Permian/Triassic transition: Nature, v. 337, p. 64–68.

Hallam, A., 1991, Why was there a delayed radiation after the end-Palaeozoic extinctions?: Historical Biology, v. 5, p. 257–262.

Hallam, A., 1994, The earliest Triassic as an anoxic event, and its relationship to the end-Paleozoic mass extinction, *in* Embry, A.F., Beauchamp, B., and Glass, D.J., eds., Pangea: Global environments and research: Canadian Society of Petroleum Geologists Memoir 17, p. 797–804.

Hallam, A., and Wignall, P.B., 1997, Mass extinctions and their aftermath: Oxford, Oxford University Press, 320 p.

Hallam, A., and Wignall, P.B., 1999, Mass extinctions and sea-level changes: Earth-Science Reviews, v. 48, p. 217–250.

Holser, W.T., and Magaritz, M., 1987, Events near the Permian-Triassic boundary: Modern Geology, v. 11, p. 155–180.

Holser, W.T., Schönlaub, H.P., Attrep, M., Jr., Boeckelmann, K., Klein, P., Magaritz, M., and Orth, C.J., 1989, A unique geochemical record at the Permian/Triassic boundary: Nature, v. 337, p. 39–44.

Hotinski, R.M., Bice, K.L., Kump, L.R., Najjar, R.G., and Arthur, M.A., 2001, Ocean stagnation and end-Permian anoxia: Geology, v. 29, p. 7–10.

Hotton, N., III, 1967, Stratigraphy and sedimentation in the Beaufort Series (Permian-Triassic), South Africa, *in* Teichert, C., and Yochelson, E.L., eds., Essays in paleontology and stratigraphy, R.C. Moore Commemorative Volume: Lawrence, University of Kansas Press, p. 390–428.

Isbell, J.L. and Askin, R.A., 1999, Search for impact at the Permian-Triassic boundary in Antarctica and Australia: Comment: Geology, v. 27, p. 859.

Isozaki, Y., 1995, Superanoxia across the Permo-Triassic boundary: Record in accreted deep-sea pelagic chert in Japan, *in* Embry, A.F., Beauchamp, B., and Glass, D.J., eds., Pangea: Global environments and research: Canadian Society of Petroleum Geologists Memoir 17, p. 805–812.

Isozaki, Y., 1997, Permo-Triassic boundary superanoxia and stratified superocean: Records from lost deep sea: Science, v. 276, p. 235–238.

Isozaki, Y., 2001, An extraterrestrial impact at the Permian-Triassic Boundary?: Comment: Science, v. 293, p. 2344a.

Jablonski, D., 1986, Causes and consequences of mass extinctions, a comparative approach, *in* Elliot, D.K., ed., Dynamics of extinction: Chichester, Wiley, p. 183–229.

Jin Yugan, Wardlaw, B.R., Glenister, B.F., and Kotlyar, G.V., 1997, Permian chronostratigraphic subdivisions: Episodes, v. 20, no. 1., p. 6–10.

Jin Yugan, Wang Wei, Wang Yue, and Cao Changqun, 1998, Prospects for global correlation of Permian sequences: Proceedings of the Royal Society of Victoria, v. 111, p. 73–83.

Jin Yugan, Wang Yue, Wang Wei, Shang Qinghua, Cao Changqun, and Erwin, D.H., 2000a, Pattern of marine mass extinction near the Permian-Triassic boundary in South China: Science, v. 289, p. 432–436.

Jin Yugan, Shang Qinghua, and Cao Changqun, 2000b, Late Permian magnetostratigraphy and its global correlation: Chinese Science Bulletin, v. 45, p. 698–704.

Jin Yugan, Zhang, J., and Shang Qinghua, 1994, Two phases of the end-Permian mass extinction, *in* Embry, A.F., Beauchamp, B., and Glass, D.J., eds., Pangea: Global environments and resources: Canadian Society of Petroleum Geologists Memoir 17, p. 813–822.

Kaiho, K., Kajiwara, Y., Nakano, T., Miura, Y., Kawahata, H., Tazaki, K., Ueshima, M., Chen Z.Q., and Shi, G.R., 2001, End-Permian catastrophe by a bolide impact: Evidence of a gigantic release of sulfur from the mantle: Geology, v. 29, p. 815–818.

Kajiwara, Y., Yamakita, S., Ishida, K., Ishiga, H., and Imai, A., 1994, Development of a largely anoxic stratified ocean and its temporary mixing at the Permian/Triassic boundary supported by the sulfur isotope record: Paleaeogeography, Palaeoecology, Palaeoecology, v. 111, p. 367–379.

Kakuwa, Y., 1996, Permian-Triassic mass extinction event recorded in bedded chert sequence in southwest Japan: Palaeogeography, Palaeoclimatology, Palaeoecology, v. 121, p. 35–51.

Kamo, S.L., Czamanske, G.K., and Krough, T.E., 1996, A minimum U-Pb age for Siberian flood basalt volcanism: Geochimica et Cosmochimica Acta, v. 60, p. 3505–3511.

Kamo, S.L., Czamnske, G.K., Amelin, Y., Fedorenko, V.A., and Trofimov, V.R., 2000, U-Pb zircon and baddeleyite and U-Th-Pb perovskite ages from Siberian flood volcanism, Maymecha-Kotuy area, Siberia: Goldschmidt 2000, Oxford, UK, European Association for Geochemistry and the Geochemical Society, Journal of Conference Abstracts, v. 5, no. 2, p. 569.

Kershaw, S., Zhang Tingshan, and Lan Guangzhi, 1999, A ?microbialite carbonate crust at the Permian-Triassic boundary in South China, and its palaeoenvironmental significance: Palaeogeography, Palaeoclimatology, Palaeoecology, v. 146, p. 1–18.

King, G.M., 1991, Terrestrial tetrapods and the end Permian event: A comparison of analyses: Historical Biology, v. 5, p. 239–255.

King, G.M., and Jenkins, I., 1997, The dicynodont *Lystrosaurus* from the Upper Permian of Zambia: Evolutionary and stratigraphic implications: Palaeontology, v. 40, p. 149–156.

Knoll, A.H., Bambach, R.K., Canfield, D.E., and Grotzinger, J.P., 1996, Comparative earth history and Late Permian mass extinction: Science, v. 273, p. 452–457.

Kozur, H.W., 1998, Some aspects of the Permian-Triassic boundary (PTB) and of the possible causes for the biotic crisis around this boundary: Palaeogeography, Palaeoclimatology, Palaeoecology, v. 143, p. 227–272.

Krull, E.S., Retallack, G.J., Campbell, H.J., and Lyon G.L. 2000, $^{13}C_{org}$ chemostratigraphy of the Permian-Triassic boundary in the Maitai Group, New Zealand: Evidence for high-latitudinal methane release: New Zealand Journal of Geology and Geophysics, v. 43, p. 21–32.

Krull, E.S., and Retallack, G.J., 2000, ^{13}C depth profiles from paleosols across the Permian-Triassic boundary: Evidence for methane release: Geological Society of America Bulletin, v. 112, p. 1459–1472.

Lehrmann, D.J., 1999, Early Triassic calcimicrobial mounds and biostromes of the Nanpanjiang Basin, South China: Geology, v. 27, p. 359–362.

Lind, E.N., Kropotov, S.V., Czamanske, G.K., Gromme, S.C., and Fedorenko, V.A., 1994, Paleomagnetism of the Siberian flood basalts of the Noril'sk area: A constrain on eruption duration: International Geological Review, v. 36, p. 1139–1150.

Looy, C.V., Brygman, W.A., Dilcher, D.L., and Visscher, H., 1999, The delayed resurgence of equatorial forests after the Permian-Triassic ecologic crisis: Proceedings of the National Academy of Sciences of the United States of America, v. 96, p. 13857–13862.

Lozovsky, V.R., 1998, The Permian-Triassic boundary in the continental series of Eurasia: Palaeogeography, Palaeoclimatology, Palaeoecology, v. 143, p. 273–283.

Lucas, S.G. and Yin, H.F., editors, 1998, The Permian-Triassic boundary and global correlations: Palaeogeography, Palaeoclimatology, Palaeoecology, Special Volume 143, no. 4, p. 195–384.

MacLeod, K.G., Smith, R.M.H., Koch, P.L., and Ward, P.D., 2000, Timing of mammal-like reptile extinctions across the Permian-Triassic boundary in South Africa: Geology, v. 28, p. 227–230.

Magaritz, M., Krishnamurthy, R.V., and Holser, W.T., 1992, Parallel trends in organic and inorganic isotopes across the Permian/Triassic boundary: American Journal of Science, v. 292, p. 727–739.

Marshall, C.R., 1994, Confidence intervals on stratigraphic ranges: Partial relaxation of the assumption of randomly distributed fossil horizons: Paleobiology, v. 20, p. 459–469.

Marshall, C.R., and Ward, P.D., 1996, Sudden and gradual molluscan extinctions in the latest Cretaceous of western European Tethys: Science, v. 274, p. 1360–1363.

Martin, E.E., and Macdougall, J.D., 1995, Sr and Nd isotopes at the Permian/Triassic boundary: A record of climate change: Chemical Geology, v. 125, p. 73–95.

Marzoli, A.R., Renne, P., Piccirillo, E.M., Ernesto, M., Bellienni, G., and De Min, A., 1999, Extensive 200-million-year-old continental flood basalts of the Central Atlantic Magmatic Province: Science, v. 284, p. 616–618.

McLoughlin, S., Lindstrom, S., and Drinnan, A.N., 1997, Gondwanan floristic and sedimentological trends during the Permian-Triassic transition: New evidence from the Amery Group, northern Prince Charles Mountains, East Antarctica: Antarctic Science, v. 9, p. 281–298.

Mei, S.L., Zhang, K., and Wardlaw, B.R., 1998, A refined succession of Chanhsingian and Griesbachian neogondolellid conodonts from the Meishan section, candidate of the global stratotype section and point of the Permian-Triassic boundary: Palaeogeography, Palaeoclimatology, Palaeoecology, v. 143, p. 213–226.

Meldahl, K.H., 1990, Sampling, species abundance, and the stratigraphic signature of mass extinction: A test using Holocene tidal flat Mollusca: Geology, v. 18, p. 890–893.

Melosh, H.J., 2000, Can impacts induce volcanic eruptions?, *in* Catastrophic events and mass extinctions: Impacts and beyond: Houston, Texas, Lunar and Planetary Institute, LPI Contribution No. 1053, p. 141–142.

Menning, M., 1995, A numerical time scale for the Permian and Triassic Periods: An integrated analysis, *in* Scholle, P.A., Peryt, T.M., and Ulmer-Scholle, D.S., eds., The Permian of northern Pangea: Berlin, Springer-Verlag, v. 1, p. 77–97.

Menning, M., and Jin Yugan, 1998, Permo-Triassic magnetostratigraphy in China: The type section near Taiyuan, Shanxi Province, North China: Comment: Geophysics Journal International, v. 133, p. 213–216.

Mertmann, D., 2000, Sequence stratigraphic subdivision of the Permian Zaluch Group (Salt Range and Trans Indus Ranges, Pakistan): International Geological Congress, 31st, Rio de Jainero, Brazil, Abstracts, CD-ROM.

Mii, H.S., Grossman, E.L., and Yancey, T.E., 1997, Stable carbon and oxygen isotopic shifts in Permian seas of West Spitsbergen: Global change or diagenetic artifact: Geology, v. 25, p. 227–230.

Min, K., Mundil, R., Renne, P.R., and Ludwig, K.R., 2000, A test for systematic errors in $^{40}Ar/^{39}Ar$ geochronology through comparison with U/Pb analysis of a 1.1-Ga rhyolite: Geochimica et Cosmochimica Acta, v. 64, p. 73–98.

Min, K., Renne, P.R., and Huff, W.D., 2001, $^{40}Ar/^{39}Ar$ dating of Ordovician K-bentonites in Laurentian and Baltoscandia: Earth and Planetary Science Letters, v. 185, p. 121–134.

Morante, R., 1996, Permian and early Triassic isotopic records of carbon and strontium in Australia and a scenario of events about the Permian-Triassic boundary: Historical Biology, v. 11, p. 289–310.

Morgan, W.J., 1981, Hot spot tracks and the opening of the Atlantic and Indian Oceans, *in* Emiliani, C., ed., The oceanic lithosphere: New York, Wiley, The Sea, v. 7, p. 443–487.

Mory, A.J., Lasky, R.P., Gilkson, A.Y., and Pirajno, F., 2000a, Woodleigh, Carnarvon Basin, Western Australia: A new 120 km diameter impact structure: Earth and Planetary Science Letters, v. 177, p. 119–128.

Mory, A.J., Lasky, R.P., Gilkson, A.Y., and Pirajno, F., 2000b, Response to "Critical comment on: A.J. Mory, et al., Woodleigh, Carnarvon Basin, Western Australia: A new 120 km diameter impact structure" by W.U. Reimold and C. Koeberl: Earth and Planetary Science Letters, v. 184, p. 359–365.

Mundil, R., Metacalfe, I., Ludwig, K.R. Renne, P.R. Oberli, F., and Nicoll, R.S., 2001, Timing of the Permian-Triassic biotic crisis: Implications from new zircon U/Pb age data (and their limitations): Earth and Planetary Science Letters, v. 187, p. 131–145.

Musashi, M., Isozaki, Y., Koike, T., and Krueulen, R., 2001, Stable carbon isotope signature in mid-Panthalassa shallow-water carbonates across the Permo-Triassic boundary: Evidence for ^{13}C-depleted superocean: Earth and Planetary Science Letters, v. 191, p. 9–20.

Newell, A.J., Tverdokhlebov, V.P., and Benton, M.J., 1999, Interplay of tectonics and climate on a transverse fluvial system, Upper Permian, southern Uralian foreland basin, Russia: Sedimentary Geology, v. 127, p. 11–29.

Newell, N.D., 1967, Revolutions in the history of life, *in* Albritton, C.C., Jr., ed., Uniformity and simplicity: Geological Society of America Special Paper 89, p. 63–91.

Ogg, J.G., and Steiner, M.B., 1991, Early Triassic magnetic polarity time scale integration of magnetostratigraphy, ammonite zonation and sequence stratigraphy from stratotype sections (Canadian Arctic archipelago): Earth and Planetary Science Letters, v. 107, p. 69–89.

Orchard, M.J., and Krystyn, L., 1998, Conodonts of the lowermost Triassic of Spiti, and a new zonation based on Neogondolella successions: Rivista Italiana di Paleontologia e Stratigrafia, v. 104, p. 341–368.

Orth, C.J., 1989, Geochemistry of the bio-event horizons, *in* Donovan, S., ed., Mass extinction: Processes and evidence: London, Belkhaven Press, p. 37–72.

Orth, C.J., Attrep, M., Jr., and Quintana, L.R., 1990, Iridium anomalies: Abundance patterns across bio-event horizons in the fossil record, *in* Sharpton, V.L., and Ward, P.D., eds., Global catastrophes in Earth history: Geological Society of America Special Paper 247, p. 45–60.

Ouyang Shu, and Utting, J., 1990, Palynology of Upper Permian and Lower Triassic rocks, Meishan, Changxing County, Zhejiang Province, China: Review of Palaeobotany and Palynology, v. 66, p. 65–103.

Palfy, J., Mortensen, J.K., Carter, E.S., Smith, P.L., and Tipper, H.W., 2000, Timing the end-Triassic mass extinction: First on land, then in the sea?: Geology, v. 28, p. 39–42.

Parrish, J.M., 1995, Geologic evidence of Permian climate, *in* Scholle, P.A., Peryt, T.M., and Ulmer-Scholle, D.S., eds., The Permian of northern Pangea: Berlin, Springer-Verlag, v. 1, p. 53–61.

Puffer, J.H., 2001, Contrasting high field strength element contents of continental flood basalts from plume versus reactivated-arc sources: Geology, v. 29, p. 675–678.

Rampino, M.R., Prokoph, A., and Adler, A., 2000, Tempo of the end-Permian event: High-resolution cyclostratigraphy at the Permian-Triassic boundary: Geology, v. 28, p. 643–646.

Rampino, M.R., and Strothers, R.B., 1988, Flood basalt volcanism during the past 250 million years: Science, v. 241, p. 663–668.

Raup, D.M., 1979, Size of the Permo-Triassic bottleneck and its evolutionary implications: Science, v. 206, p. 217–218.

Reimold, W.U., and Koeberl, C., 2000, Critical comment on: A.J. Mory et al., Woodleigh, Carnarvon Basin, Western Australia: A new 120 km diameter impact structure: Earth and Planetary Science Letters, v. 184, p. 353–367.

Reinhardt, J.W., 1988, Uppermost Permian reefs and Permo-Triassic sedimentary facies from the southeastern margin of Sichuan Basin, China: Facies, v. 18, p. 231–288.

Renne, P.R., and Basu, A.R., 1991, Rapid eruption of the Siberian traps flood basalts at the Permo-Triassic boundary: Science, v. 253, p. 176–179.

Renne, P.R., Swisher, C.C., Deino, A.L., Karner, D.B., Owens, T.L., and DePaolo, D.J., 1998a, Intercalibration of standards, absolute ages and uncertainties in $^{40}Ar/^{39}Ar$ dating: Chemical Geology, v. 145, p. 117–152.

Renne, P.R., Zhang Zichao, Richards, M.A., Black, M.T., and Basu, A.R., 1995, Synchrony and causal relations between Permian-Triassic boundary crises and Siberian flood volcanism: Science, v. 269, p. 1413–1416.

Renne, P.R., Sharp, W.D., and Becker, T.A., 1998b, $^{40}Ar/^{39}Ar$ dating of langbeinite ($K_2Mg_2(SO_4)_3$) in late Permian evaporites of the Salado Formation, southeastern New Mexico, USA: Proceedings of the Goldschmidt Conference, Abstracts, p. 1253–1254.

Renne, P.R., Steiner, M.B., Sharp, W.D., and Ludwig, K.R., 1996, $^{40}Ar/^{39}Ar$ and U/Pb SHRIMP dating of latest Permian tephras in the Midland Basin, Texas: American Geophysical Union, Fall Meeting, Abstracts, p. F794.

Retallack, G.J., 1995, Permian-Triassic life crisis on land: Science, v. 267, p. 77–80.

Retallack, G.J., 1997, Earliest Triassic origin of *Isoetes* and quillwort evolutionary radiation: Journal of Paleontology, v. 71, p. 500–521.

Retallack, G.J., 1999, Postapocalyptic greenhouse paleoclimate revealed by earliest Triassic paleosols in the Sydney Basin, Australia: Geological Society of America Bulletin, v. 111, p. 52–70.

Retallack, G.J., and Krull, E.S., 1999, Landscape ecological shift at the Permian-Triassic boundary in Antarctica: Australian Journal of Earth Sciences, v. 46, p. 785–812.

Retallack, G.J., Seyedolai, A., Krull, E.S., Holser, W.T., Ambers, C.P., and Kyte, F.T., 1998, Search for evidence of impact at the Permian-Triassic boundary in Antarctica and Australia: Geology, v. 26, p. 979–982.

Retallack, G.J., Veevers, J.J., and Morante, R., 1996, Global coal gap between Permian-Triassic extinction and Middle Triassic recovery of peat-forming plants: Geological Society of America Bulletin, v. 108, p. 195–207.

Riebesell, U., Zondervan, I., Rost, B., Tortell, P.D., Zeebe, R.E., and Morel, F.M.M., 2000, Reduced calcification of marine plankton in response to increased atmospheric CO_2: Nature, v. 407, p. 364–367.

Ross, C.A., and Ross, J.R.P., 1987, Late Paleozoic sea levels and depositional sequences, *in* Ross, C.A., and Haman, D., eds., Timing and depositional history of eustatic sequences: Constraints on seismic stratigraphy: Cushman Foundation for Foraminiferal Research Special Publication No. 2, p. 137–149.

Rubidge, B.S., 1995, Biostratigraphy of the Beaufort Group (Karoo Supergroup): Geological Survey of South Africa, Biostratigraphic Series, no. 1, p. 1–46.

Schmitz, M.D., and Bowring, S.A., 2001, U-Pb zircon and titanite systematics of the Fish Canyon Tuff: An assessment of high-precision U-Pb geochronology and its application to young volcanic rocks: Geochimica et Cosmochimica Acta, v. 65, p. 2571–2587.

Scholger, R., Mauritsch, H.J., and Bradner, R., Permian-Triassic boundary magnetostratigraphy from the southern Alps (Italy): Earth and Planetary Science Letters, v. 176, p. 495–508.

Scholle, P.A., 1995, Carbon and sulfur isotope stratigraphy of the Permian and adjacent intervals, *in* Scholle, P.A., Peryt, T.M., and Ulmer-Scholle, D.S., eds., The Permian of northern Pangea: Berlin, Springer-Verlag, v. 1, p. 133–149.

Schubert, J.K., and Bottjer, D.J., 1995, Aftermath of the Permian-Triassic mass extinction event: Paleoecology of Lower Triassic carbonates in the western USA: Palaeogeography, Palaeoclimatology, Palaeoecology, v. 116, p. 1–39.

Sepkoski, J.J., Jr., 1992, A compendium of fossil marine animal families (second edition): Milwaukee Public Museum Contributions in Biology and Geology, no. 83, p. 155.

Shao Longyi, Zhang Pengfei, Duo Jianwei, and Shen Shuzhong, 2000, Carbon isotope compositions of the Late Permian carbonate rocks in southern China: Their variations between the Wujiaping and Changxing formations: Palaeogeography, Palaeoclimatology, Palaeoecology, v. 161, p. 179–192.

Shen, Jianwei, Kwamura, T., and Yang Wanrong, 1998, Upper Permian coral reef and colonia rugose carals in northwest Huwan, South China; Facies, v. 39, p. 35–65.

Shen Shuzhong, and Shi Guirong, 2000, Wuchaipingian (early Lopingian, Permain) global brachiopod palaeobiogeography: A quantitative approach: Paleaeogeography, Palaeoecology, Palaeoecology, v. 162, p. 299–318.

Shishkin, M.A., and Rubidge, B.S., 2000, A relict rhinesuchid (Amphibia: Temnospondyli) from the Lower Triassic of South Africa: Palaeontology, v. 43, p. 653–670.

Signor, P.W., III, and Lipps, J.H., 1982, Sampling bias, gradual extinction patterns, and catastrophes in the fossil record, *in* Silver, L.T., and Schultz, P.H., eds., Geological implications of impacts of large asteroids and comets on the Earth: Geological Society of America Special Paper 190, p. 291–296.

Smith, R.M.H., 1995, Changing fluvial environments across the Permian-Triassic boundary in the Karoo Basin, South Africa and possible causes of tetrapod extinctions: Palaeogeography, Palaeoclimatology, Palaeoecology, v. 117, p. 81–95.

Stanley, S.M., and Yang, X., 1994, A double mass extinction at the end of the Paleozoic Era: Science, v. 266, p. 1340–1344.

Steiner, M., Ogg, J., Zhang, Z., and Sun, S., 1989, The late Permian/Early Triassic magnetic polarity time scale and plate motions of South China: Journal of Geophysical Research, v. 94, p. 7343–7363.

Sweet, W.C., Yang Zunyi, Dickins, J.M., and Yin Hongfu, eds., 1992, Permo-

Triassic events in the eastern Tethys: Cambridge, Cambridge University Press, 181 p.

Tanton, L.T.E., and Hager, B.H., 2000, Melt intrusion as a trigger for lithospheric foundering and the eruption of the Siberian flood basalts: Geophysical Research Letters, v. 27, p. 3937–3940.

Teichert, C., 1990, The Permian-Triassic boundary revisited, *in* Kauffman, E.G., and Walliser, O.H., eds., Extinction events in Earth history: Berlin, Springer-Verlag. p. 199–238.

Thackeray, J.F., van der Merwe, N.J., Lee-Thorp, J.A., Sillen, A., Lanham, J.L., Smith, R., Keyser, A., and Monteiro, P.M.S., 1990, Changes in carbon isotope ratios in the late Permian recorded in therapsid tooth apatite: Nature, v. 347, p. 751–753.

Tong Jinnan, Yin Hongfu, and Zhang Kexing, 1999, Permian and Triassic sequence stratigraphy and sea level changes of eastern Yangtze platform: Journal of China University of Geosciences, v. 10, p. 161–169.

Toomey, D.F., 1991, Late Permian reefs of southern Tunisia: Facies patterns and comparison with the Capitan Reef, southwestern United States: Facies, v. 25, p. 119–146.

Tosk, T.A., and Andersson, K.A., 1988, Late Early Triassic foraminifers from possible dysaerobic to anaerobic paleoenvironments of the Thaynes Formation, southeast Idaho: Journal of Foraminiferal Research, v. 18, p. 286–301.

Twitchett, R.J., and Wignall, P., 1996, Trace fossils and the aftermath of the Permo-Triassic mass extinction: Evidence from northern Italy: Palaeogeography, Palaeoclimatology, Palaeoecology, v. 124, p. 137–151.

Twitchett, R.J., Looy, C.V., Morante, R., Visscher, H., and Wignall, P., 2001, Rapid and synchronous collapse of marine and terrestrial ecosystems during the end-Permian biotic crisis: Geology, v. 29, p. 351–354.

Veevers, J.J., Conaghan, P.J., and Shaw, S.E., 1994, Turning point in Pangean environmental history at the Permian/Triassic (P/Tr) boundary, *in* Klein, G.D., ed., Pangea: Paleoclimate, tectonics, and sedimentation during accretion, zenith, and breakup of a supercontinent: Geological Society of America Special Paper 288, p. 187–196.

Visscher, H., Brinkhuis, H., Dilcher, D.L., Elsik, W.C., Eshet, Y., Looy, C.V., Rampino, M.R., and Traverse, A., 1996, The terminal Paleozoic fungal event: Evidence of terrestrial ecosystem destabilization and collapse: Proceedings of the National Academy of Sciences of the United States of America, v. 93, p. 2155–2158.

Wang, K., Geldsetzer, H.H.J., and Krouse, H.R., 1994, Permian-Triassic extinction: Organic ^{13}C evidence from British Columbia, Canada: Geology, v. 22, p. 580–584.

Wang Xiang-Dong, and Sugiyama, T., 2000, Diversity and extinction patterns of Permian coral faunas of China: Lethaia, v. 33, p. 285–294.

Wang Zhiqiang, 1993, Evolutionary ecosystem of Permian-Triassic red beds in N. China: A historical record of natural global desertification, *in* Lucas, S.G., and Morales, M., eds., The nonmarine Triassic: Bulletin of the New Mexico Museum of Natural History and Science, v. 3, p. 471–476.

Wang Zhiqiang, 2000, Vegetation of the eve of the P-T event in North China and plant survival strategies: An example of Upper Permian refugium in northwestern Shanxi, China: Acta Palaeontologica Sinica, v. 39 (supplement), p. 127–153.

Ward, P.D., Montgomery, D.R., and Smith, R., 2000, Altered river morphology in South Africa related to the Permian-Triassic extinction: Science, v. 289, p. 1740–1743.

Wignall, P.B., 2001, Large igneous provinces and mass extinction: Earth-Science Reviews, v. 53, p. 1–33.

Wignall, P.B., and Benton, M.J., 1999, Lazarus taxa and fossil abundance at times of biotic crisis: Journal of the Geological Society, London, v. 156, p. 453–456.

Wignall, P.B., and Hallam, A., 1992, Anoxia as a cause of the Permo-Triassic mass extinction: Facies evidence from northern Italy and the western United States: Palaeogeography, Palaeoclimatology, Palaeoecology, v. 93, p. 21–46.

Wignall, P.B., and Hallam, A., 1993, Griesbachian (earliest Triassic) paleoenvironmental changes in the Salt Range, Pakistan and southeast China and their bearing on the Permo-Triassic mass extinction: Palaeogeography, Palaeoclimatology, Palaeoecology, v. 101, p. 215–237.

Wignall, P.B., Hallam, A., Lai Xulong, and Yang Fengqing, 1995, Palaeoenvironmental changes across the Permian/Triassic boundary at Shangsi (N. Sichuan, China): Historical Biology, v. 10, p. 175–189.

Wignall, P.B., Morante, R., and Newton, R., 1998, The Permo-Triassic transition in Spitsbergen: $^{13}C_{org}$ chemostratigraphy, Fe and S geochemistry, facies, fauna and trace fossils: Geological Magazine, v. 135, p. 47–62.

Wignall, P.B., and Twitchett, R.J., 1996, Oceanic anoxia and the end Permian mass extinction: Science, v. 272, p. 1155–1158.

Wolf-Gladrow, D.A., Riebesell, U., Burkhardt, S., and Bijma, J., 1999, Direct effects of CO_2 concentration of growth and isotopic composition of marine plankton: Tellus, v. 51B, p. 461–476.

Woods, A.D., and Bottjer, D.J., 2000, Distribution of ammonoids in the Lower Triassic Union Wash Formation (Eastern California): Evidence for paleoceanographic conditions during recovery from the end-Permian mass extinction: Palaios, v. 15, p. 535–545.

Woods, A.D., Bottjer, D.J., Mutti, M., and Morrison, J., 1999, Lower Triassic large sea-floor carbonate cements: Their origin and a mechanism for the prolonged biotic recovery from the end-Permian mass extinction: Geology, v. 27, p. 645–648.

Xu Daoyi, Ma Shulang, Chai Zhifang, Mao Xiuying, Sun Yeyin, Zhang Qingweng, and Yang Zunyi, 1985, Abundance variation of Iridium anomalies and trace elements at the Permian/Triassic boundary at Shangsi in China: Nature, v. 314, p. 154–156.

Xu Daoyi, Zhang Qinwen, Sun Yeyin, Yan Zheng, Chai Zhifang, and He Jinwen, 1989, Astrogeological events in China: New York, Van Nostrand Reinhold, 264 p.

Yang Fengqing and Wang Hongmei, 2000, Ammonoid succession model across the Paleozoic-Mesozoic transition in South China, *in* Yin Hongfu, Dickins, J.M., Shi, G.R., and Tong Jinnan, eds., Permian-Triassic evolution of Tethys and western circum-Pacific: Amsterdam, Elsevier, p. 353–370.

Yang Zunyi, Wu Shunbao, Yin Hongfu, Xu Guirong, Zhang Kexing, and Bie Xianmei, 1993, Permo-Triassic events of South China: Geological Publishing House, Beijing, 153 p.

Yin Hongfu, Dickins, J.M., Shi Guanrong, and Tong Jinnan, eds., 2000, Permian-Triassic evolution of Tethys and western circum-Pacific: Amsterdam, Elsevier, 392 p.

Yin Hongfu and Tong Jinnan, 1998, Multidisciplinary high-resolution correlation of the Permian-Triassic boundary: Palaeogeography, Palaeoclimatology, Palaeoecology, v. 143, p. 199–212.

Zhang Kexing, Tong Jinnan, Yin Hongfu, and Wu Shunbao, 1996, Sequence stratigraphy near the Permian-Triassic boundary at Meishan section, South China, *in* Yin Hongfu, ed., The Palaeozoic-Mesozoic boundary: Candidates of the global stratotype section and point (GSSP) of the Permian-Triassic boundary: Wuhan, China, University of Geosciences Press, p. 57–64.

Zhou Lei, and Kyte, F.T, 1988, The Permian-Triassic boundary event: A geochemical study of three Chinese sections: Earth and Planetary Science Letters, v. 90, p. 411–421.

Zhou Zuren, Glenister, B.F., Furnish, W.M., and Spinosa, C., 1996, Multi-episodal extinction and ecological differentiation of Permian ammonoids: Permophiles, v. 29, p. 195–212.

Manuscript Submitted November 23, 2000; Accepted by the Society March 22, 2001

Geological Society of America
Special Paper 356
2002

Permian-Triassic boundary in the southwestern United States: Hiatus or continuity?

Walter Alvarez*
Diane O'Connor
Department of Earth and Planetary Science, University of California, Berkeley, California 94720-4767, USA

ABSTRACT

Study of the Permian-Triassic (P-Tr) mass extinction is hampered by the rarity of marine outcrops covering this stratigraphic interval, and it has long been thought that the P-Tr event record is lost because of a stratigraphic hiatus in much of the world. The existence of a P-Tr hiatus in the southwestern United States (mainly Utah and Nevada) has been so thoroughly accepted that this region has been ignored by most P-Tr researchers for years. We review the evidence for a P-Tr hiatus in the southwestern United States and find it to be unconvincing. The P-Tr boundary has long been thought to coincide with a global sea-level low, but recent work by others suggests a global sea-level rise at the time of the P-Tr mass extinction—i.e., a complete reversal of interpretation. Paleontological evidence for missing Upper Permian in the southwestern United States is weak and based on relatively outdated biostratigraphy. Recent conodont studies have suggested there may be younger Permian present than previously thought. Physical evidence of a P-Tr unconformity is notably lacking in most of the southwestern United States. We do not claim that complete P-Tr sections exist here, but we argue that the evidence for unconformity is unconvincing. We suggest that this region, and perhaps many others around the world, are worth reexamining for possibly complete, or nearly complete, records of what happened during the P-Tr mass-extinction event.

INTRODUCTION

It is perhaps surprising that 18 years after the first Snowbird Conference on the Cretaceous-Tertiary (K-T) mass extinction, in 1982, we still have little understanding of the Permian-Triassic (P-Tr) mass extinction. Yet that is indeed the case. It has been difficult to achieve an understanding of the event or events across the P-Tr boundary for a number of reasons. The P-Tr boundary is nearly four times as old as the K-T boundary and as a result, much less record remains. In particular, most seafloor older than Jurassic has been subducted, so it is not possible to recover from the P-Tr interval the kind of deep-sea sediment cores that have been so useful in study of the K-T boundary. P-Tr deep-sea sediments have been observed in only a few offscraped remnants, in most cases severely deformed and/or metamorphosed.

Age correlation is also more difficult. The planktic foraminifera whose global dispersal of planktic foraminifera in surface ocean waters makes them ideal for global correlation in K-T studies had not yet evolved in P-Tr time. Early workers primarily used ammonoids, which are now known to be abundant only in Boreal settings, inhibiting correlation with the Tethyan realm. Conodonts are proving to be much more cosmopolitan, but there are still complications, such as correlating

*E-mail: platetec@socrates.berkeley.edu

Alvarez, W., and O'Connor, D., 2002, Permian-Triassic boundary in the southwestern United States: Hiatus or continuity?, *in* Koeberl, C., and MacLeod, K.G., eds., Catastrophic Events and Mass Extinctions: Impacts and Beyond: Boulder, Colorado, Geological Society of America Special Paper 356, p. 385–393.

the conodont zones with the ammonoid zones. Late Permian conodont zonations are still being modified, and the current zonation has not yet been widely applied.

A major milestone in stratigraphic practice was recently passed when, after a 20-year international effort, the stages of the marine Permian were approved by the Subcommission on Permian Stratigraphy (Jin et al., 1997) Fig. 1. The Lower Permian (Cisuralian) type sections are in Russia, the middle Permian (Guadalupian) ones in Texas, and those of the Upper Permian (Lopingian) in China. The Permian-Triassic boundary is approaching formalization (Yang et al., 1995). Although this will be a very important step, it will not constrain the solution to the problem of the P-Tr event(s). The proposed boundary horizon is at the widely recognizable and therefore stratigraphically useful first occurrence of the conodont species *Hindeodus parvus*, placing the proposed formal boundary slightly above the mass-extinction level. The Lower Triassic stages (Silberling and Tozer, 1968) have not yet been formalized.

Most P-Tr workers have long been convinced that almost everywhere in the world the boundary is marked by a significant hiatus. This absence of stratigraphic record means that little can be learned about the P-Tr event in most places. The P-Tr boundary unconformity is thought to affect the marine record almost everywhere in the world except for a small area in the southern Alps, a few sections in Pakistan and, notably, many places in southern China. The Alpine sections have been intensely studied and the Pakistan sections are of difficult access, so most detailed P-Tr research is now being done in southern China by Chinese geologists and paleontologists, increasingly in collaboration with Western colleagues. The conodont studies in China have enabled workers in other areas to improve the correlations of their sections. On this basis, sections in Western Canada (Henderson, 1997) and Arctic Canada (Henderson and Baud, 1997) have been shown to contain the Upper Permian, contrary to previous studies. Recent work in Greenland and Spitsbergen (Twitchett et al., 2001; R. Twitchett, 2001, personal commun.) gives a similar result. In addition, work in Texas (Kozur, 1992) has indicated that the highest Permian present is considerably younger than previously believed.

The southwestern United States is among the many regions where an Upper Permian hiatus has long been accepted. In this paper we review the evidence that led to this conclusion. Although we cannot demonstrate that the southwestern United States contains complete P-Tr boundary sections, we will show why we think the evidence for a hiatus is unconvincing. Our conclusion will be that the southwestern United States appears to be a promising area for future P-Tr stratigraphic research, because it may contain complete and/or nearly complete sections. (In the following, we use the term "southwestern United States" to refer to Utah, Arizona, and Nevada, excluding Western Texas.)

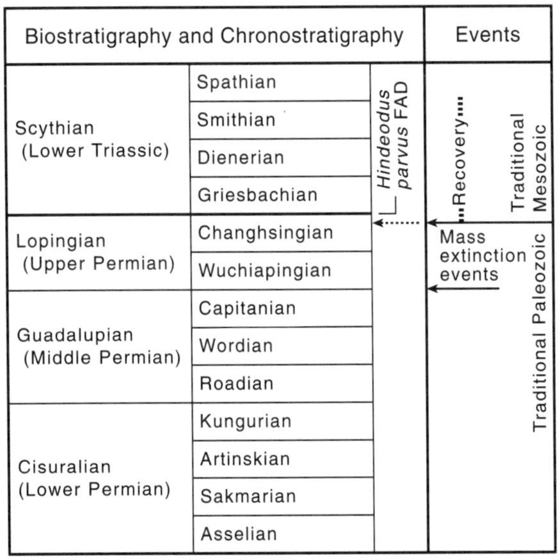

Biostratigraphy and Chronostratigraphy			Events	
Scythian (Lower Triassic)	Spathian	*Hindeodus parvus* FAD	....Recovery....	Traditional Mesozoic
	Smithian			
	Dienerian			
	Griesbachian			
Lopingian (Upper Permian)	Changhsingian		Mass extinction events	Traditional Paleozoic
	Wuchiapingian			
Guadalupian (Middle Permian)	Capitanian			
	Wordian			
	Roadian			
Cisuralian (Lower Permian)	Kungurian			
	Artinskian			
	Sakmarian			
	Asselian			

Figure 1. Stratigraphy of Permian and Lower Triassic, showing currently used series names (left column) and stage names (second column). Permian-Triassic boundary is currently placed at first appearance datum (FAD) of conodont *Hindeodus parvus*. This datum falls above the level of mass extinction, which has traditionally been used to mark Paleozoic-Mesozoic (Permian-Triassic) boundary. We make no judgement on which view is preferable, as long as distinction is kept clear. Biotic recovery from the P-Tr mass extinction continued well into Early Triassic. Recent work indicates that an earlier mass extinction took place at about the end of the Capitanian (Jin et al., 1994; Stanley and Yang, 1994).

THE SOUTHWESTERN UNITED STATES AT P-TR BOUNDARY TIME

The southwestern United States would at first seem a likely place to find useful P-Tr boundary sections. In Permian and Triassic time this region formed one sector of the western margin of Pangea. The Sonoma orogenic belt was farther west, across an unknown distance of ocean. Unfortunately, this margin presents stratigraphic difficulties because it is not as rich in fossils as many parts of the Tethyan realm. In addition, the search for P-Tr boundary sections is complicated by the fact that this margin was later involved in both compressional and extensional tectonics.

Passive-margin craton and miogeocline

Throughout the Paleozoic, thin cratonal sequences accumulated in the interior of the United States and thicker miogeoclinal sequences were deposited on the western edge of the continent. Commonly there is little difference in facies, and only sedimentary thickness distinguishes craton from miogeocline. In the late Paleozoic and early Mesozoic, the boundary between craton and miogeocline trended approximately southwest from Salt Lake City to Las Vegas. The evolution of the stratigraphy across this boundary was synthesized by Hintze (1988). We consider two areas of possible P-Tr interest (Fig. 2).

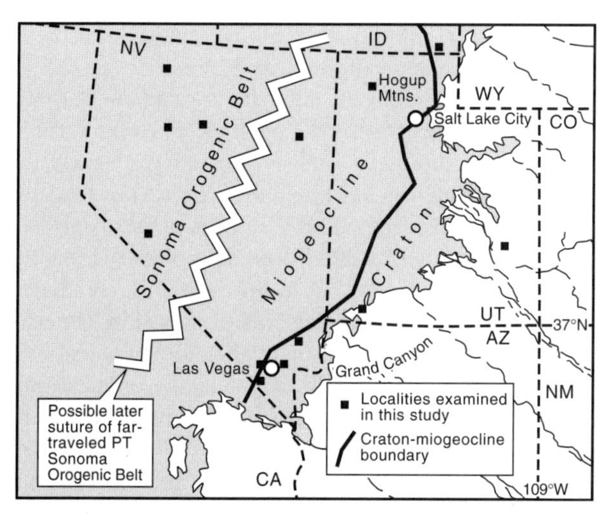

Figure 2. Setting of Permian-Triassic sedimentation in southwestern United States. The coastline and rivers are simplified after an online Early Triassic paleogeographic map by Ronald C. Blakey, University of Northern Arizona (http://vishnu.glg.nau.edu/rcb/tripaleo.html). On this kind of map, features like the coast and rivers are drawn to look realistic, but only general pattern is controlled by data; details represent artistic license. The Sonoma orogenic belt was deformed at about P-Tr time; it may have originally been far away, and have been accreted to North America later. Mtns.—Mountains, PT—Permian-Triassic.

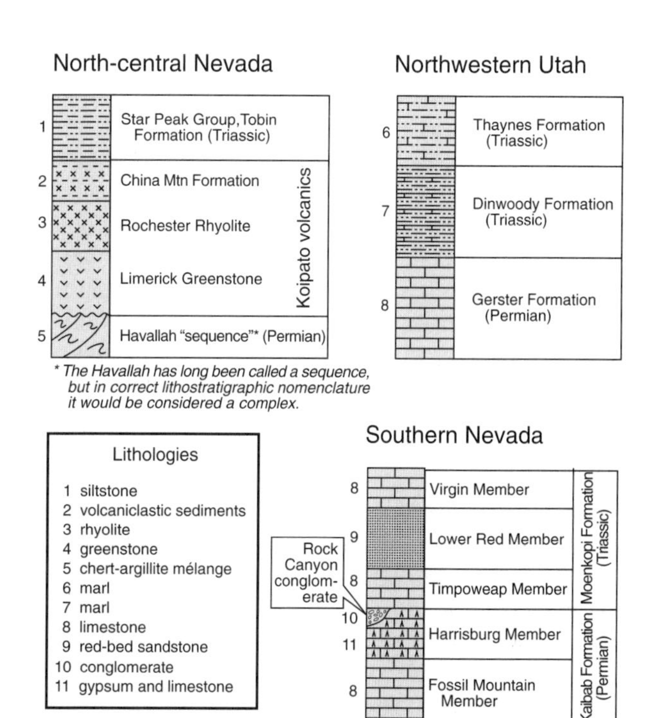

Figure 3. Simplified diagrams showing stratigraphic units in three critical locations in the southwestern United States: north-central Nevada, northwestern Utah, and southern Nevada. Mtn—Mountain.

1. Northeastern Nevada and northwestern Utah. The youngest Permian formation in Utah according to Hintze (1988, Fig. 40) is the Gerster Limestone, which outcrops in a few fault-block ranges west of Great Salt Lake in northwestern Utah and in northeast Nevada (Fig. 3). The Gerster is the highest formation in the Park City Group, which interfingers in a very complex way with the enigmatic phosphate-rich Phosphoria Formation slightly to the north. Complicated stratigraphy, tectonic disruption, and limited outcrop make the Gerster difficult to study, and the literature on this unit is limited.

The Gerster is in the thick miogeoclinal province, and its cherty, brachiopod-bearing limestone facies suggests deposition well away from shore, perhaps on a carbonate ramp. Biostratigraphy based on brachiopods (Wardlaw, 1974) and on preliminary conodont studies (Clark et al., 1979; Clark et al., 1977; Wardlaw and Collinson, 1979), has convinced students of the Gerster that it contains little, if any, Permian sediments that would now be included in the Upper Permian. However, there is reason to question this conclusion, and it is worth exploring the possibility that the Gerster continues to the very top of the Permian.

In the area of occurrence of the Gerster Limestone, the oldest Triassic unit is the Dinwoody Formation (Kummel, 1954), in which marine shale and limestone are the dominant lithologies. Exposures of Dinwoody resting on Gerster outcrop in a small area of the Hogup Mountains, on the northwest shore of Great Salt Lake (Stifel, 1964), and this looks to us like the most promising candidate for a complete or nearly complete P-Tr section in this region.

In the Hogup Mountains, a careful study of conodonts in the Dinwoody (Paull, 1982; Paull and Paull, 1982) led to the conclusion that the very lowest conodont zone of the Triassic (as understood at that time)—the portion of the *Hindeodus typicalis* zone beneath the *Isarcicella isarcica* zone—was present. In a wider study Paull and Paull (1986) found this very oldest Triassic to be present in a 270000 km^2 area that includes the Hogups. Accepting the view that the upper part of the Permian is missing due to exposure and erosion, they concluded that there had been an extremely rapid sea-level rise at the very beginning of the Triassic ("orders of magnitude greater than those suggested for eustatic sea level changes associated with continental glaciation or related to ridge spreading" [Paull and Paull, 1986, p. 243]). The resulting transgression would have swept across a top Permian that had been eroded to a very smooth, level surface, so that the basal Triassic conodont zone could be present over such a wide belt. They searched for a basal conglomerate at about 90 places in Utah, Idaho, Wyoming, and Montana; only 9 of these sections contained even modest conglomerates as possible physical evidence for unconformity. Interestingly, Paull and Paull (1982) reported 4 elements of *Hindeodus typicalis* within the top Gerster, suggestive of an Upper Permian age, according to the latest conodont zonations (Sweet, 1992).

We suggest that both of these difficulties—wide occurrence of lowermost Triassic and rarity of physical evidence for transgression—would be explained if the Gerster is complete to the very top of the Permian, and there was no exposure or significant erosion, removing the need for an instantaneous transgression. Renewed biostratigraphic study of the Gerster formation will be necessary to test this possibility.

2. Southern Nevada. On the Colorado Plateau of northwest Arizona, the highest Permian present is spectacularly exposed because the soft red beds of the Triassic Moenkopi Formation have been erosionally stripped off the resistant carbonates of the Permian Kaibab Formation (Fig. 3). The resulting topography is the Kaibab Plateau, into which the Grand Canyon is incised. This stratigraphy continues west of the Colorado Plateau into the Basin and Range of southern Nevada around Las Vegas.

Although the P-Tr units in southern Nevada are west of the Hogup Mountains (Fig. 2), they were located closer to the shoreline, because of the southwest trend of the craton-miogeocline boundary line. In this area, the Kaibab and Moenkopi both contain gypsum, indicative of near-sea-level conditions; gypsum is absent or at least very rare in the Gerster and Dinwoody.

The Kaibab-Moenkopi contact has long been considered to be an unconformity representing a substantial amount of missing time (Dake, 1920; Gilbert, 1877; McKee, 1954). To us, this appears likely to be correct, but it is worth discussing the arguments, some of which are not iron-clad, because of the slight possibility that useful P-Tr sections might be found here.

The first line of evidence is paleontological. Although the Moenkopi is considered "Early and Middle (?) Triassic" on the basis of scarce fossil evidence (Stewart et al., 1972, p. 71), the oldest part of the formation has not been more precisely dated, because of the rarity of fossils in the basal Timpoweap Member and their absence in the silty-sandy red beds of the overlying Lower Red Member (Fig. 3).

In reconnaissance work the Kaibab was thought to be of Carboniferous age (Gilbert, 1877; Lee, 1907; Walcott, 1880). (At that time the present Kaibab was considered the upper part of the Aubrey Limestone, a term no longer used.) Subsequent paleontological work by G.H. Girty moved the Kaibab to the Permian (Schimer, 1919; Schuchert, 1918), but no higher than Lower Permian or possibly Guadalupian. Most of this biostratigraphy was based on brachiopods; almost no modern conodont work has been published. Furthermore, most of the biostratigraphy has been done on the lower, more fossiliferous Fossil Mountain Member, and less has been done on the upper, sparsely fossiliferous Harrisburg Member. Again, it is possible that the Kaibab might reach the top of the Permian, but this would need further paleontological work to test.

A second argument for a Kaibab age low in the Permian comes from the recognition of Kaibab in the northwestern Utah–northeastern Nevada area, discussed above, lying below the Plympton Formation, which underlies the Gerster (Hintze, 1988, Fig. 40). The problem here is to ascertain whether this northern Kaibab is age-equivalent to the Kaibab of the Grand Canyon and southern Nevada, or is an older unit of similar facies. This has not been demonstrated because of the discontinuous nature of outcrops in the Basin and Range Province separating the two areas of the Kaibab.

The third, and strongest, argument for unconformity at the Kaibab-Moenkopi contact is the presence of recognizable erosional channels cutting the top Kaibab in a few places (Billingsley, 1998). Perhaps the most striking of these is in Timpoweap Canyon, near Hurricane, Utah (Nielson, 1991; Nielson and Johnson, 1979), where a 20-m-deep paleovalley is filled with cobble conglomerate (Fig. 3). However, this conglomerate shows up in only one of 14 sections measured in Timpoweap Canyon by Nielson and Johnson (1979); all the others show a Kaibab-Moenkopi contact with no angular discordance or other obvious indication of unconformity. The situation in Timpoweap Canyon reflects the regional nature of the contact: locally there is clear evidence for erosion, but in most places there is not. Can we conclude that the local erosion reflects a general erosion which has left little or no trace? Bissell (1969) has forcefully raised this question for the southern Nevada P-Tr contact.

Although paleovalleys cutting into the top of the Kaibab demonstrate at least local erosion, there may be places where the Kaibab was not eroded before deposition of the Moenkopi. A similar situation occurs in Greenland, where complete basinal sections lie between submarine conglomerate channels (R.J. Twitchett, 2001, personal commun.).

Tectonic complications

It is perhaps surprising that there could still be questions about the nature of the P-Tr contact in the southwestern United States, but this is a difficult region in which to do stratigraphy, for a number of reasons.

1. The Permian-Triassic Sonoma Orogeny. Tectonism in north-central Nevada (Fig. 3) at about P-Tr boundary time is attested to by a spectacular angular unconformity. Oceanic argillites and cherts of the Havallah "sequence" are intensely deformed (Brueckner and Snyder, 1985). These rocks contain radiolarians ranging in age from Mississippian to Permian (Leonardian) in a vertical arrangement that indicates multiple thrust packages (Stewart et al., 1986). The Havallah is unconformably overlain by the Koipato volcanics, including a lower greenstone unit and two rhyolite units, from the upper of which Silberling (1973) reported marine volcaniclastic deposits containing the ammonoid *Subcolumbites*, indicating a Spathian (late Early Triassic) age. The Koipato is covered by marine sedimentary rocks of the Star Peak Group. The lowest unit in this group, the Tobin Formation, also carries *Subcolumbites* (Nichols and Silberling, 1977). However, an abundance of *Lingula* and *Claraia* (Muller et al., 1951) suggests that the Tobin Formation may in part be Griesbachian or Dienerian, i.e., earliest Triassic age.

Based on the Havallah-Koipato angular unconformity, Nevada geologists have long recognized a Sonoma Orogeny, very close in age to the P-Tr boundary. This agreed well with the early twentieth century concept that orogeny and extinction were related. It also led to the expectation that nearby areas would have incomplete stratigraphy because of postorogenic erosion. This view may still be distantly echoed in the standard view of a ubiquitous P-Tr hiatus in the southwestern United States.

Before the acceptance of plate tectonics, the Sonoma orogenic belt was thought always to have lain very close to the P-Tr continental-margin carbonates of northeastern Nevada and northwestern Utah. This view is compatible with the suggestion of Collinson et al. (1976) that Triassic sedimentary units onlapped westward onto a Sonoma topographic high during a post–P-Tr transgression. In the plate-tectonic view, the Sonoma orogenic belt may have been located at the margin of North America, or may have been accreted later after transport from a distant location; this question is unresolved (Ketner, 1984; Skalbeck et al., 1989).

2. Compressional tectonics in the Cretaceous. Thick miogeoclinal units of P-Tr age, west of the craton, underwent compressional deformation during the Late Cretaceous Sevier Orogeny. A complicated thrust system, including the Canyon Range and Pavant thrusts (Villien and Kligfield, 1986), passes through western Utah, and P-Tr sections are commonly folded as a result. For example, the P-Tr contact in the Hogup Mountains is clearly folded into a broad syncline (Stifel, 1964), but our field observations suggest that imbricate thrusting is present as well. Caution is certainly needed in using long sections measured in the Permian and Triassic anywhere in this region.

3. Extensional tectonics in the Tertiary. Basin and Range extension in the Late Tertiary has further complicated the situation. P-Tr outcrops are restricted to small areas within isolated fault-block ranges. Normal faults cut the older Sevier thrusts. There is as yet no agreement on the geometry or displacement history of the normal faults, so it is not possible to make accurate reconstructions of the original positions of the various known P-Tr localities.

Summary

Because of tectonic complexities and limited outcrop, it is not presently possible, and may never be possible, to understand the P-Tr geology of the southwestern United States to the extent that can be done in broadly exposed, more straightforward areas like west Texas and the southern Alps. Nevertheless, critical and informative sections across the P-Tr boundary may exist in this area. One of the lessons of K-T research in the 1980s was that when studying a geological event, one can often learn much from a very short stratigraphic section, as long as it records the event in detail.

IS THERE REALLY A P-TR HIATUS IN THE SOUTHWESTERN UNITED STATES?

There is a contact between Permian and Triassic strata exposed in many places in the southwestern United States, but this contact is almost universally believed to represent a substantial unconformity that can be of no value in understanding the P-Tr event. This conclusion is so thoroughly accepted that we have found no paper in a major journal since 1982 addressing the P-Tr boundary question that uses information from the southwestern United States, although recent work on the subsequent Early Triassic recovery has been done here (Bottjer and Schubert, 1997; Schubert and Bottjer, 1995), and Wignall and Hallam (1992) examined sections in nearby Idaho as part of a study of anoxia. Except for the spectacular Permian reef complexes of west Texas, the western United States is almost unmentioned in the major P-Tr synthesis book by Erwin (1993).

From a careful reading of the literature on the Permian and Triassic in the southwestern United States and reconnaissance field work at 13 areas in Nevada, Utah, and southern Idaho, some with multiple sections, we have become convinced that the arguments for a major P-Tr unconformity in this area are flawed. We emphatically stress that we make no claim that the P-Tr boundary record here actually is complete. However, we believe that this area, excluded from most research on the P-Tr extinction for almost two decades, merits a new look.

Analysis of the arguments favoring a hiatus

Expectations from global sea-level history. The supposed P-Tr unconformity in the southwestern United States and throughout most of the rest of the world has commonly been attributed to erosion during a sea-level lowstand in the Late Permian: Erwin (1993, p. 145–146) stated the following.

> The standard description of the end-Permian regression includes a long-term drop in sea level beginning near the end of the Early Permian. . . . The regression continued into the Upper Permian, accelerating near the Permo-Triassic boundary where sea level reached perhaps the lowest point in the Phanerozoic. The drop is variously estimated at 210 m (Forney, 1975) to 280 m (Holser and Magaritz, 1987). During the Early Triassic transgression, sea level rose rapidly by some 220 m. It should come as no surprise that with a regression of this magnitude preservation of marine sections across the Permo-Triassic boundary was limited.

In the light of modern sequence-stratigraphic understanding of the sedimentary response to sea-level fluctuations (Miall, 1997), it is difficult a priori to accept the reality of a sea-level low marked by a near-global unconformity. Sediments removed by erosion during the Late Permian in one area will of necessity be deposited somewhere else, but those places have not yet been found. Although sediments carried as far as the deep ocean may have been subducted or intensely deformed at a trench, it should still be possible to find Upper Permian sediments deposited out on the continental margin, below the level reached by any lowered strand line. These deposits have not been identified. Furthermore, in South China, Upper Permian sediments are found in many outcrops over a broad area (Yang and Li, 1992). These widespread shallow-marine deposits were not removed by Late

Permian erosion, which argues against a major global sea-level fall at that time.

In the past few years, a different view of this eustatic event has been emerging, which places the P-Tr boundary within a transgression, rather than a regression. This is based on observations from Pakistan and the southern Alps (Hallam and Wignall, 1999), although in the southern Alps a regression does occur a few meters below the extinction horizon (R.J. Twitchett, 2000, personal commun.). The fact that such a complete reversal of interpretation could take place underlines the tentative nature of our understanding of the P-Tr event. If this new view is correct, we may expect complete marine P-Tr sections to be much more common than previously believed.

The primary evidence supporting a P-Tr unconformity in the southwestern United States comes from biostratigraphy. How well can strata of the P-Tr interval be dated and correlated?

Biostratigraphic indications of missing section

The task of dating the higher parts of the Permian in the southwestern United States has proven to be difficult—so much so that we are not persuaded by the outdated biostratigraphic argument for a P-Tr hiatus. Ammonoids and fusulinaceans have been useful elsewhere in the world for biostratigraphy around the P-Tr boundary. In the southwestern United States, fusulinaceans are rare or absent above the Lower Permian, and ammonoids are too rare to be of much use. Biostratigraphy was therefore originally based on brachiopods, and more recently on conodonts.

Brachiopods. Brachiopods give a general idea of where southwestern United States Permian formations fall within that system, and have been interpreted as indicating that Upper Permian is not present. However, examination of the data leads us to question whether a hiatus can be inferred on the basis of brachiopods.

McKee (1938, Table 15, p. 170–171) found that the pattern of abundance among 21 brachiopod species from the Kaibab Limestone more closely resembles that of the Lower Permian Leonardian of West Texas than it does that of the overlying Wordian or underlying Hessian, and specifically that *Productus bassi*, which is extremely abundant in the Kaibab, does not occur above the Leonardian in Texas. Although suggestive, this does not conclusively demonstrate the absence of Upper Permian, for two reasons. (1) The assemblage listed from the Kaibab comes from the Fossil Mountain member (McKee's β member), not from the overlying Harrisburg member (his α member). (2) The absence of *Productus bassi* above the Leonardian in Texas may be facies controlled.

Wardlaw (1974) used brachiopods to divide the Gerster Limestone into four biozones—from base upward: (1) a *Megousia* total range zone between the first appearance data and last appearance data of *Megousia leptosa*, (2) a "Transition" interval zone, (3) a *Yakovlevia* total range zone between the FAD and LAD of *Yakovlevia multistriata*, and (4) an "Upper" zone, from the LAD of *Y. multistriata* to the top of the Gerster. The appearance datums could not be directly correlated to well-dated sections elsewhere, but statistically the Gerster fauna resembles that of the Roadian-Wordian of West Texas somewhat more closely than it does that of the Capitanian. Yet the "Upper" zone is ~140–175 m thick (more than half the total Gerster) in 3 of the 10 sections. This does not eliminate the possibility that this poorly controlled zone in the upper Gerster may extend up to the top of the Upper Permian.

Articulate brachiopods are rare or absent in the southwestern United States Lower Triassic, but the inarticulate brachiopod, *Lingula*, occurs in great abundance in the lowermost Triassic, for example in the Hogup Mountains (Paull and Paull, 1982). A *Lingula* acme zone is characteristic of the basal Triassic stage, the Griesbachian, indicating that little or no section is missing at the base of the Triassic.

Conodonts. Conodonts have been very useful for long-range correlation throughout much of the Paleozoic, although they are proving to be more complicated for the Middle and Upper Permian, because many species are provincial and not useful for global correlation. Recent work in China and other Tethyan localities has, however, greatly improved the conodont biostratigraphy for the Upper Permian and across the P-Tr boundary (Mei et al., 1999, Mei et al., 1998). Uppermost Permian conodonts have been found in Arctic and western Canada in areas which were formerly believed to have P-Tr unconformities (Henderson, 1997; Henderson and Baud, 1997). In addition, Kozur (1992) has reported the presence of Upper Permian conodonts from the Glass Mountains in Texas, which aids stratigraphic correlation between the proposed type sections of the Middle Permian in Texas and the Upper Permian in China. Furthermore, the conodont *Hindeodus typicalis* has been reported from the Gerster in northwestern Utah (Paull and Paull, 1982). Because this conodont is now thought to range from the mid-Wuchiapingian into the Lower Triassic, the top of the Gerster at that locality must be younger than Guadalupian.

The absence of a particular index fossil is not evidence of a hiatus. In the present condition of biostratigraphic uncertainty, we conclude that one cannot exclude the possibility of continuous sections across the P-Tr boundary in the southwestern United States on the basis of current conodont data.

Physical evidence for unconformity

Even in the absence of a compelling biostratigraphic case, physical evidence for unconformity might be available, and this has been the topic of some discussion in the literature. We observed two different situations during field reconnaissance:

1. In northern Arizona and adjacent parts of Utah and Nevada, the P-Tr contact is locally cut by channels filled with coarse conglomerate. One of the best examples is at Timpoweap Canyon (Fig. 3) in southwestern Utah (Nielson, 1991). The standard interpretation is that these features represent uplift of an ancestral Mogollon Highland in southern Arizona leading to

subaerial erosion prior to Triassic. These conglomerates deserve reexamination, for it is possible that they might represent submarine canyon fills. Bissell (1969) has questioned whether there is convincing evidence for a P-Tr unconformity in this area.

2. In the rest of Utah and Nevada and in adjacent Idaho and Wyoming, it has been very difficult to find physical evidence for unconformity at the P-Tr contact. The resulting discomfort is reflected in many statements in the literature, e.g., in Bissell (1973, p. 330).

> Throughout [northeastern Nevada and northwestern Utah] I have searched for evidence of what we have been led to believe record a substantial unconformity separating the Triassic from Permian rocks. . . . Evidences of subaerial erosion, truncation of strata, angular discordance, channeling, and other criteria indicative of uplift and development of a substantial disconformity at the top of the Permian were not found in the basin . . . [T]he bulk of the evidence favors the interpretation that marine waters remained in the miogeosyncline during the Permian-Triassic interim. . . . Had marine waters retreated over an areally-extensive region from the Cordilleran miogeosyncline during Late Permian time and exposed a vast region to subaerial erosion, those evidences of such an event surely would be present at more than one locality. They are absent.

Boyd and Maughan (1973, p. 310–311) stated the following.

> . . . Sheldon et al. (1967) noted that widespread key beds in the top few feet of the Phosphoria Formation are in the same stratigraphic position relative to the overlying Triassic. Subsurface control is good in the shelf sequence, and parallelism of electric log correlations above and below the contact is well documented . . . over a distance of 200 miles. . . . Minor intraformational pebble beds and discontinuity surfaces have been found within both Permian . . . and Triassic sequences. At certain localities, some of these have more impressive disconformity credentials than does the Permian-Triassic contact.

The authors of these quotations were puzzled by the conflict between biostratigraphic evidence for a P-Tr hiatus and the absence of physical evidence for a hiatus. With the biostratigraphic evidence now called into question, the absence of physical evidence may turn out to have been the more significant observation.

DISCUSSION

This paper terminates with a discussion, but no conclusions. We have argued that the evidence for a ubiquitous P-Tr hiatus in the southwestern United States is not compelling, and that stratigraphically complete sections across the P-Tr boundary may be present. Testing this possibility will require renewed paleontological work, particularly on conodonts.

The hope of finding complete P-Tr marine sections is encouraged by the recent view that the P-Tr event may have occurred during a transgression, instead of a lowstand. This suggests that complete P-Tr sections may be much more common throughout the world than is generally believed.

If so, the P-Tr case would be reminiscent of what happened in early work on the K-T boundary. While doing the research and writing of the 1980 iridium paper (Alvarez et al., 1980), the Berkeley group knew of only two localities with complete K-T sections—Gubbio in Italy, and Stevns Klint in Denmark—and learned of one more, in New Zealand, before the paper was finalized. At that time the K-T was believed by many to be marked by a widespread hiatus, as Kaufmann (1979, p. 32) stated.

> Over 90 percent of exposed Cretaceous/Tertiary boundary sequences have major disconformities, paraconformities, and/or intercalations of nonmarine sediments in the critical boundary zone. Part or all of the marine Late Maastrichtian and Danian are commonly missing in continental shelf sites, and more in cratonic interior sites. A minimum of 1–3 m.y. of marine history (and biotas) is thus absent in many sequences. . . .

And yet, within a few years, dozens of complete K-T sections were known, and now the total is in the hundreds (Claeys et al., this volume).

However, locating the P-Tr mass extinction in a stratigraphic section is more difficult than was the case for the K-T boundary. In the latter case, planktic forams are widespread in pelagic sediments and they were decimated by the K-T event. The K-T mass extinction is clearly marked by a temporary disappearance of all planktic forams large enough to be seen with a hand lens. The conodonts provide the best biostratigraphic information on the P-Tr boundary, but they can only be studied with laboratory preparation and microscopy.

In the southwestern United States, the contact between the highest Permian and the lowest Triassic is often a sharp facies boundary, such as the shale-rich Dinwoody over the limestone of the Gerster, or the red siltstone of the Moenkopi over the limestone of the Kaibab. Where this contact juxtaposes facies unlikely to have been side by side, such as shale over clean limestone, Walther's Law (that facies in vertical contact should be facies that originally were laterally adjacent, because sea-level rises and falls should laterally shift the facies belts) is violated. One explanation would be that a hiatus separates the two facies. An alternative explanation would be that the stratigraphic record is continuous, but that a sudden mass extinction altered the nature of the sedimentary input by eliminating most of the carbonate-producing organisms. For this hypothesis to be accepted, it would have to be shown that little or no record is missing at a hiatus, and this will require improved biostratigraphic studies.

In closing, we stress again that we do not maintain that complete P-Tr sections exist in the southwestern United States, but we are not convinced by the arguments that they are absent. As proved to be the case with the Cretaceous-Tertiary mass extinction, there may turn out to be far more sections recording the Permian-Triassic mass extinction than are presently known.

ACKNOWLEDGMENTS

We thank Richard Twitchett for useful insights in the field, based on his P-Tr studies elsewhere, and Joshua Feinberg, Erick Staley, Sarah Yoder, and Kathy Blum for help with the fieldwork. This research has benefited from discussions with William B.N. Berry, Richard A. Muller, Frank Asaro, Daniel B. Karner, and Jonathan Levine. Constructive reviews by Richard Twitchett and Wolfgang Kiessling greatly improved this paper. We appreciate the work of Christian Koeberl in organizing the 2000 Vienna conference.

REFERENCES CITED

Alvarez, L.W., Alvarez, W., Asaro, F., and Michel, H.V., 1980, Extraterrestrial cause for the Cretaceous-Tertiary extinction: Science, v. 208, p. 1095–1108.

Billingsley, G.H., 1998, Paleo Permian/Triassic river systems of northwestern Arizona : Geological Society of America Abstracts with Programs, v. 30, no. 5, p. 4.

Bissell, H.J., 1969, Permian and Lower Triassic transition from the shelf to basin (Grand Canyon, Arizona to Spring Mountains, Nevada): Geology and natural history of the Grand Canyon region: Four Corners Geological Society, 5th Field Conference, Powell Centennial River Expedition, p. 135–169.

Bissell, H.J., 1973, Permian-Triassic boundary in the eastern Great Basin area: Canadian Society of Petroleum Geologists Memoir, v. 2, p. 318–344.

Bottjer, D.J., and Schubert, J.K., 1997, Paleoecology of Lower Triassic marine carbonates in the southwestern USA: Brigham Young University Geology Studies, v. 42, p. 15–18.

Boyd, D.W., and Maughan, E.K., 1973, Permian-Triassic boundary in the Middle Rocky Mountains: Canadian Society of Petroleum Geologists Memoir 2, p. 294–317.

Brueckner, H.K., and Snyder, W.S., 1985, Structure of the Havallah Sequence, Golconda Allochthon, Nevada: Evidence for prolonged evolution in an accretionary prism: Geological Society of America Bulletin, v. 96, p. 1113–1130.

Clark, D.L., Peterson, D.O., Stokes, W.L., Wardlaw, B., and Wilcox, J.D., 1977, Permian-Triassic sequence in northwest Utah: Geology, v. 5, p. 655–658.

Clark, D.L., Carr, T.R., Behnken, F.H., Wardlaw, B.R., and Collinson, J.W., 1979, Permian conodont biostratigraphy in the Great Basin: Brigham Young University Geology Studies, v. 26, p. 143–149.

Collinson, J.W., Kendall, C.G., and Marcantel, J.B., 1976, Permian-Triassic boundary in eastern Nevada and west-central Utah: Geological Society of America Bulletin, v. 87, p. 821–824.

Dake, C.L., 1920, The pre-Moenkopi (pre-Permian?) unconformity of the Colorado Plateau: Journal of Geology, v. 28, p. 61–74.

Erwin, D.H., 1993, The great Paleozoic crisis: New York, Columbia University Press, 327 p.

Forney, G.G., 1975, Permo-Triassic sea level change: Journal of Geology, v. 83, p. 773–779.

Gilbert, G.K., 1877, Report on the geology of the Henry Mountains: Washington D.C., Government Printing Office, 160 p.

Hallam, A., and Wignall, P.B., 1999, Mass extinctions and sea-level changes: Earth-Science Reviews, v. 48, p. 217–250.

Henderson, C.M., 1997, Uppermost Permian conodonts and the Permian-Triassic boundary in the Western Canada Sedimentary Basin: Bulletin of Canadian Petroleum Geology, v. 45, p. 693–707.

Henderson, C.M., and Baud, A., 1997, Correlation of the Permian-Triassic boundary in Arctic Canada and comparison with Meishan, China: Proceedings of the 30th International Geological Congress, v. 11, p. 143–152.

Hintze, L.F., 1988, Geologic history of Utah: Provo, Utah, Department of Geology, Brigham Young University, 203 p.

Holser, W.T., and Magaritz, M., 1987, Events near the Permian-Triassic boundary: Modern Geology, v. 11, p. 155–180.

Jin, Y., Zhang, J., and Shang, Q., 1994, Two phases of the end-Permian mass extinction: Canadian Society of Petroleum Geologists Memoir 17, p. 813–822.

Jin, Y.-G., Wardlaw, B.R., Glenister, B.F., and Kotlyar, G.V., 1997, Permian chronostratigraphic subdivisions: Episodes, v. 20, p. 10–15.

Kauffman, E.G., 1979, The ecology and biogeography of the Cretaceous-Tertiary extinction event, *in* Christensen, W.K., and Birkelund, T., eds., Proceedings of the 2nd Cretaceous-Tertiary Boundary Events Symposium: Copenhagen, University of Copenhagen, p. 29–37.

Ketner, K.B., 1984, Recent studies indicate that major structures in northeastern Nevada and the Golconda thrust in north-central Nevada are of Jurassic or Cretaceous age: Geology, v. 12, p. 483–486.

Kozur, H., 1992, Dzhulfian and early Changxingian (Late Permian) Tethyan conodonts from the Glass Mountains, West Texas: Neues Jahrbuch fuer Geologie und Palaeontologie, Abhandlungen, v. 187, p. 99–114.

Kummel, B., 1954, Triassic stratigraphy of southeastern Idaho and adjacent areas: U.S. Geological Survey Professional Paper, v. 254-H, p. 165–194.

Lee, W.T., 1907, The Iron County Coal Field, Utah: U.S. Geological Survey Bulletin, v. 316, p. 359–375.

McKee, E.D., 1938, The environment and history of deposition of the Toroweap and Kaibab Formations of northern Arizona and southern Utah: Washington D.C., Carnegie Institution of Washington, 268 p.

McKee, E.D., 1954, Stratigraphy and history of the Moenkopi Formation of Triassic age: Geological Society of America Memoir 61, p. 1–133.

Mei, S., Zhang, K., and Wardlaw, B.R., 1998, A refined succession of Changhsingian and Griesbachian neogondolellid conodonts from the Meishan section, candidate of the global stratotype section and point of the Permian-Triassic boundary: Palaeogeography, Palaeoclimatology, Palaeoecology, v. 143, p. 213–226.

Mei, S., Henderson, C.M., and Jin, Y., 1999, Permian conodont provincialism, zonation and global correlation: Permophiles, v. 35, p. 9–16.

Miall, A.D., 1997, The geology of stratigraphic sequences: Berlin, Springer-Verlag, 433 p.

Muller, S.W., Ferguson, H.G., and Roberts, R.J., 1951, Mount Tobin Quadrange, Nevada: U.S. Geological Survey Geological Quadrangle Map GQ-7, scale 1:125 000, one sheet.

Nichols, K.M., and Silberling, N.J., 1977, Stratigraphy and depositional history of the Star Peak Group (Triassic), northwestern Nevada: Geological Society of America Special Paper 178, p. 1–73.

Nielson, R.L., 1991, Petrology, sedimentology and stratigraphic implications of the Rock Canyon Conglomerate, southwestern Utah: Utah Geological Survey, Miscellaneous Publications, v. 91-7, p. 1–65.

Nielson, R.L., and Johnson, J.L., 1979, The Timpoweap Member of the Moenkopi Formation, Timpoweap Canyon, Utah: Utah Geology, v. 6, p. 17–28.

Paull, R.K., 1982, Conodont biostratigraphy of Lower Triassic rocks, Terrace Mountains, northwestern Utah: Utah Geological Association Publication, v. 10, p. 235–249.

Paull, R.K., and Paull, R.A., 1982, Permian-Triassic unconformity in the Terrace Mountains, northwestern Utah: Geology, v. 10, p. 582–587.

Paull, R.K., and Paull, R.A., 1986, Epilogue for the Permian in the western Cordillera: A retrospective view from the Triassic: Contributions to Geology, University of Wyoming, v. 24, p. 243–252.

Schimer, H.W., 1919, Permo-Triassic of northwestern Arizona: Geological Society of America Bulletin, v. 30, p. 471–498.

Schubert, J.K., and Bottjer, D.J., 1995, Aftermath of the Permian-Triassic mass extinction event: Paleoecology of Lower Triassic carbonates in the western USA: Palaeogeography, Palaeoclimatology, Palaeoecology, v. 116, p. 1–39.

Schuchert, C., 1918, On the Carboniferous of the Grand Canyon of Arizona: American Journal of Science, v. 195 (ser. 4, v. 45), p. 347–361.

Sheldon, R.P., Cressman, E.R., Cheney, T.M., and McKelvey, V.E., 1967, Middle Rocky Mountains and northeastern Great Basin, Chapter H, *in* Paleotectonic investigations of the Permian System in the United States: U.S. Geological Survey Professional Paper, v. 515, p. 153–170.

Silberling, N.J., 1973, Geologic events during Permian-Triassic time along the Pacific margin of the United States: Canadian Society of Petroleum Geologists Memoir 2, p. 345–362.

Silberling, N.J., and Tozer, E.T., 1968, Biostratigraphic classification of the marine Triassic in North America: Geological Society of America Special Paper 110, p. 1–63.

Skalbeck, J.D., Burmester, R.F., Beck, M.E., Jr., and Speed, R.C., 1989, Paleomagnetism of the Late Permian-Early Triassic Koipato Volcanics, Nevada: Implications for latitudinal displacement: Earth and Planetary Science Letters, v. 95, p. 403–410.

Stanley, S.M., and Yang, X., 1994, A double mass extinction at the end of the Paleozoic Era: Science, v. 266, p. 1340–1344.

Stewart, J.H., Murchey, B.L., Jones, D.L., and Wardlaw, B.R., 1986, Paleontologic evidence for complex tectonic interlayering of Mississippian to Permian deep-water rocks of the Golconda Allochthon in Tobin Range, north-central Nevada: Geological Society of America Bulletin, v. 97, p. 1122–1132.

Stewart, J.H., Poole, F.G., and Wilson, R.F., 1972, Stratigraphy and origin of the Triassic Moenkopi Formation and related strata in the Colorado Plateau region: U.S. Geological Survey Professional Paper, v. 691, p. 1–195.

Stifel, P.B., 1964, Geology of the Terrace and Hogup Mountains, Box Elder County, Utah [Ph.D. thesis]: Salt Lake City, University of Utah, 173 p.

Sweet, W.C., 1992, A conodont-based high-resolution biostratigraphy for the Permo-Triassic boundary interval, *in* Sweet, W.C., Yang, Z., Dickins, J.M., and Yin, H., Permo-Triassic events in the eastern Tethys: Cambridge, Cambridge University Press, p. 120–133.

Twitchett, R.J., Looy, C.V., Morante, R., Visscher, H., and Wignall, P.B., 2001, Rapid and synchronous collapse of marine and terrestrial ecosystems during the end-Permian mass extinction event: Geology, v. 29, p. 351–354.

Villien, A., and Kligfield, R.M., 1986, Thrusting and synorogenic sedimentation in central Utah: Association of Petroleum Geologists Memoir, v. 41, p. 281–307.

Walcott, C.D., 1880, The Permian and other Paleozoic Groups of the Kanab Valley, Arizona: American Journal of Science, v. 120 (ser. 3, v. 20), p. 221–225.

Wardlaw, B., and Collinson, J.W., 1979, Youngest Permian conodont faunas from the Great Basin and Rocky Mountain regions: Brigham Young University Geology Studies, v. 26, p. 151–159.

Wardlaw, B.R., 1974, The biostratigraphy and paleoecology of the Gerster Formation (Upper Permian) in Nevada and Utah [Ph.D. thesis]: Columbus, Ohio, Case Western Reserve University, 239 p.

Wignall, P.B., and Hallam, A., 1992, Anoxia as a cause of the Permian/Triassic mass extinction: Facies evidence from northern Italy and the Western United States: Palaeogeography, Palaeoclimatology, Palaeoecology, v. 93, p. 21–46.

Yang, Z., Sheng, J., and Yin, H., 1995, The Permian-Triassic boundary: The global stratotype and point (GSSP): Episodes, v. 18, p. 49–53.

Yang, Z.Y., and Li, Z.S., 1992, Permo-Triassic boundary relations in South China, *in* Sweet, W.C., Yang, Z.Y., Dickins, J.M., and Yin, H.F., eds., Permo-Triassic events in the eastern Tethys: Cambridge, Cambridge University Press, p. 9–20.

Manuscript Submitted November 15, 2000; Accepted by the Society March 22, 2001

Printed in the U.S.A.

Geological Society of America
Special Paper 356
2002

Extent, duration, and nature of the Permian-Triassic superanoxic event

Paul B. Wignall
Department of Earth Sciences, University of Leeds, Leeds LS2 9JT, UK
Richard J. Twitchett
Department of Earth Sciences, University of Bristol, Bristol, B58 IRJ, UK

ABSTRACT

The widespread development of anoxic and dysoxic deposition in marine settings occurred during the Permian-Triassic (P-Tr) transition interval. Facies varied according to paleobathymetry and paleolatitude. Thus, dark gray, uranium-enriched shales characterize deeper shelf locations over wide areas of northern Boreal seas, whereas the oceanic record consists of condensed, organic-rich, black shales. Finely laminated, pyrite-rich, micritic mudstones occur in equatorial Tethyan sections. Contemporaneous dolomitization in many shallow-marine settings provides further indirect evidence for widespread P-Tr anoxia. Similarly, common reports of unusual stromatolites in the earliest Triassic Griesbachian Stage could reflect the widespread occurrence of direct calcite precipitation from carbonate-saturated anoxic bottom waters. Oxygen-poor conditions are first recorded from the Late Permian, deep-water, accreted oceanic terranes of Japan. Such conditions vastly increased in extent in the interval between the latest Permian and the late Griesbachian, when dysaerobic facies developed in all but the shallowest of marine settings. The Panthalassa ocean was probably truly euxinic in this interval. Anoxia was never so extensive or so intense after this interval, and the superanoxic event ceased abruptly in equatorial Tethyan latitudes in the latest Griesbachian. Elsewhere, anoxia persisted at least into the Dienerian Stage in the Perigondwanan shelf sections of the Neo-Tethys, and deep-water anoxia may have persisted in Panthalassa until the middle Triassic.

INTRODUCTION

The possibility that oceanic anoxia may have been responsible for the marine Permian-Triassic (P-Tr) extinction event was first mooted by Hallam (1989, p. 445) and by Wignall (1990, p. 106–107). Their evidence included, respectively, the presence of finely laminated strata in Chinese boundary sections and the widespread occurrence of dysaerobic taxa in the earliest Triassic. The subsequent documentation of oxygen-poor deposition in both shallow shelf (Wignall and Hallam, 1992) and deep-water oceanic sections (Isozaki and Maruyama, 1992) established anoxia as a viable kill mechanism for the marine mass extinction at the end of the Permian (Hallam, 1994; Isozaki, 1994, 1997; Wignall and Twitchett, 1996; Hallam and Wignall, 1997).

Oceanic anoxic events are not infrequent occurrences in the Phanerozoic, and many are associated with extinction events, but several are not (e.g., the Selli event of the Aptian). It has been argued that the P-Tr event was especially lethal because of its development throughout almost the entire water column, from abyssal depths into the lower shoreface (Wignall and Twitchett, 1996). Isozaki's (1997) study of deep-water sediments from Japanese accreted terranes concluded that oxygen-poor deposition spanned the entire middle Permian to Middle

Wignall, P.B., and Twitchett, R.J., 2002, Extent, duration, and nature of the Permian-Triassic superanoxic event, *in* Koeberl, C., and MacLeod, K.C., eds., Catastrophic Events and Mass Extinctions: Impacts and Beyond: Boulder, Colorado, Geological Society of America Special Paper 356, p. 395–413.

Triassic, an interval of ~20 m.y. To distinguish this event from the much briefer duration (<1 m.y.) oceanic anoxic events of the Cretaceous, Isozaki (1994) coined the term "superanoxic event". The "super" epithet thus conveys the unusually long duration of the event, although it is also appropriate because of the unusually shallow water occurrence of oxygen deficiency.

Changes in oceanic oxygenation state have figured in other P-Tr extinction mechanisms. Thus, Polish workers suggested that Late Permian euxinic oceans were replaced by oxygenated oceans in the Early Triassic, a change associated with a postulated drastic decline in nutrient availability which is held responsible for the marine mass extinction (Malkowski et al., 1989). Identical changes of oceanic redox were proposed by Knoll et al. (1996), although they suggested that the release of CO_2 into shelf locations (hypercapnia poisoning) during Early Triassic oceanic ventilation (triggered by severe glaciation) was the cause of the mass extinction. A somewhat different history was proposed by several Japanese workers who postulated a brief oxygenation event at the P-Tr boundary punctuating a prolonged interval of oceanic euxinicity (Kajiwara et al., 1994; Suzuki et al., 1998).

This chapter aims to review the evidence for redox changes in the shelf seas and oceans of the world during the Late Permian–Early Triassic interval, in an attempt to resolve the conflicting proposals outlined here. We have utilized published records from all the major regions with an extensive P-Tr boundary record supplemented with new data from sections in British Columbia and Greenland. Several regions, notably the central Tethys, lack detailed facies analysis across the boundary, but the record of facies successions provides comparisons with the better documented areas. The review has enabled the global timing and extent of the superanoxic event to be defined, although much work remains to be done, particularly on the younger Early Triassic record.

INDICATORS OF ANOXIA AND DYSOXIA

The ability to determine ancient oxygen levels has improved greatly in the past decade and a plethora of independent geochemical, sedimentological, and paleoecological indices are now available (Table 1). Many of these techniques have been applied to P-Tr boundary sections.

Note that the term anoxic is used in its strict sense to refer to conditions lacking oxygen, and anaerobic applies to the strata accumulated under such conditions; dysoxic refers to oxygen-poor conditions, whereas dysaerobic is a descriptive term for strata and fossils accumulated under dysoxic conditions (Wignall, 1994). Euxinic refers to conditions with an anoxic, sulfidic lower water column.

P-TR BOUNDARY

The definition of the P-Tr boundary has been debated for more than a century. The lowest occurrence of *Otoceras boreale* was, for a long time, the favored level. However, the restriction of *O. boreale* to the Arctic region, its extreme rarity within this region, and uncertainty regarding its species-level status (Kummel, 1972) have made it difficult to correlate this first appearance datum (FAD) beyond the type section. These problems have been partially resolved by adopting the FAD of the conodont *Hindeodus parvus* as the base of the Triassic. However, it appears that this level is higher than the traditional boundary defined by *Otoceras* (e.g., Sweet, 1992; Kozur, 1994; Wignall et al., 1996). Recent work also shows that the FAD of *H. parvus* is above the palynological P-Tr transition (Twitchett et al., 2001). Problems still remain in correlating the Late Permian–Early Triassic interval, particularly within the later Permian, for which several regional stage schemes are currently in use. The stratigraphic units used in our study, and the conodont zones used to determine them, are given in Figure 1.

PERMIAN-TRIASSIC PALEOGEOGRAPHY

Global paleogeography during the P-Tr interval was dominated by three features, the Panthalassa and Tethyan oceans and the supercontinent Pangea (Fig. 2). By the end of the Permian Pangea had already begun to disintegrate due to the rifting of several small and mid-sized continents from the Perigondwanan margin of southern Pangea. The rapid northward drift of these continents, termed the Cimmerian continent (Şengör, 1984; Şengör et al., 1988), opened up a Neo-Tethyan ocean to the south as the Paleo-Tethyan ocean was subducted to the north. The Cimmerian continent is often depicted as a linear feature, but the provinciality of Permian foraminifera indicates that the Lhasa block rifted later and drifted slower (Şengör et al., 1988; Kobayashi, 1999). It is thus shown lagging behind the other Cimmerian continents in our reconstruction (Fig. 2). Two elongate seaways formed major embayments of the Pangea margin, one along the northeastern margin of Greenland and another along an axis through Pakistan to Madagascar.

BOUNDARY SECTIONS

Numerous P-Tr boundary sections occur throughout the world, although there are significant areas that lack data. Boundary sections are well represented from shelf locations around the margins of the Tethyan oceans and on the Cimmerian continent. Middle to high paleolatitude deposition is recorded in the extensive shelf seas on the southern margin of the Boreal ocean, now exposed in the sections of Arctic Canada and Spitsbergen. The western margin of the Panthalassa ocean is equally well represented in the sections of the U.S. and Canadian Rockies. Intraoceanic sediments of Panthalassa are well known; both deep-water, chert-dominated sections and seamount limestones are seen in the accreted terranes of Japan and western North America (Isozaki, 1997). The original location of these terranes within Panthalassa is unknown. Deep-water

TABLE 1. CRITERIA FOR IDENTIFYING ANCIENT OXYGEN LEVELS

Indicator	Significance for interpreting oxygen levels
Organic C enrichment	High organic C values are a common, but far from ubiquitous, feature of anoxic deposition. Factors such as carbonate dilution, sedimentation rate, and surface-water productivity can also affect organic C values and so confound any one-to-one relationship with oxygen levels.
Pyrite S/organic C ratios	High ratios are indicative of euxinic conditions (Raiswell and Berner, 1985).
DOP (degree of pyritization of Fe)	High values (>0.75) indicate anoxic deposition, although these values cannot diagnose euxinicity (Raiswell et al., 2001).
Pyrite framboid size distributions	Small populations (<6.0 mm) with low variability are diagnostic of euxinic conditions (Wilkin et al., 1996).
Th/U ratios (measured using a field portable gamma-ray spectrometer	Th/U ratios >3 typify aerobic shales, but decline below this level in anoxic settings due to enrichment with authigenic U (Langmuir, 1978; Myers and Wignall, 1987).
Fine lamination	Preservation of lamination requires the suppression of bioturbation, which can occur under high sedimentation rates and, more commonly, anoxic bottom waters.
Dysaerobic trace fossils	Small burrow diameter and low-diversity characterize dysaerobic trace fossil assemblages (Savrda and Bottjer, 1986).
Dysaerobic benthos	Dysoxic bottom waters are characterised by a low-diversity, thin-shelled benthic fauna. Bivalves with a "paper pecten" morphology are common (Wignall, 1994).

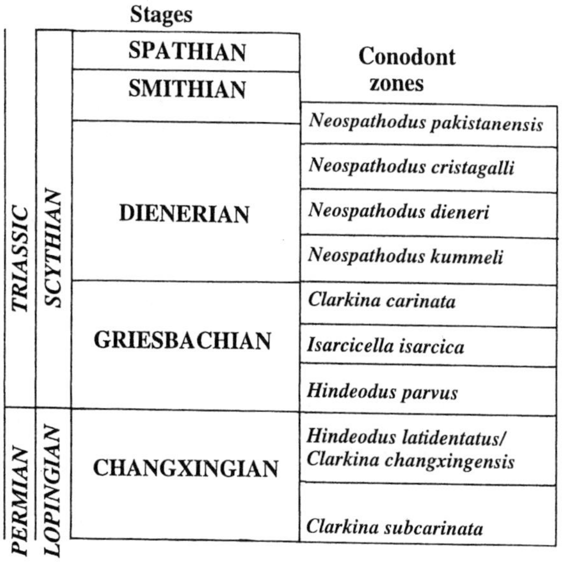

		Stages	Conodont zones
TRIASSIC	SCYTHIAN	SPATHIAN	
		SMITHIAN	
		DIENERIAN	*Neospathodus pakistanensis*
			Neospathodus cristagalli
			Neospathodus dieneri
			Neospathodus kummeli
		GRIESBACHIAN	*Clarkina carinata*
			Isarcicella isarcica
			Hindeodus parvus
PERMIAN	LOPINGIAN	CHANGXINGIAN	*Hindeodus latidentatus/ Clarkina changxingensis*
			Clarkina subcarinata

Figure 1. Stage names from around Permian-Triassic boundary interval, and associated conodont zones.

oceanic sections from Tethys are also known from localities in Sicily and Oman (Blendinger, 1988; Catalano et al., 1991).

Western Tethys

The P-Tr transition of the westernmost Tethys is recorded in the carbonate ramp facies of the Werfen Formation and equivalents that occur over a wide area of northern Italy (from Lombardy eastward to the Carnic Alps) and southern Austria (Broglio-Loriga et al., 1990). Near-identical successions are also seen in Hungary and the former Yugoslavia (Haas et al., 1988; Cassinis et al., 2000). In the more proximal developments of the Werfen Formation an oolitic grainstone unit, the Tesero Oolite Horizon, is developed at the base of the formation; this horizon passes upward into, and is interbedded with, flaggy micrites of the Mazzin Member (Fig. 3).

The P-Tr boundary, defined by the lowest occurrence of *H. parvus*, occurs in the basal Mazzin Member (Wignall et al., 1996). The last diverse Permian assemblage occurs below this in a packstone below the first oolitic grainstone of the Tesero Oolite Horizon. Some Permian taxa (calcareous algae, foraminifera, and brachiopods) range for a few meters above this level, although their preservation is poor (Wignall and Hallam, 1992). It has been argued that the base of the Tesero Oolite Horizon marks the level of an abrupt mass extinction (Rampino and Adler, 1998), although it could equally record a facies or taphonomic control due to the decline in fossil preservation in the dolomitized and recrystallized oolitic grainstones of the Tesero Oolite Horizon.

Dysoxic conditions are developed in western Tethys a short distance above the base of the Tesero Oolite Horizon, in the latest Permian, where the first Mazzin-style micrites occur interbedded with oolitic grainstones (Fig. 3). The evidence includes the presence of fine lamination, high pyrite content (Wignall and Hallam, 1992), common reworked lags of pyrite (Twitchett and Wignall, 1996), an impoverished dysaerobic assemblage dominated by *Claraia* and *Planolites* (Twitchett, 1999), well preserved, although not abundant, organic matter

Early Changxingian

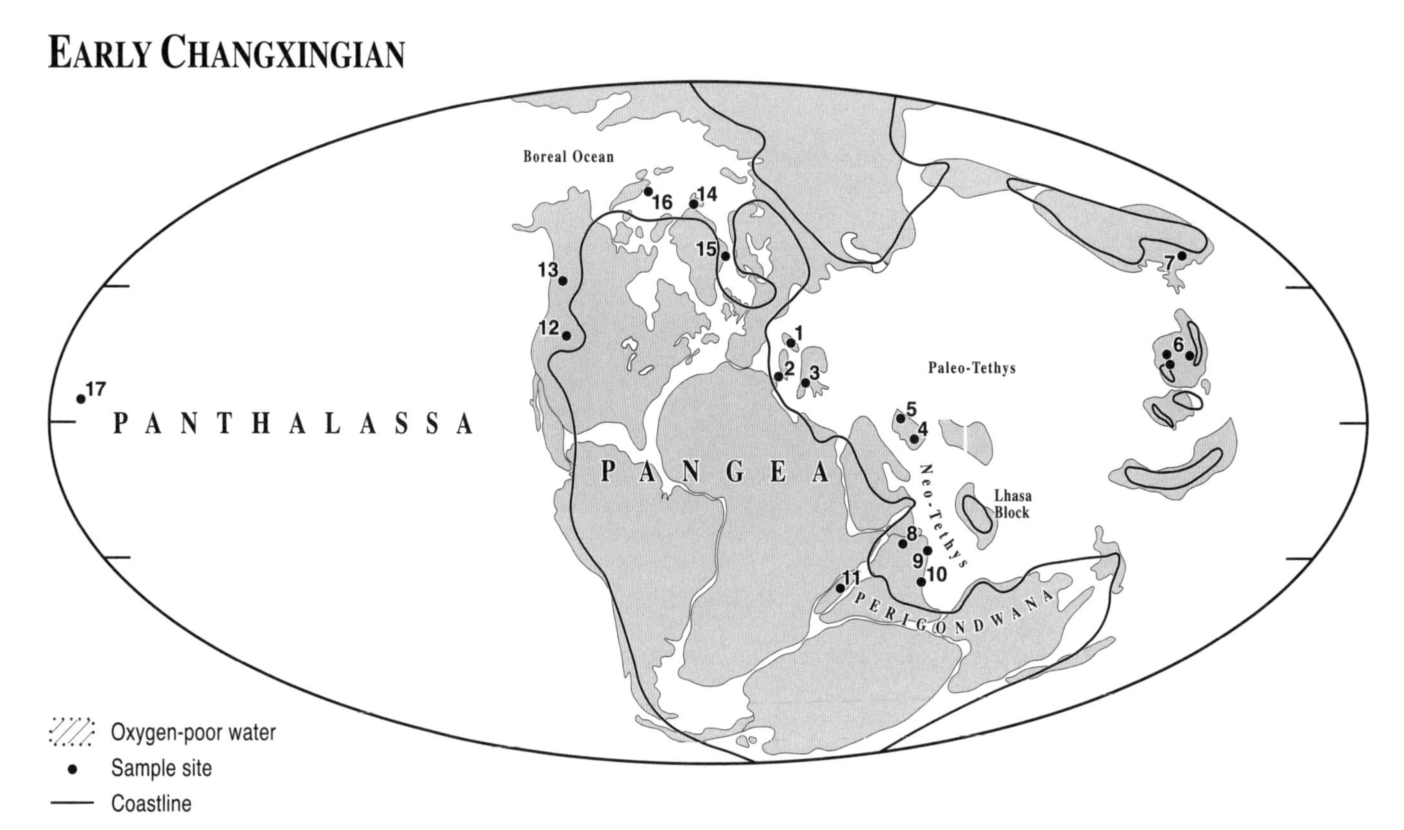

Figure 2. Global paleogeography in early Changxingian Stage, showing locations of regions discussed in text. 1. Northern Italy (Wignall and Hallam, 1992). 2. Sicily (Catalano et al., 1991). 3. Turkey (Marcoux and Baud, 1986). 4. Central Iran (Iranian-Japanese Research Group, 1981). 5. Northwestern Iran and Armenia (Teichert et al., 1973; Kozur et al., 1980). 6. Southern China (Wignall and Hallam, 1993, 1996; Wignall et al., 1995). 7. South Primorie (Zakharov, 1992). 8. Salt Range, Pakistan (Wignall and Hallam, 1993). 9. Kashmir, India (Nakazawa et al., 1975). 10. Southern Tibet (Kapoor and Tokuoka, 1985). 11. Madagascar (Treat, 1933). 12. Western United States (Wignall and Hallam, 1992). 13. British Columbia (Henderson, 1997). 14. Spitsbergen (Wignall et al., 1998). 15. East Greenland (Twitchett et al., 2001). 16. Arctic Canada (Embry, 1993). 17. Ocean-floor sediments of Panthalassa, now found in Japanese accreted terranes (Isozaki, 1994, 1997; Kajiwara et al., 1994).

(Wolbach et al., 1994), and low Th/U ratios (Wignall and Twitchett, 1996; Fig. 3).

Dysoxic Mazzin Member deposition was terminated by a regional regression in the mid-late Griesbachian, when there was the brief establishment of peritidal carbonate deposition (Andraz Horizon). Subsequent transgression in the latest Griesbachian reestablished carbonate ramp deposition (Siusi Member) over wide areas. The lower part of the Siusi Member is of similar facies to the Mazzin Member, with characteristically low Th/U ratios (Fig. 3). An abrupt transition to oxygenated deposition (higher Th/U ratios, well bioturbated sediment) occurs in the latest Griesbachian portion of the Siusi Member. The Dolomites section thus records dysoxic deposition on a carbonate ramp from the latest Permian to latest Griesbachian interval. Only the very shallow, red peritidal facies of the Andraz Horizon appear to record oxic deposition within this interval.

The Dolomites sections are the only ones to have received detailed facies analysis in the western Tethyan region. However, lithological descriptions from other areas (e.g., Baud et al., 1989) indicate that the redox changes seen in the Dolomites probably occurred throughout the region. An intriguing succession of lithological changes is seen in the Cürük dag section of the western Taurides of Turkey (Marcoux and Baud, 1986; Baud et al., 1997; Fig. 4). The Late Permian Pamucak Formation is a platform carbonate containing a diverse fauna of brachiopods and foraminifera. This is sharply overlain by an oolitic grainstone that marks the base of the Kokarkuyu Formation. This part of the succession is clearly comparable with that seen in the Dolomites. However, the overlying 12 m of section is markedly different and consists of a series of domal "stromatolites" to 40 cm high, composed of radiating fans of calcite, interbedded with cryptalgal laminites and thrombolites. This level is succeeded by "10 m of well bedded dark euxinic calcilutites" (Marcoux and Baud, 1986, p. 248) of probable Griesbachian age. Oxygenation during deposition of the stromatolites was not discussed by Baud et al. (1997), but their descriptions and illustrations are comparable to similar features described from Smithian-Spathian strata in California (Woods et al., 1999). These younger examples of calcite fans are inter-

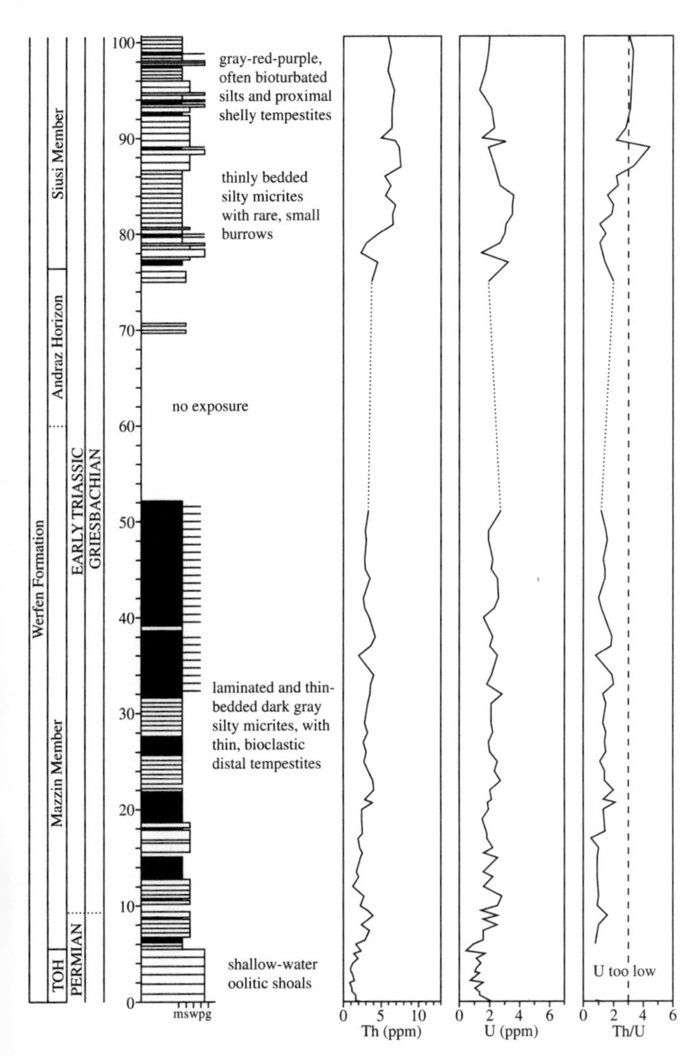

Figure 3. Permian-Triassic boundary section from l'Uomo section of western Dolomites, Italy. Shallow-water oolitic facies of Tesero Oolite Horizon (TOH) are overlain by dysaerobic, distal ramp facies of Mazzin and Siusi Members. Peritidal facies of Andraz Horizon are poorly exposed at this section. Spectral gamma ray data indicate onset of oxygen-poor deposition at base of Mazzin Member (Th/U ratios <3), in latest Permian, and rapid return of oxic conditions (Th/U ratios increase >3) in lower Siusi Member, late in Griesbachian. Key to lithologies: m, marly micrites; s, silty micrites; w, wackestones; p, packstones; g, grainstones. Black shading shows beds that are laminated.

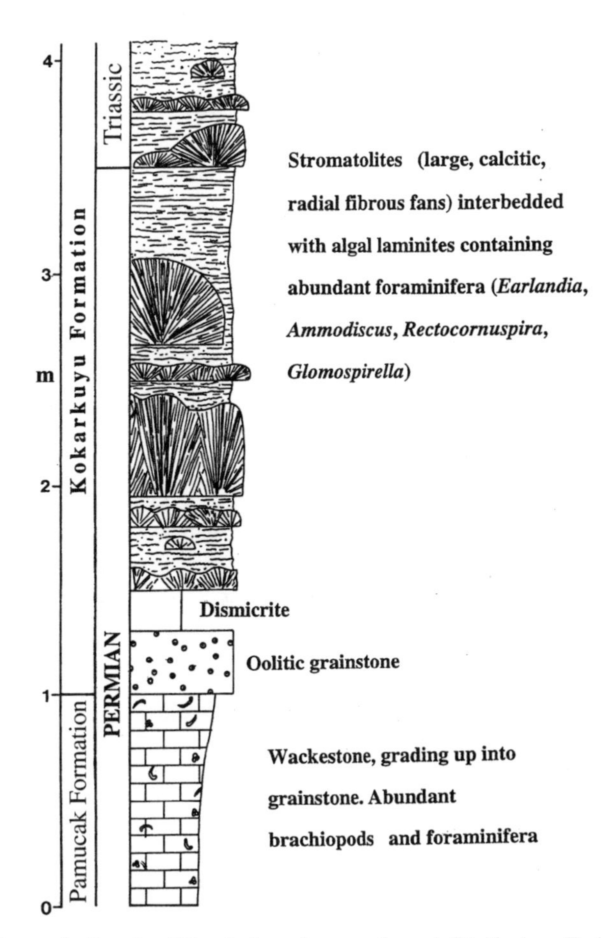

Figure 4. Permian-Triassic boundary section at Cürük dag, Turkey, based on lithological description of Marcoux and Baud (1986). Position of P-Tr boundary is based on low point of $\delta^{13}C$ curve measured at this section by Baud et al. (1997). Radial fibrous fan fabric of stromatolites is suggestive of direct precipitation from anoxic, carbonate-rich waters (Woods et al., 1999).

preted to have formed by direct precipitation from bicarbonate-rich anoxic waters. The Turkish examples may also have formed by a similar process, although the presence of an abundant, low-diversity benthic foraminiferal assemblage indicates at least some oxygenation events.

Evidence for oxygenation changes in the deeper parts of western Tethys is confined to small sections in the Sosio valley of central Sicily, where the latest Permian consists of condensed red claystones with a rich fauna of deep-water conodonts (*Clarkina* sp.), radiolarians, and benthic ostracods (Catalano et al., 1991). The ostracods are a pandemic, cold-adapted fauna that probably lived below the oceanic thermocline; Kozur (1991) termed them the paleopsychrospheric fauna. The red claystones are sharply overlain by a 2-m-thick, finely laminated, organic-rich shale containing abundant pyrite framboids and a single conodont taxon, *H. parvus* (Gullo and Kozur, 1993). The succeeding *I. isarcica* zone consists of brown-weathering, pyritic shales with thin, graded limestone beds (Wignall et al., 1996); younger Griesbachian strata are unknown in the Sosio Valley sections. This deep Tethyan section thus records a dramatic shift from fully oxic to benthos-free, euxinic-style deposition at the P-Tr boundary.

Anoxic and/or dysoxic deposition appears to have been rare in the post-Griesbachian interval of the western Tethys, although there have been relatively few detailed facies studies. There is no evidence of oxygen restriction in the peritidal to outer ramp facies of the Werfen Formation after the latest Griesbachian (Twitchett, 1999). The Spathian basinal sediments of Romania, which consist of thinly bedded, ammonoid-bearing, black, marly limestones (Cassinis et al., 2000), may be an example. Dienerian calcilutite deposition at Cürük dag was reported to be euxinic by Marcoux and Baud (1986), although the biostratigraphic control is poor. Thus, current data indicate post-Griesbachian oxygen restriction in the western Tethys was limited to a few deep, basinal locations.

Central Tethys

Central Tethyan P-Tr boundary sediments are well represented in Iran. Sections occur in the southwest (Zagros basin), in central Iran (Abadeh region), in northwestern Iran, and across the border in Armenia (Julfa region) (Teichert et al., 1973; Szabo and Kheradipur, 1978; Iranian-Japanese Research Group, 1981). All sections record a similar depositional history, as do the Transcaucasian sections of Armenia (Kozur et al., 1980). In the late Changxingian there was regional deepening and the widespread development of the Paratirolites Limestone, a red or pink, thoroughly bioturbated, nodular carbonate with common ammonoids (Teichert et al., 1973; Fig. 5). This overlies a variable succession of limestones and shales (Abadeh Formation) that contains a diverse Late Permian assemblage of brachiopods, corals, and dascyclad algae (Iranian-Japanese Research Group, 1981). The benthic fauna becomes less diverse in the Paratirolites Limestone, whereas the nektonic fauna (araxoceratid ammonoids and conodonts) becomes more abundant and diverse; this faunal change is almost certainly due to the deepening of the benthic environment.

In most regions of Iran the Paratirolites Limestone is sharply overlain by a thin development of basal Triassic shales that are variously described as being yellow-gray, olive-gray, and green-brown (Fig. 5). The overlying beds belong to the Elikah Formation, a Griesbachian-Dienerian unit that is widespread throughout Iran. This consists of finely laminated, thin-bedded limestones and interbeds of greenish shale with abundant *Claraia* and *Eumorphotis*. The Iranian-Japanese Research Group (1981) describes the limestones as vermicular due to the presence of trace fossils on bedding planes. The bioturbation intensity appears not to have been sufficient to disturb the thin interbeds (cf. Iranian-Japanese Research Group 1981, Plate 6). In central Iran the basal 2 m of the Elikah Formation consists of vermicular limestones with large stromatolites, described by the Iranian-Japanese Research Group (1981) as massive (Altiner et al., 1979, also noted that they are dolomitized).

The sections of Armenian Transcaucasia are comparable to those of Iran. Thus, a late Changxingian, nodular red limestone with *Paratirolites* is overlain by a thin claystone bed of P-Tr boundary age (Kozur et al., 1980). A 2-m-thick unit of domal stromatolites that consists of radiating fans of prismatic calcite (Baud et al., 1997) overlies the boundary clay. The common occurrence of reworked and broken crystals indicates that the fans grew on the seafloor and are not a diagenetic replacement texture. These fans are overlain by thin-bedded, black, and dark gray limestones and shales of the Karabaglar Suite. Kozur (*in* Wignall et al., 1996) reported that the shales are intensely bioturbated in the lower part (*H. parvus* zone) of this unit and that finely laminated, anoxic facies only appear in the *I. isarcica* zone. No ichnofabric descriptions are available for the youngest Griesbachian portion of the Karabaglar Suite.

The Transcaucasian and Iranian P-Tr sections are in need

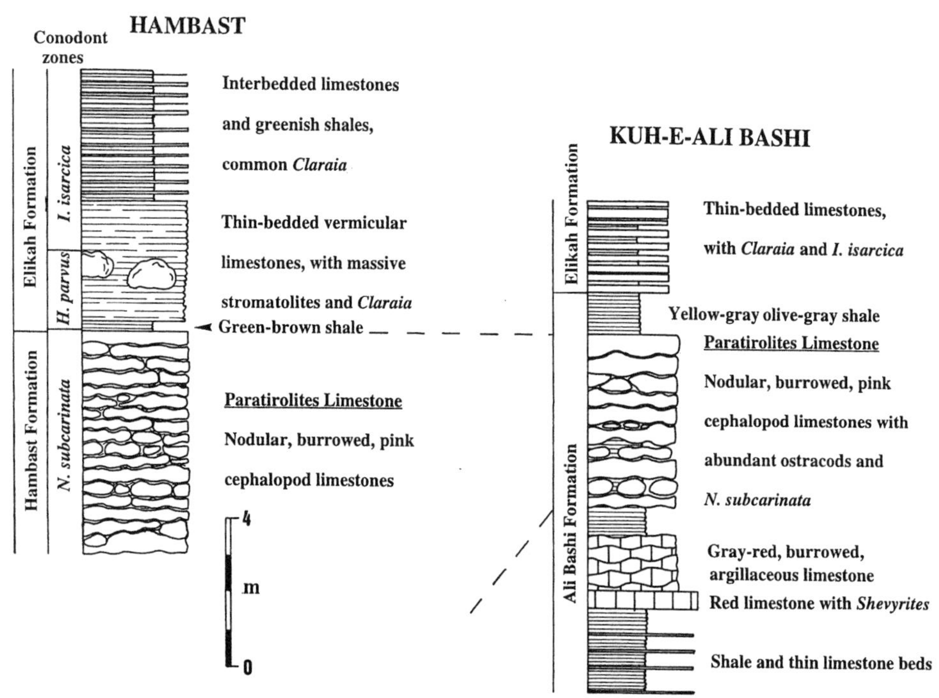

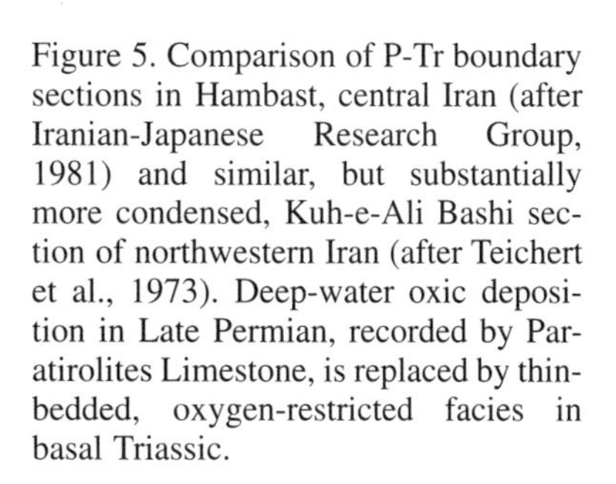
Figure 5. Comparison of P-Tr boundary sections in Hambast, central Iran (after Iranian-Japanese Research Group, 1981) and similar, but substantially more condensed, Kuh-e-Ali Bashi section of northwestern Iran (after Teichert et al., 1973). Deep-water oxic deposition in Late Permian, recorded by Paratirolites Limestone, is replaced by thin-bedded, oxygen-restricted facies in basal Triassic.

of detailed facies analysis, but the available lithological descriptions imply substantial changes in benthic oxygenation levels during the boundary interval. The Paratirolites Limestone was clearly deposited in deep, well-oxygenated conditions, but the depositional environment of the overlying shale is more enigmatic. The impoverished conodont fauna and presence of fine lamination suggest anoxic deposition, in which case the shale could be the equivalent of the *H. parvus* zone shale of Sicily. The overlying stromatolitic limestones have been ascribed a lagoonal and/or intertidal origin in all previous studies (e.g., Iranian-Japanese Research Group, 1981; Baud et al., 1989, 1997). However, stromatolites are encountered in a broad range of shallow-marine environments in the Early Triassic (Schubert and Bottjer, 1992), and their presence in the basal Triassic of central Tethys need not necessarily imply such shallow water depths. Baud et al.'s (1997) description of the Transcaucasian stromatolites suggests they may not be of cyanobacterial origin, but rather a further example of directly precipitated fans formed in anoxic conditions (Woods et al., 1999). The succeeding thinly bedded strata with trace fossils of the Elikah Formation and Karabaglar Suite are probably dysaerobic facies; it is possible that the *I. isarcica* zone was deposited in more oxygen-restricted conditions than the *H. parvus* zone. Very little information is available from the higher levels of these units, but the thinly bedded style of deposition appears to have persisted until at least the latest Griesbachian.

Central Tethyan sections are also well developed in the mountains of Oman, where depositional settings from shallow platform to slope and basin are recorded. Most of these sections still require detailed facies and biostratigraphic analysis, but some general observations can be made. The shallowest environments are recorded in the Upper Saiq and lower Mahil Formations and record deposition in the shallow subtidal to supratidal settings of the Arabian platform (e.g., Rabu et al., 1990). Unfortunately, pervasive dolomitization has affected the entire Wordian to Upper Triassic record, and little facies or faunal information has been gleaned. Nevertheless, bioturbation is present in the upper (?Changxingian) units of the Saiq Formation, suggesting oxic conditions, whereas the lower Mahil Formation (of Griesbachian-Dinerian age; A. Baud, 2001, personal commun.) is laminated to thinly bedded.

Deeper water sections are also well represented in the Oman Mountains, and provide a unique central Tethyan record of such environments. Slope deposits of Permian to Triassic age are represented by the lower members of the Maqam Formation (Watts, 1988). The uppermost Permian strata consist of partially dolomitized, cherty calcarenites that are well bioturbated and contain a diverse ichnofaunal assemblage, including *Palaeophycus, Chondrites*, and large *Rhizocorallium*. The overlying Early Triassic strata consist of laminated and thinly bedded calcilutites with decimeter- to meter-thick, erosive-based, flat-pebble conglomerates and several large (10–20 m thick) channels containing a variety of flat-pebble clasts in a finer (often ooidal) matrix (Watts and Garrison, 1986). The presence of lamination suggests that these slope sediments were deposited under anoxic conditions. The flat pebbles were derived from shallower settings and show that these areas were also subject to oxygen restriction (cf. Wignall and Twitchett, 1999). There is an increase in both the number and thickness of bioturbated horizons upward in the section (in particular within the Smithian), and an increase in the size, diversity, and depth of the trace fossil assemblage. Thus, the trace-fossil record shows that, in the slope environments of the central Tethys, the Griesbachian and Dienerian sediments were deposited under anoxic-dysoxic conditions and there was significant improvement in benthic oxygenation in the Smithian.

Eastern Tethys

The P-Tr sections of southern China provide one of the best marine records of this interval. Facies and faunal changes are best known from the condensed Meishan section of Zhejiang Province (Yin et al., 1992; Wignall and Hallam, 1993; Jin et al., 2000). There are 333 species known from the latest Permian Changxing Formation at Meishan, of which slightly more than 50% range up to the top of the formation; all but 6 species disappear in a series of beds that mark the P-Tr transition (Jin et al., 2000). The last diverse assemblage occurs in the top of bed 24, a burrowed wackestone (Fig. 6). This is sharply overlain by a lamina composed almost entirely of pyrite with an elevated chalcophile content (As, Se, Sb, and Mo) characteristic of anoxic deposition (Chai et al., 1992). The overlying beds, two

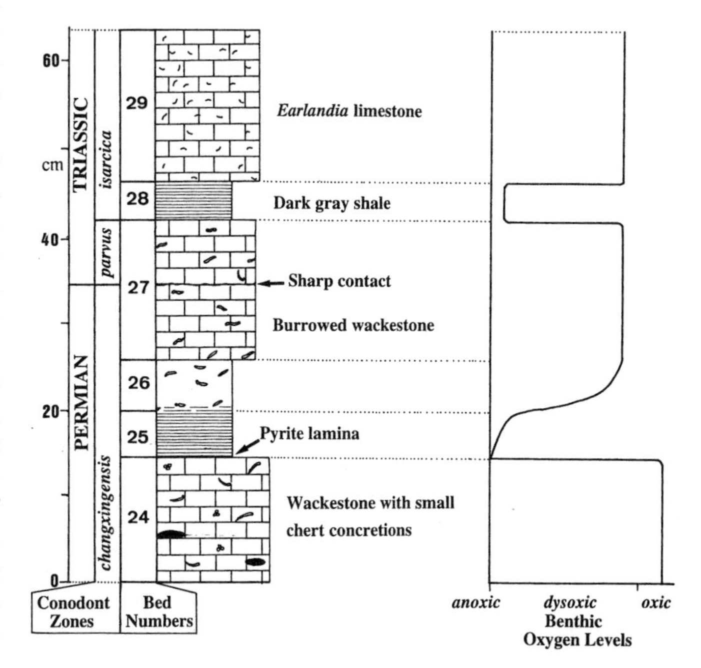

Figure 6. P-Tr boundary section at classic section D of Meishan, showing interpreted oxygenation levels (see text). Main extinction level occurs at bed 24–25 contact, which coincides with abrupt decline in oxygenation.

thin boundary clays, consist of a pale cream, finely laminated volcanic ash (bed 25) and a dark gray shale (bed 26): both beds are also enriched in chalcophile elements and contain abundant prasinophyte algae (Yin et al., 1992). Such algae are common in many Phanerozoic anoxic facies (Tyson, 1995), and their presence in the boundary clays is further evidence for oxygen-poor conditions at this level. However, intense bioturbation by shallow-penetrating *Planolites* in bed 26 suggests a progressive improvement in oxygen levels from the base of bed 25 upward (Fig. 6). The overlying bioturbated wackestone (bed 27) straddles the P-Tr boundary and, but for the presence of scattered grains of pyrite, contains little evidence for anoxic deposition. The younger Griesbachian strata at Meishan consist of frequent alternations of anoxic facies (organic-rich, dark gray shales) and dysoxic facies (thin beds of micrite and pale marls); pyrite is present throughout (Wignall and Hallam, 1993).

A similar decline in oxygenation is recorded in other sections in southern China, although the nature of the transition is dependent upon paleobathymetry. At Shangsi, in northern Sichuan, an expanded Changxingian basinal section consists of intensely burrowed, radiolarian-rich micritic limestones indicative of well-oxygenated, deep-water conditions (Wignall et al., 1995). These are succeeded by laminated, pyrite-rich, gray-green siliceous marls with thin interbeds of organic-rich shales. This transition from oxic to anoxic facies occurs at a level within the latest Changxingian. Anaerobic facies persist until the latest Griesbachian; younger strata are absent from Shangsi (Wignall et al., 1995).

Much shallower P-Tr conditions are recorded in southern Sichuan, where there are late Changxingian large sponge-algal reefs (Wignall and Hallam, 1996). The reefs are sharply overlain by a 4-m-thick bed of partially dolomitized biopelmicrites containing a fauna of foraminifera and gastropods of latest Permian age. This is overlain by the Feixianguan Formation, which consists of centimeter-scale alternations of finely laminated micrite and marls. Microburrowing (<1 mm burrow diameters) is present, but the shallow burrow depths have not disrupted the finely laminated fabric. Thus, in shallow-water settings persistent dysoxia appears to have enabled a small burrowing infauna to colonize the seafloor (Wignall and Hallam, 1996). In several sections in south Sichuan, Changxingian reefs are overlain by curious, digitate microbial (?) crusts (Kershaw et al., 1999). These have been recrystallized, making confirmation of the microbial fabric problematic, but Kershaw et al. (1999) speculated that the crusts could have formed by microbially mediated precipitation from beneath anoxic bottom waters. True calcimicrobial mounds have been recorded from an isolated carbonate platform in the Griesbachian of the Nanpanjiang basin (Lehrmann, 1999). In this case, the presence of a diverse benthic fauna within interbedded strata suggests that bottom waters are unlikely to have been anoxic in this very shallow water setting.

The anoxic and/or dysoxic facies of the Chinese P-Tr boundary beds contrast with the widespread development of oxygenated conditions in both deep- and shallow-water settings in the preceding Changxingian. However, only sparse information is available for post-Griesbachian oxygenation levels in China. Thus, the termination of the anoxic and/or dysoxic conditions is poorly defined. At Meishan and in the nearby Hushan section finely laminated anaerobic facies persist at least into the earliest Dienerian (Wignall and Hallam, 1993). Anoxia also persisted until at least the end of the Griesbachian in Sichuan (Wignall et al., 1995; Wignall and Hallam, 1996). Younger Scythian deposition (Smithian-Spathian Stages) is recorded in the marginal marine clastics and platform carbonates of the Jialing Formation Fan, (1980). Reinhardt (1988) noted that the latter facies consist of thin- to medium-bedded limestones and dolomitic marls containing very few fossils, but common burrows that are restricted to bedding planes. This could record an interval of dysoxic deposition, but more information is needed.

Further evidence for depositional conditions in the eastern Tethys comes from the diverse selection of outcrops in South Primorie (Fig. 2), including accreted oceanic terranes and epicontinental basin sediments. In the distal shelf facies of the Artemkova River basin, the shale-dominated Griesbachian strata are described as black, thin bedded, and containing carbonate nodules (Zakharov, 1992), features suggestive of anoxic deposition. Anoxia may also have been locally developed prior to this; Zakharov et al. (1995) recorded Changxingian strata from the Artemkova region as consisting of thin-bedded, gray-green pelites with abundant *Posidonia*, a typical dysaerobic bivalve. However, other Changxingian sections in South Primorie yield a diverse assemblage of benthic fossils suggesting well-oxygenated depositional conditions. In the oceanic terranes the P-Tr transition is marked by the cessation of carbonate formation on seamounts and the development of siliceous mudstone deposition (Khanchuk and Panchenko, 1997); the latter facies change is also seen in contemporaneous Japanese oceanic sediments (see following). The Gorbusha Suite of east Primorie includes thrust slices of red cherts and jaspers that yield Smithian-Spathian conodonts (Buryi, 1997). The red color indicates the presence of ferric iron, and thus oxidizing conditions in the later Scythian.

Perigondwanan margin

P-Tr boundary shelf sediments from the southern margin of the Neo-Tethys are well developed in Pakistan and northern India. In the Salt Range of Pakistan, Wignall and Hallam (1993) documented late Griesbachian–late Dienerian dysaerobic strata in the lower part of the Mittiwali Marls (alternations of organic-rich shale and finely laminated, gray-green shales with a low-diversity paper pectin fauna). Older Griesbachian and latest Changxingian strata in the Salt Range consist of dedolomitized limestones that, before their extensive diagenesis, consisted of cross-bedded calcarenites composed of abraded echinoderm and brachiopod fossils (Wignall et al., 1996), clearly not dysaerobic facies.

In the basinal setting of Kashmir oxygen-restricted deposition appears to have begun considerably earlier, because Nakazawa et al. (1975) described black shales, with common *Claraia* species, interbedded with thin limestone, ranging from the late Changxingian to the late Dienerian (Wignall et al., 1996). Similar facies developments of similar age are widespread in the Himalayan region (Kapoor and Tokuoka, 1985).

To the south of the Himalayan sections, there is a further marine Perigondwanan P-Tr record in Madagascar. In southern Madagascar the P-Tr transition is recorded in a succession consisting of interdigitating fan-delta conglomerates and shallow-marine facies (Wescott, 1988). The unusual marine strata include stromatolites as high as 2 m and laminated, calcareous siltstones. In northern Madagascar Lower Triassic strata have yielded a famous fauna of fully articulated fish (Merle and Fournier, 1910). Such exceptional preservation implies anoxic deposition, but unfortunately the precise age of the fish is not known, although it is probably attributable to the Griesbachian, given that it occurs with *Claraia griesbachi* (Treat, 1933). Although detailed analysis has not been done, there seem to be many similarities between Madagascar and East Greenland (see following), including depositional environment (rapid sedimentation in a narrow, elongate, fault-controlled embayment), faunal assemblages, and styles of preservation. Nielsen (1936) suggested that the Madagascar fish fauna is slightly younger than that of East Greenland, possibly the result of a slightly later onset of anoxic conditions in Madagascar.

Eastern Panthalassan margin

A large marine embayment marked the eastern margin of the Panthalassa ocean in the western United States (Fig. 2), and the infill is now seen in an extensive series of outcrops in Nevada, Utah, Idaho, Wyoming, and Montana. Wignall and Hallam (1992) examined Lower Triassic strata from deep-water sections in southeastern Idaho and recorded dysoxic-anoxic deposition (finely laminated, pyritic micrites and dark gray shales) in the Griesbachian portion of the Dinwoody Formation. Younger strata in the region record more oxygenated shallow-water deposition. Anoxic deposition is recorded in the Smithian-Spathian interval, but only in deep, basinal sections (Tosk and Anderson, 1988; Schubert and Bottjer, 1995). Contemporaneous anoxic, deep-water deposition is recorded by the finely laminated, siliceous mudstones of the Union Wash Formation of California (Woods and Bottjer, 2000). Woods et al. (1999) recorded directly precipitated calcite fans from this unit that are similar to "stromatolites" reported from the western and central Tethys.

Deep-water P-Tr sedimentary records are present in the Western Canadian sedimentary basin (Henderson, 1997). One of these, the section at Ursula Creek, on the shores of Williston Lake in British Columbia, was measured and the Th and U concentrations recorded using a field portable gamma ray spectrometer (Fig. 7). The Late Permian record consists of both massive and bedded radiolarian cherts of the Fantasque Formation. Values of both Th and U are low (and below the accurate detection limit of the spectrometer), reflecting the low clastic content of the sediments and probable oxic depositional conditions; this interpretation is supported by the presence of pervasive bioturbation. Toward the top of the Fantasque Formation thin interbeds of black shale also occur, indicating a transitional contact with the overlying Grayling Formation, which is dominated by this lithology (Fig. 7). The associated enrichment in Th and U reflects an increased clastic content, and the fluctuations of Th/U ratios indicate variable benthic oxygenation, also recorded by the presence of finely laminated, pyritic cherts interbedded with burrowed cherts in the topmost meters of the Fantasque Formation. The overlying Grayling and Toad Formations record persistent oxygen-poor deposition that varied in intensity. Thus, the most anoxic facies occur near the base of the formation (early Griesbachian), where Th/U values decline to 1.0 at a level that is also characterized by the presence of several thin beds of finely laminated dolomite. The same level also coincides with reported total organic carbon (TOC) values of 1.5%–2.0% (Wang et al., 1994). Wang et al. (1994) also noted low H/C ratios of 0.5, indicating that the sediments are overmature; thus the original organic carbon content was even greater. The succeeding late Griesbachian–Dienerian section records a slight improvement in oxygenation before the reestablishment of intense anoxia in the base of the Toad Formation. At Ursula Creek these conditions persisted from the Smithian Stage to the Anisian Stage.

Boreal shelf seas

Redox changes on the margin of the high-paleolatitude Boreal ocean have been investigated in the P-Tr sections of western Spitsbergen (Wignall and Twitchett, 1996; Wignall et al., 1998). Here a diverse fauna of brachiopods, sponges, and bryozoans ranges to near the top of the Kapp Starostin Formation (Ezaki et al., 1994). The loss of the siliceous sponges coincides with the disappearance of thick beds of chert, a lithological change that marks the base of the overlying Vardebukta Formation. The basal meters of this formation record the rapid decline in benthic oxygenation (bioturbation declines and then disappears, pyrite appears, and palynomorph preservation improves) until, at a level immediately below the P-Tr boundary, anoxic facies are developed with characteristically high pyrite S/TOC ratios (Wignall et al., 1998). Anoxic facies range to a level around the base of the Selmanset Member, a Dienerian shoreface sandbody (Wignall and Twitchett, 1996). Anoxic facies became extensive again in the deeper water facies of the Spitsbergen region during the Smithian-Spathian interval, when finely laminated, dark gray shales of the Tvillingodden Formation were deposited over a wide area (Mørk et al., 1982). Anoxia persisted into the Anisian in basinal locations with the deposition of the highly organic-rich, phosphatic shales of the Botneheia Formation.

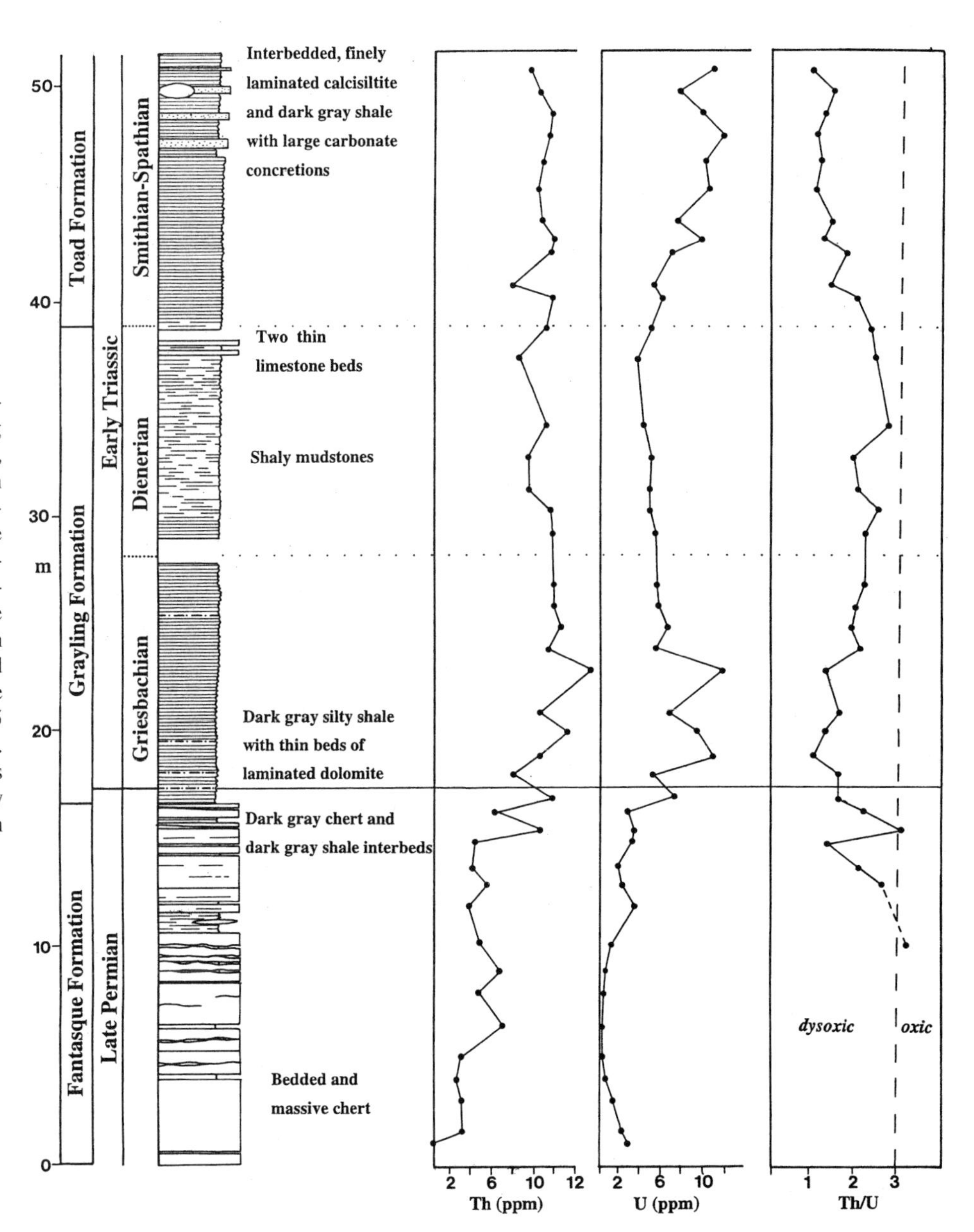

Figure 7. Sedimentary log and concentrations of Th and U, measured using field-portable gamma ray spectrometer, in Ursula Creek section of Williston Lake, British Columbia. Position of P-Tr boundary and stage boundaries are based on conodont data (J.-P. Zonneveld, 2000, personal commun.). Calculated Th/U ratio is depicted for those values that exceed accurate detection limit of instrument (>1.8 ppm for U and >2 ppm for Th). Ratios of <3.0 indicate dysoxic deposition and values <1.5 suggest intensely anoxic deposition. Oxic deposition of radiolarian cherts is gradually replaced by progressively more oxygen-restricted deposition in latest Permian and earliest Triassic.

Similar lithological changes are seen in the Lower Triassic succession of the Sverdrup basin in Arctic Canada (Embry, 1993), although it is unclear from the lithological descriptions available if the same sequence of redox changes is recorded. A microfacies study of the dark gray basinal shales (Blind Fiord Formation) of the region would be particularly useful.

A major marine embayment along the east coast of Greenland also contains a thick Permian-Triassic succession. Although the latest Permian is generally thought to be absent in this region (e.g., Seidler, 2000), conformable shale on shale contacts are known from the western margin of the Jameson Land basin toward the southern termination of the seaway (Perch-Nielsen et al., 1972; Twitchett et al., 2001). Localities in this region record the transition from upper Schuchert Dal Formation into the base of the overlying Wordie Creek Formation. The gray-green, micaceous mud-siltstones of the upper Schuchert Dal Formation contain a diverse Late Permian invertebrate fauna of brachiopods, rugose corals, ammonoids, and foraminifera. The sediments are well bioturbated and contain a diverse trace-fossil assemblage that implies a well-oxygenated benthic environment. The overlying dark gray, micaceous, and pyritic mud-siltstones of the basal Wordie Creek Formation are generally laminated, except for rare horizons containing sparse *Planolites*. These latest Changxingian sediments contain a depauperate benthic fauna of foraminifera and bedding-plane assemblages of the paper pecten *Claraia* (Twitchett et al., 2001). A well-preserved fish fauna is common throughout the lower Wordie Creek Formation (Bendix-Almgreen, 1976). Facies and paleoecological evidence therefore suggests deposition under anoxic-dysoxic conditions. This interpretation has been inves-

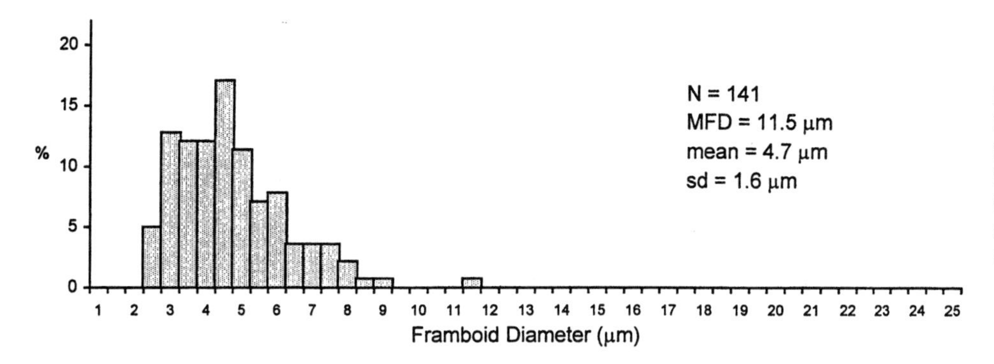

Figure 8. Example of typical pyrite framboid size-frequency distribution from silty shales of lower Wordie Creek Formation of Schuchert Dal, East Greenland. N is sample size, MFD is diameter of largest framboid, and sd is standard deviation.

tigated further by means of pyrite framboid size analysis (Fig. 8). The upper beds of the Schuchert Dal Formation are nearly devoid of pyrite with the rare exception of large framboids (~40 μm). In contrast, the shales of the lower Wordie Creek Formation contain prolific numbers of pyrite framboids with a small-sized population suggestive of euxinic conditions (Table 1). The presence of a short "tail" of larger framboids suggests transient seafloor oxygenation and brief periods of diagenetic framboid growth (Wignall and Newton, 1998). The dark gray shales of the lower Wordie Creek Formation are replaced by gray-green and purple varieties at high levels, indicating improvement in seafloor oxygenation in the region during the late Griesbachian.

Panthalassa ocean

The accreted terranes of Japan provide an extensive, if much deformed, record of deposition on the Panthalassan ocean floor (Isozaki et al., 1990). Much of the Carboniferous to Jurassic rocks are red radiolarian cherts, indicating oxic deposition, but all P-Tr boundary sections display a remarkably similar and unusual succession of lithologies. During the Late Permian, hematitic red cherts were replaced by pyritic gray cherts (Kubo et al., 1996), followed by a gradual transition to siliceous claystones and then black, organic-rich shales at the P-Tr boundary (Isozaki, 1994, 1997). The overlying Scythian succession generally records the reverse series of lithological changes: siliceous claystones overlie the black shales and are in turn overlain by gray cherts. Red cherts do not reappear until the Anisian.

Interpreting oxygen levels within this stratigraphy has proved controversial. The siliceous claystones, known as Toishi facies (Musashino, 1993), consist of alternations of 2–3-cm-thick beds of pale gray, siliceous claystone and thinner (1 cm thick) beds of chert or black shale. Fine lamination and layers and veins of crystalline pyrite are commonly seen, U is enriched, and the facies has, not unreasonably, been consistently interpreted as indicative of anoxic deposition (Musashino, 1993; Kajiwara et al., 1994; Isozaki, 1994, 1997). The interpretation of the depositional conditions of the black shale has proved much more controversial. The black shale is very rich in organic carbon (3–8 wt%), finely laminated, contains abundant framboidal pyrite, and is enriched in U, all good evidence for anoxic deposition, as suggested by Isozaki (1994, 1997). In contrast, Suzuki et al. (1998) suggested that the black shales record oxic deposition under conditions of high primary productivity, reflecting vigorous upwelling during the P-Tr boundary interval. This interpretation disregards the overwhelming sedimentary and geochemical evidence for anoxic deposition, but it highlights the salient point that the black shale is more than an order of magnitude enriched in TOC relative to the siliceous claystones (Kajiwara et al., 1994). Rather than increased productivity, this could reflect a lower sedimentation rate and thus less clastic dilution in the black shales. However, the best-dated sections suggest that 2 m of black shale accumulated during the Griesbachian-Dienerian Stages (Yamakita et al., 1999), an interval of ~3 m.y., at a sedimentation rate of 0.7 m/m.y. In other less-well-dated sections the black shale unit is as much as 15 m thick (Isozaki, 1994). In comparison, the thickness of the lower (Late Permian) unit of the siliceous claystones does not exceed 1 m (Isozaki, 2000, personal commun.), indicating sedimentation rates of ~0.2 m/m.y., using the time scale of Bowring et al. (1998); this is less than the black shale sedimentation rate. Clastic dilution therefore does not seem to explain the relatively low TOC values of the Toishi facies relative to the black shales, implying elevated organic carbon production during deposition of the shales in the earliest Triassic.

The $\delta^{34}S_{pyr}$ curve for the Japanese P-Tr sections reveals a 10‰–15‰ negative shift within the black shale from values of ~ − 10‰, which Kajiwara et al. (1994) interpreted as evidence of repeated oxidation-reduction cycling beneath an oxic water column. Thus, like Suzuki et al. (1998), they suggested that the black shale unit records a brief oxic event within a longer term anoxic event. However, significant sulfate fractionation is also encountered in euxinic environments. For example, the transition to euxinic conditions in the Black Sea in the late Pleistocene caused a large negative shift of $\delta^{34}S_{pyr}$ values, and pyrite forming in the water column today records the most extreme S fractionation of any modern environment (Calvert et al., 1996). We suggest that the $\delta^{34}S_{pyr}$ trend recorded by Kajiwara et al. (1994) is a record of oxygenation changes that are opposite to those they propose. The relatively heavy values of pyrite in the siliceous claystones either side of the black shale suggest formation in a semiclosed system (with respect to sulfate) during late diagenesis, also indicated by the occurrence of pyrite in veins and layers rather than as framboids. The diagenetic, rather than syngenetic, origin of pyrite in this facies implies dysoxic depositional conditions. The black shale, however, has the attributes of euxinic deposition.

The deep-water Japanese sections provided the basis for Isozaki's (1994, 1997) concept of an 8 m.y. superanoxic event lasting from the middle Permian to the Middle Triassic. Evidence for truly euxinic conditions is confined to the black shale that, recent conodont findings indicate, accumulated from the *parvus* zone (Yamakita et al., 1999) to some poorly dated interval in the late Griesbachian-Dienerian (Isozaki, 1997). The bounding Toishi claystones clearly record oxygen-poor deposition, but the relatively heavy $\delta^{34}S_{pyr}$ values suggest that euxinicity was not established at this time.

Carbonate facies that formed on seamounts are also incorporated in the terranes of Japan, thus providing evidence of very shallow-water conditions within the Panthalassa ocean. Sano and Nakashima's (1997) detailed facies analysis of a seamount succession from the Chichibu terrane revealed a Changxingian record consisting of massive, peloidal lime-mudstones that become progressively more heavily dolomitized toward the top. These are overlain by an upward-deepening Scythian succession that begins with gastropod-rich, carbonaceous, cryptomicrobial micrites that contain *H. parvus* and *I. isarcica* (Koike, 1996). Sano and Nakashima (1997) interpreted these sediments as a peritidal dysaerobic facies. The younger Griesbachian consists of oncolitic limestones, and the Dienerian to Spathian succession is composed of coquinas containing an appreciable diversity of bivalves but no evidence for further oxygen-restricted deposition.

P-TR SUPERANOXIC EVENT

The foregoing documentation of Late Permian to Early Triassic sections reveals comparable redox histories for many regions. For four regions in particular, the record is sufficient for spatial and temporal changes of marine oxygenation from deep- to shallow-water sites to be documented in the P-Tr interval (Fig. 9). We have also attempted to plot the spatial distribution of oxygen-poor, shallow marine waters (Figs. 10–12).

Oxygen-restricted deposition began at the end of the Middle Permian in the deep waters of the Panthalassa ocean (Isozaki, 1994, 1997), but dysaerobic facies are rarely encountered in contemporaneous epicontinental settings, except in local basin centers (e.g., the Artemkova River region of South Primorie). This situation changed dramatically in the latest Permian (late *C. changxingensis—H. latidentatus* zone) when dysoxic or anoxic deposition became widespread (Fig. 10). This transition is seen in both basinal sections (e.g., Ursula Creek, British Columbia, Guryul Ravine, Kashmir) and inner ramp or shelf sections (e.g., Dolomite sections, northern Italy, sections of southwest Sichuan, China) and coincides with the transition to black shale deposition in the accreted terranes of Japan. The deeper waters of the Panthalassa ocean may have become euxinic at this time, while seamount crests underwent dysoxic deposition. Similarly, in epicontinental settings, the deeper water settings became anoxic while updip the presence of microburrowed horizons and a dysaerobic fauna indicates that variable, but generally low, oxygen levels were prevalent (e.g., Wignall and Hallam, 1996).

The onset of anoxia-dysoxia was diachronous within some regions. Thus, in the western Tethys the ramp sections of the Dolomites first developed laminated, pyritic micrites a short, but appreciable, distance below the P-Tr boundary, while anoxic facies developed at the base of the Triassic in the deeper water Sicilian sections (Fig. 9). This implies that oxygen-poor bottom waters may have started near shore and spread offshore within equatorial Tethys. Much clearer diachroneity is seen in the ramp settings of Perigondwana, where dysaerobic facies developed in deep-water sections (Kashmir) considerably earlier than in shallow-water sections (Salt Range, Fig. 9). During the early Griesbachian the Salt Range is one of the few areas of the world not to record oxygen-poor deposition. However, by late Griesbachian time all but the shallowest water settings recorded oxygen-poor deposition; this interval can be regarded as the acme of the superanoxic event (Fig. 11).

The timing of the cessation of anoxia is much less well defined than the onset. In the few sections studied, anoxic-dysoxic deposition ceased abruptly in the latest Griesbachian (Wignall and Twitchett, 1996; Twitchett, 1999). There is no evidence for oxygen-poor waters in the Tethyan and Boreal oceans by mid-Dienerian time (Figs. 9 and 12). Within Panthalassa, the British Columbia data (Fig. 7) suggest a weakening of the intensity of anoxia in this region. Only in deeper settings on the Perigondwana margin did the event apparently continue with little appreciable change.

Subsequent changes in oxygenation are not known in any detail or with any biostratigraphic resolution, but it appears that, at least on the margins of the Panthalassa and Boreal oceans, anoxic deposition became widespread during the Smithian and Spathian Stages. However, unlike the Griesbachian event, there is no evidence that anoxia extended into shallow waters at that time, and the event does not appear to have occurred in Tethyan regions. Evidence from the ramp settings of the eastern Panthalassan margin demonstrate clearly that shallow-water environments were free of oxygen-poor waters during the Smithian-Spathian interval (Schubert and Bottjer, 1995).

DISCUSSION

Anoxia and extinction

The P-Tr superanoxic event was perhaps the most severe event of its kind in the Phanerozoic; it was both globally widespread and developed in exceptionally shallow waters. There

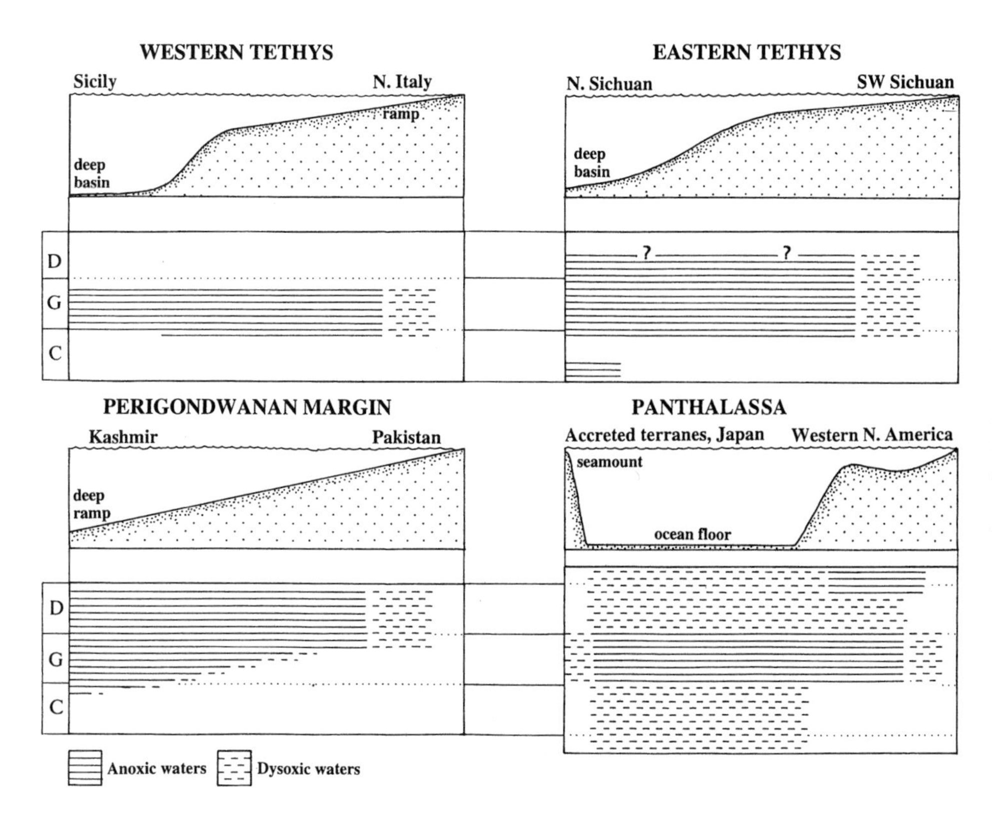

Figure 9. Plot of extent of anoxic waters and dysoxic waters (solid and dashed horizontal lines, respectively) on time-environment diagrams from selected regions, depicting near-synchronous onset of anoxia in latest Permian and much more variable (and less well defined) cessation of this superanoxic event. Stages: C, Changxingian; G, Griesbachian; D, Dienerian.

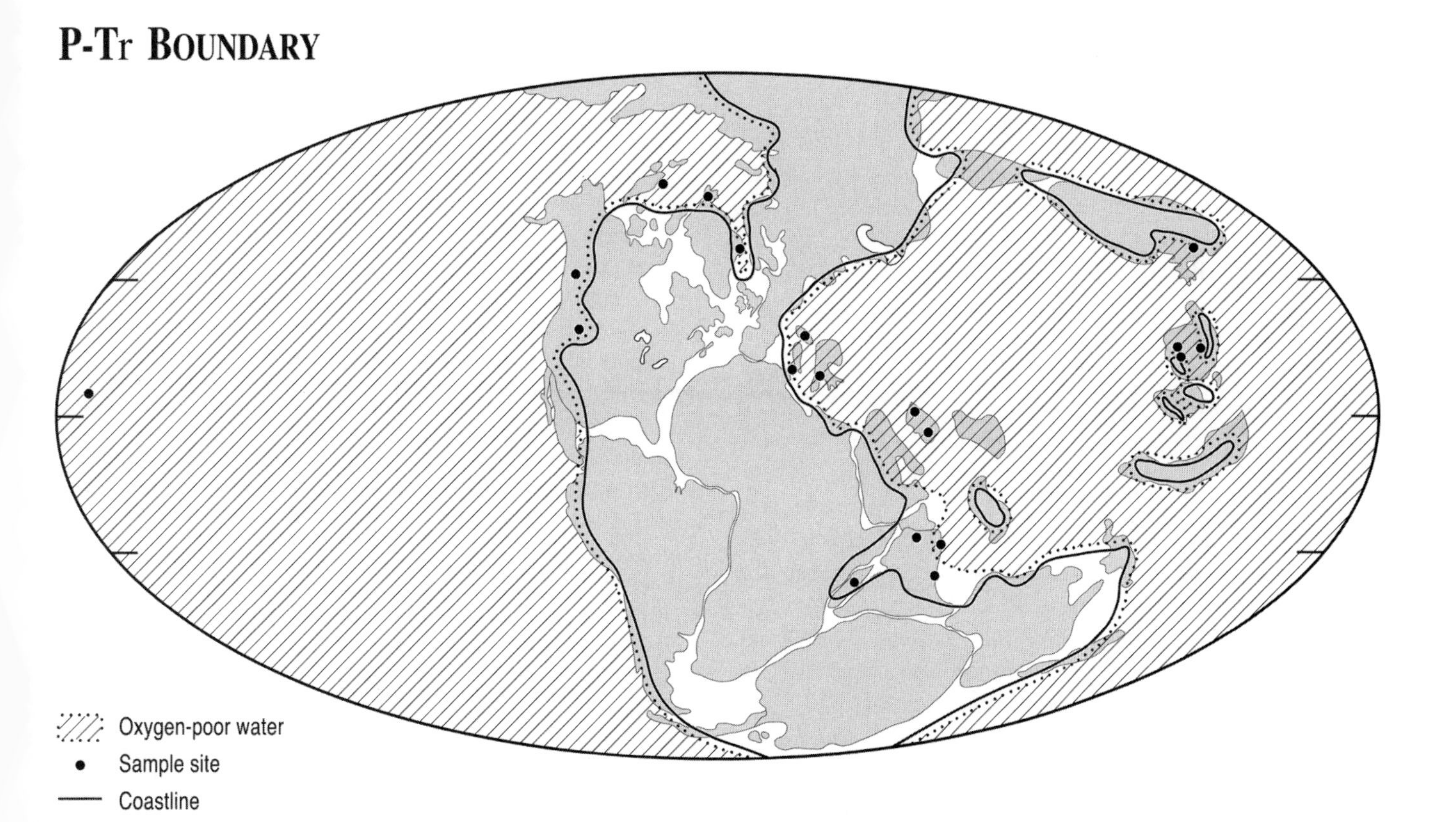

Figure 10. Extent of oxygen-poor shallow-marine conditions in Permian-Triassic (P-Tr) boundary interval (late *C. changxingensis—H. parvus* conodont zones). Note that only on southern margins of Neo-Tethys, in Perigondwanan region, did oxic deposition persist at this time.

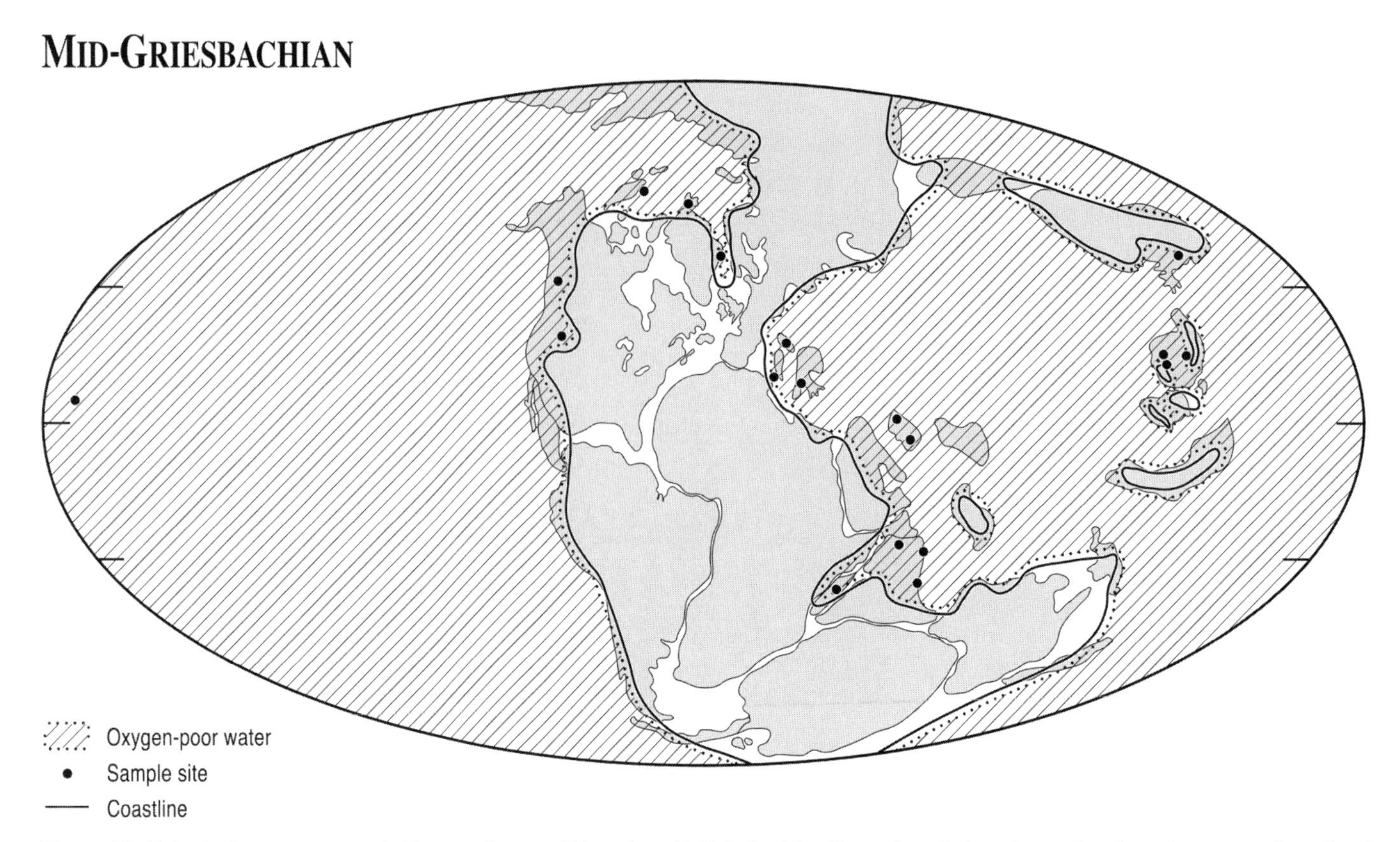

Figure 11. Extent of oxygen-poor shallow-marine conditions in mid-Griesbachian Stage (late *I. isarcica*–early *C. carinata* zones), peak of Permian-Triassic superanoxic event.

were few inhabitable marine areas at this time, a factor that suggests that it was, in a large part, responsible for the mass extinction. In comparison with the best known oceanic anoxic event, at the Cenomanian-Turonian boundary, the P-Tr event lasted considerably longer and developed in both deep- and shallow-marine sites. In comparison, the Cenomanian-Turonian event only affected deep shelf and oceanic sites (Arthur et al., 1987).

As with all kill mechanisms, the crucial factor is to demonstrate the close coincidence in timing of the event with the extinction. The best example of the coincidence between end-Permian mass extinction and the development of anoxic facies is the intensively studied Meishan section; it can also be seen in many other areas of China (Fig. 6; Wignall et al., 1995; Wignall and Hallam, 1996). The level of extinction has also been well defined in northern Italy and Spitsbergen (Wignall and Hallam, 1992; Wignall et al., 1998; Twitchett, 1999) and, although there is diversity decline in the few meters of strata before the development of anoxia in these regions, the correspondence is still close. The nearshore Pakistani sections record the onset of dysaerobic deposition considerably later than elsewhere in the world, and it is significant that abundant benthic fauna occur up to the level of this late Griesbachian facies change (Wignall and Hallam, 1993). Thus, Permian holdover taxa were able to survive in areas where the onset of oxygen-poor deposition was delayed.

Kirchner and Weil (2000) noted that origination pulses only follow extinction pulses after an appreciable lag, without an appreciable burst of diversification in the immediate aftermath of the extinction event. However, this observation depends on how the end of the extinction pulse is defined. It has been argued that the environmental stresses that caused the end-Permian extinction (i.e., marine anoxia) may have persisted throughout the Early Triassic and caused the prolonged delay in recovery (Hallam, 1991; Isozaki, 1994, 1997). In this scenario the delay is a consequence of environmental factors rather than an intrinsic property of the biological system. Our review lends credence to Hallam and Isozaki's proposition. Only during the Dienerian Stage did the extent of anoxic-dysoxic conditions contract, and it is noteworthy that the first evidence for diversity recovery occurs at this time (Wignall et al., 1998; Twitchett, 1999). The subsequent reexpansion of anoxic-dysoxic deposition in the Smithian-Spathian may account for the limited recovery seen in this interval (cf. Schubert and Bottjer, 1995). In the western Tethys, facies evidence suggests that deleterious environmental changes, unrelated to oxygenation, also prevented pre-Spathian recovery in shallow-marine settings of this region (Twitchett and Wignall, 1996; Twitchett, 1999).

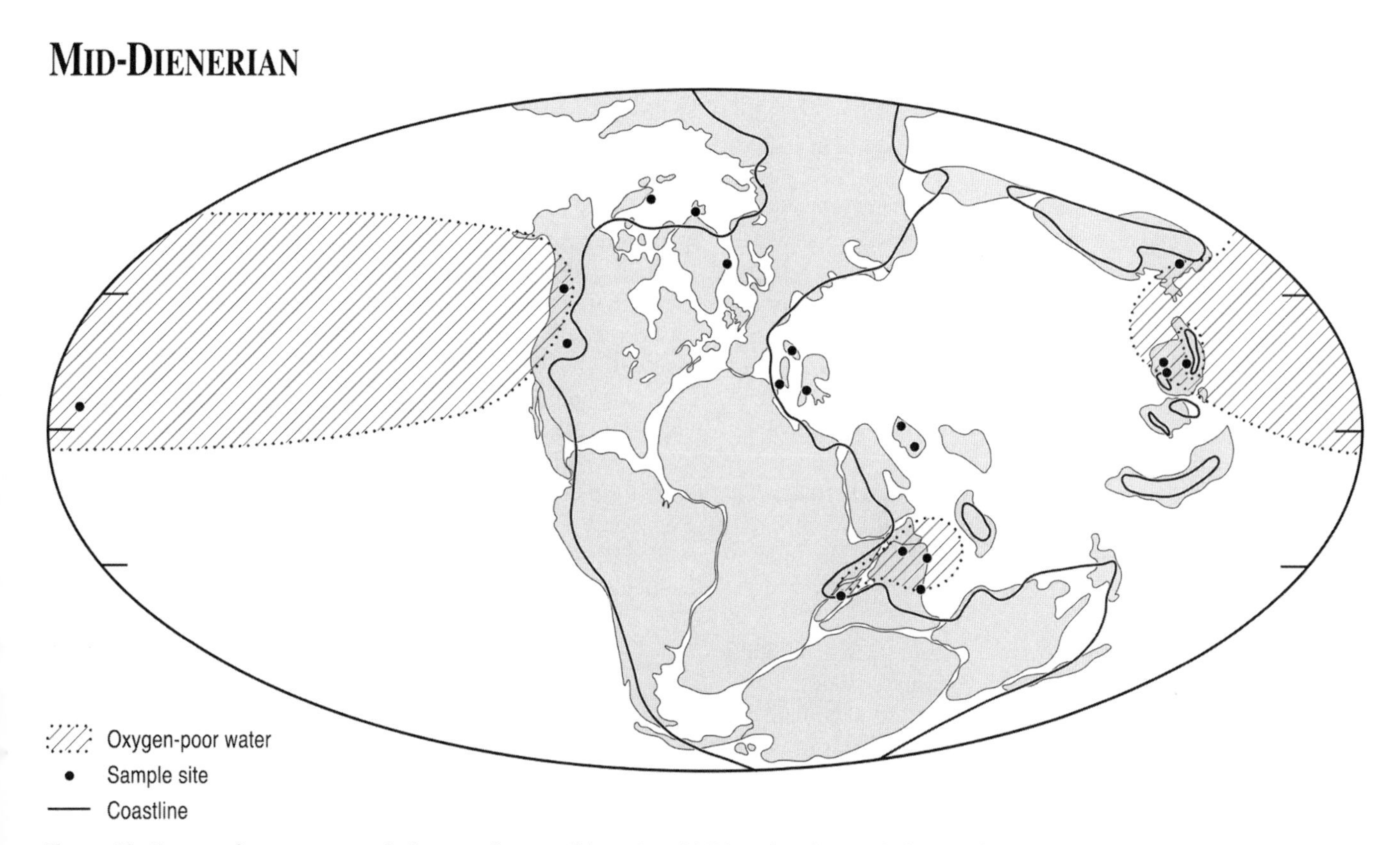

Figure 12. Extent of oxygen-poor shallow-marine conditions in mid-Dienerian Stage, during waning stages of superanoxic event. Anoxic deposition had ceased throughout Paleo-Tethys at this time.

Other extinction mechanisms

Changes of marine oxidation state have figured in other models for the end-Permian mass extinction. Knoll et al. (1996) proposed that CO_2 buildup in the euxinic waters of Late Permian oceans was followed by oxygenation and overturn of the oceans in the latest Permian. Malkowski et al. (1989) also invoked a major increase in oceanic ventilation at the P-Tr boundary, a change they suggested would cause a collapse of primary productivity. Both these models, although potentially correct in their predictions of the consequences of oceanic overturn, require oceanic redox changes that are the reverse of those recorded here.

Many workers have suggested that there was a catastrophic crash of primary productivity in the latest Permian (e.g., Malkowski et al., 1989; Holser and Magaritz, 1992; Wang et al., 1994; Wignall and Twitchett, 1996; Hallam and Wignall, 1997; Twitchett, 2001). However, modeling of nutrient dynamics in euxinic oceans suggests that the development of the P-Tr superanoxic event should have stimulated a productivity increase due to the greater availability of dissolved phosphorus in anoxic waters (van Cappellen and Ingall, 1994). Phosphorus-stimulated productivity increase is only a phenomenon of long-term euxinicity; there is a lag period of several thousand years between the onset of anoxia and buildup of phosphorus in solution (van Cappellen and Ingall, 1994). This may have occurred within oceanic settings, where the large volume of anoxic waters immured them against oxygenation events, but shallow-marine environments record much more variable anoxic-dysoxic conditions, in which the phosphorus-enhanced productivity factor never occurred. This may explain the observation that organic carbon enrichment is only seen in very deep shelf locations and oceanic settings during the P-Tr interval. Thus, the transition from organic-poor Toishi facies to organic-rich shale deposition seen in the basal Triassic of Japan (and the similar change seen in Sicily) may reflect the shallowing of euxinic waters to a point where phosphorus is leaked into surface waters. The development of dysoxic waters even over seamounts is testimony to the extreme shallowing of the oceanic chemocline at this time.

Indicators of anoxia

Many dysaerobic facies have been recorded from P-Tr boundary sections; they reflect the diverse paleobathymetries and paleolatitudes of the superanoxic event. It is interesting to note the widespread occurrence of dolomitization at that time, particularly of very shallow water strata immediately beneath the lowest dysaerobic facies. Thus, dolomitization is seen in Italy, Austria, Turkey, Oman, Pakistan, shallow-marine sections of southern China, and the Panthalassa seamounts. Recent study has suggested that dolomitization results from anaerobic microbial activity, intervals of extensive dolomitization coinciding with periods of poor oceanic oxygenation (Burns et al., 2000). Therefore, the widespread dolomitization in the P-Tr interval may be further evidence of the superanoxic event.

Stromatolites and other microbial structures are also widely reported from the earliest Triassic (e.g., the Dolomites, northern Italy, Turkey, central Iran, Armenia, southern China, Madagascar). This resurgence of typical Precambrian fossils may record the relaxation of grazing pressures in the aftermath of the end-Permian extinction (Schubert and Bottjer, 1992; Lehrmann, 1999). It may reflect the ease of carbonate precipitation by cyanobacteria in supersaturated oceans following the demise of most calcite-secreting organisms at the end of the Permian. The global development of anoxia, often in very shallow water, may have supplied bicarbonate-rich waters to the photic zone, as predicted for stratified oceans (Riding, 1993, 2000; Kempé and Kazmierczak, 1994).

Many so-called stromatolites may be inorganic precipitates from beneath bicarbonate-rich, euxinic waters. Woods et al. (1999) described examples of fans of radiating, prismatic calcite crystals from the younger Scythian record of California that closely match the descriptions of many Griesbachian stromatolites. This alternative requires support from further microfacies analysis, but it may have important consequences for many facies models, because the stromatolites are often used as evidence for shallow-water-lagoonal-peritidal interpretations, notably in Turkey and Iran. The examples of Wood et al. (1999) come from a deep-water setting, well beneath the photic zone.

Ultimate cause?

We have previously suggested that the origin of oceanic anoxia during the P-Tr crisis was probably related to a severe global warming event (Wignall and Twitchett, 1996). Evidence for warming is diverse and includes oxygen isotope fluctuations (Holser et al., 1991), data from high-latitude paleosols (Retallack, 1999; Krull and Retallack, 2000), and the appearance of warm-water marine taxa in high latitudes (Wignall et al., 1998). Release of volcanic CO_2 from Siberian volcanism has been postulated as a likely cause of the global warming (Wignall and Hallam, 1993), perhaps exacerbated by the release of methane from gas hydrates (Erwin, 1993).

The consequence of extreme global warming would be a major reduction of the equator to pole temperature gradient, the engine of oceanic circulation, and a decline in dissolved oxygen concentrations (Wignall and Twitchett, 1996). Computer modeling has lent support to this scenario (Hotinski et al., 2000, 2001). By assuming an equator to pole gradient of surface-water temperatures of only 12°C, and polar temperature of 12°C, Hotinski et al. (2001) predicted dysoxic to anoxic deep-ocean waters for a Late Permian continental configuration. This is due to sluggish circulation and the "difference in solubility of oxygen in the warmed high-latitude regions where deep waters are formed" (Hotinski et al., 2001, p. 8). Several of the detailed predictions of this computer modeling may be reflected in the geological record. Thus, the greatest intensity of oxygen restriction occurs along the North American margin (Hotinski et al., 2001, Fig. 3), a prediction that may relate to the more prolonged duration of the superanoxic event in this region (Woods and Bottjer, 2000). The weakest expression of oceanic dysoxia was in the Neo-Tethys adjacent to the Perigondwanan margin, a region where the expression of the event is relatively weak (Fig. 9).

Computer modeling confirms our suggestion that high-latitude surface-water temperatures were a crucial control of oceanic ventilation in the Late Permian. Extreme global warming at the end of this interval appears to have ensured the exceptionally widespread development of oxygen-poor waters, a phenomenon that we believe was largely responsible for the coincident marine mass extinction.

ACKNOWLEDGMENTS

The Natural Environment Research Council provided funding for the research in Greenland and British Columbia. We thank the Cambridge Arctic Shelf Programme for logistic support in Greenland, and John-Paul Zonneveld (Canadian Geological Survey) for the organization of field work in British Columbia. Iain Armstrong and Rob Newton provided invaluable field assistance in Greenland and British Columbia, respectively. We thank Aymon Baud, Leo Krystyn, and Sylvian Richoz for their organization of an Omani field trip, and Dave Bottjer, Yukio Isozaki, Ken MacLeod, Greg Retallack, and Brad Sageman for their comments on the manuscript.

REFERENCES CITED

Altiner, D., Baud, A., Guex, J., and Stampfli, G., 1979, La limite Permien-Trias dans quelques localities du Moyen Orient: Recherches stratigraphiques et micropaleontologiques: Rivista Italiana Paleontologia Stratigrafia, v. 85, p. 683–714.

Arthur, M.A., Schlanger, S.O., and Jenkyns, H.C., 1987, The Cenomanian-Turonian oceanic anoxix event. II. Palaeoceanographic controls on organic-matter production and preservation, *in* Brooks, J., and Fleet, A.J., eds., Marine petroleum source rocks: Geological Society [London] Special Publication 26, p. 401–420.

Baud, A., Cirilli, S., and Marcoux, J., 1997. Biotic response to mass extinction: The lowermost Triassic microbialites: Facies, v. 36, p. 238–242.

Baud, A., Magaritz, M., and Holser, W.T., 1989, Permian-Triassic of the Tethys: Carbon isotope studies: Geologisches Rundschau, v. 78, p. 649–677.

Bendix-Almgreen, S.E., 1976, Palaeovertebrate faunas of Greenland, *in* Escher, A., and Watt, W.S., eds., Geology of Greenland: Grønlands Geologiske Undersøgelse, p. 536–573.

Blendinger, W., 1988, Permian to Jurassic deep water sediments of the Eastern Oman Mountains: Their significance for the evolution of the Arabian margin of the south Tethys: Facies, v. 19, p. 1–32.

Bowring, S.A., Erwin, D.H., Jin, Y., Martin, M.W., Davidek, K., and Wang, W., 1998, U/Pb zircon geochronology and tempo of the end-Permian mass extinction: Science, v. 280, p. 1039–1045.

Broglio Loriga, C., Goczan, F., Haas, J., Lenner, K., Neri, C. Oravesz Sheffer, A., Posenato, R., Szabo, A., and Toth Makk, A., 1990, The Lower Triassic sequences of the Dolomites (Italy) and Transdanubian mid-mountains (Hungary) and their correlation: Memoirie delli Instituti di geologia e mineralogia dell'Universita di Padova, v. 42, p. 41–103.

Burns, S.J., McKenzie, J.A., and Vasconcelos, C., 2000, Dolomite formation and biogeochemical cycles in the Phanerozoic: Sedimentology, v. 47, p. 49–61.

Buryi, G.I., 1997, Early Triassic conodont biofacies of Primorye: Mémoirie de Géologie (Lausanne), v. 30, p. 35–44.

Calvert, S.E., Thode, H.G., Yeung, D., and Carlin, R.E., 1996, A stable isotope study of pyrite formation in the Late Pleistocene and Holocene sediments of the Black Sea: Geochimica et Cosmochimica Acta, v. 60, p. 1261–1270.

Cassinis, G., Di Stefano, P., Massari, F., Neri, C., and Venturini, C., 2000, The Permian of South Europe and its interregional correlation, *in* Yin, H., et al., eds., Permian-Triassic evolution of Tethys and western circum-Pacific: Amsterdam, Elsevier, p. 37–70.

Catalano, R., Di Stefano, P., and Kozur, H., 1991 Permian circumpacific deep-waterfaunas from the western Tethys (Sicily, Italy): New evidences for the position of the Permian Tethys: Palaeogeography, Palaeoclimatology, Palaeoecology, v. 87, p. 75–108.

Chai, C., Zhou, Y., Mao, X., Ma, S., Ma, J., Kong, P., and He, J., 1992, Geochemical constraints on the Permo-Triassic boundary event in South China, *in* Yin, H., et al., eds., Permo-Triassic events in the eastern Tethys: Cambridge, Cambridge University Press, p. 158–168.

Embry, A.F., 1993, Global sequence boundaries of the Triassic and their identification in the Western Canadian Sedimentary Basin: Bulletin of Canadian Petroleum Geology, v. 45, p. 415–433.

Erwin, D.H., 1993, The great Paleozoic crisis: Life and death in the Permian: New York, Columbia University Press, 327 p.

Ezaki, Y., Kawamura, T., and Nakamura, K., 1994, Kapp Starostin Formation in Spitsbergen: A sedimentary and faunal record of Late Permian palaeoenvironments in an Arctic region: Canadian Society of Petroleum Geologists Memoir 17, p. 647–655.

Fan, J.-S., 1980, The main features of marine Triassic sedimentary facies in southern China: Rivista Italiana Paleontologia, v. 85, p. 1125–1146.

Gullo, M., and Kozur, H., 1993, First evidence of Scythian conodonts in Sicily: Neues Jahrbuch für Geologie und Paläontologie Monatsheft, v. 1993, p. 477–488.

Haas, J., Góczán, F., Oravecz-Sheffer, A., Barbas-Stuhl, A., Majoros, G., and Bérczi-Makk, A., 1988, Permian-Triassic boundary in Hungary: Memoirie della Società Geologica Italiana, v. 34, p. 221–241.

Hallam, A., 1989, The case for sea-level change as a dominant causal factor in mass extinction of marine invertebrates: Royal Society of London Philosophical Transactions, v. B325, p. 437–455.

Hallam, A., 1991, Why was there a delayed radiation after the end-Palaeozoic extinction?: Historical Biology, v. 5, p. 257–262.

Hallam, A., 1994, The earliest Triassic as an anoxic event, and its relationship to the end-Palaeozoic mass extinction: Canadian Society of Petroleum Geologists Memoir 17, p. 797–804.

Hallam, A., and Wignall, P.B., 1997, Mass extinctions and their aftermath: Oxford, Oxford University Press, 320 p.

Henderson, C.M., 1997, Uppermost Permian conodonts and the Permian-Triassic boundary in the Western Canadian Sedimentary Basin: Bulletin of Canadian Petroleum Geology, v. 45, p. 693–707.

Holser, W.T., and Magaritz, M., 1992, Cretaceous/Tertiary and Permian/Triassic boundary events compared: Geochimica et Cosmochimica Acta, v. 56, p. 3297–3309.

Holser, W.T., Schönlaub, H.-P., Boeckelmann, K., Magaritz, M., and Orth, C.J., 1991, The Permian-Triassic of the Gartnerkofel-1 Core (Carnic Alps, Austria): Synthesis and conclusions: Abhandlungen der Geologischen Bundesanstalt, Band 45, 213–232.

Hotinski, R.M., Kump, L.R., and Najjar, R.G., 2000, Opening Pandora's Box: The impact of open system modeling on interpretations of anoxia: Paleoceanography, v. 15, p. 267–279.

Hotinski, R.M., Bice, K.L., Kump, L.R., Najjar, R.G., and Arthur, M.A., 2001, Ocean stagnation and end-Permian anoxia: Geology, v. 29, p. 7–10.

Iranian-Japanese Research Group, 1981, The Permian and Lower Triassic systems in Abadeh Region, central Iran: Memoir of the Faculty of Science, Kyoto University, v. 47, p. 61–133.

Isozaki, Y., 1994, Superanoxia across the Permo-Triassic boundary: Recorded in accreted deep-sea pelagic chert in Japan: Canadian Society of Petroleum Geologists Memoir 17, p. 805–812.

Isozaki, Y., 1997, Permo-Triassic boundary superanoxia and stratified superocean: Records from lost deep sea: Science, v. 276, p. 235–238.

Isozaki, Y., Maruyama, S., and Furuoka, F., 1990, Accreted oceanic materials in Japan: Tectonophysics, v. 181, p. 179–205.

Isozaki, Y., and Maruyana, S., 1992, Deep-sea anoxia across the Permo-Triassic boundary: Records in accreted pelagic chert in Japan: Eos (Transactions, American Geophysical Union), v. 73, p. 272.

Jin, Y.G., Wang, Y., Wang, W., Shang, Q.H., Cao, C.Q., and Erwin, D.H., 2000, Patterns of marine mass extinction near the Permian-Triassic boundary in South China: Science, v. 289, p. 432–436.

Kajiwara, Y., Yamakita, S., Ishiga, K., Ishida, H., and Imai, A., 1994, Development of a largely anoxic stratified ocean and its temporary massive mixing at the Permian/Triassic boundary supported by the sulfur isotopic record: Palaeogeography, Palaeoclimatology, Palaeoecology, v. 111, p. 367–379.

Kapoor, H.M., and Tokuoka, T., 1985, Sedimentary facies of the Permian and Triassic of the Himalaya, *in* Nakazawa, K.D., and Dickins, J.M., eds., The Tethys: Tokyo, Tokai University Press, p. 23–58.

Kempé, S., and Kazmierczak, J., 1994, The role of alkalinity in the evolution of ocean chemistry, organization of living systems, and bio-calcification processes: Bulletin de l'Institut océanographique, Monaco, Special No. 13, p. 61–116.

Kershaw, S., Zhang, T., and Lan, G., 1999, A ?microbial carbonate crust at the Permian-Triassic boundary in South China, and its palaeoenvironmental significance: Palaeogography, Palaeoclimatology, Palaeoecology, v. 146, p. 1–18.

Khanchuk, A.I., and Panchenko, I.V., 1997, Permian and Triassic rocks in terranes of the southern far east Russia: Mémoirie de Geologie (Lausanne), v. 30, p. 1–4.

Kirchner, J.W., and Weil, A., 2000, Delayed biological recovery from extinctions throughout the fossil record: Nature, v. 404, p. 177–180.

Knoll, A.H., Bambach, R.K., Canfield, D.E., and Grotzinger, J.P., 1996, Comparative Earth history and Late Permian mass extinction: Science, v. 273, p. 452–457.

Kobayashi, F., 1999, Tethyan uppermost Permian (Dzhulfian and Dorashamian) foraminiferal faunas and their palaeogeographic and tectonic significance: Palaeogeography, Palaeoclimatology, Palaeoecology, v. 150, p. 279–308.

Koike, T., 1996, The first occurrence of Griesbachian conodonts in Japan: Transactions and Proceedings of the Palaeontological Society of Japan, New Series, no. 181, p. 337–346.

Kozur, H., 1991, Permian deep-water ostracods from Sicily (Italy). 2. Biofacial evolution and remarks to the Silurian to Triassic paleopsychrospheric ostracods: Geologisch-Paläontologische Mitteilungen Innsbruck, v. 3, p. 25–38.

Kozur, H., 1994, The correlation of the Zechstein with the marine standard: Jahrbuch der Geologischen Bundesanstalt, v. 137, p. 85–103.

Kozur, H., Leven, E.Y., Lozovskiy, V.R., and Pyatakova, M.V., 1980, Subdivisions of Permian-Triassic boundary beds in Transcaucasia on the basis of conodonts: International Geological Review, v. 22, p. 361–368.

Krull, E.S., and Retallack, G.J., 2000, ^{13}C depth profiles from paleosols across the Permian-Triassic boundary: Evidence for methane release: Geological Society of America Bulletin, v. 112, p. 1459–1472.

Kubo, K., Isozaki, Y., and Matsuo, M., 1996, Colour of bedded chert and redox condition of depositional environments: ^{57}Fe Mössbauer spectroscopic study on chemical state of iron in Triassic deep-sea pelagic chert: Journal of the Geological Society of Japan, v. 102, p. 40–48.

Kummel, B., 1972, The Lower Triassic (Scythian) ammonoid *Otoceras*: Bulletin of the Museum of Comparative Zoology, v. 143, p. 365–417.

Langmuir, D., 1978, Uranium solution-mineral equilibria at low temperatures with applications to sedimentary ore deposits: Geochimica et Cosmochimica Acta, v. 42, p. 547–569.

Lehrmann, D.J., 1999, Early Triassic calcimicrobial mounds and biostromes of the Nanpanjiang basin, South China: Geology, v. 27, p. 359–362.

Malkowski, K., M. Gruszczynski, M., Hoffman, A., and Halas, A., 1989, Oceanic stable isotope composition and a scenario for the Permo-Triassic crisis: Historical Biology, v. 2, p. 289–309.

Marcoux, J., and Baud, A., 1986, The Permo-Triassic boundary in the Antalya Nappes (western Taurides, Turkey): Memoirie de la Società Geologica Italiana, v. 34, p. 243–252.

Merle, A., and Fournier, E., 1910, Sur le Trias marin du Nord de Madagascar: Bulletin de la Société Géologique France, serie 4, v. 10, p. 660–664.

Mørk, A., Knarud, R., and Worsley, D., 1982, Depositional and diagenetic environments of the Triassic and Lower Jurassic succession of Svalbard, *in* Embry, A.F., and Balkwill, H.R., eds., Arctic geology and geophysics: Canadian Sociey of Petroluem Geologists Memoir 8, p. 371–398.

Musashino, M., 1993, Chemical composition of the "Toishi-type" siliceous shale, Part 1: Bulletin of the Geological Survey of Japan, v. 44, p. 699–705.

Myers, K.J. and Wignall, P.B., 1987, Understanding Jurassic organic-rich mudrocks—New concepts using gamma-ray spectrometry and palaeoecology: Examples from the Kimmeridge Clay of Dorset and the Jet Rock of Yorkshire, *in* Leggett, J.K., and Zuffa, G.G., eds., Marine clastic sedimentology: London, Graham and Trotman, p. 172–189.

Nakazawa, K., Kapoor, H.M., Ishii, K., Bando, Y., Okimura, Y., and Tokuoka, T., 1975, The Upper Permian and Lower Triassic in Kashmir, India: Memoir of the Faculty of Science, Kyoto University, Geology and Mineralogy Series, v. 42, p. 1–106.

Nielsen, E., 1936, Some few preliminary remarks on Triassic fishes from East Greenland: Meddeleser om Grønland, v. 112, p. 1–55.

Perch-Nielsen, K., Bromley, R.G., Birkenmajer, K., and Aellen, M., 1972, Field observations in Palaeozoic sediments of Scoresby Land and northern Jameson Land: Rapport Grønlands Geoligiske Undersøgelse, v. 48, p. 39–59.

Rabu, D., Le Metour, J., Bechennec, F., Beurrier, M., Villey, M., and Boudillon-Jeudy de Grissac, C., 1990, Sedimentary aspects of the Eo-Alpine cycle on the northeast edge of the Arabian Platform (Oman Mountains), *in* Robertson, A.H.F., Searle, M.P., and Ries, A.C., eds., The geology and tectonics of the Oman region: Geological Society [London] Special Publication 49, p. 49–68.

Raiswell, R., and Berner, R.A., 1985, Pyrite formation in euxinic and semi-euxinic sediments: American Journal of Science, v. 285, p. 710–724.

Raiswell, R., Newton, R., and Wignall, P.B., 2001, A water column anoxicity indicator: Resolution of biofacies variations in the Kimmeridge Clay (Upper Jurassic, UK): Journal of Sedimentary Research, v. 71A, p. 286–294.

Rampino, M.R., and Adler, A.C., 1998, Evidence of abrupt latest Permian mass extinction of foraminifera: Results of tests for the Signor-Lipps effect: Geology, v. 26, p. 415–418.

Reinhardt, J.W., 1988, Uppermost Permian reefs and Permo-Triassic sedimentary facies from the southeastern margin of Sichuan Basin, China: Facies, v. 18, p. 231–288.

Retallack, G.J., 1999, Postapocalyptic greenhouse paleoclimate revealed by earliest Triassic paleosols in the Sydney Basin, Australia: Geological Society of America Bulletin, v. 111, p. 52–70.

Riding, R., 1993, Phanerozoic patterns of marine $CaCO_3$ precipitation: Naturwissenschaften, v. 80, p. 513–516.

Riding, R., 2000, Microbial carbonates: The geological record of calcified bacterial-algal mats and biofilms: Sedimentology, v. 47 (supplement), p. 179–214.

Sano, H., and Nakashima, K., 1997, Lowermost Triassic (Griesbachian) microbial bindstone-cementstone facies, southwest Japan: Facies, v. 36, p. 1–24.

Savrda, C.E., and Bottjer, D.J., 1986, Trace fossil model for reconstruction of paleo-oxygenation in bottom waters: Geology, v. 14, p. 3–6.

Schubert, J.K., and Bottjer, D.J., 1992, Early Triassic stromatolites as post-mass extinction disaster form: Geology, v. 20, p. 883–886.

Schubert, J.K., and Bottjer, D.J., 1995, Aftermath of the Permian-Triassic mass extinction event: Paleoecology of Lower Triassic carbonates in the western USA: Palaeogeography, Palaeoclimatology, Palaeoecology, v. 116, p. 1–40.

Seidler, L., 2000, Incised submarine canyons governing new evidence of Early Triassic rifting in East Greenland: Palaeogeography, Palaeoclimatology, Palaeoecology, v. 161, p. 267–293.

Şengör, A.M.C., 1984, The Cimmeride orogenic system and the tectonics of Eurasia: Geological Society of America Special Paper 195, p. 1–82.

Şengör, A.M.C., Altiner, D., Cin, A., Ustaömer, T., and Hsü, K.J., 1988, Origin and assembly of the Tethyside orogenic collapse at the expense of Gondwana Land, *in* Audley-Charles, M.G., and Halllam, A., eds., Gondwana and Tethys: Geological Society [London] Special Publication 37, p. 119–181.

Suzuki, N., Ishida, K., Shinomiya, Y., and Ishiga, H., 1998, High productivity in the earliest Triassic ocean: Black shales, southwest Japan: Palaeogeography, Palaeoclimatology, Palaeoecology, v. 141, p. 53–65.

Sweet, W.C., 1992, A conodont-based high-resolution biostratigraphy for the Permo-Triassic boundary interval, *in* Yin, H.-F., et al., eds., Permo-Triassic events in the eastern Tethys: Cambridge, Cambridge University Press, p. 120–133.

Szabo, F., and Kheradipur, A., 1978, Permian and Triassic stratigraphy of Zagros Basin, south-west Iran: Journal of Petroleum Geology, v. 1, p. 57–82.

Teichert, C., Kummel, B., and Sweet, W., 1973, Permian-Triassic strata, Kuh-e-Ali Bashi, northwestern Iran: Bulletin of the Museum of Comparative Zoology, v. 145, p. 359–472.

Tosk, T.A., and Anderson, K.A., 1988, Late Early Triassic foraminfers from possible dysaerobic to anaerobic paleoenvironments of the Thaynes Formation, Southeast Idaho: Journal of Foraminiferan Research, v. 18, p. 286–301.

Treat, V.-C., 1933, Paléontologie de Madagascar 19, le Permo-Trias Marin: Annales de Paléontologie, v. 22, p. 39–95.

Twitchett, R.J., 1999, Palaeoenvironments and faunal recovery after the end-Permian mass extinction: Palaeogeography, Palaeoclimatology, Palaeoecology, v. 154, p. 27–37.

Twitchett, R.J., 2001, Incompleteness of the Permian-Triassic fossil record: A consequence of productivity decline? Geological Journal, v. 36, p. 341–353.

Twitchett, R.J., and Wignall, P.B., 1996, Trace fossils and the aftermath of the Permo-Triassic mass extinction: Evidence from northern Italy: Palaeogeography, Palaeoclimatology, Palaeoecology, v. 124, p. 137–151.

Twitchett, R.J., Looy, C.V., Morante, R., Visscher, H., and Wignall, P.B., 2001, Rapid and synchronous collapse of marine and terrestrial ecosystems during the end-Permian mass extinction event: Geology, v. 29, p. 351–354.

Tyson, R.V., 1995, Sedimentary organic matter: London, Chapman and Hall, 615 p.

Van Cappellen, P., and Ingall, E.D., 1994, Benthic phosphorus regeneration, net primary production, and oceanic anoxia: A model of the coupled marine biogeochemical cycles of carbon and phosphorus: Paleoceanography, v. 9, p. 677–692.

Wang, K., Geldsetzer, H.H.J., and Krouse, H.R., 1994, Permian-Triassic extinction: Organic $\delta^{13}C$ evidence from British Columbia, Canada: Geology, v. 22, p. 580–584.

Watts, K.F., 1988, Triassic carbonate submarine fans along the Arabian platform margin, Sumeini Group, Oman: Sedimentology, v. 35, p. 43–71.

Watts, K.F., and Garrison, R.E., 1986, Sumeini Group, Oman: Evolution of a Mesozoic carbonate slope on a south Tethyan continental margin: Sedimentary Geology, v. 48, p. 107–168.

Wescott, W.A., 1988, A late Permian fan-delta system in the southern Morondava Basin, Madagascar, *in* Nemec, W., and Steel, R.J., eds., Fan deltas: Sedimentology and tectonic settings: London, Blackie and Son, p. 226–238.

Wignall, P.B., 1990, Observations on the evolution and classification of dysaerobic communities, *in* Miller, W., ed., Paleocommunity temporal dynamics: The long-term development of mutlispecies assemblies: Paleontological Society Special Publication 5, p. 99–111.

Wignall, P.B., 1994, Black shales: Oxford, Oxford University Press, 130 p.

Wignall, P.B., and Hallam, A., 1992, Anoxia as a cause of the Permian/Triassic extinction: Facies evidence from northern Italy and the western United States: Palaeogeography, Palaeoclimatology, Palaeoecology, v. 93, p. 21–46.

Wignall, P.B., and Hallam, A., 1993, Griesbachian (earliest Triassic) palaeoenvironmental changes in the Salt Range, Pakistan and southwest China and their bearing on the Permo-Triassic mass extinction: Palaeogeography, Palaeoclimatology, Palaeoecology, v. 102, p. 215–237.

Wignall, P.B., and Hallam, A., 1996, Facies change and the end-Permian mass extinction in S.E. Sichuan, China: Palaios, v. 11, p. 587–596.

Wignall, P.B., Hallam, A., Lai, X., and Yang, F., 1995, Palaeoenvironmental changes across the Permian/Triassic boundary at Shangsi (N. Sichuan, China): Historical Biology, v. 10, p. 175–189.

Wignall, P.B., Kozur, H., and Hallam, A., 1996, The timing of palaeoenvironmental changes at the Permian/Triassic (P/Tr) boundary using conodont biostratigraphy: Historical Biology, v. 12, p. 39–62.

Wignall, P.B., Morante, R., and Newton, R., 1998, The Permo-Triassic transition in Spitsbergen: $\delta^{13}C_{org.}$ chemostratigraphy, Fe and S geochemistry, facies, fauna and trace fossils: Geological Magazine, v. 135, p. 47–62.

Wignall, P.B., and Newton, R., 1998, Pyrite framboid diameter as a measure of oxygen deficiency in ancient mudrocks: American Journal of Science, v. 298, p. 537–552.

Wignall, P.B., and Twitchett, R.J., 1996, Oceanic anoxia and the end Permian mass extinction: Science, v. 272, p. 1155–1158.

Wignall, P.B., and Twitchett, R.J., 1999, Unusual intraclastic limestones in Lower Triassic carbonates and their bearing on the aftermath of the end Permian mass extinction: Sedimentology, v. 46, p. 303–316.

Wilkin, R.T., Barnes, H.L., and Brantley, S.L., 1996, The size distribution of framboidal pyrite in modern sediments: An indicator of redox conditions: Geochimica at Cosmochimica Acta, v. 60, p. 3897–3912.

Wolbach, W.S., Roegge, D.R., and Gilmour, I., 1994, The Permian-Triassic of the Gartnerkofel-1 core (Carnic Alps, Austria): Organic carbon isotope variation [Abstract]: *in* New developments regarding the K/T event and other catastrophes in Earth history: Houston, Texas, Lunar and Planetary Institute, LPI Contribution 825, p. 133–134.

Woods, A.D., and Bottjer, D.J., 2000, Distribution of ammonoids in the Lower Triassic Union Wash Formation (Eastern California): Evidence for paleoceanographic conditions during recovery from the end-Permian mass extinction: Palaios, v. 15, p. 535–545.

Woods, A.D., Bottjer, D.J., Mutti, M., and Morrison, J., 1999, Lower Triassic large sea-floor carbonate cements: Their origin and a mechanism for the prolonged biotic recovery from the end-Permian mass extinction: Geology, v. 27, p. 645–648.

Yamakita, S., Kadota, N., Kato, T., Tada, R., Ogihara, S., Tajika, E., and Hamada, Y., 1999, Confirmation of the Permian/Triassic boundary in deep-sea sedimentary rocks; earliest Triassic conodonts from black carbonaceous claystone of the Ubara section in the Tamba Belt, Southwest Japan: Journal of the Geological Society of Japan, v. 105, p. 895–898.

Yin, H.-F., Huang, S., Zhang, K., Hansen, H.J., Yang, F.Q., Ding, M., and Bie, X., 1992, The effects of volcanism on the Permo-Triassic mass extinction in South China, *in* Sweet, W.C., et al., eds., Permo-Triassic events in the eastern Tethys: Cambridge, Cambridge University Press, p. 146–157.

Zakharov, Y.D., 1992, The Permo-Triassic boundary in the southern and eastern USSR and its international correlation, *in* Sweet, W.C., et al., eds., Permo-Triassic events in the eastern Tethys: Cambridge, Cambridge University Press, p. 46–55.

Zakharov, Y.D., Kotlyar, G.V., and Oleinikov, A.V., 1995, Late Dorashamian (Late Changxingian) invertebrates of the Far East: Geology of the Pacific Ocean, v. 12, p. 216–229.

Manuscript Submitted October 10, 2000; Accepted by the Society March 22, 2001

Geological Society of America
Special Paper 356
2002

Abruptness of the end-Permian mass extinction as determined from biostratigraphic and cyclostratigraphic analyses of European western Tethyan sections

Michael R. Rampino
Earth and Environmental Science Program, New York University, 100 Washington Square East, New York, New York 10003, USA,
and NASA, Goddard Institute for Space Studies, 2880 Broadway, New York, New York 10025, USA
Andreas Prokoph
Department of Earth Sciences, University of Ottawa, 365 Nicholas Street, Ottawa, Ontario K1N 6N5, Canada
Andre C. Adler
Department of Physics, New York University, 2-4 Washington Place, New York, New York 10003, USA
Dylan M. Schwindt
Earth and Environmental Science Program, New York University, 100 Washington Square East, New York, New York 10003, USA

ABSTRACT

The Permian-Triassic (P-Tr) boundary is marked by the most severe mass extinction in the geologic record, but the time interval over which the extinctions occurred has been a subject of much debate. Published biostratigraphic data from P-Tr boundary sections in the southern Alps, Italy, were used to test the effects of sampling and species abundance on the record of the latest Permian extinction. The results for foraminiferal taxa match simulations for abrupt extinctions; the extinction level is close to the base of the Tesero horizon of the Werfen Formation. Using biostratigraphic estimates of sedimentation rates, we constrain the extinction interval to <10 k.y.

High-resolution cyclostratigraphy on a 10^4 yr scale across the P-Tr boundary in a core from the Carnic Alps (Austria) shows significant cycles in the ratio ~40:10:4.7:2.3 m, identified with Milankovitch cycles of ~412:100:40:20 k.y. (eccentricity 1 and 2, obliquity, and precession). Cycle analysis indicates continuity of deposition across the P-Tr boundary, and an average accumulation rate of ~10 cm/k.y. The results define the dramatic faunal shift across the boundary within an interval of <8 k.y., and the accompanying sharp negative global carbon isotope shift within <30 k.y., suggesting a catastrophic cause.

INTRODUCTION

The latest Permian (251 Ma) mass extinction is the most severe in the geologic record (Raup, 1979; Erwin, 1993; Stanley and Yang, 1994). Recent studies suggest that the marine and terrestrial extinctions were more abrupt than formerly thought (Bowring et al., 1998; Retallack, 1995; Rampino and Adler, 1998), and precise radiometric dating has bracketed the marine

Rampino, M.R., Prokoph, A., Adler, A.C., and Schwindt, D.M., 2002, Abruptness of the end-Permian mass extinction as determined from biostratigraphic and cyclostratigraphic analyses of European western Tethyan sections, *in* Koeberl, C., and MacLeod, K.G., eds., Catastrophic Events and Mass Extinctions: Impacts and Beyond: Boulder, Colorado, Geological Society of America Special Paper 356, p. 415–427.

extinctions and associated negative carbon isotope anomaly within <1 m.y. (Bowring et al., 1998). Many paleontological studies, however, have concluded that the disappearance of Late Permian species was gradual (see Erwin, 1993, for a review). A basic problem in using biostratigraphic last occurrences to infer patterns of extinction is that last occurrences almost always underestimate the time of extinction. Signor and Lipps (1982) showed that a random distribution of errors at the end points of biostratigraphic ranges could produce range truncations and apparent gradual decline preceding a sudden extinction boundary. Several studies have explored the ramifications of this so-called Signor-Lipps effect (Raup, 1989; Strauss and Sadler, 1989; Springer, 1990; Meldahl, 1990; Marshall, 1990, 1994).

Meldahl (1990) simulated abrupt, gradual, and stepwise modes of extinction using modern biostratigraphic data, and Marshall (1990) used statistical methods to determine the uncertainties in the ends of stratigraphic ranges of taxa (50% confidence intervals), and used these data to estimate the most likely position for a predicted abrupt extinction of the taxa involved. In order to apply these statistical methods to the Permian-Triassic (P-Tr) extinction boundary, we utilized published data on occurrences of Late Permian fusulinid and other foraminifera from several fossiliferous marine sections across the well-studied P-Tr boundary in the southern Alps. In order to quantify the time scale of the extinctions, we utilized high-resolution cyclostratigraphy at the 10^4 yr scale across the P-Tr boundary in similar strata in the Gartnerkofel-1 core (Carnic Alps, Austria). We applied spectral and wavelet time-series analysis and detected Milankovitch orbital cycles in several variables in the core. The cycles were used to estimate sediment accumulation rates, and hence the time scale, and to measure completeness across the P-Tr boundary. This chapter represents a combination of two studies (Rampino and Adler, 1998; Rampino et al., 2000) and extends those results to present a coherent picture of the extent and tempo of the P-Tr mass extinction.

PERMIAN-TRIASSIC SECTIONS IN THE SOUTHERN ALPS, ITALY

In the southern Alps of Italy, the P-Tr boundary interval is marked by the transition from the Upper Permian Bellerophon Formation to the Lower Triassic Werfen Formation (Fig. 1). The Bellerophon Formation is composed of alternating limestones, dolomites, and evaporitic facies of shallow-water origin. The distribution of fossils in the Bellerophon is controlled by facies; the richest fossil assemblages occur episodically in open lagoonal facies and shallow-marine dark limestones. Locally, the topmost section of the Bellerophon Formation consists of a sequence (0.5–2 m thick) of dark marls and limestones containing abundant fusulinids and other foraminifers, calcareous algae, and brachiopods. The Bellerophon Formation is overlain by the thin (~6 m in the study area) oolitic Tesero horizon of the Werfen Formation. There is no evidence of a significant stratigraphic gap, although the Bellerophon-Tesero boundary is apparently time transgressive in the area (Buggisch and Noe, 1986; Broglio-Loriga et al., 1986; Broglio-Loriga and Cassinis, 1992.)

The regional P-Tr boundary is placed by many workers near the base or within the lower 1–2 m of the Tesero horizon, with the last occurrences of typical Late Permian foraminifera and other characteristic Late Permian invertebrates, and the appearance of taxa considered representative of the earliest Triassic *Otoceras* zone (Broglio-Loriga et al., 1986; Sweet et al., 1992; Broglio-Loriga and Cassinis, 1992). However, the recognition of a global stratigraphic marker and criteria for associated boundary placement at the base of the Triassic are still discussed. The Mazzin Member of the Werfen Formation, a sequence of alternating marls and calcarenites that have structures suggesting a near wave-base origin, overlies the Tesero horizon (Broglio-Loriga and Cassinis, 1992).

Published data from systematic sampling of four sections, the Sass de Putia, Tesero, Monte Ruche, and Dierico, Italy, sections (Fig. 1, A–D) (Broglio-Loriga et al., 1986; Buggisch and Noe, 1986; Broglio-Loriga and Cassinis, 1992), including vertical ranges of fusulinid and other foraminifers, were used in our tests of the temporal aspects of the end-Permian extinctions. We chose to focus on foraminifera to avoid problems of sample size as much as possible (Koch and Morgan, 1988). In all cases, we used a vertical scale in meters below the local datum, chosen for convenience at the base of the Tesero horizon.

RESULTS OF STRATIGRAPHIC RANGE ANALYSIS

Using data from modern molluscan fauna collected in cores from tidal flat areas, Meldahl (1990) found that when the frequency of last occurrences is plotted versus the sampling depth sudden extinction produces a tailed distribution, the mode of last occurrences corresponding closely to the extinction boundary. Stepwise extinction simulations produce a distribution with distinct modes at each of the simulated steps, and simulations of gradual extinctions give a broad convex distribution without distinct modes (Meldahl, 1990).

Last occurrences of species versus their stratigraphic abundance (defined as the percent of stratigraphic sample intervals in which the species occurs) for simulated sudden extinctions show a hollow distribution curve. Extinction is typically not accurately recorded for species having <15% stratigraphic abundance (i.e., occurring in <15% of the sampled intervals) (Meldahl, 1990). Simulated stepped extinction events showed a series of hollow curves that marked the extinction pulses in the stepped model. The curves for simulated gradual extinctions display a distinct triangular distribution. Simulations of sudden extinctions showed an accelerating decline in the number of species, whereas gradual extinction showed a contrasting pattern of constant decline. The simulation of stepped extinctions displayed an intermediate pattern of stepped decline.

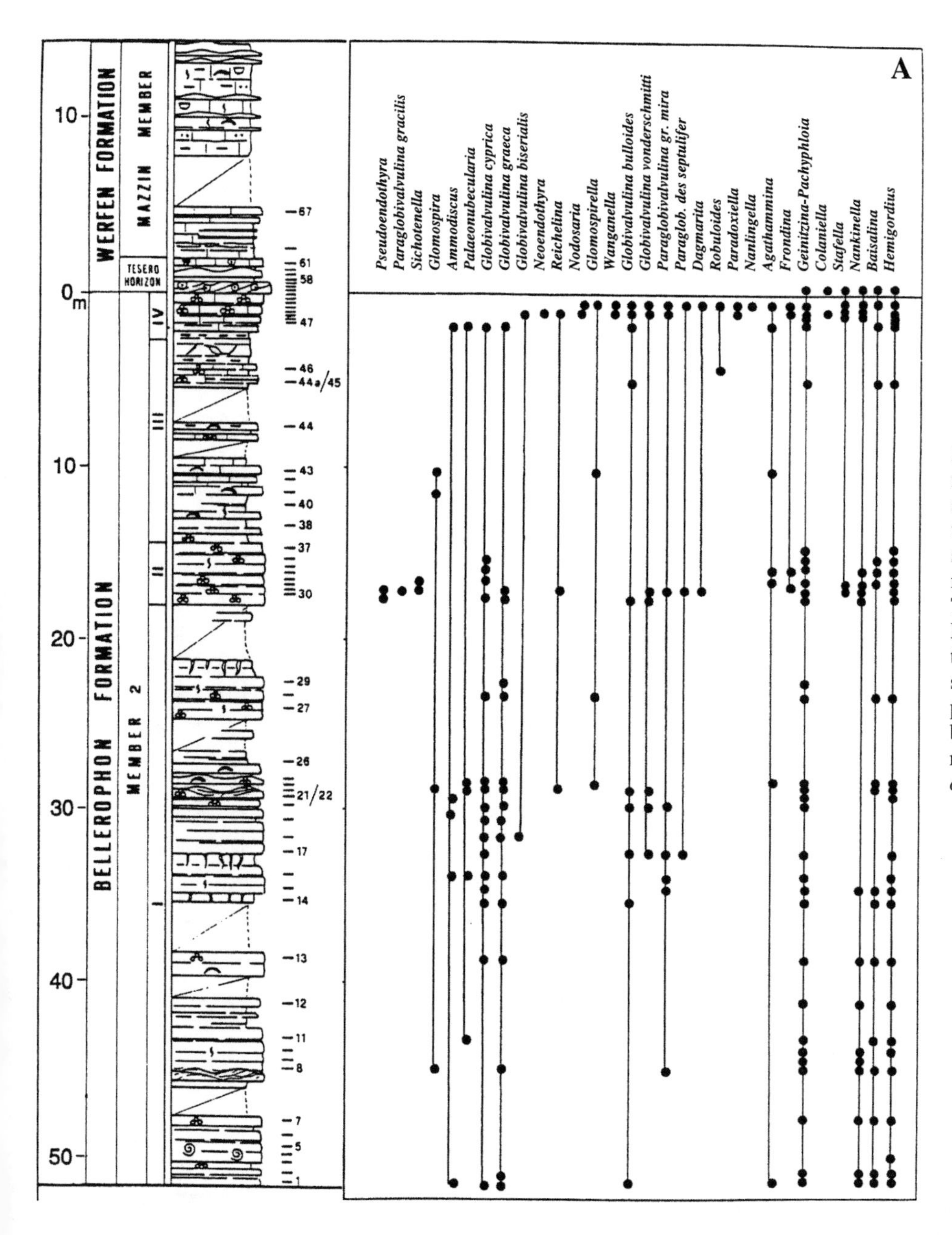

Figure 1. A: Stratigraphic ranges for foraminifera taxa in Sass de Putia, Italy, section (data, sample numbers, and lithology as in Broglio-Loriga and Cassinis, 1992). Vertical scale is in meters relative to datum chosen at base of Tesero horizon as defined in Broglio-Loriga and Cassinis (1992). Ranges for the first 11 taxa in other sections studied show that they actually did not disappear prior to the Permian-Triassic (P-Tr) boundary. In all diagrams, fossil occurrences are indicated by dots. (*Figure 1 continues on following pages.*)

The data from the Sass de Putia section consist of vertical ranges of 30 foraminifera taxa as much as 52 m below the Tesero horizon datum (Fig. 1A); the average sample spacing is ~1.2 m. Systematic sampling shows that foraminifera-rich facies occur episodically. In this section, no Permian foraminifera were found in samples taken ≥2 m above the base of the Tesero horizon (Broglio-Loriga and Cassinis, 1992). Following Meldahl (1990), we plotted the following parameters versus stratigraphic position in meters below the Tesero horizon datum for the Sass de Putia section: (1) the number of last occurrences (Fig. 2A), and (2) the last occurrence versus stratigraphic abundance (stratigraphic abundance was determined as the percent of the fossiliferous [nonbarren] samples that contained the taxon in question) (Fig. 2B).

The number of last occurrences versus depth for the Sass de Putia section (Fig. 2A) shows the typical shape produced in simulations of sudden extinctions, demonstrated to be significantly different from both stepwise and gradual extinction distributions (a significance level <0.01 in Kolmogorov-Smirnov two-sample comparisons; Meldahl, 1990). The curve of last occurrence versus stratigraphic abundance (Fig. 2B) displays the hollow shape typical of sudden extinctions. We find that only taxa having <15% stratigraphic abundance tend to have last occurrences well below the boundary datum, similar to the results of Meldahl (1990) for a sudden extinction simulation. We also note that the taxa that have last occurrences below the base of the Tesero horizon last occur in the other Italian sections (Fig. 1, B–D) at or very close to the Tesero horizon datum. The

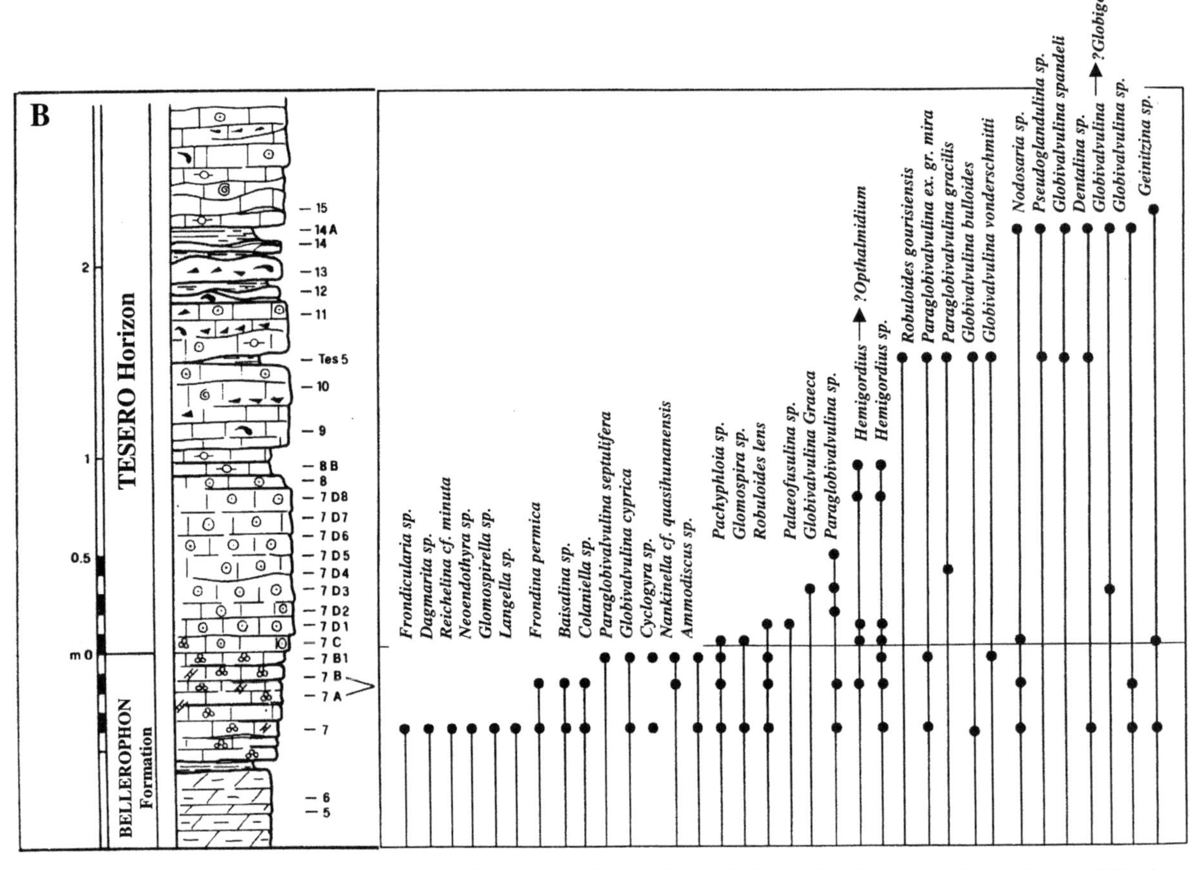

Figure 1. B: Terminations of ranges of Late Permian foraminifera in Tesero, Italy, section (data, sample numbers and lithology as in Broglio-Loriga and Cassinis, 1992).

total number of taxa recorded (diversity) versus depth over the 50 m section shows an irregular pattern that reflects episodic foraminifera-rich zones.

We used the method of Marshall (1990) to determine the 50% confidence intervals for uncertainty of the ends of stratigraphic ranges of fauna from the Sass de Putia section. We considered, but did not utilize, Marshall's (1994) method because of the relatively sparse data. Of the 30 taxa on which we had range data, however, only 18 had more than 3 occurrences and only 6 taxa had ranges determined by 15 or more occurrences, which defined error bars of 2.5 m or less (Table 1). All of these taxa last appeared between the base of the Tesero horizon and +2.4 m above it. Marshall's (1990) statistical analysis indicates that the best estimate of a putative abrupt extinction is determined by the median of the 50% confidence intervals of the taxa involved. Considering the calculated uncertainties in the upper range limit for the six taxa utilized, the best fit position for an abrupt extinction would be +3.2 m above the base of the Tesero horizon as defined in Figure 1A.

The details of the uppermost Permian biostratigraphy can be extracted from the Tesero, Monte Ruche, and Dierico sections (Fig. 1, B–D), which provide data on ranges of foraminiferal taxa within the foraminifera-rich uppermost 2–6 m of

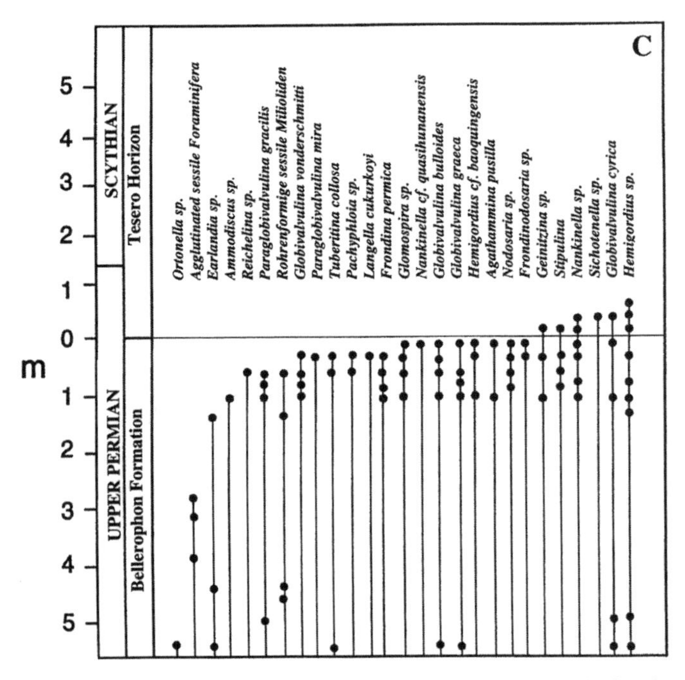

Figure 1. C: Terminations of ranges of Late Permian foraminifera in Monte Ruche, Italy, section (data from Broglio-Loriga and Cassinis, 1992).

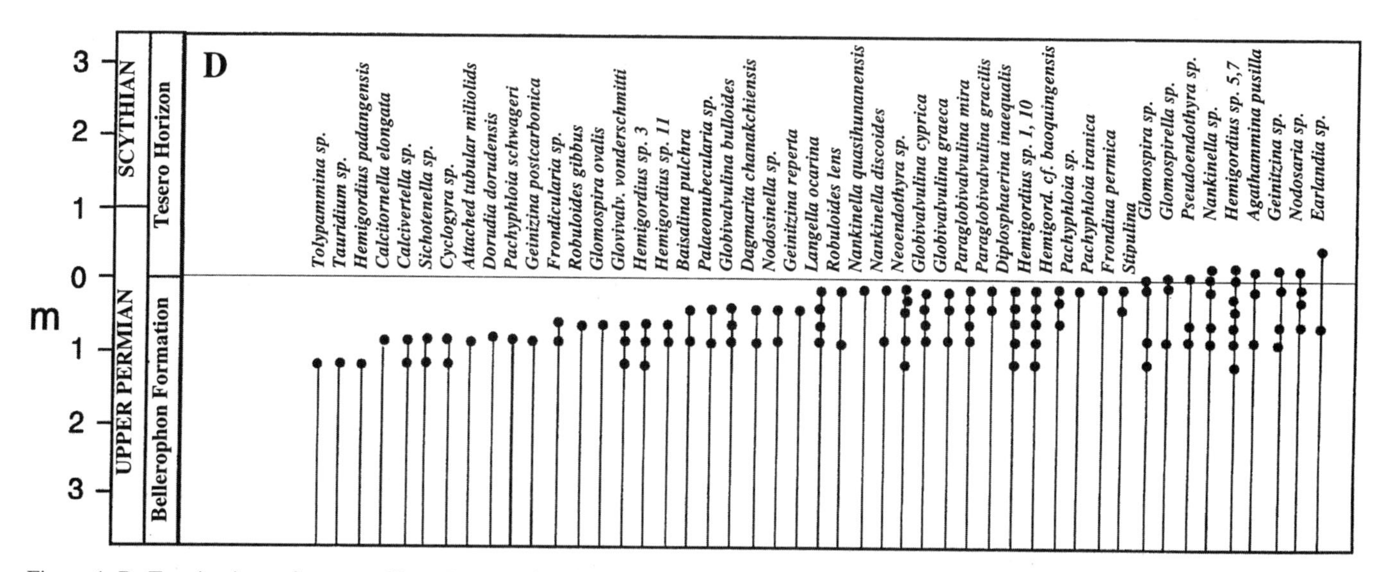

Figure 1. D: Terminations of ranges of Late Permian foraminifera from Dierico, Italy, section (data from Buggisch and Noe, 1986).

the Bellerophon Formation and within the lower Tesero horizon. Sampling intervals average from ~0.14 to 0.4 m. The total recorded number of species declines as the Tesero horizon is approached (e.g., in the Monte Ruche section the number of species drops from 32 to 22 within the meter below the Tesero horizon, and in the Dierico section the number drops from 15 to 10 species in the same interval), but the decline is episodic. In these sections, most Late Permian foraminifera last occur within the uppermost 0.2–0.6 m of the Bellerophon Formation to the lowermost 0.1–0.2 m of the Tesero horizon.

In the Tesero, Italy, section (Fig. 1B), 20 of 34 foraminifera taxa (60%) last occur within the 1 m of strata bracketing the base of the Tesero horizon, and the occurrences of Late Permian foraminifera above the lower 0.2 m of the Tesero horizon are very spotty (only 40% of the samples yielded foraminifera, and no sample had more than 8 taxa). In the Monte Ruche, Italy, section (Fig. 1C), 42 of 48 recognized taxa (88%) last occur within the last 1 m of the Bellerophon Formation, and only 5 taxa are found in the basal sample of the Tesero horizon. Only one foraminifera taxon (*Earlandia sp.*), which is known from sections elsewhere to have survived the P-Tr extinction, is found in a single sample within the lower 1 m of the Tesero horizon. In the Dierico, Italy, section (Fig. 1D), only 6 taxa of the 26 recorded (20%) persist into the Tesero horizon, and none occur more than ~1 m above the base of the unit.

The question of reworking of the typical Late Permian foraminifera that occur within the Tesero horizon in the southern Alps sections (Buggisch and Noe, 1986; Broglio-Loriga and Cassinis, 1992) can be addressed by comparing the taxa that were found in each of the four sections considered here. Only members of the genus *Hemigordius* are found ≥1 m above the base of the Tesero horizon in more than one section (Tesero and Dierico), although they last occur near the base of the Tesero horizon in the other sections studied. Late Permian taxa found within the lower 1 m of the Tesero horizon in some outcrops do not persist above the basal part of the horizon in other sections (Broglio-Loriga et al., 1986; Buggisch and Noe, 1986; Broglio-Loriga and Cassinis, 1992). Furthermore, the Permian fusulinids and other taxa found within the Tesero horizon are reported to be fragmented, abraded, and thinly coated with carbonate, and ooid nuclei in the Tesero horizon commonly contain small foraminifers (Cirilli et al., 1998). Permian-type palynomorphs found mixed with Triassic forms in the lower Tesero horizon are reported to be dark and abraded as a result of mechanical damage, indicating reworking from older deposits (Cirilli et al., 1998).

The Tesero horizon represents a transgressive deposit accompanied by an abrupt increase in hydrodynamic energy in which physical mixing of sediments is important (Cirilli et al., 1998). Such deposits, by their very nature, must involve erosion of preexisting sediments and incorporation of older sediments and fossils, and thus we would expect a mixed fauna in these high-energy deposits. In shallow-water marine sediments, a combination of relatively rapid rates of sedimentation (~3–10 cm/k.y.) and biodiffusion (~10 cm^2/yr) (Wheatcroft, 1990; Boudreau, 1994) might mix Late Permian taxa to ~1 m above a level representing their sudden truncation. These considerations suggest that the Late Permian taxa found episodically above the lowermost part of the Tesero horizon were most likely reworked, although they might represent short-term survivors of an event that apparently decimated the foraminifer population (Broglio-Loriga and Cassinis, 1992).

BIOSTRATIGRAPHIC EVIDENCE FOR SUDDEN EXTINCTION AT THE P-TR BOUNDARY

Our results suggest that the pattern of last occurrences of Late Permian foraminifera in the Alpine region is best explained by a sudden extinction event close to the transition between the Bellerophon Formation and the overlying Tesero

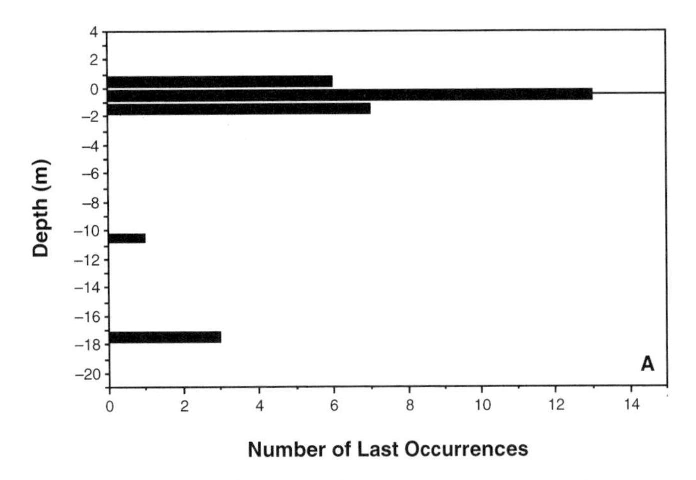

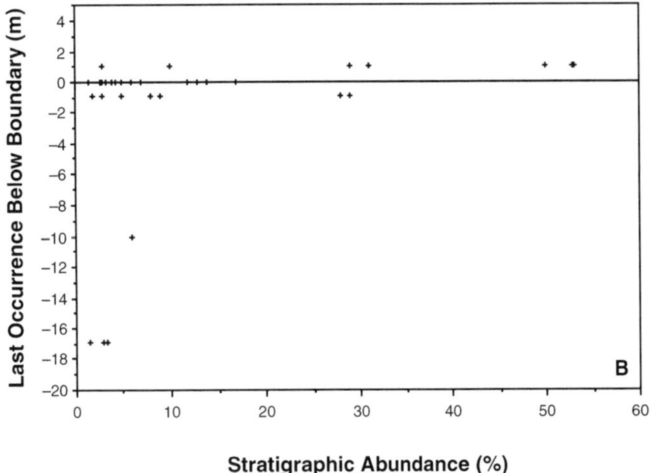

Figure 2. Results for data from Sass de Putia, Italy, section. A: Number of last occurrences of foraminifera taxa vs. depth. B: Last occurrence of foraminifera taxa vs. stratigraphic abundance. Datum is base of Tesero horizon.

horizon. We note that other components of the Late Permian benthic fauna (e.g., brachiopods) show a similar pattern of last occurrences close to that datum (Broglio-Loriga and Cassinis, 1992). Shallow-water carbonate sedimentary strata such as those of the upper Bellerophon Formation are known to have sedimentation rates in the range of ~30–100 m/m.y. (Perrodon and Masse, 1984). In the closely sampled Monte Ruche, Italy, section (Fig. 1C), 88% of Late Permian foraminifera taxa listed disappear in the last 1 m of the Bellerophon Formation, or within ≤30 k.y., based on the estimates of sedimentation rates. Furthermore, mixing could smear an instantaneous extinction over such an interval.

The oolitic facies of the Tesero horizon has been interpreted as part of a very rapid transgression (Wignall and Hallam, 1992; Cirilli et al., 1998) produced by a relative sea-level rise, and no significant break is recognized between the Bellerophon Formation and Tesero horizon (Broglio-Loriga and Cassinis, 1992; Cirilli et al., 1998). The presence of punctuated aggradational cycles forming ~2-m-thick parasequences across the P-Tr boundary (Wignall and Hallam, 1992) suggests control of sedimentation by Milankovitch-band cyclicities (most likely related to the ~23 k.y. precession cycle; see following). Such a transgression would take at most a few thousand years in the southern Alps region, and so would not greatly affect our estimates of the abruptness of the latest Permian extinction event.

The transition between the Bellerophon and Werfen Formations is marked by other evidence that supports its choice as a mass-extinction horizon. Biological evidence for a widespread drastic marine extinction includes the proliferation of acritarchs 1–2 m above the base of the Tesero horizon at Tesero, and proliferation of other stress-tolerant and/or opportunistic species (e.g., stromatolitic algal mats, lingulid brachiopods) in the lower Werfen Formation in the Alps (e.g., Buser et al., 1986) and in the earliest Triassic elsewhere (Schubert and Bottjer, 1992; Kershaw et al., 1999). A marked increase in fungal remains has been recognized in the uppermost Bellerophon Formation limestones and within the Tesero horizon (Visscher and Brugman, 1986) that apparently represents a worldwide ecological crisis event on land (Eshet et al., 1995; Visscher et al., 1996; Retallack, 1995; Ward et al., 2000).

A drastic global decrease in primary productivity and biomass in the oceans is also indicated by the carbon isotope record at the P-Tr boundary, although other explanations have been suggested (see Knoll et al., 1996; Krull and Retallack, 2000). In the Tesero section, for example, the base of the Tesero horizon is marked by a significant negative $\delta^{13}C$ anomaly (Magaritz et al., 1988) that is apparently a worldwide event (Magaritz et al., 1992; Morante et al., 1994; Wang et al., 1994). That section shows a rapid decrease in $\delta^{13}C$ of >1‰ over a thickness of <20 cm in the uppermost Bellerophon limestones, followed by further drop in $\delta^{13}C$ of 1‰ in the first 1 m of the lower Tesero horizon, and a continued drop in $\delta^{13}C$ in the upper Tesero horizon and overlying Mazzin deposits (Magaritz et al., 1988).

In the Gartnerkofel core from the Carnic Alps in Austria, a similar abrupt negative $\delta^{13}C$ anomaly is present toward the top of the 6-m-thick Tesero horizon (Magaritz and Holser, 1991). This shows that the horizon is time transgressive in the Alpine region, and that the abrupt change in carbon isotopes and associated faunal event does not everywhere correspond to the Bellerophon-Werfen Formation contact and facies change. Foraminifera are not preserved in the dolomitized Tesero oolites of the Gartnerkofel core and associated outcrops (Boeckelmann, 1991). In southern China, the negative $\delta^{13}C$ shift occurs at the boundary between the Upper Permian Changxing Formation (and its equivalents) and the overlying Triassic sediments (Jin et al., 2000). The last appearances of the similar Late Permian foraminiferal fauna in southern China (including *Palaeofusulina, Hemigordius, Nodosaria, Glomospira, Reichelina, Geinitzina, Glomospirella,* and *Pseudoglobulina*) are all recorded <50 cm from the carbon isotope anomaly (Zunyi and Lishun, 1992), supporting the conclusion that the abrupt disappearance of foraminifera seen in Italy is part of a global extinction event.

TABLE 1. CONFIDENCE INTERVALS FOR UNCERTAINTY OF THE ENDS OF STRATIGRAPHIC RANGES FOR FORAMINIFERAL TAXA, SASS DE PUTIA, ITALY

Taxa with more than 3 occurrences	Stratigraphic distance between first and last appearance (m)	Number of occurrences	Range extension calculated using the 50% confidence interval (m)
Glomospira	35	4	9.0
Ammodiscus	50	5	9.4
Paleonubecularia	41	5	7.8
*Globivalvulina cyprica**	50	17	2.2*
*Globivalvulina graeca**	50	16	2.4*
Reichelina	28	3	11.5
Glomospirella	28	4	7.3
Globlvalvulina bulloides	51	10	4.1
Globlvalvulina vonderschmitti	32	7	3.9
Paraglobivalvulina gr. mira	44	8	4.6
Paraglob. des septulifer	32	3	13.2
Agathammina	51	7	6.2
Frondina	16	4	4.3
*Geinitzina-Pachyphloia**	52	31	1.2*
Stafella	18	6	2.6
*Nankinella**	52	17	2.3*
*Baisalina**	52	18	2.2*
*Hemigordius**	52	30	1.3*

*Taxa used to determine extinction level (see text).

CYCLOSTRATIGRAPHY IN THE GARTNERKOFEL CORE

The 331 m Gartnerkofel-1 core from the Carnic Alps in Austria penetrated the P-Tr boundary succession of dolomitized limestones and interbedded thin marls and shales (bedding rhythms averaging ~20–30 cm) of shallow-marine (subtidal to supratidal) origin. These also consist of the uppermost Permian Bellerophon Formation and the overlying Lower Triassic Werfen Formation (basal oolitic Tesero horizon and the overlying Mazzin, Seis, and Campil Members) (Holser and Schönlaub, 1991) (Fig. 3). The upper 57 m in the core consists of Middle Triassic Muschelkalk conglomerate; interbedded ash-flow tuff is present from 30 to 34 m. Dolomitization of the carbonates was apparently a syndepositional and very early diagenetic process, possibly promoted by reducing conditions, and the sequence shows evidence of later pressure solution (Boeckelmann and Magaritz, 1991). Similar platform dolomites of other ages preserve original meter- to decimeter-scale cycles in the Milankovitch frequency band (Balog et al., 1999), and hence the Gartnerkofel-1 core was considered suitable for cycle analyses.

The P-Tr boundary interval in the Garnerkofel-1 core spans a 6–7 m interval marked by the last sample containing typical latest Permian fusulinids at 231 m and the first sample with a typical earliest Triassic invertebrate fauna at 224 m (the intervening Tesero horizon is devoid of fauna in the core; Jenny-Deshusses, 1991), preceded by the first occurrence of the earliest Triassic conodont *Hindeodus parvus* at 225 m (Schönlaub, 1991). A major negative carbon isotope shift of ~2.3‰ takes place within the same interval (from 228 to 224 m) in the core (Magaritz and Holser, 1991). In the nearby Reppwand outcrop section, the same faunal change takes place over only 0.8 m across the base of the Tesero horizon (Jenny-Deshusses, 1991).

Summary results of the analyses of the Garnerkofel-1 core contained the down-hole logging records of gamma-ray counts and density (Schmöller, 1991). The well bore was a cased and dry hole, logged with an OYO Geologger. The dolomites that make up the Upper Permian Bellerophon Formation and the Lower Triassic Werfen Formation show low gamma counts and predominantly high densities, whereas intercalated shaly beds give higher gamma rates. The density log represents relative densities, and values in the range of ~1.5–2 g/cm^3 or below are interpreted as indicating cavities. The continuous logs were digitized to provide 1561 gamma ray and 1555 density data points for analysis. Measurements of carbon isotope ratios on 379 samples of bulk carbonate from 57 to 330 m in the core were taken from Magaritz and Holser (1991).

WAVELET ANALYSIS

We utilized the continuous wavelet transform (CWT), which provides the amplitudes of various frequencies, and information about their time or depth dependence (Bolton et al., 1995; Prokoph and Barthelmes, 1996). The CWT of a depth-related series $f(t)$ is defined as

$$W_\psi(a,b) = \left(\frac{1}{\sqrt{a}}\right)\int f(t)\psi\left(\frac{t-b}{a}\right) dt, \quad (1)$$

where the base wavelet (which is here the Morlet wavelet) has a length that is usually much shorter than the depth series $f(t)$. W stands for the wavelet coefficients, which are given according

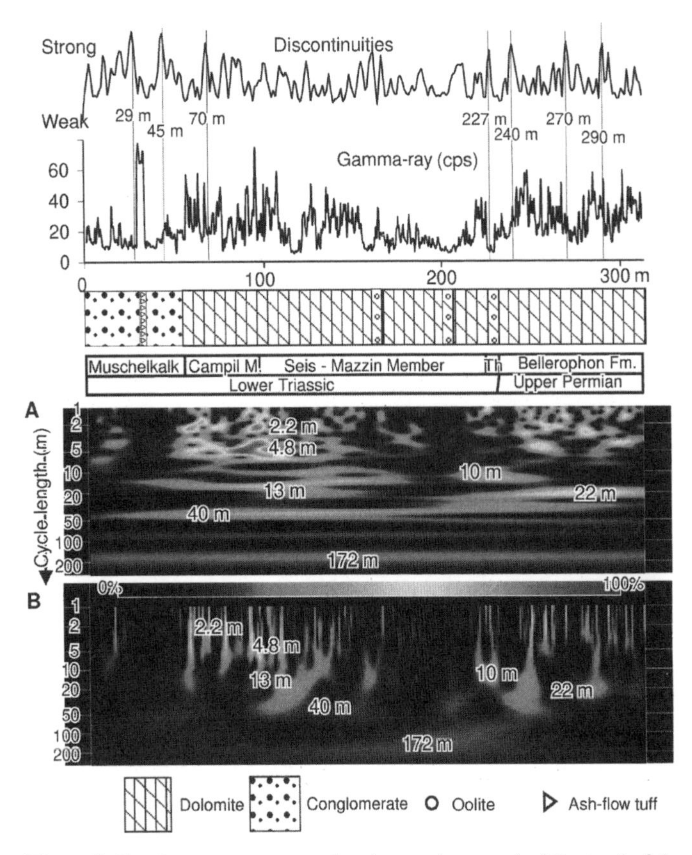

Figure 3. Continuous gamma ray log (counts/s, or cps) of Gartnerkofel-1 core (Schmöller, 1991), digitized at 0.2 m intervals, including its discontinuity curve at $l = 10$, and simplified lithology and stratigraphy. (Parameter l represents relation between periodicity and depth resolution.) A: Scalogram of wavelet coefficients for $l = 10$, with relative intensity scale; B: Scalogram for $l = 2$. Wavelet scalograms provide relative intensity of periodic signals at depth. (High gamma ray values (~70 cps) for andesitic ash-flow tuff between 30 and 34 m were set to 11.5 cps to avoid artificial event peaks.)

to a and b in depth-frequency space, and graphically represented by means of a scalogram (Prokoph and Barthelmes, 1996). The variable a is the scale factor that determines the frequencies (or scale), so that varying a gives rise to a spectrum; and b is the shift of the analysis window in depth, so that varying b represents the "sliding window" of the wavelet over $f(t)$ (Chao and Naito, 1995).

The Morlet wavelet is formed by a periodic sinusoidal function with a Gaussian envelope, and has already provided robust results in analyses of climate-related records. The Morlet mother wavelet (shifted and scaled) is defined as

$$\psi^{l}_{a,b}(t) = \pi^{-1/4}(al)^{-1/2}e^{-i2\pi\ 1/a\ (t-b)}e^{-1/2\ (t-b/al)^2}. \quad (2)$$

The parameter l represents the relation between periodicity and depth resolution. Parameters $l = 2$ and $l = 10$ were chosen, which give sufficiently precise results in resolution of depth and frequency, respectively. The most significant (major) periodicity through the succession can be used to identify variations in accumulation rate directly from the scalogram.

Discontinuities in the sequence can be identified as follows: A function $f(t)$ has a discontinuity of degree k at t_0, $k = 0, 1, \ldots$, if the kth order left and right derivatives at t_0 are different; i.e., $f(k)(t_0+) \neq f(k)(t_0-)$, the difference being the size of the discontinuity (Lee, 1989). Using the wavelet transform, a discontinuity u_b between the depth intervals b and $b - 1$ can be interpreted as

$$u_b = \sqrt{\sum_{a=1}^{n}\left(\frac{W_b(a)}{\Sigma W_b} - \frac{W_{b-1}(a)}{\Sigma W_{b-1}}\right)^2}, \quad (3)$$

where n is the total number of frequencies; u_b represents not only the short-time difference between the signals y_b and y_{b-l}, but also the longer time trend with respect to length l. The average spacing of ~0.2 m for the logging data and 0.8 m for the isotope data gives Nyquist periods of 0.4 and 1.6 m, respectively.

The Nyquist period for the isotope data can extend to 9.8 m at the maximum sampling interval (4.9 m), but is lower in the key interval around the P-Tr boundary in the Garnerkofel-1 core, with sampling at ~0.5 m intervals. In order to avoid artificial nonstationarities produced by unevenness in the sampling interval, no CWT with $l = 2$ has been carried out on the isotope data. The method is used here for depth-dependent data as demonstrated for hemipelagic sediments (Prokoph and Barthelmes, 1996). CWT allows detection of nonstationarities and periodicities in a way similar to advanced evolutionary spectral analysis (e.g., multiple taper spectral analysis; Park and Maasch, 1993).

Results of the wavelet analysis reveal that the gamma ray and density logging data are both characterized by relatively continuous cycles with periods of 10–12 m and 37–42 m (Figs. 3 and 4). In the density log, 22 periodic ~10 m cycles occur from ~270 m to 57 m depth in the core, and the P-Tr boundary interval, at ~230 m in the core, does not show any obvious disruption in this cyclicity (Fig. 4). A single truncation of the 10 m density cycles occurs at ~183 m core depth. In the gamma ray log, less stable periodicities of ~2.2 m and 4.8 m also occur in the interval from 175 to 57 m core depth (Fig. 3). Below ~270 m, a periodicity of ~22 m replaces the ~10 m period (Fig. 3) (Rampino et al., 2000).

In the carbon isotope record, periodic 8–12 m and 4.4–4.7 m cycles occur from ~250 to 160 m in the core (Fig. 5). Two strong negative carbon isotope anomalies occur at ~187 and 220 m core depth (Magaritz and Holser, 1991), producing a single 33 m pseudo-period in the wavelet scalogram. These strong carbon isotope anomalies (1.5‰–~3‰) dwarf the ~0.5‰ carbon isotope variations in the rest of the core, which show a weaker cycle of ~175 m, and cycles of 21 m, and 39 m below 250 m depth (Fig. 5).

Our wavelet analyses suggest that the interval from 57 m to 270 m in the core shows no remarkable nonstationarities in

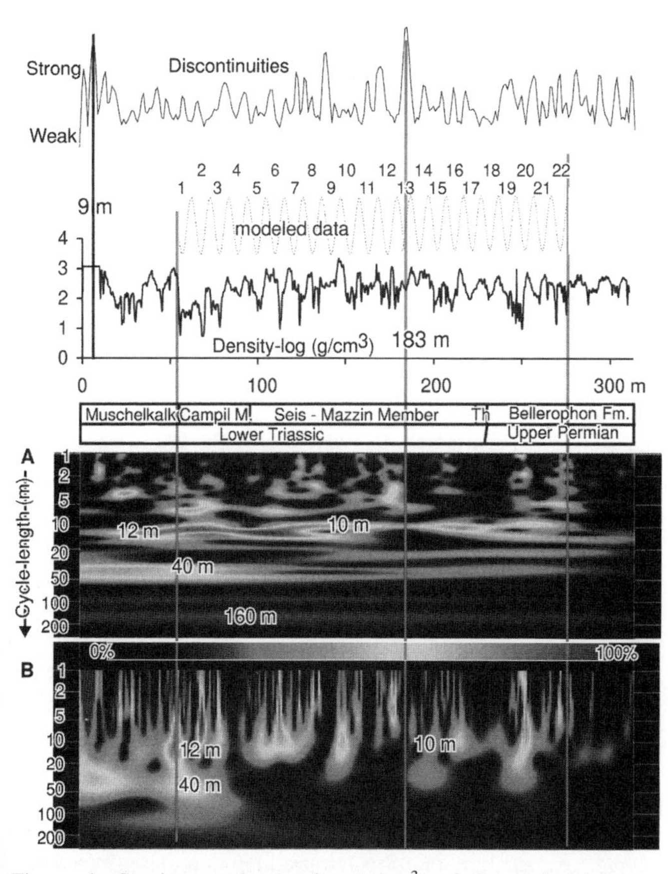

Figure 4. Continuous density log (g/cm^3) of Gartnerkofel-1 core (Schmöller, 1991), digitized at 0.2 m intervals, including its discontinuity curve, and ideal sinusoidal ~10 m cycle curve for 22 cycle intervals (dashed line). Wavelet scalogram for A is l = 10, and for B is l = 2. Parameter l represents relation between periodicity and depth resolution.

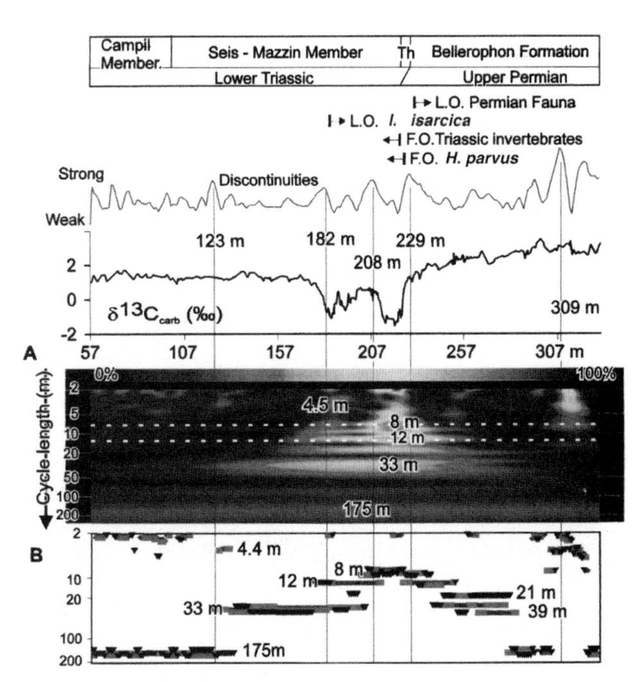

Figure 5. Nonequidistant measured δ^{13}C isotope data from Gartnerkofel-1 core (mean sample spacing = 0.8 m) (from Magaritz and Holser, 1991), including its discontinuity curve. A: Wavelet scalogram with intensity scale (see Fig. 3). B: Occurrence of three major periodicities (regardless of their absolute amplitude) throughout core (blue = first period, red = second period, black = third period). Also shown are last occurrence (L.O.) datum for Late Permian fauna, first occurrence (F.O.) datum for characteristic Early Triassic invertebrates, F.O. datum for earliest Triassic index conodont *H. parvus*, and L.O. datum for Early Triassic conodont *I. isarcica* (see text).

density or gamma ray counts (except at 183 m). In order to establish significant periods for the full time series, we utilized Fast Fourier Transform (FFT) spectral analysis (Mann and Lees, 1996), and to reduce high-frequency noise, the normalized (m = 0, d = 1) density and gamma ray data were first stacked together (Fig. 6) (Fischer et al., 1996).

The FFT power spectrum of the stacked data reveals four period bands located at ~36 m, 9.5–13 m (maximum at 11.5 m), ~4.4–5 m, and 2–2.3 m (Fig. 6). The ratios of the dominant cycles of ~40:10:4.7:2.2 m detected by FFT and CWT analyses are approximately correlative to the ratio of Milankovitch cycles calculated for the Late Permian (~412:100:37:19 k.y.), corresponding to the two eccentricity cycles, obliquity, and precession cycles (Berger et al., 1989), and suggesting control of cyclicity by orbital variations. The Milankovitch-band peaks in our stacked data are all significantly above the calculated red-noise and white-noise variances of the FFT power spectrum (Mann and Lees, 1996); red noise is considered an autoregressive (AR) (1) process with lag-one autocorrelation coefficient (Bartlett, 1966) (Fig. 6).

A stable ~2.2 m cycle in the interval from 132 to 120 m core depth (Fig. 6) indicates almost constant average sedimentation rate through the P-Tr boundary interval. Below ~270 m in the core, detection of an ~22 m cycle in the gamma ray data, and weaker 10 m and 41 m cycles in the δ^{13}C data, suggests that the sedimentation rate may have been similar to the upper part of the Bellerophon Formation at ~10 cm/k.y., the ~22 m cycle possibly expressing a multiple of the 10–11 m eccentricity cycle (Rampino et al., 2000).

TEMPO OF THE END-PERMIAN EXTINCTION

The recognition of Milankovitch-band cycles allows estimation of an average sedimentation rate of ~10 cm/k.y., which is reasonable for shallow-water carbonate sediments in regions of rapid continental-margin subsidence (Perrodon and Masse, 1984; Schwarzacher and Haas, 1986). The apparent regularity of the 10–11 m cycle over 22 cycles in the density data suggests that the sedimentation rate was rather consistent for ~2.2 m.y. from the latest Permian through the Early Triassic. The regularity also suggests that no major change in sedimentation rate took place at the P-Tr boundary (Fig. 7), despite the faunal extinctions that marked that interval. This is supported by the

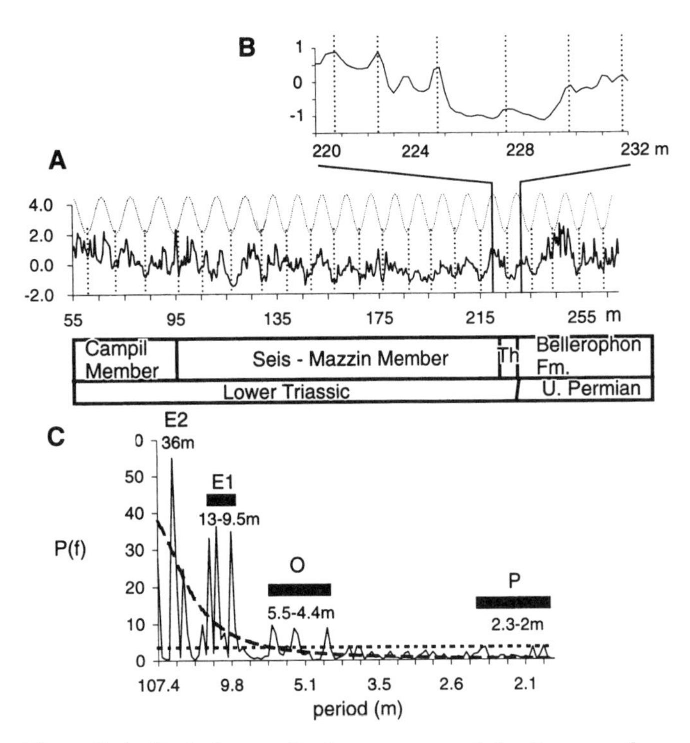

Figure 6. A: Stacked normalized gamma ray and density curves from Gartnerkofel-1 core (55–270 m depth), showing modeled ~10.5 m cycle (dotted line) fitted to stacked data, with correlation markers (dashed lines). B: Permian-Triassic boundary interval, with best-fit ~2.2 m cycle marked by dashed lines. C: Fast Fourier Transform (FFT) power spectrum for 57–270 m section of core; highlighted cycle bands interpreted as ~400 k.y. (E2), ~100 k.y. (E1), ~40 k.y. (O), and ~20 k.y. (P) Milankovitch bands, and calculated red-noise (dashed line) and white-noise (dotted line) variances of FFT power spectrum.

discontinuity analyses (Figs. 3–5), which show no evidence for a significant break in the cyclic sequence in the vicinity of the P-Tr boundary (Rampino et al., 2000)

The carbonate content of the sediments also remains high across the boundary interval, although the lowermost Triassic carbonates contain somewhat more insoluble residue and shaly interbeds, and the carbonate component changed from largely biomicrite in the Bellerophon Formation to nonbiogenic oolite and cements in the lower Tesero horizon (Holser and Schönlaub, 1991). The continuity of carbonate deposition may be explained by a shift in ocean chemistry that maintained the removal of calcium carbonate in shallow-marine waters despite a decrease in biogenic carbonate production (Caldeira and Rampino, 1993).

The time series show evidence for relatively strong discontinuities in the interval 180–190 m within the Lower Triassic, which is characterized by a thin pyrite-rich layer that also shows enriched Ir (233 ppt) (Holser and Schönlaub, 1991), and is centered on the uppermost minimum in $\delta^{13}C$. (A second Ir-peak [150 ppt] occurs near the lower $\delta^{13}C$ minimum.) The upper Ir anomaly is also marked by the last occurrences of the characteristic Early Triassic (Griesbachian) conodonts *Hindeodus parvus* and *Isarcicella isarcica* (Schönlaub, 1991) (Fig. 7).

Previous estimates of sedimentation rates for the P-Tr interval in the Garnerkofel-1 core vary widely (Schönlaub, 1991; Zeissl and Mauritsch, 1991). An independent estimate of sedimentation rate can be derived from graphic correlation studies of similar sections in the southern Alps in Italy, which suggest that the duration of sedimentation of the Tesero horizon in the southern Alps (where it averages ~6 m thick) was ~70 k.y. (Sweet et al., 1992), giving an average lowermost Triassic sedimentation rate of ~9 cm/k.y. Moreover, the Tesero horizon encompasses three of the ~2.3 m precession cycles in the Gartnerkofel core (or ~60 k.y.; Fig. 7), and similarly in sections in the Italian Alps the horizon contains three parasequences or punctuated aggradational cycles related to relative sea-level changes most likely controlled by precession cycles. At sedimentation rates of ~9–10 cm/k.y., the entire 140 m lowermost Triassic (Griesbachian) succession at Gartnerkofel (composed of the Tesero horizon and the Mazzin and Seis Members of the Werfen Formation) (Holser and Schönlaub, 1991) would have been deposited in ~1.4–1.6 m.y. This agrees with estimates of ~1.6 m.y. for the Griesbachian using recent geologic time scales (Gradstein et al., 1994).

Milankovitch periods in $\delta^{13}C$ and density in these shallow-water carbonates were most likely the result of climatically induced oscillations of sea level and climate, coupled with changes in ocean circulation and productivity, that affected sedimentation or early diagenesis (Schwarzacher and Haas, 1986; Schwarzacher, 1993). Fluctuations in gamma radiation reflect varying input of clay minerals and the presence of shaly interbeds (Schmöller, 1991). Throughout the P-Tr boundary interval in the Garnerkofel-1 core, the 100 k.y. eccentricity cycle seems to be dominant. Weaker obliquity and precession cycles are in line with the location of the Austrian section in the latest Permian, close to the equator in the western bight of the Tethys, where obliquity and precessional effects on seasonal contrast would be subdued (Schwarzacher, 1993).

Within the Lower Triassic sequence in the core (~225–57 m in the core) the 10 m and 41 m cycles most likely reflect eccentricity-forced sedimentation or early diagenesis, and approximately constant average sediment accumulation rates of the episodic storm-induced deposition that may be responsible for the ~30 cm bedding rhythms. The precession cycle is represented stratigraphically by a series of parasequences or punctuated aggradational cycles, averaging ~2 to 2.5 m in thickness, most likely controlled by relative sea-level changes (Wignall and Hallam, 1992). The primary ratio between precession and eccentricity (Park and Maasch, 1993) is altered somewhat in our data, probably because of the relatively large or nonequidistant sampling intervals, averaging effects due to diagenetic processes, and brief erosive events, which tend to preferentially alter high-frequency periodic signals in the sedimentary record (Schwarzacher, 1993).

The Gartnerkofel section can be directly correlated with

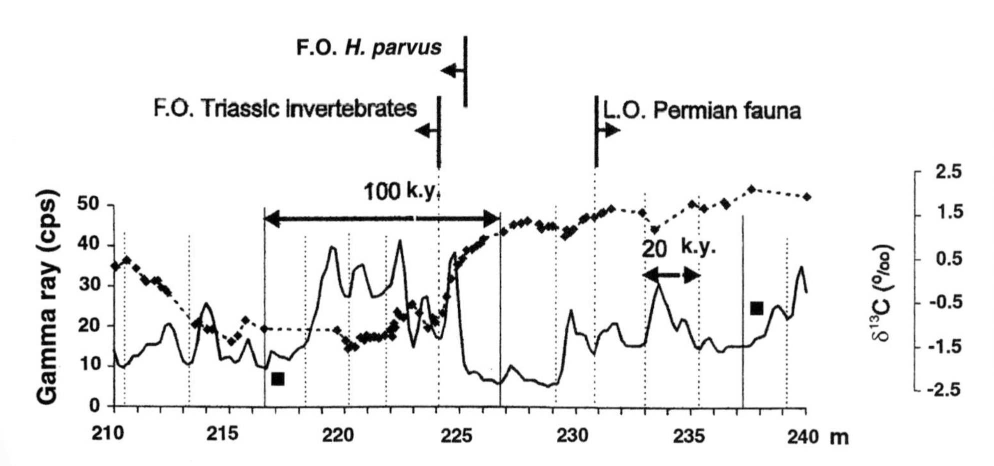

Figure 7. Gamma ray (solid line; cps, counts per second) and carbon isotope (diamonds) values for interval 210–240 m, spanning Permian-Triassic boundary in Gartnerkofel-1 core. Also shown are best-fit 100 k.y. and 20 k.y. cycles in gamma ray data, last occurrence (L.O.) datum of Late Permian fauna, and first occurrence (F.O.) datums of *H. parvus* (marker for earliest Trassic) and typical Early Triassic invertebrates.

the classic Meishan, China, P-Tr boundary section, where precise radiometric dating of volcanic ash layers by Bowring et al. (1998) has defined the end-Permian extinction pulse to an interval of <1 m.y., and the negative carbon isotope shift to a period ≤165 k.y. Using the improved resolution provided by cycle analysis of the expanded section in the Garnerkofel-1 core, we find that the abrupt faunal change takes place over ≤6 m, or <60 k.y. In the nearby Reppwand outcrop section, the last typical Permian foraminifera were sampled 0.6 m below the base of the Tesero horizon (Jenny-Deshusses, 1991), and *H. parvus* first occurs in a sample only 0.2 m above the base of the horizon, an interval of only 0.8 m (Schönlaub, 1991). At the average sedimentation rates calculated for the section, the P-Tr faunal turnover could have occurred in ≤8 k.y., close to the limit of time resolution induced by bioturbation and reworking in these sediments (e.g., Boudreau, 1994).

In the Gartnerkofel core and the Reppwand section, the major negative carbon isotope shift (which takes place over ~3 m) would have occurred in ≤30 k.y. (Fig. 5). The carbon isotope excursion, which marks the boundary worldwide in marine (Erwin, 1993; Bowring et al., 1998; Magaritz and Holser, 1991; Wang et al., 1994) and nonmarine sediments (Morante et al., 1994; MacLeod et al., 2000), would have lasted ~480 k.y. into the Early Triassic, according to our estimates. The most likely explanation for lighter δ^{13}C values in both carbonate and organic carbon fractions is greatly reduced oceanic productivity in the wake of a mass extinction of phytoplankton at the end of the Permian that would have led to a marked isotopic homogenization of the oceans (Magaritz et al., 1992; Rampino and Caldeira, 1993; Wang et al., 1994).

Such an event, termed the "Strangelove" ocean, is recorded at the Cretaceous-Tertiary boundary, and at other mass-extinction boundaries, and could occur on time scales as short as 10^3 yr (Magaritz et al., 1992). A nearly complete cessation of primary productivity in the surface ocean, such as would be associated with a mass extinction of plankton, could lead to a rapid negative shift in δ^{13}C of ~2‰–3‰ in ocean surface waters, similar to that seen in the P-Tr boundary sections in Italy, Austria, and elsewhere (Rampino et al., 2000).

Our biostratigraphic and cyclostratigraphic results indicate that the extinctions that mark the P-Tr boundary were very sudden, perhaps less than the resolution window for events in the Garnerkofel-1 core, and suggest a catastrophic cause. The biostratigraphic analyses of the Italian P-Tr sections suggest that the extinctions took place within ~1 m in the lowermost Tesero horizon, or <10 k.y. based on our estimates of average sedimentation rates in these sections. The cyclostratigraphic analyses of the Austrian section indicate that the faunal turnover that marks the boundary may have taken place in <8 k.y., and that the section records the essentially continuous transition from the Paleozoic to the Mesozoic Era.

ACKNOWLEDGMENTS

We thank K. Caldeira, D. Erwin, Y. Eshet, J. Ogg, J. Rial, M. Steiner, R. Stothers, H. Visscher, and P. Wignall for helpful discussions. C. Koeberl, J. Pálfy, and F. Sierro provided critical reviews. D. Winiarski provided technical support. Schwindt and Rampino received support from New York University Scholar and Research Challenge grants, respectively, and Prokoph was supported by a grant from the Deutsche Forschungsgemeinschaft.

REFERENCES CITED

Balog, A., Read, J.F., and Haas, J., 1999, Climate-controlled early dolomite, Late Triassic cyclic platform carbonates, Hungary: Journal of Sedimentary Research, v. 69, p. 267–282.

Bartlett, M.S., 1966, An introduction to stochastic processes: Cambridge, UK, Cambridge University Press, 384 p.

Berger, A.L., Loutre, M.F., and Dehant, V., 1989, Astronomical frequencies for pre-Quaternary palaeoclimate studies: Terra Nova, v. 1, p. 474–479.

Boeckelmann, K., 1991, The Permian-Triassic of the Gartnerkofel-1 Core and the Reppwand outcrop section (Carnic Alps, Austria): Abhandlungen der Geologischen Bundesanstalt, v. 45, p. 17–36.

Boeckelmann, K., and Magaritz, M., 1991, The Permian-Triassic of the Gartnerkofel-1 core (Carnic Alps, Austria): Dolomitization of the Permian-Triassic sequence: Abhandlungen der Geologischen Bundesanstalt, v. 45, p. 61–68.

Bolton, E.W., Maasch, K.A., and Lilly, J.M., 1995, A wavelet analysis of Plio-

Pleistocene climate indicators: A new view of periodicity evolution: Geophysical Research Letters, v. 22, p. 2753–2756.

Boudreau, B.P., 1994, Is burial velocity a master parameter for bioturbation?: Geochimica et Cosmochimica Acta, v. 58, p. 1243–1249.

Bowring, S.A., Erwin, D.H., Jin, Y.G., Martin, M.W., Davidek, K., and Wang, W., 1998, U/Pb zircon geochronology and tempo of the end-Permian mass extinction: Science, v. 280, p. 1039–1045.

Broglio-Loriga, C., and Cassinis, G., 1992, The Permo-Triassic boundary in the Southern Alps (Italy) and in adjacent Periadriatic regions, *in* Sweet, W.C., Zunyi, Y., Dickins, J.M., and Hongfu, Y., eds., Permo-Triassic events in the eastern Tethys: Cambridge, Cambridge University Press, p. 78–97.

Broglio-Loriga, C., Neri, C., Pasini, M., and Posenato, R., 1986, Marine fossil assemblages from Upper Permian to lowermost Triassic in the western Dolomites: Società Geologica Italiana, Memorie, v. 34, p. 5–44.

Buggisch, W., and Noe, S., 1986, Upper Permian and Permian-Triassic boundary of the Carnia (Bellerophon Formation, Tesero Horizon, northern Italy): Società Geologica Italiana, Memorie, v. 34, p. 91–106.

Buser, S., Grad, K., Ogorelec, B., Ramovs, A., and Sribar, L., 1986, Stratigraphical, paleontological and sedimentological characteristics of Upper Permian beds in Slovenia, NW Yugoslavia: Società Geologica Italiana, Memorie, v. 34, p. 195–210.

Caldeira, K., and Rampino, M.R., 1993, The aftermath of the K/T boundary mass extinction: Biogeochemical stabilization of the carbon cycle and climate: Paleoceanography, v. 8, p. 515–525.

Chao, B.F., and Naito, I., 1995, Wavelet analysis provides a new tool for studying Earth's rotation: Eos (Transactions of the American Geophysical Union), v. 76, p. 161, 164–165.

Cirilli, S., Pirini Radrizzani, C., Ponton, M., and Radrizzani, S., 1998, Stratigraphical and palaeoenvironmental analysis of the Permian-Triassic transition in the Badia Valley (Southern Alps, Italy): Palaeogeography, Palaeoclimatology, Palaeoecology, v. 138, p. 85–113.

Erwin, D.H., 1993, The great Paleozoic crisis: New York, Columbia University Press, 257 p.

Eshet, Y., Rampino, M.R., and Visscher, H., 1995, Fungal event and palynological record of ecological crisis and recovery across the Permian-Triassic boundary: Geology, v. 23, p. 967–970.

Fischer, D.A., Koerner, R.M., Kuivinen, K., Clausen, H.B., Johnson, S.J., Steffensen, J.P., Gundestrup, N., and Hammer, C.U., 1996, Inter-comparison of ice core ^{18}O and precipitation records from sites in Canada and Greenland over the last 3500 years, and over the last few centuries in detail using EOF techniques: NATO Advanced Study Institute Series 141, p. 297–328.

Gradstein, F.M., Agterberg, F.P., Ogg, J.G., Hardenbol, J., van Veen, P., Thierry, J., and Huang, Z., 1994, A Mesozoic time scale: Journal of Geophysical Research, v. 99, p. 24051–24074.

Holser, W.T., and Schönlaub, H.P., 1991, The Permian-Triassic of the Gartnerkofel-1 core (Carnic Alps, Austria): Synthesis and conclusions: Abhandlungen der Geologischen Bundesanstalt, v. 45, p. 213–232.

Jenny-Deshusses, C., 1991, The Permian-Triassic of the Gartnerkofel-1 core (Carnic Alps, Austria): Foraminifera and algae of the core and outcrop section: Abhandlungen der Geologischen Bundesanstalt, v. 45, p. 99–108.

Jin, Y.G., Wang, Y., Wang, W., Shang, Cao, C.Q., and Erwin, D.H., 2000, Patterns of marine mass extinction near the Permian-Triassic boundary: Science, v. 289, p. 432–436.

Kershaw, S., Zhang, T., and Lan, G., 1999, A ?microbialite carbonate crust at the Permian-Triassic boundary in South China, and its palaeoenvironmental significance: Palaeogeography, Palaeoclimatology, Palaeoecology, v. 146, p. 1–18.

Knoll, A.H., Bambach, R.K., Canfield, D.E., and Grotzinger, J.P., 1996, Comparative Earth history and Late Permian mass extinction: Science, v. 273, p. 452–457.

Koch, C.F., and Morgan, J.P., 1988, On the expected distribution of species' ranges: Paleobiology, v. 14, p. 126–138.

Krull, E.S., and Retallack, G.J., 2000, δ^{13}C depth profiles from paleosols across the Permian-Triassic boundary: Evidence for methane release: Geological Society of America Bulletin, v. 112, p. 1459–1472.

Lee, D., 1989, Discontinuity detection and curve fitting, *in* Chui, C.K., Schumaker, L.L., and Ward, J.D., eds., Approximation theory 6: New York, Academic Press, v. 2, p. 373–376.

MacLeod, K.G., Smith, R.M.H., Koch, P.L., and Ward, P.D., 2000, Timing of mammal-like reptile extinctions across the Permian-Triassic boundary in South Africa: Geology, v. 28, p. 227–230.

Magaritz, M., and Holser, W.T., 1991, The Permian-Triassic of the Gartnerkofel-1 core (Carnic Alps, Austria): Carbon and oxygen isotope variation: Abhandlungen der Geologischen Bundesanstalt, v. 45, p. 149–163.

Magaritz, M., Bär, R., Baud, A., and Holser, W.T., 1988, The carbon-isotope shift at the Permian-Triassic boundary in the Southern Alps is gradual: Nature, v. 331, p. 337–339.

Magaritz, M., Krishnamurthy, R.V., and Holser, W.T., 1992, Parallel trends in organic and inorganic carbon isotopes across the Permian/Triassic boundary: American Journal of Science, v. 292, p. 727–739.

Mann, M.S., and Lees, J.M., 1996, Robust estimation of background noise and signal detection in climatic time series: Climatic Change, v. 33, p. 409–445.

Marshall, C.R., 1990, Confidence intervals on stratigraphic ranges: Paleobiology, v. 16, p. 1–10.

Marshall, C.R., 1994, Confidence intervals on stratigraphic ranges: Partial relaxation of the assumption of randomly distributed fossil horizons: Paleobiology, v. 20, p. 459–469.

Meldahl, K.H., 1990, Sampling, species abundance, and the stratigraphic signature of mass extinction: A test using Holocene tidal flat molluscs: Geology, v. 18, p. 890–893.

Morante, R., Veevers, J.J., Andrew, A.S., and Hamilton, P.J., 1994, Determination of the Permian-Triassic boundary in Australia from carbon isotope stratigraphy: Australian Petroleum Exploration Association Journal, v. 34, p. 330–336.

Park, J., and Maasch, K.A., 1993, Plio-Pleistocene time evolution of the 100-kyr cycle in the marine paleoclimate records: Journal of Geophysical Research, v. 98, p. 447–463.

Perrodon, A., and Masse, P., 1984, Subsidence, sedimentation and petroleum systems: Journal of Petroleum Geology, v. 7, p. 5–26.

Prokoph, A., and Barthelmes, F., 1996, Detection of nonstationarities in geological time series: Wavelet transform of chaotic and cyclic sequences: Computers and Geoscience, v. 22, p. 1097–1108.

Rampino, M.R., and Adler, A.C., 1998, Evidence for abrupt latest Permian mass extinction of foraminifera: Results of tests for the Signor-Lipps effect: Geology, v. 26, p. 415–418.

Rampino, M.R., Prokoph, A., and Adler, A.C., 2000, Tempo of the end-Permian event: High-resolution cyclostratigraphy at the Permian-Triassic boundary: Geology, v. 28, p. 643–646.

Raup, D.M., 1979, Size of the Permo-Triassic bottleneck and its evolutionary implications: Science, v. 206, p. 217–218.

Raup, D.M., 1989, The case for extraterrestrial causes of extinction: Royal Society of London Philosophical Transactions, ser. B, v. 325, p. 421–435.

Retallack, G.J., 1995, Permian-Triassic life crisis on land: Science, v. 267, p. 77–80.

Schmöller, R., 1991, The Permian-Triassic of the Gartnerkofel-1 core (Carnic Alps, Austria): Remarks on the natural gamma ray log and density log Abhandlungen der Geologischen Bundesanstalt, v. 45, p. 209–211.

Schönlaub, H.P., 1991, The Permian-Triassic of the Gartnerkofel-1 core (Carnic Alps, Austria): Conodont biostratigraphy: Abhandlungen der Geologischen Bundesanstalt, v. 45, p. 79–98.

Schubert, J.K., and Bottjer, D.J., 1992, Early Triassic stromatolites as post mass extinction disaster forms: Geology, v. 20, p. 883–886.

Schwarzacher, W., 1993, Cyclostratigraphy and Milankovitch theory: Developments in Sedimentology 52: Amsterdam, Elsevier, 225 p.

Schwarzacher, W., and Haas, J., 1986, Comparative statistical analysis of som

Hungarian and Austrian Upper Triassic peritidal carbonate sequences: Acta Geologica Hungarica, v. 29, p. 175–196.

Signor, P.W., III, and Lipps, J.H., 1982, Sampling bias, gradual extinction patterns and catastrophes in the fossil record, *in* Silver, L.T., and Shultz, P.H., eds., Geological implications of impacts of large asteroids and comets on the earth: Geological Society of America Special Paper 190, p. 291–296.

Springer, M.S., 1990, The effect of random range truncations on patterns of evolution in the fossil record: Paleobiology, v. 16, p. 512–520.

Stanley, S.M., and Yang, X., 1994, Two extinction events in the Late Permian: Science, v. 266, p. 1340–1344.

Strauss, D., and Sadler, P.M., 1989, Classical confidence intervals and Bayesian probability estimates for ends of local taxon ranges: Mathematical Geology, v. 21, p. 411–427.

Sweet, W.C., Yang, Z., Dickins, J.M., and Yin, H., 1992, Permo-Triassic events in the eastern Tethys: An overview, *in* Sweet, W.C., Zunyi, Y., Dickins, J.M., and Hongfu, Y., eds., Permo-Triassic events in the eastern Tethys: Cambridge, Cambridge University Press, p. 1–8.

Visscher, H., and Brugman, W.A., 1986, The Permian-Triassic boundary in the Southern Alps: A palynological approach: Società Geologica Italiana, Memorie, v. 34, p. 121–128.

Visscher, H., Brinkhuis, H., Dilcher, D., Elsik, W.C., Eshet, Y., Looy, C., Rampino, M.R., and Traverse, A., 1996, The terminal Paleozoic fungal event: Evidence of terrestrial ecosystem destabilization and collapse: Proceedings of the National Academy of Sciences of the United States of America, v. 93, p. 2155–2158.

Wang, K., Geldsetzer, H.H.J., and Krouse, H.R., 1994, Permian-Triassic extinction: Organic ^{13}C evidence from British Columbia, Canada: Geology, v. 22, p. 580–584.

Ward, P., Montgomery, D.R., and Smith, R., 2000, Altered river morphology in South Africa related to the Permian-Triassic extinction: Science, v. 289, p. 1740–1743.

Wheatcroft, R.A., 1990, Preservation potential of sedimentary event layers: Geology, v. 18, p. 843–845.

Wignall, P.B., and Hallam, A., 1992, Anoxia as a cause of the Permian/Triassic mass extinction: Facies evidence from northern Italy and the western United States: Palaeogeography, Palaeoclimatology, Palaeoecology, v. 93, p. 21–46.

Zeissl, W., and Mauritsch, H., 1991, The Permian-Triassic of the Gartnerkofel-1 core (Carnic Alps, Austria): Magnetostratigraphy: Abhandlungen der Geologischen Bundesanstalt, v. 45, p. 193–207.

Zunyi, Y., and Lishun, L., 1992, Permo-Triassic boundary relations in South China, *in* Sweet, W.C., Zunyi, Y., Dickins, J.M., and Hongfu, Y., eds., Permo-Triassic events in the eastern Tethys: Cambridge, Cambridge University Press, p. 9–20.

Manuscript Submitted October 16, 2000; Accepted by the Society March 22, 2001

Printed in the U.S.A.

Geological Society of America
Special Paper 356
2002

Permian-Triassic boundary in the northwest Karoo basin: Current stratigraphic placement, implications for basin development models, and the search for evidence of impact

P. John Hancox
D. Brandt
W.U. Reimold
Impact Cratering Research Group, Geology Department, University of the Witwatersrand, Private Bag 3, Wits 2050, South Africa
C. Koeberl
Institute of Geochemistry, University of Vienna, Althanstrasse 14, A-1090 Vienna, Austria
J. Neveling
Council for Geoscience, Private Bag X112, Pretoria 0001, South Africa, and Department of Palaeontology, University of the Witwatersrand, Wits 2050, South Africa

ABSTRACT

Recent geological and paleontological studies of the Permian-Triassic (P-Tr) boundary sequence in the northwest of the main Karoo basin of South Africa have shown that there is some ambiguity as to the exact placement of the contact. The change in fluvial style from meandering to braided, tetrapod biostratigraphy, and geochemical signatures favor the historical placement of the boundary, in the north of the basin, above a laterally continuous horizon of brown-weathering carbonate concretions that at places contain isolated skulls of the dicynodont *Dicynodon*. At the same stratigraphic level, fine-grained sandstone lenses are regarded as the horizon from which the temnospondyl amphibian *Uranocentradon* was recovered. Two series of samples straddling the historical P-Tr boundary at the town of Senekal were analyzed geochemically and petrographically for evidence of impact. Neither section shows any enrichment of siderophile elements or iridium, or any evidence of impact deformation in the form of planar deformation features in quartz. These studies have, however, shown that there is a significantly different chemical signature above and below the historical P-Tr boundary in the north of the basin. These findings suggest that: the extinction event documented in the south of the basin may be represented by nondeposition or an unconformity in the north; that the change in fluvial style from meandering to braided may be useful for positioning the P-Tr boundary in the continental Karoo sequence; that dynamic subsidence probably played a limited role in creating distal sector accommodation space during the Late Permian; and that the sequences in the south and north of the basin are out-of-phase in the sense of a purely reciprocal stratigraphic model. Biostratigraphic studies based on wood, however, suggest that the P-Tr boundary may be higher in the sequence, in the Verkykerskop Formation, but still place the P-Tr boundary in the north of the basin at an unconformity surface.

Hancox, P.J., Brandt, D., Reimold, W.U., Koeberl, C., and Neveling, J., 2002, Permian-Triassic boundary in the northwest Karoo basin: Current stratigraphic placement, implications for basin development models, and the search for evidence of impact, *in* Koeberl, C., and MacLeod, K.G., eds., Catastrophic Events and Mass Extinctions: Impacts and Beyond: Boulder, Colorado, Geological Society of America Special Paper 356, p. 429–444.

INTRODUCTION

The Permian-Triassic (P-Tr) faunal crisis has long been considered as the greatest mass-extinction event to befall the planet (e.g., Erwin, 1990, 1993, 1994). The Permian-Triassic (P-Tr) boundary in the shallow marine realm is now fairly well documented, especially for the stratotype in southern China at the Meishan section (Chai et al., 1992; Yin et al., 1992, 1996; Zang et al., 1992, 1996; Xu and Yang, 1993; Jin et al., 2000). The correlation of the marine with the nonmarine P-Tr event is less well understood, and the continental P-Tr boundary sequence is also less well known. In the terrestrial realm a major reduction in tetrapod diversity at the family level occurred, and data on faunal and floral turnovers have shown that the terrestrial P-Tr extinction was highly catastrophic (Retallack, 1995; Ward, 2000); however, the time span and cause of the extinction are less well documented (King, 1990; Smith, 1995; Rampino et al., 2000).

Causal mechanisms for the P-Tr boundary event include environmental shifts (Thackeray et al., 1990; Stanley and Yang, 1994; Smith, 1995; Retallack, 1999), volcanism (Campbell et al., 1992; Renne et al., 1995), asteroid or comet impact (Rampino, 1992; Rampino and Haggerty, 1996; Retallack et al., 1998), and various combinations of these processes (e.g., Erwin, 1993, 1994; Bowring et al., 1998). Evidence for impact in particular is controversial, but the idea has received support from numerous works (Brandner et al., 1986; Holser and Magaritz, 1992; Retallack et al., 1998; Mory et al., 2000; Becker et al., 2001). Retallack et al. (1998) reported the occurrence of planar defects in quartz in continental P-Tr boundary sections in Australia and Antarctica, but stated that the evidence of impact at the P-Tr boundary was still ambiguous. The iridium data obtained by certain of these workers also vary considerably, with high values being reported for the Meishan section, but in most boundary sequences, an anomaly is either undetectable or of only moderate value (Yin and Tong, 1998). The unevenness of the iridium distribution may also point to an origin that is not extraterrestrial in nature. The geochemical evidence interpreted to favor an impact origin put forward by Becker et al. (2001) is also controversial and has been criticized by both Farley and Mukhopadhyay (2001) and Isozaki (2001).

Apart from these problems, fossiliferous continental P-Tr boundary sequences are fairly rare (Lozovsky, 1998) and notoriously difficult to date. Certain biological events (mass extinctions in particular) may be isochronous across a basin or region, and thus useful for correlation of P-Tr boundary sequences. Lozovsky (1998) noted that for correlation of these sequences, the most important fossils were to be found in the tetrapod faunas, and that the P-Tr boundary in the terrestrial realm is characterized by a change from Late Permian faunal communities containing *Dicynodon* (or closely allied forms) to one dominated by *Lystrosaurus*. Such a sequence has long been recognized in the Karoo basin of South Africa (Anderson and Cruickshank, 1978; King, 1990). In the south of the main Karoo basin, this sequence has been the focus of renewed activity and research in terms of its stratigraphy, paleontology (Smith, 1995; MacLeod et al., 2000; Ward et al., 2000), paleomagnetics (Kirschvink and Ward, 1998), and stable carbon and oxygen isotope signatures (Thackeray et al., 1990; MacLeod et al., 2000). Paleomagnetic studies (Kirschvink and Ward, 1998) have not been as successful as hoped for, in most part due to the remagnetization associated with Mesozoic igneous activity (Bachtadse et al., 1987).

Although much work has focused on the P-Tr boundary sequence in the south of the Karoo basin, the boundary in the north of the basin is not as well studied or understood; therefore, we focus on this understudied part of the basin.

REGIONAL GEOLOGY AND STRATIGRAPHIC OVERVIEW

The Karoo basin (Fig. 1) is a retro-foreland system that accumulated sediment from Late Carboniferous to Early Jurassic time in southwestern Gondwana (Johnson, 1991; Cole, 1992; Catuneanu et al., 1998). The sequence is highly asymmetrical in nature, being thickest in the south (proximal sector) and thinning dramatically to the north (distal sector) (Fig. 2).

The lithostratigraphic placement of the P-Tr boundary in the southern, proximal sector of the basin was a point of some contention (e.g., Hotton, 1967; Hiller and Stavrakis, 1984) until Smith (1995) showed that it could be placed within the upper part of the predominantly fluviatile Palingkloof Member of the Balfour Formation (Adelaide Subgroup). This coincides biostratigraphically with an overlap zone, which includes the last appearance datum (LAD) of *Dicynodon lacerticeps* and the first appearance datum (FAD) of *Lystrosaurus* (Smith, 1995; MacLeod et al., 2000). Historically the P-Tr boundary has been placed biostratigraphically at the contact between the *Dicynodon* and *Lystrosaurus* assemblage zones (Rubidge et al., 1995), i.e., the FAD of *Lystrosaurus*; however, Smith and Ward (2000) showed that the boundary is more correctly placed at the LAD of *Dicynodon*, and that the FAD of *Lystrosaurus* in the Karoo basin occurs within the Permian.

Although various authors believe that the Balfour and Katberg Formations are separated by an unconformity (Hotton, 1967; Groenewald, 1996), Smith (1995) and Ward et al. (2000) documented several sections in the central and southern parts of the basin where the transition from the mudrock-dominated Balfour Formation (Adelaide Subgroup) to the sandstone-dominated Katberg Formation (Tarkastad Subgroup) shows no evidence of disconformity or hiatus other than that normally associated with fluvial sequences (Smith, 1995; Smith and Ward, 2000), particularly at Lootsberg Pass (Fig. 1) and Bethulie. Smith and Ward (2000) also documented that there is little lithological change between the sections at Lootsberg and Bethulie, despite their south-north (proximal-distal) geographical separation of more than 200 km. In the south of the basin, the mass extinction that marks the P-Tr boundary takes place

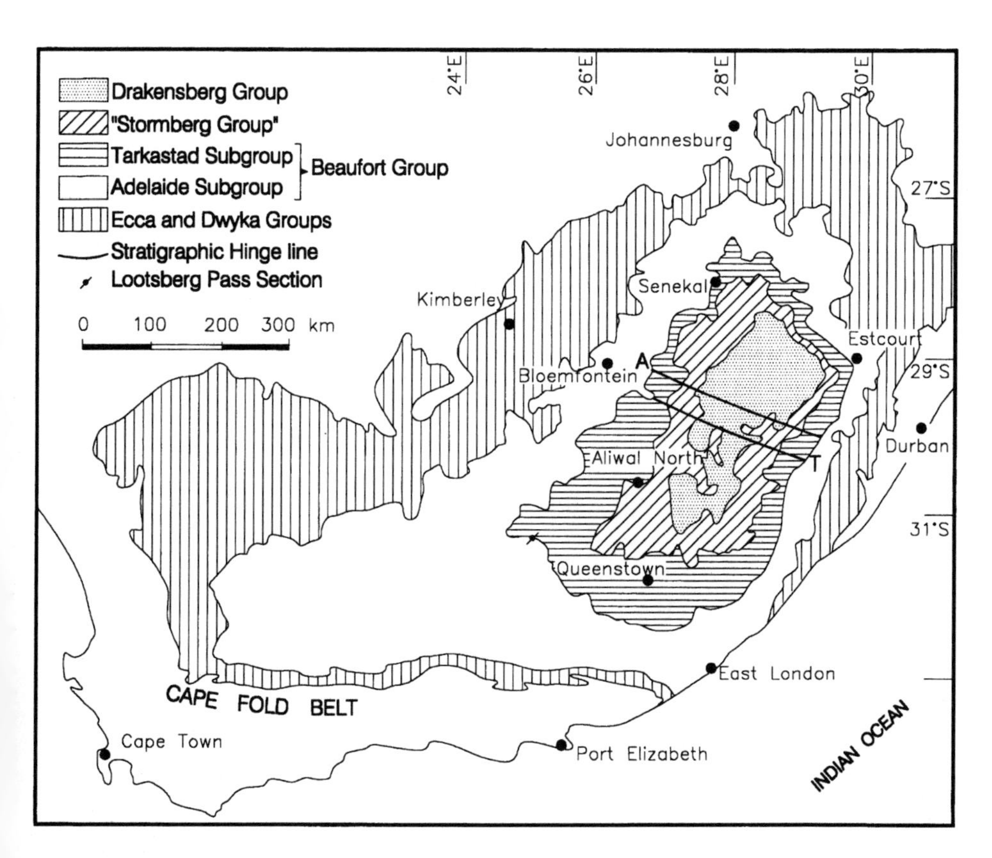

Figure 1. Geographic distribution of rocks of main Karoo basin in South Africa showing position of proposed stratigraphic hinge lines for the Adelaide (A) and Tarkastad (T) subgroups (after Catuneanu et al., 1998).

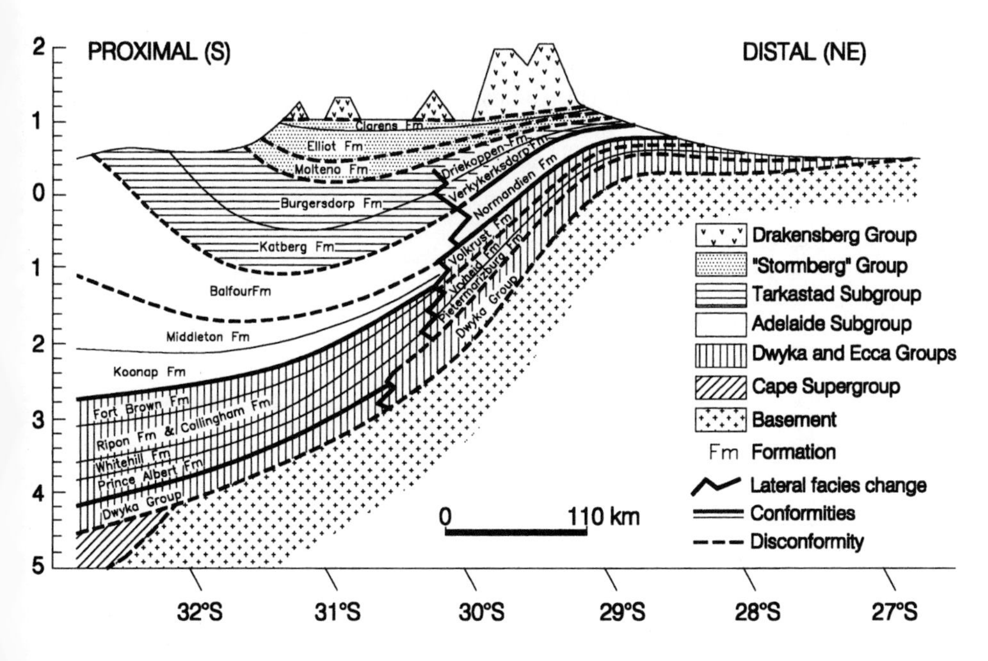

Figure 2. Proximal to distal stratigraphic changes in the Karoo retro-foreland basin between lat 27 and 30°S (after Catuneanu et al., 1998).

within a single facies and is overlain by an anomalously nonfossiliferous event bed above the LAD of *Dicynodon*. This event bed of laminated maroon mudrock occurs within a 30–40-m-thick sedimentary sequence marked by a reddening of the flood-plain rocks and a change from high- to low-sinuosity fluvial systems (Smith, 1995; Ward et al., 2000). This rapid and apparently basin-wide change from meandering to braided fluvial environments was originally linked to a tectonically driven increase in regional slope (Smith, 1995); however, Ward et al. (2000) linked it to increased sediment yield brought about by a hypothesized mass die-off of the vegetational cover. This mass extinction of plants at the P-Tr boundary is evidenced elsewhere in the world by a fungal spike (Visscher et al., 1996) that has recently been documented for the southern Karoo basin at Carlton Heights (M. Rampino, 2000, personal commun.), just north of Lootsberg Pass.

Catuneanu et al. (1998) proposed a reciprocal basin development model for the Karoo, and the P-Tr boundary sequence needs to be studied within the framework of this model. The reciprocal model postulates two distinct tectonic settings, with proximal and distal sectors separated by a hinge line, evidenced in the field by major facies changes (Figs. 1 and 2). This model states that in a basin in which dynamic subsidence plays no role, flexurally induced subsidence will be out of phase in the proximal and distal sectors (Fig. 3), and that temporally complete sequences can only be generated by stacking the out of phase proximal and distal sector stratigraphies. That is, any conformable stratigraphic sequence in the proximal sector will have a correlative unconformity surface in the distal sector, and vice versa. Catuneanu et al. (1997) and Pysklywec and Mitrovica (1999), however, also showed that in addition to static loading, dynamic loading by viscous mantle corner flow above a subducting plate might cause long-wavelength subsidence. The effect of this additional subsidence is shown in Figure 4. Pysklywec and Mitrovica (1999) further proposed that dynamic subsidence ceases in the Late Permian or Early Triassic, with the cessation of subduction of the paleo-Pacific plate.

Within the confines of the Karoo basin, both the Lootsberg and Bethulie sections occur in the proximal sector (Fig. 1). Therefore, if the P-Tr boundary sequence in the southern (proximal) sector is complete (as claimed by Smith and Ward, 2000; Ward et al., 2000), then the sequence in the northern (distal) sector should be represented by either an unconformity (including paleosol development or erosional downcutting) or by a condensed section (if dynamic subsidence plays a role in generating accommodation space). However, there are few data in the literature for these ideas to be vigorously tested, and insufficient focus has been placed on a boundary sequence in the northern (distal) sector.

HISTORICAL PLACEMENT OF THE BOUNDARY IN THE NORTH OF THE BASIN

The P-Tr boundary in the north of the Karoo basin has been placed lithostratigraphically at the base of the Tarkastad Subgroup (Kitching, 1977; Groenewald, 1989). In the northern (distal) sector this occurs at the contact between the Normandien and Verkykerskop formations (Groenewald, 1984, 1989), which are the distal lithological equivalents of the Balfour and Katberg Formations in the south of the basin (Fig. 2). Although not formally recognized by the South African Committee on Stratigraphy, the terms Normandien and Verkykerskop formation are used in this paper to distinguish these sequences from those in the southern (proximal) sector of the basin.

At the town of Senekal in the Free State (Fig. 5), the P-Tr boundary has historically been placed within an old quarry (now a nursery) to the northeast of the town (Kitching, 1977). Within this quarry the boundary is marked by a zone of brown-weathering calcareous nodules and concretionary structures that are laterally traceable over many kilometers. As for the boundary in the south of the Karoo basin, the correct documentation of the P-Tr boundary in the north requires an understanding of the temporal changes that occur in the fauna and flora across the contact. To this end a project was begun in 1994 to reevaluate the biostratigraphy of the P-Tr boundary in the north of the basin. This study focused in particular on the stratigraphic

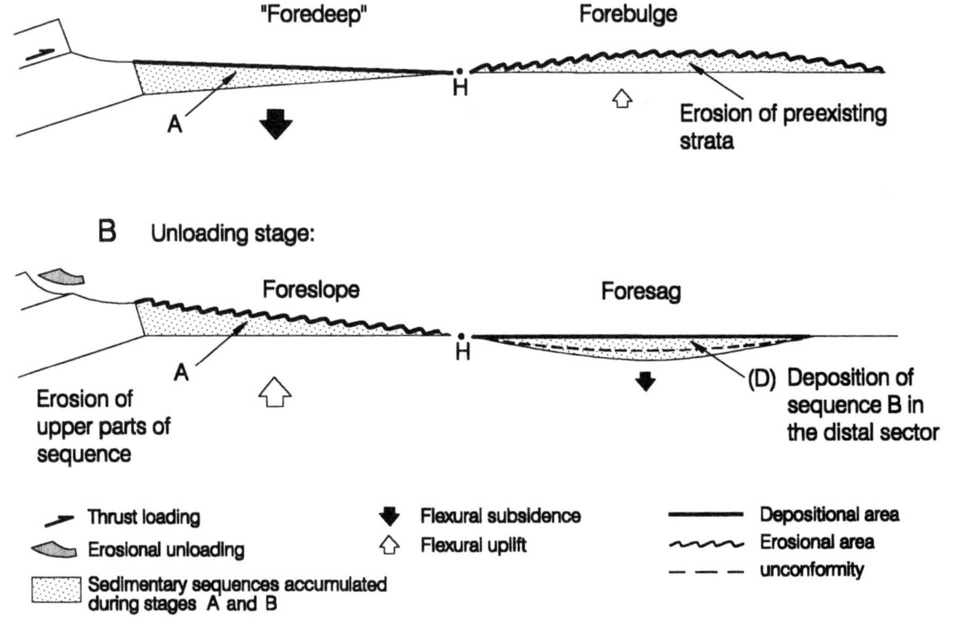

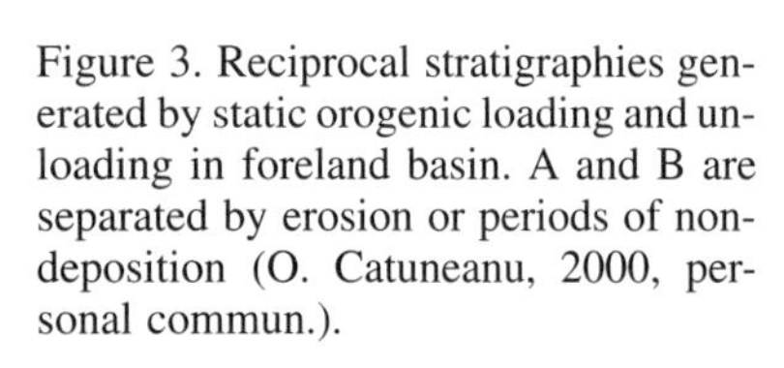
Figure 3. Reciprocal stratigraphies generated by static orogenic loading and unloading in foreland basin. A and B are separated by erosion or periods of nondeposition (O. Catuneanu, 2000, personal commun.).

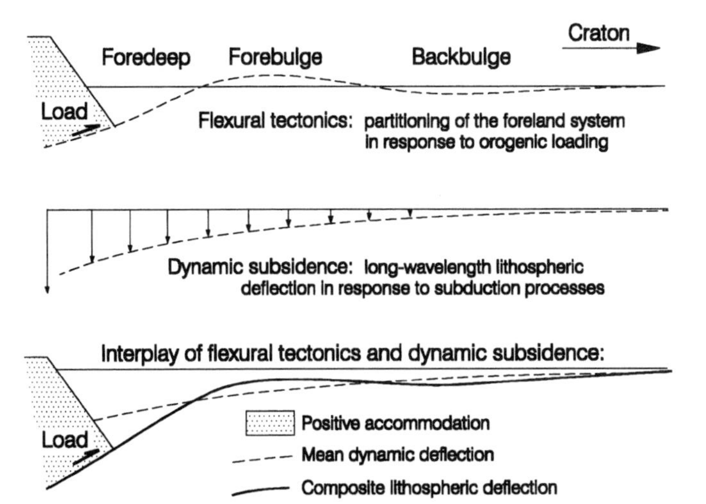

Figure 4. Effect of dynamic subsidence on generation of accommodation space in foreland basin (O. Catuneanu, 2000, personal commun.).

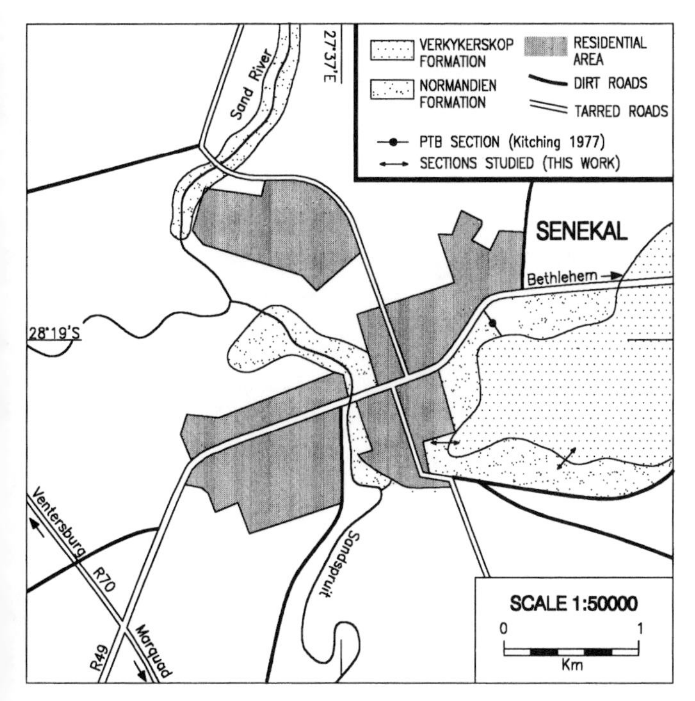

Figure 5. Map of town of Senekal showing position of historical contact section and sections studied for this project.

placement of the temnospondyl amphibian *Uranocentradon senekalensis*, known from several specimens from a single locality, originally described by Van Hoepen (1911) as being from a quarry in the hill near the town of Senekal. Most include *Uranocentradon* as a member of the Triassic *Lystrosaurus* assemblage zone fauna (Kitching, 1978; Groenewald and Kitching, 1995), making it the only amphibian genus in the Karoo basin considered to have survived the end-Permian mass extinction. However, recent studies of the locality of occurrence of this form (E. Latimer, 2000, personal commun.) favor its placement in the Permian Normandien formation, as a component of the *Dicynodon* assemblage zone fauna.

The initial study also showed that the section contained both *Dicynodon* and *Lystrosaurus* assemblage zone fossils, and by the current definition, the time span encompassing the terrestrial P-Tr boundary. This work, in turn, formed the background for an expanded study of the lithostratigraphy, paleontology, and geochemical signatures of the boundary section in the northern, distal sector of the basin. Preliminary results of this work were presented in Brandt et al. (2000) and Hancox (2000).

AIMS AND METHODOLOGY

The stratigraphy and paleontology of the P-Tr boundary section in the north of the Karoo basin have not previously been adequately documented or defined. This information is vital for determining whether the section is temporally complete in the north, and to determine how it relates to the sequence in the south. The main aim of this chapter is therefore to determine the correct placement of the P-Tr boundary in the northern (distal) sector of the Karoo basin, and to define the boundary in terms of its sedimentological, paleontological, and geochemical signatures.

To this end two sections were measured across the historical P-Tr boundary at the town of Senekal, and all lithological and paleontological parameters were noted. Fossil specimens were collected and are stored in the Bernard Price Institute for Palaeontological Research (BPI) in Johannesburg. Due to the paucity of tetrapod fossils in the sequence, the biostratigraphic significance (Bamford, 1999), and the abundance of fossil wood at all levels, and the fact that the P-Tr boundary should be marked by a major floral turnover (Visscher et al., 1996; Ward et al, 2000), it was decided to try and determine the placement of the P-Tr boundary based on the biostratigraphic occurrence of fossil wood. A number of mudrock samples were macerated for palynological analysis at the Bernard Price Institute, but to date have not proved to be productive.

Due to the fact that all previous evidence for impact at the P-Tr boundary has been disputed, and to aid in defining the P-Tr boundary in the north of the basin, the historical P-Tr boundary sequence at Senekal was sampled for geochemical and petrographic analysis. Two sections across the P-Tr boundary, south and southeast of the town of Senekal (Fig. 5), were sampled and analyzed chemically and petrographically, in particular searching for shocked quartz and for distinct chemical evidence. Samples were carefully collected from <1–2 cm intervals in these two sections. Because no significant carbonate (or other volatile-rich compounds) contents were noted in the course of cursory X-ray diffraction (XRD) analysis, sample preparation only involved drying at 40°C, careful screening for removal of any modern organic (plant) debris, and then powderization and/or homogenization with agate mortar and pestle.

The chemical analytical results are listed in Table 1 (section A) and Table 2 (section B). In these tables the sample numbers correspond to the distance (in centimeters) of a given sample from the presumed boundary level (A-0, B-0). Major element analyses were carried out on powdered samples by standard X-ray fluorescence (XRF) techniques (see Reimold et al., 1994, for details on procedures, precision, and accuracy). The concentrations of the trace elements Rb, Sr, Ba, Cs, V, Cu, Zn, Co, Ni, Cr, Nb, Y, and Zr, some of which were also analyzed by instrumental neutron activation analysis (INAA), were also determined by XRF. In Tables 1 and 2 where only some trace element data are presented, they represent XRF results. Where

TABLE 1. CHEMICAL DATA OF PROFILE A

	A − 13.0	A − 11.0	A − 8.0	A − 5.0 ≥ − 7.0	A − 5.0	A0 − 3.0	A − 2.0	A − 1.5	A − 1.0	A − 0.9	A − 0.3	A + 1.5	A + 3.5	A + 8.5
SiO_2	62.1	61.5	63.6	62.4	60.1		60.8	62.9	61.5	64.7		62.8	73.6	66.7
TiO_2	0.87	0.89	0.84	0.84	0.96		0.99	0.91	0.92	0.80		0.84	0.59	0.54
Al_2O_3	17.2	17.6	17.1	17.5	18.2		18.1	16.8	17.3	17.1		12.0	13.7	14.0
Fe_2O_3	6.09	6.22	5.85	5.73	6.52		6.12	5.98	6.22	5.06		11.9	2.57	6.73
MnO	0.11	0.11	0.10	0.10	0.23		0.24	0.12	0.59	0.10		0.14	0.33	0.11
MgO	1.79	1.65	1.45	1.52	1.72		1.62	0.84	1.38	1.47		0.39	0.31	0.44
CaO	1.21	0.86	1.24	1.29	1.27		1.14	1.54	1.03	1.23		0.97	1.18	0.85
Na_2O	2.06	1.28	1.50	1.47	2.01		1.48	1.87	2.00	1.40		2.94	2.47	3.14
K_2O	3.18	3.90	3.38	3.5	3.48		3.76	3.56	3.54	3.14		2.5	2.85	3.59
P_2O_5	0.14	0.14	0.18	0.18	0.19		0.15	0.52	0.12	0.18		0.13	0.08	0.08
LOI	5.17	5.72	4.95	5.25	5.49		5.60	4.72	5.34	4.87		5.49	2.45	3.74
Total	99.92	99.90	100.16	99.78	100.24		99.95	99.66	99.95	100.04		100.12	100.13	99.88
Sc	14.3	14.0	14.4	15.8	16.9	16.8	15.8	13.7	16.4	13.0	16.7	6.86	8.37	5.74
V	110		113	129			149			118			57	
Cr	79	75	79	81	85	84	83	73	93	75	96	63	51	35
Co	23.3	21.2	18.2	54.1	33.8	32.3	25.9	17.9	69.4	14.1	66.4	9.0	16.3	7.1
Ni	56	39	67	39	64	95	36	56	108	48	93	34	29	20
Cu	<2		7	8			3			6			<2	
Zn	101	110	105	108	132	147	112	106	121	103	121	57	58	42
As	20	16	14	20	<2	34	31	34	31	11	34	70	12	42
Se	<1	2	<1	1	1	2	<2	1	1	1	1	2	1	1
Br	0.4	0.6	0.4	0.3	0.3	0.1	0.5	0.4	0.3	0.3	0.6	0.4	0.4	0.4
Rb	183	163	176	162	182	184	167	141	167	152	148	99.9	113	105
Sr	181		239	224			223			240			383	
Y	34		39	41			38			37			20	
Zr	200	321	232	237	411	416	238	347	427	229	503	713	232	301
Nb	15		17	17			18			17			12	
Sb	4.38	3.03	4.31	5.53	<1	7.05	6.89	8.81	9.45	2.93	11.3	14.5	3.01	9.13
Cs	7.58	7.14	7.09	10.1	8.16	8.14	7.76	7.25	11.50	6.01	15.10	3.75	4.53	4.16
Ba	5410	4800	7150	8490	2210	2220	10700	2120	3320	6400	2780	6450	14100	6220
La	53.2	45.2	62.8	52.8	56.4	58.2	50.3	42.0	68.3	48.6	68.1	51.0	39.2	20.5
Ce	96.5	83.1	113	86.6	154	175	119	72.4	306	77.8	282	93.5	92.4	42.5
Nd	48.9	39.2	54.1	50.1	53.4	54.9	46.8	37.4	56.9	45.4	59.6	47.2	34.9	15.6
Sm	10.9	8.24	12.4	11.0	12.2	12.2	9.99	8.48	12.0	9.16	11.4	8.83	6.44	3.05
Eu	1.58	1.36	1.77	1.82	1.98	1.97	1.64	1.43	1.89	1.50	1.68	1.12	1.24	0.78
Gd	7.24	6.21	9.09	7.40	9.65	8.78	7.70	6.32	7.95	6.75	7.35	5.61	3.80	2.07
Tb	1.02	0.93	1.35	1.05	1.49	1.29	1.17	0.94	1.12	1.00	1.02	0.77	0.50	0.29
Tm	0.60	0.49	0.65	0.57	0.71	0.72	0.56	0.51	0.61	0.51	0.52	0.40	0.28	0.18
Yb	4.12	3.31	4.24	3.91	4.63	4.86	3.64	3.48	4.13	3.39	3.47	2.65	1.93	1.22
Lu	0.61	0.48	0.65	0.55	0.69	0.69	0.54	0.49	0.58	0.5	0.51	0.39	0.29	0.18
Hf	6.71	5.86	6.91	6.29	6.66	7.14	6.47	6.29	6.81	6.08	7.07	15.7	7.15	5.66
Ta	1.34	1.11	1.27	1.14	1.30	1.36	1.15	1.05	1.18	1.02	1.11	0.87	0.73	0.61
Ir (ppb)	<1	<1	<2	0.2	<1	<1	<1	<1	<1	<1	<1	<1	<1	<1
Ir (ppt)*	337	108	101		117	165	320		196	321	216	225	131	117
Au (ppb)	2	1.8	0.3	0.3	<10	2	1.5	1	3	1.5	2.4	<20	0.5	2.5
Th	20.3	18.1	20.0	18.3	21.2	21.3	19.1	17.1	19.8	17.2	19.7	26.0	13.0	8.03
U	17.8	15.8	22.3	22.5	16.4	18.1	18	13.1	19.8	11.8	34.2	26.1	19.3	11.3
K/U	1483	2047	1258	1291	1762	–	1734	2256	1484	2209	–	795	1226	2637
Th/U	1.14	1.15	0.90	0.81	1.29	1.18	1.06	1.31	1.00	1.46	0.58	1.00	0.67	0.71
La/Th	2.62	2.50	3.14	2.89	2.66	2.73	2.63	2.46	3.45	2.83	3.46	1.96	3.02	2.55
Zr/Hf	29.8	54.8	33.6	37.7	61.7	58.3	36.8	55.2	62.7	37.7	71.1	45.4	32.4	53.2
Hf/Ta	5.01	5.28	5.44	5.52	5.12	5.25	5.63	5.99	5.77	5.96	6.37	18.05	9.79	9.28
La/Yb	12.9	13.7	14.8	13.5	12.2	12.0	13.8	12.1	16.5	14.3	19.6	19.2	20.3	16.8
Eu/Eu*	0.54	0.58	0.51	0.62	0.56	0.58	0.57	0.60	0.59	0.58	0.56	0.49	0.77	0.95

**Note:* Determined by γ–γ coincidence spectrometry.

Blank spaces: not determined; major element data in weight %; trace element data in ppm, unless denoted otherwise. LOI is loss on ignition.

	B+15–18B	+12–15B	+12.5B	+11.5B	+10.0B	+8.0B	+6.5B	+4.4–5.0B	+4.0B	+3.0B	+2.0B	+1.0B	–OAB	–0BB	–0CB	–0.5AB	–0.5BB	–0.5CB	–1.0B	–1.5B	–2.5B	–3.5B	–4.0B	–5.0AB	–5.0BB	–8.5B	–9.0B	–10.0B	–11.0B	–12.0B	–13.0B	–14.0B	–16.0B	–17.5B	–18.0B	–25.0B	–37.0
SiO_2	72.69	73.59	74.04	73.71	73.99	73.02	64.41	72.34	72.5	72.99	72.69	67.84	63		65.61	70.03	63.36				63.76	67.62		66.94	69.62	64.28	64.26	64.75	65.86	66.93	64.12	64.82	64.31	63.82	63.52	64.42	68.91
TiO_2	0.57	0.51	0.53	0.53	0.54	0.53	0.91	0.51	0.53	0.51	0.56	0.80	1.00		0.85	0.49	0.81				0.93	0.81		0.81	0.79	0.84	0.78	0.79	0.79	0.69	0.76	0.77	0.86	0.85	0.87	0.79	0.69
Al_2O_3	13.92	13.28	13.78	13.94	13.55	13.89	16.54	15.15	14.68	13.53	14.47	15.26	17.11		16.11	13.53	15.89				15.31	15.29		15.16	14.52	16.03	16.29	16.24	15.64	14.37	16.56	16.17	16.53	16.65	16.71	16.26	14.9
Fe_2O_3	3.15	2.44	2.23	2.03	2.21	2.17	7.04	2.63	2.85	3.37	2.32	5.18	6.93	8.46	5.67	2.67	6.54	7.48	6.31	6.39	6.34	5.09		5.06	4.97	5.40	6.29	5.88	5.64	5.26	6.55	6.06	6.38	6.68	7.10	6.43	4.86
MnO	0.04	0.05	0.07	0.06	0.09	0.06	0.2	0.04	0.04	0.03	0.03	0.05	0.31		0.32	0.04	0.22				0.06	0.08		0.07	0.05	0.14	0.12	0.09	0.08	0.13	0.06	0.13	0.06	0.04	0.02	0.07	0.07
MgO	0.55	0.32	0.38	0.33	0.32	0.37	1.45	0.39	0.56	0.37	0.45	0.84	1.61		1.13	0.41	1.25				2.01	1.31		1.48	1.32	1.70	1.79	1.62	1.59	1.09	1.85	1.75	1.95	1.97	2.22	1.61	1.26
CaO	1.29	1.48	1.18	1.42	1.31	1.48	1.17	0.74	0.88	0.79	0.86	0.88	1.31		1.69	1.50	1.42				2.01	1.39		1.47	1.39	1.27	1.22	1.33	1.32	1.85	1.28	1.28	1.29	1.29	1.57	1.35	1.26
Na_2O	2.36	2.46	3.09	3.13	2.02	3.51	2.40	2.71	2.85	3.17	3.18	3.09	2.28	1.87	2.41	1.79	1.86	1.81	1.91	1.97	2.28	0		2.18	0	1.90	2.00	1.94	0	1.79	1.39	1.48	0	1.79	0	1.34	1.87
K_2O	2.14	2.40	2.19	2.13	2.30	1.49	3.28	2.74	2.65	2.82	2.90	2.29	1.75	2.84	1.76	2.24	1.98	3.55	3.43	2.37	1.81	2.79		2.69	2.39	3.01	1.98	3.19	2.82	2.74	3.07	3.04	3.15	1.73	3.20	3.09	2.90
P_2O_5	0.12	0.10	0.09	0.10	0.09	0.10	0.21	0.09	0.10	0.12	0.10	0.13	0.30		0.18	0.10	0.17				0.20	0.21		0.24	0.25	0.19	0.17	0.17	0.20	0.21	0.18	0.18	0.20	0.21	0.19	0.18	0.21
LOI	3.19	3.17	3.05	3.25	3.7	3.59	2.60	3.09	3.24	2.56	2.80	3.84	4.25		4.62	7.59	6.41				5.19	4.14		4.40	4.16	5.25	4.87	4.60	5.73	3.67	4.53	4.36	4.79	4.71	4.73	4.67	3.50
TOTAL	100.0	99.8	100.6	100.6	100.1	100.2	100.2	100.4	100.9	100.3	100.4	100.2	99.8		100.4	100.4	99.9				99.9	98.7		100.5	99.5	100.0	99.8	100.6	99.7	98.7	100.4	100.0	99.5	99.7	100.1	100.2	100.4
Sc			7.40	7.80	7.2	6.96	6.25	6.73	8.28	6.74	7.84	12.3	13.4	20.3	14.7	16.9	20.0	18.7	17.4	17.5	14.2	13.4	18.9	12.6	11.9	13.1	14.0	13.9						14.7			
V	71	58	60	50	46	50	53	56		63	60	99															100	97		81	103	103				110	74
Cr	46.0	40.0	34.8	40.3	45.2	40.0	41.0	33.5	37.7	32.0	33.0	61.0	64.8	96.6	68.3	80.7	89.3	92.5	83.4	86.2	74.1	70	92.3	66.2	64.9	65.0	67.1	71.0		67.0	82.0	78.0		70.1		83.0	63.0
Co	13.0	13.0	8.08	10.9	18.8	17.0	10.0	12.3	15.1	11.9	13.8	16.0	35.5	40.8	29.7	39.9	77.5	18.8	28.3	29.1	14.6	21.6	15.2	13.1	11.9	19.2	11.0	12.9		20.0	17.0	10.0		12.9		12.0	9.0
Ni	16	12	26	17	17	15	10	20	43	22	19	29	30	89	49	21	42	59	89	37	42	69	15	44	46	41	36	28		31	34	36		28		39	28
Cu	<2	<2	<2	<2	5	6	<2	<2		<2	<2	<2															4	<2		3	5	6				9	3
Zn	58	45	58	49	40	68	43	45	50	46	53	75	149	237	167	144	162	136	177	156	132	125	148	109	103	143	135	123		80	86	91		121		94	75
As			6.45	5.32	5.75	8.30	6.18	18.7	42.1	42.4	18.4	65.8	7.25	85.8	51.2	65.0	49.2	56.2	28.9	19.5	12.2	9.93	18.1	8.59	8.12	8.49	12	8.57						12.1			
Se			1.41	1.44	0.83	1.10	0.91	1.01	1.17	0.89	1.36	2.09	2.16	2.75	1.69	2.87	2.43	2.90	2.80	2.95	2.19	2.3	2.76	2.07	2.4	1.43	2.05	2.28						2.15			
Br			0.54	0.30	0.43	0.76	0.56	0.36	0.59	0.31	0.24	0.18	0.76	0.78	0.29	0.66	0.91	0.52	1.04	0.64	0.60	0.66	0.81	0.69	0.85	0.73	0.19	0.6						0.66			
Rb	94.0	87.0	100	93.6	77.9	93.0	79.8	102	106	91.1	106	132	142	175	131	139	145	172	177	160	142	146	181	131	125	140	154	157		125	141	151		165		151	127
Sr	288	314	296	333	334	337	335	324		313	310	263															216	182		255	212	182				184	237
Yb	23	20	18	17	18	19	19	19		16	17	22															41	34		36	40	41				41	34
Zr	159	147	138	170	192	163	165	154	228	150	132	247	366	409	335	346	407	325	351	339	336	282	337	273	298	262	205	173		196	194	185		222		216	219
Nb	10	11	11	11	11	10	10	10		11	11	15															17	14		15	17	18				17	15
Sb			1.08	0.95	0.73	0.84	0.61	1.51	2.21	2.59	1.48	2.74	1.79	6.67	4.50	4.03	4.58	5.23	2.86	2.06	1.95	1.4	1.74	1.45	1.28	1.09	1.28	1.42						1.93			
Cs			5.99	7.21	6.12	7.64	5.74	5.18	5.93	4.12	5.11	6.00	6.82	7.68	6.5	8.02	10.4	10.3	10.3	8.33	7.85	8.34	10.9	6.9	7.12	8.73	8.42	8.54						8.19			
Ba	5585	6690	6350	5004	5290	5885	5450	6170	1220	9585	7195	9560	8590	4110	5160	3190	2050	2500	2650	1500	1380	1540	1030	933	932	918	4702	4394		4774	3697	3284		702		3035	3745
La			23.4	21.2	21.3	28.2	25.2	21.5	24.5	21.9	19.8	28.6	69.5	67.2	58.3	41.5	158	32.5	65.7	96.5	69.7	58.5	70.3	45.4	77.6	52.0	55.6	52.3						51.4			
Ce			50.9	52.0	49	61.7	48.7	41.4	53.4	37.4	52.7	59.0	120	173	116	152	373	67.0	130	160	106	137	101	91.9	82.9	194	96.4	101						96.2			
Nd			23.5	24.9	20.9	25.8	20.8	20.9	24.5	15.8	22.4	26.2	40.6	61.4	52.5	44.9	125	31.8	65.1	76.0	65.8	57.9	64.9	41.4	54.6	41.1	41.4	46.4						45.5			
Sm			4.26	3.94	3.6	5.33	3.86	3.83	4.35	3.21	4.52	5.17	8.41	11.3	8.57	7.34	19.7	6.87	11.8	13.1	108	9.92	11.8	8.08	9.85	8.61	8.58	9.06						8.85			
Eu			1.03	0.99	0.9	1.23	0.97	0.93	1.11	0.82	0.97	1.12	1.87	2.23	1.98	1.59	3.55	1.32	2.16	2.53	2.07	2.17	2.27	1.69	2.11	1.81	1.69	1.72						1.57			
Gd			3.36	3.61	3.05	4.46	3.13	3.08	4.09	2.72	3.28	4.62	7.42	9.95	6.99	6.52	13.9	6.65	10.3	10.5	8.86	8.74	10.7	7.45	8.75	7.74	7.54	7.69						7.57			
Tb			0.52	0.59	0.49	0.70	0.48	0.47	0.68	0.43	0.50	0.75	1.18	1.63	1.09	1.11	2.04	1.13	1.68	1.65	1.39	1.44	1.76	1.25	1.43	1.27	1.18	1.26						1.21			
Tm			0.27	0.28	0.28	0.33	0.24	0.28	0.37	0.26	0.26	0.43	0.59	0.83	0.60	0.60	0.86	0.64	0.79	0.79	0.68	0.69	0.88	0.65	0.7	0.65	0.61	0.62						0.61			
Yb			1.78	1.82	1.88	2.17	1.59	1.84	2.44	1.68	1.73	2.96	3.88	5.51	4.07	4.06	5.45	4.31	5.13	5.15	4.45	4.52	5.84	4.32	4.56	4.31	3.97	4.10						3.93			
Lu			0.27	0.25	0.28	0.32	0.27	0.27	0.36	0.25	0.26	0.43	0.59	0.81	0.62	0.62	0.79	0.63	0.73	0.76	0.68	0.69	0.85	0.63	0.69	0.64	0.59	0.61						0.60			
Hf			5.07	5.39	5.53	4.85	3.83	3.75	4.99	4.05	3.75	8.10	7.37	7.19	6.79	6.54	7.05	7.69	7.02	7.07	6.57	7.2	7.35	6.31	6.81	5.46	5.47	5.64						5.10			
Ta			0.64	0.61	0.6	0.62	0.54	0.69	0.77	0.60	0.66	1.07	1.06	1.63	1.13	1.27	1.50	1.78	1.78	1.5	1.29	1.37	1.60	1.10	1.15	1.10	1.20	1.29						1.32			
Ir(ppb)			<1	<2	<1	<1	<1	<1	<1	<1	<1	<1	<1	<2	<2	<2	<2	<2	<1	<1	<2	<1	<1	<1	<1	<1	<2	<2						<1			
Ir(ppt)*			170		247	55	109	362	166	253	239	285	37		235	54		297	145	113	345	145	287	245	113	79	368	103									
Au			3	3	<7	<8	5	6	3	3	4	3	7	15	<9	3	26	7	5	<6	8	4	16	2	8	4	3	1						6			
Th			10.0	9.93	8.44	9.99	8.03	10.1	11.6	8.94	9.88	18.7	19.0	24.9	17.5	20.2	23.1	25.4	23.2	22	18.6	19.3	23.8	16.9	16.8	17.0	18.6	18.8						18.3			
U			3.75	3.62	2.53	5.17	2.81	4.59	3.27	4.02	3.38	7.22	4.65	7.84	5.04	7.45	10.1	9.25	7.85	6.6	5.59	5.03	7.19	4.72	5.2	4.83	4.88	5.29						5.77			
K/U			4848	4885	7547	2392	9690	4956	6727	5823	7123	2633	3124	3010	2899	2496	1627	3189	3631	2985	2688	4605		4731	3815	5173	3368	5006						2489			
Th/U			2.67	2.74	3.34	1.93	2.86	2.20	3.55	2.22	2.92	2.59	4.09	3.18	3.47	2.71	2.29	2.75	2.96	3.33	3.33	3.84	3.31	3.58	3.23	3.52	3.81	3.55						3.17			
La/Th			2.34	2.13	2.52	2.82	3.14	2.13	2.11	2.45	2.00	1.53	3.66	2.70	3.33	2.05	6.84	1.28	2.83	4.39	3.75	3.03	2.95	2.69	4.62	3.06	2.99	2.78						2.81			
Zr/Hf			27.2	31.5	34.7	33.6	43.1	41.1	45.7	37.0	35.2	30.5	49.7	56.9	49.3	52.9	57.7	42.3	50.0	47.9	51.1	39.2	45.9	43.3	43.8	48.0	37.5	30.7						43.5			
Hf/Ta			7.92	8.84	9.22	7.82	7.09	5.43	6.48	6.75	5.68	7.57	6.95	4.41	6.01	5.15	4.70	4.32	3.94	4.71	5.09	5.26	4.59	5.74	5.92	4.96	4.56	4.37						3.86			
La/Yb			13.1	11.6	11.3	13.0	15.8	11.7	10.0	13.0	11.4	9.66	17.9	12.2	14.3	10.2	29.0	7.54	12.8	18.7	15.7	12.9	12.0	10.5	17.0	12.1	14.0	12.8						13.1			
Eu/Eu*			0.83	0.80	0.83	0.77	0.85	0.83	0.80	0.85	0.77	0.70	0.72	0.64	0.78	0.70	0.66	0.60	0.60	0.66	0.20	0.71	0.62	0.67	0.69	0.68	0.64	0.63						0.59			

Note: *Determined by γ–γ coincidence spectrometry

Blank spaces: not determined; major element data in weight %; trace element data in ppm, unless otherwise denoted. LOI is loss on ignition.

both XRF and INAA could be obtained, only those data with relatively better accuracies are reported. INAA was carried out on 200 mg aliquots according to techniques described in detail by Koeberl (1993). In addition, iridium in selected samples has been analyzed by γ-γ coincidence spectrometry after irradiation with thermal neutrons, which allows the almost interference-free determination of iridium abundances at sub-part per billion (ppb) levels (Huber et al., 2000). Thin sections were prepared from all samples of the interval close to the presumed boundary layer, and studied with the aid of a petrographic microscope.

STRATIGRAPHY OF THE P-TR BOUNDARY SECTION AT SENEKAL

Lithostratigraphy

The P-Tr boundary section at the town of Senekal (Fig. 6) is contained within a ~65-m-thick sequence of intercalated mudrocks, siltstones, and sandstones of the Normandien Formation, and sandstones and minor conglomerates of the Verkykerskop Formation. The P-Tr boundary has in the past been placed at the level of a laterally continuous, brown-weathering, carbonate concretionary horizon (Kitching, 1977), which is ~1–2 m below the first sandstone attributable to the Verkykerskop Formation.

Below the level of the concretionary layer, the Normandien Formation consists of a sequence of intercalated mudrocks, siltstones, and lenticular sandstones. Although only one channel sandstone is shown at the base of the section in Figure 6, as many as three laterally discontinuous, single-storied lenticular sandstones may occur below the level of the concretionary layer. The basal sandstone in the section is a medium- to coarse-grained unit, internally stratified by horizontal (Sh) and trough cross-stratification (St). The upper two sandstones are much finer, rarely being medium grained. These channel sandstones and overbank fines are here interpreted to represent deposition in a high-sinuosity, mixed-load meandering system. The coarse nature of the lowermost sandstone would preclude it having the

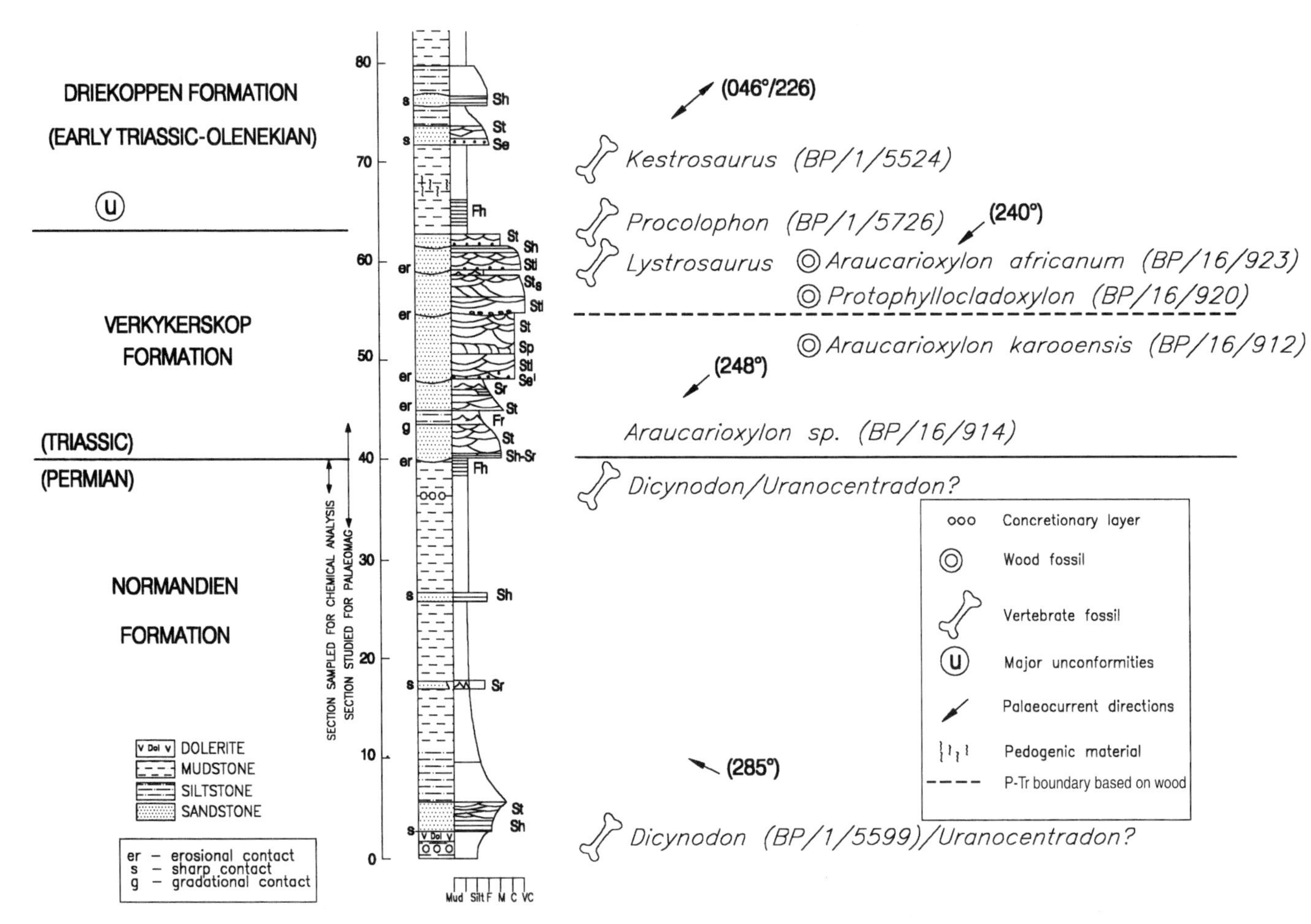

Figure 6. Generalized stratigraphic log of Permian-Triassic (P-Tr) boundary section at Senekal in northwestern Free State. Occurrence of *Lystrosaurus* documented at ~60 m is based on previous find by J.W. Kitching. Facies codes are as follows: Se, erosional scour fill; Sei, intraformational scour fill; Sh, horizontally stratified sandstone; St, trough cross-stratified sandstone; Stl, large-scale trough cross-stratified sandstone; Sts, small-scale trough-cross stratified sandstone; Sr, ripple cross-stratified sandstone; Sp, planar cross-stratified sandstone; Fh, horizontally laminated fines; Fr, ripple cross-laminated fines.

same source as the channel sandstones of the Balfour Formation in the south (which are much finer) and would favor an intrabasinal source, or cannibalization of the proximal sector strata during late-stage orogenic unloading (Fig. 3).

The succession above the concretionary layer may be formed by either a fine-grained, lenticular sandstone, or by a 1–2-m-thick sequence of horizontally laminated fines (Fh). Petrographically this unit shows lamination on a millimeter to submillimeter scale; the silt fraction is composed predominantly of quartz and kaolinite, with slightly larger diagenetic muscovite.

The contact between the Normandien and overlying Verkykerskop formations may be strongly erosive, and at places the laminated siltstone-mudrock unit has been removed. At other places the basal part of the Verkykerskop Formation may directly overlie fine-grained lenticular sandstone units of the Normandien Formation. At Senekal, the Verkykerskop Formation is composed of five laterally continuous, medium- to coarse-grained sandstone bodies, at places separated by thin units of ripple-laminated siltstones (Fr). This association is most common at the base of the sequence, whereas upsection, erosional downcutting often removes the preceding siltstone unit. Evidence of this downcutting is often documented at the base of the major sandstone units by accumulations of intraformational mud-pebble conglomerates. In general the sandstones of the Verkykerskop Formation are dominantly multistory, laterally extensive, and internally structured by horizontal stratification (Sh) and large-scale trough cross-stratification (St_l), and there is evidence of channel switching and erosional downcutting and scour. Two major episodes of extensive erosional downcutting in particular are documented in the sequence. The first occurs just below the 50 m mark (Fig. 6) and shows only evidence of intraformational scour, whereas the more extensive surface at the 55 m mark has accumulations of extraformational metaquartzite clasts. Paleocurrent evidence documents a source area to the northeast, at variance to the source area for the underlying Normandien Formation that is from the southeast. Petrographically the sandstones of the Verkykerskop Formation consist predominantly of fairly well sorted, subrounded quartz grains, and lesser feldspars (dominantly plagioclase), muscovite, and biotite. Quartz grain boundaries may show evidence of pressure solution; the sandstones in general are bound by rim cements of silica and iron oxide.

Due to their alluvial architecture and internal structure, unimodal paleocurrent directions, and the fact that they are bedload dominated, the channel sandstones of the Verkykerskop Formation are interpreted as having been deposited by large braided rivers receiving detritus from the northeast. The overall coarsening-upward nature of the sequence and sequential downcutting, evidenced by the incised erosional basal surfaces, reflect either a decrease in accommodation space through time, or an increase in sediment supply and slope.

Above the Verkykerskop Formation, the sequence changes abruptly to one dominated by laminated maroon mudrocks of the Driekoppen Formation (the northern equivalent of the Burgersdorp Formation). This abrupt facies translocation represents a major change in fluvial style and an increase in accommodation space in the distal sector (Hancox, 1998a). Although not well documented in the literature, such rapid translocations of facies may represent sequence boundaries (e.g., Rogers, 1994). The included lower *Cynognathus* assemblage zone fauna of the Driekoppen Formation is now well documented as latest Early Triassic (late Olenekian) in age (Hancox et al., 1995; Shishkin et al., 1995; Hancox, 1998a, 1998b; Damiani, 1999).

Tetrapod biostratigraphy

Tetrapod remains are scarce in the sections at Senekal; however, fossils of the dicynodont *Dicynodon* are known from two quarries in the Normandien Formation (Kitching, 1977) (Fig. 6). Of these, only a single specimen (BP/1/5599) was worthy of collection. The stratigraphic placement of the temnospondyl amphibian *Uranocentradon* has recently been reviewed and is included in Figure 6 at both of its proposed localities. We have not found *Lystrosaurus* at this section; however, Broom (1912) documented its occurrence here, and J.W. Kitching showed one of us (Hancox) where he collected this form at Senekal. To the east of the town the uppermost sandstone of the Verkykerskop Formation has produced fossils assignable to the *Lystrosaurus* assemblage zone, in particular the Triassic captorhinid reptile *Procolophon* (BP/1/5726) (Fig. 7). This sandstone is beneath the base of the late Early Triassic (late Olenekian) Driekoppen Formation and, based on the assignment of *Procolophon* to the early Scythian (early Olenekian) (Shishkin et al., 1995), the boundary between these two formations represents an unconformity surface of some magnitude.

Wood biostratigraphy

In terms of biostratigraphy of wood in the main Karoo basin (Bamford, 1999), the P-Tr boundary is believed to coincide with the LAD of *Australoxylon teixeirae* and *Araucarioxylon karooensis*, and the FAD of *Prototaxoxylon africanum*. At Senekal, fossil wood (Fig. 8) is preserved throughout the section between the last stratigraphic occurrence of *Dicynodon* and the first occurrence of *Lystrosaurus* and *Procolophon*.

To date the fossil wood data show that *Protophyllocladoxylon*, a Triassic genus (M. Bamford, 2000, personal commun.), occurs at the 55 m level (Fig. 6), below a large in situ (Fig. 8) log assigned to *Araurácarioxylon africanum*. This latter species is known to cross the P-Tr boundary in the Karoo basin. In situ specimens of *Araurácarioxylon karooensis* (and *Araurácarioxylon* sp.) occur fairly frequently below the 55 m level. On the basis of the wood data, the P-Tr boundary would therefore occur within the Verkykerskop Formation, and would tie in lithologically with an erosional unconformity marked by numerous extraformational clasts of metaquartzite at the 55 m mark.

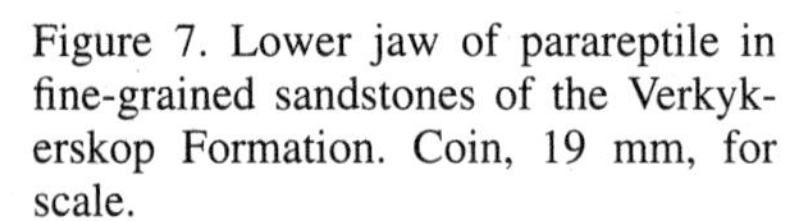

Figure 7. Lower jaw of parareptile in fine-grained sandstones of the Verkykerskop Formation. Coin, 19 mm, for scale.

Figure 8. In situ *Araucarioxylon africanum* from the upper part of the Verkykerskop Formation. Hammer, 30 cm, for scale.

SEARCH FOR EVIDENCE OF IMPACT

The two sections sampled for evidence of impact consist of horizontally laminated to massive interbedded mudrock and/or siltstone and fine sandstone, overlain by a laterally continuous, horizontally and planar cross-stratified sandstone that contains iron-oxide nodules. These nodules are mainly found in a 15 cm zone directly overlying the boundary. The sandstones display an erosive lower contact and may have scoured away a significant portion of the original clay surface. In thin section, the clay-rich siltstone shows banding on a millimeter to submillimeter scale. The silt-sized grains consist predominantly of quartz and kaolinite (confirmed by XRD), with interspersed, slightly larger grains of diagenetic muscovite. Iron-oxide staining in narrow bands (thought to have originally been organic-rich bands) is primarily responsible for darkening of certain zones. The sandstones consist mainly of well-sorted, rounded to very angular quartz grains, and minor feldspar (mostly plagioclase), biotite, and muscovite. The latter is thought to also be of diagenetic origin. Quartz grain boundaries show evidence of pressure solution, which is probably, in part, responsible for the high angularity of the grains. Microdeformation features observed in the quartz grains are only those expected for quartz sourced from a metamorphic provenance, and none of the samples studied from this interval display any petrographic evidence that could be linked to impact-generated shock metamorphism.

The chemical results listed in Table 1 for profile A clearly show that the samples from below the presumed boundary position are distinctly different from those above the boundary sample (A-0.3). In terms of major elements, Al_2O_3 and MgO in samples above the boundary are depleted, whereas Fe_2O_3 and Na_2O are significantly enriched. The trace elements Sc, V, Co, Ni, Rb, and Cs, and the rare earth elements (REE) Eu, Tb, Yb, and Lu, as well as Ta, are relatively depleted above the boundary. These samples have relatively higher concentrations of Sr and Hf. Most important, the sample from just about at the boundary (A-0.3) only shows values for Cr, Co, and Ni, which are also attained in other samples away from the boundary. No iridium enrichment in the samples from close to the boundary can be reported (compare the γ-γ coincidence spectrometric data in Table 1). This result is also evident in the profile plots of Figure 9. In fact, the few excursions of individual data for siderophile elements do not coincide with the exact position of the P-Tr boundary, nor do they always coincide for the various elements plotted. The REE patterns for the two sample traverses are shown in Figure 10. The patterns for sample profile A are distinctly different with regard to the samples from above and below the boundary. Samples from above the boundary are characterized by relatively lower abundances of the REEs and have either a lesser Eu anomaly than the samples from below the boundary, or lack such an anomaly completely. In addition, it is obvious that two samples (A-0.3 and A-1.0) from just at the boundary level are characterized by patterns that are generally similar to those for samples from below the boundary, but exhibit distinct Ce anomalies. The reason for this effect is unclear and could be the result of secondary hydrothermal processes. However, it cannot be excluded that this Ce anomaly is a primary feature of the boundary layer material.

Several samples were collected at the boundary in profile B, and the chemical data for samples from above, at, and below the boundary are compared in Table 2. With regard to the major element data, samples from above and below the boundary are not significantly different, with the notable exception of Fe_2O_3 contents that are relatively higher in most boundary and below-boundary samples. MgO is also relatively enriched in these samples. A number of trace element data, however, discriminate samples from above the boundary from those at and below the boundary rather well: enrichments for Sc, Cr, Co, Ni, Zn, As, and all REEs, Hf, Ta, and Th are noted. Boundary samples are distinguished from below-boundary samples by slightly higher Cr, distinctly higher Co, and variable Ni, including several very high values, higher Zn, variable but in some samples significantly enriched As, on average slightly enriched Zr, and marginally lower REE contents.

The siderophile element profiles plotted in Figure 9 for the B section samples illustrate that some positive excursions for Cr and Ni occur at or near the presumed boundary layer. These are not, however, accompanied by simultaneous Co excursions, and in part can be linked to Fe enrichment. This leaves the possibility that these effects could be of either primary or secondary nature. In addition, it is obvious from Table 2 that these positive excursions do not always coincide with relatively high iridium concentrations. It is also interesting to note that two of the boundary samples have high Au abundances, which coincide with other mobile elements such as Rb and As. This finding may also suggest hydrothermal alteration.

The REE patterns for section B samples (Fig. 10) indicate significant differences between samples from above, at, and below the boundary. Samples from above the boundary are characterized by relatively lower REE abundances and less pronounced Eu anomalies. Boundary, or near-boundary, samples cover a relatively wide range of REE abundances, but show the same patterns as samples from below the boundary.

With regard to both sets of samples, the REE patterns are typically light REE enriched and fractionated, as well as heavy REE depleted and relatively unfractionated. The overall ranges of abundances are similar for both sections, as is the presence or absence of pronounced Eu anomalies in samples from above and below the boundary, respectively. It can be speculated that these significant changes in chemical characteristics from below and/or at the boundary, and from above it, may be the result of different source areas for the two sequences, different environmental conditions of sedimentation, or of postdepositional chemical overprint. There is strong chemical evidence for a well-defined change at the historical P-Tr boundary at Senekal, as documented by sedimentological and tetrapod paleontological evidence.

DISCUSSION

From the work undertaken to date, it is evident that the resolution for the placement of the contact in the northern (distal) sector of the main Karoo basin is not as refined as in the south. Accepting the paleontological definition for the P-Tr boundary in the south of the basin (Smith, 1995; Ward et al., 2000), it is clear that the boundary must be between the 38 m and 60 m level in the section at Senekal (Fig. 6), because this is the stratigraphic succession between the LAD of *Dicynodon* and the FAD of *Lystrosaurus* and *Procolophon*. *Dicynodon* assemblage zone fossils dominate below the level of the base of the Verkykerskop Formation, and unlike in the south of the basin, no faunal overlap zone with *Lystrosaurus* or *Lystrosaurus* assemblage zone forms have been documented.

It is also pertinent here to address the question of the stratigraphic placement and age assignation of *Uranocentradon senekalensis*. As previously stated, most believe that this form is from the *Lystrosaurus* assemblage zone (Groenewald and Kitching, 1995); however, regardless of whether they came from the upper or lower quarry in the section at Senekal (Fig. 6), the *Uranocentradon* specimens would still be associated with a Permian *Dicynodon* assemblage zone fauna. This finding may account for the previously supposed anomalous occurrence of a large temnospondyl form in the *Lystrosaurus* assemblage zone, and it appears therefore that *Uranocentradon* may be added to the list of end-Permian casualties in the Karoo basin.

Geochemical and isotopic studies have shown that there is also a marked difference in the chemical signatures of the up-

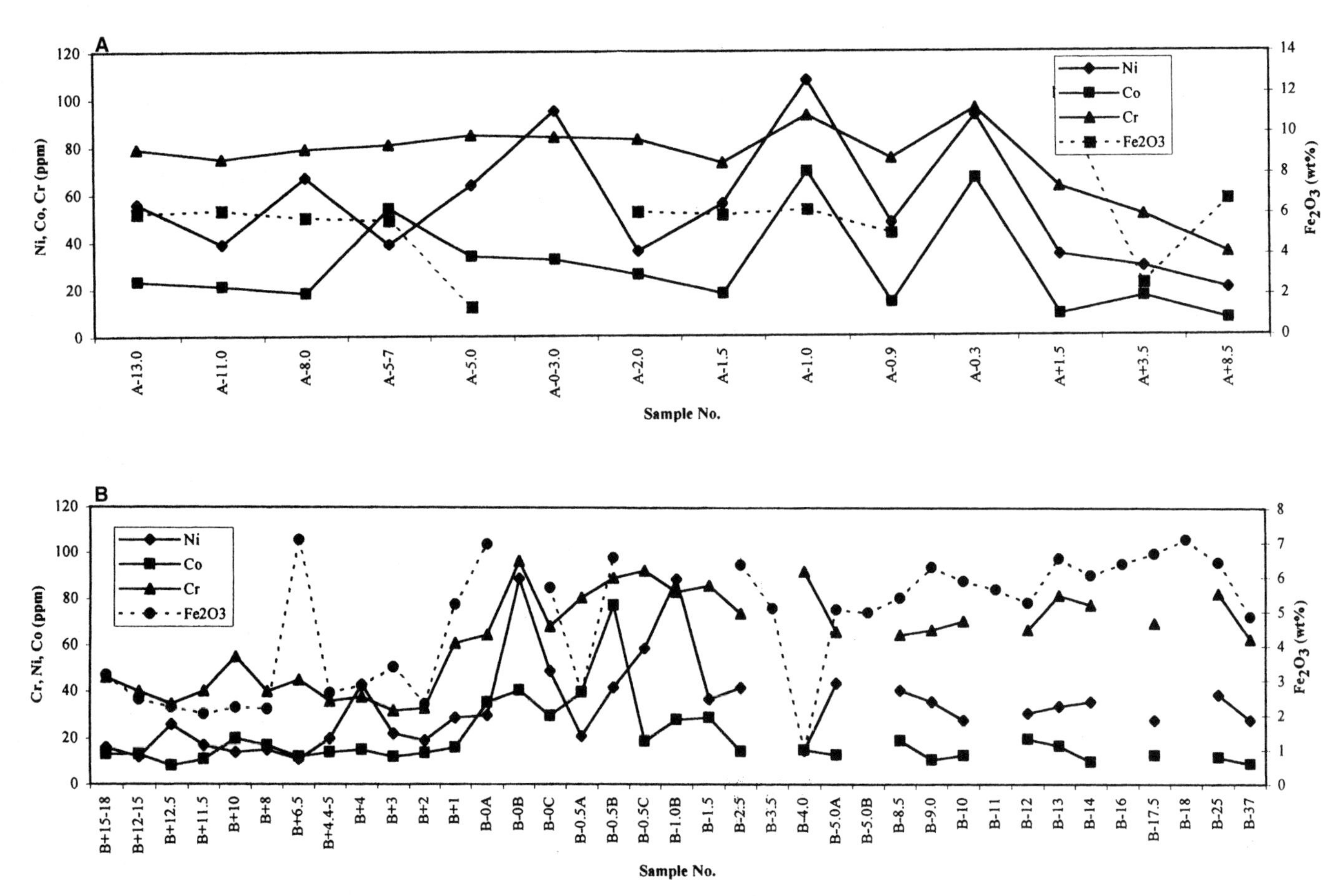

Figure 9. Profiles for several trace elements (in ppm) and Fe_2O_3 (in wt%) across sections A and B at Senekal. Sample numbers refer to sample distances (in cm) from the presumed boundary.

permost Normandien and lowermost Verkykerskop formations across the historical boundary. At present, however, no evidence of impact in terms of an iridium spike, enrichment in siderophile elements, or planar deformation features in quartz has been discovered, and therefore no unequivocal evidence of impact is known in any continental P-Tr boundary section.

Although the sequence in the north of the basin contains evidence of a change from meandering to braided river fluvial styles, as documented for the proximal sector by Smith (1995) and Ward et al. (2000), it differs in a number of significant ways. First, the change is not gradational and there is no thick (30–40 m) intermediate facies sequence of red fines with concretionary structures, and second, there is a change in source area from the southeast for the Normandien Formation, to the northeast for the Verkykerskop Formation. Ward et al. (2000) attributed this change in fluvial style to major vegetational die-off, because they believed it could no longer be tied to a period of tectonic uplift in the south of the basin. Accepting that the 246 ± 2 Ma compressional tectonic event (Hälbich, 1992) now postdates the P-Tr boundary, this would seem like a sound argument; however two additional mechanisms may be invoked to account for a steepening of regional slope and change in fluvial style across the P-Tr boundary in the south. These are early unloading stage isostatic adjustment (Catuneanu et al., 1998) and/or loss of dynamic support coupled to cessation of flat-plate subduction (Pysklywec and Mitrovica, 1999).

Recent studies of the nature of generation of accommodation space in foreland basins (e.g., Catuneanu et al., 1997, 1998) have furthermore shown that sequences in the proximal sector are related to orogenic loading stages (Fig. 3), possibly coupled to dynamic deflection from mantle flow. However, Pysklywec and Mitrovica (1999) showed that for the main Karoo basin, mantle flow-induced dynamic deflection is insufficient to account for all of the observed near-field (proximal) subsidence. In effect, this means that if the coarsening-upward sequence that encompasses the P-Tr boundary in the south of the basin is temporally complete, as suggested by Ward et al. (2000), then proximal sector load-induced subsidence must have occurred during the Late Permian in order to generate accommodation space for the deposition of the Balfour Formation. This would be a necessary part of any model trying to account for proximal sector subsidence, even in the absence of any dateable tectonic event. Although the change in fluvial style across the P-Tr boundary sequence in the south of the basin may well tie in

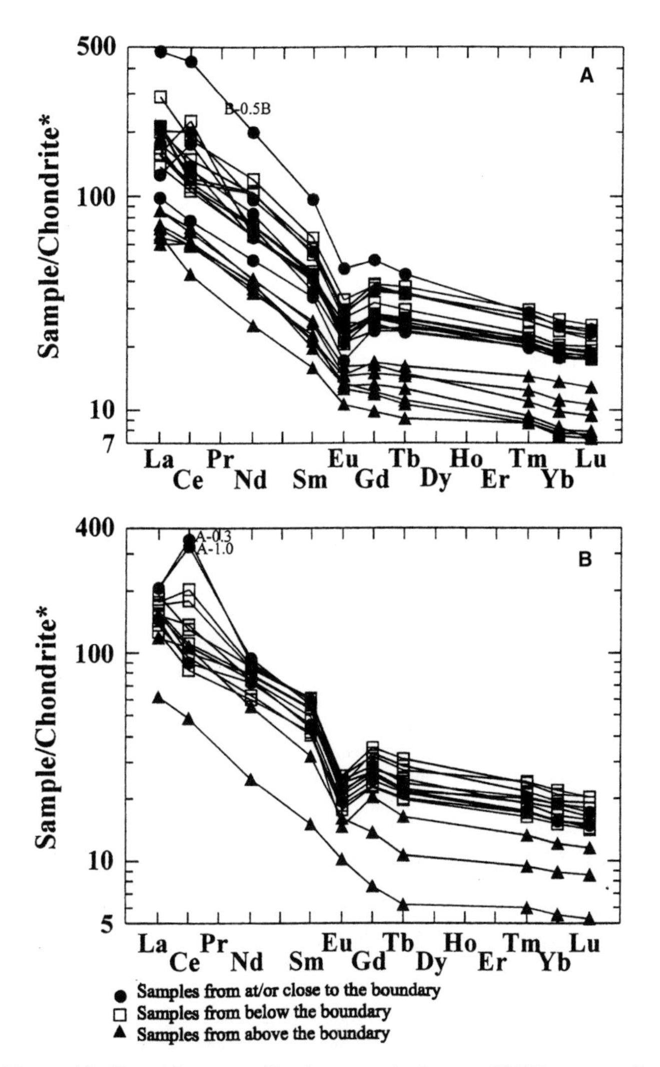

Figure 10. Chondrite-normalized rare earth element (REE) patterns for samples from profiles A and B (normalization factors after Nakamura, 1974).

with a vegetational die-off, the causal effect of this mechanism is therefore not beyond reproach, and other causal factors should not be discarded without significant further evidence. Changes from meandering to braided fluvial styles are common throughout the Beaufort Group, and are well documented in the Burgersdorp Formation (Hancox, 1998b), where no significant change in vegetation is known.

On the basis of the biostratigraphic cut-off for the P-Tr boundary on fossil wood (Bamford, 1999), the boundary in the Senekal section would be above the historical contact, and seems to coincide with an erosional unconformity at approximately the 55 m mark in Figure 7. Below this level, all the fossil wood collected to date may be assigned to *Araucarioxylon karooensis* (or *Araucarioxylon* sp.), a form believed to be restricted to the Permian in the Karoo basin (Bamford, 1999). The amount of large logs of fossilized wood in the northern P-Tr boundary section may be in keeping with the die-off of big trees and cover vegetation proposed by Ward et al. (2000); however, no fungal spike has been documented. This is contrary to the situation in the south of the basin, and may be further evidence that in the north of the basin this period of time is actually represented by nondeposition or erosion.

The data from the northern sector of the main Karoo basin allow for two possible placements of the P-Tr boundary. Biostratigraphic data based on tetrapods, the change in fluvial style, and paleocurrent direction, coupled to the change in chemical signature, would allow for the boundary to be maintained at or near its historical placement (Kitching, 1977), above the brown-weathering concretionary layer at the base of the Verkykerskop Formation. Given this scenario, the actual temporal boundary may be represented either by paleosol development (concretionary horizon), or by an erosional surface at the base of the Verkykerskop Formation. In either scenario, dynamic subduction could not have played a major role in the creation of distal sector accommodation space during proximal sector subsidence in the Late Permian. If the boundary is represented by a temporal unconformity, this is what a purely flexural reciprocal model without dynamic subsidence would predict. In this model the time missing would be equivalent to the temporal P-Tr boundary in the south of the basin. This finding is, however, somewhat at odds with the model of Catuneanu et al. (1998), who documented the contact between the Normandien and Verkykerskop formations as being conformable.

Because both the Normandien and Verkykerskop formations occur in the distal sector (foresag) within the current reciprocal model for the main Karoo basin (Catuneanu et al., 1998), it is interesting to postulate on the mechanism and timing of the generation of accommodation space in the distal sector. Such subsidence may be accounted for in two major ways, long-wavelength subduction-induced dynamic subsidence, and reciprocal flexure. Pysklywec and Mitrovica (1999) postulated that long-wavelength subduction-induced subsidence would produce considerable subsidence in the distal sector of the Karoo basin, but this subsidence would decrease significantly with the decoupling of the descending slab. Given that there is little evidence for any dynamic subsidence at the P-Tr boundary in the north of the basin, it is possible that decoupling of the paleo-Pacific plate occurred toward the end of the Permian. Reciprocal flexural tectonic settings have, however, been invoked to explain the creation of distal sector accommodation space and the complex stratigraphic tie-up between the sequences in the south and north of the Karoo basin.

Within a flexural model the documented changes in source area and paleocurrent directions for the Normandien and Verkykerskop formations could be accounted for by their position relative to the spatial configuration of the foresag. That the Normandien Formation is sourced from the southeast suggests it is most likely that the sequence represents deposition on the northerly dipping proximal part of the foresag, whereas the northeasterly sourced Verkykerskop Formation represents deposition on the southerly dipping distal foresag slope. This interpretation would be in line with the southerly shift of the hinge line between the Adelaide and Tarkastad subgroups postulated by Catuneanu et al. (1998) (Fig. 1).

CONCLUSIONS

Although at present the exact placement and nature of the boundary in the northwest of the basin are not as well understood as those in the south, the data of this study support the historical placement of the contact, and show that the temporal equivalent of the P-Tr boundary in the south of the basin is represented by a hiatal surface in the north. This fact has important implications for biostratigraphic studies of the P-Tr boundary in the Karoo basin, because such a sequence boundary creates the illusion of rapid change (extinction) across the boundary. This effectively prohibits any meaningful study of the tempo of faunal or floral change over the P-Tr boundary in the north of the Karoo basin. Because the temporal P-Tr event is not preserved, this fact also has ramifications for studies for evidence of impact or volcanism. Further studies as to the cause of the terrestrial P-Tr boundary event in the Karoo basin should therefore focus only on the P-Tr boundary sequence in the temporally complete southern (proximal) sector of the basin.

The findings of this study are entirely compatible with recent models predicting out of phase reciprocal stratigraphies, and decoupling of the descending slab toward the end of the Permian, but have shown that the current reciprocal model for the main Karoo basin (Catuneanu et al., 1998) may require slight amendment and refinement. The study has also shown that an understanding of the basinal context of a proposed P-Tr boundary sequence is as important as sedimentological, stratigraphic, or paleontological studies, and that, in future, workers should be aware of the tectonic setting of the sections they are studying.

Further studies will focus on better resolution of the boundary layer, and will include additional fossil collecting and a paleomagnetic study. A full taxonomic reassessment of both *Dicynodon* and *Lystrosaurus* would also be beneficial for understanding the temporal ranges of the different species of these forms, which may in turn allow for the amount of time missing at the northern P-Tr boundary sequences to be estimated. Stable carbon isotope work on the carbonate concretionary layer, pedogenic nodules in the intraformational lags of the Verkykerskop Formation, and on dicynodont tusks will also be undertaken to try and isolate any carbon excursion in the sequence, as documented in the south of the basin by MacLeod et al. (2000). We hope that these ongoing studies will help to identify the true nature and stratigraphic placement of the P-Tr boundary contact in the north of the main Karoo basin, which, in turn, will allow for better correlation between the proximal and distal sectors, as well as aiding the refinement of basin development models.

ACKNOWLEDGMENTS

Hancox thanks Octavian Catuneanu for his help in understanding the nature and tectonic histories of foreland basins, and for his many helpful discussions about the Permian-Triassic Karoo stratigraphy. We thank Ken MacLeod, Mike Rampino, and Roger Smith for their reviews and helpful comments on this chapter. This research is supported in part by grants from the National Research Foundation (South Africa) and the Research Council of the University of the Witwatersrand. This is University of the Witwatersrand Impact Cratering Research Group Contribution 27.

REFERENCES CITED

Anderson, J.M., and Cruickshank, A.R.I., 1978, The biostratigraphy of the Permian and Triassic. 5. A review of the classification and distribution of Permo-Triassic tetrapods: Palaeontologia Africana, v. 21, p. 15–44.

Bachtadse, V., van der Voo, R., and Hälbich, I.W., 1987, Palaeomagnetism of the Western Cape Fold Belt, South Africa, and its bearing on the Paleozoic apparent polar wander path for Gondwana: Earth and Planetary Science Letters, v. 84, p. 487–499.

Bamford, M., 1999, Permo-Triassic fossil woods from the South African Karoo Basin: Palaeontologia Africana, v. 35, p. 25–40.

Becker, L., Poreda, R.J., Hunt, A.G., Bench, T.E., and Rampino, M., 2001, Impact event at the Permian-Triassic boundary: Evidence from extraterrestrial noble gases in fullerenes: Science, v. 291, p. 1530–1533.

Bowring, S.A., Erwin, D.H., Jin, Y.G., Martin, W.W., Davidek, K., and Wang, W., 1998, U/Pb zircon geochronology and tempo of the end Permian mass extinction: Science, v. 280, p. 1039–1045.

Brandner, R., Donofrio, D.A., Krainer, K., Mostler, H., Nazarow, M.N., Resch, W., Stingl, V., and Weissert, H., 1986, Events at the Permian-Triassic boundary in the southern and northern Alps: Società Geologica Italiana, Pavia, Abstracts, p. 15.

Brandt, D., Hancox, P.J., Reimold, W.U., and Koeberl, C., 2000, Search for impact evidence at the Permo-Triassic boundary in the northeastern Free State province, South Africa, *in* Catastrophic events and mass extinctions: Impacts and beyond: Houston, Texas, Lunar and Planetary Institute, LPI Contribution No. 1053, p. 15–16.

Broom, R., 1912, Note on the temnospondylous stegocephalian, *Rhinesuchus*: Transactions of the Geological Society of South Africa, v. 14, p. 79–81.

Campbell, Y.H., Czamanske, G.K., Fedrenko, V.A., Hill, R.J., and Stepanov, V.D., 1992, Synchronism of the Siberian Traps and the Permian-Triassic boundary: Science, v. 258, p. 1760–1763.

Catuneanu, O., Beaumont, C., and Waschbusch, P., 1997, Interplay of static loads and subduction dynamics in foreland basins: Reciprocal stratigraphies and the "missing" peripheral bulge: Geology, v. 25, p. 1087–1090.

Catuneanu, O., Hancox, P.J., and Rubidge, B.S., 1998, Reciprocal flexural behaviour and contrasting stratigraphies: A new basin development model for the Karoo retroarc foreland system, South Africa: Basin Research, v. 10, p. 417–439.

Chai, Z.F., Zhou, Y.Q., Mao., X.Y., Ma, S.L., Ma, J.G., Kong, P., and He, J.W., 1992, Geochemical constraints on the Permo-Triassic event in South China, *in* Sweet, W.C., Yang, Z.Y., Dickens, J.M., and Yin, H.F., eds., Permo-Triassic events in the eastern Tethys: Cambridge, Cambridge University Press, p. 158–168.

Cole, D.I., 1992, Evolution and development of the Karoo Basin, *in* De Wit, M.J., and Ransome, I.G.D., eds., Inversion tectonics of the Cape Fold Belt, Karoo and Cretaceous Basins of southern Africa: Rotterdam, Balkema, p. 87–99.

Damiani, R.J., 1999, *Parotosuchus* (Amphibia, Temnospondyli) *in* Gondwana: Biostratigraphic and palaeobiogeographic implications: South African Journal of Science, v. 95, p. 458–460.

Erwin, D.H., 1990, The end-Permian mass extinction: Annual Review of Ecology and Systematics, v. 21, p. 69–91.

Erwin, D.H., 1993, The great Paleozoic crises: Life and death in the Permian: New York, Columbia University Press, 327 p.

Erwin, D.H., 1994, The Permo-Triassic extinction: Nature, v. 367, p. 231–236.

Farley, K.A., and Mukhopadhyay, S., 2001, An extraterrestrial impact at the Permian-Triassic boundary?: Science, v. 293, p. 2343a.

Groenewald, G.H., 1984, Stratigrafie en Sedimentologie van die Groep Beaufort in die Noordoos Vrystaat [M.S. thesis]: Johannesburg, South Africa, Rand Afrikaans University, 189 p.

Groenewald, G.H., 1989, Stratigrafie en Sedimentologie van die Groep Beaufort in die Noordoos Vrystaat: Bulletin of the Geological Survey, South Africa, v. 96, p. 1–62.

Groenewald, G.H., 1996, Stratigraphy of the Tarkastad Subgroup, Karoo Supergroup, South Africa [Ph.D. thesis]: Port Elizabeth, University of Port Elizabeth, South Africa, v. 1, 145 p.

Groenewald, G.H., and Kitching, J.W., 1995, Biostratigraphy of the Lystrosaurus Assemblage Zone, *in* Rubidge, R.B.S., ed., Reptilian biostratigraphy of the Permian-Triassic Beaufort Group (Karoo Supergroup): South African Committee for Stratigraphy, Biostratigraphic Series, v. 1, p. 35–39.

Hälbich, I.W., 1992, The Cape Fold Belt orogeny: State of the art 1970's–1980's, *in* DeWit, M.J., and Ransome, I.G.D., eds., Inversion tectonics of the Cape Fold Belt, Karoo and Cretaceous basins of southern Africa: Rotterdam, Balkema, p. 141–148.

Hancox, P.J., 1998a, The nonmarine Triassic of South Africa: Zentralblatt für Geologie und Paläontologie, Teil 1, Heft 11-12, p. 1285–1324.

Hancox, P.J., 1998b, A stratigraphic, sedimentological, and palaeoenvironmental synthesis of the Beaufort-Molteno contact in the Karoo Basin [Ph.D. thesis]: Johannesburg, University of the Witwatersrand, South Africa, v. 1, 381 p.

Hancox, P.J., 2000, The Permo-Triassic boundary in the northwest Karoo Basin, *in* Catastrophic events and mass extinctions: Impacts and beyond: Houston, Texas, Lunar and Planetary Institute, LPI Contribution No. 1053, p. 66–67.

Hancox, P.J., Shishkin, M.A., Rubidge, B.S., and Kitching, J.W., 1995, A threefold subdivision of the Cynognathus Assemblage Zone (Beaufort Group, South Africa) and its palaeogeographical implications: South African Journal of Science, v. 91, p. 143–144.

Hiller, N., and Stavrakis, N., 1984, Permo-Triassic fluvial systems in the southeastern Karoo Basin, South Africa: Palaeogeography, Palaeoclimatology, Palaeoecology, v. 45, p. 1–21.

Holser, W.T., and Magaritz, M., 1992, Cretaceous/Tertiary and Permian/Triassic boundary events compared: Geochimica et Cosmochimica Acta, v. 56, p. 3297–3309.

Hotton, N., 1967, Stratigraphy and sedimentation in the Beaufort Series (Permian-Triassic), South Africa, *in* Teichert, C., and Yochelson, E.I., eds., Essays in palaeontology and stratigraphy: Special Publication of the University of Kansas, Number 2, p. 390–427.

Huber, H., Koeberl, C., McDonald, I., Reimold, W.U., 2000, Use of the γ–γ coincidence spectrometry in the geochemical study of diamictites from South Africa: Journal of Radioanalalysis and Nuclear Chemistry, v. 244, p. 603–607.

Isozaki, Y., 2001, An extraterrestrial impact at the Permian-Triassic boundary?: Science, v. 293, p. 2343a.

Jin, Y.G., Wang, Y., Wang, W., Shang, Q.H., Cao, C.Q., and Erwin, D.H., 2000, Pattern of marine mass extinction near the Permian-Triassic boundary in South China: Science, v. 289, p. 432–436.

Johnson, M.A., 1991, Sandstone petrography, provenance, and plate tectonic setting in Gondwana context of the south-eastern Cape Karoo Basin: South African Journal of Geology, v. 94, n. 2/3, p. 137–154.

King, G.M., 1990, Dicynodonts and the end Permian event: Palaeontologia Africana, v. 27, p. 31–39.

Kirschvink, J.L., and Ward, P.D., 1998, Magnetostratigraphy of Permian/Triassic boundary sediments in the Karoo of southern Africa: Journal of African Earth Sciences, v. 27(1A), p. 124.

Kitching, J.W., 1977, The distribution of the Karroo vertebrate fauna: Bernard Price Institute for Palaeontological Research, Memoir No. 1, 131 p.

Kitching J.W., 1978, The stratigraphic distribution and occurrence of South African fossil Amphibia in the Beaufort beds: Palaeontologia Africana, v. 21, p. 101–112.

Koeberl, C., 1993, Instrumental neutron activation analysis of geochemical and cosmochemical samples: A fast and proven method for small sample analysis: Journal of Radioanalysis and Nuclear Chemistry, v. 168, p. 47–60.

Lozovsky, V.R., 1998, The Permian-Triassic boundary in the continental series of Eurasia: Palaeogeography, Palaeoclimatology, Palaeoecology, v. 143, p. 273–283.

Lucas, S.G., 1998, Global Triassic tetrapod biostratigraphy and biochronology: Palaeogeography, Palaeoclimatology, Palaeoecology, v. 143, p. 347–384.

MacLeod, K.G., Smith, R.M.H., Koch, P.L., and Ward, P.D., 2000, Timing of mammal-like reptile extinctions across the Permian Triassic boundary in South Africa: Geology, v. 28, p. 227–230.

Mory, A.J., Iasky, R.P., Glikson, A.Y., and Pirajno, F., 2000, Woodleigh, Carnarvon Basin, Western Australia: A new 120 km impact structure: Earth and Planetary Science Letters, v. 177, p. 119–128.

Nakamura, N., 1974, Determination of REE, Ba, Fe, Mg, NA, and K in carbonaceous and ordinary chondrites: Geochemica et Cosmochemica Acta, v. 38, p. 757–775.

Pysklywec, R.N., and Mitrovica, J.X., 1999, The role of subduction-induced subsidence in the evolution of the Karoo Basin: The Journal of Geology, v. 107, p. 155–164.

Rampino, M.R., 1992, A major Late Permian impact event on the Falkland Plateau: Eos (Transactions, American Geophysical Union), v. 73, p. 336.

Rampino, M.R., and Haggerty, B.M., 1996, Impact crises and mass extinctions: A working hypothesis, *in* Ryder, G., Fastovsky, D., and Gartner, S., eds., The Cretaceous-Tertiary event and other catastrophes in Earth history: Geological Society of America Special Paper 307, p. 11–30.

Rampino, M.R., Prokoph, A., and Adler, A.C., 2000, Abrupt changes at the Permian/Triassic Boundary: Tempo of events from high-resolution cyclostratigraphy, *in* Catastrophic events and mass extinctions: Impacts and beyond: Houston, Texas, Lunar and Planetary Institute, LPI Contribution No. 1053, p. 176.

Reimold, W.U., Koeberl, C., and Bishop, J., 1994, Roter Kamm impact crater, Namibia: Geochemistry of basement rocks and breccias: Geochimica et Cosmochimica Acta, v. 58, p. 2689–2710.

Renne, P.R., Zhang, Z., Richards, M.A., Black, M.T., and Basu, A.R., 1995, Synchrony and causal relations between Permian-Triassic boundary crises and Siberian flood volcanism: Science, v. 269, p. 1413–1415.

Retallack, G.J., 1995, Permian-Triassic extinction on land: Science, v. 267, p. 77–80.

Retallack, G.J., 1999, Postapocalyptic greenhouse paleoclimate revealed by earliest Triassic paleosols in the Sydney basin, Australia: Geological Society of America Bulletin, v. 111, p. 52–70.

Retallack, G.J., Seyedolali, A., Krull, E.S., Holser, W.T., Ambers, C.P., and Kyle, F.T., 1998, Search for evidence of impact at the Permo-Triassic boundary in Antarctica and Australia: Geology, v. 26, no. 11, p. 979–982.

Rogers, R.R., 1994, Nature and origin of through-going discontinuities in nonmarine foreland basin strata, Upper Cretaceous, Montana: Implications for sequence analysis: Geology, v. 22, p. 1119–1122.

Rubidge, B.S., Johnson, M.R., Kitching, J.W., Smith, R.M.H., Keyser, A.W., and Groenewald, G.H., 1995, An introduction to the biozonation of the Beaufort Group, *in* Rubidge, B.S., ed., Reptilian biostratigraphy of the Permian-Triassic Beaufort Group (Karoo Supergroup): South African Committee for Stratigraphy, Biostratigraphic Series, v. 1, p. 1–2.

Shishkin, M.A., Rubidge, B.S., and Hancox, P.J., 1995, Vertebrate biozonation of the Upper Beaufort Series of South Africa: A new look on correlation of the Triassic biotic events in Euramerica and southern Gondwana, *in*

Sun, A., and Wang, Y., eds., Sixth Symposium on Mesozoic Ecosystems and Biota: Beijing, China Ocean Press, p. 39–41.

Smith, R.M.H., 1995, Changing fluvial environments across the Permian-Triassic boundary in the Karoo Basin, South Africa and possible causes of tetrapod extinctions: Palaeogeography, Palaeoclimatology, Palaeoecology, v. 117, p. 81–104.

Smith, R.M.H., and Ward, P., 2000, Pattern of vertebrate extinctions across an event bed at the Permian/Triassic boundary in the Karoo Basin of South Africa: Palaeontological Society of Southern Africa, 11th Biennial Conference, Pretoria, Abstracts, p. 3–5.

Stanley, G.D., and Yang, X., 1994, Two extinction events in the Late Permian: Science, v. 266, p. 1340–1344.

Thackeray, J.F., Vandermerwe, N.J., Leethorp, J.A., Sillen, A., Lanham, J.L., Smith, R.M.H., Keyser, A., and Monteiro, P.M.S., 1990, Changes in carbon isotope ratios in the late Permian recorded in therapsid tooth apatite: Nature, v. 347, p. 751–753.

Van Hoepen, E.C.N., 1911, Korte voorlopige beschrijving van te Senekal gevonden stegocephalen: Annals of the Transvaal Museum, v. 3, p. 102–106.

Visscher, H., Brinkhuis, H., Dilcher, D.L., Elsik, W.C., Eshet, Y., Looy, C.V., Rampino, M.R., and Traverse, A., 1996, The terminal Paleozoic fungal event: Evidence of terrestrial ecosystem destabilization and collapse: Proceedings of the National Academy of Sciences of the United States of America, v. 93, p. 2155–2158.

Ward, P., 2000, The P/T Boundary on land: A summary, *in* Catastrophic events and mass extinctions: Impacts and beyond: Houston, Texas, Lunar and Planetary Institute, LPI Contribution No. 1053, p. 235.

Ward, P., Montgomery, D.R., and Smith, R.M.H., 2000, Altered river morphology in South Africa related to the Permian-Triassic extinction: Science, v. 289, p. 1740–1743.

Xu, D.Y., and Yan, Z., 1993, Carbon-isotope and iridium event markers near the Permian Triassic boundary in the Meishan section, Zhejiang Province, China: Palaeogeography, Palaeoclimatology, Palaeoecology, v. 104, p. 171–176.

Yin, H.F., Huang, S.J., Zhang, K.X., Hansen, H.J., Yang, F.Q., Ding, M.H., Bie, X.M., 1992, The effects of volcanism on the Permo-Triassic mass extinction in South China, *in* Sweet, W.C., Yang, Z.Y., Dickens, J.M., and Yin, H.F., eds., Permo-Triassic events in the eastern Tethys: Cambridge, Cambridge University Press, p. 146–157.

Yin, H.F., Wu, S.B., Ding, M.H., Zhang, K.X., Tong, J.N., Yang, F.Q., Lai, X.L. 1996, The Meishan section, candidate of the Global Stratotype section and Point (GSSP) of Permian-Triassic Boundary (PTB), *in* Yin, H.F., ed, The Palaeozoic-Mesozoic boundary: Candidates of the Global Stratotype Section and Point (GSSP) of the Permian-Triassic boundary: Wuhan, China University of Geosciences Press, p. 3–30.

Yin, H., and Tong, J., 1998, Multidisciplinary high-resolution correlation of the Permian-Triassic boundary: Palaeogeography, Palaeoclimatology, Palaeoecology, v. 143, p. 199–212.

Zang, Z.C., Claoué-Long, J.C., Ma, G.G., Du, S.H., 1992, Age determination of the Permian-Triassic boundary at Meishan, Changxing, Zhejiang Province: Geological Review, v. 38, no. 4, p. 372–381.

Zang, K.X., Tong, J.N., Yin, H.F., Wu, S.B., 1996, Sequence stratigraphy near the Permian-Triassic boundary at Meishan section, South China, *in* Yin, H.F., ed., The Palaeozoic-Mesozoic boundary: Candidates of the Global Stratotype Section and Point (GSSP) of the Permian-Triassic boundary: Wuhan, China University of Geosciences Press, p. 57–64.

Manuscript Submitted October 23, 2000; Accepted by the Society March 22, 2001

Printed in the U.S.A.

Geological Society of America
Special Paper 356
2002

Chemical signatures of the Permian-Triassic transitional environment in Spiti Valley, India

A.D. Shukla
N. Bhandari*
P.N. Shukla
Physical Research Laboratory, Navarangpura, Ahmedabad 380 009, India

ABSTRACT

Chemical, lithological, and paleontological data indicate that the Permian-Triassic transition record is present in a thin ferruginous band at the junction between the Permian Productus shale and Triassic limestone in the Spiti Valley, Himalaya, India. This ferruginous band consists of mixed mineral assemblages that indicate widely differing environments of origin. On the basis of chemical criteria, specifically the concentration of U and the ratios of Th/U, $(Ce/La)_N$, La/Th, and Th/Yb, and a large positive Eu anomaly, the environmental conditions during the deposition of the transitional sedimentary strata have been deduced. Anoxic conditions seem to have existed for a period of ~100–300 k.y. prior to the end-Permian, consistent with the global anoxia proposed by several workers. The presence of multiple components in the ferruginous band, development of anoxia, and the Eu anomaly suggest changes that may be related to sudden events associated with the Permian-Triassic transition.

INTRODUCTION

The Permian-Triassic (P-Tr) transition (251.4 ± 0.3 Ma; Bowring et al., 1998) is the most severe extinction event in the Phanerozoic history of the Earth. It changed the evolutionary pattern of life on the Earth in a major way. The proposed causes of the P-Tr extinction include (1) sea-level changes (Newell, 1967; Erwin, 1994), (2) Siberian volcanism and aerosol loading of the atmosphere (Renne and Basu, 1991; Renne et al., 1995), (3) anoxia and overturning of the oceans (Wignall and Hallam, 1992; Wignall and Twitchett, 1996; Knoll et al., 1996), and (4) extraterrestrial impact (Rampino, 1992; Bhandari et al., 1992; Retallack et al., 1998). Recent studies carried out on P-Tr sections of Antarctica and Australia, although not conclusive, have indicated the possible presence of impact markers, such as planar deformation features in quartz (Retallack et al., 1998). The present study is a continuation of our earlier work wherein we reported a high positive europium anomaly in the chocolate colored ferruginous band material at the P-Tr boundary in some Spiti Valley sections (Bhandari et al., 1992; Bhandari, 1998) and discussed possible terrestrial and extraterrestrial causes for its origin.

SPITI VALLEY P-TR SECTIONS

The P-Tr sections, exposed along the Spiti River and its tributaries, the Pen and Lingati Rivers, are shown in Figure 1. Good exposures occur near the villages of Attargoo, Lalung, Guling, and Losar (Ganmachidam Hill). The sections contain Permian gray-black shales (Gungri Formation of the Kuling Group) overlain by massive Triassic limestones (Mikin Formation of the Lilang Group). The shales and limestones are separated by a thin (2-5 cm) chocolate colored ferruginous, pebbly-sandy layer (Fig. 2). This layer marks the transition

*E-mail: bhandari@prl.ernet.in

Shukla, A.D., Bhandari, N., and Shukla, P.N., 2002, Chemical signatures of the Permian-Triassic transitional environment in Spiti Valley, India, *in* Koeberl, C., and MacLeod, K.G, eds., Catastrophic Events and Mass Extinctions: Impacts and Beyond: Boulder, Colorado, Geological Society of America Special Paper 356, p. 445–453.

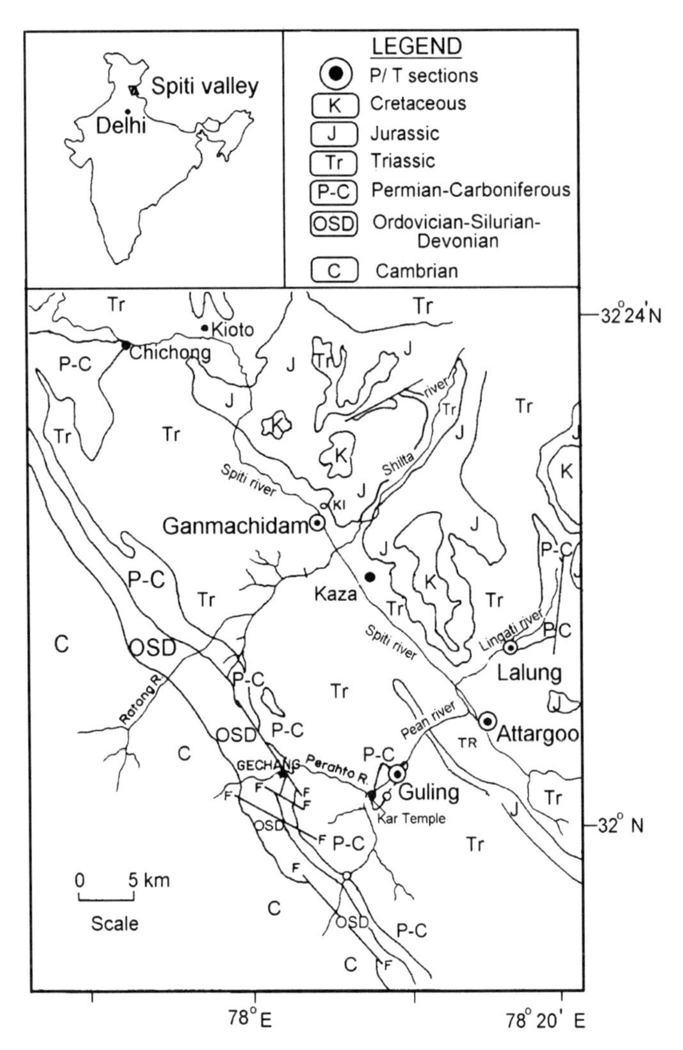

Figure 1. Location and geology of Spiti basin (after Hayden, 1904) showing exposure of Permian-Triassic (P/T) sections.

between the Permian *Productus* shale and limestone of the *Otoceras* zone, which represents the lowermost Triassic (Griesbachian). The contact with the underlying *Productus* shale as well as the overlying Triassic limestone is sharp. The Permian *Productus* shale changes color from grayish to black toward the ferruginous band, suggesting gradually diminishing oxygen availability during the deposition of the shales close to the boundary.

The presence of a ferruginous layer has also been reported in Lahaul Valley and in the Guryul ravine section in Kashmir (Singh et al., 1995); thus, the layer appears to be widespread, occurring at least on a regional scale on the Indian subcontinent. The presence of this layer is taken to represent a small subaerial or submarine hiatus (Bhatt et al., 1981; Bhargava, 1987; Srikantia and Bhargava, 1998), and may be taken as evidence of strong regression at the end of Permian. Tozer (1988), in a detailed summary of the P-Tr sections, concluded that a universal unconformity below the *Otoceras* zone signified a worldwide geological event. However, we present here new evidence that suggests that the P-Tr boundary may be present in the Spiti sections.

BIOSTRATIGRAPHY

The biostratigraphy of the P-Tr sections of the Spiti Valley is based on the work carried out by several workers, starting with the first identification of the Griesbachian (Griesbach, 1891; Hayden and von Kraft, *in* Hayden, 1904). Bhatt et al. (1981), Azmi (1982), and Srikantia and Bhargava (1998) carried out detailed studies on several sections of the Spiti region, summarized briefly in the following. They regarded the basal 1.15 m of the thick, brownish, bedded limestone, designated as the *Otoceras-Ophiceras* zone of Mikin Formation belonging to the Lilang Group, to represent the beginning of the Triassic. The Mikin Formation consists of dark gray to gray, locally cherty dolomite and shale lenses and cyclical carbonate and shale units. Its age ranges from Scythian to Anisian (Srikantia and Bhargava, 1998), representing the beginning of Triassic. The basal *Otoceras* zone is ~55 cm thick and contains the ammonoid index fossil *Otoceras woodwardi*. Several species of *Ophiceras, Glyptophiceras himalayanum*, and the conodont taxa *Hindeodus typicalis, Neogondelella carinata*, and *N. planata* are present in the *Otoceras-Ophiceras* bed, indicating that it is of Griesbachian age (sensu Tozer, 1988). This age is further defined by the occurrence of typical Dienerian conodonts such as *Neospathodus kummeli, N. nepalensis, N. dieneri, N. crystagalli*, and *N. novaehollandiae* in the basal part of the overlying *Ophiceras* zone. The contact of the Triassic limestone with the underlying *Productus* shale of Gungri Formation is marked by the thin (≤5 cm) ferruginous layer that is devoid of fossils, but is persistent with slight variation in thickness in all the P-Tr sections of the Spiti basin. This ferruginous band was interpreted by Bhatt et al. (1981) as caused by subaerial exposure, whereas Bhargava (1987) explained it as a submarine break. About 1.30 m of *Productus* shale below this layer is barren. The development of anoxia in the latest Permian might be the cause of the absence of fossils in this zone. The occurrence of the *Marginifera himalayensis, Productus gangeticus, Chonetes* cf. *lissarensis, Athyris (Cleiothyridina) gerardi, Spiriferalla rajah*, and the cephalopods *Xenaspis carbonaria* and *Cyclolobus oldhami* suggest a Dzhulfian age, possibly extending into part of the Dorashmian or Changxingian (Srikantia and Bhargava, 1998). *Cyclolobus walkeri*, which is Dzhulfian age, is known to occur 1 m below the ferruginous layer (Bhatt et al., 1980) and therefore a Dorashmian age can be assigned to the strata that contain no fossils. It has therefore been debated whether the youngest Permian is present in this section. However, *Cyclolobus walkeri* has been observed in Changxingian strata (P.B. Wignall, 2001, personal commun.; Zakharov et al., 1997), indicating that the uppermost Permian may be present in the Spiti sections. In view of this discussion it is likely that the ferruginous band represents the P-Tr boundary.

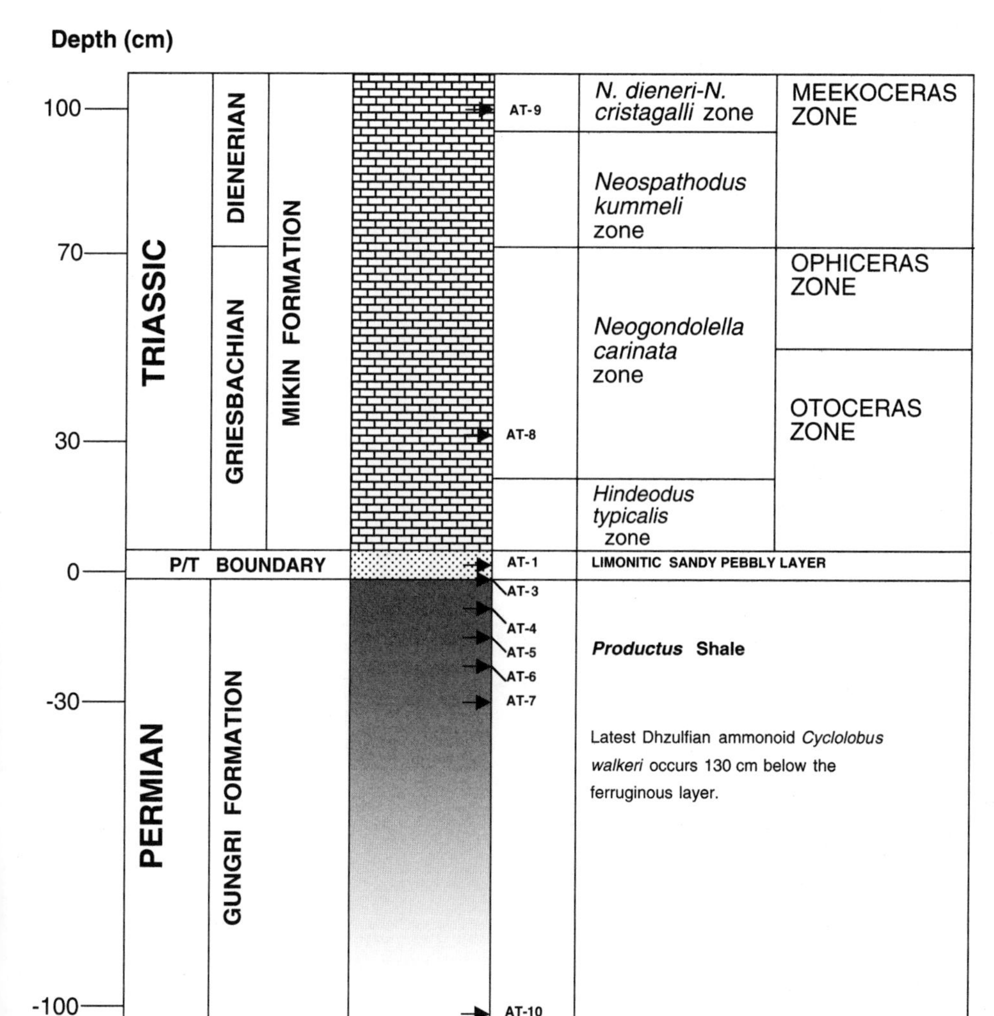

Figure 2. General lithological and biostratigraphical section across Permian-Triassic (P/T) boundary in Spiti Valley. Stratigraphic locations of Attargoo samples (AT-1 – AT-10) are shown.

EXPERIMENTAL DETAILS AND RESULTS

Several samples from the Lalung section, representing 1 m of the uppermost Permian and 0.3 m of the lowermost Triassic, were collected in 1985 (Bhandari et al., 1992). Samples from the Guling section, covering ~7 m section of the Permian *Productus* shale and ~50 cm of Triassic limestone, including the ferruginous layer, were collected in 1989. The Guling, Ganmachidam, and Attargoo sections covering ~1 m around the ferruginous band were sampled in 1997 during International Geological Correlation Program field workshop 386.

X-ray diffraction studies of the ferruginous band were made with a Philips X-ray diffractometer (model PW1730) using copper K_α source and Ni filter. The machine was operated at 35 kV with 25 mA current and the spectra were compared with standard charts to determine the minerals present. The scanning electron microscopic (SEM) studies were carried out using a Cambridge S4-10 SEM with 20 kV operating voltage, having EDAX facility with a Kevex Si (Li) detector for qualitative determination of the surface elemental composition.

We have measured the concentration of U, Th, and K in selected samples of these sections using gamma ray spectrometry, following the procedure described by Sen Gupta (1986). The samples were cleaned and kept in sealed plastic boxes for ~15 days to attain radioactive equilibrium between radium, radon, and some of its daughter isotopes. The concentrations were estimated based on the characteristic gamma rays using a standard basalt reference source (107) having known concentrations of U (5.69 ppm), Th (14.5 ppm), and K (2.63%), in which the uranium series nuclides are expected to be in secular equilibrium. The concentrations of U, Th, and K for the Guling and Attargoo sections are presented in Tables 1 and 2. The errors of measurement of these radionuclides (including systematic and statistical uncertainties) are <5%.

To understand the geochemical nature of the sedimentary strata, we also measured concentrations of several major, minor, and trace elements in the Attargoo, Guling and Lalung sections using instrumental neutron activation analysis (INAA) following the procedures described by Bhandari et al. (1994). Some samples were separated in dark brown and nodular components

TABLE 1. THE U, Th, AND K CONCENTRATIONS IN SAMPLES FROM THE GULING PERMIAN-TRIASSIC SECTION

Sample	Depth* (cm)	Nature	U (ppm)	Th (ppm)	K (%)	Th/U
GL-21	16.0	Triassic limestone	1.58	1.51	0.30	0.96
GL-20	7.0	Triassic limestone	1.71	0.92	0.03	0.54
GL-19	0	Ferruginous band	7.44	7.2	1.12	0.97
GL-18	−4.0	Permian shale	4.67	27.5	3.77	5.89
GL-17	−7.0	Permian shale	4.13	26.9	3.59	6.52
GL-16	−10.5	Permian shale	4.09	29.0	3.79	7.08
GL-15	−15.0	Permian shale	3.91	30.4	3.99	7.79
GL-14	−30.0	Permian shale	3.24	29.6	3.92	9.14
GL-13	−45.0	Permian shale	3.49	27.7	3.65	7.95
GL-12	−57.5	Permian shale	3.56	30.1	3.92	8.46
GL-11	−77.5	Permian shale	3.07	28.7	3.77	9.35
GL-10	−97.5	Permian shale	3.22	30.3	4.18	7.24
GL-9	−122.5	Permian shale	3.26	28.7	3.70	8.80
GL-8	−147.5	Permian shale	3.41	27.1	3.64	7.95
GL-7	−197.5	Permian shale	3.23	29.3	3.62	9.09
GL-6	−247.5	Permian shale	3.58	28.3	3.68	7.91
GL-5	−297.5	Permian shale	3.12	26.7	3.39	8.57
GL-4	−397.5	Permian shale	3.12	27.6	3.73	7.42
GL-3	−497.5	Permian shale	3.34	26.9	3.55	8.56
GL-2	−597.5	Permian shale	2.95	25.4	3.40	8.47
GL-1	−697.5	Permian shale	4.02	23.2	3.38	5.77

*Depth is measured relative to the base of the ferruginous band.

TABLE 2. THE U, Th, AND K CONCENTRATIONS IN SAMPLES FROM THE ATTARGOO PERMIAN-TRIASSIC SECTION

Sample	Nature	Depth* (cm)	U (ppm)	Th (ppm)	K (%)	Th/U
AT-9	Triassic limestone	100.0	6.18	19.2	3.15	3.10
AT-8	Triassic limestone	30.0	0.62	1.55	0.27	2.50
AT-1	Ferruginous band	0	6.71	11.70	1.54	1.74
AT-3	Permian shale	−2.50	6.29	24.5	3.89	3.89
AT-4	Permian shale	−7.50	6.10	18.9	3.11	3.10
AT-5	Permian shale	−12.0	3.81	27.6	4.14	7.24
AT-6	Permian shale	−20.0	3.08	30.9	4.49	10.02
AT-7	Permian shale	−30.0	3.48	25.6	3.83	7.37
AT-10	Permian shale	−100.0	2.91	26.9	4.12	9.25

*Depth is measured relative to the base of the ferruginous band.

on the basis of color and appearance (Tables 3 and 4). About 150 mg of each sample together with Allende meteorite and BCR-1 U.S. Geological Survey (USGS) standards were irradiated in the DHRUVA reactor at Bhabha Atomic Research Centre, Mumbai, for a fluence of $\sim 10^{18}$ n/cm^2. The samples were counted at different intervals of time on a large volume (148 cm^3) hyperpure Ge detector having counting efficiency relative to 7.5 cm NaI (Tl) scintillator of 45.3%, located inside a 10-cm-thick lead shield. Using standard procedures, several elements, including nine rare earth elements (REE; La, Ce, Nd, Sm, Eu, Gd, Tb, Yb, Lu), the siderophiles (Fe, Co, Cr), and lithophiles (e.g., Ca, Ba, Hf, Th, Ta, Sc) were estimated. The results are given in Tables 3 and 4. The statistical errors (1σ) in concentrations of Fe, Co, Cr, Hf, Sc, Ce, Sm, and Eu are $<1\%$; Ta, Cs, Sb, La, Tb, Yb, and Lu are $\leq 5\%$ and Sr, Ba, Rb, Zr, and Nd are up to 7%. Replicate measurements of terrestrial and meteorite standards (USGS standards BCR-1 [Columbia River Basalt], AGV-1 [andesite], and Allende meteorite [split 4, position 24], supplied by E. Jarosewich) showed that the concentrations of all the elements are reproducible within $\pm 1\%$–5% (except for Sr, Ba, Rb, and Zr, which were within $\pm 10\%$), indicating that the systematic errors from the reported values as well as the precision of measurements are within these ranges. The concentrations of various elements in the standards were taken from the published literature (e.g., Jarosewich, 1990; Kallemeyn et al., 1989; Laul, 1979)

FERRUGINOUS BAND

The chocolate colored ferruginous band that separates the Permian sedimentary rocks and the Triassic limestones is friable, unlaminated, sandy, pebbly, and devoid of any burrow markings. Macroscopic, physical examination and X-ray diffraction studies of the ferruginous layer, which is best developed near Attargoo village in the Spiti Valley (Fig. 1), shows the presence of goethite, quartz, gypsum, and feldspar. Man-

TABLE 3. CONCENTRATION OF VARIOUS ELEMENTS IN THE SAMPLES FROM THE ATTARGOO (AT) AND GULING (GL) SECTIONS

Nature#	AT-9 Lime-stone	AT-8 Lime-stone	AT-2 Lime-stone	AT-1B FB Bulk	AT-1D FB Dark	AT-4 Shale	AT-5 Shale	AT-6 Shale	AT-7 Shale	AT-10 Shale	GL-20 Lime-stone	GL-19B FB bulk	GL19 FB brown	GL19 FB dark	GL-18 Shale
Depth* (cm)	100	30	1	0	0	−7.5	−12	−20	−30	−100	9.5	0	0	0	−4.0
Fe (wt%)	0.74	1.92	5.73	23.9	37.4	3.98	2.51	0.93	5.17	3.78	0.57	2.24	35.4	12.1	4.21
Co (μg/g)	8.1	3.3	5.3	153	133	11.5	2.7	2.7	9.0	10.5	1.0	2.0	16.3	12.9	34.7
Cr	8.3	6.0	8.1	24.8	23.8	71.2	106	98.8	103	103	1.7	14.5	19.5	13	112
Zn	14.0	N.M.	15.1	854	851	74.8	14.2	N.M.	N.M.	21.9	2.8	33.1	N.M.	N.M.	N.M.
Sb	0.37	0.19	0.66	7.16	10.7	4.25	1.49	1.20	0.42	0.87	0.23	0.23	16.2	6.08	1.3
Ba	63	88	211	4945	3040	983	673	670	619	589	20	233	342	422	349
Sr	374	309	302	285	240	352	64	41	26	33	326	326	58	N.M.	201
Rb	9.2	5.5	20.1	49.6	48.3	126	173	154	68.3	252	N.M.	N.M.	N.M.	N.M.	N.M.
Zr	46.8	11.8	92.4	257	227	262	194	173	93.4	181	N.M.	N.M.	N.M.	N.M.	5.3
Hf	0.52	0.23	0.52	1.56	1.40	5.29	4.88	4.45	2.53	4.30	0.11	2.45	0.40	0.90	17.7
Ta	0.17	0.05	0.17	0.63	0.65	1.67	2.29	2.06	0.87	1.98	N.M.	N.M.	N.M.	N.M.	N.M.
Th	2.17	0.67	3.27	6.50	5.54	18.9	26.7	22.1	6.92	20.1	N.M.	N.M.	N.M.	N.M.	35.7
Sc	1.97	1.20	3.53	10.0	9.82	11.3	17.9	18.1	16.9	17.8	0.63	5.21	3.19	2.96	10.5
La	10.6	7.8	30.8	41.9	35.7	76.8	47.9	58.0	53.0	56.9	15.7	25.9	19.4	205	45.3
Ce	22.4	11.1	78.7	128	113	168	101	110	90.7	114	20.0	59.7	23.1	302	120
Nd	9.5	6.6	26.5	49.1	49.7	44.6	25.7	30.3	29.6	34.8	11.2	26.4	12.4	126	38.9
Sm	2.15	1.31	6.56	15.8	16.5	8.57	4.14	5.37	6.05	7.17	3.8	9.62	5.41	36.8	7.32
Eu	0.48	0.26	1.84	4.5	4.75	1.72	0.73	0.88	0.88	1.11	0.47	1.86	0.78	23.5	1.15
Gd	1.6	0.7	5.5	13.3	14.0	8.0	5.7	4.8	5.4	7.1	1.9	7.4	2.9	63.3	8.9
Tb	0.26	0.10	1.01	3.08	3.30	1.12	0.48	0.46	0.31	0.73	0.24	1.02	0.38	6.46	0.78
Yb	0.85	0.51	2.70	4.87	5.91	4.01	3.52	2.84	3.03	3.10	0.68	1.78	0.63	9.46	2.99
Lu	0.11	0.08	0.35	0.76	0.84	0.58	0.43	0.43	0.41	0.42	0.12	0.27	0.17	1.21	0.47
La/Th	4.9	11.6	9.4	6.5	6.4	4.1	1.8	2.6	7.7	2.8	N.M.	N.M.	8.8	N.M.	1.3
Th/Yb	2.6	1.3	1.2	1.3	0.9	4.7	7.6	7.8	2.3	6.5	N.M.	N.M.	3.5	0.1	12.0
$(Ce/La)_N$	0.97	0.66	1.18	1.41	1.47	1.01	0.97	0.88	0.79	0.92	2.0	1.07	0.55	0.68	0.98
(Eu/Eu*)†	0.76	0.74	0.91	0.94	0.93	0.63	0.46	0.52	0.46	0.47	0.47	0.65	0.54	1.47	0.43

*Depth is measured from the base of the ferruginous band.
†(Eu/Eu*) is normalized to chondrites. Eu* = (Sm + Gd)/2.
N.M.—Not measured; FB—ferruginous band

TABLE 4. CONCENTRATION OF VARIOUS ELEMENTS IN THE SAMPLES FROM THE LALUNG (LL) SECTIONS

Nature#	LL14 Lime-stone	LL13 Lime-stone	LL12 Lime-stone	LL 11B2 Clay	LL 11B1 FB	LL 11 FB bulk-1	LL 11 FB bulk-2	LL 11 FB brown	LL 11 FB dark	LL9 Shale	LL8 A Shale	LL8 B Clay	LL7 Shale	LL4 Shale	LL1 Shale
Depth* (cm)	27	17	6	0	0	0	0	0	0	−20	−30	−30	−40	−70	−100
Fe (wt%)	2.09	3.03	4.55	10.38	13.46	3.05	6.51	26.8	1.13	0.85	1.15	5.76	2.02	2.46	3.56
Co (μg/g)	1.9	2.5	8.8	72.8	38.9	15.3	44.9	40.4	11.7	2.2	2.4	38.4	7.7	23.8	12.9
Cr	N.M.	N.M.	N.M.	N.M.	N.M.	N.M.	20.2	11.8	18.4	N.M.	N.M.	N.M.	N.M.	N.M.	N.M.
Zn	N.M.	N.M.	N.M.	N.M.	N.M.	N.M.	208	420	85.3	N.M.	N.M.	N.M.	N.M.	N.M.	N.M.
Sb	N.M.	N.M.	N.M.	N.M.	N.M.	N.M.	25.7	53.8	6.9	N.M.	N.M.	N.M.	N.M.	N.M.	N.M.
Ba	39	49	97	344	322	347	252	104	222	566	508	378	504	464	450
Sr	N.M.	N.M.	N.M.	N.M.	N.M.	N.M.	489	297	654	N.M.	N.M.	N.M.	N.M.	N.M.	N.M.
Zr	N.M.	N.M.	N.M.	N.M.	N.M.	N.M.	827	401	1281	N.M.	N.M.	N.M.	N.M.	N.M.	N.M.
Hf	0.15	0.28	0.45	2.0	1.0	N.M.	1.18	0.50	0.98	4.6	4.5	3.4	4.3	3.8	4.0
Th	1.04	1.24	2.23	4.84	5.39	N.M.	N.M.	N.M.	N.M.	22.5	22.8	17.9	22.6	20.6	19.4
Sc	N.M.	N.M.	N.M.	N.M.	N.M.	N.M.	10.1	6.8	5.9	N.M.	N.M.	N.M.	N.M.	N.M.	N.M.
La	11.7	8.8	16.9	41.7	63.9	101	113	43.1	122	48.6	51.2	41.9	48.4	39.8	44.5
Ce	13.8	19.2	34.2	120	200	208	341	147	374	104	119	95.7	114	102.0	94.0
Nd	12.9	9.1	14.5	42.7	70.2	111	158	53.6	182	29.4	32.4	27.1	29.3	28.7	29.6
Sm	2.7	1.94	3.23	8.93	19.42	25.6	59.7	19.0	63.3	5.38	6.30	5.65	6.38	5.38	6.33
Eu	0.51	0.47	0.98	5.19	10.78	18.0	19.7	6.36	26.8	0.81	1.0	0.89	1.05	1.0	0.97
Gd	1.8	1.6	2.8	12.5	27.1	33.1	31.3	15.1	52.0	4.1	4.4	4.3	5.5	5.2	5.6
Tb	0.31	0.21	0.37	1.70	3.84	N.M.	5.56	2.52	6.67	0.46	0.58	0.59	0.71	0.72	0.69
Yb	0.65	0.62	1.06	3.05	2.99	9.89	9.11	3.60	10.4	2.03	1.72	1.61	2.13	1.89	1.86
Lu	0.65	0.09	0.18	0.70	0.83	1.31	1.33	0.61	1.45	0.36	0.31	0.26	0.36	0.35	0.32
La/Th	11.3	7.0	7.6	8.6	11.8	28.,6	N.M.	N.M.	N.M.	2.2	2.2	2.3	2.1	1.9	2.3
Th/Yb	1.6	2.0	2.1	1.6	1.8	0.4	N.M.	N.M.	N.M.	11.1	13.3	11.1	10.6	10.9	10.4
$(Ce/La)_N$	0.55	1.01	0.94	1.33	1.44	0.95	1.40	1.58	1.42	0.99	1.07	1.06	1.09	1.19	0.98
(Eu/Eu*)†	0.66	0.80	0.98	1.48	1.44	1.89	1.26	1.11	1.39	0.51	0.55	0.53	0.53	0.57	0.49

*Depth is measured from the base of the ferruginous band (FB).
† (Eu/Eu*) is normalized to chondrites. Eu* = (Sm + Gd)/2.

ganese in the form of oxide with minor iron has also been observed by SEM studies, indicating the presence of a marine component, whereas the presence of gypsum indicates evaporitic deposition in alkaline conditions. Mössbauer spectroscopic studies have shown the presence of nanometer-size superparamagnetic oxide and/or hydroxide phases in this layer (Verma et al., 2001), indicating formation under oxic conditions.

DISCUSSION

Oceanic anoxia is considered to have been a global phenomenon during the P-Tr transition on the basis of studies carried out on the sections of the southern Alps, Spitsbergen, East Greenland, British Columbia, and Japan (Wignall and Hallam, 1992; Wignall and Twitchett, 1996; Isozaki, 1997). We first attempt to ascertain if anoxia in the Tethyan regime was recorded. In this discussion we assume that the ferruginous layer is a reference horizon representing the P-Tr transition. Because U in the sedimentary strata remains insoluble in U^{+4} state in reducing environments and is soluble in U^{+6} state in oxic conditions, its concentration is a good indicator of anoxia (Veeh, 1967). The depth profiles of U, Th/U, and Th in the Guling and Attargoo sections are shown in Figures 3 and 4A, respectively. We confine our discussion to the Permian because it represents the same facies, and the interpretation is meaningful if lithological changes are absent. As can be seen from Figures 3 and 4A, the U concentration starts increasing (from 3 to 6.3 ppm) in the uppermost Permian and becomes highest (7.4 ppm) in the ferruginous band. From the increase in the concentration of U, we can infer that the depositional condition was progressively becoming more reducing toward the uppermost Permian. To minimize the effects of depositional or lithological changes, we also consider Th/U ratios. We find that the ratio in the lower part of the Permian section is close to the crustal value of 3.8 (Taylor and McLennan, 1985) and represents oxic conditions, but starts decreasing ~20–30 cm below the ferruginous band (P-Tr boundary) in both the sections at Attargoo as well as at Guling. The sedimentation rate of this section is not known, but Srikantia and Bhargava (1998) gave evidence that this formation mostly represents shelf sedimentation under restricted conditions, implying that the sedimentation rate was low. Assuming a plausible value of ≥1–3 mm/k.y., as is found for standing waters in deep sea (Twenhofel, 1950), the duration of anoxic conditions in the Spiti basin can be estimated to be 100–300 k.y. or less. This time span is similar to the period of catastrophic addition of light carbon for a short duration (≤165 k.y.) before the P-Tr crisis in the southern China sections (Bowring et al., 1998). Furthermore, the reducing nature of the basin is supported by the $(Ce/La)_N$ ratio (normalized to North American Shale Composite; La = 31 μg/g; Ce = 67 μg/g) measured in the Attargoo P-Tr section (Fig. 4B), and earlier work carried out at Lalung by Bhandari et al. (1992). The observations discussed here lead to the conclusion that the ocean was anoxic during the deposition of the ferruginous band. The anoxic environment can also be inferred from Figure 4 (A and B), where it is shown that the Th/U ratio is minimum and $(Ce/La)_N$ is maximum in this layer. On the basis of these observations, we infer that the sediments deposited during the latest Permian in the Spiti Valley were in a relatively reducing environment compared to the underlying Permian. The situation is similar to the P-Tr section of the Carnic Alps (Holser et al., 1989). We further compare the Th/U profile across the P-Tr boundary of the Spiti Valley with the Carnic Alps (Holser and Schönlaub, 1991) and Meishan sections (Zhou and Kyte, 1988), which have been studied in detail. In the absence of proper time scales and accurate sedimentation rates for all the sections, we compare the depth profiles to see the trend of changes. The nature of the Th/U excursion observed in the Spiti sections (Fig. 5), located at that time in the Southern Hemisphere, is similar to that observed elsewhere. We therefore conclude that the anoxia observed by various workers in different parts of the globe just before P-Tr time was also present in the Spiti Valley and that these observations support that anoxia was a global phenomenon. We have also carried out $\delta^{13}C$ measurements in or-

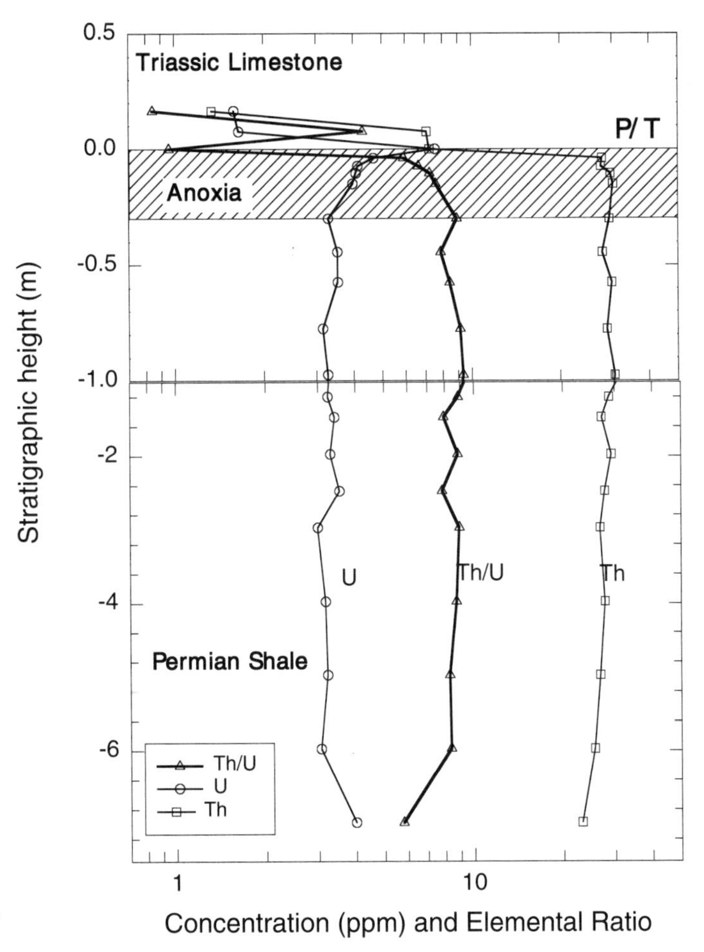

Figure 3. U, Th, and Th/U profiles in Guling section. Note change in scale at −1 m. Diagonal ruled area from ~0 to −0.3 m defines horizon of anoxia buildup. P/T—Permian-Triassic boundary.

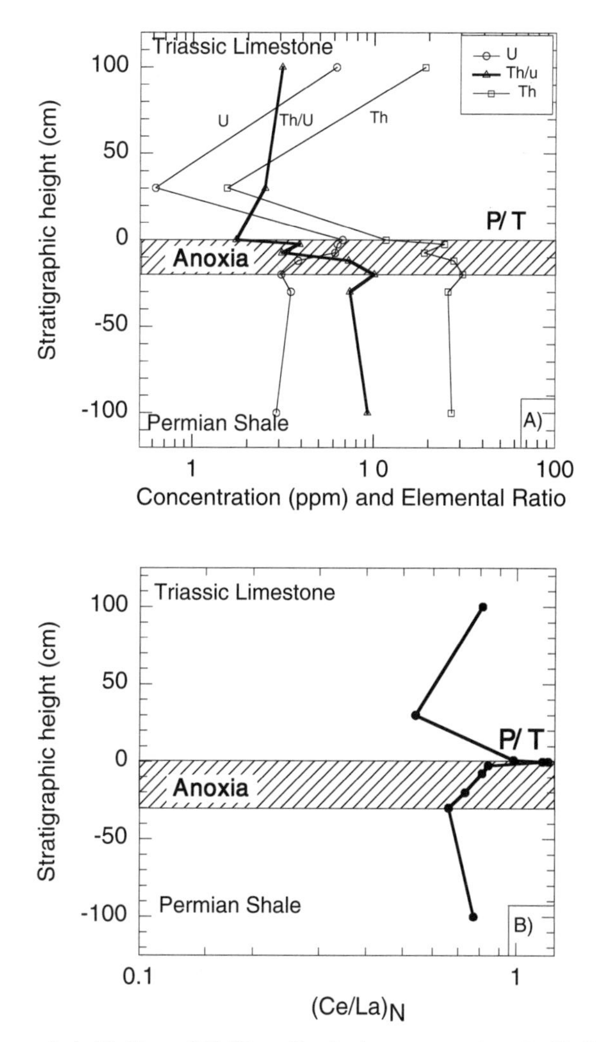

Figure 4. A: U, Th, and Th/U profiles in Attargoo section. B: $(Ce/La)_N$ profile, normalized to North American Shale Composite (La = 31 ppm, Ce = 67 ppm). Diagonal ruled area from ~0 to −30 cm defines horizon of anoxia build up. P/T—Permian-Triassic boundary.

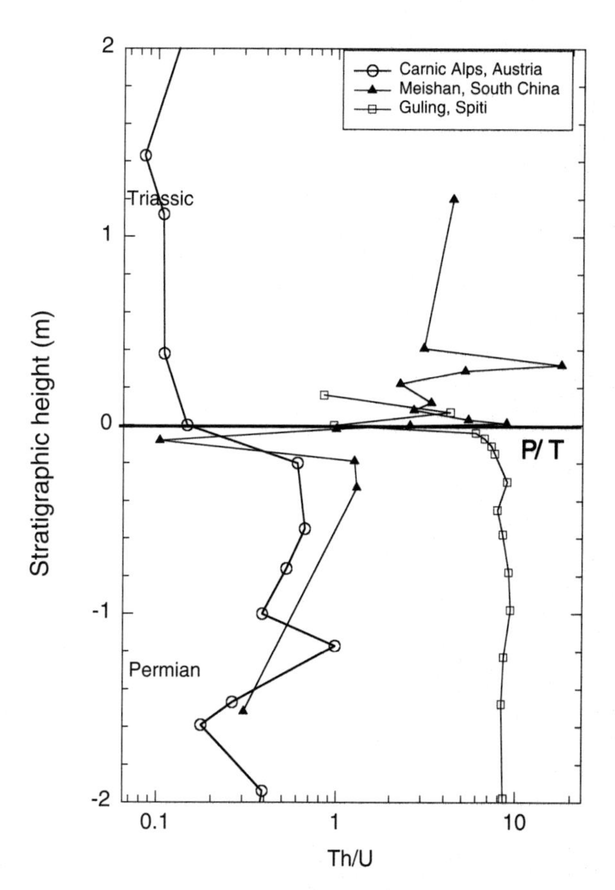

Figure 5. Comparison of Th/U profile from Guling, Carnic Alps (Holser and Schönlaub, 1991), and Meishan (Zhou and Kyte, 1988) sections. P/T—Permian-Triassic boundary.

ganic component of this section that show abrupt and large excursion, characteristic of the P-Tr boundary, further supporting that boundary sediments are present in this section.

The ferruginous band present in the P-Tr sections of the Spiti Valley has a sharp contact with the Permian black shales, indicating that its deposition was abrupt. The presence of goethite and other nanometer-size superparamagnetic particles of oxides and oxyhydroxides of iron, as determined by Mössbauer spectroscopy (Verma et al., 2001), indicate oxic environmental conditions during their formation. However, the presence of gypsum along with manganese-bearing iron-oxide phases suggests evaporitic conditions in the marine environment, which was anoxic. In nature, change from anoxic to oxic condition occurs slowly; however, the time available, based on the occurrence of both Changxingian and Griesbachian fossils in this region, was short. One possibility is that this layer was not deposited in situ, but has extraneous material.

In addition to the high U concentration, the ferruginous layer has a high REE content (Tables 3 and 4). At some locations, the ferruginous layer or the uppermost Permian shale contains dark nodules that exhibit a large positive Eu anomaly, even when normalized to chondrites (Fig. 6). The highest value ($[Eu/Eu^*]_{chon}$ = 1.9) occurs in the Lalung P-Tr section of the Spiti Valley (Bhandari et al., 1992), but these Eu-bearing nodules do not occur everywhere and their distribution seems to be patchy. From the data (Tables 3 and 4) it is evident that the dark nodular grains are major contributors to the enrichment of REEs in the ferruginous band and in the uppermost shale. The ferruginous band is also enriched in chalcophile elements (Tables 3 and 4).

Siberian and south China volcanism (Renne and Basu, 1991; Campbell et al., 1992; Erwin, 1994) occurred at about the same time as the Permian to Triassic transition. We therefore looked for volcanic contributions in the sedimentary strata of Permian and Triassic sections of the Spiti Valley. Elemental ratios such as La/Th and Th/Yb have characteristic values for various sedimentary and igneous rock types and have been used for identifying a change in the source supply of the sediments (Wang et al., 1986). The average values of these ratios in the Attargoo section for Permian shale (Table 3) are La/Th = 3.8

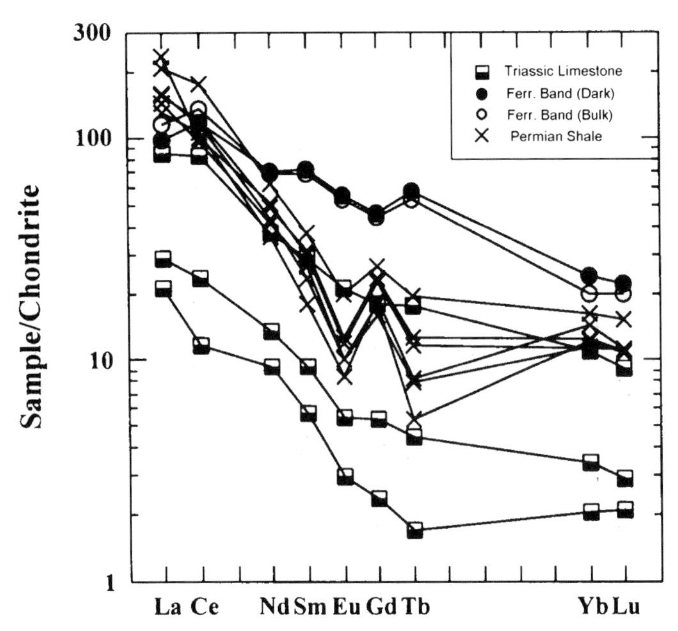

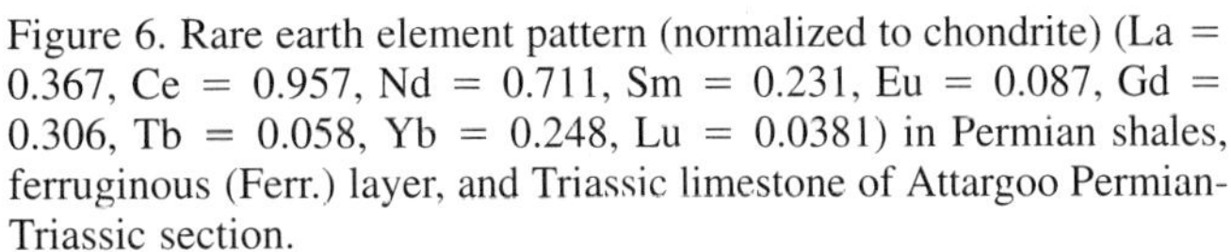
Figure 6. Rare earth element pattern (normalized to chondrite) (La = 0.367, Ce = 0.957, Nd = 0.711, Sm = 0.231, Eu = 0.087, Gd = 0.306, Tb = 0.058, Yb = 0.248, Lu = 0.0381) in Permian shales, ferruginous (Ferr.) layer, and Triassic limestone of Attargoo Permian-Triassic section.

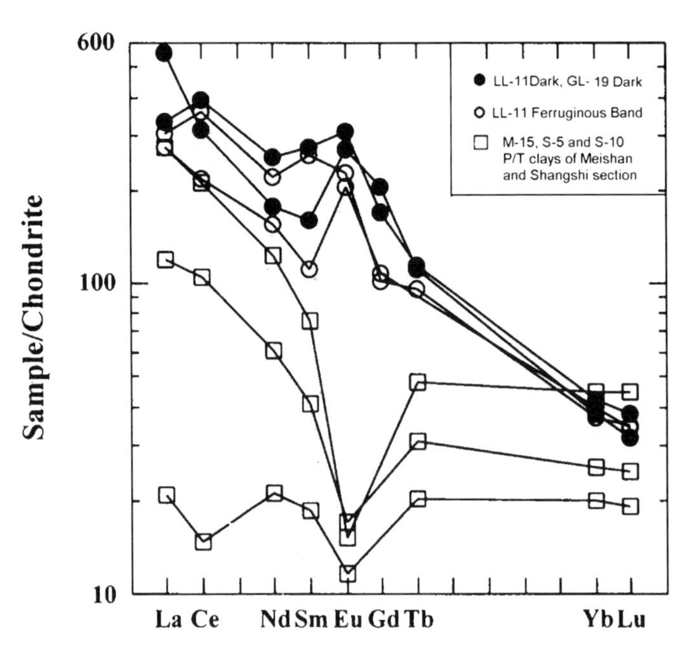

Figure 7. Comparison of rare earth element pattern observed in ferruginous band at Lalung (Spiti Valley) with three typical Permian-Triassic (P/T) clays (M-15, S-5, and S-10) from the Meishan and Shangshi sections of southern China (from Zhou and Kyte, 1988).

and Th/Yb = 5.7, similar to the values generally observed for shales (La/Th = 2.7, Th/Yb = 5.8); as expected, however, the Triassic limestones have average La/Th = 8.6 and Th/Yb = 1.7, similar to basalts (La/Th = 7; Th/Yb = 1.7), thus indicating the presence of a volcanic component in the Triassic limestones. A similar trend is found for the Lalung section (Table 4), where higher Th/Yb and lower La/Th ratios are observed for Permian shales and vice versa for Triassic limestones (Bhandari et al., 1992). This could be due to Panjal and Phe volcanism, which occupied large areas in the adjoining Kashmir and Zanskar basins in the Tethyan realm of the northwestern Himalaya (Srikantia and Bhargava, 1998).

We further compare the trace element data and diagnostic trace elemental ratios used by Zhou and Kyte (1988) to check the contribution of South China volcanism in the Spiti P-Tr ferruginous band. The REE pattern of the ferruginous band is compared with the pattern in the P-Tr clays of the Meishan and Shangshi sections in Figure 7. It appears that the ferruginous band has higher concentrations of REEs in comparison to the Chinese sections. Furthermore, in the elemental correlation diagrams (e.g., Lu vs. Hf, Hf vs. Th, Co-Th-Hf and U vs. La) used by Zhou and Kyte (1988) for identifying the acidic components, the samples of the Spiti layer are in a different field. The absence of the acidic components found in the P-Tr clays of Meishan and Shangshi indicates that this debris did not reach the Tethyan realm.

In view of the evidence of shocked quartz from Antarctic and Australian sections (Retallack et al., 1998) and possible, although inconclusive, evidence of impact craters like Woodleigh (Mory et al., 2000; Reimold and Koeberl, 2000) on the western coast of Australia, or the Falkland Plateau (Rampino, 1992) or in Brazil (Hammerschmidt and Engelhardt, 1995), it might be possible that the signatures of impact are preserved only in the Southern Hemisphere P-Tr sections because of their proximity to the proposed crater sites.

In conclusion, arguments based on available paleontological evidence and chemical signatures are presented here that suggest that the records of P-Tr transitions are preserved in the Spiti Valley. Changes in Th/U and $(Ce/La)_N$ ratios start occurring 20–30 cm below the ferruginous band, indicating anoxic conditions corresponding to a relatively short time span, similar to those observed in other P-Tr sections globally. A sudden occurrence of a ferruginous band consisting of mineral assemblages formed under highly oxidizing environments is observed where the anoxia was at a maximum; this ferruginous band could therefore be extraneous in origin. Chemical signatures of volcanic origin are observed in the overlying limestones. The results presented here thus indicate that the P-Tr transition is recorded in the sedimentary strata of the Spiti Valley.

ACKNOWLEDGMENTS

We thank R.J. Azmi for participating in the field trip and for his help in collection of samples of the Guling section. We thank H. Strauss for organizing the field trip under International Geological Correlation Program 386, during which the Attargoo section was sampled. We are grateful to P.B. Wignall, P.J. Hancox, L. Krystyn, and C. Koeberl for useful comments that helped in improving the manuscript. The assistance provided by Pranav Adhyaru in maintaining the gamma ray spectrometers and K.R. Nambiar in preparation of the manuscript is appreciated.

REFERENCES CITED

Azmi, R.J., 1982, Triassic conodont biostratigraphy of Spiti Tethys Himalaya, India: Third European Conodont Symposium (ECOS III), Institute of Mineralogy, Paleontology and Quaternary Geology, University of Lund, Lund, Sweden, Abstracts, p. 3–4.

Bhandari, N., 1998, Astronomical and terrestrial causes of physical, chemical and biological changes at geological boundaries: Proceedings of the Indian Academy of Sciences (Earth and Planetary Science Letters), v. 107, p. 251–263.

Bhandari, N., Shukla, P.N., and Azmi, R.J., 1992, Positive Europium anomaly at the Permian-Triassic boundary, Spiti, India: Geophysical Research Letters, v. 19, p.1531–1534.

Bhandari, N., Gupta, M., Pandey, J., and Shukla, P.N., 1994, Chemical profiles in K/T boundary section of Meghalaya, India: Cometary, asteroidal or volcanic: Chemical Geology, v. 113, p. 45–60.

Bhargava, O.N., 1987, Stratigraphy, microfacies and paleoenvironment of the Lilang group (Scythian-Dogger), Spiti Valley, Himachal Himalaya, India: Journal of Palaeontological Society of India, v. 25, p. 91–107.

Bhatt, D.K., Fuchs, G., Prashara, K.C., Krysten, L., Arora, R.K., and Golebiowski, R., 1980, Additional ammonoid layer in the Upper Permian sequence of Spiti: Bulletin of Indian Geological Association, v. 13, p. 57–61.

Bhatt, D.K., Joshi, V.K., and Arora, R.K., 1981, Conodonts of Otoceras beds of Spiti: Journal of Palaeontological Society of India. v. 25, p 130–134.

Bowring, S.A., Erwin, D.H., Jin, Y.G., Martin, M.W., Davidek, K., and Wang, W., 1998, U/Pb Zircon geochronology and tempo of the end-Permian mass extinction: Science, v. 280, p. 1039–1045.

Campbell, I.H., Czamanske, G.K., Fedorenko, V.A., Hill, R.I., and Stepanov, V., 1992, Synchronism of the Siberian Traps and the Permian-Triassic boundary: Science, v. 258, p.1760–1763.

Erwin, D.H., 1994, The Permo-Triassic extinction: Nature, v. 367, p. 231–235.

Griesbach, C.L.,1891, Geology of Central Himalayas: Memoirs of the Geological Society of India, v. 23, p. 1–232.

Hammerschmidt, K., and Engelhardt, W.V., 1995, $^{40}Ar/^{39}$ Ar dating of the Araguainha impact structure, Mato Grosso, Brazil: Meteoritics, v. 30, p. 227–233.

Hayden, H.H., 1904, The geology of Spiti, with parts of Bashahr and Rupshu: Memoirs of the Geological Society of India, v. 36, p. 1–121.

Holser, W.T., and Schönlaub, H.P., 1991, The Permian-Triassic boundary in the Carnic Alps of Austria (Gartnerkofel Region): Abhandlungen Der Geologischen Bundesanstalt Band, Wein, v. 45, 232 p.

Holser, W.T., Schönlaub, H.P., Attrep, M., Jr., Boeckelmann, K., Klein, P., Magaritz, M., Orth, C.J., Feninger, A., Jenny, C., Kralik, M., Mauritsch, H., Pak, E., Schramm, J.M., Stattegger, K., and Schmöller, R., 1989, A unique geochemical record at the Permian-Triassic boundary: Nature, v. 337, p. 39–44.

Isozaki, Y., 1997, Permo-Triassic boundary superanoxia and stratified superocean: Records from lost Deep-Sea: Science, v. 276, p. 235–238.

Jaresowich, E., 1990, Chemical analyses of meteorites: A compilation of stony and iron meteorite analyses: Meteoritics, v. 25, p. 323–337.

Kallemeyn, G.W., Rubin, A.E., Wang, D., and Wasson, J.T., 1989, Ordinary chondrites: Bulk compositions, classification, lithophile-element fractionations, and composition-petrologic type relationships: Geochimica et Cosmochimica Acta, v. 53, p. 2747–2767.

Knoll, A.H., Bambach, R.K., Canfield, D.E., and Grotzinger, J.P., 1996, Comparative Earth history and late Permian mass extinction: Science, v. 273, p. 452–457.

Laul, J.C., 1979, Neutron activation analysis of geological materials: Atomic Energy Review, v. 17, n. 3, p. 603–695.

Mory, A.J., Iasky, R.P., Glikson, A.Y., and Pirajno, F., 2000, Woodleigh, Carnarvon Basin, Western Australia: A new 120 km diameter impact structure: Earth and Planetary Science Letters, v. 177, p. 119–128.

Newell, N.D., 1967, Revolutions in the history of life: Geological Society of America Special Paper 89, p. 63–91.

Rampino, M.R., 1992, A major Late Permian impact event on the Falkland Plateau: Eos (Transactions, American Geophysical Union), v. 73, p. 336.

Reimold, W.U., and Koeberl, C., 2000, Critical comment on A.J. Mory et al. "Woodleigh, Carnarvon Basin, western Australia: A new 120 km diameter impact structure": Comment: Earth and Planetary Science Letters, v. 184, p. 353–357.

Renne, P.R., and Basu, A.R., 1991, Rapid eruption of the Siberian Traps flood basalts at the Permo-Triassic boundary: Science, v. 253, p. 176–179.

Renne, P.R., Zichao, Z., Richards, M.R., Black, M.T., and Basu, A.R., 1995, Synchrony and causal relations between Permian-Triassic boundary crises and Siberian flood volcanism: Science, v. 269, p. 1413–1416.

Retallack, G.J., Seyedolali, A., Krull, E.S., Holser, W.T., Ambers, C.P., and Kyte, F.T., 1998, Search for evidence of impact at the Permian-Triassic boundary in Antarctica and Australia: Geology, v. 26, p. 979–982.

Sen Gupta, D., 1986, Cosmic ray induced thermoluminescence in moon and meteorites and terrestrial dating applications [Ph. D. thesis]: Ahmedabad, India, Physical Research Laboratory, 278 p.

Singh, T., Tiwari, R.S., Vijaya, and Ram-Avtar, 1995, Stratigraphy and palynology of Carboniferous-Permian-Triassic succession in Spiti valley, Tethys Himalaya, India: Journal of the Palaeontological Society of India, v. 40, p. 55–76.

Srikantia, S.V., and Bhargava, O.N., 1998, Geology of Himachal Pradesh: Bangalore, Geological Society of India, 416 p.

Taylor, S.R., and McLennan, S.M., 1985, The continental crust: Its composition and evolution: Oxford, Blackwell, 312 p.

Tozer, E.T., 1988, Towards a definition of Permian-Triassic boundary: Episodes, v. 11, p. 251–255.

Twenhofel, W.H., 1950, Principles of sedimentation: New York, McGraw Hill, 673 p.

Veeh, H.H., 1967, Deposition of uranium from the ocean: Earth and Planetary Science Letters, v. 3, p. 145–150.

Verma, H.C., Upadhyay, C., Tripathi, R.P., Bhandari, N., and Shukla, A.D., 2001, Nano-sized iron phases at the K/T and P/T boundaries revealed by Mössbauer Spectroscopy: Lunar and Planetary Science, v. 32, Abstract #1270, CD-ROM.

Wang, Y.L., Liu, Y.-G., and Schmitt, R.A., 1986, Rare earth element geochemistry of south Atlantic deep sea sediments: Ce anomaly change at ~54 My: Geochimica et Cosmochimica Acta, v. 50, p. 1337–1355.

Wignall, P.B., and Hallam, A., 1992, Anoxia as a cause of the Permian/Triassic extinction: Facies as evidence from northern Italy and the western United States: Palaeogeography, Palaeoclimatology, Palaeoecology, v. 93, p. 21–46.

Wignall, P.B., and Twitchett, R., 1996, Oceanic anoxia and the end Permian mass extinction: Science, v. 272, p. 1155–1158.

Zakharov, Y.D., Oleinikov, A., and Kotlyar, G.V., 1997, Late Changxingian ammonoids, bivalves, and brachiopods in South Primorye, *in* Dickens, J.M., et al., eds., Late Palaeozoic and early Mesozoic circum-Pacific events and the global correlation: Cambridge, Cambridge University Press, p. 142–146.

Zhou, L., and Kyte, F.T., 1988, The Permian-Triassic boundary event: A geochemical study of three Chinese sections: Earth and Planetary Science Letters, v. 90, p. 411–421.

MANUSCRIPT SUBMITTED OCTOBER 5, 2000; ACCEPTED BY THE SOCIETY MARCH 22, 2001

Printed in the U.S.A.

Geological Society of America
Special Paper 356
2002

Synchronous record of δ¹³C shifts in the oceans and atmosphere at the end of the Permian

Mark A. Sephton*
Department of Geochemistry, Institute of Earth Sciences, Utrecht University, Budapestlaan 4, 3584 CD Utrecht, The Netherlands, and Department of Marine Biogeochemistry and Toxicology, Netherlands Institute for Sea Research, PO Box 59, 1790 AB Den Burg, Texel, The Netherlands
Cindy V. Looy
Laboratory of Palaeobotany and Palynology, Utrecht University, Budapestlaan 4, 3584 CD Utrecht, The Netherlands
Ruben J. Veefkind
Department of Geochemistry, Institute of Earth Sciences, Utrecht University, Budapestlaan 4, 3584 CD Utrecht, The Netherlands, and Department of Marine Biogeochemistry and Toxicology, Netherlands Institute for Sea Research, PO Box 59, 1790 AB Den Burg, Texel, The Netherlands
Henk Brinkhuis
Laboratory of Palaeobotany and Palynology, Utrecht University, Budapestlaan 4, 3584 CD Utrecht, The Netherlands
Jan W. De Leeuw
Department of Geochemistry, Institute of Earth Sciences, Utrecht University, Budapestlaan 4, 3584 CD Utrecht, The Netherlands, and Department of Marine Biogeochemistry and Toxicology, Netherlands Institute for Sea Research, PO Box 59, 1790 AB Den Burg, Texel, The Netherlands
Henk Visscher
Laboratory of Palaeobotany and Palynology, Utrecht University, Budapestlaan 4, 3584 CD Utrecht, The Netherlands

ABSTRACT

In conjunction with the profound ecologic crisis at the end of the Permian, the most conspicuous geochemical event is the worldwide negative shift in the carbon isotopic composition ($\delta^{13}C$) of both carbonates and sedimentary organic matter. Comparative carbon isotopic analyses of carbonates and the molecular fossils of land plant leaf cuticles from a marine Permian-Triassic transition section in the southern Alps, northeastern Italy, substantiates the concept of synchronous disturbances in oceanic and atmospheric chemistry and, therefore, verifies the primary nature of the end-Permian $\delta^{13}C$ disturbance. The $\delta^{13}C$ excursion appears to be a consequence of the ecological crisis, and the global reservoir of soil organic matter may be the only plausible source of ^{13}C-depleted carbon.

INTRODUCTION

Global extinctions at the end of the Permian reflect the most devastating ecologic crisis of Phanerozoic time (Erwin, 1993, 1994; Hallam and Wignall, 1997). The ecosystem collapse affected not only marine, but also terrestrial biota (Retallack, 1995; Smith, 1995; Visscher et al., 1996). The extinction events are accompanied by conspicuous negative shifts in the

*Corresponding author: Planetary and Space Sciences Research Institute, Open University, Milton Keynes MK7 6AA, UK
E-mail: M.A.Sephton@open.ac.uk

Sephton, M.A., Looy, C.V., Veefkind, R.J., Brinkhuis, H., De Leeuw, J.W., and Visscher, H., 2002, Synchronous record of $\delta^{13}C$ shifts in the oceans and atmosphere at the end of the Permian, *in* Koeberl, C., and MacLeod, K.G., eds., Catastrophic Events and Mass Extinctions: Impacts and Beyond: Boulder, Colorado, Geological Society of America Special Paper 356, p. 455–462.

carbon isotopic compositions of carbonates ($\delta^{13}C_{carb}$) that have been recorded in latest Permian sections throughout the world (Magaritz et al., 1988; Baud et al., 1989; Holser et al., 1989; Erwin, 1993; Bowring et al., 1998; Jin et al., 2000).

End-Permian $\delta^{13}C_{carb}$ shifts are generally interpreted as reflecting a reapportioning of carbon between the Earth's inorganic and organic reservoirs. A thorough appreciation of the nature of the isotopic trends is essential to help us understand the association between changes in end-Permian oceanic and atmospheric chemistry and the ecologic crisis. Because $\delta^{13}C_{carb}$ provides a record of oceanic bicarbonate, the $\delta^{13}C$ shifts are generally interpreted in terms of a prominent global change in the $\delta^{13}C$ values of the surface waters from which the carbonate originated. However, individual end-Permian $\delta^{13}C_{carb}$ profiles from different sections frequently display significant variations in the absolute magnitude of the $\delta^{13}C$ shifts as well as many small-scale fluctuations. These differences are likely to be secondary diagenetic effects, which can mimic primary perturbations in the global carbon cycle (Scholle, 1995).

Complementary to the $\delta^{13}C_{carb}$ record, carbon isotope profiles for both marine (Magaritz et al., 1992; Wang et al., 1995; Wignall et al., 1988; Looy, 2000; Twitchett et al., 2001) and terrestrial (Morante et al., 1994; Krull and Retallack, 2000; Krull et al., 2000) sedimentary organic matter ($\delta^{13}C_{org}$) also display negative shifts at the end of the Permian. This could imply a means of confirming the primary origin of the negative shifts in $\delta^{13}C_{carb}$. However, it appears that $\delta^{13}C_{org}$ trends may not always be a straightforward record of isotopic changes in surficial carbon reservoirs, because shifts in $\delta^{13}C_{org}$ may be caused by variations in the contributions from different sources for the organic matter (Foster et al., 1997). Most of the $\delta^{13}C_{org}$ data available originate from shallow-marine sedimentary rocks, so that $\delta^{13}C_{org}$ values can reflect the carbon isotopic composition of a mixture of marine and land-derived material. Consequently, even when $\delta^{13}C_{carb}$ and $\delta^{13}C_{org}$ values are measured in the same sections, bulk $\delta^{13}C_{org}$ data are inappropriate for verifying to what extent the end-Permian carbon isotope event represents a dramatic disturbance in the global carbon cycle. In terrestrial sedimentary strata, $\delta^{13}C$ trends for land-plant derived organic matter (Morante et al., 1994; Morante, 1996) and the tusks of vertebrates (Thackeray et al., 1990; MacLeod et al., 2000), could reflect atmospheric change associated with the ecologic crisis. However, substantiation of a suspected synchronism with the marine $\delta^{13}C_{carb}$ trends is hampered by the lack of accurate time controls when attempting to correlate terrestrial and marine Permian-Triassic sedimentary records.

A further complicating factor when interpreting $\delta^{13}C_{org}$ profiles is the significant variation among the ^{13}C content of organic compounds from individual plants. Many of the enzymatic processes involved in secondary carbon metabolism have associated isotope effects, and different metabolites can have different $\delta^{13}C$ values (O'Leary, 1981; Farquhar et al., 1989). Lipids, for example, are generally enriched in ^{12}C; the difference between $\delta^{13}C$ values for whole-leaf carbon and lipids may be as much as 10‰ (O'Leary, 1981). Dissimilarities in the relative proportions of different plant constituents contributing to and/or preserved in sedimentary rocks can also disturb $\delta^{13}C_{org}$ trends (Van Kaam-Peters et al., 1998).

To remove the uncertainty surrounding the end-Permian $\delta^{13}C$ shifts, a record is required that cannot have been produced by diagenetic alteration of carbonates, or variations in the sources of sedimentary organic matter. In this chapter, we demonstrate that the real and profound nature of the end-Permian global $\delta^{13}C$ shift can be verified by comparing shallow-marine $\delta^{13}C_{carb}$ data with compound-specific carbon isotopic signatures of *n*-alkanes ($\delta^{13}C_{alk}$). These *n*-alkanes are the chemical fossils of leaf cuticles from land plants, transported and incorporated into marine sediments. We document concurrent shifts in $\delta^{13}C_{carb}$ and $\delta^{13}C_{alk}$ values from the same samples, confirming that these data represent an approximately synchronous record of the primary chemical changes occurring in the surface ocean, atmosphere, and terrestrial biosphere during the end-Permian ecologic crisis.

CARBON ISOTOPES, KEROGEN, AND MOLECULAR FOSSILS

We established $\delta^{13}C_{carb}$ and $\delta^{13}C_{alk}$ profiles for the latest Permian sedimentary strata at Val Badia (western Dolomites, southern Alps, northeast Italy) (Fig. 1). In this section, close to the boundary between the Bellerophon Formation and the Werfen Formation, the end-Permian ecosystem collapse is evidenced by the last occurrences of a variety of foraminifera and gymnospermous pollen, as well as an abundance of fungal remains (Cirilli et al., 1998). It is now accepted that the first appearance of the conodont *Hindeodus parvus* should mark the onset of the Triassic in the marine realm (Yin et al., 1996). One of the implications of this definition is that the crisis occurs before, rather than at, the biostratigraphically defined Permian-Triassic boundary. In the western Dolomites, the first occurrences of *H. parvus* are in the basal part of the Mazzin Member of the Werfen Formation (Kozur, 1998). Unfortunately, there are no conodont records available for the Val Badia section.

The $\delta^{13}C_{carb}$ profile for Val Badia displays high Late Permian values (SD600 to SD0-6) that are followed by a sharp drop (SD0-6 to SU30) and a subsequent gradual decline into younger samples (SU30 to SU1300). The abrupt onset of the observed trend is common to coeval sections throughout the southern Alps (Magaritz et al., 1988; Holser et al., 1989) and elsewhere.

By applying gas chromatography–isotope ratio mass spectrometry (GC-IRMS) (Hayes et al., 1990; Van Kaam-Peters et al., 1997), the $\delta^{13}C_{alk}$ profile is constructed on the basis of $\delta^{13}C$ values measured for individual *n*-alkanes that are present in fractionated solvent extracts of sedimentary organic matter in the Val Badia section (Fig. 2). In the thermally mature Val Badia sedimentary rocks (mean vitrinite reflectance values 0.62%), long-chain *n*-alkanes are likely to represent the products

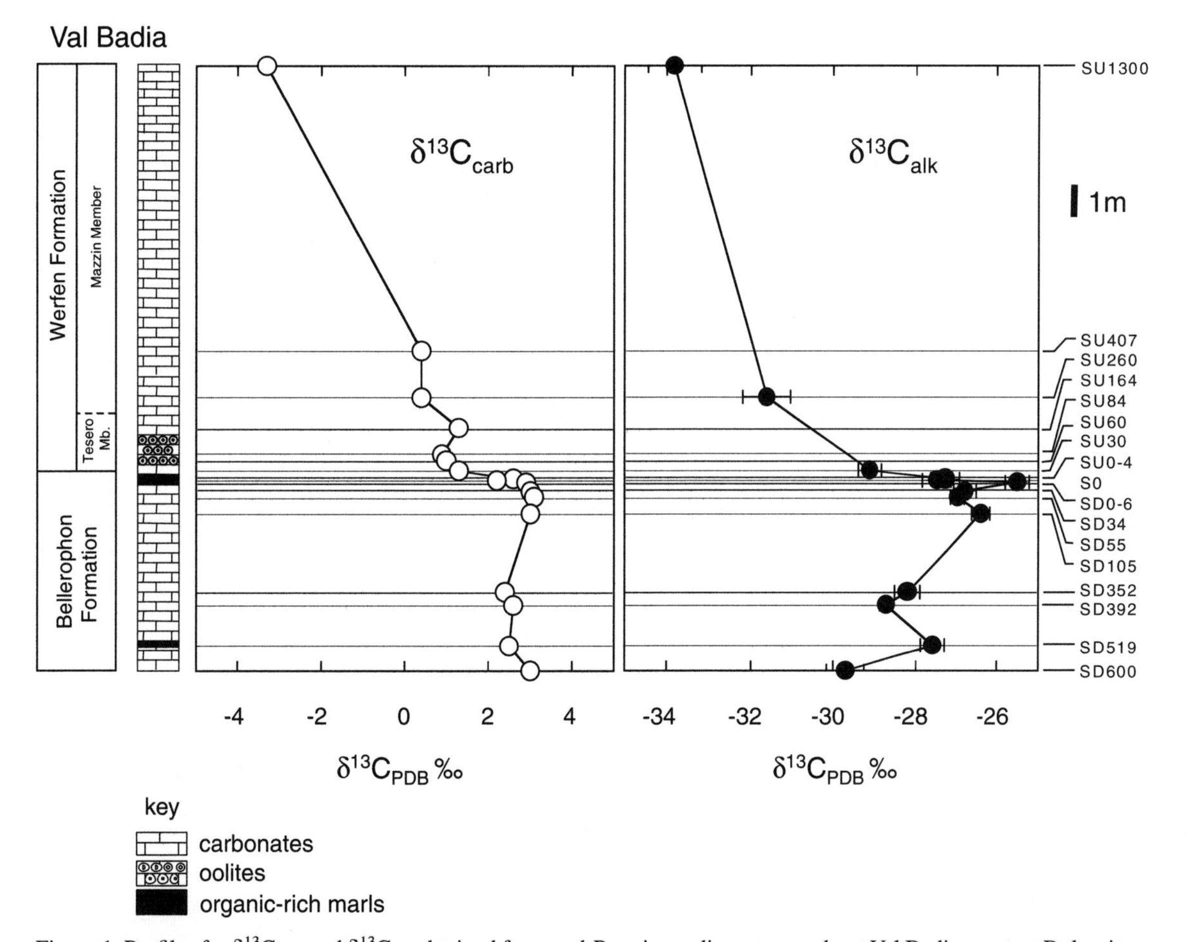

Figure 1. Profiles for $\delta^{13}C_{carb}$ and $\delta^{13}C_{alk}$ obtained from end-Permian sedimentary rocks at Val Badia, western Dolomites, northeastern Italy, illustrating isotopic harmony between these marine and terrestrial carbonaceous materials. For locality, description of lithology, and microfossil content of section, see Cirilli et al. (1998). Sample codes indicate their position (in cm) either above (SU) or below (SD) conspicuous organic-rich marl (S0) situated close to extinction event (Cirilli et al., 1998). Newly defined (conodont based) Permian-Triassic boundary is in basal part of Mazzin Member (see text). $\delta^{13}C_{carb}$ was determined by heating samples under vacuum to 400°C (30 min) before treatment with H_3PO_4 to produce CO_2. CO_2 was introduced into VG SIRA 24 isotope ratio mass spectrometry and ratio of $^{13}C/^{12}C$ measured. $\delta^{13}C_{alk}$ is an average of $\delta^{13}C$ values for C_{19} through C_{29} *n*-alkanes, and was obtained by extracting samples (as in Sephton et al., 1999), and then fractionating extracts (as in Kohnen et al., 1990) to give a saturated hydrocarbon fraction. Compound identification was accomplished (as in Sephton et al., 1999), and isotopic measurement of individual *n*-alkanes was performed (as in Van Kaam-Peters et al., 1997). Each sample was run in triplicate, except for S0, which was run in quadruplicate. Standard deviations (1σ) for data are indicated. All obtained carbon isotope ratios are expressed in usual delta notation relative to international PDB (Peedee belemnite) standard as follows: $\delta^{13}C$ (‰) $= [(^{13}C/^{12}C)_{sample}/(^{13}C/^{12}C)_{PDB} - 1] \times 10^3$. Mb—Member.

of the thermal breakdown of highly aliphatic biopolymers, selectively preserved in kerogen (Tegelaar et al., 1989a). In general, the two principal precursor macromolecules of the *n*-alkanes are cutan, a resistant polymethylenic structure in leaf cuticles (Nip et al., 1986; Tegelaar et al., 1989a, 1989b, 1991), and algaenan, the resistant cell wall material found in a variety of fresh-water and marine algae (Goth et al., 1988; Gelin et al., 1996). The *n*-alkanes from Val Badia display the unimodal distribution patterns characteristic of cutan; a dominance of algaenan would have resulted in a predominantly bimodal distribution.

Palynofacies analysis of the chemically investigated samples (Fig. 3) confirms earlier observations (Cirilli et al., 1998) that kerogens from the end-Permian sedimentary rocks at Val Badia are overwhelmingly dominated by land-derived material. Structured palynodebris is generally degraded. This category includes mainly tracheal structures, sclerenchymous structures, and cuticular structures reflecting a land plant origin. Although structureless, subordinate opaque palynodebris can also be attributed to land plants (e.g., Cope, 1981). The abundant amorphous organic matter appears to be an end product of the decomposition of the structured palynodebris. Moreover, the amorphous material is nonfluorescent and, because algal organic matter should fluoresce at the measured maturity, a

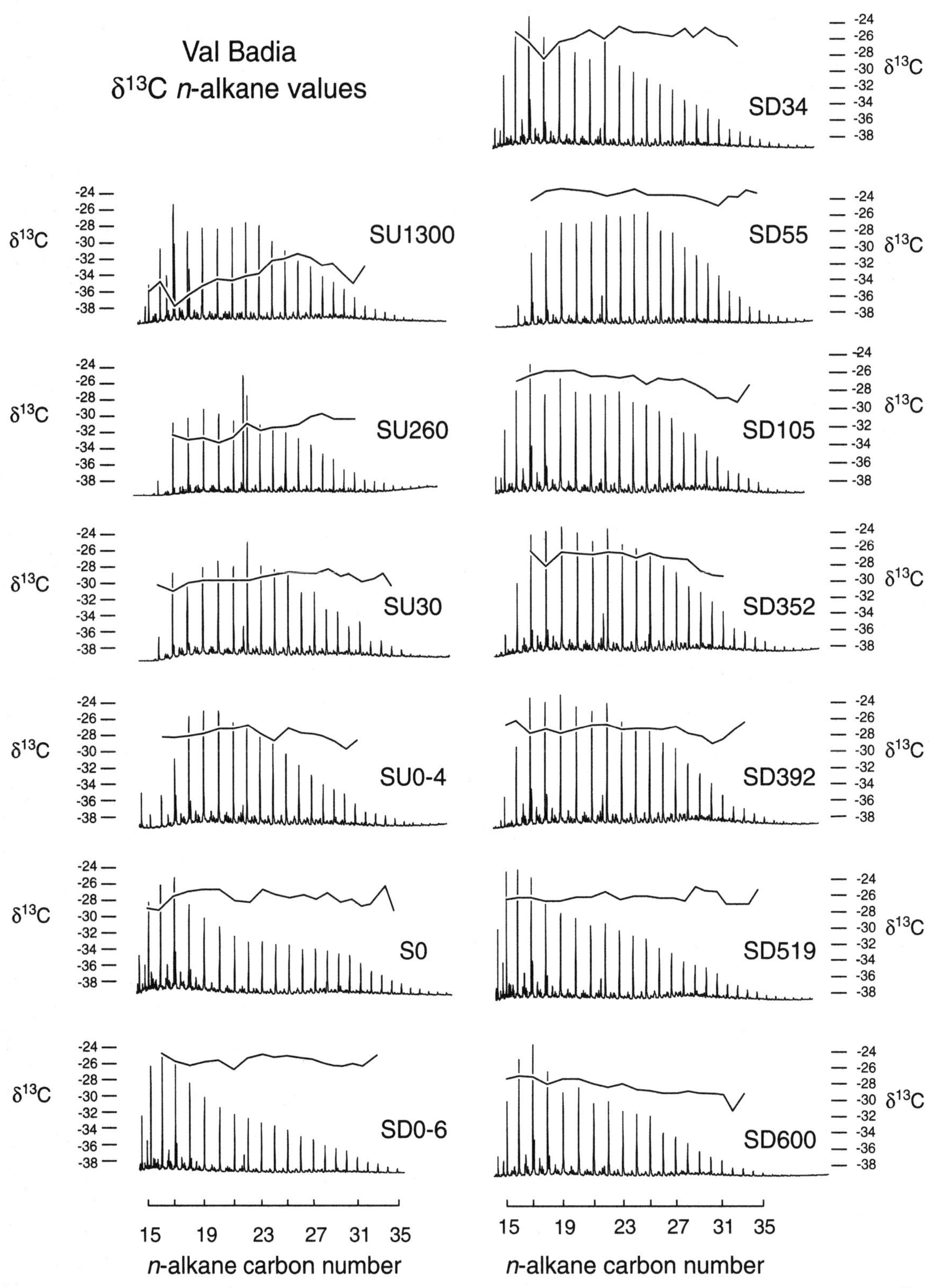

Figure 2. Gas chromatography traces of saturated hydrocarbon fractions from Val Badia sedimentary rocks. $\delta^{13}C$ values of individual *n*-alkanes are indicated, illustrating isotopic homogeneity within, but significant variation between, samples.

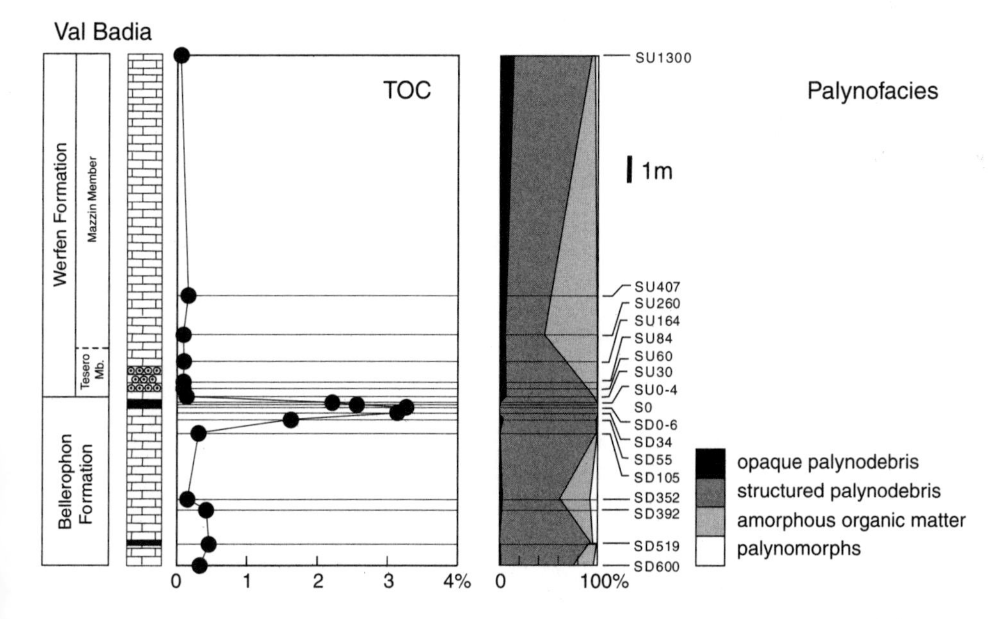

Figure 3. Profiles for percentage of total organic carbon (TOC) (left) and principal palynofacies categories (right), obtained from end-Permian sedimentary rocks at Val Badia, western Dolomites, northeastern Italy. For locality, description of lithology, and microfossil content of section, see Cirilli et al. (1998). Sample codes correspond to those of Figure 1. TOC was determined as dry-weight percentages. Sample preparation for palynofacies took place according to standard procedures of Laboratory of Palaeobotany and Palynology, Utrecht University, including sieving over a 10 µm sieve. Data for palynofacies percentage diagram were obtained by counting at least 200 organic particles per sample. Mb—Member.

marine origin is unlikely. Except for subordinate marine acritarchs in samples from the Mazzin Member, all palynomorphs (pollen, spores, fungal remains) are land derived.

To corroborate the presence of cutan rather than algaenan, we investigated the kerogen samples with flash pyrolysis–gas chromatography (Py-GC) and flash pyrolysis–gas chromatography–mass spectrometry (Py-GC-MS). Two end-member kerogens were detected on the basis of characteristic pyrolysis products; one dominated by aromatic and furan rings (Sephton et al., 1999), the other dominated by long-chain *n*-alk-1-ene/*n*-alkane doublets with a similar distribution pattern, as observed for the extracted *n*-alkanes (Fig. 4A). Comparative analysis of fossil leaf cuticles of *Ortiseia*, a common conifer genus in the Upper Permian of the western Dolomites and other parts of Europe (Poort et al., 1997), confirms that the aliphatic molecules are the pyrolysis products of cutan (Fig. 4B). Subtle variations in the distribution of the pyrolysis products are presumably due to changes in the land plant assemblage, which contributed cuticular material to the sediments.

The homogeneity of the $\delta^{13}C_{alk}$ values makes it possible to construct mean C_{19} to C_{29} *n*-alkane values that accurately represent the carbon isotopic ratios of the aliphatic hydrocarbons in each sample. Throughout the section, the $\delta^{13}C_{alk}$ values appear to be harmonious with the $\delta^{13}C_{carb}$ trend (Fig. 1). Initial high values (SD600 to SD0-6) are followed by a drop (SD0-6 to SU30) and a subsequent gradual decline (SU30 to SU1300).

The concomitant change of $\delta^{13}C_{alk}$ and $\delta^{13}C_{carb}$ to more negative values indicates that the chemical fossils of leaf cuticles from land plants contain an accessible record of the carbon isotopic expression associated with the end-Permian ecologic crisis. This is because the $\delta^{13}C$ signature of fossil land plant material primarily reflects the carbon isotopic composition of atmospheric CO_2. The relatively constant isotopic separation maintained by the $\delta^{13}C_{alk}$ and $\delta^{13}C_{carb}$ values (Fig. 5) corroborates the proposal that a single source of organic matter, i.e., land plants, has contributed to the aliphatic material (cf. Foster et al., 1997). Because secondary diagenetic effects are unlikely to produce isotopic changes in the $\delta^{13}C$ values of two sedimentary carbon-containing phases that agree in both sign and magnitude, the $\delta^{13}C_{alk}$ and $\delta^{13}C_{carb}$ profiles must reflect simultaneous primary disturbances in end-Permian atmospheric and oceanic chemistry.

COUPLING BETWEEN THE CARBON ISOTOPE EVENT AND THE ECOLOGIC CRISIS

The concurrent atmospheric and oceanic decline in $\delta^{13}C$ and its association with biological evidence of the ecosystem collapse at Val Badia support the concept of a causal relationship between the end-Permian ecologic crisis and redistribution of ^{13}C-depleted carbon between major carbon reservoirs. Yet, until recently, it was unknown whether ecosystem collapse was a consequence or a cause of changes in global carbon cycling. It is frequently emphasized that the marine extinction is coincident with the carbon isotope event. This observation (e.g., Jin et al., 2000) is likely to be influenced by the condensed nature of many of the carbonate-rich Permian-Triassic transition sections in the Tethys realm, including the boundary stratotype at Meishan, southern China. However, in the greatly expanded siliciclastic Permian-Triassic section of Jameson Land, East Greenland, detailed analysis of the sequence of isotope events and significant quantitative changes in the marine and terrestrial fossil record has now clearly revealed that the marine and terrestrial ecosystem collapse preceded the onset of the negative excursions in the $\delta^{13}C_{carb}$ and $\delta^{13}C_{org}$ record (Looy, 2000; Twitchett et al., 2001). Consequently, the cause of ^{13}C depletion cannot be the cause of the ecologic crisis. Scenarios relying on rapid overturn of anoxic deep oceans (Knoll et al., 1996) or

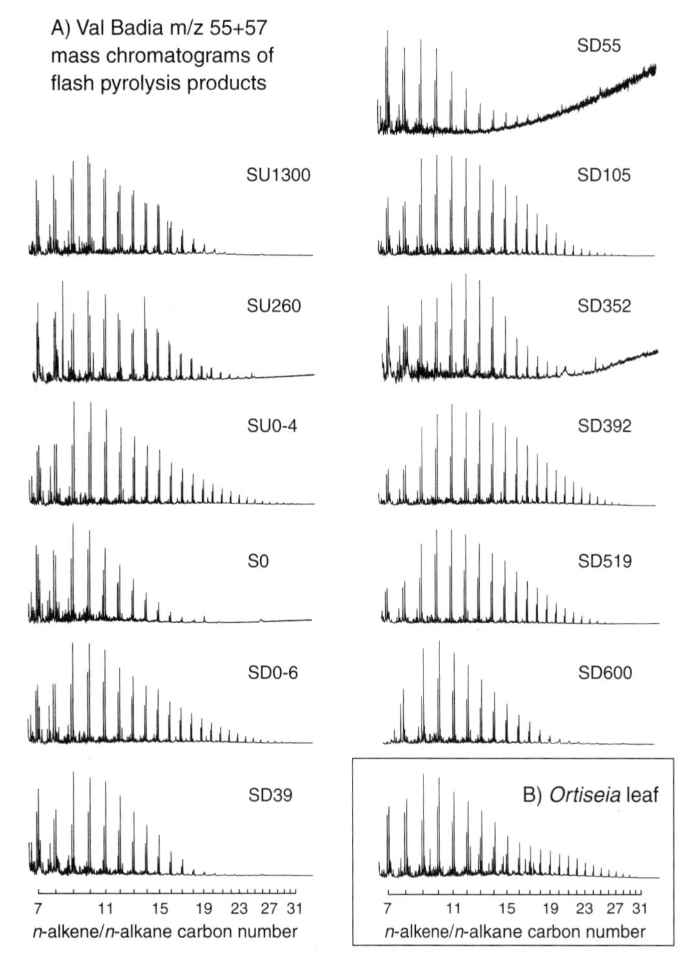

Figure 4. Summed mass chromatograms of mass/charge (m/z) 55 + 57 from Curie point (610°C) flash pyrolysis-gas chromatography-mass spectrometry (Py-GC-MS) analyses of (A) kerogens from Val Badia section and (B) isolated leaf of Late Permian conifer genus *Ortiseia*. These mass chromatograms selectively display *n*-alkene and *n*-alkane patterns in pyrolysis products, and comparisons indicate that cutan is dominant aliphatic biopolymer in Val Badia sedimentary rocks. Carbon number retention times of *n*-alkene and *n*-alkane doublets are shown. Py-GC-MS conditions were as in Sephton et al. (1999).

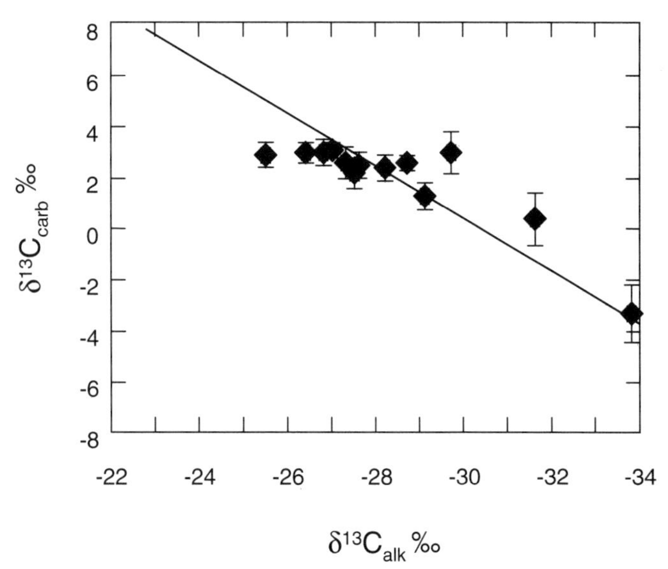

Figure 5. Cross-plots of $\delta^{13}C_{carb}$ and $\delta^{13}C_{alk}$ from Val Badia, northern Italy. Data form single population within which alkanes display relatively constant isotopic relationship with carbonate (x = y − 30.5‰). This correlation indicates that single source of organic matter (i.e., land plants) has contributed to alkane parent material throughout section.

dissociation of oceanic CH_4-hydrate (Erwin, 1993; Morante, 1996) as mechanisms for massive release of light carbon are therefore unsatisfactory.

In order to assess whether specific end-Permian carbon reservoirs could have had the necessary combination of isotopic composition and size to produce the observed perturbations, the release mechanism and origin of the ^{13}C-depleted carbon should be in harmony with both mass-balance and time-frame considerations (Spitzy and Degens, 1985). Following the verification of the primary nature for the initial ~3‰ shift in $\delta^{13}C_{carb}$ observed at Val Badia, this value may be used confidently in quantitative considerations. From estimated sedimentation rates in Greenland, it may be inferred that the initial $\delta^{13}C_{carb}$ decline had a duration of <50 k.y. (Looy, 2000; Twitchett et al., 2001). Together, these data corroborate the concept that the initial release of ^{13}C-depleted carbon into surficial environments at the end of the Permian was both massive and fast.

Considering the observation that the abrupt initial step of the end-Permian $\delta^{13}C$ excursion follows ecosystem collapse, the global reservoir of soil organic matter with mean $\delta^{13}C$ values of −25‰ may be the only plausible source of ^{13}C-depleted carbon. End-Permian accumulations could have been mobilized as extensive soil erosion followed the massive die-back of woody vegetation that characterized end-Permian ecosystem collapse on land (Visscher et al., 1996; Retallack, 1999; Ward et al., 2000). In the Val Badia section, evidence for increased soil erosion is provided by an elevated total organic carbon content, in combination with the land plant origin of palynodebris categories in the microscopic organic matter assemblage (Fig. 3). The heavily degraded nature of this buried plant debris indicates decomposition on land prior to transport into the marine depositional environment. Because rootlets deteriorating in soils are known to become inertinized and structureless (Cope, 1981), this may also apply to the opaque palynodebris. From a biogeochemical point of view, the unusually high abundance of cyclic diaryl ethers, such as dibenzofuran and its alkyl derivatives, in the Val Badia kerogen could well reflect the rapid redeposition of soil organic matter (Sephton et al., 1999).

ACKNOWLEDGMENTS

We thank D.H. Erwin and P.J. Hancox for constructive reviews. This work was supported by the Council for Earth and Life Sciences with financial aid of the Netherlands Organization for

Scientific Research. Netherlands Research School of Sedimentary Geology publication 2000 0902.

REFERENCES CITED

Baud, A., Magaritz, M., and Holser, W.T., 1989, Permian-Triassic of the Tethys: Carbon isotope studies: Geologische Rundschau, v. 78, p. 649–677.

Bowring, S.A., Erwin, D.H., Jin, Y.G., Martin, M.W., Davidek, K., and Wang, W., 1998, U/Pb zircon geochronology and tempo of the end-Permian mass extinction: Science, v. 280, p. 1039–1045.

Cirilli, S., Pirini Radrizzani, C., Ponton, M., and Radrizzani, S., 1998, Stratigraphical and palaeoenvironmental analysis of the Permian-Triassic transition in the Badia Valley (Southern Alps, Italy): Palaeogeography, Palaeoclimatology, Palaeoecology, v. 138, p. 85–113.

Cope, M.J., 1981, Products of natural burning as a component of the dispersed organic matter of sedimentary rocks, *in* Brooks, J., ed., Organic maturation studies and fossil fuel exploration: London, Academic Press, p. 89–109.

Erwin, D.H., 1993, The great Paleozoic crisis: New York, Columbia University Press, 327 p.

Erwin, D.H., 1994, The Permian-Triassic extinction: Nature, v. 367, p. 231–236.

Farquhar, G.D., Ehleringer, J.R., and Hubick, K.T., 1989, Carbon isotope discrimination and photosynthesis: Annual Review of Plant Physiology and Plant Molecular Biology, v. 40, p. 503–537.

Foster, C.B., Logan, G.A., Summons, R.E., Gorter, J.D., and Edwards, D.S., 1997, Carbon isotopes, kerogen types and the Permian-Triassic boundary in Australia: Implications for exploration: Australian Petroleum Production and Exploration Association Journal, v. 37, p. 442–459.

Gelin, F., Boogers, I., Noordeloos, A.A.M., Sinninghe Damsté, J.S., Hatcher, P.G., and De Leeuw, J.W., 1996, Novel, resistant microalgal polyethers: An important sink of organic carbon in the marine environment?: Geochimica et Cosmochimica Acta, v. 60, no. 7, p. 1275–1280.

Goth, K., De Leeuw, J.W., Puttmann, W., and Tegelaar, E.W., 1988, Origin of Messel Oil Shale kerogen: Nature, v. 336, p. 759–761.

Hallam, A., and Wignall, P.B., 1997, Mass extinctions and their aftermath: Oxford, Oxford University Press, 320 p.

Hayes, J.M., Freeman, K.H., Popp, B.N., and Hoham, C.H., 1990, Compound-specific isotopic analyses: A novel tool for reconstruction of ancient biogeochemical processes: Organic Geochemistry, v. 16, no. 4–6, p. 1115–1128.

Holser, W.T., Schonlaub, H.P., Attrep, M., Boeckelmann, K., Klein, P., Magaritz, M., Orth, C.J., Fenninger, A., Jenny, C., Kralik, M., Mauritsch, H., Pak, E., Schramm, J.M., Stattegger, K., and Schmoller, R., 1989, A unique geochemical record at the Permian-Triassic boundary: Nature, v. 337, p. 39–44.

Jin, Y.G., Wang, Y., Wang, W., Shang, Q.H., Cao, C.Q. and Erwin, D.H., 2000, Pattern of marine mass extinction near the Permian-Triassic boundary in South China: Science, v. 289, p. 432–436.

Knoll, A.H., Bambach, R.K., Canfield, D.E., and Grotzinger, J.P., 1996, Comparative Earth history and Late Permian mass extinction: Science, v. 273, p. 452–457.

Kohnen, M.E.L., Sinninghe Damsté, J.S., Rijpstra, W.I.C., and De Leeuw, J.W., 1990, Alkylthiophenes as sensitive indicators of paleoenvironmental changes: A study of a Cretaceous oil-shale from Jordan: American Chemical Society, Symposium Series, v. 429, p. 444–485.

Kozur, H.W., 1998, Some aspects of the Permian-Triassic boundary (PTB) and of the possible causes for the biotic crisis around this boundary: Palaeogeography, Palaeoclimatology, Palaeoecology, v. 143, p. 227–272.

Krull, E.S., and Retallack, G.J., 2000, ^{13}C depth profiles from paleosols across the Permian- Triassic boundary: Evidence for methane release: Geological Society of America Bulletin, v. 112, no. 9, p. 1459–1472.

Krull, E.S., Retallack, G.J., Campbell, H.J., and Lyon, G.L., 2000, $^{13}C_{(org)}$ chemostratigraphy of the Permian-Triassic boundary in the Maitai Group, New Zealand: Evidence for high-latitudinal methane release: New Zealand Journal of Geology and Geophysics, v. 43, no. 1, p. 21–32.

Looy, C.V., 2000, The Permian-Triassic biotic crisis: collapse and recovery of terrestrial ecosystems [Ph.D. thesis]: Utrecht, Utrecht University, Laboratory of Palaeobotany and Palynology, LPP Contributions Series 13, 114 p.

MacLeod, K.G., Smith, R.M.H., Koch, P.L., and Ward, P.D., 2000, Timing of mammal-like reptile extinctions across the Permian-Triassic boundary in South Africa: Geology, v. 28, no. 3, p. 227–230.

Magaritz, M., Bar, R., Baud, A., and Holser, W.T., 1988, The carbon-isotope shift at the Permian-Triassic boundary in the southern Alps is gradual: Nature, v. 331, p. 337–339.

Magaritz, M., Krishnamurthy, R.V., and Holser, W.T., 1992, Parallel trends in organic and inorganic carbon isotopes across the Permian-Triassic boundary: American Journal of Science, v. 292, no. 10, p. 727–739.

Morante, R., 1996, Permian and Early Triassic isotopic records of carbon and strontium in Australia and a scenario of events about the Permian-Triassic boundary: Historical Biology, v. 11, p. 289–310.

Morante, R., Veevers, J.J., Andrew, A.S., and Hamilton, P.J., 1994, Determination of the Permian-Triassic boundary in Australia from carbon-isotope stratigraphy: Australian Petroleum Exploration Association Journal, v. 34, p. 330–336.

Nip, M., Tegelaar, E.W., Brinkhuis, H., De Leeuw, J.W., Schenck, P.A., and Holloway, P.J., 1986, Analysis of modern and fossil plant cuticles by Curie-point Py-Gc and Curie-point Py-GC-MS–recognition of a new, highly aliphatic and resistant bio-polymer: Organic Geochemistry, v. 10, no. 4–6, p. 769–778.

O'Leary, M.H., 1981, Carbon isotope fractionation in plants: Phytogeochemistry, v. 20, no. 4, p. 367–567.

Poort, R.J., Clement-Westerhof, J.A., Looy, C.V., and Visscher, H., 1997, Aspects of Permian palaeobotany and palynology. 17. Conifer extinction in Europe at the Permian-Triassic junction: Morphology, ultrastructure and geographic/stratigraphic distribution of *Nuskoisporites dulhuntyi* (prepollen of *Ortiseia*, Walchiaceae): Review of Palaeobotany and Palynology, v. 97, p. 9–39.

Retallack, G.J., 1995, Permian-Triassic life crisis on land: Science, v. 267, p. 77–80.

Retallack, G.J., 1999, Postapocalyptic greenhouse paleoclimate revealed by earliest Triassic paleosols in the Sydney Basin, Australia: Geological Society of America Bulletin, v. 111, no.1, p. 52–70.

Scholle, P.A., 1995, Carbon and sulfur isotope stratigraphy of the Permian and adjacent intervals, *in* Scholle, P.A., Peryt, T.M., and Ulmer-Scholle, D.S., eds., The Permian of Northern Pangea: Berlin, Springer-Verlag, p. 133–149.

Sephton, M.A., Looy, C.V., Veefkind, R.J., Visscher, H., Brinkhuis, H., and De Leeuw, J.W., 1999, Cyclic diaryl ethers in a Late Permian sediment: Organic Geochemistry, v. 30, no. 4, p. 267–273.

Smith, R.M.H., 1995, Changing fluvial environments across the Permian-Triassic boundary in the Karoo Basin, South Africa and possible causes of tetrapod extinction: Palaeogeography, Palaeoclimatology, Palaeoecology, v. 117, no. 1–2, p. 81–104.

Spitzy, A., and Degens, E.T., 1985, Modelling stable isotope fluctuations through geologic time: Mitteilungen des Geologisch-Paläontologischen Institutes der Universität Hamburg, v. 59, p. 155–166.

Tegelaar, E.W., De Leeuw, J.W., Derenne, S., and Largeau, C., 1989a, A reappraisal of kerogen formation: Geochimica et Cosmochimica Acta, v. 53, no. 11, p. 3103–3106.

Tegelaar, E.W., Matthezing, R.M., Jansen, J.B.H., Horsfield, B., and De Leeuw, J.W., 1989b, Possible origin of normal-alkanes in high-wax crude oils: Nature, v. 342, p. 529–531.

Tegelaar, E.W., Kerp, H., Visscher, H., Schenck, P.A., and De Leeuw, J.W., 1991, Bias of the paleobotanical record as a consequence of variations in the chemical composition of higher vascular plant cuticles: Paleobiology, v. 17, no. 2, p. 133–144.

Thackeray, J.F., Van der Merwe, J.A., Lee-Thorp, J.A., Sillen, A., Lanham, J.L., Smith, R., Keyser, A., and Monteiro, P.M.S., 1990, Changes in carbon isotope ratios in the Late Permian recorded therapsid tooth apatite: Nature, v. 347, p. 751–753.

Twitchett, R.J., Looy, C.V., Morante, R., Visscher, H., and Wignall, P.B., 2001, Rapid and synchronous collapse of marine and terrestrial ecosystems during the end-Permian biotic crisis: Geology, v. 29, p. 351–354.

Van Kaam-Peters, H.M.E., Schouten, S., De Leeuw, J.W., and Sinninghe Damsté, J.S., 1997, A molecular and carbon isotope biogeochemical study of biomarkers and kerogen pyrolysates of the Kimmeridge Clay facies: Palaeoenvironmental implications: Organic Geochemistry, v. 27, no. 7–8, p. 399–422.

Van Kaam-Peters, H.M.E., Schouten, S., Koster, J., and Sinninghe Damsté, J.S., 1998, Controls on the molecular and carbon isotopic composition of organic matter deposited in a Kimmeridgian euxinic shelf sea: Evidence for preservation of carbohydrates through sulfurisation: Geochimica et Cosmochimica Acta, v. 62, no. 19–20, p. 3259–3283.

Visscher, H., Brinkhuis, H., Dilcher, D.L., Elsik, W.C., Eshet, Y., Looy, C.V., Rampino, M.R., and Traverse, A., 1996, The terminal Paleozoic fungal event: Evidence of terrestrial ecosystem destabilization and collapse: Proceedings of the National Academy of Science of the United States of America, v. 93, p. 2155–2158.

Wang, K., Geldsetzer, H.H.J., and Krouse, H.R., 1995, Permian-Triassic extinction: organic ^{13}C evidence from British Columbia, Canada: Geology, v. 22, p. 580–584.

Ward, P.D., Montgomery, D.R., and Smith R., 2000, Altered river morphology in South Africa related to the Permian-Triassic extinction: Science, v. 289, p. 1740–1743.

Wignall, P.B., Morante, R., and Newton, R., 1988, The Permo-Triassic transition in Spitsbergen: $^{13}C_{org}$ chemostratigraphy, Fe and S geochemistry, facies, fauna and trace fossils: Geological Magazine, v. 135, p. 47–62.

Yin, H., Sweet, W.C., Glenister, B.F., Kotlyar, G., Kozur, H., Newell, N.D., Sheng, J., Yang, Z., and Zakharov, Y.D., 1996, Recommendation of the Meishan section as global stratotype section and point for basal boundary of Triassic System: Newsletter on Stratigraphy, v. 34, no. 2, p. 81–108.

Manuscript Submitted October 5, 2000; Accepted by the Society March 22, 2001

Geological Society of America
Special Paper 356
2002

Late Ordovician extinction: A Laurentian view

William B.N. Berry
Department of Earth and Planetary Science, University of California, Berkeley, California 94720, USA
Robert L. Ripperdan
Department of Geology, University of Puerto Rico, Mayaguez, Puerto Rico 00681
Stanley C. Finney
Department of Geological Sciences, California State University, Long Beach, California 90840, USA

ABSTRACT

Integrated sedimentologic, biostratigraphic, chemostratigraphic, and tectonic analyses of Late Ordovician stratigraphic sequences on the Laurentian plate indicate that one of the most severe extinctions among marine organisms in the Phanerozoic may be linked to climate and environmental changes related to a relatively brief glacial interval. These studies indicate that most of the extinctions occurred in relation to environmental change stimulated by relative sea-level fall or rise. Although some of the extinctions took place at the onset of glaciation, more of them occurred as sea level rose with the onset of deglaciation. A significant excursion in carbon isotopic ratios coincided with a cycle of sea-level fall and rise. Deposition of siliciclastic sediments in the Late Ordovician Queenston Delta complex on the southern margin of Laurentia may have been responsible for a brief decline in atmospheric carbon dioxide concentration, which, in turn, led to global cooling and glaciation. The Queenston Delta sedimentary record suggests that the complex accumulated for only a brief interval that coincided with the duration of glaciation.

INTRODUCTION

The Late Ordovician extinction among marine organisms was the second- or third-most severe during the Phanerozoic. That extinction was preceded in the early Middle Ordovician by one of the most dramatic radiations among marine organisms in the Phanerozoic. Sepkoski (1995, p. 393) pointed out that during the early part of the Ordovician, "more than three times more biodiversity was added to the marine system than during the Early Cambrian metazoan radiation, and this diversity was accumulated at least four times more rapidly than during the Mesozoic radiations." As noted by many (Berry and Boucot, 1973; Sheehan, 1973, 1975, 1988; Sheehan et al., 1996; Brenchley, 1984, 1988; Brenchley et al., 1991, 1994; Brenchley and Marshall, 1999; Cocks and Rickards, 1988; Wilde et al., 1990), the Late Ordovician extinction may be related to continental glaciation near the south pole. Long (1993) postulated that Late Ordovician ice volume could have been similar to that of the present-day Antarctic ice sheets. Poussart et al. (1999) proposed a model for Late Ordovician climate cooling indicating that the north polar area could also have been ice covered.

Late Ordovician climate change involved both cooling from an essentially greenhouse state to icehouse conditions associated with glaciation and then a return to preglacial conditions. Inasmuch as two significant climate changes are involved, the relationship between the organism extinctions and the climate changes is interesting. The extinctions essentially ended marked diversifications in many groups of marine organisms. Also of interest is how glaciation was initiated at a time when atmospheric carbon dioxide concentration was 10–16 times present atmospheric level (Berner, 1994; Kump et al., 1999).

Potential causes of the Late Ordovician glaciation have been discussed (see Brenchley et al., 1994; Brenchley and Marshall, 1999; and Kump et al., 1999), but no consensus has been

Berry, W.B.N., Ripperdan, R.L., and Finney, S.C., 2002, Late Ordovician extinction: A Laurentian view, *in* Koeberl, C., and MacLeod, K.G., eds., Catastrophic Events and Mass Extinctions: Impacts and Beyond: Boulder, Colorado, Geological Society of America Special Paper 356, p. 463–471.

reached. Kump et al. (1999, p. 173) suggested that the "Taconic orogeny, which commenced in the late-middle Ordovician, caused a long-term decline in atmospheric pCO_2 through increased weatherability of silicate rocks." Their proposal is consistent with the several discussions of the relationship between tectonic uplift and climate change in the volume on that topic edited by Ruddiman (1997). In a summary chapter in that volume, Ruddiman et al. (1997, p. 494) concluded that "uplift-driven chemical weathering has lowered CO_2 levels" and that lowered CO_2 levels apparently have led to continental glaciation. Consideration of the latest Ordovician sedimentary rock record on the southern side of the Laurentian plate enhances the understanding of the links between tectonism, chemical weathering, and glaciation discussed by Kump et al. (1999). That depositional record is reviewed here in light of discussions of climate change influenced by tectonic uplift in Ruddiman (1997). The patterns of organism extinctions that resulted from a set of environmental changes that may be linked to probable tectonic uplift-induced climate changes are also described herein. Discussion of those patterns forms the primary focus of this contribution.

The impacts of glaciation and subsequent deglaciation within an interval of time that may have been <500 k.y. in the Late Ordovician included not only changes in sea level but also those in seawater temperature, sites of oceanic upwelling, surface ocean circulation, and sites of origination of ocean deep waters. The extinctions of marine organisms appear to result from one or more of these climate and environmental shifts. Despite intensive searches, no evidence for an associated impact has been found (Orth et al., 1986; Wilde et al., 1986; Robertson et al., 1991; Wang et al., 1992, 1993).

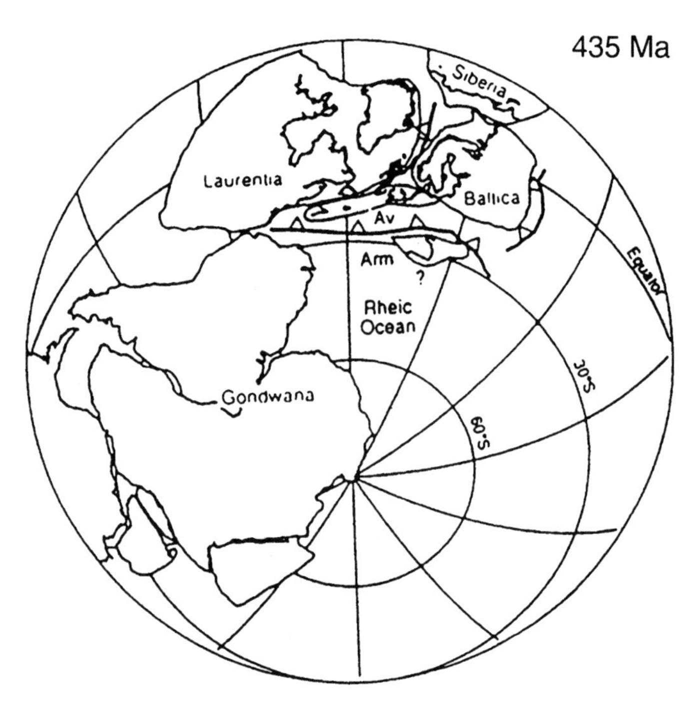

Figure 1. Latest Ordovician geography from Van Staal et al. (1998). Av—Avalon Plate, Arm—Armorican Plate.

PALEOGEOGRAPHIC SETTING

During the later part of the Ordovician, a relatively large plate, Laurentia, was within the tropics, essentially astride the equator (Fig. 1). What is today western North America was on the northern side of the Laurentian plate. The northern side of that plate was a passive tectonic margin. Gently sloping submarine ramps developed on this margin. Carbonates accumulated in shallow-marine environments along the southern or interior portions of the ramps. Depositional environments deepened northward. The change to slope along the plate margin was approximately in what is today central Nevada. Finney et al. (1997, 1999) described stratigraphic sequences that accumulated in shallow and deep shelf as well as slope environments during the Late Ordovician in the western United States.

The southern side of the plate was an active plate boundary. Plates that were 1200–1500 km long accreted with this margin throughout the Late Ordovician (Van Staal, 1994; MacNiocaill et al., 1997; Robinson et al., 1998; Van Staal et al., 1998). A number of collisions of these small plates with Laurentia in the late Middle Ordovician (Van Staal, 1994) resulted in those tectonic events collectively called the Taconic orogeny. During the latest Ordovician, plate collisions with Laurentia led indirectly to development of broad areas of fluviatile and deltaic environments in which a spectrum of siliciclastic materials accumulated that classically have been called the Queenston Delta in the central and southern Appalachians (Dennison, 1976). The sedimentology of these siliciclastic materials was analyzed by Dorsch and Driese (1995), who drew attention to the relationships of the sediments to glacio-eustatic sea-level fall and subsequent rise. Continued collisions on the southern side of Laurentia during the latest Ordovician-Early Silurian resulted indirectly in development of unconformable relationships between the Queenston deltaic complex and overlying units (Berry and Boucot, 1970; Dennison, 1976; Dorsch and Driese, 1995).

BIOSTRATIGRAPHIC ZONES AND STAGES USEFUL IN CORRELATION

Graptolite zones seem to be the biostratigraphic units most useful for Late Ordovician correlations among a number of areas. The Late Ordovician (Ashgill Series) graptolite zones commonly used for correlations are cited in Table 1 with approximate correlatives with stages that are recognized by shelly, primarily brachiopod, faunas. In light of Late Ordovician conodont-based correlations in Sweet (2000, Figs. 5–7) and graptolite-based correlations of the type Ashgill by Rickards (2001), the Hirnantian Stage appears to be only the youngest part of the Ashgill or Late Ordovician. The upper part of the North American Richmond Stage appears to be coeval with the

TABLE 1. LATE ORDOVICIAN AND EARLIEST SILURIAN (*P. ACUMINATUS*) GRAPTOLITE ZONES AND THEIR CORRELATIONS WITH LATE ORDOVICIAN AND EARLIEST SILURIAN STAGES AND SERIES

Graptolite zones	Shelly Faunal stages	North American stages	Standard series
Parakidograptus acuminatus	Rhuddanian		Llandovery
Persculptograptus persculptus			
Normalograptus extraordinarius	Hirnantian	Richmondian	
			Ashgill
Paraorthograptus pacificus			
	Rawtheyian	————	
Dicellograptus complanatus ornatus		Maysvillian	

Hirnantian. Rickards' (2001) correlations imply that the glaciation took place within the *N. extraordinarius* and, probably, in the lower *P. persculptus* zone, and that these zones are approximate correlatives of the Hirnantian Stage. Discussions with R.B. Rickards (1999, personal commun.) on the durations of these graptolite zones suggest that each of them may have been 500 k.y., or perhaps less, in duration: if so, then the glaciation may have persisted for perhaps ~500 k.y.

The Late Ordovician stratigraphic succession along Vinini Creek in central Nevada includes an interval of black, organic-rich mudstones that suggest that upwelling conditions along the slope-shelf change or platform margin led to development of and subsequent preservation of large numbers of several graptolite species in the *D. complanatus ornatus* and *P. pacificus* zones (Fig. 2). Most extinctions took place near the *P. pacificus-N. extraordinarius* zone boundary (Fig. 2). Most graptolite species that became extinct were those that lived close to the edges of an oxygen minimum zone that developed under upwelling waters (Finney and Berry, 1997; Finney et al., 1997, 1999). Graptolite survivors were those that lived in relatively more oxic waters near the ocean surface. Most of these survivors are normalograptids.

Fossil collecting in the Vinini Creek and Monitor Range stratal successions in central Nevada indicates that most conodont species persisted through nearly all of the graptolite zones cited herein, disappearing from the succession during the *P. persculptus* zone, when depositional environments became markedly shallow (Sweet, 2000). Sweet (2000) described the conodonts in these successions, their stratigraphic occurrences, and relevant correlations.

Conodont taxa considered to be basin dwellers found in the Nevada successions disappeared before those considered to have lived in shelf sea settings (Finney et al., 1997, 1999; Sweet, 2000) (Fig. 2). Barnes and Zhang reviewed global Late Ordovician conodont extinctions and noted essentially the same pattern. They stated "there is no dramatic change in conodont faunas" through the graptolite zones cited above until the *P. persculptus* zone, and they concluded that "the main conodont extinction bioevent" is "concentrated in the lower *persculptus* Zone" (Barnes and Zhang, 1999, p. 211).

Correlations of rocks bearing conodonts and benthic faunas suggest that most benthic taxa (brachiopods, corals, crinoids) disappeared as environments in which they lived shallowed during the *P. pacificus* zone. Late Ordovician Laurentian plate shelly faunal developments were discussed by Sheehan et al. (1996), Jin (1999), and Copper (1999).

Jin (1999) reviewed development of brachiopod faunas that lived in environments in the interior parts of the Laurentian plate in the Late Ordovician. He pointed out that many taxa originated in environments on the platform margins during the Middle Ordovician and migrated to interior sites as sea level rose in the Rawtheyian-Hirnantian interval (Jin, 1999). These faunas were restricted to the Laurentian plate (Jin, 1999, Fig. 2). Analysis of Laurentian Late Ordovician brachiopods led Jin (1999, p. 204) to conclude that by approximately the Rawtheyian, "the evolution of the North American epicontinental brachiopod fauna reached a climax in species diversity, abundance and geographic distribution." He also noted that an abundance of large shells of several species could be found in several sites within the interior of the Laurentian plate. During the Hirnantian interval, many of the Rawtheyian brachiopods became extinct. Jin (1999, p. 205) pointed out that during glacio-eustatic sea-level fall, marine environments across the interior of the Laurentian plate had one of at least three possible fates: (1) complete disappearance, (2) slow shrinking to hypersaline (salt, gypsum, and anhydrite) environments, or (3) changing from sites of dominantly carbonate deposition to sites of either mixed carbonate and siliciclastic or almost entirely siliciclastic deposition. Jin (1999) stated that most of the large-shelled brachiopod taxa that were so abundant in shelf sea settings in the interior of the Laurentian plate in the Rawtheyian became extinct, whereas taxa with smaller shells migrated to sites in the plate margins where they survived.

Copper (1999) reviewed Late Ordovician into Silurian faunal changes in a carbonate succession on Anticosti Island, a site that was near a margin of the Laurentian plate. His studies (Copper, 1999, Fig. 1) suggest that brachiopod faunas underwent two extinctions. The first was at the Rawtheyian-Hirnantian boundary, which probably coincided with onset of environmental conditions associated with marked development of continental glaciation and related sea-level fall. The second was at the close of the Hirnantian, during the initial phase of deglaciation and glacio-eustatic sea-level rise. Certain tabulate coral and many stromatoporoid taxa survived cooling at the

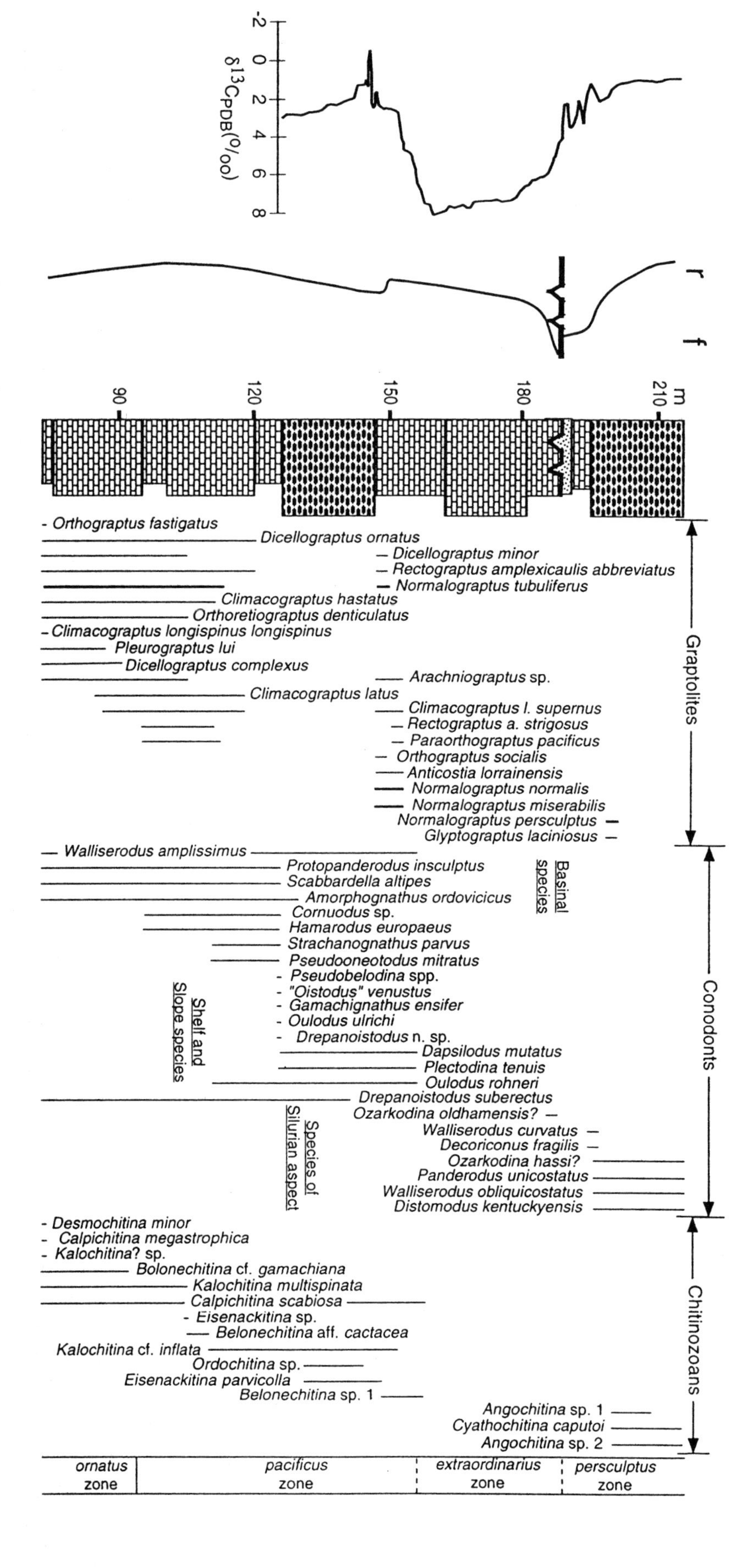

Figure 2. Carbon isotope, sea-level change, and stratigraphic range data for graptolites, conodonts, and chitinozoans illustrating relationships among organism extinctions, sea-level changes, and shift in carbon isotopic concentrations in Late Ordovician strata at Copenhagen Canyon in Monitor Range, Nevada. Data in this figure may be related to Figure 3 by stratigraphic thicknesses indicated. PDB is Peedee belemnite.

Rawtheyian-Hirnantian boundary (Copper, 1999, Fig. 2). New taxa among both stromatoporoids and tabulate corals appeared during the Hirnantian, and patch reefs developed during that interval: Copper (1999, p. 207) commented that "on Anticosti, regional diversity in the Hirnantian was higher than in the preceding Rawtheyian." A number of stromatoporoid and tabulate coral taxa became extinct with warming conditions at the end of the Hirnantian (Copper, 1999, Fig. 2). Copper (1999, p. 207) concluded from his analysis of Late Ordovician Anticosti faunas that there were "virtually no losses at the family, superfamily and suborder level in the corals (only lichenarids vanished), stromatoporoids (only aulacerids were lost), and brachiopods (even the classic Ordovician zygospirinids and catazyginids survived)."

Analyses of global patterns in developments among Late Ordovician brachiopods by Sheehan et al. (1996) are consistent with the patterns cited for the Anticosti succession. Sheehan et al. (1996, p. 479) concluded from their global study that the first major brachiopod extinction event took place "at the beginning of glaciation when sea level was depressed as much as 100 m." The consequence of lowered sea level was that faunas in shallow waters of shelf seas were eliminated. Sheehan et al. (1996, p. 479) noted that Late Ordovician glacio-eustatic sea-level fall was not as great as that during the Pleistocene, but that the impact on shelf sea benthic faunas in the Late Ordovician was far more dramatic because shelf seas were far more widespread and diverse habitats more prevalent. Hirnantian brachiopod faunas were adapted to relatively cooler ocean waters and to life in depositional environments in which siliciclastic sediments were deposited. The second significant extinction occurred at the end of the Hirnantian, when shallow shelf seas again spread widely across major plates as sea level rose during deglaciation (Sheehan et al., 1996, p. 480).

DEPOSITIONAL ENVIRONMENTS

Changes in depositional environments that reflect sea-level fall and rise through the Late Ordovician in Nevada were described by Cooper (1997; and *in* Finney et al., 1997, 1999). Cooper's analyses documented that environments shallowed in all shelf as well as basin sequences through the *P. pacificus* zone. The interior part of the shelf was exposed by the later part of the *P. pacificus* zone. Shallow-marine environments in which carbonates accumulated shifted gradually toward the shelf margin during the *P. pacificus* zone into the *N. extraordinarius* zone. Carbonates were deposited in sites that had been outer shelf and slope by the *N. extraordinarius* zone. Depositional environments began to deepen within the *P. persculptus* zone. By the latter part of that zone, sea level appears to have risen significantly (Finney et al., 1999).

CARBON ISOTOPES

Ripperdan et al. (1998) (see Figs. 2 and 3) determined the carbon isotopic signatures of marine carbonates in the central Nevada outer shelf and basin successions. The $\delta^{13}C$ value remained relatively constant during the *D. complanatus ornatus* and most of the *P. pacificus* (unit Ka of Fig. 3) zones. Near the end of the *P. pacificus* zone (border of Ka and Kb shown in Fig. 3), relatively low energy sedimentation was replaced by an ~10 m interval of coarse-grained strata. Sedimentary petrologic analyses of these strata suggest a brief interval of relatively rapid sea-level variation followed by distinct evidence of sea-level fall (within unit Kb of Fig. 3, see also Fig. 2). The carbon isotopic values recovered from the coarse-grained strata suggest a brief fall in local seawater $\delta^{13}C$ values. A sharp rise in $\delta^{13}C$ values occurred simultaneously with sea-level fall and remained high through the next 20 m of strata (most of unit Kb and Kc shown in Fig. 3), which encompassed most of the interval of sea-level lowstand. The $\delta^{13}C$ values began to fall sharply within an ~6 m interval immediately below an oxidized surface (Figs. 2 and 3) interpreted to represent the maximum extent of glacio-eustatic sea-level drawdown (Cooper, 1997; Finney et al., 1999). Analysis of the sedimentary, biotic, and carbon isotopic data indicates that $\delta^{13}C$ variations and the carbon cycle change that stimulates them were strongly associated with sea-level fluctuations during the glacio-eustatic interval. The absolute values and overall trend in $\delta^{13}C$ values from above the oxidized surface are nearly identical to $\delta^{13}C$ values in strata deposited prior to sea-level fall (unit Ka of Fig. 3). Potentially, one inference from these Nevada data is that the atmospheric carbon dioxide concentration dropped after initiation of glaciation.

Kump et al. (1999) determined similar changes in Late Ordovician carbon isotopic values and suggested that the enhanced concentration of $\delta^{13}C$ is a reflection of increased weathering of carbonate rocks that had been deposited in shallow-marine platform environments. Most of these environments were exposed during sea-level fall. This explanation is reasonable in light of Late Ordovician lithofacies development (Finney et al., 1997) evidenced by the Nevada strata.

The pattern of $\delta^{13}C$ variation found in the central Nevada succession suggests that the shift in marine carbon cycling involved transfer of surface ocean carbon into a transient reservoir during sea-level lowstand, followed by a return to prior conditions during (and after) sea-level rise. A possibility for such storage might have been the spread of cyanobacteria, eucaryotic algae, lichens, and certain nonvascular land plants (embryophytes). A.J. Boucot (1999, personal commun.), who collected Late Ordovician-Early Silurian land plant remains from a number of localities along the southern side of Laurentia, pointed out that land plants and bacteria could have spread as sea level fell and many former marine environments were exposed. Graham (1993) reviewed evidence for Late Ordovician-Silurian embryophtes and other possible Late Ordovician-Early Silurian land vegetation, suggesting that embryophytes of that time could have resembled the modern prostrate thalloid liverwort, *Ricciocarpus*. She pointed out that many cyanobacteria living in modern soils have ultraviolet-resistant sheath pigments and that many of them, as well as certain modern soil-dwelling

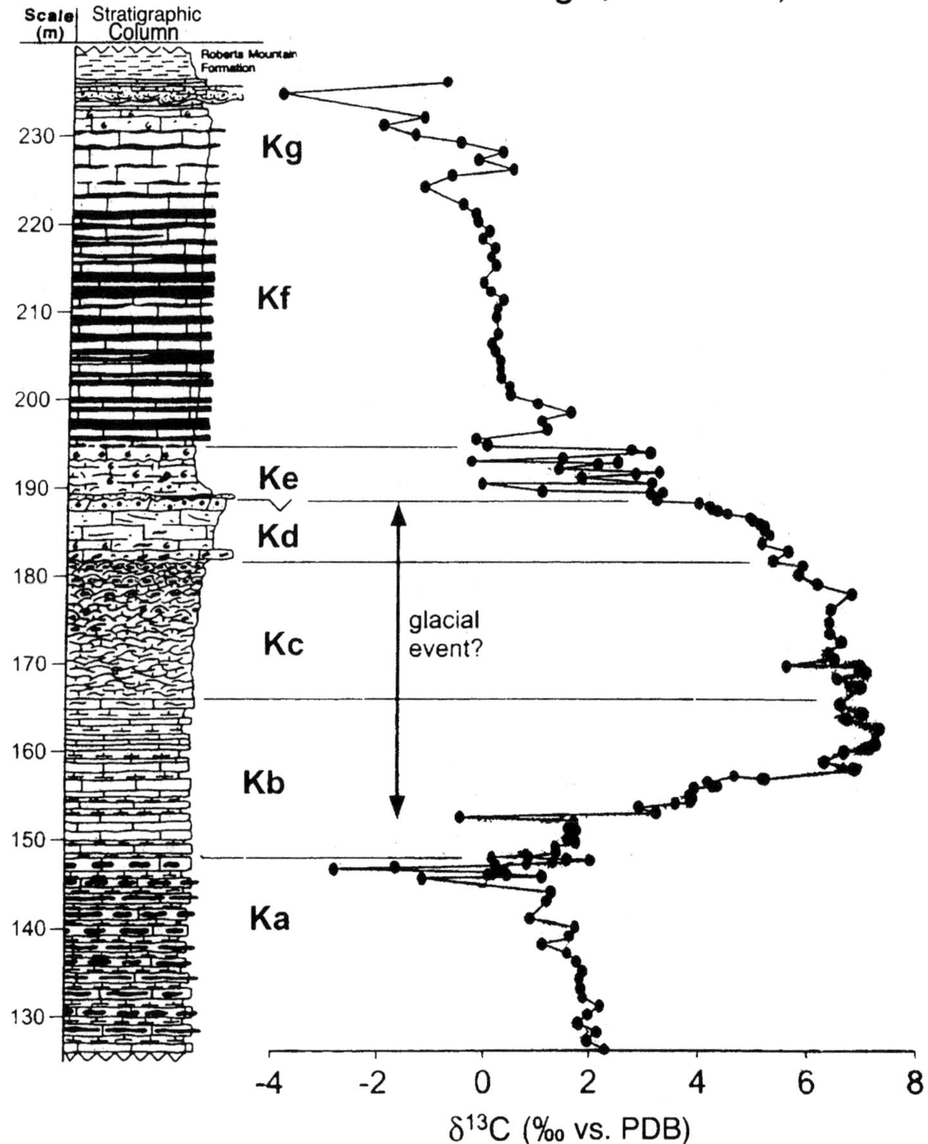

Figure 3. Details of carbon isotopic analyses in context of sedimentologic examination of stratigraphic sequence at Copenhagen Canyon, Monitor Range, Nevada (Fig. 2 indicates stratigraphic context). Units Ka and Kf are both alternating layers of cherty lime mudstone and medium-bedded lime mudstone deposited in moderate depth shelf environments. Carbon isotope ratios in these units are preglaciation and postglaciation. Kb is lime mudstone with thin shale partings that developed at times of slight sea-level rises that swept graptolites into site. Carbon isotopic values shown closely follow relative changes in sea level as determined from petrographic analysis. (PDB is Peedee belemnite.) Kc is cliff-forming, medium-bedded lime mudstone. Units Kb and Kc accumulated in shelf sea environments shallower than units Ka and Kf. Kd includes burrowed lime mudstones, wackestones, and packstones with abundant brachiopod and crinozoan fragments, and beds of ooids. It accumulated in shallow subtidal and intertidal settings. Ke includes beds of rip-up clasts, burrowed lime mudstones, and well-sorted, ripple-laminated quartz sandstones. It accumulated in intertidal to supratidal conditions during which rising sea level resulted in significant reworking of bottom sediment. Analysis of sediments suggests that maximum lowstand occurred during unit Kd and that units Kb and Kc accumulated during falling sea level. Copenhagen Canyon data suggest that shift in carbon cycling occurred in conjunction with sea-level change, although two events were not linked precisely. Unit Kg is skeletal wackestone with phosphatic grainstone at the top.

eucaryotic algae, may have had Late Ordovician-Early Silurian analogues (Graham, 1993). Postglaciation sea-level rise would have flooded tidal wetland environments on the Laurentian plate that had been colonized by organisms during glacio-eustatic lowstand. The loss of these former wetland environments would have resulted in significant losses of vegetation and bacteria and, as a consequence, loss of a transient carbon reservoir.

QUEENSTON DELTA FORMATION: AN INFLUENCE ON GLOBAL CLIMATE

Active plate margin tectonism and resultant uplifts characterized the southern side of the Laurentian plate during much of the Ordovician (Van Staal, 1994; MacNiocaill et al., 1997; Robinson et al., 1998; Van Staal et al., 1998). Analyses of geologic structures in New England and Maritime Canada indicate that a number of tectonic terranes or small plates collided with Laurentia along an area that is now New Jersey and New York, New England, and Maritime Canada from the Middle through Late Ordovician (Van Staal, 1994; MacNiocaill et al., 1997; Robinson et al., 1998; Van Staal et al., 1998). Terrane collision that commenced in the later part of the Middle Ordovician and continued into the Late Ordovician resulted in development of a topographically high land that formed a source area for the extensive fluviatile and deltaic deposits classically known as the Queenston Delta (Dennison, 1976). Dorsch and Driese

(1995) described the stratigraphy and sedimentologic aspects of these deposits.

Dorsch and Driese (1995, Fig. 1) indicated that the Juniata Formation and a unit they termed Lower Tuscarora Sandstone (these are rock units within the Queenston deltaic complex) are, at least in part, latest Ordovician in age. Sweet (2000, Figs. 5–8) indicated that latest Ordovician strata in eastern North America could be correlated with the *N. extraordinarius* and the lower part of the *P. persculptus* graptolite zones based upon his graphic correlations of Ordovician conodont successions. Both Dennison (1976) and Dorsch and Driese (1995) drew attention to a hiatus in deposition within the Queenston deltaic complex that could be related to Late Ordovician glacio-eustatic sea-level lowstand. The lower part of the strata included in the Queenston deltaic complex are thus most likely correlative with onset of glaciation as well as much of the glacial interval.

Paleomagnetism studies indicate that Laurentia and the small plates that collided with it were within the tropics (MacNiocaill et al., 1997; Robinson et al., 1998; Van Staal et al., 1998). Driese and Foreman (1992) described paleosols that formed during Late Ordovician pedogenesis of Queenston deltaic complex estuarine and upper tidal flat deposits. They suggested that these paleosols developed in subtropical climate conditions in which rainfall was high, at least seasonally (Driese and Foreman, 1992).

In his discussion of the Queenston deltaic complex of sediments, Dennison (1976) indicated that this sedimentary complex may be recognized within a belt that is ~1500 km long and ranges from 200 to ~300 km wide. That belt extends from New York southwest into Alabama (Dennison, 1976, Fig. 5). The sedimentary complex attains a maximum thickness of 400–500 m at sites in New York, Pennsylvania, and West Virginia. Dennison (1976) described the Queenston deltaic complex as including a significant component of nonmarine flood-plain, river channel, and overbank deposits as well as shallow subtidal deposits that accumulated in delta plain environments. He indicated that interior deltaic deposits accumulated in extensive mudflats exposed subaerially during glacio-eustatic sea-level fall (Dennison, 1976, p. 115).

Much of the Queenston deltaic complex is separated from superjacent strata by a regional unconformity (Berry and Boucot, 1970; Dennison, 1976; Dorsch and Driese, 1995). Uplift and erosion of former depositional sites appear to have resulted from continued tectonic uplifts driven by plate collisions.

The volume, geographic extent, source region, and lithologic aspects of the siliciclastic sediments that compose the Queenston deltaic complex (Dorsch and Driese, 1995) suggest that it was derived by chemical weathering of a tropical terrane formed by tectonic uplift. The origin and accumulation of the Queenston deltaic complex siliciclastics are consistent with discussions in Raymo and Ruddiman (1992), Kump and Arthur (1997), and Ruddiman (1997) on the relationship between chemical weathering of tectonically active mountain belts and its influence on global climate. Edmund and Huh (1997, p. 349), for example, concluded from a study of chemical weathering of orogenic regions that "tectonically active mountain belts are the loci for accelerated drawdown of atmospheric CO_2; hence their initiation and evolution have a direct influence on global climate." The volume, geographic extent, and paleogeography of the source region of the Queenston deltaic complex suggest that a similar mechanism could have been responsible for the onset of Late Ordovician glaciation.

Inasmuch as the solar constant was ~4.5% less in the Late Ordovician than at present (Berner, 1994; Gibbs et al., 1997), a modest decline in atmospheric carbon dioxide concentration brought about by chemical erosion of siliciclastics could have been enough to allow ice to accumulate at the poles and initiate development of Hirnantian ice sheets. Two factors suggest that deglaciation could have started when the source region had been eroded markedly and part of the area of sediment accumulation became a source of sediment. These factors are (1) the relatively greater proportion of mudstones and claystones in postglaciation sediments than in sediments deposited during glaciation, and (2) an unconformity between some Queenston deltaic complex sediments and superjacent strata (Dorsch and Driese, 1995).

SUMMARY

The Late Ordovician extinctions among marine organisms appear to be linked temporally with two climate changes: the first, preglacial to glacial conditions, and the second, return to nonglacial climates. Late Ordovician tectonic, depositional, and faunal records of the Laurentian plate have yielded significant evidence indicating that extinctions in the oceans took place as a consequence of environmental changes that are related to glacio-eustatic sea-level fall and subsequent rise. Late Ordovician plate margin tectonism and derivation of the Queenston deltaic siliciclastics from a tectonically uplifted source region seem to have led to glaciation. Deglaciation followed after the source of the sediments had been eroded significantly.

The Laurentian plate record suggests that many of the extinctions among marine organisms took place during warming and the environmental changes related to it. Conodonts, stromatoporoids, and certain tabulate corals did not become extinct at the onset of glaciation; rather, they underwent significant extinctions during warming and deglaciation. Graptolite extinctions occurred when plate margin upwelling ceased near the onset of sea-level fall and the loss of upwelling conditions as glaciation proceeded. Graptolite faunal diversity was enhanced when upwelling conditions and associated oxygen minimum zones redeveloped after sea level rose in the early part of the Silurian. Two major faunal turnovers, one at the onset of cooling and glaciation and one near the onset of warming and deglaciation, characterize Late Ordovician brachiopod faunal diversity changes.

The Late Ordovician Laurentian plate tectonic, stratigraphic, and faunal record provides evidence of links between coeval climate, environment, and faunal changes. Analysis of this record indicates that extensive chemical weathering of a source region for siliciclastics on the southern margin probably led to climate cooling and related environmental changes. The study of stratigraphic ranges of many species suggests that the climate shift toward warming and deglaciation, and resultant environmental changes, led to more extinctions than did the change toward cooling.

REFERENCES CITED

Barnes, C.R. and Zhang, S., 1999, Pattern of conodont extinction and recovery across the Ordovician-Silurian boundary interval: Acta Universitatis Carolinae, Geologica, v. 43, p. 211–213.

Berner, R.A., 1994, GEOCARB II: A revised model of atmospheric CO_2 over Phanerozoic time: American Journal of Science, v. 294, p. 56–91.

Berry, W.B.N. and Boucot, A.J., 1970, Correlation of the North American Silurian rocks: Geological Society of America Special Paper 102, 289 p.

Berry, W.B.N., and Boucot, A.J., 1973, Glacio-eustatic control of Late Ordovician-Early Silurian platform sedimentation and faunal changes: Geological Society of America Bulletin, v. 84, p. 275–284.

Brenchley, P.J., 1984, Late Ordovician extinctions and their relationship to the Gondwana glaciation, *in* Brenchley, P.J., ed., Fossils and climate: Chichester, UK, John Wiley & Sons, p. 291–327.

Brenchley, P.J., 1988, Environmental changes close to the Ordovician-Silurian boundary: Bulletin of the British Museum (Natural History), Geology Series, v. 43, p. 377–385.

Brenchley, P.J., and Marshall, J.D., 1999, Relative timing of critical events during late Ordovician mass extinction: New data from Oslo: Acta Universitatis Carolinae, Geologica, v. 43, p. 187–190.

Brenchley, P.J., Romano, M., Young, T.P., and Storch, P., 1991, Hirnantian glaciomarine diamictites: Evidence for the spread of glaciation and its effects on Ordovician faunas, *in* Barnes, C.R., and Williams, S.H., eds., Advances in Ordovician geology: Geological Survey of Canada Paper 90-9, p. 325–336.

Brenchley, P.J., Marshall, J.D., Carden, G.A.F., Robertson, D.B.R., Long, D.G.F., Meidla, T., Hints, L., and Anderson, T.F., 1994, Bathymetric and isotopic evidence for a short-lived Late Ordovician glaciation in a greenhouse period: Geology, v. 22, p. 295–298.

Brenchley, P.J., Carden, G.A.F., and Marshall, J.D., 1995, Environmental changes associated with the "first strike" of the Late Ordovician mass extinction: Modern Geology, v. 20, p. 69–82.

Cocks, L.R.M., and Rickards, R.B., eds., 1988, A global analysis of the Ordovician-Silurian boundary: Bulletin of the British Museum (Natural History), Geology Series, v. 43, p. 1–394.

Cooper, J.D., 1997, Late Ordovician crises: The depositional and sequence stratigraphic framework of platform to basin succession, central Nevada: Geological Society of America Abstracts with Programs, v. 29, p. 355.

Copper, P., 1999, Brachiopods during and after the Late Ordovician mass extinctions, on Anticosti Island, E. Canada: Acta Universitatis Carolinae, Geologica, v. 43, p. 207–209.

Dennison, J.D., 1976, Appalachian Queenston Delta related to eustatic sea-level drop accompanying Late Ordovician glaciation centered in Africa, *in* Bassett, M.G., ed., The Ordovician system: Proceedings of a Palaeontological Association Symposium, Birmingham, September 1974: Cardiff, University of Wales Press and National Museum of Wales, p. 107–120.

Dorsch, J., and Driese, S.G., 1995, The Taconic foredeep as sediment sink and sediment exporter: Implications for the origin of the white quartzarenite blanket (Upper Ordovician-Lower Silurian) of the central and southern Appalachians: American Journal of Science, v. 295, p. 201–243.

Driese, S.G., and Foreman, J.L., 1992, Paleopedology and paleoclimatic implications of Late Ordovician vertic paleosols, Juniata Formation, southern Appalachians: Journal of Sedimentary Petrology, v. 62, p. 71–83.

Edmond, J.M. and Huh, Youngsook, 1997, Chemical weathering yields from basement and orogenic terrains in hot and cold climates, *in* Ruddiman, W.F., ed., Tectonic uplift and climate change: New York, Plenum Press, p. 329–351.

Finney, S.C., and Berry, W.B.N., 1997, New perspectives on graptolite distributions and their use as indicators of platform margin dynamics: Geology, v. 25, p. 919–922.

Finney, S.C., Cooper, J.D., and Berry, W.B.N., 1997, Late Ordovician mass extinction: Sedimentologic, cyclostratigraphic, and biostratigraphic records from platform and basin successions, central Nevada: Brigham Young University Geology Studies, v. 42, p. 79–103.

Finney, S.C., Berry, W.B.N., Cooper, J.D., Ripperdan, P.L., Sweet, W.C., Jacobson, S.R., Soufiane, A., Achab, A., and Noble, P.J., 1999, Late Ordovician mass extinction: A new perspective from stratigraphic sections in central Nevada: Geology, v. 27, p. 215–218.

Gibbs, M.T., Barron, E.J., and Kump, L.R., 1997, An atmospheric pCO_2 threshold for glaciation in the Late Ordovician: Geology, v. 25, p. 447–450.

Graham, L.E., 1993, Origin of land plants: New York, John Wiley, 287 p.

Jin, J., 1999. Evolution and extinction of the Late Ordovician epicontinental brachiopod fauna of North America: Acta Universitatis Carolinae, Geologica, v. 43, p. 203–206.

Kump, L.R., and Arthur, M.A., 1997, Global chemical erosion during the Cenozoic: Weatherability balances the budget, *in* Ruddiman, W.F., ed., Tectonic uplift and climate change: New York, Plenum Press, p. 399–426.

Kump, L.R., Arthur, M.A., Patzkowsky, M.E., Gibbs, M.T., Pinkus, D.S., and Sheehan, P.M., 1999, A weathering hypothesis for glaciation at high atmospheric pCO_2 during the Late Ordovician: Palaeogeography, Palaeoclimatology, Palaeoecology, v. 152, p. 173–187.

Long, D.G.F., 1993, Oxygen and carbon isotopes and event stratigraphy near the Ordovician-Silurian boundary, Anticosti Island, Quebec: Palaeogeography, Paleoclimatology, Palaeoecology, v. 104, p. 49–59.

MacNiocaill, C., van der Plujm, B., and van ver Voo, R., 1997, Ordovician paleogeography and the evolution of the Iapetus Ocean: Geology, v. 25, p. 159–162.

Marshall, J.D., Brenchley, P.J., Mason, P., Wolff, G.A., Astini, R.A., Hints, L., and Meidla, T., 1997, Global isotopic events associated with mass extinction and glaciation in the Late Ordovician: Palaeogeography, Palaoeclimatology and Palaeoecology, v. 132, p. 195–210.

Orth, C.J., Gilmore, J.S., Quintana, L.R., and Sheehan, P.M., 1986, The terminal Ordovician extinction: Geochemical analysis of the Ordovician/Silurian boundary, Anticosti Island, Quebec: Geology, v. 14, p. 433–436.

Poussart, P.F., Weaver, A.J., and Barnes, C.R., 1999, Late Ordovician glaciation and high atmospheric CO_2: Addressing an apparent paradox via a coupled model approach: Acta Universitatis Carolinae, Geologica, v. 43, p. 167–169.

Raymo, M.E., and Ruddiman, W.F., 1992, Tectonic forcing of late Cenozoic climate: Nature, v. 359, p. 117–122.

Rickards, R.B., 2001, The graptolitic age of the type Ashgill Series (Ordovician), Cumbria: Yorkshire Geological Society Transactions (in press).

Ripperdan, R.L., Cooper, J.D., and Finney, S.C., 1998, High-resolution C and lithostratigraphic profiles from Copenhagen canyon, Nevada: Clues to the behavior of ocean carbon during the Late Ordovician global crisis: Mineralogical Magazine, v. 62A, p. 1279–1280.

Robertson, D.B.R., Brenchley, P.J., and Owen, A.W., 1991, Ecological disruption close to the Ordovician-Silurian boundary: Historical Biology, v. 5, p. 131–144.

Robinson, P., Tucker, R.D., Bradley, D., Berry, H.N., Jr., and Osberg, P., 1998, Paleozoic orogens in New England, USA: Geologiska Foreningen Forhandlungen, v. 120, p. 119–148.

Ruddiman, W.F., editor, 1997, Tectonic uplift and climate change: New York, Plenum Press, 489 p.

Ruddiman, W.F., Raymo, M.E., Prell, W.L. and Kutzbach, J.E., 1997, The

uplift-climate connection: A synthesis, *in* Ruddiman, W.F., ed., Tectonic uplift and climate change: New York, Plenum Press, p. 471–515.
Sepkoski, J.J., Jr., 1995, The Ordovician radiations: Diversification and extinction shown by global genus-level taxonomic data, *in* Cooper, J.D., Droser, M.L., and Finney, S.C., eds., Ordovician odyssey: Short papers for the Seventh International Symposium on the Ordovician System, Las Vegas, Nevada, USA: Fullerton, California, SEPM (Society for Sedimentary Geology), p. 393–396.
Sheehan, P.M., 1973, The relation of Late Ordovician glaciation to the Ordovician-Silurian changeover in North American brachiopod faunas: Lethaia, v. 6, p. 147–154.
Sheehan, P.M. 1975, Brachiopod synecology in a time of crisis (Late Ordovician–Early Silurian): Paleobiology, v. 1, p. 205–212.
Sheehan, P.M., 1988, Late Ordovician events and the terminal Ordovician extinction: New Mexico Bureau of Mines and Mineral Resources Memoir 44, p. 405–415.
Sheehan, P.M., Coorough, P.J., and Fastovsky, D.E., 1996, Biotic selectivity during the K/T and Late Ordovician extinction events, *in* Ryder, G., Fastovsky, D.E., and Gartner, S., eds., The Cretaceous-Tertiary event and other catastrophes in Earth history: Geological Society of America Special Paper 307, p. 477–489.
Sweet, W.C., 2000, Conodonts and biostratigraphy of Upper Ordovician strata along a shelf to basin transect in central Nevada: Journal of Paleontology, v. 74, p. 1148–1160.
Van Staal, C.R., 1994, Brunswick subduction complex in the Canadian Appalachians: Record of the Late Ordovician to Late Silurian collision between Laurentia and the Gander margin of Avalon: Tectonics, v. 13, p. 946–962.
Van Staal, C.R., Dewey, J.F., MacNiocaill, C., and McKerrow, W.S., 1998, The Cambrian-Silurian tectonic evolution of the northern Appalachians and British Caledonides: History of a complex, west and southwest Pacific-type segment of Iapetus, *in* Blundell, D.J., and Scott, A.C., eds., Lyell: The past is the key to the present: Geological Society [London] Special Publication 143, p. 199–242.
Wang, K., Chatterton, B.D.E., Attrep, M., and Orth, C.J., 1992, Iridium abundance maxima at the latest Ordovician mass extinction horizon, Yangtze Basin, China: Terrestrial or extraterrestrial?: Geology, v. 20, p. 39–42.
Wang, K., Orth, C.J., Attrep, M., Jr., Chatterton, B.D.E., Wang, X., and Li, J.-J., 1993, The great latest Ordovician extinction on the south China Plate: Chemostratigraphic studies of the Ordovician-Silurian boundary interval on the Yangtze Platform: Palaeogeography, Palaeoclimatology, Palaeoecology, v. 104, p. 61–79.
Wilde, P., Berry, W.B.N., Quinby-Hunt, M.S., Orth, C.J., Quintana, L.R., and Gilmore, J.S., 1986, Iridium abundances across the Ordovician-Silurian stratotype: Science, v. 233, p. 339–341.
Wilde, P., Quinby-Hunt, M.S., and Berry, W.B.N., 1990, Vertical advections from oxic to anoxic water from the main pycnocline as a cause of rapid extinctions or rapid radiations, *in* Kauffman, E.G., and Walliser, O.H., eds., Extinction events in Earth history: Berlin, Springer-Verlag, v. 30, p. 85–98.

Manuscript Submitted October 5, 2000; Accepted by the Society March 22, 2001

Printed in the U.S.A.

Geological Society of America
Special Paper 356
2002

Late Devonian sea-level changes, catastrophic events, and mass extinctions

Charles A. Sandberg*
U.S. Geological Survey, Box 25046, MS 939, Federal Center, Denver, Colorado 80225-0046, USA
Jared R. Morrow
Department of Earth Sciences, University of Northern Colorado, Greeley, Colorado 80639, USA
Willi Ziegler
Forschungsinstitut Senckenberg, Senckenberganlage 25, D-60325 Frankfurt am Main, Germany

ABSTRACT

Late Devonian history is explained through event stratigraphy comprising a sequence of 18 sea-level changes, catastrophic events, and mass extinctions. Generally rising sea level during the initial Frasnian Stage, beginning with the Taghanic onlap and ending with a sea-level fall and major mass extinction, was interrupted by several exceptionally rapid, very high rises of sea level. These rises may be related to a series of comet showers, as suggested by the coincidence of the Alamo Impact in Nevada and the older Amönau Event in Germany with two of the sea-level rises. The subcritical, off-platform marine Alamo Impact is demonstrated to have produced greatly different effects in deep water from those previously recorded on the carbonate platform.

The series of comet showers, most notably those around the Frasnian-Famennian boundary, evidenced by microtektites in widely separated regions, not only produced the late Frasnian mass extinction, but also induced global cooling. This cooling resulted in Southern Hemisphere glaciation. Generally falling sea level during the later Famennian Stage was interrupted by several warmer, interglacial episodes, evidenced by glacio-eustatic rises. Another, less severe mass extinction occurred during an abrupt sea-level fall near the end of the Famennian. This glacio-eustatic fall is interpreted to have resulted from a severe, terminal glacial episode.

Interpretation of Late Devonian history suggests that impacts and comet showers coincided with sea-level rises, whereas mass extinctions occurred during, not at the start of, sea-level falls.

INTRODUCTION

The Late Devonian Epoch, one of the most intensively studied of all the Paleozoic epochs, was a time of major sea-level changes and catastrophic events, some of which were impact related, and two mass extinctions, one of which was impact related. Detailed knowledge and dating of Late Devonian events resulted from a high-resolution biochronology, based primarily on conodont zonations (Sandberg and Ziegler, 1996) and supported in part by ammonoid, ostracod, and spore zonations. This detailed knowledge was gained by intensive biostratigraphic studies during the past two decades, inspired by the International Union of Geological Sciences Subcommission on Devonian Stratigraphy and by International Geological Cor-

*E-mail: sandberg@usgs.gov

Sandberg, C.A., Morrow, J.R., and Ziegler, W., 2002, Late Devonian sea-level changes, catastrophic events, and mass extinctions, *in* Koeberl, C., and MacLeod, K.G., eds., Catastrophic Events and Mass Extinctions: Impacts and Beyond: Boulder, Colorado, Geological Society of America Special Paper 356, p. 473–487.

relation Programme (IGCP) Projects on Global Bio-Events and Mass Extinctions (Kauffman and Walliser, 1990; Walliser, 1986, 1996a). The initial Late Devonian Frasnian Stage was a time of general transgression during the Taghanic onlap that began in the late Middle Devonian (Johnson et al., 1985). The final Famennian Stage was a time of general regression, probably due to Southern Hemisphere glaciation, interrupted by four major transgressions, probably related to interglacial episodes. Both the late Frasnian and late Famennian mass extinctions occurred during, not at the start of, rapid regressions that closely followed rapid transgressions. The stepwise late Frasnian mass extinction (Sandberg et al., 1988; Schindler, 1990a, 1990b; McGhee, 1996), one of the five greatest in Earth's history, is believed to have occurred as a result of environmental stresses that were related to not just one but to a series of multiple, noncritical impacts, beginning with the Alamo Impact (Morrow et al., 1998). The late Famennian mass extinction, however, is believed to have occurred at the culmination of stresses produced by alternating glacial and interglacial episodes (Sandberg et al., 1988).

CONODONT BIOCHRONOLOGY

The Late Devonian was a time of many sea-level changes, catastrophic events, and two mass extinctions, knowledge of which is enabled mainly by means of a high-resolution conodont biochronology (Sandberg and Ziegler, 1996). The Devonian Period lasted 46 m.y., and comprises 57 conodont zones. The Late Devonian alone lasted 15 m.y., and comprises 32 of these zones (Fig. 1). The older Late Devonian Frasnian Stage, 5 m.y. long, contains 10 of these zones, whereas the younger Famennian Stage, 10 m.y. long, contains the other 22 zones. Two Frasnian zones are further divisible, and we intend to formally subdivide them in a later manuscript. Herein we informally recognize three subdivisions of the Early *rhenana* Zone: an early part, a middle part encompassing the entry and life span of the opportunistic species *Palmatolepis semichatovae*, and a late part apparently lacking this species. We also recognize two important subdivisions of the *linguiformis* Zone: an early part, encompassing the entry and life span of the nominal species *Palmatolepis linguiformis*; and an upper part, apparently lacking this species but containing *Pa. praetriangularis* and common *Ancyroides ubiquitus*.

Although the large number of recognized events might be considered to be an artifact of this detailed biochronology, we believe that the opposite is true. The multiple events that occurred during the Late Devonian caused the rapid evolution of conodonts and other marine organisms. This rapid evolution was punctuated first by the late Frasnian mass extinction (the so-called F-F or Kellwasser Event) and later by the late Famennian mass extinction (the so-called D-C or Hangenberg Event). *Palmatolepis* and two other pelagic genera, *Mesotaxis* and *Siphonodella*, are the principal taxa employed by the standard Late Devonian conodont zonation (Ziegler and Sandberg, 1990). However, only one species of *Palmatolepis, Pa. praetriangularis*, survived the late Frasnian mass extinction. This single species gave rise to the earliest Famennian species *Palmatolepis triangularis* and its more than 100 descendant species and subspecies during the Famennian. Later, only one of these descendant species, *Palmatolepis gracilis*, barely survived the late Famennian mass extinction, only to die out very early in the ensuing Early Carboniferous (Mississippian) Epoch.

LATE DEVONIAN SEA-LEVEL CHANGES

The now widely accepted Late Devonian sea-level curve shown in Figure 2 was devised by Johnson et al. (1985) and improved for the western United States by Johnson and Sandberg (1989). This curve illustrates that a general transgression or sea-level rise began with the Taghanic onlap in the Middle Devonian (Givetian) Middle *varcus* Zone and continued through most of the Frasnian until just before the late Frasnian mass extinction within the *linguiformis* Zone (Fig. 1). The sea-level curve also demonstrates that a general Famennian regression occurred and was terminated by a second severe eustatic fall just before the start of the Early Carboniferous (Mississippian) Kinderhookian Stage. Sandberg et al. (1988) attributed this general Famennian regression to Southern Hemisphere glaciation that directly followed and may have resulted from the same catastrophic events that produced the late Frasnian mass extinction. This general regression was interrupted by four major sea-level rises that they attributed to interglacial episodes.

The Late Devonian sea-level curve was used by Sandberg et al. (1988, their Fig. 4) to illustrate that the late Frasnian mass extinction as well as the late Famennian mass extinction occurred during severe eustatic falls that immediately followed major eustatic rises. They also demonstrated by a series of three marine cross sections, two for the *linguiformis* Zone and one for the earliest Early *triangularis* Zone, how these sea-level changes dramatically altered not only conodont biofacies, but also sedimentation patterns. Furthermore, they documented these rapid changes of sea level sedimentologically not only at the Steinbruch Schmidt section in a deep-water, submarine-rise setting in Germany, but also at sections in inner shelf and outer shelf and slope settings in Belgium, Nevada, and Utah. Whereas it is well documented that a stepwise extinction began with the severe sea-level fall, the ultimate mass extinction took place well within this sea-level fall. The magnitude of this sea-level fall, which continued into the early Famennian, was demonstrated by a map showing that this regression caused the sea to retreat an average of 400 km westward from the Transcontinental arch in the western United States.

The occurrence of the two Late Devonian mass extinctions during sea-level falls was reemphasized by Sandberg and Ziegler (1991, 1992) at meetings of IGCP Project 216, dealing with Event Markers in Earth History and with Phanerozoic Global Bio-Events and Event Stratigraphy. This observation had been first made in relation to the Devonian-Carboniferous boundary

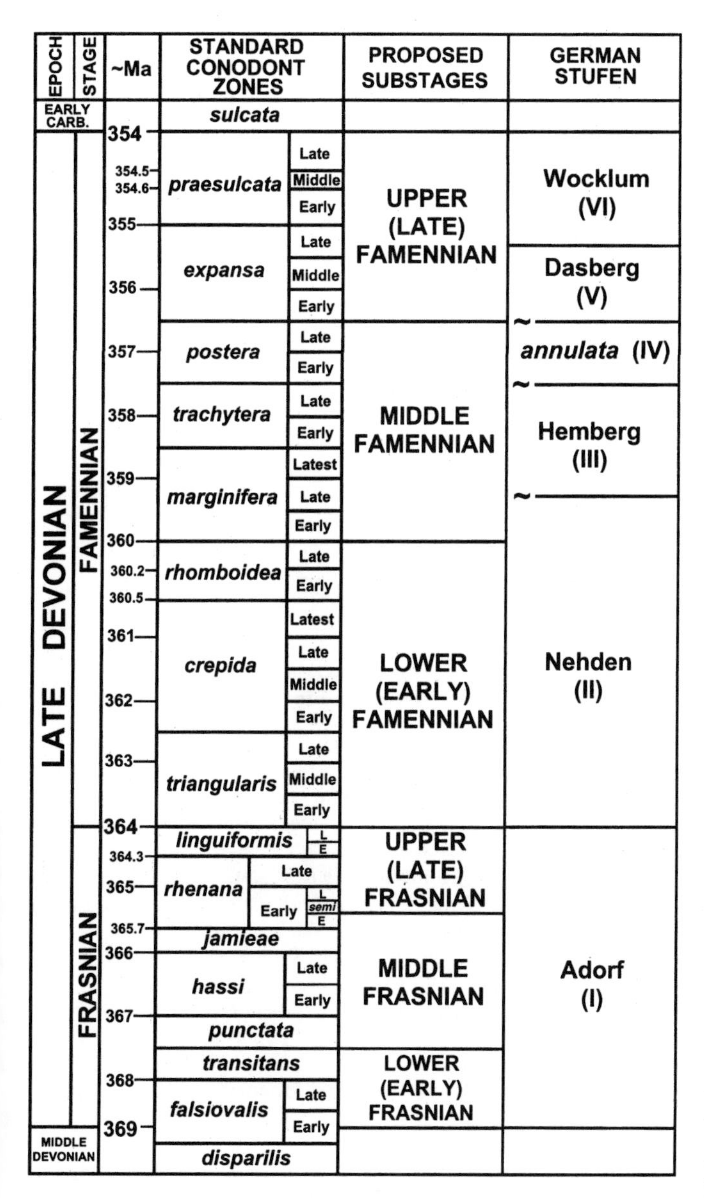

Figure 1. Late Devonian conodont zonation and approximate ages (after Sandberg and Ziegler, 1996). Substages proposed by Sandberg and Ziegler (1998); informal subdivisions of Early *rhenana* and *linguiformis* Zones are added. E is early part, L is late part, *semi* is *Palmatolepis semichatovae* interval, Carb. is Carboniferous.

(Ziegler and Sandberg, 1984). Schindler (1990a, 1990b) effectively demonstrated not only the stepwise extinction during the late Frasnian sea-level fall, but also the concurrent introduction of new shallow-water faunas prior to the final mass extinction. The finding that a sea-level fall preceded the late Frasnian mass extinction was later corroborated biostratigraphically and sedimentologically by many other workers in other parts of the world (e.g., Goodfellow et al., 1989; Geldsetzer et al., 1993 [western Canada]; Matyja and Narkiewicz, 1992, and Racki, 1998a [Poland]; Lazreq, 1992, 1999 [Morocco]; Girard and Feist, 1997 [southern France]; Ji, 1989, and Muchez et al., 1996

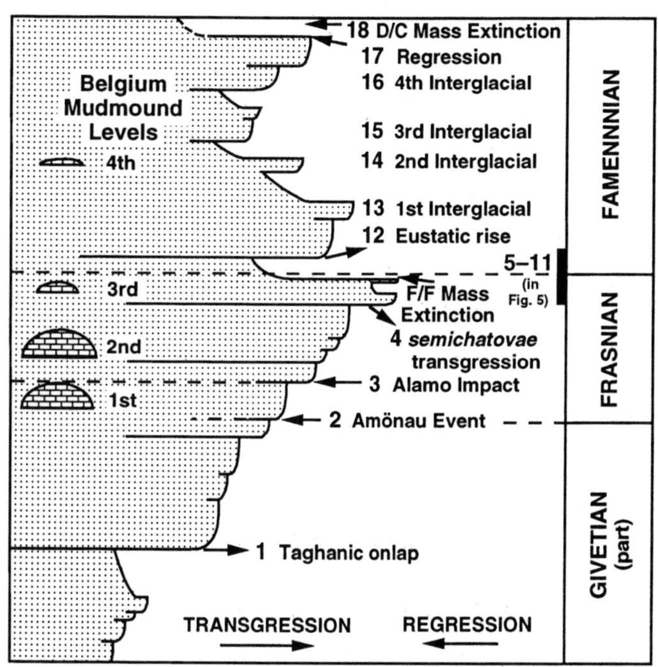

Figure 2. Late Middle to Late Devonian sea-level curve, showing positions of 18 sea-level changes, catastrophic events, and mass extinctions, which are more fully described in Table 1, and four levels of Belgium mudmounds. Events 5–11 are positioned in Figure 5 against more detailed sea-level curve for interval shown here by thick black bar. D/C is Devonian-Carboniferous, F/F is Frasnian-Famennian.

[southern China]; Schindler et al., 1998 [Germany]). Furthermore, a sequence stratigraphic study across the Frasnian-Famennian boundary (Morrow and Sandberg, 1997) substantiated the initial sedimentological analysis. In contradiction to overwhelming evidence presented by these and many other workers directly involved in Late Devonian studies, Hallam and Wignall (1999) misinterpreted, on the basis of a biased or incomplete literature survey, that the late Frasnian mass extinction had little support in sequence stratigraphic analysis and differed from the other four major mass extinctions in having occurred during a highstand or sea-level rise.

LATE DEVONIAN EVENTS

The 18 major Late Devonian sea-level changes, catastrophic events, and two mass extinctions that we now recognize are positioned against the Late Devonian sea-level curve in Figure 2 and are listed with more detail in Table 1. We emphasize that this table, although having some events in common, does not revise or supplant the tabulation of 17–19 late Middle to Late Devonian eustatic and epeirogenic events in the western United States (Sandberg et al., 1989, 1997), which improved on earlier tabulations of only 13 such events in that region and around the "Old Red Continent" (Sandberg et al., 1983, 1986). These tabulations were matched by the recording of a similar sequence of mainly eustatic Late Devonian events

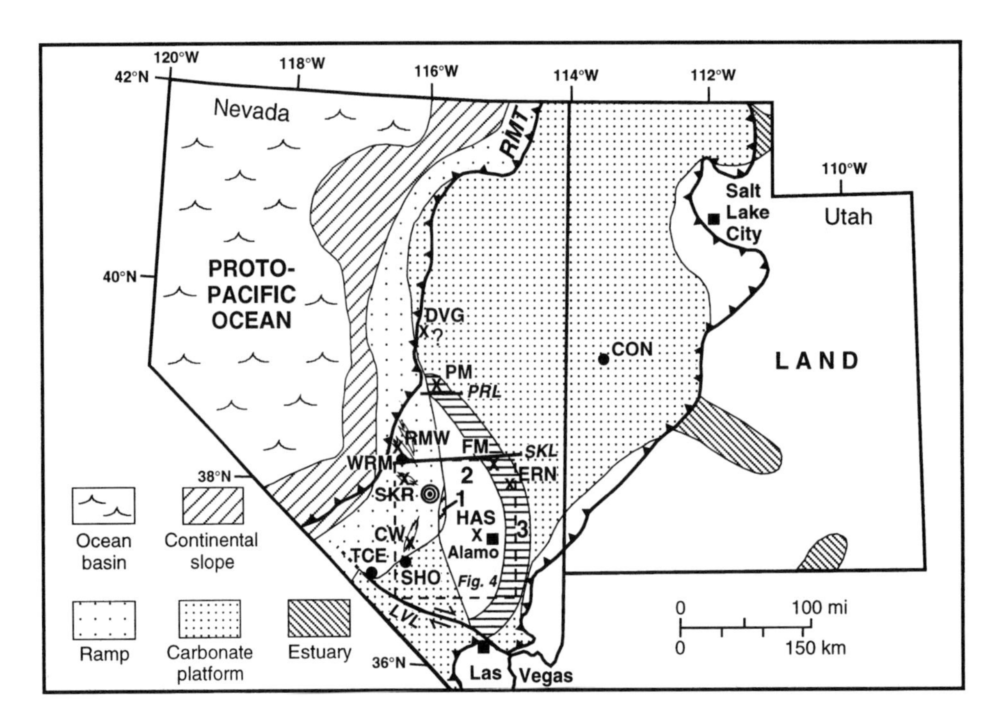

Figure 3. Paleotectonic map (partly restored) of Nevada and Utah during middle Frasnian *punctata* Zone, showing possible site of subcritical, oceanic Alamo impact (bullseye) and distribution of resulting Alamo Impact Breccia. Alamo Breccia Zones 1, 2, and 3 form semicircular pattern on ramp (Zone 1), outer carbonate platform (Zone 2), and peritidal, inner platform (Zone 3). Also shown are locations of four newly discovered deep-water channels of Alamo Breccia, some representing crater fill and some transporting breccia debris seaward. Selected major post-breccia structural features include RMT, latest Devonian to earliest Mississippian Roberts Mountains thrust, which significantly displaced Devonian transitional and oceanic facies; and three Tertiary lineaments that affected distribution of Alamo Breccia: LVL, Las Vegas lineament; SKL, Silver King lineament; and PRL, Pancake Range lineament. Important measured sections: CON, Little Mile-and-a-Half Canyon; CW, Carbonate Wash; DVG, Devils Gate; ERN, East Ridge North; FM, Fox Mountain; HAS, Hancock Summit (type locality of Alamo Breccia); PM, Portuguese Mountain; RMW, Rawhide Mountain West; SHO, Shoshone Mountain; SKR, Streuben Knob; TCE, Tarantula Canyon East; WRM, Warm Springs. Alamo Breccia localities are indicated (X) and other important data points are indicated by dots. Area of Figure 4 is shown by dashed rectangle.

in western Pomerania, Poland (Matyja, 1993). These earlier tabulations had a very different emphasis, however, from our global tabulation.

Herein, we explain more fully the 18 events summarized in Table 1 and discuss in detail the Amönau Event, Alamo Impact, and late Frasnian and late Famennian mass extinctions.

Taghanic onlap (Event 1)

The Taghanic onlap (start of transgressive-regressive [T-R] cycle IIa of Johnson et al., 1985) marked the end of faunal provincialism, which had persisted since Early Devonian time and is well documented for conodonts and brachiopods (Johnson, 1970; Klapper and Johnson, 1980; Ziegler and Lane, 1987; Johnson and Sandberg, 1989). It also marked the start of an episode of cosmopolitanism, which continued uninterrupted until the late Frasnian mass extinction. This onlap and long episode of general sea-level rise began in the Middle Devonian Middle *varcus* Zone and was coincident with the Acadian orogeny, which produced a continental collision of Europe, and probably North Africa, with the present eastern side of North America. The Taghanic onlap ended with the rapid, eustatic sea-level fall and regression that coincided with the late Frasnian mass extinction near the end of the *linguiformis* Zone (Fig. 2).

Amönau Event (Event 2)

The enigmatic Amönau Event occurred in the Rheinisches Schiefergebirge of central Germany coincident with a major pulse of the Taghanic onlap at the start of the Frasnian Stage within the late part of the Early *falsiovalis* Zone (Fig. 1). The event was recognized and named by Sandberg et al. (2000) on the basis of a reinterpretation of the Amönau Breccia. This so-called tuff breccia, containing large blocks derived from Devonian reefs mixed with basalt clasts and glass shards and found in two quarries near Wetter, Germany, was originally inter-

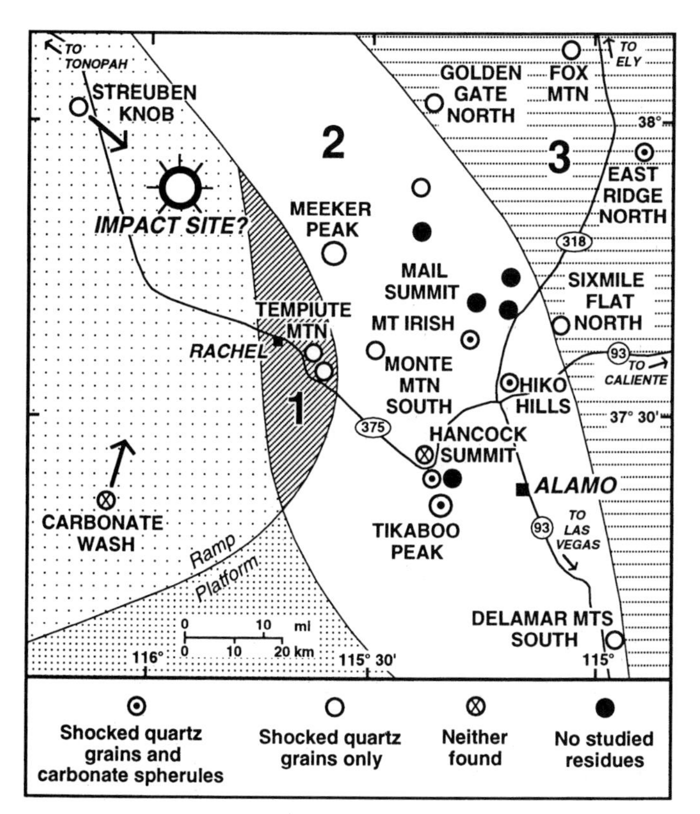

Figure 4. Northeast quadrant of distribution pattern of Alamo Breccia, southern Nevada, as indicated in Figure 3. Shown are inferred location of now buried, tectonically disturbed impact site; Zones 1, 2, and 3 of Alamo Breccia; probable direction of deep-water channels; and recovery of impact-fallout material at studied localities. Line bounding Zone 2 on west is position of Late Devonian *punctata* Zone shelf margin. Weak iridium anomaly was found at Meeker Peak. Mt—Mount, Mtn—Mountain, Mts—Mountains.

preted by Bender et al. (1984, p. 31) to have resulted from "erosional destruction of a reef during a period of volcanic activity." Sandberg et al. (2000) found that the presumed volcanic explosion at Amönau was coincident with another tuff breccia at Donsbach, 38 km to the southwest, and a bentonite bed at Blauer Bruch near Bad Wildungen, 40 km to the northeast. Thus, subaerial volcanism, which might have resulted from an impact on a submarine rise, began simultaneously across a 78-km-long transect of the Rheinisches Schiefergebirge. At the same time, a stratigraphic hiatus was produced by deposition of exhalative hematite at Martenberg, 40 km northwest of Blauer Bruch. Subsequent field work suggested that a 170-m-thick debris flow near Günterod (Huckriede, 1992) could be part of the same event. All three known breccias might represent post-impact crater fill. However, we emphasize that evidence for an impact has not yet been found because of the scarcity of exposures, except in a few quarries and stream cuts, in this region of predominantly highly weathered slate hills. The Amönau Event is meaningful, however, as a harbinger of later, well-documented catastrophic events and mass extinctions that shaped Late Devonian history.

Alamo Impact (Event 3)

The biochronologically dated, well-documented Alamo Impact in southern Nevada (Fig. 3) is especially significant because of its possible relationship to the younger Frasnian mass extinction and to several other, radiometrically dated Late Devonian impacts. The Alamo Impact has been dated by conodonts in beds below, within, and above the resulting breccia as having occurred during the early Frasnian *punctata* Zone (Fig. 1), ~3 m.y. before this mass extinction within the latest Frasnian *linguiformis* Zone (Sandberg and Warme, 1993; Sandberg and Morrow, 1998). Because of the number of other Late Devonian impacts, such as the Flynn Creek in Tennessee and the Siljan in southern Sweden, we suggest that the Alamo Impact may have been part of a comet shower or the first—or second, if the Amönau Event proves to be impact related—of a series of comet showers. These showers would explain the demise of three successively less extensive and biotically less diverse reef levels in Belgium (Fig. 2), and the progressive weakening of marine communities before their final extinction. The timing of the Alamo Impact coincides with the demise of the first reef level. Unfortunately, the other Devonian impacts have been dated by different radiometric methods and are difficult to relate to one another and to the Alamo Impact.

Even though the location of its crater is uncertain because of later tectonic dismemberment and burial beneath valley fill and volcanics, the Alamo Impact has been documented on the basis of many different criteria. Originally, the occurrence of three zones of megabreccia in a semicircular distribution pattern, 200 km in diameter and located mainly on a shallow carbonate platform, and the presence of a weak iridium anomaly were documented by Warme and Sandberg (1995). They illustrated an example of the shocked quartz grains, identified by Leroux et al. (1995), which occur abundantly within turbidites in the upper part of the megabreccia. Localities at which shocked quartz grains have been recovered in all three zones, as well in a deep-water channel farther west (Morrow et al., 1998), as reported herein, are shown in Figure 4. Impact-related injected dikes and sills within the 300 m of strata below the Alamo Breccia at Tempiute Mountain, which may be located close to the eastern crater rim, have been described and illustrated (Warme and Sandberg, 1995; Sandberg et al., 1997). Impact-fallout, zoned, carbonate spherules within the breccia were recognized by Warme and Kuehner (1998). Like the shocked quartz grains, these spherules occur in small blocks within coarse turbidites in the upper part of breccia. Shocked quartz grains were found within the turbidites, as well as within the contained blocks, and in the cores of spherules. Localities where both shocked quartz grains and carbonate spherules occur in Zones 2 and 3 are shown in Figure 4. The final line of evidence, which also permits estimation of the crater depth, is the presence in these same blocks of common, ejected Middle Ordovician and possibly older conodonts, which were blasted from strata 1.5 km or more below the Late Devonian seafloor

TABLE 1. LATE DEVONIAN EVENTS

Number	Event
18	Late Famennian mass extinction within eustatic fall; loss of many Devonian species including Lazarus fauna
17	Start of major eustatic fall at climax of Southern Hemisphere glaciation; biotic decline begins
16	Eustatic rise at start of 4th interglacial episode; Etroeungt Lazarus fauna suddenly appears in Northern Hemisphere
15	Eustatic rise at start of 3rd interglacial episode; *annulata* black basinal shales deposited in Germany
14	Eustatic rise at start of 2nd interglacial episode; Baelen mudmound formed in Belgium
13	Eustatic rise at start of 1st interglacial episode; *Cheiloceras* dark shales deposited
12	Eustatic rise, producing initial post-extinction biotic radiation
11	Carbonate-platform margins collapse due to glacioeustatic lowering of seas; widespread coarse tsunamite breccias result
10	Storms and (or) tsunami scour or remove Frasnian-Famennian boundary beds in France, Germany, and Nevada
9	Late Frasnian mass extinction within eustatic fall; widespread layer of abiotic extinction shale deposited; start of Southern Hemisphere glaciation
8	Rapidly increased shallowing; storm deposits; stepwise extinction of some deep-water species as shallow-water species move into deeper water; spore floras change
7	Eustatic fall; start of global cooling; reduction in number and diversity of deep-water faunas
6	Continued eustatic rise; stratification of water column, resulting in widespread basinal anoxia; start of rapid evolution of deep-water entomozoan ostracods
5	Eustatic rise producing dysoxia in deep basins; start of events leading to mass extinction
4	Major *semichatovae* transgression producing biotic changes and stress
3	Alamo Impact in Nevada, coinciding with eustatic rise and demise of 1st level of Belgian mudmounds
2	Amönau Event, coinciding with onset of impact-induced(?) volcanic activity in central Germany and with eustatic rise at start of Frasnian
1	Taghanic onlap at start of plate-tectonic movements in Northern Hemisphere; end of faunal provincialism

(Sandberg, 1998; Morrow et al., 1998). The recovery of rare reworked Ordovician conodonts within the groundmass of the breccia was reported by Warme and Sandberg (1995).

More complete descriptions of these lines of evidence were given by Warme and Kuehner (1998) and Morrow et al. (1998, 1999). Warme and Kuehner (1998) and Warme and Chamberlain (2000), however, interpreted the impact site to be on the shallow-water carbonate platform, rather than in deeper water off the platform margin, as interpreted by Morrow et al. (1998, 1999). Our current study of deep-water channels, several of which contain shocked quartz grains, provides further support for a deeper water impact site.

Our interpretation of the deeper water site of the subcritical Alamo Impact is best explained by reference to our partly restored *punctata* Zone paleotectonic map (Fig. 3) showing the three partial rings of Alamo Breccia in relation to the coeval platform, ramp, and continental slope. The innermost ring, Zone 1, a 130-m-thick turbidite, was deposited by crater fill that poured seaward off the Late Devonian carbonate-platform margin. The middle ring, Zone 2, deposited on the shallow carbonate platform, consists of a lower debrite of huge, seaward-sliding blocks evolving upward into several normally graded, coarse- to fine-grained turbidites resulting from the ensuing megatsunami. The outer ring, Zone 3, consists of stranded, locally isolated, high-water-line megatsunami deposits on the peritidal inner carbonate platform and adjacent Late Devonian shoreline. These rings are now truncated and/or segmented by several generally east-west-trending strike-slip lineaments, the three most critical of which are shown in Figure 3.

Warme and Kuehner (1998) and Warme and Chamberlain (2000), misinterpreting the stratigraphy of rocks enclosing the Alamo Breccia, tried to show that, if the impact were entirely on the carbonate platform, structural reinterpretation of the Timpahute Range would produce three complete rings. However, our recent study of deeper water channels and of Zone 1 show that these contemporaneous Alamo Breccia deposits overlie a different suite of older Devonian rocks than those that underlie the breccia on the carbonate platform. We now realize that the pattern of rings need not be completely circular even if the original crater and crater rim were circular. The Zone 2 ring is a shallow-water deposit, whereas the Zone 3 ring is peritidal to nonmarine. Given the early Frasnian paleogeography, which shows progressively deeper water toward the proto-Pacific Ocean basin on the west (Fig. 3), it is physically impossible for these rings to be completed by very shallow water and land to the west, even if the impact had been on the carbonate platform. In direct contradiction to the complete-ring interpretation of Warme and his coworkers, we have discovered four deep-water channels (Fig. 3), in several of which we recovered shocked quartz grains. These channels are interpreted to form a pattern radiating from the probable crater site, very much like the pattern shown by Ormö and Lindström (2000) for the Early Ordovician marine Lockne crater in Sweden. Thus, we now interpret the Alamo Impact depositional phenomena to form two different patterns, depending on paleotectonic setting, and theorize that the western rim of the crater is located farther west and downslope, perhaps below the Roberts Mountains thrust.

The northeast quadrant of the Alamo Breccia deposits (enlarged in Fig. 4) focuses on the evidence for our interpretations and shows some of the more important localities. We interpret that the Streuben Knob channel, containing shocked quartz grains, flowed craterward from the western crater rim. Conversely, we interpret that the Carbonate Wash channel, containing blocks of carbonate-platform rocks but no shocked quartz grains, represents crater fill from the shattered platform margin. We also record, for the first time, the occurrence of carbonate spherules in Zone 3 at East Ridge North.

Semichatovae *transgression (Event 4)*

The *semichatovae* transgression (Event 4) was an abrupt, but short-lived major deepening event (Fig. 2). This event corresponds to the start of T-R cycle IId of Johnson et al. (1985). This transgression carried *Palmatolepis semichatovae*, an opportunistic representative of a pelagic genus, far onto shallow carbonate platforms, normally uninhabitable for *Palmatolepis*, throughout the world. This species composes 70%–100% of *Palmatolepis* populations on the carbonate platform, but <10% of such populations in deep basins (Sandberg et al., 1989). This population migration is unique among all organisms in the Late Devonian record. Although the *semichatovae* transgression did not directly terminate the second level of Belgian Late Devonian mudmounds, it is recorded by an abrupt sedimentologic change, possibly associated with the onset of volcanism, between the second and third levels (Sandberg et al., 1992). We offer no explanation for this globally recorded, exceptionally high, very rapid eustatic rise, but ponder whether, like the Alamo Event, it could be impact related.

Late rhenana *Zone eustatic rise (Event 5)*

This eustatic rise, shown as the next sea-level spike above the *semichatovae* transgression in Figure 2, followed a eustatic fall late in the Early *rhenana* Zone and early in the Late *rhenana* Zone. It equaled Event 4 in intensity and resulted in the demise of the third level of Belgian Late Devonian mudmounds. Event 5 deepening, which produced dysoxia in most marine basins, was the start of a series of events leading to the late Frasnian mass extinction, as illustrated in the detailed Late Devonian sea-level curve (Fig. 5). As in the case of Event 4, we ponder whether Event 5 could be impact related. Regardless, the close timing of two such catastrophic floodings undoubtedly weakened and destabilized marine populations, so that they more easily succumbed to the late Frasnian mass extinction during a major eustatic fall.

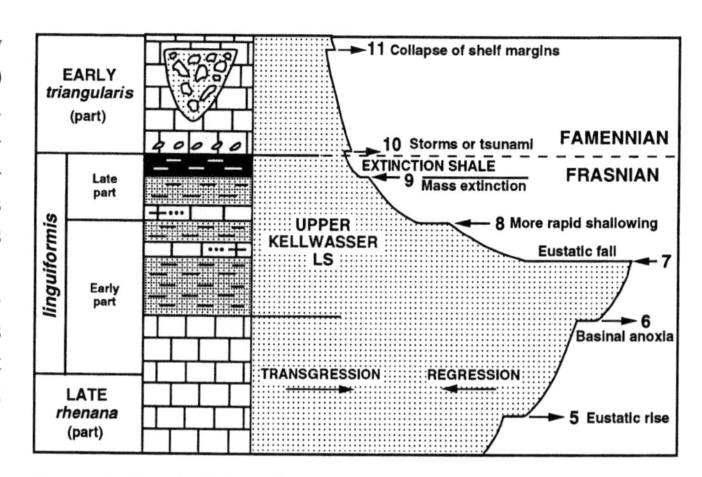

Figure 5. Detailed Late Devonian sea-level curve across Frasnian-Famennian stage boundary and generalized pattern of sedimentation related to late Frasnian mass extinction. Events 5–11 are more fully described in Table 1. Conodont zonation is at left and position of Upper Kellwasser Limestone (LS) is at right of lithologic column.

Linguiformis *Zone anoxia (Event 6)*

Deepening continued uninterrupted from the Late *rhenana* into the *linguiformis* Zone. Eventually, well within the *linguiformis* Zone, the oceans deepened so considerably that even in most epicontinental seaways, the water column became stratified into aerobic, dysaerobic, and anaerobic layers. The resulting basinal anoxia was recognized by the overstepping of black carbonaceous shale and chert over lighter colored carbonate rocks, as exemplified within several stratigraphic sections in Nevada, Belgium, and Germany (Sandberg et al., 1988). The most important, readily accessible reference sections for faunal, sedimentologic, geochemical, and geomagnetic studies of Events 6 to 9 are Devils Gate, Nevada, and Steinbruch Schmidt, Germany.

Conodont abundances decreased by at least 80% as a result of the basinal anoxia, but few if any significant deeper water extinctions occurred at that time. In fact, deep-water entomozoan ostracod populations continued to thrive and evolve (Buggisch et al., 1978; Groos-Uffenorde and Wang, 1989; Groos-Uffenorde and Schindler, 1990). Nevertheless, the anoxic events were considered to be the actual cause of the Frasnian-Famennian faunal crisis by Joachimski and Buggisch (1993), who based this interpretation on a positive $\delta^{13}C$ excursion and an increase in organic carbon burial.

However, the deepening during Event 6 changed marine conditions on the shallow shelf areas and carbonate platforms, even where obvious lithologic evidence of the event is lacking (Bratton et al., 1999). Many workers have previously used the highest occurrences of colonial and large solitary corals or of *Amphipora* biostromes in shallow-water or peritidal settings, respectively, to approximate the end of the Frasnian in the absence of definitive conodont evidence (e.g., Geldsetzer et al., 1993). However, it is more likely that these Frasnian corals and biostromes disappeared concurrently with the increased deepening in Event 6 within the *linguiformis* Zone, if they had not already been decimated or terminated earlier, concurrently with the start of deepening and demise of the third level of Belgian mudmounds in the Late *rhenana* Zone (Event 5). Similarly, extinctions of some species of shallow-water atrypid and gypidulid brachiopods, commonly attributed to the late Frasnian mass extinction (Event 9), may actually have occurred earlier, during Events 5 and 6.

Linguiformis *Zone eustatic fall pulses (Events 7 and 8)*

A severe eustatic fall (Event 7) followed by a pulse of even more rapid shallowing (Event 8) preceded the late Frasnian mass extinction terminating the *linguiformis* Zone (Fig. 5). The sedimentologic and oxygenation changes accompanying these pulses were described in detail by Sandberg et al. (1988), so only the drastic conodont faunal changes are summarized here. Stepwise extinction and genetic mutations of almost all species of the deep-water genus *Palmatolepis* accompanied the pulses of falling sea level. These changes are exemplified by the two most important taxa, *Palmatolepis linguiformis* and *Pa. rhenana rhenana. Palmatolepis linguiformis*, the entry of which defines the start of the *linguiformis* Zone (Ziegler and Sandberg, 1990), died out at Event 8. Consequently, we now informally divide the *linguiformis* Zone into early and late parts, with and without this key species (Fig. 5). Bizarre, oddly shaped mutants, illustrated by Ziegler and Sandberg (1990, 2000), preceded the demise of *Pa. rhenana rhenana* within the late part of the *linguiformis* Zone. Contemporaneous with these changes were major changes in shallower water conodont faunas, as exemplified by two important species, *Icriodus alternatus* and *Ancyroides ubiquitus. Icriodus alternatus*, which had been a minor constituent of all conodont faunas during Event 5 and totally absent from deep-water faunas during Event 6, resurged to compose 30% or more of all deep- and shallow-water faunas during Event 7. *Ancyroides ubiquitus*, which is a further aid to identification of the late part of the *linguiformis* Zone, evolved from *As. uddeni* late within the early part of the *linguiformis* Zone and survived the late Frasnian mass extinction only to die out early in the Early *triangularis* Zone.

The stepwise extinction and changes in conodont faunas during Events 7 and 8 are matched by equally significant stepwise extinctions and changes in other groups, as described by Schindler (1990a, 1990b). The stepwise extinctions involve trilobites in Event 6, gephuroceratid ammonoids and some entomozoan ostracods in Event 8, and finally homoctenids, which like *Ancyroides ubiquitus*, barely survived Event 9. Concurrently with these extinctions, a large bivalve *Buchiola* invaded shallower water strata during Event 8.

Late Frasnian mass extinction (Event 9)

Biological and geochemical studies of the late Frasnian mass extinction, one of the five greatest in the Phanerozoic record, have increased exponentially since its recognition by McLaren (1967 [personal commun.], 1970, 1982). The literature has grown so voluminous that only the more recent references and the most relevant older references can be cited here. Comprehensive bibliographies listing most of the other equally significant studies up to the date of publication were presented by Sandberg et al. (1988), McGhee (1996), and Hallam and Wignall (1999). By combining this vast amount of knowledge contained in these studies with our own investigations of the late Frasnian mass-extinction (Event 9), we can now answer with confidence the important questions of what happened, and when and where this happened. These answers pertain mainly to the marine realm. Admittedly, we know little about the effect of the extinction on the landmasses, except for some studies of palynomorphs and estuarine fish. We can, however, partly answer the question of how the mass extinction took place, but the remaining question of why remains equivocal. We attempt to explain the merits of the opposing arguments as to whether this was an Earth-bound or an impact-related event.

The late Frasnian mass extinction, or Kellwasser Crisis (Schindler, 1990a, 1990b), decimated most groups of marine organisms, as thoroughly summarized by Walliser (1996b), but apparently did not entirely wipe out any of them. Although the mass extinction affected the Earth's tropical and subtropical regions, which encompassed most of the areas of studied Devonian rocks, we know little about the temperate and polar regions. Apparently these regions as well as the deep ocean basins provided refuges for survivors or unaffected species that later repopulated the Famennian oceans. For example, although reef-building Devonian corals and large shallow-water solitary rugose corals were wiped out, small deep-water corals must have survived, because they are recorded in early Famennian rocks. Either their descendants or large rugose corals that survived in some unknown refuge produced a Lazarus fauna, containing large rugose corals late in the Famennian (Event 16). Likewise, although reef-building bulbous stromatoporoids were wiped out, descendants returned to produce large, pillow-shaped colonies in the late Famennian.

There are many other examples of the existence of survivors in other groups; we cite here only a few prime examples. Among the conodonts, all but one species of the pelagic genus *Palmatolepis* became extinct (Sandberg et al., 1988). The lone survivor *Pa. praetriangularis*, which had evolved just before the mass extinction, gave rise to the myriad of Famennian species, which in diversity far surpassed their Frasnian relatives. Regarding the mass extinction of ostracods (Lethiers and Casier, 1999), investigations of a single locality, Devils Gate, Nevada, documented their near mass extinction (Casier et al., 1996), the survival of some species (Casier and Lethiers, 1998a), and their eventual recovery (Casier and Lethiers,

1998b). The extinction of brachiopod faunas was effectively summarized by Racki (1998a). From our own studies at Devils Gate and elsewhere in Nevada, we note that although gypidulid and atrypid brachiopods were essentially wiped out, other groups such as the cyrtospiriferids and rhynchonellids were unaffected and occur profusely in post-extinction beds. A single possible atrypid survivor, *Peratos* sp., in the Famennian of Morocco was illustrated by Schindler (1990a). Likewise, possible homoctenid survivors high in the lower Famennian of South China were illustrated by You (2000), but the possibility of their reworking from the Frasnian or oldest part of the Famennian cannot be ruled out.

Having discussed what happened and where it happened, we now answer the question of when the mass extinction happened. From the conodont evidence and the international decision to locate the base of the Famennian at the start of sedimentation following the extinction, Event 9 can be dated as the final Frasnian event at the end of the late part of the *linguiformis* Zone (Fig. 5). This extinction apparently took place within a sedimentologically calculated interval of 20 k.y. or less (Sandberg et al., 1988), just prior to 364 Ma, which is the conodont biochronologic date assigned to the start of the Famennian (Sandberg and Ziegler, 1996). The mass extinction occurred during a long, catastrophic eustatic sea-level fall that immediately followed two closely spaced, rapid sea-level rises (Sandberg et al., 1988; Fig. 5).

Our findings contradict the conclusion reached by Hallam and Wignall (1999) that the late Frasnian mass extinction differed from the other four major Phanerozoic mass extinctions in having occurred during a sea-level rise rather than during a sea-level fall. Their conclusion was based on an incomplete literature survey that resulted in several incorrect assumptions. First, they assumed that *Icriodus* was restricted to shallow-water facies in the Frasnian. This is incorrect for two reasons. At least one species, *Icriodus symmetricus*, preferred moderately deep water settings and is uncommon in shallow-water settings. They assumed that the species *Icriodus alternatus*, the proliferation of which they challenged, inhabited only shallow-water settings. In fact, it also inhabited many preextinction, deep-water settings such as around the black smoker at the famous Martenberg section in Germany and the submarine rise at Steinbruch Schmidt, where it makes up 4% of youngest Late *rhenana* Zone faunas (Ziegler and Sandberg, 1990). Hallam and Wignall (1999, p. 226) stated that because of the shallow-water Frasnian occurrence of *Icriodus*, conodont workers assumed that "it had a similar facies distribution in the basal Famennian" and that the proliferation of the conodont *Icriodus* did not signal sea-level changes but rather a filling of vacant ecospace by lucky survivors. In addition to being inaccurate for several reasons, these are self-contradictory statements. The proliferation of *Icriodus* began during Events 7 and 8, not during the mass extinction (Event 9). The earliest Famennian conodont biofacies distribution of *Icriodus* in comparison to Frasnian and younger Famennian patterns was depicted by Sandberg et al. (1997, their Table 1). Sandberg et al. also showed that the "*Icriodus*" in later Famennian biofacies was not a true *Icriodus* but rather reiterative, homeomorphic shallow-water species of *Pelekysgnathus*, as documented by Sandberg and Dreesen (1984). Third, they assumed that the *Icriodus* bloom was the sole evidence on which the sea-level fall was based. They ignored the many documented sedimentologic and megafaunal changes that accompanied the *Icriodus* bloom prior to the mass extinction. The findings redocumented herein are supported by the findings of an overwhelming consensus of specialists who have studied the Frasnian-Famennian boundary globally (e.g., Goodfellow et al., 1989; Schindler, 1990a, 1990b; Lazreq, 1992, 1999; Matyja and Narkiewicz, 1992; Muchez et al., 1996; Girard and Feist, 1997; Racki, 1998a; Schindler et al., 1998).

Our scenario to partly explain how and why the late Frasnian mass extinction happened involves our sequence of Events 1–8. We agree with McGhee (1981, 1988), that the marine ecosystem had collapsed and reef and mudmound communities had been destroyed well before the mass extinction occurred. Applying the association of sea-level rises with the known Alamo Impact and enigmatic Amönau Event to other unexplained sea-level rises, we theorize that the Frasnian was a time when the Earth was subjected to a number of subcritical oceanic impacts produced by a series of comet showers. The two short, catastrophic rises that immediately preceded the mass extinction were particularly devastating to marine communities. Thus, the final event that triggered the mass extinction could have been another group of small, subcritical impacts. Following Sandberg et al. (1988), we interpret that one of these impacts could have been in the Southern Hemisphere, possibly close to Australia and China, and that this impact produced a global cooling and subsequent Southern Hemisphere glaciation during the Famennian (Caputo, 1985). A strong $\delta^{13}C$ anomaly associated with a weak iridium anomaly has been found at the Frasnian-Famennian boundary in southern China (Wang et al., 1991). Younger, early Famennian microtektites reported there by Wang (1992) were interpreted by Claeys and Casier (1994) as being possibly reworked. Evidence of another possible impact site is provided by the finding of iridium enrichments in the boundary interval in western New York State (Over et al., 1997). Even more compelling evidence for a third impact site is the finding by Claeys and Casier (1994) of microtektite-like glass precisely at the boundary layer at the Hony railroad cut, Belgium, which was previously measured and used to interpret the mass extinction by Sandberg et al. (1988). The Siljan Impact in Sweden was suggested as the possible source for the Belgian microtektites (Claeys and Casier, 1994).

Several respected investigators of the Frasnian-Famennian boundary interval disagree for different reasons with an impact origin for the extinction (e.g., Copper, 1986; McGhee et al., 1986; Joachimski and Buggisch, 2000; Racka and Racki, 2000; Racki, 1998b, 1999). Whereas we agree with Copper (1986) regarding most stepwise floral and faunal extinctions, we do not agree that continental suturing took place between the

Frasnian and Famennian, or that the diverted, cold polar currents would have been dysaerobic. According to our sequence of events, continental suturing—i.e., the Acadian orogeny—would have already occurred in Event 1, concurrent with the onset of cosmopolitanism. In addition, using the modern analog, Antarctic waters are known to be not dysaerobic, but well oxygenated and teeming with biota. McGhee et al. (1986) concluded that no geochemical evidence existed at Steinbruch Schmidt, Germany, for an impact during the Kellwasser Event. They analyzed the entire Upper Kellwasser Limestone, which incorporates Events 6–9. However, the relevant extinction layer (Event 9) at this locality is a possibly slumped or tectonically squeezed 5 cm interval. If this interval were to be properly analyzed geochemically, it should be done by means of a core hole well back from the cliff face. In all fairness, McGhee (1996) later thoroughly summarized and gave equal weight to all opposing theories, but seemingly favored an impact origin for the late Frasnian mass extinction. Although Joachimski and Buggisch (2000) tentatively reported, on the basis of $\delta^{18}O$ shifts, a 7 °C decrease in tropical surface-water temperature, this does not negate an impact. The large decrease could have been produced by the impact-related global cooling postulated by Sandberg et al. (1988) or even by an ancient El Niño-La Niña cycle. Racki (1998b, 1999) reverted to a Variscan plate tectonic explanation for the biotic crisis, but added the possibility of volcanic-hydrothermal processes, while not excluding minor cometary strikes. Regardless of the ultimate solution for the cause of Event 9, we concede that given the preexisting weakness of marine communities, the triggering mechanism for the extinction could have been relatively minor.

Violent post-extinction currents (Event 10)

At most studied sections where the Frasnian-Famennian boundary can be precisely pinpointed in carbonate-platform, shelf, and slope settings in Belgium, France, Germany, Nevada, and Utah, but where a dark-colored extinction shale is not preserved, there is convincing evidence of immediate post-extinction scour. For example, in a slope sequence at Burg Berg, Germany (Ziegler and Sandberg, 1990; new data herein), initial sampling of a bed only 5 cm thick yielded a mixed *linguiformis* and Early *triangularis* Zone conodont fauna. However, vertically slabbing this bed disclosed that it comprised three thin layers. Horizontally slicing a slab and individually analyzing each layer revealed that the lower, medium gray micrite layer and the medial, dark gray micrite layer both contained a fauna that we now assign to the early part of the *linguiformis* Zone. The medial layer is scoured and channeled by a chaotically bedded, sandy, finely conglomeratic, grayish orange wackestone upper layer yielding an Early *triangularis* Zone fauna. Detailed sedimentologic analysis of boundary beds elsewhere produces similar results. At Hony, Belgium, unusual pentagonal crinoid columnals not recorded in older beds of this middle-shelf sequence attest to the occurrence of storms that carried these remains shoreward from deeper water (Sandberg et al., 1988). At the Frasnian-Famennian global stratotype section and point (GSSP) at Coumiac, France, a stratigraphic gap was found to encompass the youngest part of the *linguiformis* Zone and oldest part of the Early *triangularis* Zone (Sandberg et al., 1988; Ziegler and Sandberg, 1996). Even in the unbroken boundary sequence of a submarine-ridge setting at Steinbruch Schmidt, the first layer deposited during the earliest part of the Early *triangularis* Zone contains pink stained synsedimentary micrite pebbles probably related to rip-up by tsunami or storm currents. Only at Devils Gate, Nevada (Sandberg et al., 1988), are thick slope sequences deposited during both the late part of the *linguiformis* Zone and the earliest part of the Early *triangularis* Zone fortuitously preserved. From our study of these and many other boundary sequences worldwide, we conclude that the immediate post-extinction world was wracked by storms and/or tsunamis. This would accord with microtektites found at the boundary layer in Belgium (Claeys and Casier, 1994) and the reworking of microtektites in southern China (Wang, 1992) into higher layers.

Collapse of carbonate-platform margins (Event 11)

The last of the seven events associated with the late Frasnian mass extinction was the collapse of carbonate-platform margins (Fig. 5), which occurred as a result of continued regression, accelerated by the lowering of oceans with the onset of Southern Hemisphere glaciation. During this regression, the Late Devonian seaway offlapped the Transcontinental arch and regressed more than 400 km westward in the western United States (Sandberg et al., 1988, Fig. 2). This drastic lowering, probably exceeding 100 m, shallowed or exposed the margins of carbonate platforms, dead Frasnian mudmounds and reefs, and submarine ridges, and caused their collapse. This catastrophic event, which is represented by debris-flow deposits and tsunamites in some regions, is also responsible for the removal by tsunamis of older deposits of the Early *triangularis* Zone in most other regions. Hence, researchers generally record that the Early *triangularis* Zone is highly condensed. This phenomenon is best exemplified by the excellent exposure at Devils Gate, Nevada (Sandberg et al., 1988, 1997). There, a thick breccia deposit, interpreted as a tsunamite, is preserved in a channel. In one measured section at this locality, the channel overlies 4.5 m of Early *triangularis* Zone deposits, but in another, only 200 m along strike, it truncates them. This tsunamite consists mainly of slabs of Early *triangularis* Zone shallow-water deposits removed from the proto-Antler forebulge on the west. Its distal edge is preserved at Coyote Knolls, Utah (Sandberg et al., 1988, 1997). Similar debris-flow breccias have been recorded in Poland (Matyja and Narkiewicz, 1992) and Morocco (Walliser et al., 1989).

Middle **triangularis** *Zone biotic radiation (Event 12)*

Post-extinction adaptive radiation of the surviving biota began during the eustatic rise (start of T-R cycle IIe of Johnson et al., 1985; Fig. 2) that occurred in the middle of the Middle *triangularis* Zone. This eustatic rise has been documented by later researchers in most regions of the world. We interpret that it occurred as the initial pulse of Southern Hemisphere glaciation waned and cold waters from a melting ice cap were gradually released to the oceans. The burgeoning of faunas is best exemplified by conodont and brachiopod populations, but little is known about the effects on other groups such as the surviving corals and stromatoporoids.

First interglacial episode (Event 13)

Early Famennian global warming caused further melting of the Southern Hemisphere ice cap and raised ocean temperatures to such a degree that the first interglacial episode ensued in the middle of the Middle *crepida* Zone (Fig. 1). This produced a major transgressive interruption of the general Famennian regression (Fig. 2) and a great sedimentologic and biotic change. This change, termed the *Cheiloceras* Event (Walliser, 1985) after the ammonoid genus, was first recognized in Germany by the overstepping of black shale over lighter colored carbonate rocks. It flooded shelf areas with deeper water, resulting in deposition of red-stained nodular cephalopod limestone in Belgium and producing similar sedimentologic changes throughout Europe and Morocco (Dreesen, 1989). In basinal sections in Germany and Morocco, especially where the intervening, lowest part of the Famennian is greatly condensed, the *Cheiloceras* black shale is recognized as a third Kellwasser Event or is included in the Kellwasser facies (e.g., Wendt and Belka, 1991). In Nevada, Event 13 resulted in the expansion of the Late Devonian Pilot basin and in the overstepping of shallow-water, carbonate-platform facies of the Guilmette Formation by the moderately deep water, cephalopod-bearing, nodular West Range Limestone.

Second interglacial episode (Event 14)

The second major transgression to interrupt the general Famennian regression resulted from the second interglacial episode in the Early *marginifera* Zone (Fig. 1). This sea-level rise resulted in deposition of the Baelen mudmound, representing the fourth level of Belgian mudmounds and the only well-documented Famennian mudmound anywhere (Dreesen et al., 1985). Unlike the Frasnian stromatoporoid- and coral-rich Belgian mudmounds, the Baelen mudmound was structurally bound by cyanobacteria and algae and overgrown by thickets of crinoids and sponges. However, the few known global occurrences of early Famennian small, deep-water corals may be related to this event. In Nevada, Event 14 resulted in the deepening and further expansion of the Pilot basin. Concurrently, the great Palliser bank and its equivalents, extending from Alberta to Nevada, developed on the outer margin of the carbonate platform (Sandberg et al., 1989).

Third interglacial episode (Event 15)

The third interglacial episode, although not as pronounced as the second, has been recognized in several regions, but it is best represented by the overstepping of black shale over carbonate rocks in Germany. There, it has been termed the *annulata* Event, after the ammonoid *Platyclymenia annulata*. Faunal changes accompanying this event were described by Walliser (1996b), who accepted its conventional dating as Late *trachytera* Zone. Ziegler and Sandberg (2000) tentatively redated the *annulata* Event as the next younger, Early *postera* Zone. In western North America, Event 15 is represented by the major transgression that resulted in deposition of the Trident Member of the Three Forks Formation and its lateral equivalents. The Trident Member and its deep-water equivalents in Nevada and California contain the ammonoid *Platyclymenia americana*, but the Trident has yielded only a nonpalmatolepid conodont fauna previously assigned roughly to the *trachytera* Zone (Sandberg et al., 1989). The age of these rocks, like the *annulata* Event, may need to be reevaluated.

Fourth interglacial episode (Event 16)

The fourth interglacial episode resulted in the final Famennian transgression, initiating T-R cycle IIf of Johnson et al. (1985), beginning in the Early *expansa* Zone. Like the transgression resulting from the second interglacial episode (Event 14), the fourth was a major eustatic event. Because of the accompanying important faunal changes, Events 14 and 16 are both employed for defining the proposed Famennian substages (Fig. 1). Event 16 introduced the shallow-water Etroeungt fauna into the Northern Hemisphere in widely separated areas, such as Utah and Arizona, northern France and southern Belgium, and the southern Urals. The Etroeungt fauna was a Lazarus fauna, characterized by the enigmatic reappearance of large clisiophyllid and caninoid corals, probably descended from supposedly extinct Frasnian rugose corals but for which there is no intervening Famennian record. As a result of extreme global warming and warm ocean currents, these corals must have been introduced from an unknown refuge, possibly in the Southern Hemisphere or in Asia. Stromatoporoids, occurring in large pillow-shaped colonies, also mysteriously reappeared at this time.

In addition to causing a rapid acceleration in conodont evolution, Event 16 also produced major changes in megafaunas, particularly the brachiopods and armored fishes. Among the brachiopods, first the highly diverse Percha fauna, including unusually large species such as *Paurorhyncha endlichi*, evolved. This fauna was followed in the next younger Middle *expansa* Zone by the diverse Louisiana Limestone fauna, which contains many forerunners of Carboniferous genera, such as *Syringothyris*. Among the fish, some of the largest known arthrodires inhabited the warm seas. The unusually large size of both the megafauna and conodonts supports the interpretation that Event 16 was a warm, tropical, interglacial episode in the Northern Hemisphere.

Middle praesulcata *Zone eustatic fall (Event 17)*

A major eustatic fall (Ziegler and Sandberg, 1984), probably associated with a resurgence of Southern Hemisphere glaciation (Caputo, 1985), began in the Middle *praesulcata* Zone and continued into the Early Carboniferous. The Middle *praesulcata* Zone eustatic fall (Event 17) and the ensuing mass extinction (Event 18) have been collectively termed the Hangenberg Event and were discussed in detail by Walliser (1996b). The initial pulse of this eustatic fall is evidenced in every studied section globally by a short stratigraphic interval wherein the pelagic siphonodellid conodont biofacies is replaced by a shallow-water protognathodid biofacies. Where initially recognized in a closely sampled, condensed carbonate sequence at the Trolp Quarry, near Graz, Austria, this gap amounts to only 5 cm. The gap is interpreted to indicate that sea level dropped so catastrophically at the start of the eustatic fall that the pelagic species *Siphonodella praesulcata* was forced to retreat from epicontinental seas to the ocean basins, and hence left no trace in a short interval of uppermost Devonian rocks of all continents. In the United States, the eustatic fall is characterized by regressive sandstones that retreated seaward from the Transcontinental arch; the upper part of the Sappington Member of the Three Forks Formation to the west and the Berea Sandstone to the east. Both units contain the diagnostic latest Famennian *Retispora lepidophyta* spore flora. The continuing eustatic fall induced a stepwise mass extinction of pelagic conodont species and the disappearance of the Etroeungt coral fauna, which reappeared again at different times during the Early Carboniferous.

Late Famennian mass extinction (Event 18)

The late Famennian (or D-C), mass extinction (Event 18) occurred during the eustatic sea-level fall that continued across the Devonian-Carboniferous boundary. As suggested by Sandberg et al. (1988), we interpret this mass extinction to be the terminal event of the Devonian Southern Hemisphere glaciation. However, Caplan and Bustin (1999, p. 148) concluded that climatic glacial cooling led to a "D-C mini-glaciation in Gondwana," thus inferring that Event 18 was an initial, not a terminal, glacial event. They also concluded that the global faunal crisis occurs "at the base of a globally extensive black, organic-rich mudrock." This statement is contradicted by the study of Ziegler and Sandberg (1984), who evaluated the most continuous D-C boundary sections in North America, Austria, Germany, the southern Urals, and South China, and showed that the black extinction shale is only locally preserved as at some localities in Germany and South China. More commonly the sea-level fall is represented by a hiatus or discontinuity or is masked within shallow-water deposits, as at the selected GSSP in the Montagne Noire, southern France (Ziegler and Sandberg, 1996).

Event 18 was not as intense as the late Frasnian mass extinction and is not included among the five major Phanerozoic mass extinctions. It caused the demise of the dominant pelagic conodont genus *Palmatolepis*; only a single species, *Palmatolepis gracilis*, survived into the Early Carboniferous. However, many shallow-water conodont genera, including *Protognathodus*, which had evolved shortly before the mass extinction, lived well into the Carboniferous. Likewise, shallow-water Carboniferous-type latest Famennian brachiopods, including the widespread *Syringothyris* fauna, were unaffected by the mass extinction and continued to flourish. Thus, it is likely that the late Famennian mass extinction affected mostly pelagic, as well as benthic and nektobenthic organisms, as suggested by Caplan and Bustin (1999). The Etroeungt coral fauna probably disappeared before the final mass extinction. As was the case in the early Famennian, following the late Frasnian mass extinction, only generally small, deep-water corals are known in the earliest Carboniferous.

CONCLUSIONS

Late Devonian geologic history comprises two very different but related parts. The initial Frasnian Stage was a time of generally rising sea level, accentuated by several catastrophic rises and punctuated by a mass extinction that occurred during a drastic, abrupt sea-level fall. The later Famennian Stage was a time of generally falling sea level, interrupted by several glacio-eustatic rises and terminated by a less severe mass extinction during another abrupt sea-level fall.

The most reasonable explanation for the catastrophic pulses of eustatic sea-level rise during the Frasnian and the ensuing late Frasnian mass extinction is a series of comet showers. This is suggested by a large number of known and possible impacts, such as the Siljan and Flynn Creek impacts. These are radiometrically dated by different methods and so cannot be positively correlated with one another. However, compelling evidence for relating the catastrophic rises to comet showers is provided by the biochronologically dated, well-documented Alamo Impact, which accompanied such a major rise. The off-platform, subcritical Alamo Impact is reinterpreted to have produced a greatly different sedimentary pattern in deep water from that previously recorded on the carbonate platform. The rises associated with the Alamo Impact shower and later comet showers altered or terminated three levels of Belgian mud-mounds, progressively diminished their size, and decimated reef communities. As a result, the global ecosystem was weakened and thus susceptible to the terminal Frasnian sea-level fall, which accompanied the mass extinction.

A logical explanation for the generally falling Famennian sea level is Southern Hemisphere glaciation resulting from global cooling that was induced by comet showers occurring just before, and possibly just after, the late Frasnian mass extinction. These showers are evidenced by microtektites found at and just above the Frasnian-Famennian boundary in widely separated areas. The four major pulses of sea-level rise that interrupted the general fall are attributable to glacio-eustatic rises during warm interglacial episodes. Each rise is associated with increased biotic abundance and diversity, resulting from the many more hospitable niches provided by warmer oceans. The mass extinction just before the end of the Famennian occurred during the sea-level fall produced by the terminal Devonian glacial episode.

ACKNOWLEDGMENTS

Peter Bender and Jens Ormö assisted our current study of the Amönau Breccia. Alan K. Chamberlain, Hans-Christian Kuehner, and, especially, John E. Warme helped immeasurably with our earlier interpretations of Alamo Breccia sedimentology and the Alamo Impact. Roland Dreesen contributed significantly to our study of Belgian mudmounds and to our understanding of Belgian Devonian event stratigraphy. Raymond C. Gutschick has collaborated with Sandberg since 1962 in studying the uppermost Famennian rocks of the western United States and introduced him to the Upper Devonian of Michigan and Indiana. Forrest G. (Barney) Poole has collaborated with Sandberg since 1968 in regional Devonian studies of the western United States and has contributed his considerable expertise on the geology of Nevada. Eberhard Schindler provided helpful comments on an early version of our manuscript. We thank Hans-Peter Schönlaub and an anonymous reviewer, whose comments helped improve our final manuscript.

Acknowledgment is made to the donors of the Petroleum Research Fund, administered by the American Chemical Society, for support of J.R. Morrow's contribution to this research.

REFERENCES CITED

Bender, P., Hühner, G., Kupfahl, H.-G., and Voutta, U., 1984, Ein Mitteldevon/Oberdevon-Profil bei Amönau auf Bl. 5018 Wetter (Hessen): Geologisches Jahrbuch Hessen, v. 112, p. 31–65.

Bratton, J.F., Berry, W.B.N., and Morrow, J.R., 1999, Anoxia pre-dates Frasnian-Famennian boundary mass extinction horizon in the Great Basin, USA: Palaeogeography, Palaeoclimatology, Palaeoecology, v. 154, p. 275–292.

Buggisch, W., Rabien, A., and Hühner, G., 1978, Biostratigraphische Parallelisierung und Faziesvergleich von oberdevonischen Becken- und Schwellen-Profilen E Dillenburg: Geologisches Jahrbuch Hessen, v. 106, p. 53–115.

Caplan, M.L., and Bustin, R.M., 1999, Devonian-Carboniferous Hangenberg mass extinction event, widespread organic-rich mudrock and anoxia: Causes and consequences: Palaeogeography, Palaeoclimatology, Palaeoecology, v. 148, p. 187–208.

Caputo, M.V., 1985, Late Devonian glaciation in South America: Palaeogeography, Palaeoclimatology, Palaeoecology, v. 51, p. 291–317.

Casier, J.-G., and Lethiers, F., 1998a, Les ostracodes survivant à l'extinction du Dévonien supérieur dans la coupe du Col de Devils Gate (Nevada, U.S.A.): Geobios, v. 30, no. 6, p. 811–821.

Casier, J.-G., and Lethiers, F., 1998b, The recovery of the ostracod fauna after the Late Devonian mass extinction: The Devils Gate Pass section example (Nevada, USA): Comptes rendus de l'Académie des Sciences, Serie 2, Sciences de la Terre et des Planetes, Earth and Planetary Sciences, v. 327, p. 501–507.

Casier, J.-G., Lethiers, F., and Claeys, P., 1996, Ostracod evidence for an abrupt mass extinction at the Frasnian/Famennian boundary (Devils Gate, Nevada, USA): Comptes rendus de l'Académie des Sciences, Serie 2a, Sciences de la Terre et des Planetes, Earth and Planetary Sciences, p. 415–422.

Claeys, P., and Casier, J.-G., 1994, Microtektite-like glass associated with the Frasnian-Famennian boundary mass extinction: Earth and Planetary Science Letters 122, p. 303–315.

Copper, P., 1986, Frasnian/Famennian mass extinction and cold-water oceans: Geology, v. 14, p. 835–839.

Dreesen, R., 1989, The "*Cheiloceras* Limestone", a Famennian (Upper Devonian) event-stratigraphic marker in Hercynian Europe and northwestern Africa: Bulletin de la Société Belge de Géologie, v. 28, no. 2, p. 127–133.

Dreesen, R., Bless, M.J.M., Conil, R., Flajs, G., and Laschet, C., 1985, Depositional environment, paleoecology and diagenetic history of the "Marbre rouge à crinoïdes de Baelen" (Late Upper Devonian, Verviers syclinorium, eastern Belgium): Annales de la Société Géologique de Belgique, v. 108, p. 311–359.

Geldsetzer, H.J., Goodfellow, W.D., and McLaren, D.J., 1993, The Frasnian-Famennian extinction event in a stable cratonic shelf setting: Trout River, Northwest Territories, Canada: Palaeogeography, Palaeoclimatology, Palaeoecology, v. 104, p. 81–95.

Girard, C., and Feist, R., 1997, Eustatic trends in conodont diversity across the Frasnian-Famennian boundary in the stratotype area, Montagne Noire, Southern France: Lethaia, v. 29, p. 329–337.

Goodfellow, W.D., Geldsetzer, H.H.J., McLaren, D.J., Orchard, M.J., and Klapper, G., 1989, The Frasnian-Famennian extinction: The current results and possible causes, *in* McMillan, N.J., Embry, A.F., and Glass, D.J., eds., Devonian of the World: Calgary, Canadian Society of Petroleum Geologists Memoir 14, v. 3, p. 9–21.

Groos-Uffenorde, H., and Wang, S.-Q., 1989, The entomozoacean succession of South China and Germany (Ostracoda, Devonian): Courier Forschungsinstitut Senckenberg, v. 110, p. 61–79.

Groos-Uffenorde, H., and Schindler, E., 1990, The effect of global events on entomozoan Ostracoda, *in* Whathey, R., and Maybury, C., eds., Ostracoda and global events: London, Chapman and Hall, p. 101–112.

Hallam, A., and Wignall, P.B., 1999, Mass extinctions and sea-level changes: Earth Sciences Reviews, v. 48, p. 217–250.

Huckriede, H., 1992, Das Barytlager von Günterrod in der Dillmulde (Rheinisches Schiefergebirge, Deutschland): Eine allochthone Scholle in givetischen Debris-Flow-Sedimenten: Geologisches Jahrbuch Hessen, v. 120, p. 117–144.

Ji, Q., 1989, On the Frasnian-Famennian mass extinction event in South China: Courier Forschungsinstitut Senckenberg, v. 117, p. 275–301.

Joachimski, M.M., and Buggisch, W., 1993, Anoxic events in the late Frasnian—Cause of the Frasnian-Famennian faunal crisis?: Geology, v. 21, p. 675–678.

Joachimski, M.M., and Buggisch, W., 2000, The Late Devonian mass extinction—Impact or earth-bound event? [abs.], *in* Catastrophic events and mass extinctions: Impacts and beyond: Houston, Texas, Lunar and Planetary Institute, LPI Contribution No. 1053, p. 83–84.

Johnson, J.G., 1970, Taghanic onlap and the end of North American Devonian provinciality: Geological Society of America Bulletin, v. 81, p. 2077–2105.

Johnson, J.G., Klapper, G., and Sandberg, C.A., 1985, Devonian eustatic fluctuations in Euramerica: Geological Society of America Bulletin, v. 96, p. 567–587.

Johnson, J.G., and Sandberg, C.A., 1989, Devonian eustatic events in the Western United States and their biostratigraphic responses, *in* McMillan, N.J.,

Embry, A.F., and Glass, D.J., eds., Devonian of the World: Canadian Society of Petroleum Geologists Memoir 14, v. 3, p. 171–179.

Kauffman, E.G., and Walliser, O.H., eds., 1990, Extinction events in Earth history, *in* Proceedings of the IGCP Project 216 "Global Biological Events in Earth History": Heidelberg, Germany, Springer-Verlag, Lecture Notes in Earth Sciences, v. 30, 432 p.

Klapper, G., and Johnson, J.G., 1980, Endemism and dispersal of Devonian conodonts: Journal of Paleontology, v. 54, p. 400–455.

Lazreq, N., 1992, The upper Devonian of M'rirt (Morocco): Courier Forschungsinstitut Senckenberg, v. 154, p. 107–123.

Lazreq, N., 1999, Biostratigraphie des conodontes du Givétien au Famennien du Maroc central: Biofaciès et événement Kellwasser: Courier Forschungsinstitut Senckenberg, v. 214, 111 p.

Leroux, H., Warme, J.E., and Doukhan, J.-C., 1995, Shocked quartz in the Alamo Breccia, southern Nevada: Evidence for a Devonian impact: Geology, v. 23, p. 1003–1006.

Lethiers, F., and Casier, J.-G., 1999, Autopsie d'une extinction biologique: Un exemple: La crise de la limite Frasnien-Famennien (364 Ma): Comptes rendus de l'Académie des Sciences, Serie 2, Sciences de la Terre et des Planetes, Earth and Planetary Sciences, no. 329, p. 303–315.

Matyja, H., 1993, Upper Devonian of Western Pomerania: Acta Geologica Polonica, v. 43, no. 1–2, 94 p.

Matyja, H., and Narkiewicz, M., 1992, Conodont biofacies succession near the Frasnian/Famennian boundary: Some Polish examples: Courier Forschungsinstitut Senckenberg, v. 154, p. 124–147.

McGhee, G.R., Jr., 1981, The Frasnian-Famennian extinctions: A search for extraterrestrial causes: Field Museum of Natural History Bulletin, v. 52, no. 7, p. 3–5.

McGhee, G.R., Jr., 1988, The Late Devonian extinction event: Evidence for abrupt ecosystem collapse: Paleobiology, v. 14, no. 3, p. 250–257.

McGhee, G.R., Jr., 1996, The Late Devonian mass extinction: The Frasnian/Famennian crisis: New York, Columbia University Press, 303 p.

McGhee, G.R., Jr., Orth, C.J., Quintana, L.R., Gilmore, J.S., and Olsen, E.J., 1986, Late Devonian "Kellwasser Event" mass-extinction horizon in Germany: No chemical evidence for a large-body impact: Geology, v. 14, p. 776–779.

McLaren, D.J., 1970, Presidential address: Time, life, and boundaries: Journal of Paleontology, v. 44, p. 801–815.

McLaren, D.J., 1982, Frasnian-Famennian extinctions: Geological Society of America Special Paper 190, p. 477–484.

Morrow, J.R., and Sandberg, C.A., 1997, Sequence stratigraphy across the F-F (mid-Late Devonian) boundary, central Great Basin; a conodont-based "reality check": Geological Society of America Abstracts with Programs, v. 29, no. 2, p. 41.

Morrow, J.R., Sandberg, C.A., Warme, J.E., and Kuehner, H.-C., 1998, Regional and possible global effects of sub-critical Late Devonian Alamo Impact Event, southern Nevada, USA: Journal of The British Interplanetary Society, v. 51, p. 451–460.

Morrow, J.R., Sandberg, C.A., and Ziegler, W., 1999, Recognition of mid-Frasnian (early Late Devonian) oceanic impacts: Alamo, Nevada, USA, and Amönau, Hessen, Germany [abs.]: European Science Foundation—Alfred Wegener Institute IMPACT Workshop, Oceanic Impacts: Mechanisms and Environmental Perturbations, Bremerhaven, Germany, Berichte zur Polarforschung 343, p. 66–69.

Muchez, P., Boulvain, F., Dreesen, R., and Hou, H.F., 1996, Sequence stratigraphy of the Frasnian-Famennian transitional strata: A comparison between South China and southern Belgium: Palaeogeography, Palaeoclimatology, Palaeoecology, v. 123, p. 289–296.

Ormö, J., and Lindström, M., 2000, When a cosmic impact strikes the sea bed: Geological Magazine, v. 137, no. 1, p. 67–80.

Over, D.J., Conaway, C.A., Katz, D.J., Goodfellow, W.D., and Gregoire, D.C., 1997, Platinum group element enrichments and possible chondritic Ru:Ir across the Frasnian-Famennian boundary, western New York State: Palaeogeography, Palaeoclimatology, Palaeoecology, v. 132, p. 399–410.

Racka, M., and Racki, G., 2000, Geochemical aspects of the Frasnian-Famennian mass extinction: The Polish example [abs.], *in* Catastrophic events and mass extinctions: Impacts and beyond: Houston, Texas, Lunar and Planetary Institute, LPI Contribution No. 1053, p. 174.

Racki, G., 1998a, The Frasnian-Famennian brachiopod extinction events: A preliminary review: Acta Palaeontologica Polonica, v. 43, no. 2, p. 395–411.

Racki, G., 1998b, Frasnian-Famennian biotic crisis: Undervalued tectonic control?: Palaeogeography, Palaeoclimatology, Palaeoecology, v. 141, p. 177–198.

Racki, G., 1999, The Frasnian-Famennian biotic crisis: How many (if any) bolide impacts?: Geologisches Rundschau, v. 87, p. 617–632.

Sandberg, C.A., 1998, Tiny teeth shed light on ancient comets: U.S. Geological Survey News Release, March 20, 1998, 2 p.

Sandberg, C.A., and Dreesen, R., 1984, Late Devonian icriodontid biofacies models and alternate shallow-water conodont zonation, *in* Clark, D.L., ed., Conodont biofacies and provincialism: Geological Society of America Special Paper 196, p. 143–178.

Sandberg, C.A., Gutschick, R.C., Johnson, J.G., Poole, F.G., and Sando, W.J., 1983, Middle Devonian to Late Mississippian geologic history of the Overthrust Belt region, western United States: Denver, Colorado, Rocky Mountain Association of Geologists, Geologic Studies of the Cordilleran Thrust Belt, v. 2, p. 691–719.

Sandberg, C.A., Gutschick, R.C., Johnson, J.G., Poole, F.G., and Sando, W.J., 1986, Middle Devonian to Late Mississippian event stratigraphy of Overthrust belt region, western United States, *in* Bless, M.J.M., and Streel, M., eds., Late Devonian events around the Old Red continent: Société Géologique de Belgique Annales, v. 109, pt. 1, Special volume "Aachen 1986", p. 205–207.

Sandberg, C.A., and Morrow, J.R., 1998, Role of conodonts in deciphering and dating Late Devonian Alamo Impact Megabreccia, southeastern Nevada, USA [abs.], *in* Bagnoli, G., ed., Seventh International Conodont Symposium (ECOS 7): Bologna-Modena, Italy, Abstracts volume, p. 93–94.

Sandberg, C.A., Morrow, J.R., and Warme, J.E., 1997, Late Devonian Alamo Impact Event, global Kellwasser Events, and major eustatic events, eastern Great Basin, Nevada and Utah: Brigham Young University Geology Studies, v. 42, pt. 1, p. 129–160.

Sandberg, C.A., Morrow, J.R., and Ziegler, W., 2000, Possible impact origin of the enigmatic early Late Devonian Amönau Breccia, Rheinisches Schiefergebirge, Germany [abs.], *in* Catastrophic events and mass extinctions: Impacts and beyond: Houston, Texas, Lunar and Planetary Institute, LPI Contribution No. 1053, p. 187.

Sandberg, C.A., Poole, F.G., and Johnson, J.G., 1989, Upper Devonian of Western United States, *in* McMillan, N.J., Embry, A.F., and Glass, D.J., eds., Devonian of the World: Canadian Society of Petroleum Geologists Memoir 14, v. 1, p. 183–220.

Sandberg, C.A., and Warme, J.E., 1993, Conodont dating, biofacies, and catastrophic origin of Late Devonian (early Frasnian) Alamo Breccia, southern Nevada: Geological Society of America Abstracts with Programs, v. 25, no. 3, p. 77.

Sandberg, C.A., and Ziegler, W., 1991, Extreme falls of sea level accompanied late Frasnian (F/F) and late Famennian (D/C) mass extinctions in the Late Devonian [abs.]: Calgary, International Union of Geological Sciences, Joint Meeting of IGCP Projects 216, 293, and 303, Event Markers in Earth History, Programs and Abstracts, p. 63.

Sandberg, C.A., and Ziegler, W., 1992, Late Devonian conodont evolution and post-crisis recoveries [abs.]: Göttingen, Germany, International Union of Geological Sciences, IGCP Project 216, Fifth International Conference on Global Bioevents, Phanerozoic Global Bio-Events and Event-Stratigraphy, Abstracts volume, p. 92.

Sandberg, C.A., and Ziegler, W., 1996, Devonian conodont biochronology in geologic time calibration: Senckenbergiana lethaea, v. 76, p. 259–265.

Sandberg, C.A., and Ziegler, W., 1998, Comments on proposed Frasnian and Famennian Subdivisions (Document submitted to IUGS Subcommission on Devonian Stratigraphy Meeting in Bologna, Italy): Subcommission on Devonian Stratigraphy Newsletter No. 15, p. 43–46.

Sandberg, C.A., Ziegler, W., Dreesen, R., and Butler, J.L., 1988, Late Frasnian mass extinction: Conodont event stratigraphy, global changes, and possible causes, *in* Ziegler, W., ed., 1st International Senckenberg Conference and 5th European Conodont Symposium (ECOS V), Contribution 1: Courier Forschungsinstitut Senckenberg, v. 102, p. 263–307.

Sandberg, C.A., Ziegler, W., Dreesen, R., and Butler, J.L., 1992, Conodont biochronology, biofacies, taxonomy, and event stratigraphy around middle Frasnian Lion Mudmound (F2h), Frasnes, Belgium: Courier Forschungsinstitut Senckenberg, v. 150, 87 p.

Schindler, E., 1990a, The late Frasnian (Upper Devonian) Kellwasser Crisis: Lecture Notes in Earth Sciences, v. 30, p. 151–159.

Schindler, E., 1990b, Die Kellwasser-Krise (hohe Frasne-Stufe, Ober-Devon): Göttinger Arbeiten Geologie und Paläontologie, Göttingen, no. 46, 115 p.

Schindler, E., Schülke, I., and Ziegler, W., 1998, The Frasnian/Famennian boundary at the Sessacker Trench section near Oberscheld (Dill Syncline), Rheinisches Schiefergebirge, Germany: Senckenbergiana lethaea, v. 77, no. 1–2, p. 243–261.

Walliser, O.H., 1985, Natural boundaries and Commission boundaries in the Devonian: Courier Forschungsinstitut Senckenberg, v. 75, p. 401–408.

Walliser, O.H., ed., 1986, Global bio-events, *in* Proceedings of the First International Meeting of the IGCP Project 216 "Global Biological Events in Earth History": Lecture Notes in Earth Sciences, v. 8, Heidelberg, Germany, Springer-Verlag, 442 p.

Walliser, O.H., ed., 1996a, Global events and event stratigraphy in the Phanerozoic: Results of international interdisciplinary cooperation in the IGCP Project 216 "Global Biological Events in Earth History": Heidelberg, Germany, Springer-Verlag, 333 p.

Walliser, O.H., 1996b, Global events in the Devonian and Carboniferous, *in* Walliser, O.H., ed., Global events and event stratigraphy in the Phanerozoic: p. 225–250.

Walliser, O.H., Groos-Uffenorde, H., and Schindler, E., 1989, On the Upper Kellwasser Horizon (boundary Frasnian/Famennian): Courier Forschungsinstitut Senckenberg, v. 110, p. 247–255.

Wang, K., 1992, Glassy microspherules (microtektites) from an Upper Devonian limestone: Science, v. 256, p. 1546–1549.

Wang, K., Orth, C.J., Attrep, M., Jr., Chatterton, B.D.E., Hou, H., and Geldsetzer, H.H.J., 1991, Geochemical evidence for a catastrophic biotic event at the Frasnian/Famennian boundary in South China: Geology, v. 19, p. 776–779.

Warme, J.E., and Chamberlain, A.K., 2000, Primary ejecta, tsunami reworking, tectonic dismemberment: Reconstructing the Late Devonian Alamo Breccia and crater, Nevada [abs.], *in* Catastrophic events and mass extinctions: Impacts and beyond: Houston, Texas, Lunar and Planetary Institute, LPI Contribution No. 1053, p. 238.

Warme, J.E., and Kuehner, H.-C., 1998, Anatomy of an anomaly: The Devonian catastrophic Alamo impact breccia of southern Nevada: International Geology Review, v. 40, p. 189–216.

Warme, J.E., and Sandberg, C.A., 1995, The catastrophic Alamo breccia of southern Nevada: Record of a Late Devonian extraterrestrial impact: Courier Forschungsinstitut Senckenberg, v. 188, p. 31–57.

Wendt, J., and Belka, Z., 1991, Age and depositional environment of Upper Devonian (early Frasnian to early Famennian) black shales and limestones (Kellwasser facies) in the eastern Anti-Atlas, Morocco: Facies, v. 25, p. 51–90.

You, X.L., 2000, Famennian tentaculitids of China: Journal of Paleontology, v. 74, no. 5, p. 969–975.

Ziegler, W., and Lane, H.R., 1987, Cycles in conodont evolution from Devonian to mid-Carboniferous, *in* Aldridge, R.J., ed., Palaeobiology of conodonts: British Micropalaeontological Society, p. 147–163.

Ziegler, W., and Sandberg, C.A., 1984, Important candidate sections for stratotype of conodont based Devonian-Carboniferous boundary, *in* Paproth, E., and Streel, M., eds., The Devonian-Carboniferous boundary: Courier Forschungsinstitut Senckenberg, v. 67, p. 231–239.

Ziegler, W., and Sandberg, C.A., 1990, The Late Devonian standard conodont zonation: Courier Forschungsinstitut Senckenberg, v. 121, 115 p.

Ziegler, W., and Sandberg, C.A., 1996, Reflexions on Frasnian and Famennian Stage boundary decisions as a guide to future deliberations: Newsletters on Stratigraphy, v. 33, p. 157–180.

Ziegler, W., and Sandberg, C.A., 2000, Utility of palmatolepids and icriodontids in recognizing Upper Devonian Series, Stage, and possible Substage boundaries, *in* Bultynck, P., ed., Subcommission on Devonian Stratigraphy: Recognition of Devonian series and stage boundaries in geological areas: Courier Forschungsinstitut Senckenberg, v. 225, p. 335–348.

MANUSCRIPT SUBMITTED OCTOBER 5, 2000; ACCEPTED BY THE SOCIETY MARCH 22, 2001

Geological Society of America
Special Paper 356
2002

Impact-generated carbonate accretionary lapilli in the Late Devonian Alamo Breccia

John E. Warme
Matthew Morgan
Department of Geology and Geological Engineering, Colorado School of Mines, Golden, Colorado 80401, USA
Hans-Christian Kuehner
Texaco North Sea Ltd., Langlands House, Hemtley Street, Aberdeen A B10 1SH, Scotland

ABSTRACT

Carbonate accretionary lapilli occur in the Late Devonian Alamo Breccia of south-central Nevada. They provide evidence for the extraterrestrial impact origin of the breccia, and help unravel the complicated events that formed it. The accretionary lapilli (Alamo lapilli) are concentrated in lapilli beds, and portions of the latter occur as reworked clasts that are isolated within the upper half of the thick breccia. The Alamo lapilli resemble volcanic accretionary lapilli reported from both silicate and carbonatite volcanoes. They are variable in size and detail, but generally exhibit a nucleus interpreted to be altered target rock, an enveloping mantle of silt- and sand-sized particles, and a very fine grained peripheral crust. Spherule composition is entirely carbonate, except for sparse shocked quartz grains incorporated into the mantle and diagenetic iron oxides. Rare preservation of undeformed bed segments shows the stratigraphy of the accretionary lapilli; they were deposited in poorly size-sorted layers with varying proportions of matrix. Their preserved form is spherical, deformed, or broken, implying varying degrees of damage before or during deposition, before bed hardening, and during catastrophic reworking and dewatering of the Alamo Breccia.

We propose that the carbonate accretionary lapilli were preserved when target carbonate formations were pulverized by impact pressure and calcinated by impact heat, creating quicklime. The lapilli evolved by adhesion of particles within the impact cloud; they were partially cemented in flight by hydration, and then precipitated as one or more beds over early ejectite, debris flows, and/or nearly contemporaneous tsunamites. The cementation process continued in the lapilli beds so that portions of them survived reworking and initial breccia settling and dewatering. Coherent, isolated fragments of lapilli beds and deformed bed masses are preserved as much as 25 m beneath the top of the breccia, indicating the thickness of rock reworked during the Alamo Event.

Warme, J.E., Morgan, M., and Kuehner, H.-C., 2002, Impact-generated carbonate accretionary lapilli in the Late Devonian Alamo Breccia, *in* Koeberl, C., and MacLeod, K.G., eds., Catastrophic Events and Mass Extinctions: Impacts and Beyond: Boulder, Colorado, Geological Society of America Special Paper 356, p. 489–504.

INTRODUCTION

Carbonate spheroidal particles (Fig. 1), interpreted as impact-generated carbonate accretionary lapilli, represent an important constituent of the Late Devonian Alamo impact breccia in Nevada. For brevity, and to distinguish them from other kinds of accretionary spherules and spherical impact products, we call them Alamo lapilli. Shocked quartz grains are dispersed in the Alamo Breccia, and are incorporated into the Alamo lapilli. The beautifully preserved lapilli and shocked quartz are the best observable physical evidence for the impact origin of the breccia, although they compose much less than 1% of the deposit.

Probably within only minutes to a day, a series of impact-related events (Alamo Event of Warme and Sandberg, 1995, 1996; Morrow et al., 1998) eroded and resedimented a thick interval of carbonate platform rock of the Guilmette Formation, which became the thick Alamo carbonate megabreccia. The Alamo lapilli occur in isolated clasts scattered within the Alamo Breccia, and provide a key to understanding the character and the timing of events that led to the Alamo Breccia formation. Alamo lapilli are interpreted to have accreted in a vapor-rich impact plume by adhesion of carbonate particles and other contemporaneous processes. Their structure resembles some kinds of armored accretionary volcanic lapilli, having a central nucleus interpreted as carbonate target rock, an enveloping mantle of sand- and silt-sized carbonate fragments, and an outer crust of very fine silt- and clay-sized carbonate particles. Alamo lapilli imply that the impact target was dominated by carbonate rock and was wet. Evidence suggests that they precipitated as one or more widespread beds. The beds were then dismembered, by tsunami erosion and/or other energetic impact processes, and redeposited as isolated clasts in the massive breccia. Preserved lapilli survived these events by rapid cementation processes that began in flight and continued in the precipitated bed(s). This early cementation caused segments of the beds to be durable enough to withstand subsequent catastrophic transportation and deposition.

Figure 1. Alamo carbonate accretionary lapilli exposed on weathered surface of lapilli bed showing sizes, proportions of nuclei, mantles, and crust, and grainy matrix in this sample. Lapilli are circular, flattened, and in various stages of breakage.

Other examples of ancient carbonate lapilli may have formed and been preserved by impact processes. Carbonate strata anywhere may contain unrecognized intervals of impact lapilli and lapillistone.

ALAMO BRECCIA

The Alamo Breccia is a widespread carbonate bed of Late Devonian age (ca. 365 Ma) that covered much of what is now south-central Nevada. The Alamo Breccia was deposited during the catastrophic Alamo Event. The Alamo Breccia was discovered during field work in 1990 as a stratigraphic anomaly in the Guilmette Formation; it is a thick mass-flow deposit that strongly contrasts with the depositional style of the shallow-water cyclic carbonate platform deposits characteristic of the Guilmette. Evidence for the impact origin of the Alamo Breccia includes its impact-generated shocked quartz grains, an iridium anomaly, conodont microfossils ejected from target rock, and the carbonate accretionary lapilli (Alamo lapilli) that are the topic of this chapter. Warme and Kuehner (1998) provided a comprehensive description and history of discovery of the Alamo Breccia and an interpretation of its genesis. Additional major publications on the Alamo Breccia include Warme and Sandberg (1995, 1996), Laroux et al. (1995), Sandberg et al. (1997), and Morrow et al. (1998).

The process of discovery, description, and interpretation of the Alamo Breccia is not complete. This contribution is a progress report that calls attention to the breccia, to the Alamo lapilli, and to the likely genetic relationships between the two. A description of the Alamo Breccia is presented first because it is the carrier of the Alamo lapilli, and interpretation of the Alamo Breccia and the Alamo lapilli are interdependent.

Geologic framework

The Alamo Breccia is recognized by us in 16 mountain ranges in south-central Nevada, and has been discovered in several more ranges by others (Sandberg et al., this volume). The general area and some key localities where Alamo lapilli are found are shown in Figure 2. The full dimensions of the Alamo Breccia are not known, and calculations vary depending upon the assumed original areal distribution of the deposit and the style of tectonic reconstruction adopted for southern Nevada since the Devonian (e.g., east-west crustal shortening, lengthening, and tectonic timing). Warme and Kuehner (1998) conservatively estimated that the Alamo Breccia covers an area of

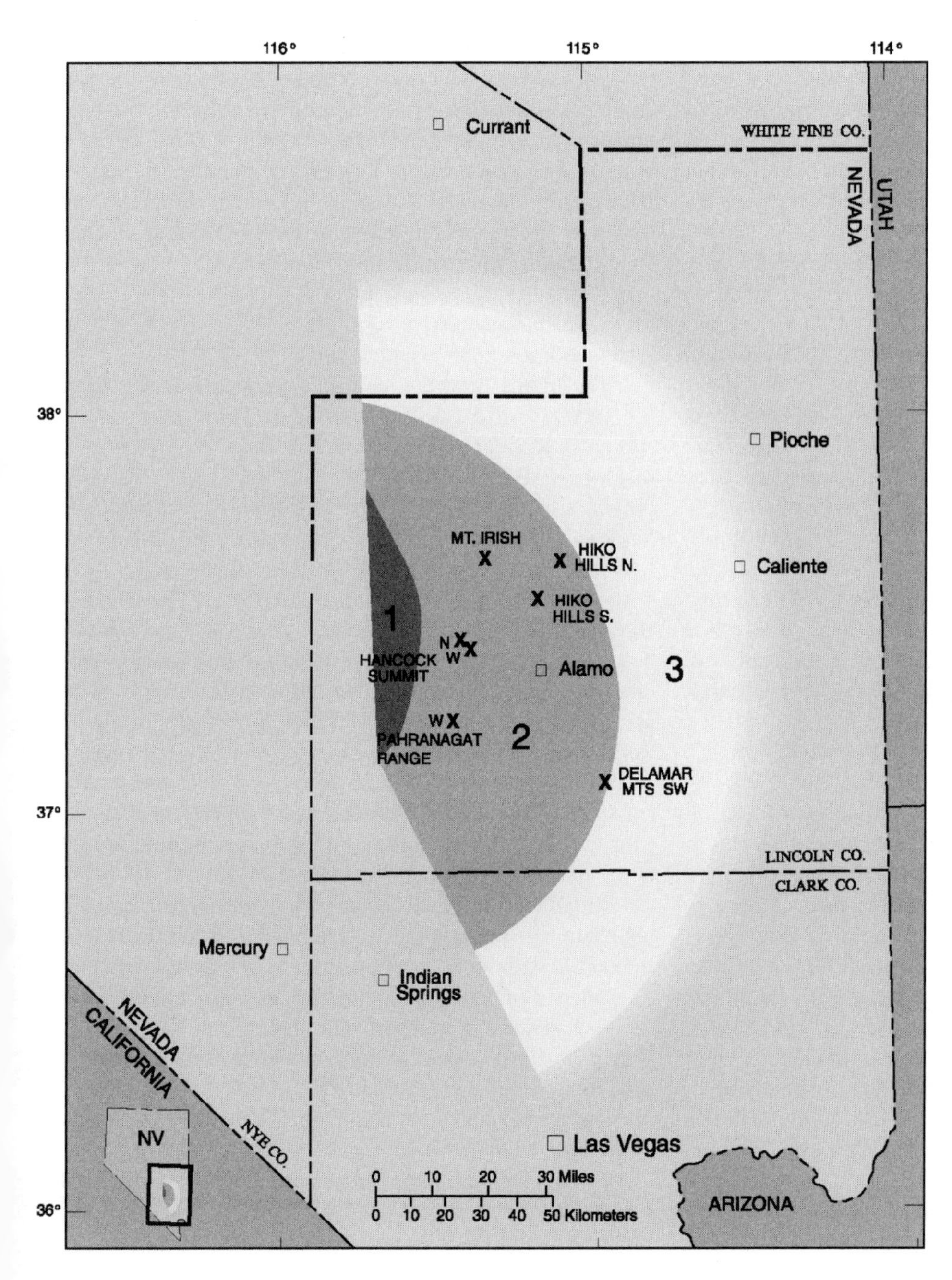

Figure 2. Index map of southern Nevada. Shading shows known distribution of Alamo Breccia. Concentric zones represent thickening of deposit and increased erosion of underlying strata from Zone 3 to Zone 1. Thicknesses of breccia: Zone 1 > 100 m; Zone 2 < 100 m; Zone 3 < 10 m. Locations cited in text are marked with X (after Warme and Kuehner, 1998). Mt.—Mount, N.–North, S.—South, Co.—County, Mts.—Mountains.

$\sim$10 × 10^3 km^2, contains a volume of $\sim$500 km^3, and has an average thickness of $\sim$50 m.

Stratigraphic position

Sandberg et al. (1997) formalized the Alamo Breccia as the Middle Member of the Guilmette Formation. The Guilmette represents the shallow-water platform facies of a north-south-trending carbonate platform that is interpreted as the western margin of the North American continent in the Late Devonian (e.g., Cook, 1983). Over much of its area, the Alamo Breccia is intercalated with the shallow-water platform carbonates of the Guilmette. The Alamo Breccia is precisely dated to have formed within a single conodont zone of the Frasnian Stage, i.e., early Late Devonian (Warme and Sandberg, 1995, 1996; Sandberg et al., 1997).

The present epiplatform distribution of the Alamo Breccia is a north-south-extending, eastward-curved, semicircle (Fig. 2). From the eastern periphery to its center the breccia overlies increasingly older sediments, having cut more deeply into the carbonate platform facies of the Guilmette. At the western limit in Figure 2 the erosion surface under the Alamo Breccia cuts completely through the Guilmette and overlies the earlier Devonian Bay State and/or Sentinel Mountain Formation (Warme and Kuehner, 1998). The Alamo Breccia increases in thickness from <1 m around the periphery of the semicircle to >100 m at the center. Recent work west of the semicircle has identified deep-water equivalents of the Alamo Breccia (Morrow et al., 2001, 2002; Morrow and Sandberg, 2001).

Post-Devonian thrusting has probably significantly telescoped Paleozoic formations of southern Nevada. Chamberlain (1999) and Warme and Chamberlain (2000) used existing and new data on formation facies and thicknesses, Alamo Breccia trends, and thrust fault distributions and styles, to propose a preliminary palinspastic reconstruction of the thrust belt across the area of the Alamo Breccia in southern Nevada. The resulting balanced cross sections yielded a more circular original distribution of the Alamo Breccia. This may be the first example where an impact deposit was used to help solve a tectonic problem. The missing crater area, and thrusted segments of the Alamo Breccia-bearing Guilmette Formation, were likely displaced hundreds of kilometers from the west. Most or all of the crater and large portions of the Alamo Breccia are probably still buried by thrust sheets, or have been uplifted and eroded away.

Composition

The Alamo Breccia at locations shown in Figure 2 is composed almost entirely of limestone or dolostone and appears to be derived directly from the interval of the carbonate platform that was catastrophically disintegrated by the impact. It is probable that the impact crater was excavated in the carbonate beds that dominate Devonian to Middle Ordovician formations. An exception is the regional, thin, late Early or early Middle Devonian quartzose Oxyoke Canyon Sandstone. However, sandstone clasts are exceedingly rare and shale clasts are absent in the Alamo Breccia. It is significant that no clasts of the distinctive Middle Ordovician Eureka Quartzite have been identified in the Breccia, thus limiting the maximum depth of crater excavation to less than ~1500 m. Shocked quartz, presumed to be sourced from the thin Oxyoke Canyon Sandstone, is dispersed as free grains in the Alamo Breccia matrix, and is incorporated into the Alamo lapilli. Conodonts reworked into the Alamo Breccia were reported by Warme and Sandberg (1995, 1996), indicating crater excavation down to Ordovician formations ~1200 m below the base of the Guilmette (Morrow et al., 1998).

Lateral zones

The present epiplatform distribution of the Alamo Breccia is divided into three semiconcentric zones (Fig. 2) defined by their thicknesses: >100 m in western Zone 1, <100 m in central Zone 2, and <10 m in eastern Zone 3. It is ~1 m thick at several localities around the Zone 3 periphery. The clasts containing Alamo lapilli are most common in the breccia in Zone 2, are sparse in Zone 1, and are rare or absent in Zone 3. The light tan-gray color of the breccia in Zone 3, similar to the Alamo lapilli clasts in Zones 1 and 2, and the presence of shocked quartz grains in the Zone 3 breccia matrix both suggest that much of the volume there may be composed of disintegrated Alamo lapilli that were reworked by tsunamis from the marine direction (west) and by massive runoff of water, condensed from the impact plume, from the land (east).

Vertical stratification

The Alamo Breccia is vertically composed of four different lithofacies, best expressed in Zone 2. From the base of the Alamo Breccia upward, they are: D—monomict breccia, C—megaslabs, B—megaclast polymict breccia, and A—graded polymict breccia (Fig. 3). The D interval is a discontinuous, basal monomict breccia that separates the overlying C megaslabs from the underlying intact bedrock. C megaslabs (slab = ~66-1050 m; see extended Udden-Wentworth classification of Blair and McPherson, 1999), usually tens of meters thick and hundreds of meters long, are parallel or subparallel to the underlying bedrock, and have not been lifted far off the basal surface of the Alamo Breccia. D and C lithofacies are the same as the D and C units of Warme and Sandberg (1995, 1996). The widespread C megaslabs and related basal D breccias were somehow generated by Alamo impact seismic waves, crustal tilting, tsunami shear, and/or other processes (Warme and Kuehner, 1998), and are still being studied.

Lithofacies B is a disorganized polymict breccia containing abundant megaclasts that were defined by Kuehner (1997) as longer than 2 m in outcrop exposure. They float within a polymict matrix of carbonate clasts. Clasts and matrix exhibit an array of shallow-water carbonate lithofacies and fossils that most commonly appear to be fragments of the Guilmette Formation. At many Zone 2 localities, and at the base of the breccia in Zone 1, megaclasts are both matrix- and clast-supported, but a fine-grained matrix is always present.

Lithofacies A contains most of the lapillistone clasts. It is composed of one or more graded beds of polymict breccia. Figure 3 shows the vertical distribution of B lithofacies, overlain by several graded beds of A lithofacies that are thinner and finer grained upward. The upper beds are generally better graded, better sorted, and more clast supported. Kuehner (1997) stretched a 2 m interval of cord horizontally across Alamo Breccia outcrops and documented clast sizes, compositions, and orientations. The outcome for one locality is shown in Figure 3. One result of his study was the discovery of multiple graded beds within the Alamo Breccia, in addition to the obvious one at the top. Ungraded or roughly graded beds of lithofacies B are overlain by several better graded intervals defined as lithofacies A. In Figure 3 we extended the A lithofacies downward to include all well-graded beds. Only the upper one comprised the A unit of previous publications (Warme and Sandberg, 1995, 1996; Warme and Kuehner, 1998). We reduced the B

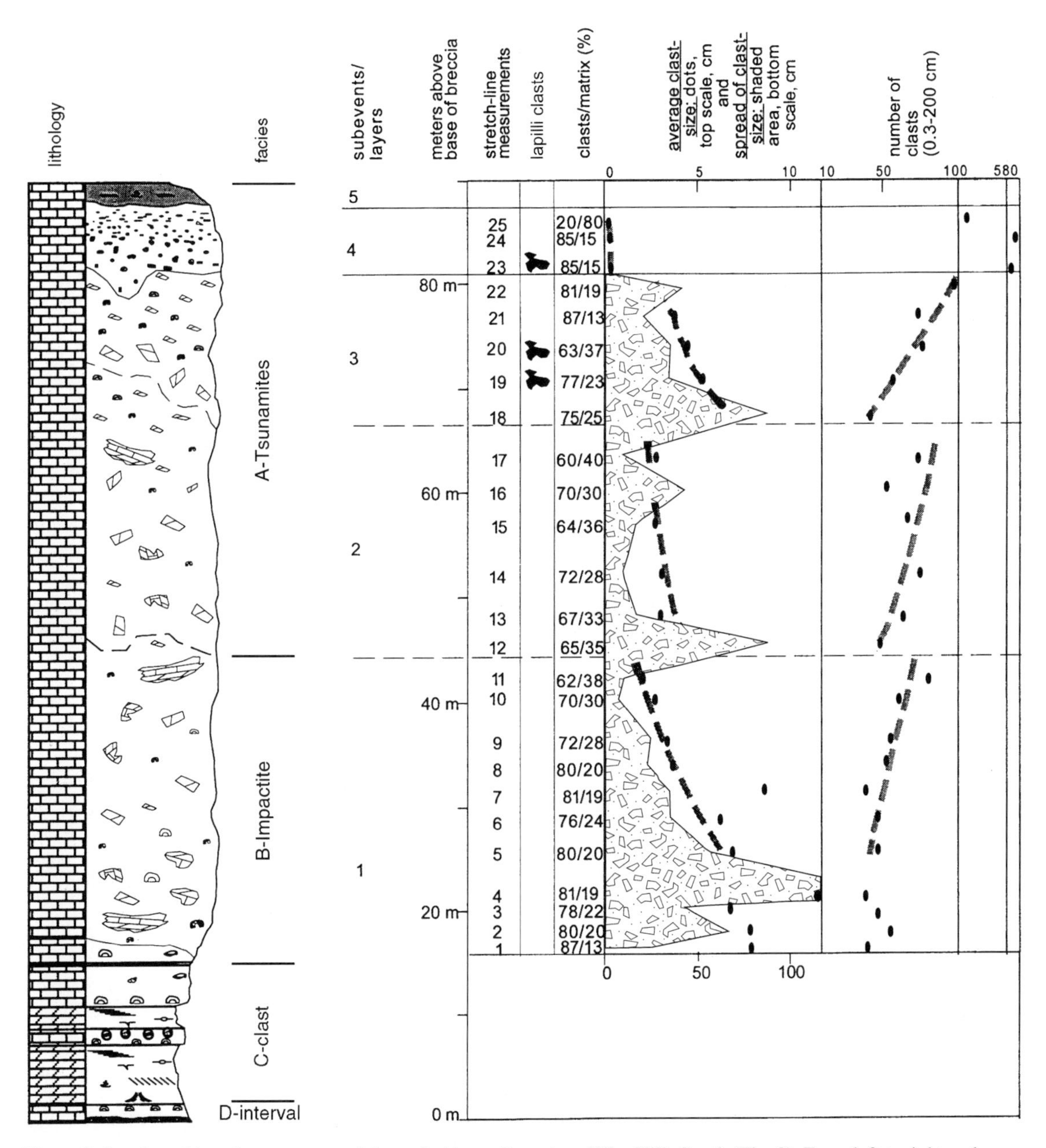

Figure 3. Stratigraphic column measured through Alamo Breccia at Hiko Hills South (Fig. 2). From left to right, columns represent distribution of: lithology, clasts, and weathering profile, A–D lithofacies, erosion and deposition events, stratigraphic position of clast-size measurements using stretch line, location of three preserved segments of spherule beds, and trends in clast/matrix ratios, average clast sizes, range of clast sizes, and number of clasts per measurement. Lithofacies D is only 10–20 cm thick and very subtle at this location. Lithofacies C clast at base is 18 m thick and overlain by five graded beds that become progressively thinner and finer grained upward. B lithofacies in first layer is highly disorganized and may represent primary ejecta. Overlying beds are interpreted as tsunamites.

lithofacies thickness to include only the more disorganized interval above the base of the Alamo Breccia, or over C megaslabs if they are present, and thus reduced the relative thickness of the B unit of previous papers. In this revised scheme, Alamo lapilli appear to be restricted to the A lithofacies.

Figure 4 shows the contact between two A lithofacies graded beds. These boundaries are rarely as flat and sharp as this example. More typically, the contacts are distorted, and in many localities they apparently have been destroyed by pervasive dewatering, loading, shearing, and other bed adjustments that we attribute to successive rapid cycles of catastrophic scour, transportation, and deposition of the Alamo Breccia during the Alamo Event. The number of vertical intervals recognized in the Alamo Breccia changes between lateral zones. In

Figure 4. Scour surface ~10 m below top of Alamo Breccia at Hiko Hills North locality, interpreted as two separate tsunamites. Surface at pencil tip separates sorted pebbles and cobbles from overlying sorted fine pebbles, granules, and sand. Pencil width is 7 mm.

Zone 2 as many as five layers are recognized (Fig. 3). In Zones 1 and 3 the deposit consists of three or fewer layers (Kuehner, 1997).

In Zone 3 the erosional base of the Alamo Breccia is ~200 m above the base of the Upper Devonian Guilmette Formation. In Zone 2 it is ~100 m from the base. In both zones the underlying and overlying facies are upward-shallowing cycles typical of carbonate platforms. In Zone 1, the Alamo Breccia is in the deeper-water platform or carbonate ramp facies of Middle Devonian dolostones.

Overlying the Alamo Breccia in Zone 1 are thick, deep-water, anoxic limestones, punctuated by graded quartzose beds that are very different from the shallow-water carbonate platform beds over the Alamo Breccia in Zones 2 and 3. Warme and Kuehner (1998) suggested that this thin-bedded facies was deposited in the newly formed crater. This analysis is a departure from the view of Morrow et al. (1998) and earlier workers, whereby these distinctive limestones were deposited in deeper water within a shelf-edge salient. Earlier paleogeographic reconstructions placed the platform margin, or shelf-slope break, to the west, along an approximate north-south line through central Nevada (e.g., Poole and Sandberg, 1977; Sandberg et al., 1989).

We interpret the significant bedrock deformation, fluidization, and fractures that characterize the underlying Middle Devonian dolostones in Zone 1 to be products of shock metamorphism, and the Alamo Breccia over the dolostones to represent deposition closest to the crater of any known outcrops (Warme and Kuehner, 1998). The dolostones are pervasively sheared, and penetrated by sedimentary dikes and sills that are most obvious along the contact with the underlying Oxyoke Canyon Sandstone, ~300 m below the base of the Alamo Breccia in Zone 1. The Oxyoke Canyon was apparently less cemented than the overlying more brittle dolostones. Beds of the sandstone were injected into the dike and sill system as folded quartzose fragments and dispersed quartz grains, some of which exhibit shock-induced parallel deformation features (PDFs).

Genesis of the Alamo Breccia

We have adopted the published marine impact model of Oberbeck et al. (1993), derived from studies of lunar and terrestrial cratering, to interpret the genesis of the Alamo Breccia (Kuehner, 1997; Kuehner and Warme, 1998). In their model an impact into shallow-marine sediments deposited ejecta nearby, which was then reworked by successive impact-generated multiple waning tsunamis. These wave systems consecutively eroded and redeposited the newly littered and damaged sea bed. We suggest that similar processes account for the puzzling distribution of the carbonate lapillistone as isolated clasts in the Alamo Breccia.

In this scenario, a Late Devonian bolide struck the carbonate platform of what is now western North America. We do not know the location of the crater. However, the damaged bedrock under the Alamo Breccia in Zone 1 indicates that Zone 1 was closer to the crater than other known outcrops.

It is possible that crustal tilting caused by impact rebound, or a seismically induced detachment surface propagated away from the crater into Zone 2, separated the C megaslabs over the fluidized D monomict breccia. Seismic shock may have induced the concentric disintegration of the carbonate platform and created some of the B lithofacies. In either case, the model includes a high-velocity ejecta curtain that spread excavated sediments from the crater across the platform, forming a sheet of ejecta that we interpret as some or all of the B lithofacies. In either case, ensuing impact-generated tsunamis eroded and deposited sediments over the initial deposit, forming the upper polymict A lithofacies, and all lithofacies of the Alamo Breccia appear to have evolved during several closely spaced events.

CARBONATE CALCAREOUS ACCRETIONARY LAPILLI AND LAPILLI BEDS

Alamo lapilli are distinctive, and were formed during the Alamo impact event; both the Alamo lapilli and their matrix contain shocked quartz grains and a modest iridium anomaly. Fine detail reveals that they formed by accretion (Fig. 5). However, their occurrence as rare isolated lapilli bed remnants, scattered between the dominant marine limestone clasts in the Alamo Breccia, is puzzling (Fig. 6). The lapilli beds were indurated early, forming lapillistone, then were broken and deformed during transportation and deposition as the Alamo Breccia accumulated (Figs. 6, 7).

Lapilli definitions, types, and nomenclature

Important comprehensive publications on volcanic pyroclastic rocks and their classifications, including lapilli, were edited by Fisher and Schmincke (1984) and Fisher and Smith (1991). According to the American Geological Institute (1980) volcanic lapilli are defined as pyroclastic aggregates of any shape and have diameters of 2–64 mm. Aggregates >64 mm are defined as volcanic bombs. Indurated beds of lapilli are lapillistone. Accretionary lapilli are volcanic pellets with concentric structure around a center or nucleus, and armored accretionary lapilli have an outer coat that is usually finer grained, more indurated, and more brittle than the interiors (e.g., Reimers, 1983b). Most lapilli reported in the literature are from silicate volcanoes (particularly hydroclastic basaltic eruptions). However, explosive carbonatite volcanoes create carbonate lapilli (Keller, 1989), which can be armored (Deans and Roberts, 1984). The only active carbonatite volcano is Oldoinyo Lengai in Tanzania. Numerous Quaternary and older examples have been reported from Africa (Deans and Roberts, 1984) and elsewhere (Deans and Seager, 1978).

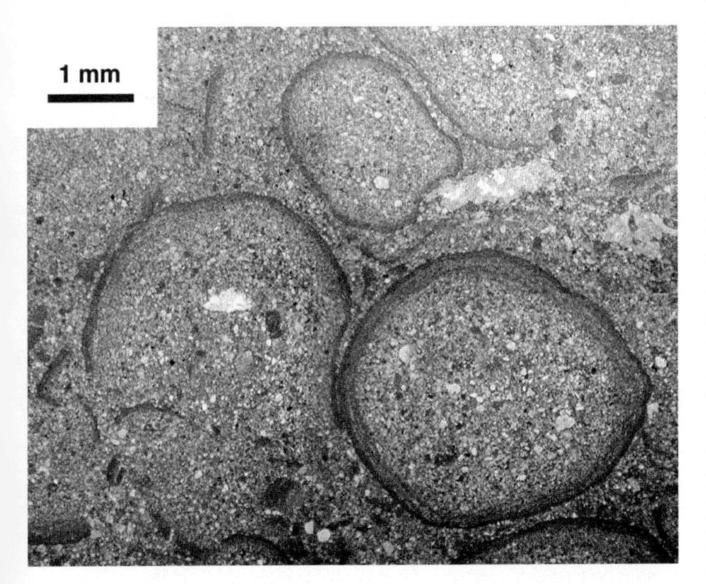

Figure 5. Photomicrograph of circular, deformed, and broken Alamo lapilli. Broken lapillis on left has exposed nucleus. Lapillis at top center is indented with fossil fragment in matrix. Broken crust fragments occur throughout matrix, which has same texture and composition as lapilli mantles.

We interpret the Alamo lapilli to have formed through the same processes as volcanic accretionary lapilli, which they closely resemble (e.g., Heiken and Wohletz, 1991). They occur in Alamo Breccia clasts with bedding characteristics of stratified silicate volcanic lapilli, including sorting and grading or reverse grading (e.g., Boulter, 1987; Ayres et al., 1991; Jones and Anhaeusser, 1993). Figure 8 shows silicate volcanic lapilli in a Devonian bed from the Carlin gold district of central Nevada. They structurally resemble Alamo lapilli (Fig. 1). Both display concentric structure and broken crusts suspended in matrix. The Alamo lapilli differ by the presence of nuclei and by their carbonate composition.

Warme and Kuehner (1998) used the term "carbonate spherules" for Alamo lapilli. Although a sphere is a "round body whose surface is at all points equidistant from the center," a spherule is "a body having the form of a sphere," and a spheroid is a solid geometrical figure "generated by rotating an ellipse about one of its axes," creating oblate or prolate versions (Random House Dictionary, 1967), these or closely related words have been given narrow definitions in the Earth sciences. For example, spheroid, spherule, spheruloid, and spherulite define specific shapes, compositions, and/or internal structures for spherical objects in sedimentary, igneous, or metamorphic rocks. At the urging of specialists who study lapilli and other spherical geologic objects, and at the risk of substituting a descriptive term for a genetic one, we have abandoned "spherule" and chosen "Alamo Breccia carbonate accretionary lapilli" or "Alamo lapilli" for brevity, to identify the unique spherical objects discussed herein.

Alamo lapilli are composed of three main layers. The literature on accretionary lapilli contains a diverse terminology for their internal layers and cover, e.g., nucleus, core, shell, layered shells, skin, cortex, rim, and armor, and some terms have been used for more than one feature. For descriptive purposes, and to separate our definitions from those used elsewhere, we use "nuclei" for lapilli that have central cores, "mantle" for accreted particles over the nucleus or in the body of nonnucleated lapilli, and "crust" for the outer shell. Proportions of these three elements are highly variable between Alamo

Figure 6. Rare, well-preserved segment of lapilli bed 5 m below top of Alamo Breccia at Hiko Hills South locality. Exposed bed is ~3 m long, as much as 70 cm thick, and is surrounded by sharp contact with common non-lapilli breccia clasts and matrix. Hammer is 35 cm long.

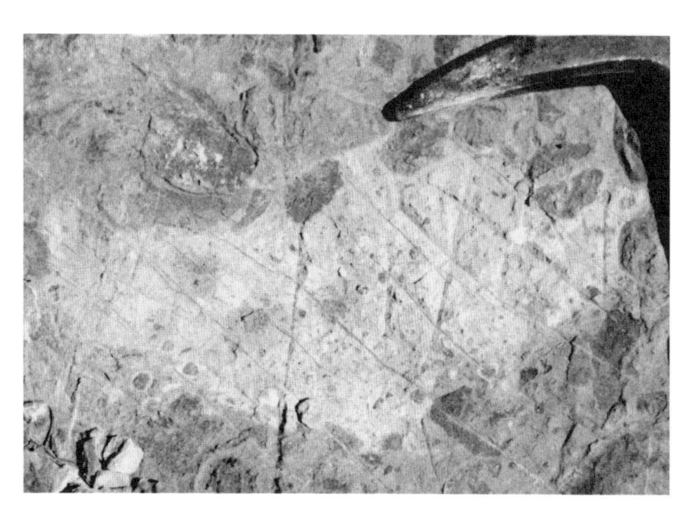

Figure 7. Light gray distorted lapilli mass in dark gray polymict tsunamite lithofacies A, ~6 m below top of Alamo Breccia at southwestern Delamar Mountains. Clast is ~30 cm across in longest dimension. Periphery is molded by penetration of breccia limestone clasts, demonstrating that mass was soft and deformed after deposition.

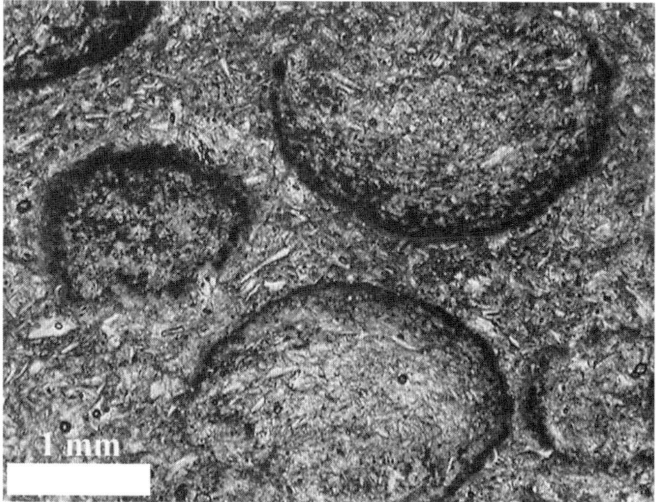

Figure 8. Volcanic accretionary lapilli from Devonian beds, Carlin Gold district, central Nevada. These lapilli are nonnucleated, but exhibit interiors that resemble matrix and have deformed and broken rims, similar to Alamo carbonate lapilli.

lapilli (Fig. 1), but they commonly occur in layers of individuals of similar size and character (Fig. 5). Alamo lapilli may or may not have distinctive nuclei. The obvious examples described here are all armored or mantled; i.e., they have well-developed crusts.

Impact lapilli

Lapilli reported from impact deposits can be silicate, which is the most common, or carbonate, or both (see following for examples). To our knowledge, no other carbonate accretionary impact lapilli have been described that are identical to Alamo lapilli, although we expect them to have formed at numerous sites over geologic time.

Spherule-shaped bodies have been described from impact sites such as the Miocene Ries crater in Germany (Graup, 1981; Newsom et al., 1990) and the Chicxulub crater in Mexico and Belize (Ocampo et al., 1996; Pope et al., 1999). Graup (1981) described and interpreted both glassy silicate lapilli and silicate accretionary lapilli from the Ries crater. He reasoned that concentric structure of clastic components, tangential arrangement of mineral fragments, and decreasing grain size from center to periphery all suggested that the Ries accretionary lapilli were formed by the same processes as volcanic lapilli. The silicate accretionary lapilli closely resemble the size and structure of the carbonate Alamo lapilli (Newsom et al., 1990). Alvarez et al. (1995) proposed a general model, and applied it to the Cretaceous-Tertiary (K-T) impact in Yucatan, whereby silicate spherules represent ejecta from the initial hot fireball. A secondary volatile-rich warm fireball hosts the formation of carbonate spherules. Pope et al. (1999) reviewed the literature on both silicate and carbonate impact accretionary spherules from several impact sites, and described accretionary dolomitic lapilli from the K-T boundary section at Albion Island, Belize. One of their published examples with an angular nucleus appears to be very similar to Alamo lapilli variants.

Alamo Breccia carbonate accretionary lapilli

Thus far the Alamo lapilli have been found in the Alamo Breccia at six general localities: Hancock Summit West, Mount Irish, Hiko Hills South, Hiko Hills North, western Pahranagat Range, and southwestern Delamar Mountains (Fig. 2). They occur imbedded in the polymict A lithofacies of the Alamo Breccia as broken and isolated beds of lapilli stone (Fig. 6), and plastically deformed bed masses (Fig. 7) that now occupy much less than 1% of the breccia volume.

Color. Alamo lapilli and spherule beds always weather to hues of very light gray or tannish gray, which contrasts with the wide range of colors from clasts of the enveloping A lithofacies of the Alamo Breccia. On weathered surfaces the Alamo lapilli are commonly lighter hued than their matrix. They are darker hued shades of gray or brownish gray on fresh surfaces. Diagenetically dolomitized beds are generally lighter gray than nonaltered limestones.

Composition. Alamo lapilli in the Alamo Breccia are almost entirely composed of carbonate mineral fragments. The obvious exceptions are the sparse but important shocked quartz grains described under the following heading. The Alamo lapilli were originally limestone. In some localities they remain limestone, and in others they are partially or entirely dolomitized. Paleozoic limestones in Nevada, including the Guilmette Formation and the Alamo Breccia Member within it, are dolomitized on many scales, including individual beds, horizontal, vertical, and inclined zones of varying thicknesses near faults or karsted surfaces, and whole mountain ranges. Some localities exhibit Alamo lapilli that vary from those with coarse calcite or faintly dolomitized nuclei to those with fully dolomitized nuclei, mantles, and crusts.

Iridium was detected in samples of combined Alamo lapilli and spherule matrix. Values of as much as 43 ± 13 parts per trillion (ppt) contrast with background values of ~10 ppt or less (C. Koeberl, 2000, personal commun.). Values near 100 ppt have been measured from the breccia matrix (Warme and Sandberg, 1995, 1996), but most matrix samples showed modest elevated levels or none.

Rare Ordovician conodonts were recovered from acidized bulk samples collected within the upper 10 m of the Breccia in several locations (Warme and Sandberg, 1995). In the Pahranagat Range, Morrow et al. (1998) collected lapilli bed samples and recovered conodonts of pre-Devonian age within them. The conodont fauna of this surprising occurrence is composed of >90% taxa as old as the Middle Ordovician. Consequently, they interpreted that sediments as deep as Ordovician age were ejected and incorporated into the lapilli beds.

Euhedral hematite grains in the Alamo lapilli and lapillistone matrix are interpreted as hydrothermal precipitates and mineral replacements. They occur in the fine-grained Alamo lapilli crusts and mantles, and rarely replace nuclei. Opaque grains that may or may not have similar origins are described in the centers of dolomite accretionary lapilli crystals at the Albion Island K-T boundary bed (Pope et al., 1999) and as unusual magnetite crystals in carbonatite beds from several locations (Deans and Seager, 1978). Similar-appearing crystals occur throughout the Alamo Breccia. They displacively stud the surfaces of shocked quartz grains dispersed in the Breccia, even occur in the grain interiors (Warme and Sandberg, 1995, 1996; Laroux et al., 1995), as well as throughout the Alamo lapilli and lapillistone matrix.

Quartz grains. Shocked quartz grains occur in three different settings related to the Alamo impact event: (1) as free grains in the matrix of the Alamo Breccia (Warme and Sandberg, 1995, 1996; Laroux et al., 1995), (2) as shocked grains in the Oxyoke Canyon Sandstone under the breccia in Zone 1 (Warme and Kuehner, 1998), and (3) as crystals in the isolated beds of Alamo lapilli reported here. Rounded quartz grains of siliciclastic origin are incorporated into the Alamo lapilli and also occur within the lapilli clast matrix (Fig. 9). The periphery of the quartz crystals is commonly damaged through dissolution and replacement by carbonate grains or the metallic oxide crystals described herein.

In a preliminary study, we documented that ~20% of the quartz grains from several Alamo lapilli clasts were shocked and exhibit planar deformation fractures (PDFs). The PDFs were measured, using a Universal Stage, to estimate pressures and temperatures at the time of lapilli formation. Applying relationships developed by Grieve et al. (1996), the results showed a range of values between weakly and strongly shocked grains. These values are represented by PDFs at {0001, 1012, 1121, 1122}, implying shock pressures in the range of 8–20 GPa. Refractive indices of these quartz crystals were also reduced by shock. According to Stoffler (1994), shock pressures of this magnitude would result in shock temperatures of ~1000 °C for porous rocks. Temperatures reached in shocked carbonates or wet carbonate target rock may be slightly lower.

Alamo lapilli structure. Alamo lapilli have a concentric structure that consists of a central nucleus of highly variable size and structure, a mantle of sand- to silt-sized grains, and a very fine grained roughly laminated crust (Figs. 1, 5, and 10). The relative proportions of these elements also vary, as shown in Figure 1. In some large Alamo lapilli, ~15–30 mm in maximum cross section, the mantles and crusts are thin and difficult to identify separately (Fig. 10).

Nuclei. Nuclei in Alamo lapilli have sharp boundaries. Some lapilli that appear nonnucleated may be exposed as small-circle cross sections away from the center. The nuclei range in size from subsand to fragments as much as 30 mm in longest dimension (Fig. 10). Larger fragments are always angular. Some nuclei are large enough to show rare macroscopic features such as bedding and fossils, but most are recrystallized to a mosaic of coarse calcite or dolomite crystals (Fig. 10). The nuclei show no evidence of plastic deformation, in contrast to the mantles and crusts.

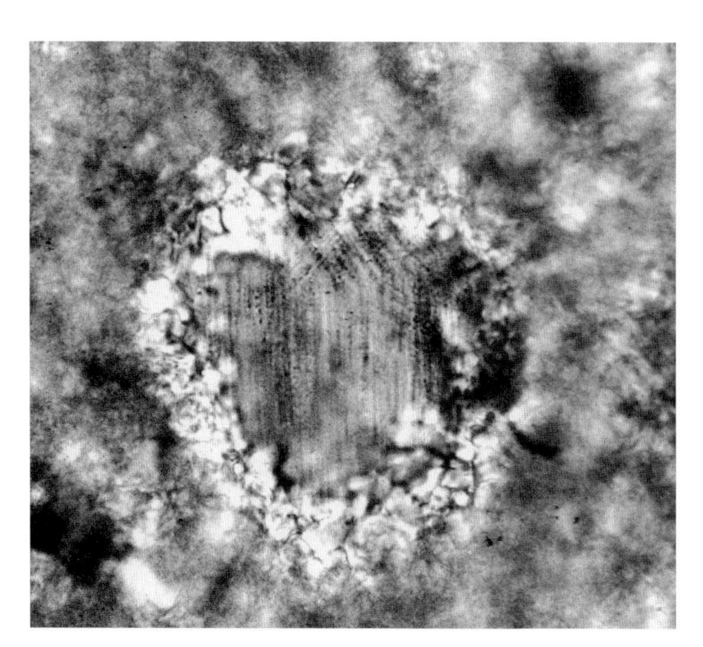

Figure 9. Shocked quartz crystal in fine-grained mantle of Alamo carbonate lapillus showing two strong directions of parallel deformation fractures and two or more weaker ones in this orientation. Grain is ~0.05 mm in diameter.

Figure 10. Very large lapillus with angular nucleus exposed on weathered surface of lapilli bed. Maximum diameters: nucleus, 20 mm; lapillus 23 mm.

Alamo lapilli beds and deformed bed masses commonly are studded with graded or irregularly dispersed angular rock fragments similar to lapilli nuclei (Fig. 11). These fragments range in diameter from a few millimeters to several centimeters. Apparently identical fragments, as much as 30 mm in exposed cross section, serve as lapilli nuclei in the same beds (Fig. 10). The isolated fragments are interpreted as nuclei of large Alamo lapilli that were stripped of their coats.

Mantles. Alamo lapilli mantles are composed of very fine grained sand- and silt-sized carbonate particles and very rare quartz grains. They commonly trend from coarser near the nucleus to finer radially outward in a single graded cycle (Figs. 1 and 5). Such outward grading is present in most examples, but is variably developed. The transition to the crust may be gradual or sharp, but the mantle is always finer grained near the crust. Rare Alamo lapilli display two cycles of grading, separated by a crust, indicating that they were recycled in some manner

Figure 11. Fragment of broken Alamo lapilli bed showing dark, angular clasts that are normally graded (smaller clasts upward) in this orientation. Many similar but smaller clasts are preserved as lapilli nuclei. Larger ones are nuclei of larger Alamo lapilli or appear to be former nuclei liberated from lapilli when they disaggregated, probably upon impact. They were heavier and accumulated first in basal portions of graded bed. Alamo lapilli with small or absent nuclei are scattered throughout bed, not graded by size. Scale in centimeters.

through the impact cloud. Alamo lapilli were preserved in various states of mantle cementation. Many were pliable (Fig. 5; see following discussion).

Crusts. In thin section the crusts are darker than mantles, and in some examples almost opaque, caused largely by very fine grained silt- to clay-sized particles. The crust is composed of a few to several concentric, discontinuous, alternations of silt- and clay-sized particles that give the appearance of fine lamellae. Crusts range in thickness from ~0.2 to 1 mm; they are typically <0.5 mm. Crusts appear more brittle than the underlying mantles, based on their tendency to crack and detach as discrete fragments that are commonly observed as matrix-supported clasts (Fig. 5) mixed with lapilli in the lapillistone beds.

Alamo lapilli sizes. Measurements of 191 Alamo lapilli were conducted on selected hand samples, thin sections, and outcrops from the Hiko Range, Western Pahranagat Range, Mount Irish Range, and the Delamar Mountains. They range in diameter from ~1 to 30 mm (average ~5 mm; Fig. 12), and show a slightly skewed distribution toward smaller lapilli. However, each bed or deformed bed mass appears to have its own array of lapilli sizes and relative proportions of nuclei, mantles, and crusts.

Alamo lapilli deformation and preservation. Generally the Alamo lapilli are circular in cross section, implying an original spherical three-dimensional shape. They also occur in various stages of deformation, breakage, and disintegration. Beds in different localities appear to have distinctive size-sorted layers and characteristic lapilli deformation such as uniform degrees of compaction. Figure 5 shows a lapilli-supported layer where some touching, unbroken individuals are deformed by adjoining ones, implying that both mantle and crust were pliable at the time of deposition and/or penecontemporaneous compaction. Other Alamo lapilli are broken, and fragments of brittle crusts and attached mantle appear to float in the bed matrix.

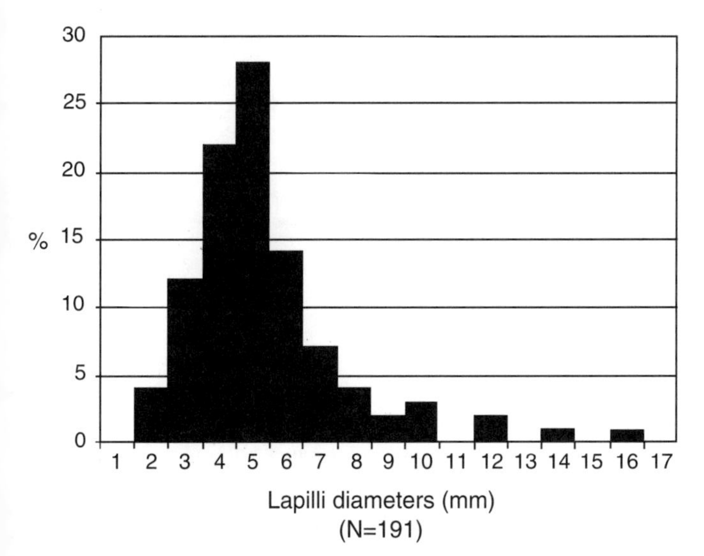

Figure 12. Histogram of 191 Alamo lapilli apparent diameters, measured in hand samples, thin sections, and on outcrops of Delamar Mountains, western Pahranagat Range, Hancock Summit North, Hiko Hills, and Mount Irish. Diameters have near normal distribution, with median and mode near 5 mm. Lapilli larger than 1 cm diameter are relatively rare.

Alamo lapilli matrix. The matrix of lapilli beds is highly variable in composition. It consists of a mixture of fine-grained particles, coarser grains, flame structures (see following), and debris embedded from the surrounding Alamo Breccia. Fine-grained matrix is commonly much more than 50% of the bed volume. Alamo lapilli can be touching (Fig. 5), dispersed, or rare to absent. We interpret that some Alamo lapilli were apparently destroyed by bombardment, dewatering, early compaction, or other processes, and became matrix of the same grain-size distribution and color as surviving Alamo lapilli (Figs. 1 and 5). The fine-grained matrix appears to represent nonaccretionary fallout, or flight-destroyed Alamo lapilli, from the impact plume. The matrix contains variable amounts of dispersed and rarely clast-supported, coarse pebble- to cobble-sized fragments. Their distribution appears disorganized or is roughly graded.

Alamo lapilli beds

Discovered fragments of Alamo lapilli beds are as much as 7 m long and 1 m thick (Fig. 6). Some of the beds have irregular and feathery edges, and underwent peripheral deformation via penetration of Alamo Breccia lithic clasts. Some beds were preserved as gently flexed, and others were squeezed and completely deformed between breccia clasts to become masses showing no original bed structure (Fig. 7). Most such occurrences are small pods <1 m in maximum exposed dimension (Fig. 7), but some stretched into taffy-like masses for 2 m or more, squeezed between Alamo Breccia clasts and matrix. These beds and distorted bed masses consist of whole, deformed, or broken Alamo lapilli, fine-grained matrix, and sand- and pebble-sized darker carbonate fragments (Fig. 11).

We do not know if one or several beds were originally deposited. If deposition of Alamo lapilli occurred more or less continuously, over perhaps several hours or a day, several beds may have accumulated between the multiple events that eroded and largely destroyed them. The variable degree of deformation suggests that some beds were more completely cemented and resistant during rip-up and transportation. Others were less cemented, ductile, and lost their original structure during transportation, deposition, and energetic penecontemporaneous dewatering, loading, and compaction. If only one bed was deposited, its lateral variability was significant because each relatively intact clast of an Alamo lapilli bed appears to have a different thickness and internal Alamo lapilli distribution. In either case, rapid cementation was necessary; such a process, allowing lapilli bed preservation, is described in the following.

The isolated clasts of Alamo lapilli in beds at several localities are graded (Fig. 11). Because the beds may have been inverted during transport, normal and/or reverse grading may have occurred, and reflect decreasing or increasing bursts of deposition from convective clouds (e.g., Fisher and Schmincke, 1984; Ayres et al., 1991).

Vertical bed distribution. Figure 3 shows an example of the stratigraphic distribution of Alamo lapilli beds in one locality, Hiko Hills South, where they occur as isolated clasts between ~5 and 20 m below the top of the breccia. Figure 6 shows one of the higher clasts at this location. At Hancock Summit West the Alamo Breccia is 55 m thick, and a deformed spherule mass is ~25 m beneath the top, in the lower part of the A lithofacies. Alamo lapilli have not been found for certain in the underlying B lithofacies near the base or in the uppermost finer grained graded bed at the top of the deposit.

"Flame" structures. Alamo lapilli beds contain enigmatic, sinuous, penetrative, carbonate ribbons. Some examples resemble common flame structures that occur in stratified rocks at bed interfaces, caused by loading and dewatering. The "flames" can penetrate deeply into the lapilli beds, but are more abundant around bed edges. They wrap around individual lapilli, as shown in the small-scale example of Figure 13. They are irregularly curved, appear stretched, and have tattered, serrated, or flame-like shapes (Warme and Kuehner, 1998). Some examples appear to carry surrounding polymict Alamo Breccia fragments into the lapilli beds. Their weathered color is dark gray to brown; fresh surfaces are much darker. In thin section, these structures are dark brown to opaque, fine outward, and contain hematite grains.

These features may be pathways created during the massive dewatering that probably occurred shortly after each depositional phase of the Alamo Breccia. In this interpretation the pathways penetrated into or through the spherule beds that blanketed and partially sealed early phases of Alamo Breccia deposition. They enveloped Alamo lapilli that were already hardened, and probably damaged and destroyed many of those that were not. Similar appearing soft-sediment deformation was described by Boulter (1987) in Archean subaqueous accretionary lapilli tuffs. They were disturbed by convolute and recumbent folding, and interpreted to be fluidized water-escape channels and forcefully injected clastic dikes.

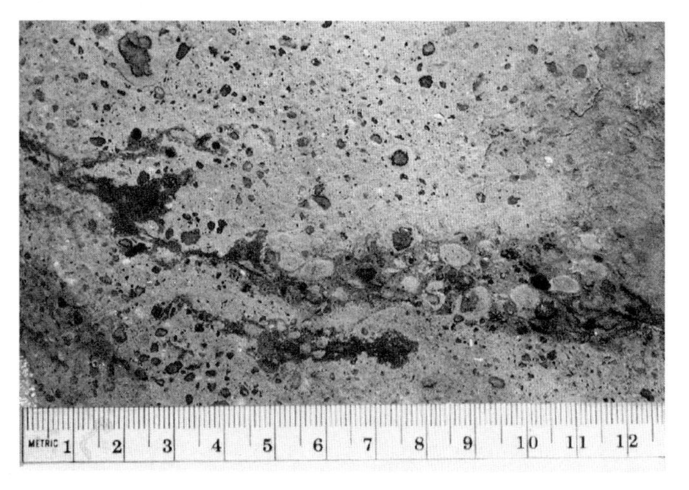

Figure 13. Detail of irregular "flame" structure penetrating lapilli bed. Darker flame matrix envelops clasts, individual lapilli, and one large circular lapillus that appears to have been better cemented than other distorted individuals. Width of photo is ~10 cm. Numbers on scale are centimeters.

The flame structures are still being investigated. An alternative interpretation is that they are carbonate equivalents of melt rock, deposited in a basal surge of melted ejecta from the impact. The attenuated stringers superficially resemble volcanic fiamme, and also resemble the stringers of siliceous melt rock in suevite of the Ries crater.

Genesis of Alamo lapilli and lapilli beds

Lapilli formation. Moore and Peck (1992) described the formation of siliceous volcanic lapilli (Fig. 8) that are structurally mimicked by the Alamo lapilli. The process they described is paraphrased in the following and applied to the proposed genesis of Alamo lapilli.

1. A hot eruptive volcanic cloud of ash and steam cools, causing condensation of water vapor from the vent and surrounding air. In Nevada, the marine Alamo Impact Event created the eruptive cloud.

2. Condensed moisture causes adhesion and rapid agglutination of ash, forming cores of accretionary lapilli. Such cores are present in some Alamo lapilli, but the nuclei are more commonly angular fragments of recrystallized carbonate rock interpreted to be derived from damaged carbonate target formations.

3. Rapid formation results in relatively coarse and unsorted ash particles in the cores and absence of concentric structure and tangential mineral orientation. In the Alamo lapilli this stage is represented by the generally disorganized inner parts of lapilli mantles.

4. More ash sticks to the falling, moist, embryonic lapilli, adhering along the largest grain surfaces and resulting in tangential arrangement of long axes (Fig. 8). Alamo equivalents are the outwardly graded mantle layers, and the very fine grained crust.

5. Rising temperatures in lower parts of the cloud reduce relative humidity and result in slower and more selective accretion of progressively finer ash. This process may account for the outward fining of mantle particles in most Alamo lapilli.

6. Multiple thin layers in the outer shell may be produced by falling through varying cloud densities or cycling through turbulent convection cells. Alamo equivalents are the crusts and the rare lapilli showing multiple mantle-crust cycles.

Preservation of accretionary lapilli

Most lapilli are volcanic particles with no characteristic structure. Structured accretionary lapilli require moisture to form. They are usually not armored (i.e., do not have crusts). However, armored lapilli are regarded as more durable and are most commonly preserved in ancient lapilli beds (e.g., Boulter, 1987; Reimer, 1983a, 1983b; Ayers et al., 1991; Jones and Anhaeusser, 1993). Possibly both types accumulated in the Alamo lapilli beds, but only the armored individuals survived or can be differentiated from the lapillistone matrix.

Reimer (1983a), Gilbert and Lane (1994), and Schumacher and Schmincke (1995) provided theory and models for the formation of silicate accretionary lapilli. Reimer (1983a) listed and explained five forces that operate to form accretionary lapilli and give them strength: electrostatic attraction, capillary pressure of pore fluids, van der Waals forces, crystallization of dissolved matter, and sintering at high temperatures. He believed that electrostatic attraction and capillary pressure were the most important forces, and that armor was formed when lapilli took a final hotter and dryer path through the impact cloud.

Both Gilbert and Lane (1994) and Schumacher and Schmincke (1995) designed innovative physical laboratory experiments to create lapilli. Gilbert and Lane concluded that growth was controlled mainly by collision and adhesion of liquid-coated particles, and by binding of grains from tension forces and secondary mineral growth. Schumacher and Schmincke (1995) showed that volcanic ash agglomeration was most effective where water content was between 15 and 25 wt% of the water-particle mixture. They believed that the most important processes for holding accreted grains together were capillary forces that create liquid bridges and electrostatic attraction. In their experiments particles >500 μm commonly acted as nuclei, and grains <350 μm were accreted around them. These sizes compare favorably with those of many Alamo lapilli.

Minerals such as sodium chloride and calcium sulfate can precipitate rapidly as lapilli desiccate during flight (Gilbert and Lane, 1994). As outlined here, we believe that calcium carbonate also precipitates from impact moisture trapped between grains wherever target rock is carbonate, such as in the southern Nevada Alamo impact area. Salt water and carbonate rock are envisioned to provide elements for almost instantaneous early cementation of Alamo lapilli. Most important, the results of recent experiments (Agrinier et al., 2001) strongly support our scenario for in-flight cementation of Alamo lapilli by $CaCO_3$.

Hardening of Alamo lapilli and lapilli beds

It has been difficult to comprehend how Alamo lapilli could be deposited in a bed that developed enough strength, within minutes or hours, to survive reworking by the violent postimpact events that formed the Alamo Breccia. The process of industrial calcining and creation of commercial concrete may provide an analogy that solves this problem and may apply to the hardening of individual lapilli, both in flight and after landing, and to rapid strengthening of the lapilli beds.

Calcining is the process of heating calcium carbonate at temperatures below its vapor point to drive off volatile components. In a heated kiln, calcium carbonate is transformed into quicklime (CaO) and carbon dioxide (CO_2) (Manahan, 1994; Karhela, 1996):

$$CaCO_3 \text{ (solid)} + \text{heat (impact)} \rightarrow CaO \text{ (solid)} + CO_2 \text{ (gas)}. \quad (1)$$

Industry standard temperatures for calcining in reaction 1 are ~900 °C. In the Alamo event these temperatures were likely exceeded, corroborated by the estimates from PDFs in shocked quartz grains of the Alamo Breccia that yield temperatures of ~1000 °C. Carbonates do not melt at these temperatures. The wet impact target area provided abundant liquid water and steam. In an exothermic reaction quicklime (CaO) reacts with water to form portlandite ($Ca[OH]_2$):

$$CaO \text{ (solid)} + H_2O \text{ (liquid)} \rightarrow Ca(OH)_2 \text{ (solid)} + \text{heat}. \quad (2)$$

During the Alamo Event this reaction occurred in the impact plume as it began to cool. With continued cooling the pliable mixture combined with previously released carbon dioxide (CO_2) to form cemented calcium carbonate ($CaCO_3$):

$$Ca(OH)_2 \text{ (solid)} + CO_2 \rightarrow CaCO_3 \text{ (solid)} + H_2O \text{ (liquid)}. \quad (3)$$

This reaction can occur quickly, and must be retarded with continued mixing and additives during commercial application. This process could explain how partially cemented and still pliable Alamo lapilli formed during the Alamo event and how, upon precipitation, they solidified quickly enough to withstand reworking by ensuing tsunamis. Impact pressure pulverized target rock, and impact heat decarbonitized and dehydrated it to form quicklime. This chemical process may cause the distinctive light color of the Alamo lapilli. During cooling of the impact plume the reaction was reversed and the quicklime reacted with water and carbon dioxide. Wet particles accreted around existing larger nuclei to form porous, pliable Alamo lapilli. Upon hydration they quickly began to cement and harden.

Agrinier et al. (2001) documented that the back-reaction to $CaCo_3$ can occur in less than 200 seconds in the temperature range of 573–973 K. Thus, the formation and growth of minerals such as sodium chloride and calcium sulfate during the flight of silicate lapilli can be augmented in carbonate lapilli with the rapid formation of strong-binding calcium carbonate. Cementation and hardening of entire beds of carbonate lapilli may continue after deposition and rapidly create a durable lapillistone capable of withstanding the following tsunamis and/or other energetic processes that reworked the Alamo Breccia.

SYNTHESIS

We propose that the Alamo impact excavated and pulverized carbonates and minor quartz sandstones beneath the Paleozoic water-saturated carbonate platform and ejected them into the atmosphere where the Alamo lapilli formed. We have divided the formation and deposition of the carbonate Alamo lapilli and lapilli beds into two phases (Fig. 14). Although speculative, this model is consistent with our current knowledge about the Alamo Breccia.

Phase 1: Formation and precipitation of Alamo lapilli

When the bolide struck, target carbonates were launched ballistically and formed a high-velocity ejecta curtain that crossed the carbonate platform and deposited material preserved as the basal lithofacies of the Alamo Breccia. Some of the target carbonates were pulverized, heated, decompressed, and dehydrated to form quicklime in the vapor-rich impact plume rising far above the impact site, duplicating the warm fireball scenario of Alvarez et al. (1995). During this process the formation of the Alamo lapilli is analogous to that of volcanic accretionary lapilli. The rising impact plume cooled, causing condensation of moisture and rapid accretion of wet particles that formed mantles around nuclei or provided cores for nucleus-free Alamo lapilli. Significant cementation occurred as quicklime reacted with water to form portlandite, which reacted with carbon dioxide to form calcium carbonate.

As the Alamo lapilli fell toward the base of the plume, higher temperatures and lower humidity reduced the size of the adhering particles, causing outward grading and development of durable crusts that armored some lapilli. Updrafts and tur-

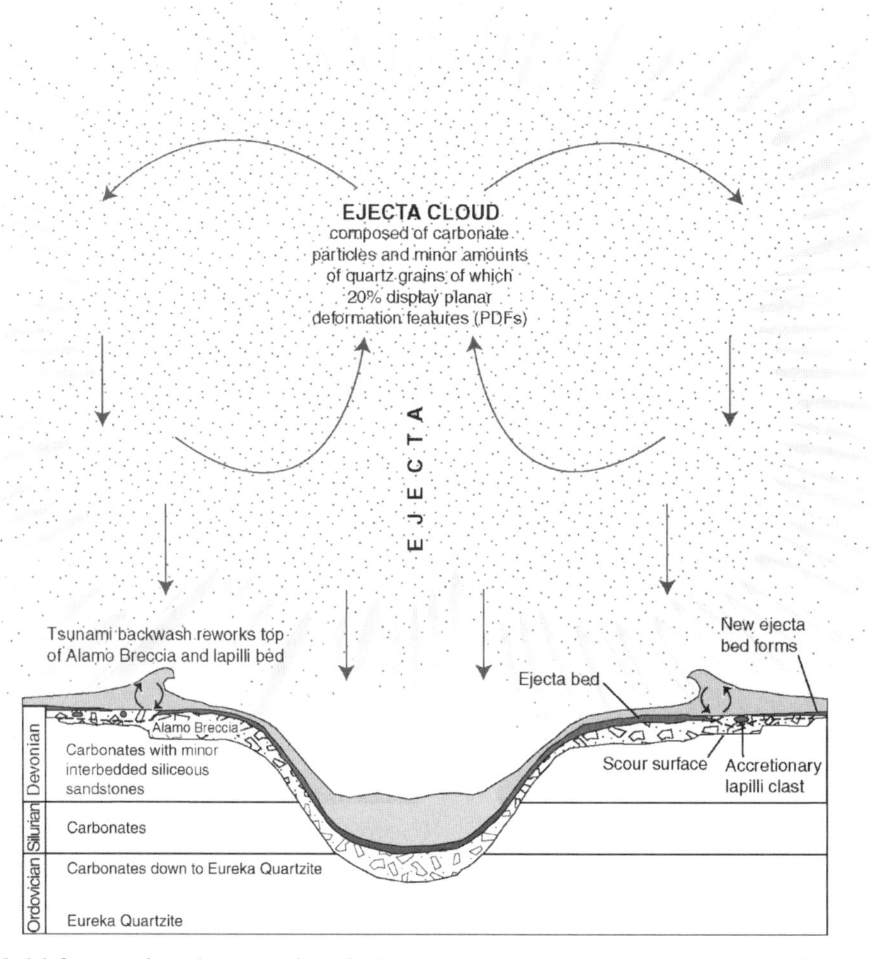

Figure 14. Model for genesis and preservation of Alamo carbonate accretionary lapilli and lapilli beds preserved in Alamo Breccia. During phase 1, the Alamo Impact excavated rocks from Devonian carbonate platform, penetrating through carbonate formations and one thin quartzose sandstone down to Middle Ordovician levels. Quartz 2 grains were deformed with parallel deformation features (PDFs). Pulverized particles were ejected in vapor-rich impact plume. In flight, particles accreted, preferably around larger nuclei, to form carbonate Alamo lapilli that precipitated as continuous lapilli beds. During phase 2, beds were quickly cemented before ensuing tsunamis fragmented and reworked them into tsunamite facies of Alamo Breccia, where they were preserved.

bulence in the plume could have recirculated some Alamo lapilli or passed them through different grain sizes in different parts of the cloud so that particles accumulated in outward-graded cycles. Alamo lapilli with larger nuclei but a missing mantle rose lower in the plume, where very fine grained particles adhered as a thin crust that directly covered the nuclei. Cooling and hydration began in the cloud, and continued upon precipitation and bed formation across the initial Alamo Breccia ejecta blanket. Alternatively, owing to the potential for $CaCo_3$ formation within 200 seconds (Agrinier et al., 2001), the lapilli beds may have formed in the impact base surge.

Shocked quartz grains were also ejected, indicating the maximum depth of excavation below the Guilmette Formation. The sparse quartz grains in the Alamo lapilli, lapilli beds, and Alamo Breccia were probably derived from the thin, earlier Devonian Oxyoke Canyon Sandstone, which occurs ~400 m below the Guilmette and contains shocked quartz grains in outcrop in Zone 1. No fragments of the distinctive milky-white Ordovician Eureka Quartzite have been identified in the breccia, limiting the maximum crater excavation to 1000–1500 m below the Guilmette Formation.

Phase 2: Reworking of Alamo lapilli and lapilli beds

The spherule bed or beds were not preserved as continuous strata. They may have been significantly loaded, thinned, partitioned, or otherwise altered by ongoing settling, dewatering, and related processes soon after deposition. They were then reworked within minutes to hours after phase 1. Most of the bed(s) disintegrated and the Alamo lapilli were destroyed and incorporated into the matrix of the tsunamite facies, where they are now largely unidentifiable. However, rapid cementation of the lapilli beds indurated them enough for fragments to survive reworking and to be incorporated as lapilli clasts into the polymict A lithofacies of the Alamo Breccia.

It is probable that precipitation of the Alamo lapilli from the impact plume commenced shortly after the impact and that spherule fallout was more or less continuous for hours. Thus far Alamo lapilli have been found only above the lowest scour surface that was created by the first tsunami. They may have accumulated below this level but were destroyed by early base surge, secondary bombardment, and bed movement, and later massive settling, compaction, and dewatering. We favor a scenario in which lapilli fallout began after the B lithofacies was deposited. The absence of Alamo lapilli in the muddy facies at the very top of the breccia indicates that fallout ceased before the last graded bed of the A lithofacies was deposited.

CONCLUSIONS

The distinctive Alamo carbonate accretionary lapilli and their enclosed grains of shocked quartz provide certainty that the Alamo Breccia is an impact deposit. The Alamo lapilli were shaped under conditions similar if not identical to those that form silicate pyroclastic accretionary lapilli, except that the physical conditions were caused by a bolide impact rather than by a volcanic explosion, and the Alamo lapilli are carbonate, which can harden more quickly than silicate. Particles composing the Alamo lapilli were pulverized by impact pressure and calcined by impact heat. The particles were moistened, accreted to nuclei to form carbonate lapilli, cemented to various degrees in flight, and then deposited over wide areas near the impact. Carbonate cement formed by devolitization of target rock to create quicklime, and rapid back-reaction to calcite that strengthened the beds so they withstood violent reworking by multiple impact-generated tsunamis and/or other processes. The result is the isolated bed fragments of distinctive carbonate lapillistone that occur within the Alamo Breccia.

ACKNOWLEDGMENTS

The work reported here was supported by grants to Kuehner from the German Academic Exchange Program, the Field Research Grant program of the Geological Society of America, and the Department of Geology and Geological Engineering at the Colorado School of Mines. Initial work on the Alamo Breccia was funded through grants from UNOCAL and National Science Foundation grant EAR-9106324.

We thank all the individuals that have propelled our understanding of the Alamo Breccia through their observations and lively discussions in the field: Alan Chamberlain for his hands-on support; Charles Sandberg and Jared Morrow for sharing their geologic knowledge of Nevada; David Roddy, Fred Hörz, and Don Lowe for crucial remarks and observations in the field; Fred Meissner for contributing ideas on carbonate cementation of the Alamo lapilli; and John Skok for rock preparation and thin sections.

Bruce Simonson, Scott Hassler, Fred Hörz, and Ken MacLeod provided valuable critiques of the draft manuscript. Simonson, Hassler, and Richard Wendtland directed us to significant literature on accretionary lapilli.

REFERENCES CITED

Agrinier, P., Deutsch, A., Schärer, U., and Martinez, I., 2001, Fast back-reactions of shock-released CO_2 from carbonates: An experimental approach: Geochimica et Cosmochimica Acta, v. 65, p. 2615–2632.

Alvarez, W., Claeys, P., and Kieffer, S.W., 1995, Emplacement of Cretaceous-Tertiary boundary shocked quartz from Chicxulub Crater: Science, v. 269, p. 930–935.

American Geological Institute, 1980, Glossary of Geology (second ed.): Falls Church, Virginia, American Geological Institute.

Ayres, L.D., Van Wagoner, N.A., and Ferreira, W.S., 1991, Voluminous shallow-water to emergent phreatomagmatic basaltic volcanoclastic rocks, Proterozoic (~1886 MA) Amisk Lake composite volcano, Flin Flon Greenstone Belt, Canada, *in* Fisher, R.V., and Smith, G.A., eds., Sedimentation in volcanic settings: SEPM (Society for Sedimentary Geology) Special Publication No. 45, p. 175–187.

Blair, T.C., and McPherson, J.G., 1999, Grain-size and textural classification of coarse sedimentary particles: Journal of Sedimentary Research, v. 69, p. 6–19.

Boulter, C.A., 1987, Subaqueous deposition of accretionary lapilli: Significance for palaeoenvironmental interpretations in Archaean greenstone belts: Precambrian Research, v. 34, p. 231–246.

Chamberlain, A.K. 1999, Structure and Devonian stratigraphy of the Timpahute Range, Nevada [Ph.D. thesis]: Golden, Colorado School of Mines, 344 p.

Cook, H.E., 1983, Introductory perspectives, basic carbonate principles, and stratigraphic and depositional models, *in* Platform margin and deepwater carbonates: Society of Economic Paleontologists and Mineralogists Short Course No. 12, p. 1:1–1:89.

Deans, T., and Roberts, B., 1984, Carbonatite tuffs and lava clasts of the Tinderet foothills, western Kenya: A study of calcified natrocarbonatites: Journal of the Geological Society of London, v. 141, p. 563–580.

Deans, T., and Seager, A.F., 1978, Stratiform magnetite crystals of abnormal morphology from volcanic carbonatites in Tanzania, Kenya, Greenland, and India: Mineralogical Magazine, v. 42, p. 463–475.

Fisher, R.V., and Schmincke, H.-U., 1984, Pyroclastic rocks: Berlin, Springer-Verlag, 472 p.

Fisher, R.V., and Smith, G.A., eds., 1991, Sedimentation in volcanic settings: Tulsa, Oklahoma, SEPM (Society for Sedimentary Geology) Special Publication No. 45, 257 p.

Gilbert, J.S., and Lane, S.J., 1994, The origin of accretionary lapilli: Bulletin of Volcanology, v. 56, p. 398–411.

Graup, G., 1981, Terrestrial chondrules, glass lapilli and accretionary lapilli from the suevite, Ries Crater, Germany: Earth and Planetary Science Letters, v. 55, p. 407–418.

Grieve, R.A., Langenhorst, F., and Stoffler, D., 1996, Shock metamorphism of quartz in nature and experiment: II. Significance in geoscience: Meteoritics and Planetary Science, v. 31, p. 6–35.

Heiken, G., and Wohletz, K., 1991, Fragmentation processes in explosive volcanic eruptions, *in* Fisher, R.V., and Smith, G.A., eds., Sedimentation in volcanic settings: SEPM (Society for Sedimentary Geology) Special Publication No. 45, p. 19–26.

Jones, I.M., and Anhaeusser, C.R., 1993, Accretionary lapilli associated with Archaean banded iron formations of the Kraaipan Group, Amalia greenstone belt, South Africa: Precambrian Research, v. 61, p. 117–136.

Karhela, T., 1996, Dynamic simulation model of rotary lime kiln [M.Sc. thesis]: Tampere University of Technology, Tampere, Finland.

Keller, J., 1989, Extrusive carbonatites and their significance, *in* Peryt, T., ed., Coated grains: Berlin, Springer-Verlag, p. 70–88.

Kuehner, H.-C., 1997, The Late Devonian Alamo impact breccia, southeastern Nevada [Ph.D. dissertation]: Golden, Colorado School of Mines, 321 p.

Laroux, H., Warme, J.E., and Doukhan, J.C., 1995, Shocked quartz in the Alamo breccia, southern Nevada: Evidence for a Devonian impact event: Geology, v. 23, no. 11, p. 1003–1006.

Manahan, S.E., 1994, Environmental Chemistry [6th edition]: Boca Raton, Florida, Lewis Publishers, 811 p.

Moore, J.G., and Peck, D.L., 1992, Accretionary lapilli in volcanic rocks of the western continental United States: Journal of Geology, v. 70, p. 182–193.

Morrow, J.R., and Sandberg, C.A., 2001, Distribution and characteristics of multi-sourced shock-metamorphosed quartz grains, Late Devonian Alamo impact, Nevada: Houston, Lunar and Planetary Institute Contribution No. 1080, 2 p.

Morrow, J.R., Sandberg, C.A., Warme, J.E., and Kuehner, H.-C., 1998, Regional and possible global effects of sub-critical Late Devonian Alamo impact event, southern Nevada, USA: Journal of the British Interplanetary Society, v. 51, p 451–460.

Morrow, J.R., Sandberg, C.A., and Poole, F.G., 2001, New evidence for deeper water site of Late Devonian Alamo impact, Nevada: Houston, Lunar and Planetary Institute Contribution No. 1080, 2 p.

Newsom, H.E., Graup, G., Iseri, D.A., Geissman, J.W., and Keil, K., 1990, The formation of the Ries Crater, West Germany: Evidence of atmospheric interactions during a larger cratering event, *in* Sharpton, V.L., and Ward, P.D., eds., Global catastrophes in Earth history: Geological Society of America Special Paper 247, p. 198–206.

Oberbeck, V.R., Marshall, J.R., and Aggarwal, H., 1993, Impacts, tillites, and the breakup of Gondwanaland: Journal of Geology, v. 101, p. 1–19.

Ocampo, A., Pope, K.O., and Fischer, A.G., 1996, Ejecta blanket deposits of the Chicxulub crater from Albion Island, Belize, *in* Ryder, G., Fastovsky, D., and Gartner, S., eds, The Cretaceous-Tertiary Event and other catastrophes in Earth history: Geological Society of America Special Paper 307, p. 75–88.

Poole, F.G., and Sandberg, C.A., 1977, Mississippian paleogeography and tectonics of the western United States, *in* Stewart, J.H., Stevens, C.H., and Fritsche, A.E., eds., Paleozoic paleogeography of the western United States: Society of Economic Paleontologists and Mineralogists, Pacific Section, Pacific Coast Paleogeography Symposium 1, p. 67–85.

Pope, K., Ocampo, A.C., Fischer, A.G., Alvarez, W., Fouke, B.W., Webster, C.L., Vega, F.J., Smit, J., Fritsche, A.E., and Claeys, P., 1999, Chicxulub impact ejecta from Albion Island, Belize: Earth and Planetary Science Letters, v. 170, p. 351–364.

Random House Dictionary of the English Language, Unabridged, 1967: New York, Random House.

Reimer, T.O., 1983a, Accretionary lapilli in volcanic ash falls: Physical factors governing their formation, *in* Peryt, T., ed., Coated grains: Berlin, Springer-Verlag, p. 56–68.

Reimer, T.O., 1983b, Accretionary lapilli and other spheroidal rocks from the Archaean Swaziland Supergroup, Barberton Mountain Land, South Africa, *in* Peryt, T., ed., Coated grains: Berlin, Springer-Verlag, p. 619–634.

Sandberg, C.A., Morrow, J.R., and Warme, J.E., 1997, Late Devonian Alamo impact event, Global Kellwasser events, and major eustatic event, Eastern Great Basin, Nevada and Utah, *in* Link, P.K., and Kowallis, B.J., Proterozoic to recent stratigraphy, tectonics, and volcanology, Utah, Nevada, Southern Idaho and Central Mexico: Brigham Young University Geology Studies, v. 42, pt. 1, p. 129–160.

Sandberg, C.A., Poole, F.G., and Johnson, J.G., 1989, Upper Devonian of western United States, *in* McMillan, N.J., Embry, A.F., and Glass, D.J., eds., Devonian of the world, v. 1, Regional synthesis: Canadian Society of Petroleum Geologists, Memoir 14, p. 183–220.

Schumacher, R., and Schmincke, H.-U., 1995, Models for the origin of accretionary lapilli: Bulletin of Volcanology, v. 56, p. 626–639.

Stoffler, D., 1984, Glasses formed by hypervelocity impact: Journal of Noncrystal solids, v. 67, p. 465–502.

Warme, J.E., and Chamberlain, A.K., 2000, Primary ejecta, tsunami reworking, tectonic dismemberment: Reconstructing the Late Devonian Alamo Breccia and crater, Nevada, *in* Catastrophic events and mass extinctions: Impacts and beyond: Houston, Lunar and Planetary Institute Contribution No. 1053, p. 238.

Warme, J.E., and Kuehner, H.-C., 1998, Anatomy of an anomaly: The Devonian catastrophic Alamo impact breccia of southern Nevada: International Geological Review, v. 40, 189–216.

Warme, J.E., and Sandberg, C.A., 1995, The catastrophic Alamo Breccia of southern Nevada: Record of a Late Devonian extraterrestrial impact: Courier Forschungsinstitut Senckenberg, v. 188, p. 31–57.

Warme, J.E., and Sandberg, C.A., 1996, Alamo megabreccia: Record of Late Devonian impact in southern Nevada: GSA Today, v. 6, no. 1, p. 1–7.

Manuscript Submitted December 1, 2000; Accepted by the Society March 22, 2001

Printed in the U.S.A.

Geological Society of America
Special Paper 356
2002

Continental Triassic-Jurassic boundary in central Pangea: Recent progress and discussion of an Ir anomaly

Paul E. Olsen*
Lamont-Doherty Earth Observatory of Columbia University, Route 9W, Palisades, New York 10964-8000, USA
Christian Koeberl
Heinz Huber
Institute of Geochemistry, University of Vienna, Althanstrasse 14, A-1090 Vienna, Austria
Alessandro Montanari
Osservatorio Geologico do Coldigioco, I-62020 Frontale di Apiro, Italy
Sarah J. Fowell
Department of Geology and Geophysics, University of Alaska, Fairbanks, Alaska 99775-5780, USA, and Lamont-Doherty Earth Observatory of Columbia University, Route 9W, Palisades, New York 10964-8000, USA
Mohammed Et-Touhami
Faculty of Sciences, Department of Geology, Université Mohamed Premier, Oujda, Morocco, and Lamont-Doherty Earth Observatory of Columbia University, Route 9W, Palisades, New York 10964-8000, USA
Dennis V. Kent
Department of Geological Sciences, Rutgers University, Piscataway, New Jersey 08854-8066, USA, and Lamont-Doherty Earth Observatory of Columbia University, Route 9W, Palisades, New York 10964-8000, USA

ABSTRACT

The Triassic-Jurassic (Tr-J) boundary marks one of the five largest mass extinctions in the past 0.5 b.y. In many of the exposed rift basins of the Atlantic passive margin of eastern North America and Morocco, the boundary is identified as an interval of stratigraphically abrupt floral and faunal change within cyclical lacustrine sequences. A comparatively thin interval of Jurassic strata separates the boundary from extensive overlying basalt flows, the best dates of which (ca. 202 Ma) are practically indistinguishable from recent dates on tuffs from marine Tr-J boundary sequences. The pattern and magnitude of the Tr-J boundary at many sections spanning more than 10° of paleolatitude in eastern North America and Morocco are remarkably similar to those at the Cretaceous-Tertiary boundary, sparking much debate on the cause of the end-Triassic extinctions, hypotheses focusing on bolide impacts and climatic changes associated with flood basalt volcanism.

Four prior attempts at finding evidence of impacts at the Tr-J boundary in these rift basin localities were unsuccessful. However, after more detailed sampling, a modest Ir anomaly has been reported (up to 285 ppt, 0.29 ng/g) in the Newark rift basin (New York, New Jersey, Pennsylvania, United States), and this anomaly is directly associated with a fern spike. A search for shocked quartz in these rift basins has thus

*E-mail: polsen@ldeo.columbia.edu

Olsen, P.E., Koeberl, C., Huber, H., Montanari, A., Fowell, S.J., Et-Touhami, M., and Kent, D.V., 2002, Continental Triassic-Jurassic boundary in central Pangea: Recent progress and discussion of an Ir anomaly, *in* Koeberl, C., and MacLeod, K.G., eds., Catastrophic Events and Mass Extinctions: Impacts and Beyond: Boulder, Colorado, Geological Society of America Special Paper 356, p. 505–522.

far been fruitless. Although both the microstratigraphy and the biotic pattern of the boundary are very similar to continental Cretaceous-Tertiary boundary sections in the western United States, we cannot completely rule out a volcanic, or other non-impact, hypothesis using data currently available.

INTRODUCTION

The Triassic-Jurassic (Tr-J) boundary (ca. 202 Ma) marks one of the five largest mass extinctions of the Phanerozoic (Sepkoski, 1997), arguably at least as large in magnitude as that at the much better known Cretaceous-Tertiary (K-T) boundary (Fig. 1). The extinctions occurred in a hothouse world during a time of extremely high CO_2 (Ekart et al., 1999) and the existence of the supercontinent of Pangea. In continental environments, the Tr-J extinctions mark the end of a regime dominated by nondinosaurian tetrapods and the beginning of the dinosaurian dominance that would last the succeeding 135 m.y.

Pangean rift basins developed during the Middle to Late Triassic along a broad zone from Greenland through the Gulf of Mexico in the ~40 m.y. preceding the Jurassic opening of the central Atlantic Ocean. Many of these rift basins preserve a detailed record of the Tr-J boundary in mostly continental environments characterized by relatively high sedimentation rates (Fig. 2). In this chapter we report on recent progress in documenting the biotic transition around the Tr-J boundary, including the recently reported evidence for an associated Ir anomaly, and the relationship between these boundary events and the Central Atlantic magmatic province.

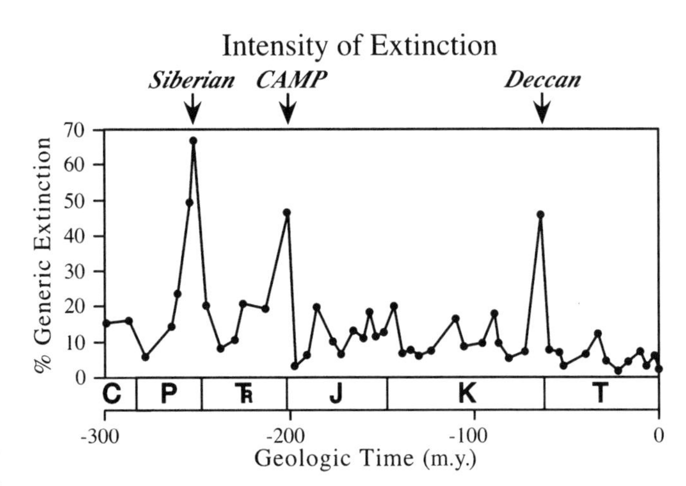

Figure 1. Generic-level extinctions of marine, shelly organisms during past 300 m.y., and distribution of giant flood basalt provinces (in italics). Modified from Sepkoski (1997) to reflect our estimate of age of Triassic-Jurassic boundary. CAMP is Central Atlantic magmatic province.

IDENTIFICATION OF THE CONTINENTAL TR-J BOUNDARY

Rift basins of central Pangea developed largely in a continental milieu. Comparison of the Tr-J boundary in the marine realm with that in continental environments is largely based on studies of Alpine marine sections correlated to other European, mostly continental, areas by palynomorphs and scant marine invertebrates (Schulz, 1967; Brugman, 1983; Beutler, 1998). Cornet and others (Cornet et al., 1973; Cornet and Traverse, 1975; Cornet, 1977; Cornet and Olsen, 1985) first identified the Tr-J boundary in the tropical Pangean basins, most notably in the Jacksonwald syncline of the Newark basin. This palynological work was augmented by subsequent studies in the Newark and other basins in eastern North America by Fowell (1994; Fowell et al., 1994; Fowell and Traverse, 1995; Olsen et al., 1990).

Within eastern North America, Olsen and Galton (1977, 1984), and Olsen and Sues (1986) identified a transition in the terrestrial tetrapod assemblages that coincides, albeit at a coarser level, with that seen in the palynomorphs. Additional vertebrate paleontological studies have refined the correlation of the Tr-J boundary in these continental rifts to the marine

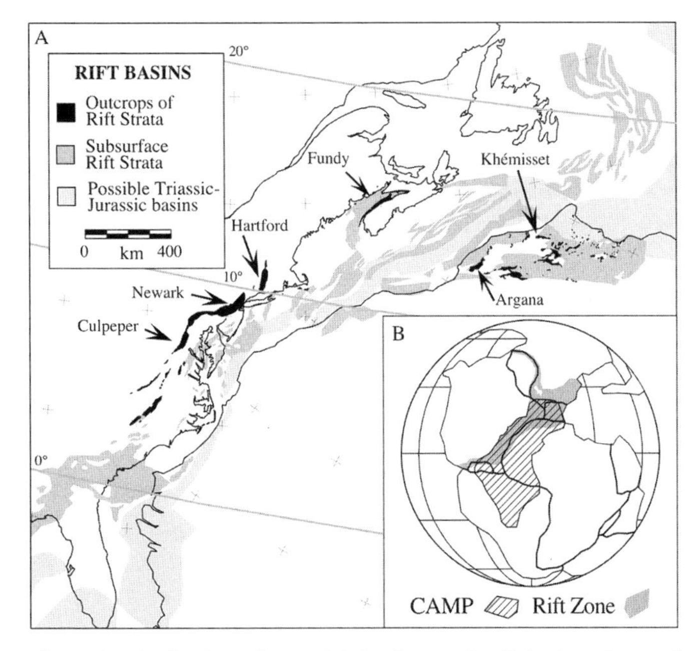

Figure 2. Distribution of central Atlantic margin rift basins of central Pangea and temporal distribution of Central Atlantic magmatic province (CAMP). A: Rift basins of central Atlantic margins of North America and western Africa in predrift coordinates (for Late Triassic) showing basins discussed in this chapter (modified from Olsen, 1997). B: Pangea during earliest Jurassic showing distribution of CAMP, overlapping much of Triassic-Jurassic rift zone (based on Olsen, 1999).

realm, in which very rare terrestrial vertebrates are found (Olsen et al., 1987; Huber et al., 1996; Lucas, 1998).

Two significant problems in the biostratigraphy of the Tr-J boundary are the interpretation of negative evidence and the presence of distinct biotic provinces in the early Mesozoic. We believe that the Tr-J boundary is marked by a catastrophic mass extinction. As in the case of the Cretaceous-Tertiary boundary, we expect the immediate postboundary biotic assemblages to consist of survivor taxa, not new appearances. However, there is a long tradition of identification of the Tr-J boundary on the basis of new appearances, notably the first appearance of the ammonite *Psiloceras planorbis* in marine sections (e.g., Page and Bloos, 1998), and this approach has been carried over to the palynological analyses (e.g., Morbey, 1975).

It is arguable that global correlation of a boundary marked by a mass extinction with a very rapid and global cause using the first appearances of taxa is inappropriate. The boundary is more appropriately marked by the last appearances of taxa and not the appearance of new taxa that might appear thousands if not millions of years in the study area after the event.

A second problem with traditional biostratigraphy across the boundary is a very strong climatic gradient (Kent and Olsen, 2000b). The central European sections that form the basis of much of the palynological and vertebrate biostratigraphy were located ~2000 km north of the tropical Pangean basins of North America on the opposite (northern) side of the northern subtropical arid belt (Kent and Muttoni, 2001; Olsen and Kent, 2000). The marked floral and faunal provinciality, probably related to the Pangean climatic belts, was reviewed by Cornet and Olsen (1985) and Olsen and Galton (1984), but the possible effects of this very strong climatic gradient on floral assemblages, particularly palynomorphs, has been either ignored or discounted in attempts at long-distance correlation.

Reliance on first appearances and the discounting of the effects of climatic gradients have led some workers to cite the absence of certain critical palynological taxa characteristic of western Europe and the abruptness of the transition as evidence of a major hiatus at the Tr-J boundary, especially in eastern North America and Greenland (e.g., Pedersen and Lund, 1980; van Veen et al., 1995; Tourani et al., 2000). It is critical to point out that there is no physical stratigraphic or sedimentologic evidence of such a hiatus in these areas; detailed cyclostratigraphic and magnetostratigraphic evidence indicates continuous deposition across the palynologically identified boundary, especially where it has been examined in the most detail in the Newark basin (Kent et al., 1995; Olsen et al., 1996a, 1996b, 2002a, 2002b).

There is no question that there are significant differences between the stratigraphic distribution of palynomorph taxa in tropical Pangean basins and in western Europe. This difference shows up very obviously in the concurrence of the rise of abundance of *Corollina* (*Classopollis*) and persistence of *Patinasporites densus* in eastern North America over the last 15 m.y. of the Triassic (Cornet, 1977); this pattern has no described European counterpart. The abrupt disappearance of *Patinasporites densus* is a signature of the Tr-J boundary in tropical Pangea. If the same kind of typological biostratigraphic philosophy were applied to the western North American continental K-T boundary, a major hiatus would be required. It is significant that the K-T boundary also exhibits dramatic climate-related floral provinciality (Sweet et al., 1990).

Radiometric dating of the Tr-J boundary provides support for the correct identification of the Tr-J boundary in the tropical Pangean rifts. Ages from lavas (and associated intrusive feeders; e.g., Ratcliffe, 1988) just above the palynological Tr-J boundary in eastern North America provide U-Pb ages of 201.3 ± 1 Ma (Gettysburg sill, Dunning and Hodych, 1990), 200.9 ± 1 Ma (Palisade sill, Dunning and Hodych, 1990), and 201.7 ± 1.3 Ma (North Mountain Basalt, Hodych and Dunning, 1992) and $^{40}Ar/^{39}Ar$ ages of 202.2 ± 1, 200.3 ± 1.2, and 201.2 ± 1.3 Ma (Palisade sill and Culpeper basin plutons, respectively; Sutter, 1988).

Until recently, these have been the only ages directly applicable to the boundary; however, Palfy et al. (2000a, 2000b) provided a date of 199.6 ± 0.3 Ma for a tuff just below the marine Tr-J boundary in British Columbia. Palfy et al. (2000a) argued that the difference between their marine date and the continental dates from eastern North America implies that the continental extinctions occurred prior to the marine extinctions. However, the differences in ages are very small, and thus we regard all these dates as indistinguishable, given reasonable, but unstated, geological and interlaboratory uncertainties. Furthermore, recent dates on the Orange Mountain Basalt of 201 ± 2.1 Ma (Hames et al., 2000) and its feeder, the Palisade sill, of 201 ± 0.6 Ma (Turrin, 2000), both from the Newark basin, are indistinguishable from the data cited by Palfy et al. (2001a). The similarity between the marine and continental dates provides powerful support for the age of the Tr-J boundary being ca. 200 Ma, which is substantially younger than in recent time scales (e.g., Harland et al., 1990; Gradstein et al., 1994; Palfy et al., this volume).

More precise dating of the boundary requires interlaboratory cross-calibration and a better understanding of the systematics of the K-Ar system in tholeiites than currently exists (e.g., Turrin, 2000). We favor a date of 202 Ma for the boundary, based on the assumption that the average of available Newark igneous dates is close to the middle of the extrusive section, and adding the duration of the older half of the interval based on Milankovitch cyclostratigraphy, then rounding to the nearest 0.5 m.y. (Olsen et al., 1996b). We believe that realistic geological uncertainties for the Newark-based radiometric boundary dates are ~2 m.y.

NEWARK BASIN BOUNDARY SECTIONS

Of all of the Tr-J Pangean rifts, the stratigraphy of the Newark basin is arguably the best known, due to more than 100 yr of field work, extensive scientific coring by the Newark Basin Coring Project (NBCP) (Olsen et al., 1996a), geotechnical coring by the Army Corps of Engineers (ACE) (Fedosh and Smoot, 1988), and petroleum industry exploration. Virtually the entire >5 km section has been recovered with redundancy in the existing >10 km of continuous core. In addition, the continental Jurassic boundary is better known in the Newark basin than elsewhere. In particular, the boundary is best known in the Jacksonwald syncline in the southwestern part of the basin (Fig. 3), because of ongoing commercial real-estate development and rich fossil content.

It has long been known that much of the Newark basin section is composed of cyclical lake deposits (Van Houten, 1962), and with the recovery of the NBCP and ACE cores, the cyclicity of virtually the entire section has been described (Olsen et al., 1996a, 1996b) (Fig. 4). Van Houten ascribed the hierarchical cyclicity in the Lockatong and Passaic Formations of the Newark basin to lake-level fluctuations controlled by

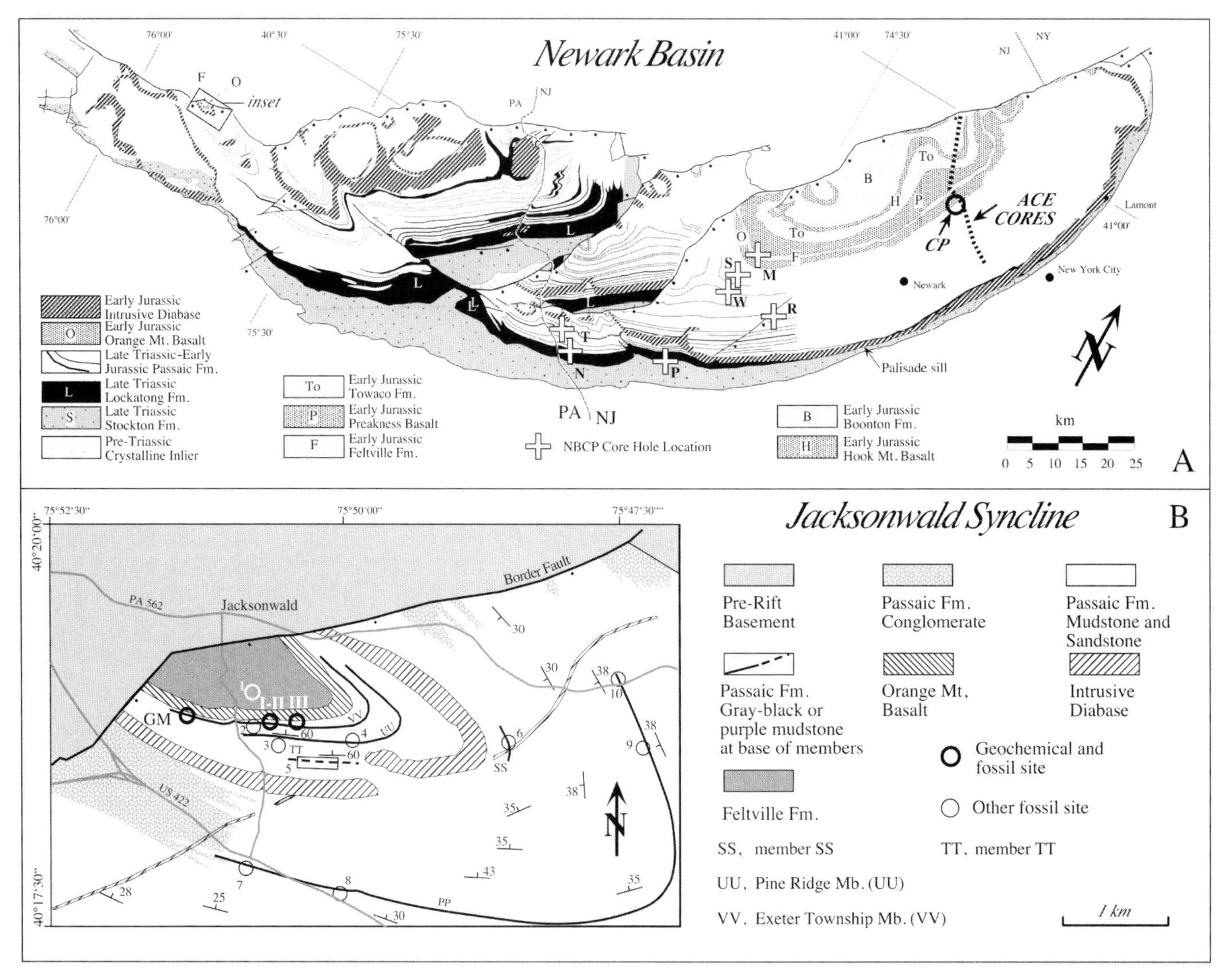

Figure 3. Position of key sections and cores within Newark basin of New Jersey, New York and Pennsylvania. A: Newark basin. Cores of Newark Basin Coring Project (NBCP): M, Martinsville; N, Nursery; P, Princeton; R, Rutgers; S, Somerset; T, Titusville; W, Weston. CP indicates Clifton-Paterson area and area in box outlines Jacksonwald syncline map of B. B: Map of Jacksonwald syncline showing positions of sections discussed in text. Locations for four sections shown in Figure 8 are as follows: GM, Grist Mills (lat 40°18′85″, long 075°51′20″); I, section I (lat 40°18′76″, long 075°50′56″); II, section II (lat 40°18′76″, long 075°50′55″); III, section III (lat 40°18′81″, long 075°50′38″). Other paleontological localities are: 1, Exeter Golf Course Estates (Feltville Formation locality for *Eubrontes giganteus*); 2, original palynological boundary sections of Cornet (1977) and Fowell (1994); 3, Wingspread footprint locality of Szajna and Silvestri (1993); 4, Pine Ridge Creek locality for pollen and footprints; 5, Pathfinder Village bone assemblage in member TT (discovery site at lat 40°18′55″, long 075°50′10″); 6, Walnut Road phytosaur tooth locality in member SS; 7, Shelbourne Square (Ames) footprint locality of Szajna and Silvestri (1993) and Silvestri and Szajna (1993); 8, Heisters Creek development footprint locality; 9, Tuplehocken Road footprint locality; 10, pollen localities OLA1 and OLA3 of Cornet (1977). Specific latitude and longitude coordinates not given here are listed in Olsen et al. (2001b).

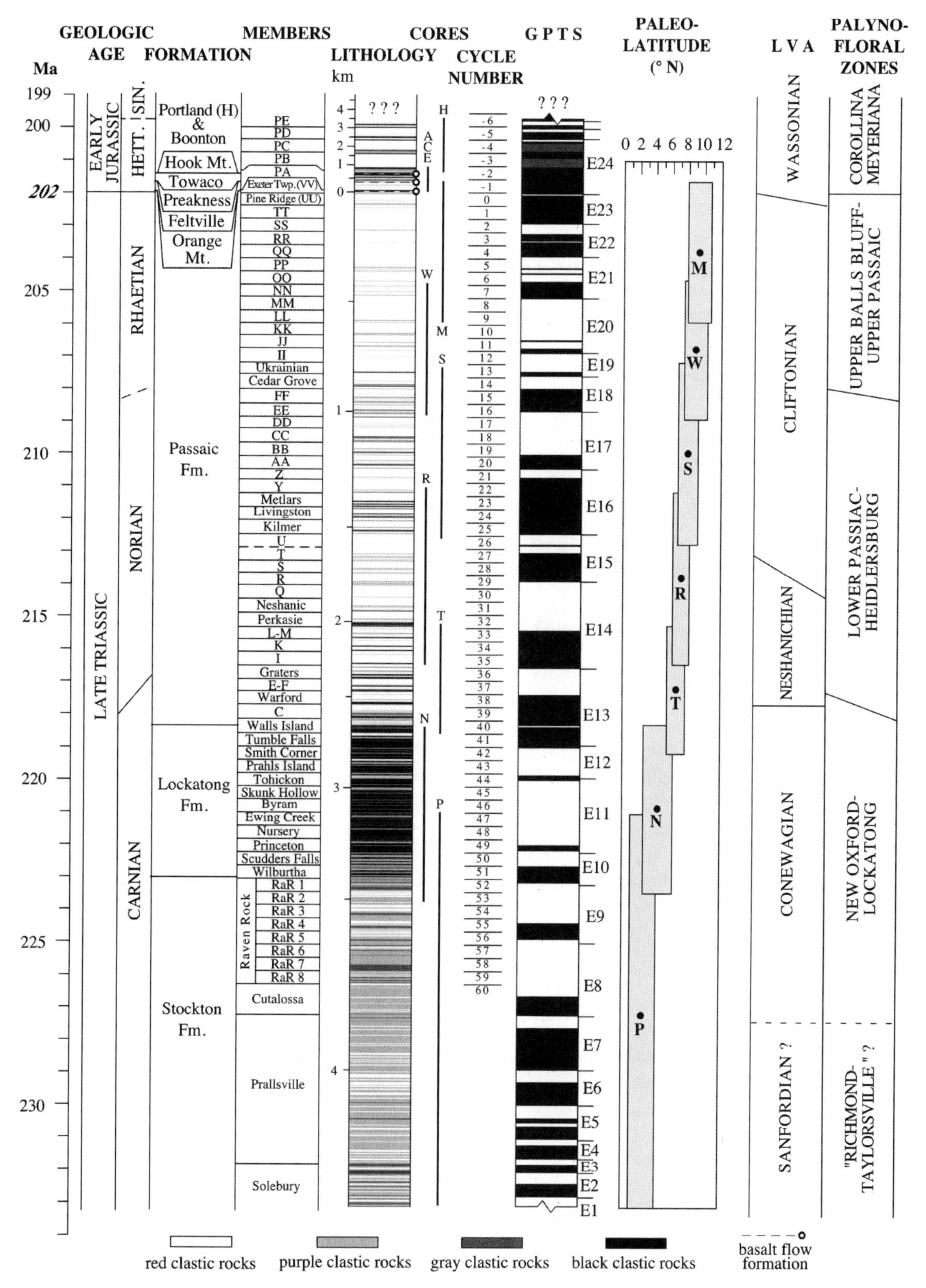

Figure 4. Time scale for Late Triassic and Early Jurassic based on geomagnetic polarity time scale (GPTS) and astronomical calibration from Newark Basin Coring Project (NBCP) Kent and Olsen, 1999a; Olsen and Kent, 1999), Army Corps of Engineers (ACE) cores (Olsen et al., 1996), and preliminary results from Hartford basin (Kent and Olsen, 1999b). Biostratigraphic data are from Huber et al. (1996), Lucas and Huber (2002), Cornet (1977), and Cornet and Olsen (1985). For GPTS, black is normal polarity, white is reversed polarity, and gray represents intervals for which there is incomplete sampling (Hartford basin section only). H, Hartford basin section; HETT, Hettangian; L.V.A., Land Mammal Ages; M, Martinsville (NBCP) core; N, Nursery (NBCP) core; P, Princeton (NBCP) cores; R, Rutgers (NBCP) cores; S, Somerset (NBCP) cores; SIN., Sinemurian; T, Titusville (NBCP) core; W, Weston Canal (NBCP) cores. Cycle number refers to 404 k.y. cycle of eccentricity with lines placed at calculated minima.

Milankovitch climate cycles (Van Houten, 1962, 1964, 1969), and his interpretation has proven applicable to the entire cyclical lacustrine sequence (Olsen et al., 1996a; Olsen and Kent, 1996, 1999). Similarly, the magnetostratigraphy of the Newark basin section has been determined from cores and outcrops (Kent et al., 1995). The magnetostratigraphy, Milankovitch cyclostratigraphy, and radiometric dates from the lavas together have provided the basis for an astronomically tuned time scale for the Late Triassic (Kent and Olsen, 1999a, 2000a) and earliest Early Jurassic (Olsen et al., 1996b; Kent and Olsen, 1999b) (Fig. 4).

Recent work on outcrops in the Hartford basin (Fig. 1) has resulted in a preliminary magnetostratigraphy and cyclostratigraphy that allows astronomical calibration of the 2 m.y. of section postdating the Jurassic lava flows in the Hartford basins that are precisely correlative with those in the Newark basin (Fig. 4) (Kent and Olsen, 1999b). This interval includes the oldest reversed polarity zones in the Early Jurassic, providing a tie to marine sections in the Paris basin as well as an upper bound to the normal polarity zone enclosing the exposed Central Atlantic magmatic province lavas.

The Jacksonwald syncline of the Newark basin has an unusually thick latest Triassic–earliest Jurassic section, marked by accumulation rates greater than anywhere else in the basin, and capped by the Orange Mountain Basalt and Feltville Formation (Fig. 3). The cyclicity and paleontological richness of the latest Triassic age sections here are better developed than anywhere else in the Newark basin. A series of largely temporary exposures created for houses over the past 15 yr have allowed detailed paleontological and cyclostratigraphic analyses of the uppermost Passaic Formation, including the Tr-J boundary (Figs. 5–7), and it is these sections that we concentrated on for the new geochemical analyses.

The boundary section exhibits strongly cyclical sediment variation; well-developed gray and black shales occur periodically (in terms of thickness) in the Milankovitch pattern typical of Newark basin lacustrine sequences. In general, gray strata produce pollen and spores and thus the boundary section is better defined biostratigraphically here than elsewhere (e.g., northern Newark basin). Over the years boundary sections have been exposed over a distance of ~2 km along strike, the easternmost (i.e., most basinward) exposures being the finest grained and having the highest proportion of gray strata. All of the Newark basin boundary sections are characterized by laterally consistent lithostratigraphy and biostratigraphy, despite the lateral changes in facies and accumulation rate (Fig. 5).

A very prominent gray shale is ~25–30 m below the Orange Mountain Basalt (Figs. 5 and 8). This unit contains palynoflorules that are dominated (60%) by *Patinasporites densus* with variable amounts (5%–20%) of *Corollina*, and other pollen and spores. This is typical of many older Late Triassic palynoflora assemblages from the Jacksonwald syncline. Above this gray shale are variegated red and gray shales and sandstones that have a lower abundance (5%–40%) of *Patinasporites densus*. Another prominent marker bed, a brown to blue-gray sandstone with abundant comminuted charcoal and wood, which we refer to as the blue-gray sandstone bed (Figs. 5, 6, and 8), is 8–12 m below the Orange Mountain Basalt. This unit generally is directly above a thin (1–10 cm) coal bed or carbonaceous shale. The highest stratigraphic occurrence of *Patinasporites densus* occurs ~1 m below the sandstone (Fig. 9). This palynoflorule (sample 6-2 of Fowell, 1994) is dominated by *Corollina* (73%) and ~10% *Patinasporites*. This is the highest assemblage that we regard as having a Triassic-aspect palynomorph assemblage.

In the 40 cm below the blue-gray sandstone, palynomorph assemblages are consistently dominated by trilete spores belonging to taxa usually attributed to ferns (Fig. 9) (first discovered by Litwin, cited *in* Smith et al., 1988). Within a couple of centimeters of the base of the blue-gray sandstone, the proportion of spores reaches a maximum of 80% (Fowell et al., 1994). *Patinasporites densus* is absent from these assemblages, and we consider these assemblages to be of earliest Jurassic age. Such fern-dominated assemblages are unknown elsewhere in Newark basin strata of Triassic age. We refer to this very anomalous high proportion of fern spores as a fern spike, in parallel with the terminology used for the K-T boundary (e.g., Tschudy et al., 1984; Nichols and Fleming, 1990). It is within this fern spore–rich interval that the Ir anomaly occurs (see following). Above the blue-gray sandstone, palynoflorules are dominated by *Corollina; Patinasporites* and other Triassic-type taxa are absent. The character of these assemblages is similar to that of the Jurassic strata overlying the basalts throughout the Newark Supergroup, and we consider them to indicate a Jurassic age for the uppermost Passaic Formation. Thus, in analogy to the K-T boundary, we hypothesize that the fern spike at least approximates the base of the Jurassic, and hence the Tr-J boundary (Fig. 7). On the basis of the cyclostratigraphically defined accumulation rate, the last Triassic aspect palynoflorule occurs, conservatively, within 25 k.y. of the base of the Orange Mountain Basalt. The interval of time between this Triassic assemblage and the lowest definitive Jurassic assemblage (Jb-6 of Cornet, 1977) is less than 10 k.y. Similarly, the Tr-J boundary, as defined here by the fern spike, occurs within 20 k.y. of the base of the Orange Mountain Basalt.

Reptile footprint taxa broadly follow the same pattern as the palynological stratigraphy as described in Olsen et al. (2002b). Footprint faunules have been recovered at many levels within the Jacksonwald syncline and there is a concentration of productive levels near the palynologically identified Tr-J boundary (Silvestri and Szajna, 1993; Szajna and Silvestri, 1996; Szajna and Hartline, 2001; Olsen et al., 2002b). Assemblages in rocks of Triassic age contain abundant *Brachychirotherium* (suchian) and more rare *Apatopus* (phytosaur), a form informally designated "new taxon A" of Szajna and Silvestri (1996) (suchian), as well as *Rhynchosauroides* (lepidosauromorph), *Batrachopus* (crocodylomorph suchian), a form referred to "new taxon B" (Szajna and Silvestri, 1996) (su-

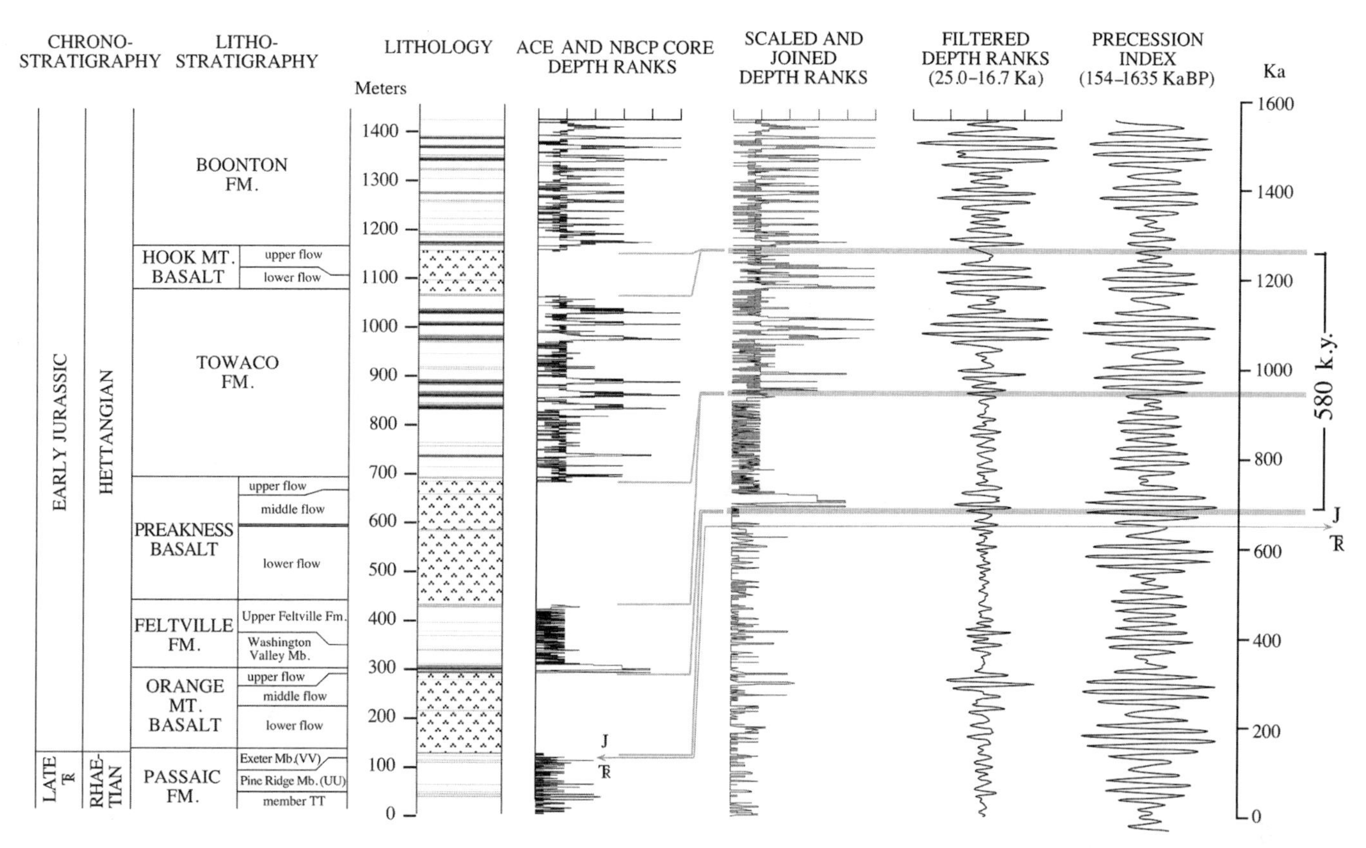

Figure 5. Cyclostratigraphic calibration of Triassic-Jurassic (Tr-J) boundary and succeeding extrusive zone flows and interbedded and overlying sedimentary strata (adapted from Olsen et al., 1996b). Depth ranks are numerical classification of sedimentary facies sequences in order of increasing interpreted relative water depth (Olsen and Kent, 1996). Comparison of depth rank curves with arbitrary segment of precession index curve indicates that it is not necessary to assume that any significant time is represented by lava flow formations and that entire flow sequence was probably deposited during interval of <600 k.y. Note also that Tr-J boundary (correlated to Jacksonwald syncline by magnetostratigraphy and lithostratigraphy; Fig. 7) is ~20 k.y. below Orange Mountain basalt. Depth rank record from strata above Preakness Basalt is based on Army Corps of Engineers Army Corps of Engineers (ACE) cores, whereas that from Passaic and Feltville Formations is based on Martinsville no. 1 core of Newark Basin Coring Project (NBCP).

chian), and abundant small- to medium-sized dinosaurian tracks usually referred to as various species of *Grallator* and *Anchisauripus* (i.e., *Eubrontes* spp. in the terminology of Olsen et al., 2002b). *Brachychirotherium, Apatopus*, and "new taxon B" have never been found in strata of Jurassic age, despite the global abundance of Early Jurassic footprint assemblages. The highest footprint assemblage with *Brachychirotherium* and "new taxon B" occurs ~11 m below the blue-gray sandstone and the fern spike (Fig. 5). Even closer to the boundary is a poorly sampled footprint-bearing level with *Rhynchosauroides, Batrachopus*, and *Grallator* and *Anchisauripus* (*Eubrontes* spp.) that is ~7 m below the blue-gray sandstone and fern spike. Although the localities are thus far not very productive, only *Grallator* and *Anchisauripus* (*Eubrontes* spp.) have been found above the blue-gray sandstone in the Jacksonwald syncline.

Abundant tetrapod bones occur in a zone ~400 m below the Orange Mountain Basalt in the Jacksonwald syncline in member TT (Fig. 3). This interval has produced numerous skeletal remains, including skulls and articulated skeletons of the procolophonid parareptile *Hypsognathus fenneri*, and the crocodylomorph cf. *Protosuchus*, as well as other as yet unidentified remains, including probable phytosaur teeth. These bone occurrences are ~800 k.y. older than the Tr-J boundary, and help define the ranges of Triassic-type taxa.

Correlation of the Jacksonwald syncline sections with the Newark basin cores is fairly straightforward, despite the muted cyclicity in the uppermost Passaic Formation in the cores. A thin but well-defined interval of reversed polarity (E23r) occurs ~17 m below the Orange Mountain Basalt in the Martinsville no. 1 core (Kent et al., 1995; Kent and Olsen, 1999a). This reversed interval is between two very thick normal polarity intervals (E23n and E24n). Magnetic polarity chron E23r has also been identified in the Jacksonwald syncline section (Olsen et al., 1996a). The pattern of prominent gray beds in the Jacksonwald syncline sections and their relationship to chron E23r are matched very closely by the relationship between very thin gray and purple bands and chron E23r in the Martinsville no. 1 core. This laterally repeated pattern permits precise outcrop and core

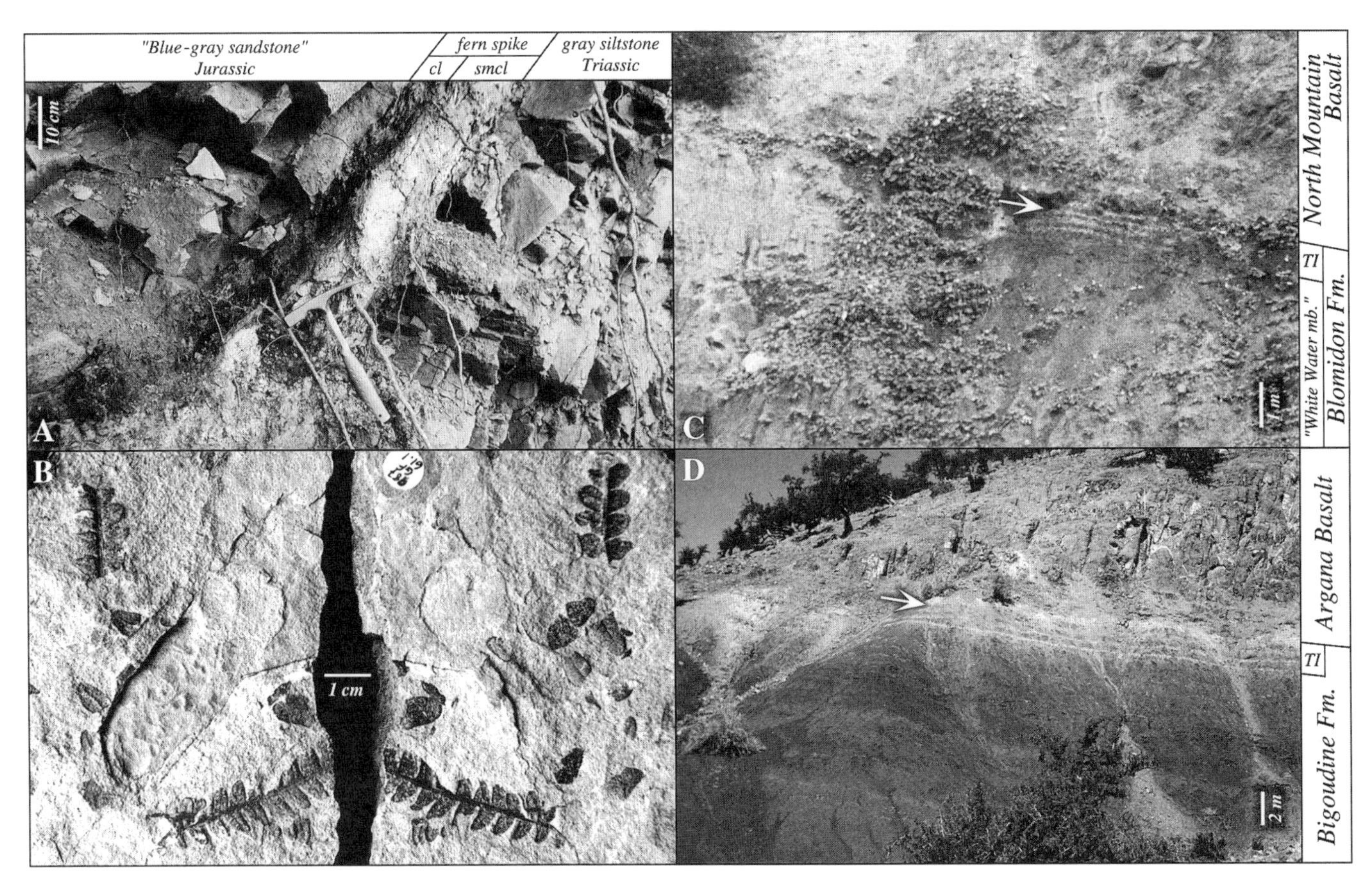

Figure 6. Photographs of sections and ferns of Triassic-Jurassic (Tr-J) boundary. A: Tr-J boundary at section I (Figs. 2 and 7). cl, coal and carbonaceous shale; smcl, smectitic claystone. Note that strata dip 60° to north (left). B: Part and counterpart of slab bearing fronds of fern *Cladophlebis* from Tr-J boundary of Fundy basin at Central Clarence (Nova Scotia Provincial Museum no. 982.GF.G1.1). C: Tr-J boundary at Partridge Island, Nova Scotia (studied by Fowell and Traverse, 1995); TI indicates palynological transition interval and arrow shows position of palynological Tr-J boundary. D: Tr-J boundary near Argana, Morocco, in Argana basin (Olsen et al., 2000).

correlation (Fig. 5) (Olsen et al., 1996a), in particular tying the Jacksonwald sections to the stratigraphy in the northern Newark basin, where the cyclostratigraphy of the bulk of the Jurassic section has been established by study of the ACE cores (Fig. 5) (Olsen et al., 1996b).

Exposures in quarries and at construction sites in the vicinity of Paterson and Clifton, New Jersey, in the same areas in which the ACE cores were drilled, have produced a series of important fossil assemblages within the upper Passaic Formation. Although there is no magnetic stratigraphy for this area of the Newark basin section, the pattern of purple intervals in the ACE cores matches that seen in the Martinsville no. 1 core, allowing lithostratigraphic correlation. On the basis of this correlation, the Martinsville no. 1 core and the ACE cores appear to have nearly the same accumulation rate for the uppermost Passaic Formation.

The uppermost few meters of the Passaic have produced an enormous number of footprints from a variety of exposures. These footprint assemblages contain only *Rhynchosauroides, Batrachopus*, and small to large *Grallator, Anchisauripus*, and *Eubrontes* (*Eubrontes* spp.). As reported in Olsen et al. (2002b), these assemblages contain the oldest examples of *Eubrontes giganteus*, a dinosaurian track ~20% larger than any older ichnospecies. At one locality within the footprint-bearing sequence, a gray lens of sandstone and shale has produced a macroflora dominated by the conifers *Brachyphyllum* and *Pagiophyllum*, and the fern *Clathropteris meniscoides*, as well as a poorly preserved palynoflorule dominated by *Corollina* and lacking Triassic-type taxa. This footprint and plant assemblage is thus of earliest Jurassic age, an interpretation supported by its stratigraphic position compared to the Martinsville no. 1 core (Fig. 5).

In close proximity to these footprint and plant localities are the exposures that yielded skeletal remains of *Hypsognathus fenneri*, including the holotype specimen (Sues et al., 2000). These occurrences are ~45 m below the Orange Mountain Basalt. On the basis of correlation with the Martinsville no. 1 core and the Jacksonwald syncline sections, these represent the youngest examples of typical Triassic osteological taxa in eastern North America. On the basis of cyclostratigraphy, the *Hypsognathus*-bearing horizons are within 500 k.y. of the palynologically defined Tr-J boundary.

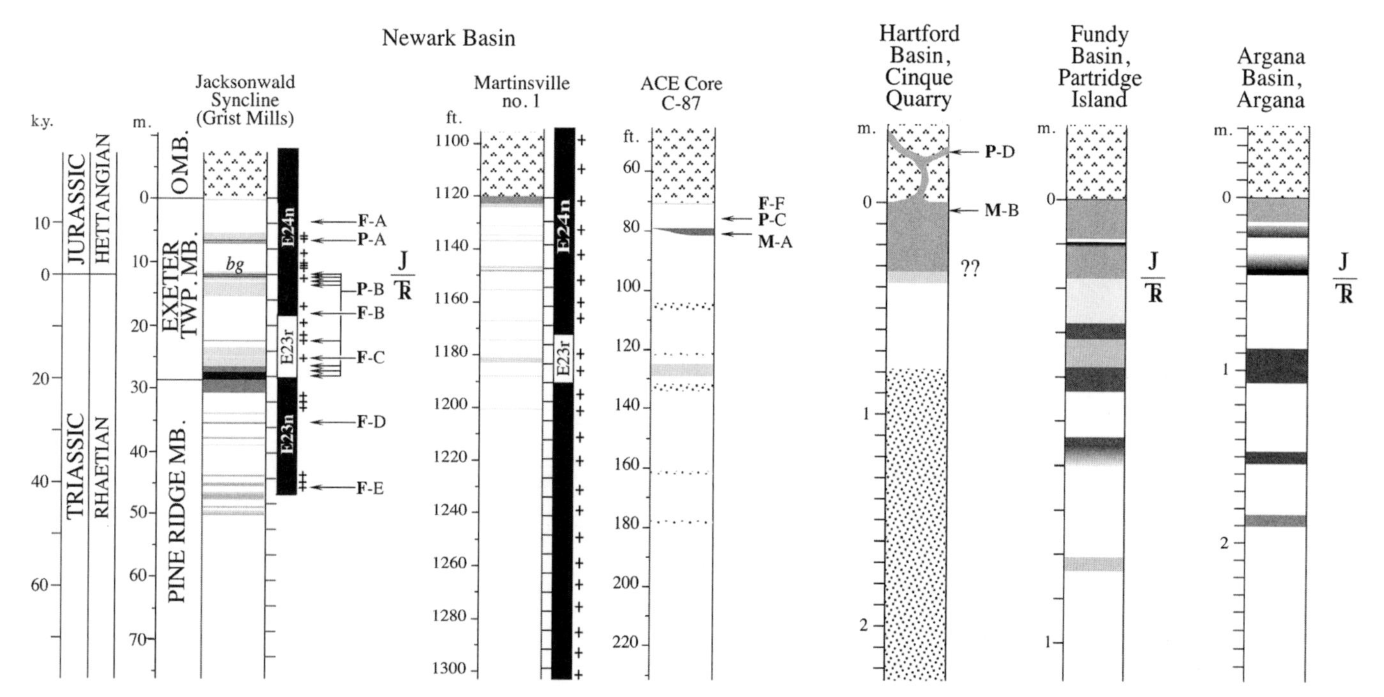

Figure 7. Detailed stratigraphy of Newark, Hartford, Fundy, and Argana basin boundaries. In all cases Triassic-Jurassic (Tr-J) boundary is based on palynomorph transitions. Newark section is from Olsen et al. (2001b). Abbreviations: F-A, highest footprint bearing level in Jacksonwald syncline with small grallatorids only (projected from Exeter Village sections); F-B, footprint assemblage with *Batrachopus* cf. *B. deweyii, Rhynchosauroides*, and small grallatorids (projected from Exeter Village sections); F-C, highest footprint level with new taxon B and *Brachychirotherium* and cf. *Apatopus*, also with *Rhynchosauroides* and small grallatorids; F-D, *Brachychirotherium*, new taxon B, *Batrachopus*, and small grallatorids; F-E, level with abundant *Brachychirotherium*, new taxon B, *Batrachopus*, and small to medium sized grallatorids; F-F, Clifton-Patterson area quarries and exposures with small to large grallatorids, including lowest occurrence of *Eubrontes giganteus*, also with *Batrachopus deweyii*, and *Rhynchosauroides*; P-A, lowest definitive Jurassic-type palynomorph level; P-B, palynomorph assemblages of Triassic aspect (lower) or dominated by spores (upper); P-C, palynomorph assemblage with *Corollina* only (poorly preserved); P-D, palynomorph assemblage with *Corollina* only in matrix between basalt pillow (Robbins, *in* Heilman, 1987); M-A, macrofossil plant assemblage dominated by *Brachyphyllum* and *Clathropteris*; M-B, macrofossil plant assemblage dominated by *Brachyphyllum* (Heilman, 1987). ACE is Army Corps of Engineers; bg indicates position of blue-gray sandstone.

Above the Orange Mountain Basalt are many horizons in the Feltville, Towaco, and Boonton Formations that have yielded very abundant tetrapod footprints of typical Connecticut Valley aspect (Olsen, 1995; Olsen et al., 2002b), as well as several well-preserved *Corollina*-dominated palynoflorules of typical Early Jurassic aspect (Cornet and Traverse, 1975; Cornet, 1977; Cornet and Olsen, 1985).

Thus, the paleontology and Milankovitch cyclostratigraphy of the upper Passaic Formation and succeeding units confines the Tr-J biological transition to within 10 k.y. based on palynology, and within 30 k.y. based on footprints (Olsen et al., 2002b). Because there are no assemblages of bones from Jurassic strata of the Newark basin, limitations based on osteological taxa are discussed in a regional context that follows.

NEWARK BASIN GEOCHEMICAL AND MINERALOGICAL ANOMALIES

The abundances and interelement ratios of the siderophile elements, such as Cr, Co, Ni, and especially the platinum group elements (PGEs) have been used to investigate the possible presence of a meteoritic component in terrestrial rocks at several geological boundaries (e.g., Alvarez et al., 1980; references *in* Montanari and Koeberl, 2000). However, the expected concentrations of PGEs in terrestrial rocks, even with extraterrestrial enrichment, are exceedingly low. For example, the addition of ~0.1% of a meteoritic (contrite) component to a crustal rock would yield an enrichment of ~0.5 ppb Ir to the crustal abundance (~0.02 ppb Ir) in the resulting impact breccia (Koeberl, 1998). Due to these low abundances, only very sensitive analytical techniques, such as Ir coincidence spectrometry (ICS) and inductively coupled plasma source–mass spectrometry (ICP-MS), after chemical preseparation of the PGEs, can be used.

There were two previous, unsuccessful, attempts to find geochemical and mineralogical anomalies at the Tr-J boundary in the Newark basin. Smith et al. (1988) looked specifically for an Ir anomaly in the same units in the Jacksonwald syncline that we examine here, but the amounts present were below their detectable limits. Mossman et al. (1998) looked for shocked quartz and Ir anomalies without reported success in the same interval.

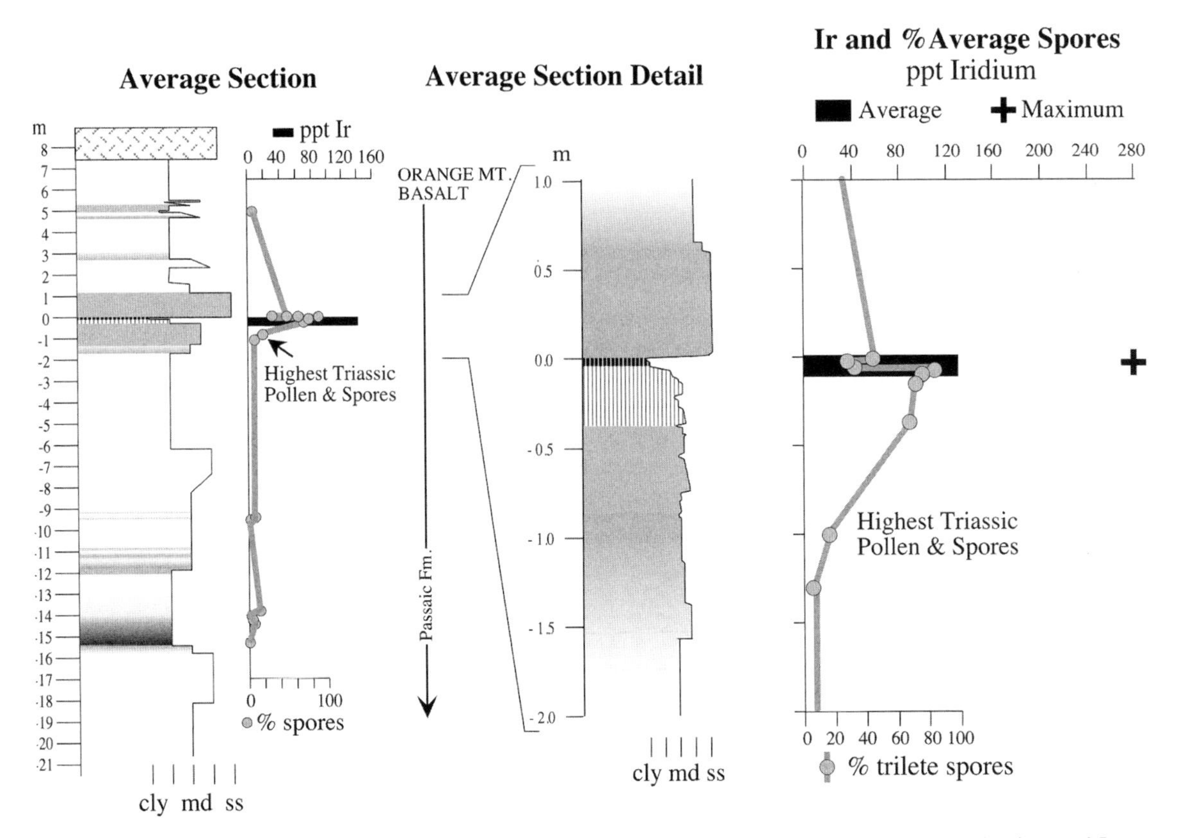

Figure 8. Summary of average Ir and pollen and spore data from Jacksonwald syncline. Color, grain size, and Ir are average values from four sections at the localities shown in Figure 3 based on Olsen et al., 2002b); pollen and spore data are from Fowell (1994). Details of the Ir anomaly are given in Olsen et al. (2002b). Spore and pollen data are from Fowell (1994).

RESULTS AND DISCUSSION

Undeterred by the previous unsuccessful attempts, Olsen et al. (2002b) examined Ir and other elemental concentrations at four sections along strike in the Jacksonwald syncline directly around the fern spike (Fig. 8). The samples show variations in Ir content from 19 to 285 ppt (0.2–0.29 ng/g) with an average maximum of 141 ppt (0.14 ng/g) (Olsen et al., 2002b). All sections except section I show a distinct Ir anomaly directly at the boundary with a distinct systematic association between Ir content and stratigraphy. The elevated levels of Ir are mostly associated with higher levels of Al in a white smectitic claystone (Smith et al., 1988), directly beneath the thin coaly layer (Figs. 6 and 7), although there is no correlation between Al and Ir in the data in general. The anomaly is directly associated with the previously identified fern spike in these sections, recalling the similar pattern at the K-T boundary in the western United States (Tschudy et al., 1984; Nichols and Fleming, 1990). It is possible that the relatively weak Ir anomaly (relative to the apparent background) seen thus far is a consequence of dilution by the coarse sampling level (~3 cm per sample) required by the very high accumulation rates (~1 m/2 k.y.) in the sampled part of the Newark basin. We can probably rule out a simple diagenetic concentration of Ir along a redox boundary because of the good correlation between Ir and stratigraphy, despite the lateral facies change from gray and black strata in the east to virtually entirely red strata in the westernmost section (Grist Mills, Fig. 3). The sample with the highest Ir content at the Grist Mills section is red (177 ppt, Olsen et al., 2002b).

In the one section (section I) that did not show a systematic association between Ir content and stratigraphy (Olsen et al., 2002b), Ir levels were highest in the blue-gray sandstone rather than in the spore-rich clays just below it. This could be because our sample of this sandstone contained mud chips eroded from a presumably Ir-enriched layer upstream. The sandstone was bulk processed and it will be necessary to run separate analyses on the sandstone and clay pebble separates to test this hypothesis.

The situation does not, unfortunately, become any clearer when considering the concentrations of other elements. Although we recognize that much more might be done with the data presented by Olsen et al. (2002b). Comparing the trends

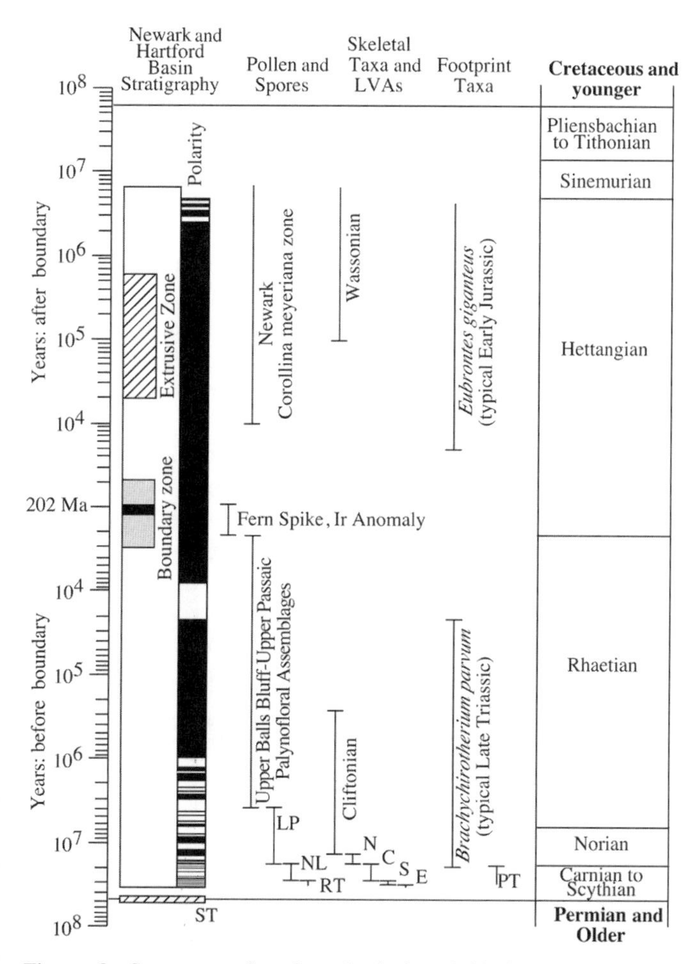

Figure 9. Summary of major physical and biotic events around Triassic-Jurassic boundary plotted on logarithmic scale. Data are primarily from Newark basin; time scale is based on Figure 4. LVAs, land vertebrate ages of Huber et al. (1996) and Lucas and Huber (2002): N, Neshanician; C, Conewagian; S, Sanfordian; and E, Economian. Pollen and spore zones are from Cornet (1977) and Cornet and Olsen (1985): LP, Lower Passaic Heidlersburg; NL, New Oxford, Lockatong; RT, Richmond Taylorsville. Footprint distribution is from Olsen et al. (2001b); PT, range of Pekin-type footprint assemblages (Olsen and Huber, 1996). Note that extrusive zone consists of lava flow formations interbedded with fossiliferous and cyclical sedimentary strata, with latter interpreted as representing nearly all of time shown. ST indicates position of Permian-Triassic Siberian Traps.

of the Ir data with those of other siderophile elements reported in Olsen et al. (2002b), such as Co, Ni, and Cr, which are often used as tracers of meteoritic components, does not yield distinct correlations.

The trace element data described by Olsen et al. (2002b) do not seem to support the idea that the Ir enrichments could be related to volcanic ash. Elements that might correlate in abundance with Ir if the source were altered volcanic material (e.g., Cs, Al, Cu, and V) reveal no significant correlations. The variations in these elements in section I are irregular and the concentrations of Cs vary with no correlation to the distinct Ir enrichment and associated Co and Ni abundance peaks. The highest Cs abundances occur for samples outside the zone of siderophile element enrichment. Thus, we consider a volcanic interpretation of the Ir enrichments unlikely, although we refrain from more detailed comment on the trace element data until more thorough sampling, especially farther from the palynologically identified boundary, can be performed.

A significant problem with the analysis by Olsen et al. (2002b) of the association between Ir abundance and the fern spike in the Jacksonwald syncline sections is that the geochemical and palynological analyses were not conducted on the same samples, or even at the same sections, and there is considerable variability from section to section in both variables. In an attempt to mitigate this problem and facilitate comparison of the Ir and pollen and spore data, Olsen et al. (2002b) averaged the Ir data and combined the spore data from all of the Jacksonwald synclines sections examined to date, using the base of the blue-gray sandstone as the correlative datum (Figs. 8 and 9). No attempt was made to account for possible lateral changes in accumulation rate.

The averaged and combined data show a strong correlation between spore percentage and Ir content Olsen et al. (2002b). However, the spore maximum is below the Ir maximum and the spore spike appears quite broad. This could reflect an actual offset between the two data sets, or it could reflect small variations in either accumulation rate or depth of erosion of the overlying blue-gray sandstone at different sections. Existing data do not permit these hypotheses to be tested. It is clear that what is needed is very detailed sampling of sections, in which splits of the same samples are subjected to palynological and geochemical analyses. In addition, Ir measurements are clearly needed from a broader stratigraphic swath around the boundary to assess background Ir levels. The available data are, however, encouraging, and suggest that there is a modest Ir anomaly at the biologically identified boundary and that it is associated with a fern spike.

Olsen et al. (2002b) also reported negative results in their search for shocked quartz. All quartz grains observed were angular, and few were clear. Most of them contained mineral and fluid inclusions. In some cases, fluid inclusions were aligned and in a very few cases they seemed to be aligned along parallel lines, but none of these features are characteristic for shock metamorphism. Besides these very few cases, no planar features or lamellae potentially representing planer deformation features were observed. Thus, as was the case for Mossman et al. (1998), the search for shocked quartz grains was not successful.

TR-J BOUNDARY IN OTHER PANGEAN CONTINENTAL RIFTS

The patterns seen in boundary sections in other continental rift basins in the Newark Supergroup and Morocco are consistent with that seen in the Newark basin, but also provide important additional information. The most southern rift known to preserve the Tr-J boundary is the Culpeper basin of Virginia.

The boundary section there was described by Fowell (1994), but currently known outcrops and exposures have not permitted detailed stratigraphic or palynological analysis.

The next basin northward for which there is significant stratigraphic information relevant to the Tr-J boundary is the Hartford basin of Connecticut and Massachusetts (Fig. 2). A section exposed in the Cinque Quarry in East Haven, Connecticut, reveals the uppermost New Haven Formation and overlying pillowed base of the Talcott Basalt (Heilman, 1987). Nearly the entire New Haven Formation is fluvial, composed of brown and minor gray coarse clastic rocks and red mudstone. The uppermost 40 cm of the New Haven Formation at the Cinque Quarry is, in contrast, gray, and may represent marginal lacustrine environments (Fig. 7). The uppermost few centimeters of gray mudstone and sandstone preserve abundant *Brachyphyllum* shoots and cones and a palynoflorule of typical Early Jurassic aspect, dominated by *Corollina* (Robbins, quoted *in* Heilman, 1987). Because there is no dispute that most of the New Haven Formation is of Late Triassic age (Cornet, 1977), the Tr-J boundary probably is either within the gray sequence below the conifer-bearing level, or closely underlying it in the red beds.

Strata interbedded with and overlying the basalts of the Hartford basin preserve a cyclostratigraphy nearly identical to that of the Newark basin, implying nearly exact synchrony of climatic and eruptive events (Olsen et al., 1996b). However, compared to the youngest formation in the Newark basin (Boonton), the Portland Formation, which overlies the youngest basalt formation in the Hartford basin (Hampden Basalt), is much thicker (~4 km) and represents a much longer time than is preserved in the Newark basin. Hence, the Portland Formation provides an important supplement to the Newark basin astronomically calibrated time scale and allows the floral and faunal change seen at the Tr-J boundary to be placed into a more extensive temporal perspective. The lower Portland Formation exhibits a cyclostratigraphy virtually identical to that of the Boonton Formation of the Newark basin. However, whereas <300 k.y. of Boonton Formation was recovered by the ACE cores, at least 3 m.y. is represented by the Portland Formation. The magnetostratigraphy of ~2 m.y. of the lower parts of the Portland was determined from outcrop samples (Kent and Olsen, 1999b). Although clearly not known at the same level of detail as the Newark basin time scale, which is based on continuous core, the Portland Formation magnetostratigraphy does delimit the maximum duration of chron E24n, the top of which is not seen in the Newark basin record (Kent and Olsen, 1999a) (Fig. 4). The polarity reversal stratigraphy of the Portland Formation also provides a critical link to the marine polarity sequence from the Paris basin (Yang et al., 1996), indicating that the upper part of the Portland Formation is of Sinemurian age. Based on the Newark basin time scale, with the addition of the Portland Formation information, the duration of the Hettangian is thus ~2 m.y., in excellent agreement with new information from U-Pb dates from marine sections in British Columbia (Palfy et al., 2000b).

Palynoflorules from the Jurassic part of the Hartford basin section show only minor changes through time (Cornet, 1977; Cornet and Olsen, 1985), predominately involving the appearance of new taxa and species-level changes in the dominant pollen taxon *Corollina*. Vertebrate footprint and bone assemblages from Hettangian and Sinemurian strata of the Hartford basin show no obvious changes at all, except perhaps an increase in the size of some footprint forms (e.g., *Anomoepus*: Olsen and Rainforth, 2002). The slow change through the Hettangian and into the Sinemurian, spanning at least 3 m.y., contrasts dramatically with the extraordinarily abrupt change seen at the Tr-J boundary, which conservatively took less than 20 k.y.

In predrift configuration, the next continental rift basins north for which there is significant information on the Tr-J boundary are those in Morocco, particularly the Argana basin (Fig. 2). The stratigraphy of the boundary section in the Argana basin differs from that in the Newark basin in that cyclical gray and black strata are limited in outcrop to a couple of meters below the Argana Basalt (Olsen et al., 2000). Palynoflorules from these very thin (<20 cm) gray and black mudstones, which are interbedded with red mudstones, show a transition from an assemblage with *Patinasporites* to one without this taxon or any other forms typical of the Triassic. Instead it is dominated by *Corollina* (Olsen et al., 2000). The physical stratigraphy of this sequence, despite its very condensed appearance, as well as that of the overlying Argana Basalt and Amsekroud Formation, is closely comparable to that of the Newark basin section, implying a nearly identical sequence and timing of events (Olsen et al., 2002a).

Of special interest is the physical stratigraphy of the Khémisset basin in Morocco. In outcrop, the stratigraphy of the sedimentary section immediately below the oldest basalt is virtually identical to that seen in the Argana basin. However, in the subsurface, these cyclical red, gray, and black mudstones pass into interbedded black, red, and white halite and potash salts (Et-Touhami, 2000). Our preliminary palynological results from outcrop sections suggest that the Tr-J boundary is in the cyclical mudstone sequence just below the oldest basalt, and that the boundary should be present within the salt in the subsurface as well, a hypothesis we are examining.

Where studied, the physical stratigraphy of the sequence of basalts and the directly underlying and overlying sediments of the other Tr-J that crop out in basins of Morocco is very similar to that in the Newark and Argana basins, albeit very condensed. Thus, we argue for correlation of biotic and tectonic events at a very fine scale across the Moroccan basins. We hypothesize that the similarity, in detail, of the sequences just below the oldest basalts in all these basins to that seen in the Argana and Newark basins demarcates the position of the Tr-J boundary. However, published interpretations of the biostratigraphy of these basins differ dramatically from that presented here (Figs. 6 and 7). According to the summary of Oujidi et al. (2000), the ages of these homotaxial sequences are dramatically different in different basins, ranging in age from Ladinian to Norian. We attribute these differences in interpretation to a lack of appreciation of floral provinciality, the importance of the fact that earliest Jurassic assemblages are characterized by survivor assemblages rather than the appearance of new taxa, and by reliance on ostracode and bivalve taxa of dubious identification and biostratigraphic utility. Our palynological work underway on various sections in Morocco will test our hypothesis.

The Fundy basin of the Maritime provinces of Canada is the next continental rift basin the north that has outcrops, and it provides the most information on the Tr-J boundary after the Newark basin. Fowell and others (Fowell and Olsen, 1993; Fowell, 1994; Fowell et al., 1994) have shown that the Tr-J boundary is preserved within the uppermost few meters of the Blomidon Formation, below the North Mountain Basalt (Fig. 6) (Kent and Olsen, 2000b). The boundary section is very condensed and closely comparable to that in the Argana basin (Fig. 6). Like the Newark basin, the palynology of the boundary is marked by an abrupt disappearance of assemblages with *Patinasporites* and their replacement by assemblages dominated by *Corollina*. Although no fern spike has been found in the section studied in detail by Fowell and Traverse (1995) at Partridge Island (Cumberland County, Nova Scotia), outcrops of the uppermost few meters of Blomidon Formation at Central Clarence (Annapolis County, Nova Scotia) have one layer that produces a macroflora consisting entirely of the fern *Cladophlebis* (Carroll et al., 1972) (Fig. 6). This is the only foliage macroflora from the Fundy basin of Nova Scotia, and that it consists entirely of ferns, at the expected position of the Tr-J boundary, suggests to us that it may represent the fern spike. Unfortunately, the exposures at Central Clarence preclude a detailed study of this section without significant excavation.

The cyclostratigraphy of the lower McCoy Brook Formation, overlying the North Mountain Basalt, is closely comparable to that of the Feltville Formation of the Newark basin and its equivalents in other rifts in the eastern United States (Olsen et al., 1996b, 2002a). A rich vertebrate assemblage has been recovered from the lower McCoy Brook Formation and is notable for the abundance of well-preserved tetrapod bones and skeletons (Olsen et al., 1987). This assemblage is the oldest known from the continental Early Jurassic that is reasonably diverse and well dated. Despite the sampling of both terrestrial and aquatic habitats, typical Triassic osseous and footprint taxa are absent. In addition, while the remains of small tetrapods are very common at multiple levels, procolophonids are absent. Thus, this assemblage helps delimit the osseous record of at least some of the tetrapod extinctions to within 300 k.y. around the boundary. Footprint assemblages from the rest of the McCoy Brook Formation outcrops are consistent with this picture and are entirely of Connecticut Valley aspect (i.e., Early Jurassic).

Anders and Asaro (reported *in* Olsen et al., 1990) examined the upper 100 m of the Blomidon Formation for shocked quartz and Ir anomalies, but were unsuccessful. Mossman et al. (1998) examined the stratigraphic region near the Tr-J boundary, including at Partridge Island, for shocked quartz and Ir. Although some planar features were seen, they concluded that none were characteristic of shocked metamorphism. They also concluded that although no distinct Ir anomaly was found, the highest Ir amounts were found in proximity to the Tr-J boundary. We stress that these results are not inconsistent with our results from the Newark basin, but that much tighter geochemical sampling directly tied to the stratigraphic interval with the biotic turnover needs to be conducted in the Fundy basin.

Thus, the emerging picture of faunal and floral change around the Tr-J boundary in eastern North America and Morocco is one of extraordinarily rapid, synchronous change over a very large area. This change, at least in the Newark basin and possibly in the Fundy basin, is associated with a geologically abrupt burst in fern abundance suggestive of major ecological disruption. It is important to note that within this context, with the exception of the appearance of the large dinosaurian ichnospecies *Eubrontes giganteus*, the earliest Jurassic floral and tetrapod assemblages consist entirely of survivor taxa with no originations and no apparent replacement of Triassic forms by Jurassic ecological vicars. It has been hypothesized that even the appearance of *Eubrontes giganteus* may represent a consequence of ecological release upon the extinction of Triassic competitor forms, largely members of the nondinosaurian Crurotarsi, such as rauisuchians and phytosaurs (Olsen et al., 2002b). It is also important to note that the Tr-J boundary marks the end of the persistent floral and faunal provinciality that characterized the Triassic, and the establishment of a nearly cosmopolitan terrestrial community (Cornet and Olsen, 1985; Olsen and Galton, 1984; Sues et al., 1994). The biological data available thus far are consistent with a catastrophic end to the Triassic comparable in magnitude and similar in pattern to that characterizing the terminal Cretaceous event. The results from the Newark basin Tr-J boundary reported by Olsen et al. (2002b) show a modest Ir anomaly associated with the biotic turnover and the fern spike that is remarkably similar to that described for the K-T boundary. We stress that these results need to be tested by much more widespread geochemical and palynological analyses, both stratigraphically and geographically.

RELATIONSHIP TO POSSIBLE IMPACT

It is clear that, at least superficially, there is a strong similarity between the K-T boundary in the western North American interior and the Tr-J boundary in eastern North America. This similarity includes the specific pattern of floral and faunal extinction, the fern spike, and the presence of an apparent Ir anomaly. The similar patterns might indicate similar cause, and an impact origin for both boundaries has been suggested (Dietz, 1986; Olsen et al., 1987, 1990). Although shocked quartz has not been reported from eastern North America, Bice et al. (1992) reported it from a Tr-J boundary section in Tuscany, and there is an additional report from the Kendelbach section in

Austria (Badjukov et al., 1987). However, in both cases the shocked quartz was identified only petrographically, which is now not considered definitive (e.g., Grieve et al., 1996; Mossman et al., 1998), and in neither case has there been a subsequent attempt at independent confirmation. Of course, if the impact site were very distant, or in oceanic crust, it is possible that shocked quartz would be very rare or absent.

Originally, the giant Manicouagan impact was suggested as a possible cause (Olsen et al., 1987) of the Tr-J mass extinctions. However, U-Pb dates from this feature by Hodych and Dunning (1992) suggest that its age is 214 ± 1 Ma, which is consistent with older $^{40}Ar/^{39}Ar$ and Rb-Sr dates from the impact, but incompatible with the dates of the basalts overlying the boundary (ca. 200 Ma), which at least at the level of uncertainty of radiometric dates, should be the age of the boundary.

RELATIONSHIP WITH CENTRAL ATLANTIC MAGMATIC PROVINCE

A remarkable aspect of the Tr-J boundary is the very close proximity in both stratigraphic thickness and time (~20 k.y.) to the oldest exposed Central Atlantic magmatic province flood basalts in eastern North America and Morocco. The Central Atlantic magmatic province tholeiites may represent the largest known (in area at least) igneous event in Earth history, covering an area of 7×10^6 km^2 (e.g., Marzoli et al., 1999) (Fig. 2). The preerosion volume of the Central Atlantic magmatic province may have been in excess of 3×10^6 km^3, making it larger than any other known continental flood basalt province. The close association between the Central Atlantic magmatic province lavas and the boundary has led to speculation that the extinctions might have been caused by climatic changes resulting from gas and aerosol emissions from the eruptions (Courtillot et al., 1994; McHone, 1996; Marzoli et al., 1999; Palfy et al., 2000a). McElwain et al. (1999) found that the stomatal density in a range of plant taxa drops significantly within the same taxa at the florally identified Tr-J boundary in Greenland (Kap Stewart Formation). This change is direct evidence suggesting a major increase in CO_2 at the boundary that they speculated might be due to Central Atlantic magmatic province volcanism. Smith et al. (1988) noted that a volcanic source could be responsible for the smectitic clay at the boundary in the Jacksonwald syncline of the Newark basin. It is also possible that an Ir anomaly, especially a modest one, could be explained by deep-seated basaltic volcanism as suggested by Olmez et al. (1986) for the K-T boundary, although this hypothesis is not supported by our geochemical analyses of associated elemental concentrations.

A significant problem with the volcanic hypothesis for the origin of the Tr-J mass extinction is that wherever the biological signature of the mass extinction and the oldest Central Atlantic magmatic province basalts have been observed in the same section, the basalts invariably postdate the extinctions, albeit by only a short time. However, the basalts that are known in superposition with the boundary amount to a small part of the Central Atlantic magmatic province; their temporal relationship with the rest of the Central Atlantic magmatic province and the boundary is unknown at the required fine level of resolution.

Olsen (1999) and Olsen et al. (2002a) pointed out that there is some slim paleomagnetic evidence for Central Atlantic magmatic province eruption occurring just prior to the boundary, consisting of very rare dikes of Central Atlantic magmatic province radiometric age (i.e., 200 Ma) with reversed magnetic polarity. Because all the Central Atlantic magmatic province basalts above the boundary are uniformly of normal polarity, this shows that some of the Central Atlantic magmatic province igneous activity occurred at a different time than the known flows. The temporally closest interval of reversed polarity is E23r, located just below the boundary (Figs. 4–7). If further research confirms the reality of the reversed dikes, a significant portion of the Central Atlantic magmatic province could easily predate or be synchronous with the boundary. For example, a huge basaltic edifice that could predate the boundary includes the massive seaward-dipping reflectors off the southeastern United States that may be part of Central Atlantic magmatic province (Holbrook and Kelemen, 1993; Olsen, 1999; Olsen et al., 2002a) and could be volumetrically as large as the rest of the province. Thus we do not reject the possibility that at least part of the Central Atlantic magmatic province, perhaps even the largest part, could have been emplaced just before or during the Tr-J mass extinctions.

IMPACT AND VOLCANISM?

Rampino and Stothers (1988) and Courtillot et al. (1994), on the basis of a compilation of published radiometric dates of igneous rocks and literature ages for geologic boundaries, showed that there is a very good correlation between major continental flood basalts and mass extinctions over the past 300 m.y. Particularly prominent are the three largest Phanerozoic mass extinctions (Permian-Triassic, Tr-J, and K-T) and their remarkably tight association with the three largest Phanerozoic continental flood basalts (Siberian, Central Atlantic magmatic province, and Deccan, respectively). It seems very difficult to dismiss this correlation as a coincidence, and it is particularly interesting that at least two (Tr-J and K-T) have evidence of an impact (the latter impact evidence is unimpeachable). Thus, we cannot dismiss the possibility that the flood basalt volcanism was perhaps somehow triggered or enhanced by an impact (e.g., Boslough et al., 1996), despite the fact that preliminary models of the energetics of impacts suggest that causing volcanism de novo with an impact would be very difficult (e.g., Melosh, 2000).

CONCLUSIONS

The biotic pattern around the continental Tr-J boundary has many similarities with the much better understood K-T boundary, including the very short duration of the extinction event, its selectivity, the composition of postboundary assemblages made up of only survivor taxa, and the presence of a regional fern spike at the microfloral extinction level. We have shown here new evidence of at least a modest Ir anomaly associated with the fern spike. A summary of the data for the Tr-J boundary based mostly on the Newark basin is shown in Figure 9. We have also shown the enormous area over which the stratigraphy of the continental boundary is remarkably consistent. Also shared with both the K-T and Permian-Triassic boundaries are temporally associated massive continental flood basalts. Unlike the K-T boundary, evidence for an impact at the Tr-J boundary is not yet conclusive. It is difficult to dismiss as coincidental the co-occurrence of mass extinctions and flood basalts, including that at the Tr-J boundary. The newly reported Ir anomaly could be consistent with either an impact or deep-seated volcanic origin, although the latter receives no support from the trace element concentrations or stratigraphic relationships reported here. The microstratigraphy is very similar to continental K-T boundary sections, and this lithological similarity is matched by a similar biotic pattern. However, without additional, more stratigraphically extensive sampling, we cannot completely rule out a volcanic or other nonimpact origin for this anomaly.

ACKNOWLEDGMENTS

Work on this project by Olsen, Et-Touhami, and Kent was funded by U.S. National Science Foundation grants EAR-98-14475 and EAR-9804851 (to Olsen and Kent) and a grant from the Lamont Climate Center to Olsen. Work in Morocco was aided by logistical support from ONAREP, for which we are very grateful. Et-Touhami was supported during work on this project by a fellowship from the Fulbright Foreign Student Program (MACECE). The laboratory work in Austria was supported by Austrian Science Foundation grant Y58-GEO (to Koeberl). We thank Robert Grantham for access to the fossil collections of the Nova Scotia Provincial Museum in Halifax. We also thank Emma Rainforth, David Mossman, and Roy Schlische for helpful reviews. This is Lamont-Doherty Earth Observatory contribution 6155.

REFERENCES CITED

Alvarez, L.W., Alvarez, W., Asaro, F., and Michel, H.V., 1980, Extraterrestrial cause for the Cretaceous Tertiary extinction: Science, v. 208, p. 1095–1108.

Badjukov, D.D., Lobitzer, H., and Nazarov, M.A., 1987, Quartz grains with planar features in the Triassic-Jurassic boundary sediments from the northern calcareous Alps, Austria: Houston, Texas, Lunar and Planetary Sciences Conference, Abstract, v. 28, p. 38.

Beutler, G., 1998, Keuper: Hallesches Jahrbuch für Geowissenschaften, Reihe B: Geologie, Paläontolgie, Mineralogie, v. 6, p. 45–58.

Bice, D.M., Newton, C.R., McCauley, S., Reiners, P.W., 1992, Shocked quartz at the Triassic-Jurassic boundary in Italy: Science, v. 255, p. 443–446.

Boslough, M.B., Chael, E.P., Truncano, T.G., Crawford, D.A., and Campbell, D.L., 1996, Axial focusing of impact energy in the earth's interior: A possible link to flood basalts and hotspots, *in* Ryder, G., Fastovsky, D., and Gartner, S., eds., The Cretaceous-Tertiary event and other catastrophes in Earth history: Geological Society of America Special Paper 307, p. 541–569.

Brugman, W.A., 1983, Permian-Triassic palynology: Utrecht, Netherlands, State University Utrecht, Laboratory of Paleobotany and Palynology, 121 p.

CANMET, 1994, Catalogue of Certified Reference Materials: Ottawa, Canadian Certified Reference Materials Project, 94-1E.

Carroll, R.L, Belt, E.S, Dineley, D.L, Baird, D, and McGregor, D.C., 1972, Excursion A59: Vertebrate Paleontology of Eastern Canada: 24th International Geological Congress, Montreal, Guidebook, p. 1–113.

Cornet, W.B., 1977, The palynostratigraphy and age of the Newark Supergroup [Ph.D. thesis]: University Park, Pennsylvania State University, 527 p.

Cornet, B., and Olsen, P.E., 1985, A summary of the biostratigraphy of the Newark Supergroup of eastern North America, with comments on early Mesozoic provinciality, *in* Weber, R., ed., Symposio Sobre Flores del Triasico Tardio su Fitografia y Paleoecologia, Memoria, Proceedings II, Latin-American Congress on Paleontology, 1984: Mexico City, Instituto de Geologia Universidad Nacional Autonoma de Mexico, p. 67–81.

Cornet, B., and Traverse, A., 1975, Palynological contributions to the chronology and stratigraphy of the Hartford Basin in Connecticut and Massachusetts: Geoscience and Man, v. 11, p. 1–33.

Cornet, B., Traverse, A., and McDonald, N.G., 1973, Fossil spores, pollen, and fishes from Connecticut indicate Early Jurassic age for part of the Newark Group: Science, v. 182, p. 1243–1247.

Courtillot, V., Jaeger, J.J., Yang, Z., Féraud, G., and Hofmann, C., 1994, The influence of continental flood basalts on mass extinctions: Where do we stand?, *in* Ryder, G., Fastovsky, D., and Gartner, S., eds., The Cretaceous-Tertiary event and other catastrophes in Earth history: Geological Society of America Special Paper 307, p. 513–525.

Dietz, R.S., 1986, Triassic-Jurassic extinction event, Newark basalts and impact-generated Bahama nexxus: Houston, Texas, Lunar and Planetary Institute, LPI Contribution No. 600, p. 1–10.

Dunning, G.R., and Hodych, J.P., 1990, U/Pb zircon and baddeleyite ages for the Palisades and Gettysburg sills of the northeastern United States: Implications for the age of the Triassic/Jurassic boundary: Geology, v. 18, p. 795–798.

Ekart, D.D., Cerling, T.E., Montanez, I.P., and Tabor, N.J., 1999, A 400 million year carbon isotope record of pedogenic carbonate: Implications for paleoatomospheric carbon dioxide: American Journal of Science, v. 299, p. 805–827.

Et-Touhami, M., 2000, Lithostratigraphy and depositional environments of lower Mesozoic evaporites and associated red beds, Khemisset basin, northwestern Morocco, *in* Bachmann, G., and Lerché, eds., Epicontinental Triassic, Volume 2: Zentralblatt für Geologie und Paläontologie, Teils, 1998, Heft 9-10, p. 1217–1241.

Fedosh, M.S., and Smoot, J.P., 1988, A cored stratigraphic section through the northern Newark Basin, New Jersey: U.S. Geological Survey Bulletin 1776, p. 19–24.

Fowell, S.J., 1994, Palynology of Triassic/Jurassic boundary sections from the Newark Supergroup of eastern North America: Implications for catastrophic extinction scenarios [Ph.D. thesis]: New York, Columbia University, 154 p.

Fowell, S.J., Cornet, B., and Olsen, P.E., 1994, Geologically rapid Late Triassic extinctions: Palynological evidence from the Newark Supergroup, *in* Klein, G.D., ed., Pangaea: Paleoclimate, tectonics and sedimentation during accretion, zenith and break-up of a supercontinent: Geological Society of America Special Paper 288, p. 197–206.

Fowell, S.J., and Olsen, P.E., 1993. Time-calibration of Triassic/Jurassic microfloral turnover, eastern North America: Tectonophysics, v. 222, p. 361–369.

Fowell, S.J., and Traverse, A., 1995, Palynology and age of the upper Blomidon Formation, Fundy Basin, Nova Scotia: Review of Palaeobotany and Palynology, v. 86, p. 211–233.

Govindaraju, K., 1989, 1989 Compilation of working values and sample descriptions for 272 geostandards: Geostandards Newsletter, v. 13, p. 1–113.

Gradstein, F.M., Agterberg, F.P., Ogg, J.G., Hardenbol. J., Van Veen, P., Thierry, J., and Huang, Z., 1994, A Mesozoic time scale: Journal of Geophysical Research, v. 9. p. 24051–24074.

Grieve, R.A.F., Langenhorst, F., and Stoeffler, D., 1996, Shock metamorphism of quartz in nature and experiment. 2. Significance in geoscience: Meteoritics, v. 31, p. 6–35.

Hames, W.E., Renne, P.R., and Ruppel, C., 2000, New evidence for geologically instantaneous emplacement of earliest Jurassic Central Atlantic magmatic province basalts on the North American margin: Geology, v. 28, p. 859–862.

Harland, W.B., Armstrong, R.L., Cox, A.V., Craig, L.E., Smith, A.G., and Smith, D.G., 1990, A geologic time scale, 1989: Cambridge, Cambridge University Press, 263 p.

Heilman, J.J., 1987, That catastrophic day in the Early Jurassic: Journal of Science Education, v. 25, p. 8–25.

Hodych, J.P., and Dunning, G.R., 1992, Did the Manicouagan impact trigger end-of-Triassic mass extinction?: Geology, v. 20, p. 51–54.

Holbrook, W.S., and Kelemen, P.B., 1993, Large igneous province on the US Atlantic margin and implications for magmatism during continental breakup: Nature, v. 364, p. 433–437.

Huber, P., Lucas, S.G., and Hunt, A.P., 1996, Vertebrate biochronology of the Newark Supergroup Triassic, eastern North America, *in* Morales, M., ed., The continental Jurassic: Museum of Northern Arizona Bulletin 60, p. 179–186.

Jarosewich, E., Clarke, R.S., Jr., and Barrows, J.N., editors, 1987, The Allende meteorite reference sample: Smithsonian Contributions to the Earth Sciences, v. 27, p. 1–49.

Kent, D.V., and Muttoni, G., 2002, Mobility of Pangea: Implications for late Paleozoic and early Mesozoic paleoclimate, *in* LeTourneau, P.M., and Olsen, P.E., eds., The Great Rift Valleys of Pangea in eastern North America, Volume 1: Tectonics, structure, and volcanism of Supercontinent breakup: New York, Columbia University Press (in press).

Kent, D.V., and Olsen, P.E., 1999a, Astronomically tuned geomagnetic polarity time scale for the Late Triassic: Journal of Geophysical Research, v. 104, p. 12831–12841.

Kent, D.V., and Olsen, P.E., 1999b, Search for the Triassic/Jurassic long normal and the J1 cusp: Eos (Transactions, American Geophysical Union) Supplement, v. 80, no. 46, p. F306.

Kent, D.V., and Olsen, P.E., 2000a, Implications of a new astronomical time scale for the Late Triassic, *in* Bachmann, G., and Lerche, I., eds., Epicontinental Triassic, Volume 3: Zentralblatt für Geologie und Paläontologie, Teil I, 1998, Heft 11/12, p. 1463–1474.

Kent, D.V., and Olsen, P.E., 2000b, Magnetic polarity stratigraphy and paleolatitude of the Triassic–Jurassic Blomidon Formation in the Fundy basin (Canada): Implications for early Mesozoic tropical climate gradients: Earth and Planetary Science Letters, v. 179, no. 2. p. 311–324.

Kent, D.V., Olsen, P.E., and Witte, W.K., 1995, Late Triassic-Early Jurassic geomagnetic polarity and paleolatitudes from drill cores in the Newark rift basin (Eastern North America): Journal of Geophysical Research, v. 100, no. B8, p. 14965–14998.

Koeberl, C., 1993, Instrumental neutron activation analysis of geochemical and cosmochemical samples: A fast and reliable method for small sample analysis: Journal of Radioanalytical and Nuclear Chemistry, v. 168, p. 47–60.

Koeberl, C., 1998, Identification of meteoritic components in impactites, *in* Grady, M.M., Hutchison, R., McCall, G.J.H., and Rothery, D.A., eds., Meteorites: Flux with time and impact effects: Geological Society [London] Special Publication 140, p. 133–153.

Koeberl, C., and Huber, H., 2000, Multiparameter–coincidence spectrometry for the determination of iridium in geological materials: Journal of Radioanalytical and Nuclear Chemistry, v. 244, p. 655–660.

Lucas, S.G., 1998, Global Triassic tetrapod biostratigraphy and biochronology, *in* The Permian-Triassic boundary and global Triassic correlations: Palaeogeography, Palaeoclimatology, Palaeoecology, v. 143, p. 345–382.

Lucas, S.G., and Huber, P., 2002, Vertebrate biostratigraphy and biochronology of the nonmarine Late Triassic, *in* LeTourneau, P.M., and Olsen, P.E., eds., Aspects of Triassic-Jurassic geoscience: New York, Columbia University Press (in press).

Marzoli, A., Renne, P.R., Piccirillo, E.M., Ernesto, M., Bellieni, G., and De-Min, A., 1999, Extensive 200-million-year-old continental flood basalts of the Central Atlantic Magmatic Province: Science, v. 284, p. 616–618.

McElwain, J.C., Beerling, D.J., and Woodward, F.I., 1999, Fossil plants and global warming at the Triassic-Jurassic boundary: Science, v. 285, p. 1386–1390.

McHone, J.G., 1996. Broad-terrane Jurassic flood basalts across northeastern North America: Geology, v. 24, p. 319–322.

Melosh, H.J., 2000, Can impacts induce volcanic eruptions?: Houston, Texas, Lunar and Planetary Institute, LPI Contribution 1053, Abstract no. 3144.

Montanari, A., and Koeberl, C., 2000, Impact stratigraphy: The Italian record: Heidelberg, Germany, Springer-Verlag, Lecture Notes in Earth Sciences, v. 93, 364 p.

Morbey, S.J., 1975, The palynostratigraphy of the Rhaetian Stage, Upper Triassic in the Kendelbachgraben, Austria: Palaeontographica, v. 152, p. 1–75.

Mossman, D.J., Grantham, R.G., and Langenhorst, F., 1998, A search for shocked quartz at the Triassic-Jurassic boundary in the Fundy and Newark basins of the Newark Supergroup: Canadian Journal of Earth Science, v. 35, p. 101–109.

Nichols, D.J., and Fleming, R.F., 1990, Plant microfossil record of the terminal Cretaceous event in the western United States and Canada, *in* Sharpton, V.L., and Ward, P.D., eds., Global catastrophes in Earth history: An interdisciplinary conference on impacts, volcanism, and mass mortality: Geological Society of America Special Paper 247, p. 445–456.

Olmez, I., Finnegan, D.L., and Zoller, W.H., 1986, Iridium emissions from Kilauea volcano: Journal of Geophysical Research, v. 91, p. 653–652.

Olsen, P.E., 1995, Paleontology and paleoenvironments of Early Jurassic age strata in the Walter Kidde Dinosaur Park (New Jersey, USA), *in* Baker, J.E.B., ed., Field Guide and Proceedings of the Twelfth Annual Meeting of the Geological Association of New Jersey, Geological Association of New Jersey: Paterson, New Jersey, William Paterson College, p. 156–190.

Olsen, P.E., 1997, Stratigraphic record of the early Mesozoic breakup of Pangea in the Laurasia-Gondwana rift system: Annual Reviews of Earth and Planetary Science, v. 25, p. 337–401.

Olsen, P.E., 1999, Giant lava flows, mass extinctions, and mantle plumes: Science, v. 284, p. 604–605.

Olsen, P.E., and Galton, P.M., 1977, Triassic-Jurassic tetrapod extinctions: Are they real?: Science, v. 197, p. 983–986.

Olsen, P.E., and Galton, P.M., 1984, A review of the reptile and amphibian assemblages from the Stormberg of Southern Africa with special emphasis on the footprints and the age of the Stormberg: Palaeontologia Africana, v. 25, p. 87–110.

Olsen, P.E., and Huber, P., 1996, The oldest Late Triassic footprint assemblage from North America (Pekin Formation, Deep River basin, North Carolina USA): Southeastern Geology, v. 38, no. 2, p. 77–90.

Olsen, P.E., and Kent, D.V., 1996, Milankovitch climate forcing in the tropics of Pangea during the Late Triassic: Palaeogeography, Palaeoclimatology and Palaeoecology, v. 122, p. 1–26.

Olsen, P.E., and Kent, D.V., 1999, Long-period Milankovitch cycles from the Late Triassic and Early Jurassic of eastern North America and their implications for the calibration of the early Mesozoic time scale and the long-term behavior of the planets: Royal Society of London Philosophical Transactions, ser. A, v. 357, p. 1761–1787.

Olsen, P.E., and Kent, D.V., 2000, High resolution early Mesozoic Pangean climatic transect in lacustrine environments, *in* Bachmann, G., and Lerche, I., eds., Epicontinental Triassic, Volume 3: Zentralblatt für Geologie und Paläontologie, Teil I, 1998, Heft 11/12, p. 1475–1496.

Olsen, P.E., and Rainforth, E., 2002, The Early Jurassic ornithischian dinosaurian ichnite *Anomoepus, in* LeTourneau, P.M., and Olsen, P.E., eds., The Great Rift Valleys of Pangea in eastern North America, Volume 2: Sedimentology, Stratigraphy, and Paleontology: New York, Columbia University Press (in press).

Olsen, P.E., and Sues, H.-D., 1986, Correlation of the continental Late Triassic and Early Jurassic sediments, and patterns of the Triassic-Jurassic tetrapod transition, *in* Padian, K., ed., The beginning of the Age of Dinosaurs: Faunal change across the Triassic-Jurassic boundary: New York, Cambridge University Press, p. 321–351.

Olsen, P.E., Shubin, N.H., and Anders, M., 1987, New Early Jurassic tetrapod assemblages constrain Triassic-Jurassic tetrapod extinction event: Science, v. 237, p. 1025–1029.

Olsen, P.E., Fowell, S.J., and Cornet, B., 1990, The Triassic-Jurassic boundary in continental rocks of eastern North America: A progress report, *in* Sharpton, V.L., and Ward, P.D., eds., Global catastrophes in Earth history: An interdisciplinary conference on impacts, volcanism, and mass mortality: Geological Society of America Special Paper 247, p. 585–593.

Olsen, P.E., Kent, D.V., Cornet, B., Witte, W.K., and Schlische, R.W., 1996a, High-resolution stratigraphy of the Newark rift basin (Early Mesozoic, Eastern North America): Geological Society of American Bulletin, v. 108, p. 40–77.

Olsen P.E., Schlische R.W., Fedosh M.S., 1996b, 580 k.y. duration of the Early Jurassic flood basalt event in eastern North America estimated using Milankovitch cyclostratigraphy, *in* Morales, M., ed., The continental Jurassic: Museum of Northern Arizona Bulletin 60, p. 11–22.

Olsen, P.E., and Kent, D.V., Fowell, S.J., Schlische, R.W., Withjack, M.O., and LeTourneau, P.M., 2000, Implications of a comparison of the stratigraphy and depositional environments of the Argana (Morocco) and Fundy (Nova Scotia, Canada) Permian-Jurassic basins, *in* Oujidi, M., and Et-Touhami, M., eds., Le Permien et le Trias du Maroc, Actes de la Première Réunion du Groupe Marocain du Permien et du Trias: Oujda, Hilal Impression, p. 165–183.

Olsen, P.E., Kent, D.V., Et-Touhami, M., and Puffer, J.H., 2002a, Cyclo-, magneto-, and bio-stratigraphic constraints on the duration of the CAMP event and its relationship to the Triassic-Jurassic boundary: American Geophysical Union Memoir (in press).

Olsen, P.E., Kent, D.V., Sues, H.-D., Koeberl, C., Huber, H., Montanari, A., Rainforth, E.C., Fowell,. S.J., Szajna, M.J., and Hartline, B.W., 2002b, Ascent of dinosaurs linked to Ir anomaly and "fern spike" at Triassic-Jurassic boundary: Science, v. 296, p. 1305–1307.

Oujidi, M., Courel, L., Benaouiss, N., El Mostaine, M., El Youssi, M., Et-Touhami, M., Ouarhache, D., Sabaoui, A., and Tourani, A., 2000, Triassic series of Morocco: Stratigraphy, palaeogeography and structuring of the southwestern peri-Tethyan platform: An overview, *in* Crasquin-Soleau, S., and Barrier, E., eds., Peri-Tethys Memoir 5: New data on peri-Tethyan sedimentary basins: Mémoires du Muséum National d'Histoire Naturelle, v. 182, p. 23–38.

Page, K.N., and Bloos, G., 1998, The base of the Jurassic system in west Somerset, South-West England: New observations on the succession of ammonite faunas of the lowest Hettangian stage: Proceedings of the Ussher Society, v. 9, p. 231–235.

Palfy, J., Mortensen, J.K., Smith, P.L., Carter, E.S., Friedman, R.M., and Tipper, H.W., 2000a, Timing the end-Triassic mass extinction: First on land, then in the sea?: Geology, v. 28, p. 39–42.

Palfy, J., Smith, P.L., and Mortensen, J.K., 2000b, A U-Pb and $^{40}Ar/^{39}Ar$ time scale for the Jurassic: Canadian Journal of Earth Sciences, v. 37, p. 923–944.

Pedersen, K.R., and Lund, J.J., 1980, Palynology of the plant-bearing Rhaetian to Hettangian Kap Stewart Formation, Scoresby Sund, East Greenland: Review of Palaeobotany and Palynology, v. 31, p. 1–69.

Rampino, M.R., and Stothers, R.B., 1988, Flood basalt volcanism during the past 250 million years: Science, v. 241, p. 663–668.

Ratcliffe, N.M., 1988, Reinterpretation of the relationship of the western extension of the Palisades Sill to the lava flows at Ladentown, New York, based on new core data: U.S. Geological Survey Bulletin 1776, p. 113–135.

Reimold, W.U., Koeberl C., and Bishop J., 1994, Roter Kamm impact crater, Namibia: Geochemistry of basement rocks and breccias: Geochimica et Cosmochimica Acta, v. 58, p. 2689–2710.

Schulz, E., 1967, Sporenpalaeontologische Untersuchungen raetoliassischer Schichten im Zentralteil des germanischen Beckens: Palaeontologische Abhandlungen, Abteilung B: Palaeobotanik v. 2, no. 3, p. 541–626.

Sepkoski, J.J., Jr., 1997, Biodiversity: Past, present, and future: Journal of Paleontology, v. 71, p. 533–539.

Silvestri, S.M., and Szajna, M.J., 1993, Biostratigraphy of vertebrate footprints in the Late Triassic section of the Newark basin, Pennsylvania: Reassessment of stratigraphic ranges, *in* Lucas, S.G., and Morales, M., eds., The nonmarine Triassic: New Mexico Museum of Natural History and Science Bulletin 3, p. 439–445.

Smith, R.C., Berkheiser, S.W., Barnes, J.H., and Hoff, D.T., 1988, Strange clay baffles geologists: Pennsylvania Geology, v. 19, p. 8–13.

Sues, H.-D., Olsen, P.E., Scott, D.M., and Spencer, P.S., 2000, Cranial osteology of *Hypsognathus fenneri*, a latest Triassic procolophonid reptile from the Newark Supergroup of eastern North America: Journal of Vertebrate Paleontology, v. 20, p. 275–284.

Sues, H.-D., Shubin, N.H., and Olsen, P.E., 1994, A new sphenodontian (Lepidosauria: Rhynchocephalia) from the McCoy Brook Formation (Lower Jurassic) of Nova Scotia, Canada: Journal of Vertebrate Paleontology, v. 14, no. 3, p. 327–340.

Sutter, J.F., 1988, Innovative approaches to the dating of igneous events in the early Mesozoic basins of the Eastern United States: U.S. Geological Survey Bulletin 1776, p. 194–200.

Sweet, A.R., Braman, D.R., and Lerbekmo, J.F., 1990, Palynofloral response to K/T boundary events: A transitory interruption within a dynamic system, *in* Sharpton, V.L., and Ward, P.D., eds., Global catastrophes in Earth history: An interdisciplinary conference on impacts, volcanism, and mass mortality: Geological Society of America Special Paper 247, p. 457–469.

Szajna, M.J., and Hartline, B.W., 2001, A new vertebrate footprint locality from the Late Triassic Passaic Formation near Birdsboro, Pennsylvania, *in* LeTourneau, P.M., and Olsen, P.E., eds., Aspects of Triassic-Jurassic geoscience: New York, Columbia University Press (in press).

Szajna, M.J., and Silvestri, S.M., 1996, A new occurrence of the ichnogenus *Brachychirotherium*: Implications for the Triassic-Jurassic extinction event, *in* Morales, M., ed., The continental Jurassic: Museum of Northern Arizona, Bulletin 60, p. 275–283.

Tourani, A., Lund, J.J., Benaouiss, N., and Gaup, R., 2000, Stratigraphy of Triassic synrift deposition in Western Morocco, *in* Bachmann, G., and Lerche, I., eds., Epicontinental Triassic, Volume 2: Zentralblatt für Geologie und Paläontologie, Teil I, 1998, Heft 11/12, p. 1193–1215.

Tschudy, R.H., Pilmore, C.L., Orth, C.J., Gilmore, J.S., and Knight, J.D., 1984, Disruption of the terrestrial plant ecosystem at the Cretaceous-Tertiary boundary, western interior: Science, v. 225, p. 1030–1032.

Turrin, B., 2000, $^{40}Ar/^{39}Ar$ mineral ages and potassium and argon systematics from the Palisade Sill, New York: Eos (Transactions, American Geophysical Union) Supplement, v. 81, no. 48, p. F1326.

Van Houten, F.B., 1962, Cyclic sedimentation and the origin of analcime-rich Upper Triassic Lockatong Formation, west central New Jersey and adjacent Pennsylvania: American Journal of Science, v. 260, p. 561–576.

Van Houten, F.B., 1964, Cyclic lacustrine sedimentation, Upper Triassic Lockatong Formation, central New Jersey and adjacent Pennsylvania, *in* Meriam, D.F., ed., Symposium on cyclic sedimentation: Kansas Geological Survey Bulletin, v. 169, p. 497–531.

Van Houten, F.B., 1969, Late Triassic Newark Group, north-central New Jersey and adjacent Pennsylvania and New York, *in* Subitzki, S., ed., Geology of selected areas in New Jersey and eastern Pennsylvania and Guidebook of Excursions, Geological Society of America, Field Trip 4, Atlantic City, New Jersey: New Brunswick, Rutgers University Press, p. 314–347.

van Veen, P.M., Fowell, S.J., and Olsen, P.E., 1995, Time calibration of Triassic/Jurassic microfloral turnover, eastern North America: Discussion and reply: Tectonophysics, v. 245, p. 93–99.

Yang, Z., Moreau, M.-G., Bucher, H., Dommergues, J.-L., and Trouiller, A., 1996, Hettangian and Sinemurian magnetostratigraphy from Paris Basin: Journal of Geophysical Research, v. 101, p. 8025–8042.

Manuscript Submitted December 14, 2000; Accepted by the Society March 22, 2001

Printed in the U.S.A.

Geological Society of America
Special Paper 356
2002

Dating the end-Triassic and Early Jurassic mass extinctions, correlative large igneous provinces, and isotopic events

József Pálfy*
Hungarian Natural History Museum, P.O. Box 137, H-1431, Budapest, Hungary
Paul L. Smith
James K. Mortensen
Department of Earth and Ocean Sciences, University of British Columbia, Vancouver, British Columbia V6T 1Z4, Canada

ABSTRACT

The end-Triassic marks one of the five biggest mass extinctions, and was followed by a well-known second-order extinction event in the Early Jurassic. Previously published geological time scales were inadequate for correlation of extinctions with other global events and to unravel their dynamics. Here we present a revised time scale based on high-precision U-Pb ages integrated with ammonoid biochronology resolved to the zone level. This compilation suggests that the end of the Triassic Period (ca. 200 Ma) coincided with peak volcanism in the Central Atlantic magmatic province and that terrestrial floral and faunal extinctions may have slightly preceded the marine biotic crisis. The $^{87}Sr/^{86}Sr$ and $\delta^{13}C$ stratigraphic records are compatible with volcanically induced global environmental change that could be the proximal cause of extinction.

The revised Early Jurassic time scale suggests that peak extinction in the early Toarcian occurred at 183 Ma. Recent isotopic dating of flood basalts from the southern Gondwanan Karoo and Ferrar provinces documents a synchronous culmination in volcanic activity at 183 ± 2 Ma. The onset of volcanism is correlative with the start of a rapid rise in seawater $^{87}Sr/^{86}Sr$ ratios. A recently recognized negative $\delta^{13}C$ anomaly, tentatively ascribed to a massive release of methane hydrate, and the subsequent widespread oceanic anoxia suggest that the environmental perturbations thought to trigger the extinction also seriously disrupted the global carbon cycle. The interval between these two extinctions is 18 m.y., significantly shorter than the hypothetical 26 m.y. periodicity of extinctions.

INTRODUCTION

The Mesozoic era is framed by the two most studied mass extinctions, the end-Permian and Cretaceous-Tertiary events. Much less research effort has been devoted to two other events that occurred in the first half of the Mesozoic, i.e., at the close of Triassic and in the Early Jurassic. The end-Triassic mass extinction is the least studied and most poorly understood event among the five major mass extinctions (Hallam, 1996a). A subsequent extinction in the Early Jurassic, a second-order event

*E-mail: palfy@paleo.nhmus.hu

Pálfy, J., Smith, P.L., and Mortensen, J.K., 2002, Dating the end-Triassic and Early Jurassic mass extinctions, correlative large igneous provinces, and isotopic events, *in* Koeberl, C., and MacLeod, K.G., eds., Catastrophic Events and Mass Extinctions: Impacts and Beyond: Boulder, Colorado, Geological Society of America Special Paper 356, p. 523–532.

close to the Pliensbachian-Toarcian boundary, is one of the better known minor events (Harries and Little, 1999). However, a common impediment to the reconstruction of these crises and identification of their causes is the inadequate knowledge of their timing, due to the poor calibration of the latest Triassic and Early Jurassic time scale.

Extraterrestrial impacts, climate changes, sea-level changes, oceanic anoxia, and flood basalt volcanism are among the most frequently cited agents that could lead to elevated extinction rates or ecosystem collapse. In each case, testing of competing hypotheses requires precise timing and correlation of events. A recent revision of the Jurassic numerical time scale employed high-precision U-Pb zircon or $^{40}Ar/^{39}Ar$ geochronology of volcanic-ash layers embedded in fossiliferous sedimentary rocks (Pálfy et al., 2000b). Herein we review the new timing limitations of early Mesozoic extinctions in order to gain new insights into their potential causes.

The ramifications of the refined time frame have been discussed elsewhere for the end-Triassic (Pálfy et al., 2000a) and early Toarcian events (Pálfy and Smith, 2000). In this chapter we compare the two events and cite evidence for the synchronicity of extinctions and pulses of flood basalt volcanism. Temporal relationships between biotic crises and flood basalt volcanism in the Central Atlantic magmatic province and the Karoo-Ferrar igneous province are analyzed using a summary of recently published isotopic dates for the two large igneous provinces. Synchrony suggests possible causal relationships, as proposed earlier (e.g., Rampino and Stothers, 1988; Courtillot, 1994). If volcanically triggered environmental perturbations were the driving force of these extinctions, then distinctive isotopic signatures are expected in the stratigraphic record and can be used to test hypotheses. Therefore we also assess the compatibility of such scenarios with recent isotopic data.

RADIOMETRIC DATING OF EARLY MESOZOIC EXTINCTIONS

The accuracy of numerical time scales commonly used in the 1990s (e.g., Harland et al., 1990; Gradstein et al., 1994; Fig. 1) is compromised by several problems: (1) they are based on a small number of isotopic ages, (2) many of the isotopic ages were produced by K-Ar and Rb-Sr dating methods, which are considered less reliable than U-Pb and $^{40}Ar/^{39}Ar$ ages, and (3) the unjustified assumption of equal duration of biochronologic units is used for interpolation and the estimation of boundary ages. Recent integrated dating around the critical extinction intervals, primarily using U-Pb geochronology and ammonite biochronology from the western North American Cordillera, is summarized here. This work led to the first interpolation-free, independent stage and zonal boundary and duration estimates (Pálfy et al. 2000b; Fig. 1). The biochronological underpinning of the time scale is a North American regional ammonoid zonal scheme correlated with the primary standard chronostratigraphy of northwestern Europe (Smith et al., 1988, 1994; Jakobs et al., 1994).

Age of the end-Triassic event

It is generally accepted that a mass extinction corresponds to the Triassic-Jurassic (Tr-J) system boundary, even though detailed documentation is hindered by a dearth of fossiliferous and continuous sections worldwide (Hallam, 1990). One of the four proposed sections for the basal Jurassic Global Stratotype Section and Point is located on Kunga Island (Queen Charlotte Islands, British Columbia). At this locality, a tuff layer in the marine sedimentary section that contains the Tr-J boundary yielded a U-Pb zircon age of 199.6 ± 0.4 Ma (Pálfy et al., 2000a). This age provides a direct estimate for the age of the Tr-J boundary because the sampled layer is immediately below the system boundary as defined by integrated radiolarian, ammonoid, and conodont biochronology. Ages quoted by published time scales are invariably older by several million years (Fig. 1). The two most widely used estimates are 208.0 ± 7.5 Ma (Harland et al., 1990) and 205.7 ± 4.0 Ma (Gradstein et al., 1994). Other time scales list 208 Ma (Palmer, 1983), 210 Ma (Haq et al., 1988), and 203 Ma (Odin, 1994) as the best boundary estimates. Several additional biostratigraphically defined U-Pb dates were recently obtained from marine island arc terranes of the North American Cordillera (Table 1). These dates were not considered in previous time scales, but they convincingly support the conclusion that the true age of the Tr-J boundary is close to 200 Ma (Pálfy et al., 2000a).

The age of the end-Triassic extinction can also be estimated from terrestrial sections in eastern North America, where precise U-Pb dates are available from volcanic units within the continental Newark Supergroup (Dunning and Hodych, 1990; Hodych and Dunning, 1992). The North Mountain Basalt was dated as 201.7 $+1.4/-1.1$ Ma, whereas the Palisades and Gettysburg sills yielded ages of 200.9 ± 1.0 Ma and 201.3 ± 1.0 Ma, respectively. On the basis of geochemical and field evidence, the Palisades sill appears to have fed the lowermost flows of the Orange Mountain Basalt (Ratcliffe, 1988). The extrusive volcanic rocks postdate the palynologically defined Tr-J boundary (Fowell and Olsen, 1993) by only 20–40 k.y., on the basis of cyclostratigraphic evidence (Olsen et al., 1996). Vertebrate extinction, as deduced from tetrapod remains (Olsen et al., 1987) and their trace fossil record (Silvestri and Szajna, 1993), is coincident with the peak in floral turnover. The three overlapping isotopic ages and their respective errors suggest that the terrestrial extinction occurred no later than 200.6 Ma. The marine event, best represented by a sharp turnover in radiolarian taxa and delimited by the U-Pb age from Kunga Island, did not occur before 200.0 Ma. Taking these dates at face value suggests that the crisis of terrestrial biota preceded that of the marine realm by at least 600 k.y. (Pálfy et al., 2000a). This is the first indication of such temporal dichotomy within a major mass extinction.

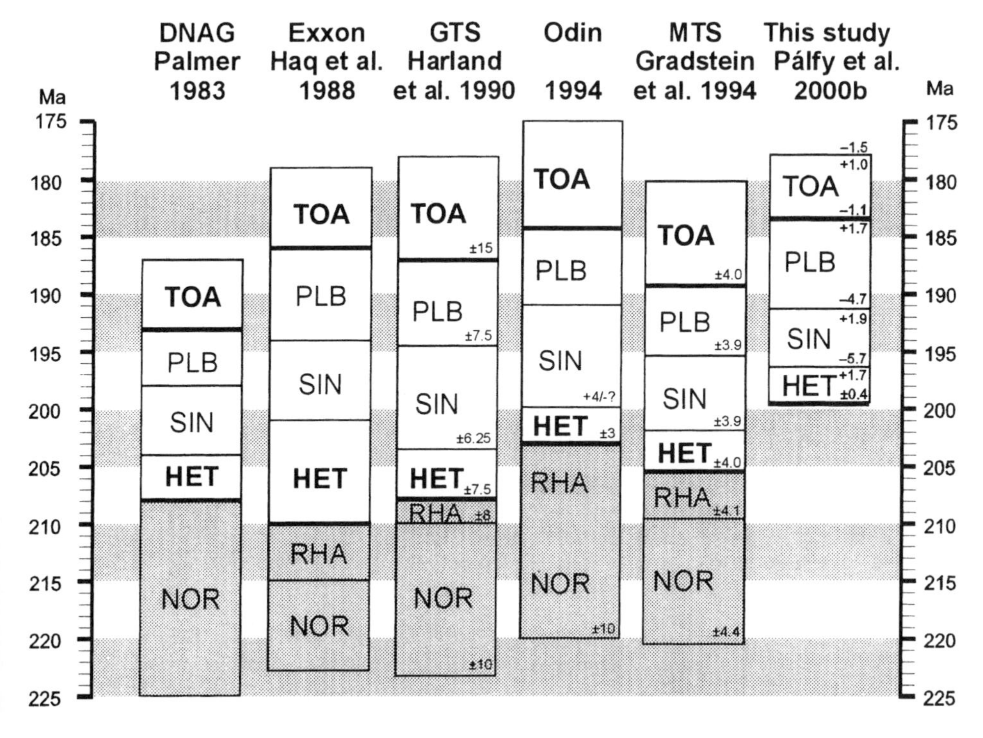

Figure 1. Comparison of Early Jurassic and latest Triassic (shaded) numerical time scales. Note different estimates for age of Triassic-Jurassic and Pliensbachian-Toarcian boundaries in previously widely used time scales vs. revised scale of Pálfy et al. (2000b). Small numbers at stage boundaries indicate stated uncertainty of estimates. Stage abbreviations: NOR, Norian; RHA, Rhaetian; HET, Hettangian; SIN, Sinemurian; PLB, Pliensbachian; TOA, Toarcian. Time-scale abbreviations: DNAG, Decade of North American Geology; GTS, geological time scale; MTS, Mesozoic time scale. Note that Rhaetian stage is not recognized by DNAG scale, whereas Odin (1994) did not estimate age of Norian-Rhaetian boundary.

TABLE 1. LIST OF RECENTLY PUBLISHED U-Pb ZIRCON DATES RELEVANT TO THE AGE OF THE TRIASSIC-JURASSIC BOUNDARY

Dated rock	Locality	U-Pb age (Ma)	Biochronologic age	
			Maximum	Minimum
Tuff in Talkeetna Formation	Puale Bay, Alaska	197.8 ± 1.0	Middle Hettangian	Late Hettangian
Tuff in Talkeetna Formation	Puale Bay, Alaska	197.8 ± 1.2/−0.4	Middle Hettangian	Late Hettangian
Tuff in Kamishak Formation	Puale Bay, Alaska	200.8 ± 2.8	Middle Hettangian	Middle Hettangian
Goldslide Porphyry (Goldslide Intrusions)	Stewart, B.C.	197.6 ± 1.9	Hettangian	Hettangian
Tuff in Hazelton Group	Stewart, B.C	199 ± 2	Hettangian	Hettangian
Biotite Porphyry (Goldslide Intrusions)	Stewart, B.C	201.8 ± 0.5	Norian	Rhaetian
Griffith Creek volcanics	Spatsizi River, British Columbia	205.8 ± 0.9	Norian	Rhaetian
Griffith Creek volcanics	Spatsizi River, British Columbia	205.8 ± 1.5/−3.1	Norian	Rhaetian

Note: References to sources of isotopic ages and their biochronologic constraints are given in Pálfy et al. (2000b). See Figure 1 for stages.

Age of the Early Jurassic event

An Early Jurassic (Pliensbachian) extinction event was first recognized from a global database of the stratigraphic ranges of marine animal families and genera resolved to stratigraphic stages (Raup and Sepkoski, 1984). An independent compilation of fossil families detected to 5% marine extinction in both the Pliensbachian and Toarcian, and 2.4%–12.8% extinction among continental organisms in the Toarcian (Benton, 1995). On the basis of detailed analysis of the fossil record of northwestern European epicontinental seas, Hallam (1986, 1996a) regarded the extinction as a regional event (Fig. 2D), the later phase of which coincided with widespread anoxia in the early Toarcian Falciferum zone (Jenkyns, 1988). Little and Benton (1995) analyzed the time distribution of global family extinctions and found that a protracted interval of five zones spanning the Pliensbachian-Toarcian stage boundary showed elevated extinction levels (Fig. 2C). However, outcrop-scale studies of the most fossiliferous sections in England and Germany displayed a clear species extinction peak correlating with the anoxic event in the Falciferum zone (Little, 1996). The global extent of the Pliensbachian-Toarcian extinction event was established through detailed studies in the Andean basin (Aberhan and Fürsich, 1997) and deep-water facies of the western Tethys (Vörö, 1993) and Japan (Hori, 1993).

Previous best estimates for the Pliensbachian-Toarcian boundary are 187.0 ± 15 Ma (Harland et al., 1990) and 189.6 ± 4.0 (Gradstein et al., 1994). This portion of our revised Jurassic time scale is built upon 14 recently obtained U-Pb ages from volcanic layers that are also dated by ammonoid biochronology in the North American Cordillera (Pálfy et al., 2000b). None of these isotopic ages had been used in earlier

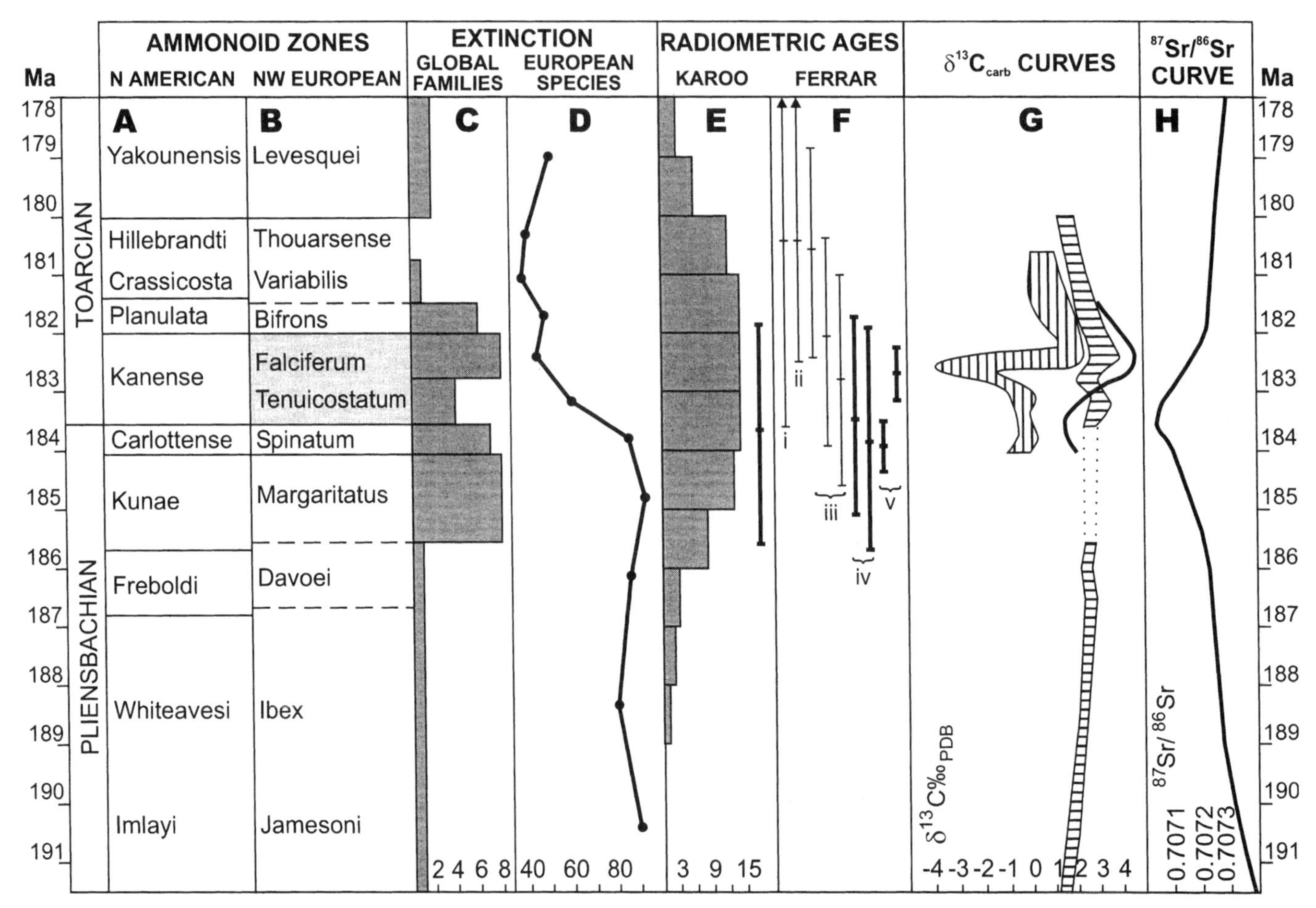

Figure 2. Correlation of marine extinction event, Karoo and Ferrar flood basalt volcanism, and carbon and strontium isotope stratigraphy in numerically calibrated ammonoid zonal chronostratigraphic framework. A: North American regional standard ammonoid zonation (Pliensbachian: Smith et al., 1988; Toarcian: Jakobs et al., 1994); lack of horizontal line between zones indicates that no numerical estimate is available for zone boundary. B: Northwest European standard ammonoid zonation; shading indicates extent of organic-rich deposits in Tenuicostatum and Falciferum zones in western Tethys and northwest Europe. C: Number of global family extinctions by zone (Little and Benton, 1995). D: Cumulative species diversity per zone, expressed in number of species of bivalves, ammonoids, rhynchonellid brachiopods, crinoids, foraminifera, and ostracods from Britain (Hallam, 1996a). E: Radiometric ages from Karoo Group (recalculated from published sources with corrected standard ages for $^{40}Ar/^{39}Ar$ geochronology and 2σ external errors [unless indicated otherwise] for valid comparison; Renne et al., 1998); age spectrum histogram of 28 $^{40}Ar/^{39}Ar$ dates with 1σ internal errors (Duncan et al., 1997); and error bar of U-Pb age (Encarnación et al., 1996). F: $^{40}Ar/^{39}Ar$ (thin lines) and U-Pb (heavy lines) ages from Ferrar Group (recalculated from published sources with corrected standard ages for $^{40}Ar/^{39}Ar$ geochronology and 2σ external errors for valid comparison; Renne et al., 1998). Error bars, from left to right: i, composite of 11 $^{40}Ar/^{39}Ar$ ages by Heimann et al. (1994); ii, composite of two $^{40}Ar/^{39}Ar$ ages by Foland et al. (1993); iii, three $^{40}Ar/^{39}Ar$ ages by Duncan et al. (1997); iv, two U-Pb ages by Encarnación et al. (1996); and v, two U-Pb ages by Minor and Mukasa (1997). G: Carbon isotope profiles: horizontal hachure, composite profile from Central Apennines, Italy (E. Morettini, 1999, personal commun.); vertical hachure, England (Jenkyns and Clayton, 1997; Hesselbo et al., 2000); solid line, composite curve from Tethyan sections (Jenkyns et al., 1991). H: Seawater $^{87}Sr/^{86}Sr$ curve simplified from Jones et al. (1994); monotonous decline of curve starts in Hettangian from values >0.7077 (modified from Pálfy and Smith, 2000).

time scales (Table 2). The density and biochronologic resolution of the isotopic age database across the Pliensbachian-Toarcian transition allows, for the first time, the estimation of zonal boundary ages for six consecutive zones. Zonal boundary ages are calculated using the chronogram method (Harland et al., 1990), except for the base of Crassicosta zone, which is directly dated in the Queen Charlotte Islands (Pálfy et al., 1997). Ammonoid provinciality warrants the use of the North American regional ammonoid zonal scale (Fig. 2A), which is correlated with the northwest European standard chronostratigraphy following Smith et al. (1988) and Jakobs et al. (1994) (Fig. 2B). Calculated best estimates for initial zonal boundaries are as follows (Fig. 2A): Kunae zone (early-late Pliensbachian boundary), 185.7 + 0.5/ − 0.6 Ma; Carlottense zone: 184.1 + 1.2/ − 1.6 Ma; Kanense zone (Pliensbachian-Toarcian boundary), 183.6 + 1.7/ − 1.1 Ma; Planulata zone, 182.0 + 3.3/ − 1.8 Ma; Crassicosta zone, 181.4 ± 1.2 Ma.

TABLE 2. LIST OF RECENTLY PUBLISHED U-Pb ZIRCON DATES USED IN ESTIMATING THE AGE OF PLIENSBACHIAN AND TOARCIAN ZONAL BOUNDARIES

Dated rock	Locality	U-Pb age (Ma)	Biochronologic age (zone or stage)	
			Maximum	Minimum
Chuchi intrusion	Chuchi property	188.5 ± 2.5	Whiteavesi	Kunae
Tuff in Laberge Group	Atlin Lake (East shore)	187.5 ± 1.0	Whiteavesi	Whiteavesi
Tuff in Hazelton Group	Todagin Mountain, Spatsizi area	185.6 ± 7.3/−0.6	Freboldi	Freboldi
Granitoid boulder in Laberge Gr.	Atlin Lake (Sloko Island)	186.6 ± 0.5/−1.0*	Sinemurian	Kunae
Tuff in Laberge Group	Atlin Lake (Copper Island)	185.8 ± 0.7	Kunae	Kunae
Tuff in Hazelton Group	Skinhead Lake	184.7 ± 0.9	Kunae	Kunae
Tuff (Nordenskiold volcanics)	Whitehorse	184.1 ± 5.8/−1.6	Kunae	Kunae
Eskay porphyry	Eskay Creek, Iskut River area	184 ± 6/−1	Carlottense	Aalenian
McEwan Creek pluton	McEwan Creek, Spatsizi area	183.2 ± 0.7	Kanense	Planulata
Tuff in Hazelton Group	Mt. Brock range, Spatsizi area	180.4 ± 11.2/−0.4	Kanense	Planulata
Tuff in Whiteaves Formation	Yakoun River, Queen Charlotte Island	181.4 ± 1.2	Crassicosta	Crassicosta
Tuff in Hazelton Group	Julian Lake	178 ± 1	Yakounensis	Aalenian
Eskay rhyolite	Eskay Creek anticline, west limb	175.1 ± 4.7	Yakounensis	Aalenian
Eskay rhyolite	Eskay Creek anticline, east limb	174.1 ± 4.5/−1.1	Yakounensis	Aalenian

*Maximum age.
References to sources of isotopic ages and their biochronologic constraints are given in Pálfy et al. (2000b).
See Figure 2 for ammonoid zones.

RADIOMETRIC DATING OF COEVAL VOLCANIC EVENTS

The flood basalt volcanism of large igneous provinces has been invoked as a cause of extinctions (e.g., Rampino and Stothers, 1988, Courtillot, 1994). Although the proposal was made on broad time correlation when the numerical ages of both the volcanic provinces and the extinction boundaries were poorly established, recent advances permit a reevaluation of temporal linkages.

Age of the Central Atlantic magmatic province

Basalts around the margins of the central Atlantic form a vast, once contiguous volcanic province, termed the Central Atlantic magmatic province. The volume of coeval volcanic rocks is estimated to be in excess of 2×10^6 km^3 (Marzoli et al., 1999); thus the Central Atlantic magmatic province ranks among the most voluminous of the Phanerozoic volcanic provinces. Tholeiitic basalts of the Central Atlantic magmatic province are now exposed along the eastern seaboard of North America, Brazil, and adjacent parts of northeastern South America, western Africa, and the Iberian Peninsula.

Recent studies significantly improved the dating of this extensive basaltic volcanism. The ^{40}Ar/^{39}Ar dating by Marzoli et al. (1999), mainly from previously undated areas in South America, documented that Central Atlantic magmatic province eruptions represent a short-lived episode that reached its peak intensity at 200 Ma. Marzoli et al. compiled a total of 41 newly reported and previously published ^{40}Ar/^{39}Ar and U-Pb dates, and calculated a mean age of 199.0 ± 2.4 Ma (2σ external error). This data set includes U-Pb dates from the Newark Supergroup discussed herein. Further ^{40}Ar/^{39}Ar dating of related rocks from the southwestern United States is fully compatible with this age distribution (Hames et al., 2000).

An additional line of evidence for a short-lived volcanic pulse is furnished by paleomagnetic studies. Kent et al. (1995) documented that extrusive rocks in the Newark Supergroup all belong to the single normal polarity chron that spans the Tr-J boundary. Similarly, paleomagnetic data from South American reported by Marzoli et al. (1999) and previous studies summarized therein all revealed normal magnetic polarity for Central Atlantic magmatic province volcanic rocks.

Age of the Karoo-Ferrar volcanic province

The Karoo province in South Africa and the Ferrar province in Antarctica are disjunct parts of a once contiguous large igneous province of Jurassic age in Gondwana. The estimated minimum volume of coeval igneous rocks is 2.5×10^6 km^3 (Encarnación et al., 1996); thus it ranks among the most voluminous flood basalt provinces of the Phanerozoic (Rampino and Stothers, 1988). Early radiometric dating, relying on the K-Ar method, was plagued with problems. A suite of whole-rock ages for the Karoo Group is distributed between 135 and 225 Ma, with apparent peaks of volcanic intensity at 193 ± 5 and 178 ± 5 Ma (Fitch and Miller, 1984). The K-Ar chronometer often yields anomalously young or old ages in disturbed systems, due to Ar loss or uptake of excess Ar, respectively. The use of ^{40}Ar/^{39}Ar and U-Pb dating, however, permits reliable determination of the true crystallization age of mafic igneous rocks. For valid comparison between dates recently obtained using different isotopic methods, we recalculated the published ages to reflect external errors (i.e., including decay constant uncertainty) at the 2σ level and the currently accepted ages of standards for ^{40}Ar/^{39}Ar dating (Renne et al., 1998).

Duncan et al. (1997) reported 28 precise ^{40}Ar/^{39}Ar plateau ages from Karoo Group basalts and dolerites in South Africa and Namibia. The ages range between 179 and 186 Ma, with a majority at 183 ± 2 Ma (Fig. 2E). A U-Pb age of 183.7 ± 1.9 Ma obtained by Encarnación et al. (1996) from a tholeiitic sheet in South Africa is in good agreement with the $^{40}Ar/^{39}Ar$ results (Fig. 2E).

Various units within the Ferrar Group in Antarctica have also been dated (Fig. 2F). From the Kirkpatrick Basalt, Foland et al. (1993) reported two nearly identical incremental heating $^{40}Ar/^{39}Ar$ ages of 180.4 ± 2.1 Ma. Also from the Kirkpatrick Basalt, Heimann et al. (1994) reported 11 $^{40}Ar/^{39}Ar$ plateau ages that form a tight cluster and permit a composite age determination of 180.3 ± 3.6 Ma. Basalts from the Kirwan Mountains, East Antarctica, yielded $^{40}Ar/^{39}Ar$ plateau ages of 180.6 ± 1.8, 182.7 ± 1.8, and 182.8 ± 1.8 Ma (Duncan et al., 1997). Concordant U-Pb ages of 183.4 ± 1.9 and 183.8 ± 1.9 were obtained by Encarnación et al. (1996) from sills within the Ferrar Group. Minor and Mukasa (1997) dated (U-Pb) two samples from the Dufek intrusion (which forms part of the Ferrar Group) as 183.9 ± 0.4 and 182.7 ± 0.5 Ma.

These radiometric ages suggest a short-lived magmatic episode, represented by coeval rocks of the Karoo and Ferrar Groups. Such brevity of volcanism is typical of most other large flood basalt provinces of the world (Coffin and Eldholm, 1994). The cluster of ages around 183 ± 2 Ma is interpreted as the peak of magmatic activity. Additional support for a short-lived magmatic episode is provided by paleomagnetic results. Hargraves et al. (1997) demonstrated that the bulk of basalts in the Karoo province erupted during a single polarity epoch.

INTERVAL BETWEEN EXTINCTIONS

Ever since the Cretaceous-Tertiary extinction was first linked to an extraterrestrial impact (Alvarez et al., 1980), other mass extinction horizons have been scrutinized for impact signatures. The case has also been made that extraterrestrial impacts were the prime causes of many or all extinctions based, in part, on their periodicity and the well-established Cretaceous-Tertiary case of synchrony (Raup and Sepkoski, 1984; Rampino and Haggerty, 1996). The end-Triassic and the Toarcian events represent two consecutive mass extinctions in a series of such events that were hypothesized to exhibit periodicity at a 26 m.y. phase length (Raup and Sepkoski, 1984). The periodicity hypothesis has been much debated and one frequently cited argument against it is the deficiency of the underlying time scale (Hoffman, 1985; Heisler and Tremaine, 1989). When updating the extinction time series analysis using the Harland et al. (1990) time scale, Sepkoski (1996) found that the end-Triassic and Pliensbachian extinctions did not closely follow the 26 m.y. pattern.

A corollary of our refined dating is a fresh look at the spacing of mass extinctions in question. The revised time scale confirms that the spacing of these two events is significantly less than 26 m.y. Accepting 199.6 ± 0.4 Ma as the best estimate for the Triassic-Jurassic boundary and 182.0 +3.3/−1.8 Ma as the end of the Falciferum zone, i.e., a time that postdates the Toarcian species extinction peak, it is evident that the two consecutive extinction events likely took place less than 18 m.y. apart.

ISOTOPIC EVIDENCE FOR SYNCHRONOUS GLOBAL ENVIRONMENTAL PERTURBATIONS

Stable isotope stratigraphy is a powerful tool for reconstructing past environmental change. Dramatic environmental changes, accompanied by disruption and reorganization of the global carbon cycle and other biogeochemical cycles, are known to occur associated with mass extinctions. There is ample evidence for pronounced isotopic excursions associated with many extinction events, most notably the end-Permian and the end-Cretaceous events (e.g., Baud et al., 1989; Zachos et al., 1989; Holser et al., 1996). Carbon isotope stratigraphy is especially useful for documenting changes in the biogeochemical cycling of carbon (Kump and Arthur, 1999), whereas oxygen isotopes are widely used for paleotemperature reconstruction. In addition, strontium isotope stratigraphy can be used as a proxy for changes in paleoclimate and global tectonics (e.g., Hodell and Woodruff, 1994).

Having established the temporal synchrony between early Mesozoic pulses of flood basalt volcanism and extinctions, isotopic signatures may provide clues to their cause and effect relationship.

Stable isotope record across the Tr-J boundary

Until recently, relatively little was known about the stable isotope history of the end-Triassic event. Among the available marine sections, early work concentrated on Kendelbachgraben (Morante and Hallam, 1996) and Lorüns (McRoberts et al., 1997) in Austria and New York Canyon in Nevada (Taylor et al., 1992). All met with mixed success. McRoberts et al. (1997) reported a small negative anomaly confined to a single sample from the boundary interval in the Lorüns section. However, at this locality the boundary is defined by bivalves; hence the stratigraphic completeness of the section cannot be proven. At Kendelbachgraben, a negative $\delta^{13}C_{carb}$ excursion is accompanied by a positive $\delta^{13}C_{org}$ and a negative $\delta^{18}O$ excursion, indicating diagenetic overprint rather than a primary signal (Hallam and Goodfellow, 1990; Morante and Hallam, 1996). The preliminary report of isotopic variations from Nevada (Taylor et al., 1992) cannot be properly evaluated. However, significant negative carbon isotope excursions were recently reported from marine Triassic-Jurassic boundary strata of widely separated localities of the eastern Pacific and western Tethys. In the Queen Charlotte Islands, western Canada, a negative $\delta^{13}C$ spike of $-1.5‰$ measured in bulk organic matter, from a level that corresponds to the radiolarian extinction (Ward et al., 2001). An even larger anomaly of up to $-3.5‰$ was detected in both bulk carbonate and organic matter in a section at Csővár, Hungary (Pálfy et al., 2001), within the Triassic-Jurassic transition defined by ammonoid and conodont biostratigraphy. Organic carbon isotopic data from a terrestrial section in Greenland also hint at a negative excursion at the boundary (McElwain et al., 1999).

The available data suggest that the terminal Triassic extinction coincided with a negative carbon isotope anomaly, similar to many other extinction events. However, more research is needed to assess the magnitude and duration of the anomaly, and to interpret the underlying changes in the global carbon cycle.

Stable isotope record of the Toarcian

Recognition of a prominent positive $\delta^{13}C$ excursion in the Falciferum zone, along with widespread organic-rich facies, is the basis for defining an early Toarcian oceanic anoxic event (Jenkyns, 1988). Originally the $\delta^{13}C$ maximum was thought to be restricted to the Falciferum zone, but in several Tethyan sections, the rise of $\delta^{13}C$ begins in the Tenuicostatum zone (Jenkyns et al., 1991; Jiménez et al., 1996; E. Morettini, 1999, personal commun.) (Fig. 2G). Organic-rich black shale deposition is also known in the Tenuicostatum zone in Spain and Italy (Jiménez et al., 1996; E. Morettini, 1999, personal commun.), and manganese-rich deposits are widespread in the Tenuicostatum to Falciferum zones (Jenkyns et al., 1991).

More recent stable isotope results of Hesselbo et al. (2000) indicate that the positive carbon isotope excursion was preceded by a short and intense negative anomaly that is equally measurable in both marine carbonate and terrestrial organic matter. Hesselbo et al. suggested that the isotopically light carbon could have entered the world ocean, the atmosphere, and the biosphere through sudden release of methane hydrate, which in turn might have been triggered by the warming of deep ocean water. The short-lived negative spike needs further investigation, because it has not been confirmed by a study on closely sampled belemnites (McArthur et al., 2000).

A $\delta^{18}O$ minimum in the Falciferum zone records a paleotemperature maximum for the Toarcian (Jenkyns and Clayton, 1997). Jenkyns and Clayton speculated that a correlation with increased CO_2 level is a strong possibility supported by the low $\delta^{13}C$ values of organic matter. The CO_2 from voluminous volcanic outgassing is a possible cause of greenhouse warming that may have been accelerated by gas-hydrate release.

Strontium isotope stratigraphy across the Tr-J boundary

The strontium isotopic evolution across the Tr-J transition is revealed by recent detailed studies from both the Jurassic (Jones et al., 1994) and the Triassic (Korte, 1999). A Late Triassic climb of the $^{87}Sr/^{86}Sr$ ratio was sharply reversed immediately prior to the Tr-J boundary (Veizer et al., 1999). The most likely event that affected the Sr isotopic composition of seawater is the initiation of volcanism in the Central Atlantic Magmatic Province, which could have simultaneously altered the global climate and the composition of exposed rocks.

Strontium isotope stratigraphy of the Pliensbachian-Toarcian

Temporal variations in the $^{87}Sr/^{86}Sr$ ratio of the Early Jurassic oceans were documented by Jones et al. (1994) (Fig. 2H) and refined by McArthur et al. (2000). Following a nearly continuous decline from the Hettangian to the Pliensbachian, the curve reaches a minimum at the Pliensbachian-Toarcian boundary and rises in the Toarcian; the steepest slope is recorded for the Falciferum zone. It is notable that the major Early Jurassic inflection appears to coincide with the inception of Karoo-Ferrar volcanism. The early Toarcian rise can be related to increased humidity and continental weathering, possibly enhanced by acid rain, under escalating greenhouse conditions triggered by volcanic emissions.

DISCUSSION

The synchrony of the end-Triassic extinction and volcanism in the Central Atlantic magmatic province, and the Early Jurassic extinction and volcanism in the Karoo-Ferrar province now appear well established. Geochemical data, primarily carbon and strontium isotope stratigraphy, support models that call on environmental perturbations triggered by volcanism as a potential factor in the biotic extinctions.

Here we discuss three key issues: extinction dynamics, carbon isotope evolution, and strontium isotope evolution. We compare the two extinctions to provide insight and to indicate where further research is most needed.

For the Early Jurassic event, the tempo of extinction was found curiously different at the species level from that at higher taxonomic ranks. Little and Benton (1995) documented that elevated family extinction level was sustained through five zones in the late Pliensbachian-early Toarcian, an interval of 4 m.y. in the revised time scale used here. However, abrupt species-level extinctions are concentrated in the early Toarcian Tenuicostatum zone. The subsequent Falciferum zone can be described as a survival interval, which in turn was followed by rapid repopulation and recovery (Harries and Little, 1999). Thus the Pliensbachian-Toarcian extinction exhibits the attributes of both press and pulse events (Erwin, 1998).

Few similarly detailed studies are available from the Tr-J boundary. Sudden extinction and major turnover are recorded among radiolarians (Carter, 1994), pollen (Fowell and Olsen, 1993), and megaflora (McElwain et al., 1999). The duration of the Rhaetian, a time of protracted decline for many fossil groups including the ammonites, bivalves, and conodonts (Hallam, 1996a), is poorly defined due to a lack of isotopic ages (see Fig. 1). Timing of the earliest Jurassic marine biotic recovery is better known. The early Hettangian (Planorbis zone) is characterized by a low-diversity fauna worldwide and is perhaps best regarded as a postextinction lag or survival period. True recovery and diversification started in the middle Hettangian within many clades (Hallam, 1996b). Hettangian U-Pb dates from Alaska (Table 1) (Pálfy et al., 1999) indicate that recovery was underway within less than 2 m.y. Whether the apparent length of the biotic crisis and the delayed rebound are artifacts of inadequate sampling (Signor and Lipps, 1982; Erwin, 1998) remains to be tested. Attributes of a mixed press and pulse event similar to the Toarcian are apparent from the available data. This pattern and the suggested temporal difference between terrestrial and marine events are consistent with extinction scenarios that invoke long-term environmental change, but with a shorter response time or lower threshold in the more vulnerable terrestrial biota.

Gradual global warming, induced by flood basalt volcanism, may trigger the sudden release of methane hydrate and a positive feedback mechanism that in turn may cause catastrophic climate change and extinctions. This is the scenario suggested by the sharp negative $\delta^{13}C$ spike for the early Toarcian (Hesselbo et al., 2000). Methane hydrate release, first invoked to account for the late Paleocene thermal maximum (Dickens et al., 1995), has also been considered for the Permian-Triassic event (Erwin, 1993; Krull and Retallack, 2000). However, it is not clear whether this model can be applied to the Tr-J extinction event. The need to improve our understanding of the carbon isotope record is underscored by three lines of evidence that suggest that the model may be applicable: (1) paleobotanical data indicate a super-greenhouse Earth at the Tr-J boundary (McElwain et al., 1999), (2) the similar press and pulse dynamics of the two extinction events, and (3) the synchronous occurrence of flood basalt volcanism and extinction.

We suggest that the end-Pliensbachian reversal and Toarcian increase of the Sr isotope ratio records the onset of Karoo-Ferrar volcanism. A similar inflection occurs in the Late Permian, although it appears to slightly predate the Siberian Traps (Martin and Macdougall, 1995). The formation of the Central Atlantic magmatic province around the Tr-J boundary coincides with a downturn of the Sr curve. Modeling suggests that continental flood basalt volcanism could alter seawater chemistry via enhanced weathering and increased riverine flux (Martin and Macdougall, 1995). We speculate that changes of opposite sense across the Tr-J boundary may reflect the low to middle paleolatitude of the Central Atlantic magmatic province (vs. the middle- to high-latitude Siberian Traps and Karoo-Ferrar provinces), whereby basalt weathering exerts greater influence on the oceanic Sr budget and explains a shift toward less radiogenic values (Taylor and Lasaga, 1999).

Advances in our understanding of the early Mesozoic extinctions are expected from further integrated paleontological and geochemical studies. We cannot escape a conclusion that flood basalt volcanism in the Central Atlantic magmatic province and the Karoo-Ferrar provinces likely played a significant role in driving the end-Triassic and Toarcian extinctions, although details of the links are only beginning to emerge.

ACKNOWLEDGMENTS

Pálfy acknowledges support from the Hungarian Scientific Research Fund (grant F23451) and the Alexander von Humboldt Foundation. Presentation of this paper at the Vienna conference was made possible by a grant-in-aid to Pálfy. Smith and Mortenesen acknowledge support of the Natural Sciences and Engineering Research Council of Canada. Reviews and comments by M. Coffin, K. MacLeod, and P.B. Wignall helped improve the manuscript.

REFERENCES CITED

Aberhan, M., and Fürsich, F.T., 1997, Diversity analysis of Lower Jurassic bivalves of the Andean Basin and the Pliensbachian-Toarcian mass extinction: Lethaia, v. 29, p. 181–195.

Alvarez, L., Alvarez, W., Asaro, F., and Michel, H., 1980, Extraterrestrial cause for the Cretaceous-Tertiary extinction: Science, v. 208, p. 1095–1108.

Baud, A., Magaritz, M., and Holser, W.T., 1989, Permian-Triassic of the Tethys: Carbon isotope studies: Geologische Rundschau, v. 78, p. 649–677.

Benton, M.J., 1995, Diversification and extinction in the history of life: Science, v. 268, p. 52–58.

Carter, E.S., 1994, Evolutionary trends in latest Norian through Hettangian radiolarians from the Queen Charlotte Islands, British Columbia: Géobios, Mémoire Spécial, v. 17, p. 111–119.

Coffin, M.F., and Eldholm, O., 1994, Large igneous provinces: Crustal structure, dimensions and external consequences: Reviews of Geophysics, v. 32, p. 1–36.

Courtillot, V., 1994, Mass extinctions in the last 300 million years: One impact and seven flood basalts?: Israel Journal of Earth Sciences, v. 43, p. 255–266.

Dickens, R.G., O'Neil, J.R., Rea, D.K., and Owen, R.M., 1995, Dissociation of oceanic methane hydrate as a cause of the carbon isotope excursion at the end of the Paleocene: Paleoceanography, v. 10, p. 965–971.

Duncan, R.A., Hooper, P.R., Rehacek, J., Marsh, J.S., and Duncan, A.R., 1997, The timing and duration of the Karoo igneous event, southern Gondwana: Journal of Geophysical Research, v. 102, p. 18 127–18 138.

Dunning, G.R., and Hodych, J.P., 1990, U/Pb zircon and baddeleyite ages for the Palisades and Gettysburg sills of the northeastern United States: Implications for the age of the Triassic/Jurassic boundary: Geology, v. 18, p. 795–798.

Encarnación, J., Fleming, T.H., Elliot, D.H., and Eales, H., 1996, Synchronous emplacement of Ferrar and Karoo dolerites and the early breakup of Gondwana: Geology, v. 24, p. 535–538.

Erwin, D.H., 1993, The great Paleozoic crisis: New York, Columbia University Press, 327 p.

Erwin, D.H., 1998, The end and the beginning: Recoveries from mass extinctions: Trends in Ecology & Evolution, v. 13, p. 344–349.

Fitch, F.J., and Miller, J.A., 1984, Dating Karoo igneous rocks by the conventional K/Ar and $^{40}Ar/^{39}Ar$ age spectrum methods, *in* Erlank, A.J., ed., Petrogenesis of the volcanic rocks of the Karoo Province: Geological Society of Africa Special Publication 13, p. 247–266.

Foland, K.A., Fleming, T.H., Heimann, A., and Elliot, D.H., 1993, Potassium-argon dating of fine-grained basalts with massive Ar-loss: Application of the $^{40}Ar/^{39}Ar$ technique to plagioclase and glass from the Kirkpatrick Basalt, Antarctica: Chemical Geology, v. 107, p. 173–190.

Fowell, S.J., and Olsen, P.E., 1993, Time calibration of Triassic/Jurassic microfloral turnover, eastern North America: Tectonophysics, v. 222, p. 361–369.

Gradstein, F.M., Agterberg, F.P., Ogg, J.G., Hardenbol, J., van Veen, P., Thierry, J., and Huang, Z., 1994, A Mesozoic time scale: Journal of Geophysical Research, v. 99, p. 24 051–24 074.

Hallam, A., 1986, The Pliensbachian and Tithonian extinction events: Nature, v. 319, p. 765–768.
Hallam, A., 1990, The end-Triassic mass extinction event, *in* Sharpton, V.L., and Ward, P.D., eds., Global catastrophes in Earth history: An interdisciplinary conference on impacts, volcanism, and mass mortality: Geological Society of America Special Paper 247, p. 577–583.
Hallam, A., 1996a, Major bio-events in the Triassic and Jurassic, *in* Walliser, O.H., ed., Global events and event stratigraphy in the Phanerozoic: Berlin, Springer-Verlag, p. 265–283.
Hallam, A., 1996b, Recovery of the marine fauna in Europe after the end-Triassic and early Toarcian mass extinctions, *in* Hart, M.B., ed., Biotic recovery from mass extinction events: Geological Society [London] Special Publication 102, p. 231–236.
Hallam, A., and Goodfellow, W.D., 1990, Facies and geochemical evidence bearing on the end-Triassic disappearance of the Alpine reef ecosystem: Historical Biology, v. 4, p. 131–138.
Hames, W.E., Renne, P.R., and Ruppel, C., 2000, New evidence for geologically instantaneous emplacement of earliest Jurassic Central Atlantic Magmatic Province basalts on the North American margin: Geology, v. 28, p. 859–862.
Haq, B.U., Hardenbol, J., and Vail, P.R., 1988, Mesozoic and Cenozoic chronostratigraphy and cycles of sea level change: Society of Economic Paleontologists and Mineralogists Special Publication 42, p. 71–108.
Hargraves, R.B., Rehacek, J., and Hooper, P.R., 1997, Paleomagnetism of the Karoo igneous rocks in South Africa: South African Journal of Geology, v. 100, p. 195–212.
Harland, W.B., Armstrong, R.L., Cox, A.V., Craig, L.E., Smith, A.G., and Smith, D.G., 1990, A geologic time scale 1989: Cambridge, Cambridge University Press, 263 p.
Harries, P.J., and Little, C.T.S., 1999, The early Toarcian (Early Jurassic) and the Cenomanian-Turonian (Late Cretaceous) mass extinctions: Similarities and contrasts: Palaeogeography, Palaeoclimatology, Palaeoecology, v. 154, p. 39–66.
Heimann, A., Fleming, T.H., Elliot, D.H., and Foland, K.A., 1994, A short interval of Jurassic continental flood basalt volcanism in Antarctica as demonstrated by $^{40}Ar/^{39}Ar$ geochronology: Earth and Planetary Science Letters, v. 121, p. 19–41.
Heisler, J., and Tremaine, S., 1989, How dating uncertainties affect the detection of periodicity in extinctions and craters: Icarus, v. 77, p. 213–219.
Hesselbo, S.P., Gröcke, D.R., Jenkyns, H.C., Bjerrum, C.J., Farrimond, P., Morgans Bell, H.S., and Green, O.R., 2000, Massive dissociation of gas hydrate during a Jurassic oceanic anoxic event: Nature, v. 406, p. 392–395.
Hodell, D.A., and Woodruff, F., 1994, Variations in the strontium isotopic ratio of seawater during the Miocene: Stratigraphic and geochemical implications: Paleoceanography, v. 9, p. 405–426.
Hodych, J.P., and Dunning, G.R., 1992, Did the Manicouagan impact trigger end-of-Triassic mass extinction?: Geology, v. 20, p. 51–54.
Hoffman, A., 1985, Patterns of family extinction: Dependence on definition and geologic time scale: Nature, v. 315, p. 659–662.
Holser, W.T., Magaritz, M., and Ripperdan, R.L., 1996, Global isotopic events, *in* Walliser, O.H., ed., Global events and event stratigraphy in the Phanerozoic: Berlin, Springer-Verlag, p. 63–88.
Hori, S.R., 1993, Toarcian oceanic event in deep-sea sediments: Geological Survey of Japan Bulletin, v. 44, p. 555–570.
Jakobs, G.K., Smith, P.L., and Tipper, H.W., 1994, An ammonite zonation for the Toarcian (Lower Jurassic) of the North American Cordillera: Canadian Journal of Earth Sciences, v. 31, p. 919–942.
Jenkyns, H.C., 1988, The early Toarcian (Jurassic) anoxic event: Stratigraphic, sedimentary, and geochemical evidence: American Journal of Science, v. 288, p. 101–151.
Jenkyns, H.C., and Clayton, C., 1997, Lower Jurassic epicontinental carbonates and mudstones from England and Wales: Chemostratigraphic signals and the early Toarcian anoxic event: Sedimentology, v. 44, p. 687–706.
Jenkyns, H.C., Géczy, B., and Marshall, J.D., 1991, Jurassic manganese carbonates of central Europe and the early Toarcian anoxic event: Journal of Geology, v. 99, p. 137–149.
Jiménez, A.P., Jiménez de Cisneros, C., Rivas, P., and Vera, J.A., 1996, The early Toarcian anoxic event in the westernmost Tethys (Subbetic): Paleogeographic and paleobiogeographic significance: Journal of Geology, v. 104, p. 399–416.
Jones, C.E., Jenkyns, H.C., and Hesselbo, S.P., 1994, Strontium isotopes in Early Jurassic seawater: Geochimica et Cosmochimica Acta, v. 58, p. 1285–1301.
Kent, D.V., Olsen, P.E., and Witte, W.K., 1995, Late Triassic-earliest Jurassic geomagnetic polarity sequence and paleolatitudes from drill cores in the Newark rift basin, eastern North America: Journal of Geophysical Research, v. 100, p. 14965–14998.
Korte, C., 1999, $^{87}Sr/^{86}Sr$-, $\delta^{18}O$- und $\delta^{13}C$-evolution des triassischen Meerwassers: geochemische und stratigraphische Untersuchungen an Conodonten und Brachiopoden: Bochumer Geologische und Geotechnische Arbeiten, v. 52, p. 1–171.
Krull, E.S., and Retallack, G.J., 2000, $\delta^{13}C$ depth profiles from paleosols across the Permian-Triassic boundary: Evidence for methane release: Geological Society of America Bulletin, v. 112, p. 1459–1472.
Kump, L.R., and Arthur, M.A., 1999, Interpreting carbon-isotope excursions: Carbonates and organic matter: Chemical Geology, v. 161, p. 181–198.
Little, C.T.S., 1996, The Pliensbachian-Toarcian (Lower Jurassic) extinction event, *in* Ryder, G., et al., eds., The Cretaceous-Tertiary event and other catastrophes in Earth history: Geological Society of America Special Paper 307, p. 505–512.
Little, C.T.S., and Benton, M.J., 1995, Early Jurassic mass extinction: A global long-term event: Geology, v. 23, p. 495–498.
Martin, E.E., and Macdougall, J.D., 1995, Sr and Nd isotopes at the Permian/Triassic boundary: A record of climate change: Chemical Geology, v. 125, p. 73–100.
Marzoli, A., Renne, P.R., Piccirillo, E.M., Ernesto, M., Bellieni, G., and De Min, A., 1999, Extensive 200-million-year-old continental flood basalts of the Central Atlantic Magmatic Province: Science, v. 284, p. 616–618.
McArthur, J.M., Donovan, D.T., Thirlwall, M.F., Fouke, B.W., and Mattey, D., 2000, Strontium isotope profile of the Early Toarcian (Jurassic) Oceanic Anoxic Event, the duration of ammonite biozones, and belemnite paleotemperatures: Earth and Planetary Science Letters, v. 179, p. 269–285.
McElwain, J.C., Beerling, D.J., and Woodward, F.I., 1999, Fossil plants and global warming at the Triassic-Jurassic boundary: Science, v. 285, p. 1386–1390.
McRoberts, C.A., Furrer, H., and Jones, D.S., 1997, Palaeoenvironmental interpretation of a Triassic-Jurassic boundary section from Western Austria based on palaeoecological and geochemical data: Palaeogeography, Palaeoclimatology, Palaeoecology, v. 136, p. 79–95.
Minor, D.R., and Mukasa, S.B., 1997, Zircon U-Pb and hornblende $^{40}Ar/^{39}Ar$ ages for the Dufek layered mafic intrusion, Antarctica: Implications for the age of the Ferrar large igneous province: Geochimica et Cosmochimica Acta, v. 61, p. 2497–2504.
Morante, R., and Hallam, A., 1996, Organic carbon isotopic record across the Triassic-Jurassic boundary in Austria and its bearing on the cause of the mass extinction: Geology, v. 24, no. 5, p. 391–394.
Odin, G.S., 1994, Geological time scale: Comptes Rendus de l'Académie des Sciences, Paris, Série 2, v. 318, p. 59–71.
Olsen, P.E., Schlische, R.W., and Fedosh, M.S., 1996, 580 ky duration of the Early Jurassic flood basalt event in eastern North America estimated using Milankovitch cyclostratigraphy, *in* Morales, M., ed., The continental Jurassic: Museum of Northern Arizona Bulletin 60, p. 11–22.
Olsen, P.E., Shubin, N.H., and Anders, M.H., 1987, New Early Jurassic tetrapod assemblages constrain Triassic-Jurassic tetrapod extinction event: Science, v. 237, p. 1025–1029.
Pálfy, J., Demény, A., Haas, J., Hetényi M., Orchard, M., and Vető, I., 2001, Carbon isotope anomaly and other geochemical changes at the Triassic-Jurassic boundary from a marine section in Hungary: Geology, v. 29, p. 1047–1050.

Pálfy, J., and Smith, P.L., 2000, Synchrony between Early Jurassic extinction, oceanic anoxic event, and the Karoo-Ferrar flood basalt volcanism: Geology, v. 28, p. 747–750.

Pálfy, J., Parrish, R.R., and Smith, P.L., 1997, A U-Pb age from the Toarcian (Lower Jurassic) and its use for time scale calibration through error analysis of biochronologic dating: Earth and Planetary Science Letters, v. 146, p. 659–675.

Pálfy, J., Smith, P.L., Mortensen, J.K., and Friedman, R.M., 1999, Integrated ammonite biochronology and U-Pb geochronology from a Lower Jurassic section in Alaska: Geological Society of America Bulletin, v. 111, p. 1537–1549.

Pálfy, J., Mortensen, J.K., Carter, E.S., Smith, P.L., Friedman, R.M., and Tipper, H.W., 2000a, Timing the end-Triassic mass extinction: First on land, then in the sea?: Geology, v. 28, p. 39–42.

Pálfy, J., Mortensen, J.K., and Smith, P.L., 2000b, A U-Pb and $^{40}Ar/^{39}Ar$ time scale for the Jurassic: Canadian Journal of Earth Sciences, v. 37, p. 923–944.

Palmer, A.R., 1983, The Decade of North American Geology 1983 geologic time scale: Geology, v. 11, p. 503–504.

Rampino, M.R., and Haggerty, B.M., 1996, Impact crises and mass extinction: A working hypothesis, *in* Ryder, G., et al., eds., The Cretaceous-Tertiary event and other catastrophes in Earth history: Geological Society of America Special Paper 307, p. 11–30.

Rampino, M.R., and Stothers, R.B., 1988, Flood basalt volcanism during the last 250 million years: Science, v. 241, p. 663–667.

Ratcliffe, N.M., 1988, Reinterpretation of the relationships of the western extension of the Palisades sill to the lava flows at Ladentown, New York, based on new core data, *in* Froelich, A.J., and Robinson, G.R., eds., Studies of the early Mesozoic basins of the eastern United States: U.S. Geological Survey Bulletin 1776, p. 113–135.

Raup, D.M., and Sepkoski, J.J., Jr., 1984, Periodicity of extinctions in the geologic past: Proceedings of the National Academy of Sciences of the United States of America, v. 81, p. 801–805.

Renne, P.R., Swisher, C.C., Deino, A.L., Karner, D.B., Owens, T.L., and DePaolo, D.J., 1998, Intercalibration of standards, absolute ages and uncertainties in $^{40}Ar/^{39}Ar$ dating: Chemical Geology, v. 145, p. 117–152.

Sepkoski, J.J., Jr., 1996, Patterns of Phanerozoic extinction: A perspective from global data bases, *in* Walliser, O.H., ed., Global events and event stratigraphy in the Phanerozoic: Berlin, Springer-Verlag, p. 35–51.

Signor, P.W., and Lipps, J.H., 1982, Sampling bias, gradual extinction patterns and catastrophes in the fossil record, *in* Silver, L.T., and Schultz, P.H., eds., Geological implications of impacts of large asteroids and comets on the Earth: Geological Society of America Special Paper 190, p. 291–296.

Silvestri, S.M., and Szajna, M.J., 1993, Biostratigraphy of vertebrate footprints in the Late Triassic section of the Newark basin, Pennsylvania: Reassessment of stratigraphic ranges, *in* Lucas, S.G., and Morales, M., eds., The nonmarine Triassic: New Mexico Museum of Natural History and Science Bulletin 3, p. 439–445.

Smith, P.L., Tipper, H.W., Taylor, D.G., and Guex, J., 1988, An ammonite zonation for the Lower Jurassic of Canada and the United States: The Pliensbachian: Canadian Journal of Earth Sciences, v. 25, p. 1503–1523.

Smith, P.L., Beyers, J.M., Carter, E.S., Jakobs, G.K., Pálfy, J., Pessagno, E., and Tipper, H.W., 1994, North America, Lower Jurassic, *in* Westermann, G.E.G., and Riccardi, A.C., eds., Jurassic taxa ranges and correlation charts for the Circum Pacific: Newsletters on Stratigraphy, v. 31, p. 33–70.

Taylor, A.S., and Lasaga, A.C., 1999, The role of basalt weathering in the Sr isotope budget of the oceans: Chemical Geology, v. 161, p. 199–214.

Taylor, D.G., Boelling, K., Holser, W.T., Magaritz, M., and Guex, J., 1992, Ammonite biostratigraphy and geochemistry of latest Triassic and earliest Jurassic strata from the Gabbs and Sunrise Formations, Nevada: Geological Society of America Abstracts with Programs, v. 24, p. 85.

Veizer, J., Ala, D., Azmy, K., Bruckschen, P., Buhl, D., Bruhn, F., Carden, G.A.F., Diener, A., Ebneth, S., Godderis, Y., Jasper, T., Korte, C., Pawellek, F., Podlaha, O.G., and Strauss, H., 1999, $^{87}Sr/^{86}Sr$, $\delta^{13}C$ and $\delta^{18}O$ evolution of Phanerozoic seawater: Chemical Geology, v. 161, p. 59–88.

Vörös, A., 1993, Jurassic brachiopods from the Bakony Mts. (Hungary): Global and local effects on changing diversity, *in* Pálfy, J., and Vörös, A., eds., Mesozoic brachiopods of Alpine Europe: Budapest, Hungarian Geological Society, p. 179–187.

Ward, P.D., Haggart, J.W., Carter, E.S., Wilbur, D., Tipper, H.W., and Evans, T., 2001, Sudden productivity collapse associated with the Triassic-Jurassic boundary mass extinction: Science, v. 292, p. 1148–1151.

Zachos, J.C., Arthur, M.A., and Dean, W.E., 1989, Geochemical evidence for suppression of pelagic marine productivity at the Cretaceous Tertiary boundary: Nature, v. 337, p. 61–64.

Manuscript Submitted November 3, 2000; Accepted by the Society March 22, 2001

Geological Society of America
Special Paper 356
2002

Sea-level changes and black shales associated with the late Paleocene thermal maximum: Organic-geochemical and micropaleontologic evidence from the southern Tethyan margin (Egypt-Israel)

Robert P. Speijer*
Thomas Wagner
Department of Geosciences, FB5, Bremen University, P.O. Box 330440, 28334 Bremen, Germany

ABSTRACT

Organic geochemistry and microfossil contents of six sections spanning the late Paleocene thermal maximum are investigated. The sections are arranged along a depth transect (~50–600 m) across an epicontinental basin covering Egypt and Israel. This study is aimed at unraveling paleoceanographic changes associated with the late Paleocene thermal maximum. In three sections (~200–600 m paleodepth), black shales, consisting of dark brown laminated marls with as much as 2.7% total organic carbon (TOC), mark the late Paleocene thermal maximum. The black shales of the deeper sites correlate with pink to gray fissile marls in the shallowest section. In the two remaining sections, this stratigraphic interval is missing. A relative sea-level fall (~30 m) immediately preceded the late Paleocene thermal maximum, during which sea-level rose again by ~20 m. This rise may have been eustatically controlled, possibly through a combination of thermal expansion of the oceanic water column and melting of unknown sources of high-altitude or polar ice caps in response to global warming. During the late Paleocene thermal maximum, the upwelling of low-oxygen intermediate Tethyan water into the epicontinental basin led to enhanced biological productivity and anoxia at the seafloor. Before and after the late Paleocene thermal maximum, upwelling and biological productivity were less intense, and seafloor dysoxia was restricted to neritic parts of the basin. The presence of similar TOC-rich beds in extensive areas in southern Asia indicates that the Tethyan continental margins may have acted as significant carbon sinks during the late Paleocene thermal maximum.

INTRODUCTION

The late Paleocene thermal maximum represents an abrupt and transient period of extreme global warmth (Zachos et al., 1993). It is marked by a ~90 k.y. period (~4.5 precession cycles; Röhl et al., 2000) of high-latitude and deep-ocean warming and is associated with major evolutionary turnovers among mammals, foraminifera, and calcareous nannofossils (Aubry et al., 1998). The onset of the late Paleocene thermal maximum coincides with the start of the most severe perturbation of the global carbon cycle during the Cenozoic (Kennett and Stott, 1991; Norris and Röhl, 1999). This perturbation, which is

*E-mail: speijer@uni-bremen.de

Speijer, R.P, and Wagner, T., 2002, Sea-level changes and black shales associated with the late Paleocene thermal maximum: Organic-geochemical and micropaleontologic evidence from the southern Tethyan margin (Egypt-Israel), *in* Koeberl, C., and MacLeod, K.G., eds., Catastrophic Events and Mass Extinctions: Impacts and Beyond: Boulder, Colorado, Geological Society of America Special Paper 356, p. 533–549.

thought to have lasted ~220 k.y., is indicated by the abrupt 3‰ negative carbon isotope excursion recorded in marine and terrestrial carbonates worldwide (Kennett and Stott, 1991; Koch et al., 1992; Bains et al., 1999; Röhl et al., 2000). The carbon isotope excursion is thought to have resulted from several massive methane-hydrate melting events, induced by deep ocean warming (Dickens et al., 1995; Bains et al., 1999; Katz et al., 1999; Norris and Röhl, 1999; Röhl et al., 2000). Before the late Paleocene thermal maximum, large methane-hydrate reservoirs within the sediment are thought to have been stable below ~1000 m depth (under high pressure and relatively low temperature). Warming of intermediate and deep waters and underlying sediments with the onset of the late Paleocene thermal maximum may have disrupted the methane-hydrate stability zone, thus leading to injection of methane into the ocean and atmosphere (Dickens et al., 1995).

A direct consequence of methane release into the water column would be an increase in oxygen consumption, especially in areas where free methane is injected (Dickens, 2000a, 2001b). During the late Paleocene thermal maximum, this effect would have been most notable at intermediate depths (middle to lower bathyal) where methane-hydrate melting is thought to have occurred. Dysoxia at the seafloor and within the water column are indicated by sedimentologic and biotic data from a number of cores and localities from this depth range (Thomas, 1989, 1998; Kaiho et al., 1996; Kelly et al., 1996; Bralower et al., 1997; Katz et al., 1999). Oxygen deficiency is also regarded as one of the main causes for the largest Cenozoic extinction (40%) of deep-sea benthic foraminifera (Thomas, 1998, and references therein). This extinction, known as the benthic extinction event, has also been demonstrated, albeit in a milder form, in neritic successions bordering the Tethys and Atlantic (Speijer et al., 1996a, 1996b; Cramer et al., 1999).

Epicontinental basins bordering the Tethys appear to have been particularly prone to severe oxygen deficiency during the late Paleocene thermal maximum. A total organic carbon (TOC) rich sapropelite unit as thick as 1 m has been recorded in numerous localities over a vast area between the Crimea and Uzbekistan (northern Tethyan margin) by Gavrilov et al. (1997). Gavrilov et al. (1997) noted TOC-rich (to 17%), laminated beds with enrichments of a suite of elements typical of deposition under suboxic to anoxic conditions. Similarly, Speijer et al. (1997) found dark sapropelic beds with micropaleontologic indications for anoxia in association with the benthic extinction event in localities in Turkmenistan (northern Tethys) and Egypt (southern Tethys). We refer to these deposits as black shales (Tyson, 1987). For their corresponding oxygen-depleted environments we adopted the nomenclature of Tyson and Pearson (1991). Chemostratigraphic and biostratigraphic studies indicate that the black shales on both Tethyan margins correlate with the onset and early part of the carbon isotope excursion and the late Paleocene thermal maximum (Speijer et al., 2000).

In this chapter we provide a paleoenvironmental synthesis of an area on the southern Tethyan margin (Egypt-Israel) in an ~2 m.y. interval comprising the late Paleocene thermal maximum. It builds upon the stratigraphic framework in Speijer et al. (2000), which was constructed through biostratigraphic and chemostratigraphic interpretation of five of the six sections on a north-south paleobathymetric transect presented here (Fig. 1). We draw the paleoenvironmental information from microbiotic (mostly benthic foraminifera) interpretations published in earlier reports (e.g., Speijer et al., 1996b, 1997), and complement these by new microbiotic, sedimentologic, and organic-petrographic and/or geochemical observations and data. Late Paleocene to early Eocene paleoenvironments of four sections (Gebel Aweina, Gebel Duwi, Ben Gurion, and Wadi Nukhl) have been intensively discussed, also by others (Table 1). For addi-

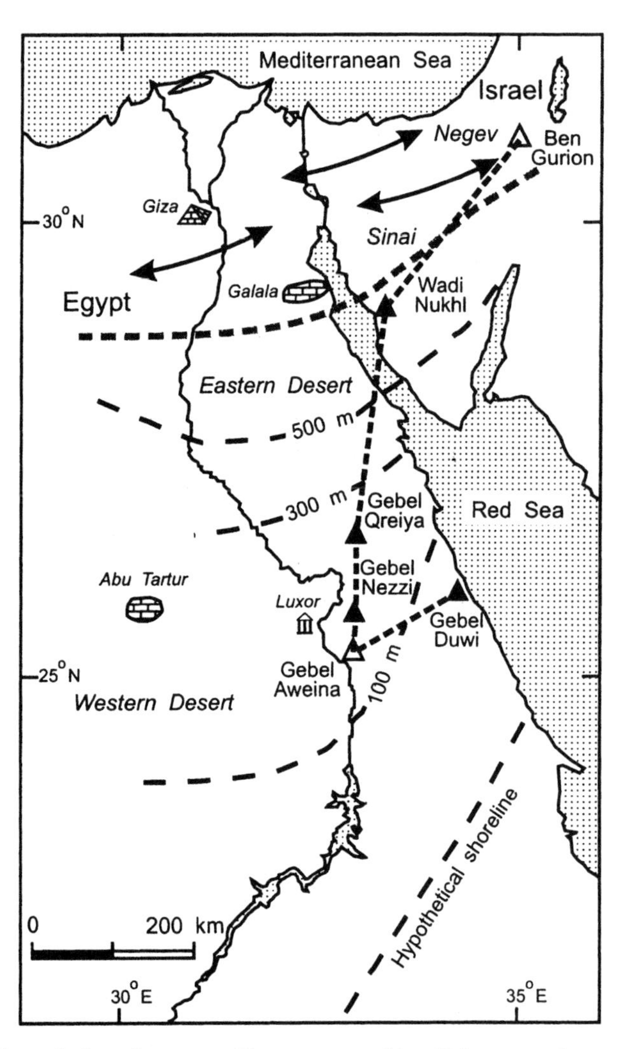

Figure 1. Location map with reconstructed late Paleocene paleogeography and paleobathymetry (no correction for opening of Red Sea rift). Filled triangles indicate sections containing subzone P5b (early part of carbon isotope excursion-late Paleocene thermal maximum); open triangles represent sections that do not contain subzone P5b. Thick dashed line separates tectonically stable shelf in south from tectonically active unstable shelf (Syrian arc) in north. Arrows indicate general orientation of double-plunging anticlines (submarine swells) of unstable shelf. Modified from Speijer et al. (2000).

TABLE 1. PREVIOUS STUDIES DEFINING PALEOENVIRONMENTAL PARAMETERS

Parameter	Ben Gurion*	Wadi Nukhl†	Gebel Qreiya	Gebel Aweina	Gebel Duwi
Depth and/or sea level	1, 2, 4, 5, 9	2, 10, 14	14	2, 6, 7	2, 6, 8, 9, 14
Oceanic circulation	1, 2, 4, 5, 9, 11, 13	2, 10, 14	14	2, 3, 6, 11	2, 6, 8, 9, 11, 14
Surface productivity	1, 5, 9, 11, 13,	2, 10, 14	14	2, 3, 6, 11	2, 6, 8, 9, 11, 14
Bottom oxygenation	1, 2, 4, 9, 11	2, 10	N.D.	2, 6, 11	2, 6, 8, 9, 11
Temperature	11, 13	N.D.	N.D.	6, 11	6, 11
Salinity	11	N.D.	N.D.	6, 11	6, 11
Climate	5, 9, 13, 11	2, 14	14	2, 6, 11	2, 6, 8, 9, 11, 14

Note: 1 = Benjamini, 1992; 2 = Speijer, 1994; 3 = Charisi and Schmitz, 1995; 4 = Speijer, 1995; 5 = Lu et al., 1996; 6 = Schmitz et al., 1996; 7 = Speijer et al., 1996a; 8 = Speijer et al., 1996b; 9 = Schmitz et al., 1997; 10 = Speijer et al., 1997; 11 = Charisi and Schmitz, 1998; 12 = Speijer and Schmitz, 1998; 13 = Bolle et al., 2000a; 14 = Bolle et al., 2000b.
*Also known as Nahal Avdat in refs. 2 and 4
†In ref. 14 the nearby Gebel Matulla.
N.D.—Not discussed previously.

tional stratigraphic studies on the sections discussed here, we refer to the references cited, and the literature cited therein, in Table 1. In addition, two less well-studied sections, Gebel Nezzi and Gebel Qreiya, provide new key data to the synthesis presented here. We focus on the relationship between the deposition of the black shales and the relative sea-level record of the area.

GEOLOGIC SETTING

The studied sections are in the eastern part of an extensive epicontinental basin that covered most of Egypt, Israel, and Jordan at the end of the Paleocene. The Egyptian sections are located on the so-called stable shelf, which underwent relatively little structural deformation during the Late Cretaceous and early Paleogene (Said, 1962). The Ben Gurion section is on the unstable shelf, also known as the Syrian arc, a fold belt that has been active since the Late Cretaceous (Shahar, 1994). Disregarding the structural highs of the Syrian arc, the basin generally deepened in a northwest direction. Microbiota indicate that the shallowest deposition (middle neritic, ~50–100 m) was at Gebel Duwi, and the deepest (upper bathyal, 500–600 m) was at Wadi Nukhl and Ben Gurion. Deposition at the Gebels Aweina, Nezzi, and Qreiya occurred at ~200 m paleodepth (Fig. 1). We use the terms neritic and bathyal as representing paleodepths less than and in excess of 200 m, respectively (Van Morkhoven et al., 1986). However, the true continental slope was situated north of the epicontinental basin studied (Mart, 1991).

STRATIGRAPHY

In Egypt, upper Paleocene to lower Eocene marls and shales are usually classified as the Esna Formation. The Esna Formation is intercalated between two more calcareous units, the upper Paleocene Tarawan Formation and the lower Eocene Thebes Formation (Said, 1990). In southern Israel, other lithostratigraphic units (Taqiye Formation, Hafir Member and Mor Formation, respectively) are employed, but the stratigraphic succession is very similar there (Benjamini, 1992). The Esna Formation and its lateral equivalents consist mostly of monotonous and fissile gray to brown-green marls and shales, containing abundant and mostly well-preserved calcareous microbiota. The fissile character relates to the generally high smectite contents of the clay fraction (Strouhal, 1993; Bolle et al., 2000a, 2000b). In the interval of major biotic changes associated with the late Paleocene thermal maximum and carbon isotope excursion, however, beds as thick as 1 m of different lithologies (black shales and foraminifera-rich calcarenitic marls) are present. These black shales should not be confused with occasional dark gray to black beds present in many Paleocene marl successions in the region (see following). The stratigraphic framework for all sections but Nezzi is based on planktic and benthic foraminifera and carbon isotopes (Speijer et al., 2000). For Nezzi, only foraminiferal data are available. Further calcareous nannofossil data pertaining explicitly to the late Paleocene thermal maximum in these sections were presented in Speijer (1995), Schmitz et al. (1996), Bolle et al. (2000b), and Monechi et al. (2000). The stratigraphic interval documented here is bracketed by the base of the Esna Formation (uppermost part of zone P4) below, and by the P5–P6 zonal boundary above (Fig. 2). We adopted biostratigraphic terminology from Berggren et al. (1995). Subdivision of zone P5 (subzones P5a–P5c), however, is after Speijer et al. (2000). Subzone P5b (*Morozovella allisonensis* total range subzone) constitutes the lower part of the stratigraphic interval spanning the carbon isotope excursion and the late Paleocene thermal maximum (black shales). This subzone is present in four of our sections. In the Gebel Aweina and Ben Gurion sections, omission surfaces, characterized by a sharp but burrowed contact between the pre-late Paleocene thermal maximum marls and the calcarenitic marls, mark discontinuities at this level.

MATERIAL AND METHODS

Samples were collected from decimeter-deep trenches or holes along the outcrop sections. Some influence of weathering cannot be ruled out (van Os et al., 1996). For Duwi, Aweina,

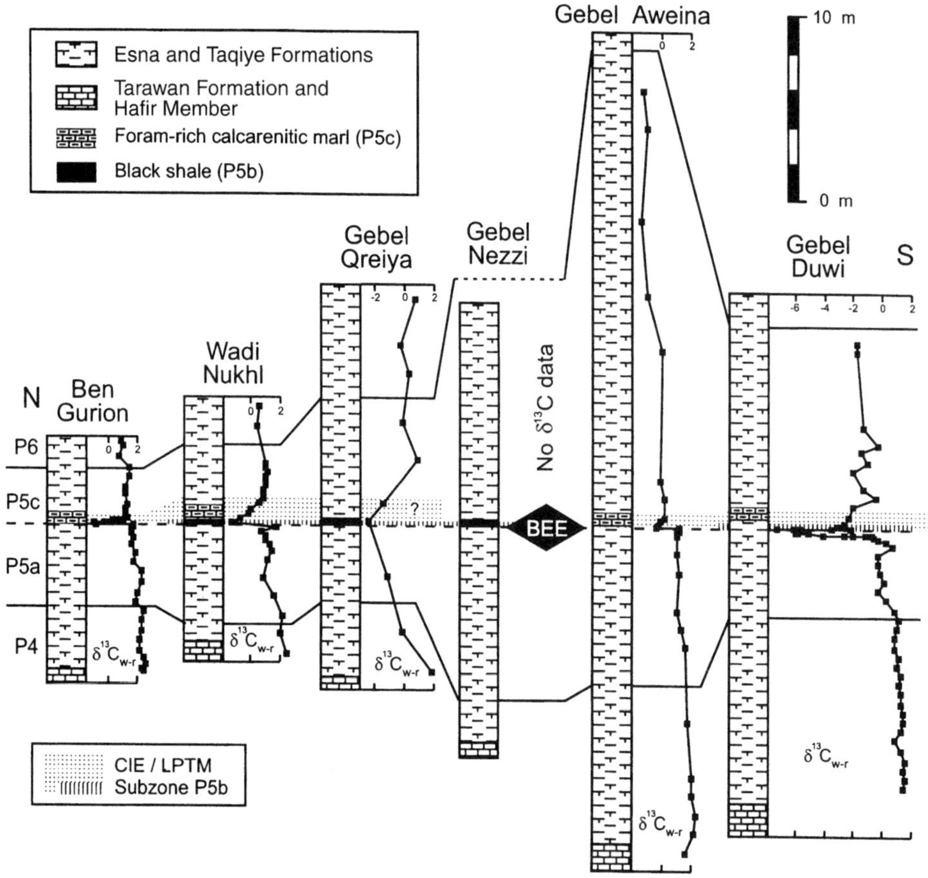

Figure 2. Stratigraphic correlation chart of lower part of Esna and Taqiye Formations along paleodepth transect. Whole-rock $\delta^{13}C$ excursions (CIE, carbon isotope excursion; w-r, whole rock) coincide with benthic foraminiferal extinction event. Initiation and termination of CIE is uncertain at Qreiya. Large isotopic shift below carbon isotope excursion in Duwi results from diagenesis (Speijer et al., 2000). Modified from Speijer et al. (2000). LPTM is late Paleocene thermal maximum.

and Ben Gurion, high-resolution (centimeter scale) sample sets across the late Paleocene thermal maximum were available. The Nukhl, Qreiya, and Nezzi sections were logged and sampled with little (Nukhl) or no prior knowledge on the late Paleocene thermal maximum (Nezzi and Qreiya, sampled by F. Hendriks and P. Luger, personal commun., 1985). Accordingly, only medium- to low-resolution (25–200 cm sample spacing) sets were available, each yielding one sample in the black shales. Herein only the lower half of the sapropelic bed in Nukhl (Speijer et al., 1997, 2000) is considered as black shale. The upper half (brown bioturbated marl) and the overlying calcarenitic marls are discussed together. All samples were processed according to standard micropaleontologic procedures. Stratigraphic and compositional data of the foraminiferal assemblages were determined qualitatively and quantitatively (e.g., Speijer et al., 1996b). In this chapter we summarize micropaleontologic data from Speijer et al. (1996a, 1996b, 1997) and Speijer and Schmitz (1998), and incorporate new observations on samples from Duwi, Nezzi, Aweina, and Qreiya.

Geochemical analyses were performed on the three sections (Nukhl, Nezzi, Qreiya) containing the peculiar black shale bed within subzone P5b, immediately above the benthic extinction event. Additional $CaCO_3$ results of Aweina, Duwi, and Ben Gurion can be found in Schmitz et al. (1996, 1997). Values of organic and inorganic carbon were measured on homogenized samples using a Leco CS-300 elemental analyzer at Bremen University (precision of measurement $\pm 3\%$). Before determination of organic carbon, calcium carbonate was removed by repetitive addition of 0.25 N HCl. The carbonate content was calculated from the difference between total and organic carbon and expressed as calcite ($CaCO_3 = [C_{total} - C_{organic}] \times 8.33$). Rock-Eval pyrolysis was run on whole-rock samples at the Alfred Wegener Institute in Bremerhaven according to the procedures described by Espitalié et al. (1985). Hydrogen indices (HI) were calculated using TOC values obtained from Leco analysis. Organic petrologic analyses were performed on resin-embedded bulk sediment samples using a Zeiss Axiophot, equipped with incident white and reflected ultraviolet light.

RESULTS

Background sedimentation

The $CaCO_3$ contents of the marls of the Esna Formation (Fig. 3) show an overall decline from the top of the Tarawan Formation (~70%) upward to just below the late Paleocene thermal maximum (10%–40%). In the calcarenitic foraminifera-rich marl beds above the benthic extinction event in Nukhl (as

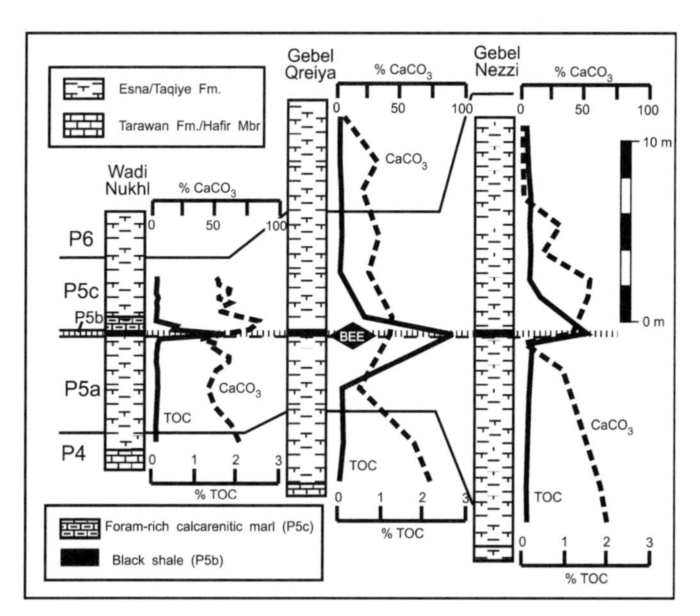

Figure 3. $CaCO_3$ and total organic carbon (TOC) profiles of lower part of Esna Formation in three sections containing late Paleocene thermal maximum black shale. Black shales are characterized by peak values in TOC content (1.5%–2.7%) and average $CaCO_3$ contents (40%–50%).

in Aweina, Ben Gurion, and Duwi; see Schmitz et al., 1996, 1997), this decline is followed by a temporary $CaCO_3$ peak (~80%). In Nezzi and Qreiya this peak is less pronounced (~50% $CaCO_3$). The temporary peaks are followed by a further general decline upward in all sections except Ben Gurion (see also Schmitz et al., 1996, 1997). Intervals marked by severe postdepositional dissolution distort this general pattern, particularly in the neritic sections. Beyond the late Paleocene thermal maximum, the TOC content of the Esna Formation is invariably low, ranging between 0% and 0.2%.

Planktic foraminifera assemblages from all sections are generally rich and diverse, mainly consisting of species of the genera *Morozovella, Acarinina, Globanomalina*, and *Subbotina*. Planktic/Benthic (P/B) ratios of well-preserved assemblages range from 70% to 95% planktics. Taphonomic processes leading to dissolution and fragmentation, however, have occasionally reduced this value to 5%–50% planktics. Consequently, there is no distinct decrease in P/B ratios from the deeper to shallower localities.

The following benthic foraminiferal data are largely from earlier papers (Speijer, 1994, 1995; Speijer et al., 1996a, 1996b, 1997; Speijer and Schmitz, 1998). Beyond the late Paleocene thermal maximum, seven main benthic assemblages can be bathymetrically ordered (Fig. 4). The nominate taxa of the main assemblages are the most common and/or characteristic ones. Before the benthic extinction event, the *Frondicularia phosphatica* assemblage is restricted to the middle neritic Duwi section, the *Angulogavelinella avnimelechi* assemblage characterizes the outer neritic localities, and the *Gavelinella beccariiformis* assemblage marks bathyal deposits. In Duwi, the *Anomalinoides aegyptiacus* assemblage gradually replaced the *F. phosphatica* assemblage just prior to the benthic extinction event. Well after the benthic extinction event, the *Valvulineria scrobiculata* assemblage succeeds the *A. aegyptiacus* assemblage there. Overall, the *Bulimina callahani* assemblage replaced the outer neritic *A. avnimelechi* assemblage and the *Nuttallides truempyi* assemblage succeeded the *G. beccariiformis* assemblage of the deepest localities. The deepest assemblages are composed of a mixture of deep-sea and neritic taxa, whereas the outer neritic assemblages contain <10% deep-sea taxa.

Other biotic components of the micropaleontologic residues are ostracodes (common in Duwi, rare elsewhere), fish scales and teeth (rare to common), radiolarians (abundant at the base of the Esna Formation in Nukhl and Ben Gurion), and rare molds of juvenile gastropods. Not a single sample yielded dinocysts, spores, or pollen (H. Brinkhuis, 1994, 1999, personal communs.).

Sedimentation during late Paleocene thermal maximum and carbon isotope excursion

Geochemistry of black shales. The late Paleocene thermal maximum black shale beds in Nukhl, Qreiya, and Nezzi are 25–50 cm thick (Table 2). They consist of dark brown marls (~40%–50% $CaCO_3$; Fig. 3; Table 3) that contain abundant phosphatic peloids (Fig. 5, A and B) and 1.5% (Nezzi) to 2.7% TOC (Qreiya). The overlying beds show a modest (0.5%) TOC enrichment. Fluorescence petrography of these beds reveals a fine wavy-laminated sedimentary texture and abundant threads of weakly fluorescent marine amorphous organic matter. This pattern is well known from many other TOC-rich shales deposited under conditions of oxygen depletion (e.g., Wignall, 1994). Larger, structured organic particles (macerals) are rare and mostly reveal a weak, brownish fluorescence. Stronger fluorescing liptinites are restricted to algae bodies, showing characteristic internal structures (Fig. 5C). Terrestrial organic matter is rare and very small ($\ll$5 μm).

Despite qualifying as typical black shales (Tyson, 1987), these beds differ from other marine black shales with regard to their thermal signature and hydrocarbon charging (e.g., Kuhnt et al., 1990; Mostafa, 1993; Mello et al., 1995; Alsharhan and Salah, 1997; Wagner and Pletsch, 1999). Rock-Eval data from the late Paleocene thermal maximum black shales and similar middle Paleocene TOC-rich units in Egypt (Table 3) reveal pyrolytic signatures indicative of preservation of extremely hydrogen-depleted organic matter (S2 yields <0.35 mg HC/g TOC corresponding to HI <30 mg HC/g TOC). These low hydrogen indices favor the presence of severely oxidized type IV kerogen, which is commonly associated with TOC-poor, oxic depositional environments (Jones, 1987) rather than with dysoxic and/or anoxic settings, where hydrocarbon source rocks accumulate.

Optical properties of the shales, however, challenge the

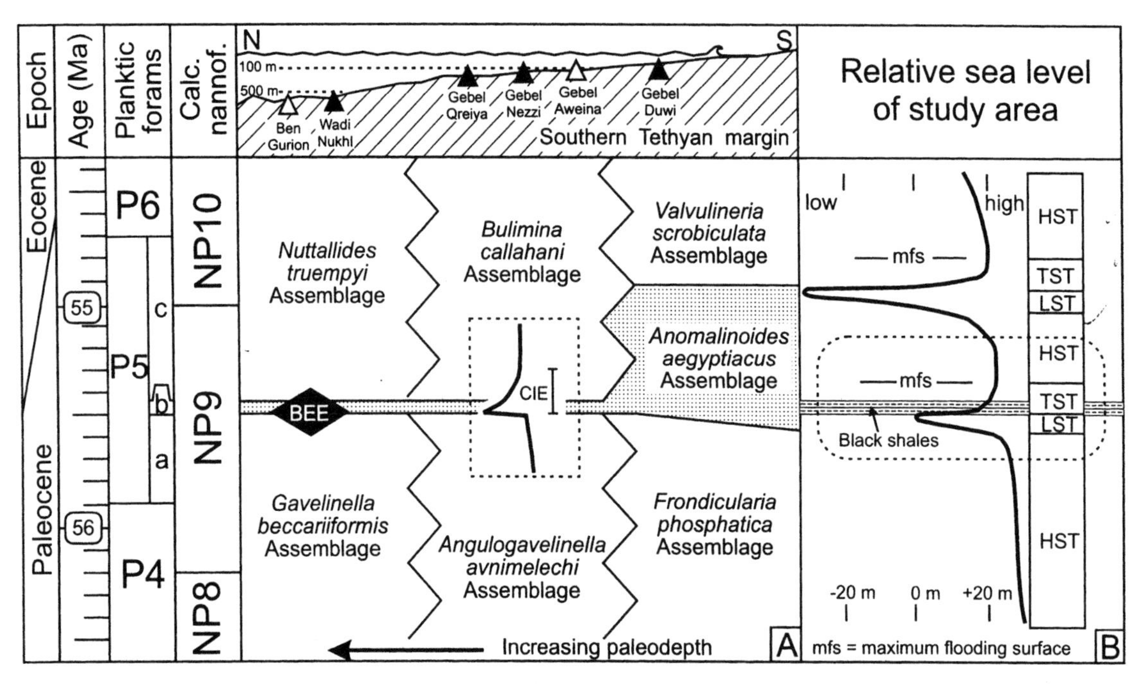

Figure 4. A: Schematic spatial and temporal distribution of main benthic foraminifera assemblages. Inset shows schematic profile and duration of carbon isotope excursion (CIE). B: Relative sea-level curve of area and sequence-stratigraphic interpretation. All data are plotted against biostratigraphic and chronostratigraphic schemes of Berggren et al. (1995). Note that estimated age of initiation of late Paleocene thermal maximum in this scheme is slightly older than astronomically calibrated age (ca. 54.95 Ma) of Norris and Röhl (1999). Subdivision of zone P5 is after Speijer et al. (2000). HST, TST, and LST are highstand, transgressive, and lowstand systems tracts, respectively.

TABLE 2. DIMENSIONS OF LATE PALEOCENE THERMAL MAXIMUM/CARBON ISOTOPE EXCURSION RELATED PARAMETERS

Unit	Ben Gurion (in cm)	Wadi Nukhl (in cm)	Gebel Qreiya (in cm)	Gebel Nezzi (in cm)	Gebel Aweina (in cm)	Gebel Duwi (in cm)
Calcarenitic marl	20	75	N.A.	N.A.	40	~60
Black shale (Zone P5b)	N.A.	~25	~40	~50	N.A.	~25*
CIE	30	110	N.D.	N.D.	30	~80

*Fissile marl correlating with black shales
Note: N.D. = No or insufficient data
N.A. = Not applicable

idea of an oxic environment, and suggest alternative explanations. In this respect, maximum temperature values (T_{max}) obtained by Rock-Eval analysis need to be considered. T_{max} depicts the temperature of maximum hydrocarbon generation from the kerogen fraction (S2), and is a direct measure of the thermal maturity of the organic matter. By definition, organic matter is considered thermally immature if T_{max} is below 435°C (<0.5%–0.6% Rm vitrinite reflectance), mature between 435 and 460°C (to 1.3% Rm), and overmature when exceeding 460°C (Tissot and Welte, 1984). T_{max} data obtained from the late Paleocene thermal maximum black shales and middle Paleocene intervals show, with the exception of the black shale from Nezzi, thermally mature to overmature organic matter. Being aware of the analytical problems of T_{max} detection in the presence of very low hydrocarbon yields (Wagner and Dupont, 1999), we consider the consistency of T_{max} signatures to be at least indicative for the organic matter maturity stage.

Micropaleontology of black shales. Planktic foraminifera, particularly *Acarinina* spp., are the most abundant component of the black shales. The marker species of subzone P5b, *Morozovella allisonensis*, is rare (Nukhl) to common (Qreiya), whereas other *Morozovella* are very rare. Benthic foraminifera are also rare, leading to unusually high P/B ratios (99% planktics) in all black shale samples. Three benthic species (*Anomalinoides aegyptiacus*, *Stainforthia farafraensis*, and *Valvulineria* sp.) make up ~95% of the benthic assemblages (Fig. 6). These three taxa are otherwise very rare in outer neritic and deeper deposits in the region, but the former two are common to abundant in middle neritic deposits of Gebel Duwi. Preservation of the foraminifera varies from poor (Nukhl; recrystal-

TABLE 3. ELEMENTAL AND PYROLYTIC DATA FROM STUDIED SEDIMENTS

Profile	Sample	Level (m)	TOC (%)	$CaCO_3$ (%)	Stot (%)	T_{max} (°C)	Hydrogen Index (mgHC/gTOC)
G. Qreiya, 271185	20	20.4	0.10	10.69	0.06	N.D.	N.D.
	21	17.9	0.11	36.37	0.16	N.D.	N.D.
	22	16.4	0.13	24.92	0.12	N.D.	N.D.
	23	13.8	0.15	37.80	0.09	N.D.	N.D.
	24	11.8	0.11	29.65	0.11	N.D.	N.D.
	25	9.4	0.65	47.66	0.09	N.D.	N.D.
	26	8.4	2.74	46.08	0.22	502	11
	28	5.4	0.16	22.94	0.08	N.D.	N.D.
	29	2.4	0.20	64.85	0.14	N.D.	N.D.
	30	0.3	0.11	77.59	0.08	N.D.	N.D.
G. Nezzi, 41285	19	22.5	0.08	0.03	0.45	N.D.	N.D.
	18	20	0.12	0.10	0.03	N.D.	N.D.
	17	18.5	0.17	1.10	0.05	N.D.	N.D.
	16	17	0.15	28.71	0.1	N.D.	N.D.
	15	15.5	0.17	15.07	0.05	N.D.	N.D.
	14	14	0.13	50.85	0.45	N.D.	N.D.
	13	13	0.38	48.93	0.09	N.D.	N.D.
	12	11	1.46	36.38	0.19	362	1
	11	10.5	0.19	0.00	0.07	N.D.	N.D.
	10	9	0.16	29.94	0.26	N.D.	N.D.
	9	8	0.14	34.25	0.09	N.D.	N.D.
	8	5.5	0.11	45.06	0.09	N.D.	N.D.
	7	2.5	0.08	58.34	0.06	N.D.	N.D.
	6	0.5	0.10	63.62	0.14	N.D.	N.D.
Wadi Nukhl	1385	9.55	0.08	48.05	0.14	N.D.	N.D.
	1384	9.3	0.08	50.88	0.15	N.D.	N.D.
	1383	9.05	0.08	51.74	0.14	N.D.	N.D.
	1382	8.8	0.06	60.39	0.08	N.D.	N.D.
	1381	8.55	0.07	59.22	0.08	N.D.	N.D.
	1380	8.3	0.10	49.36	0.04	N.D.	N.D.
	1379	8.05	0.07	61.31	0.14	N.D.	N.D.
	1378	7.8	0.08	52.63	0.13	N.D.	N.D.
	1377	7.55	0.08	53.11	0.15	N.D.	N.D.
	1376	7.3	0.07	66.48	0.16	N.D.	N.D.
	1375	7.05	0.06	82.58	0.05	N.D.	N.D.
	1374	6.8	0.55	79.61	0.05	N.D.	N.D.
	1373	6.55	0.41	73.98	1.57	N.D.	N.D.
	1372	6.3	1.55	53.29	0.22	506	7
	1371	6.05	0.15	36.49	0.41	N.D.	N.D.
	1370	5.8	0.09	51.61	0.18	N.D.	N.D.
	1369	5.55	0.09	47.21	0.13	N.D.	N.D.
	1368	5.3	0.07	53.00	0.14	N.D.	N.D.
	1367	5.05	0.07	59.48	0.13	N.D.	N.D.
	1366	4.8	0.07	59.14	0.17	N.D.	N.D.
	1364	4.3	0.10	49.71	0.14	N.D.	N.D.
	1363	3.3	0.10	43.48	0.16	N.D.	N.D.
	1362	2.3	0.08	48.71	0.12	N.D.	N.D.
	1361	1.3	0.06	61.19	0.15	N.D.	N.D.
	1360	0.3	0.06	67.84	0.18	N.D.	N.D.
G. Nezzi, 51285	21	N.A.	0.13	45.75	0.11	N.D.	N.D.
(Mid Paleocene)	19	N.A.	0.91	51.85	0.43	501	10
	17	N.A.	0.15	18.06	0.04	446	123
G. Qreiya, 271185	40	N.A.	1.03	7.13	0.12	501	17
(Mid Paleocene)	41	N.A.	2.38	46.29	3.84	467	15
	42	N.A.	0.32	6.07	0.14	509	31

Note: Mid-Paleocene data are included for comparison of T_{max} values. TOC = total organic carbon; Stot = total sulfur; T_{max} = temperature of maximum hydrocarbon generation rate obtained from the kerogen fraction (52 window) by Rock-Eval Pyrolysis; N.D. = No data; N.A. = Not applicable.

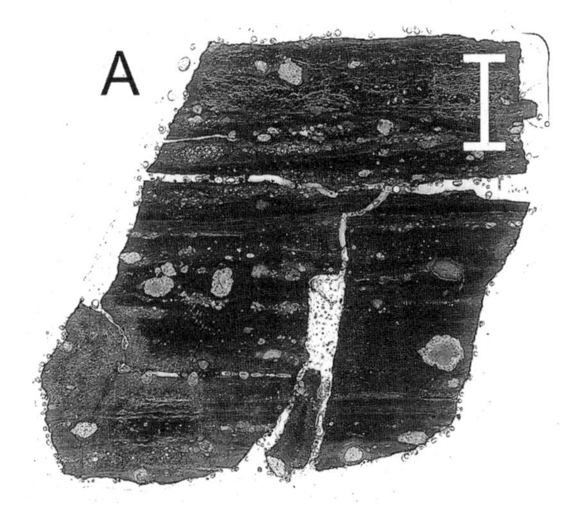

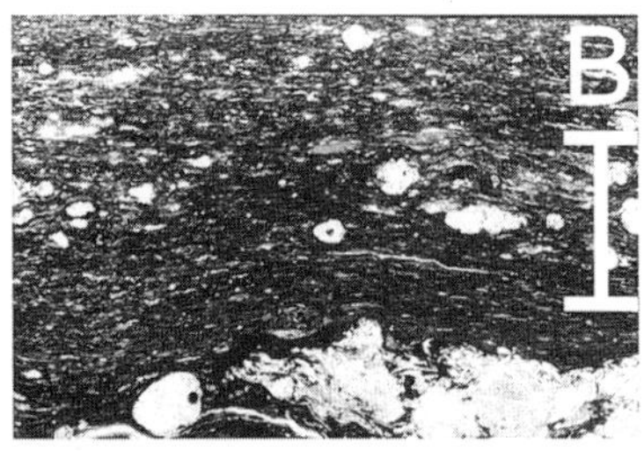

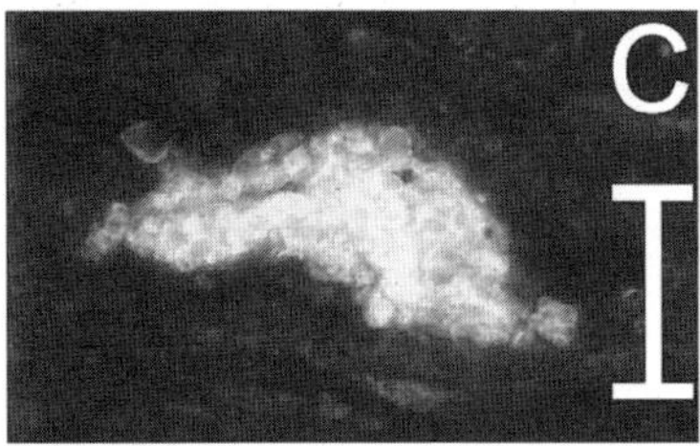

Figure 5. A: Thin section of late Paleocene thermal maximum black shale at Gebel Nezzi, showing numerous pale phosphatic peloids (coprolites) in finely laminated dark brown matrix. Scale bar is 1 cm. B: Polarized light microscope image of A, showing wavy lamination around phosphatic peloids and other grains (e.g., foraminifera), induced by compaction (Wignall, 1994). Scale bar is 400 μm. C: Fluorescence petrography of A, showing rare algal body with internal structures. Black shales at Nukhl and Qreiya are sedimentologically and petrographically very similar. Scale bar is 100 μm.

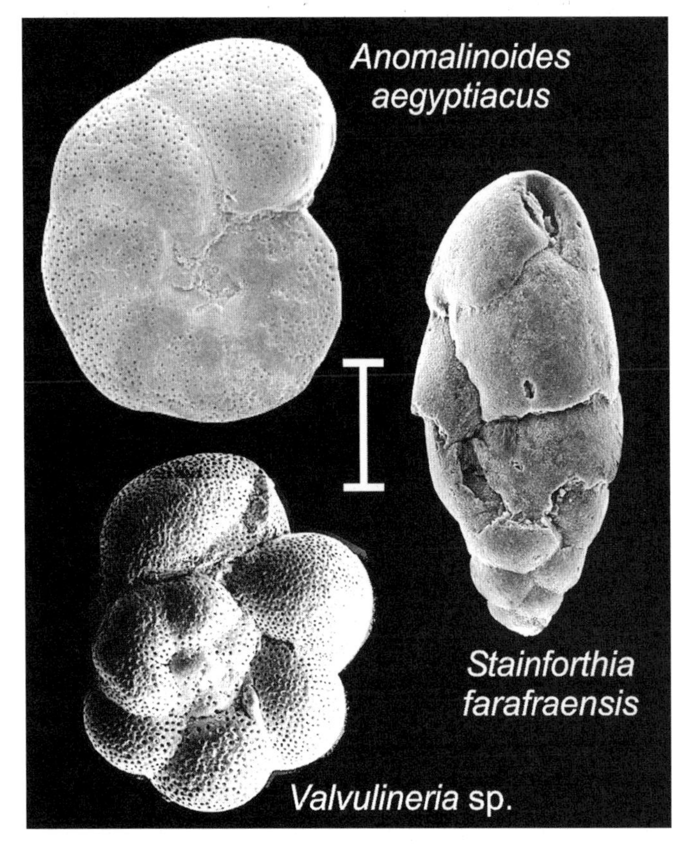

Figure 6. Dominant benthic foraminifera taxa of *Anomalinoides aegyptiacus* assemblage (Fig. 4), together composing up to 95% of benthic assemblages in late Paleocene thermal maximum black shales and in equivalent fissile marl at Duwi. Scale bar is 100 μm.

lization) to good (Qreiya). In addition to foraminifera, fish scales and bones are abundant, whereas ostracodes appeared to be absent. Palynologic residues of the black shales did not yield dinocysts, pollen, or spores (H. Brinkhuis, 1999, personal commun.).

Gebel Duwi. The Duwi section presents a special situation for the late Paleocene thermal maximum interval. Subzone P5b is 20 cm thick, and consists of fissile pink to gray marls (25% $CaCO_3$) with a modest TOC enrichment (0.5%). Despite its different appearance, the well-preserved foraminiferal residues are nearly identical to those of the black shales, both qualitatively and quantitatively. They also both contain numerous phosphatic peloids and common fish remains. The main micropaleontologic difference is the presence of rare ostracodes.

Other black shales in the region. At various levels in Paleocene successions of the region, dark gray and fissile shales to several meters thick are occasionally intercalated. These have also been referred to as black shales (e.g., Benjamini, 1992), but should not be confused with the late Paleocene thermal maximum black shales. Unlike the latter, these black shales are generally composed of strongly fissile (not laminated) shales with a very low $CaCO_3$ content (e.g., the smectite-rich P2–P3 interval of the Ben Gurion 400 section; Benjamini, 1992; Schmitz et al., 1997; Bolle et al. 2000a). Their foraminifera assemblages either correspond to the background assemblages described here or are poorly preserved and dominated by thick-shelled benthic foraminifera and/or agglutinated taxa, indicating postdepositional dissolution (e.g., Speijer and Schmitz, 1998; Speijer's data on Ben Gurion). The fissile black shales 1 m below the benthic extinction event in the Ben Gurion section contain no planktic foraminifera and larger numbers of buliminid taxa (Benjamini, 1992), and are compositionally, stratigraphically, and genetically unrelated to the late Paleocene thermal maximum black shales discussed here.

Foraminifera-rich calcarenitic marl beds. The calcarenitic marl beds at the base of subzone P5c (Nukhl, Ben Gurion, Aweina, and Duwi) represent the interval with the transition to stable $\delta^{13}C$ values and thus the upper part of the carbon isotope excursion (Fig. 2). These beds are homogeneous and contain abundant and diverse planktic foraminifera, typically more than 10 000 specimens (>125 μm) per gram sediment and with

P/B ratios >95%. Benthic foraminifera and ostracode assemblages are also relatively rich and diverse. This bed signals a gradual transition to depositional conditions resembling those prevailing prior to the late Paleocene thermal maximum.

DISCUSSION

Thermal maturation or weathering effect of late Paleocene thermal maximum black shales?

Two mechanisms are envisaged to explain the observed extremely low hydrocarbon yields and high T_{max} values of the black shales. Such pyrolytic signatures argue for severe influence of weathering effects, or for postdepositional thermal heating of the successions studied. Pyrolytic signatures of various Cretaceous and Neogene sections from the Gulf of Suez area show T_{max} values in general below 432°C and corresponding HI ranging from 300 to 676 mg HC/g TOC (Mostafa, 1993), which would suggest that material studied in this paper has been subjected to intense weathering. Alternatively, heating of the sediments related to tectonic mechanisms has to be considered in view of the geotectonic position of the studied localities close to the Red Sea rift system (Fig. 1). Following this interpretation we refer to a correlation of rank parameters recently compiled by the Shipboard Scientific Party (1998) of ODP Leg 171B in which T_{max} values of ~500°C would correspond to peak burial temperatures to 180°C. At such temperatures most oil-prone kerogen has stopped producing liquid hydrocarbons, leaving a residual organic fraction in the original sedimentary beds. The observation of a weak, mostly dark brownish fluorescence in the laminated late Paleocene thermal maximum shales is consistent with that interpretation; however, it does not rule out any influence of weathering processes. Fluorescence intensity decreases and fluorescence color shifts to higher wavelengths (from yellow to brown) with increasing thermal maturity (Tissot and Welte, 1984).

Assuming an average geothermal gradient of ~30°C/km, alteration of late Paleocene thermal maximum black shales at peak temperatures of ~180°C require a sediment overburden of ~5–6 km. Stratigraphic data, however, indicate that overburden in the Eastern Desert (Nezzi and Qreiya) probably did not exceed ~1 km (Said, 1990), which would lead to burial temperatures of only ~50°C. We speculate that rifting of the Red Sea and the Gulf of Suez during the late Cenozoic may have provided sufficient additional heat to the surrounding continental margins, including its Paleogene sedimentary cover, to cause the high peak temperatures of the black shales suggested by the pyrolytic analysis. Considering the realistic Paleocene $\delta^{13}C$ values (see also Schmitz et al., 1996), this heating apparently did not have a great influence on stable isotopic signatures.

Relative sea-level change across the late Paleocene thermal maximum

Sedimentologic and micropaleontologic data indicate fluctuations in relative sea level in an interval bracketing the late Paleocene thermal maximum in Egypt (Fig. 4). The microbiota of the neritic sections in particular enable reconstruction of the relative sea-level curve for the area. Sections in the Eastern and Western Deserts indicate maximum transgression coinciding with largest paleodepths during late Paleocene biochrons P4 and NP8 at 57 Ma (Luger, 1985; Speijer and Schmitz, 1998). This is followed by general regression and shallowing into the early Eocene. Two regressive and/or shallowing peaks are superimposed on this overall regressive trend. One large sea-level fluctuation (~70 m) within biochron P5c close to the NP9–NP10 boundary was demonstrated in the Aweina section (Speijer and Schmitz, 1998). A sea-level fluctuation of lesser magnitude, previously unnoticed in Egypt, encompasses the late Paleocene thermal maximum. In the Duwi section, just prior to the late Paleocene thermal maximum, deeper dwelling ostracoda gradually disappear, replaced by shallower taxa and indicating shallowing from 80–100 m to 50–70 m. At the onset of the late Paleocene thermal maximum this trend rapidly reverses and deeper conditions (70–100 m) were attained again (Speijer and Morsi, 2002). We also find a succession of last appearances of deeper dwelling benthic foraminifera in the Duwi section, also indicative of shallowing immediately prior to the late Paleocene thermal maximum. The subsequent deepening is not clear from the benthic foraminiferal record. With the onset of basin-wide decrease of oxygenation and increase of food levels at the seafloor, the previously established bathymetric gradient in benthic foraminifera assemblages became disrupted. A low-diversity *A. aegyptiacus* assemblage dominated at all depths studied. As normal conditions returned and benthic foraminifera assemblages again became established along a bathymetric gradient, water depths were similar to those prior to the late Paleocene thermal maximum.

Sedimentologic observations provide additional indications for relative sea-level fluctuations in the studied interval. In sections like Aweina and Ben Gurion, the top of the pre-late Paleocene thermal maximum succession is eroded and the early part of the late Paleocene thermal maximum (i.e., the black shale) is lost in a hiatus, indicated by an omission surface. In Nezzi, the black shale fills a channel (P. Luger, 1998, personal commun.). In numerous sections on Sinai, Lüning et al. (1998) also found discontinuities within the NP9–NP10 interval. Our observations on this Sinai material indicate that these discontinuities also include the late Paleocene thermal maximum. These discontinuities are likely the result of changes in the interplay between sediment input, carbonate production, redistribution, and relative sea-level change.

In a sequence stratigraphic interpretation (Fig. 4), using EXXON nomenclature (e.g., Haq et al., 1988), the beds underlying the late Paleocene thermal maximum represent the late highstand systems tract and subsequent lowstand systems tract. The black shales with numerous peloids and fish remains suggest slow sedimentation typical of a condensed sequence, representing the early part of the transgressive systems tract. The overlying, often calcarenitic, beds with high numbers of planktic foraminifera seem to result from a continuation of condensed sedimentation (reduction of input from land, but high carbonate production; see following), representing the maximum flooding surface and the transition to the next highstand systems tract. Thus, a complete short-term (third or fourth order) sea-level cycle appears to envelope the late Paleocene thermal maximum. Foraminiferal abundance and $CaCO_3$ data indicate that the next highstand systems tract was reached well before $\delta^{13}C$ values stabilized. These observations indicate that the transgressive systems tract containing the black shales at its base had a duration similar to that of the late Paleocene thermal maximum (~90 k.y.), using the temporal data of the carbon isotope excursion (~220 k.y.) by Röhl et al. (2000).

Eustatic sea-level rise at the onset of the late Paleocene thermal maximum

A very similar black shale in the same stratigraphic position has been recorded in numerous sections on the northern margin of the Tethys (Gavrilov et al., 1997; Speijer et al., 1997). It contains phosphatic concentrations and enrichments of a suite of elements typical for deposition under suboxic to anoxic conditions (Gavrilov et al., 1997). Also similar to the Egyptian record, these black shales constitute a condensed deposit during a transgressive pulse. However, this transgressive pulse is superimposed on a long-term transgressive trend (Gavrilov et al., 1997), not on a regressive one as in the Middle East. Apparently, at least one of these long-term trends does not relate to eustatic change, but rather to the general tectonic development of the individual continental margin.

Some of the best-studied neritic records across the late Paleocene thermal maximum are found in the subsurface of the New Jersey coastal plain. The investigation of numerous cores shows a complicated stratigraphic architecture of upper Paleocene and lower Eocene rocks, characterized by numerous smaller and larger gaps (Olsson and Wise, 1987, Gibson et al., 2000). The late Paleocene thermal maximum in this area is characterized by the transition from glauconite-rich clayey sands and sandy clays to clays. The two cores currently providing the best record across this interval are the Clayton core (Gibson et al., 1993) and the Bass River core (Cramer et al., 1999). Both studies agree that the lithologic change results from a relative sea-level rise coinciding with the late Paleocene thermal maximum. As in the Tethys, this sea-level rise appears to be associated with the spread of dysoxia on the shelf (Olsson and Wise, 1987; Gibson et al., 1993).

Thus, although there are no indications of a sharp sea-level fall on the Asian and North American passive continental margins, there is ample evidence of sea-level rise and/or transgression associated with the onset of the late Paleocene thermal maximum. Although proving synchronicity of sea-level fluctuations from one basin to another is usually a delicate operation, here the time control is very good and suggests that the sea-level rise observed is eustatic in origin and intimately linked to the climatic changes of the late Paleocene thermal maximum. In the Duwi section, an increase in paleodepth of ~20 m during the late Paleocene thermal maximum is estimated (Speijer and Morsi, 2002). Theoretically this value is composed of a combination of eustatic change, subsidence, compaction, and sediment input. Average subsidence (without correction for compaction) of the Paleocene section of Duwi is ~15 m/m.y. (150 m total thickness and assuming no significant change in water depth through the ~10 m.y. of the Paleocene; Said, 1990). Thus, during the ~90 k.y. of the late Paleocene thermal maximum, average subsidence led to ~1.5 m deepening, while ~0.5 m of compacted sediment accumulated, leading to a net paleodepth increase of only ~1 m. This value is negligible; assuming no major tectonic reorganization, an ~20 m eustatic rise can be inferred. There are no estimates for the amount of sea-level rise in the other regions and so this value is currently the best estimate of a possible eustatic change coinciding with the late Paleocene thermal maximum.

This coincidence also makes it plausible that both are causally linked. The current best explanation for the rapidity and magnitude of the carbon isotope excursion is the massive release of methane from deep-sea sediment reservoirs (Dickens et al., 1995; Bains et al., 1999; Katz et al., 1999; Dickens, 2000a). Bains et al. (1999) and Röhl et al. (2000) suggested that there were two or three main phases of methane release within a 20 k.y. period at the onset of the late Paleocene thermal maximum. These would be a response to deep-sea warming and may have acted as a positive feedback for climate warming during the late Paleocene thermal maximum.

An additional consequence of deep-sea warming would be the thermal expansion of the oceanic water column. This expansion would be ~1 m/°C (Schulz and Schäfer-Neth, 1997) for an expanding water column of 4 km. During the late Paleocene thermal maximum, deep ocean temperatures warmed by 4–8°C (Kennett and Stott, 1991; Zachos et al., 1993). Depending on how much of the entire water column was involved in the deep-sea warming, the resulting sea-level rise would be ~4–8 m at the most. This would account for less than half of the inferred sea-level rise. Additional sea-level rise may have been generated by the melting of polar ice caps and high-altitude ice covers. However, there is little evidence of polar ice during the Paleocene, also because of a general lack of high-latitude Paleocene deposits. The potential global waxing and waning of mountainous glaciers at that time is similarly difficult to quantify. Significant amounts of water stored in soils and lakes may have been transferred to the oceanic reservoir as a response to global warming, but these cannot have been sufficient to account for the rest of the inferred sea-level rise. However, a eustatic rise of ~20 m within <100 k.y. is difficult to explain without calling upon glacio-eustatic changes in addition to the thermal expansion of oceanic water. This shows that the sea-level record at late Paleocene thermal maximum time demands a careful examination of other continental margin records.

Formation of the black shales and foraminifera-rich marls

There are various models for the formation of black shales: a common factor in all of them is a considerable degree of oxygen deficiency at the seafloor (Wignall, 1994). For certain periods of black shale formation (e.g., in the Cretaceous), even fully anoxic conditions on an ocean-wide scale have been suggested (e.g., Arthur et al., 1987; Sinninghe Damsté and Köster, 1998). However, in many black shales, evidence of benthic life is present in the form of benthic microfauna or small burrows, i.e., the quasianaerobic biofacies (Koutsoukos and Hart, 1990; Savrda and Bottjer, 1991; Tyson and Pearson, 1991). Benthic foraminifera are the most common group of shelly benthic microbiota in recent oxygen-deficient environments and in black shales. This group has numerous representatives that appear to thrive under such conditions (e.g., Sen Gupta and Machain-Castillo, 1993; Bernhard, 1996). Experimental and field studies indicate that some species are able to survive under fully anoxic conditions (e.g., Moodley and Hess, 1992; Bernhard, 1996; Bernhard et al., 1997). However, it has not been demonstrated that reproduction can take place in the absence of oxygen, and thus long-term anoxia would lead to disappearance of all living benthic foraminifera and stop in situ shell production.

The late Paleocene thermal maximum black shales studied here show a fine wavy-laminated microtexture, indicating a general absence of bioturbating organisms and suboxic to anoxic conditions within the sediment (Wignall, 1994). The wavy character of the lamination results from compaction of the fine-grained sediment around the larger grains such as peloids and microfauna (Wignall, 1994). In situ, opportunistic benthic foraminifera are present in very low numbers in all black shales studied, thus representing a quasianaerobic biofacies. Considering that microhabitat preferences are not fixed, but rather show a dynamic response to oxygen and nutrient gradients at the top of the sediment (Jorissen et al., 1995; van der Zwaan et al., 1999), we assume that all common to abundant (in a relative sense) benthic taxa (*A. aegyptiacus, Stainforthia farafraensis,* and *Valvulineria* sp.) lived on top of the sediment. Their very low numbers (in an absolute sense) cannot be caused by postmortem destruction; the shells of planktic foraminifera are abundant and both are well preserved, especially in Qreiya (as in the fissile marls of Duwi). These black shale characteristics and inferred ecologic responses suggest that bottom-water conditions may either have been suboxic throughout or may have fluctuated between anoxic and suboxic or dysoxic. In modern environments, stable suboxic conditions tend to favor certain adapted microbiota to proliferate, leading to high-standing stocks and large numbers of tests (Bernhard et al., 1997; van der Zwaan et al., 1999). The late Paleocene thermal maximum black shales, however, yield very low numbers of opportunistic benthic foraminifera, despite the lack of postmortem loss and the somewhat condensed nature of the sediment. Thus, from a numeric perspective we exclude the possibility that the bottom water was permanently suboxic and adopt the alternative that the conditions fluctuated between anoxic and suboxic or dysoxic. We hypothesize that during extended periods of anoxia, all local benthic organisms became extinct. During brief periods of improved ventilation, some pioneer benthic taxa were able to colonize the seafloor from shallower areas, protected from severe anoxia (see Speijer et al., 1997, for details on extinction and repopulation patterns). In order to preserve lamination and to prevent pioneer taxa from producing abundant offspring, ventilation of the seafloor must have been a rare and short-lived event, for example following severe storms.

Productivity changes

It was suggested earlier that primary production and the vertical organic carbon flux in this basin increased considerably during the latest Paleocene and particularly during deposition of the black shales and calcarenitic marls associated with the late Paleocene thermal maximum and carbon isotope excursion (Speijer, 1994; Schmitz et al., 1997; Speijer et al., 1997; Charisi and Schmitz, 1998; Speijer and Schmitz, 1998). TOC enrichment of the black shales is consistent with this (e.g., Calvert and Pedersen, 1992; Bertrand and Lallier-Vergès, 1993), particularly because the bulk of the TOC is represented by amorphous organic matter of marine origin and the TOC contents are expected to have been considerably higher before thermal maturation. The domination of the benthic foraminifera *A. aegyptiacus* is thought to be typical for eutrophic conditions (Speijer et al., 1996b).

Biotic evidence for productivity changes of the pelagic ecosystem during the late Paleocene thermal maximum is highly equivocal. The only pelagic organisms in the black shales and overlying foraminifera-rich marls are calcareous nannoplankton and planktic foraminifera. Monechi et al. (2000) noted the blooming of *Coccolithus pelagicus* during the late Paleocene thermal maximum in Egypt. In the modern ocean, blooms of this taxon typify high-latitude oceans as well as mid-latitude upwelling systems (Baumann et al., 2000; Cachao and Moita, 2000). In Mediterranean Quaternary sediments, *C. pelagicus* shows abundance peaks associated with intervals characterized by sapropels (Müller, 1985). These distribution data strongly suggest that the late Paleocene thermal maximum is characterized by enhanced biological productivity through upwelling.

Among the planktic foraminifera, the genus *Acarinina*, including the species *A. africana* and *A. sibaiyaensis*, is unusually abundant in the black shales, largely at the expense of *Morozovella*. Both genera are usually considered as surface dwellers typical for oligotrophic conditions (e.g., Boersma and Premoli-Silva, 1991; Hallock et al., 1991; Berggren and Norris, 1997).

The same dramatic compositional change during the late Paleocene thermal maximum was observed in the central Pacific ocean and interpreted as an indication for extreme oligotrophy (Kelly et al., 1996, 1998). However, we consider it very unlikely that conditions resembling extreme oligotrophy of an open ocean could ever have prevailed in this marginal basin. Humid and warm conditions prevailing on the southeastern continent at this time led to enhanced kaolinite input into the basin (Bolle et al., 2000b) and would have supplied sufficient nutrients to prevent the basin from becoming extremely oligotrophic. This paradox demonstrates that the current understanding of the autecology of Paleogene planktic foraminifera is still quite poor.

Radiolaria are often associated with modern and ancient upwelling systems. In the studied region, radiolaria are fairly abundant only in the basal meters of the Esna Formation (and lateral equivalents) in Nukhl and Ben Gurion. Their disappearance prior to the late Paleocene thermal maximum has been used as an argument for a decrease in upwelling strength during the late Paleocene (Benjamini, 1992; Lu et al., 1996). However, benthic foraminifera assemblages from the pre-late Paleocene thermal maximum Esna Formation in Nukhl and Ben Gurion are uniform (Speijer, 1995), regardless of the presence of radiolaria, and suggest oligotrophic deep-water conditions throughout (Speijer et al., 1997). Whatever caused the radiolaria to bloom did not result in an enhanced flux of organic matter to the seafloor, which makes a connection to high productivity less likely. The biogenic barium record of Ben Gurion is also consistent with more or less stable paleoproductivity prior to the late Paleocene thermal maximum (Schmitz et al., 1997).

Diatoms and organic-walled cysts derived from heterotrophic dinoflagellates are other potential indicators for enhanced productivity of surface waters. However, representatives of these groups were not found in the studied sections, neither in background sediments nor in the black shales (H. Brinkhuis and J. Fenner, 1999, personal commun.). Paleocene deposits in Tunisia contain microfaunal assemblages nearly identical to those from Egypt (Speijer et al., 1996b; Kouwenhoven et al., 1997). Because these deposits also provided rich dinocyst assemblages (Bujak and Brinkhuis, 1998; H. Brinkhuis, 2001, personal commun.), we believe that the absence of organic walled dinocysts in the Egyptian sections was probably caused by oxidation or other (postdepositional) effects.

In conclusion, we stress that the TOC enrichment of the black shales was probably not merely caused by enhanced preservation through a decrease in oxygen concentrations of the ambient water, but also resulted from a higher organic carbon flux stimulated by upwelling. This view is in accordance with numerous studies that call upon a combination of higher paleoproductivity and preservation leading to organic-rich deposits (e.g., Calvert and Pedersen, 1992; Bertrand and Lallier-Vergès, 1993). Which primary producers were stimulated remains uncertain. The use of general pelagic proxies (at generic or higher systematic level) for trophic conditions for the late Paleocene is unwarranted and the ecology of the pelagic ecosystem requires a more thorough understanding prior to unequivocally determining changes in biological productivity. From this perspective, it is interesting to note that biogenic barium records suggest increased paleoproductivity during the late Paleocene thermal maximum in several Atlantic and Southern Ocean localities (Bains et al., 2000) and that one of these localities, Ocean Drilling Program Site 1051 (Norris and Röhl, 1999), is marked by the same unusual planktic foraminifera assemblage that typifies the late Paleocene thermal maximum black shale assemblages in Egypt.

Paleoceanographic model

We envisage the following paleogeographic and paleoceanographic configurations for the studied region during the late Paleocene (Fig. 7). Hemipelagic sedimentation characterized large parts of the epicontinental basin studied. The cosmopolitan character of pelagic and benthic microbiota indicates good surface and bottom connections with the Tethys ocean, despite the presence of Syrian arc swells. These swells may have modulated but not restricted surface- and intermediate-water circulation in and out of the epicontinental basin. The only significant local barrier was the Galala platform in the northern part of the Eastern Desert (Fig. 1; e.g., Kuss et al., 2000). The lack of barriers also means that prevailing surface winds (northeast trade winds) were largely unhindered by geographic obstacles and could lead to upwelling of nutrient-rich deeper waters through offshore Ekman transport (Speijer, 1994; Speijer et al., 1996b). As discussed herein, various micropaleontologic and geochemical proxies indicate upwelling and enhanced biological productivity during the late Paleocene, particularly during and to a lesser extent after the late Paleocene thermal maximum (Schmitz et al., 1996, 1997; Speijer et al., 1996b, 1997; Charisi and Schmitz, 1998; Speijer and Schmitz, 1998). Prior to the late Paleocene thermal maximum at a time of gradually falling sea level, elevated biological productivity led to an oxygen minimum zone confined to the shallowest part of the basin (Fig. 7A). Wind stress and high excess evaporation led to an inflow of well-oxygenated Tethyan intermediate water. Deeper parts of the basin were sufficiently ventilated to enable the settlement of diverse benthic communities. Clay-mineral analysis (Bolle et al., 2000a, 2000b) suggests that terrestrial climate was generally warm and arid in the northern part (southern Israel, Sinai) and alternating dry and wet in the southern part (southern Egypt). These climatic conditions could have favored the creation of restricted loci for dense saline water formation, particularly in the northern part of the studied basin. The existence of such loci, however, remains speculative (see also Schmitz et al., 1996; Charisi and Schmitz, 1998). With the onset of the late Paleocene thermal maximum (Fig. 7B), sea level rapidly rose. The inflowing intermediate water from the Tethys, like Pacific and Atlantic intermediate water masses (e.g., Bralower et al.,

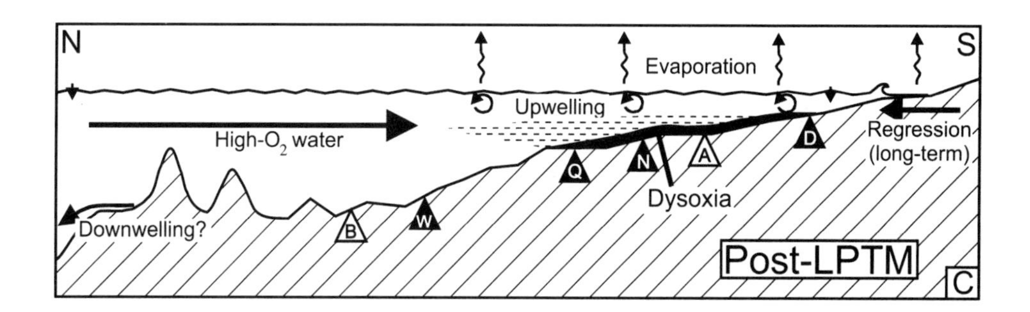

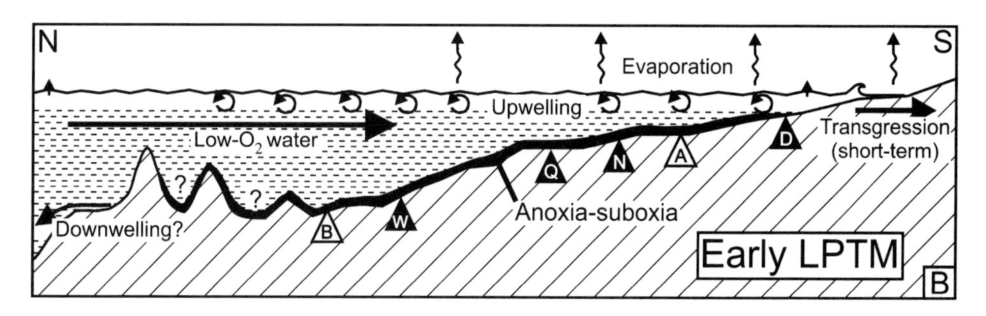

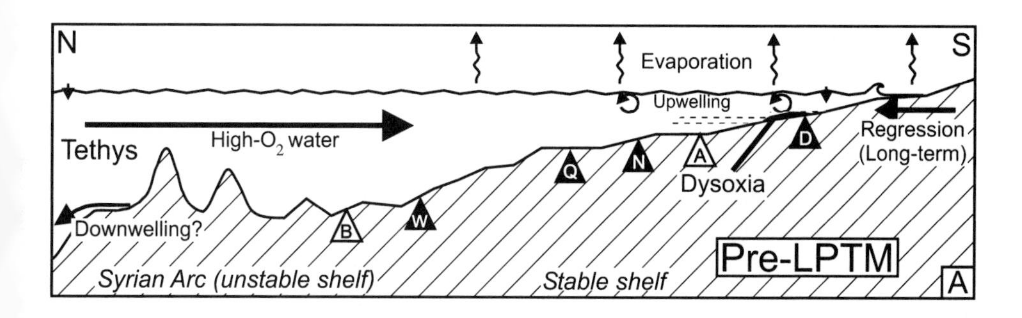

Figure 7. Postulated late Paleocene paleoceanographic conditions in basin covering Eastern Desert, Sinai, and southern Israel. Depth of swells of Syrian arc (unstable shelf) is uncertain. Passages between swells enabled exchange with deep sea Tethys. A: Reconstruction prior to late Paleocene thermal maximum (LPTM). B: Early part of LPTM. C: After LPTM. Triangles are localities studied: B, Ben Gurion; W, Wadi Nukhl; Q, Gebel Qreiya; N, Gebel Nezzi; A, Gebel Aweina; and D, Gebel Duwi.

1995; Kaiho et al., 1996; Pardo et al., 1997; Thomas, 1998; Katz et al., 1999; Dickens, 2000a, 2001c), probably contained less dissolved oxygen. Incorporation of this water into epicontinental circulation in combination with intensified upwelling and biological productivity led to severe seafloor anoxia in all studied parts of the basin. Rare ventilation events, e.g., resulting from severe storms, may have occasionally supplied sufficient oxygen for pioneer benthic foraminifera to settle briefly. Increasing humidity on the southeastern landmass brought kaolinite and additional nutrients into the basin (Bolle et al., 2000b). Following the late Paleocene thermal maximum, i.e., from the second half of the carbon isotope excursion onward, elevated productivity associated with a high carbonate production (foraminifera-rich calcarenitic marl beds) persisted through the remainder of the studied interval. At the same time, Tethyan intermediate waters became better oxygenated, so that the oxygen minimum zone (dysoxia at the seafloor) became restricted to neritic environments and bathyal sites were bathed in well-oxygenated waters (Fig. 7C). Continental climate was similar to prior to the late Paleocene thermal maximum, although more humid conditions appear to have persisted SE of Duwi (Bolle et al., 2000a, 2000b).

Carbon burial and termination of the carbon isotope excursion

The magnitude and duration of the carbon isotope excursion are related to several basic components of the global carbon cycle, particularly the mass of carbon in the ocean, and the rates and isotopic compositions of carbon fluxes to and from the ocean (Kump and Arthur, 1999; Dickens, 2001b). Recent studies have highlighted the significance of the carbon isotope excursion, stressing that it can only be explained by a carbon injection to the ocean and atmosphere on a scale similar to that projected for anthropogenic inputs from fossil-fuel combustion (Dickens et al., 1997; Norris and Röhl, 1999; Bains et al., 2000). Although many Paleogene conditions were different from those of today, the late Paleocene thermal maximum may offer an analog for how global biogeochemical cycles responded to a massive input of carbon (Dickens et al., 1997).

One particularly relevant issue is the rate at which the carbon isotope excursion returns to pre-carbon isotope excursion values, because this represents the removal of excess carbon from the ocean and atmosphere after carbon injection (Dickens, 2001b). Norris and Röhl (1999) estimated that the carbon isotope excursion lasted ~150 k.y. Assuming a Paleogene carbon cycle with masses and fluxes similar to present, this response is shorter than predicted (~200 k.y.) if inputs from rivers, volcanoes, and weathering and outputs to carbonate and organic matter remained constant after carbon injection. There are several possible explanations for the discrepancy, of which two seem particularly plausible (Dickens, 2001b): (1) the total carbon inputs and outputs were enhanced or (2) the burial of organic matter increased relative to carbonate. The high TOC contents of sediment deposited during the late Paleocene thermal maximum in the northern Tethys (Gavrilov et al., 1997; Speijer et al., 1997) and the southern Tethys are consistent with the second explanation. They may also support the first explanation if the flux of TOC increased. TOC contents increased from ~0.1% to ~2% in Egyptian black shales and to 17% in northern Tethyan black shales (Gavrilov et al., 1997), marking a 20–170 fold increase. Although sedimentation rates decreased at the same time in response to sea-level rise, it is unrealistic to explain the entire TOC increase by condensation, and thus the TOC flux increased significantly. However, at least two basic issues beyond the scope of this paper need to be resolved before carbon removal during the late Paleocene thermal maximum can be defined. First, it is uncertain whether TOC concentrations increased in other marine sediments, deep and shallow; records from other locations do not show strongly enhanced TOC contents across the late Paleocene thermal maximum. Second, Röhl et al. (2000) suggested that the duration of the carbon isotope excursion was closer to ~220 k.y., which may negate the need for enhanced carbon removal after massive carbon injection.

CONCLUSIONS

On three continental margins, the late Paleocene thermal maximum is associated with transgression and sea-level rise, suggesting a eustatic control. A possible eustatic rise of ~20 m could in part (maximum 8 m) have resulted from thermal expansion of the oceanic water column. We speculate that the remaining difference may have been caused by glacio-eustatic mechanisms. As sea-level rose, the oxygen minimum zone in the studied basin expanded and intensified, leading to anoxia at the seafloor. This was caused by a combination of inflow of oxygen-deficient Tethyan water (possibly due to methane oxidation) and higher organic carbon flux to the seafloor. TOC-rich late Paleocene thermal maximum sediments are widespread along the Tethyan continental margins, and may have played an important role in drawing down carbon that entered the oceans and atmosphere with the onset of the late Paleocene thermal maximum.

ACKNOWLEDGMENTS

Financial support was provided through grants from the Deutsche Forschungsgemeinschaft (Sp-612/1) and the Zentrale Kommissionen für Forschungsförderung und wissenschaftlichen Nachwuchs (FNK, Bremen University) to Speijer. We thank Jan Smit and an anonymous reader for insightful and constructive reviews, Gerald Dickens, Henk Brinkhuis, and Ken MacLeod for comments and improvements of parts of the text, and Peter Luger for providing samples and stratigraphic logs of the Qreiya and Nezzi sections.

REFERENCES CITED

Alsharhan, A.S., and Salah, M.G., 1997, A common source rock for Egyptian and Saudi hydrocarbons in the Red Sea: American Association of Petroleum Geologists Bulletin, v. 81, p. 1640–1659.

Arthur, M.A., Schlanger, S.O., and Jenkyns, H.C., 1987, The Cenomanian-Turonian oceanic anoxic event. 2. Palaeoceanographic controls on organic-matter production and preservation, *in* Brooks, J., and Fleet, A.J., eds., Marine petroleum source rocks: Geological Society [London] Special Publication 26, p. 401–420.

Aubry, M.-P., Berggren, W.A., and Lucas, E., 1998, Late Paleocene-early Eocene climatic and biotic events in the marine and terrestrial records: New York, Columbia University Press, 513 p.

Bains, S., Corfield, R.M., and Norris, R.D., 1999, Mechanisms of climate warming at the end of the Paleocene: Science, v. 285, p. 724–727.

Bains, S., Norris, R.D., Corfield, R.M., and Faul, K.L., 2000, Termination of global warmth at the Palaeocene/Eocene boundary through productivity feedback: Nature, v. 407, p. 171–174.

Baumann, K.H., Young, J.R., Cachao, M., and Ziveri, P., 2000, Biometric study of *Coccolithus pelagicus* and its palaeoenvironmental utility: Journal of Nannoplankton Research, v. 22, p. 82.

Benjamini, C., 1992, The Paleocene-Eocene boundary in Israel: A candidate for the boundary stratotype: Neues Jahrbuch für Geologie und Paläontologie, Abhandlungen, v. 186, p. 49–61.

Berggren, W.A., and Norris, R.D., 1997, Biostratigraphy, phylogeny and systematics of Paleocene trochospiral planktonic foraminifera: Micropaleontology, v. 43, Supplement, 116 p.

Berggren, W.A., Kent, D.V., Swisher, C.C., III, and Aubry, M.-P., 1995, A revised Cenozoic geochronology and chronostratigraphy, *in* Berggren, W.A., Kent, D.V., Aubry, M.-P., and Hardenbol, J., eds., Geochronology, time scales and global stratigraphic correlation: SEPM (Society for Sedimentary Geology), Special Publication 54, p. 129–212.

Bernhard, J.M., 1996, Microaerophilic and facultative anaerobic benthic foraminifera: A review of experimental and ultrastructural evidence: Revue de Paléobiologie, v. 15, p. 261–275.

Bernhard, J.M., and Bowser, S.S., 1999, Benthic foraminifera of dysoxic sediments: Chloroplast sequestration and functional morphology: Earth-Science Reviews, v. 46, p. 149–165.

Bernhard, J.M., Sen Gupta, B.K., and Borne, P.F., 1997, Benthic foraminiferal proxy to estimate dysoxic botom-water oxygen concentrations: Santa Barbara Basin, U.S. Pacific continental margin: Journal of Foraminiferal Research, v. 27, p. 10.

Bertrand, P., and Lallier-Vergès, E., 1993, Past sedimentary organic matter accumulation and degradation controlled by productivity: Nature, v. 364, p. 786–788.

Boersma, A., and Premoli-Silva, I., 1991, Distribution of Paleogene planktonic foraminifera: Analogies with the recent?: Palaeogeography, Palaeoclimatology, Palaeoecology, v. 83, p. 29–48.

Bolle, M.P., Pardo, A., Adatte, T., von Salis, K., and Burns, S., 2000a, Climatic evolution on the southeastern margin of the Tethys (Negev, Israel) from the Palaeocene to the early Eocene: Focus on the late Palaeocene thermal maximum: Journal of the Geological Society of London, v. 157, p. 929–941.

Bolle, M.P., Tantawy, A.A., Pardo, A., Adatte, T., Burns, S., and Kassab, A., 2000b, Climatic and environmental changes documented in the upper Paleocene to lower Eocene of Egypt: Eclogae Geologicae Helvetiae, v. 93, p. 33–51.

Bralower, T.J., Zachos, J.C., Thomas, E., Parrow, M., Paull, C.K., Kelly, D.C., Silva, I.P., Sliter, W.V., and Lohmann, K.C., 1995, Late Paleocene to Eocene paleoceanography of the equatorial Pacific Ocean: Stable isotopes recorded at Ocean Drilling Program Site 865, Allison Guyot: Paleoceanography, v. 10, p. 841–865.

Bralower, T.J., Thomas, D.J., Zachos, J.C., Hirschmann, M.M., Röhl, U., Sigurdsson, H., Thomas, E., and Whitney, D.L., 1997, High-resolution records of the late Paleocene thermal maximum and circum-Caribbean volcanism: Is there a causal link?: Geology, v. 25, p. 963–966.

Bujak, J.P., and Brinkhuis, H., 1998, Global warming and dinocyst changes across the Paleocene/Eocene Epoch boundary, *in* Aubry, M.-P., Lucas, S.G., and Berggren, W.A., eds., Late Paleocene-early Eocene climatic and biotic events in the marine and terrestrial records: New York, Columbia University Press, p. 277–295.

Cachao, M., and Moita, M.T., 2000, *Coccolithus pelagicus*, a productivity proxy related to moderate fronts off western Iberia: Marine Micropaleontology, v. 39, p. 131–155.

Calvert, S.E., and Pedersen, T.F., 1992, Organic carbon accumulation and preservation in marine sediments: How important is anoxia?, *in* Whelan, J.K., and Farrington, J.W., eds., Organic matter; productivity, accumulation, and preservation in recent and ancient sediments: New York, Columbia University Press, p. 231–263.

Charisi, S.D., and Schmitz, B., 1995, Stable ($\delta^{13}C$, $\delta^{18}O$) and strontium ($^{87}Sr/^{86}Sr$) isotopes through the Paleocene at Gebel Aweina, eastern Tethyan region: Palaeogeography, Palaeoclimatology, Palaeoecology, v. 116, p. 103–129.

Charisi, S.D., and Schmitz, B., 1998, Paleocene to early Eocene paleoceanography of the Middle East: The $\delta^{13}C$ and $\delta^{18}O$ isotopes from foraminiferal calcite: Paleoceanography, v. 13, p. 106–118.

Cramer, B.S., Aubry, M.P., Miller, K.G., Olsson, R.K., Wright, J.D., and Kent, D.V., 1999, An exceptional chronologic, isotopic, and clay mineralogic record of the latest Paleocene thermal maximum, Bass River, New Jersey, Ocean Drilling Program 174AX: Bulletin de la Société Géologique de France, v. 170, p. 883–897.

Dickens, G.R., 2000, Methane oxidation during the late Palaeocene thermal maximum: Bulletin de la Société Géologique de France, v. 171, p. 37–49.

Dickens, G.R., 2001a, Carbon addition and removal during the Late Palaeocene thermal maximum: Basic theory with a preliminary treatment of the isotope record at Ocean Drilling Program Site 1051, Blake Nose, *in* Kroon, D., Norris, R., and Klaus, A., eds., Western North Atlantic Palaeogene and Cretaceous palaeoceanography: Geological Society [London] Special Publication 183, p. 293–306.

Dickens, G.R., 2001b, On the fate of past gas: What happens to methane released from a bacterially mediated gas hydrate capacitor?: Geochemistry, Geophysics, Geosystems, v. 2, Paper number 2000GC000131 (Available at http://www.gcubed.org).

Dickens, G.R., O'Neil, J.R., Rea, D.K., and Owen, R.M., 1995, Dissociation of oceanic methane hydrate as a cause of the carbon isotope excursion at the end of the Paleocene: Paleoceanography, v. 10, p. 965–971.

Dickens, G.R., Castillo, M.M., and Walker, J.C.G., 1997, A blast of gas in the latest Paleocene: Simulating first-order effects of massive dissociation of oceanic methane hydrate: Geology, v. 25, p. 259–262.

Espitalié, J., Deroo, G., and Marquis, F., 1985, La pyrolyse Rock-Eval et ses applications: Revue de l'Institut Français du Pétrole, v. 40, p. 25–89.

Gavrilov, Y.O., Kodina, L.A., Lubchenko, I.Y., and Muzylev, N.G., 1997, The late Paleocene anoxic event in epicontinental seas of peri-Tethys and formation of the sapropelite unit: Sedimentology and geochemistry: Lithology and Mineral Resources, v. 32, p. 427–450.

Gibson, T.G., Bybell, L.M., and Owens, J.P., 1993, Latest Paleocene lithologic and biotic events in neritic deposits of southwestern New Jersey: Paleoceanography, v. 8, p. 495–514.

Gibson, T.G., Bybell, L.M., and Mason, D.B., 2000, Stratigraphic and climatic implications of clay mineral changes around the Paleocene/Eocene boundary of the northeastern US margin: Sedimentary Geology, v. 134, p. 65–92.

Hallock, P., Premoli-Silva, I., and Boersma, A., 1991, Similarities between planktonic and larger foraminiferal evolutionary trends through Paleogene paleoceanographic changes: Palaeogeography, Palaeoclimatology, Palaeoecology, v. 83, p. 49–64.

Haq, B.U., Hardenbol, J., and Vail, P., 1988, Mesozoic and Cenozoic chronostratigraphy and cycles of sea level change, *in* Wilgus, C.K., et al., eds., Sea-level changes: An integrated approach: Society of Economic Paleontologists and Mineralogists, Special Publication 42, p. 71–108.

Jones, R.W., 1987, Organic facies: Advances in petroleum geochemistry, v. 2, p. 1–90.

Jorissen, F.J., de Stigter, H.C., and Widmark, J.G.V., 1995, A conceptual model explaining benthic foraminiferal microhabits: Marine Micropaleontology, v. 26, p. 3–15.

Kaiho, K., Arinobu, T., Ishiwatari, R., Morgans, H.E.G., Okada, H., Takeda, N., Tazaki, K., Zhou, G., Kajiwara, Y., Matsumoto, R., Hirai, A., Niitsuma, N., and Wada, H., 1996, Latest Paleocene benthic foraminiferal extinction and environmental changes at Tawanui, New Zealand: Paleoceanography, v. 11, p. 447–465.

Katz, M.E., Pak, D.K., Dickens, G.R., and Miller, K.G., 1999, The source and fate of massive carbon input during the latest Paleocene thermal maximum: Science, v. 286, p. 1531–1533.

Kelly, D.C., Bralower, T.J., Zachos, J.C., Premoli Silva, I., and Thomas, E., 1996, Rapid diversification of planktonic foraminifera in the tropical Pacific (ODP Site 865) during the late Paleocene thermal maximum: Geology, v. 24, p. 423–426.

Kelly, D.C., Bralower, T.J., and Zachos, J.C., 1998, Evolutionary consequences of the latest Paleocene thermal maximum for tropical planktonic foraminifera: Palaeogeography, Palaeoclimatology, Palaeoecology, v. 141, p. 139–161.

Kennett, J.P., and Stott, L.D., 1991, Abrupt deep-sea warming, palaeoceanographic changes and benthic extinctions at the end of the Palaeocene: Nature, v. 353, p. 225–229.

Koch, P.L., Zachos, J.C., and Gingerich, P.D., 1992, Correlation between isotope records in marine and continental carbon reservoirs near the Palaeocene/Eocene boundary: Nature, v. 358, p. 319–322.

Koutsoukos, E.A.M., and Hart, M.B., 1990, Cretaceous foraminiferal morphogroup distribution patterns, palaeocommunities and trophic structures: A case study from the Sergipe Basin, Brazil: Transactions of the Royal Society of Edinburgh: Earth Sciences, v. 81, p. 221–246.

Kouwenhoven, T.J., Speijer, R.P., Van Oosterhout, C.W.M., and Van der Zwaan, G.J., 1997, Benthic foraminiferal assemblages between two major extinction events; the Paleocene El Kef section, Tunisia: Marine Micropaleontology, v. 29, p. 105–127.

Kuhnt, W., Herbin, J.P., Thurow, J., and Wiedmann, J., 1990, Distribution of Cenomanian-Turonian organic facies in the western Mediterranean and along the adjacent Atlantic margin, *in* Huc, A.Y., ed., Deposition of organic facies, American Association of Petroleum Geologists Studies in Geology, v. 30, p. 133–160.

Kump, L.R., and Arthur, M.A., 1999, Interpreting carbon-isotope excursions: Carbonates and organic matter: Chemical Geology, v. 161, p. 181–198.

Kuss, J., Scheibner, C., and Gietl, R., 2000, Carbonate to platform transition along an upper Cretaceous to lower Tertiary Syrian Arc uplift, Galala Plateaus, Eastern Desert of Egypt: GeoArabia, v. 5, p. 405–424.

Lu, G., Keller, G., Adatte, T., and Benjamini, C., 1996, Abrupt change in the upwelling system along the southern margin of the Tethys during the

Paleocene-Eocene transition event: Israel Journal of Earth Sciences, v. 44, p. 185–195.

Luger, P., 1985, Stratigraphie der marinen Oberkreide und des Alttertiärs im südwestlichen Obernil-Becken (SW-Ägypten) unter besonderer Berücksichtigung der Mikropaläontologie, Palökologie und Paläogeographie: Berliner Geowissenschaftliche Abhandlungen, Reihe A: Geologie und Paläontologie, v. 63, 151 p.

Lüning, S., Marzouk, A.M., and Kuss, J., 1998, The Paleocene of central East Sinai, Egypt: "Sequence stratigraphy" in monotonous hemipelagites: Journal of Foraminiferal Research, v. 28, p. 19–39.

Mart, Y., 1991, Some Cretaceous and early Tertiary structures along the distal continental margin of the southeastern Mediterranean: Israel Journal of Earth Sciences, v. 40, p. 77–90.

Mello, M.R., Telnaes, N., and Maxwell, J.R., 1995, The hydrocarbon source potential in the Brazilian marginal basins: A geochemical and paleoenvironmental assessment, *in* Huc, A.Y., ed., Paleogeography, paleoclimate, and source rocks: American Association of Petroleum Geologists Studies in Geology 40, p. 233–272.

Monechi, S., Angori, E., and Speijer, R.P., 2000, Upper Paleocene biostratigraphy in the Mediterranean region: Zonal markers, diachronism and preservational problems: GFF, v. 122, p. 108–110.

Moodley, L., and Hess, C., 1992, Tolerance of infaunal benthic for low and high oxygen concentrations: The Biological Bulletin, v. 183, p. 94–98.

Mostafa, A.E.D.R., 1993, Organic geochemistry of source rocks and related crude oils in the Gulf of Suez area, Egypt, 147 of Berliner Geowissenschaftliche Abhandlungen, Reihe A: Berlin, Selbstverlag Fachbereich Geowissenschaften, FU Berlin, 163 p.

Müller, C., 1985, Late Miocene to recent Mediterranean biostratigraphy and paleoenvironments based on calcareous nannoplankton, *in* Stanley, D.J., and Wezel, F.C., eds., Geological evolution of the Mediterranean Basin: New York, Springer-Verlag, p. 471–485.

Norris, R.D., and Röhl, U., 1999, Carbon cycling and chronology of climate warming during the Palaeocene/Eocene transition: Nature, v. 401, p. 775–778.

Olsson, R.K., and Wise, S.W., Jr., 1987, Upper Paleocene to middle Eocene depositional sequences and hiatuses in the New Jersey Atlantic Margin, *in* Ross, C.A., and Haman, D., eds., Timing and depositional history of eustatic sequences: Constraints on seismic stratigraphy: Cushman Foundation for Foraminiferal Research, Special Publication 24, p. 99–112.

Pardo, A., Keller, G., Molina, E., and Canudo, J.I., 1997, Planktic foraminiferal turnover across the Paleocene-Eocene transition at DSDP Site 401, Bay of Biscay, North Atlantic: Marine Micropaleontology, v. 29, p. 129–158.

Röhl, U., Bralower, T.J., Norris, R.D., and Wefer, G., 2000, New chronology for the late Paleocene thermal maximum and its environmental implications: Geology, v. 28, p. 927–930.

Said, R., 1962, The geology of Egypt: Amsterdam, Elsevier, 377 p.

Said, R., 1990, Cenozoic, *in* Said, R., ed., The geology of Egypt: Rotterdam, Balkema, p. 451–486.

Savrda, C.E., and Bottjer, D.J., 1991, Oxygen-related biofacies in marine strata: An overview and update, *in* Tyson, R.V., and Pearson, T.H., eds., Modern and ancient continental shelf anoxia: Geological Society [London] Special Publication 58, p. 201–219.

Schmitz, B., Speijer, R.P., and Aubry, M.-P., 1996, Latest Paleocene benthic extinction event on the southern Tethyan shelf (Egypt): Foraminiferal stable isotopic ($\delta^{13}C$, $\delta^{18}O$) records: Geology, v. 24, p. 347–350.

Schmitz, B., Charisi, S.D., Thompson, E.I., and Speijer, R.P., 1997, Barium, SiO_2 (excess), and P_2O_5 as proxies of biological productivity in the Middle East during the Palaeocene and the latest Palaeocene benthic extinction event: Terra Nova, v. 9, p. 95–99.

Schulz, M., and Schäfer-Neth, C., 1997, Translating Milankovitch climate forcing into eustatic fluctuations via thermal deep water expansion: A conceptual link: Terra Nova, v. 9, p. 228–231.

Sen Gupta, B.K., and Machain-Castillo, M.L., 1993, Benthic foraminifera in oxygen-poor habitats: Marine Micropaleontology, v. 20, p. 183–201.

Shahar, J., 1994, The Syrian arc system: An overview: Palaeogeography, Palaeoclimatology, Palaeoecology, v. 112, p. 125–142.

Shipboard Scientific Party, 1998, Explanatory notes, *in* Kroon, D., Norris, R.D., et al., eds., Proceedings of the Ocean Drilling Program, Part A, Initial Reports, Leg 171B: College Station, Texas, Ocean Drilling Program, p. 11–44.

Sinninghe Damsté, J.S., and Köster, J., 1998, A euxinic southern North Atlantic Ocean during the Cenomanian/Turonian oceanic anoxic event: Earth and Planetary Science Letters, v. 158, p. 165–173.

Speijer, R.P., 1994, Extinction and recovery patterns in benthic foraminiferal paleocommunities across the Cretaceous/Paleogene and Paleocene/Eocene boundaries [Ph.D. thesis]: Geologica Ultraiectina, v. 124, 191 p.

Speijer, R.P., 1995, The late Paleocene benthic foraminiferal extinction as observed in the Middle East: Bulletin de la Société Belge de Géologie, v. 103, p. 267–280.

Speijer, R.P., and Morsi, A.M., 2002, Ostracode turnover and sea-level changes associated with the Paleocene-Eocene thermal maximum: Geology, v. 30, p. 23–26.

Speijer, R.P., and Schmitz, B., 1998, A benthic foraminiferal record of Paleocene sea level and trophic/redox conditions at Gebel Aweina, Egypt: Palaeogeography, Palaeoclimatology, Palaeoecology, v. 137, p. 79–101.

Speijer, R.P., Schmitz, B., Aubry, M.-P., and Charisi, S.D., 1996a, The latest Paleocene benthic extinction event: Punctuated turnover in outer neritic foraminiferal faunas from Gebel Aweina, Egypt: Israel Journal of Earth Sciences, v. 44, p. 207–222.

Speijer, R.P., Van der Zwaan, G.J., and Schmitz, B., 1996b, The impact of Paleocene/Eocene boundary events on middle neritic benthic foraminiferal assemblages from Egypt: Marine Micropaleontology, v. 28, p. 99–132.

Speijer, R.P., Schmitz, B., and Van der Zwaan, G.J., 1997, Benthic foraminiferal extinction and repopulation in response to latest Paleocene Tethyan anoxia: Geology, v. 25, p. 683–686.

Speijer, R.P., Schmitz, B., and Luger, P., 2000, Stratigraphy of late Palaeocene events in the Middle East: Implications for low- to middle-latitude successions and correlations: Journal of the Geological Society of London, v. 157, p. 37–47.

Strouhal, A., 1993, Tongeologische Entwicklungstrends in kretazischen und tertiären Sedimenten Nordostafrikas: Regionale Fallbeispiele: Berliner Geowissenschaftliche Abhandlungen, Reihe A: Geologie und Paläontologie, v. 155, 68 p.

Thomas, E., 1989, Development of Cenozoic deep-sea benthic foraminiferal faunas in Antarctic waters, *in* Crame, J.A., ed., Origins and evolution of the Antarctic biota: Geological Society [London] Special Publication 47, p. 283–296.

Thomas, E., 1998, Biogeography of the late Paleocene benthic foraminiferal extinction, *in* Aubry, M.P., Lucas, S.G., and Berggren, W.A., eds., Late Paleocene-early Eocene climatic and biotic events in the marine and terrestrial records: New York, Columbia University Press, p. 214–243.

Tissot, B.P., and Welte, D.H., 1984, Petroleum formation and occurrence (second edition): Berlin, Springer-Verlag, 699 p.

Tyson, R.V., and Pearson, T.H., 1991, Modern and ancient continental shelf anoxia: An overview, *in* Tyson, R.V., and Pearson, T.H., eds., Modern and ancient continental shelf anoxia: Geological Society [London] Special Publication 58, p. 1–24.

Tyson, R.V. 1987, The genesis and palynofacies characteristics of marine petroleum source rocks, *in* Brooks, J., and Fleet, A.J., eds., Marine petroleum source rocks: Geological Society [London] Special Publication 26, p. 47–67.

van der Zwaan, G.J., Duijnstee, I.A.P., den Dulk, M., Ernst, S.R., Jannink, N.T., and Kouwenhoven, T.J., 1999, Benthic foraminifers: Proxies or problems? A review of paleocological concepts: Earth-Science Reviews, v. 46, p. 213–236.

Van Morkhoven, F.P.C.M., Berggren, W.A., and Edwards, A.S., 1986, Cenozoic cosmopolitan deep-water benthic foraminifera: Pau, Elf Aquitaine, Bulletin des Centres de Recherches Exploration-Production Elf-Aquitaine, Memoir 11, 421 p.

van Os, B., Middelburg, J.J., and de Lange, G.J., 1996, Extensive degradation and fractionation of organic matter during subsurface weathering: Aquatic Geochemistry, v. 1, p. 303–312.

Wagner, T., and Dupont, L.M., 1999, Terrestrial organic matter in marine sediments: Analytical approaches and eolian-marine records from the central equatorial Atlantic, *in* Fischer, G., and Wefer, G., eds., The use of proxies in paleoceanography: Examples from the South Atlantic: Heidelberg, Springer-Verlag, p. 547–574.

Wagner, T., and Pletsch, T., 1999, Tectono-sedimentary controls on Cretaceous black shale deposition along the opening equatorial Atlantic gateway (ODP Leg 159), *in* Cameron, N.R., Bate, R.H., and Clure, V.S., eds., The oil and gas habitats of the South Atlantic: Geological Society [London] Special Publication 153, p. 241–265.

Wignall, P.B., 1994, Black shales: Oxford, Oxford University Press, Oxford Monographs on Geology and Geophysics 30, 127 p.

Zachos, J.C., Lohmann, K.C., Walker, J.C.G., and Wise, S.W., 1993, Abrupt climate changes and transient climates during the Paleogene: A marine perspective: Journal of Geology, v. 101, p. 191–213.

Manuscript Submitted October 17, 2000; Accepted by the Society March 22, 2001

Printed in the U.S.A.

Geological Society of America
Special Paper 356
2002

Sedimentary record of impact events in Spain

Enrique Díaz-Martínez*
Enrique Sanz-Rubio
Jesús Martínez-Frías
Centro de Astrobiología, Consejo Superior de Investigaciones Científicas—Instituto Nacional de Técnica Aeroespacial, Carretera Torrejón-Ajalvir Kilómetro 4, 28850 Torrejón de Ardoz, Madrid, Spain

ABSTRACT

A review of the evidence of meteorite-impact events in the sedimentary record of Spain reveals that the only proven impact-related bed is the clay layer at the Cretaceous-Tertiary boundary (at Zumaya and Sopelana in the Bay of Biscay region, and at Caravaca, Agost, and Alamedilla in the Betic Cordilleras). Other deposits previously proposed as impact related can now be rejected, or are dubious and still debated. These include the Pelarda Formation, alleged to represent proximal ejecta from the Azuara structure; the Paleocene-Eocene boundary near Zumaya (western Pyrenees) and Alamedilla (Betic Cordillera); and the Arroyofrío Oolite Bed, which has been alleged as distal ejecta of an unknown Callovian-Oxfordian impact event. The scarcity of evidence for meteorite-impact events in the sedimentary record is possibly due to a lack of detailed studies. We propose several sedimentary units that could potentially be related to impact events, and where future research should focus.

INTRODUCTION

The sedimentary record of Spain presents evidence for at least one impact event, as well as a number of units of potential impact clastic origin (some of which are currently under investigation). In this chapter we summarize and review the information available about the sedimentary record of meteorite impacts in Spain (Fig. 1), most of it published in Spanish journals. In addition, we propose several stratigraphic units with potential for future research. In this contribution we attempt (1) to bring to the attention of the international community recent and ongoing research relating to the sedimentary record of impact events in Spain, (2) to review the current knowledge and interpretation of units previously proposed as related to impacts, and (3) to promote new research within selected units to evaluate the possibility of an impact origin.

DISTAL RECORD OF IMPACT EVENTS

The evidence from sedimentary units to be considered as distal impact ejecta may consist of geochemical anomalies of elements and isotopes (e.g., Ir, $^{187}Os/^{188}Os$), the presence of impact ejecta in the sediments (e.g., shocked minerals, microtektites, or spherules), or tsunami deposits (Montanari and Koeberl, 2000). Evidence for distal impact ejecta in the sedimentary record of Spain has been proposed in relation with the Dogger-Malm, Cretaceous-Tertiary (K-T), and Paleocene-Eocene boundaries, as discussed in the following. In brief, the only proven distal record of an impact event in Spain is found at the K-T boundary. Studies of the Paleocene-Eocene extinction event in Spanish sections have shown major changes in paleoceanographic conditions, the causes of which are still debated, but an impact origin remains only probable. Future work should

*E-mail: diazme@inta.es

Díaz-Martínez, E., Sanz-Rubio, E., and Martínez-Frías, J., 2002, Sedimentary record of impact events in Spain, *in* Koeberl, C., and MacLeod, K.G., eds., Catastrophic Events and Mass Extinctions: Impacts and Beyond: Boulder, Colorado, Geological Society of America Special Paper 356, p. 551–562.

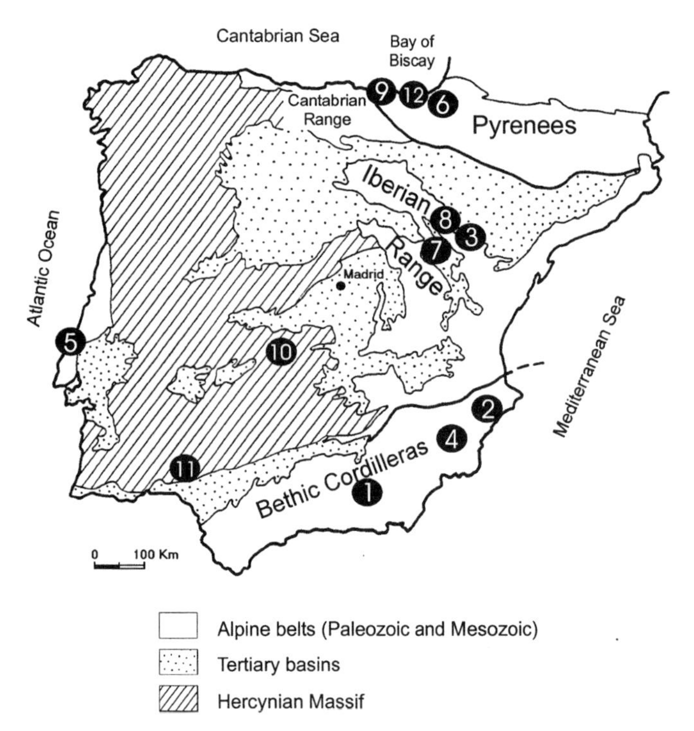

Figure 1. Localities mentioned in text: 1, Alamedilla; 2, Agost; 3, Azuara and Lécera; 4, Caravaca; 5, Nazaré; 6, Osinaga and Musquiz; 7, Pozuel del Campo; 8, Ricla; 9, Sopelana; 10, Valdelacasa and Navalpino; 11, Valverde del Camino; 12, Zumaya.

focus on high-resolution studies on marine stratigraphic sections comprising critical boundaries related to major biotic and/or climatic events (e.g., Triassic-Jurassic, Paleocene-Eocene, late Eocene).

Middle-Upper Jurassic boundary

The Middle-Upper Jurassic boundary throughout many peri-Atlantic basins is associated with a stratigraphic gap spanning at least the upper Callovian–lower Oxfordian interval (three ammonite biozones), although, in places, the missing record is much longer. On a regional scale, it is normally accepted that a major tectono-eustatic event controlled this widespread stratigraphic boundary, usually including emersion and/or condensed levels (Aurell, 1991; Aurell et al., 1994). These features are recorded, among others, in the Lusitanian basin (west-central Portugal), the Iberian, Catalonian, and Cantabrian basins (northeastern and northwestern Spain), Bourgogne and the Paris basin (France), the Jura basin (Switzerland), and the Neuquén basin (Argentina).

The Dogger-Malm boundary at Ricla and Pozuel del Campo (Iberian Range; Fig. 1) presents several features that were interpreted by Meléndez et al. (1987) as related to an impact event. This hypothesis was based on the presence of conspicuous geochemical anomalies (e.g., heavy metals and platinum group elements [PGE]), volcanic and hydrothermal activity, submarine corrosion, high concentration of iron-rich spherules, and Fe-Mn Bacterial-fungal stromatolites. According to Sepkoski (1996), the Callovian-Oxfordian interstage boundary coincides with a >20% extinction of marine fossil genera. Relatively high levels of extinction percentages are reported throughout the Middle and Upper Jurassic, although no clearly defined peak can be identified (Sepkoski, 1996). In any case, these values are higher than the percentage of extinction coinciding with other known large impact events, such as the late Eocene Chesapeake and Popigai events. No proven impact structure or impact signatures have been found at or near the Middle-Upper Jurassic boundary anywhere in the world that could be related to a large impact event (Montanari and Koeberl, 2000). Therefore, any evidence in the sedimentary record that is not unequivocal should be carefully considered before a cosmic origin is inferred.

The unit studied by Meléndez et al. (1987) is known as the Arroyofrío Oolite Bed, a thin discontinuous bed at the top of the Chelva Formation and directly below the Yátova Formation; both of these formations are shallow-marine carbonate units found at many sections throughout the eastern branch of the Iberian Range (Gómez, 1979; Aurell and Meléndez, 1990). The Arroyofrío Oolite Bed is a condensed unit, <1 m thick, consisting of wackestone and packstone with iron oolites and bioclasts. Bioclasts include ammonites, planktonic foraminifera, brachiopods, and belemnites, which were dated as mid-Callovian to early Oxfordian by Ramajo et al. (2000). Workers in Spanish basins usually interpret the Arroyofrío Oolite Bed as a result of a series of punctuated subaerial exposure and transgressive events resulting in condensed carbonate sedimentation in a shallow-marine setting near local paleogeographic highs, under the influence of local currents and regional tectonic or tectono-eustatic controls (Aurell et al., 1990, 1994; Aurell, 1991; Ramajo and Aurell, 1997; Ramajo et al., 2000).

Meléndez et al. (1987) mentioned sedimentological and biostratigraphic evidence for hardground and hiatus development, together with local (parautochthonous) resedimentation. Their studies revealed geochemical anomalies of certain siderophile elements (Fe, Mn, Ni, Co). In some cases, Pt and Ir were found in relatively high proportions. In our opinion, the high proportions of Fe + Mn/Al (indicative of hydrothermal processes), in conjunction with the evidence for submarine corrosion by acid waters, and the occasional presence of bacterial stromatolites, point the geochemical anomalies being related to shallow submarine hydrothermal vents and volcanic activity. Based on the high concentration of Ni-Fe-rich spherules found at one locality (Ricla), Meléndez et al. (1987) interpreted the volcanic and hydrothermal activity as triggered by the impact of a cosmic body. In their interpretation, the other phenomena recorded at the boundary represent the effects of such an impact. However, the evidence presented in favor of a cosmic origin for the disconformity at the Middle-Upper Jurassic boundary in the Iberian Range is not unequivocal. Discussing a recently discovered modern analogue for iron ooids and pisoids in a shallow-marine volcanic setting in Indonesia, Sturesson et al. (2000) demonstrated that iron ooids form by chemical precipitation of cryptocrystalline iron oxyhydroxides on available grains on the seafloor, from seawater enriched with Fe, Al and Si. The enrichment can be the result of hydrothermal fluids, volcanic ash falling into shallow basins, or rapid weathering of fresh volcanic rocks. More detailed research should be carried out on the geochemistry of the spherules and PGE anomalies found within the Arroyofrío Oolite Bed before a possible cosmic origin should be considered.

K-T boundary

The Spanish sedimentary record presents good examples of continuous upper Maastrichtian sedimentary sequences, such as the sections at Agost in Alicante (Groot et al., 1989), Caravaca in Murcia, and Zumaya in the Bay of Biscay region (Smit and Romein, 1985) (Fig. 1). The Agost and Caravaca sections are in the Betic Cordillera (southeastern Spain), whereas the Bay of Biscay region includes Zumaya and other remarkable sections in Spain (Sopelana, Osinaga, Musquiz) and France (Bidart, Hendaye) (Fig. 1). Most of these sections are in pelagic to hemipelagic facies and contain rich foraminiferal and nannofossil faunas and floras, insignificant amounts of macrofossils, and little or no evidence for hardgrounds or omission surfaces (Smit, 1999). High-resolution studies resulted in a magnetostratigraphic record for the Caravaca and Agost sections, but a reliable magnetostratigraphy for the Upper Cretaceous of the Bay of Biscay region has not been established (Kate and Sprenger, 1993; Moreau et al., 1994).

Sections of the Betic Cordilleras. The Agost and Caravaca sections occur in the peri-Mediterranean Alpine orogenic belt. They are among the most complete marine sections for the K-T transition, in which the K-T boundary layer provides an excellent record of the distal ejecta facies related to the Chicxulub impact (Groot et al., 1989; Martínez-Ruíz, 1994). Marl is their main lithology of both sections, which are composed calcite, quartz, and clay minerals. A clayey 2–3-mm-thick layer appears on top of the Maastrichtian, marking the K-T boundary. It is characterized by an abrupt decrease in carbonate and by an increase in clay mineral content (Ortega-Huertas et al., 1995; Martínez-Ruíz et al., 1997). PGE anomalies and spherules are confined to the boundary layer (200–400 spherules/cm^3), where spherules and smectites are the main components, and there are minor amounts of illite and kaolinite (Smit, 1990; Ortega-Huertas et al., 1995; Martínez-Ruíz et al., 1997). The composition of most of the spherules is K-feldspar and Fe-oxides, probably as a result of diagenetic alteration and replacement of precursor clinopyroxene, as interpreted from relict crystalline textures (Martínez-Ruíz et al., 1997; Smit, 1999).

The Agost section is located 1.5 km north of Agost (Alicante Province, southeastern Spain), and covers the Late Cretaceous through middle Eocene record. The K-T transition is represented by open sea deposits. The Maastrichtian record consists of light gray pelagic marls, and interbeds of calcareous marls and scarce turbiditic calcarenite beds rich in macroforaminifera (Usera et al., 2000). The K-T boundary is represented by a dark gray, 12-cm-thick clay layer that has a red-yellowish lamina, enriched in goethite and hematite, at the base (Usera et al., 2000). This lamina contains impact evidence, such as spherules, isotopic changes, and anomalies of Ir, Co, Ni, Cr, and other elements (Martínez-Ruíz et al., 1992a, 1997). Fe-oxide spherules at Agost are more abundant than K-feldspar spherules, some of the Fe-oxide spherules showing fibroradial and dendritic textures (Martínez-Ruíz et al., 1997). The Danian record comprises mainly gray marl with some interbeds of marly limestones, but toward the top of the section, reddish colors are dominant. The sedimentary continuity of the section has been demonstrated by biostratigraphic studies (Molina et al., 1996). Several models for the extinction of planktonic foraminifera have been proposed for the K-T transition at Agost, such as an almost total catastrophic mass extinction (Smit, 1990), a gradual mass extinction (Canudo et al., 1991; Pardo et al., 1996), and a catastrophic mass extinction superposed onto a less-evident gradual trend (Molina et al., 1996).

The Caravaca section is located 4 km southwest of Caravaca (Murcia Province, southeastern Spain; Fig. 1) and constitutes one of the most complete and least disturbed K-T sections in the world (Canudo et al., 1991; MacLeod and Keller, 1991). Terminal Maastrichtian–basal Paleocene sediments at Caravaca were deposited in a middle bathyal environment (200–1000 m depth), as indicated by benthic foraminiferal assemblages (Coccioni and Galeotti, 1994). Cretaceous and Tertiary lithologies of the transition are dominantly marly. The K-T boundary clay layer is a 7–10-cm-thick dark clay-marl bed. The upper part of the boundary clay layer is disturbed, and burrows several centimeters in length have been described (Arinobu et al., 1999). A 1–2-mm-thick orange basal layer rich in goethite also has high Ir and Os concentrations, in conjunction with V, Cr, Fe, Ni, Zn, and As anomalies, and a high content of small spherules (Smit and Hertogen, 1980; Smit and Klaver, 1981; Smit, 1982; Smit and ten Kate, 1982; Smit and Romein, 1985; Schmitz, 1988; Martínez-Ruíz et al., 1992b). The spherules are mainly made of K-feldspar, 0.1–0.8 mm in diameter, and were first discovered at the K-T boundary clay of the Caravaca section by Smit and Klaver (1981). Fe-oxide spherules are rare at Caravaca (Martínez-Ruíz et al., 1997).

Kaiho and Lamolda (1999) concluded, for the Caravaca section, that most planktonic foraminifera did not survive and abruptly became extinct at the K-T boundary, on the basis of stable isotope and foraminiferal abundance determinations. In addition, Arinobu et al. (1999) carried out a study of carbon isotope stratigraphy and detected a spike of the pyrosynthetic polycyclic aromatic hydrocarbons (PAHs) at the Caravaca K-T boundary. Arinobu et al. proposed that the combustion of terrestrial organic matter in massive global fires was the most probable mechanism for the origin of these PAHs.

The Alamedilla section (Granada; Fig. 1) has been described as the closest site to the Chicxulub crater (~7000 km away) with an undisturbed ejecta layer (Smit, 1999). Droplets recently found at Alamedilla were interpreted as altered tektites, indicating that tektites end microkrystites may occur together in the same ejecta layer (Smit, 1999).

Sections of the Bay of Biscay region. Stratigraphic sections around the Bay of Biscay region are valuable for testing hypotheses of K-T transition extinctions (Smit et al., 1987). The reasons for this are: (1) these sections are considered by micropaleontologists to be relatively complete (Smit et al., 1987), (2) they exhibit high sedimentation rates, resulting in increased resolution of the stratigraphy, (3) they were deposited in a pelagic, but nonturbiditic, environment, and (4) well-exposed outcrops along coastlines facilitate access for measuring sections and collecting samples. All the sections contain a conformable sequence of Upper Cretaceous and lower Tertiary marine strata that were deposited in the Basque-Cantabrian basin (Lamolda et al., 1981). This basin is part of the continental margin of northern Spain, and is mostly filled with Mesozoic rocks. The most striking geological feature of the region is the great thickness of its Mesozoic-Tertiary sequence, which exceeds 15 km (García-Mondéjar et al., 1985). This basin was one of several forming along the boundary of the European-Iberian plates during the Late Cretaceous (Ward, 1988). Although deposition of turbiditic sediments dominated from the Campanian to the early Maastrichtian, the reduction of siliciclastic material influx and basin-wide shallowing and regression during the late Maastrichtian resulted in limestone-marl rhythmites (Lamolda et al., 1981). Immediately following the K-T boundary, there was an even more dramatic reduction in siliciclastic influx into the basin, resulting in the deposition of pink coccolith limestones during the Danian (Ward, 1988). The most representative and most studied sections for the K-T boundary are Zumaya (Gipuzkoa Province) and Sopelana (Bizkaia Province) (Fig. 1). In addition, the nearby sections of Bidart and Hendaye (south of France) are also well-known for the K-T transition in the same region.

The Maastrichtian-Paleocene of the Zumaya section is the most thoroughly studied of the Bay of Biscay sections. It is the thickest, best exposed, and least faulted section (Ward, 1988). A continuous section from the lower Campanian to the Eocene is exposed along the coastal cliff west of Zumaya. The advantages of the Zumaya section are (Lamolda et al., 1988): (1) sedimentary continuity across the K-T boundary, (2) relative abundance of fossil remains through the Maastrichtian, (3) almost complete absence of turbidites in the purple marls and limestones of late Maastrichtian and early Paleocene ages, (4) thickness of and high sedimentation rate for the transitional beds, and (5) absence of tectonism affecting the transitional beds. The section has no boundary clay, but the boundary layer at Zumaya is pyritic, and therefore easy to recognize (Wiedmann, 1988). The uppermost part of the Maastrichtian is composed of several thin beds of green marls (1–5 cm thick), a sandy gray-brown bed, and then purple marls (Lamolda et al., 1988). The K-T boundary is marked by a single or multiple calcite vein (of supergenic nature) 2–3 cm thick, with gray dark shale interbeds. Ir anomalies and spherules interpreted as microkrystites altered to As-rich pyrite have been described from the Zumaya section (Smit, 1982; Smit and Romein, 1985; Schmitz et al., 1997) and from the Sopelana section (Rocchia et al., 1990). Above the calcite vein, 7–8 cm of dark gray shales occur, and 25 cm of gray marls forming the so-called boundary marls (Lamolda et al., 1988).

Paleocene-Eocene boundary

The largest extinction event having affected the deep-sea benthic foraminiferal fauna during the past 90 m.y. occurred in the latest Paleocene (Schmitz et al., 1997). The Zumaya section (Bay of Biscay region) also contains one of the most expanded and biostratigraphically complete Paleocene-Eocene transitions, deposited in a middle or lower bathyal environment (Pujalte et al., 1993). At Zumaya, the benthic extinction event closely coincides with deposition of a clay interval that indicates strong $CaCO_3$ undersaturation (Canudo et al., 1995). Approximately 15 m below the benthic extinction event, the section is dominated by gray marls that underlie a 4 m thick, mainly reddish-brown clay interval. The benthic extinction event occurs at the base of the clay interval. A transition from marls to limestone occurs above the clay interval.

High-resolution $\delta^{18}O$ and $\delta^{13}C$, calcareous nannofossil, and planktic and benthic foraminifera studies showed that, below the marl-clay transition, there is a 40–50-cm-thick interval that contains a detailed record of a gradual succession of faunal and geochemical events culminating in the benthic extinctions (Schmitz et al., 1997). There is a significant Ir anomaly (133 ppt over a background of 38 ppt) in a 1-cm-thick, gray marl layer ~40 cm below the base of the clay interval. Above the Ir anomaly, a negative gradual excursion of $\delta^{13}C$ is developed in a 40-cm-thick, glauconitic, greenish-brown marl bed. The relation of these anomalies to an impact event and its role regarding mass extinctions related to a Paleocene-Eocene (P-E) event are debatable. Schmitz et al. (1997) indicated that, if the Ir anomaly can be related to an impact event, it may not have been of any consequence for ongoing paleoceanographic changes and later mass extinctions.

Major biotic and geochemical changes have also been shown in the P-E at Alamedilla (Granada, Spain), where the transition is marked by major faunal turnover in planktic foraminifera, mass extinction in benthic foraminifera, negative $\delta^{18}O$ excursion in benthic foraminifera, negative $\delta^{13}C$ excursion in both planktic and benthic foraminifera, decrease in calcite preservation, increase in detrital flux, and changes in clay mineral composition (Lu et al., 1995). According to Montanari and Koeberl (2000), there is no relationship between the four or five impact craters known that are roughly P-E age, and the P-E benthic extinction and associated $\delta^{13}C$ shift.

PROXIMAL RECORD OF IMPACT EVENTS

The Azuara structure (41°01′N, 00°55′W, Zaragoza province, northeastern Spain), ~30 km in diameter (Fig. 2), is the

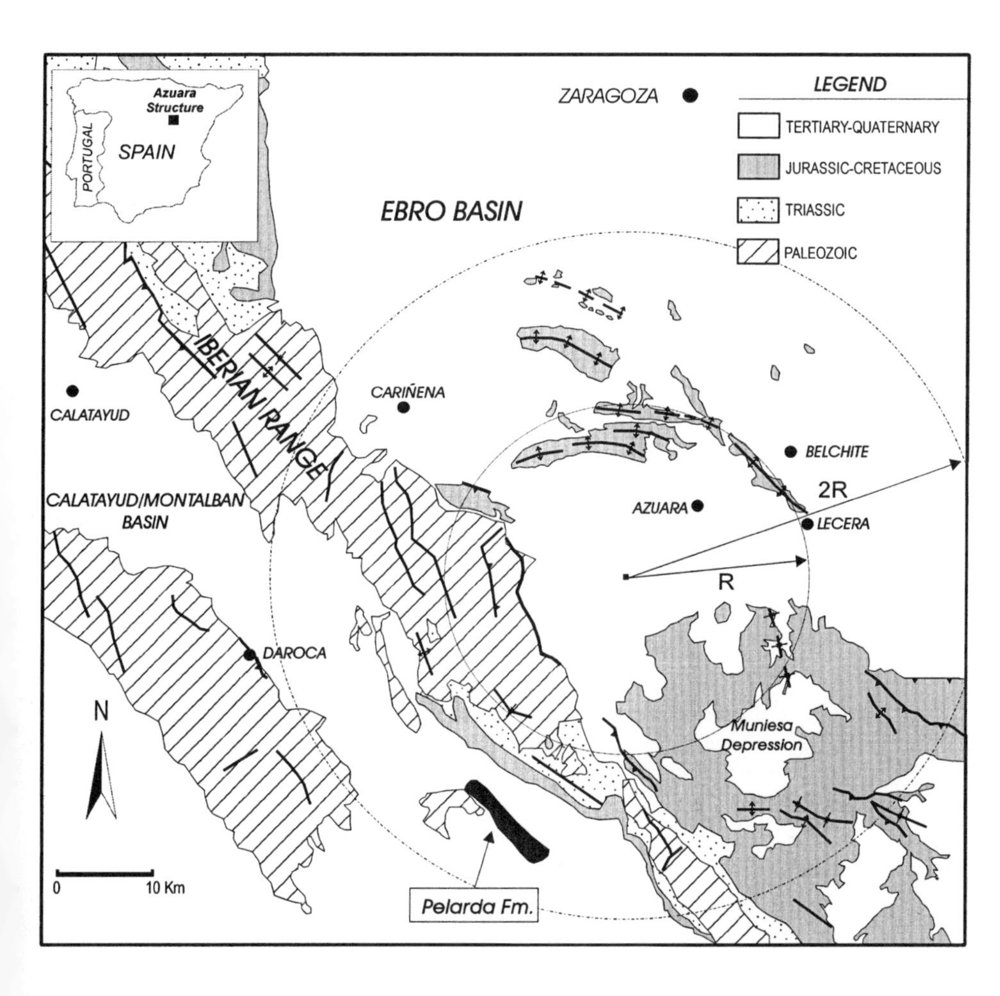

Figure 2. Geological sketch of Azuara structure and adjacent Calatayud-Montalbán and Ebro basins (modified after Cortés and Casas-Sáinz, 1996). Pelarda Formation is located to south of Azuara structure (circled in map; R). A distance of one radius from outlined structure has also been marked (2R). Fm.—Formation.

only structure on the Iberian Peninsula for which an impact origin has been formally proposed. Following its identification in the 1980s, a strong debate arose about either a tectonic or an impact-induced origin for this structure (e.g., Ernstson and Fiebag, 1992; Aurell et al., 1993). The controversy remains, although the arguments in favor of the impact hypothesis are gradually being rejected, because most of the evidence is inconclusive and allows for other interpretations. It is interesting that other meteorite-impact craters, similar in size and age to those proposed for the Azuara structure (Ries in Germany, Haughton in Canada), display numerous impact-related features (impact melts, widespread shock metamorphism) that are certainly not observed at Azuara.

The Azuara structure is located ~50 km south of Zaragoza, at the northeastern side of the Iberian Range, close to the Ebro basin (northestern Spain; Fig. 2). The present-day structure observed in the Azuara region corresponds to a sedimentary basin filled with Tertiary deposits and delimited by folds and thrusts involving Precambrian-Paleozoic basement and Mesozoic and Cenozoic supracrustal rocks. An impact origin for the Azuara structure was interpreted from evidence such as inverted stratigraphy, occurrence of megabreccias and megablocks, breccia dikes, a negative gravity anomaly, and features alleged to be indicative of high-pressure and high-temperature effects (Ernstson et al., 1985, 1999; Ernstson and Claudín, 1990; Ernstson and Fiebag, 1992). Several lines of evidence based on the sedimentary and structural evolution of the Azuara area, as well as that of the Iberian Range and the Ebro basin, were presented against the hypothesis of a meteorite impact (Aurell et al., 1993), alternative interpretations being proposed for the criteria used as evidence. For example, the inverted stratigraphy is due to Cenozoic Alpine tectonism, most breccias are due to diagenetic and/or edaphic processes (evaporite dissolution with collapse of host carbonate rocks, karst and caliche development), breccia dikes are also due to karst and paleosol development on carbonates, and the negative gravity anomaly comes from an incomplete data set restricted to the interior of the Azuara sedimentary basin, and resulting from its bowl shape. In addition, Cortés and Casas-Sainz (1996) considered that the Azuara structure is consistent with a north-south regional shortening during the Tertiary that controlled deformation both in the Variscan basement and in the Mesozoic-Tertiary sedimentary cover. They interpreted the structure as a synclinal basin located over an important depression of the Hercynian basement that is bounded by a fold and thrust arc in the northern part, and a poorly defined fold system toward the south. These interpretations refute part of the evidence put forward by Ernstson's group.

Immediately after an impact event, the impact crater is surrounded by a deposit of debris ejected as the result of the collision. Most of these ejecta are close to the crater rim, and continuous ejecta normally extend about one crater radius from the crater rim, in the case of a nonoblique impact (Melosh, 1989). The only unit proposed as probable proximal ejecta related to the Azuara structure is the Pelarda Formation (Ernstson and Claudín, 1990), located ~10 km to the south of the supposed crater rim (Fig. 2), and overlying alluvial fan deposits of the adjacent Calatayud-Montalbán basin. The origin and age of this unit are debated. Although the Pelarda Formation has been traditionally interpreted as one of the frequent Pliocene-Pleistocene aluvial sedimentary cover units present throughout the area (Instituto Tecnológico Geominero de España [ITGE], 1989, 1991), Ernstson and Claudín (1990) and Ernstson and Fiebag (1992) interpreted this formation as the remnant of an originally extensive ejecta blanket around the Azuara structure. The conglomerates and diamictites of the Pelarda Formation, which has an outcrop of ~30 km^2, are basically composed of rounded to subrounded quartzite clasts (to 1 m in diameter) eroded from the local Paleozoic basement, embedded within a mixed clayey-silty-sandy matrix, and with apparently no internal fabric (Fig. 3B). Carls and Monninger (1974) reported some Buntsandstein pebbles, but they did not observe limestone components. However, Ernstson and Claudín (1990) added to the previous work the identification of Buntsandstein megaclasts, sporadic limestone clasts, and lower Tertiary marls as clasts within the conglomerates. Striated and polished boulders and cobbles of quartzite, schist, and slate were also described by Ernstson and Claudín (1990). Plastically deformed and fractured clasts, some of them showing rotational deformation, and multiple sets of planar deformation features in quartz were also described by Ernstson and Claudín as evidence for shock deformation and metamorphism. However, the deformational features proposed as evidence for shock metamorphism are unrelated to impact metamorphism (F. Langenhorst, personal commun., 2000; see also Langenhorst and Deutsch, 1996).

On the assumption that the Azuara structure may be an impact crater, only biostratigraphic and lithostratigraphic methods help provide an age for the alleged proximal ejecta. This is because no impact melt sheet or true suevites have ever been described for the structure, and therefore radiometric methods could not be applied. Cenozoic sediments cover almost the total surface area of the structure, and there are no deep boreholes. Ernstson and Fiebag (1992) suggested a late Eocene–Oligocene age for the Pelarda Formation because the Miocene sediments are not affected by tectonics, and Eocene sediments are incorporated into some breccia dikes. However, vertebrate paleontological data for the units immediately below the Pelarda Formation suggest an age younger than early Oligocene (Olalla paleontological site, MP 21 zone; Peláez-Campomanes, 1993). These data do not exclude a Pliocene-Pleistocene age for the Pelarda Formation, which in our opinion can also be interpreted as local Pliocene-Pleistocene alluvial deposits, which are common throughout central Spain along zones of major relief.

POTENTIAL IMPACT CLASTIC BEDS

Useful criteria for the recognition of potential impact-related units in the sedimentary record are the presence of breccia or diamictite beds as probable proximal impact ejecta, and the presence of spherules as probable distal impact ejecta. This is particularly true when these deposits coincide in time with a well-dated massive extinction event, and/or when they roughly coincide with the age of a known impact event. However, once identified, the potential of the deposit to be impact related needs to be proved with unequivocal criteria characteristic of meteorite impacts: marked geochemical anomalies and shock metamorphic features (planar deformation features, diaplectic glass, high-pressure polymorphs).

Breccia and diamictite beds are identified in the Spanish sedimentary record: our review of the literature revealed that most of them have been interpreted as the result of resedimentation related to slope and/or tectonic instability, and more rarely as glacial deposits. Many of these units are clearly related to active tectonism or eustacy, although there are some that might be impact related. For these, the evidence for a strictly terrestrial origin (i.e., unrelated to cosmic impact) is not always unequivocal: some features remain to be explained, and alternative hypotheses may relate the deposits to impact events. Following is a brief review of the principal characteristics of several units in the sedimentary record of Spain that we have identified as potential impact clastic beds. Our current research is oriented toward the verification of the terrestrial or impact origin of these strata.

Vendian-Cambrian boundary

Deep-marine breccias and olistostromes known as the Fuentes Bed (Nivel de Fuentes) in the Central Iberian Zone of the Hercynian Massif broadly coincide with the Vendian-Cambrian boundary (location 10 in Fig. 1). They have been traditionally interpreted as tectonically induced strata, within the context of the Cadomian or late Pan-African orogeny (San José, 1984). The unit is present in the Montes de Toledo and Las Hurdes (Alvarez-Nava et al., 1988; Robles and Alvarez-Nava, 1988) (Fig. 1), whereas it is absent from other areas within the Central Iberian Zone (as in southern Salamanca; Nozal and Robles, 1988). Along the northeastern flank of the Valdelacasa antiform, near its type locality (town of Fuentes) in the central Montes de Toledo, the Fuentes Bed unconformably overlies the deformed deep-marine shales and sandstones of the Late Proterozoic (Riffean) Domo Extremeño Group (Alvarez-Nava et al., 1988; Pardo and Robles, 1988; Santamaría and Remacha, 1994). To the southeast of the Valdelacasa antiform, but still within it, the Fuentes Bed overlies both the Domo Extremeño Group and a remnant of the Late Proterozoic (Vendian) Ibor Group (Santamaría and Pardo, 1994) (Figs. 3, C and D, and 4). Farther to the south of the Valdelacasa antiform, in the Villarta-Navalpino antiform, the Fuentes Bed is

Figure 3. A: Microphotograph showing general aspect of Arroyofrío Oolite Bed (Dogger-Malm boundary). Cross-polarized light. Scale bar is 1 mm. B: Paleozoic quartzite boulders at southwestern part of Pelarda Formation outcrops. C: General view of Fuentes Bed megabreccia, showing characteristic chaotic aspect. D: Carbonate and quartzite boulders of Fuentes Bed embedded in plastically deformed muddy matrix. Circled hammer for scale. E: Sandstone clast in Orea Formation diamictite in Iberian Range (east of Checa, Guadalajara Province). F: Limestone clast in Orea Formation diamictite in eastern Iberian Range (northwest of Fombuena, Zaragoza Province).

known as the Navalpino Breccia and overlies shallow-marine limestones of the Ibor Group (San José, 1984).

The Fuentes Bed was first described and defined by Moreno (1974, 1975). In the Valdelacasa antiform it is a rather continuous bed, between 200 and 300 m thick (Fig. 4). Moreno (1977) interpreted it as the result of a single event representing an isochron. However, detailed sedimentologic analysis proved that it represents a series of multiple resedimentation events (slumps, debris flows, and olistostromes), with no interbeds of in situ (autochthonous) sedimentation separating them (Santamaría and Remacha, 1994). The size of the clasts varies from millimeter and centimeter size to blocks of several meters, and slabs of 20 to 30 m. The composition is highly polymictic, and consists of all the lithologies of the underlying Ibor and Domo Extremeño Groups, i.e., limestone, dolostone, shale, siltstone, sandstone, conglomerate, and graywacke. Thin sections reveal the complex character of the matrix, which also includes small lithic igneous clasts of probable volcanic origin, and highly

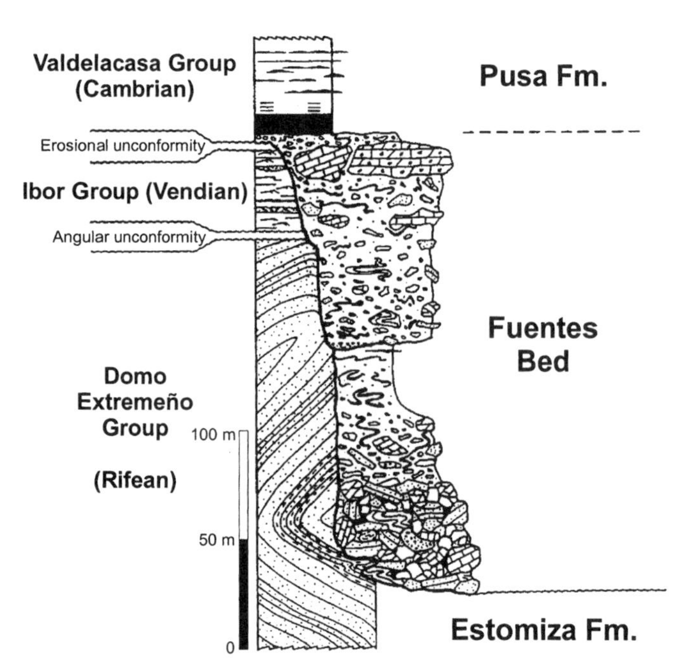

Figure 4. Schematic section of Fuentes Bed at El Membrillar (southeast Valdelacasa antiform), indicating lithostratigraphic nomenclature and approximate ages. Modified from Alvarez-Nava et al. (1988) and Santamaría and Pardo (1994). Fm.—Formation.

deformed quartz clasts, which seem to lack planar deformation features. Many terrigenous clasts display plastic deformation and partial disaggregation in the matrix, indicating their incomplete consolidation at the time of resedimentation. Carbonate clasts are thoroughly recrystallized and have been transformed to dolomite and/or magnesite during diagenesis.

Latest Ordovician

Shallow-marine, late Ashgill (Hirnantian) diamictites are present in northern Africa and western and southern Europe, along the former margin of Gondwana, and have been generally interpreted to be coeval with the north African glaciation (Fortuin 1984; Robardet and Doré, 1988). In Spain, these diamictites are known by different formation names throughout the Central Iberian Zone: pelitas con fragmentos (or fragment-bearing shales), Gualija Formation, Orea Formation, and Chavera Formation (Fortuin, 1984; Portero and Dabrio, 1988; Robardet and Doré, 1988). Apart from diamictites, these uppermost Ordovician units also consist of graywacke, shale, sandstone, and conglomerate, with a variable total thickness of as much as 200 m at some localities. The diamictites include reworked clasts and fossils recycled from underlying Ordovician units (e.g., quartz, limestone, shale, sandstone) (Fig. 3, E and F), and are overlain by a thin ubiquitous quartzite (García-Palacios et al., 1996).

The diamictites at most of the sections are resedimented (Portero and Dabrio, 1988). Evidence for a glacial origin is scarce and inconclusive: glacially striated clasts are extremely rare and dubious, whereas no striated pavements or boulder pavements have ever been found. These features remain unexplained, and detailed sedimentological and geochemical studies need to be done.

Late Devonian

Shallow-marine diamictites within the Phyllite-Quartzite Group of the Iberian Pyrite Belt (South Portuguese Zone of the Hercynian Massif) are commonly interpreted as large debris flows related to tectonism (Moreno and Sáez, 1990; Moreno et al., 1995). The age of the Phyllite Quartzite Group in the South Portuguese Zone is not well defined, but it is broadly considered to be of Late Devonian age (Moreno et al., 1995). Sedimentary facies and sequences in the Phyllite Quartzite Group represent storm-dominated shallow clastic shelf deposits, interrupted toward the top by thick (to 60 m), conspicuous beds of massive diamictites. They are common throughout the Iberian Pyrite Belt, but are particularly frequent in the Valverde del Camino antiform, 40 km north of Huelva (Fig. 1). The matrix of the diamictites is abundant (80%–90%) and muddy (shaly). Clasts within the matrix consist of partially consolidated resedimented parautochthonous sandstones of variable size, normally reaching 20 cm in size. Meter-sized slumps and isolated blocks are also common.

Moreno et al. (1995) interpreted these large debris flows as having been triggered by earthquakes, and related them to tectonic instability, in particular at the beginning of the Hercynian orogeny in the region. However, mass gravity flows can also be triggered by wave loading during strong storms and tsunami events. Mass-extinction events and meteorite-impact structures are known in the Late Devonian (Sandberg et al., 2000), and therefore other possibilities should be considered in the interpretation of the Phyllite Quartzite Group.

Triassic-Jurassic boundary

In most of Spain, the Triassic-Jurassic boundary is marked by a regional erosional unconformity and/or earliest Jurassic (Hettangian) breccias (Cortes de Tajuña Formation). In some localities, the breccias are interpreted as being related to rifting and eustacy (Aurell et al., 1992; Gallego et al., 1994; Campos et al., 1996), whereas in others they are considered to be the result of collapse after evaporite dissolution (Gómez and Goy, 1998). In particular, Campos et al. (1996) described erosion of underlying Triassic units and tectonic collapse of a shallow carbonate platform of Hettangian age developed during rift extension. Gómez and Goy (1998) identified an evaporite unit (Lécera Formation) from subsurface data in eastern Spain (Fig. 1), coinciding with the Triassic-Jurassic boundary, and consisting of 100–200 m of gypsum, anhydrite, and carbonates. This unit was found to underlie, laterally grade into and replace the Cortes de Tajuña Formation in many areas.

One of the most important extinction events of the Phanerozoic, including 47% extinction of all known marine animal genera (Sepkoski, 1996), took place near the Triassic-Jurassic boundary. At the same time, several meteorite-impact structures of intermediate and large size are known that have Late Triassic-Early Jurassic ages (Rochechouart, Manicouagan, Lake Saint Martin, Red Wing, Obolon), and impact signatures (iridium anomaly, shocked quartz) have been found with them (Montanari and Koeberl, 2000).

Cenomanian-Turonian boundary

Possible impact ejecta have recently been found north of Nazaré (Mesozoic Lusitanian basin of Portugal; Fig. 1), near the Cenomanian-Turonian boundary, which may be related to the Tore Seamount, a possible impact structure located off the coast of Portugal (Pena dos Reis et al., 1997; Monteiro et al., 1997, 1998, 1999). It may be possible to find corresponding distal ejecta in the frequently excellent exposures of shallow- and deep-marine sequences covering this same interval (Cenomanian-Turonian boundary) in Spain. Some of the Spanish sections are already well dated on the basis of calcareous nannoflora, planktic foraminifera, and ostracod biostratigraphy (Gorostidi and Lamolda, 1991; Gil et al., 1993; Paul et al., 1994; Floquet et al., 1996). The sedimentology and sequence stratigraphy for most of these units are well known (Alonso et al., 1993; Segura et al., 1993; García-Quintana et al., 1996). During the Archaeocretacea zone of the latest Cenomanian, the same event occurred both in the Basque-Cantabrian passive margin (northern Spain) and in its Iberian hinterland (Castilian proximal ramp, central and eastern Spain), connecting the Atlantic with the Tethys, as evidenced by deposition of a thin black shale bed, formation of a glauconitic and pyritic hardground, low sedimentation rate, anoxia (or hypoxia), geochemical shift, and biological extinction (Floquet et al., 1996). The potential relation of these processes to a meteorite impact has not been explored. Several other intermediate-size (15–25 km in diameter) impact craters have been identified with a Cenomanian or Turonian age: Steen River (Canada), Dellen (Sweden), and Boltysh (Ukraine) (Montanari and Koeberl, 2000).

CONCLUSIONS

1. The only proven impact-related bed in the sedimentary record of Spain is the clay layer at the K-T boundary (at Zumaya and Sopelana in the Bay of Biscay region, and at Caravaca, Agost, and Alamedilla in the Betic Cordilleras).

2. Other deposits previously proposed as impact related can now be rejected, or are dubious and still debated. These include (1) the Pelarda Formation, alleged to represent proximal ejecta from the Azuara structure, (2) the Paleocene-Eocene boundary near Zumaya (western Pyrenees) and Alamedilla (Betic Cordillera), and (3) the Arroyofrío Oolite Bed, alleged to be distal ejecta of an unknown Callovian-Oxfordian impact event.

3. The scarcity of evidence for meteorite-impact events in the sedimentary record is related to the lack of detailed studies. We propose several sedimentary units that could potentially be related to impact events, and where future research should focus.

ACKNOWLEDGMENTS

This research is supported by the Spanish Center for Astrobiology (Consejo Superior de Investigaciones Científicas—Instituto Nacional de Técnica Aeroespacial). We appreciate helpful and constructive reviews of this manuscript by Bruce Simonson and Wolf Uwe Reimold, as well as comments by editor Christian Koeberl. The IMPACT program of the European Science Foundation financed the participation of Díaz-Martínez and Sanz-Rubio in Short Courses on Impact Stratigraphy and Impact Metamorphism (2000), which suggested to us the need for this review, and greatly helped the development of ideas on prospective units.

REFERENCES CITED

Alonso, A., Floquet, M., Mas, R., and Meléndez, A., 1993, Late Cretaceous carbonate platforms: Origin and evolution, Iberian Range, Spain: American Association of Petroleum Geologists Memoir, v. 56, p. 297–313.

Álvarez-Nava, H., García-Casquero, J.L., Gil-Toja, A., Hernández-Urroz, J., Lorenzo-Alvarez, S., López-Díaz, F., Mira-López, M., Monteserín, V., Nozal, F., Pardo, M.V., Picart, J., Robles, R., Santamaría, J., and Solé, F.J., 1988, Unidades litoestratigráficas de los materiales precámbrico-cámbricos en la mitad suroriental de la Zona Centro-Ibérica [abs.]: 2 Congreso Geológico de España, Granada, Actas, v. 1, p. 19–22.

Arenillas, I., Arz, J.A., and Molina, E., 1997, El límite Cretácico/Terciario con foraminíferos planctónicos en Osinaga y Músquiz (Navarra, Pirineos): Geogaceta, v. 21, p. 25–28.

Arinobu, T., Ishiwatari, R., Kaiho, K., and Lamolda, M.A., 1999, Spike of pyrosynthetic polycyclic hydrocarbons associated with an abrupt decrease in $\delta^{13}C$ of a terrestrial biomarker at the Cretaceous-Tertiary boundary at Caravaca, Spain: Geology, v. 27, p. 723–726.

Arz, J.A., Canudo, J.I., and Molina, E., 1992, Estudio comparativo del Maastrichtiense de Zumaya (Pirineos) y Agost (Béticas) basado en el análisis cuantitativo de los foraminíferos planctónicos [abs.]: 3 Congreso Geológico de España, Salamanca, Actas, v. 1, p. 487–491.

Arz, J.A., Arenillas, I., López-Oliva, J.G., and Molina, E., 1998, Modelos de extinción de foraminíferos en el límite Cretácico/Terciario (K/T) de El Mulato (México) y Agost (España): Geogaceta, v. 23, p. 15–18.

Aurell, M., 1991, Identification of systems tracts in low-angle carbonate ramps: Examples from the Upper Jurassic of the Iberian Chain (Spain): Sedimentary Geology, v. 73, p. 101–105.

Aurell, M., and Meléndez, G., 1990, Upper Jurassic of the northeastern Iberian Chain (E Spain): A synthesis: Publicaciones del Seminario de Paleontología de Zaragoza, v. 2, p. 5–31.

Aurell, M., Meléndez, A., and Meléndez, G., 1990, Caracterización de la secuencia Oxfordiense en el sector central de la Cordillera Ibérica: Geogaceta, v. 8, p. 73–76.

Aurell, M., Meléndez, A., San Román, J., Guimerá, J., Roca, E., Salas, R., Alonso, A., and Mas, R., 1992, Tectónica sinsedimentaria distensiva en el límite Triásico-Jurásico en la Cordillera Ibérica: 3 Congreso Geológico de España, Salamanca, Actas, v. 1, p. 50–54.

Aurell, M., González, A., Pérez, A., Guimerá, J., Casas, A., and Salas, R., 1993, The Azuara impact structure (Spain): New insights from geophysical and geological investigations: Discussion: Geologische Rundschau, v. 82, p. 750–755.

Aurell, M., Fernández-López, S., and Meléndez, G., 1994, The Middle-Upper Jurassic oolitic ironstone level in the Iberian Range (Spain): Eustatic implications: Geobios, v. 17, p. 549–561.

Baceta, J.I., Pujalte, V., Orue-Etxebarria, X., Payros, A., Apellániz, E., and Núñez-Betelu, K., 1997, El Cretácico superior y Paleógeno del País Vasco: Ciclos sedimentarios y eventos biológicos en una cuenca marina profunda: Geogaceta, v. 22, p. 225–231.

Campos, S., Aurell, M., and Casas, A., 1996. Origen de las brechas de la base del Jurásico en Morata de jalón (Zaragoza): Geogaceta, v. 20, p. 887–889.

Canudo, J.I., Keller, G., and Molina, E., 1991, Cretaceous/Tertiary boundary extinction pattern and faunal turnover at Agost and Caravaca, S.E. Spain: Marine Micropaleontology, v. 17, p. 319–341.

Canudo, J.I., Keller, G., Molina, E., and Ortíz, N., 1995, Planktic foraminiferal turnover and $\delta^{13}C$ isotopes across the Paleocene-Eocene transition at Caravaca and Zumaya, Spain: Palaeogeography, Palaeoclimatolology, Palaeoecology, v. 116, p. 75–100.

Carls, P., and Monninger, W., 1974, Ein Block-Konglomerat im Tertiär der östlichten Iberischen Ketten (Spanien): Neues Jahrbuch für Geologie und Paläontologie, Abhandlungen, v. 145, p. 1–16.

Coccioni, R., and Galeotti, S., 1994, K-T boundary extinction: Geologically instantaneous or gradual event? Evidence from deep sea benthic foraminifera: Geology, v. 22, p. 779–782.

Cortés, A.L., and Casas-Sainz, A.M., 1996, Deformación alpina de zócalo y cobertera en el borde norte de la Cordillera Ibérica (Cubeta de Azuara-Sierra de Herrera): Revista de la Sociedad Geológica de España, v. 9, p. 51–66.

Ernstson, K., and Claudin, F., 1990, Pelarda Formation (Eastern Iberian Chains, NE Spain): Ejecta of the Azuara impact structure: Neues Jahrbuch für Geologie und Paläontologie, Monatshefte, v. 10, p. 581–599.

Ernstson, K., and Fiebag, J., 1992, The Azuara impact structure (Spain): New insights from geophysical and geological investigations: Geologische Rundschau, v. 81, p. 403–425.

Ernstson, K., Hammann, W., Fiebag, J., and Graup, G., 1985, Evidence of an impact origin for the Azuara structure (Spain): Earth and Planetary Science Letters, v. 74, p. 361–370.

Ernstson, K., Rampino, M.R., Anguita, F., Hiltl, M., and Siegert, I., 1999, Shock deformation of autochthonous conglomerates near the Azuara impact structure, Spain: Geological Society of America Abstracts with Programs, v. 27, no. 6, p. A122.

Floquet, M., Mathey, B., Métais, E., Emmanuel, L., Babinot, J.-F., Magniez-Jeannin, F., and Tronchetti, G., 1996, Correlation of sedimentary events during the latest Cenomanian from the Basque Basin to the castillian Ramp (northern Spain): Geogaceta, v. 20, p. 50–53.

Fortuin, A.R., 1984, Late Ordovician glaciomarine deposits (Orea Shale) in the Sierra de Albarracín, Spain: Palaeogeography, Palaeoclimatology, Palaeoecology, v. 48, p. 245–261.

Gallego, R., Aurell, M., Badenas, B., Fontana, B., and Meléndez, G., 1994, Origen de las brechas de la base del Jurásico de Leitza (Cordillera Vasco-Cantábrica oriental, Navarra): Geogaceta, v. 15, p. 26–29.

García-Mondéjar, J., Hines, F.M., Pujalte, V., and Reading, H.G., 1985, Sedimentation and tectonics in the western Basque-Cantabrian area (Northern Spain) during Cretaceous and Tertiary times, *in* Milá, M.D., and Rosell, J., eds., Excursion Guidebook: 6th European Regional Meeting, Lleida, Spain, p. 309–392.

García-Palacios, A., Gutiérrez-Marco, J.C., and Herranz, P., 1996, Edad y correlación de la "Cuarcita del Criadero" y otras unidades cuarcíticas del límite Ordovícico-Silúrico en la Zona Centroibérica meridional (España y Portugal): Geogaceta, v. 20, p. 19–22.

García-Quintana, A., Segura, M., García-Hidalgo, J.F., Ruiz, G., Gil, J., and Carenas, B., 1996, Discontinuidades estratigráficas y secuencias deposicionales del Cretácico medio (Albiense superior-Turoniense medio) en la Cordillera Ibérica central y el Sistema Central meridional: Geogaceta, v. 20, p. 119–122.

Gil, J., García, A., and Segura, M., 1993, Secuencias deposicionales del Cretácico en el flanco sur del Sistema Central: Geogaceta, v. 13, p. 43–45.

Gómez, J.J., 1979, El Jurásico en facies carbonatadas del Sector Levantino de la Cordillera Ibérica: Madrid, Universidad Complutense de Madrid, Departamento de Estratigrafía, Seminarios de Estratigrafía, Monografías, v. 4, 683 p.

Gómez, J.J., and Goy, A., 1998, Las unidades litoestratigráficas del tránsito Triásico-Jurásico en la región de Lécera (Zaragoza): Geogaceta, v. 23, p. 63–66.

Gorostidi, A., and Lamolda, M.A., 1991, El paso Cenomaniense-Turoniense de Menoyo (Alava): Variaciones de la nanoflora calcárea: Geogaceta, v. 10, p. 54–57.

Grieve, R.A.F., 1987, Terrestrial impact structures: Annual Review of Earth and Planetary Sciences, v. 15, p. 245–270.

Groot, J.J., de Jonge, R.B.G., Langereis, C.G., ten Kate, W.G.H.Z., and Smit, J., 1989, Magnetostratigraphy of the Cretaceous-Tertiary boundary at Agost (Spain): Earth and Planetary Science Letters, v. 94, p. 385–397.

Herranz, P., San José, M.A., and Vilas, L., 1977, Ensayo de correlación del Precámbrico entre los Montes de Toledo occidentales y el Valle del Matachel: Estudios Geológicos, v. 33, p. 327–342.

Instituto Tecnológico Geominero de España, 1989, Memoria Hoja número 466 (Moyuela) del Mapa Geológico de España E.: Madrid, Ministerio de Industria y Energía, Servicio de Publicaciones, 116 p., scale 1:50 000, 1 sheet.

Instituto Tecnólogico Geominero de España, 1991, Memoria Hoja número 40 (Daroca) del Mapa Geológico de España E.: Madrid, Ministerio de Industria y Energía, Servicio de Publicaciones, 239 p., scale 1:200 000, 1 sheet.

Kaiho, K., and Lamolda, M.A., 1999, Catastrophic extinction of planktonic foraminifera at the Cretaceous-Tertiary boundary evidenced by stable isotopes and foraminiferal abundance at Caravaca, Spain: Geology, v. 27, p. 355–358.

Kate, W.G.T., and Sprenger, A., 1993, Orbital cyclicities above and below the Cretaceous/Paleogene boundary at Zumaya (N Spain), Agost and Relleu: Sedimentary Geology, v. 87, p. 69–101.

Lamolda, M.A., and Mao, S., 1999, The Cenomanian-Turonian boundary event and dinocyst record at Ganuza (northern Spain): Palaeogeography, Palaeoclimatology, Palaeoecology, v. 150, p. 65–82.

Lamolda, M., Rodríguez-Lázaro, J., and Wiedmann, J., 1981, Field Guide: Excursions to Coniacion-Maastrichtian of Basque Cantabric Basin: Barcelona, Universidad Autónoma Barcelona, Publicaciones de Geología, v. 14, p. 1–53.

Lamolda, M.A., Mathey, B., and Wiedmann, J., 1988, Field-guide excursion to the Cretaceous-Tertiary boundary section at Zumaya (northern Spain), *in* Lamolda, M.A., Kauffman, E.G., and Walliser, O.H., eds., Paleontology and evolution: Extinction events 2nd International Conference on Global Bioevents. Madrid, Revista de Paleontología, no. extraordinario, p. 141–155.

Langenhorst, F., and Deutsch, A., 1996, The Azuara and Rubielos structures, Spain: Twin impact craters or Alpine thrust systems? TEM investigations on deformed quartz disprove shock origin [abs.]: Lunar and Planetary Science Conference, 27th, Houston, Texas, Lunar and Planetary Institute, v. 2, p. 725–726.

Lu, G., Keller, G., Ortiz, N., Adatte, T., and Molina, E., 1995, Faunal isotopic and sedimentary changes at the Alamedilla section, Spain: The P-E event in the Deep Tethys Basin: Geological Society of America Abstracts with Programs, v. 27, no. 6, p. A405.

Macleod, N., and Keller, G., 1991, How complete are Cretaceous/Tertiary boundary sections?: A chronostratigraphic estimate based on graphic correlation: Geological Society of America Bulletin, v. 103, p. 1439–1457.

Martínez-Ruíz, F., 1994, Geoquímica y mineralogía del tránsito Cretácico-Terciario en las Cordilleras Béticas y en la Cuenca Vasco-Cantábrica. [Ph.D. thesis]: Granada, Universidad de Granada, 280 p.

Martínez-Ruíz, F., Ortega-Huertas, M., Palomo, I., and Barbieri, M., 1992a, The geochemistry and mineralogy of the Cretaceous-Tertiary boundary at Agost (southeast Spain): Chemical Geology, v. 95, p. 265–281.

Martínez-Ruíz, F., Acquafredda, P., Palomo, I., and Ortega-Huertas, M., 1992b, New data on the spherules from the Cretaceous-Tertiary boundary layer at Caravaca (SE Spain): Geogaceta, v. 12, p. 30–32.

Martínez-Ruíz, F., Ortega-Huertas, M., Palomo, I., and Acquafredda, P., 1997, Quench textures in altered spherules from the Cretaceous-Tertiary boundary layer at Agost and Caravaca, SE Spain: Sedimentary Geology, v. 113, p. 137–147.

Meléndez, G., Sequeiros, L., Brochwicz-Lewinski, W., Gasiewicz, A., Suffzynsky, S., Szatkowski, K., Zbik, M., and Tarkowski, R., 1987, El límite Dogger-Malm en la Cordillera Ibérica: Anomalías geoquímicas y fenómenos asociados: Geogaceta, v. 2, p. 5–7.

Melosh, H.J., 1989, Impact cratering: Oxford, Oxford University Press, 245 p.

Molina, E., Arenillas, I., and Arz, J.A., 1996, The Cretaceous-Tertiary boundary mass extinction in planktonic foraminifera at Agost, Spain: Revue de Micropaléontologie, v. 39, p. 225–243.

Montanari, A., and Koeberl, C., 2000, Impact stratigraphy: The Italian record: Heidelberg, Springer-Verlag, Lecture Notes in Earth Sciences, v. 93, 364 p.

Monteiro, J.F., Ribeiro, A., Munha, J., Fonseca, P.E., Brandao Silva, J., Moita, C., and Galopim de Carvalho, A., 1997, Ejecta from meteorite impact near the Cenomanian-Turonian boundary found at North of Nazaré, Portugal [abs.]: Lunar and Planetary Science Conference, 29th, Houston, Texas, Lunar and Planetary Institute, v. 2, p. 967–968.

Monteiro, J.F., Munha, J., and Ribeiro, A., 1998, Impact ejecta horizon near the Cenomanian-Turonian boundary north of Nazaré, Portugal [abs.]: Meteoritics and Planetary Science, v. 33, p. A112–A113.

Monteiro, J.F., Ribeiro, A., and Munha, J., 1999, The Tore "sea-mount": A possible megaimpact in the deep ocean [abs.]: Reports on Polar Research, v. 343, p. 64–66.

Moreau, M.G., Cojan, I., and Ory, J., 1994, Mechanisms of remanent magnetization in marl and limestone alternations: Case study: Upper Cretaceous (Chron 31–30), Sopelana, Basque Country: Earth and Planetary Science Letters, v. 123, p. 15–37.

Moreno, C., and Sáez, R., 1990, Sedimentación marina somera en el Devónico del anticlinorio de Puebla de Guzmán, Faja Pirítica Ibérica: Geogaceta, v. 8, p. 62–64.

Moreno, C., Sierra, S., and Sáez, R, 1995. Mega-debris flows en el tránsito Devónico-Carbonífero de la Faja Pirítica Ibérica: Geogaceta, v. 17, p. 9–11.

Moreno, F., 1974, Las formaciones anteordovícicas del anticlinal de Valdelacasa: Boletín Geológico y Minero, v. 85, p. 396–400.

Moreno, F., 1975, Olistostromas, fanglomerados y slump folds: Distribución de facies en las series de tránsito Cámbrico-Precámbrico en el anticlinal de Valdelacasa (provincias de Toledo, Cáceres y Ciudad Real): Estudios Geológicos, v. 31, p. 249–260.

Moreno, F., 1977, Tectónica y sedimentación de las Series de Tránsito (Precámbrico terminal) entre el anticlinal de Valdelacasa y el Valle de Alcudia: Ausencia de Cámbrico: Studia Geologica, v. 12, p. 123–136.

Nozal, F., and Robles, R., 1988, Series y correlación de los materiales anteordovícicos en los Montes de Toledo y el sur de Salamanca [abs.]: 2 Congreso Geológico de España, v. 1, p. 139–143.

Ortega-Huertas, M., Martínez-Ruíz, F., Palomo, I., and Chamley, H., 1995, Comparative mineralogical and geochemical clay sedimentation in the Bethic Cordilleras and Basque-Cantabrian Basin areas at the Cretaceous-Tertiary boundary: Sedimentary Geology, v. 94, p. 209–227.

Pardo, M.V., and Robles, R., 1988, La discordancia basal del Grupo Valdelacasa en el Anticlinal de Valdelacasa (sector central de los Montes de Toledo) [abs.]: 2 Congreso Geológico de España, v. 2, p. 165–168.

Pardo, A., Ortiz, N., and Keller, G., 1996, Latest Maastrichtian and K/T boundary foraminiferal turnover and environmental changes at Agost (Spain), *in* MacLeod, N., and Keller, G., eds., Cretaceous-Tertiary mass extinctions: Biotic and environmental changes: New York, Norton and Company, p. 139–171.

Paul, C.R.C., Mitchell, S., Lamolda, M.A., and Gorostidi, A., 1994, The Cenomanian-Turonian boundary event in northern Spain: Geological Magazine, v. 131, p. 801–817.

Peláez-Campomanes, P., 1993, Micromamíferos del Paleógeno continental español: Sistemática, biocronología y paleoecología [Ph.D. thesis]: Madrid, Universidad Complutense de Madrid, 385 p.

Pena dos Reis, R.P., Corrochano, A., and Armenteros, I., 1997, El paleokarst de Nazaré (Cretácico Superior de la Cuenca Lusitana, Portugal): Geogaceta, v. 22, p. 149–152.

Portero, J.M., and Dabrio, C., 1988, Evolución tectosedimentaria del Ordovícico y Silúrico de los Montes de Toledo meridionales y Campo de Calatrava [abs.]: 2 Congreso Geológico de España, v. 1, p. 161–164.

Pujalte, V., Robles, S., Robador, A., Baceta, J.L., and Orue-Etxebarria, X., 1993, Shelf-to-basin Paleocene palaeogeography and depositional sequences, western Pyrenees, north Spain: Special Publication International Association of Sedimentologists, v. 18, p. 369–395.

Ramajo, J., and Aurell, M., 1997, Análisis sedimentológico de las discontinuidades y depósitos asociados del Calloviense superior-Oxfordiense medio en la Cordillera Ibérica Noroccidental: Cuadernos de Geología Ibérica, v. 22, p. 213–236.

Ramajo, J., Aurell, M., Delvene, G., and Pérez-Urresti, I., 2000, El Calloviense-Oxfordiense en el sector Oliete-Torre de las Arcas (Teruel): Geotemas, v. 1, p. 213–216.

Robardet, M., and Doré, F., 1988, The Late Ordovician diamictic formations from southwestern Europe: North-Gondwana glaciomarine deposits: Palaeogeography, Palaeoclimatology, Palaeoecology, v. 66, p. 19–31.

Robles, R., and Alvarez-Nava, H., 1988, Los materiales precámbrico-cámbricos del Domo de Las Hurdes: Existencia de tres series sedimentarias separadas por discordancias, SO de Salamanca (Zona Centro-Ibérica) [abs.]: 2 Congreso Geológico de España, v. 1, p. 185–189.

Rocchia, R., Boclet, D., Bonté, P.H., Buffetaut, E., Orue-Etxebarría, X., and Jéhanno, C., 1990, Structure de l'anomalie en iridium à la limite Crètacé-Tertiare du site de Sopelana (Pays Basque Espagnol): Comptes Rendus de l'Academie des Sciences, v. 307, p. 1217–1223.

San José, M.A. de, 1984, Los materiales anteordovícicos del Anticlinal de Navalpino (provincias de Badajoz y Ciudad Real, España central): Cuadernos de Geología Ibérica, v. 9, p. 81–117.

Sandberg, C.A., Ziegler, W., and Morrow, J.R., 2000, Late Devonian events and mass extinctions [abs.], *in* Catastrophic events and mass extinctions: Impacts and beyond: Houston, Texas, Lunar and Planetary Institute, LPI Contribution No. 1053, p. 188–189.

Santamaría, J., and Pardo, M.V., 1994, Las Megabrechas del Membrillar y su relación con el sustrato, Precámbrico-Cámbrico de la Zona Centro-Ibérica: Geogaceta, v. 15, p. 10–13.

Santamaría, J., and Remacha, E., 1994, Variaciones laterales del "Nivel de Fuentes", Precámbrico-Cámbrico de la Zona Centro-Ibérica: Geogaceta, v. 15, p. 14–16.

Schmitz, B., 1988, Origin of microlayering in worldwide distributed Ir-rich marine Cretaceous/Tertiary boundary clays: Geology, v. 16, p. 1068–1072.

Schmitz, B., Asaro, F., Molina, E., Monechi, S., von Salis, K., and Speijer, R.P., 1997, High-resolution iridium, $\delta^{13}C$, $\delta^{18}O$, foraminifera and nannofossil profiles across the latest Paleocene benthic extinction event at Zumaya, Spain: Palaeogeography, Palaeoclimatolology, Palaeoecolology, v. 133, p. 49–68.

Segura, M., García-Hidalgo, J.F., Carenas, B., and García-Quintana, A., 1993, Late Cenomanian-Early Turonian platform from central eastern Iberia, Spain: American Association of Petroleum Geologists Memoir, v. 56, p. 283–296.

Sepkoski, J.J., 1996, Patterns of Phanerozoic extinctions: A perspective from global data bases, *in* Walliser, O.H., ed., Global events and event stratigraphy: Berlin, Springer-Verlag, p. 35–52.

Smit, J., 1982, Extinction and evolution of planktonic foraminifera after a major impact at the Cretaceous/Tertiary boundary, *in* Silver, L.T., and Schultz, P.H., eds., Geological implications of impacts of large asteroids and comets on the earth: Geological Society of America Special Paper 190, p. 329–352.

Smit, J., 1990, Meteorite impact, extinctions and the Cretaceous-Tertiary boundary: Geologie en Mijnbouw, v. 69, p. 187–204.

Smit, J., 1999, The global stratigraphy of the Cretaceous-Tertiary boundary impact ejecta: Annual Review of Earth and Planetary Science, v. 27, p. 75–113.

Smit, J., and Hertogen, J., 1980, An extraterrestrial event at the Cretaceous-Tertiary boundary: Nature, v. 285, p. 198–200.

Smit, J., and Klaver, G., 1981, Sanidine spherules at the Cretaceous-Tertiary boundary indicate a large impact event: Nature, v. 292, p. 47–49.

Smit, J., and ten Kate, W.G.H.Z., 1982, Trace element patterns at the Cretaceous-Tertiary boundary: Consequences of a large impact: Cretaceous Research, v. 3, p. 307–332.

Smit, J., and Romein, A.J.T., 1985, A sequence of events across the Cretaceous-Tertiary boundary: Earth and Planetary Science Letters, v. 74, p. 155–170.

Smit, J., Klaver, G., and Van Kempen, T.M.G., 1987, Three unusually complete Spanish sections: Caravaca, Agost, Zumaya, *in* Lamolda, M.A., and Cearreta, A., eds., Paleontology and evolution: Extinction events (2nd International Conference on Global Bioevents): Bilbao, Abstracts, p. 270.

Sturesson, U., Heikoop, J.M., and Risk, M.J., 2000, Modern and Palaeozoic iron ooids: A similar volcanic origin: Sedimentary Geology, v. 136, p. 137–146.

Usera, J., Molina, E., Montoya, P., Robles, F., and Santisteban, C., 2000, Límites entre sistemas y pisos en la Provincia de Alicante, *in* Cañaveras, J.C., García del Cura, M.A., and Meléndez, A., eds., Itinerarios geológicos por la Provincia de Alicante y limítrofes: Alicante, Spain, Universidad de Alicante, p. 43–58.

Ward, P.D., 1988, Maastrichtian ammonite and inoceramid ranges from Bay of Biscay Cretaceous-Tertiary boundary sections, *in* Lamolda, M.A., Kauffman, E.G., and Walliser, O.H., eds., Paleontology and evolution: Extinction events (2nd International Conference on Global Bioevents): Madrid, Revista de Paleontología, no. extraordinario, p. 119–126.

Wiedmann, J., 1988, The Basque coastal sections of the K/T boundary: A key to understanding "mass extinction" in the fossil record, *in* Lamolda, M.A., Kauffman, E.G., and Walliser, O.H., eds., Paleontology and evolution: Extinction events (2nd International Conference on Global Bioevents): Madrid, Revista de Paleontología, no. extraordinario, p. 127–140.

Manuscript Submitted October 16, 2000; Accepted by the Society March 22, 2001

Printed in the U.S.A.

Geological Society of America
Special Paper 356
2002

Postglacial impact events in Estonia and their influence on people and the environment

Anto Raukas*
Institute of Geology at Tallinn Technical University, 7 Estonia Avenue, Tallinn 10143, Estonia

ABSTRACT

During the past decades the role of impacts of large asteroids and comets in the geological and biological evolution of Earth has been the subject of much debate. In contrast, possible environmental and social consequences of impacts of medium and small meteoroids have not been sufficiently studied. In Estonia, two groups of impact craters and four single craters, a total of 15 depressions, have been identified. The Kaali craters on Saaremaa Island were formed ca. 7500 yr ago and the Ilumetsa craters in southeast Estonia formed ca. 6600 yr ago, when the area of Estonia was already inhabited. A meteorite with a mass of several thousand tonnes could have induced remarkable environmental consequences (e.g., forest fires) and left indelible impressions in the minds of people.

INTRODUCTION

Several tens of impact craters, ranging from Neoproterozoic to Holocene in age and <100 m to >50 km in diameter, occur in both the crystalline basement and the sedimentary cover of the East European craton in the Fennoscandian-Baltic region (Puura et al., 1994). In the Baltic states 18 impact craters and a large number of meteorite falls have been recorded. The density of extraterrestrial phenomena is highest in Estonia, where impact craters have attracted the attention of scientists since the beginning of the nineteenth century.

In Estonia, 15 impact craters are known, and 5 meteorite falls have been registered in the past two centuries (Fig. 1). The number of meteorite craters per square kilometer is also higher there than in neighboring regions. The large number of identified meteorites and impact events can be explained by local conditions favoring the study of meteorites: the relatively high public awareness, the influence of Tartu University (established in 1632), and the activity of some German estate owners as amateur naturalists. Consequently, many fragments of meteorites, carefully preserved by individual collectors, found their way into scientific institutions.

The minimum size of a cosmic body that can cause a global catastrophe should be several kilometers in diameter. The diameters of the cosmic bodies responsible for the formation of the Neugrund and Kärdla craters in Estonia (Fig. 1) are estimated as ~400 and 200 m, respectively (K. Suuroja, personal commun., 2001).

Meteorites with a mass of several thousands of tonnes can also cause severe damage, if they fall on densely populated areas. To investigate possible environmental consequences of such cosmic bodies, we examine the record of Holocene meteorite falls in Estonia and the likely imprint on the memories of local people.

METEORITE FALLS IN ESTONIA WITNESSED BY LOCAL PEOPLE

The first scientifically recorded meteorite fall in Estonia occurred on 4 July 1821, when a stony meteorite of the size of a man's head fell near the village of Kaiavere (Fig. 1). According to eyewitnesses, a huge stone "whizzed from the east towards the west like an arrow. In a field at Kaiavere it hit the ground, split up with a terrifying crack and smashed into

*E-mail: Raukas@gi.ee

Raukas, A., 2002, Postglacial impact events in Estonia and their influence on people and the environment, *in* Koeberl, C., and MacLeod, K.G., eds., Catastrophic Events and Mass Extinctions: Impacts and Beyond: Boulder, Colorado, Geological Society of America Special Paper 356, p. 563–569.

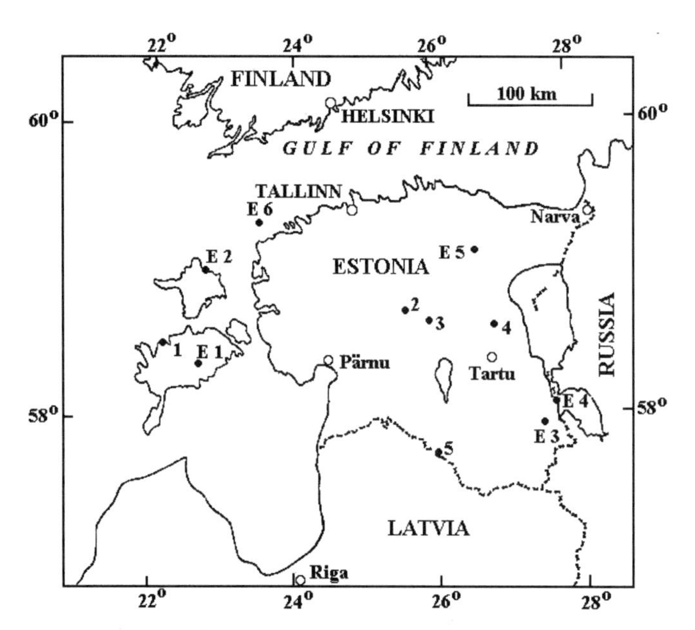

Figure 1. Distribution of impact craters and meteorite falls in Estonia. Craters: E1, Kaali; E2, Kärdla; E3, Ilumetsa; E4, Tsõõrikmäe; E5, Simuna; E6, Neugrund. Meteorite falls: 1, Kaande (Oesel); 2, Tännasilma; 3, Pilistvere; 4, Kaiavere; 5, Iigaste.

pieces" (Aaloe, 1971, p. 3). Although several people touched the still-warm meteorite pieces, not a single meteorite fragment ever reached researchers (Aaloe, 1971).

On 11 May 1855, a heavy fall of stony meteorites was visible on Kaande beach in the northwestern part of Saaremaa Island, formerly known as Oesel (Fig. 1). The fall of meteorites was accompanied by a deafening thunder-like noise and a strong whistling sound. Black tails were observed behind the falling stones. The seventh rumble was accompanied by a strong explosion. At first, the eyewitnesses thought that it was a cannon ball that had flown over their heads and hit the ground. At the fall site, people found several stones covered with black fusion crust. One of them had caused a depression, ~28.5 cm in depth and ~45 cm in diameter. Six meteorites, the largest weighing 3.5 kg, were found at the site. Others had evidently fallen into the sea (Goebel, 1856). Six days later, a meteorite fall was observed at Iigaste, south Estonia (Fig. 1, Grewingk and Schmidt, 1864).

On 8 August 1863, people harvesting rye in a field at Pilistvere (Fig. 1) witnessed a heavy fall of meteorites (Aaloe, 1964). The largest of the meteorites weighed 12.1 kg. One stone, 6876 g in weight, had fallen through the roof of the Kurla innkeeper's pigsty and rested amid the animals. On 28 June 1872, a 2.7 kg stone meteorite fell at Tännasilma (Fig. 1; Grewingk, 1874).

On 1 June 1937, a cosmic body entered the atmosphere and moved from the east-northeast to the west-southwest at an azimuth of 259° and fell at an angle of ~60°. It exploded at an estimated height of 28 km to the east of the Viru-Roela settlement (Kipper, 1937; Bronŝten, 1991). It is possible that this cosmic body formed a depression in the vicinity of Orguse Village near Simuna (Fig. 1). The diameter of the crater at the top of the mound is 8.5 m and its depth is 1.9 m. The mound around the depression is low (20–25 cm) and inconspicuous, but clearly ring-shaped and continuous. The depression was formed at a site with two distinct layers of sediment: 1.1 m of loose sand above clayey till (Pirrus, 1995). Meteoritic material has not been found at the site. Fortunately, no people were hit by the falling cosmic bodies.

POSTGLACIAL METEORITE FALLS AND THE MEMORIES OF LOCAL PEOPLE

Tsõõrikmäe crater

The oldest meteorite crater of Quaternary age in Estonia is located at Tsõõrikmägi ("ring hill" in the south Estonian dialect) in the southeastern part of the Republic (Fig. 1). Its mound is flattened, but a ring structure is well preserved. The diameter of the crater at the top of the mound is 38–40 m, and its depth from the highest point of the rim is 5.5 m. The crater is located in reddish-brown basal till. The peat in the depression is 4.5 m thick and, according to the palynological and ^{14}C data, its formation started 9500–10 000 yr ago Pirrus and Tiirmaa, 1984), when the area was probably not yet or only very sparsely inhabited. For this reason, this meteoritic event is not represented in the mythology and oral history of the people of southern Estonia.

Kaali craters

Nine craters are located in an area of 1 km^2 within the Kaali crater field (58°24′N, 22°40′E) in the southeastern part of Saaremaa Island (Figs. 1 and 2). Until the 1960s, these were the only craters in Europe that had been proven to be of impact origin. Different hypotheses about the origin of the craters were advanced between 1794 and 1937. Rauch (1794) was the first naturalist who gave a scientific opinion on the main Kaali crater, suggesting that it represented a fossil volcano. For a long time, in addition to a volcanic origin, emission of gas and steam, salt tectonics, karst formation, and several other processes were suggested to have formed the crater. The idea of a cosmic origin was first advanced by Kalkun-Kaljuvee in 1919 (*in* Kaljuvee, 1933) and proven in 1937 by Reinwald (1938), who collected the first small fragments of meteoritic iron. The chemical analysis of the pieces showed that Fe and Ni oxides amounted to 91.5 and 8.3 wt%, respectively (Spencer, 1938).

According to Yudin and Smyshlyayev (1963), the Kaali meteorite also contains 0.41 wt% Co. Mineralogical analyses of meteorite fragments (Buchwald, 1975; Yudin, 1968) showed minerals characteristic of iron meteorites, such as kamacite (α-FeNi), taenite (γ-FeNi), schreibersite-rhabdite ($[Fe, Ni, Co]_3P$),

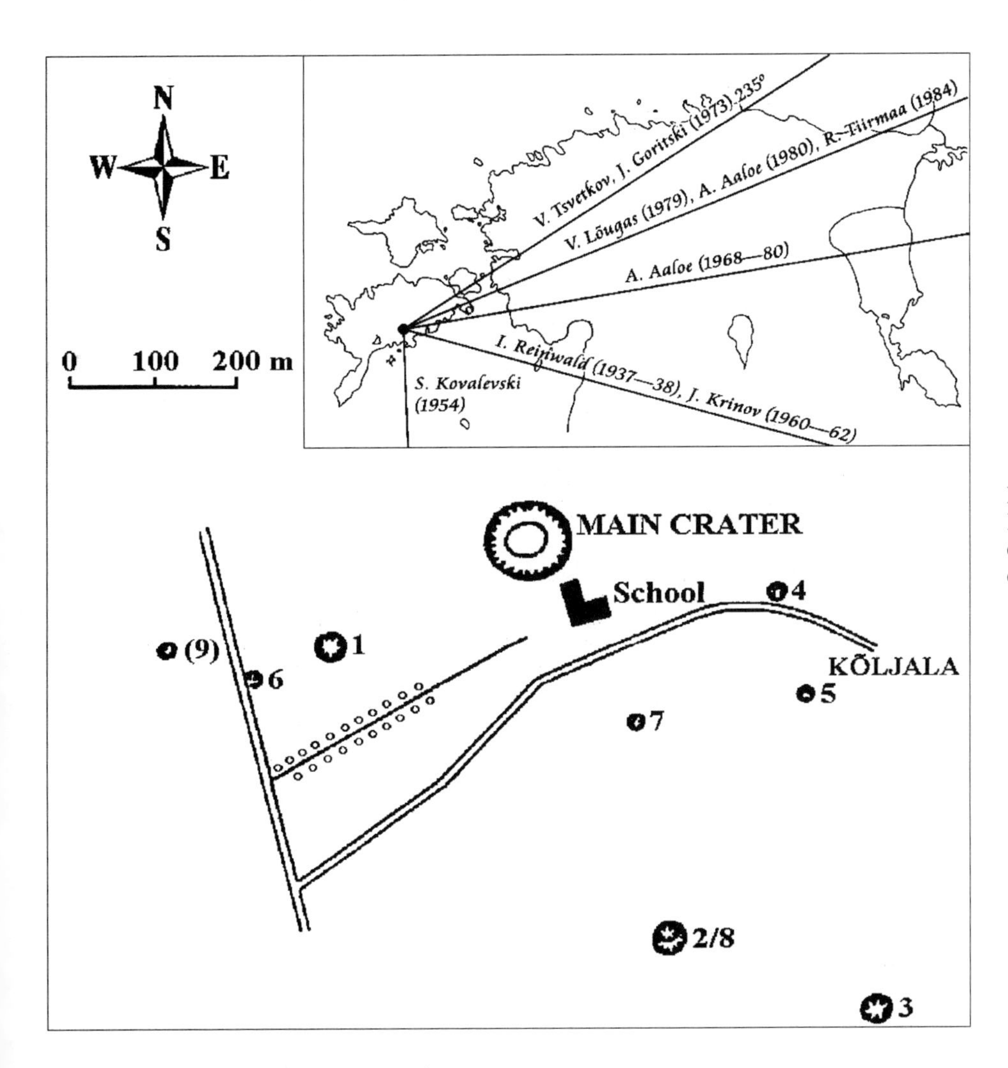

Figure 2. Location of impact craters in Kaali area and possible directions of incidence of Kaali meteorite, according to different authors (after Lõugas, 1996).

troilite (FeS), and olivine (Mg, Fe[SiO_4]). Goethite (α-Fe_2O_3) also occurs in the iron-shale crust of the meteorite, forming colloform textures. The crust locally contains pseudomorphs after schreibersite. The average content of kamacite is 96.4 vol%, and clear Widmanstätten patterns are commonly present. Taenite is rich in Ni (to 50 wt%); its average proportion is 1.8 vol%, and it occurs in the form of 1–0.15-mm-wide bands between kamacite bands. Schreibersite (average proportion 1.7 vol%) occurs throughout the meteorite, mostly in the form of rhabdite (Yudin and Smyshlyayev, 1963).

The Kaali meteorite belongs to the IA group of coarse octahedrites (Buchwald, 1975) and contains several rare elements, such as Ga (74.9–80.4 μg/g), Ge (293–305 μg/g), Ir (3.04 μg/g), Re (240 mg/g), Pt (7.7 μg/g), and Au (1.71 μg/g) (Wasson, *in* Czegka and Tiirmaa, 1998).

The main crater in the Kaali field of craters has a diameter of 105–110 m at the top of the mound and is at least 22 m deep. A natural body of water, named Lake Kaali, occupies its floor. The diameter of the lake depends on the seasonal water level and ranges from 30 to 60 m. The water depth is 1–6 m, and the maximum thickness of lake sediments is 5.8 m. The upper part of the mound consists of material ejected from the crater during the explosion and of partly overhanging dolomite layers tilted at angles of 25°–90° (Fig. 3). The uplifted bedrock complex has an average thickness of 10 m, and has been split into 9 blocks as wide as 50 m (Tiirmaa, 1997). The smaller craters, locally known as dry lakes, are 1–4-m-deep hollows with diameters of 12–14 m. These features have never aroused the interest of local people, and in the course of time came to be filled with rubbish and stones removed from fields. For this reason, several small craters may have remained undiscovered.

Because the sudden appearance of the crater lake and a new landform with uplifted and deformed dolomite blocks (Fig. 3) was mysterious, people started to use the site as a place of sacrifice. Eichwald (1854) believed that there was an ancient stronghold, where a natural karst lake, surrounded by a human-made wall, served as a well. The first archaeological finds in the east wall of the main crater date from 1976. In 1978, excavations were begun at the site of the fortification, remains of which had been discovered on the outer slope of the northeast-

Figure 3. Dolomites, uplifted due to Kaali impact, in wall of main Kaali crater. Because sudden appearance of crater lake was mysterious, people started to use the site as place of sacrifice (photo by R. Tiirmaa).

ern wall of the main crater (Lõugas, 1995). The excavations yielded surprisingly few archaeological artifacts that included some fragments of earthenware dating from the seventh century B.C. Most of the pottery fragments were from the Iron Age, the beginning of which has been locally traced to ca. 600 B.C. (Lõugas, 1980).

Lennart Meri, former President of Estonia and a well-known writer, who analyzed the Kaali catastrophe in 1976, maintained that the striking impression produced by this catastrophic event must have been long preserved in the minds of islanders (Meri, 1976). Recollections of the event were preserved in the form of tales orally passed from one generation to the next. The event found reflection in the Nordic mythology (*Kalevala, Edda*) and, probably, even in the myth about Phaethon by the Roman author Virgil (Lõugas, 1996, p. 9).

Phaethon, the son of the sun god Helios, and a woman Clymene wanted to drive the chariot of the sun through the heavens for a single day. The four, fiery white, stallions knew that an inexperienced hand was guiding them. They swerved from the well-worn path of the sun and bolted high into the sky. The chariot soon tumbled towards the Earth and began to scorch it. It started fires that destroyed woods and fields and dried up rivers and oceans. The mortals cried to the gods for help. To prevent further damage, Jupiter hurled a thunderbolt at Phaethon, the irresponsible chariot driver. The boy flew like a shooting star, his hair in flames, through the air, and fell to the earth at the mouth of the Eredanus. The latter may be interpreted as the Daugava River in Latvia, because the legend tells about amber which was formed by the tears of the sisters mourning the loss of their brother Phaethon.

Opinions differ as to the direction from south to northeast and angle of incidence of the Kaali meteorite (Fig. 2). On the basis of the size of the craters, Reinwald (1937a, 1937b) came to the conclusion that the direction of movement was from the southeast to the northwest (east-southeast–west-northwest). This idea was supported by Krinov (1962); I also support it (Raukas, *in* Raukas and Tiirmaa, 1998). It is known that a meteorite with a reasonable mass maintains its initial cosmic velocity for a longer time than smaller ones. Therefore, the larger one should be in front of the dispersal ellipse, and the smaller ones behind it (Fig. 2). The morphology of the wall of the main crater, the geophysical data available in the destruction zones of the main and secondary craters 1 and 6 (Aaloe et al., 1982), and the distribution of dispersed material in the craters and outside of the crater field (Aaloe and Tiirmaa, 1981, 1982; Tiirmaa, 1988) suggested that the meteorite probably fell from the east-northeast.

The fall of the meteorite from the northeast is strongly supported by archaeologists (Lõugas, 1996). According to a legend, Tarapita, the great god of ancient inhabitants of Saaremaa Island, dashed as a fireball with a long pile-shaped tail (pile is "vai" in Estonian) from his birthplace, in the northeast, at Ebavere—the highest hill of northern Estonia—to his new abode on Saaremaa. Probably, the Estonian "vai" gave the name to the small Votian nation to the east of Estonia ("vailased"; "vadjalased" in Estonian; "vaddalain" in Votic). In the Baltic-Finnish lexis the word "vai" (also "vaaja") has also another meaning, i.e., thunderbolt. Generally, people fear thunder: therefore, historians consider it possible that the flash of the Kaali meteorite at the beginning of Tarapita's flight, i.e., at the moment when the meteorite entered the atmosphere, rendered the name to the tribe living in the area.

Westren-Doll (*in* Lõugas, 1996, p. 121) described the event: "The reason why the god had to change his abode was the all-of-a-sudden appearance of the famous Lake Kaali in the vicinity of the Valjala stronghold. As studies have shown, Lake Kaali was formed by a falling meteorite. It could have been that the inhabitants of Virumaa (in northeast Estonia) seeing a meteorite falling on Saaremaa thought that it was their god that thundered through the air towards his new abode" (translation mine).

Unfortunately, this beautiful story is far from being realistic. The spherules in surficial layers of the Estonian mainland, which, according to Aaloe and Tiirmaa (1982) show the fall trajectory, are mostly industrial in origin and were probably transported to Saaremaa and western Estonia by winds from the industrial region of northeastern Estonia and Tallinn. Therefore, the explanation by Reinwald (1937a, 1937b) and Krinov (1962) favoring a meteorite fall from the southeast is considered more reasonable.

The energy needed for the formation of the main Kaali crater is estimated as 4×10^{19} ergs, and approximately two orders of magnitude less for the formation of the small craters. The initial velocity of the meteorite with an initial mass of 400–10 000 t (most probably ~1000 t upon entering the atmosphere) is estimated as 15–45 km/s. At the time of impact, its weight was probably 20–80 t and its velocity was 10–20 km/s (Bronšten and Stanyukovich, 1963). According to Pokrovski (1963), the diameter of the Kaali meteorite was probably 4.8 m, its mass 450 t, and its impact velocity 21 km/s. Koval (1974) suggested a somewhat lower velocity (~13 km/s) of the Kaali meteorite at impact. His calculations showed that the mass of the meteorite, which produced the main crater, must have been 40–50 t. According to Koval (1974), the meteorites that formed the small craters may have weighed between 1 and 6 t.

The explosive energy of the Hiroshima atomic bomb was 2×10^{21} ergs as calculated from TNT equivalents. Thus, the force of the Kaali meteorite (4×10^{19} ergs) could have matched a small atomic bomb. Its explosion was probably heard not only on Saaremaa and Hiiumaa Islands, but also all over mainland Estonia, in southern Sweden, and in neighboring areas. The explosion was probably followed by a weak earthquake. A witness to the event, east or west of Saaremaa Island, could see a ball of fire falling from the sky to the ground. Such an awesome event must have remained in people's memories. Archaeologists believe that the falling meteorite may have started forest fires and even destroyed some settlements on Saaremaa (Lõugas, 1996).

Age of the Kaali craters

On the basis of archaeological evidence obtained from the burning of ancient strongholds at Asva and Ridala on the island of Saaremaa, as well as on dendrochronological data, archaeologists reached the conclusion that the Kaali meteorite could not have fallen before the turn of the seventh to the eighth centuries B.C. (Lõugas, 1995). The idea was supported by the geologist Aaloe (Aaloe et al., 1975; Aaloe, 1981), who based his conclusion on radiocarbon dates of charcoal from the bottom of the twin craters 2 and 8 (2530 ± 130 ^{14}C yr B.P., TA-19 and 2660 ± 250 ^{14}C yr B.P., TA-22), and from the bottom of crater 4 (2920 ± 240 ^{14}C yr B.P., TA-769). According to Aaloe, the age of the craters is 2800 ± 100 yr. The analysis of trace elements in peat cores taken from the Piila bog (58°25′N, 22°36′E), 8 km northwest of the craters, suggested an age of 2400–2370 ^{14}C yr B.P. for the impact (Rasmussen et al., 2000).

However, the palynological analyses by Kessel (1981) showed that the basal sediments in the Kaali main crater are at least 3500 yr old. Saarse et al. (1992) obtained an age of 3390 ± 35 ^{14}C yr B.P. (Tln-1353) for the bottom sediments in the main Kaali crater. Saarse et al. believed that the Kaali meteorite fell ca. 4000 yr ago. However, there is no proof that the sediments studied actually originated from the base of the section. The time interval between crater formation and the age of the lowermost radiocarbon-dated sample from the crater lake deposit is also uncertain. It must be assumed that the craters are much older than 4000 yr.

According to Linstow (1919), the Kaali craters formed 4000–8000 yr ago. Kraus et al. (1928) placed this event at 12 000 yr ago. Reinwald (1937a, 1937b) suggested a 4000–5000 yr age for the craters. Aaloe (1958) supported this age, but later changed his opinion (Aaloe et al., 1975; Aaloe, 1981).

During our study, detailed geological and geophysical mapping and complex studies of meteorite debris and micrometeorites in the Kaali crater field were carried out. Surrounding lakes and bogs were also searched for microimpactites. In the four mires studied, including the above-mentioned Piila bog, microimpactites—mainly glassy spherules formed by melting and vaporization of meteoritic matter and target rocks—were identified only in one layer dated by palynological and radiocarbon methods as ca. 7500 yr old (Raukas et al., 1995, 1999). Microimpactites consist mainly of silica and calcium, with an admixture of iron and nickel, and several other elements (Raukas, 2000). The dates of the peat layer with microimpactites provide a basis for the conclusion that the Kaali craters were formed ca. 7500–7600 yr ago. At that time, the northern Baltic area was inhabited, and it is conceivable that a large crater-producing meteorite impact produced a striking and long-lasting impression on people.

Ilumetsa craters

In 1938, five impact craters were discovered in the course of geological mapping at Ilumetsa (58°57′N, 27°24′E) in southeast Estonia (Fig. 1). The largest of these, which has been studied in particular detail (Fig. 4), is called Põrguhaud (Hell's Grave). It is difficult to say whether the name is associated with the fall of a meteorite or with the morphology of the depression ("an entrance to Hell"). The crater has a diameter of 75–80 m. Its base is covered with a thin layer of gyttja and a 2-m-thick layer of peat. Radiocarbon dates obtained on the lowermost organic layer are 6030 ± 100 ^{14}C yr B.P. (TA-130) and 5970 ± 100 ^{14}C yr B.P. (TA-725); with palynological evidence,

Figure 4. Põrguhaud crater at Ilumetsa (photo by J. Nõlvak).

these suggest an age of ca. 6000 yr for the crater (Liiva et al., 1979). In the summer of 1996, we found glassy spherules, probably of impact origin, at a depth of 5.70 m in the Meenikunno Bog, 6 km southwest of the Ilumetsa craters. The layer was dated by the ^{14}C method, and the dates obtained are: 6542 ± 50 ^{14}C yr B.P. (Tln-2214) for the depth interval 5.6–5.7 m and 6697 ± 50 ^{14}C yr B.P. (Tln-2316) for the depth interval 5.7–5.8 m. Based on these ages and palynological evidence from the Meenikunno Bog, it can be concluded that the Ilumetsa craters were formed ca. 6600 yr ago (Raukas and Tiirmaa, 2000).

Põrguhaud is smaller than the main Kaali crater, and, accordingly, the energy needed for the formation of Põrguhaud must have been an order of magnitude less, probably $\sim 10^{18}$ ergs. Southeast Estonia was densely populated at 6500 yr ago, and the event was fixed in oral history. In Estonian mythology, the meteorite was interpreted as a creature that came to be known by various names, including "spark-tail," "treasure-bringing goblin," "demon of fortune," and "fiery flying dragon." However, in essence, this was an "enormously large and hot streak of fire, which flew through the heavens to fetch treasures for its masters."

CONCLUSIONS

There were eyewitnesses to all the meteorite falls recorded in Estonia during the nineteenth century. There is no evidence that anyone was hurt or killed in these falls. At the time of the two meteorite falls, which generated impact craters at Kaali and Ilumetsa, these areas were already inhabited. It is plausible that the impressions produced by these events were fixed in people's memories and in folklore.

ACKNOWLEDGMENTS

I thank the Austrian and Estonian Academies of Sciences for the financial support that allowed me to participate in the conference *Catastrophic Events and Mass Extinctions: Impacts and Beyond* in Vienna, July 9–12, 2000. I thank Helle Kukk and Rein Vaher for assistance with manuscript preparation; Henning Dypwik (University of Oslo) and Uwe Reimold (University of Witwatersrand, Johannesburg) for thorough and constructive reviews; and Maurice Schwartz (Western Washington University) for language improvements. Research was financed by the Estonian Science Foundation (Project 1905).

REFERENCES CITED

Aaloe, A., 1958, Kaalijärve meteoriidikraatri nr. 5 uurimised 1955. aastal [Studies of the Kaali meteorite crater no. 5 in 1955]: Tallinn, Eesti NSV Teaduste Akadeemia Geoloogia Instituudi Uurimused, v. 2, p. 105–117 (in Estonian, with English and Russian summaries).

Aaloe, A., 1964, Pilistvere meteoriidisajust [The Pilistvere meteorite shower]: Eesti Loodus, no. 4, p. 206 (in Estonian).

Aaloe, A., 1971, Kaiavere meteoriit [The Kaiavere meteorite]: Eesti Loodus, no. 8, p. 485 (in Estonian).

Aaloe, A., 1981, Erinevused Kaali kraatrite vanuse määrangutes [Discrepancies in dating the Kaali meteorite craters]: Eesti Loodus, no. 4, p. 236–237 (in Estonian, with English and Russian summaries).

Aaloe, A.O., Andra, H., and Andra, V., 1982, Determination of the direction of the meteorite rain at Kaali by means of geophysical methods: Eesti NSV Teaduste Akadeemia Toimetised, v. 31, no. 2, p. 56–61 (in Russian, with English and Estonian summaries).

Aaloe, A., Eelsalu, H., Liiva, A., and Lõugas V., 1975, Võimalusi Kaali kraatrite vanuse täpsustamiseks [On the correction of the age of the Kaali meteorite craters]: Eesti Loodus, no. 12, p. 706–709 (in Estonian, with English and Russian summaries).

Aaloe, A., and Tiirmaa, R., 1981, Pulverized and impactite meteoritic matter in the Kaali crater field: Eesti NSV Teaduste Akadeemia Toimetised, Geoloogia, v. 30, no. 1, p. 20–27 (in Russian, with English and Estonian summaries).

Aaloe, A., and Tiirmaa, R., 1982, Meteorite matter in the small craters of Kaali and their vicinity: Meteoritika, no. 41, p. 120–125 (in Russian).

Bronšten, V.A., 1991, About the Bolide 1 of June 1937 and Simuna Crater: Astronomichesky Vestnik, v. 25, no. 1, p. 104–108 (in Russian).

Bronšten, V., and Stanyukovich, K., 1963, On the fall of the Kaali meteorite: Tallinn, Eesti NSV Teaduste Akadeemia Geoloogia Instituudi Uurimused, v. 11, p. 73–83 (in Russian, with English and Estonian summaries).

Buchwald, V.F., 1975, Handbook of iron meteorite: Berkeley, University of California Press, v. 2, p. 704–707.

Czegka, W., and Tiirmaa, R., 1998, Das Holozäne Meteoritenkraterfeld von Kaali auf Saaremaa (Ösel), Estland: Aufschluss, v. 49, p. 233–252.

Eichwald, E., 1854, Die Grauwackenschichten von Liv- und Estland: Bulletin de la Société Impériale des Naturalistes de Moscou, v. 27, no. 1, p. 1–111.

Goebel, A., 1856, Untersuchung eines am 11. Mai (29 April) 1855 auf Oesel niedergefallenen Meteorsteins: Archiv für die Naturkunde Liv-, Ehst- und Kurlands, Dorpat, ser. 1, v. 1, p. 168–169.

Grewingk, C., 1874, Über einen Meteoritenfall, der am 16/28 Juni 1872, im Kreise Jerwen Estlands, auf dem Gute Allenküll bei dem zum Dorfe Tennasilm gehörigen Gesinde Sikkensaare statthatte: Sitzungsberichte der Naturforscher-Gesellschaft zu Dorpat, v. 3, p. 390–391.

Grewingk, C., and Schmidt, C., 1864, Über die Meteoritenfälle von Pillistfer, Buschof und Igast in Liv- und Kurland: Archiv für die Naturkunde Liv-, Ehst- und Kurlands, Dorpat, ser. 1, v. 3, p. 421–552.

Kaljuvee, J., 1933, Die Grossprobleme der Geologie: Tallinn (Reval), F. Wassermann, 162 p.

Kessel, H., 1981, Kui vanad on Kaali järviku põhjasetted? [How old are the bottom deposits of Lake Kaali]: Eesti Loodus, no. 4, p. 150–155 (in Estonian, with English and Russian summaries).

Kipper, A., 1937, 1937.a. 1. Juuni meteoriidist [About the Meteorite of 1 June 1937]: Eesti Loodus, no. 4, p. 150–155 (in Estonian).

Koval, V.I., 1974, About the mass and composition of the Kaali meteorite: Astronomicheskiye Vesti, v. 8, no. 3, p. 169–176 (in Russian).

Kraus, E., Meyer, R., and Wegener, A., 1928, Untersuchungen über den Krater von Sall auf Ösel: Gerlands Beiträge zur Geophysik, v. 20, p. 312–378.

Krinow, E.L., 1960, Die meteoritischen Krater Kaalijärv auf der Insel Saaremaa, Estnische SSR: Chemie der Erde, v. 20, p. 199–216.

Krinov, E.L., 1962, Meteorite craters on the surface of the earth: Meteoritika, no. 22, p. 3–30 (in Russian).

Liiva, A., Kessel, H., and Aaloe, A., 1979, Ilumetsa kraatrite vanus [The age of Ilumetsa craters]: Eesti Loodus, no. 12, p. 762–764 (in Estonian, with English and Russian summaries).

Linstow, O., 1919, Der Krater von Sall auf Oesel: Zentralblatt für Mineralogie und Geologie, v. 21/22, p. 326–339.

Lõugas, V., 1980, Archaeological excavations in the Kaali crater area: Eesti NSV Teaduste Akadeemia Toimetised, Ühiskonnateadused, v. 29, no. 4, p. 357–360.

Lõugas, V., 1995, Must auk Saaremaa ajaloos [Black hole in the history of Saaremaa]: Luup, no. 2, p. 52–56 (in Estonian).

Lõugas, V., 1996, Kaali kraatriväljal Phaethonit otsimas [Searching for Phaethon in the Kaali crater field]: Tallinn, Eesti Entsüklopeediakirjastus, 486 p. (in Estonian).

Meri, L., 1976, Hõbevalge [Silver White]: Tallinn, Eesti Raamat, 486 p. (in Estonian).

Pirrus, E., 1995, The ratio of depression and mound volumes: A promising criterion for identification of small meteorite craters, *in* Proceedings of the Estonian Academy of Sciences: Geology, v. 44, no. 3, p. 188–196.

Pirrus, E., and Tiirmaa, R., 1984, Tsõõrikmägi—ka meteoriidikraater? [Tsõõrikmägi—also a meteorite crater?]: Eesti Loodus, no. 9, p. 566–571; no. 10, p. 638–642 (in Estonian, with English and Russian summaries).

Pokrovski, G., 1963, Computation of the parameters of a meteorite according to the crater caused by its fall: Eesti NSV Teaduste Akadeemia Geoloogia Instituudi Uurimused, v. 11, p. 61–71 (in Russian, with English and Estonian summaries).

Puura, V., Lindström, M., Floden, T., Pipping, F., Motuza, G., Lehtinen, M., Suuroja, K., and Murnieks, A., 1994, Structure and stratigraphy of meteorite craters in Fennoscandia and Baltic region: A first outlook, *in* Proceedings of the Estonian Academy of Sciences: Geology, v. 43, p. 93–108.

Rasmussen, L.K., Aaby, B., and Gwozdz, R., 2000, The age of the Kaalijärv meteorite craters: Meteoritics and Planetary Science, no. 35, p. 1067–1071.

Rauch, J.E., 1794, Nachricht von der alten lettischen Burg Pilliskaln, und von mehrern ehemaligen festen Plätzen der Letten und Ehsten; auch von etlichen andern lief- und ehstländischen Merkwürdigkeiten: Neue Nordische Miscellaneen, no. 9 und 10, p. 540–541.

Raukas, A., 2000, Investigation of impact spherules: A new promising method for the correlation of Quaternary deposits: Quaternary International, v. 68–71, p. 241–252.

Raukas, A., Pirrus, R., Rajamäe, R., and Tiirmaa, E., 1995, On the age of meteorite craters at Kaali (Saaremaa Island, Estonia), *in* Proceedings of the Estonian Academy of Sciences: Geology, v. 44, p. 177–183.

Raukas, A., Pirrus, R., Rajamäe, R., and Tiirmaa, R., 1999, Tracing the age of the catastrophic impact event in sedimentary sequences around the Kaali meteorite craters on the Island of Saaremaa, Estonia: Pact 57, Rixensart, p. 435–453.

Raukas, A., and Tiirmaa, R., 1998, Estonian Holocene meteorite craters as unique natural monuments, *in* Miidel, A., ed., ProGeo '97 in Estonia, Proceedings: Tallinn, Geological Survey of Estonia, p. 53–57.

Raukas, A., and Tiirmaa, R., 2000. Dating of Ilumetsa impact craters, SE Estonia, *in* Plado, J., and Pesonen, L.J., eds., Meteorite impacts in Precambrian Shields, 4th Workshop of the European Science Foundation Impact Programme, Lappajärvi—Karikkoselkä—Sääksjärvi, Finland: Espoo, Programme and Abstracts, p. 88.

Reinwald, I., 1937a, Meteoorkraatrid Saaremaal [Meteorite craters in Saaremaa]: Looduskaitse, no. 1, p. 118–131 (in Estonian).

Reinwald, I., 1937b, Kaali järve meteoorkraatrite väli [The Kaalijärv field of meteorite craters]: Loodusvaatleja, no. 4, p. 97–102 (in Estonian).

Reinwald, I., 1938, The finding of meteoritic iron in Estonian craters: A long search richly awarded: The Sky Magazine of Cosmic News, v. 2, no. 6, p. 6–7.

Saarse, L., Rajamäe, R., Heinsalu, A., and Vassiljev, J., 1992, The biostratigraphy of sediments deposited in the Lake Kaali meteorite impact structure, Saaremaa Island, Estonia: Bulletin of the Geological Society of Finland, v. 63, p. 129–139.

Spencer, L.J., 1938, The Kaalijärv Meteorite from the Estonian Craters: Mineralogical Magazine, v. 25, p. 75–80.

Tiirmaa, R., 1988, Distribution of pulverized meteoritic matter in the Kaali crater field, *in* Proceedings of the Academy of Sciences of the Estonian SSR: Geology, v. 37, p. 43–46 (in Russian, with English and Estonian summaries).

Tiirmaa, R., 1997, Meteorite craters, *in* Raukas, A., and Teedumäe, A., eds., Geology and mineral resources of Estonia: Tallinn, Estonian Academy Publishers, p. 378–383.

Yudin, I.A., 1968, About mineralogy of Kaali meteorite: Meteoritika, no. 28, p. 44–49 (in Russian).

Yudin, I.A., and Smyshlyayev, S., 1963, Mineragraphic and chemical studies of Kaali iron meteorite: Tallinn, Eesti NSV Teaduste Akadeemia Geoloogia Instituudi Uurimused, v. 11, p. 53–59 (in Russian, with English and Estonian summaries).

Manuscript Submitted October 5, 2000; Accepted by the Society March 22, 2001

Geological Society of America
Special Paper 356
2002

Mineralogical investigations of experimentally shocked dolomite: Implications for the outgassing of carbonates

Roman Skála*
Czech Geological Survey, Geologická 6, CZ-15200 Praha 5, Czech Republic
Jana Ederová
Pavel Matějka
Institute of Chemical Technology, Technická 5, CZ-16628 Praha 6, Czech Republic
Friedrich Hörz
National Aeronautics and Space Administration, Johnson Space Center, Houston, Texas 77058, USA

ABSTRACT

The response of carbonate-bearing sediments to transient shock waves is important in understanding atmospheric CO_2 pollution caused by large-scale impacts on Earth. As a consequence, we examined polycrystalline dolomite that was experimentally shocked to pressures from 4.6 to 68.0 GPa.

Electron microprobe data, at spatial resolutions of 3–5 µm, do not reveal any decomposition products of dolomite. XRD and micro-Raman spectroscopy show distinct broadening of reflections and spectral bands, respectively, with increasing peak shock pressure. The substantial scatter in the bandwidth of the Raman spectra demonstrates considerable localization of shock-induced damage on scales of ~10–20 µm. XRD patterns reveal the presence of small quantities of MgO and calcite at pressures >60 GPa; these phases must be very fine grained (<1 µm), because they could not be resolved via microprobe. These observations suggest that dolomite decomposition begins at pressures just above 60 GPa, but it remains a relatively minor effect at even 68 GPa. However, massive outgassing occurs at slightly higher (>70 GPa) pressures, as inferred from the characteristic and reproducible failure of the metallic containers that house the dolomite target during these shock experiments. Identical containers housing volatile-free silicates routinely survive shocks as high as 90 GPa.

These relatively high threshold pressures for the outgassing of dolomite are consistent with recent experimental and theoretical efforts by a number of groups on the shock behavior of calcite. Cumulatively, these developments imply that the total amount of CO_2 released during the Cretaceous-Tertiary event could be considerably smaller than previously estimated.

INTRODUCTION

The most common rock-forming carbonates, calcite and dolomite, make up a considerable fraction of terrestrial sediments that may be sufficiently shocked during large-scale hypervelocity impacts to liberate CO_2, thereby perturbing the atmosphere and biosphere in possibly catastrophic fashion. Lange and Ahrens (1986) estimated that ~30% of all impact

*E-mail: skala@cgu.cz

Skála, R., Ederová, J., Matějka, P., and Hörz, F., 2002, Mineralogical investigations of experimentally shocked dolomite: Implications for the outgassing of carbonates, *in* Koeberl, C., and MacLeod, K.G., eds., Catastrophic Events and Mass Extinctions: Impacts and Beyond: Boulder, Colorado, Geological Society of America Special Paper 356, p. 571–585.

craters on Earth occurred in targets containing some carbonates.

Following Chatterjee et al. (1998), relatively modest temperatures are needed to decompose carbonates at ambient pressure, such as 436 °C for 1 Do → 1 Cc + 1 Pe + 1 CO_2 and/or 894 °C for 1 Cc → 1 Lm + 1 CO_2 (Do = dolomite, Cc = calcite; Pe = periclase, and Lm = lime). Such temperatures are easily attained during impact and specifically upon shock pressure release from relatively modest shock states (e.g., Gupta et al., 2000). However, the CO_2 release of carbonates under shock conditions is not well understood, in addition to a number of other reactions and phase transitions that were suggested for shock-processed carbonates. Experimental results and thermodynamic modeling seem to produce conflicting results on occasion.

The first shock-induced devolatilization studies of carbonates in the early 1980s were in part based on experiments, but mostly on thermodynamic considerations; they suggested that devolatilization of calcite commences at <15 GPa, with massive decarbonation occurring at 15 GPa (30% CO_2 loss) and increasing to 50% CO_2 loss at 20 GPa, as summarized by Lange and Ahrens (1986). However, more recent studies reveal both calcite and dolomite to be unexpectedly stable under shock conditions, and no significant outgassing is observed at pressures >40 GPa for calcite and >60 GPa for dolomite (e.g., Kotra et al., 1983; Martinez et al., 1995; Bell et al., 1998; Langenhorst et al., 1998, 2000; Skála et al., 1999, 2000). Love and Ahrens (1998) and Gupta et al. (2000) measured the in situ temperatures of calcite and concluded that devolatilization of calcite commences at ~75 GPa during shock compression and that it is complete at <100 GPa. However, Gupta et al. (2000) also concluded that the post-shock temperatures upon pressure release from modest 18 GPa shocks should suffice to initiate devolatilization of calcite, consistent with the earlier observations of Boslough et al. (1982) and Lange and Ahrens (1986).

Martinez et al. (1995) were the first to stress that purely thermodynamic considerations (e.g., Kieffer and Simonds, 1980; Vizgirda and Ahrens, 1982; Lange and Ahrens, 1986) consistently underestimate the pressure for the onset of devolatilization compared to postmortem analysis of experimentally shocked samples, a view that seems to be supported by shock-recovery experiments of others (e.g., Bell, 1997; Langenhorst et al., 1998, 2000). Such largely thermodynamic considerations also form the basis of specific calculations by, e.g., O'Keefe and Ahrens (1989), Takata and Ahrens (1994), Ivanov et al. (1996), Pope et al. (1994), Pierazzo et al. (1998), and Pierazzo and Melosh (1999) regarding the Cretaceous-Tertiary (K-T) event. Most concluded that the K-T bolide liberated sufficient CO_2 to trigger a greenhouse effect and thereby a global environmental crisis. A more quantitative understanding of the devolatilization of carbonates as a function of shock stress is thus critical to better constrain such calculations.

Additional shock effects in carbonates include complex oxidation-reduction reactions and phase changes. In the extreme, shocked carbonates may be reduced to elemental carbon, as suggested by Miura and Okamoto (1997), Miura et al. (1999), and Rietmeijer et al. (1999). Martinez et al. (1996) observed the following reaction during static compression experiments of dolomite: $CaMg(CO_3)_2$ (dolomite) → $CaCO_3$ (aragonite) + $MgCO_3$ (magnesite). They postulated that such mineral assemblage should be found in shocked carbonates from terrestrial impact structures. The mineral portlandite, $Ca(OH)_2$, was reported from the Haughton impact crater and was interpreted as a weathering product of devolatilized carbonates (Ostertag et al., 1985). However, this observation has not been confirmed by later studies (Metzler et al., 1988). Other observations also indicate reactions of carbonates with silicates in highly shocked, vesicular, carbonate-rich materials from Haughton (Martinez et al., 1994). Also, Graup (1999) observed immiscibility of impact-generated carbonate and silicate melts in the suevites from the Ries crater, and Jones et al. (2000) described crystallized calcite melts from Chicxulub ejecta. To complicate matters, the impact melts from Meteor Crater, Arizona, do not show such carbonate-silicate reactions, but all carbonate residue is completely dissolved into silicate-dominated impact melts (Hörz et al., 1998, 2000; See et al., 1999). Such observations also apply to molten dolomite and sandstone in the impact melts from Popigai (Kenkmann et al., 1999). Current interest also focuses on the instantaneous back-reaction(s) of shock-devolatilized carbonate residues with freshly released CO_2, initially theorized by Kieffer and Simonds (1980), and subsequently reported from the Haughton crater (Martinez et al., 1994). Such back-reactions were also experimentally produced by Agrinier et al. (1998, 2001), Deutsch et al. (1998), Ivanov et al. (2000), and Langenhorst et al. (2000), leading them to conclude that actual liberation of gaseous CO_2 into the atmosphere is not very efficient.

Many of these reactions and observations were derived from experimentally and naturally shocked calcite; analogous processes can be expected for dolomite. The purpose of this report is to present mineralogic and petrographic observations of experimentally shocked dolomite and to contribute to some of the above issues.

EXPERIMENTAL METHODS

The dolomite used in this study was a dense (0.04% porosity) rock composed of equigranular dolomite grains, typically 25 μm across. Unfortunately, the origin of this material is unknown, because it was acquired (for other purposes) from a local landscaper in Houston, Texas, without further documentation. The shock-recovery experiments employed a 20-mm-caliber powder propellant gun at the Johnson Space Center as detailed in, e.g., Hörz (1970) and Gibbons et al. (1975). Figure 1 illustrates the experimental set-up immediately before impact. The projectile is a polycarbonate cylinder that has a metal flyer plate of thickness T_p inserted into its front face. A three-part target assembly effectively embeds the geologic target in a metal matrix. The geologic specimens were doubly polished

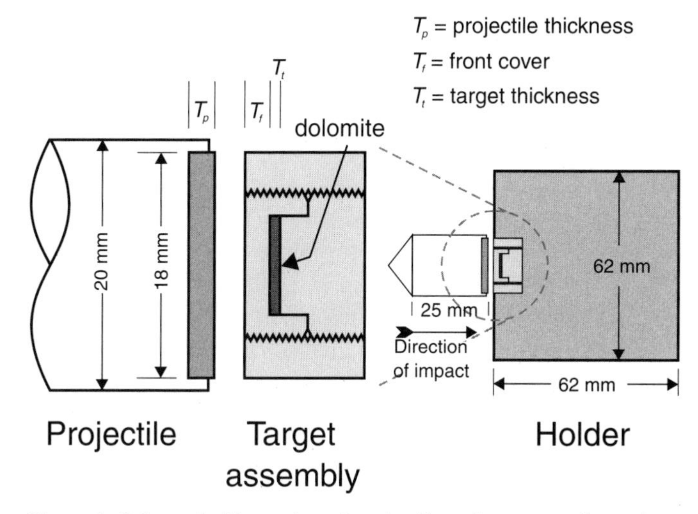

Figure 1. Schematic illustration of projectile and target configurations employed in the 20 mm powder propellant gun at Johnson Space Center. Note relatively small dimensions of the geological sample, a disc ~7 mm in diameter and 0.5 mm thick (for additional details see text).

TABLE 1. EXPERIMENTAL CONDITIONS FOR THE RECOVERY EXPERIMENTS OF DOLOMITE TARGETS

Peak pressure (GPa)	Flyer plate material	Target assembly material	Projectile velocity (km/s)
4.6	Lexan	Al2024	1.131
8.2	Lexan	SS304	1.694
16.9	Al2024	SS304	1.345
17.0	Al2024	SS304	1.345
17.6	Al2024	SS304	1.371
19.2	Al2024	SS304	1.478
20.0	Al2024	SS304	1.540
24.1	SS304	SS304	1.121
26.0	SS304	SS304	1.192
29.8	SS304	SS304	1.340
41.0	SS304	Fansteel 77	1.413
55.2	Tungsten	Fansteel 77	1.260
65.2	Tungsten	Fansteel 77	1.461
68.0	Tungsten	Fansteel 77	1.533

Note: Pressures are calculated via the impedance match method (Duvall, 1962) using the equations of state provided by Marsh (1980) for the diverse target assembly and flyer plate materials.

discs ~7.1 mm in diameter and 0.6 mm thick (T_t). The inset shows that the entire target assembly is tightly jacketed by a larger cylinder, the target holder, made from the same metal as the assembly. This massive holder is mounted inside the impact chamber by three plastic screws that break on purpose during impact, thus allowing the entire cylinder to move freely and to act as an efficient momentum trap. The most critical dimensions in Figure 1 relate to the thickness of the flyer plate (T_p), the front cover (T_f typically 0.9–0.95 T_p), and the thickness of the geologic target ($T_t < 0.5\ T_p$). Most flyer plates were 2 mm thick and made from Al 2024-T4, stainless steel 304, Fansteel 77 (a high-density W-Ni alloy), or pure W, but all experiments >60 GPa employed 1-mm-thick flyer plates of pure tungsten. These geometries permit the geologic specimen to equilibrate with the (calculated) pressure of the surrounding metal via any number ($n = 3-5$) of reverberations of the shock wave at the rock-metal interfaces (e.g., Gibbons et al., 1975). The pressure quoted refers to the projectile-assembly interface and was solved via a graphical impedance method (e.g., Duvall, 1962; Bowden et al., 2000). Pressure pulse duration varied from ~2 µs (at the low pressures) to ~0.5 µs at the highest pressures. Cooling of the geologic target is totally dominated by the radiative heat loss of the massive, metallic target holder inside the impact chamber; there were no special provisions to quench the samples. The target assembly is press fit into the sample holder and both are recovered as a single unit following an experiment; the deformed front face is then carefully machined, while being cooled with gaseous N_2, until the geologic specimen is exposed and amenable to physical quarrying and recovery. Pertinent details of the shock-recovery experiments are summarized in Table 1.

Carbon-coated, polished thin sections were made of the experimentally shocked materials and investigated compositionally using an energy-dispersive Link eXL or Link ISIS 300 X-ray spectrometer connected to a CamScan 4 scanning electron microscope at the Czech Geological Survey in Prague (CGS). Accelerating voltage was 15 kV and sample current 3 nA; spectrum acquisition time during quantitative analysis was typically 80 s. The spectra were processed by Link analytical software, which utilizes the procedure correcting for atomic number, absorption, and fluorescence (ZAF-type correction). The analyzed elements were Ca, Mg, Fe, and Mn; other elements were below the detection limit. Synthetic olivine (Mg_2SiO_4, Fe_2SiO_4, and Mn_2SiO_4) and wollastonite ($CaSiO_3$) were used as compositional standards. We did not use carbonate standards because carbonates are unstable during prolonged exposure to the electron beam. Empirical mineral formulae were calculated assuming a total of two cations per formula unit. Back-scattered electron (BSE) images were collected using a Robinson detector in the CamScan 4 scanning electron microscope. The images were acquired electronically by a personal computer employing Link ISIS 300 or Channel+ software.

Powder X-ray diffraction data were collected using a Philips X'Pert MPD diffraction system consisting of a vertical goniometer (PW3020) in the Bragg-Brentano reflecting geometry (goniometer radius = 173 mm), a generator (PW1870), and control electronics (PW3710) at the CGS. Powder patterns were taken at ambient pressure and temperature. The grain size of the powder was from 2 µm to 5 µm. The goniometer was equipped with a secondary-beam graphite monochromator and a proportional counter. Both primary and secondary Soller collimators had an axial divergence of 2.3°. The copper tube was powered at 40 kV and 40 mA. The samples were prepared as slurry mounts on a low-background silicon sample holder for the collection of accurate intensity data with the sample mass kept more or less constant at ~10 mg. Powder patterns were step-scanned in the angular range from 5° to 65° 2θ with step width 0.1° 2θ and 1.5 s counting time per step for phase identification; we scanned from 15° to 145° 2θ with step width 0.02° 2θ and ~8 s exposure per step for quantitative phase analysis and determination of peak widths, and from 29 to 33° 2θ with step width 0.015° 2θ and 10 s exposure per a step for the evaluation of intensity data for the strongest 104 reflection. For the phase identification we used the program Bede Search/Match with the International Center for Diffraction Data Powder Diffraction File 2 (ICDD PDF2) database (release 1998).

The Raman spectra were collected using a LabRam system (Dilor) equipped with external Ar^+-ion laser (Melles Griot) at the Institute of Chemical Technology in Prague. The 488 nm line was used for excitation. The power of the laser head was adjusted to 25 mW, yet a gray filter reduced this power by about one order of magnitude. A high-powered objective (Olympus; ×100) was used to focus the laser beam on the sample; the latter could be translated along arbitrary X-Y coordinates via a motorized sample stage. The scattered light was analyzed by a spectrograph with holographic grating (1800 g/mm); the slit width was 100 μm, and the confocal hole opening was 1000 μm. A Peltier-cooled charge-coupled device (CCD) detector (1024 × 256 pixels; ~ −70 °C) monitored the dispersed light. The alignment of the system was regularly checked using a sample of silicon and by measuring the grating in the zero-order position. The time of acquisition of a particular spectral window was optimized for individual sample measurements (~10 s). Five individual runs were summed to obtain a single spectrum. Three such average spectra were collected at three different locations for each individual sample. Spectra were collected over a spectral range 100–1900 cm^{-1}. The spectrometer and the positioning of a sample were controlled via personal computer (Pentium, Dell) with Labspec v. 2.08 (Dilor) software. The same software was used for data manipulation and analysis. Three spectral bands (at ~1100 cm^{-1}, 300 cm^{-1}, 176 cm^{-1}) were fitted with mixed Gaussian and Lorentzian functions to accurately calculate both their position and width.

A thermobalance Stanton-Redcroft TG-750 (at the Institute of Chemical Technology) was used for thermogravimetric (TG) measurements. The amount of sample used for individual heating runs varied between 1.9 and 2 mg. The TG curves were recorded simultaneously on a two-channel recorder and by personal computer. The heating rate was 10 °C·min^{-1} and air flow was 10 mL·min^{-1}. Differential thermogravimetric (DTG) curves were obtained by differentiating the TG curves with a graphical program.

RESULTS

Shock-recovery experiments and failure mode of target containers

The macroscopic deformation of the target assembly and its holder increases progressively with increasing peak pressure. Although ultimate mechanical failure of the entire target object is possible at very high projectile velocities and/or energies, the typical limit for the successful recovery of geologic material relates to the formation, by plastic flow, of a crater-like depression in the front face of the target holder, centered on the target assembly. The latter becomes greatly thinned and smeared along the entire crater bottom; the geologic sample is engulfed in this material flow and is substantially thinned and deformed as well. Substantial metal deformation occurs at pressures where most silicates begin to melt, and these melts ultimately escape at very high pressures through the most minute gaps and fractures represented by the original horizontal or vertical joints (although screwed or press fit) of the original assembly parts. These highly deformed joints will intercept the bottom and walls of the evolving crater, allowing the melts to escape; the latter disperse beyond recovery into the voluminous impact chamber (~1 m^3). This loss mechanism effectively limits the recovery of silicates to pressures <90 GPa using the Johnson Space Center facilities and procedures.

It is significant that this stage of target deformation was never reached in our dolomite experiments. Instead, the target failed in a different fashion and at lower pressures, ~70 GPa. The central front cover of thickness T_f (see Fig. 1) was catastrophically dislodged and found as a coherent disk inside the impact chamber. The diameter of this plate was approximately that of the dolomite sample. The original target well housing the dolomite was barely deformed, but empty; all dolomite was dispersed into the impact chamber. The vertical, machined sides and the bottom of the target well were well preserved and hardly deformed. A jagged fracture surface extends from the top of this cylindrical well to the original target surface, however. These relationships suggest that the front cover was removed violently around the periphery of the target well.

This mode of container failure is not only different from experiments employing silicates, but it is highly reproducible. It is typical of all carbonates and was observed with calcite and siderite samples as well. It is interesting that the threshold pressure for this style of target failure is mineral dependent with dolomite > calcite > siderite. The gas pressure inside the target well becomes apparently sufficient to violently dislodge the front cover. As a consequence, we interpret this specific mode of target failure to reflect the onset of significant devolatilization. For dolomite, this threshold pressure is reached at ~70 GPa. Unfortunately, we cannot specify the volume fraction of the target that outgassed at this pressure. Nevertheless, the pressure interval that separates successful dolomite recovery experiments (68 GPa) from failed ones (70 GPa) is extremely narrow, suggesting that something changes dramatically over a small pressure increment. Despite considerable efforts, we did not recover even small, physical fragments of dolomite inside the target well and inside the impact chamber at 70 GPa. This seems consistent with the notion that most of the target was rendered into gas-rich, possibly molten, materials that were finely dispersed beyond recovery.

Electron microprobe analysis and scanning electron microscopy

The electron microprobe investigations revealed that the starting material was a dolomite of elevated Ca content; Ca/Mg ratios ranged from 1.02 to 1.17. The major element composition is illustrated via a ternary plot in Figure 2 using Ca, Mg, and (Fe + Mn) as end members. Note that Figure 2 illustrates averaged compositions of ~20–30 individual analyses per sample. Individual microprobe analyses are shown in histogram form in Figure 3, after recasting the measured weight oxide percent into atoms per formula units. The purpose of Figures 2 and 3 is to illustrate the homogeneous composition of the initial dolomite, and the lack of any systematic compositional changes in the shocked products.

Accessory minerals of the starting material, identified by microprobe, were K-feldspar, quartz, and pyrite. We searched extensively in the shocked samples for the presence of pure CaO or MgO as evidence of outgassing, especially in all samples >50 GPa, yet we did not find any new, shock-induced phases via optical microscope and electron microprobe BSE images. These observations limit potentially new phases to grain-sizes <5 μm.

In samples shocked above 40 GPa, we observed droplets and/or veinlets of tungsten-bearing compounds, either in isolation or as highly localized clusters. They obviously originated from the shock-melted W-Ni alloy of the sample containers. Highly vesicular patches of frothy appearance are frequently associated with such tungsten-rich melts, and an example is illustrated in Figure 4. BSE images also reveal an increasing degree of fracturing with increasing shock pressure, as illustrated in Figure 5.

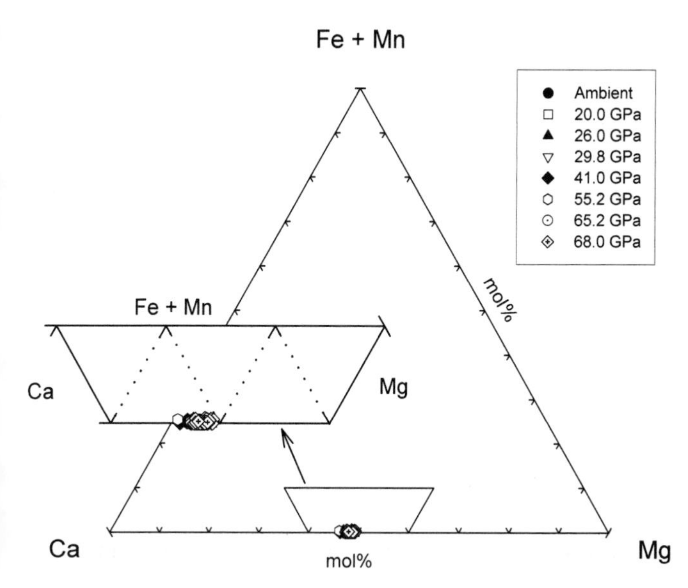

Figure 2. Average composition of unshocked and shocked dolomite plotted as molar percentages of $CaCO_3$, $MgCO_3$, and (Fe, Mn)O_3. Note tight clustering of all data with entire compositional range corresponding to ~50.3 and 53.7 mol% $CaCO_3$.

X-ray diffraction

Qualitative phase analysis of samples shocked below 30 GPa reveals the dominant nature of dolomite with subordinate and variable amounts of quartz and feldspar, totally consistent with the microprobe observations. Figure 6 displays examples of typical X-ray diffraction (XRD) patterns and matching standards from the PDF2 database. XRD peaks of quartz and feldspar broaden gradually at modest pressures and disappear completely in samples shocked above 30 GPa due to the formation of diaplectic glasses (Skála et al., 1999). Specimens shocked above 40 GPa contain variable, yet small amounts of tungsten, tungsten carbide (W_2C), and scheelite ($CaWO_4$), all representing artifacts of the sample containers. In samples shocked above 60 GPa, periclase and pure calcite have been detected. A summary of qualitative phase compositions is given in Table 2.

Qualitative modal compositions were also determined for the devolatilized residues of the thermogravimetry runs (see following); typical XRD patterns are illustrated in Figure 7. These materials contain the thermal decomposition products of dolomite, such as portlandite and periclase, together with the refractory constituents of the original target (quartz), and the reaction products of the container metal (e.g., scheelite) and/or oxidization products of this alloy (e.g., tungsten oxide). Table 3 lists all phases identified in the residues of the thermogravimetrically studied samples.

We tested two approaches to obtain the quantitative modal composition of the shock products. The first utilized the concept of reference intensity ratio (RIR; for details see Snyder and Bish, 1989). Using net intensities of the strongest peaks of each phase, we calculated the content of periclase to be ~2.5 wt% in the sample shocked to 65.2 GPa and 3.5 wt% in the sample shocked to 68.0 GPa (Skála and Hörz, 2000). These values, however, appear to be inconsistent with the CO_2 loss determined from TG analyses, suggesting that the RIR method may systematically overestimate the content of periclase. We consider potential preferred orientation in our samples the most likely source of error in our RIR calculations. Consequently, we used the Rietveld method, which can correct for this effect.

The Rietveld refinements were carried out using the program FullProf2.k (Rodríguez-Carvajal, 1990, 2000) with a graphic shell WinPLOTR (Roisnel and Rodríguez-Carvajal, 2000). The model structures of individual phases are those of the Inorganic Crystal Structure Database (ICSD), release 2000/1. Refined parameters were scale factors, background polynomial coefficients, sample displacement, unit-cell dimensions, and half-width parameters. For dolomite, the fractional atomic coordinates and preferred orientation parameters were also varied. Using these Rietveld procedures, the periclase content in the 68 GPa sample is reduced to ~0.6%, a value that is consistent with the CO_2 loss measured via TG. The quantitative

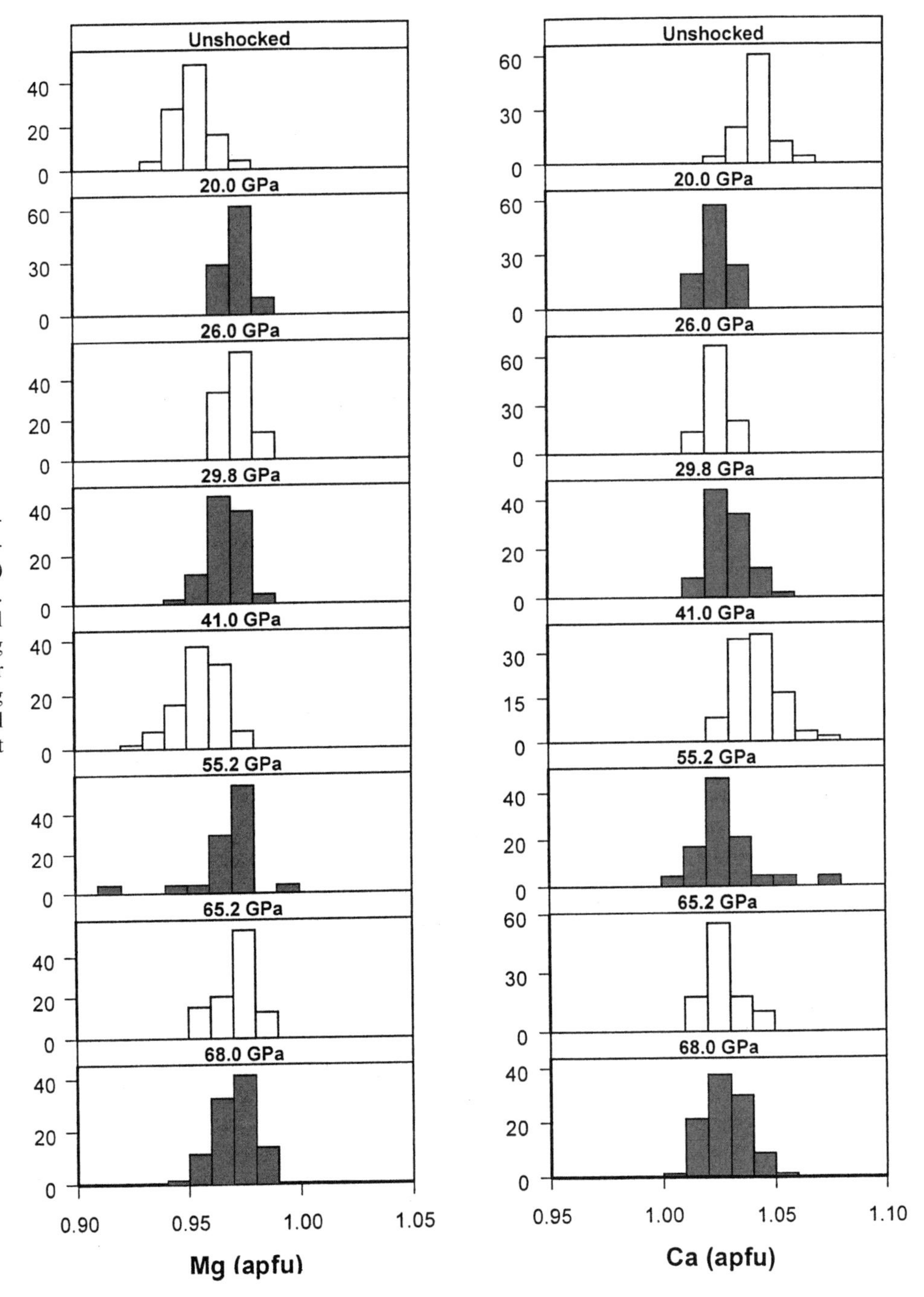

Figure 3. Histograms displaying absolute concentration of calcium and magnesium (atoms per formula units, apfu) in unshocked and shocked dolomites. Compositional variation is small at all pressures, suggesting that outgassing products like pure Mg and Ca oxides or reaction products like pure Ca or Mg carbonates cannot occupy substantial fractions of the electron beam's foot pad, ~2–3 µm in diameter.

phase compositions of all shocked materials are shown in Figure 8.

We were not able to address quantitatively the content of shock-produced, amorphous materials, such as diaplectic quartz and feldspar glasses at >30 GPa. The decreased, absolute intensity of the strongest dolomite 104 reflection implies the loss of diffracting mass, most likely by mechanical comminution to domain sizes beyond coherent X-ray diffraction (e.g., Hörz and Quaide, 1973). The modal concentrations of Figure 8 can only refer to the crystalline portions of the shock products, and they do not characterize the entire sample mass.

We also refined the dimensions of the dolomite unit cell (Table 4) and calculated theoretical densities for the samples, as illustrated in Figure 9. We observe a systematic increase of ~0.5% in the cell volume over the pressure range from 0–30 GPa. The data scatter uncomfortably at higher pressures; neglecting the 40 GPa result, most other experiments suggest a relatively invariant density over the pressure range 30–70 GPa.

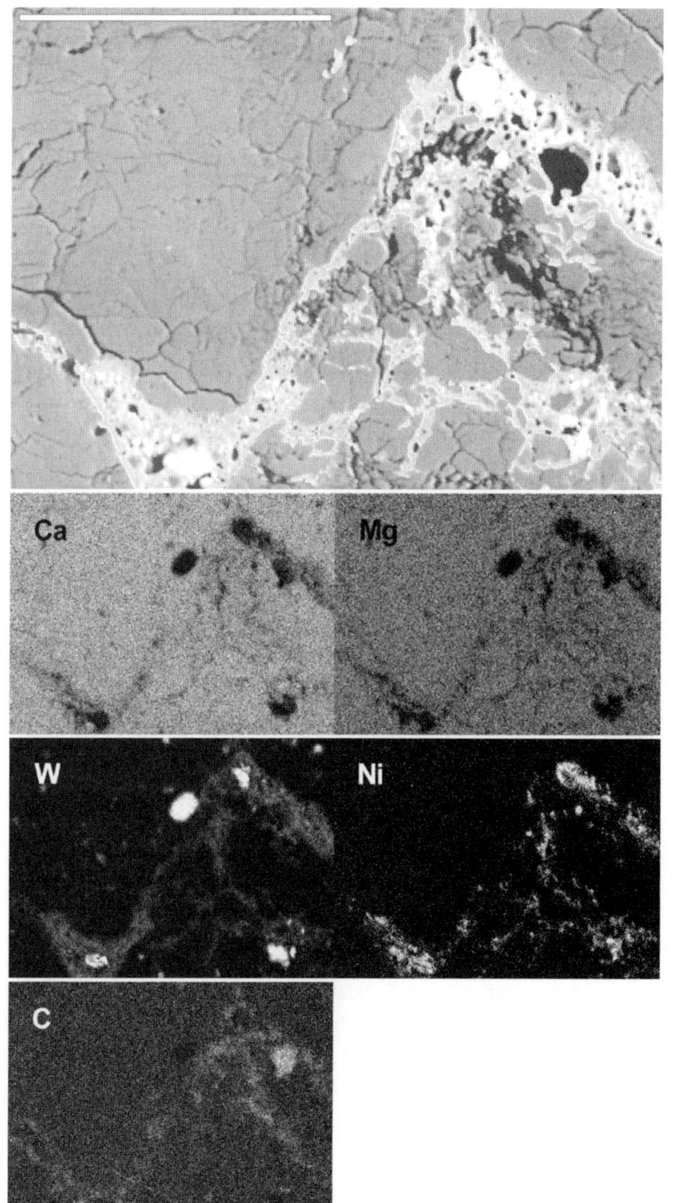

Figure 4. Typical appearance of tungsten-bearing regions in dolomite targets shocked above 40 GPa. Tungsten-bearing compounds (tungsten metal, tungsten carbide, and scheelite) form irregularly shaped patches that often contain voids and bubbles or small veinlets that crosscut dolomite (sample shocked to 41.0 GPa). Back-scattered electron image and distribution maps for Ca, Mg, W, Ni, and C shown. Scale bar ~100 μm.

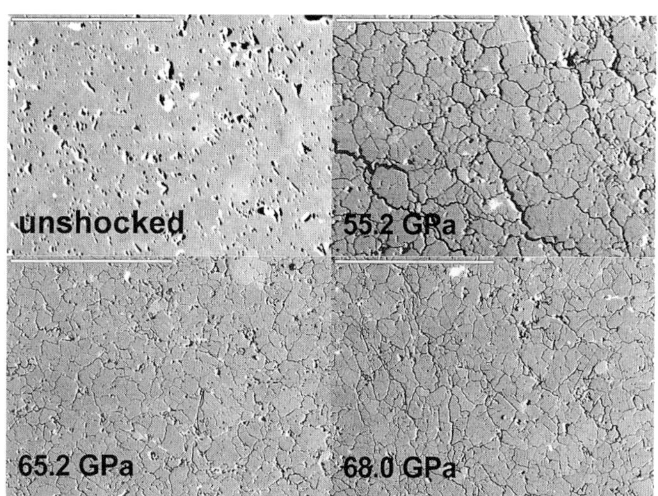

Figure 5. Back-scattered electron images of selected experimental shots. Samples shocked to 55.2 GPa (upper right), 65.2 GPa (lower left), and 68.0 GPa (lower right) are compared to unshocked material (upper left). Note gradual decrease of fragment size with increasing peak shock pressure. Scale bar ~200 μm.

Within a 1 σ error, most shocked dolomite has modestly lower crystal density than the unshocked material.

We observed a noticeable decrease in intensity of the strongest 104 reflection of dolomite in our experimental charges. We modeled the observed intensities using the centroid method of the WinPLOTR (Roisnel and Rodríguez-Carvajal, 2000) program and applying the split Pearson VII profile shape function of the program XFIT (Coelho and Cheary, 1997). The values calculated with both methods are identical within experimental error and are plotted versus peak shock pressure in Figure 10. There is a trend for systematically decreasing intensity at pressures <30 GPa, with absolute intensity changes approaching 50% and nearly linearly related to peak pressure; however, the relative intensity of the 104 peak seems invariant at pressures >30 GPa.

The changes in crystal density (Fig. 9) as well as in the intensity of the major X-ray diffraction peak (Fig. 10) indicate that the dolomite lattice deforms progressively at relatively modest pressures (<30 GPa) and that it reaches some configuration at 30–40 GPa that seems fairly stable against additional, mechanical deformation at pressures as high as 68 GPa.

Raman spectroscopy

For each sample, three individual spectra were collected by scanning sample areas of ~20–30 μm². The Raman spectra of dolomite should yield a total of eight fundamental Raman-active modes (Gillet et al., 1993); of these, four are of E_g symmetry and four are of A_g symmetry. We have observed seven of these modes in our spectra, yet three are very weak and one is weak. Consequently, we were able to fit selected spectral bands only, at about 1100, 300, and 176 cm^{-1}, corresponding to one internal (A_g, ν_1) and two lattice (E_g) modes. Typical examples of these spectral bands are shown in Figure 11. The fitted spectral band positions and bandwidths varied to some extent even within individual samples due to the heterogeneous distribution of shock deformations. As a consequence, three sampling spots were analyzed per specimen; representative values for peak positions and bandwidths were subsequently

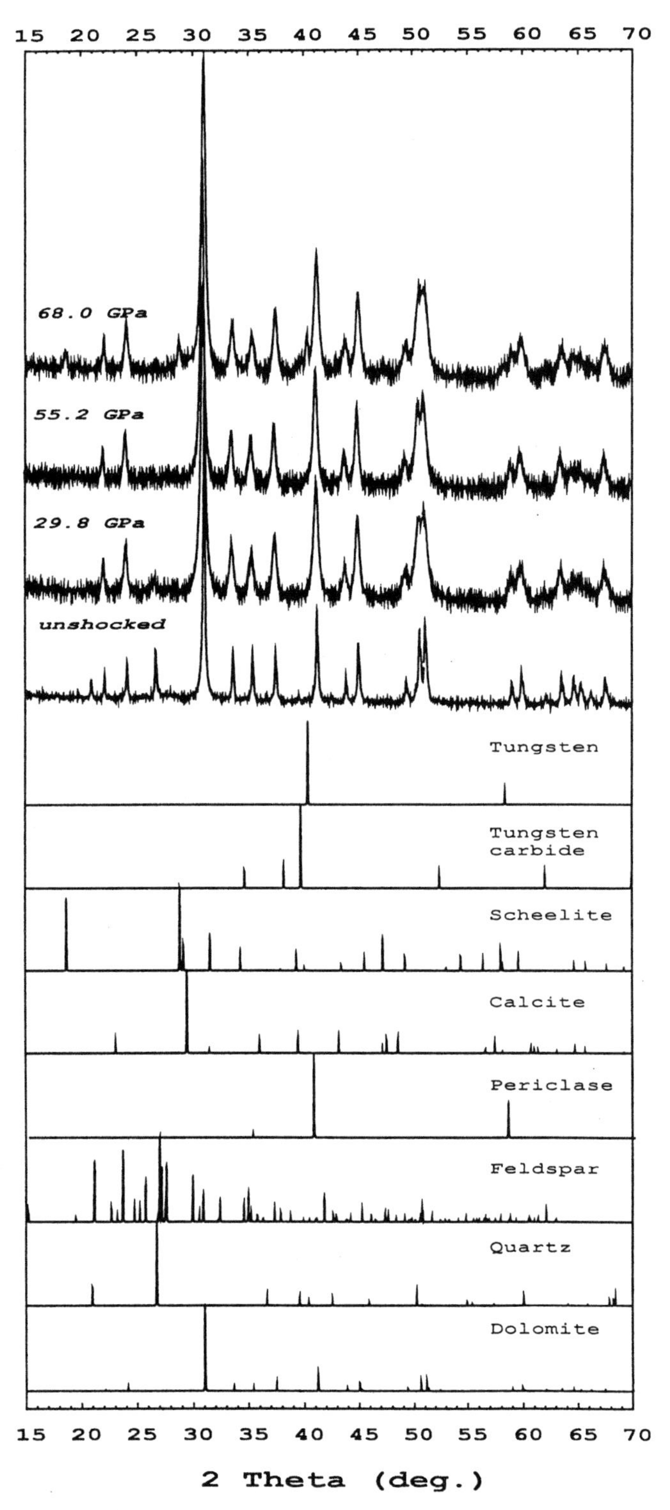

Figure 6. X-ray diffraction powder patterns of samples shocked to 29.8 GPa, 55.2 GPa, and 68.0 GPa for comparison with unshocked dolomite and standard patterns of pure phases. Note gradual broadening of dolomite peaks with increasing peak shock pressure. The intensity of quartz and feldspar reflections decreases and disappears entirely above 30 GPa, suggesting complete transformation to diaplectic glass.

obtained by averaging the two observations of closest agreement and discarding the third measurement.

Most spectra reveal three shock-induced effects: (1) substantial peak broadening with increasing peak pressure, (2) a decrease in absolute intensity of the spectral bands, and (3) a shift of the spectral band positions (Skála et al., 2000). The shifts of spectral band positions are generally in the opposite direction compared to those from static compression tests (e.g., Gillet et al., 1993). However, the absolute shifts are significantly smaller in the shocked samples, indicating extremely small changes in unit-cell dimensions. The broadening of Raman spectral bands with increasing peak shock pressure is illustrated in Figures 11 and 12. The relative broadening of Raman spectral bands relates sensitively to shock pressure. The bandwidth increases by ~20% for samples shocked at 20 GPa and by as much as 60% at >65 GPa.

Thermogravimetry

Thermogravimetry confirms and expands the insights gained from microprobe and X-ray diffraction analyses. The accessory minerals that do not decompose upon heating to ~1000°C amount to as much as 5% by weight, as illustrated in Table 5 and Figure 13. DTG curves of our samples are illustrated in Figure 14. The first DTG peak (at ~770–800°C) is associated with the reaction 1 Do → 1 Cc + 1 Pe + 1 CO_2. The second DTG peak at higher temperatures (~800–830°C) corresponds to decarbonization of the remaining calcite. The shoulder on the lower temperature slope of the first DTG peak is due to a small content of iron (Smykatz-Kloss, 1974). None of the measurable quantities (such as the absolute amount of released gas, the DTG peak temperature, or the width of the temperature interval for complete devolatilization to take place) correlates systematically with shock pressure.

INTERPRETATIONS

The modest shift of spectral band positions in the Raman spectra indicates very small changes in unit-cell dimensions of dolomite, which is confirmed by direct measurements using XRD data and the Rietveld method. The density of shocked dolomite may be as much as 0.5% smaller than the unshocked material. The magnitude of this change is consistent with the general hysteresis behavior described by Grady et al. (1976) that is associated with the transition from high- to low-density phases in shocked dolomite. The systematic spectral band broadening may be attributed to some widening in the distribution of bond lengths and angles and some residual stress and/ or strain following relaxation from the shock-compressed state. The increase of microstrain and the decrease in size of coherently diffracting domains with increasing pressure are also manifested by systematic broadening of the XRD peaks, as shown by Skála et al. (1999) for dolomite and generally by Hörz

TABLE 2. QUALITATIVE PHASE COMPOSITION IN STARTING AND RECOVERED MATERIALS DETERMINED FROM POWDER DIFFRACTION DATA

p (GPa)	Primary minerals	Minerals due to dolomite decomposition	Phases representing experimental artifacts
Unshocked	Dolomite, quartz, feldspar		
4.6	Dolomite, quartz, feldspar		
8.2	Dolomite, quartz, feldspar		
16.9	Dolomite, quartz, feldspar		
17.0	Dolomite, quartz, feldspar		
17.6	Dolomite, quartz, feldspar		
19.2	Dolomite, quartz, feldspar		
20.0	Dolomite, quartz, feldspar		
24.1	Dolomite, quartz, feldspar		
26.0	Dolomite, quartz, feldspar		
29.8	Dolomite, quartz, feldspar		
41.0	Dolomite		Tungsten, scheelite
55.2	Dolomite		
65.2	Dolomite	Periclase, calcite	Tungsten, scheelite, tungsten carbide W_2C
68.0	Dolomite	Periclase, calcite	Tungsten, scheelite, tungsten carbide W_2C

Note: The amount of feldspar did not allow more detailed determination of mineral species from XRD data but according to chemical data from EMPA the feldspar should be potassic. The gradual amorphization of feldspar and quartz resulted in their disappearing in samples recovered from shots above 30 GPa.

and Quaide (1973) for a number of experimentally shocked silicates.

The gradual decrease in intensity of the XRD reflections indicates a progressive decrease of the absolute fraction of coherently diffracting, crystalline dolomite that remains in the samples. The magnitude of intensity loss in the XRD data of the 104 reflection (Fig. 10) suggests that only slightly more than 50% of the sample coherently diffracts X-rays at pressures >40 GPa. However, the intensity decrease is not linear with pressure. A polynomial dependence best fits the observations with a relatively steep portion between 15 and 40 GPa.

Neither XRD data nor Raman spectra reveal that the breakdown products of dolomite contain magnesite + aragonite as postulated from static compression experiments by Martinez et al. (1996). Instead, calcite and periclase were found as major shock products in samples shocked above 60 GPa. They most likely are the result of thermally induced breakdown of dolomite according to the chemical reaction $CaMg(CO_3)_2$ (dolomite) → $CaCO_3$ (calcite) + MgO (periclase) + CO_2 ↑ (gaseous).

This reaction will occur above some threshold pressure, which we found to be ~60 GPa. The Rietveld data indicate that the content of periclase is no more than 0.6 wt% at 68 GPa. Such a periclase content implies that ~2.5% of the original dolomite devolatilized. An absolute upper limit of degassing is given by the thermogravimetry methods that suggest that ~7% of dolomite is devolatilized at 68 GPa. As a consequence, outgassing seems relatively modest at pressures <70 GPa.

We cannot positively exclude that the dominant phase assemblage formed during initial shock compression is aragonite + magnesite, as postulated by Martinez et al. (1996). Upon decompression and cooling of this material, the aragonite could undergo a temperature-induced phase transition to calcite, and the magnesite (Mst) may decompose to periclase and CO_2. We note that the reaction 1 Mst → 1 Pe + CO_2 starts at 400°C (and ambient pressure) following the BAYES program of Chatterjee et al. (1998).

The scatter in Raman bandwidths and intensities among individual measurements within individual samples implies that the shock deformation is heterogeneous on scales of tens of microns, i.e., on scales of component grains. The microprobe analyses (see Figs. 2 and 3), at modestly smaller spatial scales of 3–5 μm, consistently failed to resolve any products of outgassing, such as MgO or pure calcite, that are indicated by X-ray diffraction. All electron microprobe data yield MgO/CaO ratios that are consistent with dolomite. This implies that Mg- or Ca-rich decomposition products must be extremely fine grained, substantially below the spatial resolution of the electron microprobe beam (e.g., <1 μm), yet large enough to coherently diffract X-rays (e.g., >5 nm). These limitations seem consistent with the concept of thermally induced shear bands, resulting in highly localized hot spots, as the dominant devolatilization mechanism of shocked carbonates (e.g., Grady, 1980; Kondo and Ahrens, 1983; Tyburczy and Ahrens, 1986).

DISCUSSION

A number of independent analysis methods indicate that modest, yet variable amounts of carbon dioxide are being released during shock loading and associated postshock heating of dolomite from peak pressures as high as 68 GPa. Additional and somewhat indirect observations related to the failure mode of our sample containers suggest that dolomite outgasses massively at peak pressures >70 GPa. However, destruction of the sample containers at these and higher pressures prevents the recovery and characterization of the solid residues.

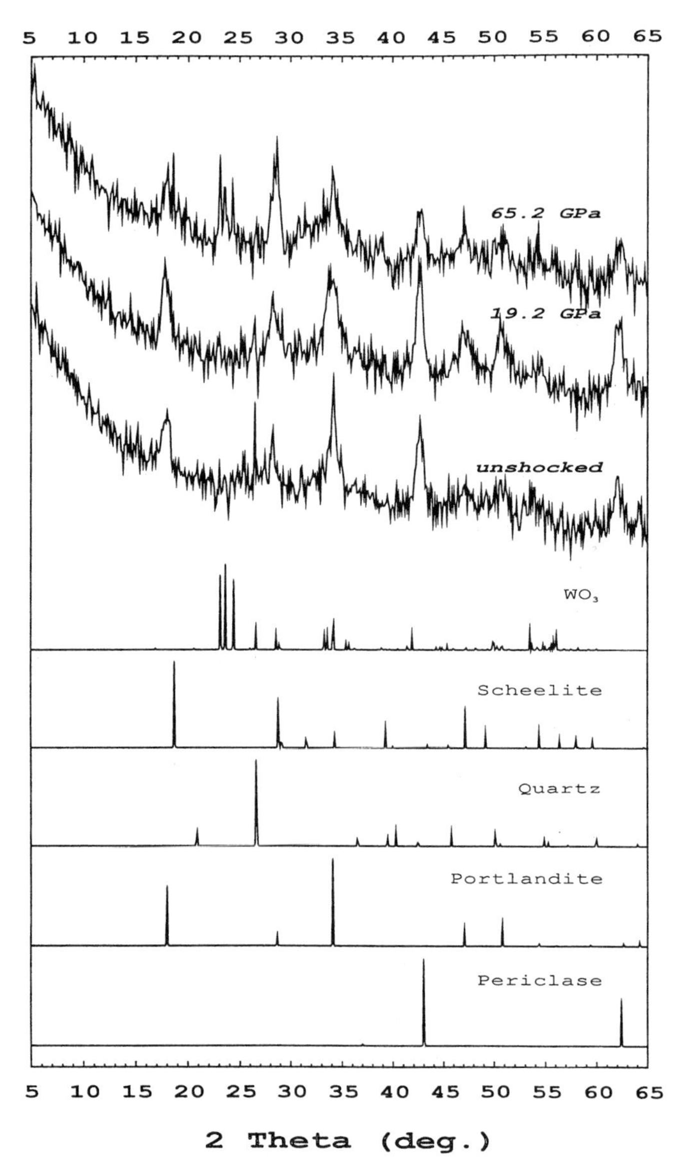

Figure 7. X-ray diffraction powder patterns of refractory residua following thermogravimetric (TG) analyses of unshocked dolomite and samples shocked to 29.8 GPa, 55.2 GPa, and 68.0 GPa, including comparisons with standard patterns of pure phases. Tungsten oxide formed during TG analysis; metallic impurities were derived from sample container.

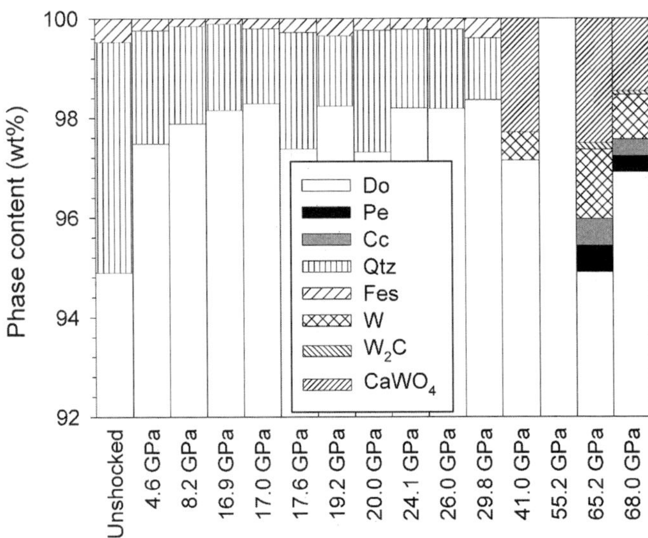

Figure 8. Quantitative phase composition of experimentally shocked dolomite samples based on X-ray powder diffraction patterns and their analysis using the Rietveld method. Note that amount of quartz and feldspar decreases at pressures of 25-30 GPa. At pressures above 30 GPa, the reaction products of the sample container material (tungsten-bearing phases) begin to appear and products of incipient dolomite vaporization (periclase and calcite) are present only at pressures above 60 GPa.

The current studies substantiate the earlier observations of Martinez et al. (1995), who found that experimentally measured devolatilization pressures are generally higher than those deduced from purely thermodynamic considerations. While there is an emerging consensus that calcite will devolatilize at substantially higher peak pressures (75 GPa; Gupta et al., 2000) than previously thought, Gupta et al. also postulated that devolatilization commences upon decompression of calcite from pressures as low as 18 GPa. We do not confirm this for dolomite, however. It is possible that the metal jackets that house our dolomite targets will modify and ultimately inhibit the devolatilization process. These containers prevent any liberated gas, once formed, from escaping efficiently and freely; gas pressure may build up and inhibit additional outgassing.

TABLE 3. QUALITATIVE PHASE COMPOSITION IN RESIDUA AFTER THERMOGRAVIMETRY

p (GPa)	Minerals due to thermal decomposition of dolomite	Refractory admixtures in dolomite	Phases representing experimental artifacts
Unshocked	Periclase, portlandite	Quartz	
8.2	Periclase, portlandite	Quartz	
19.2	Periclase, portlandite		
29.8	Periclase, portlandite	Quartz	
41.0	Periclase, portlandite		Tungsten oxide WO_3, scheelite
55.2	Periclase, portlandite		
65.2	Periclase, portlandite		Tungsten oxide WO_3, scheelite

Notes: We failed to identify other phases which we expected to occur—namely feldspar and tungsten carbide—due mainly to poor resolution of the powder patterns acquired. Presence of WO_3 can be explained by oxidation of metallic W during heating of the sample when making TG analysis.

TABLE 4. UNIT-CELL VOLUME AND THEORETICAL DENSITIES FROM RIETVELD REFINEMENT

p (GPa)	V_{cell} (Å^3)	D_x (g.cm^{-3})	ΔD_x (rel. %)
Unshocked	320.4(1)	2.879(1)	0.00(5)
4.6	320.5(1)	2.871(1)	−0.28(5)
8.2	320.5(1)	2.871(1)	−0.28(5)
17.0	320.5(2)	2.871(2)	−0.28(8)
17.6	320.6(2)	2.870(2)	−0.31(8)
29.8	321.0(3)	2.865(3)	−0.49(11)
19.2	320.6(2)	2.870(2)	−0.31(8)
20.0	320.7(2)	2.865(2)	−0.49(8)
24.1	320.7(2)	2.869(2)	−0.35(8)
26.0	320.8(2)	2.864(2)	−0.52(8)
16.9	320.7(2)	2.869(2)	−0.35(8)
55.2	320.8(2)	2.866(2)	−0.45(8)
41.0	321.0(3)	2.873(3)	−0.21(11)
65.2	320.6(2)	2.867(2)	−0.42(8)
68.0	320.7(2)	2.866(2)	−0.45(8)

Notes: $\Delta D_x = 100 \times (D_{x,s} - D_{x,u})/D_{x,u}$ where subscript *u* refers to the unshocked and the subscript *s* to the shocked state. Error values in parentheses represent 1 σ.

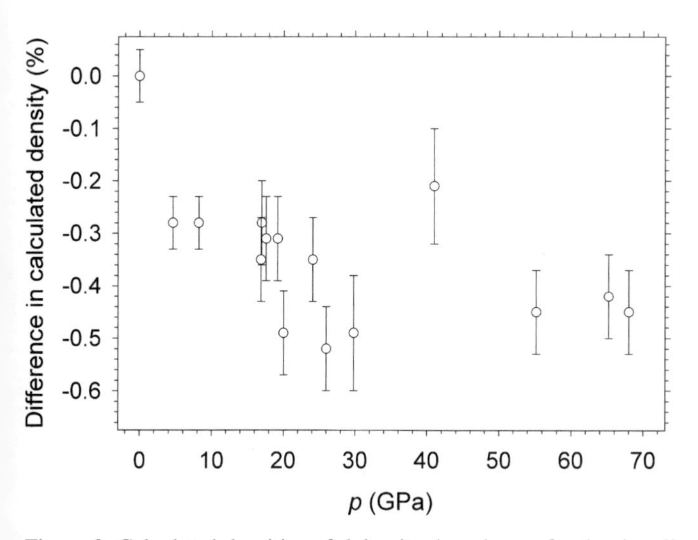

Figure 9. Calculated densities of dolomite, based on refined unit-cell dimensions, as function of increasing peak shock pressure. Note that density of all shocked samples is slightly lower than that of the starting material.

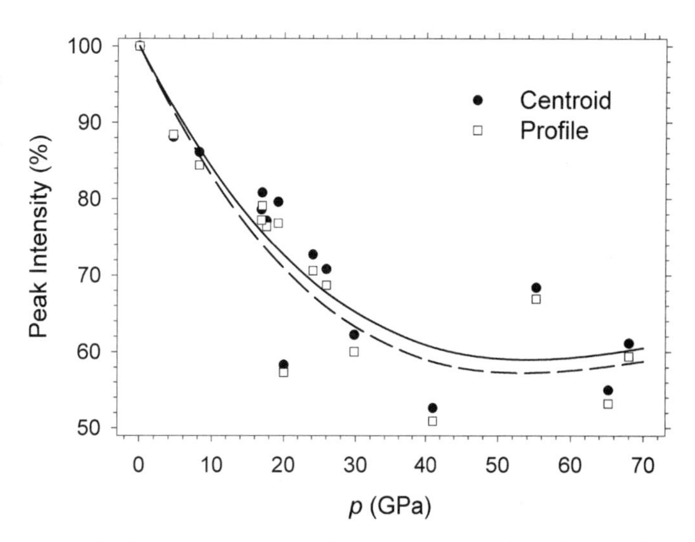

Figure 10. Decrease in the intensity of the strongest dolomite peak 104 as a function of shock pressure. Two relative intensity values were calculated, one using centroid method, the other using profile shape function (see text for details). Curves are fitted by a third-order polynomial.

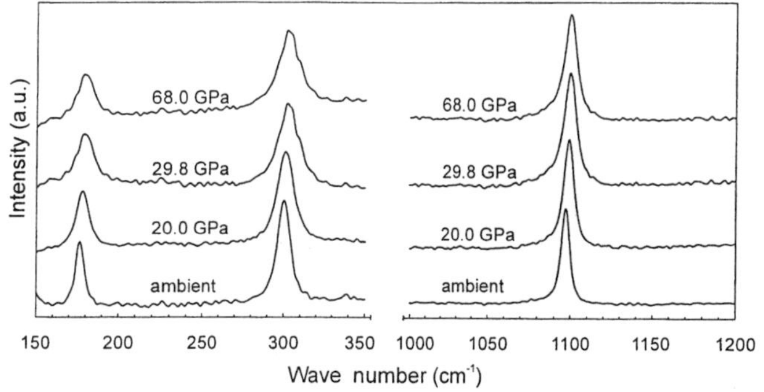

Figure 11. Partial Raman spectra of dolomite samples shocked to 20.0 GPa, 29.8 GPa, and 68.0 GPa and comparison with unshocked raw material. The most prominent shock-produced feature is a broadening of Raman bands; slightly shifted band positions are produced.

None of our samples has back-reacted with gaseous CO_2, however, and there is no evidence for secondary carbonates even at the highest pressures. Such back-reactions were described from calcite experiments at shock pressures of 80 GPa (Langenhorst et al., 1998, 2000; Deutsch et al., 1998) and from some frothy materials, obviously once molten, from the Haughton Dome crater (Martinez et al., 1994). Langenhorst et al. (2000) and Agrinier et al. (2001) propose such back-reactions to constitute effective sinks of most CO_2 that was liberated during natural events. We note that Langenhorst et al. (1998, 2000) used large, explosively driven flyer plates. Their sample containers are dimensionally scaled accordingly and are so massive that they can tolerate the high gas pressures inside the interior target well. As a consequence, back-reactions of gaseous species and their own residues are common in the experiments of Langenhorst et al. (2000), yet such reactions are absent in our experiments. Back-reactions are not significant in the melts from Meteorite Crater, Arizona, largely a mixture of dolomite residues and quartz (Horz et al., 2000), or in the melts from the Chixculub event (e.g., Sigurdson et al., 1992). While the occurrence of such back-reactions is undisputed, we suggest that Langenhorst et al. (2000) and Agrinier et al. (2001) may have overestimated their significance as an effective sink of shock-liberated gases.

All shock recovery experiments are also by necessity conducted at very small scales, both dimensionally as well as kinetically, compared to natural impacts, especially those of sufficient size to produce global consequences. The relatively short

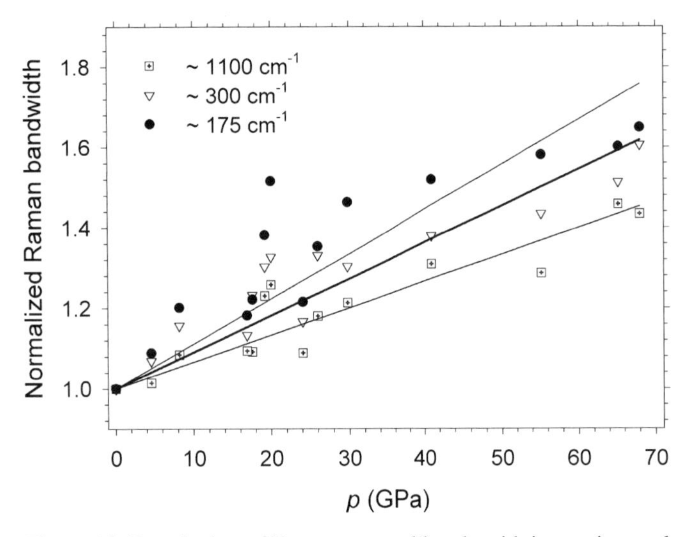

Figure 12. Broadening of Raman spectral bands with increasing peak shock pressure; each bandwidth is normalized to that of unshocked dolomite.

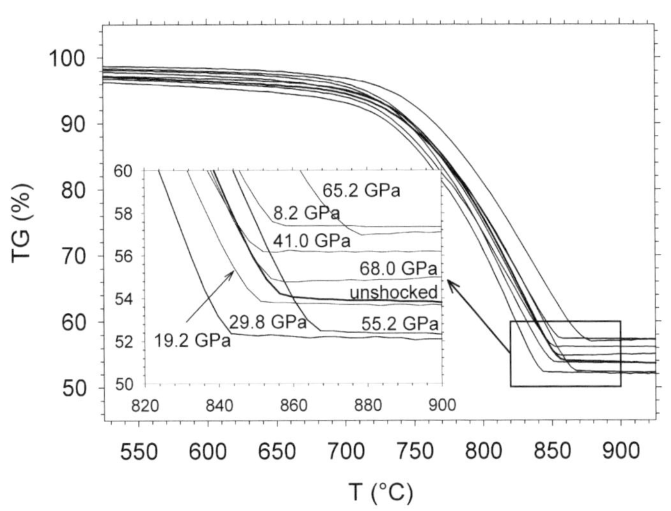

Figure 13. Thermogravimetric (TG) curves of experimentally shocked dolomites and comparison with unshocked standard. Total evolved carbon dioxide varies from 43 to 48 wt%; the measured stoichiometry of the unshocked sample should yield 47.5 wt% CO_2; the difference between this value and the measured one represents the maximum, shock-induced CO_2 loss (after correction for presence of volatile-free accessories like quartz, feldspars, and W-bearing components).

TABLE 5. RESULTS OF THERMOGRAVIMETRY ANALYSES

Peak pressure (GPa)	Sample amount (mg)	Weight loss measured (wt%)	Weight loss expected (wt%)	Difference to measured (wt%)	Difference to expected (wt%)	Difference to measured (rel%)	Difference to expected (rel%)	Peak DTG 1 (°C)	Peak DTG 2 (°C)	Decarbonization interval (K)
Unshocked	1.922	46.4	47.5	0.0	1.2	0.0	2.5	777	831	189
8.2	1.953	42.7	47.6	3.7	4.9	7.9	10.3	786	826	195
19.2	1.928	46.4	47.6	0.0	1.2	−0.1	2.5	769	812	176
29.8	1.950	48.0	47.6	−1.6	−0.4	−3.5	−0.9	769	818	179
41.0	1.941	43.8	47.5	2.5	3.7	5.4	7.8	766	823	176
55.2	1.919	47.8	47.6	−1.4	−0.2	−3.1	−0.4	779	835	186
65.2	1.953	43.0	47.6	3.3	4.6	7.2	9.6	799	835	189
68.0	1.946	45.2	47.6	1.1	2.4	2.5	5.0	778	822	183

Notes: According to Smykatz-Kloss (1974), the weight loss for stoichiometric $CaMg(CO_3)_2$ (NBS sample 88) is 47.7 wt%; differential thermagravimetric (DTG) peak for sample amounts of 1 g has maximum at 753 °C and for 100 g there are two peaks at 807 °C and 901 °C, respectively.

pulse durations of the experiment may not properly duplicate such large-scale impacts (e.g., Bowden et al., 2000), and it is possible that all experiments underestimate the volatile loss of carbonates. Owing to the small masses involved in the experiments, the experimental charges may cool much more rapidly than large masses of naturally shocked rocks. It seems obvious to postulate that the shock behavior of carbonates, especially their devolatilization, is substantially dominated by thermal effects rather than by the mechanical deformation of their crystal structure under high stresses and high strain rates. Although the latter may dominate the behavior of silicates and may be successfully reproduced at small laboratory scales, the postshock thermal environment may ultimately be decisive for the behavior of carbonates; this post-shock, thermal environment may not be simuated adequately by laboratory experiments of short pulse durations.

The detailed shock behavior, and specifically the threshold pressures for incipient and complete outgassing, of dolomite and carbonates in general is not well understood at present. A substantial discrepancy remains between observations from experimentally shocked samples and purely thermodynamical considerations, although the gap appears to close (e.g., Lange and Ahrens, 1986; Martinez et al., 1995; Langenhorst et al., 2000; Gupta et al., 2000). Current shock-recovery experiments suggest that carbonates appear surprisingly stable under shock conditions compared to major rock-forming minerals such as quartz and feldspar (e.g., Bell, 1997). We simply observe that devolatilization of dolomite is not very significant at pressures <70 GPa.

We suggest further that mineralogical evidence for the volatile-free residua of dolomite will be difficult to find in natural impact formations. Such residues are unstable at ambient

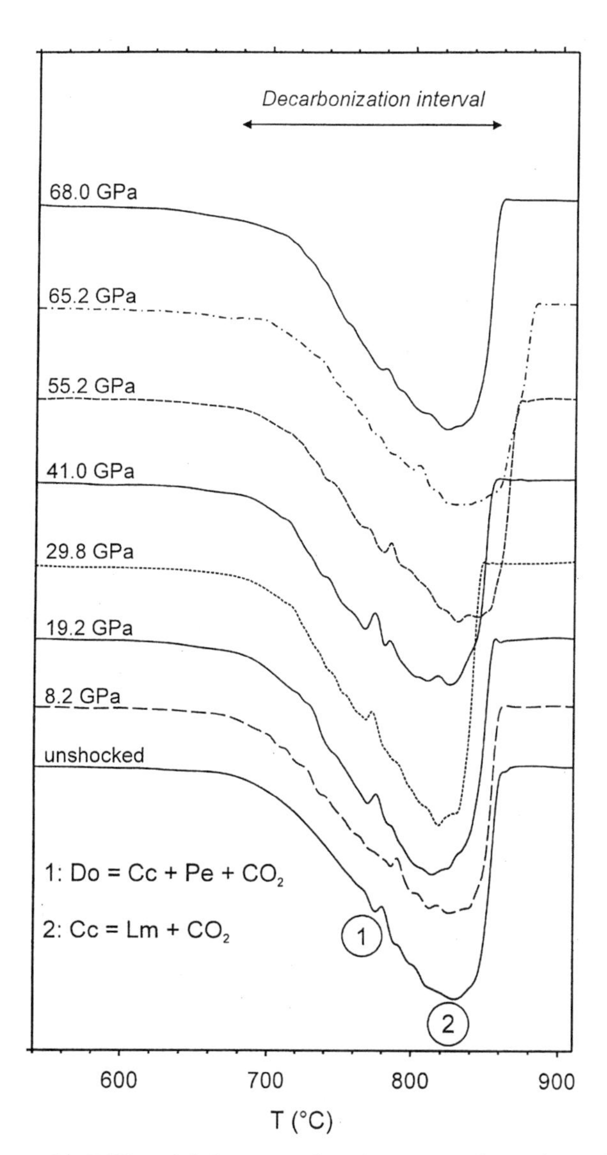

Figure 14. Differential thermogravimetric curves of experimentally shocked dolomite. The curves contain two separate, major peaks corresponding to (1) decomposition of dolomite (Do) to periclase (Pe), calcite (Cc), and CO_2, and, at higher temperature, (2) devolatilization of calcite to lime (Lm) and CO_2. The modest shoulder at lower temperatures (~720–760°C) reflects modest iron contents of the dolomite used. The decarbonization interval shown corresponds to that indicated in Table 5.

temperatures and will rapidly convert to secondary phases, the latter largely reflecting some specific, ambient environment rather than the formative conditions of their shock-produced precursor phases. Diagnostic shock evidence will thus be modified rapidly and may be difficult to prove. In addition, many terrestrial targets contain crystalline rocks or quartz, and the CaO and MgO of the devolatilized carbonate may combine with the silicate fraction into genuine silicate melts, as was the case at Meteor Crater (e.g., Hörz et al., 2000) or the K-T melt beads from Haiti (Sigurdsson et al., 1992).

CONCLUSIONS

Our findings on experimentally shocked dolomite substantiate that carbonates are surprisingly stable against shock deformation and that they devolatilize at shock stresses similar to those needed to cause shock melting of common silicates, such as quartz and feldspar. Specifically, modest outgassing of dolomite occurs at pressures as high as 70 GPa, a result consistent with the earlier observations of Martinez at al. (1995). Our results are also consistent with the growing body of evidence that calcite is similarly resistant to shock and that it undergoes substantial gas loss only at pressures >65 GPa (Bell, 1997; Langenhorst et al., 2000; Gupta et al., 2000).

These observations suggest that the production of CO_2 may have been overestimated in earlier models of the Chixculub impact. More work is needed to reduce the current discrepancies between theory and experiment. However, additional geologic information is needed as well to render Chixculub-specific calculations more realistic: e.g., the total modal abundance and stratigraphic position of carbonates (and sulfates) within the Chixculub target are not well known, nor are impact speed, incident angle, and projectile density. These parameters are crucial to determine how much carbonate was shocked beyond some threshold isobar for devolatilization and how much CO_2 was ultimately produced by the K-T impact. In addition, the absolute concentration and role of CO_2, vis-à-vis other environmental factors such as SO_x or copious amounts of dust, is poorly understood. Past estimates of the total CO_2 production and its contribution to the environmental crisis at the K-T boundary are not well defined.

ACKNOWLEDGMENTS

Skála thanks the European Science Foundation IMPACT program and the Organizing Committee of the conference in Vienna for their support to participate in the meeting. We thank F. Veselovský of the Czech Geological Survey in Prague for the careful porosity measurements of the unshocked material, and G. Hayes (Lockheed-Martin, Houston) for the skillful operation of the 20 mm gun at the Johnson Space Center. We acknowledge enlightening discussions with M.S. Bell, M. Cintala, P. Jakeš, and D. Ming, and helpful reviews by A. Deutsch and an anonymous reviewer.

REFERENCES CITED

Agrinier, P., Deutsch, A., Schärer, U., Martinez, I., and Javoy, M., 1998, On the kinetics of reaction of CO_2 with hot CaO during impact events: An experimental study: Lunar and Planetary Science, v. 29, Abstract #1217, CD-ROM.

Agrinier, P., Deutsch, A., Schärer, U., and Martinez, I., 2001, Fast back-reaction of shock-released CO_2 from carbonates: An experimental approach: Geochimica et Cosmochimica Acta, v. 65, p. 2615–2632.

Bell, M.S., 1997, Experimental shock effects in calcite, gypsum, and quartz [abs.]: Meteoritics and Planetary Science, v. 32, Supplement, p. A11.

Bell M.S., Hörz, F., and Reid A., 1998, Characterization of experimental shock effects in calcite and dolomite by x-ray diffraction: Lunar and Planetary Science, v. 29, Abstract #1422, CD-ROM.

Boslough, M.B., Ahrens, T.J., Vizgirda, J., Becker, R.H., and Epstein, S., 1982, Shock-induced devolatilization of calcite: Earth and Planetary Science Letters, v. 61, p. 166–170.

Bowden, E., Kondo, K., Ogura, T., Jones, A.P., Price, G.D., and DeCarli, P.S., 2000, Loading path effects on the shock metamorphism of porous quartz: Lunar and Planetary Science, v. 31, Abstract #1582, CD-ROM.

Chatterjee, N.D., Krüger, R., Haller, G., and Olbricht, W., 1998, The Bayesian approach to an internally consistent thermodynamic database: Theory, database, and generation of phase diagrams: Contributions to Mineralogy and Petrology, v. 133, p. 149–168.

Coelho, A.A., and Cheary, R.W., 1997, X-ray Line Profile Fitting Program, XFIT: Sydney, New South Wales, Australia, School of Physical Sciences, University of Technology, Computer program, ftp://ftp.minerals.csiro.au/pub/xtallography/koalariet

Deutsch, A., Schärer, U., and Agrinier, P., 1998, Evidence for back-reacted carbonates in distant ejecta of the Chicxulub impact event? An experimental approach: Lunar and Planetary Science, v. 29, Abstract #1386, CD-ROM.

Duvall, G.E., 1962, Concepts of shock wave propagation: Bulletin of the Seismological Society of America, v. 4, p. 869–893.

Gibbons, R.V., Morris, R.V., and Hörz, F., 1975, Petrographic and ferromagnetic resonance studies of experimentally shocked regolith analogs, *in* Proceedings of Lunar Science Conference, 6th: New York, Pergamon Press, p. 3143–3171.

Gillet, Ph., Biellmann, C., Reynard, B., and McMillan, P., 1993, Raman spectroscopic studies of carbonates. 1. High-pressure and high-temperature behaviour of calcite, magnesite, dolomite and aragonite: Physics and Chemistry of Minerals, v. 20, p. 1–18.

Grady, D.E., 1980, Shock deformation of brittle solids: Journal of Geophysical Research, v. 85, p. 913–924.

Grady, D.E., Murri, W.J., and Mahrer, K.D., 1976, Shock compression of dolomite: Journal of Geophysical Research, v. 81, p. 889–893.

Graup, G., 1999, Carbonate-silicate liquid immiscibility upon impact melting, Ries Crater, Germany: Meteoritics and Planetary Science, v. 34, p. 425–438.

Gupta, S.C., Love, S.G., and Ahrens, T.J., 2000, Shock temperatures in calcite ($CaCO_3$): Implication for shock induced decomposition, *in* Furnish, M.D., Chhabildas, L.C., and Hixson, R.S., eds., Shock compression of condensed matter: College Park, Maryland, American Institute of Physics, p. 1263–1266.

Hörz, F., 1970, A small ballistic range for impact metamorphism studies: NASA Technical Note D-5787, 18 p.

Hörz, F., and Quaide, W.L., 1973, Debye-Scherrer investigations of experimentally shocked silicates: The Moon, v. 6, p. 45–82.

Hörz, F., See, T.H., Yang, V., and Mittlefehldt, D., 1998, Major element composition of ballistically dispersed melt particles from Meteor Crater, Arizona: Lunar and Planetary Science, v. 29, Abstract #1777, CD-ROM.

Hörz, F., See, T.H., Mittlefehldt, D.W., Galindo, C., and Golden, D.C., 2000, Major element composition of the impact melts at Meteor Crater, Arizona: Lunar and Planetary Science, v. 31, Abstract #1730, CD-ROM.

ICDD PDF2 database, 1998, International Center for Diffraction Data, Newton Square, Pennsylvania, USA.

ICSD database, release 2000/1, Fachinformationszentrum Karlsruhe, Germany and National Institute of Standards and Technology: Gaithersburg, Maryland, USA.

Ivanov, B.A., Badjukov, D.D., Yakovlev, O.I., Gerasimov, M.V., Dikov, Yu.P., Pope, K.O., Ocampo, A., 1996, Degassing of sedimentary rocks due to Chicxulub impact: Hydrocode and physical simulations, *in* Ryder, G., Fastovsky, D., and Gartner, S., eds., The Cretaceous-Tertiary event and other catastrophes in Earth history: Geological Society of America Special Paper 307, p. 125–139.

Ivanov, B.A., Langenhorst, F., Deutsch, A., and Hornemann, U., 2000, How strong was shock-induced CO_2 degassing in the K/T event? [abs.], *in* Catastrophic events and mass extinctions: Impacts and beyond: Houston, Texas, Lunar and Planetary Institute, LPI Contribution No. 1053, p. 80–81.

Jones, A.P., Claeys P., and Heuschkel, S., 2000, Impact melting of carbonates from the Chicxulub crater, *in* Gilmour, I., and Koeberl, C., eds., Impacts and the early earth: Berlin, Springer-Verlag, Lecture Notes in Earth Sciences, v. 91, p. 343–362.

Kenkmann, T., Greshake, A., Schmitt, R.T., Tagle, R., Claeys, P., and Stöffler, D., 1999, Naturally shock-induced phase transformations in a dolomite-sandstone: An example from the Popigai Impact Crater, Siberia [abs.]: Lunar and Planetary Science, v. 30, Abstract #1561, CD-ROM.

Kieffer, S.W., and Simonds, C.H., 1980, The role of volatiles and the lithology in the impact cratering process: Reviews of Geophysics and Space Physics, v. 18, p. 143–181.

Kondo, K.-I., and Ahrens, T.J., 1983, Heterogeneous shock-induced thermal radiation in minerals: Physics and Chemistry of Minerals, v. 9, p. 173–181.

Kotra, R.K., See, T.H., Gibson, E.K., Hörz, F., Cintala, M.J., and Schmidt, R.S., 1983, Carbon dioxide loss in experimentally shocked calcite and limestone [abs.]: Lunar and Planetary Science, v. 14, p. 401–402.

Lange, M.A., and Ahrens, T.J., 1986, Shock-induced CO_2 loss from $CaCO_3$; implications for early planetary atmospheres: Earth and Planetary Science Letters, v. 77, p. 409–418.

Langenhorst, F., Deutsch, A., and Hornemann, U., 1998, On the shock behavior of calcite: Dynamic 85-GPa compression and multianvil decompression experiments [abs.]: Meteoritics and Planetary Science, v. 33, Supplement, p. A90.

Langenhorst, F., Deutsch, A., Ivanov, B.A., and Hornemann, U., 2000, On the shock behavior of $CaCO_3$: Dynamic loading and fast unloading experiments: Modeling mineralogical observations: Lunar and Planetary Science, v. 31, Abstract #1851, CD-ROM.

Love, S.G., and Ahrens, T.J., 1998, Measured shock temperatures in calcite and their relation to impact-melting and devolatilization: Lunar and Planetary Science, v. 29, Abstract #1206, CD-ROM.

Marsh, S.P., 1980, LASL Shock Hugoniot Data, University of California Press, Berkeley, California, 658 p.

Martinez, I., Agrinier, P., Schärer, U., and Javoy, M., 1994, A SEM-ATEM and stable isotope study of carbonates from the Haughton impact crater, Canada: Earth and Planetary Science Letters, v. 121, p. 559–574.

Martinez, I., Deutsch, A., Schärer, U., Ildefonse, Ph., Guyot, F., and Agrinier, P., 1995, Shock recovery experiments on dolomite and thermodynamical modelling of impact induced decarbonization: Journal of Geophysical Research, v. 100, p. 15465–15476.

Martinez, I., Zhang, J., and Reeder, R.J., 1996, In situ X-ray diffraction of aragonite and dolomite at high pressure and high temperature: Evidence for dolomite breakdown to aragonite and magnesite: American Mineralogist, v. 81, p. 611–624.

Metzler, A., Redeker, H.-J., Stöffler, D., and Ostertag, R., 1988, Composition of the crystalline basement and shock metamorphism of crystalline and sedimentary target rocks at the Haughton impact crater, Devon Island, Canada: Meteoritics, v. 23, p. 197–207.

Miura, Y., and Okamoto, M., 1997, Change of limestone by impacts: Source of impact induced graphite from target, *in* IGCP Project 384, Symposium Impacts and Extraterrestrial Spherules: New Tools For Global Correlation: Tallinn, Estonia, Excursion Guide and Abstracts, p. 38–40.

Miura, Y., Kedves, M.Á., and Kobayashi, H., 1999, Carbon source from limestone target by impact reaction at the K/T boundary [abs.]: International Symposium on Planetary Impact Events and Their Consequences on Earth, PIECE '99, Yamaguchi, Japan, p. 58

O'Keefe, J.D., and Ahrens, T.J., 1989, Impact production of CO_2 by the Cretaceous/Tertiary extinction bolide and the resultant heating of the earth: Nature, v. 338, p. 247–249.

Ostertag, R., Robertson, P.B., Stöffler, D., and Wöhrmeyer, C., 1985, First results of a multidisciplinary analysis of the Haughton impact crater, Devon Island, Canada, III. Petrography and shock metamorphism [abs.]: Lunar and Planetary Science, v. 16, p. 633–634.

Pierazzo, E., and Melosh, H.J., 1999, Hydrocode modeling of Chicxulub as an oblique impact event: Earth and Planetary Science Letters, v. 165, p. 163–176.

Pierazzo, E., Kring, D.A., and Melosh, H.J., 1998, Hydrocode simulation of the Chicxulub impact event and the production of climatically active gases: Journal of Geophysical Research, v. 103, p. 28607–28625.

Pope, K.O., Baines, K.H., Ocampo, A.C., and Ivanov, B.A., 1994, Impact winter and the Cretaceous/Tertiary extinctions: Results of a Chicxulub asteroid impact model: Earth and Planetary Science Letters, v. 128, p. 719–725.

Rietmeijer, F.J.M., Bunch, T.E., and Schultz, P.H., 1999, A preliminary analytical electron microscope study of experimentally shocked dolomite with emphasis on neoformed carbon phases: Lunar and Planetary Science, v. 30, Abstract #1051, CD-ROM.

Rodríguez-Carvajal, J., 1990, FULLPROF: A program for Rietveld refinement and pattern matching analysis: Abstracts of the Satellite Meeting on Powder Diffraction of the 15th Congress of the International Union of Crystallography, Toulouse, France, p. 127.

Rodríguez-Carvajal, J., 2000, FULLPROF.2k: Rietveld, Profile matching and integrated intensities refinement of X-ray and/or neutron data (powder and/or single-crystal): Computer program, Laboratoire Léon Brillouin, Centre d'Etudes de Saclay, Gif-sur-Yvette Cedex, France, ftp://charybde.saclay.cea.fr/pub/divers/fullp

Roisnel, T., and Rodríguez-Carvajal, J., 2000, WinPLOTR. March 2000 version: Computer program, Laboratoire Léon Brillouin, Centre d'Etudes de Saclay, Gif-sur-Yvette Cedex, France, available from http://www-llb.cea.fr/fullweb/winplotr/winplotr.htm

See, T.H., Galindo, C., Golden, D.C., Yang, V., Mittlefehldt, D., and Hörz, F., 1999, Major-element composition of ballistically dispersed melts from Meteor Crater, Arizona: Lunar and Planetary Science, v. 30, Abstract #1633, CD-ROM.

Sigurdsson, H., D'Hondt, S., and Carey, S., 1992, The impact of the Cretaceous Tertiary bolide on evaporite terrane and generation of major sulfuric acid aerosol: Earth and Planetary Science Letters, v. 109, p. 543–560.

Skála, R., and Hörz, F, 2000, Mineralogical studies of experimentally shocked dolomite: Implications for the outgassing of carbonates [abs.], *in* Catastrophic events and mass extinctions: Impacts and beyond: Houston, Texas, Lunar and Planetary Institute, LPI Contribution No. 1053, p. 202–203.

Skála, R., Hörz, F., and Jakeš, P., 1999, X-ray powder diffraction study of experimentally shocked dolomite: Lunar and Planetary Science, v. 30, Abstract #1327, CD-ROM.

Skála, R., Matějka, P., and Hörz, F., 2000, Experimentally shocked dolomite: A Raman spectroscopic study: Lunar and Planetary Science, v. 31, Abstract #1567, CD-ROM.

Smykatz-Kloss, W., 1974, Differential thermal analysis: Application and results in mineralogy: Berlin, Springer-Verlag, 185 p.

Snyder, R.L., and Bish, D.L., 1989, Quantitative analysis, *in* Bish, D.L., and Post, J.E., eds., Modern powder diffraction: Reviews in Mineralogy, v. 20, p. 101–144.

Takata, T., and Ahrens, T.J., 1994, Numerical simulation of impact cratering at Chicxulub and the possible causes of KT catastrophe [abs.], *in* New developments regarding the KT event and other catastrophes in Earth history: Houston, Texas, Lunar and Planetary Institute, LPI Contribution No. 825, p. 125–126.

Tyburczy, J.A., and Ahrens, T.J., 1986, Dynamic compression and volatile release of carbonates: Journal of Geophysical Research, v. 91, p. 4730–4744.

Vizgirda, J., and Ahrens, T.J., 1982, Shock compression of aragonite and implications for the equation of state of carbonates: Journal of Geophysical Research, v. 87, p. 4747–4758.

Manuscript Submitted October 14, 2000; Accepted by the Society March 22, 2001

Geological Society of America
Special Paper 356
2002

How strong was impact-induced CO_2 degassing in the Cretaceous-Tertiary event? Numerical modeling of shock recovery experiments

Boris A. Ivanov*
Institute for Dynamics of Geospheres, Russian Academy of Sciences, Moscow, Russia 117939
Falko Langenhorst*
Bayerisches Geoinstitut, Universität Bayreuth, D-95440 Bayreuth, Germany
Alexander Deutsch*
Institut für Planetologie, Universität Münster, D-48149 Münster, Germany
Ulrich Hornemann
Ernst-Mach-Institut, Am Klingelberg 1, D-79588 Efringen-Kirchen, Germany

ABSTRACT

The impact-induced release of CO_2 is one of the potentially most important mechanisms to explain drastic and rapid changes of the climate at the Cretaceous-Tertiary (K-T) boundary. General predictions, modeling, and laboratory experiments, however, yield questionable estimates on the amplitude of CO_2 outgassing. We present new results on shock recovery experiments with calcite. A new approach is used to better understand experimental results: the numerical simulation of the shock loading and following release of the samples. The calculations allow us to construct detailed pressure-temperature paths for different parts of a sample. Mineralogical observations, as well as modeling, indicate that at low and moderate pressures ($<\sim$70 GPa in the reverberation experiments), compact calcite develops deformation effects, such as dislocations and mechanical twins, whereas the major effect for release from high shock pressures is melting, not degassing. In porous or preheated calcite, shock and postshock temperatures are higher, resulting in a lowering of the threshold for melting and degassing. The presence of specific textures, e.g., bubbles and foamy aggregates, indicative of a mobilized gaseous phase in porous samples, in combination with the absence of CaO, is interpreted as the result of a rapid back reaction due to the high affinity of CO_2 to lime. To delimit kinetic effects seems to be of fundamental importance for a proper estimate of the CO_2 release from carbonate lithologies in the context of the Chicxulub impact event.

*E-mails: Ivanov, baivanov@online.ru; Langenhorst, falko.langenhorst@uni-bayreuth.de; Deutsch, deutsca@uni-muenster.de

Ivanov, B.A., Langenhorst, F., Deutsch, A., and Hornemann, U., 2002, How strong was impact-induced CO_2 degassing in the Cretaceous-Tertiary event?: Numerical modeling of shock recovery experiments, *in* Koeberl, C., and MacLeod, K.G., eds., Catastrophic Events and Mass Extinctions: Impacts and Beyond: Boulder, Colorado, Geological Society of America Special Paper 356, p. 587–594.

INTRODUCTION

The impact-related release of greenhouse gases is one of the potential mechanisms that could have triggered or accelerated the crisis of life at the Cretaceous-Tertiary (K-T) boundary via drastic and fast changes of the climate. Estimates for the total input of CO_2 into the atmosphere due to the Chicxulub impact event range from 2.6 10^{14} kg (Ivanov et al., 1996) to 10^{17} kg (Takata and Ahrens, 1994); the latter value is 10 times the amount of CO_2 that was estimated to be present in the Late Cretaceous atmosphere (Berner, 1994). Because anhydrite was present in the Chicxulub target, sulfur-bearing gases have been released in the context of the K-T impact (Pope et al., 1993, 1994; Ivanov et al., 1996). The type and amount of released gases control possible scenarios of the K-T biospheric perturbation; an excess of carbon dioxide would yield global warming, whereas sulfuric acid aerosol (produced from sulfur-bearing gases) would initiate global cooling (so-called "impact winter"; for reviews see Toon et al., 1997; Pope et al., 1997).

The shock decomposition of carbonate minerals and rocks has been intensively studied in experiments (see compilations in Agrinier et al., 2001; Skála et al., this volume). Minerals that definitely are the product of the thermal decomposition of shocked carbonates are mostly absent from samples from natural impact sites, and have also been only rarely observed in shock recovery experiments (Martinez et al., 1995; Skála et al., this volume). These observations are not well understood, and require additional investigations.

A reliable quantification of how much CO_2 entered the atmosphere at 65 Ma is not yet available for the following reasons. (1) Geologic factors such as thickness and composition of the platform sediments at the impact site are not well defined. (2) Contrasting experimental results for decarbonation in shock experiments range from ~10 to >60 GPa (see Agrinier et al., 2001; Skála et al., this volume). (3) There are unexplained differences between thermodynamical calculations and experimental data. (4) Clear criteria to identify shocked as well as back-reacted carbonates at terrestrial impact sites are lacking (Langenhorst and Deutsch, 1998). It is, however, commonly accepted that (1) carbonates decompose after pressure release due to high residual temperatures, (2) porosity controls shock and postshock temperatures, and hence, the threshold of CO_2 release, and (3) fast back reactions trap a significant part of the gaseous species (Agrinier et al., 2001). Here we present the results of dynamic experiments with shock-wave compression of calcite and recovery of shocked samples, and, in particular, focus on the numerical modeling of these experiments.

EXPERIMENTS

In order to better define shock metamorphism of calcite, especially its potential degassing behavior, we have performed systematic high-explosive shock experiments in the pressure range to 100 GPa. Variations in container geometries, sample porosities (single-crystal calcite versus calcite powder), and preheating temperatures were used to enhance shock and postshock temperatures. Precursor materials used were high-quality, fracture-free single-crystal calcite and synthetic, chemically pure $CaCO_3$ powder with grain sizes of a few micrometers. The conventional setups for shock experiments at variable preheating temperatures (to 600°C) were described in detail in Langenhorst and Deutsch (1994). A new design of the high-explosive setup with an open drill hole on the rear side of the container was tested, resulting in strong deformation and heating of the sample (Fig. 1). The subsequent numerical simulation focuses on the effects of this complicated setup on the thermal history of the calcite sample disk.

In addition, we have performed fast decompression experiments in a multianvil apparatus with controlled quenching of the sample. The use of a multianvil apparatus is a new experimental approach to exclusively simulate the pressure-temperature (*P-T*) path of shock decompression. Calcite powder, loaded to hydrostatic pressures of 25 GPa and temperatures to 3000 K, can now be decompressed and quenched to ambient conditions within a few seconds. This is approximately the time scale of unloading comparable to the Chicxulub impact. The first results for this type of experiment as well as for the recovery experiments were presented by Langenhorst et al. (2000).

Mineralogical characterization of recovered samples involved X-ray diffraction, optical, field-emission scanning

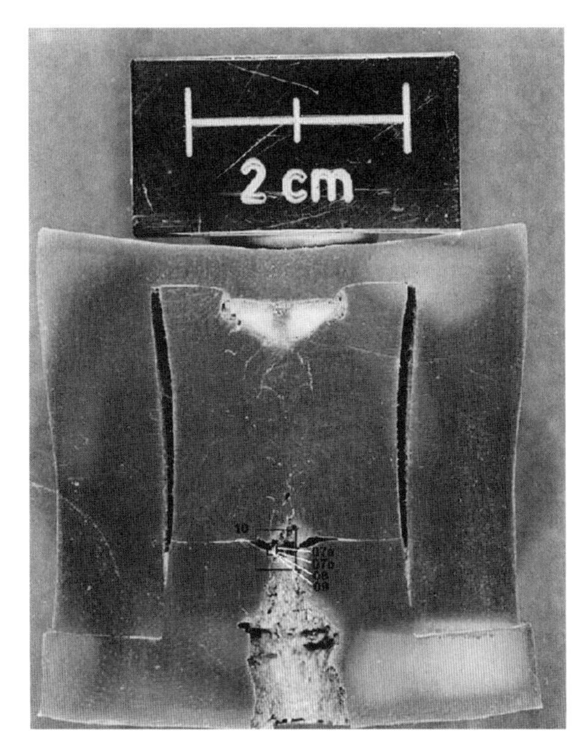

Figure 1. Sectioned ARMCO steel container with central vent hole on rear side of sample. Upper part of this hole was sealed during unloading. Final shock pressure = 73 GPa. For schematic drawing of general experimental setup, see Langenhorst and Deutsch (1994).

(SEM) and transmission electron (TEM) microscopy (Figs. 2–4). These techniques were used to detect possible decomposition and reaction products, as well as deformation effects. Mineralogical characterization of recovered, shocked calcite samples involved X-ray powder diffraction, polarizing microscopy, field-emission gun scanning and transmission electron microscopy (FEG-SEM, FEG-TEM; Bayerisches Geoinstitut, Bayreuth). These techniques were used to detect possible decomposition and reaction products, as well as deformation effects. X-ray diffraction spectra were recorded with a STOE STADIP powder diffractometer, operating with Co K_α radiation at 40 kV and 30 mA. Secondary electron images were acquired on a JEOL 6300F field emission SEM (Institute für Planetologie, Universität Münster), enabling imaging of uncoated, untreated samples at low acceleration voltages. To study the shocked samples with TEM, thin sections were prepared. After a first inspection with polarizing microscopy, these thin sections were thinned to electron transparency in an ion-milling machine, operating at 4.5 kV and an incidence angle of 13°. The microstructure and composition of shocked, thinned samples was then studied using a 200 kV Philips FEG-TEM equipped with a NORAN energy-dispersive detector.

Single-crystal calcite recovered from shock experiments with our conventional setup shows a network of irregular cracks that is most pronounced at highest pressure. Investigations with the field emission SEM (FE-SEM) reveal the presence of numerous bubbles and vesicles on the surface of cracks with diameters in the submicrometer range (Fig. 4). Using TEM we detect numerous dislocations and mechanical twins in the interior of calcite shocked at pressures of 64 and 85 GPa (Fig. 2). However, initially porous calcite samples, shocked to identical pressures, developed spherically shaped foamy aggregates, which under TEM are defect free (Fig. 3). This suggests complete shock melting, and partial, local mobilization of a gaseous species. X-ray diffraction and TEM microanalyses failed, however, to detect CaO in these samples, the expected solid residue in case of CO_2 release, which may indicate that fast back reaction has taken place. Experiments performed with the modified experimental setup yielded similar results. The shock and unloading process caused drastic deformation and melting in the sample disk, and resulted in a sealing of the container vent hole. Melting occurred in the central part of the disk, followed by crystallization into fine-grained, defect-free crystals. Deformation along shear planes has produced numerous mechanical twins in the margins of the disk. Fast decompression experiments in the multianvil apparatus corroborate these results. The TEM analysis shows a significant coarsening of grains (>10 μm) in the initially fine grained (<5 μm) calcite powder; mechanical twinning is also in the more coarse-grained calcite aggregate. Solid residues of $CaCO_3$ degassing have not been discovered in this type of experiment.

NUMERICAL SIMULATIONS OF EXPERIMENTS

The lack of unambiguous signs for degassing in the recovered single-crystal calcite samples demands a more intense investigation of the kinematic and thermodynamic processes that act during the experimental shock compression and unloading. Therefore, we performed numerical modeling of the real experiments. Previously we used hydrocodes mostly to simulate whole-scale impact crater formation (e.g., Ivanov, 1994; Ivanov et al., 1996; Ivanov and Deutsch, 1999). Here we present the first results of numerical modeling of shock compression laboratory experiments, which all were done with the reverberation technique. Details of the numerical simulations are presented in the Appendix.

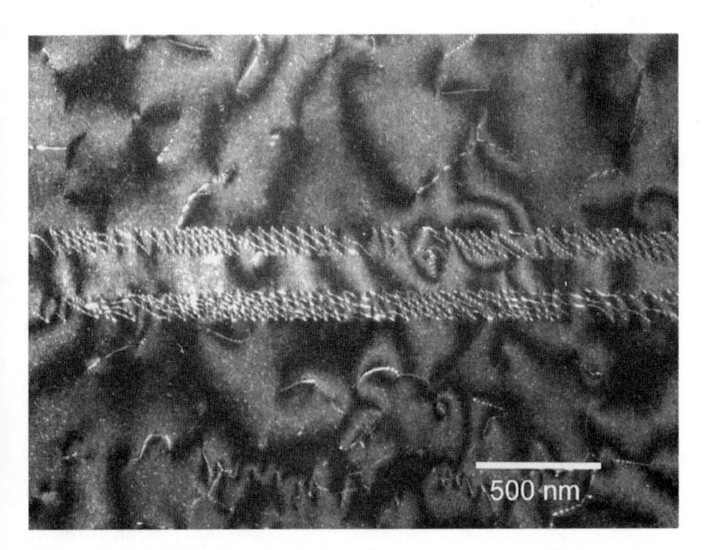

Figure 2. Weak beam transmission electron microscope image of partial and perfect dislocations and mechanical twin in single-crystal calcite shocked at 73 GPa.

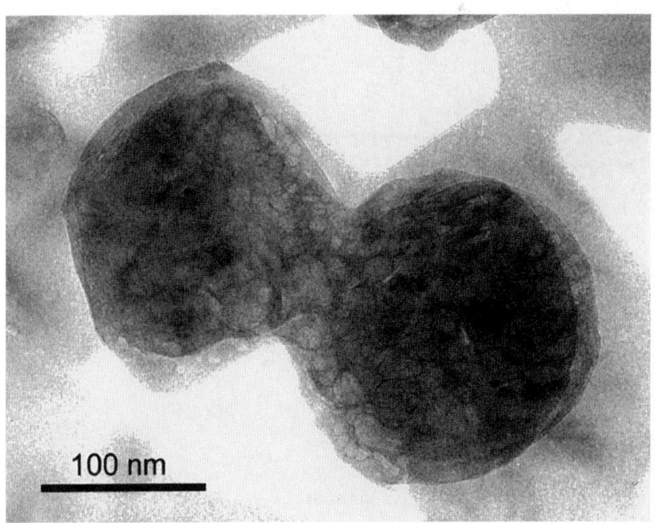

Figure 3. Bright field transmission electron microscope image of spherically shaped, foamy calcite aggregates. Starting material was porous calcite powder shocked at 85 GPa.

The *P-T* paths in the shock experiments are modeled using the heavily modified SALE (simplified arbitrary Lagrangian and Eulerian) code (Amsden et al., 1980; Ivanov et al., 1997; Ivanov and Deutsch, 1999). This code computes problems of several materials in the Eulerian mode taking into account material strength. We used the equation of state for calcite with constant Grüneisen parameters for each material ($CaCO_3$, CaO, and CO_2) as derived by Bobrovsky et al. (1976) (see Ivanov and Deutsch, 2001, for details).

The simulations take into account the geometries of the ARMCO (American Rolling Mill Corporation) iron sample container, and the calcite sample disk that is totally enclosed in this container. The upper part of the container is treated as flyer plate (Fig. 5). Model runs yield the *P-T* evolution in the experiment: during shock reverberation; mainly the first wave heats the target. For example, the first wave creates a shock pressure of 41 GPa in the calcite disk, and gives a shock temperature of 1128 K. The following reverberations compress the specimen to 85 GPa but add only 70 K to the shock temperature, obviously insufficient for decomposition after pressure release (cf. Irving and Wyllie, 1973; Martinez et al., 1995; Jones et al., 2000; Ivanov and Deutsch, 2001). For comparison, the single 85 GPa shock wave has a shock temperature of 2700 K, enough to release the sample into the decomposition field (Martinez et al., 1995; Gupta et al., 2000).

The new setup of the sample container with a vent hole at the rear side (Fig. 1) changes the situation: the lower central part of the calcite specimen is injected into the collapsing hole. We test the influence of the vent hole first in Lagrangian mode (see Appendix). The numerical modeling shows that the plastic flow of calcite at the beginning of the jet injection in the hole creates additional heating of the material. Numerical modeling in the Eulerian mode reveals a new effect of the following implosion; i.e., when calcite is injected into the vent hole, the shock wave in the iron container continues to propagate. The compressed iron begins to flow to the center of the vent hole, closing it. In the computation, this effect appears like a cylindrical implosion. Consequently, calcite, injected to the vent hole undergoes a secondary compression and heating event. The *P-T* trajectory of this secondary (implosion) heating is shown in Figure 6. Due to the solid lower boundary in our calculations, calcite cannot decompose, because there is no space for the expansion. In the real experiments, a small part of calcite has a chance to decompose by expanding into the vent hole.

EXPAND TO DECOMPOSE

The first shock experiments with carbonates revealed that the degree of decomposition depends on the ambient CO_2 pressure (Tyburczy and Ahrens, 1986). In terms of thermodynamic equilibrium, the important issue in the assessment of shock-related CO_2 release is the pressure-volume (*P-V*) relationship at the decomposition boundary. By definition, the decomposition boundary of $CaCO_3$ is defined by the loci of points in a *P-T* plane where solid or liquid $CaCO_3$ coexists in thermodynamic equilibrium with its decomposition products CaO and CO_2. Depending on the extent of decomposition, the same *P-T* points are occupied by original $CaCO_3$ (0% of the reaction) as well as by CaO and CO_2 (complete decomposition). At low pressures, expansion is required due to the large volume of CO_2, whereby the extent of decomposition (0%–100%) controls the

Figure 4. Photomicrograph of surface of calcite single crystal shocked at 73 GPa in conventional high-explosive setup (cf. Langenhorst and Deutsch, 1994). Note numerous cracks and bubbles on surface. Uncoated sample; field emission scanning electron microscope JEOL 6300F; operating conditions: 1 kV acceleration voltage, 60 pA beam current.

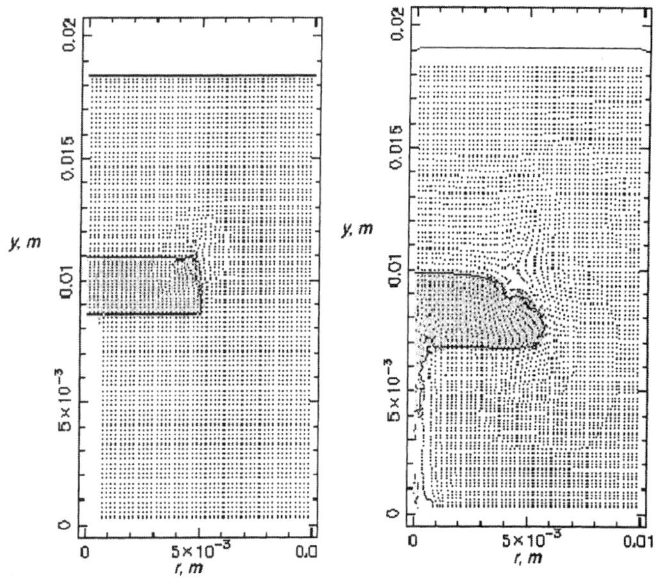

Figure 5. Time frames of numerical simulation of experiment shown in Figure 1. Calcite sample (light gray) is compressed inside ARMCO iron sample holder (dark gray dots). Left axis is axis of cylindrical symmetry. Calcite jet in vent hole can be seen in right panel.

corresponding change in specific volume. If the decomposition occurs along an isotherm, the pressure will remain constant all the time when the reaction proceeds from 0% to 100%. The sole parameter of the system that changes is volume (see Ivanov and Deutsch, 2001, for a more elaborate discussion).

Figure 6 depicts the field of calcite decomposition in the *P-V* plane according to Bobrovsky et al. (1976). The decarbonation reaction follows isotherms from 0% to 100% of decomposition. To reach a given level of decomposition, the calcite specimen needs enough volume to allow expansion of gaseous CO_2; otherwise the specimen remains close to the initial density of the calcite. At each *P-T* point at the decomposition boundary, pristine calcite coexists with its decomposition products CaO and CO_2. Consequently, in equilibrium, the percentage of calcite decomposition depends only on the volume available for the expansion of CO_2. This point is also crucial in the assessment of decarbonation in natural impacts, during release after shock compression, because the completeness of the decomposition relies on the rate of CO_2 diffusion out of the place of release. Kinetics may limit the percentage of degassing that is finally reached.

The CO_2 needs to expand dramatically to reach an observable level of decomposition: this explains why it is difficult to observe decomposition in small-scale laboratory experiments. For example, the *P-V* relations for specimens degassed at 1% and 10% of the initial mass are shown in Figure 7. At normal atmospheric pressure (1 bar = 0.0001 GPa), only 1% of decomposition occurs if carbon dioxide has a volume 40 times larger than the initial sample volume.

For the temperature range of 1800–2000 K, predicted for the calcite jet in the vent hole experiment (Figs. 5 and 6), the *P-V* diagram suggests that CO_2 should expand 10–20 times to produce total decomposition of calcite. The lack of such a large free volume may explain why we failed to observe mineralogical signatures of outgassing in this experiment. If, however, the strength of the container is lower than in the described case, carbonates may decompose totally in laboratory-scale experiments (see Skála et al., this volume).

DISCUSSION AND CONCLUSIONS

The combined experimental and modeling results of our study indicate that (1) specific designs are required to decompose carbonates in small-scale laboratory shock experiments, and (2) the major shock effect of calcite at high pressure is melting, not degassing. (3) The necessity for a large free volume to allow CO_2 expansion was realized in some single-shot experiments with carbonates that were mounted in evacuated target chambers (e.g., Tyburczy and Ahrens, 1986). It remains uncertain if such a setup is a realistic approximation of processes occurring in natural impact events. (4) At low shock pressures, activation of dislocations and mechanical twins is dominant. The presence of specific textures, e.g., bubbles and foamy aggregates, indicative of a mobilized gaseous phase

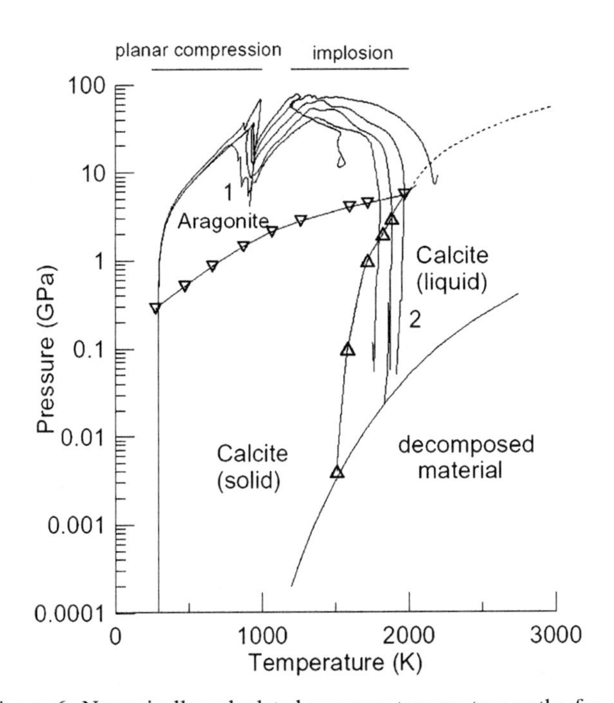

Figure 6. Numerically calculated pressure-temperature paths for selected particles in shock experiment (Fig. 1). Main phase curves for calcite shown for comparison are after Irving and Wyllie (1973) and Huang and Willie (1976). (See Ivanov and Deutsch, 2001, for review and update of calcite phase diagram.)

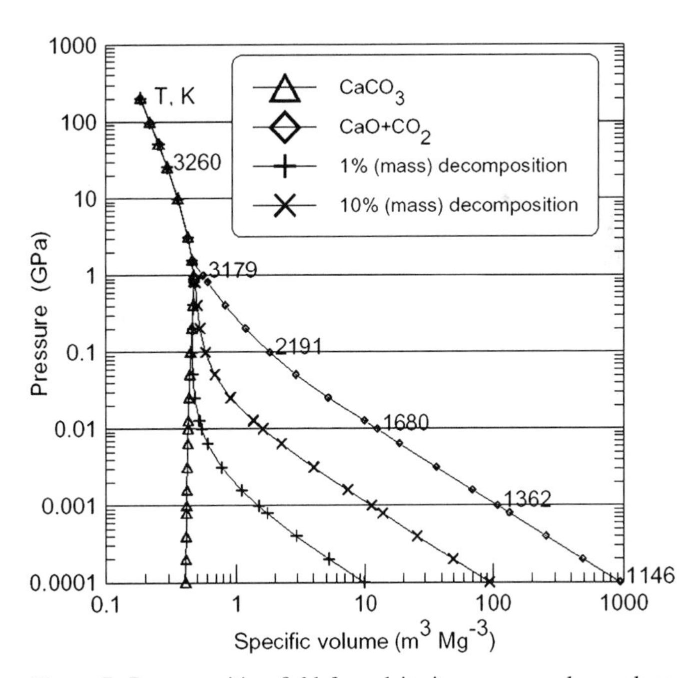

Figure 7. Decomposition field for calcite in pressure-volume plane. Decomposition reaction follows isotherms from 0% to 100% of decomposition. To reach given level of decomposition, specimen should have enough volume to expand CO_2 gas; otherwise, specimen will stay close to initial calcite density.

combined with the absence of CaO may indicate a rapid back reaction that occurred due to the high affinity of CO_2 with CaO.

The numerical modeling of experiments helps us to understand details of shock loading and release processes. Computations confirm that a pressure of 70 GPa, reached by shock reverberation, does not heat calcite to shock temperatures, and hence, postshock temperatures, required for massive CO_2 production. The vent hole in the sample holder results in the new effect of an additional heating of those parts of the calcite sample disk that are injected into this hole. This secondary heating brings the material into the thermodynamic field where thermal decomposition is possible. However, the volume available for gas expansion is too small to allow outgassing of calcite in a quantity sufficient to be detected by the employed mineralogical methods. Despite this rather unsatisfying result of the first test, the new design with the vent hole suggests a new direction for future experiments.

Skála et al. (this volume) report on the catastrophic disintegration of sample containers of dolomite shocked at ~70 GPa, and interpret this disintegration as a result of excessive vapor production of the carbonate target. This statement is enforced by the fact that in similar experiments with fluid-free and/or fluid-poor silicates, the containers did not break, allowing recovery of the samples. Our numerical modeling shows that properties of a sample holder are also affected by the shock loading. The strength of an experimental assembly is not a single material parameter, but a complex function of material parameters, changed by shock loading, and geometry of an experiment. Hence, container destruction alone is not clear-cut proof of total decomposition of the carbonate sample. Numerical modeling may help to interpret the experimental results. Regardless of what really caused the failure of the container, the pressure of 70 GPa for massive CO_2 release from dolomite (Skála et al., this volume) is larger than the threshold estimates based on thermodynamic equilibrium calculations for dolomite (Martinez et al., 1995). Conditions for effective calcite degassing in large-scale impacts, such as the Chicxulub (K-T) cratering event, need to be reevaluated in this context. The requirement of separating CaO and CO_2 in space prompts the question of whether the thermodynamic equilibrium is a valid assumption. Recent experiments with the reverse reaction (CaO + $CO_2 \Rightarrow CaCO_3$) by Agrinier et al. (2001) and Deutsch et al. (1998) show that diffusion of CO_2 into hot CaO grains occurs on time scales of ~100 s. A similar time scale may be valid in direct decomposition. To define this kinetic effect seems to be of fundamental importance for proper estimates of CO_2 production due to the Chicxulub (K-T) event. We conclude that previously published values for the amount of CO_2, which should have entered the atmosphere in the aftermath of this impact, are too high.

ACKNOWLEDGMENTS

This study was supported by German Science Foundation grants DE 401/15, HO 1446/3, LA 830/4, 436 RUS 17/20/99, and 436 RUS 17/34/00. Part of the work was done when Ivanov was a visiting professor at the Institute für Planetologie, Universität Münster. We thank F. Bartschat, T. Grund, U. Heitmann, R. Thewes, H. Heying, M. Feldhaus (Münster), and H. Schulze (Bayreuth) for technical assistance, and R. Skála, A. Jones, and C. Koeberl for constructive reviews and editing.

APPENDIX: NUMERICAL MODELING OF LABORATORY EXPERIMENTS

To test usefulness of the modified SALE (simplified arbitrary Lagrangian and Eulerian) code, three problems were calculated: (1) one-dimensional shock compression, (2) two-dimensional compression of a calcite disk inside the ARMCO iron sample holder, and (3) two-dimensional compression of a calcite disk inside the ARMCO iron sample holder with a vent hole at the back side of the sample.

The geometry of the two-dimensional computational zone reproduces the experiment with shock compression of the calcite disk, 10 mm in diameter and 3 mm thick, at a final pressure of 73 GPa.

Equation of state and mechanical parameters

Tillotson equation of state and temperature estimates. Despite the presence of modern sophisticated equations of state, such as ANEOS (analytical equation of state), the simple analytical equation of state, as proposed by Tillotson (1962) for metals, is still widely used. The Tillotson EOS does not contain an explicit expression for temperature. As for most hydrocodes, pressure is computed via density, ρ, and specific internal energy, e; the absence of the explicit temperature calculations is not a problem. However, as soon as mechanical parameters of materials are treated as a function of temperature (e.g., melting or thermal decomposition), one needs to compute temperature simultaneously with pressure. We have added temperature estimates to the Tillotson EOS in this work.

The thermal energy was computed by subtraction of the cold energy of the total specific internal energy (SIE). The cold energy, e_c, is computed with the same Tillotson EOS via integration of the usual equation for cold energy:

$$de_c/dV = -P(\rho, e_c), \tag{1}$$

where V is a specific volume ($V = 1/\rho$, where ρ is density), T is temperature, and pressure P is computed with the Tillotson EOS. Equation 1 is integrated once for each material of interest. Results are stored in the computer memory as an array of values $e_c(\rho_i)$. During the code run the EOS subroutine computes a value interpolating the e_c value for a given density ρ in the interval $\rho_{i+1} > \rho > \rho_i$. The subtraction of the cold energy from the current value of specific internal energy gives the estimate of thermal internal energy:

$$e_{th} = e - e_c. \tag{2}$$

The temperature may be estimated for a known thermal energy with a standard Debye approach. A simple estimate of the temperature may be calculated as

$$T = T_0 + (e_{th} - e_{th0})/c, \tag{3}$$

where T_0 is initial (ambient) temperature, e_{th0} is thermal energy at $T = T_0$, and c is the effective heat capacity for a temperature range of interest.

ARMCO iron. The hydrostatic (volumetric) compression of the ARMCO iron is described with the Tillotson EOS for iron (Melosh, 1989). Main parameters are: density $\rho_{0I} = 7900$ kg m^{-3}, and Tillotson EOS coefficients $A = 128$ GPa, $B = 105$ GPa, and $E_0 = 9.5$ MJ kg^{-1}.

For temperature calculations, the constant heat capacity coefficient of 0.6 kJ kg^{-1} K^{-1} is used. The strength (plastic limit) is assumed to depend on plastic deformations, ρ, and temperature in accordance with the Johnson and Cook (1983) model:

$$\sigma = (E + F\,\varepsilon^n)(1 - T*^m), \quad (4)$$

where $E = 175$ MPa, $F = 280$ MPa, $n = 0.32$, $m = 0.55$, and $T*$ is a reduced temperature showing how close is material to the melting temperature, T_{melt}:

$$T* = (T - T_0)/(T_{melt} - T_0). \quad (5)$$

Calcite. Two kinds of EOS are used: the Tillotson EOS for limestone (with parameters listed by Melosh, 1989) and the theoretical EOS for calcite derived by Bobrovsky et al. (1976). The Bobrovsky EOS includes a direct temperature estimation, and the Tillotson EOS temperature is computed in the same way as for iron, with a constant value of heat capacity of 1.0 kJ $kg^{-1}K^{-1}$.

Deviatoric stresses in iron and calcite are computed with the Johnson and Cook (1983) model for metals (JC-M) and Johnson and Holmquist (1993) model for brittle solids (JH-B). For simple tests, the constant plastic yield of 1.5 GPa for iron and 1 GPa for calcite is used.

The geometry includes the ARMCO iron sample container, 4 cm in diameter and 4 cm in height, and a 5 mm (or 10 mm) thick flyer plate (upper surface layer of the sample container). The flyer plate velocity of 3 kms^{-1} provides a pressure impulse of 73 GPa in the container. The calcite disk is situated at the depth of 5 mm (or 10 mm) under the contact container surface.

One-dimensional test. One-dimensional computations facilitate quick reconnaissance of pressure and temperature rise with a number of shock reflections in the $CaCO_3$ disk inside the ARMCO steel container. The shock reverberates several times (the first four reflections are important for shock heating; see Martinez et al., 1995). The pressure-particle velocity (*P-U*) representation is shown in Figure A1. Due to the shock reverberation, pressure increases almost twice, while the temperature rises only 70 K after the passage of the first shock wave.

The general parameters during the loading-unloading cycle are shown in Figure A2. It is possible to see four jumps of pressure, density, and temperature, followed by gradual unloading when the rarefraction wave reaches the sample position from the rear side of the flyer plate. Note that plastic strain increases at least twice during unloading. It is important to remember that more than half of a residual strain may accumulate during the unloading phase at relatively small pressures. The unloading path depends on the sample-container geometry and may be different for different experiments.

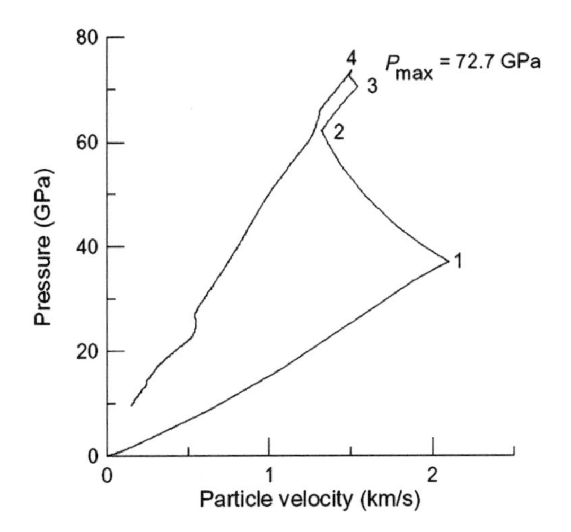

Figure A1. Calculated loading and unloading path in central point of sample in pressure-particle velocity representation. Numbers of reflections demonstrate that maximum pressure is reached by four reflected waves.

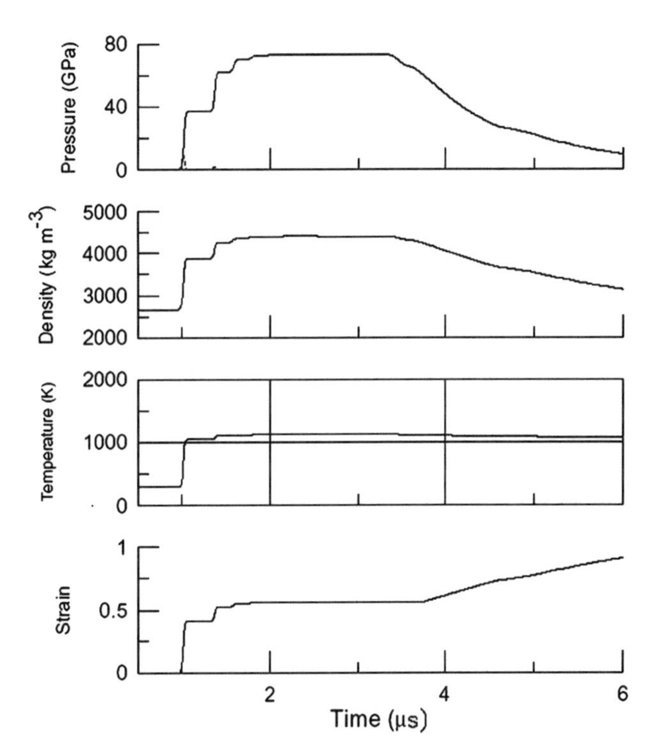

Figure A2. Pressure, density, temperature, and plastic (permanent) strain during reverberative compression of calcite sample at maximum pressure of 73 GPa. Due to shock reverberation pressure increased almost twice, while temperature rose only 70 K after first shock passage.

Figure A3 shows the pressure-temperature dependence for the central point of a sample in comparison with the thermal decomposition curve (Tyburczy and Ahrens, 1986). It seems that the average shock heating due to shock reverberation to a pressure of 73 GPa cannot heat the sample above the threshold for degassing (~1150 K at normal pressure). The iron is heated ~400 K above the preimpact temperature (293 K).

REFERENCES CITED

Agrinier, P., Deutsch, A., Schärer, U., and Martinez, I., 2001, Fast back-reactions of shock-released CO_2 from carbonates: An experimental approach: Geochimica et Cosmochimica Acta, v. 65, n. 15, p. 2615–2632.

Amsden, A.A, Ruppel, H.M, and Hirt, C.W., 1980, SALE: A simplified ALE computer program for fluid flow at all speeds: Los Alamos, New Mexico, Los Alamos National Laboratory, Report LA-8095, 101 p.

Berner, R.A., 1994, GEOCARB II: A revised model for atmospheric CO_2 over Phanerozoic time: American Journal of Science, v. 294, p. 56–91.

Bobrovsky, S.V., Gogolev, V.M., Zamyshlyaev, B.V., Lozhkina, V.P., and Rasskazov, V.V., 1976, The study of thermal decomposition influence on the spallation velocity for strong shock waves in solids: Fiziko-Technicheskie Problemy Razrabotki Poleznykh Iskopaemyh (Physical and technical problems of mining), v. 3, p. 49–57 (in Russian).

Deutsch, A., Schärer, U., and Agrinier, P., 1998, Evidence for back-reacted carbonates in distant ejecta of the Chicxulub impact event? An experimental approach [abs.]: Lunar and Planetary Science, v. 29, abstract #1386 (CD-ROM).

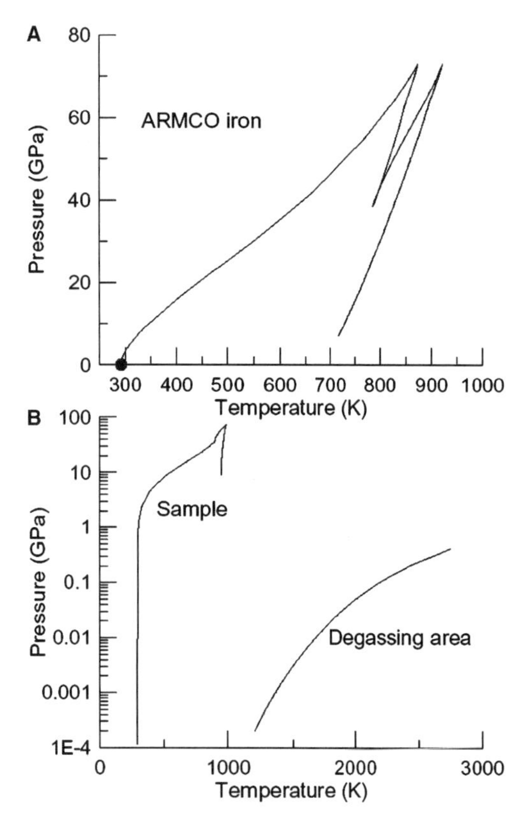

Figure A3. Pressure-temperature dependence for iron (top) and calcite (bottom). Temperature of unloaded calcite at 73 GPa is below degassing limit at pressure = 1 bar.

Gupta, S.C., Love, S.G., and Ahrens, T.J., 2000, Shock temperatures in calcite ($CaCO_3$): Implication for shock induced decomposition, *in* Furnish, M.D., Chhabildas, L.C., and Hixson, R.S., eds., Shock compression of condensed matter: Woodbury, New York, American Institute of Physics, p. 1263–1266.

Huang, W.L., and Wyllie, P.J., 1976, Melting relationships in the systems CaO-CO_2 and MgO-CO_2 to 33 kbars: Geochimica et Cosmochimica Acta, v. 40, p. 129–132.

Irving, A.J., and Wyllie, P.J., 1973, Melting relationships in CaO-CO_2 and MgO-CO_2 to 36 kilobars with comments on CO_2 in the mantle: Earth and Planetary Science Letters, v. 20, p. 220–225.

Ivanov B.A., 1994, Geomechanical models of impact cratering: Puchezh-Katunki structure, *in* Dressler, B.O., Grieve, R.A.F., and Sharpton, V.L., eds., Large meteorite impacts and planetary evolution: Geological Society of America Special Paper 293, p. 81–91.

Ivanov, B.A., and Deutsch, A., 1999, Sudbury impact event: Cratering mechanics and thermal history, *in* Dressler, B.O., and Grieve, R.A.F., eds., Large meteorite impacts and planetary evolution II: Geological Society of America Special Paper 339, p. 389–397.

Ivanov, B.A., and Deutsch, A., 2001, Calcite phase diagram in relation to shock decomposition [abs.]: Lunar and Planetary Science, v. 32, abstract #1740 (CD-ROM).

Ivanov, B.A., Badjukov, O.I., Yakovlev, M.I., Gerasimov, M.V., Dikov, Yu. P., Pope, K.O., and Ocampo, A.C., 1996, Degassing of sedimentary rocks due to Chicxulub impact: Hydrocode and physical simulations, *in* Ryder, G., Fastovsky, D., and Gartner, S., eds., The Cretaceous-Tertiary event and other catastrophes in Earth history: Geological Society of America Special Paper 307, p. 125–139.

Ivanov, B.A., DeNiem, D., and Neukum, G., 1997, Implementation of dynamic strength models into 2D hydrocodes: Applications for atmospheric breakup and impact cratering: International Journal of Impact Engineering, v. 20, p. 411–430.

Johnson, G.R., and Cook, W.C., 1983, A constitutive model and data for metals subjected to large strains, high strain rates and high temperatures: Seventh International Symposium on Ballistics, The Hague, The Netherlands, April 1983, p. 1–7.

Johnson, G.R., and Holmquist, T.J., 1993, An improved computational constitutive model for brittle materials, *in* Schmidt, S.C., Shaner, J.W., Samara, G.A., and Ross, M., eds., High-Pressure Science and Technology, Proceedings of the Joint International Association for Research and Advance of High Pressure and Technology, American Institute of Physics Press, p. 981–984.

Jones, A.P., Claeys, P., and Heuschkel, S., 2000, Impact melting of carbonates from the Chicxulub crater, *in* Gilmor, I., and Koeberl, C., eds., Impacts and the early Earth: Berlin-Heidelberg, Springer Verlag, p. 343–362.

Langenhorst, F., and Deutsch, A., 1994, Shock experiments on pre-heated alpha- and beta-quartz. 1. Optical and density data: Earth and Planetary Science Letters, v. 125, p. 407–420.

Langenhorst, F., and Deutsch, A., 1998. Mineralogy of astroblemes: Terrestrial impact craters, *in* Marfunin, A.S., ed., Advanced mineralogy, vol. 3, Mineral matter in space, mantle, ocean floor, biosphere, environmental management, jewelry, Chapter 1.10: Berlin, Springer-Verlag, p. 95–119.

Langenhorst, F., Deutsch, A., Ivanov, B.A., and Hornemann, U., 2000, On the shock behavior of $CaCO_3$: Dynamic loading and fast unloading experiments–Modeling–Mineralogical Observations [abs.]: Lunar and Planetary Science, v. 31, abstract #1851 (CD-ROM).

Martinez, I., Deutsch, A., Schärer, U., Ildephonse, P., Guyot, F., and Agrinier, P., 1995, Shock recovery experiments on dolomite and thermodynamical modeling of impact induced decarbonatation: Journal of Geophysical Research, v. 100, p. 15465–15476.

Melosh, H.J., 1989, Impact cratering: A geologic process: New York, Oxford University Press, 245 p.

Pope, K.O., Ocampo, A.C., Baines, K.H., and Ivanov, B.A., 1993, Global blackout following the K-T Chicxulub impact: Results of impact and atmospheric modeling [abs.]: Lunar and Planetary Science, v. 24, p. 1165–1166.

Pope, K.O., Baines, K.H., Ocampo, A.C., and Ivanov B.A., 1994, Impact winter and the Cretaceous/Tertiary extinctions: Results of a Chicxulub asteroid impact model: Earth and Planetary Science Letters, v. 128, p. 719–725.

Pope, K.O, Baines, K.H., Ocampo, A.C., and Ivanov, B.A., 1997, Energy, volatile production, and climatic effects of the Chicxulub–Cretaceous/Tertiary impact: Journal of Geophysical Research, v. 102, p. 21645–21664.

Takata, T., and Ahrens, T.J., 1994, Numerical simulation of impact cratering at Chicxulub and the possible causes of KT catastrophe [abs.]: LPI Contribution 825, Houston, Texas, Lunar and Planetary Institute, p. 125–126.

Tillotson, J.H., 1962, Metallic equations of state for hypervelocity impact: San Diego, California, Advanced Research Project Agency, General Atomic Report GA-3216, 141 p.

Toon, O.B., Turco, R.P., Covey, C., Zahnle, K., and Morrison, D., 1997, Environmental pertubations caused by the impacts of asteroids and comets: Reviews of Geophysics, v. 35, p. 41–78.

Tyburczy, J.A., and Ahrens, T.J., 1986, Dynamic compression and volatile release of carbonates: Journal of Geophysical Research, v. 91, p. 4730–4744.

Manuscript Submitted December 11, 2000; Accepted by the Society March 22, 2001

Geological Society of America
Special Paper 356
2002

Laboratory impact experiments versus natural impact events

Paul S. DeCarli*
*SRI International, 333 Ravenswood Avenue, Menlo Park, California 94025, USA, and
Department of Geological Sciences, University College London,
Gower Street, London WC1E 6BT, UK*
Emma Bowden
Adrian P. Jones
G. David Price
*Department of Geological Sciences, University College London,
Gower Street, London WC1E 6BT, UK*

ABSTRACT

Laboratory studies of shock metamorphism have long provided a basis for recognition of the diagnostic features of ancient terrestrial impact craters. However, there are significant differences between the range of parameters accessible in laboratory impact experiments and the conditions of large natural impact events. The basic premise of the laboratory calibrations is that peak pressure is the most significant parameter governing shock metamorphism. To show that other parameters are important, the shock formation of diamond is discussed in detail. Shock-induced phase transitions in silicates are discussed in the framework of current efforts to infer possible kinetic effects. In an effort to encourage critical examination of the literature, we call attention to the characteristics and limitations of experimental techniques. Particular emphasis is placed on the value of thoroughly documenting both the details of shock-loading experiments and the assumptions underlying shock calculations, to permit eventual reassessment of the results in the light of new information.

INTRODUCTION

This chapter is the outgrowth of a conversation in 1995 between one of us (Paul DeCarli) and Gene Shoemaker (we collaborated on research between about 1958 and 1963 before our careers diverged). The topic of conversation was the current lack of interaction between the geology-oriented investigators of shock metamorphism and the physics-oriented investigators of shock-wave propagation in solids. On the one hand, one must persuade the geologists to present their work in a form that is more useful to physicists. Experimental papers on shock metamorphism frequently do not describe the shock-loading aspects of the experiments in sufficient detail to permit faithful reproduction of the experimental conditions. Such papers do not appeal to experimental physicists, and they are also of limited usefulness to the computational physicists who engage in detailed modeling of experiments. On the other hand, one must convince the physicists that the efforts of the geologists may be important to the solution of some long-standing problems in shock physics. For example, shock-physics investigators have long been concerned that shock-wave propagation in geological materials could be modified by the kinetics of phase transformations. To date, kinetic effects have not observed to operate on the microsecond time scale of Hugoniot measurements.

The question is whether kinetic effects could be important on longer time scales, from the millisecond domain of a small terrestrial impact (Meteor Crater) to the multisecond domain of a large impact (Chicxulub). We present evidence indicating that

*E-mail: pdecarli@unix.sri.com

DeCarli, P.S., Bowden, E., Jones, A.P., and Price, G.D., 2002, Laboratory impact experiments versus natural impact events, *in* Koeberl, C., and MacLeod, K.G., eds., Catastrophic Events and Mass Extinctions: Impacts and Beyond: Boulder, Colorado, Geological Society of America Special Paper 356, p. 595–605.

the question may be answered by evaluation of the evidence contained in the shock-metamorphosed rocks of impact craters. Herein we address geologists; a paper currently in preparation for an American Physical Society meeting will address the shock physicists.

In order to properly address possible kinetic effects in natural impacts, possible discrepancies between small-scale laboratory experiments and large-scale natural impacts must be considered. In addition, we must address the commonly held view that shock-wave effects are unique and different from the effects of static high pressure. One must also confront the view that peak shock pressure is the most important parameter in the description of a shock event.

This chapter is not intended as a critical review of prior work on shock metamorphism. Such a review would have to be book length to do justice to the extensive research of the past 40 years. For the most part, the papers that were cited were chosen because they contained material that was directly relevant to our arguments; however, the cited papers contain numerous references to prior work. The emphasis of this chapter is on phase transitions, which may be governed by kinetic factors. We therefore do not discuss those shock-metamorphic features that may depend primarily on peak shock pressure.

The section on diamond is based largely on research, including hundreds of shock-recovery experiments, performed between 1958 and 1970 by one of us (DeCarli). Much of this work could not be published prior to expiration of the patents because of commercial considerations. Here we attempt to collect the information most useful to students of natural impact diamond; we do not attempt to fully describe commercial shock synthesis of diamond.

The discussion of shock-diamond synthesis has multiple purposes: it is intended to illustrate the complexity of shock synthesis and the importance of details such as loading path and the crystallinity of the carbon from which the diamond is synthesized. The agreement between static high-pressure results and shock results is noted. Shock-synthesized diamond is present in a number of large impact craters. The fact of diamond survival permits us to set bounds on postshock temperature of the surrounding rock and thus indirectly infer kinetic effects on the response of the rock. We deliberately do not cite the extensive literature on studies of natural impact diamond. We did not wish to cite any paper without relating it to the laboratory studies; we chose to cite several short papers in order to present brief examples of the relationship between nature and experiment. The cited papers contain adequate references to the main body of work on natural impact diamond.

BACKGROUND

Much of our current knowledge of shock-metamorphic effects on minerals is based on the results of laboratory shock-loading experiments, including both equation-of-state measurements and recovery experiments. Some recent studies of naturally shocked material, including both meteorites and material from impact craters, have been controversial because of implied disagreement with the results of the laboratory studies. We suggest that the disagreements may stem partially from the differences between laboratory experiments and large natural impacts.

The effective duration of a typical laboratory shock-loading experiment is of the order of a microsecond. In a large natural impact resulting in a 100-km-diameter crater, the effective high-pressure duration at a depth of 7 km will be $\sim$1 s, the approximate transit time of a relief wave from the surface. The relevance of the laboratory experiment will depend on whether kinetic effects are important to the development of particular shock-metamorphic features.

Another major difference between laboratory and nature is the postshock cooling history. Laboratory shock experiments are conducted on a small scale; samples cool within seconds. The effective cooling time of the shock-heated material in a large impact crater may exceed 10^5 yr. Shocked meteorites may have a complex history of repeated impacts.

In consideration of these differences, one might expect to find occasional discrepancies between observations on naturally shocked rocks and the results of laboratory peak-shock-pressure calibration experiments. We do not imply that laboratory shock-loading experiments are generally irrelevant nor do we denigrate the excellent work that has been done by the majority of our colleagues.

The basic premise of most shock-pressure calibration studies is that observable shock-metamorphic effects are primarily a function of peak shock pressure. This view may be correct with respect to some shock-metamorphic effects, such as the formation of planar deformation features in quartz. Other shock-metamorphic effects, in particular reconstructive phase transformations, may depend very much on other parameters. In addition to peak shock pressure, one must consider loading path, shock duration, the initial state (preshock) of the rock, the geometry of the shock event, postshock cooling history, and possibly other factors.

One must accept that the parameter space accessible to laboratory shock-loading experiments is small, in comparison to the parameter space of natural events. It may therefore be useful to also use data from static high-pressure experiments and from computational simulations in efforts to interpret the histories of impact metamorphosed rocks.

SHOCK-RECOVERY EXPERIMENTS

Figure 1 illustrates the essential components of a shock-recovery experiment. A flyer plate, accelerated by either explosives or a gun, impacts a sample-container assembly at a known (measured) velocity. The resultant pressure at the impact surface is a function of the impact velocity and the shock impedances of flyer plate and container materials. If sample and container have the same shock impedance (a matched impedance experiment), the shock wave propagates unperturbed from container through sample. Radial momentum traps and rear surface spall plates protect the sample and container from the destructive effects of rarefactions originating at free surfaces. In the ideal world, the sample container would survive without significant plastic deformation; the sample would be recovered in one piece by machining open the container. In practice, matched impedance recovery experiments are difficult because of the relatively low shock impedance of most rocks and minerals; the best impedance matches are generally low-strength materials. The usual result is that the sample and container are recovered as fragments; painstaking work under a microscope is required to pick out the sample fragments.

Some of the earliest shock-recovery experiments (DeCarli and Jamieson, 1959) were performed in a matched impedance geometry. In many experiments the sample was lost. The high impedance container geometry was developed to permit more reliable sample recovery. This shock-recovery technique was first described in 1963 (Milton and DeCarli, 1963). Initially, the sample container and momentum traps were made from ordinary low-carbon steel. The first improvement was to make the containers from austenitic stainless steel, which remains ductile over a wide range of temperatures. It then became possible to conduct experiments over a wide range of initial temperatures (from liquid helium temperature to 1000°C) at pressures to 150 GPa (limited to 100 GPa at temperatures near 1000°C). The objective of these experiments under extreme conditions was to study shock-diamond synthesis.

The great advantage of the matched impedance recovery experiment is that the loading path of the sample generally matches the loading path in a natural impact event. In an experiment with a high impedance container, the sample achieves pressure equilibrium with the container via a sequence of shock reflections. A concise description of the relevant shock-wave theory along with references to more extensive accounts can be found in Melosh (1989). The calculation of the shock history of the sample is a cumbersome numerical problem, using the nonlinear equations of state of container and sample materials and satisfying the requirement that pressure and particle velocity match at interfaces. The problem is readily solved by graphical methods. Figure 2 is a graphical calculation of the pressure of the initially transmitted shock in the sample and the pressures of subsequent shock reflections. In this example, the container material is 304 stainless steel and the sample is Coconino Sandstone. The flyer plate (impactor) is stainless steel and the impact velocity is 966 m/s. The initial shock pressure transmitted into the sandstone is ~5 GPa; essential equilibration with the 20 GPa of the container requires three or four shock reflections. The samples in matched impedance and high impedance containers can thus reach the same peak pressures, but the loading paths differ.

The increase in internal energy in a single shock is $(E_1 - E_0) = (P_0 + P_1) * (V_0 - V_1)/2$, where E, P, and V represent internal energy, pressure, and specific volume, respectively; the subscripts 0 and 1 refer to initial and final states, respectively.

The total increase in internal energy, for initial and reflected

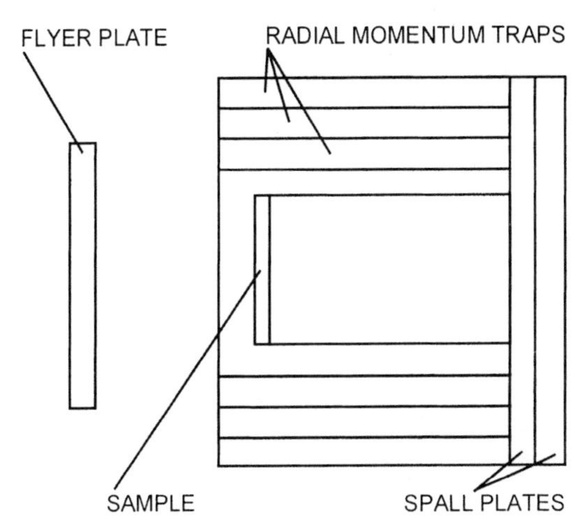

Figure 1. Illustration of essential elements of shock-recovery experiment. Disc-shaped flyer plate impacts cylindrical sample container-radial momentum trap–spall plate assembly.

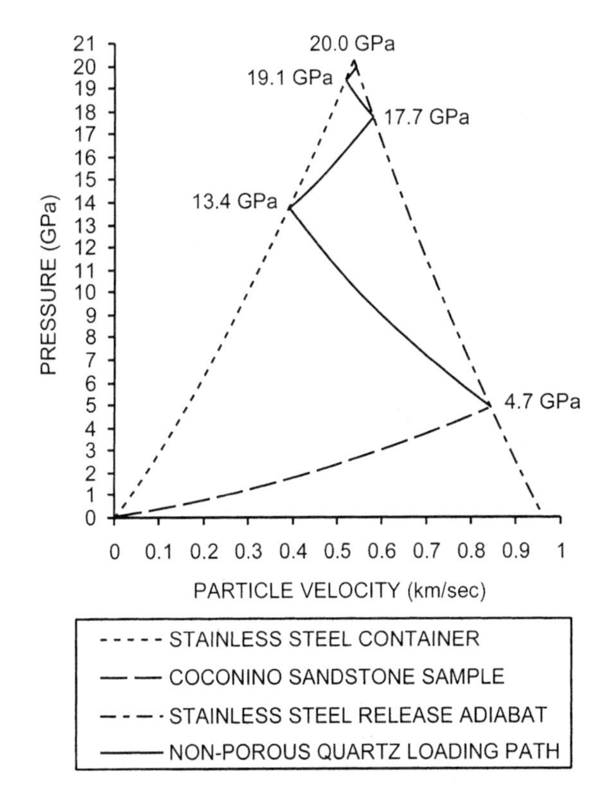

Figure 2. Graphical calculation of loading path for Coconino Sandstone sample in high-impedance (304 stainless steel) sample recovery assembly. Initial shock state, 4.7 GPa, is determined as intersection of Coconino Hugoniot with release adiabat of 304 stainless steel. First reflected shock state is determined as intersection of reflected shock Hugoniot of Coconino with Hugoniot of 304 stainless steel. Reflected shock Hugoniot of Coconino, centered on 4.7 GPa state, is approximated by Hugoniot of initially nonporous quartzite. Subsequent reflected shock states are calculated similarly.

shocks, is simply the sum of the increases calculated for each stage, as shown in Figure 3. The product of pressure and volume has the dimensions of specific energy: GPa * cm^3/g = 1000 J/g.

The total energy increase is due to both thermal and mechanical terms. The mechanical portion, which is recovered upon unloading, is equal to the area under the release adiabat. Coconino Sandstone is porous; we assume complete crush up of the porosity as a result of shock loading to 20 GPa. The path of the release adiabat is approximated by assuming release along the Hugoniot down to the intersection with the hydrostat of solid quartz; further release to ambient pressure follows the hydrostat. The thermal energy, the area enclosed by the loading path and the release adiabat, is often referred to as the waste heat. The waste heat estimate is the basis for estimating the postshock temperature of the material. The shock temperature can be estimated by calculating the adiabatic temperature increase (starting at the postshock temperature) resulting from loading along the release adiabat to the peak shock pressure. The calculation can be refined with, e.g., iterative corrections for thermal expansion, but uncertainties in specific heat ratios at high pressure can be much larger than the corrections.

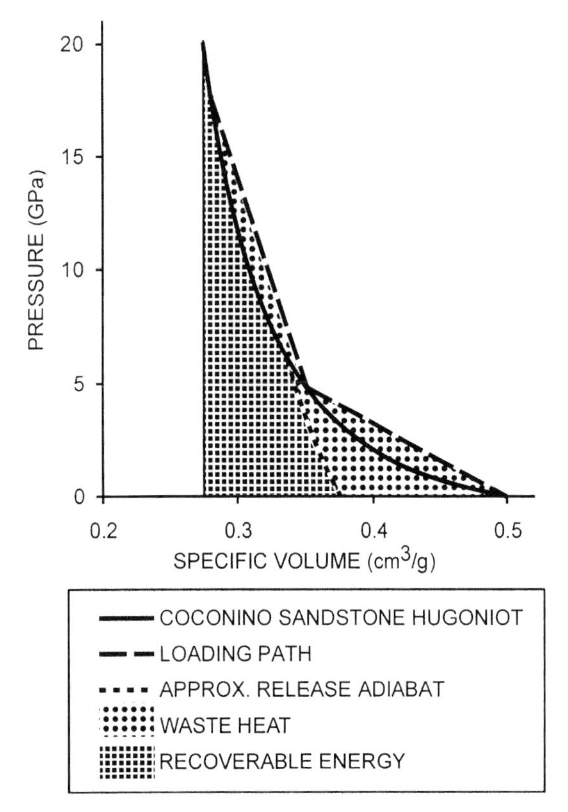

Figure 3. Graphical calculation of internal energy increase of Coconino sandstone sample shock loaded to 20 GPa via reflected shock-loading path of Figure 2. Loading path is assumed to follow Rayleigh lines, straight lines connecting successive states (pressure, volume). Internal energy increase for each stage is equal to area under Rayleigh line. Total energy increase on shock compression to given pressure is sum of internal energy increases for each shock and successive shock reflections. Recoverable energy, mechanical energy that is recovered on decompression, is equal to area under release adiabat. Net internal energy increase after decompression, often referred to as waste heat, is thus represented by area between loading path and unloading path. In this example, waste heat is ~340 J/g, corresponding to postshock temperature of ~400°C after equilibration of hotspots.

There are complications, however. It is well known that the shock compression of a porous solid results in an initial heterogeneous temperature distribution, on the scale of material heterogeneity (e.g., grain size). The evidence for thermal heterogeneity is noted in studies of the effects of natural and laboratory impacts on initially porous material. (Kieffer et al., 1976; Bowden et al., 2000). The temperature of a micrometer-sized hotspot can be thousands of degrees hotter than the average temperature. These hotspots form within nanoseconds at the shock front and thermally equilibrate more slowly, on a time scale of microseconds, in accordance with the ordinary laws of heat flow. The details of hotspot formation and thermal equilibration in shock compression of porous solids are currently being studied computationally, using micrometer cell sizes and nanosecond time steps, on a massively parallel supercomputer (Baer, 2000)

One may simply calculate the average postshock temperature after equilibration. For the example of Figure 3, the waste heat is ~340 J/g, corresponding to a postshock temperature of ~400°C. For a matched impedance experiment at the same peak pressure of 20 GPa (Fig. 4), the waste heat is 1250 J/g, corresponding to an equilibrated postshock temperature of ~1350°C. These simple calculations show that there can be a large difference between a laboratory experiment with a sample in a high impedance container and a natural impact in which the material is singly shocked to the same peak pressure.

Certain critical assumptions are made in the temperature calculation. It was assumed that the loading path followed the Rayleigh line between initial and shock Hugoniot states. The Hugoniot data of Ahrens and Gregson (1964) were used to calculate the amplitude of the initially transmitted shock and their Hugoniot data on initially nonporous quartzites were used to calculate reflected shock states. Complete crushup of porosity was assumed to calculate a release adiabat, and it was assumed that no phase transitions occurred on either loading or unloading. These appear to be reasonable assumptions, based on available data, for a microsecond-duration shock-loading event. The question addressed in the section on kinetics is whether these assumptions may be questioned in the instance of a longer duration, millisecond to second, shock-loading event.

One may ask whether a high impedance container experiment with a preheated sample might serve as a proxy for an experiment in a matched impedance recovery geometry. This was certainly not the case in our experiments on shock synthesis of diamond. Matched impedance experiments with a porous (1.57 g/cm^3) commercial graphite at ~20 GPa yielded more than 25% diamond. Experiments with the same material in high impedance containers at 20 GPa yielded only traces (~0.1%)

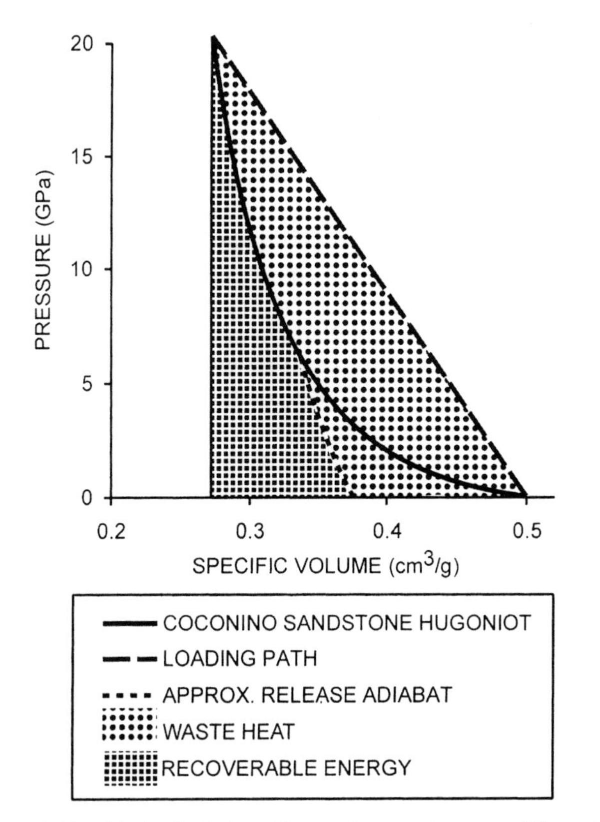

Figure 4. Graphical calculation of internal energy increase of Coconino Sandstone sample shock loaded to 20 GPa via single shock in matched impedance container assembly. In this instance sample is loaded by single shock from initial state to final state. Waste heat, represented by area between loading and unloading paths, is ~1250 J/g, corresponding to postshock temperature of ~1350°C, after equilibration of hotspots.

of diamond. The yield was not improved by preheating the sample. Eventually we determined that the important parameter was the amplitude of the shock initially transmitted into the sample; subsequent shock reflections had no effect on diamond yield.

We hope that these examples illustrate why it is important to document shock-loading details in archival experimental papers. It is not enough to list calculated pressures. The shock physicist who may want to model or simply recalculate an experiment using new Hugoniot or release data will need to know the details, including flyer plate material and dimensions, sample container material and dimensions, sample material and porosity, and the assumptions of any calculations that were made.

Two recent papers illustrate both the benefits of including sufficient detail and the importance of knowing the porosity of samples. Anderson and Ahrens (1998) reported Hugoniot measurements on pieces of the Murchison carbonaceous chondrite. They noted a disagreement between their measurements and the Hugoniot that was calculated by simple addition of the relative volumes at pressure of the major constituent minerals. The disagreement was unexpected; the agreement between calculated and measured Hugoniots is usually excellent. However, the additional details in the paper included the fact that they had not actually measured the porosity of their samples. Rather, they used mineral norms together with composition data to calculate the porosity of Murchison. Their result, 16% porosity, is significantly lower than measured values, which cluster at ~23% porosity. When we use the value of 23% for porosity, we find excellent agreement between our calculated Hugoniot and their measurements. Although this error in porosity seems minor, it has had repercussions for users of the data. Tomeoka et al. (1999) took the Hugoniot data and porosity estimate at face value in their analysis of shock-recovery experiments on Murchison. As a consequence, their waste heat calculations were affected. The paper is sufficiently detailed that we were able to calculate that their waste heat values would have been 300 J/g higher if they had used the 23% porosity value.

The experiments of Tomeoka et al. (1999) used high impedance containers; they interpreted their results in terms of equivalent matched impedance experiments. Thus, their 48.8 GPa experiment is equivalent, in terms of postshock temperature, to a 17 GPa natural event (after our correction to 23% initial porosity). The container impedance effect on postshock temperature is high in the case of porous samples, as the preceding example shows. However, it is not trivial for nonporous samples. We note that Ivanov et al. (2000) made detailed calculations indicating that, in their shock-recovery experiments on nonporous calcite, 85 GPa in reflected shock loading is equivalent to an ~41 GPa single shock event. Work by Bowden et al. (2000), in which both matched impedance containers and high impedance containers were used, experimentally verified the importance of loading path. With the notable exception of the matched impedance experiments of Hörz (1968) on quartz, most shock-recovery experiments on rocks and minerals have employed high shock impedance (iron, copper, stainless steel) sample containers. When one also considers that the maximum duration of laboratory experiment is small in comparison with natural events, it may be prudent to consider the laboratory results as qualitative rather than precise indicators of shock pressure.

SHOCK SYNTHESIS OF DIAMOND

The presence of diamond in a rock has long been considered evidence of a history of high pressure within the stability field of diamond. The presence of diamond in some meteorites was considered, at least until 1961, as evidence that the meteorites must have originated in bodies of at least lunar size, needed to provide static pressures within the stability field of diamond. Lipschutz and Anders (1961) suggested that diamonds in the Canyon Diablo iron meteorite were formed by shock compression during impact with the Earth. Working independently, DeCarli and Jamieson (1961) reported synthesis of diamond by shock compression of graphitic carbon to a peak pressure of ~30 GPa for a duration of ~1 µs. At the time it was widely believed that the duration of shock compression was too short to permit a conventional nucleation and growth mechanism to operate. To lend credibility to the publication, it was necessary to propose a diffusionless mechanism, c-axis compression of rhombohedral graphite.

In DeCarli and Jamieson (1961), no mention was made of lower or upper pressure bounds for shock synthesis of diamond, and the proposed mechanism was sheer speculation, supported only by a mention of failure to make diamond in some experiments with well-ordered hexagonal graphite. Nevertheless, various other researchers inferred that shock formation of diamond could be used as a 30 GPa calibration point.

We now have a much better understanding of the conditions for shock-wave syntheses of diamond (DeCarli, 1966, 1979, 1995, 1996, 1998). In later experiments it became clear that the amount of diamond produced far exceeded the amount of rhombohedral graphite in the starting material. That simple c-axis compression mechanism may operate, but other mechanisms must operate as well. It was also clear that the early failure to make diamond from well-crystallized hexagonal graphite was due to the fact that we had not yet explored the appropriate pressure-temperature regime.

Hundreds of experiments were performed on shock synthesis of diamond. Peak pressures spanned the range 10–150 GPa; initial sample temperatures ranged from 4 to 1300 K; peak pressure durations ranged from 0.1 to ~3 μs; starting materials included carbon black, charcoal, various grades of Acheson graphite, natural graphite, and pyrolytic graphite; and initial sample porosities ranged from 0% to ~60%. Sample container shock impedances ranged from low (ice) through matched (pressed sodium chloride) to high (stainless steel) (DeCarli, 1966)

Our interpretation of the results of all of those shock-recovery experiments appears to be completely consistent with the static high-pressure studies of Bundy and Kasper (1967) on the direct (uncatalyzed) synthesis of diamond. They compressed carbon (including both disordered carbon and well-crystallized graphite) at room temperature to pressures in the range of 10–20 GPa. The samples were heated transiently by discharge of a capacitor through the sample. Electrical measurements provided a basis for estimating the temperature history of the sample. The effective duration of the high-temperature regime was ~1 ms. With the knowledge that the compression of carbon at room temperature is reversible, one may argue that these direct conversion flash-heating experiments were analogous to shock compression. We emphasize the distinction between direct conversion diamond and the standard commercial product made by the catalytic method. Direct synthesis requires substantially higher pressures, and the fine-grained polycrystalline product is unlike either the product of catalytic synthesis or natural mantle-derived diamond.

Bundy and Kasper found that there were two distinct diamond synthesis regimes, depending on the properties of the source carbon. When the source carbon was a well-ordered dense graphite, a mixture of cubic and hexagonal (lonsdaleite) diamond could be made at pressures above ~15 GPa and simultaneous transient temperatures in the range of 1300–2000 K. The diamond is polycrystalline and the hexagonal phase has a preferred orientation, (100) parallel to (001) graphite. Although the preferred orientation implies a martensitic mechanism, Bundy and Kasper noted that it did not seem possible to convert one structure into the other by simple shear processes. Given the relatively modest temperature requirement, only about one-fourth of the absolute melting point of carbon, one may speculate that the mechanism for the transformation is nearly diffusionless.

When the source carbon was a less well-ordered graphitic carbon, purely cubic diamond could be made at pressures above ~15 GPa and temperatures above 3000 K. The diamond is polycrystalline with a small grain size, ~50 nm, and with no evidence for preferred orientation. The requisite temperature is high enough to justify speculation that the mechanism is nucleation and growth. It is obvious that the hot diamond must be quenched to lower temperature (to avoid graphitization) before the pressure is reduced into the stability field of graphite. The static data thus impose severe constraints on impact formation of diamond from natural disordered carbons, such as coal or charcoal.

The response of well-ordered graphite is particularly interesting with respect to observations of diamond in impact-metamorphosed rocks. The description of diamonds found in the Popigai impact structure (Koeberl et al., 1997) is in essential agreement with Bundy and Kasper's results and with our own studies of the product of laboratory shock synthesis from well-ordered graphite. Koeberl et al. offered some particularly convincing evidence for the impact origin of the Popigai diamonds. Can we infer criteria by which we can identify impact diamonds that are found out of context, i.e., not associated with obvious impact structures? With the exception of shock-produced nanodiamonds (2–7 nm diameter), all impact diamonds are polycrystalline. On the basis of X-ray diffraction line broadening, one would estimate the crystallite size of impact diamonds (including meteoritic diamond, the laboratory product, and diamonds found in association with terrestrial impact craters) to be <100 nm; however, lattice strain and stacking faults also contribute to line broadening. Polycrystalline mantle-derived diamonds are known, but, to the best of our knowledge, have crystallite sizes that are a minimum of several orders of magnitude larger than impact diamond. As noted by Koeberl et al. (1997), evidence for lonsdaleite is found in some, but not all, impact diamonds. Because lonsdaleite has never been found in mantle-derived diamond, the presence of lonsdaleite is an excellent shock indicator. The absence of lonsdaleite signifies nothing. Although some mantle-derived diamonds may exhibit strain birefringence, one generally expects diamond to be optically isotropic. The impact diamond shown in Figure 1 of Koeberl et al. (1997) is strongly birefringent, in agreement with our own observations on diamond made by shock from well-ordered graphite. Anyone conducting an optical search for impact diamond must be alert to birefringent diamond, but impact diamond may also be optically isotropic. El Goresy et al. (2001) performed a detailed study of the provenance of diamond in the impact-metamorphosed gneisses of the Ries impact crater. Their results in general accord with, although are much more detailed than, prior studies of direct transition diamond synthesized from well-crystallized graphite either in laboratory shock experiments or by Bundy and Kasper's quasistatic flash-heating method.

The conditions for shock synthesis of diamond and its subsequent preservation can be used to provide an estimate of shock pressure in the surrounding rock or metal. We have made simple estimates of the peak temperature in graphite inclusions in various matrices (DeCarli, 1996). Based on the static data, the peak temperature of the graphite (well crystallized) must be above ~1300 K for the conversion to diamond. However, the temperature of the diamond (and the surrounding rock) should be <2000 K to avoid rapid graphitization on release of pressure. Because graphite has a lower shock impedance that the matrix rock, the graphite will reach peak pressure via a series of shock reflections. The details of the pressure and temperature history at any point in the inclusion will depend on the exact geometry of the inclusion with respect to the shock. We simplify the matter by assuming plate-like inclusions and a shock-propagation direction normal to the plane of the flake. We also ignore possible kinetic effects on the shock response of the matrix.

For graphite in marble, the peak temperature range in the graphite of 1300–2000 K corresponds to a peak pressure range in the marble of ~30–45 GPa; the temperature of the marble on release will be <1800 K. For graphite in granite or quartzite, the 1300–2000 K graphite temperature range corresponds to the ~27–38 GPa pressure range. For graphite in nickel-iron meteorites, the 1300–2000 K range corresponds to peak pressures in the range of ~60–120 GPa; the temperature of the iron on release from 120 GPa will be <1700 K. This latter calculation is confirmed by our own data from shock-recovery experiments as well as by Trueb's (1968) study of diamond synthesized from graphite in cast iron shocked to 100 GPa. We lack specific Hugoniot and release data on the Popigai gneiss, but we estimate that it should be roughly intermediate in shock properties between the marble and granite. Although it is compositionally similar to granite, the sillimanite and cordierite constituents are slightly less compressible than quartz and feldspar. It can be seen that the presence of shock-synthesized diamond does not serve as a particularly precise piezometer. If we found an isolated polycrystalline diamond that contained lonsdaleite and was a paramorph after graphite, we could be virtually certain that this was impact diamond. Without knowing anything about the nature of the matrix in which the graphite was shocked, we could only infer that the peak shock pressure in the matrix was within the range of 15–150 GPa.

Koeberl et al. (1997) estimated that Popigai diamonds formed within the peak pressure range of 35–60 GPa, based on petrographic observations of shock phenomena in the surrounding rocks. Our estimate of 30 GPa is in essential agreement with their lower bound, but we suspect that their upper bound of 60 GPa may be ~20 GPa too high. Our crude estimates indicate that the gneiss would be at a temperature of ~3000 K after release from 60 GPa. These estimates are consistent with an extrapolation of Boslough's (1988) measurements of post-shock temperatures in silica. However, we lack the accurate Hugoniot and release adiabat data needed to make a rigorously defensible calculation of the release temperature of the gneiss. If kinetic effects were important, even such a defensible calculation, based on microsecond duration laboratory data, might grossly underestimate the actual release temperature. On the basis of current knowledge, we estimate that the peak shock pressure in the diamond-bearing gneiss was probably in the range of 30–40 GPa.

The presence of diamond in a shocked rock provides a clue to interpreting possible kinetic effects in longer duration (millisecond to second) shock events. That the diamonds survived without completely graphitizing delimits the postshock thermal environment. Because the postshock temperature of a rock is a function of the difference between the loading path and the release path, knowledge of the postshock temperature provides a basis for assessing whether the loading and release paths could have been affected by kinetics (DeCarli and Bowden, 2000) The kinetics of graphitization of single crystal diamond have been studied thoroughly (Howes, 1962; Evans and James, 1964). These data indicate that, in the absence of oxygen, a 100 mg diamond would be completely graphitized within one week at a temperature of 1960 K. At 1640 K, the extrapolated graphitization time would be ~1000 yr. The data on graphitization kinetics of natural single crystal diamond probably do not apply to the polycrystalline and strained impact diamonds associated with Popigai and other craters. It would therefore be useful to have results of graphitization studies on Popigai and similar impact diamonds.

Impact diamond may also be formed from disordered carbon precursors. Commercial shock synthesis of diamond powder is descended from the DeCarli and Jamieson experiments that used commercially available porous Acheson graphite (DeCarli, 1966; Cowan et al., 1968). In accordance with the static flash-heating results, very high temperatures (>3000 K) appear to be required. It is believed that the diamond forms in hotspots caused by shock interactions around pores (DeCarli, 1979, 1995). This diamond must be quenched before pressure release to inhibit graphitization. Because graphite is an excellent thermal conductor, the hotspots can be quenched on a microsecond time scale via conduction to the surrounding cooler graphite. Extrapolating from laboratory shock-loading experiments on commercial coke carbon, one would predict that naturally occurring charcoals, coals, and cokes would yield diamond as a result of shock loading to peak pressures in the range of 15–50 GPa.

In our experience there are two types of diamond formed by shocking disordered carbon. The first type comprises equiaxed polycrystalline particles having diameters in the micrometer to tens of micrometers range. The particles are usually black, although they may contain optically isotropic transparent regions. X-ray diffraction and electron-diffraction patterns show only cubic diamond reflections; line broadening implies crystallite size of <50 nm. This type of diamond is commercially synthesized and marketed for use in grinding and polishing compounds. In most catalogs of petrographic laboratory supplies, the shock-synthesized diamond is not identified as such, but rather as polycrystalline micron diamond. We note with approval that El Goresy et al. (2001) were careful to characterize the abrasives used to prepare samples of diamond-bearing gneisses. Single-crystal micrometer-size diamond is easily distinguished from impact diamond.

The maximum size of diamond that can be recovered from shocked disordered carbon is simply the maximum size that can be quenched before the pressure decays. We speculate that most of the diamond particles formed from disordered carbon by impact will be small, ≤10 μm. Although large impacts can provide relatively long quenching times, the natural environment of disordered carbon, e.g., charcoal in soil, is not necessarily optimal for quenching of shock-formed diamond. It has been estimated that the maximum diameter diamond that could conceivably be formed and quenched in a natural impact is ~1 cm (DeCarli, 1998). This estimate was intended to provide a firm upper limit and invoked the most favorable environment for both shock loading and postshock quenching that could be conceived. The purpose of the estimate was to test the hypothesis that carbonado, a type of natural polycrystalline diamond, is impact diamond (Smith and Dawson, 1985). Because the largest known carbonado is 10 cm diameter, it seems highly unlikely that carbonado could be impact diamond. Note that our estimate of the maximum size of impact diamond does not apply to diamond made from a well-crystallized graphite precursor, as is the case with the Popigai and Ries diamonds. Because these diamonds can be made at modest temperature, ~1300 K, in a matrix at a comparable postshock temperature, rapid quenching to avoid graphitization is unnecessary. The only limitation on diamond size is the size of precursor crystal of well-ordered graphite.

The second type of diamond made by shocking disordered carbon is nanodiamond, cubic diamond particles having diameters in the range of 2–7 nm. These diamonds strongly resemble both the nanodiamonds found in carbonaceous chondrites and the diamonds found in the detonation products of oxygen-deficient high explosives (DeCarli, 1995). One may ask whether the diamonds found in carbonaceous chondrites could have been formed by shock compression. We suggest that the possibility should be kept open until a way is found to distinguish nanodiamonds formed by shock compression and nanodiamonds formed by some other process, such as metastable growth, in the solar nebula. Purified detonation product nanodiamond is commercially marketed as Ultra Disperse Diamond; it is often used as a proxy for meteoritic nanodiamond. We surmise that the nanodiamond found in the Cretaceous-Tertiary boundary layer (Hough et al., 1997) could have been impact synthesized from a disordered form of natural carbon, such as charcoal. Of course, measurements of trapped gases provide a means of distinguishing meteoritic nanodiamond from terrestrial nanodiamond. A final note on nanodiamond: one recent study indicates that upon heating in vacuum, nanodiamond may transform to onion-like fullerenes (Mal'kov and Titiov, 1995).

The observation of purely cubic diamond intergrown with silicon carbide in Ries suevite (Hough et al., 1995) implies high local temperatures, >1800°C, to account for the reaction of carbon with silica. Survival of the diamond implies probable quenching prior to pressure decay. The inferred high temperature does not strongly limit the estimate of shock pressure. In the most likely scenario, if the source carbon were porous charcoal (or similar organic matter) in a porous silicate matrix, e.g., soil, the peak shock pressure could have been as low as 15 GPa. Less likely, if the source carbon were fully dense graphite, the peak shock pressure could have been as high as 50 GPa, given a local environment of incompressible minerals that would not have been strongly shock heated and that could serve as heat sinks. It is not necessary to assume an exotic mechanism, such as chemical vapor deposition (CVD) growth of diamond and silicon carbide in a low-pressure plasma (Hough et al., 1995). The time available at high pressure in the Ries event was of the order of 0.5 s, more than enough time for reduction of silica by hot carbon, formation of the observed diamond-silicon particles, and subsequent quenching.

PHASE TRANSFORMATIONS IN ROCKS AND MINERALS

It has long been customary to interpret discontinuities in the Hugoniot data in terms of phase transitions. McQueen (1964) summarized a decade of Los Alamos work that established that Hugoniot discontinuities could be associated with shock-induced phase transitions in pure metals and metal alloys. The rapidity of those shock-induced phase transitions was not surprising to metallurgists who were accustomed to rapid diffusionless (martensitic) phase transitions. However, some shock-wave workers speculated that shock-wave compression was fundamentally different from static compression (Alder, 1963). It was then only a short step to the interpretation of Hugoniot discontinuities in terms of extraordinarily rapid reconstructive phase transitions in silicates.

The speculation that shock-wave compression was fundamentally different from static compression was effectively refuted by Jeanloz (1980), who presented extensive evidence contradicting the interpretations of Hugoniot data on olivines, pyroxenes, and quartz in terms of very rapid reconstructive phase transitions. Jeanloz's view, which we share, is that minerals can compress to a volume nearly equal to the volume of a high-pressure phase without actually undergoing a reconstructive phase transition.

Data on the shock-wave formation of cubic diamond from disordered porous graphite can be explained by invoking hotspots. If the hotspots are hot enough, a microsecond at pressure is sufficient time for nucleation and growth of fine-grained polycrystalline diamonds. Heterogeneous compression effects are observed to occur even in initially homogeneous materials. Optical studies imply that localized hotspots may be thousands of degrees higher than the surrounding matrix (Kondo and Ahrens, 1983). These localized hotspots can account for the rapid formation of high-pressure phases via conventional nucleation and growth mechanisms; there is no need to invoke novel mechanisms peculiar to shock compression. If the duration of the shock is sufficient, the hotspots are quenched at pressure, and the high-pressure phases can be preserved. In this spirit, Grady et al. (1975) interpreted Hugoniot sound-velocity data as indicative of localized hotspots in which stishovite could be formed by an ordinary nucleation and growth process. Although the sound-velocity data appear to be well established, the interpretation is by no means exclusive; other interpretations of the data are possible. High-pressure minerals are also found in the so-called melt veins of certain meteorites (Chen et al., 1996, and references therein). The presence of these minerals can be accounted for as the result of ordinary nucleation and growth in the following scenario. Assume that the melt zone forms upon arrival of a long-duration shock wave. A variety of mechanisms can produce localized melting, including adiabatic shear (analogous to tachylite formation), shock compression of a locally porous region, and shock collisions. The temperature of the melt zone, a few tens of micrometers in width, will equilibrate with its surroundings on a time scale of a few tens of microseconds. The conditions (pressure, temperature, and time) permit nucleation and growth of high-pressure phases that can be quenched prior to pressure decay, if the effective shock duration is long enough.

It has long been known that melt veins can be made experimentally in shock-recovery experiments (Fredriksson et al., 1963). However, no one has been able to perform a shock-recovery experiment in which the duration of shock pressure was sufficient (10 μs or greater) to permit quenching of the melt vein prior to pressure decay. The presence of a high-pressure mineral in a rock is usually a good indication that the rock was subjected to a pressure within the stability field of the high-pressure phase. If there are data on the kinetics of the reversion of this high-pressure phase to its low-pressure form, the post-shock thermal history of the rock can be delimited. With enough knowledge of Hugoniots and release adiabats, the bounds on the thermal history can be used to infer an upper pressure limit to the most recent shock event.

EVIDENCE FOR KINETIC EFFECTS ON DYNAMIC PHASE TRANSITIONS

One might look for kinetic effects by the traditional method of increasing the duration of the experiment at a given pressure and temperature. This is very difficult for shock experiments, in which shock-duration scales (approximately) with the cube root of the dimensions of the experiment. A typical microsecond-duration laboratory shock experiment might use 10 kg of high explosive to shock load a 10 g rock sample contained in a precisely fabricated 50 kg recovery assembly. In order to increase the duration by a factor of 10, the amount of explosive required would be ~10 t; the cost of such an experiment would be about a million dollars.

Alternatively, one might vary temperature at a given pressure and duration to obtain evidence of kinetic effects. Two independent sets of shock-recovery experiments on preheated quartz samples show that the conversion to diaplectic glass increases with increasing temperature of the target (Gratz et al., 1993; Langenhorst et al., 1992). Subsequent work (Huffman et al., 1993; Huffman and Reimold, 1996) confirmed the temperature effect on conversion of quartz and showed a similar temperature effect on conversion of plagioclase and K-feldspar to diaplectic glass.

One also finds indirect evidence for kinetic effects by comparing Hugoniot data on an initially porous material with data on the nonporous material. The initially porous material is much hotter, when singly shocked to a given shock pressure, than the nonporous material singly shocked to the same pressure. The initially porous material would therefore be expected to have a higher volume, because of thermal expansion, than the initially nonporous material at the same pressure. Using the data of Ahrens and Gregson (1964), the Hugoniot of Coconino Sandstone (porous quartz, initially 2.0 g/cm^3) can be compared with the Hugoniot of Arkansas novaculite (nonporous quartzite, 2.65 g/cm^3). The predicted relationship, initially porous less dense than initially nonporous, holds up to ~10 GPa. At higher pressures, Coconino Sandstone becomes the denser material. One explanation of this behavior is that the sandstone, because of its higher temperature, is beginning to transform to a denser structure. At a shock pressure of 10 GPa, the internal energy content of the sandstone is ~700 J/g higher than the initially solid material.

An attempt was made to infer kinetic effects on the basis of a study of a sample of naturally shocked Coconino Sandstone from Meteor Crater (DeCarli and Bowden, 2000). Stishovite, coesite, quartz, and diaplectic glass were present; their relative proportions provided a basis for estimating the release adiabat. The problem was to account for the survival of stishovite. Using the estimated release adiabat along with the Coconino Hugoniot, it was estimated that the postshock temperature would be much too high to permit survival of stishovite. Assuming water-saturated Coconino Sandstone, as suggested by Kieffer et al. (1976), the Hugoniot and release adiabat were recalculated. The postshock temperature was still much too high. However, if a large kinetic effect on the Hugoniot was assumed, the calculated postshock temperature could be reduced to a value below 600°C, the upper limit for stishovite survival.

An assumption of kinetic effects could resolve other

apparent conflicts between postshock temperatures inferred by comparisons with results of laboratory studies and postshock temperatures inferred from the survival of heat-sensitive phases in the shocked rocks of large impact craters. There may be an additional problem in that evidence for low temperatures, the survival of heat-sensitive phases, may coexist with evidence for high temperatures, petrographic evidence of material that was once molten. It can readily be determined whether the high-pressure phases are misidentified. Another possible explanation for the conflict is that the melt structures are misidentified. We suggest that petrographers should consider the possibility that diaplectic glass that was deformed at high pressure could be misidentified as material that was molten on release of pressure. Grady et al. (1975) observed that the sound velocities of quartz and feldspar were anomalously low in the shock pressure above ~20 GPa. An alternate explanation for the data is that the shear strengths of the material became negligible. The hypothesis that kinetic effects are important in large crater-forming events implies that diaplectic glass may form at much lower pressures than observed in laboratory experiments.

There was one instance of a misleading experimental result in one of the papers cited in this section. Huffman et al. (1993) performed shock-recovery experiments on preheated Westerly granite samples in high-shock-impedance stainless steel containers. They found evidence of bulk (as opposed to tachylite) melting in samples preheated to 750°C and shocked to only 5 GPa. We calculate that the postshock temperature increase attributable to shock compression and release would have been only ~20°C. This paper is sufficiently detailed to permit reinterpretation of their results. Huffman et al. reported that some samples were reduced in thickness by as much as 50%. One may infer that one of the causes of the observed melting was the additional heating due to plastic deformation that occurred during release of pressure.

SUMMARY AND CONCLUSIONS

We have attempted to demonstrate the importance of thoroughly documenting the conditions of shock-recovery experiments. In the section on diamond we attempted to show that the conditions governing shock-induced phase transitions may be very complex. The section on phase transitions supports the view that shock-induced phase transitions follow the same rules as phase transitions under static high pressure. We have presented preliminary and indirect evidence for kinetic effects on phase transformations in long-duration shock events.

ACKNOWLEDGMENTS

We thank Tom Sharp of Arizona State University and an anonymous reviewer for their incisive comments on an earlier version of this manuscript. We thank the conference organizers for providing an excellent opportunity for interdisciplinary communication.

REFERENCES CITED

Ahrens, T.J., and Gregson, V.G., 1964, Shock compression of crustal rocks: Data for quartz, calcite, and plagioclase rocks: Journal of Geophysical Research, v. 69, p. 4839–4874.

Alder, B.J., 1963, Physics experiments with strong pressure pulses, *in* Paul, W., and Warshauer, D.W., eds., Solids under pressure: New York, McGraw-Hill, p. 385–420.

Anderson, W.W., and Ahrens, T.J., 1998, Shock wave equations of state of chondritic meteorites, *in* Schmidt, S.C., Dandekar, D.P., and Forbes, J.W., eds., Shock compression of condensed matter—1997: American Institute of Physics Conference Proceedings, v. 429, p. 115–118.

Baer, M.R., 2000, Computational modeling of heterogeneous materials at the mesoscale, *in* Furnish, M.D., Chabildas, L.C., and Hixon, R.S., eds., Shock compression of condensed matter—1999: American Institute of Physics Conference Proceedings, v. 505, p. 27–33.

Boslough, M.B., 1988, Postshock temperatures in silica: Journal of Geophysical Research, v. 93, p. 6477–6484.

Bowden, E., Kondo, K., Ogura, T., Jones, A.P., Price, G.D., and DeCarli, P.S., 2000, Loading path effects on the shock metamorphism of porous quartz [abs.]: Lunar and Planetary Science, v. 31, Abstract #1582, CD-ROM.

Bundy, F.P., and Kasper, J.S., 1967, Hexagonal diamond: A new form of carbon: Journal of Chemical Physics, v. 46, p. 3437–3446.

Chen, M., Sharp, T.G., El Goresy, A., Wopenka, B., and Xie, X., 1996, The majorite-pyrope + magnesiowustite assemblage: Constraints on the history of shock veins in chondrites: Science, v. 271, p. 1570–1573.

Cowan, G.R., Dunnington, B.W., and Holtzman, A.H., 1968, United States Patent 3,401,019.

DeCarli, P.S., 1966, United States Patent 3,238,019.

DeCarli, P.S., 1979, Nucleation and growth of diamond in shock wave experiments, *in* Timmerhaus, K.D., and Barber, M.S., eds., High pressure science and technology, Volume 1: New York, Plenum, p. 940–943.

DeCarli, P.S., 1995, Shock wave synthesis of diamond and other phases, *in* Drory, M.D., Bogy, D.B., Donley, M.S., and Field, J.E., eds., Mechanical behavior of diamond and other forms of carbon: Proceedings, Materials Research Society Symposium, Materials Research Society, v. 383, p. 21–31.

DeCarli, P.S., 1996, Were carbonados synthesized by an ancient impact?, *in* Schmidt, S.C., and Tao, W.C., eds., Shock compression of condensed matter: 1995, Proceedings, American Institute of Physics Conference: American Institute of Physics, v. 370, p. 757–760.

DeCarli, P.S.,1998, More on the possibility of impact origin of carbonado, *in* Schmidt, S.C., Dandekar, D.P., and Forbes, J.W., eds., Shock compression of condensed matter–1997: Proceedings, American Institute of Physics Conference, American Institute of Physics, v. 429, p. 681–684.

DeCarli, P.S., and Bowden, E., 2000, A bootstrap estimate of peak pressure in a sample of Meteor Crater Coconino sandstone: Meteoritics and Planetary Science, v. 35, p. A47.

DeCarli, P.S., and Jamieson, J.C., 1959, Formation of an amorphous form of quartz under shock conditions: Journal of Chemical Physics, v. 31, p. 1675–1676.

DeCarli, P.S., and Jamieson, J.C., 1961, Formation of diamond by explosive shock: Science, v. 133, p. 1821–1822.

El Goresy, A., Gillet, P., Ming Chen, Künstler, F., Graup, G., and Stähle, V., 2001, In-situ discovery of shock-induced graphite-diamond phase transition in gneisses from the Ries Crater, Germany: American Mineralogist, v. 86, p. 611–621.

Evans, T., and James, P.F., 1964, A study of the transformation of diamond to graphite: Proceedings of the Royal Society, v. A 277, p. 260–269.

Fredriksson, K., DeCarli, P.S., and Aaramae, A., 1963, Shock-induced veins in chondrites, *in* Priester, W., ed., Space Research 3, Proceedings of the 3rd International Space Science Symposium, Washington, April 30–May 9, 1962: Amsterdam, North-Holland Publishing Co., p. 974–983.

Grady, D.E., Murri, W.J., and DeCarli, P.S., 1975, Hugoniot sound velocities and phase transformations in two silicates: Journal of Geophysical Research, v. 80, p. 4857–4861.

Gratz, A.J., Nellis, W.J., Christie, J.M., Brocious, W., Swegle, J., and Cordier, P., 1992, Shock metamorphism of quartz with initial temperatures − 170 to + 1000°C: Physics and Chemistry of Minerals, v. 19, p. 267–288.

Hörz, F., 1968, Statistical measurements of deformation structures and refractive indices in experimentally shock loaded quartz, *in* French, B.M., and Short, N.M., eds., Shock metamorphism of natural materials: Baltimore, Maryland, Mono Book Corp., p. 243–253.

Hough, R.M., Gilmore, I., Pillinger, C.T., Arden, J.W., Gilkes, K.W.R., Yuan, J., and Milledge, H.J., 1995, Diamond and silicon carbide in impact melt rock from the Ries impact crater: Nature, v. 378, p. 41–44.

Hough, R.M., Langenhorst, F., Montanari, A., Pillinger, C.T., and Gilmour, I., 1997, Diamonds from the iridium-rich K-T boundary layer at Arroyo el Mimbral, Tamaulipas, Mexico: Geology, v. 25, p. 1019–1022.

Howes, V.R, 1962, Graphitization of diamond: Proceedings of the Royal Society, v. 80, p. 260–269.

Huffman, A.R., and Reimold, W.U., 1996, Experimental constraints on shock-induced microstructures in naturally deformed silicates: Tectonophysics, v. 256, p. 165–217.

Huffman, A.R., Brown, J.M., Carter, N.L., and Reimold, W.U., 1993, The microstructural response of quartz and feldspar under shock loading at variable temperature: Journal of Geophysical Research, v. 98, p. 22171–22197.

Ivanov, B.A., Langenhorst, F., Deutsch, A., and Hornemann, U., 2000, How strong was impact-induced CO_2 degassing in the K/T event? [abs.], *in* Catastrophic events and mass extinctions: Impacts and beyond: Houston, Texas, Lunar and Planetary Institute, LPI Contribution No. 1053, p. 80–81.

Jeanloz, R., 1980, Shock effects in olivine and implications for Hugoniot data: Journal of Geophysical Research, v. 85, p 3163–3176.

Kieffer, S.W., Phakey, P.P., and Christie, J.M., 1976, Shock processes in porous quartzite: Transmission electron microscope observations and theory: Contributions to Mineralogy and Petrology, v. 59, p. 41–93.

Koeberl, C., Masaitis, V.L., Shafranovsky, G.I., Gilmour, I., Langenhorst, F., and Schrauder, M., 1997, Diamonds from the Popigai impact structure, Russia: Geology, v. 25, p. 967–970.

Kondo, K., and Ahrens, T.J., 1983, Heterogeneous shock-induced thermal radiation in minerals: Physics and Chemistry of Minerals, v. 9, p. 173–181.

Langenhorst, F., Deutsch, A., Stöffler, D., and Hornemann, U., 1992, Effect of temperature on shock metamorphism of single-crystal quartz: Nature, v. 356, p. 507–509.

Lipschutz, M.E., and Anders, E., 1961, The record in meteorites. 4. Origin of diamonds in iron meteorites: Geochimica et Cosmochimica Acta, v. 24, p. 83–105.

Mal'kov, I.Yu., and Titiov, V.M., Structure and properties of detonation soot particles, *in* Schmidt, S.C., and Tao, W.C., eds., Shock compression of condensed matter—1995: Proceedings, American Institute of Physics Conference, American Institute of Physics, v. 370, p 783–786.

McQueen, R.G., 1964, Laboratory techniques for very high pressures and the behavior of metals under dynamic loading, *in* Gschneider, K.S., Jr., Hepworth, M.T., and Parlee, N.A.D., eds., Metallurgy at high pressures and high temperatures: New York, Gordon and Breach Science Publishers, p. 44–132.

McQueen, R.G., Marsh, S.P., Taylor, J.W., Fritz, J.N., and Carter, W.J., 1970, The equation of state of solids from shock wave studies, *in* Kinslow, R., ed., High-velocity impact phenomena, Appendix E: New York, Academic Press, p. 530–568.

Melosh, H.J., 1989, Impact cratering: A geologic process: New York, Oxford University Press, 245 p.

Milton, D.J., and DeCarli, P.S., 1963, Maskelynite: Formation by explosive shock: Science, v. 147, p. 144–145.

Smith, J.V., and Dawson, J.B., 1985, Carbonado: Diamond aggregates from early impacts of crustal rocks?: Geology, v. 13, p. 342–343.

Tomeoka, K., Yamahana, Y., and Sekine, T., 1999, Experimental shock metamorphism of the Murchison CM carbonaceous chondrite: Geochimica et Cosmochimica Acta, v. 63, p. 3683–3703.

Trueb, L.F., 1968, An electron-microscope study of shock-synthesized diamond: Journal of Applied Physics, v. 39, p. 4707–4716.

MANUSCRIPT SUBMITTED DECEMBER 1, 2000; ACCEPTED BY THE SOCIETY MARCH 22, 2001

Printed in the U.S.A.

Geological Society of America
Special Paper 356
2002

Comparison of the osmium and chromium isotopic methods for the detection of meteoritic components in impactites: Examples from the Morokweng and Vredefort impact structures, South Africa

Christian Koeberl*
Institute of Geochemistry, University of Vienna, Althanstrasse 14, A-1090 Vienna, Austria
Bernhard Peucker-Ehrenbrink*
Department of Marine Chemistry and Geochemistry, Woods Hole Oceanographic Institution, Woods Hole, Massachusetts 02543-1541, USA
Wolf Uwe Reimold*
Impact Cratering Research Group, Department of Geology, University of the Witwatersrand, Johannesburg 2050, South Africa
Alex Shukolyukov*
Scripps Institute of Oceanography, University of California, San Diego, California 92093-0212, USA
Günter W. Lugmair*
Scripps Institute of Oceanography, University of California, San Diego, California 92093-0212, USA, and Max-Planck-Institute for Chemistry, P.O. Box 3060, 55020 Mainz, Germany

ABSTRACT

Breccias and melt rocks found at possible meteorite impact structures on Earth may contain a minor extraterrestrial component. In the absence of evidence of shock-metamorphic effects in such rocks, the unambiguous detection of an extraterrestrial component can be of diagnostic value regarding the impact origin of a geologic structure. Previously the concentrations of siderophile elements, mainly the platinum-group elements (PGEs), or osmium isotopic studies have been used to detect such a meteoritic component. The Cr isotopic method has shown great potential for identifying the type of impactor. Here we use a combination of trace element (PGE) and Os and Cr isotopic data for rocks from the two large impact structures, Vredefort and Morokweng, both in South Africa, to compare the two isotopic methods directly for the first time. The percentage of a meteoritic component in Vredefort granophyre was too low to yield a Cr isotopic signal, but there was a distinct Os isotope effect. For Morokweng, both methods yielded positive results, the Cr isotope data suggesting that the projectile had the composition of an ordinary chondrite, probably an L chondrite. Comparison of the two methods thus confirms that the Os isotopic method is a

*E-mails: Koeberl, christian.koeberl@univie.ac.at; Peucker-Ehrenbrink, behrenbrink@whoi.edu; Reimold, 065wur@cosmos.wits.ac.za; Shukolyukov, ashukolyukov@ucsd.edu; Lugmair, lugmair@mpch-mainz.mpg.de

Koeberl, C., Peucker-Ehrenbrink, B., Reimold, W.U., Shukolyukov, A., and Lugmair, G.W., 2002, Comparison of the osmium and chromium isotopic methods for the detection of meteoritic components in impactites: Examples from the Morokweng and Vredefort impact structures, South Africa, *in* Koeberl, C., and MacLeod, K.G., eds., Catastrophic Events and Mass Extinctions: Impacts and Beyond: Boulder, Colorado, Geological Society of America Special Paper 356, p. 607–617.

more sensitive tool for detecting an extraterrestrial component, whereas the Cr isotopic method, where applicable, can provide additional information regarding the nature of the impactor.

INTRODUCTION

The verification of an extraterrestrial component in impact-derived melt rocks or breccias can provide evidence confirming an impact origin of a geologic structure. Similar approaches are of great value in the investigation of distal ejecta layers; perhaps the best example is that of the Cretaceous-Tertiary (K-T) boundary, where the discovery of an extraterrestrial component (Alvarez et al., 1980) opened the door to a reassessment of the importance of impact events in the geological history of the Earth. During impact, a small amount of the finely dispersed meteoritic melt or vapor is mixed with a much larger quantity of target rock vapor and melt, and this mixture later forms impact-melt rocks, melt breccias, or impact glass. In most cases, the contribution of meteoritic matter to these impactite lithologies is very small ($\ll 1\%$), leading to only slight chemical changes in the resulting impactites.

The detection of such small amounts of meteoritic matter within the normal upper crustal compositional signature of the target rocks is extremely difficult. Only elements that have high abundances in meteorites, but low abundances in terrestrial crustal rocks (e.g., the siderophile elements) are useful. Another complication is the existence of a variety of meteorite groups and types (the three main groups are stony meteorites, iron meteorites, and stony-iron meteorites, in order of decreasing abundance; see, e.g., Wasson, 1985; Papike, 1998, for reviews on meteorites), which have widely varying siderophile element compositions. Distinctly higher siderophile element contents in impact melts, compared to target rock abundances, can be indicative of the presence of either a chondritic or an iron meteoritic component. Achondritic projectiles (stony meteorites that underwent magmatic differentiation) are much more difficult to discern, because they have significantly lower abundances of the key siderophile elements. Furthermore, in order to reliably determine the target rock contribution of such elements, i.e., the so-called indigenous component, absolute certainty must be attained that all contributing terrestrial target rocks have been identified and their relative contributions to the melt mixture are reasonably well known.

Geochemical methods have been used to determine the presence of the traces of such an extraterrestrial component (see review in Koeberl, 1998). In the absence of actual meteorite fragments, it is necessary to chemically search for traces of meteoritic material mixed in with the target rocks in breccias and melt rocks. Meteoritic components have been identified for ~40 impact structures (see Koeberl, 1998, for a list), of the more than 160 impact structures that have so far been identified on Earth. This number reflects mostly the extent to which these structures have been studied in detail, because only a few of these impact structures were first identified by finding a meteoritic component (the majority were confirmed by the identification of shock-metamorphic effects). The identification of a meteoritic component can be achieved by determining the concentrations and interelement ratios of siderophile elements, especially the platinum group elements (PGEs), which are several orders of magnitude more abundant in meteorites than in terrestrial upper crustal rocks. This is illustrated for one of the PGEs, Ir, in Figure 1. Iridium is most often determined as a proxy for all PGEs, because it can be measured with the best detection limit of all PGEs by neutron activation analysis (which was, for a long time, the only more or less routine method for Ir measurements at abundance levels less than parts per billion in small samples).

The use of PGE abundances and ratios avoids some of the ambiguities that result if only moderately siderophile elements (e.g., Cr, Co, Ni) are used in an identification attempt. However, problems may arise if the target rocks have high abundances of siderophile elements or if the siderophile element concentrations in the impactites are very low. In such cases, the Os and Cr isotopic systems can be used to establish the presence of a meteoritic component in a number of impact-melt rocks and breccias (e.g., Koeberl and Shirey, 1997; Shukolyukov and Lugmair, 1998). In the past, PGE data were used to estimate the type or class of meteorite for the impactor (e.g., Morgan, 1978; Palme, 1982; Palme et al., 1978, 1979, 1981), but these attempts were not always successful. It is difficult to distinguish among different chondrite types based on siderophile element (or even PGE) abundances, which has led to conflicting conclusions regarding the nature of the impactor at a number of structures (see Koeberl, 1998, for details). Clearly, the identification of a meteoritic component in impactites is not a trivial problem. In this study we use a combination of trace element (PGE) analyses and the results from both Os and Cr isotopic studies to illustrate the pros and cons of each method with two case studies, the Vredefort and Morokweng impact structures of South Africa.

EXPERIMENTAL METHODS: BACKGROUND AND ANALYTICAL PROCEDURES

Osmium isotopic method

The use of the Os isotopic system is based on the formation of ^{187}Os (one of seven stable isotopes of Os) by β-decay of ^{187}Re (half-life of 42.3 ± 1.3 Ga). Meteorites (and the terrestrial mantle) have Os and other PGEs contents higher by factors

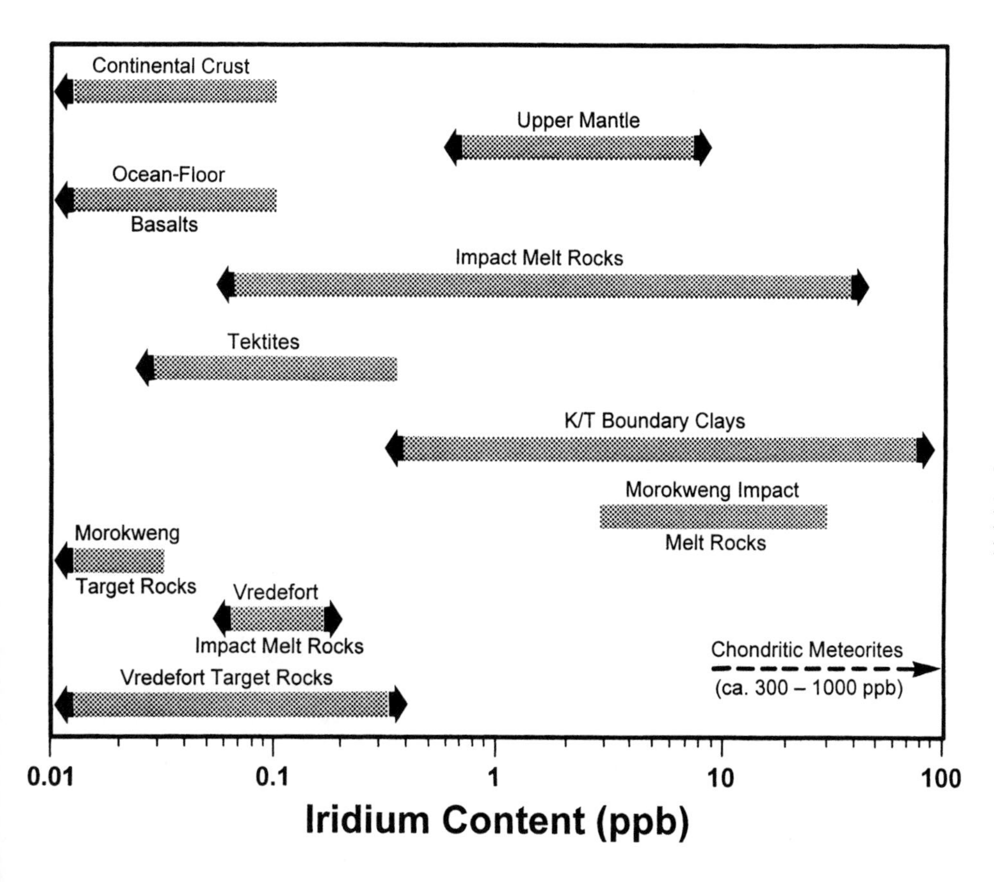

Figure 1. Approximate range of iridium concentrations in variety of terrestrial rocks, impactites, distal impact ejecta (i.e., tektites and Cretaceous-Tertiary [K-T] boundary clays), and meteorites (for data sources, see Koeberl, 1998), in comparison with data for target rocks and impact-melt rocks at Vredefort and Morokweng impact structures (data from French et al., 1989; Koeberl et al., 1997; Koeberl and Reimold, 2002).

of 10^2–10^5 than terrestrial crustal rocks. In addition, meteorites have relatively low Re and high Os abundances, resulting in Re/Os ratios $\leq$0.1, whereas the Re/Os ratio of terrestrial crustal rocks is usually no less than 10. Similar to the Rb-Sr and Sm-Nd isotopic methods, the abundance of radiogenic ^{187}Os is normalized to the abundance of nonradiogenic ^{188}Os. The use of the method is based on the fact that the ^{187}Os/^{188}Os isotopic ratios for meteorites and terrestrial crustal rocks are significantly different. As a result of the high Re and low Os concentrations in old crustal rocks, their ^{187}Os/^{188}Os ratio increases rapidly with time (average upper crustal ^{187}Os/^{188}Os = 1–1.2). In contrast, meteorites have low ^{187}Os/^{188}Os ratios of ~0.11–0.18, and, because Os is much more abundant in meteorites than Re, only small changes in the meteoritic ^{187}Os/^{188}Os ratio occur with time. In the laboratory, the abundances of Os (in some cases also of Re) and the ^{187}Os/^{188}Os isotopic ratio are measured by sensitive negative thermal ionization mass spectrometry (NTIMS) (e.g., Creaser et al., 1991).

The first application of this isotope system to impact-related materials was reported by Luck and Turekian (1983), who measured the Os isotopic composition of K-T boundary clays, which was feasible because of the relatively high Os abundances (tens of parts per billion) in these rocks. However, the first application for impact-crater studies, by Fehn et al. (1986), was hampered by a lack in sensitivity, because target rocks and typical impactites have very low (sub-parts per billion) Os abundances. This changed with the introduction of the NTIMS technique (Völkening et al., 1991), which allowed the measurement of Os isotopic ratios in samples of a few grams mass containing sub-parts per billion amounts of Os. The first successful application of this method dealt with the Bosumtwi impact structure in Ghana (Koeberl and Shirey, 1993).

As a result of the relatively high Os abundances in meteorites, the addition of even a small amount of meteoritic matter to the crustal target rocks leads to an almost complete change of the Os isotopic signature of the resulting impact melt or breccia. Contamination from an achondritic meteorite requires much higher percentages of meteoritic addition due to the much lower PGE abundances in achondrites compared to chondritic and iron meteorites. In addition, the present-day ^{187}Os/^{188}Os ratio of mantle rocks is ~0.13, which is similar to meteoritic values. Thus, contribution of a significant mantle component (e.g., ultramafic rocks) in an impact breccia or melt rock has to be excluded. The PGE abundances in typical rocks from the upper mantle are at least two orders of magnitude lower than those in (chondritic and iron) meteorites. Thus, to achieve the same disturbance of the Os isotope value, at least 100 times higher mantle contributions (i.e., 10 wt% or more) than meteoritic contributions (e.g., 0.1 wt%) would be needed. The presence of such a significant mantle component would be easily discernable from petrographic studies of the clast population in breccias, and/or from geochemical data (especially Sr and Nd isotopes) of melt rocks. Compared to the use of PGE elemental abundances and ratios, the Os isotope method is superior with respect to detection limit and selectivity, as discussed in detail (with several case histories) by Koeberl and Shirey (1997). There are, however, a few disadvantages; the laboratory procedures (sample preparation, digestion, and measurement) are complex, and the Os isotopic method does not allow determination of the projectile type.

Chromium isotopic method

This new method provides evidence not only for the presence of an extraterrestrial component in impactites, but also information regarding the type of projectiles (Shukolyukov and Lugmair, 1998). It is based on the determination of the relative abundances of ^{53}Cr, which is the daughter product of the extinct radionuclide ^{53}Mn (half-life = 3.7 m.y.). The ^{53}Cr relative abundances are measured as the deviations of the $^{53}Cr/^{52}Cr$ ratio in a sample relative to the standard terrestrial $^{53}Cr/^{52}Cr$ ratio by high-precision TIMS. These deviations are usually expressed in ε units (1 ε is 1 part in 10^4, or 0.01%). Terrestrial rocks are not expected to show (and do not reveal) any variation in $^{53}Cr/^{52}Cr$ ratio, because the homogenization of the Earth was completed long after all ^{53}Mn (which was injected into, or formed within, the solar nebula) had decayed. In contrast, most meteorite groups analyzed so far (Shukolyukov and Lugmair, 1998, 2000), such as carbonaceous, ordinary, and enstatite chondrites, primitive achondrites, and other differentiated meteorites (including the SNC meteorites, which originated from Mars) show a variable excess of ^{53}Cr relative to terrestrial samples. The range for meteorites is about +0.1 to +1.3 ε, depending on the meteorite type, except for carbonaceous chondrites, which show an apparent deficit in ^{53}Cr of ~ −0.4 ε. These differences reflect a heterogeneous distribution of ^{53}Mn in the early solar system and early Mn/Cr fractionation in the solar nebula and in meteorite parent bodies (Lugmair and Shukolyukov, 1998). The negative ε value for carbonaceous chondrites is an artifact of using the $^{54}Cr/^{52}Cr$ ratio for a second-order fractionation correction (Lugmair and Shukolyukov, 1998), because these meteorites carry a pre-solar ^{54}Cr component (Shukolyukov et al., 2000). The actual, unnormalized $^{53}Cr/^{52}Cr$ ratio is similar to that of other undifferentiated meteorites, and the apparent ^{53}Cr deficit in the carbonaceous chondrites is actually due to an excess of ^{54}Cr. However, the presence of ^{54}Cr excesses in bulk carbonaceous chondrites allows us to distinguish clearly these meteorites from the other meteorite classes.

Shukolyukov and Lugmair (1998) have used the Cr isotopic method on samples from the K-T boundary in Denmark and Spain and found that ~80% of the Cr in these samples originated from an impactor with a carbonaceous chondritic composition. Shukolyukov et al. (2000) found a similar (carbonaceous chondritic) signal for some samples from Archean spherule layers in the Barberton Mountain Land, South Africa. Shukolyukov and Lugmair (2000) reported on analyses of impact-melt rocks from the East Clearwater (Canada), Lappajärvi (Finland), and Rochechouart (France) impact structures. In all three cases, the impactors were found to have had ordinary chondritic composition.

The Cr isotopic method thus has the advantage over both the use of Os isotopes or PGE abundance ratios of being selective not only regarding the Cr source (terrestrial versus extraterrestrial), but also regarding the meteorite type. A disadvantage of this method is the complicated and time-consuming analytical procedure. In addition, a significant proportion of the Cr in an impactite, compared to the abundance in the target, has to be of extraterrestrial origin. This point is illustrated in Figure 2, which shows the detection limit of the Cr isotopic method, assuming a chondritic composition of an extraterrestrial component. This detection limit is a function of the Cr content in the (terrestrial) target rocks that were involved in the formation of the impact breccias or melt rocks. For example, if the average Cr concentration in the target is ~185 ppm (the average Cr concentration in the bulk continental crust; Taylor and McLennan, 1985), only an extraterrestrial component of >1.2% can be detected.

Analytical procedures

For the present study, we selected impact-melt rocks and target rocks from two large impact structures in South Africa, Vredefort and Morokweng (see following for details on the structures and samples). The Os isotopic analyses of the Vredefort rocks are those of Koeberl et al. (1996), where analytical details are given. We analyzed 12 samples from the Morokweng impact structure for Os isotopic composition, and the abundances of Os, Ir, and Pt were determined using methods described in detail by Ravizza and Pyle (1997) and Hassler et al.

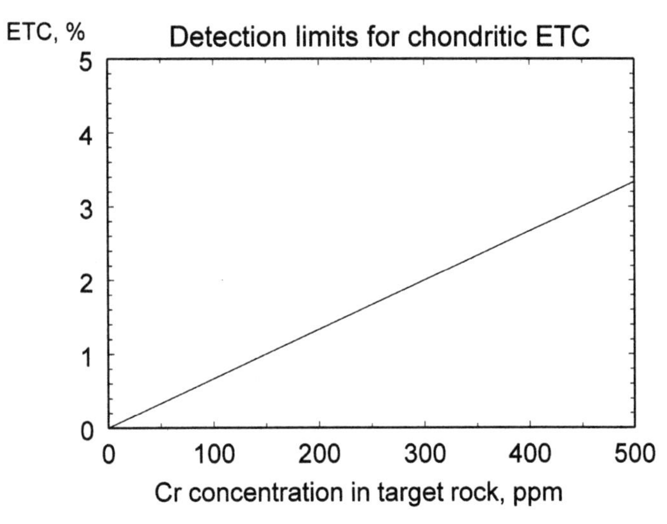

Figure 2. Detection limits for extraterrestrial component (ETC) in impactites (e.g., suevitic breccias or impact-melt rocks), using chromium isotope method. Detection limit, assuming chondritic composition of extraterrestrial component, is plotted (in percent) on y-axis as function of Cr content in terrestrial target rocks that were involved in impactite formation.

(2000). The samples were digested by NiS fire assay, followed by transfer of volatile OsO_4 by an Ar gas stream directly (without nebulizer) into the torch of a magnetic sector ICP-MS (Finnigan Element), allowing the rapid determination of the Os content and isotopic composition. The liquid residue after sparging of Os can then be used for conventional inductively coupled plasma mass spectroscopy (ICP-MS) measurement of the abundances of the PGEs (here, Ir, Pt, and Os). Due to the considerable analytical effort for the determination of the Cr isotopic data, only two melt-rock samples from each of the two impact structures were measured. The methods were described in detail by Shukolyukov and Lugmair (1998) and Shukolyukov et al. (2000).

VREDEFORT IMPACT STRUCTURE: BACKGROUND

The Vredefort structure in South Africa, centered ~140 km southwest of Johannesburg, currently has a diameter of ~100 km, which is believed to represent the central uplift of this impact structure. The entire impact structure could initially have been as large as 300 km, comprising the entire Witwatersrand basin (Henkel and Reimold, 1998). The origin of the structure has been debated during most of the twentieth century; e.g., microscopic deformation features in quartz, with an appearance similar to deformation lamellae observed in some tectonically deformed rocks, were controversial because of their unusual appearance at the optical level (cf. Grieve et al., 1990; Reimold, 1993). However, recent data confirmed the existence of impact-characteristic shock-metamorphic effects in Vredefort rocks, such as basal Brazil twins in quartz (Leroux et al., 1994) and shock-characteristic planar deformation features (PDFs) in zircon (Kamo et al., 1996), supporting an impact origin as opposed to an internal origin of the structure (see Reimold and Gibson, 1996, for a review). The Vredefort event is well dated as 2023 ± 4 Ma (Kamo et al., 1996). Dikes of granophyric rock, the so-called Vredefort granophyre, occur in the basement core of the structure and along the boundary between the core and the supracrustal rocks of the collar. Previous studies indicated that the granophyre could have formed by impact melting of granite, shale, and quartzite. A major mafic contribution is unlikely due to the scarcity of mafic clasts and because the granophyre composition can be perfectly modeled from mixing of felsic crustal and supracrustal rocks (French et al., 1989; French and Nielsen, 1990; Reimold et al., 1990; Koeberl et al., 1996; Reimold and Gibson, 1996). All these authors agreed that the Vredefort granophyre represents an impact-melt rock, probably material that gravitationally settled into impact-generated fractures in the crater basement after the impact event.

MOROKWENG IMPACT STRUCTURE: BACKGROUND

The Morokweng impact structure is centered at 23°32′ E and 26°20′ S, close to the border with Botswana, in the Northwest Province of South Africa. The structure was recognized as a circular positive magnetic anomaly of as much as 350 nT above regional background. This anomaly forms a central 30-km-diameter, near-circular area, which is surrounded by a concentric, magnetically quiet zone that is 20 km wide. Refined processing of the gravity and aeromagnetic data revealed the possible presence of a larger circular structure (Corner et al., 1997). The discovery of impact-characteristic shock-metamorphic effects in rocks from the Morokweng area (e.g., Corner et al., 1997; Koeberl et al., 1997; Hart et al., 1997) confirmed the presence of a large meteorite-impact structure. The size of the Morokweng impact structure is still debated (e.g., Reimold et al., 1999; Andreoli et al., 1999), but new evidence (Reimold et al., 2000) seems to favor a diameter of ≥70–80 km. Three boreholes in the south-central area of the aeromagnetic anomaly (MWF03, core depth 130.3 m; MWF04, 189.3 m; MWF05, 271.3 m) were sampled (see Koeberl et al., 1997; Reimold et al., 1999, for detailed information on the three cores).

All three boreholes penetrated a top layer of the Tertiary to Holocene Kalahari Group calcrete, which is directly underlain by a dark-brown melt rock having a thickness of about 125 m in borehole MWF05. Only in borehole MWF05 was the lower contact of the melt rock intersected. In this hole, granitic rocks were reached at 225 m depth, whereas the other holes were terminated while still within the melt-rock unit. Most of the melt rock appears fresh and homogeneous, except for a large number of lithic clasts. The clast population includes many gabbro fragments, but microscopic studies reveal that felsic, clearly granitoid-derived, clasts are dominant. The granites drilled below the melt body in core MWF05 are locally brecciated and pervasively recrystallized. Some primary minerals are preserved and display shock deformation in the form of PDFs in quartz, plagioclase, alkali feldspar, and K-feldspar (Koeberl et al., 1997; Reimold et al., 1999). Some thin (<10 cm wide) breccia veins occur in the granitoids of drill core MWF05, which have since been identified as injections of melt, or as partly recrystallized, locally produced, cataclastic material. It is not clear yet whether granitoid basement has been reached at the bottom of drill core MWF05, but it is possible that a (mega?) breccia zone below the melt rock and above the basement was intersected. SHRIMP ion probe dating of zircons from the melt rock yielded an age of 146.2 ± 1.5 Ma, which is indistinguishable from that of the Jurassic-Cretaceous boundary (Koeberl et al., 1997; Reimold et al., 1999).

In 1999, we were able to sample and study a 3.5-km-long KHK-1 drill core obtained by Anglogold Limited on the farm Kelso 351 (~23°12′E, 26°40′S), ~38–40 km southwest of the presumed center of the impact structure. The rocks recovered from this core are considered to be representative of the Morokweng target rocks, especially the more mafic rocks that were not well represented among the (predominantly granitoid) clast population of the three central Morokweng drill cores. Preliminary stratigraphic results for the KHK-1 core were presented by Reimold and Koeberl (1999), and further details on the drill core stratigraphy, as well as first results of chemical and chronological analysis of drill core samples, were reported by Reimold et al. (2000). The top 599.15 m (not recovered) represent dolomite and chert of the 2.25–2.5 Ga Transvaal Supergroup. This package continues until 889.1 m, to the contact with a gabbroic intrusion. Some dolomite follows, above a package of arenitic metasediments (to 1200 m). A series of mafic volcanic flows was intersected to a depth of 1417.8 m, where a 1.2 m cataclastic breccia that exhibited no shock-deformation features was observed. The breccia is underlain by more metasediment, including some diamictite also lacking shock-deformation features. Felsic volcanics and a felsic granophyre follow until 1784.2 m, below which more metasediment, including two thin diamictite bands, was encountered. Below this depth, a thick package of felsic granophyre was intersected to the depth of 2669.4 m, where it is terminated by a gabbroic intrusion. We did not find any quartz clasts with PDFs in this felsic granophyre. A thick gabbro intrusion follows until 3012 m depth, where locally pegmatoidal, but mostly micropegmatoidal, granitoids continue until the final depth of the borehole (3420 m). Whereas no evidence of shock metamorphism could be detected in any of the granophyric rocks of the drill core, there is ample evidence that at least some of the granophyric material is of secondary origin, formed from melting of a granitoid precursor.

RESULTS: METEORITIC COMPONENT AT VREDEFORT

Vredefort granophyre has been interpreted either as an igneous intrusion or as impact-melt rock that was injected into fractures in the floor of the impact structure (French et al., 1989; French and Nielsen, 1990). In addition, French et al. (1989) analyzed the Ir content of seven Granophyre samples and reported a range of 57–130 ppt Ir (Fig. 1). These authors also analyzed a suite of country rocks and found that the Vredefort granophyre is enriched in Ir by a factor of ~20–50 compared to granitic country rocks (target rocks?). However, they also found that some Witwatersrand shale samples have Ir contents of 160–330 ppt (Fig. 1). French et al. (1989) suggested that all Ir in the Vredefort granophyre could be explained by admixture of Ir from shales and similar rocks during impact melting. However, this would require that at least one-third of the Vredefort granophyre composition be derived from shale, which contradicts mixing calculations of French and Nielsen (1990), who only found a 10% shale contribution. All other relevant country rocks have Ir contents of 3–62 ppt. Thus, it seems that there is excess Ir in the Vredefort granophyre; nevertheless, this ambiguity made the Vredefort granophyre, and Vredefort in general, an ideal target for an Os isotopic study.

Such a study was undertaken by Koeberl et al. (1996) in order to (1) search for a meteoritic component in Vredefort granophyre, and (2) to confirm that the granophyre represents an impact-melt rock. The Os abundances in the granophyre range from 0.11 to 1.11 ppb, which is significantly higher than the average of the source-rock values, indicating a distinct enrichment of Os in most of the granophyre samples compared to the country rocks. In addition, the $^{187}Os/^{188}Os$ ratios of the granophyre samples are significantly lower than those of the supracrustal country rocks (Table 1). The $^{187}Re/^{188}Os$ and $^{187}Os/^{188}Os$ ratios of the Vredefort granophyre scatter about a 2 Ga isochron, the majority of the initial $^{187}Os/^{188}Os$ ratios (at 2 Ga) ranging from 0.13 to 0.22 (Fig. 3). These values overlap the meteoritic data range and indicate that all measured granophyre samples contain some meteoritic Os. In addition, the Re-Os isotopic composition of the granophyre is significantly different from that of any of the target rocks. Shale samples were also analyzed and yielded values of 37–162 ppt Os, but the Os isotopic compositions are significantly different from those of the granophyre samples.

Thus, a meteoritic component in the Vredefort granophyre is the only explanation that is in agreement with these observations. This conclusion is supported by some enrichments in Cr, Co, Ni, and Ir in the granophyre compared to the country rocks, although these enrichments are not as unambiguous as the Re-Os data. Assuming chondritic meteorite Os abundances (~500 ppb), Koeberl et al. (1996) concluded that the Vredefort granophyre contains ≤0.2% of a chondritic component. However, the meteoritic component is not homogeneously distributed, probably due to a nugget effect, as is evident from a spread in $^{187}Re/^{188}Os$ ratios, in agreement with observations from other impact-melt rocks (cf. Koeberl, 1998). Meteorites contain ~10 times less Re than Os, indicating that the Re contribution from the meteoritic material to the Vredefort granophyre was ≤30%, subordinate to the Os, which is almost exclusively of meteoritic origin. The almost constant Re abundances found in the granophyre samples indicate homogenization during the impact. Thus, the Vredefort granophyre is a mixture of a large amount of low-Os, high-Re material (crustal rocks) with a small contribution of high-Os, low-Re meteoritic material. The Os isotope results showed that all analyzed Vredefort granophyre samples contain some meteoritic Os, confirming that the Vredefort granophyre is an impact-melt rock.

To confirm the findings from the Os isotope study, and for comparison of the two isotopic techniques, we analyzed two Vredefort granophyre samples for their Cr isotopic composition. In contrast to the Os isotope study, the Cr isotope study did not reveal the presence of extraterrestrial Cr: the two Vredefort granophyre samples yielded $^{53}Cr/^{52}Cr$ values of -0.01 ± 0.06 ε and $+0.03 \pm 0.03$ ε ($2\sigma_{mean}$), which is indistinguishable from the terrestrial mean value (0 ε) (Table 2). The average Cr content of the Vredefort granophyre is ~420 ppm (Koeberl et al., 1996; Reimold et al., 1999), whereas Cr contents in the various target rocks range from 6 to 750 ppm; a possible average Cr content is ~300 ppm. Based on the Cr data, the upper limit for a chondritic component in these samples is ~2% (Fig. 2), which is consistent with the better defined limit (~0.2%) provided by the Os isotopes. An abundance of terrestrial Cr masks the cosmic Cr and, in this case, the sensitivity of the Cr

TABLE 1. COMPOSITION OF VREDEFORT GRANOPHYRE AND COUNTRY ROCK AND MOROKWENG DRILL CORE SAMPLES

Sample	Type	Ir (ppb)	Pt (ppb)	Os (ppb)	$^{187}Os/^{188}Os$
Vredefort samples					
AV81-22	Granophyre	0.13	n.d.	n.d.	n.d.
AV81-59	Granophyre	0.094	3.0	n.d.	n.d.
AV81-63	Granophyre	0.105	1.6	n.d.	n.d.
BG-7/2	Granophyre	n.d.	n.d.	0.111	0.558
BG-9	Granophyre	n.d.	n.d.	0.228	0.410
BG-4/2	Granophyre	n.d.	n.d.	0.356	0.286
BG-4/1	Granophyre	n.d.	n.d.	1.11	0.196
AV81-35	Alkali granite	0.0011	<0.5	n.d.	n.d.
AV81-5	Ventersdorp Lava	0.016	n.d.	n.d.	n.d.
AV81-12	Witwatersrand shale	0.290	n.d.	n.d.	n.d.
UP-63	Ventersdorp Lava	n.d.	n.d.	0.0158	6.272
SNE	Witwatersrand shale	n.d.	n.d.	0.162	1.013
S1/2	Witwatersrand shale	n.d.	n.d.	0.0368	1.572
Morokweng drill core samples					
MO-15	Melt rock	9.6	18.1	7.87	0.1316
MO-20	Melt rock	7.7	16.4	6.31	0.1341
MO-35cl	Clast	11.2	25.1	8.11	0.1449
MO-37a	Melt rock	4.0	11.0	3.45	0.1319
MO-48	Melt rock	10.4	20.6	8.84	0.1320
MO-63a	Breccia	0.029	<0.040	0.34	0.2027
MO-65	Granite	0.016	0.907	0.29	0.2417
MO-69b	Breccia	0.004	<0.040	0.035	1.383
MO-70	Breccia	0.016	n.d.	0.036	0.1283
1006	Mafic rock	0.014	0.0059	0.042	9.533
1536.7	Mafic rock	<0.002	0.184	0.032	12.41
2617.8	Mafic rock	<0.002	<0.040	0.009	0.6423

Note: Vredefort Os content and Os isotope data from Koeberl et al. (1996); Ir and Pt data from French et al. (1989). All Morokweng data (this work) by isotope dilution magnetic sector (inductively coupled plasma mass spectroscopy) (Finnigan Element). Os analyses done by sparging of OsO_4 in Ar directly into the ICP-MS (Hassler et al., 2000). n.d. = not determined.

isotopic method is not sufficient to resolve the meteoritic component.

RESULTS: METEORITIC COMPONENT AT MOROKWENG

With the exception of a few obviously altered melt rock samples, the Morokweng melt body is extremely homogeneous in composition. Variations for major elements do not exceed 2–5 relative%. The Morokweng melt rock contains relatively high proportions of CaO (on average, 3.41 wt%), MgO (3.70 wt%), and Fe_2O_3 (5.87 wt%), and an average SiO_2 content of 65.75 wt%. Siderophile elements are consistently enriched (the variation between samples is less than a factor of 2) in the melt-rock samples in comparison with rocks of such major element composition (granodioritic to dioritic), with average values of 440 ppm for Cr, 50 ppm for Co, 780 ppm for Ni, and 32 ppb for Ir. No variation with depth and no differences between drill cores were found (Koeberl et al., 1997; Reimold et al., 1999). Koeberl et al. (1997) noted that it was highly unlikely that mafic to ultramafic country rocks, or other mantle-derived sources, were responsible for these high siderophile element and Ir concentrations. The abundances of the PGEs in the various Morokweng impact-melt rock samples analyzed were found to have almost chondritic ratios, and, depending on the values used for normalization, ~2–5 wt% of a chondritic component is present in the melt rocks (Koeberl et al., 1997). In contrast to melt rocks from most other impact structures (Koeberl, 1998), the meteoritic component has a high abundance and is uniformly distributed in the Morokweng impact-melt rocks. To confirm the presence of this component, we performed Cr and Os isotopic studies, the preliminary results of which were reported in abstract form by Shukolyukov et al. (1999) and Koeberl et al. (2000), respectively.

We measured 12 samples from the MWF03, MWF04, MWF05, and KHK-1 cores by magnetic sector ICP-MS for the Os isotopic composition and the abundances of Ir, Os, and Pt. The samples included four impact-melt rocks, one melt-saturated clast within a melt rock, three vein breccia samples (Reimold et al., 1999), a granite clast, and three mafic target rocks. Stratigraphic and petrographic information on these samples was given by Koeberl et al. (1997) and Reimold et al. (1999, 2000). The analytical results are given in Table 1. The elemental abundances of the three PGEs measured in this study agree well with data from Koeberl et al. (1997), Andreoli et al. (1999), and McDonald et al. (2001). Iridium abundances in the melt-rock samples range up to 11.2 ppb (Fig. 1), and there are corresponding high abundances of Os and Pt. PGE abundances in granitic and mafic target rocks are very low; Ir values in the mafic rocks are <2–14 ppt, and 4–29 ppt in the felsic rocks or vein breccias, and abundances of Os and Pt are somewhat higher (Table 1). The target rock values measured here are below the detection limit of all other studies cited above. The Ir and Os data are almost identical to data obtained by NiS fire assay and/or neutron activation analysis (Koeberl et al., 1997; Koeberl and Reimold, 2002), but Pt data in the present data set are somewhat lower. It is interesting that the Ir abundances measured by instrumental neutron activation analysis in two different laboratories (Koeberl et al., 1997; Koeberl and Reimold, 2002; Andreoli et al., 1999) are higher by a factor of two or more compared to the ICP-MS data. The chondrite-normalized abundance patterns of the PGEs (and other siderophile elements), however, have similar shapes, at elevated absolute abundances. The reasons for this discrepancy are unknown, but other comparative studies (neutron activation versus ICP-MS) also found a similar result (Dai et al., 2001). Nevertheless, the chondrite-normalized abundance patterns of the impact-melt rock PGE data given in Table 1 are flat and indicate the presence of ~2% of a meteoritic component, whereas the normalized patterns of the various felsic and mafic target rocks are highly fractionated (for a more detailed discussion, see Koeberl and Reimold, 2002).

The results for Os show high contents of as much as ~9 ppb in the impact-melt rock samples, and correspondingly low, but remarkably uniform, isotopic ratios ($^{187}Os/^{188}Os$ 0.1316–0.1341). The breccias show a much wider variation in both isotope ratio and Os abundance. One of the breccias has a low Os abundance, but a fairly low isotope ratio. Considering that this breccia (MO70) is of granitic composition, and other granites or granitic breccias may have lower PGE contents, it is possible that this sample contains a small, but dominating, extraterrestrial Os component. Figure 4 shows the Os content versus the Os isotopic composition. The MO70 breccia sample plots close to the hyperbolic mixing line that connects the meteorite and target rock fields. The impact-melt rock samples plot close to this line at higher meteorite contribution values. The mafic rock samples from the KHK-1 deep drill core, representing the mafic contribution to the Morokweng impact-melt rocks, have very low Os abundances (9–42 ppt) and high isotope ratios ($^{187}Os/^{188}Os$ to 12). This result clearly indicates that the mafic target rocks of the area did not contribute measurably

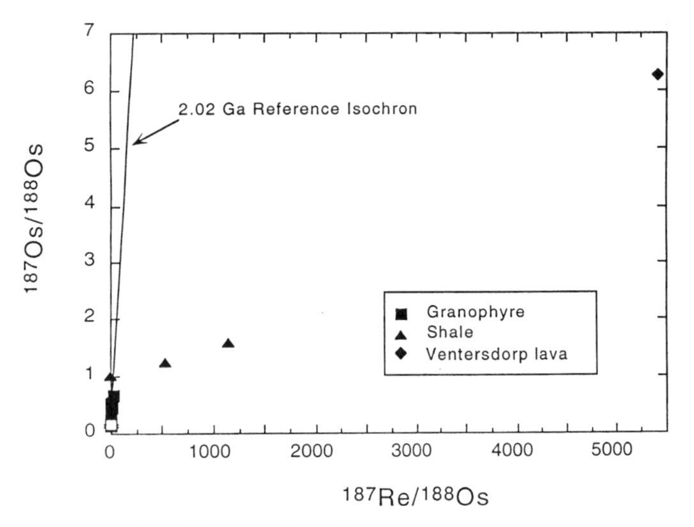

Figure 3. $^{187}Os/^{188}Os$ vs. $^{187}Re/^{188}Os$ diagram for Vredefort granophyre samples and some country rocks that are assumed to be representative of target rocks, compared to field for chondrites and iron meteorites (after Koeberl et al., 1996). All but one of Vredefort granophyre samples plot close to 2 Ga reference isochron, which intersects meteorite data array. This shows that Vredefort granophyre samples have narrow range of initial $^{187}Os/^{188}Os$ (at 2 Ga) ratios that overlap range of meteoritic initial ratios, which, in absence of any significant mantle rock component in Vredefort granophyre (Koeberl et al., 1996), confirms presence of meteoritic component in these samples, and that Vredefort granophyre is impact-melt rock.

TABLE 2. CHROMIUM ISOTOPE DATA FOR VREDEFORT GRANOPHYRE AND MOROKWENG DRILL CORE SAMPLES

Sample	Type	Cr (ppm)	$^{53}Cr/^{52}Cr$ (ε)
Vredefort samples			
AV81-22	Granophyre	394	n.d.
BG-9	Granophyre	419	−0.01 ± 0.06
BG-4/2	Granophyre	429	+0.03 ± 0.03
AV81-35	Alkali granite	2.7	n.d.
UP-63	Ventersdorp Lava	223	n.d.
SNE	Witwatersrand shale	142	n.d.
S1	Witwatersrand shale	138	n.d.
Morokweng drill core samples			
MO-15	Melt rock	359	+0.24 ± 0.04
MO-48	Melt rock	408	+0.27 ± 0.03

Note: Vredefort Cr content from Koeberl et al. (1996); Morokweng Cr content from Koeberl and Reimold (2002). n.d. = not determined.

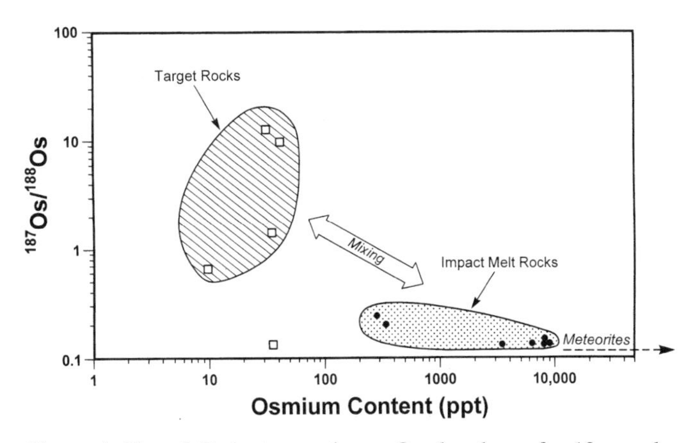

Figure 4. Plot of Os isotope ratio vs. Os abundance for 12 samples from Morokweng impact structure, South Africa. Target rocks (including those from deep drill core outside of structure) define low Os abundance, radiogenic field, whereas impact-melt rocks have clear meteoritic component (high Os concentration, nonradiogenic). Mixing between two components follows mixing hyperbola (or set of hyperbolas) that passes close to vein breccia point just below target rock field.

to the high siderophile element abundances observed in the melt rocks. It is evident that the Os isotope data confirm the presence of a meteoritic component in these melt rocks.

Because the meteoritic component in the Morokweng impact-melt rocks seems to be fairly abundant, we performed a Cr isotopic study. This study had the following goals: first, we wanted to confirm the extraterrestrial nature of the siderophile element anomaly indicated by the PGE data of Koeberl et al. (1997) and Hart et al. (1997), and second, to use the potential of the Cr isotope signal to provide information about the type of projectile involved. The results (Table 2) indicated that about half of the Cr in the melt rock is of extraterrestrial origin. It was also possible to show that only an ordinary chondritic source, rather than a carbonaceous chondritic source, can explain the Cr isotope data. A similar conclusion, based only on PGE abundance data, was obtained by Koeberl et al. (1997), who concluded that the PGE data fit best an ordinary chondrite source (probably H-chondrite). However, using the Cr isotope data together with trace element data (Fig. 5; Shukolyukov et al., 1999), we conclude that the Morokweng bolide was most likely of L-chondritic composition. This interpretation agrees well with the conclusions based on PGE abundances (McDonald et al., 2001).

CONCLUSIONS

It is clear that there are advantages and disadvantages of both the Os and Cr isotopic methods for the determination of a meteoritic component in impact breccias, relative to each other and in comparison with PGE abundance measurements. PGE abundance studies may indicate the presence of an extraterrestrial component, and the normalized patterns can give an indication of the impactor type. However, it is not possible to distinguish between terrestrial and extraterrestrial PGEs. This determination is possible in principle by both the Os and Cr isotopic methods. In the Os isotopes, it is necessary to distinguish between a possible mantle signal and an extraterrestrial signal, because the Os isotopic ratios for mantle and meteoritic material are very similar. There is a significant difference in the total abundance of Os in mantle rocks (~1–4 ppb Os) and meteorites (typical chondrites have ~400–800 ppb Os). Thus, at least 100 times more mantle than meteoritic material needs to be added to normal crustal rocks (e.g., in an impact breccia) to result in the same Os isotopic ratio of the bulk rock. Detailed field investigations and petrographic studies, if necessary combined with trace element and Rb-Sr and/or Sm-Nd isotopic analyses, will show if significant amounts of ultramafic materials are present.

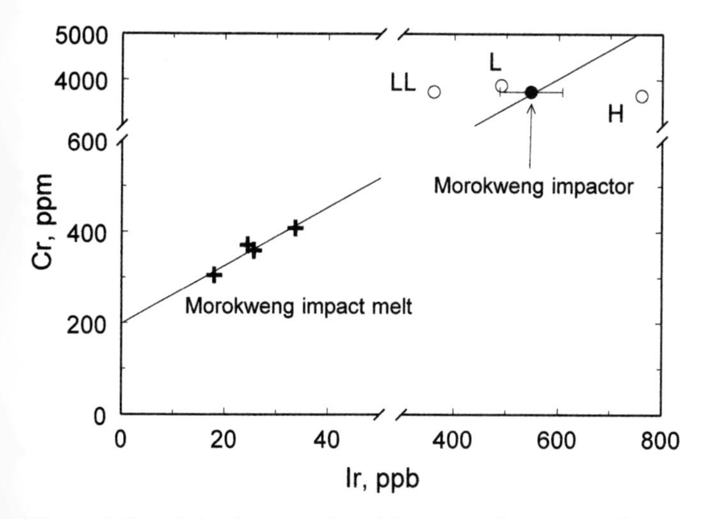

Figure 5. Correlation between Cr and Ir contents in impact-melt rock samples from Morokweng impact structure. Position of projectile point was determined using concentration of Ir (from Koeberl et al., 1997; Koeberl and Reimold, 2002) and proportion of cosmic Cr (obtained from Cr isotopic composition) in samples MO15 and MO48. Cr and Ir concentrations for ordinary chondrites are from Wasson and Kallemeyn (1988).

However, the Cr isotopic composition allows the distinction between terrestrial and extraterrestrial Cr, and there is a difference between some meteorite groups (Shukolyukov and Lugmair, 1998). For example, ordinary chondrites (types H and L) have isotopic characteristics that are slightly different from those of enstatite chondrites and are, even more significantly, different from those of carbonaceous chondrites. Thus, assuming that a certain (fairly significant) percentage of the chromium in an impact melt of an ejecta layer is of extraterrestrial origin, it should be possible by measuring the Cr isotopic composition to determine not only that an extraterrestrial component is present, but also which meteorite type might have been involved. This determination is not possible for iron meteorites, which do not carry significant amounts of Cr, but basaltic achondrites (such as the eucrites, which have very low PGE abundances and, thus, low Os) have high Cr abundances. Therefore, the Cr isotopic method may be the only reasonable way to identify achondritic impactors. The analytical effort for Cr isotope measurements is substantial. The differences in isotopic composition are very small, requiring extremely precise and time-consuming measurements, severely limiting the number of samples that can be measured. In addition, a substantial amount of chromium has to be of extraterrestrial origin to show an effect in the Cr isotopic composition that is larger than the precision of the measurement. Thus, while being more selective than the Os isotopic method, the Cr isotopic method is less sensitive.

ACKNOWLEDGMENTS

Supported by the Austrian Fonds zur Förderung der wissenschaftlichen Forschung, Y58-GEO (Koeberl), the U.S. National Science Foundation (Peucker-Ehrenbrink), the National Research Foundation of South Africa (Reimold), and NASA grant 5-8172 (Shukolyukov and Lugmair). We thank Anglogold Limited and their Consulting Geologist G. Cantello for permission to study the KHK-1 borehole and to publish these results. We also thank M. Humayun, F.T. Kyte, and K. MacLeod for helpful suggestions on the manuscript, and Kyte for the hint that a eucrite signal could be detected by using the Cr isotopic method. This is Wits Impact Cratering Research Group contribution 16, and Woods Hole Oceanographic Institution contribution 10410.

REFERENCES CITED

Alvarez, L.W., Alvarez, W., Asaro, F., and Michel, H.V., 1980, Extraterrestrial cause for the Cretaceous-Tertiary extinction: Science, v. 208, p. 1095–1108.

Andreoli, M.A.G., Ashwal, L.D., Hart, R.J., and Huizenga, J.M., 1999, A Ni- and PGE-enriched quartz norite impact melt complex in the Late Jurassic Morokweng impact structure, South Africa, *in* Dressler, B.O., and Sharpton, V.L., eds., Large meteorite impacts and planetary evolution 2, Geological Society of America Special Paper 339, p. 91–108.

Corner, B., Reimold, W.U., Brandt, D., and Koeberl, C., 1997, Morokweng impact structure, Northwest Province, South Africa: Geophysical imaging and some preliminary shock petrographic studies: Earth and Planetary Science Letters, v. 146, p. 351–364.

Creaser, R.A., Papanastassiou, D.A., and Wasserburg, G.J., 1991, Negative thermal ion mass spectrometry of osmium, rhenium and iridium: Geochimica et Cosmochimica Acta, v. 55, p. 397–401.

Dai, X., Koeberl, C., and Fröschl, H., 2001, Determination of PGEs in impact melt breccias using NAA and USN-ICP-MS after anion exchange preconcentration: Analytica Chimica Acta, v. 436, p. 79–85.

Fehn, U., Teng, R., Elmore, D., and Kubik, P.W., 1986, Isotopic composition of osmium in terrestrial samples determined by accelerator mass spectrometry: Nature, v. 323, p. 707–710.

French, B.M., and Nielsen, R.L., 1990, Vredefort bronzite granophyre: Chemical evidence for an origin as meteorite impact melt: Tectonophysics, v. 171, p. 119–138.

French, B.M., Orth, C.J., and Quintana, C.R., 1989, Iridium in the Vredefort bronzite granophyre: Impact melting and limits on a possible extraterrestrial component, *in* Proceedings, 19th Lunar and Planetary Science Conference: New York, Cambridge University Press, p. 733–744.

Grieve, R.A.F., Coderre, J.M., Robertson, P.B., and Alexopoulos, J., 1990, Microscopic planar deformation features in quartz of the Vredefort structure: Anomalous but still suggestive of an impact origin: Tectonophysics, v. 171, p. 185–200.

Hart, R.J., Andreoli, M.A.G., Tredoux, M., Moser, D., Ashwal, L.D., Eide, E.A., Webb, S.J., and Brandt, D., 1997, Late Jurassic age for the Morokweng impact structure, southern Africa: Earth and Planetary Science Letters, v. 147, p. 25–35.

Hassler, D.R., Peucker-Ehrenbrink, B., and Ravizza, G.E., 2000, Rapid determination of Os isotopic composition by sparging OsO_4 into a magnetic-sector ICP-MS: Chemical Geology, v. 166, p. 1–14.

Henkel, H., and Reimold, W.U., 1998, Integrated geophysical modelling of a giant, complex impact structure: Anatomy of the Vredefort Structure, South Africa: Tectonophysics, v. 287, p. 1–20.

Kamo, S.L., Reimold, W.U., Krogh, T.E., and Colliston, W.P., 1996, A 2.023 Ga age for the Vredefort impact event and a first report of shock metamorphosed zircons in pseudotachylitic breccias and Granophyre: Earth and Planetary Science Letters, v. 144, p. 369–388.

Koeberl, C., 1998, Identification of meteoritical components in impactites, *in* Grady, M.M., Hutchison, R., McCall, G.J.H., and Rothery, D.A., eds., Meteorites: Flux with time and impact effects: Geological Society [London] Special Publication 140, p. 133–152.

Koeberl, C., and Reimold, W.U., 2002, Geochemistry and petrography of impact breccias and target rocks from the 145 Ma Morokweng Impact Structure, South Africa: Geochimica et Cosmochimica Acta, v. 66, in press.

Koeberl, C., and Shirey, S.B., 1993, Detection of a meteoritic component in Ivory Coast tektites with rhenium-osmium isotopes: Science, v. 261, p. 595–598.

Koeberl, C., and Shirey, S.B., 1997, Re-Os systematics as a diagnostic tool for the study of impact craters and distal ejecta: Palaeogeography, Palaeoclimatology, Palaeoecology, v. 132, p. 25–46.

Koeberl, C., Reimold, W.U., and Shirey, S.B., 1996, A Re-Os isotope and geochemical study of the Vredefort Granophyre: Clues to the origin of the Vredefort structure, South Africa: Geology, v. 24, p. 913–916.

Koeberl, C., Armstrong, R.A., and Reimold, W.U., 1997, Morokweng, South Africa: A large impact structure of Jurassic-Cretaceous boundary age: Geology, v. 25, p. 731–734.

Koeberl, C., Peucker-Ehrenbrink, B., and Reimold, W.U., 2000, Meteoritic component in impact melt rocks from the Morokweng, South Africa, impact structure: An Os isotopic study: Lunar and Planetary Science, v. 31, Abstract #1595, CD-ROM.

Leroux, H., Reimold, W.U., and Doukhan, J.C., 1994, A T.E.M. investigation of shock metamorphism in quartz from the Vredefort dome, South Africa: Tectonophysics, v. 230, p. 223–239.

Luck, J.M., and Turekian, K.K., 1983, Osmium-187/Osmium-186 in manganese nodules and the Cretaceous-Tertiary boundary: Science, v. 222, p. 613–615.

Lugmair, G.W., and Shukolyukov, A., 1998, Early solar system timescales according to ^{53}Mn-^{53}Cr systematics: Geochimica et Cosmochimica Acta, v. 62, p. 2863–2886.

McDonald, I., Andreoli M.A.G., Hart, R.J., and Tredoux, M., 2001, Platinum-group elements in the Morokweng impact structure, South Africa: Evidence for the impact of a large ordinary chondrite projectile at the Jurassic-Cretaceous boundary: Geochimica et Cosmochimica Acta, v. 65, p. 299–309.

Morgan, J.W., 1978, Lunar crater glasses and high-magnesium australites: Trace element volatilization and meteoritic contamination: Proceedings of the 9th Lunar and Planetary Science Conference: New York, Pergamon Press, p. 2713–2730.

Palme, H., 1982, Identification of projectiles of large terrestrial impact craters and some implications for the interpretation of Ir-rich Cretaceous/Tertiary boundary layers, *in* Silver, L.T., and Schultz, P.H., eds., Geological implications of impacts of large asteroids and comets on Earth: Geological Society of America Special Paper 190, p. 223–233.

Palme, H., Janssens, M.-J., Takahasi, H., Anders, E., and Hertogen, J., 1978, Meteorite material at five large impact craters: Geochimica et Cosmochimica Acta, v. 42, p. 313–323.

Palme, H., Göbel, E., and Grieve, R.A.F., 1979, The distribution of volatile and siderophile elements in the impact melt of East Clearwater (Quebec): Proceedings of the 10th Lunar and Planetary Science Conference: New York, Pergamon Press, p. 2465–2492.

Palme, H., Grieve, R.A.F., and Wolf, R., 1981, Identification of the projectile at the Brent crater, and further considerations of projectile types at terrestrial craters: Geochimica et Cosmochimica Acta, v. 45, p. 2417–2424.

Papike, J., editor, 1998, Planetary materials: Reviews in Mineralogy, v. 36, 915 p.

Ravizza, G., and Pyle, D., 1997, PGE and Os isotopic analyses of single sample aliquots with NiS fire assay preconcentration: Chemical Geology, v. 141, p. 251–268.

Reimold, W.U., 1993, A review of the geology of and deformation related to the Vredefort Structure, South Africa: Journal of Geological Education, v. 41, p. 106–117.

Reimold, W.U., and Gibson, R.L., 1996, Geology and evolution of the Vredefort impact structure, South Africa: Journal of African Earth Sciences, v. 23, p. 125–162.

Reimold, W.U., and Koeberl, C., 1999, The deep borehole into the Morokweng impact structure [abs.]: Meteoritics and Planetary Science, v. 34, p. A97.

Reimold, W.U., Horsch, H., and Durrheim, R.J., 1990, The "Bronzite"-Granophyre from the Vredefort structure: A detailed analytical study and re-

flections on the origin of one of Vredefort's enigmas: Proceedings of the 20th Lunar and Planetary Science Conference: Houston, Texas, Lunar and Planetary Institute, p. 433–450.
Reimold, W.U., Koeberl, C., Brandstätter, F., Kruger, F.J., Armstrong, R.A., and Bootsman, C., 1999, The Morokweng impact structure, South Africa: Geologic, petrographic, and isotopic results, and implications for the size of the structure, *in* Dressler, B.O., and Sharpton, V.L., eds., Large meteorite impacts and planetary evolution 2: Geological Society of America Special Paper 339, p. 61–90.
Reimold, W.U., Armstrong, R.A., and Koeberl, C., 2000, New results from the deep borehole at Morokweng, North West Province, South Africa: Constraints on the size of the J/K boundary age impact structure: Lunar and Planetary Science, v. 31, Abstract #1074, CD-ROM.
Shukolyukov, A., and Lugmair, G.W., 1998, Isotopic evidence for the Cretaceous-Tertiary impactor and its type: Science, v. 282, p. 927–929.
Shukolyukov, A., and Lugmair, G.W., 2000, Extraterrestrial matter on Earth: Evidence from the Cr isotopes [abs.], *in* Catastrophic events and mass extinctions: Impacts and beyond: Houston, Texas, Lunar and Planetary Institute, LPI Contribution No. 1053, p. 197–198.
Shukolyukov, A., Lugmair, G.W., Koeberl, C., and Reimold, W.U., 1999, Chromium in the Morokweng impact melt rocks: Isotope evidence for extraterrestrial component and type of the impactor [abs.]: Meteoritics and Planetary Science, v. 34, p. A107–A108.
Shukolyukov, A., Kyte, F.T., Lugmair, G.W., Lowe, D.R., and Byerly, G.R., 2000, The oldest impact deposits on earth: First confirmation of an extraterrestrial component. *in* Gilmour, I., and Koeberl, C., eds., Impacts and the early earth: Heidelberg, Germany, Springer-Verlag, Lecture Notes in Earth Sciences, v. 91, p. 99–116.
Taylor S.R., and McLennan, S.M., 1985, The continental crust: Its composition and evolution: Oxford, Blackwell, p. 57–72.
Völkening, J., Walczyk, T., and Heumann, K.G., 1991, Osmium isotope ratio determinations by negative thermal ionization mass spectrometry: International Journal of Mass Spectrometry and Ion Processes, v. 105, p. 147–159.
Wasson, J.T., 1985, Meteorites: New York, W.H. Freeman and Company, 267 p.
Wasson, J.T., and Kallemeyn, G., 1988, Compositions of chondrites: Philosophical Transactions of the Royal Society of London, ser. A, v. 325, p. 535–544.

MANUSCRIPT SUBMITTED OCTOBER 24, 2000; ACCEPTED BY THE SOCIETY MARCH 22, 2001

Printed in the U.S.A.

Geological Society of America
Special Paper 356
2002

Numerical modeling of the formation of large impact craters

Boris A. Ivanov*
Natalia A. Artemieva*
*Institute for Dynamics of Geospheres, Russian Academy of Sciences,
Leninsky Prospect 38-6, Moscow 117939, Russia*

ABSTRACT

Herein we present numerical simulations of Chicxulub-scale impact events. We used the three-dimensional version of the SOVA (solid, vapor, air) computer code in hydrodynamic approximation (no strength) to model the initial stage of an oblique impact, and the two-dimensional version of the SALE (simplified arbitrary Lagrangian and Eulerian) code for modeling of the crater collapse. Our estimates of the melt production are slightly higher than previously published estimates. In the case of a 15-km-diameter projectile, the melt zone may reach the crust-mantle boundary, but the mantle is not melted, because of its higher melting entropy. A surprising result is the considerable deviation of the transient cavity volume, obtained in our simulations, from traditional predictions of the experimental scaling law. It seems that high-velocity (>10 km/s) oblique impacts have almost the same cratering efficiency as vertical ones, while laboratory impacts (<5 km/s) follow the experimental scaling law. We estimate the volume of impact melt for a Chicxulub-size crater to be ~40 000 to 50 000 km^3, for projectile diameters estimated from the experimental scaling law. If all the melt was deposited inside the crater, this volume would be large enough to create a melt pool with a diameter of 100 km and a depth of 6 km. Ejection of melt outside the crater rim decreases the impact-melt body thickness. Implementation of acoustic fluidization into the SALE code allows us to reproduce Chicxulub as a peak-ring crater.

INTRODUCTION

The importance of the Cretaceous-Tertiary impact event, which formed the Chicxulub crater, for the terrestrial biosphere is widely accepted. This crater is a unique example of a large and relatively young impact structure, with a well-preserved subsurface structure, in contrast to two older large impact structures (Vredefort and Sudbury). Seismic, gravity, and magnetic data define a structure ~180 km in diameter (Penfield and Camargo, 1991). The Chicxulub structure consists of a centrally located positive gravity anomaly (~40 km diameter) and another ring-shaped anomaly (~100 km diameter), which is defined by two concentric negative anomaly troughs with diameters of ~70 and 120 km, reflecting the peak-ring structure (Hildebrand et al., 1991). Additional analysis of the gravity data detected a 104 ± 6 km diameter ring, a discontinuous 150 ± 16 km ring, and a 204 ± 16 km diameter rim (Sharpton et al., 1993). The seismic data reveal two deformation zones with normal faults at radial distances of 55–65 and 85–98 km and monoclines at the distances of 120–135 km (Snyder et al., 1999), as well as the general subsurface basin structure (Morgan et al., 1997; Morgan and Warner, 1999).

Numerical modeling of complex phenomena, such as impact crater collapse, still meets many problems (e.g., see

*E-mails: Ivanov, baivanov@idg.chph.ras.ru; Artemieva, art@idg.chph.ras.ru

Ivanov, B.A., and Artemieva, N.A., 2002, Numerical modeling of the formation of large impact craters, *in* Koeberl, C., and MacLeod, K.G., eds., Catastrophic Events and Mass Extinctions: Impacts and Beyond: Boulder, Colorado, Geological Society of America Special Paper 356, p. 619–630.

O'Keefe and Ahrens, 1993, 1999; Melosh and Ivanov, 1999). Herein we report some new results on attempts to quantitatively reproduce impact-crater morphology and subsurface structure.

Another important aspect of crater formation is the obliquity (angle) of impact. Most previous attempts to numerically simulate the formation of the Chicxulub crater have been done with two-dimensional hydrocodes (e.g., Roddy et al., 1987; Ivanov et al., 1996; O'Keefe and Ahrens, 1999). In this approach one assumes axial symmetry for the cratering flow. This assumption results in the inevitable restriction to simulate the vertical impact only. The most probable angle α of impact is 45°, the probability decreasing as $\sin 2\alpha$ (Shoemaker, 1962). Consequently, the probability of a strictly vertical impact of comets and asteroids on Earth and other planetary bodies is zero, while one-half of all impacts are in the range from 30–60°.

This issue was addressed by Pierazzo and Crawford (1998) and Pierazzo and Melosh (1999, 2000a, 2000b, 2000c), who primarily studied the initial stage of the impact and paid attention to the melt production from target rocks and the fate of the projectile. They used the same projectile size for all impact angles. To calculate the melt production, their results can be scaled for any projectile size, provided the actual geological structure is not very important. However, gravity adds an additional scale for modeling for the late stage of the transient cavity growth and its subsequent collapse. The amount of the melt within the final crater is the most geophysically interesting aspect. Pierazzo and Melosh (2000a) used the well-known scaling law of Schmidt and Housen (1987) for the calculation of a vertical impact, and some corrections regarding impact angle after Chapman and McKinnon (1986). The corrections were obtained from laboratory experiments (Gault and Wedekind, 1978) for impacts into dry sand with laboratory impact velocities. Extrapolation of these scaling laws to planetary-size impacts should be justified from numerical experiments. To check the available scaling law for oblique impacts, we make several test calculations, modeling oblique impacts with different projectile sizes, and keeping the gravity and crustal thickness constant. The obtained results also provide some insight into the significance of using two-dimensional simulations with axial symmetry for the late stages of cratering.

TRANSIENT CAVITY GROWTH AFTER AN OBLIQUE IMPACT

Hydrocode and equation of state in use

To solve hydrodynamic equations in a three-dimensional grid of computational cells we use the SOVA (solid, vapor, air) code, developed by Shuvalov (1999; Shuvalov et al., 1999). The hydrocode allows the modeling of multidimensional and multimaterial flows with strong shock waves. Each cell of the target initially is marked by a massless tracer particle. These tracers follow the material flow through the mesh, recording maximum pressure of a given Lagrangian (material) point. By using the peak shock pressure we are able to determine the volume of original target that undergoes melting or vaporization.

In the current version the SOVA code can treat only viscous stresses, and the modeling is carried out in a hydrodynamic regime with no elastic and plastic stresses. This means that the projectile, crust, and mantle behave as a heavy fluid. However, in large-scale impact events the effect of gravity is so important that it is possible to ignore the finite strength of rocks during the initial stage of cratering. The absence of strength, however, prevents estimates of the final crater shape. However, as the first step in three-dimensional modeling of large-scale impact events, the hydrodynamic approximation seems to be a useful and necessary step before using more sophisticated models.

The hydrocode allows us to describe only three materials with different properties, and for this reason the target lithology is simplified to two layers: crust and mantle. The sedimentary layer, with a thickness of 3 km, is excluded from consideration, because it does not change substantially the crater growth and final crater morphology (but this layer is of great importance as a possible source of climatically active gases). The thickness of crust is estimated as 33 km, in accordance with the geophysical data (Morgan et al., 1997). The increase of temperature with depth z corresponds to the real continental thermal gradient. The lithosphere is in hydrostatic equilibrium, i.e., the pressure gradient dP/dz is proportional to the target density $\rho_t(z)$.

The ANEOS (analytical equation of state; Thompson and Lauson, 1972), with input data for granite and dunite as proposed by Pierazzo et al. (1997), is used to prepare tables and to describe the material pressure and temperature versus density and internal energy of the projectile and target materials in a wide range of both parameters (from a cold solid state to a hot plasma). The EOS of air is used to reproduce the behavior of the terrestrial atmosphere.

Estimates of projectile size

Using the value for the final crater diameter in the range 180 (Snyder et al., 1999) to 195 km (Morgan et al., 1997), we can estimate the transient crater diameter, D_{tr}, using the Croft's (1985) scaling rule:

$$D_{tr} = D_{sc}^{0.15} \cdot D^{0.85}, \tag{1}$$

where D_{sc} is the value of the crater diameter, corresponding to the transition from simple to complex craters. This value depends on the target rocks and planet gravity and equals about 4 km in the case of crystalline rocks on Earth. Equation 1 estimates the transient cavity diameter in the range of 100–110 km.

The dependence of the transient crater diameter on the parameters of the impact may be described by a scaling law, as suggested by Schmidt and Housen (1987):

$$D_{tr} = 1.16\ (\rho_P/\rho_t)^{1/3}\ D_P^{\ 0.78}\ v^{0.43}\ g^{-0.22}, \qquad (2)$$

where ρ_P and ρ_t are densities of the projectile and the target, D_P is the projectile diameter, v is the impact velocity, and g is surface gravity acceleration. Laboratory studies of oblique impacts by Gault and Wedekind (1978) established that cratering efficiency decreases with sinα for particulate targets (sand), where gravity largely controls the limit of crater growth. As Chapman and McKinnon (1986) pointed out, impact-angle effects can be simply incorporated into existing scaling relations (equation 2), if the vertical component of impact, v sinα, is used:

$$D_{tr} = 1.16\ (\rho_P/\rho_t)^{1/3}\ D_P^{\ 0.78}\ (v\ \sin\alpha)^{0.43}\ g^{-0.22}. \qquad (3)$$

Consequently, for the same target/projectile density ratio and for a constant value of impact velocity, a crater of a constant diameter is created by projectiles, whose size, $D_P(\alpha)$, increase while impact angle, α, decreases:

$$D_P(\alpha) = D_P(90°)(\sin\ \alpha^{-0.55}). \qquad (4)$$

Using equations 1 and 2, we obtain, in the case of vertical impact of dunite projectile (ρ_P = 3.32 g/cm^3) with a velocity of 20 km/s into the granite target (ρ_t = 2.63 g/cm^3), a result of D_P = 14–18 km. However, standard simulations of the Cretaceous-Tertiary event use a smaller value of 10 km diameter (Roddy et al., 1987; Pierazzo and Crawford, 1998). This diameter of the asteroid was first defined by Alvarez et al. (1980) on the basis of Ir anomaly analysis. To compare our results with previous publications, we also use this smaller value for the projectile diameter. Consequently, craters for most of our simulations should be slightly smaller than the Chicxulub crater. For oblique impacts, projectile diameters increase, according to equation 4, to 12 km (for an impact angle of 45°) and 15 km (30°). In several computer runs, we used large projectiles with D_P = 15 km to study the influence of a finite crust thickness on the melt production.

With these assumptions we make several three-dimensional hydrocode runs to estimate the melt production, and to trace the evolution of the transient cavity volume and shape.

Melt production

The computational region is described in the following with Cartesian coordinates x, y, z. The trajectory plane corresponds to $y = 0$. The vertical coordinate z presents depth ($z < 0$) or height ($z > 0$) in respect to the original ground surface ($z = 0$). The x coordinate measures distance in the direction, parallel to the trajectory plane. The projectile touches the surface at the point $x = 0$, $y = 0$, $z = 0$, moving from right to left (from positive x to negative x).

Early evolution of the transient cavity is shown in Figure 1 for the 30° impact. The transient cavity grows mostly downward in respect to the point of impact: 10 s after impact, crater

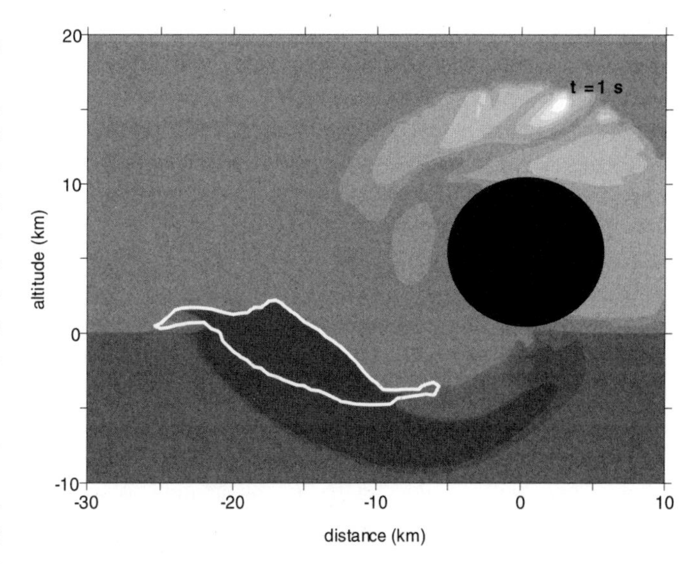

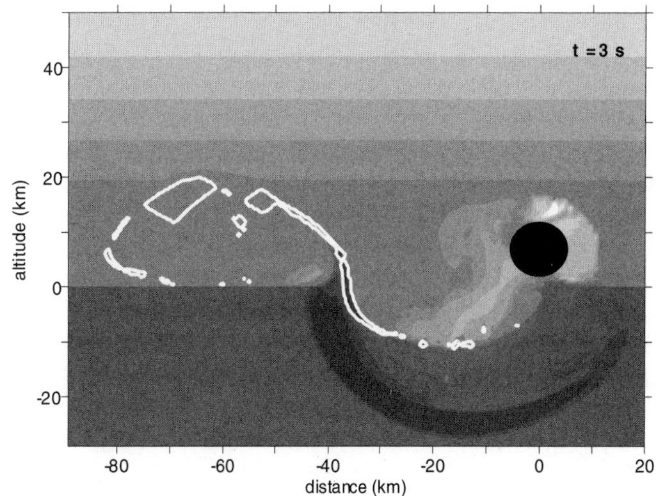

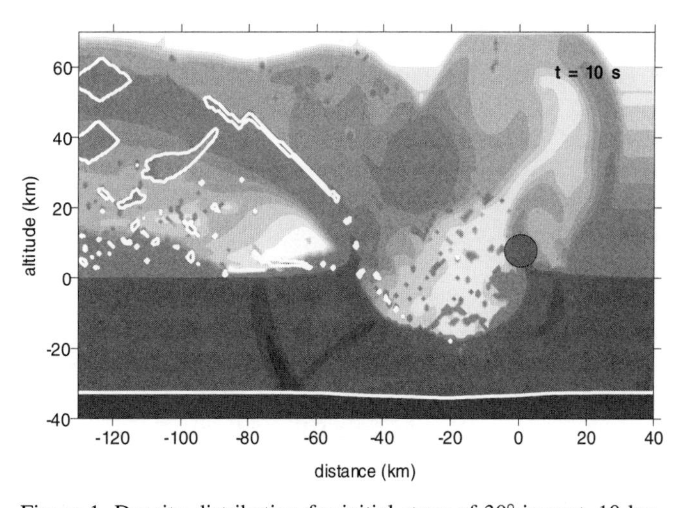

Figure 1. Density distribution for initial stage of 30° impact. 10-km-diameter projectile touches surface at point $x = 0$, $z = 0$ with a velocity of 20 km/s. White line limits projectile material in atmosphere and indicates crust-mantle boundary. Crust is modeled as granite, projectile and mantle are modeled as dunite. At 10 s after impact (bottom plate), shock wave reflected from crust-mantle boundary is visible.

depth reaches 18 km, its width exceeds 30 km, and the length (in the direction of the flight) is ~50 km. However, at ~40 s the transient cavity has the shape of an ellipsoid of revolution, with a width roughly equal to the crater depth.

Table 1 presents the main results for impact melt production, listing the impact melt/projectile volume ratio, V_m/V_{pr}, for the impacts of dunite projectiles into granite targets. The melting shock pressure for granite is assumed to be 50 GPa. This value depends, in principle, on the target porosity and water content, but is suitable for the present estimates (see Pierazzo et al., 1997). Analysis of initial tracer positions shows that even for a vertical impact the melting zone is above a depth of 30 km and is even shallower in the case of an oblique impact. Some part of the melt (~20%–30%) is ejected from the crater during the transient cavity growth, but to estimate the final volume and final position of melt within the crater, one needs to model the final crater formation collapse. This problem is discussed later in this paper.

We used a low spatial resolution of 10 cells per projectile radius (cppr); however, this allowed us to compute reasonably well both impact melting and transient cavity evolution. An estimate of the influence of computational spatial resolution on the melt production shows that ~20% of the melt is lost in comparison with a two times higher resolution of 20 cppr (for the impact velocity of 20 km/s), and this melt deficiency only slightly depends on the impact angle. This result of the melt production decrease with the spatial resolution decrease differs from that of Pierazzo et al. (1997), where losses are higher. Table 2 compares results obtained by Pierazzo and Melosh (2000a) and in our study, listing volumes of target material compressed above a given level of shock pressure, P_0. One can see that the SOVA hydrocode with ANEOS gives a systematically larger volume of compressed material than the hydrocode and EOS used by Pierazzo and Melosh (2000a). The reason for this discrepancy needs additional study. The difference is probably due to an increased number of tracer particles (~100 000) in our simulations (in contrast, only 1000 particles were used by Pierazzo and Melosh (2000a)).

In two additional runs we compute the impact-melt production for larger projectile with diameter D_P = 15 km (impact angles 90° and 45°), which is close to estimates with equations 1–3 for Chicxulub. Figure 2 illustrates the melt-volume geometry for these runs. In contrast to a smaller projectile (D_P = 10 km in the case of vertical impact, and D_P = 12 km for 45° impact), for the large projectile the melt zone in crust reaches the Moho. For dunite, onset of the shock melt pressure is ~120 GPa (Pierazzo et al., 1997), and the mantle is still unmelted. Shock reflection at the crust-mantle boundary creates a local zone (a collar) of a crust material compressed above the nominal onset value of 50 GPa. However, in this zone the maximum shock pressure (formally recorded in the numerical modeling) is reached by double compression (primary shock wave + reflected shock wave). Consequently, the entropy jump is smaller than for the compression in a single shock. The exact value of melted material in this case needs further study. In our present calculations the near-boundary zone increases the melt/projectile ratio by ~10% in the case of vertical impact, and ~5% for a 45° impact.

TABLE 1. IMPACT MELT–PROJECTILE VOLUME RATIO, V_m/V_p

Impact velocity (km/s)	Impact angle 30°	45°	90°
40	43.6	—	72.7
20	13.1	19	23.2
20	—	21.8 (12)*	29.6 (20)*
11	—	3.8	5.2

*Default resolution is 10 cppr (cells per projectile radius). Higher spatial resolution is listed in parentheses.

TABLE 2. VOLUME OF THE TARGET MATERIAL COMPRESSED TO PRESSURE P HIGHER P_0

Pressure P_0 (GPa)	Impact angle 45° P&M* (km³)	Impact angle 45° This work (km³)	Impact angle 90° P&M* (km³)	Impact angle 90° This work (km³)
250	475	657	738	1061
100	3220	4092	3903	5231
50	8987	10584	10488	14850

*P&M—Pierazzo and Melosh (2000a).

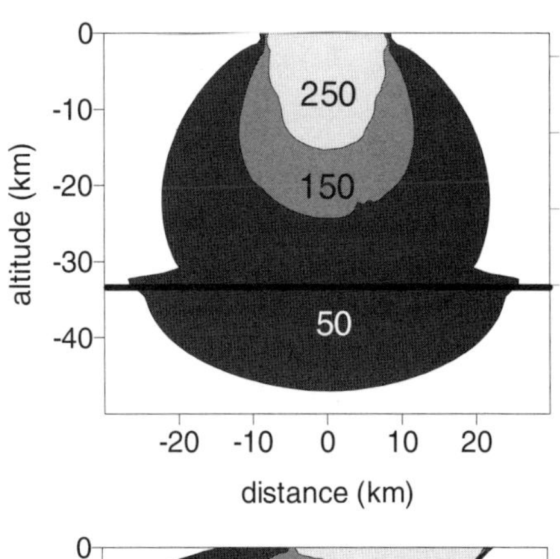

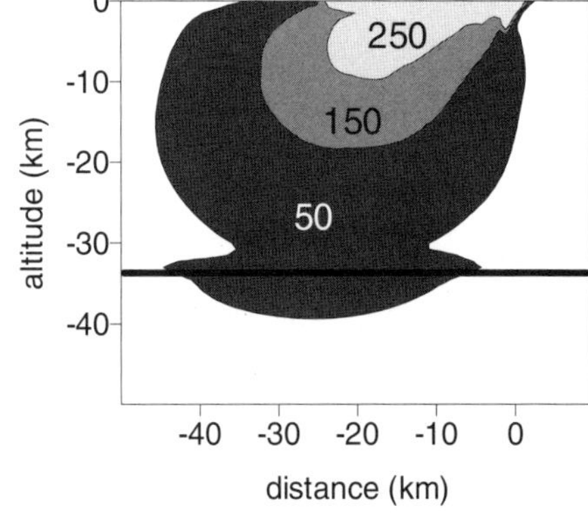

Figure 2. Peak shock pressure contours in plane of impact (y = 0) for angles of 90° (top) and 45° (bottom). Projectile diameter is 15 km. Crust-mantle boundary at depth of 33 km is shown by thick line. Crust material compressed above 50 GPa is melted after pressure release.

Transient cavity growth

One of the goals of the presented modeling is to study the growth and evolution of a transient cavity after an oblique impact. The current version of the SOVA hydrocode does not take deviatoric stress into account. Consequently, the code treats solid material as a liquid (however, with a proper density). Hence, available three-dimensional calculations model properly only the high-pressure stage of cratering. However, large-scale impact craters are created in the so-called gravity regime, where the transient crater growth is controlled with a gravity field (e.g., Melosh, 1989). Initially, the projectile transfers its kinetic energy to the target in the form of kinetic and internal energy. Later, the kinetic energy, transmitted to the target, is converted to heat by work against material strength and internal friction, as well as to the potential energy of a transient cavity and ejection of material. For large craters, the potential energy may be larger than the work of plastic (strength) stresses. Consequently, the hydrodynamic approach gives a good approximation for the maximum size of a transient cavity. In our two-dimensional calculations (see next section), we analyze the crater growth and collapse with a full description of material strength.

In this work we have made several long-term runs to estimate the maximum volume of a transient cavity. Three runs have been continued until central uplift formation. Figure 3 illustrates the crater profile evolution for a 45° impact angle. The transient cavity grows mostly forward in respect to the point of impact at $x = 0$, producing, however, a transient cavity close to a hemisphere at ~40 s after the impact.

Figure 4 shows the shape of the transient cavity in the planes $z = 0$ and $y = 0$ at this time. The velocity field (Fig. 4B) illustrates that at the moment when the depth of the transient cavity stops to grow, target material near the surface still continues to move upward and outward. This indicates that the transient cavity collapse starts not in immovable material, but in a flow field with velocities to 0.5 km/s. The comparison of kinetic energy below the plane $z = 0$ with the potential energy of the transient cavity shows that for a Chicxulub-scale impact event, the kinetic energy does not drop below ~80% of the potential energy. This is in dramatic contrast with small-scale events that we have modeled for comparison. For example, in modeling a 10 km impact crater, the minimum kinetic energy of a flow field is only 30% of the cavity potential energy. This result shows that crater-collapse modeling with initial conditions (immobile material) may predict the final crater shape improperly.

Figure 5 shows a snapshot for the later time (138 s); shading highlights isotherms below the growing crater. The dark shade in the central mound corresponds to the position of impact melt. One can see that the flow field still "remembers" the direction of impact. However the general picture appears more or less symmetrical.

Transient cavity scaling for oblique impacts

Initial projectile diameters have been estimated from equations 3 and 4. However, first three-dimensional code runs have shown that the transient cavity size for oblique impacts is too large; the computed cavity volume is much larger than for the

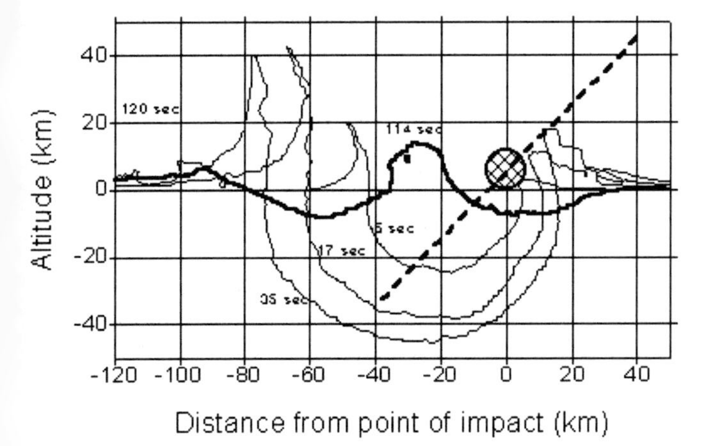

Figure 3. Evolution of transient cavity for 45° impact. Dark circle shows projectile position at moment of first contact with target.

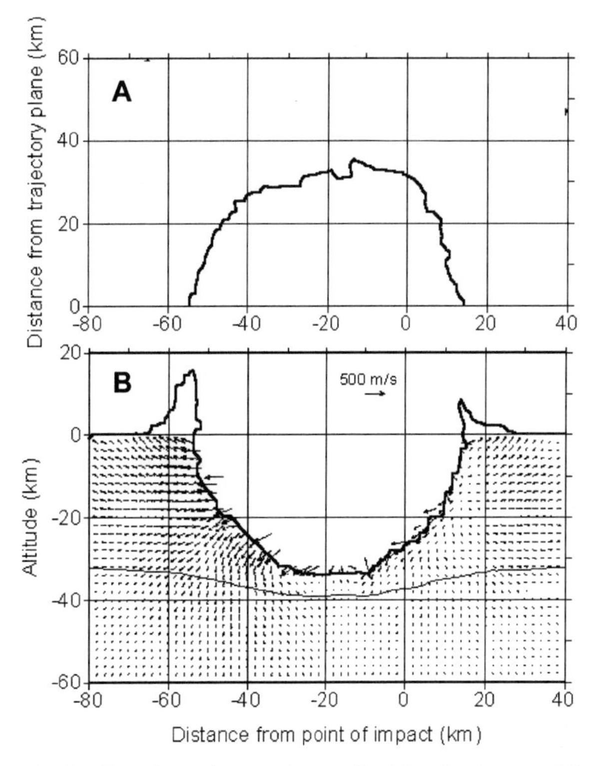

Figure 4. Profiles of growing transient cavity 40 s after impact of dunite projectile (diameter $D_P = 10$ km, impact velocity $v = 20$ km/s, impact angle $\alpha = 45°$). A: Cavity profile in x-y plane at $z = 0$ km (preimpact surface). B: Vertical cross section in bilateral symmetry plane $y = 0$. Velocity field in B demonstrates onset of crater collapse. Effective center of velocity field is shifted ~20 km downrange (left from point of impact).

vertical impact. The derivation of a more appropriate scaling law is the matter of ongoing calculations. However, we discuss some first results here.

The Schmidt and Housen scaling law (equation 4) may be rewritten for the transient cavity volume, V_{tr} (e.g., Pierazzo and Melosh, 2000a):

$$V_{tr} = 0.28\ (\rho_p \rho_t)\ g^{-0.65}\ D_{pr}^{2.35}\ (v \sin \alpha)^{1.3}. \quad (5)$$

The comparison of three-dimensional hydrocode model results with equation 5 is shown in Figure 6 (for a check of the reliability of our results, we compare with a model of a Bosumtwi-scale crater with $D = 10$ km; see Ivanov and Artemieva, 2001). For a vertical impact, the results of the model are in a good agreement with equation 5. However, for lower angles, equation 5 underestimates the maximum transient volume by a factor of 1.6 at $\alpha = 45°$ and by a factor of ~2 at $\alpha = 30°$. A review of published results shows that, for strength craters in metals, a deviation of crater parameters from simple relations similar to those of equation 5 is well established from experiments (Burchell and Mackay, 1998) and from industrial hydrocode modeling (Hayhurst et al., 1995). The deviation increases for larger impact velocities. To test this idea for large-scale gravity craters, we computed the initial stage of the impact for various impact angles and velocities.

Although several runs (Fig. 6) continue to the maximum volume of a transient cavity, it is a time-consuming procedure. To save computational time, we find that the maximum kinetic energy, transferred below the initial target surface, K_{tarmax}, correctly predicts the final transient crater volume. At the same time, the value of K_{tarmax} is reached 10–20 times sooner than the maximum crater volume (4–5 s for a Chicxulub-scale impact in comparison to 40–60 s). Figure 7 compares the values of K_{tarmax}, normalized to the initial projectile kinetic energy, K_0 for various impact angles and velocities. The three-dimensional numerical simulations show that for vertical impact the energy that is available for the crater excavation decreases with the impact velocity, while for oblique impacts the ratio K_{tarmax}/K_0 decreases with the velocity more slowly, or is almost constant for $\alpha = 30°$. The result is that for a lower impact velocity the cratering efficiency decreases with the impact angle in correspondence with equation 5, while for larger impact velocity (20 km/s), the difference between vertical and oblique impact is not as pronounced. The deviation of our results for $v = 20$ km/s shown in Figure 6 seems to be the consequence of the high-velocity effect.

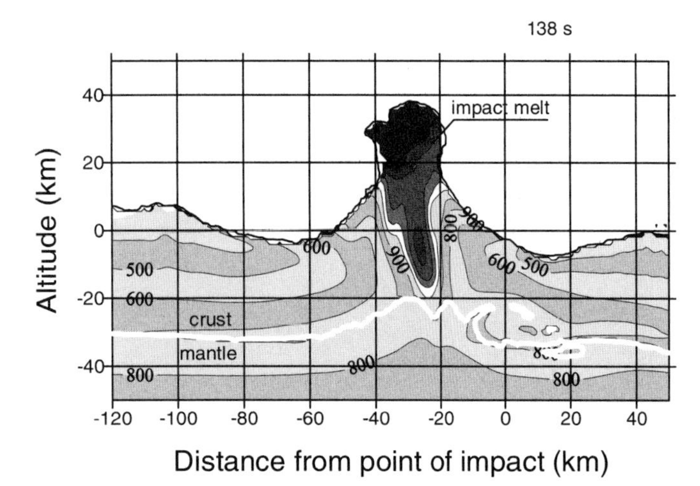

Figure 5. Temperature field in target during phase of maximum overshot. Isotherms are shown every 100 K. Below 1000 K, 100 K strips are shaded alternately dark and light gray. Above 1100 K darker shading shows position of impact melt. Thick white line shows crust-mantle material boundary. Nonphysical oscillations on this boundary are artifacts, because pure hydrodynamics approximation is used.

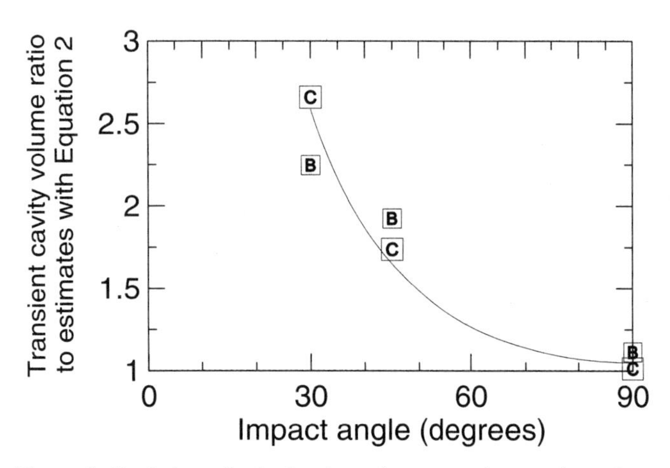

Figure 6. Deviation of calculated maximum transient cavity volume estimated from equation 5 for Bosumtwi (B) and Chicxulub-sized craters (C) simulations at 20 km/s.

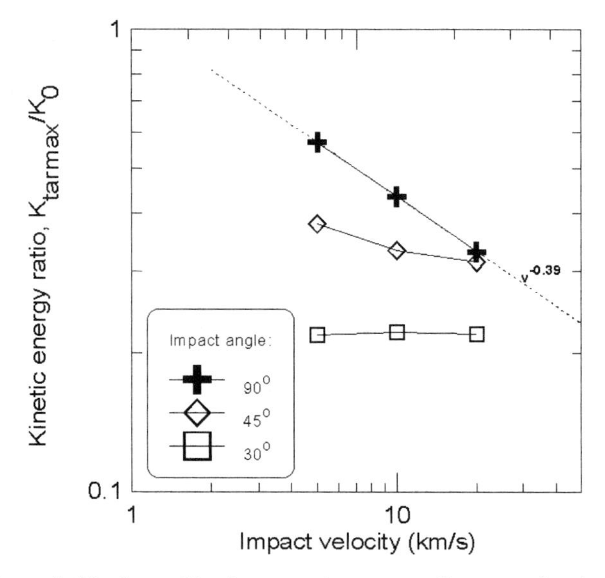

Figure 7. Maximum kinetic energy in a target, K_{tarmax}, as fraction of initial projectile kinetic energy, K_0, for various impact angles and velocities. For vertical impact K_{tarmax} decreases with impact velocity as $v^{-0.39}$.

The velocity effect discussed here may depend on the properties of projectile and target material. Here we briefly outline the possible direction of a future, more detailed, study. Intuitively it can be assumed that an oblique impact has the same efficiency if a projectile penetrates deep into the target during the compression stage. At this stage the projectile transfers its energy to the target. An oblique impact may have very low cratering efficiency if the projectile shears over the surface, producing jets. The latter situation was modeled by Schultz and Gault (1990), who showed that for impact angles of <15°, the ricochet energy reaches 70%–90% of the initial projectile energy.

The first step to quantify the velocity dependence for oblique impact efficiency is to compare the time of projectile penetration into the target with the time of the passage of the shock wave to the rear of the projectile. If the projectile penetrates the target surface before the shock wave reached its rear side, the chances for decapitation (in the sense of Schultz and Gault, 1990) are low. In this case most of the projectile energy is delivered below the target surface, despite the oblique nature of the impact. However, shock waves in a projectile can change the direction of the projectile motion, and part of the projectile never strikes the target surface (i.e., it ricochets), or it strikes it downrange of the main crater. Tentatively we name these two regimes the high efficiency and low efficiency regimes.

To estimate critical angles and velocities for various material, we use a model for planar impacts of various materials with various velocities; the simplest equation of state for solids is in the form: $U = C + Su$, where U is the velocity of the shock wave, u is the material velocity behind the shock, and S and C are empirically determined parameters that describe the target and the projectile materials (for details, see Melosh, 1989, p. 54). It is easy to obtain an analytical solution for this simplified EOS. Table 3 summarizes our planar impact estimates for various projectile-target pairs. The table shows critical planar impact velocity, v_{cr}, when the shock wave reaches the rear of the projectile just at the moment when the rear is at the level of the target surface. For higher impact velocities, the projectile penetrates deeper into the target at the end of compression, whereas for lower ones some part of the projectile is above the target surface after compression. To obtain results for an oblique impact, we define a critical angle for each pair of target-projectile materials and impact velocity. If the vertical component of the impact velocity $v \sin\alpha$ is higher than v_{cr}, then an oblique impact has a high cratering efficiency, close to that of a vertical impact: It is obvious from Table 3 that all laboratory experiments produce impacts with low efficiency: the laboratory impact velocity limit of 5 km/s is below the critical velocity for practically all materials in use (aluminum, iron, quartz). In contrast, planetary impacts (20 km/s, granite-dunite) are effective if the impact angle is above 30°. These preliminary estimates confirm the idea about oblique impact cratering efficiency (Fig. 7), and the results, obtained earlier, are not surprising now. We should not use the scaling law of equation 3, but we should use about the same projectile size for an oblique impact.

The result of the higher cratering efficiency of oblique impacts with $v > v_{cr}$ is that one should check once more the melt/transient cavity volume ratio, as reported by Pierazzo and Melosh (2000a), who used a low-velocity scaling law (equation 6) to estimate the transient cavity volume. Surprisingly, our results, compared to those of Pierazzo and Melosh (2000a), give both larger impact melt volumes and transient cavity volumes. Consequently, the ratio of impact melt/transient cavity volume, V_m/V_{tr}, is nearly constant within the range of impact angles from 45° to 90° (Table 4), and is decreasing at smaller angles.

TWO-DIMENSIONAL SIMULATIONS OF THE CHICXULUB-SCALE CRATER COLLAPSE

The numerical modeling of oblique impacts gives valuable information about the initial stages of crater formation. However, we are not able yet to compute the entire sequence of crater-forming processes with the three-dimensional code. For this reason, we are forced to use less-sophisticated two-

TABLE 4. IMPACT MELT PRODUCTION FOR A CHICXULUB-SCALE IMPACT EVENT

Impact angle	Projectile diameter (km)	Transient cavity volume V_{tr} (km³)	Melt volume V_m (km³)	V_m/V_{tr}
90°	10	82400	12160	0.148
45°	12	137400	19790	0.144
30°	15	240000	25840	0.107

TABLE 3. VERTICAL COMPONENT OF THE IMPACT VELOCITY CRITICAL FOR HIGH CRATERING EFFICIENCY AT OBLIQUE IMPACT

	Target material					
Projectile material	Iron (km/s)	Aluminum (km/s)	Granite (km/s)	Calcite (km/s)	Ice (km/s)	Dunite (km/s)
Iron	18.10	8.54	7.56	7.98	5.08	8.90
Aluminum	44.94	16.84	14.68	15.82	8.92	17.00
Granite	21.78	11.06	9.70	10.28	6.02	11.62
Calcite	—	14.24	12.08	13.12	6.74	14.72
Ice	—	—	—	—	6.42	—
Dunite	15.84	11.78	11.04	11.40	8.68	12.02

dimensional numerical simulations to estimate the final crater morphology and the final melt distribution. In this section we report our results for the Chicxulub-scale crater-formation modeling.

O'Keefe and Ahrens (1993, 1999) presented calculations with standard rock mechanical properties. They claimed that the main causes of the gravity crater modification (the gravity collapse) are thermal softening of rocks (O'Keefe and Ahrens, 1993) and accumulated rock damage (O'Keefe and Ahrens, 1999). In addition to published standard models, we use acoustic fluidization (Melosh and Ivanov, 1999) with appropriate scaling for various crater size. Acoustic fluidization gives an elegant phenomenological model, which allows a temporary decrease in rock friction and facilitates transient cavity collapse, and restores normal fragmented rock properties at the end of collapse, producing the definite final crater shape (see the review by Melosh and Ivanov, 1999). This feature of the model is the most attractive for the crater formation analysis. However, the model is not the ultimate answer to understanding the formation of a complex crater.

Scaling of parameters for the acoustic fluidization model

The block model used here is a version of the general acoustic fluidization model proposed by Melosh (1979, 1983). The formulation of the block model version (Ivanov and Kostuchenko, 1997) and its comparison with Melosh's original model were presented by Melosh and Ivanov (1999). According to the block model, the low-frequency oscillations of rock fragments (blocks) periodically decrease contact forces between adjacent blocks; hence the friction is decreased and slip occurs if oscillations are strong enough. When shear stress, τ, is larger than the so called Bingham limit, the motion of a system of oscillating blocks may be presented as viscous motion with an effective kinematic viscosity:

$$\nu_{lim} = c_{af}\, h^2/T, \tag{6}$$

where h is the characteristic block size, T is the period of oscillations, and c_{af} is a numerical coefficient in the range from 4 to 8 (depending of internal model assumptions). Another important parameter of the block model is the decay time of the block oscillations, T_{dec}. This value is closely connected to the quality factor Q, which is the ratio of the energy stored per cycle to the energy lost over the same period:

$$Q = T/T_{dec}. \tag{7}$$

For the case of Puchezh-Katunki impact crater, with a diameter of ~40 km (Masaitis and Pevzner, 1999), suitable values are $\nu_{lim} = 8.8 \cdot 10^4$ m^2/s and $T_{dec} = 45$ s (Ivanov and Kostuchenko, 1997). This choice of model parameters allowed us to reproduce the real crater morphology. Keeping viscosity and decay time constant, various block size h and oscillations period T are possible according to equation 6, as well as various Q values (Q is proportional to T). These values are summarized in Table 5.

To scale the values of T, h, and T_{dec} for another crater size we use the simple idea that blocks oscillate due to the presence of a soft breccia between adjacent blocks, which dampens the movements. Under this assumption we can write the equation for the oscillating motion of a block (height h, density ρ) "jumping" at the breccia layer (thickness h_b, density ρ_b, characteristic sound speed c_b, compressibility module $\rho_b\, c_b^2$):

$$M\, d^2 y/dt^2 + \rho_b\, c_b^2\, y/h_b = 0, \tag{8}$$

where y is a coordinate in the direction of the oscillating motion, and M is the block mass per unit area of a contact ($M = \rho h$). The equation gives a standard description of harmonic oscillations with the period

$$T = 2\pi/\omega = 2\pi\, h/c_b\, [(\rho/\rho_b)(h_b/h)]^{1/2}. \tag{9}$$

The values of the speed of sound c_b for the ratios $h_b/h = 0.1$ and 0.2 with a typical density ratio $\rho/\rho_b = 2$ are given in Table 5 for various oscillation periods T.

In the simple linear model of scaling, tested in our work, the values of c_b and $(\rho/\rho_b)(h_b/h)$ are assumed to be constant for craters of different sizes. Hence, the oscillation period T is proportional to the characteristic block size h.

The scaling law used here may be presented as a flow chart.

1. For a given crater rim diameter, the transient crater diameter is calculated with Croft's model (equation 1).

2. The characteristic block size, h, is estimated as a fraction of the transient crater diameter, D_{tr}.

3. The characteristic oscillation period, T, is assumed to be proportional to h (equation 9).

4. The effective viscosity is calculated with equation 6.

5. The block oscillation decay time for a given Q is proportional to T. Hence, T_{dec} is assumed to be proportional to the transient crater diameter.

From this description one can easily see the strong and weak points of the model. For example, the assumed breccia compressibility should increase with the lithostatic pressure. In the upper pressure limit, the breccia is compressed to the density of intact rocks, thus, block oscillations are not possible. This and many other phenomena should be implemented in the model in the future. Here we used only one simple restriction:

TABLE 5. BLOCK MODEL PARAMETERS FOR THE PUCHEZH-KATUNKI CRATER

Quality factor Q	Block size h (m)	Oscillation period T (s)	Sound speed c_b (m/s)	
			$h_b/h = 0.1$	$h_b/h = 0.2$
10	316	4.50	197.1	278.8
50	141	0.90	440.8	623.4
100	100	0.45	623.4	881.7
200	71	0.23	881.7	1246.8

Note: T is period of oscillations through m/s

if the pressure is above a limiting value, P_{max}, the strength of the oscillations is set to zero.

For specific calculations one can adjust the model parameters so they better fit the observations. Table 6 displays the actual parameters used in the calculations of Ivanov and Artemieva (2001) for the Bosumtwi crater (D = 10 km), and the ones used here for the Chicxulub impact structure (D = 180 km).

Final crater shape

The best test for any model of the formation of a complex crater is the quantitative reproduction of impact-crater morphology and subsurface structure (Ivanov and Kostuchenko, 1997; Ivanov, 1998; Melosh and Ivanov, 1999; O'Keefe and Ahrens, 1993, 1999). Here we present results from numerical modeling with the heavily modified two-dimensional SALE (simplified arbitrary Lagrangian and Eulerian) code (see the original version by Amsden et al., 1980, and modifications in Ivanov et al., 1997; Ivanov and Deutsch, 1999). The code has been modified to allow computations in the Eulerian mode for the complex target rocks with different material strengths. Lagrangian description of the rocks is a convenient tool to reproduce the crater collapse, including shear localization, slumping, and faults (Ivanov, 1998; Kenkmann et al., 2000). The simulated shear faults may be compared with geological and geophysical data to tune the numerical model, as demonstrated by Kenkmann et al. (2000). In the case of Chicxulub, it will soon be possible to do so, because the necessary field data are just now becoming available (cf. Morgan et al., this volume; Snyder et al., 1999). The Lagrangian method does not allow us to model the initial stages of the impact, where flows with strong deformations arise. Calculations in the modified Eulerian mode of the SALE hydrocode allow us to compute the complete crater formation sequence, including the impact and shock-wave generation prior to the growth of the transient cavity. However, the phenomenon of shear localization is not as straightforward in this model, due to natural smoothing of the calculated deformation during the advection of materials through the Eulerian grid.

Here we used a slow impact velocity (12 km/s) to avoid massive vaporization of rocks (equations of state used in the SALE code are not useful for an exact description of vaporized and partially vaporized materials). The size of projectile was chosen by trial and error to create a transient crater 40 km deep. Target and projectile materials are the same as in three-dimensional simulations, but no atmosphere is simulated (vacuum above the target). To simulate temporary dynamic friction reduction, we implemented the acoustic fluidization model into the SALE code. The growth of the transient cavity is similar to our three-dimensional calculations for vertical impacts and to the results obtained for Sudbury (Ivanov and Deutsch, 1999). The cavity is close to a hemisphere; the floor starts to uplift before the end of ejection; dynamic overshoot of the central uplift takes place well above the preimpact surface; and the central uplift collapse results in the double ring crater formation. Figure 8 shows the final crater and crust-mantle boundary profiles for two hydrocode runs: (1) with dry friction, thermal softening, and acoustic fluidization with parameters calculated from a scaling law (see Table 6, Chicxulub-1); (2) as in (1), but with artificially enhanced acoustic fluidization to check the sensitivity of the model (Table 6, Chicxulub-2). Both runs produce a double ringed crater.

Similar to three-dimensional calculations, in our two-dimensional calculations we use tracer particles embedded into the computational grid to record the peak shock pressure. After the run the initial and final coordinates of the tracer particles allow to find the location of the melt. Figure 9 illustrates the geometry of the maximum pressure field with respect to the initial tracer particle positions (left panel) and the final particle positions (400 s after impact). To construct Figure 9, every tracer particle position is drawn with a light gray dot if it

TABLE 6. PARAMETERS OF THE BLOCK MODEL USED FOR SIMULATIONS

	Bosumtwi	Chicxulub-1	Chicxulub-2
Effective viscosity (m^2/s)	1.7×10^4	2.6×10^5	3×10^4
Oscillation period (s)	0.1	0.4	0.4
Oscillation decay time (s)	25	160	100
Limiting pressure P_{max} (GPa)	25	25	25

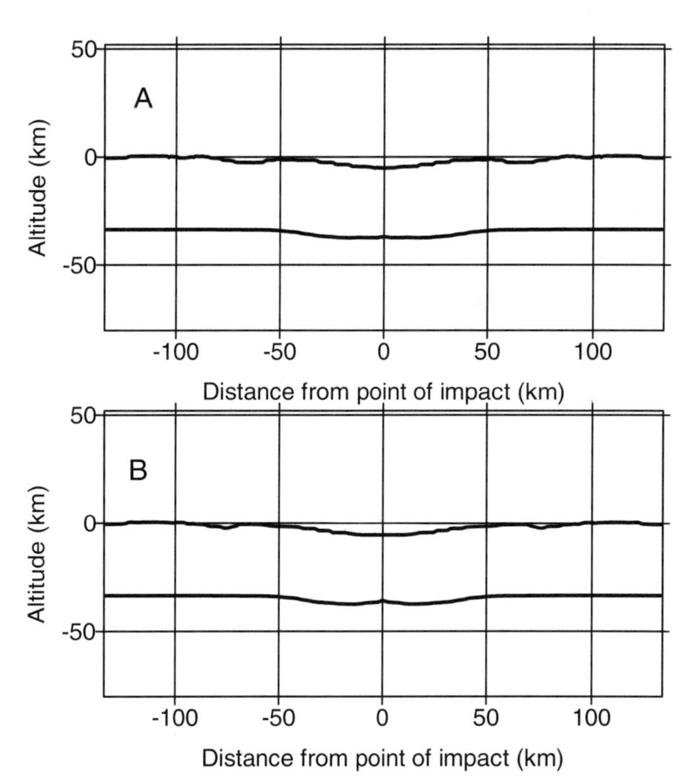

Figure 8. Crater and crust-mantle boundary profiles for numerical models. A: With dry friction, thermal softening, and acoustic fluidization. B: With artificially enhanced acoustic fluidization. Both runs produce a peak-ring crater.

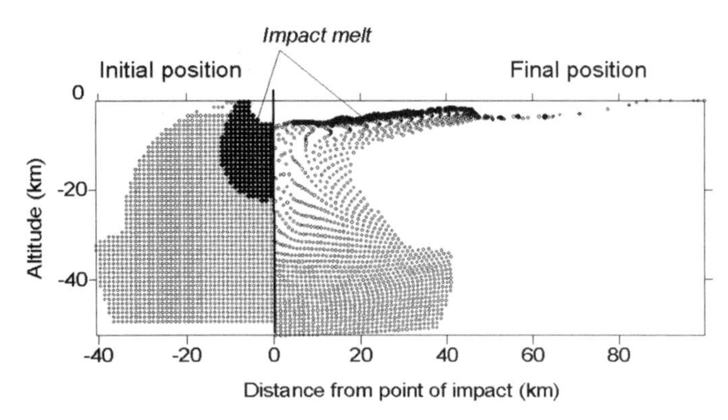

Figure 9. Initial (negative distance) and final (positive distance) positions of tracers for vertical impact of projectile with diameter of 16 km and impact velocity of 12 km/s. Gray dots show material compressed to shock pressures in range 10–50 GPa (shock metamorphism); black dots show impact melt compressed above 50 GPa. Initial depth of melt zone is ~22 km. Melt occupies central depression of collapsed uplift with diameter of 90 km.

recorded the maximum shock pressure in the range 10–50 Gpa, and with a dark dot if it was shocked to a pressure in the range 50–110 GPa. The latter pressure interval corresponds to granite melting after pressure release (Pierazzo et al., 1997). In the final position (right panel in Fig. 9), most of tracer particles from the melt zone are deposited on the top of the central uplift depression in a circle with a diameter of 90 km. The SALE code is not as exact as the SOVA that was used for three-dimensional modeling (so called first-order accuracy hydrocode versus second-order accuracy). Therefore, the volume of impact melt calculated with SALE may differ at low resolution (8 cppr) from volumes calculated with SOVA at high resolution (20 cppr). However, the current version of SALE used here allows us to compute the final positions of particles, including impact melt. Combining the advantages of 2 sets of computations (three-dimensional SOVA and two-dimensional SALE), we obtain a melt sheet with a diameter of 90 km (Fig. 9), and the volume of melt of $V_m \sim 48\,000$ km^3, for a vertical impact of a 15-km-diameter projectile. The maximum thickness of the melt sheet is estimated as ~6 km, provided that all melt remains in the crater. Ejection of melt outside the crater decreases this estimate. This calculated value of the dimensions of a possible melt sheet seems agree with previously published geological models (Kring, 1995; Grieve and Therriault, 2000), where an interior melt sheet with a diameter of 100 km and thickness of 3–7 km was suggested.

DISCUSSION

The formation of large (100–300 km in diameter) terrestrial impact craters is a complex natural phenomenon still under investigation. This work was meant as a test for the relatively new three-dimensional SOVA code, which is capable of calculating oblique impacts. The results for the initial stage of an impact (projectile penetration and shock melting), computed with the SOVA code, are in good agreement with previously published models (Pierazzo and Melosh, 1999, 2000a). In addition, we computed the subsequent cratering stages, i.e., the transient cavity growth to the beginning of its gravitational collapse. We find that for relatively high impact velocities (20 km/s) the efficiency of the cratering excavation, measured in terms of the maximum volume of a transient cavity, for a 45° impact appears to be comparable with that of a vertical impact. This confirms previously published data showing that the angle dependence of the crater excavation efficiency varies with the impact velocity.

The possible outcome may influence the interplanetary comparison of craters. For example, the critical velocity (Table 3) for a dunite projectile impacting the granite target is ~11 km/s. The average impact velocity on Mercury (~23 km/s) is twice as large as on Mars (10–12 km/s). Consequently, the angle dependence of the crater excavation efficiency may be different on Mars and Mercury: lower velocity impacts on Mars can be better approximated with the experimental scaling law (equations. 3 and 6), whereas oblique impacts on Mercury may be more efficient due to higher impact velocity. This problem needs further study.

Acoustic fluidization is important for the explanation of the formation of a complex crater. In the case of large craters, comparable in diameter to Chicxulub, the implementation of acoustic fluidization allows us to model the peak-ring morphology, similar to the model based on seismic data (Morgan et al., 1997, and this volume). However, with the simple two-layer model of the target (granite crust over dunite mantle), it is impossible to reproduce the measured magnetic and gravity anomalies. Similar results of the crater collapse modeling were published by Morgan et al. (2000). Our approach involves the complete computation of the process, starting with the compression stage of cratering (in terms of Melosh, 1989) and shock-wave generation.

The modeled position of melt bodies and highly shocked material will, in the future, allow us to compare models with the magnetic anomaly data. Currently the resolution of our computations for the final phase of cratering is not high enough (the final crater relief is comparable with the size of computational cells).

CONCLUSIONS

At the beginning of the crater collapse, the transient crater shape is more or less symmetric for impact angles of between 90° (vertical impact) and 30°, provided the impact velocity is high enough (~20 km/s and more).

We calculated a wide range of impact conditions for Chicxulub-scale events, and in no case did the impact melting zone enter the mantle.

Numerical modeling confirms the possibility of the existence of an impact melt sheet for a Chicxulub-size terrestrial crater, with a diameter of 100 km and a thickness of 3–6 km.

Collapse of a Chicxulub-size transient cavity results in the formation of a peak-ring crater.

These and other characteristic features found from numerical modeling may be used for the verification of numerical models by comparison with observational geophysical and geological data.

ACKNOWLEDGMENTS

We acknowledge the personnel of the DLR (Deutschen Zentrum für Luft- und Raumfahrt) Institute of Space Sensor Technology and Planetary Exploration (Berlin, Adlershof) and the Institute for Geophysics, University of Kiel, for technical support with the three-dimensional numerical modeling. Artemieva was supported by a mobility grant from the European Science Foundation Impact program during her visit to the University of Kiel. J. Morgan and H. Henkel made very important comments and suggestions as reviewers. We also appreciate the improvement of the style of the manuscript by J. Morgan and the very valuable scientific and linguistic editing by C. Koeberl.

REFERENCES CITED

Alvarez, L.W., Alvarez, W., Asaro, F., and Michel, H.V., 1980, Extraterrestrial cause for the Cretaceous-Tertiary extinction: Science, v. 208, p. 1095–1108.

Amsden, A.A., Ruppel, H.M., and Hirt, C.W., 1980, SALE: A Simplified ALE Computer Program for Fluid Flow at All Speeds: Los Alamos, New Mexico, LA-8095 Report, Los Alamos National Laboratories, 101 p.

Burchell, M.J., and Mackay, N.G., 1998, Crater ellipticity in hypervelocity impacts on metals: Journal of Geophysical Research, v. 103, p. 22761–22774.

Chapman, C.R., and McKinnon, W.B., 1986, Cratering of planetary satellites, *in* Burns, J.A., and Matthews, M.S., eds., Satellites: Tucson, University of Arizona Press, p. 492–580.

Croft, S.K., 1985, The modification stage of basin formation: Conditions of ring formation, *in* Schultz, P.H., and Merrill, R.B., eds., Multi-ring basins, *in* Proceedings, 12th Lunar and Planetary Science Conference 12A: New York, Pergamon, p. 227–257.

Gault, D.E., and Wedekind, J.A., 1978, Experimental studies of oblique impact, *in* Proceedings, 9th Lunar and Planetary Science Conference: New York, Pergamon, p. 3843–3875.

Grieve, R., and Therriault, A., 2000, Vredefort, Sudbury, Chicxulub: Three of a kind?: Annual Reviews of Earth and Planetary Sciences, v. 28, p. 305–338.

Ivanov, B.A., 1998, Large impact crater formation: Thermal softening and acoustic fluidization: Meteoritics and Planetary Science, v. 33, Supplement, p. A76.

Ivanov, B.A., and Artemieva, N.A., 2001, Transient cavity scaling for oblique impacts: Lunar and Planetary Science, v. 32, Abstract #1327, CD-ROM.

Ivanov, B.A., and Deutsch, A., 1999, Sudbury impact event: Cratering mechanics and thermal history, *in* Dressler, B., and Grieve, R.A.F., eds., Large meteorite impacts and planetary evolution II: Geological Society of America Special Paper 339, p. 389–397.

Ivanov, B.A., and Kostuchenko, V.N., 1997, Block oscillation model for impact crater collapse: Lunar and Planetary Science, v. 28 Abstract #1655, CD-ROM.

Ivanov, B.A., Kocharyan, G.G., Kostuchenko, V.N., Kirjakov A.F., and Pevzner, L.A., 1996, Puchezh-Katunki impact crater: Preliminary data on recovered core block structure [abs]: Lunar and Planetary Science, v. 27, p. 589–590.

Ivanov, B.A., DeNiem D., and Neukum, G., 1997, Implementation of dynamic strength models into 2D hydrocodes: Applications for atmospheric breakup and impact cratering: International Journal of Impact Engineering, v. 20, p. 411–430.

Hayhurst, C.J., Ranson, H.J., Gardner, D.J., and Birnbaum, N.K., 1995, Modelling of microparticle hypervelocity oblique impacts on thick targets: International Journal of Impact Engineering, v. 17, p. 375–386.

Hildebrand, A.R., Penfield, G.T., Kring, D.A., Pilkington, M., Camargo, A., Jacobsen, S.B., and Boynton, W.V., 1991, A possible Cretaceous-Tertiary boundary impact crater on the Yucatan peninsula, Mexico: Geology, v. 19, p. 867–871.

Kenkmann, T., Ivanov, B.A., and Stöffler, D., 2000, Identification of ancient impact structures: Low-angle faults and fault patterns of crater basements, *in* Gilmore, I., and Koeberl C., eds., Impacts and the early Earth: Heidelberg, Germany, Springer-Verlag, Lecture Notes in Earth Sciences, v. 91, p. 279–307.

Kring, D.A., 1995, The dimension of the Chicxulub impact crater and impact melt sheet: Journal of Geophysical Research, v. 100, p. 979–986.

Masaitis, V.L., and Pevzner, L.A., eds., 1999, Deep drilling in the Puchezh-Katunki impact structure: St. Petersburg VSEGEI (Karpinsky Institute), 392 p. (in Russian).

Melosh, H.J., 1979, Acoustic fluidization: A new geologic process?: Journal of Geophysical Research, v. 84, p. 7513–7520.

Melosh, H.J., 1983, Acoustic fluidization: American Scientist, v. 71, p. 158–165.

Melosh, H.J., 1989, Impact cratering: A geologic process: New York, Oxford University Press, 245 p.

Melosh, H.J., and Ivanov, B.A., 1999, Impact crater collapse: Annual Review of Earth and Planetary Science, v. 27, p. 385–415.

Morgan, J., Warner, M., and the Chicxulub Working Group, 1997, Size and morphology of the Chicxulub impact crater: Nature, v. 390, p. 472–476.

Morgan, J., and Warner, M., 1999, Morphology of the Chicxulub impact crater: Peakring crater or multi-ring basin?, *in* Dressler, B., and Grieve, R.A.F., eds., Large meteorite impacts and planetary evolution II: Geological Society of America Special Paper 339, p. 281–290.

Morgan, J.V., Warner, M.R., Collins, G.S., Melosh, H.J., and Christeson, G.L., 2000, Peak-ring formation in large impact craters: Geophysical constraints from Chicxulub: Earth and Planetary Science Letters, v. 183, p. 347–354.

O'Keefe, J.D., and Ahrens, T.J., 1993, Planetary cratering mechanics: Journal of Geophysical Research, v. 98, p. 17011–17028.

O'Keefe, J.D., and Ahrens, T.J., 1999, Complex craters: Relationships of stratigraphy and rings to impact conditions: Journal of Geophysical Research, v. 104, p. 27091–27104.

Penfield, G.T., and Camargo Z., 1991, Interpretation of geophysical cross-sections on the north flank of the Chicxulub impact structure [abs.]: Lunar and Planetary Science, v. 22, p. 1051.

Pierazzo, E., and Crawford, D.A., 1998, Modelling Chicxulub as an oblique impact event: Results of hydrocode simulations [abs.]: Lunar and Planetary Science, v. 29, Abstract #1704, CD-ROM.

Pierazzo, E., and Melosh, H.J., 1999, Hydrocode modeling of Chicxulub as an oblique impact event: Earth and Planetary Science Letters, v. 165, p. 163–176.

Pierazzo, E., and Melosh, H.J., 2000a, Melt production in oblique impacts: Icarus, v. 144, p. 252–261.

Pierazzo, E., and Melosh, H.J., 2000b, Understanding oblique impacts from experiments, observations, and modeling: Annual Reviews of Earth and Planetary Science, v. 28, p. 141–167.

Pierazzo, E., and Melosh, H.J., 2000c, Hydrocode modeling of oblique impacts: The fate of the projectile: Meteoritics and Planetary Science, v. 35, p. 117–130.

Pierazzo, E., Vickery A.M., and Melosh H.J., 1997, A re-evaluation of impact melt production: Icarus, v. 127, p. 498–423.

Roddy, D.J., Schuster, S.H., Rosenblatt, M., Grant, L.B., Hassig, P.J., and Kreyenhagen, K.N., 1987, Computer simulations of large asteroid impacts into oceanic and continental sites: Preliminary results on cratering, and ejecta dynamics: International Journal of Impact Engineering, v. 5, p. 525–541.

Schultz, P.H., and Gault, D.E., 1990, Prolonged global catastrophes from oblique impacts, *in* Sharpton, V.L., and Ward, P.D., eds., Global catastrophes in Earth history: An interdisciplinary conference on impacts, volcanism, and mass mortality: Geological Society of America Special Paper 247, p. 239–261.

Schmidt, R.M., and Housen K.R., 1987, Some recent advances in the scaling of impact and explosion cratering: International Journal of Impact Engineering, v. 5, p. 543–560.

Sharpton, V.L., Burke, K., Hall, S.A., Lee, S., Marin, L.E., Suarez, G., Quezada-Muneton, J.M., and Urrutia-Fucugauchi, J., 1993, Chicxulub impact basin: Gravity characteristics and implications for basin morphology and deep structure [abs.]: Lunar and Planetary Science, v. 24, p. 1283–1284.

Shoemaker, E.M., 1962, Interpretation of lunar craters, *in* Kopal, Z., ed., Physics and astronomy of the moon: New York, Academic Press, p. 283–359.

Shuvalov, V.V., 1999, 3D hydrodynamic code SOVA for interfacial flows, application to thermal layer effect: Shock Waves, v. 9, p. 381–390.

Shuvalov, V.V., Artemieva, N.A., and Kosarev, I.B., 1999, 3D hydrodynamic code SOVA for multimaterial flows, application to Shoemaker-Levy 9 comet impact problem: International Journal of Impact Engineering, v. 23, p. 847–858.

Snyder, D.B., Hobbs R.W., and the Chicxulub working group, 1999, Ringed structural zones with deep roots formed by the Chicxulub impact: Journal of Geophysical Research, v. 104, p. 10743–10755.

Thompson, S.L., and Lauson, H.S., 1972, Improvements in the Chart D radiation-hydrodynamic code 3: Revised analytic equation of state: Albuquerque, New Mexico, Report SC-RR-71 0714, Sandia Laboratories, 119 p.

Manuscript Submitted December 9, 2000; Accepted by the Society March 22, 2001

Printed in the U.S.A.

Geological Society of America
Special Paper 356
2002

Multiple stages of condensation in impact-produced vapor clouds

Detlef de Niem*
*Deutsches Zentrum für Luft- und Raumfahrt (DLR),
Institute of Space Sensor Technology and Planetary Exploration,
Rutherfordstraße 2, D-12489 Berlin, Germany*

ABSTRACT

The condensation of vapor formed by the impact of an asteroid or comet onto a planetary surface is investigated using the classical theory of homogeneous nucleation. Multiple nucleation stages (multiple peaks of the formation rate of critical nuclei of the condensed phase) appear during the long-term evolution of the vapor plume. Several times the nucleation rate rises to a maximum in a very short time interval, and subsequently drops to zero. Each of these events is followed by a phase of growth of supercritical clusters into macroscopic droplets. For a Chicxulub-scale impact, the first nucleation maximum occurs typically a few hundreds of a second, the second a few minutes, and the third more than one hour after the vapor becomes saturated (this last event may be suppressed for part of the matter subject to interaction with the atmosphere or due to irradiation and ionization in outer space). Each event produces a different size scale and number density of droplets; ultimately the size distribution is multimodal. The number of new-generation droplets formed per molecular unit of the gas is increasing from stage to stage, by roughly three orders of magnitude as compared with a previous generation. Thermal characteristics of millimeter-sized spherules of the first generation are a temperature of formation around ~3900 K and several minutes of growth at temperatures above 2000 K. More abundant small (~10 μm) spherules form later, at a greater distance from the impact site, at temperatures below 2000 K.

INTRODUCTION

In hypervelocity impacts of asteroids or comets on the surface of a planetary body, depending on the impact velocity, a significant amount of the projectile and target material may be vaporized and expand as a dense gas cloud. In simulations of asteroid impacts (e.g., see Melosh and Pierazzo, 1997, Pierazzo et al., 1997, 1998), a thermodynamic equilibrium equation of state was used for the liquid-vapor region, so the kinetics of condensation is not treated.

For questions related to the final amount of condensed material, the condensate size distribution, and the thermodynamic conditions during the formation of droplets, a different approach is necessary. The amount of condensed material in the form of micrometer-sized grains is an important parameter for the climatic effects of large-scale impacts. In studies of the kinetics of condensation in impact-produced vapors (e.g. the pioneering work of Raizer, 1960, or O'Keefe and Ahrens, 1981), some analytical gas-dynamic solutions were used for the hydrodynamics, and the solution of the kinetic equations was approximated only in a simple way. In contrast, here differential equations for the moments of the size distribution of condensate

*E-mail: detlef.deniem@dlr.de

de Niem, D., 2002, Multiple stages of condensation in impact-produced vapor clouds, *in* Koeberl, C., and MacLeod, K.G., eds., Catastrophic Events and Mass Extinctions: Impacts and Beyond: Boulder, Colorado, Geological Society of America Special Paper 356, p. 631–644.

droplets are obtained and solved numerically, using a simplified hydrodynamic background solution. Previously, for atmospereless bodies such as the Moon or asteroids, hydrodynamics were approximated by a solution of Zel'dovich and Raizer (1967), valid only for a perfect gas. In this work, a similar solution is found for a van der Waals gas, which is the simplest model for studying phase transitions in dense real gases; it is possible to approximate hydrodynamics with the help of this solution to study the kinetics of vapor condensation. To investigate effects of an atmosphere on the propagation of the vapor cloud, numerical hydrodynamic simulations are required. In this case the lowest temperatures and primary condensation sites are found in the release wave behind the atmospheric shock. Formation of critical nuclei of the liquid phase begins in a narrow layer, initially shaped like a mushroom. Later, the region containing condensate droplets encompasses larger parts of the original vapor plume, while the shock wave detaches, pushing away parts of the upper atmosphere and creating a storm in the near-surface layer. This process is investigated with the help of numerical hydrodynamics in cylindrical geometry, again assuming a van der Waals equation of state. However, no detailed consideration of the release of gases in the course of crater formation is made. It is the vapor of the projectile that led to the global identification of the Cretaceous-Tertiary boundary through isotopic anomalies, and the condensation of this material is investigated here.

As the vapor cloud cools adiabatically, locally it crosses the phase boundary, and further expansion leads to supercooling. To determine the thermodynamic conditions where the gas reaches the phase boundary, only the specific entropy of the material produced in the shock wave has to be known locally. The detailed hydrodynamic history is less important, if a perfect adiabatic expansion is assumed. From the point of view of kinetics of condensation, these local thermodynamic conditions and the knowledge of the volume expansion rate are sufficient to study further evolution of the vapor in a Lagrangean formulation of the kinetic and energy equations. Condensation is not complete if an adiabatic expansion is fast enough (i.e., that the rates of formation and evaporation of molecular clusters are driven away from equilibrium), and the final state is one of thermodynamic nonequilibrium characterized by a nonzero mass fraction of remaining vapor (Raizer, 1960; Zel'dovich and Raizer, 1967). This nonequilibrium thermodynamic state would also be characterized by nonequilibrium chemical compositions. Because density and pressure during the expansion decrease by many orders of magnitude, it is difficult to treat the process with conventional hydrodynamic algorithms. A possible way to derive the necessary hydrodynamic background variables for the nucleation kinetics is to obtain them from simulations without kinetics of phase transitions. This is not self-consistent, but nucleation kinetics is described by differential equations that are very difficult to combine with any numerical hydrodynamic algorithm. Herein I simplify the hydrodynamic treatment and focus on the condensation kinetics.

Zel'dovich and Raizer (1967) characterized the condensation process in the following way. (1) The vapor adiabat reaches the phase boundary, at which point the vapor becomes saturated, and further expansion leads to supercooling. (2) At some critical degree of supercooling, the nucleation rate becomes high enough that a large number of critical nuclei of the new phase forms. (3) In the course of nucleation, the gas pressure decreases, the supercooling becomes subcritical, and formation of new clusters is inhibited. (4) In the meantime, newly formed clusters grow into droplets of macroscopic dimensions. (5) At some point, the flux of molecules at the surface of the droplets is so low that condensation stops; a non-zero mass fraction of gas remains.

Herein I show that this is an oversimplified scenario. Several nucleation events appear at different times and spatial scales, even for the most simple hydrodynamic background solution. The supercooling ratio, which measures the deviation from local thermodynamic equilibrium, follows a complicated oscillating pattern. This results in the formation of several generations of droplets, changing the overall characteristics of the condensates. Each time a new generation appears, the overall root-mean-square size of droplets decreases, because the older and large droplets are outnumbered by finer, newly formed droplets. Thermodynamic conditions at the time of the various nucleation events are different, so marked differences in chemical composition among the generations should also be expected. Because the pressure in the various nucleation events differs by many orders of magnitude, the presence of an atmosphere may inhibit the oscillatory evolution of the supercooling and suppress very low-pressure stages of nucleation that would otherwise occur in a free expansion into space. Part of the vapor plume flows along nearly ballistic trajectories at high altitude, before returning to the surface, over ~0.5–2 h, long enough to complete several nucleation stages.

HYDRODYNAMICS OF THE VAPOR CLOUD

Evolution of the dense core: An analytical solution

An analytical hydrodynamic solution describes the evolution of the dense core of the vapor cloud, using the van der Waals equation of state (EOS). This particular solution does not treat the interaction of the shock with a surrounding atmosphere, but it is indicative of the initial dynamics and will serve as a background solution to study the kinetics of phase transitions. I compare it to more realistic numerical simulations. To study quenching into a metastable thermodynamic state, no Maxwell construction is carried out in the two-phase region. Later, in the treatment of the kinetic equations, an energy equation is used that is valid for a nonequilibrium two-phase mixture, including a consistent description of the phase boundary, so the approximations made here are in effect for the initial dynamics, but not for the subsequent kinetics of condensation. The solution derived in this section is valid while the matter is in the gas phase or at temperatures above the critical point. As the phase boundary is approached, the solution continues to be a valid approximation when the vapor remains in a metastable state and even later, as long as the condensed mass fraction remains small. The velocity early approaches an asymptotic value, so the solution is simpler while the mass fraction of the condensed phase is larger. Thereafter, only the time dependence of the density, in a Lagrangean frame, or equivalently, the divergence of the velocity field is necessary to solve the kinetic and energy equations.

Assuming hemispherical symmetry, and that the radial velocity is given by a Hubble-like law:

$$v_r = \dot{R}\eta, \tag{1}$$

the mass density is of the form

$$\rho(t,r) = \frac{\mu(\eta)}{R^3}, \tag{2}$$

where t is the time, r the radial coordinate, $\eta = r/R(t)$ is a dimensionless variable, and $R(t)$ is a time-dependent scale factor. The mass interior to a shell $\eta =$ constant is conserved. The function $\mu(\eta)$ is left undetermined by the continuity equation (see Zel'dovich and Raizer, 1967). Here, a solution with constant μ is selected, so the total mass inside a hemispere with radius $R(t)$ is $M = (2\pi/3)\mu$. The pressure is obtained from the van der Waals EOS

$$P = \frac{R_g T\rho}{1 - b\rho} - a\rho^2, \tag{3}$$

where T is the temperature, R_g is the gas constant, and the van der Waals parameters a and b are related to the pressure and temperature at the critical point, P_c and T_c, respectively.

$$P_c = \frac{a}{27b^2}, \quad T_c = \frac{8a}{27bR_g}, \tag{4}$$

(e. g., see Guggenheim, 1957). The internal energy for a van der Waals gas with constant specific heat c_V is given by $\varepsilon = c_V T - a\rho$, then the first law of thermodynamics leads to a specific entropy s of the form

$$\ln s = (p + a\rho^2)b^\gamma \left(\frac{1}{b\rho} - 1\right)^\gamma + const, \tag{5}$$

where the adiabatic exponent γ is defined by

$$\gamma := \frac{R_g + c_V}{c_V}. \tag{6}$$

The energy equation is equivalent to vanishing partial time derivative of the specific entropy at constant η because of the special ansatz for the velocity. Using equation 5, this leads to

$$P = \frac{A(\eta)}{b^\gamma}\left(\frac{1}{b\rho} - 1\right)^{-\gamma} - a\rho^2, \tag{7}$$

with an undetermined function of the entropy, $A(\eta)$. After some algebra, Euler's equation for the radial velocity reduces to

$$\frac{d^2R}{dt^2} = -\frac{2}{\rho R}\frac{\partial P}{\partial \eta^2}. \tag{8}$$

Inserting the pressure from equation 7, the solution for the unknown function of entropy is $A(\eta) = C(1 - \eta^2)$, where C is a constant. The pressure at the outer boundary $r = R(t)$ equals $(-a\rho^2)$, whereas the temperature is zero there (this follows from equation 7 and the EOS). It is impossible to apply a zero-pressure boundary condition for a van der Waals gas; however, this is not considered as a major deficit of the analytical solution. Effects at the boundary are not correctly modeled by the present solution; conditions in the interior roughly agree with a numerical solution, however. At the boundary the vapor is separated by a very low temperature contact discontinuity from the shocked atmospheric air; this supplies a small but nonzero boundary value for the temperature. Integration of equation 8, inserting the pressure from equation 7 gives a first integral

$$\frac{\dot{R}^2}{2} = -\frac{2C(\gamma - 1)}{3b^{(\lambda-1)}}\left(\frac{R^3}{b\mu} - 1\right)^{-(\gamma-1)} + \frac{1}{2}\dot{R}_\infty^2. \tag{9}$$

The first term on the right side vanishes for $\rho \rightarrow 0$, so $\dot{R}_\infty$ is the asymptotic velocity at the boundary $\eta = 1$. C is determined by the initial conditions at the center. For nonnegative right side in equation 9 at initial time it follows that

$$1 - \frac{4(1 - b\rho_0)(P_0 + a\rho_0^2)}{3(1 - \gamma)\rho_0\dot{R}_\infty^2} \geq 0, \tag{10}$$

where an index zero denotes initial values at the center. If the equality sign holds, the gas is initially at rest. This leads to a maximum possible central temperature $T_{0,\max}$

$$\begin{aligned} R_g T_{0,\max} &= (\gamma - 1)(\varepsilon_0 + a\rho_0) \\ &= \frac{3(\gamma - 1)}{4}\dot{R}_\infty^2 = \frac{5}{2}\left(\frac{E}{M} + a\rho\right), \end{aligned} \tag{11}$$

where the last identity on the right side of equation 10 is valid for the total energy E and the density ρ at arbitrary time. A derivation is lengthy but straightforward, using the first integral (equation 9). The other two equalities follow, inserting the EOS and the relation for the internal energy. Because the density is decreasing with time, the total energy reaches an asymptotic value of $(3/10)M\dot{R}_\infty^2$. Another relation can be found for the mass-averaged squared sound speed at the initial time:

$$\overline{c_s^2(t_0)}: = \frac{2\pi}{M}\int_0^R r^2\rho c_s^2 dr = \frac{2\gamma(\gamma - 1)(\varepsilon_0 + a\rho_0)}{5(1 - b\rho_0)^2} - 2a\rho_0. \quad (12)$$

A derivation of this identity is lengthy, so it is only sketched. The squared sound speed of a van der Waals gas is

$$c_s^2 = \left(\frac{\partial P}{\partial \rho}\right)_\varepsilon + \frac{P}{\rho^2}\left(\frac{\partial P}{\partial \varepsilon}\right)_\rho = \frac{\gamma(\gamma - 1)(\varepsilon + a\rho)}{(1 - b\rho)^2} - 2a\rho. \quad (13)$$

Because ρ is only time-dependent, the problem in equation 12 is reduced to that of the mass average of internal energy, because averages of all other parts are proportional to the total mass. The total energy is already known (see equation 11). The remaining problem is an average of the kinetic energy. The latter can be evaluated with the help of the expression for velocity (equation 1), the energy integral (equation 9), and (equation 11). Knowing ε_0 and $\overline{c_s^2(t_0)}$, the asymptotic velocity $\dot{R}_\infty$ and the initial density ρ_0 can be determined from equations 11 and 12.

The model is aimed at studying the vaporization and dispersion of the projectile, and not detailed enough to investigate degassing of the target. Therefore, material parameters are derived for a projectile consisting only of dunite to apply this solution to the Chicxulub impact. The van der Waals parameters for dunite corresponding to a total critical pressure of 6.4926 GPa and a critical temperature of 13070 K, and a mean molar mass of 35.173 g are given by $a = 620.207$ m^5 kg^{-1}s^{-2} and $b = 5.948068\ 10^{-5}$ m^3 kg^{-1}. (This is using equation 4 and assuming that in the gas phase, Mg_2SiO_4 transforms into $2MgO + SiO + O_2$, so the molar mass of the gas mixture is 25% of the molar mass of Mg_2SiO_4, to match the correct number of translational degrees of freedom in the expression for the pressure, in the ideal gas limit.) A value of $\gamma = 1.307938$ for the adiabatic exponent corresponds to a specific heat at constant volume c_V of 27 kJ mole^{-1}K^{-1}, with the help of equation 6. For comparison, a mixture of gases of diatomic molecules would have a classical value of $c_V = 7/2R_g = 29.1$ kJ mole^{-1}K^{-1} corresponding to $\gamma = 1.4$. Initial values $\varepsilon_0 = 3.71\ 10^7$ J kg^{-3} and $\sqrt{\overline{c_s^2(t_0)}} = 3.3$ km s^{-1} from Melosh and Pierazzo (1997) lead to an initial density of $\rho_0 = 6990.987$ kg m^{-3} and a central temperature of $T_0 = 53979$ K. These values correspond to a highly compressed projectile. Therefore, an initial radius of $R_0 = 5 \times (3320\ \text{kg m}^{-3}/\rho_0)^{1/3}$ km is chosen to match the initial mass of a projectile with a diameter of 10 km at the normal density of dunite of 3320 kg m^{-3}.

Figure 1 (A, B, and D) shows the evolution of the core density, the central pressure, and the time derivative of the scale factor, $\dot{R}$, with respect to time. In Figure 1C, the temperature is shown with respect to density, with the phase boundary indicated. Note that densities at the liquid part of the coexistence curve are unrealistic, due to the approximation by the van der Waals EOS. For these parameters, the saturation curve is reached at $t_1 = 8.07$ s at a pressure of $P_1 = 2.258\ 10^6$ Pa; in the center of the vapor cloud, the temperature is $T_1 = 3937$ K. In a spherically symmetric numerical solution (see Figs. 2 and 3), with homogeneous initial conditions, the temperature and pressure corresponding to the point that saturation is reached at the center of the cloud are nearly identical, because of constancy of the entropy, whereas the time of t_1 at ~6.88 s is somewhat earlier. This time difference to a numerical solution is due to the different initial pressure distribution for the latter (see next section), whereas the thermodynamic conditions are matched, because the temperature and pressure values at saturation are determined by the initial specific entropy only. The subsequent evolution of a shell of gas that has crossed the coexistence curve can only be studied with the help of nucleation theory; the semianalytical solution is valid to this moment, for pressure and temperature. The velocity is close to the asymptotic value at t_1, so the evolution of the total density and the rate of volume expansion predicted by the this solution are useful as an input for the condensation kinetics.

A different type of solution was proposed by Zel'dovich and Raizer (1967), valid for expansion of a perfect gas into vacuum; this solution has no generalization for the van der Waals EOS, however. The main difference between both solutions is that the value of the asymptotic velocity at the outer boundary is $\dot{R}_\infty^2 = (10/3 + 4/[\gamma - 1])\ E/M$ for the case of Zel'dovich and Raizer, leading to a value of $\dot{R}_\infty$ approximately two times larger for the same E/M than in the present solution. In an atmosphere, the velocity at this boundary is determined by the conditions at the contact discontinuity traveling behind the shock wave in the surrounding air, and the speed of the atmospheric shock is spatially inhomogeneous due to the vertical density gradient in the atmosphere (see following). The present solution was not developed to match the speed of an expansion into a vacuum.

Numerical solution in hemispherical geometry

For a more realistic treatment of the hydrodynamics, a state of the art finite-volume Godunov-type scheme was selected, which allows the use of equations of state of the Mie-Grüneisen form, for which the van der Waals EOS is a subcase. Roe's Riemann solver (Roe, 1981) has been modified, taking the van der Waals EOS into account; second-order accuracy is obtained with the help of wave-propagation methods (LeVeque, 1997). The program is based on the publicly available CLAWPACK software (see LeVeque, 1997).

The initial conditions in the simulation correspond to homogeneous distribution of pressure $P_0 = 1.22387 \times 10^{11}$ Pa and density $\rho_0 = 6990.987$ kg m^{-3}, with an initial radius of the cloud of $R_0 = 5 \times (3320\ \text{kg m}^{-3}/\rho_0)^{1/3}$ km. A homogeneous atmosphere with $P = 10^5$ Pa and $\rho = 1.5$ kg m^{-3} is

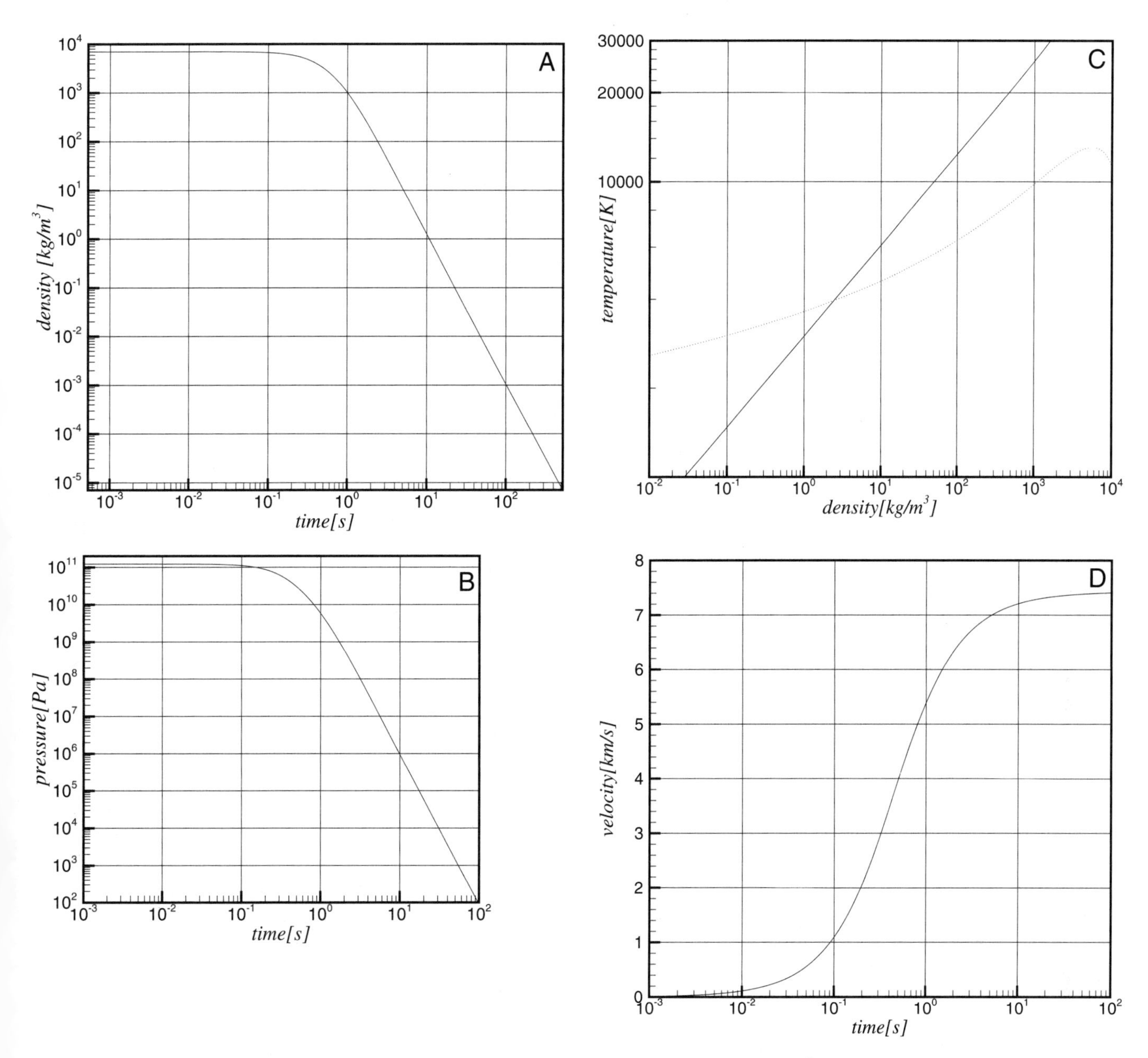

Figure 1. A: Semianalytical solution for expansion of van der Waals gas. Central density is in kg/m^3 vs. time. B: Semianalytical solution for expansion of van der Waals gas. Central pressure is in Pa vs. time. C: Semianalytical solution for expansion of van der Waals gas: Central temperature in K vs. density in kg/m^3; coexistence curve is indicated (dotted). Intersection of thermodynamic path with phase boundary occurs at time of ~8.07 s. Note that liquid densities at coexistence curve (right of critical point) are unrealistic, because van der Waals parameters are fitted to match critical pressure and temperature. D: Semi-analytical solution for expansion of van der Waals gas. Velocity (in km/s) at the outer boundary of core vs. time.

modeled, using the same EOS for air (air has γ ~1.2–1.3 in a wide range of temperatures and pressures [see Zel'dovich and Raizer, 1967], so this is a compromise). The resulting initial atmospheric temperature is 289.7 K. Because no thermal processes in the atmospheric gas are investigated here, it is only necessary to apply the appropriate counter-pressure; the temperature is not relevant.

Figure 2 shows the distribution of temperature at the time t_1 of saturation in the center. Also shown is the saturation temperature, T_s (for each cell, the saturation temperature, corresponding to the local value of pressure, was determined; using the van der Waals EOS, no approximation by a more simple form of the phase boundary is made).

A contact discontinuity marks the boundary between

dunite vapor and the air swept up by the atmospheric shock. This contact discontinuity is a cold zone: to the left, the temperature decreases outward; from the center, the density jumps downward, and the temperature rises afterward, reaching a high value in the atmospheric shock. Metastability of the vapor first appears near the contact discontinuity, later in the core. The metastable zone initially forms a thin layer following the shock. Later, at t_1 ~6.88 s, material in the center arrives at the phase boundary. The saturation temperature in the undisturbed atmosphere (right of the shock discontinuity) is larger than the atmospheric temperature, due to the description by the same van der Waals EOS; this is irrelevant, however, for the thermodynamic conditions in the dunite vapor. The spatial resolution of discontinuities with the second-order accurate Godunov-type method is sufficient (see Fig. 3). The temperature distribution near the narrow contact zone can only be resolved with state of the art hydrodynamic algorithms. Newman et al. (1999), using a realistic atmospheric model and advanced hydrodynamic software, had to modify the EOS for the projectile to exclude condensation effects, because condensation "causes the computation to grind to a halt" (Newman et al., 1999, p. 233).

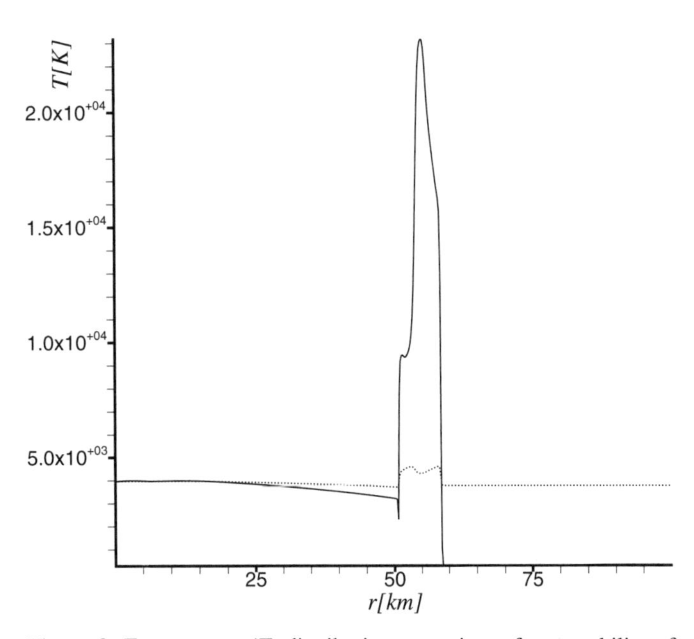

Figure 2. Temperature (T) distribution at t_1, time of metastability of cloud core. Saturation temperature (dotted) is shown for comparison. Cold zone can be recognized at ~50 km distance from center, marking contact discontinuity between evaporated projectile material and shocked atmospheric air.

Simulation with exponential atmosphere

The hydrodynamic algorithm has been applied in cylindrical geometry to study the effects of an atmospheric density gradient, to obtain a more realistic flow pattern. The initial conditions are the same as in the spherically symmetric solution, with the exception that the constant-density atmosphere is replaced by an isothermal temperature, T = 289.7 K, and exponential atmosphere with a scale height of 10 km and a surface pressure of 10^5 Pa. Far from being a realistic atmospheric model, this was introduced to investigate the effects due to an atmospheric density gradient.

Figure 4 shows contours of the degree of supercooling $\theta = (T_s - T)/T_s$, which is a measure of the deviation from

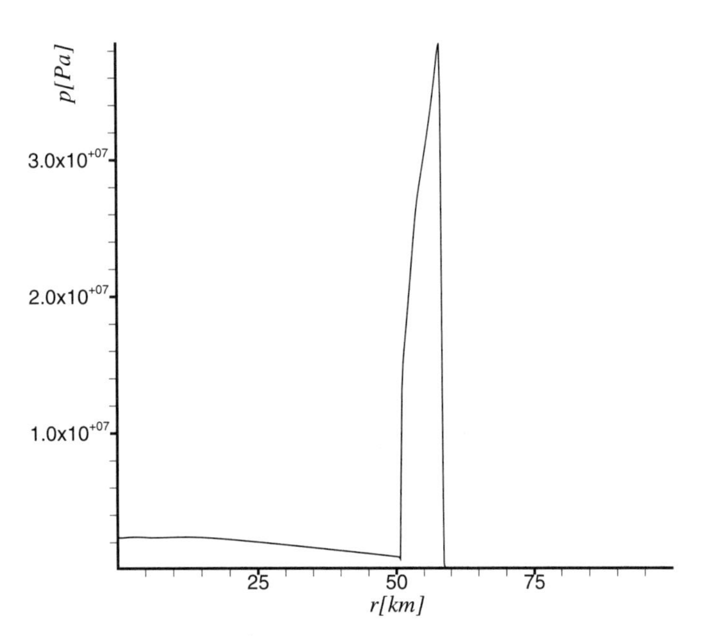

Figure 3. Pressure (P) distribution at t_1, time of metastability of cloud core: shock wave propagates through atmospheric air (right), followed by flattened distribution of pressure in region of evaporated projectile.

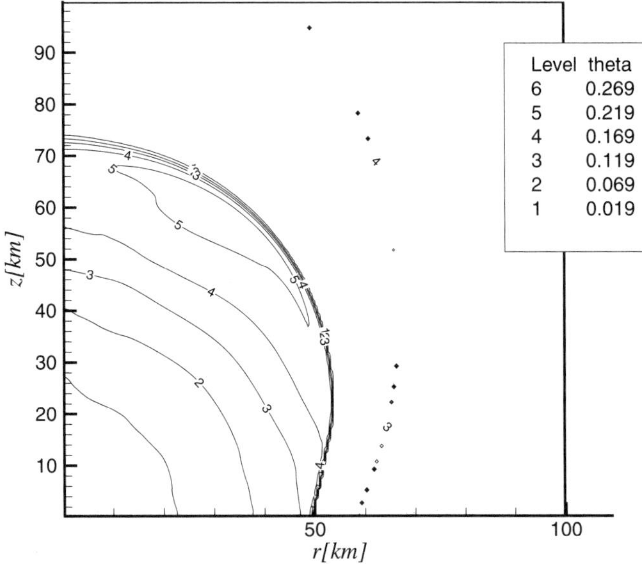

Figure 4. Contours of supercooling parameter θ at t_1, time of metastability in core. Outer boundary of contours marks region where condensation occurs.

thermodynamic equilibrium, at the onset of metastabilty in the cloud center, at $t_1 = 6.75$ s. The hydrodynamic simulation is not able to describe kinetics of condensation, so the maximum value of ~0.3 for the supercooling close to the contact discontinuity is meaningless, because formation of the condensed phase begins at lower values. However, the spatial shape of the metastable region is informative. Contours for large values of θ are arranged in an onion-shell pattern, with individual contours shaped like a mushroom, for values where they do not extend to the ground. Artifacts mark the location of the shock front (the front is visible in pressure contours; see Fig. 5) and are caused by the one-fluid hydrodynamic treatment (the atmosphere is described by the same van der Waals EOS).

Figures 5 through 7 display logarithmic contours of pressure, density, and temperature, at the moment of saturation in the center, t_1, respectively. The central pressure and temperature are $P_1 = 2.386 \times 10^6$ Pa and $T_1 = 3965$ K, at this time, which compare to the analytical results of $P_1 = 2.258 \times 10^6$ Pa and $T_1 = 3937$ K. Exact agreement with the analytical values cannot be expected because the hydrodynamic evolution of the inhomogeneous vapor cloud is not perfectly adiabatic.

The expansion of the shock is not spherically symmetric at distances larger than a scale height. Therefore, the portion of the atmosphere that is pushed apart by the shock is less than that contained in a hemisphere with radius equal to the thickness of the atmosphere. The elongated shape of the shock front in the vertical direction can be explained with the help of Kompaneets' (1960) theory: the numerical results obtained here qualitatively agree with those of Newman et al. (1999). Note the excellent resolution of the shock front by the second-order Godunov-type method based on LeVeque's (1997) work. The largest values of pressure and density are found in the shock front near the surface (see Figs. 5 and 6). With available numerical hydrodynamic methods, treatment of condensation of the impact vapor cloud is only possible as far as local thermodynamic equilibrium is valid. Even such a treatment faces severe difficulties with state of the art hydrodynamic algorithms, because these are based on the assumption of hyperbolicity (reality of the speed of sound) and effectively come to a halt if borders of thermodynamic stability are reached (Newman et al., 1999).

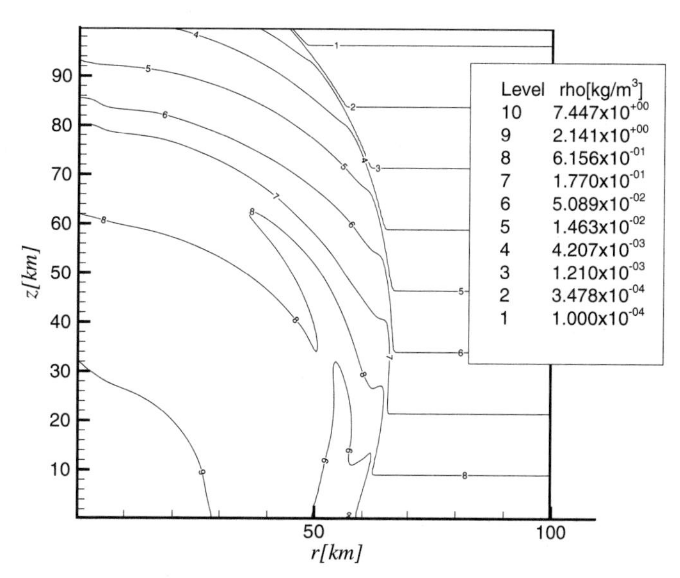

Figure 6. Contours of density at t_1, time of metastability in core; values are in kg/m^3. Exponential density gradient in undisturbed atmosphere (right of shock) is visible as equally spaced contours (logarithmic levels).

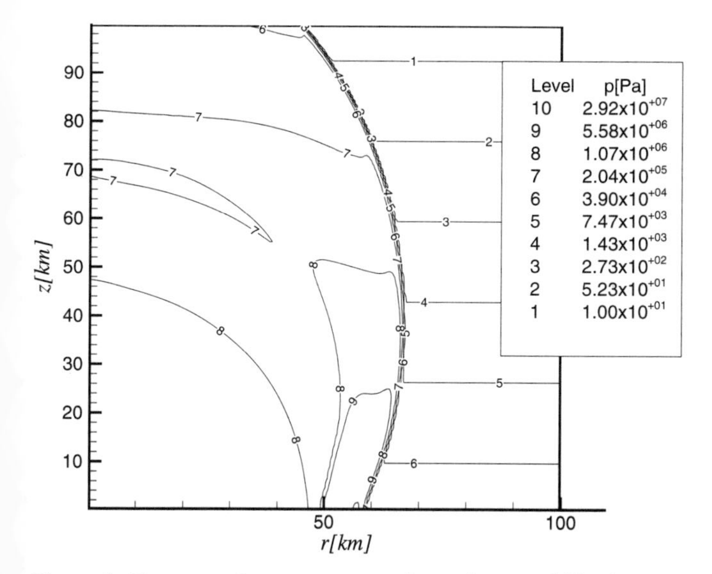

Figure 5. Contours of pressure, at t_1, time of metastability in core. Values are in Pa. Note excellent resolution of shock front due to second-order accurate Godunov-type method.

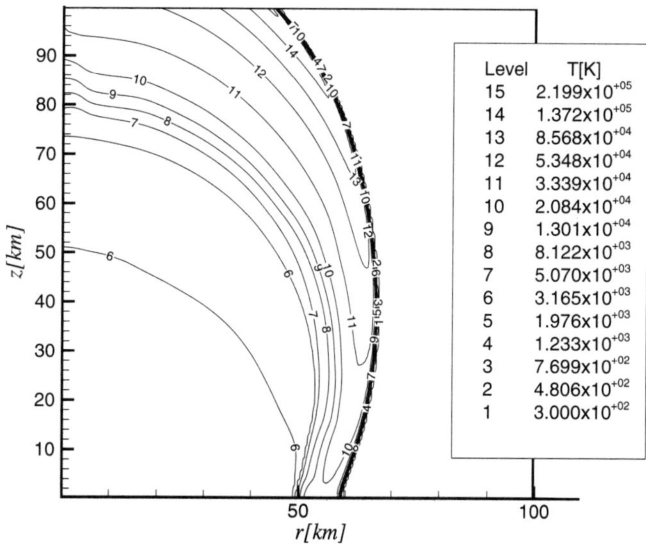

Figure 7. Contours of temperature, in K, at t_1, time of metastability in core. Shock acceleration by exponential density gradient causes large temperatures reached at high altitude, in front.

KINETIC THEORY AND CONDENSATION

Homogeneous nucleation rate and droplet model

In a situation as complicated as a planetary impact, condensation of gases proceeds by homogeneous and heterogeneous nucleation, depending if early-stage evaporation of projectile and target material or later-stage degassing is investigated. The initial temperatures and densities during the evaporation of the projectile are so high that solid aerosols or others type of solid nucleation centers are absent; therefore homogeneous nucleation of molecular clusters will initially occur. If the vapor cloud is not subject to instabilities, the large outward-directed pressure gradient prevents diffusion of solid particles into the dense vapor cloud. If the cloud is at low density, the velocity of expansion is large enough that dust produced later in the formation of the crater cannot penetrate into upper parts of the cloud. This does not exclude more complicated heterogeneous nucleation processes during recondensation of evaporated target rocks. Only homogeneous nucleation is investigated here, also because it is the isotopic signature of the projectile material that led to a global identification of the Cretaceous-Tertiary boundary.

The Becker-Döring-Zel'dovich nucleation theory describes the kinetics of a first-order phase transition (e.g., melting, solidification, evaporation, condensation) as a hierarchy of clustering reactions for attachment or evaporation of single monomers (Zel'dovich, 1942; Frenkel, 1946; see also Abraham, 1974, for an introduction). Following the work of Raizer (1960), who investigated the fate of condensation products of an iron meteorite expanding into vacuum, few have studied the problem of condensation in impacts (e.g., O'Keefe and Ahrens, 1981). The steady-state solution for the nucleation rate gives the number of critical nuclei per molecule of the vapor formed per unit of time:

$$I = \pi r_{cr}^2 n_b v_{th} \kappa \exp\left(-\frac{4\pi\sigma r_{cr}^2}{3k_B T}\right), \qquad (14)$$

with

$$v_{th} = \left(\frac{8k_B T}{\pi m}\right)^{1/2}, \quad \kappa = \sqrt{\frac{\sigma}{k_B T}\frac{\Omega_a}{2\pi r_{cr}^2}}, \qquad (15)$$

where k_B is Boltzmann's constant, n_b and v_{th} and are the number density and the thermal velocity of the vapor, respectively; m is the molecular mass of the monomer; Ω_a is the molecular volume of the liquid, and σ is the surface tension. The radius of critical clusters,

$$r_{cr} = \frac{2\sigma\Omega_a}{\lambda m \theta}, \qquad (16)$$

is a function of the degree of supercooling, θ, defined by

$$\theta = \frac{T_s - T}{T_s}, \qquad (17)$$

where λ is the enthalpy of evaporation per unit mass; T_s is the temperature of saturated vapor at given mass density and liquid fraction. The critical cluster radius r_{cr} is given in the so-called capillarity approximation here, where the surface tension is identical to its macroscopic value. The critical cluster radius is located at an unstable maximum of the free energy of formation, ΔF; liquid droplets larger than r_{cr} grow, whereas smaller droplets evaporate. The dimensionless Zel'dovich (or statistical) prefactor κ measures the width of the extremum in the size distribution of subcritical nuclei. The so-called dynamical prefactor in Equation 14 is the rate of collision of vapor molecules with the surface of a critical drop. The exponential is the probability of thermal fluctuations leading to formation of critical nuclei, dominating the expression. Small changes in the supercooling ratio change the rate by several orders of magnitude. The precise form of the prefactors in Equation 14 is relatively less important than an accurate model for the energy of formation of critical droplets in view of the changes over many orders of magnitude caused by the exponential.

Recent theories for the nucleation rate rely on Monte-Carlo simulations or molecular dynamics and do not lead to simple analytical expressions (see Senger et al., 1999). Such models are too ambitious in a dynamical situation such as an impact, because they require complicated time-consuming numerical calculations for each set of thermodynamic parameters. However, a generalization for more complicated droplet models is straightforward: The free energy of formation, ΔF, is expressed in terms of the number of molecules in the droplet n; the Zel'dovich factor κ is then given by

$$\kappa^2 = -\frac{1}{2\pi k_B T}\left(\frac{\partial^2 \Delta F(n_{cr})}{\partial n_{cr}^2}\right). \qquad (18)$$

Here n_{cr} is related to the radius of a spherical droplet by $n_{cr} = (4\pi/3)r_{cr}^3/\Omega_a$, and the exponential in Equation 13 has to be replaced by

$$\exp\left(-\frac{\Delta F(n_{cr})}{k_B T}\right),$$

and the critical droplet size is found from $\partial\Delta F(n_{cr})/\partial n_{cr} = 0$.

Furthermore, a model for the growth of droplets larger than r_{cr} is required because the nucleation rate I only describes their formation. Following Frenkel (1946), the number n of molecules in a droplet of radius r changes according to

$$\frac{dn}{dt} = 4\pi r^2 \frac{\rho_b v_{th}}{m}\left[1 - \exp\left(\frac{1}{k_B T}\frac{\partial\Delta F(n)}{\partial n}\right)\right], \qquad (19)$$

where the forefactor is the rate of collision of vapor molecules with the surface of the droplet, and the expression in brackets is caused by the difference between forward and backward processes. Raizer (1960) further approximated the exponential:

$$\frac{\partial \Delta F(n)}{\partial n} \cong -\lambda\theta\rho_a\Omega_a = -\lambda m\theta, \quad (20)$$

neglecting the contribution of surface energy to the free energy of formation for $r > r_{cr}$. The formulation used here allows the incorporation of general droplet models; see McClurg and Flagan (1998) for a discussion of such models based on statistical mechanics and partition functions.

Growth of supercritical droplets: Kinetic equation

The degree of condensation $x(t)$ is the ratio of the number of molecules in the liquid state to their total number in the metastable vapor (Zel'dovich and Raizer, 1967):

$$x(t) = \int_{t_I}^{t} I(t')n(t, t')dt', \quad (21)$$

where t_1 is the time where the coexistence curve is crossed and $n(t, t')$ is a solution of Equation 19 with initial conditions $n(t = t', t') = n_{cr}(t')$. The integrand of Equation (21) depends on the history each thermodynamic variable and of $x(t)$ itself in a complicated way. One can reduce the evaluation of the integral to a system of ordinary differential equations, introducing the droplet radius $r(t, t')$ corresponding to the number $n(t, t')$. This radius fulfills a differential equation, where the right side is only dependent on thermodynamic state variables, but independent of $r(t)$:

$$\frac{dr}{dt} = \frac{\rho_b}{\rho_a} v_{th}\left[1 - \exp\left(-\frac{\lambda m\theta}{k_B T}\right)\right], \quad (22)$$

where Equation 19 has been used together with the assumption of constant specific molecular volume of the liquid phase Ω_a. Relaxing constancy of Ω_a would add a term linear in $r(t)$ to the right side. Others, e.g., Deguchi (1980) or Chigai et al. (1999), neglected the exponential on the right side of Equation 22, but this term is significant, because the exponent is small. To derive this term, it has to be assumed that growing droplets stay at the same temperature as the surrounding vapor. This is a good approximation in the beginning of macroscopic growth; later, when no temperature equilibrium is maintained, the growth is stopped due to the low density of surrounding gas. The growth rate immediately after formation must be correctly described; omission of this term in the early growth phase would result in an overestimate of the final sizes.

The expression $n(t, t')$ in the integrand of Equation 21 is written in terms of the droplet radius

$$n(t, t') = \frac{4\pi}{3\Omega_a} r^3(t, t'), \quad n(t, t) = \frac{4\pi}{3\Omega_a} r_{cr}^3(t), \quad (23)$$

where the second equality describes the fact that all nuclei are born with radius equal to the critical one. Repeated time differentiation of Equation 21 leads to a system of differential equations. The right side in Equation 21 depends on time through the upper bound of the integral as well as through the first argument of $n(t, t')$ in the integrand. The time derivative of the integrand is obtained with the help of Equations 22 and 23. The resulting system is

$$\frac{dx_0}{dt} = I(t)\frac{4\pi r_{cr}^3(t)}{3\Omega_a} + x_1(t)\frac{dr}{dt}, \quad (24)$$

$$\frac{dx_1}{dt} = I(t)\frac{4\pi r_{cr}^2(t)}{\Omega_a} + x_2(t)\frac{dr}{dt}, \quad (25)$$

$$\frac{dx_2}{dt} = I(t)\frac{8\pi r_{cr}(t)}{\Omega_a} + x_3(t)\frac{dr}{dt}, \quad (26)$$

$$\frac{dx_3}{dt} = I(t)\frac{8\pi}{\Omega_a}, \quad (27)$$

with the variables x_i, $(i = 0.3)$ defined as:

$$x_0(t) := x(t), \quad (28)$$

$$x_1(t) := \frac{4\pi}{\Omega_a}\int_{t_I}^{t} I(t')r^2(t, t')dt', \quad (29)$$

$$x_2(t) := \frac{8\pi}{\Omega_a}\int_{t_I}^{t} I(t')r(t, t')dt', \quad (30)$$

$$x_3(t) := \frac{8\pi}{\Omega_a}\int_{t_I}^{t} I(t')dt', \quad (31)$$

and the initial conditions $x_i(t_1) = 0$, $(i = 0.3)$. t_1 is the time where the vapor becomes saturated. The variable $x_3(t)$ is proportional to the total number of droplets per molecule of the vapor, formed until t, the quotient $x_2(t)/x_3(t)$ is the mean size, $2x_1(t)/x_3(t)$ is the mean squared size of droplets formed up to a time t. These moment equations (named such because x_i are related to the moments of the size distribution) were derived by Deguchi (1980) originally. Closure of the hierarchy of moment equation results because of the simplifying assumption (Equation 20), leading to the expression for the macroscopic growth rate (Equation 22). This rate is independent of the radius (otherwise, there would be an infinite hierarchy of differential equations). All time derivatives appearing in the moment Equations 24–27 have to be interpreted as Lagrangean (convective) in the hydrodynamic context. The temperature is found from an energy equation (an extension of that used by previous authors) for a van der Waals gas-liquid mixture with constant specific heats $c_{V,1}$, $c_{V,2}$ for the gaseous and liquid phases

$$[c_{v,1}(1 - x) + c_{v,2}x]\frac{dT}{dt} = \frac{R_g T \rho_1}{(1 - b\rho_1)\rho^2}\frac{d\rho}{dt} + \left[(c_{v,1} - c_{v,2})T + \frac{a(\rho_2 - \rho_1)^2}{\rho_2}\right]\frac{dx}{dt}. \quad (32)$$

This form of the energy equation is derived in the Appendix. The gas density ρ_1 has to be evaluated from

$$\rho_1 = (1 - x)\rho\left[1 - x\frac{\rho}{\rho_2(T)}\right], \quad (33)$$

and $\rho_2(T)$ is the liquid density at the coexistence curve, a function of temperature only. At small condensed-matter fractions or at low densities, this can be approximated by $\rho_1 \sim (1 - x)\rho$. Although effects of nonideality of the vapor are only important if the coexistence curve is reached at high densities, this form of the energy equation has been used throughout in the numerical solution.

The energy Equation 32 together with the moment Equations 24–27 form a closed system if only the total density $\rho(t)$ is known along a Lagrangean trajectory, e.g., from a background solution. The velocity field is not required then. One can obtain ρ and the rate of volume expansion $(-1/\rho)d\rho/dt$ directly from a numerical simulation (using tracer particles), or supply a semianalytical solution for ρ because all other thermodynamic variables are determined by ordinary differential equations, in a Lagrangean description. Only the volume expansion rate, or, equivalently, the velocity divergence is required, because this leads to a solution of the continuity equation for the density.

In this work, the background solution is the simple analytical model; however, it only enters through the volume expansion rate $(-1/\rho)d\rho/dt$ and the initial conditions at the time of saturation, t_1. The use of a background solution neglects the back-reaction of condensation on hydrodynamics for times later than t_1. The possible feedback of the condensation kinetics consists in a change of the pressure gradient, modifying the velocity field, and consequently the volume expansion rate. Condensation is a late-time phenomenon, but at late times, pressure forces in the Euler equation are small (the pressure is nearly equal to the saturation pressure, decreasing exponentially with temperature, because the thermodynamic path closely follows the phase boundary [see Fig. 8D]; the temperature and total density decrease according to power laws, asymptotically, so the ratio of the pressure acceleration to gravity is very small). Consequently, matter above the atmosphere moves along gravity-dominated, nearly ballistic trajectories. This trend will continue until the vapor cloud recollapses, falling back onto the atmosphere, and dust particles decouple from the low-density gas.

Raizer (1960) and his followers did not derive the system of Equations 24–27, but approximated the solution of Equation 21 at the time of onset of nucleation, made the approximation that some asymptotic velocity was reached at t_1, and used the asymptotic dependence $\rho \sim t^{(-3)}$ in the treatment of the energy equation (written in the ideal gas limit). Here, the evolution of the degree of condensation and the moments of the droplet size distribution is calculated in detail, and a more realistic energy equation is used. Moreover, the theoretical model developed here can easily be adapted to a situation where the volume expansion rate or, equivalently, the Lagrangean evolution of the density is obtained from a realistic hydrodynamic simulation, with the help of tracer particles.

RESULTS

Although the analytical solution for the isentropic expansion of a van der Waals gas in spherical symmetry contains large simplifications, it is exploited here, but only to describe the history of the scale factor, or equivalently, the volume expansion rate. All other variables follow from the more accurate Lagrangean form of the energy equation and from the moment equations. A disturbing feature of this analytical solution is that different layers of material would start with different specific entropies, varying spatially like $s \sim \ln(1 - \eta^2)$, and the initial temperature varies like $T \sim (1 - \eta^2)$. In contrast, material that is initially below the critical point is not present in the numerical simulation, because here the initial temperature and entropy are homogeneous inside the projectile. An impact with 20 km s^{-1} velocity leads to temperatures significantly above the critical point for more realistic EOS for minerals (in the planar impact approximation). As far as the material expands only adiabatically in the numerical solution, the point of intersection with the coexistence curve is the same for all parts of the cloud with the same initial value of the pressure. A feature found in realistic numerical simulations of impacts is that an isobaric core develops, extending to ~0.7 projectile radii for 20 km/s impact velocity (smaller than the projectile size, at a state of compression to roughly twice the normal density; see Pierazzo et al., 1997, their Fig. 10). A homogeneous distribution of entropy and temperature is not unrealistic as an initial condition for the vapor cloud. During adiabatic expansion, the main difference is in the saturation time t_1, and the value of the volume expansion rate at saturation reached in different parts of the vapor cloud originally in the isobaric core, not in thermodynamic conditions. The central part reached saturation at a time approximately a factor of two later than the material near the contact discontinuity, in the simplified hydrodynamic simulations performed here.

There is some limiting adiabat, for which the coexistence curve is intersected at the critical point. This limiting adiabat divides the $\rho - T$ diagram into two parts, where material can intersect the coexistence curve coming from either the gas or the liquid phase. In the latter case, matter has to stay at densities

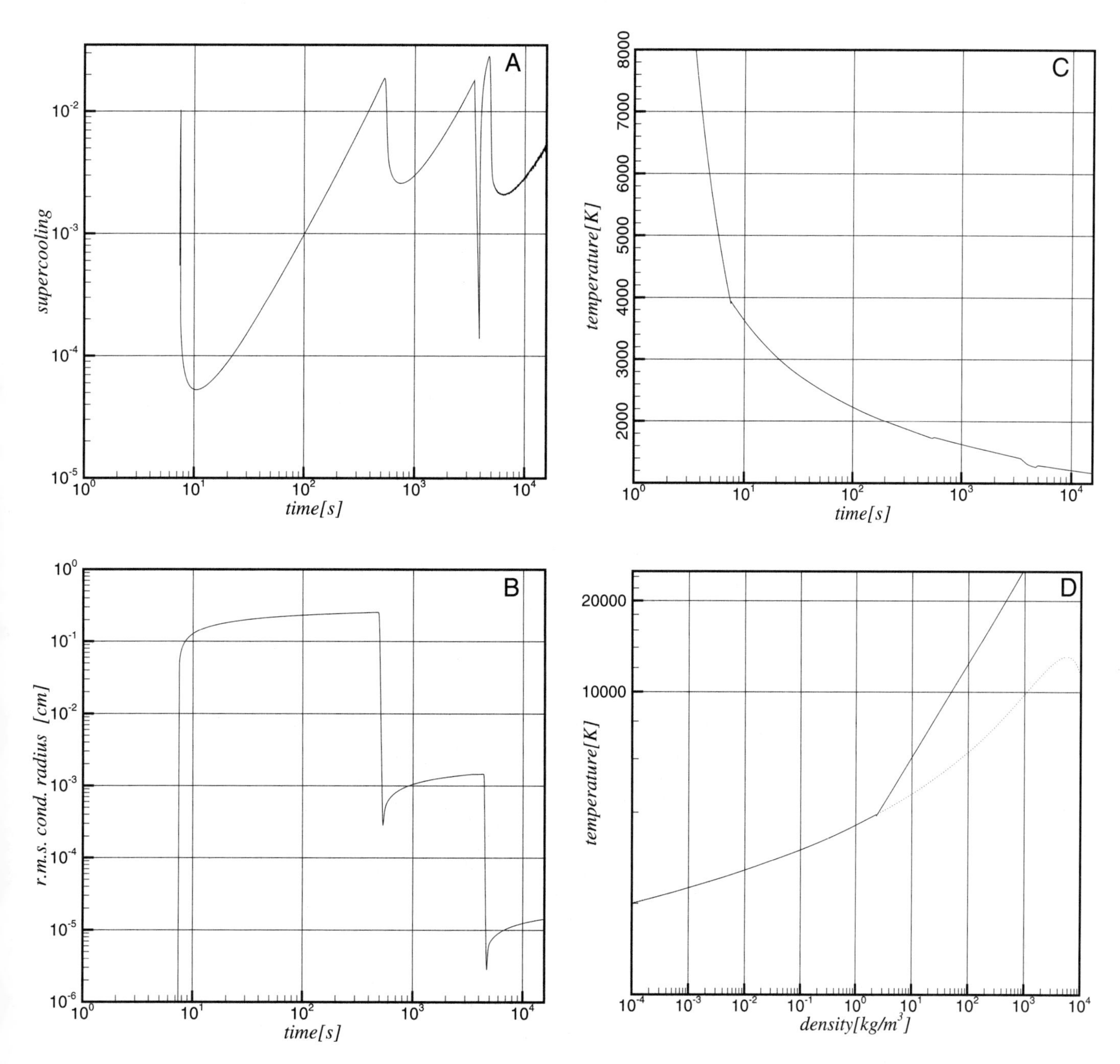

Figure 8. A: Condensation history of layer at $\eta = 0$, using semianalytical solution for total density. Supercooling parameter θ vs. time t. B: Condensation history of layer at $\eta = 0$, using semianalytical solution for total density. Root-mean-square droplet radius (in cm) vs. time. C: Condensation history of layer at $\eta = 0$, using semianalytical solution for total density. Temperature, in K, vs. time. D: Condensation history of layer at $\eta = 0$, using semianalytical solution for total density. Temperature, in K, vs. density, in kg/m^3. Phase boundary of van der Waals gas is indicated (dotted).

larger than the critical for long enough time until reaching a temperature below the critical point. If the temperature decreases below the critical point, a much larger reduction in density is necessary because $\rho \sim T^{1/(\gamma-1)}$ (in the ideal gas limit), and the adiabatic index is $\sim$1.3, here; i.e., a reduction of density by a factor of $\sim$10 is necessary to achieve a temperature decrease by a factor of two. Consequently, no evaporating liquid was observed. It is important to use a correct description of the gas phase, especially of the adiabatic exponent in the expanded state, for more complicated rock equations of state. A seemingly minor change of the adiabatic exponent can have drastic consequences for the fate of the shocked projectile during expansion.

Using the analytical solution for the Lagrangean evolution

of the total mass density, the second-order differential equation for the scale factor $R(t)$, the energy equation, and the moment Equations 24–27 have been integrated numerically with the help of a seventh- to eighth-order Runge-Kutta method (part of the optimal control software package TOMP of Kraft, 1994, subroutine rk87). As a representative example, the evolution at the center $\eta = 0$ is followed, because the specific entropy at this point in the analytical solution is the same as in the numerical hydrodynamic solution. Initial values for central pressure and temperature correspond to the Chicxulub-size impact. A surface tension of $\sigma = 0.3$ Nm^{-1} has been used, which is appropriate for silicates (Melosh, 1989). Figure 8A shows the history of the supercooling parameter θ, Figure 8B shows the root-mean-square droplet radius $\sqrt{\overline{r^2}} = \sqrt{2x_1/x_3}$, Figure 8C shows the temperature with respect to time, and Figure 8D shows a plot of T versus total mass density ρ. Several nucleation events (peaks in the θ-history) appear, repeating regularly on a logarithmic time scale. The degree of supercooling follows a complicated oscillating pattern. In Figure 8B, the interpretation is that several generations of droplets form, with very different final dimensions. Each time a new generation appears, the root-mean-square radius of droplets decreases, because older and larger droplets are outnumbered by finer, newly formed ones. Growth of already formed larger droplets is not halted when a new generation appears; this is implied in the moment equations. This is not explicitly visible here, because the moments do not allow us to reconstruct the size distribution. The rate of mass deposition of the condensed phase is proportional to the surface area, at any scale (see Equation 19). This means that the dominance of a small-scale generation in the size distribution does not imply that it is more abundant in terms of mass. Figure 8B shows that the root-mean-square size drops by approximately three orders of magnitude at the time when a new generation appears. This implies that the new generation is approximately three orders of magnitude more abundant. The distribution of droplet sizes is characterized by several maxima that correspond to repeated nucleation events and subsequent periods of macroscopic growth. The condensation temperature strongly changes between different generations, so later generations could be distinguished by chemical composition. A certain amount of reheating is visible in the temperature history (nonmonotonous changes) each time a peak of the supercooling is reached. This effect is contained in the last term on the right side of the energy Equation 32. Larger droplets should record the history of thermodynamic conditions in their layers in terms of composition, because they participated during the growth of the next small-scale generation. For the initial values and parameters used in this work, the largest droplets begin to form at a temperature of ~3900 K, whereas the second generation begins with temperatures of ~1700 K, and a third at ~1300 K (see Fig. 8C). The large droplets formed at ~3900 K stay in the liquid state for several minutes, until their atmospheric reentry. The droplet radius at the end of the first growth phase is ~3 mm; ~9 min later, another nucleation event follows (this occurs on ballistic trajectories for material leaving the atmosphere). The second nucleation event leads to final droplet radii of ~15 μm, and is eventually finished after ~1.5 h of ballistic flight. For material staying in the lower atmosphere, further stages of nucleation, except the first, are suppressed, because of the low pressures and densities required. Material on long-distance ballistic trajectories may go through a sequence of at least two nucleation stages. The present simple model is unable to predict the global distribution of the condensation products, but it is clear that the second-generation droplets are distributed worldwide, whereas the largest droplets are found closer to the impact site, because their time of formation is earlier, and they decouple from the gas flow due to their higher surface/mass ratio.

It is clear that the asymptotic form of the expansion rate $(-1/\rho)(d\rho/dt) = 3/t$ following from the spherical model is inappropriate for a geometry of ballistic Keplerian orbits, characteristic for the late, gravity-dominated phase of the vapor expansion. Therefore, a more detailed model will need to include such effects.

CONCLUSIONS

In the impact of a large asteroid at 20 km/s, vapor condensation is a nonequilibrium phenomenon, characterized by a hierarchic size distribution where the smallest sized droplets, originating late in the event, are the most abundant. The reason for such a multimodal size distribution is the repeating, nonmonotonous character of condensation in rapid adiabatic expansion. This is visible in the history of the supercooling parameter, measuring deviation from local thermodynamic equilibrium (see Fig. 8A). Small-scale (~10 μm) condensate particles are more abundant in number by three orders of magnitude as compared to millimeter-sized droplets. Although these small-scale droplets dominate in number, the first generation is larger in terms of mass. The largest droplets originate during high-temperature condensation, beginning with temperatures of ~3900 K, whereas a smaller generation begins to form when the temperature is ~1700 K. Large droplets stay at temperatures in excess of 2000 K during a time interval of at least ~2 min for conditions applicable in the Chicxulub impact.

I did not address the question of the chemical composition of condensates. This would have required extending the theory of nucleation to the case of a number of different chemically interacting species undergoing phase transitions. Another point is that classical nucleation theory is derived assuming that the monomer and n-mer cluster gases are ideal and noninteracting (e.g., see Abraham, 1974), which is questionable for those parts of the vapor cloud where nucleation starts at high densities. A general formulation of nucleation theory for arbitrary EOS is not available. These complications affect the largest droplets originating from material that is shocked to lower entropies such that the thermodynamic path intersects the phase boundary closer to the critical point. However, the majority of droplets form at low densities in the upper atmosphere. Nucleation theory is a field of ongoing research and subject to revisions. The steady nucleation rate is very sensitive with respect to the surface tension parameter. Size effects for the surface tension of small critical clusters are debated (McGraw and Laaksonen 1996). Measurements of the macroscopic surface tension for common minerals are mostly absent. Such uncertainties have been discussed by Chigai et al. (1999), who found that predictions of macroscopically observable quantities by the moment Equations 24–27 were insensitive to the precise value of the nucleation rate and surface tension.

ACKNOWLEDGMENTS

This work was supported by the Deutsches Zentrum für Luft- und Raumfahrt (DLR). I thank E. Pierazzo and D. Ebel for helpful discussions.

APPENDIX: GENERALIZED ENERGY EQUATION

The energy equation has to take into account nonideality of the vapor. Here it is derived for arbitrary EOS; the final result is specialized to a van der Waals gas-liquid mixture.

The specific entropy of the pure gaseous and liquid phases is given by

$$Tds_i = c_{V,i}dT + T\left(\frac{\partial P}{\partial T}\right)_{vi} dV_i, \tag{34}$$

where $V_i := 1/\rho_i$ is the specific volume, and indices $i = 1{,}2$ denote the gaseous and liquid phases, respectively. For general equations of state, the energy equation follows from the vanishing of the Lagrangean time derivative of the entropy, or $Tds/dt = 0$, and the additivity of specific entropy and volume $s = (1 - x)s_1 + xs_2$, and $V = (1 - x)V_1 + xV_2$:

$$\begin{aligned} Tds = {} & [c_{V,1}(1 - x) + c_{V,2}x]dT \\ & + \left[\varepsilon_2 - \varepsilon_1 - \Delta\mu + \left(T\left(\frac{\partial P}{\partial T}\right)_{V_1} - P\right)(V_1 - V_2)\right]dx \\ & + T\left(\frac{\partial P}{\partial T}\right)_{V_1} dV + xT\left[\left(\frac{\partial P}{\partial T}\right)_{V_2} - \left(\frac{\partial P}{\partial T}\right)_{V_1}\right]dV_2. \end{aligned} \tag{35}$$

For the sake of completeness, a nucleation barrier $\Delta\mu$ is included for a general gas-liquid nonequilibrium mixture, and both phases coexist at equal pressure P (in thermodynamic equilibrium, both chemical potentials would be equal, the nucleation barrier is of the order of $H_V\theta$, where H_V is the specific enthalpy of evaporation). To a very good degree of approximation, the liquid density ρ_2 is only a function of temperature and equal to the respective value at the coexistence curve. Now the formulas can be specialized for the case of a van der Waals gas. The derivative of the liquid density with respect to T along the coexistence curve (which follows after tedious algebra using the equation of state and the Clausius-Clapeyron equation) is:

$$\frac{d\rho_2}{dT} = \frac{\rho_2(1 - b\rho_2)}{T(1 - b\rho_{1,c} - 2b\rho_2)}\left[1 + \frac{(c_{V,2} - c_{V,1})T\rho_1}{a(\rho_2 - \rho_{1,c})^2}\right], \tag{36}$$

where $\rho_{1,C}$ is the gas density at the coexistence curve, for given temperature T. For the calculation of the EOS pressure, the actual gas density ρ_1 has to be found from

$$\rho_1 = (1 - x)\rho\left[1 - x\frac{\rho}{\rho_2(T)}\right], \tag{37}$$

if one does not make the approximation $\rho_1 \cong (1 - x)\rho$ of Zeldovich and Raizer (1967). If the degree of condensation x is large only at late times where ρ is many orders of magnitude below ρ_2, the difference of the right side in Equation 37 to $(1 - x)\rho$ is not very important. The partial derivative of pressure with respect to temperature at constant specific volume is given by

$$T\left(\frac{\partial P}{\partial T}\right)_{V_i} = \frac{R_g T\rho_i}{(1 - b\rho_i)} = P + a\rho_i^2, \tag{38}$$

using the equation of state. This leads to the following form of the last term in Equation 35:

$$-x\frac{a\rho_2}{T}\left[1 - \left(\frac{\rho_1}{\rho_2}\right)^2\right]\frac{(1 - b\rho_2)}{(1 - b\rho_{1,c} - 2b\rho_2)}\left[1 + \frac{(c_{V,2} - c_{V,1})T\rho_1}{a(\rho_2 - \rho_{1,c})^2}\right]dT. \tag{39}$$

This term can be viewed as a small correction to $xc_{V,2}dT$ in the first term in Equation 35 ($a\rho_2$ is of the order of the binding energy, but $b\rho_2$ is very close to one; far from the critical point; only the product $a\rho_2(1 - b\rho_2)$ matters). If the variability of the liquid density ρ_2 and the nucleation barrier are neglected, one arrives at the following form of the energy equation:

$$\begin{aligned} [c_{V,1}(1 - x) + c_{V,2}x]\frac{dT}{dt} = {} & \frac{R_g T\rho_1}{(1 - b\rho_1)\rho^2}\frac{d\rho}{dt} \\ & + \left[(c_{V,1} - c_{V,2})T + \frac{a(\rho_2 - \rho_1)^2}{\rho_2}\right]\frac{dx}{dt}, \end{aligned} \tag{40}$$

where the gas density ρ_1 has to be found from Equation 37 for known x, and ρ and $\rho_2(T)$ have to be calculated from the coexistence curve of the van der Waals gas.

REFERENCES CITED

Abraham, F.F., 1974, Homogeneous nucleation theory: Advances in theoretical chemistry, Supplement 1: New York, Academic Press, 263 p.

Chigai T., Yamamoto, T., and Kozasa, T., 1999, Formation conditions of presolar TiC core-graphite mantle spherules in the Murchison Meteorite: Astrophysical Journal, v. 510, p. 999–1010.

Deguchi, S., 1980, Grain formation in cool stellar envelopes: Astrophysical Journal, v. 236, p. 567–576.

Frenkel, Ya.I., 1946, Kinetic theory of liquids: Oxford, Clarendon Press, 488 p.

Guggenheim, E.A., 1957, Thermodynamics: An advanced treatment for chemists and physicists: Amsterdam, North-Holland Publishing Company, p. 117 and p. 163.

Kompaneets, A.S., 1960, A point explosion in an inhomogeneous atmosphere: Soviet Physics Dokladyi, v. 5, p. 46–48 (English translation).

Kraft, D., 1994, Fortran modules for optimal control calculations: Association of Computing Machinery Transactions on Mathematical Software, v. 20, n. 3, p. 262–281.

LeVeque, R.J. 1997, Wave propagation algorithms for multi-dimensional hyperbolic systems: Journal of Computational Physics, v. 131, p. 327–353.

McClurg, R.B., and Flagan, R.C., 1998, Critical comparison of droplet models in homogeneous nucleation theory: Journal of Colloid and Interface Science, v. 201, p. 194–199.

McGraw, R., and Laaksonen, R., 1996, Interfacial curvature free energy, the Kelvin relation, and vapor-liquid nucleation rate: Journal of Chemical Physics, v. 106, n. 12, p. 5284–5287.

Melosh, H.J., and Pierazzo, E. 1997, Impact vapor plume expansion with realistic geometry and equation of state [abs.]: Lunar and Planetary Science, v. 28, p. 935.

Melosh, H.J., 1989, Impact cratering: A geological process: New York, Oxford University Press, p. 68–71.

Newman, W.I., Symbalisty, E.M.D., Ahrens, T.J., and Johnes, E.M., 1999, Impact erosion of planetary atmospheres: Some surprising results: Icarus, v. 138, p. 224–240.

O'Keefe, J.D., and Ahrens, T.J., 1981, The interaction of the Cretaceous-Tertiary extinction bolide with the atmosphere, ocean, and solid earth: Conference on large body impacts and terrestrial evolution, Snowbird, Utah, p. 71.

Pierazzo, E., Vickery, A.M., and Melosh, H.J., 1997, A reevaluation of impact melt production: Icarus, v. 127, p. 408–423.

Pierazzo, E., Kring, D.A., and Melosh, H.J., 1998, Hydrocode simulation of the Chicxulub impact event and the production of climatically active gases: Journal of Geophysical Research, v. 103, p. 28607–28626.

Raizer, Yu.P., 1960, Condensation of a cloud of vaporized matter expanding in a vacuum: Soviet Physics JETP, v. 37, p. 1229–1235.

Roe, P.L., 1981, Approximate Riemann solvers, parameter vectors, and difference schemes: Journal of Computational Physics, v. 43, p. 357–372.

Senger, B., Schaaf, P., Cort, D.S., Bowles, R., Voegel, J.-C., and Reiss, H., 1999: A molecular theory of the homogeneous nucleation rate. I. Formulation and fundamental issues: Journal of Chemical Physics, v. 110, n. 13, p. 6421–6431.

Zel'dovich, Ya.B., 1942, Theory of the formation of a new phase: Zhurnal Eksperimental'noi i Theoreticeskoi Fiziki, n. 12, p. 525–538.

Zel'dovich, Ya.B., and Raizer Yu., P., 1967, Physics of shock waves and high-temperature hydrodynamic phenomena: New York, Academic Press, p. 571–597.

Manuscript Submitted October 5, 2000; Accepted by the Society March 22, 2001

Printed in the U.S.A.

Geological Society of America
Special Paper 356
2002

On the completeness of the discovery rate of the potentially hazardous asteroids

Eric W. Elst*
Royal Observatory at Uccle, Ringlaan 3, B-1180 Uccle, Belgium

ABSTRACT

An investigation was made on the discovery completeness of the potentially hazardous asteroids in function of the absolute magnitude (*H*). There are some indications that the most dangerous objects of that kind have already been discovered. From a study of the orbital elements of the known potentially hazardous asteroids (272 objects, September 2000), the ephemerides, the minimum orbital intersection distances, and the encounter conditions, it follows that there is, at least for the near future, no reason to fear a disastrous hit of one of these objects on the Earth. Almost complete sky coverage during the past four years by LINEAR (Lincoln Laboratory Experimental Test System, New Mexico) and other large asteroid-search programs has resulted, for the limiting magnitude of $V = 19.0$, in nearly discovery completeness of the potentially hazardous asteroids below $H = 18.5$.

INTRODUCTION

During the past decade there has been understandable interest in the potentially hazardous asteroids. The fear of a disastrous hit on the Earth with global consequences for human life has been supported by several new findings of impact structures on Earth (e.g., Fennoscandian impact craters; Pesonen, 1996). It has become clear that not only the Moon and several planets of the solar system, but also Earth, have collided with smaller bodies (asteroids and comets). Tunguska, in Siberia, has been revisited by the Russians as well as by several international groups; new-found impact places, such as Chicxulub (Yucatan, Mexico), have been thoroughly investigated by international teams. The asteroids (such as Mathilda, Gaspra, Eros), which have been recently photographed by spacecrafts (e.g., NEAR; Near Earth Asteroid Rendezvous, Jet Propulsion Laboratory, Pasadena, California, United States), have many craters on their surfaces.

During the past decade, several international conferences have drawn the interest of scientists of different disciplines, and have focused the attention of the public on impact asteroids, sometimes with the consequence of creating undue concern. Furthermore, the discovery of a few potentially hazardous asteroids which were thought to be on a nearly collisional course with the Earth, such as 1997 XF11 (Marsden, 1998) and 1999 AN10 (Milani et al., 1999), have not been reassuring. Therefore, an investigation of the potentially dangerous objects that already have been discovered during the past, and of objects that still may be discovered, could be very useful.

MAIN BELT ASTEROIDS, COLLISIONS, AND RESONANCE DYNAMICS

Main belt asteroids are not considered to be dangerous, because their orbits are safely stable, at least for longer periods than the history of mankind. However, the main belt may be the reservoir of new objects that, due to internal collisions in the belt, could find their way toward the inner parts of the planetary system (e.g., Ipatov, 1995). However, the collisions do not lead directly to the generation of new potentially hazardous asteroids, but they may supply new objects (fragments of the parental bodies) that are partially injected into resonant, chaotic

*E-mail: ericelst@oma.be

Elst, E.W., 2002, On the completeness of the discovery rate of the potentially hazardous asteroids, *in* Koeberl, C., and MacLeod, K.G., eds., Catastrophic Events and Mass Extinctions: Impacts and Beyond: Boulder, Colorado, Geological Society of America Special Paper 356, p. 645–650.

orbits undergoing large variations of the eccentricity and becoming therefore amenable to planetary encounters (Menichella et al., 1996). Farinella et al. (1993) estimated the fraction of main belt fragments that ending up into either the 3/1 mean motion resonance with Jupiter or the ν6 secular resonance, which are the two most effective dynamical routes from the main belt to the inner planet zone. It is interesting that, depending on the detailed assumptions adopted on the collisional physics, in particular the ejection velocity distribution of the fragments, only 1%–4% of these fragments may reach the routes. Froeschlé and Morbidelli (1993) developed the resonant proper element algorithm, which allows us to identify the dynamical nature of the resonant objects. This algorithm is a powerful tool for studying the mechanism of the fragment and/or meteorite transport to the inner solar system.

During the past decade, there has been a tremendous increase in the effort to discover new asteroids at several observatories. Asteroid discoverers admire the ambition of van Houten et al. (1970): during the 1960s and 1970s, when there was hardly an interest in asteroids, they discovered thousands of asteroids, on plates taken by Tom Gehrels at Palomar (Leiden-Palomar Surveys,Table 1). However, the main impulse to observe potentially hazardous asteroids came from events like the Shoemaker-Levy 9 comet impact on Jupiter in 1996 and from the theoretical work done by several authors such as Milani (1993), Ferraz-Mello (1993), Froeschlé and Morbidelli (1993), Farinella et al. (1993), Rabinowitz (1994), Menichella et al. (1996), and Steel (1996).

Since the general introduction of the charge coupled devices (CCDs), mounted on all kind of telescopes, the rate of new discoveries of asteroids has grown immensely (Tables 2 and 3). From this we infer that the amount of asteroids in the main belt (and at other places in the solar system) may be limitless. Therefore we first focus our attention on the main belt objects.

TABLE 1. MINOR PLANET DISCOVERERS: NUMBERED OBJECTS, SEPTEMBER 2000

Discoveries	Codiscoveries	Period	Name(s)
1538		1997–2000	LINEAR
1166		1960–1977	C.J. van Houten,*
981	23	1986–1998	E.W. Elst
705	1	1991–2000	T. Kobayashi
539	5	1977–1994	E. Bowell
523	144	1975–1989	S.J. Bus
487	15	1966–1986	N.S. Chernykh
483		1987–2000	S. Ueda, H. Kaneda
422	60	1965–1993	H. Debehogne
420	5	1989–1993	H.E. Holt
392		1914–1957	K. Reinmuth
383		1987–1997	K. Endate,**
336	177	1973–1995	E.F. Helin
293		1973–1995	C + E Shoemaker,***
286		1985–2000	Spacewatch
274		1992–1993	UESAC
239	15	1966–1992	L.I. Chernykh
234		1978–1990	A. Mrkos
228	20	1891–1932	M.F. Wolf
200	80	1961–1995	F. Börngen
188	10	1975–1996	R.H. Naught

Notes: *: van Houten-Groeneveld and T. Gehrels.**: and K. Watanabe.***: D.H. Levy and H.E. Holt.
LINEAR: Lincoln Laboratory Experimental Test System, New Mexico.
UESAC: Uppsala-ESO asteroids and comets searches.
Source: Minor Planet Center (2000).

Completeness of main belt objects

Figure 1 is a plot of the number of all (discovered) asteroids as a function of the semimajor axis (in astronomical units, AU). Asteroids are accumulating at well-defined distances (from the Sun), indicating the presence of so-called asteroid families, i.e., asteroids with almost the same orbital elements. However, if we want to study asteroid families in a more appropriate way, it will be necessary to use proper elements. Asteroid families become clearly visible in the proper element diagrams (semimajor axes versus inclinations or eccentricities; e.g., see Kne-

TABLE 2. LEADING OBSERVING PROGRAMS: ASTROMETRIC POSITIONS (1997–1999)

Program	Number	Telescope	Investigator(s)
LINEAR	725277	1.0-m reflector	G.H. Stokes
Spacewatch	126143	0.91-m reflector	R.S. McMillan
LONEOS	103493	0.59-m Schmidt	E.L.G. Bowell
NEAT	43333	1.0-m reflector	E.F. Helin
Xinglong	35240	0.60-m Schmidt	J. Zhu
ESO	33744	1.0-m Schmidt	E.W. Elst,*
ODAS	27742	0.9-m Schmidt	A. Maury, G. Hahn
Klet	19764	0.57-m reflector	J. Tichá
Catalina	17792	0.41-m Schmidt	S.M. Larson
Oizumi	13929	0.25-m reflector	T. Kobayashi
Visnjan	13295	0.41-m reflector	K. Korlevic
Ondrejov	10300	0.65-m reflector	P. Pravec
Woomera	8843	0.30-m reflector	F.B. Zoltowski
Prescott	7659	0.46-m reflector	P. Comba

Notes: *: C.-I. Lagerkvist. LINEAR: Lincoln Laboratory Experimental Test System, New Mexico.
LONEOS: Lowell Observatory Near Earth Objects, Arizona. NEAT: Near Earth Asteroids, Haleakala, Maui. ESO: European Southern Observatory, Chile. ODAS: Observatoire de la Côte d'Azur, Asteroid Searches.
Source: IAU Commission 20 (1999).

TABLE 3. MINOR PLANET CENTER ARCHIVE STATISTICS: ASTROMETRIC POSITIONS (SEPTEMBER 2000)

Date	Numbered objects	Unnumbered objects
2000 September 13	2145798	3051281
2000 July 26	2008481	2862252
2000 June 21	1941688	2713750
2000 May 23	1874284	2541569
2000 April 18	1799778	2281176
2000 March 20	1654324	2061381
2000 February 22	1584702	1913075
2000 January 24	1509331	1795549
1999 December 22	1419893	1575129
1999 November 23	1358402	1513800
1999 October 26	1287819	1479916
1999 September 28	1210373	1421197

Source: Minor Planet Center (2000)

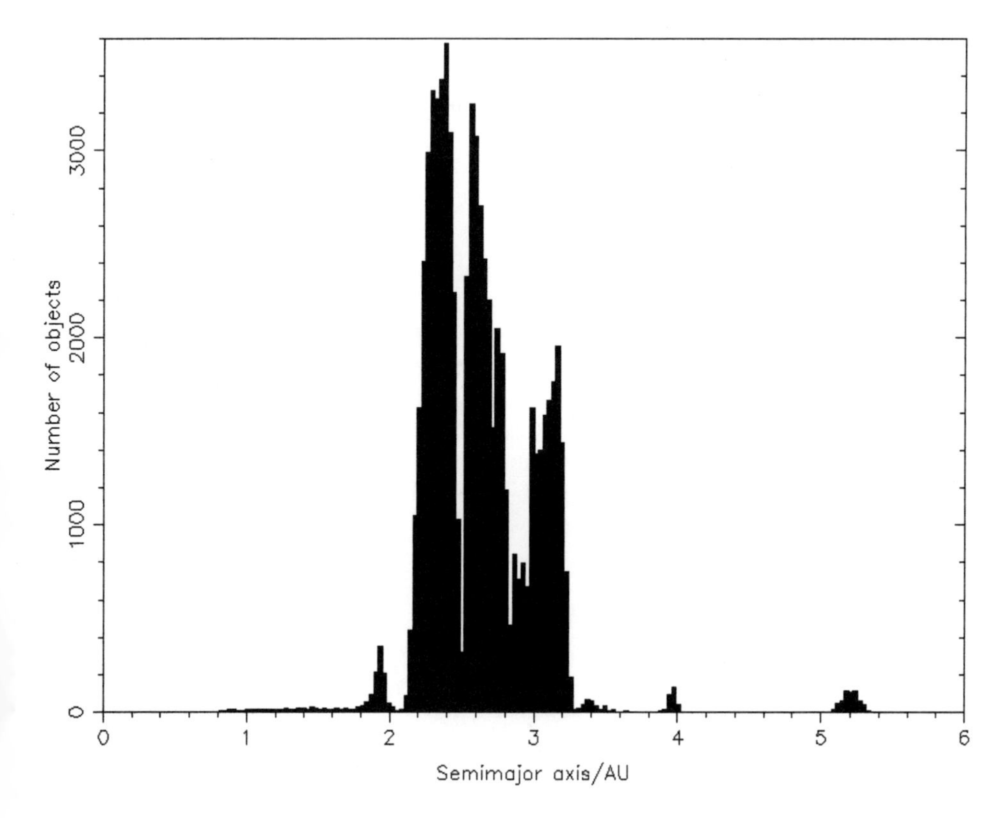

Figure 1. Accumulation of asteroids at particular places in main belt from 1 to 5.2 astronomical units (AU). Different peaks correspond with presence of families, such as Flora group, from 2.10 to 2.30 AU, the Pallas group from 2.50 to 2.82 AU, Koronis group from 2.83 to 2.91 AU, and Themis group from 3.08 to 3.24 AU. Hungaria and Hilda asteroids are indicated at 1.9 and 4.0 AU, respectively. Trojans are accumulating at 5.2 AU. It is clear that height of peak is measure for relative completeness below certain limiting magnitude. *Source:* Minor Planet Center (2000), various lists and plots.

zevic and Milani, 1993). The first peak (Fig. 1) corresponds neatly with the Flora-group of asteroids, whereas the two smaller peaks, respectively at 1.9 and 4 AU, represent the Hungaria-resonance group and Hilda-family. It is clear that the peaks (which are here intermingled with the presence of other families) are a measure for relative completeness of certain asteroids at well-defined distances from the Sun. The peaks may move to higher values, depending on how far the search for new asteroids extends (indicated by the *V*-magnitude). Sky coverage by the larger survey programs (LINEAR, Spacewatch [Arizona], LONEOS [Lowell Observatory Near Earth Objects, Arizona], NEAT [Near Earth Asteroids, Haleakala, Maui] and Catalina Sky Survey [Arizona]) is currently between *V*19.0 and *V*19.5.

A better and more reliable measure for completeness is given by the cumulative distribution curves of asteroids, because the location of the inflection point indicates very well the rate of completeness, expressed in absolute magnitude, for a selected group of asteroids (Nakos, 2000, personal commun.). We therefore look at the cumulative distribution curve of the Jupiter Trojans (Fig. 2), which we use only to demonstrate the phenomenon of completeness. Conclusions drawn from the Trojans should not be extended to the potentially hazardous asteroids, because Trojans, characterized by low eccentricities and sometimes high inclinations, move in relatively stable orbits, which are only slightly perturbed (Milani, 1993). Potentially hazardous asteroids, however, show in general large eccentricities, making them observable only during a short time (a few weeks or even less) when they are in the vicinity of the Earth. However, because sky coverage of the observations has been extended significantly during the past years, there is good hope that all objects, to a well-defined and not too faint absolute magnitude, will be discovered during the following decade. The cumulative distribution curve for the Trojans (Fig. 2) with the inflection point at absolute magnitude, $H = 11.5$, indicates that completeness has been reached for objects with $H < 11.5$. There may be some small scatterings, depending on the inclination of a particular object. High inclination asteroids (with sometimes very high absolute magnitudes) tend to be discovered much later than their ecliptic counterparts (during the past asteroid searches have been mostly done near the ecliptic), because these objects are moving most of the time outside the ecliptic (e.g., see Elst, 1996). The trend toward approaching fainter absolute magnitudes showed up in my investigation (discoveries) of new Trojans from 1986 to 1997 (Table 4). Averaging the absolute magnitude per year, we obtain the following result: 1986, 8.6 (1 object); 1987, 10.3 (3 objects); 1989, 10.3 (1 object); 1990, 10.1 (1 object); 1991, 10.9 (1 object); 1993, 10.6 (3 objects); 1994, 10.6 (3 objects); 1996, 11.3 (13 objects); 1997, 10.5 (2 objects). 1996 GE19 and 1997 JG15 are high-inclination objects.

Potentially hazardous asteroids

The known 272 potentially hazardous asteroids (minor planets with the greatest potential for close approaches to the

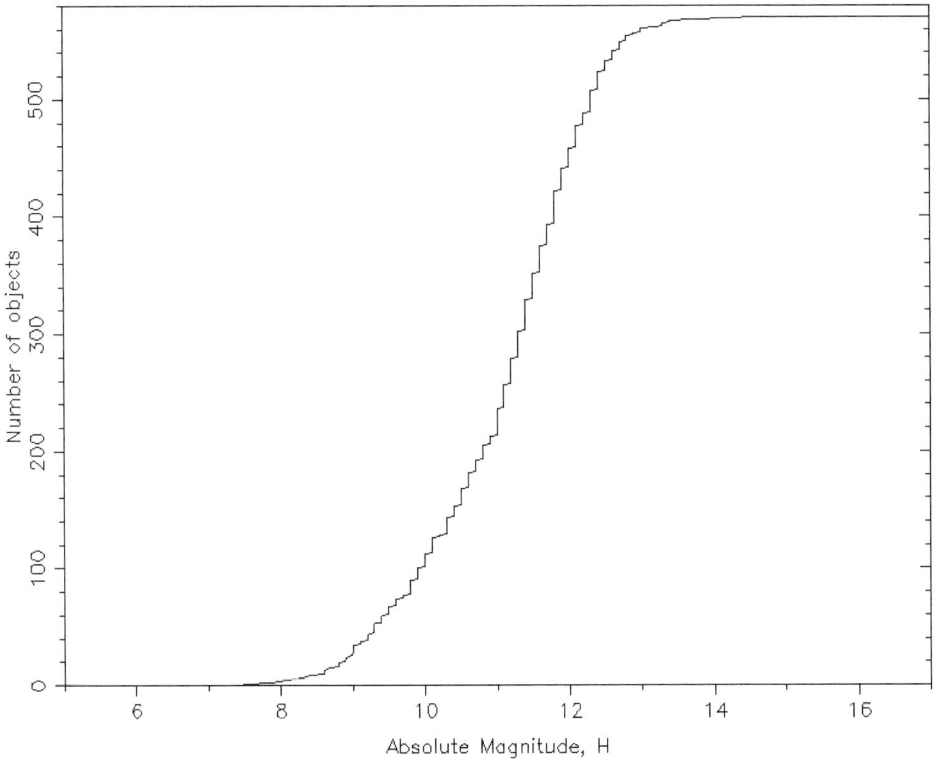

Figure 2. Cumulative distribution of Jupiter Trojans in function of absolute magnitude *H*. If at certain epoch full sky coverage has been accomplished, we may assume that all asteroids with that particular absolute magnitude, for searches below limiting magnitude, have been discovered. *Source:* Minor Planet Center (2000), various lists and plots.

TABLE 4. DISCOVERIES OF TROJANS AT THE OBSERVATORIES OF HAUTE PROVENCE, CAUSSOLS, AND EUROPEAN SOUTHERN OBSERVATORY, CHILE

Number	Name	Designation	Ln	H	Observatory
5254	Ulysses	1986 VG1	L4	8.8	OHP
7815	Dolon	1987 QN	L5	10.4	ESO
4060	Deipylos	1987 YT1	L4	8.9	ESO
12238		1987 YU1	L4	10.8	ESO
4501	Eurypylos	1989 CJ3	L4	10.3	ESO
11509		1990 VL6	L5	10.1	ESO
16560		1991 VZ5	L5	10.9	ESO
11552		1993 BD4	L5	10.6	Caussols
11554		1993 BZ12	L5	10.5	ESO
		1994 CX13	L5	11.2	ESO
		1994 CR18	L5	10.0	ESO
		1994 ES6	L5	10.5	ESO
12444		1996 GE19	L4	10.8	ESO
		1996 HV9	L5	11.1	ESO
9430		1996 HU10	L5	10.5	ESO
		1996 HN17	L5	11.0	ESO
		1996 HF19	L5	12.5	ESO
		1996 HZ25	L5	13.0	ESO
		1996 TN39	L4	11.4	ESO
13650		1996 TN49	L4	11.9	ESO
13184		1996 TS49	L4	10.8	ESO
		1996 TC51	L4	11.3	ESO
13185		1996 TH52	L4	11.3	ESO
		1996 TD57	L4	10.5	ESO
		1996 TA58	L4	11.4	ESO
		1997 JG15	L5	10.5	ESO
12052		1997 JB16	L5	10.6	ESO

Notes: Ln: Lagrange Libration Point. H: absolute magnitude. *Source:* Minor Planet Circulars (1986–1997)

Earth) include those objects with *H* brighter than or equal to $V = 22.0$. Specification of the encounter conditions is given by the value of *M* (perpendicular distance to the Earth orbit when the object is at the Earth's distance from the Sun) and the value of *N* (the distance of the object's nodal points from the Earth's orbit). There are 46 potentially hazardous asteroids that have been permanently numbered (Table 5).

We now look at the distribution of all near Earth objects, including the 272 potentially hazardous asteroids, in function of the absolute magnitude (Fig. 3). A prominent peak appears at $H = 18.5$, indicating that for that particular magnitude nearly all near Earth objects have been discovered (not taking into account those objects with very high inclinations, which may be discovered later). This value ($H = 18.5$) corresponds to a diameter of 530–1200 m. Potentially hazardous asteroids have a relatively short dynamic lifetime. Farinella et al. (1993) estimated the collisional lifetime of main belt asteroids: 100-km-sized bodies approach the age of the solar system, whereas the age of the smaller bodies is roughly proportional to the square root of size. Farinella's model predicts also a flux of ~100 fragments larger than 1 km in diameter per 1 m.y. into the resonant routes, which appears to be the same order as the loss rate of the near Earth objects/m.y.

CONCLUSIONS

We may conclude from the present investigation that searches for potentially hazardous asteroids have reached the

TABLE 5. LIST OF THE NUMBERED POTENTIALLY HAZARDOUS ASTEROIDS

Number	Name	Designation	H
12923		1999 GK4	16.1
16960		1998 QS52	14.3
12538		1998 OH	16.1
13651		1997 BR	17.6
8566		1996 EN	16.5
7482		1994 PC1	16.8
5693		1993 EA	17.0
5604		1992 FE	16.4
6489	Golevka	1991 JX	19.2
9856		1991 EE	17.4
7822		1991 CS	17.4
5189		1990 UQ	17.3
4953		1990 MU	14.1
8014		1990 MF	18.7
11500		1989 UR	18.4
6239	Minos	1989 QF	17.9
4769	Castalia	1989 PB	16.9
7335		1989 JA	17.0
4581	Asclepius	1989 FC	20.4
4179	Toutatis	1989 AC	15.30
7753		1988 XB	18.6
6037		1988 EG	18.7
4450	Pan	1987 SY	17.2
4486	Mithra	1987 SB	15.6
4034		1986 PA	18.1
14827		1986 JK	18.3
3362	Khufu	1984 QA	18.3
3671	Dionysus	1984 KD	16.3
3200	Phaethon	1983 TB	14.6
3757		1982 XB	18.95
3361	Orpheus	1982 HR	19.03
4660	Nereus	1982 DB	18.2
3908	Nyx	1980 PA	17.4
4015	Wilson-Harrington	1979 VA	15.99
2135	Aristaeus	1977 HA	17.94
2340	Hathor	1976 UA	19.2
2102	Tantalus	1975 YA	16.2
1981	Midas	1973 EA	15.5
5011	Ptah	6743 P-L	17.1
2061	Anza	1960 UA	16.56
4183	Cuno	1959 LM	14.4
1620	Geographos	1951 RA	15.60
1566	Icarus	1949 MA	16.9
2201	Oljato	1947 XC	15.25
2101	Adonis	1936 CA	18.7
1862	Apollo	1932 HA	16.25

Notes: N: Number of the asteroid. H: Absolute magnitude.
Source: Minor Planet Center (2000)

1-km-sized objects, without discovering a single object that is on its way to Earth (orbital calculations by Milani, 2000). We should still have time to discover new fragments larger than 1 km that will be injected into resonant routes during the next 1 m.y. Therefore we shouldn't be too worried about globally disastrous events in the near future, because 1-km-sized objects are exactly the size limit for objects with global consequences, if they should hit the Earth. However, objects such as Mithra (Elst et al., 1989), Alinde, Taranis, and Toutatis may be a short list of names to remember. Impacts of 3–5-km-diameter asteroids like those, even in the oceans, would be disastrous (Andrews, 1997).

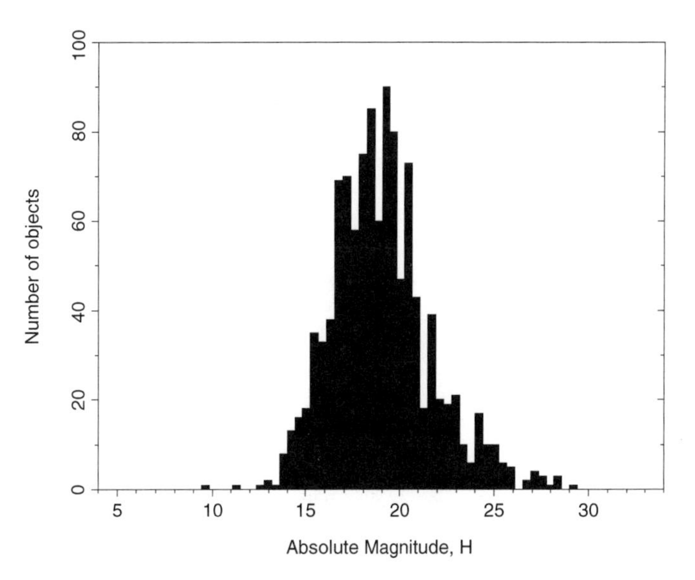

Figure 3. Distribution of near Earth asteroids (NEAs; including potentially hazardous asteroids) in function of absolute magnitude *H*. Depending on how far searches have proceeded, this distribution may move to higher values, moving at same time to fainter absolute magnitudes. From this plot it becomes clear that we are approaching near completeness for NEAs with absolute magnitude below 18.5. *Source:* Minor Planet Center (2000), various lists and plots.

REFERENCES CITED

Andrews, A.D., 1997, Asteroids and doomsday: Irish Astronomical Journal, v. 24, no. 1, p. 6–8.

Elst, E.W., Ivanova, V., and Shkodrov, V., 1989, On the discovery of the Apollo object 1987 SB: Astronomische Nachrichten, v. 310, p. 157–158.

Elst, E.W., 1996, Discovery and rediscovery of Trojan asteroids: Earth, Moon and Planets, v. 72, p. 275–277.

Farinella, P., Froeschlé, C., and Gonczi, R., 1993, Meteorite delivery and transport: Asteroids, comets, meteors: Dordrecht, Kluwer, p. 205–222.

Ferraz-Mello S., 1993, Kirkwood gaps and resonant groups: Asteroids, comets, meteors: Dordrecht, Kluwer, p. 175–188.

Froeschlé, C., and Morbidelli, A., 1993, The secular resonances in the solar system: Asteroids, comets, meteors: Dordrecht, Kluwer, p. 189–204.

Ipatov, S.I., 1995, Migration of small bodies to the earth: Solar System Research, v. 29, p. 261–286.

Knezevic, Z., and Milani, A., 1993, Asteroid proper elements: The big picture: Asteroids, comets, meteors: Dordrecht, Kluwer, p. 143–158.

Marsden, B.G., 1998, International Astronomical Union (IAU) Circular 6879.

Menichella M., Paolicchi, and P., Farinella, P., 1996, The main belt as a source of near-Earth asteroids: Earth, Moon and Planets, v. 72, p. 133–149.

Michel, P., Froeschlé, and C., Farinella, P., 1996, Dynamical evolution of NEAs: Close encounters, secular perturbations and resonances: Earth, Moon and Planets, v. 72, p. 151–164.

Milani, A., Chesley S.R., and Valsecchi G.B., 1999, Close approaches of asteroid 1999 AN10: Resonant and non-resonant returns: Astronomy and Astrophysics, v. 346, p. L65–L68.

Milani, A., 1993, The dynamics of the Trojan asteroids: Asteroids, comets, meteors: Dordrecht, Kluwer, p. 159–174.

Milani, A., 2000, Encounter conditions, Personal Homepage of Andrea Milani: http://copernico.dm.unipi.it/~milani/homemilani.html.

Minor Planet Center, 2000, Various lists and plots: http://cfa-www.harvard.edu/iau/mpc/html.

Pesonen, L.J., 1996, The impact cratering record of Fennoscandia: Earth, Moon and Planets, v. 72, p 377–393.
Rabinowitz, D.L., 1994, A near-Earth asteroid belt: Icarus, v. 111, p. 364–377.
Steel, D.I., 1996, The limitations of NEO-uniformitarianism: Earth, Moon and Planets, v. 72, p. 279–292.
Van Houten, C.J., van Houten-Groeneveld, I., Herget, P., and Gehrels, T., 1970, The Palomar-Leiden survey of faint minor planets: Astronomy and Astrophysics, Supplement Series, v. 2, p. 339–448.

Manuscript Submitted October 10, 2000; Accepted by the Society March 22, 2001

Printed in the U.S.A.

Geological Society of America
Special Paper 356
2002

Possible sources of Earth crossers

Nina A. Solovaya*
Astronomical Institute of the Slovak Academy of Sciences, Dúbravská cesta 9, 842 28 Bratislava, Slovak Republic, and Celestial Mechanics Department, Sternberg State Astronomical Institute, University Prospect 13, 119 899 Moscow, Russia
Eduard M. Pittich*
Astronomical Institute of the Slovak Academy of Sciences, Dúbravská cesta 9, 842 28 Bratislava, Slovak Republic

ABSTRACT

We discuss the orbital evolution of solar system objects that can be sources of Earth-crossing bodies. One of the candidates of the source of Earth crossers is a group of high-inclination asteroids within the main asteroid belt. The results of their orbital evolution, obtained by numerical integration using a dynamical model of the solar system consisting of all planets, showed that, during the 200 k.y. interval that was investigated, bodies with initial inclinations within 40°–120° and eccentricities within 0.2–0.4 periodically change their inclinations, eccentricities, and perihelion distances within a rather wide range. During this time, they can reach the vicinity of the Sun and cross the Earth orbit. These bodies could potentially be dangerous for the Earth.

We studied the dynamical behavior of known Solar and Heliospheric Observatory (SOHO) comets, assuming their orbits to be near parabolic. The parabolic orbits of these comets, as derived from coronographs data covering short near perihelion part of the orbits, are not precisely determined. However, it is possible to make assumptions about their near-parabolic eccentricities. Sungrazing comets discovered from the Earth with better determined orbits support these assumptions. The equations of motion of the model SOHO comets with near-parabolic eccentricities were also numerically integrated within an interval of 200 k.y.; the bodies show periodic changes in their eccentricities, inclinations, and perihelion distances. Some of these bodies come periodically to the vicinity of the Earth and can be candidates for collisions.

INTRODUCTION

It is known that the impact of a large asteroid or comet on the Earth would have catastrophic consequences. In the history of the Earth there are many examples of craters that formed from such collisions, and there are similar features on the surface of the Moon and satellites of other planets. We know of at least one collision with a large body that had devastating consequences for the biological evolution on Earth, that at the Cretaceous-Tertiary boundary, ca. 65 Ma (e.g., Pillmore and Miggins, 2000).

Our knowledge about asteroids is not very old. The first asteroid discovery, Ceres, was by G. Piazzi, January 1, 1801, at Palermo. The first asteroid (Apollo) with an orbit reaching the Earth's orbit was found in 1932. Due to secular motion of the perihelion and nodes, some time in the future the Apollo orbit may cross the Earth's orbit (e.g., Williams, 1969, 1971), and a collision of this asteroid and the Earth would be possible.

*E-mails: Solovaya, solov@sai.msu.ru; Pittich, pittich@savba.sk

Solovaya, N.A., and Pittich, E.M., 2002, Possible sources of Earth crossers, *in* Koeberl, C., and MacLeod, K.G., eds., Catastrophic Events and Mass Extinctions: Impacts and Beyond: Boulder, Colorado, Geological Society of America Special Paper 356, p. 651–657.

The first ten asteroids crossing the Earth's orbit were discovered before the 1970s. In 1973, an observation program to search for Earth-crossing asteroids started at Palomar Observatory. Until the end of the twentieth century, studies searching from different observatories have discovered thousands of asteroids in the solar system. The majority of them are concentrated in the main asteroid belt between the orbits of Mars and Jupiter. More than 400 asteroids orbit the Sun with perihelion distances (the minimum distance from the Sun) of <1.33 AU (astronomical units). Their sizes vary from 41 km (Ganymed) to a few meters (asteroid 1991 VG) (Shor, 1996).

The other small bodies that are potentially dangerous for the Earth are comets. According to Bailey (1990), 10% of all craters on the Earth and Moon surfaces originate from collision of comets with these bodies. The unique appearance of comets in the sky attracted the attention of people throughout history. The first preserved record of an observed comet seems to be from 239 B.C., the famous Halley comet. The rapid motion and large size of comets in the sky, in contrast with asteroids, made their existence known much earlier than asteroids. The discoveries of the SOHO (a space mission, the Solar and Heliospheric Observatory) comets during the past five years showed an extensive population of comets that have orbits that bring them very close to the Sun; they are generally called sungrazers. This raises the question of whether some of them could be Earth crossers.

The SOHO comets are bodies with radii <1000 m. According to the distribution function of cometary radii, most of them are probably as small as 100 m in radius. According to the estimate of Brandt et al. (1997), 16–800 objects with nuclear radii of 250 m have a perihelion distance between of 1 and 2 AU. Due to their small size, the SOHO comets are very faint objects on the sky, and they are detectable only when they become bright enough, which happens either when the comets are near the Sun, or when they pass close to the Earth.

By the end of the year 2000, 301 such comets were known (Gregory and Myers, 2001; Biesecker, 2001). From the Earth, only 14 sungrazing comets were discovered within the time period of A.D. 1106–1970. In contrast to the comets discovered by the SOHO spacecraft, these were very bright objects, some of them visible during daylight, e.g., comet C/1106, or C/1882, the famous September comet (e.g., Vsekhsvyatskij, 1956). All other sungrazers were detected using the coronographs on the satellite P78-1 (6 comets) SMM (Solar Maximum Mission, 10), and SOHO (271) spacecrafts. The first spacecraft operated from 1979 to 1983, the second from 1987 to 1989, and the third from December 1995.

Generally, comets move around the Sun on periodic or aperiodic orbits with prograde or retrograde orientation without any preference of direction or shape of the orbit. At the end of 1999, more than 1100 comets were known: 258 have an orbital period shorter than 30 yr, 24 have orbits between 30 and 200 yr, 243 have orbits longer than 200 yr, 151 have near-hyperbolic orbits, and more than 400 have parabolic and near-parabolic orbits (Marsden and Williams, 1999). About 50% of all known cometary orbits, except the sungrazers, have their perihelion inside the Earth's orbit. However, because their nodes are far from the Earth's orbit, only a few of them can cross the Earth orbit and be dangerous.

The orbits of sungrazers differ from those of other known comets: they have inclinations that are higher than 90°, and many have very similar orbits. The medians of inclinations (i) and perihelion distances (q) calculated from the orbital data in Gregory and Myers (2001) and Biesecker (2001), are 145° and 0.0055 AU, respectively. The imprecise determination of their orbits is due to the limited observation period shortly before the perihelion passage. Therefore, their orbits have been calculated only in the first approximation as parabolic ones.

HILL CRITERION FOR UNSTABLE ASTEROID ORBITS

The origin of near-Earth asteroids has been the subject of a long-standing debate, starting with Öpik (1951). Today it seems likely that these asteroids almost surely have diverse origins. Some of them may have been derived from old Mars crossers, some from separate regions of the main asteroid belt, and some from the cometary population (Shoemaker et al., 1979, 1990). This idea is supported by the taxonomy and mineralogical classification of these objects. It shows that practically all taxonomy classes, except one or two low-albedo ones, are present among the near-Earth asteroids (e.g., Lupishko, 2000). The stability of asteroid orbits was discussed by many (e.g., Carusi and Valsecchi, 1979; Everhart, 1979; Greenberg and Scholl, 1979). Some of the orbits are stable on a time scale of $\sim 10^9$ yr; others can be chaotic with regular and irregular retrograde orbits. The main asteroid belt regions bordered by low-order commensurabilities and by secular resonances are probable sources of near-Earth asteroids, e.g., as shown by Williams (1969, 1971) and Wetherill (1979).

All known asteroids move in the prograde direction, as do the planets. Retrograde asteroids have not yet been observed. Most of the known asteroids have orbital inclinations no higher than 20°. The largest inclination observed for an asteroid is 64° (asteroid 2102). However, the major asteroid surveys are limited by low ecliptic latitude, which limits knowledge of the high-inclination population (Gehrels, 1979).

The investigation of the orbital evolution of asteroids and comets, and the study of the transfer mechanism among different types of orbits, including trans-Neptunian orbits through the main asteroid belt, long-period and short-period cometary orbits, and Earth-crossing orbits, is important for understanding many cosmogenic and geophysical problems. In celestial mechanics the investigation of the motion of two bodies with large masses and one massless body is called the restricted three-body problem. The masses of asteroids and comets are very small in comparison to the Sun and the planets: in the solar system, the Sun and Jupiter have the dominant masses. Therefore, in a first approximation we can study the motion of massless asteroids under the influence of the Sun and Jupiter only.

In the classic three-body problem (Sun-Jupiter-asteroid), the only known integral of the restricted problem can be obtained directly from the differential equations of motion of three bodies. This integral is known as the Jacobian integral. Hill applied the Jacobian integral for considering the question of stability of motion (Hill, 1902; Chebotarev, 1965). The Jacobian integral establishes the connection between the constant of integration C and velocity and coordinates of an asteroid. If C_1 represents the value of the Jacobian constant in a singular point and C represents the value of that constant for the asteroid, then Hill's stability criterion is defined by the condition: $C > C_1$. This means that an asteroid inside the sphere of influence of the Sun with a value of C of the Jacobian integral is trapped there indefinitely. This criterion is valid both in two and three dimensions. In our work we took model asteroids from the main asteroid belt, located at the boundary of the Hill stability criterion in the restricted three-body problem, and investigated their orbital evolution under the gravitational influence of all planets.

LONG-TERM EVOLUTION OF HIGH-INCLINATION ASTEROIDS

A dynamic model of the solar system, with all planets, but massless asteroids, was used. The initial semimajor axis of the asteroids, a_0, was calculated from the equation of the Jacobian integral of the restricted three-body problem Sun-Jupiter-asteroid. The Jacobian integral establishes the connection between the initial values of coordinates and velocities, and allows us to define regions where a motion with the given initial conditions may take place. In the Sun-centered sidereal coordinate system the Jacobian integral has the form (Solovaya et al., 1992)

$$C = [1/(2a) + \sqrt{p} \cos I] + 1/2\mu^2 + \mu[2K(\kappa_2)/(\pi\nu) - R \cos b \cos(l - l_j)], \quad (1)$$

where $\nu^2 = (\rho^2 + 3\alpha^2)^2$, $\alpha = (1 - \mu)e_j$, and $\kappa_2{}^2 = [4\alpha^2 - (\rho^2 + 3\alpha^2 - \nu^2)/2]/\nu^2$; R is the Sun to asteroid distance, ρ is the Jupiter to asteroid distance, $\mu = 0.0009538$ is the ratio of Jupiter's mass to the sum of the masses of Sun and Jupiter, e_j is the eccentricity of Jupiter's orbit, l and b are the longitude and latitude of an asteroid, l_j is the longitude of Jupiter, a, I, and p are the osculating elements and the parameter of an asteroid orbit, and $K(\kappa_2)$ is the elliptic integral. When $e_j = 0$, then $K(\kappa_2) = \pi/2$. C is the Jacobian constant.

It is known that if the Jacobian constant C for an asteroid is higher than $C(L_1)$, then the asteroid's orbit is stable according to Hill's criterion. $C(L_1)$ is the Jacobian constant for the inner equilibrium point, L_1. It represents the boundary of the Hill stability region. Asteroids with similar values of the Jacobian constants have been observed for many years. Their orbits are well determined. The initial values of the semimajor axis for the model asteroids were calculated from Equation 1 with $C(L_1) = 1.538$. This value of the Jacobian constant is valid for the system Sun-Jupiter-asteroid when the eccentricity value of Jupiter is maximum, i.e., $e_j = 0.062$. Equation 1 was solved for i_0 equal to 40°, 60°, 80°, 100°, 120°, and 140°, and for e_0 equal to 0.0, 0.2, and 0.4. The initial value of argument of perihelion ω_0 of the asteroids was selected from the equation

$$\cos \Theta_\pi = \cos \omega \cos l_{j\pi} + \sin \omega \sin l_{j\pi} \cos i \quad (2)$$

at $\Theta_\pi = 180°$ and varied with the step of 60° from 0° to 360°. The initial value of the ascending node was $\Omega_0 = 90°$. The initial epoch of the numerical integration was March 25, 1991 UT (Universal Time). For this moment l_j was taken from Abalakin (1989).

The investigation of the long-period evolution of the model asteroids under the gravitational influence of all planets of the solar system may reveal whether the asteroids have Earth-crossing orbits. Therefore, we performed a numerical integration of the equation of motion of the model asteroids for the period of 200 k.y., using Everhart's (1985) integrator. We found that within a period of no longer than 20 k.y. some of the model asteroids crossed the Earth orbit and that their perihelions are very close to the Sun. This migration of orbits between the main asteroid belt orbits and Earth crossers is periodically repeated during the entire investigated period of 200 k.y. The behavior of the osculating orbital element, perihelion distance q, is shown in Figure 1 for 20 k.y. This short period enables us to see more details of the changes of orbital elements due to migration. For very low eccentricities the variations of all elements are negligible. Therefore, the perihelion distance q is not shown in Figure 1.

Bodies with $q \leq 1$, for which the heliocentric distance r is equal to 1 when they are in the ascending or descending node, cross the Earth's orbit. At favorable geometrical conditions they can collide with the Earth. These asteroidal orbits can be determined from the equation for calculation of their heliocentric distance in the problem of two bodies

$$r = q(1 + e)/(1 + e \cos \nu), \quad (3)$$

where v is the true anomaly. For the conditions that lead to collision with the Earth, i.e., $r = 1$ AU, $\nu = -\omega$ for the ascending node, or $\nu = 180° - \omega$ for the descending node, we obtain the argument of perihelion ω from Equation 3:

$$\omega = \arccos\{1/e[+q(1 + e) - 1]\} \quad (4)$$

for the ascending node and

$$\omega = \arccos\{1/e[-q(1 + e) + 1]\} \quad (5)$$

for the descending one.

The asteroids fulfilling the conditions given in Equations 4 and 5 are designated in Figure 1 by solid lines. The number of such cases cannot be neglected.

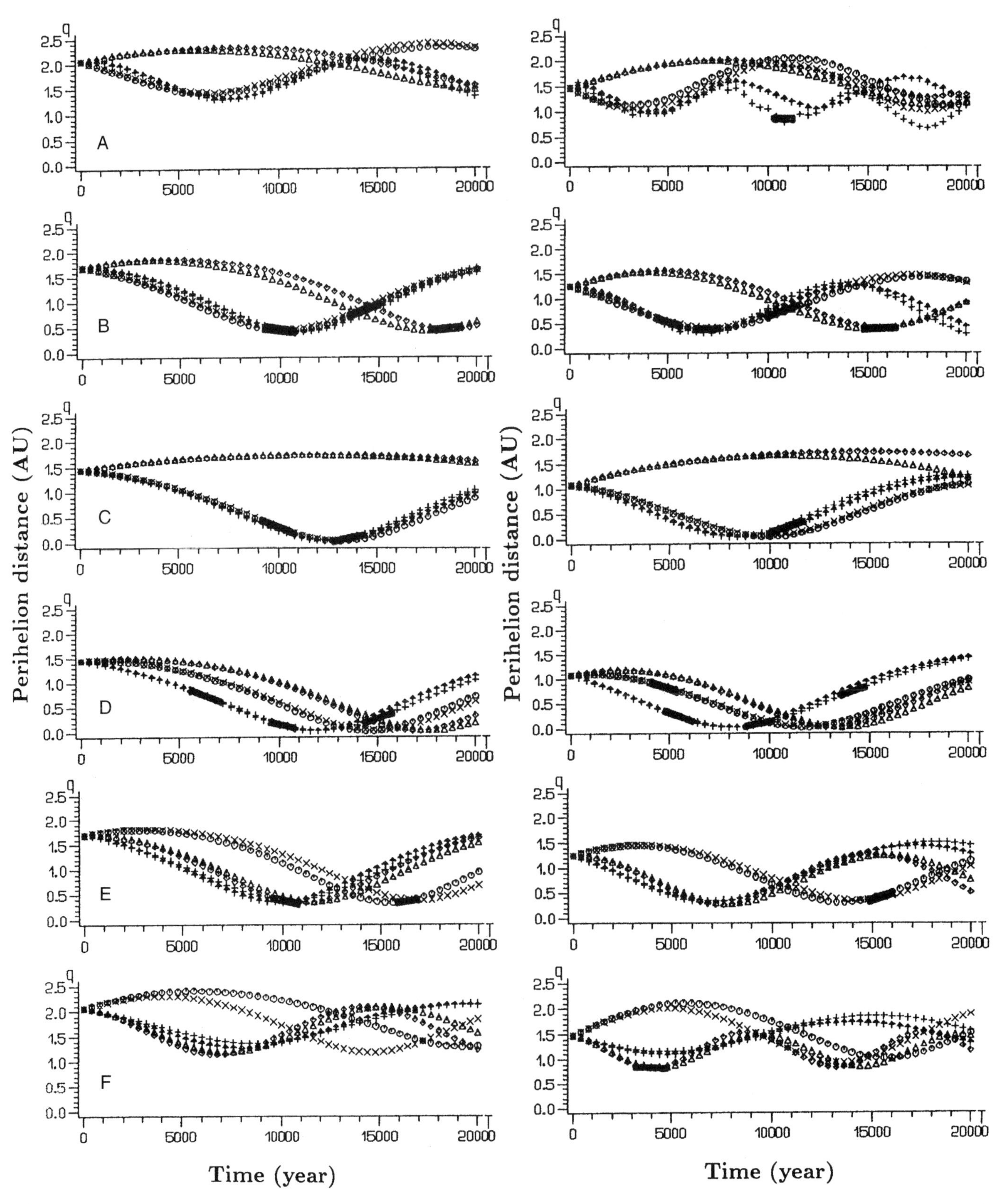

Figure 1. Time evolution of the perihelion distance q of the model asteroids within 20,000 years; initial values are: inclination $i_0 = 40°$ (a), $i_0 = 60°$ (b), $i_0 = 80°$ (c), $i_0 = 100°$ (d), $i_0 = 120°$ (e), $i_0 = 140°$ (f), eccentricity $e_0 = 0.2$ (left panels), $e_0 = 0.4$ (right panels), and argument of perihelion ω_0 for ⦶ — $\omega_0 = 0°$, △ — $\omega_0 = 60°$, + — $\omega_0 = 120°$, × — $\omega_0 = 180°$, ◇ — $\omega_0 = 240°$, and ⍏ — $\omega_0 = 300°$. The solid lines indicate sections of orbits when the nodes are close to the Earth's orbit.

LONG-TERM EVOLUTION OF ORBITS SIMILAR TO THOSE OF SOHO COMETS

To investigate the evolution of model orbits similar to those of SOHO comets within a period of 200 k.y., we used initial orbital elements, the values of which are medians of elements of retrograde orbits of the observed SOHO comets with different eccentricities, *e*. We studied the numerical integration of model orbits with the following initial parameters: argument of perihelion $\omega_0 = 80°$, longitude of the ascending node $\Omega_0 = 0°$, inclination $i_0 = 145°$, perihelion distance $q_0 = 0.0055$ AU, and eccentricity e_0 ranging from 0.99 to 0.9999999.

The long-term numerical integration of the differential equations was done for an interval of 200 k.y. from March 25, 1991, using Everhart's (1985) integrator, as described here. The initial date for the integration was chosen arbitrarily and has no

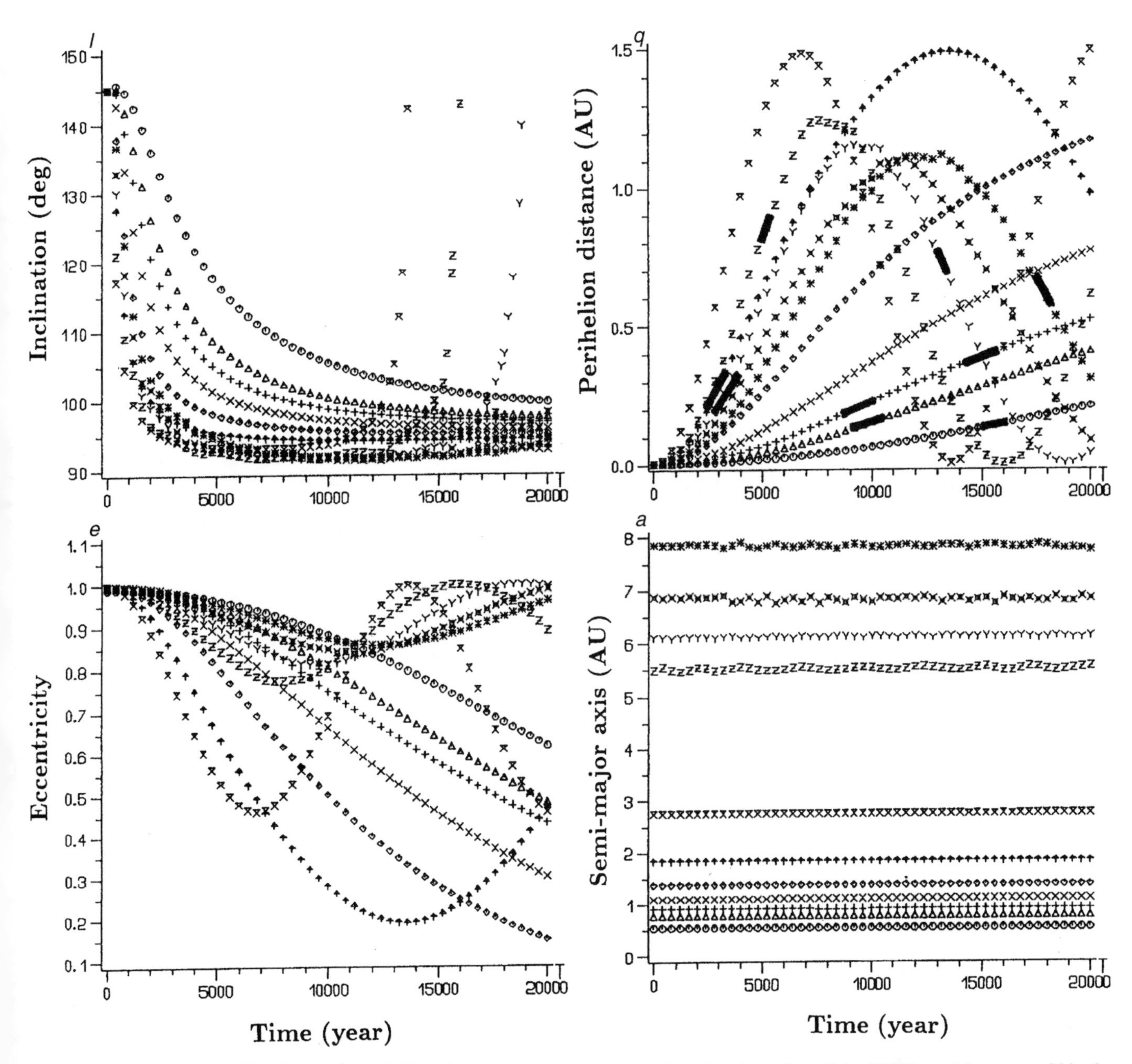

Figure 2. Time evolution of inclination i, perihelion distance q, eccentricity e, and semi-major axis a of the SOHO model comets within the next 20,000 years; initial values are: inclination $i_0 = 145°$, perihelion distance $q_0 = 0.055$ AU, argument of perihelion $\omega_0 = 80°$, and eccentricity e_0 for ⦶ — $e_0 = 0.9900$, △ — $e_0 = 0.9930$, + — $e_0 = 0.9940$, × — $e_0 = 0.9950$, ◇ — = 0.9960, ⯐ — $e_0 = 0.9970$, ⧖ — $e_0 = 0.9980$, Z — $e_0 = 0.9990$, Y — $e_0 = 0.9991$, ⌘ — $e_0 = 0.9992$, and ✳ — $e_0 = 0.9993$. The solid lines indicate sections of orbits when the nodes are close to the Earth's orbit.

influence on the calculation results. The same initial date was chosen for the numerical integration of the asteroids. The behavior of the osculating elements; i.e., inclination i, eccentricity e, perihelion distance q, and semimajor axis a, resulting from the study of the orbital evolution of the SOHO model orbits within the period of 20 k.y. for the eccentricities smaller than 0.9994, is shown in Figure 2.

CONCLUSIONS

We studied the orbital evolution of two groups of small bodies in the solar system that can be in Earth-crossing orbits. The first group includes asteroids from the main asteroid belt, which orbits close to the boundary of the Hill stability criterion. Asteroids with high-inclination orbits can under the influence of the gravitational forces of other planets, periodically changing their orbital eccentricities and inclinations in such a way that they can become Earth crossers for a certain period.

The results of the numerical integration (Fig. 1) show that, within the investigated period, the semimajor axes a of the orbits change little, without any secular perturbations. However, the eccentricities and inclinations of the asteroid orbits show large long-period perturbations. Consequently, they change within a wide range. Because the perihelion distance is a function of the eccentricity, $q = a(1 - e)$, the perihelion distances change together with the eccentricities. For eccentricities close to zero the variations of all elements are negligible.

When the initial eccentricity e_0 is equal to 0.2 and 0.4, and for initial inclinations of $i_0 = 60°$, 80°, 100°, and 120°, the eccentricities change periodically within a wide range and the perihelion distances are <1 AU. Asteroids moving in these orbits can cross the Earth's orbits in their nodes. When an asteroid and the Earth are at the same time near one of the asteroid nodes, the probability of collisions is high.

In the group of model asteroids investigated here, collisions can occur depending on the initial values of eccentricity e_0, inclination i_0, and the argument of perihelion ω_0. Figure 1 shows that a high collision probability is not rare. For example, asteroids with initial values of $e_0 = 0.2$, $i_0 = 100°$, and $\omega_0 = 120°$ can collide with a period of ~5 k.y. (see Fig. 1D, left), and asteroids with initial values of $e_0 = 0.4$, $i_0 = 60°$, and $\omega_0 = 0°$ can collide even more often than at 5 k.y. intervals (Fig. 1B, right).

The second group investigated here includes sungrazing comets with very small perihelion distances and eccentricities close to 1. The results of our investigation are shown in Figure 2. The evolution of the orbits of the sungrazers is similar to the evolution of the orbits of high-inclination asteroids. Within the investigated period the semimajor axes a of the sungrazer orbits have small periodic perturbations, but the eccentricities and inclinations of their orbits change within a wide range. Therefore, their perihelion distances also vary widely, from close to 0 AU to 1.5 AU.

The comets that come close to the Earth's orbit and can be candidates for collisions are shown in Figure 2 by solid lines. We see that the possibility of the collisions of sungrazers with the Earth is not negligible. For example, within 20 k.y. the Earth can collide with sungrazing comets moving in orbits with eccentricities from 0.9900 to 0.9993, if the Earth and the comet are both close to the orbital node of the sungrazing comet. For eccentricities higher than 0.9993, the orbits are long periodical, close to parabolic ones. Comets with such orbits reach the Kuiper belt and the more distant outer part of the solar system with heliocentric distances, depending on their eccentricities. These comets do not come close to the Earth's orbit.

ACKNOWLEDGMENTS

This work was supported by the Slovak Academy of Sciences Grant VEGA 1005. The computations were done on ORIGIN 2000 computer of the computer center of the Slovak Academy of Sciences. We thank both reviewers for their helpful comments, and Christian Koeberl for valuable suggestions and for helping in the preparation of the manuscript.

REFERENCES CITED

Abalakin, V.K., ed., 1989, Astronomical almanach on 1991: Leningrad, Science, 693 p. [in Russian].

Bailey, M.E., 1990, Comet crater versus asteroid craters: Manchester, England, University of Manchester, Department of Astronomy, Preprint, 17 p.

Biesecker, D., 2001, LASCO sungrazers: http://sungrazer.nascom.nasa.gov/

Brandt, J., Randall, C., Stewart. I., A'Hearn, M., Fernandez, Y., and Schleicher, D., 1997, The lost tribe of small comets: American Astronomical Society Meeting, v. 191, p. 33.03.

Carusi, A., and Valsecchi, G.B., 1979, Numerical simulations of close encounters between Jupiter and minor bodies, *in* Gehrels, T., ed., Asteroids: Tucson, University of Arizona Press, p. 391–416.

Chebotarev, G.A., 1965, Analytical and numerical methods in celestial mechanics: Moscow, Nauka, 365 p.

Everhart, E., 1979, Chaotic orbits in the solar system, *in* Gehrels, T., ed., Asteroids: Tucson, University of Arizona Press, p. 283–288.

Everhart, E., 1985, An efficient integrator that used Gauss-Radau spacing, *in* Carusi, A., and Valsecchi, G.B., eds., Proceedings of the International Astronomical Union, Colloquium 83, Dynamics of comets: Their origin and evolution: Dordrecht, The Netherlands, Reidel, p. 185–202.

Gehrels, T., 1979, The asteroids: History, surveys, techniques, and future work, *in* Gehrels, T., ed., Asteroids: Tucson, University of Arizona Press, p. 3–24.

Greenberg, R., and Scholl, H., 1979, Resonances in the asteroid belt, *in* Gehrels, T., ed., Asteroids: Tucson, University of Arizona Press, p. 310–333.

Gregory, S.E., and Myers, D.C., 2001, Pre-LASCO sungrazers: http://sungrazer.nascom.nasa.gov/

Hill, G.W., 1902, Illustration of periodic solutions in the problem of three bodies: Astronomical Journal, v. 22, p. 93–97.

Lupishko, D.F., 2000, Physical properties of near-Earth asteroids as principal impactors onto the Earth [abs.], *in* Catastrophic events and mass extinctions: Impacts and beyond: Houston, Texas, Lunar and Planetary Institute, LPI Contribution No. 1053, p. 118–119.

Marsden, B.G., and Williams, G.V., 1999, Catalogue of cometary orbits 1999: Cambridge, Minor Planet Center, Harvard-Smithsonian Astrophysical Observatory, 127 p.

Öpik, E.J., 1951, Collision probabilities with the planet and distribution of interplanetary matter: Proceedings of the Royal Irish Academy, v. 54A, p. 165–199.

Pillmore, C.L., and Miggins, D.P., 2000, A new $^{40}Ar/^{39}Ar$ age determination on the K/T boundary interval [abs.], *in* Catastrophic events and mass extinctions: Impacts and beyond: Houston, Texas, Lunar and Planetary Institute, LPI Contribution No. 1053, p. 166.

Shoemaker, E.M., Williams, J.G., Helin, E.F., and Wolfe, R.F., 1979, Earth-crossing asteroids: Orbital classes, collision rates with Earth, and origin, *in* Gehrels, T., ed., Asteroids: Tucson, University of Arizona Press, p. 253–282.

Shoemaker, E.M., Wolfe R.F., and Shoemaker, C.S., 1990, Asteroids and comet flux in the neighborhood of Earth, *in* Sharpton, V.L., and Ward, P.D., eds., Global catastrophes in Earth history: Geological Society of America Special Paper 247, p. 155–170.

Shor, V.A, 1996, History of origin of problem of asteroid-comet dangerous, *in* Sokolskij, A.G., ed., Asteroid-comet dangerous: St. Petersburg, Institute of Theoretical Astronomy, Russian Academy of Sciences, p. 9–21 [in Russian].

Solovaya, N.A., Gerasimov, I.A., and Pittich, E.M., 1992, 3-D orbital evolution model of outer asteroid belt, *in* Harris, A.W., and Bowell, E. eds., Asteroids, comets, meteors 1991: Lunar and Planetary Institute, Houston, Texas, p. 565–568.

Solovaya, N.A., and Pittich, E.M., 1996, Orbital stability of high inclination asteroids, *in* Ferraz-Mello, S., Morando, B., and Arlot, J.-E., eds., Proceedings of the International Astronomical Union, Symposium 172, Dynamics, ephemerides and astrometry of the solar system: Dordrecht, The Netherlands, Reidel, Proceedings of the International Astronomical Union, Symposium 172, p. 187–192.

Vsekhsvyatskij, S.K., 1956, Physical characteristics of comets: Moscow, Physical and Mathematical Publishing House, 575 p.

Wetherill, G.W., 1979, Steady state populations of Apollo-Amor objects: Icarus, v. 37, p. 96–112.

Williams, J.G., 1969, Secular perturbations in the solar system [Ph.D. thesis]: Los Angeles, University of California, 270 p.

Williams, J.G., 1971, Proper elements, families, and belt boundaries, *in* Gehrels, T., ed., Physical studies of minor planets: Washington, D.C., NASA SP-267, p. 177–181.

Manuscript Submitted October 5, 2000; Accepted by the Society March 22, 2001

Geological Society of America
Special Paper 356
2002

Measurement of the lunar impact record for the past 3.5 b.y. and implications for the Nemesis theory

Richard A. Muller*
Department of Physics and Lawrence Berkeley Laboratory, 50-5032 LBL, University of California, Berkeley, California 94720, USA

ABSTRACT

Measurements of the ages of 155 lunar spherules from the Apollo 14 site suggest that the solar system impact rate over the past 3.5 b.y. first gradually declined, and then increased starting at 0.4 Ga, back to the level it had been 3 Ga. A possible explanation is offered in terms of the Nemesis theory, which postulated a solar companion star. A sudden change in the orbit of that star at 0.4 Ga transformed a circular orbit (which does not trigger comet showers) into an eccentric orbit (which does). The Nemesis theory is speculative but viable; contrary to prior assertions, the orbit is sufficiently stable to account for the data.

INTRODUCTION

A new approach to estimating past impact rates has been developed by our group at Berkeley: rather than measuring the ages of identified craters, the ages of small glass droplets called spherules that are produced from unidentified impact craters are measured (Muller, 1993). Most lunar spherules are formed in impacts; others are primarily from pyroclastic eruptions. From analysis of the spherule ages, it was concluded that most of the spherules come from separate craters (Culler et al., 2000; Muller et al., 2000a). Even though we do not know the crater from which a spherule originates, the distribution of spherule ages gives us information about the rate at which craters were formed.

Even 1 g of lunar soil, collected from the Apollo missions, typically contains more than 100 spherules in the size range of 150–250 μm diameter. At the Apollo 14 landing site, the spherules contain sufficient potassium to allow $^{40}Ar/^{39}Ar$ measurement of their age to a mean accuracy of 221 m.y.; half of the spherules had age uncertainties <132 m.y. Larger spherules, which have higher total potassium contents, yielded the most accurate ages.

The original goal of the experiment was to test the Nemesis theory (Davis et al., 1984), which predicts that the impact rate will have a large component from periodic comet showers, events in which multiple impacts occur within short (1–2 m.y.) time spans. Moreover, the theory predicts that the comet showers will recur whenever the cause (Nemesis, the hypothetical companion star to the Sun) reaches perihelion, which is once per orbit. The spacing between comet showers would be regular, with a 26–28 m.y. spacing, matching the period seen in paleontological extinctions; however, due to perturbations in the orbit from passing stars, there are expected variations of several million years in the timing of the showers, as pointed out by Davis et al. (1984). Herein I review the status of the Nemesis theory, including the critical issue of the stability of the Nemesis orbit.

In March 2000, Culler et al. (2000) reported the measurement of the chemical composition and ages of 155 spherules from the Apollo 14 site. Based on the belief that the soil sample at our site was well mixed, we (Culler et al., 2000) argued that the observed spherule age distribution reflects the cratering rate on the surface of the Moon and the Earth. Additional aspects of our study of these issues are discussed in this chapter.

The spherule age distribution is shown in Figure 1. At the bottom of the plot are 155 Gaussian curves, one for each

*E-mail: muller@physics.berkeley.edu; web site: http://muller.lbl.gov

Muller, R.A., 2002, Measurement of the lunar impact record for the past 3.5 b.y. and implications for the Nemesis theory, *in* Koeberl, C., and MacLeod, K.G., eds., Catastrophic Events and Mass Extinctions: Impacts and Beyond: Boulder, Colorado, Geological Society of America Special Paper 356, p. 659–665.

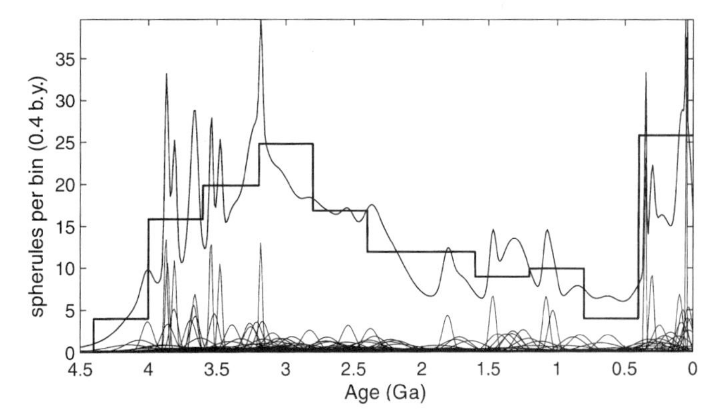

Figure 1. Distribution of spherule ages found at Apollo 14 site. Histogram shows number of spherules found in each 0.4 b.y. age bin. At bottom of plot, age measurement of each spherule is depicted by 155 separate Gaussian curves. Area of each Gaussian is constant (representing one spherule each), but width represents 1 standard deviation uncertainty in age. Continuous curve is sum of these Gaussians (ideogram). Ideogram is best estimate, in statistical sense, of age distribution of spherules. Note, however, that sharp spikes in ideogram are artifacts of spherules with highly precise ages.

spherule. The areas of each Gaussian are equal, but the width (root mean square deviation) for each is set equal to the 1 standard deviation error in the age uncertainty for that spherule. The sum of these individual Gaussian curves, the ideogram, is the smooth curve. The ideogram represents the best estimate, in a statistical sense, for the spherule age distribution. Note, however, that the sharp peaks are artifacts of several ages that were determined with high precision. The uncertainties in the ages were determined by a least-squares analysis of the isotope correlation plots, and represent the 1 standard deviation in the ages of those fits.

The spherule age distribution shows a broad peak near 3 Ga (3000 Ma). The smaller number for older ages is likely due to the relatively young age of the Apollo 14 landing site, estimated as 3.85 Ga. All spherules produced at the Apollo 14 site must be younger than this. The older spherules are presumed to be secondary, i.e., created and first deposited at distant sites and subsequently transported to the Apollo 14 site from secondary impacts.

There is a gradual decrease in spherule number from 3 to 0.4 Ga. This is consistent with previously reported estimates of a decrease in cratering rates during the same interval (BVSP, 1981; Ryder et al., 1991). The decrease could be due to gradual reduction in the number of impactors in the solar system as asteroids and comets are ejected by Jupiter's gravitational perturbation or deflected into the Sun.

The most surprising feature of Figure 1 is the sharp increase in the number of spherules with ages between 0 and 0.4 Ga. A recent increase of a factor of two had previously been suggested on the basis of measurements on the Earth and crater counting on the Moon (Grieve and Shoemaker, 1994; McEwen et al., 1997; Shoemaker et al., 1990). In the Berkeley data, the rate in the last 0.4 b.y. increased by a factor of 3.7 $\pm$ 1.2 (compared to the preceding 0.8 b.y.), back to the level that it had been 3 b.y. earlier.

There are several systematics that could create a difference between the spherule age distribution and the lunar impact rate. The most important of these is the fact that the lunar surface is mixed over time (lunar "gardening"). Older spherules are more likely to be buried, so it is more likely that we would find recent spherules at the top of the soil (where our sample was collected). This possibility is discussed in the next section.

SPHERULE AGES AND ANALYSIS

At the beginning of this study, two 1 g samples of lunar soil from the lunar missions Apollo 11 and Apollo 12 were obtained. The spherules were separated from the soil and the ages measured at the Berkeley Geochronology Center using the $^{40}Ar/^{39}Ar$ isochron technique. Poor precision was achieved because of the low potassium levels; several individual age uncertainties were >1 b.y. However, based on a chemical analysis of these spherules, we (Culler and Muller, 1999) deduced that most of the spherules came from local impacts. It was known that the Fra Mauro Formation at the Apollo 14 landing site was enriched in potassium by a factor of 5–10 compared to that at other Apollo landing sites. We requested and obtained samples from this location, and found that the spherules did have the higher potassium content expected from a local origin: with this higher potassium, higher precision ages were obtained. The median age uncertainty (1σ) for our 155 Apollo 14 spherules was 0.13 b.y.

A statistical analysis was done to estimate the number of craters represented by the 155 spherules. To understand the method used, assume for the moment that every spherule had an age that differed significantly from that of every other spherule. Then we could conclude that the spherules all came from different craters. Of course, this condition was not met, as can be seen in the overlapping Gaussian plots in Figure 1. However, we could still estimate the independence of the spherules from the ratio of spherules that occur within 1σ versus between 1σ and 2σ. If the spherules are truly independent, then these two numbers should be equal. To the extent that they are different (i.e., there are more close ages), we can conclude that there are multiple spherules from the same age. Using this method, we estimated that ~146 of the spherules all came from separate craters (Muller et al., 2000b). This was consistent with a previous conclusion that most of the spherules were from different craters based on a chemical analysis of spherules (Culler and Muller, 1999).

The samples were collected from the lunar surface, and this creates a potential bias in favor of recent ages. The lunar soil is constantly overturned through impacts of meteoroids, a process referred to as lunar gardening. However, shallow soil is gardened more rapidly than deep soil. We might expect that old spherules would be uniformly mixed down to the base of the regolith, but recent spherules would be found primarily at the surface. This would result in an excess of recent spherules in our samples, which is what was found.

To minimize the systematics of lunar gardening, we chose a sample that had been collected in the vicinity of a very recent (25 Ma) impact that formed Cone crater. Calculations (Heiken et al., 1991; McGetchin et al., 1973) and seismic data (Chao, 1973) imply the presence of a 10–50-cm-thick ejecta layer from Cone crater at the location where the sample was collected. Because the sample was scooped from the surface of the regolith, it may consist almost entirely of Cone crater ejecta, which is thought to be well mixed. Because Cone crater is much deeper than the local regolith (which is ~8 m deep), these spherules may contain an approximately uniform sample of the cratering history of the landing site.

Despite the conclusion that the best model was that no gardening correction was necessary, several standard gardening corrections were applied to see their potential effect on the conclusions. The corrections consisted of power-law models in which the density of spherules at any given depth is assumed to depend on age, with young spherules concentrated near the surface, and older spherules more uniformly distributed. However, none of the gardening corrections yielded acceptable models in the sense that they suppressed the recent (0.4 Ga) peak, and yet had a cratering rate between 0.4 and 3 Ga that was compatible with previous limits based on crater counting (BVSP, 1981; Grieve and Shoemaker, 1994; Shoemaker et al., 1990). The failure of the gardening corrections can be traced to the fact they all assumed a smooth correction, and any smooth function that suppresses the recent peak results in an unacceptably high cratering rate in the region near 3 Ga.

Lunar gardening at one special location need not follow a simple power law, because it could be significantly affected by a few events. The standard gardening corrections are meant to reflect average changes, not the gardening at every location on the lunar surface. Additional information about gardening can be obtained, in principle, by measuring cosmic ray exposure ages, and such work is in progress.

The specific possibility that the 0.4 Ga increase could be due to spherules coming directly from Cone crater was examined, and we discussed this analysis in Culler et al. (2000). Of the 11 spherules that had ages compatible with that of Cone crater age (25 Ma), 3 are black, 3 are yellow, 3 are orange, 1 is green, and 1 is white. This diversity of color suggests that they did not all come from the same (Cone) crater. Moreover, removal of all 11 spherules compatible with the Cone crater age leaves 15 spherules in the 0–0.4 Ga bin, still a significant increase in the past 0.4 b.y., although the statistical significance of the effect is reduced from just over three standard deviations (3.7 ± 1.2) to just under two (2.1 ± 1.2). The first of these results is considered to be more accurate, because the spherule colors suggest that the 11 spherules discounted were not from Cone crater, as the reduced statistical significance requires.

Hörz (2000) suggested that the increase in spherules in the past 4 b.y. does not reflect an increase in cratering, but simply an increase in the efficiency of spherule production. His model is consistent with the data, but requires several ad hoc assumptions. Nevertheless, until measurements are made at other locations, it cannot be ruled out. For a discussion, see Muller et al. (2000a).

ROBUSTNESS

The data were analyzed in many different ways to see if our conclusions are robust to the method of analysis, and to seek correlations that could help us to understand systematic biases.

The use of the ideogram to represent the spherule age data avoids potential biases due to binning. However, because the ideogram is unfamiliar to many scientists, the data were also analyzed using the histogram method with different size bins: 400, 200, 100, and 50 b.y. The 0.4 b.y. increase is robust under these different binnings. In Figure 2 the histogram uses 50 m.y. bins. With very large 1 b.y. bins, as shown in Figure 3, there is no significant increase in cratering in the last bin; see the discussion in Muller et al. (2000a). In this plot, the recent increase is averaged with the period of low impact rate that immediately preceded it, and they cancel. Thus, without the age precision afforded by the measurements, the variations in the cratering rate (assuming it is real) would have been missed. Figure 4 shows the age distribution of the 58 spherules that had age uncertainties <0.1 b.y. In this plot, the 0.4 b.y. increase is still strong.

In the initial sorting of spherules, each was visually inspected and assigned a color in one of six categories: black or opaque (85 total spherules), yellow (36), gray (13), orange or red (10), green (9), and white (2). These variations in color are thought to relate to the concentration of titanium and to a lesser extent iron in the glass of the spherules. To search for possible biases, the age distribution of each color group was plotted separately. The 0.4 Ga increase was present (within the limited

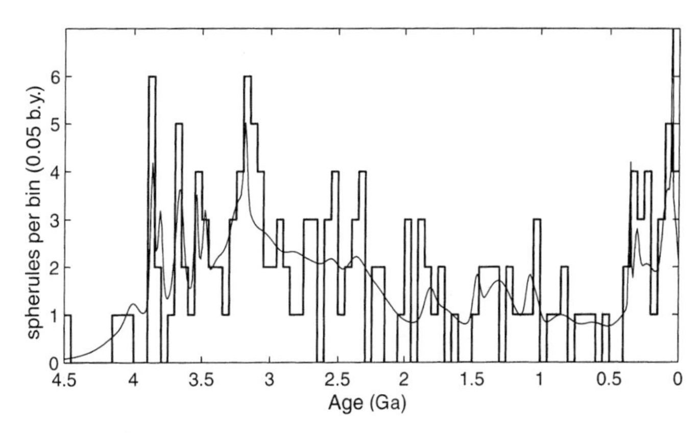

Figure 2. Distribution of spherule ages, plotted in 0.05 b.y. (50 m.y.) bins. Ideogram is identical to that in Figure 1. Plot illustrates that recent 0.4 Ga (400 Ma) increase is not artifact of choice of bin size.

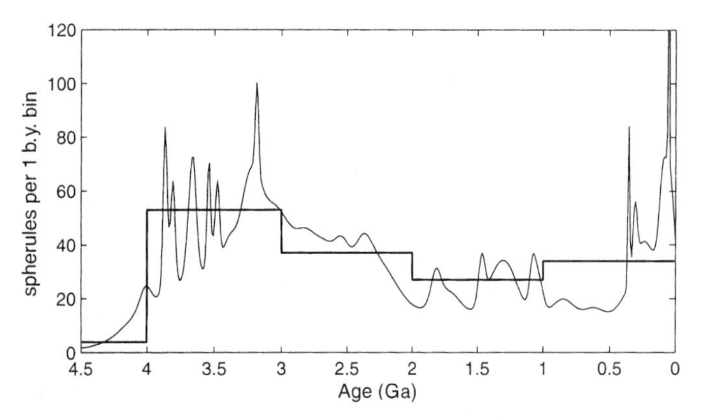

Figure 3. Distribution of spherule ages, plotted in 1 b.y. (1000 m.y.) bins. This plot shows no statistically significant increase in last 1 b.y. vs. previous period 3 to 1 b.y. Recent 0.4 Ga peak disappears because it is averaged with preceding bins, which were low. This plot illustrates that with poor age resolution, recent increase would have been missed. Ideogram is identical to that in Figure 1.

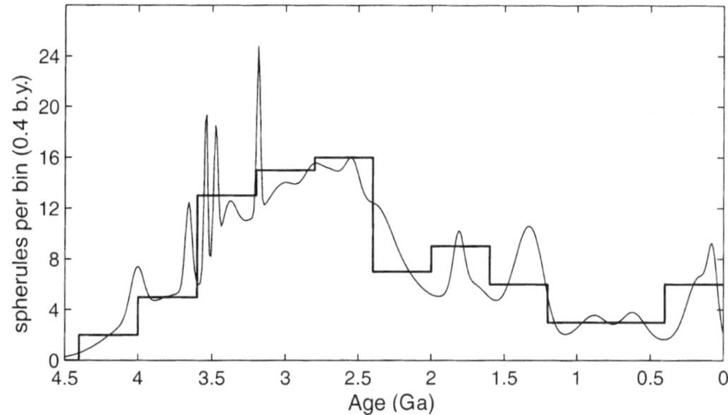

Figure 5. Distribution of spherule ages for black and opaque spherules. In our search for systematic anomalies, this plot deviated most from others. Number of events in past 0.4 b.y. is only 6, well below peak level at 3 Ga of 16. Based on Figure 1, we would have expected 13.3 events in this bin. Probability of obtaining 6 or few events when 13.3 are expected is ~2%. Plot shows small increase in final bin, and anomaly could be presence of unusual number of black and opaque spherules near age 3 Ga.

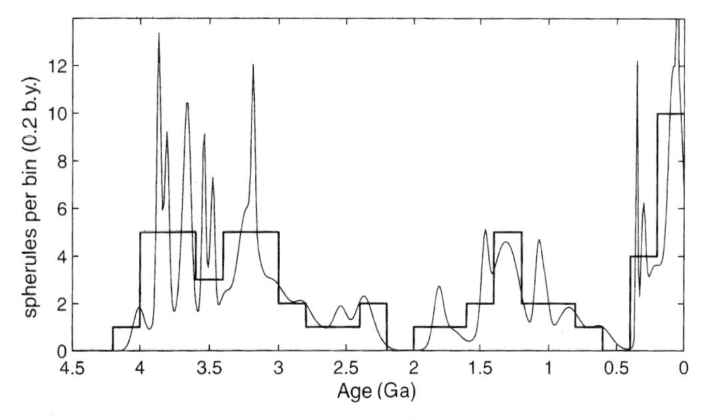

Figure 4. Distribution of spherule ages for most accurately dated spherules, those with age uncertainties (1 standard deviation) of 0.1 b.y. (100 m.y.) or better. Ideogram also contains only this subset of spherules. Recent increase is still evident.

statistics) in all groups except the black and opaque spherules. The age distribution for these spherules is shown in Figure 5. There is a slight increase present in the most recent bin of this plot, but it appears to be significantly less than in the other plots. I do not have an explanation for this discrepancy: the estimate of the probability of observing the 6 events in this last bin, when 13.3 were expected (based on the number of black and opaque spherules), is 2%. Perhaps it is simply that if there are 50 different plots (about this many were generated), to find one that is deviant at the 2% level is expected. It is also possible that there was a more efficient mechanism for generation of black and opaque spherules at 3 Ga. The statistics of the last three bins are very low, but they are compatible with a rise in the last 0.4 b.y. when compared to that of the previous 0.8 b.y.

The most likely model is that the spherule age distribution is proportional to the cratering rate near the Apollo 14 site. We do not know the size distribution that contributes to the spherules, and whether this matches the same craters that are measured in the crater counting methods. However, if we specifically degrade our age resolution, the results are consistent with those previously reported (Muller et al., 2000a). A final conclusion about the solar system cratering rate based on lunar spherule ages should wait until additional lunar sites are measured. Such measurements are underway. For the remainder of this chapter, the 0.4 Ga increase is assumed to be real.

NEMESIS THEORY

The Nemesis theory postulates that there is a companion star to the Sun, orbiting at a distance of ~3 light years, with a period of 26 m.y. If this orbit has an eccentricity >~0.5, then it passes close enough to the Oort comet cloud to trigger a comet shower once per orbit. Such periodic showers could lead to periodic extinctions of life on Earth, and to periodic increases in the cratering rate on the Moon.

The 3 light year orbit is larger than known for any double star system, and there has been considerable speculation that the orbit is unstable. The original paper (Davis et al., 1984) stated that the orbit had a stability time constant of about 1 b.y., and that small orbit to orbit variations of a few million years should be expected. Many subsequent analyses of the Nemesis orbit stability have been published, and these are summarized in the following.

A detailed study of the Nemesis orbit stability was published by Hut (1984), based on extensive computer simulations of the effects of passing stars, using a realistic distribution of star masses and velocities. Hut concluded that the Nemesis orbit was unstable unless it was close to the plane of the galaxy. For this orbit, however, the stability he found agreed with the estimate of ~1 b.y. in the original Nemesis paper. Hut also reiterated a surprising discovery about the lifetime of the orbit, i.e., that it decreased linearly with time. This is in contrast to other well-known behavior, such as the lifetime of radioactive particles. For radioactive particles the lifetime is independent of time; e.g., the ^{14}C nuclei that remain after 100 k.y. still decay with the same 5.7 k.y. half-life as the original nuclei. For Nemesis, the behavior is different. At formation, the expected lifetime of Nemesis would have been 5–6 b.y. Now that 4–5 b.y. have passed, the residual lifetime is only 1 b.y. Of course, this is just the average behavior, and the actual behavior of Nemesis (shown in specific simulations in Hut, 1984) could show sudden changes from the nearby passage of one massive star. (The 0.4 Ga increase is interpreted herein in terms of such a change.) The importance of the Hut calculation is that if estimates of the present lifetime of the Nemesis orbit are less than the lifetime of the solar system, it does not mean the star could not have survived for that period. This subtlety has been missed by many.

The Nemesis theory was based on the idea that passages of the solar companion star Nemesis through the Oort comet cloud would trigger comet showers. The same issue of *Nature* that contained the Hut article, contained several other articles on the same subject. One was by Hills (1984), who had originally discovered the possibility of "comet showers" (Hills, 1981). Hills (1984) showed that in its present orbit, the stability of Nemesis would be about 1 b.y., the same value obtained by Hut (1984).

Torbett and Smoluchowski (1984) also analyzed the stability of the Nemesis orbit. They assumed (incorrectly, as argued herein) that a stability time of 4–5 b.y. is needed for the present Nemesis orbit, in order for it to have survived. They also raised the possibility that passing giant molecular clouds, much more massive than stars, would dominate the orbital perturbations, and make the Nemesis orbit even more unstable. We (Morris and Muller, 1986) pointed out in a subsequent paper that their molecular cloud calculation had ignored the fact that giant molecular clouds are not only massive but also very large and diffuse, and thus only part of their mass (effectively that between Nemesis and the Sun) contributes to the tidal field that disrupts the orbit. When the diffuse nature of these clouds is taken into account, their effect on the stability is less than that of passing stars.

The stability of the Nemesis orbit was also analyzed by Clube and Napier (1984). Unfortunately, this article confused the parameters of the Nemesis theory with those of a similar theory by Whitmire and Jackson (1984), which had been published simultaneously with the Nemesis theory. Whitmire and Jackson also postulated a companion star to the Sun; however, they assumed a very small star, and this required them to postulate an extremely eccentric orbit, one that would be truly unstable. In contrast, the eccentricity assumed in the Nemesis theory was taken to be the median for stochastic orbits, i.e., 0.7. Clube and Napier argued (correctly) that the highly eccentric orbit is unstable, but they mistakenly ascribed that to the Nemesis theory. They also invoked the giant molecular cloud perturbations, and also ignored (as did Torbett and Smoluchowski, 1984) the large size of these clouds that reduces the effect they have on the Nemesis orbit (Morris and Muller, 1986). They stated that if the effect of molecular clouds were ignored, then the residual Nemesis stability would still be only 1 b.y., and they considered that insufficient. This estimate is the same as that determined by Hut, and it agrees with the estimate in the original Nemesis paper. It appears that Clube and Napier were confusing the present expected lifetime with the past expected lifetime.

The confusion about the stability of the Nemesis orbit was made worse by an editorial comment that appeared in the same issue of *Nature* as the articles by Hut, Hills, Torbett and Smoluchowski, and Clube and Napier. The comment was by Bailey (1984, p. 602), and it was titled *Nemesis for Nemesis*. Bailey stated: "the Nemesis proposal is extended and shown, in fact, to be quite incapable of producing the strictly periodic sequence for which is was originally designed." This was a misreading of the original Nemesis paper (Davis et al., 1984), which explicitly pointed out that the period would not be constant but would have orbit to orbit variations of several million years. Bailey's comment also characterized the paper by Hut (1984), as a "near retraction" of the Nemesis theory. Yet Hut (1984, personal commun.) considers his paper to be a vindication of the original Nemesis calculations, not a retraction. M. Bailey (1984, personal commun.) later said that he never wrote the words "near retraction," but that they had been inserted by the editor at *Nature*. An informal and unscientific survey of astronomers who discredit the Nemesis orbit (taken by me, over the following decade) showed that none of them had read the Hut paper. This is not surprising; why bother to read a paper when, according to the accompanying comment, it amounts to a near retraction?

The strongest argument against the Nemesis idea is not any difficulty with its orbit. It is the fact that the theory predicts that most of the mass extinctions of Raup and Sepkoski (1984, 1986; Sepkoski, 1989) should have been caused by impacts, and yet little evidence has been adduced since 1984 in favor of this conclusion. If we ever understand the origins of the other extinctions, and they are shown to be unrelated to impacts, then the Nemesis theory loses its only reason for existence. However, the Nemesis theory makes definite predictions, and thus is falsifiable. We should be able to find the star (the search has been stalled by telescope difficulties), and there should be a periodicity in impacts that can be seen in the crater data. Unfortunately, the age accuracy achieved thus far in the lunar spherule project is insufficient to see the expected 26 m.y. cycle.

NEMESIS INTERPRETATION OF THE 0.4 GA INCREASE

I assume for the purposes of the following discussion that the increase in the number of lunar spherules after 0.4 Ga reflects an actual increase in cratering rates. The Nemesis hypothesis can readily accommodate the increase. Hut's theory of Nemesis stability (Hut, 1984) described the average behavior of the orbit, which was an ensemble of many individual simulations. In reality, the orbit is very unlikely to have followed the average behavior. Let us hypothesize, therefore, that the orbit was relatively circular during the period 2–0.4 Ga. As shown in Figure 6A, the orbit does not enter the Oort comet cloud, but remains outside. In such an orbit, comet showers would not be triggered, and the rate of impacts on the Earth would reflect a background level from distant comets and asteroids.

If a passing star gave a major perturbation to Nemesis at 0.4 Ga, the orbit could have been perturbed into the postulated eccentricity of 0.7, as shown in Figure 6B. This would require a relatively close encounter with a passing star, probably <1 light year; however, the calculations of Hills (1981) and the simulations of Hut (1984) show that such an encounter is likely to happen sometime during the life of the solar system. If we postulate that it occurred at 0.4 Ga, from then on, the impact flux on the Earth would have two sources: a steady-state background, and a component from periodic comet showers. This could accommodate the observed increase.

Ultimately, the existence of Nemesis must be confirmed by direct observation. In the original theory, we assumed that Nemesis is a red dwarf star, and should be readily visible from the Earth. The Hipparcos satellite, unfortunately, surveyed only about 25% of the known candidates. Future parallex surveys, if they reach stars as dim as 10th magnitude, should find the star if it is there, or prove (by lack of discovery) that it is not there.

FUTURE SPHERULE MEASUREMENTS

The spherule method has had an initial success that suggests significant potential for future measurements. The increase in the past 0.4 b.y. has important implications, so it must be confirmed (or shown to be wrong) by making measurements at a different lunar site. This requires another high-potassium site, and/or the use of larger spherules. Use of such spherules may also allow sufficient improvement in the age determination that we could directly test the prediction that there is a significant component of impacts that occur in showers. A search for periodicity could yield evidence for or against the Nemesis hypothesis. In addition, measurements in the lunar highlands could potentially determine older cratering rates, and answer lingering questions about the existence and intensity of the late heavy bombardment.

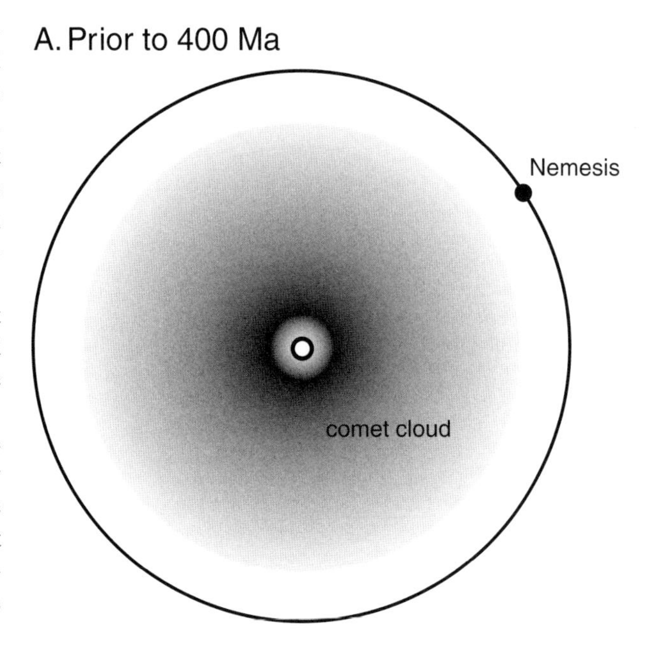

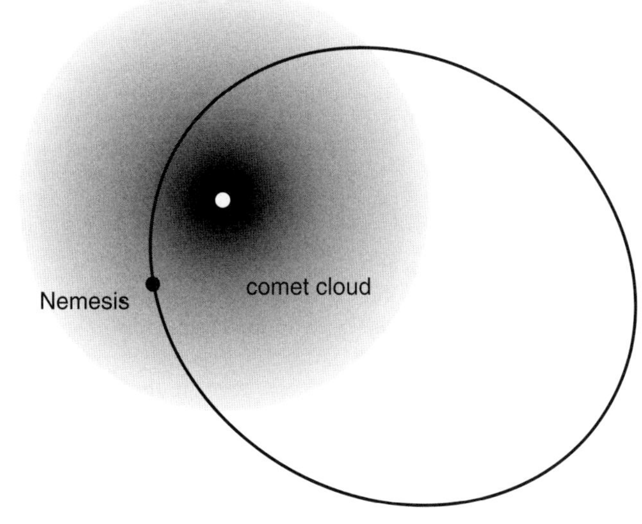

Figure 6. Hypothesized orbits of Nemesis. A: Circular orbit with 26 m.y. period, hypothesized as orbit prior to 0.4 Ga. B: Present orbit, with eccentricity 0.7. Change in eccentricity is attributed to close passage of star, as in calculations of Hut (1984). Inner Oort comet cloud is depicted in gray. Clear region near center in (A) represents depletion in orbits that pass close to Sun; size of depleted region is exaggerated for purposes of illustration. Depletion of these comets is result of perturbations of Jupiter. There is no depleted region in B, because of recent perihelion of Nemesis that perturbed comets into this region of space. Comets fill inner "eye" until they are kicked out (or into Sun) by gravitational perturbations of Jupiter. Expected duration of comet shower is few million years (Muller, 1985).

ACKNOWLEDGMENTS

I am very grateful to my collaborators, Paul Renne and Tim Culler, for their many contributions. (They allowed me to write this chapter without their coauthorship so that I could speculate freely on Nemesis.) I thank the referees for their detailed comments and very helpful suggestions. This work could not have been done without the support of the Ann and Gordon Getty Foundation.

REFERENCES CITED

Bailey, M., 1984, Nemesis for nemesis: Nature, v. 311, p. 602.

BVSP, 1981, Basaltic volcanism on the terrestrial planets: New York, Pergamon, 1286 p.

Chao, E.C.T., 1973, Geologic implications of the Apollo 14 Fra Mauro breccias and comparison with ejecta from the Ries crater, Germany: U.S. Geological Survey Journal of Research, v. 1, p. 1–18.

Clube, S., 1987, The origin of dust in the solar system: Royal Society of London Philosophical Transactions, v. 323, p. 421–436.

Clube, S.V.M., and Napier, W.M., 1984, Terrestrial catastrophism: Nemesis or Galaxy: Nature, v. 311, p. 635–636.

Culler, T.S., Becker, T.A., Muller, R.A., and Renne, P.R., 2000, Lunar impact history from ^{40}Ar-^{39}Ar dating of glass spherules: Science, v. 287, p. 1785–1788.

Culler, T.S., and Muller, R.A., 1999, Use of surface features and chemistry to determine the origin of fourteen Apollo 11 glass spherules: Lawrence Berkeley Laboratory Report LBNL-45703, 23 p.

Davis, M., Hut, P., and Muller, R.A., 1984, Extinction of species by periodic comet showers: Nature, v. 308, p. 715–717.

Grieve, R.A.F., and Shoemaker, E.M., 1994, The record of past impacts on Earth, *in* Gehrels, T., ed., Hazards due to comets and asteroids: Tucson, University of Arizona Press, p. 417–462.

Heiken, G.H., Vaniman, D.T., and Frenchx, B.M., 1991, Lunar sourcebook: Cambridge, UK, Cambridge University Press, 736 p.

Hills, J.G., 1981, Comet showers and the steady-state infall of comets from the Oort cloud: Astronomical Journal, v. 86, p. 1730–1740.

Hills, J.G., 1984, Dynamical constraints on the mass and perihelion distance of Nemesis and the stability of its orbit: Nature, v. 311, p. 636–638.

Hörz, F., 2000, Time-variable cratering rates?: Science, v. 288, p. 2095a.

Hut, P., 1984, How stable is an astronomical clock that can trigger mass extinctions on Earth?: Nature, v. 311, p. 636–640.

McEwen, A.S., Moore, J.M., and Shoemaker, E.M., 1997, The Phanerozoic impact cratering rate: Evidence from the farside of the moon: Journal of Geophysical Research, v. 102, p. 9231–9242.

McGetchin, T.R., Settle, M., and Head, J.W., 1973, Radial thickness variation in impact crater ejecta: Implications for lunar basin deposits: Earth and Planetary Science Letters, v. 20, p. 226–236.

Morris, D., and Muller, R.A., 1986, Tidal gravitational forces: The infall of "new" comets and comet showers: Icarus, v. 65, p. 1–12.

Muller, R.A., 1993, Cratering rates from lunar spherules: Lawrence Berkeley Laboratory Report LBL-34168, 7 p.

Muller, R.A., 1985, Evidence for a solar companion star, *in* Papagiannis, M.D., ed., The search for extraterrestrial life: Recent developments: New York, International Astronomical Union, 430 p.

Muller, R.A., Becker, T.A., Culler, T.S., Karner, D.B., and Renne, P.R., 2000a, Time-variable cratering rates?: Science, v. 288, n. 23, p. 2095a.

Muller, R.A., Becker, T.A., Culler, T.S., and Renne, P.R., 2000b, Solar system impact rates measured from lunar spherule ages, *in* Peucker-Ehrenbrink, B., and Schmitz, B., eds., Accretion of extraterrestrial matter throughout Earth's history: New York, Kluwer Publishers, 466 p.

Raup, D., and Sepkoski, J., 1986, Periodic extinction of families and genera: Science, v. 231, p. 833–836.

Raup, D., and Sepkoski, J., 1984, Periodicity of extinctions in the geologic past: Proceedings of the National Academy of Sciences of the United States of America, v. 81, p. 801–805.

Ryder, G., Bogard, D., and Garrison, D., 1991, Probable age of Autolycus and calibration of lunar stratigraphy: Geology, v. 19, p. 143–146.

Sepkoski, J.J., 1989, Periodicity in extinction and the problem of catastrophism in the history of life: Geological Society [London] Journal, v. 146, p. 7–19.

Shoemaker, E.M., Wolfe, R.F., and Shoemaker, C.S., 1990, Asteroid and comet flux in the neighborhood of Earth, *in* Sharpton, V.L., and Ward, P., eds., Global catastrophes in Earth history: Geological Society of America Special Paper 247, p. 155–170.

Torbett, M.V., and Smoluchowski, R., 1984, Orbital stability of the unseen solar companion linked to periodic extinction events: Nature, v. 311, p. 641–642.

Whitmire, D.P., and Jackson, A.A., 1984, Are periodic mass extinctions driven by a distant solar companion?: Nature, v. 308, p. 713–715.

Manuscript Submitted November 23, 2000; Accepted by the Society March 22, 2001

Printed in the U.S.A.

Geological Society of America
Special Paper 356
2002

Role of the galaxy in periodic impacts and mass extinctions on the Earth

Michael R. Rampino
Earth and Environmental Science Program, New York University, 100 Washington Square East, New York, New York 10003 USA, and NASA, Goddard Institute for Space Studies, 2880 Broadway, New York, New York 10025, USA

ABSTRACT

Impacts of large comets and asteroids on the Earth are energetic enough to cause mass extinction of species. Several studies have concluded that large impact craters and extinctions show a correlation, and that both records display periods in the range of ~31 ± 5 m.y. This might be the result of periodic or quasiperiodic showers of Oort cloud comets with a similar cycle. One candidate for a pacemaker for such periodic comet showers is the Sun's vertical oscillation through the plane of the galaxy, with a half-period over the past 250 m.y. estimated in the range of ~33 ± 7 m.y. Thus, major events in the history of life on the Earth may be partly related to the dynamics of the galaxy. This result could have significant astrobiological implications for other planetary systems with outer comet clouds.

INTRODUCTION

In 1984, Raup and Sepkoski first presented statistical analyses suggesting that mass extinctions of life showed an ~26–30 m.y. period, and thus might have a common cause. Alvarez et al. (1980) had reported evidence for a link between the major mass extinction at the end of the Cretaceous Period (65 Ma) and the impact of a large asteroid or comet, and thus it became apparent that periodic impacts could be the cause of periodic extinctions. In response to those findings, Rampino and Stothers (1984a) suggested a model in which the periodic or quasiperiodic extinction events were related to impacts from showers of comets caused by the periodic passage of the solar system through the central plane of the Milky Way Galaxy.

In discussing the impact and extinction issue, David Raup (1986, p. 117) observed that, "More fundamental is the question of whether extinctions and impacts are really regularly spaced in time. This gives rise to some difficult problems in the statistical analyses of time series, an area fraught with problems and pitfalls. It is no wonder, therefore, that good statisticians disagree on the validity of the basic claims of periodicity. When it is all over, we may know whether there are large scale cycles in the solar system or galactic environment that have strongly influenced the history of the earth and of life—or we may have found one more example of scientists seeing cycles where none exist."

Most studies have started with the reported periodicities and the purported correlation between impact and mass extinction in order to provide a rationale for proposing pacemakers for periodic comet showers, including companion stars, unknown planets, and galactic dynamics (e.g., see Smoluchowski et al., 1986). By contrast, here I begin with the dynamics of the galaxy, and make predictions about the expected consequences for the Oort comet cloud and for comet impacts on the inner planets.

GALACTIC DYNAMICS AND CYCLES

The Milky Way is a large spiral galaxy, ~30 kiloparsecs (kpc) in diameter. The galaxy contains ~400 billion stars, gas, and dust clouds, with a total mass of ~4×10^{11} M_o (solar masses). The disk of the galaxy is highly flattened; the radius

Rampino, M.R., 2002, Role of the galaxy in periodic impacts and mass extinctions on the Earth, *in* Koeberl, C., and MacLeod, K.G., eds., Catastrophic Events and Mass Extinctions: Impacts and Beyond: Boulder, Colorado, Geological Society of America Special Paper 356, p. 667–678.

is ~15 kpc and the thickness is only ~0.5–0.8 kpc. The Sun and planets are located ~8.5 kpc from the galactic center, and only 10–30 pc above the central plane of the disk (Bailey et al., 1990; Reed, 1997).

Within the galaxy, the Sun and planets move in a quasielliptical orbit between ~8.4 and 9.7 kpc from the galactic center, with a full period of revolution of about 250 ± 15 m.y., and a radial apogalactic to perigalactic epicycle of 170 ± 10 m.y. (Bailey et al., 1990; Matese et al., 1995). The solar system is currently close to and moving toward the perigalactic position (Matese et al., 1995).

The solar system also moves perpendicular to the galactic plane in a harmonic fashion, with an estimated period of 63 ± 11 m.y. (1/2 period of 32 ± 5 m.y.) and an amplitude of 71 ± 22 pc out of the plane (the full ranges given here are the result of uncertainties in the mass density in the galactic disk; see following) (Stothers, 1998). The Sun and planets passed through the galactic plane recently (in the past 2–3 m.y.) moving "upward." The combined effect of the two solar motions has been called the "Galactic Carrousel" (Fig. 1).

For the galactic periodicity models, the critical parameter is the time between plane crossings (the half-period of vertical oscillations) which is given by $P_{1/2} = (\pi/4G\rho_0)^{1/2}$, where ρ_0 is the mean volume density of matter near the galactic plane, G is the universal gravitational constant, and $P_{1/2}$ is the half period of vertical oscillation. If ρ_0 is given in $M_o pc^{-3}$ (Stothers, 1998), then this formula reduces to $P_{1/2} = 13.2\rho_0^{1/2}$.

The observed visible matter in the plane is 0.10 ± 0.01 $M_o pc^{-3}$ (Stothers, 1998), and some additional dark matter is commonly believed to be present, which is critical to the periodicity question. Determination of the volume density of matter comes from analysis of the positions and velocities of tracer stars in the direction perpendicular to the galactic mid-plane. Stothers (1998) utilized all 28 published determinations of local mass density to estimate an average value of ρ_0 of 0.15 ± 0.01 $M_o pc^{-3}$, implying ~30% dark matter (probably in the form of cold interstellar clouds, if present) (see also Gould et al., 1996).

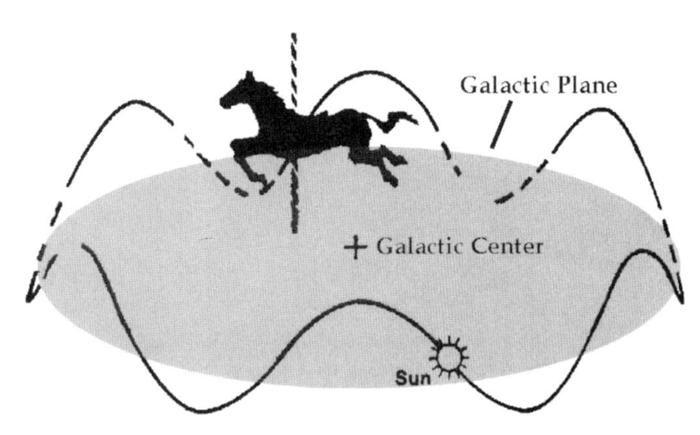

Figure 1. "Galactic carrousel" effect. Combined vertical oscillation of solar system perpendicular to galactic plane, and revolution of solar system around galaxy.

Stothers' (1998) result would give a half-period of 34 ± 1 m.y. for the solar system in our neighborhood, or 37 ± 3 m.y. over the past 250 m.y., taking into consideration the variations in galactic position.

Two recent studies using the improved astrometry made possible by the Hipparcos satellite, however, produced lower, but still mutually incompatible estimates of ρ_0. Crezé et al. (1998), using ~3000 A- to F-type stars closer than 125 pc and more luminous than $M_v = 2.5$, determined a total disk matter density of only 0.076 ± 0.015 $M_o pc^{-3}$ (M_v = visual magnitude). Pham (1997) used 10 000 F stars from the Hipparcos survey and found a total disk density of 0.11 ± 0.1 $M_o pc^{-3}$. If correct, these estimates, which contain little or no dark matter, would give considerably longer $P_{1/2}$ values of 47 m.y. (Crezé et al., 1998) and 40 m.y. (Pham, 1997), which would not match the observed geological periods.

By contrast, Bahcall et al. (1992), using a set of K III stars, found a significant dark-matter component, giving a total disk density of 0.26 $M pc^{-3}$, and therefore a $P_{1/2}$ of only 26 m.y. Thus, the question of the mass density in the galactic disk, which controls the periodicity and is thus critical in determining if the proposed model fits the geologic and paleontologic periods, is still unsettled.

In the Rampino and Stothers (1984a, 1986) model, the half-cycle of vertical oscillation modulated the probability of encounters with molecular clouds that could perturb the Oort comet cloud and cause comet showers. The distribution of clouds in the galactic disk suggests that an encounter would be more likely as the Sun passes through the galactic mid-plane region, and hence the encounters would be quasiperiodic, with a period equal to the time between plane crossings (Rampino and Stothers, 1986).

The original model was criticized on the grounds that the amplitude of the Sun's present excursions from the galactic plane versus the scale height for the perturbing interstellar clouds was insufficient to produce a quasiperiodic modulation of comets that would show up over the background of nonperiodic impacts (Thaddeus and Chanan, 1985). Numerical simulations using the best available astronomical data, however, suggested that this modulation should be detectable in the terrestrial record of cratering (Stothers, 1985).

Another criticism of the original model was that the most recent extinction recognized by Raup and Sepkoski (1984) (in the mid-Miocene, ca. 11 Ma) and the most recent plane crossing in the past few million years seemed out of phase. Rampino and Stothers (1984a) originally argued that there could be enough scatter in the cloud encounters to explain the apparent offset. Subsequent work, however, has identified a significant extinction event in the Pliocene (ca. 2.3 Ma), although both the mid-Miocene and Pliocene extinctions were minor events as compared with the end-Cretaceous (65 Ma) or even the late Eocene (35 Ma) extinctions (Raup and Sepkoski, 1986; Sepkoski, 1995). Thus, an impact-extinction event predicted for the present plane-crossing episode might be in the future.

Tidal forces produced by the overall gravitational field of the galaxy can also cause perturbations of cometary orbits. Because these forces vary with the changing position of the solar system in the galaxy, they also provide a mechanism for periodic variation in the flux of Oort cloud comets into the inner solar system (Bailey et al., 1990; Matese at al., 1995). The cycle time and degree of modulation also depend critically on the mass distribution in the galactic disk.

Matese et al. (1995), using the range of estimates of galactic mass density, calculated periods ranging from 44 to 28 m.y., with a best-fit estimate of ~33 m.y., and peak to trough Oort cloud currents of 2.5 to 1 to 4 to 1. The standard deviations of the flux peaks were 4–5 m.y. in most model cases, and the times of peak comet flux lagged the times of galactic plane crossing by ~1–2 m.y. (Valtonen et al., 1995). The amplitude of the comet pulses and the length of the cycle interval are also modulated somewhat by the epicyclic motion of the Sun about the galactic center, with a period of about 170 ± 10 m.y. (Matese et al., 1995) (Fig. 2).

Because the largest terrestrial impactors are most likely comets (Shoemaker et al., 1990), the largest impact craters should preferentially show the galactic modulation of comet flux. If large comets commonly break up, however, then the smaller impacts may also show a periodic signal. A comet shower in the late Eocene (ca. 35 Ma) is suggested by enriched extraterrestrial ^{3}He in sediments covering ~2–3 m.y., apparently in dust from a pulse of comets or comet breakup in the inner solar system. This interval brackets the dates of two known large impacts (and several less-well-dated smaller craters) (Farley et al., 1998). Similar pulses of increased comet flux could also explain the stepped nature of some extinction events (e.g., the Cenomanian-Turonian boundary at 93 Ma) and clusters of similar-age craters and impact layers reported from the geologic record (Hut et al., 1987; Shoemaker and Wolfe, 1986; Montanari et al., 1998).

Thus, galactic models predict periodic to quasiperiodic disturbances of the Oort cloud comets with periods ranging from ~28 to 44 m.y. (using the entire range of estimates of mass density in the galactic plane). The comet showers produced by these perturbations could last for as much as 9 m.y. (a significant portion of the 1/2 period of oscillation). Mass extinctions should be correlated with the largest cometary impactors during the period of increased flux of comets, and hence should follow the galactic half-period.

Having approached the problem from the point of galactic dynamics, the evidence for impact as a cause of mass extinction and the time structure of the impact and mass extinction records are now examined.

LARGE IMPACTS AND MASS EXTINCTIONS

The major mass extinction that marks the Cretaceous-Tertiary (K-T) boundary (65 Ma) coincided with the impact of a comet or asteroid ~10 km in diameter that created the ~180-km-diameter Chicxulub impact structure in the Yucatán region of Mexico (e.g., Hildebrand et al., 1995; Pope et al., 1998) and a global fallout layer (Alvarez et al., 1980). This large impact (energy equivalent to ~10^8 Mt of TNT) is calculated to have caused a severe global catastrophe, primarily related to dense clouds of fine ejecta, production of nitric oxides and acid rain, and smoke clouds from fires (Toon et al., 1997). A particularly severe effect of such large impactors is the predicted global distribution of fires from the global reentry of debris that

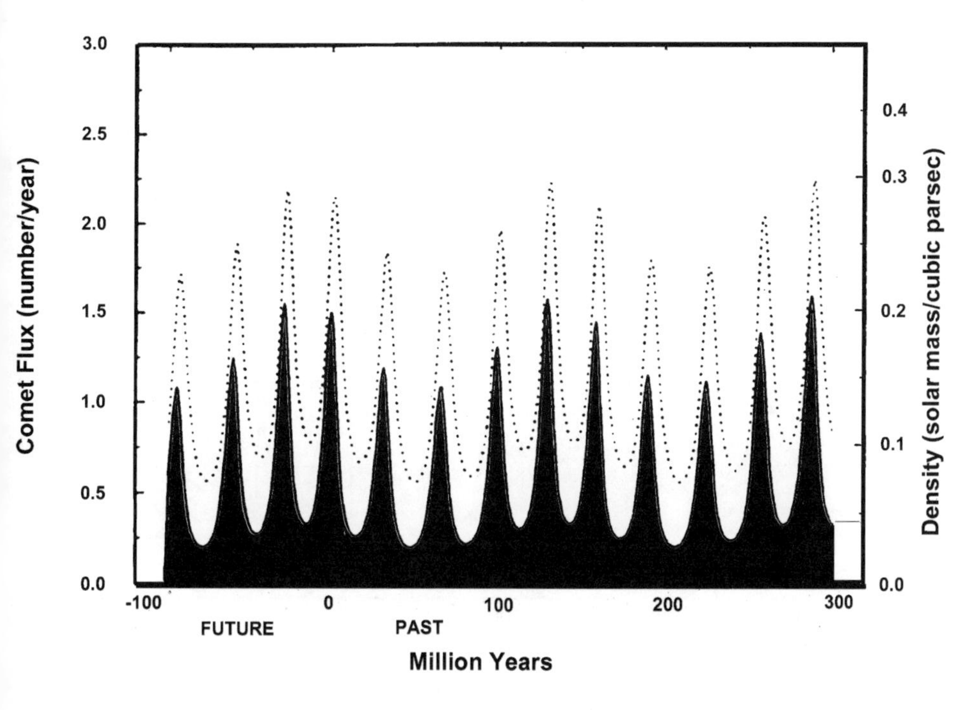

Figure 2. Time dependence of Jupiter-dominated current of new comets (comets per year with $q < 5.8$ astronomical units) (filled black) from outer Oort cloud due to adiabatically changing galactic tides based on galactic model (q = perihelion distance). Variations in galactic density at solar location are also shown (dotted line). Model was run for interval ranging from 100 m.y. in the future to 300 m.y. in past (after Matese et al., 1995).

generated an initial global heat pulse from atmospheric friction (Melosh et al., 1990).

Other effects of large impacts include tsunamis, ballistic debris flows, enhanced greenhouse effect from atmospheric water vapor derived from an ocean impact or CO_2 released by impact into carbonate rocks, and cooling and acid rain from large amounts of sulfuric acid aerosols derived from calcium-sulfate target rocks (Toon et al., 1997).

The cratering record of the inner planets and the expected times between collisions of bodies of various sizes with the Earth based on observations of earth-crossing asteroids and comets predict that in an ~100 m.y. period, the Earth should be hit by one ~10-km-diameter (~10^{24} J, 10^8 Mt) object (most likely a comet), and several asteroids or comets larger than a few kilometers in diameter (~10^{23} J or 10^7 Mt TNT equivalent) (Shoemaker et al., 1990).

A $\geq 10^8$ Mt, or first-order, impact event would produce a more severe and widespread environmental disaster (including a ≥200-km-diameter crater, a global heat pulse, and resulting fires) than an ~10^7 Mt second-order event (Melosh et al., 1990; Toon et al., 1997). Using the first-order K-T impact as a standard, these results lead to the prediction of ~5 major mass extinctions, and ~25 ± 5 less-severe pulses of extinction during the Phanerozoic Eon (the past 540 m.y.) resulting from large impacts.

The independent paleontological record of extinctions for that interval clearly shows 5 major mass extinctions and ~20 less-severe extinction pulses, in agreement with the estimates for impact-induced extinctions (Fig. 3; note that the higher overall extinction rates during the period prior to ca. 500 Ma may be an artifact of relatively poor knowledge of the fossil record, e.g., Sepkoski, 1995).

Proposed "kill curve" relationships between mass extinctions and impacts of various magnitudes (Fig. 4; Raup, 1992) can be compared with data representing all of the largest known (>80 km diameter) impact craters with well-defined ages that overlap the ages of mass-extinction boundaries (when the full dating uncertainties in both are taken into consideration) (Table 1). The observed points in Figure 4 agree with the predicted curves within the errors permitted by the geologic data, supporting at least a first-order relationship between large impacts and mass extinctions.

Matsumoto and Kubotani (1996) compared the rate of formation of large craters with the mass-extinction record. They found a statistical correlation between rate of formation of large craters and extinction events in the past 300 m.y. (see also Stothers, 1988; Yabushita, 1998). Moreover, there is some evidence that smaller craters tend to match times of stratigraphic stage boundaries defined by lesser faunal changes (Stothers, 1993; Rampino and Schwindt, 1999).

Some correlations are still problematic. For example, the ~100-km-diameter Manicouagan impact structure has been dated by U-Pb methods as 214 ± 1 Ma. The age of the Triassic-Jurassic boundary (Rhaetian-Hettangian), as evidenced by a palynological break in the Newark basin (Fowell et al., 1994), suggests a revision in the age of that boundary from ca. 205.7 ± 4.0 Ma (Gradstein and Ogg, 1996) to a younger age of ca. 201 Ma. Thus, Manicouagan seems to be too old to correlate with the end of the Rhaetian (with 76% marine species extinction) as had earlier been proposed. The crater might, however,

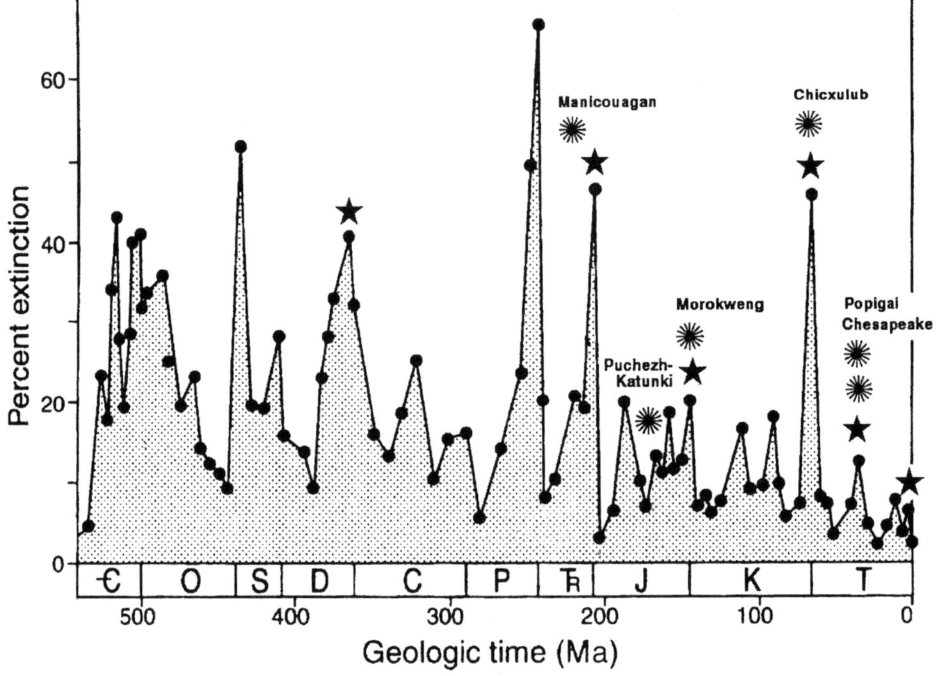

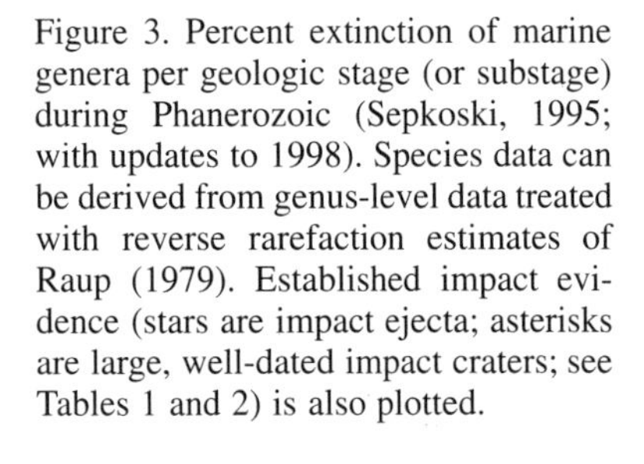

Figure 3. Percent extinction of marine genera per geologic stage (or substage) during Phanerozoic (Sepkoski, 1995; with updates to 1998). Species data can be derived from genus-level data treated with reverse rarefaction estimates of Raup (1979). Established impact evidence (stars are impact ejecta; asterisks are large, well-dated impact craters; see Tables 1 and 2) is also plotted.

be related to the Norian-Rhaetian boundary (~40% species extinction) dated as ca. 209.6 ± 4.1 Ma, or perhaps the Carnian-Norian boundary at 220.7 ± 4.4 Ma (~42% species extinction) (Fig. 3). This would support suggested modifications of the original kill curve, with a possible threshold level for significant global mass extinctions at impacts that create craters somewhat larger than 100 km in diameter (~10^7 Mt events) (Poag, 1997; also see Montanari et al., 1998).

The Puchezh-Katunki crater, previously dated as 220 ± 10 Ma and tentatively correlated with the Carnian-Norian boundary (220.7 ± 4.4 Ma), has been redated as 167 ± 3 Ma. This new date suggests a possible correlation with the end-Bajocian event at 171 ± 4 Ma (~15% species extinction) (Fig. 3).

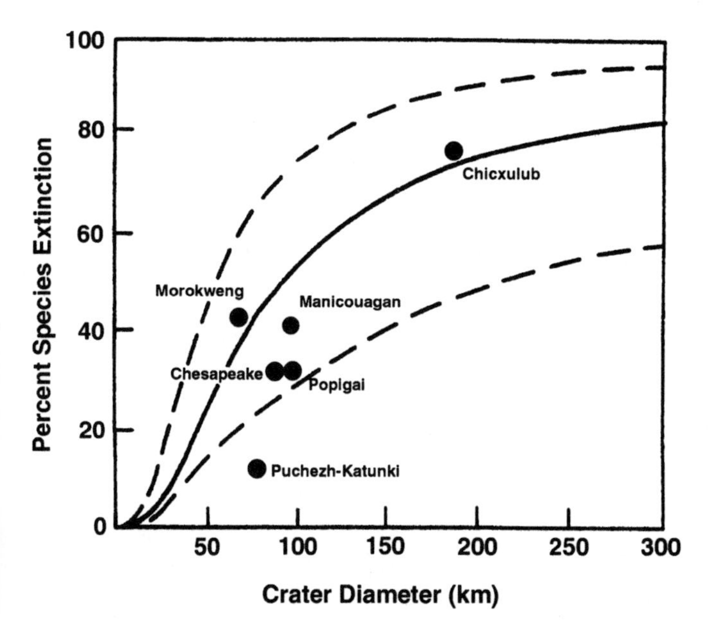

Figure 4. Proposed kill curve for Phanerozoic marine species plotted against estimated size of impact craters associated with extinctions of various magnitudes (assuming that two are related) (solid line with dashed estimated error lines; after Raup, 1992). Largest well-dated impact craters with ages overlapping mass extinction times are plotted for comparison (see Table 1). New dating of extinction record suggests that Manicouagan crater (214 ± 1 Ma) may be associated with end-Norian extinction (209.6 ± 4.1 Ma) event (~40% species kill), and that Puchezh-Katunki crater (167 ± 3 Ma) might be correlated with extinction event at end of Bajocian stage (171 ± 4 Ma). Kill curve proposed by Poag (1997) suggests step function in curve, with rapid rise in global species kill from very low values, at crater size somewhat greater than 100 km in diameter.

TABLE 1. LARGE DATED IMPACT CRATERS WITH ASSOCIATED MARINE EXTINCTIONS

Name	Diameter (km)	Age (Ma)	Extinction	Species (%)
Popigai	100	35.7 ± 0.8	Late Eocene	30
Chesapeake	90	35.2 ± 0.3	Late Eocene	30
Chicxulub	180	64.98 ± 0.5	Cretaceous-Tertiary	76
Morokweng	>70?	145.0 ± 0.8	Jurassic-Cretaceous	42
Puchezh-Katunki	80	167 ± 3	end-Bajocian?	15
Manicouagan	100	214.0 ± 1.0	end-Norian	40
			or	
			end-Carnian	42

Note: Craters after Montanari and Koeberl (2000); extinctions after Sepkoski (1995).

Another possible problem with the comet impact model, however, is that some recent studies of Cr isotopes in impact melts suggest that several large impact craters (e.g., Morokweng) were created by normal chondritic impactors (McDonald et al., 2001; Koeberl et al., this volume), not Oort cloud comets.

IMPACTS AND EXTINCTION EVENTS: EVIDENCE

Shocked minerals (including, e.g., shocked quartz, stishovite, shocked zircons), impact glass (microtektites and tektites), microspherules with structures indicating high-temperature origin, and Ni-rich spinels are considered evidence of impact. Although the K-T boundary is considered by most workers to be the only mass extinction convincingly tied to evidence for a large impact, these materials have been reported in stratigraphic horizons close to the times of five other recorded extinction pulses (Table 2) (Rampino et al., 1997; Grieve, 1997). Recent preliminary reports of shocked quartz(?) from the Permian-Triassic (P-T) boundary (ca. 251 Ma) in Antarctica and Australia (Retallack et al., 1998), and a Triassic-Jurassic boundary site in eastern North America (Mossman et al., 1998) require further data for confirmation. Fullerenes with trapped rare gases that show carbonaceous chondrite-like abundance ratios at the P-T boundary in China and Japan have been interpreted as evidence for a large impact event (Becker et al., 2001).

Thus far, at least 6 of the ~25 extinction peaks in Figure 3 (Pliocene, ca. 2.3 Ma; late Eocene, ca. 35 Ma; Cretaceous-Tertiary, 65 Ma; Jurassic-Cretaceous [J-K], ca. 144 Ma; Late

TABLE 2. STRATIGRAPHIC EVIDENCE OF IMPACT DEBRIS AT OR NEAR EXTINCTION EVENTS

Age	Evidence	Sources
Pliocene (2.3 Ma)	Microkrystites, microtektites	Margolis et al. (1991)
Late Eocene (35 Ma)	Microtektites (multiple), tektites, microspherules, shocked quartz, iridium	Montanari and Koeberl (2000)
Cretaceous-Tertiary (65 Ma)	Microtektites, tektites, shocked minerals, stishovite, Ni-rich spinels, iridium	Montanari and Koeberl (2000)
Jurassic-Cretaceous (ca. 142 Ma)	Shocked quartz	Dypvik et al. (1996)
Late Triassic (ca. 201 Ma)	Shocked quartz (multiple)	Bice et al. (1992)
Late Devonian (ca. 368–365 Ma)	Microtektites (multiple)	McGhee (1996)

Triassic, ca. 201 Ma; Late Devonian, ca. 365 Ma) are associated with stratigraphic evidence of impact ejecta (Table 2), and three (late Eocene, K-T and J-K) are associated with large (≥100 km diameter) dated impact craters, with two other possibilities: one of the Late Triassic extinctions (with Manicouagan) and the end Bajocian event (with the new age for Puzech-Katunki of 167 ± 3 Ma) (Fig. 3). Several other extinction events are associated with possible stratigraphic evidence of impact consisting of iridium concentrations somewhat elevated with respect to background values (see tables in Rampino et al., 1997; Grieve, 1997).

Clusters of impacts could explain the difficulty in global correlation of some geologic time boundaries. For example, the Jurassic-Cretaceous boundary does not yet have an internationally accepted definition (Gradstein and Ogg, 1996), stemming in part from the difficulty of correlating faunal changes in the northern (Boreal and sub-Boreal) and equatorial (Tethyan) regions of the time. Dypvik et al. (1996) found that the ejecta layer from the 40-km-diameter Mjølnir impact structure in the Barents Sea can be traced in seismic-reflection profiles to an Upper Jurassic-Lower Cretaceous section in a drill hole near the structure. The ejecta unit contains shocked quartz grains and an Ir anomaly of as much as ~1 ppb with chondritic siderophile element ratios, and is ~3 m below the occurrence of ammonites of the *Hectoroceras kochi* zone of the Boreal faunal realm, hence is placed very close to the regional J-K boundary at the top of the Volgian Stage (recently dated as 142.0 ± 2.6 Ma).

This agrees with the earlier report of an Ir anomaly in a Jurassic-Cretaceous sequence in northern Siberia (Zhakarov et al., 1993). The Ir anomaly, with chondritic siderophile element ratios and abundant pyritic spherules, occurs in a thin, laminated phosphatic limestone layer ~4 m below the local base of the same ammonite zone, and thus is at the same stratigraphic level as the ejecta layer from the Mjølnir impact, ~2300 km distant.

In the Tethyan realm, the J-K boundary has been chosen at a slightly older level of faunal change, at the base of the Berriasian Stage (144.2 ± 2.6 Ma). The top of the Volgian is correlated with the slightly younger lower-middle Berriasian boundary (Sey and Kalacheva, 1996). In the sub-Boreal realm, the British Portlandian-Purbeckian Stages span the J-K boundary interval, and the lower-middle Berriasian boundary has been correlated with the boundary between the lower and middle Purbeck. In England, this zone is marked by an unusual shell-rich marine intercalation called the Cinder Beds (Hallam et al., 1991). This would be a promising zone to search for further evidence of the Mjølnir impact.

The age of the larger (>70 km diameter) Morokweng impact structure in South Africa (145.0 ± 0.8 Ma) (Koeberl et al., 1997) matches the timing of the slightly older Tethyan J-K boundary. The smaller Gosses Bluff impact in Australia (142.5 ± 0.5 Ma) also dates from the same interval. These data suggest that several impacts and extinction events over a period of at least several million years (ca. 145–142 Ma) were involved in the worldwide Jurassic-Cretaceous transition.

IMPACTS AND REGIONAL EVENTS

Raup (1992) suggested the hypothesis that all extinction episodes, from the largest mass extinctions to the faunal events that define stages and substages in the stratigraphic record, could be related to impacts. For example, collisions with impactors of 2–3 km diameter (with energies in the 10^6 Mt range, or third-order events), producing impact craters from ~30–60 km in diameter, should occur about every 1 m.y., or ~600 such impacts should have occurred during the 545 m.y. of the Phanerozoic. About 30% of these impacts (~200) would have hit the continents, which means that North America, for example (representing ~15% of total continental area) would have undergone ~30 such impacts, or ~1/20 m.y.: 10 such impact structures between ~30 and 60 km in diameter are known (Table 3). Calculations suggest that these energetic events should have left a significant imprint on regional geologic records (Adushkin and Nemchinov, 1994).

For example, the scaling law for severe blast devastation (fatal overpressures) around impacts, $R_s = 8E_k^{1/3}$, where R_s is the radius of blast effects in kilometers, and E_k is the kinetic energy of the impactor in megatons. The scaling law for fire ignition by impacts is $R_f = 3E_r^{1/2}$, where R_f is the ignition radius in kilometers and E_r is the thermal radiation energy in megatons (the ratio of thermal radiation to kinetic energy for an impact is ~20%). In the case of an ~10^6 Mt impact event, the radius of blast devastation could be ~800 km, and the radius of fire ignition >1000 km. The radius of severe seismic effects of such an impact (and the resulting magnitude 9 earthquake) is estimated as ~1000 km.

A possible example of the regional environmental effects of such an impact involves the 35-km-diameter Manson impact structure in northwest Iowa (74.1 ± 0.1 Ma), estimated to have resulted from an ~10^6 Mt impact (Izett et al., 1993) (Fig. 5). A search for Manson ejecta in Upper Cretaceous marine rocks to the west led to the discovery of shocked minerals in a wide-

TABLE 3. NORTH AMERICAN IMPACT STRUCTURES BETWEEN ~25 AND 60 km IN DIAMETER

Impact structure	Approximate diameter* (km)	Age (Ma)
Beaverhead	60?	~600
Charlevoix	54	357 ± 15
Slate Islands	30	~450
Clearwater West	36	290 ± 20
Clearwater East	26	290 ± 20
Saint Martin	40	220 ± 32
Carswell	39	115 ± 10
Manson	35	74.1 ± 0.1
Montagnais	45	50.5 ± 0.8
Mistastin	28	38 ± 4
Haughton	24	23 ± 1

*Montanari and Koeberl (2000).

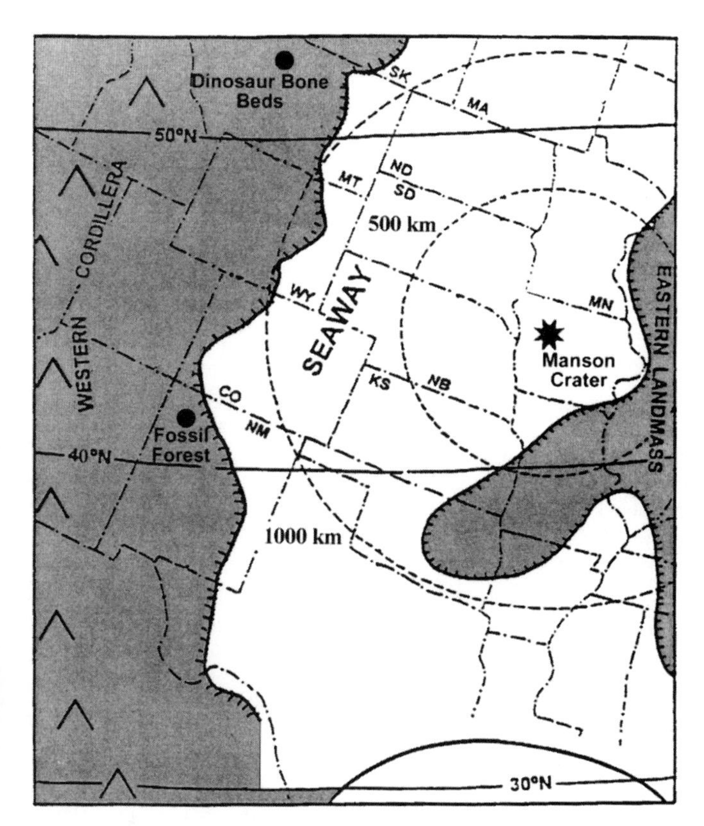

Figure 5. Location of Manson impact crater (74.1 ± 0.1 Ma) (star) and radii of blast destruction and fire at ~1000 km (according to formulae in Aduskin and Nemchimov, 1994). Locations of dinosaur kill sites in Montana and Fossil Forest site in New Mexico are shown by dots (see text).

spread layer in southeastern South Dakota, which has been interpreted as a probable tsunami deposit.

There has been much discussion as to whether dinosaurs in western North America were declining prior to the end of the Cretaceous Period. The confusion may be the result of a regional extinction caused by the Manson impact. Coincident with the timing of the Manson event, nine genera of dinosaurs disappeared in western North America, followed by the immigration and rapid evolution of other dinosaur taxa and by major changes in marine vertebrates and mammals in the same region (Russell, 1993).

Catastrophic dinosaur bone beds representing mass mortality, some indicating extremely violent conditions and containing herds of >10 000 individuals, occur in rocks dated as ca. 74 Ma in Montana ~1400 km from Manson (Rampino et al., 1997) (Fig. 5). This would correspond to part of the zone of instantaneous death or the zone of slow death of Cutler and Behrensmeyer (1996), where both enhanced preservation (rapid burial) and unusual enrichment of bones (destruction of large herds) could lead to impact-induced bone beds. A late Campanian fossil forest site in northwestern New Mexico might also be the result of destruction related to the Manson impact (Rampino et al., 1997). Thus, a number of puzzling features of the Late Cretaceous extinctions in western North America might be explained by regionally devastating effects of the Manson impact ca. 74 Ma.

The Azuara impact structure in Spain (Ernstson et al., 1985) is a similar ~40-km-diameter crater, tentatively dated as ca. 30–40 Ma (late Eocene to early Oligocene). (Ernstson et al. [2000] suggested the presence of a twin impact structure, the Rubielos de la Cérida structure, also ~40 km in diameter, to the east.) The Pellarda Formation and related deposits in northeastern Spain might represent debris-flow diamictites produced by the impact.

The aftermath of the Azuara impact should have been widespread environmental devastation in Europe. A drastic change occurred in the region's fauna and flora ca. 30–35 Ma, an event so severe that it was named the Grande Coupure (the Great Cut) by French paleontologists. More than 60% of the native western European species suddenly disappeared, to be subsequently replaced by immigrant species from eastern Europe and Asia, with a burst of evolution in surviving rodents and carnivores.

The ~60-km-diameter Charlevoix structure in southern Quebec is dated as 357 ± 15 Ma. At the Devonian-Mississippian boundary (ca. 354 Ma), the Appalachian region in Pennsylvania and Maryland was covered by the unusual Spechty Kopf Formation of rapidly deposited polymictic diamictite grading upward into pebbly mudstones, sands, and laminites (see McGhee, 1996). The Spechty Kopf sediments crop out over a distance of 400 km, and the region is ~1000 km from the Charlevoix structure.

These results and other examples in the literature suggest that impact-derived deposits are more widespread in the geologic record than previously thought, and that continued examination of unusual and neglected sedimentary formations (e.g., breccias, diamictites, debris-flow deposits, tempestites, seismites) will produce more evidence of impacts as an important factor in sedimentation.

PERIODIC MASS EXTINCTIONS?

Raup and Sepkoski (1984, 1986) reported a statistically significant 26 m.y. periodicity in extinction time series for the past 250 m.y., with a secondary periodicity of 30 m.y. Periods of ~26–31 m.y. have been derived using various subsets of extinction events (family and genus levels), different geologic time scales, and various methods of time-series analysis. Rampino and Stothers (1986) reported detection of a similar ~29 m.y. component in the record of extinctions of nonmarine vertebrates. The interpretation of these results has been a subject of considerable ongoing debate (see Rampino and Haggerty, 1996a).

Analysis of 21 extinctions over the longer record going back to 515 Ma resulted in a spectrum with the highest peak at 27.3 m.y. (Rampino and Haggerty, 1996a, 1996b), and truncation analysis of that extinction time series revealed that all records longer than 144 m.y. exhibited a major spectral peak at ~26–27 m.y. In the longer records, the second- or third-highest peak is between 35 and 37 m.y.

Using the 11 mass extinctions over the past 250 m.y., Rampino and Stothers (1998) performed linear time-series analyses either allowing the phase to be a free parameter, or with fixed phasing, in which case they only used trial starting epochs in the range 0 ± 1 Ma (the time of most recent galactic plane crossing). When the phase is allowed to vary, only one high spectral peak at 27 m.y. is apparent. When the phase is fixed at the present time, however, the highest peak shifts to 28 m.y. and two somewhat smaller peaks at 32 and 35 m.y. become considerably more prominent. Napier (1998) analyzed the Rampino and Caldeira (1993) dataset of geologic events (which includes the extinctions) and found that the 25–27 m.y. periodicity detected earlier (see also Liritzis, 1993) was in agreement with periods expected from the galactic oscillation hypothesis.

The three most severe mass extinctions (end-Ordovician, ca. 435 Ma; end-Permian, ca. 250 Ma; and end-Cretaceous, 65 Ma) are separated by ~180 m.y. The solar system undergoes a perigalactic revolution cycle with an estimated period of ~170 ± 10 m.y. (Matese et al., 1985), and this cycle might also modulate the flux of Oort cloud comets (Bailey et al., 1990). Matese and Whitmire (1996) found evidence for such a perturbation of the present Oort comet cloud by the galactic radial tide, due to the entire distribution of matter interior to the solar orbit.

PERIODIC IMPACTS?

The initial time-series analyses of impact craters (Rampino and Stothers, 1984a, 1984b; Alvarez and Muller, 1984), and a number of subsequent studies (Shoemaker and Wolfe, 1986; Rampino and Stothers, 1986; Stothers, 1988; Yabushita 1991, 1992, 1996a, 1996b) found evidence for a possible period in impact crater ages ranging from 28 to 36 m.y. Although a peak at ~30 m.y. commonly shows up in statistical analyses, the significance of the peak has been questioned (e.g., Grieve and Shoemaker, 1994; Grieve and Pesonen, 1996). For example, Montanari et al. (1998) detected peaks of ~33 m.y. and 64 m.y. (the multiple?), but concluded that the peaks were most likely artifacts of the statistical analysis.

The differences in the formal periods derived from analyses of extinctions and cratering might at first seem problematic. However, several studies have concluded that the observed differences in the formal periodicity are to be expected, taking into consideration problems in dating and the likelihood that both records would be mixtures of periodic and random events (Stothers, 1988, 1989; Trefil and Raup, 1987; Fogg, 1989).

Considering that most Earth-crossing asteroids are in the ≤1 km size range (Shoemaker et al., 1990), we might expect that for impact craters on Earth ≤20 km in diameter, the presumably random asteroid flux would dominate over any signal from periodic comets in that size range. If this is true, the cratering period should show up preferentially in the largest craters. In a test of this idea, Matese et al. (1998) found a best-fitting period of 36 ± 2 m.y., using only the nine largest well-dated craters, when the phase was fixed to the time of most recent galactic plane crossing.

Linear time-series analyses were performed (see Rampino, 1998; Rampino and Stothers, 1998) on sets of various sized craters using the revised list of 34 impact craters shown in Table 4. As Table 5 shows, two peaks were detected, a narrow peak

TABLE 4. IMPACT CRATERS

Crater	Location	Diameter (km)	Age (Ma)
Zhamanshin	Kazakhstan	13.5	1.0 ± 0.1
Bosumtwi	Ghana	10.5	1.03 ± 0.02
El'gygytgyn	Russia	18	3.5 ± 0.5
Bigach	Kazakhstan	5 ± 3	6 ± 3
Karla	Russia	10	5 ± 1
Ries	Germany	24	15.1 ± 1
Haughton	Canada	24	23.4 ± 1
Chesapeake	USA	90	35.2 ± 0.3
Popigai	Russia	100	35.7 ± 0.2
Wanapitei	Canada	7.5	37 ± 2
Mistastin	Canada	28	38 ± 4
Logoisk	Belarus	15	30 ± 5
Ragozinka	Russia	9	46 ± 3
Kamensk	Russia	25	49.15 ± 0.18
Montagnais	Canada	45	50.5 ± 0.8
Marquez	USA	13	58 ± 2
Chicxulub	Mexico	195	64.98 ± 0.05
Kara, Ust-Kara	Russia	65	70.3 ± 2.2
Manson	USA	35	74.1 ± 0.1
Lappajarvi	Finland	23	77.3 ± 0.4
Dellen	Sweden	19	89 ± 2.7
Steen River	Canada	25	95 ± 7
Boltysh	Ukraine	24	95 ± 10
Avak	USA	12	100 ± 5
Carswell	Canada	39	115 ± 10
Mien	Sweden	9	121 ± 2.3
Tookoonooka	Australia	55	128 ± 5
Gosses Bluff	Australia	22	142.5 ± 0.8
Mjølnir	Barents Sea	40	142.6 ± 2.6
Morokweng	South Africa	≥70	145.0 ± 0.8
Puchezh-Katunki	Russia	80	167 ± 3
Rochechouart	France	23	214 ± 8
Manicouagan	Canada	100	214 ± 1
Araguainha	Brazil	40	247 ± 5.5

Note: After Montanari and Koeberl (2000). (Diameter ≥ 5 km and age estimates ≤ 250 Ma with maximum error of ±10 Ma.)

TABLE 5. RESULTS OF SPECTRAL ANALYSES OF IMPACT CRATER AGES

Diameter (km)	Number of craters	Phase	Highest peak (m.y.)	Second-Highest peak (m.y.)
≥5	31	Free	30	35
		Fixed at 0	30	35
≥35	11	Free	35	None
		Fixed at 0	35	None
≥90	5	Free	36	29
		Fixed at 0	36	30

Note: Ages from Rampino and Stothers (1998).

at 30 ± 0.5 m.y. and a broader peak at 35 ± 2 m.y. In the cases where only the largest craters were utilized, the highest peak in the period spectrum is located at ~36 m.y., in agreement with the results of Matese et al. (1998).

In an earlier study designed to test the effects of the errors in the ages of the craters on the periods detected, Stothers (1988) showed that the dating errors alone were capable of shifting the dominant period between 30 and 35 m.y. We note, however, that the periods detected in the mass extinction and cratering records are statistically significant (at the 5% level) only when they are treated as periods (and phases) known a priori. Thus, it is still difficult to say whether the best period in cratering is closer to 30 or 36 m.y., and the width of the comet flux pulses (~8 m.y.) in the galactic models might make it difficult to determine significance in any case.

COMET SHOWERS IN THE GEOLOGIC RECORD: EVIDENCE PRO AND CON

Theoretical modeling suggests that gravitational perturbation of the Oort cloud should greatly enhance the flux of active comets over a period of several million years. Disintegration and sublimation of the cometary nuclei during orbital evolution would increase the dust flux in the inner solar system. Because the dust is swept into the Sun on a lifetime that is short or comparable to the average lifetime of a long-period comet (~600 k.y.), the period of most probable cometary impacts should be associated with an increase in interplanetary dust particle flux.

The best evidence for a comet shower in the geologic record is the work of Farley et al. (1998) on $^3He/^4He$ ratios in late Eocene sediments from Italy, where an increase in extraterrestrial ^3He accretion rate is temporally correlated with indicators of impact. The late Eocene sequence contains impact evidence in the form of Ir anomalies, shocked quartz, microspherules, and microtektites (Table 2), and two large well-dated impact craters, Popigai (35.7 ± 0.2 Ma) and Chesapeake (35.2 ± 0.3 Ma) (Table 4).

By contrast, at the K-T boundary, recent work by Mukhopadhyay et al. (2000) suggests that the ^3He accretion rate was constant from 65.4 to 63.9 Ma at the Italian K-T boundary site, which seems to rule out a comet shower at that time. They suggested that the K-T impactor was a lone comet or an asteroid. Their results, however, are based on constant sedimentation rates across the boundary, which contradicts previous studies that have found evidence for a significant reduction in sedimentation rates across the K-T boundary.

Minor increases (a factor <2) in the extraterrestrial ^3He accretion rate are suggested between 70.5 and 68 Ma, and ca. 66 Ma, 1 m.y. prior to the K-T boundary. Furthermore, two relatively large craters of Maastrichtian-Campanian age are known, Kara Ust-Kara and Manson (Table 4). A comet shower during this period, ca. 70 Ma, would fit an ~35–36 m.y. galactic cycle (ca. 0 Ma, 35 Ma, 70 Ma). Further studies of ^3He accretion in the upper Maastrichtian and across the K-T boundary at other sites in the world are necessary to resolve the issue of a comet shower of late Maastrichtian to K-T age.

NONGALACTIC MODEL FOR IMPACTS AND EXTINCTIONS

A galactic explanation for cyclical impacts and mass extinctions has an advantage in being attractive in the sense that it would make comet impacts a predictable effect of the large-scale dynamics of the galaxy. The idea predicts that the largest impactors, those that would be associated with severe mass extinctions, are comets derived from the Oort comet cloud.

There are two other known reservoirs of potential Earth impacts in the solar system, the Kuiper Belt of large comets and the Main Belt asteroids, centered at 3.0 astronomical units between the orbits of Mars and Jupiter. If Kuiper Belt objects or asteroids are the main culprits in terrestrial mass extinctions, then it is difficult to see how they could be either cyclical or related to galactic motions (but see Matese and Whitmire, 1986). There is some recent evidence based on Cr isotopes, however, that several of the large craters of the past 200 m.y. (Morokweng, Clearwater, Rochechouart) were produced by asteroids of normal chondritic composition (Koeberl et al., this volume; Shukolyukov and Lugmair, 2000; McDonald et al., 2001). Furthermore, Veverka et al. (2000) determined that Eros, a 38-km-diameter near Earth object, is of L chondritic composition.

The parent asteroid of L chondrite meteorites (which represent 38% of present meteorite falls) underwent a major collision and may have been catastrophically fragmented ca. 540-440 Ma, as evidenced by $^{40}Ar/^{39}Ar$ ages of heavily shocked and degassed L chondrites (Haack et al, 1996; Bogard, 1995). Recent dating of lunar glass spherules suggests that the impact cratering rate on the Moon (and presumably on the Earth) increased by a factor of 3.7 ± 1.2 in the past 400 m.y. (Culler et al., 2000), and McEwen et al. (1997) provided evidence for a similar increase in impacts from crater counts on the moon for the past 500 m.y. This near coincidence in timing between the increased flux of impactors and the fragmentation of the L chondrite parent asteroid suggests that a significant portion of large Earth impactors could have been created by that breakup event.

Thus, catastrophic collisions in the asteroid belt that effectively deliver objects of various sizes to the inner solar system might have caused the increase in impactor flux in the past 400–500 m.y. and could be responsible for many mass-extinction events. There are clear astrobiological implications here. A galactic trigger for impacts on the Earth could also work in other planetary systems containing an outer comet cloud. If comet clouds are an integral part of planetary system evolution, then inner planets of other stars should also undergo periodic comet showers. These would be of similar cyclicity if the cycle is controlled by the vertical oscillation of stars with respect to the galactic plane.

By contrast, if asteroids were important in causing mass extinctions, then only those solar systems with an asteroid belt in which random collisions can take place would have inner planets that were bombarded with a frequency similar to that of the Earth. Asteroid belts may be a rare occurrence, and thus many planetary systems would probably lack the special conditions that created an increased flux of impactors. If impact-induced mass extinctions are an important driver in evolution, then the many planetary systems that would lack asteroid belts may lack a critical impetus for the evolution of complex life.

CONCLUSIONS

Galactic dynamics and solar system structure predict that gravitational perturbations of the Oort comet cloud should cause increases in the comet flux in the inner solar system (comet showers) with a period or quasiperiod close to the half-period of the solar system's vertical oscillation about the galactic plane, as given by the simple formula $P_{1/2} = 13.2\ \rho_0^{1/2}$. According to recent calculations of the mass density in the plane region (ρ_0) utilizing the current range of estimates of the total mass of the galactic disk (including dark matter), the half-period could range from ~26–40 m.y.

Geologic data on mass extinctions of life and evidence of large impacts on the Earth are thus far consistent with a quasiperiodic modulation of the flux of Oort cloud comets with a mean period of ~30 or 36 m.y. Multiple large impacts over periods of several million years, a possible signature of comet showers, are indicated at a number of critical periods, including the late Eocene, Late Jurassic, and Late Devonian. Studies of ^{3}He accretion in the late Eocene also support a comet shower. One problem is that recent analysis of ^{3}He accretion rates across the K-T boundary failed to detect evidence of a comet shower at that time.

Discrepancies in the periodicities detected in the extinctions and cratering may be the result of a combination of dating errors, mixtures of periodic and nonperiodic events, the small data sets, and real irregularities in the underlying cycle. Further astronomical and geological studies (including the discovery and accurate age dating of additional impact craters) should help to clarify and refine both the expected astronomical cycle times and the periodicities detectable in the geologic record.

ACKNOWLEDGMENTS

I thank K. Caldeira, T. Gehrels, J. Matese, A. Montanari, Y. Miura, W. Napier, R. Stothers, and S. Yabushita for discussions and criticism. R. Muller and J. Raitala provided critical reviews. D. Winiarski provided technical support. Jack Sepkoski kindly sent me an updated version of his extinction data set in 1998. This chapter is dedicated to his memory.

REFERENCES CITED

Adushkin, V.V., and Nemchinov, I.V., 1994, Consequences of impacts of cosmic bodies on the surface of the Earth, *in* Gehrels, T., ed., Hazards due to asteroids and comets: Tucson, University of Arizona Press, p. 721–778.

Alvarez, L.W., Alvarez, W., Asaro, F., and Michel, H.V., 1980, Extraterrestrial cause of the Cretaceous/Tertiary extinction: Science, v. 208, p. 1095–1108.

Alvarez, W., and Muller, R.A., 1984, Evidence from crater ages for periodic impacts on the Earth: Nature, v. 308, p. 718–720.

Bahcall, J.N., Flynn, C., and Gould, A., 1992, Local dark matter from a carefully selected sample: Astrophysical Journal, v. 389, p. 234–250.

Bailey, M.E., Clube, S.V.M., and Napier, W.M., 1990, The origin of comets: Oxford, Pergamon, 577 p.

Becker, L., Poreda, R.J., Hunt, A.G., Bunch, T.E., and Rampino, M.R., 2001, Impact event at the Permian-Triassic boundary: Evidence from extraterrestrial noble gases in fullerenes: Science, v. 291, p. 1530–1533.

Bice, D.M., Newton, C.R., McCauley, S., Reiners, P.W., and McRoberts, C.A., 1992, Shocked quartz at the Triassic-Jurassic boundary in Italy: Science, v. 259, p. 443–446.

Bogard, D.D., 1995, Impact ages of meteorites: A synthesis: Meteoritics, v. 30, p. 244–268.

Crezé, M., Chereul, E., Bienaymé, O., and Pichon, C., 1998, The distribution of nearby stars in phase space mapped by Hipparcos. 1. The potential well and local dynamical mass: Astronomy and Astrophysics, v. 329, p. 920–936.

Culler, T.S., Becker, T.A., Muller, R.A., and Renne, P.R., 2000, Lunar impact history from ^{40}Ar/^{39}Ar dating of glass spherules: Science, v. 287, p. 1785–1788.

Cutler, A.H., and Behrensmeyer, A.K., 1996, Models of vertebrate mass mortality events at the K/T boundary, *in* Ryder, G., Fastovsky, D., and Gartner, S., eds., The Cretaceous-Tertiary event and other catastrophes in Earth history: Boulder, Colorado, Geological Society of America Special Paper 307, p. 375–379.

Dypvik, H., Gudlaugsson, S.T., Tsikalas, F., Attrep, M., Jr., Ferrell, R.E., Jr., Krinsley, D.H., Mørk, A., Faleide, J.I., and Nagy, J., 1996, Mjølnir structure: An impact crater in the Barents Sea: Geology, v. 24, p. 779–782.

Ernstson, K., Hammann, W., Fiebag, J., and Graup, G., 1985, Evidence of an impact origin for the Azuara structure (Spain): Earth and Planetary Science Letters, v. 74, p. 361–370.

Ernstson, K., Rampino, M.R., and Hiltl, M., 2001, Cratered cobbles in Triassic Buntsandstein conglomerates in northeastern Spain: An indicator of shock deformation in the vicinity of large impacts: Geology, v. 29, p. 11–14.

Farley, K.A., Montanari, A., Shoemaker, E.M., and Shoemaker, C.S., 1998, Geochemical evidence for a comet shower in the Late Eocene: Science, v. 280, p. 1250–1253.

Fogg, M.J., 1989, The relevance of the background impact flux to cyclic impact/mass extinction hypotheses: Icarus, v. 79, p. 382–395.

Fowell, S.J., Cornet, B., and Olsen, P.E., 1994, Geologically rapid Late Triassic extinctions: Palynological evidence from the Newark Supergroup, *in* Klein, G.D., ed., Pangea: Paleoclimate, tectonics, and sedimentation during accretion, zenith, and breakup of a supercontinent: Boulder, Colorado, Geological Society of America Special Paper 288, p. 197–206.

Gould, A., Bahcall, J.N., and Flynn, C., 1996, Disk M dwarf luminosity function from Hubble Space Telescope star counts: Astrophysical Journal, v. 465, p. 759–768.

Gradstein, F.M., and Ogg, J.G., 1996, A Phanerozoic time scale: Episodes, v. 19, nos. 1 and 2, insert.

Grieve, R.A.F., 1996, Chesapeake Bay and other terminal Eocene impacts: Meteoritics and Planetary Science, v. 31, p. 166–167.

Grieve, R.A.F., 1997, Extraterrestrial impact events: The record in the rocks and the stratigraphic column: Palaeoclimatology, Palaeogeography, Palaeoecology, v. 132, p. 5–23.

Grieve, R.A.F., and Pesonen, L.J., 1996, Terrestrial impact craters: Their spatial and temporal distribution and impacting bodies: Earth, Moon, and Planets, v. 72, p. 357–376.

Grieve, R.A.F., and Shoemaker, E.M., 1994, The record of past impacts on Earth, *in* Gehrels, T., ed., Hazards due to comets and asteroids: Tucson, University of Arizona Press, p. 417–462.

Haack, H., Farinella, P., and Scott, E.R.D., 1996, Meteoritic, asteroidal, and theoretical constraints on the 500 Ma disruption of the L chondrite parent body: Icarus, v. 119, 182–191.

Hallam, A., Grose, J.A., and Ruffell, A.H., 1991, Paleoclimatic significance of changes in clay mineralogy across the Jurassic-Cretaceous boundary in England and France: Palaeogeography, Palaeoclimatology, Palaeoecology, v. 81, p. 173–187.

Heisler, J., and Tremaine, S., 1989, How dating uncertainties affect the detection of periodicity in extinctions and craters: Icarus, v. 77, p. 213–219.

Hildebrand, A.R., Pilkington, M., Connors, M., Ortiz-Aleman, C., and Chavez, R.E., 1995, Size and structure of the Chicxulub Crater revealed by horizontal gravity gradients and cenotes: Nature, v. 376, p. 415–417.

Hut, P., Alvarez, W., Elder, W.P., Hansen, T., Kauffman, E.G., Keller, G., Shoemaker, E.M., and Weissman, P.R., 1987, Comet showers as a cause of mass extinctions: Nature, v. 329, p. 118–126.

Izett, G.A., Cobban, W.A., Obradovich, J.D., and Kunk, M.J., 1993, The Manson impact structure: $^{40}Ar/^{39}Ar$ age and its distal impact ejecta in the Pierre Shale in southeastern South Dakota: Science, v. 262, p. 729–732.

Koeberl, C., Armstrong, R.A., and Reimold, W.U., 1997, Morokweng, South Africa: A large impact structure of Jurassic/Cretaceous boundary age: Geology, v. 25, p. 731–734.

Liritzis, I., 1993, Cyclicity in terrestrial upheavals during the Phanerozoic eon: Quarterly Journal of the Royal Astronomical Society, v. 34, p. 251–260.

Margolis, S.V., Claeys, P., and Kyte, F.T., 1991, Microtektites, microkrystites, and spinels from a Late Pliocene asteroid impact in the Southern Ocean: Science, v. 251, p. 1594–1597.

Matese, J.J., Whitman, P.G., Innanen, K.A., and Valtonen, M.J., 1995, Periodic modulation of the Oort cloud comet flux by the adiabatically changing galactic tide: Icarus, v. 116, p. 255–268.

Matese, J.J., Whitman, P.G., Innanen, K.A., and Valtonen, M.J., 1998, Variability of the Oort comet flux: Can it be manifest in the cratering record?: Highlights in Astronomy, v. 11A, p. 252–256.

Matese, J.J., and Whitmire D., 1986, Planet X and the origins of the shower and steady-state flux of short period comets: Icarus, v. 65, p. 37–50.

Matese, J., and Whitmire, D., 1996, Tidal imprint of distant galactic matter on the Oort comet cloud: Astrophysical Journal, v. 472, p. L41–L43.

Matsumoto, M., and Kubotani, H., 1996, A statistical test for correlation between crater formation rate and mass extinctions: Monthly Notices of the Royal Astronomical Society, v. 282, p. 1407–1412.

McDonald, I., Andreoli, M.A.G., Hart, R.J., and Tredoux, M., 2001, Platinum-group elements in the Morokweng impact structure, South Africa: Evidence for the impact of a large ordinary chondrite projectile at the Jurassic-Cretaceous boundary: Geochimica et Cosmochimica Acta, v. 65, p. 299–309.

McEwen, A.S., Moore, J.M., Shoemaker, E.M., 1997, The Phanerozoic impact cratering rate: Evidence from the farside of the moon: Journal of Geophysical Research, v. 102, p. 9231–9242.

McGhee, G., 1996, Late Devonian extinctions: New York, Columbia University Press, 378 p.

Melosh, H.J., Schneider, N.M., Zahnle, K.J., and Latham, D., 1990, Ignition of global wildfires at the Cretaceous/Tertiary boundary: Nature, v. 343, p. 251–254.

Montanari, A., and Koeberl, C., 2000, Impact stratigraphy: Heidelberg, Springer-Verlag, 364 p.

Montanari, A., Campo Bagatin, A., and Farinella, P., 1998, Earth cratering record and impact energy flux in the last 150 Ma: Planetary and Space Science, v. 46, p. 271–281.

Mossman, D.J., Grantham, R.G., and Langenhorst, F., 1998, A search for shocked quartz at the Triassic-Jurassic boundary in the Fundy and Newark basins of the Newark Supergroup: Canadian Journal of the Earth Sciences, v. 35, p. 101–109.

Mukhopadhyay, S., Farley, K.A., and Montanari, A., 2001, A 35 Myr record of helium in pelagic limestones from Italy: Implications for interplanetary dust accretion from the early Maastrichtian to the middle Eocene: Geochimica et Cosmochimica Acta, v. 65, p. 653–669.

Napier, W.M., 1998, NEOs and impacts: The galactic connection: Celestial Mechanics and Dynamical Astronomy, v. 69, p. 59–75.

Pham, H.A., 1997, Estimation of the local mass density from an F-star sample observed by Hipparcos, *in* Perryman, M.A.C., and Bernacca, P.L., eds., HIPPARCOS Venice '97, European Space Agency Special Publication 402: Noordwijk, The Netherlands, European Space Agency Publication Division, p. 559–561.

Poag, C.W., 1997, Roadblocks on the kill curve: Testing the Raup hypothesis: Palaios, v. 12, p. 582–590.

Pope, K.O., D'Hondt, S.L., and Marshall, C.R., 1998, Meteorite impact and the mass extinction of species at the Cretaceous/Tertiary boundary: Proceedings of the National Academy of Sciences of the United States of America, v. 95, p. 11 028–11 029.

Rampino, M.R., 1998, The galactic theory of mass extinctions: An update: Celestial Mechanics and Dynamical Astronomy, v. 69, p. 49–58.

Rampino, M.R., and Caldeira, K., 1993, Major episodes of geologic change: Correlations, time structure and possible causes: Earth and Planetary Science Letters, v. 114, p. 215–227.

Rampino, M.R., and Haggerty, B.M., 1996a, The "Shiva Hypothesis": Impacts, mass extinctions, and the galaxy: Earth, Moon and Planets, v. 72, p. 441–460.

Rampino, M.R., and Haggerty. B.M., 1996b, Impact crises and mass extinctions: A working hypothesis, *in* Ryder, G., Fastovsky, D., and Gartner, S., eds., The Cretaceous-Tertiary event and other catastrophes in Earth history: Geological Society of America Special Paper 307, p. 11–30.

Rampino, M.R., and Schwindt, D.M., 1999, Comment on "The age of the Kara impact structure, Russia" by Trieloff et al.: Meteoritics and Planetary Science, v. 34, p. 1–2.

Rampino, M.R., and Stothers, R.B., 1984a, Terrestrial mass extinctions, cometary impacts and the sun's motion perpendicular to the galactic plane: Nature, v. 308, p. 709–712.

Rampino, M.R., and Stothers, R.B., 1984b, Geological rhythms and cometary impacts: Science, v. 226, p. 1427–1431.

Rampino, M.R., and Stothers, R.B., 1986, Geologic periodicities and the galaxy, *in* Smoluchowski, R., Bahcall, J.N., and Matthews, M.S., eds., The galaxy and the solar system: Tucson, University of Arizona Press, p. 241–259.

Rampino, M.R., and Stothers, R.B., 1998, Mass extinctions, comet impacts, and the galaxy: Highlights in Astronomy, v. 11A, p. 246–251.

Rampino, M.R., Haggerty, B.M., and Pagano, T.C., 1997, A unified theory of impact crises and mass extinctions: Quantitative tests: Annals of the New York Academy of Sciences, v. 822, p. 403–431.

Raup, D.M., 1979, Size of the Permian-Triassic bottleneck and its evolutionary implications: Science, v. 206, p. 217–218.

Raup, D.M., 1986, Extinctions in the geologic past, *in* Elliott, D.K., ed., Dynamics of extinction: New York, John Wiley and Sons, p. 109–119.

Raup, D.M., 1992, Large-body impact and extinction in the Phanerozoic: Paleobiology, v. 18, p. 80–88.

Raup, D.M., and Sepkoski, J.J., Jr., 1984, Periodicity of extinctions in the geologic past: Proceedings of the National Academy of Sciences of the United States of America, v. 81, p. 801–805.

Raup, D.M., and Sepkoski, J.J., Jr., 1986, Periodic extinctions of families and genera: Science, v. 231, p. 833–836.

Reed, B.C., 1997, The sun's displacement from the galactic plane: Limits from the distribution of OB-star latitudes: Publications of the Astronomical Society of the Pacific, v. 109, p. 1145–1148.

Retallack, G.J., Seyedolali, A., Krull, E.S., Holser, W.T., Ambers, C.P., and

Kyte, F.T., 1998, Search for evidence of impact at the Permian-Triassic boundary in Antarctica and Australia: Geology, v. 26, p. 979–982.

Russell, D.A., 1993, Manson extinctions: Science, v. 262, p. 1956–1957.

Sey, I.I., and Kalacheva, E.D., 1996, Jurassic/Cretaceous boundary in the Boreal Realm (biostratigraphy and Boreal-Tethyan correlations): International Subcommission on Jurassic Stratigraphy, Newsletter No. 24, p. 50–53.

Sepkoski, J.J., Jr., 1995, Patterns of Phanerozoic extinction: A perspective from global data bases, *in* Walliser, O.H., ed., Global events and event stratigraphy in the Phanerozoic: Berlin, Springer, p. 35–51.

Shoemaker, E.M., and Wolfe, R.F., 1986, Mass extinctions, crater ages, and comet showers, *in* Smoluchowski, R., Bahcall, J.N., and Matthews, M.S., eds., The galaxy and the solar system: Tucson, University of Arizona Press, p. 338–386.

Shoemaker, E.M., Wolfe, R.F., and Shoemaker, C.S., 1990, Asteroid and comet flux in the neighborhood of Earth: Geological Society of America Special Paper 247, p. 155–170.

Shukolyukov, A., and Lugmair, G.W., 2000, Extraterrestrial matter on Earth: Evidence from the Cr isotopes [abs.], *in* Catastrophic events and mass extinctions: Impacts and beyond: Houston, Texas, Lunar and Planetary Institute, LPI Contribution No. 1053, p. 197–198.

Smoluchowski, R., Bahcall, J.N., and Matthews, M.S., editors, 1986, The galaxy and the solar system: Tucson, University of Arizona Press, 483 p.

Stothers, R.B., 1985, Terrestrial record of the solar system's oscillation about the galactic plane: Nature, v. 317, p. 338–341.

Stothers, R.B., 1988, Structure of Oort's comet cloud inferred from terrestrial impact craters: Observatory, v. 108, p. 1–9.

Stothers, R.B., 1989, Structure and dating errors in the geologic time scale and periodicity in mass extinctions: Geophysical Research Letters, v. 16, p. 119–122.

Stothers, R.B., 1993, Impact cratering at geologic stage boundaries: Geophysical Research Letters, v. 20, p. 887–890.

Stothers, R.B., 1998, Galactic disc dark matter, terrestrial impact cratering and the law of large numbers: Monthly Notices of the Royal Astronomical Society, v. 300, p. 1098–1104.

Thaddeus, P., and Chanan, G., 1985, Cometary impacts, molecular clouds, and the motion of the Sun perpendicular to the galactic plane: Nature, v. 314, p. 73–75.

Toon, O.B., Zahnle, K., Morrison, D., Turco, R.P., and Covey, C., 1997, Environmental perturbations caused by the impacts of asteroids and comets: Reviews of Geophysics, v. 35, p. 41–78.

Trefil, J.S., and Raup, D.M., 1987, Numerical simulations and the problem of periodicity in the cratering record: Earth and Planetary Science Letters, v. 82, p. 159–164.

Valtonen, M.J., Zheng, J.Q., Matese, J.J., and Whitman, P.G., 1995, Near-Earth populations of bodies coming from the Oort Cloud and their impacts with planets: Earth, Moon, and Planets, v. 71, p. 219–223.

Veverka, J., Robinson, M., Thomas, P., Murchie, S., Bell, J.F., III, Izenberg, N., Chapman, C., Harch, A., Bell, M., Carcich, B., Cheng, A., Clark, B., Doningue, D., Dunham, D., Farquhar, R., Gaffey, M.J., Hawkins, E., Joseph, J., Kirk, R., Li, H., Lucey, P., Malin, M., Martin, P., McFadden, L., Merline, W.J., Miller, J.K., Owen, W.M., Jr., Peterson, C., Prockter, L., Warren, J., Wellnitz, D., Williams, B.G., and Yeomans, D.K., 2000, NEAR at Eros: Imaging and spectral results: Science, v. 289, p. 2088–2097.

Yabushita, S., 1991, A statistical test for periodicity hypothesis in the crater formation rate: Monthly Notices of the Royal Astronomical Society, v. 250, p. 481–485.

Yabushita, S., 1992, Periodicity in the crater formation rate and implications for astronomical modeling, *in* Clube, S.V.M., Yabushita, S., and Henrard, J., eds., Dynamics and evolution of minor bodies with galactic and geological implications: Dordrecht, The Netherlands, Kluwer, p. 161–178.

Yabushita, S., 1996a, Are cratering and probably related geological records periodic?: Earth, Moon, and Planets, v. 72, p. 343–356.

Yabushita, S., 1996b, Statistical tests of a periodicity hypothesis for crater formation rate-2: Monthly Notices of the Royal Astronomical Society, v. 279, p. 727–732.

Yabushita, S., 1998, A statistical test of correlations and periodicities in the geological records: Celestial Mechanics and Dynamical Astronomy, v. 69, p. 31–48.

Zhakarov, V.A., Lapukhov, A.S., and Shenfil, O.V., 1993, Iridium anomaly at the Jurassic-Cretaceous boundary in northern Siberia: Russian Journal of Geology and Geophysics, v. 34, p. 83–90.

Manuscript Submitted October 30, 2000; Accepted by the Society March 22, 2001

Printed in the U.S.A.

Geological Society of America
Special Paper 356
2002

Solar system linked to a gigantic interstellar cloud during the past 500 m.y.: Implications for a galactic theory of terrestrial catastrophism

Carlos A. Olano*
Instituto Argentino de Radioastronomía, C.C. 5, 1894 Villa Elisa, Argentina

ABSTRACT

Modern investigations have revealed that the galactic environment of the Sun may have played a role in climatic and geological changes on the Earth through time. Here I present the results of a model of the local system of gas and stars, in which the Sun is gravitationally linked to a gigantic interstellar cloud during the past 500 m.y. The Sun would have been captured by the supercloud during the cloud's process of formation ca. 500 Ma, remaining linked to it. The original supercloud, with a mass of $\sim 2 \times 10^7$ solar masses, has been forming stars and evolving in the solar neighborhood under the influence of the pressure of external gas and shear due to the galactic differential rotation. In our model, Gould's belt, the local arm, and the Sirius supercluster, the main subsystems of the local system, would have been formed within the postulated supercloud. We propose a variation of the Rampino and Stothers mechanism to explain the 26 m.y. quasiperiodicity of mass extinctions and the abrupt increase in the production rate of lunar craters over the past ~ 400 m.y. We suggest that perturbations on the Oort comet cloud caused by encounters with intermediate-sized clouds of the supercloud about the extremes of the Sun motion perpendicular to the galactic plane could generate periodic comet showers on the Earth, in agreement with the apparent cyclicity in the mass extinction and geological records.

INTRODUCTION

Several investigators have speculated on the role that external influences, such as the explosion of a nearby supernova (e.g., Leitch and Vasisht, 1998), the penetration of a dense interstellar cloud (McCrea, 1975; Yabushita and Allen, 1989; Zank and Frisch, 1999) or the impact of a large asteroid or comet (Urey, 1973; Hoyle and Wickramasinghe, 1978), might have played in the geological and biological evolution of the Earth (van den Bergh, 1989, 1994; Toon et al., 1994).

The discoveries of the iridium anomaly at the Cretaceous-Tertiary boundary (Alvarez et al., 1980), a possible 26 m.y. periodicity in the occurrence of mass extinctions (Raup and Sepkoski, 1986), and a possible periodicity in impacts on Earth in approximate correspondence with mass extinctions (Alvarez and Muller, 1984; Rampino and Stothers, 1984; Stothers, 1998), have provided an empirical basis for astrophysical theories of terrestrial catastrophism. The similarity in the periods suggests that the mass extinctions were caused by periodic cometary showers. However, the reality of the 26 m.y. periodicity has been criticized by some (see Jetsu and Pelt, 2000, and references therein). Despite this, periodic cometary showers, as

*Member of the Carrera del Investigador Científico of CONICET, Buenos Aires, Argentina.

Olano, C.A., 2002, Solar system linked to a gigantic interstellar cloud during the past 500 m.y.: Implications for a galactic theory of terrestrial catastrophism, *in* Koeberl, C., and MacLeod, K.G., eds., Catastrophic Events and Mass Extinctions: Impacts and Beyond: Boulder, Colorado, Geological Society of America Special Paper 356, p. 679–683.

defined by time-series analysis of the terrestrial cratering record, are viable (Stothers and Rampino, 1990; Yabushita, 1992; Grieve and Shoemaker, 1994). Moreover, there are some cases where the connection between major impact events and mass extinctions can be more or less firmly established (see Rampino, this volume; Rampino and Haggerty, 1994).

To explain these observations, two major competing hypotheses have been proposed. The first, advocated by Rampino and Stothers (1984, 1986, 2000), introduced the idea that the oscillatory motion of the solar system about the plane of the galaxy would allow for large clouds of gas and dust to perturb the Oort comet cloud, leading to cyclical comet showers. The second hypothesis invoked a companion star to the Sun as the perturber of the Oort cloud (Whitmire and Jackson, 1984; Davis et al., 1984). However, both hypotheses have certain difficulties (Clube and Napier, 1986; Thaddeus, 1986; Tremaine, 1986; Bailey et al., 1987).

I present a model of the local system of gas and stars (Olano, 2001) that leads to the conclusion that the solar system has been attached to a gigantic interstellar cloud during the past 500 m.y. In view of this result, I suggest a variation of the Rampino and Stothers mechanism that could resolve its original difficulties.

MODEL FOR THE LOCAL SYSTEM OF GAS AND STARS

The local galactic environment within 1 kpc (kiloparsec; 3.26×10^3 light years) of the Sun comprises a massive, expanding ring of interstellar matter (Olano, 1982) associated with star-forming molecular clouds and a group of relatively young stars of spectral type OB known as Gould's belt. This structure is likely the result of the disintegration of a formerly bound supercloud. Gould's belt is a discrete system that forms part of the so-called Orion or local arm of the galaxy, which is thought to be a material interarm branch or spur. Another characteristic of the solar neighborhood is the existence of at least three superclusters of stars near the Sun: Sirius, Pleiades, and Hyades (Palouš, 1986).

The central thesis of this paper is that Gould's belt, the local arm, and the superclusters were formed in different epochs within a supercloud of 2×10^7 solar masses and 500 pc of radius. This supercloud was moving initially almost ballistically in the galactic field until an encounter with a major spiral arm initiated a braking process. The estimated mass for the supercloud is a typical mass for the largest clouds of the galaxy, and most of the mass in the interstellar medium is in the form of giant cloud complexes containing 10^5 to 10^7 solar masses (Elmegreen and Elmegreen, 1983, 1987). The stars of the older generations (i.e., the Sirius supercluster, the age of which is ca. 500 Ma) tended to conserve the kinematics of the pre-braking phase of the supercloud, while the gas and early star complexes reflect the recent kinematics resulting from the braking process. I calculated back in time the epicyclic galactic orbits of the Sirius supercluster and the postulated supercloud, starting from their current state as initial conditions and taking into account the action of a frictional force on the gas. From the condition that the Sirius supercluster and the supercloud shared the same orbits before the separation of gas and stars due to the braking of the gas, I determined the model's free parameters. The main observational evidence supporting our hypothesis is that the supercloud's track derived from the model coincides with a large tunnel in the distribution of local interstellar matter, toward the galactic longitude of 240° (Heiles, 1998) (Fig. 1).

The model results suggest that interaction of the shock of the galactic spiral density wave with the supercloud converted the gas of the central regions of the supercloud into a flattened disk, the precursor of Gould's belt, and gas in the outer supercloud into an expanding superring, the precursor of the local arm. The inclination of Gould's belt and the galactic longitude of its nodal line can be explained by the model.

In order to study the evolution of the configuration and kinematics of the local superring under the galactic force field and the frictional force, I assumed that the superring initially had a cylindrical form, with one axis perpendicular to the galactic plane and centered on the position that the supercloud's center had 100 m.y. ago. The superring was moving with the combination of the velocity of its gravitational center, the velocity of rotation about the gravity center (both conserved from the supercloud), and a velocity of radial expansion with respect to the ring center. The superring model provides a good fit to the configuration of the local arm and to the kinematics of the interstellar matter associated with the local arm, i.e., Lindblad's feature C/H (Lindblad, 1967) (Fig. 2).

RESULTS AND CONCLUSIONS

An interesting prediction of the model is that the Sun has been gravitationally bound to the supercloud, rotating in sense contrary to the supercloud's rotation. With the values obtained for the model's parameters, I calculated the Sun's orbit with respect to the supercloud's gravitational center for the past 100 m.y., the time in which the braking force acted on the supercloud (Fig. 3). A plausible explanation for the capture of the Sun by the supercloud is that while the Sun was passing through an extended concentration of gas, ca. 500 Ma, this gas was suddenly accelerated and organized into a supercloud, making the difference between the velocity of the center of mass of the supercloud and the Sun's velocity relative to the regional standard of rest lower than the escape velocity. This could explain why the Sun's current velocity of ~11 km/s with respect to the local standard of rest is significantly lower than the average of 60 km/s typical for similar G-type stars in the galaxy (Clube and Napier, 1982; Bailey, 1983).

SUPERCLOUD AND TERRESTRIAL CATASTROPHISM

The idea that the Sun may have been linked strongly to a gigantic interstellar cloud for a long time has interesting impli-

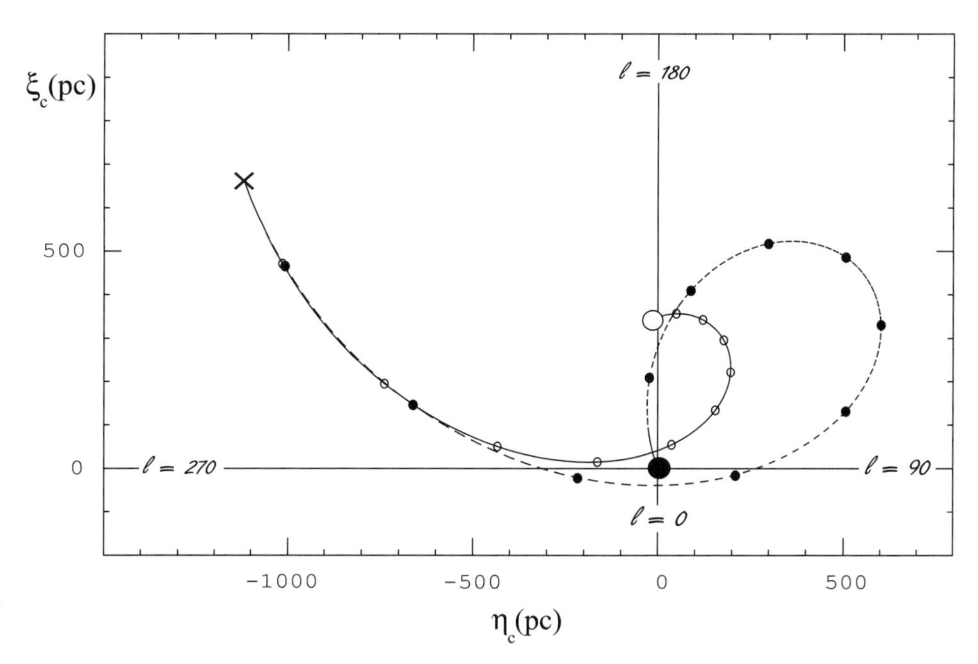

Figure 1. Epicycle orbits of supercloud's center (full line and open circles) and centroid of Sirius supercluster (broken line and filled circles). ξ_c axis points continuously toward anticenter (galactic longitude = 180°) and η_c axis in direction of galactic rotation (galactic longitude = 90°). Coordinate system (ξ_c, η_c) has its origin at local standard of rest. We use subscript c to differentiate this coordinate system from that of Figure 3. Small circles mark positions for each 10^7 yr. Larger circles indicate present positions, and cross indicates meeting point (100 Ma). Present position of Sirius supercluster is very close to Sun's position. Galactic longitudes 0°, 90°, 180°, and 270°, denoted by letter l, are indicated; pc is parsec.

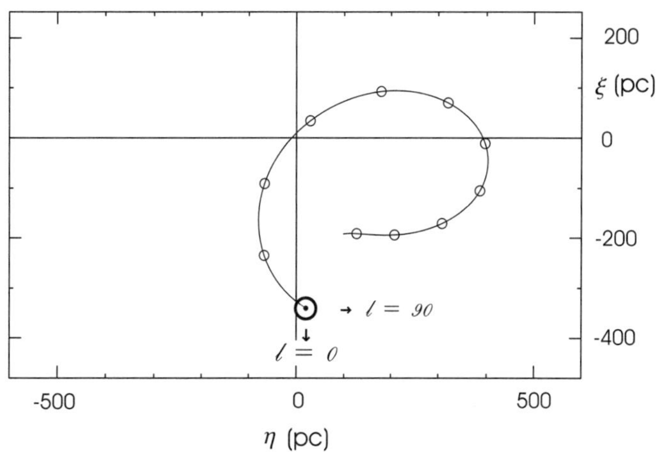

Figure 3. Epicyclic orbit of sun with respect to supercloud's center in past 100 m.y. Sun symbol indicates Sun's present position. Coordinate system (ξ, η) has same orientation as system (ξ_c, η_c) (see Fig. 1), but with origin at gravitational center of supercloud. Open circles mark Sun's positions each 10^7 yr. Galactic longitudes 0° and 90°, denoted by letter l, are indicated; pc is parsec.

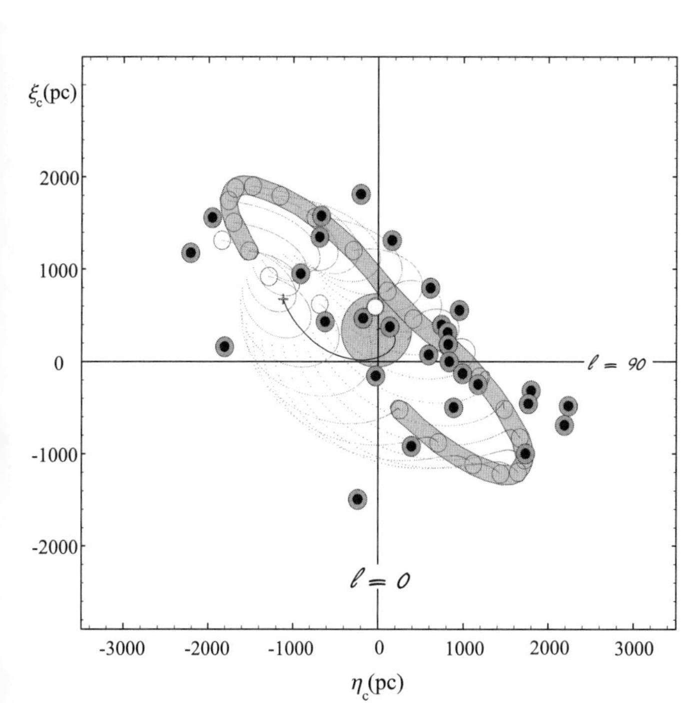

Figure 2. Evolution of local system during past 100 m.y. Dotted lines indicate trajectories of outer parts of supercloud (i.e., superring), and circles represent their present positions. Open circles represent back side of superring, and shaded stripe on circles outlines denser sector of superring. Full line shows trajectory of supercloud's nucleus. Its present position is sketched by large shaded circle. Locations of associations of stars of spectral type O-B2, projected onto galactic plane, are shown by shaded circles with central solid circles. Coordinate origin coincides with Sun's present position. Galactic longitudes 0° and 90°, denoted by letter l, are indicated; pc is parsec. See Figure 1.

cations for terrestrial impacts and mass extinctions (Clube, 1988; Napier, 1988; Bailey et al., 1990). The Oort cloud is the bridge between the solar system and the supercloud and would have been disturbed repeatedly during the past 500 m.y. by the Sun's encounters with substructures in the supercloud and the tidal forces caused by the galactic disk and the supercloud, increasing the flux of comets at the Earth (Clube and Napier, 1986).

The Sun and the supercloud would have oscillated about the galactic plane as two linked oscillators. The vertical oscillation period of the Sun with respect to the supercloud's midplane is given approximately by $[\pi/G(\rho + \rho_0)]^{1/2}$, where ρ and ρ_0 are the density of the supercloud and the background density of the galactic disk in the solar neighborhood, respectively. The estimate of ρ is ~0.165 solar masses/pc^3 and the estimate of ρ_0 is ~0.12 solar masses/pc^3 (Olano, 2001). Thus the Sun crosses the midplane of the supercloud every 26 m.y. The periodic passages of the Sun through the midplane of the supercloud, where the probability of an encounter with a dense substructure is higher, could explain the 26 m.y. periodicity found in the cratering and paleontological records (Raup and Sepkoski, 1986; Shoemaker and Wolfe, 1986) (Fig. 4). This is similar to the mechanism proposed by Rampino and Stothers (1984, 1986, 2000), except that the galactic plane is here replaced by the supercloud's plane. The amplitude of the Sun's oscillation in the direction perpendicular to the galactic plane, however, is not large enough compared with the probable thickness of the supercloud's disk to modulate the comet flux. However, according to our model, the times in which the Sun crossed the supercloud's midplane do not coincide with the dates of the main mass-extinction episodes (see Fig. 4). Although the period, P, of these crossing times is similar to that of the extinctions (~26 m.y.), there is a phase difference of 1/2P. Figure 4 shows the positions of the Sun at the times of episodes of extinction during the past 100 m.y.: middle Miocene (11 Ma), late Eocene (33 Ma), Maastrichtian (65 Ma), and Cenomanian (91 Ma) (Shoemaker and Wolfe, 1986, and references therein). Note that our model shows that the dates of these extinctions are in approximate correspondence with the times of the largest elongations of the Sun's vertical oscillations relative to the supercloud midplane. This could be explained by the fact that at these extreme positions the Oort cloud would have been exposed to stronger perturbations, because the component of the velocity of the Sun in direction perpendicular to the galactic plane is there small, and hence the duration of the encounters with substructures of the supercloud would be relatively long (Olano, 1993).

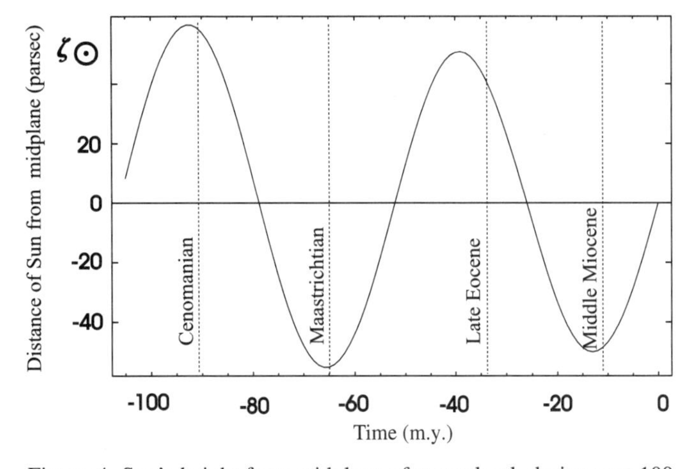

Figure 4. Sun's height from midplane of supercloud, during past 100 m.y. $\zeta\odot$ axis is perpendicular to galactic plane and points toward galactic north pole. Positions of Sun above and below midplane of supercloud at times of mass-extinction events (vertical shaded lines) are shown.

A comet shower produced by a star passing close to the Sun (10^4 astronomical units) can be great enough that several comets would hit the Earth during the shower (Hills, 1981). A penetrating encounter with an intermediate-size cloud (5×10^3 solar masses) can cause a temporary increase in the influx rate of Oort cloud comets comparable to that caused by a close stellar passage, if the Oort cloud is heavily concentrated (Fernández and Ip, 1991). A reasonable model for the structure of the supercloud might be to suppose it consists in clouds of 5×10^3 solar masses and 15 pc of radius. Therefore the supercloud, the total mass of which is estimated as 2×10^7 solar masses, comprises 4000 intermediate-size clouds. Representing the supercloud with a disk of 500 pc of radius and 200 pc of thickness, the cloud density is $\sim 3 \times 10^{-5}$ clouds/pc^3. With an impact parameter of 15 pc and a mean encounter velocity of 10 km/s, the Sun meets a cloud on average every 5 m.y. We argue that the 26 m.y. periodicity of comet showers is due to the fact that the stronger perturbations of the Oort cloud occur near the extremes of the Sun's vertical motion. This is a consequence of the fact that the magnitude of the impulsive perturbation on the Oort cloud is inversely proportional to the encounter velocity (Weissman, 1991), and the encounter velocity is minimal at the extremes of the Sun's vertical motion. That is, the catastrophic collisions did not occur at the center of the supercloud, but at the edges. I will present a detailed analysis of the perturbations at the Oort comet cloud caused by encounters with gas substructures of the proposed supercloud, and by the tidal forces of the supercloud and galactic disk, in the future.

Culler et al. (2000) discovered that the production rate of lunar craters increased dramatically in the past 400 m.y., reaching the level that it had ~3000 m.y. earlier (see also Muller, 2000). The theory of the capture of the Sun by a supercloud ca. 500 Ma could explain this observation. Furthermore, the supercloud theory might eventually provide a key for understanding the origin and evolution of the Oort comet cloud. It is interesting to note that the capture of the Sun by the supercloud was roughly coincident with the Cambrian explosion of complex life on Earth.

ACKNOWLEDGMENTS

I thank M.R. Rampino for his constructive criticism and helpful comments. This work was supported in part by the Consejo Nacional de Investigaciones Científicas y Técnicas (CONICET) under project PIP-0608/98.

REFERENCES CITED

Alvarez, L.W., Alvarez, W., Asaro, F., and Michel, H.V., 1980, Extraterrestrial cause for the Cretaceous-Tertiary extinction: Experimental results and theoretical interpretation: Science, v. 208, p. 1095–1108.

Alvarez, W., and Muller, R.A., 1984, Evidence from crater ages for periodic impacts on the Earth: Nature, v. 308, p. 718–720.
Bailey, M.E., 1983, The structure and evolution of the solar system comet cloud: Monthly Notices of the Royal Astronomical Society, v. 204, p. 603–633.
Bailey, M.E., Clube, S.V.M., and Napier, W.M., 1990, The origin of comets: New York, Pergamon Press, 577 p.
Bailey, M.E., Wilkinson, D.A., and Wolfendale, A.W., 1987, Can episodic comet showers explain the 30-Myr cyclicity in the terrestrial record?: Monthly Notices of the Royal Astronomical Society, v. 227, p. 863–885.
Clube, S.V.M., 1988, The catastrophic role of giant comets, *in* Clube, S.V.M., ed., Catastrophes and evolution: Cambridge University Press, p. 81–112.
Clube, S.V.M., and Napier, W.M., 1982, Spiral arms, comets and terrestrial catastrophism: The Quarterly Journal of the Royal Astronomical Society, v. 23, p. 45–46.
Clube, S.V.M., and Napier, W.M., 1986, Giant comets and the galaxy: Implications of the terrestrial record, *in* Smoluchowski, R., Bahcall, J.N., and Matthews, M.S., eds., The galaxy and the solar system: Tucson, University of Arizona Press, p. 260–285.
Culler, T.S., Becker, T.A., Muller, R.A., and Renne, P.R., 2000, Lunar impact history from $^{40}Ar/^{39}Ar$ dating of glass spherules: Science, v. 287, p. 1785–1788.
Davis, M., Hut, P., and Muller, R.A., 1984, Extinction of species by periodic comet showers: Nature, v. 308, p. 715–717.
Elmegreen, B.G., and Elmegreen, D.M., 1983, Regular string of HII regions and superclouds in spiral galaxies: Clues to the origin of cloudy structure: Monthly Notices of the Royal Astronomical Society, v. 203, p. 31–45.
Elmegreen, B.G., and Elmegreen, D.M., 1987, HI superclouds in the inner galaxy: Astrophysical Journal, v. 320, p. 182–198.
Fernández, J.A., and Ip, W.-H., 1991, Statistical and evolutionary aspects of cometary orbits, *in* Newburn, R.L., Jr., Neugebauer, M., and Rahe, J., eds., Comets in the post-Halley era: Dordrecht, Netherlands, Kluwer Academic Publishers, v. 1, p. 487–535.
Grieve, R.A.F., and Shoemaker, E.M., 1994, The record of past impacts on Earth, *in* Gehrels, T., ed., Hazards due to comets and asteroids: Tucson, University of Arizona Press, p. 417–462.
Heiles, C., 1998, Whence the local bubble, Gum, Orion? GSH 238 + 00 + 09, a nearby major superbubble toward galactic longitude 238°: Astrophysical Journal, v. 498, p. 689–703.
Hills, J.G., 1981, Comet showers and the steady-state infall of comets from the Oort cloud: Astronomical Journal, v. 86, p. 1730–1740.
Hoyle, F., and Wickramasinghe, C., 1978, Comets, ice ages and ecological catastrophes: Astrophysics and Space Science, v. 53, p. 523–526.
Jetsu, L., and Pelt, J., 2000, Spurious periods in the terrestrial impact crater record: Astronomy and Astrophysics, v. 353, p. 409–418.
Leitch, E.M., and Vasisht, G., 1998, Mass extinctions and the sun's encounters with spiral arms: New Astronomy, v. 3, p. 51–56.
Lindblad, P.O., 1967, 21-cm observations in the region of the galactic anticentre: Bulletin of the Astronomical Institute of the Netherlands, v. 19, p. 34–73.
McCrea, W.H., 1975, Ice ages and the galaxy: Nature, v. 255, p. 607–609.
Muller, R.A., 2000, A 3.5 billion year record of the solar system impacts, and the recent (0.4 Gyr) increase, *in* Catastrophic events and mass extinctions: Impacts and beyond: Houston, Texas, Lunar and Planetary Institute, LPI Contribution No. 1053 p. 146–147.
Napier, W.M., 1988, Terrestrial catastrophism and galactic cycles, *in* Clube, S.V.M., ed., Catastrophes and evolution: Cambridge, UK, Cambridge University Press, p. 133–167.
Olano, C.A., 1982, On a model of local gas related to Gould's belt: Astronomy and Astrophysics, v. 112, p. 195–208.
Olano, C.A., 1993, Hacia una teoría galáctica del catastrofismo terrestre: Boletín de la Asociación Argentina de Astronomía, v. 38, p. 11–24.
Olano, C.A., 2001, The origin of the local system of gas and stars: Astronomical Journal, v. 121, p. 295–308.
Palouš, J., 1986, The local velocity field in the last billion years, *in* Smoluchowski, R., Bahcall, J.N., and Matthews, M.S., eds., The galaxy and the solar system: Tucson, University of Arizona Press, p. 47–57.
Rampino, M.R., and Haggerty, B.M., 1994, Extraterrestrial impacts and mass extinctions of life, *in* Gehrels, T., ed., Hazards due to comets and asteroids: Tucson, University of Arizona Press, p. 827–857.
Rampino, M.R., and Stothers, R.B., 1984, Terrestrial mass extinctions, cometary impacts and the sun's motion perpendicular to the galactic plane: Nature, v. 308, p. 709–712.
Rampino, M.R., and Stothers, R.B., 1986, Geologic periodicities and the galaxy, *in* Smoluchowski, R., Bahcall, J.N., and Matthews, M.S., eds., The galaxy and the solar system: Tucson, University of Arizona Press, p. 241–259.
Rampino, M.R., and Stothers, R.B., 2000, Periodic comet showers, mass extinctions, and the galaxy, *in* Catastrophics events and mass extinctions: Impacts and beyond: Houston, Texas, Lunar and Planetary Institute, LPI Contribution No. 1053, p. 175.
Raup, D.M., and Sepkoski, J.J., Jr., 1986, Periodic extinction of families and genera: Science, v. 231, p. 833–836.
Shoemaker, E.M., and Wolfe, R.F., 1986, Mass extinctions, crater ages and comet showers, *in* Smoluchowski, R., Bahcall, J.N., and Matthews, M.S., eds., The galaxy and the solar system: Tucson, University of Arizona Press, p. 338–386.
Stothers, R.B., 1998, Galactic disc dark matter, terrestrial impact cratering and the law of large numbers: Monthly Notices of the Royal Astronomical Society, v. 300, p. 1098–1104.
Stothers, R.B., and Rampino, M.R., 1990, Periodicity in flood basalts, mass extinctions and impacts: A statistical view and a model, *in* Sharpton V.L., and Ward, P.D., Global catastrophes in Earth history: Geological Society of America Special Paper 247, p. 9–18.
Thaddeus, P., 1986, Molecular clouds and periodic events in the geologic past, *in* Smoluchowski, R., Bahcall, J.N., and Matthews, M.S., eds., The galaxy and the solar system: Tucson, University of Arizona Press, p. 61–68.
Toon, O.B., Zahnle, K., Turco, R.P., and Covey, C., 1994, Environmental perturbations caused by asteroid impacts, *in* Gehrels, T., ed., Hazards due to comets and asteroids: Tucson, University of Arizona Press, p. 791–826.
Tremaine, S., 1986, Is there evidence for a solar companion star?, *in* Smoluchowski, R., Bahcall, J.N., and Matthews, M.S., eds., The galaxy and the solar system: Tucson, University of Arizona Press, p. 409–416.
Urey, H.C., 1973, Cometary collisions and geological periods: Nature, v. 242, p. 32–33.
Van den Bergh, S., 1989, Life and death in the inner solar system: Publications of the Astronomical Society of the Pacific, v. 101, p. 500–509.
Van den Bergh, S., 1994, Astronomical catastrophes in Earth history: Publications of the Astronomical Society of the Pacific, v. 106, p. 689–695.
Weissman, P.R., 1991, Dynamical history of the Oort cloud, *in* Newburn, R.L., Jr., Neugebauer, M., and Rahe, J., eds., Comets in the post-Halley era: Dordrecht, Netherlands, Kluwer Academic Publishers, v. 1, p. 463–486.
Whitmire, D.P., and Jackson, A.A., 1984, Are periodic mass extinctions driven by a distant solar companion?: Nature, v. 308, p. 713–715.
Yabushita, S., and Allen, A.J., 1989, On the effect of accreted interstellar matter on the terrestrial environment: Monthly Notices of the Royal Astronomical Society, v. 238, p. 1465–1478.
Yabushita, S., 1992, Periodicity and decay of craters over the past 600 Myr: Earth, Moon, and Planets, v. 58, p. 57–63.
Zank, G.P., and Frisch, P.C., 1999, Consequences of a change in the galactic environment of the sun: Astrophysical Journal, v. 518, p. 965–973.

Manuscript Submitted October 5, 2000; Accepted by the Society March 22, 2001

Printed in the U.S.A.

Geological Society of America
Special Paper 356
2002

Grazing meteoroids could ignite continental-scale fires

Vladimir V. Svetsov
*Institute for Dynamics of Geospheres, Russian Academy of Sciences,
Leninskiy Prospekt 38-6, 117334, Moscow, Russia*

ABSTRACT

Meteoroids captured by the Earth into orbits with perigees larger than the Earth's radius can graze the atmosphere and travel a long distance, from one to several thousand kilometers, at low altitudes. The dynamics of large grazing asteroids, from 0.5 to 2 km in size, is studied using equations of motion in the gravitational field and pancake models for atmospheric drag and dissipation of energy. A numerical model is developed for calculation of radiation effects on the Earth's surface. It is found that large grazing meteoroids probably traveled a long distance over land several times during the Phanerozoic and could have caused continental-scale wildfires.

INTRODUCTION

Large impacts can cause various environmental stresses with dramatic effects on the Earth's biosphere (Gilmour et al., 1989; Toon et al., 1994; Rampino and Haggerty, 1994). The record of mass extinctions derived from paleontological studies shows that 5 major and ~20 smaller extinction pulses have happened during the past 540 m.y. (Sepkoski, 1995), but only some of these seem to be associated with geological and stratigraphic evidence of major impacts (Rampino et al., 1997).

The size and energy of impactors are commonly accepted as critical parameters controlling impact-induced mass extinctions (Morrison et al., 1994). The mass extinction at the Cretaceous-Tertiary (K-T) boundary was most likely caused by the 10–15 km impactor that produced the Chicxulub crater (Alvarez et al., 1980; Smit, 1994). Large amounts of soot discovered at the K-T boundary suggest that global fires developed after the impact (Wolbach et al., 1985, 1988, 1990; Anders et al., 1986; Vencatesan and Dahl, 1989; Gilmour and Anders, 1989). Such wildfires can give rise to severe environmental stresses, releasing large amounts of heat, soot, aerosols, CO_2, and other pyrotoxins that by far exceed those produced by the impactor.

The worldwide distribution of iridium at the K-T boundary indicates that the impact debris moved along ballistic trajectories away from the impact site (Argyle, 1989). Schultz and Gault (1982) were the first to argue for the importance of the reentry component in the global atmospheric heating and subsequent biospheric stresses. The more recent studies (Melosh et al., 1990; Zahnle, 1990) have shown that the thermal flux released from the reentering impact ejecta is likely to be the chief cause of the forest ignition over the Earth. However, the estimated thermal radiation flux at the Earth's surface is near the lowest limit required for ignition of wood (Melosh et al., 1990). This means that smaller impactors, 1–2 km in size, are not effective ignitors, because they produce an energy flux on the ground that is 2–3 orders of magnitude lower (Toon et al., 1994).

Vertical impact trajectories are not very effective in energy partition from the viewpoint of maximum effect on the environment. Much of impactor preentry kinetic energy is transferred to the planetary surface (O'Keefe and Ahrens, 1982); a smaller portion of the energy is converted to radiation of a fireball and kinetic energy of high-velocity reentering ejecta. However, a typical stony 1-km-diameter impactor has a kinetic energy of 10^{20} J, which is sufficient to set fires over the total surface of land covered by forests and grass.

Schultz and Gault (1982, 1990) suggested that hypervelocity ricochet after a large oblique impact, a downrange plume, a great number of smaller Tunguska-scale ricochet clusters, and orbital injection of debris could be the most damaging to the biosphere. Schultz and D'Hondt (1996) argued that the Chicxulub crater is likely to have been caused by a low-angle

Svetsov, V.V., 2002, Grazing meteoroids could ignite continental-scale fires, *in* Koeberl, C., and MacLeod, K.G., eds., Catastrophic Events and Mass Extinctions: Impacts and Beyond: Boulder, Colorado, Geological Society of America Special Paper 356, p. 685–694.

impactor. They found that thermal radiation from the downrange ballistic debris could be more than an order of magnitude higher than the average values estimated by Melosh et al. (1990) for a hemispherical ejecta vapor plume. However, in oblique impacts, most of the meteoroid energy can be lost in the atmosphere prior to the impact. In this chapter I focus on the problem of entry phenomena, fragmentation, energy release in the atmosphere, and radiation impulse caused by large meteoroids with shallow trajectories.

An impactor severely dispersed in the atmosphere can produce a strong effect on the environment. If the trajectory has a low inclination angle or goes beyond the solid Earth and grazes the atmosphere, the meteoroid intercepts more atmospheric mass and can lose a significant fraction of its energy under aerodynamic load. Its fragments can start motion along other Earth-bounded trajectories ending at distant points on the Earth surface. Meteoroids with grazing or low-angle trajectories can release their energy over significant distances (1000 km) comparable with the size of a continent. Therefore, grazing impactors could be very dangerous to the biosphere.

Despite the rarity of grazing events, it is likely that they have happened during geologic history. Small meteoroids, e.g., the daytime fireball of August 10, 1972, registered over the United States (Ceplecha, 1994), and the fireball of October 13, 1990, over Czechoslovakia and Poland (Borovicka and Ceplecha, 1992), grazed the atmosphere and returned to space with reduced speeds. The Rio Cuarto crater field in Argentina (Schultz and Lianza, 1992; Schultz et al., 1994) and several elongated craters discovered recently in South Australia (Haines et al., 1999) are signatures of larger impactors having landed at very low angles. No charcoal layer approaching the magnitude of that at the K-T boundary has been found elsewhere (Wolbach et al., 1990). However, carbon layers of smaller magnitude have been detected regionally and it is possible that continental-scale wildfires are the cause.

Grazing impacts of relatively small (2–40 m) high-strength bodies were studied by Hills and Goda (1997), who investigated the probability that such objects could be captured into elliptical orbits about the Earth. They took into account only meteoroid dynamics. Here large (0.5–2 km) grazers are considered as a possible cause of fires and environmental stresses. I use similar dynamical models and develop another for calculations of radiation energy at the Earth's surface.

SPECIAL FEATURES OF LARGE GRAZING IMPACTS

Grazers are only a small fraction of all meteoroids striking the Earth, but this fraction is not negligibly small. The following relation for a body moving around the Earth can readily be derived from the laws of angular momentum end energy conservation:

$$p^2V_\infty^2 = V_\infty^2R^2 + V_e^2\,RR_E, \tag{1}$$

where p is the collision parameter (a distance from the Earth's center to a rectilinear trajectory in the absence of gravitational attraction), R is the orbit perigee, $R_E = 6400$ km is the Earth's radius, V_∞ is the impactor's velocity at infinity, and $V_e = 11.2$ km/s is the Earth's escape velocity. Because the flux of impactors is homogeneous and proportional to p^2 far from the Earth, it is easy to calculate the probability of grazing impacts.

The probability that the closest approach distance of a trajectory to the Earth's surface is within an interval from 0 to ΔR is equal to

$$\frac{\Delta R}{R_E}\frac{1 + 2(V_\infty/V_e)^2}{1 + (V_\infty/V_e)^2}. \tag{2}$$

Assuming that the grazers are meteoroids with theoretical perigees from 0 to 60 km above the Earth's surface (in the absence of any atmospheric dissipation), one can calculate that they compose from 1% to 2% of the total number of impactors hitting the Earth. The relative proportion of grazers increases with impactor velocity, reaching 2% for long-period comets. In the range of the most probable impact velocities the relative amount of grazers varies only slightly about 1.5%.

The chance that a 10 km impactor approached the Earth along a grazing trajectory is very small for the recent geological past. Smaller, 1 km asteroids have typical time intervals between the impact of ~0.2 m.y. (Shoemaker, 1983; Chapman and Morrison, 1994), and, therefore, 25–50 such grazers have collided with the Earth during the Phanerozoic.

The atmospheric density ρ varies with height h in the vertical direction approximately as

$$\rho = \rho_0\exp(-h/H), \tag{3}$$

where H is the atmospheric scale height (7–8 km), and ρ_0 is the density at a certain altitude. It is easy to show by geometry that the atmospheric density varies with length x in the tangential direction (perpendicular to the vertical) as

$$\rho = \rho_0\exp(-x^2/2R_EH). \tag{4}$$

The scale length in this tangential direction (on which the atmospheric density varies e-fold) is equal to $2(2R_EH)^{1/2}$. This is ~600 km, i.e., ~100 times larger than H. A meteoroid strongly decelerates when it intercepts an atmospheric mass equal to its own mass. The intercepted mass is proportional to the atmospheric scale length, the meteoroid's cross-sectional area, and the atmospheric density at an altitude of deceleration. Therefore, a 1 km grazing impactor will significantly decelerate at the same altitudes where 10 m vertical impactors are atmospherically braked. These heights are typically above 20–30 km.

Cometary impactors can be considered as strengthless bodies when they enter the atmosphere. Large stony meteoroids are also weak. Their strength is much lower than that of meteorite samples, and, moreover, 1 km bodies are likely to be gravitationally bounded rubble piles rather than solid objects (Love and Ahrens, 1996; Melosh and Ryan, 1997). The low average densities measured for some asteroids strengthen the rubble-pile hypothesis. For example, the density of the asteroid Mathilde was found to be remarkably low, 1.3 g/cm^3 (Veverka et al., 1999). Other evidence supporting this hypothesis comes from the measured rotational rates of asteroids. The rotation period distribution abruptly truncates at the point where rubble piles would begin to fly apart (Harris, 1996).

Such strengthless or weak meteoroids are very likely to fragment into a great number of pieces in the atmosphere. A conglomerate of fragments enlarges its cross section under aerodynamic load and decelerates more effectively than a solid impactor. Several models have been suggested for calculating the lateral expansion (pancaking) of fragmented fluid-like meteoroids in the atmosphere (Zahnle, 1992; Hills and Goda, 1993; Chyba et al., 1993; Korycansky et al., 2000). These so-called pancake models, applicable to sufficiently large meteoroids, can be used for calculations of the motion and atmospheric dispersion of large grazing impactors.

If the size of an object moving horizontally in an exponential atmosphere is about or larger than the characteristic scale height H, hydrodynamic flow around the body differs from the flow around a smaller (or vertically moving) meteoroid because the shock wave accelerates upward in the direction of diminishing atmospheric density. The critical sizes of meteoroids can be estimated in the following way. The meteoroid decelerates and heats air in the shock wave. The amount of energy released per unit length of trajectory is (Bronshten, 1983)

$$E = \frac{1}{2} C_D \pi r^2 \rho_c V^2, \tag{5}$$

where C_D is the drag coefficient, r is the meteoroid radius, V is its velocity, and ρ_c is the atmospheric density at the flight height. In a rough approximation, the process of energy release and shock-wave propagation can be treated as a linear explosion.

A strong point explosion in an exponential atmosphere was considered by Zel'dovich and Raizer (1967). If the explosion energy is large enough, the upward propagating shock wave accelerates while the downward shock slows down. Some portion of the atmospheric gas can acquire high velocities, sufficient to escape from the Earth's gravitational field (Ahrens et al., 1989). Hydrodynamic simulations (Jones and Kodis, 1982) show that explosion geometry (a disk instead of a point) only slightly influences the flow developed in the atmosphere after a lapse of time. By analogy with a point explosion, the shock wave upward velocity U in a linear explosion can be estimated (Zel'dovich and Raizer, 1967) as

$$U = \lambda \left(\frac{E}{\rho_c}\right)^{1/2} \frac{\exp(z/2H)}{z}, \tag{6}$$

where z is the distance to the shock front, and λ is a numerical constant of the order of unity. Upward acceleration begins at a distance $z = 2H$ if the shock wave is strong, i.e., U far exceeds the speed of sound. This gives the following criterion of the flow with an accelerating shock wave:

$$M \frac{r}{2H} \gg 1, \tag{7}$$

where M is the Mach number.

The upward-propagating shock wave goes to infinity during a finite time, and at this point the flow field becomes similar to a jet expanding into vacuum. The time of the atmospheric rupture (Zel'dovich and Raizer, 1967) is

$$\tau = \int_0^\infty \frac{dz}{U} = \frac{16}{\lambda \pi^{1/2}} \frac{H^2}{rV}. \tag{8}$$

If $\tau = r/V$, the velocity field becomes directed upward just behind the body. The critical meteoroid radius can be estimated as

$$r_c = H \frac{4}{\lambda^{1/2} \pi^{1/4}}. \tag{9}$$

I have carried out three-dimensional hydrodynamic simulations for an object with a cubic shape moving horizontally in an exponential atmosphere. The Mach number was 60 (a typical value for meteors), the characteristic scale height was 8 km, and the body size was equal to 5, 10, and 15 km. The results of simulations are demonstrated in Figure 1. The shock wave around a 5 km body accelerates upward, but the volume of heated gas does not go far from the trajectory. Beginning from a size of 10 km, the major portion of the heated gas acquires significant upward velocities. A flow field in the cross section behind the 10 km body is shown in Figure 2. The velocity field is similar to that in the expanding vapor plume produced by a large vertical impactor on the Earth's surface. This suggests that, in principle, grazing meteoroids can set fires by two different mechanisms.

Wood on the ground can be ignited by thermal radiation emitted by the air heated in the shock wave. This mechanism dominates for grazing objects smaller than 5–10 km. The radiating gas is located in the wake around the trajectory and has sufficient time to emit a major portion of its internal energy.

Another mechanism of ignition can possibly be operative for objects larger than 5–10 km. These objects can be swarms of fragments from the disruption of initially smaller meteoroids. Hot air and vapor expand in the vacuum behind the object and then move along ballistic trajectories and return to the atmosphere. The air and condensed vapor can act similarly to the high-velocity reentering ejecta in the case of the Chicxulub impactor, heating the atmosphere at high altitudes over an area around the trajectory. This mechanism, however, turns out to be inefficient.

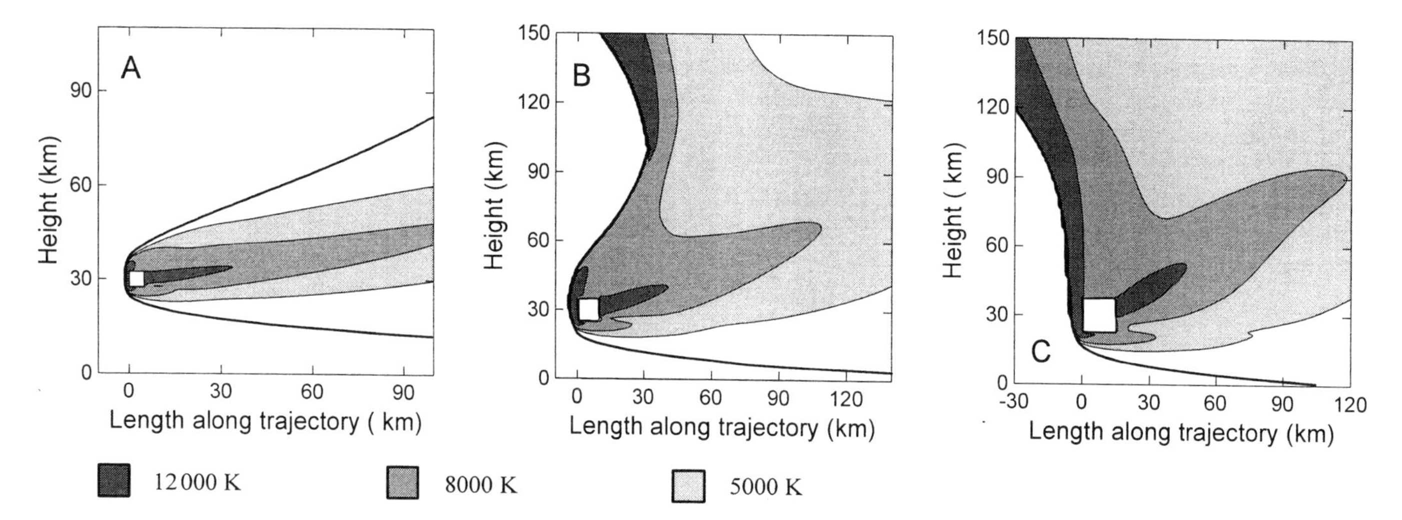

Figure 1. Meteoroids of cubic shape moving parallel to Earth's surface in exponential atmosphere at speed $V = 20$ km/s. Sizes of bodies are 5 km (A), 10 km (B), and 15 km (C). Shock wave and isolines of specific internal energy of air are shown in vertical plane through trajectory. Shaded areas correspond to following values of internal energy relative to stagnation value $V^2/2$: 0.05 (temperature above ~5000 K), 0.2 (temperature above ~8000 K), and 0.5 (temperature above ~12 000 K).

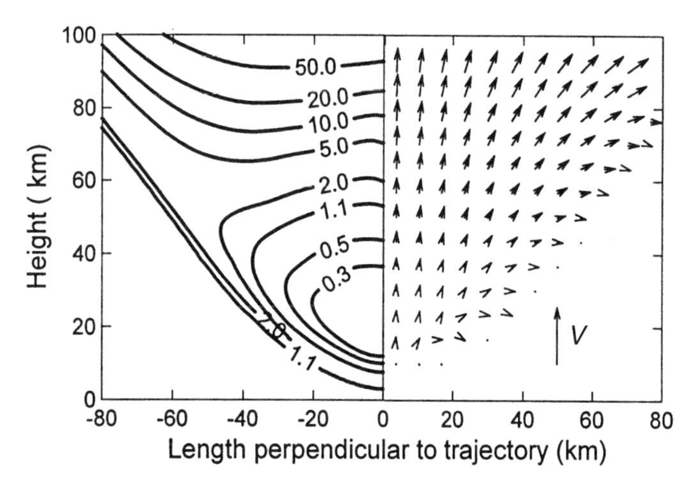

Figure 2. Isolines of air density relative to atmospheric density at given altitude (to left) and velocity field (to right) in vertical plane perpendicular to trajectory at distance of 130 km behind 10 km meteoroid shown in Figure 1 B; meteoroid's velocity V is shown by arrow as scale.

I have made calculations of thermal energy produced on the Earth's surface via the first mechanism and made some estimates for the second reentry mechanism.

NUMERICAL MODELS

The motion of a body in the Earth's gravitational field is governed by the well-known equations of mechanics. To specify the meteoroid's initial orbit one can choose two parameters: V_∞, the meteoroid velocity far from the Earth prior to gravitational acceleration, and h_0, the minimum approach of a meteoroid to the Earth's surface in the absence of the atmosphere (perigee minus Earth's radius). This is a hyperbolic trajectory; the meteoroid's positions and velocities in the absence of the atmosphere at any time instant are determined by simple universal formulas following from the equations of motion in a central gravitational field.

The atmosphere decelerates the impactor. The drag force acting on it in the direction tangential to the trajectory diminishes the meteoroid's velocity V according to the following drag equation (Bronshten, 1983),

$$m\,\frac{dV}{dl} = -\,\frac{1}{2}\,C_D A \rho V, \tag{10}$$

where l is the length along the trajectory, m is the meteoroid mass, C_D is the drag coefficient depending on the body's shape, A is the meteoroid cross-sectional area perpendicular to the trajectory, and ρ is the atmospheric density. The atmospheric drag and dispersion of fragments were taken into account in the calculations when the meteoroid height above the Earth's surface was smaller than 100 km. The atmosphere was assumed to be spherical with a tabulated density in the vertical direction.

The calculations of motion were made in the following way. The trajectory below 100 km altitude was divided to small segments. The choice of the segment lengths Δl is dictated by the necessary accuracy; Δl was chosen smaller where deceleration was greater. At every computational step, the meteoroid velocity is known at the beginning point of a segment. First, I compute the meteoroid velocity at the end point of a segment as it would be in the absence of any atmospheric drag, using the analytical formulas of mechanics. Then this velocity is diminished by an increment $\Delta V = -C_D A \rho V \Delta l / 2m$ following from drag equation 10. From the beginning of the next segment the body moves along another trajectory, which is defined by the new velocity. This splitting method for calculations of meteoroid dynamics provides exact solutions in the absence of atmospheric drag and gives reasonable accuracy for the motion with atmospheric deceleration.

The meteoroid in flight is treated as a strengthless, strongly fragmented, fluid-like object. The ram pressure on its leading surface causes a relatively slow movement of fragments in the direction perpendicular to the trajectory; the object gradually enlarges its cross-sectional area. This process is taken into account using a pancake model proposed by Hills and Goda (1993). According to this model, the meteoroid's radius r grows along the trajectory as

$$\frac{dr}{dl} = \left(\frac{\rho}{\rho_m}\right)^{1/2}, \quad (11)$$

where ρ_m is the meteoroid density. The meteoroid shape is assumed to be cylindrical, so that $m = 2\pi\rho_m r^3$.

Two modifications of this model have been made. First, the pancake model of Hills and Goda (1993) allows infinite enlargement of the meteoroid cross section, which will not happen in reality. Numerical simulations based on hydrocodes show that the object cannot be flattened as a genuine pancake (Svetsov et al., 1995; Crawford, 1997; Korycansky et al., 2000). Instead, it represents a swarm of fragments and vapor of restricted lateral size. Strengthless objects involving particles and vapor can also change their shape, diminishing the drag (Schultz, 1992). The reduction in drag coefficient can be formally treated as a reduction in cross-sectional area. Korycansky et al. (2000) heuristically restricted the pancaking rate for the case of a vertical impact terminated at the planetary surface. Here I restricted the maximum radius of a flying dispersed object, not allowing it to be larger than five preentry radii (r_0).

The maximum size of dispersed objects is large, 5 km for a 1 km intruder. It is comparable with the characteristic scale height, and therefore the atmospheric density and stagnation pressure change across the leading surface of the meteoroid. To take into account this factor, another modification of the model has been done in the following way. The lateral expansion of a fragmented meteoroid is allowed to be different at its edges. The lateral expansion in a given direction is assumed to be equal to $(\rho/\rho_m)^{1/2}$, as in equation 11, but here ρ is the atmospheric density at the edge of the flattened meteoroid and depends on the direction. The meteoroid is assumed to move as a whole, but the shape of its cross section is distorted, being elongated downward.

The initially circular leading surface is divided into some number of sectors. At every computational step along the trajectory, the length of each sector $r(\varphi)(0 \leq \varphi < \pi)$ is modified according to equation 11. Instead of ρA, I calculate and substitute to equation 10 an integral

$$\bar{\rho}\bar{A} = 2\int_0^{\pi} d\varphi \int_0^{r(\varphi)} \rho r' dr'. \quad (12)$$

To calculate the radiation energy on the ground I treat each segment of the trajectory as a point source emitting from a unit length the radiation energy

$$E(l) = \frac{1}{2}\alpha C_D \bar{\rho}\bar{A} V^2, \quad (13)$$

where α is the coefficient of conversion of the kinetic energy to radiation energy.

The radiation flux on the Earth surface is inversely proportional to a squared distance from the source and exponentially diminishes with visibility. The total radiation energy absorbed by a unit area on the Earth's surface is calculated as an integral along the trajectory

$$e = \int \frac{E\cos\Psi\exp(-\tau)}{4\pi S^2} dl, \quad (14)$$

where S is the distance from the unit area to the point on the trajectory, ψ is the angle between the normal to the surface and the direction to the trajectory point, and τ is the optical thickness along a ray. In calculations, the Earth's surface is covered by a grid. For every point of this computational grid the integration in equation 14 is made along a visible part of the trajectory for $\cos\psi > 0$. It was assumed that the unit surface area absorbing the radiation is oriented to the nearest point of the trajectory.

The optical thickness is computed as an integral along the line connecting the surface point with the radiation source

$$\tau = \int_0^s \frac{ds}{L}. \quad (15)$$

The visibility L is assumed to be infinite above 10 km and inversely proportional to atmospheric density below this altitude, being equal to some constant value L_0 at sea level.

After the absorbed thermal energy is calculated over the Earth's surface seen from the trajectory, the area of initial ignition can be determined assuming that dry forest materials are ignited when e is above 50 J/cm^2, the value defined from nuclear-weapon tests (Glasstone and Dolan, 1977).

The model has several free parameters that are not well determined and may be different for different impactors. Moreover, the process of meteoroid pancaking and deceleration may be to some extent different for identical impactors due to development of hydrodynamic instabilities on the surface of fragmented meteoroids (Korycansky et al., 2000). The purpose, however, is to predict probable radiation effects of typical grazing encounters and, therefore, average values for some of these parameters can be chosen. I have chosen $\rho_m = 2.2$ g/cm^3, a density of carbonaceous chondrites, and $C_D = 1.7$, an exact value of the drag coefficient for a cylinder. Because the fireball created by a 1 km moving meteoroid is comparable in size with those of powerful nuclear explosions, I have taken $\alpha = 0.5$, a reasonable approximation for such explosions (Glasstone and

Dolan, 1977; Svetsov, 1995). The best atmospheric visibility at sea level is 20–40 km. Clouds can entirely screen the ground. However, at least some meteoroids will fly over continents during fair weather, and I chose $L_0 = 20$ km to assess the effect produced by these more dangerous grazers.

The rate and degree of meteoroid pancaking are the most uncertain input parameters. To assess their influence on results I have carried out additional test calculations, taking ρ in equations 10 and 11 equal to its value at the height of the trajectory, i.e., assuming the rate of lateral expansion to be the same in all directions. I also varied the maximum radius of the swarm of fragments from 2.5 to 10 initial radii. Another test has been made with the pancake model of Chyba et al. (1993).

In the main series of calculations I varied the initial meteoroid diameter from 0.5 to 2 km and its velocity at infinity V_∞ from 5 to 25 km/s. Velocity $V_\infty = 15$ km/s corresponds to the mean impact velocity of asteroids.

RESULTS

Let us consider output dynamical characteristics of a 1 km grazing asteroid. Figure 3 shows the actual closest approach distance to the Earth's surface as a function of its theoretical closest approach, h_0. Below some critical h_0, the asteroid decelerates in the atmosphere and strikes the surface at a low speed. Above this h_0, the atmosphere only slightly influences the grazer and the actual closest approach distance is almost the same as the theoretical closest approach. The critical values of h_0 grow with diminishing V_∞. The meteoroid energy dissipated in the atmosphere is shown in Figure 4. Below the critical values of h_0 almost all the preentry kinetic energy is released in the air; these grazers do not create craters.

Figure 5 predicts the fate of a 1 km grazing asteroid as a function of h_0 and V_∞. In a zone above the upper curve, with high values of h_0 and V_∞, the meteoroid escapes the Earth's gravitational field with a diminished velocity. Below another curve the meteoroid strikes or, more likely, disintegrates and evaporates after a relatively short passage in the atmosphere. If the meteoroid parameters are between the curves, it makes from half a loop to several revolutions around the planet before full deceleration. In all the cases, the atmospheric drag ultimately leads to full deceleration of all the fragments. The zone with capture into bound orbits around the Earth narrows with increasing V_∞. Slow meteoroids have a better chance of making several orbits around the Earth.

A 1 km grazer has dynamics quite similar to that of the smaller, 2–20 m grazing meteoroids studied by Hills and Goda (1997). The main difference is in critical values of h_0, which are 10–20 km smaller for 1 km meteoroids. However, large

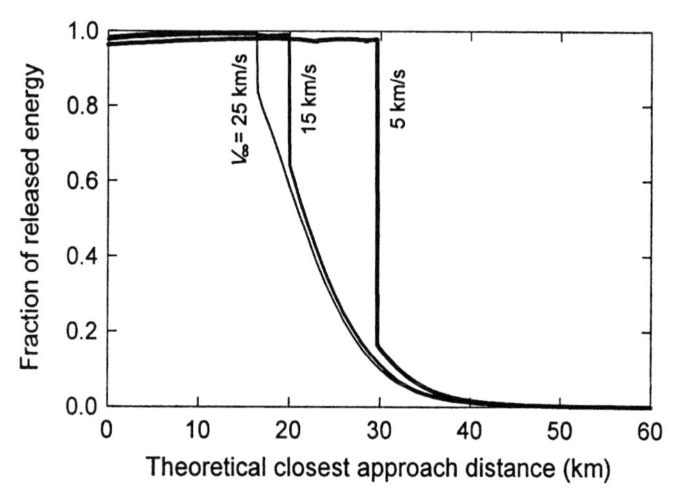

Figure 4. Fraction of preentry kinetic energy released in atmosphere by 1 km grazer as function of h_0 for three velocities V_∞.

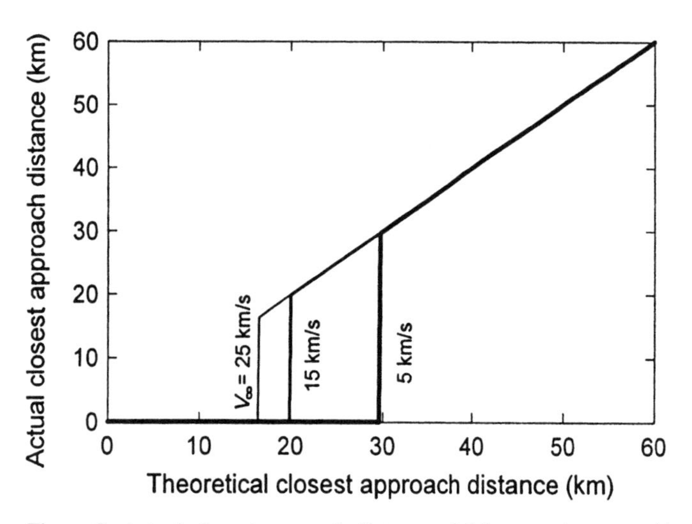

Figure 3. Actual closest approach distance of 1 km grazing asteroid decelerating in atmosphere as function of h_0, theoretical closest approach distance without any atmospheric drag. V_∞ is velocity at infinity.

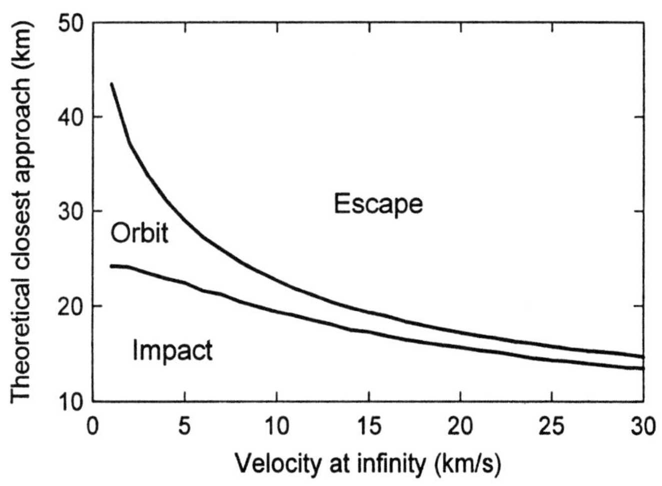

Figure 5. Three regions of meteoroid parameters where 1-km-diameter meteoroid either escapes Earth, strikes, or is captured to bound orbit and makes from half to several loops around Earth.

grazing meteoroids produce a significant radiation effect at the Earth's surface.

Figure 6 shows the length of an area that could be ignited by 0.5, 1, and 2 km grazers. This ignition length is close to the length of an arc traveled by the meteoroid in the atmosphere below 70 km. The maximum length of the ignition area can reach 5000–6000 km, approximately the Earth's radius. The meteoroids captured into elliptical orbits can set fires in two areas located as far as the Earth's diameter from each other; however, those are very rare events. The probable length of initial ignition is from 1000 to 3000 km for most grazers.

Figure 7 shows the area of ignition as a function of h_0 for several values of V_∞. The grazing meteoroids set fires along a surface strip with a typical width varying from ~100 to ~1000 km. The probable area of initial ignition is ~1% of the total land area. The fires, however, can develop and spread over a larger territory comparable in size with a continent. The largest area is ignited by objects persisting in orbit about the Earth, but to provide a global effect, asteroids or comets must be aimed at a very narrow gap of h_0, some kilometers in size.

The width of the ignition area is sensitive to visibility and weather, but the length depends on visibility only slightly due to the very high level of thermal energy delivered to the surface just below the trajectory. This thermal energy for 1 km grazers is shown in Figure 8 for several h_0. An estimate of thermal

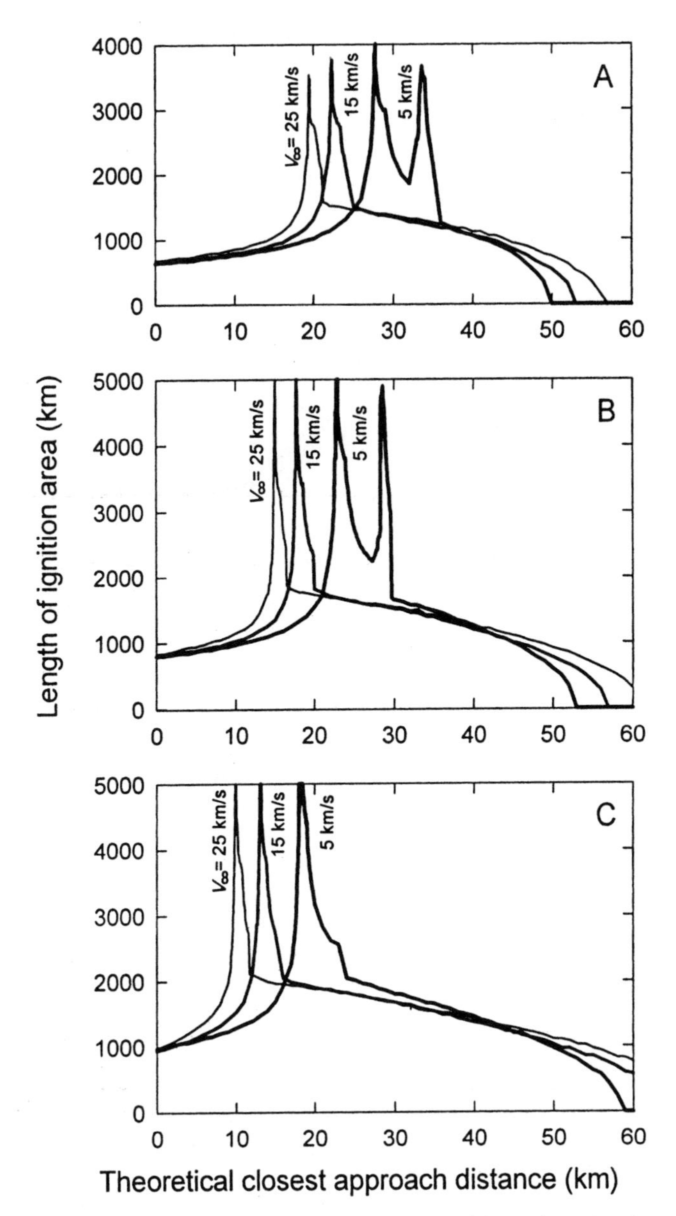

Figure 6. Length of ignited area as function of h_0 and V_∞ for three asteroid diameters: 0.5 km (A), 1 km (B), and 2 km (C).

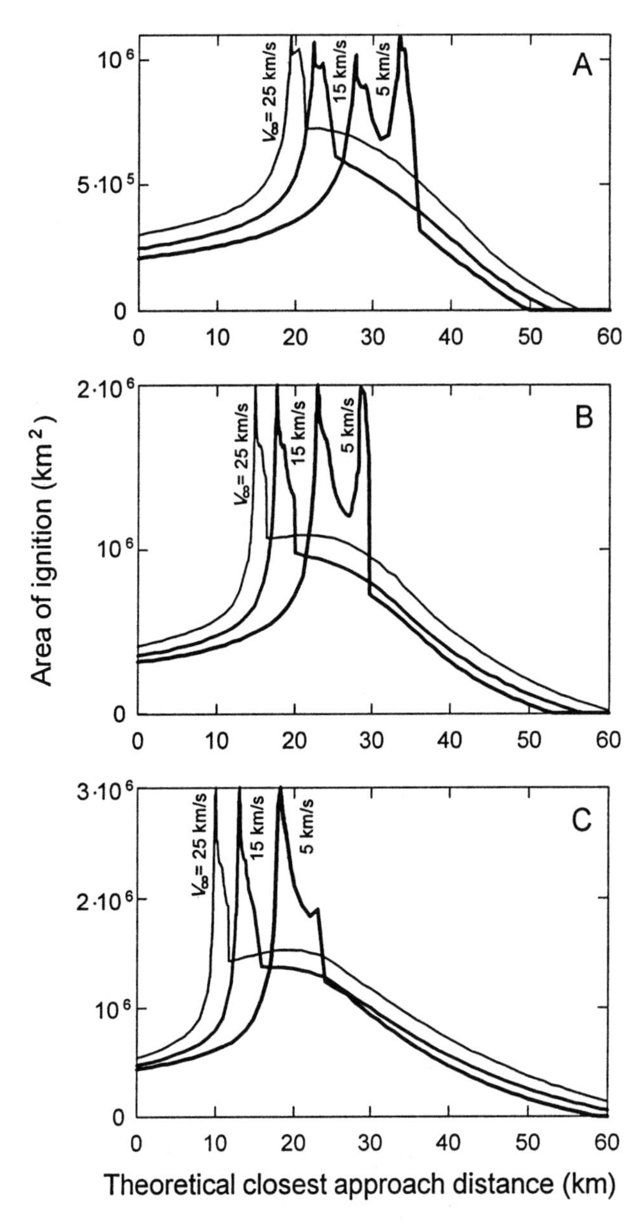

Figure 7. Area of forest ignition produced by meteoroids with diameters 0.5 km (A), 1 km (B), and 2 km (C). V_∞ is meteoroid's velocity at infinity.

energy required to fuse or vaporize soils or rocks can be obtained from the equation of thermal conduction. If a flat material surface is exposed to constant heat flux q, the temperature of the surface can be easily derived from the well-known solution of the heat-conduction equation as follows:

$$\lambda T = qbt^{1/2}. \quad (16)$$

Here λ is the heat conduction coefficient, and b^2 is the temperature conduction coefficient; $b^2 = \lambda\rho_m{}^{-1}c^{-1}$, where ρ_m is the material density and c is specific heat capacity. The typical values of these parameters for rocks are $\lambda = 2.5$ W/m K, $c = 1$ J/g K. For the case considered here, the time of heating is of the order of duration of radiation impulse from powerful nuclear explosions. For kilometer-sized fireballs this time will be ~10 s (Glasstone and Dolan, 1977).

Specific energy $e = qt$ required to heat the surface to temperature T can be estimated as

$$e = \frac{\lambda T t^{1/2}}{b}. \quad (17)$$

For a typical temperature of rock fusion 1800 K the critical value of e is ~1 kJ/cm^2. The energy required to vaporize rocks or soils is approximately twice this value. Thus the maximum energies shown in Figure 8 are well above the critical values required to melt and vaporize rocks. A thickness of a melted layer is limited by the energy absorbed by the surface and by the depth of heating by thermal conduction, which is equal to $2bt^{1/2}$ (Zel'dovich and Raizer, 1967). This thickness is smaller than 1 cm, and rocks bearing marks of such fusion will thereby differ from impact glasses or fulgurites.

The computational tests have shown that most of the input parameters do not significantly influence the main results. If I change the meteoroid density, drag coefficient, or the model of pancaking, approximately the same length and width of the ignition band are obtained at some other values of h_0 or V_∞. The results are most sensitive to restrictions on the maximum meteoroid radius r_{max}. Figure 9 demonstrates the length of ignition area for $r_{max}/r_0 = 2.5$ and $r_{max}/r_0 = 10$. In comparison to the case of restriction by 5 initial radii, demonstrated in the preceding figures, the output values are shifted to other h_0. Thus, the probable values of ignition bands obtained here seem to be determined with reasonable accuracy.

Let us estimate the radiation effect from a 2-km-diameter grazing asteroid that can be produced by the second mechanism, via reentry of condensed material. Evidently, this effect is more probable for soft-stone or cometary impactors entering the atmosphere at high speeds. A stripped mass m_v of meteoroid vapor is roughly equal to the encountered atmospheric mass (Crawford, 1997):

$$\frac{dm_v}{dl} = \frac{1}{2}\pi r^2 \rho. \quad (18)$$

The results obtained by Melosh et al. (1990) for the impact at the K-T boundary suggest that a significant radiation effect can occur if the reentering mass deposition is ~1 g/cm^2. (As mentioned, reentry of low-angle hypervelocity debris yields a greater effect [Schultz and D'Hondt, 1996], but the flight along a grazing trajectory is closer to the case of expanding cloud considered by Melosh et al. [1990].) Assuming that a swarm of debris from a 2 km meteoroid has a diameter of 10 km and flies at an altitude of 15 km where the atmospheric density is 10^{-4} g/cm^3, one can calculate that the 1 g/cm^2 mass flux could be deposited in the atmosphere if the vapor mass m_v were uniformly distributed over the band with a width of 400 km. For the angle of the expanding vapor jet equal to 45° (Fig. 2), this can be achieved only if the velocity of the material is low, ~1–2 km/s. However, the reentering material cannot produce a significant radiation effect at such low speeds. On the other hand,

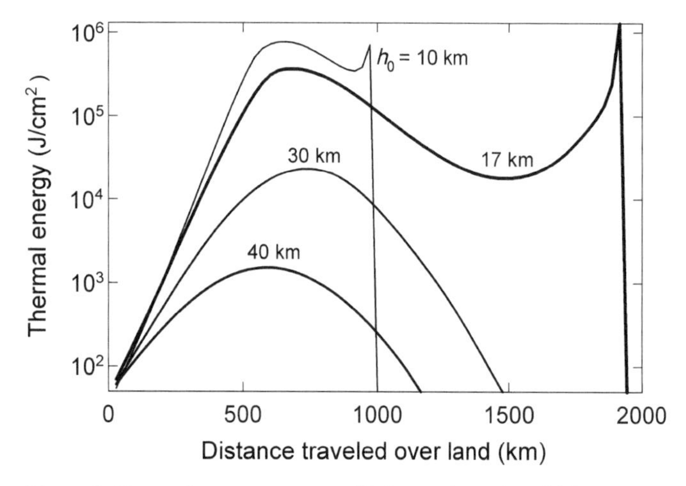

Figure 8. Thermal energy on ground below trajectories of 1 km grazers with velocity $V_\infty = 15$ km/s as function of distance traveled over Earth's surface for several values of h_0.

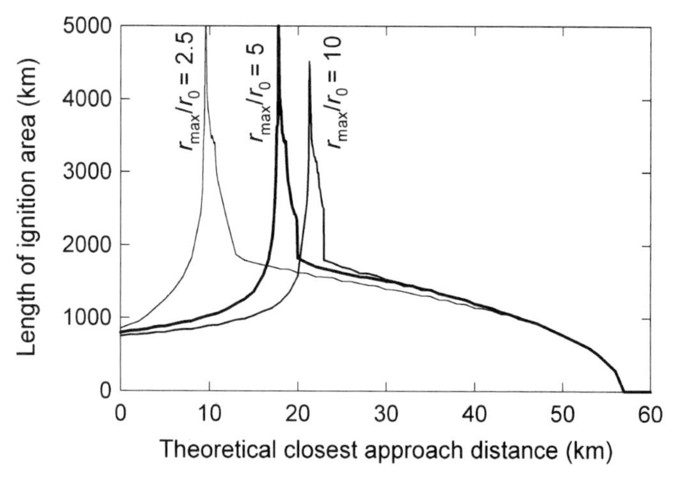

Figure 9. Length of ignited area for 1 km asteroid with velocity $V_\infty = 15$ km/s obtained for assumptions that ratio of object's maximum radius (r_{max}) to its initial radius (r_0) cannot exceed 2.5, 5, and 10.

at higher speeds the condensed material will be distributed over a very large area and the radiation flux will be too low. Thus, the second mechanism of ignition is not effective, although it can enhance, to some extent, the radiation flux in the vicinity of the trajectory due to the central part of a jet directed upward and re-entering close to the trajectory.

CONCLUSIONS

The estimates show that during the Phanerozoic several grazing meteoroids, $\sim$1 km in size, could very likely have flown over land during good weather and ignited forests or other flammable materials over areas from 10^6 to 2×10^6 km^2. The length of the ignition band could be from 1000 to 3000 km. The area of initial ignition could further develop into continental-scale wildfires. Such probable events should be taken into account in assessments of the causes of regional or global catastrophes.

Soft landing of meteoroid remnants could happen during a grazing impact event, but full meteoroid ablation, similar to the case of the Tunguska bolide, seems more probable due to the high degree of dispersion and intense radiation. The absence of craters and meteorites hinders the identification of such grazing impacts. However, traces of melted rocks on the Earth's surface in a long and narrow strip under the trajectory could be discovered. It is possible that such traces could be also left by low-angle impacts, e.g., in the area of the Rio Cuarto crater field (Schultz and Lianza, 1992; Schultz et al., 1994).

ACKNOWLEDGMENTS

I thank P.W. Haines and P.H. Schultz for helpful comments and reviews of the manuscript.

REFERENCES CITED

Ahrens, T.J., O'Keefe, J.D., and Lange, M.A., 1989, Formation of atmospheres during accretion of the terrestrial planets, *in* Atreya, S.K., Pollack, J.B., and Matthews, M.S., eds., Origin and evolution of planetary and satellite atmospheres: Tucson, University of Arizona Press, p. 328–385.

Alvarez, L.W., Alvarez, W., Asaro, F., and Michel, H.V., 1980, Extraterrestrial cause for the Cretaceous-Tertiary extinction: Science, v. 208, p. 1095–1108.

Anders, E., Wolbach, W.S., and Lewis, R.C., 1986, Cretaceous extinctions and wildfires: Science, v. 234, p. 261–264.

Argyle, E., 1989, The global fallout signature of the K-T bolide impact: Icarus, v. 77, p. 220–222.

Borovicka, J., and Ceplecha, Z., 1992, Earth-grazing fireball of October 13, 1990: Astronomy and Astrophysics, v. 257, p. 323–328.

Bronshten, V.A., 1983, Physics of meteoric phenomena: Dordrecht, The Netherlands, Reidel, 356 p.

Ceplecha, Z., 1994, Earth-grazing daylight fireball of August 10, 1972: Astronomy and Astrophysics, v. 283, p. 287–288.

Chapman, C.R., and Morrison, D., 1994, Impacts on the earth by asteroids and comets: Assessing the hazard: Nature, v. 367, p. 33–40.

Chyba, C.F., Thomas, P.J., and Zahnle, K.J., 1993, The 1908 Tunguska explosion: Atmospheric disruption of a stony asteroid: Nature, v. 361, p. 40–44.

Crawford, D.A., 1997, Comet Shoemaker-Levy 9 fragment size estimates: How big was the parent body?, *in* Remo, J.L., ed., Near-Earth objects: Annals of the New York Academy of Sciences, v. 822, p. 155–173.

Gilmour, I., and Anders, E., 1989, Cretaceous-Tertiary boundary event: Evidence for a short time scale: Geochimica et Cosmochimica Acta, v. 53, p. 503–511.

Gilmour, I., Wolbach, W.S., and Anders, E., 1989, Major wildfires at the Cretaceous-Tertiary boundary, *in* Clube, S.V.M., ed., Catastrophes and evolution: Astronomical foundations: Cambridge, Cambridge University Press, p. 195–213.

Glasstone, S., and Dolan, P.J., 1977, Effects of nuclear weapons: Washington, D.C., U.S. Department of Defense and Energy, 653 p.

Haines, P.W., Therriault, A.M., and Kelley, S.P., 1999, Evidence for Mid-Cenozoic(?), low-angle multiple impacts in South Australia: Meteoritics and Planetary Science, v. 34, p. A49–A50.

Harris, A.W., 1996, The rotation rates of very small asteroids: Evidence for "rubble pile" structure: Lunar and Planetary Science Conference, 27th, Houston, Texas, Lunar and Planetary Institute, p. 493–494.

Hills, J.G., and Goda, M.P., 1993, The fragmentation of small asteroids in the atmosphere: Astronomical Journal, v. 105, p. 1114–1144.

Hills, J.G., and Goda, M.P., 1997, Meteoroids captured into Earth orbit by grazing atmospheric encounters: Planetary and Space Science, v. 45, p. 595–602.

Jones, E.M., and Kodis, J.W., 1982, Atmospheric effects of large body impacts: The first few minutes, *in* Silver, L.T., and Schultz, P.H., eds., Geological implications of impacts of large asteroids and comets on the earth: Geological Society of America Special Paper 190, p. 175–186.

Korycansky, D.G., Zahnle, K.J., and Mac Low, M.-M., 2000, High-resolution calculations of asteroid impacts into the Venusian atmosphere: Icarus, v. 146, p. 387–403.

Love, S.G., and Ahrens, T.J., 1996 Catastrophic impacts on gravity dominated asteroids: Icarus, v. 124, p. 141–155.

Melosh, H.J., and Ryan, E.V., 1997, Asteroids: Shattered but not dispersed: Icarus, v. 129, p. 562–564.

Melosh, H.J., Schneider, N.M., Zahnle, K.J., and Latham, D., 1990, Ignition of global wildfires at the Cretaceous/Tertiary boundary: Nature, v. 343, p. 251–254.

Morrison, D., Chapman, C.R., and Slovic, P., 1994, The impact hazard, *in* Gehrels, T., ed., Hazards due to comets and asteroids: Tucson, University of Arizona Press, p. 59–91.

O'Keefe, J.D., and Ahrens, T.J., 1982, Cometary and meteorite swarm impact on planetary surfaces: Journal of Geophysical Research, v. 87, p. 6668–6680.

Rampino, M.R., and Haggerty, B.M., 1994, Extraterrestrial impacts and mass extinctions of life, *in* Gehrels, T., ed., Hazards due to comets and asteroids: Tucson, University of Arizona Press, p. 827–857.

Rampino, M.R., Haggerty, B.M., and Pagano, T.C., 1997, A unified theory of impact crises and mass extinctions: Quantitative tests, *in* Remo, J.L., ed., Near-Earth objects: Annals of the New York Academy of Sciences, v. 822, p. 403–431.

Schultz, P.H., 1992, Atmospheric effects on ejecta emplacement and crater formation on Venus from Magellan: Journal of Geophysical Research, v. 97, p. 16183–16248.

Schultz, P.H., and D'Hondt, S., 1996, Cretaceous-Tertiary (Chicxulub) impact angle and its consequences: Geology, v. 24, p. 963–967.

Schultz, P.H., and Gault, D.E., 1982, Impact ejecta dynamics in an atmosphere: Experimental results and extrapolations, *in* Silver, L.T., and Schultz, P.H., eds., Geological implications of impacts of large asteroids and comets on the earth: Geological Society of America Special Paper 190, p. 153–174.

Schultz, P.H., and Gault, D.E., 1990, Prolonged global catastrophes from oblique impacts, *in* Sharpton, V.L., and Ward, P.E, eds., Global catastrophes in Earth history: Geological Society of America Special Paper 247, p. 239–261.

Schultz, P.H., and Lianza, R.E., 1992, Recent grazing impacts on the earth recorded in the Rio Cuarto crater field, Argentina: Nature, v. 355, p. 234–237.

Schultz, P.H., Koeberl, C., Bunch, T., Grant, J., and Collins, W., 1994, Ground truth for oblique impact processes: New insight from the Rio Cuarto, Argentina, crater field: Geology, v. 22, p. 889–892.

Sepkoski, J.J., Jr., 1995, Patterns of Phanerozoic extinction: A perspective from global data bases, *in* Walliser, O.H., ed., Global events and event stratigraphy in the Phanerozoic: Berlin, Springer, p. 35–51.

Shoemaker, E.M., 1983, Asteroid and comet bombardment of the earth: Annual Reviews of Earth and Planetary Sciences, v. 11, p. 461–494.

Smit, J., 1994, Extinctions at the Cretaceous-Tertiary boundary: The link to the Chicxulub impact, *in* Gehrels, T., ed., Hazards due to comets and asteroids: Tucson, University of Arizona Press, p. 859–878.

Svetsov, V.V., 1995, Explosions in the lower and middle atmosphere: The spherically symmetrical stage: Combustion, Explosion and Shock Waves, v. 30, p. 696–707.

Svetsov, V.V., Nemtchinov, I.V., and Teterev, A.V., 1995, Disintegration of large meteoroids in Earth's atmosphere: Theoretical models: Icarus, v. 116, p. 131–153, Errata, v. 120, p. 443.

Toon, O.B., Zahnle, K., Turco, R.P., and Covey, C., 1994, Environmental perturbations caused by asteroid impacts, *in* Gehrels, T., ed., Hazards due to comets and asteroids: Tucson, University of Arizona Press, p. 791–826.

Vencatesan, M.I., and Dahl, J., 1989, Organic geochemical evidence for global fires at the Cretaceous/Tertiary boundary: Nature, v. 338, p. 57–60.

Veverka, J., Thomas, P., Harch, A., Clark, B., Bell, J.F., III, Carcich, B., Joseph, J., Murchie, S., Izenberg, N., Chapman, C., Merline, W., Malin, M., McFadden, L., and Robinson, M., 1999, NEAR encounter with asteroid 253 Mathilde: Overview: Icarus, v. 140, p. 3–16.

Wolbach, W.S., Lewis, R., and Anders, E., 1985, Cretaceous extinctions: Evidence for wildfires and search for meteoritic material: Science, v. 230, p. 167–170.

Wolbach, W.S., Gilmour, I., Anders, E., Orth, C.J., and Brooks, R.R., 1988, Global fire at the Cretaceous-Tertiary boundary: Nature, v. 334, p. 665–669.

Wolbach, W.S., Gilmour, I., and Anders, E., 1990, Major wildfires at the Cretaceous/Tertiary boundary, *in* Sharpton, V.L., and Ward, P.E., eds., Global catastrophes in Earth history: Geological Society of America Special Paper 247, p. 391–400.

Zahnle, K.J., 1990, Atmospheric chemistry by large impacts, *in* Sharpton, V.L., and Ward, P.E., eds., Global catastrophes in Earth history: Geological Society of America Special Paper 247, p. 271–288.

Zahnle, K.J., 1992, Airburst origin of dark shadows on Venus: Journal of Geophysical Research, v. 97, p. 10243–10255.

Zel'dovich, Ya.B., and Raizer, Yu.P., 1967, Physics of shock waves and high-temperature hydrodynamic phenomena: New York, Academic Press, 916 p.

Manuscript Submitted October 5, 2000; Accepted by the Society March 22, 2001

Printed in the U.S.A.

Geological Society of America
Special Paper 356
2002

Atmospheric erosion and radiation impulse induced by impacts

V.V. Shuvalov*
N.A. Artemieva
Institute for Dynamics of Geospheres, Leninsky Prospekt, 38, Building 6, Moscow 119334, Russia

ABSTRACT

Two effects of Chicxulub-scale impacts are considered: radiation impulse and impact-induced atmospheric erosion. Detailed numerical simulations show that radiation of the Chicxulub impact plume could be responsible for igniting wildfires over ~3%–10% of the Earth's surface, close to the impact point. Atmospheric erosion is calculated based on detailed numerical modeling of all stages of the impact, including the flight through the atmosphere, cratering, and plume evolution. The mass of escaping air is found to be smaller than predicted by previous investigations. Preliminary results of three-dimensional numerical simulations show that oblique impacts are more effective from the viewpoint of atmospheric erosion.

INTRODUCTION

Impacts of large cosmic bodies played a great role in the evolution of the Earth (Melosh, 1989). An individual impact could cause local and global changes in the atmosphere, hydrosphere, and solid Earth, could influence the Earth's biosphere, and could even lead to mass extinctions (Silver and Schultz, 1982; Sharpton and Ward, 1990; Toon et al., 1997). The time scale of these effects is $\sim10^3$ yr or less; they may be referred to as short term in the history of Earth. Impacts as a whole controlled the formation and evolution of the solid Earth, its hydrosphere, and atmosphere during the period of heavy bombardment 4.5–3.5 b.y. before present and possibly later. These are long-term effects (~1 m.y.). In this chapter we consider two effects resulting from large impacts: the radiation impulse (short term) and atmospheric erosion (long term).

Light impulse and following global wildfires resulting from the Chicxulub-scale impact are considered as a probable reason for global changes in the biosphere (Toon et al., 1997). Radiation is emitted both by the entering impactor and by the hot air-vapor cloud, or plume, expanding through the Earth's atmosphere just after the impact. Moreover, Schultz and Gault (1982, 1990) and Melosh et al. (1990) proposed that global wildfires would be caused by the radiation generated due to ejecta reentry.

The impacts of large cosmic bodies (>~100 m) could substantially influence the evolution of planetary atmospheres (Lange and Ahrens, 1982; Matsui and Abe, 1986; Ahrens et al., 1989; Melosh and Vickery, 1989; Vickery and Melosh, 1990; Zahnle et al., 1992). The growth of the atmospheric mass was defined by the release of volatiles in the impact process. At the same time some portion of atmospheric gas could be accelerated by the shock wave resulting from the expansion of vapor plume and escape from the Earth. The influence of a single impact can be negligible, but a great number of impacts in the history of the Earth are believed to have affected the evolution of the atmosphere.

In order to study radiation effects and atmospheric erosion we performed detailed numerical simulations of the main stages of large impacts, including the flight through the atmosphere, cratering, plume expansion, and shock-wave propagation through the ambient air. The SOVA (solid air vapor) multi-

*E-mail: shuvalov@idg.chph.ras.ru

Shuvalov, V.V., and Artemieva, N.A., 2002, Atmospheric erosion and radiation impulse induced by impacts, *in* Koeberl, C., and MacLeod, K.G., eds., Catastrophic Events and Mass Extinctions: Impacts and Beyond: Boulder, Colorado, Geological Society of America Special Paper 356, p. 695–703.

material hydrocode (Shuvalov, 1999a) is used to model the impacts of 10–30-km-diameter cosmic bodies (both comets and asteroids) with velocities ranging from 20 to 50 km/s. SOVA is an Eulerian material response code with some Lagrangian features. It allows a consideration of strong hydrodynamic flows with accurate description of the boundaries between different materials (e.g., soil, gas, water). The code is similar in concept to the CTH hydrocode (McGlaun et al., 1990), which is widely used in the United States. ANEOS (analytical equation of state, Thompson and Lauson, 1972) and Tillotson (1962) equations of state are used to describe the thermodynamical properties of water and solids. Detailed tables of thermodynamical and optical properties of air (Kuznetsov, 1965) and H-chondrite vapor (Kosarev et al., 1996) are applied to calculate the radiation fluxes.

The computational grid consists of 300 × 1000 cells in vertical and horizontal directions. The space step is 250 m for the 10 km impactor and 750 m for the 30 km impactor (20 cells across projectile radius) for the central 200 × 500 cell region around the point of impact and increases progressively far from the center.

The simulations do not include material strength; therefore, the stress tensor is considered to be spherical. This, however, does not affect the results of the simulations because the strength of the target becomes important only at the late stage of the crater formation (Melosh and Ivanov, 1999). During that stage a nonevaporated cold target material is ejected at a low (<1 km/s) velocity, which does not contribute to the plume radiation and atmospheric erosion.

RADIATION EFFECTS

The first estimates of the radiation produced in impact events were based on an analogy between impacts and nuclear explosions (Nemtchinov and Svetsov, 1991). The difference between a meteor and nuclear-like explosion was emphasized by Schultz and Gault (1982) and Jones and Kodis (1982). Recent investigations (Boslough and Crawford, 1997; Shuvalov, 1999b) show that a great difference between an atmospheric meteor and a nuclear-like explosion can appear due to the wake created during impactor flight through the atmosphere. The influence of the wake was underestimated in the previous works.

In this study the temperature and density distributions obtained in the course of hydrodynamical numerical impact simulations are used to calculate radiation intensities and radiation fluxes on the ground surface at different distances from the impact site at different moments of time. We consider a 10-km-diameter stony asteroid impacting at the velocity of 20 km/s against a granite target (Alvarez et al., 1980). The radiation transfer does not considerably influence the impact hydrodynamics itself and is ignored in the calculations. A more detailed description of the numerical procedure can be found in Shuvalov (2001).

Figure 1 shows the view of the impact plume from the side by displaying the intensity shading of the brightness, expressed in terms of effective temperature T_{eff}. The value of T_{eff} is defined from the relation $\sigma T_{eff}^4 = I$, where σ is the Stefan-Boltzmann constant, $I = \int_0^\infty I_{\varepsilon} d\varepsilon$ is the total radiation intensity, and I_ε is the spectral radiation intensity along the horizontal line crossing the plume. The source image is obtained without taking into account the absorption of light by the cold undisturbed atmosphere. This view is a characteristic of the light source, and does not depend on the distance from the observation point. The effective temperature is low at low altitudes (below 30 km), because the dense cold material ejected from the boundary of the growing cavity screens this part of the plume.

The value of T_{eff} varies in a narrow range from ~1.5×10^3 K to 2.5×10^3 K. This temperature is close to the temperature of condensation. Earlier, within first several seconds after the collision and during the impactor flight through the atmosphere, the fireball is much brighter (T_{eff} ~$20–30 \times 10^3$ K), but it is very small. The duration of this bright phase is very short compared to the total emission time. Therefore, most of the radiation is emitted at a later stage, 1–5 min after the collision. Almost all the radiation is emitted by the impactor and ground vapor and/or two-phase vapor-solid mixture. The radiation of the shock-compressed air is negligible in comparison with the total emitted radiation, although the brightest initial peak results mainly from the air emission.

The average real plume temperature (not effective radiation temperature) remains almost constant (close to the temperature of vapor-liquid transition) for a long time due to the energy release resulting from continuing condensation. The upper boundary of the radiating volume is controlled by adiabatic and radiative cooling of the vapor, and by the decrease of its density and optical thickness.

It follows from the numerical simulations that a rough estimate of the radiation impulse can be obtained by assuming that the plume temperature equals ~2×10^3 K, its characteristic scale is ~1000 km, and the height of the upper boundary is 1200 km. The time scale for light emission is 5–10 min.

The radiation intensity at the Earth's surface depends on several factors, such as geometrical divergence, curvature of the Earth's surface and its topography, absorption by the ambient (cold) atmospheric gas, and absorption and scattering by aerosol particles suspended in the air. Among these factors the most undefined is aerosol absorption and scattering. Typical estimates of the radiation-free path L_r in the atmosphere are ~40 km for very good sunny days and 10 km for rainy conditions (Glasstone and Dolan, 1977). Figure 2 shows the time-integrated incident radiation energy per unit surface versus distance from the impact site. In this calculation it is assumed that the surface being radiated is optimally oriented (from the viewpoint of maximum radiation). A critical value of the incident radiation energy for ignition is estimated to be ~20–100 J/cm^2

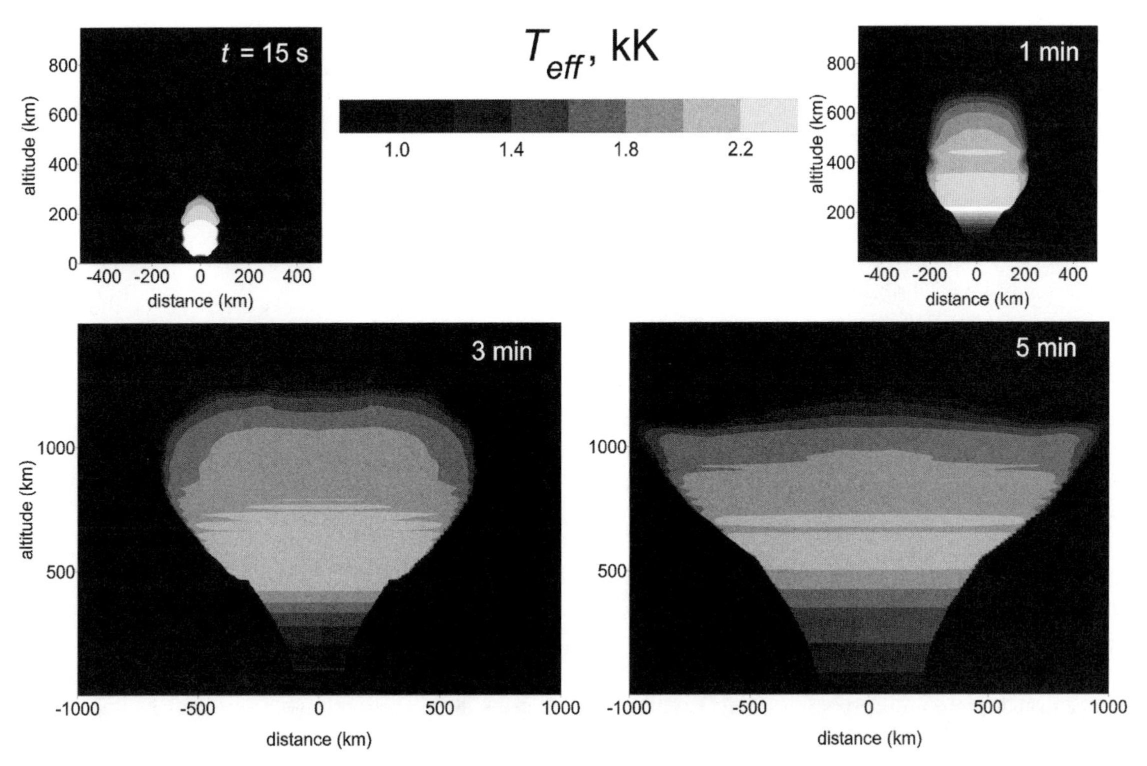

Figure 1. Views of impact plume from side (distributions of effective radiation temperature) at different moments of time.

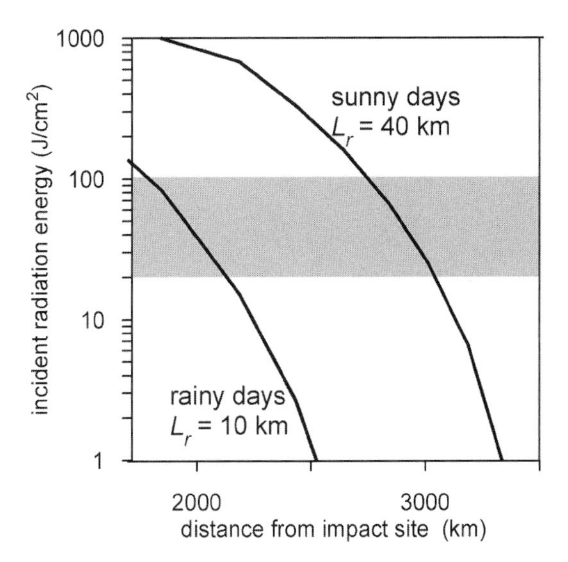

Figure 2. Incident radiation energy vs. distance from impact site, using Glasstone and Dolan (1977) values for radiation-free path in atmosphere. L_r is radiation-free path in atmosphere.

depending on forest type, moisture, and other conditions (Glasstone and Dolan, 1977).

Almost all the radiation is emitted in the infrared region with the exception of the short initial peak. The maximum of the radiation spectra falls to 0.5–0.7 eV.

It follows from Figure 2 that wildfires could arise at distances of 2000–3000 km from the impact site. A two-fold increase or decrease of the impactor size can extend this range to 1500–4000 km.

Some peculiarities can be induced by the angle of impact (Schultz and Gault, 1982, 1990; Schultz, 1996). Recent investigations (Pierazzo and Melosh, 1999, 2000) have shown that the impact angle can influence considerably the fate of the projectile and the degassing of the sedimentary layer. However, numerical simulations of the light flashes generated by meteoroids impacting the Moon (Nemtchinov et al., 1998) have shown that the difference in the total radiation impulse between a vertical and an inclined impact does not exceed a factor of two. The radiation is larger for the vertical impact because a larger part of impactor kinetic energy is transformed into heat. Such variations cannot change the area of forest ignition significantly. In experiments by Schultz (1996) oblique impacts were observed to produce a lower temperature but more vaporized mass, thereby leading to a greater total luminosity. This effect is not very surprising for the impacts with small (5–7 km/s) velocities, because the mass of produced vapor is small as compared with the projectile mass. Moreover, in a low-velocity vertical impact, high-temperature vapor is screened by low-temperature ejecta. It is not the case for impacts with velocities exceeding 10–15 km/s, where the vaporized mass strongly increases (Pierazzo et al., 1998).

This work suggests that the direct radiation induced by the Chicxulub impact expansion plume could be responsible for igniting wildfires over ~3%–10% of the Earth's surface, close to the impact point. This is an upper estimate because this area was partially covered by water.

Melosh et al. (1990) suggested that the global wildfires could be caused by ejecta reentry. In their calculations the fall-back of the impact-created debris was treated like a large number of small meteoroid impacts. However, these calculations were based on arbitrary assumptions of the ejecta velocity distribution. In this study this distribution was a direct result of numerical simulations of the impact. A simple consideration shows that all the mass ejected with velocities of <5 km/s falls within the area affected by direct plume radiation. The proportion of high-velocity ejecta (velocity >5 km/s) appears to be small: several percent of the impactor mass (see Fig. 3). Melosh et al. (1990) assumed that the mass of ejecta with velocities of 5–10 km/s was $\sim 5 \cdot 10^{15}$ kg (i.e., 3–5 times larger than the mass of 10 km stony impactor). Under this assumption the average irradiance from the reentering ejecta was near the lower limit required for ignition of solid wood. It follows from the present investigation that the mass of high-velocity ejecta ($\sim 10^{14}$ kg) is less by almost two orders of magnitude.

Therefore, in our opinion the process of ejecta reentry is probably not responsible for global wildfires. However, more detailed investigation is needed. In particular, the interaction of the reentering ejecta aerosol cloud with the Earth's atmosphere probably cannot be described as a sum of independent micro-impacts (airbursts). Correspondingly, the irradiance from the reentering ejecta cannot be estimated as a sum of independent micrometeors (shooting stars).

Another scenario may be considered to explain a great amount of soot, which was produced after the Cretaceous-Tertiary (K-T) impact (Wolbach et al., 1990; Tinus and Roddy, 1990). The local wildfires and other mechanisms detailed in Toon et al. (1997) could lead to a global mortality of forests and other plants. Dead forests are known to be much more subject to ignition than living forests. Moreover, in dry forests the fraction of totally burned mass of wood increases considerably (several times), leading to an increase of the total mass of soot produced. The ignition of dead forests could result from the natural thunderstorm activity enhanced by strong atmospheric perturbations due to the impact and its consequences. There is no evidence of instant global fires and extinction. The period (or periods) of wildfires could continue for several years. In other words, a great amount of soot could be produced by a large number of local wildfires (maybe not instantaneous) in the dead forests. In this case the global mortality of forests was not the result of global wildfires, but rather wildfires and K-T soot could result from the forest mortality.

Similar ideas were proposed in Argyle (1986). The main objection against the long time ignition is that carbon does not overlie the iridium layer (Wolbach et al., 1990). The data show that carbon appears even in the lowermost 0.3 cm of the 0.6 cm K-T boundary clay at Woodside Creek and Sumbar (Wolbach et al., 1988). This fact gives an upper estimate of the wildfires duration, i.e., the time of sedimentation. In Eder and Preisinger (1987) this time is estimated to be ~1 yr. However, the time of sedimentation, which is defined by Stoke's law, strongly depends on the sizes of the particles. These sizes are poorly known values. Even a three-fold decrease in particle sizes increases the time of sedimentation by an order of magnitude. In other words, there are no strong data rejecting the long-time ignition. This problem needs more detailed investigation.

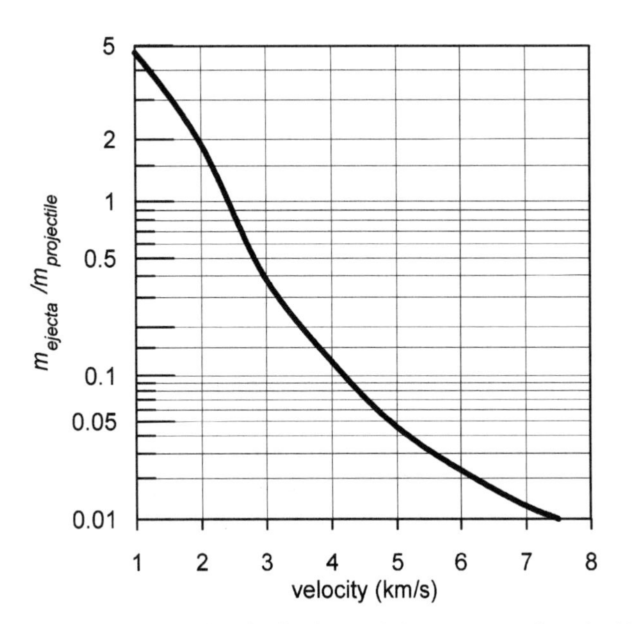

Figure 3. Ejecta velocity distribution (relative mass, *m*, ejected with velocity exceeding given one).

ATMOSPHERIC EROSION

Cameron (1983) was the first to suggest that planetary atmospheric erosion could result from impact events. The monumental importance of this process was justified by Ahrens et al. (1989) and Ahrens (1993). The main results concerning impact-induced atmospheric erosion were obtained with the use of a simplified sector blow-off model (Vickery and Melosh, 1990). This approximation uses the Zel'dovich and Raizer (1967) solution for the expansion of the vapor plume and the momentum balance between the expanding gas and the mass of vapor plus overlying atmosphere. This theory predicts that the escaping material is contained within a cone, its angle depending on the impact energy. For a sufficiently energetic impact all the air in the hemisphere above a plane tangent to the point of impact will be ejected. The blow-off model was widely used to study evolution of planetary atmospheres (Zahnle et al., 1990, 1992; Evans et al., 1995). Newman et al. (1999) showed that atmospheric stratification violates an isotropy of the flow induced by impact. Using more sophisticated analytical model based on the solution of Kompaneets (1960) and detailed numerical simulations, Newman et al. concluded that atmospheric escape region looks like a narrow vertical column rather than a cone. The more important result is that the mass of escaping air proves to be considerably less (by a factor of five for Chicxulub impact) than predicted by the blow-off model.

However, both models do not take into account the influence of the atmospheric perturbations created during impactor flight through the atmosphere and the influence of cratering process on the air ejection. The significance of these factors was outlined in Vickery (1994) and Svetsov (1996). In this study we use direct numerical simulations of the impact to calculate atmospheric erosion induced by impacts of large (10–30 km in diameter) cosmic bodies.

Projectile penetration into the planetary atmosphere results in formation of a wake, i.e., a rarefied hot channel with the density, that is 1–2 orders of magnitude lower than the ambient air density. In the case under consideration the total mass ejected from the crater is much greater than the mass of air within the cone (or column) of ejection. Therefore, the main part of ejecta is not exposed to atmospheric drag. However, a leading, very hot, rarefied and fast, portion of vapor undergoes considerable drag and expands within the rarefied wake preferentially (see Fig. 4). In some sense the wake is like an empty tube through which the hot rarefied vapor moves to high altitudes. The shock wave generated by expansion plume in the ambient air propagates preferentially along the wake as well. Thus, the main part of the fast ejecta moves through the rarefied wake and can accelerate only a small mass of air filling the wake. This effect depends on impactor size. For a 10-km-diameter projectile the presence of the wake leads to the decrease of the atmospheric erosion as compared with the flow without the wake. As the impact energy increases (a 30-km-diameter projectile) the hot and fast mass of the plume also increases, and atmospheric drag becomes too small to influence the plume evolution. In this case the presence of the wake does not influence atmospheric erosion.

The mass of escaping atmospheric air m_{esc} is determined as the mass of air that reaches the altitude of 300 km having a velocity exceeding 11 km/s at the moment of 20 s. Several runs were continued to the time of 60 s; this did not lead to increase

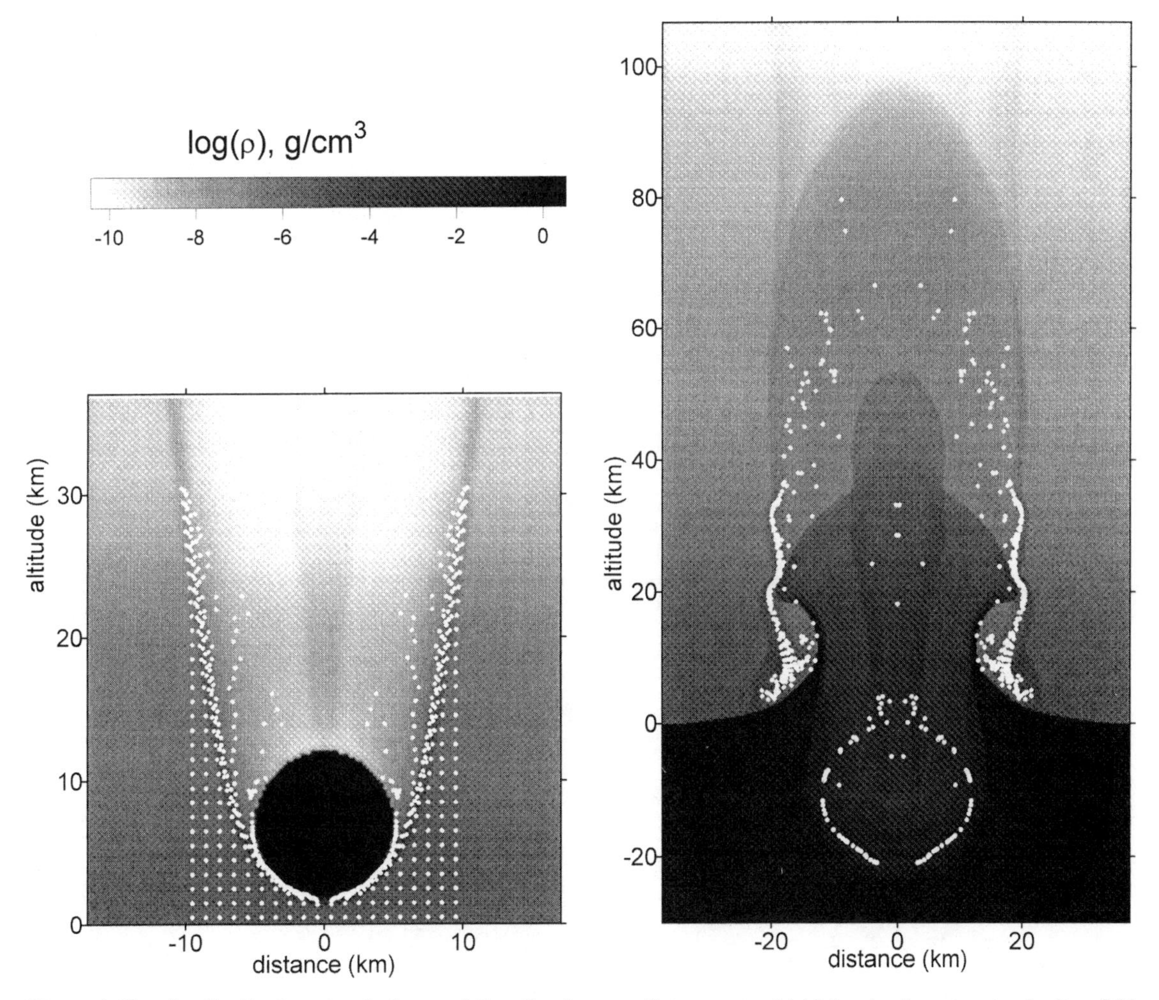

Figure 4. Density distributions just before and 5 s after impact of stony asteroid 10 km in diameter at velocity of 20 km/s. Log (ρ) is logarithm of density (measured in g/cm^3). White circles show positions of tracer particles initially distributed in cylinder (R < 10 km, Z < 30 km). R—radius; Z—height.

of escaping mass. Some results of numerical simulations are summarized in Table 1.

As expected, the atmospheric erosion increases with the increase of impact velocity, and the comet impacts are more effective than the asteroid impacts. The influence of the target (water or rocks) is found to be small. The presence of the wake leads to a decrease of the value of m_{esc} for the 10 km impactors.

Apart from the total escaping mass it is interesting to know how much air is removed from each altitude. It is important from the viewpoint of possible changes in the chemical composition of the atmosphere, induced by impacts, because the relative concentrations of different components change with the altitude. To follow the motion of different atmospheric layers and to determine the initial distribution of the escaping air (atmospheric escape region), 15 000 tracer particles were initially distributed below 100 km. Several typical distributions are shown in Figure 5.

TABLE 1. VERTICAL IMPACT RESULTS

Impactor	Diameter (km)	Velocity (km/s)	m_{esc}/m_{imp}
Asteroid	30	20	4.7×10^{-5}
Asteroid	10	20	1.6×10^{-4}
Asteroid	10	30	2.5×10^{-4}
Asteroid	10	50	8.8×10^{-4}
Asteroid*	10	20	2.5×10^{-4}
Asteroid**	10	20	1.8×10^{-4}
Comet	30	50	6.5×10^{-3}
Comet	10	20	1.3×10^{-3}
Comet	10	30	9.7×10^{-3}
Comet	10	50	2.6×10^{-2}

Note: m_{esc} is the mass of escaping air; m_{ipm} is the impactor mass.
*The run without taking into account the wake.
**Impact into the ocean 4 km deep.

At very high impact energies (diameter $D = 30$ km, velocity $V = 50$ km/s) the atmospheric escape region looks like a cone. As the impact energy decreases the cone gradually transforms into a column. However, the air from the bottom part of this cone or column (which has the highest density and contains the most part of cone or column air mass) does not escape from the Earth. This effect cannot be explained in the frame of the blow-off model (Vickery and Melosh, 1990) and even more complicated model of Newman et al. (1999).

Figure 4 illustrates the fate of the air surrounding the point of impact. Just before a meteoroid strikes the ground surface this air mass is contained in the shock-compressed layer surrounding the projectile. Some portion of this gas is forced into the target, mixes with melt, and is further ejected with rather small velocity (less than the escape velocity). The other part of the air surrounding the impactor spreads along the ground surface. The dense curtain of ejecta separates this air from the wake and upward-expanding plume. These peculiarities of the flow (typical for all the impacts under consideration) considerably decrease the atmospheric erosion as compared to both the Melosh and Vickery and Newman et al. models.

Only the results for a vertical impact are presented in Table 1. It is not clear in advance how impact angle can influence the mass of escaping air. Artem'eva and Shuvalov (1994) showed that the role of the wake becomes less significant for an oblique impact. However, a hypervelocity jet directed along the target surface is known to be formed after the oblique impact (Schultz

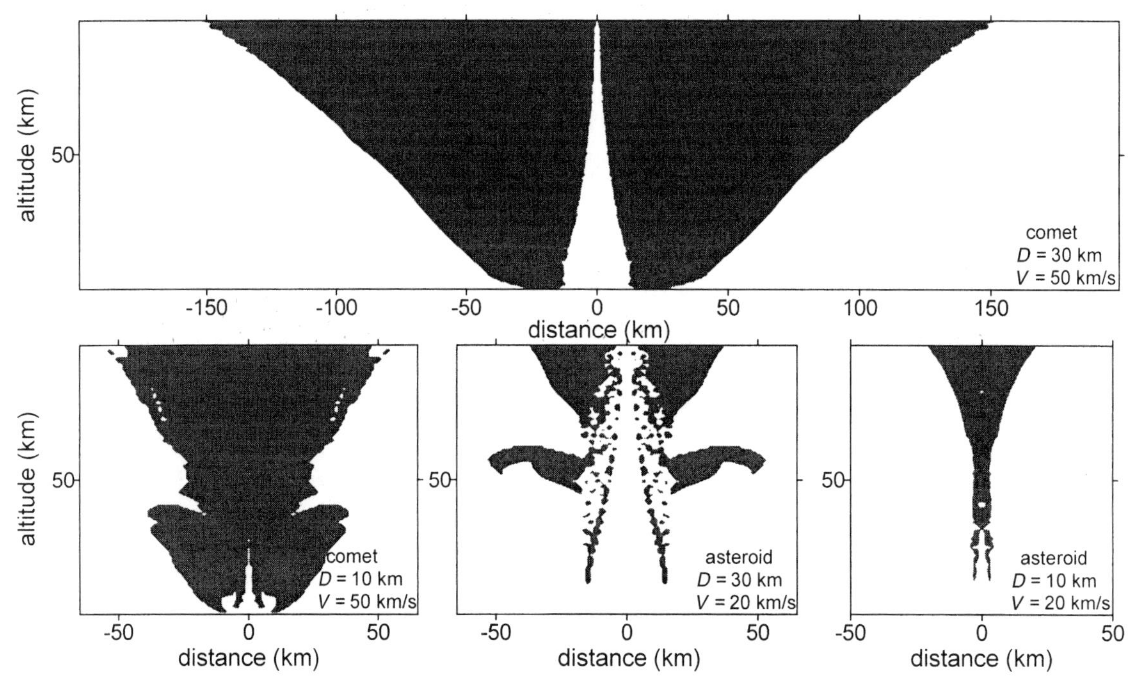

Figure 5. Dark gray marks volumes occupied by escaping air before asteroid impact (atmospheric erosion region). *D* is diameter, *V* is velocity.

and Gault, 1990; Schultz, 1996; Artemieva and Shuvalov, 2001). This jet can accelerate a great mass of air distributed in the shock-compressed layer surrounding the projectile, i.e., the air from the apex of the atmospheric escape cone (column). As a result the total mass of escaping air could increase considerably.

To test this idea we perform numerical simulations of the oblique (45°) impact of a stony body 10 km in diameter with the velocity of 20 km/s. The results of these simulations are shown in Figure 6, which is similar to Figure 4, illustrating a vertical impact of the same projectile. The mass of escaping air is proved to be an order of magnitude greater than that in a vertical impact. The atmospheric escape region looks like a cone rather than a column.

However, these results can be considered as preliminary and qualitative ones only. Three-dimensional numerical simulations are much more expensive than the two-dimensional ones. For this reason we use a crude computational grid (~10 cells across impactor radius) in the three-dimensional run. To obtain more reliable results numerical simulations should be repeated with a higher spatial resolution; numerical simulations of the impact with various impact angles (from near vertical to grazing) are also of great interest.

CONCLUSIONS

The results of detailed numerical simulations show that the direct radiation induced by the Chicxulub impact could ignite wildfires over ~3%–10% of the Earth's surface. Almost all the radiation is emitted in the infrared region with exception of the short initial peak. The maximum of radiation spectrum falls to 0.5–0.7 eV.

We conclude that at least the vertical impacts apparently could not decrease the mass of the Earth's atmosphere. The ratio of escaping air mass to the impactor mass does not exceed the value of 2%–3%, which is considerably less than the mass of volatiles delivered by impactors (Chyba, 1990) and released during the impact (Pierazzo et al., 1998). It follows from our simulations that previous investigations (Vickery and Melosh, 1990; Newman et al., 1999) overestimated atmospheric erosion.

Most likely the atmospheric erosion could not result in the deficit of xenon (Zahnle, 1993), because the relative escaping mass increases with altitude in all runs.

Oblique impacts are believed to be more effective from the viewpoint of atmospheric erosion. However, for stony impactors the escaping mass to impactor mass ratio does not exceed the value of 1%, even for an oblique impact.

Atmospheric erosion induced by impacts can be more effective on planets with lower gravity. A rough estimate of the erosion of the hypothetical primitive Martian atmosphere of 1 bar can be obtained by replacing the Earth's escape velocity, 11 km/s, by the Martian one, 5 km/s. Our simulations show that the ratio of m_{esc}/m_{imp} for Mars is 3–10 times larger than that for the Earth.

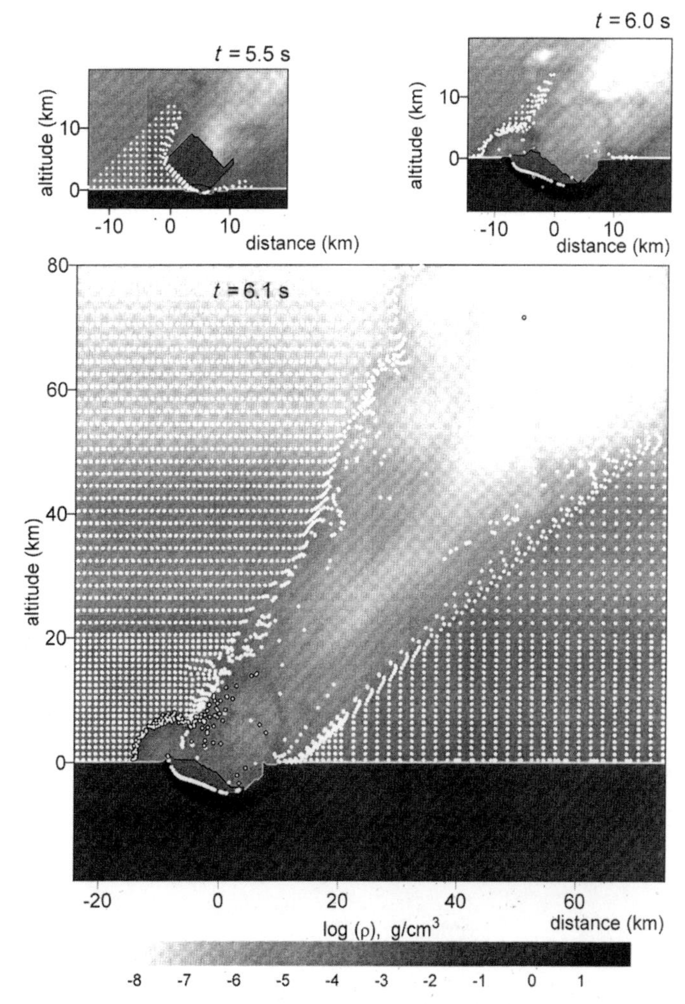

Figure 6. Density distributions just before and 5 s after impact of stony asteroid 10 km in diameter at velocity of 20 km/s. Log (ρ) is logarithm of density (measured in g/cm³). Dark tracers have velocity exceeding 11 km/s. Later some other tracers can reach escape velocity.

ACKNOWLEDGMENTS

We are grateful to P. Schultz and D. Kring for their accurate and helpful reviews. We appreciate the European Science Foundation (ESF, program IMPACT) for the financial support that allowed us to attend the conference in Vienna. Special thanks to C. Koeberl for his hospitality and highly efficient organizing work. This study was supported by Russian Fund of Basic Research (grant 00-05-81152 Bel2000_a).

REFERENCES CITED

Ahrens, T.J., 1993, Impact erosion of terrestrial planetary atmospheres: Annual Review of Earth and Planetrary Sciences, v. 21, p. 525–555.

Ahrens, T.J., O'Keefe, J.D., and Lange, M.A., 1989, Formation of atmospheres during accretion of the terrestrial planets, *in* Atreya, S.K., Pollack, J.B., and Matthews, M.S., eds., Origin and evolution of planetary and satellite atmospheres: Tucson, University of Arizona Press, p. 328–385.

Alvarez, L.W., Alvarez, W., Asaro, F., and Michel, H.V., 1980, Extraterrestrial cause for the Cretaceous-Tertiary extinction: Science, v. 208, p. 1095–1108.

Argyle, E., 1986, Cretaceous extinctions and wildfires: Science, v. 234, p. 261.

Artem'eva, N.A., and Shuvalov, V.V., 1994, Oblique impact: Atmospheric effects [abs.]: Lunar and Planetary Science Conference, 25th, Houston, Texas, Lunar and Planetary Institute, part 1, p. 39–40.

Artemieva, N.A., and Shuvalov, V.V., 2001, Extraterrestrial material deposition after the impacts into continental and oceanic sites: Geological and biological effects of impact events, *in* Buffetaut, E., and Koeberl, C., eds., ESF Impact: Berlin, Germany, Springer Verlag, p. 249–263.

Boslough, M.B., and Crawford, D.A., 1997, Shoemaker-Levy 9 and plume-forming collisions on Earth, *in* Remo, J.L., ed., Near-Earth objects: New York, New York Academy of Sciences, p. 236–282.

Cameron, A.G.W., 1983, Origin of the atmospheres of the terrestrial planets: Icarus, v. 56, p. 195–201.

Chyba, C.F., 1990, Impact delivery and erosion of planetary oceans in the early inner solar system: Nature, v. 343, no. 9, p. 129–133.

Eder, J., and Preisinger, A., 1987, Zeitstructur globaler Ereignisse, veranschaulicht an der Kreide-Tertiar-Grenze: Naturwissenschaften, v. 74, p. 35–37.

Evans, N.J., Ahrens, T.J., and Gregoire, D.C., 1995, Fractionation of ruthenium from iridium at the Cretaceous-Tertiary boundary: Earth and Planetary Sciences Letters, v. 134, p. 141–153.

Glasstone, S., and Dolan, P.J., 1977, The effects of nuclear weapons: Washington, D.C., U.S. Government Printing Office, 653 p.

Jones, E.M., and Kodis, J.W., 1982, Atmospheric effects of large body impacts: The first few minutes, *in* Silver, L.T., and Schultz, P.H., eds., Geological implications of impacts of large asteroids and comets on the earth: Geological Society of America Special Paper 190, p. 175–186.

Kompaneets, A.S., 1960, A point explosion in an inhomogeneous atmosphere: Soviet Physics Doklads English Translation, v. 5, p. 46–48.

Kosarev, I.B., Loseva, T.V., and Nemtchinov, I.V., 1996, Optical properties of vapor and ablation of large chondritic and icy bodies in the earth's atmosphere: Solar System Research, v. 30, no. 4, p. 265–278.

Kuznetsov, N.M., 1965, Thermodynamic functions and shock adiabats for air at high temperatures [in Russian]: Moscow, Mashinostroyenie, 464 p.

Lange, M.A., and Ahrens, T.J., 1982, The evolution of an impact-generated atmosphere: Icarus, v. 51, p. 96–120.

Matsui, T., and Abe, Y., 1986, Evolution of an impact-induced atmosphere and magma ocean on the accreting earth: Nature, v. 319, p. 303–305.

McGlaun, J.M., Thompson, S.L., and Elrick, M.G., 1990, CTH: A three-dimensional shock wave physics code: International Journal of Impact Engineering, v. 10, p. 351–360.

Melosh, H.J., 1989, Impact cratering: A geologic process: New York, Oxford University Press, 245 p.

Melosh, H.J., and Ivanov, B.A., 1999, Impact crater collapse: Annual Review of Earth and Planetary Sciences, v. 27, p. 385–425.

Melosh, H.J., Schneider, N.M., Zahnle, K.J., and Lathan, D., 1990, Ignition of global wildfires at the Cretaceous/Tertiary boundary: Nature, v. 343, p. 251–254.

Melosh, H.J., and Vickery, A., 1989, Impact erosion of the primordial atmosphere of Mars: Nature, v. 338, p. 487–492.

Nemtchinov, I.V., Shuvalov, V.V., Artem'eva, N.A., Ivanov, B.A., Kosarev, I.B., and Trubetskaya, I.A., 1998, Light flashes caused by meteoroid impacts on the lunar surface: Solar System Research, v. 32, no. 2, p. 99–114.

Nemtchinov, I.V., and Svetsov, V.V., 1991, Global consequences of radiation impulse caused by comet impact: Advanced Space Review, v. 11, p. (6)95–(6)97.

Newman, W.I., Symbalisty, E.M.D., Ahrens T.J., and Jones, E.M., 1999, Impact erosion of planetary atmospheres: Some surprising results: Icarus, v. 138, p. 224–240.

Pierazzo, E., Kring, D.A., and Melosh, H.J., 1998, Hydrocode simulations of the Chicxulub impact event and the production of climatically active gases: Journal of Geophysical Research, v. 103, no. E12, p. 28607–28625.

Pierazzo, E., and Melosh, H.J., 1999, Hydrocode modeling of Chicxulub as an oblique impact event: Earth and Planetary Science Letters, v. 165, p. 163–176.

Pierazzo, E., and Melosh, H.J., 2000, Hydrocode modeling of oblique impacts: The fate of the projectile: Meteoritics and Planetary Science, v. 35, p. 117–130.

Schultz, P.H., 1996, Effect of impact angle on vaporization: Journal of Geophysical Research, v. 101, no. E9, p. 21117–21136.

Schultz, P.H., and Gault, D.E., 1982, Impact ejecta dynamics in an atmosphere: Experimental results and extrapolations, *in* Silver, L.T., and Schultz, P.H., eds., Geological implications of impacts of large asteroids and comets on the earth: Geological Society of America Special Paper 190, p. 153–174.

Schultz, P.H., and Gault, D.E., 1990, Prolonged global catastrophes from oblique impacts, *in* Sharpton, V.L., and Ward, P.D., eds., Global catastrophes in Earth history: An interdisciplinary conference on impacts, volcanism and mass mortality: Geological Society of America Special Paper 247, p. 239–261.

Sharpton, V.L., and Ward, P.D., editors, 1990, Global catastrophes in Earth history: An interdisciplinary conference on impacts, volcanism, and mass mortality: Geological Society of America Special Paper 247, 631 p.

Shuvalov, V.V., 1999a, 3D hydrodynamic code SOVA for interfacial flows, application to thermal layer effect: Shock Waves, v. 9, no. 6, p. 381–390.

Shuvalov, V.V., 1999b, Atmospheric plumes created by meteoroids impacting the earth: Journal of Geophysical Research, v. 104, no. E3, p. 5877–5890.

Shuvalov, V.V., 2001, Radiation effects of the Chicxulub impact event. Geological and biological effects of impact events, *in* Buffetaut, E., and Koeberl, C., eds., ESF Impact: Berlin, Germany, Springer Verlag, p. 237–247.

Silver, L.T., and Schultz, P.H., editors, 1982, Geological implications of impacts of large asteroids and comets on the earth: Geological Society of America Special Paper 190, 528 p.

Svetsov, V.V., 1996, Total ablation of the debris from the 1908 Tunguska explosion: Nature, v. 383, p. 697–699.

Thompson, S.L., and Lauson, H.S., 1972, Improvements in the Chart D radiation-hydrodynamic CODE 3: Revised analytic equations of state: Albuquerque, New Mexico, Sandia National Laboratory, Report SC-RR-71 0714, 119 p.

Tillotson, J.H., 1962, Metallic equations of state for hypervelocity impact: General Atomic Report GA-3216, 137 p.

Tinus, R.W., and Roddy, D.J., Effects of global atmospheric perturbations on forest ecosystems in the Northern Temperate Zone: Predictions of seasonal depressed-temperature kill mechanisms, biomass production, and wildlife soot emissions, *in* Sharpton, V.L., and Ward, P.D., eds., Global catastrophes in Earth history: An interdisciplinary conference on impacts, volcanism, and mass mortality: Geological Society of America Special Paper 247, p. 77–86.

Toon, O.B., Turco, R.P., and Covey, C., 1997, Environmental perturbations caused by the impacts of asteroids and comets: Reviews of Geophysics, v. 35, no. 1, p. 41–78.

Vickery, A.M., 1994, Impact erosion of atmospheres [abs.]: Lunar and Planetary Science Conference, 25th, Houston, Texas, Lunar and Planetary Institute, p. 1437–1438.

Vickery, A.M., and Melosh, H.J., 1990, Atmospheric erosion and impactor retention in large impacts with application to mass extinctions, *in* Sharpton, V.L., and Ward, P.D., eds., Global catastrophes in Earth history: An interdisciplinary conference on impacts, volcanism, and mass mortality: Geological Society of America Special Paper 247, p. 289–300.

Wolbach, W.S., Gilmour, I., Anders, E., Orth, C.J., and Brooks, R.R., 1988, A global fire at the Cretaceous-Tertiary boundary: Nature, v. 334, p. 665–669.

Wolbach, W.S., Gilmour, I., and Anders, E., 1990, Major wildfires at the Cretaceous/Tertiary boundary, *in* Sharpton, V.L., and Ward, P.D., eds., Global catastrophes in Earth history: An interdisciplinary conference on impacts, volcanism, and mass mortality: Geological Society of America Special Paper 247, p. 391–400.

Zahnle, K.J., 1993, Xenological constraints on the impact erosion of the early martian atmosphere: Journal of Geophysical Research, v. 98, no. 6, p. 10899–10913.

Zahnle, K., Pollack, J.B., and Kasting, J.F., 1990, Mass fractionation of noble gases in diffusion-limited hydrodynamic hydrogen escape: Icarus, v. 84, p. 503–527.

Zahnle, K., Pollack, J.B., and Gripspoon, D., 1992, Impact-generated atmospheres over Titan, Ganymede and Callisto: Icarus, v. 95, p. 1–23.

Zel'dovich, Ya.B., and Raizer, Yu.P., 1967, Physics of shock waves and high-temperature hydrodynamic phenomena: New York, Academic Press, 916 p.

Manuscript Submitted October 10, 2000; Accepted by the Society March 22, 2001

Printed in the U.S.A.

Geological Society of America
Special Paper 356
2002

Toxins produced by meteorite impacts and their possible role in a biotic mass extinction

M.V. Gerasimov*
Space Research Institute, Russian Academy of Science, Profsoyuznaya Street, 84/32, Moscow, 117997, Russian Federation

ABSTRACT

Chemical contamination of ecosystems due to the production of toxic gases by large impacts is a possible cause of biotic mass extinctions. Some globally distributed toxins are hazardous even at low concentrations. This extinction mechanism has not been studied in detail due to the lack of chemical models of the effects of an impact event. This chapter presents experimentally derived information about the chemistry of impacts and considers the possible effect of chemical contamination on the biosphere, with special reference to the Chicxulub impact and the Cretaceous-Tertiary mass extinction. The Chicxulub impact released CO, SO_2, CS_2, H_2S, and other gases that spread from the point of the impact, dissolved in the atmosphere, and may have produced lethal concentrations over large areas. The global concentrations of many gases were probably orders of magnitude above permissible exposure levels for human beings and may have resulted in long-term mortal effects by carcinogenesis, mutagenesis, or teratogenesis. Production of toxins during a large impact may be a sufficient mechanism for some extinctions.

INTRODUCTION

Impacts of large asteroids or comets can produce global catastrophes on the Earth. Such catastrophes may involve temporal changes in the composition and density of an atmosphere, in the climatic and oceanic circulation, and other changes that may eventually equilibrate by endogenic processes to a regular state. Some changes may have no recovery to the former state, for example the extinction of biota. Among the different factors of an impact that are significant for the survivability of life forms are physical (e.g., the action of shock waves, expansion and luminosity of the fireball, ejecta, tsunami, high-amplitude earthquakes), chemical (e.g., the damage of the chemical equilibrium of the atmosphere by the release of impact-generated gaseous components and chemical reactions between atmospheric gases and vapor cloud, formation of toxic components, acidic rains, change of pH in the ocean water, formation of aerosols), and climatic effects. The significance of physical and chemical effects changes with the distance from the center of an impact (shown in Fig. 1). Near the crater there is a zone of immediate total extermination of all forms of life mainly due to physical factors. Physical factors are reduced at greater distances from the impact, where there is a zone with immediate selective extermination. This means that some life forms survive while others are killed. At even greater distances physical factors do not have mortal effects. Chemical factors at large distances can still be significant, because chemical products can be transported long distances and some of the globally distributed toxins are dangerous even at low concentrations. Chemical changes in the environment can cause slow extermination of some life forms that are sensitive to such changes. If an impact is large enough to cause climate changes, then the entire planet may become a hazardous environment for some life forms.

The significance of chemical changes and the production

*E-mail: mgerasim@mx.iki.rssi.ru

Gerasimov, M.V., 2002, Toxins produced by meteorite impacts and their possible role in a biotic mass extinction, *in* Koeberl, C., and MacLeod, K.G., eds., Catastrophic Events and Mass Extinctions: Impacts and Beyond: Boulder, Colorado, Geological Society of America Special Paper 356, p. 705–716.

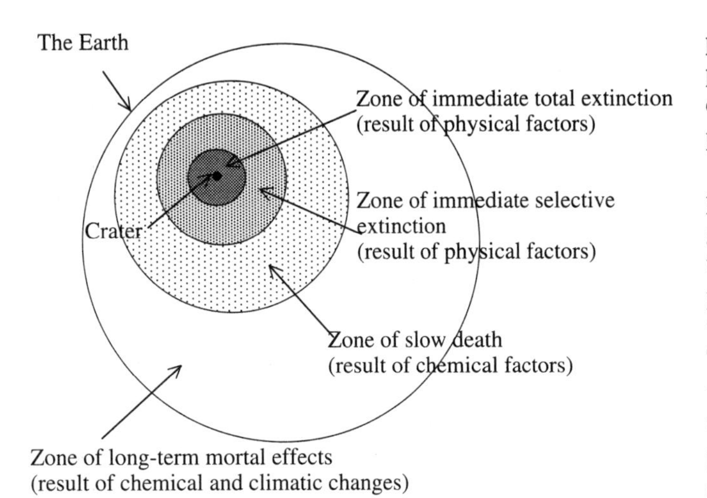

Figure 1. Schematic zoning of catastrophic impact effects.

of toxins during an impact have received limited coverage in the literature and are not well understood. Fegley et al. (1986) investigated theoretically the possible production of nitrous oxides and hydrogen cyanide in an air shock wave due to the passage of a large asteroid through the Earth's atmosphere. It was shown that the chemical effect of acidic rains due to the production of nitrous oxides in a shock wave could be sufficient to raise the acidity of the ocean and to cause biospheric trauma. The impact-induced degassing of volatile-bearing minerals was investigated experimentally using a shock-wave-loading technique. Lange and Ahrens (1986) and Tyburczy and Ahrens (1986) investigated the shock loading of calcite and reported on the release of CO_2. Tyburczy and Ahrens (1993) investigated the shock-induced devolatilization of anhydrite and proposed its simple decomposition scheme with release of SO_2. The forming gas phase in the experiments was not investigated in detail, and CO_2 and SO_2 were considered as the main results of impact degassing of carbonates and sulfates. Theoretical estimates of gas and aerosol emissions generated by the Chicxulub impact were made by Siret et al. (2000), who argued for an insignificant SO_2 emission due to the formation of aerosols in the vapor plume. A detailed investigation of gases formed during an impact simulated with a laser pulse was reported by Gerasimov and Mukhin (1984) and Mukhin et al. (1989). The possible effect of released gases on the biosphere was not considered. This chapter focuses on experimentally derived information about the chemistry of impacts and its possible effect on the biosphere as the mechanism of extinction.

FORMATION OF THE CHEMICAL COMPOSITION OF GASES DURING AN IMPACT

It is not a simple task to determine the gas output of an impact event. The general approach can be to determine the difference between the gaseous inventory before and after an impact within the undisturbed vicinity of the crater. The main problem here is how to determine when the production of impact gases ceases, because degassing of hot rocks and chemical changes in the products of an impact proceed long after the physical displacement of the crater material has stopped.

The volatile components that are processed during an impact are a combination of the volatiles of the impactor, the target, and the atmospheric gases. The formation of gas species that are released during an impact is a multistage and complex process. The decomposition of volatile-bearing minerals starts with the passage of a shock wave through the colliding material. The release of gases into an expanding vapor-gas cloud starts with the decrease of the pressure in a rarefaction wave when the partial pressure of vapor and gases exceeds the ambient pressure. Atmospheric gases involved in the impact are first subjected to a strong air shock wave during the passage of an impactor through the atmosphere. During the cratering stage the atmospheric gases are mixed with the impact vapor plume and participate in plume chemistry. An impact-generated cloud is mainly composed of volatile gas components, siliceous vapor, condensed particles, and ejecta. During its expansion the composition of gases can vary due to chemical reactions between degassed and admixed atmospheric gas molecules and, in addition, due to reactions of gases with condensed rock material and ejecta. It is important that the condensation of silicates proceeds with the formation of small nanoscale nuclei, which have enormous surface area and increase the role of surface and catalytic reactions.

The integral composition of the plume, at any time, is a combination of the individual chemical compositions of all of its parts, which are defined by the history of their pressure, temperature, and elemental composition. Gerasimov et al. (1999) showed that at the onset of an impact-induced vaporization the pressure and temperature are high enough to provide thermodynamical equilibrium conditions. Equilibrium conditions are terminated after a certain degree of expansion of the vapor-gas cloud when quenching of the chemical composition of expanding gases begins. Further changes in the composition of evolved gases can occur due to interactions with the planetary surface and atmosphere. It is reasonable to consider the chemical composition of gases at the point of quenching as the composition of primary impact-generated gases and their further chemical changes by reactions with the environment as secondary evolution.

EXPERIMENT

The complexity of the chemistry of an expanding multicomponent gas-vapor cloud implied that an experimental approach to its investigation is more appropriate than theoretical modeling. The chemical composition of gases formed in an impact-generated expanding vapor cloud was investigated experimentally more than a decade ago (Gerasimov and Mukhin, 1984; Mukhin et al., 1989). The simulation of impact vaporization was done by use of a laser pulse system. The laser pulse installation is shown schematically in Figure 2. Principal aspects of the simulation were described by Gerasimov et al. (1999). Basically, the procedure is as follows: a sample is mounted in a closed cell and a powerful laser beam is focused on its surface. The luminous energy of a laser pulse was 200–600 J and the duration of a pulse was ~1 ms. The beam was focused into a spot 2–5 mm in diameter to provide power densities in the range 10^5–10^7 W/cm^2. The cell was filled by helium at atmospheric pressure. The gases produced by the laser pulse were concentrated in a cold trap by purging the cell with a steady flow of helium gas. The cold trap had two loops (see Fig. 2B): one was filled by glass spheres for freezing out most of the gases, and the other was filled by an activated coal for retention of hydrogen and oxygen. After the purging of ~10 cell volumes of helium the cold trap was closed, then heated, and the gases by the same flow of helium (which was also used as a carrier gas for gas chromatography) were transmitted into the inlet of a gas chromatograph-mass spectrometer. In some experiments the cell was filled by hydrogen to perform vaporization in reduced environment.

Among samples used for experiments were: natural milky quartz, basalt, pyroxene, peridotite, picritic gabbro dolerite, and meteorites (ordinary L5 chondrite Tsarev, and carbonaceous C3O chondrite Kainsaz).

Chromatographic analyses were performed using two columns. The first was filled with Porapac Q (P_Q) and was used for the separation of most of components, but it did not separate the most volatile components, including H_2, O_2, N_2, CH_4, and CO. The second was filled by molecular sieves (type 5A) to separate these volatile components, but it totally absorbed the other. To make analyses of a full range of gas components from a single pulse, two switches were used (see Fig. 2B). These switches permitted the carrier gas to pass in three different modes: (1) a dual-column mode involving successive passage through P_Q and 5A columns; (2) a single 5A column mode; and (3) a single P_Q column mode. The analytical cycle was the following: from the cold trap the entire mixture was passed into P_Q column while the columns were in the dual-column mode; after a calibrated period of time, the unresolved portion of volatile gases were passed into the 5A column and then the columns were switched into the second single 5A column mode (other gases were sealed during this period in the P_Q column). After full analysis of the volatile gases on 5A column, the system was switched into the third single P_Q column mode for the analysis of all other gases. The gas chromatograph-mass spectrometer instrument also permitted the acquisition of the mass spectrum of every individual component for its reliable identification. An example of the gas chromatogram is presented in Figure 3.

Experiments in a neutral (helium) environment gave infor-

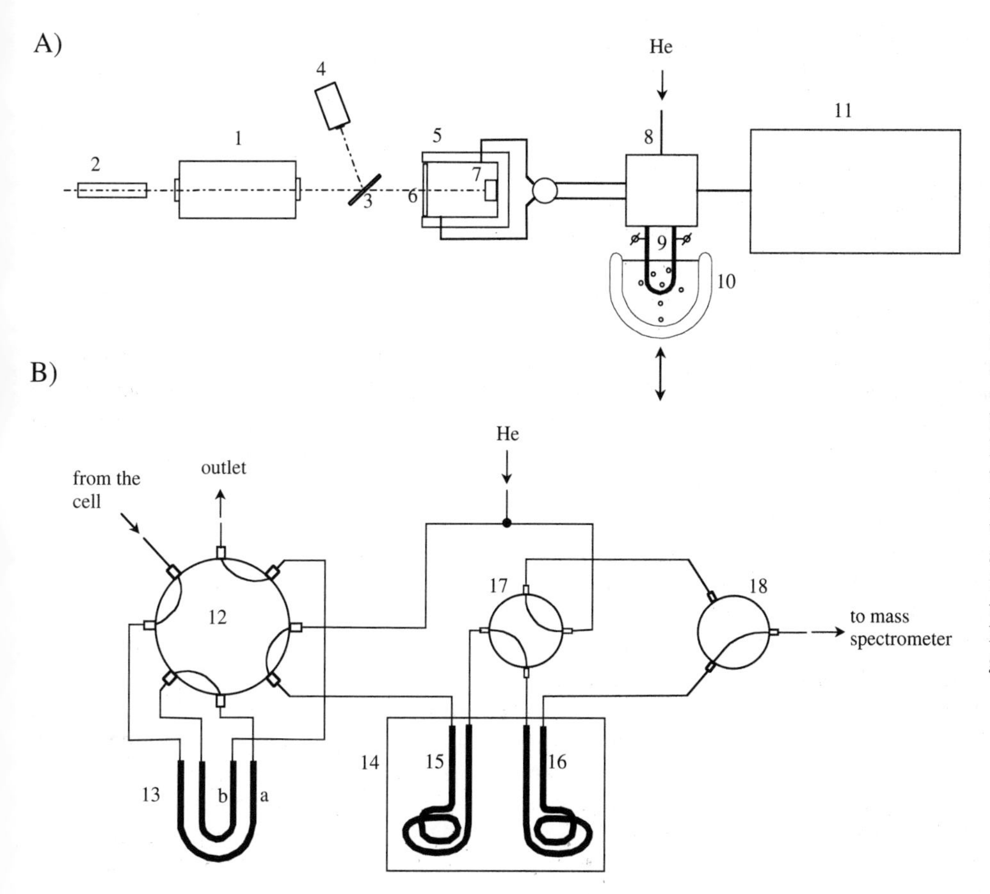

Figure 2. A: Schematic diagram of laser pulse system for simulation of impact vaporization chemistry. 1, pulsed neodymium glass laser; 2, He-Ne laser for adjustment of optical axis; 3, glass plate; 4, calorimeter; 5, sample cell; 6, quartz window; 7, sample; 8, valve assemblage; 9, cold trap; 10, liquid nitrogen Dewar; 11, chromatograph-mass spectrometer. B: Schematic diagram of cold trap and valve assemblage. 12, cold trap valve; 13, cold trap (a, loop with glass spherules; b, loop with activated coal); 14, chromatographic heater; 15, column with Porapac Q; 16, column with molecular sieves 5A; 17 and 18, chromatographic column valves.

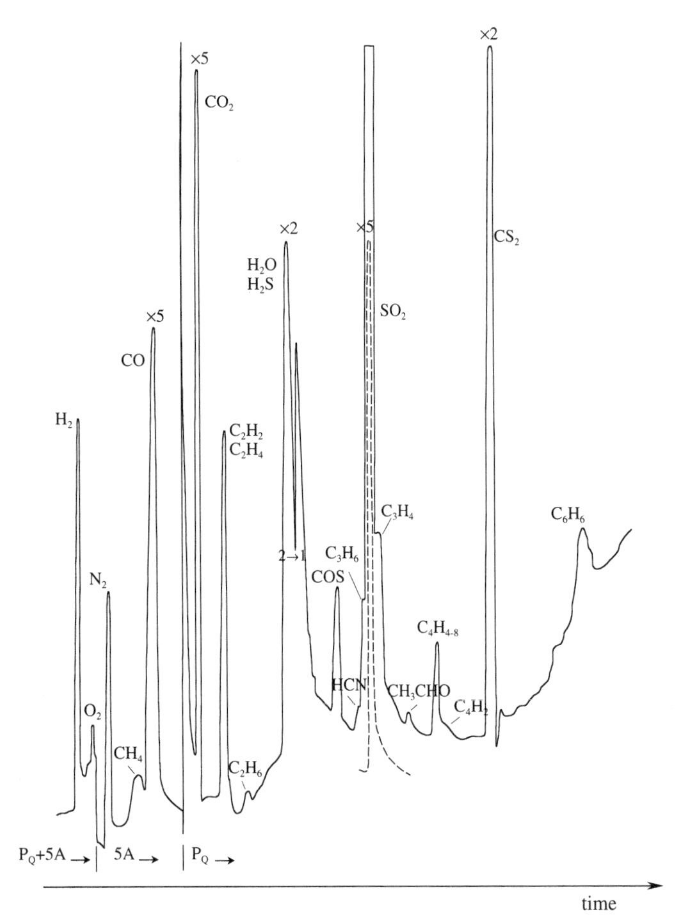

Figure 3. Chromatogram of gases from vaporization experiment of ordinary chondrite (L5 Tsarev) in helium. Labels P_Q + 5A, 5A, and P_Q indicate work time of chromatographic columns: successive passage through Porapac Q and molecular sieves 5A columns; and switching gas mixture passage through only 5A and P_Q columns. Sign ×n indicates n-fold reduction of sensitivity of recorder.

mation about the gas mixture produced after the quenching, and secondary reactions between the products of vaporization. Experiments in a reduced (hydrogen) environment gave information about the gas mixture formed by interactions with ambient gas. The detailed description of the experimental procedure shows that it was possible to measure only gases that survived the entire experimental cycle. The most active components (e.g., radicals, atomic hydrogen and oxygen), which could also be a result of quenching of the expanding gas-vapor cloud, may have been lost in the analytical system.

These experiments reveal a rather complex chemistry for the gas-vapor cloud and the formation of gas mixtures, which usually were composed (see Fig. 3) of H_2, O_2, N_2, CO, CO_2, H_2O, HCN, SO_2, H_2S, CS_2, COS, and hydrocarbons from CH_4 to C_6H_6 (Gerasimov and Mukhin, 1984; Mukhin et al., 1989). Many of these components are hazardous to living organisms at low concentrations. Typical results for the analyses of gases released during laser pulse vaporization of various samples are presented in Table 1.

In general, the amount of gases released was to a certain degree proportional to the concentration of the component elements in the starting material. Figure 4 shows the release of H_2 and the sum of CO + CO_2 (main carbon-containing components) normalized to the energy output of the pulse. For a certain sample, the dispersion of measured quantities of released gases per unit of pulse energy was within an order of magnitude, which is not surprising taking into account the inhomogeneity of samples, the partitioning of volatiles between the gas and condensed phases, and the physical instabilities of the pulse processes. Nevertheless, Figure 4 shows a steady growth of the normalized release of volatiles from volatile-poor (e.g., quartz) to volatile-rich samples (e.g., carbonaceous chondrite).

All the measured gas components were built from the elements O, H, C, N, and S. The mixtures of gases that formed in the laser pulse experiments were composed of both reduced and oxidized gas components, and included the simultaneous presence of noticeable quantities of H_2 and O_2, representing a non-equilibrium assemblage for normal conditions. Figure 5 shows the measured quantities of H_2 versus O_2 in experiments with terrestrial samples and meteorites in helium. H_2 and O_2 were a noticeable part of the gas mixture and amounted in some cases to as much as 80% of its volume. The change in the proportion of H_2 to O_2 indicates the change in redox conditions in the resulting gas mixture: from oxidized conditions in case of experiments with quartz, basalt, and pyroxene to reduced conditions in case of experiments with peridotite, gabbro, and meteorites.

Despite the lack of O_2 in the gas mixtures after laser pulse experiments with meteorites, it has been shown in experiments with a time-of-flight mass spectrometer (measuring the composition of gas components in the expanding vapor cloud) that even for meteorites, O_2 and O were the most abundant species (Gerasimov et al., 1987). Oxygen was generated by the thermal decomposition of rocks. Thermodynamic calculations show that the partial pressure of molecular and atomic oxygen in the vapor cloud could be about one-third of the total pressure. After quenching of the expanding vapor cloud, the oxygen reacted in a closed cell during the oxidation of condensed siliceous material, which filled the interior of the cell with heavy smog. Only a small portion of oxygen in the vapor cloud was left in the gas phase as O_2 in experiments with terrestrial samples. In the case of meteorites, all of the released oxygen was consumed and even some portion of small stable background oxygen was absorbed. This indicates that oxygen very efficiently interacted with condensed material.

Carbon was distributed among a wide range of components, including CO, CO_2, hydrocarbons, HCN, and CS_2. Figure 6 shows the pattern of the volume ratios of carbon containing gases normalized to CO_2 in experiments in helium and hydrogen atmospheres. In general, the pattern is rather stable because the dispersion of ratios for most of components did not exceed an order of magnitude. There was an increase in the production of some hydrocarbons in the experiments in a H_2

TABLE 1. TYPICAL CHEMICAL COMPOSITION OF GAS MIXTURES OBTAINED IN EXPERIMENTS WITH DIFFERENT SAMPLES IN HELIUM AND HYDROGEN ENVIRONMENTS

Sample	Release of gases*	Composition of gas mixtures (mol%)										
		H_2	O_2	N_2	CH_4	CO	CO_2	hc	SO_2	H_2S	COS	CS_2
		Experiments in helium environment										
Quartz	3.3	<10	77.8	<2.8	<0.3	7.8	14.4	≤0.05	N.D.	N.D.	N.D.	N.D.
Basalt	17.0	12.4	37.2	0.9	1.0	16.7	29.1	2.7	<0.3	<0.3	<0.3	<0.3
Pyroxene	8.5	≤12	69.5	1.8	0.6	15.9	8.5	3.4	<0.3	<0.3	<0.3	<0.3
Peridotite	15.7	46.1	12.8	<1.0	0.4	24.1	15.9	0.7	N.D.	N.D.	N.D.	N.D.
Gabbro	153.0	62.2	~0.03	0.2	0.1	8.2	4.1	0.5	23.7	0.34	0.1	0.6
Tsarev(L5)	52.0	56.6	abs.	1.2	0.3	23.4	8.2	1.4	5.0	0.5	0.2	2.9
Kainsaz (CO3)	142.0	45.7	abs.	0.3	0.6	41.3	10.5	1.4	<0.01	<0.01	0.07	<0.01
		Experiments in hydrogen environment										
Pyroxene	16.5	N.D.	abs.	N.D.	17.1	62.6	11.8	7.4	<0.1	N.D.	<0.1	<0.1
Gabbro	25.3	N.D.	abs.	N.D.	4.4	44.0	2.6	4.0	<0.1	43.1	<0.1	<0.1
Tsarev (L5)	50.5	N.D.	abs.	N.D.	2.3	10.1	4.2	2.2	<0.04	81.2	<0.04	<0.04
Kainsaz (CO3)	119.2	N.D.	abs.	N.D.	2.9	59.8	6.6	2.7	<0.01	27.7	0.2	<0.01

Note: hc—hydrocarbons; N.D.—no data; abs.—detected absorption of oxygen.
*In units $\times 10^{-5}$ cm^3 STP/J (STP—standard temperature and pressure)

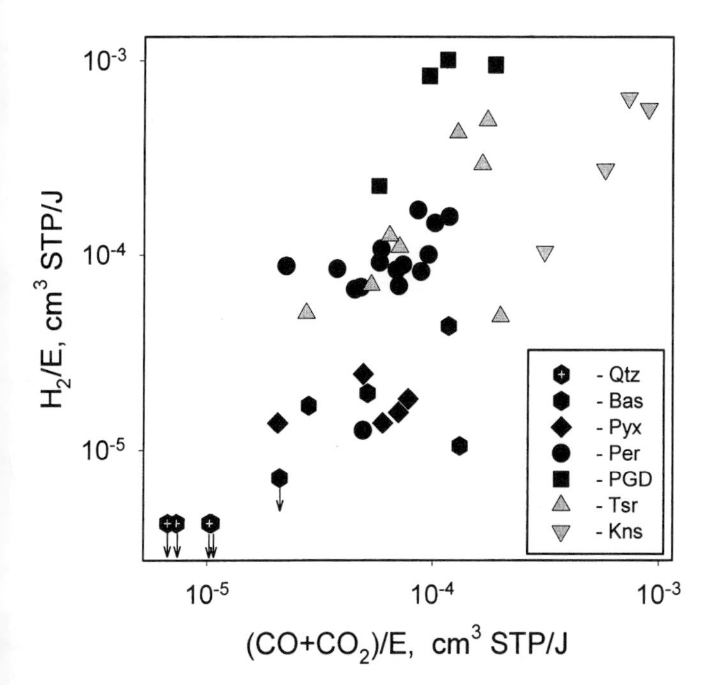

Figure 4. Release of H_2 and sum of CO + CO_2 per unit of laser pulse energy from samples: Qrz, quartz; Bas, basalt; Pyx, pyroxene; Per, peridotite; PGD, picritic gabbro dolerite; Tsr, ordinary chondrite (L5) Tsarev; Kns, carbonaceous chondrite (CO3) Kainsaz. STP is standard temperature and pressure. Arrows mean upper limit.

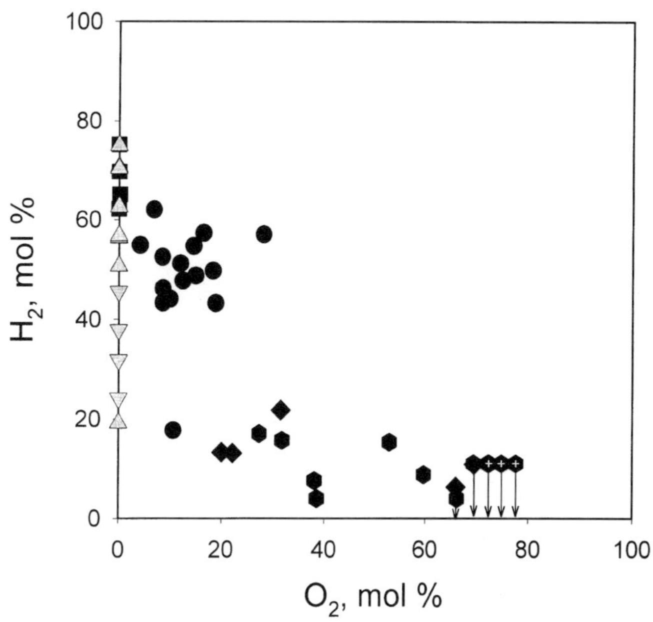

Figure 5. Concentration of H_2 vs. O_2 in released gas mixtures. Symbols as in Figure 4.

atmosphere, the most pronounced effect being observed for methane and benzene. CO and CO_2 were the main carbon-containing components in the experiments with all samples and in both helium and hydrogen atmospheres. The ratio of CO/CO_2 did not depend strongly on the redox state of the atmosphere, but instead on sample type. Figure 7 shows the CO/CO_2 ratio versus the power density of a laser pulse, which is proportional to the temperature of vaporization. All terrestrial rocks did not show any significant growth of CO/CO_2 ratio with the change of power density. The ratio was mostly within its dispersion for individual samples. Their CO/CO_2 ratio was close to unity, which is consistent with thermodynamically calculated values. Meteorites had noticeably higher values for the CO/CO_2 ratio, which is consistent with the deficit of oxygen in their vapor clouds, shown in experiments by the lack of a residual O_2 component. This observation indicates that the formation of the carbon-containing gases probably occurred in an internal region of the gas cloud, without significant mixing with ambient gas. It is important that the concentration of hydrocarbons produced is orders of magnitude larger than that calculated for a gas-phase equilibrium. The stability of the pattern

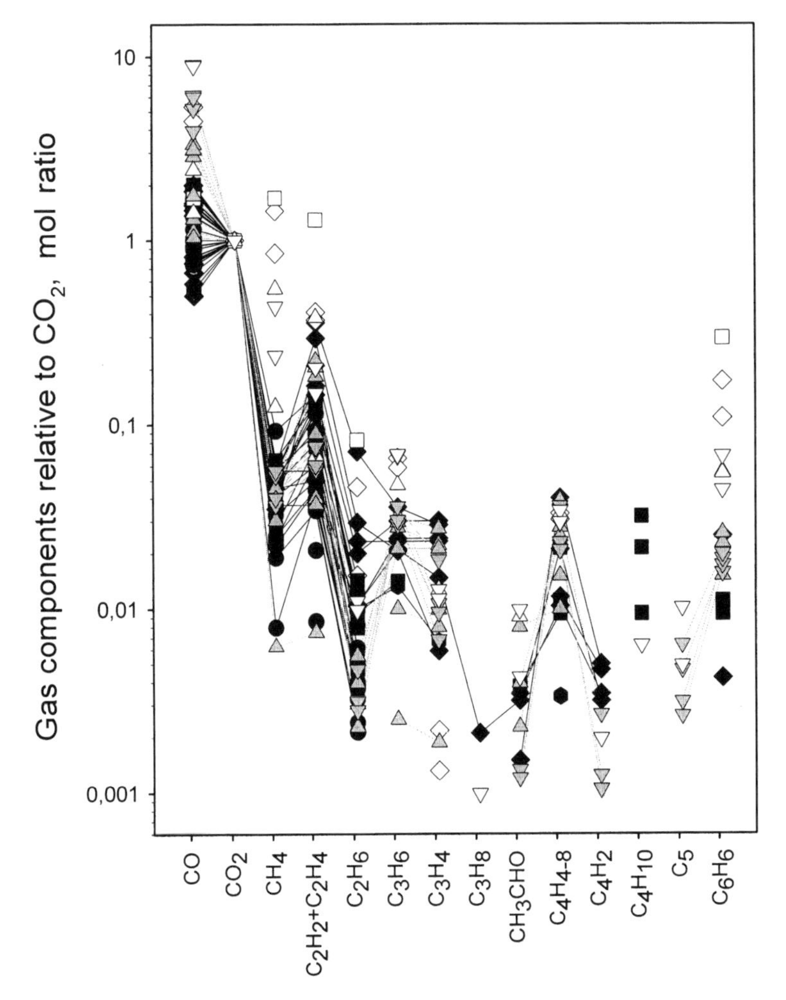

Figure 6. Pattern of carbon-containing gases relative to CO_2 in experimentally obtained gas mixtures. Symbols as in Figure 4. Open symbols indicate experiments in hydrogen environment.

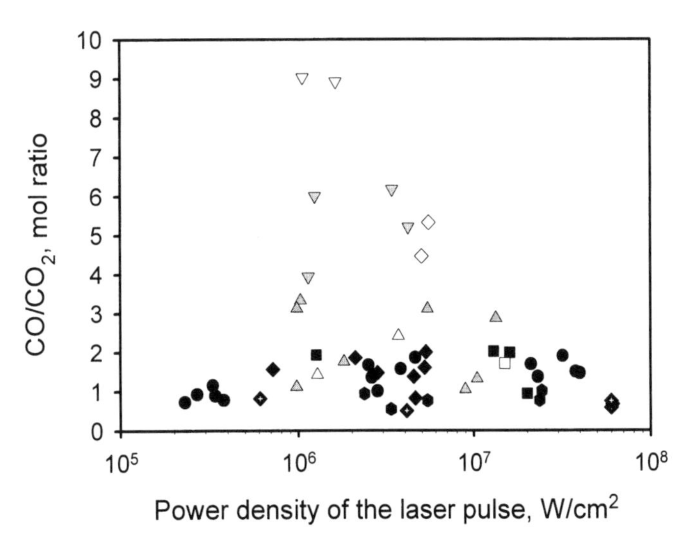

Figure 7. Ratio CO/CO_2 vs. power density of laser pulses in experimentally obtained gas mixtures. Symbols as in Figure 6.

of hydrocarbons and their high concentration compared to gas equilibrium conditions strongly argues for the role of reactions on the surface of nanophase condensed particles, probably due to Fischer-Tropsch type of catalytic reactions, involving CO and H_2. Analyses of the gases in laser pulse experiments could measure only light hydrocarbons that remained as gases at room temperature. Heavier hydrocarbons were adsorbed, mainly on the surface of dispersed condensed nanoparticles. To investigate the possible formation of heavier hydrocarbons during laser pulse experiments, we performed laser pulse vaporization experiments on pyroxene in a methane atmosphere. An atmosphere of methane was the source of sufficient quantities of carbon and hydrogen and provided reduced conditions favorable for the formation of organics. After the laser pulse, the condensed material that formed was collected and put in hexane to dissolve any organic components. The extracted mixture was concentrated and analyzed using the gas chromatography—mass spectrometry technique. A chromatogram of the extracted products is shown in Figure 8. A wide range of polycyclic aromatic hydrocarbons (PAH) was identified, ranging from naphthalene to benzpyrenes. This experiment implies that the syn-

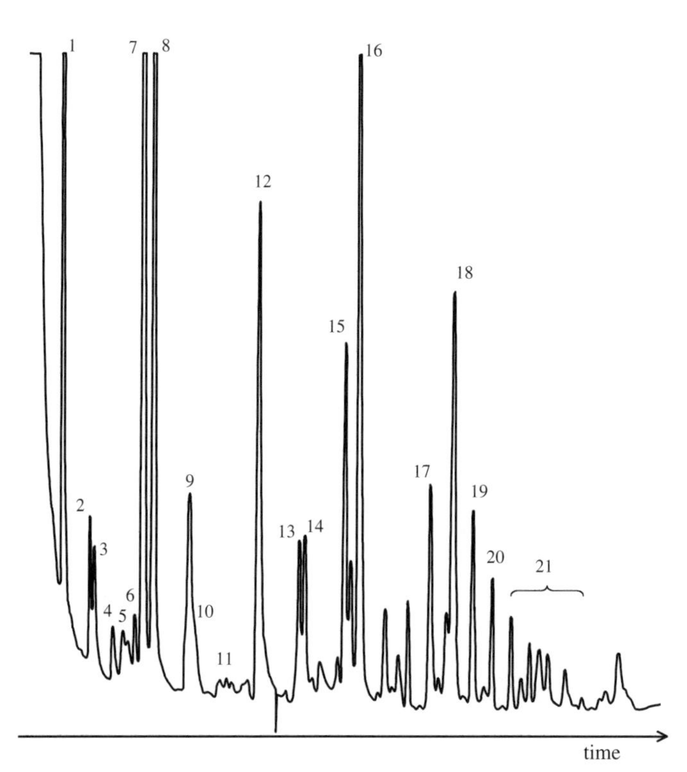

Figure 8. Chromatogram of polycyclic aromatic hydrocarbons extracted from condensed phase in experiment with vaporization of pyroxene in CH_4 atmosphere. 1, naphthalene; 2 and 3, monometylnaphthalene; 4–8, acenaphthene, biphenyl, dimethylnaphthalene; 9, fluorene; 10, monomethylbiphenil; 11, monomethylfluorene; 12, antracene, phenanthrene; 13, ?$C_{15}H_{10}$; 14, dibuthyl phthalate; 15 and 16, pyrene, fluoranthene, benzacenaphthene; 17–20, naphthacene, 1,2 benzanthracene, hryzene, threephenylene, 3,4 benzphenanthrene; 21, benzpyrenes.

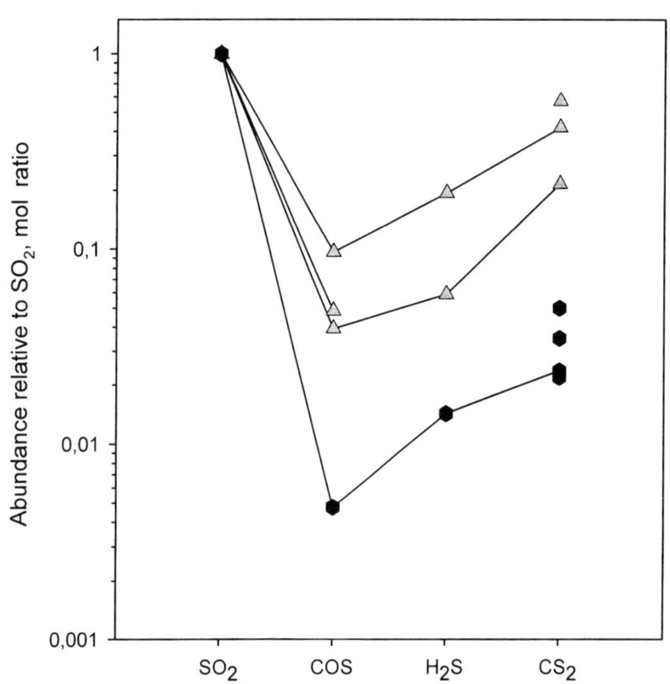

Figure 9. Pattern of sulfur-containing gases relative to SO_2 in experimentally obtained gas mixtures. Symbols as in Figure 4.

thesis of complex organic components during simulated impact vaporization is an important process.

Sulfur was an abundant component only for three samples: carbonaceous and ordinary chondrites, and gabbro. Among the sulfur-containing gases identified were SO_2, COS, H_2S, and CS_2. The pattern of released sulfur containing gases was not as stable as that of carbon. The most pronounced difference was the change in the pattern when hydrogen and helium environments were compared. While the main component in the helium atmosphere was SO_2, in the hydrogen atmosphere it was H_2S. This indicated that the formation of the resultant sulfuric gases proceeded in the region of a vapor cloud that was well mixed with ambient gas that took part in chemical reactions. Another difference was the very weak release of sulfur-containing gases in experiments with carbonaceous chondrite in a helium atmosphere (only the release of COS approached the normal level), while in hydrogen there was a normally high release of H_2S. Laser pulse experiments with ordinary chondrite and gabbro samples in helium atmosphere have shown similar patterns of sulfuric gases (Fig. 9). The production of SO_2 in the case of gabbro was higher compared to the ordinary chondrite experiment, but the ratios between other gases remained the same. Unexpectedly, the sulfur chemistry of the carbonaceous chondrite experiment was different from the ordinary chondrite and gabbro experiments. It could be expected that a higher production of COS and CS_2 components would result due to the higher concentration of carbon in the system. Hence, the only reasonable explanation of the results is that there was a sufficiently higher trapping efficiency of carbonaceous material in respect to SO_2 and CS_2 gases.

Additional information on the chemistry of sulfur during simulated impact vaporization was obtained by an X-ray photoelectron spectroscopic (XPS) analysis of the condensed material. XPS analysis of the condensed material after vaporization of calcite + gypsum + silicate targets was interpreted as an indication of the formation of SO_3 and H_2SO_4 in the vapor cloud (Ivanov et al., 1996; Gerasimov et al., 1994, 1997). SO_3 was not detected by gas chromatography and mass spectrometry probably due to being absorbed by the condensate. The loss of SO_3 in the analytical system by its reaction with H_2O, resulting in formation of H_2SO_4 and further reaction with still walls, can also not be excluded.

Nitrogen was a minor element in the system. Two nitrogen-containing components were measured in gas mixtures: N_2 and HCN. In some experimental runs with carbonaceous chondrite a minor amount of CH_3CN was detected. The measured quantities of N_2 in experiments with terrestrial rocks were at the detection limit level. It was more easily detected in case of meteorite experiments, which usually have a nitrogen concentration of 10–45 ppm (Kothari and Goel, 1974). The measured ratios of HCN/N_2 were in the range of 0.1–0.8. The experimental system was unsuitable for the analysis of nitrogen oxides and their formation was expected, but not measured in the experiment.

POSSIBLE RELEASE OF GASEOUS COMPONENTS IN THE CHICXULUB IMPACT EVENT

The amount of gases released into the atmosphere depended on the composition and dimension of the projectile and impacted target rocks. To evaluate a possible gaseous output during a large impact it is interesting to do this for the Chicxulub impact, which is the main candidate for the K-T impact, as an example. The main problem in extrapolating the results of the laser pulse impact simulation experiments to large impacts is to account for differences in scale.

There is certain evidence indicating that the results of the simulation experiments can be applied to the evaluation of the composition of gases formed during a large-scale impact. During an impact the main component of the vapor cloud is a siliceous material, which condenses into nanoscale nuclei throughout the expanding cloud. This condensation process causes the relationship between pressure and temperature in the cloud to be close to vapor-solid equilibrium conditions. It was shown (Gerasimov et al., 1999) that in an expanding cloud that contains condensing solid material, the relative concentrations C_A, C_B, and C_{AB}, which correspond to components A, B, and AB in a reaction $A + B \leftrightarrow AB$ depend on temperature T as shown in this expression:

$$\frac{C_A \cdot C_B}{C_{AB}} \propto \sqrt{T} \cdot \exp\left(\frac{\lambda - E_r}{RT}\right), \quad (1)$$

where λ is a specific heat of vaporization of silicates, E_r is the effective heat of the reaction, and R is the universal gas constant. Because λ and E_r for interested reactions have comparable values, the exponent has a weak dependence on temperature and it favors a small change in the composition of gas mixture during an expansion. In contrast, the quenching of chemical reactions has a strong dependency on temperature and was evaluated as ~3000 K for laser pulse simulation experiments and ~2000 K for a several-kilometer-size impact (Gerasimov et al., 1999). Taking into account the weak dependence of the products of reactions on temperature, we can assume that there is little change in the composition of a gaseous phase forming in an expanding vapor cloud as temperatures change across this range. Thermodynamic calculations of the chemical composition of the system SiO_2–S–C–H_2O–N (SiO_2 97.3 wt%, S 2%, C 0.2%, H_2O 0.5%, N 50 ppm) with pressure and temperature varying according to the saturation of SiO_2 vapor are shown in Figure 10. The weak dependence of the products of chemical reactions on temperature was considered the main reason for the independence of the composition of gases from the variation within two orders of magnitude of the luminosity density of the laser pulse (see Fig. 7). The stability of the pattern of carbon-containing gases and to some extent that of the sulfur-containing gases is also evidence of the stability of the composition of a gas mixture forming during the expansion of a vapor cloud with varying scales.

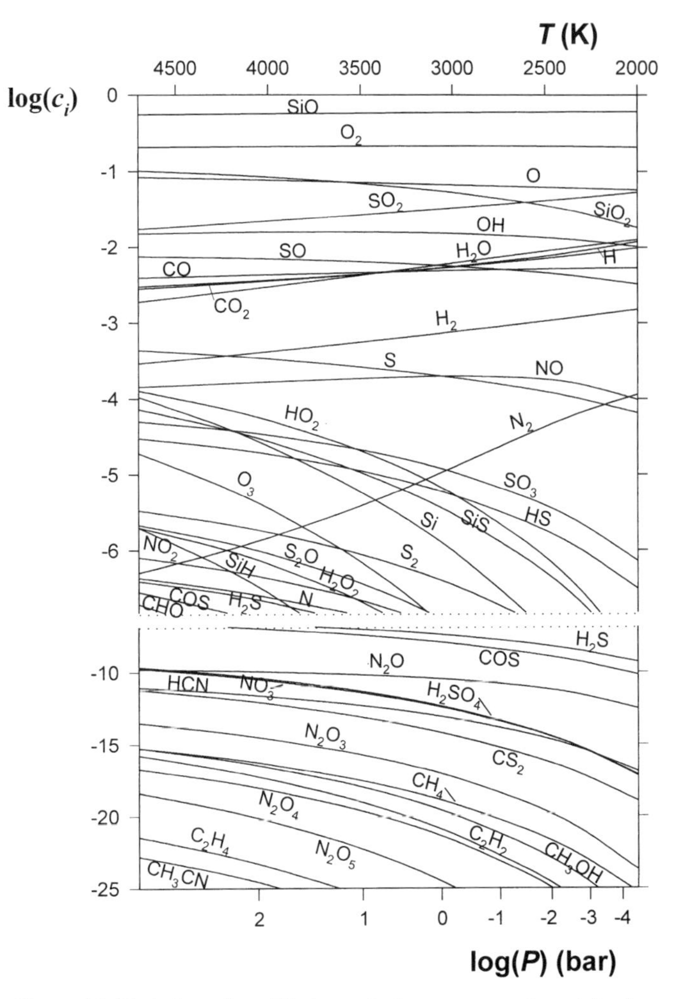

Figure 10. Variation of equilibrium relative concentrations (c_i) of gas components in system SiO_2–S–C–H_2O–N (SiO_2 97.3 wt%, S 2%, C 0.2%, H_2O 0.5%, N 50 ppm) with pressure (P) and temperature (T) varying according to SiO_2 vapor-melt equilibrium. Lower half of graph shows only components discussed in text. Calculation code and thermodynamic data were same as those discussed by Fegley et al. (1986). Modified from Gerasimov et al. (1999).

To evaluate the significance of released gases in the perturbations of the environment during a large impact we need to know first the absolute amounts of all gas components produced by this impact. To do this we make the basic assumption that during an impact the forming gas mixtures are qualitatively the same and the ratios between gas components within carbon- and sulfur-containing groups are similar to those measured in the laser pulse simulation experiments. The following volume ratios were used for the evaluation of the forming gases: $CO:CO_2:CH_4:C_2H_2:C_2H_4:C_2H_6:CH_3CHO:C_3H_6:C_3H_4:C_4:C_5$:

C_6H_6 = 1:1:0.04:0.05:0.05:0.01:0.005:0.02:0.01:0.02:0.005: 0.02 and SO_2:COS:H_2S:CS_2 = 1:0.03:0.05:0.1. These ratios are related to experimental data obtained in helium atmosphere and defined by the internal chemistry of the vapor plume. In the Chicxulub impact event, an increase in the amount of hydrogen in the plume could be expected due to involvement of ocean water, and this hydrogen, in turn, could affect the ratios for sulfur-containing gases in favor of H_2S. Nevertheless, the helium experiment ratios seem to be adequate, because gas mixtures in experiments with ordinary chondrite and gabbro samples in helium had ~60% hydrogen and the starting gabbro sample had 3.4 wt% of water.

The amount of degassed carbon and sulfur during the Chicxulub impact was estimated to have been in the range of 2.4–8.1 × 10^{17} g (Pierazzo et al., 1998) and 3–38 × 10^{16} g (Ivanov et al., 1996), respectively. Based on the experimentally derived ratios, the amounts of released different gas components calculated for the Chicxulub impact are presented in Table 2. Possible amounts of released gases are presented as two models; the first model used minimum values of degassed carbon and sulfur, and the second model used their maximum values. The difference between these two models does not define the possible range of minimum and maximum values for each gas component, because experimentally derived ratios between gases vary within an order of magnitude, thereby decreasing the minimum and increasing the maximum values. These models were built mainly to illustrate possible variations associated with changes in the output of volatiles in an impact. The amount of HCN was calculated based on normal nitrogen concentration of ~40 ppm (Kothari and Goel, 1974) in a meteoritic impactor with a mass of 10^{19} g and conversion of 20% of nitrogen into HCN. A possible admixture of atmospheric N_2 into the vapor plume was not accounted for because there are no reliable models that evaluate the mixing of a plume with an atmosphere. A simple estimation shows that an admixture of the air from a cylindrical layer corresponding to a crater diameter ~180 km and a thickness of 1 km involves an order of magnitude more nitrogen into the system than was delivered by the impactor. Table 2 also does not include gases, which could be a result of an air shock wave chemistry producing nitrogen oxides and hydrogen cyanide (Fegley et al., 1986). Additional effects will increase the production of gases during an impact.

TABLE 2. ESTIMATED AMOUNTS OF GASES THAT COULD BE PRODUCED DURING THE CHICXULUB IMPACT

Gas components	Degassed by the Chicxulub impact (g)	
	Model 1	Model 2
CO_2	3.4 × 10^{17}	1.1 × 10^{18}
CO	2.2 × 10^{17}	7.2 × 10^{17}
SO_2	4.7 × 10^{16}	5.9 × 10^{17}
CS_2	5.6 × 10^{15}	7.1 × 10^{16}
H_2S	1.3 × 10^{15}	1.6 × 10^{16}
COS	1.3 × 10^{15}	1.7 × 10^{16}
CH_4	4.9 × 10^{15}	1.6 × 10^{16}
C_2H_2	1.0 × 10^{16}	3.3 × 10^{16}
C_2H_4	1.1 × 10^{16}	3.6 × 10^{16}
C_2H_6	2.3 × 10^{15}	7.7 × 10^{15}
CH_3CHO	1.7 × 10^{15}	5.6 × 10^{15}
C_3H_6	6.5 × 10^{15}	2.2 × 10^{16}
C_3H_4	3.1 × 10^{15}	1.0 × 10^{16}
C_4	8.3 × 10^{15}	2.8 × 10^{16}
C_5	2.5 × 10^{15}	8.5 × 10^{15}
C_6H_6	1.2 × 10^{16}	4.0 × 10^{16}
HCN	1.5 × 10^{14}	1.5 × 10^{14}

Note: See the description of models 1 and 2 in the text.

TOXIC EFFECTS OF RELEASED GASES

Table 2 gives a list of relatively stable gas components; those are the ones that survived the experimental process, as the most aggressive components were apparently lost in the analytical system. Nevertheless, the injection of large quantities of the listed components into the atmosphere could have resulted in a serious perturbation of the ecosystem and in unsuitable living conditions for some life forms.

The exposure of any living organism to toxins could be the result of eating or respiration of poisoned substances, or the absorption of toxins through the skin. Some toxins can be removed from the body after exposure, but some toxins could accumulate in an organism. The mortal action of toxins could be a result of, e.g., asphyxia, paralyses of different systems in the body, and general illness. There could also be long-term effects such as the development of cancer or chronic illnesses, mutations, and the damage of reproduction functions. Some characteristics of toxic action of impact gases can be derived from the *Encyclopedia of Toxicology* (Wexler, 1998). Carbon dioxide causes anoxia and exerts a direct toxic effect to the heart. Carbon monoxide displaces oxygen in hemoglobin, causing anoxia; extended exposure to this gas causes damage to the reproductive system, and death of neuronal cells. Hydrocarbons have general irritants, cause allergies, and long-term exposure to certain types of them causes narcosis, carcinogenesis, mutagenesis, and teratogenesis. Sulfur dioxide produces heart, lung, and kidney damage. Hydrogen sulfide causes eye irritation, brain-tissue necrosis, changes in the liver, hyperemia, ataxia, anorexia, sudden loss of consciousness, and cardiac arrest. Carbon disulfide causes organic brain damage, peripheral nervous system damage, neurobehavioral dysfunction, strong eye irritation, skin blistering, and reproductive disorders. Carbonyl sulfide causes eye irritation and acts on the central nervous system to produce respiratory paralysis. Hydrogen cyanide is highly toxic and causes paralysis of the respiratory center.

A critical parameter for any living organism to survive is the concentration of gas components in the environment and the time of exposure to this concentration. While the plume was expanding from the center of an impact, the concentration of injected gases was decreasing with distance. Gases could be removed from an atmosphere due to geochemical processes such as weathering and dissolution in the ocean; however, most of the geochemical processes occur over longer time scales compared to the lateral spreading of an impact generated cloud and its consequence on living beings, so the proposed model considers only the effect of the decrease of concentration of gas components with the spreading of a cloud from the center of an impact. The model considers an ideal mixing of released gases in a portion of modern atmosphere defined by a projection of the cloud on the Earth's surface, the relative concentrations of gas components being constant vertically and laterally within this cloud. The mixing ratio of a gas component c_i is determined as

$$c_i = \frac{M_i}{M_a(S) + \sum_i M_i}, \quad (2)$$

where $M_a(S)$ is the mass of modern Earth's atmosphere over a planetary surface S, M_i is the mass of released gas component i. To evaluate the significance of the role of toxic gases in the extinction of life forms, it is interesting to see in what portion of an atmosphere it is necessary to dissolve injected gases such that their concentration will drop below the lethal concentration or permissible exposure limit. Lethal concentration defines a concentration of a toxic gas component above which some organisms (e.g., human beings and many animals) will die after contact with this gas for few minutes or hours. The permissible exposure limit defines the concentration of a toxic gas component below which living beings will be safe for an unlimited period of time. Concentrations of toxic gases lower than lethal concentration, but greater than permissible exposure limit, are hazardous for many organisms, because they can result in long-term, potentially lethal, effects, such as cancer, mutations, and teratogenesis. Different organisms have different resistance to different toxic gases and therefore there cannot be a simple model for the entire biosphere. Moreover, for some living forms the injection of a certain gas can result in the opposite effect; i.e., gases that are mortal for one organism can be favorable for another (e.g., for some forms of bacteria). Permissible exposure limits were defined mainly for human beings and there are a variety of standards that have been established by different countries and for different situations (e.g., limits for exposure at the workplace, limits for injected temporal exposure, limits for permanent over life exposure). Permissible exposure limits were used in the model, which were accepted in the USSR for continuous lifetime exposure (Grushko, 1986). Lethal concentration values used in the model of Wexler (1998) were defined for humans and animals. Table 3 shows the concentration of released gas components if mixed with the total terrestrial atmosphere in terms of permissible exposure limits, and the percentage of the total Earth's surface covered by an expanding cloud, when the concentration of a certain released gas component by mixing with the atmosphere is reduced to its lethal concentration value.

Table 3 shows that some gases have a noticeably high toxic effect for human beings and consequently for animals who have analogous reactions to toxic gases. The most significant effect is from the release of carbon monoxide and sulfur-containing gases. The expansion of a cloud over >10% of the Earth's territory could be lethal by direct action of released toxic gases: 10% of the Earth's territory is an area with diameter ~8000 km, which is equal to the dimensions of a large continent. Table 3 also shows that even after dissolution of released gases from a Chicxulub-like impact into the entire Earth's atmosphere, the concentration of many components will exceed permissible exposure limits by several orders of magnitude. Among these components are carbon monoxide, sulfur-containing gases, and hydrocarbons that can result in a long period mortality due to the damage of reproduction function, carcinogenesis, mutagenesis, and teratogenesis.

Hydrogen cyanide in the model for the entire atmosphere has a concentration three times higher than its permissible exposure limit. An account of nitrogen that could be admixed from an atmosphere could increase its concentration by orders of magnitude, making it a significant component.

Table 3 does not include any information for polycyclic aromatic hydrocarbons (PAH) because their rate of production in a Chicxulub-like impact could not be derived from the available experimental data. The significance of PAHs even at a low rate of production can be associated with the possibility of synthesis of some species that are highly toxic even at low concentration. For example Figure 8 shows the possibility of synthesis of benzpyrenes, among which 1,2-benzpyrene is known to be an extremely dangerous toxin, having the permissible exposure limit mixing ratio of 10^{-12} (Grushko, 1986). Some PAH can be accumulated in a body, increasing the harmful effect over time.

DISCUSSION AND CONCLUSIONS

The model used amounts of carbon (Pierazzo et al., 1998) and sulfur (Ivanov et al., 1996), which were calculated for the Chicxulub impact as a result of degassing of evaporites from target rocks. Model 2 used values of degassed carbon and sulfur that corresponded to impactor concentrations of ~8 wt% and ~4 wt% if divided by the mass of the assumed impactor (~10^{19} g). If we consider the material of meteorites as a possible source for the impactor, the concentration of sulfur is ~2 wt% and the concentration of carbon is 0.1–0.5 wt% for ordinary and 1–2 wt% for carbonaceous chondrites (Jarosewich, 1990). An impact of a two times larger (by weight) meteorite can produce the same output of sulfur-containing gases as in model 2, and a four times larger carbonaceous chondrite-like asteroid can produce the same output of carbon-containing gases without involving target volatiles. In the case of an impact of a comet, a smaller impactor can produce the same amount of toxic gases as in model 2.

The model presented here shows that the amount and chemical composition of impact-generated gases in a Chicxulub-like impact could be significant factors in the extinction of advanced life forms over large territories. The main effect of extinction

TABLE 3. PROBABLE TOXIC EFFECT OF GASES RELEASED BY THE CHICXULUB IMPACT

Gas component	PEL* ppm	Over globe mixing, in numbers of PEL		LC§ ppm	Dimension of zone with concentration ≥ LC, % of Earth's surface	
		Model 1	Model 2		Model 1	Model 2
CO_2	500	0.14	0.45	100000	0.05	0.17
CO	1	43	143	1000	4.3	14.3
COS				1000	0.01	0.3
SO_2	5	2	24	1000	0.9	12
CS_2	0.005	223	2820	220	0.5	6.4
H_2S	0.008	31	395	100	0.24	3
HCN	0.01†	3	3	100†	0.02	
C_2H_4	3	0.7	2			
CH_3CHO	0.01	34	113			
C_3H_6	3	0.4	1.4			
C_6H_6	0.1	24	80	32000		

*Permissible exposure limit from Grushko (1986).
†From Bobkov and Smirnov (1970).
§Lethal concentration from Wexler (1998).

derives from the release of carbon monoxide and sulfur-containing gases. This means that the proposed model is not critically dependent on the stability of the forming pattern of the gases. Carbon monoxide and carbon dioxide are the major carbon-containing gases and their ratio is rather stable. The ratio could be shifted slightly in favor of carbon dioxide if lower temperatures of quenching are considered for large-scale impacts (see Fig. 10) but it also could be shifted in favor of carbon monoxide if the material of meteoritic impactor was a significant part in the plume where carbon-containing gases were formed. The uncertainty in the pattern of sulfur-containing gases is not critical because all of them have high toxicity and replacement of one by another will not significantly change the general toxic effect.

The model considered only the individual action of toxins while they were released as mixture. The question of whether there could be a cumulative toxic effect of the mixture of gases, when the action of some toxins could amplify the action of another by suppression of the resistive abilities of living systems, must be considered individually for certain forms of life, but it is beyond the scope of this chapter. Nevertheless, these effects, if they exist, can increase the territory affected by the strong toxic effect of the released gases. If we consider a cumulative action of gases, then as much as 40% of the Earth's territory in model 2 could be covered by a cloud with lethal concentration.

The cloud, which expands infinitely from the center of an impact and ultimately dissolves in the atmosphere, has a constantly decreasing concentration of released gas components, and its toxic action is limited by a certain territory and time of expansion. One could expect that the expansion of a cloud will be stopped with the loss of its dynamic forces and the cloud later will be transported by normal atmospheric circulation, covering long distances without sufficient loss of the concentration of toxins. Such a scenario looks realistic and must be considered in models of atmospheric circulation.

The model highlights the significance of hydrocarbons and PAHs that are forming during an impact. While their direct action was not very significant, their long-term action could have a noticeable effect due to carcinogenesis, mutagenesis, and teratogenesis. The effect could be more pronounced for PAHs rather than for light hydrocarbons despite higher rate of production of the latter, because PAHs are not volatile and would be concentrated by precipitation on the Earth's surface in the living zone of living organisms. To explain the role of PAHs in biotic extinctions, it is necessary to have the information about the efficiency of their production at conditions such as those of the Chicxulub impact, and this needs further investigation.

The model presented here does not account for the chemistry of chlorine, which could also be an important volatile element in the plume, because salty ocean water was involved in the Chicxulub impact. Components such as phosgene ($COCl_2$) or dioxins could be synthesized in the vapor plume during an impact, but there are no data on their production efficiency. Phosgene is known to be an efficient war gas and dioxins are extremely dangerous components even at very low concentrations.

Another possible effect of an impact that needs to be investigated is the local change in the oxygen budget in the atmosphere. Experiments have indicated that large portions of molecular oxygen were mobilized into the vapor cloud due to the thermal decomposition of rocks, and later this oxygen was almost totally used up for the condensation of solid material. Condensation of meteoritic material resulted in the absorption of additional portions of molecular oxygen from the atmosphere, while condensation of terrestrial rocks resulted in the evolution of molecular oxygen into the atmosphere. The degree of absorption of molecular oxygen by condensing meteoritic material was not investigated, and therefore it is not possible to evaluate a possible effect of local oxygen depletion in the atmosphere. Nevertheless, any depletion of molecular oxygen could affect the local ecosystem.

The injection of a variety of gases into an atmosphere during an impact could promote other mechanisms of extinction. The formation of aerosols of sulfuric acid particles from sulfur dioxide and water vapor in the upper atmosphere (Pope et al., 1994) or their direct synthesis in the vapor plume (Ivanov et al., 1996) could damage the solar radiation balance and hence cause climatic change. Sulfur- and chlorine-containing gases could result in damage to the ozone layer. The dissolution of gases in the ocean could result in a change of the water pH, and affect the aquatic biota. Additional disturbances in the ecosystem are expected to enhance the toxic effect of evolved gases due to suppression of resistive capabilities of living systems.

ACKNOWLEDGMENTS

I thank E.N. Safonova for the discussion and for the help with the preparation of the paper. I also thank Kevin Pope and Mark Sephton for reading an earlier version of the paper and for their valuable suggestions that led to additions and improvements to the manuscript. The research was supported in part by INTAS grant 97-11084.

REFERENCES CITED

Bobkov, S.S. and Smirnov, S.K., 1970, Sinilnaya kislota [in Russian]: Moskva, Khimiya, 176 p.

Fegley, B., Jr., Prinn, R.G., Hartman, H., and Watkins, G.H., 1986, Chemical effects of large impacts on the earth's primitive atmosphere: Nature, v. 319, no. 6051, p. 305–308.

Gerasimov, M.V., and Mukhin, L.M., 1984, Studies of the chemical composition of gaseous phase released from laser pulse evaporated rocks and meteorite materials [abs.]: Lunar and Planetary Science Conference, 15th, Houston, Texas, Lunar and Planetary Institute, p. 298–299.

Gerasimov, M.V., Satovsky, B.L., and Mukhin, L.M., 1987, Mass-spectrometrical analyses of gases originated during impulsive evaporation of meteorites and terrestrial rocks [abs.]: Lunar and Planetary Science Conference, 18th, Houston, Texas, Lunar and Planetary Institute, p. 322–323.

Gerasimov, M.V., Dikov, Yu.P., Yakovlev, O.I., and Wlotzka, F., 1994, High-temperature vaporization of gypsum and anhydrite: Experimental results [abs.]: Lunar and Planetary Science Conference, 25th, Houston, Texas, Lunar and Planetary Institute, p. 413–414.

Gerasimov M.V., Dikov Yu.P., Yakovlev O.I., and Wlotzka F., 1997, Experimental investigation of the chemistry of vaporization of targets in relation to the Chicxulub impact [abs.]: Houston, Texas, Lunar and Planetary Institute, LPI Contribution No. 922, p. 15–16.

Gerasimov, M.V., Ivanov, B.A., Yakovlev, O.I., and Dikov, Yu.P., 1999, Physics and Chemistry of Impacts, *in* Ehrenfreund, P., Krafft, K., Kochan, H., and Pirronello, V., eds., Laboratory astrophysics and space research, Astrophysics and Space Science Library, Volume 236: Dordrecht, Kluwer Academic Publishers, p. 279–330.

Grushko, Ya.M., 1986, Toxic organic species in industrial pollution into atmosphere [in Russian]: Leningrad, Khimiya, 207 p.

Ivanov, B.A., Badukov, D.D., Yakovlev, O.I., Gerasimov, M.V., Dikov, Yu.P., Pope, K.O., and Ocampo, A.C., 1996, Degassing of sedimentary rocks due to Chicxulub impact: Hydrocode and physical simulations, *in* Ryder, G., Fastovsky, D., and Gartner, S., eds., The Cretaceous-Tertiary event and other catastrophes in Earth history: Geological Society of America Special Paper 307, p. 125–139.

Jarosewich, E., 1990, Chemical analyses of meteorites: Meteoritics, v. 25, p. 323–337.

Kothari, B.K., and Goel, P.S., 1974, Total nitrogen in meteorites: Geochimica et Cosmochimica Acta, v. 38, p. 1493–1507.

Lange, M.A., and Ahrens, T.J., 1986, Shock-induced CO_2 loss from $CaCO_3$: Implications for early planetary atmospheres: Earth and Planetary Science Letters, v. 77, p. 409–418.

Mukhin, L.M., Gerasimov, M.V., and Safonova, E.N., 1989, Origin of precursors of organic molecules during evaporation of meteorites and mafic terrestrial rocks: Nature, v. 340, p. 46–48.

Pierazzo, E., Kring, D.A., and Melosh, H.J., 1998, Hydrocode simulation of the Chicxulub impact event and the production of climatically active gases: Journal of Geophysical Research, v. 103, p. 28 607–28 625.

Pope, K.O., Bains, K.H., Ocampo, A.C., and Ivanov, B.A., 1994, Impact winter and the Cretaceous/Tertiary extinctions: Results of a Chicxulub asteroid impact model: Earth and Planetary Science Letters, v. 128, p. 719–725.

Siret, D., Robin, E., Guern, F.Le., and Cheynet, B., 2000, Condensation in an impact vapor plume: Estimates of gas and aerosol emissions generated by the Chicxulub impact [abs.]: Catastrophic events and mass extinctions: Impacts and beyond: Houston, Texas, Lunar and Planetary Institute, LPI Contribution No. 1053, p. 201.

Tyburczy, J.A., and Ahrens, T.J., 1986, Dynamic compression and volatile release of carbonates: Journal of Geophysical Research, v. 91, p. 4730–4744.

Tyburczy, J.A., and Ahrens, T.J., 1993, Impact-induced devolatilization of $CaSO_4$ anhydrite and implications for K/T extinctions: Preliminary results [abs.]: Lunar and Planetary Science Conference, 24th, Houston, Texas, Lunar and Planetary Institute, p. 1449–1450.

Wexler, Ph., editor, 1998, Encyclopedia of toxicology: London, Academic Press, v. 1–3 (v. 1, 605 p.; v. 2, 614 p.; v. 3, 486 p.).

Manuscript Submitted November 23, 2000; Accepted by the Society March 22, 2001

Geological Society of America
Special Paper 356
2002

Modeling long-term climatic effects of impacts: First results

Thomas Luder
Willy Benz
Thomas F. Stocker
Physics Institute, University of Bern, Sidlerstrasse 5, CH-3012 Bern, Switzerland

ABSTRACT

Catastrophic impacts of asteroids or comets on Earth are events that have occurred in the past and will happen in the future. While climate perturbations induced by an impact-generated dust cloud have been studied in recent years, the studies have been restricted to the first year following the event. The aim of this study is to assess the long-term response of the climate including ocean circulation following the impact of a 5-km-diameter asteroid.

We modeled the dust evolution, calculated radiation transfer through dust layers, and used a zonally averaged, coupled ocean-atmosphere model to obtain the climatic response. Within less than one year, mean sea surface temperatures (SST) dropped by ~2 °C. Even though most of the dust had been removed from the atmosphere after several months, SST deviations remained larger than 0.1 °C for a period of 20 yr following impact. In the first year, ocean temperatures below 500 m depth did not change. Although 60 yr later the main temperature deviations were still located above 1000 m depth, temperature changes occurred throughout the ocean. After 2000 yr the oceans showed slightly increased temperatures. A zone below the surface of the northern Atlantic exhibited increased temperatures by 3 °C. Precipitations decreased within months to about half of normal, but recovered after 1 yr. We did not observe glaciation effects.

INTRODUCTION

Temperature drops and darkness lasting for months are some of the outcomes triggered by impacts of asteroids and comets on the Earth. The wide range of effects (see review by Toon et al., 1997) starts with an air blast and seismic waves; the destruction scale depends on the impactor size, but ranges from local for 1 km bolides to continental for 10 km objects. Ocean impacts induce tsunamis, which threaten coastal regions (Hills et al., 1994; Ward and Asphaug, 2000). It is also presumed that part of the global land biomass catches fire (Wolbach et al., 1990) due to the impact ejecta energy deposition into the atmosphere (Melosh et al., 1990). Chemical effects such as the release of nitric oxide (Zahnle, 1990), worldwide acid rain (Lewis et al., 1982; Prinn and Fegley, 1987), and destruction of the ozone layer have also been invoked.

The globally distributed ejecta in the stratosphere perturbs the radiation budget of air, land, and water on Earth. Larger fragments settle rapidly, whereas submicrometer dust particles remain in the stratosphere for months (Toon et al., 1982). If the impact releases sulfur (Pope et al., 1994, 1997; Kring et al., 1996; Pierazzo et al., 1998), sulfuric acid aerosols could be found in the atmosphere for years. Because of their long residence times in the air and their interaction with radiation, small dust particles and, if released, sulfuric acid aerosols disturb the climate. The effect of impact-induced dust on global tempera-

Luder, T., Benz, W., and Stocker, T.F., 2002, Modeling long-term climatic effects of impacts: First results, *in* Koeberl, C., and MacLeod, K.G., eds., Catastrophic Events and Mass Extinctions: Impacts and Beyond: Boulder, Colorado, Geological Society of America Special Paper 356, p. 717–729.

tures has been investigated with one-dimensional radiative-convective atmospheric models (Toon et al., 1982) and with three-dimensional atmospheric general circulation models (Covey et al., 1990, 1994). The focus has been on changes in the atmosphere, whereby the underlying surface has been taken into account as a heat reservoir. Typically a zero heat capacity was chosen for continental areas. Oceans were represented with either infinite or finite heat capacities. While in the former case these simulations are appropriate only as long as the sea temperatures have not reacted to the perturbation, in the latter case the sea is approximated by static oceans, and influences and changes in the circulation are neglected. Therefore, the time span integrated by these models is usually restricted to 1 yr.

To consider climatic changes on time scales of centuries to millennia, we use a low-order coupled ocean-atmosphere model for paleoclimate studies, coupled to a multiscattering radiative transfer and a dust evolution simulation code. We discuss assumptions and present our model, and then report on climatic results.

The results do not necessarily reflect the Cretaceous-Tertiary (K-T) event, because this study investigates climatic effects of an impact with the present oceanic configuration and the present climate, and because the amount of dust considered corresponds to a smaller impactor (5 km) than the K-T bolide. We regard this work as a contribution to the exploration of the consequences of a collision with an asteroid or comet today. However, we think that similar effects might have occurred after real impacts in the past.

ASSUMPTIONS AND MODEL

Organization of the model

Model simulations of the K-T ejecta trajectories give evidence of a nonspherical symmetric distribution of the dust around the globe (Argyle, 1988; Durda et al., 1997). However, for simplicity, we assume in this study that the dust has spread out globally and that the distribution does not depend on geographical location. Thus horizontal motions of dust are not allowed in the model. The temporal development of the dust layer in the stratosphere is modeled with a one-dimensional dust evolution code that takes into account the settling and the coagulation processes of dust particles. A multiscattering radiation transfer code calculates solar and infrared energy fluxes through the dust layer specified by the dust evolution model. The energy fluxes depend on a set of parameters, including the solar position and the temperature of the underlying ocean. An ocean circulation model simulates the response of the climate to the perturbed energy fluxes.

Initial dust distribution after the impact

The plume of melted and vaporized target and impactor expands at velocities of the order of Earth's escape velocity and pervades the atmosphere. Once these particles have left the vicinity of the atmosphere, their orbits become pure Keplerian. Some of this material will leave the Earth, and some moves at altitudes of hundreds of kilometer around the globe. Then it reenters the atmosphere from above. This takes place at all locations around the Earth only hours after the ejection (Melosh et al., 1990). Thus, compared to the lifetime of dust particles in the atmosphere, the dust is assumed to be distributed both instantaneously and globally. Therefore, our model starts with the dust already spread worldwide and uniformly.

A good guess of the size of ejected particles is taken from the size distribution of volcanic dust (Toon et al., 1982). Farlow et al. (1981) measured sizes of aerosols collected in the stratosphere between and after the eruptions of Mount St. Helens, and fitted the results with log-normal size distributions. They found mode radii r_0 from 0.48 μm to 0.94 μm. For aerosols in the stratosphere 1 yr after an average volcanic eruption, Jaenike (1988) gave a log-normal size distribution with a mode radius of $\sim$0.217 μm. This value is smaller than the radii from Farlow et al. (1981) because the size distribution by Farlow et al. (1981) refers to samples of fresh volcanic aerosols, while the data given by Jaenike (1988) represent an evolved population. In 1 yr the larger particles have moved out of the stratosphere faster than smaller particles; therefore, the size distribution shifts to the smaller sizes.

In their investigation of the evolution of an impact-induced dust cloud, Toon et al. (1982) used a log-normal particle size distribution with $r_0 = 0.5$ μm. They stated a low sensitivity of the cloud evolution to changes in the mode radius r_0. In our model the amount of dust in the stratosphere after the impact is insensitive to the value of r_0 in the interval between 0.5 μm and 0.9 μm. We conclude that the precise value is not critical. We use a log-normal size distribution with mode radius r_0 of 0.7 μm, which is in the range of the values given by Farlow et al. (1981) for the fresh volcanic aerosols.

At the larger end of the size distribution we neglect all particles with diameter larger than 4 μm for several reasons. (1) Farlow et al. (1981) found no particles larger than 3 μm in diameter. (2) Because the fall speed of large particles is high, they are quickly removed from the atmosphere by settling and washout before they have an influence on the climate (Toon et al., 1997). (3) Because of the steepness of the size distribution, their number is smaller by several orders of magnitude than the particles with the typical radius of 0.7 μm. To appraise the total initial amount of dust, we use the estimate by Toon et al. (1997) that the mass of submicrometer dust reaching the stratosphere equals a fraction $\alpha = 30\%$ of the impactor's mass, which is a small fraction of the total ejecta. Let R, d, and ρ denote the radius of the Earth, the impactor diameter, and its density, respectively. Then the mass of the impactor and the surface area of the Earth are $\pi \cdot \rho \cdot d^3/6$ and $4 \cdot \pi \cdot R^2$. Hence, if the dust is distributed worldwide, the column density of dust is given by

$$L = \alpha \cdot \rho \cdot d^3/(24 \cdot R^2) \quad (1)$$

(Fig. 1). In the climate simulation of this study we use a dust load of $L = 100$ g/m^2, which corresponds to a bolide with diameter of ~5 km and a density of 3000 kg/m^3.

For simplicity we assume that the number density of the particles does not depend on height at the beginning. Therefore, our model distributes the dust homogeneously along the vertical axis. Toon et al. (1982) injected the initial dust in two scenarios, between 12 km and 42 km, and between 66 km and 88 km. They found that the optical depth of an evolving dust layer is only slightly sensitive to the initial height of the particles. Covey et al. (1990) inserted the dust at altitudes from ~15 to 30 km. We set the initial lower and upper boundaries of the dust layer at $z = 20$ km and $z = 100$ km, respectively. Variations in the upper boundary do not change the evolution of the dust layer, because the Stoke's settling velocities of dust particles at these high levels are of ~100 m/s. To move from 100 km height to 80 km or 60 km takes <1 h or ~1 d, respectively, for all particles. Because total dust evolution and climatic change occur on time scales of weeks to months and even years, these initial differences in dust layer are not important. However, fall velocities of micrometer particles in the lower stratosphere are ~1 mm/s. Thus, it is not probable that impact-formed particles can be found at the lower boundary of the stratosphere around the globe shortly after they have penetrated the atmosphere from above. Therefore, we do not set the lower initial dust boundary at the tropopause but higher, at 20 km.

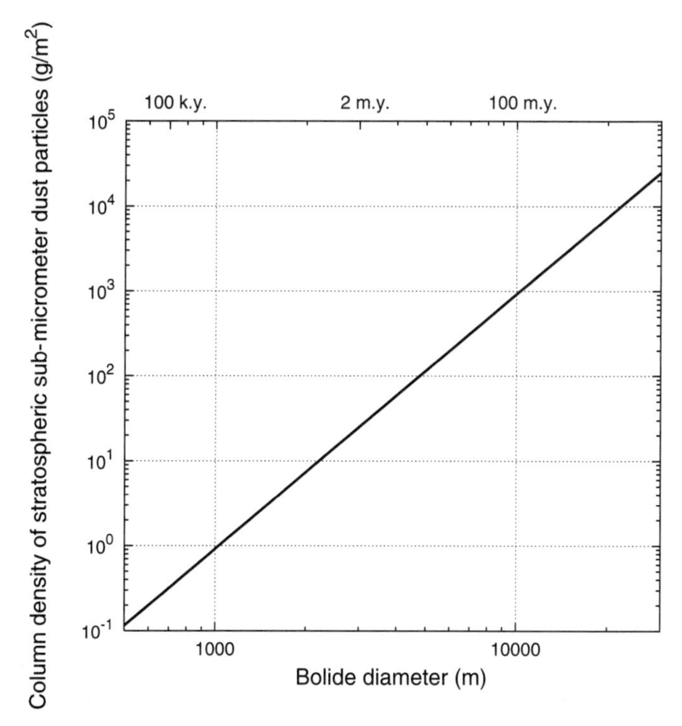

Figure 1. Amount of globally distributed submicrometer dust particles lofted by impacts into stratosphere from equation 1 (see text), assuming impactor density of 3000 kg/m^3. Column density of 1 g/m^2 corresponds to global dust mass of 5.1×10^{14} g. For comparison, density of sheet of paper is ~50–100 g/m^2. Upper abscissa indicates time intervals between impacts of bodies of given size or larger (from Fig. 1 in Morrison et al., 1994).

In our simulations, dust grains are assumed spherical, which is a crude simplification of their natural shapes. However, this assumption greatly simplifies the calculation of scattering and absorption cross sections of electromagnetic waves.

Dust evolution model

The dust is specified by the particle size distribution as a function of time and height in the atmosphere. The distribution evolves with time due to coagulation and sedimentation processes, which are the main mechanisms that alter the local number of particles (Toon et al., 1997).

We assume that all particles fall with a velocity that is given by the equilibrium of gravity and drag. The Stoke's drag with Cunningham slip correction factor (Otto et al., 1999) is determined by particle size, assuming spherical shape, and local parameters of the atmosphere. Thus the fall velocities depend on, among other factors, particle diameter, air temperature, and density. Large particles have a larger volume to cross-section ratio than small particles; therefore their settling speed is larger. Because the drag force increases with number density of the air molecules, and air density decreases exponentially with height, the fall speed is much smaller in lower parts of the atmosphere than in the upper parts. Thus all particles fall fast as they enter the atmosphere from above, and they slow down as they reach denser, lower atmospheric levels.

Particles collide due to different settling speeds and due to individual Brownian motion. If a collision occurs, there is a chance that the particles stick to each other. The probability is large for small relative velocities and vice versa. In the model, we use a sticking probability of unity, if the relative speed is smaller than a specific critical sticking velocity. Otherwise the particles do not stick. The critical velocity depends on the particles' radii, masses, and material constants. Analytical expressions for the sticking velocity are from Dominik and Tielens (1997), Chokshi et al. (1993), and Wurm and Blum (1998). Destruction of particles due to high-speed collisions and the influence of liquid aerosols on the coagulation of dust particles are not allowed for in the model.

The effect of coagulation is a reduction in the number of small particles and an increasing number of large ones. Because large particles settle faster than small ones, coagulation reduces the residence time of a dust layer in the atmosphere significantly. Figure 2 shows the evolution of the amount of dust in the stratosphere for five initial values of the column dust density. To simulate the sedimentation and the coagulation of the dust, we do not follow single particles but propagate the particle size distribution along the vertical axis. In the model, the atmosphere from 0 to 100 km height is divided into 1000 cells.

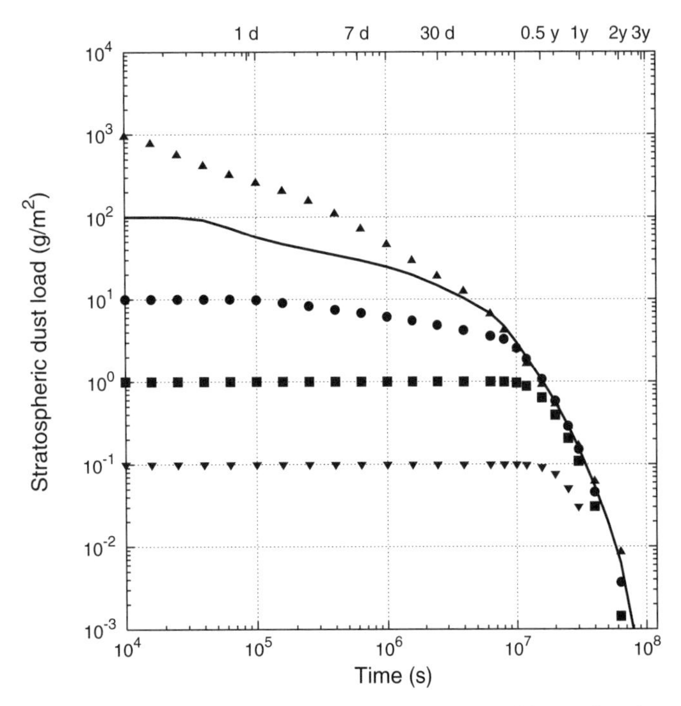

Figure 2. Total column density (load) of dust above 12 km as function of time for five initial dust loads. Solid line represents nominal case used in climate simulation. Initial loads of 1, 10, 100, or 1000 g/m^2 correspond to bolide diameters of ~1, 2, 5, or 10 km, respectively (Fig. 1). Due to coagulation, dust loads converge.

The dust population in every cell is represented by 100 particle volume bins.

Because the particles of all sizes drop down according to their local fall velocities, the numbers of particles in every bin obey advection equations, which are solved with an upwind difference scheme. Because all bins in every cell exchange particles due to coagulation, all advection equations are coupled. The length of the time step is determined by the stipulation that every particle of every size bin has to fulfill the equivalent of a local Courant criterion.

Radiation transfer calculations through the dust layer

As far as radiation transfer is concerned, a dust layer in the atmosphere can be compared with water clouds. In contrast to soot particles, dust grains and water droplets are reflective. Consequently, parts of an electromagnetic wave impinging on a dust grain are not only absorbed but also scattered in all directions. In general, scattering dominates over absorption for visible radiation (Ramaswamy and Kiehl, 1985). Thus substantial fractions of the energy of an incident solar ray are reflected many times from particle to particle in the dust layer. Therefore, the radiation field of visible light consists of the direct, damped solar beam as well as diffuse light. Because the absorption probability for an electromagnetic wave by a dust particle is small but not zero, the wave is absorbed before it has been scattered infinite times and so not only the direct solar beam but also the diffuse light are damped. However, the penetration depth of the diffuse radiation into the layer is larger than of the direct beam. Consequently, on an overcast day, while the Sun is not visible, the environment is well illuminated due to the diffuse light from the cloud cover.

Model. A single scattering model is a simplification of the radiation transfer through dust. Whereas Toon et al. (1982) used a multiscattering method to calculate the implications of a dust layer on radiation fluxes, we believe that Covey et al. (1994) neglected multiscattering effects.

In our multiscattering code, we first calculate the scattering phase function, extinction, absorption, and scattering cross sections of single spherical particles using Mie theory (Hansen and Travis, 1974; Liou, 1980; Goody and Yung, 1989). Complex indices of refraction depending on wavelength for dust-like aerosols are from Jaenike (1988). Second, we allow for the multiple scattering effects with the Doubling and Adding method (Goody and Yung, 1989). This technique starts with an optically thin sublayer of the dust cloud. In that layer, multiple scattering effects are improbable due to its tenuity. Therefore the transmission, absorption, and scattering of the thin layer can be calculated based on the radiative properties of single dust particles. Then, the Doubling and Adding method piles up two such thin layers and determines the transmission, absorption, and scattering of the resulting doubled layer. Multiple scattering between the sublayers is now taken into account explicitly. By repeating this process with the doubled layer as input, we obtain a layer with a four-fold thickness. Further repetitions of the doubling step lead quickly to layers with a large optical depth.

The energy absorbed by the dust particles has to be reemitted as infrared radiation. As a part of the total radiation field, it contributes to the net energy flux, and therefore infrared radiation has to be taken into account. We determine the temperature profile in the dust layer assuming that dust particles are in energetic equilibrium with the local radiation. The wavelengths considered range from 0.15 to 81.37 μm, divided into 63 finite bands. The wavelength range is chosen such that all typical spectra are covered: 99.99%, 99.67%, and 98.66% of the energies in the spectra of black bodies with temperatures of 6000 K, 300 K, and 200 K, respectively, are within this range. The results of radiation calculations depend on the number of bands, if it is below 40. However, more than 60 bands provide enough resolution on the wavelength axis.

For the radiative transfer calculations, we take the distribution of the dust from the dust evolution code. The lifetime of aerosols in the troposphere is much smaller than in the stratosphere because of rainout; therefore, we neglect all dust particles that have left the stratosphere and entered the troposphere. As concerns radiation, we assume that all absorbing gaseous constituents of the atmosphere are situated below the dust particles in the stratosphere. This splitting approach is justified by the fact that <20% of the mass of the real atmosphere is located above the tropopause. Thus in our model the radiation leaving the dust layer at its lower end does not impinge directly on the ocean surface, but is transferred through the gaseous part of the atmosphere. Due to the dust, the spectra of this light is shifted from solar toward the infrared frequencies; therefore we have to compute spectral air transmission coefficients, which depend on wavelength. These calculations are performed with MODTRAN, a "Moderate Resolution Transmittance" code, which is able to calculate atmospheric transmittance and radiance for frequencies from 0 to 1.5×10^{15} Hz at a resolution of 6×10^{10} Hz (Kneizys et al., 1988; Berk et al., 1989).

The radiation field in the stratospheric dust layer is determined by the boundary conditions. At the upper end we assume that the Sun is the only source of energy, and that all upward energy fluxes leave the system and are lost to space. With this, we only have to specify the position of the Sun with respect to zenith. At the lower end of the dust layer we do not allow for heat conduction between the dust layer and the underlying troposphere. Therefore, the boundary conditions at the lower end involve solar and infrared fluxes, which are given by the reflectivity and the radiative emissions of the atmospheric and oceanic cells below. These parameters are computed and specified by the model of the ocean circulation (see following).

Results of radiation transfer calculations. As an example we present calculated radiative energy fluxes between a planar gray body with temperature and emissivity of 295 K and 0.95, respectively, and an overlying dust layer. The Sun is located above the layer at 36° and 64° below zenith in Figure 3 (A and B, respectively). Figure 3C represents nighttime. The radiation fluxes shown are the body's loss of energy due to its infrared emission, the gain due to the absorption of solar energy and infrared radiation from the dust, and the net flux. As expected, the solar energy absorbed decreases with the amount of dust. While column densities below 0.1 g/m^2 do not lead to a significant change in the amount of solar radiation, dust loads above 10 g/m^2 reduce the solar light by more than a factor of 10 (Fig. 3, A and B). The infrared radiation fluxes at the surface of the body are of the same magnitude as the solar flux (Fig. 3, A, B, and C). The infrared energy absorbed by the body increases with the number of dust particles having a high temperature and being optically close to the body. (Low-temperature particles and particles optically far from the body contribute only to a small extent to the infrared radiation at the lower boundary of the dust layer.) Therefore, the infrared energy increases with the thickness of the dust layer, until the optical depth of the cloud is so large that the particles at the upper boundary of the layer cannot contribute anymore to the infrared field at the lower boundary, and the infrared energy does not increase further with dust load. In the limit of dust layers with large optical depth (dust load larger than ~300 g/m^2), the particles at the lower boundary are decoupled from the upper boundary, in particular from the solar radiation. At the surface of the body a radiative equilibrium attunes such that the infrared emission of the dust equals the emission and the absorption of the body.

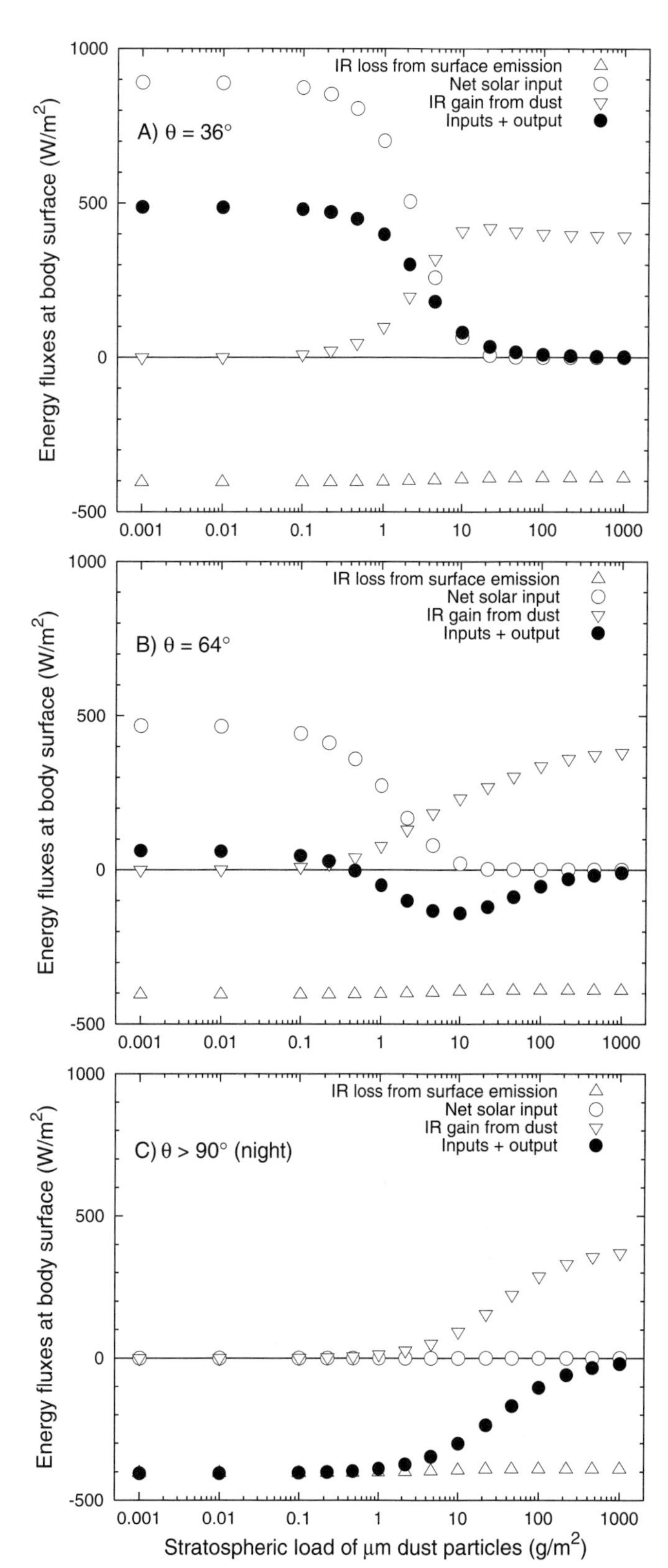

Figure 3. Examples of energy flux calculations with radiation transfer model between surface of gray body and dust layer as function of dust load. Positive energy fluxes represent energy input for body, negative values express energy output. Sun is situated above dust layer at $\theta = 36°$ and $\theta = 64°$ below zenith in A and B. C shows nighttime situation. Temperature, albedo for visible light, and infrared (IR) emissivity of body are 295 K, 0.2, and 0.95, respectively. Note minimum of total net flux at 10 g/m^2 dust load in B and insulation effect due to infrared emission of dust. See text for possible consequences for climate change.

Therefore, the body is radiatively insulated against space. Its radiative energy balance becomes zero.

The magnitude of this insulation due to the dust is visible in the nighttime example of Figure 3C. The emission of the body is ~400 W/m^2. If no dust layer covers the body, this energy is radiated to space. However, a dust load of ~1000 g/m^2 absorbs this flux and radiates it back to the body. Consequently we expect that diurnal terrestrial temperature variations are reduced during the presence of an impact-induced cloud.

Furthermore, the insulation leads to temperature increases in the polar regions, where the infrared emission is larger than the solar input. Therefore, with a dust layer, the reduction in solar radiation is smaller than the flux of infrared energy, which is not emitted to space but reabsorbed by the polar surface.

The combination of reduction of solar input and insulation due to dust causes an interesting effect. In the example of Figure 3B, the net energy balance of the body is not a monotonous function of amount of dust. The surface of the body undergoes a maximal energy loss of ~150 W/m^2 at an intermediate dust load of 10 g/m^2. For smaller and larger amounts of dust, the solar and the dust infrared input increase the net energy input of the body. The reason for this phenomenon is the low position of the sun, implying a smaller penetration depth than in the high solar inclination case. Therefore, at 10 g/m^2 dust load the solar energy at the surface is already strongly reduced. Furthermore, the dust particles heated by the Sun appear only in upper parts of the layer. More infrared radiation is emitted directly to space, and less energy reaches the surface. For dust loads larger than 10 g/m^2, the energy loss of the surface is smaller, because the magnitude of the insulation effect increases.

The consequence of a maximal energy loss effect at 10 g/m^2 might be that very large asteroid impacts lead to smaller climatic perturbations than medium-sized bolides, which produce a dust load of ~10 g/m^2. From equation 1 or Figure 1 we deduce that an asteroid of 1–3 km in diameter is in that range. This presumption will be investigated in a later study.

Thermohaline ocean circulation model

The climatic relevance of the ocean circulation originates in the heat capacity of seawater, which is enormous compared to air. To provide an order of magnitude, we assume that the temperature in the oceans is uniform. Then it would take the Sun ~1 yr to increase the temperature of the water from 0 to 1 K. Because the ocean flows circulate cold and warm waters, they are conveyors of large vertical and horizontal energy fluxes. The heat content of the air is much smaller: a layer of ~3 m of ocean water corresponds to the total heat capacity of the atmosphere.

If less energy reaches the ocean surface as a result of an impact, the temperatures of the sea surface layer and the atmosphere decrease. However, because circulation exchanges waters between the upper layer of the ocean and underlying parts by advection, the atmosphere is, over time, brought into thermal and radiative contact with large masses of water. Therefore, the variations of air and surface water temperatures are smaller than in the case of a resting ocean. Consequently the atmospheric and climatic response depends on the circulation of the oceans.

If the investigations of the climatic changes induced by an impact are to be extended to time scales of hundreds of years, the ocean circulation has to be taken into account, not only because of its implications at the surface. The perturbations may induce a change in the ocean circulation patterns and thereby change global energy fluxes, leading to climatic effects that last longer than the residence time of the dust in the atmosphere. A hypothetical example is a decrease of the Gulf Stream, which would cause a temperature drop in Europe.

Model. To investigate the long-term climatic effects, we use a two-dimensional, zonally averaged dynamic ocean circulation model (Wright and Stocker, 1992; Stocker et al., 1992). It represents the Atlantic, Indian, and Pacific basins in their present-day shape, connected with the Southern Ocean. A simple thermodynamic sea-ice model is included.

The oceanic spatial grid consists of 14 vertical and 9–14 horizontal cells, depending on the basin, with a meridional width of between 7.5° and 15°. In this model, the governing equations are written in spherical coordinates and include hydrostatic, Boussinesq, and rigid-lid approximations. Momentum equations are balances between Coriolis forces, horizontal pressure gradients, and zonal wind stress. Time-dependence enters via the equations for temperature and salinity, including vertical and meridional advection, diffusion, and convection. Horizontal and vertical eddy diffusivities are taken as constant (1000 m^2/s and 4×10^{-5} m^2/s, respectively).

With present-day conditions this model shows the vertical current in the northern Atlantic and in the Southern Oceans, and upwelling of waters toward the surface in the Pacific and the Indian Oceans. Observed temperature and salinity patterns as well as meridional fluxes of heat and freshwater are well reproduced.

The atmospheric part does not contain a circulation model. Fluxes of energy and water within the atmosphere, between ocean and air, and between continents and oceans are parameterized in terms of surface air temperature. The model does not take into account climatic processes on land such as snow. Before the dust injection, the model oceans are in a steady state that represents the present climatic situation. At time $t = 0$ the given dust load is injected. Thereafter, the simulation progresses as the dust cloud evolves. After the column density of dust has fallen below the 0.001 g/m^2 level, the climate model runs for additional 2000 yr.

Restrictions

Our model does not render all of the complex mechanisms that may influence the climate after an impact. We do not take into account smoke and soot particles, which are incorporated into the atmosphere by fires ignited in the aftermath of an impact. If this hypothesis is correct, then these aerosols are produced preferentially over the continents and incorporated into the atmosphere from the ground. Hence, an investigation of the climatic effects of soot and smoke has to consider the global transport of such particles in the air with three-dimensional convective models of the atmosphere, which is not in the scope of this work.

On the one hand, the radiative effects of soot particles would be a further reduction of the atmospheric transmissivity and thus a potential decrease in energy flux at the Earth's surface. On the other hand, the atmospheric albedo would be decreased due to the low reflectivity of soot. Therefore soot enhances the conversion of solar light into infrared energy. This effect could increase temperatures. We ignore variations in the densities of some air constituents, such as H_2O, CO_2, CH_4, N_2O, SO_2, SO_3, and O_3. In part, the amounts of change in these gases depend on the compositions of the projectile and the ground at the impact site. Therefore, studies about the climatic relevance of these atmospheric changes may refer to a specific impact. The results of the work by Pope et al. (1997) about the K-T event at Chicxulub can be applied probably only with caution to a generic impact location.

RESULTS

We run the climate code with a scenario of an initial load of 100 g/m^2. The global mass of dust is 5×10^{13} kg, which corresponds to an impactor ~5 km in diameter and 3000 kg/m^3 in density (Fig. 1). We discuss temperature deviations, the response of the Atlantic thermohaline circulation, and change in precipitation.

Global average of ocean temperatures

The mean ocean temperature shows the global long-term behavior of the climate model (Fig. 4). The mean ocean temperature drops by 0.05 °C in the first year after the impact. It recovers to the preimpact value ~65 yr later. It increases further until 130 yr after impact. A decrease in the next 40 yr is followed by a long-term increase: 2000 yr after impact, the mean ocean temperature is 0.05 °C higher than before the impact.

The mean temperature provides a broad overview of how the modeled climate system reacts to the dust coverage. In the months with a dust-loaded atmosphere the changed radiation fluxes immediately displace the climate from its previous steady state. One year after the impact, when most of the dust particles have settled, and the energetic boundary conditions are almost equal to the conditions before the impact, the climate is not in equilibrium. During several hundreds of years the system then passes through transient states, until it comes to an equilibrium

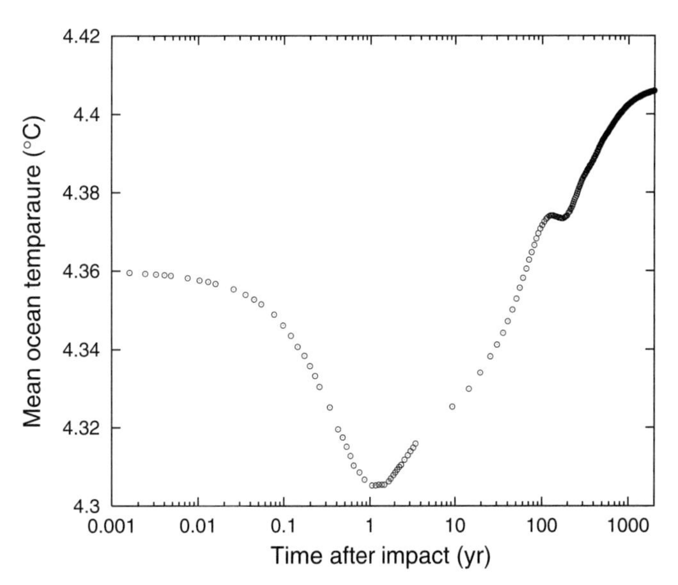

Figure 4. Mean global ocean temperature. Increase after 2000 yr is mainly due to higher temperature zone in northern Atlantic.

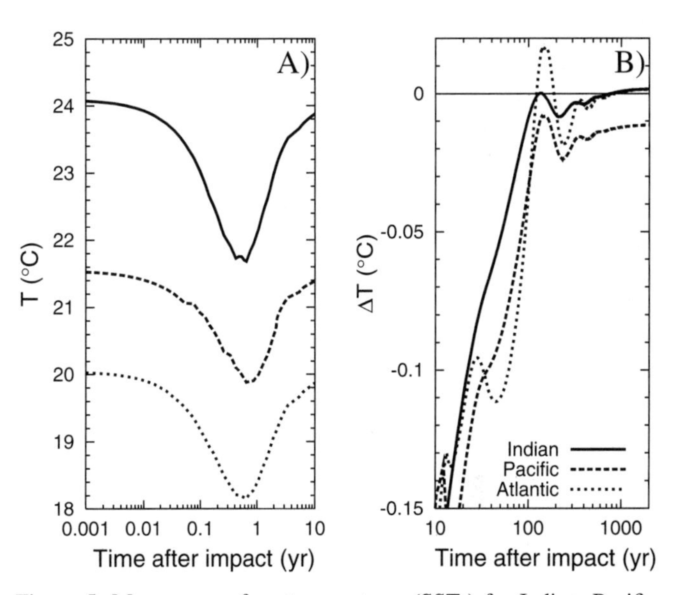

Figure 5. Mean sea surface temperatures (SSTs) for Indian, Pacific, and Atlantic Oceans in first 10 yr after impact. Maximal excursions are ~ −2.4, −1.7, and −1.9 °C, and occur between 7 and 8 months after impact. B: Deviations of SSTs with respect to values before impact from 10 to 2000 yr after impact. They are smaller than 0.1 °C for first time after 20–40 yr.

~2000 yr after the impact. This new steady state differs from the one before the impact, pointing to a hysteresis effect.

The relevance of the mean ocean temperature for biological individuals is low because different parts of an ocean respond with different amplitudes to the perturbation. Therefore, we present temperatures with spatial resolution in the following sections.

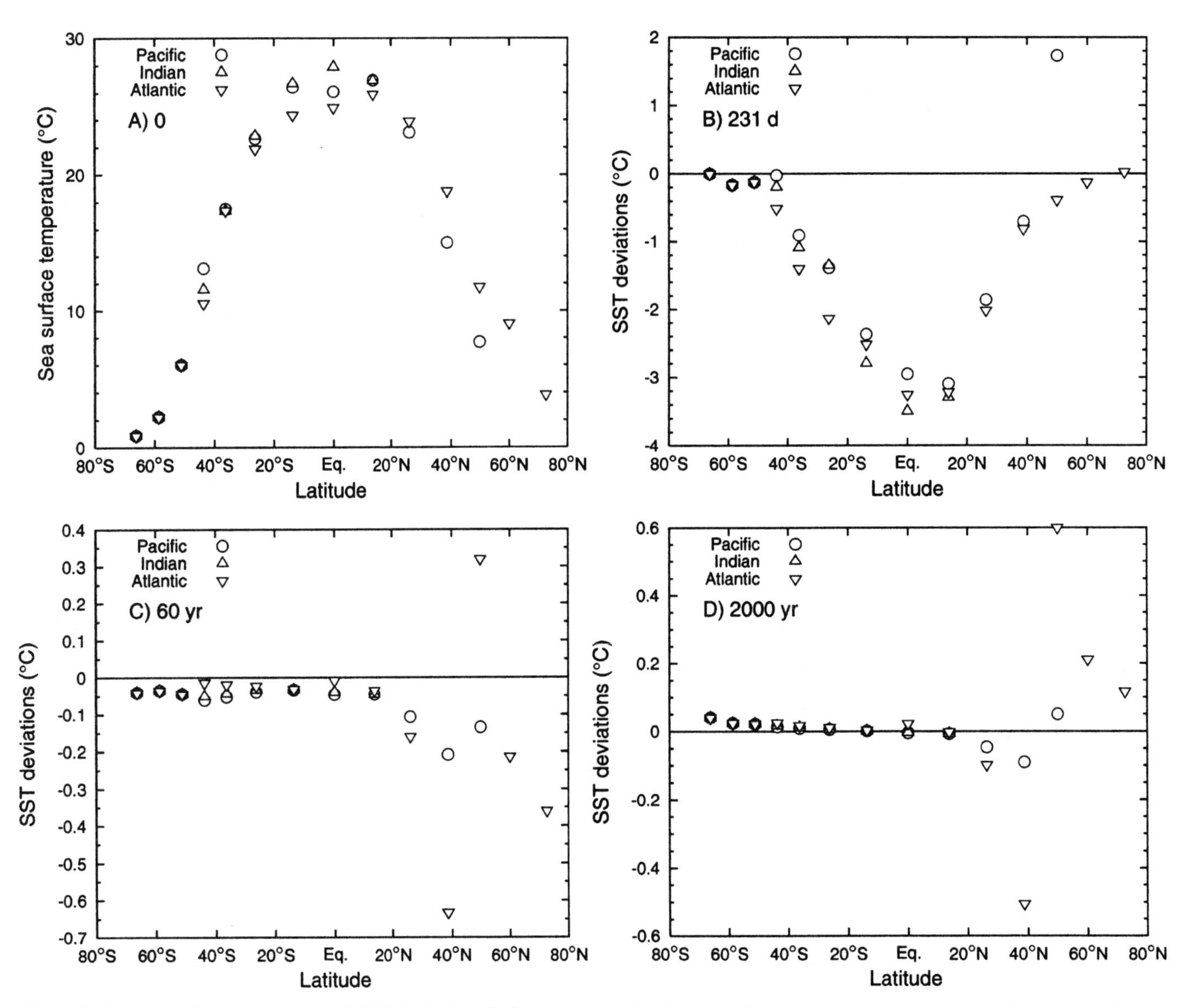

Figure 6. A: Sea surface temperatures (SSTs) in Pacific, Indian, and Atlantic Oceans before impact as function of latitude. B: Deviations of SST 231 d after impact with respect to preimpact values. C: Deviations 60 yr after impact. D: Deviations 2000 yr after impact. Identical values of all three basins in three southernmost cells come from strong coupling of oceans due to southern circumpolar current. Interpretation of deviations in northern Atlantic is given in text.

Sea surface temperatures

The mean sea surface temperatures (SSTs) of the Indian, Pacific, and Atlantic Oceans drop sharply following the impact (Fig. 5A). Peak amplitudes of ~ -2 °C are reached after 7–8 months, after which time the temperatures recover. However, 2 yr after the impact, the deviations are still more than -0.1 °C (Fig. 5B), followed by small variations on a time scale of 200 yr.

Figure 6 (A–D) shows SSTs in the Indian, Pacific, and Atlantic as a function of latitude before and after the impact. Figure 6B reflects the SST 7–8 months after impact, when the deviations are maximal. By that time, the dust has been diluted by a factor of more than 100, reaching a load of 1 g/m^2 in absolute terms (Fig. 2). The largest coolings of more than 3 °C appear at the equator. Because the radiative energy losses are much smaller near the poles than at the equator, the drops of SST at high latitude are not as large as at low latitudes at this early time. In the northern Pacific the surface temperatures have increased by 1.7 °C. However, the waters in the northern Atlantic will show a further decrease in temperature, because the thermohaline circulation transports less energy than before from the cooled equatorial regions into the polar zone.

After 3 yr the dust column density has fallen to such low levels that the radiation transfer has recovered its initial value. Nevertheless, as is shown by the evolution of the mean sea

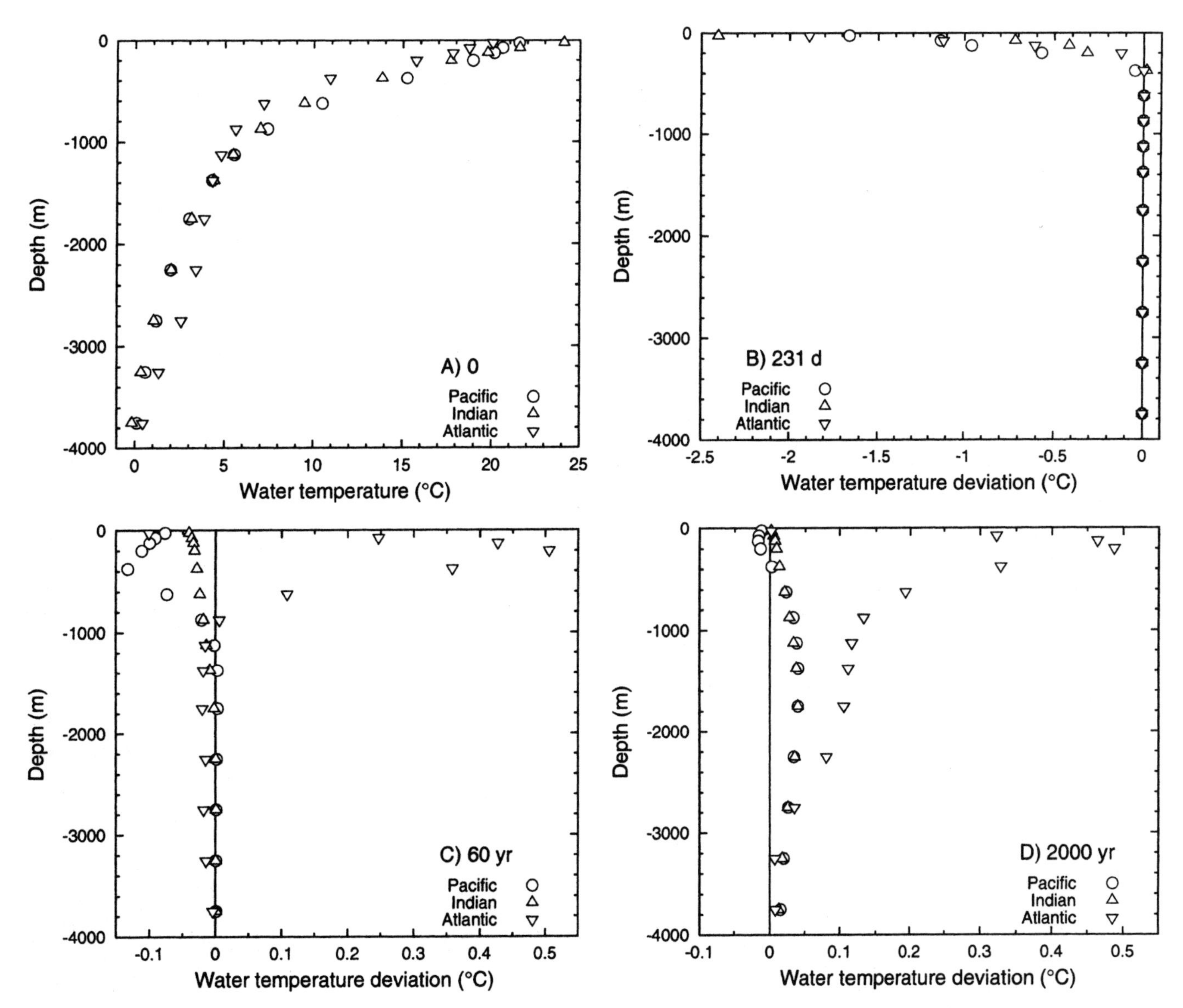

Figure 7. A: Water temperature profiles in Pacific, Indian, and Atlantic Oceans before impact. B: Deviations of water temperatures 231 d after impact with respect to preimpact values. C: Deviations 60 yr after impact. D: Deviations 2000 yr after impact.

temperature (Fig. 4), the oceans do not return to their previous state: 60 yr after impact, they are in an intermediate state. The SST deviations are still all negative except in one Atlantic cell at lat 50°N (Fig. 6C). We try to explain this behavior in the following section about the Atlantic circulation. Typical deviations with respect to preimpact temperatures are ~ -0.1 °C, but northern regions of the Atlantic show larger changes.

In the following 2000 yr, the state of the oceans converges to a new climatic equilibrium. After that time, the SSTs of the Southern Hemisphere are higher than before the impact, while the surface temperatures of the Atlantic and Pacific from lat 20°S to 45°S latitude are lower (Fig. 6D). However, in the new steady state the surfaces of Pacific and Atlantic north from lat 45°S are warmer than before the impact.

Sea temperature profiles

The horizontal currents in the uppermost tens of meters of the oceans are affected by atmospheric winds at the surface, resulting in a typical water velocity of 1 m/s. In deeper layers, horizontal and vertical motions of the water are driven by density gradients, which depend on temperature and salinity. The water and heat exchanges induced by these currents occur on time scales much longer than the atmospheric time scales. While the residence time of water at the surface is some months, it amounts to ~100 yr for the uppermost 100 m, and the age of deep water is ~1000 yr (von Reden et al., 1997). Therefore, the vertical transport of temperature anomalies is slow compared with the lifetime of the bolide-induced dust cloud.

Figure 7A shows water temperatures as function of depth in the equilibrium state of the model before the impact. Months after the impact the temperature deviations are largest at the surface (Fig. 7B). They are −2.4 °C, −1.9 °C, and −1.7 °C for Indian, Atlantic, and Pacific, respectively, and reduce to ~−0.5 °C at a depth of 200 m. Below 400 m, the water temperatures have not changed: 60 yr after impact (Fig. 7C), temperature deviations appear down to 4 km depth, although the largest changes occur only in the uppermost 700 m. At the surface, all three basins are colder than before the impact. The cooling increases with depth in the Pacific down to 400 m. From there to the bottom, the deviations are still negative but small. In the Indian Ocean, they are negative and small at all depths. A zone in the Atlantic down to 800 m exhibits warmer temperatures with a peak value of +0.5 °C at 100 m. After 2000 yr (Fig. 7D) the temperatures have risen above those preceding the impact, except in the uppermost 200 m of the Pacific. It is interesting that the increase of temperature by 0.5 °C in the Atlantic 100 m below the surface remains.

Atlantic circulation and temperature

Figure 8 shows the Atlantic stream function, the contours of which are parallel to the motion of the water. The flow is dominated by a basin-wide overturn circulation. In a layer from the surface to a depth of 500–1000 m, it transports warm water from southern and equatorial to northern regions, where the temperature differences of the warm water and the cold atmosphere induce high evaporation rates. Consequently, the ocean temperature drops and salinity increases. Both effects increase the density of the water, which in turn sinks to deep sea levels in the northern Atlantic. Below 1500 m depth the water flows southward and upwells again. South of the equator, a local second circulation rotates in the opposite direction. It transports water from the equator to lat 30°S, where the water is advected back to the equator at a depth of a few hundred meters. A symmetrical third local current in the Northern Hemisphere is superimposed over the large first basin-wide circulation. At the surface, the net effects of these three currents are (1) ascending water at the equator from several 100 m below, (2) subsiding water from 20°N to 40°N and 20°S to 40°S latitude in both hemispheres, (3) a fast horizontal transport of surface water from the equator to 20°N and 20°S latitude, and a slow horizontal surface motion in the northern Atlantic from 20°N to the poles. Due to the circulation, the impact-induced cooling in the surface water layers are advected vertically.

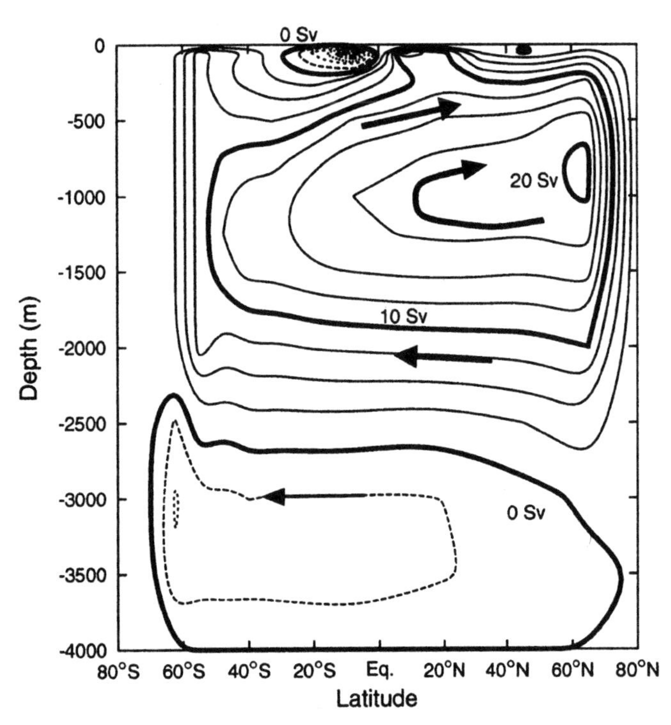

Figure 8. Latitude-depth section of stream function in Atlantic before impact. Unit is 1 Sv (Sverdrup) = 10^6 m^3/s. Motion of water is parallel to isolines of stream function. Contour interval is 2.5 Sv. Amount of water flowing between two contours is equal to difference of values of stream function. Positive and negative values are represented by solid lines and dashed lines, respectively.

Figure 9 shows the steady-state temperatures in the Atlantic before the impact. Temperature deviations 231 d after the impact are given in Figure 10. At 20°S and 30°N latitude, the deviations have advanced further downward than at the equator, which is explained by the equatorial upwelling and the subsiding waters from 20°N to 40°N and 20°S to 40°S latitude. There are nearly no changes in subsurface temperatures in the polar regions, because there the surface temperature changes are much smaller than at the equator.

The Atlantic temperatures of the steady state 2000 yr later show an increase at 200 m below the surface at lat 40°N with maximum temperature deviation of more than +3 °C (Fig. 11). In the steady state before the impact, potential instabilities in the northern Atlantic generated mixing of water from vertically adjacent cells. This mixing process convected warm water from a depth of 100 m toward the surface. However, the instabilities between lat 20°N and 45°N have disappeared 2000 yr after the impact. Thus the vertical mixing and the related warm water transport to the surface stop in this region, which is a significant change in the thermohaline circulation. Therefore, we observe the strongly negative temperature deviation at the Atlantic surface at lat 40°N (Fig. 6D). However, the potential instabilities and the vertical mixing have not disappeared north from lat 45°N. Because of the increased temperature in the subsurface cell in the northern Atlantic (Fig. 11), the mixing transports warmer water to the surface. Hence, the mixing gives rise to the peak in the SST (Fig. 6D) at lat 50°N.

The basin-wide pattern of the circulation remains, although during the 2000 year transition the strength of the Atlantic overturn circulation (Fig. 8) shows an oscillation with a period of 200 yr. This oscillation is model related (Aeberhardt et al., 2000) and may not reflect a natural behavior of the ocean.

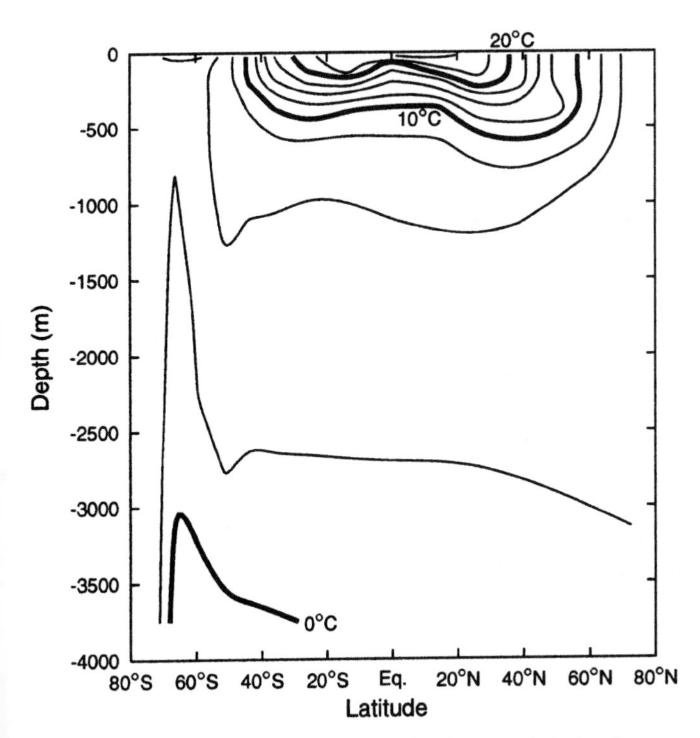

Figure 9. Temperature-latitude-depth section in Atlantic before impact. Contour interval is 2.5 °C.

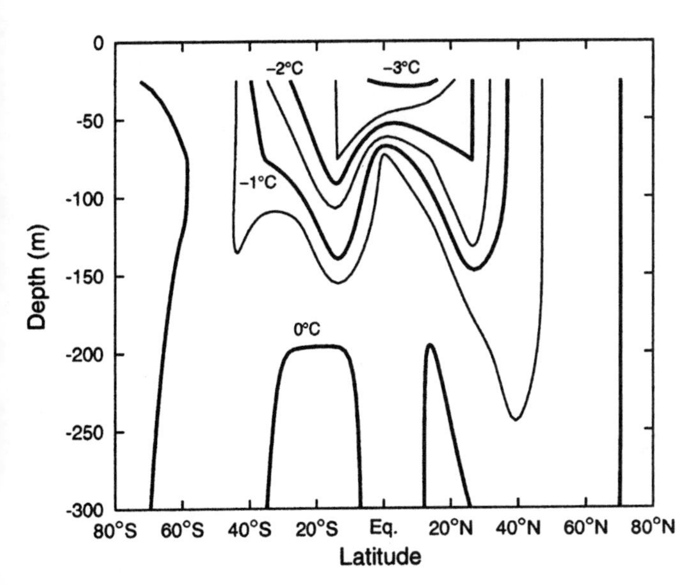

Figure 10. Deviations of temperature field in uppermost 300 m of Atlantic 231 d after impact with respect to values in equilibrium state before impact (Fig. 9). Largest cooling appears near equator at surface. Below 250 m deviations are smaller than 0.5 °C. Contour interval is 0.5 °C.

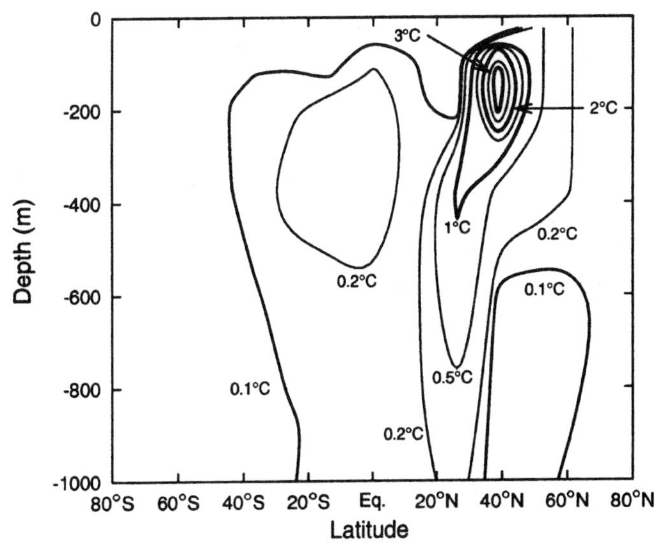

Figure 11. Deviations of temperature field in uppermost 1000 m of Atlantic 2000 yr after impact with respect to values before impact (Fig. 9). Except for cell with increased temperatures at 40° N latitude, differences are below 0.5 °C.

Air temperatures over the oceans

Due to its smaller heat capacity, the temperature of the air over the ocean changes much more rapidly than the water temperature following a change in input energy. Figure 12 shows deviations of air temperatures in three latitudinal bands near the north pole, the equator, and the south pole. In the first 50 d, the polar air temperatures increase by 5 °C (north) and 6 °C (south). This response is probably caused by the radiative insulation provided by the dust, which is most effective for low solar positions or during nighttime (Fig. 3, B and C), and therefore in the polar regions. The equatorial air temperature is increased during ~2 months; however, the peak of more than 1 °C is reached 9 d after the impact. The following phase of colder equatorial temperature lasts for more than 5 months, before the minimum value is reached at a time corresponding to the lowest SSTs (Fig. 5). The lowest temperature is 3 °C below the preimpact value. These results compare with the climate simulations by Covey et al. (1994), who used a general atmospheric circulation model coupled to a thermodynamic model of the upper mixed layer of the oceans, to consider climatic effects of dust generated by an impact. Their simulation started with a dust load being 50 times larger than the one considered here, which corresponds to an impactor diameter of 17.5 km according to equation 1 and an assumed density of 3000 kg/m^3. They investigated large-particle and small-particle scenarios; the small-particle scenario compares with the dust particle sizes used in this study. For the temperatures of air over oceans at the end of 1 yr, coolings of 6 °C and 4 °C were found for the large-particle and small-particle scenario, respectively. In the polar regions, their simulation showed increased temperatures between 0 °C and more than 10 °C compared to their control case in the period from 10 to 20 days after impact. After 1 yr, positive and negative temperature deviations occur in Antarctica as well as in the arctic region.

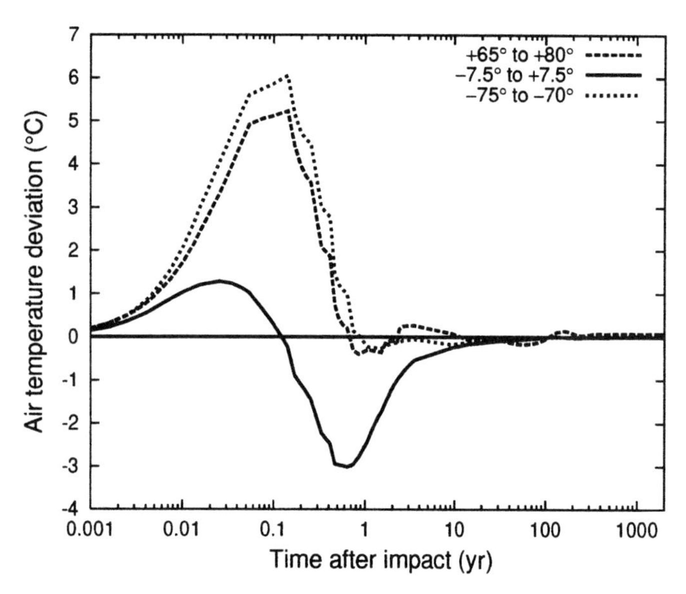

Figure 12. Deviations of air temperatures over sea in three latitudinal bands: (A) 75°S to 70°S, (B) 7.5°S to 7.5°N, and (C) 65°N to 80°N. Amplitude of cooling in equatorial region is ~3 °C. Air temperatures in polar latitudinal bands increase by 5 °C and 6 °C, respectively, within first 50 d and return to preimpact values in <1 yr. Temperatures before impact are −6.4 °C, 26.7 °C, and −4.5 °C for southern, equatorial, and northern regions, respectively.

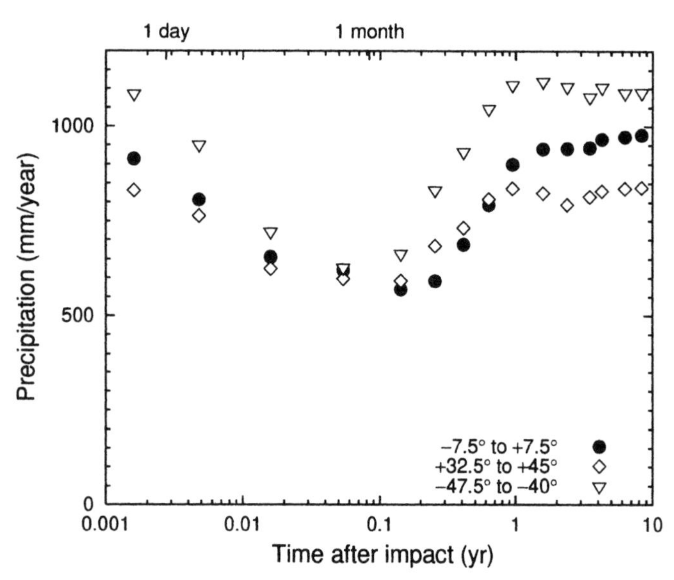

Figure 13. Precipitations after impact in three latitude bands.

Precipitation

Covey et al. (1994) noted a collapse of the hydrological cycle in their simulations. They reported that the globally averaged precipitation decreases by 90% after 3 months and remains below 50% after 1 yr. We find a smaller decrease in precipitation. Figure 13 shows precipitation rates as function of time after the impact in three latitude bands located at the equator and in the Northern and Southern Hemispheres. The reduction of precipitation is <50%, and the minimum is reached in the first 2 months. After 1 yr, deviations of precipitation rates are within 10% of original value. The determination of evaporation and precipitation rates was given in Stocker et al. (1992).

CONCLUSIONS

In this chapter we investigated impact-induced climate changes over 2000 yr by using a model that includes dust settling and coagulation, radiation transfer, and ocean-atmosphere coupling. The amount of dust injected into the atmosphere corresponds to an impactor with a diameter of 5 km, assuming a density of 3000 kg/m^3.

In the first year, equatorial SSTs dropped by −3 °C, while the polar SSTs showed much smaller deviations: 500 m below the surface of the oceans, temperatures remained unchanged. Air temperatures dropped (−3 °C) with the exception of the polar regions (+6 °C). Precipitation decreased in 3 months to 50%, but recovered within 1 yr.

The climate is a system with more than one equilibrium. Once it has been pushed out of one equilibrium state due to the large amount of dust in the atmosphere, it will settle over time into a new equilibrium that does not necessarily coincide with the initial one. The reasons are the large oceanic heat inertia and the complexity of the system. Therefore, the climatic response operates on time scales different from the short-term radiative forcing due to dust studied here. The time scales of some of the climatic excursions, which are initiated by the perturbed atmospheric energy fluxes, range from fractions of 1 yr to hundreds of years. For example, 2 yr after impact, deviations of air and SSTs were still ~−1 °C, and after 60 yr the amplitudes were reduced to 0.1 °C. On long time scales, the ocean circulation advected the temperature anomalies from the surface to the depth. Because such transport is slow, there is a considerable time lag between the surface changes and response of deeper zones. The major deviations were still restricted to depths above 1000 m 60 yr after the impact. On average, the deep-sea temperatures started changing only after ~100 yr. The climate took more than 2000 yr to reach a new steady state.

Both ocean circulation and temperatures changed as part of the global climate changes. During the first millennium after the impact the strength of the Atlantic overturning circulation showed a damped oscillation. Therefore the transport of surface waters toward the bottom was not constant, which may disturb global cycles such as the carbon and the oxygen cycles. After 2000 yr, the main new feature in the circulation pattern was a weaker convection zone in the northern Atlantic.

An extreme climatic scenario is the glaciation of vast areas. However, we did not observe growth or formation of sea ice shields. At the poles the cooling was too small, and in the middle and low latitudes the ocean temperatures did not drop below freezing.

ACKNOWLEDGMENTS

We thank Karen Bice and David Kring for their helpful reviews. This work has been supported in part by the Swiss National Science Foundation.

REFERENCES CITED

Aeberhardt, M., Blatter, M., and Stocker, T.F., 2000, Variability on the century time scale and regime changes in a stochastic forced zonally averaged ocean-atmosphere model: Geophysical Research Letters, v. 27, p. 1303–1306.

Argyle, E., 1988, The global fallout signature of the K/T bolide impact: Icarus, v. 77, p. 220–222.

Berk, A., Bernstein, L.S., and Robertson, D.C., 1989, MODTRAN: A Moderate Resolution Model for LOWTRAN 7: Hanscom Air Force Base, Massachusetts, Air Force Geophysics Laboratory Technical Report GL-TR-89–0122, 11 p.

Chokshi, A., Tielens, A.G.G.M., and Hollenbach, D., 1993, Dust coagulation: Astrophysical Journal, v. 407, p. 806–819.

Covey, C., Ghan, S.J., and Walton, J.J., 1990, Global environmental effects of impact-generated aerosols: Results from a general circulation model, *in* Sharpton, V., and Ward, P., eds., Global catastrophes in Earth history: Geological Society of America Special Paper 247, p. 263–270.

Covey, C., Thompson, S.L., Weissman, P.R., and MacCracken, M.C., 1994, Global climatic effects of atmospheric dust from an asteroid or comet impact on Earth: Global and Planetary Change, v. 9, p. 263–273.

Dominik, C., and Tielens, A.G.G.M., 1997, The physics of dust coagulation and the structure of dust aggregates in space: Astrophysical Journal, v. 480, p. 647–673.

Durda, D.D., Kring, D.A., Pierazzo, E., and Melosh, H.J., 1997, Model calculations of the proximal and globally distributed distal ejecta from the Chicxulub impact crater [abs.]: Lunar and Planetary Science Conference, 28th, Houston, Texas, Lunar and Planetary Institute, p. 315–316.

Farlow, N.H., Oberbeck, V.R., Snetsinger, K.G., Ferrey, G.V., Polkowski, G., and Hayes, D.M., 1981, Size distribution and mineralogy of ash particles in the stratosphere from eruptions of Mount St. Helens: Science, v. 211, p. 832–834.

Goody, R.M., and Yung, Y.L., 1989, Atmospheric radiation, theoretical basis (second edition): New York, Oxford University Press, 519 p.

Hansen, J.E., and Travis, L.D., 1974, Light scattering in planetary atmospheres: Space Science Reviews, v. 16, p. 527–610.

Hills, J.G., Nemchinov, I.V., Popov, S.P., and Teterev, A.V., 1994, Tsunami generated by small asteroid impacts, *in* Gehrels, T., ed., Hazards due to comets and asteroids: Tucson, University of Arizona Press, p. 779–789.

Hoffman, P.F., Kaufman, A.J., Halverson, G.P., and Schrag, D.P., 1998, A Neoproterozoic snowball Earth: Science, v. 281, p. 1342–1346.

Jaenike, R., 1988, Properties of atmospheric aerosols, *in* Fischer, G., ed., Landolt-Boernstein, New Series: Numerical data and functional relationships in science and technology, Group 5: Geophysics and space research, Volume 4: Meteorology, Sub-Volume B: Physical and chemical properties of the air: Berlin, Springer-Verlag, p. 405–428.

Kneizys, F.X., Shettle, E.P., Abreu, L.W., Chetwynd, J.H., Anderson, G.P., Gallery, W.O., Selby, J.E.A., and Clough, S.A., 1988, Users Guide to LOWTRAN 7: Hanscom Air Force Base, Massachusetts, Air Force Geophysics Laboratory, AFGL-TR-88-0177, 137 p.

Kring, A.H., Melosh, H.J., and Hunten, D.M., 1996, Impact-induced perturbations of atmospheric sulfur: Earth and Planetary Science Letters, v. 140, p. 201–212.

Lewis, J.S., Watkins, G.H., Hartmann, H., and Prinn, R.G., 1982, Chemical consequences of major impact events on Earth, *in* Silver, L.T., and Schultz, P.H., eds., Geological implications of impacts of large asteroids and comets on the Earth: Geological Society of America Special Paper 190, p. 215–221.

Liou, K.N., 1980, An introduction to atmospheric radiation: New York, Academic Press, 392 p.

Melosh, H.J., Schneider, N.M., Zahnle, K.J., and Latham, D., 1990, Ignition of global wildfires at the Cretaceous/Tertiary boundary: Nature, v. 343, p. 251–254.

Morrison, D., Chapman, C.R., and Slovic, P., 1994, The impact hazard, *in* Gehrels, T., ed., Hazards due to comets and asteroids: Tucson, University of Arizona Press, p. 59–91.

Otto, E., Fissan, H., Park, S.H., and Lee, K.W., 1999, The log-normal size distribution theory of brownian aerosol coagulation for the entire particle size range. 2. Analytical solution using Dahneke's coagulation kernel: Journal of Aerosol Science, v. 30, p. 17–34.

Pierazzo, E., Kring, D.A., and Melosh, H.J., 1998, Hydrocode simulation of the Chicxulub impact event and the production of climatically active gases: Journal of Geophysical Research, v. 103, p. 28 607–28 625.

Pope, K.O., Baines, K.H., Ocampo, A.C., and Ivanov, B.A., 1994, Impact winter and the Cretaceous/Tertiary extinctions: Results of a Chicxulub asteroid impact model: Earth and Planetary Science Letters, v. 128, p. 719–725.

Pope, K.O., Baines, K.H., Ocampo, A.C., and Ivanov, B.A., 1997, Energy, volatile production, and climatic effects of the Chicxulub Cretaceous/Tertiary impact: Journal of Geophysical Research, v. 102, p. 21 645–21 664.

Prinn, R.G., and Fegley, B., Jr., 1987, Bolide impacts, acid rain, and biospheric traumas at the Cretaceous-Tertiary boundary: Earth and Planetary Science Letters, v. 83, p. 1–15.

Ramaswamy, V., and Kiehl, J.T., 1985, Sensitivities of the radiative forcing due to large loadings of smoke and dust aerosols: Journal of Geophysical Research, v. 90, p. 5597–5613.

von Reden, K.F., McNichol, A.P., Peden, J.C., Elder, K.L., Gagnon, A.R., and Schneider, R.J., 1997, AMS measurements of the ^{14}C distribution in the Pacific Ocean: Nuclear Instruments and Methods in Physics Research B, v. 123, p. 438–442.

Stocker, T.F., Wright, D.G., and Mysak, L.A., 1992, A zonally averaged, coupled ocean-atmosphere model for paleoclimate studies: Journal of Climate, v. 5, p. 773–797.

Toon, O.B., Pollack, J.B., Ackermann, T.P., Turco, R.P., McKay, C.P., and Liu, M.S., 1982, Evolution of an impact-generated dust cloud and its effects on the atmosphere, *in* Silver, L.T., and Schultz, P.H., eds., Geological implications of impacts of large asteroids and comets on the Earth: Geological Society of America Special Paper 190, p. 187–200.

Toon, O.B., Zahnle, K.J., Morrison, D., Turco, R.P., and Covey, C., 1997, Environmental perturbations caused by the impacts of asteroids and comets: Reviews of Geophysics, v. 35, p. 41–78.

Ward, S.N., and Asphaug, E., 2000, Asteroid impact tsunami: A probabilistic hazard management: Icarus, v. 145, p. 64–78.

Williams, D.M., Kasting, J.F., and Frakes, L.A., 1998, Low-latitude glaciation and rapid changes in the earth's obliquity explained by obliquity-oblateness feedback: Nature, v. 396, p. 453–455.

Wolbach, W.S., Gilmour, I., and Anders, E., 1990, Major wildfires at the Cretaceous/Tertiary boundary, *in* Sharpton, V., and Ward, P., eds., Global catastrophes in Earth history: Geological Society of America Special Paper 247, p. 391–400.

Wright, D.G., and Stocker, T.F., 1992, Sensitivities of a zonally averaged global ocean circulation model: Journal of Geophysical Research, v. 97, p. 12 707–12 730.

Wurm, G., and Blum, J., 1998, Experiments on preplanetary dust aggregation: Icarus, v. 132, p. 125–136.

Zahnle, K.J., 1990 Atmospheric chemistry by large impacts, *in* Sharpton, V., and Ward, P., eds., Global catastrophes in Earth history: Geological Society of America Special Paper 247, p. 271–288.

Manuscript Submitted October 10, 2000; Accepted by the Society March 22, 2001

Printed in the U.S.A.

Index

(Italic page numbers indicate major references.)

A

Abadeh Formations
  fossils, 400
Acadian orogeny, 482
accommodation space, 441
acid rain, 341, 343
acoustic fluidization, *626*
acoustic horizons, 80
Acraman impact structure, 26
adaptive radiation, *482*
adiabatic exponent
  equation, 633
age dating. *See also* radiometric dating
  argon-argon, *371,* 527
  Carnavon basin, 373
  uranium-lead, *371, 524,* 527
Agost section
  K-T boundary, *553*
air temperature
  impact effects, *727*
Alamedilla section, 553
  P-E boundary, 551, *554,* 559
Alamo Breccia, *478, 490*
  genesis, *494*
  lapilli, 489, *490*
Alamo Impact Event, 473, 477, *477,* 481, 484, 489, 490, 493, 500
Alamo lapilli, *497*
Alamo Wash, 309
  palynology, 310
Alamoan fauna, 332
  San Juan basin, 307
Albion Island, 140
  impact deposit, 126
  lapilli, 497
algae
  Bellerophon Formation, 416
alkanes
  biomarkers, *356*
  Val Badia, *457*
Allende chondrite, 34
Allende meteorite, 33, 448
  nanoparticles, 349
Alvarez hypothesis, 8, 22, 190
American Commission on Stratigraphic Nomenclature, 314
ammonites, 266
  Bjala, 236
  Hornerstown Formation, 295
  Navesink Formation, 294
ammonoids
  Hunan Province, *367*
  Koipate volcanics, 388
  P-Tr boundary, *367*
  P-Tr boundary, 378
Amönau Breccia, 476
Amönau Event, *476,* 481
  Germany, 473
Amsekroud Formation, 516
Ancón Formation, *111, 117,* 127, 138
  deposition, 120
Ancora borehole, 98
  spherule, 99
angle of impact
  bolide, *620*
Anglo-Paris basin, 272
angular momentum law, 686
anhydrite
  in suevite, 49
Animas Formation, 309, 310
Anjar, 202
  fullerenes, 346
  iron phases, *207, 208,* 210
  Mössbauer spectrum, 204
  volcano-sedimentary sequence, *346*
annulata Event, 483
anoxia
  Dogie Creek, 343
  early Toarcian, 529
  late Paleocene, 543
  nitrogen isotopes, *356*
  P-Tr boundary, *372, 374,* 377, *395,* 403, *409, 450*
  Panthalassa, 395
  Perigondwanan shelf, 395
  Salt Range, 375
Antarctica
  Permian swamp, 368
Anticosti Island
  brachiopods, *465*
apatite, 330
  isomorphic substitution, 328
Aquia Formation, 296
Argana basin
  Tr-J boundary, 516
argument of perihelion
  equation, 653
Arroyofrío Oolite Bed, 551, 552, 559
Artemkova River basin, 402
asteroids, 646, *646,* 652, 656. *See also* Earth crossers
  Eltanin, 26, 29, 31
  Eros, 9
  grazing, 685, 690
  hazardous, *645*
  impacts, *717*
  near-Earth, *652*
  spectral reflectance, 23
Atlantic circulation model, *726*
atmospheric general circulation models, 718
atomic absorption spectrometry
  Blake Nose, 191
Attargoo section
  rare earth elements, 451
  Spiti Valley, 447
automatic powder diffractometer, 214
Aztecs, 2
  myth, 3
Azuara impact structure, 551, 554, *555,* 559, 673

B

back reactions, *581,* 587
Bahama platform, 89, 110
Bahía Honda allochthon, 111, 126, 127
Baja
  mass wasting, 80
Balfour Formation
  Karoo basin, 440
  P-Tr boundary, *430*
Baptornis, 304
Barberton Greenstone Belt, 26
  spherules, 35
barium
  biogenic, 544
Barrel Spring, 309
  K-T boundary, *319*
  palynology, 310
  palynomorphs, *318,* 333
Basin and Range, 388
Basque-Cantabrian basin
  turbidite, 554
Bass River, 98
  core, 542
  foraminifera, 100
  oxygen isotopes, 104
  spherules, 61, 99
  strontium isotopes, 100, 104, *104*
bathymetry
  Moncada Formation, 120
Bay of Biscay
  K-T boundary, *554*
Bay State Formation, 492
Beaufort Group, 441
Becker-Döring-Zel'dowich nucleation theory, 638
belemnites
  Navesink Formation, 293
Bellerophon Formation, 419
  carbon isotopes, 456
  carbonate component, 424
  foraminifera, 419
  fossils, 416
  Gartnerkofel core, 421
Beloc
  deposition, *183*
  isotopes, *179*
  lithology, *167*
  sample preparation, *164*
  sections, *164*
  shocked quartz, 64
  spherules, *167*
Beltic Cordillera
  K-T boundary, *553*
Ben Gurion section, 535
  radiolaria, 544
benthic extinction event, 534
  Nukhl, 536
benzpyrenes
  toxins, 714
Bermuda Rise, 79, 91
  drill holes, 84
  lithostratigraphy, *84*
  mass wasting, *80*
  slope failure, 93
  turbidites, 89, 284
Bethulie section
  Karoo basin, 432

Betonnie Tsosie Wash, 314
Bidart section
  K-T boundary, 554
biochronology
  dinosaurs, *331*
  Late Devonian, *474*
biogenic particle flux, 169
biomagnetochronology
  Gulf of Mexico, *260*
  Mexico, 253
  Tunisia, 253
biomarker
  Brownie Butte, *357*
  hydrocarbon, *356*
  Phosphoria Formation, 375
  Raton basin, *357*
biopolymers, 457
biosphere
  chemical contamination, 705
biostratigraphy
  Ancora borehole, *101*
  Bass River borehole, *101*
  Bjala, *236*
  dinoflagellate, 295
  New Jersey coastal plain, *103*
  P-Tr boundary, *364*
  planktonic, 294
  Spiti Valley, *446*
  Tr-J boundary, *507*
biostromes, 378
biotic recovery
  early Triassic, *378*
bioturbation, 91
  Bjala, 236
  Changxing Formation, 402
biozones
  Bochil, 256
  El Kef, *256*
  Gulf of Mexico, 261
  La Ceiba, *256*
birds
  Hornerstown Formation, 295
  K-T boundary, *303*
bivalves
  Navesink Formation, 293
  Red Bank Formation, 294
Bjala
  geology, *215*
  K-T boundary, *213*
  section, 233
black shales
  geochemistry, *537*
  Paleocene, *540, 543*
Blake Nose
  deformation, *81,* 88
  K-T boundary, 197, 279
  mass wasting, 279
  submarine slope failure, 61
Blake Plateau, 79, 89
  mass wasting, *80*
  sediment, 283
  slope failure, 88
  slumping, 93
Blauer Bruch
  bentonite, 477
blocking temperature, 206, *207*
Blomidon formation
  Tr-J boundary, 517
Bochil, 253
  shocked quartz, 64
  stratigraphy, *254*
bolide, 11
Boltzmann's constant, 638
bony fish
  Hornerstown Formation, 295
Boonton Formation
  cyclostratigraphy, 516
  tetrapod footprints, 513
Botsumtwi crater
  model parameters, 627
boundary clay, 11, 265. *See also* clay layer
  Bjala, 226, *240*
  Caravaca section, 553
  Gulf of Mexico, 261
  Meishan, 372
  Spain, 559
  Sugarite site, 338
  Wachapo Mountains, 372
  Zumaya section, 554
boundary sections
  P-Tr boundary, *396*
  Passaic Formation, *510*
boundary sediments
  P-Tr boundary, *400, 402*
Bourgogne basin, 552
brachiopods, 298
  Anticosti Island, *465*
  Bellerophon Formation, 416
  extinction, 367, 481
  Gerster Limestone, 390
  Hogup Mountains, 390
  Inversand Pit, 296
  Kaibab Limestone, 390
  Late Ordivician, *467*
  minimalist organisms, 297
  Navesink Formation, 293, 295
  P-Tr boundary, 378, *390*
Brazos River section, 298
breccia
  Bochil, *254*
  Campeche bank, 60, 64
  ejecta, 64
  impact, 39, 41, 48
  polymict, 30, 262
  Tr-J boundary, 558
  Yucatan platform, *60*
Brownie Butte
  biomarkers, *357*
  hopanes, *358*
  lithology, *352*
  organic geochemistry, *352*
  sulfur isotopes, *338*
bulk rock composition
  Bjala, *242*
Bunte Breccia, 60
Burgersdorp Formation
  depositional environment, 441

C

Cacarajicara Formation, 61, 70, 121, 122, *125,* 126
  deposition, *140*
  lithostratigraphy, *127*
  Middle Calcarenite Member, *133*
  paleocurrents, *138*
  Upper Lime Mudstone Member, *138*
Cache Creek terrane, 372
Calatayud-Montalbán basin, 556
calcareous sandstone complex
  Moncada Formation, *112*
calcimicrobial mound, 378
calcining equation, 501
calcite
  decomposition, 591
  shock experiments, *588*
calcium carbonate
  Esna Formation, 536
Campeche bank
  breccia, 60, 64
CamScan 4 scanning electron microscope, 573
CamScan CS 44 scanning electron microscope, 266
Canadian margin
  impact, 93
Cantabrian basin, 552
Canyon Diablo
  diamond synthesis, *599*
  sulfur isotopes, 338
Canyon Range thrust, 389
Cape Fold Belt
  uplift, 368
Capitanian extinction phase, 366
Caravaca
  chromium isotopes, 34
  section, *553*
  spinel, 29
carbon cycle, 355
  mass extinction, 528
carbon disulfide
  toxicity, 713
carbon isotopes, 455
  Bellerophon Formation, 456
  Beloc, 179
  Bjala, 221, *248,* 249
  Brownie Butte, 351, *353*
  central Nevada, *467*
  early Toarcian, *529*
  El Kef, 256
  Gartnerkofel core, 421, 425
  Kendelbachgraben, 528
  late Devonian, 481
  late Paleocene, 534, *537,* 542, *545*
  Little Ben Sandstone, 369
  Mendez Formation, *153*
  P-Tr boundary, 377, 416, 420, *420, 456*
  Paleocene, 540
  Panthallasa, 369
  Queen Charlotte Islands, 528
  Raton basin, 351, *353*
  Spiti Valley, *450*
  Tr-J boundary, *528, 529*

Val Badia, *456,* 459
Werfen Formation, 456
Woodside Creek, 356
York Canyon, 355
carbon monoxide
toxicity, 713
carbon nitrogen ratio, 356
carbon shift
P-Tr boundary, *375*
carbon to sulfur ratio, 338
carbonate
Sichuan Province, 374
carbonate compensation depth, 79, 80, 84
mass wasting, *92*
carbonate platforms
collapse, *482*
Carbonate Wash channel, 479
carbonyl sulfide
toxicity, 713
Caribbean
iridium anomaly, 164
shocked quartz, 164
spherule, 164
Carlin gold district, 495
Carnavon basin, 373
Carnic Alps
thorium uranium ratio, *450*
Catalonian basin, 552
cathodoluminescence, 267
Caudipteryx, 304
cementation
Alamo lapilli beds, *499*
Central America, 164
Central Atlantic magmatic province, 523, *527,* 529
Tr-J boundary, *518,* 530
Changxing Formation, *401*
carbon shift, 420
marine extinction, 363
reef complex, 374
Charlevoix structure, 673
Chavera Formation, 558
Cheiloceras Event, 483
Chelva Formation, 552
Chesapeake Bay impact, 33, 93, 675
Chichibu terrane, 406
Chicxulub crater, 28, 29, *39, 41,* 55, 669
coesite, 47
ejecta sequence, *56*
gravity data, *42*
magnetic anomaly, *43*
model parameters, 627
mulit-ring basin, 41
multi-ring, 45
peak ring, 42, 45
projectile, 634
seismic reflection, *43*
shocked quartz, 47
suevite, 47, *48,* 64
velocity data, 43
Chicxulub impact, 8, 65, 97, 164, 356
Cacarajicara Formation, 140
gases, 712
palynology, 356
chlorine, 715
Chlorite
Bjala, 242
chondrichthyians
Hornerstown Formation, 295
Mount Laurel Formation, 293
chondrite, 22
chromium isotopes, 21, 24, 607, *610*
impactor, 201
Morokweng, *615*
Vredefort, *612*
chronozone
Bjala, *216*
Cimmerian continent, 396
Cinque Quarry, 516
clams
Hornerstown Formation, 295
clay layer. *See also* boundary clay
Bjala, 241, 249
El Kef, 256
claystone
Moncada Formation, *115*
Clayton core, 542
Clayton Formation, 98
climate
Bjala, *225, 245,* 248
impact, *717,* 728
Kazakstan, 248
Late Ordivician, *463*
clinoptilolite
Beloc, 168
clinopyroxene
spherules, 28
coagulation
dust, 718
coal, 360
coalification
nitrogen isotopes, *355*
Coconino Sandstone
Hugoniot, 603
shock experiments, *597*
codex, 1
Telleriano-Remensis, 2
Vaticanus A, 2
coelacanth
Mount Laurel Formation, 293
coesite, 48
Chicxulub impact structure, 47, 50
formation temperature and pressure, 51
optical identification, 49
collision
asteroids, 645, 675
Laurentia, 468
Colorado Plateau, 388
comet showers, *484,* 668, *675,* 682
comets, *652*
Hale-Bopp, 16
Halley, 23
Shoemaker-Levy 9, 10, 11
sungrazing, 656
condensation
droplet model, *638*
kinetics, *639*
nucleation events, 632
temperature, 642
vapor, 642
Cone crater, 661
conifers
extinction, 368
conodonts
Alamo Breccia, 490, 497
Dinwoody formation, 387
extinction, 367, 480
Garnerkofel-1 core, 421
Glass Mountains, 390
Late Devonian, 479
Monitor Range, *465*
P-Tr boundary, *364,* 372, *390,* 391
Pahranagat Range, 497
Sosio valley, 399
Vinni Creek, *465*
zones, 386, *474*
corals
Anticosti Island, 465
Corpman crater, 45
Cortes de Tajuña Formation, 558
coupled ocean-atmosphere model, 718
Courant-Friedrichs-Lewys condition, 73
Coxquihui
iridium anomaly, 158
spherule, 158
crater
formation, *41*
cratering, 695
inner planets, 670
craton
western United States, *386*
critical velocity
impact, 628
crocodiles
Hornerstown Formation, 295
Navesink Formation, 293
cross-bedding, 89
Csõvár
carbon isotopes, *529*
Cuban fold belt, *110*
Cuban island arc, 126
Culpepper basin, 515
Cürük Dag section, *398*
cyclostratigraphy
Boonton Formation, 516
Feltville Formation, 517
Gartnerkofel core, *421*
McCoy Brook Formation, 517
P-Tr boundary, 415
Portland Formation, 516

D

darkness
impact, 717
database
K-T boundary, 55, 56, 65, 66
debris flow
Gulf of Mexico, 58
decarbonation, *572*
Deccan flood basalt, 346
K-T boundary, 376
Deccan Traps, 103, 104, 105

Deccan volcanism, 158
decomposition
  calcite, 591
decompression experiments
  multianvil apparatus, *588*
Deep Sea Drilling Program, 80
degassing
  Beloc, *167*
  calcite, *588*
Delamar Mountains
  Alamo lapilli, 499
Delaware basin
  age, 371
Delmarva Peninsula, 294
deposition
  Ancón Formation, 120
  Beloc, *183*
  Cacarajicara Formation, *140*
  Fish Clay, *269*
  Moncada Formation, *118*
depositional environment
  Karoo basin, *440*
  Nevada, *467*
devolatilization, 583
  dolomite, *574*
dewatering
  Alamo Breccia, 500
diagenesis
  Beloc, 168
  Blake Nose, *193,* 197
diamictite, *558*
  Pelarda Formation, 556
diamond
  Popigai impact structure, 600
  Ries impact crater, *601*
  synthesis, 596, *599*
Dierico section
  biostratigraphy, 416, 418
dinoflagellates, 85, 265
  adapters, *272*
  distribution, *269*
  DSDP 387, 89
  Fish Clay, *266*
  Hornerstown Formation, 295
  Inversand Pit, 295
  Navesink Formation, 295
  opportunists, *272*
  preservation, 266
  resting time, *271*
  Stevns Klint, *266,* 268, *272*
  survival, 273
dinosaurs
  Barrel Spring area, 326
  eggs, 307
  Fruitland Formation, 321
  geochemistry, *325*
  Hunter Wash, 326
  Kimbeto site, *324*
  Kirtland Formation, 321
  Navesink Formation, 293, 295
  Ojo Alamo Sandstone, *309, 315,* 321
  Pot Mesa site, 324
  San Juan basin, 309, *331*
  survival, *333*
  trace elements, *326*
Dinwoody Formation, 387, 403
dioxins, 715
disconformity
  Cacarajicara Formation, 125
  K-T, *183*
diversity
  New Jersey coastal plain, *292*
  recovery, 408
Dogger-Malm boundary
  impact event, 552
Dogie Creek
  sulfur isotopes, *338, 342*
dolomite
  detection, *575*
  shocking methods, *572*
dolomitization
  P-Tr boundary, 395, 409, 410
Domo Extremeño Group, 556
Donsbach tuff breccia, 477
downcutting
  Karoo basin, *437*
Driekoppen Formation, 437
droplet model
  condensation, *638*
  kinetic equation, *639*
droplet radius equation, 639
DSDP Hole 398D, 61
  slumping, 93
DSDP Site 1052
  structure, 88
DSDP Site 385
  K-P boundary, *83*
  stratigraphy, *84*
DSDP Site 386, 61
  K-P boundary, *83*
  stratigraphy, *85*
  turbidite, 89, *90,* 93
DSDP Site 387, 61
  K-P boundary, *83*
  stratigraphy, *85*
  turbidite, 89
DSDP Site 398D
  stratigraphy, *86*
DSDP Site 576
  meteorites, 30
DSDP Site 577
  spherules, 28, 29
DSDP Site 603
  K-P boundary, *83*
  stratigraphy, *84,* 89
  turbidite, 89, 91
DSDP Site 605
  K-P boundary, *82*
  structure, 88
Dufek intrusion
  age, 528
dunite
  Chicxulub impact, 634
dust
  distribution
    K-T boundary, *718*
  evolution code, 720
  interplanetary, *23, 34*
  layer, *718*
  size, *718*
Duwi section
  sea level, 542
dysoxia, 396
  late Paleocene, 534
  western Tethys, 397

E

Earth crossers, *651. See also* asteroids
East Greenland
  palynomorphs, 364
East Yucatán
  impact ejecta, 126
echinoids, 298
  Bjala, 236
Eckman transport, 544
effective kinematic viscosity
  equation, 626
effective temperature equation, 696
ejecta
  Alamo Impact, 495
  Blake Nose, 283
  Chicxulub crater, *56*
  distribution, 65
  ODP Site 1049, *279,* 288
El Caribe
  iridium anomaly, 158
  spherule, 158
El Guayal
  shocked quartz, 64
El Kef, 242, 253
  Global Stratotype Section and Point, 254, 255, 261
El Mimbral
  section, 260
  spherules, 196
El Mulato section, 260
electron impact ionization mass spectra, 346
electron microprobe
  shocked dolomite, *575*
electron microscopy
  Blake Nose, 191
Elikah Formation
  Iran, 400, 401
Elles, 242
Eltanin
  asteroid, 29
  impact, 29
  oceanic event, 30
Emeishan flood basalt, 377
energy
  kinetic, 11
energy conservation law, 686
energy dispersive x-ray, 243
energy equation, 639, 692
entropy
  equation, 633
environmental scanning electron microscope, 243
ephemerides, 645
equations
  adiabatic exponent, 633
  argument of perihelion, 653
  calcining, 501

continuous wavelet transform, 421
discontinuity equation, 422
droplet radius, 639
effective kinematic viscosity, 626
effective temperature, 696
energy equation, 639, 692
entropy, 633
Eulers equation for the radial velocity, 633
excess concentration, 170
galactic periodicity, 668
harmonic oscillations, 626
heat conduction, 692
mass conservation, 72
mass density, 633
mixing ratio, 714
momentum, 72
Morlet mother wavelet, 422
Navier-Stokes, 71
nucleation rate, 638
oscillating motion of a block, 626
quality factor, 626
radial velocity, 633
radiation energy, 687
ratio of excess to bulk concentrations, 170
reduced sulfate, 342
scaling law, 621
sulfate reduction rate, 342
transient crater diameter, 620
Zel'dovich factor, 638
equilibrium
thermodynamic, 592
Eros
asteroids, 9
erosion
atmospheric, *698*
Bjala, 242
Esna Formation
lithology, *535*
Etroeungt fauna, 483
Eulers equation for the radial velocity, 633
Eureka Quartzite, 492, 503
europium anomaly, *177,* 179
Beloc, 184
Lalung section, 451
Spiti Valley, 445
evaporites
Spain, 558
evolved gas analysis
Beloc, *182*
excess concentration equation, 170
experimental shocking, 571
extinction, 223, 226. *See also* mass extinction
abruptness, *416*
Early Jurassic, *525*
end-Triassic, 505
Late Ordovician, 469
peaks, 671
radiolaria, 528
rate, 266, 288
Tr-J boundary, *529*
extraterrestrial component
sediment, *21*

F

fallout layer
Raton basin, 333
Fannin drill hole, 312
Fantasque Formation, 403
faunal turnover
Late Ordovician, 469
fecal pellets, 105
Feixianguan Formation, *402*
Feltville Formation, 510
cyclostratigraphy, 517
tetrapod footprints, *512*
fern spike
Jacksonwald syncline, 514
Passaic Formation, 510
Tr-J boundary, 519
Ferrar Group
age, *528*
Ferrar province, 523
ferruginous band
P-Tr boundary, 445
Spiti Valley, *448*
thorium uranium ratio, 450
X-ray diffraction, 447
Finnigan MAT 251, 214
Finnigan MAT Delta S/GC isotope ratio mass spectrometer, 353
fires
global, 685
first law of thermodynamics, 633
Fischer-Tropsch type catalytic reactions, 710
Fish Canyon Tuff
age, *371*
Fish Clay
cyst abundance, 268
deposition, *269*
dinoflagellate cysts, 265, *266*
pyrite spherules, 337
fish fauna
Madagascar, 403
fish scales
Egypt, 537
flame structures
Alamo lapilli, 500
flash pyrolysis-gas chromatography, 459
flightless birds
K-T boundary, *305*
flood basalt
Deccan, 346, 376
Emeishan, 377
Siberian, 363, 373, *375,* 377
flood basalts
Tr-J boundary, 518, 519
Flora group
asteroids, 647
fluid-escape structure
Cacarajicara Formation, 126
fluorescence
black shales, 541
flux
extraterrestrial, 32
helium, 33
Flynn Creek impact, 484
food chain, 297
footprints
Passaic Formation, *512*
foraminifera, 61, 84, 85, 100, 105, 226
Bass River, 100, 294
Bellerophon Formation, 416, 419
Bjala, 223, 236, 237, *245*
black shales, 543
Blake Nose, 192
diversity, 291
DSDP Hole 398D, 86, 88
extinction, 146, *278*
Gebel Duwi, 538, 540
late Paleocene, *538*
Loma Cerca, *152*
mass extinction, *232*
Mendez Formation, 145
Monte Ruche, 419, 420
New Jersey Coastal plain, 103
ODP Site 1049, 279, 282, 288
Paleocene, *537*
paleoslope model, 100
size, *89, 282*
species richness, 266
Tesero horizon, *419*
formations
Abadeh Formations, 400
Amsekroud Formation, 516
Ancon Formation, 127, 138
Animas Formation, 309
Aquia Formation, 296
Balfour Formation, 430, 440
Bay State Formation, 492
Bellerophon Formation, 416, 419, 456
Blomidon Formation, 517
Boonton Formation, 513
Burgersdorp Formation, 441
Cacarajicara Formation, 61, 70, 121, 122, *125,* 126
Changxing Formation, 374, *401,* 420
Chavera Formation, 558
Chelva Formation, 552
Clayton Formation, 98
Coconino Sandstone, 597, 603
Dinwoody Formation, 387, 403
Driekoppen Formation, 437
Elikah Formation, 400, 401
Esna Formation, *535*
Eureka Quartzite, 492, 503
Fantasque Formation, 403
Feixianguan Formation, *402*
Feltville Formation, 510, *513,* 517
Fruitland Formation, 311
Gerster Formation, *387,* 388, 391
Grayling Formation, 403
Green River Formation, 356, 359
Gualija Formation, 558
Guilmette Formation, 483, 490, 491, 503

formations (*continued*)
Hornerstown Formation, 294
Juniata Formation, 469
Kaibab Formation, 38, 391
Kapp Starostin Formation, 403
Katberg Formation, 430
Kirtland Shale, 310, 314
Kokarkuyu Formation, 398
Lockaton Formation, 508
Lower Tuscarora Sandstone, 469
Maqam Formation, *401*
McCoy Brook Formation, 517
Mendez Formation, 145, *146*
Messel Shale, 359
Moenkopi Formation, 388, 391
Moncada Formation, 75, 109
Moreno Formation, 127
Mount Laurel, 293
Navesink Formation, 97, 293, 294
New Haven Formation, 516
Normandien Formation, 432, *436*
Ojo Alamo Sandstone, *307*
Orea Formation, 558
Oxyoke Canyon Sandstone, 492, 494, 503
Pamecak Formation, 398
Paratirolites Limestons, 400
Passaic Formation, 508
Pelarda Formation, 551, 556, 559, 673
Peñalver Formation, 61, 70, 126, 140
Phosphoria Formation, 387
Plympton Formation, 388
Polier Formation, 127, 140
Portland Formation, 516
Prairie Bluff Chalk, 98
Quartermaster Formation, 371
Red Bank, 293
Salado Formation, 371
San Cayetano Formation, 140
Schuchert Dal Formation, *404*
Sentinel Mountain Formation, 492
Spechty Kopf Formation, 673
Talcott Basalt, 516
Tarawan Formation, 535
Thaynes Formation, 379
Thebes Formation, 535
Three Forks Formation, 483, 484
Tinton Formation, 293
Toad Formation, 403
Tobin Formation, 388
Towaco Formation, 513
Tvillingodden Formation, 403
Union Wash Formation, 379, 403
Vardebukta Formation, 403
Verkykerskop Formation, 429, 432, 436
Vincentown Formation, 294, 296
Werfen Formation, 397, 416, 456
West Range Limestone, 483
Wordie Creek Formation, *404*
Yátova Formation, 552
Fossil Mountain Member
Kaibab Formation, 388
fossil wood
P-Tr boundary, 441
fossils
abundance, *282*
Baptornis, 304
Bellerophon Formation, 416
Beloc, 167
Bjala, *220, 240*
Bochil, *255*
brachiopods, 298
Cacarajicara Formation, *127*
Caudipteryx, 304
clay layer 2, 241
dinoflagellates, 85, 89
Dinwoody Formation, 387
DSDP Site 385, *84*
echinoids, 236, 298
fecal pellets, 105
foraminifera, 61, 84, 85, 105, 145, 192, 223, 226, 236, 237, 245, *278,* 279
Gargantauvis philoinos, 304
Gargantuavis, 303
Gastornis, 304
Gastornithidae, 303
Gebel Duwi section, *540*
globotruncanids, 241
graptolites, *464*
guembelitrids, 241
Gulf of Mexico, *261*
Gungri Formation, *446*
hadrosaur, *311*
hedbergellids, 241
Hesperornis, 304
heterohelicids, 241
Hornerstown Formation, *295*
Karoo basin, *368, 433, 439*
kerogen assemblage, 85
La Ceiba, 254
Meishan, *367*
Mikin Formation, *446*
molluscs, 266, 298
Mount Laurel Formation, *293*
nannofossils, 61, 85, 110, 192, 223, 226, 236, 237, 279
nautiloids, 298
New Jersey coastal plain, 292
Ojo Alamo Formation, *322*
Paleocene, *537*
Passaic Formation, *510*
Patagopteryx deferrariisi, 304
Pithonelloideae, 271
plant debris, 89
pollen, 89
Red Bank Formation, *294*
rugoglobigerinids, 241
Sass de Putia section, *417*
sponges, 298
tetrapods, *437*
Val Badia, *456*
Werfen Formation, 416
Fourier transform infrared spectroscopy, 346
Fox Hills Formation, 294
Fra Mauro Formation
lunar, 660
Frasnian-Famennian boundary, *482*
Fruitland Formation
dinosaurs, 321
palynology, 311
Fuentes Bed
Vendian-Cambrian boundary, *556*
fullerenes, *27*
K-T boundary, *345*
origin, *349*
P-Tr boundary, 373
Fundy basin
Tr-J boundary, 517
fungal remains
P-Tr boundary, *368*
fungal spike, 376
Karoo basin, 431
P-Tr boundary, 377
furan rings, 459
fusulinids
Garnerkofel-1 core, 421

G

Galactic Carrousel, 668
galactic periodicity
equation, 668
Galala platform, 544
gamma ray
counts, 421
spectrometry, 447
Ganmachidam section
Spiti Valley, 447
Gargantauvis philoinos, 304
Gartnerkofel core
carbon isotope anomaly, 420
cyclostratigraphy, *421*
iridium anomaly, 373
sedimentation rate, *424*
gas chromatography, *353*
carbon isotopes, 456
gas plume
composition, 706
Gasbuggy drill core, 312
gases
Chicxulub impact event, 712
toxic, *713*
Gastornis, 304
Gastornithidae, 303
gastropods
Egypt, 537
Hornerstown Formation, 295
P-Tr boundary, 378
Gebel Duwi
deposition, 535
fossils, *540*
Gebels Aweina
deposition, 535
genesis
Alamo lapilli, *500*
geochronology
uranium-lead, 524
Gerster Limestone, *387,* 388, 391
brachiopods, 390
Geulhemmerberg
K-T boundary section, 269
organic compounds, 357

glaciation
Late Ordovician, *463,* 469
Southern Hemisphere, 474, 482, 484
glauconite, *293*
Global Stratotype Section and Point
El Kef, 254
global warming, 104, 105, *530. See also* greenhouse effect
earliest Triassic, 372
Famennian, *483*
P-Tr boundary, 377, 410
goethite, 204, *208,* 209
Meghalaya, 206
Mössbauer spectrum, 205
ODP Site 1049, 279
Turkmenistan, 206
gold, 27
Gorbusha Suite, 402
Gosses Bluff impact, 672
Gould's belt, 679, 680
graptolites
Late Ordovician, *464*
gravitational perturbation
Oort comet cloud, *675*
gravity flows
Bochil, 255
Gulf of Mexico, 58
Grayling Formation, 403
Green River Formation
hopanes, 359
nitrogen isotopes, 356
greenhouse effect, 375, 376. *See also* global warming
Tr-J boundary, 529
greenhouse gases, 225, 588
greigite, 222
Grist Mills section
iridium, 514
Gualija Formation, 558
Guaniguanico terrane, 110, 121, 126, 131
Guatemala
spherule, 146
Guayal section, 254
Gubbio, 202
iron phases, 207, 210
Mössbauer spectrum, 204
particle size, *207*
Guilmette Formation, 483, 490, 491, 503
Gulf of Mexico
biomagnetochronology, *260*
debris flow, 58
fossils, *261*
gravity flows, 58
K-T boundary, *254*
shocked quartz, *64*
spherules, 61
tsunami deposits, 28, *58, 70*
Guling section
Spiti Valley, 447
Gungri Formation
fossils, *446*
Günterod
debris flow, 477
Guryal ravine
ferruginous layer, 446
gypsum, 388

H

hadrosaur
Ojo Alamo Formation, *316*
Ojo Alamo Sandstone, 307, *311,* 325, 333
Haiti
platinum group element anomaly, 248
Halley's comet, 2, 652
Hamersley basin
spherule, 26
Hancock Summit West
Alamo Breccia, 500
Hangenberg Event, 484
hardground, 552
harmonic oscillations
equation, 626
Hartford basin
correlation, 510
Tr-J boundary, *516*
Haughton impact crater, 40
portlandite, 572
Havallah sequence
radiolarians, 388
Havallah-Koipato angular unconformity, 388
heat conduction
equation, 692
helium isotopes, 21, 24, *32*
hematite, *208,* 209
Alamo lapilli, 497
hemispherical geometry
numerical solution, 634
Hendaye section
K-T boundary, 554
Hercynian Massif, 556
Hesperornis, 304
Hewlett Packard 5890 gas chromatograph, 353
hiatus
Gulf of Mexico, 262
Hunter Wash area, 325
Mexico, 260
P-Tr boundary, 385, *389*
Queenston deltaic complex, 469
high-resolution electron-impact ionization mass spectrometry, 345
Hiko Hills
Alamo lapilli, 499, 500
Hilda family
asteroids, 647
Hipparcos satellite, 664, 668
Hiroshima atomic bomb
explosive energy, 567
Hogup Mountains
brachiopods, 390
Dinwoody Formation, 387
homogenite, 140
Peñalver Formation, 122
Hony railroad cut
microtektite, 481
hopanes, 352, 360
Brownie Butte, *358*
Green River Formation, 359
Messel shale, 359
Raton basin, *358*
Hornerstown Formation, 294
hotspots
shock metamorphism, 598
Hugoniot
Coconino Sandstone, 603
discontinuities, 602
Hunan Province
ammonoids, *367*
Hungaria resonance group
asteroids, 647
Hunter Wash, 309
ash beds, 332
dinosaur bones, 326
hiatus, 325
paleomagnetism, 331
Hut calculation, 663
Hyades supercluster, 680
hydrocarbons
toxicity, 713
hydrocode
two-dimensional, 620
hydrodynamics
algorithm, 636
vapor cloud, *632*
hydrogen cyanide
impact generated, 706
toxicity, 713
hydrogen sulfide
toxicity, 713
hydrothermal vents
submarine, 552
hyperfine magnetic field, 202, 204
hypervelocity jet, *700*

I

Iberian Abyssal Plain, 79, *85*
Iberian basin, 552
Iberian Peninsula
Azuara structure, *555*
Iberian Pyrite Belt, 558
Ibor Group, 556
ice age, 17
ice sheet
Antarctic, 100
ignition area
grazing meteoroids, 691
Iigaste
meteorites, 564
Illawarra mixed polarity superchron, 370
illite
Bjala, 242
Blake Nose, 193
Ilumetsa craters, 563, *567*
impact, *18*
angle, 697
breccia, *615* (*See also* suevite)
Canadian margin, 93
Chesapeake Bay, 93
civilization-ending, *12*

impact (*continued*)
 climate changes, *717*
 craters, *39*
 darkness, 717
 dust, 11
 ejecta
  K-T boundary, 559
  Moncada Formation, *118*
  ODP Site 1049, 278
 event
  Dogger-Malm boundary, 552
  geochemical anomalies, 551
  numerical simulations, 619
  survivors, *287*
 frequency, 10
 glass, 48, 671 (*See also* microtektite; tektite)
 hazard, 8, 9
 kinetic energy, 11
 lethality, *7*
 melt
  Meteor Crater, 572
  modeling, 628
  numerical modeling, 622
  Popigai, 572
 nuclear, 11
 ocean, *31*
 periodicity, *674*
 prediction, 8, 16
 risk, *13*
 spherule, *28*
 structures
  Acraman, 26
  Carnavon basin, 373
  Woodleigh, 373
 vaporization, *631*
impactites
 extraterrestrial, *610*
impactors, *8*. *See also* impact
 bolide phase, 11
 ejecta plume phase, 11
 K-T, 13
 low angle, *686*
 radiation, 695
inductively coupled plasma-mass spectrometry, 236, 513
 Blake Nose, 191
inoceramids
 Mount Laurel Formation, 293
Institute of High Pressure Physics, 49
instrumental neutron activation analysis, 326
interstellar cloud, 679
 giant, 680
Inversand Pit, 291, 294
 glauconite, 294
ion-exchange uptake, 328
iridium anomaly, 27, 55, 56, *62,* 65, 201, 231, 255
 Alamo Breccia, 490
 Alamo Impact, 477
 Alamo lapilli, 497
 Anjar, 346, 349
 Beloc, 146, 174, 175, 183, 184
 Bjala, *222,* 236, 240, 241, 248
 Caravaca section, 553
 Caribbean, 164
 Central America, 164
 Coxquihui, 158
 distribution, *62*
 El Caribe, 158
 El Kef, 256
 Gartnerkofel core, 373
 Grist Mills section, 514
 Gulf of Mexico, 261
 iron phases, *207*
 Jacksonwald syncline, *514*
 Mendez Formation, 146
 Mexico, 248
 Moncada Formation, 110, *118*
 New Jersey passive margin, 97
 New York State, 481
 Newark basin, 517
 ODP 1049, 282
 ODP Site 1049, 278, 284
 P-Tr boundary, 372
 Passaic Formation, 510
 Tesero Oolite, 373
 Tr-J boundary, 506, 519
 tracer, 21, 22
 Vredefort, 612
 Zumaya section, 554
iridium coincidence spectrometry, 513
iron
 magnetically ordered phases, 204
 mineralogy, 201, *207, 208*
iron sulfur relationship, *342*
iron-oxide nodules
 Karoo basin, *438*
isomer shift, 202, 204
isotopes. *See also* carbon isotopes; oxygen isotopes
 Beloc, *179*
 Bjala, *245*
 DSDP Hole 398D, 86
 foraminifera, 100
 helium, *32*
 high latitudes, 232
 Mendez Formation, *153*
 ODP Site 1049, 284

J

Jacalteca
 myth, 5
Jacksonwald syncline
 tetrapods, 511
 Tr-J boundary, 508, 510
Jacobian integral equation, *653*
Jameson Land
 basin, 404
 mass extinction, *459*
Jan Mayen hotspot, 376
Japan
 sedimentation rates, *405*
Juniata Formation, 469
Jupiter Trojans, 647
Jura basin, 552
Jurassic Global Stratotype Section, 524

K

K-T boundary
 Barrel Spring, *319*
 Blake Nose, 279
 database, 55
 flightless birds, *305*
 Gulf of Mexico, *254*
 regression, 57
 Spain, *553*
Kaali craters, 563, *564*
 age, *567*
Kaali meteorite, *564*
Kaande beach
 meteorites, *564*
Kaiavere
 meteor, 563
Kaibab Limestone, 388, 391
 brachiopods, 390
Kalahara Group calcrete
 Morokweng, 611
kaolinite
 Bjala, 242
 Blake Nose, 193
 Tunisia, 248
Kapp Starostin Formation
 fossils, 403
Kara Ust-Kara crater, 675
Karabaglar Suite, 400, 401
Kärdla crater, 563
Karoo basin
 boundary sequence, 425
 braided streams, 379
 lithostratigraphy, *436*
 P-Tr boundary, *432, 433*
 province, 523
 stratigraphy, *430*
 tetrapods, *437*
Karoo Group basalt, *528*
Karoo-Ferrar volcanism, 529
 strontium isotopes, 530
Katberg Formation, 430
Kazakstan
 climate, 248
Kellwasser Crisis, 480
Kendelbach, 517
 carbon isotopes, 528
kerogens, 85, 457, 459
 black shale, 537, 541
Khémisset basin
 stratigraphy, *516*
Kiaman paleomagnetic superchron, 376
Kimbeto Member
 dinosaurs, *324,* 326
 Ojo Alamo Sandstone, 310
kinetic effects
 phase transitions, *603*
kinetic equation
 droplet model, *639*
Kirtland Formation, 310, 314
 dinosaurs, 321
 geologic history, *330*
 K-T boundary, *321*
 paleomagnetism, *331*
 palynomorphs, *321*

trace elements, *328*
uranium, *329*
Klenova, 45
Koipate volcanics
ammonoids, 388
Kokarkuyu Formation
stromatolites, 398
KTbase, *56,* 65, 66
Kuiper Belt, 656, 675
Kunga Island
Jurassic Global Stratotype Section, 524

L

La Ceiba, 253
K-T boundary, 261
stratigraphy, *254*
La Güira Member
Ancón Formation, 111
La Lajilla
spherule, 196
La Sierrita, 145, *146*
lithology, *152*
lag deposit
Hornerstown Formation, 296
Lagrangian
evolution, 641
method, 627
Lahaul Valley
ferruginous layer, 446
Lalung section
Spiti Valley, 447
Lambert graben, 379
landslide
model, 72
lapilli. *See also* spherule
Alamo Breccia, 489, *490*
Albion Island, 497
Ries crater, 496
Las Hurdas
Fuentes Bed, 556
last occurrences
fossils, *416*
late Paleocene
paleoceanographic model, *544*
late *Rhenana* Zone eustatic rise, *479*
Laurentia
collision, 464, 468
Laurentian plate
paleoenvironment, *465*
Lazarus lineage, 378, 483
Leco CS-300 elemental analyzer, 536
lethality
impact, *7*
Lingati River, 445
Linguiformis Zone
anoxia, *479*
eustatic fall, *480*
Link eXL X-ray spectrometer, 573
Link ISIS 300 X-ray spectrometer, 573
lithology
Alamo Breccia, *492*
Bjala-1, *236*
Brownie Butte, *352*
Cacarajicara Formation, *127*
Karoo basin, *436*
Morokweng, 612
Normandien Formation, *436*
Ojo Alamo Sandstone, *309, 311*
Raton basin, *352*
Stevns Klint, *267*
Verkykerskop Formation, *437*
Lockatong Formation
cyclicity, 508
Loma Cerca
Mendez Formation, 146
paleoclimate, *157*
sediment accumulation rate, 151
lonsdaleite
shock indicator, 600
Lootsberg section
Karoo basin, 432
Lorentzian peak, 204
Los Organos belt, 110, 111, 121, 127
Lower Breccia Member
Cacarajicara Formation, *128, 141*
Lower Red Member
Moenkopi Formation, 388
Lower Tuscarora Sandstone, 469
Luda Kamchia unit, 233
lunar gardening, 661
Lusitanian basin, 552
lycopsids
early Triassic, 378

M

macrofossil
New Jersey coastal plain, 291
Madagascar
P-Tr sections, 403
maghemite
Mössbauer spectrum, 205
magic angle sample spinning nuclear magnetic resonance, 49
magnetism, 88
Bjala, *218*
Chicxulub crater, *43*
Morokweng, 611
Ojo Alamo Sandstone, *332*
P-Tr boundary, *370*
Paris basin, 516
polarity zone, 213
Portland Formation, 516
resonance spectroscopy, 345
Main Belt asteroids, 675
major elements
Beloc, *169*
Manicouagan crater, 40, 670
peak ring, 41
Manicouagan impact, 518
Manson crater, 672, 675
Maqam Formation, *401*
marl
mining, 291, 294
Paleocene, *543*
Martinsville core
reversed polarity, 511
mass conservation equation, 72
mass density equation, 633
mass extinction, 8, *17,* 79, 105, 158, 231, 256, 298. *See also* extinction
ammonites, 298
Bellerophon Formation, *420*
birds, 303
Bjala, 245
carbon cycle, 528
Devonian, *474*
diversity, *297*
end-Triassic, *523*
foraminifera, 105, 248, 253, 254, 262
Gulf of Mexico, 261
Jameson Land, *459*
K-T boundary, 190, 391, 669
Late Devonian, 476, 482
Late Famennian, *484*
Late Frasnian, *480*
Late Ordovician, *463*
New Jersey coastal plain, 292
ODP Site 1049, *287*
P-E boundary, 554
P-Tr boundary, *366, 372,* 385, 391, *409, 415,* 430, *445*
Pangea, 377
pelagic conodonts, 484
Permian, *363,* 408
slope failure, *93*
southern China, 377
Tesero Oolite Horizon, 397
Tr-J boundary, 506, 524, 559
Werfen Formation, *420*
mass flow, *90*
ODP 1049, *283*
mass spectrometry, *353*
mass wasting, 55. *See also* turbidite
Blake Nose, 279
carbonate compensation depth, *92*
Chicxulub impact, *80,* 93
Iberia, *91*
K-P boundary, *87*
western North Atlantic, *83*
Massignano quarry, 33
matched impedance recovery experiments, 597
matrix
Alamo lapilli, *499*
Maya, 2
Maymecha-Kotuy area
age, 376
Mazzin Member
Werfen Formation, 397, 398, 416
McCoy Brook Formation
cyclostratigraphy, 517
Meenikunno Bog
spherules, 568
megabreccia
Alamo Impact, 477
megatsunami, 262
megaturbidite
impact-induced, 110
Meghalaya
iron phases, *207, 208,* 210
Mössbauer spectrum, 206
section, 202

Meishan
  age, *371*
  anoxic facies, 408
  ash beds, 376
  boundary clay, 372
  fossils, *367*
  P-Tr boundary, *367,* 425
  strontium isotopes, 370, 373
  sulfur isotopes, 373
  type section, 371
melt rocks, 39, 47
melt sheet, 41
Mendez Formation, 61, 145, *146*
  Loma Cerca, 146
Mesa Portales, 321
  palynology, 311
Messel shale
  hopanes, 359
Meteor Crater
  shocked sandstone, 603
Meteor crater, 40
Meteorite Crater, 581
meteorites
  DSDP Site 576, 30
  Eltanin oceanic event, 30
  Estonia, *563, 564*
meteoroids
  grazing, 693
  pancake model, *687*
methane, 542
  hydrate, *534*
  P-Tr boundary, *375*
  recycling, 359
methanogenesis, 370
Mexico
  hiatus, 260
  iridium anomaly, 248
microimpactite
  Piila bog, 567
microkrystites
  Alamedilla section, 554
Micromass Optima mass spectrometer, 100
microspherules, 671
microtektite, 28, 164, 254, 481. *See also* tektite
  Frasnian-Famennian boundary, 485
  Late Devonian, 473
mid-Atlantic rise, 79
Middle Calcarenite Member
  Cacarajicara Formation, *133, 141*
Mie theory, 720
Mikin Formation
  fossils, *446*
Milankovitch cycle, 213, 223, 225, 415, 420, 423, 424, 507
  Bjala, *218*
  Gartnerkofel-1 core, 368
  Newark basin, 510
Milky Way, 667
mineralogy
  Beloc, *167*
  La Sierrita, *152*
  Moncada Formation, *116*
minimalist organisms
  brachiopods, 297
  sponges, 297
miogeocline
  western United States, *386*
Mississipi embayment, 69
mixing ratio equation, 714
Mjølnir crater, 40, 672
Moenkopi Formation, 391
  red beds, 388
molluscs, 266, 298
  Hornerstown Formation, 295
  Navesink Formation, 295
moment equations, *639*
momentum equation, 72
Moncada Formation, 75, 109, *111, 115*
Monitor Range
  conodonts, *465*
Monmouth Group, 293
Monte Ruche Formation
  biostratigraphy, 416, 418
  foraminifera, 419
Monte-Carlo simulations, 638
Montes de Toledo
  Fuentes Bed, 556
Moreno Formation, 127
Morocco
  rift basins, *516*
  Tr-J boundary, 515
Morokweng
  drill cores, 611
  impact structure, 610, 611, 672
mosasaurs, 293, 295
Mössbauer spectroscopy
  K-T boundary clays, 201, *202*
  particle size, *207*
  sample preparation, *202*
  Spiti Valley, *450*
Mössbauer spectrum, 202
Mount Irish Range
  Alamo lapilli, 499
Mount Laurel Formation, 293
Mount St. Helens
  volcanic dust, 718
Mulhouse basin
  derogens, 356
multi-ring basins, 41
multianvil apparatus
  decompression experiments, *588*
Murchison meteorite
  helium, 373
Muschelkalk conglomerate, 421
mutation rate, 226
myth
  Aztec, *3*
  Jacalteca, 5
  Nahua, 5
  Popoloca, 5
  Tonotac, 5
  Tzotzil, 5

N

Naashoibito Member
  definition, *312*
  Kirtland Shale, 310
Nacimiento Formation
  palynomorphs, *317*
Nahua
  myth, 5
nannofossils, 61, 86, 110, 150, 242
  Bjala, 236, 237
  Blake Nose, 192
  calcareous, 85
  extinction, 146
  ODP Site 1049, 279, 281, 282
  plankton, 223, 226
nanodiamond, 602
nanoparticles
  Allende meteorite, 349
  vapor condensate, 210
Nanpanjiang basin
  biostromes, 378
  calcimicrobial mounds, 378
NASA, 10
nautiloids, 298
Navalpino Breccia, *557*
Navesink Formation, 97, 293, 294
  deposition, 104
  fossils, 293
  shell bed, 295
  strontium isotopes, 100
  unconformity, 105
Navier-Stokes equations, 71
Near Earth Asteroid, 9, 10, 14
Nemesis
  orbit, *662*
  theory, 659, *662*
Neugrund crater, 563
Neuquén basin, 552
Nevada
  depositional environment, *467*
New Haven Formation, 516
New Jersey coastal plain, 98
  drilling project, 98
  margin slope failure, 88
  slumping, 93
  thermal maximum, 542
New Jersey transect
  reflector, 88
Newark basin
  coring project, 508
  palynology, 507
  platinum group elements, *513*
  siderophile elements, *513*
  stratigraphy, *507*
  Supergroup, 510, 515, *524*
Nezzi
  deposition, 535
Ni-rich spinel, 671
Niobrara Chalk, 304
nitric acid, 12
nitrogen
  impact vaporization, 711
nitrogen isotopes
  anoxia, *356*
  Brownie Butte, 351, *353*
  coalification, *355*
  Green River Formation, 356
  Raton basin, 351, *353*

nitrous oxide
impact generated, 706
nonlinear long-wave theory, *71*
nonoparticles
vapor condensate, 209
Nordic mythology
meteorite, 566
Noril'sk gabbro
age, 376
Normandien Formation, 432, 436
North America
palynology, 354
North American regional ammonoid zonal scale, 524, 526
nuclear war, 16
nuclear winter, 12
nucleation, 642
condensation, 632
homogeneous, *638*
rate equation, 638
Nukhl section
radiolaria, 544

O

oblique impacts
hypothesis, 64
numerical modeling, 625
transient cavity scaling, *623*
ocean circulation model, *722*
Ocean Drilling Program, 80
Ocean Drilling Program Site 1049
K-P boundary, 81
Ocean Drilling Program Site 1049A, 190
Ocean Drilling Program Site 1049B, 190
Ocean Drilling Program Site 1049C, 190
Ocean Drilling Program Site 1050
K-P boundary, 81
Ocean Drilling Program Site 1052
K-P boundary, 81
Ocean Drilling Program Site 689
spherules, 28
ocean temperature
impact effects, *723*
profile, *725*
Ojo Alamo Arroyo, 307
Ojo Alamo Sandstone, *307*
age, 310
basal contact, *314*
dinosaurs, 321
geologic history, *330*
hadrosaur, 307, *311*
lithology, *309*
paleomagnetism, *331,* 333
palynology, *310, 317*
rare earth elements, *329*
reworking, *325*
trace elements, *328*
type area, 311
uranium, *329*
Oldoinyo Lengai volcano
carbonatite, 495
Oman Mountains, 401
Oort comet cloud, 26, 33, 34, 662, 668, 681
flux, 674
gravitational perturbation, *675*
optical identification
coesite, 49
stishovite, 49
Orange Mountain Basalt, 510, *524*
age, 507
orbit
evolution, 652, *655*
Nemesis, *662*
Sun, 668
Orea Formation, 558
organic carbon
terrestrial, 352
organic geochemistry
Brownie Butte, *352*
Raton basin, *352*
Orgueil chondrite, 34
origin
spherules, *155*
orogeny
western Cuba, 126
osmium isotopes, 21, 22, 24, *32,* 607, *608*
K-T boundary clay, 609
Morokweng, 610, *614*
Vredefort, *612*
ostracods
Egypt, 537
extinction, 480
Gebel Duwi section, 540
Late Devonian, 479
Sosio valley, 399
outgassing
dolomite, 571, 579
overturn
oceanic, *374*
oxygen budget, 715
oxygen isotopes, 98. *See also* isotopes
Bass River, 104
Beloc, 180
Bjala, 221, 249
Loma Cerca, 158
Mendez Formation, *153*
Navesink Formation, 101, 103
Raton Basin, 32
Stevns Klint, 32
Tr-J boundary, *529*
oxygenation
Sosio valley, 399
southern China, *402*
Oxyoke Canyon Sandstone, 492, 494, 503
oyster
Hornerstown Formation, 295
Navesink Formation, 293

P

P-Tr boundary
anoxia, *372*
biostratigraphy, *364*
boundary event, 430
boundary sections, *396*
boundary sediments, *402*
carbon isotopes, *369*
conodonts, *364*
definition, 396
fungal remains, *368*
Karoo basin, *432, 433, 442*
magnetostratigraphy, *370*
mass extinction, *366, 409*
proposed formal boundary, 386
sea level, *371*
southern China, *401*
southwestern United States, *387*
strontium isotopes, *370*
sulfur isotopes, *370*
Pahranagat Range
conodonts, 497
paleobathymetry
Gulf of Mexico, *71*
paleoceanographic model
late Paleocene, *544*
paleoclimate
Loma Cerca, *157*
paleocurrent
Cacarajicara Formation, *138*
Moncada Formation, *119*
paleoenvironment, 57
Bjala, *245*
Laurentian plate, *465*
paleogeography
P-Tr boundary, *396*
paleomagnetism
Bjala, 214, *216*
Kirtland Formation, *331*
Laurentia, 469
New Jersey Coastal plain, *103*
Ojo Alamo Sandstone, *331*
paleoslope model
foraminifera, 100
paleosols
Early Triassic, 370
Palisade sill
age, 507
palladium
Beloc, 176, 184
Bjala, 241, 248
enrichment, 27
palygorskite
Blake Nose, 197
palynology
Alamo Wash, 310
Barrel Spring, 310, *317,* 333
Blomidon Formation, 517
Chicxulub impact, 356
East Greenland, 364
index, 334
Kimbeto site, 325
Kirtland Formation, *321*
Nacimiento Formation, *317*
Newark basin, 507
North America, 354
Ojo Alamo Sandstone, *310, 317*
Passaic Formation, *510*
Raton basin, 334
southern Israel, 368
Val Badia, *457*
westen Alps, 368

Pamecak Formation
  platform carbonate, 398
Pan-African basement, 56
pancake model
  meteoroids, *687*
Pangea, 396
  mass extinction, 377
  movement, 372
  rift basins, 506
Panjal volcanism, 452
Panthalassa
  anoxia, 395
  carbon isotopes, 369
  ocean, 396, 403, *405*
Paratirolites Limestons
  deposition, 401
  fossils, 400
Paris basin, 304, 510, 552
  magnetostratigraphy, 516
Park City Group, *387*
particle size
  iron phases, *207*
Passaic Formation
  boundary section, *510*
  cyclicity, 508
  iridium anomaly, 510
Patagopteryx, 305
Patagopteryx deferrariisi, 304
Pavant thrust, 389
peak ring, 39, 40
  Chicxulub, 41, 45
  Manicouagan crater, 41
  Popogai crater, 40
peak shock pressure, 596
Pelarda Formation, 551, 556, 559, 673
Pen River, 445
Peñalver Formation, 61, 70, 126, 140
  homogenite, 122
Perigondwanan shelf
  anoxia, 395
Perkin Elmer 5100 ZL spectrometer, 192
Permian-Carboniferous reversed polarity
    superchron, 370
Pernambuco region, 62
perturbation
  atmospheric, 699
Petriccio
  spherules, 29
PGE abundance
  spherule, *26*
phase transformation
  minerals, *602*
  shock metamorphism, 596
Phe volcanism, 452
Phillips X'Pert MPD diffraction system,
    573
phillipsite
  Beloc, 168
phosgene, 715
Phosphoria Formation, 387
  biomarker, 375
Phyllite-Quartzite Group, 558
Piila bog
  microimpactite, 567
Pilistvere
  meteorites, 564
Pilot basin
  expansion, 483
Pinar del Rio Province, 110
Pinar fault, 111
Pinos terrane, 126
pisolith
  Cacarajicara Formation, 140
Pithonelloideae
  disappearance, 273
  distribution, 271
planar deformation features
  Cacarajicara Formation, 138
  Morokweng, 611
  Pelarda Formation, 556
  shocked quartz, 497
  Vredefort, 611
plane lamination, 89
platinum anomaly
  Beloc, 174
platinum group elements, 231
  anomaly, 248
  Beloc, 163, *174,* 180
  chondrite normalized, 174
  Haiti, 248
  impact structures, *608*
  Newark basin, *513*
Pleiades supercluster, 680
plesiosaurs
  Navesink Formation, 293
Pliensbachian-Toarcian boundary, *525*
plume expansion, 695
Plympton Formation, 388
polarity
  magnetic, 226
  reversal, 213, *219,* 511
Polier Formation, 127, 140
polycrystalline micron diamond, 602
polycyclic aromatic hydrocarbons, 710,
    715
Pons Formation, 112, 131
Popigai
  crater, 33, 40, 42, 675
  diamonds, 600, 601
  impact melt, 572
Popoloca
  myth, 5
Põrguhaud crater, 567
  oral history, 568
Portland Formation, 516
  cyclostratigraphy, 516
portlandite
  Haughton impact crater, 572
Portugal
  slope failure, 93
postshock cooling history
  shock metamorphism, 596
Pot Mesa site
  dinosaurs, *324*
  Ojo Alamo Formation, *324*
Poty quarry, 62
  shocked quartz, 63
Prairie Bluff Chalk, 98
precipitation
  impact effects, *728*
predator
  duraphagous, 298
prediction
  impact, 8
Princeton Mine, 27
productivity
  Bjala, 245, *248*
  late Paleocene, *543*
  P-Tr boundary, 420
Productus shale, 445
protognathodid biofacies, 484
pseudotachylytes
  complex craters, 40
Puchezh-Katunki crater, 670
pyrite
  framboidal, 372
pyrite spherules
  Fish Clay, 337
  Stevns Klint, 337
pyrosynthetic polycyclic aromatic
    hydrocarbons
  Caravaca, 553

Q

Qreiya
  deposition, 535
quadrupole splitting, 202, 204
quality factor
  equation, 626
Quartermaster Formation
  age, 371
Queen Charlotte Islands
  carbon isotopes, 528
Queenston delta, 464, *468*

R

radial velocity
  equation, 633
radiation
  energy equation, 687
  grazing meteoroids, *691*
  impact, *696*
radiative-convective atmospheric models,
    718
radiolaria
  Ben Gurion section, 544
  Egypt, 537
  Esna Formation, 544
  extinction, 528
  Havallah sequence, 388
  Nukhl section, 544
  P-Tr boundary, 372
  Sosio valley, 399
radiometric dating. *See also* age dating
  New Jersey coastal plain, 294, 295
  Tr-J boundary, *507*
Raman spectra, 49, 50, 571, 573, *577*
Rancocas Group, 294
rare earth elements
  Beloc, *177*
  Blake Nose, 193, *196*

chondrite normalized, *182*
ferruginous band, *451*
Kirtland Formation, *329*
Ojo Alamo Sandstone, *329*
Spiti Valley, *448*
ratio of excess to bulk concentrations
equation, 170
Raton basin, 48
biomarkers, *357*
fallout layer, 333
hopanes, *358*
lithology, *352*
organic geochemistry, *352*
oxygen isotopes, 32
palynomorphs, 334
receding wave, 69
tsunami, *73*
Red Bank Formation, 293
red beds
Moenkopi Formation, 388
red layer
El Kef, 261, 262
Red Sea rift system, 541
reentry, 692
reference intensity ratio, 575
reflectors, 89
chaotic, 93
K-P boundary, *81*
New Jersey transect, 88
refugia, *13,* 16, 17
regression
K-T boundary, 57
Reppwand outcrop, 421
foraminifera, 425
reptiles
footprints, 510
resting time
dinoflagellate cysts, *271*
reversed polarity, 513, 519
Martinsville no. 1 core, 511
reworking
Alamo lapilli, *503*
ODP Site 1049, 284
Ojo Alamo Sandstone, *325*
rhenium osmium ratio, 609
rhythmites
Basque-Cantabrian basin, 554
ricochet
hypervelocity, 685
Ries crater, 40, 60
diamond, 601
lapilli, 496
suevite, 500, 572, 602
Rietveld method, 575
rifting
Gulf of Suez, 541
Red Sea, 541
rim wave, 70
tsunami, 70, *73*
Rio Cuarto crater field, 686, 693
Rio Grande embayment
rushing wave, 69
Rock-Eval pyrolysis, 536
Rosario belt, 121, 126, 127
Rubielos de la Cérida structure, 673
rushing wave, 69
tsunami, *73*

S

Saaremaa Island
craters, *564*
Salado Formation
age, 371
SALE hydrocode, 627, 628
Salt Range
boundary sediments, 402
marine facies, 379
sequence, 364
sample preparation
Anjar, *346*
Bjala, *214, 233*
Blake Nose, *190*
Brownie Butte, *338, 352*
Dogie Creek, *338*
Egyptian sections, *535*
Karoo basin, *433*
lunar spherules, *660*
Mössbauer spectroscopy, *202*
ODP Site 1049, *278*
Raton basin, *352*
Spiti Valley, *447*
Stevns Klint, *266*
San Cayetano Formation, 140
San Juan basin
dinosaurs, 311, *331*
sandstone
Moncada Formation, *115*
sapropelite, 534
Sasayama section, 373
Sass de Putia section, 416
fossils, *417*
scaling law, *626*
equation, 621
scanning electron microscopy
shocked dolomite, *575*
schlieren, 145
Schuchert Dal Formation, *404*
sea level
Late Devonian, 473, *474,* 475
late Paleocene, *541*
P-Tr boundary, *371, 389*
sediment
El Kef, 262
extraterrestrial component, *21*
rhythmic, 225, 233
sedimentation rate
Gartnerkofel core, *424*
Japan, *405*
seismic reflection
Chicxulub crater, *43*
semichatovae transgression, *479*
Sentinel Mountain Formation, 492
Sevier Orogeny, 389
Shangshi boundary section
helium, 373
shark teeth
Navesink Formation, 293, 295
Red Bank Formation, 294
shock
diamonds, 595
history, 597
impact, 637
loading experiments, 596
melting
calcite, 589
pressure, 497
recovery, 587
shock experiments
calcite, *588*
container failure, 574
shock metamorphism, 50, 51, 494
experiments, *596*
kinetic effects, 595, 596
laboratory studies, 595
Vredefort, 611
shock wave, 70, 596, 695
induced tsunami, 70
shocked minerals, 671
shocked quartz, *63*
Alamo Breccia, 490, 497
Alamo Impact, 477
Alamo lapilli, 497
Antarctica, 452
Australia, 452
Beloc, 64, 146
Bjala, 221
Bochil, 64
Cacarajicara Formation, 125, *138*
Caribbean, 164
Central America, 164
Chicxulub crater, 55
Chicxulub impact structure, 47
El Kef, 256
Gulf of Mexico, *64*
La Ceiba, 254
ODP Site 1049, 277, 279, 288
Oxyoke Canyon Sandstone, 497
P-Tr boundary, 373
planar deformation fractures, 497
Poty quarry, 63
size, 65
Streuben Knob channel, 479
shocked zircon, 56
Shoemaker-Levy 9 comet, 646
SHRIMP ion probe dating, 611
Siberian flood basalts, 373, *375,* 377
P-Tr boundary, 378
Siberian Traps, 375
age, *371*
Sichuan Province
carbonate, 374
siderophile elements
Newark basin, *513*
Signor-Lipps effect, 245, 416
Siljan impact, 481, 484
siphonodellid conodont biofacies, 484
Sirius supercluster, 679, 680
slope failure, 88, 93
slumping, 93
smectite
Beloc, 168, *183*
Bjala, *242*

smectite (*continued*)
Blake Nose, 193
Jacksonwald syncline, 518
ODP Site 1049, 279
Stevns Klint, 269
Tunisia, 248
soft-sediment deformation
ODP 1049, 279
Sohm Abyssal Plain, 91
SOHO comets, 652
Sohryngkew River section, 202
solar insolation, 225
Sonoma orogenic belt, 386, 389
Sonoma Orogeny, *388*
Sopelana section
K-T boundary, 554
South Pole
cosmic spherules, 28
southern China
mass extinction, 377
oxygenation, *402*
P-Tr research, 386
P-Tr sections, *401*
SOVA multi-material hydrocode, 695
space debris, 10
Spaceguard Survey, *13*
Spacewatch Program, 10
Spain
K-T boundary, 551
Spathian basinal sediments
ammonoids, 399
Spearman rank correlation matrix, 180
Spechty Kopf Formation, 673
species richness
Bjala, *237*
Stevns Klint, 266
spherules, 27, 262. *See also* lapilli
Alamo Impact, 477
Ancora borehole, 99
Barberton Greenstone Belt, 35
Bass River, 99, 104
Beloc, *146, 167, 180*
Blake Nose, 189, *190, 193,* 197
Cacarajicara Formation, 138, *139*
Caribbean, 164
Central America, 164
clinopyroxene, 28
Cone crater, *661*
cosmic, 21, *22, 28*
Coxquihui, 158
DSDP Site 387, 85
DSDP Site 577, 28, 29
DSDP Site 603, *84*
El Caribe, 158
El Kef, 256
El Mimbral, 146, 196
Estonia, 567
Guatemala, 146
Gulf of Mexico, 61
Hamersley basin, 26
impact, *28*
La Ceiba, 254
La Lajilla, 196
La Sierrita, *148*
lunar, *659*
Meenikunno Bog, 568
Mendez Formation, 145, *148, 156*
ODP Site 1049, 189, 279, *283*
ODP Site 689, 28
Petriccio, 29
PGE abundance, *26*
spinel-bearing, *29*
Stevns Klint, 337
Yucatan, 497
spherulite
spinel-bearing, 35
spinel
Carvaca, 29
K-T boundry, *222*
magnesioferrite, 221
nickel rich, 146, *219*
ODP Site 1049, 281, 282
spherules, *29*
tracers, 30
Spiti Valley
biostratigraphy, *446*
sponges, 298
minimalist organisms, 297
Star Peak Group, 388
Stara Planica zone, 233
Steinbruch Schmidt section, 474
stepwise extinction, 475, *480*
Stevns Klint
boundary section, 265
Ce anomaly, 179
chromium isotopes, 34
lithology, *267*
oxygen isotopes, 32
pyrite spherules, 337
stishovite, 48
Strangelove ocean, 425
stratigraphy
Bermuda Rise, *84*
Bochil, *254*
DSDP Site 385, *84*
DSDP Site 386, *85*
DSDP Site 387, *85*
DSDP Site 603, *84,* 89
Karoo basin, *430*
Khémisset basin, *516*
La Ceiba, *254*
Newark basin, *507*
Streuben Knob channel
shocked quartz, 478
stromatolites
Armenian Transcaucasia sections, 400
Madagascar, 403
P-Tr boundary, 410
Triassic, 395
stromatoporoids
Anticosti Island, 465
strontium isotopes
Bass River, 100, 104, *104*
Early Jurassic oceans, 529
Hornerstown Formation, 294
K-Tr boundary, *370*
Meishan, 370, 373
Navesink sequence, 100
New Jersey Coastal Plain, 103
stratigraphy, 528
Tr-J boundary, 529, 530
submarine slope failure
Blake Nose, 61
subsidence
long-wavelength subduction-induced dynamic subsidence, 441
reciprocal flexure, 441
Sudbury
breccia, 28
crater, 27, 345
fullerenes, 348
helium, 373
suevite, 60. *See also* breccia, impact
Chicxulub crater, 47, *48,* 64
mineralogy, 49
Ries crater, 500, 572
Sugarite site
boundary claystone, 338
sulfur isotopes, 338
Suisi Member
Werfen Formation, 398
sulfate aerosols, 376
sulfides, 337
sulfur
impact vaporization, 711
microbiological utilization, 337
sulfur dioxide
toxicity, 713
sulfur isotopes
above K-T boundary, *342*
below K-T boundary, *340*
Brownie Butte, *338*
Dogie Creek, *338, 342*
Japan, *405*
K-T boundary, *337*
Meishan, 373
P-Tr boundary, *370*
Sugarite Site, 338
sulfuric acid, 12
sungrazers, 652
superanoxic event
P-Tr boundary, *406*
supercloud, 682
supercooling parameter, 642
supercooling ratio, 632
superparamagnetism, 201, 206, 207, 451
natural fine particle, 202
origin, *208*
superring, *680*
survival
dinosaurs, *332*
survivor taxa, 507
Tr-J boundary, 517
Sverdup basin
lithology, 404
Sydney basin
Permian swamp, 368
synthesis
complex organic components, 711
Syrian arc, 535

T

Taconic orogeny, 464
Taghanic onlap, 474, *476*

Talcott Basalt, 516
Tambor volcano, 12
Tarawan Formation, 535
tectonics
  P-Tr boundary, *386*
tektite, 84, 104. *See also* impact, glass; microtektite
  Alamedilla section, 554
  ODP Site 1049, 277, 283, 288
terrigenous flux, 169
Tesero Oolite
  biostratigraphy, 416, 418
  carbonate component, 424
  foraminifera, *419*
  iridium anomaly, 373
  Werfen Formation, 397
Tethyan margin
  black shales, 534
Tethyan ocean, 396
tetrapods
  footprints, 513
  Jacksonwald syncline, 511
  Karoo basin, *437*
Thaynes Formation, 379
Thebes Formation, 535
theory of homogenous nucleation, 631
thermal maturation
  late Paleocene, *541*
thermal maximum
  late Paleocene, *533,* 535, *537, 542,* 544
thermogravimetry
  shocked dolomite, *578*
Three Forks Formation, 483, 484
tidal forces
  galactic, 669
Timpahute Range, *478*
Timpoweap Canyon
  conglomerate, 388, 390
Timpoweap Member
  Moenkopi Formation, 388
Tinton Formation, 293
Toad Formation, 403
Tobin Formation, 388
Tonotac
  myth, 5
Tore Seamount, 559
Torino scale, 14
Toshi claystones, 406
total organic carbon, *546*
  black shales, 544
Towaco Formation
  tetrapod footprints, 513
toxins
  impact generated, 705
Tr-J boundary, 505
  age, *524*
  carbon isotopes, *528*
  Central Atlantic magmatic province, *518*
  Hartford basin, *516*
  radiometric dating, *507*
trace elements
  Beloc, *169*
  dinosaur bones, *326*
  Karoo basin, *434, 439*
  Kirtland Formation, *328*
  Newark basin, *515*
  Ojo Alamo Sandstone, *328*
tracers, *24,* 79. *See also* isotopes
  chemical, *24*
  extraterrestrial, *21*
  isotope, *31*
  spinel, 30
Transcaucasia sections, 400
  stromatolites, 400
transgression
  Hornerstown Formation, 294
transient cavity, *621, 623*
transient crater diameter equation, 620
Transvaal Supergroup
  core, 612
Tsõõrikmäe crater, *564*
tsunami, 97, 105
  Cacarajicara Formation, 140
  Chicxulub impact, 79, 98, 105
  crater generated, 70, 75
  Gulf of Mexico, 28, *58, 70*
  landslide generated, 70, *74,* 75
  model, *69, 70*
  Moncada Formation, 110
  receding wave, *73*
  rim wave, 70, *73*
  rushing wave, *73*
  wave rhythm, 121, 122
tsunamites
  Gulf of Mexico, *61*
Tunguska
  bolide, 693
  event, 10, 31
  impact, 15
Tunisia
  kaolinite, 248
  smectite, 248
turbidite, 125
  Basque-Cantabrian basin, 554
  Bermuda Rise, 89, 284
  Cacarajicara Formation, 140
  DSDP Site 386, 89, *90,* 93
  DSDP Site 387, 89
  DSDP Site 603, 84, 89, 91
  La Ceiba, 254
  ODP Site 1049, 288
Turkmenistan, 202
  iron phases, *207*
  ultrafine particles, 206
turtles
  Hornerstown Formation, 295
  Navesink Formation, 293, 295
Tvillingodden Formation, 403
Tzotzil
  myth, 5

U

Ultra Disperse Diamond, 602
ultrafine particles
  Turkmenistan, 206
unconformity, 294
  Clayton Formation, 98
  DSDP Site 385, *84*
  Loma Cerca, 150
  Navesink sequence, *100,* 105
  Ojo Alamo Sandstone, *315*
  Queenston deltaic complex, 469
  Tr-J boundary, 558
Union Wash Formation, 379, 403
Upper Lime Mudstone Member
  Cacarajicara Formation, *138, 141*
Ural Mountains, 368
uranium
  Japan, *405*
  Kirtland Formation, *329*
  Ojo Alamo Sandstone, *329*
  P-Tr boundary, 450
Ursell parameter, 73
Ursula Creek section
  thorium and uranium, 403

V

Val Badia
  carbon isotopes, *456*
  fossils, *456*
Valdelacasa antiform, 556, 557
van der Waals equation, 632
vapor cloud
  chemical composition, *706*
  condensation, 631
  experimental, *707*
  hydrodynamics, *632*
vaporization
  impact, *631,* 707
  kinetics, *632*
  nanoparticles, 209, 210
  projectile, 634
Vardebukta Formation, 403
Varian 3400 gas chromatograph, 353
Varian INOVA 500 NMR spectrometer, 347
velocity effect
  numerical modeling, 625
Verkykerskop Formation, 429, 432, 436
VG AUTOSPEC mass spectrometer, 346
VG Prism, 214
VG Prism II ratio mass spectrometer, 150, 236
VG SIRA 24 mass spectrometer, 353
Vienna Peedee Belemnite standard, 150
Vigo Seamount, 86
Vincentown Formation, 294, 296
Vinni Creek
  conodonts, *465*
  stratigraphy, 465
volcanism, 16
Von Neumann method, 73
Vredefort crater, 40, 49, 610, *611*
Vredefort granophyre, *612*

W

Wachapo Mountains
  boundary clay, 372
Wadi Nukhl
  deposition, 535

warming trend
  New Jersey coastal plain, 103
weakly interacting massive particles, 374
Werfen Formation
  carbon isotopes, 456
  carbonate ramp facies, 397
  fossils, 416
  Gartnerkofel core, 421
West Range Limestone, 483
Western Pahranagat Range
  Alamo lapilli, 499
wildfires
  Chicxulub impact, 698, 700
  global, 695
Witwatersrand basin, 611
wood biostratigraphy
  Karoo basin, *437*
Woodleigh crater, 373, 452
Woodside Creek
  carbon isotopes, 356
  nitrogen isotopes, 356
Wordie Creek Formation, *404*

X

x-ray diffraction, 236
  K-T boundary clays, 204
  Meghalaya, 206
  shocked dolomite, *575*

Y

Yátova Formation, 552
Yenisei-Khtanga trough, 376
York Canyon
  carbon isotopes, 355
  coal, 354
Yucatan basin, 110
Yucatan platform, 121

Z

Zel'dovich factor
  equation, 638
zeolites
  Beloc, 168, 180
  Blake Nose, 196, 197
  ODP Site 1049, 279
zircon
  shocked, 56
zoning
  Alamo Breccia, *492*
Zumaya
  K-T boundary, *554*
  P-E boundary, 551, *554,* 559